FORMULAS FROM GEOMETRY

Triangle

$h = a \sin \theta$

$\text{Area} = \dfrac{1}{2}bh$

(Law of Cosines)

$c^2 = a^2 + b^2 - 2ab \cos \theta$

Right Triangle

(Pythagorean Theorem)

$c^2 = a^2 + b^2$

Equilateral Triangle

$h = \dfrac{\sqrt{3}s}{2}$

$\text{Area} = \dfrac{\sqrt{3}s^2}{4}$

Parallelogram

$\text{Area} = bh$

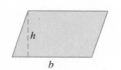

Trapezoid

$\text{Area} = \dfrac{h}{2}(a + b)$

Circle

$\text{Area} = \pi r^2$

$\text{Circumference} = 2\pi r$

Sector of Circle

(θ in radians)

$\text{Area} = \dfrac{\theta r^2}{2}$

$s = r\theta$

Circular Ring

(p = average radius,
w = width of ring)

$\text{Area} = \pi(R^2 - r^2)$
$\quad\quad\; = 2\pi pw$

Sector of Circular Ring

(p = average radius,
w = width of ring,
θ in radians)

$\text{Area} = \theta pw$

Ellipse

$\text{Area} = \pi ab$

$\text{Circumference} \approx 2\pi\sqrt{\dfrac{a^2 + b^2}{2}}$

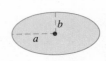

Cone

(A = area of base)

$\text{Volume} = \dfrac{Ah}{3}$

Right Circular Cone

$\text{Volume} = \dfrac{\pi r^2 h}{3}$

$\text{Lateral Surface Area} = \pi r\sqrt{r^2 + h^2}$

Frustum of Right Circular Cone

$\text{Volume} = \dfrac{\pi(r^2 + rR + R^2)h}{3}$

$\text{Lateral Surface Area} = \pi s(R + r)$

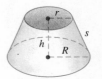

Right Circular Cylinder

$\text{Volume} = \pi r^2 h$

$\text{Lateral Surface Area} = 2\pi rh$

Sphere

$\text{Volume} = \dfrac{4}{3}\pi r^3$

$\text{Surface Area} = 4\pi r^2$

Wedge

(A = area of upper face,
B = area of base)

$A = B \sec \theta$

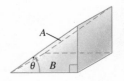

Calculus

CALCULUS

WITH ANALYTIC GEOMETRY

Alternate Fourth Edition

ROLAND E. LARSON

ROBERT P. HOSTETLER

The Pennsylvania State University
The Behrend College

with the assistance of
BRUCE H. EDWARDS
University of Florida
and
DAVID E. HEYD
The Pennsylvania State University
The Behrend College

D. C. HEATH AND COMPANY
Lexington, Massachusetts   Toronto

Chapter Opener Photo Credits

Chapter 1: Portfield-Chickering/Photo Researchers.
Chapter 2: Ray Ellis/Photo Researchers.
Chapter 3: Dan Helms/duomo.
Chapter 4: Bohdan Hrynewych/Stock, Boston.
Chapter 5: NASA/Johnson Space Center.
Chapter 6: Focus on Sports.
Chapter 7: Lionel Delevingne/Stock, Boston.
Chapter 8: Mark Antman/The Image Works.
Chapter 9: NASA/Johnson Space Center.
Chapter 10: Focus on Sports.
Chapter 11: Kent and Donna Dannen/Photo Researchers.
Chapter 12: NASA/Johnson Space Center.
Chapter 13: Steven Sutton/duomo.
Chapter 14: Ellis Herwig/Stock, Boston.
Chapter 15: Ford Motor Company.
Chapter 16: AP/Wide World Photos.
Chapter 17: NASA/Johnson Space Center.
Chapter 18: Cary Wolinsky/Stock, Boston.

Acquisitions Editors: Mary Lu Walsh and Ann Marie Jones

Developmental Editor: Cathy Cantin

Production Editor: Laurie Johnson

Designer: Sally Steele

Production Coordinator: Michael O'Dea

Photo Researcher: Wendy Johnson

Composition: Jonathan Peck Typographers

Technical Art: Folium

Cover: Martucci Studio

Preface

We wrote this text, *Calculus with Analytic Geometry, Alternate Fourth Edition,* guided by two primary objectives that have crystallized over many years of teaching calculus. For the student, our objective was to explain the basic concepts in a precise, readable manner and to clearly define and demonstrate calculus. For the instructor, our objective was to design a comprehensive teaching instrument that uses proven pedagogical techniques, thus freeing the instructor to make the most efficient use of classroom time.

Changes in the Alternate Fourth Edition

In the Alternate Fourth Edition, all examples, theorems, definitions, and prose have been revised—or at least considered for revision. Many new examples, exercises, and applications have been added to the text. The major changes in the Alternate Fourth Edition, listed by chapter, are as follows:

- The basic organization of Chapter 1 is the same as in the Alternate Third Edition. Section 1.4 contains a new summary of Equations of Lines, and Section 1.5 contains a new summary of the graphs of basic functions.
- Chapter 2 begins with a new introduction to the tangent line problem. Section 2.5 has a new example dealing with the ε-δ definition applied to a quadratic function.
- The "Four-Step Process" terminology has been eliminated from Section 3.1 in favor of simply applying the definition of the derivative. The introduction to the Chain Rule has been rewritten.
- Section 4.3 on the First Derivative Test has been reorganized. The section on business and economics applications has been moved to the end of the chapter.
- At the request of several users of the text, Chapter 5 has been reorganized. In an effort to get to the Fundamental Theorem of Calculus earlier in the chapter, we moved the section on integration by substitution to Section 5.5. The material on sigma notation was condensed and now appears as the first part of Section 5.2. The introduction of the natural logarithmic function was moved to Chapter 7.

- In Chapter 6, we moved the material on arc length and surface area to Section 6.4, preceding the section on work.
- Chapter 7 has been reorganized. At the request of several users, the chapter now begins with an introduction to the natural exponential function. The natural logarithmic function is then introduced as the inverse of the natural exponential function.
- The tabular method of repeated integration by parts was added to Section 9.2.
- At the request of many users of the Alternate Third Edition, the introduction to Taylor polynomials, which was in the first section of Chapter 10, has been moved to Section 10.7.
- In Chapter 11, the material on classifying conics was moved from Section 11.3 to Section 11.4.
- In Chapter 12, the section on area and arc length in polar coordinates now occurs before the section on polar equations for conics and Kepler's Laws.
- Chapter 13 was completely rewritten. At the request of several users, we now treat vectors and vector-valued functions *in the plane* separately from the corresponding topics in space.
- Chapter 14 was also rewritten. It now introduces the geometry of space, and then goes on to generalize two-dimensional vectors and vector-valued functions to three dimensions.
- New material on computer graphics in three dimensions was added to Section 15.1. A new discussion of continuity of a function of three variables was added to Section 15.2. The material on complex zeros of polynomial functions was deleted from Section 15.9, and a new example on least squares regression was added. In Section 15.10, the method of Lagrange multipliers was rewritten to agree with standard presentation.
- In Chapter 16, the introduction to Jacobians was rewritten and moved from the fourth section to the end of the chapter. The section that introduces cylindrical and spherical coordinates was moved from Chapter 13 to Chapter 16, just before integration in these two nonrectangular coordinate systems is discussed.
- The table of contents for Chapter 17 is the same as in the Alternate Third Edition, but the chapter was substantially rewritten. Section 17.1 now formally defines inverse square fields, has five new examples, and introduces divergence of a vector field. In Section 17.2, the introduction to line integrals was rewritten, and two new examples were added. In Section 17.3, Theorem 17.7 is new. In Section 17.5, Examples 2 and 5 were rewritten, and material on Gauss's Law was added.

Features

Order of Topics   The eighteen chapters readily adapt to either semester or quarter systems. In each system both differentiation *and* integration can be introduced in the first course of the sequence. There is some flexibility in the order and depth in which the chapters can be covered. For instance, much of the precalculus material in Chapter 1 can be used as individual review. The ε-δ discussion of limits in Chapter 2 can be given minimal coverage. Sections

6.6 and 6.7 can be given later coverage in the course. Chapter 10 can be covered at any time after Chapter 9. For instructors wishing to introduce trigonometric functions earlier in the course, the syllabus for an early trigonometry course in the Instructor's Guide suggests an alternative sequence of the topics in this text.

Definitions and Theorems   Special care has been taken to state the definitions and theorems simply, without sacrificing accuracy.

Proofs   We have chosen to include only those proofs that we have found to be both instructive and within the grasp of a beginning calculus student. Moreover, in presenting proofs, we have found that extensive detail often obscures rather than illuminates. For this reason, many of the proofs are presented in outline form, with an emphasis on the essence of the argument. (See the proof of the Product Rule in Section 3.4.) In some cases, we have included a more complete discussion of proofs in Appendix A. (See the proof of the Chain Rule in Section 3.5.)

Graphics   The Alternate Fourth Edition has over 2325 figures. Of these, 1080 are in the examples and exposition, over 650 are in the exercise sets, and over 590 are in the odd-numbered answers. The new art package in the Alternate Fourth Edition was computer-generated for accuracy. Designed with additional colors used systematically, this improved art program will help students visualize mathematical concepts, particularly in the presentation of complex, three-dimensional material. For example, brown is used for both the axes and primary graphs, blue is used for all planes in three-dimensional space, and red is used for all primary three-dimensional surfaces.

Exercises   Over 1000 new problems have been added, so the text now contains nearly 8200 exercises. The exercises are graded, progressing from skill-development problems to more challenging problems involving applications and proofs. Many exercise sets begin with a group of exercises that provides the graphs of the functions involved. Review exercises are included at the end of each chapter.

Examples   The text contains nearly 1000 examples, each titled for easy reference. Many of the examples include side comments in red that clarify the steps of the solution.

Computer/Calculator Exercises   Another new feature in the Alternate Fourth Edition is the addition of exercises requiring the use of a computer or graphics calculator. See, for instance, Exercises 69 and 70 in Section 1.3 or Exercises 57 and 58 in Section 8.2.

Numerical Methods   With the increasing power and accessibility of computers, numerical techniques are becoming more widely used. This edition reflects this trend by introducing numerical integration earlier in the text and by adding a tabular method of integration by parts. Calculators or computers are useful in these areas, as well as with topics such as limits, Newton's Method, and Taylor polynomials.

Chapter Introductions   As a new feature in the Alternate Fourth Edition, each chapter begins with a chapter overview and a special motivational application.

Applications   We have tried to choose applications that reflect variety and realism, and that require a minimal knowledge of other fields. The Alternate Fourth Edition contains over 2000 different applications.

Enhanced Presentation   As a new feature, the Alternate Fourth Edition has been designed with a functional use of four colors that strengthens the text as a pedagogical tool. In the page design, each color is used consistently to aid both reading and reference. For example, all theorems and definitions are highlighted by brown boxes, and equation side comments are given in red.

Summaries   Many sections have summaries that identify core ideas and procedures—see Sections 3.4, 4.10, 8.6, 10.6, and 13.6. In some instances, an entire section summarizes the preceding topics—see Sections 4.6 and 9.1.

Historical Notes   Throughout the text, we include several short biographical notes about prominent mathematicians. These are designed to help students gain an appreciation for both the people involved in the development of calculus and the nature of the problems that calculus was designed to solve.

Remarks   The text contains many special instructional notes to students in the form of "Remarks." These notes appear after definitions, theorems, or examples and are designed to give additional insight, help avoid common errors, or describe generalizations.

Supplements

For Students

- The *Study and Solutions Guide* by David E. Heyd contains detailed solutions to several representative problems from each exercise set. In the text, these exercises are identified in the exercise sets by blue numbers. The solutions to these exercises are given in greater detail than the examples in the text, with special care taken to show the algebra involved. In addition, the *Study and Solutions Guide* contains a review of algebra.

For Instructors

- A *Complete Solutions Guide* by Dianna L. Zook is available in three volumes. This guide contains brief solutions to every exercise in the text.
- An *Instructor's Guide* by Ann R. Kraus contains sample tests for each chapter in the text as well as suggestions for classroom instruction.
- A package of color transparencies of figures from the text is available.

For Students and Instructors

- *Calculus Applications in Engineering and Science* by Stuart Goldenberg and Harvey Greenwald contains solved examples and exercises (with answers to the odd-numbered ones) covering applications in engineering, physics, chemistry, biology, and other fields.

Software

A library of software products for this text is available for the IBM-PC, Macintosh, and Apple II. To make the most of the software products described below, instructors or students may choose to solve appropriate exercises from the text and those provided in a User Manual as well as analyze functions of the user's own selection. All of the following products can be used for classroom demonstration and have been class tested.

- *Math Lab Calculus*, The Math Lab (IBM-PC, Apple II)
  Offering both two-dimensional and three-dimensional computer graphics and numerical computations, the package consists of one disk and a User Manual with 80 lab assignments. These assignments are presented in worksheet format and are keyed to the text. Additionally, all computer/calculator exercises in the text can be solved using this software.
- *Math Utilities* (updated version 4.0), Bridge Software (IBM-PC)
  This is a program of three powerful graphing packages: CURVES, SURFS (Surfaces), and DIFFS (Differential Equations). Each package includes one disk and a User Manual. These enhanced graphing packages allow the user to save graphs and offer labelling and annotating capabilities for creating classroom handouts.
- *Graphitti*, George Best (IBM-PC, Macintosh)
  This software graphs both two- and three-dimensional figures as well as offers some features that help teach concepts through simulation, such as rotation of a curve about an axis. A lab package that accompanies the software provides assignments in a worksheet format keyed to the text.
- *Computer Activities for Calculus*, Technology Training Associates (IBM-PC)
  This package includes three disks and a User Manual. Each easy-to-use disk has two units of tutorials for self-study and the Grapher, a graphing tool for two-dimensional figures that is designed to handle a wide variety of elementary functions.
- *TrueBASIC CALCULUS*, TrueBASIC (IBM-PC) and *TrueBASIC MULTIVARIATE CALCULUS: MACFUNCTION* (Macintosh)
  Each package includes one disk and a User Manual. *TrueBASIC CALCULUS* performs numerical routines and two-dimensional graphing for topics in the first two semesters of calculus. *MACFUNCTION* is a tool for plotting and examining three-dimensional graphs of functions.

Computerized Testing

- *HeathTest Plus for Calculus* (IBM-PC, Apple II, Macintosh)
  Instructors can produce chapter tests, mid-terms, and final exams easily and accurately. Instructors can also edit existing questions or add new ones as

IBM-PC is a registered trademark of International Business Machines Corporation. Apple and Macintosh are the registered trademarks of Apple Computer, Inc. TrueBASIC is the registered trademark of TrueBASIC, Inc.

desired, or preview questions on screen and add them to a test with a single keystroke. The software supports graphics and offers both multiple-choice and open-ended questions. A User Manual and a printed test item file are available.

- *HeathTest for Calculus* (IBM-PC)

  This is an algorithm-based program that generates tests, quizzes, or worksheets in a multiple-choice format. A User Manual with printed test items accompanies the software.

Acknowledgments

We would like to thank the many people who have helped us at various stages of this project during the past seventeen years. Their encouragement, criticisms, and suggestions have been invaluable to us.

Fourth Edition Focus Group   Homer F. Bechtell, University of New Hampshire; K. Elayn Gay, University of New Orleans; Hideaki Kaneko, Old Dominion University; Judith A. Palagallo, University of Akron; John Tweed, Old Dominion University

Fourth Edition Reviewers   Keith Bergeron, United States Air Force Academy; Jorge Cossio, Miami-Dade Community College; Rosario Diprizio, Oakton Community College; Ali Hajjafar, University of Akron; Ransom Van B. Lynch, Phillips Exeter Academy; Bennet Manvel, Colorado State University; Duff A. Muir, United States Air Force Academy; Charlotte J. Newsom, Tidewater Community College; Terry J. Newton, United States Air Force Academy; Wayne J. Peeples, University of Texas; Jorge A. Perez, LaGuardia Community College; Barry J. Sarnacki, United States Air Force Academy; George W. Schultz, St. Petersburg Junior College; Frank Soler, De Anza College; Michael Steuer, Nassau Community College; John Tweed, Old Dominion University; Jay Wiestling, Palomar College; August J. Zarcone, College of Dupage; Li Fong, Johnson County Community College

Third Edition Reviewers   Dennis Albér, Palm Beach Junior College; Garret J. Etgen, University of Houston; William R. Fuller, Purdue University; Timothy J. Kearns, Boston College; Norbert Lerner, State University of New York at Cortland; Robert L. Maynard, Tidewater Community College; Barbara L. Osofsky, Rutgers University; Jean E. Rubin, Purdue University; Lawrence A. Trivieri, Mohawk Valley Community College; J. Philip Smith, Southern Connecticut State University

Second Edition Reviewers   Harry L. Baldwin, Jr., San Diego City College; Phillip A. Ferguson, Fresno City College; Thomas M. Green, Contra Costa College; Arnold J. Insel, Illinois State University; William J. Keane, Boston College; David C. Lantz, Colgate University; Richard E. Shermoen, Washburn University; Thomas W. Shilgalis, Illinois State University; Florence A. Warfel, University of Pittsburgh

First Edition Reviewers   Paul W. Davis, Worcester Polytechnic Institute; Eric R. Immel, Georgia Institute of Technology; Frank T. Kocher, Jr., Pennsylvania State University; Joseph F. Krebs, Boston College; Maurice L. Monahan, South Dakota State University; Robert A. Nowlan, Southern Connecticut State University; N. James Schoonmaker, University of Vermont; Bert K. Waits, Ohio State University

We would also like to thank users of the Third Edition who answered a questionnaire concerning changes they wanted in the new edition. They are George Anderson, Rhode Island College; Frank P. Battles, Massachusetts Maritime Academy; Derek I. Bloomfield, Orange County Community College; Karen J. Edwards, Paul Smith's College; Theodore Hanley, State University of New York at Utica/Rome; Peter Herron, Suffolk County Community College; Ann M. Joyce, Chestnut Hill College; Arthur Kaufman, College of Staten Island; Alan Levine, Franklin and Marshall College; James Magliano, Union County College; Frank Morgan, Castleton State College; Raymond Pluta, Castleton State College; M. Susan Richman, Pennsylvania State University; Carmen Vlad, Pace University; Christopher White, Castleton State College.

A special thanks to all the people at D. C. Heath and Company who worked with us in the development of the Fourth Edition, especially Ann Marie Jones, Mathematics Acquisitions Editor; Cathy Cantin, Developmental Editor; Laurie Johnson, Production Editor; Sally Steele, Designer; Carolyn Johnson, Editorial Assistant; Mike O'Dea, Production Manager.

Several other people also worked on this project: David E. Heyd wrote the *Study and Solutions Guide* and solved the exercises; Dianna L. Zook wrote the *Complete Solutions Manual*; Ann R. Kraus wrote the *Instructor's Guide*; Timothy R. Larson prepared the art; Linda L. Kifer proofread the galleys; Linda M. Bollinger proofread the galleys and typed the supplements; Helen Medley solved the exercises and performed an accuracy check for the text; and Kathleen Evanoff, Randall Hammond, and Paula Sibeto solved the exercises and assisted with the production of the supplements package.

A special note of thanks goes to the over 500,000 students who have used earlier editions of the text.

On a personal level, we are grateful to our wives, Deanna Gilbert Larson and Eloise Hostetler, for their love, patience, and support. Also, a special thanks goes to R. Scott O'Neil.

If you have suggestions for improving this text, please feel free to write to us. Over the past seventeen years we have received many useful comments from both instructors and students, and we value these very much.

Roland E. Larson
Robert P. Hostetler

Calculus

To accommodate the different methods of teaching calculus, D. C. Heath also offers the two texts described below. Each has its own supplements package. The following is a brief discussion of how each book differs from CALCULUS WITH ANALYTIC GEOMETRY, Alternate Fourth Edition.

Calculus with Analytic Geometry, Fourth Edition
Larson/Hostetler/Edwards

This text is also designed for a three-semester course. All six trigonometric functions are reviewed in Chapter 1, then used throughout the text. Additionally, this text offers a different treatment of the following topics: limits, applications of integration, exponential and logarithmic functions, and vectors.

Calculus with Analytic Geometry, Third Edition, Part I
Larson/Hostetler

This single-variable text is designed for a two-semester course. All six trigonometric functions are reviewed in Chapter 1, then used throughout the text. Additionally, this text offers a different treatment of the following topics: units, applications of integration, and exponential and logarithmic functions.

Contents

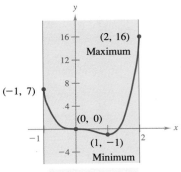

$f(x) = 3x^4 - 4x^3$

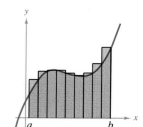

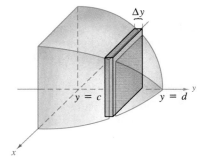

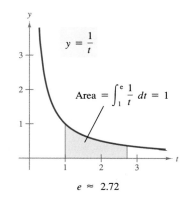

$$y = \frac{1}{t}$$

$$\text{Area} = \int_1^e \frac{1}{t}\,dt = 1$$

$e \approx 2.72$

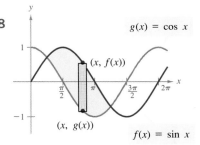

$g(x) = \cos x$

$(x, f(x))$

$(x, g(x))$

$f(x) = \sin x$

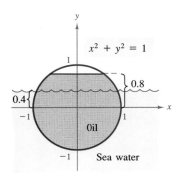

$x^2 + y^2 = 1$

Oil

Sea water

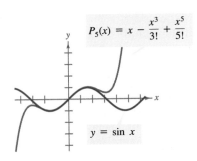

$$P_5(x) = x - \frac{x^3}{3!} + \frac{x^5}{5!}$$

$y = \sin x$

Ellipse   Hyperbola

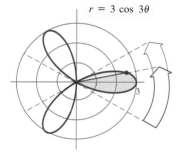

$r = 3 \cos 3\theta$

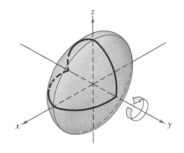

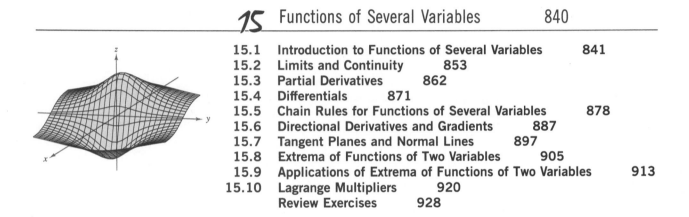

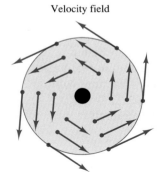

Velocity field

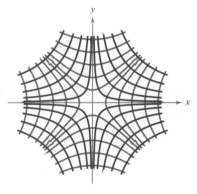

What Is Calculus?

We begin to answer this question by saying that calculus is the reformulation of elementary mathematics through the use of a limit process. If limit processes are unfamiliar to you, then this answer is, at least for now, somewhat less than illuminating. From an elementary point of view, we may think of calculus as a "limit machine" that generates new formulas from old. Actually, the study of calculus involves three distinct stages of mathematics: *precalculus mathematics* (the length of a line segment, the area of a rectangle, and so forth), the *limit process*, and new *calculus* formulations (derivatives, integrals, and so forth).

Some students try to learn calculus as if it were simply a collection of new formulas. This is unfortunate. When students reduce calculus to the memorization of differentiation and integration formulas, they miss a great deal of understanding, self-confidence, and satisfaction.

On the following two pages we have listed some familiar precalculus concepts coupled with their more powerful calculus versions. Throughout this text, our goal is to show you how precalculus formulas and techniques are used as building blocks to produce the more general calculus formulas and techniques. Don't worry if you are unfamiliar with some of the "old formulas" listed on the following two pages—we will be reviewing all of them.

As you proceed through this text, we suggest that you come back to this discussion repeatedly. Try to keep track of where you are relative to the three stages involved in the study of calculus. For example, the first three chapters break down as follows: precalculus (Chapter 1), the limit process (Chapter 2), and new calculus formulas (Chapter 3). This cycle is repeated many times on a smaller scale throughout the text. We wish you well in your venture into calculus.

WITHOUT CALCULUS	WITH DIFFERENTIAL CALCULUS
value of $f(x)$ when $x = c$	limit of $f(x)$ as x approaches c
slope of a line	slope of a curve
secant line to a curve	tangent line to a curve
average rate of change between $t = a$ and $t = b$	instantaneous rate of change at $t = c$
curvature of a circle	curvature of a curve
height of a curve when $x = c$	maximum height of a curve on an interval
tangent plane to a sphere	tangent plane to a surface
direction of motion along a straight line	direction of motion along a curved line

WITHOUT CALCULUS	WITH INTEGRAL CALCULUS
area of a rectangle	area under a curve
work done by a constant force	work done by a variable force
center of a rectangle	centroid of a region
length of a line segment	length of an arc
surface area of a cylinder	surface area of a solid of revolution
mass of a solid of constant density	mass of a solid of variable density
volume of a rectangular solid	volume of a region under a surface
sum of a finite number of terms $a_1 + a_2 + \cdots + a_n = S$	sum of an infinite number of terms $a_1 + a_2 + a_3 \cdots = S$

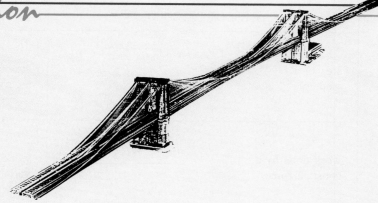

The Brooklyn Bridge was designed by the American civil engineer John A. Roebling (1806–1869). When it was completed in 1883, it was the longest suspension bridge in the world—with a span of 1595 feet. Other famous suspension bridges include the George Washington Bridge with a span of 3500 feet (completed in 1931) and the Golden Gate Bridge with a span of 4200 feet (completed in 1937).

Equation for a Suspension Bridge

A suspension bridge is supported by cables hung on two towers. The distance between the two towers is called the *span* of the bridge, and the height of each tower is called the *cable sag*. The weight of the roadbed is uniformly distributed along the cables, causing each cable to take the form of a parabola.

The Brooklyn Bridge has a span of 1595 feet between its supporting 275-foot towers. To find an equation for one of the parabolic cables between the towers, we consider the origin, (0, 0), to occur at the vertex (the lowest point) of the parabola. This implies that the cable meets the towers at the points

$$\left(-\frac{1595}{2}, 275\right) \quad \text{and} \quad \left(\frac{1595}{2}, 275\right).$$

Substituting the three points (0, 0), (−797.5, 275), and (797.5, 275) into the general equation of a parabola

$$y = ax^2 + bx + c$$

we obtain the equation

$$y = \frac{275}{(797.5)^2}x^2 = \frac{44}{101{,}761}x^2.$$

Chapter Overview

This first chapter contains a review of basic algebra and analytic geometry. The more familiar you are with the material in this chapter, the more successful you will be in calculus.

Section 1.1 reviews the properties of the real numbers and the real number line. The next two sections review the fundamental concepts of plane analytic geometry, the Cartesian Plane, and graphs of equations in two variables.

Section 1.4 discusses the slope of a line—this concept is critical in calculus. This section begins by showing how the slope of a line is related to the *average rate of change* of one variable with respect to another.

The concept of a **function** is also critical in calculus, and we review several fundamental ideas related to functions in Section 1.5. For instance, this section reviews the graphs of such basic functions as

$$f(x) = x \qquad f(x) = x^2$$
$$f(x) = x^3 \qquad f(x) = \sqrt{x}$$
$$f(x) = |x| \qquad f(x) = \frac{1}{x}.$$

Familiarity with the graphs of these functions will help you in later chapters.

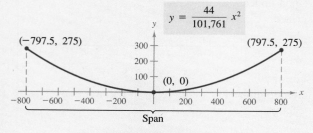

$$y = \frac{44}{101{,}761}x^2$$

(−797.5, 275)   (797.5, 275)

(0, 0)

Span

The Brooklyn Bridge

See Exercise 60, Section 1.3.

The Cartesian Plane and Functions

1.1 Real Numbers and the Real Line

Real numbers ▪ The real line ▪ Order and inequalities ▪ Absolute value ▪ Distance on the real line ▪ Intervals on the real line

In this first chapter we will lay the foundation for studying calculus. We assume that you have a good working knowledge of basic algebra. This is essential for the study of calculus.

The real line

The Real Line

FIGURE 1.1

To represent the set of real numbers we use a coordinate system called the **real line** or *x*-axis (Figure 1.1). The real number corresponding to a particular point on the real line is called the **coordinate** of the point. As Figure 1.1 shows, it is customary to identify those points whose coordinates are integers.

The point on the real line corresponding to zero is called the **origin** and is denoted by 0. The **positive direction** (to the right) is denoted by an arrowhead and indicates the direction of increasing values of *x*. Numbers to the right of the origin are **positive;** numbers to the left of the origin are **negative.** We use the term **nonnegative** to describe a number that is either positive or zero. Similarly, the term **nonpositive** is used to describe a number that is either negative or zero.

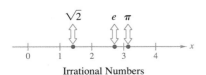

One-to-one correspondence between real numbers and points on the real line.

FIGURE 1.2

Each point on the real line corresponds to one and only one real number, and each real number corresponds to one and only one point on the real line. This type of relationship is called a **one-to-one correspondence.**

Each of the four points in Figure 1.2 corresponds to a real number that can be expressed as the ratio of two integers. $\left(\text{Note that } 4.5 = \frac{9}{2} \text{ and } -2.6 = -\frac{13}{5}.\right)$ We call such numbers **rational.** Rational numbers can be represented either by *terminating decimals* such as $\frac{2}{5} = 0.4$, or by *repeating decimals* such as $\frac{1}{3} = 0.333 \ldots = 0.\overline{3}$.

Real numbers that are not rational are called **irrational.** They cannot be represented as terminating or repeating decimals. To represent an irrational number, we usually resort to a decimal approximation. For example, $\sqrt{2} \approx 1.4142135623$, $\pi \approx 3.1415926535$, and $e \approx 2.7182818284$. (See Figure 1.3.)

$\sqrt{2}$ e π

Irrational Numbers

FIGURE 1.3

Order and inequalities

One important property of real numbers is that they are **ordered.**

DEFINITION OF ORDER ON THE REAL LINE	If a and b are real numbers, then a is **less than** b if $b - a$ is positive. We denote this order by the **inequality** $$a < b.$$ The symbol $a \leq b$ means that a is **less than or equal to** b. The statement b is **greater than** a is equivalent to saying a is less than b.

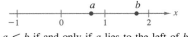

$a < b$ if and only if a lies to the left of b.

FIGURE 1.4

Geometrically, $a < b$ if and only if a lies to the *left* of b on the real line. (See Figure 1.4.) For example, $1 < 2$ because 1 lies to the left of 2 on the real line.

The following properties are often used to work with inequalities. Similar properties are obtained if $<$ is replaced by $\leq$ and $>$ is replaced by $\geq$.

THEOREM 1.1 PROPERTIES OF INEQUALITIES	1. If $a < b$ and $b < c$, then $a < c$. 2. If $a < b$ and $c < d$, then $a + c < b + d$. 3. If $a < b$ and k is any real number, then $a + k < b + k$. 4. If $a < b$ and $k > 0$, then $ak < bk$. 5. If $a < b$ and $k < 0$, then $ak > bk$.

REMARK   Note that we *reverse the inequality* when we multiply by a negative number. For example, if $x < 3$, then $-4x > -12$. This principle also applies to division by a negative number. Thus, if $-2x > 4$, then $x < -2$.

When three real numbers a, b, and c are ordered such that $a < b$ and $b < c$, we say that b is **between** a and c and we write $a < b < c$.

Occasionally it is convenient to use set notation to describe collections of real numbers. A **set** is a collection of elements. For example, the two major sets we have been discussing are the set of real numbers and the set of points on the real line. Often, we will restrict our interest to a **subset** of one of these two sets, in which case it is convenient to use **set notation** of the form

$$\{x: \text{condition on } x\}.$$

The set of all x such that a certain condition is true

For example, we can describe the set of positive real numbers as $\{x: 0 < x\}$. The **union** of two sets A and B is the set of elements that are members of A *or* B or both. This union is denoted by $A \cup B$. The **intersection** of two sets A and B is the set of elements that are members of A *and* B. This intersection is denoted by $A \cap B$. Two sets are called **disjoint** if they have no elements in common.

The most common sets we work with are subsets of the real line called **intervals.** For example, the **open** interval $(a, b) = \{x: a < x < b\}$ is the set of all real numbers greater than a and less than b, where a and b are called the **endpoints** of the interval. Note that the endpoints are not included in an open interval. Intervals that include their endpoints are called **closed** and are denoted by $[a, b] = \{x: a \leq x \leq b\}$. The nine basic types of intervals on the real line are shown in Table 1.1. The first four are called **bounded intervals** and the remaining five are called **unbounded intervals.**

TABLE 1.1   Intervals on the Real Line

	Interval notation	Set notation	Graph
Open interval	(a, b)	$\{x: a < x < b\}$	
Closed interval	$[a, b]$	$\{x: a \leq x \leq b\}$	
Half-open intervals	$[a, b)$	$\{x: a \leq x < b\}$	
	$(a, b]$	$\{x: a < x \leq b\}$	
Infinite intervals	$(-\infty, a]$	$\{x: x \leq a\}$	
	$(-\infty, a)$	$\{x: x < a\}$	
	(b, ∞)	$\{x: b < x\}$	
	$[b, \infty)$	$\{x: b \leq x\}$	
	$(-\infty, \infty)$	$\{x: x \text{ is a real number}\}$	

REMARK   We use the symbols ∞ and $-\infty$ to refer to positive and negative infinity. These symbols do not denote real numbers; they merely enable us to describe unbounded conditions more concisely. For instance, the interval $[b, \infty)$ is unbounded to the right since it includes *all* real numbers that are greater than or equal to b.

EXAMPLE 1   Intervals on the real line

Describe the intervals on the real line that correspond to the temperature ranges (in degrees Celsius) for water in the following two states.
(a) liquid   (b) gas

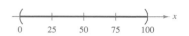

(a) Temperature range of water
     (in degrees Celsius)

(b) Temperature range of steam
     (in degrees Celsius)

FIGURE 1.5

SOLUTION

(a) Since water is in a liquid state at temperatures greater than $0°$ and less than $100°$, we have the interval

$$(0, 100) = \{x: 0 < x < 100\}$$

as shown in Figure 1.5(a).

(b) Since water is in a gaseous state (steam) at temperatures greater than or equal to $100°$, we have the interval

$$[100, \infty) = \{x: 100 \leq x\}$$

as shown in Figure 1.5(b).

In calculus we are frequently asked to solve inequalities involving variable expressions such as $2x - 5 < 7$. We say that a is a **solution** of this inequality if the inequality is true when a is substituted for x. The set of all values of x that satisfy the inequality is called the **solution set** of the inequality.

EXAMPLE 2   Solving an inequality

Find the solution set of the inequality $2x - 5 < 7$.

SOLUTION

Using the properties in Theorem 1.1, we have

$$2x - 5 < 7$$
$$2x - 5 + 5 < 7 + 5 \qquad \text{Add 5 to both sides}$$
$$2x < 12$$
$$\tfrac{1}{2}(2x) < \tfrac{1}{2}(12) \qquad \text{Multiply both sides by } \tfrac{1}{2}$$
$$x < 6.$$

Thus, the interval representing the solution is $(-\infty, 6)$.

REMARK   In Example 2, all five inequalities listed as steps in the solution have the same solution set and are called **equivalent.**

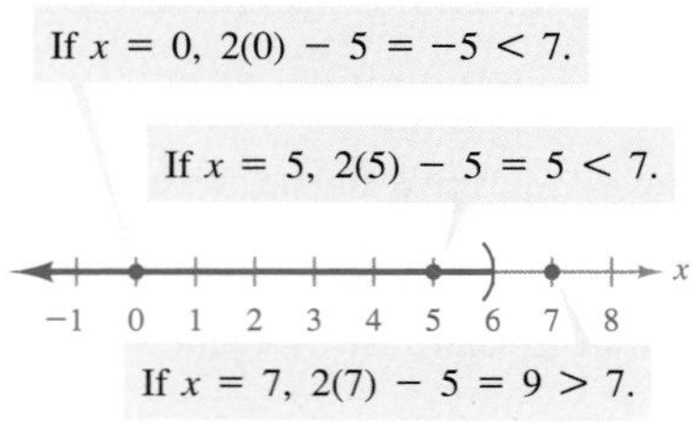

Once you have solved an inequality, check some x-values in your solution interval to see whether they satisfy the original inequality. You also might check some values outside your solution interval to verify that they do not satisfy the inequality. For example, Figure 1.6 shows that when $x = 0$ or $x = 5$ the inequality is satisfied, but when $x = 7$ the inequality is not satisfied.

If $x = 0$, $2(0) - 5 = -5 < 7$.

If $x = 5$, $2(5) - 5 = 5 < 7$.

If $x = 7$, $2(7) - 5 = 9 > 7$.

FIGURE 1.6

EXAMPLE 3   Finding the intersection of two solution sets

Find the intersection of the solution sets of the inequalities

$$-3 \leq 2 - 5x \qquad \text{and} \qquad 2 - 5x \leq 12.$$

SOLUTION

We could solve both inequalities and then find the intersection of the resulting solution sets. However, since the expression $2 - 5x$ occurs on the left side of one inequality and the right side of the other, it is convenient to work with both inequalities at the same time.

$$-3 \le \quad 2 - 5x \quad \le 12$$
$$-3 - 2 \le 2 - 5x - 2 \le 12 - 2 \qquad \text{Subtract 2}$$
$$-5 \le \quad -5x \quad \le 10$$
$$\frac{-5}{-5} \ge \quad \frac{-5x}{-5} \quad \ge \frac{10}{-5} \qquad \text{Divide by } -5 \text{ and reverse the inequality}$$
$$1 \ge \quad x \quad \ge -2$$

Thus, the interval representing the solution is $[-2, 1]$, as shown in Figure 1.7.

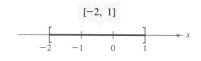

$[-2, 1]$

FIGURE 1.7

The inequalities in Examples 2 and 3 involve first-degree polynomials. For inequalities involving polynomials of higher degree we use the fact that a polynomial can change signs *only* at its real zeros (a **zero** of a polynomial is a number at which the value of the polynomial is zero). Between two consecutive real zeros a polynomial must be entirely positive or entirely negative. This means that when the real zeros of a polynomial are put in order, they divide the real line into **test intervals** in which the polynomial has no sign changes. That is, if a polynomial has the factored form

$$(x - r_1)(x - r_2) \cdots (x - r_n), \qquad r_1 < r_2 < r_3 < \cdots < r_n$$

then the test intervals are

$$(-\infty, r_1), (r_1, r_2), \ldots, (r_{n-1}, r_n), \text{ and } (r_n, \infty).$$

For example, the polynomial

$$x^2 - x - 6 = (x - 3)(x + 2)$$

can change signs only at $x = -2$ and $x = 3$.

EXAMPLE 4   Solving an inequality involving a quadratic

Find the solution set of the inequality $x^2 < x + 6$.

SOLUTION

$$x^2 < x + 6 \qquad \text{Given}$$
$$x^2 - x - 6 < 0 \qquad \text{Polynomial form}$$
$$(x - 3)(x + 2) < 0 \qquad \text{Factor}$$

Thus, the polynomial $x^2 - x - 6$ has $x = -2$ and $x = 3$ as its zeros, and we can solve the inequality by testing the sign of $x^2 - x - 6$ in each of the following open intervals.

$$(-\infty, -2), \qquad (-2, 3), \qquad (3, \infty)$$

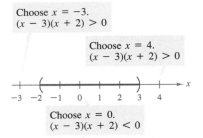

Choose $x = -3$.
$(x - 3)(x + 2) > 0$

Choose $x = 4$.
$(x - 3)(x + 2) > 0$

Choose $x = 0$.
$(x - 3)(x + 2) < 0$

FIGURE 1.8

To test an interval, we choose an arbitrary number in the interval and compute the sign of each factor of $x^2 - x - 6$. For example, for any x in the open interval $(-\infty, -2)$, the factors $(x - 3)$ and $(x + 2)$ are both negative. Consequently, the product (of two negative numbers) is positive and the inequality is *not* satisfied in the interval $(-\infty, -2)$. We suggest that you use the testing format shown in Figure 1.8. Since the inequality $(x - 3)(x + 2) < 0$ is satisfied only for values of x in the center interval, we conclude that the solution set is the open interval $(-2, 3)$.

Absolute value and distance

The **absolute value** of a real number a is denoted by $|a|$ and is defined as follows.

DEFINITION OF ABSOLUTE VALUE

If a is a real number, then the **absolute value** of a is

$$|a| = \begin{cases} a, & \text{if } a \geq 0 \\ -a, & \text{if } a < 0. \end{cases}$$

The absolute value of a number can never be negative. For example, let $a = -4$. Then, since $-4 < 0$, we have

$$|a| = |-4| = -(-4) = 4.$$

Remember that the symbol $-a$ does not necessarily mean that $-a$ is negative. Theorems 1.2 and 1.3 contain some useful properties of absolute value.

**THEOREM 1.2
OPERATIONS WITH
ABSOLUTE VALUE**

If a and b are real numbers and n is a positive integer, then the following properties are true.

1. $|ab| = |a||b|$ 2. $\left|\dfrac{a}{b}\right| = \dfrac{|a|}{|b|}, b \neq 0$

3. $|a| = \sqrt{a^2}$ 4. $|a^n| = |a|^n$

REMARK   You are asked to prove these properties in Exercises 67–71.

**THEOREM 1.3
INEQUALITIES AND
ABSOLUTE VALUE**

If a and b are real numbers and k is positive, then the following properties are true.

1. $-|a| \leq a \leq |a|$
2. $|a| \leq k$ if and only if $-k \leq a \leq k$.
3. $k \leq |a|$ if and only if $k \leq a$ or $a \leq -k$.
4. Triangle Inequality: $|a + b| \leq |a| + |b|$

Properties 2 and 3 are also true if $\leq$ is replaced by $<$.

PROOF   We give a proof of Property 4 and leave the proofs of the first three properties as exercises (see Exercises 72–74). Using Property 1, we have $-|a| \leq a \leq |a|$ and $-|b| \leq b \leq |b|$. Adding these two inequalities produces

$$-(|a| + |b|) \leq a + b \leq |a| + |b|.$$

Now, by Property 2 (using $k = |a| + |b|$), we can conclude that

$$|a + b| \leq |a| + |b|.$$

EXAMPLE 5   Solving an inequality involving an absolute value

Sketch the solution set of $|x - 3| \leq 2$.

SOLUTION

Using Property 2 of Theorem 1.3, we have

$$-2 \leq \quad x - 3 \quad \leq 2$$
$$-2 + 3 \leq x - 3 + 3 \leq 2 + 3$$
$$1 \leq \quad x \quad \leq 5.$$

Thus, the solution set is the closed interval $[1, 5]$, as shown in Figure 1.9.

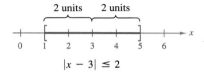

FIGURE 1.9

EXAMPLE 6   A two-interval solution set

Find the solution set of $3 < |x + 2|$.

SOLUTION

Using Property 3 of Theorem 1.3, we have

$$3 < x + 2 \quad \text{or} \quad x + 2 < -3$$
$$1 < x \quad \text{or} \quad x < -5.$$

Thus, the solution set consists of the union of the disjoint intervals $(-\infty, -5)$ and $(1, \infty)$, as shown in Figure 1.10.

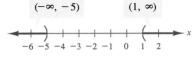

FIGURE 1.10

Examples 5 and 6 illustrate the general results shown in Figure 1.11. Note that if $d > 0$, the solution set for the inequality $|x - a| \leq d$ consists of a *single* interval, while the solution set for the inequality $|x - a| \geq d$ consists of *two* disjoint intervals.

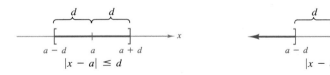

FIGURE 1.11

The **distance between two points** a and b on the real line is given by

$$d = |a - b|.$$

The **directed distance from** a **to** b is $b - a$ and the **directed distance from** b **to** a is $a - b$, as shown in Figure 1.12.

FIGURE 1.12

EXAMPLE 7   Distance on the real line

(a) The distance between -3 and 4 is given by

$$|4 - (-3)| = |7| = 7 \qquad \text{or} \qquad |-3 - 4| = |-7| = 7.$$

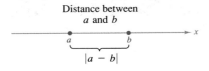

FIGURE 1.13

(See Figure 1.13.)

(b) The directed distance from -3 to 4 is $4 - (-3) = 7$.

(c) The directed distance from 4 to -3 is $-3 - 4 = -7$.   ▭

To find the **midpoint** of an interval with endpoints a and b, we simply find the average value of a and b. That is,

$$\text{midpoint of interval } (a, b) = \frac{a + b}{2}.$$

To show that this is the midpoint, you need only show that $(a + b)/2$ is equidistant from a and b.

EXERCISES for Section 1.1

In Exercises 1–10, determine whether the real number is rational or irrational.

1. 0.7

2. -3678

3. $\dfrac{3\pi}{2}$

4. $3\sqrt{2} - 1$

5. $4.345\overline{1451}$

6. $\dfrac{22}{7}$

*7. $\sqrt[3]{64}$

8. $0.8177817\overline{177}$

9. $4\dfrac{5}{8}$

10. $(\sqrt{2})^3$

*A blue number indicates that a detailed solution can be found in the *Study and Solutions Guide*.

In Exercises 11–14, express the repeating decimal as a ratio of integers using the following procedure. Let $x = 0.6363\ldots$. Then $100x = 63.6363\ldots$. Subtracting the first equation from the second produces $99x = 63$ or $x = \frac{63}{99} = \frac{7}{11}$.

11. $0.36\overline{36}$

12. $0.318\overline{18}$

13. $0.297\overline{297}$

14. $0.9900\overline{9900}$

15. Given $a < b$, determine which of the following are true.

(a) $a + 2 < b + 2$

(b) $5b < 5a$

(c) $5 - a > 5 - b$

(d) $\dfrac{1}{a} < \dfrac{1}{b}$

(e) $(a - b)(b - a) > 0$

(f) $a^2 < b^2$

16. For $A = \{x: 0 < x\}$, $B = \{x: -2 \le x \le 2\}$, and $C = \{x: x < 1\}$, find the indicated interval.
   (a) $A \cup B$ (b) $A \cap B$
   (c) $B \cap C$ (d) $A \cup C$
   (e) $A \cap B \cap C$

In Exercises 17 and 18, complete the table by filling in the appropriate interval notation, set notation, and graph on the real line.

17.

Interval notation	Set notation	Graph
$(-\infty, -4]$		
	$\{x: 3 \le x \le \frac{11}{2}\}$	
$(-1, 7)$		

18.

Interval notation	Set notation	Graph
	$\{x: 10 < x\}$	
$(\sqrt{2}, 8]$		
	$\{x: \frac{1}{3} < x \le \frac{22}{7}\}$	

In Exercises 19–44, solve the inequality and graph the solution on the real line.

19. $x - 5 \ge 7$ **20.** $2x > 3$
21. $4x + 1 < 2x$ **22.** $2x + 7 < 3$
23. $2x - 1 \ge 0$ **24.** $3x + 1 \ge 2x + 2$
25. $-4 < 2x - 3 < 4$ **26.** $0 \le x + 3 < 5$
27. $\frac{3}{4}x > x + 1$ **28.** $-1 < -\frac{x}{3} < 1$
29. $\frac{x}{2} + \frac{x}{3} > 5$ **30.** $x > \frac{-1}{x}$
31. $|x| < 1$ **32.** $\frac{x}{2} - \frac{x}{3} > 5$
33. $\left|\frac{x-3}{2}\right| \ge 5$ **34.** $\left|\frac{x}{2}\right| > 3$
35. $|x - a| < b$ **36.** $|x + 2| < 5$

37. $|2x + 1| < 5$ **38.** $|3x + 1| \ge 4$
39. $\left|1 - \frac{2}{3}x\right| < 1$ **40.** $|9 - 2x| < 1$
41. $x^2 \le 3 - 2x$ **42.** $x^4 - x \le 0$
43. $x^2 + x - 1 \le 5$ **44.** $2x^2 + 1 < 9x - 3$

In Exercises 45–48, find the directed distance from a to b, the directed distance from b to a, and the distance between a and b.

45.

46.

47. (a) $a = 126$, $b = 75$
    (b) $a = -126$, $b = -75$
48. (a) $a = 9.34$, $b = -5.65$
    (b) $a = \frac{16}{5}$, $b = \frac{112}{75}$

In Exercises 49–52, find the midpoint of the given interval.

49.

50.

51. (a) $[7, 21]$ (b) $[8.6, 11.4]$
52. (a) $[-6.85, 9.35]$ (b) $[-4.6, -1.3]$

In Exercises 53–58, use absolute values to define each interval (or pair of intervals) on the real line.

53.

54.

55.

56.

57. (a) All numbers that are at most 10 units from 12.
    (b) All numbers that are at least 10 units from 12.

58. (a) y is at most 2 units from a.
    (b) y is less than δ units from c.

59. The balance in an account after t years is given by

$$A = P + Prt$$

where P dollars is the initial investment and r is the simple interest rate (in decimal form). In order for an investment of $1000 to attain a balance that is greater than $1250 in two years, what should the interest rate be?

60. In the manufacture and sale of a certain product, the revenue for selling x units is

$$R = 115.95x$$

and the cost of producing x units is

$$C = 95x + 750.$$

In order for a profit to be realized, R must be greater than C. For what values of x will this product return a profit?

61. A utility company has a fleet of vans. The annual operating cost of each van is estimated to be

$$C = 0.32m + 2300$$

where C is measured in dollars and m is measured in miles. If the company wants the annual operating cost of each van to be less than $10,000, then m must be less than what value?

62. The heights, h, of two-thirds of the members of a certain population satisfy the inequality

$$\left| \frac{h - 68.5}{2.7} \right| \leq 1$$

where h is measured in inches. Determine the interval on the real line in which these heights lie.

63. To determine if a coin is fair (has an equal probability of landing tails up or heads up), an experimenter tosses it 100 times and records the number of heads, x. Through statistical theory, the coin is declared unfair if

$$\left| \frac{x - 50}{5} \right| \geq 1.645.$$

For what values of x will the coin be declared unfair?

64. The estimated daily production, p, at a refinery is given by

$$|p - 2,250,000| < 125,000$$

where p is measured in barrels of oil. Determine the high and low production levels.

In Exercises 65 and 66, determine which of the two given real numbers is greater.

65. (a) π or $\dfrac{355}{113}$ (b) π or $\dfrac{22}{7}$

66. (a) $\dfrac{224}{151}$ or $\dfrac{144}{97}$ (b) $\dfrac{73}{81}$ or $\dfrac{6427}{7132}$

In Exercises 67–74, prove the given property.

67. $|ab| = |a||b|$

68. $|a - b| = |b - a|$ [Hint: Use Exercise 67 and the fact that $(a - b) = (-1)(b - a)$.]

69. $\left| \dfrac{a}{b} \right| = \dfrac{|a|}{|b|}$, $b \neq 0$

70. $|a| = \sqrt{a^2}$

71. $|a^n| = |a|^n$, $n = 1, 2, 3, \ldots$

72. $-|a| \leq a \leq |a|$

73. $|a| \leq k$ if and only if $-k \leq a \leq k$, $k > 0$.

74. $k \leq |a|$ if and only if $k \leq a$ or $a \leq -k$, $k > 0$.

1.2 The Cartesian Plane

The Cartesian plane ▪ The Distance Formula ▪ The Midpoint Formula ▪ Equations of circles ▪ Completing the square

Just as real numbers can be represented by points on the real line, we can represent ordered pairs of real numbers by points in a plane. An **ordered pair** (x, y) of real numbers has x as its *first* member and y as its *second* member. The model for representing ordered pairs is called the **rectangular coordinate system,** or the **Cartesian plane.** It is developed by considering two real lines intersecting at right angles (Figure 1.14).

The horizontal real line is usually called the **x-axis,** and the vertical real line is usually called the **y-axis.** Their point of intersection is called the **origin.** The two axes divide the plane into four parts called **quadrants.**

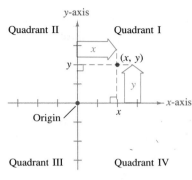

The Cartesian Plane

FIGURE 1.14

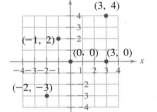

FIGURE 1.15

We identify each point in the plane by an ordered pair (x, y) of real numbers x and y, called **coordinates** of the point. The number x represents the directed distance from the y-axis to the point, and y represents the directed distance from the x-axis to the point (Figure 1.14). For the point (x, y), the first coordinate is called the x-coordinate or **abscissa,** and the second coordinate is called the y-coordinate or **ordinate.** For example, Figure 1.15 shows the location of the points $(-1, 2)$, $(3, 4)$, $(0, 0)$, $(3, 0)$, and $(-2, -3)$ in the Cartesian plane.

REMARK Note that we use an ordered pair (a, b) to denote either a point in the plane *or* an open interval on the real line. As the nature of the problem clarifies whether a point in the plane or an open interval is being discussed, there should be no confusion.

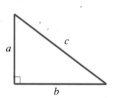

Pythagorean Theorem:
$a^2 + b^2 = c^2$

FIGURE 1.16

The Distance and Midpoint Formulas

Section 1.1 defined the distance between two points x_1 and x_2 on the real line. We will now find the distance between two points in the plane. Recall from the Pythagorean Theorem that, for a right triangle with hypotenuse c and sides a and b, we have the relationship $a^2 + b^2 = c^2$. Conversely, if $a^2 + b^2 = c^2$, then the triangle is a right triangle (Figure 1.16).

Suppose we want to determine the distance d between the two points (x_1, y_1) and (x_2, y_2) in the plane. With these two points, a right triangle can be formed, as shown in Figure 1.17. The length of the vertical side of the triangle is $|y_2 - y_1|$. Similarly, the length of the horizontal side is $|x_2 - x_1|$. By the Pythagorean Theorem, it follows that

$$d^2 = |x_2 - x_1|^2 + |y_2 - y_1|^2$$
$$d = \sqrt{|x_2 - x_1|^2 + |y_2 - y_1|^2}.$$

Replacing $|x_2 - x_1|^2$ and $|y_2 - y_1|^2$ by the equivalent expressions $(x_2 - x_1)^2$ and $(y_2 - y_1)^2$, we obtain

$$d = \sqrt{(x_2 - x_1)^2 + (y_2 - y_1)^2}.$$

We choose the positive square root for d because the distance *between* two points is not a directed distance. We have therefore established the following theorem.

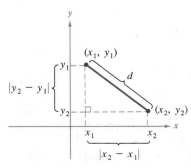

Distance Between Two Points

FIGURE 1.17

THEOREM 1.4 **DISTANCE FORMULA**	The distance d between the points (x_1, y_1) and (x_2, y_2) in the plane is given by $$d = \sqrt{(x_2 - x_1)^2 + (y_2 - y_1)^2}.$$

EXAMPLE 1   Finding the distance between two points

Find the distance between the points $(-2, 1)$ and $(3, 4)$.

SOLUTION

Applying the Distance Formula, we have

$$\begin{aligned} d &= \sqrt{[3 - (-2)]^2 + (4 - 1)^2} \\ &= \sqrt{(5)^2 + (3)^2} \\ &= \sqrt{25 + 9} \\ &= \sqrt{34} \approx 5.83. \end{aligned}$$

EXAMPLE 2   Verifying a right triangle

Plot the points $(2, 1)$, $(4, 0)$, and $(5, 7)$ and use the Distance Formula to show that the three points form the vertices of a right triangle.

SOLUTION

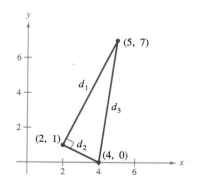

FIGURE 1.18

Figure 1.18 shows the triangle formed by the three points. Moreover, the three sides of the triangle have the following lengths.

$$\begin{aligned} d_1 &= \sqrt{(5 - 2)^2 + (7 - 1)^2} = \sqrt{9 + 36} = \sqrt{45} \\ d_2 &= \sqrt{(4 - 2)^2 + (0 - 1)^2} = \sqrt{4 + 1} = \sqrt{5} \\ d_3 &= \sqrt{(5 - 4)^2 + (7 - 0)^2} = \sqrt{1 + 49} = \sqrt{50} \end{aligned}$$

Since $d_1{}^2 + d_2{}^2 = 45 + 5 = 50 = d_3{}^2$, we can apply the Pythagorean Theorem to conclude that the triangle must be a right triangle.

The formula for the midpoint of a line segment in the plane is similar to that for an interval on the real line. The proof is left as an exercise (see Exercise 63).

THEOREM 1.5 **MIDPOINT FORMULA**	The midpoint of the line segment joining the points (x_1, y_1) and (x_2, y_2) is $$\left(\frac{x_1 + x_2}{2}, \frac{y_1 + y_2}{2}\right).$$

EXAMPLE 3 Finding the midpoint of a line segment

Find the midpoint of the line segment joining the points $(-5, -3)$ and $(9, 3)$.

SOLUTION

By the Midpoint Formula, the midpoint is

$$\left(\frac{-5 + 9}{2}, \frac{-3 + 3}{2}\right) = (2, 0).$$

(See Figure 1.19.)

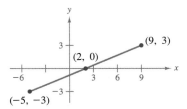

FIGURE 1.19

EXAMPLE 4 Finding points at a specified distance from a given point

Find x so that the distance between $(x, 3)$ and $(2, -1)$ is 5.

SOLUTION

Using the Distance Formula, we have

$$d = 5 = \sqrt{(x - 2)^2 + (3 + 1)^2}$$
$$25 = (x^2 - 4x + 4) + 16$$
$$0 = x^2 - 4x - 5$$
$$0 = (x - 5)(x + 1).$$

Therefore, $x = 5$ or $x = -1$, and we conclude that there are two solutions. That is, both of the points $(5, 3)$ and $(-1, 3)$ lie 5 units from the point $(2, -1)$, as shown in Figure 1.20.

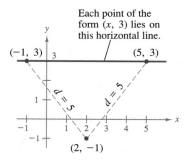

FIGURE 1.20

Circles

One straightforward application of the Distance Formula is in developing an equation for a circle in the plane.

DEFINITION OF A CIRCLE IN THE PLANE	Let (h, k) be a point in the plane and let $r > 0$. The set of all points (x, y) such that r is the distance between (h, k) and (x, y) is called a **circle.** The point (h, k) is the **center** of the circle, and r is the **radius** (see Figure 1.21).

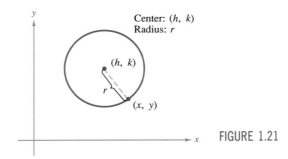

FIGURE 1.21

We can use the Distance Formula to write an equation for the circle with center (h, k) and radius r as

$$[\text{distance between } (h, k) \text{ and } (x, y)] = r$$
$$\sqrt{(x - h)^2 + (y - k)^2} = r.$$

By squaring both sides of this equation, we obtain the **standard form of the equation of a circle,** as indicated in the following theorem.

THEOREM 1.6 STANDARD FORM OF THE EQUATION OF A CIRCLE	The point (x, y) lies on the circle of radius r and center (h, k) if and only if $$(x - h)^2 + (y - k)^2 = r^2.$$

It follows from Theorem 1.6 that the standard form of the equation of a circle with center at the origin, $(h, k) = (0, 0)$, is

$$x^2 + y^2 = r^2.$$

If $r = 1$, then the graph of this equation is called the **unit circle.**

EXAMPLE 5 Finding the equation of a circle

The point $(3, 4)$ lies on a circle whose center is at $(-1, 2)$, as shown in Figure 1.22. Find an equation for the circle.

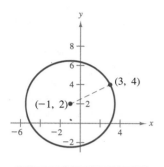

$(x + 1)^2 + (y - 2)^2 = 20$

FIGURE 1.22

SOLUTION

The radius of the circle is the distance between $(-1, 2)$ and $(3, 4)$. Thus,

$$r = \sqrt{[3 - (-1)]^2 + (4 - 2)^2} = \sqrt{16 + 4} = \sqrt{20}.$$

Therefore, the standard form of the equation of this circle is

$$[x - (-1)]^2 + (y - 2)^2 = (\sqrt{20})^2$$
$$(x + 1)^2 + (y - 2)^2 = 20.$$

By squaring and simplifying, the equation $(x - h)^2 + (y - k)^2 = r^2$ can be written in the following **general form of the equation of a circle.**

$$Ax^2 + Ay^2 + Cx + Dy + F = 0, \quad A \neq 0$$

To convert such an equation to the standard form

$$(x - h)^2 + (y - k)^2 = p$$

we use a process called **completing the square.** If $p > 0$, then the graph of the equation is a circle. If $p = 0$, then the graph is the single point (h, k). Finally, if $p < 0$, then the equation has no graph.

EXAMPLE 6   Completing the square

Sketch the graph of the circle whose general equation is

$$4x^2 + 4y^2 + 20x - 16y + 37 = 0.$$

SOLUTION

To complete the square, first divide by 4 so that the coefficients of x^2 and y^2 are both 1.

$$4x^2 + 4y^2 + 20x - 16y + 37 = 0 \qquad \text{General form}$$

$$x^2 + y^2 + 5x - 4y + \frac{37}{4} = 0 \qquad \text{Divide by 4}$$

$$(x^2 + 5x + \quad) + (y^2 - 4y + \quad) = -\frac{37}{4} \qquad \text{Group terms}$$

$$\left(x^2 + 5x + \frac{25}{4}\right) + (y^2 - 4y + 4) = -\frac{37}{4} + \frac{25}{4} + 4 \qquad \begin{array}{l}\text{Complete the}\\ \text{square by}\\ \text{adding } \frac{25}{4} \text{ and}\\ \text{4 to both sides}\end{array}$$

$$\underbrace{\qquad}_{(\text{half})^2} \qquad \underbrace{\qquad}_{(\text{half})^2}$$

$$\left(x + \frac{5}{2}\right)^2 + (y - 2)^2 = 1 \qquad \text{Standard form}$$

Note that we complete the square by adding the square of half the coefficient of x *and* the square of half the coefficient of y to both sides of the equation. Therefore, the circle is centered at $\left(-\frac{5}{2}, 2\right)$, and its radius is 1, as shown in Figure 1.23.

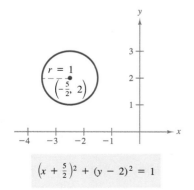

$\left(x + \frac{5}{2}\right)^2 + (y - 2)^2 = 1$

FIGURE 1.23

Pierre de Fermat

René Descartes

We have now introduced some fundamental concepts of *analytic geometry*. Because these concepts are in common use today, it is easy to overlook their revolutionary character. At the time analytic geometry was being developed by Pierre de Fermat (1601–1655) and René Descartes (1596–1650), the two major branches of mathematics—geometry and algebra—were largely independent of each other. Circles belonged to geometry and equations belonged to algebra. The coordination of the points on a circle and the solutions of an equation belongs to what is now called analytic geometry. It is important to become skilled in analytic geometry so that you may move easily between geometry and algebra. For instance, in Example 5, we are given a geometric description of a circle and are asked to find an algebraic equation for the circle. Thus, we are moving from geometry to algebra. Similarly, in Example 6 we are given an algebraic equation and asked to sketch a geometric picture. In this case, we are moving from algebra to geometry. These two examples illustrate the two most common problems in analytic geometry.

1. Given a graph, find its equation.

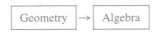

2. Given an equation, find its graph.

In the next two sections, we will look more closely at these two types of problems.

EXERCISES for Section 1.2

In Exercises 1–6, (a) plot the points, (b) find the distance between the points, and (c) find the midpoint of the line segment joining the points.

1. (2, 1), (4, 5) **2.** (−3, 2), (3, −2)

3. $\left(\frac{1}{2}, 1\right)$, $\left(-\frac{3}{2}, -5\right)$ **4.** $\left(\frac{2}{3}, -\frac{1}{3}\right)$, $\left(\frac{5}{6}, 1\right)$

5. (1, $\sqrt{3}$), (−1, 1) **6.** (−2, 0), (0, $\sqrt{2}$)

In Exercises 7–10, show that the given points form the vertices of the indicated polygon. (A rhombus is a quadrilateral whose sides are all of the same length.)

Vertices	*Figure*
7. (4, 0), (2, 1), (−1, −5)	Right triangle
8. (1, −3), (3, 2), (−2, 4)	Isosceles triangle
9. (0, 0), (1, 2), (2, 1), (3, 3)	Rhombus
10. (0, 1), (3, 7), (4, 4), (1, −2)	Parallelogram

In Exercises 11–14, use the Distance Formula to determine whether the given points are collinear (lie on the same line).

11. (0, −4), (2, 0), (3, 2)
12. (0, 4), (7, −6), (−5, 11)
13. (−2, 1), (−1, 0), (2, −2)
14. (−1, 1), (3, 3), (5, 5)

In Exercises 15 and 16, find x so that the distance between the points is 5.

15. (0, 0), (x, −4) **16.** (2, −1), (x, 2)

In Exercises 17 and 18, find y so that the distance between the points is 8.

17. (0, 0), (3, y) **18.** (5, 1), (5, y)

In Exercises 19 and 20, find the relationship between x and y so that (x, y) is equidistant from the two given points.

19. $(4, -1), (-2, 3)$ **20.** $\left(3, \frac{5}{2}\right), (-7, -1)$

21. Use the Midpoint Formula to find the three points that divide the line segment joining (x_1, y_1) and (x_2, y_2) into four equal parts.

22. Use the result of Exercise 21 to find the points that divide the line segment joining the given points into four equal parts.
(a) $(1, -2), (4, -1)$ (b) $(-2, -3), (0, 0)$

In Exercises 23 and 24, complete the square for each expression.

23. (a) $x^2 + 5x$ (b) $x^2 + 8x + 7$
24. (a) $4x^2 - 4x - 39$ (b) $5x^2 + x$

In Exercises 25–30, match the given equation with its graph. [Graphs are labeled (a)–(f).]

25. $x^2 + y^2 = 1$
26. $(x - 1)^2 + (y - 3)^2 = 4$
27. $(x - 1)^2 + y^2 = 0$
28. $\left(x + \frac{1}{2}\right)^2 + \left(y - \frac{3}{4}\right)^2 = \frac{1}{4}$
29. $(x + 3)^2 + (y - 1)^2 = 16$
30. $x^2 + (y - 1)^2 = 1$

(a)
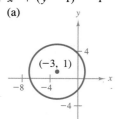

(b)

(c)

(d)
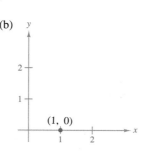

(e)
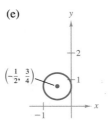

(f)
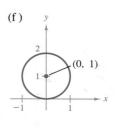

In Exercises 31–40, write the equation of the specified circle in general form.

31. Center: $(0, 0)$; radius: 3
32. Center: $(0, 0)$; radius: 5
33. Center: $(2, -1)$; radius: 4
34. Center: $(-4, 3)$; radius: $\frac{5}{8}$
35. Center: $(-1, 2)$; point on circle: $(0, 0)$
36. Center: $(3, -2)$; point on circle: $(-1, 1)$
37. Endpoints of diameter: $(2, 5), (4, -1)$
38. Endpoints of diameter: $(1, 1), (-1, -1)$
39. Points on circle: $(0, 0), (0, 8), (6, 0)$
40. Points on circle: $(1, -1), (2, -2), (0, -2)$

In Exercises 41–48, write the given equation (of a circle) in standard form and sketch its graph.

41. $x^2 + y^2 - 2x + 6y + 6 = 0$
42. $x^2 + y^2 - 2x + 6y - 15 = 0$
43. $x^2 + y^2 - 2x + 6y + 10 = 0$
44. $3x^2 + 3y^2 - 6y - 1 = 0$
45. $2x^2 + 2y^2 - 2x - 2y - 3 = 0$
46. $4x^2 + 4y^2 - 4x + 2y - 1 = 0$
47. $16x^2 + 16y^2 + 16x + 40y - 7 = 0$
48. $x^2 + y^2 - 4x + 2y + 3 = 0$

49. Find an equation for the path of a communications satellite in a circular orbit 22,000 miles above the earth. (Assume that the radius of the earth is 4000 miles.)
50. Find the equation of the circle passing through the points $(1, 2), (-1, 2)$, and $(2, 1)$.
51. Find the equation of the circle passing through the points $(4, 3), (-2, -5)$, and $(5, 2)$.
52. Find the equations of the circles passing through the points $(4, 1)$ and $(6, 3)$ and having radius $\sqrt{10}$.

In Exercises 53–56, sketch the set of all points satisfying the given inequality.

53. $x^2 + y^2 - 4x + 2y + 1 \leq 0$
54. $x^2 + y^2 - 4x + 2y + 1 > 0$
55. $(x + 3)^2 + (y - 1)^2 < 9$
56. $(x - 1)^2 + \left(y - \frac{1}{2}\right)^2 > 1$

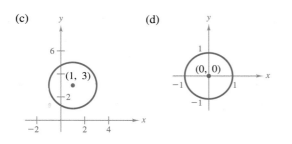

57. Prove that

$$\left(\frac{2x_1 + x_2}{3}, \frac{2y_1 + y_2}{3}\right)$$

is one of the points of trisection of the line segment joining (x_1, y_1) and (x_2, y_2). Also, find the midpoint of the line segment joining

$$\left(\frac{2x_1 + x_2}{3}, \frac{2y_1 + y_2}{3}\right)$$

and (x_2, y_2) to find the second point of trisection.

58. Use the results of Exercise 57 to find the points of trisection of the line segment joining the following points.
  (a) $(1, -2)$ and $(4, 1)$ (b) $(-2, -3)$ and $(0, 0)$

59. Prove that the line segments joining the midpoints of the opposite sides of a quadrilateral bisect each other.
60. Prove that the midpoint of the hypotenuse of a right triangle is equidistant from each of the three vertices.
61. Prove that an angle inscribed in a semicircle is a right angle.
62. Prove that the perpendicular bisector of a chord of a circle passes through the center of the circle.
63. Prove the Midpoint Formula (Theorem 1.5).

1.3 Graphs of Equations

The graph of an equation ■ Point-plotting method ■ Intercepts of a graph ■ Symmetry of a graph ■ Points of intersection ■ Mathematical models

Using a graph to show how two quantities are related is common. News magazines frequently show graphs that compare the gross national product or the unemployment rate to the time of year. Industries and businesses use graphs to report their monthly production and sales statistics. The value of such graphs is that they provide a geometric picture of the way one quantity changes with respect to another.

Frequently a relationship between two quantities is expressed as an equation. For instance, degrees on the Fahrenheit scale are related to degrees on the Celsius scale by the equation $F = \frac{9}{5}C + 32$. In this section we introduce a basic procedure for sketching the graph of such an equation.

TABLE 1.2

x	0	1	2	3	4
y	7	4	1	-2	-5

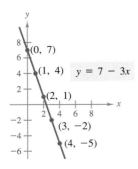

Solution points
of $y = 7 - 3x$

FIGURE 1.24

The graph of an equation

Consider the equation $3x + y = 7$. If $x = 2$ and $y = 1$, the equation is satisfied and we call the point $(2, 1)$ a **solution point** of the equation. Of course, there are other solution points, such as $(1, 4)$ and $(0, 7)$. We can construct a **table of values** for x and y by choosing arbitrary values for x and determining the corresponding values for y. To determine the values for y, it is convenient to write the equation in the form

$$y = 7 - 3x.$$

Thus, $(0, 7)$, $(1, 4)$, $(2, 1)$, $(3, -2)$, and $(4, -5)$ are all solution points of the equation $3x + y = 7$, as shown in Table 1.2. Actually, there are infinitely many solution points of this equation, and the set of all such points is called the **graph** of the equation, as shown in Figure 1.24.

REMARK Even though we refer to the sketch shown in Figure 1.24 as the graph of $y = 7 - 3x$, it really represents only a *portion* of the graph. The entire graph would extend beyond the page.

DEFINITION OF THE GRAPH OF AN EQUATION IN TWO VARIABLES	The **graph of an equation** involving two variables x and y is the set of all points in the plane that are solution points of the equation.

EXAMPLE 1   The point-plotting method

Sketch the graph of the equation $y = x^2 - 2$.

SOLUTION

First, we make a table of values (Table 1.3) by choosing several convenient values of x and calculating the corresponding values of $y = x^2 - 2$. Next, we locate these points in the plane, as in Figure 1.25(a). Finally, we connect the points by a *smooth curve*, as shown in Figure 1.25(b). This particular graph is called a **parabola.** It is one of the conic sections we will study in Chapter 11.

TABLE 1.3

x	-2	-1	0	1	2	3
y	2	-1	-2	-1	2	7

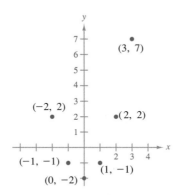

(a) Plot several points.

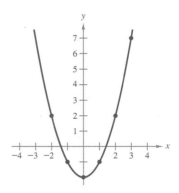

(b) Connect points with a smooth curve.

FIGURE 1.25

We call this method of sketching a graph the *point-plotting method*. It consists of three basic steps.

1. Make up a table of several solution points of the equation.
2. Plot these points in the plane.
3. Connect the points with a smooth curve.

In later chapters, we will discuss more sophisticated graphing techniques. In the meantime, when using the point-plotting method, we must plot a sufficient number of points to reveal the basic shape of the graph. With too few solution points, we can grossly misrepresent the graph of a given equation. For instance, how would you connect the four points shown in Figure 1.26? Without additional points or more information about the equation, any one of the three graphs shown in Figure 1.27 would be reasonable.

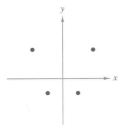

FIGURE 1.26

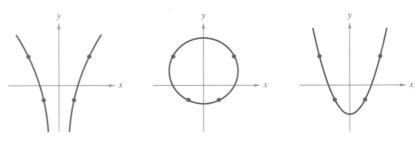

FIGURE 1.27

Two types of solution points that are especially useful are those having zero as either their x- or y-coordinate. Such points are called **intercepts,** because they are the points at which the graph intersects the x- or y-axis. Specifically, the point $(a, 0)$ is called an ***x*-intercept** of the graph of an equation if it is a solution point of the equation. Such points can be found by letting y be zero and solving the equation for x. Similarly, the point $(0, b)$ is called a ***y*-intercept** of the graph of an equation if it is a solution point of the equation. Such points can be found by letting x be zero and solving the equation for y.

REMARK   Some texts denote the x-intercept as the x-coordinate of the point $(a, 0)$ rather than the point itself. Unless it is necessary to make a distinction, we will use the term *intercept* to mean either the point or the coordinate.

It is possible for a graph to have no intercepts, or it might have several. For instance, consider the four graphs shown in Figure 1.28.

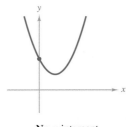

No x-intercept
One y-intercept

Three x-intercepts
One y-intercept

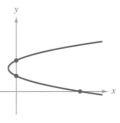

One x-intercept
Two y-intercepts

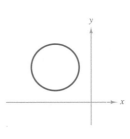

No intercepts

FIGURE 1.28

EXAMPLE 2   Finding *x*- and *y*-intercepts

Find the *x*- and *y*-intercepts for the graphs of the following equations.

(a) $y = x^3 - 4x$ (b) $y^2 - 3 = x$

SOLUTION

(a) Let $y = 0$. Then, $0 = x(x^2 - 4)$ has solutions $x = 0$ and $x = \pm 2$.

  x-intercepts: $(0, 0)$, $(2, 0)$, $(-2, 0)$

  Let $x = 0$. Then $y = 0$.

  y-intercept: $(0, 0)$

  (See Figure 1.29.)

(b) Let $y = 0$. Then $-3 = x$.

  x-intercept: $(-3, 0)$

  Let $x = 0$. Then $y^2 - 3 = 0$ has solutions $y = \pm\sqrt{3}$.

  y-intercepts: $(0, \sqrt{3})$, $(0, -\sqrt{3})$

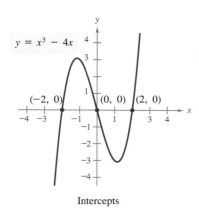

Intercepts

FIGURE 1.29

EXAMPLE 3   Sketching the graph of an equation

Sketch the graph of the equation $x^2 + 4y^2 = 16$.

SOLUTION

Sometimes it helps to rewrite an equation before calculating solution points. For example, if we rewrite the equation $x^2 + 4y^2 = 16$ as

$$x = \pm\sqrt{16 - 4y^2} = \pm 2\sqrt{4 - y^2}$$

then we can easily determine several solution points by choosing values for *y* and calculating the corresponding values for *x*. (Note that $x = \pm 2\sqrt{4 - y^2}$ is defined only when $|y| \le 2$.) By plotting these points and connecting them with a smooth curve, we create the graph shown in Figure 1.30. This particular graph is called an **ellipse.** It is one of the conic sections we will study in Chapter 11.

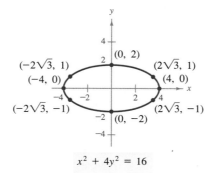

FIGURE 1.30 $x^2 + 4y^2 = 16$

Symmetry of a graph

The graphs shown in Figures 1.25(b) and 1.30 are said to be **symmetric** with respect to the y-axis. This means that if the Cartesian plane were folded along the y-axis, the portion of the graph to the left of the y-axis would coincide with the portion to the right of the y-axis. Another way to describe this symmetry is to say that the graph is a reflection of itself with respect to the y-axis. Symmetry with respect to the x-axis can be described similarly.

Knowing that a graph has symmetry *before* attempting to sketch it is helpful because then we need only half as many solution points as we would otherwise. We define three basic types of symmetry, as shown in Figure 1.31.

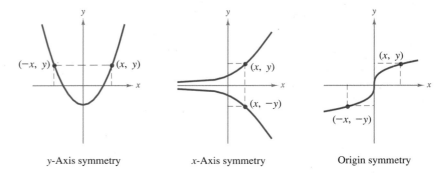

FIGURE 1.31 y-Axis symmetry x-Axis symmetry   Origin symmetry

DEFINITION OF SYMMETRY

A graph is said to be **symmetric with respect to the y-axis** if, whenever (x, y) is a point on the graph, $(-x, y)$ is also a point on the graph.

A graph is said to be **symmetric with respect to the x-axis** if, whenever (x, y) is a point on the graph, $(x, -y)$ is also a point on the graph.

A graph is said to be **symmetric with respect to the origin** if, whenever (x, y) is a point on the graph, $(-x, -y)$ is also a point on the graph.

REMARK   Note that a graph is symmetric with respect to the origin if a rotation of $180°$ (about the origin) leaves the graph unchanged.

Suppose we apply the definition of symmetry to the graph of the equation shown in Figure 1.25(b).

$$y = x^2 - 2 \qquad \text{Given equation}$$
$$y = (-x)^2 - 2 \qquad \text{Replace } x \text{ by } -x$$
$$y = x^2 - 2 \qquad \text{Equivalent equation}$$

Since substituting $-x$ for x produces an equivalent equation, it follows that if (x, y) is a solution point of the given equation, then $(-x, y)$ must also be a solution point. Therefore, the graph of $y = x^2 - 2$ is symmetric with respect to the y-axis.

A similar test can be made for symmetry with respect to the x-axis or the origin. These three tests are summarized as follows.

TESTS FOR SYMMETRY

1. The graph of an equation in x and y is symmetric with respect to the y-axis if replacing x by $-x$ yields an equivalent equation.
2. The graph of an equation in x and y is symmetric with respect to the x-axis if replacing y by $-y$ yields an equivalent equation.
3. The graph of an equation in x and y is symmetric with respect to the origin if replacing x by $-x$ *and* y by $-y$ yields an equivalent equation.

EXAMPLE 4 Testing for origin symmetry

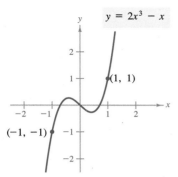

Origin Symmetry

FIGURE 1.32

Show that the graph of $y = 2x^3 - x$ is symmetric with respect to the origin.

SOLUTION

We apply the test for origin symmetry as follows.

$$y = 2x^3 - x \qquad \text{Given equation}$$
$$-y = 2(-x)^3 - (-x) \qquad \text{Replace } x \text{ by } -x \text{ and } y \text{ by } -y$$
$$-y = -2x^3 + x$$
$$y = 2x^3 - x \qquad \text{Equivalent equation}$$

Since the replacement produces an equivalent equation, we conclude that the graph of $y = 2x^3 - x$ is symmetric with respect to the origin, as shown in Figure 1.32.

EXAMPLE 5 Using symmetry to sketch a graph

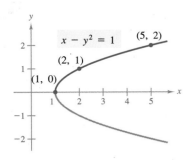

First, plot the points above the x-axis, then use symmetry to complete the graph.

FIGURE 1.33

Sketch the graph of $x - y^2 = 1$.

SOLUTION

The graph is symmetric with respect to the x-axis since replacing y by $-y$ yields

$$x - (-y)^2 = 1$$
$$x - y^2 = 1.$$

This means that the graph below the x-axis is a mirror image of the graph above the x-axis. Hence, we first sketch the graph above the x-axis and then reflect it to obtain the entire graph, as shown in Figure 1.33.

Points of intersection

Since each point of a graph is a solution point of its corresponding equation, a **point of intersection** of two graphs is simply a solution point that satisfies both equations. Moreover, the points of intersection of two graphs can be found by solving the equations simultaneously.

EXAMPLE 6   Finding points of intersection

Find all points of intersection of the graphs of

$$x^2 - y = 3 \qquad \text{and} \qquad x - y = 1.$$

SOLUTION

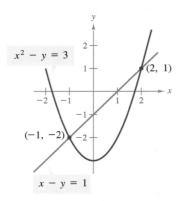

$x^2 - y = 3$

$(2, 1)$

$(-1, -2)$

$x - y = 1$

Two Points of Intersection

FIGURE 1.34

It is helpful to begin by making a sketch of each equation on the *same* coordinate plane, as shown in Figure 1.34. Having done this, it appears that the two graphs have two points of intersection. To find these two points, we proceed as follows.

$$y = x^2 - 3 \qquad \text{Solve first equation for } y$$
$$y = x - 1 \qquad \text{Solve second equation for } y$$
$$x^2 - 3 = x - 1 \qquad \text{Equate } y\text{-values}$$
$$x^2 - x - 2 = 0$$
$$(x - 2)(x + 1) = 0 \qquad \text{Solve for } x$$

The corresponding values of y are obtained by substituting $x = 2$ and $x = -1$ into either of the original equations. For instance, if we choose the equation $y = x - 1$, then the values of y are 1 and -2, respectively. Therefore, the two points of intersection are $(2, 1)$ and $(-1, -2)$.   ▭

Mathematical models

In applications we frequently use equations to form **mathematical models** of real-world phenomena. In developing a mathematical model to represent actual data, we strive for two (often conflicting) goals: accuracy and simplicity. That is, we want the model to be simple enough to be workable, yet accurate enough to produce meaningful results. Our next example describes a typical mathematical model.

EXAMPLE 7   A mathematical model

The median income (between 1955 and 1985) for married couples in the United States is given in Table 1.4. A mathematical model* for these data is given by

$$y = 0.033286t^2 - 0.130718t + 5.05716$$

where y represents the median income in thousands of dollars and t represents the year, with $t = 0$ corresponding to 1955. Using a graph, compare the data with the model and use the model to predict the median income for 1990.

*This model was developed using a procedure called the method of least squares. For a discussion of this method, see Section 15.9, Exercise 21.

TABLE 1.4

Year	1955	1960	1965	1970	1975	1980	1985
Income (in 1000s)	4.6	5.9	7.3	10.5	14.9	23.1	31.1

SOLUTION

Table 1.5 and Figure 1.35 compare the values given by the model with the actual values.

TABLE 1.5

t	0	5	10	15	20	25	30
y	5.1	5.2	7.1	10.6	15.8	22.6	31.1
Actual (income)	4.6	5.9	7.3	10.5	14.9	23.1	31.1

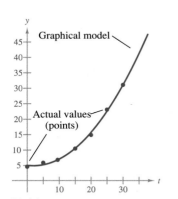

Model:
$y = 0.033286t^2 - 0.130718t + 5.05716$

FIGURE 1.35

To predict the median income for 1990, we let $t = 35$ and calculate y as follows.

$$y = 0.033286(35^2) - 0.130718(35) + 5.05716 \approx 41.3$$

Thus, we estimate that the 1990 median income will be $41,300. ▭

EXERCISES for Section 1.3

In Exercises 1–6, match the given equation with its graph. [Graphs are labeled (a)–(f).]

1. $y = x - 2$

2. $y = -\dfrac{1}{2}x + 2$

3. $y = x^2 + 2x$

4. $y = \sqrt{9 - x^2}$

5. $y = 4 - x^2$

6. $y = x^3 - x$

(a)
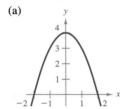

(b)
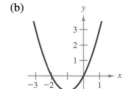

(c)
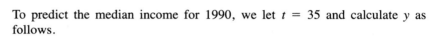

(d)
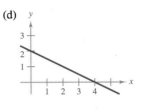

(e)
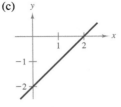

(f)
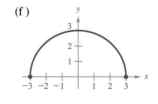

In Exercises 7–16, find the intercepts.

7. $y = 2x - 3$

8. $y = (x - 1)(x - 3)$

9. $y = x^2 + x - 2$

10. $y^2 = x^3 - 4x$

11. $y = x^2\sqrt{9 - x^2}$

12. $xy = 4$

13. $y = \dfrac{x - 1}{x - 2}$

14. $y = \dfrac{x^2 + 3x}{(3x + 1)^2}$

15. $x^2 y - x^2 + 4y = 0$

16. $y = 2x - \sqrt{x^2 + 1}$

In Exercises 17–26, check for symmetry with respect to both axes and to the origin.

17. $y = x^2 - 2$

18. $y = x^4 - x^2 + 3$

19. $x^2 y - x^2 + 4y = 0$

20. $x^2 y - x^2 - 4y = 0$

21. $y^2 = x^3 - 4x$

22. $xy^2 = -10$

23. $y = x^3 + x$

24. $xy = 1$

25. $y = \dfrac{x}{x^2 + 1}$

26. $y = x^3 + x - 3$

In Exercises 27–46, sketch the graph of each equation. Identify the intercepts and test for symmetry.

27. $y = x$

28. $y = x - 2$

29. $y = x + 3$

30. $y = 2x - 3$

31. $y = -3x + 2$

32. $y = -\dfrac{1}{2}x + 2$

33. $y = \dfrac{1}{2}x - 4$

34. $y = x^2 + 3$

35. $y = 1 - x^2$

36. $y = 2x^2 + x$

37. $y = -2x^2 + x + 1$

38. $y = x^3 - 1$

39. $y = x^3 + 2$

40. $y = \sqrt{9 - x^2}$

41. $x^2 + 4y^2 = 4$

42. $9x^2 + y^2 = 9$

43. $y = (x + 2)^2$

44. $x = y^2 - 4$

45. $y = \dfrac{1}{x}$

46. $y = 2x^4$

In Exercises 47–56, find the points of intersection of the graphs of the equations; check your results.

47. $x + y = 2,\ 2x - y = 1$

48. $2x - 3y = 13,\ 5x + 3y = 1$

49. $x + y = 7,\ 3x - 2y = 11$

50. $x^2 + y^2 = 25,\ 2x + y = 10$

51. $x^2 + y^2 = 5,\ x - y = 1$

52. $x^2 + y = 4,\ 2x - y = 1$

53. $y = x^3,\ y = x$

54. $y = x^4 - 2x^2 + 1,\ y = 1 - x^2$

55. $y = x^3 - 2x^2 + x - 1,\ y = -x^2 + 3x - 1$

56. $x = 3 - y^2,\ y = x - 1$

In Exercises 57 and 58, find the sales necessary to break even ($R = C$) for the given cost C of x units and the given revenue R obtained by selling x units.

57. $C = 8650x + 250{,}000$
   $R = 9950x$

58. $C = 5.5\sqrt{x} + 10{,}000$
   $R = 3.29x$

In Exercises 59–62, determine whether the points lie on the graph of the given equation.

59. Equation: $2x - y - 3 = 0$
   Points: $(1, 2),\ (1, -1),\ (4, 5)$

60. Equation: $y = \dfrac{44}{101{,}761}x^2$
   Points: $(0, 0),\ (-797.5, 275),\ (797.5, 275)$

61. Equation: $x^2 y - x^2 + 4y = 0$
   Points: $\left(1, \dfrac{1}{5}\right),\ \left(2, \dfrac{1}{2}\right),\ (-1, -2)$

62. Equation: $x^2 - xy + 4y = 3$
   Points: $(0, 2),\ \left(-2, -\dfrac{1}{6}\right),\ (3, -6)$

63. For what values of k does the graph of $y = kx^3$ pass through the given point?
   (a) $(1, 4)$ (b) $(-2, 1)$
   (c) $(0, 0)$ (d) $(-1, -1)$

64. For what values of k does the graph of $y^2 = 4kx$ pass through the given point?
   (a) $(1, 1)$ (b) $(2, 4)$
   (c) $(0, 0)$ (d) $(3, 3)$

65. The Consumer Price Index (CPI) for selected years is given in the following table.

Year	1970	1975	1980	1985	1987
CPI	116.3	161.2	246.8	322.2	333.9

A mathematical model for the CPI during this time period is

$$y = 0.1t^2 + 11.9t + 111.4$$

where y represents the CPI and t represents the year, with $t = 0$ corresponding to 1970.
   (a) Use a graph to compare the CPI with the model.
   (b) Use the model to predict the CPI for 1995.

66. From the model in Exercise 65, we obtain the model

$$V = \dfrac{1000}{t^2 + 119t + 1114}$$

where V represents the purchasing power of the dollar (in terms of constant 1967 dollars) and t represents the year, with $t = 0$ corresponding to 1970. Use the model to complete the following table.

t	0	5	10	15	20	25
V						

67. The farm population in the United States as a percentage of the total population for selected years is given in the following table.

Year	1950	1960	1970	1980	1985
Percentage	15.3	8.7	4.8	2.7	2.2

A mathematical model for these data is given by

$$y = \frac{1000}{11t + 27}$$

where y represents the percentage and t represents the year, with $t = 0$ corresponding to 1950.
(a) Use a graph to compare the actual percentage with that given by the model.
(b) Use the model to predict the farm percentage of the population in 1995.

68. The average number of acres per farm in the United States for selected years is given in the following table.

Year	1950	1960	1970	1980	1985
Number of acres	213	297	374	427	446

A mathematical model for these data is given by

$$y = -0.08t^2 + 9.69t + 211.79$$

where y represents the average acreage and t represents the year, with $t = 0$ corresponding to 1950.
(a) Use a graph to compare the actual number of acres per farm with that given by the model.
(b) Use the model to predict the average number of acres per farm in the United States in 1995.

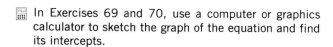 In Exercises 69 and 70, use a computer or graphics calculator to sketch the graph of the equation and find its intercepts.

69. $y = \dfrac{1}{16}(x^5 - 6x^4 + 9x^3 + 32)$

70. $y = \dfrac{5}{x^2 + 1} - 1$

71. Prove that if a graph is symmetric with respect to the x-axis and to the y-axis, then it is symmetric with respect to the origin. Give an example to show that the converse is not true.

72. Prove that if a graph is symmetric with respect to one axis and the origin, then it is symmetric with respect to the other axis also.

1.4 Lines in the Plane

The slope of a line ▪ Equations of lines ▪ Sketching the graph of a line ▪ Parallel lines ▪ Perpendicular lines

In Chapter 3, you will see that one of the primary problems in calculus is measuring (instantaneous) rates of change. In this section, we discuss the noncalculus version of this problem—measuring an *average* rate of change. We begin with an example.

Consider an automobile that is traveling at a *constant* rate on a straight highway. At 2:00 P.M. the car has traveled 20 miles from a particular city, and at 4:00 P.M. the car has traveled 132 miles, as shown in Figure 1.36. How fast is the car traveling?

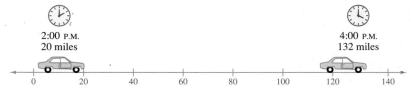

FIGURE 1.36

To measure the rate in *miles per hour*, we divide the distance traveled by the elapsed time.

$$\text{rate (mph)} = \frac{\text{distance (miles)}}{\text{time (hours)}} = \frac{132 - 20}{4 - 2} = \frac{112}{2} = 56 \text{ mph}$$

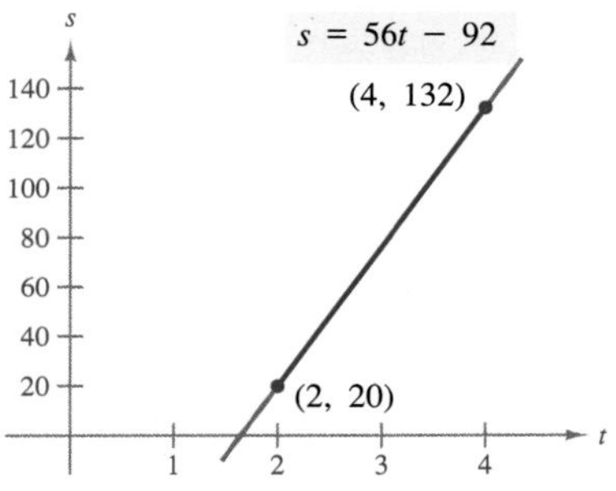

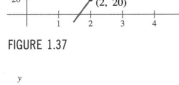

FIGURE 1.37

If we let s be the distance from the city in miles and t be the time in hours, then the distance is related to the time by the **linear equation**

$$s = 56t - 92, \qquad 2 \le t \le 4.$$

The graph of the equation $s = 56t - 92$ is a line (in this text, we use the term *line* to mean *straight line*), as shown in Figure 1.37. For every unit that t increases, the distance s increases 56 units. Mathematically, we say that this line has a *slope* of 56.

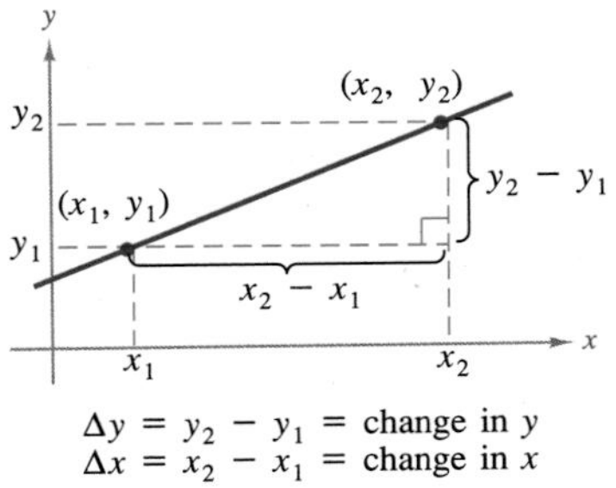

The slope of a line

By the **slope** of a (nonvertical) line, we mean the number of units a line rises (or falls) vertically for each unit of horizontal change from left to right. For instance, consider the two points (x_1, y_1) and (x_2, y_2) on the line in Figure 1.38. As we move from left to right along this line, a vertical change of $\Delta y = y_2 - y_1$ units corresponds to a horizontal change of $\Delta x = x_2 - x_1$ units. (Δ is the Greek uppercase letter *delta*, and the symbols Δy and Δx are read "delta y" and "delta x.") We use the ratio of Δy to Δx to define the slope of a line as follows.

$$\Delta y = y_2 - y_1 = \text{change in } y$$
$$\Delta x = x_2 - x_1 = \text{change in } x$$

FIGURE 1.38

DEFINITION OF THE SLOPE OF A LINE	The **slope** m of a nonvertical line passing through the points (x_1, y_1) and (x_2, y_2) is $$m = \frac{\Delta y}{\Delta x} = \frac{y_2 - y_1}{x_2 - x_1}, \quad x_1 \ne x_2.$$

REMARK   Note that

$$\frac{y_2 - y_1}{x_2 - x_1} = \frac{-(y_1 - y_2)}{-(x_1 - x_2)} = \frac{y_1 - y_2}{x_1 - x_2}.$$

Hence, it does not matter in which order we subtract *as long as* we are consistent and both "subtracted coordinates" come from the same point.

EXAMPLE 1   The slope of a line passing through two points

(a) The slope of the line containing $(-2, 0)$ and $(3, 1)$ is

$$m = \frac{1 - 0}{3 - (-2)} = \frac{1}{3 + 2} = \frac{1}{5}.$$

(b) The slope of the line containing $(-1, 2)$ and $(2, 2)$ is

$$m = \frac{2 - 2}{2 - (-1)} = \frac{0}{3} = 0.$$

(c) The slope of the line containing $(0, 4)$ and $(1, -1)$ is

$$m = \frac{-1 - 4}{1 - 0} = \frac{-5}{1} = -5.$$

(d) We do not define the slope of the vertical line containing $(3, 4)$ and $(3, 1)$.

(See Figure 1.39.)

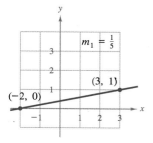

If m is positive, the line rises.

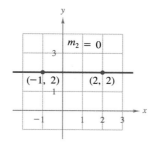

If m is zero, the line is horizontal.

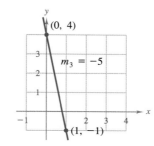

If m is negative, the line falls.

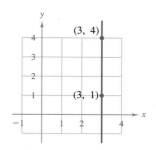

If the line is vertical, the slope is undefined.

FIGURE 1.39

REMARK   Note that we do not define the slope of a vertical line.

It is important to realize that *any* two points on a nonvertical line can be used to calculate its slope. This can be verified from the similar triangles shown in Figure 1.40. (See Exercise 77.) (Recall that the ratios of corresponding sides of similar triangles are equal.)

Equations of lines

If we know the slope of a line and one point on the line, how can we determine the equation of the line? Figure 1.40 leads us to the answer to this question. If (x_1, y_1) is a point lying on a line of slope m and (x, y) is any *other* point on the line, then

$$\frac{y - y_1}{x - x_1} = m.$$

This equation, involving the two variables x and y, can be rewritten in the form

$$y - y_1 = m(x - x_1)$$

which is called the **point-slope equation of a line.**

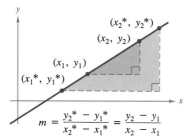

$$m = \frac{y_2{}^* - y_1{}^*}{x_2{}^* - x_1{}^*} = \frac{y_2 - y_1}{x_2 - x_1}$$

Any two points on a line can be used to determine its slope.

FIGURE 1.40

THEOREM 1.7 **POINT-SLOPE EQUATION** **OF A LINE**	The equation of the line with slope m passing through the point (x_1, y_1) is given by $$y - y_1 = m(x - x_1).$$

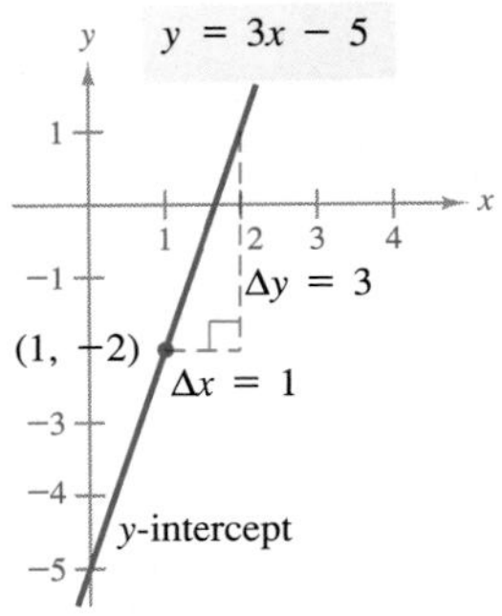

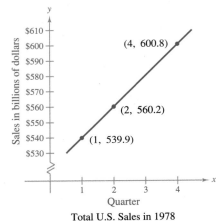

FIGURE 1.41

EXAMPLE 2   The point-slope equation of a line

Find an equation of the line that has a slope of 3 and passes through the point $(1, -2)$.

SOLUTION

$$y - y_1 = m(x - x_1) \qquad \text{Point-slope form}$$
$$y - (-2) = 3(x - 1)$$
$$y + 2 = 3x - 3$$
$$y = 3x - 5$$

(See Figure 1.41.)

EXAMPLE 3   An application: total U.S. sales

The total U.S. sales (including inventories) during the first two quarters of 1978 were 539.9 and 560.2 billion dollars, respectively. Assuming a *linear growth pattern*, estimate the total sales during the fourth quarter of 1978.

SOLUTION

Referring to Figure 1.42, we let $(1, 539.9)$ and $(2, 560.2)$ be two points on the line representing total U.S. sales. We let x represent the quarter and y represent the sales in billions of dollars. The slope of the line passing through these two points is

$$m = \frac{560.2 - 539.9}{2 - 1} = 20.3.$$

Thus, the equation of the line is

$$y - y_1 = m(x - x_1)$$
$$y - 539.9 = 20.3(x - 1)$$
$$y = 20.3(x - 1) + 539.9$$
$$y = 20.3x + 519.6.$$

Now, using this linear model, we estimate the fourth quarter sales ($x = 4$) to be

$$y = (20.3)(4) + 519.6 = 600.8 \text{ billion dollars.}$$

(In this particular case, the estimate proves to be quite good. The actual fourth quarter sales in 1978 were 600.5 billion dollars.)

FIGURE 1.42

Total U.S. Sales in 1978

REMARK   The estimation method illustrated in Example 3 is called **linear extrapolation.** Note that the estimated point does not lie between the two given points. When the estimated point lies between the two given points, we call the procedure **linear interpolation.**

Sketching the graph of a line

In Section 1.2, we mentioned that many problems in analytic geometry can be classified in two basic categories: (1) Given a graph, what is its equation? and (2) Given an equation, what is its graph? The point-slope equation of a line fits in the first category. However, this form is *not* particularly useful for solving problems in the second category. The form that is best suited to sketching the graph of a line is called the **slope-intercept** form for the equation of a line.

THEOREM 1.8 THE SLOPE-INTERCEPT EQUATION OF A LINE	The graph of the equation $$y = mx + b$$ is a line having a *slope* of m and a *y-intercept* at $(0, b)$.

EXAMPLE 4   Sketching lines in the plane

Sketch the graphs of the following linear equations.

(a) $y = 2x + 1$ (b) $y = 2$ (c) $3y + x - 6 = 0$

SOLUTION

(a) Since $b = 1$, the y-intercept occurs at $(0, 1)$, and since the slope is $m = 2$, we know that this line rises 2 units for each unit it moves to the right. (See Figure 1.43(a).)

(b) Since $b = 2$, the y-intercept occurs at $(0, 2)$, and since the slope is $m = 0$, we know that the line is horizontal. That is, it doesn't rise or fall. (See Figure 1.43(b).)

(c) We begin by writing the equation in slope-intercept form.

$$3y + x - 6 = 0$$
$$3y = -x + 6$$
$$y = -\frac{1}{3}x + 2$$

Thus, the y-intercept occurs at $(0, 2)$ and the slope is $m = -\frac{1}{3}$. This means that the line falls 1 unit for every 3 units it moves to the right. (See Figure 1.43(c).)

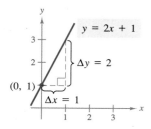

(a) $m = 2$, line rises.

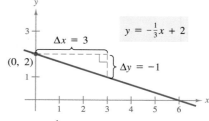

(b) $m = 0$, line is horizontal.   (c) $m = -\frac{1}{3}$, line falls.

FIGURE 1.43

Since the slope of a vertical line is not defined, its equation cannot be written in the slope-intercept form. However, the equation of *any* line can be written in the **general form**

$$Ax + By + C = 0$$

where A and B are not *both* zero. For instance, the vertical line given by $x = a$ can be represented by the general form $x - a = 0$. We summarize the five most common forms of equations of lines in the following list.

SUMMARY OF EQUATIONS OF LINES

1. General form: $Ax + By + C = 0$
2. Vertical line: $x = a$
3. Horizontal line: $y = b$
4. Point-slope form: $y - y_1 = m(x - x_1)$
5. Slope-intercept form: $y = mx + b$

Parallel and perpendicular lines

The slope of a line is a convenient tool for determining whether two lines are parallel or perpendicular. This is seen in the following two theorems.

THEOREM 1.9 PARALLEL LINES

Two distinct nonvertical lines are parallel if and only if their slopes are equal.

THEOREM 1.10 PERPENDICULAR LINES

Two nonvertical lines are perpendicular if and only if their slopes are related by the following equation.

$$m_1 = -\frac{1}{m_2}$$

PROOF We will prove only one direction of the theorem and leave the other direction as an exercise (see Exercise 78). Let us assume that we are given two nonvertical perpendicular lines L_1 and L_2 with slopes m_1 and m_2. For simplicity's sake let these two lines intersect at the origin, as shown in Figure 1.44. The vertical line $x = 1$ will intersect L_1 and L_2 at the respective points $(1, m_1)$ and $(1, m_2)$. Since the triangle formed by these two points and the origin is a right triangle, we can apply the Pythagorean Theorem and conclude that

$$\left(\begin{array}{c}\text{distance between} \\ (0,0) \text{ and } (1, m_1)\end{array}\right)^2 + \left(\begin{array}{c}\text{distance between} \\ (0,0) \text{ and } (1, m_2)\end{array}\right)^2 = \left(\begin{array}{c}\text{distance between} \\ (1, m_1) \text{ and } (1, m_2)\end{array}\right)^2.$$

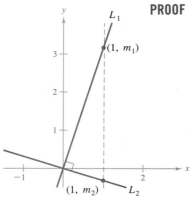

The slopes of perpendicular lines are negative reciprocals of each other.

FIGURE 1.44

Using the Distance Formula, we have

$$(\sqrt{1 + m_1{}^2})^2 + (\sqrt{1 + m_2{}^2})^2 = (\sqrt{0^2 + (m_1 - m_2)^2})^2$$
$$1 + m_1{}^2 + 1 + m_2{}^2 = (m_1 - m_2)^2$$
$$2 + m_1{}^2 + m_2{}^2 = m_1{}^2 - 2m_1m_2 + m_2{}^2$$
$$2 = -2m_1m_2$$
$$-\frac{1}{m_2} = m_1.$$

EXAMPLE 5   Finding parallel and perpendicular lines

Find an equation for the line that passes through the point $(2, -1)$ and is
(a) parallel to the line $2x - 3y = 5$
(b) perpendicular to the line $2x - 3y = 5$.

SOLUTION

Writing the equation $2x - 3y = 5$ in slope-intercept form, we have

$$y = \frac{2}{3}x - \frac{5}{3}.$$

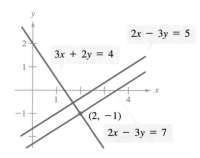

FIGURE 1.45

Therefore, the given line has a slope of $m = \frac{2}{3}$.

(a) The line through $(2, -1)$ that is parallel to the given line has an equation of the form

$$y - (-1) = \frac{2}{3}(x - 2)$$
$$3(y + 1) = 2(x - 2)$$
$$2x - 3y = 7.$$

(See Figure 1.45.) (Note the similarity to the original equation.)

(b) Using the negative reciprocal of the slope of the given line, we find the slope of a line perpendicular to the given line to be $-\frac{3}{2}$. Therefore, the line through the point $(2, -1)$ that is perpendicular to the given line has the equation

$$y - (-1) = -\frac{3}{2}(x - 2)$$
$$2(y + 1) = -3(x - 2)$$
$$3x + 2y = 4.$$

(See Figure 1.45.) ▭

EXERCISES for Section 1.4

In Exercises 1–6, estimate the slope of the given line from its graph.

1.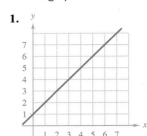

2.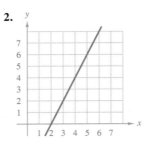

3.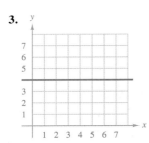

4.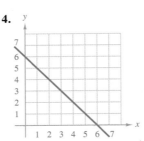

5.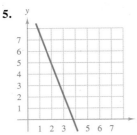

6.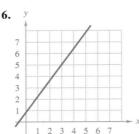

In Exercises 7–12, plot the given pair of points and find the slope of the line passing through them.

7. $(3, -4)$, $(5, 2)$

8. $(-2, 1)$, $(4, -3)$

9. $\left(\frac{1}{2}, 2\right)$, $(6, 2)$

10. $(2, 1)$, $(2, 5)$

11. $(1, 2)$, $(-2, 4)$

12. $\left(\frac{7}{8}, \frac{3}{4}\right)$, $\left(\frac{5}{4}, -\frac{1}{4}\right)$

In Exercises 13–16, use the given point on the line and the slope of the line to find three additional points that the line passes through. (The solution is not unique.)

Point	Slope	Point	Slope
13. $(2, 1)$	$m = 0$	**14.** $(-3, 4)$	m undefined
15. $(1, 7)$	$m = -3$	**16.** $(-2, -2)$	$m = 2$

In Exercises 17–20, find the slope and y-intercept (if possible) of the line specified by the given equation.

17. $x + 5y = 20$

18. $6x - 5y = 15$

19. $x = 4$

20. $y = -1$

In Exercises 21–26, find an equation for the line that passes through the given points, and sketch the graph of the line.

21. $(2, 1)$, $(0, -3)$

22. $(-3, -4)$, $(1, 4)$

23. $(0, 0)$, $(-1, 3)$

24. $(-3, 6)$, $(1, 2)$

25. $(1, -2)$, $(3, -2)$

26. $\left(\frac{7}{8}, \frac{3}{4}\right)$, $\left(\frac{5}{4}, -\frac{1}{4}\right)$

In Exercises 27–32, find an equation of the line that passes through the given point and has the indicated slope. Sketch the line.

Point	Slope	Point	Slope
27. $(0, 3)$	$m = \frac{3}{4}$	**28.** $(-1, 2)$	m undefined
29. $(0, 0)$	$m = \frac{2}{3}$	**30.** $(-2, 4)$	$m = -\frac{3}{5}$
31. $(0, 2)$	$m = 4$	**32.** $(0, 4)$	$m = 0$

33. Find an equation of the vertical line with x-intercept at 3.

34. Show that the line with intercepts $(a, 0)$ and $(0, b)$ has the following equation.

$$\frac{x}{a} + \frac{y}{b} = 1, \quad a \neq 0, b \neq 0$$

In Exercises 35–40, use the result of Exercise 34 to write an equation of the indicated line.

35. x-intercept: $(2, 0)$
y-intercept: $(0, 3)$

36. x-intercept: $(-3, 0)$
y-intercept: $(0, 4)$

37. x-intercept: $\left(-\frac{1}{6}, 0\right)$
y-intercept: $\left(0, -\frac{2}{3}\right)$

38. x-intercept: $\left(-\frac{2}{3}, 0\right)$
y-intercept: $(0, -2)$

39. Point on line: $(1, 2)$
x-intercept: $(a, 0)$
y-intercept: $(0, a)$
$(a \neq 0)$

40. Point on line: $(-3, 4)$
x-intercept: $(a, 0)$
y-intercept: $(0, a)$
$(a \neq 0)$

In Exercises 41–46, write an equation of the line through the given point (a) parallel to the given line and (b) perpendicular to the given line.

Point	Line
41. $(2, 1)$	$4x - 2y = 3$
42. $(-3, 2)$	$x + y = 7$
43. $\left(\dfrac{7}{8}, \dfrac{3}{4}\right)$	$5x + 3y = 0$
44. $(-6, 4)$	$3x + 4y = 7$
45. $(2, 5)$	$x = 4$
46. $(-1, 0)$	$y = -3$

In Exercises 47–52, sketch the graph of the equation.

47. $y = -3$ **48.** $x = 4$
49. $2x - y - 3 = 0$ **50.** $x + 2y + 6 = 0$
51. $y = -2x + 1$ **52.** $y - 1 = 3(x + 4)$

In Exercises 53 and 54, find an equation of the line determined by the points of intersection of the graphs of the parabolas.

53. $y = x^2$
   $y = 4x - x^2$

54. $y = x^2 - 4x + 3$
   $y = -x^2 + 2x + 3$

In Exercises 55 and 56, determine whether the three given points are collinear (lie on the same straight line).

55. $(-2, 1), (-1, 0), (2, -2)$
56. $(0, 4), (7, -6), (-5, 11)$

In Exercises 57–60, refer to the triangle in the accompanying figure.

FIGURE FOR 57–60

57. Find the coordinates of the point of intersection of the perpendicular bisectors of the sides.
58. Find the coordinates of the point of intersection of the medians.
59. Find the coordinates of the point of intersection of the altitudes.
60. Show that the points of intersection of Exercises 57, 58, and 59 are collinear.

61. Find an equation of the line giving the relationship between the temperature in degrees Celsius C and degrees Fahrenheit F. Use the fact that water freezes at 0° Celsius (32° Fahrenheit) and boils at 100° Celsius (212° Fahrenheit).

62. Use the result of Exercise 61 to complete the following table.

C		$-10°$	$10°$			$177°$
F	$0°$			$68°$	$90°$	

63. A company reimburses its sales representatives $95 per day for lodging and meals plus 25¢ per mile driven. Write a linear equation giving the daily cost C to the company in terms of x, the number of miles driven.

64. A manufacturing company pays its assembly line workers $9.50 per hour *plus* an additional piecework rate of $0.75 per unit produced. Find a linear equation for the hourly wages W in terms of x, the number of units produced per hour.

65. A small business purchases a piece of equipment for $875. After 5 years the equipment will be obsolete and have no value. Write a linear equation giving the value y of the equipment during the 5 years it will be used. (Let t represent the time in years.)

66. A company constructs a warehouse for $825,000. It has an estimated useful life of 25 years, after which its value is expected to be $75,000. Use straight-line depreciation to write a linear equation giving the value y of the warehouse during its 25 years of useful life. (Let t represent the time in years.)

67. A real estate office handles an apartment complex with 50 units. When the rent is $380 per month, all 50 units are occupied. However, when the rent is $425, the average number of occupied units drops to 47. Assume that the relationship between the monthly rent p and the demand x is linear. (Note: Here we use the term *demand* to refer to the number of occupied units.)
   (a) Write a linear equation giving the quantity demanded x in terms of the rent p.
   (b) (Linear extrapolation) Use this equation to predict the number of units occupied if the rent is raised to $455.
   (c) (Linear interpolation) Predict the number of units occupied if the rent is lowered to $395.

68. The number of subscribers to cable TV for the years 1980 and 1986 were 16 million and 37.5 million, respectively. Assume that the relationship between the year t and the number of subscribers y is linear.
   (a) Write the equation giving the number of subscribers y in terms of t. (Let $t = 0$ represent 1980.)
   (b) [Linear extrapolation] Use this equation to estimate the number of subscribers in 1990.
   (c) [Linear interpolation] Estimate the number of subscribers in 1985.
   (d) What information is given by the slope of the line in part (a)?

In Exercises 69–74, find the distance between the given point and line (or two lines) using the following formula for the distance between the point (x_1, y_1) and the line $Ax + By + C = 0$.

$$\frac{|Ax_1 + By_1 + C|}{\sqrt{A^2 + B^2}}$$

69. Point: $(0, 0)$; line: $4x + 3y = 10$
70. Point: $(2, 3)$; line: $4x + 3y = 10$
71. Point: $(-2, 1)$; line: $x - y - 2 = 0$
72. Point: $(6, 2)$; line: $x = -1$
73. Lines: $x + y = 1$, $x + y = 5$
74. Lines: $3x - 4y = 1$, $3x - 4y = 10$

75. Prove that the diagonals of a rhombus intersect at right angles.

76. Prove that the figure formed by connecting consecutive midpoints of the sides of any quadrilateral is a parallelogram.

77. Prove that if the points (x_1, y_1) and (x_2, y_2) lie on the same line as $(x_1{}^*, y_1{}^*)$ and $(x_2{}^*, y_2{}^*)$, then

$$\frac{y_2{}^* - y_1{}^*}{x_2{}^* - x_1{}^*} = \frac{y_2 - y_1}{x_2 - x_1}.$$

Assume $x_1 \neq x_2$ and $x_2{}^* \neq x_1{}^*$.

78. Complete the proof of Theorem 1.10. That is, prove that if the slopes of two nonvertical lines are negative reciprocals of each other, then the lines are perpendicular.

1.5   Functions

Definition of function ▪ Function notation ▪ The graph of a function ▪ Transformations of graphs ▪ Classifications of functions ▪ Combinations of functions

Many common relationships involve two variables in such a way that the value of one of the variables depends on the value of the other. For example, the sales tax on an item depends on its selling price. The distance an object moves in a given time depends on its speed.

Consider the relationship between the area of a circle and its radius. This relationship can be expressed by the equation $A = \pi r^2$, where the value of A depends on the choice of r. We refer to A as the **dependent variable** and to r as the **independent variable.**

Of particular interest are relationships such that to every value of the independent variable there corresponds *one and only one* value of the dependent variable. We call this type of correspondence a **function.**

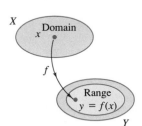

FIGURE 1.46

DEFINITION OF A FUNCTION

A **function** f from a set X into a set Y is a correspondence that assigns to each element x in X exactly one element y in Y. We call y the **image** of x under f and denote it by $f(x)$. The **domain** of f is the set X, and the **range** consists of all images of elements in X. (See Figure 1.46.)

If to each value in its range there corresponds exactly one value in its domain, the function is said to be **one-to-one.** Moreover, if the range of f consists of all of Y, then the function is said to be **onto.**

In the first twelve chapters of this text, we work with functions whose domains and ranges are sets of real numbers. We call such functions **real-valued functions of a real variable.** Other types of functions will be introduced in Chapters 13–17.

Functions can be specified in a variety of ways. We will, however, concentrate primarily on functions that are given by equations involving the dependent and independent variables. To evaluate a function described by an equation, we generally isolate the dependent variable on the left side of the equation. For instance, the equation $x + 2y = 1$, written as

$$y = \frac{1 - x}{2}$$

describes y as a function of x, and we can denote this function as

$$f(x) = \frac{1 - x}{2}.$$

This function notation has the advantage of clearly identifying the dependent variable as $f(x)$ while at the same time telling us that x is the independent variable and that the function itself will be called "f." The symbol $f(x)$ is read "f of x." The $f(x)$ notation also allows us to be less wordy. Instead of asking, "What is the value of y that corresponds to $x = 3$?" we can ask, "What is $f(3)$?" In general, to denote the value of the dependent variable when $x = a$, we use the symbol $f(a)$. For example, the value of f when $x = 3$ is

$$f(3) = \frac{1 - (3)}{2} = \frac{-2}{2} = -1.$$

In an equation that defines a function, the role of the variable x is simply that of a placeholder. For instance, the function given by

$$f(x) = 2x^2 - 4x + 1$$

can be described by the form

$$f(\ ) = 2(\ )^2 - 4(\ ) + 1$$

where parentheses are used instead of x. Therefore, to evaluate $f(-2)$, we simply place -2 in each set of parentheses.

$$f(-2) = 2(-2)^2 - 4(-2) + 1 = 2(4) + 8 + 1 = 17$$

REMARK   Although we generally use f as a convenient function name and x as the independent variable, we also can use other symbols. For instance, the following equations all define the same function.

$$f(x) = x^2 - 4x + 7 \qquad f(t) = t^2 - 4t + 7 \qquad g(s) = s^2 - 4s + 7$$

EXAMPLE 1 Evaluating a function

For the function f defined by $f(x) = x^2 + 7$, evaluate the following.

(a) $f(3a)$ (b) $f(b - 1)$ (c) $\dfrac{f(x + \Delta x) - f(x)}{\Delta x}$, $\Delta x \neq 0$

SOLUTION

(a) $f(3a) = (3a)^2 + 7$ Replace x with $3a$

    $= 9a^2 + 7$

(b) $f(b - 1) = (b - 1)^2 + 7$ Replace x with $b - 1$

      $= b^2 - 2b + 1 + 7$

      $= b^2 - 2b + 8$

$f(x) = x^2 + 7$

(c) $\dfrac{f(x + \Delta x) - f(x)}{\Delta x} = \dfrac{[(x + \Delta x)^2 + 7] - [x^2 + 7]}{\Delta x}$

      $= \dfrac{x^2 + 2x\Delta x + (\Delta x)^2 + 7 - x^2 - 7}{\Delta x}$

      $= \dfrac{2x\Delta x + (\Delta x)^2}{\Delta x}$

      $= \dfrac{\Delta x(2x + \Delta x)}{\Delta x}$

      $= 2x + \Delta x$ ▭

REMARK   The ratio in Example 1(c) is called a difference quotient and has a special significance in calculus. We will say more about this in Chapter 3.

The domain of a function may be described explicitly, or it may be described *implicitly* by an equation used to define the function. (The implied domain is the set of all real numbers for which the equation is defined.) For example, the function given by

$$f(x) = \frac{1}{x^2 - 4}, \quad 4 \le x \le 5$$

has an explicitly defined domain given by $\{x: 4 \le x \le 5\}$. On the other hand, the function given by

$$g(x) = \frac{1}{x^2 - 4}$$

has an implied domain which is the set $\{x: x \ne \pm 2\}$. Another common type of implied domain is that used to avoid even roots of negative numbers. For example, the function given by

$$f(x) = \sqrt{x + 2}$$

has the implied domain $\{x: x \ge -2\}$.

EXAMPLE 2   Finding the domain and range of a function

Determine the domain and range for the function of x defined by

$$f(x) = \sqrt{x - 1}.$$

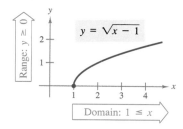

FIGURE 1.47

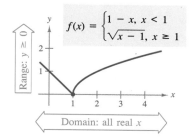

FIGURE 1.48

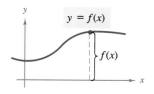

The Graph of a Function

FIGURE 1.49

SOLUTION

Since $\sqrt{x - 1}$ is not defined for $x - 1 < 0$ (that is, for $x < 1$), we must have $x \geq 1$. Therefore, the domain is the interval $[1, \infty)$.

To find the range, we observe that $f(x) = \sqrt{x - 1}$ is never negative. Moreover, as x takes on the various values in the domain, $f(x)$ takes on all nonnegative values and we find the range to be the interval $[0, \infty)$.

The graph of the function is shown in Figure 1.47.   ▭

EXAMPLE 3   A function defined by more than one equation

Determine the domain and range for the function of x given by

$$f(x) = \begin{cases} 1 - x, & \text{if } x < 1 \\ \sqrt{x - 1}, & \text{if } x \geq 1. \end{cases}$$

SOLUTION

Since f is defined for $x < 1$ and $x \geq 1$, the domain of the function is the entire set of real numbers.

On the portion of the domain for which $x \geq 1$, the function behaves as in Example 2. For $x < 1$, the value of $1 - x$ is positive, and therefore the range of the function is the interval $[0, \infty)$. (See Figure 1.48.)   ▭

REMARK   Note that the function given in Example 2 is one-to-one whereas the function given in Example 3 is not one-to-one.

The graph of a function

As you study this section, remember that the graph of the function $y = f(x)$ consists of all points $(x, f(x))$ as shown in Figure 1.49, where

> $x =$ the directed distance from the y-axis
> $f(x) =$ the directed distance from the x-axis.

Since, by the definition of a function, there is exactly one y-value for each x-value, it follows that a vertical line can intersect the graph of a function of x at most once. This observation provides us with a convenient visual test for functions. For example, in part (a) of Figure 1.50, we see that the graph does not define y as a function of x since a vertical line intersects the graph twice.

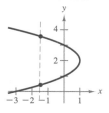

(a) Not a function of x

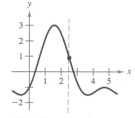

(b) A function of x

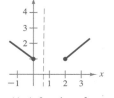

(c) A function of x

FIGURE 1.50

Vertical Line Test for Functions

Figure 1.51 shows the graphs of six basic functions. You need to know these graphs well.

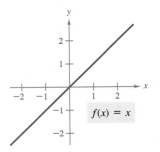

(a) Identity function

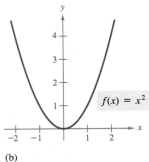

(b)

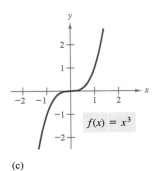

(c)

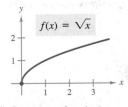

(d) Square root function

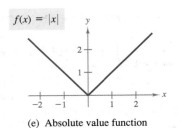

(e) Absolute value function

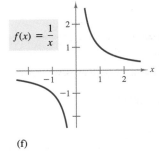

(f)

FIGURE 1.51

Function notation lends itself well to describing transformations of graphs in the plane. Some families of graphs all have the same basic shape. For example, consider the graph of $y = x^2$, as shown in Figure 1.52. Now compare this graph to those shown in Figure 1.53.

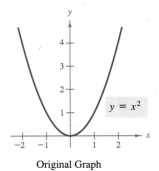

Original Graph

FIGURE 1.52

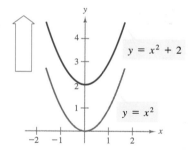

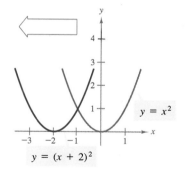

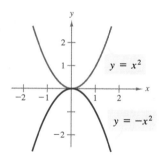

(a) Vertical shift upward   (b) Horizontal shift to the left   (c) Reflection

FIGURE 1.53

Each of the graphs in Figure 1.53 is a **transformation** of the graph of $y = x^2$. The three basic types of transformations illustrated by these graphs are (1) vertical shifts, (2) horizontal shifts, and (3) reflections.

BASIC TYPES OF TRANSFORMATIONS ($c > 0$)		
	Original graph:	$y = f(x)$
	Horizontal shift c units to the **right:**	$y = f(x - c)$
	Horizontal shift c units to the **left:**	$y = f(x + c)$
	Vertical shift c units **downward:**	$y = f(x) - c$
	Vertical shift c units **upward:**	$y = f(x) + c$
	Reflection (about the x-axis):	$y = -f(x)$

Classifications and combinations of functions

The modern notion of a function is derived from the efforts of many seventeenth- and eighteenth-century mathematicians. Of particular note was Leonhard Euler (1707–1783), to whom we are indebted for the function notation $y = f(x)$. By the end of the eighteenth century, mathematicians and scientists had concluded that most real-world phenomena can be represented by mathematical models taken from a basic collection of functions called **elementary functions.** Elementary functions are divided into three categories: (1) algebraic, (2) logarithmic and exponential, and (3) trigonometric. We will study the logarithmic and exponential functions in Chapter 7 and the trigonometric functions in Chapter 8.

The most common type of algebraic function is a **polynomial function**

$$f(x) = a_n x^n + a_{n-1}x^{n-1} + \cdots + a_2 x^2 + a_1 x + a_0, \quad a_n \neq 0$$

where the positive integer n is the **degree** of the polynomial function. The numbers a_i are called **coefficients,** with a_n the **leading coefficient** and a_0 the **constant term** of the polynomial function. It is common practice to use subscript notation for coefficients of general polynomial functions, but for polynomial functions of low degree we often use the following simpler forms.

Leonhard Euler

Zeroth degree:	$f(x) = a$	Constant function
First degree:	$f(x) = ax + b$	Linear function
Second degree:	$f(x) = ax^2 + bx + c$	Quadratic function
Third degree:	$f(x) = ax^3 + bx^2 + cx + d$	Cubic function

Although the graph of a polynomial function can have several turns, eventually the graph will rise or fall without bound as x moves to the right or left. Whether the graph of

$$f(x) = a_n x^n + a_{n-1} x^{n-1} + \cdots + a_2 x^2 + a_1 x + a_0$$

eventually rises or falls can be determined by the function's degree (odd or even) and by the leading coefficient a_n, as indicated in Figure 1.54. Note that the dashed portions of the graphs indicate that the **leading coefficient test** determines *only* the right and left behavior of the graph.

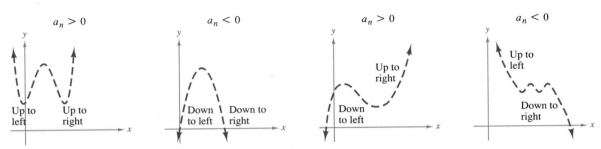

Graphs of polynomial functions of even degree   Graphs of polynomial functions of odd degree

FIGURE 1.54   Leading Coefficient Test for Polynomial Functions

Just as a rational number can be written as the quotient of two integers, a **rational function** can be written as the quotient of two polynomials. Specifically, a function f is **rational** if it has the form

$$f(x) = \frac{p(x)}{q(x)}, \quad q(x) \neq 0$$

where $p(x)$ and $q(x)$ are polynomials.

Polynomial functions and rational functions are two examples of a larger class of functions called **algebraic functions**. An algebraic function is one that can be expressed as a finite number of sums, differences, multiples, quotients, and radicals involving x^n. For example, the following functions are algebraic.

$$f(x) = \sqrt{x + 1} \quad \text{and} \quad g(x) = x + \frac{1}{\sqrt[3]{x + 1}}$$

Functions that are not algebraic are called **transcendental.** For instance, the trigonometric functions discussed in Section 8.1 are transcendental.

Two functions can be combined in various ways to create new functions. For example, if

$$f(x) = 2x - 3 \quad \text{and} \quad g(x) = x^2 + 1$$

we can form the following functions.

$$f(x) + g(x) = (2x - 3) + (x^2 + 1) = x^2 + 2x - 2 \qquad \text{Sum}$$
$$f(x) - g(x) = (2x - 3) - (x^2 + 1) = -x^2 + 2x - 4 \qquad \text{Difference}$$
$$f(x)g(x) = (2x - 3)(x^2 + 1) = 2x^3 - 3x^2 + 2x - 3 \qquad \text{Product}$$
$$\frac{f(x)}{g(x)} = \frac{2x - 3}{x^2 + 1} \qquad \text{Quotient}$$

We can combine two functions in yet another way to form what is called a **composite function.**

DEFINITION OF COMPOSITE FUNCTION	Let f and g be functions. The function given by $(f \circ g)(x) = f(g(x))$ is called the **composite** of f with g. The domain of $f \circ g$ is the set of all x in the domain of g such that $g(x)$ is in the domain of f.

It is important to realize that the composite of f with g may not be equal to the composite of g with f. This is illustrated in the following example.

EXAMPLE 4   Composition of functions

Given $f(x) = 2x - 3$ and $g(x) = x^2 + 1$, find $f \circ g$ and $g \circ f$.

SOLUTION

Since $f(x) = 2x - 3$, we have

$$(f \circ g)(x) = f(g(x)) = 2(g(x)) - 3 = 2(x^2 + 1) - 3 = 2x^2 - 1$$

and since $g(x) = x^2 + 1$, we have

$$(g \circ f)(x) = g(f(x)) = (f(x))^2 + 1 = (2x - 3)^2 + 1 = 4x^2 - 12x + 10.$$

Note that $(f \circ g)(x) \neq (g \circ f)(x)$. ▭

In Section 1.3, we defined an x-intercept of a graph to be a point $(a, 0)$ at which the graph crosses the x-axis. If the graph represents a function f, then the number a is called a **zero** of f. In other words, the zeros of a function f are the solutions of the equation $f(x) = 0$. For example, the function $f(x) = x - 4$ has a zero at $x = 4$ because $f(4) = 0$.

In Section 1.3 we also discussed different types of symmetry. In the terminology of functions, we say that a function is **even** if its graph is symmetric with respect to the y-axis, and a function is **odd** if its graph is symmetric with respect to the origin. Thus, the symmetry tests in Section 1.3 yield the following test for even and odd functions.

THEOREM 1.11 TEST FOR EVEN AND ODD FUNCTIONS	The function $y = f(x)$ is **even** if $f(-x) = f(x)$. The function $y = f(x)$ is **odd** if $f(-x) = -f(x)$.

REMARK   Except for such trivial cases as the constant function $f(x) = 0$, the graph of a function cannot have symmetry with respect to the x-axis because it then would fail the vertical line test for the graph of a function.

EXAMPLE 5   Even and odd functions

Determine whether the following functions are even, odd, or neither. In each case find the zeros of the function.

(a) $f(x) = x^3 - x$ (b) $g(x) = x^2 + 1$

SOLUTION

(a) This function is odd since

$$f(-x) = (-x)^3 - (-x) = -x^3 + x = -(x^3 - x) = -f(x).$$

The zeros of f are found as follows.

$$x^3 - x = 0 \qquad \text{Let } f(x) = 0$$
$$x(x^2 - 1) = x(x - 1)(x + 1) = 0 \qquad \text{Factor}$$
$$x = 0, 1, -1 \qquad \text{Zeros of } f$$

(b) This function is even since

$$g(-x) = (-x)^2 + 1 = x^2 + 1 = g(x).$$

It has no zeros since $x^2 + 1$ is positive for all x. (See Figure 1.55.)

REMARK   Each of the functions in Example 5 is either even or odd. However, some functions such as

$$f(x) = x^2 + x + 1$$

are neither even nor odd.

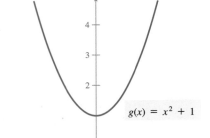

(a) Odd function

(b) Even function

FIGURE 1.55

EXERCISES for Section 1.5

1. Given $f(x) = 2x - 3$, find the following.
   (a) $f(0)$ (b) $f(-3)$
   (c) $f(b)$ (d) $f(x - 1)$
2. Given $f(x) = x^2 - 2x + 2$, find the following.
   (a) $f\left(\dfrac{1}{2}\right)$ (b) $f(-1)$
   (c) $f(c)$ (d) $f(x + \Delta x)$
3. Given $f(x) = \sqrt{x + 3}$, find the following.
   (a) $f(-2)$ (b) $f(6)$
   (c) $f(c)$ (d) $f(x + \Delta x)$

4. Given $f(x) = 1/\sqrt{x}$, find the following.
   (a) $f(2)$ (b) $f\left(\dfrac{1}{4}\right)$
   (c) $f(x + \Delta x)$ (d) $f(x + \Delta x) - f(x)$
5. Given $f(x) = |x|/x$, find the following.
   (a) $f(2)$ (b) $f(-2)$
   (c) $f(x^2)$ (d) $f(x - 1)$
6. Given $f(x) = |x| + 4$, find the following.
   (a) $f(2)$ (b) $f(-2)$
   (c) $f(x^2)$ (d) $f(x + \Delta x) - f(x)$

7. Given $f(x) = x^2 - x + 1$, find

$$\frac{f(2 + \Delta x) - f(2)}{\Delta x}.$$

8. Given $f(x) = 1/x$, find

$$\frac{f(1 + \Delta x) - f(1)}{\Delta x}.$$

9. Given $f(x) = x^3$, find

$$\frac{f(x + \Delta x) - f(x)}{\Delta x}.$$

10. Given $f(x) = 3x - 1$, find

$$\frac{f(x) - f(1)}{x - 1}.$$

11. Given $f(x) = 1/\sqrt{x - 1}$, find

$$\frac{f(x) - f(2)}{x - 2}.$$

12. Given $f(x) = x^3 - x$, find

$$\frac{f(x) - f(1)}{x - 1}.$$

In Exercises 13–22, find the domain and range of the given function, and sketch its graph.

13. $f(x) = 4 - x$ **14.** $f(x) = \frac{1}{3}x$

15. $f(x) = 4 - x^2$ **16.** $g(x) = \frac{4}{x}$

17. $h(x) = \sqrt{x - 1}$ **18.** $f(x) = \frac{1}{2}x^3 + 2$

19. $f(x) = \sqrt{9 - x^2}$ **20.** $h(x) = \sqrt{25 - x^2}$

21. $f(x) = |x - 2|$ **22.** $f(x) = \frac{|x|}{x}$

In Exercises 23–28, use the vertical line test to determine whether y is a function of x.

23. $y = x^2$ **24.** $y = x^3 - 1$

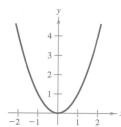   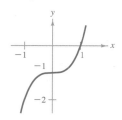

25. $x - y^2 = 0$ **26.** $x^2 + y^2 = 9$

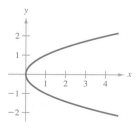   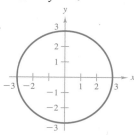

27. $\sqrt{x^2 - 4} - y = 0$ **28.** $x - xy + y + 1 = 0$

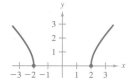   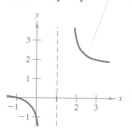

In Exercises 29–36, determine whether y is a function of x.

29. $x^2 + y^2 = 4$ **30.** $x = y^2$
31. $x^2 + y = 4$ **32.** $x + y^2 = 4$
33. $2x + 3y = 4$ **34.** $x^2 + y^2 - 4y = 0$
35. $y^2 = x^2 - 1$ **36.** $x^2y - x^2 + 4y = 0$

37. Use the graph of $f(x) = \sqrt{x}$ to sketch the graph of each of the following.
  (a) $y = \sqrt{x} + 2$ (b) $y = -\sqrt{x}$
  (c) $y = \sqrt{x - 2}$ (d) $y = \sqrt{x + 3}$
  (e) $y = \sqrt{x - 4}$ (f) $y = 2\sqrt{x}$

38. Use the graph of $f(x) = 1/x$ to sketch the graph of each of the following.
  (a) $y = \frac{1}{x} - 1$ (b) $y = \frac{1}{x + 1}$
  (c) $y = \frac{1}{x - 1}$ (d) $y = -\frac{1}{x}$
  (e) $y = \frac{4}{x}$ (f) $y = -\frac{1}{x} + 2$

39. Use the graph of $f(x) = x^2$ to determine a formula for the indicated function.
  (a)   (b)

        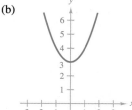

40. Use the graph of $f(x) = |x|$ to determine a formula for the indicated function.

(a)

(b)

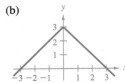

41. Given $f(x) = \sqrt{x}$ and $g(x) = x^2 - 1$, find the following.
   (a) $f(g(1))$
   (b) $g(f(1))$
   (c) $g(f(0))$
   (d) $f(g(-4))$
   (e) $f(g(x))$
   (f) $g(f(x))$

42. Given $f(x) = 1/x$ and $g(x) = x^2 - 1$, find the following.
   (a) $f(g(2))$
   (b) $g(f(2))$
   (c) $f\left(g\left(\frac{1}{\sqrt{2}}\right)\right)$
   (d) $g\left(f\left(\frac{1}{\sqrt{2}}\right)\right)$
   (e) $g(f(x))$
   (f) $f(g(x))$

In Exercises 43–46, find the composite functions $(f \circ g)$ and $(g \circ f)$. What is the domain of each function? Are the two composite functions equal?

43. $f(x) = x^2,$ $g(x) = \sqrt{x}$
44. $f(x) = x^3,$ $g(x) = \sqrt[3]{x}$
45. $f(x) = x + 1,$ $g(x) = \dfrac{1}{x}$
46. $f(x) = x^2 - 1,$ $g(x) = x$

In Exercises 47–50, find the (real) zeros of the given function.

47. $f(x) = x^2 - 9$
48. $f(x) = x^3 - x$
49. $f(x) = \dfrac{3}{x - 1} + \dfrac{4}{x - 2}$
50. $f(x) = a + \dfrac{b}{x}$

In Exercises 51–54, determine whether the function is even, odd, or neither.

51. $f(x) = 4 - x^2$
52. $f(x) = \sqrt[3]{x}$
53. $f(x) = x(4 - x^2)$
54. $f(x) = 4x - x^2$

55. Show that the following function is odd.

$$f(x) = a_{2n+1}x^{2n+1} + \cdots + a_3x^3 + a_1x$$

56. Show that the following function is even.

$$f(x) = a_{2n}x^{2n} + a_{2n-2}x^{2n-2} + \cdots + a_2x^2 + a_0$$

57. Show that the product of two even (or two odd) functions is even.
58. Show that the product of an odd function and an even function is odd.

In Exercises 59–62, express the indicated values as functions of x.

59. R and r

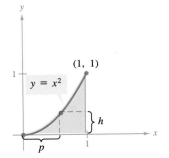

60. R and r

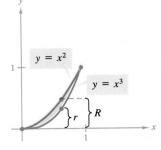

61. h and p

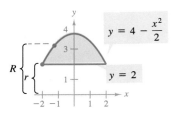

62. h and p

63. A rectangle has a perimeter of 100 feet (see figure). Express the area A of the rectangle as a function of x.

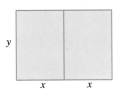

64. A rancher has 200 feet of fencing to enclose two adjacent rectangular corrals (see figure). Express the area A of the enclosures as a function of x.

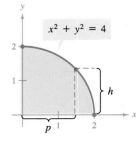

65. An open box is to be made from a square piece of material 12 inches on a side, by cutting equal squares from each corner and turning up the sides (see figure). Express the volume V as a function of x.

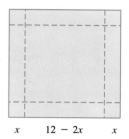

x $12 - 2x$ x

66. A rectangle is bounded by the x-axis and the semicircle $y = \sqrt{25 - x^2}$ (see figure). Write the area A of the rectangle as a function of x.

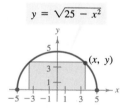

$y = \sqrt{25 - x^2}$

67. A rectangular package with square cross sections has a combined length and girth (perimeter of a cross section) of 108 inches. Express the volume V as a function of x (see figure).

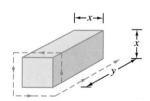

68. A closed box with a square base of side x has a surface area of 100 square feet (see figure). Express the volume V of the box as a function of x.

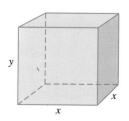

69. A man is in a boat 2 miles from the nearest point on the coast. He is to go to a point Q, 3 miles down the coast and 1 mile inland (see figure). He can row at 2 mph and walk at 4 mph. Express the total time T of the trip as a function of x.

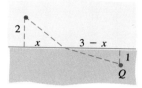

70. The portion of the vertical line through the point $(x, 0)$ that lies between the x-axis and the graph of $y = \sqrt{x}$ is revolved about the x-axis. Express the area A of the resulting disk as a function of x (see figure).

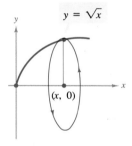

$y = \sqrt{x}$

In Exercises 71 and 72, use a computer or graphics calculator to (a) sketch the graph of f, (b) find the zeros of f, and (c) determine the domain of f.

71. $f(x) = x\sqrt{9 - x^2}$

72. $f(x) = 2\left(\pi x^2 + \dfrac{6}{x}\right)$

REVIEW EXERCISES for Chapter 1

In Exercises 1–4, sketch the interval(s) defined by the given inequality.

1. $|x - 2| \le 3$

2. $|3x - 2| \le 0$

3. $4 < (x + 3)^2$

4. $\dfrac{1}{|x|} < 1$

In Exercises 5 and 6, find the midpoint of the given interval.

5. $\left[\dfrac{7}{8}, \dfrac{10}{4} \right]$

6. $\left[-1, \dfrac{3}{2} \right]$

7. Find the midpoints of the sides of the triangle whose vertices are $(1, 4)$ $(-3, 2)$, and $(5, 0)$.

8. Find the vertices of the triangle whose sides have midpoints $(0, 2)$, $(1, -1)$, and $(2, 1)$.

In Exercises 9–12, determine the radius and center of the given circle and sketch its graph.

9. $x^2 + y^2 + 6x - 2y + 1 = 0$

10. $4x^2 + 4y^2 - 4x + 8y = 11$

11. $x^2 + y^2 + 6x - 2y + 10 = 0$

12. $x^2 - 6x + y^2 + 8y = 0$

13. Determine the value of c so that the given circle has a radius of 2.
$$x^2 - 6x + y^2 + 8y = c$$

14. Find an equation in x and y such that the distance between (x, y) and $(-2, 0)$ is twice the distance between (x, y) and $(3, 1)$.

15. Find an equation for the circle whose center is $(1, 2)$ and whose radius is 3. Then determine whether the following points are inside, outside, or on the circle.
(a) $(1, 5)$ (b) $(0, 0)$
(c) $(-2, 1)$ (d) $(0, 4)$

16. Find an equation for the circle whose center is $(2, 1)$ and whose radius is 2. Then determine whether the following points are inside, outside, or on the circle.
(a) $(1, 1)$ (b) $(4, 2)$
(c) $(0, 1)$ (d) $(3, 1)$

In Exercises 17–20, sketch the graph of the given equation.

17. $y = \dfrac{-x + 3}{2}$

18. $y = 1 + \dfrac{1}{x}$

19. $y = 7 - 6x - x^2$

20. $y = 6x - x^2$

In Exercises 21 and 22, determine whether the given points lie on the same straight line.

21. $(-1, 3)$, $(2, 9)$, $(3, 1)$

22. $(2, 5)$, $(4, 10)$, $(6, 20)$

In Exercises 23–26, use the slope and y-intercept to sketch the graph of the given line.

23. $4x - 2y = 6$

24. $0.02x + 0.15y = 0.25$

25. $-\dfrac{1}{3}x + \dfrac{5}{6}y = 1$

26. $51x + 17y = 102$

27. Find equations of the lines passing through $(-2, 4)$ and having the following characteristics.
(a) Slope of $\frac{7}{16}$
(b) Parallel to the line $5x - 3y = 3$
(c) Passing through the origin
(d) Parallel to the y-axis

28. Find equations of the lines passing through $(1, 3)$ and having the following characteristics.
(a) Slope of $-\frac{2}{3}$
(b) Perpendicular to the line $x + y = 0$
(c) Passing through the point $(2, 4)$
(d) Parallel to the x-axis

29. The midpoint of a line segment is $(-1, 4)$. If one end of the line segment is $(2, 3)$, find the other end.

30. Find the point that is equidistant from $(0, 0)$, $(2, 3)$, and $(3, -2)$.

In Exercises 31 and 32, find the point(s) of intersection of the graphs of the given equations.

31. $3x - 4y = 8$, $x + y = 5$

32. $x - y + 1 = 0$, $y - x^2 = 7$

In Exercises 33–38, find a formula for the given function and find the domain.

33. The value v of a farm at $850 per acre, with buildings, livestock, and equipment worth $300,000, is a function of the number of acres a.

34. The value v of wheat at $3.25 per bushel is a function of the number of bushels b.

35. The surface area s of a cube is a function of the length of an edge x.

36. The surface area s of a sphere is a function of the radius r.

37. The distance d traveled by a car at a speed of 45 miles per hour is a function of the time traveled t.

38. The area a of an equilateral triangle is a function of the length of one of its sides x.

39. The sum of two positive numbers is 500. Let one of the numbers be x, and express the product P of the two numbers as a function of x.

40. The product of two positive numbers is 120. Let one of the numbers be x, and express the sum of the two numbers as a function of x.

In Exercises 41–46, sketch the graph of the given equation and use the vertical line test to determine whether the equation expresses y as a function of x.

41. $x^2 - y = 0$

42. $x^2 + 4y^2 = 16$

43. $x - y^2 = 0$

44. $x^3 - y^2 + 1 = 0$

45. $y = x^2 - 2x$

46. $y = 36 - x^2$

47. Given $f(x) = 1 - x^2$ and $g(x) = 2x + 1$, find the following.

   (a) $f(x) + g(x)$
   (b) $f(x) - g(x)$

   (c) $f(x)g(x)$
   (d) $\dfrac{f(x)}{g(x)}$

   (e) $f(g(x))$
   (f) $g(f(x))$

48. Given $f(x) = 2x - 3$ and $g(x) = \sqrt{x + 1}$, find the following.

   (a) $f(x) + g(x)$
   (b) $f(x) - g(x)$

   (c) $f(x)g(x)$
   (d) $\dfrac{f(x)}{g(x)}$

   (e) $f(g(x))$
   (f) $g(f(x))$

49. Consider a plane flying at a constant rate on a direct route between two cities. The distance s (in miles) it has traveled in t hours is given by $s = 560t$.

   (a) Sketch the graph of this equation for $t \geq 0$.
   (b) What information is given by the slope of the line?

50. Find an equation of the line that bisects the acute angle formed by the lines $y = \sqrt{3}x$ and $y = 2$.

51. Sales representatives for a certain company are required to use their own cars for transportation. The cost to the company is $150 per day for lodging and meals, plus $0.30 per mile driven. Write a linear equation expressing the daily cost C to the company in terms of x, the number of miles driven.

52. A contractor purchases a piece of equipment for $36,500 that uses an average of $9.25 per hour for fuel and maintenance. The equipment operator is paid $13.50 per hour, and customers are charged $30 per hour.

   (a) Write an equation for the cost C of operating this equipment t hours.
   (b) Write an equation for the revenue R derived from t hours of use.
   (c) Find the break-even point for this equipment by finding the time at which $R = C$.

Chapter 2 Application

On construction sites, otherwise harmless objects such as metal bolts can be life-threatening hazards. If a metal bolt is dropped from the 100th floor of a skyscraper (approximately 1000 feet above ground), its velocity at ground level will be about 250 feet/second.

Velocity of a Free-Falling Object

A classic problem in calculus concerns the velocity of a free-falling object. For instance, consider an object that is dropped from a height of 25 feet above the earth's surface. We let $s(t)$ represent the height (in feet) of the object at time t (measured in seconds). Assuming that the only force acting on the object is that due to gravity, and neglecting air resistance, the height is given by the position function

$$s(t) = \frac{1}{2}gt^2 + v_0 t + s_0$$

where $g = -32$ ft/sec^2 is the acceleration due to gravity, $v_0 = 0$ is the initial velocity, and $s_0 = 25$ is the initial height. Thus, for this object, the position function is

$$s(t) = -16t^2 + 25.$$

Suppose we want to determine the velocity of the object at time $t = 1$ second. When $t = 1$, the height is $s(1) = 9$ feet and its *average* velocity during the time interval $[0, 1]$ is $(9 - 25)/(1 - 0) = -16$ ft/sec. Similarly, the average velocity during the time interval $[0.5, 1]$ is $[s(1) - s(0.5)]/(1 - 0.5) = (9 - 21)/0.5 = -24$ ft/sec.

In general, the average velocity during the time interval $[t, 1]$ is $[s(1) - s(t)]/(1 - t)$. To find the *instantaneous* velocity when $t = 1$, we let t approach 1 and evaluate the resulting limit.

$$
\begin{aligned}
v &= \lim_{t \to 1} \frac{s(1) - s(t)}{1 - t} = \lim_{t \to 1} \frac{9 - (-16t^2 + 25)}{1 - t} \\
&= \lim_{t \to 1} \frac{16(t^2 - 1)}{1 - t} \\
&= \lim_{t \to 1} \frac{16(t + 1)(t - 1)}{-(t - 1)} \\
&= \lim_{t \to 1} -16(t + 1) \\
&= -32 \text{ ft/sec}
\end{aligned}
$$

See Exercise 51, Section 2.2.

Chapter Overview

The concept of the **limit** of a function is the primary idea that distinguishes calculus from algebra and analytic geometry. Section 2.1 begins with a brief discussion of the way a limit will be used later (in Chapter 3) to solve the *tangent line problem*. The section then gives an informal description of the idea of the limit

$$\lim_{x \to a} f(x) = L.$$

This is followed by a strategy for finding limits and a discussion of properties of limits. In this section, it is important that you become familiar with several types of functions whose limits are easily found. For instance, the limit of $f(x) = x^2$ as x approaches 2 is simply $f(2) = 4$.

In Section 2.2, we use the properties discussed in Section 2.1 to find limits that are not so straightforward. This section also introduces the concept of a one-sided limit.

Section 2.3 introduces the notion of continuity. Informally, when we say that a function f is *continuous* on an interval (a, b), we mean that the graph of f has no holes, gaps, or jumps on the interval.

Section 2.4 discusses infinite limits and vertical asymptotes. (Limits at infinity and horizontal asymptotes are discussed later in the text, in Section 4.5.)

The last section in the chapter presents a more theoretical definition of the limit of a function—the so-called "ε-δ definition."

Limits and Their Properties

2.1 An Introduction to Limits

The tangent line problem ▪ Informal definition of limits ▪ Limits that fail to exist ▪ A strategy for finding limits ▪ Limits of algebraic functions

The notion of a limit is fundamental to the study of calculus. Thus, it is important to acquire a good working knowledge of limits before moving on to other topics in calculus.

The tangent line problem

To give you some idea of the way we will be using limits, we begin this section with a brief introduction to a classic problem in calculus, called the *tangent line problem*. In this problem, we are given a function f and a point P on its graph and asked to find an equation of the tangent line to the graph at the point P, as shown in Figure 2.1. To answer this question, we must first define what we mean when we say that a line is tangent to a curve at a point.

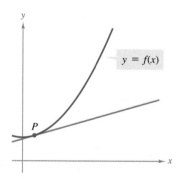

The tangent line to the graph of f at P

FIGURE 2.1

Except for cases involving a vertical tangent line, the problem of finding the **tangent line** at a point P is equivalent to finding the *slope* of the tangent line at P. We can approximate this slope by using a line through the point of tangency P and a second point on the curve, as shown in Figure 2.2(a). We call such a line a **secant line.** If $P = (c, f(c))$ is the point of tangency and $Q = (c + \Delta x, f(c + \Delta x))$ is a second point on the graph of f, then the slope of the secant line through these two points is given by

$$m_{\text{sec}} = \frac{f(c + \Delta x) - f(c)}{c + \Delta x - c} = \frac{f(c + \Delta x) - f(c)}{\Delta x}.$$

The idea here is that by letting the point Q approach the point P, the slope of the secant line will approach the slope of the tangent line, as indicated in Figure 2.2(b). When such a "limiting position" exists, we say that the slope of the tangent line is equal to the **limit** of the slope of the secant line. (Much more will be said about this important problem in Chapter 3.)

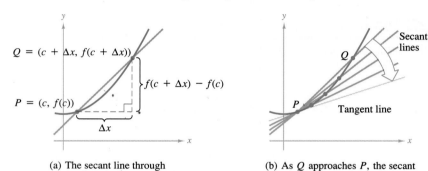

FIGURE 2.2

(a) The secant line through
$(c, f(c))$ and $(c + \Delta x, f(c + \Delta x))$

(b) As Q approaches P, the secant lines approach the tangent line.

Introduction to limits

Suppose we are asked to sketch the graph of the function f given by

$$f(x) = \frac{x^3 - 1}{x - 1}, \quad x \neq 1.$$

For all values other than $x = 1$, we can use standard curve-sketching techniques. However, at $x = 1$, we are not sure what to expect. To get an idea of the behavior of the graph of f near $x = 1$, we use two sets of x-values—one set that approaches 1 from the left and one set that approaches 1 from the right, as shown in Table 2.1.

TABLE 2.1

	x approaches 1 from the left					x approaches 1 from the right				
x	0.5	0.75	0.9	0.99	0.999	1	1.001	1.01	1.1	1.25	1.5
$f(x)$	1.750	2.313	2.710	2.970	2.997	?	3.003	3.030	3.310	3.813	4.750
		$f(x)$ approaches 3					$f(x)$ approaches 3				

When we plot these points, it appears that the graph of f is a parabola that has a gap at the point $(1, 3)$, as shown in Figure 2.3. Although x cannot equal 1, we can move arbitrarily close to 1, and as a result, $f(x)$ moves arbitrarily close to 3. Using limit notation, we say that the *limit of $f(x)$ as x approaches 1 is 3* and write

$$\lim_{x \to 1} f(x) = 3.$$

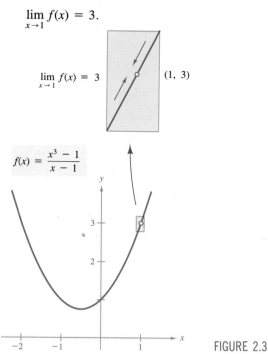

$$\lim_{x \to 1} f(x) = 3 \qquad (1, 3)$$

$$f(x) = \frac{x^3 - 1}{x - 1}$$

FIGURE 2.3

This discussion leads to the following informal definition of a limit.

INFORMAL DEFINITION OF A LIMIT	If $f(x)$ becomes arbitrarily close to a unique number L as x approaches c from either side, then we say that the **limit** of $f(x)$, as x approaches c, is L, and we write $$\lim_{x \to c} f(x) = L.$$

Throughout this text, when we write

$$\lim_{x \to c} f(x) = L$$

we imply two statements—the limit **exists** *and* the limit is L. Some functions do not have a limit as $x \to c$, but those that do cannot have two different limits as $x \to c$. In other words, *if the limit of a function exists, it is unique.*

For many functions, we can estimate the value of the limit L by using a calculator and evaluating the function at several points near c, as we did in Table 2.1.

EXAMPLE 1   Using a calculator to estimate a limit

Evaluate $f(x) = x/(\sqrt{x + 1} - 1)$ at several points near $x = 0$, and use the result to estimate the limit

$$\lim_{x \to 0} \frac{x}{\sqrt{x + 1} - 1}.$$

SOLUTION

Table 2.2 lists the values of $f(x)$ for several x-values near 0.

TABLE 2.2

		x approaches 0 from the left				*x* approaches 0 from the right			
x	-0.1	-0.01	-0.001	-0.0001	0	0.0001	0.001	0.01	0.1
$f(x)$	1.9487	1.9950	1.9995	1.9999	?	2.0001	2.0005	2.0050	2.0488

| | | $f(x)$ approaches 2 | $f(x)$ approaches 2 | |

From the results shown in this table, we estimate the limit to be 2. The graph of f is shown in Figure 2.4.   ▭

In Example 1, note that the function is undefined at $x = 0$ and yet $f(x)$ appears to be approaching a limit as $x \to 0$. This often happens and it is important to realize that *the existence or nonexistence of $f(x)$ at $x = c$ has no bearing on the existence of the limit of $f(x)$ as x approaches c.* This is further demonstrated in the next example.

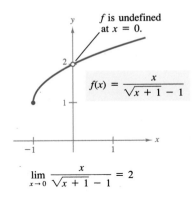

f is undefined at $x = 0$.

$$f(x) = \frac{x}{\sqrt{x + 1} - 1}$$

$$\lim_{x \to 0} \frac{x}{\sqrt{x + 1} - 1} = 2$$

FIGURE 2.4

EXAMPLE 2   Finding a limit

Find the limit of $f(x)$ as x approaches 2, where f is defined by

$$f(x) = \begin{cases} 1, & x \neq 2 \\ 0, & x = 2. \end{cases}$$

SOLUTION

Since $f(x) = 1$ for all x other than $x = 2$, and since the value of $f(2)$ is immaterial, we conclude that the limit is 1, as shown in Figure 2.5.   ▭

REMARK   Note in the first three figures that we use an open "dot" in a graph to mean that the point is not part of the graph, whereas a solid dot means that the point is part of the graph.

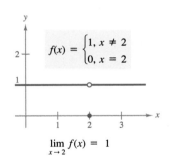

$$f(x) = \begin{cases} 1, & x \neq 2 \\ 0, & x = 2 \end{cases}$$

$$\lim_{x \to 2} f(x) = 1$$

FIGURE 2.5

Limits that fail to exist

In Examples 1 and 2 we looked at two limits that do exist. We can learn a great deal about what it means for a limit to exist by looking at some situations for which a limit does not exist.

EXAMPLE 3   Behavior that differs from the right and left

Show that the following limit does not exist.

$$\lim_{x \to 0} \frac{|x|}{x}$$

SOLUTION

We consider the function $f(x) = |x|/x$. From Figure 2.6, we see that for positive x, $|x|/x = 1$, and for negative x, $|x|/x = -1$. This means that no matter how close we get to zero, we will have both positive and negative x-values that yield $f(x) = 1$ and $f(x) = -1$. Specifically, if δ (the lowercase Greek letter *delta*) is a positive number, then for x-values satisfying the inequality $0 < |x| < \delta$, we can classify the values of $|x|/x$ as follows.

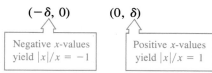

$$(-\delta, 0) \qquad (0, \delta)$$

| Negative x-values yield $|x|/x = -1$ | Positive x-values yield $|x|/x = 1$ |

This implies that the limit does not exist.

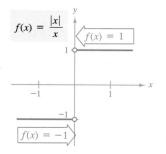

$f(x) = \dfrac{|x|}{x}$

$f(x) = 1$

$f(x) = -1$

$\lim\limits_{x \to 0} f(x)$ does not exist.

FIGURE 2.6

EXAMPLE 4   Unbounded behavior

Discuss the existence of the following limit.

$$\lim_{x \to 0} \frac{1}{x^2}$$

SOLUTION

We let $f(x) = 1/x^2$. From Figure 2.7, we see that as x approaches 0 from either the right or the left, $f(x)$ increases without bound. This means that by choosing x close enough to 0, we can force $f(x)$ to be as large as we want. For instance, $f(x)$ will be larger than 100 if we choose x that is within $\frac{1}{10}$ of 0. That is,

$$0 < |x| < \frac{1}{10} \quad \Longrightarrow \quad f(x) = \frac{1}{x^2} > 100.$$

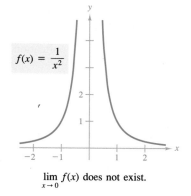

$f(x) = \dfrac{1}{x^2}$

$\lim\limits_{x \to 0} f(x)$ does not exist.

FIGURE 2.7

Similarly, we can force $f(x)$ to be larger than 1,000,000, as follows.

$$0 < |x| < \frac{1}{1000} \quad \Longrightarrow \quad f(x) = \frac{1}{x^2} > 1,000,000$$

Since $f(x)$ is not approaching a real number L as x approaches 0, we say that the limit does not exist.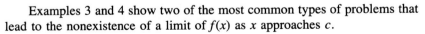

Examples 3 and 4 show two of the most common types of problems that lead to the nonexistence of a limit of $f(x)$ as x approaches c.

1. $f(x)$ approaches a different number from the right side of c than it approaches from the left side.
2. $f(x)$ increases or decreases without bound as x approaches c.

There are many other interesting functions that have unusual limit behavior. An often cited one is the *Dirichlet function*

$$f(x) = \begin{cases} 0, & \text{if } x \text{ is rational} \\ 1, & \text{if } x \text{ is irrational.} \end{cases}$$

This function has *no limit* at any real number c.

In the early development of calculus, the definition of function was much more restricted than it is today, and "functions" such as the Dirichlet function would not have been considered. The modern definition of a function was given by the German mathematician Peter Gustav Dirichlet (1805–1859). Dirichlet made many contributions to mathematics and, together with Cauchy, Riemann, and Weierstrauss, developed much of the rigor of modern calculus.

Peter Gustav Dirichlet

A strategy for finding limits

We pointed out that the limit of $f(x)$ as $x \to c$ does not depend upon the value of f at $x = c$. *However*, if it happens that the limit is precisely $f(c)$, then we say that the limit can be evaluated by **direct substitution.** That is,

$$\lim_{x \to c} f(x) = f(c). \qquad \text{Substitute for } x$$

Such *well-behaved* functions are said to be **continuous at** c and we will examine this concept more closely in Section 2.3. An important application of direct substitution is shown in the following theorem.

THEOREM 2.1 FUNCTIONS THAT AGREE AT ALL BUT ONE POINT	Let c be a real number and $f(x) = g(x)$ for all $x \neq c$ in an open interval containing c. If the limit of $g(x)$ as $x \to c$ exists, then the limit of $f(x)$ also exists, and $$\lim_{x \to c} f(x) = \lim_{x \to c} g(x).$$

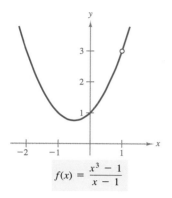

$$f(x) = \frac{x^3 - 1}{x - 1}$$

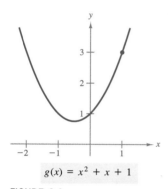

$$g(x) = x^2 + x + 1$$

FIGURE 2.8

EXAMPLE 5   Two functions that agree at all but one point

Show that the functions

$$f(x) = \frac{x^3 - 1}{x - 1} \quad \text{and} \quad g(x) = x^2 + x + 1$$

have the same values for all x other than $x = 1$.

SOLUTION

By factoring the numerator of f, we have

$$f(x) = \frac{x^3 - 1}{x - 1} = \frac{(x - 1)(x^2 + x + 1)}{x - 1}.$$

Thus, for $x \neq 1$, we can cancel like factors to obtain

$$f(x) = \frac{(x - 1)(x^2 + x + 1)}{x - 1} = x^2 + x + 1 = g(x), \quad x \neq 1.$$

Therefore, for all points other than $x = 1$, the functions f and g are identical. This is shown graphically in Figure 2.8.

Note that in Figure 2.8 there is no gap in the graph of the polynomial function g, and we will see in Theorem 2.4 that the limit of $g(x)$ as x approaches 1 is simply

$$\lim_{x \to 1} g(x) = \lim_{x \to 1} (x^2 + x + 1) = 1^2 + 1 + 1 = 3 = g(1).$$

Thus, by applying Theorem 2.1, we can conclude that the limit of $f(x)$ as $x \to 1$ is also 3.

This discussion identifies the following basic strategy for finding limits.

STRATEGY FOR FINDING LIMITS

1. Learn to recognize which limits can be evaluated by direct substitution. (Several types will be discussed in the remainder of this section.)
2. If the limit of $f(x)$ as $x \to c$ *cannot* be evaluated by direct substitution, try to find a function g that agrees with f for all x other than $x = c$. (Choose g so that the limit of $g(x)$ *can* be evaluated by direct substitution.)
3. Apply Theorem 2.1 to conclude that

$$\lim_{x \to c} f(x) = \lim_{x \to c} g(x) = g(c).$$

We devote the remainder of this section to the study of limits that can be evaluated by direct substitution. In the next section, we will look at techniques for evaluating limits for which direct substitution fails.

Limits of algebraic functions

We begin with a look at the limits of some basic functions.

THEOREM 2.2 **SOME BASIC LIMITS**	If b and c are real numbers and n is a positive integer, then the following properties are true. 1. $\lim_{x \to c} b = b$ 2. $\lim_{x \to c} x = c$ 3. $\lim_{x \to c} x^n = c^n$

EXAMPLE 6   Evaluating a limit

Applying Theorem 2.2, we have the following.

(a) $\lim_{x \to 2} 3 = 3$ (b) $\lim_{x \to 2} x^2 = 2^2 = 4$

By combining Theorem 2.2 with the following theorem, we can find limits for a wide variety of algebraic functions. (The proofs of Properties 4 and 5 of Theorem 2.3 are given in Appendix A.)

THEOREM 2.3 **PROPERTIES OF LIMITS**	Let b and c be real numbers, and n a positive integer, and let f and g be functions that each have a limit as $x \to c$. 1. Scalar multiple: $\lim_{x \to c} [b\, f(x)] = b\left[\lim_{x \to c} f(x)\right]$ 2. Sum or difference: $\lim_{x \to c} [f(x) \pm g(x)] = \lim_{x \to c} f(x) \pm \lim_{x \to c} g(x)$ 3. Product: $\lim_{x \to c} [f(x)g(x)] = \left[\lim_{x \to c} f(x)\right]\left[\lim_{x \to c} g(x)\right]$ 4. Quotient: $\lim_{x \to c} \dfrac{f(x)}{g(x)} = \dfrac{\lim_{x \to c} f(x)}{\lim_{x \to c} g(x)}$, provided $\lim_{x \to c} g(x) \neq 0$ 5. Power: $\lim_{x \to c} [f(x)]^n = \left[\lim_{x \to c} f(x)\right]^n$

EXAMPLE 7   The limit of a polynomial

Find the following limit.

$$\lim_{x \to 2} (4x^2 + 3)$$

SOLUTION

Using the results of Example 6 and the properties listed in Theorem 2.3, we have

$$\lim_{x \to 2} (4x^2 + 3) = \lim_{x \to 2} 4x^2 + \lim_{x \to 2} 3 \qquad \text{Property 2}$$

$$= 4\left[\lim_{x \to 2} x^2\right] + \lim_{x \to 2} 3 \qquad \text{Property 1}$$

$$= 4(4) + 3 = 19.$$

Note that in Example 7, the limit (as $x \to 2$) of the *polynomial function* $p(x) = 4x^2 + 3$ is simply the value of p at $x = 2$.

$$\lim_{x \to 2} p(x) = p(2) = 4(2^2) + 3 = 19$$

This *direct substitution* property is valid for all polynomial functions, as stated in the following theorem.

**THEOREM 2.4
LIMIT OF A POLYNOMIAL
FUNCTION**

If p is a polynomial function and c is a real number, then

$$\lim_{x \to c} p(x) = p(c).$$

PROOF   Let the polynomial function p be given by

$$p(x) = a_n x^n + \cdots + a_1 x + a_0.$$

Repeated applications of the sum and scalar multiple properties produce

$$\lim_{x \to c} p(x) = a_n \left[\lim_{x \to c} x^n \right] + \cdots + a_1 \left[\lim_{x \to c} x \right] + \lim_{x \to c} a_0.$$

Finally, using Properties 1, 2, and 3 of Theorem 2.2, we obtain

$$\lim_{x \to c} p(x) = a_n c^n + \cdots + a_1 c + a_0 = p(c).$$

**THEOREM 2.5
LIMIT OF A RATIONAL FUNCTION**

If r is a rational function given by $r(x) = p(x)/q(x)$, and c is a real number such that $q(c) \neq 0$, then

$$\lim_{x \to c} r(x) = r(c) = \frac{p(c)}{q(c)}.$$

PROOF   By Theorem 2.4 we know that for the polynomial functions p and q, we have

$$\lim_{x \to c} p(x) = p(c) \qquad \text{and} \qquad \lim_{x \to c} q(x) = q(c).$$

Moreover, since $q(c) \neq 0$, we can apply Property 4 of Theorem 2.3 to conclude that

$$\lim_{x \to c} r(x) = \lim_{x \to c} \frac{p(x)}{q(x)} = \frac{\lim_{x \to c} p(x)}{\lim_{x \to c} q(x)} = \frac{p(c)}{q(c)} = r(c).$$

EXAMPLE 8   The limit of a rational function

Find the following limit.

$$\lim_{x \to 1} \frac{x^2 + x + 2}{x + 1}$$

SOLUTION

Since the denominator is not zero when $x = 1$, we can apply Theorem 2.5 to obtain

$$\lim_{x \to 1} \frac{x^2 + x + 2}{x + 1} = \frac{1^2 + 1 + 2}{1 + 1} = \frac{4}{2} = 2.$$

Polynomial functions and rational functions constitute two of the three basic types of algebraic functions. The following theorem deals with the limit of a third type of algebraic function—one that involves radicals. A proof of this theorem is given in Appendix A.

THEOREM 2.6
LIMIT OF A FUNCTION INVOLVING A RADICAL

If $c > 0$ and n is *any* positive integer, or if $c < 0$ and n is an *odd* positive integer, then

$$\lim_{x \to c} \sqrt[n]{x} = \sqrt[n]{c}.$$

The next theorem involves the limit of a composite function. Its proof is also given in Appendix A.

THEOREM 2.7
LIMIT OF A COMPOSITE FUNCTION

If f and g are functions such that

$$\lim_{x \to c} g(x) = L \qquad \text{and} \qquad \lim_{x \to L} f(x) = f(L)$$

then

$$\lim_{x \to c} f(g(x)) = f(L).$$

To use Theorem 2.7 to find the limit of a composite function such as

$$\lim_{x \to 0} \sqrt{x^2 + 4}$$

we let $g(x) = x^2 + 4$, where $\lim_{x \to 0} (x^2 + 4) = g(0) = 4$ and $f(x) = \sqrt{x}$ where $\lim_{x \to 4} \sqrt{x} = 2$. Then it follows that

$$\lim_{x \to 0} f(g(x)) = \lim_{x \to 0} \sqrt{x^2 + 4} = 2.$$

This procedure is demonstrated further in Example 9.

EXAMPLE 9   The limit of a composite function

Find the following limit.

$$\lim_{x \to 3} \sqrt[3]{2x^2 - 10}$$

SOLUTION

Since

$$\lim_{x \to 3} (2x^2 - 10) = 2(3^2) - 10 = 8$$

we have

$$\lim_{x \to 3} \sqrt[3]{2x^2 - 10} = \sqrt[3]{8} = 2.$$

EXERCISES for Section 2.1

In Exercises 1–6, complete the table and use the result to estimate the given limit.

1. $\lim_{x \to 2} \dfrac{x - 2}{x^2 - x - 2}$

x	1.9	1.99	1.999	2.001	2.01	2.1
$f(x)$						

2. $\lim_{x \to 2} \dfrac{x - 2}{x^2 - 4}$

x	1.9	1.99	1.999	2.001	2.01	2.1
$f(x)$						

3. $\lim_{x \to 0} \dfrac{\sqrt{x + 3} - \sqrt{3}}{x}$

x	−0.1	−0.01	−0.001
$f(x)$			

x	0.001	0.01	0.1
$f(x)$			

4. $\lim_{x \to -3} \dfrac{\sqrt{1 - x} - 2}{x + 3}$

x	−3.1	−3.01	−3.001
$f(x)$			

x	−2.999	−2.99	−2.9
$f(x)$			

*5. $\lim_{x \to 3} \dfrac{[1/(x + 1)] - (1/4)}{x - 3}$

x	2.9	2.99	2.999	3.001	3.01	3.1
$f(x)$						

6. $\lim_{x \to 4} \dfrac{[x/(x + 1)] - (4/5)}{x - 4}$

x	3.9	3.99	3.999	4.001	4.01	4.1
$f(x)$						

*A blue number indicates that a detailed solution can be found in the *Study and Solutions Guide*.

In Exercises 7–12, use the given graph to find the limit (if it exists).

7. $\lim\limits_{x \to 3} (4 - x)$

8. $\lim\limits_{x \to 1} (x^2 + 2)$

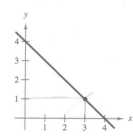

9. $\lim\limits_{x \to 2} f(x)$

$f(x) = \begin{cases} 4 - x, & x \neq 2 \\ 0, & x = 2 \end{cases}$

10. $\lim\limits_{x \to 1} f(x)$

$f(x) = \begin{cases} x^2 + 2, & x \neq 1 \\ 1, & x = 1 \end{cases}$

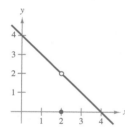

11. $\lim\limits_{x \to 5} \dfrac{|x - 5|}{x - 5}$

12. $\lim\limits_{x \to 3} \dfrac{1}{x - 3}$

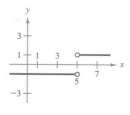
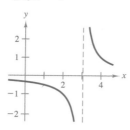

In Exercises 13–26, find the indicated limit.

13. $\lim\limits_{x \to 2} x^2$

14. $\lim\limits_{x \to -3} (3x + 2)$

15. $\lim\limits_{x \to 0} (2x - 1)$

16. $\lim\limits_{x \to 1} (-x^2 + 1)$

17. $\lim\limits_{x \to 2} (-x^2 + x - 2)$

18. $\lim\limits_{x \to 1} (3x^3 - 2x^2 + 4)$

19. $\lim\limits_{x \to 3} \sqrt{x + 1}$

20. $\lim\limits_{x \to 4} \sqrt[3]{x + 4}$

21. $\lim\limits_{x \to -4} (x + 3)^2$

22. $\lim\limits_{x \to 0} (2x - 1)^3$

23. $\lim\limits_{x \to 2} \dfrac{1}{x}$

24. $\lim\limits_{x \to -3} \dfrac{2}{x + 2}$

25. $\lim\limits_{x \to -1} \dfrac{x^2 + 1}{x}$

26. $\lim\limits_{x \to 3} \dfrac{\sqrt{x + 1}}{x - 4}$

27. If $\lim\limits_{x \to c} f(x) = 2$ and $\lim\limits_{x \to c} g(x) = 3$, find the following.

(a) $\lim\limits_{x \to c} [5g(x)]$

(b) $\lim\limits_{x \to c} [f(x) + g(x)]$

(c) $\lim\limits_{x \to c} [f(x)g(x)]$

(d) $\lim\limits_{x \to c} \dfrac{f(x)}{g(x)}$

28. If $\lim\limits_{x \to c} f(x) = \frac{3}{2}$ and $\lim\limits_{x \to c} g(x) = \frac{1}{2}$, find the following.

(a) $\lim\limits_{x \to c} [4f(x)]$

(b) $\lim\limits_{x \to c} [f(x) + g(x)]$

(c) $\lim\limits_{x \to c} [f(x)g(x)]$

(d) $\lim\limits_{x \to c} \dfrac{f(x)}{g(x)}$

29. If $\lim\limits_{x \to c} f(x) = 4$, find the following.

(a) $\lim\limits_{x \to c} [f(x)]^3$

(b) $\lim\limits_{x \to c} \sqrt{f(x)}$

(c) $\lim\limits_{x \to c} [3f(x)]$

(d) $\lim\limits_{x \to c} [f(x)]^{3/2}$

30. If $\lim\limits_{x \to c} f(x) = 27$, find the following.

(a) $\lim\limits_{x \to c} \sqrt[3]{f(x)}$

(b) $\lim\limits_{x \to c} \dfrac{f(x)}{18}$

(c) $\lim\limits_{x \to c} [f(x)]^2$

(d) $\lim\limits_{x \to c} [f(x)]^{2/3}$

In Exercises 31 and 32, use a computer or graphics calculator to sketch the graph of the function f and find the specified limit (if it exists).

31. $f(x) = \dfrac{\sqrt{x + 5} - 3}{x - 4}$, $\quad \lim\limits_{x \to 4} f(x)$

32. $f(x) = \dfrac{x - 3}{x^2 - 4x + 3}$, $\quad \lim\limits_{x \to 3} f(x)$

33. If $f(2) = 4$, can we conclude anything about

$$\lim\limits_{x \to 2} f(x)?$$

Give reasons for your answer.

34. If

$$\lim\limits_{x \to 2} f(x) = 4,$$

can we conclude anything about $f(2)$? Give reasons for your answer.

35. Find two functions f and g such that

$$\lim\limits_{x \to 0} f(x) \quad \text{and} \quad \lim\limits_{x \to 0} g(x)$$

do not exist, but

$$\lim\limits_{x \to 0} [f(x) + g(x)]$$

does exist.

2.2 Techniques for Evaluating Limits

Cancellation technique ▪ Rationalization technique ▪ One-sided limits

In Section 2.1, we catalogued several types of limits that can be evaluated by *direct substitution*. We now look at some techniques for reducing other limits to this form. In the first example we use the Factor Theorem from algebra that states that for a polynomial function p, $p(c) = 0$ if and only if $(x - c)$ is a factor of $p(x)$. We can use this result in evaluating the limit of a rational function, as follows. If

$$r(x) = \frac{p(x)}{q(x)}$$

and

$$\lim_{x \to c} p(x) = p(c) = 0 \qquad \text{and} \qquad \lim_{x \to c} q(x) = q(c) = 0$$

then we may be able to evaluate the limit of $r(x)$ as x approaches c by cancelling the common factor $(x - c)$ out of the numerator and the denominator, as shown in the following example.

EXAMPLE 1   Cancellation technique

Find the following limit.

$$\lim_{x \to -3} \frac{x^2 + x - 6}{x + 3}$$

SOLUTION

Although we are taking the limit of a rational function $p(x)/q(x)$, we *cannot* apply Theorem 2.5, because the limit of the denominator is zero.

$$\lim_{x \to -3} \frac{x^2 + x - 6}{x + 3}$$

$$\boxed{\lim_{x \to -3} (x^2 + x - 6) = p(-3) = 0}$$

Direct substitution fails

$$\boxed{\lim_{x \to -3} (x + 3) = q(-3) = 0}$$

However, since the limit of the numerator is also zero, we know that the numerator and denominator have a *common factor* of $(x + 3)$. Thus, for all $x \neq -3$, we can cancel this factor to obtain

$$\frac{x^2 + x - 6}{x + 3} = \frac{(x + 3)(x - 2)}{x + 3} = x - 2, \qquad x \neq -3.$$

Finally, by Theorem 2.1, it follows that

$$\lim_{x \to -3} \frac{x^2 + x - 6}{x + 3} = \lim_{x \to -3} (x - 2) = -5.$$

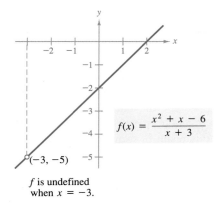

$$f(x) = \frac{x^2 + x - 6}{x + 3}$$

f is undefined
when $x = -3$.

FIGURE 2.9

This result is shown graphically in Figure 2.9. Note that the graph of the function *f* coincides with the graph of the function $g(x) = x - 2$, except that the graph of *f* has a hole at the point $(-3, -5)$. ▭

In Example 1, direct substitution produced the meaningless fractional form $0/0$. We call such an expression an **indeterminate form** because we cannot (from the form alone) determine the limit. When you try to evaluate a limit and encounter this form, remember that you must rewrite the fraction so that the new denominator does not have zero as its limit. One way to do this is to *cancel like factors*, as shown in Example 1 and further demonstrated in the next example.

EXAMPLE 2   Cancellation technique

Find the following limit.

$$\lim_{x \to 1} \frac{x - 1}{x^3 - x^2 + x - 1}$$

SOLUTION

By direct substitution, we obtain the indeterminate form $0/0$.

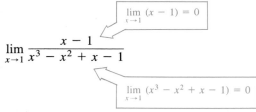

$$\lim_{x \to 1} \frac{x - 1}{x^3 - x^2 + x - 1}$$

Since the limit as *x* approaches 1 is zero for both the numerator *and* denominator and both are polynomials, we know that $(x - 1)$ is a common factor. Thus, for all $x \ne 1$, we can cancel this factor to obtain

$$\frac{x - 1}{x^3 - x^2 + x - 1} = \frac{x - 1}{(x - 1)(x^2 + 1)} = \frac{1}{x^2 + 1}, \qquad x \ne 1.$$

Now, by Theorem 2.1, it follows that

$$\lim_{x \to 1} \frac{x - 1}{x^3 - x^2 + x - 1} = \lim_{x \to 1} \frac{1}{x^2 + 1} = \frac{1}{2}.$$

This result is shown graphically in Figure 2.10. ▭

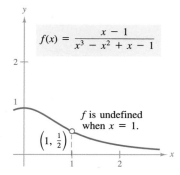

$$f(x) = \frac{x - 1}{x^3 - x^2 + x - 1}$$

f is undefined
when $x = 1$.

$\left(1, \frac{1}{2}\right)$

FIGURE 2.10

REMARK In Example 2 the factorization of the denominator can be obtained by dividing by $(x - 1)$ or by grouping as follows.

$$x^3 - x^2 + x - 1 = x^2(x - 1) + (x - 1)$$
$$= (x - 1)(x^2 + 1)$$

A second way to find the limit of a function for which direct substitution yields the indeterminate form 0/0 is to use a *rationalization technique*. This technique can be used to rationalize either the numerator or the denominator. For instance, in the next example we begin by rationalizing the numerator—then we cancel common factors in the rationalized form.

EXAMPLE 3   Rationalization technique

Find the following limit.

$$\lim_{x \to 0} \frac{\sqrt{x + 1} - 1}{x}$$

SOLUTION

By direct substitution, we obtain the indeterminate form 0/0.

$$\lim_{x \to 0} (\sqrt{x + 1} - 1) = 0$$

$$\lim_{x \to 0} \frac{\sqrt{x + 1} - 1}{x}$$

$$\lim_{x \to 0} x = 0$$

In this case, we rewrite the fraction by rationalizing (eliminating the radical in) the numerator.

$$\frac{\sqrt{x + 1} - 1}{x} = \left(\frac{\sqrt{x + 1} - 1}{x} \right) \left(\frac{\sqrt{x + 1} + 1}{\sqrt{x + 1} + 1} \right) = \frac{(x + 1) - 1}{x(\sqrt{x + 1} + 1)}$$

$$= \frac{x}{x(\sqrt{x + 1} + 1)} = \frac{1}{\sqrt{x + 1} + 1}, \quad x \neq 0$$

Therefore, by Theorem 2.1, we have

$$\lim_{x \to 0} \frac{\sqrt{x + 1} - 1}{x} = \lim_{x \to 0} \frac{1}{\sqrt{x + 1} + 1} = \frac{1}{1 + 1} = \frac{1}{2}.$$

Table 2.3 reinforces our conclusion that this limit is $\frac{1}{2}$.

TABLE 2.3

	x approaches 0 from the left					x approaches 0 from the right			
x	−0.25	−0.1	−0.01	−0.001	0	0.001	0.01	0.1	0.25
$\dfrac{\sqrt{x + 1} - 1}{x}$	0.5359	0.5132	0.5013	0.5001	?	0.4999	0.4988	0.4881	0.4721

$f(x)$ approaches $\frac{1}{2}$	$f(x)$ approaches $\frac{1}{2}$

REMARK   The rationalization technique for evaluating limits is based on multiplication by a convenient form of 1. In Example 3, the convenient form is

$$1 = \frac{\sqrt{x + 1} + 1}{\sqrt{x + 1} + 1}.$$

One-sided limits

In Section 2.1, we saw that one way in which a limit can fail to exist is when a function approaches a different value from the right side of c than it approaches from the left side of c. To further investigate this type of behavior, we first need to look at a different type of "limit," called a **one-sided limit.** For example, when we talk about the **limit from the right,** we mean that x approaches c from values greater than c. We denote this by

$$\lim_{x \to c^+} f(x) = L. \qquad \text{Limit from the right}$$

Similarly, the **limit from the left** means that x approaches c from values less than c. We denote this by

$$\lim_{x \to c^-} f(x) = L. \qquad \text{Limit from the left}$$

One-sided limits are useful in taking limits of functions involving radicals. For instance, if n is an even integer, then

$$\lim_{x \to 0^+} \sqrt[n]{x} = 0.$$

EXAMPLE 4   A one-sided limit

Find the limit of $f(x) = \sqrt{x - 1}$ as x approaches 1 from the right.

SOLUTION

As indicated in Figure 2.11, the limit as x approaches 1 from the right is given by

$$\lim_{x \to 1^+} \sqrt{x - 1} = 0. \qquad \blacksquare$$

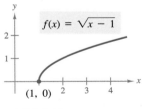

$f(x) = \sqrt{x - 1}$

$(1, 0)$

FIGURE 2.11

Another use of one-sided limits is that of investigating the behavior of **step functions.** One common type of step function is the **greatest integer function,** $[\![x]\!]$, defined by

$$[\![x]\!] = \text{greatest integer } n \text{ such that } n \le x.$$

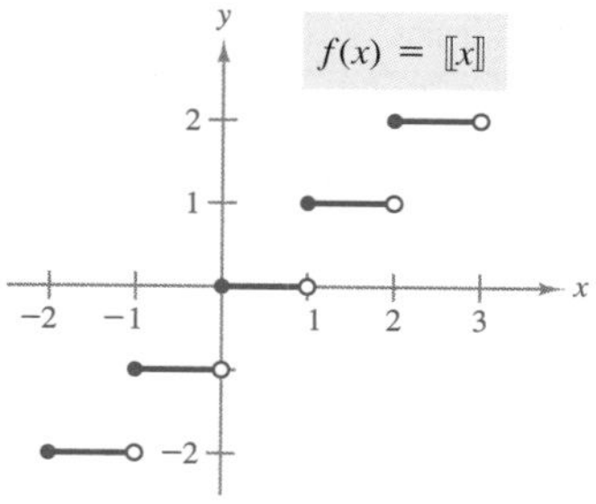

$f(x) = [\![x]\!]$

Greatest Integer Function

FIGURE 2.12

EXAMPLE 5 The greatest integer function

Find the limit of $f(x) = [\![x]\!]$ as x approaches 0 from the left and from the right. (See Figure 2.12.)

SOLUTION

The limit as x approaches 0 *from the left* is given by

$$\lim_{x \to 0^-} [\![x]\!] = \lim_{x \to 0^-} (-1) = -1$$

and the limit as x approaches 0 *from the right* is given by

$$\lim_{x \to 0^+} [\![x]\!] = \lim_{x \to 0^+} (0) = 0.$$

In Figure 2.12 we see that the greatest integer function approaches a different number from the right of 0 than it approaches from the left of 0. In such cases we say that the (two-sided) limit *does not exist*. The following theorem makes this more explicit. The proof of this theorem follows directly from the definitions of limits and one-sided limits.

THEOREM 2.8
THE EXISTENCE OF A LIMIT

If f is a function and c and L are real numbers, then the limit of $f(x)$ as x approaches c is L if and only if

$$\lim_{x \to c^-} f(x) = L \qquad \text{and} \qquad \lim_{x \to c^+} f(x) = L.$$

Theorem 2.8 is particularly useful for determining that a limit does not exist as demonstrated in Example 6.

EXAMPLE 6 Comparing the limits from the left and right

Evaluate the following limit.

$$\lim_{x \to 1} f(x), \quad \text{where } f(x) = \begin{cases} 2x - x^3, & x < 1 \\ 2x^2 - 2, & x \geq 1 \end{cases}$$

SOLUTION

Since f is defined differently for $x < 1$ than for $x \geq 1$, we consider the following one-sided limits.

$$\lim_{x \to 1^-} f(x) = \lim_{x \to 1^-} (2x - x^3) = 2 - 1 = 1$$

$$\lim_{x \to 1^+} f(x) = \lim_{x \to 1^+} (2x^2 - 2) = 2 - 2 = 0$$

Because these one-sided limits are not equal, we conclude that the limit of $f(x)$ as $x \to 1$ *does not exist*. This is shown graphically in Figure 2.13.

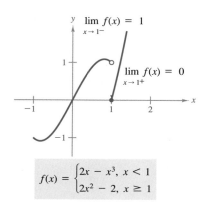

$\lim_{x \to 1^-} f(x) = 1$

$\lim_{x \to 1^+} f(x) = 0$

$$f(x) = \begin{cases} 2x - x^3, & x < 1 \\ 2x^2 - 2, & x \geq 1 \end{cases}$$

FIGURE 2.13

EXERCISES for Section 2.2

In Exercises 1–4, use the graph to visually determine the limit, if it exists.

1. $g(x) = \dfrac{-2x^2 + x}{x}$

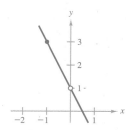

2. $h(x) = \dfrac{x^2 - 3x}{x}$

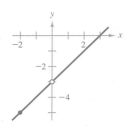

(a) $\lim\limits_{x \to 0} g(x)$

(b) $\lim\limits_{x \to -1} g(x)$

(a) $\lim\limits_{x \to -2} h(x)$

(b) $\lim\limits_{x \to 0} h(x)$

3. $g(x) = \dfrac{x^3 - x}{x - 1}$

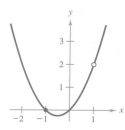

4. $f(x) = \dfrac{1}{x - 1}$

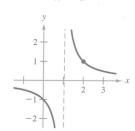

(a) $\lim\limits_{x \to 1} g(x)$

(b) $\lim\limits_{x \to -1} g(x)$

(a) $\lim\limits_{x \to 1} f(x)$

(b) $\lim\limits_{x \to 2} f(x)$

In Exercises 5–22, find the limit (if it exists).

5. $\lim\limits_{x \to -1} \dfrac{x^2 - 1}{x + 1}$

6. $\lim\limits_{x \to -1} \dfrac{2x^2 - x - 3}{x + 1}$

7. $\lim\limits_{x \to 3} \dfrac{x - 3}{x^2 - 9}$

8. $\lim\limits_{x \to -1} \dfrac{x^3 + 1}{x + 1}$

9. $\lim\limits_{x \to -2} \dfrac{x^3 + 8}{x + 2}$

10. $\lim\limits_{\Delta x \to 0} \dfrac{(x + \Delta x)^2 - x^2}{\Delta x}$

11. $\lim\limits_{\Delta x \to 0} \dfrac{2(x + \Delta x) - 2x}{\Delta x}$

12. $\lim\limits_{\Delta x \to 0} \dfrac{(x + \Delta x)^3 - x^3}{\Delta x}$

13. $\lim\limits_{\Delta x \to 0} \dfrac{(x + \Delta x)^2 - 2(x + \Delta x) + 1 - (x^2 - 2x + 1)}{\Delta x}$

14. $\lim\limits_{\Delta x \to 0} \dfrac{(1 + \Delta x)^3 - 1}{\Delta x}$

15. $\lim\limits_{x \to 5} \dfrac{x - 5}{x^2 - 25}$

16. $\lim\limits_{x \to 2} \dfrac{2 - x}{x^2 - 4}$

17. $\lim\limits_{x \to 1} \dfrac{x^2 + x - 2}{x^2 - 1}$

18. $\lim\limits_{x \to 0} \dfrac{\sqrt{2 + x} - \sqrt{2}}{x}$

19. $\lim\limits_{x \to 0} \dfrac{\sqrt{3 + x} - \sqrt{3}}{x}$

20. $\lim\limits_{x \to 0} \dfrac{[1/(x + 4)] - (1/4)}{x}$

21. $\lim\limits_{x \to 0} \dfrac{[1/(2 + x)] - (1/2)}{x}$

22. $\lim\limits_{x \to 3} \dfrac{\sqrt{x + 1} - 2}{x - 3}$

In Exercises 23–26, use a computer or calculator to complete a table of values near $x = c$ to estimate the limit. Then, find the limit by analytic methods and compare the result to your estimated limit.

23. $\lim\limits_{x \to 0} \dfrac{\sqrt{x + 2} - \sqrt{2}}{x}$

24. $\lim\limits_{x \to 1} \dfrac{1 - x}{\sqrt{5 - x^2} - 2}$

25. $\lim\limits_{x \to 0} \dfrac{[1/(2 + x)] - (1/2)}{x}$

26. $\lim\limits_{x \to 2} \dfrac{x^5 - 32}{x - 2}$

In Exercises 27–32, use the graph to determine the following visually.

(a) $\lim\limits_{x \to c^+} f(x)$ (b) $\lim\limits_{x \to c^-} f(x)$ (c) $\lim\limits_{x \to c} f(x)$

27.

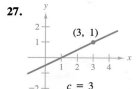

28.

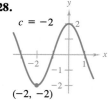

29.

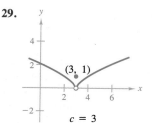

30.

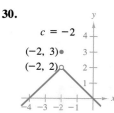

31.

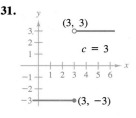

32.

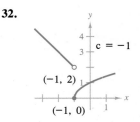

In Exercises 33–50, find the limit (if it exists).

33. $\lim\limits_{x \to 5^+} \dfrac{x - 5}{x^2 - 25}$

34. $\lim\limits_{x \to 2^+} \dfrac{2 - x}{x^2 - 4}$

35. $\lim\limits_{x \to 2^+} \dfrac{x}{\sqrt{x^2 - 4}}$

36. $\lim\limits_{x \to 4^-} \dfrac{\sqrt{x} - 2}{x - 4}$

37. $\lim\limits_{\Delta x \to 0^+} \dfrac{2(x + \Delta x) - 2x}{\Delta x}$

38. $\lim\limits_{\Delta x \to 0^+} \dfrac{(x + \Delta x)^2 + (x + \Delta x) - (x^2 + x)}{\Delta x}$

39. $\lim\limits_{\Delta x \to 0^+} \dfrac{\dfrac{1}{x + \Delta x} - \dfrac{1}{x}}{\Delta x}$

40. $\lim\limits_{x \to 1^-} \dfrac{x^2 - 2x + 1}{x - 1}$

41. $\lim\limits_{x \to 0} \dfrac{|x|}{x}$

42. $\lim\limits_{x \to 2} \dfrac{|x - 2|}{x - 2}$

43. $\lim\limits_{x \to 3} f(x), \quad f(x) = \begin{cases} \dfrac{x + 2}{2}, & x \le 3 \\ \dfrac{12 - 2x}{3}, & x > 3 \end{cases}$

44. $\lim\limits_{x \to 2} f(x), \quad f(x) = \begin{cases} x^2 - 4x + 6, & x < 2 \\ -x^2 + 4x - 2, & x \ge 2 \end{cases}$

45. $\lim\limits_{x \to 1} f(x), \quad f(x) = \begin{cases} x^3 + 1, & x < 1 \\ x + 1, & x \ge 1 \end{cases}$

46. $\lim\limits_{x \to 1} f(x), \quad f(x) = \begin{cases} x, & x \le 1 \\ 1 - x, & x > 1 \end{cases}$

47. $\lim\limits_{x \to 3^-} 2[\![x - 3]\!]$

48. $\lim\limits_{x \to 1^+} [\![2x]\!]$

49. $\lim\limits_{x \to 1} \left(\left[\!\!\left[\dfrac{x}{4} \right]\!\!\right] + x \right)$

50. $\lim\limits_{x \to 2} \left[\!\!\left[\dfrac{x - 1}{2} \right]\!\!\right]$

In Exercises 51 and 52, use the position function $s(t) = -16t^2 + 25$ giving the height (in feet) of a free-falling object. As discussed in the Chapter 2 Application, the velocity at time $t = a$ is given by

$$\lim_{t \to a} \frac{s(a) - s(t)}{a - t}.$$

51. Find the velocity when $t = 0.5$ seconds.

52. Find the velocity when $t = 1.1$ seconds.

In Exercises 53 and 54, use a computer or graphics calculator to sketch the graph of the function f and find the specified limit (if it exists).

53. $f(x) = \dfrac{x^2 + 2x - 8}{x^2 + 6x + 8}, \qquad \lim\limits_{x \to -4} f(x)$

54. $f(x) = \dfrac{x^2 - 5x + 6}{x^2 - 4x + 4}, \qquad \lim\limits_{x \to 2} f(x)$

2.3 Continuity

Continuity at a point ▪ Continuity on an open interval ▪ Continuity on a closed interval ▪ Properties of continuity ▪ Intermediate Value Theorem

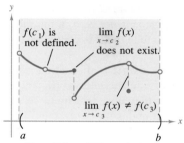

$f(c_1)$ is not defined.

$\lim\limits_{x \to c_2} f(x)$ does not exist.

$\lim\limits_{x \to c_3} f(x) \ne f(c_3)$

Three Points of Discontinuity

FIGURE 2.14

In mathematics the term *continuous* has much the same meaning as it does in our everyday usage. To say that a function is continuous at $x = c$ means that there is no interruption in the graph of f at c. That is, its graph is unbroken at c and there are no holes, jumps, or gaps. For example, Figure 2.14 identifies three values of x at which the graph of f is not continuous. At all other points of the interval (a, b), the graph of f is uninterrupted and we say it is **continuous** at such points. Thus, it appears that the continuity of a function at $x = c$ can be destroyed by any one of the following conditions:

1. The function is not defined at $x = c$.
2. The limit of $f(x)$ does not exist at $x = c$.
3. The limit of $f(x)$ exists at $x = c$, but is not equal to $f(c)$.

This brings us to the following definition.

DEFINITION OF CONTINUITY

Continuity at a Point: A function f is called **continuous at** c if the following three conditions are met.

1. $f(c)$ is defined
2. $\lim\limits_{x \to c} f(x)$ exists
3. $\lim\limits_{x \to c} f(x) = f(c)$

Continuity on an Open Interval: A function is called **continuous on an open interval** (a, b) if it is continuous at each point in the interval. A function that is continuous on the entire real line $(-\infty, \infty)$ is called **everywhere continuous.**

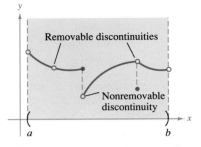

Removable discontinuities

Nonremovable discontinuity

FIGURE 2.15

A function f is said to be **discontinuous at** c if f is defined on an open interval containing c (except possibly at c) and f is not continuous at c. Discontinuities fall into two categories: **removable** and **nonremovable.** A discontinuity at $x = c$ is called removable if f can be made continuous by appropriately defining (or redefining) f at $x = c$. For example, in Figure 2.15, the function f has two removable discontinuities and one nonremovable discontinuity.

In Section 2.1 we studied several types of functions that meet the three conditions for continuity. In Example 1 we use this knowledge of limits to examine two functions that are continuous and one that is not.

EXAMPLE 1 Testing for continuity

Determine whether the following functions are continuous on the given interval.

(a) $f(x) = \dfrac{1}{x}, \quad (0, 1)$

(b) $f(x) = \dfrac{x^2 - 1}{x - 1}, \quad (0, 2)$

(c) $f(x) = x^3 - x, \quad (-\infty, \infty)$

SOLUTION

The graphs of these three functions are shown in Figure 2.16.
(a) Since f is a rational function whose denominator is not zero in the interval $(0, 1)$, we can apply Theorem 2.5 to conclude that f is continuous on $(0, 1)$.
(b) Since f is undefined at $x = 1$, we conclude that it is discontinuous at $x = 1$ and continuous for all other values of x in the interval $(0, 2)$.
(c) Since polynomial functions are defined over the entire real line, we can apply Theorem 2.4 to conclude that f is continuous on $(-\infty, \infty)$. ▭

REMARK In part (b) of Example 1 the discontinuity at $x = 1$ is *removable*. Specifically, by defining $f(1)$ to be 2, we would obtain a function that is continuous on $(0, 2)$.

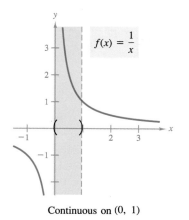

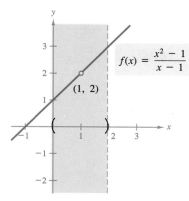

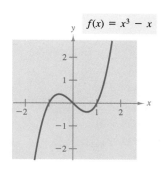

Continuous on $(0, 1)$ Discontinuous at $x = 1$ Continuous on $(-\infty, \infty)$

FIGURE 2.16

Each of the intervals in Example 1 is open. To discuss continuity on a closed interval, we use the concept of one-sided limits, as defined in the previous section.

DEFINITION OF CONTINUITY ON A CLOSED INTERVAL

A function f is **continuous on the closed interval** $[a, b]$ if it is continuous on the open interval (a, b) and

$$\lim_{x \to a^+} f(x) = f(a) \qquad \text{and} \qquad \lim_{x \to b^-} f(x) = f(b).$$

The function f is called **continuous from the right at a** and **continuous from the left at b**.

Similar definitions can be made to cover continuity on half-open intervals of the form $(a, b]$ and $[a, b)$, or on infinite intervals. For example, the function $f(x) = \sqrt{x}$ is continuous on the infinite interval $[0, \infty)$, as discussed in Example 2.

EXAMPLE 2   Continuity at the endpoint of an interval

Discuss the continuity of the function given by $f(x) = \sqrt{x}$.

SOLUTION

The domain of f is $[0, \infty)$. At all points in the interval $(0, \infty)$ the continuity of f follows from Theorems 2.2 and 2.6. Moreover, at the left endpoint of the domain, we have

$$\lim_{x \to 0^+} \sqrt{x} = 0 = f(0). \qquad \text{Limit from the right}$$

Therefore, f is continuous from the right at $x = 0$, and we conclude that f is continuous on its entire domain, as shown in Figure 2.17.

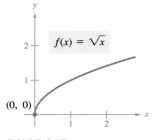

FIGURE 2.17

$$g(x) = \begin{cases} 5 - x, -1 \le x \le 2 \\ x^2 - 1, \ 2 < x \le 3 \end{cases}$$

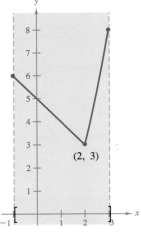

FIGURE 2.18

EXAMPLE 3 Continuity on a closed interval

Discuss the continuity of

$$g(x) = \begin{cases} 5 - x, & -1 \le x \le 2 \\ x^2 - 1, & 2 < x \le 3. \end{cases}$$

SOLUTION

From our work in Section 2.1, we know that the polynomial functions given by $5 - x$ and $x^2 - 1$ are continuous for all real x. Thus, to conclude that g is continuous on the entire interval $[-1, 3]$, we need only check the behavior of g when $x = 2$. By taking the one-sided limits when $x = 2$, we see that

$$\lim_{x \to 2^-} g(x) = \lim_{x \to 2^-} (5 - x) = 3 \qquad \text{Limit from the left}$$

and

$$\lim_{x \to 2^+} g(x) = \lim_{x \to 2^+} (x^2 - 1) = 3. \qquad \text{Limit from the right}$$

Since these two limits are equal, we can apply Theorem 2.8 to conclude that

$$\lim_{x \to 2} g(x) = g(2) = 3.$$

Thus, g is continuous at $x = 2$, and consequently it is continuous on the entire interval $[-1, 3]$. The graph of g is shown in Figure 2.18. ▭

Properties of continuity

In Section 2.1, we looked at several properties of limits. Each property yields a corresponding property pertaining to the continuity of a function. The proof of Theorem 2.9 follows directly from Theorem 2.3.

THEOREM 2.9
PROPERTIES OF CONTINUITY

If b is a real number and f and g are continuous at $x = c$, then the following functions are also continuous at c.

1. Scalar multiple: bf 2. Sum and difference: $f \pm g$

3. Product: fg 4. Quotient: $\dfrac{f}{g}$, if $g(c) \ne 0$

The following summary lists some common types of functions that are continuous at every point in their domain.

1. Polynomial functions: $p(x) = a_n x^n + a_{n-1} x^{n-1} + \cdots + a_1 x + a_0$

2. Rational functions: $r(x) = \dfrac{p(x)}{q(x)}, \quad q(x) \ne 0$

3. Radical functions: $f(x) = \sqrt[n]{x}$

By combining Theorem 2.9 with this summary, we can conclude that a wide variety of elementary functions are continuous at every point in their domains. For example, the following function is continuous at every point in its domain.

$$f(x) = \frac{x^2 + 1}{\sqrt{x}}$$

The next theorem, which is a consequence of Theorem 2.7, allows us to determine the continuity of a *composite* function, such as $f(x) = \sqrt{x^2 + 1}$. A proof of this theorem is given in Appendix A.

**THEOREM 2.10
CONTINUITY OF A
COMPOSITE FUNCTION**

If g is continuous at c and f is continuous at $g(c)$, then the composite function given by $(f \circ g)(x) = f(g(x))$ is continuous at c.

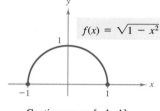

$f(x) = \sqrt{1 - x^2}$

Continuous on $[-1, 1]$

REMARK   Note that one consequence of the theorem is that if f and g satisfy the given conditions, we can determine the limit of $f(g(x))$ as x approaches c to be

$$\lim_{x \to c} f(g(x)) = f\left(\lim_{x \to c} g(x)\right) = f(g(c)).$$

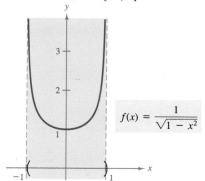

$f(x) = \dfrac{1}{\sqrt{1 - x^2}}$

Continuous on $(-1, 1)$

EXAMPLE 4   Testing for continuity

Find the intervals for which the three functions shown in Figure 2.19 are continuous.

SOLUTION

(a) The function $f(x) = \sqrt{1 - x^2}$ is continuous on the *closed* interval $[-1, 1]$.
(b) The function $f(x) = 1/\sqrt{1 - x^2}$ is continuous on the *open* interval $(-1, 1)$. (Note that f is *undefined* for all x such that $|x| \geq 1$.)
(c) At $x = \pm 1$, the limits from the right and left are zero. Thus, the function $f(x) = |x^2 - 1|$ is continuous on the entire real line, that is, the interval $(-\infty, \infty)$.

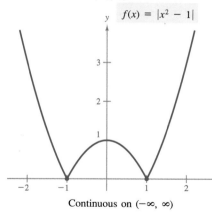

$f(x) = |x^2 - 1|$

Continuous on $(-\infty, \infty)$

Intermediate Value Theorem

We conclude this section with an important theorem concerning the behavior of functions that are continuous on a closed interval.

FIGURE 2.19

THEOREM 2.11 **INTERMEDIATE VALUE THEOREM**	If f is continuous on $[a, b]$ and k is any number between $f(a)$ and $f(b)$, then there is at least one number c in $[a, b]$ such that $f(c) = k$.

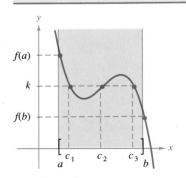

f is continuous.
(For k, there exist 3 c's.)

FIGURE 2.20

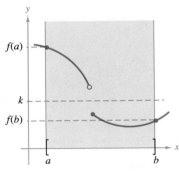

f is not continuous.
(For k, there are no c's.)

FIGURE 2.21

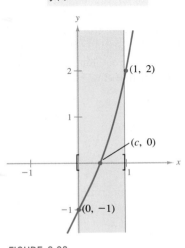

$f(x) = x^3 + 2x - 1$

FIGURE 2.22

By referring to a text on advanced calculus, you will find that a proof of this theorem is based on a property of real numbers called *completeness*. The Intermediate Value Theorem states that for a continuous function f, if x takes on all values between a and b, then $f(x)$ must take on all values between $f(a)$ and $f(b)$. As a simple example of this theorem, consider a person's height. Suppose that a boy is 5 feet tall on his thirteenth birthday and 5 feet 7 inches tall on his fourteenth birthday. Then, for any height h between 5 feet and 5 feet 7 inches, there must have been a time t when his height was exactly h. This seems reasonable because we believe that normal human growth is continuous and that a person's height could not abruptly change from one value to another.

The Intermediate Value Theorem guarantees the existence of *at least one* number c in the closed interval $[a, b]$. There may, of course, be more than one number c such that $f(c) = k$, as shown in Figure 2.20. A discontinuous function might not possess the intermediate value property. For example, the graph of the discontinuous function shown in Figure 2.21 jumps over the horizontal line given by $y = k$, and for this function there is no value of c in $[a, b]$ such that $f(c) = k$.

The Intermediate Value Theorem often can be used to locate the zeros of a function that is continuous on a closed interval. Specifically, if f is continuous on $[a, b]$ and $f(a)$ and $f(b)$ differ in sign, then the Intermediate Value Theorem guarantees the existence of at least one zero of f in the closed interval $[a, b]$.

EXAMPLE 5 An application of the Intermediate Value Theorem

Use the Intermediate Value Theorem to show that the polynomial function $f(x) = x^3 + 2x - 1$ has a zero in the interval $[0, 1]$.

SOLUTION

Since

$$f(0) = 0^3 + 2(0) - 1 = -1 \qquad f(0) < 0$$

and

$$f(1) = 1^3 + 2(1) - 1 = 2 \qquad f(1) > 0$$

we can apply the Intermediate Value Theorem to conclude that there must be some c in $[0, 1]$ such that $f(c) = 0$, as shown in Figure 2.22. ▭

REMARK Note that the Intermediate Value Theorem tells us that (at least) one c exists in the interval (a, b). However, the theorem does not give us a method for finding c. Such theorems are called **existence theorems**.

EXERCISES for Section 2.3

In Exercises 1–6, find the points of discontinuity (if any).

1. $f(x) = -\dfrac{x^3}{2}$

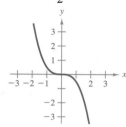

2. $f(x) = \dfrac{x^2 - 1}{x}$

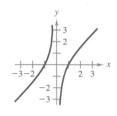

3. $f(x) = \dfrac{x^2 - 1}{x + 1}$

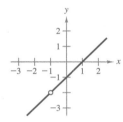

4. $f(x) = \dfrac{1}{x^2 - 4}$

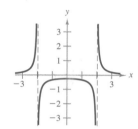

5. $f(x) = \begin{cases} x, & x < 1 \\ 2, & x = 1 \\ 2x - 1, & x > 1 \end{cases}$

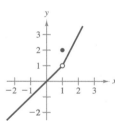

6. $f(x) = \dfrac{[\![x]\!]}{2} + x$

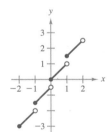

In Exercises 7–24, find the discontinuities (if any) for the given function. Which of the discontinuities are removable?

7. $f(x) = x^2 - 2x + 1$

8. $f(x) = \dfrac{1}{x^2 + 1}$

9. $f(x) = \dfrac{1}{x - 1}$

10. $f(x) = \dfrac{x}{x^2 - 1}$

11. $f(x) = \dfrac{x}{x^2 + 1}$

12. $f(x) = \dfrac{x - 3}{x^2 - 9}$

13. $f(x) = \dfrac{x + 2}{x^2 + 3x - 10}$

14. $f(x) = \dfrac{x - 1}{x^2 + x - 2}$

15. $f(x) = \begin{cases} x, & x \le 1 \\ x^2, & x > 1 \end{cases}$

16. $f(x) = \begin{cases} -2x + 3, & x < 1 \\ x^2, & x \ge 1 \end{cases}$

17. $f(x) = \begin{cases} \dfrac{x}{2} + 1, & x \le 2 \\ 3 - x, & x > 2 \end{cases}$

18. $f(x) = \begin{cases} -2x, & x \le 2 \\ x^2 - 4x + 1, & x > 2 \end{cases}$

19. $f(x) = \dfrac{|x + 2|}{x + 2}$

20. $f(x) = \dfrac{|x - 3|}{x - 3}$

21. $f(x) = \begin{cases} |x - 2| + 3, & x < 0 \\ x + 5, & x \ge 0 \end{cases}$

22. $f(x) = \begin{cases} 3 + x, & x \le 2 \\ x^2 + 1, & x > 2 \end{cases}$

23. $f(x) = [\![x - 1]\!]$

24. $f(x) = x - [\![x]\!]$

In Exercises 25–30, discuss the continuity of the composite function $h(x) = f(g(x))$.

25. $f(x) = x^2$, $g(x) = x - 1$

26. $f(x) = \dfrac{1}{\sqrt{x}}$, $g(x) = x - 1$

27. $f(x) = \dfrac{1}{x - 1}$, $g(x) = x^2 + 5$

28. $f(x) = \sqrt{x}$, $g(x) = x^2$

29. $f(x) = \dfrac{1}{x}$, $g(x) = \dfrac{1}{x - 1}$

30. $f(x) = \dfrac{1}{\sqrt{x}}$, $g(x) = \dfrac{1}{x}$

In Exercises 31–34, sketch the graph of the given function to determine any points of discontinuity.

31. $f(x) = \dfrac{x^2 - 16}{x - 4}$

32. $f(x) = \dfrac{x^3 - 8}{x - 2}$

33. $f(x) = [\![x]\!] - x$

34. $f(x) = \dfrac{|x^2 - 1|}{x}$

In Exercises 35–38, find the interval(s) for which the function is continuous.

35. $f(x) = \dfrac{x^2}{x^2 - 36}$

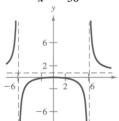

36. $f(x) = x\sqrt{x + 3}$

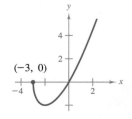

37. $f(x) = \dfrac{x}{x^2 + 1}$

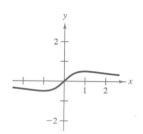

38. $f(x) = \dfrac{x + 1}{\sqrt{x}}$

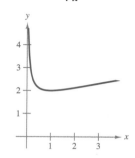

In Exercises 39 and 40, prove that the given function has a zero in the indicated interval.

Function	Interval
39. $f(x) = x^2 - 4x + 3$	[2, 4]
40. $f(x) = x^3 + 3x - 2$	[0, 1]

In Exercises 41 and 42, use the Intermediate Value Theorem to approximate the zero of the given function in the interval [0, 1]. (a) Begin by locating the zero in a subinterval of length 0.1. (b) Refine your approximation by locating the zero in a subinterval of length 0.01.

41. $f(x) = x^3 + x - 1$ **42.** $f(x) = x^3 + 3x - 2$

In Exercises 43–46, verify the applicability of the Intermediate Value Theorem in the indicated interval and find the value of c guaranteed by the theorem.

43. $f(x) = x^2 + x - 1$, [0, 5], $f(c) = 11$
44. $f(x) = x^2 - 6x + 8$, [0, 3], $f(c) = 0$
45. $f(x) = x^3 - x^2 + x - 2$, [0, 3], $f(c) = 4$

$$f(x) = \dfrac{x^2 + x}{x - 1}, \left[\dfrac{5}{2}, 4\right], f(c) = 6$$

47. Determine the constant a so that the following function is continuous on the entire real line.

$$f(x) = \begin{cases} x^3, & x \le 2 \\ ax^2, & x > 2 \end{cases}$$

48. Determine the constants a and b so that the following function is continuous on the entire real line.

$$f(x) = \begin{cases} 2, & x \le -1 \\ ax + b, & -1 < x < 3 \\ -2, & x \ge 3 \end{cases}$$

49. Is the function $f(x) = \sqrt{1 - x^2}$ continuous at $x = 1$? Give the reason for your answer.

50. A union contract guarantees a 9 percent annual salary increase for five years. For an initial salary of $28,500, the salary S is given by

$$S = 28,500(1.09)^{[\![t]\!]}$$

where $t = 0$ corresponds to 1985. Sketch a graph of this function and discuss its continuity.

51. A dial-direct long distance call between two cities costs $1.04 for the first two minutes and $0.36 for each additional minute or fraction thereof. Use the greatest integer function to write the cost C of a call in terms of the time t (in minutes). Sketch a graph of this function and discuss its continuity.

52. The number of units in inventory in a small company is given by

$$N(t) = 25\left(2\left[\!\left[\dfrac{t + 2}{2}\right]\!\right] - t\right)$$

where t is the time in months. Sketch the graph of this function and discuss its continuity. How often must this company replenish its inventory?

53. Use a computer or graphics calculator to sketch the graph of the function f and determine whether it is continuous on the entire real line.

$$f(x) = \begin{cases} 2x - 4, & x \le 3 \\ x^2 - 2x, & x > 3 \end{cases}$$

54. At 8:00 A.M. on Saturday a man begins running up the side of a mountain to his weekend campsite. On Sunday at 8:00 A.M. he runs back down the mountain. It takes him 20 minutes to run up, but only 10 minutes to run down. At some point on the way down he realizes that he passed the same place at the exact same time on Saturday. Prove that he is correct. [Hint: Let $s(t)$ and $r(t)$ be the position functions for the run up and down, respectively, and apply the Intermediate Value Theorem to the function $f(t) = s(t) - r(t)$.]

55. Prove Theorem 2.9.

56. Prove that if f is continuous and has no zeros on $[a, b]$, then either

$$f(x) > 0 \quad \text{for all } x \text{ in } [a, b]$$

or

$$f(x) < 0 \quad \text{for all } x \text{ in } [a, b].$$

57. The Dirichlet function f is defined as follows.

$$f(x) = \begin{cases} 0, & x \text{ is rational} \\ 1, & x \text{ is irrational} \end{cases}$$

Show that this function is discontinuous at every real number.

2.4 Infinite Limits

Infinite limits ▪ Vertical asymptotes

In this section we look at another important way in which a limit can fail to exist. We begin with an example. Let f be the function given by

$$f(x) = \frac{3}{x-2}.$$

From Figure 2.23 and Table 2.4, we can see that $f(x)$ *decreases without bound* as x approaches 2 from the left, and $f(x)$ *increases without bound* as x approaches 2 from the right. Symbolically, we write

$$\lim_{x \to 2^-} \frac{3}{x-2} = -\infty \quad \text{and} \quad \lim_{x \to 2^+} \frac{3}{x-2} = \infty.$$

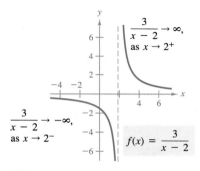

FIGURE 2.23

TABLE 2.4

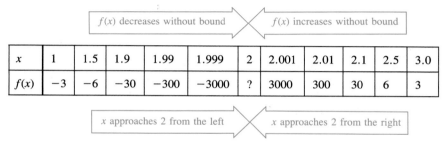

			$f(x)$ decreases without bound				$f(x)$ increases without bound				
x	1	1.5	1.9	1.99	1.999	2	2.001	2.01	2.1	2.5	3.0
$f(x)$	−3	−6	−30	−300	−3000	?	3000	300	30	6	3
			x approaches 2 from the left				x approaches 2 from the right				

In general, the types of limits where $f(x)$ increases or decreases without bound as x approaches c are called **infinite limits.**

DEFINITION OF INFINITE LIMITS

The statement

$$\lim_{x \to c} f(x) = \infty$$

means that $f(x)$ *increases* without bound as x approaches c. The statement

$$\lim_{x \to c} f(x) = -\infty$$

means that $f(x)$ *decreases* without bound as x approaches c.

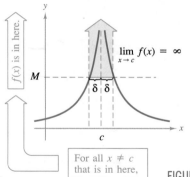

FIGURE 2.24

REMARK The equal sign in the statement $\lim f(x) = \infty$ does not mean that the limit exists! On the contrary, it tells us how the limit *fails to exist* by denoting the unbounded behavior of $f(x)$ as x approaches c. Thus, when we say "the limit of $f(x)$ is infinite as x approaches c" we really mean that "the limit does not exist *and f* has an infinite discontinuity at $x = c$."

To say that $f(x)$ increases without bound as $x \to c$ means that for each $M > 0$, there exists an open interval I containing c such that $f(x) > M$ for all x in I (other than $x = c$), as illustrated in Figure 2.24. A similar interpretation is given to define what it means for $f(x)$ to decrease without bound.

Infinite limits from the right and left are defined similarly. The four possible one-sided infinite limits are

$$\lim_{x \to c^-} f(x) = -\infty, \qquad \lim_{x \to c^-} f(x) = \infty \qquad \text{Infinite limits from the left}$$

$$\lim_{x \to c^+} f(x) = -\infty, \qquad \lim_{x \to c^+} f(x) = \infty \qquad \text{Infinite limits from the right}$$

If $f(x) \to \infty$ (or $f(x) \to -\infty$) from the left or from the right, we say that f has an **infinite discontinuity** at $x = c$.

EXAMPLE 1   Determining infinite limits from a graph

Use Figure 2.25 to determine the limit of each function as $x \to 1$ from the left and from the right.

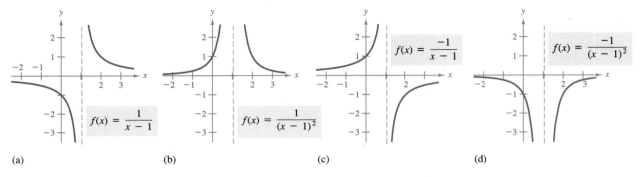

(a)   (b)   (c)   (d)

FIGURE 2.25

SOLUTION

(a) $\displaystyle \lim_{x \to 1^-} \frac{1}{x - 1} = -\infty$ and $\displaystyle \lim_{x \to 1^+} \frac{1}{x - 1} = \infty$

(b) $\displaystyle \lim_{x \to 1} \frac{1}{(x - 1)^2} = \infty$ Limit from both sides is ∞

(c) $\displaystyle \lim_{x \to 1^-} \frac{-1}{x - 1} = \infty$ and $\displaystyle \lim_{x \to 1^+} \frac{-1}{x - 1} = -\infty$

(d) $\displaystyle \lim_{x \to 1} \frac{-1}{(x - 1)^2} = -\infty$ Limit from both sides is $-\infty$

If it were possible to extend the graphs in Figure 2.25 up and down toward infinity, you would see that each graph becomes arbitrarily close to the vertical line $x = 1$. We call this line a **vertical asymptote** of the graph of f.

DEFINITION OF VERTICAL ASYMPTOTE	If $f(x)$ approaches infinity (or negative infinity) as x approaches c from the right or left, then we call the line $x = c$ a **vertical asymptote** of the graph of f.

REMARK   It should be clear that if a function f has a vertical asymptote at $x = c$, then f is *discontinuous* at c.

In Example 1, note that each of the functions is a *quotient* and that the vertical asymptote occurs at a number where the denominator is zero (and the numerator is not zero). The following theorem generalizes this observation. (A proof of the theorem is given in Appendix A.)

THEOREM 2.12
VERTICAL ASYMPTOTES

Let f and g be continuous on an open interval containing c. If $f(c) \neq 0$, $g(c) = 0$, and there exists an open interval containing c such that $g(x) \neq 0$ for all $x \neq c$ in the interval, then the graph of the function given by

$$h(x) = \frac{f(x)}{g(x)}$$

has a vertical asymptote at $x = c$.

EXAMPLE 2   Finding vertical asymptotes

Determine all vertical asymptotes for the graphs of the following functions.

(a) $f(x) = \dfrac{1}{2(x + 1)}$ (b) $f(x) = \dfrac{x^2 + 1}{x^2 - 1}$

SOLUTION

(a) When $x = -1$, the denominator is zero and the numerator is not zero. Hence, by Theorem 2.12, we conclude that $x = -1$ is a vertical asymptote, as shown in Figure 2.26(a).

(b) By factoring the denominator as $(x^2 - 1) = (x - 1)(x + 1)$, we see that the denominator is zero at $x = -1$ and $x = 1$. Moreover, since the numerator is not zero at these two points, we apply Theorem 2.12 to conclude that the graph of f has the two vertical asymptotes shown in Figure 2.26(b).

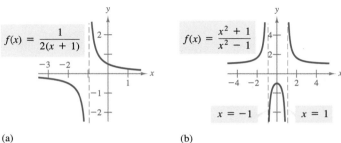

FIGURE 2.26   (a)   (b)

Theorem 2.12 requires the value of the numerator at $x = c$ to be nonzero. If both the numerator and denominator are zero at $x = c$, then we obtain the *indeterminate form* $0/0$ and we cannot determine the limit behavior at $x = c$ without further investigation. One case for which we can determine the limit is when the numerator and denominator are polynomials. We do this by cancelling common factors, as shown in the next example.

EXAMPLE 3 A rational function with common factors

Determine all vertical asymptotes of the graph of

$$f(x) = \frac{x^2 + 2x - 8}{x^2 - 4}.$$

SOLUTION

By factoring both the numerator and the denominator, we have

$$f(x) = \frac{x^2 + 2x - 8}{x^2 - 4} = \frac{(x + 4)(x - 2)}{(x + 2)(x - 2)} = \frac{x + 4}{x + 2}, \quad x \neq 2.$$

Now, at all x-values other than $x = 2$, the graph of f coincides with the graph of $g(x) = (x + 4)/(x + 2)$. Thus, we apply Theorem 2.12 to g to conclude that there is a vertical asymptote at $x = -2$, as shown in Figure 2.27. Note that $x = 2$ is *not* a vertical asymptote. ▭

Since the graph of f in Example 3 has a vertical asymptote at $x = -2$, we know that the limit as $x \to -2$ from the right (or the left) is either ∞ or $-\infty$. But without looking at the graph, how can we determine the following limits?

$$\lim_{x \to -2^-} \frac{x^2 + 2x - 8}{x^2 - 4} = -\infty \quad \text{and} \quad \lim_{x \to -2^+} \frac{x^2 + 2x - 8}{x^2 - 4} = \infty$$

Once you have determined that the graph of a function has a vertical asymptote at a particular x-value, we suggest that you settle the question of whether $f(x)$ approaches positive or negative infinity, using the graphical approach outlined in the following example.

EXAMPLE 4 Determining infinite limits

Find the following limits.

$$\lim_{x \to 1^-} \frac{x^2 - 3x}{x - 1} \quad \text{and} \quad \lim_{x \to 1^+} \frac{x^2 - 3x}{x - 1}$$

SOLUTION

1. Factor both numerator and denominator. (Cancel any common factors.)

$$f(x) = \frac{x^2 - 3x}{x - 1} = \frac{x(x - 3)}{x - 1}$$

2. The (remaining) factors of the numerator yield the x-intercepts of the function. In this case, the x-intercepts occur at $(0, 0)$ and $(3, 0)$.
3. The (remaining) factors of the denominator yield the vertical asymptotes of the function. In this case, a vertical asymptote occurs at $x = 1$.

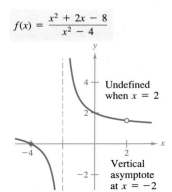

$$f(x) = \frac{x^2 + 2x - 8}{x^2 - 4}$$

Undefined when $x = 2$

Vertical asymptote at $x = -2$

FIGURE 2.27

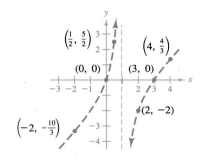

(a)

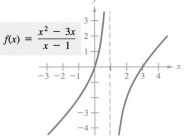

(b)

FIGURE 2.28

4. Determine a few additional points on the graph, making certain that at least one point between each intercept and asymptote is chosen. Then plot the information obtained in the first four steps, as shown in Figure 2.28(a).

x	-2	$\frac{1}{2}$	2	4
$f(x)$	$-\frac{10}{3}$	$\frac{5}{2}$	-2	$\frac{4}{3}$

5. Finally, we complete the sketch as shown in Figure 2.28(b) and conclude that

$$\lim_{x \to 1^-} \frac{x^2 - 3x}{x - 1} = \infty \quad \text{and} \quad \lim_{x \to 1^+} \frac{x^2 - 3x}{x - 1} = -\infty. \qquad \square$$

We conclude this section with a theorem involving the limits of sums, products, and quotients of functions.

THEOREM 2.13
PROPERTIES OF INFINITE LIMITS

If c and L are real numbers and f and g are functions such that

$$\lim_{x \to c} f(x) = \infty \quad \text{and} \quad \lim_{x \to c} g(x) = L$$

then the following properties are true.

1. Sum or difference: $\lim_{x \to c} [f(x) \pm g(x)] = \infty$

2. Quotient: $\lim_{x \to c} \dfrac{g(x)}{f(x)} = 0$

3. Product: $\lim_{x \to c} [f(x)g(x)] = \infty, \quad L > 0$

   $\lim_{x \to c} [f(x)g(x)] = -\infty, \quad L < 0$

Similar properties hold for one-sided limits and for functions for which the limit of $f(x)$ as x approaches c is $-\infty$.

EXAMPLE 5   Determining limits

Find the following limits.

(a) $\displaystyle\lim_{x \to 0} \left(1 + \frac{1}{x^2} \right)$ (b) $\displaystyle\lim_{x \to 1^-} \frac{x^2 + 1}{1/(x - 1)}$

SOLUTION

(a) Since $\lim\limits_{x\to 0} (1) = 1$ and $\lim\limits_{x\to 0} (1/x^2) = \infty$, we can apply Property 1 of Theorem 2.13 to conclude that

$$\lim_{x\to 0}\left(1 + \frac{1}{x^2}\right) = \infty.$$

(b) Since $\lim\limits_{x\to 1^-} (x^2 + 1) = 2$ and $\lim\limits_{x\to 1^-} [1/(x - 1)] = -\infty$, we can apply Property 2 of Theorem 2.13 to conclude that

$$\lim_{x\to 1^-} \frac{x^2 + 1}{1/(x - 1)} = 0.$$

∎

EXERCISES for Section 2.4

In Exercises 1 and 2, determine whether $f(x)$ approaches ∞ or $-\infty$ as x approaches -2 from the left and from the right.

1. $f(x) = \dfrac{1}{(x + 2)^2}$

2. $f(x) = \dfrac{1}{x + 2}$

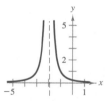

In Exercises 3–6, determine whether $f(x)$ approaches ∞ or $-\infty$ as x approaches -3 from the left and from the right.

3. $f(x) = \dfrac{1}{x^2 - 9}$

4. $f(x) = \dfrac{x}{x^2 - 9}$

5. $f(x) = \dfrac{x^3}{x^2 - 9}$

6. $f(x) = \dfrac{x^2}{x^2 - 9}$

In Exercises 7 and 8, find the vertical asymptotes of the given function.

7. $f(x) = \dfrac{x^2 - 2}{x^2 - x - 2}$

8. $f(x) = \dfrac{x^3}{x^2 - 1}$

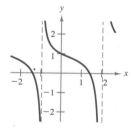

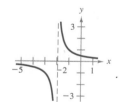

In Exercises 9–18, find the vertical asymptotes (if any) of the given function.

9. $f(x) = \dfrac{1}{x^2}$

10. $f(x) = \dfrac{4}{(x - 2)^3}$

11. $f(x) = \dfrac{x^2}{x^2 + x - 2}$

12. $f(x) = \dfrac{2 + x}{1 - x}$

13. $f(x) = \dfrac{x^3}{x^2 - 4}$

14. $f(x) = \dfrac{-4x}{x^2 + 4}$

15. $f(x) = 1 - \dfrac{4}{x^2}$

16. $f(x) = \dfrac{-2}{(x - 2)^2}$

17. $f(x) = \dfrac{x}{x^2 + x - 2}$

18. $f(x) = \dfrac{1}{(x + 3)^4}$

In Exercises 19–22, determine whether the given function has a vertical asymptote or a removable discontinuity at $x = -1$.

19. $f(x) = \dfrac{x^2 - 1}{x + 1}$

20. $f(x) = \dfrac{x^2 - 6x - 7}{x + 1}$

21. $f(x) = \dfrac{x^2 + 1}{x + 1}$

22. $f(x) = \dfrac{x - 1}{x + 1}$

In Exercises 23–32, find the indicated limit.

23. $\lim\limits_{x\to 2^+} \dfrac{x - 3}{x - 2}$

24. $\lim\limits_{x\to 1^+} \dfrac{2 + x}{1 - x}$

25. $\lim\limits_{x\to 4} \dfrac{x^2}{x^2 - 16}$

26. $\lim\limits_{x\to 4} \dfrac{x^2}{x^2 + 16}$

27. $\lim\limits_{x\to 0^-} \left(1 + \dfrac{1}{x}\right)$

28. $\lim\limits_{x\to 0^-} \left(x^2 - \dfrac{1}{x}\right)$

29. $\lim\limits_{x\to 1} \dfrac{x^2 - x}{(x^2 + 1)(x - 1)}$

30. $\lim\limits_{x\to 1} \dfrac{x^3 - 1}{x^2 + x + 1}$

31. $\lim\limits_{x\to 1^+} \dfrac{x^2 + x + 1}{x^3 - 1}$

32. $\lim\limits_{x\to 0^-} \dfrac{x^2 - 2x}{x^3}$

In Exercises 33–38, find the indicated limit (if it exists), given that

$$f(x) = \frac{1}{(x - 4)^2} \quad \text{and} \quad g(x) = x^2 - 5x.$$

33. $\lim\limits_{x \to 4} f(x)$

34. $\lim\limits_{x \to 4} g(x)$

35. $\lim\limits_{x \to 4} [f(x) + g(x)]$

36. $\lim\limits_{x \to 4} [f(x)g(x)]$

37. $\lim\limits_{x \to 4} \left[\dfrac{f(x)}{g(x)} \right]$

38. $\lim\limits_{x \to 4} \left[\dfrac{g(x)}{f(x)} \right]$

39. The cost in dollars of removing p percent of the air pollutants from the stack emission of a utility company that burns coal to generate electricity is

$$C = \frac{80,000p}{100 - p}, \quad 0 \le p < 100.$$

(a) Find the cost of removing 15 percent.
(b) Find the cost of removing 50 percent.
(c) Find the cost of removing 90 percent.
(d) Find the limit of C as $x \to 100^-$.

40. The cost in millions of dollars for the federal government to seize x percent of a certain illegal drug as it enters the country is given by

$$C = \frac{528x}{100 - x}, \quad 0 \le x < 100.$$

(a) Find the cost of seizing 25 percent.
(b) Find the cost of seizing 50 percent.
(c) Find the cost of seizing 75 percent.
(d) Find the limit of C as $x \to 100^-$.

41. A 25-foot ladder is leaning against a house, as shown in the figure. If the base of the ladder is pulled away from the house at a rate of 2 feet per second, the top will move down the wall at a rate of

$$r = \frac{2x}{\sqrt{625 - x^2}} \text{ ft/sec.}$$

(a) Find the rate when x is 7 feet.
(b) Find the rate when x is 15 feet.
(c) Find the limit of r as $x \to 25^-$.

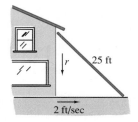

FIGURE FOR 41

42. Coulomb's Law states that the force F of a point charge q_1 on a point charge q_2, when the charges are r units apart, is proportional to the product of the charges and inversely proportional to the square of the distance between them. If a point particle with a charge of $+1$ is placed on a line between two particles 5 units apart, each with a charge of -1, the net force on the particle with a positive charge is given by

$$F = -\frac{k}{x^2} + \frac{k}{(x - 5)^2}, \quad 0 < x < 5$$

where x is the distance shown in the figure. Sketch the graph of F.

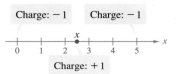

FIGURE FOR 42

43. Use a computer or graphics calculator to sketch the graph of the function

$$f(x) = \frac{1}{x^2 - 25}$$

and find $\lim\limits_{x \to 5^-} f(x)$.

44. Find functions f and g such that

$$\lim\limits_{x \to c} f(x) = \infty \quad \text{and} \quad \lim\limits_{x \to c} g(x) = \infty$$

but

$$\lim\limits_{x \to c} [f(x) - g(x)] \ne 0.$$

2.5 ε-δ Definition of Limits

Formal definition of limit ▪ Proofs of limit theorems ▪ Formal definition of infinite limit ▪ The Squeeze Theorem

Augustin-Louis Cauchy

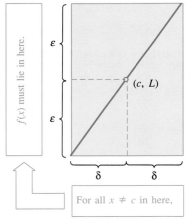

$$\lim_{x \to c} f(x) = L$$

We begin this section with another look at our informal description of a limit given at the beginning of this chapter. If $f(x)$ becomes arbitrarily close to a single number L as x approaches c from either side, then we say that the **limit** of $f(x)$, as x approaches c, is L, and we write

$$\lim_{x \to c} f(x) = L.$$

At first glance, this description looks fairly technical—so why do we call it informal? The answer lies in the exact meaning of the two phrases

"$f(x)$ becomes arbitrarily close to L"   and   "x approaches c."

The first person to assign a mathematically rigorous meaning to these two phrases was Augustin-Louis Cauchy (1789–1857). His ε-δ definition of a limit is the standard used today. (ε is the lowercase Greek letter *epsilon* and δ is the lowercase Greek *delta*.)

In Figure 2.29, let ε represent a (small) positive number. Then the phrase "$f(x)$ becomes arbitrarily close to L" means that $f(x)$ lies in the interval $(L - \varepsilon, L + \varepsilon)$. In terms of absolute value, we write this as

$$|f(x) - L| < \varepsilon.$$

Similarly, the phrase "x approaches c" means that there exists a positive number δ such that x lies in either the interval $(c - \delta, c)$ or the interval $(c, c + \delta)$. This fact can be concisely expressed by the double inequality

$$0 < |x - c| < \delta.$$

The first inequality, $0 < |x - c|$, expresses the fact that $x \neq c$, while the second, $|x - c| < \delta$, says that x is within a distance δ of c.

This brings us to the following formal ε-δ definition of a **limit.**

FIGURE 2.29

DEFINITION OF LIMIT

Let f be a function defined on an open interval containing c (except possibly at c) and let L be a real number. The statement

$$\lim_{x \to c} f(x) = L$$

means that for each $\varepsilon > 0$ there exists a $\delta > 0$ such that

$$|f(x) - L| < \varepsilon \qquad \text{whenever} \qquad 0 < |x - c| < \delta.$$

REMARK   Note that the inequality $0 < |x - c| < \delta$ implies that $x \neq c$. In other words, the value of the limit of $f(x)$ as $x \to c$ does not depend on the value of $f(x)$ at c.

Note that the definition has an *order* to it: "For each $\varepsilon > 0$ there exists a $\delta > 0$." First we are given an ε-value, then we must find an appropriate δ-value. We do not require any specific δ to work for more than one choice of ε. Furthermore, the number δ is not unique, for if a specific δ works, then any smaller positive number will also work. To gain a better understanding of the relationship between ε and δ, consider the graphical description shown in Figure 2.30.

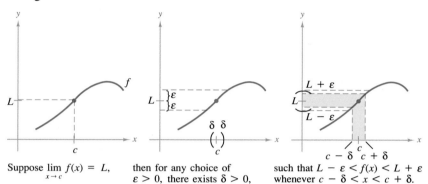

FIGURE 2.30

Suppose $\lim_{x \to c} f(x) = L$,   then for any choice of $\varepsilon > 0$, there exists $\delta > 0$,   such that $L - \varepsilon < f(x) < L + \varepsilon$ whenever $c - \delta < x < c + \delta$.

The next three examples illustrate some ways to determine a δ-value for a given ε.

EXAMPLE 1   Finding a δ for a given ε

Given the limit

$$\lim_{x \to 3} (2x - 5) = 1$$

find δ such that $|(2x - 5) - 1| < 0.01$ whenever $0 < |x - 3| < \delta$.

SOLUTION

In this problem, we are working with a given value of ε, namely $\varepsilon = 0.01$. To find an appropriate δ, we try to establish a connection between the two absolute values $|(2x - 5) - 1|$ and $|x - 3|$. By simplifying the first absolute value, we get

$$|(2x - 5) - 1| = |2x - 6| = 2|x - 3|.$$

In other words, the inequality $|(2x - 5) - 1| < 0.01$ is equivalent to $2|x - 3| < 0.01$, and we have

$$|x - 3| < \frac{0.01}{2} = 0.005.$$

Thus, we choose $\delta = 0.005$. This choice works because $0 < |x - 3| < 0.005$ implies that

$$|(2x - 5) - 1| = 2|x - 3| < 2(0.005) = 0.01$$

as indicated in Figure 2.31.

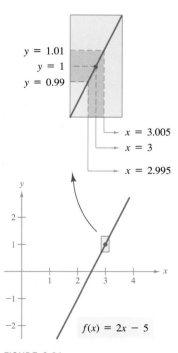

$y = 1.01$
$y = 1$
$y = 0.99$

$x = 3.005$
$x = 3$
$x = 2.995$

$f(x) = 2x - 5$

FIGURE 2.31

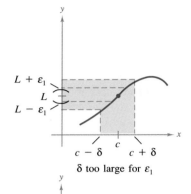

δ too large for ε_1

δ_1 sufficiently small for ε_1

FIGURE 2.32

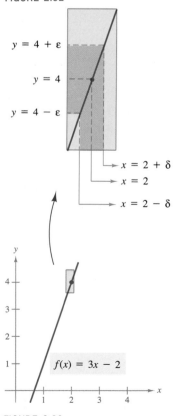

FIGURE 2.33

REMARK   Note in Example 1 that 0.005 is the *largest* value of δ that will guarantee $|(2x - 5) - 1| < 0.01$ whenever $0 < |x - 3| < \delta$. Any *smaller* positive value of δ would, of course, also work.

In Example 1, we found a δ-value for a *particular* ε. Figure 2.32 shows how a new ε-value, ε_1, can require a new choice for δ. The original value for δ is now too large, and we must find a smaller value δ_1 that works for ε_1. In other words, finding a δ-value for a particular ε does not prove the existence of the limit. To do that, we must prove that we can find δ for *any* ε, as demonstrated in the next example.

EXAMPLE 2   Using the ε-δ definition of a limit

Use the ε-δ definition of a limit to prove that

$$\lim_{x \to 2} (3x - 2) = 4.$$

SOLUTION

We are required to show that for each $\varepsilon > 0$, there exists a $\delta > 0$ such that

$$|(3x - 2) - 4| < \varepsilon \qquad \text{whenever} \qquad 0 < |x - 2| < \delta.$$

Since our choice for δ depends on ε, we try to establish a connection between the absolute values $|(3x - 2) - 4|$ and $|x - 2|$. By simplifying the first absolute value, we get

$$|(3x - 2) - 4| = |3x - 6| = 3|x - 2|.$$

Thus, the inequality $|(3x - 2) - 4| < \varepsilon$ requires $3|x - 2| < \varepsilon$, so that we have

$$|x - 2| < \frac{\varepsilon}{3}.$$

Finally, we choose $\delta = \varepsilon/3$. This choice works because

$$0 < |x - 2| < \delta = \frac{\varepsilon}{3}$$

implies that

$$|(3x - 2) - 4| = 3|x - 2| < 3\left(\frac{\varepsilon}{3}\right) = \varepsilon$$

as shown in Figure 2.33.   ▭

EXAMPLE 3   Using the ε-δ definition of a limit

Use the ε-δ definition of a limit to prove that

$$\lim_{x \to 2} x^2 = 4.$$

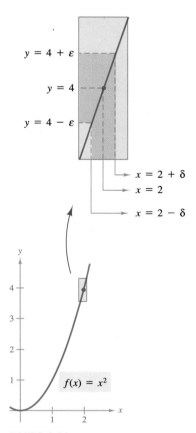

FIGURE 2.34

SOLUTION

We must show that for each $\varepsilon > 0$, there exists a $\delta > 0$ such that

$$|x^2 - 4| < \varepsilon \quad \text{whenever} \quad 0 < |x - 2| < \delta.$$

We begin by writing $|x^2 - 4| = |x - 2||x + 2|$. For all x in the interval $(1, 3)$, we know that $|x + 2| < 5$. Thus, letting δ be the minimum of $\varepsilon/5$ and 1, it follows that whenever $0 < |x - 2| < \varepsilon/5$, we have

$$|x^2 - 4| = |x - 2||x + 2| < \left(\frac{\varepsilon}{5}\right)(5) = \varepsilon$$

as shown in Figure 2.34.

Example 4 uses the ε-δ definition to prove the nonexistence of a limit.

EXAMPLE 4   A limit that does not exist

Show that the limit of $f(x)$ as x approaches 0 *does not* exist for the function

$$f(x) = \begin{cases} 1, & x \le 0 \\ 2, & x > 0. \end{cases}$$

SOLUTION

Assume that the limit of $f(x)$ as $x \to 0$ exists and is equal to L. We choose $\varepsilon = \frac{1}{2}$, which is less than the jump discontinuity at $x = 0$, as shown in Figure 2.35. Then, if there exists a $\delta > 0$ such that

$$|f(x) - L| < \frac{1}{2} \quad \text{whenever} \quad 0 < |x - 0| < \delta$$

the limit from the left requires

$$|1 - L| < \frac{1}{2} \quad \text{whenever} \quad -\delta < x < 0$$

and the limit from the right requires

$$|2 - L| < \frac{1}{2} \quad \text{whenever} \quad 0 < x < \delta.$$

Together these statements imply that

$$-\frac{1}{2} < L - 1 < \frac{1}{2} \quad \text{and} \quad -\frac{1}{2} < L - 2 < \frac{1}{2}$$

or

$$\frac{1}{2} < L < \frac{3}{2} \quad \text{and} \quad \frac{3}{2} < L < \frac{5}{2}.$$

But no single value for L can satisfy both of these inequalities. Therefore, our assumption that the limit exists is false, and we conclude that the limit of $f(x)$ as $x \to 0$ does not exist.

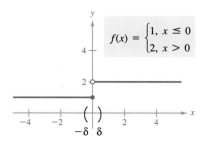

FIGURE 2.35

In Example 4 we made use of one-sided limits. For future reference we provide the formal ε-δ definition of one-sided limits.

FORMAL DEFINITION OF ONE-SIDED LIMITS

1. $\lim\limits_{x \to c^-} f(x) = L$ if for each $\varepsilon > 0$, there exists a $\delta > 0$ such that

$$|f(x) - L| < \varepsilon \qquad \text{whenever} \qquad -\delta < x - c < 0.$$

2. $\lim\limits_{x \to c^+} f(x) = L$ if for each $\varepsilon > 0$, there exists a $\delta > 0$ such that

$$|f(x) - L| < \varepsilon \qquad \text{whenever} \qquad 0 < x - c < \delta.$$

There are several equivalent forms of the limit statement $\lim\limits_{x \to c} f(x) = L$. In the future we will use whichever form is most convenient for the task at hand. Try using the ε-δ definition of a limit to establish the equivalence of the following three forms.

1. $\lim\limits_{x \to c} f(x) = L$
2. $\lim\limits_{x \to c} [f(x) - L] = 0$
3. $\lim\limits_{h \to 0} f(c + h) = L$

Proofs of limit theorems

Your appreciation of the ε-δ definition of a limit should grow as you see how useful it is in proving properties of limits. In the next two examples, we demonstrate the type of argument needed to establish some of the basic theorems identified in Section 2.1.

EXAMPLE 5 Using the ε-δ definition to prove Theorem 2.1

Prove Theorem 2.1: Let c be a real number and $f(x) = g(x)$ for all $x \neq c$ in an open interval containing c. If the limit of $g(x)$ as $x \to c$ exists, then the limit of $f(x)$ also exists, and

$$\lim_{x \to c} f(x) = \lim_{x \to c} g(x).$$

SOLUTION

Assume that the limit of $g(x)$ as $x \to c$ is L. Then, by the definition of limit, for each $\varepsilon > 0$ there exists a $\delta > 0$ such that $f(x) = g(x)$ in the open intervals $(c - \delta, c)$ and $(c, c + \delta)$, and

$$|g(x) - L| < \varepsilon \qquad \text{whenever} \qquad 0 < |x - c| < \delta.$$

However, since $f(x) = g(x)$ for all x in the open interval other than $x = c$, it follows that

$$|f(x) - L| < \varepsilon \qquad \text{whenever} \qquad 0 < |x - c| < \delta.$$

Thus, we conclude that the limit of $f(x)$ as $x \to c$ is also L. ▭

EXAMPLE 6   Using the ε-δ definition to prove Theorem 2.2

Prove that if c is a real number, then

$$\lim_{x \to c} x = c.$$

SOLUTION

We need to show that for each $\varepsilon > 0$ there exists a $\delta > 0$ such that

$$|x - c| < \varepsilon \qquad \text{whenever} \qquad 0 < |x - c| < \delta.$$

The right-hand inequality is similar to the left-hand one, and we simply choose $\delta = \varepsilon$.   ▭

EXAMPLE 7   Using the ε-δ definition to prove Theorem 2.3

Prove that if the functions f and g have limits as $x \to c$, then

$$\lim_{x \to c} [f(x) + g(x)] = \lim_{x \to c} f(x) + \lim_{x \to c} g(x).$$

SOLUTION

Assume that

$$\lim_{x \to c} f(x) = L \qquad \text{and} \qquad \lim_{x \to c} g(x) = K.$$

Choose $\varepsilon > 0$. Then, since $\varepsilon/2 > 0$, we know that there exists $\delta_1 > 0$ and $\delta_2 > 0$ such that

$$0 < |x - c| < \delta_1 \qquad \text{implies} \qquad |f(x) - L| < \frac{\varepsilon}{2}$$

and

$$0 < |x - c| < \delta_2 \qquad \text{implies} \qquad |g(x) - K| < \frac{\varepsilon}{2}.$$

If δ is the smaller of δ_1 and δ_2, then

$$0 < |x - c| < \delta \quad \text{implies} \quad |f(x) - L| < \frac{\varepsilon}{2} \quad \text{and} \quad |g(x) - K| < \frac{\varepsilon}{2}.$$

Finally, we apply the Triangle Inequality to conclude that

$$|[f(x) + g(x)] - (L + K)| < |f(x) - L| + |g(x) - K| < \frac{\varepsilon}{2} + \frac{\varepsilon}{2} = \varepsilon.$$

Hence

$$0 < |x - c| < \delta \quad \Longrightarrow \quad |f(x) + g(x) - (L + K)| < \varepsilon$$

which implies that

$$\lim_{x \to c} [f(x) + g(x)] = L + K = \lim_{x \to c} f(x) + \lim_{x \to c} g(x).$$

In Section 2.4, we discussed vertical asymptotes and their relationship to infinite limits. The following definition formalizes the definition of an infinite limit.

FORMAL DEFINITION OF INFINITE LIMITS

The statement

$$\lim_{x \to c} f(x) = \infty$$

means that for each $M > 0$ there exists a $\delta > 0$ such that $f(x) > M$ whenever $0 < |x - c| < \delta$.

The statement

$$\lim_{x \to c} f(x) = -\infty$$

means that for each $N < 0$ there exists a $\delta > 0$ such that $f(x) < N$ whenever $0 < |x - c| < \delta$.

To define the **infinite limit from the left,** we replace

$$0 < |x - c| < \delta \quad \text{by} \quad c - \delta < x < c.$$

To define the **infinite limit from the right,** we replace

$$0 < |x - c| < \delta \quad \text{by} \quad c < x < c + \delta.$$

EXAMPLE 8 Using the M-δ definition to prove Theorem 2.13

Prove that if c and L are real numbers and f and g are functions such that

$$\lim_{x \to c} f(x) = \infty \quad \text{and} \quad \lim_{x \to c} g(x) = L$$

then

$$\lim_{x \to c} [f(x) + g(x)] = \infty.$$

SOLUTION

To show that the limit of $f(x) + g(x)$ is infinite, we choose $M > 0$, and we need to find $\delta > 0$ such that

$$[f(x) + g(x)] > M \quad \text{whenever} \quad 0 < |x - c| < \delta.$$

For simplicity's sake, we assume L is positive, and let $M_1 = M + 1$. Since the limit of $f(x)$ is infinite, there exists δ_1 such that

$$f(x) > M_1 \qquad \text{whenever} \qquad 0 < |x - c| < \delta_1.$$

Also, since the limit of $g(x)$ is L, there exists δ_2 such that

$$|g(x) - L| < 1 \qquad \text{whenever} \qquad 0 < |x - c| < \delta_2.$$

By letting δ be the smaller of δ_1 and δ_2, we conclude that $0 < |x - c| < \delta$ implies

$$f(x) > M + 1 \qquad \text{and} \qquad |g(x) - L| < 1.$$

The second of these two inequalities implies that $g(x) > L - 1$, and by adding this to the first inequality, we have

$$f(x) + g(x) > M + L > M.$$

Thus,

$$\lim_{x \to c} [f(x) + g(x)] = \infty. \qquad \blacksquare$$

Later in the text when we encounter the indeterminate form $0/0$ in a limit involving trigonometric functions, it will often require a considerable amount of ingenuity to determine the limit. The following theorem is helpful. It concerns the limiting behavior of a function that is squeezed between two other functions, each of which has the same limit at a given point, as shown in Figure 2.36.

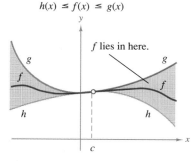

$$h(x) \le f(x) \le g(x)$$

f lies in here.

The Squeeze Theorem

FIGURE 2.36

THEOREM 2.14
THE SQUEEZE THEOREM

If $h(x) \le f(x) \le g(x)$ for all x in an open interval containing c, except possibly at c itself, and if

$$\lim_{x \to c} h(x) = L = \lim_{x \to c} g(x) \qquad \text{then} \qquad \lim_{x \to c} f(x) = L.$$

PROOF For $\varepsilon > 0$ there exist δ_1 and δ_2 such that

$$|h(x) - L| < \varepsilon \qquad \text{whenever} \qquad 0 < |x - c| < \delta_1$$

and

$$|g(x) - L| < \varepsilon \qquad \text{whenever} \qquad 0 < |x - c| < \delta_2.$$

Let δ be the smaller of δ_1 and δ_2. Then, if $0 < |x - c| < \delta$, it follows that $|h(x) - L| < \varepsilon$ and $|g(x) - L| < \varepsilon$, which implies that

$$-\varepsilon < h(x) - L < \varepsilon \qquad \text{and} \qquad -\varepsilon < g(x) - L < \varepsilon$$
$$L - \varepsilon < h(x) \qquad \text{and} \qquad g(x) < L + \varepsilon.$$

Now, since $h(x) \le f(x) \le g(x)$, it follows that

$$L - \varepsilon < f(x) < L + \varepsilon$$
$$|f(x) - L| < \varepsilon.$$

Therefore,

$$\lim_{x \to c} f(x) = L.$$

EXERCISES for Section 2.5

In Exercises 1–6, find the limit L and then find $\delta > 0$ such that $|f(x) - L| < 0.01$ whenever $0 < |x - c| < \delta$.

1. $\lim\limits_{x \to 2} (3x + 2)$

2. $\lim\limits_{x \to 4} \left(4 - \dfrac{x}{2}\right)$

3. $\lim\limits_{x \to 2} (x^2 - 3)$

4. $\lim\limits_{x \to 5} \sqrt{x - 4}$

5. $\lim\limits_{x \to 2} \dfrac{x^2 - 3x + 2}{x - 2}$

6. $\lim\limits_{x \to -2} \dfrac{x^2 + 5x + 6}{x + 2}$

In Exercises 7–22, find the indicated limit L. Then use the definitions given in this section to verify the limit.

7. $\lim\limits_{x \to 2} (x + 3)$

8. $\lim\limits_{x \to -3} (2x + 5)$

9. $\lim\limits_{x \to 6} 3$

10. $\lim\limits_{x \to 2} (-1)$

11. $\lim\limits_{x \to 0} \sqrt[3]{x}$

12. $\lim\limits_{x \to 3} |x - 3|$

13. $\lim\limits_{x \to 0} x^2$

14. $\lim\limits_{x \to 2} x^3$

15. $\lim\limits_{x \to 2} \dfrac{x^2 + x - 6}{x - 2}$

16. $\lim\limits_{x \to -10} \dfrac{x^2 + 10x}{x + 10}$

17. $\lim\limits_{x \to 2} \dfrac{1}{x}$

18. $\lim\limits_{x \to -1} \dfrac{3}{x + 2}$

19. $\lim\limits_{x \to 2} (x^2 - 2)$

20. $\lim\limits_{x \to -1} x^2$

21. $\lim\limits_{x \to 0^+} \sqrt{x}$

22. $\lim\limits_{x \to 3^-} f(x), \quad f(x) = \begin{cases} 2, & x \le 3 \\ 0, & x > 3 \end{cases}$

In Exercises 23–26, find the infinite limit and use the definition given in this section to verify your result.

23. $\lim\limits_{x \to -1^+} \dfrac{1}{x + 1}$

24. $\lim\limits_{x \to -1^-} \dfrac{1}{x + 1}$

25. $\lim\limits_{x \to 2} \dfrac{1}{(x - 2)^2}$

26. $\lim\limits_{x \to 0} \dfrac{1}{x^2}$

In Exercises 27 and 28, prove that the given function is continuous at the specified value of x.

27. $f(x) = x^2, \quad x = 3$

28. $f(x) = 4 - 3x, \quad x = 1$

In Exercises 29 and 30, use the Squeeze Theorem to find $\lim\limits_{x \to c} f(x)$.

29. $c = 0$
$4 - x^2 \le f(x) \le 4 + x^2$

30. $c = a$
$b - |x - a| \le f(x) \le b + |x - a|$

31. Prove that if $\lim\limits_{x \to c} f(x)$ exists, then the limit must be unique. [Hint: Let $\lim\limits_{x \to c} f(x) = L_1$ and $\lim\limits_{x \to c} f(x) = L_2$ and prove that $L_1 = L_2$.]

32. Prove that if $\lim\limits_{x \to c} f(x)$ exists and $\lim\limits_{x \to c} [f(x) + g(x)]$ does not exist, then $\lim\limits_{x \to c} g(x)$ does not exist.

33. Prove that if $\lim\limits_{x \to c} f(x) = 0$, then $\lim\limits_{x \to c} |f(x)| = 0$.

34. Prove that if $\lim\limits_{x \to c} |f(x)| = 0$, then $\lim\limits_{x \to c} f(x) = 0$. [Note: This is the converse of Exercise 33.]

35. Prove that if $\lim\limits_{x \to c} f(x) = L$, then $\lim\limits_{x \to c} |f(x)| = |L|$. [Hint: Use the inequality $\big| |f(x)| - |L| \big| \le |f(x) - L|$.]

36. Find a function f to show that the converse of Exercise 35 is not true.

REVIEW EXERCISES for Chapter 2

In Exercises 1–20, find the given limit (if it exists).

1. $\lim\limits_{x \to 2} (5x - 3)$

2. $\lim\limits_{x \to 2} (3x + 5)$

3. $\lim\limits_{x \to 2} (5x - 3)(3x + 5)$

4. $\lim\limits_{x \to 2} \dfrac{3x + 5}{5x - 3}$

5. $\lim\limits_{t \to 3} \dfrac{t^2 + 1}{t}$

6. $\lim\limits_{t \to 3} \dfrac{t^2 - 9}{t - 3}$

7. $\lim\limits_{t \to -2} \dfrac{t + 2}{t^2 - 4}$

8. $\lim\limits_{x \to 0} \dfrac{\sqrt{4 + x} - 2}{x}$

9. $\lim\limits_{x \to 0} \dfrac{[1/(x + 1)] - 1}{x}$

10. $\lim\limits_{s \to 0} \dfrac{(1/\sqrt{1 + s}) - 1}{s}$

11. $\lim\limits_{x \to -1} \dfrac{x^3 + 1}{x + 1}$

12. $\lim\limits_{x \to -2} \dfrac{x^2 - 4}{x^3 + 8}$

13. $\lim\limits_{x \to 0^+} \left(x - \dfrac{1}{x^3}\right)$

14. $\lim\limits_{x \to 2} \dfrac{1}{\sqrt[3]{x^2 - 4}}$

15. $\lim\limits_{x \to -2^-} \dfrac{2x^2 + x + 1}{x + 2}$

16. $\lim\limits_{x \to 1/2} \dfrac{2x - 1}{6x - 3}$

17. $\lim\limits_{x \to -1} \dfrac{x + 1}{x^3 + 1}$

18. $\lim\limits_{x \to -1} \dfrac{x + 1}{x^4 - 1}$

19. $\lim\limits_{x \to 1} \dfrac{x^2 - 2x + 1}{x + 1}$

20. $\lim\limits_{x \to -1^+} \dfrac{x^2 - 2x + 1}{x + 1}$

21. Estimate the limit

$$\lim\limits_{x \to 1^+} \dfrac{\sqrt{2x + 1} - \sqrt{3}}{x - 1}$$

by completing the following table.

x	1.1	1.01	1.001	1.0001
$f(x)$				

22. Estimate the limit

$$\lim_{x \to 1^+} \frac{1 - \sqrt[3]{x}}{x - 1}$$

by completing the following table.

x	1.1	1.01	1.001	1.0001
$f(x)$				

23. Evaluate the limit in Exercise 21 by rationalizing the numerator.

24. Evaluate the limit in Exercise 22 by rationalizing the numerator. [Use $a^3 - b^3 = (a - b)(a^2 + ab + b^2)$.]

In Exercises 25–30, determine whether the given limit statement is true or false.

25. $\lim\limits_{x \to 0} \dfrac{|x|}{x} = 1$

26. $\lim\limits_{x \to 0} x^3 = 0$

27. $\lim\limits_{x \to 2} f(x) = 3$, where $f(x) = \begin{cases} 3, & x \le 2 \\ 0, & x > 2 \end{cases}$

28. $\lim\limits_{x \to 3} f(x) = 1$, where

$$f(x) = \begin{cases} x - 2, & x \le 3 \\ -x^2 + 8x - 14, & x > 3 \end{cases}$$

29. $\lim\limits_{x \to 0^+} \sqrt{x} = 0$

30. $\lim\limits_{x \to 0} \sqrt[3]{x} = 0$

In Exercises 31–38, determine the intervals on which the given function is continuous.

31. $f(x) = [\![x + 3]\!]$

32. $f(x) = \dfrac{3x^2 - x - 2}{x - 1}$

33. $f(x) = \begin{cases} \dfrac{3x^2 - x - 2}{x - 1}, & x \ne 1 \\ 0, & x = 1 \end{cases}$

34. $f(x) = \begin{cases} 5 - x, & x \le 2 \\ 2x - 3, & x > 2 \end{cases}$

35. $f(x) = \dfrac{1}{(x - 2)^2}$

36. $f(x) = \sqrt{\dfrac{x + 2}{x}}$

37. $f(x) = \dfrac{3}{x + 1}$

38. $f(x) = \dfrac{x + 1}{2x + 2}$

39. Determine the value of c so that the following function is continuous on the entire real line.

$$f(x) = \begin{cases} x + 3, & x \le 2 \\ cx + 6, & x > 2 \end{cases}$$

40. Determine the values of b and c so that the following function is continuous on the entire real line.

$$f(x) = \begin{cases} x + 1, & 1 < x < 3 \\ x^2 + bx + c, & |x - 2| \ge 1 \end{cases}$$

41. A sum of $5000 is deposited in a savings plan that pays 12 percent compounded semiannually. The amount in the account after t years is given by

$$A = 5000(1.06)^{[\![2t]\!]}.$$

Sketch a graph of this function and discuss its continuity.

42. A sum of $1000 is deposited in a savings plan that pays 14 percent compounded annually. The amount in the account after t years is given by

$$A = 1000(1.14)^{[\![t]\!]}.$$

Sketch a graph of this function and discuss its continuity.

Chapter 3 Application

High-diver Greg Louganis is shown in mid-air after diving from a 10-meter platform. Louganis is considered by many to be the greatest diver of all time. In 1976, at the age of 16, he won an Olympic silver medal for the United States in the platform competition. Later, in 1984 and again in 1988, he won both Olympic gold medals in diving—one in the platform competition and one in the springboard competition.

The Velocity of a High-Diver

A diver jumps from a 30-foot platform. The initial velocity of the diver is 5 feet per second (upward). When will the diver hit the water, and how fast will the diver be traveling at that time? To answer these questions, we use the classical model for the position of a free-falling object. This model assumes that air resistance is negligible and is given by

$$s(t) = \frac{1}{2}gt^2 + v_0 t + h_0$$

where $s(t)$ is the height of the diver in feet, t is the time in seconds, $g = -32$ ft/sec^2 is the acceleration due to gravity, v_0 is the initial velocity, and h_0 is the initial height. Using $v_0 = 5$ and $h_0 = 30$, the position function becomes

$$s(t) = -16t^2 + 5t + 30.$$

To find the time when the diver hits the water, we let $s(t) = 0$ and solve for t. The solution of this equation is $t \approx 1.53$ seconds. The velocity of the diver is given by the derivative of the position function

$$v(t) = s'(t) = -32t + 5$$

where $v(t)$ is measured in feet per second. Thus, when $t \approx 1.53$ seconds, the velocity is $v(1.53) \approx -44$ feet per second (or approximately 30 miles per hour).

Diver's Velocity at Water for Varying Platform Heights
(Ten meters is standard for Olympic competition.)

Height (in feet)	30	40	50	60
Velocity (in ft/sec)	−44.1	−50.8	−56.8	−62.2

See Exercise 51, Section 3.3.

Chapter Overview

Calculus has two fundamental operations—differentiation and integration. In this chapter we introduce the first of these.

Section 3.1 begins by taking a detailed look at the problem of finding the **tangent line** to the graph of a function at a specified point. The section also shows how the **derivative** of a function can be used to find the slope of the tangent line at a point.

Section 3.2 presents the second primary use of the derivative—that of finding the instantaneous rate of change of one variable with respect to another. For example, the velocity of an object moving along a straight line can be found by differentiating the position function of the object.

In the first two sections of the chapter, the emphasis is on the *use* of derivatives. In these two sections, the actual *calculation* of derivatives is done by finding a limit, and this can be quite tedious. In Sections 3.3–3.5, we streamline the differentiation process by presenting several "differentiation rules." It is important that you memorize each rule, so that you can find derivatives efficiently.

Section 3.6 introduces a procedure called **implicit differentiation.** This procedure can be used to find derivatives of functions that are implied by equations. For example, implicit differentiation can be applied to the circle $x^2 + y^2 = 1$ to find the slope of the tangent line at a given point on the circle.

The chapter closes by showing how differentiation can be used in applications involving rates of change (Section 3.7).

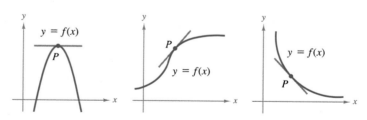

Differentiation

3.1 The Derivative and the Tangent Line Problem

Tangent line to a curve ▪ Vertical tangent lines ▪ The derivative of a function ▪ Differentiability and continuity

Isaac Newton

Calculus grew out of four major problems that European mathematicians were working on during the seventeenth century.

1. The tangent line problem (Sections 2.1 and 3.1)
2. The velocity and acceleration problem (Section 3.2)
3. The minimum and maximum problem (Section 4.1)
4. The area problem (Section 5.2)

Each problem involves the notion of a limit, and we could introduce calculus with any of the four problems.

Recall that we gave a brief introduction to the tangent line problem in Section 2.1. Although partial solutions to this problem were given by Pierre de Fermat (1601–1655), René Descartes (1596–1650), Christian Huygens (1629–1695), and Isaac Barrow (1630–1677), credit for the first general solution usually is given to Isaac Newton (1642–1727) and Gottfried Leibniz (1646–1716). Newton's work on the problem stemmed from his interest in optics and light refraction.

The tangent line problem

What do we mean when we say that a line is tangent to a curve at a point? For a circle, we can characterize the tangent line at point P as the line that is perpendicular to the radial line at point P, as shown in Figure 3.1. For a general curve, however, the problem is more difficult. For example, how would you define the tangent lines shown in Figure 3.2?

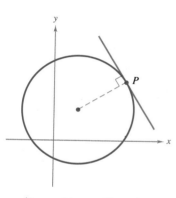

Tangent Line to a Circle

FIGURE 3.1

Tangent Line to a Curve at a Point

FIGURE 3.2

95

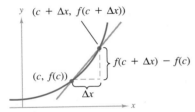

The secant line through $(c, f(c))$
and $(c + \Delta x, f(c + \Delta x))$

FIGURE 3.3

Essentially, the problem of finding the tangent line at a point P boils down to the problem of finding the *slope* of the tangent line at P. We can approximate this slope using a line through the point of tangency and a second point on the curve, as shown in Figure 3.3. We call such a line a **secant line***. If $(c, f(c))$ is the point of tangency and $(c + \Delta x, f(c + \Delta x))$ is a second point on the graph of f, then the slope of the secant line through these two points is given by

$$m_{\text{sec}} = \frac{f(c + \Delta x) - f(c)}{c + \Delta x - c}$$

$$= \frac{f(c + \Delta x) - f(c)}{\Delta x}.$$ Slope of secant line

The right-hand side of this equation is called a **difference quotient.** The denominator Δx is called the **change in** x, and the numerator $\Delta y = f(c + \Delta x) - f(c)$ is called the **change in** y.

The beauty of this procedure is that we can obtain more and more accurate approximations to the slope of the tangent line by choosing a sequence of points closer and closer to the point of tangency, as shown in Figure 3.4.

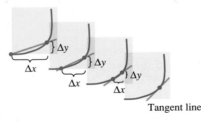

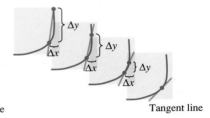

As $\Delta x \to 0$ from the left, the secant lines approach the tangent line.

As $\Delta x \to 0$ from the right, the secant lines approach the tangent line.

FIGURE 3.4

DEFINITION OF TANGENT LINE WITH SLOPE m	If f is defined on an open interval containing c and the limit $$\lim_{\Delta x \to 0} \frac{\Delta y}{\Delta x} = \lim_{\Delta x \to 0} \frac{f(c + \Delta x) - f(c)}{\Delta x} = m$$ exists, then we call the line passing through $(c, f(c))$ with slope m the **tangent line** to the graph of f at the point $(c, f(c))$.

We often refer to the slope of the tangent line to the graph of f at the point $(c, f(c))$ as simply the **slope of the graph of** f **at** $x = c$.

*This use of the word *secant* comes from the Latin *secare*, meaning to cut, and not from a reference to the trigonometric function of the same name.

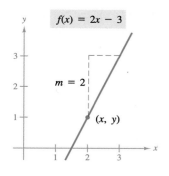

$f(x) = 2x - 3$

FIGURE 3.5

EXAMPLE 1 The slope of the graph of a linear function

Find the slope of the graph of $f(x) = 2x - 3$ at the point $(2, 1)$.

SOLUTION

$$
\lim_{\Delta x \to 0} \frac{f(2 + \Delta x) - f(2)}{\Delta x} = \lim_{\Delta x \to 0} \frac{[2(2 + \Delta x) - 3] - [2(2) - 3]}{\Delta x}
$$

$$
= \lim_{\Delta x \to 0} \frac{4 + 2\Delta x - 3 - 4 + 3}{\Delta x}
$$

$$
= \lim_{\Delta x \to 0} \frac{2\Delta x}{\Delta x} = \lim_{\Delta x \to 0} 2 = 2
$$

Therefore, the slope of f at $(2, 1)$ is $m = 2$, as shown in Figure 3.5. ▭

REMARK   Note in Example 1 that the limit definition of the slope of f agrees with our usual definition of the slope of a line as discussed in Section 1.4.

We know that the graph of a linear function has the same slope at any point. This is not true of nonlinear functions, as can be seen in the following example.

EXAMPLE 2 Tangent lines to the graph of a nonlinear function

Find the slope of the tangent line to the graph of $f(x) = x^2 + 1$ at the points $(0, 1)$ and $(-1, 2)$, as shown in Figure 3.6.

SOLUTION

To solve this problem, we let $(x, f(x))$ represent an arbitrary point on the graph of f. Then the slope of the tangent line at $(x, f(x))$ is given by

$$
\lim_{\Delta x \to 0} \frac{f(x + \Delta x) - f(x)}{\Delta x} = \lim_{\Delta x \to 0} \frac{x^2 + 2x(\Delta x) + (\Delta x)^2 + 1 - x^2 - 1}{\Delta x}
$$

$$
= \lim_{\Delta x \to 0} \frac{2x(\Delta x) + (\Delta x)^2}{\Delta x}
$$

$$
= \lim_{\Delta x \to 0} (2x + \Delta x) = 2x.
$$

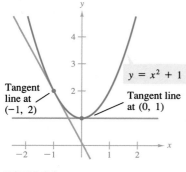

Tangent line at $(-1, 2)$

Tangent line at $(0, 1)$

$y = x^2 + 1$

FIGURE 3.6

Therefore, the slope at *any* point $(x, f(x))$ on the graph of f is given by $m = 2x$. At the point $(0, 1)$, the slope is $m = 2(0) = 0$, and at the point $(-1, 2)$, the slope is $m = 2(-1) = -2$. ▭

REMARK   In Example 2 note that x is held constant in the limit process (as $\Delta x \to 0$).

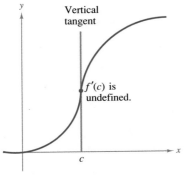

Vertical tangent

$f'(c)$ is undefined.

FIGURE 3.7

Our definition of a tangent line to a curve does not cover the possibility of a vertical tangent line. For vertical tangent lines, we give the following definition. If f is continuous at c and

$$\lim_{\Delta x \to 0} \left| \frac{f(c + \Delta x) - f(c)}{\Delta x} \right| = \infty$$

then the vertical line passing through $(c, f(c))$ is called a **vertical tangent line** to the graph of f. For example, the function shown in Figure 3.7 has a vertical tangent line at $x = c$. If the domain of f is a closed interval $[a, b]$, then we extend the definition of a vertical tangent line to include the endpoints by considering continuity and limits from the right (for $x = a$) and from the left (for $x = b$).

The derivative of a function

We have now arrived at a crucial point in the study of calculus. The limit used to define the slope of a tangent line is also used to define one of the two fundamental operations of calculus—**differentiation.**

DEFINITION OF THE DERIVATIVE OF A FUNCTION

The **derivative** of f at x is given by

$$f'(x) = \lim_{\Delta x \to 0} \frac{f(x + \Delta x) - f(x)}{\Delta x}$$

provided the limit exists.

The process of finding the derivative of a function is called **differentiation.** A function is called **differentiable** at x if its derivative exists at x and **differentiable on an open interval** (a, b) if it is differentiable at every point in the interval.

In addition to $f'(x)$, read "f prime of x," other notations are used to denote the derivative of $y = f(x)$. The most common are

$$f'(x), \qquad \frac{dy}{dx}, \qquad y', \qquad \frac{d}{dx}[f(x)], \qquad D_x[y]. \qquad \text{Notation for derivatives}$$

The notation dy/dx is read as "the derivative of y *with respect to* x." Using limit notation, we have

$$\frac{dy}{dx} = \lim_{\Delta x \to 0} \frac{\Delta y}{\Delta x} = \lim_{\Delta x \to 0} \frac{f(x + \Delta x) - f(x)}{\Delta x} = f'(x).$$

EXAMPLE 3   Finding the derivative by the limit process

Find the derivative of $f(x) = x^3 + 2x$.

SOLUTION

$$f'(x) = \lim_{\Delta x \to 0} \frac{f(x + \Delta x) - f(x)}{\Delta x}$$

$$= \lim_{\Delta x \to 0} \frac{(x + \Delta x)^3 + 2(x + \Delta x) - (x^3 + 2x)}{\Delta x}$$

$$= \lim_{\Delta x \to 0} \frac{3x^2\Delta x + 3x(\Delta x)^2 + (\Delta x)^3 + 2\Delta x}{\Delta x}$$

$$= \lim_{\Delta x \to 0} \frac{\cancel{\Delta x}[3x^2 + 3x\Delta x + (\Delta x)^2 + 2]}{\cancel{\Delta x}}$$

$$= \lim_{\Delta x \to 0} [3x^2 + 3x\Delta x + (\Delta x)^2 + 2]$$

$$= 3x^2 + 2$$

Remember that the derivative $f'(x)$ gives us a formula for finding the slope of the tangent line at the point $(x, f(x))$ on the graph of f. This is illustrated in the following example.

EXAMPLE 4 Using the derivative to find the slope at a point

Find $f'(x)$ for $f(x) = \sqrt{x}$, and use the result to find the slope of the graph of f at the points $(1, 1)$ and $(4, 2)$. Discuss the behavior of f at $(0, 0)$.

SOLUTION

We use the procedure for rationalizing numerators, as discussed in Section 2.3.

$$f'(x) = \lim_{\Delta x \to 0} \frac{f(x + \Delta x) - f(x)}{\Delta x}$$

$$= \lim_{\Delta x \to 0} \frac{\sqrt{x + \Delta x} - \sqrt{x}}{\Delta x}$$

$$= \lim_{\Delta x \to 0} \left(\frac{\sqrt{x + \Delta x} - \sqrt{x}}{\Delta x}\right)\left(\frac{\sqrt{x + \Delta x} + \sqrt{x}}{\sqrt{x + \Delta x} + \sqrt{x}}\right)$$

$$= \lim_{\Delta x \to 0} \frac{(x + \Delta x) - x}{\Delta x(\sqrt{x + \Delta x} + \sqrt{x})}$$

$$= \lim_{\Delta x \to 0} \frac{\cancel{\Delta x}}{\cancel{\Delta x}(\sqrt{x + \Delta x} + \sqrt{x})}$$

$$= \lim_{\Delta x \to 0} \frac{1}{\sqrt{x + \Delta x} + \sqrt{x}} = \frac{1}{2\sqrt{x}}$$

Therefore, at the point $(1, 1)$ the slope is $f'(1) = \frac{1}{2}$, and at the point $(4, 2)$ the slope is $f'(4) = \frac{1}{4}$, as shown in Figure 3.8. At the point $(0, 0)$ the slope is undefined, since substituting $x = 0$ in $f'(x)$ produces division by zero. Moreover, because the limit of $f'(x)$ as $x \to 0$ from the right is infinite, the graph of f has a vertical tangent line at $(0, 0)$.

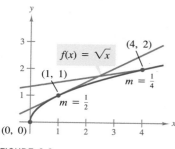

FIGURE 3.8

In many applications, it is convenient to use a variable other than x as the independent variable. Example 5 shows a function that uses t as the independent variable.

EXAMPLE 5 Finding the derivative of a function

Find the derivative with respect to t for the function $y = 2/t$.

SOLUTION

Considering $y = f(t)$, we obtain the following.

$$\frac{dy}{dt} = \lim_{\Delta t \to 0} \frac{f(t + \Delta t) - f(t)}{\Delta t}$$

$$= \lim_{\Delta t \to 0} \frac{\dfrac{2}{t + \Delta t} - \dfrac{2}{t}}{\Delta t} = \lim_{\Delta t \to 0} \frac{\dfrac{2t - 2(t + \Delta t)}{t(t + \Delta t)}}{\Delta t}$$

$$= \lim_{\Delta t \to 0} \frac{-2\Delta t}{\Delta t(t)(t + \Delta t)}$$

$$= \lim_{\Delta t \to 0} \frac{-2}{t(t + \Delta t)} = -\frac{2}{t^2} \qquad \blacksquare$$

Differentiability and continuity

There is a close relationship between differentiability and continuity. The following alternate limit form of the derivative is useful in investigating this relationship.

THEOREM 3.1 **ALTERNATE FORM** **OF THE DERIVATIVE**	The derivative of f at c is given by $$f'(c) = \lim_{x \to c} \frac{f(x) - f(c)}{x - c}$$ provided this limit exists.

PROOF The derivative of f at c is given by

$$f'(c) = \lim_{\Delta x \to 0} \frac{f(c + \Delta x) - f(c)}{\Delta x}.$$

In Figure 3.9, we can see that if $x = c + \Delta x$, then $x \to c$ as $\Delta x \to 0$. Thus, if we replace $c + \Delta x$ by x, we can write

$$f'(c) = \lim_{\Delta x \to 0} \frac{f(c + \Delta x) - f(c)}{\Delta x} = \lim_{x \to c} \frac{f(x) - f(c)}{x - c}.$$

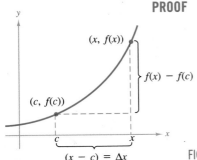

FIGURE 3.9

Note that the existence of the limit in this alternate form requires that the one-sided limits

$$\lim_{x \to c^-} \frac{f(x) - f(c)}{x - c}$$

and

$$\lim_{x \to c^+} \frac{f(x) - f(c)}{x - c}$$

exist and are equal. For convenience, we refer to these one-sided limits as the **derivatives from the left and from the right,** respectively. Keep in mind, however, that if these one-sided limits are not equal at c, then the derivative does not exist at c.

The next two examples will give you a sense of the relationship between differentiability and continuity.

EXAMPLE 6 A function whose one-sided derivatives are different

The function $f(x) = |x - 2|$ shown in Figure 3.10 is continuous at $x = 2$. However, the one-sided limits

$$\lim_{x \to 2^-} \frac{f(x) - f(2)}{x - 2} = \lim_{x \to 2^-} \frac{|x - 2| - 0}{x - 2} = -1 \qquad \text{Derivative from the left}$$

and

$$\lim_{x \to 2^+} \frac{f(x) - f(2)}{x - 2} = \lim_{x \to 2^+} \frac{|x - 2| - 0}{x - 2} = 1 \qquad \text{Derivative from the right}$$

are not equal. Therefore, f is not differentiable at $x = 2$ and the graph of f does not have a tangent line at the point $(2, 0)$. ▭

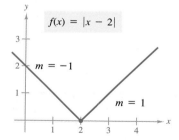

$f(x) = |x - 2|$

$m = -1$

$m = 1$

Not differentiable at $x = 2$, since the one-sided derivatives are not equal.

FIGURE 3.10

EXAMPLE 7 A function with a vertical tangent

The function $f(x) = x^{1/3}$ is continuous at $x = 0$, as shown in Figure 3.11. However, since the following limit is infinite,

$$\lim_{x \to 0} \frac{f(x) - f(0)}{x - 0} = \lim_{x \to 0} \frac{x^{1/3} - 0}{x}$$

$$= \lim_{x \to 0} \frac{1}{x^{2/3}} = \infty$$

we conclude that the tangent line is vertical at $x = 0$. Therefore, f is not differentiable at $x = 0$. ▭

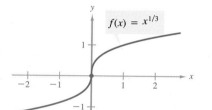

$f(x) = x^{1/3}$

Not differentiable at $x = 0$, since f has a vertical tangent at $x = 0$.

FIGURE 3.11

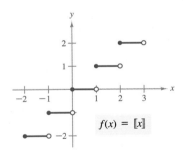

$f(x) = [\![x]\!]$

Not differentiable at $x = 0$,
since f is not continuous at $x = 0$.

FIGURE 3.12

In Examples 6 and 7, we saw that a function is not differentiable at a point at which its graph has a sharp turn *or* a vertical tangent. Differentiability can also be destroyed by a discontinuity. For instance, the greatest integer function $f(x) = [\![x]\!]$ is not continuous at $x = 0$ and hence is not differentiable at $x = 0$ (see Figure 3.12). Another way of saying this is that if a function is differentiable at a point, then it is continuous at the point. We formalize this result in the following theorem.

THEOREM 3.2 **DIFFERENTIABILITY** **IMPLIES CONTINUITY**	If f is differentiable at $x = c$, then f is continuous at $x = c$.

PROOF To prove that f is continuous at $x = c$, we will show that $f(x)$ approaches $f(c)$ as $x \to c$. To do this, we use the differentiability of f at $x = c$ and consider the following limit.

$$\lim_{x \to c} [f(x) - f(c)] = \lim_{x \to c} \left[(x - c)\left(\frac{f(x) - f(c)}{x - c} \right) \right]$$

$$= \left[\lim_{x \to c} (x - c) \right] \left[\lim_{x \to c} \frac{f(x) - f(c)}{x - c} \right]$$

$$= (0)[f'(c)] = 0$$

Since the difference $[f(x) - f(c)]$ approaches zero as $x \to c$, we conclude that

$$\lim_{x \to c} f(x) = f(c).$$

Therefore, f is continuous at $x = c$.

EXERCISES for Section 3.1

In Exercises 1 and 2, trace the curve on another piece of paper and sketch the tangent line at the point (x, y).

1.
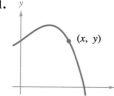

2.
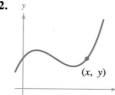

In Exercises 3 and 4, estimate the slope of the curve at the point (x, y).

3. (a)
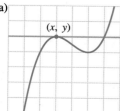
(b)
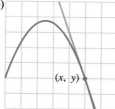

4. (a)
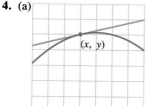
(b)
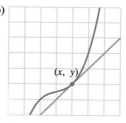

In Exercises 5–14, use the definition of the derivative to find $f'(x)$.

5. $f(x) = 3$
6. $f(x) = 3x + 2$
7. $f(x) = -5x$
8. $f(x) = 1 - x^2$
9. $f(x) = 2x^2 + x - 1$
10. $f(x) = \sqrt{x - 4}$
11. $f(x) = \dfrac{1}{x - 1}$
12. $f(x) = \dfrac{1}{x^2}$
13. $f(t) = t^3 - 12t$
14. $f(t) = t^3 + t^2$

In Exercises 15–20, find the equation of the tangent line to the graph of f at the indicated point. Then verify your answer by sketching both the graph of f and the tangent line.

Function	Point of Tangency
15. $f(x) = x^2 + 1$	$(2, 5)$
16. $f(x) = x^2 + 2x + 1$	$(-3, 4)$

Function	Point of Tangency
17. $f(x) = x^3$	$(2, 8)$
18. $f(x) = x^3$	$(-2, -8)$
19. $f(x) = \sqrt{x + 1}$	$(3, 2)$
20. $f(x) = \dfrac{1}{x + 1}$	$(0, 1)$

In Exercises 21–26, use the alternate form of the derivative (Theorem 3.1) to find the derivative at $x = c$ (if it exists).

21. $f(x) = x^2 - 1, \ c = 2$
22. $f(x) = x^3 + 2x, \ c = 1$
23. $f(x) = x^3 + 2x^2 + 1, \ c = -2$
24. $f(x) = \dfrac{1}{x}, \ c = 3$
25. $f(x) = (x - 1)^{2/3}, \ c = 1$
26. $f(x) = |x - 2|, \ c = 2$

In Exercises 27–36, find every point at which the function is differentiable.

27. $f(x) = |x + 3|$
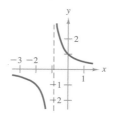

28. $f(x) = |x^2 - 9|$
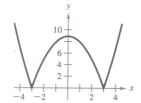

29. $f(x) = \dfrac{1}{x + 1}$

30. $f(x) = \dfrac{2x}{x - 1}$
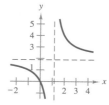

31. $f(x) = (x - 3)^{2/3}$
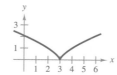

32. $f(x) = x^{2/5}$
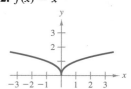

33. $f(x) = \sqrt{x} - 1$

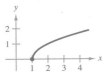

34. $f(x) = \dfrac{x^2}{x^2 - 4}$

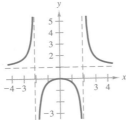

35. $f(x) = \begin{cases} 4 - x^2, & 0 < x \\ x^2 - 4, & x \le 0 \end{cases}$

36. $f(x) = \begin{cases} x^2 - 2x, & x > 1 \\ x^3 - 3x^2 + 3x, & x \le 1 \end{cases}$

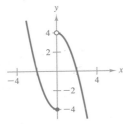

FIGURE FOR 35

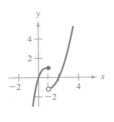

FIGURE FOR 36

In Exercises 37–40, find the derivatives from the left and from the right at $x = 1$ (if they exist). Is the function differentiable at $x = 1$?

37. $f(x) = \sqrt{1 - x^2}$

38. $f(x) = \begin{cases} x - 1, & x \le 1 \\ (x - 1)^2, & x > 1 \end{cases}$

39. $f(x) = \begin{cases} (x - 1)^3, & x \le 1 \\ (x - 1)^2, & x > 1 \end{cases}$

40. $f(x) = \begin{cases} x, & x \le 1 \\ x^2, & x > 1 \end{cases}$

In Exercises 41 and 42, find an equation of the line that is tangent to the graph of f and parallel to the given line.

Function	Line
41. $f(x) = x^3$	$3x - y + 1 = 0$
42. $f(x) = \dfrac{1}{\sqrt{x}}$	$x + 2y - 6 = 0$

In Exercises 43 and 44, find the equations of the two tangent lines to the graph of f that pass through the indicated point.

43. $f(x) = 4x - x^2$

44. $f(x) = x^2$

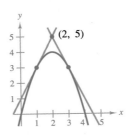

FIGURE FOR 43

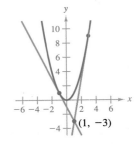

FIGURE FOR 44

45. Assume $f'(c) = 3$. Find $f'(-c)$ for the following conditions.
  (a) f is an odd function.
  (b) f is an even function.

46. Sketch the graph of f and f' on the same set of axes for each of the following.
  (a) $f(x) = x^2$ (b) $f(x) = x^3$

In Exercises 47–50, determine whether the statement is true or false.

47. The slope of the graph of $y = x^2$ is different at every point on the curve.

48. If a function is continuous at a point, then it is differentiable at that point.

49. If a function is differentiable at a point, then it is continuous at that point.

50. If a function has derivatives from both the right and the left at a point, then it is differentiable at that point.

 In Exercises 51 and 52, use a computer or graphics calculator to sketch the graph of f over the interval $[-2, 2]$ and complete the following table.

x	-2	-1.5	-1	-0.5	0	0.5	1	1.5	2
$f(x)$									
$f'(x)$									

51. $f(x) = \dfrac{1}{4}x^3$

52. $f(x) = \dfrac{4}{x}$

3.2 Velocity, Acceleration, and Other Rates of Change

Straight-line motion ▪ Average velocity ▪ Instantaneous velocity ▪ Acceleration ▪ Higher-order derivatives ▪ Other rates of change

We have seen how the derivative is used to determine slope. We now consider another use—to determine the rate of change in one variable with respect to another. Applications involving rates of change can be found in a wide variety of fields. A few examples are population growth rates, production rates, the rate of water flow, velocity, and acceleration.

Straight-line motion

A common use of a rate of change is to describe the motion of an object moving in a straight line. Such motion is called **rectilinear motion.** It is customary to use either a horizontal or vertical line with a designated origin to represent the line of motion. Movement to the right (or upward) is considered to be in the **positive direction,** and movement to the left (or down) is considered to be in the **negative direction.**

The function s that gives the position (relative to the origin) of an object as a function of time t is called a **position function.** If, over a period of time Δt, the object changes its position by the amount

$$\Delta s = s(t + \Delta t) - s(t) \qquad \text{Change in distance}$$

then, by the familiar formula

$$\text{rate} = \frac{\text{distance}}{\text{time}}$$

the **average rate of change** in distance with respect to time is given by

$$\frac{\text{change in distance}}{\text{change in time}} = \frac{\Delta s}{\Delta t}.$$

We call this average rate of change the **average velocity.**

DEFINITION OF AVERAGE VELOCITY

If $s(t)$ gives the position at time t of an object moving in a straight line, then the **average velocity** of the object over the interval $[t, t + \Delta t]$ is given by

$$\text{average velocity} = \frac{\Delta s}{\Delta t} = \frac{s(t + \Delta t) - s(t)}{\Delta t}.$$

EXAMPLE 1 Finding average velocities for a falling object

If a free-falling object is dropped from a height of 100 feet, its height s at time t is given by the position function $s = -16t^2 + 100$, where s is measured in feet and t is measured in seconds. Find the average rate of change of the height over the following intervals.

(a) [1, 2]   (b) [1, 1.5]   (c) [1, 1.1]

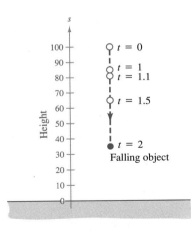

TABLE 3.1

s	84	80.64	64	36
t	1	1.1	1.5	2

FIGURE 3.13

SOLUTION

Using the position equation, we determine the heights at $t = 1$, 1.1, 1.5, and 2, as shown in Table 3.1 and Figure 3.13.

(a) For the interval [1, 2], the object falls from a height of 84 feet to a height of 36 feet. Thus, the average rate of change is

$$\frac{\Delta s}{\Delta t} = \frac{36 - 84}{2 - 1} = \frac{-48}{1} = -48 \text{ ft/sec.}$$

(b) For the interval [1, 1.5], the average rate of change is

$$\frac{\Delta s}{\Delta t} = \frac{64 - 84}{1.5 - 1} = \frac{-20}{0.5} = -40 \text{ ft/sec.}$$

(c) For the interval [1, 1.1], the average rate of change is

$$\frac{\Delta s}{\Delta t} = \frac{80.64 - 84}{1.1 - 1} = \frac{-3.36}{0.1} = -33.6 \text{ ft/sec.}$$

REMARK   Note that the average velocities in Example 1 are *negative*, indicating that the object is moving downward.

Suppose that in Example 1 we wanted to find the velocity at the instant, $t = 1$ second. We call this the **instantaneous velocity,** or simply the **velocity** of the object when $t = 1$. Just as we approximate the slope of the tangent line by calculating the slope of the secant line, we can approximate the velocity at $t = 1$ by calculating the average velocity over a small interval $[1, 1 + \Delta t]$, as shown in Table 3.2.

TABLE 3.2

Δt	1	0.5	0.1	0.01	0.001	0.0001
$\dfrac{\Delta s}{\Delta t}$	-48	-40	-33.6	-32.16	-32.016	-32.0016

From this table, it seems reasonable to conclude that the velocity when $t = 1$ is -32 feet per second. We will verify this conclusion after presenting the following definition.

DEFINITION OF (INSTANTANEOUS) VELOCITY	If $s = s(t)$ is the position function for an object moving along a straight line, then the **velocity** of the object at time t is given by $$v(t) = \lim_{\Delta t \to 0} \frac{s(t + \Delta t) - s(t)}{\Delta t} = s'(t).$$

REMARK   Note that velocity is given by the *derivative* of the position function.

EXAMPLE 2   Using the derivative to find velocity

Find the velocity at $t = 1$ and $t = 2$ of a free-falling object whose position function is

$$s(t) = -16t^2 + 100$$

where s is measured in feet and t is measured in seconds.

SOLUTION

Using the limit definition of the derivative, we find the velocity function to be

$$
\begin{aligned}
v(t) = s'(t) &= \lim_{\Delta t \to 0} \frac{s(t + \Delta t) - s(t)}{\Delta t} \\
&= \lim_{\Delta t \to 0} \frac{-16(t + \Delta t)^2 + 100 - (-16t^2 + 100)}{\Delta t} \\
&= \lim_{\Delta t \to 0} \frac{-32t\Delta t - 16(\Delta t)^2}{\Delta t} \\
&= \lim_{\Delta t \to 0} (-32t - 16\Delta t) \\
&= -32t.
\end{aligned}
$$

Therefore, the velocity at $t = 1$ is $v(1) = -32$ ft/sec and the velocity at $t = 2$ is $v(2) = -64$ ft/sec.

Sometimes there is confusion about the terms "speed" and "velocity." We define **speed** as the absolute value of velocity; as such, it is always nonnegative. Thus, speed indicates only how fast an object is moving, whereas velocity indicates both the speed and the direction of motion relative to a given coordinate system.

Just as we can obtain the velocity function by differentiating the position function, we can obtain the **acceleration** function by differentiating the velocity function.

DEFINITION OF ACCELERATION

If s is the position function for an object moving along a straight line, then the **acceleration** of the object at time t is given by

$$a(t) = v'(t)$$

where $v(t)$ is the velocity at time t.

REMARK   If the time t is expressed in seconds and the distance s is expressed in feet, then velocity will be expressed in feet per second (ft/sec) and acceleration will be given in feet per second per second (ft/sec^2).

EXAMPLE 3   Finding acceleration as the derivative of the velocity

Find the acceleration of a free-falling object whose position function is

$$s(t) = -16t^2 + 100.$$

SOLUTION

From Example 2, we know the velocity function for this object is

$$v(t) = -32t.$$

Thus, the acceleration is given by

$$a(t) = v'(t) = -32 \text{ ft/sec}^2.$$

The acceleration found in Example 3 is called the **acceleration due to gravity,** denoted by g, and its exact value depends on one's location on the earth. The *standard value* of g is -32.174 feet per second per second (or -9.81 meters per second per second).

In general, the position of a free-falling object (neglecting air resistance) under the influence of gravity can be represented by the equation

$$s(t) = \frac{1}{2}gt^2 + v_0 t + s_0$$

where s_0 is the initial height of the object, v_0 is the initial velocity of the object, and g is the acceleration due to gravity. Considering the acceleration due to the earth's gravity to be $g = -32$ feet per second per second, we obtain the position function

$$s(t) = -16t^2 + v_0 t + s_0.$$

Remember that for free-falling objects, we consider the velocity to be positive for upward motion and negative for downward motion.

EXAMPLE 4 Finding the acceleration of a moving object

Suppose that the velocity of an automobile starting from rest is given by

$$v = \frac{80t}{t + 5} \text{ ft/sec.}$$

Make a table to compare the velocity and acceleration of the automobile when $t = 0, 5, 10, \ldots , 60$ seconds.

SOLUTION

The position of the car is shown in Figure 3.14. The acceleration at time t is given by

$$
\begin{aligned}
a = \frac{dv}{dt} &= \lim_{\Delta t \to 0} \frac{1}{\Delta t}\left[\frac{80(t + \Delta t)}{(t + \Delta t) + 5} - \frac{80t}{t + 5}\right] \\
&= \lim_{\Delta t \to 0} \frac{80}{\Delta t}\left[\frac{5\Delta t}{(t + \Delta t + 5)(t + 5)}\right] \\
&= \lim_{\Delta t \to 0} \frac{400}{(t + \Delta t + 5)(t + 5)} = \frac{400}{(t + 5)^2} \text{ ft/sec}^2.
\end{aligned}
$$

Table 3.3 compares the velocity and acceleration of the automobile at five-second intervals during its first minute of travel.

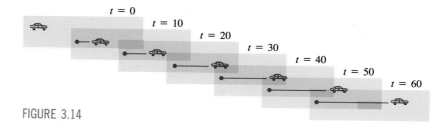

$t = 0$ $t = 10$ $t = 20$ $t = 30$ $t = 40$ $t = 50$ $t = 60$

FIGURE 3.14

TABLE 3.3

t	0	5	10	15	20	25	30	35	40	45	50	55	60
v	0	40	53.3	60.0	64.0	66.7	68.6	70.0	71.1	72.0	72.7	73.3	73.8
$\dfrac{dv}{dt}$	16	4	1.78	1.00	0.64	0.44	0.33	0.25	0.20	0.16	0.13	0.11	0.09

Note from Table 3.3 that the acceleration approaches zero as the velocity levels off. This observation should agree with your experience of riding in an accelerating car—you do not feel the velocity, but you feel the acceleration. In other words, you feel changes in velocity.

Higher-order derivatives

To derive the acceleration function from the position function, we need to differentiate the position function *twice*.

$$s(t) \qquad \text{Position function}$$
$$v(t) = s'(t) \qquad \text{Velocity function}$$
$$a(t) = v'(t) = s''(t) \qquad \text{Acceleration function}$$

We call $a(t)$ the **second derivative** of $s(t)$ and denote it by $s''(t)$.

The second derivative is an example of a **higher-order derivative.** We can define derivatives of any positive integer order. For instance, the **third derivative** is the derivative of the second derivative. We denote higher-order derivatives as follows.

First derivative: $\quad y', \quad f'(x), \quad \dfrac{dy}{dx}, \quad \dfrac{d}{dx}[f(x)], \quad D_x(y)$

Second derivative: $\quad y'', \quad f''(x), \quad \dfrac{d^2y}{dx^2}, \quad \dfrac{d^2}{dx^2}[f(x)], \quad D_x^2(y)$

Third derivative: $\quad y''', \quad f'''(x), \quad \dfrac{d^3y}{dx^3}, \quad \dfrac{d^3}{dx^3}[f(x)], \quad D_x^3(y)$

Fourth derivative: $\quad y^{(4)}, \quad f^{(4)}(x), \quad \dfrac{d^4y}{dx^4}, \quad \dfrac{d^4}{dx^4}[f(x)], \quad D_x^4(y)$

$$\vdots$$

nth derivative: $\quad y^{(n)}, \quad f^{(n)}(x), \quad \dfrac{d^ny}{dx^n}, \quad \dfrac{d^n}{dx^n}[f(x)], \quad D_x^n(y)$

Other rates of change

Velocity and acceleration are only two examples of rates of change. In general, we can use the derivative to measure the rate of change of any variable with respect to another [provided the two variables are related by a differentiable function $y = f(x)$]. When determining the rate of change of one variable with respect to another, we must be careful to distinguish between average and instantaneous rates of change. The distinction between these two rates of change is comparable to the distinction between the slope of the secant line through two points on a curve and the slope of the tangent line at one point on a curve.

REMARK   In future work with the derivative, we will use "rate of change" to mean "instantaneous rate of change."

EXAMPLE 5   Finding the average rate of change over an interval

The concentration of a drug in a patient's bloodstream is monitored over 10-minute intervals for two hours. Find the average rates of change (in milligrams per minute) over the time intervals [0, 10], [0, 20], and [100, 110] for the concentrations in Table 3.4.

TABLE 3.4

t (min)	0	10	20	30	40	50	60	70	80	90	100	110	120
C (mg)	0	2	17	37	55	73	89	103	111	113	113	103	68

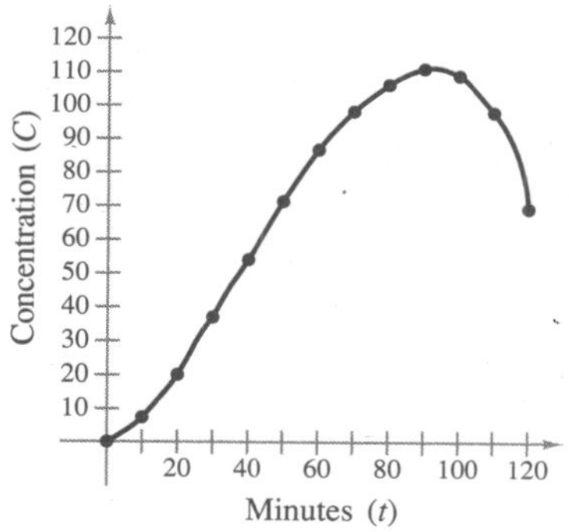

Drug Concentration in Bloodstream

FIGURE 3.15

SOLUTION

For the interval [0, 10], the average rate of change is

$$\frac{\Delta C}{\Delta t} = \frac{2 - 0}{10 - 0} = \frac{2}{10} = 0.2 \text{ mg/min.}$$

For the interval [0, 20], the average rate of change is

$$\frac{\Delta C}{\Delta t} = \frac{17 - 0}{20 - 0} = \frac{17}{20} = 0.85 \text{ mg/min.}$$

For the interval [100, 110], the average rate of change is

$$\frac{\Delta C}{\Delta t} = \frac{103 - 113}{110 - 100} = \frac{-10}{10} = -1 \text{ mg/min.}$$

Note in Figure 3.15 that the average rate of change is positive when the concentration increases and negative when the concentration decreases. ▭

To conclude this section, we give a summary concerning the derivative and its interpretations.

INTERPRETATIONS OF THE DERIVATIVE

If the function given by $y = f(x)$ is differentiable at x, then its derivative

$$\frac{dy}{dx} = f'(x) = \lim_{\Delta x \to 0} \frac{f(x + \Delta x) - f(x)}{\Delta x}$$

denotes both

1. the *slope* of the graph of f at x and
2. the *instantaneous rate of change* in y with respect to x.

EXERCISES for Section 3.2

In Exercises 1–6, find the average rate of change of the given function over the indicated interval. Compare this average rate of change to the instantaneous rates of change at the endpoints of the interval.

Function	Interval
1. $f(t) = 2t + 7$	[1, 2]
2. $f(t) = 3t - 1$	$\left[0, \frac{1}{3}\right]$

Function	Interval
3. $f(x) = \dfrac{1}{x + 1}$	[0, 3]
4. $f(x) = \dfrac{-1}{x}$	[1, 2]
5. $f(t) = t^2 - 3$	[2, 2.1]
6. $f(x) = x^2 - 6x - 1$	[−1, 3]

7. The height s at time t of a silver dollar dropped from the World Trade Center is given by $s(t) = -16t^2 + 1350$, where s is measured in feet and t is measured in seconds $[s'(t) = -32t]$.
   (a) Find the average velocity on the interval $[1, 2]$.
   (b) Find the instantaneous velocity when $t = 1$ and $t = 2$.
   (c) How long will it take the dollar to hit the ground?
   (d) Find the velocity of the dollar when it hits the ground.

8. An automobile's velocity starting from rest is given by

$$v = \frac{100t}{2t + 15}$$

   where v is measured in feet per second. Find the acceleration at the following times.
   (a) 5 seconds   (b) 10 seconds   (c) 20 seconds

In Exercises 9–14, use the following position and velocity functions for free-falling objects.

$$s(t) = -16t^2 + v_0 t + s_0$$
$$s'(t) = -32t + v_0$$

9. A projectile is shot upward from the surface of the earth with an initial velocity of 384 feet per second. What is its velocity after 5 seconds? After 10 seconds?

10. Repeat Exercise 9 for an initial velocity of 256 feet per second.

11. A pebble is dropped from a height of 600 feet. Find the pebble's velocity when it hits the ground.

12. A ball is thrown straight down from the top of a 220-foot building with an initial velocity of -22 feet per second. What is its velocity after 3 seconds? What is its velocity after falling 108 feet?

13. To estimate the height of a building, a stone is dropped from the top of the building into a pool of water at ground level. How high is the building if the splash is seen 6.8 seconds after the stone is dropped?

14. A ball is dropped from a height of 100 feet. One second later another ball is dropped from a height of 75 feet. Which ball hits the ground first?

15. A car is traveling at a rate of 66 feet per second (45 miles per hour) when the brakes are applied. The position function for the car is $s(t) = -8.25t^2 + 66t$, where s is measured in feet and t is measured in seconds. Use this function to complete the following table.

t	0	1	2	3	4
$s(t)$					
$v(t)$					
$a(t)$					

16. The position function for an object is given by $s(t) = 10t^2$, $0 \le t \le 10$, where s is measured in feet and t is measured in seconds. Use this function to complete the following table.

t	0	2	4	6	8	10
$s(t)$						
$v(t)$						
$a(t)$						

17. Find the average velocity of the car in Exercise 15 during each given time interval.

18. Find the average velocity of the object in Exercise 16 during each given time interval.

In Exercises 19–24, find the indicated derivative.

Given	*Find*
19. $f'(x) = x^2$	$f''(x)$
20. $f''(x) = x^3$	$f'''(x)$
21. $f''(x) = 2 - \dfrac{2}{x}$	$f'''(x)$
22. $f'''(x) = 2\sqrt{x} - 1$	$f^{(4)}(x)$
23. $f^{(4)}(x) = 2x + 1$	$f^{(6)}(x)$
24. $f(x) = 2x^2 - 2$	$f''(x)$

25. The annual inventory cost for a certain manufacturer is given by

$$C = \frac{1,008,000}{Q} + 6.3Q$$

   where Q is the order size when the inventory is replenished. Find the change in annual cost when Q is increased from 350 to 351 and compare this with the rate of change

$$\frac{dC}{dQ} = -\frac{1,008,000}{Q^2} + 6.3$$

   when $Q = 350$.

26. A car is driven 15,000 miles a year and gets x miles per gallon. Assume that the average fuel cost is $1.10 per gallon. Find the annual cost of fuel C as a function of x and use this function to complete the following table.

x	10	15	20	25	30	35	40
C							
$\dfrac{dC}{dx}$							

27. At 0° Celsius, the equation for heat loss H from the body in kilocalories per square meter per hour is

$$H(v) = 33(10\sqrt{v} - v + 10.45)$$

where v is the wind speed in meters per second. The rate of change of H is given by

$$\frac{dH}{dv} = 33\left(\frac{5}{\sqrt{v}} - 1\right).$$

Find the rate at which H is changing when (a) $v = 2$ and (b) $v = 5$.

28. When a guitar string is plucked, it vibrates with a frequency of

$$F = 200\sqrt{T}$$

where F is measured in vibrations per second and the tension T is measured in pounds. Find the rate of change of the frequency when (a) $T = 4$ and (b) $T = 9$.

29. Newton's Law of Gravitation states that the gravitational force between two (point) particles of masses m_1 and m_2, respectively, is

$$F = K\frac{m_1 m_2}{r^2}, \quad r > 0$$

where r is the distance between the particles. Find the rate of change of force between the particles with respect to r and explain why the rate of change is negative.

30. Newton's Law of Cooling states that the rate of change of the temperature of an object is proportional to the difference between the object's temperature T and the temperature T_a of the surrounding medium. Write an equation for this law.

31. The height of an object projected upward from an initial height of 100 feet above the ground is given by

$$s = -16t^2 + 27t + 100$$

where t is the time in seconds. Use a computer or graphics calculator to sketch the graph of the position function and find the time when the object reaches ground level. Find the velocity of the object when it reaches ground level.

3.3 Differentiation Rules for Powers, Constant Multiples, and Sums

Constant Rule ▪ Power Rule ▪ Constant Multiple Rule ▪ Sum and Difference Rules ▪ Applications of the derivative

In Sections 3.1 and 3.2 we used the limit definition to find derivatives. In this and the next two sections we introduce several "differentiation rules" that allow us to find derivatives without the *direct* use of the limit definition.

**THEOREM 3.3
CONSTANT RULE**

The derivative of a constant is zero.

$$\frac{d}{dx}[c] = 0, \quad c \text{ is a real number}$$

PROOF Let $f(x) = c$. Then, by the limit definition of the derivative,

$$\frac{d}{dx}[c] = f'(x) = \lim_{\Delta x \to 0} \frac{f(x + \Delta x) - f(x)}{\Delta x} = \lim_{\Delta x \to 0} \frac{c - c}{\Delta x} = 0.$$

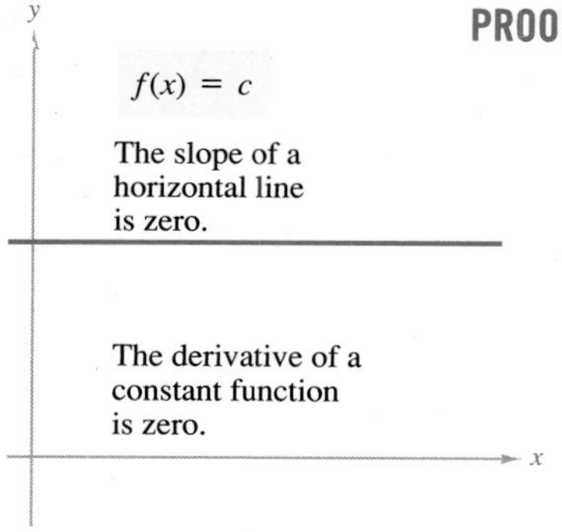

$f(x) = c$

The slope of a horizontal line is zero.

The derivative of a constant function is zero.

REMARK Note in Figure 3.16 that the Constant Rule is equivalent to saying that the slope of a horizontal line is zero.

FIGURE 3.16

EXAMPLE 1 The derivative of a constant function

Function	*Derivative*
(a) $y = 7$	$\dfrac{dy}{dx} = 0$
(b) $f(x) = 0$	$f'(x) = 0$
(c) $s'(t) = -3$	$s''(t) = 0$

Before proving the next rule, we review the procedure for expanding a binomial. Recall that

$$(x + \Delta x)^2 = x^2 + 2x\Delta x + (\Delta x)^2$$
$$(x + \Delta x)^3 = x^3 + 3x^2\Delta x + 3x(\Delta x)^2 + (\Delta x)^3$$

and the general binomial expansion for a positive integer n is

$$(x + \Delta x)^n = x^n + nx^{n-1}\Delta x + \underbrace{\frac{n(n-1)x^{n-2}}{2}(\Delta x)^2 + \cdots + (\Delta x)^n}_{(\Delta x)^2 \text{ is a factor of these terms}}.$$

This binomial expansion is used in proving a special case of the following theorem.

**THEOREM 3.4
POWER RULE**

If n is a rational number, then

$$\frac{d}{dx}[x^n] = nx^{n-1}.$$

PROOF Here, we prove only the case in which n is a positive integer greater than 1. In Exercise 59, we ask you to prove the case for $n = 1$. Section 3.4 (Example 7) has a proof for the case in which n is a negative integer, and Section 3.5 (Example 5) has a proof for the case in which n is rational. (In Chapter 7, the Power Rule will be extended to cover irrational values of n.) If n is a positive integer greater than 1, then the binomial expansion produces

$$\frac{d}{dx}[x^n] = \lim_{\Delta x \to 0} \frac{(x + \Delta x)^n - x^n}{\Delta x}$$

$$= \lim_{\Delta x \to 0} \frac{x^n + nx^{n-1}(\Delta x) + [n(n-1)x^{n-2}/2](\Delta x)^2 + \cdots + (\Delta x)^n - x^n}{\Delta x}$$

$$= \lim_{\Delta x \to 0} \left[nx^{n-1} + \left(\frac{n(n-1)x^{n-2}}{2} \right)(\Delta x) + \cdots + (\Delta x)^{n-1} \right]$$

$$= nx^{n-1} + 0 + \cdots + 0 = nx^{n-1}.$$

In the Power Rule, the case in which $n = 1$ is worth memorizing as a separate differentiation rule. That is,

$$\frac{d}{dx}[x] = 1.$$

This rule is consistent with the fact that the slope of the line given by $y = x$ is 1.

EXAMPLE 2 Applying the Power Rule

Function	Derivative
(a) $f(x) = x^3$	$f'(x) = 3x^2$
(b) $y = \dfrac{1}{x^2}$	$\dfrac{dy}{dx} = \dfrac{d}{dx}[x^{-2}] = (-2)x^{-3} = -\dfrac{2}{x^3}$

In part (b) of Example 2, note that *before* differentiating we rewrote $1/x^2$ as x^{-2}. Rewriting is the first step in *many* differentiation problems.

Given:	Rewrite:	Differentiate:	Simplify:
$y = \dfrac{1}{x^2}$	$y = x^{-2}$	$\dfrac{dy}{dx} = (-2)x^{-3}$	$\dfrac{dy}{dx} = -\dfrac{2}{x^3}$

THEOREM 3.5
CONSTANT MULTIPLE RULE

If f is a differentiable function and c is a real number, then

$$\frac{d}{dx}[cf(x)] = cf'(x).$$

PROOF

$$\frac{d}{dx}[cf(x)] = \lim_{\Delta x \to 0} \frac{cf(x + \Delta x) - cf(x)}{\Delta x}$$

$$= \lim_{\Delta x \to 0} c\left[\frac{f(x + \Delta x) - f(x)}{\Delta x}\right]$$

$$= c\left[\lim_{\Delta x \to 0} \frac{f(x + \Delta x) - f(x)}{\Delta x}\right] = cf'(x)$$

Informally, the Constant Multiple Rule states that constants can be factored out of the differentiation process.

$$\frac{d}{dx}[cf(x)] = c\frac{d}{dx}[f(x)] = cf'(x)$$

This rule is often overlooked, especially when the constant appears in the denominator, as follows.

$$\frac{d}{dx}\left[\frac{f(x)}{c}\right] = \frac{d}{dx}\left[\left(\frac{1}{c}\right)f(x)\right] = \left(\frac{1}{c}\right)\frac{d}{dx}[f(x)] = \left(\frac{1}{c}\right)f'(x)$$

EXAMPLE 3 Applying the Constant Multiple Rule with the Power Rule

Function	*Derivative*
(a) $y = \dfrac{2}{x}$	$\dfrac{dy}{dx} = \dfrac{d}{dx}[2x^{-1}] = 2\dfrac{d}{dx}[x^{-1}] = 2(-1)x^{-2} = -\dfrac{2}{x^2}$
(b) $f(t) = \dfrac{4t^2}{5}$	$f'(t) = \dfrac{d}{dt}\left[\dfrac{4}{5}t^2\right] = \dfrac{4}{5}\dfrac{d}{dt}[t^2] = \dfrac{4}{5}(2t) = \dfrac{8}{5}t$

$\Box$

REMARK   It is helpful to see that the Constant Multiple and Power Rules can be combined into one rule. The combination rule is $D_x[cx^n] = cnx^{n-1}$.

Although the two functions in the next example are very simple, errors are frequently made in differentiating functions involving a constant multiple of the first power of x. Keep in mind that $D_x[cx] = c$.

EXAMPLE 4 Applying the Constant Multiple Rule

(a) $\dfrac{d}{dx}\left[-\dfrac{3x}{2}\right] = -\dfrac{3}{2}$ (b) $\dfrac{d}{dx}[3\pi x] = 3\pi$ $\Box$

The Sum Rule given in the following theorem tells us that we can differentiate the sum of two functions separately and add the resulting derivatives.

THEOREM 3.6 **SUM AND DIFFERENCE RULES**	The derivative of the sum (or difference) of two differentiable functions is the sum (or difference) of their derivatives. $$\frac{d}{dx}[f(x) + g(x)] = f'(x) + g'(x) \qquad \text{Sum Rule}$$ $$\frac{d}{dx}[f(x) - g(x)] = f'(x) - g'(x) \qquad \text{Difference Rule}$$

PROOF   A proof of the Sum Rule follows directly from Theorem 2.3 (the limit of a sum). The Difference Rule can be proved in a similar way.

$$\frac{d}{dx}[f(x) + g(x)] = \lim_{\Delta x \to 0} \frac{[f(x + \Delta x) + g(x + \Delta x)] - [f(x) + g(x)]}{\Delta x}$$

$$= \lim_{\Delta x \to 0} \frac{f(x + \Delta x) + g(x + \Delta x) - f(x) - g(x)}{\Delta x}$$

$$= \lim_{\Delta x \to 0} \left[\frac{f(x + \Delta x) - f(x)}{\Delta x} + \frac{g(x + \Delta x) - g(x)}{\Delta x}\right]$$

$$= \lim_{\Delta x \to 0} \frac{f(x + \Delta x) - f(x)}{\Delta x} + \lim_{\Delta x \to 0} \frac{g(x + \Delta x) - g(x)}{\Delta x}$$

$$= f'(x) + g'(x)$$

The Sum and Difference Rules can be extended to cover the derivative of any finite number of functions. For instance, if

$$F(x) = f(x) + g(x) - h(x) - k(x)$$

then

$$F'(x) = f'(x) + g'(x) - h'(x) - k'(x).$$

EXAMPLE 5 Applying the Sum and Difference Rules

Function	Derivative
(a) $f(x) = x^3 - 4x + 5$	$f'(x) = 3x^2 - 4$
(b) $g(x) = -\dfrac{x^4}{2} + 3x^3 - 2x$	$g'(x) = -2x^3 + 9x^2 - 2$

Parentheses can play an important role in the Power Rule and the Constant Multiple Rule, as shown in Example 6.

EXAMPLE 6 Using parentheses when differentiating

Given Function	Rewrite	Differentiate	Simplify
(a) $y = \dfrac{5}{2x^3}$	$y = \dfrac{5}{2}(x^{-3})$	$y' = \dfrac{5}{2}(-3x^{-4})$	$y' = -\dfrac{15}{2x^4}$
(b) $y = \dfrac{5}{(2x)^3}$	$y = \dfrac{5}{8}(x^{-3})$	$y' = \dfrac{5}{8}(-3x^{-4})$	$y' = -\dfrac{15}{8x^4}$
(c) $y = \dfrac{7}{3x^{-2}}$	$y = \dfrac{7}{3}(x^2)$	$y' = \dfrac{7}{3}(2x)$	$y' = \dfrac{14x}{3}$
(d) $y = \dfrac{7}{(3x)^{-2}}$	$y = 63(x^2)$	$y' = 63(2x)$	$y' = 126x$

When differentiating functions involving radicals, we rewrite the function in terms of rational exponents, as shown in the next example.

EXAMPLE 7 Differentiating a function involving a radical

Function	Derivative
(a) $y = 2\sqrt{x}$	$\dfrac{dy}{dx} = \dfrac{d}{dx}[2x^{1/2}] = 2\left(\dfrac{1}{2}x^{-1/2}\right) = x^{-1/2} = \dfrac{1}{\sqrt{x}}$
(b) $y = \dfrac{1}{2\sqrt[3]{x^2}}$	$\dfrac{dy}{dx} = \dfrac{d}{dx}\left[\dfrac{1}{2}x^{-2/3}\right] = \dfrac{1}{2}\left(-\dfrac{2}{3}\right)x^{-5/3} = -\dfrac{1}{3x^{5/3}}$

Applications of the derivative

The first two sections of this chapter included two important applications of the derivative—the slope of a curve and rate of change. We conclude this section with two examples of these applications.

EXAMPLE 8   Using the derivative to find the slope of a curve

Find the slope of the graph of $f(x) = x^3 - 3x$ at the following points.

(a) $(-2, -2)$ (b) $(0, 0)$ (c) $(1, -2)$

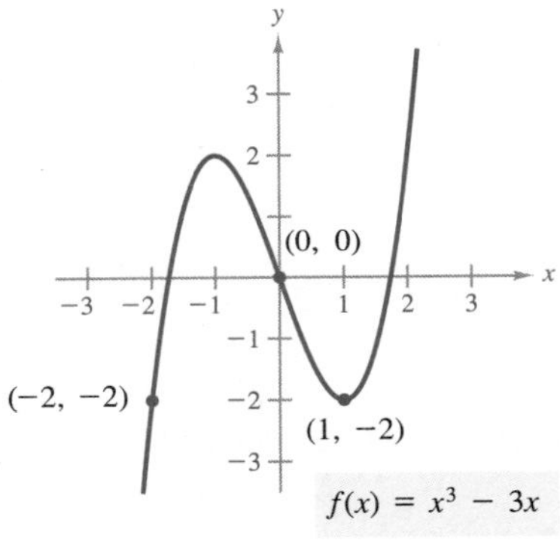

SOLUTION

The derivative of f is $f'(x) = 3x^2 - 3$. Therefore, the slopes at the indicated points are as follows.
(a) At $x = -2$, the slope is

$$f'(-2) = 3(-2)^2 - 3 = 12 - 3 = 9.$$

(b) At $x = 0$, the slope is

$$f'(0) = 3(0)^2 - 3 = -3.$$

(c) At $x = 1$, the slope is

$$f'(1) = 3(1)^2 - 3 = 0.$$

(See Figure 3.17.)

FIGURE 3.17

$f(x) = x^3 - 3x$

EXAMPLE 9   Using the derivative to find velocity

At time $t = 0$, a diver jumps from a diving board that is 32 feet above the water. The position of the diver is given by

$$s(t) = -16t^2 + 16t + 32$$

where s is measured in feet and t is measured in seconds. (See Figure 3.18.)
(a) When does the diver hit the water?
(b) What is the diver's velocity at impact?

SOLUTION

(a) To find the time at which the diver hits the water, we let $s = 0$ and solve for t.

$$-16t^2 + 16t + 32 = 0$$
$$-16(t^2 - t - 2) = 0$$
$$-16(t + 1)(t - 2) = 0$$
$$t = -1 \text{ or } 2$$

The solution $t = -1$ doesn't make sense, so we conclude that the diver hits the water at $t = 2$ seconds.

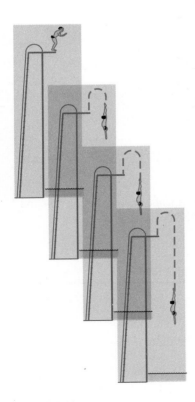

FIGURE 3.18

(b) The velocity at time t is given by the derivative

$$s'(t) = -32t + 16.$$

Therefore, the velocity at time $t = 2$ is

$$s'(2) = -32(2) + 16 = -48 \text{ ft/sec.}$$

REMARK   In Figure 3.18, note that the diver moves upward for the first half-second. This corresponds to the fact that the velocity is positive for $0 < t < \frac{1}{2}$.

EXERCISES for Section 3.3

In Exercises 1 and 2, find the slope of the tangent line to $y = x^n$ at the point (1, 1).

1. (a) $y = x^{1/2}$

(b) $y = x^{3/2}$

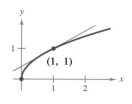

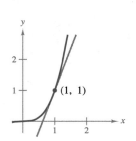

(c) $y = x^2$

(d) $y = x^3$

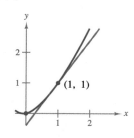

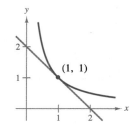

2. (a) $y = x^{-1/2}$

(b) $y = x^{-1}$

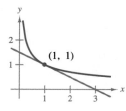

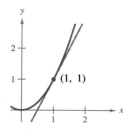

(c) $y = x^{-3/2}$

(d) $y = x^{-2}$

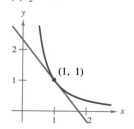

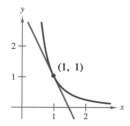

In Exercises 3–12, find the derivative of the given function.

3. $y = 3$ **4.** $f(x) = -2$
5. $f(x) = x + 1$ **6.** $g(x) = 3x - 1$
7. $g(x) = x^2 + 4$ **8.** $y = t^2 + 2t - 3$
9. $f(t) = -2t^2 + 3t - 6$ **10.** $y = x^3 - 9$
11. $s(t) = t^3 - 2t + 4$ **12.** $f(x) = 2x^3 - x^2 + 3x$

In Exercises 13–18, find the value of the derivative of the given function at the indicated point.

Function	Point
13. $f(x) = \dfrac{1}{x}$	(1, 1)
14. $f(t) = 3 - \dfrac{3}{5t}$	$\left(\dfrac{3}{5}, 2\right)$
15. $f(x) = -\dfrac{1}{2} + \dfrac{7}{5}x^3$	$\left(0, -\dfrac{1}{2}\right)$
16. $y = 3x\left(x^2 - \dfrac{2}{x}\right)$	(2, 18)
17. $y = (2x + 1)^2$	(0, 1)
18. $f(x) = 3(5 - x)^2$	(5, 0)

In Exercises 19–30, find $f'(x)$.

19. $f(x) = x^2 - \dfrac{4}{x}$

20. $f(x) = x^2 - 3x - 3x^{-2}$

21. $f(x) = x^3 - 3x - \dfrac{2}{x^4}$

22. $f(x) = \dfrac{2x^2 - 3x + 1}{x}$

23. $f(x) = \dfrac{x^3 - 3x^2 + 4}{x^2}$

24. $f(x) = (x^2 + 2x)(x + 1)$

25. $f(x) = x(x^2 + 1)$

26. $f(x) = x + \dfrac{1}{x^2}$

27. $f(x) = x^{4/5}$

28. $f(x) = x^{1/3} - 1$

29. $f(x) = \sqrt[3]{x} + \sqrt[5]{x}$

30. $f(x) = \dfrac{1}{\sqrt[3]{x^2}}$

In Exercises 31–36, complete the table, using Example 6 as a model.

Function	Rewrite	Derivative	Simplify
31. $y = \dfrac{1}{3x^3}$			
32. $y = \dfrac{2}{3x^2}$			
33. $y = \dfrac{1}{(3x)^3}$			
34. $y = \dfrac{\pi}{(3x)^2}$			
35. $y = \dfrac{\sqrt{x}}{x}$			
36. $y = \dfrac{4}{x^{-3}}$			

In Exercises 37 and 38, find an equation of the tangent line to the given function at the indicated point.

37. $y = x^4 - 3x^2 + 2$, $(1, 0)$

38. $y = x^3 + x$, $(-1, -2)$

In Exercises 39–42, determine the point(s) (if any) at which the given function has a horizontal tangent line.

39. $y = x^4 - 3x^2 + 2$

40. $y = x^3 + x$

41. $y = \dfrac{1}{x^2}$

42. $y = x^2 + 1$

43. Sketch the graphs of the two equations $y = x^2$ and $y = -x^2 + 6x - 5$, and sketch the two lines that are tangent to both graphs. Find the equations of these lines.

44. Show that the graphs of the two equations $y = x$ and $y = 1/x$ have tangent lines that are perpendicular to each other at their points of intersection.

45. The area of a square with sides of length s is given by $A = s^2$. Find the rate of change of the area with respect to s when $s = 4$.

46. The volume of a cube with sides of length s is given by $V = s^3$. Find the rate of change of the volume with respect to s when $s = 4$.

47. A company finds that charging p dollars per unit produces a monthly revenue

$$R = 12{,}000p - 1{,}000p^2, \quad 0 \le p \le 12.$$

(Note that the revenue is zero when $p = 12$ since no one is willing to pay that much.) Find the rate of change of R with respect to p when p has the following values.
(a) $p = 1$ (b) $p = 4$
(c) $p = 6$ (d) $p = 10$

48. Suppose that the profit P obtained in selling x units of a certain item each week is given by

$$P = 50\sqrt{x} - 0.5x - 500, \quad 0 \le x \le 8000.$$

Find the rate of change of P with respect to x when
(a) $x = 900$ (b) $x = 1600$
(c) $x = 2500$ (d) $x = 3600$
(Note that the eventual decline in profit occurs because the only way to sell larger quantities is to decrease the price per unit.)

49. Suppose that the effectiveness E of a painkilling drug t hours after entering the bloodstream is given by

$$E = \frac{1}{27}(9t + 3t^2 - t^3), \quad 0 \le t \le 4.5.$$

Find the rate of change of E with respect to t when
(a) $t = 1$ (b) $t = 2$
(c) $t = 3$ (d) $t = 4$

50. In a certain chemical reaction, the amount in grams Q of a substance produced in t hours is given by the equation

$$Q = 16t - 4t^2, \quad 0 < t \le 2.$$

Find the rate in grams per hour at which the substance is being produced when t has the following values.
(a) $t = \dfrac{1}{2}$ (b) $t = 1$ (c) $t = 2$

51. In the Chapter 3 Application, we developed the model

$$s(t) = -16t^2 + 5t + 30$$

for the height (in feet) of a diver at time t, where t is measured in seconds. Find the velocity of the diver when the diver hits the water.

52. In Exercise 51, the position equation corresponding to a diving platform of height h is

$$s(t) = -16t^2 + 5t + h.$$

What is the velocity of the diver when the diver hits the water for $h = 40$ ft, $h = 50$ ft, and $h = 60$ ft?

53. An astronaut standing on the moon throws a rock into the air. The height of the rock is given by

$$s = -\frac{27}{10}t^2 + 27t + 6$$

where s is measured in feet and t is measured in seconds. Find the acceleration of the rock and compare it with the acceleration due to gravity on the earth.

54. A ball is thrown upward from ground level, and its height at any time is given by

$$s(t) = -16t^2 + 48t$$

where s is measured in feet and t is measured in seconds.

(a) Find expressions for the velocity and acceleration of the ball.

(b) Find the time when the ball is at its highest point by finding the time when the velocity is zero.

(c) Find the height at the time given in part (b).

In Exercises 55 and 56, determine whether the given statement is true or false.

55. If $f(x) = g(x) + c$, then $f'(x) = g'(x)$.

56. If $y = \pi^2$, then $y' = 2\pi$.

57. Use a computer or graphics calculator to sketch the graph of f and f' over the interval $[0, 3]$ if

$$f(x) = 4.1x^3 - 12x^2 + 2.5x.$$

Determine the points (if any) at which f has horizontal tangents.

58. Use a computer or graphics calculator to find an equation of the tangent line to the graph of $f(x) = 2/x$, which passes through the point $(5, 0)$. Sketch the graph of f and the tangent line on the same set of coordinate axes. [Hint: To find the point of tangency on the graph of f, solve the equation $f'(x) = (0 - y)/(5 - x)$.]

59. Prove that $D_x[x] = 1$.

3.4 Differentiation Rules for Products and Quotients

Product Rule ▪ Quotient Rule

In Section 3.3 we saw that the derivative of a sum or difference of two functions is simply the sum or difference of their derivatives. The rules for the derivative of a product or quotient of two functions are not as simple, and you may find the results surprising.

THEOREM 3.7 PRODUCT RULE	The derivative of the product of two differentiable functions is given by the first function times the derivative of the second plus the second function times the derivative of the first.

$$\frac{d}{dx}[f(x)g(x)] = f(x)g'(x) + g(x)f'(x)$$

PROOF   Some mathematical proofs, such as the proof of the Sum Rule in the previous section, are straightforward. Others involve clever steps that may appear unmotivated to a reader. This proof involves such a step—adding and subtracting the same quantity. Once you see this step (shown in color), you should be able to justify the limits shown in the other steps.

$$\frac{d}{dx}[f(x)g(x)] = \lim_{\Delta x \to 0} \frac{f(x + \Delta x)g(x + \Delta x) - f(x)g(x)}{\Delta x}$$

$$= \lim_{\Delta x \to 0} \frac{f(x + \Delta x)g(x + \Delta x) - f(x + \Delta x)g(x) + f(x + \Delta x)g(x) - f(x)g(x)}{\Delta x}$$

$$= \lim_{\Delta x \to 0} \left[f(x + \Delta x)\frac{g(x + \Delta x) - g(x)}{\Delta x} + g(x)\frac{f(x + \Delta x) - f(x)}{\Delta x} \right]$$

$$= \lim_{\Delta x \to 0} \left[f(x + \Delta x)\frac{g(x + \Delta x) - g(x)}{\Delta x} \right] + \lim_{\Delta x \to 0} \left[g(x)\frac{f(x + \Delta x) - f(x)}{\Delta x} \right]$$

$$= \lim_{\Delta x \to 0} f(x + \Delta x) \cdot \lim_{\Delta x \to 0} \frac{g(x + \Delta x) - g(x)}{\Delta x} + \lim_{\Delta x \to 0} g(x) \cdot \lim_{\Delta x \to 0} \frac{f(x + \Delta x) - f(x)}{\Delta x}$$

$$= f(x)g'(x) + g(x)f'(x)$$

The Product Rule can be extended to cover products involving more than two terms. For example, if f, g, and h are differentiable functions of x, then

$$\frac{d}{dx}[f(x)g(x)h(x)] = f'(x)g(x)h(x) + f(x)g'(x)h(x) + f(x)g(x)h'(x).$$

EXAMPLE 1   Differentiation with the Product Rule

Find the derivative of $f(x) = (3x - 2x^2)(5 + 4x)$.

SOLUTION

Using the Product Rule, we have

$$f'(x) = \overbrace{(3x - 2x^2)}^{\text{(first)}}\overbrace{\frac{d}{dx}[5 + 4x]}^{\binom{\text{derivative}}{\text{of second}}} + \overbrace{(5 + 4x)}^{\text{(second)}}\overbrace{\frac{d}{dx}[3x - 2x^2]}^{\binom{\text{derivative}}{\text{of first}}}$$

$$= (3x - 2x^2)(4) + (5 + 4x)(3 - 4x)$$

$$= (12x - 8x^2) + (15 - 8x - 16x^2)$$

$$= -24x^2 + 4x + 15.$$

REMARK   Note that the derivative of a product of two functions is not (in general) given by the product of the derivatives of the two functions. To see this, try comparing the product of the derivatives of $f(x) = 3x - 2x^2$ and $g(x) = 5 + 4x$ with the derivative found in Example 1.

In Example 1, we had the option of finding the derivative with or without the Product Rule. To find the derivative without the Product Rule, we can write

$$D_x[(3x - 2x^2)(5 + 4x)] = D_x[-8x^3 + 2x^2 + 15x]$$
$$= -24x^2 + 4x + 15.$$

After we introduce the Chain Rule in the next section, the benefit of the Product Rule will become more apparent.

EXAMPLE 2 Using the Product Rule

Find the derivative of $y = (1 + x^{-1})(x - 1)$.

SOLUTION

$$D_x[(1\text{st})(2\text{nd})] = (1\text{st})(D_x[2\text{nd}]) + (2\text{nd})(D_x[1\text{st}])$$

$$f'(x) = (1 + x^{-1})\frac{d}{dx}[x - 1] + (x - 1)\frac{d}{dx}[1 + x^{-1}]$$

$$= (1 + x^{-1})(1) + (x - 1)(-x^{-2})$$

$$= 1 + \frac{1}{x} - \frac{x - 1}{x^2}$$

$$= \frac{x^2 + x - x + 1}{x^2} = \frac{x^2 + 1}{x^2}$$

EXAMPLE 3 Comparing the Product Rule and the Constant Multiple Rule

Find the derivative of the following.

(a) $y = \sqrt{x}g(x)$ (b) $y = \sqrt{2}g(x)$

SOLUTION

(a) Using the Product Rule, we have

$$\frac{dy}{dx} = \sqrt{x}\left(\frac{d}{dx}[g(x)]\right) + g(x)\left(\frac{d}{dx}[\sqrt{x}]\right)$$

$$= \sqrt{x}g'(x) + g(x)\left(\frac{1}{2}x^{-1/2}\right)$$

$$= \sqrt{x}g'(x) + g(x)\frac{1}{2\sqrt{x}}.$$

(b) Using the Constant Multiple Rule, we have

$$\frac{dy}{dx} = \sqrt{2}g'(x).$$

REMARK In Example 3, notice that we use the Product Rule when both factors of the product are variable, and we use the Constant Multiple Rule when one of the factors is a constant.

THEOREM 3.8
QUOTIENT RULE

The derivative of the quotient of two differentiable functions is given by the denominator times the derivative of the numerator minus the numerator times the derivative of the denominator divided by the square of the denominator.

$$\frac{d}{dx}\left[\frac{f(x)}{g(x)}\right] = \frac{g(x)f'(x) - f(x)g'(x)}{[g(x)]^2}, \quad g(x) \neq 0$$

PROOF   As in the proof of the Product Rule, a key step in this proof involves adding and subtracting the same quantity.

$$\frac{d}{dx}\left[\frac{f(x)}{g(x)}\right] = \lim_{\Delta x \to 0} \frac{\dfrac{f(x + \Delta x)}{g(x + \Delta x)} - \dfrac{f(x)}{g(x)}}{\Delta x}$$

$$= \lim_{\Delta x \to 0} \frac{g(x)f(x + \Delta x) - f(x)g(x + \Delta x)}{\Delta x g(x)g(x + \Delta x)}$$

$$= \lim_{\Delta x \to 0} \frac{g(x)f(x + \Delta x) - f(x)g(x) + f(x)g(x) - f(x)g(x + \Delta x)}{\Delta x g(x)g(x + \Delta x)}$$

$$= \frac{\displaystyle\lim_{\Delta x \to 0} \frac{g(x)[f(x + \Delta x) - f(x)]}{\Delta x} - \lim_{\Delta x \to 0} \frac{f(x)[g(x + \Delta x) - g(x)]}{\Delta x}}{\displaystyle\lim_{\Delta x \to 0} [g(x)g(x + \Delta x)]}$$

$$= \frac{g(x)\left[\displaystyle\lim_{\Delta x \to 0} \dfrac{f(x + \Delta x) - f(x)}{\Delta x}\right] - f(x)\left[\displaystyle\lim_{\Delta x \to 0} \dfrac{g(x + \Delta x) - g(x)}{\Delta x}\right]}{\displaystyle\lim_{\Delta x \to 0} [g(x)g(x + \Delta x)]}$$

$$= \frac{g(x)f'(x) - f(x)g'(x)}{[g(x)]^2}$$

EXAMPLE 4   Using the Quotient Rule

Differentiate

$$y = \frac{2x^2 - 4x + 3}{2 - 3x}.$$

SOLUTION

$$y' = \frac{(2 - 3x)\dfrac{d}{dx}[2x^2 - 4x + 3] - (2x^2 - 4x + 3)\dfrac{d}{dx}[2 - 3x]}{(2 - 3x)^2}$$

$$= \frac{(2 - 3x)(4x - 4) - (2x^2 - 4x + 3)(-3)}{(2 - 3x)^2}$$

$$= \frac{(-12x^2 + 20x - 8) - (-6x^2 + 12x - 9)}{(2 - 3x)^2}$$

$$= \frac{-6x^2 + 8x + 1}{(2 - 3x)^2}$$

REMARK   Note the use of parentheses in Example 4. A liberal use of parentheses is recommended for *all* types of differentiation problems. For instance, with the Quotient Rule, it is a good idea to enclose all factors and derivatives in parentheses and to pay special attention to the subtraction required in the numerator.

When we introduced differentiation rules in the previous section, we emphasized the need for rewriting *before* differentiating. The next example illustrates this point with the Quotient Rule.

EXAMPLE 5   Rewriting before differentiating

$$y = \frac{3 - (1/x)}{x + 5} \qquad \text{Given function}$$

$$= \frac{(3x - 1)/x}{x + 5} = \frac{3x - 1}{x(x + 5)} = \frac{3x - 1}{x^2 + 5x} \qquad \text{Rewrite}$$

$$\frac{dy}{dx} = \frac{(x^2 + 5x)(3) - (3x - 1)(2x + 5)}{(x^2 + 5x)^2} \qquad \text{Quotient Rule}$$

$$= \frac{(3x^2 + 15x) - (6x^2 + 13x - 5)}{(x^2 + 5x)^2}$$

$$= \frac{-3x^2 + 2x + 5}{(x^2 + 5x)^2} \qquad \text{Simplify} \qquad \square$$

Not every quotient needs to be differentiated by the Quotient Rule. For example, the quotients in the next example can each be considered as the product of a constant times a function of x. In such cases it is more convenient to use the Constant Multiple Rule than the Quotient Rule.

EXAMPLE 6   Differentiating quotients with the Constant Multiple Rule

Given Function	*Rewrite*	*Differentiate*	*Simplify*
(a) $y = \dfrac{x^2 + 3x}{6}$	$y = \dfrac{1}{6}(x^2 + 3x)$	$y' = \dfrac{1}{6}(2x + 3)$	$y' = \dfrac{2x + 3}{6}$
(b) $y = \dfrac{5x^4}{8}$	$y = \dfrac{5}{8}x^4$	$y' = \dfrac{5}{8}(4x^3)$	$y' = \dfrac{5}{2}x^3$
(c) $y = \dfrac{-3(3x - 2x^2)}{7x}$	$y = -\dfrac{3}{7}(3 - 2x)$	$y' = -\dfrac{3}{7}(-2)$	$y' = \dfrac{6}{7}$
(d) $y = \dfrac{9}{5x^2}$	$y = \dfrac{9}{5}(x^{-2})$	$y' = \dfrac{9}{5}(-2x^{-3})$	$y' = -\dfrac{18}{5x^3}$

$\square$

In Section 3.3, we claimed that the Power Rule, $D_x[x^n] = nx^{n-1}$, is valid for any rational number n, but we proved only the case where n is a positive integer. In the next example, we prove the rule for the case where n is a negative integer.

EXAMPLE 7 Proof of the Power Rule for negative integers

Use the Quotient Rule to prove the Power Rule for the case when n is a negative integer.

SOLUTION

If n is a negative integer, then there exists a positive integer k such that $n = -k$. Thus, by the Quotient Rule, we have

$$\frac{d}{dx}[x^n] = \frac{d}{dx}\left[\frac{1}{x^k}\right]$$
$$= \frac{x^k(0) - (1)(kx^{k-1})}{(x^k)^2}$$
$$= \frac{0 - kx^{k-1}}{x^{2k}}$$
$$= -kx^{-k-1} = nx^{n-1}.$$

The summary in Table 3.5 shows that much of the work required to obtain the simplified form of a derivative occurs *after* differentiating. Note that two characteristics of a simplified form are the absence of negative exponents and the combining of like terms.

TABLE 3.5

	$f'(x)$ after differentiating	$f'(x)$ after simplifying
Example 1	$(3x - 2x^2)(4) + (5 + 4x)(3 - 4x)$	$-24x^2 + 4x + 15$
Example 2	$(1 - x^{-1})(1) + (x - 1)(-x^{-2})$	$\dfrac{x^2 + 1}{x}$
Example 4	$\dfrac{(2 - 3x)(4x - 4) - (2x^2 - 4x + 3)(-3)}{(2 - 3x)^2}$	$\dfrac{-6x^2 + 8x + 1}{(2 - 3x)^2}$
Example 5	$\dfrac{(x^2 + 5x)(3) - (3x - 1)(2x + 5)}{(x^2 + 5x)^2}$	$\dfrac{-3x^2 + 2x + 5}{(x^2 + 5x)^2}$

We conclude this section with a summary of the differentiation rules studied so far. To become skilled at differentiation, we recommend that you memorize each rule.

SUMMARY OF DIFFERENTIATION

General Differentiation Rules

Let u and v be differentiable functions of x.

Constant Multiple Rule: $\dfrac{d}{dx}[cu] = cu'$

Sum or Difference Rule: $\dfrac{d}{dx}[u \pm v] = u' \pm v'$

Product Rule: $\dfrac{d}{dx}[uv] = uv' + vu'$

Quotient Rule: $\dfrac{d}{dx}\left[\dfrac{u}{v}\right] = \dfrac{vu' - uv'}{v^2}$

Derivatives of Algebraic Functions

Constant Rule: $\dfrac{d}{dx}[c] = 0$

Power Rule: $\dfrac{d}{dx}[x^n] = nx^{n-1}$ $\dfrac{d}{dx}[x] = 1$

EXERCISES for Section 3.4

In Exercises 1–8, find $f'(x)$ and $f'(c)$.

Function	Value of c
1. $f(x) = \dfrac{1}{3}(2x^3 - 4)$	$c = 0$
2. $f(x) = \dfrac{5 - 6x^2}{7}$	$c = 1$
3. $f(x) = 5x^{-2}(x + 3)$	$c = 1$
4. $f(x) = (x^2 - 2x + 1)(x^3 - 1)$	$c = 1$
5. $f(x) = (x^3 - 3x)(2x^2 + 3x + 5)$	$c = 0$
6. $f(x) = (x - 1)(x^2 - 3x + 2)$	$c = 0$
7. $f(x) = (x^5 - 3x)\left(\dfrac{1}{x^2}\right)$	$c = -1$
8. $f(x) = \dfrac{x + 1}{x - 1}$	$c = 2$

In Exercises 9–24, differentiate the given function.

9. $f(x) = \dfrac{3x - 2}{2x - 3}$

10. $f(x) = \dfrac{x^3 + 3x + 2}{x^2 - 1}$

11. $f(x) = \dfrac{3 - 2x - x^2}{x^2 - 1}$

12. $f(x) = x^4\left(1 - \dfrac{2}{x + 1}\right)$

13. $f(x) = \dfrac{x + 1}{\sqrt{x}}$

14. $f(x) = \sqrt[3]{x}(\sqrt{x} + 3)$

15. $h(t) = \dfrac{t + 1}{t^2 + 2t + 2}$

16. $h(x) = (x^2 - 1)^2$

17. $h(s) = (s^3 - 2)^2$

18. $f(x) = \left(\dfrac{x^2 - x - 3}{x^2 + 1}\right)(x^2 + x + 1)$

19. $g(x) = \left(\dfrac{x + 1}{x + 2}\right)(2x - 5)$

20. $f(x) = (x^2 - x)(x^2 + 1)(x^2 + x + 1)$

21. $f(x) = (3x^3 + 4x)(x - 5)(x + 1)$

22. $f(x) = \dfrac{x^2 + c^2}{x^2 - c^2}$, c is a constant

23. $f(x) = \dfrac{c^2 - x^2}{c^2 + x^2}$

24. $f(x) = \dfrac{x(x^2 - 1)}{x + 3}$

In Exercises 25–30, complete the table without using the Quotient Rule (see Example 6).

Function	Rewrite	Derivative	Simplify
25. $y = \dfrac{x^2 + 2x}{x}$			
26. $y = \dfrac{4x^{3/2}}{x}$			
27. $y = \dfrac{7}{3x^3}$			
28. $y = \dfrac{4}{5x^2}$			
29. $y = \dfrac{3x^2 - 5}{7}$			
30. $y = \dfrac{x^2 - 4}{x + 2}$			

In Exercises 31–34, find the second derivative of the given function.

31. $f(x) = 4x^{3/2}$

32. $f(x) = \dfrac{x^2 + 2x - 1}{x}$

33. $f(x) = \dfrac{x}{x - 1}$

34. $f(x) = x + \dfrac{32}{x^2}$

In Exercises 35–38, find an equation of the tangent line to the graph of the given function at the indicated point.

Function	Point
35. $f(x) = \dfrac{x}{x - 1}$	$(2, 2)$
36. $f(x) = (x - 1)(x^2 - 2)$	$(0, 2)$
37. $f(x) = (x^3 - 3x + 1)(x + 2)$	$(1, -3)$
38. $f(x) = \dfrac{(x - 1)}{(x + 1)}$	$\left(2, \dfrac{1}{3}\right)$

In Exercises 39 and 40, determine the point(s) at which the graph of the function has a horizontal tangent.

39. $f(x) = \dfrac{x^2}{x - 1}$

40. $f(x) = \dfrac{x^2}{x^2 + 1}$

41. The function

$$f(t) = \frac{t^2 - t + 1}{t^2 + 1}$$

measures the percentage of the normal level of oxygen in a pond, where t is the time in weeks after organic waste is dumped into the pond. Find the rate of change of f with respect to t when (a) $t = 0.5$, (b) $t = 2$, and (c) $t = 8$.

42. Boyle's Law states that if the temperature of a gas remains constant, its pressure is inversely proportional to its volume. Use the derivative to show that the rate of change of the pressure is inversely proportional to the square of the volume.

43. A population of 500 bacteria is introduced into a culture and grows in number according to the equation

$$P(t) = 500\left(1 + \frac{4t}{50 + t^2}\right)$$

where t is measured in hours. Find the rate at which the population is growing when $t = 2$.

44. Find $h'(c)$ where $h(x) = f(x)g(x)$ and f and g are differentiable functions such that $f'(c) = g'(c) = 0$.

45. The velocity of a particle in straight-line motion is given by

$$v(t) = \frac{760t}{4t^2 + 25}, \quad 0 \leq t \leq 4.$$

Use a computer or graphics calculator to sketch the graph of $v(t)$ and $a(t)$. Determine the acceleration of the particle when $t = 2$.

3.5 The Chain Rule

The Chain Rule ▪ The General Power Rule ▪ Simplifying derivatives

We have yet to discuss one of the most powerful rules in differential calculus—the **Chain Rule.** This differentiation rule deals with composite functions and adds a surprising versatility to the rules we discussed in the two previous sections. For example, compare the following functions. Those on the left can be differentiated without the Chain Rule, and those on the right are best done with the Chain Rule.

Without the Chain Rule	*With the Chain Rule*
$y = x^2 + 1$	$y = \sqrt{x^2 + 1}$
$y = x + 1$	$y = (x + 1)^{-1/2}$
$y = 3x + 2$	$y = (3x + 2)^5$

We begin with a physical example that should give you an intuitive idea about the nature of the Chain Rule. Basically, the rule says that if y changes dy/du times as fast as u, and u changes du/dx times as fast as x, then y changes $(dy/du)(du/dx)$ times as fast as x.

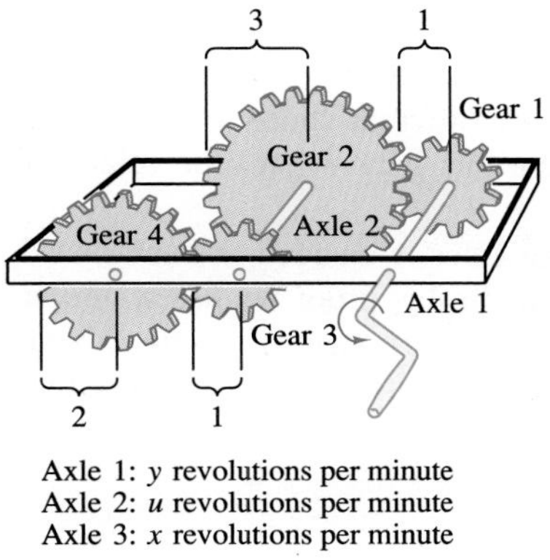

Axle 1: *y* revolutions per minute
Axle 2: *u* revolutions per minute
Axle 3: *x* revolutions per minute

FIGURE 3.19

EXAMPLE 1   The derivative of a composite function

A set of gears is constructed as shown in Figure 3.19 so that the second and third gears are on the same axle. As the first axle revolves, it drives the second axle, which in turn drives the third axle. Let *y*, *u*, and *x* represent the number of revolutions per minute of the first, second, and third axles, respectively. Find dy/du, du/dx, and dy/dx, and show that

$$\frac{dy}{dx} = \left(\frac{dy}{du}\right)\left(\frac{du}{dx}\right).$$

SOLUTION

Since the circumference of the second gear is 3 times that of the first, the first axle must make 3 revolutions to turn the second axle once. Similarly, the second gear must make 2 revolutions to turn the third axle once, and we have

$$\frac{dy}{du} = 3 \quad \text{and} \quad \frac{du}{dx} = 2.$$

Combining these two results, we know that the first axle must make 6 revolutions to turn the third axle once. Thus, we can write

$$\frac{dy}{dx} = \left(\frac{dy}{du}\right)\left(\frac{du}{dx}\right) = (3)(2) = 6.$$

In other words, the rate of change of *y* with respect to *x* is the product of the rate of change of *y* with respect to *u* and the rate of change of *u* with respect to *x*.   ▭

Example 1 illustrates a simple case of the Chain Rule. The general use is stated as follows.

**THEOREM 3.9
CHAIN RULE**

If $y = f(u)$ is a differentiable function of *u* and $u = g(x)$ is a differentiable function of *x*, then $y = f(g(x))$ is a differentiable function of *x* and

$$\frac{dy}{dx} = \frac{dy}{du} \cdot \frac{du}{dx}$$

or, equivalently,

$$\frac{d}{dx}[f(g(x))] = f'(g(x))g'(x).$$

PROOF   Let $F(x) = f(g(x))$. Then, using the alternate form of the derivative (Theorem 3.1), we need to show that for $x = c$

$$F'(c) = f'(g(c))g'(c).$$

An important consideration in this proof is the behavior of g as x approaches c. A problem occurs if there are values of x, other than c, such that $g(x) = g(c)$. In Appendix A we show how to use the differentiability of f and g to overcome this problem. For now, we assume that $g(x) \neq g(c)$ for values of x other than c.

In the proof of the Product and Quotient Rules, we added and subtracted the same quantity to obtain the desired form. In this proof we use a similar technique—multiplying and dividing by the same (nonzero) quantity. Note that since g is differentiable, it is also continuous, and it follows that $g(x) \to g(c)$ as $x \to c$.

$$F'(c) = \lim_{x \to c} \frac{f(g(x)) - f(g(c))}{x - c}$$

$$= \lim_{x \to c} \left[\frac{f(g(x)) - f(g(c))}{g(x) - g(c)} \cdot \frac{g(x) - g(c)}{x - c} \right], \quad g(x) \quad g(c)$$

$$= \left[\lim_{x \to c} \frac{f(g(x)) - f(g(c))}{g(x) - g(c)} \right] \left[\lim_{x \to c} \frac{g(x) - g(c)}{x - c} \right]$$

$$= f'(g(c))g'(c)$$

When applying the Chain Rule, it is helpful to think of the composite function $f \circ g$ as having two parts—an *inside* and an *outside*, as follows.

$$y = f(g(x)) = \overbrace{f(u)}^{\text{outside}}$$
$$\underbrace{u = g(x)}_{\text{inside}}$$

The next example illustrates two composite functions.

EXAMPLE 2 Decomposition of a composite function

$y = f(g(x))$	$u = g(x)$	$y = f(u)$
(a) $y = \dfrac{1}{x + 1}$	$u = x + 1$	$y = \dfrac{1}{u}$
(b) $y = \sqrt{3x^2 - x + 1}$	$u = 3x^2 - x + 1$	$y = \sqrt{u}$

EXAMPLE 3 Using the Chain Rule

Find dy/dx for $y = (x^2 + 1)^3$.

SOLUTION

To apply the Chain Rule, we identify the inside function u as follows.

$$y = \underbrace{(x^2 + 1)}_{u}^3 = u^3$$

Now, by the Chain Rule, we have

$$\frac{dy}{dx} = \underbrace{3(x^2 + 1)^2}_{\frac{dy}{du}}\underbrace{(2x)}_{\frac{du}{dx}} = 6x(x^2 + 1)^2.$$

The function in Example 3 is an instance of one of the most common types of composite functions, $y = [u(x)]^n$. The rule for differentiating such functions is called the **General Power Rule,** and it is a special case of the Chain Rule.

**THEOREM 3.10
GENERAL POWER RULE**

If $y = [u(x)]^n$, where u is a differentiable function of x and n is a rational number, then

$$\frac{dy}{dx} = n[u(x)]^{n-1}\frac{du}{dx}$$

or, equivalently,

$$\frac{d}{dx}[u^n] = nu^{n-1}u'.$$

PROOF Since $y = u^n$, we apply the Chain Rule to obtain

$$\frac{dy}{dx} = \left(\frac{dy}{du}\right)\left(\frac{du}{dx}\right) = \frac{d}{du}[u^n]\frac{du}{dx}.$$

Now, by the (simple) Power Rule in Section 3.3, we have $D_u[u^n] = nu^{n-1}$, and it follows that

$$\frac{dy}{dx} = nu^{n-1}\frac{du}{dx}.$$

EXAMPLE 4 Applying the General Power Rule

Find the derivative of $f(x) = (3x - 2x^2)^3$.

SOLUTION

If we let $u = 3x - 2x^2$, then $f(x) = (3x - 2x^2)^3 = u^3$, and by the General Power Rule the derivative is

$$f'(x) = \overset{n}{\overbrace{3}}\overset{u^{n-1}}{\overbrace{(3x - 2x^2)^2}}\overset{u'}{\overbrace{\frac{d}{dx}[3x - 2x^2]}}$$

$$= 3(3x - 2x^2)^2(3 - 4x) = (9 - 12x)(3x - 2x^2)^2.$$

In Section 3.3 we claimed that the Power Rule is valid for *any* rational number *n*. At this point, however, we have given proofs for only two cases— positive integer powers (Section 3.3) and negative integer powers (Section 3.4). Using the Chain Rule, we can now prove the case for rational powers.

EXAMPLE 5 The Power Rule for rational exponents

Use the Chain Rule to show that for a rational number n,

$$\frac{d}{dx}[x^n] = nx^{n-1}.$$

SOLUTION

Let n be a rational number with denominator m. Then nm is an integer. We let $u = x^n$ and consider the equation

$$(x^n)^m = x^{nm}, \quad \text{where } m \text{ and } nm \text{ are integers.}$$

Using the Chain Rule on the left and the Power Rule (for integers) on the right, we differentiate both sides of this equation to obtain

$$\overbrace{m}^{} \overbrace{(x^n)^{m-1}}^{u^{m-1}} \overbrace{\frac{d}{dx}[x^n]}^{u'} = nmx^{nm-1}$$

$$\frac{d}{dx}[x^n] = nx^{nm-1}x^{-nm+n}$$

$$= nx^{n-1}.$$

EXAMPLE 6 Differentiating functions involving radicals

Find the derivative of

$$y = \sqrt[3]{(x^2 + 2)^2}.$$

SOLUTION

We rewrite the function as $y = (x^2 + 2)^{2/3}$. Then, by the Power Rule (with $u = x^2 + 2$), we have

$$y' = \overbrace{\frac{2}{3}}^{n} \overbrace{(x^2 + 2)^{-1/3}}^{u^{n-1}} \overbrace{(2x)}^{u'} = \frac{4x}{3\sqrt[3]{x^2 + 2}}.$$

EXAMPLE 7   Differentiating quotients with constant numerators

Differentiate

$$g(t) = \frac{-7}{(2t - 3)^2}.$$

SOLUTION

If we rewrite the function as $g(t) = -7(2t - 3)^{-2}$, then the General Power Rule produces

$$g'(t) = \underbrace{(-7)}_{\substack{\text{Constant} \\ \text{Multiple Rule}}}\overbrace{(-2)}^{n}\overbrace{(2t - 3)^{-3}}^{u^{n-1}}\overbrace{(2)}^{u'} = 28(2t - 3)^{-3} = \frac{28}{(2t - 3)^3}.$$

Simplifying derivatives

The next three examples illustrate some useful techniques for simplifying the "raw derivatives" of functions involving products, quotients, and composites.

EXAMPLE 8   Simplification by factoring out the least powers

Differentiate $f(x) = x^2\sqrt{1 - x^2}$.

SOLUTION

$$
\begin{aligned}
f(x) &= x^2(1 - x^2)^{1/2} & &\text{Rewrite} \\
f'(x) &= x^2\frac{d}{dx}[(1 - x^2)^{1/2}] + (1 - x^2)^{1/2}\frac{d}{dx}[x^2] & &\text{Product Rule} \\
&= x^2\left[\frac{1}{2}(1 - x^2)^{-1/2}(-2x)\right] + (1 - x^2)^{1/2}(2x) & &\text{Power Rule} \\
&= -x^3(1 - x^2)^{-1/2} + 2x(1 - x^2)^{1/2} \\
&= x(1 - x^2)^{-1/2}[-x^2(1) + 2(1 - x^2)] & &\text{Factor} \\
&= \frac{x(2 - 3x^2)}{\sqrt{1 - x^2}} & &\text{Simplify}
\end{aligned}
$$

EXAMPLE 9   Simplifying the derivative of a quotient

Differentiate

$$f(x) = \frac{x}{\sqrt[3]{x^2 + 4}}.$$

SOLUTION

$$f(x) = \frac{x}{(x^2 + 4)^{1/3}}$$ Rewrite

$$f(x) = \frac{(x^2 + 4)^{1/3}(1) - x(1/3)(x^2 + 4)^{-2/3}(2x)}{(x^2 + 4)^{2/3}}$$ Quotient Rule

$$= \frac{1}{3}(x^2 + 4)^{-2/3}\left[\frac{3(x^2 + 4) - (2x^2)(1)}{(x^2 + 4)^{2/3}}\right]$$ Factor

$$= \frac{x^2 + 12}{3(x^2 + 4)^{4/3}}$$ Simplify ▭

EXAMPLE 10 Differentiating a quotient raised to a power

Differentiate

$$y = \left(\frac{3x - 1}{x^2 + 3}\right)^2.$$

SOLUTION

$$\frac{dy}{dx} = 2\overbrace{\left(\frac{3x - 1}{x^2 + 3}\right)}^{u^{n-1}}\overbrace{\frac{d}{dx}\left[\frac{3x - 1}{x^2 + 3}\right]}^{u'}$$

$$= \left[\frac{2(3x - 1)}{x^2 + 3}\right]\left[\frac{(x^2 + 3)(3) - (3x - 1)(2x)}{(x^2 + 3)^2}\right]$$

$$= \frac{2(3x - 1)(3x^2 + 9 - 6x^2 + 2x)}{(x^2 + 3)^3}$$

$$= \frac{2(3x - 1)(-3x^2 + 2x + 9)}{(x^2 + 3)^3}$$

Try finding y' using the Quotient Rule on $y = (3x - 1)^2/(x^2 + 3)^2$ and compare the results. ▭

EXERCISES for Section 3.5

In Exercises 1–6, complete the table using Example 2 as a model.

$y = f(g(x))$	$u = g(x)$	$y = f(u)$

1. $y = (6x - 5)^4$

2. $y = \dfrac{1}{\sqrt{x + 1}}$

3. $y = \sqrt{x^2 - 1}$

4. $y = \left(\dfrac{3x}{2}\right)^2$

5. $y = (x^2 - 3x + 4)^6$

6. $y = (5x - 2)^{3/2}$

In Exercises 7–44, find the derivative.

7. $y = (2x - 7)^3$

8. $y = (3x^2 + 1)^4$

9. $g(x) = 3(9x - 4)^4$

10. $f(x) = 2(x^2 - 1)^3$

11. $y = \dfrac{1}{x - 2}$

12. $s(t) = \dfrac{1}{t^2 + 3t - 1}$

13. $f(t) = \left(\dfrac{1}{t - 3}\right)^2$

14. $y = -\dfrac{4}{(t + 2)^2}$

15. $f(x) = \dfrac{3}{x^3 - 4}$

16. $f(x) = \dfrac{1}{(x^2 - 3x)^2}$

17. $f(x) = x^2(x - 2)^4$

18. $f(x) = x(3x - 9)^3$

19. $f(t) = \sqrt{1 - t}$

20. $g(x) = \sqrt{3 - 2x}$

21. $s(t) = \sqrt{t^2 + 2t - 1}$

22. $y = \sqrt[3]{3x^3 + 4x}$

23. $y = \sqrt[3]{9x^2 + 4}$

24. $g(x) = \sqrt{x^2 - 2x + 1}$

25. $y = 2\sqrt{4 - x^2}$

26. $f(x) = -3\sqrt[4]{2 - 9x}$

27. $f(x) = (9 - x^2)^{2/3}$

28. $f(t) = (9t + 2)^{2/3}$

29. $y = \dfrac{1}{\sqrt{x + 2}}$

30. $g(t) = \sqrt{\dfrac{1}{t^2 - 2}}$

31. $y = x\sqrt{1 - x^2}$

32. $y = x^2\sqrt{9 - x^2}$

33. $y = \dfrac{x}{\sqrt{x^2 + 1}}$

34. $y = \dfrac{x^2}{\sqrt{x^2 + 9}}$

35. $y = \dfrac{\sqrt{x} + 1}{x^2 + 1}$

36. $f(x) = \dfrac{x + 1}{2x - 3}$

37. $f(t) = \dfrac{3t + 2}{t - 1}$

38. $y = \sqrt{\dfrac{2x}{x + 1}}$

39. $g(t) = \dfrac{3t^2}{\sqrt{t^2 + 2t - 1}}$

40. $f(x) = \sqrt{x}(2 - x)^2$

41. $y = \sqrt{\dfrac{x + 1}{x}}$

42. $y = (t^2 - 9)\sqrt{t + 2}$

43. $s(t) = \dfrac{-2(2 - t)\sqrt{1 + t}}{3}$

44. $g(x) = \sqrt{x - 1} + \sqrt{x + 1}$

In Exercises 45 and 46, find an equation of the tangent line to the graph of f at the given point.

Function	*Point*
45. $f(x) = \sqrt{3x^2 - 2}$	$(3, 5)$
46. $f(x) = x\sqrt{x^2 + 5}$	$(2, 6)$

In Exercises 47–50, find the second derivative of the given function.

47. $f(x) = 2(x^2 - 1)^3$

48. $f(x) = \dfrac{1}{x - 2}$

49. $f(x) = \sqrt{x^2 + x + 1}$

50. $f(t) = \dfrac{\sqrt{t^2 + 1}}{t}$

51. Let u be a differentiable function of x. Use the fact that $|u| = \sqrt{u^2}$ to prove that

$$\dfrac{d}{dx}[|u|] = u'\dfrac{u}{|u|}, \quad u \neq 0.$$

In Exercises 52–54, use the result of Exercise 51 to find the derivative of the given function.

52. $f(x) = |x^2 - 4|$

53. $f(x) = |x^3 + x|$

54. $f(x) = \left|\dfrac{4}{x}\right|$

55. *(Doppler effect)* The frequency F of a fire truck siren heard by a stationary observer is given by

$$F = \dfrac{132,400}{331 \pm v}$$

where $\pm v$ represents the velocity of the accelerating fire truck (see figure). Find the rate of change of F with respect to v when

(a) the fire truck is approaching at a velocity of 30 m/s [use $-v$], and then when

(b) the fire truck is moving away at a velocity of 30 m/s [use $+v$].

$$F = \dfrac{132,400}{331 + v} \qquad F = \dfrac{132,400}{331 - v}$$

Doppler Effect

56. The speed S of blood that is r centimeters from the center of an artery is given by $S = C(R^2 - r^2)$, where C is a constant, R is the radius of the artery, and S is measured in centimeters per second. Suppose a drug is administered and the artery begins dilating at the rate of dR/dt. At a constant distance r, find the rate at which S changes with respect to t for $C = 1.76 \times 10^5$, $R = 1.2 \times 10^{-2}$, and $dR/dt = 10^{-5}$.

3.6 Implicit Differentiation

Implicit and explicit functions ▪ Implicit differentiation

So far, our equations involving two variables were generally expressed in the **explicit form** $y = f(x)$. That is, one of the two variables was explicitly given in terms of the other. For example,

$$y = 3x - 5, \qquad s = -16t^2 + 20t, \qquad u = 3w - w^2$$

all are written in explicit form, and we say that y, s, and u are functions of x, t, and w, respectively.

However, many functions are not given explicitly and are implied only by a given equation. For instance, the function $y = 1/x$ is defined **implicitly** by the equation $xy = 1$. Suppose that you were asked to find dy/dx in this equation. As it turns out, this is a relatively simple task, and you probably would begin by solving the equation for y.

Implicit form	*Explicit form*	*Derivative*
$xy = 1$	$y = \dfrac{1}{x} = x^{-1}$	$\dfrac{dy}{dx} = -x^{-2} = -\dfrac{1}{x^2}$

This procedure works well whenever we can easily solve for the function explicitly. However, we cannot use this procedure in cases in which we are unable to solve for y as a function of x. For instance, how would you find dy/dx in the equation

$$x^2 - 2y^3 + 4y = 2$$

where it is very difficult to express y as a function of x explicitly? To do this, we use a procedure called **implicit differentiation,** in which we assume y is a differentiable function of x.

To understand how to find dy/dx implicitly you must realize that the differentiation is taking place *with respect to x*. This means that when we differentiate terms involving x alone, we can differentiate as usual. *But* when we differentiate terms involving y, we must apply the Chain Rule because we are assuming that y is defined implicitly as a function of x. Study the next example carefully. Note in particular how the Chain Rule is used to introduce the dy/dx terms.

EXAMPLE 1   Applying the Chain Rule

Differentiate the following with respect to x.

(a) $3x^2$ (b) $2y^3$ (c) $x + 3y$ (d) xy^2

SOLUTION

(a) $\dfrac{d}{dx}[3x^2] = 6x$

(b) $\dfrac{d}{dx}[\overbrace{2y^3}^{u^n}] = \overbrace{2(3)y^2 \dfrac{dy}{dx}}^{nu^{n-1}\ u'} = 6y^2 \dfrac{dy}{dx}$ Chain Rule

(c) $\dfrac{d}{dx}[x + 3y] = 1 + 3\dfrac{dy}{dx}$

(d) $\dfrac{d}{dx}[xy^2] = x\dfrac{d}{dx}[y^2] + y^2\dfrac{d}{dx}[x]$ Product Rule

$\qquad\qquad = x\left(2y\dfrac{dy}{dx}\right) + y^2(1)$ Chain Rule

$\qquad\qquad = 2xy\dfrac{dy}{dx} + y^2$

For equations involving x and y, we suggest the following procedure for finding dy/dx implicitly.

IMPLICIT DIFFERENTIATION

Given an equation involving x and y, and assuming y is a differentiable function of x, we can find dy/dx as follows.

1. Differentiate both sides of the equation *with respect to x.*
2. Collect all terms involving dy/dx on the left side of the equation and move all other terms to the right side of the equation.
3. Factor dy/dx out of the left side of the equation.
4. Solve for dy/dx by dividing both sides of the equation by the left-hand factor that does not contain dy/dx.

EXAMPLE 2   Implicit differentiation

Find dy/dx given that $y^3 + y^2 - 5y - x^2 = -4$.

SOLUTION

1. Differentiate both sides of the equation with respect to x.

$$\frac{d}{dx}[y^3 + y^2 - 5y - x^2] = \frac{d}{dx}[-4]$$

$$\frac{d}{dx}[y^3] + \frac{d}{dx}[y^2] - \frac{d}{dx}[5y] - \frac{d}{dx}[x^2] = \frac{d}{dx}[-4]$$

$$3y^2\frac{dy}{dx} + 2y\frac{dy}{dx} - 5\frac{dy}{dx} - 2x = 0$$

2. Collect the dy/dx terms on the left side of the equation.

$$3y^2\frac{dy}{dx} + 2y\frac{dy}{dx} - 5\frac{dy}{dx} = 2x$$

3. Factor dy/dx out of the left side of the equation.

$$\frac{dy}{dx}(3y^2 + 2y - 5) = 2x$$

4. Solve for dy/dx by dividing by $(3y^2 + 2y - 5)$.

$$\frac{dy}{dx} = \frac{2x}{3y^2 + 2y - 5}$$

REMARK   In Example 2, note that implicit differentiation can produce an expression for dy/dx that contains both x and y.

To see how we can use an *implicit derivative*, consider the graph of the equation $y^3 + y^2 - 5y - x^2 = -4$ as shown in Figure 3.20. The derivative found in Example 2 gives us a formula for the slope of the tangent line at a point on this graph. The slopes at several points on the graph are as follows.

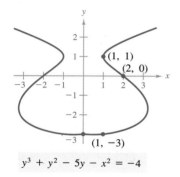

$y^3 + y^2 - 5y - x^2 = -4$

FIGURE 3.20

Point on Graph	Slope of Graph
(2, 0)	$-\dfrac{4}{5}$
(1, −3)	$\dfrac{1}{8}$
x = 0	0
(1, 1)	undefined

When using implicit differentiation, keep in mind that it is meaningless to solve for dy/dx in an equation that has no solution points. (For example, $x^2 + y^2 = -4$ has no solution points.) If a segment of a graph can be represented by a differentiable function, then dy/dx will have meaning as the slope at each point on the segment. Recall that a function is not differentiable at (1) points with vertical tangents and (2) points of discontinuity. The following example illustrates this idea.

EXAMPLE 3 Representing a graph by differentiable functions

If possible, represent the graphs of the following equations by differentiable functions. (See Figure 3.21.)

(a) $x^2 + y^2 = 0$ (b) $x^2 + y^2 = 1$ (c) $x + y^2 = 1$

SOLUTION

(a) The graph of this equation is a single point. Therefore, it does not define y as a differentiable function of x.

(b) The graph of this equation is the unit circle, centered at (0, 0). The upper semi-circle is given by the differentiable function

$$y = \sqrt{1 - x^2}, \quad -1 < x < 1$$

and the lower semi-circle is given by

$$y = -\sqrt{1 - x^2}, \quad -1 < x < 1.$$

At the point (−1, 0) and (1, 0), the slope of the graph is undefined.

(c) The upper half of this parabola is given by the differentiable function

$$y = \sqrt{1 - x}, \quad x < 1$$

and the lower half is given by

$$y = -\sqrt{1 - x}, \quad x < 1.$$

At the point (1, 0), the slope of the graph is undefined. ▭

EXAMPLE 4 Finding the slope of a curve implicitly

Determine the slope of the tangent line to the graph of $x^2 + 4y^2 = 4$ at the point $(\sqrt{2}, -1/\sqrt{2})$. (See Figure 3.22.)

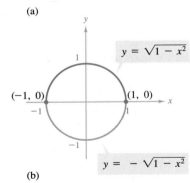

$x^2 + y^2 = 0$

(0, 0)

(a)

$y = \sqrt{1 - x^2}$

(−1, 0)   (1, 0)

$y = -\sqrt{1 - x^2}$

(b)

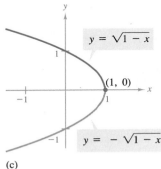

$y = \sqrt{1 - x}$

(1, 0)

$y = -\sqrt{1 - x}$

(c)

FIGURE 3.21

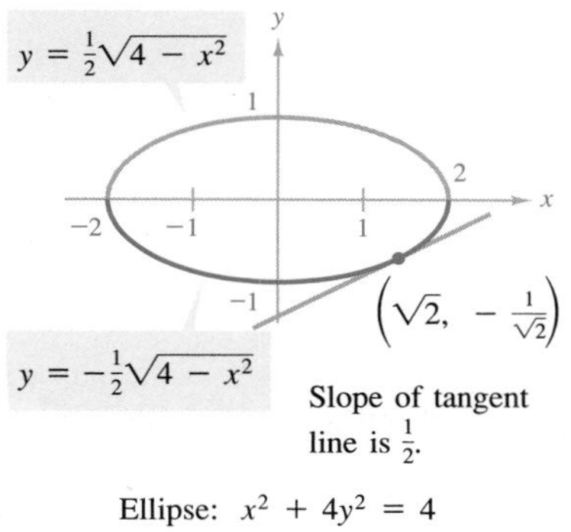

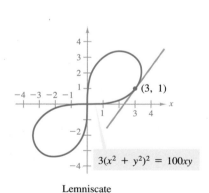

$y = \frac{1}{2}\sqrt{4 - x^2}$

$y = -\frac{1}{2}\sqrt{4 - x^2}$

$\left(\sqrt{2}, -\frac{1}{\sqrt{2}}\right)$

Slope of tangent line is $\frac{1}{2}$.

Ellipse: $x^2 + 4y^2 = 4$

FIGURE 3.22

SOLUTION

Implicit differentiation of the equation $x^2 + 4y^2 = 4$ with respect to x yields

$$2x + 8y\frac{dy}{dx} = 0$$

$$\frac{dy}{dx} = \frac{-2x}{8y} = \frac{-x}{4y}.$$

Therefore, at $(\sqrt{2}, -1/\sqrt{2})$, the slope is

$$\frac{dy}{dx} = \frac{-\sqrt{2}}{-4/\sqrt{2}} = \frac{1}{2}.$$

REMARK   To see the benefit of implicit differentiation, try doing Example 4 using the explicit function $y = -\frac{1}{2}\sqrt{4 - x^2}$. The graph of this function is the lower half of the ellipse.

EXAMPLE 5   Finding the slope of a curve implicitly

Determine the slope of the graph of $3(x^2 + y^2)^2 = 100xy$ at the point $(3, 1)$.

SOLUTION

$$\frac{d}{dx}[3(x^2 + y^2)^2] = \frac{d}{dx}[100xy]$$

$$3(2)(x^2 + y^2)\left(2x + 2y\frac{dy}{dx}\right) = 100\left[x\frac{dy}{dx} + y(1)\right]$$

$$12y(x^2 + y^2)\frac{dy}{dx} - 100x\frac{dy}{dx} = 100y - 12x(x^2 + y^2)$$

$$[12y(x^2 + y^2) - 100x]\frac{dy}{dx} = 100y - 12x(x^2 + y^2)$$

$$\frac{dy}{dx} = \frac{100y - 12x(x^2 + y^2)}{-100x + 12y(x^2 + y^2)} = \frac{25y - 3x(x^2 + y^2)}{-25x + 3y(x^2 + y^2)}$$

Now, at the point $(3, 1)$ the slope of the graph is

$$\frac{dy}{dx} = \frac{25(1) - 3(3)(3^2 + 1^2)}{-25(3) + 3(1)(3^2 + 1^2)} = \frac{25 - 90}{-75 + 30} = \frac{-65}{-45} = \frac{13}{9}$$

as shown in Figure 3.23. This graph is called a **lemniscate.**

$(3, 1)$

$3(x^2 + y^2)^2 = 100xy$

Lemniscate

FIGURE 3.23

EXAMPLE 6   Finding a differentiable function

Find dy/dx implicitly for the equation $4x - y^3 + 12y = 0$, and use Figure 3.24 to find the largest interval of the form $-a < y < a$ such that y is a differentiable function of x.

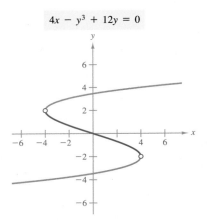

$$4x - y^3 + 12y = 0$$

FIGURE 3.24

SOLUTION

$$4x - y^3 + 12y = 0$$
$$4 - 3y^2y' + 12y' = 0$$
$$y'(-3y^2 + 12) = -4$$
$$y' = \frac{4}{3(y^2 - 4)}$$

From Figure 3.24 we see that the largest interval about the origin for which y is a differentiable function of x is $-2 < y < 2$. ▭

In implicit differentiation, the form of the derivative can often be simplified or changed to a single variable by an appropriate use of the *original* equation. A similar technique can be used to find and simplify higher-order derivatives obtained implicitly. This is demonstrated in the next example.

EXAMPLE 7 Finding the second derivative implicitly

Given $x^2 + y^2 = 25$, find y''.

SOLUTION

Differentiating each term with respect to x, we obtain

$$2x + 2yy' = 0$$
$$2yy' = -2x$$
$$y' = \frac{-2x}{2y} = -\frac{x}{y}.$$

Differentiating a second time with respect to x yields

$$y'' = -\frac{(y)(1) - (x)(y')}{y^2}$$

$$= -\frac{y - (x)(-x/y)}{y^2} \qquad \text{Substitute } y' = -x/y$$

$$= -\frac{y^2 + x^2}{y^3}$$

$$= -\frac{25}{y^3}. \qquad \text{Substitute } x^2 + y^2 = 25 \qquad ▭$$

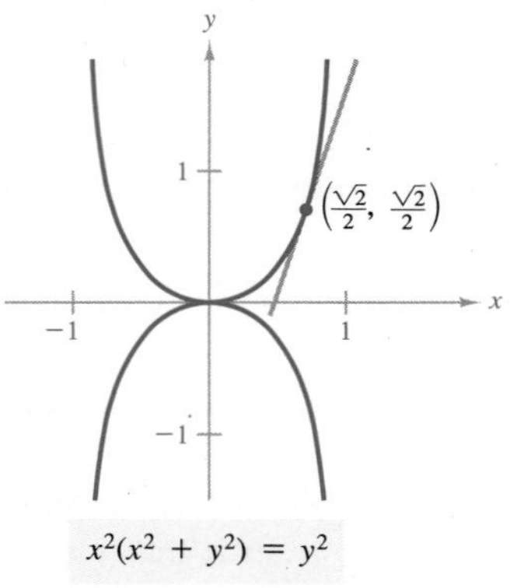

EXAMPLE 8 Finding a tangent line to a graph

Find the tangent line to the **kappa curve** given by

$$x^2(x^2 + y^2) = y^2$$

at the point $(\sqrt{2}/2, \sqrt{2}/2)$ as shown in Figure 3.25.

$$x^2(x^2 + y^2) = y^2$$

FIGURE 3.25

SOLUTION

By rewriting and differentiating implicitly, we have

$$x^4 + x^2y^2 - y^2 = 0$$
$$4x^3 + x^2(2yy') + 2xy^2 - 2yy' = 0$$
$$2y(x^2 - 1)y' = -2x(2x^2 + y^2)$$
$$y' = \frac{x(2x^2 + y^2)}{y(1 - x^2)}.$$

Now, at the point $(\sqrt{2}/2, \sqrt{2}/2)$, the slope is

$$m = \frac{(\sqrt{2}/2)[2(1/2) + (1/2)]}{(\sqrt{2}/2)[1 - (1/2)]} = \frac{3/2}{1/2} = 3$$

and the equation of the tangent line at this point is

$$y - \frac{\sqrt{2}}{2} = 3\left(x - \frac{\sqrt{2}}{2}\right)$$
$$y = 3x - \sqrt{2}.$$

The graph in Example 8 is called the *kappa curve* because it resembles the Greek letter kappa. The general solution for the tangent line to this particular curve was first discovered by the English mathematician Isaac Barrow (1630–1677). Barrow and Newton were contemporaries and corresponded frequently regarding their work in the early development of calculus.

Isaac Barrow

EXERCISES for Section 3.6

In Exercises 1–16, find dy/dx by implicit differentiation and evaluate the derivative at the indicated point.

Equation	Point
1. $x^2 + y^2 = 16$	$(3, \sqrt{7})$
2. $x^2 - y^2 = 16$	$(4, 0)$
3. $xy = 4$	$(-4, -1)$
4. $x^2 - y^3 = 0$	$(1, 1)$
5. $x^{1/2} + y^{1/2} = 9$	$(16, 25)$
6. $x^3 + y^3 = 8$	$(0, 2)$
7. $x^3 - xy + y^2 = 4$	$(0, -2)$
8. $x^2y + y^2x = -2$	$(2, -1)$
9. $y^2 = \dfrac{x^2 - 9}{x^2 + 9}$	$(3, 0)$
10. $(x + y)^3 = x^3 + y^3$	$(-1, 1)$
11. $x^3y^3 - y = x$	$(0, 0)$
12. $\sqrt{xy} = x - 2y$	$(4, 1)$
13. $x^{2/3} + y^{2/3} = 5$	$(8, 1)$
14. $x^3 + y^3 = 2xy$	$(1, 1)$
15. $x^3 - 2x^2y + 3xy^2 = 38$	$(2, 3)$
16. $x^3 - y^3 = x - y$	$(1, 1)$

In Exercises 17–20, find the slope of the tangent line to the graph at the indicated point.

17. Witch of Agnesi:
$(x^2 + 4)y = 8$
Point: $(2, 1)$

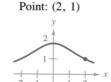

18. Cissoid:
$(4 - x)y^2 = x^3$
Point: $(2, 2)$

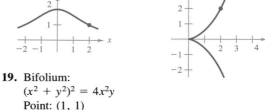

19. Bifolium:
$(x^2 + y^2)^2 = 4x^2y$
Point: $(1, 1)$

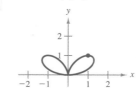

20. Folium of Descartes:

$x^3 + y^3 - 6xy = 0$

Point: $\left(\dfrac{4}{3}, \dfrac{8}{3}\right)$

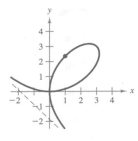

In Exercises 21–24, sketch the graph of the given equation. Then find dy/dx implicitly and explicitly and show that the two results are equivalent.

21. $x^2 + y^2 = 16$
22. $x^2 + y^2 - 4x + 6y + 9 = 0$
23. $9x^2 + 16y^2 = 144$
24. $4y^2 - x^2 = 4$

In Exercises 25–30, find d^2y/dx^2 in terms of x and y.

25. $x^2 + xy = 5$ **26.** $x^2y^2 - 2x = 3$
27. $x^2 - y^2 = 16$ **28.** $1 - xy = x - y$
29. $y^2 = x^3$ **30.** $y^2 = 4x$

In Exercises 31 and 32, find equations for the tangent line and normal line to the given circle at the indicated points. (The **normal line** at a point is perpendicular to the tangent line at the point.)

31. $x^2 + y^2 = 25$, $(4, 3)$ and $(-3, 4)$
32. $x^2 + y^2 = 9$, $(0, 3)$ and $(2, \sqrt{5})$

In Exercises 33 and 34, find the points at which the graph of the given equation has a vertical or horizontal tangent line.

33. $25x^2 + 16y^2 + 200x - 160y + 400 = 0$
34. $4x^2 + y^2 - 8x + 4y + 4 = 0$

In Exercises 35–38, sketch the intersecting graphs of the given equations and show that they are **orthogonal.** [Two graphs are orthogonal if at their point(s) of intersection, their tangent lines are perpendicular to each other.]

35. $2x^2 + y^2 = 6$ and $y^2 = 4x$
36. $y^2 = x^3$ and $2x^2 + 3y^2 = 5$
37. $x + y = 0$ and $x^2 + y^2 = 4$
38. $x^3 = 3(y - 1)$ and $x(3y - 29) = 3$

39. Two circles of radius 4 are tangent to the graph of $y^2 = 4x$ at the point $(1, 2)$. Find the equations for these two circles.

40. Show that the normal line (the line perpendicular to the tangent line) at any point on the circle $x^2 + y^2 = r^2$ passes through the origin.

In Exercises 41 and 42, use a computer to sketch the graph of the equation. Find the equation of the tangent line to the graph at the specified point and sketch its graph.

Equation	Point
41. $\sqrt{x} + \sqrt{y} = 3$	$(4, 1)$
42. $x^3 + y^3 - 6xy = 0$	$\left(\dfrac{4}{3}, \dfrac{8}{3}\right)$

43. Show that

$$\frac{dy}{dx} = \frac{-(x - h)}{y - k}$$

for the circle

$$(x - h)^2 + (y - k)^2 = r^2.$$

3.7 Related Rates

Related rate problems

We have seen how the Chain Rule can be used to find dy/dx implicitly. Another important use of the Chain Rule is to find the rates of change of two or more related variables that are changing with respect to time.

For example, when water is drained out of a conical tank (see Figure 3.26), the volume V, the radius r, and the height h of the water level are all functions of time t. Knowing that these variables are related by the equation

$$V = \frac{\pi}{3}r^2h$$

we can differentiate implicitly with respect to t to obtain the **related rate equation**

$$\frac{dV}{dt} = \frac{\pi}{3}\left[r^2\frac{dh}{dt} + h\left(2r\frac{dr}{dt}\right)\right]$$

$$= \frac{\pi}{3}\left[r^2\frac{dh}{dt} + 2rh\frac{dr}{dt}\right].$$

From this equation we see that the rate of change of V is related to the rates of change of both h and r. This is further demonstrated in the following example.

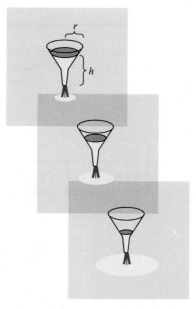

Volume is related to radius and height.

FIGURE 3.26

EXAMPLE 1   Two rates that are related

Suppose x and y are both differentiable functions of t and are related by the equation $y = x^2 + 3$. Find dy/dt when $x = 1$, given that $dx/dt = 2$ when $x = 1$.

SOLUTION

Using the Chain Rule, we can differentiate both sides of the equation *with respect to t*.

$$y = x^2 + 3 \qquad \text{Given equation}$$

$$\frac{d}{dt}[y] = \frac{d}{dt}[x^2 + 3] \qquad \text{Differentiate with respect to } t$$

$$\frac{dy}{dt} = 2x\frac{dx}{dt} \qquad \text{Chain Rule}$$

Now, when $x = 1$ and $dx/dt = 2$, we have

$$\frac{dy}{dt} = 2(1)(2) = 4.$$

In Example 1, we are *given* the following mathematical model.

Given equation: $y = x^2 + 3$

Given rate: $\dfrac{dx}{dt} = 2$ when $x = 1$

Find: $\dfrac{dy}{dt}$ when $x = 1$

Now we look at an example in which we must *create* the model from a verbal description.

FIGURE 3.27

EXAMPLE 2   An application involving related rates

A pebble is dropped into a calm pond, causing ripples in the form of concentric circles, as shown in Figure 3.27. The radius r of the outer ripple is increasing at a constant rate of 1 foot per second. When this radius is 4 feet, at what rate is the total area A of the disturbed water increasing?

SOLUTION

The variables r and A are related by the equation for the area of a circle, $A = \pi r^2$. To solve this problem, we must remember that the rate of change of the radius r is given by its derivative, dr/dt. Thus, the problem can be summarized by the following model.

Given equation: $A = \pi r^2$

Given rate: $\dfrac{dr}{dt} = 1$ when $r = 4$

Find: $\dfrac{dA}{dt}$ when $r = 4$

With this information, we proceed as in Example 1.

$$\frac{d}{dt}[A] = \frac{d}{dt}[\pi r^2]$$

$$\frac{dA}{dt} = 2\pi r \frac{dr}{dt}$$

Thus, when $r = 4$, we have

$$\frac{dA}{dt} = 2\pi(4)(1) = 8\pi \ \text{ft}^2/\text{sec}.$$

Using the procedure shown in Example 2, we suggest the following four solution steps for related rate problems.

PROCEDURE FOR SOLVING RELATED RATE PROBLEMS	1. Assign symbols to all *given* quantities and *quantities to be determined*. Make a sketch and label the quantities if feasible. 2. Write an equation involving the variables whose rates of change either are given or are to be determined. 3. Using the Chain Rule, implicitly differentiate both sides of the equation *with respect to time t*. 4. Substitute into the resulting equation all known values for the variables and their rates of change. Then solve for the required rate of change.

REMARK   Be sure to complete Step 3 *before* starting Step 4. A common error is to introduce specific values for rates that are variable too early in the solution process.

Table 3.6 shows the mathematical models for some common rates of change which can be used in the first step of the solution to related rate problems.

TABLE 3.6

Verbal statement	Mathematical model
The velocity of a car after traveling one hour is 50 miles per hour.	x = distance traveled $\dfrac{dx}{dt} = 50$ when $t = 1$
Water is being pumped into a swimming pool at the rate of 10 cubic feet per minute.	V = volume of water in pool $\dfrac{dV}{dt} = 10$ ft³/min
A gear is revolving at the rate of 25 revolutions per minute (1 rev = 2π rad).	θ = angle of revolution $\dfrac{d\theta}{dt} = 25(2\pi)$ rad/min

EXAMPLE 3 An inflating balloon

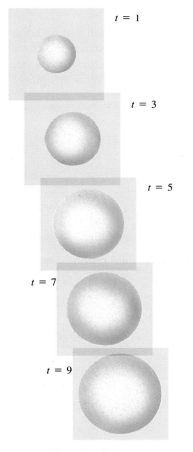

$t = 1$

$t = 3$

$t = 5$

$t = 7$

$t = 9$

Expanding Balloon

FIGURE 3.28

Air is being pumped into a spherical balloon at the rate of 4.5 cubic inches per minute, as shown in Figure 3.28. Find the rate of change of the radius when the radius is 2 inches.

SOLUTION

1. We let V be the volume of the balloon and let r be its radius. Since the volume is increasing at the rate of 4.5 in³/min, we know that at time t the rate of change of the volume is $dV/dt = \frac{9}{2}$. Thus, the problem can be stated as follows.

   Given: $\dfrac{dV}{dt} = \dfrac{9}{2}$ (constant rate)

   Find: $\dfrac{dr}{dt}$ when $r = 2$

2. To find the rate of change of the radius, we must find an equation that relates the radius r to the volume V. This is given by the formula for the volume of a sphere.

$$V = \frac{4}{3}\pi r^3$$

3. Now we implicitly differentiate *with respect to t* to obtain

$$\frac{dV}{dt} = 4\pi r^2 \frac{dr}{dt}$$

$$\frac{dr}{dt} = \frac{1}{4\pi r^2}\left(\frac{dV}{dt}\right).$$

4. Finally, when $r = 2$, the rate of change of the radius is

$$\frac{dr}{dt} = \frac{1}{16\pi}\left(\frac{9}{2}\right) \approx 0.09 \text{ in/min.}$$

In Example 3, note that the volume is increasing at a *constant* rate but the radius is increasing at a *variable* rate. Thus, when we say that two rates are related, we do not mean that they are proportional. In this particular case, the radius is growing more and more slowly as *t* increases. This is illustrated in Figure 3.28 and Table 3.7.

TABLE 3.7

t	1	3	5	7	9	11
V	4.5	13.5	22.5	31.5	40.5	49.5
r	1.02	1.48	1.75	1.96	2.13	2.28
dr/dt	0.34	0.16	0.12	0.09	0.08	0.07

EXAMPLE 4 The velocity of an airplane tracked by radar

An airplane is flying at an elevation of 6 miles on a flight path that will take it directly over a radar tracking station. Let *s* represent the distance (measured in miles) between the radar station and the plane. If *s* is decreasing at a rate of 400 miles per hour when *s* is 10 miles, what is the velocity of the plane?

SOLUTION

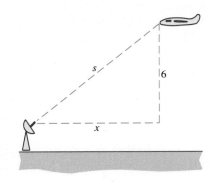

FIGURE 3.29

1. We label the distances *x* and *s* in Figure 3.29.

$$\text{Given: } \quad \frac{ds}{dt} = -400 \quad \text{when} \quad s = 10$$

$$\text{Find: } \quad \frac{dx}{dt} \quad \text{when} \quad s = 10$$

2. We use the Pythagorean Theorem to form an equation relating *s* and *x*.

$$x^2 + 6^2 = s^2$$

3. Differentiating implicitly with respect to *t*, we have

$$2x\frac{dx}{dt} = 2s\frac{ds}{dt}$$

$$\frac{dx}{dt} = \frac{s}{x}\left(\frac{ds}{dt}\right).$$

4. To find *dx/dt*, we must first find *x* when *s* = 10.

$$x = \sqrt{s^2 - 36} = \sqrt{100 - 36} = \sqrt{64} = 8$$

Finally, when $s = 10$, we have

$$\frac{dx}{dt} = \frac{10}{8}(-400) = -500 \text{ mph.}$$

Since the velocity is -500 miles per hour, the *speed* is 500 miles per hour.

EXAMPLE 5 Gravel falling in a conical pile

Gravel is falling in a conical pile at the rate of 100 cubic feet per minute. Find the rate of change of the height of the pile when the height is 10 feet. (Assume that the coarseness of the gravel is such that the radius of the cone is equal to its height.)

SOLUTION

1. We label the height and radius on the cone in Figure 3.30. Note that the given rate is a change in *volume*.

   Given: $\dfrac{dV}{dt} = 100$ (constant rate)

   Find: $\dfrac{dh}{dt}$ when $h = 10$

2. The volume of a cone is given by the formula

   $$V = \frac{\pi r^2 h}{3}$$

   and since we are given that $r = h$, this equation simplifies to

   $$V = \frac{\pi}{3}h^3.$$

3. Implicit differentiation with respect to t yields the following.

   $$\frac{dV}{dt} = \frac{\pi}{3}3h^2 \frac{dh}{dt}$$

   $$\frac{dh}{dt} = \frac{1}{\pi h^2}\left(\frac{dV}{dt}\right)$$

4. Finally, when $h = 10$ we have

   $$\frac{dh}{dt} = \frac{1}{100\pi}(100) = \frac{1}{\pi} \approx 0.318 \text{ ft/min.}$$

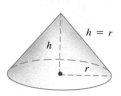

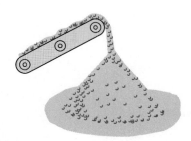

$h = r$

FIGURE 3.30

EXAMPLE 6   Tracking an accelerating object

A television camera at ground level is filming the lift-off of a space shuttle that is rising vertically according to the position equation $s = 50t^2$, where s is measured in feet and t is measured in seconds. The camera is 2000 feet from the launch pad. Find the rate of change in the distance between the camera and the base of the shuttle 10 seconds after lift-off. (Assume that the camera and the base of the shuttle are level with each other when $t = 0$.)

SOLUTION

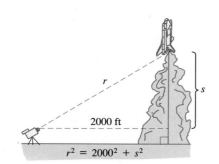

$r^2 = 2000^2 + s^2$

FIGURE 3.31

1. We let r be the distance between the camera and the base of the shuttle, as shown in Figure 3.31. Then we can find the velocity of the rocket by differentiating s with respect to t to obtain $ds/dt = 100t$. Thus, we have the following model.

   Given: $\dfrac{ds}{dt} = 100t = $ velocity

   Find: $\dfrac{dr}{dt}$ when $t = 10$

2. Using Figure 3.31 we relate s and r by the equation
   $$r^2 = 2000^2 + s^2.$$

3. Implicit differentiation with respect to t yields
   $$2r \frac{dr}{dt} = 2s \frac{ds}{dt}$$
   $$\frac{dr}{dt} = \frac{s}{r} \cdot \frac{ds}{dt} = \frac{s}{r}(10t).$$

4. Now, when $t = 10$, we know that $s = 50(10^2) = 5000$, and we have
   $$r = \sqrt{2000^2 + 5000^2} = 1000\sqrt{29}.$$

   Finally, the rate of change of r when $t = 10$ is
   $$\frac{dr}{dt} = \frac{5000}{1000\sqrt{29}}(100)(10) = 928.48 \text{ ft/sec.}$$

EXERCISES for Section 3.7

In Exercises 1–4, assume that x and y are both differentiable functions of t and find the indicated values of dy/dt and dx/dt.

Equation	Find		Given
1. $y = \sqrt{x}$	(a) $\dfrac{dy}{dt}$ when $x = 4$		$\dfrac{dx}{dt} = 3$
	(b) $\dfrac{dx}{dt}$ when $x = 25$		$\dfrac{dy}{dt} = 2$
2. $y = x^2 - 3x$	(a) $\dfrac{dy}{dt}$ when $x = 3$		$\dfrac{dx}{dt} = 2$
	(b) $\dfrac{dx}{dt}$ when $x = 1$		$\dfrac{dy}{dt} = 5$
3. $xy = 4$	(a) $\dfrac{dy}{dt}$ when $x = 8$		$\dfrac{dx}{dt} = 10$
	(b) $\dfrac{dx}{dt}$ when $x = 1$		$\dfrac{dy}{dt} = -6$

Equation	Find	Given

4. $x^2 + y^2 = 25$ (a) $\dfrac{dy}{dt}$ when $x = 3$, $y = 4$ $\qquad \dfrac{dx}{dt} = 8$

(b) $\dfrac{dx}{dt}$ when $x = 4$, $y = 3$ $\qquad \dfrac{dy}{dt} = -2$

5. The radius r of a circle is increasing at a rate of 2 inches per minute. Find the rate of change of the area when (a) $r = 6$ inches and (b) $r = 24$ inches.

6. The radius r of a sphere is increasing at a rate of 2 inches per minute. Find the rate of change of the volume when (a) $r = 6$ inches and (b) $r = 24$ inches.

7. Let A be the area of a circle of radius r that is changing with respect to time. If dr/dt is constant, is dA/dt constant? Explain why or why not.

8. Let V be the volume of a sphere of radius r that is changing with respect to time. If dr/dt is constant, is dV/dt constant? Explain why or why not.

9. A spherical balloon is inflated with gas at the rate of 20 cubic feet per minute. How fast is the radius of the balloon increasing at the instant the radius is (a) 1 foot and (b) 2 feet?

10. The formula for the volume of a cone is

$$V = \frac{1}{3}\pi r^2 h.$$

Find the rate of change of the volume if dr/dt is 2 inches per minute and $h = 3r$ when (a) $r = 6$ inches and (b) $r = 24$ inches.

11. At a sand and gravel plant, sand is falling off a conveyor and onto a conical pile at the rate of 10 cubic feet per minute. The diameter of the base of the cone is approximately three times the altitude. At what rate is the height of the pile changing when it is 15 feet high?

12. A conical tank (with vertex down) is 10 feet across the top and 12 feet deep. If water is flowing into the tank at the rate of 10 cubic feet per minute, find the rate of change of the depth of the water the instant it is 8 feet deep.

13. All edges of a cube are expanding at the rate of 3 centimeters per second. How fast is the volume changing when each edge is (a) 1 centimeter and (b) 10 centimeters?

14. The conditions are the same as in Exercise 13. Now measure how fast the *surface* area is changing when each edge is (a) 1 centimeter and (b) 10 centimeters.

15. A point is moving along the graph of $y = x^2$ so that dx/dt is 2 centimeters per minute. Find dy/dt when (a) $x = 0$ and (b) $x = 3$.

16. The conditions are the same as in Exercise 15, but now measure the rate of change of the distance between the point and the origin.

17. A point is moving along the graph of $y = 1/(1 + x^2)$ so that $dx/dt = 2$ centimeters per minute. Find dy/dt for the following values of x.
(a) $x = -2$ $\qquad$ (b) $x = 0$
(c) $x = 2$ $\qquad$ (d) $x = 10$

18. A point is moving along the graph of $y = x^3$ so that $dx/dt = 2$ centimeters per minute. Find dy/dt for the following values of x.
(a) $x = -2$ $\qquad$ (b) $x = 1$
(c) $x = 0$ $\qquad$ (d) $x = 3$

19. A swimming pool is 40 feet long, 20 feet wide, 4 feet deep at the shallow end, and 9 feet deep at the deep end (see figure). Water is being pumped into the pool at 10 cubic feet per minute, and there is 4 feet of water at the deep end.
(a) What percentage of the pool is filled?
(b) At what rate is the water level rising?

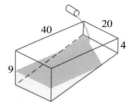

20. A trough is 12 feet long and 3 feet across the top (see figure). Its ends are isosceles triangles with an altitude of 3 feet. If water is being pumped into the trough at 2 cubic feet per minute, how fast is the water level rising when it is 1 foot deep?

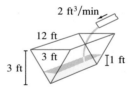

21. A ladder 25 feet long is leaning against the wall of a house (see figure). The base of the ladder is pulled away from the wall at a rate of 2 feet per second. How fast is the top moving down the wall when the base of the ladder is (a) 7 feet, (b) 15 feet, and (c) 24 feet from the wall?

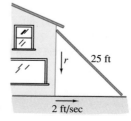

22. A construction worker pulls a 16-foot plank up the side of a building under construction by means of a rope tied to the end of the plank (see figure). Assume the opposite end of the plank follows a path perpendicular to the wall of the building and the worker pulls the rope at the rate of 0.5 feet per second. How fast is the end of the plank sliding along the ground when it is 8 feet from the wall of the building?

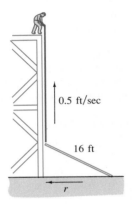

0.5 ft/sec

16 ft

r

23. Consider the right triangle formed by the moving ladder, the side of the house, and the ground in Exercise 21. When the base is 7 feet from the wall, find the rate at which the area of the triangle is changing.

24. A boat is pulled in by means of a winch on the dock 12 feet above the deck of the boat (see figure). The winch pulls in rope at the rate of 4 feet per second. Determine the speed of the boat when there is 13 feet of rope out. What happens to the speed of the boat as it gets closer to the dock?

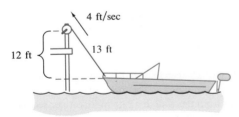

4 ft/sec

12 ft

13 ft

25. An air traffic controller spots two planes at the same altitude converging on a point as they fly at right angles to each other (see figure). One plane is 150 miles from the point and is moving at 450 miles per hour. The other plane is 200 miles from the point and has a speed of 600 miles per hour.

(a) At what rate is the distance between the planes decreasing?

(b) How much time does the traffic controller have to get one of the planes on a different flight path?

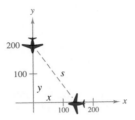

26. The point $(0, y)$ moves along the y-axis at a constant rate of R feet per second, while the point $(x, 0)$ moves along the x-axis at a constant rate of r feet per second. Find an expression for the rate of change of the distance between the two points.

27. A baseball diamond has the shape of a square with sides 90 feet long (see figure). A player 30 feet from third base is running at a speed of 28 feet per second. At what rate is the player's distance from home plate changing?

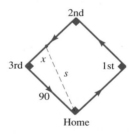

2nd

3rd   *x*   1st

          s

90

Home

28. For the baseball diamond in Exercise 27, suppose the player is running from first to second at a speed of 28 feet per second. Find the rate at which the distance from home plate is changing when the player is 30 feet from second.

29. A man 6 feet tall walks at a rate of 5 feet per second away from a light that is 15 feet above the ground (see figure). When he is 10 feet from the base of the light,
(a) at what rate is the tip of his shadow moving?
(b) at what rate is the length of his shadow changing?

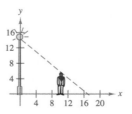

16
12
8
4

4   8   12 16 20

30. An airplane is flying at an altitude of 6 miles and passes directly over a radar antenna (see figure). When the plane is 10 miles away ($s = 10$), the radar detects that the distance s is changing at a rate of 240 miles per hour. What is the speed of the plane?

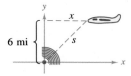

31. As a spherical raindrop falls, it reaches a layer of dry air and begins to evaporate at a rate that is proportional to its surface area ($S = 4\pi r^2$). Show that the radius decreases at a constant rate.

32. The combined electrical resistance R of R_1 and R_2, connected in parallel, is given by

$$\frac{1}{R} = \frac{1}{R_1} + \frac{1}{R_2}$$

where R, R_1, and R_2 are measured in ohms. R_1 and R_2 are increasing at rates of 1 and 1.5 ohms per second, respectively. At what rate is R changing when $R_1 = 50$ ohms and $R_2 = 75$ ohms?

33. When a certain polyatomic gas undergoes adiabatic expansion, its pressure p and volume v satisfy the equation

$$pv^{1.3} = k$$

where k is a constant. Find the relationship between the related rates dp/dt and dv/dt.

REVIEW EXERCISES for Chapter 3

In Exercises 1–24, find the derivative of the given function.

1. $f(x) = x^3 - 3x^2$

2. $f(x) = \dfrac{2x^3 - 1}{x^2}$

3. $f(x) = x^{1/2} - x^{-1/2}$

4. $f(x) = \dfrac{x + 1}{x - 1}$

5. $g(t) = \dfrac{2}{3t^2}$

6. $h(x) = \dfrac{2}{(3x)^2}$

7. $f(x) = \sqrt{x^3 + 1}$

8. $f(x) = \sqrt[3]{x^2 - 1}$

9. $f(x) = (3x^2 + 7)(x^2 - 2x + 3)$

10. $f(x) = \left(x^2 + \dfrac{1}{x}\right)^5$

11. $f(s) = (s^2 - 1)^{5/2}(s^3 + 5)$

12. $h(\theta) = \dfrac{\theta}{(1 - \theta)^3}$

13. $f(x) = \dfrac{-2x^2}{x - 1}$

14. $f(x) = \dfrac{6x - 5}{x^2 + 1}$

15. $f(x) = \dfrac{x^2 + x - 1}{x^2 - 1}$

16. $f(t) = t^2(t - 1)^5$

17. $f(x) = -2(1 - 4x^2)^2$

18. $f(x) = [(x - 2)(x + 4)]^2$

19. $f(x) = \dfrac{1}{4 - 3x^2}$

20. $f(x) = \dfrac{9}{3x^2 - 2x}$

21. $g(x) = \dfrac{2x}{\sqrt{x + 1}}$

22. $g(x) = x\sqrt{x^2 + 1}$

23. $f(t) = \sqrt{t + 1}\,\sqrt[3]{t + 1}$

24. $y = \sqrt{3x}(x + 2)^3$

In Exercises 25–30, find the second derivative of the given function.

25. $f(x) = \sqrt{x^2 + 9}$

26. $h(x) = x\sqrt{x^2 - 1}$

27. $f(t) = \dfrac{t}{(1 - t)^2}$

28. $h(x) = x^2 + \dfrac{3}{x}$

29. $g(x) = \dfrac{6x - 5}{x^2 + 1}$

30. $f(x) = (3x^2 + 7)(x^2 - 2x + 3)$

In Exercises 31–36, use implicit differentiation to find dy/dx.

31. $x^2 + 3xy + y^3 = 10$

32. $x^2 + 9y^2 - 4x + 3y - 7 = 0$

33. $y\sqrt{x} - x\sqrt{y} = 16$

34. $y^2 + x^2 - 6y - 2x - 5 = 0$

35. $y^2 - x^2 = 25$

36. $y^2 = (x - y)(x^2 + y)$

In Exercises 37–42, find the equation of the tangent line and the normal line to the graph of the given equation at the indicated point.

Equation	*Point*
37. $y = (x + 3)^3$	$(-2, 1)$
38. $y = (x - 2)^2$	$(2, 0)$
39. $x^2 + y^2 = 20$	$(2, 4)$
40. $x^2 - y^2 = 16$	$(5, 3)$
41. $y = \sqrt[3]{(x - 2)^2}$	$(3, 1)$
42. $y = \dfrac{2x}{1 - x^2}$	$(0, 0)$

43. Find the points on the graph of

$$f(x) = \frac{1}{3}x^3 + x^2 - x - 1$$

at which the slope is (a) -1, (b) 2, and (c) 0.

44. Find the points on the graph of

$$f(x) = x^2 + 1$$

at which the slope is (a) -1, (b) 0, and (c) 1.

In Exercises 45–48, find the derivative of the given function by using the definition of the derivative.

45. $f(x) = \dfrac{1}{x^2}$

46. $f(x) = \dfrac{x+1}{x-1}$

47. $f(x) = \sqrt{x+2}$

48. $f(x) = \dfrac{1}{\sqrt{x}}$

In Exercises 49 and 50, derive the equations for the velocity and acceleration of a particle having the given position function.

49. $s = t + \dfrac{1}{t+1}$

50. $s = \dfrac{1}{t^2 + 2t + 1}$

51. Suppose that the temperature T of food placed in a freezer is given by the equation

$$T = \frac{700}{t^2 + 4t + 10}$$

where t is the time in hours. Find the rate of change of T with respect to t at the following times.

(a) $t = 1$ (b) $t = 3$ (c) $t = 5$ (d) $t = 10$

52. The emergent velocity v of a liquid flowing from a hole in the bottom of a tank is given by $v = \sqrt{2gh}$, where g is the acceleration due to gravity (32 ft/s²) and h is the depth of the liquid in the tank. Find the rate of change of v with respect to h when (a) $h = 9$ and (b) $h = 4$.

53. What is the smallest initial velocity that is required to throw a stone to the top of a 49-foot silo?

54. A bomb is dropped from an airplane at an altitude of 14,400 feet. How long will it take to reach the ground? (Even though it will not be a vertical fall because of the motion of the plane, the time will be the same as that for a vertical fall.) The plane is moving at 600 miles per hour. How far will the bomb move horizontally after it is released from the plane?

55. A ball is thrown and follows a path described by $y = x - 0.02x^2$.

(a) Sketch a graph of the path.

(b) Find the total horizontal distance the ball was thrown.

(c) For what x-value does the ball reach its maximum height? (Use the symmetry of the path.)

(d) Find the equation that gives the instantaneous rate of change in the height of the ball with respect to the horizontal change and evaluate this equation at $x = 0, 10, 25, 30, 50$.

(e) What is the instantaneous rate of change of the height when the ball reaches its maximum height?

56. The path of a projectile thrown at an angle of 45° with level ground is given by

$$y = x - \frac{32}{v_0{}^2}(x^2)$$

where the initial velocity is v_0 feet per second.

(a) Sketch the path followed by the projectile.

(b) Find the x-coordinate of the point where the projectile strikes the ground. Use the symmetry of the path of the projectile to locate the x-coordinate of the point where the projectile reaches its maximum height.

(c) What is the instantaneous rate of change of the height when the projectile is at its maximum height?

57. The path of a projectile, thrown at an angle of 45° with the ground, is given by

$$y = x - \frac{32}{v_0{}^2}(x^2)$$

where the initial velocity is v_0 feet per second. Show that doubling the initial velocity of the projectile multiplies both the maximum height and the range by a factor of 4.

58. Use the equation given in Exercise 57 to find the maximum height and range of a projectile thrown with an initial velocity of 70 feet per second.

59. A point moves along the curve $y = \sqrt{x}$ in such a way that the y-value is increasing at the rate of 2 units per second. At what rate is x changing for the following values?

(a) $x = \dfrac{1}{2}$ (b) $x = 1$ (c) $x = 4$

60. The same conditions exist as in Exercise 59. Find the rate the distance between (x, y) and the origin is changing for the following.

(a) $x = \dfrac{1}{2}$ (b) $x = 1$ (c) $x = 4$

61. The cross section of a 5-foot trough is an isosceles trapezoid with a 2-foot lower base, a 3-foot upper base, and an altitude of 2 feet. Water is running into the trough at the rate of 1 cubic foot per minute. How fast is the water level rising when the water is 1 foot deep?

62. The geometric mean of x and $x + n$ is

$$g = \sqrt{x(x + n)}$$

and the arithmetic mean is

$$a = \frac{x + (x + n)}{2}.$$

Show that $dg/dx = a/g$.

63. For $y = 1/x$, show that

$$y^{(n)} = \frac{(-1)^n n!}{x^{n+1}}.$$

[Note: $n! = n(n - 1)(n - 2) \cdots (3)(2)(1)$.]

64. Sketch the graph of $f(x) = 4 - |x - 2|$.
    (a) Is f continuous at $x = 2$?
    (b) Is f differentiable at $x = 2$? Why or why not?

65. Sketch the graph of

$$f(x) = \begin{cases} x^2 + 4x + 2, & x < -2 \\ 1 - 4x - x^2, & x \geq -2. \end{cases}$$

    (a) Is f continuous at $x = -2$?
    (b) Is f differentiable at $x = -2$? Why or why not?

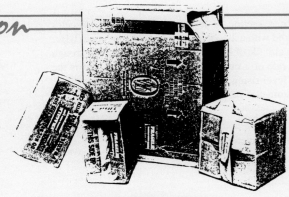

Container construction involves, among other things, consideration of the optimum use of construction material to hold a given amount. For instance, of all rectangular containers holding a given volume, a cube has the least surface area, and would therefore require the least construction material. In commercial packaging, this fact is often discounted in favor of other considerations such as the attractiveness of the packaging.

Construction of a Cylindrical Container

One of the most important applications of the derivative involves the problem of optimization—finding the least amount of material, the greatest strength, the greatest profit, and so on. As an example, consider the problem of finding the dimensions of a (right circular) cylindrical container that uses the least material. Suppose that the container is to hold 12 fluid ounces. To minimize the material used to construct the cylinder, we minimize its surface area, which is given by

$$S = 2\pi r^2 + 2\pi rh$$

where r is the radius of the circular ends of the container and h is the height of the container. Before we can apply calculus techniques to find the minimum surface area, it is helpful to rewrite this equation in terms of a single independent variable. To do that, we use the fact that the volume of the container is 12 fl. oz. = 21.66 in² = $\pi r^2 h$. Thus, we have $h = 21.66/\pi r^2$ and it follows that

$$S = 2\pi r^2 + 2\pi r\left(\frac{21.66}{\pi r^2}\right) = 2\pi\left(r^2 + \frac{21.66}{\pi r}\right).$$

The derivative of S with respect to r is

$$\frac{dS}{dr} = 2\pi\left(2r - \frac{21.66}{\pi r^2}\right).$$

Finally, by solving the equation $dS/dr = 0$, we find that the minimum surface area occurs when $r \approx 1.5$ inches and $h \approx 3.0$ inches.

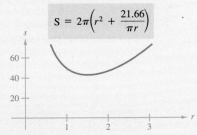

$$S = 2\pi\left(r^2 + \frac{21.66}{\pi r}\right)$$

The minimum of S occurs when $\frac{dS}{dr} = 0$. See Section 4.7, Exercise 31.

Chapter Overview

This chapter discusses several applications of the derivative of a function. These applications fall into three basic categories—curve sketching, optimization, and approximation techniques.

The chapter begins by showing how the derivative can be used to locate the minimum and maximum values of a function on a closed interval. Many of the results in the chapter depend on two important theorems called *Rolle's Theorem* and the *Mean Value Theorem*, and these are presented in Section 4.2. In Sections 4.3 through 4.5, we refine our curve-sketching techniques by showing how the first derivative can be used to determine whether a function is increasing or decreasing, and how the second derivative can be used to determine whether a function is concave upward or downward. We also present techniques for locating horizontal asymptotes of functions. Section 4.6 summarizes all of the curve-sketching techniques presented in the first four chapters of the text.

Section 4.7 is an important one because it shows how calculus can be used to solve optimization problems that occur in science and engineering. (Section 4.10 discusses optimization problems that occur in business and economics.)

Section 4.8 presents a numerical technique called *Newton's Method*. This method uses the derivative of a function to approximate the zeros of the function. Section 4.9 introduces the concept of the *differential* of a function. Finally, Section 4.10 discusses several applications of calculus in business and economics.

4

Applications of Differentiation

4.1 Extrema on an Interval

Extrema of a function ▪ Relative extrema ▪ Critical numbers

In calculus much effort is devoted to determining the behavior of a function f on an interval I. For instance, we are interested in several questions: Does f have a maximum value on I? Does it have a minimum value? Where is the function increasing? Where is it decreasing? In this chapter we will show how the derivative can be used to answer these questions. We will also show why these questions are important to applications.

We begin by looking at the minimum and maximum values of a function on an interval.

DEFINITION OF EXTREMA

Let f be defined on an interval I containing c.

1. $f(c)$ is the **minimum of f on I** if $f(c) \leq f(x)$ for all x in I.
2. $f(c)$ is the **maximum of f on I** if $f(c) \geq f(x)$ for all x in I.

The minimum and maximum of a function on an interval are called the **extreme values,** or **extrema,** of the function on the interval.

REMARK   The minimum and maximum of a function on an interval are sometimes called the **absolute minimum** and **absolute maximum** on the interval, respectively.

A function need not have a minimum or a maximum on an interval. For instance, Figure 4.1, p. 156, shows three possibilities. By comparing the first and second graphs, we see that the function $f(x) = x^2 + 1$ has both a minimum and a maximum on the closed interval $[-1, 2]$, but does not have a maximum on the open interval $(-1, 2)$. Moreover, in the third graph, we see that a discontinuity (at $x = 0$) can affect the existence of an extremum on an interval. This suggests the following theorem, which identifies conditions that guarantee the existence of both a minimum and a maximum of a function on an interval. (A proof of this theorem is not within the scope of this text.)

THEOREM 4.1
THE EXTREME VALUE THEOREM

If f is continuous on a closed interval $[a, b]$, then f has both a minimum and a maximum on the interval.

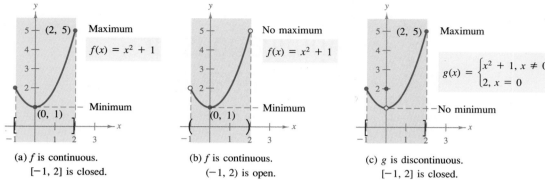

(a) f is continuous.
[−1, 2] is closed.

(b) f is continuous.
(−1, 2) is open.

(c) g is discontinuous.
[−1, 2] is closed.

FIGURE 4.1

REMARK   Note that the Extreme Value Theorem (like the Intermediate Value Theorem) is an *existence theorem* because it tells of the existence of minimum and maximum values but does not show how to find these values.

From Figure 4.1 we can see that extrema can occur at interior points or endpoints of an interval. Extrema that occur at the endpoints are called **endpoint extrema.**

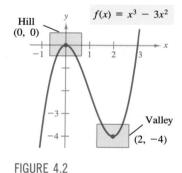

FIGURE 4.2

Relative extrema

In Figure 4.2, the graph of $f(x) = x^3 - 3x^2$ has a **relative maximum** at the point (0, 0) and a **relative minimum** at the point (2, −4). Informally, we can think of a relative maximum occurring on a "hill" on the graph, and a relative minimum occurring in a "valley" on the graph. Such hills and valleys can occur in two ways. If the hill (or valley) is smooth and rounded, then the graph has a horizontal tangent line at the high point (or low point). On the other hand, if the hill (or valley) is sharp and peaked, then the graph represents a function that is not differentiable at the high point (or low point).

DEFINITION OF RELATIVE EXTREMA	1. If there is an open interval on which $f(c)$ is a maximum, then $f(c)$ is called a **relative maximum** of f. 2. If there is an open interval on which $f(c)$ is a minimum, then $f(c)$ is called a **relative minimum** of f.

REMARK   The plural of relative maximum is relative maxima, and the plural of relative minimum is relative minima.

In Example 1, we examine the derivative of a function at several *given* relative extrema. (We will say much more about *finding* the relative extrema of a function in Section 4.3.)

EXAMPLE 1   The value of the derivative at a relative extrema

Find the value of the derivative at each of the relative extrema shown in Figure 4.3.

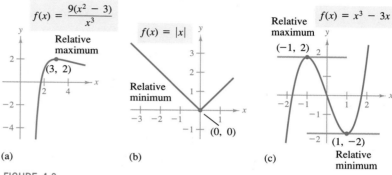

(a)   (b)   (c)

FIGURE 4.3

SOLUTION

(a) The derivative of this function is

$$f'(x) = \frac{x^3(18x) - (9)(x^2 - 3)(3x^2)}{(x^3)^2} = \frac{9(9 - x^2)}{x^4}.$$

Thus, at the point (3, 2), the value of the derivative is $f'(3) = 0$.

(b) At $x = 0$, the derivative of $f(x) = |x|$ *does not exist* because the following one-sided limits differ.

$$\lim_{x \to 0^-} \frac{f(x) - f(0)}{x - 0} = \lim_{x \to 0^-} \frac{|x|}{x} = -1 \qquad \text{Limit from the left}$$

$$\lim_{x \to 0^+} \frac{f(x) - f(0)}{x - 0} = \lim_{x \to 0^+} \frac{|x|}{x} = +1 \qquad \text{Limit from the right}$$

(c) The derivative of this function is $f'(x) = 3x^2 - 3$. Thus, at the point $(-1, 2)$, the value of the derivative is $f'(-1) = 3(-1)^2 - 3 = 0$, and at the point $(1, -2)$ the value of the derivative is $f'(1) = 3(1)^2 - 3 = 0$.

Note in Example 1 that at a relative extremum the derivative is either zero or undefined. We call the x-values at these special points **critical numbers.**

DEFINITION OF CRITICAL NUMBER

If f is defined at c, then c is called a **critical number** of f if $f'(c) = 0$ or if f' is undefined at c.

Figure 4.4 illustrates the two types of critical numbers given in this definition.

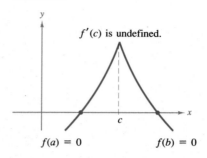

$f'(c)$ is undefined.

$f(a) = 0$ $f(b) = 0$

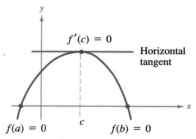

$f'(c) = 0$
Horizontal tangent

FIGURE 4.4 $f(a) = 0$ c $f(b) = 0$

THEOREM 4.2 **RELATIVE EXTREMA OCCUR ONLY** **AT CRITICAL NUMBERS**	If f has a relative minimum or relative maximum at $x = c$, then c is a critical number of f.

PROOF

Case 1: If f is *not* differentiable at $x = c$, then by definition c is a critical number of f and we have nothing to prove.

Case 2: If f is differentiable at $x = c$, then $f'(c)$ must be either positive, negative, or zero. Suppose $f'(c)$ is positive, Then we have

$$f'(c) = \lim_{x \to c} \frac{f(x) - f(c)}{x - c} > 0.$$

But this implies that there exists an interval (a, b) containing c such that

$$\frac{f(x) - f(c)}{x - c} > 0, \quad \text{for all } x \neq c \text{ in } (a, b).$$

Since this quotient is positive, the signs of the denominator and numerator must agree; this produces the following inequalities for x-values in the interval (a, b).

Left of c: $x < c$ and $f(x) < f(c)$ ⟹ $f(c)$ is not a relative minimum
Right of c: $x > c$ and $f(x) > f(c)$ ⟹ $f(c)$ is not a relative maximum

Thus, our assumption that $f'(c) > 0$ contradicts the fact that $f(c)$ is a relative extremum. Assuming that $f'(c) < 0$ produces a similar contradiction, we are left with only one possibility, namely $f'(c) = 0$.

Theorem 4.2 tells us that the relative extrema of a function can occur *only* at the critical numbers of the function. By combining this observation with our knowledge of endpoint extrema, we obtain the following guidelines for finding extrema on a closed interval.

GUIDELINES FOR FINDING EXTREMA ON A CLOSED INTERVAL

To find the extrema of a continuous function f on a closed interval $[a, b]$, we suggest the following steps.

1. Find the critical numbers of f.
2. Evaluate f at each critical number in (a, b).
3. Evaluate f at each endpoint of $[a, b]$.
4. The least of these values is the minimum, and the greatest is the maximum.

EXAMPLE 2   Finding extrema on a closed interval

Find the extrema of $f(x) = 3x^4 - 4x^3$ on the interval $[-1, 2]$.

SOLUTION

To find the critical numbers, we differentiate to obtain

$$f'(x) = 12x^3 - 12x^2 = 0 \qquad \text{Set } f'(x) = 0$$
$$12x^2(x - 1) = 0 \qquad \text{Factor}$$
$$x = 0, 1. \qquad \text{Critical numbers}$$

Since f' is defined for all x, we conclude that these are the only critical numbers of f. Finally, by evaluating f at these two critical numbers and at the endpoints of $[-1, 2]$, we determine that the maximum is $f(2) = 16$ and the minimum is $f(1) = -1$, as indicated in Table 4.1. The graph of f is shown in Figure 4.5.

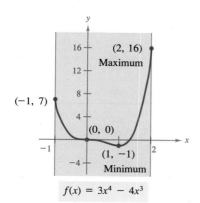

$f(x) = 3x^4 - 4x^3$

FIGURE 4.5

TABLE 4.1

Left endpoint	Critical number	Critical number	Right endpoint
$f(-1) = 7$	$f(0) = 0$	$f(1) = -1$ Minimum	$f(2) = 16$ Maximum

REMARK   Note in Figure 4.5 that the critical number $x = 0$ does not yield a relative minimum or a relative maximum. This tells us that the converse of Theorem 4.2 is not true. In other words, the critical numbers of a function need not produce relative extrema.

EXAMPLE 3   Finding extrema on a closed interval

Find the extrema of $f(x) = 2x - 3x^{2/3}$ on the interval $[-1, 3]$.

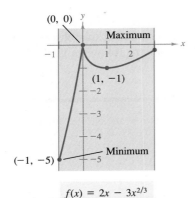

$f(x) = 2x - 3x^{2/3}$

FIGURE 4.6

SOLUTION

Differentiating produces

$$f'(x) = 2 - \frac{2}{x^{1/3}} = 2\left(\frac{x^{1/3} - 1}{x^{1/3}}\right).$$

This gives us the following critical numbers.

$$x = 1 \qquad f'(1) = 0$$
$$x = 0 \qquad f' \text{ is undefined}$$

Finally, by evaluating f at these two points and at the endpoints of the interval, we conclude that the minimum is $f(-1) = -5$ and the maximum is $f(0) = 0$, as indicated in Table 4.2. The graph of f is shown in Figure 4.6.

TABLE 4.2

Left endpoint	Critical number	Critical number	Right endpoint
$f(-1) = -5$ Minimum	$f(0) = 0$ Maximum	$f(1) = -1$	$f(3) = 6 - 3\sqrt[3]{9} \approx -0.24$

EXERCISES for Section 4.1

In Exercises 1–6, find the value of the derivative (if it exists) at the indicated extrema.

1. $f(x) = \dfrac{x^2}{x^2 + 4}$

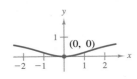

2. $f(x) = -x^2 + 4x$

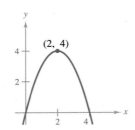

3. $f(x) = x + \dfrac{32}{x^2}$

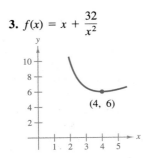

4. $f(x) = -3x\sqrt{x + 1}$

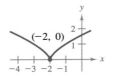

5. $f(x) = (x + 2)^{2/3}$

6. $f(x) = 4 - |x|$

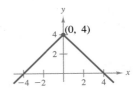

In Exercises 7–18, locate the absolute extrema of the given function on the indicated interval.

Function	Interval
7. $f(x) = 2(3 - x)$	$[-1, 2]$
8. $f(x) = \dfrac{2x + 5}{3}$	$[0, 5]$
9. $f(x) = -x^2 + 3x$	$[0, 3]$
10. $f(x) = x^2 + 2x - 4$	$[-1, 1]$
11. $f(x) = x^3 - 3x^2$	$[-1, 3]$
12. $f(x) = x^3 - 12x$	$[0, 4]$
13. $f(x) = 3x^{2/3} - 2x$	$[-1, 1]$
14. $g(x) = \sqrt[3]{x}$	$[-1, 1]$
15. $h(t) = 4 - |t - 4|$	$[1, 6]$

16. $g(t) = \dfrac{t^2}{t^2 + 3}$ $[-1, 1]$

17. $h(s) = \dfrac{1}{s - 2}$ $[0, 1]$

18. $h(t) = \dfrac{t}{t - 2}$ $[3, 5]$

19. Explain why the function $f(x) = 1/x^2$ has a maximum on $[1, 2]$ but not on $(0, 2]$.

20. Explain why the function $y = 1/(x + 1)$ has a minimum on $[0, 2]$, but not on $[-2, 0]$.

In Exercises 21–24, determine from the graph whether f possesses a minimum in the interval (a, b).

21. (a)   (b)

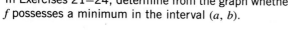

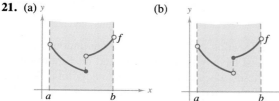

22. (a)   (b)

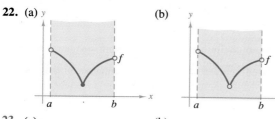

23. (a)   (b)

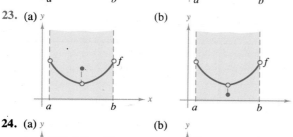

24. (a)   (b)

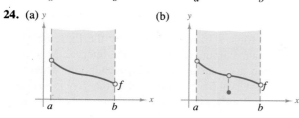

In Exercises 25 and 26, locate the absolute extrema of the function (if any exist) over the indicated interval.

25. $f(x) = 2x - 3$
    (a) $[0, 2]$
    (b) $[0, 2)$
    (c) $(0, 2]$
    (d) $(0, 2)$

26. $f(x) = 5 - x$
    (a) $[1, 4]$
    (b) $[1, 4)$
    (c) $(1, 4]$
    (d) $(1, 4)$

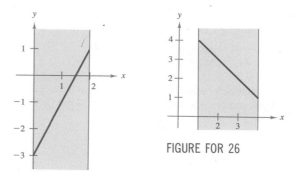

FIGURE FOR 25

FIGURE FOR 26

In Exercises 27–30, find the maximum value of $|f''(x)|$ on the indicated interval. (This value is used in the error estimate for the Trapezoidal Rule, as discussed in Section 5.6.)

Function	Interval
27. $f(x) = \dfrac{1}{x^2 + 1}$	$[0, 3]$
28. $f(x) = \dfrac{1}{x^2 + 1}$	$\left[\dfrac{1}{2}, 3\right]$
29. $f(x) = \sqrt{1 + x^3}$	$[0, 2]$
30. $f(x) = x^3(3x^2 - 10)$	$[0, 1]$

In Exercises 31–34, find the maximum value of $|f^{(4)}(x)|$ on the indicated interval. (This value is used in the error estimate for Simpson's Rule, as discussed in Section 5.6.)

Function	Interval
31. $f(x) = 15x^4 - \left(\dfrac{2x - 1}{2}\right)^6$	$[0, 1]$
32. $f(x) = x^5 - 5x^4 + 20x^3 + 600$	$\left[0, \dfrac{3}{2}\right]$
33. $f(x) = (x + 1)^{2/3}$	$[0, 2]$
34. $f(x) = \dfrac{1}{x^2 + 1}$	$[-1, 1]$

35. The formula for the power output P of a battery is given by

$$P = VI - RI^2$$

where V is the electromotive force in volts, R is the resistance, and I is the current. Find the current (measured in amperes) that corresponds to a maximum value of P in a battery for which $V = 12$ volts and $R = 0.5$ ohms. (Assume that a 15-amp fuse bounds the output in the interval $0 \leq I \leq 15$.)

36. A retailer has determined that the cost C for ordering and storing x units of a certain product is

$$C = 2x + \frac{300{,}000}{x}, \quad 0 \le x \le 300.$$

Find the order size that will minimize cost if the delivery truck can bring a maximum of 300 units per order.

In Exercises 37 and 38, use a computer or graphics calculator to sketch the function f and find its critical numbers over the specified interval.

Function	Interval
37. $f(x) = 3.2x^5 + 5x^3 - 3.5x$	[0, 1]
38. $f(x) = \frac{4}{3}x\sqrt{3 - x}$	[0, 3]

4.2 Rolle's Theorem and the Mean Value Theorem

Rolle's Theorem ▪ Mean Value Theorem ▪ Applications

The Extreme Value Theorem (Section 4.1) tells us that a continuous function on a closed interval $[a, b]$ must have both a minimum and maximum on the interval. However, both of these values can occur at the endpoints. We now present a theorem called **Rolle's Theorem,** named after the French mathematician Michel Rolle (1652–1719), that implies (under certain conditions) the existence of an extreme value in the interior of a closed interval.

THEOREM 4.3
ROLLE'S THEOREM

Let f be continuous on the closed interval $[a, b]$ and differentiable on the open interval (a, b). If

$$f(a) = f(b)$$

then there is at least one number c in (a, b) such that $f'(c) = 0$.

PROOF Let $f(a) = d = f(b)$.

Case 1: If $f(x) = d$ for all x in $[a, b]$, then f is constant on the interval, and by Theorem 3.3 $f'(x) = 0$ for all x in (a, b).

Case 2: If $f(x) > d$ for some x in (a, b), then, by the Extreme Value Theorem, we know that f has a maximum at some c in the interval. Moreover, since $f(c) > d$, this maximum does not occur at either endpoint. Therefore, f has a maximum in the *open* interval (a, b). This implies that $f(c)$ is a *relative* maximum, and by Theorem 4.2 we know c is a critical number of f. Finally, since f is differentiable at c, we can conclude that $f'(c) = 0$.

Case 3: If $f(x) < d$ for some x in (a, b), then we can use an argument similar to that in Case 2.

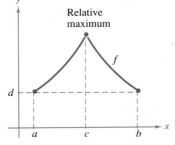

f has a critical number in (a, b).

FIGURE 4.7

If we drop the differentiability requirement from Rolle's Theorem, then f will still have a critical number in (a, b), but it need not yield a horizontal tangent. This is shown in Figure 4.7 and stated in the following corollary to Rolle's Theorem.

COROLLARY TO ROLLE'S THEOREM	Let f be continuous on the closed interval $[a, b]$. If $f(a) = f(b)$, then f has a critical number in the open interval (a, b).

EXAMPLE 1 An application of Rolle's Theorem

Find the two x-intercepts of $f(x) = x^2 - 3x + 2$ and show that $f'(x) = 0$ at some point between the two intercepts.

SOLUTION

Note that f is differentiable on the entire real line. Setting $f(x)$ equal to zero, we have

$$x^2 - 3x + 2 = 0$$
$$(x - 1)(x - 2) = 0.$$

Thus, $f(1) = f(2) = 0$, and from Rolle's Theorem we know that there exists c in the interval $(1, 2)$ such that $f'(c) = 0$. To find c we solve the equation

$$f'(x) = 2x - 3 = 0$$

and determine that $f'(x) = 0$ when $x = \frac{3}{2}$. This x-value lies in the open interval $(1, 2)$, as shown in Figure 4.8. ▭

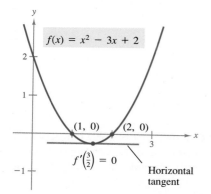

FIGURE 4.8

Rolle's Theorem states that if f satisfies the conditions of the theorem, then there must be *at least* one point between a and b at which the derivative is zero. There may of course be more than one such point, as illustrated in the next example.

EXAMPLE 2 An application of Rolle's Theorem

Let $f(x) = x^4 - 2x^2$. Find all c in the interval $(-2, 2)$ such that $f'(c) = 0$.

SOLUTION

Since $f(-2) = 8 = f(2)$ and f is differentiable, Rolle's Theorem guarantees the existence of at least one c in $(-2, 2)$ such that $f'(c) = 0$. Setting the derivative equal to zero produces

$$f'(x) = 4x^3 - 4x = 0$$
$$4x(x^2 - 1) = 0$$
$$x = 0, 1, -1.$$

Thus, in the interval $(-2, 2)$, the derivative is zero at each of these three x-values, as shown in Figure 4.9. ▭

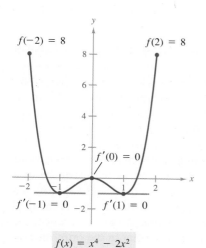

FIGURE 4.9

The Mean Value Theorem

Rolle's Theorem can be used to prove another well-known theorem in calculus—the **Mean Value Theorem.**

THEOREM 4.4 **THE MEAN VALUE THEOREM**	If f is continuous on the closed interval $[a, b]$ and differentiable on the open interval (a, b), then there exists a number c in (a, b) such that $$f'(c) = \frac{f(b) - f(a)}{b - a}.$$

PROOF   Refer to Figure 4.10. The equation of the secant line containing the points $(a, f(a))$ and $(b, f(b))$ is given by

$$y = \left[\frac{f(b) - f(a)}{b - a} \right](x - a) + f(a).$$

Let $g(x)$ be the difference between $f(x)$ and y. Then

$$g(x) = f(x) - y = f(x) - \left[\frac{f(b) - f(a)}{b - a} \right](x - a) - f(a).$$

Now, by evaluating g at a and b, we see that $g(a) = 0 = g(b)$. Furthermore, since f is differentiable, g is also differentiable, and we can apply Rolle's Theorem to the function g. Thus, there exists a point c in (a, b) such that $g'(c) = 0$. This means that

$$0 = g'(c) = f'(c) - \frac{f(b) - f(a)}{b - a}.$$

Therefore, there exists a point c in (a, b) such that

$$f'(c) = \frac{f(b) - f(a)}{b - a}.$$

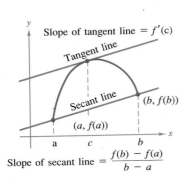

Slope of tangent line $= f'(c)$

Tangent line

Secant line

$(b, f(b))$

$(a, f(a))$

Slope of secant line $= \dfrac{f(b) - f(a)}{b - a}$

FIGURE 4.10

REMARK   The "mean" in the Mean Value Theorem refers to the mean (or average) rate of change of f in the interval $[a, b]$.

Although the Mean Value Theorem can be used directly in problem solving, it is used more often to prove other theorems. In fact, some people consider this to be the most important theorem in calculus. It was proved by a famous French mathematician Joseph-Louis Lagrange (1736–1813), and it is closely related to the Fundamental Theorem of Calculus discussed in Chapter 5. For now, you can get an idea of the versatility of this theorem by looking at the results stated in Exercises 27–32 in this section.

The Mean Value Theorem has implications for both basic interpretations of the derivative. Geometrically, the theorem guarantees the existence of a tangent line that is parallel to the secant line through the points $(a, f(a))$ and $(b, f(b))$, as shown in Figure 4.10. Example 3 illustrates this geometrical interpretation of the Mean Value Theorem. In terms of rates of change, the Mean Value Theorem tells us that there must be a point in the open interval (a, b) at which the instantaneous rate of change is equal to the average rate of change over the interval $[a, b]$. This is illustrated in Example 4.

Joseph-Louis Lagrange

EXAMPLE 3 A tangent line application of the Mean Value Theorem

Given $f(x) = 5 - (4/x)$, find all c in the interval $(1, 4)$ such that

$$f'(c) = \frac{f(4) - f(1)}{4 - 1}.$$

SOLUTION

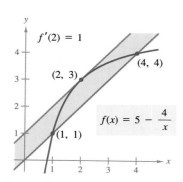

FIGURE 4.11

The slope of the secant line through $(1, f(1))$ and $(4, f(4))$ is

$$\frac{f(4) - f(1)}{4 - 1} = \frac{4 - 1}{4 - 1} = 1.$$

Since f satisfies the conditions of the Mean Value Theorem, there exists at least one c in $(1, 4)$ such that $f'(c) = 1$. Solving the equation $f'(x) = 1$ yields the following.

$$f'(x) = \frac{4}{x^2} = 1 \quad \Longrightarrow \quad x = \pm 2$$

Finally, in the interval $(1, 4)$ we choose $c = 2$, as shown in Figure 4.11.

EXAMPLE 4 A rate of change application of the Mean Value Theorem

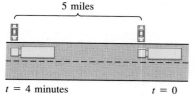

FIGURE 4.12

Two stationary patrol cars equipped with radar are 5 miles apart on a highway, as shown in Figure 4.12. As a truck passes the first patrol car, its speed is clocked at 55 miles per hour. Four minutes later, when the truck passes the second patrol car, its speed is clocked at 50 miles per hour. Prove that the truck must have exceeded the speed limit (of 55 miles per hour) at some time during the four minutes.

SOLUTION

We let $t = 0$ be the time (in hours) when the truck passes the first patrol car. Then, the time when the truck passes the second patrol car is

$$t = \frac{4}{60} = \frac{1}{15} \text{ hr.}$$

Now, if we let $s(t)$ represent the distance (in miles) traveled by the truck, we have $s(0) = 0$ and $s\left(\frac{1}{15}\right) = 5$. Therefore, the average velocity of the truck over the five-mile stretch of highway is given by

$$\text{average velocity} = \frac{s(1/15) - s(0)}{(1/15) - 0} = \frac{5}{1/15} = 75 \text{ mph.}$$

Assuming that the position function is differentiable, we can apply the Mean Value Theorem to conclude that the truck must have been traveling at a rate of 75 miles per hour sometime during the four minutes.

REMARK   A useful alternative form of the Mean Value Theorem is as follows: If f is continuous on $[a, b]$ and differentiable on (a, b), then there exists a number c in (a, b) such that $f(b) = f(a) + (b - a)f'(c)$.

When working the exercises for this section, keep in mind that polynomial functions and rational functions are differentiable at all points in their domains.

EXERCISES for Section 4.2

In Exercises 1 and 2, state why Rolle's Theorem does not apply to the function even though there exist a and b such that $f(a) = f(b) = 0$.

1. $f(x) = 1 - |x - 1|$ **2.** $f(x) = \dfrac{x^2 - 4}{x^2}$

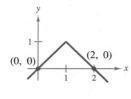

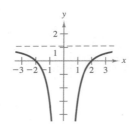

In Exercises 3–12, determine whether Rolle's Theorem can be applied to f on the indicated interval. If Rolle's Theorem can be applied, find all values of c in the interval such that $f'(c) = 0$.

Function	Interval
3. $f(x) = x^2 - 2x$	$[0, 2]$
4. $f(x) = x^2 - 3x + 2$	$[1, 2]$
5. $f(x) = (x - 1)(x - 2)(x - 3)$	$[1, 3]$
6. $f(x) = (x - 3)(x + 1)^2$	$[-1, 3]$
7. $f(x) = |x| - 1$	$[-1, 1]$
8. $f(x) = 3 - |x - 3|$	$[0, 6]$
9. $f(x) = x^{2/3} - 1$	$[-8, 8]$
10. $f(x) = x - x^{1/3}$	$[0, 1]$
11. $f(x) = \dfrac{x^2 - 2x - 3}{x + 2}$	$[-1, 3]$
12. $f(x) = \dfrac{x^2 - 1}{x}$	$[-1, 1]$

In Exercises 13–20, apply the Mean Value Theorem to f on the indicated interval. In each case, find all values of c in the interval (a, b) such that

$$f'(c) = \frac{f(b) - f(a)}{b - a}.$$

Function	Interval
13. $f(x) = x^2$	$[-2, 1]$
14. $f(x) = x(x^2 - x - 2)$	$[-1, 1]$
15. $f(x) = x^{2/3}$	$[0, 1]$
16. $f(x) = \dfrac{x + 1}{x}$	$\left[\dfrac{1}{2}, 2\right]$
17. $f(x) = \dfrac{x}{x + 1}$	$\left[-\dfrac{1}{2}, 2\right]$
18. $f(x) = \sqrt{x - 2}$	$[2, 6]$
19. $f(x) = x^3$	$[0, 1]$
20. $f(x) = x^3 - 2x$	$[0, 2]$

21. The height of a ball t seconds after it is thrown is given by

$$f(t) = -16t^2 + 48t + 32.$$

(a) Verify that $f(1) = f(2)$.
(b) According to Rolle's Theorem, what must be the velocity at some time in the interval $[1, 2]$?

22. The ordering and transportation cost C of components used in a manufacturing process is approximated by

$$C(x) = 10\left(\frac{1}{x} + \frac{x}{x + 3}\right)$$

where C is measured in thousands of dollars and x is the order size in hundreds.

(a) Verify that $C(3) = C(6)$.
(b) According to Rolle's Theorem, the rate of change of cost must be zero for some order size in the interval $[3, 6]$. Find that order size.

23. The height of an object t seconds after it was dropped from a height of 500 feet is given by

$$s(t) = -16t^2 + 500.$$

(a) Find the average velocity of the object during the first 3 seconds.
(b) Use the Mean Value Theorem to verify that at some time during the first three seconds of fall the instantaneous velocity equals the average velocity. Find that time.

24. A company introduces a new product for which the number of units sold S is given by

$$S(t) = 200\left(5 - \frac{9}{2 + t}\right)$$

where t is the time in months.

(a) Find the average rate of change of $S(t)$ during the first year.

(b) During what month does $S'(t)$ equal its average rate of change during the first year?

25. Given the function

$$f(x) = \frac{1}{x - 4}$$

show that for the interval $(2, 6)$ there exists no real number c such that

$$f'(c) = \frac{f(6) - f(2)}{6 - 2}.$$

State whether this contradicts the Mean Value Theorem and give the reason for your answer.

26. Prove the Corollary to Rolle's Theorem.

If $a > 0$ and n is any integer, prove that the polynomial function

$$p(x) = x^{2n+1} + ax + b$$

cannot have two real roots.

28. Let p be a *nonconstant* polynomial function.

(a) Prove that between any two consecutive zeros of p', there is at most one zero of p.

(b) If p has three distinct zeros in the interval $[a, b]$, prove that $p''(c) = 0$ for some real number c in (a, b).

29. Prove that if $f'(x) = 0$ for all x in an interval (a, b), then f is constant on the interval.

30. Let $p(x) = Ax^2 + Bx + C$. Prove that for any interval $[a, b]$, the value c guaranteed by the Mean Value Theorem is the midpoint of the interval.

31. Prove that if two functions f and g have the same derivatives on an interval, then they must differ only by a constant on the interval. [Hint: Let $h(x) = f(x) - g(x)$ and use the result of Exercise 29.]

32. Prove that if f is differentiable on $(-\infty, \infty)$ and $f'(x) < 1$ for all real numbers, then f has at most one fixed point. A **fixed point** of a function f is a real number c such that $f(c) = c$.

33. Use a computer or graphics calculator to sketch the graph of $f(x) = \sqrt{x}$ over the interval $[1, 9]$.

(a) Find an equation of the secant line to the graph of f passing through the points $(1, f(1))$ and $(9, f(9))$. Sketch the graph of the secant line on the same axes as the graph of f.

(b) Find the value of c in the interval $(1, 9)$ such that

$$f'(c) = \frac{f(9) - f(1)}{9 - 1}.$$

Find the equation of the tangent line to the graph of f at the point $(c, f(c))$ and sketch its graph on the same axes as the graph of f. Note that the secant line and tangent line are parallel.

4.3 Increasing and Decreasing Functions and the First Derivative Test

Increasing and decreasing functions ▪ The First Derivative Test ▪ Strictly monotonic functions

We now know that the derivative is useful in *locating* the relative extrema of a function. In this section we will show that the derivative also can be used to *classify* relative extrema as either relative minima or relative maxima. We begin by defining what is meant when we say a function increases (or decreases) on an interval.

DEFINITION OF INCREASING AND DECREASING FUNCTIONS

A function f is said to be **increasing** on an interval if for any two numbers x_1 and x_2 in the interval,

$$x_1 < x_2 \quad \text{implies} \quad f(x_1) < f(x_2).$$

A function f is said to be **decreasing** on an interval if for any two numbers x_1 and x_2 in the interval,

$$x_1 < x_2 \quad \text{implies} \quad f(x_1) > f(x_2).$$

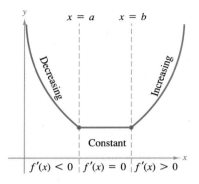

FIGURE 4.13

From this definition we can see that a function is increasing if its graph moves up as x moves to the right and a function is decreasing if its graph moves down as x moves to the right. For example, the function in Figure 4.13 is decreasing on the interval $(-\infty, a)$, is constant on the interval (a, b), and is increasing on the interval (b, ∞).

The derivative is useful in determining whether a function is increasing or decreasing on an interval. Specifically, as Figure 4.13 shows, a positive derivative implies that the graph slopes upward and the function is increasing. Similarly, a negative derivative implies that the function is decreasing. Finally, a zero derivative on an entire interval implies that the function is constant on the interval.

THEOREM 4.5
TEST FOR INCREASING OR DECREASING FUNCTIONS

Let f be a function that is differentiable on the interval (a, b).

1. If $f'(x) > 0$ for all x in (a, b), then f is increasing on (a, b).
2. If $f'(x) < 0$ for all x in (a, b), then f is decreasing on (a, b).
3. If $f'(x) = 0$ for all x in (a, b), then f is constant on (a, b).

PROOF To prove the first case, we assume that $f'(x) > 0$ for all x in the interval (a, b) and let $x_1 < x_2$ be any two points in the interval. By the Mean Value Theorem, we know there exists a number c such that $x_1 < c < x_2$, and

$$f'(c) = \frac{f(x_2) - f(x_1)}{x_2 - x_1}.$$

Since $f'(c) > 0$ and $x_2 - x_1 > 0$, we know that $f(x_2) - f(x_1) > 0$, which implies that

$$f(x_1) < f(x_2).$$

Thus, f is increasing on the interval. The second case has a similar proof (see Exercise 49), and the third case was given as Exercise 29 in Section 4.2.

To apply Theorem 4.5, note that for a continuous function on an interval (a, b), $f'(x)$ can change sign only at its critical numbers. This suggests the following guidelines. These guidelines are also valid if the interval (a, b) is replaced by an interval of the form $(-\infty, b)$, (a, ∞), or $(-\infty, \infty)$.

GUIDELINES FOR FINDING INTERVALS ON WHICH A FUNCTION IS INCREASING OR DECREASING

Let f be continuous on the interval (a, b). To find the open intervals on which f is increasing or decreasing, we suggest the following steps.

1. Locate the critical numbers of f in (a, b), and use these numbers to determine test intervals.
2. Determine the sign of $f'(x)$ at one value in each of the test intervals.
3. Use Theorem 4.5 to decide whether f is increasing or decreasing on each interval.

REMARK These guidelines require that f be *continuous* on the interval (a, b). They must be modified for functions with points of discontinuity. (See Example 4.)

EXAMPLE 1 Determining intervals on which *f* is increasing or decreasing

Find the open intervals on which

$$f(x) = x^3 - \frac{3}{2}x^2$$

is increasing or decreasing.

SOLUTION

Note that *f* is continuous on the entire real line. To determine the critical numbers of *f*, we set $f'(x)$ equal to zero.

$$f'(x) = 3x^2 - 3x = 0 \qquad \text{Let } f'(x) = 0$$
$$3(x)(x - 1) = 0 \qquad \text{Factor}$$
$$x = 0, 1 \qquad \text{Critical numbers}$$

Since there are no points for which f' is undefined, we conclude that $x = 0$ and $x = 1$ are the only critical numbers. Table 4.3 summarizes the testing of the three intervals determined by these critical numbers.

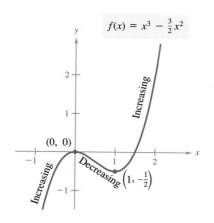

$$f(x) = x^3 - \frac{3}{2}x^2$$

Increasing

(0, 0)

Increasing

Decreasing $\left(1, -\frac{1}{2}\right)$

FIGURE 4.14

TABLE 4.3

Interval	$-\infty < x < 0$	$0 < x < 1$	$1 < x < \infty$
Test value	$x = -1$	$x = \frac{1}{2}$	$x = 2$
Sign of $f'(x)$	$f'(-1) = 6 > 0$	$f'\left(\frac{1}{2}\right) = -\frac{3}{4} < 0$	$f'(2) = 6 > 0$
Conclusion	increasing	decreasing	increasing

The graph of *f* is shown in Figure 4.14. Note that the test values in Table 4.3 were chosen for convenience; other points could have been used.

The First Derivative Test

Once we have determined the intervals on which a function is increasing or decreasing, we can locate the relative extrema of the function easily. For instance, in Figure 4.14 (Example 1), the function $f(x) = x^3 - \frac{3}{2}x^2$ has a relative maximum at the point (0, 0) because *f* is increasing immediately to the left of $x = 0$ and decreasing immediately to the right of $x = 0$. Similarly, *f* has a relative minimum at the point $\left(1, -\frac{1}{2}\right)$ because *f* is decreasing immediately to the left of $x = 1$ and increasing immediately to the right of $x = 1$. The following theorem, called the First Derivative Test, makes this more explicit.

THEOREM 4.6
THE FIRST DERIVATIVE TEST

Let c be a critical number of a function f that is continuous on an open interval I containing c. If f is differentiable on the interval, except possibly at c, then $f(c)$ can be classified as follows.

1. If f' changes from negative to positive at c, then $f(c)$ is a **relative minimum** of f.
2. If f' changes from positive to negative at c, then $f(c)$ is a **relative maximum** of f.
3. If f' does not change signs at c, then $f(c)$ is neither a relative minimum nor a relative maximum.

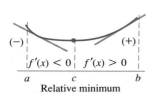

$f'(x) < 0 \quad f'(x) > 0$

$a \qquad c \qquad b$
Relative minimum

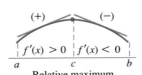

$(+) \qquad (-)$

$f'(x) > 0 \quad f'(x) < 0$

$a \qquad c \qquad b$
Relative maximum

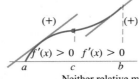

$(+) \qquad (+)$

$f'(x) > 0 \quad f'(x) > 0$

$a \qquad c \qquad b$

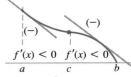

$(-) \qquad (-)$

$f'(x) < 0 \quad f'(x) < 0$

$a \qquad c \qquad b$

Neither relative minimum nor maximum

PROOF We prove the first case and leave the other two cases as exercises (see Exercise 50). Assume that f' changes from negative to positive at c. Then there exist a and b in I such that

$$f'(x) < 0 \text{ for all } x \text{ in } (a, c)$$

and

$$f'(x) > 0 \text{ for all } x \text{ in } (c, b).$$

By Theorem 4.5, f is decreasing on (a, c) and increasing on (c, b). Therefore, $f(c)$ is a minimum of f on the open interval (a, b) and, consequently, a relative minimum of f.

To apply the First Derivative Test, we suggest the same tabular format used in Example 1.

EXAMPLE 2 Applying the First Derivative Test

Use the First Derivative Test to find all relative maxima and minima for the function given by

$$f(x) = 2x^3 - 3x^2 - 36x + 14.$$

SOLUTION

$$f'(x) = 6x^2 - 6x - 36 = 0 \qquad \text{Let } f'(x) = 0$$
$$6(x^2 - x - 6) = 0$$
$$6(x - 3)(x + 2) = 0$$
$$x = -2, 3 \qquad \text{Critical numbers}$$

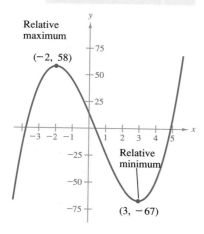

$f(x) = 2x^3 - 3x^2 - 36x + 14$

Relative
maximum

$(-2, 58)$

Relative
minimum

$(3, -67)$

FIGURE 4.15

Table 4.4 shows a practical format for applying the First Derivative Test.

TABLE 4.4

Interval	$-\infty < x < -2$	$-2 < x < 3$	$3 < x < \infty$
Test value	$x = -3$	$x = 0$	$x = 4$
Sign of $f'(x)$	$f'(-3) > 0$	$f'(0) < 0$	$f'(4) > 0$
Conclusion	increasing	decreasing	increasing

From Table 4.4, we conclude that a relative maximum occurs at $x = -2$ and a relative minimum occurs at $x = 3$. A graph of the function is shown in Figure 4.15. ▬

Note that in Examples 1 and 2 the given functions are differentiable on the entire real line. For such functions, the only critical numbers are those for which $f'(x) = 0$. In Example 3, we look at a function that has two types of critical numbers—those for which $f'(x) = 0$ and those for which f' is undefined.

EXAMPLE 3   Applying the First Derivative Test

Find the relative extrema of $f(x) = (x^2 - 4)^{2/3}$.

SOLUTION

We begin by noting that f is continuous on the entire real line. The derivative of f,

$$f'(x) = \frac{2}{3}(x^2 - 4)^{-1/3}(2x) = \frac{4x}{3(x^2 - 4)^{1/3}}$$

is zero when $x = 0$ and undefined when $x = \pm 2$. Thus, the critical numbers are $x = -2$, $x = 0$, and $x = 2$. Table 4.5 summarizes the testing of the four test intervals determined by these three critical numbers.

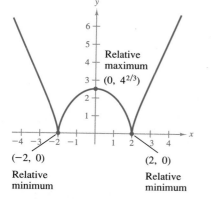

$f(x) = (x^2 - 4)^{2/3}$

Relative
maximum
$(0, 4^{2/3})$

$(-2, 0)$
Relative
minimum

$(2, 0)$
Relative
minimum

FIGURE 4.16

TABLE 4.5

Interval	$-\infty < x < -2$	$-2 < x < 0$	$0 < x < 2$	$2 < x < \infty$
Test value	$x = -3$	$x = -1$	$x = 1$	$x = 3$
Sign of $f'(x)$	$f'(-3) < 0$	$f'(-1) > 0$	$f'(1) < 0$	$f'(3) > 0$
Conclusion	decreasing	increasing	decreasing	increasing

Finally, applying the First Derivative Test, we conclude that f has a relative minimum at the point $(-2, 0)$, a relative maximum at the point $(0, 4^{2/3})$, and another relative minimum at the point $(2, 0)$, as shown in Figure 4.16. ▬

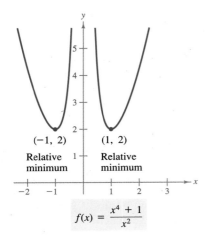

$$f(x) = \frac{x^4 + 1}{x^2}$$

(−1, 2) Relative minimum

(1, 2) Relative minimum

FIGURE 4.17

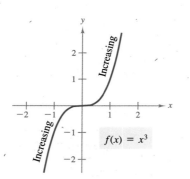

$$f(x) = x^3$$

Increasing

Increasing

Strictly monotonic function

FIGURE 4.18

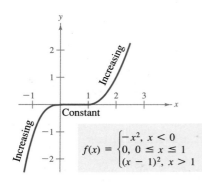

Increasing

Increasing

Constant

$$f(x) = \begin{cases} -x^2, & x < 0 \\ 0, & 0 \le x \le 1 \\ (x - 1)^2, & x > 1 \end{cases}$$

Not a strictly monotonic function

FIGURE 4.19

If a function has points of discontinuity, then these points should be used along with the critical numbers to determine the test intervals. This is demonstrated in Example 4.

EXAMPLE 4   The First Derivative Test and points of discontinuity

Find the relative extrema of $f(x) = (x^4 + 1)/x^2$.

SOLUTION

Using the Quotient Rule, we find the derivative of f to be

$$f'(x) = \frac{2(x^4 - 1)}{x^3} = \frac{2(x^2 + 1)(x - 1)(x + 1)}{x^3}.$$

Since $f'(x)$ is zero at $x = \pm 1$ and f is discontinuous at $x = 0$, we use these three x-values to determine the test intervals.

$$x = -1 \quad \text{and} \quad x = 1 \qquad \text{Critical numbers}$$
$$x = 0 \qquad\qquad\qquad\qquad \text{Point of discontinuity}$$

Table 4.6 summarizes the testing of the four intervals determined by these three x-values.

TABLE 4.6

Interval	$-\infty < x < -1$	$-1 < x < 0$	$0 < x < 1$	$1 < x < \infty$
Test value	$x = -2$	$x = -\dfrac{1}{2}$	$x = \dfrac{1}{2}$	$x = 2$
Sign of $f'(x)$	$f'(-2) < 0$	$f'\left(-\dfrac{1}{2}\right) > 0$	$f'\left(\dfrac{1}{2}\right) < 0$	$f'(2) > 0$
Conclusion	decreasing	increasing	decreasing	increasing

Finally, applying the First Derivative Test, we conclude that f has one relative minimum at the point $(-1, 2)$ and another at the point $(1, 2)$, as shown in Figure 4.17.

Strictly monotonic functions

A function is called **strictly monotonic** on an interval if it is either increasing on the entire interval or decreasing on the entire interval. For instance, the function $f(x) = x^3$ is strictly monotonic on the entire real line because it is increasing on the entire real line, as shown in Figure 4.18. The function shown in Figure 4.19 is not *strictly* monotonic on the entire real line because it is constant on the interval [0, 1].

Strictly monotonic functions play a special role in calculus. (We will say more about this in Chapter 7.)

EXERCISES for Section 4.3

In Exercises 1–6, identify the open intervals on which the function is increasing or decreasing.

1. $f(x) = x^2 - 6x + 8$

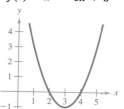

2. $y = -(x + 1)^2$

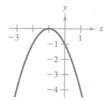

3. $y = \dfrac{x^3}{4} - 3x$

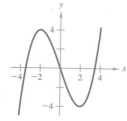

4. $f(x) = x^4 - 2x^2$

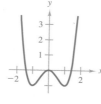

5. $f(x) = \dfrac{1}{x^2}$

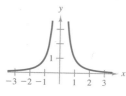

6. $y = \dfrac{x^2}{x + 1}$

In Exercises 7–26, find the critical numbers of f (if any), find the open intervals on which f is increasing or decreasing, and locate all relative extrema.

7. $f(x) = -2x^2 + 4x + 3$ **8.** $f(x) = x^2 + 8x + 10$

9. $f(x) = x^2 - 6x$ **10.** $f(x) = (x - 1)^2(x + 2)$

11. $f(x) = 2x^3 + 3x^2 - 12x$

12. $f(x) = (x - 3)^3$

13. $f(x) = x^3 - 6x^2 + 15$ **14.** $f(x) = x^4 - 2x^3$

15. $f(x) = (x - 1)^{2/3}$ **16.** $f(x) = (x - 1)^{1/3}$

17. $f(x) = x^{1/3} + 1$ **18.** $f(x) = x^{2/3}(x - 5)$

19. $f(x) = x + \dfrac{1}{x}$ **20.** $f(x) = \dfrac{x}{x + 1}$

21. $f(x) = \dfrac{x^2}{x^2 - 9}$ **22.** $f(x) = \dfrac{x + 3}{x^2}$

23. $f(x) = \dfrac{x^5 - 5x}{5}$ **24.** $f(x) = x^4 - 32x + 4$

25. $f(x) = \dfrac{x^2 - 2x + 1}{x + 1}$ **26.** $f(x) = \dfrac{x^2 - 3x - 4}{x - 2}$

In Exercises 27 and 28, determine whether the given function is strictly monotonic on the indicated interval.

27. $f(x) = x^2$
   (a) $(-\infty, \infty)$ (b) $(-\infty, 0)$ (c) $(0, \infty)$

28. $f(x) = x^3 - x$
   (a) $(-1, 0)$ (b) $\left(-1, -\dfrac{1}{2}\right)$ (c) $(-1, 1)$

29. The height (in feet) of a ball at time t (in seconds) is given by the position function

$$s(t) = 96t - 16t^2.$$

Find the open interval on which the ball is moving up and the open interval on which it is moving down. What is the maximum height of the ball?

30. Repeat Exercise 29 using the position function

$$s(t) = -16t^2 + 64t.$$

31. Coughing forces the trachea (windpipe) to contract, which affects the velocity v of the air passing through the trachea. Suppose the velocity of the air during coughing is

$$v = k(R - r)r^2$$

where k is a constant, R is the normal radius of the trachea, and r is the radius during coughing. What radius will produce the maximum air velocity?

32. The concentration C of a certain chemical in the bloodstream t hours after injection into muscle tissue is given by

$$C = \dfrac{3t}{27 + t^3}.$$

When is the concentration greatest?

33. After a drug is administered to a patient, the drug concentration in the patient's bloodstream over a two-hour period is given by

$$C = 0.29483t + 0.04253t^2 - 0.00035t^3$$

where C is measured in milligrams and t is the time in minutes. Find the open interval on which C is increasing or decreasing.

34. A fast-food restaurant sells x hamburgers to make a profit P given by

$$P = 2.44x - \dfrac{x^2}{20,000} - 5000, \quad 0 \le x \le 35,000.$$

Find the open interval on which P is increasing or decreasing.

35. After birth, an infant normally will lose weight for a few days and then start gaining. A model for the average weight W of infants over the first two weeks following birth is

$$W = 0.033t^2 - 0.3974t + 7.3032.$$

Find the open intervals on which W is increasing or decreasing.

36. The electric power P in watts in a direct-current circuit with two resistors R_1 and R_2 connected in series is

$$P = \frac{vR_1R_2}{(R_1 + R_2)^2}$$

where v is the voltage. If v and R_1 are held constant, what resistance R_2 produces maximum power?

37. The resistance R of a certain type of resistor is given by

$$R = \sqrt{0.001T^4 - 4T + 100}$$

where R is measured in ohms and the temperature T is measured in degrees Celsius. What temperature produces a minimum resistance for this type of resistor?

38. Consider the functions $f(x) = x$ and $g(x) = x^3$ on the interval $(0, 1)$.
  (a) Prove that $f(x) > g(x)$. [Hint: Show that $h(x) > 0$ where $h = f - g$.]
  (b) Sketch the graphs of f and g on the same set of axes.

39. Find a, b, c, and d so that the function given by

$$f(x) = ax^3 + bx^2 + cx + d$$

has a relative minimum at $(0, 0)$ and a relative maximum at $(2, 2)$.

40. Find a, b, and c so that the function given by

$$f(x) = ax^2 + bx + c$$

has a relative maximum at $(5, 20)$ and passes through the point $(2, 10)$.

In Exercises 41–46, assume that f is differentiable for all x. The sign of f' is as follows.

$f'(x) > 0$ on $(-\infty, -4)$

$f'(x) < 0$ on $(-4, 6)$

$f'(x) > 0$ on $(6, \infty)$

In each exercise, supply the appropriate inequality for the indicated value of c.

Function	Sign of $g'(c)$
41. $g(x) = f(x) + 5$	$g'(0)$ ___ 0
42. $g(x) = 3f(x) - 3$	$g'(-5)$ ___ 0
43. $g(x) = -f(x)$	$g'(-6)$ ___ 0
44. $g(x) = -f(x)$	$g'(0)$ ___ 0
45. $g(x) = f(x - 10)$	$g'(0)$ ___ 0
46. $g(x) = f(x - 10)$	$g'(8)$ ___ 0

In Exercises 47 and 48, use a computer or graphics calculator (a) to sketch the graph of f and f' on the same coordinate axes over the specified interval, (b) to find the critical numbers of f, and (c) to find the interval(s) on which f' is positive and the interval(s) on which it is negative. Note the behavior of f in relation to the sign of f'.

Function	Interval
47. $f(x) = 2x\sqrt{9 - x^2}$	$[-3, 3]$
48. $f(x) = 10(5 - \sqrt{x^2 - 3x + 16})$	$[0, 5]$

49. Prove the second case of Theorem 4.5.
50. Prove the second and third cases of Theorem 4.6.

4.4 Concavity and the Second Derivative Test

Concavity ▪ Points of inflection ▪ The Second Derivative Test

We have already seen that locating the intervals in which a function f increases or decreases helps to determine its graph. In this section we show that, by locating the intervals in which f' increases or decreases, we can determine where the graph of f is *curving upward* or *curving downward*. We refer to this notion of curving upward or downward as **concavity.**

DEFINITION OF CONCAVITY	Let f be differentiable on an open interval. We say that the graph of f is **concave upward** if f' is increasing on the interval and **concave downward** if f' is decreasing on the interval.

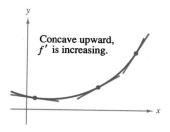

(a) The graph of f lies above its tangent lines.

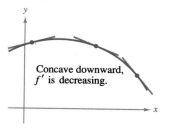

(b) The graph of f lies below its tangent lines.

FIGURE 4.20

A comparable definition of concavity is given in the two statements that follow. (See Appendix A for a proof of the equivalence of the two definitions.)

1. Let f be differentiable at c. The graph of f is **concave upward** at $(c, f(c))$ if the graph of f lies *above* the tangent line at $(c, f(c))$ on some open interval containing c. (See Figure 4.20(a).)
2. Let f be differentiable at c. The graph of f is **concave downward** at $(c, f(c))$ if the graph of f lies *below* the tangent line at $(c, f(c))$ on some open interval containing c. (See Figure 4.20(b).)

To find the open intervals on which the graph of a function f is concave upward or downward, we need to find the intervals on which f' is increasing or decreasing. For instance, the graph of $f(x) = \frac{1}{3}x^3 - x$ is concave downward on the open interval $(-\infty, 0)$ because f' is decreasing there. (See Figure 4.21.) Similarly, the graph of f is concave upward on the interval $(0, \infty)$ because f' is increasing on $(0, \infty)$.

The following theorem shows how to use the *second* derivative of a function f to determine intervals on which the graph of f is concave upward or downward. A proof of this theorem follows directly from Theorem 4.5 and the definition of concavity.

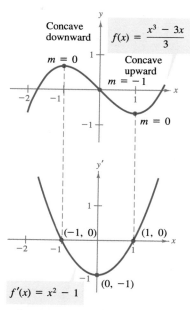

f' is decreasing. f' is increasing.

FIGURE 4.21

Let f be a function whose second derivative exists on an open interval I.

1. If $f''(x) > 0$ for all x in I, then the graph of f is concave upward.
2. If $f''(x) < 0$ for all x in I, then the graph of f is concave downward.

REMARK   A third conclusion to Theorem 4.7 is that if $f''(x) = 0$ for all x in I, then f is linear. Note, however, that we do not define concavity for straight lines. In other words, a straight line is neither concave upward nor concave downward.

We suggest the following guidelines for applying Theorem 4.7. First, locate the x-values at which $f''(x) = 0$ or f'' is undefined. Second, use these x-values to determine test intervals. Finally, test the sign of $f''(x)$ in each of the test intervals. We illustrate the procedure in Example 1.

EXAMPLE 1   Determining concavity

Determine the open intervals on which the graph of $f(x) = 6(x^2 + 3)^{-1}$ is concave upward or downward.

SOLUTION

We begin by observing that f is continuous on the entire real line. Next, we find the second derivative of f.

$$f'(x) = (-6)(2x)(x^2 + 3)^{-2} = \frac{-12x}{(x^2 + 3)^2}$$

$$f''(x) = \frac{(x^2 + 3)^2(-12) - (-12x)(2)(2x)(x^2 + 3)}{(x^2 + 3)^4} = \frac{36(x^2 - 1)}{(x^2 + 3)^3}$$

Since $f''(x) = 0$ when $x = \pm 1$ and f'' is defined on the entire real line, we test f'' in the intervals $(-\infty, -1)$, $(-1, 1)$, and $(1, \infty)$. The results are shown in Table 4.7 and Figure 4.22.

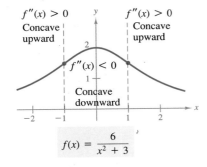

$$f(x) = \frac{6}{x^2 + 3}$$

FIGURE 4.22

TABLE 4.7

Interval	$-\infty < x < -1$	$-1 < x < 1$	$1 < x < \infty$
Test value	$x = -2$	$x = 0$	$x = 2$
Sign of $f''(x)$	$f''(-2) > 0$	$f''(0) < 0$	$f''(2) > 0$
Conclusion	Concave upward	Concave downward	Concave upward

The function given in Example 1 is continuous on the entire real line. If a function has one or more points of discontinuity, then these points should be used (along with the points at which $f''(x)$ is zero or undefined) to form the test intervals. This is illustrated in Example 2.

EXAMPLE 2 Determining concavity for a discontinuous function

Determine the open intervals in which the graph of

$$f(x) = \frac{x^2 + 1}{x^2 - 4}$$

is concave upward or downward.

SOLUTION

$$f'(x) = \frac{(x^2 - 4)(2x) - (x^2 + 1)(2x)}{(x^2 - 4)^2} = \frac{-10x}{(x^2 - 4)^2}$$

$$f''(x) = -10\frac{(x^2 - 4)^2(1) - (x)(2)(2x)(x^2 - 4)}{(x^2 - 4)^4} = \frac{10(3x^2 + 4)}{(x^2 - 4)^3}$$

There are no points at which f'' is zero, but at $x = \pm2$ the function f is discontinuous, so we test for concavity in the intervals $(-\infty, -2)$, $(-2, 2)$, and $(2, \infty)$, as shown in Table 4.8. The graph of f is shown in Figure 4.23.

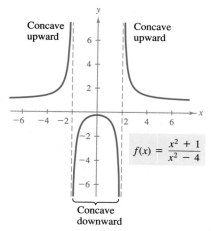

Concave upward   Concave upward

$$f(x) = \frac{x^2 + 1}{x^2 - 4}$$

Concave downward

FIGURE 4.23

TABLE 4.8

Interval	$-\infty < x < -2$	$-2 < x < 2$	$2 < x < \infty$
Test value	$x = -3$	$x = 0$	$x = 3$
Sign of $f''(x)$	$f''(-3) > 0$	$f''(0) < 0$	$f''(3) > 0$
Conclusion	Concave upward	Concave downward	Concave upward

Points of inflection

The graph in Figure 4.22 has two points at which the concavity changes. If the tangent line to the graph exists at such a point, the point is called a **point of inflection.** Three types of points of inflection are shown in Figure 4.24.

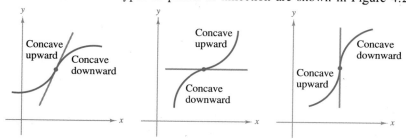

FIGURE 4.24

The graph *crosses* its tangent line at a point of inflection.

DEFINITION OF POINT OF INFLECTION

Let f be a function whose graph has a tangent line at $(c, f(c))$. The point $(c, f(c))$ is called a **point of inflection** if the concavity of f changes from upward to downward (or vice versa).

REMARK   Note in Figure 4.24 that the graph crosses its tangent line at a point of inflection.

Since a point of inflection occurs where the concavity of a graph changes, it must be true that the sign of f'' changes at such points. Thus, to locate possible points of inflection, we need only determine the values of x for which $f''(x) = 0$ or for which f'' is undefined. This is similar to the procedure for locating relative extrema of f.

THEOREM 4.8 POINTS OF INFLECTION	If $(c, f(c))$ is a point of inflection of the graph of f, then either $f''(c) = 0$ or f'' is undefined at $x = c$.

EXAMPLE 3   Finding points of inflection

Determine the points of inflection and discuss the concavity of the graph of $f(x) = x^4 - 4x^3$.

SOLUTION

Differentiating twice produces

$$f'(x) = 4x^3 - 12x^2$$
$$f'' = 12x^2 - 24x = 12x(x - 2).$$

Possible points of inflection occur at $x = 0$ and $x = 2$. By testing the intervals determined by these x-values, we conclude that they both yield points of inflection. A summary of this testing is shown in Table 4.9, and the graph of f is shown in Figure 4.25.

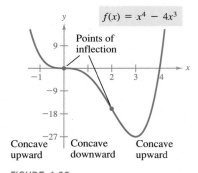

FIGURE 4.25

TABLE 4.9

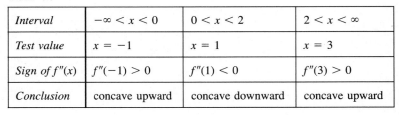

Interval	$-\infty < x < 0$	$0 < x < 2$	$2 < x < \infty$
Test value	$x = -1$	$x = 1$	$x = 3$
Sign of $f''(x)$	$f''(-1) > 0$	$f''(1) < 0$	$f''(3) > 0$
Conclusion	concave upward	concave downward	concave upward

It is possible for the second derivative to be zero at a point that is *not* a point of inflection. For instance, the graph of $f(x) = x^4$ is shown in Figure 4.26. The second derivative is zero when $x = 0$, but the point $(0, 0)$ is not a point of inflection because the graph of f is concave upward to the left and to the right of $x = 0$.

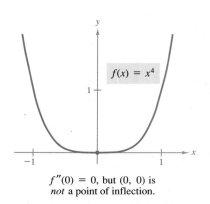

$f''(0) = 0$, but $(0, 0)$ is *not* a point of inflection.

FIGURE 4.26

The Second Derivative Test

The second derivative sometimes can be used to perform a simple test for relative maxima and minima. The test is based on the fact that if f is a function such that $f'(c) = 0$ and there exists an open interval containing c on which the graph of f is concave upward, then $f(c)$ must be a relative minimum of f. Similarly, if f is a function such that $f'(c) = 0$ and there exists an open interval containing c on which the graph of f is concave downward, then $f(c)$ must be a relative maximum of f. (See Figure 4.27.)

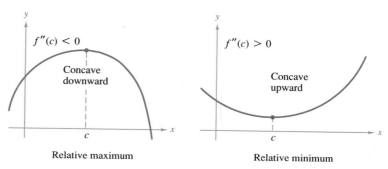

FIGURE 4.27

THEOREM 4.9 **SECOND DERIVATIVE TEST**	Let f be a function such that $f'(c) = 0$ and the second derivative of f exists on an open interval containing c. 1. If $f''(c) > 0$, then $f(c)$ is a relative minimum. 2. If $f''(c) < 0$, then $f(c)$ is a relative maximum. 3. If $f''(c) = 0$, then the test fails.

PROOF   This theorem follows from Theorem 4.7. We outline a proof of the first case. (The rest of the proof is left to you.) If $f''(c) > 0$, then f is concave upward in some interval I containing c. This implies that the graph of f lies above its tangent lines in I. Since $f'(c) = 0$, the tangent line must be horizontal at $(c, f(c))$. Thus, we can conclude that $f(c)$ is a minimum of f in the interval I, and consequently $f(c)$ must be a relative minimum of f.

REMARK   Be sure to understand that if $f''(c) = 0$, the Second Derivative Test does not apply. In such cases we can use the First Derivative Test.

EXAMPLE 4   Using the Second Derivative Test

Find the relative extrema for $f(x) = -3x^5 + 5x^3$.

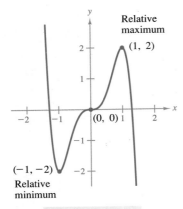

$f(x) = -3x^5 + 5x^3$

FIGURE 4.28

SOLUTION

We begin by finding the critical numbers of f.

$$f'(x) = -15x^4 + 15x^2 = 15x^2(1 - x^2) = 0$$

$$x = -1, 0, 1 \qquad \text{Critical numbers}$$

Using $f''(x) = 30(-2x^3 + x)$, we apply the Second Derivative Test as follows.

Point	Sign of f''		Conclusion
$(-1, -2)$	$f''(-1) = 30 > 0$	⟹	Relative minimum
$(1, 2)$	$f''(1) = -30 < 0$	⟹	Relative maximum
$(0, 0)$	$f''(0) = 0$	⟹	Test fails

Since the Second Derivative Test fails at $(0, 0)$, we use the First Derivative Test and observe that f increases to the left and right of $x = 0$. Thus, $(0, 0)$ is neither a relative minimum nor a relative maximum (even though the graph has a horizontal tangent line at this point). The graph of f is shown in Figure 4.28. ▭

EXERCISES for Section 4.4

In Exercises 1–6, find the open intervals on which the graph of the given function is concave upward and those on which it is concave downward.

1. $y = x^2 - x - 2$

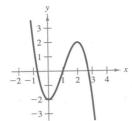

2. $f(x) = \dfrac{24}{x^2 + 12}$

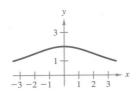

3. $y = -x^3 + 3x^2 - 2$

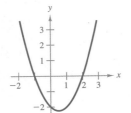

4. $f(x) = \dfrac{x^2 - 1}{2x + 1}$

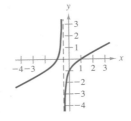

5. $f(x) = \dfrac{x^2 + 1}{x^2 - 1}$

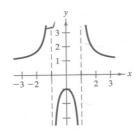

6. $y = \dfrac{-3x^5 + 40x^3 + 135x}{270}$

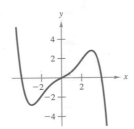

In Exercises 7–18, identify all relative extrema. Use the Second Derivative Test where applicable.

7. $f(x) = 6x - x^2$

8. $f(x) = x^2 + 3x - 8$

9. $f(x) = (x - 5)^2$

10. $f(x) = -(x - 5)^2$

11. $f(x) = x^3 - 3x^2 + 3$ **12.** $f(x) = 5 + 3x^2 - x^3$
13. $f(x) = x^4 - 4x^3 + 2$
14. $f(x) = x^3 - 9x^2 + 27x - 26$
15. $f(x) = x^{2/3} - 3$ **16.** $f(x) = \sqrt{x^2 + 1}$
17. $f(x) = x + \dfrac{4}{x}$ **18.** $f(x) = \dfrac{x}{x - 1}$

In Exercises 19–34, sketch the graph of the given function and identify all relative extrema and points of inflection.

19. $f(x) = x^3 - 12x$ **20.** $f(x) = x^3 + 1$
21. $f(x) = x^3 - 6x^2 + 12x - 8$
22. $f(x) = 2x^3 - 3x^2 - 12x + 8$
23. $f(x) = \dfrac{1}{4}x^4 - 2x^2$ **24.** $f(x) = 2x^4 - 8x + 3$
25. $f(x) = x(x - 4)^3$ **26.** $f(x) = x^3(x - 4)$
27. $f(x) = x^2 + \dfrac{1}{x^2}$ **28.** $f(x) = \dfrac{x^2}{x^2 - 1}$
29. $f(x) = x\sqrt{x + 3}$ **30.** $f(x) = x\sqrt{x + 1}$
31. $f(x) = \dfrac{x}{x^2 - 4}$ **32.** $f(x) = \dfrac{1}{x^2 - x - 2}$
33. $f(x) = \dfrac{x - 2}{x^2 - 4x + 3}$ **34.** $f(x) = \dfrac{x + 1}{x^2 + x + 1}$

35. Sketch the graph of a function f having the following characteristics.
(a) $f(2) = f(4) = 0$ (b) $f(0) = f(2) = 0$
$f'(x) < 0$ if $x < 3$ $f'(x) > 0$ if $x < 1$
$f'(3)$ is undefined $f'(1) = 0$
$f'(x) > 0$ if $x > 3$ $f'(x) < 0$ if $x > 1$
$f''(x) < 0$, $x \ne 3$ $f''(x) < 0$
36. Sketch the graph of a function f having the following characteristics.
(a) $f(2) = f(4) = 0$ (b) $f(0) = f(2) = 0$
$f'(x) > 0$ if $x < 3$ $f'(x) < 0$ if $x < 1$
$f'(3)$ is undefined $f'(1) = 0$
$f'(x) < 0$ if $x > 3$ $f'(x) > 0$ if $x > 1$
$f''(x) > 0$, $x \ne 3$ $f''(x) > 0$

In Exercises 37–40, trace the given graph of f. On the same set of axes sketch the graph of f' and f''.

37.
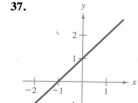

38.
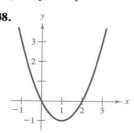

39.
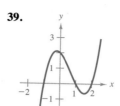

40.
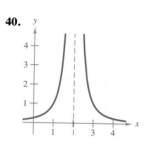

41. Find the inflection points (if any) and sketch the graph of $f(x) = (x - c)^n$ for $n = 1, 2, 3,$ and 4. What conclusion can be made about the value of n and the existence of an inflection point on the graph of f?
42. (a) Sketch the graph of $f(x) = \sqrt[3]{x}$ and identify the inflection point.
(b) Does $f''(x)$ exist at the inflection point?
43. Find a cubic polynomial function that has a relative maximum at $(3, 3)$, a relative minimum at $(5, 1)$, and a point of inflection at $(4, 2)$.
44. Find a cubic polynomial function that has a relative maximum at $(2, 4)$, a relative minimum at $(4, 2)$, and a point of inflection at $(3, 3)$.
45. Show that the point of inflection of

$$f(x) = x(x - 6)^2$$

lies midway between the relative extrema of f.
46. Prove that a cubic function with three distinct real zeros has a point of inflection whose x-coordinate is the average of the three zeros.
47. The deflection D of a particular beam of length L is given by

$$D = 2x^4 - 5Lx^3 + 3L^2x^2$$

where x is the distance from one end of the beam. Find the value of x that yields the maximum deflection.
48. The equation

$$E = \frac{kqx}{(x^2 + a^2)^{3/2}}$$

gives the electric field intensity on the axis of a uniformly charged ring, where q is the total charge on the ring, k is a constant, and a is the radius of the ring. At what value of x is E maximum?
49. A manufacturer has determined that the total cost C of operating a certain facility is given by

$$C = 0.5x^2 + 15x + 5000$$

where x is the number of units produced. At what level of production will the average cost per unit be minimum? (The average cost per unit is given by C/x.)

50. The total cost C for ordering and storing x units is

$$C = 2x + \frac{300,000}{x}.$$

What order size will produce a minimum cost?

In Exercises 51 and 52, use a computer or graphics calculator to (a) sketch the graphs of f, f', and f'' on the same coordinate axes over the specified interval, (b) find any critical numbers and the coordinates of the relative extrema of f, and (c) find any points of inflection of the graph of f.

Function	Interval
51. $f(x) = \dfrac{2x}{x^2 - x + 1}$	$[-2, 2]$
52. $f(x) = \dfrac{4x^2}{x^2 + 2}$	$[-3, 3]$

4.5 Limits at Infinity

Limits at infinity ▪ Horizontal asymptotes

In this section we take a further look at the behavior of a function on *infinite* intervals. Consider the graph of

$$f(x) = \frac{3x^2}{x^2 + 1}$$

as shown in Figure 4.29. Table 4.10 suggests that the value of $f(x)$ approaches 3 as x increases without bound ($x \to \infty$). Similarly, $f(x)$ approaches 3 as $x \to -\infty$. We denote these **limits at infinity** by

$$\lim_{x \to -\infty} f(x) = 3 \qquad \text{and} \qquad \lim_{x \to \infty} f(x) = 3.$$

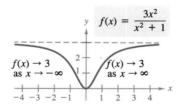

FIGURE 4.29

TABLE 4.10

	x decreases without bound								x increases without bound	
x	$-\infty \leftarrow$	-100	-10	-1	0	1	10	100	$\to \infty$
$f(x)$	3	$\leftarrow$	2.9997	2.97	1.5	0	1.5	2.97	2.9997	$\to 3$
		$f(x)$ approaches 3				$f(x)$ approaches 3				

To say that a statement is true as x increases *without bound* means that for any (large) real number M, the statement is true for *all* x in the interval $\{x\colon x > M\}$. The following definition uses this concept.

DEFINITION OF LIMITS AT INFINITY

The statement

$$\lim_{x \to \infty} f(x) = L$$

EPSILON
SMALL
NUMBER

means that for each $\varepsilon > 0$ there exists an $M > 0$ such that $|f(x) - L| < \varepsilon$ whenever $x > M$. Similarly, the statement

$$\lim_{x \to -\infty} f(x) = L$$

means that for each $\varepsilon > 0$ there exists an $N < 0$ such that $|f(x) - L| < \varepsilon$ whenever $x < N$.

REMARK   Remember that when we write

$$\lim_{x \to -\infty} f(x) = L \qquad \text{or} \qquad \lim_{x \to \infty} f(x) = L$$

we mean that the limit exists *and* the limit is equal to the number L.

The definition of a limit at infinity is illustrated graphically in Figure 4.30. In this figure, note that for a given positive number ε there exists a postive number M such that, *to the right of $x = M$*, the graph of f will lie between the horizontal lines given by $y = L \pm \varepsilon$.

In Figure 4.30, the graph of f approaches the line $y = L$ as x increases without bound. We call the line $y = L$ a **horizontal asymptote** of the graph of f.

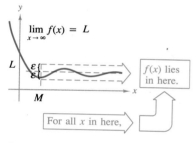

FIGURE 4.30

DEFINITION OF HORIZONTAL ASYMPTOTE

If

$$\lim_{x \to -\infty} f(x) = L \qquad \text{or} \qquad \lim_{x \to \infty} f(x) = L$$

then the line $y = L$ is called a **horizontal asymptote** of the graph of f.

REMARK   From this, it follows that the graph of a *function* of x can have at most two horizontal asymptotes—one to the right and one to the left.

Limits at infinity have many of the same properties of limits discussed in Section 2.1. For example, if $\lim\limits_{x \to \infty} f(x)$ and $\lim\limits_{x \to \infty} g(x)$ both exist, then

$$\lim_{x \to \infty} [f(x) + g(x)] = \lim_{x \to \infty} f(x) + \lim_{x \to \infty} g(x)$$

and

$$\lim_{x \to \infty} [f(x)g(x)] = \left(\lim_{x \to \infty} f(x) \right)\left(\lim_{x \to \infty} g(x) \right).$$

Similar properties hold for limits at $-\infty$.

When evaluating limits at infinity, the following theorem is helpful. (A proof of this theorem is given in Appendix A.)

THEOREM 4.10 LIMITS AT INFINITY	If r is a positive rational number and c is any real number, then
	$$\lim_{x \to \infty} \frac{c}{x^r} = 0.$$
	Furthermore, if x^r is defined when $x < 0$, then $\lim_{x \to -\infty} (c/x^r) = 0.$

EXAMPLE 1   Evaluating a limit at infinity

Find the following limit.

$$\lim_{x \to \infty} \left(5 - \frac{2}{x^2} \right)$$

SOLUTION

Using Theorem 4.10, we obtain

$$\lim_{x \to \infty} \left(5 - \frac{2}{x^2} \right) = \lim_{x \to \infty} 5 - \lim_{x \to \infty} \frac{2}{x^2} = 5 - 0 = 5.$$

EXAMPLE 2   Evaluating a limit at infinity

Find the following limit.

$$\lim_{x \to \infty} \frac{2x - 1}{x + 1}$$

SOLUTION

Note that both the numerator and the denominator approach infinity as x approaches infinity.

$$\lim_{x \to \infty} \frac{2x - 1}{x + 1} \qquad \begin{array}{l} \lim_{x \to \infty} (2x - 1) \to \infty \\[2ex] \lim_{x \to \infty} (x + 1) \to \infty \end{array}$$

To resolve this difficulty, we divide both the numerator and the denominator by x. After this division, the limit may be evaluated as follows.

$$\lim_{x \to \infty} \frac{2x - 1}{x + 1} = \lim_{x \to \infty} \frac{2 - (1/x)}{1 + (1/x)} = \frac{\displaystyle\lim_{x \to \infty} 2 - \lim_{x \to \infty} \frac{1}{x}}{\displaystyle\lim_{x \to \infty} 1 + \lim_{x \to \infty} \frac{1}{x}} = \frac{2 - 0}{1 + 0} = 2$$

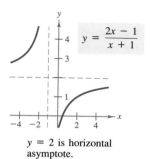

$$y = \frac{2x - 1}{x + 1}$$

$y = 2$ is horizontal asymptote.

FIGURE 4.31

Thus, the line $y = 2$ is a horizontal asymptote to the right. By taking the limit as $x \to -\infty$, we can see that $y = 2$ is also a horizontal asymptote to the left. The graph of this function is shown in Figure 4.31.

REMARK   We can test the reasonableness of the limit found in Example 2 by evaluating $f(x)$ for a few large positive values of x. For instance, $f(100) \approx 1.9703$, $f(1000) \approx 1.9970$, and $f(10000) \approx 1.9997$.

In Example 2, our first attempt to evaluate the limit resulted in the **indeterminate form** ∞/∞. We were able to resolve the difficulty by rewriting the given expression in an equivalent form. Specifically, we divided both the numerator and the denominator by x. In general, we suggest dividing by the highest power of x in the *denominator*. This is illustrated in the following example.

EXAMPLE 3   A comparison of three rational functions

Find the following limits.

(a) $\displaystyle\lim_{x \to \infty} \frac{2x + 5}{3x^2 + 1}$ (b) $\displaystyle\lim_{x \to \infty} \frac{2x^2 + 5}{3x^2 + 1}$ (c) $\displaystyle\lim_{x \to \infty} \frac{2x^3 + 5}{3x^2 + 1}$

SOLUTION

(a) We divide both the numerator and the denominator by x^2 and obtain

$$\lim_{x \to \infty} \frac{2x + 5}{3x^2 + 1} = \lim_{x \to \infty} \frac{(2/x) + (5/x^2)}{3 + (1/x^2)} = \frac{0 + 0}{3 + 0} = \frac{0}{3} = 0.$$

(b) In this case, we also divide by x^2 and obtain

$$\lim_{x \to \infty} \frac{2x^2 + 5}{3x^2 + 1} = \lim_{x \to \infty} \frac{2 + (5/x^2)}{3 + (1/x^2)} = \frac{2 + 0}{3 + 0} = \frac{2}{3}.$$

(c) Again, we divide by x^2 and obtain

$$\lim_{x \to \infty} \frac{2x^3 + 5}{3x^2 + 1} = \lim_{x \to \infty} \frac{2x + (5/x^2)}{3 + (1/x^2)}.$$

$$\lim_{x \to \infty} \left[2x + \frac{5}{x^2} \right] \to \infty$$

$$\lim_{x \to \infty} \left[3 + \frac{1}{x^2} \right] \to 3$$

In this case, we conclude that the limit *does not exist* because the numerator increases without bound while the modified denominator approaches 3.

It is instructive to compare the three rational functions in Example 3. In part (a) the degree of the numerator is *less* than the degree of the denominator and the limit of the rational function is zero. In part (b) the degrees of the

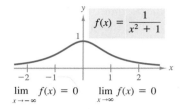

$$f(x) = \frac{1}{x^2 + 1}$$

$$\lim_{x \to -\infty} f(x) = 0 \qquad \lim_{x \to \infty} f(x) = 0$$

FIGURE 4.32

Maria Agnesi

numerator and the denominator are *equal* and the limit is simply the ratio of the two leading coefficients 2 and 3. Finally, in part (c) the degree of the numerator is *greater* than that of the denominator and the limit does not exist. This seems reasonable when we realize that for large values of x the highest-powered term of the rational function is the most "influential" in determining the limit. For instance, the limit as x approaches infinity of the function given by

$$f(x) = \frac{1}{x^2 + 1}$$

is zero since the denominator overpowers the numerator as x increases or decreases without bound, as shown in Figure 4.32.

The function shown in Figure 4.32 is a special case of a type of curve studied by the Italian mathematician Maria Gaetana Agnesi (1717–1783). The general form of this function is $f(x) = a^3/(x^2 + a^2)$ and, through a mistranslation of the Italian word *vertéré*, the curve has come to be known as the witch of Agnesi. Agnesi's work with this curve first appeared in a comprehensive text on calculus that was published in 1748.

In Figure 4.32, we can see that the function $f(x) = 1/(x^2 + 1)$ approaches the same horizontal asymptote to the right and to the left. That is,

$$\lim_{x \to -\infty} f(x) = 0 = \lim_{x \to \infty} f(x).$$

This is always the case with rational functions. Functions that are not rational, however, may approach different horizontal asymptotes to the right and to the left, as shown in Example 4.

EXAMPLE 4   A function with two horizontal asymptotes

Determine the following limits.

(a) $\displaystyle\lim_{x \to \infty} \frac{3x - 2}{\sqrt{2x^2 + 1}}$ (b) $\displaystyle\lim_{x \to -\infty} \frac{3x - 2}{\sqrt{2x^2 + 1}}$

SOLUTION

(a) For $x > 0$, we have $x = \sqrt{x^2}$. Thus, dividing both the numerator and the denominator by x produces

$$\frac{3x - 2}{\sqrt{2x^2 + 1}} = \frac{\dfrac{3x - 2}{x}}{\dfrac{\sqrt{2x^2 + 1}}{\sqrt{x^2}}} = \frac{3 - \dfrac{2}{x}}{\sqrt{\dfrac{2x^2 + 1}{x^2}}} = \frac{3 - \dfrac{2}{x}}{\sqrt{2 + \dfrac{1}{x^2}}}$$

and we can take the limit as follows.

$$\lim_{x \to \infty} \frac{3x - 2}{\sqrt{2x^2 + 1}} = \lim_{x \to \infty} \frac{3 - \dfrac{2}{x}}{\sqrt{2 + \dfrac{1}{x^2}}} = \frac{3 - 0}{\sqrt{2 + 0}} = \frac{3}{\sqrt{2}}$$

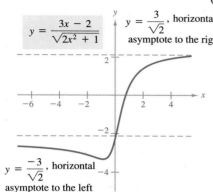

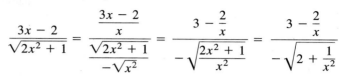

FIGURE 4.33

(b) For $x < 0$, we have $x = -\sqrt{x^2}$. Thus, dividing both the numerator and the denominator by x produces

$$\frac{3x - 2}{\sqrt{2x^2 + 1}} = \frac{\dfrac{3x - 2}{x}}{\dfrac{\sqrt{2x^2 + 1}}{-\sqrt{x^2}}} = \frac{3 - \dfrac{2}{x}}{-\sqrt{\dfrac{2x^2 + 1}{x^2}}} = \frac{3 - \dfrac{2}{x}}{-\sqrt{2 + \dfrac{1}{x^2}}}$$

and we can take the limit as follows.

$$\lim_{x \to -\infty} \frac{3x - 2}{\sqrt{2x^2 + 1}} = \lim_{x \to -\infty} \frac{3 - \dfrac{2}{x}}{-\sqrt{2 + \dfrac{1}{x^2}}} = \frac{3 - 0}{-\sqrt{2} + 0} = -\frac{3}{\sqrt{2}}$$

The graph of $f(x) = (3x - 2)/\sqrt{2x^2 + 1}$ is shown in Figure 4.33.

There are many examples of asymptotic behavior in the physical sciences. For instance, the following example describes the asymptotic recovery of oxygen in a pond.

EXAMPLE 5 An application involving oxygen levels

Suppose that $f(t)$ measures the level of oxygen in a pond, where $f(t) = 1$ is the normal (unpolluted) level and the time t is measured in weeks. When $t = 0$, organic waste is dumped into the pond, and as the waste material oxidizes, the amount of oxygen in the pond is given by

$$f(t) = \frac{t^2 - t + 1}{t^2 + 1}.$$

What percentage of the normal level of oxygen exists in the pond after 1 week? After 2 weeks? After 10 weeks? What is the limit as t approaches infinity?

SOLUTION

When $t = 1$, 2, and 10, the levels of oxygen are as follows.

$$f(1) = \frac{1^2 - 1 + 1}{1^2 + 1} = \frac{1}{2} = 50\% \qquad \text{1 week}$$

$$f(2) = \frac{2^2 - 2 + 1}{2^2 + 1} = \frac{3}{5} = 60\% \qquad \text{2 weeks}$$

$$f(10) = \frac{10^2 - 10 + 1}{10^2 + 1} = \frac{91}{101} \approx 90.1\% \qquad \text{10 weeks}$$

To take the limit as t approaches infinity, we divide the numerator and the denominator by t^2 to obtain

$$\lim_{t \to \infty} \frac{t^2 - t + 1}{t^2 + 1} = \lim_{t \to \infty} \frac{1 - (1/t) + (1/t^2)}{1 + (1/t^2)} = \frac{1 - 0 + 0}{1 + 0} = 1 = 100\%.$$

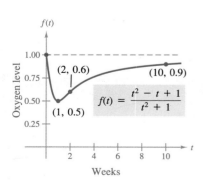

FIGURE 4.34

(See Figure 4.34.)

EXERCISES for Section 4.5

In Exercises 1–8, match the given function to one of the graphs (a)–(h), using horizontal asymptotes as an aid.

1. $f(x) = \dfrac{3x^2}{x^2 + 2}$

2. $f(x) = \dfrac{2x}{\sqrt{x^2 + 2}}$

3. $f(x) = \dfrac{x}{x^2 + 2}$

4. $f(x) = 2 + \dfrac{x^2}{x^4 + 1}$

5. $f(x) = \dfrac{-6x}{\sqrt{4x^2 + 5}}$

6. $f(x) = 5 - \dfrac{1}{x^2 + 1}$

7. $f(x) = \dfrac{4}{x^2 + 1}$

8. $f(x) = \dfrac{2x^2 - 3x + 5}{x^2 + 1}$

(a)
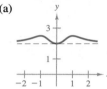

(b)
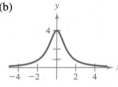

(c)
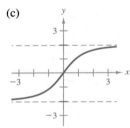

(d)
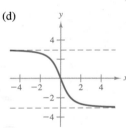

(e)
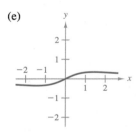

(f)
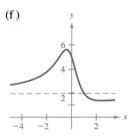

(g)
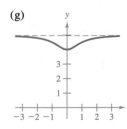

(h)
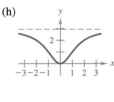

In Exercises 9–24, find the indicated limit.

9. $\lim\limits_{x \to \infty} \dfrac{2x - 1}{3x + 2}$

10. $\lim\limits_{x \to \infty} \dfrac{5x^3 + 1}{10x^3 - 3x^2 + 7}$

11. $\lim\limits_{x \to \infty} \dfrac{x}{x^2 - 1}$

12. $\lim\limits_{x \to \infty} \dfrac{2x^{10} - 1}{10x^{11} - 3}$

13. $\lim\limits_{x \to -\infty} \dfrac{5x^2}{x + 3}$

14. $\lim\limits_{x \to \infty} \dfrac{x^3 - 2x^2 + 3x + 1}{x^2 - 3x + 2}$

15. $\lim\limits_{x \to \infty} \left(2x - \dfrac{1}{x^2}\right)$

16. $\lim\limits_{x \to \infty} (x + 3)^{-2}$

17. $\lim\limits_{x \to -\infty} \left(\dfrac{2x}{x - 1} + \dfrac{3x}{x + 1}\right)$

18. $\lim\limits_{x \to \infty} \left(\dfrac{2x^2}{x - 1} + \dfrac{3x}{x + 1}\right)$

19. $\lim\limits_{x \to -\infty} \dfrac{x}{\sqrt{x^2 - x}}$

20. $\lim\limits_{x \to \infty} \dfrac{x}{\sqrt{x^2 + 1}}$

21. $\lim\limits_{x \to \infty} \dfrac{2x + 1}{\sqrt{x^2 - x}}$

22. $\lim\limits_{x \to -\infty} \dfrac{-3x + 1}{\sqrt{x^2 + x}}$

23. $\lim\limits_{x \to \infty} \dfrac{x^2 - x}{\sqrt{x^4 + x}}$

24. $\lim\limits_{x \to \infty} \dfrac{2x}{\sqrt{4x^2 + 1}}$

In Exercises 25–28, find the indicated limit. [Hint: Treat the expression as a fraction whose denominator is 1, and rationalize the numerator.]

25. $\lim\limits_{x \to -\infty} (x + \sqrt{x^2 + 3})$

26. $\lim\limits_{x \to \infty} (2x - \sqrt{4x^2 + 1})$

27. $\lim\limits_{x \to \infty} (x - \sqrt{x^2 + x})$

28. $\lim\limits_{x \to -\infty} (3x + \sqrt{9x^2 - x})$

In Exercises 29–40, sketch the graph of the given equation. As a sketching aid, examine each equation for intercepts, symmetry, and asymptotes.

29. $y = \dfrac{2 + x}{1 - x}$

30. $y = \dfrac{x - 3}{x - 2}$

31. $y = \dfrac{x^2}{x^2 + 9}$

32. $y = \dfrac{x^2}{x^2 - 9}$

33. $xy^2 = 4$

34. $x^2y = 4$

35. $y = \dfrac{2x}{1 - x}$

36. $y = \dfrac{2x}{1 - x^2}$

37. $y = 2 - \dfrac{3}{x^2}$

38. $y = 1 + \dfrac{1}{x}$

39. $y = \dfrac{x^3}{\sqrt{x^2 - 4}}$

40. $y = \dfrac{x}{\sqrt{x^2 - 4}}$

In Exercises 41–44, complete the given table and estimate the limit of $f(x)$ as x approaches infinity. Then find the limit analytically and compare your results.

41. $f(x) = \dfrac{x + 1}{x\sqrt{x}}$

x	10^0	10^1	10^2	10^3	10^4	10^5	10^6
$f(x)$							

42. $f(x) = x - \sqrt{x(x-1)}$

x	10^0	10^1	10^2	10^3	10^4	10^5	10^6
$f(x)$							

43. $f(x) = 2x - \sqrt{4x^2 + 1}$

x	10^0	10^1	10^2	10^3	10^4	10^5	10^6
$f(x)$							

44. $f(x) = x^2 - x\sqrt{x(x-1)}$

x	10^0	10^1	10^2	10^3	10^4	10^5	10^6
$f(x)$							

45. A business has a cost of $C = 0.5x + 500$ for producing x units. The average cost per unit is given by $\bar{C} = C/x$. Find the limit of $\bar{C}$ as x approaches infinity.

46. According to the theory of relativity, the mass m of a particle depends on its velocity v. That is,

$$m = \frac{m_0}{\sqrt{1 - (v^2/c^2)}}$$

where m_0 is the mass when the particle is at rest and c is the speed of light. Find the limit of the mass as v approaches c.

47. The efficiency of an internal combustion engine is defined to be

$$\text{efficiency (\%)} = 100\left[1 - \frac{1}{(v_1/v_2)^c}\right]$$

where v_1/v_2 is the ratio of the uncompressed gas to the compressed gas and c is a constant dependent upon the engine design. Find the limit of the efficiency as the compression ratio approaches infinity.

48. Verify that each of the following functions has two horizontal asymptotes.

(a) $f(x) = \dfrac{|x|}{x+1}$ 　　(b) $f(x) = \dfrac{2x}{\sqrt{x^2+1}}$

49. Use a computer or graphics calculator to sketch the graph of $f(x) = 3x/\sqrt{4x^2 + 1}$, and from the sketch locate any horizontal or vertical asymptotes.

50. Prove that if

$$p(x) = a_n x^n + \cdots + a_1 x + a_0$$
$$q(x) = b_m x^m + \cdots + b_1 x + b_0$$

then

$$\lim_{x \to \infty} \frac{p(x)}{q(x)} = \begin{cases} 0, & n < m \\ \dfrac{a_n}{b_m}, & n = m \\ \pm\infty, & n > m. \end{cases}$$

4.6 A Summary of Curve Sketching

Summary of curve-sketching techniques

It would be difficult to overstate the importance of curve sketching in mathematics. Descartes' introduction of this concept contributed significantly to the rapid advances in calculus that began during the mid-seventeenth century. In the words of Lagrange, "As long as algebra and geometry traveled separate paths their advance was slow and their applications limited. But when these two sciences joined company, they drew from each other fresh vitality and thenceforth marched on at a rapid pace toward perfection."

Today, government, science, industry, business, education, and the social and health sciences all make widespread use of graphs to describe and predict relationships between variables. We have seen, however, that sketching a graph sometimes can require considerable ingenuity.

So far, we have discussed several concepts that are useful in sketching the graph of a function.

- Domain and Range   (Section 1.5)
- *x*-intercepts and *y*-intercepts   (Section 1.3)
- Symmetry   (Section 1.3)
- Points of discontinuity   (Section 2.3)
- Vertical asymptotes   (Section 2.4)
- Horizontal asymptotes   (Section 4.5)
- Points of nondifferentiability   (Section 3.1)
- Relative extrema   (Section 4.3)
- Concavity   (Section 4.4)
- Points of inflection   (Section 4.4)

In this section we give several examples that incorporate these concepts into an effective procedure for sketching the graph of a function. The following list of suggestions for curve sketching should be helpful.

SUGGESTIONS FOR SKETCHING THE GRAPH OF A FUNCTION

1. Make a rough preliminary sketch that includes any easily determined intercepts and asymptotes.
2. Locate the *x*-values where $f'(x)$ and $f''(x)$ are either zero or undefined.
3. Test the behavior of f at and between each of these *x*-values.
4. Sharpen the accuracy of the final sketch by plotting the relative extrema, the points of inflection, and a few points between.

REMARK   Note in these guidelines the importance of *algebra* (as well as calculus) for solving the equations $f(x) = 0$, $f'(x) = 0$, and $f''(x) = 0$.

EXAMPLE 1   Sketching the graph of a rational function

Sketch the graph of the function given by

$$f(x) = \frac{2(x^2 - 9)}{x^2 - 4}.$$

SOLUTION

First derivative: $f'(x) = \dfrac{20x}{(x^2 - 4)^2}$

Second derivative: $f''(x) = -\dfrac{20(3x^2 + 4)}{(x^2 - 4)^3}$

x-intercepts: $(-3, 0)$, $(3, 0)$

y-intercept: $\left(0, \dfrac{9}{2}\right)$

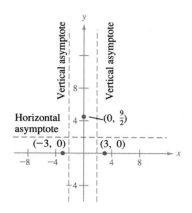

FIGURE 4.35

Vertical asymptotes: $x = -2$, $x = 2$

Horizontal asymptote: $y = 2$

Critical number: $x = 0$

Possible points of inflection: None

Points of discontinuity: $x = -2$, $x = 2$

Symmetry: With respect to y-axis

Test intervals: $(-\infty, -2)$, $(-2, 0)$, $(0, 2)$, $(2, \infty)$

The intercepts and asymptotes can be used to give preliminary clues about the graph of f, as shown in Figure 4.35. Then, we complete the graph as shown in Figure 4.36 by using a few additional points and the conclusions summarized in Table 4.11.

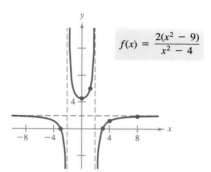

$$f(x) = \frac{2(x^2 - 9)}{x^2 - 4}$$

FIGURE 4.36

TABLE 4.11

	$f(x)$	$f'(x)$	$f''(x)$	*Shape of graph*
$-\infty < x < -2$		$-$	$-$	decreasing, concave down
$x = -2$	undefined	undefined	undefined	vertical asymptote
$-2 < x < 0$		$-$	$+$	decreasing, concave up
$x = 0$	$\frac{9}{2}$	0	$+$	relative minimum
$0 < x < 2$		$+$	$+$	increasing, concave up
$x = 2$	undefined	undefined	undefined	vertical asymptote
$2 < x < \infty$		$+$	$-$	increasing, concave down

A computer software package that generates the graph of a function would be helpful in duplicating the graphs given in this section. If you have such a package, try making changes in each function to see how the changes affect the graph. For instance, how would the graph of

$$f(x) = \frac{2(x^2 - 4)}{x^2 - 9}$$

compare to the graph of the function given in Example 1?

EXAMPLE 2 Sketching the graph of a rational function

Sketch the graph of the function given by

$$f(x) = \frac{x^2 - 2x + 4}{x - 2}.$$

SOLUTION

First derivative: $f'(x) = \dfrac{x(x - 4)}{(x - 2)^2}$

Second derivative: $f''(x) = \dfrac{8}{(x - 2)^3}$

x-intercepts: None

y-intercept: $(0, -2)$

Vertical asymptote: $x = 2$

Horizontal asymptotes: None

Critical numbers: $x = 0,\ x = 4$

Possible points of inflection: None

Points of discontinuity: $x = 2$

Test intervals: $(-\infty, 0),\ (0, 2),\ (2, 4),\ (4, \infty)$

The analysis of the graph of f is shown in Table 4.12, and the graph is shown in Figure 4.37.

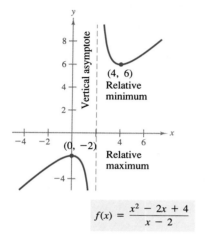

$$f(x) = \frac{x^2 - 2x + 4}{x - 2}$$

FIGURE 4.37

TABLE 4.12

	$f(x)$	$f'(x)$	$f''(x)$	*Shape of graph*
$-\infty < x < 0$		$+$	$-$	increasing, concave down
$x = 0$	-2	0	$-$	relative maximum
$0 < x < 2$		$-$	$-$	decreasing, concave down
$x = 2$	undefined	undefined	undefined	vertical asymptote
$2 < x < 4$		$-$	$+$	decreasing, concave up
$x = 4$	6	0	$+$	relative minimum
$4 < x < \infty$		$+$	$+$	increasing, concave up

Although the graph of the function in Example 2 has no horizontal asymptote, it does have a slant asymptote. The graph of a rational function (having no common factors) has a **slant asymptote** if the degree of the numerator exceeds the degree of the denominator by one. To find the slant asymptote, we use division to rewrite the rational function as the sum of a first-degree polynomial and another rational function.

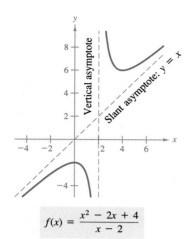

$$f(x) = \frac{x^2 - 2x + 4}{x - 2}$$

FIGURE 4.38

$$f(x) = \frac{x^2 - 2x + 4}{x - 2} = x + \frac{4}{x - 2}$$

> $y = x$ is a slant asymptote

In Figure 4.38 note that the graph of f approaches the slant asymptote $y = x$ as x approaches $-\infty$ or ∞.

EXAMPLE 3 Sketching a rational function with a slant asymptote

Sketch the graph of the function given by

$$f(x) = \frac{-x^3 + x^2 + 4}{x^2} = \frac{(-x + 2)(x^2 + x + 2)}{x^2}.$$

SOLUTION

First derivative: $f'(x) = -\dfrac{x^3 + 8}{x^3}$

Second derivative: $f''(x) = \dfrac{24}{x^4}$

x-intercept: $(2, 0)$

y-intercept: None

Vertical asymptote: $x = 0$

Slant asymptote: $y = -x + 1$ since $f(x) = -x + 1 + \dfrac{4}{x^2}$

Critical number: $x = -2$

Possible points of inflection: None

Points of discontinuity: $x = 0$

Test intervals: $(-\infty, -2), (-2, 0), (0, \infty)$

The analysis of the graph of f is shown in Table 4.13, and the graph is shown in Figure 4.39.

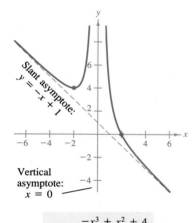

$$f(x) = \frac{-x^3 + x^2 + 4}{x^2}$$

FIGURE 4.39

TABLE 4.13

	$f(x)$	$f'(x)$	$f''(x)$	*Shape of graph*
$-\infty < x < -2$		$-$	$+$	decreasing, concave up
$x = -2$	4	0	$+$	relative minimum
$-2 < x < 0$		$+$	$+$	increasing, concave up
$x = 0$	undefined	undefined	undefined	vertical asymptote
$0 < x < \infty$		$-$	$+$	decreasing, concave up

EXAMPLE 4   Sketching the graph of a function involving a radical

Sketch the graph of the function given by

$$f(x) = \frac{x}{\sqrt{x^2 + 2}}.$$

SOLUTION

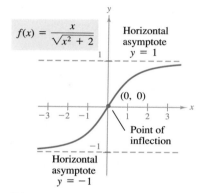

$$f(x) = \frac{x}{\sqrt{x^2 + 2}}$$

Horizontal asymptote $y = 1$

(0, 0)

Point of inflection

Horizontal asymptote $y = -1$

FIGURE 4.40

First derivative: $f'(x) = \dfrac{2}{(x^2 + 2)^{3/2}}$

Second derivative: $f''(x) = -\dfrac{6x}{(x^2 + 2)^{5/2}}$

x-intercept: (0, 0)

y-intercept: (0, 0)

Vertical asymptotes: None

Horizontal asymptotes: $y = 1$ (to the right), $y = -1$ (to the left)

Critical numbers: None

Possible points of inflection: $x = 0$

Points of discontinuity: None

Symmetry: With respect to origin

Test intervals: $(-\infty, 0)$, $(0, \infty)$

The analysis of the graph of f is shown in Table 4.14, and the graph is shown in Figure 4.40.

TABLE 4.14

	$f(x)$	$f'(x)$	$f''(x)$	*Shape of graph*
$-\infty < x < 0$		$+$	$+$	increasing, concave up
$x = 0$	0	$\dfrac{1}{\sqrt{2}}$	0	point of inflection
$0 < x < \infty$		$+$	$-$	increasing, concave down

REMARK   Although the function in Example 4 is defined over the entire real line, a square root often indicates a restricted domain. For example,

$$f(x) = \sqrt{x^2 - 4}$$

has $(-\infty, -2] \cup [2, \infty)$ as its domain.

EXAMPLE 5   Sketching the graph of a function involving cube roots

Sketch the graph of the function given by

$$f(x) = 2x^{5/3} - 5x^{4/3}.$$

SOLUTION

First derivative: $f'(x) = \dfrac{10}{3}x^{1/3}(x^{1/3} - 2)$

Second derivative: $f''(x) = \dfrac{20(x^{1/3} - 1)}{9x^{2/3}}$

x-intercepts: $(0, 0)$, $\left(\dfrac{125}{8}, 0\right)$

y-intercept: $(0, 0)$

Vertical asymptotes: None

Horizontal asymptotes: None

Critical numbers: $x = 0$, $x = 8$

Possible points of inflection: $x = 0$, $x = 1$

Points of discontinuity: None

Test intervals: $(-\infty, 0)$, $(0, 1)$, $(1, 8)$, $(8, \infty)$

The analysis of the graph of f is shown in Table 4.15, and the graph is shown in Figure 4.41.

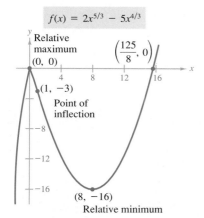

$f(x) = 2x^{5/3} - 5x^{4/3}$

Relative maximum $(0, 0)$

$\left(\dfrac{125}{8}, 0\right)$

$(1, -3)$

Point of inflection

$(8, -16)$

Relative minimum

FIGURE 4.41

TABLE 4.15

	$f(x)$	$f'(x)$	$f''(x)$	*Shape of graph*
$-\infty < x < 0$		+	−	increasing, concave down
$x = 0$	0	0	undefined	relative maximum
$0 < x < 1$		−	−	decreasing, concave down
$x = 1$	−3	−	0	point of inflection
$1 < x < 8$		−	+	decreasing, concave up
$x = 8$	−16	0	+	relative minimum
$8 < x < \infty$		+	+	increasing, concave up

REMARK   Although the function in Example 5 is differentiable over the entire real line, cube roots often indicate that there are points where the derivative is undefined. For example, $f(x) = x^{1/3}$ has a vertical tangent at $(0, 0)$.

EXAMPLE 6 Sketching the graph of a polynomial function

Sketch the graph of the function given by

$$f(x) = x^4 - 12x^3 + 48x^2 - 64x$$
$$= x(x - 4)^3.$$

SOLUTION

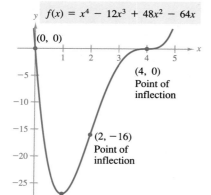

FIGURE 4.42

First derivative: $f'(x) = 4(x - 1)(x - 4)^2$

Second derivative: $f''(x) = 12(x - 4)(x - 2)$

x-intercepts: $(0, 0)$, $(4, 0)$

y-intercept: $(0, 0)$

Vertical asymptotes: None

Horizontal asymptotes: None

Critical numbers: $x = 1$, $x = 4$

Possible points of inflection: $x = 2$, $x = 4$

Points of discontinuity: None

Test intervals: $(-\infty, 1)$, $(1, 2)$, $(2, 4)$, $(4, \infty)$

The analysis of the graph of f is shown in Table 4.16, and the graph is shown in Figure 4.42.

TABLE 4.16

	$f(x)$	$f'(x)$	$f''(x)$	*Shape of graph*
$-\infty < x < 1$		$-$	$+$	decreasing, concave up
$x = 1$	-27	0	$+$	relative minimum
$1 < x < 2$		$+$	$+$	increasing, concave up
$x = 2$	-16	$+$	0	point of inflection
$2 < x < 4$		$+$	$-$	increasing, concave down
$x = 4$	0	0	0	point of inflection
$4 < x < \infty$		$+$	$+$	increasing, concave up

REMARK   The fourth degree polynomial function in Example 6 has one relative minimum and no relative maxima. In general, a polynomial function of degree n can have *at most* $n - 1$ relative extrema. Moreover, polynomial functions of even degree must have *at least* one relative extremum.

EXERCISES for Section 4.6

In Exercises 1–26, sketch the graph of the given function. Choose a scale that allows all relative extrema and points of inflection to be identified on the sketch.

1. $y = x^3 - 3x^2 + 3$

2. $y = -\frac{1}{3}(x^3 - 3x + 2)$

3. $y = 2 - x - x^3$

4. $y = x^3 + 3x^2 + 3x + 2$

5. $f(x) = 3x^3 - 9x + 1$

6. $f(x) = (x + 1)(x - 2)(x - 5)$

7. $f(x) = -x^3 + 3x^2 + 9x - 2$

8. $f(x) = \frac{1}{3}(x - 1)^3 + 2$

9. $y = 3x^4 + 4x^3$

10. $y = 3x^4 - 6x^2$

11. $f(x) = x^4 - 4x^3 + 16x$

12. $f(x) = x^4 - 8x^3 + 18x^2 - 16x + 5$

13. $f(x) = x^4 - 4x^3 + 16x - 16$

14. $f(x) = x^5 + 1$

15. $y = x^5 - 5x$

16. $y = (x - 1)^5$

17. $y = |2x - 3|$

18. $y = |x^2 - 6x + 5|$

19. $y = \frac{x^2}{x^2 + 3}$

20. $y = \frac{x}{x^2 + 1}$

21. $y = x\sqrt{4 - x}$

22. $y = x\sqrt{4 - x^2}$

23. $y = 3x^{2/3} - 2x$

24. $y = 3x^{2/3} - x^2$

25. $f(x) = \frac{x}{\sqrt{x^2 + 7}}$

26. $f(x) = \frac{4x}{\sqrt{x^2 + 15}}$

In Exercises 27–36, sketch the graph of the given function. In each case label the intercepts, relative extrema, points of inflection, and asymptotes.

27. $y = \frac{1}{x - 2} - 3$

28. $y = \frac{x^2 + 1}{x^2 - 2}$

29. $y = \frac{2x}{x^2 - 1}$

30. $y = \frac{x^2 - 6x + 12}{x - 4}$

31. $f(x) = \frac{x + 2}{x}$

32. $f(x) = x + \frac{32}{x^2}$

33. $f(x) = \frac{x^2 + 1}{x}$

34. $f(x) = \frac{x^3}{x^2 - 1}$

35. $y = \frac{x^3}{2x^2 - 8}$

36. $y = \frac{2x^2 - 5x + 5}{x - 2}$

In Exercises 37–42, determine the conditions on the coefficients of

$$f(x) = ax^3 + bx^2 + cx + d$$

such that the graph of f will resemble the given graph.

37.

38.

39.   **40.**

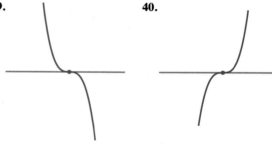

41.   **42.**

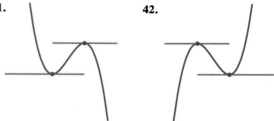

In Exercises 43–48, use the given graph (of f' or f'') to sketch a graph of the function f. [Hint: The solutions are not unique.]

43.   **44.**

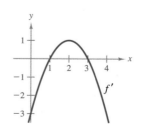

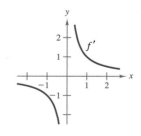

45.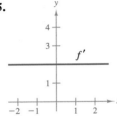

46.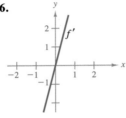

47.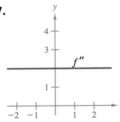

48.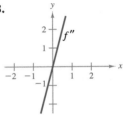

4.7 Optimization Problems

Applied minimum and maximum problems

One of the most common applications of calculus involves the determination of minimum and maximum values. Consider how frequently we hear or read terms like greatest profit, least cost, least time, greatest voltage, optimum size, least area, greatest strength, or greatest distance. Before outlining a general method of solution for such problems, we present an example.

EXAMPLE 1 Finding the maximum volume

A manufacturer wants to design an open box having a square base and a surface area of 108 square inches. What dimensions will produce a box with maximum volume?

SOLUTION

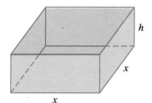

Open Box with Square Base:
$S = x^2 + 4xh = 108$

FIGURE 4.43

Since the box (in Figure 4.43) has a square base, its volume is

$$V = x^2h. \qquad \text{Primary equation}$$

(We call this the primary equation because it gives a formula for the quantity we wish to optimize.) Furthermore, since the box is open at the top, its surface area is

$$S = (\text{area of base}) + (\text{area of four sides})$$
$$S = x^2 + 4xh = 108. \qquad \text{Secondary equation}$$

Since V is to be maximized, we express it as a function of just one variable. To do this, we solve the equation $x^2 + 4xh = 108$ for h in terms of x to obtain

$$h = \frac{108 - x^2}{4x}.$$

Substituting in the equation for volume produces

$$V = x^2h = x^2\left(\frac{108 - x^2}{4x}\right) = 27x - \frac{x^3}{4}. \qquad \text{Function of one variable}$$

Before we try to find which x-value will yield a maximum value of V, we should determine the *feasible domain*. That is, what values of x make sense in this problem? We know that x should be nonnegative and that the area of the base ($A = x^2$) is at most 108. Thus, we have

$$0 \le x \le \sqrt{108}. \qquad \text{Feasible domain}$$

Now, to maximize V we find the critical numbers as follows.

$$\frac{dV}{dx} = 27 - \frac{3x^2}{4} = 0$$

$$3x^2 = 108 \implies x = \pm 6 \qquad \text{Critical numbers}$$

Evaluating V at the critical numbers in the domain and at the endpoints of the domain produces

$$V(0) = 0, \qquad V(6) = 108, \qquad \text{and} \qquad V(\sqrt{108}) = 0.$$

We conclude that V is maximum when $x = 6$ and the dimensions of the box are 6 in. $\times$ 6 in. $\times$ 3 in.

Some students have trouble with applied problems because they are too eager to use formulas. For instance, in Example 1, you should realize that there are infinitely many open boxes having 108 square inches of surface area. You might begin by asking yourself which basic shape would seem to yield a maximum volume. Should the box be tall, squat, or more cubical? You might even try calculating a few volumes, as shown in Figure 4.44, to see if you can get a better feeling for what the optimum dimensions should be. Remember that you are not ready to begin solving a problem until you have clearly identified what the problem is.

Example 1 illustrates the following five-step procedure for solving applied minimum and maximum problems.

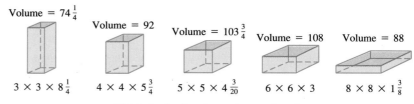

$$\text{Volume} = 74\tfrac{1}{4} \qquad \text{Volume} = 92 \qquad \text{Volume} = 103\tfrac{3}{4} \qquad \text{Volume} = 108 \qquad \text{Volume} = 88$$

$$3 \times 3 \times 8\tfrac{1}{4} \qquad 4 \times 4 \times 5\tfrac{3}{4} \qquad 5 \times 5 \times 4\tfrac{3}{20} \qquad 6 \times 6 \times 3 \qquad 8 \times 8 \times 1\tfrac{3}{8}$$

FIGURE 4.44

Which size box has the maximum volume?

PROCEDURES FOR SOLVING APPLIED MINIMUM AND MAXIMUM PROBLEMS	1. Assign symbols to all *given* quantities and quantities *to be determined*. When feasible, make a sketch. 2. Write a **primary equation** for the quantity that is to be maximized (or minimized). (A review of several useful formulas from geometry is given inside the back cover of the text.) 3. Reduce the primary equation to one having a single independent variable. This may involve the use of **secondary equations** relating the independent variables of the primary equation. 4. Determine the domain of the primary equation. That is, determine the values for which the stated problem makes sense. 5. Determine the desired maximum or minimum value by the techniques discussed in Sections 4.1 through 4.4

REMARK   When performing step 5, recall that to determine the maximum or minimum value of a continuous function f on a closed interval, we compare the values of f at its critical numbers to the values of f at the endpoints of the interval.

EXAMPLE 2   Finding the minimum distance

Find the points on the graph of $y = 4 - x^2$ that are closest to the point $(0, 2)$.

SOLUTION

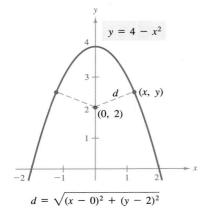

$$d = \sqrt{(x - 0)^2 + (y - 2)^2}$$

FIGURE 4.45

Figure 4.45 indicates that there are two points at a minimum distance from the point $(0, 2)$. The distance between the point $(0, 2)$ and a point (x, y) on the graph of $y = 4 - x^2$ is given by

$$d = \sqrt{(x - 0)^2 + (y - 2)^2}.$$ Primary equation

Using the secondary equation $y = 4 - x^2$, we rewrite the primary equation as

$$d = \sqrt{x^2 + (4 - x^2 - 2)^2} = \sqrt{x^4 - 3x^2 + 4}.$$

Since d is smallest when the expression inside the radical is smallest, we need only find the critical numbers of $f(x) = x^4 - 3x^2 + 4$. Note that the domain of f is the entire real line. Moreover, differentiation yields

$$f'(x) = 4x^3 - 6x = 2x(2x^2 - 3) = 0$$

$$x = 0, \ \sqrt{\frac{3}{2}}, \ -\sqrt{\frac{3}{2}}.$$

The First Derivative Test verifies that $x = 0$ yields a relative maximum, while both $x = \sqrt{3/2}$ and $x = -\sqrt{3/2}$ yield a minimum distance. Hence, the closest points are $(\sqrt{3/2}, 5/2)$ and $(-\sqrt{3/2}, 5/2)$.   ▭

EXAMPLE 3   Finding the minimum area

A rectangular page is to contain 24 square inches of print. The margins at the top and bottom of the page are each $1\frac{1}{2}$ inches. The margins on each side are 1 inch. What should the dimensions of the page be so that the least amount of paper is used?

SOLUTION

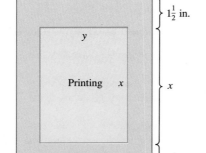

$$A = (x + 3)(y + 2)$$

FIGURE 4.46

Letting A be the area to be minimized, we have, from Figure 4.46,

$$A = (x + 3)(y + 2).$$ Primary equation

The printed area inside the margins is given by

$$24 = xy.$$ Secondary equation

Solving this equation for y produces

$$y = \frac{24}{x}$$

and the primary equation becomes

$$A = (x + 3)\left(\frac{24}{x} + 2\right) = 30 + 2x + \frac{72}{x}.$$ Function of one variable

Since x must be positive, we are interested only in values of A when $x > 0$. Now, to find the minimum area, we differentiate with respect to x and obtain

$$\frac{dA}{dx} = 2 - \frac{72}{x^2} = 0 \quad \Longrightarrow \quad x^2 = 36.$$

Thus, the critical numbers are $x = \pm 6$. Since -6 is outside the domain, we choose $x = 6$. The First Derivative Test confirms that A is a minimum at this point. Therefore, $y = \frac{24}{6} = 4$ and the dimensions of the page should be

$$x + 3 = 9 \text{ inches} \qquad \text{by} \qquad y + 2 = 6 \text{ inches}. \qquad \blacksquare$$

EXAMPLE 4   Finding the minimum length

Two posts, one 12 feet high and the other 28 feet high, stand 30 feet apart. They are to be stayed by two wires, attached to a single stake, running from ground level to the top of each post. Where should the stake be placed to use the least wire?

SOLUTION

Let W be the wire length to be minimized. Using Figure 4.47, we have

$$W = y + z. \qquad \text{Primary equation}$$

In this problem, rather than solving for y in terms of z (or vice versa), we solve for both y and z in terms of a third variable x, as shown in Figure 4.47.

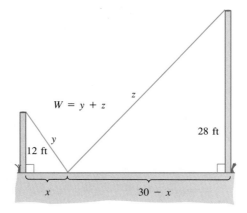

FIGURE 4.47

From the Pythagorean Theorem, we have

$$x^2 + 12^2 = y^2 \implies y = \sqrt{x^2 + 144}$$
$$(30 - x)^2 + 28^2 = z^2 \implies z = \sqrt{x^2 - 60x + 1684}.$$

Thus, W is given by

$$W = y + z = \sqrt{x^2 + 144} + \sqrt{x^2 - 60x + 1684}, \quad 0 \le x \le 30.$$

Differentiating W with respect to x yields

$$\frac{dW}{dx} = \frac{x}{\sqrt{x^2 + 144}} + \frac{x - 30}{\sqrt{x^2 - 60x + 1684}}.$$

By letting $dW/dx = 0$, we obtain

$$x^2(x^2 - 60x + 1684) = (30 - x)^2(x^2 + 144)$$

which simplifies to

$$320(x - 9)(2x + 45) = 0.$$

Since $x = -22.5$ is not in the interval $[0, 30]$ and

$$W(0) \approx 53.04, \qquad W(9) = 50, \qquad \text{and} \qquad W(30) \approx 60.31$$

we conclude that the wire should be staked at 9 feet from the 12-foot pole.

In each of the first four examples, the extreme value occurs at a critical number. Although this happens often, remember that an extreme value can occur at an endpoint of an interval. The next example illustrates this.

EXAMPLE 5 An endpoint maximum

Four feet of wire is to be used to form a square and a circle. How much of the wire should be used for the square and how much should be used for the circle to enclose the maximum total area?

SOLUTION

From Figure 4.48, the total area is given by

$$A = \text{(area of square)} + \text{(area of circle)}$$
$$A = x^2 + \pi r^2. \qquad \text{Primary equation}$$

Since the total amount of wire is 4 feet, we have

$$4 = \text{(perimeter of square)} + \text{(circumference of circle)}$$
$$4 = 4x + 2\pi r.$$

Thus, $r = 2(1 - x)/\pi$, and by substituting in the primary equation we have

$$A = x^2 + \pi \left[\frac{2(1 - x)}{\pi}\right]^2 = x^2 + \frac{4(1 - x)^2}{\pi} = \frac{1}{\pi}[(\pi + 4)x^2 - 8x + 4].$$

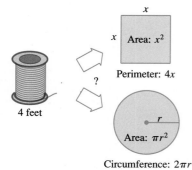

x

x Area: x^2

Perimeter: $4x$

?

Area: πr^2

r

Circumference: $2\pi r$

$4x + 2\pi r = 4$
$A = x^2 + \pi r^2$

4 feet

FIGURE 4.48

The feasible domain is $0 \le x \le 1$. Since $dA/dx = [2(\pi + 4)x - 8]/\pi$, the only critical number is $x = 4/(\pi + 4) \approx 0.56$. Therefore, since

$$A(0) \approx 1.273, \qquad A(0.56) \approx 0.56, \qquad \text{and} \qquad A(1) = 1$$

we conclude that the maximum area occurs when $x = 0$. That is, *all* the wire is used for the circle. ▭

Let's review the primary equations developed in the first five examples. As applications go, these five examples are fairly simple, and yet the resulting primary equations are quite complicated.

$$V = 27x - \frac{x^3}{4} \qquad\qquad W = \sqrt{x^2 + 144} + \sqrt{x^2 - 60x + 1684}$$

$$d = \sqrt{x^4 - 3x^2 + 4} \qquad A = \frac{1}{\pi}[(\pi + 4)x^2 - 8x + 4]$$

$$A = 30 + 2x + \frac{72}{x}$$

You must expect that real applications often involve equations that are at least as complicated as these five. Remember that one of the main goals of this course is to learn to use calculus to analyze equations that initially seem formidable.

EXERCISES for Section 4.7

In Exercises 1–6, find two positive numbers satisfying the given requirements.

1. The sum is 110 and the product is maximum.
2. The sum is S and the product is maximum.
3. The product is 192 and the sum is minimum.
4. The product is 192 and the sum of the first plus three times the second is minimum.
5. The second number is the reciprocal of the first and their sum is minimum.
6. The sum of the first and twice the second is 100 and the product is maximum.

In Exercises 7 and 8, find the length and width of a rectangle of maximum area for the given perimeter.

7. Perimeter: 100 feet   8. Perimeter: P units

In Exercises 9 and 10, find the length and width of a rectangle of minimum perimeter for the given area.

9. Area: 64 square feet   10. Area: A square feet

In Exercises 11 and 12, find the point on the graph of the function closest to the given point.

Function	Point
11. $f(x) = \sqrt{x}$	$(4, 0)$
12. $f(x) = x^2$	$\left(2, \frac{1}{2}\right)$

13. In an autocatalytic chemical reaction, the product formed is a catalyst for the reaction. If Q is the amount of the original substance and x is the amount of catalyst formed, then the rate of chemical reaction is given by

$$\frac{dQ}{dx} = kx(Q - x).$$

For what value of x will the rate of chemical reaction be greatest?

14. Show that among all positive numbers x and y with $x^2 + y^2 = r^2$, the sum $x + y$ is largest when $x = y$.

15. A dairy farmer plans to fence in a rectangular pasture adjacent to a river. The pasture must contain 180,000 square meters in order to provide enough grass for the herd. What dimensions would require the least amount of fencing if no fencing is needed along the river?

16. A rancher has 200 feet of fencing with which to enclose two adjacent rectangular corrals (see figure). What dimensions should be used so that the enclosed area will be a maximum?

17. An open box is to be made from a square piece of material, 12 inches on a side, by cutting equal squares from each corner and turning up the sides (see figure). Find the volume of the largest box that can be made in this manner.

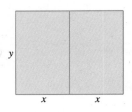

FIGURE FOR 16

FIGURE FOR 17

18. (a) Solve Exercise 17 given that the square of material is s inches on a side.
(b) If the dimensions of the square piece of material are doubled, how does the volume change?

19. An open box is to be made from a rectangular piece of material by cutting equal squares from each corner and turning up the sides. Find the dimensions of the box of maximum volume if the material has dimensions of 2 feet by 3 feet.

20. A net enclosure for practicing golf is open at one end (see figure). Find the dimensions that require the least amount of netting if the volume of the enclosure is to be $83\frac{1}{3}$ cubic meters.

21. A Norman window is constructed by adjoining a semicircle to the top of an ordinary rectangular window (see figure). Find the dimensions of a Norman window of maximum area if the total perimeter is 16 feet.

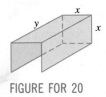

FIGURE FOR 20

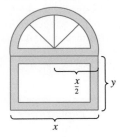

FIGURE FOR 21

22. A physical fitness room consists of a rectangular region with a semicircle on each end. If the perimeter of the room is to be a 200-meter running track, find the dimensions that will make the area of the rectangular region as large as possible.

23. A rectangle is bounded by the x- and y-axes and the graph of $y = (6 - x)/2$ (see figure). What length and width should the rectangle have so that its area is a maximum?

24. A right triangle is formed in the first quadrant by the x- and y-axes and a line through the point $(1, 2)$ (see figure). Find the vertices of the triangle so that the length of the hypotenuse is minimum.

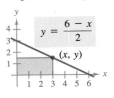

FIGURE FOR 23

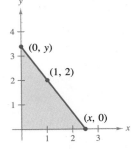

FIGURE FOR 24

25. A right triangle is formed in the first quadrant by the x- and y-axes and a line through the point $(2, 3)$. Find the vertices of the triangle so that its area is minimum.

26. Find the dimensions of the largest isosceles triangle that can be inscribed in a circle of radius 4 (see figure).

27. A rectangle is bounded by the x-axis and the semicircle $y = \sqrt{25 - x^2}$ (see figure). What length and width should the rectangle have so that its area is a maximum?

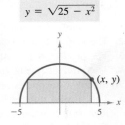

FIGURE FOR 26

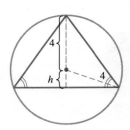

FIGURE FOR 27

28. Find the dimensions of the largest rectangle that can be inscribed in a semicircle of radius r (see Exercise 27).

29. Find the dimensions of the trapezoid of greatest area that can be inscribed in a semicircle of radius r.

30. A page is to contain 30 square inches of print. The margins at the top and bottom of the page are 2 inches. The margins on each side are only 1 inch. Find the dimensions of the page so that the least paper is used.

31. A right circular cylinder is to be designed to hold 12 fluid ounces of a soft drink and to use a minimum of material in its construction. Find the required dimensions for the container. [1 fl oz ≈ 1.80469 in.^{3}]

32. Rework Exercise 31 if the volume of the cylinder is V_0 cubic units.

33. A rectangular package to be sent by a postal service can have a maximum combined length and girth (perimeter of a cross section) of 108 inches (see figure). Find the dimensions of the package of maximum volume that can be sent. (Assume the cross section is square.)

34. Rework Exercise 33 for a cylindrical package. (The cross sections are circular.)

35. Find the volume of the largest right circular cone that can be inscribed in a sphere of radius r (see figure).

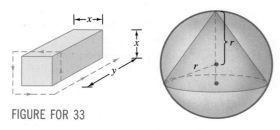

FIGURE FOR 33

FIGURE FOR 35

36. Find the volume of the largest right circular cylinder that can be inscribed in a sphere of radius r.

37. A solid is formed by adjoining two hemispheres to each end of a right circular cylinder. The total volume of the figure is 12 cubic inches. Find the radius of the cylinder that produces the minimum surface area.

38. An industrial tank of the shape described in Exercise 37 must have a volume of 3000 cubic feet. If the construction cost of the hemispherical ends is twice as much per square foot of surface area as the sides, find the dimensions that will minimize cost.

39. The combined perimeter of an equilateral triangle and a square is 10. Find the dimensions of the triangle and square that produce a minimum total area.

40. The combined perimeter of a circle and a square is 16. Find the dimensions of the circle and square that produce a minimum total area.

41. Ten feet of wire is to be used to form an isosceles right triangle and a circle. How much of the wire should be used for the circle if the total area enclosed is to be (a) minimum and (b) maximum?

42. Twenty feet of wire is to be used to form two figures. In each of the following cases, how much should be used for each figure so that the total enclosed area is a maximum?

(a) equilateral triangle and square
(b) square and regular pentagon
(c) regular pentagon and hexagon

43. A wooden beam has a rectangular cross section of height h and width w (see figure). The strength S of the beam is directly proportional to the width and the square of the height. What are the dimensions of the strongest beam that can be cut from a round log of diameter 24 inches? [Hint: $S = kh^2w$, where k is the proportionality constant.]

44. The illumination from a light source is directly proportional to the strength of the source and inversely proportional to the square of the distance from the source. Two light sources of intensities I_1 and I_2 are d units apart. At what point on the line segment joining the two sources is the illumination least?

45. A man is in a boat 2 miles from the nearest point on the coast. He is to go to a point Q, 3 miles down the coast and 1 mile inland (see figure). If he can row at 2 miles per hour and walk at 4 miles per hour, toward what point on the coast should he row in order to reach point Q in the least time?

46. The conditions are the same as in Exercise 45 except that the man can row at 4 miles per hour (see figure). How does this change the solution?

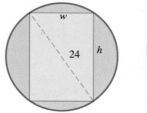

FIGURE FOR 43

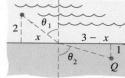

FIGURE FOR 45–46

In Exercises 47 and 48, use a computer or graphics calculator to sketch the graphs of the primary equation and its first derivative in the given applied extrema problem. From the graphs, find the required extrema.

47. Find the dimensions of the rectangle of maximum area that can be inscribed in a semicircle of radius 10.

48. Find the length of the longest pipe that can be carried level around a right-angle corner if the two intersecting corridors are of width 5 feet and 8 feet, respectively.

4.8 Newton's Method

Newton's Method ▪ Algebraic solutions of polynomial equations

In the first seven sections of this chapter, we frequently needed to find the zeros of a function. Until now our functions have been chosen carefully so that elementary algebraic techniques suffice for finding the zeros. For instance,

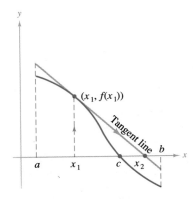

The *x*-intercept of the tangent line approximates the zero of *f*.

FIGURE 4.49

the zeros of $f(x) = x^3 - 2x^2 - x + 2$ can be found easily by factoring. In applied problems we frequently encounter functions whose zeros are more difficult to find. For instance, the zeros of a function as simple as

$$f(x) = x^3 - x + 1$$

cannot be found by elementary algebraic methods. (More is said about this at the end of this section.)

To introduce Newton's Method for approximating the real zeros of a function, we consider a function f that is continuous on the interval $[a, b]$ and differentiable on the interval (a, b). If $f(a)$ and $f(b)$ differ in sign, then by the Intermediate Value Theorem f must possess at least one zero in the interval (a, b). Suppose we estimate this zero to occur at $x = x_1$, as shown in Figure 4.49. Newton's method is based on the assumption that the graph of f and the tangent line at $(x_1, f(x_1))$ both cross the *x*-axis at *about* the same point. Since we can easily calculate the *x*-intercept for this tangent line, we use it as our second (and, we hope, better) estimate for the zero of f. The tangent line passes through the point $(x_1, f(x_1))$ with a slope of $f'(x_1)$. In point-slope form, the equation of the tangent line is therefore

$$y - f(x_1) = f'(x_1)(x - x_1)$$
$$y = f'(x_1)(x - x_1) + f(x_1).$$

Letting $y = 0$ and solving for x, we have

$$x = x_1 - \frac{f(x_1)}{f'(x_1)}.$$

Thus, from our initial guess we arrive at a new estimate

$$x_2 = x_1 - \frac{f(x_1)}{f'(x_1)}.$$

We may improve upon x_2 and calculate yet a third estimate

$$x_3 = x_2 - \frac{f(x_2)}{f'(x_2)}.$$

Repeated application of this process is called **Newton's Method.**

NEWTON'S METHOD FOR APPROXIMATING THE ZEROS OF A FUNCTION

Let $f(c) = 0$, where f is differentiable on an open interval containing c. Then, to approximate c, we use the following steps.

1. Make an initial estimate x_1 that is "close" to c. (A graph is helpful.)
2. Determine a new approximation

$$x_{n+1} = x_n - \frac{f(x_n)}{f'(x_n)}.$$

3. If $|x_n - x_{n+1}|$ is less than the desired accuracy, let x_{n+1} serve as the final approximation. Otherwise, return to step 2 and calculate a new approximation.

Each successive application of this procedure is called an **iteration.**

For many functions, just a few iterations of Newton's Method will produce approximations having very small errors. To demonstrate this, we begin with an example in which we know the exact zero.

EXAMPLE 1 Using Newton's Method

Calculate three iterations of Newton's Method to approximate a zero of $f(x) = x^2 - 2$. Use $x_1 = 1$ as the initial guess.

SOLUTION

Since $f(x) = x^2 - 2$, we have $f'(x) = 2x$ and the iterative process is given by the formula

$$x_{n+1} = x_n - \frac{f(x_n)}{f'(x_n)} = x_n - \frac{x_n^2 - 2}{2x_n}.$$

The calculations for three iterations are shown in Table 4.17.

TABLE 4.17

n	x_n	$f(x_n)$	$f'(x_n)$	$\dfrac{f(x_n)}{f'(x_n)}$	$x_n - \dfrac{f(x_n)}{f'(x_n)}$
1	1.000000	−1.000000	2.000000	−0.500000	1.500000
2	1.500000	0.250000	3.000000	0.083333	1.416667
3	1.416667	0.006945	2.833334	0.002451	1.414216
4	1.414216				

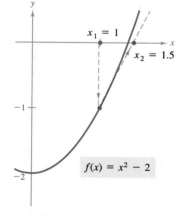

FIGURE 4.50

Of course, in this case we know that the two zeros of the function are $\pm\sqrt{2}$. To six decimal places, $\sqrt{2} = 1.414214$. Thus, after only three iterations of Newton's Method, we have obtained an approximation that is within 0.000002 of an actual root. The first iteration of this process is shown in Figure 4.50. ▬

REMARK   From the calculations in Example 1, you can see that Newton's Method has become more practical with the access to calculators and computers.

EXAMPLE 2 Using Newton's Method

Use Newton's Method to approximate the zeros of

$$f(x) = 2x^3 + x^2 - x + 1.$$

Continue the iterations until two successive approximations differ by less than 0.0001.

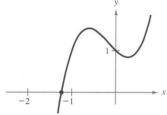

$$f(x) = 2x^3 + x^2 - x + 1$$

FIGURE 4.51

SOLUTION

We begin by sketching a graph of f and observing that it has only one zero, which occurs near $x = -1.2$, as shown in Figure 4.51. Now, we differentiate f to form the iterative formula

$$x_{n+1} = x_n - \frac{f(x_n)}{f'(x_n)} = x_n - \frac{2x_n{}^3 + x_n{}^2 - x_n + 1}{6x_n{}^2 + 2x_n - 1}.$$

The calculations are shown in Table 4.18.

TABLE 4.18

n	x_n	$f(x_n)$	$f'(x_n)$	$\dfrac{f(x_n)}{f'(x_n)}$	$x_n - \dfrac{f(x_n)}{f'(x_n)}$
1	-1.20000	0.18400	5.24000	0.03511	-1.23511
2	-1.23511	-0.00771	5.68276	-0.00136	-1.23375
3	-1.23375	0.00001	5.66533	0.00000	-1.23375
4	-1.23375				

Thus, we estimate the zero of f to be -1.23375, since two successive approximations differ by less than the required 0.0001. ▭

When, as in Examples 1 and 2, the approximations approach a limit, we say that the sequence

$$x_1, x_2, x_3, \ldots, x_n, \ldots$$

converges. Moreover, if the limit is c, then it can be shown that c must be a zero of f.

It is important to realize that Newton's Method does not always yield a convergent sequence. In such cases, we say that the method fails. One way this can happen is shown in Figure 4.52. Since Newton's Method involves division by $f'(x_n)$, it is clear that the method will fail if the derivative is zero for any x_n in the sequence. When you encounter this problem, you can usually overcome it by choosing a different value for x_1.

Another way Newton's Method can fail is illustrated in the next example.

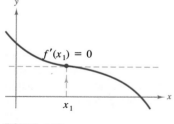

FIGURE 4.52

EXAMPLE 3 An example in which Newton's Method fails

Using $x_1 = 0.1$, show that Newton's Method fails to converge for the function $f(x) = x^{1/3}$.

SOLUTION

Since $f'(x) = \frac{1}{3}x^{-2/3}$, the iterative formula is

$$x_{n+1} = x_n - \frac{f(x_n)}{f'(x_n)} = x_n - \frac{x_n{}^{1/3}}{\frac{1}{3}x_n{}^{-2/3}} = x_n - 3x_n = -2x_n.$$

The calculations are given in Table 4.19. This table and Figure 4.53 indicate that x_n increases in magnitude as $n \to \infty$, and thus the limit of the sequence does not exist.

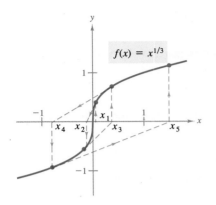

FIGURE 4.53

TABLE 4.19

n	x_n	$f(x_n)$	$f'(x_n)$	$\dfrac{f(x_n)}{f'(x_n)}$	$x_n - \dfrac{f(x_n)}{f'(x_n)}$
1	0.10000	0.46416	1.54720	0.30000	−0.20000
2	−0.20000	−0.58480	0.97467	−0.60000	0.40000
3	0.40000	0.73681	0.61401	1.20000	−0.80000
4	−0.80000	−0.92832	0.38680	−2.40000	1.60000

REMARK   In Example 3, the initial estimate $x_1 = 0.1$ fails to produce a convergent sequence. Try showing that Newton's Method also fails for every other choice of x_1 (other than the actual zero).

It can be shown that a sufficient condition to produce convergence of Newton's Method to a zero of f is that

$$\left| \frac{f(x)f''(x)}{[f'(x)]^2} \right| < 1$$

on an open interval containing the zero. For instance, in Example 1 this test would yield $f(x) = x^2 - 2, f'(x) = 2x, f''(x) = 2$, and

$$\left| \frac{f(x)f''(x)}{[f'(x)]^2} \right| = \left| \frac{(x^2 - 2)(2)}{4x^2} \right| = \left| \frac{1}{2} - \frac{1}{x^2} \right|. \qquad \text{Example 1}$$

On the interval $(1, 3)$ this quantity is less than 1 and therefore the convergence of Newton's Method is guaranteed. On the other hand, in Example 3, we have $f(x) = x^{1/3}, f'(x) = \frac{1}{3}x^{-2/3}, f''(x) = -\frac{2}{9}x^{-5/3}$, and

$$\left| \frac{f(x)f''(x)}{[f'(x)]^2} \right| = \left| \frac{x^{1/3}(-2/9)(x^{-5/3})}{(1/9)(x^{-4/3})} \right| = 2 \qquad \text{Example 3}$$

which is not less than 1 for any value of x.

EXAMPLE 4   Using Newton's Method to find a point of intersection

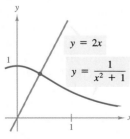

FIGURE 4.54

Estimate the point of intersection of the graphs of

$$y = \frac{1}{x^2 + 1} \qquad \text{and} \qquad y = 2x$$

as shown in Figure 4.54. Use Newton's Method and continue the iterations until two successive approximations differ by less than 0.0001.

SOLUTION

We let $2x = 1/(x^2 + 1)$. Since this implies that $2x(x^2 + 1) - 1 = 0$, we need to find the zero of the function given by

$$f(x) = 2x(x^2 + 1) - 1 = 2x^3 + 2x - 1.$$

Thus, the iterative formula for Newton's Method takes the form

$$x_{n+1} = x_n - \frac{f(x_n)}{f'(x_n)} = x_n - \frac{2x_n^3 + 2x_n - 1}{6x_n^2 + 2}$$

The calculations, shown in Table 4.20, begin with a guess of $x_1 = 0.5$.

TABLE 4.20

n	x_n	$f(x_n)$	$f'(x_n)$	$\dfrac{f(x_n)}{f'(x_n)}$	$x_n - \dfrac{f(x_n)}{f'(x_n)}$
1	0.50000	0.25957	3.50000	0.07143	0.42857
2	0.42857	0.01458	3.10204	0.00470	0.42387
3	0.42387	0.00006	3.07801	0.00002	0.42385

Thus, we approximate the point of intersection to occur when $x = 0.42385$. ▭

Algebraic solutions of polynomial equations

At the beginning of this section we mentioned that the zeros of the cubic function $f(x) = x^3 - x + 1$ cannot be found by *elementary* alebraic methods. This particular function happens to have only one real zero, and using more advanced algebraic techniques we can determine this value to be

$$x = -\sqrt[3]{\frac{3 - \sqrt{23/3}}{6}} - \sqrt[3]{\frac{3 + \sqrt{23/3}}{6}}.$$

Since the *exact* solution is written in terms of square roots and cube roots, we call it a **solution by radicals.**

The determination of radical solutions to a polynomial equation is one of the fundamental problems of algebra. The earliest result in this category is the well-known Quadratic Formula, and it dates back at least to Babylonian times. The general formula for the zeros of a cubic function was developed much later. In the sixteenth century an Italian mathematician, Jerome Cardan, published a method for finding radical solutions to cubic and quartic equations. Then, for 300 years the problem of finding a general quintic formula remained open. Finally, in the nineteenth century, the problem was answered independently by two young mathematicians. Niels Henrik Abel (1802–1829), a Norwegian mathematician, and Evariste Galois (1811–1832), a French mathematician, proved that it is not possible to solve a *general* fifth (or higher) degree polynomial equation by radicals. Of course, we can solve particular fifth-degree equations such as $x^5 - 1 = 0$, but Abel and Galois were able to show that no general *radical* solution exists.

Neils Henrik Abel

Evariste Galois

EXERCISES for Section 4.8

In Exercises 1–4, complete one iteration of Newton's Method for the given function using the indicated initial guess.

Function	Initial Guess
1. $f(x) = x^2 - 3$	$x_1 = 1.7$
2. $f(x) = 3x^2 - 2$	$x_1 = 1$
3. $f(x) = 3x^3 - 2$	$x_1 = 1$
4. $f(x) = x^3 - x^2 - 2x - 2$	$x_1 = 2.5$

In Exercises 5–10, approximate the indicated zero(s) of the function. Use Newton's Method and continue the process until two successive approximations differ by less than 0.001.

5. $f(x) = x^3 + x - 1$ **6.** $f(x) = x^5 + x - 1$

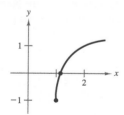 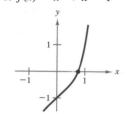

7. $f(x) = 3\sqrt{x - 1} - x$

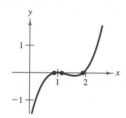

8. $f(x) = x^3 - 3.9x^2 + 4.79x - 1.881$

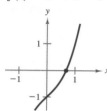

9. $f(x) = x^4 - 10x^2 - 11$ **10.** $f(x) = x^3 + 3$

 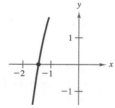

In Exercises 11–14, apply Newton's Method to approximate the x-value of the indicated point(s) of intersection of the two graphs. Continue the process until two successive approximations differ by less than 0.001.

11. $f(x) = 2x + 1$;   **12.** $f(x) = 3 - x$;

$g(x) = \sqrt{x + 4}$ $g(x) = \dfrac{1}{x^2 + 1}$

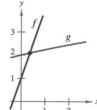 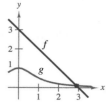

13. $f(x) = \dfrac{4}{x}$;   **14.** $f(x) = x^3$;

$g(x) = x^2 + 1$ $g(x) = x^2 + 2$

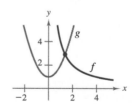 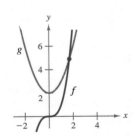

In Exercises 15–18, apply Newton's Method, using the indicated initial guess, and explain why the method fails.

15. $y = 2x^3 - 6x^2 + 6x - 1$

$x_1 = 1$

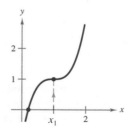

16. $y = 4x^3 - 12x^2 + 12x - 3$

$x_1 = \dfrac{3}{2}$

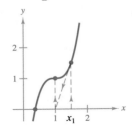

17. $y = -x^3 + 3x^2 - x + 1$ **18.** $f(x) = \sqrt[3]{x - 1}$

    $x_1 = 1$ $x_1 = 2$

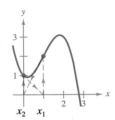   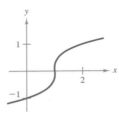

In Exercises 19 and 20, use Newton's Method to obtain a general rule for approximating the required radical.

19. $\sqrt{a}$ [Hint: Consider $f(x) = x^2 - a$.]
20. $\sqrt[n]{a}$ [Hint: Consider $f(x) = x^n - a$.]

In Exercises 21–24, use the results of Exercises 19 and 20 to approximate the indicated radical to three decimal places.

21. $\sqrt{7}$ **22.** $\sqrt{5}$
23. $\sqrt[4]{6}$ **24.** $\sqrt[3]{15}$

25. Use Newton's Method to show that the equation

$$x_{n+1} = x_n(2 - ax_n)$$

can be used to approximate $1/a$ if x_1 is an initial guess of the reciprocal of a. Note that this method of approximating reciprocals uses only the operations of multiplication and subtraction. [Hint: Consider $f(x) = (1/x) - a$.]

26. Use the result of Exercise 25 to approximate the indicated reciprocals to three decimal places.

(a) $\dfrac{1}{3}$ (b) $\dfrac{1}{11}$

In Exercises 27–30, we review some typical problems from the previous sections of this chapter. In each case, use Newton's Method to approximate the solution.

27. Find the point on the graph of $f(x) = 4 - x^2$ that is closest to the point $(1, 0)$.
28. Find the point on the graph of $f(x) = x^2$ that is closest to the point $(4, -3)$.
29. A woman is in a boat 2 miles from the nearest point on the coast (see figure). She is to go to a point Q, which is 3 miles down the coast and 1 mile inland. She can row at 3 miles per hour and walk at 4 miles per hour. Toward what point on the coast should she row in order to reach Q in the least time?

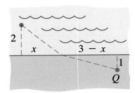

30. The concentration C of a certain chemical in the bloodstream t hours after injection into muscle tissue is given by

$$C = \frac{3t^2 + t}{50 + t^3}$$

When is the concentration greatest?

31. Use a computer or graphics calculator to approximate all the real zeros of

$$f(x) = \frac{1}{4}x^3 - 3x^2 + \frac{3}{4}x - 2$$

by Newton's Method. Sketch the graph of the function in order to make the initial estimate of a zero.

4.9 Differentials

Differentials ▪ Error propagation ▪ Differential formulas

We previously defined the derivative as the limit $(\Delta x \to 0)$ of the ratio $\Delta y / \Delta x$, and it seemed natural to retain the quotient symbolism for the limit itself. Thus, we denoted the derivative by

$$\frac{dy}{dx} = \lim_{\Delta x \to 0} \frac{\Delta y}{\Delta x}$$

even though we did not think of dy/dx as the quotient of the two separate quantities dy and dx. In this section we give separate meanings to dy and dx in such a way that their quotient, when $dx \neq 0$, is equal to the derivative of y with respect to x. We do this in the following manner.

DEFINITION OF DIFFERENTIALS

Let $y = f(x)$ represent a function that is differentiable in an open interval containing x. The **differential of x** (denoted by dx) is any nonzero real number. The **differential of y** (denoted by dy) is given by

$$dy = f'(x)\, dx.$$

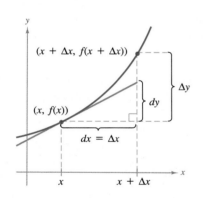

FIGURE 4.55

Note that in this definition dx can have any nonzero value. In most applications of differentials, however, we choose dx to be small and we denote this choice by $dx = \Delta x$.

One use of differentials is to approximate the change in $f(x)$ that corresponds to a change in x as shown in Figure 4.55. We denote this change by

$$\Delta y = f(x + \Delta x) - f(x).$$

The following example compares the values of dy and Δy.

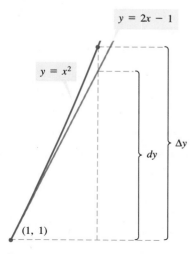

FIGURE 4.56

EXAMPLE 1 Comparing Δy and dy

Let $y = x^2$. Find dy when $x = 1$ and $dx = 0.01$. Compare this value to Δy when $x = 1$ and $\Delta x = 0.01$.

SOLUTION

Since $y = f(x) = x^2$, we have $f'(x) = 2x$, and the differential dy is given by

$$dy = f'(x)\, dx = f'(1)(0.01) \qquad \text{\small Differential of } y$$
$$= 2(0.01) = 0.02.$$

Now, since $\Delta x = 0.01$, the change in y is

$$\Delta y = f(x + \Delta x) - f(x) = f(1.01) - f(1) \qquad \text{\small Change in } y$$
$$= (1.01)^2 - 1^2 = 0.0201.$$

Figure 4.56 shows the geometric comparison of dy and Δy.

For the function in Example 1, we can see in Table 4.21 that *dy* approximates Δy more and more closely as *dx* approaches zero.

TABLE 4.21

$dx = \Delta x$	dy	Δy	$\Delta y - dy$
0.1000	0.2000	0.21000000	0.01000000
0.0100	0.0200	0.02010000	0.00010000
0.0010	0.0020	0.00200100	0.00000100
0.0001	0.0002	0.00020001	0.00000001

REMARK   Note that for the sequence shown in Table 4.21, as Δx decreases by tenths, $\Delta y - dy$ decreases by hundredths.

This type of approximation is called a **tangent line approximation** because we are using the tangent line at a point to approximate the graph of the function near the point. (Actually, this is not the first time we have encountered such an approximation. Recall from Section 4.8 that Newton's Method uses tangent lines to approximate an *x*-intercept of the graph of a function.)

The validity of using a tangent line to approximate a curve stems from the limit definition of the tangent line. That is, the existence of the limit

$$f'(x) = \lim_{\Delta x \to 0} \frac{f(x + \Delta x) - f(x)}{\Delta x}$$

implies that when Δx is close to zero (within δ units of zero), then $f'(x)$ is close to the difference quotient (within ε units) and we have

$$\frac{f(x + \Delta x) - f(x)}{\Delta x} \approx f'(x), \quad \Delta x \neq 0$$

$$f(x + \Delta x) - f(x) \approx f'(x)\,\Delta x$$

$$\Delta y \approx f'(x)\,\Delta x.$$

Substituting *dx* for Δx produces

$$\Delta y \approx f'(x)\,dx = dy, \quad dx \neq 0.$$

We formalize this result in the following theorem. The proof follows directly from the definition of the derivative.

THEOREM 4.11
THE RELATIVE SIZE
OF *dy* AND Δy

Let $y = f(x)$ be differentiable at x such that $f'(x) \neq 0$. If $dx = \Delta x$, then

$$\lim_{\Delta x \to 0} \frac{\Delta y}{dy} = 1$$

where $dy = f'(x)\,dx$ and $\Delta y = f(x + \Delta x) - f(x)$.

Error propagation

Physicists and engineers tend to make liberal use of the approximation of Δy by dy. One way this occurs in practice is in the estimation of errors propagated by physical measuring devices. For example, if we let x represent the measured value of a variable and let $x + \Delta x$ represent the exact value, then Δx is the *error in measurement*. Finally, if the measured value x is used to compute another value $f(x)$, then the difference between $f(x + \Delta x)$ and $f(x)$ is called the **propagated error.**

$$\underbrace{f(x + \overbrace{\Delta x}^{\substack{\text{measurement}\\\text{error}}})}_{\substack{\text{exact}\\\text{value}}} - \underbrace{f(x)}_{\substack{\text{measured}\\\text{value}}} = \underbrace{\Delta y}_{\substack{\text{propagated}\\\text{error}}}$$

EXAMPLE 2   Estimation of error

The radius of a ball bearing is measured to be 0.7 inch, as shown in Figure 4.57. If the measurement is correct to within 0.01 inch, estimate the propagated error in the volume V of the ball bearing.

SOLUTION

The formula for the volume of a sphere is $V = \frac{4}{3}\pi r^3$, where r is the radius of the sphere. Thus, we have

$$r = 0.7 \qquad \text{Measured radius}$$

and

$$-0.01 \le \Delta r \le 0.01. \qquad \text{Possible error}$$

To approximate the propagated error in the volume, we differentiate V to obtain $dV/dr = 4\pi r^2$ and write

$$\Delta V \approx dV = 4\pi r^2 \, dr = 4\pi(0.7)^2(\pm 0.01) \approx \pm 0.06158 \text{ in}^3. \qquad \blacksquare$$

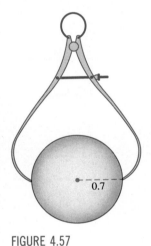

FIGURE 4.57

Would you say that the propagated error in Example 2 is large or small? The answer is best given in *relative* terms by comparing dV to V. We obtain the ratio

$$\frac{dV}{V} = \frac{4\pi r^2 \, dr}{\frac{4}{3}\pi r^3} = \frac{3 \, dr}{r} \approx \frac{3}{0.7}(\pm 0.01) \approx \pm 0.0429$$

which is called the **relative error.** The corresponding **percentage error** is

$$\frac{dV}{V}(100) \approx 4.29\%.$$

We can use the definition of a differential to rewrite each of the derivative rules in **differential form.** For example, suppose u and v are differentiable functions of x. By the definition of differentials, we have $du = u'\,dx$ and $dv = v'\,dx$. Therefore, we can write the differential form of the Product Rule as follows.

$$d[uv] = \frac{d}{dx}[uv]\ dx \qquad \text{Differential of } uv$$

$$= [uv' + vu']\ dx \qquad \text{Product Rule}$$

$$= uv'\ dx + vu'\ dx$$

$$= u\ dv + v\ du$$

The following summary lists the differential forms corresponding to several differentiation rules.

GENERAL DIFFERENTIAL FORMULAS

Let u and v be differentiable functions of x.

Constant multiple: $d[cu] = c\,du$
Sum or difference: $d[u \pm v] = du \pm dv$
Product: $d[uv] = u\,dv + v\,du$
Quotient: $d\left[\dfrac{u}{v}\right] = \dfrac{v\,du - u\,dv}{v^2}$

In the next example, we compare the derivatives and differentials of several functions.

EXAMPLE 3 Finding differentials

Function	Derivative	Differential
(a) $y = x^2$	$\dfrac{dy}{dx} = 2x$	$dy = 2x\,dx$
(b) $y = 2x^3 + x$	$\dfrac{dy}{dx} = 6x^2 + 1$	$dy = (6x^2 + 1)\,dx$
(c) $y = \sqrt{2x + 1}$	$\dfrac{dy}{dx} = \dfrac{1}{\sqrt{2x + 1}}$	$dy = \left(\dfrac{1}{\sqrt{2x + 1}}\right)dx$
(d) $y = \dfrac{1}{x}$	$\dfrac{dy}{dx} = -\dfrac{1}{x^2}$	$dy = -\dfrac{dx}{x^2}$

The notation in Example 3 is called the **Leibniz notation** for derivatives and differentials, named after the German mathematician Gottfried Wilhelm Leibniz (1646–1716). The beauty of this notation is that it provides us with an easy way to remember several important calculus formulas by making it seem as though the formulas were derived from algebraic manipulations of differentials. We will encounter several instances of this later in the text (in substitutions, inverse functions, parametric equations, and polar coordinates).

Gottfried Wilhelm Leibniz

For now, we simply compare the *Chain Rule* in Leibniz notation to other possible notations.

$$\frac{dy}{dx} = \frac{dy}{du}\frac{du}{dx}$$ 　　　　Leibniz notation

$$f'(g(x)) = f'(u)g'(x), \quad \text{where } u = g(x)$$ 　　Function notation

$$D_x[y] = D_u[y]D_x[u]$$ 　　　　Operator notation

The Leibniz form of the rule probably would appear to be true to a student in elementary algebra. Of course, it appears to be true for the *wrong reason*— that we have simply cancelled the *du*'s. Even so, the notation's many advantages overshadow the potential problem of drawing valid conclusions from invalid arguments.

EXAMPLE 4 Differential of a composite function

Use the Chain Rule to find *dy* for $y = (3x^2 + x)^4$.

SOLUTION

$$y = f(x) = (3x^2 + x)^4$$
$$f'(x) = 4(6x + 1)(3x^2 + x)^3$$
$$dy = f'(x)\,dx = 4(6x + 1)(3x^2 + x)^3\,dx$$

EXAMPLE 5 Differential of a composite function

Use the Chain Rule to find *dy* for $y = \sqrt{x^2 + 1}$.

SOLUTION

$$y = f(x) = (x^2 + 1)^{1/2}$$
$$f'(x) = \frac{1}{2}(2x)(x^2 + 1)^{-1/2} = \frac{x}{\sqrt{x^2 + 1}}$$

$$dy = f'(x)\,dx = \frac{x}{\sqrt{x^2 + 1}}\,dx$$

Differentials can be used to approximate function values. To do this for the function given by $y = f(x)$, we make use of the formula

$$f(x + \Delta x) \approx f(x) + dy = f(x) + f'(x)\,dx$$

which is derived from the approximation $\Delta y = f(x + \Delta x) - f(x) \approx dy$. The key to using this formula is choosing a value for *x* that makes the calculations easier. For instance, to approximate $\sqrt{101}$ we would let $x = 100$ and $\Delta x = 1$. This is illustrated in the next example.

EXAMPLE 6 Approximating function values

Use differentials to approximate $\sqrt{16.5}$.

SOLUTION

We use the function $f(x) = \sqrt{x}$ and choose $x = 16$. Then $dx = 0.5$, and we obtain

$$f(x + \Delta x) \approx f(x) + f'(x)\, dx = \sqrt{x} + \frac{1}{2\sqrt{x}}\, dx$$

$$\sqrt{16.5} \approx \sqrt{16} + \frac{1}{2\sqrt{16}}(0.5) = 4 + \left(\frac{1}{8}\right)\!\left(\frac{1}{2}\right)$$

$$= 4 + \frac{1}{16} = 4.0625.$$

REMARK The use of differentials to approximate function values diminished with the availability of calculators and computers. Using a calculator, we obtain the value

$$\sqrt{16.5} \approx 4.0620$$

which indicates the accuracy of the differential method in Example 6.

EXERCISES for Section 4.9

In Exercises 1–10, find the differential dy of the given function.

1. $y = 3x^2 - 4$
2. $y = 2x^{3/2}$
3. $y = 4x^3$
4. $y = 3$
5. $y = \dfrac{x + 1}{2x - 1}$
6. $y = \dfrac{x}{x + 5}$
7. $y = \sqrt{x}$
8. $y = \sqrt{x^2 - 4}$
9. $y = x\sqrt{1 - x^2}$
10. $y = \sqrt{x} + \dfrac{1}{\sqrt{x}}$

In Exercises 11–16, let $x = 2$ and use the given function and value of $\Delta x = dx$ to complete the table.

$dx = \Delta x$	dy	Δy	$\Delta y - dy$	$\dfrac{dy}{\Delta y}$
1.000				
0.500				
0.100				
0.010				
0.001				

11. $y = x - 1$
12. $y = 2x$
13. $y = x^2$
14. $y = \dfrac{1}{x^2}$
15. $y = x^5$
16. $y = \sqrt{x}$

17. The area of a square of side x is given by $A(x) = x^2$.
  (a) Compute dA and ΔA in terms of x and Δx.
  (b) Use the accompanying figure to identify the region whose area is dA.
  (c) Use the accompanying figure to identify the region whose area is $\Delta A - dA$.

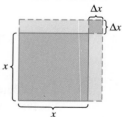

18. The measurement of the side of a square is found to be 12 inches, with a possible error of $\frac{1}{64}$ inch. Use differentials to approximate the possible error in computing the area of the square.

19. The measurement of the radius of the end of a log is found to be 14 inches, with a possible error of $\frac{1}{4}$ inch. Use differentials to approximate the possible error in computing the area of the end of the log.

20. The measurement of the edge of a cube is found to be 12 inches, with a possible error of 0.03 inch. Use differentials to approximate the maximum possible error in computing the following.
    (a) the volume of the cube
    (b) the surface area of the cube

21. The measurement of a side of a square is found to be 15 centimeters.
    (a) Approximate the percentage error in computing the area of the square if the possible error in measuring the side is 0.05 centimeters.
    (b) Estimate the maximum allowable percentage error in measuring the side if the error in computing the area cannot exceed 2.5%.

22. The measurement of the circumference of a circle is found to be 56 centimeters.
    (a) Approximate the percentage error in computing the area of the circle if the possible error in measuring the circumference is 1.2 centimeters.
    (b) Estimate the maximum allowable percentage error in measuring the circumference if the error in computing the area cannot exceed 3%.

23. The radius of a sphere is claimed to be 6 inches, with a possible error of 0.02 inch. Use differentials to approximate the maximum possible error in calculating (a) the volume of the sphere and (b) the surface area of the sphere. (c) What is the relative error in parts (a) and (b)?

24. The profit P for a company is given by

$$P = (500x - x^2) - \left(\frac{1}{2}x^2 - 77x + 3000\right).$$

Approximate the change in profit as production changes from $x = 115$ to $x = 120$ units. Approximate the percentage change in profit when x changes from $x = 115$ to $x = 120$ units.

25. The period of a pendulum is given by

$$T = 2\pi\sqrt{\frac{L}{g}}$$

where L is the length of the pendulum in feet, g is the acceleration due to gravity, and T is the time in seconds. Suppose that the pendulum has been subjected to an increase in temperature so that the length increases by $\frac{1}{2}$ percent.
    (a) Find the approximate percentage change in the period.
    (b) Using the result of part (a), find the approximate error in this pendulum clock in one day.

26. Suppose a pendulum of length L is to be used to find the acceleration of gravity at a given point on the earth's surface. Use the formula

$$T = 2\pi\sqrt{\frac{L}{g}}$$

to approximate the percentage error in the value of g if you can identify the period T to within 0.1 percent of its true value.

27. A current of I amps passes through a resistor of R ohms. **Ohms Law** states that the voltage E applied to the resistor is given by

$$E = IR.$$

If the voltage is constant, show that the magnitude of the relative error in R caused by a change in I is equal in magnitude to the relative error in I.

28. The cost in dollars of removing $p\%$ of the air pollutants in the stack emission of a utility company that burns coal to generate electricity is

$$C = \frac{80,000p}{100 - p}, \quad 0 \le p < 100.$$

Use differentials to approximate the increase in cost if the government requires the utility company to remove 2% more of the pollutants and p is currently
    (a) 40%   (b) 75%

29. Show that if $y = f(x)$ is a differentiable function, then

$$\Delta y - dy = \varepsilon\, \Delta x$$

where $\varepsilon \to 0$ as $\Delta x \to 0$.

4.10 Business and Economics Applications

Marginals ▪ Demand function

In Section 3.8 we mentioned that one of the most common ways to measure change is with respect to time. In this section we study some important rates of change in economics that are not measured with respect to time. For example, economists refer to **marginal profit, marginal revenue,** and **marginal cost** as the rates of change of the profit, revenue, and cost with respect to the number of units produced or sold.

We begin with a summary of some basic terms and formulas.

SUMMARY OF BUSINESS TERMS AND FORMULAS

Basic Terms

x is the number of units produced (or sold)
p is the price per unit
R is the total revenue from selling *x* units
C is the total cost of producing *x* units
$\overline{C}$ is the average cost per unit
P is the total profit from selling *x* units

Basic Formulas

$$R = xp$$

$$\overline{C} = C/x$$
$$P = R - C$$

The **break-even point** is the number of units for which $R = C$.

Marginals

$$\frac{dR}{dx} = \text{marginal revenue} \approx \text{the } extra \text{ revenue for selling one additional unit}$$

$$\frac{dC}{dx} = \text{marginal cost} \approx \text{the } extra \text{ cost of producing one additional unit}$$

$$\frac{dP}{dx} = \text{marginal profit} \approx \text{the } extra \text{ profit for selling one additional unit}$$

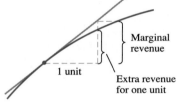

1 unit — Marginal revenue — Extra revenue for one unit

Revenue Function

FIGURE 4.58

REMARK In this summary, note that marginals can be used to approximate the *extra* revenue, cost, or profit associated with selling or producing one additional unit. This is illustrated graphically for marginal revenue in Figure 4.58.

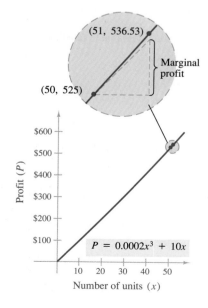

FIGURE 4.59

EXAMPLE 1   Using marginals as approximations

A manufacturer determines that the profit derived from selling *x* units of a certain item is given by $P = 0.0002x^3 + 10x$.
(a) Find the marginal profit for a production level of 50 units.
(b) Compare this to the actual gain in profit obtained by increasing the production from 50 to 51 units. (See Figure 4.59.)

SOLUTION

(a) The marginal profit is given by

$$\frac{dP}{dx} = 0.0006x^2 + 10.$$

When $x = 50$, the marginal profit is

$$\frac{dP}{dx} = (0.0006)(50)^2 + 10 = \$11.50. \qquad \text{Marginal profit}$$

(b) For $x = 50$ and 51, the actual profit is

$$P = (0.0002)(50)^3 + 10(50) = 25 + 500 = \$525.00$$
$$P = (0.0002)(51)^3 + 10(51) = 26.53 + 510 = \$536.53.$$

Thus, the additional profit obtained by increasing the production level from 50 to 51 units is

$$536.53 - 525.00 = \$11.53. \qquad \text{Extra profit for one unit} \quad \blacksquare$$

The profit function in Example 1 is unusual in that the profit continues to increase as long as the number of units sold increases. In practice, it is more common to encounter situations in which sales can be increased only by lowering the price per item, and such reductions in price ultimately will cause the profit to decline. The number of units x that consumers are willing to purchase at a given price per unit p is defined as the **demand function**

$$p = f(x). \qquad \text{Demand function}$$

EXAMPLE 2 Finding the demand function

A business sells 2000 items per month at a price of \$10 each. It is predicted that monthly sales will increase by 250 items for each \$0.25 reduction in price. Find the demand function corresponding to this prediction.

SOLUTION

From the given prediction, x increases 250 units each time p drops \$0.25 from the original cost of \$10. This is described by the equation

$$x = 2000 + 250\left(\frac{10 - p}{0.25}\right) = 12{,}000 - 1000p$$

or

$$p = 12 - \frac{x}{1000}. \qquad \text{Demand function} \quad \blacksquare$$

EXAMPLE 3 Finding the marginal revenue

A fast-food restaurant has determined that the monthly demand for their hamburgers is given by

$$p = \frac{60{,}000 - x}{20{,}000}.$$

Find the increase in revenue per hamburger for monthly sales of 20,000 hamburgers. In other words, find the marginal revenue when $x = 20{,}000$.

SOLUTION

Since the total revenue is given by $R = xp$, we have

$$R = xp = x\left(\frac{60{,}000 - x}{20{,}000}\right) = \frac{1}{20{,}000}(60{,}000x - x^2)$$

and the marginal revenue is

$$\frac{dR}{dx} = \frac{1}{20{,}000}(60{,}000 - 2x).$$

Finally, when $x = 20{,}000$, the marginal revenue is

$$\frac{dR}{dx} = \frac{1}{20{,}000}[60{,}000 - 2(20{,}000)] = \frac{20{,}000}{20{,}000} = \$1/\text{unit}.$$

REMARK Note that the demand function in Example 3 is typical in that a high demand corresponds to a low price, as seen in Figure 4.60 and Table 4.22.

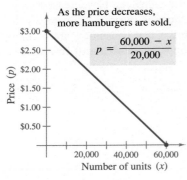

As the price decreases, more hamburgers are sold.

$$p = \frac{60{,}000 - x}{20{,}000}$$

Price (p)

Number of units (x)

FIGURE 4.60

TABLE 4.22

p	$\$0.00$	$\$0.50$	$\$1.00$	$\$1.50$	$\$2.00$	$\$2.50$	$\$3.00$
x	60,000	50,000	40,000	30,000	20,000	10,000	0

EXAMPLE 4 Finding the marginal profit

Suppose that in Example 3 the cost of producing x hamburgers is

$$C = 5000 + 0.56x.$$

Find the total profit and the marginal profit for (a) 20,000, (b) 24,400, and (c) 30,000 units.

SOLUTION

Since $P = R - C$, we use the revenue function in Example 3 to obtain

$$P = \frac{1}{20{,}000}(60{,}000x - x^2) - 5{,}000 - 0.56x$$

$$= 2.44x - \frac{x^2}{20{,}000} - 5{,}000.$$

Thus, the marginal profit is

$$\frac{dP}{dx} = 2.44 - \frac{x}{10{,}000}.$$

Table 4.23 shows the total profit and the marginal profit for the three indicated demands.

Maximum profit occurs
when $dP/dx = 0$.

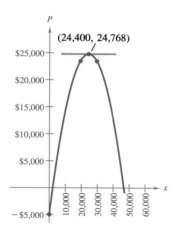

$$P = 2.44x - \frac{x^2}{20,000} - 5,000$$

FIGURE 4.61

TABLE 4.23

Demand	20,000	24,400	30,000
Profit	$23,800	$24,768	$23,200
Marginal Profit	$0.44	$0.00	−$0.56

REMARK   In Example 4 we can see that when more than 24,400 hamburgers are sold, the marginal profit is negative. This means that increasing production beyond this point will *reduce* rather than increase profit. Figure 4.61 shows that the maximum profit corresponds to the point where the marginal profit is zero.

It is worth noting that many problems in economics are minimum and maximum problems. For such problems, the procedure used in Section 4.7 is an appropriate model to follow.

EXAMPLE 5   Finding the maximum profit

In marketing a certain item, a business has discovered that the demand for the item is represented by

$$p = \frac{50}{\sqrt{x}}.$$ Demand function

The cost of producing x items is given by $C = 0.5x + 500$. Find the price per unit that yields a maximum profit.

SOLUTION

From the given cost function, we have

$$P = R - C = xp - (0.5x + 500).$$ Primary equation

Substituting for p (from the demand function), we have

$$P = x\left(\frac{50}{\sqrt{x}}\right) - (0.5x + 500) = 50\sqrt{x} - 0.5x - 500.$$

Setting the marginal profit equal to zero, we obtain

$$\frac{dP}{dx} = \frac{25}{\sqrt{x}} - 0.5 = 0$$

$$\sqrt{x} = \frac{25}{0.5} = 50 \implies x = 2500.$$

Finally, we conclude that the maximum profit occurs when the price is

$$p = \frac{50}{\sqrt{2500}} = \frac{50}{50} = \$1.00.$$

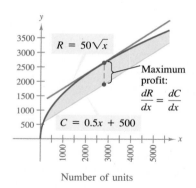

FIGURE 4.62

REMARK To find the maximum profit in Example 5, we differentiated the profit function, $P = R - C$, and set dP/dx equal to zero. From the equation

$$\frac{dP}{dx} = \frac{dR}{dx} - \frac{dC}{dx} = 0$$

it follows that the maximum profit occurs when the marginal revenue is equal to the marginal cost, as shown in Figure 4.62.

To study the effect of production levels on cost, economists use the average cost function $\overline{C}$ defined as

$$\overline{C} = \frac{C}{x}. \qquad \text{Average cost function}$$

A use of this function is illustrated in the next example.

EXAMPLE 6 Minimizing the average cost

A company estimates that the cost (in dollars) of producing x units of a certain product is given by $C = 800 + 0.04x + 0.0002x^2$. Find the production level that minimizes the average cost per unit.

SOLUTION

Substituting from the given equation for C produces

$$\overline{C} = \frac{C}{x} = \frac{800 + 0.04x + 0.0002x^2}{x} = \frac{800}{x} + 0.04 + 0.0002x.$$

Setting the derivative $d\overline{C}/dx$ equal to zero yields

$$\frac{d\overline{C}}{dx} = -\frac{800}{x^2} + 0.0002 = 0$$

$$x^2 = \frac{800}{0.0002} = 4,000,000 \quad \Longrightarrow \quad x = 2,000 \text{ units.}$$

(See Figure 4.63.)

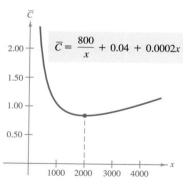

FIGURE 4.63

EXERCISES for Section 4.10

In Exercises 1–4, find the number of units x that produce a maximum revenue R.

1. $R = 900x - 0.1x^2$ **2.** $R = 600x^2 - 0.02x^3$

3. $R = \dfrac{1,000,000x}{0.02x^2 + 1800}$ **4.** $R = 30x^{2/3} - 2x$

In Exercises 5–8, find the number of units x that produce the minimum average cost per unit $\overline{C}$.

5. $C = 0.125x^2 + 20x + 5000$
6. $C = 0.001x^3 - 5x + 250$

7. $C = 3000x - x^2\sqrt{300 - x}$

8. $C = \dfrac{2x^3 - x^2 + 5000x}{x^2 + 2500}$

In Exercises 9–12, find the price per unit p that produces the maximum profit P.

Cost function	Demand function
9. $C = 100 + 30x$	$p = 90 - x$
10. $C = 2400x + 5200$	$p = 6000 - 0.4x^2$

Cost function	Demand function

11. $C = 4000 - 40x + 0.02x^2$ $p = 50 - \dfrac{x}{100}$

12. $C = 35x + 2\sqrt{x-1}$ $p = 40 - \sqrt{x-1}$

13. A manufacturer of lighting fixtures has daily production costs of

$$C = 800 - 10x + \frac{1}{4}x^2.$$

How many fixtures x should be produced each day to minimize costs?

14. The profit for a certain company is given by

$$P = 230 + 20s - \frac{1}{2}s^2$$

where s is the amount (in hundreds of dollars) spent on advertising. What amount of advertising gives the maximum profit?

15. A manufacturer of radios charges $90 per unit when the average production cost per unit is $60. To encourage large orders from distributors, the manufacturer will reduce the charge by $0.10 per unit for each unit ordered in excess of 100 (for example, there would be a charge of $88 per radio for an order size of 120). Find the largest order the manufacturer should allow so as to realize maximum profit.

16. A real estate office handles 50 apartment units. When the rent is $540 per month, all units are occupied. However, on the average, for each $30 increase in rent, one unit becomes vacant. Each occupied unit requires an average of $36 per month for service and repairs. What rent should be charged to realize the most profit?

17. A power station is on one side of a river that is $\frac{1}{2}$ mile wide, and a factory is 6 miles downstream on the other side. It costs $6 per foot to run power lines overland and $8 per foot to run them underwater. Find the most economical path for the transmission line from the power station to the factory.

18. An offshore oil well is 1 mile off the coast. The refinery is 2 miles down the coast. If laying pipe in the ocean is twice as expensive as on land, what path should the pipe follow in order to minimize the cost?

19. Assume that the amount of money deposited in a bank is proportional to the square of the interest rate the bank pays on this money. Furthermore, the bank can reinvest this money at 12 percent. Find the interest rate the bank should pay to maximize profit. (Use the simple interest formula.)

20. Prove that the average cost is minimum at the value of x where the average cost equals the marginal cost.

In Exercises 21 and 22, use the given cost function to find the value of x where the average cost is minimum. For that value of x, show that the marginal cost and average cost are equal (see Exercise 20).

21. $C = 2x^2 + 5x + 18$

22. $C = x^3 - 6x^2 + 13x$

23. A given commodity has a demand function given by $p = 100 - \frac{1}{2}x^2$ and a total cost function of $C = 4x + 375$.
  (a) What price gives the maximum profit?
  (b) What is the average cost per unit if production is set to give maximum profit?

24. When a wholesaler sold a certain product at $25 per unit, sales were 800 units each week. After a price increase of $5, the average number of units sold dropped to 775 per week. Assume that the demand function is linear and find the price that will maximize the total revenue.

25. The ordering and transportation cost C of the components used in manufacturing a certain product is given by

$$C = 100\left(\frac{200}{x^2} + \frac{x}{x+30}\right), \quad 1 \le x$$

where C is measured in thousands of dollars and x is the order size in hundreds. Find the order size that minimizes the cost. [Hint: Use Newton's Method.]

26. A company estimates that the cost in dollars of producing x units of a certain product is given by the model

$$C = 800 + 0.4x + 0.02x^2 + 0.0001x^3.$$

Find the production level that minimizes the average cost per unit. [Hint: Use Newton's Method.]

27. The revenue R for a company selling x units is given by

$$R = 900x - 0.1x^2.$$

Use differentials to approximate the change in revenue if sales increase from $x = 3000$ to $x = 3100$ units.

28. The profit P for a company selling x units is given by

$$P = (500x - x^2) - \left(\frac{1}{2}x^2 - 77x + 3000\right).$$

Use differentials to approximate the change and percentage change in profit as production changes from $x = 175$ to $x = 180$ units.

In Exercises 29–32, find η for the given demand function at the indicated x-value. Is the demand elastic, inelastic, or neither at the indicated x-value?

Demand Function	Quantity Demanded
29. $p = 400 - 3x$	$x = 20$
30. $p = 5 - 0.03x$	$x = 100$
31. $p = 400 - 0.5x^2$	$x = 20$
32. $p = \dfrac{500}{x + 2}$	$x = 23$

The relative responsiveness of consumers to a change in the price of an item is called the **price elasticity of demand.** If $p = f(x)$ is a differentiable demand function, then the price elasticity of demand is given by

$$\eta = \frac{p/x}{dp/dx}.$$

For a given price, if $|\eta| < 1$, the demand is **inelastic,** and if $|\eta| > 1$, the demand is **elastic.**

33. If the demand equation is given by $xp^m = a$, where a is a constant and $m > 1$, show that $\eta = -m$. (In other words, in terms of approximate price changes, a 1 percent increase in price results in an m percent decrease in quantity demanded.)

REVIEW EXERCISES for Chapter 4

In Exercises 1–24, make use of domain, range, symmetry, asymptotes, intercepts, relative extrema, or points of inflection to obtain an accurate graph of the given function.

1. $f(x) = 4x - x^2$ **2.** $f(x) = 4x^3 - x^4$

3. $f(x) = x\sqrt{16 - x^2}$ **4.** $f(x) = x + \dfrac{4}{x^2}$

5. $f(x) = \dfrac{x + 1}{x - 1}$ **6.** $f(x) = x^2 + \dfrac{1}{x}$

7. $f(x) = x^3 + x + \dfrac{4}{x}$ **8.** $f(x) = x^3(x + 1)$

9. $f(x) = (x - 1)^3(x - 3)^2$
10. $f(x) = (x - 3)(x + 2)^3$
11. $f(x) = (5 - x)^3$ **12.** $f(x) = (x^2 - 4)^2$
13. $f(x) = x^{1/3}(x + 3)^{2/3}$
14. $f(x) = (x - 2)^{1/3}(x + 1)^{2/3}$

15. $f(x) = x^3 + \dfrac{243}{x}$ **16.** $f(x) = \dfrac{2x}{1 + x^2}$

17. $f(x) = \dfrac{4}{1 + x^2}$ **18.** $f(x) = \dfrac{x^2}{1 + x^4}$

19. $f(x) = |x^2 - 9|$ **20.** $f(x) = |9 - x^2|$
21. $f(x) = |x^3 - 3x^2 + 2x|$
22. $f(x) = |x - 1| + |x - 3|$

23. $f(x) = \dfrac{1}{|x - 1|}$ **24.** $f(x) = \dfrac{x - 1}{1 + 3x^2}$

25. Find the maximum and minimum points on the graph of

$$x^2 + 4y^2 - 2x - 16y + 13 = 0.$$

(a) without using calculus and (b) using calculus.

26. Consider the function $f(x) = x^n$ for positive integer values of n.
(a) For what values of n does the function have a relative minimum at the origin?

(b) For what values of n does the function have a point of inflection at the origin?

In Exercises 27–32, find the point(s) guaranteed by the Mean Value Theorem for the indicated interval.

Function	Interval
27. $f(x) = \dfrac{2x + 3}{3x + 2}$	$1 \le x \le 5$
28. $f(x) = \dfrac{1}{x}$	$1 \le x \le 4$
29. $f(x) = x^{2/3}$	$1 \le x \le 8$
30. $f(x) = |x^2 - 9|$	$0 \le x \le 2$
31. $f(x) = x - \dfrac{1}{x}$	$1 \le x \le 4$
32. $f(x) = \sqrt{x} - 2x$	$0 \le x \le 4$

33. Can the Mean Value Theorem be applied to the function $f(x) = 1/x^2$ on the interval $[-2, 1]$?
34. If $f(x) = 3 - |x - 4|$, verify that $f(1) = f(7)$ and yet $f'(x)$ is not equal to zero for any x in $[1, 7]$. Does this contradict Rolle's Theorem?
35. For the function

$$f(x) = Ax^2 + Bx + C$$

determine the value of c guaranteed by the Mean Value Theorem on the interval $[x_1, x_2]$.
36. Demonstrate the result of Exercise 35 for

$$f(x) = 2x^2 - 3x + 1$$

on the interval $[0, 4]$.
37. At noon ship A was 100 miles due east of ship B. Ship A is sailing west at 12 miles per hour, and ship B is sailing south at 10 miles per hour. At what time will

the ships be nearest to each other, and what will this distance be?

38. The cost C of producing x units per day is

$$C = \frac{1}{4}x^2 + 62x + 125$$

and the price p per unit is

$$p = 75 - \frac{1}{3}x.$$

(a) What daily output produces maximum profit?
(b) What daily output produces minimum average cost?
(c) Find the elasticity of demand.

39. Find the maximum profit if the demand equation is

$$p = 36 - 4x$$

and the total cost is

$$C = 2x^2 + 6.$$

40. Find the dimensions of the rectangle of maximum area, with sides parallel to the coordinate axes, that can be inscribed in the ellipse given by

$$\frac{x^2}{144} + \frac{y^2}{16} = 1.$$

41. A right triangle in the first quadrant has the coordinate axes as sides, and the hypotenuse passes through the point $(1, 8)$. Find the vertices of the triangle so that the length of the hypotenuse is minimum.

42. The wall of a building is to be braced by a beam that must pass over a parallel fence 5 feet high and 4 feet from the building. Find the length of the shortest beam that can be used.

43. For groups of 80 or more, a charter bus company determines the rate per person according to the following formula.

Rate $= \$8.00 - \$0.05(n - 80), \quad n \geq 80$

What number of passengers will give the bus company maximum revenue?

44. Show that the greatest area of any rectangle inscribed in a triangle is one-half that of the triangle.

45. Three sides of a trapezoid have the same length s. Of all such possible trapezoids, show that the one of maximum area has a fourth side of length $2s$.

46. The cost of fuel for running a locomotive is proportional to the $\frac{3}{2}$ power of the speed $(s^{3/2})$ and is $\$50$ per hour for a speed of 25 miles per hour. Other fixed costs amount to an average of $\$100$ per hour. Find the speed that will minimize the cost per mile.

47. Find the length of the longest pipe that can be carried level around a right-angle corner if the two intersecting corridors are of width 4 feet and 6 feet.

48. Rework Exercise 47, given corridors of width a feet and b feet.

49. The cost of inventory depends on ordering cost and storage cost, according to the following inventory model.

$$C = \left(\frac{Q}{x}\right)s + \left(\frac{x}{2}\right)r$$

Determine the order size that will minimize the cost, assuming that sales occur at a constant rate, Q is the number of units sold per year, r is the cost of storing one unit for one year, s is the cost of placing an order, and x is the number of units per order.

50. The demand and cost equations for a certain product are

$$p = 600 - 3x$$

and

$$C = 0.3x^2 + 6x + 600$$

where p is the price per unit, x is the number of units, and C is the cost of producing x units. If t is the excise tax per unit, the profit for producing x units is $P = xp - C - xt$. Find the maximum profit for (a) $t = 5$, (b) $t = 10$, and (c) $t = 20$.

51. Approximate, to three decimal places, the zero of

$$f(x) = x^3 - 3x - 1$$

in the interval $[-1, 0]$.

52. Approximate, to three decimal places, the x-value of the points of intersection of the equations $y = x^4$ and $y = x + 3$.

53. Approximate, to three decimal places, the real root of

$$x^3 + 2x + 1 = 0.$$

54. The diameter of a sphere is measured to be 18 inches with a maximum possible error of 0.05 inch. Use the differential to approximate the possible error in the surface area and the volume of the sphere.

55. If a 1 percent error is made in measuring the edge of a cube, approximately what percentage error will be made in calculating the surface area and the volume of the cube?

56. A company finds that the demand for its commodity is given by

$$p = 75 - \frac{1}{4}x.$$

If x changes from 7 to 8, find the corresponding change in p. Compare the value of Δp and dp.

Chapter 5 Application

Voyager 1 and Voyager 2 were launched by the United States in 1977 to observe Jupiter and Saturn. Voyager 1 reached Jupiter in 1979 and Saturn in 1980. Voyager 2 also passed by Jupiter (1979) and Saturn (1981) and then continued on to observe the outer planets. In 1986, Voyager 2 transmitted stunning pictures of the rings of Uranus.

Escape Velocity from the Earth

The *escape velocity* from the earth is the minimum initial velocity that will allow an object, projected straight upward, to escape the earth's gravitational pull. If y is the distance from the center of the earth to the projected object, then the object will continue to move away from the earth provided that, for all t, the velocity of the object is positive. That is,

$$v = \frac{dy}{dt} > 0.$$

From Newton's Law of gravitation, we can conclude that the acceleration of the object is given by

$$\frac{dv}{dt} = -\frac{GM}{y^2}$$

where G is the gravitational constant and M is the mass of the earth. Using the Chain Rule, we have $dv/dt = (dv/dy)(dy/dt) = v(dv/dy)$. Thus, we can write

$$\int v \, dv = -GM \int \frac{1}{y^2} dy$$

$$\frac{1}{2}v^2 = \frac{GM}{y} + C.$$

Solving for the constant of integration C produces

$$v^2 = v_0{}^2 + 2GM\left(\frac{1}{y} - \frac{1}{R}\right)$$

where v_0 is the initial velocity of the object and R is the radius of the earth. Since $y > 0$, it follows that

$$v^2 > v_0{}^2 - \frac{2GM}{R}$$

which implies that v will be positive provided that

$$v_0 > \sqrt{\frac{2GM}{R}} \approx 24{,}995 \text{ mi/hr.}$$

See Exercise 46. Section 5.1.

Chapter Overview

In this chapter we introduce the second fundamental operation of calculus—**antidifferentiation** or **integration.** The chapter begins by looking at some basic rules for finding antiderivatives. For example, the family of antiderivatives of $f(x) = 2x$ is given by

$$F(x) = \int 2x \, dx = x^2 + C$$

where C is the constant of integration. Antidifferentiation has many uses. For instance, if the velocity function of an object is known, then antidifferentiation can be used to find its position function.

In Section 5.2, we introduce the problem of finding the area of a plane region. For instance, the area of the region bounded by the graphs of $f(x) = x^2$, $y = 0$, $x = 0$, and $x = 1$ is given by the **definite integral**

$$\text{area} = \int_0^1 x^2 \, dx.$$

Section 5.3 discusses several properties of definite integrals, and in Section 5.4, we present the **Fundamental Theorem of Calculus.** This theorem shows how antiderivatives can be used to solve definite integrals. The chapter closes by looking at two methods of approximating definite integrals—the Trapezoidal Rule and Simpson's Rule.

Integration

5.1 Antiderivatives and Indefinite Integration

Antiderivatives ▪ Notation for antiderivatives ▪ Basic integration rules ▪ Initial conditions and particular solutions

Until now, our study of calculus has been concerned primarily with this problem: *given a function, find its derivative*. Now, however, we will study the inverse problem: *given the derivative of a function, find the original function*. For example, suppose you were asked to find a function F that has the following derivative.

$$F'(x) = 3x^2$$

From your knowledge of derivatives, you would probably say that

$$F(x) = x^3 \quad \text{because} \quad \frac{d}{dx}[x^3] = 3x^2.$$

We call the function F an **antiderivative** of F'. For convenience, the phrase *$F(x)$ is an antiderivative of $f(x)$* is used synonymously with *F is an antiderivative of f*. For instance, we say that x^3 is an antiderivative of $3x^2$.

DEFINITION OF ANTIDERIVATIVE

A function F is called an **antiderivative** of the function f if for every x in the domain of f

$$F'(x) = f(x).$$

In this definition, we call F *an* antiderivative of f, rather than *the* antiderivative of f. To see why, consider that $F_1(x) = x^3$, $F_2(x) = x^3 - 5$, and $F_3(x) = x^3 + 97$ are all antiderivatives of $f(x) = 3x^2$. This suggests that for any constant C, the function given by $F(x) = x^3 + C$ is an antiderivative of f. This result is part of the following theorem.

229

**THEOREM 5.1
REPRESENTATION OF
ANTIDERIVATIVES**

If F is an antiderivative of f on an interval I, then G is an antiderivative of f on the interval I if and only if G is of the form

$$G(x) = F(x) + C, \quad \text{for all } x \text{ in } I$$

where C is a constant.

PROOF   There are two directions in this theorem. The proof of one direction is straightforward. That is, if $F'(x) = f(x)$ and C is a constant, then

$$G'(x) = \frac{d}{dx}[F(x) + C]$$

$$= F'(x) + 0$$

$$= f(x).$$

The proof of the other direction requires a little more work, but it can be accomplished with the Mean Value Theorem, by defining a function H such that

$$H(x) = G(x) - F(x).$$

If H is not constant on the interval I, then there must exist a and b ($a < b$) in the interval such that $H(a) \neq H(b)$. Moreover, because H is differentiable on $[a, b]$, we can apply the Mean Value Theorem to conclude that there exists some c in (a, b) such that

$$H'(c) = \frac{H(b) - H(a)}{b - a}.$$

Since $H(b) \neq H(a)$, it follows that $H'(c) \neq 0$. However, since $G'(c) = F'(c)$, we know that

$$H'(c) = G'(c) - F'(c) = 0$$

and we have a contradiction. Consequently, our assumption that $H(x)$ is not constant must be false, and we conclude that $H(x) = C$. Therefore, $G(x) - F(x) = C$ and we conclude that $G(x) = F(x) + C$.

The point of Theorem 5.1 is that we can represent the entire family of antiderivatives of a function by adding a constant to a *known* antiderivative. For example, knowing that $D_x[x^2] = 2x$, we can represent the family of *all* antiderivatives of $f(x) = 2x$ by $G(x) = x^2 + C$, where C is a constant. We call G the **general antiderivative** of f and $G(x) = x^2 + C$ the **general solution** of the equation $G'(x) = 2x$.

Notation for antiderivatives

If $y = F(x)$ is an antiderivative of f, then we say $F(x)$ is a solution of the equation

$$\frac{dy}{dx} = f(x).$$

When solving such an equation, it is convenient to write it in the equivalent differential form

$$dy = f(x)\, dx.$$

The operation of finding all solutions (the general antiderivative of f) of this equation is called **antidifferentiation** (or **integration**) and is denoted by an integral sign $\int$. The general solution of the equation $dy = f(x)\, dx$ is denoted by

Variable of integration

$$y = \int f(x)\, dx = F(x) + C.$$

Integrand   Constant of integration

We read $\int f(x)\, dx$ as the *antiderivative of f with respect to x*. Thus, the differential dx serves to identify x as the variable of integration. The term **indefinite integral** is a synonym for antiderivative.

DEFINITION OF INTEGRAL NOTATION FOR ANTIDERIVATIVES

The notation

$$\int f(x)\, dx = F(x) + C$$

where C is an arbitrary constant, means that F is an antiderivative of f. That is, $F'(x) = f(x)$ for all x in the domain of f.

The inverse nature of integration and differentiation can be verified by substituting $F'(x)$ for $f(x)$ in this definition to obtain

$$\int F'(x)\, dx = F(x) + C. \qquad \text{Integration is the inverse of differentiation}$$

Moreover, if $\int f(x)\, dx = F(x) + C$, then

$$\frac{d}{dx}\left[\int f(x)\, dx\right] = f(x). \qquad \text{Differentiation is the inverse of integration}$$

This allows us to obtain integration formulas directly from differentiation formulas, as shown in the following theorem.

THEOREM 5.2
BASIC INTEGRATION RULES

Differentiation formula	*Integration formula*
$\dfrac{d}{dx}[C] = 0$	$\displaystyle\int 0 \, dx = C$
$\dfrac{d}{dx}[kx] = k$	$\displaystyle\int k \, dx = kx + C$
$\dfrac{d}{dx}[kf(x)] = kf'(x)$	$\displaystyle\int kf(x) \, dx = k \int f(x) \, dx$
$\dfrac{d}{dx}[f(x) \pm g(x)] = f'(x) \pm g'(x)$	$\displaystyle\int [f(x) \pm g(x)] \, dx = \int f(x) \, dx \pm \int g(x) \, dx$

Power Rule

$$\frac{d}{dx}[x^n] = nx^{n-1} \qquad \int x^n \, dx = \frac{x^{n+1}}{n+1} + C, \; n \neq -1$$

REMARK Be sure you understand that the Power Rule for integration has the restriction that $n \neq -1$. The evaluation of $\int 1/x \, dx$ must wait until the introduction of the natural logarithm function in Chapter 7.

EXAMPLE 1 Applying the basic integration rules

$$\int 3x \, dx = 3 \int x \, dx \qquad \text{Constant Multiple Rule}$$

$$= 3 \int x^1 \, dx \qquad \text{Rewrite } (x = x^1)$$

$$= 3\left(\frac{x^2}{2}\right) + C \qquad \text{Power Rule } (n = 1)$$

$$= \frac{3}{2}x^2 + C \qquad \text{Simplify}$$

When evaluating indefinite integrals, a strict application of the basic integration rules tends to produce complicated constants of integration. For instance, in Example 1, we could have written

$$\int 3x \, dx = 3 \int x \, dx = 3\left[\frac{x^2}{2} + C\right] = \frac{3}{2}x^2 + 3C.$$

However, since C represents *any* constant, it is both cumbersome and unnecessary to write $3C$ as the constant of integration, and we choose the simpler form, $\frac{3}{2}x^2 + C$.

EXAMPLE 2 Applying the basic integration rules

Evaluate $\int (3x^2 + 2x) \, dx$.

SOLUTION

$$\int (3x^2 + 2x)\, dx = \int 3x^2\, dx + \int 2x\, dx \qquad \text{Sum Rule}$$

$$= 3\int x^2\, dx + 2\int x\, dx \qquad \text{Constant Multiple Rule}$$

$$= 3\left(\frac{x^3}{3}\right) + 2\left(\frac{x^2}{2}\right) + C \qquad \text{Power Rule}$$

$$= x^3 + x^2 + C$$

In Examples 1 and 2, note that the general pattern of integration is similar to that of differentiation.

Given Integral $\longrightarrow$ Rewrite $\longrightarrow$ Integrate $\longrightarrow$ Simplify

This pattern is further demonstrated in the next example.

EXAMPLE 3   Rewriting before integrating

	Given integral	Rewrite	Integrate	Simplify
(a)	$\int \dfrac{1}{x^3}\, dx$	$\int x^{-3}\, dx$	$\dfrac{x^{-2}}{-2} + C$	$-\dfrac{1}{2x^2} + C$
(b)	$\int \sqrt{x}\, dx$	$\int x^{1/2}\, dx$	$\dfrac{x^{3/2}}{3/2} + C$	$\dfrac{2}{3}x^{3/2} + C$

REMARK   Remember that we can check our answer to an antidifferentiation problem by differentiating. For instance, in Example 3(b), we can check that $\frac{2}{3}x^{3/2}$ is the correct antiderivative by differentiating to obtain $D_x\left[\frac{2}{3}x^{3/2}\right] = \left(\frac{2}{3}\right)\left(\frac{3}{2}\right)x^{1/2} = \sqrt{x}$.

The basic integration rules listed in Theorem 5.2 allow us to integrate *any* polynomial function. This is demonstrated in the next example.

EXAMPLE 4   Integrating polynomial functions

(a) $\int 1\, dx = x + C$

(b) $\int (x + 2)\, dx = \int x\, dx + \int 2\, dx = \dfrac{x^2}{2} + C_1 + 2x + C_2$

$$= \dfrac{x^2}{2} + 2x + C$$

We normally omit the third step by combining the two constants of integration, C_1 and C_2, to form a single constant, C.

(c) $\int (3x^4 - 5x^2 + x)\, dx = 3\left(\dfrac{x^5}{5}\right) - 5\left(\dfrac{x^3}{3}\right) + \dfrac{x^2}{2} + C$

$$= \dfrac{3}{5}x^5 - \dfrac{5}{3}x^3 + \dfrac{1}{2}x^2 + C$$

REMARK The integral in Example 3(a) is usually simplified to the form

$$\int 1\, dx = \int dx.$$

EXAMPLE 5 Rewriting before integrating

$\displaystyle\int \dfrac{x + 1}{\sqrt{x}}\, dx = \int \left(\dfrac{x}{\sqrt{x}} + \dfrac{1}{\sqrt{x}}\right) dx$

$\displaystyle\qquad\qquad = \int (x^{1/2} + x^{-1/2})\, dx \qquad$ Rewrite with fractional exponents

$\displaystyle\qquad\qquad = \dfrac{x^{3/2}}{3/2} + \dfrac{x^{1/2}}{1/2} + C \qquad$ Integrate

$\displaystyle\qquad\qquad = \dfrac{2}{3}x^{3/2} + 2x^{1/2} + C \qquad$ Simplify

REMARK When integrating quotients, do not integrate the numerator and denominator separately. This is no more valid in integration than it is in differentiation. For instance, in Example 4, be sure you understand that

$$\int \dfrac{x + 1}{\sqrt{x}}\, dx \neq \dfrac{\int (x + 1)\, dx}{\int \sqrt{x}\, dx}.$$

EXAMPLE 6 Rewriting before integrating

Evaluate

$$\int \dfrac{3x^2 - 4}{x^2}\, dx.$$

SOLUTION

$\displaystyle\int \dfrac{3x^2 - 4}{x^2}\, dx = \int (3 - 4x^{-2})\, dx$

$\displaystyle\qquad\qquad = 3x - 4\left(\dfrac{x^{-1}}{-1}\right) + C$

$\displaystyle\qquad\qquad = 3x + \dfrac{4}{x} + C$

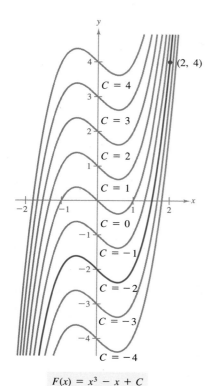

$$F(x) = x^3 - x + C$$

FIGURE 5.1

Initial conditions and particular solutions

We have mentioned already that the equation $y = \int f(x)\, dx$ has many solutions (each differing from the others by a constant). This means that the graphs of any two antiderivatives of f are vertical translations of each other. For example, Figure 5.1 shows the graphs of several antiderivatives of the form

$$y = \int (3x^2 - 1)\, dx = x^3 - x + C \qquad \text{General solution}$$

for various integer values of C. Each of these antiderivatives is a solution of the equation

$$\frac{dy}{dx} = 3x^2 - 1.$$

In many applications of integration, we are given enough information to determine a **particular solution.** To do this we need only know the value of $F(x)$ for one value of x. (This information is called an **initial condition.**) For example, in Figure 5.1, only one curve passes through the point (2, 4). To find this curve, we use the following information.

$$F(x) = x^3 - x + C \qquad \text{General solution}$$
$$F(2) = 4 \qquad \text{Initial condition}$$

By using the initial condition in the general solution, we determine that $F(2) = 8 - 2 + C = 4$, which implies that $C = -2$. Thus, we obtain

$$F(x) = x^3 - x - 2. \qquad \text{Particular solution}$$

EXAMPLE 7   Finding a particular solution

Find the general solution of the equation

$$F'(x) = \frac{1}{x^2}$$

and find the particular solution that satisfies the initial condition $F(1) = 0$.

SOLUTION

To find the general solution, we integrate to obtain

$$F(x) = \int \frac{1}{x^2}\, dx = \int x^{-2}\, dx = \frac{x^{-1}}{-1} + C = -\frac{1}{x} + C.$$

Now, since $F(1) = 0$, we can solve for C as follows.

$$F(1) = -\frac{1}{1} + C = 0 \;\Longrightarrow\; C = 1$$

Thus, the particular solution, as shown in Figure 5.2, is

$$F(x) = -\frac{1}{x} + 1.$$

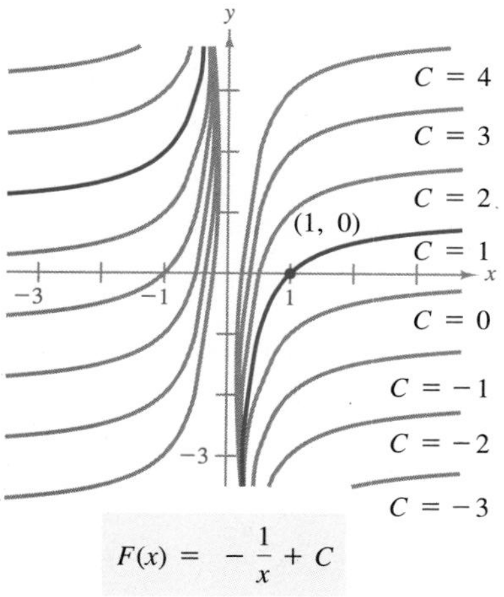

$$F(x) = -\frac{1}{x} + C$$

FIGURE 5.2

So far in this section we have been using x as the variable of integration. In applications, it is often convenient to use a different variable. For instance, in the following example involving *time*, we use t as the variable of integration.

EXAMPLE 8   An application involving gravity

A ball is thrown upward with an initial velocity of 64 feet per second from an initial height of 80 feet, as shown in Figure 5.3. (a) Find the position function giving the height s as a function of the time t. (b) When does the ball hit the ground?

SOLUTION

(a) We let $t = 0$ represent the initial time. Then the two given initial conditions can be written as follows.

$$s(0) = 80 \qquad \text{Initial height is 80 ft}$$
$$s'(0) = 64 \qquad \text{Initial velocity is 64 ft/sec}$$

Assuming the acceleration due to gravity to be -32 feet per second per second, we can write

$$s''(t) = -32$$

and, by integrating, we obtain

$$s'(t) = \int s''(t)\, dt = \int -32\, dt = -32t + C_1.$$

Using the initial velocity, we obtain $s'(0) = 64 = -32(0) + C_1$, which implies that $C_1 = 64$. Similarly, by integrating $s'(t)$, we have

$$s(t) = \int s'(t)\, dt = \int (-32t + 64)\, dt = -16t^2 + 64t + C_2.$$

Now, using the initial height, we obtain

$$s(0) = 80 = -16(0^2) + 64(0) + C_2$$

which implies that $C_2 = 80$. Therefore, the position function is

$$s(t) = -16t^2 + 64t + 80.$$

(b) Using the position function found in part (a), we can find the time that the ball hits the ground by solving the equation $s(t) = 0$.

$$s(t) = -16t^2 + 64t + 80 = 0$$
$$-16(t + 1)(t - 5) = 0$$
$$t = -1, 5$$

Since t must be positive, we conclude that the ball hits the ground 5 seconds after it was thrown.

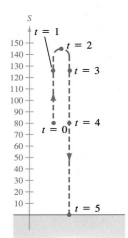

FIGURE 5.3

REMARK In Example 8, note that the position function has the form

$$s(t) = \frac{1}{2}gt^2 + v_0 t + s_0$$

where $g = -32$, v_0 is the initial velocity, and s_0 is the initial height, as presented earlier in Section 3.2.

Before you begin the exercise set for this section, be sure you realize that one of the most important steps in integration is *rewriting the integrand in a form that fits the basic integration rules*. To further illustrate this point, we list several additional examples in Table 5.1.

TABLE 5.1

Given	Rewrite	Integrate	Simplify
$\displaystyle\int \frac{2}{\sqrt{x}}\,dx$	$\displaystyle 2\int x^{-1/2}\,dx$	$\displaystyle 2\left(\frac{x^{1/2}}{1/2}\right) + C$	$4x^{1/2} + C$
$\displaystyle\int (t^2 + 1)^2\,dt$	$\displaystyle\int (t^4 + 2t^2 + 1)\,dt$	$\displaystyle\frac{t^5}{5} + 2\left(\frac{t^3}{3}\right) + t + C$	$\displaystyle\frac{1}{5}t^5 + \frac{2}{3}t^3 + t + C$
$\displaystyle\int \frac{x^3 + 3}{x^2}\,dx$	$\displaystyle\int (x + 3x^{-2})\,dx$	$\displaystyle\frac{x^2}{2} + 3\left(\frac{x^{-1}}{-1}\right) + C$	$\displaystyle\frac{1}{2}x^2 - \frac{3}{x} + C$
$\displaystyle\int \sqrt[3]{x}(x - 4)\,dx$	$\displaystyle\int (x^{4/3} - 4x^{1/3})\,dx$	$\displaystyle\frac{x^{7/3}}{7/3} - 4\left(\frac{x^{4/3}}{4/3}\right) + C$	$\displaystyle\frac{3}{7}x^{4/3}(x - 7) + C$

EXERCISES for Section 5.1

In Exercises 1–6, complete the table using Table 5.1 as a model.

Given	Rewrite	Integrate	Simplify

1. $\displaystyle\int \sqrt[3]{x}\,dx$

2. $\displaystyle\int \frac{1}{x^2}\,dx$

3. $\displaystyle\int \frac{1}{x\sqrt{x}}\,dx$

4. $\displaystyle\int x(x^2 + 3)\,dx$

5. $\displaystyle\int \frac{1}{2x^3}\,dx$

6. $\displaystyle\int \frac{1}{(2x)^3}\,dx$

In Exercises 7–26, evaluate the indefinite integral and check your result by differentiation.

7. $\displaystyle\int (x^3 + 2)\,dx$

8. $\displaystyle\int (x^2 - 2x + 3)\,dx$

9. $\displaystyle\int (x^{3/2} + 2x + 1)\,dx$

10. $\displaystyle\int \left(\sqrt{x} + \frac{1}{2\sqrt{x}}\right)\,dx$

11. $\displaystyle\int \sqrt[3]{x^2}\,dx$

12. $\displaystyle\int (\sqrt[4]{x^3} + 1)\,dx$

13. $\displaystyle\int \frac{1}{x^3}\,dx$

14. $\displaystyle\int \frac{1}{x^4}\,dx$

15. $\displaystyle\int \frac{1}{4x^2}\,dx$

16. $\displaystyle\int (2x + x^{-1/2})\,dx$

17. $\displaystyle\int \frac{x^2 + x + 1}{\sqrt{x}}\,dx$

18. $\displaystyle\int \frac{x^2 + 1}{x^2}\,dx$

19. $\displaystyle\int (x + 1)(3x - 2)\,dx$

20. $\displaystyle\int (2t^2 - 1)^2\,dt$

21. $\int \frac{t^2 + 2}{t^2} \, dt$

22. $\int (1 - 2y + 3y^2) \, dy$

23. $\int y^2 \sqrt{y} \, dy$

24. $\int (1 + 3t)t^2 \, dt$

25. $\int dx$

26. $\int 3 \, dt$

In Exercises 27–30, find the equation of the curve, given the derivative and the indicated point on the curve.

27. $\dfrac{dy}{dx} = 2x - 1$

28. $\dfrac{dy}{dx} = 2(x - 1)$

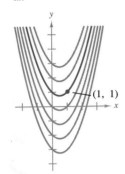

(1, 1)

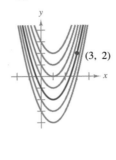

(3, 2)

29. $\dfrac{dy}{dx} = 3x^2 - 1$

30. $\dfrac{dy}{dx} = -\dfrac{1}{x^2}$

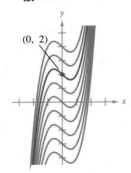

(0, 2)

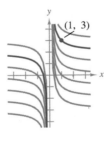

(1, 3)

In Exercises 31–34, find $y = f(x)$ satisfying the given conditions.

31. $f''(x) = 2, f'(2) = 5, f(2) = 10$
32. $f''(x) = x^2, f'(0) = 6, f(0) = 3$
33. $f''(x) = x^{-3/2}, f'(4) = 2, f(0) = 0$
34. $f''(x) = x^{-3/2}, f'(1) = 2, f(9) = -4$

In Exercises 35–39, use $a(t) = -32$ ft/s² as the acceleration due to gravity. (Neglect air resistance.)

35. An object is dropped from a balloon that is stationary at 1600 feet above the ground. Express its height above the ground as a function of t. How long does it take the object to reach the ground?

36. A ball is thrown vertically upward from the ground with an initial velocity of 60 feet per second. How high will the ball go?

37. With what initial velocity must an object be thrown upward (from ground level) to reach a maximum height of 550 feet (approximate height of the Washington Monument)?

38. Show that the height above the ground of an object thrown upward from a point s_0 feet above the ground with an initial velocity of v_0 feet per second is given by the function

$$f(t) = -16t^2 + v_0 t + s_0.$$

39. A balloon, rising vertically with a velocity of 16 feet per second, releases a sandbag at the instant when the balloon is 64 feet above the ground.
(a) How many seconds after its release will the bag strike the ground?
(b) With what velocity will it reach the ground?

40. Assume that a fully loaded plane starting from rest has a constant acceleration while moving down the runway. Find this acceleration if the plane requires, on the average, 0.7 miles of runway and a speed of 160 miles per hour before lifting off.

41. The maker of a certain automobile advertises that the automobile will accelerate from 15 to 50 miles per hour in high gear in 13 seconds. Assuming constant acceleration, compute
(a) the acceleration in feet per second per second
(b) the distance the car travels in the given time.

42. A car traveling at 45 miles per hour was brought to a stop, at constant deceleration, 132 feet from where the brakes were applied. How far had the car moved when its speed was reduced to (a) 30 miles per hour and (b) 15 miles per hour? (c) Draw the real number line from 0 to 132, and plot the points found in parts (a) and (b). What conclusions can you draw?

43. At the instant the traffic light turns green, an automobile that has been waiting at an intersection starts ahead with a constant acceleration of 6 feet per second per second. At the same instant, a truck traveling with a constant velocity of 30 feet per second overtakes and passes the car.
(a) How far beyond its starting point will the automobile overtake the truck?
(b) How fast will the automobile be traveling?

44. A ball is released from rest and rolls down an inclined plane with constant acceleration. In 4 seconds the ball travels 100 centimeters. What is its acceleration in centimeters per second per second?

45. Galileo Galilei (1564–1642) stated the following proposition concerning falling objects: The time in which any space is traversed by a body starting from rest and uniformly accelerated is equal to the time in which that same space would be traversed by the same body moving at a uniform speed whose value is the mean of the highest speed and the speed just before acceleration began. Use the techniques of this section to verify the statement.

46. In the Chapter 5 Application, we developed the equation

$$\int v \, dv = -GM \int \frac{1}{y^2} \, dy$$

where v is the velocity of an object projected from the earth, y is the distance from the center of earth, G is the gravitational constant, and M is the mass of the earth. Show that v and y are related by the equation

$$v^2 = v_0^2 + 2GM \left(\frac{1}{y} - \frac{1}{R} \right)$$

where v_0 is the initial velocity of the object and R is the radius of the earth.

47. If marginal cost is constant, show that the cost function is a straight line.

48. The marginal cost for production is

$$\frac{dC}{dx} = 2x - 12$$

and the fixed costs are $50. Find the total cost function and the average cost function.

In Exercises 49 and 50, find the revenue and demand functions for the given marginal revenue.

49. $\frac{dR}{dx} = 100 - 5x$

50. $\frac{dR}{dx} = 10 - 6x - 2x^2$

5.2 Area

Sigma notation ▪ Area ▪ The area of a plane region ▪ Upper and lower sums

In the previous section, we introduced the concept of antidifferentiation. We now look at a new problem—that of finding the area of a region in the plane. At first glance, these two ideas seem unrelated, but we will discover in Section 5.4 that they are closely related by an important theorem called the Fundamental Theorem of Calculus.

One method for finding the area of a region in the plane involves the sum of many terms. Thus, we begin this section by introducing a concise notation for sums. This notation is called **sigma notation** because it uses the uppercase Greek letter sigma, written as Σ.

DEFINITION OF SIGMA NOTATION

The sum of n terms $a_1, a_2, a_3, \ldots, a_n$ is written as

$$\sum_{i=1}^{n} a_i = a_1 + a_2 + a_3 + \cdots + a_n$$

where i is called the **index of summation,** a_i is called the **ith term** of the sum, and the **upper and lower bounds of summation** are n and 1, respectively.

REMARK   The upper and lower bounds of summation must be constant *with respect to the index of summation.* However, the lower bound does not have to be 1. Any integer value less than (or equal to) the upper bound is legitimate.

EXAMPLE 1   Examples of sigma notation

(a) $\displaystyle\sum_{i=1}^{6} i = 1 + 2 + 3 + 4 + 5 + 6$

(b) $\displaystyle\sum_{j=3}^{7} j^2 = 3^2 + 4^2 + 5^2 + 6^2 + 7^2$

(c) $\displaystyle\sum_{k=1}^{n} \frac{1}{n}(k^2 + 1) = \frac{1}{n}(1^2 + 1) + \frac{1}{n}(2^2 + 1) + \cdots + \frac{1}{n}(n^2 + 1)$

(d) $\displaystyle\sum_{i=1}^{n} f(x_i)\Delta x = f(x_1)\Delta x + f(x_2)\Delta x + \cdots + f(x_n)\Delta x$ ▭

Although any variable can be used as the index of summation, we usually prefer i, j, and k. Note from Example 1 that the index of summation does not appear in the terms of the expanded sum.

To verify the following two properties of summation, we suggest that you write them in expanded form and apply the associative and commutative properties of addition and the distributive property of addition over multiplication.

THEOREM 5.3
SUMMATION PROPERTIES

1. $\displaystyle\sum_{i=1}^{n} ka_i = k \sum_{i=1}^{n} a_i, \quad k$ is a constant

2. $\displaystyle\sum_{i=1}^{n} [a_i \pm b_i] = \sum_{i=1}^{n} a_i \pm \sum_{i=1}^{n} b_i$

The next theorem gives some useful formulas for sums of powers. A proof of this theorem is discussed in Appendix A.

THEOREM 5.4
SUMMATION FORMULAS

1. $\displaystyle\sum_{i=1}^{n} c = cn$

2. $\displaystyle\sum_{i=1}^{n} i = \frac{n(n + 1)}{2}$

3. $\displaystyle\sum_{i=1}^{n} i^2 = \frac{n(n + 1)(2n + 1)}{6}$

4. $\displaystyle\sum_{i=1}^{n} i^3 = \frac{n^2(n + 1)^2}{4}$

The importance of Theorem 5.4 is that it provides us with a simple method of evaluating certain sums. For example,

$$\sum_{i=1}^{100} i = 1 + 2 + \cdots + 100 = \frac{100(101)}{2} = 5050.$$

This is illustrated further in the next example.

EXAMPLE 2   Evaluating a sum

Evaluate the sum

$$\sum_{i=1}^{n} \frac{i + 1}{n^2}$$

for $n = 10, 100, 1000,$ and $10,000.$

SOLUTION

Applying Theorems 5.3 and 5.4, we have

$$\sum_{i=1}^{n} \frac{i+1}{n^2} = \frac{1}{n^2} \sum_{i=1}^{n} (i + 1) = \frac{1}{n^2} \left[\sum_{i=1}^{n} i + \sum_{i=1}^{n} 1 \right] = \frac{1}{n^2} \left[\frac{n(n+1)}{2} + n \right]$$

$$= \frac{1}{n^2} \left[\frac{n^2 + 3n}{2} \right] = \frac{1}{2} + \frac{3}{2n} = \frac{n+3}{2n}.$$

Now, with the formula

$$\sum_{i=1}^{n} \frac{i+1}{n^2} = \frac{n+3}{2n}$$

we can find the sum simply by substituting the appropriate values of n as shown in Table 5.2.

TABLE 5.2

n	10	100	1,000	10,000
$\sum_{i=1}^{n} \dfrac{i+1}{n^2} = \dfrac{n+3}{2n}$	0.65000	0.51500	0.50150	0.50015

In Table 5.2, note that the sum appears to approach a limit as n increases. Although our discussion of limits at infinity in Section 4.5 applied to a variable x, where x can be any real number, many of the same results hold true for limits involving the variable n, where n is restricted to positive integer values. Thus, to find the limit of $(n + 3)/2n$ as n approaches infinity, we can write

$$\lim_{n \to \infty} \frac{n+3}{2n} = \frac{1}{2}.$$

This is further illustrated in Example 3.

EXAMPLE 3   Finding the limit of a sum

Let $s(n)$ be defined by

$$s(n) = \sum_{i=1}^{n} \left(2 + \frac{i}{n} \right)^2 \left(\frac{1}{n} \right)$$

and find the limit of $s(n)$ as $n \to \infty.$

SOLUTION

Using the summation formulas given in Theorem 5.4, we can write $s(n)$ as follows.

$$
\begin{aligned}
s(n) &= \sum_{i=1}^{n} \left(2 + \frac{i}{n}\right)^2 \left(\frac{1}{n}\right) \\
&= \sum_{i=1}^{n} \left(\frac{4n^2 + 4ni + i^2}{n^2}\right)\left(\frac{1}{n}\right) \\
&= \frac{1}{n^3} \sum_{i=1}^{n} (4n^2 + 4ni + i^2) \\
&= \frac{1}{n^3}\left[\sum_{i=1}^{n} 4n^2 + \sum_{i=1}^{n} 4ni + \sum_{i=1}^{n} i^2\right] \\
&= \frac{1}{n^3}\left[4n^3 + 4n\left(\frac{n(n+1)}{2}\right) + \frac{n(n+1)(2n+1)}{6}\right] \\
&= \frac{38n^3 + 15n^2 + n}{6n^3}
\end{aligned}
$$

Thus,

$$
\lim_{n \to \infty} s(n) = \lim_{n \to \infty} \frac{38n^3 + 15n^2 + n}{6n^3} = \frac{19}{3}.
$$

Rectangle: $A = bh$

FIGURE 5.4

Triangle: $A = \frac{1}{2}bh$

FIGURE 5.5

Area

In Euclidean geometry, the simplest type of plane region is a rectangle. Although we often say that the *formula* for the area of a rectangle is $A = bh$, as shown in Figure 5.4, it is actually more proper to say that this is the *definition* of the **area of a rectangle.**

From this definition, we can develop formulas for the areas of many other plane regions. For example, to determine the area of a triangle, we can form a rectangle whose area is twice that of the triangle, as shown in Figure 5.5. Once we know how to find the area of a triangle, we can determine the area of any polygon by subdividing the polygon into triangular regions as shown in Figure 5.6.

FIGURE 5.6   Parallelogram   Hexagon   Polygon

When we move from polygons to more general plane regions, finding area becomes more difficult. The ancient Greeks were able to determine formulas for the area of some general regions (principally those bounded by conics) by the *exhaustion* method. The clearest description of this method was given by Archimedes (287–212 B.C.). Essentially, the method is a limiting process in which the area is squeezed between two polygons—one inscribed in the region and one circumscribed about the region. The process we use to determine the area of a plane region is similar to that used by Archimedes.

The area of a plane region

We introduce the general problem of finding the area of a region in the plane with an example.

EXAMPLE 4   Approximating the area of a plane region

Use the five rectangles in Figures 5.7(a) and 5.7(b) to find *two* approximations for the area of the region lying between the graph of $f(x) = -x^2 + 5$ and the *x*-axis between $x = 0$ and $x = 2$.

SOLUTION

(a) Since f is nonnegative, we can find the height of the five rectangles shown in Figure 5.7(a) by evaluating $f(x)$ at the right endpoint of each of the following intervals.

$$\left[0, \frac{2}{5}\right], \left[\frac{2}{5}, \frac{4}{5}\right], \left[\frac{4}{5}, \frac{6}{5}\right], \left[\frac{6}{5}, \frac{8}{5}\right], \left[\frac{8}{5}, \frac{10}{5}\right]$$

Evaluate $f(x)$ at the right endpoints of these intervals.

The width of each rectangle is $\frac{2}{5}$, and the right endpoints are given by

$$0 + \frac{2}{5}i, \quad i = 1, 2, \ldots, 5.$$

Thus, the sum of the areas of the five rectangles is

$$\sum_{i=1}^{5} f\left(\overbrace{\frac{2i}{5}}^{\text{height}}\right)\overbrace{\left(\frac{2}{5}\right)}^{\text{width}} = \sum_{i=1}^{5}\left[-\left(\frac{2i}{5}\right)^2 + 5\right]\left(\frac{2}{5}\right) = \frac{2}{5}\sum_{i=1}^{5}\frac{-4i^2 + 125}{25}$$

$$= \frac{2}{125}[121 + 109 + 89 + 61 + 25] = \frac{162}{25} = 6.48.$$

Each of the five rectangles lies inside the given region. Hence, we conclude that the area of the region is *greater* than 6.48.

(b) To find the sum of the areas of the five rectangles in Figure 5.7(b) we use the same basic procedure, except we evaluate f at the left endpoints of the intervals that are given by

$$0 + \frac{2}{5}(i - 1), \quad i = 1, 2, \ldots, 5$$

and obtain

$$\sum_{i=1}^{5} f\left(\overbrace{\frac{2i - 2}{5}}^{\text{height}}\right)\overbrace{\left(\frac{2}{5}\right)}^{\text{width}} = \sum_{i=1}^{5}\left[-\left(\frac{2i - 2}{5}\right)^2 + 5\right]\left(\frac{2}{5}\right) = \frac{202}{25} = 8.08.$$

From Figure 5.7(b), we see that the area of the given region is *less* than 8.08. By combining the results of parts (a) and (b), we conclude that

$$6.48 < (\text{area of region}) < 8.08.$$

Archimedes

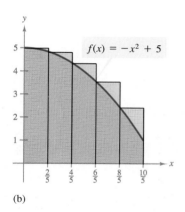

(a)

(b)

FIGURE 5.7

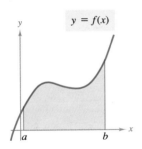

"Region under a Curve"

FIGURE 5.8

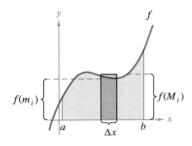

FIGURE 5.9

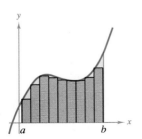

Area of inscribed rectangles
is less than area of region.

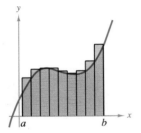

Area of circumscribed rectangles
is greater than area of region.

FIGURE 5.10

REMARK By increasing the number of rectangles used in Example 4, we can obtain closer and closer approximations for the area of the region. For instance, using 25 rectangles of width $\frac{2}{25}$ each, we could conclude that

$$7.17 < \text{(area of region)} < 7.49.$$

We now generalize the procedure described in Example 4. To begin, we look at a plane region bounded above by the graph of a nonnegative, continuous function $y = f(x)$, as shown in Figure 5.8. The region is bounded below by the x-axis, and the left and right boundaries of the region are the vertical lines $x = a$ and $x = b$. We subdivide the interval $[a, b]$ into n subintervals, each of width $\Delta x = (b - a)/n$, as shown in Figure 5.9. The endpoints of the intervals are as follows.

$$\overbrace{a + 0(\Delta x)}^{a\,=\,x_0} < \overbrace{a + 1(\Delta x)}^{x_1} < \overbrace{a + 2(\Delta x)}^{x_2} < \cdots < \overbrace{a + n(\Delta x)}^{x_n\,=\,b}$$

Since f is continuous, the Extreme Value Theorem guarantees the existence of a minimum and a maximum value of $f(x)$ in *each* subinterval.

$$f(m_i) = \text{minimum value of } f(x) \text{ in } i\text{th subinterval}$$
$$f(M_i) = \text{maximum value of } f(x) \text{ in } i\text{th subinterval}$$

Thus, we can define an **inscribed rectangle** lying *inside* the ith subregion and a **circumscribed rectangle** extending *outside* the ith subregion. We use the minimum value as the height of the inscribed rectangle and the maximum value as the height of the circumscribed rectangle, as shown in Figure 5.9. The areas of these two rectangles are related as follows.

$$\begin{pmatrix} \text{area of inscribed} \\ \text{rectangle} \end{pmatrix} = f(m_i)\Delta x \le f(M_i)\Delta x = \begin{pmatrix} \text{area of circumscribed} \\ \text{rectangle} \end{pmatrix}$$

Summing these areas produces

$$\textbf{lower sum} = s(n) = \sum_{i=1}^{n} f(m_i)\Delta x \qquad \text{Area of inscribed rectangles}$$

$$\textbf{upper sum} = S(n) = \sum_{i=1}^{n} f(M_i)\Delta x. \qquad \text{Area of circumscribed rectangles}$$

From Figure 5.10, we can see that the lower sum $s(n)$ is less than or equal to the upper sum $S(n)$, and we have $s(n) \le S(n)$. Moreover, the actual area of the region lies between these two sums. That is,

$$s(n) \le \text{(area of region)} \le S(n).$$

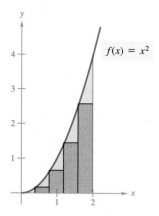

Inscribed rectangles

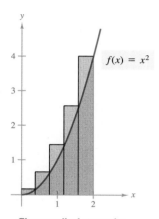

Circumscribed rectangles

FIGURE 5.11

EXAMPLE 5 Finding upper and lower sums for a region

Find the upper and lower sums for the region bounded by the graph of

$$f(x) = x^2$$

and the x-axis between $x = 0$ and $x = 2$.

SOLUTION

To begin, we partition the interval $[0, 2]$ into n subintervals, each of length

$$\Delta x = \frac{b - a}{n} = \frac{2 - 0}{n} = \frac{2}{n}.$$

Figure 5.11 shows the endpoints of the subintervals and several inscribed and circumscribed rectangles. Since f is increasing on the interval $[0, 2]$, the minimum value on each subinterval occurs at the left endpoint, and the maximum value occurs at the right endpoint.

Left endpoints	Right endpoints
$m_i = 0 + (i - 1)\left(\dfrac{2}{n}\right) = \dfrac{2(i - 1)}{n}$	$M_i = 0 + i\left(\dfrac{2}{n}\right) = \dfrac{2i}{n}$

Using the left endpoints, the lower sum is

$$
\begin{aligned}
s(n) &= \sum_{i=1}^{n} f(m_i)\Delta x = \sum_{i=1}^{n} f\left[\frac{2(i - 1)}{n}\right]\left(\frac{2}{n}\right) \\
&= \sum_{i=1}^{n} \left[\frac{2(i - 1)}{n}\right]^2\left(\frac{2}{n}\right) = \sum_{i=1}^{n} \left(\frac{8}{n^3}\right)(i^2 - 2i + 1) \\
&= \frac{8}{n^3}\left[\sum_{i=1}^{n} i^2 - 2\sum_{i=1}^{n} i + \sum_{i=1}^{n} 1\right] \\
&= \frac{8}{n^3}\left(\frac{n(n + 1)(2n + 1)}{6} - 2\left[\frac{n(n + 1)}{2}\right] + n\right) \\
&= \frac{4}{3n^3}(2n^3 - 3n^2 + n) \\
&= \frac{8}{3} - \frac{4}{n} + \frac{4}{3n^2}.
\end{aligned}
$$

Using the right endpoints, the upper sum is

$$
\begin{aligned}
S(n) &= \sum_{i=1}^{n} f(M_i)\Delta x = \sum_{i=1}^{n} f\left(\frac{2i}{n}\right)\left(\frac{2}{n}\right) \\
&= \sum_{i=1}^{n} \left(\frac{2^2 i^2}{n^2}\right)\left(\frac{2}{n}\right) = \sum_{i=1}^{n} \left(\frac{8}{n^3}\right)i^2 \\
&= \frac{8}{n^3}\left[\frac{n(n + 1)(2n + 1)}{6}\right] \\
&= \frac{4}{3n^3}(2n^3 + 3n^2 + n) \\
&= \frac{8}{3} + \frac{4}{n} + \frac{4}{3n^2}.
\end{aligned}
$$

Although it is true that (for any value of n) the lower sum in Example 5 is less than the upper sum,

$$s(n) = \frac{8}{3} - \frac{4}{n} + \frac{4}{3n^2} < \frac{8}{3} + \frac{4}{n} + \frac{4}{3n^2} = S(n)$$

we can see that the difference between these two sums lessens as n increases. In fact, if we take the limit as $n \to \infty$, both the upper and the lower sums approach $\frac{8}{3}$.

$$\lim_{n \to \infty} s(n) = \lim_{n \to \infty} \left[\frac{8}{3} - \frac{4}{n} + \frac{4}{3n^2} \right] = \frac{8}{3} \qquad \text{Lower sum limit}$$

$$\lim_{n \to \infty} S(n) = \lim_{n \to \infty} \left[\frac{8}{3} + \frac{4}{n} + \frac{4}{3n^2} \right] = \frac{8}{3} \qquad \text{Upper sum limit}$$

The next theorem shows that the equivalence of the limit (as $n \to \infty$) of the upper and lower sums is not mere coincidence. It is true for all functions that satisfy the conditions stated in the following theorem. The proof of this theorem is best left to a course in advanced calculus.

**THEOREM 5.5
LIMIT OF THE LOWER
AND UPPER SUMS**

Let f be continuous and nonnegative on the interval $[a, b]$. The limits as $n \to \infty$ of both the lower and upper sums exist and are equal to each other. That is,

$$\lim_{n \to \infty} s(n) = \lim_{n \to \infty} \sum_{i=1}^{n} f(m_i)\Delta x = \lim_{n \to \infty} S(n) = \lim_{n \to \infty} \sum_{i=1}^{n} f(M_i)\Delta x$$

where $\Delta x = (b - a)/n$ and $f(m_i)$ and $f(M_i)$ are the minimum and maximum values of f on the ith subinterval.

From this theorem, we can deduce an important result. Since the same limit is attained for both the minimum value $f(m_i)$ and the maximum value $f(M_i)$, it follows from the Squeeze Theorem (Theorem 2.14) that the choice of x in the ith subinterval does not affect the limit. This means that we are free to choose an *arbitrary* x-value in the ith subinterval, and we do this in the following *definition of the area of a region in the plane*.

**DEFINITION OF AREA OF A
REGION IN THE PLANE**

Let f be continuous and nonnegative on the interval $[a, b]$. The **area** of the region bounded by the graph of f, the x-axis, and the vertical lines $x = a$ and $x = b$ is

$$\text{area} = \lim_{n \to \infty} \sum_{i=1}^{n} f(c_i)\Delta x, \quad x_{i-1} \le c_i \le x_i$$

where $\Delta x = (b - a)/n$.

We can use this definition in two ways: (1) to *compute exact areas*, and (2) to *approximate areas*. Actually, with the techniques we have available now, there are only a few types of functions for which we can find exact areas. We are limited to the lower-degree polynomial functions covered by the summation formulas given in Theorem 5.4. In Section 5.4, however, we will extend this list greatly by a procedure described in an amazing theorem called the *Fundamental Theorem of Calculus*.

EXAMPLE 6 Finding area by the limit definition

Find the area of the region bounded by the graph of $f(x) = x^3$, the x-axis, and the vertical lines $x = 0$ and $x = 1$, as shown in Figure 5.12.

SOLUTION

We begin by noting that f is continuous and nonnegative on the interval $[0, 1]$. Next, we partition the interval $[0, 1]$ into n equal subintervals, each of width $\Delta x = 1/n$. According to the definition of area, we can choose any x-value in the ith subinterval, and for the sake of convenience we choose the right endpoint $c_i = i/n$.

$$\text{area} = \lim_{n \to \infty} \sum_{i=1}^{n} f(c_i)\Delta x = \lim_{n \to \infty} \sum_{i=1}^{n} \left(\frac{i}{n}\right)^3 \left(\frac{1}{n}\right)$$

$$= \lim_{n \to \infty} \frac{1}{n^4} \sum_{i=1}^{n} i^3 = \lim_{n \to \infty} \frac{n^2(n+1)^2}{4n^4}$$

$$= \lim_{n \to \infty} \left(\frac{1}{4} + \frac{1}{2n} + \frac{1}{4n^2}\right) = \frac{1}{4} \qquad \square$$

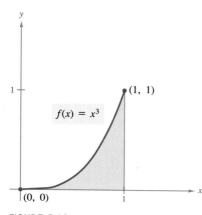

$f(x) = x^3$

$(1, 1)$

$(0, 0)$

FIGURE 5.12

EXAMPLE 7 Finding area by the limit definition

Find the area of the region bounded by the graph of $f(x) = 4 - x^2$, the x-axis, and the vertical lines $x = 1$ and $x = 2$, as shown in Figure 5.13.

SOLUTION

The function f is continuous and nonnegative on the interval $[1, 2]$, and so we begin by partitioning the interval into n equal subintervals, each of length $\Delta x = 1/n$. Choosing the right endpoint, $c_i = 1 + (i/n)$, of each subinterval, we obtain the following.

$$\text{area} = \lim_{n \to \infty} \sum_{i=1}^{n} f(c_i)\Delta x$$

$$= \lim_{n \to \infty} \sum_{i=1}^{n} \left[4 - \left(1 + \frac{i}{n}\right)^2\right]\left(\frac{1}{n}\right)$$

$$= \lim_{n \to \infty} \sum_{i=1}^{n} \left(3 - \frac{2i}{n} - \frac{i^2}{n^2}\right)\left(\frac{1}{n}\right)$$

$$= \lim_{n \to \infty} \left[\frac{1}{n}\sum_{i=1}^{n} 3 - \frac{2}{n^2}\sum_{i=1}^{n} i - \frac{1}{n^3}\sum_{i=1}^{n} i^2\right]$$

$$= \lim_{n \to \infty} \left[3 - \left(1 + \frac{1}{n}\right) - \left(\frac{1}{3} + \frac{1}{2n} + \frac{1}{6n^2}\right)\right]$$

$$= 3 - 1 - \frac{1}{3} = \frac{5}{3} \qquad \square$$

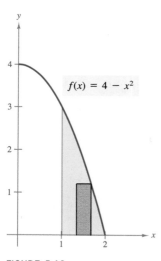

$f(x) = 4 - x^2$

FIGURE 5.13

In the next example we look at a region that is bounded by the y-axis (rather than the x-axis).

EXAMPLE 8 A region bounded by the y-axis

Find the area of the region bounded by the graph of $f(y) = y^2$ and the y-axis for $0 \le y \le 1$, as shown in Figure 5.14.

SOLUTION

When f is a continuous, nonnegative function of y, we still can use the same basic procedure illustrated in Example 6. We partition the interval $[0, 1]$ into n equal subintervals, each of width $\Delta y = 1/n$. Using the upper endpoints $c_i = i/n$, we obtain the following.

$$\text{area} = \lim_{n \to \infty} \sum_{i=1}^{n} f(c_i)\Delta y = \lim_{n \to \infty} \sum_{i=1}^{n} \left(\frac{i}{n}\right)^2\left(\frac{1}{n}\right) = \lim_{n \to \infty} \frac{1}{n^3} \sum_{i=1}^{n} i^2$$

$$= \lim_{n \to \infty} \frac{n(n + 1)(2n + 1)}{6n^3}$$

$$= \lim_{n \to \infty} \left(\frac{1}{3} + \frac{1}{2n} + \frac{1}{6n^2}\right) = \frac{1}{3}$$

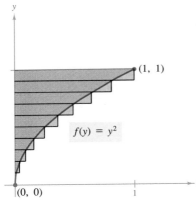

FIGURE 5.14

EXERCISES for Section 5.2

In Exercises 1–8, find the given sum.

1. $\displaystyle\sum_{i=1}^{5} (2i + 1)$

2. $\displaystyle\sum_{i=1}^{6} 2i$

3. $\displaystyle\sum_{k=0}^{4} \frac{1}{k^2 + 1}$

4. $\displaystyle\sum_{j=3}^{5} \frac{1}{j}$

5. $\displaystyle\sum_{k=1}^{4} c$

6. $\displaystyle\sum_{n=1}^{10} \frac{3}{n + 1}$

7. $\displaystyle\sum_{i=1}^{4} [(i - 1)^2 + (i + 1)^3]$

8. $\displaystyle\sum_{k=2}^{5} (k + 1)(k - 3)$

In Exercises 9–18, use sigma notation to write the given sum.

9. $\dfrac{1}{3(1)} + \dfrac{1}{3(2)} + \dfrac{1}{3(3)} + \cdots + \dfrac{1}{3(9)}$

10. $\dfrac{5}{1 + 1} + \dfrac{5}{1 + 2} + \dfrac{5}{1 + 3} + \cdots + \dfrac{5}{1 + 15}$

11. $\left[2\left(\dfrac{1}{8}\right) + 3\right] + \left[2\left(\dfrac{2}{8}\right) + 3\right] + \cdots + \left[2\left(\dfrac{8}{8}\right) + 3\right]$

12. $\left[1 - \left(\dfrac{1}{4}\right)^2\right] + \left[1 - \left(\dfrac{2}{4}\right)^2\right] + \cdots + \left[1 - \left(\dfrac{4}{4}\right)^2\right]$

13. $\left[\left(\dfrac{1}{6}\right)^2 + 2\right]\left(\dfrac{1}{6}\right) + \cdots + \left[\left(\dfrac{6}{6}\right)^2 + 2\right]\left(\dfrac{1}{6}\right)$

14. $\left[\left(\dfrac{1}{n}\right)^2 + 2\right]\left(\dfrac{1}{n}\right) + \cdots + \left[\left(\dfrac{n}{n}\right)^2 + 2\right]\left(\dfrac{1}{n}\right)$

15. $\left[\left(\dfrac{2}{n}\right)^3 - \dfrac{2}{n}\right]\left(\dfrac{2}{n}\right) + \cdots + \left[\left(\dfrac{2n}{n}\right)^3 - \dfrac{2n}{n}\right]\left(\dfrac{2}{n}\right)$

16. $\left[1 - \left(\dfrac{2}{n} - 1\right)^2\right]\left(\dfrac{2}{n}\right) + \cdots + \left[1 - \left(\dfrac{2n}{n} - 1\right)^2\right]\left(\dfrac{2}{n}\right)$

17. $\left[2\left(1 + \dfrac{3}{n}\right)^2\right]\left(\dfrac{3}{n}\right) + \cdots + \left[2\left(1 + \dfrac{3n}{n}\right)^2\right]\left(\dfrac{3}{n}\right)$

18. $\left(\dfrac{1}{n}\right)\sqrt{1 - \left(\dfrac{0}{n}\right)^2} + \cdots + \left(\dfrac{1}{n}\right)\sqrt{1 - \left(\dfrac{n - 1}{n}\right)^2}$

In Exercises 19–24, use the properties of sigma notation and summation formulas to evaluate the given sum.

19. $\displaystyle\sum_{i=1}^{20} 2i$

20. $\displaystyle\sum_{i=1}^{10} i(i^2 + 1)$

21. $\displaystyle\sum_{i=1}^{20} (i - 1)^2$

22. $\displaystyle\sum_{i=1}^{15} (2i - 3)$

23. $\displaystyle\sum_{i=1}^{15} \frac{1}{n^3}(i - 1)^2$

24. $\displaystyle\sum_{i=1}^{10} (i^2 - 1)$

In Exercises 25–30, find the limit of $s(n)$ as $n \to \infty$.

25. $s(n) = \left(\dfrac{4}{3n^3}\right)(2n^3 + 3n^2 + n)$

26. $s(n) = \left(\dfrac{8}{3} + \dfrac{4}{n} + \dfrac{4}{3n^2}\right)$

27. $s(n) = \dfrac{81}{n^4}\left[\dfrac{n^2(n+1)^2}{4}\right]$

28. $s(n) = \dfrac{64}{n^3}\left[\dfrac{n(n+1)(2n+1)}{6}\right]$

29. $s(n) = \dfrac{18}{n^2}\left[\dfrac{n(n+1)}{2}\right]$

30. $s(n) = \dfrac{1}{n^2}\left[\dfrac{n(n+1)}{2}\right]$

In Exercises 31–36, use the properties of sigma notation to find a formula for the given sum of n terms. Then use the formula to find the limit as $n \to \infty$.

31. $\displaystyle\lim_{n\to\infty} \sum_{i=1}^{n} \dfrac{1}{n^3}(i-1)^2$

32. $\displaystyle\lim_{n\to\infty} \sum_{i=1}^{n} \left(1 + \dfrac{2i}{n}\right)^2\left(\dfrac{2}{n}\right)$

33. $\displaystyle\lim_{n\to\infty} \sum_{i=1}^{n} \dfrac{16i}{n^2}$

34. $\displaystyle\lim_{n\to\infty} \sum_{i=1}^{n} \left(\dfrac{2i}{n}\right)\left(\dfrac{2}{n}\right)$

35. $\displaystyle\lim_{n\to\infty} \sum_{i=1}^{n} \left(1 + \dfrac{2i}{n}\right)^3\left(\dfrac{2}{n}\right)$

36. $\displaystyle\lim_{n\to\infty} \sum_{i=1}^{n} \left(1 + \dfrac{i}{n}\right)\left(\dfrac{2}{n}\right)$

In Exercises 37–42, use the upper and lower sums to approximate the area of the given region using the indicated number of (equal) subintervals.

37. $y = \sqrt{x}$

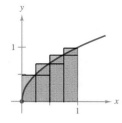

38. $y = \sqrt{x+1}$

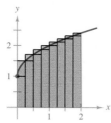

39. $y = \dfrac{1}{x}$

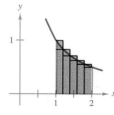

40. $y = \dfrac{1}{x-2}$

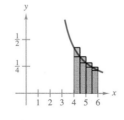

41. $y = \sqrt{1-x^2}$

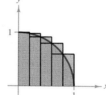

42. $y = \sqrt{x+1}$

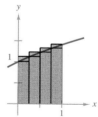

43. Consider the triangle of area 2 bounded by the graphs of $y = x$, $y = 0$, and $x = 2$.
(a) Sketch the graph of the region.
(b) Divide the interval $[0, 2]$ into n equal subintervals and show that the endpoints are

$$0 < 1\left(\frac{2}{n}\right) < \cdots < (n-1)\left(\frac{2}{n}\right) < n\left(\frac{2}{n}\right).$$

(c) Show that $s(n) = \displaystyle\sum_{i=1}^{n}\left[(i-1)\left(\frac{2}{n}\right)\right]\left(\frac{2}{n}\right)$.

(d) Show that $S(n) = \displaystyle\sum_{i=1}^{n}\left[i\left(\frac{2}{n}\right)\right]\left(\frac{2}{n}\right)$.

(e) Complete the following table.

n	5	10	50	100
$s(n)$				
$S(n)$				

(f) Show that $\displaystyle\lim_{n\to\infty} s(n) = \lim_{n\to\infty} S(n) = 2$.

44. Consider the trapezoid of area 4 bounded by the graphs of $y = x$, $y = 0$, $x = 1$, and $x = 3$.
(a) Sketch the graph of the region.
(b) Divide the interval $[1, 3]$ into n equal subintervals and show that the endpoints are

$$1 < 1 + 1\left(\frac{2}{n}\right) < \cdots < 1 + (n-1)\left(\frac{2}{n}\right) < 1 + n\left(\frac{2}{n}\right).$$

(c) Show that $s(n) = \displaystyle\sum_{i=1}^{n}\left[1 + (i-1)\left(\frac{2}{n}\right)\right]\left(\frac{2}{n}\right)$.

(d) Show that $S(n) = \displaystyle\sum_{i=1}^{n}\left[1 + i\left(\frac{2}{n}\right)\right]\left(\frac{2}{n}\right)$.

(e) Complete the following table.

n	5	10	50	100
$s(n)$				
$S(n)$				

(f) Show that $\displaystyle\lim_{n\to\infty} s(n) = \lim_{n\to\infty} S(n) = 4$.

In Exercises 45–54, use the limit process to find the area of the region between the graph of the function and the x-axis over the given interval. Sketch the region.

Function	Interval
45. $y = -2x + 3$	$[0, 1]$
46. $y = 3x - 4$	$[2, 5]$
47. $y = x^2 + 2$	$[0, 1]$
48. $y = 1 - x^2$	$[-1, 1]$
49. $y = 2x^2$	$[1, 3]$
50. $y = 2x^2 - x + 1$	$[0, 2]$
51. $y = 1 - x^3$	$[0, 1]$
52. $y = 2x - x^3$	$[0, 1]$
53. $y = x^2 - x^3$	$[-1, 1]$
54. $y = x^2 - x^3$	$[-1, 0]$

In Exercises 55 and 56, use the limit process to find the area of the region between the graph of the function and the y-axis over the given interval. Sketch the region.

Function	Interval
55. $f(y) = 3y$	$[0, 2]$
56. $f(y) = y^2$	$[0, 3]$

In Exercises 57–60, use the **Midpoint Rule**

$$\text{area} \approx \sum_{i=1}^{n} f\left(\frac{x_i + x_{i-1}}{2}\right)\Delta x$$

with $n = 4$ to approximate the area of the region bounded by the graph of the given function and the x-axis over the indicated interval.

Function	Interval
57. $f(x) = x^2 + 3$	$[0, 2]$
58. $f(x) = x^2 + 4x$	$[0, 4]$
59. $f(x) = \sqrt{x} - 1$	$[1, 2]$
60. $f(x) = \dfrac{1}{x^2 + 1}$	$[0, 2]$

61. The game commission in a certain state introduces 50 deer into some newly acquired state game land. It is believed that the size of the herd will increase according to the model

$$N = \frac{10(5 + 3t)}{1 + 0.04t}$$

where t is the time in years. Find the number in the herd when $t = 5$, 10, and 25 years. What is the limit of N as t approaches infinity?

62. (Learning Curve) Psychologists have developed models to predict performance as a function of the number of trials for a certain task. One such model is

$$P = \frac{b + \theta a(n - 1)}{1 + \theta(n - 1)}$$

where P is the percentage of correct responses after n trials and a, b, and θ are constants depending upon the actual learning situation. Find the limit of P as n approaches infinity.

63. Use a computer or calculator to complete the following table of upper and lower sums for $f(x) = \sqrt{x}$ over $[0, 4]$.

n	4	8	12	16	20
$s(n)$					
$S(n)$					

5.3 Riemann Sums and the Definite Integral

Riemann sums ▪ The definite integral ▪ Properties of definite integrals

In the definition of area in Section 5.2, we required that the partitions have subintervals of *equal width*. This was done only for computational convenience. We begin this section with an example to show that it is not necessary to have subintervals of equal width.

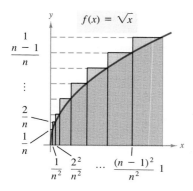

$f(x) = \sqrt{x}$

FIGURE 5.15

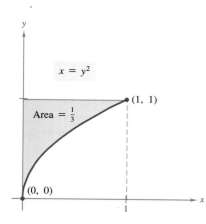

$x = y^2$

Area = $\frac{1}{3}$

(1, 1)

(0, 0)

FIGURE 5.16

Georg Friedrich Riemann

EXAMPLE 1   A partition with subintervals of unequal widths

Consider the region bounded by the graph of $f(x) = \sqrt{x}$ and the x-axis for $0 \le x \le 1$, as shown in Figure 5.15. Evaluate the limit

$$\lim_{n \to \infty} \sum_{i=1}^{n} f(c_i)\Delta x_i$$

where c_i is the right endpoint of the partition given by $x_i = i^2/n^2$ and Δx_i is the width of the ith interval.

SOLUTION

The width of the ith interval is given by

$$\Delta x_i = \frac{i^2}{n^2} - \frac{(i-1)^2}{n^2} = \frac{i^2 - i^2 + 2i - 1}{n^2} = \frac{2i - 1}{n^2}.$$

Thus, we have

$$\begin{aligned}
\lim_{n \to \infty} \sum_{i=1}^{n} f(c_i)\Delta x_i &= \lim_{n \to \infty} \sum_{i=1}^{n} \sqrt{\frac{i^2}{n^2}}\left(\frac{2i-1}{n^2}\right) \\
&= \lim_{n \to \infty} \frac{1}{n^3} \sum_{i=1}^{n} (2i^2 - i) \\
&= \lim_{n \to \infty} \frac{1}{n^3}\left[2\frac{n(n+1)(2n+1)}{6} - \frac{n(n+1)}{2}\right] \\
&= \lim_{n \to \infty} \frac{4n^3 + 3n^2 - n}{6n^3} = \frac{2}{3}.
\end{aligned}$$

The two regions shown in Figures 5.15 and 5.16 can be united to form a square whose area is 1. Furthermore, we already know from Example 8 in Section 5.2 that the region in Figure 5.16 has an area of $\frac{1}{3}$. Thus, it is reasonable to conclude that the region in Figure 5.15 has an area of $\frac{2}{3}$. This agrees with the limit found in Example 1, even though we used a partition having subintervals of unequal widths. The reason we were able to use this particular partition is that as n increases, the *width of the largest subinterval approaches zero*. This is a key feature of the development of the definite integral.

Riemann sums

In the previous section we used the limit of a sum to define the area of a special type of region in the plane. Finding area by this means is only one of *many* applications involving the limit of a sum. We will soon see that a similar approach can be used to determine quantities as diverse as arc length, average value, centroids, volumes, work, and surface areas. The following development is named after Georg Friedrich Bernhard Riemann (1826–1866). Although the definite integral had been defined and used long before the time of Riemann, he generalized the concept to cover a broader category of functions.

In the following definition of a Riemann sum, note that the function f has no restrictions other than being defined on the interval $[a, b]$. (In the previous section, we assumed f to be nonnegative because we were dealing with the area under a curve.)

DEFINITION OF A RIEMANN SUM

Let f be defined on the closed interval $[a, b]$, and let Δ be an arbitrary partition of $[a, b]$,

$$a = x_0 < x_1 < x_2 < \cdots < x_{n-1} < x_n = b$$

where Δx_i is the width of the ith subinterval. If c_i is *any* point in the ith subinterval, then the sum

$$\sum_{i=1}^{n} f(c_i)\Delta x_i, \quad x_{i-1} \leq c_i \leq x_i$$

is called a **Riemann sum** of f for the partition Δ.

REMARK   The sums in Section 5.2 are examples of Riemann sums, but there are more general Riemann sums than those covered there.

For a given partition Δ, we call the width of the largest subinterval the **norm** of the partition, and we denote it by $\|\Delta\|$. If every subinterval is of equal width, then we call the partition **regular** and denote the norm by

$$\|\Delta\| = \Delta x = \frac{b - a}{n}. \qquad \text{Regular partition}$$

For a general partition, the norm is related to the number of subintervals of $[a, b]$ in the following way.

$$\frac{b - a}{\|\Delta\|} \leq n \qquad \text{General partition}$$

Thus, we can see that the number of subintervals in a partition approaches infinity as the norm of the partition approaches zero.

$$\|\Delta\| \to 0 \quad \text{implies that} \quad n \to \infty$$

REMARK   The converse of this statement is not true. For example, let Δ_n be the partition of the interval $[0, 1]$ given by

$$0 < \frac{1}{2^n} < \frac{1}{2^{n-1}} < \cdots < \frac{1}{8} < \frac{1}{4} < \frac{1}{2} < 1.$$

For any positive value of n, the norm of the partition Δ_n is $\frac{1}{2}$. Thus, letting n approach infinity does not force $\|\Delta\|$ to approach zero. In a regular partition, the statements $\|\Delta\| \to 0$ and $n \to \infty$ are equivalent.

To define the definite integral, we use the following limit.

$$\lim_{\|\Delta\| \to 0} \sum_{i=1}^{n} f(c_i)\Delta x_i = L$$

To say that this limit exists means that for $\varepsilon > 0$ there exists a $\delta > 0$ such that for every partition with $\|\Delta\| < \delta$ it follows that

$$\left| L - \sum_{i=1}^{n} f(c_i)\Delta x_i \right| < \varepsilon.$$

(This must be true for any choice of c_i in the ith subinterval of Δ.)

DEFINITION OF THE DEFINITE INTEGRAL

If f is defined on the closed interval $[a, b]$ and the limit of a Riemann sum of f exists, then we say f is **integrable** on $[a, b]$ and we denote the limit by

$$\lim_{\|\Delta\| \to 0} \sum_{i=1}^{n} f(c_i)\Delta x_i = \int_a^b f(x)\,dx.$$

The limit is called the **definite integral** of f from a to b. The number a is the **lower limit** of integration, and the number b is the **upper limit** of integration.

It is not a coincidence that the notation for definite integrals is similar to that used for indefinite integrals. We will see why in the next section when we discuss the Fundamental Theorem of Calculus. For now it is important to see that definite integrals and indefinite integrals are different entities. A definite integral is a *number*, whereas an indefinite integral is a *family of functions*.

A sufficient condition for f to be integrable on $[a, b]$ is given in the following theorem.

THEOREM 5.6 CONTINUITY IMPLIES INTEGRABILITY

If a function f is continuous on the closed interval $[a, b]$, then f is integrable on $[a, b]$.

EXAMPLE 2   Evaluating a definite integral as a limit

Evaluate the definite integral

$$\int_{-2}^{1} 2x\,dx.$$

SOLUTION

Since $f(x) = 2x$ is continuous on the interval $[-2, 1]$, we know it is integrable. Moreover, the definition of integrability implies that any partition whose norm approaches zero can be used to determine the limit. For computational convenience, we define Δ by subdividing $[-2, 1]$ into n subintervals of equal width

$$\Delta x_i = \Delta x = \frac{b - a}{n} = \frac{3}{n}.$$

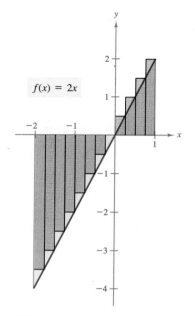

$f(x) = 2x$

FIGURE 5.17

Choosing c_i as the right endpoint of each subinterval, we have

$$c_i = a + i(\Delta x) = -2 + \frac{3i}{n}.$$

Thus, the definite integral is given by

$$\int_{-2}^{1} 2x \, dx = \lim_{\|\Delta\| \to 0} \sum_{i=1}^{n} f(c_i)\Delta x_i = \lim_{n \to \infty} \sum_{i=1}^{n} f(c_i)\Delta x$$

$$= \lim_{n \to \infty} \sum_{i=1}^{n} 2\left(-2 + \frac{3i}{n}\right)\left(\frac{3}{n}\right)$$

$$= \lim_{n \to \infty} \frac{6}{n} \sum_{i=1}^{n} \left(-2 + \frac{3i}{n}\right)$$

$$= \lim_{n \to \infty} \frac{6}{n}\left[-2n + \frac{3}{n}\left(\frac{n(n+1)}{2}\right)\right]$$

$$= \lim_{n \to \infty} \left(-12 + 9 + \frac{9}{n}\right)$$

$$= -3.$$

Note that since the particular definite integral in Example 2 is negative, it *does not* represent the area of the region shown in Figure 5.17. Definite integrals can be positive, negative, or zero. For a definite integral to be interpreted as an area (as defined in Section 5.2), the function f must satisfy the conditions given in the following theorem.

THEOREM 5.7
THE DEFINITE INTEGRAL AS
THE AREA OF A REGION

If f is continuous and nonnegative on the closed interval $[a, b]$, then the area of the region bounded by the graph of f, the x-axis, and the vertical lines $x = a$ and $x = b$ is given by

$$\text{area} = \int_{a}^{b} f(x) \, dx.$$

PROOF

We can obtain a proof by using the definition of the area of a region with Theorem 5.6. From Theorem 5.6, we know that f is integrable on $[a, b]$. This means that the definite integral of f from a to b can be written as the limit of a Riemann sum. When such a limit exists, we can use any convenient partition so long as its norm approaches zero. Thus, we choose a regular partition and conclude that the limit is equal to the area of the given region.

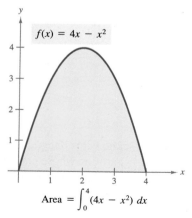

$f(x) = 4x - x^2$

$\text{Area} = \int_{0}^{4} (4x - x^2) \, dx$

FIGURE 5.18

As an example of Theorem 5.7, consider the region bounded by the graph of $f(x) = 4x - x^2$ and the x-axis, as shown in Figure 5.18. Since f is continuous and nonnegative on the closed interval $[0, 4]$, the area of the region is

$$\text{area} = \int_{0}^{4} (4x - x^2) \, dx.$$

From geometry we know several formulas for areas of plane regions. Since at this stage in our study it is not particularly easy to evaluate definite integrals using Riemann sums, it is a good idea to learn to recognize the definite integrals that correspond to areas of common geometric figures. We list some of these in the next example.

EXAMPLE 3   Areas of common geometric figures

Sketch the region corresponding to each of the following definite integrals. Then evaluate each integral using a geometric formula.

(a) $\displaystyle\int_{1}^{3} 4\,dx$

(b) $\displaystyle\int_{0}^{3} (x + 2)\,dx$

(c) $\displaystyle\int_{-2}^{2} \sqrt{4 - x^2}\,dx$

SOLUTION

A sketch of each region is shown in Figure 5.19.

(a) This region is a rectangle of height 4 and width 2. Thus,

$$\int_{1}^{3} 4\,dx = 4(2) = 8.$$

(b) This region is a trapezoid with an altitude of 3 and parallel bases of lengths 2 and 5. The formula for the area of a trapezoid is $\frac{1}{2}h(b_1 + b_2)$, and so we have

$$\int_{0}^{3} (x + 2)\,dx = \frac{1}{2}(3)(2 + 5) = \frac{21}{2}.$$

(c) This region is a semicircle of radius 2. Thus, the area is $\frac{1}{2}\pi r^2$, and we have

$$\int_{-2}^{2} \sqrt{4 - x^2}\,dx = \frac{1}{2}\pi(2^2) = 2\pi.$$

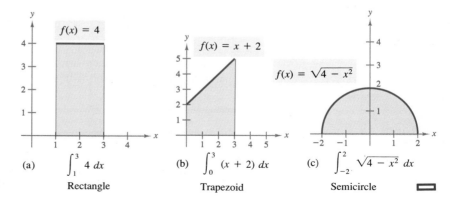

FIGURE 5.19

(a) $\displaystyle\int_{1}^{3} 4\,dx$

Rectangle

(b) $\displaystyle\int_{0}^{3} (x + 2)\,dx$

Trapezoid

(c) $\displaystyle\int_{-2}^{2} \sqrt{4 - x^2}\,dx$

Semicircle

Properties of definite integrals

In defining the definite integral of f on the interval $[a, b]$, we assumed that $a < b$. For the future, it will be convenient to extend this definition somewhat. Geometrically, the following two special definitions seem reasonable. For instance, it makes sense to define the area of a region of zero width and finite height to be zero.

DEFINITION OF TWO SPECIAL DEFINITE INTEGRALS	1. If f is defined at $x = a$, then $$\int_a^a f(x)\, dx = 0.$$ 2. If f is integrable on $[a, b]$, then $$\int_b^a f(x)\, dx = -\int_a^b f(x)\, dx.$$

EXAMPLE 4   Evaluating definite integrals

Evaluate the following definite integrals.

(a) $\displaystyle\int_2^2 \sqrt{x^2 + 1}\, dx$ (b) $\displaystyle\int_3^0 (x + 2)\, dx$

SOLUTION

(a) Since the integrand is defined at $x = 2$, and since the upper and lower limits of integration are equal, we have

$$\int_2^2 \sqrt{x^2 + 1}\, dx = 0.$$

(b) This integral is the same as that given in Example 3(b) except that the upper and lower limits are interchanged. Since the integral in Example 3(b) has a value of $\frac{21}{2}$, we have

$$\int_3^0 (x + 2)\, dx = -\int_0^3 (x + 2)\, dx = -\frac{21}{2}. \qquad \blacksquare$$

Sometimes it is convenient to break a definite integral into two (or more) parts, as shown in the following theorem.

THEOREM 5.8 ADDITIVE INTERVAL PROPERTY	If f is integrable on the three closed intervals determined by a, b, and c, then $$\int_a^b f(x)\, dx = \int_a^c f(x)\, dx + \int_c^b f(x)\, dx.$$

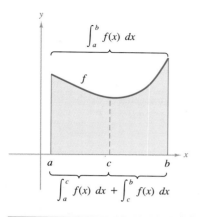

FIGURE 5.20

This theorem is valid for any integrable function and for any three numbers a, b, and c. However, rather than give a formal proof for the general case, it seems more instructive to give a geometric argument for the case in which $a < c < b$ and f is continuous and nonnegative. Figure 5.20 shows that the larger region can be divided at $x = c$ into two subregions whose intersection is a line segment. Since the line segment has zero area, it follows that the area of the larger region is equal to the sum of the areas of the two smaller regions.

Since the definite integral is defined as the limit of a sum, it inherits the two properties of summation given in Theorem 5.3.

THEOREM 5.9
PROPERTIES OF DEFINITE INTEGRALS

If f and g are integrable on $[a, b]$ and k is a constant, then the following properties are true.

1. $\displaystyle\int_a^b kf(x)\, dx = k\int_a^b f(x)\, dx$

2. $\displaystyle\int_a^b [f(x) \pm g(x)]\, dx = \int_a^b f(x)\, dx \pm \int_a^b g(x)\, dx$

REMARK   Property 2 in Theorem 5.9 can be extended to cover any finite number of functions. For example,

$$\int_a^b [f(x) + g(x) + h(x)]\, dx = \int_a^b f(x)\, dx + \int_a^b g(x)\, dx + \int_a^b h(x)\, dx.$$

EXAMPLE 5   Evaluation of a definite integral

Evaluate

$$\int_1^3 (-x^2 + 4x - 3)\, dx$$

using the following values:

$$\int_1^3 x^2\, dx = \frac{26}{3}, \qquad \int_1^3 x\, dx = 4, \qquad \text{and} \qquad \int_1^3 dx = 2.$$

SOLUTION

Using Theorem 5.9, we have

$$\int_1^3 (-x^2 + 4x - 3)\, dx = \int_1^3 (-x^2)\, dx + \int_1^3 4x\, dx + \int_1^3 (-3)\, dx$$

$$= -\int_1^3 x^2\, dx + 4\int_1^3 x\, dx - 3\int_1^3 dx$$

$$= -\left(\frac{26}{3}\right) + 4(4) - 3(2) = \frac{4}{3}.$$

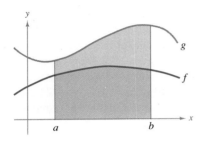

If f and g are continuous on the closed interval $[a, b]$ and $0 \le f(x) \le g(x)$ for $a \le x \le b$, then the following properties are true. First, the area of the region bounded by the graph of f and the x-axis (between a and b) must be nonnegative; second, this area must be less than or equal to the area of the region bounded by the graph of g and the x-axis (between a and b), as shown in Figure 5.21. These two results are generalized in the following theorem. (A proof of this theorem is given in Appendix A.)

FIGURE 5.21

THEOREM 5.10
PRESERVATION OF INEQUALITY

1. If f is integrable and nonnegative on the closed interval $[a, b]$, then
$$0 \le \int_a^b f(x) \, dx.$$

2. If f and g are integrable on the closed interval $[a, b]$ and $f(x) \le g(x)$ for every x in $[a, b]$, then
$$\int_a^b f(x) \, dx \le \int_a^b g(x) \, dx.$$

EXERCISES for Section 5.3

In Exercises 1–10, set up a definite integral that yields the area of the given region. (Do not evaluate the integral.)

1. $f(x) = 3$

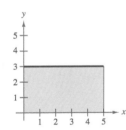

2. $f(x) = 4 - 2x$

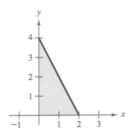

5. $f(x) = 4 - x^2$

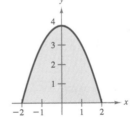

6. $f(y) = (y - 2)^2$

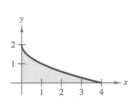

3. $f(x) = 4 - |x|$

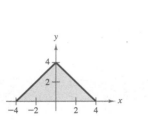

4. $f(x) = x^2$

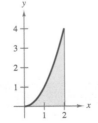

7. $g(y) = y^3$

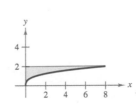

8. $f(x) = (x^2 + 1)^3$

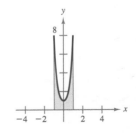

9. $f(x) = \sqrt{x + 1}$

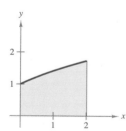

10. $f(x) = \dfrac{1}{x^2 + 1}$

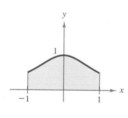

In Exercises 25–30, evaluate the definite integral by the limit definition.

25. $\displaystyle\int_4^{10} 6 \, dx$

26. $\displaystyle\int_{-2}^3 x \, dx$

27. $\displaystyle\int_{-1}^1 x^3 \, dx$

28. $\displaystyle\int_0^1 x^3 \, dx$

29. $\displaystyle\int_1^2 (x^2 + 1) \, dx$

30. $\displaystyle\int_1^2 4x^2 \, dx$

In Exercises 11–20, sketch the region whose area is indicated by the given definite integral. Then use a geometric formula to evaluate the integral.

11. $\displaystyle\int_0^3 4 \, dx$

12. $\displaystyle\int_{-a}^a 4 \, dx$

13. $\displaystyle\int_0^4 x \, dx$

14. $\displaystyle\int_0^4 \dfrac{x}{2} \, dx$

15. $\displaystyle\int_0^2 (2x + 5) \, dx$

16. $\displaystyle\int_0^5 (5 - x) \, dx$

17. $\displaystyle\int_{-1}^1 (1 - |x|) \, dx$

18. $\displaystyle\int_{-a}^a (a - |x|) \, dx$

19. $\displaystyle\int_{-3}^3 \sqrt{9 - x^2} \, dx$

20. $\displaystyle\int_{-r}^r \sqrt{r^2 - x^2} \, dx$

21. Given $\int_0^5 f(x) \, dx = 10$ and $\int_5^7 f(x) \, dx = 3$, find the following.

(a) $\displaystyle\int_0^7 f(x) \, dx$

(b) $\displaystyle\int_5^0 f(x) \, dx$

(c) $\displaystyle\int_5^5 f(x) \, dx$

(d) $\displaystyle\int_0^5 3f(x) \, dx$

22. Given $\int_0^3 f(x) \, dx = 4$ and $\int_3^6 f(x) \, dx = -1$, find the following.

(a) $\displaystyle\int_0^6 f(x) \, dx$

(b) $\displaystyle\int_6^3 f(x) \, dx$

(c) $\displaystyle\int_4^4 f(x) \, dx$

(d) $\displaystyle\int_3^6 -5f(x) \, dx$

23. Given $\int_2^6 f(x) \, dx = 10$ and $\int_2^6 g(x) \, dx = -2$, find the following.

(a) $\displaystyle\int_2^6 [f(x) + g(x)] \, dx$

(b) $\displaystyle\int_2^6 [g(x) - f(x)] \, dx$

(c) $\displaystyle\int_2^6 2g(x) \, dx$

(d) $\displaystyle\int_2^6 3f(x) \, dx$

24. Given $\int_{-1}^1 f(x) \, dx = 0$ and $\int_0^1 f(x) \, dx = 5$, find the following.

(a) $\displaystyle\int_{-1}^0 f(x) \, dx$

(b) $\displaystyle\int_0^1 f(x) \, dx - \int_{-1}^0 f(x) \, dx$

(c) $\displaystyle\int_{-1}^1 3f(x) \, dx$

(d) $\displaystyle\int_0^1 3f(x) \, dx$

In Exercises 31 and 32, use Example 1 as a model to evaluate the limit

$$\lim_{n \to \infty} \sum_{i=1}^n f(c_i)\Delta x_i$$

over the given region bounded by the graphs of the given equations.

31. $f(x) = \sqrt{x}, \; y = 0, \; x = 0, \; x = 2$
[Hint: Let $c_i = 2i^2/n^2$.]

32. $f(x) = \sqrt[3]{x}, \; y = 0, \; x = 0, \; x = 1$
[Hint: Let $c_i = i^3/n^3$.]

In Exercises 33–36, express the given limit as a definite integral on the interval $[a, b]$ where c_i is any point in the ith subinterval.

Limit	Interval
33. $\displaystyle\lim_{\|\Delta\| \to 0} \sum_{i=1}^n (3c_i + 10)\Delta x_i$	$[-1, 5]$
34. $\displaystyle\lim_{\|\Delta\| \to 0} \sum_{i=1}^n 6c_i(4 - c_i)^2 \Delta x_i$	$[0, 4]$
35. $\displaystyle\lim_{\|\Delta\| \to 0} \sum_{i=1}^n \sqrt{c_i^2 + 4} \, \Delta x_i$	$[0, 3]$
36. $\displaystyle\lim_{\|\Delta\| \to 0} \sum_{i=1}^n \left(\dfrac{3}{c_i^2}\right)\Delta x_i$	$[1, 3]$

37. Prove that if f is a continuous function on a closed interval $[a, b]$, then

$$\left| \int_a^b f(x) \, dx \right| \le \int_a^b |f(x)| \, dx.$$

38. Determine whether the function

$$f(x) = \dfrac{1}{x - 4}$$

is integrable on the interval $[3, 5]$, and give a reason for your answer.

39. Determine whether the function

$$f(x) = \begin{cases} 1, & x \text{ is rational} \\ 0, & x \text{ is irrational} \end{cases}$$

is integrable on the interval [0, 1] and give a reason for your answer.

40. Give an example of a function that is integrable on the interval [−1, 1], but not continuous on [−1, 1].

5.4 The Fundamental Theorem of Calculus

The Fundamental Theorem of Calculus ▪ The Mean Value Theorem for Integrals ▪ Average value of a function on an interval ▪ The Second Fundamental Theorem of Calculus

We have now looked at the two major branches of calculus: differential calculus (introduced with the tangent line problem) and integral calculus (introduced with the area problem). Initially, there seems no reason to assume that these two problems are related. But there is a very close connection. This connection was discovered independently by Isaac Newton and Gottfried Leibniz, and consequently these two men are usually credited with discovering calculus. The connection is stated in a theorem that is appropriately called the **Fundamental Theorem of Calculus.**

Roughly, the theorem tells us that differentiation and (definite) integration are inverse operations, in roughly the same sense as division and multiplication are inverse operations. To see how Newton and Leibniz might have anticipated this relationship, let's look at the approximations used in the development of the two operations. In Figure 5.22(a) we use the *quotient* $\Delta y/\Delta x$ (the slope of the secant line) to approximate the slope of the tangent line at (x, y). Similarly, in Figure 5.22(b) we use the *product* $\Delta y \Delta x$ (the area of a rectangle) to approximate the area under the curve. Thus, at least in the primitive approximation stage, the two operations appear to have an inverse relationship. The Fundamental Theorem of Calculus tells us that the limit processes (used to define the derivative and the definite integral) preserve this inverse relationship.

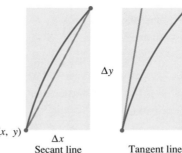

Secant line   Tangent line   Rectangle   Region under a curve

FIGURE 5.22   (a) Slope $\approx \dfrac{\Delta y}{\Delta x}$ (b) Area $\approx \Delta y \, \Delta x$

THEOREM 5.11 **THE FUNDAMENTAL THEOREM OF CALCULUS**	If a function f is continuous on the closed interval $[a, b]$, then $$\int_a^b f(x) \, dx = F(b) - F(a)$$ where F is any function that $F'(x) = f(x)$ for all x in $[a, b]$.

PROOF The key to the proof is in writing the difference $F(b) - F(a)$ in a convenient form. Let Δ be the following partition of $[a, b]$.

$$a = x_0 < x_1 < x_2 < \cdots < x_{n-1} < x_n = b$$

By pairwise subtraction and addition of like terms, we can write

$$F(b) - F(a) = F(x_n) - F(x_{n-1}) + F(x_{n-1}) - \cdots - F(x_1) + F(x_1) - F(x_0)$$

$$= \sum_{i=1}^{n} [F(x_i) - F(x_{i-1})].$$

Now, by the Mean Value Theorem, we know that there exists a number c_i in the ith subinterval such that

$$F'(c_i) = \frac{F(x_i) - F(x_{i-1})}{x_i - x_{i-1}}.$$

Since $F'(c_i) = f(c_i)$, we let $\Delta x_i = x_i - x_{i-1}$ and write

$$F(b) - F(a) = \sum_{i=1}^{n} f(c_i)\Delta x_i.$$

This important equation tells us that by applying the Mean Value Theorem we can always find a collection of c_i's such that the *constant* $F(b) - F(a)$ is a Riemann sum of f on $[a, b]$. Taking the limit (as $\|\Delta\| \to 0$), we have

$$F(b) - F(a) = \int_a^b f(x)\, dx.$$

Three comments are in order regarding the Fundamental Theorem of Calculus. First, *provided we can find* an antiderivative of f, we now have a way to evaluate a definite integral without having to use the limit of a sum. Second, in applying this theorem it is helpful to use the notation

$$\int_a^b f(x)\, dx = \left[F(x) \right]_a^b = F(b) - F(a).$$

For instance, we write

$$\int_1^3 x^3\, dx = \left[\frac{x^4}{4} \right]_1^3 = \frac{3^4}{4} - \frac{1^4}{4} = \frac{81}{4} - \frac{1}{4} = 20.$$

Third, we observe that the constant of integration C can be dropped from the antiderivative, because

$$\int_a^b f(x)\, dx = \left[F(x) + C \right]_a^b$$

$$= [F(b) + C] - [F(a) + C]$$

$$= F(b) - F(a).$$

EXAMPLE 1   Evaluating a definite integral

(a) $\displaystyle\int_{1}^{2} (x^2 - 3)\, dx = \left[\frac{x^3}{3} - 3x\right]_{1}^{2} = \left(\frac{8}{3} - 6\right) - \left(\frac{1}{3} - 3\right) = -\frac{2}{3}$

(b) $\displaystyle\int_{1}^{4} 3\sqrt{x}\, dx = 3\int_{1}^{4} x^{1/2}\, dx = 3\left[\frac{x^{3/2}}{3/2}\right]_{1}^{4} = 2(4)^{3/2} - 2(1)^{3/2} = 14$ ▭

EXAMPLE 2   A definite integral involving absolute value

Evaluate

$$\int_{0}^{2} |2x - 1|\, dx.$$

SOLUTION

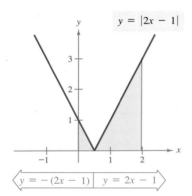

$y = |2x - 1|$

$y = -(2x - 1)$ | $y = 2x - 1$

FIGURE 5.23

From Figure 5.23 and the definition of absolute value, we note that

$$|2x - 1| = \begin{cases} -(2x - 1), & x < \dfrac{1}{2} \\ 2x - 1, & x \geq \dfrac{1}{2}. \end{cases}$$

Hence, we rewrite the integral in two parts, as follows.

$$\begin{aligned}
\int_{0}^{2} |2x - 1|\, dx &= \int_{0}^{1/2} -(2x - 1)\, dx + \int_{1/2}^{2} (2x - 1)\, dx \\
&= \left[-x^2 + x\right]_{0}^{1/2} + \left[x^2 - x\right]_{1/2}^{2} \\
&= \left(-\frac{1}{4} + \frac{1}{2}\right) - (0 + 0) + (4 - 2) - \left(\frac{1}{4} - \frac{1}{2}\right) = \frac{5}{2}
\end{aligned}$$

▭

EXAMPLE 3   Using the Fundamental Theorem to find area

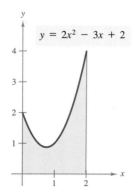

$y = 2x^2 - 3x + 2$

FIGURE 5.24

Find the area of the region bounded by the graph of $y = 2x^2 - 3x + 2$, the x-axis, and the vertical lines $x = 0$ and $x = 2$, as shown in Figure 5.24.

SOLUTION

$$\begin{aligned}
\text{area} &= \int_{0}^{2} (2x^2 - 3x + 2)\, dx \\
&= \left[\frac{2x^3}{3} - \frac{3x^2}{2} + 2x\right]_{0}^{2} \\
&= \frac{16}{3} - 6 + 4 = \frac{10}{3}
\end{aligned}$$

▭

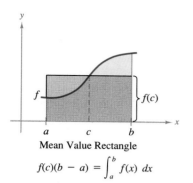

Mean Value Rectangle

$$f(c)(b - a) = \int_a^b f(x)\ dx$$

The Mean Value Theorem for Integrals

In Section 5.2, we observed that the area of a region under a curve is greater than the area of an inscribed rectangle and less than the area of a circumscribed rectangle. The Mean Value Theorem for Integrals states that somewhere "between" the inscribed and circumscribed rectangles there is a rectangle whose area is precisely equal to the area of the region under the curve, as shown in Figure 5.25.

FIGURE 5.25

THEOREM 5.12 MEAN VALUE THEOREM FOR INTEGRALS	If f is continuous on the closed interval $[a, b]$, then there exists a number c in the open interval (a, b) such that $$\int_a^b f(x)\ dx = f(c)(b - a).$$

PROOF   *Case 1:* If f is constant on the interval $[a, b]$, the result is trivial, since c can be any point in (a, b).

Case 2: If f is not constant on $[a, b]$, then by the Extreme Value Theorem we choose $f(m)$ and $f(M)$ to be the minimum and maximum values of f on $[a, b]$. Since $f(m) \le f(x) \le f(M)$ for all x in $[a, b]$, we conclude from Theorem 5.10 that

$$\int_a^b f(m)\ dx \le \int_a^b f(x)\ dx \le \int_a^b f(M)\ dx.$$

This inequality is depicted graphically in Figure 5.26.

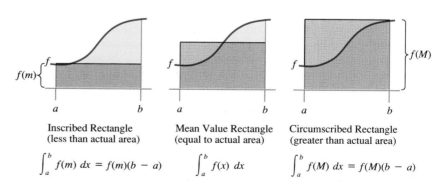

Inscribed Rectangle (less than actual area)	Mean Value Rectangle (equal to actual area)	Circumscribed Rectangle (greater than actual area)
$\int_a^b f(m)\ dx = f(m)(b - a)$	$\int_a^b f(x)\ dx$	$\int_a^b f(M)\ dx = f(M)(b - a)$

FIGURE 5.26

Thus, we have

$$f(m)(b - a) \le \int_a^b f(x)\ dx \le f(M)(b - a)$$

$$f(m) \le \frac{1}{b - a} \int_a^b f(x)\ dx \le f(M).$$

Finally, by the Intermediate Value Theorem we conclude that there exists some c in (a, b) such that

$$f(c) = \frac{1}{b - a} \int_a^b f(x)\, dx$$

$$f(c)(b - a) = \int_a^b f(x)\, dx.$$

REMARK   Note that the Mean Value Theorem for Integrals does not specify how to determine c. It merely guarantees the existence of at least one number c in the interval.

The value of $f(c)$, given in the Mean Value Theorem for Integrals, is called the **average value** of f on the interval $[a, b]$.

DEFINITION OF THE AVERAGE VALUE OF A FUNCTION ON AN INTERVAL

If f is continuous on $[a, b]$, then the **average value** of f on this interval is given by

$$\frac{1}{b - a} \int_a^b f(x)\, dx.$$

To see why we call this the average value of f, suppose that we partition $[a, b]$ into n subintervals of equal width $\Delta x = (b - a)/n$. If c_i is any point in the ith subinterval, then the arithmetic average (or mean) of the function values at the c_i's is given by

$$a_n = \frac{1}{n}[f(c_1) + f(c_2) + \cdots + f(c_n)]. \qquad \text{Average of } f(c_i), \ldots, f(c_n)$$

By multiplying and dividing by $(b - a)$, we can write the average as

$$a_n = \frac{1}{n} \sum_{i=1}^{n} f(c_i)\left(\frac{b - a}{b - a}\right)$$

$$= \frac{1}{b - a} \sum_{i=1}^{n} f(c_i)\left(\frac{b - a}{n}\right)$$

$$= \frac{1}{b - a} \sum_{i=1}^{n} f(c_i)\Delta x.$$

Finally, taking the limit as $n \to \infty$ produces the average value of f on the interval $[a, b]$, as given in the above definition.

This development of the average value of a function on an interval is only one of many practical uses of definite integrals to represent summation processes. In Chapter 6 we will study other applications, such as volume, arc length, centers of mass, and work.

EXAMPLE 4   Finding the average value of a function

Find the average value of $f(x) = 3x^2 - 2x$ on the interval $[1, 4]$.

SOLUTION

The average value is given by

$$\frac{1}{b-a} \int_a^b f(x)\,dx = \frac{1}{3} \int_1^4 (3x^2 - 2x)\,dx = \frac{1}{3}\left[x^3 - x^2 \right]_1^4$$

$$= \frac{1}{3}[64 - 16 - (1-1)] = \frac{48}{3} = 16.$$

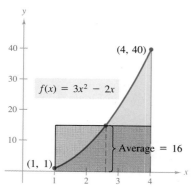

$f(x) = 3x^2 - 2x$

$(4, 40)$

Average $= 16$

$(1, 1)$

FIGURE 5.27

REMARK   Note in Figure 5.27 that the area of the region is equal to the area of the rectangle whose height is the average value.

The Second Fundamental Theorem of Calculus

When we defined the definite integral of f on the interval $[a, b]$, we used the constant b as the upper limit of integration and x as the variable of integration. We now look at a slightly different situation in which the variable x is used as the upper limit of integration. To avoid the confusion of using x in two different ways, we temporarily switch to using t as the variable of integration. (Remember that the definite integral is *not* a function of its variable of integration. Moreover, any variable can be used.)

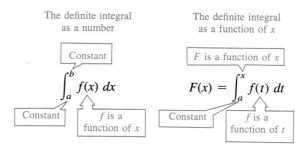

The definite integral as a number

Constant

$$\int_a^b f(x)\,dx$$

Constant f is a function of x

The definite integral as a function of x

F is a function of x

$$F(x) = \int_a^x f(t)\,dt$$

Constant f is a function of t

EXAMPLE 5   The definite integral as a function

Evaluate the function

$$F(x) = \int_0^x (3 - 3t^2)\,dt$$

at $x = 0, \dfrac{1}{4}, \dfrac{1}{2}, \dfrac{3}{4},$ and 1.

SOLUTION

We could evaluate five different definite integrals, one for each of the given upper limits. However, it is much simpler to fix x (as a constant) temporarily and apply the Fundamental Theorem once, to obtain

$$\int_0^x (3 - 3t^2)\, dt = \left[3t - t^3 \right]_0^x = [3x - x^3] - [3(0) - 0^3] = 3x - x^3.$$

Now, using $F(x) = 3x - x^3$ we have the result shown in Figure 5.28. ▭

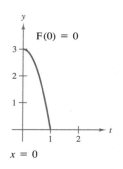 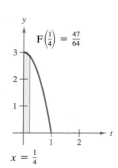 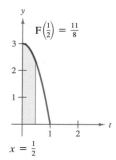 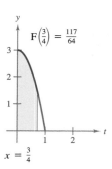 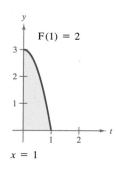

FIGURE 5.28

In Example 5, note that the derivative of F is the original integrand (with only the variable changed). That is,

$$\frac{d}{dx}[F(x)] = \frac{d}{dx}[3x - x^3] = \frac{d}{dx}\left[\int_0^x (3 - 3t^2)\, dt \right] = 3 - 3x^2.$$

We generalize this result in the following theorem, called the **Second Fundamental Theorem of Calculus.**

THEOREM 5.13
THE SECOND FUNDAMENTAL THEOREM OF CALCULUS

If f is continuous on an open interval I containing a, then for every x in the interval,

$$\frac{d}{dx}\left[\int_a^x f(t)\, dt \right] = f(x).$$

PROOF We define F as

$$F(x) = \int_a^x f(t)\, dt.$$

Then, by the definition of the derivative, we have

$$F'(x) = \lim_{\Delta x \to 0} \frac{F(x + \Delta x) - F(x)}{\Delta x}$$

$$= \lim_{\Delta x \to 0} \frac{1}{\Delta x}\left[\int_a^{x+\Delta x} f(t)\, dt - \int_a^x f(t)\, dt \right]$$

$$= \lim_{\Delta x \to 0} \frac{1}{\Delta x}\left[\int_a^{x+\Delta x} f(t)\, dt + \int_x^a f(t)\, dt \right]$$

$$= \lim_{\Delta x \to 0} \frac{1}{\Delta x}\left[\int_x^{x+\Delta x} f(t)\, dt \right].$$

Now, from the Mean Value Theorem for Integrals, we know there exists a number c in the interval $[x, x + \Delta x]$ such that the integral in the above expression is equal to $f(c)\Delta x$. Moreover, since $x \leq c \leq x + \Delta x$, it follows that $c \to x$ as $\Delta x \to 0$. Thus, we have

$$F'(x) = \lim_{\Delta x \to 0} \left[\frac{1}{\Delta x} f(c)\Delta x \right] = \lim_{\Delta x \to 0} f(c) = f(x).$$

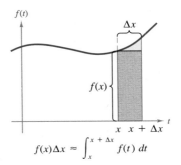

$f(t)$

Δx

$f(x)$

$x \quad x + \Delta x$

$f(x)\Delta x \approx \int_x^{x + \Delta x} f(t)\, dt$

FIGURE 5.29

REMARK   Using the area model for definite integrals, we can view the approximation

$$f(x)\Delta x \approx \int_x^{x+\Delta x} f(t)\, dt$$

as saying that the area of the rectangle of height $f(x)$ and width Δx is approximately equal to the area of the region lying between the graph of f and the x-axis on the interval $[x, x + \Delta x]$, as shown in Figure 5.29.

Note that the Second Fundamental Theorem of Calculus tells us that if a function is continuous, then we can be sure that it has an antiderivative. This antiderivative need not, however, be an elementary function. (Recall the discussion of elementary functions in Section 1.5.)

EXAMPLE 6   Applying the Second Fundamental Theorem of Calculus

Evaluate

$$\frac{d}{dx} \int_0^x \sqrt{t^2 + 1}\, dt.$$

SOLUTION

Note that $f(t) = \sqrt{t^2 + 1}$ is continuous on the entire real line. Thus, using the Second Fundamental Theorem of Calculus, we can write

$$\frac{d}{dx} \int_0^x \sqrt{t^2 + 1}\, dt = \sqrt{x^2 + 1}. \qquad \blacksquare$$

EXERCISES for Section 5.4

In Exercises 1–24, evaluate the definite integral.

1. $\displaystyle\int_0^1 2x\, dx$

2. $\displaystyle\int_2^7 3\, dv$

3. $\displaystyle\int_{-1}^0 (x - 2)\, dx$

4. $\displaystyle\int_2^5 (-3v + 4)\, dv$

5. $\displaystyle\int_{-1}^1 (t^2 - 2)\, dt$

6. $\displaystyle\int_0^3 (3x^2 + x - 2)\, dx$

7. $\displaystyle\int_0^1 (2t - 1)^2\, dt$

8. $\displaystyle\int_{-1}^1 (t^3 - 9t)\, dt$

9. $\displaystyle\int_1^2 \left(\frac{3}{x^2} - 1 \right) dx$

10. $\displaystyle\int_0^1 (3x^3 - 9x + 7)\, dx$

11. $\displaystyle\int_1^2 (5x^4 + 5)\, dx$

12. $\displaystyle\int_{-3}^3 v^{1/3}\, dv$

13. $\displaystyle\int_{-1}^1 (\sqrt[3]{t} - 2)\, dt$

14. $\displaystyle\int_{-2}^{-1} \sqrt{\frac{-2}{x}}\, dx$

15. $\displaystyle\int_1^4 \frac{u - 2}{\sqrt{u}}\, du$

16. $\displaystyle\int_{-2}^{-1} \left(u - \frac{1}{u^2} \right) du$

17. $\displaystyle\int_0^1 \frac{x - \sqrt{x}}{3}\, dx$

18. $\displaystyle\int_0^2 (2 - t)\sqrt{t}\, dt$

19. $\int_{-1}^{0} (t^{1/3} - t^{2/3}) \, dt$ **20.** $\int_{-8}^{-1} \frac{x - x^2}{2\sqrt[3]{x}} \, dx$

21. $\int_{-1}^{1} |x| \, dx$ **22.** $\int_{0}^{3} |2x - 3| \, dx$

23. $\int_{0}^{4} |x^2 - 4x + 3| \, dx$ **24.** $\int_{-1}^{1} |x^3| \, dx$

In Exercises 25–30, determine the area of the indicated region.

25. $y = x - x^2$ **26.** $y = -x^2 + 2x + 3$

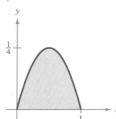   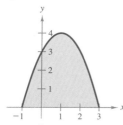

27. $y = 1 - x^4$ **28.** $y = \dfrac{1}{x^2}$

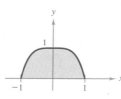   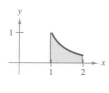

29. $y = \sqrt[3]{2x}$ **30.** $y = (3 - x)\sqrt{x}$

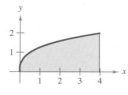   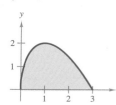

In Exercises 31–34, find the area of the region bounded by the graphs of the given equations.

31. $y = 3x^2 + 1$, $x = 0$, $x = 2$, $y = 0$
32. $y = 1 + \sqrt{x}$, $x = 0$, $x = 4$, $y = 0$
33. $y = x^3 + x$, $x = 2$, $y = 0$
34. $y = -x^2 + 3x$, $y = 0$

In Exercises 35–38, find the values of c guaranteed by the *Mean Value Theorem for Integrals* for the given function over the specified interval.

Function	Interval
35. $f(x) = x^3$	$[0, 2]$
36. $f(x) = \dfrac{9}{x^3}$	$[1, 3]$

Function	Interval
37. $f(x) = -x^2 + 4x$	$[0, 3]$
38. $f(x) = \sqrt{x}$	$[1, 9]$

In Exercises 39–42, sketch the graph of the given function over the specified interval. Find the average value of the function over the interval and all values of x where the function equals its average value.

Function	Interval
39. $f(x) = 4 - x^2$	$[-2, 2]$
40. $f(x) = \dfrac{x^2 + 1}{x^2}$	$\left[\dfrac{1}{2}, 2\right]$
41. $f(x) = x - 2\sqrt{x}$	$[0, 4]$
42. $f(x) = \dfrac{1}{(x - 3)^2}$	$[0, 2]$

In Exercises 43–48, (a) integrate to find F as a function of x and (b) demonstrate the Second Fundamental Theorem of Calculus by differentiating the result of part (a).

43. $F(x) = \int_{0}^{x} (t + 2) \, dt$ **44.** $F(x) = \int_{0}^{x} t(t^2 + 1) \, dt$

45. $F(x) = \int_{8}^{x} \sqrt[3]{t} \, dt$ **46.** $F(x) = \int_{4}^{x} \sqrt{t} \, dt$

47. $F(x) = \int_{1}^{x} \frac{1}{t^2} \, dt$ **48.** $F(x) = \int_{0}^{x} t^{3/2} \, dt$

In Exercises 49–52, use the Second Fundamental Theorem of Calculus to find $F'(x)$.

49. $F(x) = \int_{-2}^{x} (t^2 - 2t + 5) \, dt$

50. $F(x) = \int_{1}^{x} \sqrt[4]{t} \, dt$

51. $F(x) = \int_{-1}^{x} \sqrt{t^4 + 1} \, dt$

52. $F(x) = \int_{1}^{x} \frac{t^2}{t^2 + 1} \, dt$

53. The volume V in liters of air in the lungs during a 5-second respiratory cycle is approximated by the model

$$V = 0.1729t + 0.1522t^2 - 0.0374t^3$$

where t is the time in seconds. Approximate the average volume of air in the lungs during one cycle.

54. The velocity v of the flow of blood at a distance r from the central axis of an artery of radius R is given by

$$v = k(R^2 - r^2)$$

where k is the constant of proportionality. Find the average rate of flow of blood along a radius of the artery. (Use zero and R as the limits of integration.)

55. The air temperature during a period of 12 hours is given by the model

$$T = 53 + 5t - 0.3t^2, \quad 0 \le t \le 12$$

where t is measured in hours and T in degrees Fahrenheit (see figure). Find the average temperature during (a) the first 6 hours of the period and (b) the entire period.

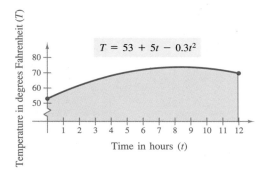

5.5 Integration by Substitution

u-substitution ▪ Pattern recognition ▪ Change of variables ▪ The General Power Rule for Integration ▪ Change of variables for definite integrals ▪ Integration of even and odd functions

In this section we demonstrate techniques for integrating composite functions. We will split the discussion into two parts—*pattern recognition* and *change of variables*. Both techniques involve a **u-substitution.** With pattern recognition we perform the substitution mentally, and with change of variables we write the substitution steps.

The role of substitution in integration is comparable to the role of the Chain Rule in differentiation. Recall that for differentiable functions given by $y = F(u)$ and $u = g(x)$, the Chain Rule states that

$$\frac{d}{dx}[F(g(x))] = F'(g(x))g'(x).$$

From our definition of an antiderivative, it follows that

$$\int F'(g(x))g'(x)\,dx = F(g(x)) + C = F(u) + C.$$

We state these results in the following theorem.

THEOREM 5.14 **ANTIDIFFERENTIATION OF** **A COMPOSITE FUNCTION**	Let f and g be functions such that $f \circ g$ and g' are continuous on an interval I. If F is an antiderivative of f on I, then $$\int f(g(x))g'(x)\,dx = F(g(x)) + C.$$

Pattern recognition

There are several techniques for applying substitution, each differing slightly from the other. However, you should remember that the goal is the same with every technique—*we are trying to find an antiderivative of the integrand.*

Note that the statement of Theorem 5.14 doesn't tell how to distinguish between $f(g(x))$ and $g'(x)$ in the integrand. As you become more experienced at integration, your skill in doing this will increase. Of course, part of the key is familiarity with derivatives.

Examples 1 and 2 show how to apply the theorem *directly,* by recognizing the presence of $f(g(x))$ and $g'(x)$. Note that the composite function in the integrand has an *outside function f* and an *inside function g.* Moreover, the derivative $g'(x)$ is present as a factor of the integrand.

$$\underset{\text{Inside} \qquad \text{Derivative} \atop \text{of inside}}{\overset{\text{Outside}}{\int f(g(x))g'(x)\ dx}} = F(g(x)) + C$$

EXAMPLE 1   Recognizing the $f(g(x))g'(x)$ pattern

Evaluate $\int (x^2 + 1)^2(2x)\ dx$.

SOLUTION

Letting $g(x) = x^2 + 1$, we have $g'(x) = 2x$ and $f(g(x)) = [g(x)]^2$, and we recognize that the integrand follows the $f(g(x))g'(x)$ pattern. Moreover, by the Power Rule, we know that $F(g(x)) = \frac{1}{3}[g(x)]^3$ is an antiderivative of f. Thus, by Theorem 5.14, we have

$$\int \underset{[g(x)]^2}{(x^2 + 1)^2} \underset{g'(x)}{(2x)}\ dx = \underset{F(g(x))}{F(x^2 + 1)} + C = \underset{\frac{1}{3}[g(x)]^3}{\frac{(x^2 + 1)^3}{3}} + C.$$

Try using the Chain Rule to check that the derivative of $\frac{1}{3}(x^2 + 1)^3 + C$ is the integrand of the original integral.   ▭

EXAMPLE 2   Recognizing the $f(g(x))g'(x)$ pattern

Evaluate $\int 5\sqrt{5x + 1}\ dx$.

SOLUTION

By letting $g(x) = 5x + 1$, we have $g'(x) = 5$, and we recognize that the integrand follows the $f(g(x))g'(x)$ pattern. Moreover, by the Power Rule, we know that

$$F(g(x)) = \frac{[g(x)]^{3/2}}{3/2} = \frac{2}{3}[g(x)]^{3/2}$$

is an antiderivative of f. Thus, by Theorem 5.3, we have

$$\int \underbrace{(5x + 1)^{1/2}}_{[g(x)]^{1/2}} \underbrace{(5)}_{g'(x)} \, dx = \underbrace{F(5x + 1)}_{F(g(x))} + C = \underbrace{\frac{2}{3}(5x + 1)^{3/2}}_{\frac{2}{3}[g(x)]^{3/2}} + C.$$

You can check this answer by differentiating $\frac{2}{3}(5x + 1)^{3/2} + C$ to obtain the original integrand. ▭

Both of the integrands in Examples 1 and 2 fit the $f(g(x))g'(x)$ pattern exactly—we only had to recognize the pattern. We can extend this technique considerably with the Constant Multiple Rule

$$\int kf(x) \, dx = k \int f(x) \, dx.$$

Many integrands contain the essential part (the variable part) of $g'(x)$, but are missing a constant multiple. In such cases we can multiply and divide by the necessary constant multiple, as shown in the following example.

EXAMPLE 3   Multiplying and dividing by a constant

Evaluate $\int x(x^2 + 1)^2 \, dx$.

SOLUTION

This is similar to the integral given in Example 1, except that there is a missing factor of 2. Recognizing that $2x$ is the derivative of $x^2 + 1$, we let $g(x) = x^2 + 1$ and supply the $2x$ as follows.

$$\int x(x^2 + 1)^2 \, dx = \int (x^2 + 1)^2 \left(\frac{1}{2}\right)(2x) \, dx \qquad \text{Multiply and divide by 2}$$

$$= \frac{1}{2} \int (x^2 + 1)^2 (2x) \, dx \qquad \text{Constant Multiple Rule}$$

$$= \frac{1}{2} \int [g(x)]^2 g'(x) \, dx$$

$$= \frac{1}{2} \frac{[g(x)]^3}{3} + C \qquad \text{Integrate}$$

$$= \frac{(x^2 + 1)^3}{6} + C \qquad\qquad\qquad ▭$$

REMARK   Be sure you see that the *Constant* Multiple Rule applies only to constants. You cannot multiply and divide by a variable and then move the variable outside the integral sign. For instance,

$$\int (x^2 + 1)^2 \, dx \neq \frac{1}{2x} \int (x^2 + 1)^2 (2x) \, dx.$$

After all, if it were legitimate to move variable quantities outside the integral sign, you could move the entire integrand out and simplify the whole process! But the result would be incorrect.

Change of variables

The integration technique used in Examples 1 through 3 depends on the ability to recognize (or create) integrands of the form $f(g(x))g'(x)$. With a formal **change of variables,** we completely rewrite the integral in terms of u and du (or any other convenient variable). Although this procedure involves more written steps, it is useful for complicated integrands. The change of variable technique uses the Leibniz notation for the differential. That is, if $u = g(x)$, we write $du = g'(x) \, dx$, and the integral in Theorem 5.14 takes the form

$$\int f(g(x))g'(x) \, dx = \int f(u) \, du = F(u) + C.$$

We illustrate the procedure in the next several examples.

EXAMPLE 4   *Change of variable*

Evaluate $\int \sqrt{2x - 1} \, dx$.

SOLUTION

First, we let u be the inner function, $u = 2x - 1$. Then, we obtain

$$du = 2 \, dx. \qquad \text{\small Solve for } du$$

Now, since $\sqrt{2x - 1} = \sqrt{u}$ and $dx = du/2$, we substitute to obtain

$$\int \sqrt{2x - 1} \, dx = \int \sqrt{u} \left(\frac{du}{2} \right) \qquad \text{\small Integral in terms of } u$$

$$= \frac{1}{2} \int u^{1/2} \, du$$

$$= \frac{1}{2} \left(\frac{u^{3/2}}{3/2} \right) + C \qquad \text{\small Antiderivative in terms of } u$$

$$= \frac{1}{3} u^{3/2} + C.$$

Back-substitution of $u = 2x - 1$ yields

$$\int \sqrt{2x - 1} \, dx = \frac{1}{3}(2x - 1)^{3/2} + C. \qquad \text{\small Antiderivative in terms of } x \qquad \blacksquare$$

EXAMPLE 5 *Change of variable*

Evaluate $\int x\sqrt{2x - 1}\, dx$.

SOLUTION

As in the previous example, we let $u = 2x - 1$ and obtain $dx = du/2$. Since the integrand contains a factor of x, we must also solve for x in terms of u as follows.

$$u = 2x - 1 \implies x = \frac{u + 1}{2} \qquad \text{Solve for } x \text{ in terms of } u$$

Thus, the integral becomes

$$\int x\sqrt{2x - 1}\, dx = \int \left(\frac{u + 1}{2}\right)u^{1/2}\left(\frac{du}{2}\right) = \frac{1}{4}\int (u^{3/2} + u^{1/2})\, du$$

$$= \frac{1}{4}\left[\frac{u^{5/2}}{5/2} + \frac{u^{3/2}}{3/2}\right] + C.$$

Back-substitution of $u = 2x - 1$ yields

$$\int x\sqrt{2x - 1}\, dx = \frac{1}{10}(2x - 1)^{5/2} + \frac{1}{6}(2x - 1)^{3/2} + C. \qquad \blacksquare$$

To complete the change of variable in Example 5, we solved for x in terms of u. Sometimes this is very difficult. Fortunately it is not always necessary, as shown in the next example.

EXAMPLE 6 *Change of variable*

Evaluate $\int x\sqrt{x^2 - 1}\, dx$.

SOLUTION

Since $\sqrt{x^2 - 1} = (x^2 - 1)^{1/2}$, we let $u = x^2 - 1$. Then

$$du = (2x)\, dx.$$

Now, since $x\, dx$ is part of the given integral, we write

$$\frac{du}{2} = x\, dx.$$

Substituting u and $du/2$ in the given integral yields

$$\int x\sqrt{x^2 - 1}\, dx = \int u^{1/2}\frac{du}{2} = \frac{1}{2}\int u^{1/2}\, du$$

$$= \frac{1}{2}\left(\frac{u^{3/2}}{3/2}\right) + C = \frac{1}{3}u^{3/2} + C.$$

Back-substitution of $u = x^2 - 1$ yields

$$\int x\sqrt{x^2 - 1}\, dx = \frac{1}{3}(x^2 - 1)^{3/2} + C. \qquad \blacksquare$$

We summarize the steps used for integration by substitution in the following guidelines.

GUIDELINES FOR INTEGRATION BY SUBSTITUTION

1. Choose a substitution $u = g(x)$. Usually, it is best to choose the *inner* part of a composite function, such as a quantity raised to a power.
2. Compute $du = g'(x)\, dx$.
3. Rewrite the integral in terms of the variable u.
4. Evaluate the resulting integral in terms of u.
5. Replace u by $g(x)$ to obtain an antiderivative in terms of x.

The General Power Rule for Integration

One of the most common u-substitutions involves quantities in the integrand that are raised to a power. Because of the importance of this type of substitution, we give it a special name—the **General Power Rule.** A proof of this rule follows directly from the (simple) Power Rule for integration, together with Theorem 5.14.

**THEOREM 5.15
THE GENERAL POWER
RULE FOR INTEGRATION**

If g is a differentiable function of x, then

$$\int [g(x)]^n g'(x)\, dx = \frac{[g(x)]^{n+1}}{n+1} + C, \quad n \neq -1.$$

Equivalently, if $u = g(x)$, then

$$\int u^n\, du = \frac{u^{n+1}}{n+1} + C, \quad n \neq -1.$$

$\int 3(3x-1)^4\, dx$

Study the following example carefully to see the variety of integrals that can be evaluated with the General Power Rule.

EXAMPLE 7   Substitution and the General Power Rule

(a) $\displaystyle \int 3(3x-1)^4\, dx = \int \overbrace{(3x-1)^4}^{u^4}\overbrace{(3)\, dx}^{du} = \overbrace{\frac{(3x-1)^5}{5}}^{u^5/5} + C$

(b) $\displaystyle \int (2x+1)(x^2+x)\, dx = \int \overbrace{(x^2+x)^1}^{u^1}\overbrace{(2x+1)\, dx}^{du} = \overbrace{\frac{(x^2+x)^2}{2}}^{u^2/2} + C$

(c) $\displaystyle \int 3x^2\sqrt{x^3-2}\, dx = \int \overbrace{(x^3-2)^{1/2}}^{u^{1/2}}\overbrace{(3x^2)\, dx}^{du} = \overbrace{\frac{(x^3-2)^{3/2}}{3/2}}^{u^{3/2}/(3/2)} + C$

(d) $\displaystyle\int \frac{-4x}{(1-2x^2)^2}\,dx = \int \overbrace{(1-2x^2)^{-2}}^{u^{-2}}\overbrace{(-4x)\,dx}^{du} = \overbrace{\frac{(1-2x^2)^{-1}}{-1}}^{u^{-1}/(-1)} + C$ ▢

Some integrals whose integrand involves a quantity raised to a power cannot be evaluated by the General Power Rule. Consider the two integrals

$$\int x(x^2+1)^2\,dx \quad\text{and}\quad \int (x^2+1)^2\,dx.$$

Can both be evaluated by the General Power Rule? We evaluated the first integral in Example 3 by letting $u = x^2 + 1$, from which we obtained $du = 2x\,dx$. However, in the second integral the substitution $u = x^2 + 1$ fails, since the integrand lacks the critical factor x needed for du. Fortunately, *for this particular integral,* we can expand the integrand into the polynomial form

$$(x^2+1)^2 = x^4 + 2x^2 + 1$$

and use the Power Rule to integrate each term.

Change of variables for definite integrals

When a definite integral involves a u-substitution, it is often convenient to determine the limits of integration for the variable u rather than to convert the antiderivative back to the variable x and evaluate at the original limits. This change of variables is stated explicitly in the next theorem. The proof follows from Theorem 5.14 combined with the Fundamental Theorem of Calculus.

THEOREM 5.16 **CHANGE OF VARIABLES FOR** **DEFINITE INTEGRALS**	If the function $u = g(x)$ has a continuous derivative on the closed interval $[a, b]$ and f has an antiderivative over the range of g, then $$\int_a^b f(g(x))g'(x)\,dx = \int_{g(a)}^{g(b)} f(u)\,du.$$

EXAMPLE 8   Change of variables

Evaluate

$$\int_0^1 x(x^2+1)^3\,dx.$$

SOLUTION

To evaluate this integral, we let $u = x^2 + 1$. Then we have

$$u = x^2 + 1 \implies du = 2x\,dx.$$

Before substituting, we determine the new upper and lower limits of integration.

Lower limit	Upper limit
When $x = 0$, $u = 0^2 + 1 = 1$.	When $x = 1$, $u = 1^2 + 1 = 2$.

Now, we substitute to obtain

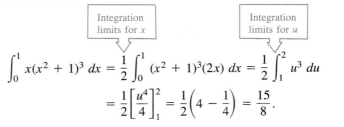

$$\int_0^1 x(x^2 + 1)^3 \, dx = \frac{1}{2} \int_0^1 (x^2 + 1)^3 (2x) \, dx = \frac{1}{2} \int_1^2 u^3 \, du$$

$$= \frac{1}{2} \left[\frac{u^4}{4} \right]_1^2 = \frac{1}{2} \left(4 - \frac{1}{4} \right) = \frac{15}{8}.$$

EXAMPLE 9 Change of variables

Evaluate

$$A = \int_1^5 \frac{x}{\sqrt{2x - 1}} \, dx.$$

SOLUTION

To evaluate this integral, we let $u = \sqrt{2x - 1}$. Then

$$u^2 = 2x - 1 \quad \Longrightarrow \quad x = \frac{u^2 + 1}{2} \quad \Longrightarrow \quad dx = u \, du.$$

Before substituting, we determine the new upper and lower limits of integration.

Lower limit	Upper limit
When $x = 1$, $u = \sqrt{2 - 1} = 1$.	When $x = 5$, $u = \sqrt{10 - 1} = 3$.

Now, we substitute to obtain

$$\int_1^5 \frac{x}{\sqrt{2x - 1}} \, dx = \int_1^3 \frac{1}{2} \left(\frac{u^2 + 1}{u} \right) u \, du = \frac{1}{2} \int_1^3 (u^2 + 1) \, du$$

$$= \frac{1}{2} \left[\frac{u^3}{3} + u \right]_1^3 = \frac{1}{2} \left(9 + 3 - \frac{1}{3} - 1 \right) = \frac{16}{3}.$$

REMARK Geometrically, we can interpret the equation

$$\int_1^5 \frac{x}{\sqrt{2x - 1}} \, dx = \int_1^3 \frac{u^2 + 1}{2} \, du$$

to mean that the two *different* regions shown in Figures 5.30 and 5.31 have the *same* area.

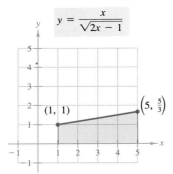

Region before substitution

FIGURE 5.30

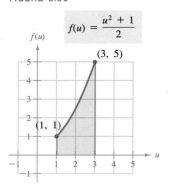

Region after substitution

FIGURE 5.31

When evaluating definite integrals by substitution, do not be surprised if the upper limit of integration of the u-variable form is smaller than the lower limit. If this happens, don't rearrange the limits. Simply evaluate as usual. For example, after substituting $u = \sqrt{1-x}$ in the integral

$$\int_0^1 x^2(1-x)^{1/2}\, dx$$

we have $u = \sqrt{1-1} = 0$ when $x = 1$, and $u = \sqrt{1-0} = 1$ when $x = 0$. Thus, the correct u-variable form of this integral is

$$-2\int_1^0 (1-u^2)^2 u^2\, du.$$

Integration of even and odd functions

Even by making a change of variables, integration often is not a simple task. Occasionally, you will be able to simplify the evaluation of a definite integral (over an interval that is symmetric to the origin) by recognizing the integrand to be an even or odd function. This result is described in the next theorem.

THEOREM 5.17 **INTEGRATION OF EVEN** **AND ODD FUNCTIONS**	Let f be integrable on the closed interval $[-a, a]$. 1. If f is an *even* function, then $$\int_{-a}^a f(x)\, dx = 2\int_0^a f(x)\, dx.$$ 2. If f is an *odd* function, then $$\int_{-a}^a f(x)\, dx = 0.$$

PROOF We prove Property 1 and leave the proof of Property 2 to you. Since f is even, we know that $f(x) = f(-x)$. Now we use Theorem 5.14 with the substitution $u = -x$ to obtain

$$\int_{-a}^0 f(x)\, dx = \int_a^0 f(-u)(-du) = -\int_a^0 f(u)\, du$$
$$= \int_0^a \int_0^a f(u)\, du = \int_0^a f(x)\, dx.$$

Finally, using Theorem 5.8, we have

$$\int_{-a}^a f(x)\, dx = \int_{-a}^0 f(x)\, dx + \int_0^a f(x)\, dx$$
$$= \int_0^a f(x)\, dx + \int_0^a f(x)\, dx = 2\int_0^a f(x)\, dx.$$

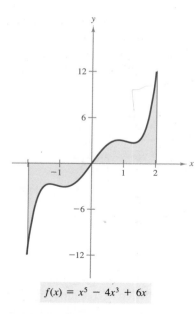

$f(x) = x^5 - 4x^3 + 6x$

FIGURE 5.32

EXAMPLE 10 Integration of an odd function

Evaluate

$$\int_{-2}^{2} (x^5 - 4x^3 + 6x) \, dx.$$

SOLUTION

By letting $f(x) = x^5 - 4x^3 + 6x$, we have

$$f(-x) = (-x)^5 - 4(-x)^3 + 6(-x) = -x^5 + 4x^3 - 6x = -f(x).$$

Thus, f is an odd function, and since $[-2, 2]$ is symmetric about the origin, we can apply Theorem 5.17 to conclude that

$$\int_{-2}^{2} (x^5 - 4x^3 + 6x) \, dx = 0.$$

REMARK From Figure 5.32, we see that the two regions on either side of the y-axis have the same area. However, since one lies below the x-axis and one lies above, integration produces a cancellation effect. (We will say more about finding the area of a region below the x-axis in Section 6.1.)

EXERCISES for Section 5.5

In Exercises 1–4, complete the table by identifying u and du for the given integral.

$$\int f(g(x))g'(x) \, dx \qquad u = g(x) \qquad du = g'(x) \, dx$$

1. $\int (5x^2 + 1)^2(10x) \, dx$

2. $\int x^2\sqrt{x^3 + 1} \, dx$

3. $\int \frac{x}{\sqrt{x^2 + 1}} \, dx$

4. $\int (x^3 + 3) \, 3x^2$

In Exercises 5–28, evaluate the indefinite integral and check the result by differentiation.

5. $\int (1 + 2x)^4(2) \, dx$

6. $\int (x^2 - 1)^3(2x) \, dx$

7. $\int \sqrt{9 - x^2}(-2x) \, dx$

8. $\int (1 - 2x^2)^3(-4x) \, dx$

9. $\int x^2(x^3 - 1)^4 \, dx$

10. $\int x(4x^2 + 3)^3 \, dx$

11. $\int 5x\sqrt[3]{1 - x^2} \, dx$

12. $\int u^3\sqrt{u^4 + 2} \, du$

13. $\int \frac{x^2}{(1 + x^3)^2} \, dx$

14. $\int \frac{x^2}{(16 - x^3)^2} \, dx$

15. $\int \frac{4x}{\sqrt{16 - x^2}} \, dx$

16. $\int \frac{10x^2}{\sqrt{1 + x^3}} \, dx$

17. $\int \frac{x + 1}{(x^2 + 2x - 3)^2} \, dx$

18. $\int \frac{x - 4}{\sqrt{x^2 - 8x + 1}} \, dx$

19. $\int \left(1 + \frac{1}{t}\right)^3\left(\frac{1}{t^2}\right) \, dt$

20. $\int \frac{1}{(3x)^2} \, dx$

21. $\int \frac{1}{\sqrt{2x}} \, dx$

22. $\int \frac{1}{2\sqrt{x}} \, dx$

23. $\int \frac{x^2 + 3x + 7}{\sqrt{x}} \, dx$

24. $\int \frac{t + 2t^2}{\sqrt{t}} \, dt$

25. $\int t^2\left(t - \frac{2}{t}\right) \, dt$

26. $\int \left(\frac{t^3}{3} + \frac{1}{4t^2}\right) \, dt$

27. $\int (9 - y)\sqrt{y} \, dy$

28. $\int 2\pi y(8 - y^{3/2}) \, dy$

In Exercises 29–38, evaluate the indefinite integral by the method shown in Example 5.

29. $\int x\sqrt{x + 2} \, dx$

30. $\int x\sqrt{2x + 1} \, dx$

31. $\int x^2\sqrt{1 - x} \, dx$

32. $\int x^3\sqrt{x + 2} \, dx$

33. $\int \frac{x^2 - 1}{\sqrt{2x - 1}} \, dx$

34. $\int \frac{2x - 1}{\sqrt{x + 3}} \, dx$

35. $\int \dfrac{-x}{(x+1) - \sqrt{x+1}}\, dx$

36. $\int t\sqrt[3]{t-4}\, dt$

37. $\int \dfrac{x}{\sqrt{2x+1}}\, dx$ **38.** $\int (x+1)\sqrt{2-x}\, dx$

In Exercises 39–50, evaluate the definite integral.

39. $\displaystyle\int_{-1}^{1} x(x^2 + 1)^3\, dx$ **40.** $\displaystyle\int_{0}^{1} x\sqrt{1-x^2}\, dx$

41. $\displaystyle\int_{0}^{4} \dfrac{1}{\sqrt{2x+1}}\, dx$ **42.** $\displaystyle\int_{0}^{2} \dfrac{x}{\sqrt{1+2x^2}}\, dx$

43. $\displaystyle\int_{1}^{9} \dfrac{1}{\sqrt{x}(1+\sqrt{x})^2}\, dx$ **44.** $\displaystyle\int_{0}^{2} x\sqrt[3]{4+x^2}\, dx$

45. $\displaystyle\int_{1}^{2} (x-1)\sqrt{2-x}\, dx$ **46.** $\displaystyle\int_{0}^{4} \dfrac{x}{\sqrt{2x+1}}\, dx$

47. $\displaystyle\int_{3}^{7} x\sqrt{x-3}\, dx$ **48.** $\displaystyle\int_{0}^{1} \dfrac{1}{\sqrt{x}+\sqrt{x+1}}\, dx$

49. $\displaystyle\int_{0}^{7} x\sqrt[3]{x+1}\, dx$ **50.** $\displaystyle\int_{-2}^{6} x^2\sqrt[3]{x+2}\, dx$

51. Use the fact that

$$\int_{0}^{2} x^2\, dx = \frac{8}{3}$$

to evaluate the following definite integrals without using the Fundamental Theorem of Calculus.

(a) $\displaystyle\int_{-2}^{0} x^2\, dx$ **(b)** $\displaystyle\int_{-2}^{2} x^2\, dx$

(c) $\displaystyle\int_{0}^{2} -x^2\, dx$ **(d)** $\displaystyle\int_{-2}^{0} 3x^2\, dx$

52. Find the equation of the function f whose graph passes through the point $\left(0, \frac{7}{3}\right)$ and whose derivative is $f'(x) = x\sqrt{1 - x^2}$.

53. A lumber company is seeking a model that yields the average weight loss W per log as a function of the number of days of drying time t. The model is to be reliable up to 100 days after the log is cut. Based on the weight loss during the first 30 days, it was determined that

$$\frac{dW}{dt} = \frac{12}{\sqrt{16t+9}}.$$

(a) Find W as a function of t. Note that no weight loss occurs until the tree is cut.

(b) Find the total weight loss after 100 days.

54. The marginal cost for a certain commodity has been determined to be

$$\frac{dC}{dx} = \frac{12}{\sqrt[3]{12x+1}}.$$

(a) Find the cost function if $C = 100$ when $x = 13$.

(b) Graph the marginal cost function and the cost function on the same set of axes.

5.6 Numerical Integration

The Trapezoidal Rule ▪ Simpson's Rule

Occasionally, we encounter functions for which we cannot find antiderivatives. Of course, that may be due to a lack of cleverness on our part. On the other hand, some elementary functions simply do not possess antiderivatives that are elementary functions. For example, there is no elementary function that has either of the following functions as its derivative.

$$\sqrt[3]{x}\sqrt{1-x} \qquad \sqrt{1-x^3}$$

If we wish to evaluate a definite integral involving a function whose antiderivative we cannot find, then the Fundamental Theorem of Calculus cannot be applied, and we must resort to an approximation technique. We describe two such techniques in this section.

The Trapezoidal Rule

One way to approximate a definite integral is by the use of n trapezoids, as shown in Figure 5.33. In the development of this method, we assume that f is continuous and positive on the interval $[a, b]$, and thus the definite integral $\int_{a}^{b} f(x)\, dx$ represents the area of the region bounded by the graph of f and the x-axis, from $x = a$ to $x = b$.

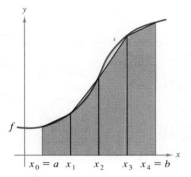

FIGURE 5.33

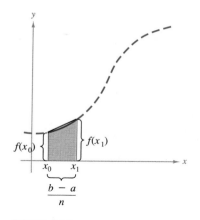

FIGURE 5.34

First, we partition the interval $[a, b]$ into n equal subintervals, each of width $\Delta x = (b - a)/n$, such that

$$a = x_0 < x_1 < x_2 < \cdots < x_n = b.$$

We then form trapezoids for each subinterval, as shown in Figure 5.34. The areas of these trapezoids are as follows.

$$\text{area of first trapezoid} = \left[\frac{f(x_0) + f(x_1)}{2}\right]\left(\frac{b - a}{n}\right)$$

$$\text{area of second trapezoid} = \left[\frac{f(x_1) + f(x_2)}{2}\right]\left(\frac{b - a}{n}\right)$$

$$\vdots$$

$$\text{area of } n\text{th trapezoid} = \left[\frac{f(x_{n-1}) + f(x_n)}{2}\right]\left(\frac{b - a}{n}\right)$$

Finally, the sum of the areas of the n trapezoids is

$$\text{area} = \left(\frac{b - a}{n}\right)\left[\frac{f(x_0) + f(x_1)}{2} + \frac{f(x_1) + f(x_2)}{2} + \cdots + \frac{f(x_{n-1}) + f(x_n)}{2}\right]$$

$$= \left(\frac{b - a}{2n}\right)[f(x_0) + \underbrace{f(x_1) + f(x_1)}_{2f(x_1)} + \underbrace{f(x_2) +}_{2f(x_2)} \cdots + \underbrace{f(x_{n-1})}_{2f(x_{n-1})} + f(x_n)]$$

$$= \left(\frac{b - a}{2n}\right)[f(x_0) + 2f(x_1) + 2f(x_2) + \cdots + 2f(x_{n-1}) + f(x_n)].$$

Letting $\Delta x = (b - a)/n$, we can take the limit as $n \to \infty$, to obtain

$$\lim_{n \to \infty} \left(\frac{b - a}{2n}\right)[f(x_0) + 2f(x_1) + \cdots + 2f(x_{n-1}) + f(x_n)]$$

$$= \lim_{n \to \infty} \left[\frac{[f(a) - f(b)]\Delta x}{2} + \sum_{i=1}^{n} f(x_i)\Delta x\right]$$

$$= \lim_{n \to \infty} \frac{[f(a) - f(b)](b - a)}{2n} + \lim_{n \to \infty} \sum_{i=1}^{n} f(x_i)\Delta x$$

$$= 0 + \int_a^b f(x)\, dx.$$

This brings us to the following theorem, which we call the Trapezoidal Rule.

THEOREM 5.18
THE TRAPEZOIDAL RULE

Let f be continuous on $[a, b]$. The Trapezoidal Rule for approximating $\int_a^b f(x)\, dx$ is given by

$$\int_a^b f(x)\, dx \approx \frac{b - a}{2n}[f(x_0) + 2f(x_1) + 2f(x_2) + \cdots + 2f(x_{n-1}) + f(x_n)].$$

Moreover, as $n \to \infty$, the right-hand side approaches $\int_a^b f(x)\, dx$.

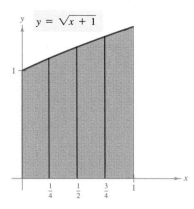

$y = \sqrt{x + 1}$

Four subintervals

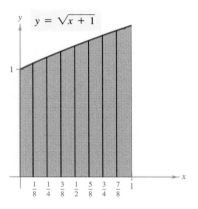

$y = \sqrt{x + 1}$

Eight subintervals

FIGURE 5.35

EXAMPLE 1 Approximation with the Trapezoidal Rule

Use the Trapezoidal Rule to approximate the definite integral

$$A = \int_0^1 \sqrt{x + 1} \; dx.$$

Compare the results for $n = 4$ and $n = 8$, as shown in Figure 5.35.

SOLUTION

When $n = 4$, we have $\Delta x = 1/4$, and by the Trapezoidal Rule we have

$$A \approx \frac{1}{8}\left[\sqrt{1} + 2\sqrt{\frac{5}{4}} + 2\sqrt{\frac{6}{4}} + 2\sqrt{\frac{7}{4}} + \sqrt{2} \right]$$

$$= \frac{1}{8}\left[1 + 2\left(\frac{\sqrt{5}}{2}\right) + 2\left(\frac{\sqrt{6}}{2}\right) + 2\left(\frac{\sqrt{7}}{2}\right) + \sqrt{2} \right]$$

$$\approx 1.2182.$$

When $n = 8$, we have $\Delta x = 1/8$, and

$$A \approx \frac{1}{16}\left[\sqrt{1} + 2\sqrt{\frac{9}{8}} + 2\sqrt{\frac{10}{8}} \right.$$

$$+ 2\sqrt{\frac{11}{8}} + 2\sqrt{\frac{12}{8}} + 2\sqrt{\frac{13}{8}}$$

$$\left. + 2\sqrt{\frac{14}{8}} + 2\sqrt{\frac{15}{8}} + \sqrt{2} \right]$$

$$\approx 1.2188.$$

For this particular integral, we could have found an antiderivative and determined that the exact area of the region is $\frac{2}{3}(2^{3/2} - 1) \approx 1.2190$. ▭

It is interesting to compare the Trapezoidal Rule for approximating definite integrals to the Midpoint Rule given in Section 5.2 (Exercises 57–60). For the Trapezoidal Rule we average the functional values at the endpoints of the subintervals, but for the Midpoint Rule we take the functional value of the subinterval midpoints.

Midpoint Rule   *Trapezoidal Rule*

$$\int_a^b f(x) \; dx \approx \sum_{i=1}^{n} f\left(\frac{x_i + x_{i-1}}{2}\right)\Delta x \qquad \int_a^b f(x) \; dx \approx \sum_{i=1}^{n} \left(\frac{f(x_i) + f(x_{i-1})}{2}\right)\Delta x$$

There are two important points that should be made concerning the Trapezoidal Rule (or the Midpoint Rule). First, the approximation tends to become more accurate as n increases. For instance, in Example 1, if $n = 16$, the

Trapezoidal Rule yields an approximation of 1.2189. Second, although we could have used the Fundamental Theorem to evaluate the integral in Example 1, this theorem cannot be used to evaluate an integral such as

$$\int_0^1 \sqrt{x^3 + 1} \, dx$$

(even though it looks simple) because $\sqrt{x^3 + 1}$ has no elementary antiderivative. Yet, the Trapezoidal Rule can be applied readily to this integral.

Simpson's Rule

One way to view the trapezoidal approximation of a definite integral is to say that on each subinterval we approximate f by a *first*-degree polynomial. In Simpson's Rule, named after the English mathematician Thomas Simpson (1710–1761), we take this procedure one step further and approximate f by *second*-degree polynomials.

Before presenting Simpson's Rule, we give a theorem for evaluating integrals of polynomials of degree 2 (or less).

THEOREM 5.19
INTEGRAL OF
$p(x) = Ax^2 + Bx + C$

If $p(x) = Ax^2 + Bx + C$, then

$$\int_a^b p(x) \, dx = \left(\frac{b - a}{6}\right)\left[p(a) + 4p\left(\frac{a + b}{2}\right) + p(b)\right].$$

PROOF

$$\int_a^b p(x) \, dx = \int_a^b (Ax^2 + Bx + C) \, dx$$

$$= \left[\frac{Ax^3}{3} + \frac{Bx^2}{2} + Cx\right]_a^b$$

$$= \frac{A(b^3 - a^3)}{3} + \frac{B(b^2 - a^2)}{2} + C(b - a)$$

$$= \left(\frac{b - a}{6}\right)[2A(a^2 + ab + b^2) + 3B(b + a) + 6C]$$

Now, by expanding and collecting terms, the expression inside the brackets becomes

$$\underbrace{(Aa^2 + Ba + C)}_{p(a)} + \underbrace{4\left\{A\left(\frac{b + a}{2}\right)^2 + B\left(\frac{b + a}{2}\right) + C\right\}}_{4p\left(\frac{a + b}{2}\right)} + \underbrace{(Ab^2 + Bb + C)}_{p(b)}$$

and we have

$$\int_a^b p(x) \, dx = \left(\frac{b - a}{6}\right)\left[p(a) + 4p\left(\frac{a + b}{2}\right) + p(b)\right].$$

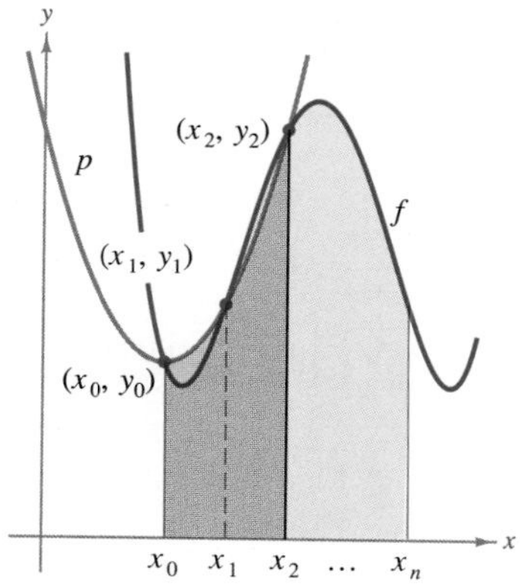

To develop Simpson's Rule for approximating a definite integral, we again partition the interval $[a, b]$ into n equal subintervals, each of width $\Delta x = (b - a)/n$. This time, however, we require n to be even, and group the subintervals into pairs such that

$$a = \underbrace{x_0 < x_1 < x_2}_{[x_0, x_2]} < \underbrace{x_3 < x_4 <}_{[x_2, x_4]} \cdots < \underbrace{x_{n-2} < x_{n-1} < x_n}_{[x_{n-2}, x_n]} = b.$$

Then on each (double) subinterval $[x_{i-2}, x_i]$ we approximate f by a polynomial p of degree less than or equal to 2. For example, on the subinterval $[x_0, x_2]$, we choose the polynomial of least degree passing through the points (x_0, y_0), (x_1, y_1), and (x_2, y_2), as shown in Figure 5.36. Now, using p as an approximation for f on this subinterval, we have

$$\int_{x_0}^{x_2} f(x)\, dx \approx \int_{x_0}^{x_2} p(x)\, dx = \frac{x_2 - x_0}{6}\left[p(x_0) + 4p\left(\frac{x_2 + x_0}{2}\right) + p(x_2)\right]$$

$$= \frac{2[(b - a)/n]}{6}[p(x_0) + 4p(x_1) + p(x_2)]$$

$$= \frac{b - a}{3n}[f(x_0) + 4f(x_1) + f(x_2)].$$

Repeating this procedure on the entire interval $[a, b]$ produces the following theorem.

$$\int_{x_0}^{x_2} p(x)\, dx \approx \int_{x_0}^{x_2} f(x)\, dx$$

FIGURE 5.36

THEOREM 5.20
SIMPSON'S RULE (n is even)

Let f be continuous on $[a, b]$. Simpson's Rule for approximating $\int_a^b f(x)\, dx$ is given by

$$\int_a^b f(x)\, dx \approx \frac{b - a}{3n}[f(x_0) + 4f(x_1) + 2f(x_2) + 4f(x_3) + \cdots + 4f(x_{n-1}) + f(x_n)].$$

Moreover, as $n \to \infty$, the right-hand side approaches $\int_a^b f(x)\, dx$.

REMARK Note that the coefficients in Simpson's Rule have the following pattern.

1 4 2 4 2 4 . . . 4 2 4 1

In Example 1 we used the Trapezoidal Rule to estimate $\int_0^1 \sqrt{x + 1}\, dx$. In the next example we see how well Simpson's Rule works for the same integral.

EXAMPLE 2 Approximation with Simpson's Rule

Use Simpson's Rule to approximate the definite integral

$$A = \int_0^1 \sqrt{x + 1}\, dx.$$

Compare the results for $n = 4$ and $n = 8$.

SOLUTION

When $n = 4$, we have $\Delta x = \frac{1}{4}$, and by Simpson's Rule we have

$$A \approx \frac{1}{12}\left[\sqrt{1} + 4\sqrt{\frac{5}{4}} + 2\sqrt{\frac{6}{4}} + 4\sqrt{\frac{7}{4}} + \sqrt{2}\right]$$

$$= \frac{1}{12}\left[1 + 4\left(\frac{\sqrt{5}}{2}\right) + 2\left(\frac{\sqrt{6}}{2}\right) + 4\left(\frac{\sqrt{7}}{2}\right) + \sqrt{2}\right] \approx 1.2189.$$

When $n = 8$, we have $\Delta x = \frac{1}{8}$, and

$$A \approx \frac{1}{24}\left[\sqrt{1} + 4\sqrt{\frac{9}{8}} + 2\sqrt{\frac{10}{8}} + 4\sqrt{\frac{11}{8}}\right.$$

$$\left. + 2\sqrt{\frac{12}{8}} + 4\sqrt{\frac{13}{8}} + 2\sqrt{\frac{14}{8}} + 4\sqrt{\frac{15}{8}} + \sqrt{2}\right]$$

$$\approx 1.2190.$$

In Examples 1 and 2 we were able to calculate the exact values of the integrals and compare those values to our approximations to see how close they were. In practice, of course, we would not bother with an approximation if it were possible to evaluate the integral exactly. However, if we must use an approximation technique, it is important to know how accurate we can expect the approximation to be. The following theorem, which we list without proof, gives the formulas for estimating the error involved in the use of Simpson's Rule and the Trapezoidal Rule.

THEOREM 5.21
ERROR IN THE TRAPEZOIDAL AND SIMPSON'S RULES

If f has a continuous second derivative on $[a, b]$, then the error E in approximating $\int_a^b f(x)\, dx$ by the Trapezoidal Rule is

$$E \le \frac{(b-a)^3}{12n^2}[\max |f''(x)|], \quad a \le x \le b. \qquad \text{Trapezoidal Rule}$$

Moreover, if f has a continuous fourth derivative on $[a, b]$, then the error E in approximating $\int_a^b f(x)\, dx$ by Simpson's Rule is

$$E \le \frac{(b-a)^5}{180n^4}[\max |f^{(4)}(x)|], \quad a \le x \le b. \qquad \text{Simpson's Rule}$$

Theorem 5.21 states that the errors generated by the Trapezoidal Rule and Simpson's Rule have upper bounds dependent on the extreme values of $f''(x)$ and $f^{(4)}(x)$, respectively, in the interval $[a, b]$. Furthermore, it is evident from this theorem that these errors can be made arbitrarily small by *increasing* n, provided that f'' and $f^{(4)}$ are continuous and therefore bounded in $[a, b]$. The next example shows how to determine a value of n that will bound the error within a predetermined **tolerance**.

EXAMPLE 3   The approximate error in the Trapezoidal Rule

Use the Trapezoidal Rule to estimate the value of

$$\int_0^1 \sqrt{1 + x^2} \, dx.$$

Determine n so that the approximation error is less than 0.01.

SOLUTION

Letting $f(x) = \sqrt{1 + x^2}$, we have

$$f'(x) = x(1 + x^2)^{-1/2}$$

$$f''(x) = x\left(-\frac{1}{2}\right)(2x)(1 + x^2)^{-3/2} + (1 + x^2)^{-1/2}$$

$$= (1 + x^2)^{-3/2}$$

which implies that the maximum value of $|f''(x)|$ on the interval [0, 1] is $|f''(0)| = 1$. Thus, by Theorem 5.21, we can write

$$E \le \frac{(b - a)^3}{12n^2}|f''(0)| \le \frac{1}{12n^2}(1) = \frac{1}{12n^2}.$$

To obtain an error E that is less than 0.01, we must choose n so that $(1/12n^2) \le 1/100$. Thus,

$$100 \le 12n^2 \quad \Longrightarrow \quad 2.89 \approx \sqrt{\frac{100}{12}} \le n.$$

Therefore, we choose $n = 3$ (since n must be greater than or equal to 2.89) and apply the Trapezoidal Rule, as shown in Figure 5.37, to obtain

$$\int_0^1 \sqrt{1 + x^2} \, dx$$

$$\approx \frac{1}{6}\left[\sqrt{1 + 0^2} + 2\sqrt{1 + (1/3)^2} + 2\sqrt{1 + (2/3)^2} + \sqrt{1 + 1^2}\right]$$

$$\approx 1.154.$$

Thus, with an error no larger than 0.01, we know that

$$1.144 \le \int_0^1 \sqrt{1 + x^2} \, dx \le 1.164.$$

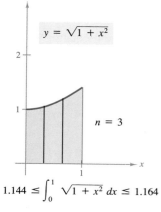

$$1.144 \le \int_0^1 \sqrt{1 + x^2} \, dx \le 1.164$$

FIGURE 5.37

EXAMPLE 4   The approximate error in Simpson's Rule

Use Simpson's Rule to estimate the value of

$$\int_1^3 \frac{1}{x} \, dx.$$

Determine n so that the approximation error is less than 0.01.

SOLUTION

According to Theorem 5.21, the error in Simpson's Rule involves the fourth derivative. Hence, by successive differentiation, we have

$$f(x) = x^{-1}$$
$$f'(x) = -x^{-2}$$
$$f''(x) = 2x^{-3}$$
$$f'''(x) = -6x^{-4}$$
$$f^{(4)}(x) = 24x^{-5}.$$

Since, in the interval $[1, 3]$, the maximum of $|f^{(4)}(x)|$ is $|f^{(4)}(1)| = 24$, we have

$$E \leq \frac{(b - a)^5}{180n^4}|f^{(4)}(1)| = \frac{32}{180n^4}|f^{(4)}(1)|$$
$$= \frac{32}{180n^4}(24) = \frac{64}{15n^4}.$$

Now by choosing n so that $(64/15n^4) < 1/100$, we have

$$\frac{6400}{15} < n^4 \quad \Longrightarrow \quad 4.54 < n.$$

Therefore, we choose $n = 6$, as shown in Figure 5.38, and obtain

$$\int_1^3 \frac{1}{x}\,dx$$
$$\approx \frac{2}{18}\left[\frac{1}{1} + 4\left(\frac{1}{4/3}\right) + 2\left(\frac{1}{5/3}\right) + 4\left(\frac{1}{6/3}\right) + 2\left(\frac{1}{7/3}\right) + 4\left(\frac{1}{8/3}\right) + \frac{1}{3}\right]$$
$$\approx 1.0989$$

and we conclude that

$$1.0889 \leq \int_1^3 \frac{1}{x} \leq 1.1089.$$

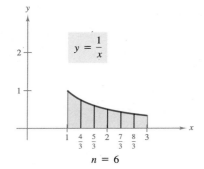

$n = 6$

FIGURE 5.38

REMARK   You may wonder why we introduced the Trapezoidal Rule, since for a fixed n Simpson's Rule usually gives a more accurate approximation. The main reason is that its error can be estimated more easily than the error involved in Simpson's Rule. For instance, if

$$f(x) = \sqrt{x}\sqrt[3]{x + 1}$$

then to estimate the error in Simpson's Rule we would need to find the fourth derivative of f—a huge task! Therefore, we may prefer to use the Trapezoidal Rule, even if we have to use a larger n to obtain the desired accuracy.

EXERCISES for Section 5.6

In Exercises 1–10, use the Trapezoidal Rule and Simpson's Rule to approximate the value of the definite integral for the indicated value of n. Round the answer to four decimal places and compare the results with the exact value of the definite integral.

1. $\int_0^2 x^2\,dx$

2. $\int_0^1 \left(\frac{x^2}{2}+1\right)dx$

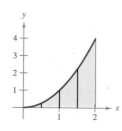

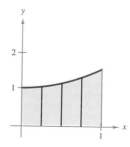

3. $\int_0^2 x^3\,dx$

4. $\int_1^2 \frac{1}{x^2}\,dx$

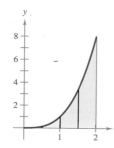

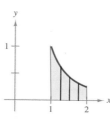

5. $\int_0^2 x^3\,dx,\ n=8$

6. $\int_0^8 \sqrt[3]{x}\,dx,\ n=8$

7. $\int_4^9 \sqrt{x}\,dx,\ n=8$

8. $\int_1^3 (4-x^2)\,dx,\ n=4$

9. $\int_1^2 \frac{1}{(x+1)^2}\,dx,\ n=4$

10. $\int_0^2 x\sqrt{x^2+1}\,dx,\ n=4$

In Exercises 11–20, approximate each integral using (a) the Trapezoidal Rule and (b) Simpson's Rule.

11. $\int_0^4 \frac{1}{x+1}\,dx,\ n=4$

12. $\int_0^4 \sqrt{1+x^2}\,dx,\ n=4$

13. $\int_0^2 \sqrt{1+x^3}\,dx,\ n=2$

14. $\int_0^2 \frac{1}{\sqrt{1+x^3}}\,dx,\ n=4$

15. $\int_0^1 \sqrt{x}\sqrt{1-x}\,dx,\ n=4$

16. $\int_0^1 \frac{1}{x^2+1}\,dx,\ n=2$

17. $\int_{-2}^2 \frac{1}{x^2+1}\,dx,\ n=8$

18. $\int_{-1}^1 x\sqrt{x+1}\,dx,\ n=4$

19. $\int_1^7 \frac{\sqrt{x-1}}{x}\,dx,\ n=6$

20. $\int_2^5 \frac{1}{1+\sqrt{x-1}}\,dx,\ n=6$

In Exercises 21–24, find the maximum possible error in approximating the given integral by (a) the Trapezoidal Rule and (b) Simpson's Rule with $n=4$.

21. $\int_0^2 x^3\,dx$

22. $\int_0^2 x^4\,dx$

23. $\int_0^1 \frac{1}{x+1}\,dx$

24. $\int_0^1 \frac{1}{x^2+1}\,dx$

In Exercises 25–28, find n so that the error in the approximation of the definite integral is less than 0.00001 using (a) the Trapezoidal Rule and (b) Simpson's Rule.

25. $\int_1^3 \frac{1}{x}\,dx$

26. $\int_0^1 \frac{1}{1+x}\,dx$

27. $\int_0^2 \sqrt{1+x}\,dx$

28. $\int_0^2 (x+1)^{2/3}\,dx$

29. Use Simpson's Rule with $n=6$ to approximate π correct to five decimal places using the equation

$$\pi = \int_0^1 \frac{4}{1+x^2}\,dx.$$

30. Prove that Simpson's Rule is exact when approximating the integral of a cubic polynomial function, and demonstrate the result for

$$\int_0^1 x^3\,dx,\ n=2.$$

In Exercises 31 and 32, use the Trapezoidal Rule to estimate the number of square feet of land in a given lot where x and y are measured in feet, as shown in the accompanying figures. In each case the land is bounded by a stream and two straight roads that meet at right angles.

31.

x	y
0	125
100	125
200	120
300	112
400	90
500	90
600	95
700	88
800	75
900	35
1000	0

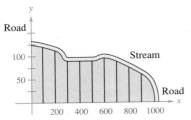

32.

x	y
0	75
10	81
20	84
30	76
40	67
50	68
60	69
70	72
80	68
90	56
100	42
110	23
120	0

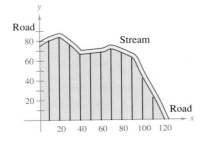

33. To estimate the surface area of a pond, a surveyor takes several measurements, as shown in the accompanying figure. Use (a) the Trapezoidal Rule and (b) Simpson's Rule to estimate the surface area of this pond. [The measurements are given in feet.]

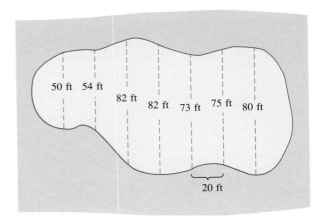

34. The following table lists several physical measurements gathered in an experiment. Assume the function $y = f(x)$ is continuous and approximate the integral $\int_0^2 f(x)\, dx$ using (a) the Trapezoidal Rule and (b) Simpson's Rule.

x	0.00	0.25	0.50	0.75	1.00
y	4.32	4.36	4.58	5.79	6.14

x	1.25	1.50	1.75	2.00
y	7.25	7.64	8.08	8.14

35. Use a computer or calculator to complete the following table of numerical approximations to the integral

$$\int_0^4 \sqrt{2 + 3x^2}\, dx.$$

n	Trapezoidal Rule	Simpson's Rule
4		
8		
12		
16		
20		

36. Use a computer or calculator and Simpson's Rule (with $n = 10$) to approximate t to three decimal places in the integral equation

$$\int_0^t \sqrt{1 + x^3}\, dx = 2.$$

REVIEW EXERCISES for Chapter 5

In Exercises 1–18, find the indefinite integral.

1. $\displaystyle\int \frac{2}{3\sqrt[3]{x}}\, dx$

2. $\displaystyle\int \frac{2}{\sqrt[3]{3x}}\, dx$

3. $\displaystyle\int (2x^2 + x - 1)\, dx$

4. $\displaystyle\int \frac{x^3 - 2x^2 + 1}{x^2}\, dx$

5. $\displaystyle\int \frac{(1 + x)^2}{\sqrt{x}}\, dx$

6. $\displaystyle\int x^2\sqrt{x^3 + 3}\, dx$

7. $\displaystyle\int \frac{x^2}{\sqrt{x^3 + 3}}\, dx$

8. $\displaystyle\int \frac{x^2 + 2x}{(x + 1)^2}\, dx$

9. $\displaystyle\int (x^2 + 1)^3\, dx$

10. $\displaystyle\int \sqrt{2 - 5x}\, dx$

11. $\displaystyle\int x(x^2 + 1)^3\, dx$

12. $\displaystyle\int x^2\sqrt{4 - x^3}\, dx$

13. $\displaystyle\int \frac{x}{(x^2 + 1)^3}\, dx$

14. $\displaystyle\int \frac{x}{\sqrt{25 - 9x^2}}\, dx$

15. $\displaystyle\int x^2\sqrt{x + 5}\, dx$

16. $\displaystyle\int x\sqrt{x + 5}\, dx$

17. $\displaystyle\int \frac{x^3 + 1}{x^2}\, dx$

18. $\displaystyle\int \left(x + \frac{1}{x}\right)^2\, dx$

19. Write in sigma notation the sum of the following.
   (a) the first ten positive odd integers
   (b) the cubes of the first n positive integers
   (c) $6 + 10 + 14 + 18 + \cdots + 42$

20. Evaluate the following sums for $x_1 = 2$, $x_2 = -1$, $x_3 = 5$, $x_4 = 3$, and $x_5 = 7$.

   (a) $\displaystyle\frac{1}{5}\sum_{i=1}^{5} x_i$

   (b) $\displaystyle\sum_{i=1}^{5} \frac{1}{x_i}$

   (c) $\displaystyle\sum_{i=1}^{5} (2x_i - x_i^2)$

   (d) $\displaystyle\sum_{i=2}^{5} (x_i - x_{i-1})$

In Exercises 21–30, use the Fundamental Theorem of Calculus to evaluate the definite integral.

21. $\displaystyle\int_0^4 (2 + x)\, dx$

22. $\displaystyle\int_{-1}^1 (t^2 + 2)\, dt$

23. $\displaystyle\int_{-1}^1 (4t^3 - 2t)\, dt$

24. $\displaystyle\int_3^6 \frac{x}{3\sqrt{x^2 - 8}}\, dx$

25. $\displaystyle\int_0^3 \frac{1}{\sqrt{1 + x}}\, dx$

26. $\displaystyle\int_0^1 x^2(x^3 + 1)^3\, dx$

27. $\displaystyle\int_4^9 x\sqrt{x}\, dx$

28. $\displaystyle\int_1^2 \left(\frac{1}{x^2} - \frac{1}{x^3}\right) dx$

29. $\displaystyle 2\pi \int_0^1 (y + 1)\sqrt{1 - y}\, dy$

30. $\displaystyle 2\pi \int_{-1}^0 x^2\sqrt{x + 1}\, dx$

31. Find the function f whose derivative is $f'(x) = -2x$ and whose graph passes through the point $(-1, 1)$.

32. A function f has a second derivative $f''(x) = 6(x - 1)$. Find the function if its graph passes through the point $(2, 1)$ and at that point is tangent to the line given by $3x - y - 5 = 0$.

33. An airplane taking off from a runway travels 3600 feet before lifting off. If it starts from rest, moves with constant acceleration, and makes the run in 30 seconds, with what speed does it lift off?

34. The speed of a car traveling in a straight line is reduced from 45 to 30 miles per hour in a distance of 264 feet. Find the distance in which the car can be brought to rest from 30 miles per hour, assuming the same constant acceleration.

35. A ball is thrown vertically upward from ground level with an initial velocity of 96 feet per second.
  (a) How long will it take it to rise to its maximum height?
  (b) What is the maximum height?
  (c) When is the velocity of the ball one-half the initial velocity?
  (d) What is the height of the ball when its velocity is one-half the initial velocity?

36. Repeat Exercise 35 for an initial velocity of 128 feet per second.

37. Consider the region bounded by $y = mx$, $y = 0$, $x = 0$, and $x = b$.
  (a) Find the upper and lower sum to approximate the area of the region when $\Delta x = b/4$.
  (b) Find the upper and lower sum to approximate the area of the region when $\Delta x = b/n$.
  (c) Find the area of the region by letting n approach infinity in both sums of part (b). Show that in each case you obtain the formula for the area of a triangle.
  (d) Find the area of the region by using the Fundamental Theorem of Calculus.

38. (a) Find the area of the region bounded by the graphs of $y = x^3$, $y = 0$, $x = 1$, and $x = 3$ by the limit definition.
  (b) Find the area of the given region by using the Fundamental Theorem of Calculus.

In Exercises 39–44, sketch the graph of the region whose area is given by the integral and find the area.

39. $\int_1^3 (2x - 1)\, dx$

40. $\int_0^2 (x + 4)\, dx$

41. $\int_3^4 (x^2 - 9)\, dx$

42. $\int_{-1}^2 (-x^2 + x + 2)\, dx$

43. $\int_0^1 (x - x^3)\, dx$

44. $\int_0^1 \sqrt{x}(1 - x)\, dx$

In Exercises 45–48, find the average value of the function over the given interval. Find the values of x where the function assumes its mean value and sketch the graph of the function.

Function	Interval
45. $f(x) = \dfrac{1}{\sqrt{x - 1}}$	$[5, 10]$
46. $f(x) = x^3$	$[0, 2]$
47. $f(x) = x$	$[0, 4]$
48. $f(x) = x^2 - \dfrac{1}{x^2}$	$[1, 2]$

In Exercises 49 and 50, use Simpson's Rule with ($n = 4$) to approximate the definite integral.

49. $\int_1^2 \dfrac{1}{1 + x^3}\, dx$

50. $\int_0^1 \dfrac{x^{3/2}}{3 - x^2}\, dx$

51. Suppose that gasoline is increasing in price according to the equation

$$p = 1 + 0.1t + 0.02t^2$$

where p is the dollar price per gallon and $t = 0$ represents the year 1983. If an automobile is driven 15,000 miles a year and gets M miles per gallon, then the annual fuel cost is

$$C = \frac{15,000}{M} \int_t^{t+1} p\, dt.$$

Find the annual fuel cost for the years (a) 1985 and (b) 1990.

In Exercises 52 and 53, the function

$$f(x) = kx^n(1 - x)^m, \qquad 0 \le x \le 1$$

where $n, m > 0$ and k is a constant, can be used to represent various probability distributions. If k is chosen so that

$$\int_0^1 f(x)\, dx = 1$$

the probability that x will fall between a and b is given by

$$P_{a,b} = \int_a^b f(x)\, dx.$$

52. The probability of recall in a certain experiment is found to be

$$p_{a,b} = \int_a^b \frac{15}{4} x\sqrt{1-x}\ dx$$

where x represents the percentage of recall (see figure).

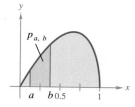

FIGURE FOR 52

(a) For a randomly chosen individual, what is the probability that he or she will recall between 50% and 75% of the material?

(b) What is the median percentage recall? That is, for what value of b is it true that the probability from 0 to b is 0.5?

53. The probability of finding between a and b percentage of iron in ore samples taken from a certain region is given by

$$p_{a,b} = \int_a^b \frac{1155}{32} x^3(1-x)^{3/2}\ dx.$$

(See figure.) What is the probability that a sample will contain between
(a) 0% and 25%? (b) 50% and 100%?

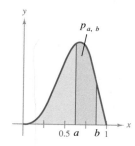

FIGURE FOR 53

Chapter 6 Application

American football, in its modern form, is a twentieth-century invention. In the 1800s a rough, soccer-like game was played with a "round football." In 1905, at the request of President Theodore Roosevelt, the Intercollegiate Athletic Association (which became the NCAA in 1910) was formed. With the introduction of the forward pass in 1906, the shape of the ball was altered to make it easier to grip.

Chapter Overview

Integration has a wide variety of applications. For each of the applications presented in this chapter, we will begin with a known formula, such as the area of a rectangular region, the volume of a circular disc, or the work done by a constant force. Then we will show how the limit of a sum gives rise to new formulas that involve integration.

The chapter begins by showing how to use a definite integral to find the area of a region bounded by two curves. Section 6.2 presents a technique for finding the volume of a solid of revolution by the *disc method*. The section closes by showing that the disc method is a special case of using integration to find the volume of a solid with known cross sections. Section 6.3 presents an alternate way to find the volume of a solid of revolution—using the *shell method*.

Section 6.4 discusses the length of a curve and the surface area of a surface of revolution. Section 6.5 shows how integration can be used to find the work done by a variable force, and Section 6.6 shows how integration can be used to find fluid pressure. For instance, integration can be used to find the water pressure against the side of a dam.

The last section in the chapter discusses moments, centers of mass, and centroids. All three of these concepts have many uses in engineering and physics, and several are discussed in the section.

Volume of a Football

A regulation-size football has the following dimensions.

> Length: 11.00–11.25 in.
> Girth: 21.25–21.50 in.

A mathematical model that falls within these dimensions is given by revolving the graph of

$$f(x) = -0.0944x^2 + 3.4, \quad -5.5 \le x \le 5.5$$

about the x-axis. The resulting solid is called a *solid of revolution* and its volume is given by

$$\text{volume} = \pi \int_{-5.5}^{5.5} [f(x)]^2 \, dx$$

$$= \pi \int_{-5.5}^{5.5} [-0.0944x^2 + 3.4]^2 \, dx$$

$$\approx 232 \text{ cubic inches.}$$

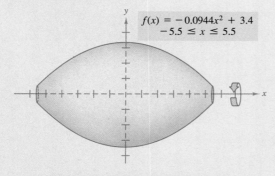

$$f(x) = -0.0944x^2 + 3.4$$
$$-5.5 \le x \le 5.5$$

Football-shaped solid formed by revolving a parabolic segment about the x-axis.

See Exercise 39, Section 6.2

Applications of Integration

6.1 Area of a Region Between Two Curves

Area of a region between two curves ▪ Points of intersection of two curves ▪ Representative elements of integration

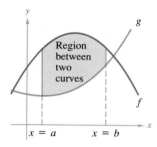

FIGURE 6.1

With a few modifications we can extend the application of definite integrals from the area of a region *under* a curve to the area of a region *between* two curves. If, as in Figure 6.1, the graphs of both f and g lie above the x-axis, we can geometrically interpret the area of the region between the graphs as the area of the region under the graph of g subtracted from the area of the region under the graph of f as shown in Figure 6.2.

Although the graphs of f and g are shown above the x-axis in Figure 6.1, this is not necessary. The same integrand $[f(x) - g(x)]$ can be used as long as f and g are continuous and $g(x) \le f(x)$ on the interval $[a, b]$. This result is summarized in the following theorem.

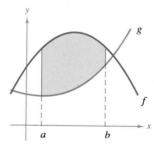
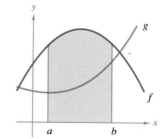

FIGURE 6.2

$$\begin{array}{ccccc} \text{(Area of region between } f \text{ and } g) & = & \text{(Area of region under } f) & - & \text{(Area of region under } g) \\ \int_a^b [f(x) - g(x)] \, dx & = & \int_a^b f(x) \, dx & - & \int_a^b g(x) \, dx \end{array}$$

**THEOREM 6.1
AREA OF A REGION BETWEEN
TWO CURVES**

If f and g are continuous on $[a, b]$ and $g(x) \le f(x)$ for all x in $[a, b]$, then the area of the region bounded by the graphs of f and g and the vertical lines $x = a$ and $x = b$ is

$$A = \int_a^b [f(x) - g(x)] \, dx.$$

PROOF

We partition the interval $[a, b]$ into n subintervals, each of width Δx, and sketch a **representative rectangle** of width Δx and height $f(x_i) - g(x_i)$, where x_i is in the ith interval, as shown in Figure 6.3. The area of this representative rectangle is

$$\Delta A_i = (\text{height})(\text{width}) = [f(x_i) - g(x_i)]\Delta x.$$

By adding the areas of the n rectangles and taking the limit as $\|\Delta\| \to 0$ ($n \to \infty$), we have

$$\lim_{n \to \infty} \sum_{i=1}^{n} [f(x_i) - g(x_i)]\Delta x.$$

Since f and g are continuous on $[a, b]$, $f - g$ is also continuous on this interval and the limit exists. Therefore, the area A of the given region is

$$A = \lim_{n \to \infty} \sum_{i=1}^{n} [f(x_i) - g(x_i)]\Delta x$$

$$= \int_a^b [f(x) - g(x)]\ dx.$$

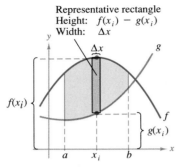

Representative rectangle
Height: $f(x_i) - g(x_i)$
Width: Δx

FIGURE 6.3

It is important to realize that the area formula in Theorem 6.1 depends *only* on the continuity of f and g and the assumption that $g(x) \le f(x)$. The graphs of f and g can be placed anywhere with respect to the x-axis, as illustrated in Figure 6.4.

Representative rectangles are used throughout this chapter in various applications of integration. A vertical rectangle (of width Δx) implies integration with respect to x, while a horizontal rectangle (of width Δy) implies integration with respect to y.

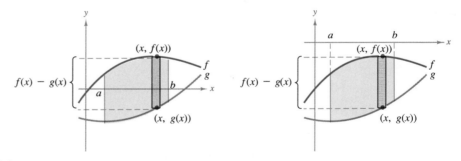

FIGURE 6.4 Height of representative rectangle is $f(x) - g(x)$ regardless of the relative position of the x-axis.

EXAMPLE 1 Finding the area of a region between two curves

Find the area of the region bounded by the graphs of $y = x^2 + 2$, $y = -x$, $x = 0$, and $x = 1$.

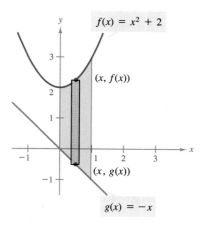

$f(x) = x^2 + 2$

$(x, f(x))$

$(x, g(x))$

$g(x) = -x$

FIGURE 6.5

SOLUTION

If we let $g(x) = -x$ and $f(x) = x^2 + 2$, then $g(x) \le f(x)$ for all x in $[0, 1]$, as shown in Figure 6.5. Thus, the area of the representative rectangle is

$$\Delta A = [f(x) - g(x)]\Delta x = [(x^2 + 2) - (-x)]\Delta x.$$

Therefore, by Theorem 6.1, we find the area of the region to be

$$A = \int_a^b [f(x) - g(x)]\, dx = \int_0^1 [(x^2 + 2) - (-x)]\, dx$$

$$= \left[\frac{x^3}{3} + \frac{x^2}{2} + 2x \right]_0^1$$

$$= \frac{1}{3} + \frac{1}{2} + 2 = \frac{17}{6}.$$

In Example 1, the graphs of $f(x) = x^2 + 2$ and $g(x) = -x$ do not intersect, and the values of a and b are given explicitly. A more common problem involves the area of a region bounded by two *intersecting* graphs, where the values of a and b must be calculated.

EXAMPLE 2 A region lying between two intersecting graphs

Find the area of the region bounded by the graphs of $f(x) = 2 - x^2$ and $g(x) = x$.

SOLUTION

From Figure 6.6, we see that the graphs of f and g have two points of intersection. To find the x-coordinates of these points, we set $f(x)$ and $g(x)$ equal to each other and solve for x.

$$2 - x^2 = x \qquad \text{Set } f(x) \text{ equal to } g(x)$$

$$-x^2 - x + 2 = 0$$

$$-(x + 2)(x - 1) = 0 \qquad x = -2 \text{ and } x = 1$$

Thus, we have $a = -2$ and $b = 1$. Since $g(x) \le f(x)$ on the interval $[-2, 1]$, the representative rectangle has an area of

$$\Delta A = [f(x) - g(x)]\Delta x = [(2 - x^2) - x]\Delta x$$

and the area of the region is

$$A = \int_{-2}^1 [(2 - x^2) - x]\, dx = \left[-\frac{x^3}{3} - \frac{x^2}{2} + 2x \right]_{-2}^1$$

$$= \left(-\frac{1}{3} - \frac{1}{2} + 2 \right) - \left(\frac{8}{3} - 2 - 4 \right)$$

$$= \frac{9}{2}.$$

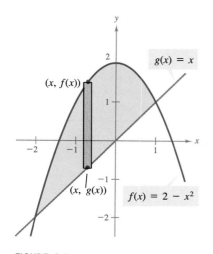

y

$g(x) = x$

$(x, f(x))$

$(x, g(x))$

$f(x) = 2 - x^2$

FIGURE 6.6

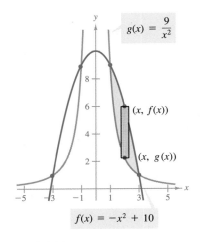

$g(x) = \dfrac{9}{x^2}$

$(x, f(x))$

$(x, g(x))$

$f(x) = -x^2 + 10$

FIGURE 6.7

EXAMPLE 3 A region lying between two intersecting graphs

The graphs of

$$f(x) = -x^2 + 10 \qquad \text{and} \qquad g(x) = \frac{9}{x^2}$$

intersect four times, bounding two regions of equal areas, as shown in Figure 6.7. Find the area of one of these regions.

SOLUTION

To find the points of intersection of the graphs of $f(x) = -x^2 + 10$ and $g(x) = 9/x^2$, we set the two functions equal to each other and solve for x.

$$-x^2 + 10 = \frac{9}{x^2} \qquad \text{Set } f(x) \text{ equal to } g(x)$$

$$-x^4 + 10x^2 - 9 = 0$$

$$-(x^2 - 9)(x^2 - 1) = 0 \qquad x = -3, -1, 1, 3$$

We choose to find the area of the region in the first quadrant. Thus, $a = 1$ and $b = 3$. Since $-x^2 + 10 \geq (9/x^2)$ on the interval $[1, 3]$, the area of the region is

$$A = \int_1^3 \left[(-x^2 + 10) - \frac{9}{x^2} \right] dx$$

$$= \left[-\frac{x^3}{3} + 10x + \frac{9}{x} \right]_1^3$$

$$= (-9 + 30 + 3) - \left(-\frac{1}{3} + 10 + 9 \right)$$

$$= \frac{16}{3}.$$

If two curves intersect at *more* than two points, then to find the area of the region between the curves, we must find all points of intersection and check to see which curve is above the other in each interval determined by these points.

EXAMPLE 4 Curves that intersect at more than two points

Find the area of the region between the graphs of $f(x) = 3x^3 - x^2 - 10x$ and $g(x) = -x^2 + 2x$.

SOLUTION

Solving for x in the equation $f(x) = g(x)$ produces

$$f(x) - g(x) = (3x^3 - x^2 - 10x) - (-x^2 + 2x) = 0$$

$$3x^3 - 12x = 0$$

$$3x(x^2 - 4) = 0.$$

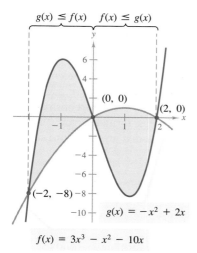

$g(x) \leq f(x)$ $f(x) \leq g(x)$

(0, 0)

(2, 0)

(−2, −8)

$g(x) = -x^2 + 2x$

$f(x) = 3x^3 - x^2 - 10x$

FIGURE 6.8

Thus, the two graphs intersect when $x = -2$, 0, and 2. In Figure 6.8, we see that $g(x) \leq f(x)$ in the interval $[-2, 0]$. However, the two graphs switch at the origin, and $f(x) \leq g(x)$ in the interval $[0, 2]$. Hence, we need two integrals—one for the interval $[-2, 0]$ and one for $[0, 2]$.

$$A = \int_{-2}^{0} [f(x) - g(x)]\, dx + \int_{0}^{2} [g(x) - f(x)]\, dx$$

$$= \int_{-2}^{0} (3x^3 - 12x)\, dx + \int_{0}^{2} (-3x^3 + 12x)\, dx$$

$$= \left[\frac{3x^4}{4} - 6x^2 \right]_{-2}^{0} + \left[\frac{-3x^4}{4} + 6x^2 \right]_{0}^{2}$$

$$= -(12 - 24) + (-12 + 24) = 24$$

If the graph of a function of y is a boundary of a region, it is often convenient to use representative rectangles that are *horizontal* and find the area by integrating with respect to y. In general, to determine the area between two curves, we use

$$A = \int_{x_1}^{x_2} \underbrace{[(\text{top curve}) - (\text{bottom curve})]}_{\text{in variable } x}\, dx \qquad \text{Vertical rectangles}$$

$$A = \int_{y_1}^{y_2} \underbrace{[(\text{right curve}) - (\text{left curve})]}_{\text{in variable } y}\, dy \qquad \text{Horizontal rectangles}$$

where (x_1, y_1) and (x_2, y_2) are either adjacent points of intersection of the two curves involved or points on the specified boundary lines.

EXAMPLE 5   Horizontal representative rectangles

Find the area of the region bounded by the graphs of $x = 3 - y^2$ and $y = x - 1$.

SOLUTION

We consider $g(y) = 3 - y^2$ and $f(y) = y + 1$. These two curves intersect when $y = -2$ and $y = 1$, as shown in Figure 6.9. Since $f(y) \leq g(y)$ on this interval, we have

$$\Delta A = [g(y) - f(y)]\Delta y = [(3 - y^2) - (y + 1)]\Delta y.$$

Hence, the area is

$$A = \int_{-2}^{1} [(3 - y^2) - (y + 1)]\, dy$$

$$= \int_{-2}^{1} (-y^2 - y + 2)\, dy = \left[\frac{-y^3}{3} - \frac{y^2}{2} + 2y \right]_{-2}^{1}$$

$$= \left(-\frac{1}{3} - \frac{1}{2} + 2 \right) - \left(\frac{8}{3} - 2 - 4 \right) = \frac{9}{2}.$$

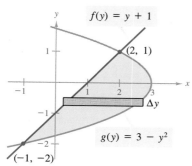

$f(y) = y + 1$

(2, 1)

Δy

$g(y) = 3 - y^2$

(−1, −2)

Horizontal rectangles
(Integration with respect to y)

FIGURE 6.9

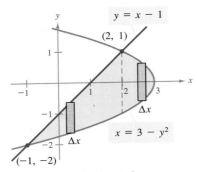

Vertical rectangles
(Integration with respect to *x*)

FIGURE 6.10

REMARK Note in Example 5 that by integrating with respect to *y* we need only one integral. If we had integrated with respect to *x*, we would have needed two integrals, as shown in Figure 6.10.

In this section, we developed the integration formula for the area between two curves by using a rectangle as the *representative element*. For each new application in the remaining sections of this chapter, we will construct an appropriate representative element using precalculus formulas you already know. Each integration formula then will be obtained by summing these representative elements.

For example, in this section we developed the area formula as follows.

$$A = (\text{height})(\text{width}) \longrightarrow \Delta A = [f(x) - g(x)] \, \Delta x \longrightarrow A = \int_a^b [f(x) - g(x)] \, dx$$

EXAMPLE 6   An application

The total consumption of petroleum fuel for transportation in the United States from 1960 to 1979 followed a growth pattern described by the equation

$$f(t) = 0.000433t^2 + 0.0962t + 2.76, \quad -10 \le t \le 9$$

where $f(t)$ is measured in billions of barrels and t in years, with $t = 0$ corresponding to January 1, 1970. When crude oil prices increased dramatically in the late 1970s, the growth pattern for consumption changed and began following the pattern described by the model

$$g(t) = -0.00831t^2 + 0.152t + 2.81, \quad 9 \le t \le 16$$

as shown in Figure 6.11. Find the total amount of fuel saved from 1979 through 1985 as a result of fuel being consumed at the post-1979 rate rather than at the pre-1979 rate.

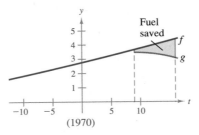

FIGURE 6.11

f: Pre-1979 consumption rate
g: Post-1979 consumption rate

SOLUTION

Since the graph of the pre-1979 model lies above the post-1979 graph on the interval [9, 16], the amount of gasoline saved is given by the following integral.

$$\int_9^{16} [\overbrace{(0.000433t^2 + 0.0962t + 2.76)}^{f(t)} - \overbrace{(-0.00831t^2 + 0.152t + 2.81)}^{g(t)}] \, dt$$

$$= \int_9^{16} (0.008743t^2 - 0.0558t - 0.05) \, dt$$

$$= \left[\frac{0.008743t^3}{3} - \frac{0.0558t^2}{2} - 0.05t \right]_9^{16}$$

$$\approx 4.58 \text{ billion barrels}$$

Therefore, approximately 4.58 billion barrels of fuel were saved. (At 42 gallons per barrel, about 200 billion gallons were saved!)

EXERCISES for Section 6.1

In Exercises 1–6, find the area of the given region.

1. $f(x) = x^2 - 6x$
  $g(x) = 0$

2. $f(x) = x^2 + 2x + 1$
  $g(x) = 2x + 5$

5. $f(x) = 3(x^3 - x)$
  $g(x) = 0$

6. $f(x) = (x - 1)^3$
  $g(x) = x - 1$

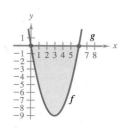

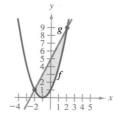

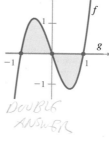

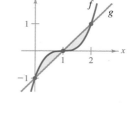

3. $f(x) = x^2 - 4x + 3$
  $g(x) = -x^2 + 2x + 3$

4. $f(x) = x^2$
  $g(x) = x^3$

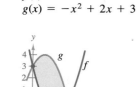

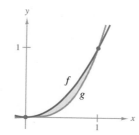

In Exercises 7–24, sketch the region bounded by the graphs of the given functions and find the area of the region.

7. $f(x) = x^2 - 4x$, $g(x) = 0$
8. $f(x) = 3 - 2x - x^2$, $g(x) = 0$
9. $f(x) = x^2 + 2x + 1$, $g(x) = 3x + 3$
10. $f(x) = -x^2 + 4x + 2$, $g(x) = x + 2$
11. $y = x$, $y = 2 - x$, $y = 0$

12. $y = \dfrac{1}{x^2}$, $y = 0$, $x = 1$, $x = 5$

13. $f(x) = 3x^2 + 2x$, $g(x) = 8$

14. $f(x) = x(x^2 - 3x + 3)$, $g(x) = x^2$

15. $f(x) = x^3 - 2x + 1$, $g(x) = -2x$, $x = 1$

16. $f(x) = \sqrt[3]{x}$, $g(x) = x$

17. $f(x) = \sqrt{3x + 1}$, $g(x) = x + 1$

18. $f(x) = x^2 + 5x - 6$, $g(x) = 6x - 6$

19. $y = x^2 - 4x + 3$, $y = 3 + 4x - x^2$

20. $y = x^4 - 2x^2$, $y = 2x^2$

21. $f(y) = y^2$, $g(y) = y + 2$

22. $f(y) = y(2 - y)$, $g(y) = -y$

23. $f(y) = y^2 + 1$, $g(y) = 0$, $y = -1$, $y = 2$

24. $f(y) = \dfrac{y}{\sqrt{16 - y^2}}$, $g(y) = 0$, $y = 3$

In Exercises 25–28, use integration to find the area of the triangle having the given vertices.

25. $(0, 0)$, $(4, 0)$, $(4, 4)$ **26.** $(0, 0)$, $(4, 0)$, $(6, 4)$

27. $(0, 0)$, $(a, 0)$, (b, c) **28.** $(2, -3)$, $(4, 6)$, $(6, 1)$

In Exercises 29 and 30, find b so that the line $y = b$ divides the region bounded by the graphs of the two equations into two regions of equal area.

29. $y = 9 - x^2$, $y = 0$ **30.** $y = 9 - |x|$, $y = 0$

31. The graphs of $y = x^4 - 2x^2 + 1$ and $y = 1 - x^2$ intersect at three points. However, the area between the curves *can* be found by a single integral. Explain why this is so, and write an integral for this area.

32. The area of the region bounded by the graphs of $y = x^3$ and $y = x$ *cannot* be found by the single integral

$$\int_{-1}^{1} (x^3 - x)\, dx.$$

Explain why this is so. Use symmetry to write a single integral that does represent the area.

In Exercises 33 and 34, find the area of the region bounded by the graph of the function and the tangent line to the graph at the specified point.

Function	Point
33. $f(x) = x^3$	$(1, 1)$
34. $f(x) = \sqrt[3]{x - 1}$	$(2, 1)$

In Exercises 35 and 36, use a computer or calculator and Simpson's Rule (with $n = 4$) to approximate the area of the region bounded by the graphs of the given equations.

35. $y = \sqrt{1 + x^3}$, $y = \dfrac{1}{2}x + 2$, $x = 0$

36. $y = \sqrt{x + x^2}$, $y = 0$, $x = 0$, $x = 1$

In Exercises 37 and 38, evaluate the given limit and sketch the graph of the region whose area is given by the limit.

37. $\displaystyle \lim_{\|\Delta\| \to 0} \sum_{i=1}^{n} (x_i - x_i^2)\, \Delta x$

where $x_i = i/n$ and $\Delta x = 1/n$

38. $\displaystyle \lim_{\|\Delta\| \to 0} \sum_{i=1}^{n} (4 - x_i^2)\, \Delta x$

where $x_i = -2 + (4i/n)$ and $\Delta x = 4/n$

In Exercises 39 and 40, two models R_1 and R_2 are given for revenue (in billions of dollars) for a large corporation. The model R_1 gives projected annual revenues from 1990 to 1995, with $t = 0$ corresponding to 1990, and R_2 gives projected revenues if there is a decrease in growth of corporate sales over the period. Approximate the total reduction in revenue if corporate sales are actually closer to the model R_2.

39. $R_1 = 7.21 + 0.58t$
$R_2 = 7.21 + 0.45t$

40. $R_1 = 7.21 + 0.26t + 0.02t^2$
$R_2 = 7.21 + 0.1t + 0.01t^2$

In Exercises 41–44, find the consumer surplus and producer surplus for the given supply and demand curves. The consumer surplus and producer surplus are represented by the areas shown in the accompanying figure.

Demand function	Supply function
41. $p_1(x) = 50 - 0.5x$	$p_2(x) = 0.125x$
42. $p_1(x) = 1000 - 0.4x^2$	$p_2(x) = 42x$
43. $p_1(x) = \dfrac{10,000}{\sqrt{x + 100}}$	$p_2(x) = 100\sqrt{0.05x + 10}$
44. $p_1(x) = \sqrt{25 - 0.1x}$	$p_2(x) = \sqrt{9 + 0.1x} - 2$

In Exercises 45 and 46, use Simpson's Rule on a computer or calculator (with $n = 10$) to approximate the area of the region bounded by the graphs of the given equations.

45. $y = \sqrt{\dfrac{x^3}{4 - x}},\ y = 0,\ x = 3$

46. $y = x\sqrt{\dfrac{4 - x}{4 + x}},\ y = 0,\ x = 4$

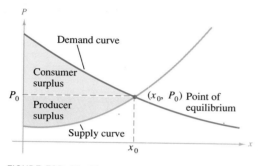

FIGURE FOR 41–44

6.2 Volume: The Disc Method

Solid of revolution ▪ The disc method ▪ The washer method ▪ Solids with known cross sections

In Chapter 5, we mentioned that area is only *one* of the many applications of the definite integral. Another important application is its use in finding the volume of a three-dimensional solid. In this section we consider a particular type of three-dimensional solid—one whose cross sections are similar. We begin with solid figures having *circular* (or annular) cross sections. Such solids are called **solids of revolution** and are used commonly in engineering and manufacturing. Some examples are axles, funnels, pills, bottles, and pistons.

If a region in the plane is revolved about a line, the resulting solid is called a **solid of revolution,** and the line is called the **axis of revolution.** The simplest such solid is a right circular cylinder or **disc,** which is formed by revolving a rectangle about an axis adjacent to one of the sides of the rectangle as shown in Figure 6.12. The volume of such a disc is given by

volume of disc $= \pi R^2 w$

where R is the radius of the disc and w is the width.

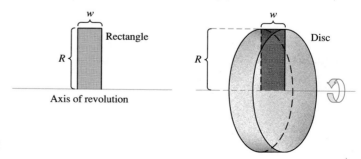

FIGURE 6.12

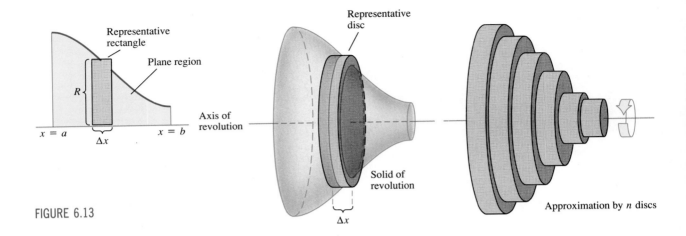

FIGURE 6.13

To see how to use the volume of a disc to find the volume of a general solid of revolution, consider a solid of revolution formed by revolving the plane region in Figure 6.13 about the indicated axis. To determine the volume of this solid, we consider a representative rectangle in the plane region. When this rectangle is revolved about the axis of revolution, it generates a representative disc whose volume is

$$\Delta V = \pi R^2 \Delta x.$$

If we approximate the volume of the solid by n such discs of width Δx and radius $R(x_i)$, we have

$$\text{volume of solid} \approx \sum_{i=1}^{n} \pi [R(x_i)]^2 \Delta x$$

$$= \pi \sum_{i=1}^{n} [R(x_i)]^2 \Delta x.$$

By taking the limit as $\|\Delta\| \to 0$ $(n \to \infty)$, we have

$$\text{volume of solid} = \lim_{n \to \infty} \pi \sum_{i=1}^{n} [R(x_i)]^2 \Delta x$$

$$= \pi \int_{a}^{b} [R(x)]^2 \, dx.$$

Schematically, the disc method looks like this.

Known precalculus formula		*Representative element*		*New integration formula*

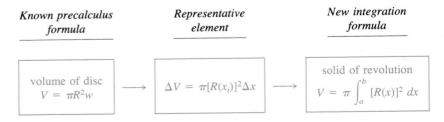

A similar formula can be derived if the axis of revolution is vertical.

| THE DISC METHOD | To find the volume of a solid of revolution with the **disc method,** use one of the following as indicated in Figure 6.14. |

Horizontal axis of revolution

volume $= V = \pi \displaystyle\int_a^b [R(x)]^2 \, dx$

Vertical axis of revolution

volume $= V = \pi \displaystyle\int_c^d [R(y)]^2 \, dy$

REMARK   In Figure 6.14 note that we can determine the variable of integration by placing a representative rectangle in the *plane* region "perpendicular" to the axis of revolution. If the width of the rectangle is Δx, we integrate with respect to x, and if the width of the rectangle is Δy, we integrate with respect to y.

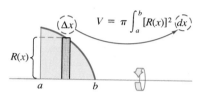

FIGURE 6.14

Horizontal axis of revolution

Vertical axis of revolution

The simplest application of the disc method involves a plane region bounded by the graph of f and the x-axis. If the axis of revolution is the x-axis, then the radius $R(x)$ is simply $f(x)$, as shown in Example 1.

EXAMPLE 1   Finding the volume of a solid of revolution: the disc method

Find the volume of the solid formed by revolving the region bounded by the graph of $f(x) = \sqrt{3x - x^2}$ and the x-axis ($0 \le x \le 3$) about the x-axis.

SOLUTION

From the representative rectangle in Figure 6.15, we see that the radius of this solid is given by

$$R(x) = f(x) = \sqrt{3x - x^2}$$

and it follows that its volume is

$$V = \pi \int_a^b [R(x)]^2 \, dx = \pi \int_0^3 (\sqrt{3x - x^2})^2 \, dx$$

$$= \pi \int_0^3 (3x - x^2) \, dx = \pi \left(\frac{3x^2}{2} - \frac{x^3}{3} \right) \Big]_0^3 = \frac{9\pi}{2}.$$

REMARK   Note in Example 1 that the problem was solved *without* referring to the three-dimensional portion of the sketch in Figure 6.15. In general, to set up an integral to find the volume of a solid of revolution, a sketch of the plane region is more useful than a sketch of the solid, since the radius is more easily visualized in the plane region.

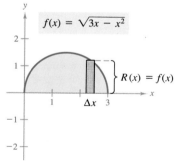

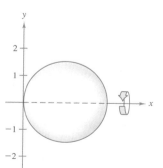

FIGURE 6.15

EXAMPLE 2   Finding the volume of a solid of revolution: the disc method

Find the volume of the solid formed by revolving the region bounded by $f(x) = 2 - x^2$ and $g(x) = 1$ about the line $y = 1$, as shown in Figure 6.16.

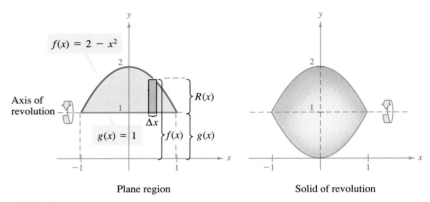

Plane region   Solid of revolution

FIGURE 6.16

SOLUTION

We begin by finding the points of intersection of f and g.

$$2 - x^2 = 1 \qquad \text{Set } f(x) \text{ equal to } g(x)$$
$$x^2 = 1$$
$$x = \pm 1$$

Now, to find the radius, we subtract $g(x)$ from $f(x)$ to obtain

$$R(x) = f(x) - g(x) = (2 - x^2) - 1 = 1 - x^2.$$

Finally, we integrate between -1 and 1 to find the volume.

$$V = \pi \int_a^b [R(x)]^2 \, dx = \pi \int_{-1}^{1} (1 - x^2)^2 \, dx$$
$$= \pi \int_{-1}^{1} (1 - 2x^2 + x^4) \, dx$$
$$= \left[\pi \left(x - \frac{2x^3}{3} + \frac{x^5}{5} \right) \right]_{-1}^{1} = \frac{16\pi}{15} \approx 3.35 \qquad \blacksquare$$

The washer method

The disc method can be extended to cover solids of revolution with a hole by replacing the representative disc with a representative **washer.** The washer is formed by revolving a rectangle about an axis, shown in Figure 6.17. If r and R are the inner and outer radii of the washer and w is the width of the washer, then the volume is given by

$$\text{volume of washer} = \pi(R^2 - r^2)w.$$

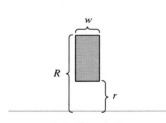

Axis of revolution

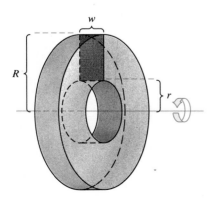

FIGURE 6.17

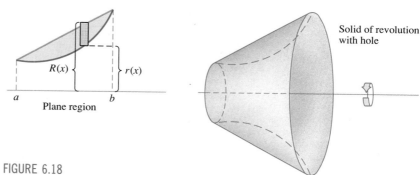

FIGURE 6.18

Now, suppose that a region is bounded by an **outer radius** $R(x)$ and an **inner radius** $r(x)$, as shown in Figure 6.18. If this region is revolved about its axis of revolution, then the volume of the resulting solid is given by

$$V = \pi \int_a^b [R(x)]^2 \, dx - \pi \int_a^b [r(x)]^2 \, dx = \pi \int_a^b ([R(x)]^2 - [r(x)]^2) \, dx.$$

Note that the integral involving the inner radius represents the volume of the hole and is *subtracted* from the integral involving the outer radius.

EXAMPLE 3   The washer method

Find the volume of the solid formed by revolving the region bounded by the graphs of $y = \sqrt{x}$ and $y = x^2$ about the x-axis, as shown in Figure 6.19.

SOLUTION

From Figure 6.19, we have

$$R(x) = \sqrt{x} \qquad \text{Outer radius}$$
$$r(x) = x^2. \qquad \text{Inner radius}$$

Now, integrating between 0 and 1, we have

$$V = \pi \int_a^b ([R(x)]^2 - [r(x)]^2) \, dx = \pi \int_0^1 (x - x^4) \, dx$$
$$= \pi \left[\frac{x^2}{2} - \frac{x^5}{5} \right]_0^1 = \frac{3\pi}{10}. \qquad \blacksquare$$

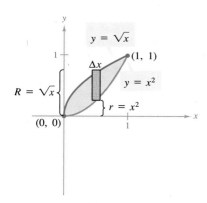

Plane region

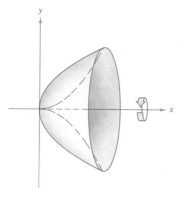

Solid of revolution

FIGURE 6.19

In each example so far, the axis of revolution has been *horizontal* and we integrated with respect to x. In the next example, the axis of revolution is *vertical* and (using the disc method) we must integrate with respect to y. In this example, we need two separate integrals to compute the volume.

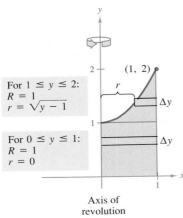

For $1 \leq y \leq 2$:
$R = 1$
$r = \sqrt{y - 1}$

For $0 \leq y \leq 1$:
$R = 1$
$r = 0$

Axis of
revolution

FIGURE 6.20

EXAMPLE 4 Two-integral case, integrating with respect to y

Find the volume of the solid formed by revolving the region bounded by the graphs of $y = x^2 + 1$, $y = 0$, $x = 0$, and $x = 1$ about the y-axis, as shown in Figure 6.20.

SOLUTION

We use two integrals because we have no single formula to represent the inner radius r. When $0 \leq y \leq 1$, then $r = 0$, but when $1 \leq y \leq 2$, r is determined by the equation $y = x^2 + 1$, which implies that $r = \sqrt{y - 1}$. Since $R = 1$, the volume is

$$V = \pi \int_0^1 (1^2 - 0^2) \, dy + \pi \int_1^2 [1^2 - (\sqrt{y - 1})^2] \, dy$$

$$= \pi \int_0^1 1 \, dy + \pi \int_1^2 (2 - y) \, dy$$

$$= \pi y \Big]_0^1 + \pi \left[2y - \frac{y^2}{2} \right]_1^2 = \pi \left(1 + 4 - 2 - 2 + \frac{1}{2} \right) = \frac{3\pi}{2}.$$

Note that the first integral $\pi \int_0^1 1 \, dy$ represents the volume of a right circular cylinder of radius 1 and height 1. This portion of the volume could have been determined without using calculus. ▭

EXAMPLE 5 An application

A manufacturer drills a hole through the center of a metal sphere of radius 5 inches, as shown in Figure 6.21(a). The hole has a radius of 3 inches. What is the volume of the resulting metal ring?

SOLUTION

We imagine the ring to be generated by a segment of the circle whose equation is $x^2 + y^2 = 25$, as shown in Figure 6.21(b). Since the radius of the hole is 3 inches, we let $y = 3$ and solve the equation $x^2 + y^2 = 25$ to determine that the limits of integration are $x = \pm 4$. Thus, we have

$$r(x) = 3 \qquad \text{and} \qquad R(x) = \sqrt{25 - x^2}$$

and the volume is given by

$$V = \pi \int_a^b ([R(x)]^2 - [r(x)]^2) \, dx$$

$$= \pi \int_{-4}^4 [(\sqrt{25 - x^2})^2 - (3)^2] \, dx = \pi \int_{-4}^4 (16 - x^2) \, dx$$

$$= \pi \left[16x - \frac{x^3}{3} \right]_{-4}^4 = \frac{256\pi}{3} \text{ in}^3.$$

▭

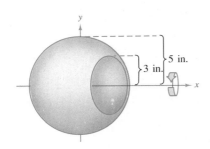

(a) Solid of revolution

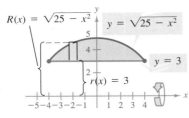

$R(x) = \sqrt{25 - x^2}$

$y = \sqrt{25 - x^2}$

$y = 3$

$r(x) = 3$

(b) Plane region

FIGURE 6.21

Solids with known cross sections

With the disc method, we can find the volume of a solid having a circular cross section whose area is $\Delta A = \pi R^2$. We can generalize this method to solids of any shape, as long as we know a formula for the area of an arbitrary cross section. Some common cross sections are squares, rectangles, triangles, semicircles, and trapezoids.

VOLUMES OF SOLIDS WITH KNOWN CROSS SECTIONS

1. For cross sections of area $A(x)$, taken perpendicular to the x-axis,

$$\text{volume} = \int_a^b A(x)\ dx.$$

2. For cross sections of area $A(y)$, taken perpendicular to the y-axis,

$$\text{volume} = \int_c^d A(y)\ dy.$$

(See Figure 6.22.)

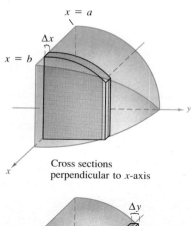

Cross sections perpendicular to x-axis

Cross sections perpendicular to y-axis

FIGURE 6.22

EXAMPLE 6 Triangular cross sections

Find the volume of the solid whose base is the area bounded by the lines

$$f(x) = 1 - \frac{x}{2}, \qquad g(x) = -1 + \frac{x}{2}, \qquad \text{and} \qquad x = 0$$

and whose cross sections perpendicular to the x-axis are equilateral triangles, as shown in Figure 6.23.

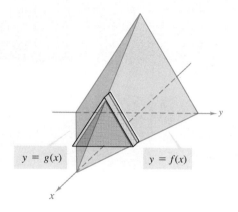

Cross sections are equilateral triangles.

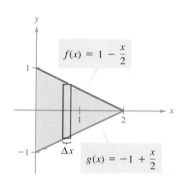

Triangular base in xy-plane

FIGURE 6.23

SOLUTION

For each triangular cross section, we have

$$\text{base} = \left(1 - \frac{x}{2}\right) - \left(-1 + \frac{x}{2}\right) = 2 - x \qquad \text{Length of base}$$

$$\text{area} = \frac{\sqrt{3}}{4}(\text{base})^2 \qquad\qquad \text{Area of equilateral triangle}$$

$$A(x) = \frac{\sqrt{3}}{4}(2 - x)^2. \qquad\qquad \text{Area of cross section}$$

Since x ranges from 0 to 2, the volume of the solid is

$$V = \int_a^b A(x) \, dx$$

$$= \int_0^2 \frac{\sqrt{3}}{4}(2 - x)^2 \, dx$$

$$= -\frac{\sqrt{3}}{4}\left[\frac{(2 - x)^3}{3}\right]_0^2$$

$$= \frac{2\sqrt{3}}{3}.$$

EXAMPLE 7   An application to geometry

Prove that the volume of a pyramid with a square base is $V = \frac{1}{3}hB$, where h is the height of the pyramid and B is the area of the base.

SOLUTION

In Figure 6.24, we intersect the pyramid with a plane parallel to the base at height y to form a square cross section whose sides are of length b'. Using similar triangles, we can show that

$$\frac{b'}{b} = \frac{h - y}{h}$$

or

$$b' = \frac{b}{h}(h - y)$$

where b is the length of the sides of the base of the pyramid. Thus,

$$A(y) = (b')^2$$

$$= \frac{b^2}{h^2}(h - y)^2.$$

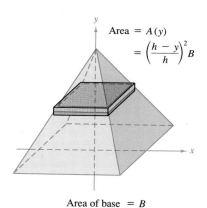
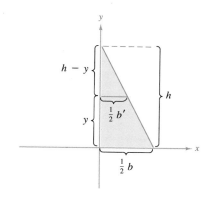

FIGURE 6.24

Integrating between 0 and h, we have

$$\text{volume} = \int_0^h \frac{b^2}{h^2}(h-y)^2 \, dy$$

$$= \frac{b^2}{h^2}\int_0^h (h-y)^2 \, dy$$

$$= -\left(\frac{b^2}{h^2}\right)\frac{(h-y)^3}{3}\Big]_0^h$$

$$= \frac{b^2}{h^2}\left(\frac{h^3}{3}\right)$$

$$= \frac{b^2 h}{3} = \frac{1}{3}hB.$$

An interesting generalization of the formula in Example 7 is that the volume of *any cone* can be determined by the formula $V = \frac{1}{3}hB$, where B is the area of the base and h is the height of the cone, as shown in Figure 6.25.

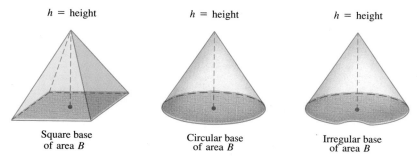

FIGURE 6.25

EXERCISES for Section 6.2

In Exercises 1–8, find the volume of the solid formed by revolving the given region about the *x*-axis.

1. $y = -x + 1$

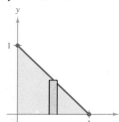

2. $y = 4 - x^2$

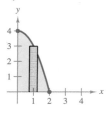

3. $y = \sqrt{4 - x^2}$

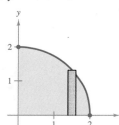

4. $y = x^2$

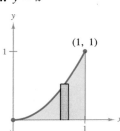

5. $y = \sqrt{x}$

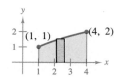

6. $y = \sqrt{4 - x^2}$

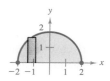

7. $y = x^2,\ y = x^3$

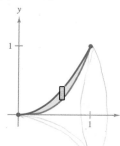

8. $y = 2,\ y = 4 - \dfrac{x^2}{4}$

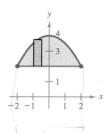

In Exercises 9–12, find the volume of the solid formed by revolving the given region about the *y*-axis.

9. $y = x^2$

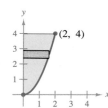

10. $y = \sqrt{16 - x^2}$

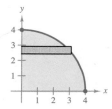

11. $y = x^{2/3}$

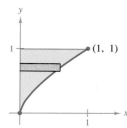

12. $x = -y^2 + 4y$

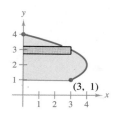

In Exercises 13–16, find the volume of the solid generated by revolving the region bounded by the graphs of the given equations about the indicated lines.

13. $y = \sqrt{x},\ y = 0,\ x = 4$
   (a) the *x*-axis   (b) the *y*-axis
   (c) the line $x = 4$ (d) the line $x = 6$

14. $y = 2x^2,\ y = 0,\ x = 2$
   (a) the *y*-axis   (b) the *x*-axis
   (c) the line $y = 8$ (d) the line $x = 2$

15. $y = x^2,\ y = 4x - x^2$
   (a) the *x*-axis   (b) the line $y = 6$

16. $y = 6 - 2x - x^2,\ y = x + 6$
   (a) the *x*-axis   (b) the line $y = 3$

In Exercises 17–20, find the volume of the solid generated by revolving the region bounded by the graphs of the given equations about the *x*-axis.

17. $y = x\sqrt{4 - x^2},\ y = 0$

18. $y = 2x^{2/3} - x,\ y = 0$

19. $y = \dfrac{1}{x},\ y = 0,\ x = 1,\ x = 4$

20. $y = \dfrac{3}{x + 1},\ y = 0,\ x = 0,\ x = 8$

In Exercises 21–24, find the volume of the solid generated by revolving the region bounded by the graphs of the given equations about the line $y = 4$.

21. $y = x$, $y = 3$, $x = 0$
22. $y = x^2$, $y = 4$
23. $y = \dfrac{1}{x^2}$, $y = 0$, $x = 1$, $x = 4$
24. $y = \sqrt{x}$, $y = 0$, $0 \le x \le 4$

In Exercises 25–28, find the volume of the solid generated by revolving the region bounded by the graphs of the given equations about the line $x = 6$.

25. $y = x$, $y = 0$, $y = 4$, $x = 6$
26. $y = 6 - x$, $y = 0$, $y = 4$, $x = 0$
27. $x = y^2$, $x = 4$
28. $xy^2 = 12$, $y = 2$, $y = 6$, $x = 6$

29. The region bounded by the parabola $y = 4x - x^2$ and the x-axis is revolved about the x-axis. Find the volume of the resulting solid.

30. If the equation of the parabola in Exercise 29 were changed to $y = 4 - x^2$, would the volume of the solid generated be different? Why or why not?

31. The upper half of the ellipse $9x^2 + 25y^2 = 225$ is revolved about the x-axis to form a prolate spheroid (shaped like a football). Find the volume of the spheroid.

32. The right half of the ellipse $9x^2 + 25y^2 = 225$ is revolved about the y-axis to form an oblate spheroid (shaped like an M&M candy). Find the volume of the spheroid.

33. If the portion of the line $y = \frac{1}{2}x$ lying in the first quadrant is revolved about the x-axis, a cone is generated. Find the volume of the cone extending from $x = 0$ to $x = 6$.

34. Use the disc method to verify that the volume of a right circular cone is $\frac{1}{3}\pi r^2 h$, where r is the radius of the base and h is the height.

35. Use the disc method to verify that the volume of a sphere of radius r is $\frac{4}{3}\pi r^3$.

36. A sphere of radius r is cut by a plane h $(h < r)$ units above the equator. Find the volume of the solid (spherical segment) above the plane.

37. A cone with a base of radius r and height H is cut by a plane to parallel to and h units above the base. Find the volume of the solid (frustrum of a cone) below the plane.

38. The region bounded by $y = \sqrt{x}$, $y = 0$, $x = 0$, and $x = 4$ is revolved about the x-axis. (a) Find the value of x in the interval $[0, 4]$ that divides the solid into two parts of equal volume. (b) Find the values of x in the interval $[0, 4]$ that divide the solid into three parts of equal volume.

39. In the Chapter 6 Application, we showed that a solid of revolution having the form of a regulation-size football can be generated by revolving the graph of

$$f(x) = -0.0944x^2 + 3.4, \quad -5.5 \le x \le 5.5$$

about the x-axis. (The x and y values are measured in inches.) Show that the volume of this solid is approximately 232 cubic inches.

40. A tank on a water tower is a sphere of radius 50 feet. Determine the depth of the water when the tank is filled to 21.6 percent of its total capacity.

41. A tank on the wing of a jet is formed by revolving the region bounded by the graph of $y = \frac{1}{8}x^2\sqrt{2 - x}$ and the x-axis about the x-axis (see figure), where x and y are measured in meters. Find the volume of the tank.

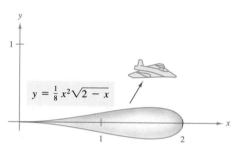

$y = \frac{1}{8}x^2\sqrt{2 - x}$

FIGURE FOR 41

In Exercises 42 and 43, use Simpson's Rule (with $n = 4$) to approximate the volume of the solid generated by revolving the region bounded by the graphs of the given equations about the x-axis.

42. $y = \sqrt[3]{x^2 - 1}$, $y = 0$, $x = 1$, $x = 3$
43. $y = \sqrt[4]{x^2 + 1}$, $y = 0$, $x = 0$, $x = 2$

44. Match each of the following integrals with the solid whose volume it represents, and give the dimensions of each solid.
(a) right circular cylinder   (b) ellipsoid
(c) sphere   (d) right circular cone
(e) torus

___ $\pi \displaystyle\int_0^h \left(\frac{rx}{h}\right)^2 dx$

___ $\pi \displaystyle\int_0^h r^2\, dx$

___ $\pi \displaystyle\int_{-r}^r (\sqrt{r^2 - x^2})^2\, dx$

___ $\pi \displaystyle\int_{-b}^b \left(a\sqrt{1 - \frac{x^2}{b^2}}\right)^2 dx$

___ $\pi \displaystyle\int_{-r}^r [(R + \sqrt{r^2 - x^2})^2 - (R - \sqrt{r^2 - x^2})^2]\, dx$

45. Find the volume of the solid whose base is bounded by the circle $x^2 + y^2 = 4$, with the indicated cross sections taken perpendicular to the x-axis.

(a) squares   (b) equilateral triangles

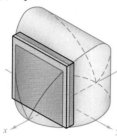
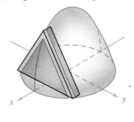

(c) semicircles   (d) isosceles right triangles

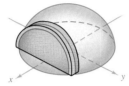
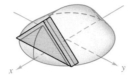

46. Find the volume of the solid whose base is bounded by the graphs of $y = x + 1$ and $y = x^2 - 1$, with the indicated cross sections taken perpendicular to the x-axis.

(a) squares   (b) rectangles of height 1

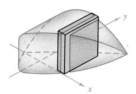
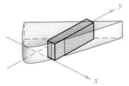

(c) semiellipses of height 2 ($A = \pi ab$)   (d) equilateral triangles

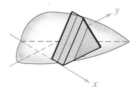

47. The base of a solid is bounded by $y = x^3$, $y = 0$, and $x = 1$. Find the volume of the solid for the following cross sections (taken perpendicular to the y-axis).

(a) squares
(b) semicircles
(c) equilateral triangles
(d) trapezoids for which $h = b_1 = \frac{1}{2}b_2$, where b_1 and b_2 are the lengths of the upper and lower bases
(e) semiellipses whose heights are twice the lengths of their bases

48. Find the volume of the solid of intersection (the solid common to both) of the two right circular cylinders of radius r whose axes meet at right angles (see figure).

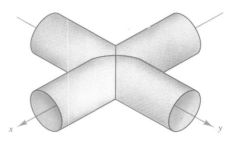

Two intersecting cylinders

Solid of intersection

FIGURE FOR 48

49. A wedge is cut from a right circular cylinder of radius r inches by a plane through the diameter of the base, making an angle of 45° with the plane of the base (see figure). Find the volume of the wedge.

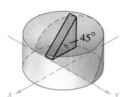

FIGURE FOR 49

50. *Cavalieri's Theorem*   Prove that if two solids have equal altitudes and all plane sections parallel to their bases and at equal distances from their bases have equal areas, then the solids have the same volume.

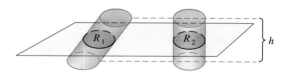

Area of R_1 = area of R_2

FIGURE FOR 50

In Exercises 51 and 52, use Simpson's Rule on a computer or calculator (with $n = 10$) to approximate the volume of the solid generated when the region bounded by the graphs of the given equations is revolved about the x-axis.

51. $y = \dfrac{10}{\sqrt{x} + 2}$, $y = 0$, $x = 0$, $x = 4$

52. $y = \dfrac{8x}{9 + x^2}$, $y = 0$, $x = 0$, $x = 5$

6.3 Volume: The Shell Method

The shell method ▪ Comparison of disc and shell methods

In this section, we look at an alternate method for finding the volume of a solid of revolution, a method that uses cylindrical shells. We will compare the advantages of the two methods later in this section.

To introduce the **shell method,** consider a representative rectangle as shown in Figure 6.26, where w = width of the rectangle, h = height of the rectangle, and p = distance between axis of revolution and *center* of the rectangle. When this rectangle is revolved about its axis of revolution, it forms a cylindrical shell (or tube) of thickness w. To find the volume of this shell we consider two cylinders. The radius of the larger cylinder corresponds to the outer radius of the shell, and the radius of the smaller cylinder corresponds to the inner radius of the shell. Since p is the average radius of the shell, we know the outer radius is $p + (w/2)$ and the inner radius is $p - (w/2)$. Thus, the volume of the shell is given by the difference.

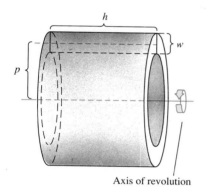

FIGURE 6.26

$$\text{volume of shell} = (\text{volume of cylinder}) - (\text{volume of hole})$$

$$= \pi\left(p + \frac{w}{2}\right)^2 h - \pi\left(p - \frac{w}{2}\right)^2 h$$

$$= \pi\left(p^2 + pw + \frac{w^2}{4} - p^2 + pw - \frac{w^2}{4}\right)h$$

$$= \pi 2phw = 2\pi(\text{average radius})(\text{height})(\text{thickness})$$

We use this formula to find the volume of a solid of revolution as follows. Assume that the plane region in Figure 6.27 is revolved about a line to form the indicated solid. If we consider a horizontal rectangle of width Δy, then as the plane region is revolved about a line parallel to the x-axis, the rectangle generates a representative shell whose volume is

$$\Delta V = 2\pi[p(y)h(y)]\,\Delta y.$$

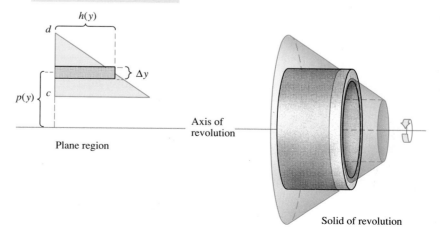

FIGURE 6.27

Now, if we approximate the volume of the solid by n such shells of thickness Δy, height $h(y_i)$, and average radius $p(y_i)$, we have

$$\text{volume of solid} \approx \sum_{i=1}^{n} 2\pi[p(y_i)h(y_i)]\,\Delta y = 2\pi\sum_{i=1}^{n}[p(y_i)h(y_i)]\,\Delta y.$$

By taking the limit as $\|\Delta\| \to 0$ $(n \to \infty)$, we have

$$\text{volume of solid} = \lim_{n\to\infty} 2\pi\sum_{i=1}^{n}[p(y_i)h(y_i)]\,\Delta y = 2\pi\int_{c}^{d}[p(y)h(y)]\,dy.$$

THE SHELL METHOD

To find the volume of a solid of revolution with the **shell method,** use one of the following, as shown in Figure 6.28.

Horizontal axis of revolution

$$\text{volume} = V = 2\pi\int_{c}^{d} p(y)h(y)\,dy$$

Vertical axis of revolution

$$\text{volume} = V = 2\pi\int_{a}^{b} p(x)h(x)\,dx$$

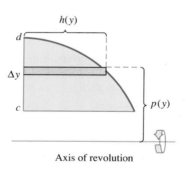

FIGURE 6.28

Axis of revolution

Axis of revolution

EXAMPLE 1 Using the shell method to find volume

Find the volume of the solid of revolution formed by revolving the region bounded by $y = x - x^3$ and the x-axis $(0 \leq x \leq 1)$ about the y-axis.

SOLUTION

Since the axis of revolution is vertical, we use a vertical representative rectangle, as shown in Figure 6.29. The width Δx indicates that x is the variable of integration. The distance from the center of the rectangle to the axis of revolution is $p(x) = x$, and the height of the rectangle is $h(x) = x - x^3$. Since x ranges from 0 to 1, the volume of the solid is

$$V = 2\pi\int_{a}^{b} p(x)h(x)\,dx = 2\pi\int_{0}^{1} x(x - x^3)\,dx = 2\pi\int_{0}^{1}(-x^4 + x^2)\,dx$$

$$= 2\pi\left[-\frac{x^5}{5} + \frac{x^3}{3}\right]_{0}^{1} = \frac{4\pi}{15}.$$

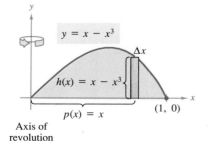

Axis of revolution

FIGURE 6.29

EXAMPLE 2   Using the shell method to find volume

Find the volume of the solid of revolution formed by revolving the region bounded by

$$y = \frac{1}{(x^2 + 1)^2}$$

and the x-axis $(0 \le x \le 1)$ about the y-axis.

SOLUTION

Since the axis of revolution is vertical, we use a vertical representative rectangle, as shown in Figure 6.30. The width Δx indicates that x is the variable of integration. The distance from the center of the rectangle to the axis of revolution is $p(x) = x$, and the height of the rectangle is $h(x) = 1/(x^2 + 1)^2$. Since x ranges from 0 to 1, the volume of the solid is

$$V = 2\pi \int_a^b p(x)h(x)\,dx = 2\pi \int_0^1 \frac{x}{(x^2 + 1)^2}\,dx$$

$$= \left[-\frac{\pi}{x^2 + 1} \right]_0^1 = \frac{\pi}{2}.$$

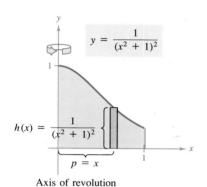

$y = \dfrac{1}{(x^2 + 1)^2}$

$h(x) = \dfrac{1}{(x^2 + 1)^2}$

$p = x$

Axis of revolution

FIGURE 6.30

Comparison of disc and shell methods

The disc and shell methods can be distinguished as follows. For the disc method, the representative rectangle is always *perpendicular* to the axis of revolution, whereas for the shell method, the representative rectangle is always *parallel* to the axis of revolution, as shown in Figure 6.31.

Often, one method is more convenient to use than the other. The following example illustrates a case in which the shell method is preferable.

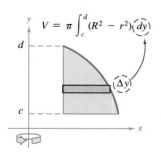

$$V = \pi \int_c^d (R^2 - r^2)\,dy$$

(a) Vertical axis
     of revolution

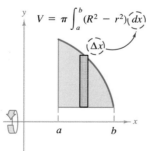

$$V = \pi \int_a^b (R^2 - r^2)\,dx$$

Horizontal axis
of revolution

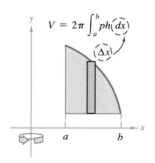

$$V = 2\pi \int_a^b ph\,dx$$

(b) Vertical axis
     of revolution

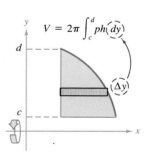

$$V = 2\pi \int_c^d ph\,dy$$

Horizontal axis
of revolution

FIGURE 6.31

EXAMPLE 3 Shell method preferable

Find the volume of the solid formed by revolving the region bounded by the graphs of $y = x^2 + 1$, $y = 0$, $x = 0$, and $x = 1$ about the y-axis.

SOLUTION

In Example 4 in the previous section, we found that the disc method requires two integrals to determine the volume of this solid. See Figure 6.32(a).

$$V = \pi \int_0^1 (1^2 - 0^2)\, dy + \pi \int_1^2 [1^2 - (\sqrt{y-1})^2]\, dy \qquad \text{Disc method}$$

$$= \pi \int_0^1 1\, dy + \pi \int_1^2 (2 - y)\, dy$$

$$= \pi y \Big]_0^1 + \pi \left[2y - \frac{y^2}{2} \right]_1^2$$

$$= \pi \left(1 + 4 - 2 - 2 + \frac{1}{2} \right) = \frac{3\pi}{2}$$

From Figure 6.32(b), we can see that the shell method requires only one integral to find the volume.

$$V = 2\pi \int_a^b p(x)h(x)\, dx \qquad \text{Shell method}$$

$$= 2\pi \int_0^1 x(x^2 + 1)\, dx$$

$$= 2\pi \left[\frac{x^4}{4} + \frac{x^2}{2} \right]_0^1$$

$$= 2\pi \left(\frac{3}{4} \right) = \frac{3\pi}{2}$$

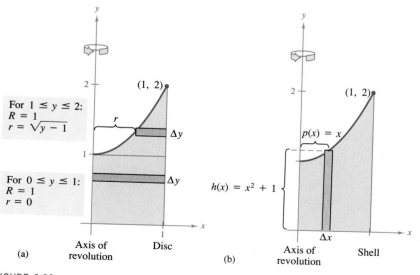

For $1 \le y \le 2$:
$R = 1$
$r = \sqrt{y-1}$

For $0 \le y \le 1$:
$R = 1$
$r = 0$

$h(x) = x^2 + 1$

$p(x) = x$

(a)   Axis of revolution   Disc

(b)   Axis of revolution   Shell

FIGURE 6.32

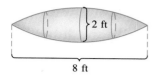

FIGURE 6.33

EXAMPLE 4   Disc method preferable

A pontoon is to be made in the shape shown in Figure 6.33. The pontoon is designed by rotating the graph of

$$y = 1 - \frac{x^2}{16}, \quad -4 \leq x \leq 4$$

about the x-axis, where x and y are measured in feet. Find the volume of the pontoon.

SOLUTION

In Figure 6.34 we compare the shell and disc methods. Using the shell method, we see that Δy is the width of the rectangle, $p(y) = y$, and $h(y) = 2x$. Solving for x in the equation $y = 1 - (x^2/16)$, we obtain

$$x = 4\sqrt{1 - y}$$

or

$$h(y) = 2x = 8\sqrt{1 - y}.$$

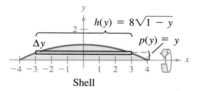
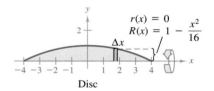

FIGURE 6.34

Hence, the shell method results in the integral

$$V = 2\pi \int_0^1 8y\sqrt{1 - y}\, dy. \qquad \text{Shell method}$$

Although this is not a particularly difficult integral, its evaluation does require a u-substitution. By contrast, the disc method yields the relatively simple integral

$$V = \pi \int_{-4}^4 \left(1 - \frac{x^2}{16}\right)^2 dx$$

$$= \pi \int_{-4}^4 \left(1 - \frac{x^2}{8} + \frac{x^4}{256}\right) dx \qquad \text{Disc method}$$

$$= \pi \left[x - \frac{x^3}{24} + \frac{x^5}{1280}\right]_{-4}^4$$

$$= \frac{64\pi}{15}$$

$$\approx 13.4 \text{ ft}^3.$$

Note that for the shell method in Example 4 we had to solve for x in terms of y in the equation $y = 1 - (x^2/16)$. Sometimes solving for x is very difficult (or even impossible). In such cases we must use a vertical rectangle (of width Δx), thus making x the variable of integration. The position (horizontal or vertical) of the axis of revolution then determines the method to be used. This is illustrated in Example 5.

EXAMPLE 5   Shell method necessary

Find the volume of the solid formed by revolving the region bounded by the graphs of $y = x^3 + x + 1$, $y = 1$, and $x = 1$ about the line $x = 2$, as shown in Figure 6.35.

SOLUTION

In the equation $y = x^3 + x + 1$, we cannot easily solve for x in terms of y. (See the discussion at the end of Section 4.8.) Therefore, the variable of integration must be x, and we choose a vertical representative rectangle. Since the rectangle is parallel to the axis of revolution, we use the shell method and obtain

$$V = 2\pi \int_a^b p(x)h(x) \, dx = 2\pi \int_0^1 (2 - x)(x^3 + x + 1 - 1) \, dx$$

$$= 2\pi \int_0^1 (-x^4 + 2x^3 - x^2 + 2x) \, dx$$

$$= 2\pi \left[-\frac{x^5}{5} + \frac{x^4}{2} - \frac{x^3}{3} + x^2 \right]_0^1$$

$$= 2\pi \left(-\frac{1}{5} + \frac{1}{2} - \frac{1}{3} + 1 \right) = \frac{29\pi}{15}.$$

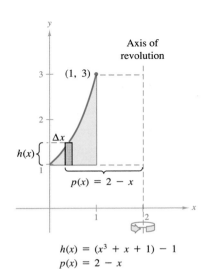

$h(x) = (x^3 + x + 1) - 1$
$p(x) = 2 - x$

FIGURE 6.35

EXERCISES for Section 6.3

In Exercises 1–20, use the shell method to find the volume of the solid generated by revolving the given plane region about the indicated line.

1. $y = x$

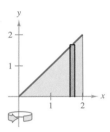

2. $y = 1 - x$

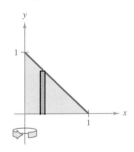

3. $y = x$

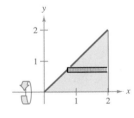

4. $y = 2 - x$

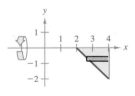

5. $y = \sqrt{x}$

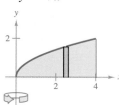

6. $y = x^2 + 4$

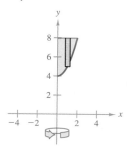

7. $y = x^2$, $y = 0$, $x = 2$, about the y-axis
8. $y = x^2$, $y = 0$, $x = 4$, about the y-axis
9. $y = x^2$, $y = 4x - x^2$, about the y-axis
10. $y = x^2$, $y = 4x - x^2$, about the line $x = 2$
11. $y = x^2$, $y = 4x - x^2$, about the line $x = 4$
12. $y = \dfrac{1}{x}$, $x = 1$, $x = 2$, $y = 0$, about the x-axis
13. $y = 4x - x^2$, $y = 0$, about the line $x = 5$
14. $x + y^2 = 9$, $x = 0$, about the x-axis
15. $y = 4x - x^2$, $x = 0$, $y = 4$, about the y-axis
16. $y = 4 - x^2$, $y = 0$, about the y-axis
17. $y = \sqrt{x}$, $y = 0$, $x = 4$, about the line $x = 6$
18. $y = 2x$, $y = 4$, $x = 0$, about the y-axis
19. $x = y - y^2$, $x = 0$, about the line $y = 2$
20. $x = y - y^2$, $x = 0$, about the line $y = -4$

In Exercises 21–24, use the disc *or* shell method to find the volume of the solid generated by revolving the region bounded by the graphs of the given equations about the specified line.

21. $y = x^3$, $y = 0$, $x = 2$
  (a) the x-axis   (b) the y-axis
  (c) the line $x = 4$ (d) the line $y = 8$
22. $y = \dfrac{1}{x^3}$, $y = 0$, $x = 1$, $x = 2$

  (a) the x-axis   (b) the y-axis
  (c) the line $x = 4$
23. $x^{1/2} + y^{1/2} = a^{1/2}$, $x = 0$, $y = 0$
  (a) the x-axis   (b) the y-axis
  (c) the line $x = a$
24. $x^{2/3} + y^{2/3} = a^{2/3}$ (hypocycloid)
  (a) the x-axis   (b) the y-axis

25. A solid is generated by revolving the region bounded by $y = \frac{1}{2}x^2$ and $y = 2$ about the y-axis. A hole, centered along the axis of revolution, is drilled through this solid so that one-quarter of the volume is removed. Find the diameter of the hole.
26. A solid is generated by revolving the region bounded by $y = \sqrt{9 - x^2}$ and $y = 0$ about the y-axis. A hole, centered along the axis of revolution, is drilled through this solid so that one-third of the volume is removed. Find the diameter of the hole.

27. Let a sphere of radius r be cut by a plane, thus forming a segment of height h. Show that the volume of this segment is $\frac{1}{3}\pi h^2(3r - h)$.

In Exercises 28 and 29, use Simpson's Rule (with $n = 4$) to approximate the volume of the solid. The solid is generated by revolving the region bounded by the graphs of the given equations about the y-axis.

28. $y = \sqrt{1 - x^3}$, $y = 0$, $x = 0$
29. $x^{4/3} + y^{4/3} = 1$

30. A pond is approximately circular with a diameter of 400 feet (see figure). Starting at the center, the depth of the water is measured every 25 feet and recorded in the accompanying table. Use Simpson's Rule to approximate the volume of water in the pond.

x	0	25	50	75	100
Depth	20	19	19	17	15

x	125	150	175	200
Depth	14	10	6	0

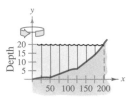

Distance from center   FIGURE FOR 30

31. A storage shed has a circular base of diameter 80 feet (see figure). Starting at the center, the interior height is measured every 10 feet and recorded in the accompanying table. Use Simpson's Rule to approximate the volume of the building.

x	0	10	20	30	40
Height	50	46	40	20	0

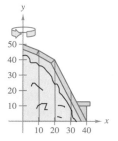

FIGURE FOR 31

32. Match each of the following integrals with the solid whose volume it represents and give the dimensions of each solid.

——— $2\pi \displaystyle\int_0^r hx\,dx$

——— $2\pi \displaystyle\int_0^r hx\left(1 - \dfrac{x}{r}\right)dx$

——— $2\pi \displaystyle\int_0^r 2x\sqrt{r^2 - x^2}\,dx$

——— $2\pi \displaystyle\int_0^b 2ax\sqrt{1 - \dfrac{x^2}{b^2}}\,dx$

——— $2\pi \displaystyle\int_{-r}^r (R - x)(2\sqrt{r^2 - x^2})\,dx$

(a) right circular cone
(b) torus
(c) sphere
(d) right circular cylinder
(e) ellipsoid

33. A **torus** is formed by revolving the circle $x^2 + y^2 = 1$ about the vertical line $x = 2$, as shown in the figure. Find the volume of this "doughnut-shaped" solid. [Hint: The integral $\int_{-1}^1 \sqrt{1 - x^2}\,dx$ represents the area of a semicircle.]

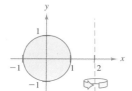

FIGURE FOR 33

34. A hole is cut through the center of a sphere of radius r. The height of the remaining spherical ring is h, as shown in the figure. Show that the volume of the ring is $V = \pi h^3/6$. [Note that the volume is independent of r.]

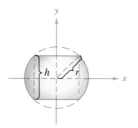

FIGURE FOR 34

In Exercises 35 and 36, use Simpson's Rule on a computer or calculator (with $n = 10$) to approximate the volume of the solid generated when the region bounded by the graphs of the given equations is revolved about the y-axis.

35. $y = \sqrt[3]{(x - 2)^2(x - 6)^2}$, $y = 0$, $x = 2$, $x = 6$

36. $y = \dfrac{2}{1 + \sqrt{x}}$, $y = 0$, $x = 1$, $x = 4$

6.4 Arc Length and Surfaces of Revolution

Arc length ▪ Surface of revolution ▪ Area of a surface of revolution

Christian Huygens

In this section the summation character of the definite integral is used to find the arc length of a plane curve and the area of a surface of revolution. In both cases, we approximate an arc (a segment of a curve) by straight line segments whose lengths are given by the familiar distance formula

$$d = \sqrt{(x_2 - x_1)^2 + (y_2 - y_1)^2}.$$

If a segment of a curve has a finite arc length, we say that it is **rectifiable.** The problem of calculating the length of a curve has motivated a wide range of mathematical work. Some early contributions to the problem were made by the Dutch mathematician Christian Huygens (1629–1695), who invented the pendulum clock. Another pioneer in the work of rectifiable curves was James Gregory (1638–1675), a Scottish mathematician. Both men played important roles in the early development of calculus.

We will see in developing the formula for arc length that a sufficient condition for the graph of a function f to be rectifiable between $(a, f(a))$ and $(b, f(b))$ is that f' is continuous on $[a, b]$. Such a function is said to be **continuously differentiable** on $[a, b]$ and its graph on the interval $[a, b]$ is called a **smooth curve.**

Assume that a function given by $y = f(x)$ is continuously differentiable on the interval $[a, b]$, and let s denote the length of its graph on this interval. We approximate the graph of f by n line segments whose endpoints are determined by the partition

$$a = x_0 < x_1 < x_2 < \cdots < x_n = b$$

as shown in Figure 6.36.

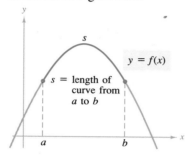
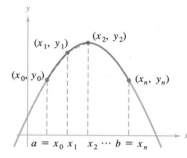

FIGURE 6.36

By letting $\Delta x_i = x_i - x_{i-1}$ and $\Delta y_i = y_i - y_{i-1}$, we approximate the total arc length by

$$s \approx \sum_{i=1}^{n} \sqrt{(\Delta x_i)^2 + (\Delta y_i)^2}.$$

Taking the limit as $\|\Delta\| \to 0$, $(n \to \infty)$, we obtain

$$s = \lim_{n \to \infty} \sum_{i=1}^{n} \sqrt{(\Delta x_i)^2 + (\Delta y_i)^2} = \lim_{n \to \infty} \sum_{i=1}^{n} \sqrt{1 + \left(\frac{\Delta y_i}{\Delta x_i}\right)^2} (\Delta x_i).$$

Since $f'(x)$ exists for each x in (x_{i-1}, x_i), the Mean Value Theorem guarantees the existence of c_i in (x_{i-1}, x_i) such that

$$f(x_i) - f(x_{i-1}) = f'(c_i)(x_i - x_{i-1}) \implies \frac{\Delta y_i}{\Delta x_i} = f'(c_i).$$

Moreover, since f' is continuous on $[a, b]$, we know that $\sqrt{1 + [f'(x)]^2}$ is also continuous (and hence integrable) on $[a, b]$ and we have

$$s = \lim_{n \to \infty} \sum_{i=1}^{n} \sqrt{1 + [f'(c_i)]^2} (\Delta x_i) = \int_a^b \sqrt{1 + [f'(x)]^2} \, dx.$$

We call s the **arc length** of f between a and b.

DEFINITION OF ARC LENGTH

If the function given by $y = f(x)$ represents a smooth curve on the interval $[a, b]$, then the **arc length** of f between a and b is given by

$$s = \int_a^b \sqrt{1 + [f'(x)]^2} \, dx.$$

Similarly, for a smooth curve given by $x = g(y)$, the **arc length** of g between c and d is given by

$$s = \int_c^d \sqrt{1 + [g'(y)]^2} \, dy.$$

Definite integrals representing arc length often are very difficult to evaluate. In this section we present a few examples. In Chapter 9, with more advanced integration techniques, we will be able to tackle more difficult arc length problems.

Since the definition for arc length can be applied to a linear function, we should check to see that this new definition agrees with the standard distance formula for the length of a line segment. We do this in Example 1.

EXAMPLE 1 The length of a line segment

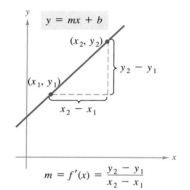

$y = mx + b$

(x_2, y_2)

$y_2 - y_1$

(x_1, y_1)

$x_2 - x_1$

$m = f'(x) = \dfrac{y_2 - y_1}{x_2 - x_1}$

FIGURE 6.37

Find the arc length from (x_1, y_1) to (x_2, y_2) on the graph of $f(x) = mx + b$, as shown in Figure 6.37.

SOLUTION

Since

$$m = f'(x)$$

$$= \frac{y_2 - y_1}{x_2 - x_1}$$

it follows that

$$s = \int_{x_1}^{x_2} \sqrt{1 + [f'(x)]^2}\, dx$$

$$= \int_{x_1}^{x_2} \sqrt{1 + \left(\frac{y_2 - y_1}{x_2 - x_1}\right)^2}\, dx$$

$$= \left[\sqrt{\frac{(x_2 - x_1)^2 + (y_2 - y_1)^2}{(x_2 - x_1)^2}}(x) \right]_{x_1}^{x_2}$$

$$= \sqrt{\frac{(x_2 - x_1)^2 + (y_2 - y_1)^2}{(x_2 - x_1)^2}}(x_2 - x_1)$$

$$= \sqrt{(x_2 - x_1)^2 + (y_2 - y_1)^2}$$

which is the formula for the distance between two points in the plane. ▭

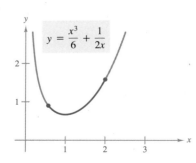

$y = \dfrac{x^3}{6} + \dfrac{1}{2x}$

FIGURE 6.38

EXAMPLE 2 Finding arc length

Find the arc length of the graph of

$$f(x) = \frac{x^3}{6} + \frac{1}{2x}$$

on the interval $\left[\frac{1}{2}, 2\right]$, as shown in Figure 6.38.

SOLUTION

Since

$$f'(x) = \frac{3x^2}{6} - \frac{1}{2x^2} = \frac{1}{2}\left(x^2 - \frac{1}{x^2}\right)$$

the arc length is

$$\begin{aligned}
s &= \int_a^b \sqrt{1 + \left(\frac{dy}{dx}\right)^2}\, dx \\
&= \int_{1/2}^2 \sqrt{1 + \left[\frac{1}{2}\left(x^2 - \frac{1}{x^2}\right)\right]^2}\, dx \\
&= \int_{1/2}^2 \sqrt{\frac{1}{4}\left(x^4 + 2 + \frac{1}{x^4}\right)}\, dx \\
&= \int_{1/2}^2 \frac{1}{2}\left(x^2 + \frac{1}{x^2}\right) dx \\
&= \left[\frac{1}{2}\left(\frac{x^3}{3} - \frac{1}{x}\right)\right]_{1/2}^2 \\
&= \frac{1}{2}\left(\frac{13}{6} + \frac{47}{24}\right) = \frac{99}{48} = \frac{33}{16}.
\end{aligned}$$

EXAMPLE 3   Finding arc length

Find the arc length of the graph of $(y - 1)^3 = x^2$ on the interval $[0, 8]$, as shown in Figure 6.39.

SOLUTION

We can solve for either x or y in the equation $(y - 1)^3 = x^2$. For this example we choose to solve for x, and we obtain $x = \pm(y - 1)^{3/2}$. By choosing the positive value for x, we have

$$\frac{dx}{dy} = \frac{3}{2}(y - 1)^{1/2}.$$

Therefore, the arc length is

$$\begin{aligned}
s &= \int_c^d \sqrt{1 + \left(\frac{dx}{dy}\right)^2}\, dy = \int_1^5 \sqrt{1 + \left[\frac{3}{2}(y - 1)^{1/2}\right]^2}\, dy \\
&= \int_1^5 \sqrt{\frac{9}{4}y - \frac{5}{4}}\, dy \\
&= \frac{1}{2}\int_1^5 \sqrt{9y - 5}\, dy \\
&= \frac{1}{18}\left[\frac{(9y - 5)^{3/2}}{3/2}\right]_1^5 = \frac{1}{27}(40^{3/2} - 4^{3/2}) \\
&= \frac{8}{27}(10^{3/2} - 1) \approx 9.0734.
\end{aligned}$$

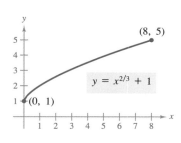

FIGURE 6.39

$y = x^{2/3} + 1$

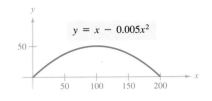

FIGURE 6.40

EXAMPLE 4   An application of arc length

Approximate the distance traveled by a projectile that travels along the path given by

$$y = x - 0.005x^2$$

where x and y are measured in feet, as shown in Figure 6.40.

SOLUTION

Since

$$\frac{dy}{dx} = 1 - 0.01x = 1 - \frac{x}{100}$$

we can use the following arc length to find the distance traveled.

$$
\begin{aligned}
s &= \int_a^b \sqrt{1 + \left(\frac{dy}{dx}\right)^2}\, dx \\
&= \int_0^{200} \sqrt{1 + \left(1 - \frac{x}{100}\right)^2}\, dx \\
&= \int_0^{200} \sqrt{2 - \frac{x}{50} + \frac{x^2}{10,000}}\, dx
\end{aligned}
$$

Using Simpson's Rule (with $n = 10$), we can approximate this distance to be 229.56 feet.　▭

Surfaces of revolution

In Sections 6.2 and 6.3 integration was used to calculate the volume of a solid of revolution. We now look at a procedure for finding the area of a surface of revolution.

DEFINITION OF SURFACE OF REVOLUTION	If the graph of a continuous function is revolved about a line, the resulting surface is called a **surface of revolution.**

To find the area of a surface of revolution, we use the formula for the lateral surface area of the frustum of a right circular cone. Consider the line segment in Figure 6.41, where

L = length of line segment

r_1 = radius at left end of line segment

r_2 = radius at right end of line segment.

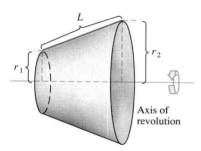

FIGURE 6.41

When the line segment is revolved about its axis of revolution, it forms a frustum of a right circular cone, with

$$S = 2\pi r L \qquad \text{Lateral surface area of frustum}$$

$$r = \frac{1}{2}(r_1 + r_2). \qquad \text{Average radius of frustum}$$

(In Exercise 36 you are asked to verify this formula for S.)

Now, suppose the graph of a function f, having a continuous derivative on the interval $[a, b]$, is revolved about the x-axis to form a surface of revolution, as shown in Figure 6.42. Let Δ be a partition of $[a, b]$, with subintervals of width Δx_i. Then the line segment of length

$$\Delta L_i = \sqrt{\Delta x_i^2 + \Delta y_i^2}$$

generates a frustum of a cone.

By the Intermediate Value Theorem, a point d_i exists such that $r_i = f(d_i)$ is the average radius of this frustum. Finally, the lateral surface area, ΔS_i, of the frustum is given by

$$\begin{aligned}
\Delta S_i &= 2\pi r_i \Delta L_i \\
&= 2\pi f(d_i)\sqrt{\Delta x_i^2 + \Delta y_i^2} \\
&= 2\pi f(d_i)\sqrt{1 + \left(\frac{\Delta y_i}{\Delta x_i}\right)^2}\, \Delta x_i.
\end{aligned}$$

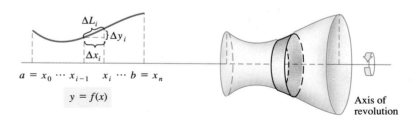

FIGURE 6.42

By the Mean Value Theorem, a point c_i exists in (x_{i-1}, x_i) such that

$$f'(c_i) = \frac{f(x_i) - f(x_{i-1})}{x_i - x_{i-1}} = \frac{\Delta y_i}{\Delta x_i}.$$

Therefore, $\Delta S_i = 2\pi f(d_i)\sqrt{1 + [f'(c_i)]^2}\, \Delta x_i$, and the total surface area can be approximated by

$$S \approx 2\pi \sum_{i=1}^{n} f(d_i)\sqrt{1 + [f'(c_i)]^2}\, \Delta x_i.$$

It can be shown that the limit of the right side as $\|\Delta\| \to 0$ (or $n \to \infty$) is

$$S = 2\pi \int_a^b f(x)\sqrt{1 + [f'(x)]^2}\, dx.$$

In a similar manner, it follows that if the graph of f is revolved about the y-axis, then S is given by

$$S = 2\pi \int_a^b x\sqrt{1 + [f'(x)]^2}\ dx.$$

In both formulas for S, we can regard the products $2\pi f(x)$ and $2\pi x$ as the circumference of the circle traced by a point (x, y) on the graph of f as it is revolved about the x- or y-axis (Figures 6.43 and 6.44). In one case the radius is $r = f(x)$, and in the other case the radius is $r = x$. Moreover, by appropriately adjusting r, we can generalize this formula for surface area to cover *any* horizontal or vertical axis of revolution, as indicated in the following definition.

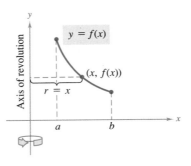

FIGURE 6.43   FIGURE 6.44

DEFINITION OF THE AREA OF A SURFACE OF REVOLUTION

If $y = f(x)$ has a continuous derivative on the interval $[a, b]$, then the area S of the surface of revolution formed by revolving the graph of f about a horizontal or vertical axis is

$$S = 2\pi \int_a^b r(x)\sqrt{1 + [f'(x)]^2}\ dx$$

where $r(x)$ is the distance between the graph of f and the axis of revolution.

REMARK   If $x = g(y)$ on the interval $[c, d]$, then the surface area is

$$S = 2\pi \int_c^d r(y)\sqrt{1 + [g'(y)]^2}\ dy$$

where $r(y)$ is the distance between the graph of g and the axis of revolution.

EXAMPLE 5   The area of a surface of revolution

Find the area of the surface formed by revolving the graph of $f(x) = x^3$ on the interval $[0, 1]$ about the x-axis, as shown in Figure 6.45.

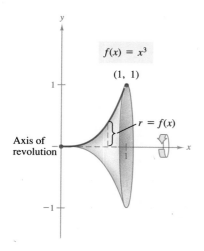

FIGURE 6.45

SOLUTION

The distance between the x-axis and the graph of f is $r(x) = f(x)$, and since $f'(x) = 3x^2$, the surface area is given by

$$S = 2\pi \int_a^b r(x)\sqrt{1 + [f'(x)]^2}\, dx = 2\pi \int_0^1 x^3\sqrt{1 + (3x^2)^2}\, dx$$

$$= \frac{2\pi}{36} \int_0^1 (36x^3)(1 + 9x^4)^{1/2}\, dx$$

$$= \frac{\pi}{18}\left[\frac{(1 + 9x^4)^{3/2}}{3/2}\right]_0^1$$

$$= \frac{\pi}{27}(10^{3/2} - 1) \approx 3.563.$$ ▭

EXAMPLE 6 The area of a surface of revolution

Find the area of the surface formed by revolving the graph of $f(x) = x^2$ on the interval $[0, \sqrt{2}]$ about the y-axis, as shown in Figure 6.46.

SOLUTION

In this case, the distance between the graph of f and the y-axis is $r(x) = x$, and since $f'(x) = 2x$, the surface area is

$$S = 2\pi \int_a^b r(x)\sqrt{1 + [f'(x)]^2}\, dx = 2\pi \int_0^{\sqrt{2}} x\sqrt{1 + (2x)^2}\, dx$$

$$= \frac{2\pi}{8} \int_0^{\sqrt{2}} (1 + 4x^2)^{1/2}(8x)\, dx$$

$$= \frac{2\pi}{8}\left[\frac{(1 + 4x^2)^{3/2}}{3/2}\right]_0^{\sqrt{2}}$$

$$= \frac{\pi}{6}[(1 + 8)^{3/2} - (1)^{3/2}] = \frac{13\pi}{3}.$$ ▭

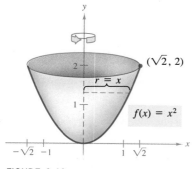

FIGURE 6.46

EXERCISES for Section 6.4

In Exercises 1 and 2, find the distance between the given points by (a) using the distance formula and (b) determining the equation of the line through the points and using the formula for arc length.

1. $(0, 0)$, $(5, 12)$ **2.** $(1, 2)$, $(7, 10)$

In Exercises 3–8, find the arc length of the graph of the given function over the indicated interval.

Function	Interval
3. $y = \dfrac{2}{3}x^{3/2} + 1$	$[0, 1]$

Function	Interval
4. $y = x^{3/2} - 1$	$[0, 4]$
5. $y = \dfrac{x^4}{8} + \dfrac{1}{4x^2}$	$[1, 2]$
6. $y = \dfrac{3}{2}x^{2/3}$	$[1, 8]$
7. $y = \dfrac{x^5}{10} + \dfrac{1}{6x^3}$	$[1, 2]$
8. $y = 2 - \dfrac{2}{3}x$	$[0, 3]$

In Exercises 9–14, find a definite integral that represents the arc length of the curve over the indicated interval. (Do not evaluate the integral.)

Function	Interval
9. $y = x^2 + x - 2$	$[-2, 1]$
10. $y = \dfrac{1}{x + 1}$	$[0, 1]$
11. $y = 4 - x^2$	$[0, 2]$
12. $y = \sqrt[3]{x}$	$[-8, 8]$
13. $x = \dfrac{1}{y^2}$	$[1, 2]$
14. $x = \sqrt{a^2 - y^2}$	$\left[0, \dfrac{a}{2}\right]$

In Exercises 15–18, use Simpson's Rule (with $n = 4$) to approximate the arc length of the function over the indicated interval.

Function	Interval
15. $y = \dfrac{1}{x}$	$[1, 3]$
16. $y = x^2$	$[0, 1]$
17. $y = x^3$	$[0, 2]$
18. $y = \sqrt{x}$	$[1, 3]$

19. A fleeing object leaves the origin and moves up the y-axis (see figure). At the same time, a pursuer leaves the point $(1, 0)$ and moves always toward the fleeing object. If the pursuer's speed is twice that of the fleeing object, the equation of the path is

$$y = \frac{1}{3}(x^{3/2} - 3x^{1/2} + 2).$$

How far has the fleeing object traveled when it is caught?

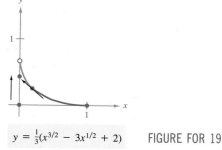

$y = \frac{1}{3}(x^{3/2} - 3x^{1/2} + 2)$ FIGURE FOR 19

20. In Exercise 19, show that the pursuer has traveled twice as far as the fleeing object.
21. Find the arc length from $(0, 3)$ clockwise to $(2, \sqrt{5})$ along the circle $x^2 + y^2 = 9$.
22. Find the arc length from $(-3, 4)$ clockwise to $(4, 3)$ along the circle $x^2 + y^2 = 25$. Show that the result is one-fourth of the circumference of the circle.

In Exercises 23–26, find the area of the surface of revolution generated by revolving the given plane curve about the x-axis.

Function	Interval
23. $y = \dfrac{x^3}{3}$	$[0, 3]$
24. $y = \sqrt{x}$	$[1, 4]$
25. $y = \dfrac{x^3}{6} + \dfrac{1}{2x}$	$[1, 2]$
26. $y = \dfrac{x}{2}$	$[0, 6]$

In Exercises 27 and 28, find the area of the surface of revolution generated by revolving the given plane curve over the indicated interval about the y-axis.

Function	Interval
27. $y = \sqrt[3]{x} + 2$	$[1, 8]$
28. $y = 4 - x^2$	$[0, 2]$

29. A right circular cone is generated by revolving the region bounded by $y = hx/r$, $y = h$, and $x = 0$ about the y-axis. Verify that the lateral surface area of the cone is

$$S = \pi r \sqrt{r^2 + h^2}.$$

30. A sphere of radius r is generated by revolving the graph of $y = \sqrt{r^2 - x^2}$ about the x-axis. Verify that the surface area of the sphere is $4\pi r^2$.
31. Find the area of the zone of a sphere formed by revolving the graph of $y = \sqrt{9 - x^2}$, $0 \le x \le 2$, about the y-axis.
32. Find the area of the zone of a sphere formed by revolving the graph of $y = \sqrt{r^2 - x^2}$, $0 \le x \le a$, about the y-axis. Assume that $a < r$.
33. An ornamental light bulb is designed by revolving the graph of

$$y = \frac{1}{3}x^{1/2} - x^{3/2}, \quad 0 \le x \le \frac{1}{3}$$

about the x-axis, where x and y are measured in feet (see figure). Find the surface area of the bulb and use the result to approximate the amount of glass needed to make the bulb. (Assume the glass is 0.015 inch thick.)

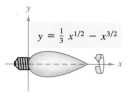

FIGURE FOR 33

34. Given a circular sector with radius L and central angle θ (see figure) show that its area is given by

$$S = \frac{1}{2}L^2\theta.$$

35. By joining the straight line edges of the sector of Exercise 34, a right circular cone is formed (see figure) and the lateral surface area of the cone is the same as the area of the sector. Show that the area is given by $S = \pi r L$ where r is the radius of the base of the cone. [Hint: The arc length of the sector equals the circumference of the base of the cone.]

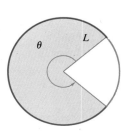

FIGURE FOR 34

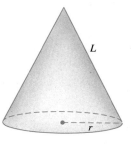

FIGURE FOR 35

36. Use the result of Exercise 35 to verify that the formula for the lateral surface area of the frustum of a cone with slant height L and radii r_1 and r_2 (see figure) is given by $S = \pi(r_1 + r_2)L$.

[Note: This formula was used to develop the integral for finding the surface area of a surface of revolution.]

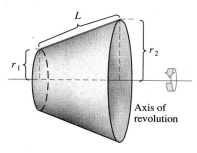

FIGURE FOR 36

In Exercises 37 and 38, use Simpson's Rule on a computer or calculator (with $n = 10$) to approximate the surface area of a vase generated by revolving the region bounded by the graphs of the given equations about the y-axis.

37. $x = \dfrac{y}{27}(2y^2 - 19y + 51)$, $x = 0$, $y = 0.5$, $y = 6$

38. $x = \dfrac{18}{y^2 + 9}$, $x = 0$, $y = 0$, $y = 6$

6.5 Work

Work done by a constant force ▪ Work done by a variable force

The concept of work is important to scientists and engineers for determining the energy needed to perform various jobs. For instance, it is useful to know the amount of work done when a crane lifts a steel girder, when a spring is compressed, when a rocket is propelled into the air, or when a truck pulls a load along a highway.

In general, we say that **work** is done by a force when it moves an object. If the force applied to the object is *constant*, we have the following definition of work.

DEFINITION OF WORK DONE BY A CONSTANT FORCE

If an object is moved a distance D in the direction of an applied constant force F, then the **work** W done by the force is defined as $W = FD$.

There are many types of forces—centrifugal, electromotive, and gravitational, to name a few. A **force** can be thought of as a *push* or a *pull*; a force changes the state of rest or state of motion of a body. For gravitational forces on the earth, we commonly use units of measure corresponding to the weight of an object, as illustrated in the following example.

EXAMPLE 1   Work done by a constant force

Determine the work done in lifting a 150-pound object 4 feet.

SOLUTION

The magnitude of the required force F is the weight of the object, as shown in Figure 6.47. Thus, the work done in lifting the object 4 feet is

$$\text{work} = W = FD = 150(4) = 600 \text{ ft} \cdot \text{lb}.$$

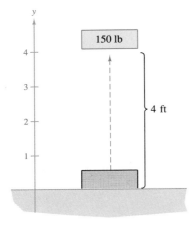

FIGURE 6.47

REMARK   In the U.S. system of measurement, work is typically expressed in foot-pounds (ft · lb), inch-pounds, or foot-tons. In the centimeter-gram-second (C-G-S) system, the basic unit of force is the **dyne**—the force required to produce an acceleration of one centimeter per second per second in a gram mass. In this system work is typically expressed in dyne-centimeters (erg) or newton-meters (joule), where one joule $= 10^7$ ergs.

Work done by a variable force

In Example 1, the force involved is *constant*. If a *variable* force is applied to an object, then calculus is needed to determine the work done, because the amount of force changes as the object changes positions. For instance, the force required to compress a spring increases as the spring is compressed.

Suppose that an object is moved along a straight line from $x = a$ to $x = b$ by a continuously varying force $F(x)$. Let Δ be a partition that divides the interval $[a, b]$ into n subintervals determined by

$$a = x_0 < x_1 < x_2 < \cdots < x_n = b$$

and let $\Delta x_i = x_i - x_{i-1}$. For each i, choose c_i such that $x_{i-1} \le c_i \le x_i$. Then at c_i the force is given by $F(c_i)$. Since F is continuous, we conclude (assuming that Δx_i is small) that the work done in moving the object through the ith subinterval is approximated by the increment $\Delta W_i = F(c_i) \Delta x_i$, as shown in Figure 6.48. By adding the work done in each subinterval, we can approximate the total work done as the object moves from a to b by

$$W \approx \sum_{i=1}^{n} \Delta W_i = \sum_{i=1}^{n} F(c_i)\Delta x_i.$$

Taking the limit of this sum as $\|\Delta\| \to 0$, $(n \to \infty)$, we have

$$W = \lim_{n \to \infty} \sum_{i=1}^{n} F(c_i)\Delta x_i = \int_a^b F(x) \, dx.$$

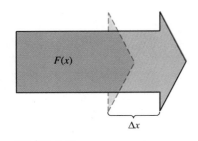

FIGURE 6.48

We therefore define work done by a variable force as follows.

DEFINITION OF WORK DONE BY A VARIABLE FORCE

If an object is moved along a straight line by a continuously varying force $F(x)$, then the **work** W done by the force as the object is moved from $x = a$ to $x = b$ is given by

$$W = \lim_{n \to \infty} \sum_{i=1}^{n} \Delta W_i = \int_{a}^{b} F(x)\ dx.$$

Emilie de Breteuil

The remaining examples in this section use some well-known physical laws. The discovery of many of these laws occurred during the same period in which calculus was being developed. In fact, during the seventeenth and eighteenth centuries, there was little difference between physicists and mathematicians. One such physicist-mathematician was Emilie de Breteuil (1706–1749). Breteuil was instrumental in synthesizing the work of many·other scientists including Newton, Leibniz, Huygens, Kepler, and Descartes. Her physics text *Institutions* was widely used for many years.

The following three laws of physics were developed by Robert Hooke (1635–1703), Isaac Newton (1642–1727), and Charles Coulomb (1736–1806). Although Hooke and Coulombe are known primarily for their work in physics, they were also gifted mathematicians.

1. **Hooke's Law:** The force F required to compress or stretch a spring (within its elastic limits) is proportional to the distance d that the spring is compressed or stretched from its original length. That is,

   $$F = kd$$

   where the constant of proportionality k depends on the specific nature of the spring.

2. **Law of Universal Gravitation:** The force F of attraction between two particles of mass m_1 and m_2 is proportional to the product of the masses and inversely proportional to the square of the distance d between the two particles. That is,

   $$F = k\frac{m_1 m_2}{d^2}.$$

   (If m_1 and m_2 are given in grams and d in centimeters, then F will be in dynes for a value of $k = 6.670 \times 10^{-8}$.)

3. **Coulomb's Law:** The force between two charges q_1 and q_2 in a vacuum is proportional to the product of the charges and inversely proportional to the square of the distance d between the two charges. That is,

   $$F = k\frac{q_1 q_2}{d^2}.$$

   (If q_1 and q_2 are given in electrostatic units and d in centimeters, then F will be in dynes for a value of $k = 1$.)

EXAMPLE 2 An application of Hooke's Law

A force of 750 pounds compresses a spring 3 inches from its natural length of 15 inches. Find the work done in compressing the spring an additional 3 inches.

SOLUTION

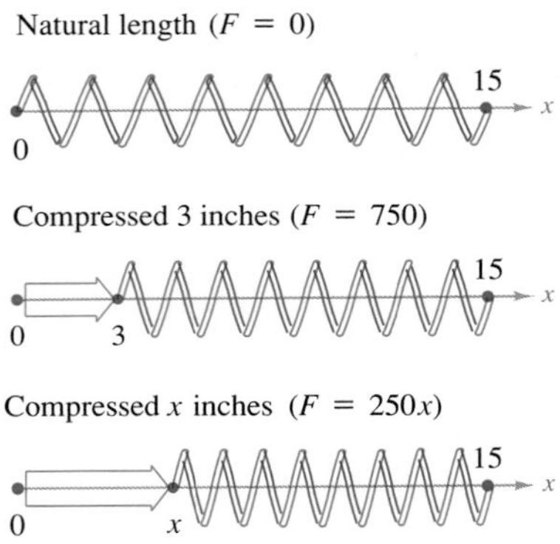

Natural length ($F = 0$)

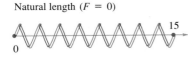

Compressed 3 inches ($F = 750$)

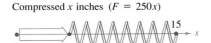

Compressed x inches ($F = 250x$)

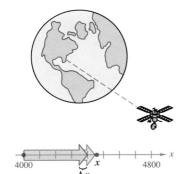

FIGURE 6.49

Using Hooke's Law, the force $F(x)$ required to compress the spring x units (from its natural length) is given by $F(x) = kx$. Using the given data, we have

$$F(3) = 750 = (k)(3)$$

and thus $k = 250$ and $F(x) = 250x$, as shown in Figure 6.49. To find the increment of work, we assume that the force required to compress the spring over a small increment Δx is nearly constant. Thus, the increment of work is

$$\Delta W = (\text{force})(\text{distance increment}) = (250x)\Delta x.$$

Now, since the spring is compressed from $x = 3$ to $x = 6$ inches less than its natural length, the work required is

$$W = \int_{3}^{6} 250x \, dx = 125x^2 \Big]_{3}^{6} = 4500 - 1125 = 3375 \text{ in} \cdot \text{lb.}$$

Note that we do not integrate from $x = 0$ to $x = 6$ because we were asked to determine the work done in compressing the spring an *additional* 3 inches (not including the first 3 inches).

EXAMPLE 3 An application involving gravitational force

If a space module weighs 15 tons on the surface of the earth, how much work is done in propelling the module to a height of 800 miles above the earth, as shown in Figure 6.50? (Do not consider the effect of air resistance or the weight of the propellant.)

SOLUTION

Since the weight of a body varies inversely as the square of its distance from the center of the earth, the force $F(x)$ exerted by gravity is

$$F(x) = \frac{C}{x^2}.$$

Since the module weighs 15 tons on the surface of the earth and the radius of the earth is approximately 4000 miles, we obtain

$$15 = \frac{C}{(4000)^2} \quad \Longrightarrow \quad C = 240{,}000{,}000.$$

Space module is moved 800 miles above the earth.

FIGURE 6.50

Thus, the increment of work is

$$\Delta W = \text{(force)(distance increment)}$$
$$= \frac{240,000,000}{x^2} \Delta x.$$

Finally, since the module is propelled from $x = 4000$ to $x = 4800$ miles, the total work done is

$$W = \int_{4000}^{4800} \frac{240,000,000}{x^2} \, dx = \frac{-240,000,000}{x} \Big]_{4000}^{4800}$$
$$= -50,000 + 60,000 = 10,000 \text{ mile-tons}$$
$$\approx 1.056 \times 10^{11} \text{ ft} \cdot \text{lb}.$$

The solutions to Examples 2 and 3 conform to our development of work as the summation of increments in the form

$$\Delta W = \text{(force)(distance increment)} = F(\Delta x).$$

An equally useful way to formulate the increment of work is

$$\Delta W = \text{(force increment)(distance)} = (\Delta F)x.$$

This second interpretation of ΔW is useful in problems involving the movement of nonrigid substances such as fluids or chains, as seen in the next two examples.

EXAMPLE 4 Work required to move a liquid

SOLUTION

A spherical tank of radius 8 feet is half full of oil that weighs 50 pounds per cubic foot. Find the work required to pump the oil out through a hole in the top of the tank.

We consider the oil to be subdivided into discs of thickness Δy and radius x, as shown in Figure 6.51. Since the increment of force for each disc is given by its weight, we have

$$\Delta F = \text{weight} = \left(\frac{50 \text{ lb}}{\text{ft}^3}\right)(\text{volume}) = 50(\pi x^2 \Delta y).$$

Now, for a circle of radius 8 and center at $(0, 8)$, we have

$$x^2 + (y - 8)^2 = 8^2$$
$$x^2 = 16y - y^2$$

and we can write the force increment as

$$\Delta F = 50(\pi x^2 \Delta y) = 50\pi(16y - y^2)\Delta y.$$

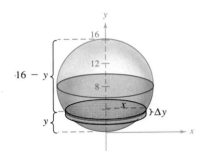

FIGURE 6.51

In Figure 6.51, we see that a disc y feet from the bottom of the tank must be moved a distance of $(16 - y)$ feet. Therefore, the increment of work is

$$\Delta W = \Delta F(16 - y) = 50\pi(16y - y^2)\Delta y(16 - y)$$
$$= 50\pi(256y - 32y^2 + y^3)\Delta y.$$

Finally, since the tank is half full, y ranges from 0 to 8, and the work required to empty the tank is

$$W = \int_0^8 50\pi(256y - 32y^2 + y^3)\, dy = 50\pi\left[128y^2 - \frac{32}{3}y^3 + \frac{y^4}{4}\right]_0^8$$
$$= 50\pi\left(\frac{11264}{3}\right) \approx 589{,}782 \text{ ft} \cdot \text{lb.}$$

EXAMPLE 5 Work done in lifting a chain

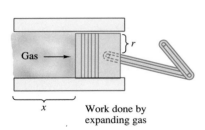

FIGURE 6.52

A 20-foot chain, weighing 5 pounds per foot, is lying coiled on the ground. How much work is required to raise one end of the chain to a height of 20 feet so that it is fully extended, as shown in Figure 6.52?

SOLUTION

Imagine that the chain is divided into small sections, each of length Δy. Then the weight of each section is the increment of force

$$\Delta F = (\text{weight}) = \left(\frac{5 \text{ lb}}{\text{ft}}\right)(\text{length}) = 5\Delta y.$$

Since a typical section (initially on the ground) is raised to a height of y, we conclude that the increment of work is

$$\Delta W = (\text{force increment})(\text{distance}) = (5\,\Delta y)y = 5y\Delta y.$$

Finally, since y ranges from 0 to 20, the total work is

$$W = \int_0^{20} 5y\, dy = \frac{5y^2}{2}\Big]_0^{20} = \frac{5(400)}{2} = 1000 \text{ ft} \cdot \text{lb.}$$

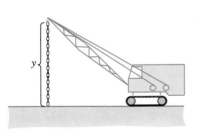

Gas →

Work done by
expanding gas

FIGURE 6.53

In the next example we consider a piston of radius r in a cylindrical casing, as shown in Figure 6.53. As the gas in the cylinder expands, the piston moves and work is done. If p represents the pressure of the gas (in pounds per square foot) against the piston head and V represents the volume of the gas (in cubic feet), then the work increment involved in moving the piston Δx feet is

$$\Delta W = (\text{force})(\text{distance increment}) = F(\Delta x) = p(\pi r^2)\Delta x = p\Delta V.$$

Thus, as the volume of the gas expands from V_0 to V_1, the work done in moving the piston is

$$W = \int_{V_0}^{V_1} p\, dV.$$

Assuming the pressure of the gas to be inversely proportional to its volume, we have $p = k/V$ and the integral for work becomes

$$W = \int_{V_0}^{V_1} \frac{k}{V}\, dV.$$

EXAMPLE 6   Work done by a gas

A quantity of gas with an initial volume of 1 cubic foot and pressure of 500 pounds per square foot expands to a volume of 2 cubic feet. Find the work done by the gas. (Assume the pressure is inversely proportional to the volume.)

SOLUTION

Since $p = k/V$ and $p = 500$ when $V = 1$, we have $k = 500$. Thus, the work is

$$W = \int_{V_0}^{V_1} \frac{k}{V}\, dV = \int_1^2 \frac{500}{V}\, dV.$$

Using Simpson's Rule (see Example 4, Section 5.6) we can approximate the work to be

$$W \approx 346.6 \text{ ft} \cdot \text{lb.}$$

EXERCISES for Section 6.5

1. Determine the work done in lifting a 100-pound bag of sugar 10 feet.
2. Determine the work done by a hoist in lifting a 2400-pound car 6 feet.
3. A force of 25 pounds is required to slide a cement block on a plank in a construction project. The plank is 12 feet long. Determine the work done in sliding the block along the length of the plank.
4. The locomotive of a freight train pulls its cars with a constant force of 9 tons while traveling at a constant rate of 55 miles per hour on a level track. How many foot-pounds of work does the locomotive do in a distance of one-half mile?
5. A force of 5 pounds compresses a 15-inch spring a total of 4 inches. How much work is done in compressing the spring 7 inches?
6. How much work is done in compressing the spring in Exercise 5 from a length of 10 inches to a length of 6 inches?
7. A force of 60 pounds stretches a spring 1 foot. How much work is done in stretching the spring from 9 inches to 15 inches?
8. A force of 200 pounds stretches a spring 2 feet on a mechanical device for driving fence posts. Find the work done in stretching the spring the required 2 feet.

9. A force of 15 pounds stretches a spring 6 inches in an exercise machine. Find the work done in stretching the spring 1 foot from its natural position.
10. An overhead garage door has two springs, one on each side of the door. A force of 15 pounds is required to stretch each spring 1 foot. Because of the pulley system, the springs only stretch one-half the distance the door travels. Find the work done by the pair of springs if the door moves a total of 8 feet and the springs are at their natural length when the door is open.
11. A rectangular tank with a base 4 feet by 5 feet and a height of 4 feet is filled with water (see figure). (The water weighs 62.4 pounds per cubic foot.) How much work is done in pumping water out over the top edge in order to empty
    (a) half of the tank?
    (b) all of the tank?

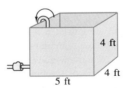

5 ft

4 ft

4 ft

FIGURE FOR 11

12. Repeat Exercise 11 for a tank filled with gasoline that weighs 42 pounds per cubic foot.

13. A cylindrical water tank 12 feet high with a radius of 8 feet is buried so that the top of the tank is 3 feet below ground level (see figure). How much work is done in pumping a full tank of water up to ground level?

14. Suppose the tank in Exercise 13 is located on a tower so that the bottom of the tank is 20 feet above the level of a stream (see figure). How much work is done in filling the tank half full of water through a hole in the bottom, using water from the stream?

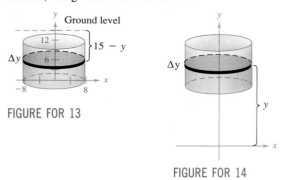

FIGURE FOR 13

FIGURE FOR 14

15. A hemispherical tank of radius 6 feet is positioned so that its base is circular. How much work is required to fill the tank with water through a hole in the base if the water source is at the base?

16. Suppose the tank in Exercise 15 is inverted and the top 2 feet of water is pumped out through a hole in the top. How much work is done?

17. An open tank has the shape of a right circular cone (see figure). The tank is 8 feet across the top and is 6 feet high. How much work is done in emptying the tank by pumping the water over the top edge?

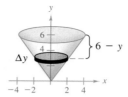

FIGURE FOR 17

18. If water is pumped in through the bottom of the tank in Exercise 17, how much work is done to fill the tank
    (a) to a depth of 2 feet?
    (b) from a depth of 4 feet to a depth of 6 feet?

19. A cylindrical gasoline tank 3 feet in diameter and 4 feet long is carried on the back of a truck and is used to fuel tractors in the field. The axis of the tank is horizontal. Find the work done to pump the entire contents of the full tank into a tractor if the opening on the

tractor tank is 5 feet above the top of the tank in the truck. Assume gasoline weighs 42 pounds per cubic foot. (Hint: Evaluate one integral by a geometric formula and the other by observing that the integrand is an odd function.)

20. The top of a cylindrical storage tank for gasoline at a service station is 4 feet below ground level. The axis of the tank is horizontal and its diameter and length are 5 feet and 12 feet, respectively. Find the work done in pumping the entire contents of the full tank to a height of 3 feet above ground level. (Hint: Evaluate one integral by a geometric formula and the other by observing that the integrand is an odd function.)

21. Neglecting air resistance and the weight of the propellant, determine the work done in propelling a 4-ton satellite to a height of (a) 200 miles and (b) 400 miles above the earth.

22. Use the information from Exercise 21 to write the work W of the propulsion system as a function of the height h of the satellite above the earth. Find the limit (if it exists) of W as h approaches infinity.

23. Neglecting air resistance, determine the work done in propelling a 10-ton satellite to a height of (a) 11,000 miles and (b) 22,000 miles above the earth.

24. If a lunar module weighs 12 tons on the surface of the earth, how much work is done in propelling the module from the surface of the moon to a height of 50 miles? Consider the radius of the moon to be 1100 miles and its force of gravity to be one-sixth that of the earth's.

25. Two electrons repel each other with a force that varies inversely as the square of the distance between them. If one electron is fixed at the point $(2, 4)$, find the work done in moving a second electron from $(-2, 4)$ to $(1, 4)$.

26. The force generated by a press in a manufacturing process is given by

$$F(x) = 10,000\sqrt{1 + x^5}, \quad 0 \le x \le 4$$

where x is the distance in feet the press moves. Use Simpson's Rule (with $n = 8$) to approximate the work done through one cycle of the press.

In Exercises 27–30, consider a 15-foot chain hanging from a winch 15 feet above ground level. Find the work done by the winch in winding up the required amount of chain, if the chain weighs 3 pounds per foot.

27. Wind up the entire chain.

28. Wind up one-third of the chain.

29. Run the winch until the bottom of the chain is at the 10-foot level.

30. Wind up the entire chain with a 100-pound load attached.

In Exercises 31 and 32, consider a 15-foot hanging chain that weighs 3 pounds per foot. Find the work done in lifting the chain vertically to the required position.

31. Take the bottom of the chain and raise it to the 15-foot level, leaving the chain doubled and still hanging vertically (see figure).

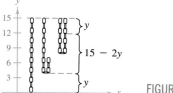

FIGURE FOR 31

32. Repeat Exercise 31 raising the bottom of the chain to the 12-foot level.

In Exercises 33 and 34, consider a demolition crane with a 500-pound ball suspended from a 40-foot cable that weighs 1 pound per foot.

33. Find the work required to wind up 15 feet of the apparatus.

34. Find the work required to wind up all 40 feet of the apparatus.

In Exercises 35 and 36, find the work done by the gas for the given volumes and pressures. Assume the pressure is inversely proportional to the volume. (Use Example 6 as a model.)

35. A quantity of gas with an initial volume of 2 cubic feet and pressure of 1000 pounds per square foot expands to a volume of 3 cubic feet.

36. A quantity of gas with an initial volume of 1 cubic foot and pressure of 2000 pounds per square foot expands to a volume of 4 cubic feet.

37. In a manufacturing process an object is moved linearly 5 feet with a variable force given by

$$F(x) = 100x\sqrt{125 - x^3}, \quad 0 \le x \le 5$$

where F is given in pounds and x gives the position of the unit in feet. Use Simpson's Rule on a computer or calculator (with $n = 10$) to approximate the work done to move one unit through the given process.

6.6 Fluid Pressure and Fluid Force

Fluid pressure ▪ Force exerted by a fluid

Swimmers know that the deeper an object is submerged in a fluid, the greater the pressure on the object. We define **pressure** to be the force per unit of area over the surface of a body. For example, since a ten-foot column of water (one-inch square) weighs 4.3 pounds, we say that the *fluid* pressure at a depth of 10 feet of water is 4.3 pounds per square inch.* At 20 feet, this would increase to 8.6 pounds per square inch, and in general the pressure is proportional to the depth of the object in the fluid, as indicated in the following definition.

DEFINITION OF FLUID PRESSURE

The **pressure** on an object at depth h in a liquid is given by

$$\text{pressure} = P = wh$$

where w is the weight of the liquid per unit of volume.

*The total pressure on an object in ten feet of water would also include the pressure due to the earth's atmosphere. At sea level, atmospheric pressure is approximately 14.7 pounds per square inch.

Blaise Pascal

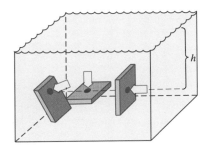

FIGURE 6.54

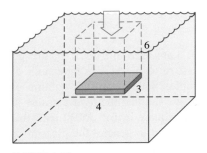

FIGURE 6.55

Listed below are some commonly used weights of fluids in pounds per cubic foot.

Water	62.4
Ethyl alcohol	49.4
Gasoline	41.0–43.0
Kerosene	51.2
Mercury	849.0
Glycerin	78.6
Sea water	64.0

When calculating fluid pressure, we use an important (and rather surprising) physical law called **Pascal's Principle,** named after the French mathematician Blaise Pascal (1623–1662). Pascal is well known for his work in many areas of mathematics and physics and also for his influence on Leibniz. Although much of Pascal's work in calculus was intuitive and lacked the rigor of modern mathematics, he nevertheless anticipated many important results. Pascal's Principle states that the pressure exerted by a fluid at a depth h is transmitted equally *in all directions*. For example, in Figure 6.54, the pressure at the indicated depth is the same for all three objects.

Since fluid pressure is given in terms of force per unit area ($P = F/A$), the fluid force on a *submerged horizontal* surface of area A is given by

$$\text{fluid force} = F = PA = (\text{pressure})(\text{area}).$$

EXAMPLE 1   Fluid force on a horizontal surface

Find the fluid force on a rectangular metal sheet 3 feet by 4 feet that is submerged in 6 feet of water, as shown in Figure 6.55.

SOLUTION

Since the weight of water is 62.4 pounds per cubic foot and the sheet is submerged in 6 feet of water, the fluid pressure is

$$P = (62.4)(6) = 374.4 \text{ lb/ft}^2.$$

Now, since the total area of the sheet is $A = (3)(4) = 12$ square feet, the fluid force is given by

$$F = PA = (374.4)(12) = 4492.8 \text{ lb.}$$

REMARK   The answer to Example 1 is independent of the size of the body of water. The fluid force would be the same in a swimming pool or a lake.

In Example 1, the fact that the sheet is rectangular and horizontal means that we do not need the methods of calculus to solve the problem. We now look at a surface that is submerged vertically in a fluid. This problem is more difficult because the pressure is not constant over the surface.

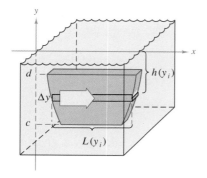

FIGURE 6.56

Suppose a vertical plate is submerged in a fluid of weight w (per unit of volume), as shown in Figure 6.56. We wish to determine the total force against *one side* of this region from depth c to depth d. First, we subdivide the interval $[c, d]$ into n subintervals, each of width Δy. Now consider the representative rectangle of width Δy and length $L(y_i)$, where y_i is in the ith subinterval. The force against this representative rectangle is

$$\Delta F_i = w(\text{depth})(\text{area}) = wh(y_i)L(y_i)\Delta y.$$

The force against n such rectangles is

$$\sum_{i=1}^{n} \Delta F_i = w \sum_{i=1}^{n} h(y_i)L(y_i)\Delta y.$$

Note that w is considered to be constant and is factored out of the summation. Therefore, taking the limit as $\|\Delta\| \to 0$ ($n \to \infty$), we find the total force against the region to be

$$F = w \lim_{n \to \infty} \sum_{i=1}^{n} h(y_i)L(y_i)\Delta y = w \int_{c}^{d} h(y)L(y) \, dy.$$

DEFINITION OF FORCE EXERTED BY A FLUID

The **force F exerted by a fluid** of constant weight w (per unit of volume) against a submerged vertical plane region from $y = c$ to $y = d$ is given by

$$F = w \int_{c}^{d} h(y)L(y) \, dy$$

where $h(y)$ is the depth of the fluid at y and $L(y)$ is the horizontal length of the region at y.

EXAMPLE 2   Fluid force on a vertical surface

A vertical gate in a dam has the shape of an isosceles trapezoid 8 feet across the top and 6 feet across the bottom, with a height of 5 feet, as shown in Figure 6.57(a). What is the fluid force against the gate if the top of the gate is 4 feet below the surface of the water?

SOLUTION

In setting up a mathematical model for this problem, we are at liberty to locate the *x*- and *y*-axes in a number of different ways. A convenient approach is to let the *y*-axis bisect the gate and place the *x*-axis at the surface of the water, as shown in Figure 6.57(b). Thus, the depth of the water at y is

$$\text{depth} = h(y) = -y.$$

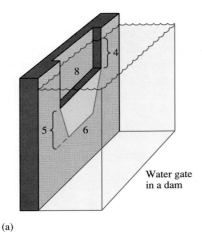

Water gate
in a dam

(a)

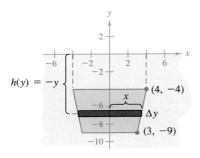

$h(y) = -y$

(b)

FIGURE 6.57

To find the length $L(y)$ of the region at y, we find the equation of the line forming the right side of the gate. Since this line passes through the two points $(3, -9)$ and $(4, -4)$, its equation is

$$y - (-9) = \frac{-4 - (-9)}{4 - 3}(x - 3) = 5(x - 3)$$

$$y = 5x - 24 \implies x = \frac{y + 24}{5}.$$

From Figure 6.57(b) we can see that the length of the region at y is given by

$$\text{length} = 2x$$
$$= \frac{2}{5}(y + 24)$$
$$= L(y).$$

Finally, by integrating from $y = -9$ to $y = -4$, we find the fluid force to be

$$F = w \int_c^d h(y)L(y)\, dy$$

$$= 62.4 \int_{-9}^{-4} (-y)\left(\frac{2}{5}\right)(y + 24)\, dy$$

$$= -62.4\left(\frac{2}{5}\right) \int_{-9}^{-4} (y^2 + 24y)\, dy$$

$$= -62.4\left(\frac{2}{5}\right)\left[\frac{y^3}{3} + 12y^2\right]_{-9}^{-4}$$

$$= -62.4\left(\frac{2}{5}\right)\left(\frac{-1675}{3}\right)$$

$$= 13{,}936 \text{ lb.} \qquad \blacksquare$$

REMARK In Example 2, we let the x-axis coincide with the surface of the water. This was convenient, but arbitrary. In choosing a coordinate system to represent a physical situation, you should consider various possibilities. Often, you can simplify the calculations in a problem by locating the coordinate system so as to take advantage of special characteristics of the problem, such as symmetry.

EXAMPLE 3 Fluid force on a vertical surface

A circular observation window on a marine science ship has a radius of 1 foot, and the center of the window is 8 feet below water level, as shown in Figure 6.58(a). What is the fluid force on the window?

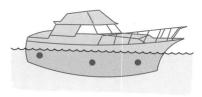

(a)

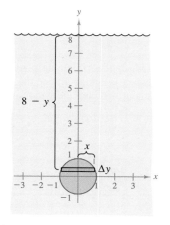

(b)

FIGURE 6.58

SOLUTION

To take advantage of symmetry, we locate a coordinate system so that the origin coincides with the center of the window, as shown in Figure 6.58(b). The depth at y is then given by

$$\text{depth} = h(y) = 8 - y.$$

The horizontal length of the window is $2x$, and we can use the equation for the circle, $x^2 + y^2 = 1$, to solve for x as follows.

$$\text{length} = 2x = 2\sqrt{1 - y^2} = L(y)$$

Finally, since y ranges from -1 to 1, we have, using the weight of sea water as 64 pounds per cubic foot,

$$F = w \int_c^d h(y)L(y)\, dy = 64 \int_{-1}^1 (8 - y)(2)\sqrt{1 - y^2}\, dy.$$

Initially it looks as if this integral would be difficult to solve. However, if we break the integral into two parts and apply symmetry, the solution is simple:

$$F = 64(16) \int_{-1}^1 \sqrt{1 - y^2}\, dy - 64(2) \int_{-1}^1 y\sqrt{1 - y^2}\, dy.$$

The second integral is zero (since the integrand is odd and the limits of integration are symmetric to the origin). Moreover, by recognizing that the first integral represents the area of a semicircle of radius 1, we have

$$F = 64(16)\left(\frac{\pi}{2}\right) - 64(2)(0) = 512\pi \approx 1608.5 \text{ lb.}$$

EXAMPLE 4 Fluid force on a vertical surface

A swimming pool is 2 feet deep at one end and 10 feet deep at the other, as shown in Figure 6.59(a). The pool is 40 feet long and 30 feet wide with vertical sides. Find the fluid force against one of the 40-foot sides.

SOLUTION

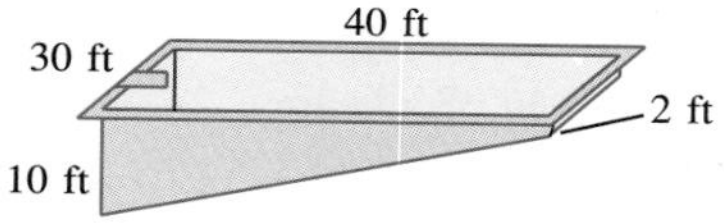

By placing the x- and y-axes as shown in Figure 6.59(b), we see that the depth at y is

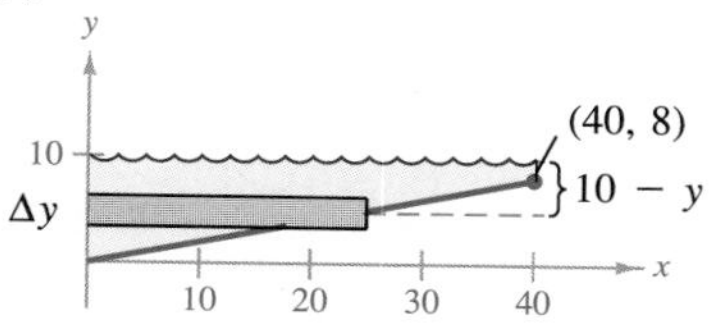

$$\text{depth} = h(y) = 10 - y.$$

The equation representing the base of the side is given by

$$y = mx + b = \frac{1}{5}x \quad \Longrightarrow \quad x = 5y.$$

However, we must note that the equation $x = 5y$ is only valid for $0 \le y \le 8$. When y is between 8 and 10, x is a constant 40. Thus, the length of the side of the pool is

$$\text{length} = L(y) = \begin{cases} 5y, & 0 \le y \le 8 \\ 40, & 8 \le y \le 10. \end{cases}$$

30 ft · 40 ft · 2 ft · 10 ft

(a)

10 · Δy · (40, 8) · $\}10 - y$ · 10 · 20 · 30 · 40

(b)

FIGURE 6.59

Therefore, the fluid force on the side of the pool is given by the two integrals

$$F = \int_0^8 62.4(10 - y)(5y) \, dy + \int_8^{10} 62.4(10 - y)(40) \, dy$$

$$= 312 \int_0^8 (10y - y^2) \, dy + 2496 \int_8^{10} (10 - y) \, dy$$

$$= 312\left[5y^2 - \frac{y^3}{3} \right]_0^8 + 2496\left[10y - \frac{y^2}{2} \right]_8^{10}$$

$$= 312\left(\frac{448}{3}\right) + 2496(2) = 51{,}584 \text{ lb.}$$

REMARK In Example 4, note that the 30-foot width of the swimming pool is a "red herring." That is, the width of the pool is unnecessary information, and the fluid force against the 40-foot sides can be determined without knowing the width of the pool.

EXERCISES for Section 6.6

In Exercises 1 and 2, find the fluid force on the top side of the metal sheet of given area submerged horizontally in 5 feet of water.

1. 3 square feet   **2.** 18 square feet

In Exercises 3 and 4, find the **buoyant force** of a rectangular solid of given dimensions submerged in water so that the top side is parallel to the surface of the water. The buoyant force is the difference between the fluid forces on the top and bottom sides of the solid.

3. 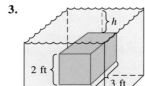   **4.**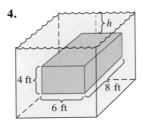

In Exercises 5–10, find the fluid force on the indicated vertical side of a tank. Assume that the tank is full of water.

5. Rectangle

6. Triangle

7. Trapezoid

8. Semicircle

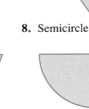

9. Parabola

$$y = x^2$$

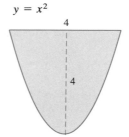

10. Semiellipse

$$y = -\frac{1}{2}\sqrt{36 - 9x^2}$$

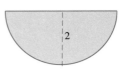

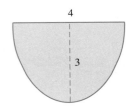

In Exercises 11–14, find the fluid force on the given vertical plates submerged in water.

11. Square

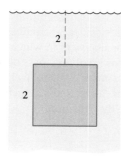

12. Square

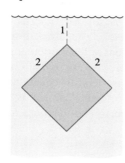

13. Triangle

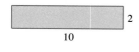

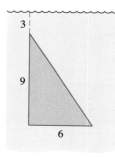

14. Rectangle

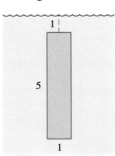

In Exercises 15–18, the given vertical plate is the side of a form for poured concrete that weighs 140.7 pounds per cubic foot. Determine the fluid force on the plate.

15. Rectangle

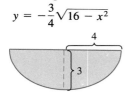

16. Rectangle

17. Semiellipse

$$y = -\frac{3}{4}\sqrt{16 - x^2}$$

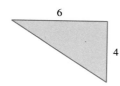

18. Triangle

19. A cylindrical gasoline tank is placed so that the axis of the cylinder is horizontal. Find the fluid force on a circular end of the tank if the tank is half full, assuming that the diameter is 3 feet and the gasoline weighs 42 pounds per cubic foot.

20. Repeat Exercise 19 for a tank that is full. (Evaluate one integral by a geometric formula and the other by observing that the integrand is an odd function.)

21. A circular plate of radius r feet is submerged vertically in a tank of fluid that weighs w pounds per cubic foot. The center of the circle is k feet below the surface of the fluid. Show that the fluid force on the surface of the circle is given by

$$F = wk(\pi r^2).$$

(Evaluate one integral by a geometric formula and the other by observing that the integrand is an odd function.)

22. A rectangular plate of height h feet and base b feet is submerged vertically in a tank of fluid that weighs w pounds per cubic foot. The center is k feet below the surface of the fluid, where $h \le k/2$. Show that the fluid force on the surface of the rectangle is given by

$$F = wkhb.$$

23. A porthole on a vertical side of a submarine (submerged in sea water) is 1 foot square. Find the fluid force on the porthole, assuming that the center of the square is 15 feet below the surface.

24. Repeat Exercise 23 for a circular porthole that has a diameter of 1 foot. The center is 15 feet below the surface.

25. A swimming pool is 20 feet wide, 40 feet long, 4 feet deep at one end, and 8 feet deep at the other. The bottom is an inclined plane. Find the fluid force on each of the vertical walls.

26. The vertical cross section of an irrigation canal is modeled by

$$f(x) = \frac{5x^2}{x^2 + 4}$$

where x is measured in feet and $x = 0$ corresponds to the center of the canal. Use Simpson's Rule (with $n = 6$) to approximate the fluid force against a vertical gate used to stop the flow of water if the water is 3 feet deep.

In Exercises 27 and 28, use Simpson's Rule on a computer or calculator (with $n = 10$) to approximate the fluid force on the vertical plate bounded by the x-axis and the top half of the graph of the given equation. Assume that the base of the plate is 12 feet beneath the surface of the water.

27. $x^{2/3} + y^{2/3} = 4^{2/3}$

28. $\dfrac{x^2}{28} + \dfrac{y^2}{16} = 1$

6.7 Moments, Centers of Mass, and Centroids

Mass ▪ Moments ▪ Center of mass ▪ Centroid ▪ Theorem of Pappus

In this section we look at several important applications of integration that are related to **mass.** Mass is a measure of a body's resistance to changes in motion, and is independent of the particular gravitational system in which the body is located. However, because so many applications involving mass occur on the earth's surface, we tend to equate an object's mass with its *weight*. This is not technically correct. Weight is a type of force and as such is dependent on gravity. Force and mass are related by the equation

force = (mass)(acceleration).

In Table 6.1 we list some commonly used measures of mass and force, together with their conversion factors.

TABLE 6.1

System of measurement	Measure of mass	Measure of force
U.S.	slug	pound = (slug)(ft/sec^2)
International	kilogram	newton = (kilogram)(m/sec^2)
C-G-S	gram	dyne = (gram)(cm/sec^2)

Conversions:	
1 pound = 4.448 newtons	1 slug = 14.59 kilograms
1 newton = 0.2248 pound	1 kilogram = 0.06854 slug
1 dyne = 0.02248 pound	1 gram = 0.00006854 slug

EXAMPLE 1   Mass on the surface of the earth

Find the mass (in slugs) of an object whose weight at sea level is one pound.

SOLUTION

Using 32 feet per second per second as the acceleration due to gravity, we have

$$\text{mass} = \frac{\text{force}}{\text{acceleration}} = \frac{1 \text{ lb}}{32 \text{ ft/sec}^2} = 0.03125 \frac{\text{lb}}{\text{ft/sec}^2} = 0.03125 \text{ slug}.$$

Because many applications occur on the earth's surface, this amount of mass is called a **pound mass.**

The moment of a mass

We shall consider two types of moments of a mass—the **moment about a point** and the **moment about a line.** To define these two moments, we consider an idealized situation in which a mass m is concentrated at a point. If x is the distance between this point-mass and another point P, then the **moment of m about the point P** is given by

$$\text{moment} = mx$$

and x is called the **length of the moment arm.**

The concept of moment can be demonstrated simply by a seesaw, as illustrated by Figure 6.60. Suppose a child of mass 20 kilograms sits 2 meters to the left of fulcrum P, and an older child of mass 30 kilograms sits 2 meters to the right of P. From experience, we know that the seesaw would begin to rotate clockwise, moving the larger child down. This rotation occurs because the moment produced by the child on the left is less than the moment produced by the child on the right:

$$\text{left moment} = (20)(2) = 40 \text{ kg} \cdot \text{m}$$
$$\text{right moment} = (30)(2) = 60 \text{ kg} \cdot \text{m}.$$

To balance the seesaw, the two moments must be equal. For example, if the larger child moved to a position $\frac{4}{3}$ meters from the fulcrum, then the seesaw would balance, since both children would produce a moment of 40 kilogram-meters.

To generalize this situation we introduce a coordinate line on which the origin corresponds to the fulcrum, as shown in Figure 6.61. Suppose several point masses are located on the x-axis. The measure of the tendency of this system to rotate about the origin is called the **moment about the origin,** and it is defined to be the sum of the n products $m_i x_i$.

$$M_0 = m_1 x_1 + m_2 x_2 + \cdots + m_n x_n$$

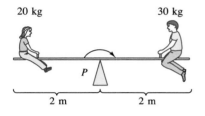

20 kg   30 kg

P

2 m   2 m

FIGURE 6.60

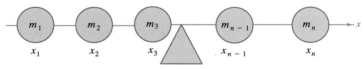

FIGURE 6.61

If $x_1 m_1 + x_2 m_2 + \cdots + x_n m_n = 0$, then the system is in equilibrium.

If M_0 is zero, then the system is said to be in **equilibrium.**

For a system that is not in equilibrium, we define the **center of mass** to be the point $\bar{x}$ at which the fulcrum could be relocated to attain equilibrium. If the system were translated $\bar{x}$ units, each coordinate x_i would become $(x_i - \bar{x})$, and since the moment of the translated system is zero, we have

$$\sum_{i=1}^{n} m_i(x_i - \bar{x}) = \sum_{i=1}^{n} m_i x_i - \sum_{i=1}^{n} m_i \bar{x} = 0.$$

Solving for $\bar{x}$, we have

$$\bar{x} = \frac{\displaystyle\sum_{i=1}^{n} m_i x_i}{\displaystyle\sum_{i=1}^{n} m_i} = \frac{\text{moment of system about origin}}{\text{total mass of system}}$$

DEFINITION OF THE MOMENT AND CENTER OF MASS OF A LINEAR SYSTEM

Let the point masses $m_1, m_2, \ldots, m_n$ be located at $x_1, x_2, \ldots, x_n$, respectively.

1. The **moment about the origin** is $M_0 = m_1 x_1 + m_2 x_2 + \cdots + m_n x_n$.
2. The **center of mass** is $\bar{x} = \dfrac{M_0}{m}$

where $m = m_1 + m_2 + \cdots + m_n$ is the **total mass** of the system.

EXAMPLE 2 The center of mass of a linear system

Find the center of mass of the linear system shown in Figure 6.62.

$$m_1 = 10, \quad m_2 = 15, \quad m_3 = 5, \quad m_4 = 10$$
$$x_1 = -5, \quad x_2 = 0, \quad x_3 = 4, \quad x_4 = 7$$

FIGURE 6.62

SOLUTION

The moment about the origin is given by

$$M_0 = m_1 x_1 + m_2 x_2 + m_3 x_3 + m_4 x_4$$
$$= 10(-5) + 15(0) + 5(4) + 10(7) = -50 + 0 + 20 + 70 = 40.$$

Since the total mass of the system is $m = 10 + 15 + 5 + 10 = 40$, the center of mass is

$$\bar{x} = \frac{M_0}{m} = \frac{40}{40} = 1.$$

Rather than define the moment of a *mass*, we could define the moment of a *force*. In this context, the center of mass is called the **center of gravity.** Suppose that a system of point masses $m_1, m_2, \ldots, m_n$ is located at $x_1, x_2, \ldots, x_n$. Then, since force = (mass)(acceleration), the total force of the system is

$$F = m_1 a + m_2 a + \cdots + m_n a = ma.$$

The **torque** (moment) about the origin is given by

$$T_0 = (m_1 a)x_1 + (m_2 a)x_2 + \cdots + (m_n a)x_n = M_0 a$$

and the **center of gravity** is

$$\frac{T_0}{F} = \frac{M_0 a}{ma} = \frac{M_0}{m} = \bar{x}.$$

Therefore, the center of gravity and the center of mass have the same location.

Two-dimensional systems

We can extend the concept of moment to two dimensions by considering a system of masses located in the xy-plane at the points (x_1, y_1), (x_2, y_2), $\ldots$, (x_n, y_n) as shown in Figure 6.63. Rather than defining a single moment (with respect to the origin), we define two moments—one with respect to the x-axis and one with respect to the y-axis.

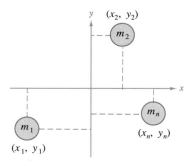

FIGURE 6.63

DEFINITION OF THE MOMENTS AND CENTER OF MASS OF A TWO-DIMENSIONAL SYSTEM	Let the point masses $m_1, m_2, \ldots, m_n$ be located at (x_1, y_1), (x_2, y_2), $\ldots$, and (x_n, y_n), respectively.

1. The **moment about the y-axis** is $M_y = m_1 x_1 + m_2 x_2 + \cdots + m_n x_n$.
2. The **moment about the x-axis** is $M_x = m_1 y_1 + m_2 y_2 + \cdots + m_n y_n$.
3. The **center of mass** $(\bar{x}, \bar{y})$ (or **center of gravity**) is given by

$$\bar{x} = \frac{M_y}{m} \qquad \text{and} \qquad \bar{y} = \frac{M_x}{m}$$

where $m = m_1 + m_2 + \cdots + m_n$ is the **total mass** of the system.

The moment of a system of masses in the plane can be taken about any horizontal or vertical line. In general, the moment about a line is the sum of the product of the masses and the *directed distances* from the points to the line.

$$\text{Moment} = m_1(y_1 - b) + m_2(y_2 - b) + \cdots + m_n(y_n - b) \qquad \begin{array}{l}\text{Horizontal line} \\ y = b\end{array}$$

$$\text{Moment} = m_1(x_1 - a) + m_2(x_2 - a) + \cdots + m_n(x_n - a) \qquad \begin{array}{l}\text{Vertical line} \\ x = a\end{array}$$

EXAMPLE 3 The center of mass of a two-dimensional system

Find the center of mass of a system of point masses $m_1 = 6$, $m_2 = 3$, $m_3 = 2$, and $m_4 = 9$, located at $(3, -2)$, $(0, 0)$, $(-5, 3)$, and $(4, 2)$, as shown in Figure 6.64.

SOLUTION

$$
\begin{aligned}
m &= 6 \quad\;\; + 3 \quad + 2 \quad\;\; + 9 \;= 20 &&\text{Mass} \\
M_y &= 6(3) \quad + 3(0) + 2(-5) + 9(4) = 44 &&\text{Moment about } y\text{-axis} \\
M_x &= 6(-2) + 3(0) + 2(3) \quad + 9(2) = 12 &&\text{Moment about } x\text{-axis}
\end{aligned}
$$

Therefore,

$$
\bar{x} = \frac{M_y}{m} = \frac{44}{20} = \frac{11}{5} \quad\text{and}\quad \bar{y} = \frac{M_x}{m} = \frac{12}{20} = \frac{3}{5}
$$

and we conclude that the center of mass is $\left(\frac{11}{5}, \frac{3}{5}\right)$.

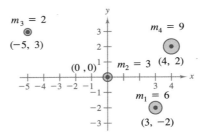

FIGURE 6.64

The center of mass of a planar lamina

In the preceding discussion we assumed the total mass of a system to be distributed at discrete points in the plane (or on a line). We now consider a thin flat plate of material of uniform density called a **planar lamina. Density** is a measure of mass per unit of volume, such as grams per cubic centimeter. We denote density by ρ, the lowercase Greek letter rho. Intuitively, we think of the center of mass $(\bar{x}, \bar{y})$ of a lamina as its balancing point. For example, the center of mass of a circular lamina is located at the center of the circle, and the center of mass of a rectangular lamina is located at the center of the rectangle, as shown in Figure 6.65.

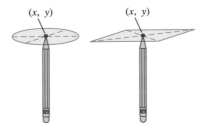

FIGURE 6.65

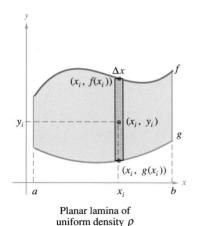

Planar lamina of
uniform density ρ

FIGURE 6.66

Consider an irregularly shaped planar lamina of uniform density ρ, bounded by the graphs of $y = f(x)$, $y = g(x)$, and $a \leq x \leq b$, as shown in Figure 6.66. The mass of this region is given by

$$m = \text{(density)(area)} = \rho \int_a^b [f(x) - g(x)] \, dx = \rho A$$

where A is the area of the region. To find the center of mass of this lamina, we partition the interval $[a, b]$ into n equal subintervals of width Δx. If x_i is the center of the ith subinterval, then we can approximate the portion of the lamina lying in the ith subinterval by a rectangle whose height is $h = f(x_i) - g(x_i)$. Since the density of the rectangle is ρ, we know that its mass is

$$m_i = \text{(density)(area)} = \underbrace{\rho}_{\text{density}} \underbrace{[f(x_i) - g(x_i)]}_{\text{height}} \underbrace{\Delta x}_{\text{width}}.$$

Now, considering this mass to be located at the center (x_i, y_i) of the rectangle, we know that the directed distance from the x-axis to (x_i, y_i) is $y_i = [f(x_i) + g(x_i)]/2$. Thus, the moment of m_i about the x-axis is

$$\text{moment} = \text{(mass)(distance)} = m_i y_i = \rho[f(x_i) - g(x_i)]\Delta x \left[\frac{f(x_i) + g(x_i)}{2}\right].$$

Summing these moments and taking the limit as $n \to \infty$, we have the moment about the x-axis defined as

$$M_x = \rho \int_a^b \left[\frac{f(x) + g(x)}{2}\right][f(x) - g(x)] \, dx.$$

For the moment about the y-axis, the directed distance from the y-axis to (x_i, y_i) is x_i and we have

$$M_y = \rho \int_a^b x[f(x) - g(x)] \, dx.$$

**DEFINITION OF MOMENTS
AND CENTER OF MASS
OF A PLANAR LAMINA**

Let f and g be continuous functions such that $f(x) \geq g(x)$ on $[a, b]$ and consider the planar lamina of uniform density ρ bounded by the graphs of $y = f(x)$, $y = g(x)$, $a \leq x \leq b$.

1. The **moments about the x- and y-axes** are

$$M_x = \rho \int_a^b \left[\frac{f(x) + g(x)}{2}\right][f(x) - g(x)] \, dx$$

$$M_y = \rho \int_a^b x[f(x) - g(x)] \, dx.$$

2. The **center of mass** $(\bar{x}, \bar{y})$ is given by $\bar{x} = \dfrac{M_y}{m}$ and $\bar{y} = \dfrac{M_x}{m}$

where $m = \rho \int_a^b [f(x) - g(x)] \, dx$ is the mass of the lamina.

REMARK   Note that the integrals for both moments can be formed by inserting into the integral for mass the directed distance from the axis (or line) about which the moment is taken to the center of the representative rectangle.

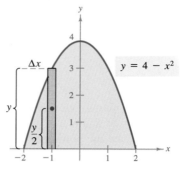

FIGURE 6.67

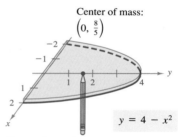

Center of mass:
$\left(0, \frac{8}{5}\right)$

$y = 4 - x^2$

Center of mass is balancing point.

FIGURE 6.68

EXAMPLE 4   The center of mass of a planar lamina

Find the center of mass of the lamina of uniform density ρ bounded by the graph of $f(x) = 4 - x^2$ and the x-axis.

SOLUTION

First, since the center of mass lies on the axis of symmetry, we know that $\bar{x} = 0$. Moreover, the mass of the lamina is given by

$$m = \rho \int_{-2}^{2} (4 - x^2)\, dx = \rho \left[4x - \frac{x^3}{3} \right]_{-2}^{2} = \frac{32\rho}{3}.$$

To find the moment about the x-axis, we place a representative rectangle in the region, as shown in Figure 6.67. The distance from the x-axis to the center of this rectangle is

$$y_i = \frac{f(x)}{2} = \frac{4 - x^2}{2}.$$

Since the mass of the representative rectangle is

$$\rho f(x) \Delta x = \rho(4 - x^2) \Delta x$$

we have

$$M_x = \rho \int_{-2}^{2} \frac{4 - x^2}{2} (4 - x^2)\, dx = \frac{\rho}{2} \int_{-2}^{2} (16 - 8x^2 + x^4)\, dx$$

$$= \frac{\rho}{2} \left[16x - \frac{8x^3}{3} + \frac{x^5}{5} \right]_{-2}^{2} = \frac{256\rho}{15}$$

and $\bar{y}$ is given by

$$\bar{y} = \frac{M_x}{m} = \frac{256\rho/15}{32\rho/3} = \frac{8}{5}.$$

Thus, the center of mass (the balancing point) of the lamina is $\left(0, \frac{8}{5}\right)$, as shown in Figure 6.68.   ▭

The density ρ in Example 4 is a common factor of both the moments and the mass, and as such cancels out of the quotients representing the coordinates of the center of mass. Thus, the center of mass of a lamina of *uniform* density depends only on the shape of the lamina and not on its density. For this reason, the point $(\bar{x}, \bar{y})$ is sometimes called the center of mass of a *region* in the plane, or the **centroid** of the region. In other words, to find the centroid of a region in the plane, we simply assume that the region has a constant density of $\rho = 1$, and compute the corresponding center of mass.

EXAMPLE 5   The centroid of a plane region

Find the centroid of the region bounded by the graphs of $f(x) = 4 - x^2$ and $g(x) = x + 2$.

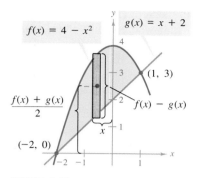

$f(x) = 4 - x^2$

$g(x) = x + 2$

$(1, 3)$

$\dfrac{f(x) + g(x)}{2}$

$f(x) - g(x)$

$(-2, 0)$

FIGURE 6.69

SOLUTION

The two graphs intersect at the points $(-2, 0)$ and $(1, 3)$, as shown in Figure 6.69. Thus, the area of the region is given by

$$A = \int_{-2}^{1} [f(x) - g(x)]\, dx = \int_{-2}^{1} (2 - x - x^2)\, dx = \frac{9}{2}.$$

The centroid $(\bar{x}, \bar{y})$ of the region has coordinates

$$\bar{x} = \frac{1}{A} \int_{-2}^{1} x[(4 - x^2) - (x + 2)]\, dx = \frac{2}{9} \int_{-2}^{1} (-x^3 - x^2 + 2x)\, dx$$

$$= \frac{2}{9} \left[-\frac{x^4}{4} - \frac{x^3}{3} + x^2 \right]_{-2}^{1} = -\frac{1}{2}$$

$$\bar{y} = \frac{1}{A} \int_{-2}^{1} \left[\frac{(4 - x^2) + (x + 2)}{2} \right] [(4 - x^2) - (x + 2)]\, dx$$

$$= \frac{2}{9}\left(\frac{1}{2}\right) \int_{-2}^{1} (-x^2 + x + 6)(-x^2 - x + 2)\, dx$$

$$= \frac{1}{9} \int_{-2}^{1} (x^4 - 9x^2 - 4x + 12)\, dx$$

$$= \frac{1}{9} \left[\frac{x^5}{5} - 3x^3 - 2x^2 + 12x \right]_{-2}^{1} = \frac{12}{5}.$$

Thus, the centroid of the region is $(\bar{x}, \bar{y}) = \left(-\frac{1}{2}, \frac{12}{5}\right)$.

For simple plane regions you may be able to find the centroid without resorting to integration. Example 6 presents such a case.

EXAMPLE 6 The centroid of a simple plane region

Find the centroid of the region shown in Figure 6.70(a).

SOLUTION

By superimposing a coordinate system on the region, as indicated in Figure 6.70(b), we locate the centroids of the three rectangles as

$$\left(\frac{1}{2}, \frac{3}{2}\right), \quad \left(\frac{5}{2}, \frac{1}{2}\right), \quad \text{and} \quad (5, 1).$$

Now, we can calculate the centroid as follows.

$$A = \text{area of region} = 3 + 3 + 4 = 10$$

$$\bar{x} = \frac{(1/2)(3) + (5/2)(3) + (5)(4)}{10} = \frac{29}{10} = 2.9$$

$$\bar{y} = \frac{(3/2)(3) + (1/2)(3) + (1)(4)}{10} = \frac{10}{10} = 1$$

Thus, the centroid of the region is $(2.9, 1)$.

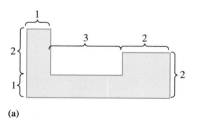

(a)

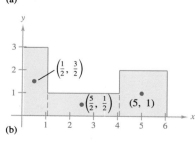

(b)

FIGURE 6.70

The final topic in this section is a useful theorem credited to Pappus of Alexandria (ca. 300 A.D.), a Greek mathematician whose eight-volume *Mathematical Collection* is a record of much of classical Greek mathematics. We delay the proof of this theorem until Section 16.4 (Exercise 43).

THEOREM 6.2
THE THEOREM OF PAPPUS

Let R be a region in a plane and let L be a line in the same plane such that L does not intersect the interior of R, as shown in Figure 6.71. If r is the distance between the centroid of R and the line, then the volume V of the solid of revolution formed by revolving R about the line is given by

$$V = 2\pi rA$$

where A is the area of R. (Note that $2\pi r$ is the distance traveled by the centroid as the region is revolved about the line.)

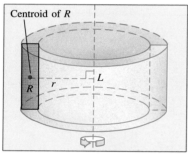

Centroid of R

Volume $= 2\pi r$(area of region R)

FIGURE 6.71

The Theorem of Pappus can be used to find the volume of a torus, as shown in the following example. Recall that a **torus** is a doughnut-shaped solid formed by revolving a circular region about a line that lies in the same plane as the circle (but does not intersect the circle).

EXAMPLE 7   Finding volume by the Theorem of Pappus

Find the volume of the torus formed by revolving the circular region bounded by $(x - 2)^2 + y^2 = 1$ about the y-axis, as shown in Figure 6.72(a).

SOLUTION

From Figure 6.72(b) we see that the centroid of the circular region is $(2, 0)$. Thus, the distance between the centroid and the axis of revolution is $r = 2$. Since the area of the circular region is $A = \pi$, the volume of the torus is given by

$$V = 2\pi rA = 2\pi(2)(\pi) = 4\pi^2 \approx 39.5.$$

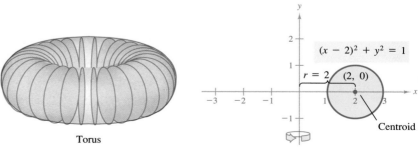

Torus

$(x - 2)^2 + y^2 = 1$

$r = 2$ $(2, 0)$

Centroid

FIGURE 6.72   (a)   (b)

EXERCISES for Section 6.7

In Exercises 1–4, find the center of the given point masses lying on the *x*-axis.

1. $m_1 = 6$, $m_2 = 3$, $m_3 = 5$
$x_1 = -5$, $x_2 = 1$, $x_3 = 3$
2. $m_1 = 7$, $m_2 = 4$, $m_3 = 3$, $m_4 = 8$
$x_1 = -3$, $x_2 = -2$, $x_3 = 5$, $x_4 = 6$
3. $m_1 = 1$, $m_2 = 1$, $m_3 = 1$, $m_4 = 1$, $m_5 = 1$
$x_1 = 7$, $x_2 = 8$, $x_3 = 12$, $x_4 = 15$, $x_5 = 18$
4. $m_1 = 12$, $m_2 = 1$, $m_3 = 6$, $m_4 = 3$, $m_5 = 11$
$x_1 = -3$, $x_2 = -2$, $x_3 = -1$, $x_4 = 0$, $x_5 = 4$

5. Notice in Exercise 3 that $\bar{x}$ is the arithmetic mean of the *x*-coordinates. Translate each point mass to the right 5 units and compare the resulting center of mass with that obtained in Exercise 3.
6. Translate each point mass in Exercise 4 to the left 3 units and compare the resulting center of mass with that obtained in Exercise 4.

In Exercises 7–10, find the center of mass of the given system of point masses.

7.

m_i	5	1	3
(x_i, y_i)	$(2, 2)$	$(-3, 1)$	$(1, -4)$

8.

m_i	10	2	5
(x_i, y_i)	$(1, -1)$	$(5, 5)$	$(-4, 0)$

9.

m_i	3	4
(x_i, y_i)	$(-2, -3)$	$(-1, 0)$

m_i	2	1	6
(x_i, y_i)	$(7, 1)$	$(0, 0)$	$(-3, 0)$

10.

m_i	4	2	5/2	5
(x_i, y_i)	$(2, 3)$	$(-1, 5)$	$(6, 8)$	$(2, -2)$

In Exercises 11–14, introduce an appropriate coordinate system and find the coordinates of the center of mass of the given planar lamina.

11.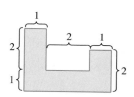

12.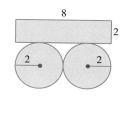

13.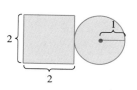

14.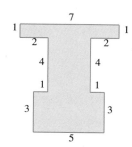

15. Suppose that the circular lamina in Exercise 11 has twice the density of the square lamina, and find the resulting center of mass.
16. Suppose that the square lamina in Exercise 11 has twice the density of the circular lamina, and find the resulting center of mass.

In Exercises 17–28, find M_x, M_y, and $(\bar{x}, \bar{y})$ for the laminas of uniform density ρ bounded by the graphs of the given equations.

17. $y = \sqrt{x}$, $y = 0$, $x = 4$
18. $y = x^2$, $y = 0$, $x = 4$
19. $y = \sqrt{x}$, $y = x$, $x \geq 0$
20. $y = x^2$, $y = x^3$
21. $y = -x^2 + 4x + 2$, $y = x + 2$
22. $y = \sqrt{3x} + 1$, $y = x + 1$
23. $x = 4 - y^2$, $x = 0$
24. $x = 2y - y^2$, $x = 0$
25. $x = -y$, $x = 2y - y^2$
26. $x = y + 2$, $x = y^2$
27. $y = x^{2/3}$, $y = 0$, $x = 8$
28. $y = x^{2/3}$, $y = 4$

In Exercises 29–34, find the centroid of the region bounded by the graphs of the given equations.

29. $y = \sqrt{1 - x^2}$, $y = 0$
30. $y = x^3$, $y = x$, $0 \le x \le 1$
31. $y = x^2$, $y = x$
32. $y = 8 - 2x$, $y = 0$, $0 \le x \le 4$
33. $y = 2x + 4$, $y = 0$, $0 \le x \le 3$
34. $y = x^2 - 4$, $y = 0$

35. Find the centroid of the triangular region with vertices $(-a, 0)$, $(a, 0)$, and (b, c), as shown in the figure. Show that it is the point of intersection of the medians of the triangle.

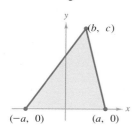

FIGURE FOR 35

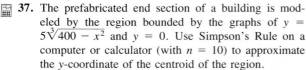

FIGURE FOR 36

36. Find the centroid of the **parabolic spandrel** shown in the figure.
37. The prefabricated end section of a building is modeled by the region bounded by the graphs of $y = 5\sqrt[3]{400 - x^2}$ and $y = 0$. Use Simpson's Rule on a computer or calculator (with $n = 10$) to approximate the y-coordinate of the centroid of the region.
38. Use Simpson's Rule on a computer or calculator (with $n = 8$) to approximate the y-coordinate of the centroid of the region bounded by the graphs $y = 8/(x^2 + 4)$,

$y = 0$, $x = -2$, and $x = 2$. The graph of the curve is called the **Witch of Agnesi,** and the exact y-coordinate of the centroid is $(\pi + 2)/2\pi$.

In Exercises 39–42, use the Theorem of Pappus to find the volume of the solid of revolution.

39. Torus formed by revolving the circle $(x - 5)^2 + y^2 = 16$ about the y-axis.
40. Torus formed by revolving the circle $x^2 + (y - 3)^2 = 4$ about the x-axis.
41. Solid formed by revolving the region bounded by the graphs of $y = x$, $y = 4$, and $x = 0$ about the x-axis.
42. Solid formed by revolving the region bounded by the graphs of $y = \sqrt{x - 1}$, $y = 0$, and $x = 5$ about the y-axis.

In Exercises 43 and 44, use the **Second Theorem of Pappus,** which is stated as follows. If a segment of a plane curve C is revolved about an axis that does not intersect the curve (except possibly at its endpoints), then the area S of the resulting surface of revolution is given by the product of the length of C times the distance d traveled by the centroid of C.

43. A sphere is formed by revolving the graph of $y = \sqrt{r^2 - x^2}$ about the x-axis. Use the formula for surface area, $S = 4\pi r^2$, to find the centroid of the semicircle $y = \sqrt{r^2 - x^2}$.
44. A torus is formed by revolving the graph of $(x - 1)^2 + y^2 = 1$ about the y-axis. Find the surface area of the torus.

REVIEW EXERCISES for Chapter 6

In Exercises 1–14, sketch the region bounded by the graphs of the given equations and determine the area of the region.

1. $y = \dfrac{1}{x^2}$, $y = 0$, $x = 1$, $x = 5$
2. $y = \dfrac{1}{x^2}$, $y = 4$, $x = 5$
3. $y = \dfrac{x}{(x^2 + 1)^2}$, $y = 0$, $x = 1$
4. $y = 1 - \dfrac{x}{2}$, $y = x - 2$, $y = 1$
5. $x = y^2 - 2y$, $x = 0$

6. $x = y^2 - 2y$, $x = -1$, $y = 0$
7. $y = x$, $y = x^3$
8. $x = y^2 + 1$, $x = y + 3$
9. $y = x^2 - 8x + 3$, $y = 3 + 8x - x^2$
10. $y = x^2 - 4x + 3$, $y = x^3$, $x = 0$
11. $y = \sqrt{x - 1}$, $y = 2$, $y = 0$, $x = 0$
12. $y = \sqrt{x - 1}$, $y = \dfrac{x - 1}{2}$
13. $\sqrt{x} + \sqrt{y} = 1$, $y = 0$, $x = 0$
14. $y = x^4 - 2x^2$, $y = 2x^2$

In Exercises 15–22, find the volume of the solid generated by revolving the plane region bounded by the given equations about the indicated line.

15. $y = x$, $y = 0$, $x = 4$
  (a) the x-axis   (b) the y-axis
  (c) the line $x = 4$ (d) the line $x = 6$
16. $y = \sqrt{x}$, $y = 2$, $x = 0$
  (a) the x-axis   (b) the line $y = 2$
  (c) the y-axis   (d) the line $x = -1$
17. $\dfrac{x^2}{16} + \dfrac{y^2}{9} = 1$
  (a) the y-axis (oblate spheroid)
  (b) the x-axis (prolate spheroid)
18. $\dfrac{x^2}{a^2} + \dfrac{y^2}{b^2} = 1$
  (a) the y-axis (oblate spheroid)
  (b) the x-axis (prolate spheroid)
19. $y = \dfrac{1}{(x^2 + 1)^2}$, $y = 0$, $x = 0$, $x = 1$,
  revolved about the y-axis
20. $y = (x + 1)^{3/2}$, $y = 0$, $x = -1$, $x = 1$,
  revolved about the x-axis
21. $y = -x^2 + 6x - 5$, $y = 0$
  (a) the x-axis   (b) the y-axis
22. $x^3 + 1$, $y = 2$, $x = 0$, revolved about the x-axis

23. Find the work done in stretching a spring from its natural length of 10 inches to a length of 15 inches, if a force of 4 pounds is needed to stretch it 1 inch from its natural position.
24. Find the work done in stretching a spring from its natural length of 9 inches to double that length, if a force of 50 pounds is required to hold the spring at double its natural length.
25. A water well has an 8-inch casing (diameter) and is 175 feet deep. If the water is 25 feet from the top of the well, determine the amount of work done in pumping it dry, assuming that no water enters the well while it is being pumped.
26. Repeat Exercise 25, assuming that water enters the well at the rate of 4 gallons per minute and the pump works at the rate of 12 gallons per minute. How many gallons are pumped in this case?
27. A chain 10 feet long weighs 5 pounds per foot and is suspended from a platform 20 feet above the ground. How much work is required to raise the entire chain to the 20-foot level?
28. A windlass, 200 feet above ground level on the top of a building, uses a cable weighing 4 pounds per foot. Find the work done in winding up the cable if
  (a) one end is at ground level.
  (b) there is a 300-pound load attached to the end of the cable.

29. A swimming pool is 5 feet deep at one end and 10 feet deep at the other, and the bottom is an inclined plane. The length and width of the pool are 40 feet and 20 feet, respectively. If the pool is full of water, what is the fluid force on each of the vertical walls?
30. Show that the fluid force against any vertical region in a liquid is the product of the weight per cubic volume of the liquid, the area of the region, and the depth of the centroid of the region.
31. Using the result of Exercise 30, find the fluid force on one side of a vertical circular plate of radius 4 feet that is submerged in water so that its center is 5 feet below the surface.
32. How much must the water level be raised to double the fluid force on one side of the plate in Exercise 31?

In Exercises 33–36, find the centroid of the region bounded by the graphs of the given equations.

33. $\sqrt{x} + \sqrt{y} = \sqrt{a}$, $x = 0$, $y = 0$
34. $y = x^2$, $y = 2x + 3$
35. $y = a^2 - x^2$, $y = 0$
36. $y = x^{2/3}$, $y = \dfrac{1}{2}x$

37. Find an integral for the arc length of the circle $x^2 + y^2 = 4$ from $(-\sqrt{3}, 1)$ clockwise to $(\sqrt{3}, 1)$.
38. Find the length of the graph of

$$y = \frac{1}{6}x^3 + \frac{1}{2x}$$

  from $x = 1$ to $x = 3$.
39. Use integration to find the lateral surface area of a right circular cone of height 4 and radius 3.
40. A gasoline tank is an oblate spheroid generated by revolving the region bounded by the graph of

$$\frac{x^2}{16} + \frac{y^2}{9} = 1$$

  about the y-axis, where x and y are measured in feet. Find the depth of the gasoline in the tank when it is filled to one-fourth its capacity.
41. Find the area of the region bounded by $y = x\sqrt{x + 1}$ and $y = 0$.
42. The region defined in Exercise 41 is revolved around the x-axis. Find the volume of the solid generated.
43. Find the volume of the solid generated by revolving the region defined in Exercise 41 about the y-axis.
44. The region bounded by $y = 2\sqrt{x}$, $y = 0$, and $x = 3$ is revolved around the x-axis. Find the surface area of the solid generated.
45. Find the arc length of the graph of $f(x) = \frac{4}{5}x^{5/4}$ from $x = 0$ to $x = 4$.

Chapter 7 Application

The Gateway Arch, St. Louis, was designed by Eero Saarinen and completed in 1965. It is over 60 stories high, is covered with over 900 tons of quarter-inch stainless steel, and has a total weight of over 16,000 tons. It is the tallest national monument of its kind—75 feet taller than the Washington Monument and 175 feet taller than the Statue of Liberty.

The Catenary

When a cable is suspended between two points, it forms a curve called a catenary. The equation for a catenary is of the form

$$y = a\left(\frac{e^{x/a} + e^{-x/a}}{2}\right).$$

Telephone cables and clotheslines take this shape. Such a free-hanging cable has all of its internal forces in equilibrium. Because of this, a catenary (turned upside down) makes a perfect arch—all of its thrust passes through the legs to the base of the arch.

The Gateway Arch in St. Louis, Missouri is one example of an arch built in the form of a catenary. Its equation is

$$y \approx 757.71 - 127.71\left(\frac{e^{x/127.71} + e^{-x/127.71}}{2}\right).$$

The height of the Gateway Arch is the same as the distance between its two legs—630 feet. To see this, note that $y \approx 630$ when $x = 0$, and $y \approx 0$ when $x = \pm 315$.

$$y \approx 757.71 - 127.71\left(\frac{e^{x/127.71} + e^{-x/127.71}}{2}\right)$$

Catenary

(0, 630)

(−315, 0)   (315, 0)

See Exercise 43, Section 7.1.

Chapter Overview

At this point in the text, we have discussed only one basic class of elementary functions—algebraic functions. This chapter introduces the second basic class of elementary functions—exponential and logarithmic functions. (In Chapter 8, we will look at the third basic class of elementary functions—trigonometric and inverse trigonometric functions.)

Section 7.1 introduces exponential functions of the form $f(x) = a^x$, where the *base a* is a positive real number ($a \neq 1$). The most important base for exponential functions is the number $e \approx 2.71828$. This number is used as the base for the **natural exponential function**

$$f(x) = e^x.$$

Differentiation and integration rules for this function are discussed in Section 7.2. (Section 7.1 also discusses compound interest and several other applications involving exponential functions.)

Section 7.3 discusses inverse functions, and this material is then used in Section 7.4 to develop properties of the **natural logarithmic function**

$$f^{-1}(x) = \ln x.$$

Differentiation and integration rules for logarithmic functions are discussed in Sections 7.5 and 7.6. Then, in Section 7.7, we show how the exponential function can be used to solve problems involving growth and decay.

In Section 7.8, we revisit a problem that was introduced in Chapter 2—determination of a limit involving an indeterminate form. To do this, we introduce a new technique called **L'Hôpital's Rule.**

7

Exponential and Logarithmic Functions

7.1 Exponential Functions

Exponential functions ▪ The natural number e ▪ Applications of exponential functions

You are familiar with the behavior of functions such as $f(x) = x^2$, $g(x) = \sqrt{x} = x^{1/2}$, and $h(x) = 1/x = x^{-1}$ that involve a variable raised to a constant power. We now look at a different type of function, called an **exponential function,** that involves a *constant raised to a variable power*. A simple example is

$$f(x) = 2^x.$$

We already know how to evaluate 2^x for *rational* values of x. For instance,

$$2^0 = 1, \quad 2^2 = 2 \cdot 2 = 4, \quad 2^{-1} = \frac{1}{2}, \quad \text{and} \quad 2^{1/2} = \sqrt{2} \approx 1.414214.$$

For *irrational* values of x, we can define 2^x by considering a sequence of rational numbers that approach x and taking a limit. A full discussion of this process would not be appropriate here, but the general idea is as follows. Suppose we want to define the number $2^{\sqrt{2}}$. Since $\sqrt{2} = 1.414214 \ldots$, we consider the following numbers (which are of the form 2^r, where r is rational).

$$2^1 = 2$$
$$2^{1.4} = 2.6390158 \ldots$$
$$2^{1.41} = 2.6573716 \ldots$$
$$2^{1.414} = 2.6647496 \ldots$$
$$2^{1.4142} = 2.6651190 \ldots$$
$$2^{1.41421} = 2.6651375 \ldots$$
$$2^{1.414214} = 2.6651449 \ldots$$

By choosing rational powers of 2 that are closer and closer to $\sqrt{2}$, we would approach the following limit.

$$2^{\sqrt{2}} = 2.6651441 \ldots$$

357

In practice, of course, we use a calculator or computer to approximate numbers such as $2^{\sqrt{2}}$.

In general, we can use any positive base $a \neq 1$ for exponential functions. Thus, the **exponential function** with base a is written as

$$f(x) = a^x.$$

Exponential functions, even with irrational values of x, obey all of the familiar properties of exponents, as given in the following theorem, which we list without proof.

THEOREM 7.1 **PROPERTIES OF EXPONENTS**	Let a and b be positive real numbers, and let x and y be any real numbers. Then the following properties are true.

1. $a^0 = 1$ 2. $a^x a^y = a^{x+y}$

3. $\dfrac{a^x}{a^y} = a^{x-y}$ 4. $(a^x)^y = a^{xy}$

5. $(ab)^x = a^x b^x$ 6. $\left(\dfrac{a}{b}\right)^x = \dfrac{a^x}{b^x}$

7. $a^{-x} = \dfrac{1}{a^x}$

EXAMPLE 1   Using properties of exponents

(a) $(2^2)(2^3) = 2^{2+3} = 2^5$

(b) $\dfrac{2^2}{2^3} = 2^{2-3} = 2^{-1} = \dfrac{1}{2}$

(c) $(3^x)^3 = 3^{3x}$

(d) $\left(\dfrac{1}{3}\right)^{-x} = (3^{-1})^{-x} = 3^x$ ▭

Example 2 shows the graphs of three different exponential functions.

EXAMPLE 2   Sketching the graph of an exponential function

Sketch the graphs of the following exponential functions.

$$f(x) = 2^x, \qquad g(x) = \left(\dfrac{1}{2}\right)^x = 2^{-x}, \qquad h(x) = 3^x$$

SOLUTION

Table 7.1 lists several values for these functions, and Figure 7.1 shows their graphs.

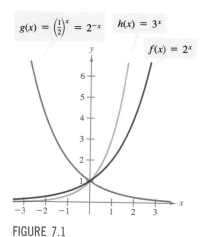

FIGURE 7.1

TABLE 7.1

x	-3	-2	-1	0	1	2	3	4
2^x	$\frac{1}{8}$	$\frac{1}{4}$	$\frac{1}{2}$	1	2	4	8	16
2^{-x}	8	4	2	1	$\frac{1}{2}$	$\frac{1}{4}$	$\frac{1}{8}$	$\frac{1}{16}$
3^x	$\frac{1}{27}$	$\frac{1}{9}$	$\frac{1}{3}$	1	3	9	27	81

The shapes of the graphs shown in Figure 7.1 are typical of the exponential functions a^x and a^{-x}, where $a > 1$, as shown in Figure 7.2. Moreover, a

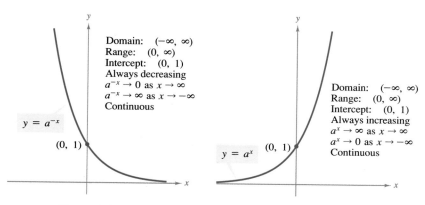

FIGURE 7.2

Characteristics of the Exponential Functions a^{-x} and a^x $(a > 1)$

summary of general characteristics of exponential functions is given in the following list.

PROPERTIES OF EXPONENTIAL FUNCTIONS	If $a > 1$ is a real number, then the exponential function $f(x) = a^x$ has the following properties.

1. The domain of $f(x) = a^x$ is $(-\infty, \infty)$, and the range is $(0, \infty)$.
2. The function $f(x) = a^x$ is continuous, increasing, and one-to-one on its entire domain.
3. The graph of $f(x) = a^x$ is concave upward on its entire domain.
4. The x-axis is a horizontal asymptote (to the left) of the graph of $f(x) = a^x$. Moreover,

$$\lim_{x \to -\infty} a^x = 0 \quad \text{and} \quad \lim_{x \to \infty} a^x = \infty.$$

REMARK   The exponential function $f(x) = a^{-x}$, where $a > 1$ has the same properties except that its graph is decreasing and the x-axis is a horizontal asymptote to the right. That is,

$$\lim_{x \to -\infty} a^{-x} = \infty \quad \text{and} \quad \lim_{x \to \infty} a^{-x} = 0.$$

EXAMPLE 3 Sketching the graph of an exponential function

Sketch the graph of

$$f(x) = 3^{-x} - 1.$$

SOLUTION

We begin by noting that the line $y = -1$ is a horizontal asymptote (to the right) of the graph of f. This is true because

$$\lim_{x \to \infty} (3^{-x} - 1) = \lim_{x \to \infty} 3^{-x} - \lim_{x \to \infty} 1$$

$$= \lim_{x \to \infty} \frac{1}{3^x} - \lim_{x \to \infty} 1$$

$$= 0 - 1$$

$$= -1.$$

Moreover, by evaluating $f(x)$ for several values of x, as shown in Table 7.2, and then plotting the corresponding points, we obtain the graph shown in Figure 7.3.

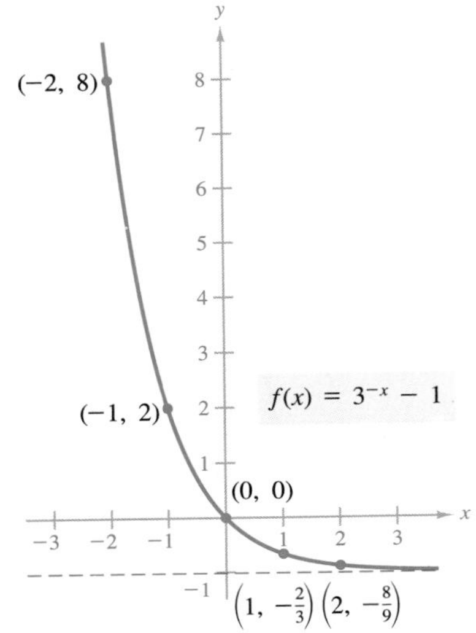

$(-2, 8)$

$(-1, 2)$ $f(x) = 3^{-x} - 1$

$(0, 0)$

$\left(1, -\frac{2}{3}\right)\left(2, -\frac{8}{9}\right)$

FIGURE 7.3

TABLE 7.2

x	-2	-1	0	1	2
$3^{-x} - 1$	8	2	0	$-\frac{2}{3}$	$-\frac{8}{9}$

The natural number *e*

We have introduced exponential functions using an unspecified base a. In calculus, the natural (or convenient) choice for a base is the irrational number e, whose decimal approximation is

$$e \approx 2.71828182846.$$

This choice may seem anything but natural. However, the convenience of this particular base will become apparent as we develop the rules for differentiating and integrating exponential functions. In the development of these rules, we will use the following definition of e.

DEFINITION OF THE NATURAL NUMBER *e*	The natural number e is defined by the following limit. $$e = \lim_{x \to 0} (1 + x)^{1/x}$$

At this point in the text, we are not in a position to prove that the limit in this definition exists. However, the values shown in Table 7.3 suggest that as x approaches 0, the values of $(1 + x)^{1/x}$ approach $e \approx 2.71828$. Figure 7.4 further reinforces this conclusion by showing the graph of $f(x) = (1 + x)^{1/x}$.

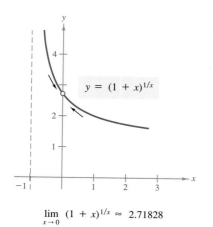

$$\lim_{x \to 0} (1 + x)^{1/x} \approx 2.71828$$

FIGURE 7.4

TABLE 7.3

x	$(1 + x)^{1/x}$
-0.1	2.8680
-0.01	2.7320
-0.001	2.7196
-0.0001	2.7184
0.0001	2.7181
0.001	2.7169
0.01	2.7048
0.1	2.5937

The graph of the natural exponential function

$$f(x) = e^x$$

is discussed in Example 4.

EXAMPLE 4   The graph of $f(x) = e^x$

Sketch the graph of $f(x) = e^x$.

SOLUTION

To begin, we evaluate $f(x) = e^x$ for several values of x, as shown in Table 7.4. Then, by plotting the points indicated in the table, we obtain the graph shown in Figure 7.5.

TABLE 7.4

x	-2	-1	0	1	2
e^x	$\dfrac{1}{e^2} \approx 0.135$	$\dfrac{1}{e} \approx 0.368$	1	$e \approx 2.718$	$e^2 \approx 7.389$

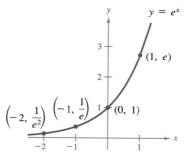

FIGURE 7.5

Applications of exponential functions

We conclude this section with two applications that should give you greater insight into the usefulness of the natural base e.

The first application deals with compound interest. Suppose P dollars are deposited in a savings account at an annual interest rate of r (where r is expressed in decimal form). If the accumulated interest is deposited into the account, what is the balance A in the account at the end of t years? The answer depends on the number of times the interest is compounded, according to the formula

$$A = P\left(1 + \frac{r}{n}\right)^{nt}$$

where n is the number of compoundings per year. For instance, the balances for a deposit of $1000 at 8 percent interest compounded n times a year for one year are shown in Table 7.5.

TABLE 7.5

1	(annually)	$1080.00
2	(semiannually)	$1081.60
4	(quarterly)	$1082.43
12	(monthly)	$1083.00
365	(daily)	$1083.28

It may surprise you to discover that as n increases, the balance A approaches a limit. In the following development, we let $x = r/n$. Then, $x \to 0$ as $n \to \infty$, and we obtain

$$A = \lim_{n \to \infty} P\left(1 + \frac{r}{n}\right)^n$$
$$= P\left[(1 + x)^{1/x}\right]^r$$
$$= Pe^r.$$

We call this limit the balance after one year of **continuous compounding.** Thus, for a deposit of $1000 at 8 percent interest, compounded continuously, the balance at the end of one year would be

$$A = 1000e^{0.08} \approx \$1083.29.$$

We summarize these results as follows.

SUMMARY OF COMPOUND INTEREST FORMULAS

Let P = amount of deposit, t = number of years, A = balance after t years, and r = annual interest rate (decimal form).

1. Compounded n times per year: $A = P\left(1 + \frac{r}{n}\right)^{nt}$

2. Compounded continuously: $A = Pe^{rt}$

EXAMPLE 5 Comparing continuous compounding to quarterly compounding

A deposit of $2500 is made in an account that pays an annual interest rate of 10 percent. Find the balance in the account at the end of 5 years if the interest is compounded (a) quarterly and (b) continuously.

SOLUTION

(a) The balance after quarterly compounding is

$$A = P\left(1 + \frac{r}{n}\right)^{nt} = 2500\left(1 + \frac{0.1}{4}\right)^{4(5)}$$

$$= 2500(1.025)^{20} = \$4096.54.$$

(b) The balance after continuous compounding is

$$A = Pe^{rt} = 2500[e^{0.1(5)}] = 2500e^{0.5} = \$4121.80.$$

EXAMPLE 6 An application to biology

A bacterial culture is growing according to the *logistics growth function*

$$y = \frac{1.25}{1 + 0.25e^{-0.4t}}, \quad 0 \le t$$

where y is the weight of the culture in grams and t is the time in hours. Find the weight of the culture after (a) 0 hours, (b) 1 hour, and (c) 10 hours. (d) What is the limit as t approaches infinity?

SOLUTION

(a) When $t = 0$,

$$y = \frac{1.25}{1 + 0.25e^0} = \frac{1.25}{1.25} = 1 \text{ g.}$$

(b) When $t = 1$,

$$y = \frac{1.25}{1 + 0.25e^{-0.4}} \approx 1.071 \text{ g.}$$

(c) When $t = 10$,

$$y = \frac{1.25}{1 + 0.25e^{-4}} \approx 1.244 \text{ g.}$$

(d) Finally, taking the limit as t approaches infinity, we have

$$\lim_{t \to \infty} \frac{1.25}{1 + 0.25e^{-0.4t}} = \frac{1.25}{1 + 0} = 1.25 \text{ g.}$$

(See Figure 7.6.)

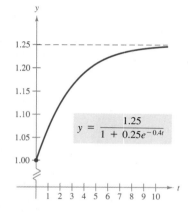

FIGURE 7.6

EXERCISES for Section 7.1

In Exercises 1 and 2, evaluate each expression.

1. (a) $25^{3/2}$ (b) $81^{1/2}$
    (c) 3^{-2} (d) $27^{-1/3}$

2. (a) $64^{1/3}$ (b) 5^{-4}

    (c) $\left(\dfrac{1}{8}\right)^{1/3}$ (d) $\left(\dfrac{1}{4}\right)^{3}$

In Exercises 3–6, use the properties of exponents to simplify each expression.

3. (a) $(5^2)(5^3)$ (b) $(5^2)(5^{-3})$

    (c) $\dfrac{5^3}{25^2}$ (d) $\left(\dfrac{1}{4}\right)^{2} 2^6$

4. (a) $(4^2)^3$ (b) $(6^4)^{1/2}$
    (c) $[(8^{-1})(8^{2/3})]^3$ (d) $(32^{3/2})(4^2)$

5. (a) $e^2(e^4)$ (b) $(e^3)^4$

    (c) $(e^3)^{-2}$ (d) $\dfrac{e^5}{e^3}$

6. (a) $\left(\dfrac{1}{e}\right)^{-2}$ (b) $\left(\dfrac{e^5}{e^2}\right)^{-1}$

    (c) e^0 (d) $\dfrac{1}{e^{-3}}$

In Exercises 7–16, solve for x.

7. $3^x = 81$ **8.** $5^{x+1} = 125$

9. $\left(\dfrac{1}{3}\right)^{x-1} = 27$ **10.** $\left(\dfrac{1}{5}\right)^{2x} = 625$

11. $4^3 = (x + 2)^3$ **12.** $18^2 = (5x - 7)^2$

13. $x^{3/4} = 8$ **14.** $(x + 3)^{4/3} = 16$

15. $e^{-2x} = e^5$ **16.** $e^x = 1$

In Exercises 17 and 18, compare the given number to e.

17. (a) $\dfrac{271,801}{99,990}$ (b) $\dfrac{299}{110}$

18. (a) $1 + 1 + \dfrac{1}{2} + \dfrac{1}{6} + \dfrac{1}{24}$

    (b) $1 + 1 + \dfrac{1}{2} + \dfrac{1}{6} + \dfrac{1}{24} + \dfrac{1}{120} + \dfrac{1}{720} + \dfrac{1}{5040}$

In Exercises 19–30, sketch the graph of the given function.

19. $y = 3^x$ **20.** $y = 3^{x-1}$

21. $y = \left(\dfrac{1}{3}\right)^{x}$ **22.** $y = 2^{x^2}$

23. $f(x) = 3^{-x^2}$ **24.** $f(x) = 3^{|x|}$

25. $h(x) = e^{x-2}$ **26.** $g(x) = -e^{x/2}$

27. $y = e^{-x^2}$ **28.** $y = e^{-x/2}$

29. $f(x) = \dfrac{2}{1 + e^{-x/4}}$ **30.** $h(x) = \dfrac{10}{1 + e^{-x}}$

In Exercises 31 and 32, find the amount of an investment of P dollars invested at r percent for t years if the interest is compounded (a) annually, (b) semiannually, (c) monthly, (d) daily, and (e) continuously.

31. $P = \$1000$, $r = 10\%$, $t = 10$ years
32. $P = \$2500$, $r = 12\%$, $t = 20$ years

In Exercises 33 and 34, find the investment that would be required at r percent compounded continuously to yield an amount of $\$100,000$ in (a) 1 year, (b) 10 years, (c) 20 years, and (d) 50 years.

33. $r = 12\%$ **34.** $r = 9\%$

35. The demand function for a certain product is given by

$$p = 5000\left(1 - \frac{4}{4 + e^{-0.002x}}\right) \quad \text{(see figure).}$$

Find the price of the product if the quantity demanded is (a) $x = 100$ units and (b) $x = 500$ units.

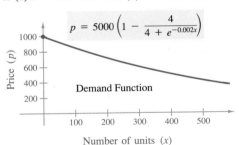

FIGURE FOR 35

36. The demand function for a certain product is given by

$$p = 500 - 0.5e^{0.004x}$$

(see figure). Find the price of the product if the quantity demanded is (a) $x = 1000$ units and (b) $x = 1500$ units.

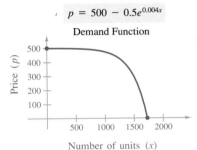

FIGURE FOR 36

37. The average time between incoming calls at a switchboard is 3 minutes. If a call has just come in, the probability that the next call will come within the next t minutes is given by

$$P(t) = 1 - e^{-t/3}.$$

Find (a) $P\left(\frac{1}{2}\right)$, (b) $P(2)$, and (c) $P(5)$.

38. A certain automobile gets 28 mi/gal at speeds of up to 50 mi/hr. At speeds of over 50 mi/hr, the number of miles per gallon drops at the rate of 12% for each 10 mi/hr. If s is the speed (in miles per hour) and y is the miles per gallon, then

$$y = 28e^{0.6-0.012s}, \quad s \geq 50.$$

Use this function to complete the following table.

Speed (s)	50	55	60	65	70
Miles per gallon (y)					

39. The population of a bacterial culture is given by the logistics growth function

$$y = \frac{850}{1 + e^{-0.2t}}$$

where y is the number of bacteria and t is the time in days.

(a) Find the limit of this function as t approaches infinity.

(b) Sketch the graph of this function.

40. The yield V (in millions of cubic feet per acre) for a forest at age t years is given by

$$V = 6.7e^{-48/t}.$$

(a) Find the volume per acre when $t = 20$ years and $t = 50$ years.

(b) Find the limiting volume of wood per acre as t approaches infinity.

(c) Sketch the graph of this function.

41. In a group project in learning theory, a mathematical model for the proportion P of correct responses after n trials was found to be

$$P = \frac{0.83}{1 + e^{-0.2n}}.$$

(a) Find the proportion of correct responses after $n = 10$ trials.

(b) Find the limiting proportion of correct responses as n approaches infinity.

42. In a typing class, the average number N of words per minute typed after t weeks of lessons was found to be

$$N = \frac{157}{1 + 5.4e^{-0.12t}}.$$

(a) Find the average number of words per minute after $t = 10$ weeks.

(b) Find the limiting number of words per minute as t approaches infinity.

43. In the Chapter 7 Application we introduced the following equation of the catenary for the Gateway Arch:

$$y \approx 757.71 - 127.71\left(\frac{e^{x/127.71} + e^{-x/127.71}}{2}\right).$$

Show that the height of the Gateway Arch is the same as the distance between its two legs.

44. Given the function

$$f(x) = \frac{2}{1 + e^{1/x}}$$

use a computer or graphics calculator to (a) sketch the graph of f, (b) find any horizontal asymptotes, and (c) find $\lim_{x \to 0} f(x)$ (if it exists).

7.2 Differentiation and Integration of Exponential Functions

Differentiation of exponential functions ■ Integration of exponential functions

In Section 7.1 we claimed that the natural base e is the most convenient base for exponential functions. One reason for this claim is that the natural exponential function $f(x) = e^x$ is its own derivative. To prove this, consider the following.

$$f'(x) = \lim_{\Delta x \to 0} \frac{f(x + \Delta x) - f(x)}{\Delta x}$$

$$= \lim_{\Delta x \to 0} \frac{e^{x + \Delta x} - e^x}{\Delta x}$$

$$= \lim_{\Delta x \to 0} \frac{e^x[e^{\Delta x} - 1]}{\Delta x}$$

Now, the definition of e,

$$e = \lim_{\Delta x \to 0} (1 + \Delta x)^{1/\Delta x}$$

tells us that for small values of Δx, we have $e \approx (1 + \Delta x)^{1/\Delta x}$, which implies that $e^{\Delta x} \approx 1 + \Delta x$. Replacing $e^{\Delta x}$ by this approximation produces the following.

$$\begin{aligned} f'(x) &= \lim_{\Delta x \to 0} \frac{e^x[e^{\Delta x} - 1]}{\Delta x} \\ &= \lim_{\Delta x \to 0} \frac{e^x[(1 + \Delta x) - 1]}{\Delta x} \\ &= \lim_{\Delta x \to 0} \frac{e^x(\Delta x)}{\Delta x} \\ &= e^x \end{aligned}$$

We summarize this result, along with its "Chain Rule version," in Theorem 7.2.

THEOREM 7.2
DERIVATIVE OF NATURAL
EXPONENTIAL FUNCTION

Let u be a differentiable function of x.

1. $\dfrac{d}{dx}[e^x] = e^x$ 2. $\dfrac{d}{dx}[e^u] = e^u \dfrac{du}{dx}$

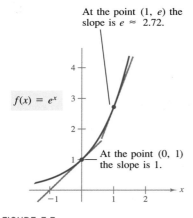

At the point $(1, e)$ the slope is $e \approx 2.72$.

$f(x) = e^x$

At the point $(0, 1)$ the slope is 1.

FIGURE 7.7

REMARK We can interpret this result geometrically by saying that the slope of the graph of $f(x) = e^x$ at any point (x, e^x) is equal to the y-coordinate of the point, as shown in Figure 7.7.

EXAMPLE 1 Differentiating exponential functions

(a) $\dfrac{d}{dx}[e^{2x-1}] = \dfrac{du}{dx} e^u = 2e^{2x-1}$ $u = 2x - 1$

(b) $\dfrac{d}{dx}[e^{-3/x}] = \dfrac{du}{dx} e^u = \left(\dfrac{3}{x^2}\right)e^{-3/x} = \dfrac{3e^{-3/x}}{x^2}$ $u = -\dfrac{3}{x}$ $\blacksquare$

EXAMPLE 2 Locating relative extrema

Find the relative extrema of $f(x) = xe^x$.

SOLUTION

The derivative of f is given by

$$\begin{aligned} f'(x) &= x(e^x) + e^x(1) \qquad \text{Product Rule} \\ &= e^x(x + 1). \end{aligned}$$

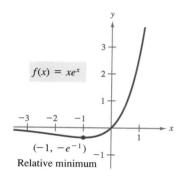

$f(x) = xe^x$

$(-1, -e^{-1})$
Relative minimum

FIGURE 7.8

Now, since e^x is never zero, the derivative is zero only when $x = -1$. Moreover, by the First Derivative Test, we can determine that this corresponds to a relative minimum, as shown in Figure 7.8. Since $f'(x) = e^x(x + 1)$ is defined for all x, there are no other critical points.　▭

EXAMPLE 3　The normal probability density function

Show that the graph of the *normal probability density function*

$$f(x) = \frac{1}{\sqrt{2\pi}} e^{-x^2/2}$$

has points of inflection when $x = \pm 1$.

SOLUTION

To locate possible points of inflection, we find the x-values for which the second derivative is zero.

$$f'(x) = \frac{1}{\sqrt{2\pi}}(-x)e^{-x^2/2}$$

$$f''(x) = \frac{1}{\sqrt{2\pi}}[(-x)(-x)e^{-x^2/2} + (-1)e^{-x^2/2}] \qquad \text{Product Rule}$$

$$= \frac{1}{\sqrt{2\pi}}(e^{-x^2/2})(x^2 - 1)$$

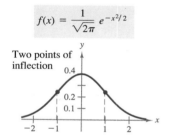

$f(x) = \dfrac{1}{\sqrt{2\pi}}\, e^{-x^2/2}$

Two points of inflection

Bell-Shaped Curve Given By Normal Probability Density Function

FIGURE 7.9

Therefore, $f''(x) = 0$ when $x = \pm 1$, and we can apply the techniques of Chapter 4 to conclude that these values yield the two points of inflection shown in Figure 7.9.　▭

REMARK　The general form of a normal probability density function is given by

$$f(x) = \frac{1}{\sigma\sqrt{2\pi}} e^{-x^2/2\sigma^2}$$

where σ is the standard deviation (σ is the lowercase Greek letter sigma). By following the procedure of Example 3, we can show that the bell-shaped curve of this function has points of inflection when $x = \pm\sigma$.

Integration of exponential functions

Each of the differentiation formulas for exponential functions has its corresponding integration formula, as shown next.

THEOREM 7.3 INTEGRATION RULES FOR EXPONENTIAL FUNCTIONS	Let u be a differentiable function of x. 1. $\displaystyle\int e^x \, dx = e^x + C$　　2. $\displaystyle\int e^u \, du = e^u + C$

EXAMPLE 4 Integrating exponential functions

Evaluate $\int e^{3x+1} \, dx$.

SOLUTION

Considering $u = 3x + 1$, we have $du = 3 \, dx$. Then

$$\int e^{3x+1} \, dx = \frac{1}{3} \int e^{3x+1}(3) \, dx = \frac{1}{3} \int e^u \, du$$

$$= \frac{1}{3} e^u + C$$

$$= \frac{e^{3x+1}}{3} + C.$$

REMARK   In Example 4 we introduced the missing *constant* factor 3 to create $du = 3 \, dx$. However, remember that you cannot introduce a missing *variable* factor in the integrand. For instance,

$$\int e^{-x^2} \, dx \neq \frac{1}{x} \int e^{-x^2}(x \, dx).$$

EXAMPLE 5   Integrating exponential functions

Evaluate $\int 5xe^{-x^2} \, dx$.

SOLUTION

If we let $u = -x^2$, then $du = -2x \, dx$, which implies that $x \, dx = -du/2$. Thus, we have

$$\int 5xe^{-x^2} \, dx = \int 5e^{-x^2}(x \, dx) = \int 5e^u \left(-\frac{du}{2} \right)$$

$$= -\frac{5}{2} \int e^u \, du$$

$$= -\frac{5}{2} e^u + C$$

$$= -\frac{5}{2} e^{-x^2} + C.$$

EXAMPLE 6   Integrating exponential functions

(a) $\displaystyle \int \frac{e^{1/x}}{x^2} \, dx = - \int \underbrace{e^{1/x}}_{e^u} \underbrace{\left(-\frac{1}{x^2} \right) \, dx}_{du} = -e^{1/x} + C \qquad u = \frac{1}{x}$

(b) $\displaystyle \int (1 + e^x)^2 \, dx = \int (1 + 2e^x + e^{2x}) \, dx = x + 2e^x + \frac{1}{2}e^{2x} + C$

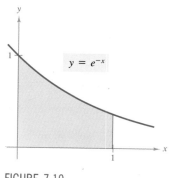

FIGURE 7.10

EXAMPLE 7 Finding the area of a region in the plane

Find the area of the region bounded by the graph of $f(x) = e^{-x}$ and the x-axis, for $0 \le x \le 1$.

SOLUTION

The region is shown in Figure 7.10, and its area is given by

$$\text{area} = \int_0^1 f(x)\, dx = \int_0^1 e^{-x}\, dx$$

$$= \left[-e^{-x}\right]_0^1$$

$$= -e^{-1} - (-1)$$

$$= 1 - \frac{1}{e} \approx 0.632.$$

EXERCISES for Section 7.2

In Exercises 1–6, find the slope of the tangent line to the given exponential function at the point $(0, 1)$.

1. $y = e^{3x}$

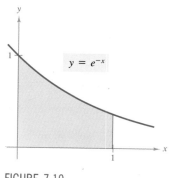

(0, 1)

2. $y = e^{2x}$

(0, 1)

3. $y = e^x$

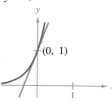

(0, 1)

4. $y = e^{-3x}$

(0, 1)

5. $y = e^{-2x}$

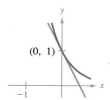

(0, 1)

6. $y = e^{-x}$

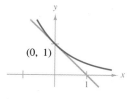

(0, 1)

In Exercises 7–24, find the derivative of the given function.

7. $y = e^{2x}$
8. $y = e^{1-x}$
9. $y = e^{-2x+x^2}$
10. $y = e^{-x^2}$
11. $f(x) = e^{1/x}$
12. $f(x) = e^{-1/x^2}$
13. $g(x) = e^{\sqrt{x}}$
14. $g(x) = e^{x^3}$
15. $f(x) = (x + 1)e^{3x}$
16. $y = x^2 e^{-x}$
17. $f(x) = \dfrac{e^{x^2}}{x}$
18. $f(x) = \dfrac{e^{x/2}}{\sqrt{x}}$
19. $y = (e^{-x} + e^x)^3$
20. $y = (1 - e^{-x})^2$
21. $f(x) = \dfrac{2}{e^x + e^{-x}}$
22. $f(x) = \dfrac{e^x - e^{-x}}{2}$
23. $y = xe^x - e^x$
24. $y = x^2 e^x - 2xe^x + 2e^x$

In Exercises 25 and 26, use implicit differentiation to find dy/dx.

25. $xe^y - 10x + 3y = 0$ **26.** $e^{xy} + x^2 - y^2 = 10$

In Exercises 27–30, find the second derivative of the exponential function.

27. $f(x) = 2e^{3x} + 3e^{-2x}$ **28.** $f(x) = 5e^{-x} - 2e^{-5x}$
29. $g(x) = (1 + 2x)e^{4x}$ **30.** $g(x) = (3 + 2x)e^{-3x}$

In Exercises 31–34, find the extrema and the points of inflection (if any exist) and sketch the graph of the function.

31. $f(x) = \dfrac{2}{1 + e^{-x}}$ **32.** $f(x) = \dfrac{e^x - e^{-x}}{2}$
33. $f(x) = x^2 e^{-x}$ **34.** $f(x) = xe^{-x}$

35. Find an equation of the line normal to the graph of $y = e^{-x}$ at $(0, 1)$.

36. Find the point on the graph of $y = e^{-x}$ where the normal line to the curve will pass through the origin.

37. Find the area of the largest rectangle that can be inscribed under the curve $y = e^{-x^2}$ in the first and second quadrants.

38. Find, to three decimal places, the value of x such that $e^{-x} = x$. [Use Newton's Method.]

39. The yield V (in millions of cubic feet per acre) for a forest stand at age t is given by

$$V = 6.7e^{(-48.1)/t}$$

where t is measured in years. Find the rate at which the yield is changing when $t = 20$ and $t = 60$ years.

40. The average typing speed (in the number of words per minute) after t weeks of lessons is given by

$$N = \frac{157}{1 + 5.4e^{-0.12t}}.$$

Find the rate at which typing speed is changing when $t = 5$ and $t = 25$ weeks.

41. In a group project in learning theory, a mathematical model for the proportion P of correct responses after n trials was found to be

$$P = \frac{0.83}{1 + e^{-0.2n}}.$$

Find the rate at which P is changing after $n = 3$ and $n = 10$ trials.

42. A lake is stocked with 500 fish, and their population increases according to the **logistics curve**

$$p(t) = \frac{10,000}{1 + 19e^{-t/5}}$$

where t is measured in months (see figure). At what rate is the fish population changing at the end of 1 month and at the end of 10 months? After how many months is the population increasing most rapidly?

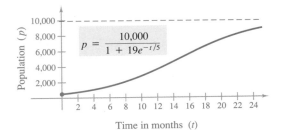

Time in months (t)

In Exercises 43–62, evaluate the integral.

43. $\displaystyle\int_0^1 e^{-2x}\,dx$

44. $\displaystyle\int_1^2 e^{1-x}\,dx$

45. $\displaystyle\int_0^2 (x^2 - 1)e^{x^3 - 3x + 1}\,dx$

46. $\displaystyle\int x^2 e^{x^3}\,dx$

47. $\displaystyle\int \frac{e^{-x}}{(1 + e^{-x})^2}\,dx$

48. $\displaystyle\int \frac{e^{2x}}{(1 + e^{2x})^2}\,dx$

49. $\displaystyle\int xe^{ax^2}\,dx$

50. $\displaystyle\int_0^{\sqrt{2}} xe^{-(x^2/2)}\,dx$

51. $\displaystyle\int_1^3 \frac{e^{3/x}}{x^2}\,dx$

52. $\displaystyle\int (e^x - e^{-x})^2\,dx$

53. $\displaystyle\int e^{-x}(1 + e^{-x})^2\,dx$

54. $\displaystyle\int e^{2x}(1 - 3e^{2x})^{-2}\,dx$

55. $\displaystyle\int e^x\sqrt{1 - e^x}\,dx$

56. $\displaystyle\int e^x(e^x - e^{-x})\,dx$

57. $\displaystyle\int \frac{e^x + e^{-x}}{\sqrt{e^x - e^{-x}}}\,dx$

58. $\displaystyle\int \frac{2e^x - 2e^{-x}}{(e^x + e^{-x})^2}\,dx$

59. $\displaystyle\int \frac{5 - e^x}{e^{2x}}\,dx$

60. $\displaystyle\int \frac{e^{2x} + 2e^x + 1}{e^x}\,dx$

61. $\displaystyle\int_{-2}^0 (3^3 - 5^2)\,dx$

62. $\displaystyle\int (3 - x)e^{(3-x)^2}\,dx$

In Exercises 63 and 64, find a function f that satisfies the given conditions.

63. $f''(x) = \dfrac{1}{2}(e^x + e^{-x})$
$f(0) = 1, f'(0) = 0$

64. $f''(x) = x + e^{-2x}$
$f(0) = 3, f'(0) = -\dfrac{1}{2}$

In Exercises 65–68, find the area of the region bounded by the graphs of the given equations.

65. $y = e^x,\ y = 0,\ x = 0,\ x = 5$
66. $y = e^{-x},\ y = 0,\ x = a,\ x = b$
67. $y = xe^{-(x^2/2)},\ y = 0,\ x = 0,\ x = \sqrt{2}$
68. $y = e^{-2x} + 2,\ y = 0,\ x = 0,\ x = 2$

In Exercises 69 and 70, find the volume of the solid generated by revolving the region bounded by the graphs of the given equations about the x-axis.

69. $y = e^x,\ y = 0,\ x = 0,\ x = 1$
70. $y = e^{-x/2},\ y = 0,\ x = 0,\ x = 4$

71. Given $e^x \geq 1$ for $x \geq 0$, it follows that

$$\int_0^x e^t\,dt \geq \int_0^x 1\,dt.$$

Perform this integration to derive the inequality $e^x \geq 1 + x$ for $x \geq 0$.

72. Integrate each term of the following inequalities in a manner similar to that of Exercise 71 to obtain each succeeding inequality for $x \geq 0$. Then evaluate both sides of each inequality when $x = 1$.

(a) $e^x \geq 1 + x$

(b) $e^x \geq 1 + x + \dfrac{x^2}{2}$

(c) $e^x \geq 1 + x + \dfrac{x^2}{2} + \dfrac{x^3}{6}$

(d) $e^x \geq 1 + x + \dfrac{x^2}{2} + \dfrac{x^3}{6} + \dfrac{x^4}{24}$

In Exercises 73 and 74, find the maximum possible error if the integral is approximated by (a) the Trapezoidal Rule and (b) Simpson's Rule.

73. $\displaystyle\int_0^1 e^{x^3}\, dx, \quad n = 4$

74. $\displaystyle\int_0^1 e^{-x^2}\, dx, \quad n = 4$

In Exercises 75 and 76, use the standard normal probability density function

$$f(z) = \frac{1}{\sqrt{2\pi}} e^{-z^2/2}.$$

The probability that z is in the interval $[a, b]$ is the area of the region bounded by $y = f(z)$, $y = 0$ on the interval $a \leq z \leq b$ and is denoted by $\Pr(a \leq z \leq b)$. Approximate the given probability. (Use Simpson's Rule with $n = 6$.)

75. $\Pr(0 \leq z \leq 1)$ **76.** $\Pr(0 \leq z \leq 2)$

77. Use Simpson's Rule on a computer (with $n = 10$) to approximate the volume of the solid generated when the region bounded by the graphs of

$$y = \frac{2}{1 + e^{1/x}},$$

$y = 0$, $x = 1$, and $x = 3$ is revolved about the y-axis.

7.3 Inverse Functions

Inverse functions ▪ Existence of an inverse function ▪ Derivative of an inverse function

When we introduced the notion of a composite function in Section 1.5, we noted that composition is not commutative. That is, it is not necessarily true that $f(g(x))$ and $g(f(x))$ are equal. We now look at a special case for which composition is commutative—when f and g are inverses of each other.

DEFINITION OF INVERSE FUNCTION

A function g is the **inverse** of the function f if

$$f(g(x)) = x \quad \text{for each } x \text{ in the domain of } g$$

and

$$g(f(x)) = x \quad \text{for each } x \text{ in the domain of } f.$$

We denote g by f^{-1} (read "f inverse").

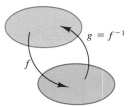

Domain of f = Range of g
Domain of g = Range of f

FIGURE 7.11

REMARK Although the notation used to denote an inverse function resembles *exponential notation*, it is a different use of -1 as a superscript. That is, in general, $f^{-1}(x) \neq 1/f(x)$.

Here are some important observations about this definition.

1. If g is the inverse of f, then f is also the inverse of g.
2. The domain of f^{-1} is equal to the range of f (and vice versa), as indicated in Figure 7.11.
3. A function need not possess an inverse, but if it does, the inverse is unique. (See Exercise 51.)

To understand the concept of an inverse function, it is helpful to think of f^{-1} as undoing what has been done by f. For example, subtraction can be used to undo addition, and division can be used to undo multiplication. Use the definition of an inverse function to check the following inverses.

1. $f(x) = x + c$ and $f^{-1}(x) = x - c$

2. $f(x) = cx$ and $f^{-1}(x) = \dfrac{x}{c}, \quad c \neq 0$

EXAMPLE 1   Verifying inverse functions

Show that the following functions are inverses of each other.

$$f(x) = 2x^3 - 1 \quad \text{and} \quad g(x) = \sqrt[3]{\frac{x + 1}{2}}$$

SOLUTION

First, note that both composite functions exist, since the domain and range of both f and g consist of the set of all real numbers. The composite of f with g is given by

$$f(g(x)) = 2\left(\sqrt[3]{\frac{x + 1}{2}}\right)^3 - 1$$

$$= 2\left(\frac{x + 1}{2}\right) - 1 = x + 1 - 1 = x.$$

The composite of g with f is given by

$$g(f(x)) = \sqrt[3]{\frac{(2x^3 - 1) + 1}{2}}$$

$$= \sqrt[3]{\frac{2x^3}{2}} = \sqrt[3]{x^3} = x.$$

Since $f(g(x)) = g(f(x)) = x$, we conclude that f and g are inverses of each other. (See Figure 7.12). ◻

In Figure 7.12 the graphs of f and f^{-1} appear to be mirror images of each other with respect to the line $y = x$. We say that the graph of f^{-1} is a **reflection** of the graph of f in the line $y = x$. This idea is generalized in the following theorem.

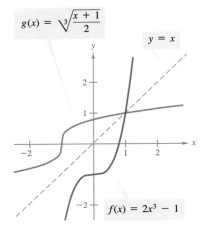

$g(x) = \sqrt[3]{\dfrac{x + 1}{2}}$

$y = x$

$f(x) = 2x^3 - 1$

FIGURE 7.12

THEOREM 7.4 **REFLECTIVE PROPERTY** **OF INVERSE FUNCTIONS**	The graph of f contains the point (a, b) if and only if the graph of f^{-1} contains the point (b, a).

PROOF   If (a, b) is on the graph of f, then $f(a) = b$ and we have

$$f^{-1}(b) = f^{-1}(f(a)) = a.$$

Thus, (b, a) is on the graph of f^{-1}, as shown in Figure 7.13. A similar argument will prove the theorem in the other direction.

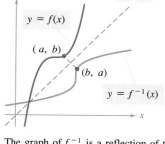

The graph of f^{-1} is a reflection of the graph of f in the line $y = x$.

FIGURE 7.13

Not every function has an inverse, and Theorem 7.4 suggests a graphical test for those that do. It is called the **horizontal line test** for an inverse function, and it follows directly from the vertical line test for functions together with the reflective property of the graphs of f and f^{-1}. The test states that a function f has an inverse if and only if every horizontal line intersects the graph of f at most once. The following theorem formally states why the horizontal line test is valid. (Recall from Section 4.3 that a function is *strictly monotonic* if it is either increasing on its entire domain or decreasing on its entire domain.)

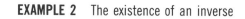

THEOREM 7.5
THE EXISTENCE OF AN
INVERSE FUNCTION

1. A function possesses an inverse if and only if it is one-to-one.
2. If f is strictly monotonic on its entire domain, then it is one-to-one and, hence, possesses an inverse.

PROOF   We leave the proof of the first part as an exercise (see Exercise 53). To prove the second part, recall from Section 1.5 that f is one-to-one if for x_1, x_2 in its domain

$$f(x_1) = f(x_2) \implies x_1 = x_2.$$

The *contrapositive* of this implication is logically equivalent and it states that

$$x_1 \neq x_2 \implies f(x_1) \neq f(x_2).$$

Now, choose x_1 and x_2 in the given interval. If $x_1 \neq x_2$, and since f is strictly monotonic, it follows that either $f(x_1) < f(x_2)$ or $f(x_1) > f(x_2)$. In either case, $f(x_1) \neq f(x_2)$. Thus, f is one-to-one on the interval.

EXAMPLE 2   The existence of an inverse

Determine which of the following functions has an inverse.

(a) $f(x) = x^3 + x - 1$ (b) $f(x) = x^3 - x + 1$

SOLUTION

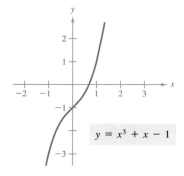

$y = x^3 + x - 1$

FIGURE 7.14

(a) From the graph of f given in Figure 7.14 it appears that f is increasing over its entire domain. To verify this, we note that the derivative, $f'(x) = 3x^2 + 1$, is positive for all real values of x. Therefore, f is strictly monotonic and it must have an inverse.

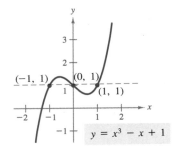

FIGURE 7.15

(b) From the graph given in Figure 7.15 we can see that the function does not pass the horizontal line test. In other words, it is not one-to-one. For instance, f has the same value when $x = -1$, 0, and 1.

$$f(-1) = f(1) = f(0) = 1 \qquad \text{Not one-to-one}$$

Therefore, by Theorem 7.5, f does not have an inverse. ▭

REMARK Often it is easier to prove that a function has an inverse than to find the inverse. For instance, it would be difficult algebraically to determine the inverse of the function in Example 2(a).

GUIDELINES FOR FINDING THE INVERSE OF A FUNCTION	1. Use Theorem 7.5 to determine whether the function given by $y = f(x)$ has an inverse. 2. Solve for x as a function of y: $x = g(y) = f^{-1}(y)$. 3. Define the domain of f^{-1} to be the range of f. 4. Verify that $f(f^{-1}(x)) = x$ and $f^{-1}(f(x)) = x$.

To avoid the confusion that could arise from using y as the independent variable for f^{-1}, it is customary to write f^{-1} as a *function of x* simply by interchanging the variables x and y after solving for x. This is illustrated in the next example.

EXAMPLE 3 Finding the inverse of a function

Find the inverse of the function given by $f(x) = \sqrt{2x - 3}$.

SOLUTION

By Theorem 7.5, this function has an inverse because it is increasing on its entire domain, as shown in Figure 7.16. To find an equation for this inverse, we let $y = f(x)$ and solve for x in terms of y.

$$\sqrt{2x - 3} = y \qquad \text{Let } y = f(x)$$
$$2x - 3 = y^2$$
$$x = \frac{y^2 + 3}{2} \qquad \text{Solve for } x$$
$$f^{-1}(y) = \frac{y^2 + 3}{2}$$

Since the range of f is $[0, \infty)$, we define this interval to be the domain of f^{-1}. Finally, using x as the independent variable, we have

$$f^{-1}(x) = \frac{x^2 + 3}{2}, \quad 0 \le x. \qquad \text{Determine domain}$$ ▭

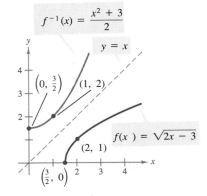

FIGURE 7.16

REMARK   Remember that any letter can be used to represent the independent variable. Thus,

$$f^{-1}(y) = \frac{y^2 + 3}{2}, \quad f^{-1}(x) = \frac{x^2 + 3}{2}, \quad \text{and} \quad f^{-1}(s) = \frac{s^2 + 3}{2}$$

all represent the same function.

Theorem 7.5 is useful in the following type of problem. Suppose you are given a function that is *not* one-to-one on its domain. By restricting the domain to an interval on which the function is strictly monotonic, you can conclude that the new function *is* one-to-one on the restricted domain. The next example illustrates this procedure.

EXAMPLE 4   Finding an interval on which a function is one-to-one

Show that the function $f(x) = x^3 - 3x$ is not one-to-one on the entire real line. Then show that $[-1, 1]$ is the largest interval, centered at the origin, for which f is strictly monotonic.

SOLUTION

It is clear that f is not one-to-one since different x-values yield the same y-value. For instance, $f(-\sqrt{3}) = f(0) = f(\sqrt{3}) = 0$. Moreover, f is decreasing on the open interval $(-1, 1)$ since its derivative

$$f'(x) = 3x^2 - 3$$

is negative there. Finally, since the left and right endpoints correspond to relative extrema of f, we can conclude that f is decreasing on the closed interval $[-1, 1]$ *and* that in any larger interval containing the origin the function would not be strictly monotonic. (See Figure 7.17.)   ▭

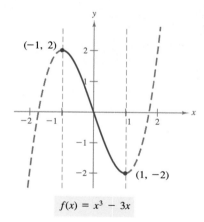

$$f(x) = x^3 - 3x$$

FIGURE 7.17

The next two theorems discuss the derivative of an inverse function. The reasonableness of Theorem 7.6 follows from the reflective property of inverse functions as shown in Figure 7.13, and a formal proof of the theorem is given in Appendix A.

THEOREM 7.6
CONTINUITY AND
DIFFERENTIABILITY OF
INVERSE FUNCTIONS

Let f be a function that possesses an inverse.

1. If f is continuous on its domain, then f^{-1} is continuous on its domain.
2. If f is increasing on its domain, then f^{-1} is increasing on its domain.
3. If f is decreasing on its domain, then f^{-1} is decreasing on its domain.
4. If f is differentiable at c and $f'(c) \neq 0$, then f^{-1} is differentiable at $f(c)$.

THEOREM 7.7 THE DERIVATIVE OF AN INVERSE FUNCTION	If f is differentiable on its domain and possesses an inverse function g, then the derivative of g is given by $$g'(x) = \frac{1}{f'(g(x))}, \quad f'(g(x)) \neq 0.$$

PROOF From Theorem 7.6 we know that g is differentiable. Using the Chain Rule we differentiate both sides of the equation $x = f(g(x))$ to obtain

$$1 = f'(g(x))\frac{d}{dx}[g(x)].$$

Since $f'(g(x)) \neq 0$, we can divide by this quantity to obtain

$$\frac{d}{dx}[g(x)] = \frac{1}{f'(g(x))}.$$

Geometrically, Theorem 7.7 tells us that the graphs of inverse functions have reciprocal slopes at the points (a, b) and (b, a), as illustrated in the next example. The equation

$$\frac{dy}{dx} = \frac{1}{dx/dy}$$

where $y = g(x)$, provides an easy way to remember this reciprocal relationship.

EXAMPLE 5 Graphs of inverse functions have reciprocal slopes

Let $f(x) = x^2$ (for $x \geq 0$) and let $f^{-1}(x) = \sqrt{x}$. Show that the slopes of the graphs of f and f^{-1} are reciprocals at the following points.

(a) $(2, 4)$ and $(4, 2)$ (b) $(3, 9)$ and $(9, 3)$

SOLUTION

The derivatives of f and f^{-1} are given by

$$f'(x) = 2x \quad \text{and} \quad (f^{-1})'(x) = \frac{1}{2\sqrt{x}}.$$

(a) At $(2, 4)$ the slope of the graph of f is $f'(2) = 2(2) = 4$. At $(4, 2)$, the slope of the graph of f^{-1} is

$$(f^{-1})'(4) = \frac{1}{2\sqrt{4}} = \frac{1}{2(2)} = \frac{1}{4}.$$

(b) At $(3, 9)$ the slope of the graph of f is $f'(3) = 2(3) = 6$. At $(9, 3)$, the slope of the graph of f^{-1} is

$$(f^{-1})'(9) = \frac{1}{2\sqrt{9}} = \frac{1}{2(3)} = \frac{1}{6}.$$

Thus, in both cases, the slopes are reciprocals, as shown in Figure 7.18.

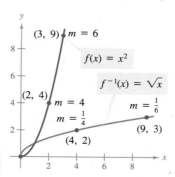

FIGURE 7.18

REMARK   In Example 5, note that f^{-1} is not differentiable at $(0, 0)$. This is consistent with Theorem 6.10 because the derivative of f at $(0, 0)$ is zero.

EXERCISES for Section 7.3

In Exercises 1–8, (a) show that f and g are inverse functions by showing that $f(g(x)) = x$ and $g(f(x)) = x$, and (b) graph f and g on the same set of coordinate axes.

1. $f(x) = x^3$ $g(x) = \sqrt[3]{x}$

2. $f(x) = x^{-1}$ $g(x) = x^{-1}$

3. $f(x) = 5x + 1$ $g(x) = \dfrac{1}{5}(x - 1)$

4. $f(x) = 3 - 4x$ $g(x) = \dfrac{1}{4}(3 - x)$

5. $f(x) = \sqrt{x - 4}$ $g(x) = x^2 + 4, \; x \geq 0$

6. $f(x) = 9 - x^2, \; x \geq 0$ $g(x) = \sqrt{9 - x}$

7. $f(x) = 1 - x^3$ $g(x) = \sqrt[3]{1 - x}$

8. $f(x) = x^{-2}, \; x > 0$ $g(x) = x^{-1/2}, \; x > 0$

In Exercises 9–22, find the inverse of f. Then graph both f and f^{-1}.

9. $f(x) = 2x - 3$ **10.** $f(x) = 3x$

11. $f(x) = x^5$ **12.** $f(x) = x^3 + 1$

13. $f(x) = \sqrt{x}$ **14.** $f(x) = x^2, \; x \geq 0$

15. $f(x) = \sqrt{4 - x^2}, \; 0 \leq x$

16. $f(x) = \sqrt{x^2 - 4}, \; x \geq 2$

17. $f(x) = \sqrt[3]{x - 1}$ **18.** $f(x) = 3\sqrt[5]{2x - 1}$

19. $f(x) = x^{2/3}, \; x \geq 0$ **20.** $f(x) = x^{3/5}$

21. $f(x) = \dfrac{x}{\sqrt{x^2 + 7}}$ **22.** $f(x) = \dfrac{x + 2}{x}$

In Exercises 23 and 24, use the graph of the function f to complete the table and sketch the graph of f^{-1}.

23.

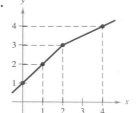

x	1	2	3	4
$f^{-1}(x)$				

24.

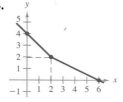

x	0	2	4
$f^{-1}(x)$			

In Exercises 25–28, use the functions

$$f(x) = \frac{1}{8}x - 3 \quad \text{and} \quad g(x) = x^3$$

to find the indicated value.

25. $(f^{-1} \circ g^{-1})(1)$ **26.** $(g^{-1} \circ f^{-1})(-3)$

27. $(f^{-1} \circ f^{-1})(6)$ **28.** $(g^{-1} \circ g^{-1})(-4)$

In Exercises 29–34, use the horizontal line test to determine whether the function is one-to-one on its entire domain and therefore has an inverse.

29. $f(x) = \dfrac{3}{4}x + 6$

30. $f(x) = 5x - 3$

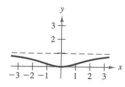

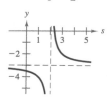

31. $f(x) = \dfrac{x^2}{x^2 + 4}$

32. $h(s) = \dfrac{1}{s - 2} - 3$

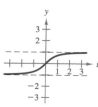

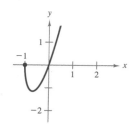

33. $g(t) = \dfrac{t}{\sqrt{t^2 + 1}}$

34. $f(x) = 3x\sqrt{x + 1}$

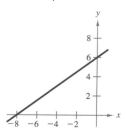

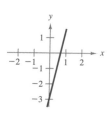

In Exercises 35–38, use the derivative to determine whether the given function is strictly monotonic on its entire domain and therefore has an inverse.

35. $f(x) = (x + a)^3 + b$

36. $f(x) = \dfrac{x^4}{4} - 2x^2$

37. $f(x) = x^3 - 6x^2 + 12x$

38. $f(x) = 2 - x - x^3$

In Exercises 39–42, show that f is strictly monotonic on the given interval and therefore has an inverse on that interval.

Function	Interval
39. $f(x) = (x - 4)^2$	$[4, \infty)$
40. $f(x) = |x + 2|$	$[-2, \infty)$
41. $f(x) = \dfrac{4}{x^2}$	$(0, \infty)$
42. $f(x) = x^3 - x$	$[1, \infty)$

In Exercises 43–46, show that the slopes of the graphs of f and f^{-1} are reciprocals at the given points.

Functions	Point
43. $f(x) = x^3$	$\left(\dfrac{1}{2}, \dfrac{1}{8}\right)$
$f^{-1}(x) = \sqrt[3]{x}$	$\left(\dfrac{1}{8}, \dfrac{1}{2}\right)$
44. $f(x) = 3 - 4x$	$(1, -1)$
$f^{-1}(x) = \dfrac{3 - x}{4}$	$(-1, 1)$
45. $f(x) = \sqrt{x - 4}$	$(5, 1)$
$f^{-1}(x) = x^2 + 4$	$(1, 5)$
46. $f(x) = \dfrac{1}{1 + x^2}$	$\left(1, \dfrac{1}{2}\right)$
$f^{-1}(x) = \sqrt{\dfrac{1 - x}{x}}$	$\left(\dfrac{1}{2}, 1\right)$

In Exercises 47 and 48, the derivative of the function has the same sign for all x in its domain, but the function is not strictly monotonic. Explain why.

47. $f(x) = \dfrac{1}{x}$

48. $f(x) = \dfrac{x}{x^2 - 4}$

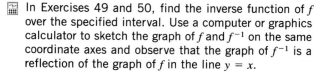 In Exercises 49 and 50, find the inverse function of f over the specified interval. Use a computer or graphics calculator to sketch the graph of f and f^{-1} on the same coordinate axes and observe that the graph of f^{-1} is a reflection of the graph of f in the line $y = x$.

Function	Interval
49. $f(x) = \dfrac{x}{x^2 - 4}$	$(-2, 2)$
50. $f(x) = 2 - \dfrac{3}{x^2}$	$(0, 10)$

51. Prove that if a function has an inverse, then the inverse is unique.

52. Prove that if f has an inverse, then $(f^{-1})^{-1} = f$.

53. Prove that a function has an inverse if and only if it is one-to-one.

54. Prove that if f and g are one-to-one functions, then $(f \circ g)^{-1}(x) = (g^{-1} \circ f^{-1})(x)$.

7.4 Logarithmic Functions

The natural logarithmic function ▪ Properties of the natural logarithmic function ▪ Logarithms to other bases

Because the natural exponential function $f(x) = e^x$ is continuous and increasing on the entire real line, it must possess an inverse function. We call this inverse function the **natural logarithmic function.** The domain of the natural logarithmic function is the set of positive real numbers.

DEFINITION OF THE NATURAL LOGARITHMIC FUNCTION	Let x be a positive real number. The **natural logarithmic function,** denoted by $\ln x$, is defined as follows. $\qquad \ln x = b \qquad$ if and only if $\qquad e^b = x$. ($\ln x$ is read as "el-en of x" or as the "natural log of x.")

This definition tells us that a logarithmic equation can be written in an equivalent exponential form, and vice versa. Here are a few examples.

Logarithmic form	*Exponential form*
$\ln 1 = 0$	$e^0 = 1$
$\ln e = 1$	$e^1 = e$
$\ln e^{-1} = -1$	$e^{-1} = \dfrac{1}{e}$

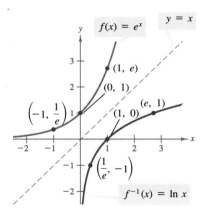

FIGURE 7.19

Because the function $f^{-1}(x) = \ln x$ is defined to be the inverse of $f(x) = e^x$, it follows that the graph of the natural logarithmic function is a reflection of the graph of the natural exponential function, as shown in Figure 7.19. Several other properties of the natural logarithmic function also follow directly from its definition as the inverse of the natural exponential function, and these are summarized in the following list.

PROPERTIES OF THE NATURAL LOGARITHMIC FUNCTION	The natural logarithmic function has the following properties. 1. The domain of $f(x) = \ln x$ is $(0, \infty)$, and the range is $(-\infty, \infty)$. 2. The function $f(x) = \ln x$ is continuous, increasing, and one-to-one on its entire domain. 3. The graph of $f(x) = \ln x$ is concave downward on its entire domain. 4. The y-axis is a vertical asymptote of the graph of $f(x) = \ln x$. Moreover, $\qquad\displaystyle \lim_{x \to 0^+} \ln x = -\infty \qquad$ and $\qquad \lim_{x \to \infty} \ln x = \infty$.

Recall from Section 7.3 that inverse functions possess the property that

$$f(f^{-1}(x)) = x \qquad \text{and} \qquad f^{-1}(f(x)) = x.$$

Because $f(x) = e^x$ and $f^{-1}(x) = \ln x$ are inverses of each other, we can conclude that

$$\ln e^x = x \quad \text{and} \quad e^{\ln x} = x.$$

These two properties are demonstrated in Example 1.

EXAMPLE 1   Applying the inverse properties of ln x and e^x

Simplify the following expressions.

(a) $\ln e^{2x}$

(b) $e^{\ln 3x}$

SOLUTION

(a) Because the natural logarithmic function is the inverse of the natural exponential function, we have

$$\ln e^{2x} = 2x.$$

(b) By the same reasoning, we have

$$e^{\ln 3x} = 3x.$$

In Section 7.1, we saw that to multiply two exponential expressions (having the same base) we add their exponents. For instance,

$$e^x e^y = e^{x+y}.$$

The logarithmic version of this property states that the natural logarithm of the product of two numbers is equal to the sum of the natural logs of the numbers. That is,

$$\ln xy = \ln x + \ln y.$$

This property and the properties dealing with the natural log of a quotient and the natural log of a power are given in the following theorem.

THEOREM 7.8
PROPERTIES OF LOGARITHMS

1. $\ln xy = \ln x + \ln y$

2. $\ln \dfrac{x}{y} = \ln x - \ln y$

3. $\ln x^y = y \ln x$

PROOF   We prove the first property and leave the proofs of the other two properties as exercises (see Exercises 71 and 72). Let

$$u = \ln x \quad \text{and} \quad v = \ln y.$$

The corresponding exponential forms of these two equations are $e^u = x$ and $e^v = y$. Multiplying x and y produces $xy = e^u e^v = e^{u+v}$, and the corresponding logarithmic form of $xy = e^{u+v}$ is

$$\ln xy = u + v.$$

Hence, $\ln xy = \ln x + \ln y$.

EXAMPLE 2   Expanding logarithmic expressions

(a) $\ln \dfrac{10}{9} = \ln 10 - \ln 9$ Property 2

(b) $\ln \sqrt{3x + 2} = \ln (3x + 2)^{1/2} = \dfrac{1}{2} \ln (3x + 2)$ Property 3

(c) $\ln \dfrac{xy}{5} = \ln (xy) - \ln 5 = \ln x + \ln y - \ln 5$ Properties 2 and 1

(d) $\ln \dfrac{(x - 3)^2}{x\sqrt[3]{x - 1}} = \ln (x - 3)^2 - \ln (x\sqrt[3]{x - 1})$

$$= 2 \ln (x - 3) - [\ln x + \ln (x - 1)^{1/3}]$$
$$= 2 \ln (x - 3) - \ln x - \ln (x - 1)^{1/3}$$
$$= 2 \ln (x - 3) - \ln x - \frac{1}{3} \ln (x - 1)$$

EXAMPLE 3   Condensing logarithmic expressions

(a) $\ln x + 2 \ln y = \ln x + \ln y^2 = \ln xy^2$

(b) $\ln (x + 1) - \dfrac{1}{2} \ln x - \ln (x^2 - 1) = \ln (x + 1) - [\ln \sqrt{x} + \ln (x^2 - 1)]$

$$= \ln (x + 1) - \ln [\sqrt{x}(x^2 - 1)]$$
$$= \ln \left[\frac{x + 1}{\sqrt{x}(x^2 - 1)} \right]$$
$$= \ln \left[\frac{1}{\sqrt{x}(x - 1)} \right]$$

The inverse properties of exponents and logarithms can be used to solve equations involving exponential or logarithmic expressions. This is demonstrated in Example 4.

EXAMPLE 4   Solving exponential and logarithmic equations

Solve for x in each of the following equations.

(a) $\ln (3x - 4) = 3$ (b) $y = e^{2x-5}$

SOLUTION

(a) $\ln (3x - 4) = 3$

$\quad e^{\ln (3x-4)} = e^3$

$\quad\quad 3x - 4 = e^3$

$\quad\quad\quad 3x = e^3 + 4$

$\quad\quad\quad\quad x = \dfrac{1}{3}(e^3 + 4) \approx 8.0285$

(b) $\quad\quad\quad y = e^{2x-5}$

$\quad\quad \ln y = \ln e^{2x-5}$

$\quad\quad \ln y = 2x - 5$

$\quad 5 + \ln y = 2x$

$\dfrac{1}{2}(5 + \ln y) = x$

$\blacksquare$

EXAMPLE 5   An application to compound interest

If P dollars are deposited at 8 percent interest compounded continuously, how long will it take for the original deposit to double?

SOLUTION

To represent a doubling of the deposit we write

$\quad Pe^{0.08t} = 2P.$

Thus, we have

$\quad e^{0.08t} = 2$

$\quad 0.08t = \ln 2$

$\quad\quad t = \dfrac{\ln 2}{0.08} \approx 8.66.$

Therefore, the balance will double by the end of 8 years and 8 months.

$\blacksquare$

Logarithms to other bases

Just as the natural logarithmic function was defined as the inverse of the natural exponential function, we can define the logarithmic function to any positive base $a \neq 1$ to be the inverse of the exponential function $f(x) = a^x$. That is,

$$y = \log_a x \quad \text{if and only if} \quad a^y = x.$$

If $a = e$, then the function given by $\log_e x = \ln x$ is simply the natural logarithmic function. Logarithmic functions to bases other than e share many of the properties of the natural logarithmic function. For instance, the following properties are true.

$$\log_a xy = \log_a x + \log_a y$$

$$\log_a \frac{x}{y} = \log_a x - \log_a y$$

$$\log_a x^y = y \log_a x$$

If $a = 10$, then the function given by $\log_{10} x$ is called the **common logarithmic function.** Common logarithms were first introduced by the Scottish mathematician John Napier (1550–1617). At that time, logarithms were used primarily as computational aids. With the creation of modern calculating devices, this use of logarithmic functions has disappeared.

John Napier

EXAMPLE 6 Properties of logarithms

(a) $\log_{10} 10^3 = 3$

(b) $\log_2 1 = 0$

(c) $\log_{10} x^2 y = 2 \log_{10} x + \log_{10} y$

(d) $\log_2 \dfrac{1}{x} = \log_2 x^{-1} = -\log_2 x$ ▭

EXAMPLE 7 Sketching the graph of a logarithmic function

Sketch the graph of the function given by $f(x) = 5 \log_{10} x$.

SOLUTION

We begin by noting that the domain of this function is the set of positive real numbers. Moreover, since

$$\lim_{x \to 0^+} f(x) = \lim_{x \to 0^+} 5 \log_{10} x = -\infty$$

it follows that the y-axis is a vertical asymptote of the graph of f. Finally, by plotting the points shown in Table 7.6, we obtain the graph shown in Figure 7.20.

FIGURE 7.20

TABLE 7.6

x	0.1	1	5	10
$5 \log_{10} x$	-5	0	3.495	5

▭

Most calculators have both natural and common logarithmic keys (often denoted by ln x and log x). Suppose, however, that you are asked to evaluate a logarithm to a base other than e or 10. In such cases, you can use the following change-of-base formula.

$$\log_a x = \frac{\log_b x}{\log_b a}$$ Change-of-base formula

With this formula, we can use a calculator to evaluate expressions such as $\log_2 14$. That is,

$$\log_2 14 = \frac{\ln 14}{\ln 2} \approx \frac{2.63906}{0.693147} \approx 3.8074.$$

EXERCISES for Section 7.4

In Exercises 1–6, write the logarithmic equation as an exponential equation and vice versa.

1. (a) $2^3 = 8$ (b) $3^{-1} = \dfrac{1}{3}$

2. (a) $27^{2/3} = 9$ (b) $16^{3/4} = 8$

3. (a) $\log_{10} 0.01 = -2$ (b) $\log_{0.5} 8 = -3$

4. (a) $e^0 = 1$ (b) $e^2 = 7.389\ldots$

5. (a) $\ln 2 = 0.6931\ldots$ (b) $\ln 8.4 = 2.128\ldots$

6. (a) $\ln 0.5 = -0.6931\ldots$
  (b) $49^{1/2} = 7$

In Exercises 7–14, solve for x (or b).

7. (a) $\log_{10} 1000 = x$ (b) $\log_{10} 0.1 = x$

8. (a) $\log_4 \dfrac{1}{64} = x$ (b) $\log_5 25 = x$

9. (a) $\log_3 x = -1$ (b) $\log_2 x = -4$

10. (a) $\log_b 27 = 3$ (b) $\log_b 125 = 3$

11. (a) $\log_{27} x = -\dfrac{2}{3}$ (b) $\ln e^x = 3$

12. (a) $e^{\ln x} = 4$ (b) $\ln x = 2$

13. (a) $x^2 - x = \log_5 25$
  (b) $3x + 5 = \log_2 64$

14. (a) $\log_3 x + \log_3 (x - 2) = 1$
  (b) $\log_{10} (x + 3) - \log_{10} x = 1$

In Exercises 15–20, sketch the graph of the function.

15. $f(x) = 3 \ln x$ **16.** $f(x) = -2 \ln x$

17. $f(x) = \ln 2x$ **18.** $f(x) = \ln |x|$

19. $f(x) = \ln (x - 1)$ **20.** $f(x) = 2 + \ln x$

In Exercises 21–24, show that the given functions are inverses of each other by sketching their graphs on the same coordinate axes.

21. $f(x) = e^{2x}$, $g(x) = \ln \sqrt{x}$

22. $f(x) = e^x - 1$, $g(x) = \ln (x + 1)$

23. $f(x) = e^{x-1}$, $g(x) = 1 + \ln x$

24. $f(x) = e^{x/3}$, $g(x) = \ln x^3$

In Exercises 25–30, apply the inverse properties of $\ln x$ and e^x to simplify the given expression.

25. $\ln e^{x^2}$ **26.** $\ln e^{2x-1}$

27. $e^{\ln (5x+2)}$ **28.** $-1 + \ln e^{2x}$

29. $e^{\ln \sqrt{x}}$ **30.** $-8 + e^{\ln x^3}$

In Exercises 31 and 32, use the properties of logarithms and the fact that $\ln 2 \approx 0.6931$ and $\ln 3 \approx 1.0986$ to approximate the given logarithm.

31. (a) $\ln 6$ (b) $\ln \dfrac{2}{3}$

  (c) $\ln 81$ (d) $\ln \sqrt{3}$

32. (a) $\ln 0.25$ (b) $\ln 24$

  (c) $\ln \sqrt[3]{12}$ (d) $\ln \dfrac{1}{72}$

In Exercises 33–42, use the properties of logarithms to write each as a sum, difference, or multiple of logarithms.

33. $\ln \dfrac{2}{3}$ **34.** $\ln (xyz)$

35. $\ln \dfrac{xy}{z}$ **36.** $\ln \sqrt{a - 1}$

37. $\ln \sqrt{2^3}$ **38.** $\ln \dfrac{1}{5}$

39. $\ln \left(\dfrac{x^2 - 1}{x^3}\right)^3$ **40.** $\ln 3e^2$

41. $\ln z(z - 1)^2$ **42.** $\ln \dfrac{1}{e}$

In Exercises 43–48, write each expression as a logarithm of a single quantity.

43. $\ln (x - 2) - \ln (x + 2)$

44. $3 \ln x + 2 \ln y - 4 \ln z$ $\ln x^3 + \ln y^2 + \ln z^{-4}$

45. $\dfrac{1}{3}[2 \ln (x + 3) + \ln x - \ln (x^2 - 1)]$

46. $2[\ln x - \ln (x + 1) - \ln (x - 1)]$

47. $2 \ln 3 - \dfrac{1}{2} \ln (x^2 + 1)$

48. $\dfrac{3}{2}[\ln (x^2 + 1) - \ln (x + 1) - \ln (x - 1)]$

In Exercises 49–60, solve for x or t.

49. $e^{\ln x} = 4$

50. $e^{\ln x^2} - 9 = 0$

51. $\ln x = 0$

52. $2 \ln x = 4$

53. $e^{x+1} = 4$

54. $e^{-0.5x} = 0.075$

55. $500e^{-0.11t} = 600$

56. $e^{-0.0174t} = 0.5$

57. $5^{2x} = 15$

58. $2^{1-x} = 6$

59. $500(1.07)^t = 1000$

60. $1000\left(1 + \dfrac{0.07}{12}\right)^{12t} = 3000$

61. A deposit of $1000 is made into a fund with an annual interest rate of 11%. Find the time for the investment to double if the interest is compounded
   (a) annually. (b) monthly.
   (c) daily. (d) continuously.

62. A deposit of $1000 is made into a fund with an annual interest rate of $10\frac{1}{2}$%. Find the time for the investment to triple if the interest is compounded
   (a) annually. (b) monthly.
   (c) daily. (d) continuously.

63. Complete the following table for the time t necessary for P dollars to triple if interest is compounded continuously at the rate r.

r	2%	4%	6%	8%	10%	12%
t						

64. The demand function for a certain product is given by

$$p = 500 - 0.5e^{0.004x}.$$

Find the quantity x demanded for a price of (a) $p = \$350$ and (b) $p = \$300$.

65. Use a calculator to demonstrate that

$$\frac{\ln x}{\ln y} \neq \ln \frac{x}{y} = \ln x - \ln y$$

by completing the following table.

x	y	$\dfrac{\ln x}{\ln y}$	$\ln \dfrac{x}{y}$	$\ln x - \ln y$
1	2			
3	4			
10	5			
4	0.5			

66. There are 25 prime numbers less than 100. The **Prime Number Theorem** states that if $p(x)$ is the number of primes less than x, then the ratio of $p(x)$ to $x/\ln x$ approaches 1 as x approaches infinity. Compute $x/\ln x$ for $x = 1000$, $x = 1{,}000{,}000$, and $x = 1{,}000{,}000{,}000$. Then compute the ratio of $p(x)$ to $x/\ln x$ given that $p(1000) = 168$, $p(10^6) = 78{,}498$, and $p(10^9) = 50{,}847{,}478$.

In Exercises 67 and 68, show that $f = g$ by using a computer or graphics calculator to sketch the graph of f and g on the same coordinate axes. (Assume $x > 0$.)

67. $f(x) = \ln \dfrac{x^2}{4}$

$g(x) = 2 \ln x - \ln 4$

68. $f(x) = \ln \sqrt{x(x^2 + 1)}$

$g(x) = \dfrac{1}{2}[\ln x + \ln (x^2 + 1)]$

In Exercises 69 and 70, evaluate the logarithm using the change of base formula. Do each problem twice. The first time use common logarithms, and the second time use natural logarithms. Round your answer to three decimal places.

69. (a) $\log_3 7$ (b) $\log_7 4$
   (c) $\log_{1/2} 10$ (d) $\log_4 0.55$

70. (a) $\log_9 0.4$ (b) $\log_2 0.125$
   (c) $\log_{15} 1250$ (d) $\log_{1/3} 0.015$

71. Prove that

$$\ln \frac{x}{y} = \ln x - \ln y.$$

72. Prove that $\ln x^y = y \ln x$.

7.5 Logarithmic Functions and Differentiation

Differentiation of the natural logarithmic function ▪ Logarithmic differentiation ▪ Bases other than *e*

The derivative of the natural logarithmic function is given in the following theorem.

THEOREM 7.9 **DERIVATIVE OF THE NATURAL** **LOGARITHMIC FUNCTION**	Let u be a differentiable function of x. $$\frac{d}{dx}[\ln x] = \frac{1}{x}, \quad x > 0 \qquad \frac{d}{dx}[\ln u] = \frac{1}{u}\frac{du}{dx} = \frac{u'}{u}, \quad u > 0$$

PROOF   We can prove the first part of the theorem by using the fact that the natural logarithmic function is the inverse of the natural exponential function $f(x) = e^x$. Using Theorem 7.7, with $f^{-1}(x) = \ln x$ and $f'(x) = e^x$, we have

$$\frac{d}{dx}[\ln x] = \frac{1}{f'(f^{-1}(x))} = \frac{1}{f'(\ln x)} = \frac{1}{e^{\ln x}} = \frac{1}{x}.$$

The second part of the theorem is simply the Chain Rule version of the first part.

So far in our development of the natural logarithmic function, it would have been difficult to predict its intimate relationship to the rational function $1/x$. Hidden relationships such as this not only illustrate the joy of mathematical discovery, they also give us logical alternatives in constructing a mathematical system. An alternative that Theorem 7.9 provides is that we could have developed the natural logarithmic function as the *antiderivative* of $1/x$, rather than as the inverse of e^x. If you are interested in pursuing this alternate development of $\ln x$, we suggest that you consult other calculus texts in your school's library.

EXAMPLE 1   Differentiation of logarithmic functions

(a) $\dfrac{d}{dx}[\ln (2x)] = \dfrac{u'}{u} = \dfrac{2}{2x} = \dfrac{1}{x}$ $u = 2x$

(b) $\dfrac{d}{dx}[\ln (x^2 + 1)] = \dfrac{u'}{u} = \dfrac{2x}{x^2 + 1}$ $u = x^2 + 1$

(c) $\dfrac{d}{dx}[x \ln x] = x\left(\dfrac{d}{dx}[\ln x]\right) + (\ln x)\left(\dfrac{d}{dx}[x]\right)$ Product Rule

$\qquad\qquad = x\left(\dfrac{1}{x}\right) + (\ln x)(1) = 1 + \ln x$ ▭

The properties of logarithms can be used to simplify the work involved in differentiating complicated logarithmic functions. This is demonstrated in Examples 2–4.

EXAMPLE 2   Logarithmic properties as an aid to differentiation

Differentiate

$$f(x) = \ln \sqrt{x + 1}.$$

SOLUTION

Since

$$f(x) = \ln \sqrt{x + 1} = \ln (x + 1)^{1/2} = \frac{1}{2} \ln (x + 1)$$

we have

$$f'(x) = \frac{1}{2}\left(\frac{1}{x + 1}\right) = \frac{1}{2(x + 1)}.$$

EXAMPLE 3   Logarithmic properties as an aid to differentiation

Differentiate

$$f(x) = \ln [x\sqrt{1 - x^2}].$$

SOLUTION

Since

$$f(x) = \ln [x\sqrt{1 - x^2}] = \ln x + \ln (1 - x^2)^{1/2}$$

$$= \ln x + \frac{1}{2} \ln (1 - x^2)$$

we have

$$f'(x) = \frac{1}{x} + \frac{1}{2}\left(\frac{-2x}{1 - x^2}\right) = \frac{1}{x} - \frac{x}{1 - x^2}$$

$$= \frac{1 - x^2 - x^2}{x(1 - x^2)}$$

$$= \frac{1 - 2x^2}{x(1 - x^2)}.$$

EXAMPLE 4   Logarithmic properties as an aid to differentiation

Differentiate

$$f(x) = \ln \frac{x(x^2 + 1)^2}{\sqrt{2x^3 - 1}}.$$

SOLUTION

Since

$$f(x) = \ln \frac{x(x^2 + 1)^2}{\sqrt{2x^3 - 1}} = \ln x + 2 \ln (x^2 + 1) - \frac{1}{2} \ln (2x^3 - 1)$$

we have

$$f'(x) = \frac{1}{x} + 2\left(\frac{2x}{x^2 + 1}\right) - \frac{1}{2}\left(\frac{6x^2}{2x^3 - 1}\right)$$

$$= \frac{1}{x} + \frac{4x}{x^2 + 1} - \frac{3x^2}{2x^3 - 1}.$$

REMARK   In Examples 2, 3, and 4, be sure you see the great benefit in applying logarithmic properties *before* differentiating. For instance, consider the difficulty of direct differentiation of the function given in Example 4.

Logarithmic differentiation

On occasion, it is convenient to use logarithms as an aid in differentiating *non*logarithmic functions. We call this procedure **logarithmic differentiation,** and we illustrate its use in Examples 5 and 6.

LOGARITHMIC DIFFERENTIATION

To differentiate the function $y = u$, use the following steps.

1. Take the natural logarithm of both sides: $\ln y = \ln u$
2. Use logarithmic properties to rid $\ln u$ of as many products, quotients, and exponents as possible.
3. Differentiate *implicitly*: $\dfrac{y'}{y} = \dfrac{d}{dx} [\ln u]$
4. Solve for y': $y' = y \dfrac{d}{dx} [\ln u]$
5. Substitute for y and simplify: $y' = u \dfrac{d}{dx} [\ln u]$

EXAMPLE 5   Logarithmic differentiation

Find the derivative of $y = x\sqrt{x^2 + 1}$.

SOLUTION

We begin by taking the natural logarithms of both sides of the equation. Then, we apply logarithmic properties and differentiate implicitly. Finally, we solve for y'.

$$y = x\sqrt{x^2 + 1}$$

$$\ln y = \ln [x\sqrt{x^2 + 1}]$$ Take log of both sides

$$= \ln x + \frac{1}{2} \ln (x^2 + 1)$$ Logarithmic properties

$$\frac{y'}{y} = \frac{1}{x} + \frac{1}{2}\left(\frac{2x}{x^2 + 1}\right)$$ Differentiate

$$= \frac{x^2 + 1 + x^2}{x(x^2 + 1)}$$

$$y' = y\left[\frac{2x^2 + 1}{x(x^2 + 1)}\right]$$ Solve for y'

$$y' = x\sqrt{x^2 + 1}\left[\frac{2x^2 + 1}{x(x^2 + 1)}\right]$$ Substitute for y

$$= \frac{2x^2 + 1}{\sqrt{x^2 + 1}}$$ Simplify

EXAMPLE 6   Logarithmic differentiation

Find the derivative of

$$y = \frac{(x - 2)^2}{\sqrt{x^2 + 1}}.$$

SOLUTION

$$\ln y = \ln \frac{(x - 2)^2}{\sqrt{x^2 + 1}}$$ Take ln of both sides

$$\ln y = 2 \ln (x - 2) - \frac{1}{2} \ln (x^2 + 1)$$ Logarithmic properties

$$\frac{y'}{y} = 2\left(\frac{1}{x - 2}\right) - \frac{1}{2}\left(\frac{2x}{x^2 + 1}\right)$$ Differentiate

$$= \frac{2}{x - 2} - \frac{x}{x^2 + 1}$$

$$y' = y\left(\frac{2}{x - 2} - \frac{x}{x^2 + 1}\right)$$ Solve for y'

$$= \frac{(x - 2)^2}{\sqrt{x^2 + 1}}\left[\frac{x^2 + 2x + 2}{(x - 2)(x^2 + 1)}\right]$$ Substitute for y

$$= \frac{(x - 2)(x^2 + 2x + 2)}{(x^2 + 1)^{3/2}}$$ Simplify

Since the natural logarithm is undefined for negative numbers, we often encounter expressions of the form $\ln |u|$. The following theorem tells us that we can differentiate functions of the form $y = \ln |u|$ as if the absolute value sign were not present.

| THEOREM 7.10 DERIVATIVE INVOLVING ABSOLUTE VALUE | If u is a differentiable function of x such that $u \neq 0$, then $$\frac{d}{dx}\left[\ln|u|\right] = \frac{u'}{u}.$$ |

PROOF If $u > 0$, then $|u| = u$, and the result follows from Theorem 7.9. If $u < 0$, then $|u| = -u$, and we have

$$\frac{d}{dx}\left[\ln|u|\right] = \frac{d}{dx}\left[\ln(-u)\right] = \frac{-u'}{-u} = \frac{u'}{u}.$$

EXAMPLE 7 Derivative involving absolute value

Find the derivative of $f(x) = \ln|2x - 1|$.

SOLUTION

Using Theorem 7.10, we let $u = 2x - 1$ and write

$$\frac{d}{dx}\left[\ln|2x - 1|\right] = \frac{u'}{u} = \frac{2}{2x - 1}.$$

EXAMPLE 8 Finding relative extrema

Locate the relative extrema of $y = \ln(x^2 + 2x + 3)$.

SOLUTION

Differentiating y, we obtain

$$\frac{dy}{dx} = \frac{2x + 2}{x^2 + 2x + 3}.$$

Now, since $dy/dx = 0$ when $x = -1$, we apply the First Derivative Test and conclude that the point $(-1, \ln 2)$ is a relative minimum. Since there are no other critical points, we conclude that this is the only relative extremum (see Figure 7.21).

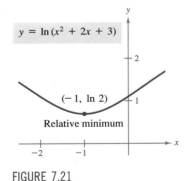

$y = \ln(x^2 + 2x + 3)$

$(-1, \ln 2)$

Relative minimum

FIGURE 7.21

Bases other than e

To differentiate exponential and logarithmic functions to other bases, you have three options: (1) convert the functions to base e and differentiate using the rules for natural exponential and logarithmic functions, (2) use logarithmic differentiation, or (3) use the following rules for bases other than e.

THEOREM 7.11 **DERIVATIVES FOR BASES** **OTHER THAN** e	Let a be a positive real number $(a \neq 1)$ and let u be a differentiable function of x. 1. $\dfrac{d}{dx}[a^x] = (\ln a)a^x$ 2. $\dfrac{d}{dx}[a^u] = (\ln a)a^u \dfrac{du}{dx}$ 3. $\dfrac{d}{dx}[\log_a x] = \dfrac{1}{(\ln a)x}$ 4. $\dfrac{d}{dx}[\log_a u] = \dfrac{1}{(\ln a)u}\dfrac{du}{dx}$

PROOF

We begin by writing a^x in the form $a^x = (e^{\ln a})^x = e^{(\ln a)x}$. Then, to prove the first property, we let $u = (\ln a)x$ and differentiate to obtain

$$\frac{d}{dx}[a^x] = \frac{d}{dx}[e^{(\ln a)x}] = e^u \frac{du}{dx} = e^{(\ln a)x}(\ln a) = (\ln a)a^x.$$

To prove the third property, we use the change of base formula and write

$$\frac{d}{dx}[\log_a x] = \frac{d}{dx}\left[\frac{1}{\ln a}\ln x\right] = \frac{1}{\ln a}\left(\frac{1}{x}\right) = \frac{1}{(\ln a)x}.$$

The second and fourth properties follow from the Chain Rule.

EXAMPLE 9 Differentiating functions to other bases

Find the derivatives of the following.

(a) $y = 2^x$ (b) $y = \log_2 (x^2 + 1)$

SOLUTION

(a) $y' = \dfrac{d}{dx}[2^x] = (\ln 2)2^x$

(b) $y' = \dfrac{1}{(\ln 2)(x^2 + 1)}(2x) = \dfrac{1}{\ln 2} \cdot \dfrac{2x}{x^2 + 1}$ ▭

When we introduced the Power Rule, $D_x[x^n] = nx^{n-1}$, in Chapter 3, we required the exponent n to be a rational number. We now extend the rule to cover any real value of n. Try to prove this theorem using logarithmic differentiation.

THEOREM 7.12 **THE POWER RULE FOR** **REAL EXPONENTS**	Let n be any real number and let u be a differentiable function of x. 1. $\dfrac{d}{dx}[x^n] = nx^{n-1}$ 2. $\dfrac{d}{dx}[u^n] = nu^{n-1}\dfrac{du}{dx}$

In the next example, we differentiate four different types of functions. Each function uses a different differentiation formula, depending on whether the base and exponent are constants or variables.

EXAMPLE 10 Comparing variables and constants

(a) $\dfrac{d}{dx}[e^e] = 0$ Constant Rule

(b) $\dfrac{d}{dx}[e^x] = e^x$ Exponential Rule

(c) $\dfrac{d}{dx}[x^e] = ex^{e-1}$ Power Rule

(d) $y = x^x$ Logarithmic differentiation

$\ln y = x \ln x$

$\dfrac{y'}{y} = x\left(\dfrac{1}{x}\right) + (\ln x)(1) = 1 + \ln x$

$y' = y(1 + \ln x) = x^x(1 + \ln x)$

EXERCISES for Section 7.5

In Exercises 1–4, find the slope of the tangent line to the given logarithmic function at the point $(1, 0)$.

1. $y = \ln x^3$

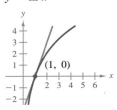

2. $y = \ln x^{3/2}$

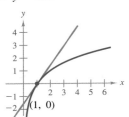

3. $y = \ln x^2$

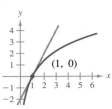

4. $y = \ln x^{1/2}$

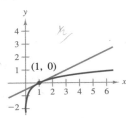

In Exercises 5–38, find dy/dx.

5. $y = \ln x^2$

6. $y = \ln (x^2 + 3)$

7. $y = \ln \sqrt{x^4 - 4x}$

8. $y = \ln (1 - x)^{3/2}$

9. $y = (\ln x)^4$

10. $y = x \ln x$

11. $y = \ln (x\sqrt{x^2 - 1})$

12. $y = \ln \left(\dfrac{x}{x + 1}\right)$

13. $y = \ln \left(\dfrac{x}{x^2 + 1}\right)$

14. $y = \dfrac{\ln x}{x}$

15. $y = \dfrac{\ln x}{x^2}$

16. $y = \ln (\ln x)$

17. $y = \ln (\ln x^2)$

18. $y = \ln \sqrt{\dfrac{x - 1}{x + 1}}$

19. $y = \ln \sqrt{\dfrac{x + 1}{x - 1}}$

20. $y = \ln \sqrt{x^2 - 4}$

21. $y = \ln \left(\dfrac{\sqrt{4 + x^2}}{x}\right)$

22. $y = \ln (x + \sqrt{4 + x^2})$

23. $y = \dfrac{-\sqrt{x^2 + 1}}{x} + \ln (x + \sqrt{x^2 + 1})$

24. $y = \dfrac{-\sqrt{x^2 + 4}}{2x^2} - \dfrac{1}{4}\ln \left(\dfrac{2 + \sqrt{x^2 + 4}}{x}\right)$

25. $y = 4^x$

26. $y = 2^{-x}$

27. $y = 5^{x-2}$

28. $y = x(7^{-3x})$

29. $y = x^2 2^x$

30. $y = 2^{x^2} 3^{-x}$

31. $y = \log_3 x$

32. $y = \log_{10} 2x$

33. $y = \log_2 \left(\dfrac{x^2}{x - 1}\right)$

34. $y = \log_3 \left(\dfrac{x\sqrt{x - 1}}{2}\right)$

35. $y = \log_5 \sqrt{x^2 - 1}$

36. $y = \log_{10} \left(\dfrac{x^2 - 1}{x}\right)$

37. $y = \ln |x^2 - 1|$

38. $y = \ln \left|\dfrac{x + 5}{x}\right|$

In Exercises 39–48, find dy/dx using logarithmic differentiation.

39. $y = x\sqrt{x^2 - 1}$

40. $y = \sqrt{(x - 1)(x - 2)(x - 3)}$

41. $y = \dfrac{x^2\sqrt{3x - 2}}{(x - 1)^2}$

42. $y = \sqrt[3]{\dfrac{x^2 + 1}{x^2 - 1}}$

43. $y = \dfrac{x(x - 1)^{3/2}}{\sqrt{x + 1}}$

44. $y = \dfrac{(x + 1)(x + 2)}{(x - 1)(x - 2)}$

45. $y = x^{2/x}$

46. $y = x^{x-1}$

47. $y = (x - 2)^{x+1}$

48. $y = (1 + x)^{1/x}$

In Exercises 49 and 50, show that the given function is a solution to the differential equation.

Function	Differential equation
49. $y = 2 \ln x + 3$	$x(y'') + y' = 0$
50. $y = x \ln x - 4x$	$x + y - xy' = 0$

In Exercises 51 and 52, find dy/dx by using implicit differentiation.

51. $x^2 - 3 \ln y + y^2 = 10$
52. $\ln xy + 5x = 30$

In Exercises 53 and 54, find an equation of the tangent line to the graph of the equation at the given point.

Equation	Point
53. $y = 3x^2 - \ln x$	$(1, 3)$
54. $x^2 + \ln (x + 1) + y^2 = 4$	$(0, 2)$

In Exercises 55–60, find any relative extrema and inflection points, and sketch the graph of the function.

55. $y = \dfrac{x^2}{2} - \ln x$ **56.** $y = x - \ln x$

57. $y = x (\ln x)$ **58.** $y = \dfrac{\ln x}{x}$

59. $y = \dfrac{x}{\ln x}$ **60.** $y = x^2 (\ln x)$

In Exercises 61 and 62, use Newton's Method to approximate, to three decimal places, the x-coordinate of the point of intersection of the graphs of the two equations.

61. $y = \ln x$ **62.** $y = \ln x$
   $y = -x$ $y = 3 - x$

63. Apply the Mean Value Theorem to the function $f(x) = \ln x$ on the closed interval $[1, e]$. Find the value of c in the open interval $(1, 3)$ such that

$$f'(c) = \frac{f(e) - f(1)}{e - 1}.$$

64. Show that

$$f(x) = \frac{\ln x^n}{x}$$

is a decreasing function for $x > e$ and $n > 0$.

65. A person walking along a dock drags a boat by a 10-foot rope. The boat travels along a path known as a *tractrix* (see figure). The equation of the path is

$$y = 10 \ln \left(\frac{10 + \sqrt{100 - x^2}}{x} \right) - \sqrt{100 - x^2}.$$

What is the slope of this path at the following x-values?
(a) $x = 10$ (b) $x = 5$

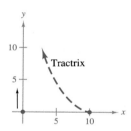

FIGURE FOR 65

7.6 Logarithmic Functions and Integration

The Log Rule for integration

The differentiation rules

$$\frac{d}{dx} [\ln |x|] = \frac{1}{x} \quad \text{and} \quad \frac{d}{dx} [\ln |u|] = \frac{u'}{u}$$

allow us to patch up the hole in our General Power Rule for integration. Recall from Section 5.5 that

$$\int u^n \, du = \frac{u^{n+1}}{n + 1} + C$$

provided $n \neq -1$. Having the differentiation formulas for logarithmic functions, we are now in a position to evaluate $\int u^n \, du$ for $n = -1$, as stated in the following theorem.

THEOREM 7.13 LOG RULE FOR INTEGRATION	Let u be a differentiable function of x.

$$1. \int \frac{1}{x}\, dx = \ln |x| + C \qquad 2. \int \frac{1}{u}\, du = \ln |u| + C$$

The second formula can also be written as

$$\int \frac{u'}{u}\, dx = \ln |u| + C.$$

EXAMPLE 1 Using the Log Rule for integration

$$\int \frac{2}{x}\, dx = 2 \int \frac{1}{x}\, dx = 2 \ln |x| + C = \ln x^2 + C \qquad \square$$

REMARK In Example 1, the absolute value is unnecessary in the final form of the antiderivative because x^2 cannot be negative.

EXAMPLE 2 Using the Log Rule with a change of variables

Evaluate

$$\int \frac{1}{2x - 1}\, dx.$$

SOLUTION

Let $u = 2x - 1$; then $du = 2\, dx$. After multiplying and dividing by 2, we have

$$\int \frac{1}{2x - 1}\, dx = \frac{1}{2} \int \left(\frac{1}{2x - 1} \right) 2\, dx$$

$$= \frac{1}{2} \int \frac{1}{u}\, du$$

$$= \frac{1}{2} \ln |u| + C$$

$$= \frac{1}{2} \ln |2x - 1| + C. \qquad \square$$

In Example 3 we use the alternate form of the Log Rule, $\int u'/u\, dx = \ln |u| + C$. This form of the Log Rule is convenient, especially for simpler integrals. To apply this rule, look for quotients in which the numerator is the derivative of the denominator.

EXAMPLE 3   Finding area with the Log Rule

Find the area of the region bounded by the graph of $y = x/(x^2 + 1)$, the x-axis, and the line $x = 3$.

SOLUTION

From Figure 7.22 we see that the area of the region is given by the definite integral

$$\int_0^3 \frac{x}{x^2 + 1}\, dx.$$

Letting $u = x^2 + 1$, we have $u' = 2x$. Thus, to apply the Log Rule, we multiply and divide by 2 and write

$$\int_0^3 \frac{x}{x^2 + 1}\, dx = \frac{1}{2} \int_0^3 \frac{2x}{x^2 + 1}\, dx \qquad \int \frac{u'}{u}\, dx = \ln |u| + C$$

$$= \frac{1}{2} \left[\ln (x^2 + 1) \right]_0^3$$

$$= \frac{1}{2}(\ln 10 - \ln 1) = \frac{1}{2} \ln 10 \approx 1.151.$$ ▭

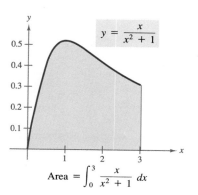

$$y = \frac{x}{x^2 + 1}$$

Area $= \int_0^3 \frac{x}{x^2 + 1}\, dx$

FIGURE 7.22

EXAMPLE 4   Recognizing quotient forms of the Log Rule

(a) $\displaystyle \int \frac{3x^2 + 1}{x^3 + x}\, dx = \ln |x^3 + x| + C \qquad u = x^3 + x$

(b) $\displaystyle \int \frac{x + 1}{x^2 + 2x}\, dx = \frac{1}{2} \int \frac{2x + 2}{x^2 + 2x}\, dx \qquad u = x^2 + 2x$

$$= \frac{1}{2} \ln |x^2 + 2x| + C$$ ▭

Integrals to which the Log Rule can be applied often appear in disguised form. For instance, if a rational function has a *numerator of degree greater than or equal to that of the denominator*, division may reveal a form to which we can apply the Log Rule. The next example illustrates this.

EXAMPLE 5   Using long division before integrating

Evaluate

$$\int \frac{x^2 + x + 1}{x^2 + 1}\, dx.$$

SOLUTION

First, by dividing we have

$$\frac{x^2 + x + 1}{x^2 + 1} = 1 + \frac{x}{x^2 + 1}.$$

Now, we integrate to obtain

$$\int \frac{x^2 + x + 1}{x^2 + 1} \, dx = \int \left(1 + \frac{x}{x^2 + 1}\right) dx$$

$$= \int dx + \frac{1}{2} \int \frac{2x}{x^2 + 1} \, dx$$

$$= x + \frac{1}{2} \ln (x^2 + 1) + C.$$

REMARK Try differentiating the antiderivative found in Example 5 to check that you obtain the original integrand.

The next example gives another instance in which the use of the Log Rule is disguised. In this case, a change of variables helps us recognize the Log Rule.

EXAMPLE 6 *Change of variables with the Log Rule*

Evaluate

$$\int \frac{2x}{(x + 1)^2} \, dx.$$

SOLUTION

Let $u = x + 1$; then $du = dx$ and $x = u - 1$.

$$\int \frac{2x}{(x + 1)^2} \, dx = \int \frac{2(u - 1)}{u^2} \, du$$

$$= 2 \int \left[\frac{u}{u^2} - \frac{1}{u^2}\right] du$$

$$= 2 \int \frac{du}{u} - 2 \int u^{-2} \, du$$

$$= 2 \ln |u| - 2\left(\frac{u^{-1}}{-1}\right) + C$$

$$= 2 \ln |u| + \frac{2}{u} + C$$

$$= 2 \ln |x + 1| + \frac{2}{x + 1} + C \qquad \text{Substitute for } u$$

As you study the two different methods shown in Examples 5 and 6, be aware that both methods involve rewriting a disguised integrand so that it fits one or more of the basic integration formulas. In Chapters 8 and 9 we will

devote much time to integration techniques. To master these techniques you must recognize the "form-fitting" nature of integration. In this sense integration is not nearly as straightforward as differentiation. Differentiation takes the form "Here is the question; what is the answer?" Integration is more like "Here is the answer; what is the question?" We suggest the following general approach to integration.

GUIDELINES FOR INTEGRATION

1. Memorize a basic list of integration formulas. (At this point our list consists only of the Power Rule, the Exponential Rule, and the Log Rule. By the end of Chapter 8, we will have expanded this list to nineteen basic formulas.)
2. Find an integration formula that resembles all or part of the integrand, and by trial and error find a choice of u that will make the integrand conform to the formula.
3. If you cannot find a u-substitution that works, try altering the integrand. You might try a trigonometric identity, division, addition, and subtraction of the same quantity—be creative.

EXAMPLE 7   *u*-Substitution and the Log Rule

Evaluate

$$\int \frac{1}{x \ln x} \, dx.$$

SOLUTION

Since neither the Power Rule, the Exponential Rule, nor the Log Rule apply directly, we consider a u-substitution. There are three basic choices for u. The choices $u = x$ and $u = x \ln x$ fail to fit the u'/u form of the Log Rule. However, the third choice does fit. Letting $u = \ln x$, we have $u' = 1/x$ and write

$$\int \frac{1}{x \ln x} \, dx = \int \frac{(1/x)}{\ln x} \, dx = \int \frac{u'}{u} \, dx = \ln |u| + C$$

$$= \ln |\ln x| + C.$$

EXAMPLE 8   *u*-Substitution and the Log Rule

Evaluate

$$\int \frac{1}{\sqrt{x} + 1} \, dx.$$

SOLUTION

Since neither the Power Rule, the Exponential Rule, nor the Log Rule apply to the integral as given, we consider the substitution $u = \sqrt{x}$. Then,

$$u^2 = x \quad \text{and} \quad 2u \, du = dx.$$

Substitution in the original integral yields

$$\int \frac{1}{\sqrt{x}+1}\,dx = \int \frac{1}{u+1}(2u\,du) = 2\int \frac{u}{u+1}\,du.$$

Since the degree of the numerator is equal to the degree of the denominator, we divide u by $(u+1)$ to obtain

$$\int \frac{1}{\sqrt{x}+1}\,dx = 2\int \left(1 - \frac{1}{u+1}\right)du$$

$$= 2(u - \ln|u+1|) + C$$

$$= 2\sqrt{x} - 2\ln(\sqrt{x}+1) + C.$$

In Section 6.5, Example 6, we used Simpson's Rule to find the work done by an expanding gas. Now, with the Log Rule, we can evaluate this work using the Fundamental Theorem of Calculus.

EXAMPLE 9   An application

A quantity of gas with an initial volume of 1 cubic foot and pressure of 500 pounds per square foot expands to a volume of 2 cubic feet. Find the work done by the gas. (Assume the pressure is inversely proportional to the volume.)

SOLUTION

Since $p = k/V$ and $p = 500$ when $V = 1$, $k = 500$. Thus, the work is

$$W = \int_{V_0}^{V_1} \frac{k}{V}\,dV = \int_1^2 \frac{500}{V}\,dV = 500\left[\ln V\right]_1^2$$

$$= 500(\ln 2 - \ln 1) \approx 346.6 \text{ ft} \cdot \text{lb.}$$

Occasionally, an integrand involves an exponential function to a base other than e. When this occurs, there are two options: (1) convert to base e using the formula $a^x = e^{(\ln a)x}$ and then integrate, or (2) integrate directly, using the integration formula

$$\int a^x\,dx = \left(\frac{1}{\ln a}\right)a^x + C$$

which follows from Theorem 7.11.

EXAMPLE 10   Integrating an exponential function to another base

Evaluate $\int 2^x\,dx$.

SOLUTION

$$\int 2^x\,dx = \frac{1}{\ln 2}2^x + C$$

EXERCISES for Section 7.6

In Exercises 1–34, evaluate each integral.

1. $\displaystyle\int \frac{1}{x+1}\,dx$

2. $\displaystyle\int \frac{1}{x-5}\,dx$

3. $\displaystyle\int \frac{1}{3-2x}\,dx$

4. $\displaystyle\int \frac{1}{6x+1}\,dx$

5. $\displaystyle\int \frac{x}{x^2+1}\,dx$

6. $\displaystyle\int \frac{x^2}{3-x^3}\,dx$

7. $\displaystyle\int \frac{x^2-4}{x}\,dx$

8. $\displaystyle\int \frac{x+5}{x}\,dx$

9. $\displaystyle\int_1^e \frac{\ln x}{2x}\,dx$

10. $\displaystyle\int_e^{e^2} \frac{1}{x(\ln x)}\,dx$

11. $\displaystyle\int_1^e \frac{(1+\ln x)^2}{x}\,dx$

12. $\displaystyle\int_0^1 \frac{x-1}{x+1}\,dx$

13. $\displaystyle\int_0^2 \frac{x^2-2}{x+1}\,dx$

14. $\displaystyle\int \frac{1}{(x+1)^2}\,dx$

15. $\displaystyle\int \frac{1}{\sqrt{x}+1}\,dx$

16. $\displaystyle\int \frac{x+3}{x^2+6x+7}\,dx$

17. $\displaystyle\int \frac{x^2+2x+3}{x^3+3x^2+9x}\,dx$

18. $\displaystyle\int \frac{(\ln x)^2}{x}\,dx$

19. $\displaystyle\int \frac{1}{x^{2/3}(1+x^{1/3})}\,dx$

20. $\displaystyle\int \frac{1}{x\ln(x^2)}\,dx$

21. $\displaystyle\int \frac{1}{1+\sqrt{x}}\,dx$

22. $\displaystyle\int \frac{1-\sqrt{x}}{1+\sqrt{x}}\,dx$

23. $\displaystyle\int \frac{\sqrt{x}}{\sqrt{x}-3}\,dx$

24. $\displaystyle\int_0^2 \frac{1}{1+\sqrt{2x}}\,dx$

25. $\displaystyle\int \frac{\sqrt{x}}{1-x\sqrt{x}}\,dx$

26. $\displaystyle\int \frac{2x}{(x-1)^2}\,dx$

27. $\displaystyle\int \frac{x(x-2)}{(x-1)^3}\,dx$

28. $\displaystyle\int \frac{x\sqrt{x}}{1+x^2\sqrt{x}}\,dx$

29. $\displaystyle\int 3^x\,dx$

30. $\displaystyle\int 4^{-x}\,dx$

31. $\displaystyle\int_{-1}^2 2^x\,dx$

32. $\displaystyle\int 2^3\,dx$

33. $\displaystyle\int x5^{x^2}\,dx$

34. $\displaystyle\int (3-x)7^{(3-x)^2}\,dx$

In Exercises 35 and 36, find the area of the indicated region.

35. $y = \dfrac{x^2+4}{x}$

36. $y = \dfrac{x+5}{x}$

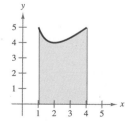

In Exercises 37–40, find the volume of the solid generated by revolving the region bounded by the graphs of the given equations about the indicated axis.

37. $y = \dfrac{1}{\sqrt{x+1}}$, $y = 0$, $x = 0$, $x = 3$, about the x-axis

38. $y = \dfrac{1}{x(x+1)}$, $y = 0$, $x = 1$, $x = 3$, about the y-axis

39. $xy = 1$, $y = 0$, $x = 1$, $x = 4$, about the line $y = 4$

40. $y = \dfrac{1}{1+\sqrt{x-2}}$, $y = 0$, $x = 2$, $x = 6$, about the y-axis

41. Find the area of the region bounded by the graphs of $y = 3^x$, $y = 0$, $x = 0$, and $x = 3$.

42. Find the centroid of the region bounded by the graphs of $y = 1/x$, $y = 0$, $x = 1$, and $x = 4$.

In Exercises 43 and 44, find the work done by the gas for the given volumes and pressures. Assume the pressure is inversely proportional to the volume. (Use Example 9 as a model.)

43. A quantity of gas with an initial volume of 2 cubic feet and an initial pressure of 1000 pounds per square foot expands to a volume of 3 cubic feet.

44. A quantity of gas with an initial volume of 1 cubic foot and an initial pressure of 2000 pounds per square foot expands to a volume of 4 cubic feet.

45. A population of bacteria is changing at the rate of

$$\frac{dP}{dt} = \frac{3000}{1+0.25t}$$

where t is the time in days. Assuming that the initial population (when $t = 0$) is 1000, write an equation that gives the population at any time t, and then find the population when $t = 3$ days.

46. The demand equation for a product is given by

$$p = \frac{90{,}000}{400+3x}.$$

Find the *average* price p on the interval $40 \le x \le 50$.

47. Find the time required for an object to cool from 300° to 250° if that time is given by

$$t = \frac{10}{\ln 2} \int_{250}^{300} \frac{1}{T-100}\,dT.$$

48. Find the average value of the function $f(x) = 1/x$ on the interval $1 \leq x \leq 5$.

49. Use Simpson's Rule with $n = 4$ to show that

$$\int_1^{2.7} \frac{1}{t}\,dt < 1 < \int_1^{2.8} \frac{1}{t}\,dt.$$

Use this inequality to show that $\ln 2.7 < \ln e < \ln 2.8$ and therefore $2.7 < e < 2.8$.

50. Use a computer and Simpson's Rule with $n = 10$ to approximate the integral

$$\int_1^x \frac{1}{t}\,dt$$

and compare the result with $\ln x$ for the specified value of x.

(a) $x = 3$ (b) $x = 8.7$

7.7 Growth and Decay

Growth and decay ▪ Applications

One practical application of exponential functions involves mathematical models in which the rate of change of a variable y is proportional to the value of y. If y is a function of time t, then we can write

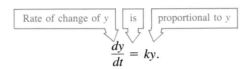

$$\frac{dy}{dt} = ky.$$

The general solution to this **differential equation** is given in the following theorem.

THEOREM 7.14 **LAW OF EXPONENTIAL GROWTH AND DECAY**	If y is a differentiable function of t such that $y > 0$ and $dy/dt = ky$, for some constant k, then $$y = Ce^{kt}.$$ C is called the **initial value** of y, and k is called the **constant of proportionality.** **Exponential growth** is indicated by $k > 0$, and **exponential decay** by $k < 0$.

PROOF   To solve the differentiable equation $y' = ky$, we use a technique called **separation of variables.** (We discuss the technique in detail in Chapter 18.)

$$y' = ky$$

$$\frac{y'}{y} = k \qquad\qquad \text{Divide by } y \text{ since } y \neq 0$$

$$\int \frac{y'}{y}\,dt = \int k\,dt \qquad\qquad \text{Integrate with respect to } t$$

$$\int \frac{1}{y}\,dy = \int k\,dt \qquad\qquad dy = y'\,dt$$

$$\ln y = kt + C_1$$

$$y = e^{kt+C_1} = e^{C_1}e^{kt} \qquad \text{Solve for } y$$

$$y = Ce^{kt} \qquad\qquad \text{Let } C = e^{C_1}$$

Example 1 involves radioactive decay, which is measured in terms of **half-life**—the number of years required for half of the atoms in a sample of radioactive material to decay. For example, the half-lives of some common radioactive isotopes are as follows.

Uranium (U^{238})	4,510,000,000 years
Plutonium (Pu^{230})	24,360 years
Carbon (C^{14})	5,730 years
Radium (Ra^{226})	1,620 years
Einsteinium (Es^{254})	276 days
Nobelium (No^{257})	23 seconds

EXAMPLE 1   Radioactive decay

A sample contains 1 gram of radium. How much radium will remain after 1000 years? [Use a half-life of 1620 years, as shown in Figure 7.23.]

SOLUTION

Let y represent the mass (in grams) of radium in the sample. Since the rate of decay is proportional to y, we apply the Law of Exponential Decay to conclude that y is of the form $y = Ce^{kt}$, where t is measured in years. We are given the following values for the function y: $y = 1$ when $t = 0$, and $y = \frac{1}{2}$ when $t = 1620$. Applying the first of these two conditions, we have

$$y = 1 = Ce^{k(0)}. \qquad t = 0$$

Thus, $C = 1$. Now, by applying the second condition, we obtain

$$y = \frac{1}{2} = e^{k(1620)}. \qquad t = 1620$$

To solve for k, we take the natural log of both sides to obtain $\ln \frac{1}{2} = 1620k$. Therefore, $k = \left(\ln \frac{1}{2}\right)/1620 \approx -0.0004279$, and the equation for y as a function of time is

$$y = Ce^{kt} \approx e^{-0.0004279t}.$$

Finally, when $t = 1000$ years, the amount of radium remaining is

$$y \approx e^{-0.0004279(1000)} \approx 0.652 \text{ g.} \qquad t = 1000$$

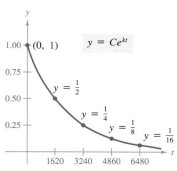

$y = Ce^{kt}$

$(0, 1)$

$y = \frac{1}{2}$

$y = \frac{1}{4}$

$y = \frac{1}{8}$

$y = \frac{1}{16}$

Radioactive Half-life of 1620 Years

FIGURE 7.23

REMARK   Example 1 demonstrates that finding exponential decay (or growth) functions basically involves solving for the constants C and k. Furthermore, we can see in Example 1 that it is easy to solve for C when we are given the value at $t = 0$.

In the next example, we demonstrate a procedure for solving for C and k in the growth equation $y = Ce^{kt}$ when we do not know the value of y at $t = 0$.

EXAMPLE 2   Population growth

Suppose that an experimental population of fruit flies increases according to the law of exponential growth. If there are 100 flies after the second day of the experiment and 300 flies after the fourth day, how many flies were in the original population?

SOLUTION

Let $y = Ce^{kt}$ be the number of flies at time t, where t is measured in days. Since $y = 100$ when $t = 2$ and $y = 300$ when $t = 4$, we have

$$100 = Ce^{2k} \quad \text{and} \quad 300 = Ce^{4k}.$$

From the first equation, we have $C = 100e^{-2k}$. Substituting this value into the second equation, we have $300 = 100e^{-2k}e^{4k} = 100e^{2k}$. Now we can solve for k as follows.

$$\frac{300}{100} = e^{2k} \quad \Longrightarrow \quad \ln 3 = 2k \quad \Longrightarrow \quad k = \frac{\ln 3}{2} \approx 0.5493$$

Therefore, the equation for exponential growth is $y = Ce^{0.5493t}$. To solve for C, we can reapply the condition $y = 100$ when $t = 2$ and obtain

$$100 = Ce^{0.5493(2)} \quad \Longrightarrow \quad C = 100e^{-1.0986} \approx 33.$$

Thus, the original population (when $t = 0$) consisted of approximately $y = C = 33$ flies, as indicated in Figure 7.24.   ▭

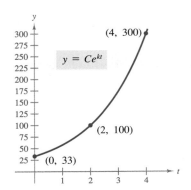

FIGURE 7.24

EXAMPLE 3   Continuously compounded interest

Money is deposited in an account for which the interest is compounded continuously. If the balance doubles in 6 years, what is the annual percentage rate?

SOLUTION

The formula for the balance A using continuously compounded interest is given by the exponential growth function

$$A = Pe^{rt}$$

where P is the original deposit, r is the annual rate (in decimal form), and t is the time in years. When $t = 6$, we know that $A = 2P$, as shown in Figure 7.25. Thus, we have

$$2P = Pe^{6r} \quad \Longrightarrow \quad 2 = e^{6r} \quad \Longrightarrow \quad \ln 2 = 6r$$

and we conclude that the annual percentage rate is

$$r = \frac{1}{6} \ln 2 \approx 0.1155 = 11.55\%.$$ ▭

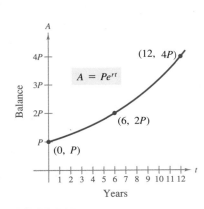

FIGURE 7.25

EXAMPLE 4   Declining sales

Four months after it stopped advertising, a manufacturing company noticed that its sales had dropped from 100,000 units per month to 80,000 units per month. If the sales follow an exponential pattern of decline, what will they be after another two months?

SOLUTION

We use the exponential model $y = Ce^{kt}$, where t is measured in months, as shown in Figure 7.26. From the initial condition ($t = 0$), we know that $C = 100{,}000$. Moreover, since $y = 80{,}000$ when $t = 4$, we have

$$80{,}000 = 100{,}000e^{4k} \quad \Longrightarrow \quad 0.8 = e^{4k} \quad \Longrightarrow \quad \ln (0.8) = 4k.$$

Thus, $k = [\ln (0.8)]/4 \approx -0.0558$. Finally, after two more months ($t = 6$), we can expect the monthly sales rate to be

$$y = 100{,}000e^{-0.0558(6)} \approx 71{,}500 \text{ units.}$$

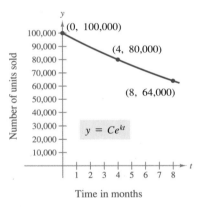

FIGURE 7.26

In each of the four examples describing exponential growth or decay, we did not actually have to solve the differential equation $y' = ky$. (We did that once in the proof of Theorem 7.14.) In the last example in this section, we demonstrate a problem whose solution involves the separation of variables technique used in proving the Law of Exponential Growth. The example involves **Newton's Law of Cooling,** which states that the rate of change in the temperature of an object is proportional to the difference between the object's temperature and the temperature of the surrounding medium.

EXAMPLE 5   Newton's Law of Cooling

Let y represent the temperature (in degrees Fahrenheit) of an object in a room whose temperature is kept at a constant 60°. If the object cools from 100° to 90° in 10 minutes, how much longer will it take for its temperature to decrease to 80°?

SOLUTION

From Newton's Law of Cooling, we know that the rate of change in y is proportional to the difference between y and 60. This gives us

$$\frac{dy}{dt} = k(y - 60), \quad 80 \le y \le 100$$

$$\left(\frac{1}{y - 60}\right)\frac{dy}{dt} = k$$

$$\int \frac{1}{y - 60}\, dy = \int k\, dt \qquad \text{Separation of variables}$$

$$\ln |y - 60| = kt + C.$$

Since $y > 60$, we can delete the absolute value signs. Now, using $y = 100$ when $t = 0$, we have $C = \ln 40$, which implies that

$$kt = \ln (y - 60) - \ln 40 = \ln \left(\frac{y - 60}{40}\right).$$

Furthermore, since $y = 90$ when $t = 10$, we have $k = \frac{1}{10} \ln \frac{3}{4}$, and thus

$$t = \left(\frac{10}{\ln (3/4)}\right) \ln \left(\frac{y - 60}{40}\right).$$

Finally, when $y = 80$,

$$t = \frac{10 \ln (1/2)}{\ln (3/4)} \approx 24.09 \text{ min.}$$

Therefore, it will require approximately 14.09 *more* minutes for the object to cool to a temperature of 80°.

EXERCISES for Section 7.7

In Exercises 1–4, find the exponential function $y = Ce^{kt}$ that passes through the two given points.

1.
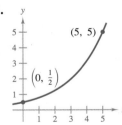

2.
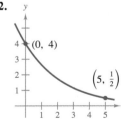

3.
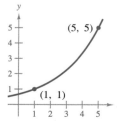

4.
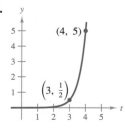

In Exercises 5–10, complete the table for a savings account in which interest is compounded continuously.

	Initial investment	Annual rate	Time to double	Amount after 10 yrs	Amount after 25 yrs
5.	$1,000	12%			
6.	$20,000	$10\frac{1}{2}$%			
7.	$750		$7\frac{3}{4}$ years		
8.	$10,000		5 years		
9.	$500			$1,292.85	
10.	$2,000				$6,008.33

11. The management at a certain factory has found that a worker can produce at most 30 units in a day. The learning curve for the number of units N produced per day after a new employee has worked t days is given by

$$N = 30(1 - e^{kt})$$

After 20 days on the job, a particular worker produced 19 units.

(a) Find the learning curve for this worker.

(b) How many days should pass before this worker is producing 25 units per day?

12. If in Exercise 11 the management requires a new employee to produce at least 20 units per day after 30 days on the job, find

(a) the learning curve that describes this minimum requirement,

(b) the number of days before a minimal achiever is producing 25 units per day.

13. The sales S (in thousands of units) of a new product after it is on the market t years is given by

$$S = Ce^{k/t}.$$

(a) Find S as a function of t if 5000 units have been sold after 1 year and the saturation point for the market is 30,000 (that is, $\lim_{t \to \infty} S = 30$).

(b) How many units have been sold after 5 years?

(c) Sketch a graph of this sales function.

14. The number of sales S (in thousands of units) of a new product after it has been on the market t years is given by

$$S = 30(1 - e^{kt}).$$

(a) Find S as a function of t if 5000 units have been sold after 1 year.

(b) How many units will saturate this market?

(c) How many units will have been sold after 5 years?

(d) Sketch a graph of this sales function.

15. A certain type of bacteria increases continuously at a rate proportional to the number present. If there are 100 present at a given time and 300 present 5 hours later, how many will there be 10 hours after the initial given time?

16. Given the conditions of Exercise 15, how long would it take for the number of bacteria to double?

17. In 1970 the population of a town was 2500, and in 1980 it was 3350. Assuming the population increases continuously at a constant rate proportional to the existing population, estimate the population in the year 2000.

18. Given the conditions of Exercise 17, how many years would be necessary for the population to double?

In Exercises 19–24, complete the table for the given radioactive isotope.

Isotope	Half-life (in yrs)	Initial quantity	Amount after 1,000 yrs	Amount after 10,000 yrs
19. Ra^{226}	1,620	10 grams		
20. Ra^{226}	1,620		1.5 grams	
21. C^{14}	5,730			2 grams
22. C^{14}	5,730	3 grams		
23. Pu^{230}	24,360		2.1 grams	
24. Pu^{230}	24,360			0.4 gram

25. Radioactive radium has a half-life of approximately 1620 years. What percentage of a given amount remains after 100 years?

26. If radioactive material decays continuously at a rate proportional to the amount present, find the half-life of the material if after 1 year 99.57 percent of an initial amount still remains.

27. C^{14} dating assumes that the carbon dioxide on earth today has the same radioactive content as it did centuries ago. If this is true, then the amount of C^{14} absorbed by a tree that grew several centuries ago should be the same as the amount of C^{14} absorbed by a tree growing today. A piece of ancient charcoal contains only 15% as much of the radioactive carbon as a piece of modern charcoal. How long ago was the tree burned to make the ancient charcoal? [The half-life of C^{14} is 5730 years.]

28. On the Richter Scale, the magnitude R of an earthquake of intensity I is given by

$$R = \frac{\ln I - \ln I_0}{\ln 10}$$

where I_0 is the minimum intensity used for comparison. Assume $I_0 = 1$.

(a) Find the intensity of the 1906 San Francisco earthquake when $R = 8.3$.

(b) Find the factor by which the intensity is increased if the Richter Scale measurement is doubled.

(c) Find dR/dI.

29. Using Newton's Law of Cooling (see Example 5), determine the reading on a thermometer 5 minutes after it is taken from a room at 72° Fahrenheit to the outdoors where the temperature is 20°, if the reading dropped to 48° after 1 minute.

30. An object in a room at 70° cools from 350° to 150° in 45 minutes. Using Newton's Law of Cooling, find the time necessary for the object to cool to 80°.

31. Using Newton's Law of Cooling, determine the outdoor temperature if a thermometer is taken from a room where the temperature is 68° to the outdoors, where after $\frac{1}{2}$ minute and 1 minute the thermometer reads 53° and 42°, respectively.

32. When an object is removed from a furnace and placed in an environment with a constant temperature of 90°, its core temperature is 1500°. One hour after it is removed, the core temperature is 1120°. Find the core temperature 5 hours after the unit is removed.

33. Atmospheric pressure P (measured in millimeters of mercury) decreases exponentially with increasing altitude x (measured in meters). The pressure is 760 millimeters of mercury at sea level ($x = 0$) and 672.71 millimeters of mercury at an altitude of 1000 meters. Find the pressure at an altitude of 3000 meters.

34. Because of a slump in the economy, a company finds that its annual revenues have dropped from $742,000 in 1988 to $632,000 in 1990. If the revenue is following an exponential pattern of decline, what is the expected revenue for 1991? (Let $t = 0$ represent 1988.)

7.8 Indeterminate Forms and L'Hôpital's Rule

Indeterminate forms ▪ L'Hôpital's Rule

In Chapter 2, we described the forms $0/0$ and ∞/∞ as *indeterminate*, because they do not guarantee that a limit exists, nor do they indicate what the limit is, if one does exist. When we encountered one of these indeterminate forms, we attempted to rewrite the expression by using various algebraic techniques, as illustrated by the examples in Table 7.7.

TABLE 7.7

Indeterminate form	Limit	Algebraic technique
$0/0$	$\lim\limits_{x \to -1} \dfrac{2x^2 - 2}{x + 1} = \lim\limits_{x \to -1} 2(x - 1) = -4$	Divide numerator and denominator by $(x + 1)$.
∞/∞	$\lim\limits_{x \to \infty} \dfrac{3x^2 - 1}{2x^2 + 1} = \lim\limits_{x \to \infty} \dfrac{3 - (1/x^2)}{2 + (1/x^2)} = \dfrac{3}{2}$	Divide numerator and denominator by x^2.

Occasionally, we can extend these algebraic techniques to find limits of transcendental functions. For instance, the limit

$$\lim_{x \to 0} \frac{e^{2x} - 1}{e^x - 1}$$

produces the indeterminate form $0/0$. Factoring and then dividing, we have

$$\lim_{x \to 0} \frac{e^{2x} - 1}{e^x - 1} = \lim_{x \to 0} \frac{(e^x + 1)(e^x - 1)}{e^x - 1}$$

$$= \lim_{x \to 0} (e^x + 1) = 2.$$

However, not all indeterminate forms can be evaluated by algebraic manipulation. This is particularly true when *both* algebraic and transcendental functions are involved. For instance, the limit

$$\lim_{x \to 0} \frac{e^{2x} - 1}{x}$$

produces the indeterminate form 0/0. Dividing the numerator and the denominator by x to obtain

$$\lim_{x \to 0} \left[\frac{e^{2x}}{x} - \frac{1}{x} \right]$$

merely produces another indeterminate form, $\infty - \infty$. Of course, we could use a calculator to estimate this limit, as shown in Table 7.8. From the table, the limit appears to be 2. (This limit will be verified in Example 1.)

TABLE 7.8

x	-1.0	-0.1	-0.01	-0.001	0	0.001	0.01	0.1	1.0
$\dfrac{e^{2x} - 1}{x}$	0.865	1.813	1.980	1.998	?	2.002	2.020	2.214	6.389

Guillaume L'Hôpital

To find the limit shown in Table 7.8, we use a theorem called **L'Hôpital's Rule.** This theorem states that under certain conditions the limit of the quotient $f(x)/g(x)$ is determined by the limit of $f'(x)/g'(x)$. This theorem is named after the French mathematician Guillaume François Antoine De L'Hôpital (1661–1704), who published the first calculus text in 1696. To prove this theorem, we use a more general result called the **Extended Mean-Value Theorem,** which states the following. If f and g are differentiable on an open interval (a, b) and continuous on $[a, b]$ such that $g(x) \neq 0$ for any x in (a, b), then there exists a point c in (a, b) such that

$$\frac{f'(c)}{g'(c)} = \frac{f(b) - f(a)}{g(b) - g(a)}.$$

We prove the Extended Mean-Value Theorem and L'Hôpital's Rule in Appendix A.

THEOREM 7.15
L'HÔPITAL'S RULE

Let f and g be functions that are differentiable on an open interval (a, b) containing c, except possibly at c itself. If the limit of $f(x)/g(x)$ as x approaches c produces the indeterminate form 0/0 or ∞/∞, then

$$\lim_{x \to c} \frac{f(x)}{g(x)} = \lim_{x \to c} \frac{f'(x)}{g'(x)}$$

provided the limit on the right exists (or is infinite).

REMARK   The indeterminate form ∞/∞ comes in four forms: ∞/∞, $(-\infty)/\infty$, $\infty/(-\infty)$, and $(-\infty)/(-\infty)$. L'Hôpital's Rule can be applied to each of these forms.

Students occasionally use L'Hôpital's Rule incorrectly by applying the Quotient Rule to $f(x)/g(x)$. Be sure you see that the rule involves $f'(x)/g'(x)$, not the derivative of $f(x)/g(x)$.

EXAMPLE 1   Indeterminate form 0/0

Evaluate

$$\lim_{x \to 0} \frac{e^{2x} - 1}{x}.$$

SOLUTION

Since direct substitution results in the indeterminate form 0/0, we apply L'Hôpital's Rule to obtain

$$\lim_{x \to 0} \frac{e^{2x} - 1}{x} = \lim_{x \to 0} \frac{\dfrac{d}{dx}[e^{2x} - 1]}{\dfrac{d}{dx}[x]} = \lim_{x \to 0} \frac{2e^{2x}}{1} = 2. \qquad \blacksquare$$

REMARK   In writing the string of equations in Example 1, we actually do not know that the first limit is equal to the second until we have shown that the second limit exists. In other words, if the second limit had not existed, it would not have been permissible to apply L'Hôpital's Rule.

Another form of L'Hôpital's Rule states that if the limit of $f(x)/g(x)$ as x approaches ∞ (or $-\infty$) produces the indeterminate form 0/0 or ∞/∞, then

$$\lim_{x \to \infty} \frac{f(x)}{g(x)} = \lim_{x \to \infty} \frac{f'(x)}{g'(x)}$$

provided the limit on the right exists. We illustrate this form of L'Hôpital's Rule in Example 2.

EXAMPLE 2   Indeterminate form ∞/∞

Evaluate

$$\lim_{x \to \infty} \frac{\ln x}{x}.$$

SOLUTION

Since direct substitution results in the indeterminate form ∞/∞, we apply L'Hôpital's Rule to obtain

$$\lim_{x \to \infty} \frac{\ln x}{x} = \lim_{x \to \infty} \frac{\dfrac{d}{dx}[\ln x]}{\dfrac{d}{dx}[x]} = \lim_{x \to \infty} \frac{1}{x} = 0. \qquad \blacksquare$$

Occasionally it is necessary to apply L'Hôpital's Rule more than once to remove an indeterminate form. This is illustrated in the next example.

EXAMPLE 3 Applying L'Hôpital's Rule more than once

Evaluate

$$\lim_{x \to -\infty} \frac{x^2}{e^{-x}}.$$

SOLUTION

Since direct substitution results in the indeterminate form ∞/∞, we apply L'Hôpital's Rule to obtain

$$\lim_{x \to -\infty} \frac{x^2}{e^{-x}} = \lim_{x \to -\infty} \frac{\dfrac{d}{dx}[x^2]}{\dfrac{d}{dx}[e^{-x}]} = \lim_{x \to -\infty} \frac{2x}{-e^{-x}}.$$

Since this limit results in the indeterminate form $(-\infty)/(-\infty)$, we apply L'Hôpital's Rule again to obtain

$$\lim_{x \to -\infty} \frac{2x}{-e^{-x}} = \lim_{x \to -\infty} \frac{\dfrac{d}{dx}[2x]}{\dfrac{d}{dx}[-e^{-x}]} = \lim_{x \to -\infty} \frac{2}{e^{-x}} = 0. \qquad \blacksquare$$

In addition to the forms $0/0$ and ∞/∞, there are other indeterminate forms such as $0 \cdot \infty$, 1^∞, ∞^0, 0^0, and $\infty - \infty$. For example, consider the following four limits that lead to the indeterminate form $0 \cdot \infty$.

$$\underbrace{\lim_{x \to 0} (x)\left(\frac{1}{x}\right)}_{\text{limit is } 1}, \qquad \underbrace{\lim_{x \to 0} (x)\left(\frac{2}{x}\right)}_{\text{limit is } 2}, \qquad \underbrace{\lim_{x \to \infty} (x)\left(\frac{1}{e^x}\right)}_{\text{limit is } 0}, \qquad \underbrace{\lim_{x \to \infty} (e^x)\left(\frac{1}{x}\right)}_{\text{limit is } \infty}$$

Since each limit is different, it is clear that the form $0 \cdot \infty$ is indeterminate in the sense that it does not determine the value (or even the existence) of the limit. The following examples indicate methods for evaluating these forms. Basically, we attempt to convert each of these forms to those for which L'Hôpital's Rule is applicable.

EXAMPLE 4 Indeterminate form $0 \cdot \infty$

Evaluate

$$\lim_{x \to \infty} e^{-x}\sqrt{x}.$$

SOLUTION

Since direct substitution produces the indeterminate form $0 \cdot \infty$, we rewrite the limit to fit the form $0/0$ or ∞/∞. In this case, we choose the second form and write

$$\lim_{x \to \infty} e^{-x}\sqrt{x} = \lim_{x \to \infty} \frac{\sqrt{x}}{e^x}.$$

Now, by L'Hôpital's Rule, we have

$$\lim_{x \to \infty} \frac{\sqrt{x}}{e^x} = \lim_{x \to \infty} \frac{1/(2\sqrt{x})}{e^x} = \lim_{x \to \infty} \frac{1}{2\sqrt{x}\, e^x} = 0.$$

If rewriting a limit in one of the forms $0/0$ or ∞/∞ does not seem to work, try the other form. For instance, in Example 4 we could have written the limit as

$$\lim_{x \to \infty} e^{-x}\sqrt{x} = \lim_{x \to \infty} \frac{e^{-x}}{x^{-1/2}}$$

which yields the indeterminate form $0/0$. However, in applying L'Hôpital's Rule to this limit, we obtain

$$\lim_{x \to \infty} \frac{e^{-x}}{x^{-1/2}} = \lim_{x \to \infty} \frac{-e^{-x}}{-1/(2x^{3/2})}$$

which also yields the indeterminate form $0/0$. Moreover, since the quotient seems to be getting more complicated, we abandon this approach and try the ∞/∞ form, as shown in Example 4.

The indeterminate forms 1^∞, ∞^0, and 0^0 arise from limits of functions that have a variable base and a variable exponent. When we encountered this type of function in Section 7.5, we used logarithmic differentiation to find the derivative. We use a similar procedure when taking limits, as indicated in the next example.

EXAMPLE 5   Indeterminate form 1^∞

Evaluate

$$\lim_{x \to \infty} \left(1 + \frac{1}{x} \right)^x.$$

SOLUTION

Since direct substitution yields the indeterminate form 1^∞, we proceed as follows. We assume the limit exists, and we represent it by

$$y = \lim_{x \to \infty} \left(1 + \frac{1}{x} \right)^x.$$

Now, taking the natural logarithm of both sides, we have

$$\ln y = \ln \left[\lim_{x \to \infty} \left(1 + \frac{1}{x} \right)^x \right]$$

and using the fact that the natural logarithmic function is continuous, we write

$$\ln y = \lim_{x \to \infty} \left[x \ln \left(1 + \frac{1}{x} \right) \right] \qquad \text{Indeterminate form: } \infty \cdot 0$$

$$= \lim_{x \to \infty} \left[\frac{\ln \left[1 + (1/x) \right]}{1/x} \right] \qquad \text{Indeterminate form: } 0/0$$

$$= \lim_{x \to \infty} \left[\frac{(-1/x^2)(1/[1 + (1/x)])}{-1/x^2} \right] \qquad \text{L'Hôpital's Rule}$$

$$= \lim_{x \to \infty} \frac{1}{1 + (1/x)} = 1.$$

Finally, since $\ln y = 1$, we know that $y = e$ and we conclude that

$$\lim_{x \to \infty} \left(1 + \frac{1}{x} \right)^x = e. \qquad \blacksquare$$

L'Hôpital's Rule can also be applied to one-sided limits, as demonstrated in Examples 6 and 7.

EXAMPLE 6   Indeterminate form 0^0

Evaluate

$$\lim_{x \to 0^+} x^x.$$

SOLUTION

Since direct substitution produces the indeterminate form 0^0, we proceed as follows.

$$y = \lim_{x \to 0^+} x^x \qquad \text{Indeterminate form: } 0^0$$

$$\ln y = \ln \left[\lim_{x \to 0^+} x^x \right] \qquad \text{Take log of both sides}$$

$$= \lim_{x \to 0^+} \left[\ln (x^x) \right] \qquad \text{Continuity}$$

$$= \lim_{x \to 0^+} \left[x \ln x \right] \qquad \text{Indeterminate form: } 0 \cdot (-\infty)$$

$$= \lim_{x \to 0^+} \frac{\ln x}{1/x} \qquad \text{Indeterminate form: } -\infty/\infty$$

$$= \lim_{x \to 0^+} \frac{1/x}{-1/x^2} \qquad \text{L'Hôpital's Rule}$$

$$= \lim_{x \to 0^+} -x = 0.$$

Now, since $\ln y = 0$, we conclude that $y = e^0 = 1$, and it follows that

$$\lim_{x \to 0^+} x^x = 1. \qquad \blacksquare$$

REMARK   When evaluating complicated limits like the one in Example 6, it is helpful to check the reasonableness of the solution with a calculator. For instance, the calculations shown in Table 7.9 are consistent with our conclusion that x^x approaches 1 as x approaches 0 from the right.

TABLE 7.9

x	1.0	0.1	0.01	0.001	0.0001	0.00001
x^x	1.0000	0.7943	0.9550	0.9931	0.9991	0.9999

EXAMPLE 7   Indeterminate form $\infty - \infty$

Evaluate

$$\lim_{x \to 1^+} \left(\frac{1}{\ln x} - \frac{1}{x - 1} \right).$$

SOLUTION

Since direct substitution yields the indeterminate form $\infty - \infty$, we try to rewrite the expression to produce a form to which we can apply L'Hôpital's Rule. In this case, we combine the two fractions to obtain

$$\lim_{x \to 1^+} \left(\frac{1}{\ln x} - \frac{1}{x - 1} \right) = \lim_{x \to 1^+} \left[\frac{x - 1 - \ln x}{(x - 1) \ln x} \right].$$

Now, since direct substitution produces the indeterminate form $0/0$, we can apply L'Hôpital's Rule to obtain

$$\lim_{x \to 1^+} \left(\frac{1}{\ln x} - \frac{1}{x - 1} \right) = \lim_{x \to 1^+} \left[\frac{1 - (1/x)}{(x - 1)(1/x) + \ln x} \right]$$

$$= \lim_{x \to 1^+} \left[\frac{x - 1}{x - 1 + x \ln x} \right].$$

This limit also yields the indeterminate form $0/0$, so we apply L'Hôpital's Rule again to obtain

$$\lim_{x \to 1^+} \left(\frac{1}{\ln x} - \frac{1}{x - 1} \right) = \lim_{x \to 1^+} \left[\frac{1}{1 + x(1/x) + \ln x} \right] = \frac{1}{2}. \qquad \blacksquare$$

We have identified the forms $0/0$, ∞/∞, $\infty - \infty$, $0 \cdot \infty$, 0^0, 1^∞, and ∞^0 as *indeterminate*. There are similar forms that you should recognize as *determinate*, such as

$$\infty + \infty \to \infty \qquad -\infty - \infty \to -\infty \qquad 0^\infty \to 0 \qquad 0^{-\infty} \to \infty.$$

(You are asked to verify two of these in Exercises 37 and 38.)

In each of the examples so far in this section, we have used L'Hôpital's Rule to find a limit that exists. L'Hôpital's Rule can also be used to conclude that a limit is infinite, and this is demonstrated in the last example.

EXAMPLE 8   An infinite limit

Evaluate

$$\lim_{x \to \infty} \frac{e^x}{x}.$$

SOLUTION

Direct substitution produces the indeterminate form ∞/∞, therefore we apply L'Hôpital's Rule to obtain

$$\lim_{x \to \infty} \frac{e^x}{x} = \lim_{x \to \infty} \frac{e^x}{1} = \lim_{x \to \infty} e^x = \infty.$$

Now, since $e^x \to \infty$ as $x \to \infty$, we conclude that the limit of e^x/x as $x \to \infty$ is also infinite.

As a final comment, we remind you that L'Hôpital's Rule can be applied only to quotients leading to the indeterminate forms $0/0$ or ∞/∞. For instance, the following application of L'Hôpital's Rule is *incorrect*.

$$\lim_{x \to 0} \frac{e^x}{x} \overset{?}{=} \lim_{x \to 0} \frac{e^x}{1} = 1 \qquad \text{Incorrect use of L'Hôpital's Rule}$$

The reason this application is incorrect is that, even though the limit of the denominator is 0, the limit of the numerator is 1—which means that the hypotheses of L'Hôpital's Rule have not been satisfied.

EXERCISES for Section 7.8

In Exercises 1–26, evaluate each limit, using L'Hôpital's Rule if necessary.

1. $\displaystyle \lim_{x \to 2} \frac{x^2 - x - 2}{x - 2}$

2. $\displaystyle \lim_{x \to -1} \frac{x^2 - x - 2}{x + 1}$

3. $\displaystyle \lim_{x \to 0} \frac{\sqrt{4 - x^2} - 2}{x}$

4. $\displaystyle \lim_{x \to 2^-} \frac{\sqrt{4 - x^2}}{x - 2}$

5. $\displaystyle \lim_{x \to 0} \frac{e^x - (1 - x)}{x}$

6. $\displaystyle \lim_{x \to 0^+} \frac{e^x - (1 + x)}{x^3}$

7. $\displaystyle \lim_{x \to 0^+} \frac{e^x - (1 + x)}{x^n}, \quad n = 1, 2, 3, \ldots$

8. $\displaystyle \lim_{x \to 1} \frac{\ln x}{x^2 - 1}$

9. $\displaystyle \lim_{x \to \infty} \frac{\ln x}{x}$

10. $\displaystyle \lim_{x \to \infty} \frac{e^x}{x}$

11. $\displaystyle \lim_{x \to \infty} \frac{3x^2 - 2x + 1}{2x^2 + 3}$

12. $\displaystyle \lim_{x \to \infty} \frac{x - 1}{x^2 + 2x + 3}$

13. $\displaystyle \lim_{x \to \infty} \frac{x^2 + 2x + 3}{x - 1}$

14. $\displaystyle \lim_{x \to \infty} \frac{x^2}{e^x}$

15. $\displaystyle \lim_{x \to 0^+} x^2 \ln x$

16. $\displaystyle \lim_{x \to 0} \left(\frac{1}{x} - \frac{1}{x^2} \right)$

17. $\displaystyle \lim_{x \to 2} \left(\frac{8}{x^2 - 4} - \frac{x}{x - 2} \right)$

18. $\displaystyle \lim_{x \to 2} \left(\frac{1}{x^2 - 4} - \frac{\sqrt{x - 1}}{x^2 - 4} \right)$

19. $\lim\limits_{x\to\infty} \dfrac{x}{\sqrt{x^2+1}}$

20. $\lim\limits_{x\to 1^+} \left(\dfrac{3}{\ln x} - \dfrac{2}{x-1}\right)$

21. $\lim\limits_{x\to 0^+} x^{1/x}$

22. $\lim\limits_{x\to 0^+} (e^x + x)^{1/x}$

23. $\lim\limits_{x\to\infty} x^{1/x}$

24. $\lim\limits_{x\to\infty} \left(1 + \dfrac{1}{x}\right)^x$

25. $\lim\limits_{x\to\infty} (1+x)^{1/x}$

26. $\lim\limits_{x\to\infty} \dfrac{\sqrt{x}}{\sqrt{x-1}}$

In Exercises 27–32, use L'Hôpital's Rule to determine the comparative rates of increase of the functions

$$f(x) = x^m$$
$$g(x) = e^{nx}$$
$$h(x) = (\ln x)^n$$

where $n > 0$, $m > 0$, and $x \to \infty$. The limits obtained in these exercises suggest that $(\ln x)^n$ tends toward infinity more slowly than x^m, which in turn tends toward infinity more slowly than e^{nx}.

27. $\lim\limits_{x\to\infty} \dfrac{x^2}{e^{5x}}$

28. $\lim\limits_{x\to\infty} \dfrac{x^3}{e^{2x}}$

29. $\lim\limits_{x\to\infty} \dfrac{(\ln x)^3}{x}$

30. $\lim\limits_{x\to\infty} \dfrac{(\ln x)^2}{x^3}$

31. $\lim\limits_{x\to\infty} \dfrac{(\ln x)^n}{x^m}$, where $0 < n, m$

32. $\lim\limits_{x\to\infty} \dfrac{x^m}{e^{nx}}$, where $0 < n, m$

In Exercises 33 and 34, L'Hôpital's Rule is used *incorrectly*. Describe the error.

33. $\lim\limits_{x\to 0} \dfrac{e^{2x}-1}{e^x} = \lim\limits_{x\to 0} \dfrac{2e^{2x}}{e^x} = \lim\limits_{x\to 0} 2e^x = 2$

34. $\lim\limits_{x\to\infty} \dfrac{e^{-x}}{1+e^{-x}} = \lim\limits_{x\to\infty} \dfrac{-e^{-x}}{-e^{-x}} = \lim\limits_{x\to\infty} 1 = 1$

35. Complete the following table to show that x eventually "overpowers" $(\ln x)^4$.

x	10	10^2	10^4	10^6	10^8	10^{10}
$\dfrac{(\ln x)^4}{x}$						

36. Complete the following table to show that e^x eventually "overpowers" x^5.

x	1	5	10	20	30	40	50	100
$\dfrac{e^x}{x^5}$								

37. Prove that if $f(x) \geq 0$, $\lim\limits_{x\to a} f(x) = 0$, and $\lim\limits_{x\to a} g(x) = \infty$, then

$$\lim\limits_{x\to a} f(x)^{g(x)} = 0.$$

38. Prove that if $f(x) \geq 0$, $\lim\limits_{x\to a} f(x) = 0$, and $\lim\limits_{x\to a} g(x) = -\infty$, then

$$\lim\limits_{x\to a} f(x)^{g(x)} = \infty.$$

39. Use the results of Exercises 37 and 38 to evaluate the following limits.
(a) $\lim\limits_{x\to 0} (e^x - 1)^{1/x^2}$
(b) $\lim\limits_{x\to 0} (e^x - 1)^{-1/x^2}$

40. The Gamma Function $\Gamma(n)$ is defined in terms of the integral of the function

$$f(x) = x^{n-1}e^{-x}, \quad n > 0.$$

Show that for any fixed value of n, the limit of $f(x)$ as x approaches infinity is zero.

41. The velocity of an object falling through a resisting medium such as air or water is given by

$$v = \dfrac{32}{k}\left(1 - e^{-kt} + \dfrac{v_0 k e^{-kt}}{32}\right)$$

where v_0 is the initial velocity, t is the time, and k is the resistance constant of the medium. Use L'Hôpital's Rule to find the formula for the velocity of a falling body in a vacuum by fixing v_0 and t and letting k approach zero. (Assume the downward direction is positive.)

42. The formula for the amount A in a savings account compounded n times a year for t years at an interest rate of r and an initial deposit of P is

$$A = P\left(1 + \dfrac{r}{n}\right)^{nt}.$$

Use L'Hôpital's Rule to show that the limiting formula as the number of compoundings per year becomes infinite is

$$A = Pe^{rt}.$$

(Note that this limiting formula is used for continuous compounding of interest.)

In Exercises 43–46, find any asymptotes and relative extrema that may exist and sketch the graph of the function. [Hint: Some of the limits required in finding the asymptotes have been found in previous exercises.]

43. $y = x^{1/x}, \quad x > 0$

44. $y = x^x, \quad x > 0$

45. $y = 2xe^{-x}$

46. $y = \dfrac{\ln x}{x}$

⌨ In Exercises 47 and 48, use a computer or graphics calculator to (a) sketch the graph of the given function, and (b) find the required limit (if it exists).

47. $\displaystyle\lim_{x \to 3} \frac{x - 3}{\ln (2x - 5)}$

48. $\displaystyle\lim_{x \to \infty} \frac{x^3}{e^{2x}}$

⌨ **49.** Use a computer or graphics calculator to sketch the graph of the function

$$f(x) = \frac{x^{k-1}}{k}$$

for $k = 1, 0.1,$ and 0.01. Also, evaluate $\displaystyle\lim_{k \to 0} \frac{x^k - 1}{k}$.

REVIEW EXERCISES for Chapter 7

In Exercises 1–6, (a) find the inverse, f^{-1}, of the given function, (b) sketch the graphs of f and f^{-1} on the same axes, and (c) verify that $f^{-1}[f(x)] = f[f^{-1}(x)] = x$.

1. $f(x) = \dfrac{1}{2}x - 3$

2. $f(x) = 5x - 7$

3. $f(x) = \sqrt{x + 1}$

4. $f(x) = x^3 + 2$

5. $f(x) = x^2 - 5, \ x \ge 0$

6. $f(x) = \sqrt[3]{x + 1}$

In Exercises 7 and 8, the function does not have an inverse. Give a restriction on the domain so that the restricted function has an inverse, and then find the inverse.

7. $f(x) = 2(x - 4)^2$

8. $f(x) = |x - 2|$

In Exercises 9–12, solve the given equation for x.

9. $e^{\ln x} = 3$

10. $\ln x + \ln (x - 3) = 0$

11. $\log_3 x + \log_3 (x - 1) - \log_3 (x - 2) = 2$

12. $\log_x 125 = 3$

In Exercises 13–34, find dy/dx.

13. $y = \ln \sqrt{x}$

14. $y = \ln \dfrac{x(x - 1)}{x - 2}$

15. $y = x\sqrt{\ln x}$

16. $y = \ln [x(x^2 - 2)^{2/3}]$

17. $y \ln x + y^2 = 0$

18. $\ln (x + y) = x$

19. $\ln y = x \ln x$

20. $\ln y = x \ln \sqrt{x^2 + 1}$

21. $y = \dfrac{1}{b^2}\left[\ln (a + bx) + \dfrac{a}{a + bx}\right]$

22. $y = \dfrac{1}{b^2}[a + bx - a \ln (a + bx)]$

23. $y = -\dfrac{1}{a} \ln \left(\dfrac{a + bx}{x}\right)$

24. $y = -\dfrac{1}{ax} + \dfrac{b}{a^2} \ln \left(\dfrac{a + bx}{x}\right)$

25. $y = \ln (e^{-x^2})$

26. $y = \ln \left(\dfrac{e^x}{1 + e^x}\right)$

27. $y = x^2 e^x$

28. $y = e^{-x^2/2}$

29. $y = \sqrt{e^{2x} + e^{-2x}}$

30. $y = x^{2x+1}$

31. $y = 3^{x-1}$

32. $y = (4e)^x$

33. $ye^x + xe^y = xy$

34. $y = \dfrac{x^2}{e^x}$

In Exercises 35–38, find the derivative of the given function. (Assume that a is constant.)

35. $y = x^a$

36. $y = a^x$

37. $y = x^x$

38. $y = a^a$

In Exercises 39–56, find the indefinite integral.

39. $\displaystyle\int \frac{1}{7x - 2} \, dx$

40. $\displaystyle\int \frac{x}{x^2 - 1} \, dx$

41. $\displaystyle\int \frac{1}{x \ln (3x)} \, dx$

42. $\displaystyle\int \frac{\ln \sqrt{x}}{x} \, dx$

43. $\displaystyle\int \frac{x^2 + 3}{x} \, dx$

44. $\displaystyle\int \frac{x^3 + 1}{x^2} \, dx$

45. $\displaystyle\int \frac{x^2}{x^3 - 1} \, dx$

46. $\displaystyle\int \left(x + \frac{1}{x}\right)^2 \, dx$

47. $\displaystyle\int \frac{1}{x\sqrt{\ln x}} \, dx$

48. $\displaystyle\int \frac{x + 2}{2x + 3} \, dx$

49. $\displaystyle\int xe^{-3x^2} \, dx$

50. $\displaystyle\int \frac{e^{1/x}}{x^2} \, dx$

51. $\displaystyle\int \frac{e^{4x} - e^{2x} + 1}{e^x} \, dx$

52. $\displaystyle\int \frac{e^{2x} - e^{-2x}}{e^{2x} + e^{-2x}} \, dx$

53. $\displaystyle\int \frac{e^x}{e^x - 1} \, dx$

54. $\displaystyle\int x^2 e^{x^3+1} \, dx$

55. $\displaystyle\int xe^{-x^2/2} \, dx$

56. $\displaystyle\int \frac{x - 1}{3x^2 - 6x - 1} \, dx$

In Exercises 57–60, evaluate the definite integral.

57. $\displaystyle\int_1^4 \frac{x + 1}{x} \, dx$

58. $\displaystyle\int_1^4 \frac{\ln x}{x} \, dx$

59. $\displaystyle\int_3^4 \frac{1}{x - 2} \, dx - \int_3^4 \frac{1}{x + 2} \, dx$

60. $\displaystyle\int_0^2 xe^{-x^2} \, dx$

In Exercises 61 and 62, sketch the graph of the region whose area is given by the integral, and find the area.

61. $\int_0^2 4e^{-2x}\,dx$ **62.** $\int_0^1 \frac{1}{x+1}\,dx$

In Exercises 63 and 64, find the average value of the function over the given interval. Find the values of x where the function assumes its mean value and sketch the graph of the function.

Function	*Interval*
63. $f(x) = \dfrac{1}{x-1}$	[5, 10]
64. $f(x) = e^{-x}$	[0, 2]

In Exercises 65 and 66, find the area of the region bounded by the graphs of the equations.

65. $y = xe^{-x^2}$, $y = 0$, $x = 0$, $x = 4$
66. $y = 3e^{-x/2}$, $y = 0$, $x = 0$, $x = 4$

67. A deposit of $500 earns interest at the rate of 5% compounded continuously. Find its value after each of the following time periods.
(a) 1 year
(b) 10 years
(c) 100 years

68. A deposit earns interest at the rate r compounded continuously and doubles in 10 years. Find r.

69. How large a deposit, at 7% interest compounded continuously, must be made to obtain a balance of $10,000 in 15 years?

70. A deposit of $2500 is made in a savings account at an annual interest rate of 12% compounded continuously. Find the average balance in this account during the first five years.

71. A population is growing continuously at the rate of $2\frac{1}{2}\%$ per year. Find the time necessary for the population to (a) double and (b) triple in size.

72. Under ideal conditions the rate of change of the air pressure, with respect to the height above sea level, is proportional to the pressure at the given height. The air pressure is 30 inches at sea level and 15 inches at 18,000 feet. Find the air pressure at 35,000 feet.

In Exercises 73 and 74, use the following model for human memory

$$p(t) = 80e^{-0.5t} + 20$$

where $p(t)$ is the percentage retained after t weeks (see figure).

73. At what rate is information being retained after (a) 1 week? (b) 2 weeks?

74. Find the average percentage retained during (a) the first 2 weeks and (b) the second 2 weeks.

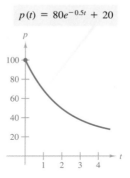

$$p(t) = 80e^{-0.5t} + 20$$

FIGURE FOR 73 and 74

In Exercises 75 and 76, the **exponential density function** is given by $P = (1/\mu)e^{-t/\mu}$, where t is the time in minutes and μ is the average time between successive events. The definite integral

$$\int_{t_1}^{t_2} \frac{1}{\mu} e^{-t/\mu}\,dt$$

gives the probability that the elapsed time before the next occurrence lies between t_1 and t_2 units of time.

75. Trucks arrive at a terminal at an average rate of 3 per hour ($\mu = 20$ minutes is the average time between arrivals). If a truck has just arrived, find the probability that the next arrival will be:
(a) within 10 minutes
(b) within 30 minutes
(c) between 15 and 30 minutes
(d) within 60 minutes

76. The average time between incoming calls at a switchboard is 3 minutes. Complete the accompanying table to show the probabilities of time elapsed between incoming calls.

Time elapsed between calls in minutes	0–2	2–4	4–6	6–8	8–10
Probability					

77. Two numbers between 0 and 10 are chosen at random. The probability that their product is less than n ($0 < n < 100$) is given by

$$P = \frac{1}{100}\left(n + \int_{10/n}^{10} \frac{n}{x}\, dx\right).$$

(a) What is the probability that the product is less than 25?

(b) What is the probability that the product is less than 50?

78. A solution of a certain drug contains 500 units per milliliter when it is prepared. After 40 days it contains 300 units per milliliter. Assuming that the rate of decomposition is proportional to the amount present, find an equation giving the amount A after t days.

79. An object is projected horizontally with an initial velocity of v_0 feet per second. Assume that the air resistance is proportional to the velocity of the object. Show that the distance traveled in t seconds is given by

$$s = \frac{v_0}{k}(1 - e^{-kt})$$

where k is a constant.

80. Assume that an object falling from rest encounters air resistance that is proportional to its velocity. Considering the acceleration due to gravity to be -32 feet per second per second, the net change in velocity is given by

$$\frac{dv}{dt} = kv - 32.$$

Find the velocity of the object as a function of time.

81. Solve Exercise 80 for an object that is projected downward with an initial velocity of 20 feet per second.

82. For the falling object in Exercise 80, find the limit of the velocity as t approaches infinity.

83. Integrate the velocity function found in Exercise 80 to find the position function s.

84. A certain automobile gets 28 miles per gallon of gasoline for speeds up to 50 miles per hour. Over 50 miles per hour the miles per gallon drop at the rate of 12% for each 10 miles per hour. If s is the speed and y is the miles per gallon, find y as a function of s by solving the differential equation

$$\frac{dy}{ds} = -0.012y, \quad s > 50.$$

In Exercises 85–92, use L'Hôpital's Rule to evaluate the given limit.

85. $\lim\limits_{x \to 1} \dfrac{(\ln x)^2}{x - 1}$

86. $\lim\limits_{x \to k} \dfrac{\sqrt[3]{x} - \sqrt[3]{k}}{x - k}$

87. $\lim\limits_{x \to \infty} \dfrac{e^{2x}}{x^2}$

88. $\lim\limits_{x \to 1^+} \left(\dfrac{2}{\ln x} - \dfrac{2}{x - 1}\right)$

89. $\lim\limits_{x \to \infty} (\ln x)^{2/x}$

90. $\lim\limits_{x \to 1} (x - 1)^{\ln x}$

91. $\lim\limits_{n \to \infty} 1000\left(1 + \dfrac{0.09}{n}\right)^n$

92. $\lim\limits_{x \to \infty} xe^{-x^2}$

Chapter 8 Application

Honey bees build their combs of pure wax, which can be produced only by the workers. Both sides of the combs are made up of geometrically perfect hexagonal cells and are constructed in a way that minimizes the amount of construction material. There are over 5 million honey bee colonies in the United States, producing over 250 million pounds of honey each year.

Construction of a Honeycomb

Honey bees use hexagonal cells to construct storage compartments. Each cell has a hexagonal base and three rhombic upper faces that meet the altitude of the cell at an angle θ. The *volume* of each cell is given by

$$V = \frac{3\sqrt{3}}{2} s^2 h$$

and is independent of the angle θ. On the other hand, the *surface area* of each cell, given by

$$S = 6hs + \frac{3s^2}{2} \left(\frac{\sqrt{3} - \cos \theta}{\sin \theta} \right)$$

does depend on the angle θ. The angle that minimizes the surface area (and thus the amount of wax needed to build the honeycomb) is given by solving the equation

$$\frac{dS}{d\theta} = \frac{3s^2}{2} \left(\frac{1 - \sqrt{3} \cos \theta}{\sin^2 \theta} \right) = 0.$$

The solution of this equation is $\cos \theta = 1/\sqrt{3}$ or $\theta \approx 54.74°$.

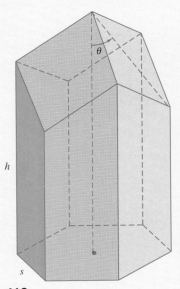

Hexagonal cell used in the construction of a honeycomb

See Exercise 82, Section 8.3.

Chapter Overview

In this chapter, we introduce the third basic class of elementary functions—trigonometric functions and inverse trigonometric functions. The first two sections in the chapter begin with a review of the definitions and graphs of the six trigonometric functions.

$$f(x) = \sin x \qquad f(x) = \cos x$$
$$f(x) = \tan x \qquad f(x) = \cot x$$
$$f(x) = \sec x \qquad f(x) = \csc x$$

Section 8.2 also discusses limits involving these six functions.

In Section 8.3, we look at differentiation rules for trigonometric functions. The section also gives some examples showing how calculus can be used to find minimum and maximum values of trigonometric functions.

Section 8.4 deals with integrals involving trigonometric functions. As you have come to expect by this time, the process of finding antiderivatives is not as straightforward as that of finding derivatives, so you should spend some extra time practicing these skills.

Sections 8.5 and 8.6 introduce the **inverse trigonometric functions.** Integration involving these functions is aided by a procedure called *completing the square*, and we illustrate this in Section 8.6.

The last section in the chapter contains a brief introduction to the **hyperbolic functions.** Hyperbolic functions are actually defined in terms of exponential functions, but we introduce them in this chapter because they have properties that are similar to trigonometric functions.

Trigonometric Functions and Inverse Trigonometric Functions

8.1 Review of Trigonometric Functions

Angles and degree measure ▪ Radian measure ▪ The trigonometric functions ▪ Evaluation of trigonometric functions ▪ Solving trigonometric equations

The concept of an angle is central to the study of trigonometry. As shown in Figure 8.1, an **angle** has three parts: an **initial ray,** a **terminal ray,** and a **vertex** (the point of intersection of the two rays). We say that an angle is in **standard position** if its initial ray coincides with the positive x-axis and its vertex is at the origin. We assume that you are familiar* with the degree measure of an angle. It is common practice to use θ (the Greek lowercase letter theta) to represent both an angle and its measure. We classify angles between $0°$ and $90°$ as **acute** and angles between $90°$ and $180°$ as **obtuse.** Positive angles are measured *counterclockwise* beginning with the initial ray. Negative angles are measured *clockwise*. For instance, Figure 8.2 shows an angle whose measure is $-45°$. We cannot assign a measure to an angle merely by knowing where its initial and terminal rays are located. To measure an angle, we must also know how the terminal ray was revolved. For example, Figure 8.2 shows that the angle measuring $-45°$ has the same terminal ray as the angle measuring $315°$. We call such angles **coterminal.**

An angle larger than $360°$ is one whose terminal ray has revolved more than one full revolution counterclockwise. Figure 8.3 shows an angle measuring more than $360°$. Similarly, we can generate angles whose measure is less than $-360°$ by revolving a terminal ray more than one full revolution clockwise.

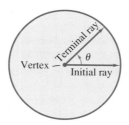

Standard Position of an Angle

FIGURE 8.1

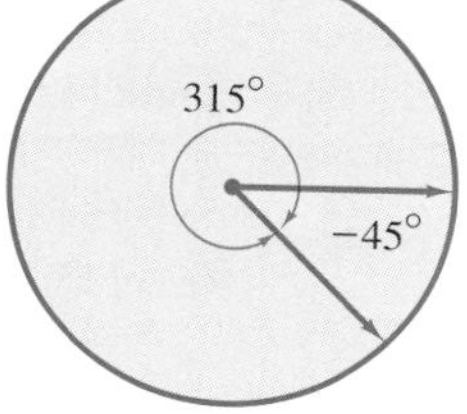

Coterminal Angles

FIGURE 8.2

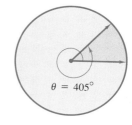

FIGURE 8.3

*For a more complete review of trigonometry, see *Algebra and Trigonometry*, 2nd Edition, by Larson and Hostetler (Lexington, Mass., D. C. Heath and Company, 1989).

419

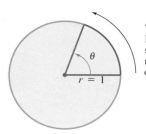

The arc length of the sector is the radian measure of θ.

Unit Circle

FIGURE 8.4

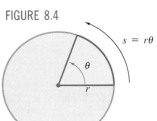

Circle of Radius r

FIGURE 8.5

Radian measure

A second way to measure angles is by radian measure. We will see in Section 8.2 that in calculus this measure is preferable to degree measure. To assign a radian measure to an angle θ, we consider θ to be the central angle of a circular sector of radius 1, as shown in Figure 8.4. The **radian measure** of θ is then defined to be the length of the arc of this sector. Recall that the total circumference of a circle is $2\pi r$. Thus, the circumference of a **unit circle** (that is, a circle of radius 1) is simply 2π, and we may conclude that the radian measure of an angle measuring 360° is 2π. In other words, $360° = 2\pi$ radians.

Using radian measure, we have a simple formula for the length s of a circular arc of radius r, as shown in Figure 8.5.

$$\text{arclength} = s = r\theta \qquad \theta \text{ measured in radians}$$

It is helpful to memorize the conversions of the common angles pictured in Figure 8.6. For other angles, you can use one of the following conversion rules.

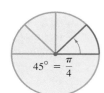

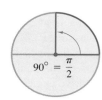
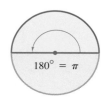
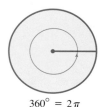

$$30° = \frac{\pi}{6} \qquad 45° = \frac{\pi}{4} \qquad 60° = \frac{\pi}{3} \qquad 90° = \frac{\pi}{2} \qquad 180° = \pi$$

$$360° = 2\pi$$

FIGURE 8.6   Radian and Degree Measure for Several Common Angles

---

CONVERSION RULES

$$180° = \pi \text{ radians}$$

Degrees $\Longrightarrow$ **Radians**   **Radians** $\Longrightarrow$ **Degrees**

$$1° = \frac{\pi}{180} \text{ radians} \qquad\qquad 1 \text{ radian} = \frac{180°}{\pi}$$

EXAMPLE 1   Conversions between degrees and radians

(a) $135° = (135 \text{ deg})\left(\dfrac{\pi \text{ rad}}{180 \text{ deg}}\right) = \dfrac{3\pi}{4} \text{ radians}$

(b) $40° = (40 \text{ deg})\left(\dfrac{\pi \text{ rad}}{180 \text{ deg}}\right) = \dfrac{2\pi}{9} \text{ radians}$

(c) $-\dfrac{\pi}{2} \text{ radians} = \left(-\dfrac{\pi}{2} \text{ rad}\right)\left(\dfrac{180 \text{ deg}}{\pi \text{ rad}}\right) = -90°$

(d) $\dfrac{9\pi}{2} \text{ radians} = \left(\dfrac{9\pi}{2} \text{ rad}\right)\left(\dfrac{180 \text{ deg}}{\pi \text{ rad}}\right) = 810°$

The trigonometric functions

There are two common approaches to the study of trigonometry. In one case the trigonometric functions are defined as ratios of two sides of a right triangle. In the other case these functions are defined in terms of a point on the terminal side of an angle in standard position. The first approach is the one generally used in surveying, navigation, and astronomy, where a typical problem involves a fixed triangle having three of its six parts (sides and angles) known and three to be determined. The second approach is the one normally used in physics, electronics, and biology, where the periodic nature of the trigonometric functions is emphasized. We define the six trigonometric functions, **sine, cosine, tangent, cotangent, secant,** and **cosecant** (abbreviated as sin, cos, etc.) from both viewpoints, as follows.

DEFINITION OF THE SIX TRIGONOMETRIC FUNCTIONS

Right triangle definitions, where $0 < \theta < \pi/2$. (Refer to Figure 8.7.)

$$\sin \theta = \frac{\text{opp.}}{\text{hyp.}} \qquad \csc \theta = \frac{\text{hyp.}}{\text{opp.}}$$

$$\cos \theta = \frac{\text{adj.}}{\text{hyp.}} \qquad \sec \theta = \frac{\text{hyp.}}{\text{adj.}}$$

$$\tan \theta = \frac{\text{opp.}}{\text{adj.}} \qquad \cot \theta = \frac{\text{adj.}}{\text{opp.}}$$

Circular function definitions, where θ is any angle. (Refer to Figure 8.8.)

$$\sin \theta = \frac{y}{r} \qquad \csc \theta = \frac{r}{y}$$

$$\cos \theta = \frac{x}{r} \qquad \sec \theta = \frac{r}{x}$$

$$\tan \theta = \frac{y}{x} \qquad \cot \theta = \frac{x}{y}$$

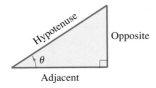

FIGURE 8.7

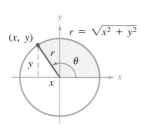

FIGURE 8.8

The following formulas are direct consequences of the definitions.

$$\csc \theta = \frac{1}{\sin \theta} \qquad \sec \theta = \frac{1}{\cos \theta} \qquad \cot \theta = \frac{1}{\tan \theta}$$

$$\tan \theta = \frac{\sin \theta}{\cos \theta} \qquad \cot \theta = \frac{\cos \theta}{\sin \theta}$$

Furthermore, since

$$\sin^2 \theta + \cos^2 \theta = \left(\frac{y}{r}\right)^2 + \left(\frac{x}{r}\right)^2 = \frac{x^2 + y^2}{r^2} = \frac{r^2}{r^2} = 1$$

we can readily obtain the Pythagorean Identity

$$\sin^2 \theta + \cos^2 \theta = 1.$$

Note that we use $\sin^2 \theta$ to mean $(\sin \theta)^2$. Additional trigonometric identities are listed next. (ϕ is the lowercase Greek letter phi.)

TRIGONOMETRIC IDENTITIES

Pythagorean identities:

$$\sin^2 \theta + \cos^2 \theta = 1$$
$$\tan^2 \theta + 1 = \sec^2 \theta$$
$$\cot^2 \theta + 1 = \csc^2 \theta$$

Sum or difference of two angles:

$$\sin (\theta \pm \phi) = \sin \theta \cos \phi \pm \cos \theta \sin \phi$$
$$\cos (\theta \pm \phi) = \cos \theta \cos \phi \mp \sin \theta \sin \phi$$
$$\tan (\theta \pm \phi) = \frac{\tan \theta \pm \tan \phi}{1 \mp \tan \theta \tan \phi}$$

Double angle formulas:

$$\sin 2\theta = 2 \sin \theta \cos \theta$$
$$\cos 2\theta = 2 \cos^2 \theta - 1$$
$$\qquad\quad = 1 - 2 \sin^2 \theta = \cos^2 \theta - \sin^2 \theta$$

Reduction formulas:

$$\sin (-\theta) = -\sin \theta$$
$$\cos (-\theta) = \cos \theta$$
$$\tan (-\theta) = -\tan \theta$$
$$\sin \theta = -\sin (\theta - \pi)$$
$$\cos \theta = -\cos (\theta - \pi)$$
$$\tan \theta = \tan (\theta - \pi)$$

Half angle formulas:

$$\sin^2 \theta = \frac{1}{2}(1 - \cos 2\theta)$$

$$\cos^2 \theta = \frac{1}{2}(1 + \cos 2\theta)$$

Law of cosines:

$$a^2 = b^2 + c^2 - 2bc \cos A$$

Law of Cosines

REMARK All angles in the remainder of this text are measured in radians unless stated otherwise. For example, when we write sin 3, we mean the sine of three radians, and when we write sin 3°, we mean the sine of three degrees.

Evaluation of trigonometric functions

There are two common methods of evaluating trigonometric functions: (1) decimal approximations with a calculator (or a table of trigonometric values) and (2) exact evaluations using trigonometric identities and formulas from geometry. We demonstrate the second method first.

EXAMPLE 2 Evaluating trigonometric functions

Evaluate the sine, cosine, and tangent of $\pi/3$.

SOLUTION

We begin by drawing the angle $\theta = \pi/3$ in the standard position, as shown in Figure 8.9. Then, since $60° = \pi/3$ radians, we obtain an equilateral triangle whose sides have a length of 1 and with θ as one of its angles. Since the altitude of this triangle bisects its base, we know that $x = \frac{1}{2}$. Now, using the Pythagorean Theorem, we have

$$y = \sqrt{r^2 - x^2} = \sqrt{1 - \left(\frac{1}{2}\right)^2} = \sqrt{\frac{3}{4}} = \frac{\sqrt{3}}{2}.$$

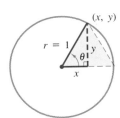

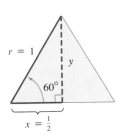

FIGURE 8.9

Thus,

$$\sin \frac{\pi}{3} = \frac{y}{r} = \frac{\sqrt{3}/2}{1} = \frac{\sqrt{3}}{2}$$

$$\cos \frac{\pi}{3} = \frac{x}{r} = \frac{1/2}{1} = \frac{1}{2}$$

$$\tan \frac{\pi}{3} = \frac{y}{x} = \frac{\sqrt{3}/2}{1/2} = \sqrt{3}.$$

The degree and radian measure of several common angles are given in Table 8.1 with the corresponding values of the sine, cosine, and tangent.

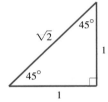

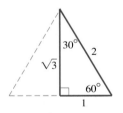

TABLE 8.1   Common First Quadrant Angles

Degrees	0	30°	45°	60°	90°
Radians	0	$\dfrac{\pi}{6}$	$\dfrac{\pi}{4}$	$\dfrac{\pi}{3}$	$\dfrac{\pi}{2}$
sin θ	0	$\dfrac{1}{2}$	$\dfrac{\sqrt{2}}{2}$	$\dfrac{\sqrt{3}}{2}$	1
cos θ	1	$\dfrac{\sqrt{3}}{2}$	$\dfrac{\sqrt{2}}{2}$	$\dfrac{1}{2}$	0
tan θ	0	$\dfrac{\sqrt{3}}{3}$	1	$\sqrt{3}$	undefined

In Figure 8.8, note that r is always positive. Thus, the quadrant signs of x and y determine the quadrant signs of the various trigonometric functions, as indicated in Figure 8.10. To extend the use of Table 8.1 to angles in quadrants other than the first quadrant, we can use the concept of a **reference angle** (as shown in Figure 8.11), together with the appropriate quadrant sign.

Quadrant II	Quadrant I
sin θ: +	sin θ: +
cos θ: −	cos θ: +
tan θ: −	tan θ: +
Quadrant III	Quadrant IV
sin θ: −	sin θ: −
cos θ: −	cos θ: +
tan θ: +	tan θ: −

FIGURE 8.10

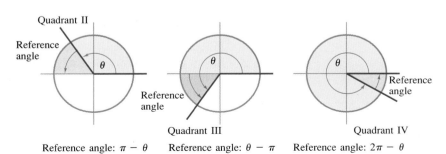

FIGURE 8.11

EXAMPLE 3 Evaluating trigonometric functions with reference angles

Function	Quadrant	Sign	Reference angle	Value
(a) $\sin \dfrac{3\pi}{4}$	II	$+$	$\pi - \dfrac{3\pi}{4} = \dfrac{\pi}{4}$	$\sin \dfrac{3\pi}{4} = \sin \dfrac{\pi}{4}$ $= \dfrac{\sqrt{2}}{2}$
(b) $\tan 330°$	IV	$-$	$360° - 330° = 30°$	$\tan 330° = -\tan 30°$ $= -\dfrac{\sqrt{3}}{3}$
(c) $\cos \dfrac{7\pi}{6}$	III	$-$	$\dfrac{7\pi}{6} - \pi = \dfrac{\pi}{6}$	$\cos \dfrac{7\pi}{6} = -\cos \dfrac{\pi}{6}$ $= -\dfrac{\sqrt{3}}{2}$

EXAMPLE 4 Trigonometric identities and calculators

(a) Using the reduction formula $\sin(-\theta) = -\sin \theta$, we have

$$\sin\left(-\frac{\pi}{3}\right) = -\sin \frac{\pi}{3} = -\frac{\sqrt{3}}{2}.$$

(b) Using the reciprocal formula $\sec \theta = 1/\cos \theta$, we have

$$\sec 60° = \frac{1}{\cos 60°} = \frac{1}{1/2} = 2.$$

(c) Using a calculator, we have

$$\cos(1.2) \approx 0.3624.$$

Remember that 1.2 is given in *radian* measure, so your calculator must be set in radian mode.

Solving trigonometric equations

In Examples 2, 3, and 4 we looked at techniques for evaluating trigonometric functions for given values of θ. In the next two examples, we look at the reverse problem. That is, if we are given the value of a trigonometric function, how can we solve for θ? For example, consider the equation

$$\sin \theta = 0.$$

We know $\theta = 0$ is one solution. But this is not the only solution. Any one of the following values of θ is also a solution.

$$\ldots, -3\pi, -2\pi, -\pi, 0, \pi, 2\pi, 3\pi, \ldots$$

We can write this infinite solution set as $\{n\pi: n \text{ is an integer}\}$.

EXAMPLE 5   Solving a trigonometric equation

Solve for θ in the following equation.

$$\sin \theta = -\frac{\sqrt{3}}{2}$$

SOLUTION

To solve the equation, we make two observations: the sine is negative in Quadrants III and IV, and $\sin (\pi/3) = \sqrt{3}/2$. By combining these two observations, we conclude that we are seeking values of θ in the third and fourth quadrants that have a reference angle of $\pi/3$. In the interval $[0, 2\pi]$, the two angles fitting these criteria are

$$\theta = \pi + \frac{\pi}{3} = \frac{4\pi}{3}$$

and

$$\theta = 2\pi - \frac{\pi}{3} = \frac{5\pi}{3}.$$

Finally, we can add $2n\pi$ to either of these angles to obtain the solution set

$$\theta = \frac{4\pi}{3} + 2n\pi, \quad n \text{ is an integer}$$

or

$$\theta = \frac{5\pi}{3} + 2n\pi, \quad n \text{ is an integer}.$$

EXAMPLE 6   Solving a trigonometric equation

Solve the following equation for θ.

$$\cos 2\theta = 2 - 3 \sin \theta, \quad 0 \le \theta \le 2\pi$$

SOLUTION

Using the double angle identity $\cos 2\theta = 1 - 2 \sin^2 \theta$, we obtain the following polynomial (in $\sin \theta$).

$$1 - 2 \sin^2 \theta = 2 - 3 \sin \theta$$
$$0 = 2 \sin^2 \theta - 3 \sin \theta + 1$$
$$0 = (2 \sin \theta - 1)(\sin \theta - 1)$$

If $2 \sin \theta - 1 = 0$, we have $\sin \theta = 1/2$ and $\theta = \pi/6$ or $\theta = 5\pi/6$. If $\sin \theta - 1 = 0$, we have $\sin \theta = 1$ and $\theta = \pi/2$. Thus, for $0 \le \theta \le 2\pi$, there are three solutions to the given equation.

$$\theta = \frac{\pi}{6}, \quad \frac{5\pi}{6}, \quad \text{or} \quad \frac{\pi}{2}$$

EXERCISES for Section 8.1

In Exercises 1 and 2, determine two coterminal angles (one positive and one negative) for the given angle. Give your answers in degrees.

1. (a)   (b)

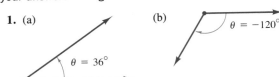

$\theta = 36°$ $\theta = -120°$

2. (a) $\theta = 300°$ (b) $\theta = -420°$

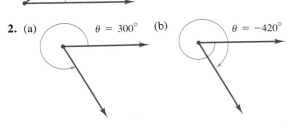

In Exercises 3 and 4, determine two coterminal angles (one positive and one negative) for the given angle. Give your answers in radians.

3. (a) $\theta = \dfrac{\pi}{9}$ (b) $\theta = \dfrac{4\pi}{3}$

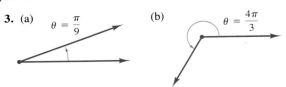

4. (a) $\theta = -\dfrac{9\pi}{4}$ (b) $\theta = \dfrac{8\pi}{9}$

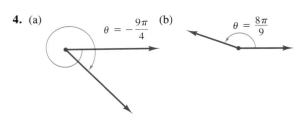

In Exercises 5 and 6, express the given angle in radian measure as a multiple of π.

5. (a) 30°   (b) 150°
    (c) 315°   (d) 120°

6. (a) −20°   (b) −240°
    (c) −270°   (d) 144°

In Exercises 7 and 8, express the given angle in degree measure.

7. (a) $\dfrac{3\pi}{2}$ (b) $\dfrac{7\pi}{6}$

    (c) $-\dfrac{7\pi}{12}$ (d) $\dfrac{\pi}{9}$

8. (a) $\dfrac{7\pi}{3}$ (b) $-\dfrac{11\pi}{30}$

    (c) $\dfrac{11\pi}{6}$ (d) $\dfrac{34\pi}{15}$

9. Let r represent the radius of a circle, θ the central angle (measured in radians), and s the length of the arc subtended by the angle. Use the relationship $\theta = s/r$ to complete the following table.

r	8 ft	15 in.	85 cm		
s	12 ft			96 in.	8642 mi
θ		1.6	$\dfrac{3\pi}{4}$	4	$\dfrac{2\pi}{3}$

10. The minute hand on a clock is $3\frac{1}{2}$ inches long (see figure). Through what distance does the tip of the minute hand move in 25 minutes?

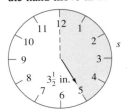

FIGURE FOR 10

11. A man bends his elbow through 75°. The distance from his elbow to the top of his index finger is $18\frac{3}{4}$ inches (see figure).
(a) Find the radian measure of this angle.
(b) Find the distance the tip of the index finger moves.

$18\frac{3}{4}$ in.   75°   FIGURE FOR 11

12. A tractor tire, 5 feet in diameter, is partially filled with a liquid ballast for additional traction. To check the air pressure, the tractor operator rotates the tire until the valve stem is at the top so that the liquid will not enter the gauge. On a given occasion, the operator notes that the tire must be rotated 80° to have the stem in the proper position.
(a) Find the radian measure of this rotation.
(b) How far must the tractor be moved to get the valve stem in the proper position?

In Exercises 13 and 14, determine all six trigonometric functions for the given angle θ.

13. (a)

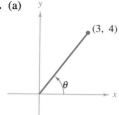

(b)

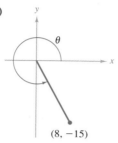

14. (a)

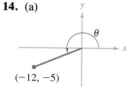

(b)

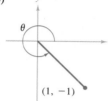

In Exercises 15 and 16, determine the quadrant in which θ lies.

15. $\sin \theta < 0$ and $\cos \theta < 0$
16. $\sin \theta > 0$ and $\cos \theta < 0$

In Exercises 17–22, find the indicated trigonometric functions from the given one. (Assume $0 < \theta < \pi/2$.)

17. Given: $\sin \theta = \dfrac{1}{2}$

Find: $\csc \theta$

18. Given: $\sin \theta = \dfrac{1}{3}$

Find: $\tan \theta$

19. Given: $\cos \theta = \dfrac{4}{5}$

Find: $\cot \theta$

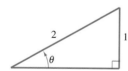

20. Given: $\sec \theta = \dfrac{13}{5}$

Find: $\cot \theta$

21. Given: $\cot \theta = \dfrac{15}{8}$

Find: $\sec \theta$

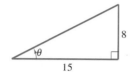

22. Given: $\tan \theta = \dfrac{1}{2}$

Find: $\sin \theta$

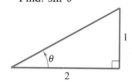

In Exercises 23–26, evaluate the sine, cosine, and tangent of the given angles *without* using a calculator.

23. (a) $60°$ (b) $\dfrac{2\pi}{3}$

(c) $\dfrac{\pi}{4}$ (d) $\dfrac{5\pi}{4}$

24. (a) $-\dfrac{\pi}{6}$ (b) $150°$

(c) $-\dfrac{\pi}{2}$ (d) $\dfrac{\pi}{2}$

25. (a) $225°$ (b) $-225°$
(c) $300°$ (d) $330°$

26. (a) $750°$ (b) $510°$

(c) $\dfrac{10\pi}{3}$ (d) $\dfrac{17\pi}{3}$

In Exercises 27–30, use a calculator to evaluate the given trigonometric functions to four significant digits.

27. (a) $\sin 10°$ (b) $\csc 10°$
28. (a) $\sec 225°$ (b) $\sec 135°$
29. (a) $\tan \dfrac{\pi}{9}$ (b) $\tan \dfrac{10\pi}{9}$
30. (a) $\cot (1.35)$ (b) $\tan (1.35)$

In Exercises 31–34, find two values of θ corresponding to the given functions. List the measure of θ in radians $(0 \leq \theta < 2\pi)$. Do not use a calculator.

31. (a) $\cos \theta = \dfrac{\sqrt{2}}{2}$ (b) $\cos \theta = -\dfrac{\sqrt{2}}{2}$
32. (a) $\sec \theta = 2$ (b) $\sec \theta = -2$
33. (a) $\tan \theta = 1$ (b) $\cot \theta = -\sqrt{3}$
34. (a) $\sin \theta = \dfrac{\sqrt{3}}{2}$ (b) $\sin \theta = -\dfrac{\sqrt{3}}{2}$

In Exercises 35–42, solve the given equation for θ $(0 \leq \theta < 2\pi)$. For some of the equations, you should use the trigonometric identities listed in this section.

35. $2 \sin^2 \theta = 1$ **36.** $\tan^2 \theta = 3$
37. $\tan^2 \theta - \tan \theta = 0$ **38.** $2 \cos^2 \theta - \cos \theta = 1$
39. $\sec \theta \csc \theta = 2 \csc \theta$ **40.** $\sin \theta = \cos \theta$
41. $\cos^2 \theta + \sin \theta = 1$ **42.** $\cos (\theta/2) - \cos \theta = 1$

In Exercises 43–46, solve for x, y, or r as indicated.

43. Solve for y.

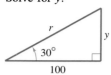

44. Solve for x.

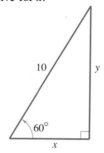

45. Solve for x.

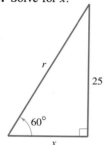

46. Solve for r.

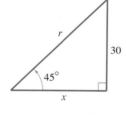

47. A 20-foot ladder leaning against the side of a house makes a 75° angle with the ground (see figure). How far up the side of the house does the ladder reach?

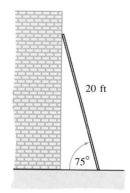

FIGURE FOR 47

48. A biologist wants to know the width w of a river in order to properly set instruments to study the pollutants in the water. From point A, the biologist walks downstream 100 feet and sights to point C. For this sighting, it is determined that $\theta = 50°$ (see figure). How wide is the river?

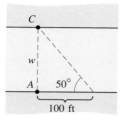

FIGURE FOR 48

49. From a 150-foot observation tower on the coast, a Coast Guard officer sights a boat in difficulty. The angle of depression of the boat is 4° (see figure). How far is the boat from the shoreline?

FIGURE FOR 49

50. A ramp $17\frac{1}{2}$ feet in length rises to a loading platform that is $3\frac{1}{2}$ feet off the ground (see figure). Find the angle that the ramp makes with the ground.

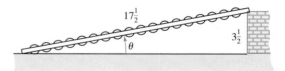

FIGURE FOR 50

8.2 Graphs and Limits of Trigonometric Functions

Graphs of trigonometric functions ▪ Limits of trigonometric functions

One of the first things we notice about the graphs of all six trigonometric functions is that they are periodic. We call a function f **periodic** if there exists a nonzero number p such that $f(x + p) = f(x)$ for all x in the domain of f. The smallest such positive value of p is called the **period** of f. Both the sine and cosine functions have a period of 2π and by plotting several values in the interval $0 \le x \le 2\pi$, we obtain the graphs shown in Figure 8.12.

Maximum Minimum

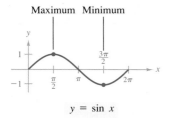

$$y = \sin x$$

Maximum Minimum Maximum

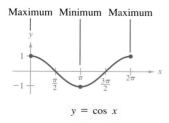

$$y = \cos x$$

FIGURE 8.12

Note in Figure 8.12 that the maximum value of sin x is 1 and the minimum value is -1. Figure 8.13 shows the graphs of all six trigonometric functions. Familiarity with these six basic graphs will serve as a valuable aid in sketching the graphs of more complicated trigonometric functions.

The graph of the function $y = a \sin bx$ oscillates between $-a$ and a and hence has an **amplitude** of $|a|$. Furthermore, since

$$bx = 0 \text{ when } x = 0 \qquad \text{and} \qquad bx = 2\pi \text{ when } x = \frac{2\pi}{b}$$

we may conclude that the function $y = a \sin bx$ has a period of $2\pi/|b|$. Table 8.2 summarizes the amplitudes and periods for some general types of trigonometric functions.

TABLE 8.2 Periods and Amplitudes of Trigonometric Functions

Function	Period	Amplitude
$y = a \sin bx \quad$ or $\quad y = a \cos bx$	$\dfrac{2\pi}{\|b\|}$	$\|a\|$
$y = a \tan bx \quad$ or $\quad y = a \cot bx$	$\dfrac{\pi}{\|b\|}$	not applicable
$y = a \sec bx \quad$ or $\quad y = a \csc bx$	$\dfrac{2\pi}{\|b\|}$	not applicable

Domain: all reals
Range: $[-1, 1]$
Period: 2π

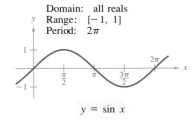

$$y = \sin x$$

Domain: all reals
Range: $[-1, 1]$
Period: 2π

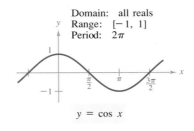

$$y = \cos x$$

Domain: all $x \neq (2n - 1)\dfrac{\pi}{2}$
Range: $(-\infty, \infty)$
Period: π

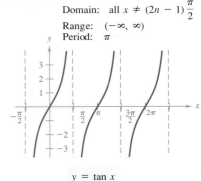

$$y = \tan x$$

Domain: all $x \neq n\pi$
Range: $(-\infty, -1]$ and $[1, \infty)$
Period: 2π

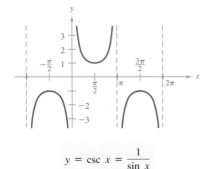

$$y = \csc x = \frac{1}{\sin x}$$

Domain: all $x \neq (2n - 1)\dfrac{\pi}{2}$
Range: $(-\infty, -1]$ and $[1, \infty)$
Period: 2π

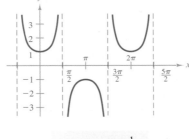

$$y = \sec x = \frac{1}{\cos x}$$

Domain: all $x \neq n\pi$
Range: $(-\infty, \infty)$
Period: π

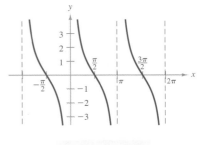

$$y = \cot x = \frac{1}{\tan x}$$

FIGURE 8.13

Graphs of the Six Trigonometric Functions

EXAMPLE 1 Sketching the graph of a trigonometric function

Sketch the graph of $f(x) = 3 \cos 2x$.

SOLUTION

The graph of $f(x) = 3 \cos 2x$ has the following characteristics.

amplitude: 3   period: $\dfrac{2\pi}{2} = \pi$

Using the basic shape of the graph of the cosine function, we sketch one period of the function on the interval $[0, \pi]$, using the following pattern.

maximum: $(0, 3)$ minimum: $\left(\dfrac{\pi}{2}, -3\right)$ maximum: $(\pi, 3)$

Then, by continuing this pattern, we sketch several cycles of the graph, as shown in Figure 8.14.

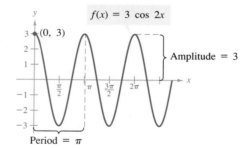

FIGURE 8.14

The discussion of horizontal shifts, vertical shifts, and reflections given in Section 1.5 can be applied to the graphs of trigonometric functions. For instance, Figure 8.15 shows three different shifted (or reflected) graphs of sine functions.

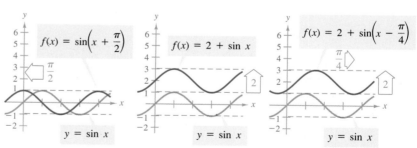

FIGURE 8.15   Horizontal shift to the left   Vertical shift upward   Horizontal and vertical shift

EXAMPLE 2 Sketching the graph of a trigonometric function

Sketch the graph of

$$f(x) = 2 \sin\left(3x - \dfrac{\pi}{2}\right).$$

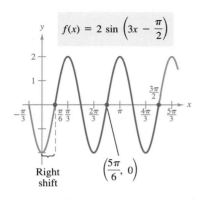

FIGURE 8.16

SOLUTION

We write

$$f(x) = 2 \sin \left[3\left(x - \frac{\pi}{6} \right) \right]$$

and observe that the graph of f has the following characteristics.

amplitude: 2   period: $\dfrac{2\pi}{3}$ right shift: $\dfrac{\pi}{6}$

Since the shift is $\pi/6$, we start one cycle at $x = \pi/6$. Then, since the period is $2\pi/3$, this cycle ends at $x = (\pi/6) + (2\pi/3) = 5\pi/6$. Two complete cycles are shown in Figure 8.16. ▭

To sketch the graph of a function that combines an algebraic function with a trigonometric function, it is helpful to use a technique called **addition of ordinates** as demonstrated in the next example.

EXAMPLE 3 Addition of ordinates

Sketch the graph of $f(x) = x + \cos x$ on the interval $0 \le x \le 2\pi$.

SOLUTION

We first make an accurate sketch of the graphs of $y = x$ and $y = \cos x$ on the same coordinate plane and then geometrically add the ordinates (y-values) for each x. This addition is aided by the use of a compass to measure the displacement of one graph from the x-axis and then mark off an equal displacement from the other graph, as shown in Figure 8.17.

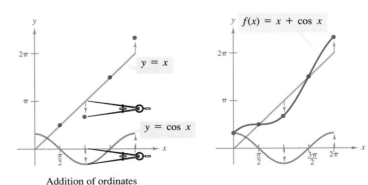

FIGURE 8.17   Addition of ordinates ▭

Limits of trigonometric functions

In the next section we will show how the derivatives of the trigonometric functions can assist us in sketching more complicated graphs. However, before we discuss these derivatives, we need to obtain some results regarding the limits of trigonometric functions. In Chapter 2, we saw that the limits of many algebraic functions can be evaluated by direct substitution. The following theorem (which we present without proof) tells us that each of the six basic trigonometric functions also possesses this desirable quality.

**THEOREM 8.1
LIMITS OF TRIGONOMETRIC
FUNCTIONS**

If c is a real number *in the domain of the given trigonometric function*, then the following properties are true.

1. $\lim\limits_{x \to c} \sin x = \sin c$ 2. $\lim\limits_{x \to c} \cos x = \cos c$

3. $\lim\limits_{x \to c} \tan x = \tan c$ 4. $\lim\limits_{x \to c} \cot x = \cot c$

5. $\lim\limits_{x \to c} \sec x = \sec c$ 6. $\lim\limits_{x \to c} \csc x = \csc c$

REMARK   From Theorem 8.1, it follows that each of the six trigonometric functions is *continuous* at every point in its domain.

EXAMPLE 4   Limits involving trigonometric functions

(a) By Theorem 8.1, we have

$$\lim_{x \to 0} \sin x = \sin (0) = 0.$$

(b) By Theorem 8.1, and Property 3 of Theorem 2.3, we have

$$\lim_{x \to \pi} (x \cos x) = \left[\lim_{x \to \pi} x\right]\left[\lim_{x \to \pi} \cos x\right] = \pi \cos (\pi) = -\pi. \quad \blacksquare$$

EXAMPLE 5   A limit involving trigonometric functions

Find the following limit.

$$\lim_{x \to 0} \frac{\tan x}{\sin x}$$

SOLUTION

Direct substitution yields the indeterminate form $0/0$. However, by using the fact that $\tan x = (\sin x)/(\cos x)$, we can rewrite the function as

$$\frac{\tan x}{\sin x} = \frac{(\sin x)/(\cos x)}{\sin x} = \frac{\sin x}{(\cos x)(\sin x)} = \frac{1}{\cos x}.$$

Thus, by Theorem 2.1, we have

$$\lim_{x \to 0} \frac{\tan x}{\sin x} = \lim_{x \to 0} \frac{1}{\cos x} = \frac{1}{1} = 1. \quad \blacksquare$$

REMARK   After we learn how to differentiate trigonometric functions, we can handle problems like the one in Example 5 with L'Hôpital's Rule.

EXAMPLE 6   A limit that does not exist

Discuss the existence of the limit.

$$\lim_{x \to 0} \sin \left(\frac{1}{x} \right)$$

SOLUTION

We let $f(x) = \sin (1/x)$. In Figure 8.18, we see that as x approaches 0, $f(x)$ oscillates between -1 and 1. Therefore, the limit does not exist because no matter how small we choose δ, it is possible to choose x_1 and x_2 within δ units of 0 such that $\sin (1/x_1) = 1$ and $\sin (1/x_2) = -1$, as indicated in Table 8.3.

TABLE 8.3

x	$\dfrac{2}{\pi}$	$\dfrac{2}{3\pi}$	$\dfrac{2}{5\pi}$	$\dfrac{2}{7\pi}$	$\dfrac{2}{9\pi}$	$\dfrac{2}{11\pi}$	$x \to 0$
$\sin \dfrac{1}{x}$	1	-1	1	-1	1	-1	Limit does not exist.

EXAMPLE 7   Testing for continuity

Find the intervals for which the two functions shown in Figure 8.19 are continuous.

SOLUTION

(a) The tangent function is undefined at $x = (\pi/2) + n\pi$. At all other points it is continuous. Thus, $f(x) = \tan x$ is continuous on the open intervals

$$\dots, \left(-\frac{3\pi}{2}, -\frac{\pi}{2} \right), \left(-\frac{\pi}{2}, \frac{\pi}{2} \right), \left(\frac{\pi}{2}, \frac{3\pi}{2} \right), \dots$$

(b) This function is similar to that in Example 6 except that the oscillations are damped by the factor x. Using the Squeeze Theorem, we have

$$-|x| \leq x \sin \frac{1}{x} \leq |x|, \quad x \neq 0$$

and we conclude that the limit as $x \to 0$ is zero. Thus, f is continuous on the entire real line.

In the next section, we will see that the following two important limits are useful in determining the derivatives of trigonometric functions.

$f(x) = \sin \dfrac{1}{x}$

$\lim_{x \to 0} f(x)$ does not exist.

FIGURE 8.18

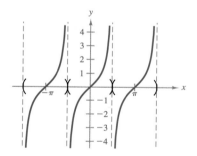

$f(x) = \tan x$

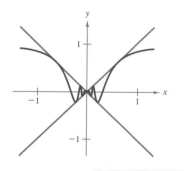

$$f(x) = \begin{cases} x \sin \dfrac{1}{x}, & x \neq 0 \\ 0, & x = 0 \end{cases}$$

FIGURE 8.19

THEOREM 8.2 TWO SPECIAL TRIGONOMETRIC LIMITS	1. $\lim\limits_{x \to 0} \dfrac{\sin x}{x} = 1$	2. $\lim\limits_{x \to 0} \dfrac{1 - \cos x}{x} = 0$

PROOF We prove the first limit and leave the proof of the second as an exercise (see Exercise 65). (Note that direct substitution produces the indeterminate form $0/0$ in both cases.) To avoid the confusion of two different uses of x, we present the proof using the variable θ, where θ is an acute positive angle (measured in radians). Figure 8.20 shows a circular section that is squeezed

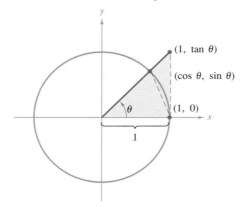

FIGURE 8.20

between two triangles. The areas of the three figures below satisfy the inequality that follows.

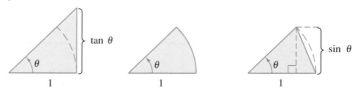

$$\text{Area of triangle} \geq \text{Area of sector} \geq \text{Area of triangle}$$

$$\frac{\tan \theta}{2} \geq \frac{\theta}{2} \geq \frac{\sin \theta}{2}$$

Multiplying each expression by $2/\sin \theta$ produces

$$\frac{1}{\cos \theta} \geq \frac{\theta}{\sin \theta} \geq 1$$

and taking reciprocals and reversing the inequalities, we have

$$\cos \theta \leq \frac{\sin \theta}{\theta} \leq 1.$$

Since $\cos \theta = \cos (-\theta)$ and $(\sin \theta)/\theta = [\sin (-\theta)]/(-\theta)$, we can conclude that this inequality is valid for all nonzero θ in the open interval $(-\pi/2, \pi/2)$. Finally, since

$$\lim\limits_{\theta \to 0} \cos \theta = 1 \qquad \text{and} \qquad \lim\limits_{\theta \to 0} 1 = 1$$

we can apply the Squeeze Theorem to conclude that

$$\lim_{\theta \to 0} \frac{\sin \theta}{\theta} = 1.$$

EXAMPLE 8   A limit involving trigonometric functions

Find the following limit.

$$\lim_{x \to 0} \frac{\tan x}{x}$$

SOLUTION

Direct substitution yields the indeterminate form 0/0. To solve this problem, we write $\tan x = (\sin x)/(\cos x)$ and obtain

$$\lim_{x \to 0} \frac{\tan x}{x} = \lim_{x \to 0} \left(\frac{\sin x}{x} \right)\left(\frac{1}{\cos x} \right).$$

Now, since

$$\lim_{x \to 0} \frac{\sin x}{x} = 1 \qquad \text{and} \qquad \lim_{x \to 0} \frac{1}{\cos x} = 1$$

we have

$$\lim_{x \to 0} \frac{\tan x}{x} = \left[\lim_{x \to 0} \frac{\sin x}{x} \right]\left[\lim_{x \to 0} \sec x \right] = (1)(1) = 1. \qquad \blacksquare$$

EXAMPLE 9   A limit involving trigonometric functions

Find the following limit.

$$\lim_{x \to 0} \frac{\sin 2x}{x}$$

SOLUTION

Direct substitution gives the indeterminate form 0/0. To solve this problem, we rewrite the limit as

$$\lim_{x \to 0} \frac{\sin 2x}{x} = 2\left[\lim_{x \to 0} \frac{\sin 2x}{2x} \right].$$

Now, by letting $y = 2x$ and observing that $x \to 0$ if and only if $y \to 0$, we can write

$$\lim_{x \to 0} \frac{\sin 2x}{x} = 2\left[\lim_{x \to 0} \frac{\sin 2x}{2x} \right] = 2\left[\lim_{y \to 0} \frac{\sin y}{y} \right] = 2[1] = 2. \qquad \blacksquare$$

EXERCISES for Section 8.2

In Exercises 1–10, determine the period and amplitude of the given function.

1. $y = 2 \sin 2x$

2. $y = \dfrac{1}{2} \sin \pi x$

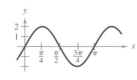

3. $y = \dfrac{5}{2} \cos \dfrac{\pi x}{2}$

4. $y = \dfrac{3}{2} \cos \dfrac{x}{2}$

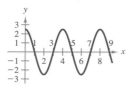

5. $y = -2 \sin \dfrac{x}{3}$

6. $y = -\cos \dfrac{2x}{3}$

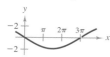

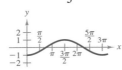

7. $y = -2 \sin 10x$

8. $y = \dfrac{1}{2} \cos \dfrac{2x}{3}$

9. $y = 3 \sin 4\pi x$

10. $y = \dfrac{2}{3} \cos \dfrac{\pi x}{10}$

In Exercises 11–14, find the period of the given function.

11. $y = 5 \tan 2x$

12. $y = 7 \tan 2\pi x$

13. $y = \sec 5x$

14. $y = \csc 4x$

In Exercises 15–28, sketch the graph of the given function.

15. $y = \sin \dfrac{x}{2}$

16. $y = 2 \cos 2x$

17. $y = -2 \sin 6x$

18. $y = \cos 2\pi x$

19. $y = -\sin \dfrac{2\pi x}{3}$

20. $y = 2 \tan x$

21. $y = \csc \dfrac{x}{2}$

22. $y = \tan 2x$

23. $y = 2 \sec 2x$

24. $y = \csc 2\pi x$

25. $y = \sin (x + \pi)$

26. $y = \cos \left(2x - \dfrac{\pi}{3} \right)$

27. $y = 1 + \cos \left(x - \dfrac{\pi}{2} \right)$

28. $y = -1 + \sin \left(x + \dfrac{\pi}{2} \right)$

In Exercises 29–32, use addition of ordinates to sketch the graph of the given function.

29. $y = x + \cos x$

30. $y = x + 2 \sin x$

31. $f(x) = \sin x + \sin 2x$

32. $f(x) = \sin x + \cos 2x$

In Exercises 33–50, determine the limit (if it exists).

33. $\displaystyle \lim_{x \to 0} \frac{\sin x}{5x}$

34. $\displaystyle \lim_{x \to 0} \frac{3(1 - \cos x)}{x}$

35. $\displaystyle \lim_{\theta \to 0} \frac{\sec \theta - 1}{\theta \sec \theta}$

36. $\displaystyle \lim_{\theta \to 0} \frac{\cos \theta \tan \theta}{\theta}$

37. $\displaystyle \lim_{x \to 0} \frac{\sin^2 x}{x}$

38. $\displaystyle \lim_{\phi \to \pi} \phi \sec \phi$

39. $\displaystyle \lim_{x \to \pi/2} \frac{\cos x}{\cot x}$

40. $\displaystyle \lim_{x \to \pi/4} \frac{1 - \tan x}{\sin x - \cos x}$

41. $\displaystyle \lim_{t \to 0} \frac{\sin^2 t}{t^2}$ $\left[\text{Hint: Find } \displaystyle \lim_{t \to 0} \left(\frac{\sin t}{t} \right)^2. \right]$

42. $\displaystyle \lim_{t \to 0} \frac{\sin 3t}{t}$ $\left[\text{Hint: Find } \displaystyle \lim_{t \to 0} 3\left(\frac{\sin 3t}{3t} \right). \right]$

43. $\displaystyle \lim_{x \to 0} \frac{\sin 2x}{\sin 3x}$

$\left[\text{Hint: Find } \displaystyle \lim_{x \to 0} \left(\frac{2 \sin 2x}{2x} \right) \left(\frac{3x}{3 \sin 3x} \right). \right]$

44. $\displaystyle \lim_{x \to 0} \frac{\tan^2 x}{x}$

45. $\displaystyle \lim_{h \to 0} \frac{(1 - \cos h)^2}{h}$

46. $\displaystyle \lim_{h \to 0} (1 + \cos 2h)$

47. $\displaystyle \lim_{x \to \pi} \cot x$

48. $\displaystyle \lim_{x \to \pi/2} \sec x$

49. $\displaystyle \lim_{x \to 0^+} \frac{2}{\sin x}$

50. $\displaystyle \lim_{x \to \pi/2^+} \frac{-2}{\cos x}$

In Exercises 51–56, find the discontinuities (if any) for the given function. Which of the discontinuities are removable?

51. $f(x) = x + \sin x$

52. $f(x) = \cos \dfrac{\pi x}{2}$

53. $f(x) = \csc 2x$

54. $f(x) = \tan \dfrac{\pi x}{2}$

55. $f(x) = \begin{cases} \csc \dfrac{\pi x}{6}, & |x - 3| < 2 \\ 2, & |x - 3| \geq 2 \end{cases}$

56. $f(x) = \begin{cases} \tan \dfrac{\pi x}{4}, & |x| < 1 \\ x, & |x| \geq 1 \end{cases}$

In Exercises 57 and 58, use a computer or graphics calculator to sketch the graph of the given functions on the same coordinate axes where x is in the interval [0, 2].

57. (a) $y = \dfrac{4}{\pi} \sin \pi x$

(b) $y = \dfrac{4}{\pi} \left(\sin \pi x + \dfrac{1}{3} \sin 3\pi x \right)$

58. (a) $y = \dfrac{1}{2} - \dfrac{4}{\pi^2} \cos \pi x$

(b) $y = \dfrac{1}{2} - \dfrac{4}{\pi^2} \left(\cos \pi x + \dfrac{1}{9} \cos 3\pi x \right)$

In Exercises 59 and 60, use a computer or graphics calculator to sketch the graph of the function f and find the specified limit (if it exists).

59. $f(x) = \dfrac{\sin 5x}{\sin 2x}$, $\lim\limits_{x \to 0} f(x)$

60. $f(x) = \dfrac{1 - \cos 3x}{2x}$, $\lim\limits_{x \to 0} f(x)$

61. Use a computer or graphics calculator to sketch the graph of

$$f(x) = \frac{\sin x}{x}, \quad x \neq 0.$$

Complete a table showing the values of $f(x)$ for $x = \pm 0.1$, $x = \pm 0.01$, and $x = \pm 0.001$.

62. Use a computer or graphics calculator to sketch the graph of the function f and determine if it is continuous on the entire real line.

$$f(x) = \begin{cases} \dfrac{\cos x - 1}{x}, & x < 0 \\ 5x, & x \geq 0 \end{cases}$$

63. Sales S, in thousands of units, of a seasonal product are given by

$$S = 58.3 + 32.5 \cos \frac{\pi t}{6}$$

where t is the time in months (with $t = 1$ corresponding to January and $t = 12$ corresponding to December). Use a computer to sketch the graph of S and determine the months when sales exceed 75,000 units.

64. When tuning a piano, a technician strikes a tuning fork for the A above middle C and sets up a wave motion that can be approximated by

$$y = 0.001 \sin 880\pi t$$

where t is the time in seconds.
(a) What is the period p of this function?
(b) What is the frequency f of this note? ($f = 1/p$)
(c) Sketch the graph of this function.

65. Prove the second part of Theorem 8.2.

8.3 Derivatives of Trigonometric Functions

Derivatives of sine and cosine functions ▪ Derivatives of other trigonometric functions ▪ Applications

In the previous section, we discussed the following limits.

$$\lim_{\Delta x \to 0} \frac{\sin \Delta x}{\Delta x} = 1 \quad \text{and} \quad \lim_{\Delta x \to 0} \frac{1 - \cos \Delta x}{\Delta x} = 0$$

These two limits are crucial in the proofs of the derivatives of the sine and cosine functions. The derivatives of the other four trigonometric functions follow easily from these two.

**THEOREM 8.3
DERIVATIVES OF
SINE AND COSINE**

$$\frac{d}{dx}[\sin x] = \cos x \qquad \frac{d}{dx}[\cos x] = -\sin x$$

PROOF We prove the first of these two rules and leave the proof of the second to you.

$$\frac{d}{dx}[\sin x] = \lim_{\Delta x \to 0} \frac{\sin(x + \Delta x) - \sin x}{\Delta x}$$

$$= \lim_{\Delta x \to 0} \frac{\sin x \cos \Delta x + \cos x \sin \Delta x - \sin x}{\Delta x}$$

$$= \lim_{\Delta x \to 0} \frac{\cos x \sin \Delta x - \sin x(1 - \cos \Delta x)}{\Delta x}$$

$$= \lim_{\Delta x \to 0} \left[\cos x \left(\frac{\sin \Delta x}{\Delta x} \right) - \sin x \left(\frac{1 - \cos \Delta x}{\Delta x} \right) \right]$$

$$= \cos x \left[\lim_{\Delta x \to 0} \frac{\sin \Delta x}{\Delta x} \right] - \sin x \left[\lim_{\Delta x \to 0} \frac{1 - \cos \Delta x}{\Delta x} \right]$$

$$= (\cos x)(1) - (\sin x)(0)$$

$$= \cos x$$

This differentiation formula is shown graphically in Figure 8.21. Note that for each x the *slope* of the sine curve determines the *value* of the cosine curve.

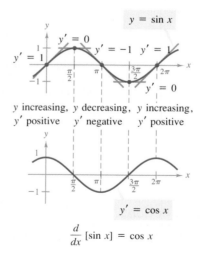

FIGURE 8.21

When taking the derivatives of trigonometric functions, the standard differentiation rules still apply—the Sum Rule, the Constant Multiple Rule, the Product Rule, and so on.

EXAMPLE 1 Derivatives involving sines and cosines

Function	Derivative
(a) $y = 3 \sin x$	$y' = 3 \cos x$
(b) $y = x + \cos x$	$y' = 1 - \sin x$
(c) $y = \dfrac{\sin x}{2} = \dfrac{1}{2} \sin x$	$y' = \dfrac{1}{2} \cos x = \dfrac{\cos x}{2}$

EXAMPLE 2 A derivative involving the Product Rule

Find the derivative of $y = 2x \cos x - 2 \sin x$.

SOLUTION

$$\frac{dy}{dx} = \overbrace{(2x)\left(\frac{d}{dx}[\cos x]\right) + (\cos x)\left(\frac{d}{dx}[2x]\right)}^{\text{Product Rule}} - \overbrace{2\frac{d}{dx}[\sin x]}^{\text{Constant Multiple Rule}}$$

$$= (2x)(-\sin x) + (\cos x)(2) - 2(\cos x)$$

$$= -2x \sin x$$

EXAMPLE 3 Using the derivative to find the slope of a curve

Find the slope of the graph of $f(x) = 2 \cos x$ at the following points.

(a) $\left(-\dfrac{\pi}{2}, 0\right)$ (b) $\left(\dfrac{\pi}{3}, 1\right)$ (c) $(\pi, -2)$

SOLUTION

The derivative of f is $f'(x) = -2 \sin x$. Therefore, the slopes at the indicated points are as follows.

(a) At $x = -\dfrac{\pi}{2}$, the slope is

$$f'\left(-\frac{\pi}{2}\right) = -2 \sin\left(-\frac{\pi}{2}\right) = -2(-1) = 2.$$

(b) At $x = \dfrac{\pi}{3}$, the slope is

$$f'\left(\frac{\pi}{3}\right) = -2 \sin\frac{\pi}{3} = -2\left(\frac{\sqrt{3}}{2}\right) = -\sqrt{3}.$$

(c) At $x = \pi$, the slope is

$$f'(\pi) = -2 \sin \pi = -2(0) = 0.$$

(See Figure 8.22.)

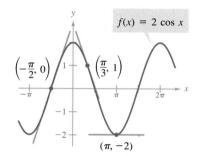

$f(x) = 2 \cos x$

$\left(-\dfrac{\pi}{2}, 0\right)$ $\left(\dfrac{\pi}{3}, 1\right)$

$(\pi, -2)$

FIGURE 8.22

Derivatives of the tangent, cotangent, secant, and cosecant functions

Knowing the derivatives of the sine and cosine functions, we can use the Quotient Rule to find the derivatives of the four remaining trigonometric functions.

THEOREM 8.4 **DERIVATIVES OF TANGENT,** **COTANGENT, SECANT,** **AND COSECANT**	$\dfrac{d}{dx}[\tan x] = \sec^2 x$ $\dfrac{d}{dx}[\sec x] = \sec x \tan x$	$\dfrac{d}{dx}[\cot x] = -\csc^2 x$ $\dfrac{d}{dx}[\csc x] = -\csc x \cot x$

PROOF   We give a proof for the derivative of the tangent function and leave the proofs of the remaining three formulas as exercises (see Exercise 88). Considering $\tan x = (\sin x)/(\cos x)$ and applying the Quotient Rule, we obtain

$$\frac{d}{dx}[\tan x] = \frac{(\cos x)(\cos x) - (\sin x)(-\sin x)}{\cos^2 x}$$

$$= \frac{\cos^2 x + \sin^2 x}{\cos^2 x} = \frac{1}{\cos^2 x} = \sec^2 x.$$

EXAMPLE 4   Differentiating trigonometric functions

Function	*Derivative*
(a) $y = x - \tan x$	$\dfrac{dy}{dx} = 1 - \sec^2 x$
(b) $y = x \sec x$	$y' = x(\sec x \tan x) + (\sec x)(1)$ $= (\sec x)(1 + x \tan x)$

Because of the abundance of trigonometric identities, the derivative of a trigonometric function can take many forms. This presents a challenge when you are trying to match your answer to one given in the back of the text. To help, we will occasionally list more than one form of the answer to trigonometric problems. The next example illustrates the possible diversity of trigonometric forms.

EXAMPLE 5   Different forms of a derivative

Differentiate the function

$$y = \frac{1 - \cos x}{\sin x} = \csc x - \cot x$$

in two forms and show that the derivatives are equal.

SOLUTION

For the first form, we can write

$$y = \frac{1 - \cos x}{\sin x}$$

$$y' = \frac{(\sin x)(\sin x) - (1 - \cos x)(\cos x)}{\sin^2 x}$$

$$= \frac{\sin^2 x + \cos^2 x - \cos x}{\sin^2 x}$$

$$= \frac{1 - \cos x}{\sin^2 x}.$$

For the second form, we can write

$$y = \csc x - \cot x$$

$$y' = -\csc x \cot x + \csc^2 x.$$

To show that these two derivatives are equal, we write

$$\frac{1 - \cos x}{\sin^2 x} = \frac{1}{\sin^2 x} - \frac{\cos x}{\sin^2 x}$$

$$= \frac{1}{\sin^2 x} - \left(\frac{1}{\sin x}\right)\left(\frac{\cos x}{\sin x}\right)$$

$$= \csc^2 x - \csc x \cot x.$$

With the Chain Rule, we can extend the six trigonometric differentiation rules to cover composite functions and we summarize the "Chain Rule Versions" of the six basic formulas as follows.

DERIVATIVES OF TRIGONOMETRIC FUNCTIONS

$$\frac{d}{dx}[\sin u] = \cos u \frac{du}{dx} \qquad \frac{d}{dx}[\cos u] = -\sin u \frac{du}{dx}$$

$$\frac{d}{dx}[\tan u] = \sec^2 u \frac{du}{dx} \qquad \frac{d}{dx}[\cot u] = -\csc^2 u \frac{du}{dx}$$

$$\frac{d}{dx}[\sec u] = \sec u \tan u \frac{du}{dx} \qquad \frac{d}{dx}[\csc u] = -\csc u \cot u \frac{du}{dx}$$

EXAMPLE 6 Applying the Chain Rule to trigonometric functions

Function	Derivative
(a) $y = \sin 2x$	$y' = \cos 2x \dfrac{d}{dx}[2x] = (\cos 2x)(2) = 2\cos 2x$
(b) $y = \cos (x - 1)$	$y' = -\sin (x - 1)$
(c) $y = \tan e^x$	$y' = e^x \sec^2 e^x$

Be sure that you understand the mathematical conventions regarding parentheses and trigonometric functions. For instance, in part (a) of Example 6 we write sin $2x$ to mean sin $(2x)$. The next example shows the effect of different placements of parentheses.

EXAMPLE 7 Parentheses and trigonometric functions

Function	*Derivative*
(a) $y = \cos 3x^2 = \cos (3x^2)$	$y' = (-\sin 3x^2)(6x) = -6x \sin 3x^2$
(b) $y = (\cos 3)x^2$	$y' = (\cos 3)(2x) = 2x \cos 3$
(c) $y = \cos (3x)^2 = \cos (9x^2)$	$y' = (-\sin 9x^2)(18x) = -18x \sin 9x^2$
(d) $y = \cos^2 3x = (\cos 3x)^2$	$y' = (2 \cos 3x)D_x[\cos 3x]$
	$\quad = 2(\cos 3x)(-\sin 3x)(3)$
	$\quad = -6 \cos 3x \sin 3x$

EXAMPLE 8 Differentiating a composite function

Differentiate

$$f(t) = \sqrt{\sin 4t}.$$

SOLUTION

First we write

$$f(t) = (\sin 4t)^{1/2}.$$

Then, by the Power Rule, we have

$$f'(t) = \frac{1}{2}(\sin 4t)^{-1/2} \frac{d}{dt} [\sin 4t]$$

$$= \frac{1}{2}(\sin 4t)^{-1/2}(4 \cos 4t)$$

$$= \frac{2 \cos 4t}{\sqrt{\sin 4t}}.$$

Applications

In the remainder of this section, we review some applications of the derivative in the context of trigonometric functions. We begin with an application to minimum and maximum values of a function.

EXAMPLE 9 Finding extrema on a closed interval

Find the extrema of $f(x) = 2 \sin x - \cos 2x$ on the interval $[0, 2\pi]$.

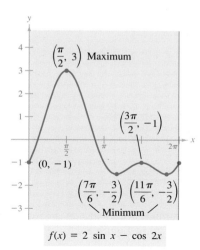

$f(x) = 2 \sin x - \cos 2x$

FIGURE 8.23

SOLUTION

This function is differentiable for all real x, so we can find all critical numbers by setting $f'(x)$ equal to zero, as follows.

$$f'(x) = 2 \cos x + 2 \sin 2x = 0$$
$$2 \cos x + 4 \cos x \sin x = 0 \qquad \text{sin } 2x = 2 \cos x \sin x$$
$$2(\cos x)(1 + 2 \sin x) = 0 \qquad \text{Factor}$$

By setting the two factors equal to zero and solving for x in the interval $[0, 2\pi]$, we have the following.

$$\cos x = 0 \quad \Longrightarrow \quad x = \frac{\pi}{2}, \frac{3\pi}{2} \qquad \text{Critical numbers}$$

$$\sin x = -\frac{1}{2} \quad \Longrightarrow \quad x = \frac{7\pi}{6}, \frac{11\pi}{6} \qquad \text{Critical numbers}$$

Finally, by evaluating f at these four critical numbers and at the endpoints of the interval, we conclude that the maximum is $f(\pi/2) = 3$ and the minimum occurs at *two* points, $f(7\pi/6) = -3/2$ and $f(11\pi/6) = -3/2$, as indicated in Table 8.4. The graph is shown in Figure 8.23.

TABLE 8.4

Left endpoint	Critical number	Critical number	Critical number	Critical number	Right endpoint
$f(0) = -1$	$f\left(\dfrac{\pi}{2}\right) = 3$ Maximum	$f\left(\dfrac{7\pi}{6}\right) = -\dfrac{3}{2}$ Minimum	$f\left(\dfrac{3\pi}{2}\right) = -1$	$f\left(\dfrac{11\pi}{6}\right) = -\dfrac{3}{2}$ Minimum	$f(2\pi) = -1$

EXAMPLE 10 Applying the First Derivative Test

Use the First Derivative Test to find all relative maxima and minima for the function given by

$$f(x) = \frac{x}{2} - \sin x$$

on the interval $(0, 2\pi)$.

SOLUTION

$$f'(x) = \frac{1}{2} - \cos x = 0 \quad \Longrightarrow \quad \cos x = \frac{1}{2}$$

Therefore, in the interval $(0, 2\pi)$, the critical numbers are $x = \pi/3$ and $x = 5\pi/3$. Table 8.5 summarizes the application of the First Derivative Test to these critical numbers.

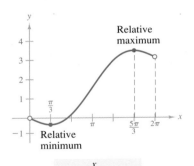

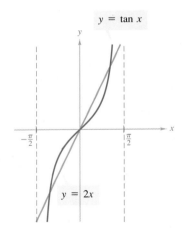

$$f(x) = \frac{x}{2} - \sin x$$

FIGURE 8.24

TABLE 8.5

Interval	$0 < x < \dfrac{\pi}{3}$	$\dfrac{\pi}{3} < x < \dfrac{5\pi}{3}$	$\dfrac{5\pi}{3} < x < 2\pi$
Test value	$x = \dfrac{\pi}{4}$	$x = \pi$	$x = \dfrac{7\pi}{4}$
Sign of $f'(x)$	$f'\left(\dfrac{\pi}{4}\right) < 0$	$f'(\pi) > 0$	$f'\left(\dfrac{7\pi}{4}\right) < 0$
Conclusion	decreasing	increasing	decreasing

From Table 8.5 we conclude that a relative minimum occurs at $x = \pi/3$ and a relative maximum occurs at $x = 5\pi/3$, as shown in Figure 8.24.

EXAMPLE 11 Using Newton's Method to find a point of intersection

Estimate the point of intersection of the graphs of $y = \tan x$ and $y = 2x$, as shown in Figure 8.25. Use Newton's Method and continue the iterations until two successive approximations differ by less than 0.0001.

SOLUTION

We let $\tan x = 2x$. Since this implies that $2x - \tan x = 0$, we need to find the zeros of the function given by

$$f(x) = 2x - \tan x.$$

Thus, the iterative formula for Newton's Method takes the form

$$x_{n+1} = x_n - \frac{f(x_n)}{f'(x_n)} = x_n - \frac{2x_n - \tan x_n}{2 - \sec^2 x_n}.$$

The calculations are shown in Table 8.6, beginning with an initial guess of $x_1 = 1.25$.

TABLE 8.6

n	x_n	$f(x_n)$	$f'(x_n)$	$\dfrac{f(x_n)}{f'(x_n)}$	$x_n - \dfrac{f(x_n)}{f'(x_n)}$
1	1.25000	-0.50957	-8.05751	0.06324	1.18676
2	1.18676	-0.10110	-5.12374	0.01973	1.16703
3	1.16703	-0.00653	-4.47832	0.00146	1.16557
4	1.16557	-0.00003	-4.43435	0.00001	1.16556

Thus, we approximate the point of intersection to occur when $x = 1.16556$.

EXAMPLE 12   The velocity of a piston

For the engine shown in Figure 8.26, a 7-inch connecting rod is fastened to a crank of radius 3 inches. The crank shaft rotates counterclockwise at a constant rate of 200 revolutions per minute. Find the velocity of the piston when $\theta = \pi/3$.

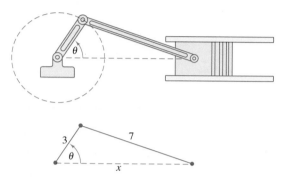

FIGURE 8.26

SOLUTION

1. We label the distances as shown in Figure 8.26. Furthermore, since a complete revolution corresponds to 2π radians, we can determine that $d\theta/dt = 200(2\pi) = 400\pi$ rad/min.

   Given: $\dfrac{d\theta}{dt} = 400\pi$ Constant rate

   Find: $\dfrac{dx}{dt}$ when $\theta = \dfrac{\pi}{3}$

2. From the Law of Cosines (see Figure 8.27), we have

   $$7^2 = 3^2 + x^2 - 2(3)(x) \cos\theta.$$

3. Implicit differentiation with respect to t yields the following.

   $$0 = 2x\frac{dx}{dt} - 6\left[-x \sin\theta\left(\frac{d\theta}{dt}\right) + \cos\theta\left(\frac{dx}{dt}\right)\right]$$

   $$(6 \cos\theta - 2x)\frac{dx}{dt} = 6x \sin\theta\left(\frac{d\theta}{dt}\right)$$

   $$\frac{dx}{dt} = \frac{6x \sin\theta}{6 \cos\theta - 2x}\left(\frac{d\theta}{dt}\right)$$

4. When $\theta = \pi/3$, we can solve for x as follows.

   $$7^2 = 3^2 + x^2 - 2(3)(x) \cos\frac{\pi}{3}$$

   $$49 = 9 + x^2 - 6x\left(\frac{1}{2}\right)$$

   $$0 = x^2 - 3x - 40$$

   $$0 = (x - 8)(x + 5)$$

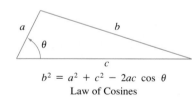

$$b^2 = a^2 + c^2 - 2ac \cos\theta$$
Law of Cosines

FIGURE 8.27

Choosing the positive solution, we have $x = 8$. Finally, when $\theta = \pi/3$ and $x = 8$, the velocity of the piston is

$$\frac{dx}{dt} = \frac{6(8)(\sqrt{3}/2)}{6(1/2) - 16}(400\pi) = \frac{9600\pi\sqrt{3}}{-13} \approx -4018 \text{ in./min.} \quad \square$$

REMARK   Note that the velocity in Example 12 is negative because the piston is moving to the left.

EXERCISES for Section 8.3

In Exercises 1–44, find the derivative of the given function.

1. $y = x^2 - \dfrac{1}{2}\cos x$

2. $y = 5 + \sin x$

3. $y = \dfrac{1}{x} - 3 \sin x$

4. $g(t) = \pi \cos t$

5. $f(x) = 4\sqrt{x} + 3 \cos x$

6. $f(x) = 2 \sin x + 3 \cos x$

7. $f(t) = t^2 \sin t$

8. $f(x) = \dfrac{\sin x}{x}$

9. $g(t) = \dfrac{\cos t}{t}$

10. $f(\theta) = (\theta + 1) \cos \theta$

11. $y = \tan x - x$

12. $y = x + \cot x$

13. $y = 5x \csc x$

14. $y = \dfrac{\sec x}{x}$

15. $f(\theta) = -\csc \theta - \sin \theta$

16. $h(x) = x \sin x + \cos x$

17. $g(t) = t^2 \sin t + 2t \cos t - 2 \sin t$

18. $h(\theta) = 5 \sec 3\theta + \tan 3\theta$

19. $f(x) = \sin \pi x \cos \pi x$

20. $f(x) = \tan 2x \cot 2x$

21. $y = \dfrac{1 + \csc x}{1 - \csc x}$

22. $y = \dfrac{\sin \theta}{1 - \cos \theta}$

23. $y = \cos 3x$

24. $y = \sin 2x$

25. $y = 3 \tan 4x$

26. $y = 2 \cos \dfrac{x}{2}$

27. $y = \sin \pi x$

28. $y = \sec x^2$

29. $y = \dfrac{1}{4} \sin^2 x$

30. $y = 5 \cos^2 \pi x$

31. $y = \dfrac{1}{4} \sin^2 2x$

32. $y = 5 \cos (\pi x)^2$

33. $y = \sqrt{\sin x}$

34. $y = \csc^2 4x$

35. $y = \sec^3 2x$

36. $y = x^2 \sin \dfrac{1}{x}$

37. $y = \ln |\csc x - \cot x|$

38. $y = \ln |\sec x + \tan x|$

39. $y = e^x(\sin x + \cos x)$

40. $y = \tan^2 e^x$

41. $y = e^{\tan x}$

42. $y = \ln |\cot x|$

43. $y = \ln |\tan x|$

44. $y = \ln |\sin x|$

In Exercises 45–50, use implicit differentiation to find dy/dx and evaluate the derivative at the indicated point.

Equation	Point
45. $\sin x + \cos 2y = 2$	$\left(\dfrac{\pi}{2}, 0\right)$
46. $2 \sin x \cos y = 1$	$\left(\dfrac{\pi}{4}, \dfrac{\pi}{4}\right)$
47. $\tan (x + y) = x$	$(0, 0)$
48. $\cot y = x - y$	$\left(\dfrac{\pi}{2}, \dfrac{\pi}{2}\right)$
49. $x \cos y = 1$	$\left(2, \dfrac{\pi}{3}\right)$
50. $x = \sec \dfrac{1}{y}$	$\left(\sqrt{2}, \dfrac{\pi}{4}\right)$

In Exercises 51 and 52, show that the function satisfies the differential equation.

51. $y = 2 \sin x + 3 \cos x$
$y'' + y = 0$

52. $y = e^x(\cos \sqrt{2}x + \sin \sqrt{2}x)$
$y'' - 2y' + 3y = 0$

In Exercises 53 and 54, find the slope of the tangent line to the given sine function at the origin. Compare this value to the number of complete cycles in the interval $[0, 2\pi]$.

53. (a) $y = \sin x$ (b) $y = \sin 2x$

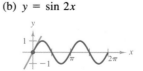

54. (a) $y = \sin 3x$ (b) $y = \sin \dfrac{x}{2}$

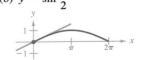

In Exercises 55 and 56, find an equation of the tangent line to the graph of the function at the indicated point.

Function	*Point*
55. $f(x) = \tan x$	$\left(-\dfrac{\pi}{4}, -1\right)$
56. $f(x) = \sec x$	$\left(\dfrac{\pi}{3}, 2\right)$

In Exercises 57–64, evaluate each limit, using L'Hô-pital's Rule when necessary.

57. $\displaystyle\lim_{x \to 0} \frac{\sin 2x}{\sin 3x}$

58. $\displaystyle\lim_{x \to \pi} \frac{\sin x}{x - \pi}$

59. $\displaystyle\lim_{x \to 0} \frac{x - \tan x}{x - \sin x}$

60. $\displaystyle\lim_{\theta \to 0} \frac{1 - \cos \theta}{\theta}$

61. $\displaystyle\lim_{\theta \to 0} \frac{1 - \cos 2\theta}{4\theta^2}$

62. $\displaystyle\lim_{x \to \pi/4} (\tan 2x - \sec 2x)$

63. $\displaystyle\lim_{x \to \infty} x \sin \frac{1}{x}$

64. $\displaystyle\lim_{x \to 0} \frac{1 - e^x}{\sin x}$

In Exercises 65–68, sketch the graph of each function on the indicated interval, making use of relative extrema and points of inflection.

Function	*Interval*
65. $f(x) = 2 \sin x + \sin 2x$	$[0, 2\pi]$
66. $f(x) = 2 \sin x + \cos 2x$	$[0, 2\pi]$
67. $f(x) = x - \sin x$	$[0, 4\pi]$
68. $f(x) = \cos x - x$	$[0, 4\pi]$

In Exercises 69 and 70, sketch the graph of the function on the interval $[-\pi, \pi]$. In each case use Newton's Method to approximate the critical number to two decimal places.

69. $f(x) = x \sin x$

70. $f(x) = x \cos x$

71. The height of a weight oscillating on a spring is given by the equation $y = \frac{1}{3} \cos 12t - \frac{1}{4} \sin 12t$, where y is measured in inches and t is measured in seconds.
(a) Calculate the height and velocity of the weight when $t = \pi/8$ seconds.
(b) Show that the maximum displacement of the weight is $\frac{5}{12}$ inches.
(c) Find the period P of y. Find the frequency f (number of oscillations per second) if $f = 1/P$.

72. The general equation giving the height of an oscillating weight attached to a spring is

$$y = A \sin \left(\sqrt{\frac{k}{m}} t\right) + B \cos \left(\sqrt{\frac{k}{m}} t\right)$$

where k is the spring constant and m is the mass of the weight.

(a) Show that the maximum displacement of the weight is $\sqrt{A^2 + B^2}$.
(b) Show that the frequency (number of oscillations per second) is $(1/2\pi)\sqrt{k/m}$. How is the frequency changed if the stiffness k of the spring is increased? How is the frequency changed if the mass m of the weight is increased?

73. A component is designed to slide a block of steel of weight W across a table and into a chute (see figure). The motion of the block is resisted by a frictional force proportional to its net weight. (Let k be the constant of proportionality.) Find the minimum force F needed to slide the block and find the corresponding value of θ. [Hint: $F \cos \theta$ is the force in the direction of motion and $F \sin \theta$ is the amount of force tending to lift the block. Therefore, the net weight of the block is $W - F \sin \theta$.]

74. When light waves traveling in a transparent medium strike the surface of a second transparent medium, they tend to "bend." This tendency is called refraction, and is given by **Snell's Law of Refraction,**

$$\frac{\sin \theta_1}{v_1} = \frac{\sin \theta_2}{v_2}$$

where θ_1 and θ_2 are the magnitudes of the angles shown in the accompanying figure, and v_1 and v_2 are the velocities of light in the two media. Show that light waves traveling from P to Q follow the path of minimum time.

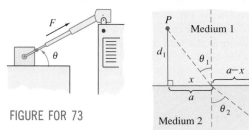

FIGURE FOR 73

FIGURE FOR 74

75. A sector with central angle θ is cut from a circle of radius 12 inches, and the resulting edges are brought together to form a cone (see figure). Find the magnitude of θ so that the volume of the cone is maximum.

76. The cross sections of an irrigation canal are isosceles trapezoids, where the length of the three sides is 8 feet (see figure). Determine the angle of elevation θ of the sides so that the area of the cross section is maximum.

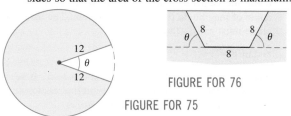

FIGURE FOR 76

FIGURE FOR 75

77. An airplane flies at an altitude of 5 miles toward a point directly over an observer (see figure). The speed of the plane is 600 miles per hour. Find the rate at which the angle of elevation θ is changing for the following angles.

(a) $\theta = 30°$ (b) $\theta = 60°$ (c) $\theta = 75°$.

78. A fish is reeled in at a rate of 1 foot per second from a bridge 15 feet above the water (see figure). At what rate is the angle between the line and the water changing when 25 feet of line is out? (Assume the fish stays near the surface of the water.)

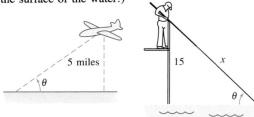

FIGURE FOR 77

FIGURE FOR 78

79. A wheel of radius 1 foot revolves at a rate of 10 revolutions per second. A dot is painted at a point P on the rim of the wheel (see figure). Find the rate of horizontal movement of the dot for the following angles.

(a) $\theta = 0°$ (b) $\theta = 30°$ (c) $\theta = 60°$

80. A patrol car is parked 50 feet from a long warehouse (see figure). The light on the car turns at a rate of 30 revolutions per minute. How fast is the light beam moving along the wall when the beam makes the following angles with the line perpendicular from the light to the wall?

(a) $\theta = 30°$ (b) $\theta = 60°$ (c) $\theta = 70°$

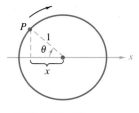

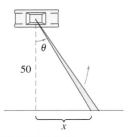

FIGURE FOR 79

FIGURE FOR 80

81. Cars on a certain roadway travel on a circular arc of radius r. In order not to rely on friction alone to overcome the centrifugal force, the road is banked at an angle of magnitude θ from the horizontal. The banking angle must satisfy the equation

$$rg \tan \theta = v^2$$

where v is the velocity of the cars and $g = 32$ ft/sec^2 is the acceleration due to gravity. Find the relationship between the related rates dv/dt and $d\theta/dt$.

82. In the Chapter 8 Application, we showed that the surface area of a cell in a honeycomb is given by

$$S = 6hs + \frac{3s^2}{2}\left(\frac{\sqrt{3} - \cos \theta}{\sin \theta}\right)$$

where h and s are positive constants and θ is the angle at which the upper faces meet the altitude of the cell. Find the angle θ ($\pi/6 \le \theta \le \pi/2$) that minimizes the surface area S.

In Exercises 83–86, use a computer (a) to sketch the graph of f and f' on the same coordinate axes over the specified interval, (b) to find the critical numbers of f, and (c) to find the interval(s) on which f' is positive and the interval(s) on which it is negative. Note the behavior of f in relation to the sign of f'.

Function	Interval
83. $f(t) = t^2 \sin t$	$[0, 2\pi]$
84. $f(x) = \dfrac{x}{2} + \cos \dfrac{x}{2}$	$[0, 4\pi]$
85. $f(x) = \sin x - \dfrac{1}{3}\sin 3x + \dfrac{1}{5}\sin 5x$	$[0, \pi]$
86. $f(x) = \sqrt{2x}\,\sin x$	$[0, 2\pi]$

87. For $f(x) = \sec^2 x$ and $g(x) = \tan^2 x$, show that $f'(x) = g'(x)$.

88. Derive the following differentiation rules.

(a) $D_x[\sec x] = \sec x \tan x$

(b) $D_x[\csc x] = -\csc x \cot x$

(c) $D_x[\cot x] = -\csc^2 x$

8.4 Integrals of Trigonometric Functions

Integrals of trigonometric functions ▪ Applications

Corresponding to each trigonometric differentiation formula is an integration formula. For instance, the differentiation formula

$$\frac{d}{dx}[\cos u] = -\sin u \frac{du}{dx}$$

corresponds to the integration formula

$$\int \sin u \; du = -\cos u + C.$$

The following list summarizes all six integration formulas corresponding to the derivatives of the basic trigonometric functions.

THEOREM 8.5 BASIC TRIGONOMETRIC INTEGRATION FORMULAS	*Integration Formula*	*Differentiation Formula*
	$\int \cos u \; du = \sin u + C$	$\frac{d}{dx}[\sin u] = \cos u \frac{du}{dx}$
	$\int \sin u \; du = -\cos u + C$	$\frac{d}{dx}[\cos u] = -\sin u \frac{du}{dx}$
	$\int \sec^2 u \; du = \tan u + C$	$\frac{d}{dx}[\tan u] = \sec^2 u \frac{du}{dx}$
	$\int \sec u \tan u \; du = \sec u + C$	$\frac{d}{dx}[\sec u] = \sec u \tan u \frac{du}{dx}$
	$\int \csc^2 u \; du = -\cot u + C$	$\frac{d}{dx}[\cot u] = -\csc^2 u \frac{du}{dx}$
	$\int \csc u \cot u \; du = -\csc u + C$	$\frac{d}{dx}[\csc u] = -\csc u \cot u \frac{du}{dx}$

EXAMPLE 1 Integration of trigonometric functions

(a) $\displaystyle \int 2 \cos x \; dx = 2 \int \cos x \; dx = 2 \sin x + C$ $u = x$

(b) $\displaystyle \int 3x^2 \sin x^3 \; dx = \int \underbrace{\sin x^3}_{\sin u}\underbrace{(3x^2)}_{du} \; dx = -\cos x^3 + C$ $u = x^3$

(c) $\displaystyle \int \sec^2 3x \; dx = \frac{1}{3} \int \underbrace{(\sec^2 3x)}_{\sec^2 u}\underbrace{(3)}_{du} \; dx = \frac{1}{3} \tan 3x + C$ $u = 3x$

The integrals in Example 1 are easily recognized as fitting one of the basic integration formulas in Theorem 8.5. However, because of the variety of trigonometric identities, it often happens that an integrand that fits one of the basic formulas will come in a disguised form. This is illustrated in the next two examples.

EXAMPLE 2 Using a trigonometric identity

Evaluate $\int \tan^2 x \, dx$.

SOLUTION

$$\int \tan^2 x \, dx = \int (-1 + \sec^2 x) \, dx = -x + \tan x + C$$

EXAMPLE 3 Using a trigonometric identity

Evaluate $\int (\csc x + \sin x)(\csc x) \, dx$.

SOLUTION

$$\int (\csc x + \sin x)(\csc x) \, dx = \int (\csc^2 x + 1) \, dx = -\cot x + x + C$$

REMARK We will say more about the use of trigonometric identities to evaluate trigonometric integrals in Section 9.3.

In addition to using trigonometric identities, another useful technique in evaluating trigonometric integrals is u-substitution, as illustrated in the next example.

EXAMPLE 4 Integration by u-substitution

Evaluate

$$\int \frac{\sec^2 \sqrt{x}}{\sqrt{x}} \, dx.$$

SOLUTION

Let $u = \sqrt{x}$. Then we have

$$u = \sqrt{x} \quad \Longrightarrow \quad du = \frac{1}{2\sqrt{x}} \, dx \quad \Longrightarrow \quad 2du = \frac{1}{\sqrt{x}} \, dx.$$

Thus,

$$\int \frac{\sec^2 \sqrt{x}}{\sqrt{x}} \, dx = \int \sec^2 \sqrt{x} \left(\frac{1}{\sqrt{x}} \, dx \right) = \int \sec^2 u(2du)$$

$$= 2 \int \sec^2 u \, du = 2 \tan u + C = 2 \tan \sqrt{x} + C. \quad \blacksquare$$

One of the most common u-substitutions involves quantities in the integrand that are raised to a power, as illustrated in the next two examples.

EXAMPLE 5   Integration by u-substitution

Evaluate $\int \sin^2 3x \cos 3x \, dx$.

SOLUTION

Since $\sin^2 3x = (\sin 3x)^2$, we let $u = \sin 3x$. Then

$$du = (\cos 3x)(3) \, dx \quad \Longrightarrow \quad \frac{du}{3} = \cos 3x \, dx.$$

Substituting u and $du/3$ in the given integral yields

$$\int \sin^2 3x \cos 3x \, dx = \int u^2 \frac{du}{3} = \frac{1}{3} \int u^2 \, du$$

$$= \frac{1}{3} \left(\frac{u^3}{3} \right) + C = \frac{1}{9} \sin^3 3x + C. \quad \blacksquare$$

EXAMPLE 6   Substitution and the Power Rule

(a) $\displaystyle \int \frac{\sin x}{\cos^2 x} \, dx = - \int \overbrace{(\cos x)^{-2}}^{u^{-2}} \overbrace{(-\sin x) \, dx}^{du} = - \overbrace{\frac{(\cos x)^{-1}}{-1}}^{u^{-1}/(-1)} + C$

$\qquad = \sec x + C$

(b) $\displaystyle \int 4 \cos^2 4x \sin 4x \, dx = - \int \overbrace{(\cos 4x)^2}^{u^2} \overbrace{(-4 \sin 4x) \, dx}^{du}$

$\qquad = - \overbrace{\frac{(\cos 4x)^3}{3}}^{u^3/3} + C$

(c) $\displaystyle \int \frac{\sec^2 x}{\sqrt{\tan x}} \, dx = \int \overbrace{(\tan x)^{-1/2}}^{u^{-1/2}} \overbrace{(\sec^2 x) \, dx}^{du} = \overbrace{\frac{(\tan x)^{1/2}}{1/2}}^{u^{1/2}/(1/2)} + C \quad \blacksquare$

In Theorem 8.5 we listed six trigonometric integration formulas—the six that correspond directly to differentiation rules. We now complete our set of basic trigonometric integration formulas by making use of the Log Rule.

EXAMPLE 7   The antiderivative of the tangent

Evaluate $\int \tan x \, dx$.

SOLUTION

This integral doesn't seem to fit any formulas on our basic list. However, by a trigonometric identity, we obtain the following quotient form.

$$\int \tan x \, dx = \int \frac{\sin x}{\cos x} \, dx.$$

Now, knowing that $D_x[\cos x] = -\sin x$, we consider $u = \cos x$ and write

$$\int \tan x \, dx = -\int \frac{(-\sin x)}{\cos x} \, dx$$

$$= -\int \frac{u'}{u} \, dx$$

$$= -\ln |u| + C$$

$$= -\ln |\cos x| + C.$$

In Example 7, we used a trigonometric identity together with the Log Rule to derive an integration formula for the tangent function. In the next example we use a rather unusual step (multiplying and dividing by the same quantity) to derive an integration formula for the secant function.

EXAMPLE 8   Antiderivative of the secant

Evaluate $\int \sec x \, dx$.

SOLUTION

Consider the following procedure.

$$\int \sec x \, dx = \int \sec x \left(\frac{\sec x + \tan x}{\sec x + \tan x} \right) dx$$

$$= \int \frac{\sec^2 x + \sec x \tan x}{\sec x + \tan x} \, dx$$

Now, letting u be the denominator of this quotient, we have

$$u = \sec x + \tan x \quad \Longrightarrow \quad u' = \sec x \tan x + \sec^2 x.$$

Therefore, we conclude that

$$\int \sec x \, dx = \int \frac{\sec^2 x + \sec x \tan x}{\sec x + \tan x} \, dx = \int \frac{u'}{u} \, dx = \ln |u| + C$$

$$= \ln |\sec x + \tan x| + C.$$

With Examples 7 and 8 we now have integration formulas for $\sin x$, $\cos x$, $\tan x$, and $\sec x$. We leave the derivations for $\csc x$ and $\cot x$ to you and summarize all six trigonometric formulas in the following theorem.

THEOREM 8.6
INTEGRALS OF THE SIX BASIC
TRIGONOMETRIC FUNCTIONS

$$\int \sin u \, du = -\cos u + C \qquad \qquad \int \cos u \, du = \sin u + C$$

$$\int \tan u \, du = -\ln |\cos u| + C \qquad \int \cot u \, du = \ln |\sin u| + C$$

$$\int \sec u \, du = \ln |\sec u + \tan u| + C \qquad \int \csc u \, du = -\ln |\csc u + \cot u| + C$$

REMARK As you memorize these formulas, note that the three formulas on the right follow the pattern set by the three on the left.

EXAMPLE 9   Integrating trigonometric functions

Evaluate

$$\int_0^{\pi/4} \sqrt{1 + \tan^2 x} \, dx.$$

SOLUTION

Since $1 + \tan^2 x = \sec^2 x$, we have

$$\int_0^{\pi/4} \sqrt{1 + \tan^2 x} \, dx = \int_0^{\pi/4} \sqrt{\sec^2 x} \, dx$$

$$= \int_0^{\pi/4} \sec x \, dx$$

$$= \left[\ln |\sec x + \tan x| \right]_0^{\pi/4}$$

$$= \ln (\sqrt{2} + 1) - \ln (1) \approx 0.8814.$$

REMARK Note in Example 9 that $\sec x > 0$ for $0 \le x \le \pi/4$ and thus it is valid to replace $\sqrt{\sec^2 x}$ by $\sec x$.

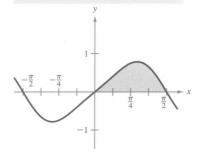

$f(x) = \sin^3 x \cos x + \sin x \cos x$

$$\text{Area} = \int_0^{\pi/2} f(x)\,dx$$

FIGURE 8.28

Applications

EXAMPLE 10 Finding the area of a plane region

Find the area of the region bounded by the graph of

$$f(x) = \sin^3 x \cos x + \sin x \cos x$$

and the x-axis over the interval $0 \le x \le \pi/2$, as shown in Figure 8.28.

SOLUTION

Using the Power Rule, we find the area of this region to be

$$A = \int_0^{\pi/2} (\sin^3 x \cos x + \sin x \cos x)\,dx = \left[\frac{\sin^4 x}{4} + \frac{\sin^2 x}{2}\right]_0^{\pi/2}$$

$$= \frac{1}{4} + \frac{1}{2} = \frac{3}{4}.$$

EXAMPLE 11 A region lying between two intersecting graphs

The sine and cosine curves intersect infinitely many times bounding regions of equal areas, as shown in Figure 8.29. Find the area of one of these regions.

SOLUTION

To find the points of intersection of the graphs of $f(x) = \sin x$ and $g(x) = \cos x$, we set the two functions equal to each other and solve for x.

$$\sin x = \cos x \qquad \text{Set } f(x) \text{ equal to } g(x)$$

$$\frac{\sin x}{\cos x} = 1$$

$$\tan x = 1 \qquad x = \frac{\pi}{4}, \frac{5\pi}{4}; \quad 0 \le x \le 2\pi$$

Thus, $a = \pi/4$ and $b = 5\pi/4$. Since $\sin x \ge \cos x$ on the interval $[\pi/4, 5\pi/4]$, the area of the region is

$$A = \int_{\pi/4}^{5\pi/4} [\sin x - \cos x]\,dx = \left[-\cos x - \sin x\right]_{\pi/4}^{5\pi/4}$$

$$= \left(\frac{\sqrt{2}}{2} + \frac{\sqrt{2}}{2}\right) - \left(-\frac{\sqrt{2}}{2} - \frac{\sqrt{2}}{2}\right)$$

$$= 2\sqrt{2}.$$

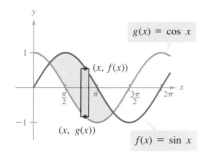

$g(x) = \cos x$

$(x, f(x))$

$(x, g(x))$

$f(x) = \sin x$

FIGURE 8.29

EXAMPLE 12 Finding the volume of a solid of revolution

Find the volume of the solid formed by revolving the region bounded by the graph of $f(x) = \sqrt{\sin x}$ and the x-axis $(0 \le x \le \pi)$ about the x-axis.

SOLUTION

From the representative rectangle in Figure 8.30, we see that the radius of this solid is given by

$$R(x) = f(x) = \sqrt{\sin x}$$

and it follows that its volume is

$$V = \pi \int_0^\pi [R(x)]^2 \, dx = \pi \int_0^\pi (\sqrt{\sin x})^2 \, dx$$

$$= \pi \int_0^\pi \sin x \, dx = -\pi \cos x \Big]_0^\pi$$

$$= \pi(1 + 1) = 2\pi.$$

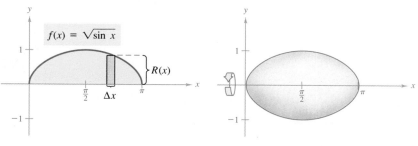

FIGURE 8.30

Plane region   Solid of revolution

EXAMPLE 13   Finding the average value of a function on an interval

The electromotive force E of a particular electrical circuit is given by

$$E = 3 \sin 2t$$

where E is measured in volts and t is measured in seconds. Find the average value of E as t ranges from 0 to 0.5 second.

SOLUTION

The average value of E is given by

$$\frac{1}{b - a} \int_a^b f(t) \, dt = \frac{1}{0.5 - 0} \int_0^{0.5} 3 \sin 2t \, dt$$

$$= 6 \int_0^{0.5} \sin 2t \, dt.$$

Letting $u = 2t$ implies that $du = 2 \, dt$. Thus, we have

$$\text{average value} = 6\left(\frac{1}{2}\right) \int_0^{0.5} (\sin 2t)(2) \, dt$$

$$= \Big[3(-\cos 2t) \Big]_0^{0.5}$$

$$= 3[-\cos (1) + 1]$$

$$\approx 1.379 \text{ volts.}$$

EXAMPLE 14 Finding arc length

Find the arc length of the graph of $y = \ln (\cos x)$ from $x = 0$ to $x = \pi/4$, as shown in Figure 8.31.

SOLUTION

Since

$$y' = -\frac{\sin x}{\cos x} = -\tan x$$

the arc length is given by

$$s = \int_a^b \sqrt{1 + (y')^2} \, dx = \int_0^{\pi/4} \sqrt{1 + \tan^2 x} \, dx = \int_0^{\pi/4} \sqrt{\sec^2 x} \, dx$$

$$= \int_0^{\pi/4} \sec x \, dx = \left[\ln |\sec x + \tan x| \right]_0^{\pi/4}$$

$$= \ln (\sqrt{2} + 1) - \ln (1) \approx 0.8814.$$

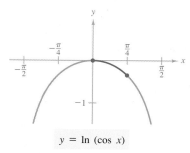

$y = \ln (\cos x)$

FIGURE 8.31

EXERCISES for Section 8.4

In Exercises 1–36, evaluate the given integral.

1. $\displaystyle\int (2 \sin x + 3 \cos x) \, dx$

2. $\displaystyle\int (t^2 - \sin t) \, dt$

3. $\displaystyle\int (1 - \csc t \cot t) \, dt$

4. $\displaystyle\int (\theta^2 + \sec^2 \theta) \, d\theta$

5. $\displaystyle\int (\sec^2 \theta - \sin \theta) \, d\theta$

6. $\displaystyle\int \sec y(\tan y - \sec y) \, dy$

7. $\displaystyle\int \sin 2x \, dx$

8. $\displaystyle\int \cos 6x \, dx$

9. $\displaystyle\int x \cos x^2 \, dx$

10. $\displaystyle\int x \sin x^2 \, dx$

11. $\displaystyle\int \sec^2 \frac{x}{2} \, dx$

12. $\displaystyle\int \csc^2 \frac{x}{2} \, dx$

13. $\displaystyle\int \frac{\csc^2 x}{\cot^3 x} \, dx$

14. $\displaystyle\int \frac{\sin x}{\cos^2 x} \, dx$

15. $\displaystyle\int \cot^2 x \, dx$

16. $\displaystyle\int \csc 2x \cot 2x \, dx$

17. $\displaystyle\int \tan^4 x \sec^2 x \, dx$

18. $\displaystyle\int \sqrt{\cot x} \csc^2 x \, dx$

19. $\displaystyle\int \cot \pi x \, dx$

20. $\displaystyle\int \tan 5x \, dx$

21. $\displaystyle\int \csc 2x \, dx$

22. $\displaystyle\int \sec \frac{x}{2} \, dx$

23. $\displaystyle\int \frac{\sec^2 x}{\tan x} \, dx$

24. $\displaystyle\int \frac{\tan^2 2x}{\sec 2x} \, dx$

25. $\displaystyle\int \frac{\sec x \tan x}{\sec x - 1} \, dx$

26. $\displaystyle\int \frac{\sin x}{1 + \cos x} \, dx$

27. $\displaystyle\int \frac{\cos t}{1 + \sin t} \, dt$

28. $\displaystyle\int \frac{\sin^2 x - \cos^2 x}{\cos x} \, dx$

29. $\displaystyle\int \frac{1 - \cos \theta}{\theta - \sin \theta} \, d\theta$

30. $\displaystyle\int \frac{1 - \sin^2 \theta}{\cos^2 \theta} \, d\theta$

31. $\displaystyle\int e^x \cos e^x \, dx$

32. $\displaystyle\int e^{\sin x} \cos x \, dx$

33. $\displaystyle\int e^{-x} \tan (e^{-x}) \, dx$

34. $\displaystyle\int e^{\sec x} \sec x \tan x \, dx$

35. $\displaystyle\int (\sin 2x + \cos 2x)^2 \, dx$

36. $\displaystyle\int (\csc 2\theta - \cot 2\theta)^2 \, d\theta$

In Exercises 37–44, evaluate the definite integral.

37. $\displaystyle\int_0^{\pi/2} \cos \frac{2x}{3} \, dx$

38. $\displaystyle\int_0^{\pi/2} \sin 2x \, dx$

39. $\displaystyle\int_{\pi/2}^{2\pi/3} \sec^2 \frac{x}{2} \, dx$

40. $\displaystyle\int_{\pi/2}^{\pi/2} (x + \cos x) \, dx$

41. $\displaystyle\int_{\pi/12}^{\pi/4} \csc 2x \cot 2x \, dx$

42. $\displaystyle\int_0^{\pi/8} \sin 2x \cos 2x \, dx$

43. $\displaystyle\int_0^1 \sec (1 - x) \tan (1 - x) \, dx$

44. $\displaystyle\int_0^{\pi/4} (\sec x)^3 (\sec x \tan x) \, dx$

In Exercises 45–50, determine the area of the given region.

45. $y = \cos \dfrac{x}{2}$

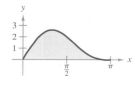

46. $y = x + \sin x$

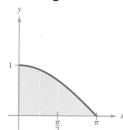

47. $y = 2 \sin x + \sin 2x$

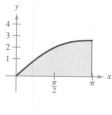

48. $y = \sin x + \cos 2x$

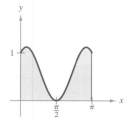

49. $f(x) = 2$
$\quad\; g(x) = \sec x$

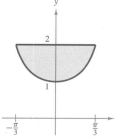

50. $f(x) = 2 \sin x$
$\quad\; g(x) = \tan x$

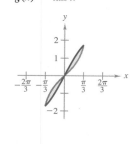

In Exercises 51–54, find the volume of the solid generated by revolving the region bounded by the graphs of the given equations about the *x*-axis.

51. $y = \sqrt{\sin x}$, $y = 0$, $0 \le x \le \pi$
52. $y = \sqrt{\cos x}$, $y = 0$, $0 \le x \le \pi/2$
53. $y = \sec x$, $y = 0$, $x = 0$, $x = \pi/4$
54. $y = \csc x$, $y = 0$, $x = \pi/6$, $x = 5\pi/6$

In Exercises 55 and 56, find the arc length of the graphs of the given function over the indicated interval. Use Simpson's Rule with $n = 4$ to approximate the definite integral.

Function	Interval
55. $y = \sin x$	$[0, \pi]$
56. $y = 4 \cos \dfrac{x}{2}$	$[0, \pi]$

57. The sales of a seasonal product are given by the model

$$S = 74.50 + 43.75 \sin \frac{\pi t}{6}$$

where S is measured in thousands of units and t is the time in months, with $t = 1$ corresponding to January. Find the average sales during the following time periods.
(a) First quarter $\quad (0 \le t \le 3)$
(b) Second quarter $\quad (3 \le t \le 6)$
(c) Entire year $\quad (0 \le t \le 12)$

58. The minimum stockpile level of gasoline in the United States can be approximated by the model

$$Q = 217 + 13 \cos \frac{\pi(t - 3)}{6}$$

where Q is measured in millions of barrels of gasoline and t is the time in months, with $t = 1$ corresponding to January. Find the average minimum level given by this model during the following time periods.
(a) First quarter $\quad (0 \le t \le 3)$
(b) Second quarter $\quad (3 \le t \le 6)$
(c) Entire year $\quad (0 \le t \le 12)$

59. The oscillating current in an electrical circuit is given by

$$I = 2 \sin (60\pi t) + \cos (120\pi t)$$

where I is measured in amperes and t is measured in seconds. Find the average current for the following time intervals.
(a) $0 \le t \le 1/60$
(b) $0 \le t \le 1/240$
(c) $0 \le t \le 1/30$

60. A horizontal plane is ruled with parallel lines 2 inches apart. If a 2-inch needle is randomly tossed onto the plane, it can be shown that the probability of the needle touching a line is given by

$$P = \frac{2}{\pi} \int_0^{\pi/2} \sin \theta \, d\theta$$

where θ is the acute angle between the needle and any one of the parallel lines. Find this probability.

61. Evaluate $\int \sin x \cos x \, dx$ two ways. First, make the substitution $u = \sin x$, and second, integrate by letting $u = \cos x$. Explain the difference in the results.

62. The graphs of the sine and cosine are symmetric to the origin and the *y*-axis respectively. Use this fact to aid in the evaluation of the following definite integrals.

(a) $\displaystyle\int_{-\pi/4}^{\pi/4} \sin x \, dx$ $\qquad$ (b) $\displaystyle\int_{-\pi/4}^{\pi/4} \cos x \, dx$

(c) $\displaystyle\int_{-\pi/2}^{\pi/2} \cos x \, dx$ $\qquad$ (d) $\displaystyle\int_{-1.32}^{1.32} \sin 2x \, dx$

(e) $\displaystyle\int_{-\pi/2}^{\pi/2} \sin x \cos x \, dx$

In Exercises 63–66, show the equivalence of each pair of formulas.

63. $\displaystyle\int \tan x \, dx = -\ln |\cos x| + C$

$\displaystyle\int \tan x \, dx = \ln |\sec x| + C$

64. $\displaystyle\int \cot x \, dx = \ln |\sin x| + C$

$\displaystyle\int \cot x \, dx = -\ln |\csc x| + C$

65. $\displaystyle\int \sec x \, dx = \ln |\sec x + \tan x| + C$

$\displaystyle\int \sec x \, dx = -\ln |\sec x - \tan x| + C$

66. $\displaystyle\int \csc x \, dx = \ln |\csc x - \cot x| + C$

$\displaystyle\int \csc x \, dx = -\ln |\csc x + \cot x| + C$

In Exercises 67–72, approximate the given integral using (a) the Trapezoidal Rule and (b) Simpson's Rule.

67. $\displaystyle\int_0^{\sqrt{\pi/2}} \cos x^2 \, dx, \; n = 4$

68. $\displaystyle\int_0^{\sqrt{\pi/4}} \tan x^2 \, dx, \; n = 4$

69. $\displaystyle\int_0^1 \sin x^2 \, dx, \; n = 2$

70. $\displaystyle\int_0^{\pi} \sqrt{x} \sin x \, dx, \; n = 4$

71. $\displaystyle\int_0^{\pi/4} x \tan x \, dx, \; n = 4$

72. $\displaystyle\int_0^{\pi/2} \sqrt{1 + \cos^2 x} \, dx, \; n = 2$

73. Use a computer and Simpson's Rule with $n = 10$ to approximate

$$\int_0^4 \sin \sqrt{x} \, dx.$$

74. Use a computer and Simpson's Rule with $n = 10$ to approximate t to three decimal places in the integral equation

$$\int_0^t \sin \sqrt{x} \, dx = 2.$$

8.5 Inverse Trigonometric Functions and Differentiation

Inverse trigonometric functions ▪ Derivatives of inverse trigonometric functions ▪ Review of basic differentiation formulas

We begin this section with a rather startling statement: *None of the six basic trigonometric functions has an inverse*. This statement follows from the fact that all six functions are periodic, and hence not one-to-one. It seems, then, that we have little to do in this section, but of course that is not the case. Rather, we need to examine the six basic trigonometric functions to see if we can redefine their domains in such a way that they will have inverses on the *restricted domains*.

We begin by looking at the sine function.

EXAMPLE 1   Finding an interval on which the sine function is one-to-one

Show that the sine function $f(x) = \sin x$ is not one-to-one on the entire real line. Then show that $[-\pi/2, \pi/2]$ is the largest interval, centered at the origin, for which f is strictly monotonic.

SOLUTION

It is clear that f is not one-to-one since many different x-values yield the same y-value. For instance,

$$\sin (0) = 0 = \sin (\pi).$$

Moreover, *f* is increasing on the open interval $(-\pi/2, \pi/2)$ since its derivative

$$f'(x) = \cos x$$

is positive there. Finally, since the left and right endpoints correspond to relative extrema of the sine, we can conclude that *f* is increasing on the closed interval $[-\pi/2, \pi/2]$ *and* that in any larger interval the function would not be strictly monotonic. (See Figure 8.32.)

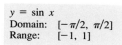

$y = \sin x$
Domain: $[-\pi/2, \pi/2]$
Range: $[-1, 1]$

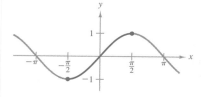

FIGURE 8.32

From Example 1, we can apply Theorem 7.5 to conclude that the *restricted* sine function, whose domain is $[-\pi/2, \pi/2]$, has an inverse function. On this interval, we define the inverse sine function to be

$$y = \arcsin x \qquad \text{if and only if} \qquad \sin y = x$$

where $-1 \le x \le 1$ and $-\pi/2 \le \arcsin x \le \pi/2$.

Under suitable restrictions, each of the six trigonometric functions is one-to-one and so possesses an inverse, as indicated in the following definition. (The term iff is used to represent the phrase "if and only if.")

DEFINITION OF INVERSE TRIGONOMETRIC FUNCTIONS			
	Function	*Domain*	*Range*
	$y = \mathbf{arcsin}\ x$ iff $\sin y = x$	$-1 \le x \le 1$	$-\dfrac{\pi}{2} \le y \le \dfrac{\pi}{2}$
	$y = \mathbf{arccos}\ x$ iff $\cos y = x$	$-1 \le x \le 1$	$0 \le y \le \pi$
	$y = \mathbf{arctan}\ x$ iff $\tan y = x$	$-\infty < x < \infty$	$-\dfrac{\pi}{2} < y < \dfrac{\pi}{2}$
	$y = \mathbf{arccot}\ x$ iff $\cot y = x$	$-\infty < x < \infty$	$0 < y < \pi$
	$y = \mathbf{arcsec}\ x$ iff $\sec y = x$	$|x| \ge 1$	$0 \le y \le \pi, y \ne \dfrac{\pi}{2}$
	$y = \mathbf{arccsc}\ x$ iff $\csc y = x$	$|x| \ge 1$	$-\dfrac{\pi}{2} \le y \le \dfrac{\pi}{2}, y \ne 0$

REMARK   The term arcsin *x* is read as the "inverse sine of *x*" or sometimes the "angle whose sine is *x*." An alternate notation for the inverse sine function is $\sin^{-1} x$.

The graphs of these six inverse functions are shown in Figure 8.33. (Compare these to the graphs of the six trigonometric functions given in Section 8.2.)

Domain: [−1, 1]
Range: [−π/2, π/2]

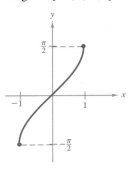

$y = \arcsin x$

Domain: (−∞, −1] and [1, ∞)
Range: [−π/2, 0) and (0, π/2]

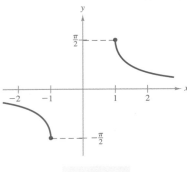

$y = \text{arccsc } x$

Domain: (−∞, ∞)
Range: (−π/2, π/2)

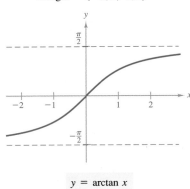

$y = \arctan x$

Domain: [−1, 1]
Range: [0, π]

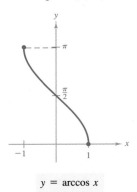

$y = \arccos x$

Domain: (−∞, −1] and [1, ∞)
Range: [0, π/2) and (π/2, π]

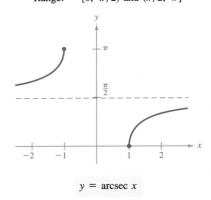

$y = \text{arcsec } x$

Domain: (−∞, ∞)
Range: (0, π)

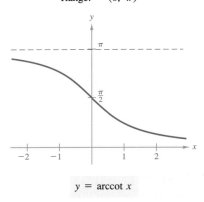

$y = \text{arccot } x$

FIGURE 8.33

When evaluating inverse trigonometric functions, remember that they denote *angles in radian measure*.

EXAMPLE 2 Evaluating inverse trigonometric functions

Evaluate the following.

(a) $\arcsin\left(-\dfrac{1}{2}\right)$ (b) $\arccos 0$

(c) $\arctan \sqrt{3}$ (d) $\arcsin (0.3)$

SOLUTION

(a) By definition, $y = \arcsin\left(-\frac{1}{2}\right)$ implies that $\sin y = -\frac{1}{2}$. In the interval $[-\pi/2, \pi/2]$, we choose $y = -\pi/6$. Therefore,

$$\arcsin\left(-\frac{1}{2}\right) = -\frac{\pi}{6}.$$

(b) By definition, $y = \arccos 0$ implies that $\cos y = 0$. In the interval $[0, \pi]$, we choose $y = \pi/2$. Therefore,

$$\arccos 0 = \frac{\pi}{2}.$$

(c) By definition, $y = \arctan \sqrt{3}$ implies that $\tan y = \sqrt{3}$. In the interval $(-\pi/2, \pi/2)$, we choose $y = \pi/3$. Therefore,

$$\arctan \sqrt{3} = \frac{\pi}{3}.$$

(d) By using a calculator set in *radian mode*, we obtain

$$\arcsin (0.3) \approx 0.3047.$$

Inverse functions possess the properties

$$f(f^{-1}(x)) = x \quad \text{and} \quad f^{-1}(f(x)) = x.$$

When applying these properties to inverse trigonometric functions, remember that the trigonometric functions possess inverses only in restricted domains. For x-values outside these domains, these two properties do not hold. For example,

$$\arcsin (\sin \pi) = \arcsin 0 = 0 \neq \pi.$$

INVERSE PROPERTIES	If $-1 \le x \le 1$ and $-\pi/2 \le y \le \pi/2$, then
	
	$\quad\sin (\arcsin x) = x \quad$ and $\quad \arcsin (\sin y) = y.$
	
	If $-\pi/2 < y < \pi/2$, then
	
	$\quad\tan (\arctan x) = x \quad$ and $\quad \arctan (\tan y) = y.$
	
	If $|x| \ge 1$ and $0 \le y < \pi/2$ or $\pi/2 < y \le \pi$, then
	
	$\quad\sec (\text{arcsec } x) = x \quad$ and $\quad \text{arcsec } (\sec y) = y.$

REMARK   Similar properties hold for the other three inverse trigonometric functions.

Notice how we use one of these inverse properties to solve the equation in the next example.

EXAMPLE 3 Solving an equation

Solve for x in the equation arctan $(2x - 3) = \pi/4$.

SOLUTION

$$\text{arctan } (2x - 3) = \frac{\pi}{4}$$

$$\tan [\text{arctan } (2x - 3)] = \tan \frac{\pi}{4}$$

$$2x - 3 = 1$$

$$x = 2$$

There are some important types of problems in calculus in which we evaluate expressions like sec (arctan x). To solve this type of problem, it helps to use right triangles, as demonstrated in the next example.

EXAMPLE 4 Using right triangles

(a) Given $y = \text{arcsin } x$, where $0 < y < \pi/2$, find cos y.
(b) Given $y = \text{arcsec } (\sqrt{5}/2)$, find tan y.

SOLUTION

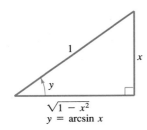

$y = \text{arcsin } x$

FIGURE 8.34

(a) Since y is the angle whose sine is x, we form a right triangle having an acute angle $y = \text{arcsin } x$, as shown in Figure 8.34. Therefore,

$$\cos y = \cos (\text{arcsin } x) \qquad \cos y = \frac{\text{adj.}}{\text{hyp.}}$$
$$= \sqrt{1 - x^2}.$$

It can also be shown that for $-\pi/2 < y \leq 0$, $y = \text{arcsin } x$ implies $\cos y = \sqrt{1 - x^2}$.

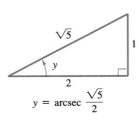

$y = \text{arcsec } \dfrac{\sqrt{5}}{2}$

FIGURE 8.35

(b) Since y is the angle whose secant is $\sqrt{5}/2$, we can sketch this angle as part of a triangle, as shown in Figure 8.35. Therefore,

$$\tan y = \tan \left[\text{arcsec } \left(\frac{\sqrt{5}}{2} \right) \right] \qquad \tan y = \frac{\text{opp.}}{\text{adj.}}$$
$$= \frac{1}{2}.$$

Derivatives of inverse trigonometric functions

In Section 7.5 we saw that the derivative of the *transcendental* function $f(x) = \ln x$ is the *algebraic* function $f'(x) = 1/x$. We will now see that the derivatives of the inverse trigonometric functions also are algebraic, even though the inverse trigonometric functions are themselves transcendental.

The following theorem lists the derivative of each of the six inverse trigonometric functions. Note that the derivatives of arccos u, arccot u, and arccsc u are the *negatives* of the derivatives of arcsin u, arctan u, and arcsec u, respectively.

THEOREM 8.7
DERIVATIVES OF THE INVERSE
TRIGONOMETRIC FUNCTIONS

Let u be a differentiable function of x.

$$\frac{d}{dx}[\arcsin u] = \frac{u'}{\sqrt{1 - u^2}} \qquad \frac{d}{dx}[\arccos u] = \frac{-u'}{\sqrt{1 - u^2}}$$

$$\frac{d}{dx}[\arctan u] = \frac{u'}{1 + u^2} \qquad \frac{d}{dx}[\text{arccot } u] = \frac{-u'}{1 + u^2}$$

$$\frac{d}{dx}[\text{arcsec } u] = \frac{u'}{|u|\sqrt{u^2 - 1}} \qquad \frac{d}{dx}[\text{arccsc } u] = \frac{-u'}{|u|\sqrt{u^2 - 1}}$$

PROOF   We prove the first of these formulas and leave the proofs of the others as an exercise (see Exercise 67). Let $f(x) = \sin x$ and $f^{-1}(x) = g(x) = \arcsin x$. Then, by using Theorem 7.7, we have

$$\frac{d}{dx}[g(x)] = \frac{1}{f'[g(x)]} = \frac{1}{\cos(\arcsin x)}.$$

Now, by using the result of Example 3(a), we have

$$\frac{d}{dx}[\arcsin x] = \frac{1}{\cos(\arcsin x)} = \frac{1}{\sqrt{1 - x^2}}.$$

If u is a differentiable function of x, then the Chain Rule gives us

$$\frac{d}{dx}[\arcsin u] = \frac{u'}{\sqrt{1 - u^2}}, \quad \text{where } u' = \frac{du}{dx}.$$

There is no common agreement regarding the definition of arcsec x (or arccsc x) for negative values of x. When we defined the range of the arcsecant, we chose to preserve the reciprocal identity

$$\text{arcsec } x = \arccos \frac{1}{x}.$$

For example, to evaluate arcsec x when $x = -2$, we write arcsec $(-2) = \arccos(-0.5) \approx 2.09$.

One of the consequences of the definition of the inverse secant function given in this text is that its graph has a positive slope at every x-value in its domain. (See Figure 8.33.) This accounts for the absolute value sign in the formula for the derivative of arcsec x.

EXAMPLE 5   Differentiating inverse trigonometric functions

(a) $\dfrac{d}{dx}[\arctan (3x)] = \dfrac{3}{1 + (3x)^2}$ $\qquad u = 3x$

$\qquad\qquad\qquad = \dfrac{3}{1 + 9x^2}$

(b) $\dfrac{d}{dx}[\arcsin \sqrt{x}] = \dfrac{(1/2)x^{-1/2}}{\sqrt{1 - x}}$ $\qquad u = \sqrt{x}$

$\qquad\qquad\qquad = \dfrac{1}{2\sqrt{x}\sqrt{1 - x}}$

$\qquad\qquad\qquad = \dfrac{1}{2\sqrt{x - x^2}}$

(c) $\dfrac{d}{dx}[\operatorname{arcsec} e^{2x}] = \dfrac{2e^{2x}}{e^{2x}\sqrt{(e^{2x})^2 - 1}}$ $\qquad u = e^{2x}$

$\qquad\qquad\qquad = \dfrac{2e^{2x}}{e^{2x}\sqrt{e^{4x} - 1}}$

$\qquad\qquad\qquad = \dfrac{2}{\sqrt{e^{4x} - 1}}$

Note that we omit the absolute value sign because $e^{2x} > 0$. ▭

EXAMPLE 6   A derivative that can be simplified

Differentiate $y = \arcsin x + x\sqrt{1 - x^2}$.

SOLUTION

$y' = \dfrac{1}{\sqrt{1 - x^2}} + x\left(\dfrac{1}{2}\right)(-2x)(1 - x^2)^{-1/2} + \sqrt{1 - x^2}$

$\quad = \dfrac{1}{\sqrt{1 - x^2}} - \dfrac{x^2}{\sqrt{1 - x^2}} + \sqrt{1 - x^2}$

$\quad = \sqrt{1 - x^2} + \sqrt{1 - x^2}$

$\quad = 2\sqrt{1 - x^2}$ ▭

EXAMPLE 7   Graphing an inverse trigonometric function

Sketch the graph of

$\qquad y = (\arctan x)^2$.

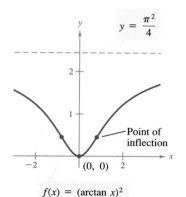

$$y = \frac{\pi^2}{4}$$

Point of inflection

$(0, 0)$

$f(x) = (\arctan x)^2$

FIGURE 8.36

SOLUTION

From the derivative

$$y' = 2(\arctan x)\left(\frac{1}{1 + x^2}\right) = \frac{2 \arctan x}{1 + x^2}$$

we can see that the only critical number is $x = 0$. By the First Derivative Test, this value corresponds to a relative minimum. From the second derivative

$$y'' = \frac{(1 + x^2)\left(\dfrac{2}{1 + x^2}\right) - (2 \arctan x)(2x)}{(1 + x^2)^2} = \frac{2(1 - 2x \arctan x)}{(1 + x^2)^2}$$

we see that points of inflection occur when $2x \arctan x = 1$. Using Newton's Method, we find that these points occur when $x \approx \pm 0.765$. Finally, the graph has a horizontal asymptote at $y = \pi^2/4$ since

$$\lim_{x \to \pm\infty} (\arctan x)^2 = \frac{\pi^2}{4}.$$

The graph is shown in Figure 8.36.　▭

We conclude this section with an example describing an application involving the derivative of an inverse trigonometric function.

EXAMPLE 8　An application

A photographer is taking a picture of a four-foot painting hung in an art gallery. The camera lens is one foot below the lower edge of the painting, as shown in Figure 8.37. How far should the camera be from the painting to maximize the angle subtended by the camera lens?

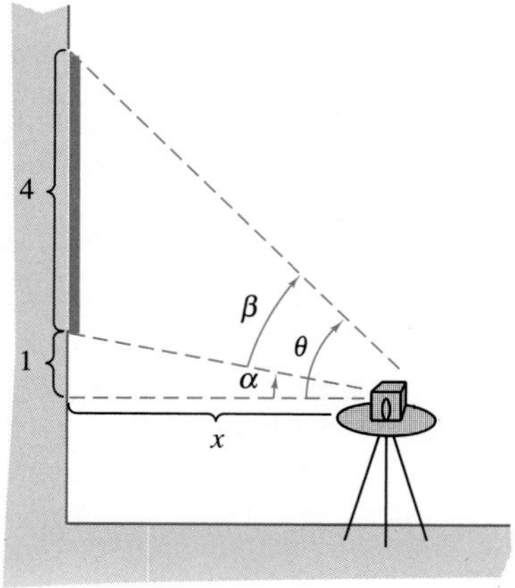

SOLUTION

In Figure 8.37, we let β be the angle we wish to maximize. Then we have

$$\beta = \theta - \alpha = \operatorname{arccot} \frac{x}{5} - \operatorname{arccot} x.$$

Differentiating, we have

$$\frac{d\beta}{dx} = \frac{-1/5}{1 + (x^2/25)} - \frac{-1}{1 + x^2}$$

$$= \frac{-5}{25 + x^2} + \frac{1}{1 + x^2}$$

$$= \frac{4(5 - x^2)}{(25 + x^2)(1 + x^2)}.$$

Since $d\beta/dx = 0$ when $x = \sqrt{5}$, we conclude from the First Derivative Test that this distance yields a maximum value for β.　▭

FIGURE 8.37

Galileo Galilei

Review of basic differentiation formulas

In the 1600s, Europe was ushered into the scientific age by such great thinkers as Descartes, Galileo, Huygens, Newton, and Kepler. These men believed that nature is governed by basic laws—laws that can, for the most part, be written in terms of mathematical equations. One of the most influential publications during this period was written by Galileo Galilei (1564–1642). It is called *Dialogue on the Great World Systems,* and it has become a classic description of modern scientific thought.

As mathematics developed during the past few hundred years, a list of elementary functions has proven sufficient for modeling most phenomena in physics, chemistry, biology, engineering, economics, and a variety of other fields. An **elementary function** is one that can be formed as the sum, product, quotient, or composition of functions from the following list.

Algebraic functions	*Transcendental functions*
Polynomial functions	Logarithmic functions
Rational functions	Exponential functions
Functions involving radicals	Trigonometric functions
	Inverse trigonometric functions

There are isolated cases of the use of other types of functions, but this list is fairly comprehensive. Moreover, with the differentiation formulas introduced in this section, we can differentiate *any* elementary function. For convenience, we summarize these differentiation rules here.

BASIC DIFFERENTIATION RULES FOR ELEMENTARY FUNCTIONS

1. $\dfrac{d}{dx}[cu] = cu'$

2. $\dfrac{d}{dx}[u \pm v] = u' \pm v'$

3. $\dfrac{d}{dx}[uv] = uv' + vu'$

4. $\dfrac{d}{dx}\left[\dfrac{u}{v}\right] = \dfrac{vu' - uv'}{v^2}$

5. $\dfrac{d}{dx}[c] = 0$

6. $\dfrac{d}{dx}[u^n] = nu^{n-1}u'$

7. $\dfrac{d}{dx}[x] = 1$

8. $\dfrac{d}{dx}[|u|] = \dfrac{u}{|u|}(u'), \ u \neq 0$

9. $\dfrac{d}{dx}[\ln u] = \dfrac{u'}{u}$

10. $\dfrac{d}{dx}[e^u] = e^u u'$

11. $\dfrac{d}{dx}[\sin u] = (\cos u)u'$

12. $\dfrac{d}{dx}[\cos u] = -(\sin u)u'$

13. $\dfrac{d}{dx}[\tan u] = (\sec^2 u)u'$

14. $\dfrac{d}{dx}[\cot u] = -(\csc^2 u)u'$

15. $\dfrac{d}{dx}[\sec u] = (\sec u \tan u)u'$

16. $\dfrac{d}{dx}[\csc u] = -(\csc u \cot u)u'$

17. $\dfrac{d}{dx}[\arcsin u] = \dfrac{u'}{\sqrt{1 - u^2}}$

18. $\dfrac{d}{dx}[\arccos u] = \dfrac{-u'}{\sqrt{1 - u^2}}$

19. $\dfrac{d}{dx}[\arctan u] = \dfrac{u'}{1 + u^2}$

20. $\dfrac{d}{dx}[\text{arccot } u] = \dfrac{-u'}{1 + u^2}$

21. $\dfrac{d}{dx}[\text{arcsec } u] = \dfrac{u'}{|u|\sqrt{u^2 - 1}}$

22. $\dfrac{d}{dx}[\text{arccsc } u] = \dfrac{-u'}{|u|\sqrt{u^2 - 1}}$

EXERCISES for Section 8.5

In Exercises 1–10, evaluate the given expression.

1. $\arcsin \dfrac{1}{2}$ **2.** $\arcsin 0$

3. $\arccos \dfrac{1}{2}$ **4.** $\arccos 0$

5. $\arctan \dfrac{\sqrt{3}}{3}$ **6.** $\text{arccot}\ (-1)$

7. $\text{arccsc}\ \sqrt{2}$ **8.** $\arcsin\ (-0.39)$

9. $\text{arcsec}\ 1.269$ **10.** $\arctan\ (-3)$

In Exercises 11–16, evaluate the given expression without a calculator. [Hint: Make a sketch of a right triangle, as illustrated in Example 4.]

11. (a) $\sin \left(\arcsin \dfrac{1}{2} \right)$ (b) $\cos \left(2 \arcsin \dfrac{1}{2} \right)$

12. (a) $\tan \left(\arccos \dfrac{\sqrt{2}}{2} \right)$ (b) $\cos \left(\arcsin \dfrac{5}{13} \right)$

13. (a) $\sin \left(\arctan \dfrac{3}{4} \right)$ (b) $\sec \left(\arcsin \dfrac{4}{5} \right)$

14. (a) $\tan\ (\text{arccot}\ 2)$ (b) $\cos\ (\text{arcsec}\ \sqrt{5})$

15. (a) $\cot \left[\arcsin \left(-\dfrac{1}{2} \right) \right]$ (b) $\csc \left[\arctan \left(-\dfrac{5}{12} \right) \right]$

16. (a) $\sec \left[\arctan \left(-\dfrac{3}{5} \right) \right]$ (b) $\tan \left[\arcsin \left(-\dfrac{5}{6} \right) \right]$

In Exercises 17–26, write the given expression in algebraic form.

17. $\tan\ (\arctan x)$ **18.** $\sin\ (\arccos x)$

19. $\cos\ (\arcsin 2x)$ **20.** $\sec\ (\arctan 3x)$

21. $\sin\ (\text{arcsec}\ x)$ **22.** $\cos\ (\text{arccot}\ x)$

23. $\tan \left(\text{arcsec}\ \dfrac{x}{3} \right)$ **24.** $\sec\ [\arcsin (x - 1)]$

25. $\csc \left(\arctan \dfrac{x}{\sqrt{2}} \right)$ **26.** $\cos \left(\arcsin \dfrac{x - h}{r} \right)$

In Exercises 27 and 28, fill in the blank.

27. $\arctan \dfrac{9}{x} = \arcsin\ (\rule{1cm}{0.4pt})$

28. $\arcsin \dfrac{\sqrt{36 - x^2}}{6} = \arccos\ (\rule{1cm}{0.4pt})$

In Exercises 29 and 30, verify each identity.

29. (a) $\text{arccsc}\ x = \arcsin \dfrac{1}{x},\ |x| \geq 1$

(b) $\text{arccot}\ x = \arctan \dfrac{1}{x},\ x > 0$

30. (a) $\arcsin\ (-x) = -\arcsin x,\ |x| \leq 1$

(b) $\arccos\ (-x) = \pi - \arccos x,\ |x| \leq 1$

In Exercises 31–34, sketch the graph of the function.

31. $f(x) = \arcsin\ (x - 1)$ **32.** $f(x) = \arctan x + \dfrac{\pi}{2}$

33. $f(x) = \text{arcsec}\ 2x$ **34.** $f(x) = \arccos \dfrac{x}{4}$

In Exercises 35–38, solve the given equation for x.

35. $\arcsin\ (3x - \pi) = \dfrac{1}{2}$ **36.** $\arctan 2x = -1$

37. $\arcsin \sqrt{2x} = \arccos \sqrt{x}$

38. $\arccos x = \text{arcsec}\ x$

In Exercises 39–58, find the derivative of the given function.

39. $f(x) = \arcsin 2x$ **40.** $f(x) = \arcsin x^2$

41. $f(x) = 2 \arcsin\ (x - 1)$ **42.** $f(x) = \arccos \sqrt{x}$

43. $f(x) = 3 \arccos \dfrac{x}{2}$ **44.** $f(x) = \arctan \sqrt{x}$

45. $f(x) = \arctan 5x$ **46.** $f(x) = x \arctan x$

47. $f(x) = \arccos \dfrac{1}{x}$ **48.** $f(x) = \text{arcsec}\ 2x$

49. $f(x) = \arcsin x + \arccos x$

50. $f(x) = \text{arcsec}\ x + \text{arccsc}\ x$

51. $h(t) = \sin\ (\arccos t)$ **52.** $g(t) = \tan\ (\arcsin t)$

53. $f(t) = \dfrac{1}{\sqrt{6}} \arctan \dfrac{\sqrt{6}t}{2}$

54. $f(x) = \dfrac{1}{2} \left(\dfrac{1}{2} \ln \dfrac{x + 1}{x - 1} - \arctan x \right)$

55. $f(x) = \dfrac{1}{2} \left(\dfrac{1}{2} \ln \dfrac{x + 1}{x - 1} + \arctan x \right)$

56. $f(x) = \dfrac{1}{2} (x\sqrt{1 - x^2} + \arcsin x)$

57. $f(x) = x \arcsin x + \sqrt{1 - x^2}$

58. $f(x) = x \arctan 2x - \dfrac{1}{4} \ln\ (1 + 4x^2)$

In Exercises 59 and 60, find the point of inflection of the graph of the given function.

59. $f(x) = \arcsin x$ **60.** $f(x) = \text{arccot}\ 2x$

In Exercises 61 and 62, find any relative extrema of the given function.

61. $f(x) = \text{arcsec}\ x - x$ **62.** $f(x) = \arcsin x - 2x$

In Exercises 63 and 64, find the point of intersection of the graphs of the given functions.

63. $y = \arccos x$, $y = \arctan x$
64. $y = \arcsin x$, $y = \arccos x$

65. A small boat is being pulled toward a dock that is 10 feet above the water. The rope is being pulled in at a rate of 1.5 feet per second. Find the rate at which the angle the rope makes with the horizontal is changing when 20 feet of rope is out.

66. An observer is standing 300 feet from the point at which a balloon is released. The balloon rises at a rate of 5 feet per second. How fast is the angle of elevation of the observer's line of sight increasing when the balloon is 100 feet high?

67. Verify the following differentiation formulas.

(a) $\dfrac{d}{dx}[\arctan u] = \dfrac{u'}{1 + u^2}$

(b) $\dfrac{d}{dx}[\text{arcsec } u] = \dfrac{u'}{|u|\sqrt{u^2 - 1}}$

(c) $\dfrac{d}{dx}[\arccos u] = \dfrac{-u'}{\sqrt{1 - u^2}}$

(d) $\dfrac{d}{dx}[\text{arccot } u] = \dfrac{-u'}{1 + u^2}$

(e) $\dfrac{d}{dx}[\text{arccsc } u] = \dfrac{-u'}{|u|\sqrt{u^2 - 1}}$

68. Show that the function

$$f(x) = \arcsin\left(\frac{x - 2}{2}\right) - 2\arcsin\frac{\sqrt{x}}{2}$$

is constant for $0 \le x \le 4$.

8.6 Inverse Trigonometric Functions: Integration and Completing the Square

Integrals involving inverse trigonometric functions ▪ Completing the square ▪ Review of basic integration formulas

The derivatives of the six inverse trigonometric functions occur in three pairs. In each pair the derivative of one function is the negative of the other. For example,

$$\frac{d}{dx}[\arcsin x] = \frac{1}{\sqrt{1 - x^2}} \quad \text{and} \quad \frac{d}{dx}[\arccos x] = -\frac{1}{\sqrt{1 - x^2}}.$$

When listing the *antiderivative* that corresponds to each of the inverse trigonometric functions, we need use only one member from each pair. For example, we choose to use arcsin x as the antiderivative of $1/\sqrt{1 - x^2}$, rather than $-\arccos x$. The next theorem gives one antiderivative formula for each of the three pairs.

**THEOREM 8.8
INTEGRALS INVOLVING INVERSE
TRIGONOMETRIC FUNCTIONS**

Let u be a differentiable function of x, and let $a > 0$.

$$\int \frac{du}{\sqrt{a^2 - u^2}} = \arcsin\frac{u}{a} + C$$

$$\int \frac{du}{a^2 + u^2} = \frac{1}{a}\arctan\frac{u}{a} + C$$

$$\int \frac{du}{u\sqrt{u^2 - a^2}} = \frac{1}{a}\text{arcsec}\frac{|u|}{a} + C$$

PROOF We prove the first formula and leave the remaining proofs as an exercise (see Exercise 52). Let $y = \arcsin(u/a)$. Then

$$y' = \frac{1}{\sqrt{1 - (u/a)^2}}\left(\frac{u'}{a}\right) = \frac{u'}{a\sqrt{(a^2 - u^2)/a^2}} = \frac{u'}{\sqrt{a^2 - u^2}}.$$

Consequently, by the definition of an antiderivative, we have

$$\int \frac{du}{\sqrt{a^2 - u^2}} = \int \frac{u'}{\sqrt{a^2 - u^2}} \, dx = \arcsin \frac{u}{a} + C.$$

EXAMPLE 1 Integration with inverse trigonometric functions

(a) $\displaystyle\int \frac{dx}{\sqrt{4 - x^2}} = \arcsin \frac{x}{2} + C$

(b) $\displaystyle\int \frac{dx}{2 + 9x^2} = \frac{1}{3} \int \frac{3 \, dx}{(\sqrt{2})^2 + (3x)^2}$ $\qquad u = 3x, \, a = \sqrt{2}$

$$= \frac{1}{3\sqrt{2}} \arctan \frac{3x}{\sqrt{2}} + C$$

The integrals in Example 1 are fairly straightforward applications of integration formulas. Unfortunately, this is not typical. The inverse trigonometric integration formulas can be disguised in many ways. In this section we give some typical ways in which these integrals can be disguised, and we illustrate techniques for removing the disguise.

EXAMPLE 2 Integration by substitution

Evaluate

$$\int \frac{dx}{\sqrt{e^{2x} - 1}}.$$

SOLUTION

Because this integral doesn't fit any of the three inverse trigonometric formulas as it stands, we try the substitution $u = e^x$.

$$u = e^x \implies du = e^x \, dx \implies dx = \frac{du}{e^x} = \frac{du}{u}$$

Now, we have

$$\int \frac{dx}{\sqrt{(e^x)^2 - 1}} = \int \frac{du/u}{\sqrt{u^2 - 1}} = \int \frac{du}{u\sqrt{u^2 - 1}}.$$

This integral now fits the arcsecant formula, and we have

$$\int \frac{dx}{\sqrt{e^{2x} - 1}} = \operatorname{arcsec} \frac{|u|}{1} + C = \operatorname{arcsec} e^x + C.$$

A useful integration technique is to rewrite the integrand as the sum of two quotients.

EXAMPLE 3   Rewriting the integrand as the sum of two quotients

Evaluate

$$\int \frac{x + 2}{\sqrt{4 - x^2}} \, dx.$$

SOLUTION

This integral does not appear to fit any of our integration formulas, but by splitting the integrand into two parts we obtain

$$\int \frac{x + 2}{\sqrt{4 - x^2}} \, dx = \int \frac{x}{\sqrt{4 - x^2}} \, dx + \int \frac{2}{\sqrt{4 - x^2}} \, dx.$$

Now, the first integral on the right can be evaluated with the Power Rule, and the second integral will yield an inverse sine function. Therefore, we write

$$\int \frac{x + 2}{\sqrt{4 - x^2}} \, dx = -\frac{1}{2} \int (4 - x^2)^{-1/2}(-2x) \, dx + 2 \int \frac{1}{\sqrt{4 - x^2}} \, dx$$

$$= -\frac{1}{2}\left[\frac{(4 - x^2)^{1/2}}{1/2} \right] + 2 \arcsin \frac{x}{2} + C$$

$$= -\sqrt{4 - x^2} + 2 \arcsin \frac{x}{2} + C.$$

Another disguised use of integration formulas occurs when the integrand is an improper rational function. (That is, the integrand is a rational function in which the degree of the numerator is greater than or equal to the degree of the denominator.) For such functions, we use long division to write the integrand as the sum of a polynomial and a proper rational function.

EXAMPLE 4   Integrating an improper rational function

Evaluate

$$\int \frac{3x^3 - 2}{x^2 + 4} \, dx.$$

SOLUTION

Since the degree of the numerator is greater than the degree of the denominator, we divide to obtain

$$\int \frac{3x^3 - 2}{x^2 + 4} \, dx = \int \left(3x - \frac{12x + 2}{x^2 + 4} \right) dx$$

$$= \int 3x \, dx - 6 \int \frac{2x}{x^2 + 4} \, dx - 2 \int \frac{dx}{x^2 + 4}$$

$$= \frac{3x^2}{2} - 6 \ln (x^2 + 4) - \arctan \frac{x}{2} + C.$$

Completing the square

Completing the square helps when quadratic functions are involved in the integrand. For example, the quadratic $x^2 + bx + c$ can be written as

$$x^2 + bx + c = x^2 + bx + \left(\frac{b}{2}\right)^2 - \left(\frac{b}{2}\right)^2 + c = \left(x + \frac{b}{2}\right)^2 + \left(\frac{4c - b^2}{4}\right).$$

Thus, we have written $x^2 + bx + c$ as the sum or difference of two squares. Notice that the key step in completing the square is to *add and subtract* $(b/2)^2$.

EXAMPLE 5 Completing the square

Evaluate

$$\int \frac{dx}{x^2 - 4x + 7}.$$

SOLUTION

By completing the square, we obtain

$$x^2 - 4x + 7 = (x^2 - 4x + 4) - 4 + 7 = (x - 2)^2 + 3 = u^2 + a^2.$$

Therefore,

$$\int \frac{dx}{x^2 - 4x + 7} = \int \frac{dx}{(x - 2)^2 + 3} \qquad u = x - 2, a = \sqrt{3}$$

$$= \frac{1}{\sqrt{3}} \arctan \frac{x - 2}{\sqrt{3}} + C. \qquad \blacksquare$$

In Example 5, we completed the square for a polynomial whose leading coefficient is 1. If the leading coefficient is not 1, we suggest factoring before completing the square, as demonstrated in the next example.

EXAMPLE 6 Completing the square when the leading coefficient is not 1

Evaluate

$$\int \frac{dx}{2x^2 - 8x + 10}.$$

SOLUTION

To complete the square, we write

$$2x^2 - 8x + 10 = 2(x^2 - 4x + 5)$$
$$= 2(x^2 - 4x + 4 - 4 + 5)$$
$$= 2[(x - 2)^2 + 1]$$
$$= 2(u^2 + a^2).$$

Therefore, we have

$$\int \frac{dx}{2x^2 - 8x + 10} = \frac{1}{2} \int \frac{dx}{(x-2)^2 + 1}$$

$$= \frac{1}{2} \int \frac{du}{u^2 + 1} \qquad u = x - 2, \ a = 1$$

$$= \frac{1}{2} \arctan u + C$$

$$= \frac{1}{2} \arctan (x - 2) + C.$$

In the next example, notice how we complete the square when the coefficient of x^2 is negative.

EXAMPLE 7 Completing the square for a negative leading coefficient

Find the area of the region bounded by the graph of $f(x) = 1/\sqrt{3x - x^2}$, the x-axis, and the lines $x = \frac{3}{2}$ and $x = \frac{9}{4}$.

SOLUTION

From Figure 8.38, we see that the area is given by

$$\text{area} = \int_{3/2}^{9/4} \frac{1}{\sqrt{3x - x^2}} \, dx.$$

To complete the square, we write

$$3x - x^2 = -(x^2 - 3x) = -\left[x^2 - 3x + \left(\frac{3}{2}\right)^2 - \left(\frac{3}{2}\right)^2 \right]$$

$$= \left(\frac{3}{2}\right)^2 - \left(x - \frac{3}{2}\right)^2$$

$$= a^2 - u^2.$$

Therefore,

$$\int_{3/2}^{9/4} \frac{dx}{\sqrt{3x - x^2}} = \int_{3/2}^{9/4} \frac{dx}{\sqrt{(3/2)^2 - [x - (3/2)]^2}}$$

$$= \left[\arcsin \frac{x - (3/2)}{3/2} \right]_{3/2}^{9/4}$$

$$= \arcsin \frac{1}{2} - \arcsin 0 = \frac{\pi}{6} - 0 = \frac{\pi}{6}.$$

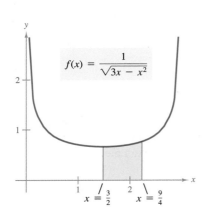

$$f(x) = \frac{1}{\sqrt{3x - x^2}}$$

$$x = \frac{3}{2} \qquad x = \frac{9}{4}$$

FIGURE 8.38

Review of basic integration formulas

We now have completed our list of the **basic integration formulas.** It is important to memorize each formula in the list.

BASIC INTEGRATION FORMULAS

1. $\displaystyle\int kf(u)\,du = k\int f(u)\,du$

2. $\displaystyle\int [f(u) \pm g(u)]\,du = \int f(u)\,du \pm \int g(u)\,du$

3. $\displaystyle\int du = u + C$

4. $\displaystyle\int u^n\,du = \frac{u^{n+1}}{n+1} + C,\ n \neq -1$

5. $\displaystyle\int \frac{du}{u} = \ln|u| + C$

6. $\displaystyle\int e^u\,du = e^u + C$

7. $\displaystyle\int \sin u\,du = -\cos u + C$

8. $\displaystyle\int \cos u\,du = \sin u + C$

9. $\displaystyle\int \tan u\,du = -\ln|\cos u| + C$

10. $\displaystyle\int \cot u\,du = \ln|\sin u| + C$

11. $\displaystyle\int \sec u\,du = \ln|\sec u + \tan u| + C$

12. $\displaystyle\int \csc u\,du = -\ln|\csc u + \cot u| + C$

13. $\displaystyle\int \sec^2 u\,du = \tan u + C$

14. $\displaystyle\int \csc^2 u\,du = -\cot u + C$

15. $\displaystyle\int \sec u \tan u\,du = \sec u + C$

16. $\displaystyle\int \csc u \cot u\,du = -\csc u + C$

17. $\displaystyle\int \frac{du}{\sqrt{a^2 - u^2}} = \arcsin \frac{u}{a} + C$

18. $\displaystyle\int \frac{du}{a^2 + u^2} = \frac{1}{a}\arctan \frac{u}{a} + C$

19. $\displaystyle\int \frac{du}{u\sqrt{u^2 - a^2}} = \frac{1}{a}\operatorname{arcsec} \frac{|u|}{a} + C$

You can learn quite a bit about the nature of integration by comparing this list to the summary of differentiation formulas given in the previous section. For differentiation, we now have formulas that allow us to differentiate *any* elementary function. For integration, this is certainly not true. The integration formulas listed above are primarily those that we happened upon when we were developing differentiation formulas. We do not have a formula for the antiderivative of a general product or quotient, the natural logarithmic function, or the inverse trigonometric functions. More importantly, we cannot apply any of the formulas in this list unless we can create the proper *du* corresponding to the *u* in the formula. The point is that we need to work more on integration techniques, which we will do in Chapter 9. To give you a better feeling for the integration problems that we *can* and *cannot* do with the techniques and formulas we now have, we conclude this section with two examples.

EXAMPLE 8   Comparing integration problems

Evaluate as many of the following integrals as you can using the formulas and techniques we have studied so far in the text.

(a) $\displaystyle\int \frac{dx}{x\sqrt{x^2 - 1}}$ (b) $\displaystyle\int \frac{x\,dx}{\sqrt{x^2 - 1}}$ (c) $\displaystyle\int \frac{dx}{\sqrt{x^2 - 1}}$

SOLUTION

(a) We *can* evaluate this integral (it fits the arcsecant formula).

$$\int \frac{dx}{x\sqrt{x^2 - 1}} = \text{arcsec } |x| + C$$

(b) We *can* also evaluate this integral (it fits the Power Rule).

$$\int \frac{x \, dx}{\sqrt{x^2 - 1}} = \frac{1}{2} \int (x^2 - 1)^{-1/2}(2x) \, dx$$

$$= \frac{1}{2}\left(\frac{(x^2 - 1)^{1/2}}{1/2}\right) + C$$

$$= \sqrt{x^2 - 1} + C$$

(c) We *cannot* evaluate this integral using our present techniques. (You should scan the list of Basic Integration Formulas to verify this conclusion.)

EXAMPLE 9 Comparing integration problems

Evaluate as many of the following integrals as you can using the formulas and techniques we have studied so far in the text.

(a) $\displaystyle\int \frac{dx}{x \ln x}$ (b) $\displaystyle\int \frac{\ln x \, dx}{x}$ (c) $\displaystyle\int \ln x \, dx$

SOLUTION

(a) We *can* evaluate this integral (it fits the log formula).

$$\int \frac{dx}{x \ln x} = \int \frac{1/x}{\ln x} \, dx = \ln |\ln x| + C$$

(b) We *can* also evaluate this integral (it fits the Power Rule).

$$\int \frac{\ln x}{x} \, dx = \int \left(\frac{1}{x}\right)(\ln x)^1 \, dx = \frac{(\ln x)^2}{2} + C$$

(c) We *cannot* evaluate this integral using our present techniques.

REMARK   Note in Examples 8 and 9 that it is the *simplest* function that we cannot yet integrate.

EXERCISES for Section 8.6

In Exercises 1–26, evaluate the given integral.

1. $\displaystyle\int_0^{1/6} \frac{1}{\sqrt{1 - 9x^2}} \, dx$

2. $\displaystyle\int_0^1 \frac{dx}{\sqrt{4 - x^2}}$

3. $\displaystyle\int_0^{\sqrt{3}/2} \frac{1}{1 + 4x^2} \, dx$

4. $\displaystyle\int_{\sqrt{3}}^3 \frac{1}{9 + x^2} \, dx$

5. $\displaystyle\int \frac{1}{x\sqrt{4x^2 - 1}} \, dx$

6. $\displaystyle\int \frac{1}{4 + (x - 1)^2} \, dx$

7. $\displaystyle\int \frac{x^3}{x^2 + 1}\, dx$

8. $\displaystyle\int \frac{x^4 - 1}{x^2 + 1}\, dx$

9. $\displaystyle\int \frac{1}{\sqrt{1 - (x + 1)^2}}\, dx$

10. $\displaystyle\int \frac{t}{t^4 + 16}\, dt$

11. $\displaystyle\int \frac{t}{\sqrt{1 - t^4}}\, dt$

12. $\displaystyle\int \frac{1}{x\sqrt{x^4 - 4}}\, dx$

13. $\displaystyle\int \frac{\arctan x}{1 + x^2}\, dx$

14. $\displaystyle\int \frac{1}{(x - 1)\sqrt{(x - 1)^2 - 4}}\, dx$

15. $\displaystyle\int_0^{1/\sqrt{2}} \frac{\arcsin x}{\sqrt{1 - x^2}}\, dx$

16. $\displaystyle\int_0^{1/\sqrt{2}} \frac{\arccos x}{\sqrt{1 - x^2}}\, dx$

17. $\displaystyle\int_{-1/2}^{0} \frac{x}{\sqrt{1 - x^2}}\, dx$

18. $\displaystyle\int_{-\sqrt{3}}^{0} \frac{x}{1 + x^2}\, dx$

19. $\displaystyle\int \frac{e^x}{\sqrt{1 - e^{2x}}}\, dx$

20. $\displaystyle\int \frac{\cos x}{\sqrt{4 - \sin^2 x}}\, dx$

21. $\displaystyle\int \frac{1}{9 + (x - 3)^2}\, dx$

22. $\displaystyle\int \frac{x + 1}{x^2 + 1}\, dx$

23. $\displaystyle\int \frac{1}{\sqrt{x}(1 + x)}\, dx$

24. $\displaystyle\int_1^2 \frac{1}{3 + (x - 2)^2}\, dx$

25. $\displaystyle\int_{\pi/2}^{\pi} \frac{\sin x}{1 + \cos^2 x}\, dx$

26. $\displaystyle\int \frac{e^{2x}}{4 + e^{4x}}\, dx$

In Exercises 27–40, evaluate the given integral. (Complete the square, if necessary.)

27. $\displaystyle\int_0^2 \frac{dx}{x^2 - 2x + 2}$

28. $\displaystyle\int_{-3}^{-1} \frac{dx}{x^2 + 6x + 13}$

29. $\displaystyle\int \frac{2x}{x^2 + 6x + 13}\, dx$

30. $\displaystyle\int \frac{2x - 5}{x^2 + 2x + 2}\, dx$

31. $\displaystyle\int \frac{1}{\sqrt{-x^2 - 4x}}\, dx$

32. $\displaystyle\int \frac{1}{\sqrt{-x^2 + 2x}}\, dx$

33. $\displaystyle\int \frac{x + 2}{\sqrt{-x^2 - 4x}}\, dx$

34. $\displaystyle\int \frac{x - 1}{\sqrt{x^2 - 2x}}\, dx$

35. $\displaystyle\int_2^3 \frac{2x - 3}{\sqrt{4x - x^2}}\, dx$

36. $\displaystyle\int \frac{1}{(x - 1)\sqrt{x^2 - 2x}}\, dx$

37. $\displaystyle\int \frac{x}{x^4 + 2x^2 + 2}\, dx$

38. $\displaystyle\int \frac{x}{\sqrt{9 + 8x^2 - x^4}}\, dx$

39. $\displaystyle\int \frac{1}{\sqrt{-16x^2 + 16x - 3}}\, dx$

40. $\displaystyle\int \frac{1}{(x - 1)\sqrt{9x^2 - 18x + 5}}\, dx$

In Exercises 41–44, use substitution to evaluate the given integral.

41. $\displaystyle\int \frac{\sqrt{x - 1}}{x}\, dx$

42. $\displaystyle\int \frac{\sqrt{x - 2}}{x + 1}\, dx$

43. $\displaystyle\int \sqrt{e^t - 3}\, dt$

44. $\displaystyle\int \frac{1}{t\sqrt{t} + \sqrt{t}}\, dt$

In Exercises 45–48, find the area of the region bounded by the graphs of the given equations.

45. $y = \dfrac{1}{1 + x^2}$, $y = 0$, $x = 0$, $x = 1$

46. $y = \dfrac{1}{\sqrt{4 - x^2}}$, $y = 0$, $x = 0$, $x = 1$

47. $y = \dfrac{1}{x^2 - 2x + 5}$, $y = 0$, $x = 1$, $x = 3$

48. $y = \dfrac{1}{\sqrt{3 + 2x - x^2}}$, $y = 0$, $x = 0$, $x = 2$

49. An object is projected upward with an initial velocity of 500 feet per second. If the air resistance is proportional to the square of the velocity, we obtain the equation

$$\frac{dv}{dt} = -(32 + kv^2)$$

where 32 feet per second is the acceleration due to gravity and k is a constant. Find the velocity (on the upward flight) as a function of time by solving the equation

$$\int \frac{dv}{32 + kv^2} = -\int dt.$$

50. A weight of mass m is attached to a spring and oscillates with simple harmonic motion (see figure). By Hooke's Law, we can determine that

$$\int \frac{dy}{\sqrt{A^2 - y^2}} = \int \sqrt{\frac{k}{m}}\, dt$$

where A is the maximum displacement, t is the time, and k is a constant. Find y as a function of t, given that $y = 0$ when $t = 0$.

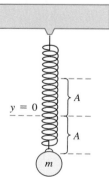

FIGURE FOR 50

51. (a) Show that

$$\int_0^1 \frac{4}{1 + x^2}\, dx = \pi.$$

(b) Approximate the number π using Simpson's Rule (with $n = 6$) and the integral of part (a).

52. Verify the following rules by differentiating.

(a) $\int \frac{du}{a^2 + u^2} = \frac{1}{a} \arctan \frac{u}{a} + C$

(b) $\int \frac{du}{u\sqrt{u^2 - a^2}} = \frac{1}{a} \operatorname{arcsec} \frac{|u|}{a} + C$

In Exercises 53–56, determine which of the given integrals can be evaluated using the Basic Integration Formulas we have studied so far.

53. (a) $\int \frac{1}{\sqrt{1 - x^2}} \, dx$ (b) $\int \frac{x}{\sqrt{1 - x^2}} \, dx$

(c) $\int \frac{1}{x\sqrt{1 - x^2}} \, dx$

54. (a) $\int e^{x^2} \, dx$ (b) $\int x \, e^{x^2} \, dx$

(c) $\int \frac{1}{x^2} e^{1/x} \, dx$

55. (a) $\int \sqrt{x - 1} \, dx$ (b) $\int x\sqrt{x - 1} \, dx$

(c) $\int \frac{x}{\sqrt{x - 1}} \, dx$

56. (a) $\int \frac{1}{1 + x^4} \, dx$ (b) $\int \frac{x}{1 + x^4} \, dx$

(c) $\int \frac{x^3}{1 + x^4} \, dx$

8.7 Hyperbolic Functions

Hyperbolic functions ▪ Differentiation and integration of hyperbolic functions ▪ Inverse hyperbolic functions ▪ Differentiation and integration involving inverse hyperbolic functions

Johann Heinrich Lambert

In this section we look briefly at a special class of exponential functions called the **hyperbolic functions.** The first person to publish a comprehensive study on hyperbolic functions was Johann Heinrich Lambert (1728–1777), a Swiss-German mathematician and colleague of Euler.

The name, *hyperbolic function,* arose from comparing the area of a semicircular region, as shown in Figure 8.39, to the area of a region under a hyperbola, as shown in Figure 8.40. The integral for the semicircular region involves an *inverse trigonometric (circular) function:*

$$\int_{-1}^{1} \sqrt{1 - x^2} \, dx = \frac{1}{2}\left[x\sqrt{1 - x^2} + \arcsin x \right]_{-1}^{1} = \frac{\pi}{2} \approx 1.571.$$

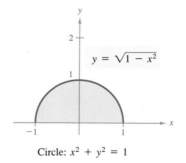

Circle: $x^2 + y^2 = 1$

FIGURE 8.39

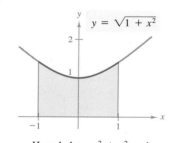

Hyperbola: $-x^2 + y^2 = 1$

FIGURE 8.40

The integral for the hyperbolic region involves an inverse hyperbolic function:

$$\int_{-1}^{1} \sqrt{1 + x^2} \, dx = \frac{1}{2}\left[x\sqrt{1 + x^2} + \sinh^{-1} x \right]_{-1}^{1} \approx 2.296.$$

This is only one of many ways in which the hyperbolic functions are similar to the trigonometric functions.

DEFINITION OF THE HYPERBOLIC FUNCTIONS

$$\sinh x = \frac{e^x - e^{-x}}{2} \qquad \cosh x = \frac{e^x + e^{-x}}{2} \qquad \tanh x = \frac{\sinh x}{\cosh x}$$

$$\operatorname{csch} x = \frac{1}{\sinh x}, x \neq 0 \qquad \operatorname{sech} x = \frac{1}{\cosh x} \qquad \coth x = \frac{1}{\tanh x}, x \neq 0$$

REMARK   We read sinh x as "the hyperbolic sine of x," cosh x as "the hyperbolic cosine of x," and so on.

The graphs of the six hyperbolic functions and their domains and ranges are shown in Figure 8.41. Note that the graphs of sinh x and cosh x can be obtained by *addition of ordinates* using the exponential functions $f(x) = \frac{1}{2}e^x$ and $g(x) = \frac{1}{2}e^{-x}$.

Domain: $(-\infty, \infty)$
Range: $(-\infty, \infty)$

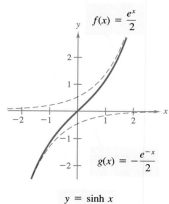

$y = \sinh x$

Domain: $(-\infty, \infty)$
Range: $[1, \infty)$

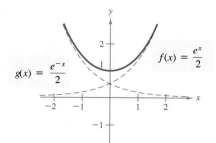

$y = \cosh x$

Domain: $(-\infty, \infty)$
Range: $(-1, 1)$

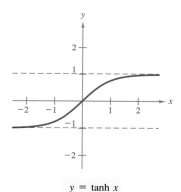

$y = \tanh x$

Domain: $(-\infty, 0)$ and $(0, \infty)$
Range: $(-\infty, 0)$ and $(0, \infty)$

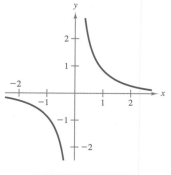

$y = \operatorname{csch} x = \dfrac{1}{\sinh x}$

Domain: $(-\infty, \infty)$
Range: $(0, 1]$

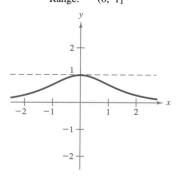

$y = \operatorname{sech} x = \dfrac{1}{\cosh x}$

Domain: $(-\infty, 0)$ and $(0, \infty)$
Range: $(-\infty, -1)$ and $(1, \infty)$

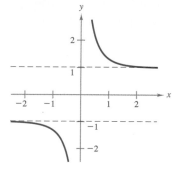

$y = \coth x = \dfrac{1}{\tanh x}$

FIGURE 8.41

Many of the trigonometric identities have corresponding *hyperbolic identities*. For instance,

$$\cosh^2 x - \sinh^2 x = \left(\frac{e^x + e^{-x}}{2}\right)^2 - \left(\frac{e^x - e^{-x}}{2}\right)^2$$

$$= \frac{e^{2x} + 2 + e^{-2x}}{4} - \frac{e^{2x} - 2 + e^{-2x}}{4} = \frac{4}{4} = 1$$

and

$$2 \sinh x \cosh x = 2\left(\frac{e^x - e^{-x}}{2}\right)\left(\frac{e^x + e^{-x}}{2}\right) = \frac{e^{2x} - e^{-2x}}{2} = \sinh 2x.$$

In the following list, notice the similarity of hyperbolic identities to trigonometric identities.

HYPERBOLIC IDENTITIES

$$\cosh^2 x - \sinh^2 x = 1 \qquad \sinh (x + y) = \sinh x \cosh y + \cosh x \sinh y$$

$$\tanh^2 x + \operatorname{sech}^2 x = 1 \qquad \sinh (x - y) = \sinh x \cosh y - \cosh x \sinh y$$

$$\coth^2 x - \operatorname{csch}^2 x = 1 \qquad \cosh (x + y) = \cosh x \cosh y + \sinh x \sinh y$$

$$\cosh (x - y) = \cosh x \cosh y - \sinh x \sinh y$$

$$\sinh^2 x = \frac{-1 + \cosh 2x}{2} \qquad \sinh 2x = 2 \sinh x \cosh x$$

$$\cosh^2 x = \frac{1 + \cosh 2x}{2} \qquad \cosh 2x = \cosh^2 x + \sinh^2 x$$

Since the hyperbolic functions are written in terms of e^x and e^{-x}, we can easily derive formulas for their derivatives. In the following theorem, we list these derivatives with the corresponding integration formulas.

**THEOREM 8.9
DERIVATIVES AND INTEGRALS
OF HYPERBOLIC FUNCTIONS**

Let u be a differentiable function of x.

$$\frac{d}{dx}[\sinh u] = (\cosh u)u' \qquad \int \cosh u \, du = \sinh u + C$$

$$\frac{d}{dx}[\cosh u] = (\sinh u)u' \qquad \int \sinh u \, du = \cosh u + C$$

$$\frac{d}{dx}[\tanh u] = (\operatorname{sech}^2 u)u' \qquad \int \operatorname{sech}^2 u \, du = \tanh u + C$$

$$\frac{d}{dx}[\coth u] = -(\operatorname{csch}^2 u)u' \qquad \int \operatorname{csch}^2 u \, du = -\coth u + C$$

$$\frac{d}{dx}[\operatorname{sech} u] = -(\operatorname{sech} u \tanh u)u' \qquad \int \operatorname{sech} u \tanh u \, du = -\operatorname{sech} u + C$$

$$\frac{d}{dx}[\operatorname{csch} u] = -(\operatorname{csch} u \coth u)u' \qquad \int \operatorname{csch} u \coth u \, du = -\operatorname{csch} u + C$$

PROOF

We give proofs for the derivatives of the hyperbolic sine and tangent, and leave the proofs of two of the remaining formulas as exercises (see Exercises 73 and 74).

$$\frac{d}{dx}[\sinh x] = \frac{d}{dx}\left[\frac{e^x - e^{-x}}{2}\right] = \frac{e^x + e^{-x}}{2} = \cosh x$$

$$\frac{d}{dx}[\tanh x] = \frac{d}{dx}\left[\frac{\sinh x}{\cosh x}\right] = \frac{\cosh x(\cosh x) - \sinh x(\sinh x)}{\cosh^2 x}$$

$$= \frac{1}{\cosh^2 x} = \text{sech}^2 x$$

EXAMPLE 1   Differentiation of hyperbolic functions

(a) $\dfrac{d}{dx}[\sinh (x^2 - 3)] = 2x \cosh (x^2 - 3)$

(b) $\dfrac{d}{dx}[\ln (\cosh x)] = \dfrac{\sinh x}{\cosh x} = \tanh x$

(c) $\dfrac{d}{dx}[x \sinh x - \cosh x] = x \cosh x + \sinh x - \sinh x = x \cosh x$ ▭

EXAMPLE 2   Finding relative extrema

Find the relative extrema of $f(x) = (x - 1) \cosh x - \sinh x$.

SOLUTION

Letting $f'(x) = 0$, we have

$$f'(x) = (x - 1) \sinh x + \cosh x - \cosh x = (x - 1) \sinh x = 0.$$

Thus, the critical numbers are $x = 1$ and $x = 0$. By the Second Derivative Test, we can verify that the point $(0, -1)$ yields a relative maximum and the point $(1, -\sinh 1)$ yields a relative minimum, as shown in Figure 8.42.

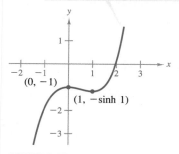

FIGURE 8.42 ▭

When a uniform flexible cable, such as a telephone wire, is suspended from two points, it takes the shape of a **catenary,** as shown in Figure 8.43. The equation for such a curve is shown in Example 3.

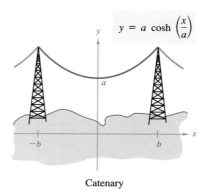

$$y = a \cosh \left(\frac{x}{a}\right)$$

Catenary   FIGURE 8.43

EXAMPLE 3   An application of hyperbolic functions

Power cables are suspended between two towers, forming the catenary in Figure 8.43. The equation for this catenary is

$$y = a \cosh \frac{x}{a}.$$

The distance between the two towers is $2b$. Find the slope of the catenary at the point where the cable meets the right-hand tower.

SOLUTION

Differentiating, we have

$$y' = a\left(\frac{1}{a}\right) \sinh \frac{x}{a} = \sinh \frac{x}{a}.$$

At b, the slope (from the left) is given by $m = \sinh (b/a)$.   ▭

EXAMPLE 4   Integrating a hyperbolic function

Evaluate $\int \cosh 2x \sinh^2 2x \, dx$.

SOLUTION

If we let $u = \sinh 2x$, then by the Power Rule it follows that

$$\int \cosh 2x \sinh^2 2x \, dx = \frac{1}{2} \int (\sinh 2x)^2 (2 \cosh 2x) \, dx$$

$$= \frac{1}{2} \left[\frac{(\sinh 2x)^3}{3} \right] + C$$

$$= \frac{\sinh^3 2x}{6} + C.$$ ▭

Unlike trigonometric functions, hyperbolic functions are *not* periodic. In fact, by looking back at Figure 8.41, we can see that four of the six hyperbolic functions are actually one-to-one (the hyperbolic sine, tangent, cosecant, and cotangent). Thus, we can apply Theorem 7.5 to conclude that these four functions have inverse functions. The other two (the hyperbolic cosine and secant) are one-to-one if their domains are restricted to the positive real numbers, and for this restricted domain they also have inverse functions. Since the hyperbolic functions are defined in terms of exponential functions, it is not surprising to find that the inverse hyperbolic functions can be written in terms of logarithmic functions, as shown in Theorem 8.10.

**THEOREM 8.10
INVERSE HYPERBOLIC
FUNCTIONS**

	Domain
$\sinh^{-1} x = \ln(x + \sqrt{x^2 + 1})$	$(-\infty, \infty)$
$\cosh^{-1} x = \ln(x + \sqrt{x^2 - 1})$	$[1, \infty)$
$\tanh^{-1} x = \dfrac{1}{2} \ln \dfrac{1 + x}{1 - x}$	$(-1, 1)$
$\coth^{-1} x = \dfrac{1}{2} \ln \dfrac{x + 1}{x - 1}$	$(-\infty, -1) \cup (1, \infty)$
$\operatorname{sech}^{-1} x = \ln \dfrac{1 + \sqrt{1 - x^2}}{x}$	$(0, 1]$
$\operatorname{csch}^{-1} x = \ln \dfrac{1 + \sqrt{1 + x^2}}{|x|}$	$(-\infty, 0) \cup (0, \infty)$

PROOF   The proof of this theorem is a straightforward application of the properties of the exponential and logarithmic functions. For example, if

$$f(x) = \sinh x = \frac{e^x - e^{-x}}{2}$$

and

$$g(x) = \ln(x + \sqrt{x^2 + 1}),$$

then

$$f(g(x)) = \frac{(x + \sqrt{x^2 + 1}) - [1/(x + \sqrt{x^2 + 1})]}{2}$$

$$= \frac{2x(x + \sqrt{x^2 + 1})}{2(x + \sqrt{x^2 + 1})} = x.$$

A similar argument can show that $g(f(x)) = x$, and we can conclude that g is the inverse function of f.

The graphs of the inverse hyperbolic functions are shown in Figure 8.44.

Domain: $[1, \infty)$
Range: $[0, \infty)$

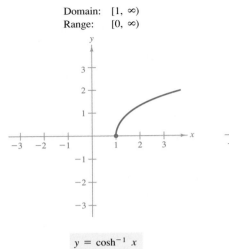

$$y = \cosh^{-1} x$$

Domain: $(-1, 1)$
Range: $(-\infty, \infty)$

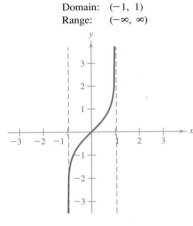

$$y = \tanh^{-1} x$$

Domain: $(-\infty, \infty)$
Range: $(-\infty, \infty)$

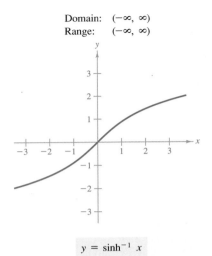

$$y = \sinh^{-1} x$$

Domain: $(0, 1]$
Range: $[0, \infty)$

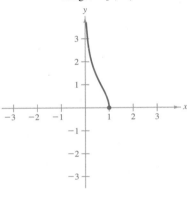

$$y = \operatorname{sech}^{-1} x$$

Domain: $(-\infty, -1)$ and $(1, \infty)$
Range: $(-\infty, 0)$ and $(0, \infty)$

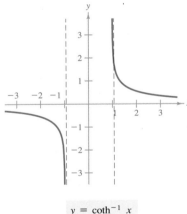

$$y = \coth^{-1} x$$

Domain: $(-\infty, 0)$ and $(0, \infty)$
Range: $(-\infty, 0)$ and $(0, \infty)$

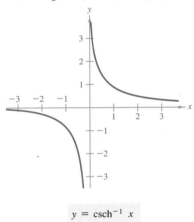

$$y = \operatorname{csch}^{-1} x$$

FIGURE 8.44

The inverse hyperbolic secant can be used to define a curve called a *tractrix* or *pursuit curve*. This is described in the following example.

EXAMPLE 5   An application: the tractrix

A person is holding a rope that is tied to a boat, as shown in Figure 8.45. As the person walks along the dock, the boat travels along a **tractrix**, given by the equation

$$y = a \operatorname{sech}^{-1} \frac{x}{a} - \sqrt{a^2 - x^2}$$

where a is the length of the rope. If $a = 20$ feet, find the distance the person must walk to bring the boat 5 feet from the dock.

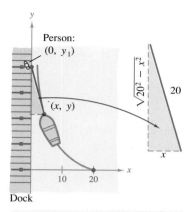

$$y = 20 \, \text{sech}^{-1} \left(\frac{x}{20}\right) - \sqrt{20^2 - x^2}$$

FIGURE 8.45

SOLUTION

In Figure 8.45, we see that the distance the person has walked is given by

$$y_1 = y + \sqrt{20^2 - x^2} = \left(20 \, \text{sech}^{-1} \frac{x}{20} - \sqrt{20^2 - x^2} \right) + \sqrt{20^2 - x^2}$$

$$= 20 \, \text{sech}^{-1} \frac{x}{20}.$$

When $x = 5$, this distance is

$$y_1 = 20 \, \text{sech}^{-1} \frac{5}{20} = 20 \ln \frac{1 + \sqrt{1 - (1/4)^2}}{1/4}$$

$$= 20 \ln (4 + \sqrt{15}) \approx 41.27 \text{ ft.}$$

The derivatives of the inverse hyperbolic functions resemble the derivatives of the inverse trigonometric functions. We list these in Theorem 8.11 with the corresponding integration formulas (listed in logarithmic form).

**THEOREM 8.11
DIFFERENTIATION AND
INTEGRATION INVOLVING
INVERSE HYPERBOLIC
FUNCTIONS**

Let u be a differentiable function of x.

$$\frac{d}{dx}[\sinh^{-1} u] = \frac{u'}{\sqrt{u^2 + 1}} \qquad\qquad \frac{d}{dx}[\cosh^{-1} u] = \frac{u'}{\sqrt{u^2 - 1}}$$

$$\frac{d}{dx}[\tanh^{-1} u] = \frac{u'}{1 - u^2} \qquad\qquad \frac{d}{dx}[\coth^{-1} u] = \frac{u'}{1 - u^2}$$

$$\frac{d}{dx}[\text{sech}^{-1} u] = \frac{-u'}{u\sqrt{1 - u^2}} \qquad\qquad \frac{d}{dx}[\text{csch}^{-1} u] = \frac{-u'}{|u|\sqrt{1 + u^2}}$$

$$\int \frac{du}{\sqrt{u^2 \pm a^2}} = \ln (u + \sqrt{u^2 \pm a^2}) + C$$

$$\int \frac{du}{a^2 - u^2} = \frac{1}{2a} \ln \left| \frac{a + u}{a - u} \right| + C$$

$$\int \frac{du}{u\sqrt{a^2 \pm u^2}} = -\frac{1}{a} \ln \frac{a + \sqrt{a^2 \pm u^2}}{|u|} + C$$

PROOF   As with the previous theorem, we can prove these differentiation and integration formulas with a straightforward application of the properties of exponential and logarithmic functions. For example,

$$\frac{d}{dx}[\tanh^{-1} x] = \frac{d}{dx}\left[\frac{1}{2} \ln \frac{1 + x}{1 - x} \right]$$

$$= \frac{1}{2}\left[\frac{d}{dx}[\ln (1 + x) - \ln (1 - x)] \right]$$

$$= \frac{1}{2}\left[\frac{1}{1 + x} - \frac{-1}{1 - x} \right] = \frac{1}{2}\left[\frac{2}{1 - x^2} \right] = \frac{1}{1 - x^2}.$$

In Exercises 75–77, you are asked to verify some of the other formulas given in this theorem.

EXAMPLE 6 Differentiation of inverse hyperbolic functions

For the tractrix given in Example 5, show that the boat is always pointing toward the person.

SOLUTION

For a point (x, y) on the tractrix, the slope of the graph gives us the direction of the boat, as shown in Figure 8.45. This slope is given by

$$
\begin{aligned}
y' &= \frac{d}{dx}\left[20 \; \text{sech}^{-1} \frac{x}{20} - \sqrt{20^2 - x^2} \right] \\
&= -20\left(\frac{1}{20}\right)\left(\frac{1}{(x/20)\sqrt{1 - (x/20)^2}}\right) - \left(\frac{1}{2}\right)\left(\frac{-2x}{\sqrt{20^2 - x^2}}\right) \\
&= \frac{-20^2}{x\sqrt{20^2 - x^2}} + \frac{x}{\sqrt{20^2 - x^2}} \\
&= -\frac{\sqrt{20^2 - x^2}}{x}.
\end{aligned}
$$

However, from Figure 8.45, we can see that the slope of the line segment connecting the point $(0, y_1)$ with the point (x, y) is also

$$
m = -\frac{\sqrt{20^2 - x^2}}{x}.
$$

Thus, we can conclude that the boat is always pointing toward the person. (It is because of this property that a tractrix is called a *pursuit curve*.) ▭

EXAMPLE 7 Integration using inverse hyperbolic functions

$$
\begin{aligned}
\int \frac{dx}{x\sqrt{4 - 9x^2}} &= \int \frac{3 \; dx}{(3x)\sqrt{4 - 9x^2}} \qquad a = 2, u = 3x \\
&= \int \frac{du}{u\sqrt{a^2 - u^2}} \\
&= -\frac{1}{a} \ln \frac{a + \sqrt{a^2 - u^2}}{|u|} + C \\
&= -\frac{1}{2} \ln \frac{2 + \sqrt{4 - 9x^2}}{|3x|} + C \qquad\qquad ▭
\end{aligned}
$$

EXAMPLE 8   Integration using inverse hyperbolic functions

Evaluate

$$\int_0^1 \frac{dx}{5 - 4x^2}.$$

SOLUTION

Since the denominator is of the form $a^2 - u^2$, we let $u = 2x$ and $a = \sqrt{5}$. Then $du = 2\,dx$, and we have

$$\int_0^1 \frac{dx}{5 - 4x^2} = \frac{1}{2} \int_0^1 \frac{2\,dx}{(\sqrt{5})^2 - (2x)^2}$$

$$= \frac{1}{2} \frac{1}{2\sqrt{5}} \left[\ln \frac{\sqrt{5} + 2x}{\sqrt{5} - 2x} \right]_0^1$$

$$= \frac{1}{4\sqrt{5}} \ln \frac{\sqrt{5} + 2}{\sqrt{5} - 2} \approx 0.3228.$$

EXERCISES for Section 8.7

In Exercises 1–6, evaluate the given function. If the function value is not a rational number, give the answer to three-decimal-place accuracy.

1. (a) $\sinh 3$ (b) $\tanh (-2)$
2. (a) $\cosh 0$ (b) $\operatorname{sech} 1$
3. (a) $\operatorname{csch} (\ln 2)$ (b) $\coth (\ln 5)$
4. (a) $\sinh^{-1} 0$ (b) $\tanh^{-1} 0$
5. (a) $\cosh^{-1} 2$ (b) $\operatorname{sech}^{-1} \frac{2}{3}$
6. (a) $\operatorname{csch}^{-1} 2$ (b) $\coth^{-1} 3$

In Exercises 7–12, verify the given identity.

7. $\tanh^2 x + \operatorname{sech}^2 x = 1$
8. $\cosh^2 x = \dfrac{1 + \cosh 2x}{2}$
9. $\sinh (x + y) = \sinh x \cosh y + \cosh x \sinh y$
10. $\sinh 2x = 2 \sinh x \cosh x$
11. $\sinh 3x = 3 \sinh x + 4 \sinh^3 x$
12. $\cosh x + \cosh y = 2 \cosh \dfrac{x + y}{2} \cosh \dfrac{x - y}{2}$

In Exercises 13–34, find y' and simplify.

13. $y = \sinh (1 - x^2)$
14. $y = \coth 3x$
15. $y = \ln (\sinh x)$
16. $y = \ln (\cosh x)$
17. $y = \ln \left(\tanh \dfrac{x}{2} \right)$
18. $y = x \sinh x - \cosh x$

19. $y = \dfrac{1}{4} \sinh 2x - \dfrac{x}{2}$
20. $y = x - \coth x$
21. $y = \arctan (\sinh x)$
22. $y = e^{\sinh x}$
23. $y = x^{\cosh x}$
24. $y = \operatorname{sech}^2 3x$
25. $y = (\cosh x - \sinh x)^2$
26. $y = \operatorname{sech} (x + 1)$
27. $y = \cosh^{-1} (3x)$
28. $y = \tanh^{-1} \dfrac{x}{2}$
29. $y = \sinh^{-1} (\tan x)$
30. $y = \operatorname{sech}^{-1} (\cos 2x)$, $0 < x < \dfrac{x}{4}$
31. $y = \coth^{-1} (\sin 2x)$
32. $y = (\operatorname{csch}^{-1} x)^2$
33. $y = 2x \sinh^{-1} (2x) - \sqrt{1 + 4x^2}$
34. $y = x \tanh^{-1} x + \ln \sqrt{1 - x^2}$

In Exercises 35–62, evaluate the integral.

35. $\displaystyle\int \sinh (1 - 2x)\,dx$
36. $\displaystyle\int \frac{\cosh \sqrt{x}}{\sqrt{x}}\,dx$
37. $\displaystyle\int \cosh^2 (x - 1) \sinh (x - 1)\,dx$
38. $\displaystyle\int \frac{\sinh x}{1 + \sinh^2 x}\,dx$
39. $\displaystyle\int \frac{\cosh x}{\sinh x}\,dx$
40. $\displaystyle\int \operatorname{sech}^2 (2x - 1)\,dx$
41. $\displaystyle\int x \operatorname{csch}^2 \frac{x^2}{2}\,dx$
42. $\displaystyle\int \operatorname{sech}^3 x \tanh x\,dx$
43. $\displaystyle\int \frac{\operatorname{csch} (1/x) \coth (1/x)}{x^2}\,dx$

44. $\displaystyle\int \sinh^2 x \, dx$

45. $\displaystyle\int_0^4 \frac{1}{25 - x^2} \, dx$ **46.** $\displaystyle\int_0^4 \frac{1}{\sqrt{25 - x^2}} \, dx$

47. $\displaystyle\int_0^{\sqrt{2}/4} \frac{2}{\sqrt{1 - 4x^2}} \, dx$ **48.** $\displaystyle\int \frac{2}{x\sqrt{1 + 4x^2}} \, dx$

49. $\displaystyle\int \frac{x}{x^4 + 1} \, dx$ **50.** $\displaystyle\int \frac{\cosh x}{\sqrt{9 - \sinh^2 x}} \, dx$

51. $\displaystyle\int \frac{1}{\sqrt{1 + e^{2x}}} \, dx$ **52.** $\displaystyle\int \frac{e^x}{1 - e^{2x}} \, dx$

53. $\displaystyle\int \frac{1}{\sqrt{x}\sqrt{1 + x}} \, dx$ **54.** $\displaystyle\int \frac{\sqrt{x}}{\sqrt{1 + x^3}} \, dx$

55. $\displaystyle\int \frac{1}{(x - 1)\sqrt{x^2 - 2x + 2}} \, dx$

56. $\displaystyle\int \frac{-1}{4x - x^2} \, dx$

57. $\displaystyle\int \frac{1}{1 - 4x - 2x^2} \, dx$

58. $\displaystyle\int \frac{1}{(x + 1)\sqrt{2x^2 + 4x + 8}} \, dx$

59. $\displaystyle\int \frac{1}{\sqrt{80 + 8x - 16x^2}} \, dx$

60. $\displaystyle\int \frac{1}{(x - 1)\sqrt{-4x^2 + 8x - 1}} \, dx$

61. $\displaystyle\int \frac{x^3 - 21x}{5 + 4x - x^2} \, dx$ **62.** $\displaystyle\int \frac{1 - 2x}{4x - x^2} \, dx$

In Exercises 63 and 64, find any relative extrema and points of inflection of the function. Sketch the graph.

63. $f(x) = x \cosh x - \sinh x$
64. $f(x) = x - \tanh x$

In Exercises 65 and 66, show that the given equation satisfies the differential equation.

65. $y = a \sinh x$, $y''' - y' = 0$
66. $y = a \cosh x$, $y'' - y = 0$

In Exercises 67 and 68, use the equation of the tractrix

$$y = a \, \text{sech}^{-1} \frac{x}{a} - \sqrt{a^2 - x^2}.$$

67. Find dy/dx.
68. Let L be the tangent line at the point P to the tractrix. If L intersects the y-axis at the point Q, show that the distance between P and Q is a.

69. Suppose that two chemicals A and B combine in a 3-to-1 ratio to form a compound. The amount of compound x being produced at any time t is proportional to the unchanged amounts of A and B remaining in the solution. Thus, if 3 kilograms of A is mixed with 2 kilograms of B, we have

$$\frac{dx}{dt} = k\left(3 - \frac{3x}{4}\right)\left(2 - \frac{x}{4}\right) = \frac{3k}{16}(x^2 - 12x + 32).$$

If 1 kilogram of the compound is formed after 10 minutes, find the amount formed after 20 minutes by solving the integral

$$\int \frac{3k}{16} \, dt = \int \frac{dx}{x^2 - 12x + 32}.$$

70. Electric wires suspended between two towers form a catenary (see figure) modeled by the equation

$$y = 60 \cosh \frac{x}{60}, \quad -60 \le x \le 60$$

where x and y are measured in feet. Find the length of the suspended cable if the towers are 120 feet apart.

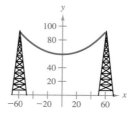

FIGURE FOR 70

71. Repeat Exercise 70, if the equation for the catenary is given by

$$y = a \cosh \frac{x}{a}, \quad -b \le x \le b$$

where a is measured in feet and the distance between the towers is $2b$ feet.

72. A barn is 100 feet long and 40 feet wide (see figure). A cross section of the roof is the inverted catenary

$$y = 31 - 20 \cosh \frac{x}{20}.$$

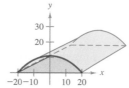

FIGURE FOR 72

Find the number of cubic feet of storage space in the barn.

In Exercises 73–77, verify the given derivative formula.

73. $\dfrac{d}{dx}[\cosh x] = \operatorname{sech}^2 x$

74. $\dfrac{d}{dx}[\operatorname{sech} x] = -\operatorname{sech} x \tanh x$

75. $\dfrac{d}{dx}[\cosh^{-1} x] = \dfrac{1}{\sqrt{x^2 - 1}}$

76. $\dfrac{d}{dx}[\sinh^{-1} x] = \dfrac{1}{\sqrt{x^2 + 1}}$

77. $\dfrac{d}{dx}[\operatorname{sech}^{-1} x] = \dfrac{-1}{x\sqrt{1 - x^2}}$

REVIEW EXERCISES for Chapter 8

In Exercises 1–24, find dy/dx.

1. $y = \dfrac{\sin x}{x^2}$

2. $y = \csc 3x + \cot 3x$

3. $y = -x \tan x$

4. $y = \dfrac{\cos (x - 1)}{x - 1}$

5. $y = \dfrac{1}{4} \sin 4x + x$

6. $y = x \cos x - \sin x$

7. $y = \dfrac{1}{2}x - \dfrac{1}{4} \sin 2x$

8. $y = \dfrac{1}{7} \sec^7 x - \dfrac{1}{5} \sec^5 x$

9. $y = 2 \csc^3 (\sqrt{x})$

10. $y = \dfrac{1}{2} \sec 2x$

11. $y = \tan \sqrt{1 - x}$

12. $y = \dfrac{1}{2}e^{\sin 2x}$

13. $y = \tan (\arcsin x)$

14. $y = \arctan (x^2 - 1)$

15. $y = x \operatorname{arcsec} x$

16. $y = \dfrac{1}{2} \arctan e^{2x}$

17. $y = x(\arcsin x)^2 - 2x + 2\sqrt{1 - x^2} \arcsin x$

18. $y = \sqrt{x^2 - 4} - 2 \operatorname{arcsec} \dfrac{x}{2}$

19. $y = \left(\dfrac{x^2 + 1}{2}\right) \arctan x$

20. $x \sin y = y \cos x$

21. $x = 2 + \sin y$

22. $\sin (x + y) = x$

23. $\cos x^2 = xe^y$

24. $\cos (x + y) = x$

In Exercises 25–30, find the second derivative of the function.

25. $f(x) = \cot x$

26. $g(t) = \sin^2 t$

27. $h(x) = \dfrac{\cos x}{x}$

28. $f(x) = x \tan x$

29. $f(x) = \arcsin 2x$

30. $f(x) = \arctan \dfrac{x}{2}$

In Exercises 31–50, evaluate the indefinite integral.

31. $\displaystyle\int \dfrac{\cos x}{1 + \sin^2 x}\, dx$

32. $\displaystyle\int \sin^3 x \cos x\, dx$

33. $\displaystyle\int \dfrac{\arctan 2x}{1 + 4x^2}\, dx$

34. $\displaystyle\int \dfrac{\cos x}{\sqrt{\sin x}}\, dx$

35. $\displaystyle\int \tan^n x \sec^2 x\, dx,\ n \neq -1$

36. $\displaystyle\int \dfrac{\sin \theta}{\sqrt{1 - \cos \theta}}\, d\theta$

37. $\displaystyle\int \dfrac{e^{-2x}}{1 + e^{-2x}}\, dx$

38. $\displaystyle\int \dfrac{x - 1}{3x^2 - 6x - 1}\, dx$

39. $\displaystyle\int \dfrac{1}{e^{2x} + e^{-2x}}\, dx$

40. $\displaystyle\int \dfrac{\tan (1/x)}{x^2}\, dx$

41. $\displaystyle\int x \tan x^2\, dx$

42. $\displaystyle\int \sec 2x \tan 2x\, dx$

43. $\displaystyle\int \dfrac{x}{\sqrt{1 - x^4}}\, dx$

44. $\displaystyle\int \dfrac{1}{3 + 25x^2}\, dx$

45. $\displaystyle\int \dfrac{x}{16 + x^2}\, dx$

46. $\displaystyle\int \dfrac{1}{16 + x^2}\, dx$

47. $\displaystyle\int \dfrac{\arctan (x/2)}{4 + x^2}\, dx$

48. $\displaystyle\int \dfrac{\arcsin x}{\sqrt{1 - x^2}}\, dx$

49. $\displaystyle\int \dfrac{4 - x}{\sqrt{4 - x^2}}\, dx$

50. $\displaystyle\int x \sin 3x^2\, dx$

In Exercises 51 and 52, show that the given equation satisfies the differential equation.

51. $y = [a + \ln |\cos x|] \cos x + (b + x) \sin x$
$y'' + y = \sec x$

52. $y = a \cos 2x + b \sin 2x + 2x^2 - 1$
$y'' + 4y = 8x^2$

In Exercises 53 and 54, find an equation of the tangent line to the graph of the equation at the specified point.

Function	Point
53. $y = x \cos x$	$\left(\dfrac{\pi}{2}, 0\right)$
54. $y = \arctan x$	$\left(1, \dfrac{\pi}{4}\right)$

55. Find the area of the largest rectangle that can be inscribed between the graph of $y = \cos x$ and the x-axis.

56. A certain lawn sprinkler is constructed in such a way that $d\theta/dt$ is constant, where θ ranges between 45° and 135° (see figure). The distance the water travels horizontally is given by

$$x = \frac{v^2 \sin 2\theta}{32}$$

where v is the speed of the water. Find dx/dt and explain why this model of lawn sprinkler does not water evenly. Which portion of the lawn receives the most water?

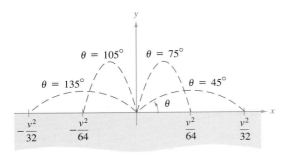

Water Sprinkler: $45° \leq \theta \leq 135°$

FIGURE FOR 56

57. A hallway of width 6 feet meets a hallway of width 9 feet at right angles. Find the length of the longest pipe that can be carried horizontally around this corner. [Hint: If L is the length of the pipe, show that

$$L = 6 \csc \theta + 9 \csc\left(\frac{\pi}{2} - \theta\right)$$

where θ is the angle between the pipe and the wall of the narrower hallway.]

58. Rework Exercise 57 if one of the hallways is of width a and the other is of width b.

59. Let θ be the angle of displacement (from the vertical) of a pendulum that is L feet long. Find the maximum rate of change of θ if

$$\theta(t) = A \sin \frac{32t}{L} + B \cos \frac{32t}{L}$$

where t is measured in seconds.

60. A buoy oscillates in simple harmonic motion ($y = A \cos \omega t$) as waves move past. At a given time it is noted that the buoy moves a total of 3.5 feet from its low point to its high point, and that it returns to the high point every 10 seconds.
   (a) Write an equation describing the motion of the buoy if it is at its high point at $t = 0$.
   (b) Determine the velocity of the buoy as a function of t.

61. Determine whether there exist any values of x in the interval $[0, 2\pi)$ such that the rate of change of $f(x) = \sec x$ and the rate of change of $g(x) = \csc x$ are equal.

62. For a person at rest, the velocity v, in liters per second, of air flow during a respiratory cycle is

$$v = 0.85 \sin \frac{\pi t}{3}$$

where t is the time in seconds. Find the volume, in liters, of air inhaled during one cycle by integrating this function over the interval $[0, 3]$.

63. The temperature in degrees Fahrenheit is given by

$$T = 72 + 12 \sin\left[\frac{\pi(t - 8)}{12}\right]$$

where t is the time in hours, with $t = 0$ representing midnight. Suppose the hourly cost of cooling a house is $0.10 per degree.
   (a) Find the cost C of cooling this house if its thermostat is set at 72° by evaluating the integral

$$C = 0.1 \int_8^{20} \left[72 + 12 \sin \frac{\pi(t - 8)}{12} - 72 \right] dt.$$

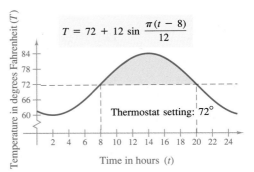

FIGURE FOR 63(a)

   (b) Find the savings in resetting the thermostat to 78° by evaluating the integral

$$C = 0.1 \int_{10}^{18} \left[72 + 12 \sin \frac{\pi(t - 8)}{12} - 78 \right] dt.$$

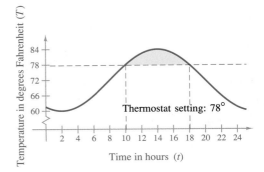

FIGURE FOR 63(b)

64. A fertilizer manufacturer finds that national sales of fertilizer roughly follow the seasonal pattern

$$F(t) = 100,000\left[1 + \sin\frac{2\pi(t - 60)}{365}\right]$$

where F is measured in pounds and t is the time in days, with $t = 1$ representing January 1. The manufacturer wants to set up a schedule to produce a uniform amount each day. What should this amount be?

In Exercises 65 and 66, find the volume of the solid generated by revolving the plane region bounded by the graphs of the given equations about the indicated line.

65. $y = \dfrac{1}{x^4 + 1}$, $y = 0$, $x = 0$, $x = 1$,

revolved about the y-axis

66. $y = \dfrac{1}{\sqrt{1 + x^2}}$, $y = 0$, $x = -1$, $x = 1$,

revolved about the x-axis

In Exercises 67–70, find dy/dx.

67. $y = \cosh(x^2 + 1)$ **68.** $y = \ln(\tanh 4x)$

69. $y = \tanh^{-1}\dfrac{x}{2}$ **70.** $y = \sinh^{-1} 3x$

In Exercises 71–74, evaluate the given integral.

71. $\displaystyle\int \frac{1}{x^2}\sinh\frac{1}{x}\,dx$ **72.** $\displaystyle\int x\,\mathrm{sech}\,x^2\,dx$

73. $\displaystyle\int \frac{1}{\sqrt{4x^2 + 1}}\,dx$ **74.** $\displaystyle\int \frac{x}{1 - x^4}\,dx$

In Exercises 75 and 76, verify the given identity.

75. $\coth^2 x - \mathrm{csch}^2 x = 1$

76. $\sinh^2 x = \dfrac{-1 + \cosh 2x}{2}$

77. Use a computer to sketch the graph of the function

$$f(x) = \sec\frac{\pi x}{6}$$

and find $\displaystyle\lim_{x \to 3^+} f(x)$.

78. A function f is defined as follows.

$$f(x) = \frac{\tan 2x}{x}, \quad x \neq 0$$

(a) Find $\displaystyle\lim_{x \to 0} f(x)$ (if it exists).

(b) Can the function be defined so that it is continuous at $x = 0$?

The space shuttle Columbia, after four test flights, made its first operational flight on November 16, 1982. It stayed in orbit five days. The second of the space shuttles, the *Challenger,* made its maiden flight on April 4, 1983. On September 29, 1988, the *Discovery* was launched. It returned four days later to Edwards Air Force Base in California.

Velocity of a Rocket

If a rocket whose initial mass is m (including fuel) is fired vertically at time $t = 0$, then its velocity at time t is given by

$$v = gt + u \ln \frac{m}{m - rt}$$

where u is the expulsion speed of the fuel, r is the rate at which the fuel is consumed, and $g = -32$ is the acceleration due to gravity.

The position equation for the rocket can be found by integrating the velocity with respect to t to obtain the equation

$$s = \frac{gt^2}{2} + u\left[t + \left(t - \frac{m}{r}\right) \ln \frac{m}{m - rt}\right].$$

For example, consider a rocket for which $m = 40{,}000$ lb, $u = 10{,}000$ ft/sec, and $r = 300$ lb/sec. If the rocket contains 36,000 pounds of fuel, then it will accelerate for 120 seconds and have a velocity of

$$v = -32t + 10{,}000 \ln \frac{40{,}000}{40{,}000 - 300t}, \quad 0 \le t \le 120$$

where v is measured in feet per second. Thus, after 120 seconds, this rocket will be traveling at 19,186 ft/sec and will have attained a height of 662,590 feet.

Chapter Overview

This chapter begins by reviewing the nineteen basic integration formulas developed in earlier chapters. Then, in Sections 9.2–9.6, we look at a variety of integration techniques: (1) integration by parts, (2) integration of powers of trigonometric functions, (3) trigonometric substitution, (4) partial fractions, (5) integration by tables, and (6) integration of rational functions of sine and cosine. Success in these new techniques depends on mastery of the basic integration formulas given in Section 9.1.

In the last section, we introduce the concept of an **improper integral.** Improper integrals come in two forms: (1) one or both of the limits of integration can be infinite such as in the improper integral

$$\int_1^\infty \frac{1}{x}\, dx$$

and (2) the integrand can have a finite number of infinite discontinuities such as in the improper integral

$$\int_0^1 \frac{1}{x^2}\, dx.$$

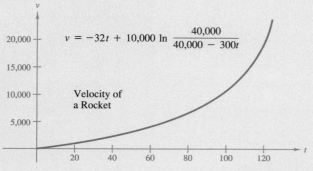

$$v = -32t + 10{,}000 \ln \frac{40{,}000}{40{,}000 - 300t}$$

Velocity of a Rocket

See Exercise 73, Section 9.2.

490

Integration Techniques and Improper Integrals

9.1 Basic Integration Formulas

Basic integration formulas ▪ Fitting integrands to basic formulas

In this chapter we study several integration techniques that greatly expand the set of integrals to which the basic integration formulas can be applied.

BASIC INTEGRATION FORMULAS

1. $\int kf(u)\, du = k \int f(u)\, du$

2. $\int [f(u) \pm g(u)]\, du = \int f(u)\, du \pm \int g(u)\, du$

3. $\int du = u + C$

4. $\int u^n\, du = \dfrac{u^{n+1}}{n+1} + C,\ n \neq -1$

5. $\int \dfrac{du}{u} = \ln|u| + C$

6. $\int e^u\, du = e^u + C$

7. $\int \sin u\, du = -\cos u + C$

8. $\int \cos u\, du = \sin u + C$

9. $\int \tan u\, du = -\ln|\cos u| + C$

10. $\int \cot u\, du = \ln|\sin u| + C$

11. $\int \sec u\, du = \ln|\sec u + \tan u| + C$

12. $\int \csc u\, du = -\ln|\csc u + \cot u| + C$

13. $\int \sec^2 u\, du = \tan u + C$

14. $\int \csc^2 u\, du = -\cot u + C$

15. $\int \sec u \tan u\, du = \sec u + C$

16. $\int \csc u \cot u\, du = -\csc u + C$

17. $\int \dfrac{du}{\sqrt{a^2 - u^2}} = \arcsin \dfrac{u}{a} + C$

18. $\int \dfrac{du}{a^2 + u^2} = \dfrac{1}{a} \arctan \dfrac{u}{a} + C$

19. $\int \dfrac{du}{u\sqrt{u^2 - a^2}} = \dfrac{1}{a} \operatorname{arcsec} \dfrac{|u|}{a} + C$

A major step in solving any integration problem is recognizing the proper basic integration formula to be used. To do this, you must memorize the basic formulas and practice using them. In this first section we review the use of basic integration formulas.

One of the challenging things about integration is the fact that slight differences in the integrand can lead to very different solution techniques.

EXAMPLE 1   A comparison of three similar integrals

Evaluate the following integrals.

(a) $\displaystyle\int \frac{4}{x^2 + 9}\, dx$ (b) $\displaystyle\int \frac{4x}{x^2 + 9}\, dx$ (c) $\displaystyle\int \frac{4x^2}{x^2 + 9}\, dx$

SOLUTION

(a) Considering the Arctangent Rule, we let $u = x$ and $a = 3$. Then

$$\int \frac{4}{x^2 + 9}\, dx = 4\int \frac{1}{x^2 + 3^2}\, dx = 4\left[\frac{1}{3} \arctan \frac{x}{3}\right] + C$$

$$= \frac{4}{3} \arctan \frac{x}{3} + C.$$

(b) Here the Arctangent Rule does not apply because the numerator contains a factor of x. Considering the Log Rule, we let $u = x^2 + 9$. Then, $du = 2x\, dx$, and we have

$$\int \frac{4x}{x^2 + 9}\, dx = 2\int \frac{2x\, dx}{x^2 + 9} = 2\int \frac{du}{u}$$

$$= 2 \ln |u| + C$$

$$= 2 \ln (x^2 + 9) + C.$$

(c) Since the degree of the numerator is equal to the degree of the denominator, we first divide to obtain

$$\frac{4x^2}{x^2 + 9} = 4 - \frac{36}{x^2 + 9}.$$

Thus,

$$\int \frac{4x^2}{x^2 + 9}\, dx = \int \left(4 - \frac{36}{x^2 + 9}\right) dx = \int 4\, dx - 36\int \frac{1}{x^2 + 9}\, dx$$

$$= 4x - 36\left[\frac{1}{3} \arctan \frac{x}{3}\right] + C$$

$$= 4x - 12 \arctan \frac{x}{3} + C. \quad \blacksquare$$

Notice in part (c) of Example 1 that some preliminary algebra was required before applying the rules for integration, and that subsequently more than one formula was needed to evaluate the resulting integral. This is also the case in the next example.

EXAMPLE 2   Using two basic formulas to solve a single integral

Evaluate

$$\int \frac{x + 3}{\sqrt{4 - x^2}} \, dx.$$

SOLUTION

We rewrite the integral as the sum of two integrals and then apply the Power Rule and the Arcsine Rule as follows.

$$\int \frac{x + 3}{\sqrt{4 - x^2}} \, dx = \int \frac{x}{\sqrt{4 - x^2}} \, dx + \int \frac{3}{\sqrt{4 - x^2}} \, dx$$

$$= -\frac{1}{2} \int \underbrace{(4 - x^2)^{-1/2}}_{u^n} \underbrace{(-2x) \, dx}_{du} + 3 \int \frac{1}{\underbrace{\sqrt{2^2 - x^2}}_{a = 2, \, u = x}} \, dx$$

$$= -\frac{1}{2} \int u^{-1/2} \, du + 3 \int \frac{1}{\sqrt{a^2 - u^2}} \, du$$

$$= -\frac{1}{2}\left(\frac{u^{1/2}}{1/2}\right) + 3 \, \arcsin \frac{u}{a} + C$$

$$= -(4 - x^2)^{1/2} + 3 \, \arcsin \frac{x}{2} + C$$

When the integrand is a quotient whose numerator is a sum (or difference), you should consider breaking the quotient into two or more parts, as in Example 2. The next example further demonstrates this useful procedure.

EXAMPLE 3   Breaking a quotient into two parts

Evaluate

$$\int \frac{1 + \cos(e^{-2x})}{e^{2x}} \, dx.$$

SOLUTION

As it stands, the integrand does not appear to fit any basic formula. However, since the numerator consists of a sum, we break the integrand into two fractions and proceed as follows.

$$\int \frac{1 + \cos e^{-2x}}{e^{2x}}\, dx = \int \frac{dx}{e^{2x}} + \int \frac{\cos e^{-2x}}{e^{2x}}\, dx$$

$$= \int e^{-2x}\, dx + \int (\cos e^{-2x})(e^{-2x})\, dx$$

$$\underset{u = -2x}{\underbrace{\phantom{e^{-2x}}}} \qquad \underset{u = e^{-2x}}{\underbrace{\phantom{\cos e^{-2x}}}}$$

$$= -\frac{1}{2} \int e^{-2x}(-2)\, dx - \frac{1}{2} \int (\cos e^{-2x})(-2e^{-2x})\, dx$$

$$= -\frac{1}{2} \int e^{u}\, du - \frac{1}{2} \int \cos u\, du$$

$$= -\frac{1}{2} e^{u} - \frac{1}{2} \sin u + C$$

$$= -\frac{1}{2} e^{-2x} - \frac{1}{2} \sin e^{-2x} + C$$

Formulas 17, 18, and 19 all have expressions involving the sum or difference of two squares:

$$a^2 - [f(x)]^2 \qquad a^2 + [f(x)]^2 \qquad [f(x)]^2 - a^2.$$

When you encounter such an expression, consider the substitution $u = f(x)$, as shown in the next example.

EXAMPLE 4   A substitution involving $a^2 - u^2$

Evaluate

$$\int \frac{x^2}{\sqrt{16 - x^6}}\, dx.$$

SOLUTION

The radical in the denominator can be written in the form $\sqrt{a^2 - u^2} = \sqrt{4^2 - (x^3)^2}$, and we try the substitution $u = x^3$. Then $du = 3x^2\, dx$, and we have

$$\int \frac{x^2}{\sqrt{16 - x^6}}\, dx = \frac{1}{3} \int \frac{3x^2\, dx}{\sqrt{16 - (x^3)^2}}$$

$$= \frac{1}{3} \int \frac{du}{\sqrt{4^2 - u^2}} = \frac{1}{3} \arcsin \frac{u}{4} + C$$

$$= \frac{1}{3} \arcsin \frac{x^3}{4} + C.$$

Also, when applying Formulas 17, 18, and 19, recall from Section 8.6 that completing the square can be helpful. For instance,

$$\int \frac{dx}{x^2 + 2x + 5} = \int \frac{dx}{(x + 1)^2 + 2^2}$$

$$= \frac{1}{2} \arctan \frac{x + 1}{2} + C.$$

Surprisingly, two of the most commonly overlooked integration formulas are the Log Rule and the Power Rule. Notice in the next two examples how these two integration formulas can be disguised.

EXAMPLE 5   A disguised form of the Log Rule

Evaluate

$$\int \frac{1}{1 + e^x} \, dx.$$

SOLUTION

The integral does not appear to fit any of the basic formulas. However, the quotient form suggests the Log Rule. If we let $u = 1 + e^x$, then $du = e^x \, dx$. We can obtain the required du by adding and subtracting e^x in the numerator as follows.

$$\int \frac{1}{1 + e^x} \, dx = \int \frac{1 + e^x - e^x}{1 + e^x} \, dx \qquad \text{Add and subtract } e^x \text{ in numerator}$$

$$= \int \left(\frac{1 + e^x}{1 + e^x} - \frac{e^x}{1 + e^x} \right) dx$$

$$= \int dx - \int \frac{e^x \, dx}{1 + e^x}$$

$$= \int dx - \int \frac{du}{u}$$

$$= x - \ln |u| + C$$

$$= x - \ln (1 + e^x) + C$$

In this chapter, you will see that there is often more than one way to solve an integration problem. For instance, in Example 5, we could multiply the numerator and denominator by e^{-x} to obtain an integral of the form $-\int (u'/u) \, dx$. See if you can get the same answer by this procedure. (Be careful; the answer will appear in a different form.)

EXAMPLE 6   A disguised form of the Power Rule

Evaluate $\int (\cot x)[\ln (\sin x)] \, dx$.

SOLUTION

Again, this integral does not appear to fit any of the basic fomulas. However, considering the two primary choices for u ($u = \cot x$ or $u = \ln \sin x$), we see that the second choice is the appropriate one, since

$$u = \ln \sin x \quad \Longrightarrow \quad du = \frac{\cos x}{\sin x} \, dx = \cot x \, dx.$$

Thus, we have

$$\int (\cot x)[\ln (\sin x)] \, dx = \int u \, du = \frac{u^2}{2} + C$$
$$= \frac{1}{2}[\ln (\sin x)]^2 + C.$$

REMARK   In Example 6, try *checking* that the derivative of $\frac{1}{2}[\ln (\sin x)]^2 + C$ is the integrand of the original integral.

Trigonometric identities can often be used to fit integrals to one of the basic integration formulas. The last example in this review section demonstrates this procedure.

EXAMPLE 7   Using trigonometric identities

Evaluate $\int \tan^2 2x \, dx$.

SOLUTION

By looking over the list of basic integration formulas, we see that $\tan^2 u$ is not on the list, but $\sec^2 u$ is on the list. This observation suggests the trigonometric identity $\tan^2 u = \sec^2 u - 1$. If we let $u = 2x$, then $du = 2 \, dx$ and

$$\int \tan^2 2x \, dx = \frac{1}{2} \int \tan^2 u \, du = \frac{1}{2} \int (\sec^2 u - 1) \, du$$
$$= \frac{1}{2} \int \sec^2 u \, du - \frac{1}{2} \int du = \frac{1}{2} \tan u - \frac{u}{2} + C$$
$$= \frac{1}{2} \tan 2x - x + C.$$

We conclude with a summary of some common procedures for fitting integrands to the basic integration formulas.

PROCEDURES FOR FITTING INTEGRANDS TO BASIC FORMULAS

Technique	*Example*
Expand (numerator)	$(1 + e^x)^2 = 1 + 2e^x + e^{2x}$
Separate numerator	$\dfrac{1 + x}{x^2 + 1} = \dfrac{1}{x^2 + 1} + \dfrac{x}{x^2 + 1}$
Complete the square	$\dfrac{1}{\sqrt{2x - x^2}} = \dfrac{1}{\sqrt{1 - (x - 1)^2}}$
Divide if rational function is improper	$\dfrac{x^2}{x^2 + 1} = 1 - \dfrac{1}{x^2 + 1}$
Add and subtract terms in numerator	$\dfrac{2x}{x^2 + 2x + 1} = \dfrac{2x + 2 - 2}{x^2 + 2x + 1}$
	$= \dfrac{2x + 2}{x^2 + 2x + 1} - \dfrac{2}{(x + 1)^2}$
Use trigonometric identities	$\cot^2 x = \csc^2 x - 1$
Multiply and divide by Pythagorean conjugate	$\dfrac{1}{1 + \sin x} = \left(\dfrac{1}{1 + \sin x}\right)\left(\dfrac{1 - \sin x}{1 - \sin x}\right)$
	$= \dfrac{1 - \sin x}{1 - \sin^2 x} = \dfrac{1 - \sin x}{\cos^2 x}$
	$= \sec^2 x - \dfrac{\sin x}{\cos^2 x}$

REMARK Watch out for this common error when fitting integrands to basic fomulas.

$$\frac{1}{x^2 + 1} \neq \frac{1}{x^2} + \frac{1}{1} \qquad \text{Do } not \text{ separate denominators}$$

EXERCISES for Section 9.1

In Exercises 1–50, evaluate the indefinite integral.

1. $\displaystyle\int (3x - 2)^4 \, dx$

2. $\displaystyle\int \frac{2}{(t - 9)^2} \, dt$

3. $\displaystyle\int (-2x + 5)^{3/2} \, dx$

4. $\displaystyle\int x\sqrt{4 - 2x^2} \, dx$

5. $\displaystyle\int \left[v + \frac{1}{(3v - 1)^3}\right] dv$

6. $\displaystyle\int \frac{2t - 1}{t^2 - t + 2} \, dt$

7. $\displaystyle\int \frac{t^2 - 3}{-t^3 + 9t + 1} \, dt$

8. $\displaystyle\int \frac{2x}{x - 4} \, dx$

9. $\displaystyle\int \frac{x^2}{x - 1} \, dx$

10. $\displaystyle\int \frac{x + 1}{\sqrt{x^2 + 2x - 4}} \, dx$

11. $\displaystyle\int \left(\frac{1}{3x - 1} - \frac{1}{3x + 1}\right) dx$

12. $\displaystyle\int \frac{e^x}{1 + e^x} \, dx$

13. $\displaystyle\int t \sin t^2 \, dt$

14. $\displaystyle\int \sec 4u \, du$

15. $\displaystyle\int \cos x \, e^{\sin x} \, dx$

16. $\displaystyle\int e^{5x} \, dx$

17. $\displaystyle\int \frac{(1 + e^t)^2}{e^t} \, dt$

18. $\displaystyle\int \frac{1 + \sin x}{\cos x} \, dx$

19. $\displaystyle\int \sec 3x \tan 3x \, dx$

20. $\displaystyle\int \frac{1}{\sqrt{x}(1 - 2\sqrt{x})} \, dx$

21. $\displaystyle\int \frac{2}{e^{-x} + 1} \, dx$

22. $\displaystyle\int \frac{1}{2e^x - 3} \, dx$

23. $\displaystyle\int \frac{1}{1 - \cos x} \, dx$

24. $\displaystyle\int \frac{1}{\sec x - 1} \, dx$

25. $\displaystyle\int \frac{2t - 1}{t^2 + 4} \, dt$

26. $\displaystyle\int \frac{2}{(2t - 1)^2 + 4} \, dt$

27. $\displaystyle\int \frac{3}{t^2 + 1} \, dt$

28. $\displaystyle\int \frac{3}{\sqrt{1 - t^2}} \, dt$

29. $\displaystyle\int \frac{1}{x\sqrt{x^2 - 4}} \, dx$

30. $\displaystyle\int \frac{-2x}{\sqrt{x^2 - 4}} \, dx$

31. $\displaystyle\int \frac{-1}{\sqrt{1 - (2t - 1)^2}} \, dt$

32. $\displaystyle\int \frac{1}{4 + 3x^2} \, dx$

33. $\int \csc \pi x \cot \pi x \, dx$

34. $\int \dfrac{1}{x\sqrt{4x^2 - 1}} \, dx$

35. $\int \dfrac{t}{\sqrt{1 - t^4}} \, dt$

36. $\int \dfrac{\sin x}{\sqrt{\cos x}} \, dx$

37. $\int \dfrac{\sec^2 x}{4 + \tan^2 x} \, dx$

38. $\int \tan^2 2x \, dx$

39. $\int \dfrac{\tan (2/t)}{t^2} \, dt$

40. $\int \dfrac{e^{1/t}}{t^2} \, dt$

41. $\int (1 + 2x^2)^2 \, dx$

42. $\int (2 - x^2)^2 \, dx$

43. $\int x\left(1 + \dfrac{1}{x}\right)^3 \, dx$

44. $\int (x + x^2)^3 \, dx$

45. $\int (1 + e^x)^2 \, dx$

46. $\int \left(\dfrac{e^x + e^{-x}}{2}\right)^2 \, dx$

47. $\int \dfrac{3}{\sqrt{6x - x^2}} \, dx$

48. $\int \dfrac{1}{(x - 1)\sqrt{4x^2 - 8x + 3}} \, dx$

49. $\int \dfrac{4}{4x^2 + 4x + 65} \, dx$

50. $\int \dfrac{1}{\sqrt{2 - 2x - x^2}} \, dx$

In Exercises 51–60, evaluate the definite integral.

51. $\displaystyle\int_0^1 xe^{-x^2} \, dx$

52. $\displaystyle\int_0^\pi \sin^2 t \cos t \, dt$

53. $\displaystyle\int_1^e \dfrac{1 - \ln x}{x} \, dx$

54. $\displaystyle\int_1^2 \dfrac{x - 2}{x} \, dx$

55. $\displaystyle\int_0^4 \dfrac{2x}{\sqrt{x^2 + 9}} \, dx$

56. $\displaystyle\int_0^{\pi/4} \cos 2x \, dx$

57. $\displaystyle\int_{\pi/4}^{\pi/2} \cot x \, dx$

58. $\displaystyle\int_0^4 \dfrac{1}{\sqrt{25 - x^2}} \, dx$

59. $\displaystyle\int_1^2 \dfrac{1}{2x\sqrt{4x^2 - 1}} \, dx$

60. $\displaystyle\int_0^{2/\sqrt{3}} \dfrac{1}{4 + 9x^2} \, dx$

In Exercises 61 and 62, find the area of the region bounded by the graph(s) of the given equation(s).

61. $y^2 = x^2(1 - x^2)$

62. $y = \sin 2x, \; y = 0, \; x = 0, \; x = \dfrac{\pi}{2}$

63. The graphs of $f(x) = x$ and $g(x) = ax^2$ intersect at the points $(0, 0)$ and $(1/a, 1/a)$. Find a so that the area of the region bounded by the graphs of these two functions is $\frac{2}{3}$.

64. Find the volume of the solid generated by revolving the region bounded by $y = e^{-x^2}$, $y = 0$, $x = 0$, and $x = 1$ about the y-axis.

65. The region bounded by $y = e^{-x^2}$, $y = 0$, $x = 0$, and $x = b$ is revolved around the y-axis. Find b so that the volume of the generated solid is $\frac{4}{3}$ cubic units.

66. Compute the average value of each of the functions over the indicated interval.
    (a) $f(x) = \sin nx, \; 0 \leq x \leq \pi/n$, n is a positive integer
    (b) $f(x) = \dfrac{1}{1 + x^2}, \; -3 \leq x \leq 3$

67. Find the x-coordinate of the centroid of the region bounded by the graphs of
    $$y = \dfrac{5}{\sqrt{25 - x^2}}, \; y = 0, \; x = 0, \; x = 4.$$

68. Find the area of the surface formed by revolving the graph of $y = 2\sqrt{x}$ on the interval $[0, 9]$ about the x-axis.

In Exercises 69 and 70, use Simpson's Rule on a computer or calculator (with $n = 12$) to approximate the arc length of the curve over the indicated interval.

Equation	Interval
69. $y = \tan \pi x$	$[0, \frac{1}{4}]$
70. $y = x^{2/3}$	$[1, 8]$

9.2 Integration by Parts

Integration by parts ▪ Tabular method

The first new integration technique we present in this chapter is called **integration by parts**. This technique applies to a wide variety of functions and is particularly useful for integrands involving a *product* of algebraic and transcendental functions. For instance, integration by parts works well with integrals like $\int x \ln x \, dx$, $\int x^2 e^x \, dx$, and $\int e^x \sin x \, dx$.

Integration by parts is based on the formula for the derivative of a product

$$\dfrac{d}{dx}[uv] = u\dfrac{dv}{dx} + v\dfrac{du}{dx} = uv' + vu'$$

where both u and v are differentiable functions of x. If u' and v' are continuous, then we can integrate both sides of this equation to obtain

$$uv = \int uv'\, dx + \int vu'\, dx = \int u\, dv + \int v\, du.$$

By rewriting this equation, we obtain the following theorem.

THEOREM 9.1
INTEGRATION BY PARTS

If u and v are functions of x and have continuous derivatives, then

$$\int u\, dv = uv - \int v\, du.$$

This formula expresses the original integrand in terms of another integral. Depending on the choices for u and dv, it may be easier to evaluate the second integral than the original one. Since the choices of u and dv are critical in the integration by parts process, we provide the following guidelines.

GUIDELINES FOR INTEGRATION BY PARTS

1. Try letting dv be the most complicated portion of the integrand that fits a basic integration formula. Then u will be the remaining factor(s) of the integrand.
2. Try letting u be the portion of the integrand whose derivative is a simpler function than u. Then dv will be the remaining factor(s) of the integrand.

EXAMPLE 1   Integration by parts

Evaluate $\int xe^x\, dx$.

SOLUTION

To apply integration by parts, we want to write the integral in the form $\int u\, dv$. There are several ways to do this.

$$\int \underbrace{(x)}_{u}\underbrace{(e^x\, dx)}_{dv}, \qquad \int \underbrace{(e^x)}_{u}\underbrace{(x\, dx)}_{dv}, \qquad \int \underbrace{(1)}_{u}\underbrace{(xe^x\, dx)}_{dv}, \qquad \int \underbrace{(xe^x)}_{u}\underbrace{(dx)}_{dv},$$

Following our guidelines, we choose the first option because the derivative of $u = x$ is simpler than x *and* $dv = e^x\, dx$ is the most complicated portion of the integrand that fits a basic integration formula. Thus, we have

$$dv = e^x\, dx \quad \Longrightarrow \quad v = \int dv = \int e^x\, dx = e^x$$
$$u = x \quad \Longrightarrow \quad du = dx.$$

Now, by the integration by parts formula, we have

$$\int u\, dv = uv - \int v\, du$$
$$\int x\, e^x\, dx = xe^x - \int e^x\, dx = xe^x - e^x + C.$$

Note in Example 1 that it is not necessary to include a constant integration when solving $v = \int e^x \, dx = e^x + C_1$. To illustrate this, we replace v by $v + C_1$ in the general formula to obtain

$$\int u \, dv = u(v + C_1) - \int (v + C_1) \, du = uv + C_1 u - \int C_1 \, du - \int v \, du$$

$$= uv + C_1 u - C_1 u - \int v \, du = uv - \int v \, du.$$

EXAMPLE 2 Integration by parts

Evaluate $\int x^2 \ln x \, dx$.

SOLUTION

In this case x^2 is more easily integrated than $\ln x$. Furthermore, the derivative of $\ln x$ is simpler than $\ln x$. Therefore, we let $dv = x^2 \, dx$.

$$dv = x^2 \, dx \quad \Longrightarrow \quad v = \int x^2 \, dx = \frac{x^3}{3}$$

$$u = \ln x \quad \Longrightarrow \quad du = \frac{1}{x} \, dx$$

Therefore, we have

$$\int x^2 \ln x \, dx = \frac{x^3}{3} \ln x - \int \left(\frac{x^3}{3}\right)\left(\frac{1}{x}\right) dx$$

$$= \frac{x^3}{3} \ln x - \frac{1}{3} \int x^2 \, dx$$

$$= \frac{x^3}{3} \ln x - \frac{x^3}{9} + C.$$

One unusual application of integration by parts involves integrands consisting of a single factor, such as $\int \ln x \, dx$ or $\int \arcsin x \, dx$. In such cases, we let $dv = dx$, as illustrated in the next example.

EXAMPLE 3 An integrand with a single term

Evaluate $\int_0^1 \arcsin x \, dx$.

SOLUTION

Letting $dv = dx$, we have

$$dv = dx \quad \Longrightarrow \quad v = \int dx = x$$

$$u = \arcsin x \quad \Longrightarrow \quad du = \frac{1}{\sqrt{1 - x^2}} \, dx.$$

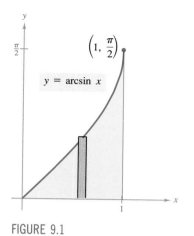

$\left(1, \dfrac{\pi}{2}\right)$

$y = \arcsin x$

FIGURE 9.1

Therefore, we have

$$\int \arcsin x \, dx = x \arcsin x - \int \frac{x}{\sqrt{1-x^2}} \, dx$$

$$= x \arcsin x + \frac{1}{2} \int (1-x^2)^{-1/2}(-2x) \, dx$$

$$= x \arcsin x + \sqrt{1-x^2} + C.$$

Now, using this antiderivative, we evaluate the definite integral as follows.

$$\int_0^1 \arcsin x \, dx = \left[x \arcsin x + \sqrt{1-x^2} \right]_0^1 = \frac{\pi}{2} - 1$$

The area represented by this definite integral is shown in Figure 9.1.   ▭

It may happen that an integral requires repeated application of the integration by parts formula. This is demonstrated in the next example.

EXAMPLE 4   Repeated application of integration by parts

Evaluate $\int x^2 \sin x \, dx$.

SOLUTION

We may consider x^2 and $\sin x$ to be equally easy to integrate. However, the derivative of x^2 becomes simpler, whereas the derivative of $\sin x$ does not. Therefore, we let $u = x^2$ and write

$$dv = \sin x \, dx \quad \Longrightarrow \quad v = \int \sin x \, dx = -\cos x$$

$$u = x^2 \quad \Longrightarrow \quad du = 2x \, dx$$

and it follows that

$$\int x^2 \sin x \, dx = -x^2 \cos x + \int 2x \cos x \, dx.$$

Now, we apply integration by parts to the new integral. We let $u = 2x$ and write

$$dv = \cos x \, dx \quad \Longrightarrow \quad v = \int \cos x \, dx = \sin x$$

$$u = 2x \quad \Longrightarrow \quad du = 2 \, dx$$

and it follows that

$$\int 2x \cos x \, dx = 2x \sin x - \int 2 \sin x \, dx = 2x \sin x + 2 \cos x + C.$$

Combining these two results, we have

$$\int x^2 \sin x \, dx = -x^2 \cos x + 2x \sin x + 2 \cos x + C.$$ ▭

When making repeated applications of integration by parts, you need to be careful not to interchange the substitutions in successive applications. For instance, in Example 4 our first substitution was $u = x^2$ and $dv = \sin x \, dx$. If, in the second application, we had switched the substitution to

$$dv = 2x \, dx \quad \Longrightarrow \quad v = \int 2x \, dx = x^2$$

$$u = \cos x \quad \Longrightarrow \quad du = -\sin x \, dx$$

we would have obtained

$$\int x^2 \sin x \, dx = -x^2 \cos x + \int 2x \cos x \, dx$$

$$= -x^2 \cos x + x^2 \cos x + \int x^2 \sin x \, dx$$

$$= \int x^2 \sin x \, dx$$

thus undoing the previous integration and returning to the *original* integral.

When making repeated applications of integration by parts, you should also watch for the appearance of a *constant multiple* of the original integral. This is illustrated in the next example.

EXAMPLE 5   Repeated application of integration by parts

Evaluate $\int e^x \cos 2x \, dx$.

SOLUTION

Our guidelines fail to help with a choice of u and dv, so we arbitrarily choose $dv = e^x \, dx$ and $u = \cos 2x$. (You might try verifying that the choice of $dv = \cos 2x \, dx$ and $u = e^x$ works equally well.)

$$dv = e^x \, dx \quad \Longrightarrow \quad v = \int e^x \, dx = e^x$$

$$u = \cos 2x \quad \Longrightarrow \quad du = -2 \sin 2x \, dx$$

Thus, it follows that

$$\int e^x \cos 2x \, dx = e^x \cos 2x + 2 \int e^x \sin 2x \, dx.$$

Making the same type of substitutions for the next application of integration by parts, we have

$$dv = e^x \, dx \quad \Longrightarrow \quad v = \int e^x \, dx = e^x$$

$$u = \sin 2x \quad \Longrightarrow \quad du = 2 \cos 2x \, dx$$

and it follows that

$$\int e^x \sin 2x \, dx = e^x \sin 2x - 2 \int e^x \cos 2x \, dx.$$

Therefore, we have

$$\int e^x \cos 2x \, dx = e^x \cos 2x + 2e^x \sin 2x - 4 \int e^x \cos 2x \, dx.$$

Now, since the right-hand integral is a constant multiple of the original integral, we add it to the left side of the equation to obtain

$$5 \int e^x \cos 2x \, dx = e^x \cos 2x + 2e^x \sin 2x$$

$$\int e^x \cos 2x \, dx = \frac{1}{5} e^x \cos 2x + \frac{2}{5} e^x \sin 2x + C. \qquad \text{Divide by 5} \quad \blacksquare$$

The integral in the next example is an important one. In Section 9.4 (Example 7) we will see that it is used to find the arc length of a parabolic segment.

EXAMPLE 6 Integration by parts

Evaluate $\int \sec^3 x \, dx$.

SOLUTION

The most complicated portion of the integrand that can be easily integrated is $\sec^2 x$. Letting $dv = \sec^2 x \, dx$ and $u = \sec x$, we have

$$dv = \sec^2 x \, dx \quad \Longrightarrow \quad v = \int \sec^2 x \, dx = \tan x$$

$$u = \sec x \quad \Longrightarrow \quad du = \sec x \tan x \, dx.$$

Therefore, we have

$$\int \sec^3 x \, dx = \sec x \tan x - \int \sec x \tan^2 x \, dx$$

$$= \sec x \tan x - \int \sec x (\sec^2 x - 1) \, dx$$

$$= \sec x \tan x - \int \sec^3 x \, dx + \int \sec x \, dx$$

$$2 \int \sec^3 x \, dx = \sec x \tan x + \int \sec x \, dx \qquad \text{Collect like integrals}$$

$$\int \sec^3 x \, dx = \frac{1}{2} \sec x \tan x + \frac{1}{2} \ln \left| \sec x + \tan x \right| + C. \qquad \blacksquare$$

Since we developed the integration by parts formula from the Product Rule for derivatives, we would expect many of our examples of this technique to involve a product. However, integration by parts is also useful in cases where the integrand is a quotient, as demonstrated in the next example.

EXAMPLE 7   An integrand involving a quotient

Evaluate

$$\int \frac{xe^x}{(x+1)^2}\, dx.$$

SOLUTION

Since $1/(x+1)^2$ is easily integrated, we make the following choices.

$$dv = \frac{dx}{(x+1)^2} \implies v = \int \frac{dx}{(x+1)^2} = -\frac{1}{x+1}$$

$$u = xe^x \implies du = (xe^x + e^x)\, dx = e^x(x+1)\, dx$$

Thus, we have

$$\int \frac{xe^x}{(x+1)^2}\, dx = xe^x\left(\frac{-1}{x+1}\right) - \int (x+1)e^x\left(\frac{-1}{x+1}\right) dx$$

$$= -\frac{xe^x}{x+1} + \int e^x\, dx$$

$$= -\frac{xe^x}{x+1} + e^x + C = \frac{e^x}{x+1} + C.$$

EXAMPLE 8   An application of integration by parts

Find the centroid of the region bounded by the graph of $y = \sin x$ and the x-axis, $0 \le x \le \pi/2$.

SOLUTION

Using the formulas presented in Section 6.7, together with Figure 9.2, we have

$$\text{area} = A = \int_0^{\pi/2} \sin x\, dx = -\cos x\Big]_0^{\pi/2} = 1$$

$$\bar{y} = \frac{1}{A} \int_0^{\pi/2} \frac{\sin x}{2} (\sin x)\, dx$$

$$= \frac{1}{4} \int_0^{\pi/2} (1 - \cos 2x)\, dx \qquad \text{Half-angle formula}$$

$$= \frac{1}{4}\left[x - \frac{\sin 2x}{2}\right]_0^{\pi/2} = \frac{\pi}{8}$$

$$\bar{x} = \frac{1}{A} \int_0^{\pi/2} x \sin x\, dx.$$

Now, using *integration by parts* on this integral, we let $dv = \sin x\, dx$, $u = x$, and obtain $v = -\cos x$ and $du = dx$. Thus, we have

$$\int x \sin x\, dx = -x \cos x + \int \cos x\, dx = -x \cos x + \sin x + C.$$

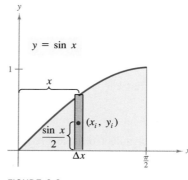

FIGURE 9.2

Now, we determine $\bar{x}$ to be

$$\bar{x} = \left[-x \cos x + \sin x \right]_0^{\pi/2} = 1.$$

Therefore, we conclude that the centroid of the region is $(1, \pi/8)$. ▭

As you gain experience in using integration by parts, your skill in determining u and dv will increase. In the following summary, we list several common integrals with suggestions for the choice of u and dv.

SUMMARY OF COMMON INTEGRALS USING INTEGRATION BY PARTS

1. $\displaystyle \int x^n e^{ax} \, dx, \qquad \int x^n \sin ax \, dx, \qquad \int x^n \cos ax \, dx$

   Let $u = x^n$ and $dv = e^{ax} \, dx$, $\sin ax \, dx$, or $\cos ax \, dx$. (Examples 1, 4)

2. $\displaystyle \int x^n \ln x \, dx, \qquad \int x^n \arcsin ax \, dx, \qquad \int x^n \arctan ax \, dx$

   Let $u = \ln x$, $\arcsin ax$, or $\arctan ax$ and $dv = x^n \, dx$. (Examples 2, 3)

3. $\displaystyle \int e^{ax} \sin bx \, dx, \qquad \int e^{ax} \cos bx \, dx$

   Let $u = \sin bx$ or $\cos bx$ and $dv = e^{ax} \, dx$. (Example 5)

Tabular method for repeated use of integration by parts

In problems involving repeated applications of integration by parts, a tabular method, illustrated in Example 9, can help to organize work. The method works well for integrals of the form

$$\int x^n \sin ax \, dx, \qquad \int x^n \cos ax \, dx, \qquad \text{and} \qquad \int x^n e^{ax} \, dx.$$

EXAMPLE 9   Using the Tabular Method

Evaluate $\int x^2 \sin 4x \, dx$.

SOLUTION

We begin as usual by letting $u = x^2$ and $dv = v' \, dx = \sin 4x \, dx$. Next, we create a table consisting of three columns as follows.

Alternate signs	u and its derivative	v' and its antiderivatives
+	x^2	$\sin 4x$
−	$2x$	$-\dfrac{1}{4} \cos 4x$
+	2	$-\dfrac{1}{16} \sin 4x$
−	0	$\dfrac{1}{64} \cos 4x$

Differentiate until you obtain 0 as a derivative.

Finally, the solution is given by multiplying the *signed* products of the diagonal entries to obtain

$$\int x^2 \sin 4x \, dx = -\frac{1}{4}x^2 \cos 4x + \frac{1}{8}x \sin 4x + \frac{1}{32} \cos 4x + C. \quad \blacksquare$$

EXERCISES for Section 9.2

In Exercises 1–34, evalute the given integral. (Note: Solve by the simplest method—not all require integration by parts.)

1. $\displaystyle\int xe^{2x} \, dx$

2. $\displaystyle\int x^2 e^{2x} \, dx$

3. $\displaystyle\int xe^{x^2} \, dx$

4. $\displaystyle\int x^2 e^{x^3} \, dx$

5. $\displaystyle\int xe^{-2x} \, dx$

6. $\displaystyle\int \frac{x}{e^x} \, dx$

7. $\displaystyle\int x^3 e^x \, dx$

8. $\displaystyle\int \frac{e^{1/t}}{t^2} \, dt$

9. $\displaystyle\int x^3 \ln x \, dx$

10. $\displaystyle\int \ln x \, dx$

11. $\displaystyle\int t \ln (t + 1) \, dt$

12. $\displaystyle\int \frac{1}{x(\ln x)^3} \, dx$

13. $\displaystyle\int (\ln x)^2 \, dx$

14. $\displaystyle\int \ln 3x \, dx$

15. $\displaystyle\int \frac{(\ln x)^2}{x} \, dx$

16. $\displaystyle\int \frac{\ln x}{x^2} \, dx$

17. $\displaystyle\int \frac{xe^{2x}}{(2x + 1)^2} \, dx$

18. $\displaystyle\int \frac{x^3 e^{x^2}}{(x^2 + 1)^2} \, dx$

19. $\displaystyle\int (x^2 - 1)e^x \, dx$

20. $\displaystyle\int \frac{\ln 2x}{x^2} \, dx$

21. $\displaystyle\int x\sqrt{x - 1} \, dx$

22. $\displaystyle\int x^2\sqrt{x - 1} \, dx$

23. $\displaystyle\int \frac{x^2}{\sqrt{2 + 3x}} \, dx$

24. $\displaystyle\int \frac{x}{\sqrt{2 + 3x}} \, dx$

25. $\displaystyle\int x \cos x \, dx$

26. $\displaystyle\int x^2 \cos x \, dx$

27. $\displaystyle\int x \sec^2 x \, dx$

28. $\displaystyle\int \theta \sec \theta \tan \theta \, d\theta$

29. $\displaystyle\int \arcsin 2x \, dx$

30. $\displaystyle\int \arccos x \, dx$

31. $\displaystyle\int \arctan x \, dx$

32. $\displaystyle\int \arctan \frac{x}{2} \, dx$

33. $\displaystyle\int e^{2x} \sin x \, dx$

34. $\displaystyle\int e^x \cos 2x \, dx$

In Exercises 35–40, evaluate the definite integral.

35. $\displaystyle\int_0^\pi x \sin 2x \, dx$

36. $\displaystyle\int_0^1 x \arcsin x^2 \, dx$

37. $\displaystyle\int_0^1 e^x \sin x \, dx$

38. $\displaystyle\int_0^1 x^2 e^x \, dx$

39. $\displaystyle\int_0^{\pi/2} x \cos x \, dx$

40. $\displaystyle\int_0^1 \ln (1 + x^2) \, dx$

In Exercises 41–46, use the tabular method for repeated applications of integration by parts to evaluate the given integral.

41. $\displaystyle\int x^2 e^{2x} \, dx$

42. $\displaystyle\int x^4 e^{-x} \, dx$

43. $\displaystyle\int x^3 \sin x \, dx$

44. $\displaystyle\int x^3 \cos 2x \, dx$

45. $\displaystyle\int x \sec^2 x \, dx$

46. $\displaystyle\int x^2(x - 2)^{3/2} \, dx$

47. Integrate $\int 2x\sqrt{2x - 3} \, dx$
(a) by parts, letting $dv = \sqrt{2x - 3} \, dx$
(b) by substitution, letting $u = \sqrt{2x - 3}$.

48. Integrate $\int x\sqrt{4 + x} \, dx$
(a) by parts, letting $dv = \sqrt{4 + x} \, dx$
(b) by substitution, letting $u = \sqrt{4 + x}$.

49. Integrate

$$\int \frac{x^3}{\sqrt{4 + x^2}} \, dx$$

(a) by parts, letting $dv = \dfrac{x}{\sqrt{4 + x^2}} \, dx$
(b) by substitution, letting $u = \sqrt{4 + x^2}$.

50. Integrate $\int x\sqrt{4 - x} \, dx$
(a) by parts, letting $dv = \sqrt{4 - x} \, dx$
(b) by substitution, letting $u = \sqrt{4 - x}$.

In Exercises 51–56, use integration by parts to verify the given formula. (Assume n is a positive integer.)

51. $\displaystyle\int x^n \sin x \, dx = -x^n \cos x + n \int x^{n-1} \cos x \, dx$

52. $\displaystyle\int x^n \cos x \, dx = x^n \sin x - n \int x^{n-1} \sin x \, dx$

53. $\displaystyle\int x^n \ln x \, dx = \frac{x^{n+1}}{(n + 1)^2}[-1 + (n + 1) \ln x] + C$

54. $\displaystyle\int x^n e^{ax} \, dx = \frac{x^n e^{ax}}{a} - \frac{n}{a} \int x^{n-1} e^{ax} \, dx$

55. $\displaystyle\int e^{ax} \sin bx \, dx = \frac{e^{ax}(a \sin bx - b \cos bx)}{a^2 + b^2} + C$

56. $\displaystyle\int e^{ax} \cos bx \, dx = \frac{e^{ax}(a \cos bx + b \sin bx)}{a^2 + b^2} + C$

In Exercises 57–60, evaluate the given integral by using the appropriate formula from Exercises 51–56.

57. $\int x^3 \ln x \, dx$

58. $\int x^2 \cos x \, dx$

59. $\int e^{2x} \cos 3x \, dx$

60. $\int x^3 e^{2x} \, dx$

In Exercises 61–64, find the area of the region bounded by the graphs of the given equations.

61. $y = xe^{-x}$, $y = 0$, $x = 4$

62. $y = \dfrac{1}{9}xe^{-x/3}$, $y = 0$, $x = 0$, $x = 3$

63. $y = e^{-x} \sin \pi x$, $y = 0$, $x = 0$, $x = 1$

64. $y = x \sin x$, $y = 0$, $x = 0$, $x = \pi$

65. Given the region bounded by the graphs of $y = \ln x$, $y = 0$, and $x = e$, find the following.
(a) the area of the region
(b) the volume of the solid generated by revolving the region about the x-axis
(c) the volume of the solid generated by revolving the region about the y-axis
(d) the centroid of the region

66. Find the volume of the solid generated by revolving the region bounded by $y = e^x$, $y = 0$, $x = 0$, and $x = 1$ about the y-axis.

67. A model for the ability M of a child to memorize, measured on a scale from 0 to 10, is given by

$$M = 1 + 1.6t \ln t, \quad 0 < t \le 4$$

where t is the child's age in years. Find the average value of this function
(a) between the child's first and second birthdays
(b) between the child's third and fourth birthdays.

68. A company sells a seasonal product, and the model for the daily revenue from the product is

$$R = 410.5t^2 e^{-t/30} + 25,000, \quad 0 \le t \le 365$$

where t is the time in days.
(a) Find the average daily receipts during the first quarter, $0 \le t \le 91$.
(b) Find the average daily receipts during the fourth quarter, $274 \le t \le 365$.

In Exercises 69 and 70, find the present value P of a continuous income flow of $c(t)$ dollars per year if

$$P = \int_0^{t_1} c(t)e^{-rt} \, dt$$

where t_1 is time in years and r is the annual interest rate compounded continuously.

69. $c(t) = 100,000 + 4000t$, $r = 9\%$, $t_1 = 10$

70. $c(t) = 30,000 + 500t$, $r = 7\%$, $t_1 = 5$

71. A string stretched between two points $(0, 0)$ and $(0, 2)$ is plucked by displacing the string h units at its midpoint. The motion of the string is modeled by a **Fourier Sine Series** whose coefficients are given by

$$b_n = h \int_0^1 x \sin \frac{n\pi x}{2} \, dx + h \int_1^2 (-x + 2) \sin \frac{n\pi x}{2} \, dx.$$

Evaluate b_n.

72. A damping force affects the vibration of a spring so that the displacement of the spring is given by

$$y = e^{-4t} (\cos 2t + 5 \sin 2t).$$

Find the average value of y on the interval from $t = 0$ to $t = \pi$.

73. In the Chapter 9 Application, we developed the following equation for the velocity of a rocket,

$$v(t) = -32t + 10,000 \ln \frac{40,000}{40,000 - 300t}$$

where v is measured in ft/sec. Use integration by parts to find the position function for the rocket. (The initial height is $s(0) = 0$.) What is the height of the rocket when $t = 120$ seconds?

74. The velocity (in ft/sec) of a rocket whose initial mass is m (including fuel) is given by

$$v = gt + u \ln \frac{m}{m - rt}$$

where u is the expulsion speed of the fuel, r is the rate at which the fuel is consumed, and $g = -32$ is the acceleration due to gravity. Find the position equation for a rocket for which $m = 50,000$ lb, $u = 12,000$ ft/sec, and $r = 400$ lb/sec. What is the height of the rocket when $t = 100$ seconds? (Assume it was fired from ground level and is moving straight up.)

75. Find the fallacy in the following argument that $0 = 1$.

$$dv = dx \quad \Longrightarrow \quad v = x$$

$$u = \frac{1}{x} \quad \Longrightarrow \quad du = -\frac{1}{x^2} \, dx$$

$$0 + \int \frac{dx}{x} = \left(\frac{1}{x}\right)(x) - \int \left(-\frac{1}{x^2}\right)(x) \, dx$$

$$= 1 + \int \frac{dx}{x}$$

Hence $0 = 1$.

76. Is there a fallacy in the following evaluation of $\int \ln (x + 5) \, dx$?

$$dv = dx \quad \Longrightarrow \quad v = x + 5$$

$$u = \ln (x + 5) \quad \Longrightarrow \quad du = \frac{1}{x + 5} \, dx$$

$$\int \ln (x + 5) \, dx = (x + 5) \ln (x + 5) - \int dx$$

$$= (x + 5) \ln (x + 5) - x + C$$

In Exercises 77 and 78, use Simpson's Rule on a computer or calculator (with $n = 12$) to approximate the area of the region bounded by the graphs of the given equations.

77. $y = \sqrt[3]{x}\sqrt{4 - x}$, $y = 0$, $x = 0$

78. $y = \sin (\pi x)^2$, $y = 0$, $x = 0$, $x = 1$

9.3 Trigonometric Integrals

Integrals involving powers of sine and cosine ▪ Integrals involving powers of secant and tangent ▪ Integrals involving sine-cosine products with different angles

In this section we introduce techniques for evaluating integrals of the form

$$\int \sin^m x \cos^n x \, dx \quad \text{and} \quad \int \sec^m x \tan^n x \, dx$$

where either m or n is a positive integer. To find antiderivatives for these forms, we try to break them into combinations of trigonometric integrals to which we can apply the Power Rule. For instance, we can evaluate $\int \sin^5 x \cos x \, dx$ with the Power Rule by letting $u = \sin x$. Then, $du = \cos x \, dx$ and we have

$$\int \sin^5 x \cos x \, dx = \int u^5 \, du = \frac{u^6}{6} + C = \frac{\sin^6 x}{6} + C.$$

Similarly, to evaluate $\int \sec^4 x \tan x \, dx$ by the Power Rule, we let $u = \sec x$. Then, $du = \sec x \tan x \, dx$ and we have

$$\int \sec^4 x \tan x \, dx = \int \sec^3 x \, (\sec x \tan x) \, dx = \frac{\sec^4 x}{4} + C.$$

To break up $\int \sin^m x \cos^n x \, dx$ into forms to which we can apply the Power Rule, we use the identities

$$\sin^2 x + \cos^2 x = 1 \qquad \text{Pythagorean identity}$$

$$\sin^2 x = \frac{1 - \cos 2x}{2} \qquad \text{Half-angle identity for } \sin^2 x$$

$$\cos^2 x = \frac{1 + \cos 2x}{2} \qquad \text{Half-angle identity for } \cos^2 x$$

as indicated in the following guidelines.

INTEGRALS INVOLVING SINE AND COSINE

1. If the power of the sine is odd and positive, save one sine factor and convert the remaining factors to cosine. Then, expand and integrate.

$$\int \overbrace{\sin^{2k+1} x}^{\text{odd}} \cos^n x \, dx = \int \overbrace{(\sin^2 x)^k}^{\text{convert to cosine}} \overbrace{\cos^n x}^{} \overbrace{\sin x \, dx}^{\text{save for } du}$$

$$= \int (1 - \cos^2 x)^k \cos^n x \sin x \, dx$$

2. If the power of the cosine is odd and positive, save one cosine factor and convert the remaining factors to sine. Then, expand and integrate.

$$\int \sin^m x \overbrace{\cos^{2k+1} x}^{\text{odd}} \, dx = \int \sin^m x \overbrace{(\cos^2 x)^k}^{\text{convert to sine}} \overbrace{\cos x \, dx}^{\text{save for } du}$$

$$= \int \sin^m x \, (1 - \sin^2 x)^k \cos x \, dx$$

3. If the powers of *both* the sine and cosine are even and nonnegative, make repeated use of the identities

$$\sin^2 x = \frac{1 - \cos 2x}{2} \qquad \text{and} \qquad \cos^2 x = \frac{1 + \cos 2x}{2}$$

to convert the integrand to odd powers of the cosine. Then, proceed as in case 2.

EXAMPLE 1   Power of sine is odd and positive

Evaluate $\int \sin^3 x \cos^4 x \, dx$.

SOLUTION

Expecting to use the Power Rule with $u = \cos x$, we *save one sine factor* to form du and convert the remaining sine factors to cosines, as follows.

$$\int \sin^3 x \cos^4 x \, dx = \int \sin^2 x \cos^4 x \, (\sin x) \, dx \qquad \text{Save } \sin x$$

$$= \int (1 - \cos^2 x) \cos^4 x \sin x \, dx \qquad \text{Identity for } \sin^2 x$$

$$= \int (\cos^4 x - \cos^6 x) \sin x \, dx$$

$$= \int \cos^4 x \sin x \, dx - \int \cos^6 x \sin x \, dx$$

$$= -\int \cos^4 x \, (-\sin x) \, dx + \int \cos^6 x \, (-\sin x) \, dx$$

$$= -\frac{\cos^5 x}{5} + \frac{\cos^7 x}{7} + C \qquad \text{Power Rule}$$

EXAMPLE 2   Power of cosine is odd and positive

Evaluate $\int \sin^2 x \cos^5 x \, dx$.

SOLUTION

Since the power of the cosine is odd and positive, we have

$$
\begin{aligned}
\int \sin^2 x \cos^5 x \, dx &= \int \sin^2 x \cos^4 x \cos x \, dx & \text{Save } \cos x \\
&= \int \sin^2 x \, (\cos^2 x)^2 \cos x \, dx \\
&= \int \sin^2 x \, (1 - \sin^2 x)^2 \cos x \, dx & \text{Identity for } \cos^2 x \\
&= \int \sin^2 x \, (1 - 2\sin^2 x + \sin^4 x) \cos x \, dx \\
&= \int [\sin^2 x - 2\sin^4 x + \sin^6 x] \cos x \, dx \\
&= \frac{\sin^3 x}{3} - \frac{2\sin^5 x}{5} + \frac{\sin^7 x}{7} + C. & \text{Power Rule}
\end{aligned}
$$

In Examples 1 and 2, *both* of the powers m and n happened to be positive integers. However, the method will work as long as either m *or* n is odd and positive. For instance, in the next example the power of the sine is 3, but the power of the cosine is $-\frac{1}{2}$.

EXAMPLE 3   Power of sine is odd and positive

Evaluate

$$
\int_0^{\pi/3} \frac{\sin^3 x}{\sqrt{\cos x}} \, dx.
$$

SOLUTION

Since the power of the sine is odd and positive, we have

$$
\begin{aligned}
\int_0^{\pi/3} \frac{\sin^3 x}{\sqrt{\cos x}} \, dx &= \int_0^{\pi/3} \frac{(1 - \cos^2 x)(\sin x)}{\sqrt{\cos x}} \, dx \\
&= \int_0^{\pi/3} [(\cos x)^{-1/2} \sin x - (\cos x)^{3/2} \sin x] \, dx \\
&= \int_0^{\pi/3} [-(\cos x)^{-1/2}(-\sin x) + (\cos x)^{3/2}(-\sin x)] \, dx \\
&= \left[-\frac{(\cos x)^{1/2}}{1/2} + \frac{(\cos x)^{5/2}}{5/2} \right]_0^{\pi/3} \\
&= \left(-\frac{2}{\sqrt{2}} + \frac{2}{5(2^{5/2})} + 2 - \frac{2}{5} \right) = \frac{32 - 19\sqrt{2}}{20} \approx 0.256.
\end{aligned}
$$

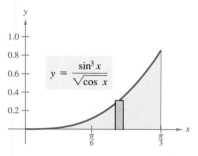

$$y = \frac{\sin^3 x}{\sqrt{\cos\ x}}$$

FIGURE 9.3

The region whose area is represented by this integral is shown in Figure 9.3.

EXAMPLE 4   Powers of sine and cosine are even and nonnegative

Evaluate $\int \cos^4 x\ dx$.

SOLUTION

Since m and n are both even and nonnegative ($m = 0$), we replace $\cos^4 x$ by $[(1 + \cos 2x)/2]^2$. Then, we have

$$\int \cos^4 x\ dx = \int \left(\frac{1 + \cos 2x}{2}\right)^2 dx \qquad \text{Identity for } \cos^2 x$$

$$= \int \left[\frac{1}{4} + \frac{\cos 2x}{2} + \frac{\cos^2 2x}{4}\right] dx$$

$$= \int \left[\frac{1}{4} + \frac{\cos 2x}{2} + \frac{1}{4}\left(\frac{1 + \cos 4x}{2}\right)\right] dx \qquad \text{Identity for } \cos^2 2x$$

$$= \frac{3}{8} \int dx + \frac{1}{4} \int 2 \cos 2x\ dx + \frac{1}{32} \int 4 \cos 4x\ dx$$

$$= \frac{3x}{8} + \frac{\sin 2x}{4} + \frac{\sin 4x}{32} + C.$$

In Example 4, suppose that we were to evaluate the *definite* integral from 0 to $\pi/2$. We would obtain

$$\int_0^{\pi/2} \cos^4 x\ dx = \left[\frac{3x}{8} + \frac{\sin 2x}{4} + \frac{\sin 4x}{32}\right]_0^{\pi/2} = \frac{3\pi}{16}.$$

Note that the only term that contributes to the solution is $3x/8$. This observation is generalized in the following formulas developed by John Wallis (1616–1703). Wallis did much of his work prior to Newton and Leibniz, and he influenced the thinking of both of these men.

John Wallis

WALLIS'S FORMULAS	If n is odd ($n \geq 3$), then

$$\int_0^{\pi/2} \cos^n x \, dx = \left(\frac{2}{3}\right)\left(\frac{4}{5}\right)\left(\frac{6}{7}\right) \cdots \left(\frac{n-1}{n}\right).$$

If n is even ($n \geq 2$), then

$$\int_0^{\pi/2} \cos^n x \, dx = \left(\frac{1}{2}\right)\left(\frac{3}{4}\right)\left(\frac{5}{6}\right) \cdots \left(\frac{n-1}{n}\right)\left(\frac{\pi}{2}\right).$$

These formulas are also valid if $\cos^n x$ is replaced by $\sin^n x$. (You are asked to prove both formulas in Exercises 85 and 86.)

Integrals involving powers of secant and tangent

To help you evaluate integrals of the form $\int \sec^m x \tan^n x \, dx$ we provide the following guidelines.

INTEGRALS INVOLVING SECANTS AND TANGENTS	1. If the power of the secant is even and positive, save a secant squared factor and convert the remaining factors to tangents. Then, expand and integrate.

$$\int \underset{\text{even}}{\overbrace{\sec^{2k} x}} \tan^n x \, dx = \int \underset{\text{convert to tangents}}{\overbrace{(\sec^2 x)^{k-1}}} \tan^n x \, \underset{\text{save for } du}{\overbrace{\sec^2 x}} \, dx$$

$$= \int (1 + \tan^2 x)^{k-1} \tan^n x \sec^2 x \, dx$$

2. If the power of the tangent is odd and positive, save a secant-tangent factor and convert the remaining factors to secants. Then, expand and integrate.

$$\int \sec^m x \, \underset{\text{odd}}{\overbrace{\tan^{2k+1} x}} \, dx = \int \sec^{m-1} x \, \underset{\text{convert to secants}}{\overbrace{(\tan^2 x)^k}} \, \underset{\text{save for } du}{\overbrace{\sec x \tan x}} \, dx$$

$$= \int \sec^{m-1} x \, (\sec^2 x - 1)^k \sec x \tan x \, dx$$

3. If there are no secant factors and the power of the tangent is even and positive, convert a tangent squared factor to secants; then expand and repeat if necessary.

$$\int \tan^n x \, dx = \int \tan^{n-2} x \, \underset{\text{convert to secants}}{\overbrace{(\tan^2 x)}} \, dx = \int \tan^{n-2} x \, (\sec^2 x - 1) \, dx$$

$$= \int \tan^{n-2} x \, (\sec^2 x) \, dx - \int \tan^{n-2} x \, dx$$

4. If the integral is of the from $\int \sec^m x \, dx$, where m is odd and positive, use integration by parts as illustrated in Example 6 in the previous section.
5. If none of the first four cases apply, try converting to sines and cosines.

EXAMPLE 5 Power of tangent is odd and positive

Evaluate

$$\int \frac{\tan^3 x}{\sqrt{\sec x}} \, dx.$$

SOLUTION

Since the power of the tangent is odd and positive, we write

$$\int \frac{\tan^3 x}{\sqrt{\sec x}} \, dx = \int (\sec x)^{-1/2} \tan^3 x \, dx$$

$$= \int (\sec x)^{-3/2} (\tan^2 x)(\sec x \tan x) \, dx \qquad \text{Save } \sec x \tan x$$

$$= \int (\sec x)^{-3/2} (\sec^2 x - 1)(\sec x \tan x) \, dx$$

$$= \int [(\sec x)^{1/2} - (\sec x)^{-3/2}](\sec x \tan x) \, dx$$

$$= \frac{2}{3}(\sec x)^{3/2} + 2(\sec x)^{-1/2} + C. \qquad \text{Power Rule}$$

EXAMPLE 6 Power of secant is even and positive

Evaluate $\int \sec^4 3x \tan^3 3x \, dx$.

SOLUTION

Since the power of the secant is even, we save a secant squared factor to create du. If $u = \tan 3x$, then

$$du = 3 \sec^2 3x \, dx$$

and we write

$$\int \sec^4 3x \tan^3 3x \, dx = \int \sec^2 3x \tan^3 3x \, (\sec^2 3x) \, dx \qquad \text{Save } \sec^2 3x$$

$$= \int (1 + \tan^2 3x) \tan^3 3x \, (\sec^2 3x) \, dx$$

$$= \frac{1}{3} \int [\tan^3 3x + \tan^5 3x](3 \sec^2 3x) \, dx$$

$$= \frac{1}{3}\left[\frac{\tan^4 3x}{4} + \frac{\tan^6 3x}{6} \right] + C. \qquad \text{Power Rule}$$

REMARK   In Example 6, the power of the tangent is odd and positive. Thus, we could have evaluated the integral with the procedure described in case 2. In Exercise 61, you are asked to show that the results obtained by these two procedures differ only by a constant.

EXAMPLE 7   Power of tangent is even

Evaluate

$$\int_0^{\pi/4} \tan^4 x \; dx.$$

SOLUTION

Since there are no secant factors, we convert a tangent squared factor to secants.

$$\int \tan^4 x \; dx = \int \tan^2 x \, (\tan^2 x) \; dx = \int \tan^2 x \, (\sec^2 x - 1) \; dx$$

$$= \int \tan^2 x \, \sec^2 x \; dx - \int \tan^2 x \; dx$$

$$= \int \tan^2 x \, \sec^2 x \; dx - \int (\sec^2 x - 1) \; dx$$

$$= \frac{\tan^3 x}{3} - \tan x + x + C$$

Thus, the definite integral (representing the area shown in Figure 9.4) has the following value.

$$\int_0^{\pi/4} \tan^4 x \; dx = \left[\frac{\tan^3 x}{3} - \tan x + x \right]_0^{\pi/4} = \frac{\pi}{4} - \frac{2}{3} \approx 0.119$$

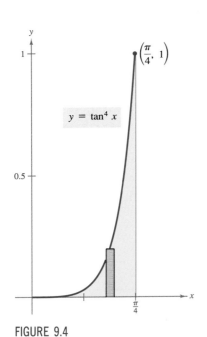

$\left(\dfrac{\pi}{4}, 1 \right)$

$y = \tan^4 x$

FIGURE 9.4

For integrals involving powers of cotangents and cosecants, we follow a strategy similar to that used for powers of tangents and secants, as illustrated in the next example.

EXAMPLE 8   Powers of cotangent and cosecant

Evaluate $\int \csc^4 x \cot^4 x \; dx$.

SOLUTION

Since the power of the cosecant is even, we write

$$\int \csc^4 x \cot^4 x \, dx = \int \csc^2 x \cot^4 x \, (\csc^2 x) \, dx$$

$$= \int (1 + \cot^2 x) \cot^4 x \, (\csc^2 x) \, dx$$

$$= -\int (\cot^4 x + \cot^6 x)(-\csc^2 x) \, dx$$

$$= -\left(\frac{\cot^5 x}{5} + \frac{\cot^7 x}{7}\right) + C$$

$$= -\frac{\cot^5 x}{5} - \frac{\cot^7 x}{7} + C.$$

EXAMPLE 9   Converting to sines and cosines

Evaluate

$$\int \frac{\sec x}{\tan^2 x} \, dx.$$

SOLUTION

Since the first two cases do not apply, we try converting the integrand to sines and cosines. In this case, we are able to integrate the resulting powers of sine and cosine as follows.

$$\int \frac{\sec x}{\tan^2 x} \, dx = \int \left(\frac{1}{\cos x}\right)\left(\frac{\cos x}{\sin x}\right)^2 dx = \int (\sin x)^{-2} (\cos x) \, dx$$

$$= -(\sin x)^{-1} + C = -\csc x + C$$

Integrals involving sine-cosine products with different angles

Occasionally, we encounter integrals involving the product of sines and cosines of two *different* angles. In such instances we use the following product-to-sum identities.

$$\sin mx \sin nx = \frac{1}{2}(\cos [(m - n)x] - \cos [(m + n)x])$$

$$\sin mx \cos nx = \frac{1}{2}(\sin [(m - n)x] + \sin [(m + n)x])$$

$$\cos mx \cos nx = \frac{1}{2}(\cos [(m - n)x] + \cos [(m + n)x])$$

EXAMPLE 10   Using product-to-sum identities

Evaluate $\int \sin 5x \cos 4x \, dx$.

SOLUTION

Considering the second product-to-sum identity, we write

$$\int \sin 5x \cos 4x \, dx = \frac{1}{2} \int (\sin 9x + \sin x) \, dx$$

$$= \frac{1}{2} \left[-\frac{\cos 9x}{9} - \cos x \right] + C$$

$$= -\frac{\cos 9x}{18} - \frac{\cos x}{2} + C.$$

As you review this section, concentrate on the *general pattern* followed in evaluating trigonometric integrals.

EXERCISES for Section 9.3

In Exercises 1–50, evaluate the given integral.

1. $\int \cos^3 x \sin x \, dx$

2. $\int \cos^3 x \sin^2 x \, dx$

3. $\int \sin^5 2x \cos 2x \, dx$

4. $\int \sin^3 x \, dx$

5. $\int \sin^5 x \cos^2 x \, dx$

6. $\int \cos^3 \frac{x}{3} \, dx$

7. $\int \cos^2 3x \, dx$

8. $\int \sin^2 2x \, dx$

9. $\int \sin^4 \pi x \, dx$

10. $\int \cos^4 \frac{x}{2} \, dx$

11. $\int \sin^2 x \cos^2 x \, dx$

12. $\int \sin^2 \frac{x}{2} \cos^2 \frac{x}{2} \, dx$

13. $\int x \sin^2 x \, dx$ (Integration by parts)

14. $\int x^2 \sin^2 x \, dx$ (Integration by parts)

15. $\int \sec 3x \, dx$

16. $\int \sec^2 (2x - 1) \, dx$

17. $\int \sec^4 5x \, dx$

18. $\int \sec^6 \frac{x}{2} \, dx$

19. $\int \sec^3 \pi x \, dx$ (Integration by parts)

20. $\int \sec^5 \pi x \, dx$ (Integration by parts)

21. $\int \tan^3 (1 - x) \, dx$

22. $\int \tan^2 x \, dx$

23. $\int \tan^5 \frac{x}{4} \, dx$

24. $\int \tan^3 \frac{\pi x}{2} \sec^2 \frac{\pi x}{2} \, dx$

25. $\int \sec^2 x \tan x \, dx$

26. $\int \csc^2 3x \cot 3x \, dx$

27. $\int \tan^2 x \sec^2 x \, dx$

28. $\int \tan^5 2x \sec^2 2x \, dx$

29. $\int \sec^5 \pi x \tan \pi x \, dx$

30. $\int \sec^4 (1 - x) \tan (1 - x) \, dx$

31. $\int \sec^6 4x \tan 4x \, dx$

32. $\int \sec^2 \frac{x}{2} \tan \frac{x}{2} \, dx$

33. $\int \sec^3 x \tan x \, dx$

34. $\int \tan^3 3x \, dx$

35. $\int \tan^3 3x \sec 3x \, dx$

36. $\int \sqrt{\tan x} \sec^4 x \, dx$

37. $\int \cot^3 2x \, dx$

38. $\int \tan^4 \frac{x}{2} \sec^4 \frac{x}{2} \, dx$

39. $\int \csc^4 \theta \, d\theta$

40. $\int \tan^3 t \sec^3 t \, dt$

41. $\int \frac{\cot^2 t}{\csc t} \, dt$

42. $\int \frac{\cot^3 t}{\csc t} \, dt$

43. $\int \sin 3x \cos 2x \, dx$

44. $\int \cos 3\theta \cos (-2\theta) \, d\theta$

45. $\int \sin \theta \sin 3\theta \, d\theta$

46. $\int x \tan^2 x^2 \, dx$

47. $\int \frac{1}{\sec x \tan x} \, dx$

48. $\int \frac{\sin^2 x - \cos^2 x}{\cos x} \, dx$

49. $\int (\tan^4 t - \sec^4 t) \, dt$

50. $\int \frac{1 - \sec t}{\cos t - 1} \, dt$

In Exercises 51–60, evaluate the given definite integral.

51. $\int_{-\pi}^{\pi} \sin^2 x \, dx$

52. $\int_0^{\pi/4} \tan^2 x \, dx$

53. $\int_0^{\pi/4} \tan^3 x \, dx$

54. $\int_0^{\pi/4} \sec^2 t \sqrt{\tan t} \, dt$

55. $\int_0^{\pi/2} \frac{\cos t}{1 + \sin t} \, dt$

56. $\int_{-\pi}^{\pi} \sin 3\theta \cos \theta \, d\theta$

57. $\int_0^{\pi/4} \sin 2\theta \sin 3\theta \, d\theta$

58. $\int_0^{\pi/2} (1 - \cos \theta)^2 \, d\theta$

59. $\int_{-\pi/2}^{\pi/2} \cos^3 x \, dx$

60. $\int_{-\pi/2}^{\pi/2} (\sin^2 x + 1) \, dx$

In Exercises 61 and 62, find the indefinite integral in two different ways and show that the results differ only by a constant.

61. $\int \sec^4 3x \tan^3 3x \, dx$

62. $\int \sec^2 x \tan x \, dx$

In Exercises 63 and 64, find the area of the region bounded by the graphs of the given equations.

63. $y = \sin^2 \pi x$, $y = 0$, $x = 0$, $x = 1$

64. $y = \tan^2 x$, $y = \dfrac{4x}{\pi}$, $x = -\dfrac{\pi}{4}$, $x = \dfrac{\pi}{4}$

In Exercises 65 and 66, find the volume of the solid generated by revolving the region bounded by the graphs of the given equations about the *x*-axis.

65. $y = \tan x$, $y = 0$, $x = -\dfrac{\pi}{4}$, $x = \dfrac{\pi}{4}$

66. $y = \sin x \cos^2 x$, $y = 0$, $x = 0$, $x = \dfrac{\pi}{2}$

In Exercises 67 and 68, for the region bounded by the graphs of the given equations, find (a) the volume of the solid formed by revolving the region about the *x*-axis and (b) the centroid of the region.

67. $y = \sin x$, $y = 0$, $x = 0$, $x = \pi$

68. $y = \cos x$, $y = 0$, $x = 0$, $x = \dfrac{\pi}{2}$

In Exercises 69–74, verify Wallis's Formulas for the given integral.

69. $\int_0^{\pi/2} \cos^3 x \, dx = \dfrac{2}{3}$

70. $\int_0^{\pi/2} \cos^5 x \, dx = \dfrac{8}{15}$

71. $\int_0^{\pi/2} \cos^7 x \, dx = \dfrac{16}{35}$

72. $\int_0^{\pi/2} \sin^2 x \, dx = \dfrac{\pi}{4}$

73. $\int_0^{\pi/2} \sin^4 x \, dx = \dfrac{3\pi}{16}$

74. $\int_0^{\pi/2} \sin^6 x \, dx = \dfrac{5\pi}{32}$

In Exercises 75–78, use integration by parts to verify the given reduction formula.

75. $\int \sin^n x \, dx = -\dfrac{\sin^{n-1} x \cos x}{n} + \dfrac{n-1}{n} \int \sin^{n-2} x \, dx$

76. $\int \cos^n x \, dx = \dfrac{\cos^{n-1} x \sin x}{n} + \dfrac{n-1}{n} \int \cos^{n-2} x \, dx$

77. $\int \cos^m x \sin^n x \, dx = -\dfrac{\cos^{m+1} x \sin^{n-1} x}{m+n} + \dfrac{n-1}{m+n} \int \cos^m x \sin^{n-2} x \, dx$

78. $\int \sec^n x \, dx = \dfrac{1}{n-1} \sec^{n-2} x \tan x + \dfrac{n-2}{n-1} \int \sec^{n-2} x \, dx$

In Exercises 79–84, find the indefinite integral by using the appropriate formula from Exercises 75–78.

79. $\int \sin^5 x \, dx$

80. $\int \cos^4 x \, dx$

81. $\int \cos^6 x \, dx$

82. $\int \sin^4 x \cos^2 x \, dx$

83. $\int \sin^2 \pi x \cos^2 \pi x \, dx$

84. $\int \sin^4 \dfrac{2\pi x}{5} \, dx$

85. (Wallis's Formula) Use the result of Exercise 76 to prove that if *n* is odd ($n \geq 3$), then
$$\int_0^{\pi/2} \cos^n x \, dx = \left(\frac{2}{3}\right)\left(\frac{4}{5}\right)\left(\frac{6}{7}\right) \cdots \left(\frac{n-1}{n}\right).$$

86. (Wallis's Formula) Use the result of Exercise 76 to prove that if *n* is even ($n \geq 2$), then
$$\int_0^{\pi/2} \cos^n x \, dx = \left(\frac{1}{2}\right)\left(\frac{3}{4}\right)\left(\frac{5}{6}\right) \cdots \left(\frac{n-1}{n}\right)\left(\frac{\pi}{2}\right).$$

In Exercises 87 and 88, use Simpson's Rule on a computer or calculator (with $n = 12$) to approximate the volume of the solid generated by revolving the region bounded by the graphs of the given equations about the *y*-axis.

87. $y = \sqrt{x} \sin \dfrac{\pi x}{2}$, $y = 0$, $x = 0$, $x = 2$

88. $y = \dfrac{10}{\sqrt{xe^x}}$, $y = 0$, $x = 1$, $x = 4$

89. The **inner product** of two functions *f* and *g* on [*a*, *b*] is given by $\langle f, g \rangle = \int_a^b f(x)g(x) \, dx$. Two distinct functions *f* and *g* are said to be **orthogonal** if $\langle f, g \rangle = 0$.
(a) Show that the functions given by $f_n(x) = \cos nx$, $n = 0, 1, 2, \ldots$ form an orthogonal family on [0, π].
(b) Show that the functions given by
$$f_n(x) = \sin nx, \quad n = 1, 2, 3, \ldots$$
$$g_n(x) = \cos nx, \quad n = 0, 1, 2, \ldots$$
form an orthogonal family on [−π, π].

9.4 Trigonometric Substitution

Trigonometric substitution

Now that we can evaluate integrals involving powers of trigonometric functions, we can use the method of **trigonometric substitution** to evaluate integrals involving the radicals

$$\sqrt{a^2 - u^2}, \qquad \sqrt{a^2 + u^2}, \qquad \text{and} \qquad \sqrt{u^2 - a^2}.$$

Our objective with trigonometric substitution is to eliminate the radical in the integrand. We do this with the Pythagorean identities

$$\cos^2 \theta = 1 - \sin^2 \theta$$
$$\sec^2 \theta = 1 + \tan^2 \theta$$
$$\tan^2 \theta = \sec^2 \theta - 1.$$

For example, if $a > 0$, we let $u = a \sin \theta$, where $-\pi/2 \leq \theta \leq \pi/2$. Then

$$\sqrt{a^2 - u^2} = \sqrt{a^2 - a^2 \sin^2 \theta} = \sqrt{a^2(1 - \sin^2 \theta)}$$
$$= \sqrt{a^2 \cos^2 \theta} = a \cos \theta.$$

Note that $\cos \theta \geq 0$, since $-\pi/2 \leq \theta \leq \pi/2$.

TRIGONOMETRIC SUBSTITUTION ($a > 0$)

1. For integrals involving $\sqrt{a^2 - u^2}$, let $u = a \sin \theta$. Then $\sqrt{a^2 - u^2} = a \cos \theta$ where $-\pi/2 \leq \theta \leq \pi/2$.

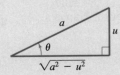

2. For integrals involving $\sqrt{a^2 + u^2}$, let $u = a \tan \theta$. Then $\sqrt{a^2 + u^2} = a \sec \theta$ where $-\pi/2 < \theta < \pi/2$.

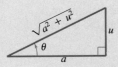

3. For integrals involving $\sqrt{u^2 - a^2}$, let $u = a \sec \theta$. Then $\sqrt{u^2 - a^2} = \pm a \tan \theta$ where $0 \leq \theta < \pi/2$ or $\pi/2 < \theta \leq \pi$. Use the positive value if $u > a$ and the negative value if $u < -a$.

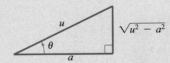

EXAMPLE 1   Trigonometric substitution: $u = a \sin \theta$

Evaluate

$$\int \frac{dx}{x^2 \sqrt{9 - x^2}}.$$

SOLUTION

First, we observe that none of the basic integration formulas in Section 9.1 apply. To use trigonometric substitution, we observe that $\sqrt{9 - x^2}$ is of the form $\sqrt{a^2 - u^2}$, with the corresponding triangle shown in Figure 9.5. Hence, we use the substitution

$$x = a \sin \theta = 3 \sin \theta$$

which implies that

$$dx = 3 \cos \theta \, d\theta, \qquad \sqrt{9 - x^2} = 3 \cos \theta, \qquad \text{and} \qquad x^2 = 9 \sin^2 \theta.$$

Therefore,

$$\int \frac{dx}{x^2 \sqrt{9 - x^2}} = \int \frac{3 \cos \theta \, d\theta}{(9 \sin^2 \theta)(3 \cos \theta)}$$

$$= \frac{1}{9} \int \frac{d\theta}{\sin^2 \theta}$$

$$= \frac{1}{9} \int \csc^2 \theta \, d\theta$$

$$= -\frac{1}{9} \cot \theta + C$$

$$= -\frac{1}{9} \left(\frac{\sqrt{9 - x^2}}{x} \right) + C \qquad \text{Substitute for } \cot \theta$$

$$= -\frac{\sqrt{9 - x^2}}{9x} + C.$$

Note that we used the triangle in Figure 9.5 to convert from θ's back to x's as follows.

$$\cot \theta = \frac{\text{adj}}{\text{opp}} = \frac{\sqrt{9 - x^2}}{x}$$

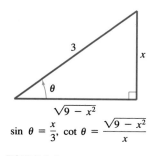

$$\sin \theta = \frac{x}{3}, \quad \cot \theta = \frac{\sqrt{9 - x^2}}{x}$$

FIGURE 9.5

In Section 8.7, we saw how the inverse hyperbolic functions can be used to evaluate the integrals

$$\int \frac{du}{\sqrt{u^2 \pm a^2}}, \qquad \int \frac{du}{a^2 - u^2}, \qquad \text{and} \qquad \int \frac{du}{u \sqrt{a^2 \pm u^2}}.$$

You can also evaluate these integrals using trigonometric substitution. This is illustrated in the next example.

EXAMPLE 2   Trigonometric substitution: $u = a \tan \theta$

Evaluate

$$\int \frac{dx}{\sqrt{4x^2 + 1}}.$$

SOLUTION

We let $u = 2x$, $a = 1$, and $2x = \tan \theta$, as shown in Figure 9.6. Then,

$$dx = \frac{1}{2} \sec^2 \theta \, d\theta \qquad \text{and} \qquad \sqrt{4x^2 + 1} = \sec \theta .$$

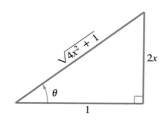

$\sec \theta = \sqrt{4x^2 + 1}, \ \tan \theta = 2x$

FIGURE 9.6

Therefore,

$$\int \frac{1}{\sqrt{4x^2 + 1}} \, dx = \frac{1}{2} \int \frac{\sec^2 \theta \, d\theta}{\sec \theta} = \frac{1}{2} \int \sec \theta \, d\theta$$

$$= \frac{1}{2} \ln |\sec \theta + \tan \theta| + C$$

$$= \frac{1}{2} \ln \left| \sqrt{4x^2 + 1} + 2x \right| + C.$$

We can extend the use of trigonometric substitution to cover integrals involving expressions such as $(a^2 - u^2)^{n/2}$ by writing the expression as

$$(a^2 - u^2)^{n/2} = (\sqrt{a^2 - u^2})^n.$$

This procedure is demonstrated in the next example.

EXAMPLE 3   Trigonometric substitution: rational powers

Evaluate

$$\int \frac{dx}{(x^2 + 1)^{3/2}}.$$

SOLUTION

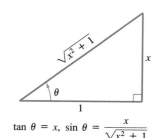

$\tan \theta = x, \ \sin \theta = \dfrac{x}{\sqrt{x^2 + 1}}$

FIGURE 9.7

We write $(x^2 + 1)^{3/2}$ in the form $(\sqrt{x^2 + 1})^3$, and let $a = 1$ and $u = x = \tan \theta$, as shown in Figure 9.7. Then,

$$dx = \sec^2 \theta \, d\theta \qquad \text{and} \qquad \sqrt{x^2 + 1} = \sec \theta.$$

Therefore,

$$\int \frac{dx}{(x^2 + 1)^{3/2}} = \int \frac{dx}{(\sqrt{x^2 + 1})^3} = \int \frac{\sec^2 \theta \, d\theta}{\sec^3 \theta}$$

$$= \int \frac{d\theta}{\sec \theta} = \int \cos \theta \, d\theta = \sin \theta + C$$

$$= \frac{x}{\sqrt{x^2 + 1}} + C. \qquad \text{Substitute for } \sin \theta$$

For definite integrals, it is often convenient to determine the integration limits for θ and thus avoid converting back to x. You might want to review this procedure in Section 5.5, Examples 8 and 9.

EXAMPLE 4   Converting the limits of integration

Evaluate

$$\int_{\sqrt{3}}^{2} \frac{\sqrt{x^2 - 3}}{x}\, dx.$$

SOLUTION

Since $\sqrt{x^2 - 3}$ has the form $\sqrt{u^2 - a^2}$, we consider $u = x$, $a = \sqrt{3}$, and let $x = \sqrt{3}\sec\theta$, as shown in Figure 9.8. Then,

$$dx = \sqrt{3}\sec\theta\tan\theta\,d\theta \qquad \text{and} \qquad \sqrt{x^2 - 3} = \sqrt{3}\tan\theta.$$

To determine the upper and lower limits of integration, we use the substitution $x = \sqrt{3}\sec\theta$ as follows.

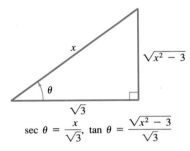

$$\sec\theta = \frac{x}{\sqrt{3}}, \quad \tan\theta = \frac{\sqrt{x^2 - 3}}{\sqrt{3}}$$

FIGURE 9.8

Lower limit	*Upper limit*
When $x = \sqrt{3}$, $\sec\theta = 1$	When $x = 2$, $\sec\theta = \dfrac{2}{\sqrt{3}}$
and $\theta = 0$.	and $\theta = \dfrac{\pi}{6}$.

Therefore, we have

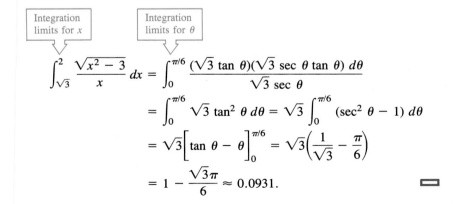

$$\int_{\sqrt{3}}^{2} \frac{\sqrt{x^2 - 3}}{x}\, dx = \int_{0}^{\pi/6} \frac{(\sqrt{3}\tan\theta)(\sqrt{3}\sec\theta\tan\theta)\, d\theta}{\sqrt{3}\sec\theta}$$

$$= \int_{0}^{\pi/6} \sqrt{3}\tan^2\theta\, d\theta = \sqrt{3}\int_{0}^{\pi/6}(\sec^2\theta - 1)\, d\theta$$

$$= \sqrt{3}\Big[\tan\theta - \theta\Big]_{0}^{\pi/6} = \sqrt{3}\left(\frac{1}{\sqrt{3}} - \frac{\pi}{6}\right)$$

$$= 1 - \frac{\sqrt{3}\pi}{6} \approx 0.0931.$$

REMARK   In Example 4, we leave as an exercise the alternate procedure of converting back to variable x and evaluating the antiderivative at the original limits of integration. (See Exercise 47.)

When using trigonometric substitution to evaluate definite integrals, you must be careful to check that the values of θ lie in the intervals discussed at the beginning of this section. For instance, if in Example 4 we had been asked to evaluate the definite integral

$$\int_{-2}^{-\sqrt{3}} \frac{\sqrt{x^2 - 3}}{x}\, dx$$

then using $u = x$ and $a = \sqrt{3}$ in the interval $[-2, -\sqrt{3}]$ would imply that $u < -a$. Thus, when determining the upper and lower limits of integration, we would have to choose θ such that $\pi/2 < \theta \le \pi$. In this case the integral would be evaluated as follows.

$$\int_{-2}^{-\sqrt{3}} \frac{\sqrt{x^2 - 3}}{x}\, dx = \int_{5\pi/6}^{\pi} \frac{(-\sqrt{3}\tan\theta)(\sqrt{3}\sec\theta\tan\theta)\, d\theta}{\sqrt{3}\sec\theta}$$

$$= \int_{5\pi/6}^{\pi} -\sqrt{3}\tan^2\theta\, d\theta$$

$$= -\sqrt{3}\int_{5\pi/6}^{\pi}(\sec^2\theta - 1)\, d\theta$$

$$= -\sqrt{3}\Big[\tan\theta - \theta\Big]_{5\pi/6}^{\pi}$$

$$= -\sqrt{3}\left[(0 - \pi) - \left(-\frac{1}{\sqrt{3}} - \frac{5\pi}{6}\right)\right]$$

$$= -1 + \frac{\sqrt{3}\pi}{6} \approx -0.0931$$

We can further expand the range of problems to which trigonometric substitution applies by employing the technique of completing the square. (To review this procedure, see Examples 5, 6, and 7 in Section 8.6.) We further demonstrate the technique of completing the square in the next example.

EXAMPLE 5   Completing the square

Evaluate

$$\int \frac{dx}{(x^2 - 4x)^{3/2}}, \quad x > 4.$$

SOLUTION

Completing the square, we have

$$x^2 - 4x = x^2 - 4x + 4 - 4 = (x - 2)^2 - 2^2 = u^2 - a^2.$$

We let $a = 2$ and $u = x - 2 = 2\sec\theta$ as shown in Figure 9.9. (Note that $x > 4$ implies that $u > a$.) Then

$$dx = 2\sec\theta\tan\theta\, d\theta \quad \text{and} \quad \sqrt{(x - 2)^2 - 2^2} = 2\tan\theta.$$

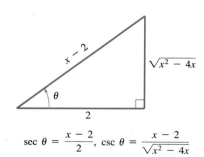

$$\sec\theta = \frac{x - 2}{2}, \quad \csc\theta = \frac{x - 2}{\sqrt{x^2 - 4x}}$$

FIGURE 9.9

Therefore,

$$\int \frac{dx}{(x^2 - 4x)^{3/2}} = \int \frac{2 \sec \theta \tan \theta \, d\theta}{(2 \tan \theta)^3} = \frac{2}{8} \int \frac{\sec \theta}{\tan^2 \theta} \, d\theta$$

$$= \frac{1}{4} \int (\sin \theta)^{-2}(\cos \theta) \, d\theta = -\frac{1}{4} \csc \theta + C$$

$$= -\frac{x - 2}{4\sqrt{x^2 - 4x}} + C. \qquad \text{Substitute for csc } \theta \qquad \blacksquare$$

In Section 9.5, we will encounter integrals involving rational functions. Some of these important integral forms can be solved using trigonometric substitution as illustrated in the next example.

EXAMPLE 6   Trigonometric substitution for rational functions

Evaluate

$$\int \frac{dx}{(x^2 + 1)^2}.$$

SOLUTION

Letting $a = 1$ and $u = x = \tan \theta$ as shown in Figure 9.10, we have

$$dx = \sec^2 \theta \, d\theta \qquad \text{and} \qquad x^2 + 1 = \sec^2 \theta.$$

Therefore,

$$\int \frac{dx}{(x^2 + 1)^2} = \int \frac{\sec^2 \theta \, d\theta}{\sec^4 \theta}$$

$$= \int \cos^2 \theta \, d\theta$$

$$= \frac{1}{2} \int (1 + \cos 2\theta) \, d\theta$$

$$= \frac{\theta}{2} + \frac{\sin 2\theta}{4} + C$$

$$= \frac{\theta}{2} + \frac{1}{2} \sin \theta \cos \theta + C$$

$$= \frac{\arctan x}{2} + \frac{1}{2}\left(\frac{x}{\sqrt{x^2 + 1}}\right)\left(\frac{1}{\sqrt{x^2 + 1}}\right) + C$$

$$= \frac{1}{2}\left(\arctan x + \frac{x}{x^2 + 1}\right) + C. \qquad \blacksquare$$

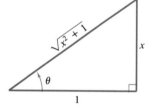

$\tan \theta = x, \ \sin \theta = \dfrac{x}{\sqrt{x^2 + 1}},$

$\cos \theta = \dfrac{1}{\sqrt{x^2 + 1}}$

FIGURE 9.10

Trigonometric substitution can be used to evaluate the three integrals listed in the following theorem. We will have occasion to use these integrals several times in the remainder of the text, and when we do, we will simply refer to this theorem. If you prefer not to memorize these formulas, you should remember that they can be generated by trigonometric substitution.

THEOREM 9.2
SPECIAL INTEGRATION
FORMULAS ($a > 0$)

1. $\int \sqrt{a^2 - u^2} \, du = \frac{1}{2}\left(a^2 \arcsin \frac{u}{a} + u\sqrt{a^2 - u^2}\right) + C$

2. $\int \sqrt{u^2 - a^2} \, du = \frac{1}{2}(u\sqrt{u^2 - a^2} - a^2 \ln |u + \sqrt{u^2 - a^2}|) + C, \quad u > a$

3. $\int \sqrt{u^2 + a^2} \, du = \frac{1}{2}(u\sqrt{u^2 + a^2} + a^2 \ln |u + \sqrt{u^2 + a^2}|) + C$

PROOF   We outline the proof of part 1 and leave the second and third parts as an exercise (see Exercise 65). For the first integral, we let $u = a \sin \theta$. Then $du = a \cos \theta \, d\theta$ and $\sqrt{a^2 - u^2} = a \cos \theta$ and we have

$$\int \sqrt{a^2 - u^2} \, du = \int a^2 \cos^2 \theta \, d\theta = a^2 \int \frac{1 + \cos 2\theta}{2} \, d\theta$$

$$= \frac{a^2}{2}\left(\theta + \frac{1}{2}\sin 2\theta\right) + C = \frac{a^2}{2}(\theta + \sin \theta \cos \theta) + C.$$

From Figure 9.11, it follows that

$$\int \sqrt{a^2 - u^2} \, du = \frac{a^2}{2}\left[\arcsin \frac{u}{a} + \left(\frac{u}{a}\right)\left(\frac{\sqrt{a^2 - u^2}}{a}\right)\right] + C$$

$$= \frac{a^2}{2}\left[\frac{a^2 \arcsin (u/a) + u\sqrt{a^2 - u^2}}{a^2}\right] + C$$

$$= \frac{1}{2}\left[a^2 \arcsin \frac{u}{a} + u\sqrt{a^2 - u^2}\right] + C.$$

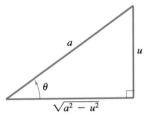

$\sin \theta = \dfrac{u}{a}, \ \cos \theta = \dfrac{\sqrt{a^2 - u^2}}{a}$

FIGURE 9.11

There are many practical applications involving trigonometric substitution, as illustrated in the last two examples.

EXAMPLE 7   An application involving arc length

Find the arc length of the graph of $f(x) = \frac{1}{2}x^2$ from $x = 0$ to $x = 1$, as shown in Figure 9.12.

SOLUTION

From the formula for arc length s, we have

$$s = \int_0^1 \sqrt{1 + [f'(x)]^2} \, dx = \int_0^1 \sqrt{1 + x^2} \, dx.$$

Letting $a = 1$ and $x = \tan \theta$, we have

$$dx = \sec^2 \theta \, d\theta \quad \text{and} \quad \sqrt{1 + x^2} = \sec \theta.$$

Moreover, the upper and lower limits are as follows.

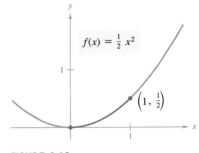

FIGURE 9.12

Lower limit	Upper limit
When $x = \tan \theta = 0,$	When $x = \tan \theta = 1,$
$\theta = 0.$	$\theta = \dfrac{\pi}{4}.$

Therefore, the arc length is given by

$$s = \int_0^1 \sqrt{1 + x^2}\, dx = \int_0^{\pi/4} \sec^3 \theta\, d\theta \qquad \text{Example 6, Section 9.2}$$

$$= \frac{1}{2}\Big[\sec \theta \tan \theta + \ln |\sec \theta + \tan \theta| \Big]_0^{\pi/4}$$

$$= \frac{1}{2}[\sqrt{2} + \ln (\sqrt{2} + 1)] \approx 1.148.$$

REMARK Try using Theorem 9.2 to evaluate the integral in Example 7.

EXAMPLE 8 A comparison of two fluid forces

A sealed barrel of oil (weighing 48 pounds per cubic foot) is floating in sea water (weighing 64 pounds per cubic foot), as shown in Figures 9.13 and 9.14. (Note that the barrel is not completely full of oil—on its side, the top 0.2 feet of the barrel is empty.) Compare the fluid force against one end of the barrel from the inside and from the outside.

SOLUTION

In Figure 9.14 we locate the coordinate system with the origin at the center of the circle given by $x^2 + y^2 = 1$. Then, to find the fluid force against an end of the barrel *from the inside*, we integrate between -1 and 0.8 (using a weight of $w = 48$) to obtain

$$F_{\text{inside}} = 48 \int_{-1}^{0.8} (0.8 - y)(2)\sqrt{1 - y^2}\, dy$$

$$= 76.8 \int_{-1}^{0.8} \sqrt{1 - y^2}\, dy - 96 \int_{-1}^{0.8} y\sqrt{1 - y^2}\, dy.$$

To find the fluid force *from the outside*, we integrate between -1 and 0.4 (using a weight of $w = 64$) to obtain

$$F_{\text{outside}} = 64 \int_{-1}^{0.4} (0.4 - y)(2)\sqrt{1 - y^2}\, dy$$

$$= 51.2 \int_{-1}^{0.4} \sqrt{1 - y^2}\, dy - 128 \int_{-1}^{0.4} y\sqrt{1 - y^2}\, dy.$$

We leave the details of integration for you to complete in Exercises 61 and 62. Intuitively, would you say that the force from the oil (the inside) or the force from the sea water (the outside) is greater? By evaluating these two integrals, you can determine that

$$F_{\text{inside}} \approx 121.3 \text{ lb} \qquad \text{and} \qquad F_{\text{outside}} \approx 93.0 \text{ lb}.$$

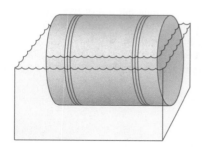

FIGURE 9.13

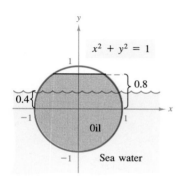

FIGURE 9.14

EXERCISES for Section 9.4

In Exercises 1–4, evaluate the indefinite integral using the substitution $x = 5 \sin \theta$.

1. $\displaystyle \int \frac{1}{(25 - x^2)^{3/2}} \, dx$

2. $\displaystyle \int \frac{1}{x^2 \sqrt{25 - x^2}} \, dx$

3. $\displaystyle \int \frac{\sqrt{25 - x^2}}{x} \, dx$

4. $\displaystyle \int \frac{x^2}{\sqrt{25 - x^2}} \, dx$

In Exercises 5–8, evaluate the indefinite integral using the substitution $x = 2 \sec \theta$.

5. $\displaystyle \int \frac{1}{\sqrt{x^2 - 4}} \, dx$

6. $\displaystyle \int \frac{\sqrt{x^2 - 4}}{x} \, dx$

7. $\displaystyle \int x^3 \sqrt{x^2 - 4} \, dx$

8. $\displaystyle \int \frac{x^3}{\sqrt{x^2 - 4}} \, dx$

In Exercises 9–12, evaluate the indefinite integral using the substitution $x = \tan \theta$.

9. $\displaystyle \int x \sqrt{1 + x^2} \, dx$

10. $\displaystyle \int \frac{x^3}{\sqrt{1 + x^2}} \, dx$

11. $\displaystyle \int \frac{1}{(1 + x^2)^2} \, dx$

12. $\displaystyle \int \frac{x^2}{(1 + x^2)^2} \, dx$

In Exercises 13–44, evaluate the indefinite integral.

13. $\displaystyle \int \frac{x}{\sqrt{x^2 + 9}} \, dx$

14. $\displaystyle \int \frac{1}{\sqrt{25 - x^2}} \, dx$

15. $\displaystyle \int_0^2 \sqrt{16 - 4x^2} \, dx$

16. $\displaystyle \int_0^2 x\sqrt{16 - 4x^2} \, dx$

17. $\displaystyle \int \frac{1}{\sqrt{x^2 - 9}} \, dx$

18. $\displaystyle \int \frac{t}{(1 - t^2)^{3/2}} \, dt$

19. $\displaystyle \int_0^{\sqrt{3}/2} \frac{t^2}{(1 - t^2)^{3/2}} \, dt$

20. $\displaystyle \int_0^{\sqrt{3}/2} \frac{1}{(1 - t^2)^{5/2}} \, dt$

21. $\displaystyle \int \frac{\sqrt{1 - x^2}}{x^4} \, dx$

22. $\displaystyle \int \frac{\sqrt{4x^2 + 9}}{x^4} \, dx$

23. $\displaystyle \int \frac{1}{x\sqrt{4x^2 + 9}} \, dx$

24. $\displaystyle \int \frac{1}{(x^2 + 3)^{3/2}} \, dx$

25. $\displaystyle \int \frac{x}{(x^2 + 3)^{3/2}} \, dx$

26. $\displaystyle \int \frac{1}{x\sqrt{4x^2 + 16}} \, dx$

27. $\displaystyle \int e^{2x}\sqrt{1 + e^{2x}} \, dx$

28. $\displaystyle \int \frac{1}{\sqrt{4x - x^2}} \, dx$

29. $\displaystyle \int (x + 1)\sqrt{x^2 + 2x + 2} \, dx$

30. $\displaystyle \int \frac{x^2}{\sqrt{2x - x^2}} \, dx$

31. $\displaystyle \int \frac{x}{\sqrt{x^2 + 4x + 8}} \, dx$

32. $\displaystyle \int \frac{x}{\sqrt{x^2 - 6x + 5}} \, dx$

33. $\displaystyle \int \frac{x^2}{\sqrt{x^2 + 10x + 9}} \, dx$

34. $\displaystyle \int (x^2 + 2x + 11)^{3/2} \, dx$

35. $\displaystyle \int e^x \sqrt{1 - e^{2x}} \, dx$

36. $\displaystyle \int \frac{\sqrt{1 - x}}{\sqrt{x}} \, dx$

37. $\displaystyle \int \frac{1}{4 + 4x^2 + x^4} \, dx$

38. $\displaystyle \int \frac{x^3 + x + 1}{x^4 + 2x^2 + 1} \, dx$

39. $\displaystyle \int \sqrt{4 + 9x^2} \, dx$

40. $\displaystyle \int \sqrt{1 + x^2} \, dx$

41. $\displaystyle \int \frac{x^2}{\sqrt{x^2 - 1}} \, dx$

42. $\displaystyle \int x^2\sqrt{x^2 - 4} \, dx$

43. $\displaystyle \int \operatorname{arcsec} 2x \, dx$

44. $\displaystyle \int x \arcsin x \, dx$

In Exercises 45 and 46, evaluate the given integral using (a) the given integration limits and (b) the limits obtained by trigonometric substitution.

45. $\displaystyle \int_0^3 \frac{x^3}{\sqrt{x^2 + 9}} \, dx$

46. $\displaystyle \int_0^{5/3} \sqrt{25 - 9x^2} \, dx$

47. Find the value of the definite integral of Example 4 by converting back to the variable x and evaluating the antiderivative at the original limits.

48. The field strength H of a magnet of length $2L$ on a particle r units from the center of the magnet is given by

$$H = \frac{2mL}{(r^2 + L^2)^{3/2}}$$

where $\pm m$ are the poles of the magnet (see figure). Find the average field strength as the particle moves from 0 to R units from the center by evaluating the integral

$$\frac{1}{R} \int_0^R \frac{2mL}{(r^2 + L^2)^{3/2}} \, dr.$$

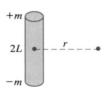

FIGURE FOR 48

In Exercises 49 and 50, find the fluid force on the circular observation window of radius 1 foot in a vertical wall of a large water-filled tank at a fish hatchery.

49. The center of the window is 3 feet below the surface of the water (see figure).

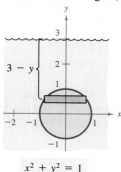

$x^2 + y^2 = 1$ FIGURE FOR 49

50. The center of the window is d feet below the surface of the water ($d > 1$).

In Exercises 51 and 52, use trigonometric substitution to find the area enclosed by the graph of the given conic.

51. $x^2 + y^2 = r^2$ **52.** $\dfrac{x^2}{2^2} + \dfrac{y^2}{3^2} = 1$

In Exercises 53 and 54, find the volume of the torus generated by revolving the region bounded by the graph of the given circle about the y-axis.

53. $(x - 3)^2 + y^2 = 1$ (see figure)
54. $(x - h)^2 + y^2 = r^2$, $h > r$

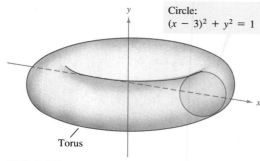

Circle:
$(x - 3)^2 + y^2 = 1$

Torus

FIGURE FOR 53

In Exercises 55 and 56, find the arc length of the given plane curve over the indicated interval.

Function	Interval
55. $y = \ln x$	$[1, 5]$
56. $y = x^2$	$[0, 3]$

In Exercises 57 and 58, find the distance that a projectile travels when it follows the path given by the graph of the equation.

Function	Interval
57. $y = x - 0.005x^2$	$[0, 200]$
58. $y = x - \dfrac{x^2}{72}$	$[0, 72]$

59. Find the surface area of the solid generated by revolving the region bounded by the graphs of $y = x^2$, $y = 0$, $x = 0$, and $x = \sqrt{2}$ about the x-axis.

60. Find the centroid of the region bounded by the graphs of $y = 3/\sqrt{x^2 + 9}$, $y = 0$, $x = -4$, and $x = 4$.

61. Evaluate the first integral for the fluid force given in Example 8.

$$F = 48 \int_{-1}^{0.8} (0.8 - y)(2)\sqrt{1 - y^2}\, dy$$

62. Evaluate the second integral for the fluid force given in Example 8.

$$F = 64 \int_{-1}^{0.4} (0.4 - y)(2)\sqrt{1 - y^2}\, dy$$

In Exercises 63 and 64, use Simpson's Rule on a computer or calculator (with $n = 12$) to approximate the centroid of the region bounded by the graphs of the given equations.

63. $y = \dfrac{4}{\ln x}$, $y = 0$, $x = 2$, $x = 10$

64. $y = 2\cos x^2$, $y = 0$, $x = 0$, $x = 1$

65. Use trigonometric substitution to verify the second and third integration formulas given in Theorem 9.2.

66. Prove that the area of the region enclosed by the ellipse

$$\frac{x^2}{a^2} + \frac{y^2}{b^2} = 1, \ a > 0, \ b > 0$$

is πab.

9.5 Partial Fractions

Partial fractions (linear factors) ▪ Partial fractions (quadratic factors)

John Bernoulli

This section examines a procedure for decomposing a rational function into simpler rational functions to which we can apply the basic integration formulas. We call this procedure the **method of partial fractions.** This technique was introduced in 1702 by John Bernoulli (1667–1748), a Swiss mathematician who was instrumental in the early development of calculus. John Bernoulli was a professor at the University of Basel and taught many outstanding students, the most famous of whom was Leonhard Euler.

To introduce the method of partial fractions, let's consider the integral of $1/(x^2 - 5x + 6)$. To evaluate this integral *without* partial fractions, we complete the square and use trigonometric substitution (see Figure 9.15) to obtain the following.

$$\int \frac{1}{x^2 - 5x + 6}\, dx = \int \frac{dx}{(x - 5/2)^2 - (1/2)^2} \qquad a = \frac{1}{2}, \; x - \frac{5}{2} = \frac{1}{2}\sec\theta$$

$$= \int \frac{(1/2)\sec\theta \tan\theta\, d\theta}{(1/4)\tan^2\theta} \qquad dx = \frac{1}{2}\sec\theta\tan\theta\, d\theta$$

$$= 2\int \csc\theta\, d\theta$$

$$= 2\ln|\csc\theta - \cot\theta| + C$$

$$= 2\ln\left|\frac{2x - 5}{2\sqrt{x^2 - 5x + 6}} - \frac{1}{2\sqrt{x^2 - 5x + 6}}\right| + C$$

$$= 2\ln\left|\frac{x - 3}{\sqrt{x^2 - 5x + 6}}\right| + C$$

$$= 2\ln\left|\frac{\sqrt{x - 3}}{\sqrt{x - 2}}\right| + C$$

$$= \ln\left|\frac{x - 3}{x - 2}\right| + C$$

$$= \ln|x - 3| - \ln|x - 2| + C$$

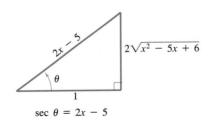

$$\sec\theta = 2x - 5$$

FIGURE 9.15

Now, to see the benefit of the partial fraction method, suppose we had observed that

$$\frac{1}{x^2 - 5x + 6} = \frac{1}{x - 3} - \frac{1}{x - 2}. \qquad \text{Partial fraction decomposition}$$

Then we could evaluate the integral easily as follows.

$$\int \frac{1}{x^2 - 5x + 6}\, dx = \int \left(\frac{1}{x - 3} - \frac{1}{x - 2}\right) dx$$

$$= \ln|x - 3| - \ln|x - 2| + C$$

This method is clearly preferable to trigonometric substitution. However, its use depends on the ability to factor the denominator, $x^2 - 5x + 6$, and to find the **partial fractions** $1/(x - 3)$ and $-1/(x - 2)$.

Recall from algebra that every polynomial with real coefficients can be factored into linear and irreducible quadratic factors.* For instance, the polynomial $x^5 + x^4 - x - 1$ can be written as

$$x^5 + x^4 - x - 1 = (x - 1)(x + 1)^2(x^2 + 1)$$

where $(x - 1)$ is a linear factor, $(x + 1)^2$ is a repeated linear factor, and $(x^2 + 1)$ is an irreducible quadratic factor. Using this factorization, we can write the partial fraction decomposition of the rational expression

$$\frac{N(x)}{x^5 + x^4 - x - 1}$$

where $N(x)$ is a polynomial of degree less than 5, as follows.

$$\frac{N(x)}{(x - 1)(x + 1)^2(x^2 + 1)} = \frac{A}{x - 1} + \frac{B}{x + 1} + \frac{C}{(x + 1)^2} + \frac{Dx + E}{x^2 + 1}$$

We summarize the steps for partial fraction decomposition as follows.

DECOMPOSITION OF $N(x)/D(x)$ INTO PARTIAL FRACTIONS	

1. **Divide if improper:** If $N(x)/D(x)$ is an improper fraction (that is, if the degree of the numerator is greater than or equal to the degree of the denominator), then divide the denominator into the numerator to obtain

$$\frac{N(x)}{D(x)} = \text{(a polynomial)} + \frac{N_1(x)}{D(x)}$$

where the degree of $N_1(x)$ is less than the degree of $D(x)$. Then apply steps 2, 3, and 4 to the proper rational expression $N_1(x)/D(x)$.

2. **Factor denominator:** Completely factor the denominator into factors of the form

$$(px + q)^m$$

and

$$(ax^2 + bx + c)^n$$

where $ax^2 + bx + c$ is irreducible.

3. **Linear factors:** For *each* factor of the form $(px + q)^m$, the partial fraction decomposition must include the following sum of m fractions.

$$\frac{A_1}{(px + q)} + \frac{A_2}{(px + q)^2} + \cdots + \frac{A_m}{(px + q)^m}$$

4. **Quadratic factors:** For *each* factor of the form $(ax^2 + bx + c)^n$, the partial fraction decomposition must include the following sum of n fractions.

$$\frac{B_1x + C_1}{ax^2 + bx + c} + \frac{B_2x + C_2}{(ax^2 + bx + c)^2} + \cdots + \frac{B_nx + C_n}{(ax^2 + bx + c)^n}$$

*For a review of factorization techniques, see *Precalculus*, 2nd edition, by Larson and Hostetler (Lexington, Mass.: D. C. Heath and Company, 1989).

Linear factors

Algebraic techniques for determining the constants in the numerators are demonstrated in the following examples.

EXAMPLE 1 Distinct linear factors

Write the partial fraction decomposition for

$$\frac{1}{x^2 - 5x + 6}.$$

SOLUTION

Since

$$x^2 - 5x + 6 = (x - 3)(x - 2)$$

we include one partial fraction for each factor and write

$$\frac{1}{x^2 - 5x + 6} = \frac{A}{x - 3} + \frac{B}{x - 2}$$

where A and B are to be determined. Multiplying this equation by the lowest common denominator, $(x - 3)(x - 2)$, leads to the **basic equation**

$$1 = A(x - 2) + B(x - 3). \qquad \text{Basic equation}$$

Since this equation is to be true for all x, we can substitute *convenient* values for x to obtain equations in A and B. These values are the ones that make particular factors zero.

To solve for A, we let $x = 3$ and obtain

$$1 = A(3 - 2) + B(3 - 3) \qquad \text{Let } x = 3 \text{ in basic equation.}$$
$$1 = A(1) + B(0)$$
$$A = 1.$$

To solve for B, we let $x = 2$ and obtain

$$1 = A(2 - 2) + B(2 - 3) \qquad \text{Let } x = 2 \text{ in basic equation.}$$
$$1 = A(0) + B(-1)$$
$$B = -1.$$

Therefore, the decomposition is

$$\frac{1}{x^2 - 5x + 6} = \frac{1}{x - 3} - \frac{1}{x - 2}$$

as indicated at the beginning of this section. ▭

REMARK   The substitutions for x in Example 1 were chosen for their convenience in determining values for A and B. We chose $x = 2$ so as to eliminate the term $A(x - 2)$, and $x = 3$ was chosen to eliminate the term $B(x - 3)$. The goal is to make *convenient* substitutions whenever possible.

EXAMPLE 2 Repeated linear factors

Evaluate

$$\int \frac{5x^2 + 20x + 6}{x^3 + 2x^2 + x} \, dx.$$

SOLUTION

Since

$$x^3 + 2x^2 + x = x(x^2 + 2x + 1) = x(x + 1)^2$$

we include one fraction for *each power* of x and $(x + 1)$ and write

$$\frac{5x^2 + 20x + 6}{x(x + 1)^2} = \frac{A}{x} + \frac{B}{x + 1} + \frac{C}{(x + 1)^2}.$$

Multiplying by the least common denominator, $x(x + 1)^2$, leads to the *basic equation*

$$5x^2 + 20x + 6 = A(x + 1)^2 + Bx(x + 1) + Cx. \qquad \text{Basic equation}$$

To solve for A, let $x = 0$. This eliminates the B and C terms and yields

$$6 = A(1) + 0 + 0$$
$$A = 6.$$

To solve for C, let $x = -1$. This eliminates the A and B terms and yields

$$5 - 20 + 6 = 0 + 0 - C$$
$$C = 9.$$

We have exhausted the most convenient choices for x, so to find the value of B we use *any other value* for x along with the calculated values of A and C. Thus, using $x = 1$, $A = 6$, and $C = 9$, we have

$$5 + 20 + 6 = A(4) + B(2) + C$$
$$= 6(4) + 2B + 9$$
$$B = -1.$$

Therefore, it follows that

$$\int \frac{5x^2 + 20x + 6}{x(x + 1)^2} \, dx = \int \left(\frac{6}{x} - \frac{1}{x + 1} + \frac{9}{(x + 1)^2} \right) dx$$

$$= 6 \ln |x| - \ln |x + 1| + 9 \frac{(x + 1)^{-1}}{-1} + C$$

$$= \ln \left| \frac{x^6}{x + 1} \right| - \frac{9}{x + 1} + C.$$

REMARK Note that it is necessary to make as many substitutions for x as there are unknowns (A, B, C, . . .) to be determined. For instance, in Example 2 we made three substitutions ($x = -1$, $x = 0$, and $x = 1$) to solve for C, A, and B, respectively.

Quadratic factors

When using the method of partial fractions with *linear* factors, a convenient choice of x immediately yields a value for one of the coefficients. With *quadratic* factors, a system of linear equations will typically have to be solved, regardless of the choice of x.

EXAMPLE 3   Distinct linear and quadratic factors

Evaluate

$$\int \frac{2x^3 - 4x - 8}{(x^2 - x)(x^2 + 4)}\, dx.$$

SOLUTION

Since

$$(x^2 - x)(x^2 + 4) = x(x - 1)(x^2 + 4)$$

we include one partial fraction for each factor and write

$$\frac{2x^3 - 4x - 8}{x(x - 1)(x^2 + 4)} = \frac{A}{x} + \frac{B}{x - 1} + \frac{Cx + D}{x^2 + 4}.$$

Multiplying by the least common denominator, $x(x - 1)(x^2 + 4)$, yields the *basic equation*

$$2x^3 - 4x - 8 = A(x - 1)(x^2 + 4) + Bx(x^2 + 4) + (Cx + D)(x)(x - 1).$$

To solve for A, let $x = 0$ and obtain

$$-8 = A(-1)(4) + 0 + 0 \quad \Longrightarrow \quad A = 2.$$

To solve for B, let $x = 1$ and obtain

$$-10 = 0 + B(5) + 0 \quad \Longrightarrow \quad B = -2.$$

At this point C and D are yet to be determined. We can find these remaining constants by choosing two other values for x and solving the resulting system of linear equations.

If $x = -1$, then, since $A = 2$ and $B = -2$, we have

$$-6 = (2)(-2)(5) + (-2)(-1)(5) + (-C + D)(-1)(-2)$$
$$2 = -C + D.$$

If $x = 2$, then we have

$$0 = (2)(1)(8) + (-2)(2)(8) + (2C + D)(2)(1)$$
$$8 = 2C + D.$$

Solving this system of two equations with two unknowns, we have

$$\begin{array}{rl} -C + D = 2 & \\ \underline{2C + D = 8} & \\ -3C \quad\quad = -6 & \text{Subtract second equation from first} \end{array}$$

which yields $C = 2$. Consequently, $D = 4$, and it follows that

$$\int \frac{2x^3 - 4x - 8}{x(x - 1)(x^2 + 4)} \, dx$$

$$= \int \left(\frac{2}{x} - \frac{2}{x - 1} + \frac{2x}{x^2 + 4} + \frac{4}{x^2 + 4} \right) dx$$

$$= 2 \ln |x| - 2 \ln |x - 1| + \ln (x^2 + 4) + 2 \arctan \frac{x}{2} + C.$$

When integrating rational expressions, keep in mind that for *improper* rational expressions like

$$\frac{N(x)}{D(x)} = \frac{2x^3 + x^2 - 7x + 7}{x^2 + x - 2}$$

you first must divide to obtain

$$\frac{N(x)}{D(x)} = 2x - 1 + \frac{-2x + 5}{x^2 + x - 2}.$$

The proper rational expression is then decomposed into its partial fractions by the usual methods.

In Examples 1, 2, and 3, we began the solution of the basic equation by substituting values of x that made the linear factors zero. This method works well when the partial fraction decomposition involves *only* linear factors. However, if the decomposition involves a quadratic factor, then an alternate procedure is often more convenient. Both methods are outlined in the following summary.

GUIDELINES FOR SOLVING THE BASIC EQUATION

Linear factors

1. Substitute the *roots* of the distinct linear factors into the basic equation.
2. For repeated linear factors use the coefficients determined in part 1 to rewrite the basic equation. Then substitute *other* convenient values of x and solve for the remaining coefficients.

Quadratic factors

1. Expand the basic equation.
2. Collect terms according to powers of x.
3. Equate the coefficients of like powers to obtain a system of linear equations involving A, B, C, and so on.
4. Solve the system of linear equations.

The second procedure for solving the basic equation is demonstrated in the next two examples.

EXAMPLE 4   Repeated quadratic factors

Evaluate

$$\int \frac{8x^3 + 13x}{(x^2 + 2)^2} \, dx.$$

SOLUTION

We include one partial fraction for each power of $(x^2 + 2)$ and write

$$\frac{8x^3 + 13x}{(x^2 + 2)^2} = \frac{Ax + B}{x^2 + 2} + \frac{Cx + D}{(x^2 + 2)^2}.$$

Multiplying by the least common denominator, $(x^2 + 2)^2$, yields the *basic equation*

$$8x^3 + 13x = (Ax + B)(x^2 + 2) + Cx + D.$$

Expanding the basic equation and collecting like terms, we have

$$8x^3 + 13x = Ax^3 + 2Ax + Bx^2 + 2B + Cx + D$$
$$8x^3 + 13x = Ax^3 + Bx^2 + (2A + C)x + (2B + D).$$

Now, we can equate the coefficients of like terms on opposite sides of the equation.

Using the known values $A = 8$ and $B = 0$, we have

$$13 = 2A + C = 2(8) + C \quad \Longrightarrow \quad C = -3$$
$$0 = 2B + D = 2(0) + D \quad \Longrightarrow \quad D = 0.$$

Finally, we conclude that

$$\int \frac{8x^3 + 13x}{(x^2 + 2)^2} \, dx = \int \left(\frac{8x}{x^2 + 2} + \frac{-3x}{(x^2 + 2)^2} \right) dx$$

$$= 4 \ln (x^2 + 2) + \frac{3}{2(x^2 + 2)} + C.$$

EXAMPLE 5   Repeated quadratic factors

Evaluate

$$\int \frac{x^2}{(x^2 + 1)^2} \, dx.$$

SOLUTION

We write

$$\frac{x^2}{(x^2 + 1)^2} = \frac{Ax + B}{x^2 + 1} + \frac{Cx + D}{(x^2 + 1)^2}$$

and multiply by $(x^2 + 1)^2$ to obtain

$$x^2 = (Ax + B)(x^2 + 1) + Cx + D$$
$$= Ax^3 + Bx^2 + (A + C)x + (B + D).$$

Therefore, by equating coefficients we obtain

$$A = 0, \quad B = 1, \quad C = 0, \quad D = -1$$

and we have

$$\int \frac{x^2}{(x^2 + 1)^2}\, dx = \int \left[\frac{1}{x^2 + 1} - \frac{1}{(x^2 + 1)^2} \right] dx.$$

The first integral can be evaluated by the Arctangent Rule, and the second integral was solved in Section 9.4 (see Example 6). Therefore, we have

$$\int \frac{x^2\, dx}{(x^2 + 1)^2} = \int \frac{dx}{x^2 + 1} - \int \frac{dx}{(x^2 + 1)^2}$$

$$= \arctan x - \frac{1}{2}\left(\arctan x + \frac{x}{x^2 + 1} \right) + C$$

$$= \frac{1}{2}\left(\arctan x - \frac{x}{x^2 + 1} \right) + C.$$

■

The final example shows how to combine the two methods effectively to solve a basic equation.

EXAMPLE 6   Linear and quadratic factors

Evaluate

$$\int \frac{3x + 4}{x^3 - 2x - 4}\, dx.$$

SOLUTION

Since

$$x^3 - 2x - 4 = (x - 2)(x^2 + 2x + 2)$$

we write

$$\frac{3x + 4}{x^3 - 2x - 4} = \frac{A}{x - 2} + \frac{Bx + C}{x^2 + 2x + 2}$$

$$3x + 4 = A(x^2 + 2x + 2) + (Bx + C)(x - 2) \qquad \text{Basic equation}$$

$$3x + 4 = (A + B)x^2 + (2A - 2B + C)x + (2A - 2C).$$

If $x = 2$, then in the basic equation we have $10 = 10A$ and $A = 1$. Now, from the expanded equation, we have the system

$$
\begin{aligned}
A + B &= 0 \\
2A - 2B + C &= 3 \\
2A - 2C &= 4.
\end{aligned}
$$

Using $A = 1$, we have $B = -1$ and $C = -1$. Therefore,

$$
\begin{aligned}
\int \frac{3x + 4}{x^3 - 2x - 4}\, dx &= \int \left(\frac{A}{x - 2} + \frac{Bx + C}{x^2 + 2x + 2} \right) dx \\
&= \int \left(\frac{1}{x - 2} + \frac{-x - 1}{x^2 + 2x + 2} \right) dx \\
&= \ln |x - 2| - \frac{1}{2} \ln (x^2 + 2x + 2) + C. \qquad \square
\end{aligned}
$$

Before we conclude this section, here are a few things you should remember. First, it is not necessary to use the partial fractions technique on all rational functions. For instance, the following integral is evaluated more easily by the Log Rule.

$$
\int \frac{x^2 + 1}{x^3 + 3x - 4}\, dx = \frac{1}{3} \int \frac{3x^2 + 3}{x^3 + 3x - 4}\, dx = \frac{1}{3} \ln |x^3 + 3x - 4| + C
$$

Second, if the integrand is not in reduced form, reducing it may eliminate the need for partial fractions, as shown in the following integral.

$$
\begin{aligned}
\int \frac{x^2 - x - 2}{x^3 - 2x - 4}\, dx &= \int \frac{(x + 1)(x - 2)}{(x - 2)(x^2 + 2x + 2)}\, dx = \int \frac{x + 1}{x^2 + 2x + 2}\, dx \\
&= \frac{1}{2} \ln |x^2 + 2x + 2| + C
\end{aligned}
$$

Finally, partial fractions can be used with some quotients involving transcendental functions. For instance, the substitution $u = \sin x$ allows us to write

$$
\begin{aligned}
\int \frac{\cos x}{\sin x(\sin x - 1)}\, dx &= \int \frac{du}{u(u - 1)} = \int \left(-\frac{1}{u} + \frac{1}{u - 1} \right) du \\
&= -\ln |u| + \ln |u - 1| + C = \ln \left| \frac{\sin x - 1}{\sin x} \right| + C.
\end{aligned}
$$

EXERCISES for Section 9.5

In Exercises 1–30, evaluate the indefinite integral.

1. $\displaystyle \int \frac{1}{x^2 - 1}\, dx$

2. $\displaystyle \int \frac{1}{4x^2 - 9}\, dx$

3. $\displaystyle \int \frac{3}{x^2 + x - 2}\, dx$

4. $\displaystyle \int \frac{x + 1}{x^2 + 4x + 3}\, dx$

5. $\displaystyle \int \frac{5 - x}{2x^2 + x - 1}\, dx$

6. $\displaystyle \int \frac{3x^2 - 7x - 2}{x^3 - x}\, dx$

7. $\displaystyle \int \frac{x^2 + 12x + 12}{x^3 - 4x}\, dx$

8. $\displaystyle \int \frac{x^3 - x + 3}{x^2 + x - 2}\, dx$

9. $\displaystyle \int \frac{2x^3 - 4x^2 - 15x + 5}{x^2 - 2x - 8}\, dx$

10. $\displaystyle \int \frac{x + 2}{x^2 - 4x}\, dx$

11. $\displaystyle \int \frac{4x^2 + 2x - 1}{x^3 + x^2}\, dx$

12. $\displaystyle \int \frac{2x - 3}{(x - 1)^2}\, dx$

13. $\int \dfrac{x^4}{(x-1)^3}\,dx$

14. $\int \dfrac{4x^2-1}{(2x)(x^2+2x+1)}\,dx$

15. $\int \dfrac{3x}{x^2-6x+9}\,dx$

16. $\int \dfrac{6x^2+1}{x^2(x-1)^3}\,dx$

17. $\int \dfrac{x^2-1}{x^3+x}\,dx$

18. $\int \dfrac{x}{x^3-1}\,dx$

19. $\int \dfrac{x^2}{x^4-2x^2-8}\,dx$

20. $\int \dfrac{2x^2+x+8}{(x^2+4)^2}\,dx$

21. $\int \dfrac{x}{16x^4-1}\,dx$

22. $\int \dfrac{x^2-4x+7}{x^3-x^2+x+3}\,dx$

23. $\int \dfrac{x^2+x+2}{(x^2+2)^2}\,dx$

24. $\int \dfrac{x^3}{(x^2-4)^2}\,dx$

25. $\int \dfrac{x^2+5}{x^3-x^2+x+3}\,dx$

26. $\int \dfrac{x^2+x+3}{x^4+6x^2+9}\,dx$

27. $\int \dfrac{6x^2-3x+14}{x^3-2x^2+4x-8}\,dx$

28. $\int \dfrac{x(2x-9)}{x^3-6x^2+12x-8}\,dx$

29. $\int \dfrac{2x^2-2x+3}{x^3-x^2-x-2}\,dx$

30. $\int \dfrac{x^3-6x^2+x-4}{1-x^4}\,dx$

In Exercises 31–36, evaluate the definite integral.

31. $\int_0^1 \dfrac{3}{2x^2+5x+2}\,dx$

32. $\int_3^4 \dfrac{1}{x^2-4}\,dx$

33. $\int_1^2 \dfrac{x+1}{x(x^2+1)}\,dx$

34. $\int_1^5 \dfrac{x-1}{x^2(x+1)}\,dx$

35. $\int_2^3 \dfrac{x^2-x+2}{x^3-x^2+x-1}\,dx$

36. $\int_0^1 \dfrac{x^2-x}{x^2+x+1}\,dx$

In Exercises 37–42, find the indefinite integral by using the indicated substitution.

37. $\int \dfrac{\sin x}{\cos x\,(\cos x-1)}\,dx$; let $u=\cos x$

38. $\int \dfrac{\sin x}{\cos x+\cos^2 x}\,dx$; let $u=\cos x$

39. $\int \dfrac{3\cos x}{\sin^2 x+\sin x-2}\,dx$; let $u=\sin x$

40. $\int \dfrac{\sec^2 x}{\tan x(\tan x+1)}\,dx$; let $u=\tan x$

41. $\int \dfrac{e^x}{(e^x-1)(e^x+4)}\,dx$; let $u=e^x$

42. $\int \dfrac{e^x}{(e^{2x}+1)(e^x-1)}\,dx$; let $u=e^x$

In Exercises 43–46, use the method of partial fractions to verify the given indefinite integral.

43. $\int \dfrac{1}{x(a+bx)}\,dx = \dfrac{1}{a}\ln\left|\dfrac{x}{a+bx}\right| + C$

44. $\int \dfrac{1}{a^2-x^2}\,dx = \dfrac{1}{2a}\ln\left|\dfrac{a+x}{a-x}\right| + C$

45. $\int \dfrac{x}{(a+bx)^2}\,dx = \dfrac{1}{b^2}\left(\dfrac{a}{a+bx}+\ln|a+bx|\right) + C$

46. $\int \dfrac{1}{x^2(a+bx)}\,dx = -\dfrac{1}{ax}-\dfrac{b}{a^2}\ln\left|\dfrac{x}{a+bx}\right| + C$

47. Find the area of the region bounded by the graphs of $y=7/(16-x^2)$ and $y=1$.

48. Find the centroid of the region bounded by the graphs of $y=2x/(x^2+1)$, $y=0$, $x=0$, and $x=3$.

49. Find the volume of the solid generated by revolving the region in Exercise 48 about the x-axis.

50. In Section 7.7 the exponential growth equation was derived from the assumption that the rate of growth is proportional to the existing quantity. In practice, there often exists some upper limit L past which growth cannot occur. In such cases, we assume the rate of growth to be proportional not only to the existing quantity, but also to the difference between the existing quantity y and the upper limit L. That is,

$$\dfrac{dy}{dt}=ky(L-y).$$

In integral form we can express this relationship as

$$\int \dfrac{dy}{y(L-y)} = \int k\,dt.$$

(a) Use partial fractions to evaluate the integral on the left, and solve for y as a function of t, where y_0 is the initial quantity.

(b) The graph of the function y is called a **logistics curve** (see figure). Show that the rate of growth is a maximum at the point of inflection, and that this occurs when $y=L/2$.

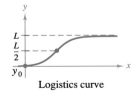

Logistics curve

FIGURE FOR 50

51. A single infected individual enters a community of n susceptible individuals. Let x be the number of newly infected individuals at time t. The common **epidemic model** assumes that the disease spreads at a rate proportional to the product of the total number infected and the number not yet infected. Thus

$$\frac{dx}{dt} = k(x + 1)(n - x)$$

and we obtain

$$\int \frac{1}{(x + 1)(n - x)}\, dx = \int k\, dt.$$

Solve for x as a function of t.

52. In a chemical reaction, one unit of compound Y and one unit of compound Z are converted into a single unit of compound X. If x is the amount of compound X formed, and the rate of formation of X is proportional to the product of the amounts of unconverted compounds Y and Z, then

$$\frac{dx}{dt} = k(y_0 - x)(z_0 - x)$$

where y_0 and z_0 are the initial amounts of substances Y and Z, respectively. From the above equation we obtain

$$\int \frac{1}{(y_0 - x)(z_0 - x)}\, dx = \int k\, dt.$$

Do the two integrations and solve for x in terms of t.

53. Use the result of Exercise 52 to find x as $t \to \infty$ if (a) $y_0 < z_0$ and (b) $y_0 > z_0$. If $y_0 = z_0$, what is the limit of x as $t \to \infty$.

In Exercises 54 and 55, use Simpson's Rule on a computer or calculator (with $n = 12$) to approximate the definite integral.

54. $\displaystyle \int_1^4 \frac{4}{\sqrt{x} + \sqrt[3]{x}}\, dx$

55. $\displaystyle \int_{-\pi}^{\pi} \sqrt{4 + 3 \sin^2 x}\, dx$

9.6 Integration by Tables and Other Integration Techniques

Integration by tables ▪ Reduction formulas ▪ Substitution for rational functions of sine and cosine

So far in this chapter we have discussed a number of integration techniques to use with the basic integration formulas. Certainly we have not considered every method for finding antiderivatives, but we have considered the most important ones.

But merely knowing *how* to use the various techniques is not enough. You also need to know *when* to use them. Integration is first and foremost a problem of recognition. That is, you must recognize which formula or technique to apply to obtain an antiderivative. Frequently, a slight alteration of an integrand will require a different integration technique, as shown below.

$$\int x \ln x\, dx = \frac{x^2}{2} \ln x - \frac{x^2}{4} + C \qquad \text{Integration by parts}$$

$$\int \frac{\ln x}{x}\, dx = \frac{(\ln x)^2}{2} + C \qquad \text{Power Rule}$$

$$\int \frac{1}{x \ln x}\, dx = \ln |\ln x| + C \qquad \text{Log Rule}$$

Integration by tables requires considerable thought and insight and often involves substitution. Many people find a table of integrals to be a valuable supplement to the integration techniques discussed in this chapter. A table of common integrals can be found in Appendix D of this text.

Each integration formula in the table in this text can be developed using one or more of the techniques we have studied. We encourage you to verify several of the formulas. For instance, Formula 4

$$\int \frac{u}{(a + bu)^2}\, du = \frac{1}{b^2}\left(\frac{a}{a + bu} + \ln |a + bu|\right) + C \qquad \text{Formula 4}$$

can be verified using the method of partial fractions, and Formula 19

$$\int \frac{\sqrt{a + bu}}{u} \, du = 2\sqrt{a + bu} + a \int \frac{du}{u\sqrt{a + bu}} \qquad \text{Formula 19}$$

can be verified using integration by parts.

The next four examples demonstrate the use of integration tables. Note that the integrals in the tables in this text are classified according to forms involving the following: u^n, $(a + bu)$, $(a + bu + cu^2)$, $\sqrt{a + bu}$, $(a^2 \pm u^2)$, $\sqrt{u^2 \pm a^2}$, $\sqrt{a^2 - u^2}$, trigonometric functions, inverse trigonometric functions, exponential functions, and logarithmic functions.

EXAMPLE 1 Integration by tables

Evaluate

$$\int \frac{dx}{x\sqrt{x - 1}}.$$

SOLUTION

Since the expression inside the radical is linear, we consider forms involving $\sqrt{a + bu}$.

$$\int \frac{du}{u\sqrt{a + bu}} = \frac{2}{\sqrt{-a}} \arctan \sqrt{\frac{a + bu}{-a}} + C \qquad \text{Formula 17 } (a < 0)$$

We let $a = -1$, $b = 1$, and $u = x$. Then $du = dx$, and we write

$$\int \frac{dx}{x\sqrt{x - 1}} = 2 \arctan \sqrt{x - 1} + C.$$

EXAMPLE 2 Integration by tables

Evaluate $\int x\sqrt{x^4 - 9} \, dx$.

SOLUTION

Since the radical has the form $\sqrt{u^2 - a^2}$, we consider the following formula.

$$\int \sqrt{u^2 - a^2} \, du = \frac{1}{2}(u\sqrt{u^2 - a^2} - a^2 \ln |u + \sqrt{u^2 - a^2}|) + C \quad \text{Formula 26}$$

We let $u = x^2$ and $a = 3$. Then $du = 2x \, dx$, and we write

$$\int x\sqrt{x^4 - 9} \, dx = \frac{1}{2} \int \sqrt{(x^2)^2 - 3^2} \, (2x) \, dx$$

$$= \frac{1}{4}(x^2\sqrt{x^4 - 9} - 9 \ln |x^2 + \sqrt{x^4 - 9}|) + C.$$

EXAMPLE 3   Integration by tables

Evaluate

$$\int \frac{x}{1 + e^{-x^2}} \, dx.$$

SOLUTION

Of the forms involving e^u, we consider the following.

$$\int \frac{du}{1 + e^u} = u - \ln (1 + e^u) + C \qquad \text{Formula 84}$$

We let $u = -x^2$. Then, $du = -2x \, dx$, and we write

$$\int \frac{x}{1 + e^{-x^2}} \, dx = -\frac{1}{2} \int \frac{-2x \, dx}{1 + e^{-x^2}}$$

$$= -\frac{1}{2}[-x^2 - \ln (1 + e^{-x^2})] + C$$

$$= \frac{1}{2}[x^2 + \ln (1 + e^{-x^2})] + C.$$

EXAMPLE 4   Integration by tables

Evaluate

$$\int \frac{\sin 2x}{2 + \cos x} \, dx.$$

SOLUTION

Substituting $2 \sin x \cos x$ for $\sin 2x$, we have

$$\int \frac{\sin 2x}{2 + \cos x} \, dx = 2 \int \frac{\sin x \cos x}{2 + \cos x} \, dx.$$

A check of the forms involving $\sin u$ or $\cos u$ shows that none of those listed applies. Therefore, we consider forms involving $a + bu$. For example,

$$\int \frac{u \, du}{a + bu} = \frac{1}{b^2}(bu - a \ln |a + bu|) + C. \qquad \text{Formula 3}$$

We let $a = 2$, $b = 1$, and $u = \cos x$. Then, $du = -\sin x \, dx$ and we write

$$2 \int \frac{\sin x \cos x}{2 + \cos x} \, dx = -2 \int \frac{\cos x \, (-\sin x \, dx)}{2 + \cos x}$$

$$= -2(\cos x - 2 \ln |2 + \cos x|) + C$$

$$= -2 \cos x + 4 \ln |2 + \cos x| + C.$$

Reduction formulas

You will notice that a number of integrals in the integration tables have the form

$$\int f(x)\ dx = g(x) + \int h(x)\ dx.$$

Such integration formulas are called **reduction formulas** because they reduce a given integral to the sum of a function and a simpler integral. We demonstrate the use of a reduction formula in the next two examples.

EXAMPLE 5   Use of a reduction formula

Evaluate

$$\int x^3 \sin x\ dx.$$

SOLUTION

From the integration table, we have the following three formulas.

$$\int u \sin u\ du = \sin u - u \cos u + C \qquad \text{Formula 52}$$

$$\int u^n \sin u\ du = -u^n \cos u + n \int u^{n-1} \cos u\ du \qquad \text{Formula 54}$$

$$\int u^n \cos u\ du = u^n \sin u - n \int u^{n-1} \sin u\ du \qquad \text{Formula 55}$$

Using Formula 54 followed by Formula 55, we have

$$\int x^3 \sin x\ dx = -x^3 \cos x + 3 \int x^2 \cos x\ dx$$

$$= -x^3 \cos x + 3\left(x^2 \sin x - 2 \int x \sin x\ dx\right).$$

Now, by Formula 52, we have

$$\int x^3 \sin x\ dx = -x^3 \cos x + 3x^2 \sin x + 6x \cos x - 6 \sin x + C.$$

EXAMPLE 6   Use of a reduction formula

Evaluate

$$\int \frac{\sqrt{3 - 5x}}{2x}\ dx.$$

SOLUTION

From the integration table, we have the following two formulas.

$$\int \frac{du}{u\sqrt{a + bu}} = \frac{1}{\sqrt{a}} \ln \left| \frac{\sqrt{a + bu} - \sqrt{a}}{\sqrt{a + bu} + \sqrt{a}} \right| + C \qquad \text{Formula 17 } (0 < a)$$

$$\int \frac{\sqrt{a + bu}}{u} du = 2\sqrt{a + bu} + a \int \frac{du}{u\sqrt{a + bu}} \qquad \text{Formula 19}$$

Using Formula 19 with $a = 3$, $b = -5$, and $u = x$, it follows that

$$\frac{1}{2} \int \frac{\sqrt{3 - 5x}}{x} dx = \frac{1}{2}\left(2\sqrt{3 - 5x} + 3 \int \frac{dx}{x\sqrt{3 - 5x}} \right)$$

$$= \sqrt{3 - 5x} + \frac{3}{2} \int \frac{dx}{x\sqrt{3 - 5x}}.$$

Now, by Formula 17, with $a = 3$, $b = -5$, and $u = x$, we conclude that

$$\int \frac{\sqrt{3 - 5x}}{2x} dx = \sqrt{3 - 5x} + \frac{3}{2}\left(\frac{1}{\sqrt{3}} \ln \left| \frac{\sqrt{3 - 5x} - \sqrt{3}}{\sqrt{3 - 5x} + \sqrt{3}} \right| \right) + C$$

$$= \sqrt{3 - 5x} + \frac{\sqrt{3}}{2} \ln \left| \frac{\sqrt{3 - 5x} - \sqrt{3}}{\sqrt{3 - 5x} + \sqrt{3}} \right| + C. \quad \blacksquare$$

Rational functions of sine and cosine

Example 4 involves a rational expression of $\sin x$ and $\cos x$. If you are unable to find an integral of this form in the integration tables, use the following special substitution to convert the trigonometric expression to a standard rational expression.

SUBSTITUTION FOR RATIONAL FUNCTIONS OF SINE AND COSINE

For integrals involving rational functions of sine and cosine, the substitution

$$u = \frac{\sin x}{1 + \cos x} = \tan \frac{x}{2}$$

yields

$$\cos x = \frac{1 - u^2}{1 + u^2}, \qquad \sin x = \frac{2u}{1 + u^2}, \qquad \text{and} \qquad dx = \frac{2\, du}{1 + u^2}.$$

PROOF From the substitution for u, it follows that

$$u^2 = \frac{\sin^2 x}{(1 + \cos x)^2} = \frac{1 - \cos^2 x}{(1 + \cos x)^2} = \frac{1 - \cos x}{1 + \cos x}.$$

Solving for $\cos x$ in this equation, we have

$$\cos x = \frac{1 - u^2}{1 + u^2}.$$

To find sin x, we write $u = \sin x/(1 + \cos x)$ as

$$\sin x = u(1 + \cos x) = u\left(1 + \frac{1 - u^2}{1 + u^2}\right) = \frac{2u}{1 + u^2}.$$

Finally, to find dx, we consider $u = \tan (x/2)$. Then $\arctan u = x/2$ and

$$dx = \frac{2\,du}{1 + u^2}.$$

EXAMPLE 7   Substitution for rational functions of sine and cosine

Evaluate

$$\int \frac{dx}{1 + \sin x - \cos x}.$$

SOLUTION

Let $u = \sin x/(1 + \cos x)$. Then

$$\int \frac{dx}{1 + \sin x - \cos x} = \int \frac{2\,du/(1 + u^2)}{1 + [2u/(1 + u^2)] - [(1 - u^2)/(1 + u^2)]}$$

$$= \int \frac{2\,du}{(1 + u^2) + 2u - (1 - u^2)} = \int \frac{2\,du}{2u + 2u^2}$$

$$= \int \frac{du}{u(1 + u)} = \int \frac{1}{u}\,du - \int \frac{1}{1 + u}\,du \qquad \text{Partial fractions}$$

$$= \ln |u| - \ln |1 + u| + C = \ln \left|\frac{u}{1 + u}\right| + C$$

$$= \ln \left|\frac{\sin x/(1 + \cos x)}{1 + [\sin x/(1 + \cos x)]}\right| + C$$

$$= \ln \left|\frac{\sin x}{1 + \sin x + \cos x}\right| + C.$$

EXERCISES for Section 9.6

In Exercises 1–52, use the integration tables at the end of the text to evaluate the given integral.

1. $\displaystyle\int \frac{x^2}{1 + x}\,dx$

2. $\displaystyle\int \frac{x}{\sqrt{1 + x}}\,dx$

3. $\displaystyle\int \frac{1}{x^2\sqrt{1 - x^2}}\,dx$

4. $\displaystyle\int x \sin x\,dx$

5. $\displaystyle\int x^2 \ln x\,dx$

6. $\displaystyle\int \text{arcsec } 2x\,dx$

7. $\displaystyle\int \frac{1}{x^2\sqrt{x^2 - 4}}\,dx$

8. $\displaystyle\int \frac{\sqrt{x^2 - 4}}{x}\,dx$

9. $\displaystyle\int xe^{x^2}\,dx$

10. $\displaystyle\int \frac{x}{\sqrt{9 - x^4}}\,dx$

11. $\displaystyle\int \frac{2x}{(1 - 3x)^2}\,dx$

12. $\displaystyle\int \frac{1}{x^2 + 2x + 2}\,dx$

13. $\displaystyle\int e^x \arccos e^x\,dx$

14. $\displaystyle\int \frac{\theta^2}{1 - \sin \theta^3}\,d\theta$

15. $\displaystyle\int x^3 \ln x\,dx$

16. $\displaystyle\int \cot^3 \theta\,d\theta$

17. $\displaystyle\int \frac{x^2}{(3x - 5)^2}\,dx$

18. $\displaystyle\int \frac{1}{2x^2(2x - 1)^2}\,dx$

19. $\displaystyle\int \frac{x}{1 - \sec x^2}\,dx$

20. $\displaystyle\int \frac{e^x}{1 - \tan e^x}\,dx$.

21. $\int \dfrac{\cos x}{1 + \sin^2 x}\, dx$

22. $\int \dfrac{1}{t[1 + (\ln t)^2]}\, dt$

23. $\int \dfrac{1}{1 + e^{2x}}\, dx$

24. $\int \dfrac{1}{\sqrt{x}(1 + 2\sqrt{x})}\, dx$

25. $\int \dfrac{\cos \theta}{3 + 2 \sin \theta + \sin^2 \theta}\, d\theta$

26. $\int x^2 \sqrt{2 + 9x^2}\, dx$

27. $\int \dfrac{1}{x^2 \sqrt{2 + 9x^2}}\, dx$

28. $\int \sqrt{3 + x^2}\, dx$

29. $\int e^x \sqrt{1 + e^{2x}}\, dx$

30. $\int \dfrac{1}{\sqrt{x}(x - 4)^{3/2}}\, dx$

31. $\int \sin^4 2x\, dx$

32. $\int \dfrac{\cos^3 \sqrt{x}}{\sqrt{x}}\, dx$

33. $\int \dfrac{1}{\sqrt{x}(1 - \cos \sqrt{x})}\, dx$

34. $\int \dfrac{1}{1 - \tan 5x}\, dx$

35. $\int t^4 \cos t\, dt$

36. $\int \sqrt{x} \arctan x^{3/2}\, dx$

37. $\int x \operatorname{arcsec}(x^2 + 1)\, dx$

38. $\int (\ln x)^3\, dx$

39. $\int \dfrac{\ln x}{x(3 + 2 \ln x)}\, dx$

40. $\int \dfrac{e^x}{(1 - e^{2x})^{3/2}}\, dx$

41. $\int \dfrac{\sqrt{2 - 2x - x^2}}{x + 1}\, dx$

42. $\int \dfrac{1}{(x^2 - 6x + 10)^2}\, dx$

43. $\int \dfrac{x}{x^4 - 6x^2 + 10}\, dx$

44. $\int (2x - 3)^2 \sqrt{(2x - 3)^2 + 4}\, dx$

45. $\int \dfrac{x}{\sqrt{x^4 - 6x^2 + 5}}\, dx$

46. $\int \dfrac{\cos x}{\sqrt{\sin^2 x + 1}}\, dx$

47. $\int \dfrac{x^3}{\sqrt{4 - x^2}}\, dx$

48. $\int \sqrt{\dfrac{3 - x}{3 + x}}\, dx$

49. $\int \dfrac{1}{x^{3/2}\sqrt{1 - x}}\, dx$

50. $\int x\sqrt{x^2 + 2x}\, dx$

51. $\int \dfrac{e^{3x}}{(1 + e^x)^3}\, dx$

52. $\int \sec^5 \theta\, d\theta$

In Exercises 53–58, verify the given integration formula.

53. $\int \dfrac{u^2}{(a + bu)^2}\, du =$
$$\dfrac{1}{b^3}\left(bu - \dfrac{a^2}{a + bu} - 2a \ln |a + bu|\right) + C$$

54. $\int \dfrac{u^n}{\sqrt{a + bu}}\, du =$
$$\dfrac{2}{(2n + 1)b}\left(u^n\sqrt{a + bu} - na \int \dfrac{u^{n-1}}{\sqrt{a + bu}}\, du\right)$$

55. $\int \dfrac{1}{(u^2 \pm a^2)^{3/2}}\, du = \dfrac{\pm u}{a^2\sqrt{u^2 \pm a^2}} + C$

56. $\int u^n \cos u\, du = u^n \sin u - n \int u^{n-1} \sin u\, du$

57. $\int \arctan u\, du = u \arctan u - \ln \sqrt{1 + u^2} + C$

58. $\int (\ln u)^n\, du = u(\ln u)^n - n \int (\ln u)^{n-1}\, du$

In Exercises 59–68, evaluate the given integral.

59. $\int \dfrac{1}{2 - 3 \sin \theta}\, d\theta$

60. $\int \dfrac{\sin \theta}{1 + \cos^2 \theta}\, d\theta$

61. $\int_0^{\pi/2} \dfrac{1}{1 + \sin \theta + \cos \theta}\, d\theta$

62. $\int_0^{\pi/2} \dfrac{1}{3 - 2 \cos \theta}\, d\theta$

63. $\int \dfrac{\sin \theta}{3 - 2 \cos \theta}\, d\theta$

64. $\int \dfrac{\sin \theta}{1 + \sin \theta}\, d\theta$

65. $\int \dfrac{\cos \sqrt{\theta}}{\sqrt{\theta}}\, d\theta$

66. $\int \dfrac{\sin \theta}{(\cos \theta)(1 + \sin \theta)}\, d\theta$

67. $\int \dfrac{1}{\sin \theta \tan \theta}\, d\theta$

68. $\int \dfrac{1}{\sec \theta - \tan \theta}\, d\theta$

69. A hydraulic cylinder on an industrial machine pushes a steel block a distance of x feet ($0 \le x \le 5$), where the variable force required is

$$F(x) = 2000xe^{-x} \text{ lb.}$$

Find the work done in pushing the block the full 5 feet through the machine.

70. Repeat Exercise 69, using a force of

$$F(x) = \dfrac{500x}{\sqrt{26 - x^2}} \text{ lb.}$$

9.7 Improper Integrals

Improper integrals (infinite limits of integration) ▪ Improper integrals (infinite discontinuities)

The definition of the definite integral $\int_a^b f(x)\,dx$ requires that the interval $[a, b]$ be finite. Furthermore, the Fundamental Theorem of Calculus, by which we have been evaluating definite integrals, requires that f be continuous on $[a, b]$. In this section we discuss a limit procedure for evaluating integrals that do not satisfy these requirements because either (1) one or both of the limits of integration are infinite, or (2) f has a finite number of infinite discontinuities in the interval $[a, b]$. Integrals that possess either property are called **improper integrals.** Note that we say that a function f has an **infinite discontinuity** at c if, *from the right or left*,

$$\lim_{x \to c} f(x) = \infty$$

or

$$\lim_{x \to c} f(x) = -\infty.$$

EXAMPLE 1   Examples of improper integrals

(a) The integrals

$$\int_1^\infty \frac{dx}{x}$$

and

$$\int_{-\infty}^\infty \frac{dx}{x^2 + 1}$$

are improper because one or both of the limits of integration are infinite.

(b) The integrals

$$\int_1^5 \frac{dx}{\sqrt{x - 1}}$$

and

$$\int_{-2}^2 \frac{dx}{(x + 1)^2}$$

are improper because the integrands have infinite discontinuities somewhere in the interval of integration. $1/\sqrt{x - 1}$ has an infinite discontinuity at $x = 1$, and $1/(x + 1)^2$ has an infinite discontinuity at $x = -1$.

Improper integrals with infinite limits of integration

To get an idea of how we can evaluate an improper integral, consider the integral

$$\int_1^b \frac{dx}{x^2} = \left[-\frac{1}{x} \right]_1^b = -\frac{1}{b} + 1 = 1 - \frac{1}{b}$$

which we can interpret as the area of the shaded region shown in Figure 9.16. Taking the limit as $b \to \infty$, we have

$$\int_1^\infty \frac{dx}{x^2} = \lim_{b \to \infty} \left[\int_1^b \frac{dx}{x^2} \right] = \lim_{b \to \infty} \left(1 - \frac{1}{b} \right) = 1$$

and we can interpret this improper integral as the area of the *unbounded* region between the graph of $f(x) = 1/x^2$ and the x-axis (to the right of $x = 1$). We define improper integrals with infinite limits of integration as follows.

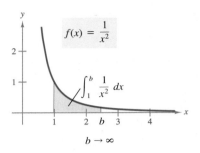

$f(x) = \frac{1}{x^2}$

$\int_1^b \frac{1}{x^2}\, dx$

$b \to \infty$

FIGURE 9.16

DEFINITION OF IMPROPER INTEGRALS WITH INFINITE LIMITS OF INTEGRATION

1. If f is continuous on the interval $[a, \infty)$, then

$$\int_a^\infty f(x)\, dx = \lim_{b \to \infty} \int_a^b f(x)\, dx.$$

2. If f is continuous on the interval $(-\infty, b]$, then

$$\int_{-\infty}^b f(x)\, dx = \lim_{a \to -\infty} \int_a^b f(x)\, dx.$$

3. If f is continuous on the interval $(-\infty, \infty)$, then

$$\int_{-\infty}^\infty f(x)\, dx = \int_{-\infty}^c f(x)\, dx + \int_c^\infty f(x)\, dx$$

where c is any real number.

In each case, if the limit exists, then the improper integral is said to **converge;** otherwise, the improper integral **diverges.** In the third case, the improper integral on the left diverges if either of the improper integrals on the right diverges.

EXAMPLE 2   An improper integral that diverges

Evaluate

$$\int_1^\infty \frac{dx}{x}.$$

SOLUTION

$$\int_1^\infty \frac{dx}{x} = \lim_{b \to \infty} \int_1^b \frac{dx}{x} = \lim_{b \to \infty} \left[\ln x \right]_1^b = \lim_{b \to \infty} (\ln b - 0) = \infty$$

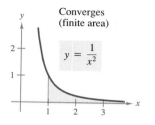

FIGURE 9.17

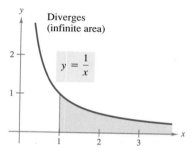

FIGURE 9.18

A comparison of the improper integral $\int_1^\infty (1/x^2)\, dx$, which converges to 1, and $\int_1^\infty (1/x)\, dx$, which diverges, suggests the rather unpredictable nature of improper integrals. The functions $f(x) = 1/x^2$ and $g(x) = 1/x$ have similar graphs, as shown in Figures 9.17 and 9.18. Yet, the shaded region shown in Figure 9.17 has a finite area, whereas the shaded region shown in Figure 9.18 has an infinite area.

EXAMPLE 3   Evaluating improper integrals

(a) $\displaystyle \int_0^\infty e^{-x}\, dx = \lim_{b \to \infty} \int_0^b e^{-x}\, dx = \lim_{b \to \infty} \left[-e^{-x} \right]_0^b = \lim_{b \to \infty} (-e^{-b} + 1) = 1$

(b) $\displaystyle \int_0^\infty \frac{1}{x^2 + 1}\, dx = \lim_{b \to \infty} \int_0^b \frac{1}{x^2 + 1}\, dx = \lim_{b \to \infty} \left[\arctan x \right]_0^b$

$$= \lim_{b \to \infty} \arctan b = \frac{\pi}{2}$$

EXAMPLE 4   An improper integral that diverges

Evaluate

$$\int_0^\infty \sin x\, dx.$$

SOLUTION

$$\int_0^\infty \sin x\, dx = \lim_{b \to \infty} \int_0^b \sin x\, dx = \lim_{b \to \infty} \left[-\cos x \right]_0^b = \lim_{b \to \infty} (1 - \cos b)$$

Since $\cos b$ does not approach a limit as b approaches infinity, we conclude that the given improper integral diverges.

REMARK   Note that the improper integrals in Examples 2 and 4 diverge for different reasons. In Example 2, the improper integral $\int_1^\infty (1/x)\, dx$ diverges because the region (to the right of 1) between $y = 1/x$ and the x-axis has an infinite area. In Example 4, the improper integral $\int_0^\infty \sin x\, dx$ diverges because of the oscillation of the sine function. This type of divergence is sometimes called *divergence by oscillation*.

In the following example, we illustrate a technique for using L'Hôpital's Rule to evaluate an improper integral.

EXAMPLE 5   Using L'Hôpital's Rule with an improper integral

Evaluate

$$\int_1^\infty (1 - x)e^{-x}\, dx.$$

SOLUTION

Using integration by parts, with $dv = e^{-x} dx$ and $u = (1 - x)$, we have

$$\int u\, dv \quad = \quad uv \quad - \quad \int v\, du$$

$$\int (1 - x)e^{-x}\, dx = -e^{-x}(1 - x) - \int e^{-x}\, dx$$
$$= -e^{-x} + xe^{-x} + e^{-x} + C = xe^{-x} + C.$$

Therefore, we have

$$\int_1^\infty (1 - x)e^{-x}\, dx = \lim_{b \to \infty} \left[xe^{-x} \right]_1^b = \left(\lim_{b \to \infty} \frac{b}{e^b} \right) - \frac{1}{e}.$$

Now, using L'Hôpital's Rule on the right-hand limit, we have

$$\lim_{b \to \infty} \frac{b}{e^b} = \lim_{b \to \infty} \frac{1}{e^b} = 0.$$

Finally, we conclude that

$$\int_1^\infty (1 - x)e^{-x}\, dx = -\frac{1}{e}.$$

(See Figure 9.19.)

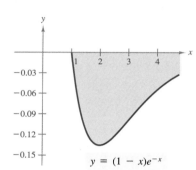

$$y = (1 - x)e^{-x}$$

FIGURE 9.19

EXAMPLE 6   An integral with infinite upper and lower limits of integration

Evaluate

$$\int_{-\infty}^\infty \frac{e^x}{1 + e^{2x}}\, dx.$$

SOLUTION

Note that the integrand is continuous on $(-\infty, \infty)$. To evaluate this integral, we break it into two parts, choosing $c = 0$ as a convenient value.

$$\int_{-\infty}^\infty \frac{e^x}{1 + e^{2x}}\, dx = \int_{-\infty}^0 \frac{e^x}{1 + e^{2x}}\, dx + \int_0^\infty \frac{e^x}{1 + e^{2x}}\, dx$$

$$= \lim_{b \to -\infty} \left[\arctan e^x \right]_b^0 + \lim_{b \to \infty} \left[\arctan e^x \right]_0^b$$

$$= \lim_{b \to -\infty} \left[\frac{\pi}{4} - \arctan e^b \right] + \lim_{b \to \infty} \left[\arctan e^b - \frac{\pi}{4} \right]$$

$$= \frac{\pi}{4} - 0 + \frac{\pi}{2} - \frac{\pi}{4}$$

$$= \frac{\pi}{2}$$

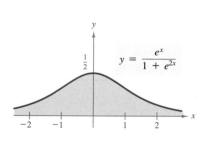

$$y = \frac{e^x}{1 + e^{2x}}$$

FIGURE 9.20

(See Figure 9.20.)

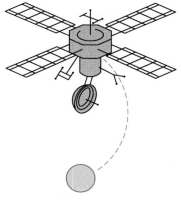

FIGURE 9.21

EXAMPLE 7   An application involving an improper integral

In Example 3 of Section 6.5, we determined that it would require 10,000 mile-tons of work to propel a 15-ton space module to a height of 800 miles above the earth. How much work is required to propel this same module an unlimited distance away from the earth's surface? (See Figure 9.21.)

SOLUTION

At first you might think that an infinite amount of work would be required. But if this were the case, then it would be impossible to send rockets into outer space. Since this has been done, the work required must be finite, and we can determine the work in the following manner. Using the integral of Example 3, Section 6.5, we replace the upper bound of 4800 miles by ∞ and write

$$W = \int_{4000}^{\infty} \frac{240,000,000}{x^2} \, dx$$

$$= \lim_{b \to \infty} \left[-\frac{240,000,000}{x} \right]_{4000}^{b}$$

$$= \lim_{b \to \infty} \left[-\frac{240,000,000}{b} + \frac{240,000,000}{4000} \right]$$

$$= 60,000 \text{ mile-tons}$$

$$\approx 6.336 \times 10^{11} \text{ ft} \cdot \text{lb.}$$

Improper integrals with infinite discontinuities

The second basic type of improper integral is one that has an infinite discontinuity *at or between* the limits of integration.

DEFINITION OF IMPROPER INTEGRALS WITH AN INFINITE DISCONTINUITY	1. If f is continuous on the interval $[a, b)$ and has an infinite discontinuity at b, then $$\int_a^b f(x) \, dx = \lim_{c \to b^-} \int_a^c f(x) \, dx.$$

2. If f is continuous on the interval $(a, b]$ and has an infinite discontinuity at a, then

$$\int_a^b f(x) \, dx = \lim_{c \to a^+} \int_c^b f(x) \, dx.$$

3. If f is continuous on the interval $[a, b]$, except for some c in (a, b) at which f has an infinite discontinuity, then

$$\int_a^b f(x) \, dx = \int_a^c f(x) \, dx + \int_c^b f(x) \, dx.$$

In each case, if the limit exists, then the improper integral is said to **converge;** otherwise, the improper integral **diverges.** In the third case, the improper integral on the left diverges if either of the improper integrals on the right diverges.

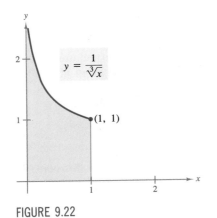

FIGURE 9.22

EXAMPLE 8   An improper integral with an infinite discontinuity

Evaluate

$$\int_0^1 \frac{dx}{\sqrt[3]{x}}.$$

SOLUTION

The integrand has an infinite discontinuity at $x = 0$, as shown in Figure 9.22. To evaluate this integral, we write

$$\int_0^1 x^{-1/3}\, dx = \lim_{b \to 0^+} \left[\frac{x^{2/3}}{2/3} \right]_b^1$$

$$= \lim_{b \to 0^+} \frac{3}{2}(1 - b^{2/3}) = \frac{3}{2}.$$

EXAMPLE 9   An improper integral that diverges

Evaluate

$$\int_0^2 \frac{dx}{x^3}.$$

SOLUTION

Since the integrand has an infinite discontinuity at $x = 0$, we write

$$\int_0^2 \frac{dx}{x^3} = \lim_{b \to 0^+} \left[-\frac{1}{2x^2} \right]_b^2 = \lim_{b \to 0^+} \left[-\frac{1}{8} + \frac{1}{2b^2} \right] = \infty.$$

Thus, we conclude that this improper integral diverges.

EXAMPLE 10   An improper integral with an interior discontinuity

Evaluate

$$\int_{-1}^2 \frac{dx}{x^3}.$$

SOLUTION

This integral is improper because the integrand has an infinite discontinuity at the interior point $x = 0$, as shown in Figure 9.23. Thus, we write

$$\int_{-1}^2 \frac{dx}{x^3} = \int_{-1}^0 \frac{dx}{x^3} + \int_0^2 \frac{dx}{x^3}.$$

From Example 9 we know that the second integral diverges. Therefore, the original improper integral also diverges.

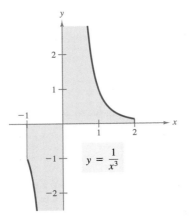

FIGURE 9.23

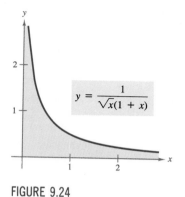

$$y = \frac{1}{\sqrt{x}(1 + x)}$$

FIGURE 9.24

REMARK   Remember to check for infinite discontinuities at interior points as well as endpoints when determining whether an integral is improper. For instance, if we had not recognized that the integral in Example 10 was improper, we would have obtained the *incorrect* result

$$\int_{-1}^{2} \frac{dx}{x^3} \stackrel{?}{=} \left[\frac{-1}{2x^2}\right]_{-1}^{2} = -\frac{1}{8} + \frac{1}{2} = \frac{3}{8}. \qquad \text{Incorrect evaluation}$$

The integral in the next example is improper for *two* reasons. One limit of integration is infinite, and the integrand has an infinite discontinuity at the other limit of integration, as shown in Figure 9.24.

EXAMPLE 11   A doubly improper integral

Evaluate

$$\int_{0}^{\infty} \frac{dx}{\sqrt{x}(x + 1)}.$$

SOLUTION

To evaluate this integral, we split it at a convenient point (say, $x = 1$) and write

$$\int_{0}^{\infty} \frac{dx}{\sqrt{x}(x + 1)} = \int_{0}^{1} \frac{dx}{\sqrt{x}(x + 1)} + \int_{1}^{\infty} \frac{dx}{\sqrt{x}(x + 1)}$$

$$= \lim_{b \to 0^+} \left[2 \arctan \sqrt{x}\right]_{b}^{1} + \lim_{c \to \infty} \left[2 \arctan \sqrt{x}\right]_{1}^{c}$$

$$= 2\left(\frac{\pi}{4}\right) - 0 + 2\left(\frac{\pi}{2}\right) - 2\left(\frac{\pi}{4}\right) = \pi. \qquad \blacksquare$$

EXAMPLE 12   An improper integral involving L'Hôpital's Rule

Evaluate

$$\int_{0}^{1} \ln x \, dx.$$

SOLUTION

Using integration by parts, with $dv = dx$ and $u = \ln x$, we have

$$\int_{0}^{1} \ln x \, dx = \lim_{b \to 0^+} \left[x \ln x - x\right]_{b}^{1} = \lim_{b \to 0^+} [0 - 1 - b \ln b + b].$$

By L'Hôpital's Rule, we have

$$\lim_{b \to 0^+} b \ln b = \lim_{b \to 0^+} \frac{\ln b}{1/b} = \lim_{b \to 0^+} \frac{1/b}{-1/b^2} = \lim_{b \to 0^+} -b = 0.$$

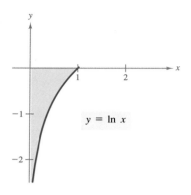

FIGURE 9.25

Finally, we conclude that

$$\int_0^1 \ln x \, dx = \lim_{b \to 0^+} [0 - 1 - b \ln b + b] = 0 - 1 - 0 + 0 = -1.$$

(See Figure 9.25.)

EXAMPLE 13   An application involving arc length

Use the formula for arc length to show that the circumference of the circle $x^2 + y^2 = 1$ is 2π.

SOLUTION

To simplify the work, we consider the quarter circle given by $y = \sqrt{1 - x^2}$, where $0 \le x \le 1$. The arc length of this quarter circle is given by

$$s = \int_0^1 \sqrt{1 + (y')^2} \, dx$$

$$= \int_0^1 \sqrt{1 + \left(\frac{-x}{\sqrt{1 - x^2}}\right)^2} \, dx$$

$$= \int_0^1 \frac{dx}{\sqrt{1 - x^2}}.$$

This integral is improper, since it has an infinite discontinuity at $x = 1$. Thus, we write

$$s = \int_0^1 \frac{dx}{\sqrt{1 - x^2}} = \lim_{b \to 1^-} \left[\arcsin x \right]_0^b = \frac{\pi}{2} - 0 = \frac{\pi}{2}.$$

Finally, multiplying by 4, we conclude that the circumference of the circle is $4s = 2\pi$, as shown in Figure 9.26.

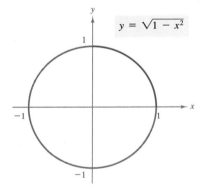

FIGURE 9.26

We conclude this section with a useful theorem describing the convergence or divergence of a common type of improper integral. The proof of this theorem is left as an exercise (see Exercise 33).

**THEOREM 9.3
A SPECIAL TYPE OF IMPROPER
INTEGRAL**

$$\int_1^\infty \frac{dx}{x^p} = \begin{cases} \dfrac{1}{p - 1}, & \text{if } p > 1 \\ \text{diverges}, & \text{if } p \le 1 \end{cases}$$

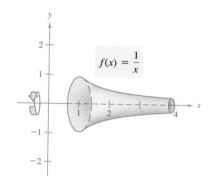

FIGURE 9.27

EXAMPLE 14 An application involving a solid of revolution

The solid formed by revolving the *unbounded* region lying between the graph of $f(x) = 1/x$ and the x-axis ($x \geq 1$) is called **Gabriel's Horn.** (See Figure 9.27.) Show that this solid has a finite volume and an infinite surface area.

SOLUTION

Using the disc method, with Theorem 9.3, we determine the volume to be

$$V = \pi \int_1^\infty \left(\frac{1}{x}\right)^2 dx$$

$$= \pi\left(\frac{1}{2-1}\right) = \pi.$$

The surface area is given by

$$S = 2\pi \int_1^\infty f(x)\sqrt{1 + [f'(x)]^2}\, dx$$

$$= 2\pi \int_1^\infty \frac{1}{x}\sqrt{1 + \frac{1}{x^4}}\, dx$$

$$= 2\pi \int_1^\infty \frac{\sqrt{1 + x^4}}{x^3}\, dx.$$

Now, using the substitution $u = x^2$ together with Formula 30 in the integration table, we have

$$S = \lim_{b\to\infty} \pi \left[\frac{-\sqrt{1 + x^4}}{x^2} + \ln\left|x^2 + \sqrt{1 + x^4}\right|\right]_1^b.$$

We leave it up to you to show that this limit is infinite.

EXERCISES for Section 9.7

In Exercises 1–32, determine the divergence or convergence of the given improper integral. Evaluate the integral if it converges.

1. $\int_0^4 \frac{1}{\sqrt{x}}\, dx$

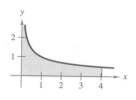

2. $\int_3^4 \frac{1}{\sqrt{x-3}}\, dx$

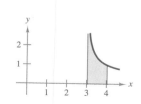

3. $\int_0^2 \frac{1}{(x-1)^{2/3}}\, dx$

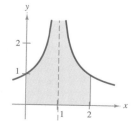

4. $\int_0^2 \frac{1}{(x-1)^2}\, dx$

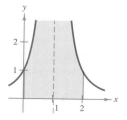

5. $\int_0^\infty e^{-x}\,dx$

6. $\int_{-\infty}^0 e^{2x}\,dx$

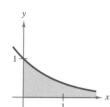

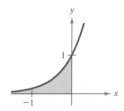

7. $\int_{-\infty}^0 xe^{-2x}\,dx$

8. $\int_0^\infty xe^{-x}\,dx$

9. $\int_0^\infty x^2 e^{-x}\,dx$

10. $\int_0^\infty (x-1)e^{-x}\,dx$

11. $\int_1^\infty \dfrac{1}{x^2}\,dx$

12. $\int_1^\infty \dfrac{1}{\sqrt{x}}\,dx$

13. $\int_0^\infty e^{-x}\cos x\,dx$

14. $\int_0^\infty e^{-ax}\sin bx\,dx,\ a > 0$

15. $\int_{-\infty}^\infty \dfrac{1}{1+x^2}\,dx$

16. $\int_0^\infty \dfrac{x^3}{(x^2+1)^2}\,dx$

17. $\int_0^\infty \dfrac{1}{e^x + e^{-x}}\,dx$

18. $\int_0^\infty \dfrac{e^x}{1+e^x}\,dx$

19. $\int_0^\infty \cos \pi x\,dx$

20. $\int_0^\infty \sin \dfrac{x}{2}\,dx$

21. $\int_0^1 \dfrac{1}{x^2}\,dx$

22. $\int_0^1 \dfrac{1}{x}\,dx$

23. $\int_0^8 \dfrac{1}{\sqrt[3]{8-x}}\,dx$

24. $\int_0^e \ln x\,dx$

25. $\int_0^1 x \ln x\,dx$

26. $\int_0^{\pi/2} \sec \theta\,d\theta$

27. $\int_0^{\pi/2} \tan \theta\,d\theta$

28. $\int_0^2 \dfrac{1}{\sqrt{4-x^2}}\,dx$

29. $\int_2^4 \dfrac{1}{\sqrt{x^2-4}}\,dx$

30. $\int_0^2 \dfrac{1}{4-x^2}\,dx$

31. $\int_0^2 \dfrac{1}{\sqrt[3]{x-1}}\,dx$

32. $\int_0^2 \dfrac{1}{(x-1)^{4/3}}\,dx$

33. Determine all values of p for which the following integral converges.

$$\int_1^\infty \dfrac{1}{x^p}\,dx$$

34. Determine all values of p for which the following integral converges.

$$\int_0^1 \dfrac{1}{x^p}\,dx$$

35. Use mathematical induction to verify that the following integral converges for any positive integer n.

$$\int_0^\infty x^n e^{-x}\,dx$$

36. Given continuous functions f and g such that $0 \le f(x) \le g(x)$ on the interval (a, ∞), prove the following.
(a) If $\int_a^\infty g(x)\,dx$ converges, then $\int_a^\infty f(x)\,dx$ converges.
(b) If $\int_a^\infty f(x)\,dx$ diverges, then $\int_a^\infty g(x)\,dx$ diverges.

In Exercises 37–46, use the results of Exercises 33–36 to determine whether the improper integral converges.

37. $\int_0^1 \dfrac{1}{x^3}\,dx$

38. $\int_0^1 \dfrac{1}{\sqrt[3]{x}}\,dx$

39. $\int_1^\infty \dfrac{1}{x^3}\,dx$

40. $\int_0^\infty x^4 e^{-x}\,dx$

41. $\int_1^\infty \dfrac{1}{x^2+5}\,dx$

42. $\int_2^\infty \dfrac{1}{\sqrt{x-1}}\,dx$

43. $\int_2^\infty \dfrac{1}{\sqrt[3]{x(x-1)}}\,dx$

44. $\int_1^\infty \dfrac{1}{\sqrt{x}(x+1)}\,dx$

45. $\int_0^\infty e^{-x^2}\,dx$

46. $\int_2^\infty \dfrac{1}{\sqrt{x}\ln x}\,dx$

47. Given the region bounded by the graphs of $y = 1/x^2$ and $y = 0$ $(x \ge 1)$, find the following.
(a) the area of the region
(b) the volume of the solid generated by revolving the region about the x-axis
(c) the volume of the solid generated by revolving the region about the y-axis

48. Given the region bounded by the graphs of $y = e^{-x}$ and $y = 0$ $(x \ge 0)$, find the following.
(a) the area of the region
(b) the volume of the solid generated by revolving the region about the y-axis

49. Sketch the graph of the hypocycloid of four cusps

$$x^{2/3} + y^{2/3} = 1$$

and find its perimeter.

50. The region bounded by

$$(x - 2)^2 + y^2 = 1$$

is revolved about the y-axis to form a torus. Find the surface area of this torus.

51. The **Gamma Function** $\Gamma(n)$ is defined by

$$\Gamma(n) = \int_0^\infty x^{n-1} e^{-x}\, dx, \quad n > 0.$$

Evaluate this function at $n = 1$, 2, and 3. Then use integration by parts to show that $\Gamma(n + 1) = n\Gamma(n)$.

52. A 5-ton rocket is fired from the surface of the earth into outer space. How much work is required to overcome the earth's gravitational force? How far has the rocket traveled when half the total work has occurred?

In Exercises 53 and 54, use the following definitions. A nonnegative function f is called a **probability density function** if

$$\int_{-\infty}^\infty f(t)\, dt = 1.$$

The **probability** that t lies between a and b is given by

$$P(a \le t \le b) = \int_a^b f(t)\, dt.$$

The **expected value** of t is given by

$$E(t) = \int_{-\infty}^\infty t f(t)\, dt.$$

53. For

$$f(t) = \begin{cases} \dfrac{1}{7} e^{-t/7}, & t \ge 0 \\ 0, & t < 0 \end{cases}$$

(a) show that f is a probability density function.
(b) find $P(0 \le t \le 5)$.
(c) find $E(t)$.

54. For

$$f(t) = \begin{cases} \dfrac{2}{5} e^{-2t/5}, & t \ge 0 \\ 0, & t < 0 \end{cases}$$

(a) show that f is a probability density function.
(b) find $P(0 \le t \le 3)$.
(c) find $E(t)$.

In Exercises 55 and 56, find the capitalized cost C of an asset (a) for $n = 5$ years, (b) for $n = 10$ years, and (c) forever. The **capitalized cost** is given by

$$C = C_0 + \int_0^n c(t) e^{-rt}\, dt$$

where C_0 is the original investment, t is the time in years, r is the annual rate compounded continuously, and $c(t)$ is annual cost of maintenance.

55. $C_0 = \$650{,}000$, $c(t) = \$25{,}000$, $r = 0.12$
56. $C_0 = \$650{,}000$, $c(t) = \$25{,}000(1 + 0.08t)$, $r = 0.12$

57. Find the value of the following integral used in electromagnetic theory.

$$P = k \int_1^\infty \frac{1}{(a^2 + x^2)^{3/2}}\, dx$$

REVIEW EXERCISES for Chapter 9

In Exercises 1–40, evaluate the given integral.

1. $\displaystyle\int \frac{x^2}{x^2 + 2x - 15}\,dx$

2. $\displaystyle\int \frac{\sqrt{x^2 - 9}}{x}\,dx$

3. $\displaystyle\int \frac{1}{1 - \sin\theta}\,d\theta$

4. $\displaystyle\int x^2 \sin 2x\,dx$

5. $\displaystyle\int e^{2x}\sin 3x\,dx$

6. $\displaystyle\int (x^2 - 1)e^x\,dx$

7. $\displaystyle\int \frac{\ln(2x)}{x^2}\,dx$

8. $\displaystyle\int 2x\sqrt{2x - 3}\,dx$

9. $\displaystyle\int \sqrt{4 - x^2}\,dx$

10. $\displaystyle\int \frac{\sqrt{4 - x^2}}{2x}\,dx$

11. $\displaystyle\int \frac{-12}{x^2\sqrt{4 - x^2}}\,dx$

12. $\displaystyle\int \tan\theta \sec^4\theta\,d\theta$

13. $\displaystyle\int \sec^4 \frac{x}{2}\,dx$

14. $\displaystyle\int \sec\theta \cos 2\theta\,d\theta$

15. $\displaystyle\int \frac{9}{x^2 - 9}\,dx$

16. $\displaystyle\int \frac{\sec^2\theta}{\tan\theta(\tan\theta - 1)}\,d\theta$

17. $\displaystyle\int \frac{x^2 + 2x}{x^3 - x^2 + x - 1}\,dx$

18. $\displaystyle\int \frac{4x - 2}{3(x - 1)^2}\,dx$

19. $\displaystyle\int \frac{3x^3 + 4x}{(x^2 + 1)^2}\,dx$

20. $\displaystyle\int \sqrt{\frac{x - 2}{x + 2}}\,dx$

21. $\displaystyle\int \frac{16}{\sqrt{16 - x^2}}\,dx$

22. $\displaystyle\int \frac{\sin\theta}{1 + 2\cos^2\theta}\,d\theta$

23. $\displaystyle\int \frac{e^x}{4 + e^{2x}}\,dx$

24. $\displaystyle\int \frac{x}{x^2 - 4x + 8}\,dx$

25. $\displaystyle\int \frac{x}{x^2 + 4x + 8}\,dx$

26. $\displaystyle\int \frac{3}{2x\sqrt{9x^2 - 1}}\,dx$

27. $\displaystyle\int \theta \sin\theta \cos\theta\,d\theta$

28. $\displaystyle\int \frac{\csc\sqrt{2x}}{\sqrt{x}}\,dx$

29. $\displaystyle\int (\sin\theta + \cos\theta)^2\,d\theta$

30. $\displaystyle\int \cos 2\theta(\sin\theta + \cos\theta)^2\,d\theta$

31. $\displaystyle\int \frac{x^{1/4}}{1 + x^{1/2}}\,dx$

32. $\displaystyle\int \sqrt{1 - \cos x}\,dx$

33. $\displaystyle\int \sqrt{1 + \cos x}\,dx$

34. $\displaystyle\int \ln\sqrt{x^2 - 1}\,dx$

35. $\displaystyle\int \ln(x^2 + x)\,dx$

36. $\displaystyle\int x \arcsin 2x\,dx$

37. $\displaystyle\int \cos x \ln(\sin x)\,dx$

38. $\displaystyle\int e^x \arctan e^x\,dx$

39. $\displaystyle\int \frac{x^4 + 2x^2 + x + 1}{(x^2 + 1)^2}\,dx$

40. $\displaystyle\int \sqrt{1 + \sqrt{x}}\,dx$

In Exercises 41–44, evaluate the given integral using the indicated methods.

41. $\displaystyle\int \frac{1}{x^2\sqrt{4 + x^2}}\,dx$

   (a) trigonometric substitution
   (b) substitution: $x = 2/u$

42. $\displaystyle\int \frac{1}{x\sqrt{4 + x^2}}\,dx$

   (a) trigonometric substitution
   (b) substitution: $u^2 = 4 + x^2$

43. $\displaystyle\int \frac{x^3}{\sqrt{4 + x^2}}\,dx$

   (a) trigonometric substitution
   (b) substitution: $u^2 = 4 + x^2$
   (c) by parts: $dv = (x/\sqrt{4 + x^2})\,dx$

44. $\displaystyle\int x\sqrt{4 + x}\,dx$

   (a) trigonometric substitution
   (b) substitution: $u^2 = 4 + x$
   (c) substitution: $u = 4 + x$
   (d) by parts: $dv = \sqrt{4 + x}\,dx$

In Exercises 45 and 46, evaluate each definite integral to two-decimal-place accuracy.

45. (a) $\displaystyle\int_0^1 e^x\,dx$

  (b) $\displaystyle\int_0^1 xe^x\,dx$

  (c) $\displaystyle\int_0^1 xe^{x^2}\,dx$

  (d) $\displaystyle\int_0^1 e^{x^2}\,dx$

46. (a) $\displaystyle\int_0^{\pi/2} \cos x\,dx$

  (b) $\displaystyle\int_0^{\pi/2} \cos^2 x\,dx$

  (c) $\displaystyle\int_0^{\pi/2} \cos x^2\,dx$

  (d) $\displaystyle\int_0^{\pi/2} \cos\sqrt{x}\,dx$

In Exercises 47 and 48, approximate to two decimal places the arc length of the curve over the given interval.

Function	Interval
47. $y = \sin x$	$[0, \pi]$
48. $y = \sin^2 x$	$[0, \pi]$

In Exercises 49 and 50, find the centroid of the region bounded by the graphs of the given equations.

49. $y = \sqrt{1 - x^2}$, $y = 0$

50. $(x - 1)^2 + y^2 = 1$, $(x - 4)^2 + y^2 = 4$

51. Approximate

$$\int_2^\infty \frac{1}{x^5 - 1} \, dx$$

using the inequality

$$\frac{1}{x^5} + \frac{1}{x^{10}} + \frac{1}{x^{15}} < \frac{1}{x^5 - 1} < \frac{1}{x^5} + \frac{1}{x^{10}} + \frac{2}{x^{15}}$$

for $x \geq 2$.

52. Let

$$I_n = \int \frac{x^{2n-1}}{(x^2 + 1)^{n+3}} \, dx, \quad n \geq 1.$$

Prove that

$$I_n = \left(\frac{n - 1}{n + 2} \right) I_{n-1}$$

and then evaluate the following.

(a) $\displaystyle \int_0^\infty \frac{x}{(x^2 + 1)^4} \, dx$

(b) $\displaystyle \int_0^\infty \frac{x^3}{(x^2 + 1)^5} \, dx$

(c) $\displaystyle \int_0^\infty \frac{x^5}{(x^2 + 1)^6} \, dx$

53. Verify the reduction formula

$$\int \tan^n x \, dx = \frac{1}{n - 1} \tan^{n-1} x - \int \tan^{n-2} x \, dx.$$

54. Show that

$$\int_x^1 \frac{1}{1 + t^2} \, dt = \int_1^{1/x} \frac{1}{1 + t^2} \, dt$$

by evaluating each integral and then using the identity

$$\arctan x + \arctan \frac{1}{x} = \frac{\pi}{2}, \quad x > 0.$$

Chapter 10 Application

A golf ball, after it is hit, continuously encounters air resistance, and its speed decreases. This means that, even on a calm day, the path of a golf ball is not parabolic. Typically, a golf ball that is hit with an iron will travel from 70–90 yards (9-iron) to 165–190 yards (2-iron).

Trajectories and Air Resistance

Neglecting air resistance, a projectile fired from ground level follows a parabolic trajectory given by

$$y = (\tan \theta)x + \frac{gx^2}{2v_0^2 \cos^2 \theta}$$

where v_0 is the initial speed, θ is the angle of projection, and $g = -32$ ft/sec is the acceleration due to gravity. If *air resistance is considered*, the trajectory is given by

$$y = \left(\tan \theta - \frac{g}{kv_0 \cos \theta}\right)x - \frac{g}{k^2} \ln \left(1 - \frac{kx}{v_0 \cos \theta}\right)$$

where k is the drag factor caused by the air resistance. Using the power series representation

$$\ln (1 + x) = x - \frac{x^2}{2} + \frac{x^3}{3} - \frac{x^4}{4} + \cdots, \qquad -1 < x < 1$$

we can rewrite the equation for the trajectory (considering air resistance) as

$$y = (\tan \theta)x + \frac{gx^2}{2v_0^2 \cos^2 \theta} + \frac{kgx^3}{3v_0^3 \cos^3 \theta} + \frac{k^2gx^4}{4v_0^4 \cos^4 \theta} + \cdots.$$

Note that the first two terms of this power series give the equation for the parabolic trajectory. Thus, when k is small (little air resistance), the trajectory is nearly parabolic.

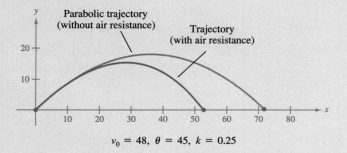

Parabolic trajectory (without air resistance)

Trajectory (with air resistance)

$v_0 = 48, \ \theta = 45, \ k = 0.25$

See Exercise 45, Section 10.10.

Chapter Overview

This chapter is divided into two basic parts. The first six sections discuss **infinite sequences** and **infinite series.** In particular, we are interested in the convergence or divergence of infinite sequences and series. For example,

$$\frac{1}{2}, \frac{1}{4}, \frac{1}{8}, \frac{1}{16}, \cdots, \frac{1}{2^n}, \cdots$$

is an infinite sequence that converges to 0. By adding the terms of this sequence, we obtain an infinite series that converges to 1. That is,

$$\sum_{n=1}^{\infty} \left(\frac{1}{2}\right)^n = \frac{1}{2} + \frac{1}{4} + \frac{1}{8} + \frac{1}{16} + \cdots = 1.$$

The last four sections in the chapter discuss **Taylor** and **Maclaurin polynomials** and power series. The nth Maclaurin polynomial for a function f is given by

$$P_n(x) = f(0) + f'(0)x + \frac{f''(0)}{2!}x^2 + \cdots + \frac{f^{(n)}(0)}{n!}x^n.$$

This polynomial can be used as an approximation for the function f—good approximations occur when x is close to zero or when n is large.

Taylor or Maclaurin "polynomials" with an infinite number of terms are called Taylor or Maclaurin **series,** and often these power series are precisely equal to the function f. In such cases, we say that the series **converges** to $f(x)$.

558

Infinite Series

10.1 Sequences

Sequences ▪ Limit of a sequence ▪ Pattern recognition for sequences ▪ Monotonic sequences ▪ Bounded sequences

In mathematics the word "sequence" is used in much the same way as in ordinary English. When we say that a collection of objects or events is *in sequence* we usually mean that the collection is ordered so that it has an identified first member, second member, third member, and so on.

We define a sequence mathematically as a *function whose domain is the set of positive integers*. Although a sequence is a function, we usually represent sequences by subscript notation, rather than the standard function notation. For instance, in the sequence

$$
\begin{array}{ccccccc}
1, & 2, & 3, & 4, & \ldots, & n, & \ldots \\
\downarrow & \downarrow & \downarrow & \downarrow & & \downarrow & \\
a_1, & a_2, & a_3, & a_4, & \ldots, & a_n, & \ldots
\end{array}
$$

1 is mapped onto a_1, 2 is mapped onto a_2, and so on. We call a_n the **nth term** of the sequence and we denote the sequence by $\{a_n\}$.

DEFINITION OF A SEQUENCE	A **sequence** $\{a_n\}$ is a function whose domain is the set of positive integers. The functional values $a_1, a_2, a_3, \ldots, a_n, \ldots$ are called the **terms** of the sequence.

REMARK   Occasionally, it is convenient to begin a sequence with a_0, so that the terms of the sequence become

$$a_0, a_1, a_2, a_3, a_4, \ldots, a_n, \ldots.$$

EXAMPLE 1 Listing the terms of a sequence

(a) For the sequence $\{a_n\} = \{3 + (-1)^n\}$, the first four terms are

$$3 + (-1)^1, \, 3 + (-1)^2, \, 3 + (-1)^3, \, 3 + (-1)^4, \, \ldots$$
$$2, 4, 2, 4, \ldots.$$

(b) For the sequence $\{b_n\} = \{2n/(1 + n)\}$, the first four terms are

$$\frac{2 \cdot 1}{1 + 1}, \frac{2 \cdot 2}{1 + 2}, \frac{2 \cdot 3}{1 + 3}, \frac{2 \cdot 4}{1 + 4}, \, \cdots$$
$$\frac{2}{2}, \frac{4}{3}, \frac{6}{4}, \frac{8}{5}, \, \cdots.$$

(c) For the sequence $\{c_n\} = \{n^2/(2^n - 1)\}$, the first four terms are

$$\frac{1^2}{2^1 - 1}, \frac{2^2}{2^2 - 1}, \frac{3^2}{2^3 - 1}, \frac{4^2}{2^4 - 1}, \, \cdots$$
$$\frac{1}{1}, \frac{4}{3}, \frac{9}{7}, \frac{16}{15}, \, \cdots.$$

In this chapter our primary interest is in sequences whose terms approach a limiting value. Such sequences are said to **converge.** For instance, the sequence $\{1/2^n\}$

$$\frac{1}{2}, \frac{1}{4}, \frac{1}{8}, \frac{1}{16}, \frac{1}{32}, \, \cdots$$

converges to 0 as indicated in the following definition.

DEFINITION OF THE LIMIT OF A SEQUENCE

The **limit** of a sequence $\{a_n\}$ is L, written as

$$\lim_{n \to \infty} a_n = L$$

if for each $\varepsilon > 0$, there exists $M > 0$ such that $|a_n - L| < \varepsilon$ whenever $n > M$. Sequences that have a (finite) limit are said to **converge,** and sequences that do not have a limit are said to **diverge.**

$y = a_n$

For $n > M$ the terms of the sequence all lie within ε units of L.

Graphically, this definition says that eventually (for $n > M$) the terms of a sequence that converges to L will lie within the band between the lines

$$y = L + \varepsilon \qquad \text{and} \qquad y = L - \varepsilon$$

as illustrated in Figure 10.1.

If a sequence $\{a_n\}$ agrees with a function f at every positive integer, and if $f(x)$ approaches a limit as $x \to \infty$, then the sequence must converge to the same limit. This result is stated formally in the following theorem.

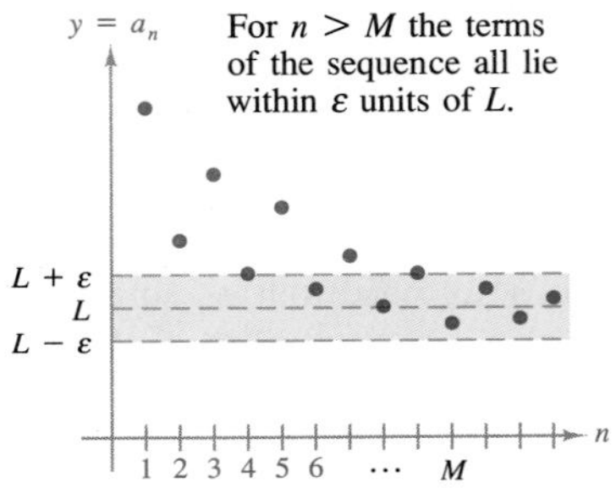

FIGURE 10.1

THEOREM 10.1 **LIMIT OF A SEQUENCE**	Let f be a function of a real variable such that $$\lim_{x \to \infty} f(x) = L.$$ If $\{a_n\}$ is a sequence such that $f(n) = a_n$ for every positive integer n, then $$\lim_{n \to \infty} a_n = L.$$

EXAMPLE 2   Finding the limit of a sequence

Find the limit of the sequence whose nth term is

$$a_n = \left(1 + \frac{1}{n}\right)^n$$

SOLUTION

From Example 5 in Section 7.8, we know that

$$\lim_{x \to \infty} \left(1 + \frac{1}{x}\right)^x = e.$$

Therefore, we can apply Theorem 10.1 to conclude that

$$\lim_{n \to \infty} a_n = \lim_{n \to \infty} \left(1 + \frac{1}{n}\right)^n = e.$$

     The following properties of limits of sequences parallel those given for limits of functions of a real variable in Section 4.5.

THEOREM 10.2 **PROPERTIES OF LIMITS** **OF SEQUENCES**	If $\lim\limits_{n \to \infty} a_n = L$ and $\lim\limits_{n \to \infty} b_n = K$, then the following properties are true. 1. $\lim\limits_{n \to \infty} (a_n \pm b_n) = L \pm K$ 2. $\lim\limits_{n \to \infty} ca_n = cL,$ c is any real number 3. $\lim\limits_{n \to \infty} (a_n b_n) = LK$ 4. $\lim\limits_{n \to \infty} \dfrac{a_n}{b_n} = \dfrac{L}{K},$ $b_n \neq 0$ and $K \neq 0$

EXAMPLE 3   Determining the convergence or divergence of a sequence

Determine the convergence or divergence of the following sequences.

   (a) $\{a_n\} = \{3 + (-1)^n\}$ (b) $\{b_n\} = \left\{\dfrac{n}{1 - 2n}\right\}$

SOLUTION

(a) Since the sequence $\{a_n\} = \{3 + (-1)^n\}$ has terms

$$2, 4, 2, 4, \ldots$$

that oscillate between 2 and 4, the limit does not exist, and we conclude that the sequence diverges.

(b) For $\{b_n\}$, we can divide the numerator and denominator by n to obtain

$$\lim_{n \to \infty} \frac{n}{1 - 2n} = \lim_{n \to \infty} \left[\frac{1}{(1/n) - 2} \right] = -\frac{1}{2}$$

and we conclude that the sequence converges to $-\frac{1}{2}$. ▭

Theorem 10.1 opens up the possibility of using L'Hôpital's Rule to determine the limit of a sequence, as demonstrated in the next example.

EXAMPLE 4 Using L'Hôpital's Rule to determine convergence

Show that the sequence whose nth term is $a_n = n^2/(2^n - 1)$ converges.

SOLUTION

We consider the function of a real variable

$$f(x) = \frac{x^2}{2^x - 1}.$$

Then, applying L'Hôpital's Rule twice, we have

$$\lim_{x \to \infty} \frac{x^2}{2^x - 1} = \lim_{x \to \infty} \frac{2x}{(\ln 2)2^x} = \lim_{x \to \infty} \frac{2}{(\ln 2)^2 2^x} = 0.$$

Since $f(n) = a_n$ for every positive integer, we can apply Theorem 10.1 to conclude that

$$\lim_{n \to \infty} \frac{n^2}{2^n - 1} = 0.$$ ▭

To simplify some of the formulas developed in this chapter, we use the symbol $n!$ (read "n factorial"), as given in the following definition.

DEFINITION OF n FACTORIAL

Let n be a positive integer; then n **factorial** is given by

$$n! = 1 \cdot 2 \cdot 3 \cdot 4 \cdots (n - 1) \cdot n.$$

Zero factorial is given by $0! = 1$.

REMARK   From this definition, we see that $0! = 1$, $1! = 1$, $2! = 1 \cdot 2 = 2$, $3! = 1 \cdot 2 \cdot 3 = 6$, and so on. Factorials follow the same conventions for order of operation as exponents. That is, just as $2x^3$ and $(2x)^3$ imply different orders of operations, $2n!$ and $(2n)!$ imply the following orders:

$$2n! = 2(n!) = 2(1 \cdot 2 \cdot 3 \cdot 4 \cdots n)$$

and

$$(2n)! = 1 \cdot 2 \cdot 3 \cdot 4 \cdots n \cdot (n + 1) \cdots 2n.$$

Another useful limit theorem that can be rewritten for sequences is the Squeeze Theorem of Section 2.5.

THEOREM 10.3 **SQUEEZE THEOREM** **FOR SEQUENCES**	If $$\lim_{n \to \infty} a_n = L = \lim_{n \to \infty} b_n$$ and there exists an integer N such that $a_n \le c_n \le b_n$ for all $n > N$, then $$\lim_{n \to \infty} c_n = L.$$

The usefulness of the Squeeze Theorem is seen in Example 5.

EXAMPLE 5   Using the Squeeze Theorem

Show that the following sequence converges and find its limit.

$$\{c_n\} = \left\{ (-1)^n \frac{1}{n!} \right\}$$

SOLUTION

To apply the Squeeze Theorem, we must find two convergent sequences that can be related to the given factorial sequence. Two possibilities are $a_n = -1/2^n$ and $b_n = 1/2^n$, both of which converge to zero. By comparing the term $n!$ with 2^n, we see that

$$n! = 1 \cdot 2 \cdot 3 \cdot 4 \cdot 5 \cdot 6 \cdots n = 24 \cdot 5 \cdot 6 \cdots n$$

and

$$2^n = 2 \cdot 2 \cdot 2 \cdot 2 \cdot 2 \cdot 2 \cdots 2 = 16 \cdot \underbrace{2 \cdot 2 \cdots 2}_{n - 4 \text{ factors}}.$$

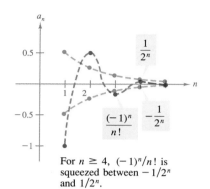

For $n \geq 4$, $(-1)^n/n!$ is squeezed between $-1/2^n$ and $1/2^n$.

FIGURE 10.2

This implies that for $n \geq 4$, $2^n < n!$, and we have

$$\frac{-1}{2^n} \leq (-1)^n\frac{1}{n!} \leq \frac{1}{2^n}$$

as illustrated in Figure 10.2. Therefore, by the Squeeze Theorem it follows that

$$\lim_{n \to \infty} (-1)^n\frac{1}{n!} = 0.$$

Table 10.1 lists the first six terms of each of the three sequences $\{a_n\}$, $\{b_n\}$, and $\{c_n\}$.

TABLE 10.1

n	1	2	3	4	5	6
$a_n = -\dfrac{1}{2^n}$	$-\dfrac{1}{2}$	$-\dfrac{1}{4}$	$-\dfrac{1}{8}$	$-\dfrac{1}{16}$	$-\dfrac{1}{32}$	$-\dfrac{1}{64}$
$b_n = (-1)^n\dfrac{1}{n!}$	$-\dfrac{1}{1}$	$\dfrac{1}{2}$	$-\dfrac{1}{6}$	$\dfrac{1}{24}$	$-\dfrac{1}{120}$	$\dfrac{1}{720}$
$c_n = \dfrac{1}{2^n}$	$\dfrac{1}{2}$	$\dfrac{1}{4}$	$\dfrac{1}{8}$	$\dfrac{1}{16}$	$\dfrac{1}{32}$	$\dfrac{1}{64}$

In Example 5 the sequence $\{c_n\}$ has both positive and negative terms. For this sequence, it happens that the sequence of absolute values, $\{|c_n|\}$, also converges to zero. You can show this by the Squeeze Theorem using the inequality

$$0 \leq \frac{1}{n!} \leq \frac{1}{2^n}, \quad n \geq 4.$$

In such cases it is often convenient to consider the sequence of positive terms, then apply Theorem 10.4, which tells us that if the positive term sequence converges to zero, the original signed sequence also must converge to zero.

THEOREM 10.4
ABSOLUTE VALUE THEOREM

For the sequence $\{a_n\}$, if

$$\lim_{n \to \infty} |a_n| = 0 \quad \text{then} \quad \lim_{n \to \infty} a_n = 0.$$

PROOF Consider the two sequences $\{|a_n|\}$ and $\{-|a_n|\}$. Since both of these sequences converge to 0 and since

$$-|a_n| \leq a_n \leq |a_n|$$

we can apply the Squeeze Theorem and conclude that $\{a_n\}$ converges to 0.

The result in Example 5 suggests something about the rate at which $n!$ increases $n \to \infty$. From Figure 10.2 we can see that both $1/2^n$ and $1/n!$ approach zero as $n \to \infty$. Yet $1/n!$ approaches 0 so much faster than $1/2^n$ does that

$$\lim_{n \to \infty} \frac{1/n!}{1/2^n} = \lim_{n \to \infty} \frac{2^n}{n!} = 0.$$

In fact, it can be shown that for any fixed number k,

$$\lim_{n \to \infty} \frac{k^n}{n!} = 0.$$

This means that *the factorial function grows faster than any exponential function*. We will find this fact to be very useful as we work with limits of sequences.

Pattern recognition for sequences

Sometimes the first several terms of a sequence are listed without the nth term, or the terms may be generated by some rule that does not explicitly identify the nth term of the sequence. In such cases, we are required to discover a *pattern* in the sequence and to describe the nth term. Once the nth term is specified, we can discuss the convergence or divergence of the sequence. This is demonstrated in the next two examples.

EXAMPLE 6   Finding the nth term of a sequence

Find a sequence $\{a_n\}$ whose first five terms are

$$\frac{2}{1}, \frac{4}{3}, \frac{8}{5}, \frac{16}{7}, \frac{32}{9}, \ldots$$

and then determine whether the particular sequence you have chosen converges or diverges.

SOLUTION

First, we note that the numerators are successive powers of 2, and the denominators form the sequence of positive odd integers. Then, by comparing a_n with n, we have the following pattern.

$$1, \quad 2, \quad 3, \quad 4, \quad 5, \quad \ldots$$
$$\downarrow \quad \downarrow \quad \downarrow \quad \downarrow \quad \downarrow$$
$$\frac{2}{1}, \frac{4}{3}, \frac{8}{5}, \frac{16}{7}, \frac{32}{9}, \ldots$$
$$\frac{2^1}{1}, \frac{2^2}{3}, \frac{2^3}{5}, \frac{2^4}{7}, \frac{2^5}{9}, \ldots$$

Using the fact that the odd integer denominators are generated by $2n - 1$, we can conclude that $a_n = 2^n/(2n - 1)$. Now, using L'Hôpital's Rule to evaluate the limit of $f(x) = 2^x/(2x - 1)$, we obtain

$$\lim_{x \to \infty} \frac{2^x}{2x - 1} = \lim_{x \to \infty} \frac{2^x(\ln 2)}{2} = \infty \quad \Longrightarrow \quad \lim_{n \to \infty} \frac{2^n}{2n - 1} = \infty.$$

Hence, the sequence *diverges*.

Without a specific rule for generating the terms of a sequence or some knowledge of the context in which the terms of the sequence are obtained, it is not possible to determine the convergence or divergence of the sequence merely from its first several terms. For instance, although the first three terms of the four sequences given next are identical, the first two sequences converge to 0, the third sequence converges to $\frac{1}{9}$, and the fourth one diverges.

$$\{a_n\}: \frac{1}{2}, \frac{1}{4}, \frac{1}{8}, \frac{1}{16}, \cdots, \frac{1}{2^n}, \cdots$$

$$\{b_n\}: \frac{1}{2}, \frac{1}{4}, \frac{1}{8}, \frac{1}{15}, \cdots, \frac{6}{(n + 1)(n^2 - n + 6)}, \cdots$$

$$\{c_n\}: \frac{1}{2}, \frac{1}{4}, \frac{1}{8}, \frac{7}{62}, \cdots, \frac{n^2 - 3n + 3}{9n^2 - 25n + 18}, \cdots$$

$$\{d_n\}: \frac{1}{2}, \frac{1}{4}, \frac{1}{8}, 0, \cdots, \frac{-n(n + 1)(n - 4)}{6(n^2 + 3n - 2)}, \cdots$$

The process of determining an nth term from the pattern observed in the first several terms of a sequence is an example of *inductive reasoning*.

EXAMPLE 7   Finding the nth term of a sequence

Determine an nth term for a sequence whose first five terms are

$$-\frac{2}{1}, \frac{8}{2}, -\frac{26}{6}, \frac{80}{24}, -\frac{242}{120} \cdots$$

and then decide if your sequence converges or diverges.

SOLUTION

We observe that the numerators are one less than 3^n. Hence, we reason that the numerators are given by the rule $3^n - 1$. If we factor the denominators, we have

$$1 = 1$$
$$2 = 1 \cdot 2$$
$$6 = 1 \cdot 2 \cdot 3$$
$$24 = 1 \cdot 2 \cdot 3 \cdot 4$$
$$120 = 1 \cdot 2 \cdot 3 \cdot 4 \cdot 5 \cdots.$$

This suggests that the denominators are represented by $n!$. Finally, since the signs alternate, we can write the nth term as

$$a_n = (-1)^n \left(\frac{3^n - 1}{n!} \right).$$

From Theorem 10.4 and the discussion about the growth of $n!$, it follows that

$$\lim_{n \to \infty} |a_n| = \lim_{n \to \infty} \frac{3^n - 1}{n!} = 0 = \lim_{n \to \infty} a_n$$

and we conclude that $\{a_n\}$ converges to 0.

Monotonic sequences

So far we have determined the convergence of a sequence by finding its limit. Even if we cannot determine the limit of a particular sequence, it still may be useful to know whether the sequence converges. Theorem 10.5 identifies a test for convergence of sequences without involving the determination of the limit. First, we look at some preliminary definitions.

DEFINITION OF A MONOTONIC SEQUENCE	A sequence $\{a_n\}$ is **monotonic** if its terms are nondecreasing $$a_1 \le a_2 \le a_3 \le \cdots \le a_n \le \cdots$$ or if its terms are nonincreasing $$a_1 \ge a_2 \ge a_3 \ge \cdots \ge a_n \ge \cdots.$$

EXAMPLE 8   Determining if a sequence is monotonic

Determine whether the sequence having the given nth term is monotonic.

(a) $a_n = 3 + (-1)^n$ (b) $b_n = \dfrac{2n}{1 + n}$ (c) $c_n = \dfrac{n^2}{2^n - 1}$

SOLUTION

(a) This sequence alternates between 2 and 4. Therefore, it is not monotonic.
(b) This sequence is monotonic because each successive term is larger than its predecessor. To see this, we compare the terms b_n and b_{n+1} as follows. (Note that because n is positive we can multiply both sides of the inequality by $(1 + n)$ and $(2 + n)$ without reversing the inequality sign.)

$$b_n = \frac{2n}{1 + n} \overset{?}{<} \frac{2(n + 1)}{1 + (n + 1)} = b_{n+1}$$
$$2n(2 + n) \overset{?}{<} (1 + n)(2n + 2)$$
$$4n + 2n^2 \overset{?}{<} 2 + 4n + 2n^2$$
$$0 < 2$$

Since the final inequality is valid, we can reverse the steps to conclude that the original inequality is also valid.

(c) This sequence is not monotonic since the second term is larger than the first term, and smaller than the third. (Note that if we drop the first term, the remaining sequence $c_2, c_3, c_4, \ldots$ is monotonic.) Figure 10.3 graphically illustrates these three sequences. ▭

REMARK   In Example 8(b), another way to see that the sequence is monotonic is to consider the derivative of the corresponding differentiable function $f(x) = 2x/(1 + x)$. Since

$$f'(x) = \frac{2}{(1 + x)^2}$$

is positive for all x, we can conclude that f is increasing. This implies that $\{a_n\}$ is increasing.

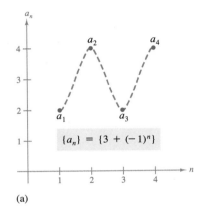

(a)

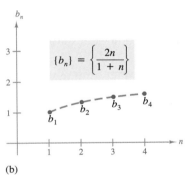

(b)

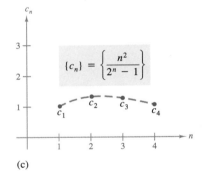

(c)

FIGURE 10.3

DEFINITION OF A BOUNDED SEQUENCE

A sequence $\{a_n\}$ is **bounded** if there is a positive real number M such that $|a_n| \leq M$ for all n. We call M an **upper bound** for the sequence.

In Figure 10.3, all three sequences are bounded. That is,

$$|3 + (-1)^n| \leq 4, \qquad \left|\frac{2n}{1 + n}\right| \leq 2, \qquad \text{and} \qquad \left|\frac{n^2}{2^n - 1}\right| \leq \frac{4}{3}.$$

One important property of the real numbers is that they are **complete.** Informally, this means that there are no holes or gaps on the real number line. (The set of rational numbers does not have the completeness property.) The completeness axiom for real numbers can be used to conclude that if a sequence has an upper bound, then it must have a **least upper bound** (an upper bound that is smaller than all other upper bounds for the sequence). For example, the least upper bound of the sequence $\{a_n\} = \{n/(n + 1)\}$,

$$\frac{1}{2}, \frac{2}{3}, \frac{3}{4}, \frac{4}{5}, \ldots, \frac{n}{n + 1}, \ldots$$

is 1. We use the completeness axiom in the proof of Theorem 10.5.

THEOREM 10.5
BOUNDED MONOTONIC
SEQUENCES

If a sequence $\{a_n\}$ is bounded and monotonic, then it converges.

PROOF

We give the proof for a nondecreasing sequence, as shown in Figure 10.4, and leave the nonincreasing case as an exercise (see Exercise 72). For the sake of simplicity, we assume that each term in the sequence is positive. Then, since the sequence is bounded, there must exist an upper bound M such that

$$a_1 \leq a_2 \leq a_3 \leq \cdots \leq a_n \leq \cdots \leq M.$$

From the completeness axiom, it follows that there is a least upper bound L such that

$$a_1 \leq a_2 \leq a_3 \leq \cdots \leq a_n \leq \cdots \leq L.$$

Now, we will show that $\{a_n\}$ converges to L. For $\varepsilon > 0$, it follows that $L - \varepsilon < L$, and therefore $L - \varepsilon$ cannot be an upper bound for the sequence. Consequently, at least one term of $\{a_n\}$ is greater than $L - \varepsilon$. That is, $L - \varepsilon < a_N$ for some positive integer N. Since the terms of $\{a_n\}$ are nondecreasing we conclude that $a_N \leq a_n$ for $n > N$. We now know that

$$L - \varepsilon < a_N < a_n \leq L < L + \varepsilon, \quad \text{for every } n > N.$$

It follows that $|a_n - L| < \varepsilon$ for $n > N$, which by definition means that $\{a_n\}$ converges to L.

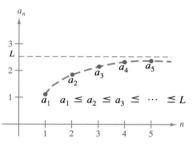

Every bounded nondecreasing
sequence converges.

FIGURE 10.4

EXAMPLE 9   Bounded and monotonic sequences

(a) The sequence

$$\{a_n\} = \left\{\frac{1}{n}\right\}$$

is both bounded and monotonic and thus, by Theorem 10.5, must converge.

(b) The divergent sequence

$$\{b_n\} = \left\{\frac{n^2}{n + 1}\right\}$$

is monotonic, but not bounded.

(c) The divergent sequence

$$\{c_n\} = \{(-1)^n\}$$

is bounded, but not monotonic.

EXERCISES for Section 10.1

In Exercises 1–8, write out the first five terms of the sequence with the given nth term.

1. $a_n = 2^n$

2. $a_n = \dfrac{n}{n+1}$

3. $a_n = \left(-\dfrac{1}{2}\right)^n$

4. $a_n = \sin\dfrac{n\pi}{2}$

5. $a_n = \dfrac{3^n}{n!}$

6. $a_n = 5 - \dfrac{1}{n} + \dfrac{1}{n^2}$

7. $a_n = \dfrac{(-1)^{n(n+1)/2}}{n^2}$

8. $a_n = \dfrac{3n!}{(n-1)!}$

In Exercises 9–24, write an expression for the nth term of the sequence.

9. $1, 4, 7, 10, \ldots$

10. $3, 7, 11, 15, \ldots$

11. $-1, 2, 7, 14, 23, \ldots$

12. $1, \dfrac{1}{4}, \dfrac{1}{9}, \dfrac{1}{16}, \ldots$

13. $\dfrac{2}{3}, \dfrac{3}{4}, \dfrac{4}{5}, \dfrac{5}{6}, \ldots$

14. $2, \dfrac{3}{3}, \dfrac{4}{5}, \dfrac{5}{7}, \dfrac{6}{9}, \ldots$

15. $2, -1, \dfrac{1}{2}, -\dfrac{1}{4}, \dfrac{1}{8}, \ldots$

16. $\dfrac{1}{2}, \dfrac{1}{3}, \dfrac{2}{9}, \dfrac{4}{27}, \dfrac{8}{81}, \ldots$

17. $2, 1+\dfrac{1}{2}, 1+\dfrac{1}{3}, 1+\dfrac{1}{4}, 1+\dfrac{1}{5}, \ldots$

18. $1+\dfrac{1}{2}, 1+\dfrac{3}{4}, 1+\dfrac{7}{8}, 1+\dfrac{15}{16}, 1+\dfrac{31}{32}, \ldots$

19. $\dfrac{1}{2 \cdot 3}, \dfrac{2}{3 \cdot 4}, \dfrac{3}{4 \cdot 5}, \dfrac{4}{5 \cdot 6}, \ldots$

20. $1, \dfrac{1}{2}, \dfrac{1}{6}, \dfrac{1}{24}, \dfrac{1}{120}, \ldots$

21. $1, \dfrac{1}{1 \cdot 3}, \dfrac{1}{1 \cdot 3 \cdot 5}, \dfrac{1}{1 \cdot 3 \cdot 5 \cdot 7}, \ldots$

22. $2, -4, 6, -8, 10, \ldots$

23. $1, -1, -1, 1, 1, -1, -1, \ldots$

24. $1, x, \dfrac{x^2}{2}, \dfrac{x^3}{6}, \dfrac{x^4}{24}, \dfrac{x^5}{120}, \ldots$

In Exercises 25–46, determine the convergence or divergence of the sequence with the given nth term. If the sequence converges, find its limit.

25. $a_n = \dfrac{n+1}{n}$

26. $a_n = \dfrac{1}{n^{3/2}}$

27. $a_n = (-1)^n\left(\dfrac{n}{n+1}\right)$

28. $a_n = \dfrac{n-1}{n} - \dfrac{n}{n-1}$
$n \geq 2$

29. $a_n = \dfrac{3n^2 - n + 4}{2n^2 + 1}$

30. $a_n = \dfrac{\sqrt{n}}{\sqrt{n}+1}$

31. $a_n = \dfrac{n^2 - 1}{n+1}$

32. $a_n = 1 + (-1)^n$

33. $a_n = \dfrac{1 + (-1)^n}{n}$

34. $a_n = \dfrac{\ln(n^2)}{n}$

35. $a_n = \cos\dfrac{n\pi}{2}$

36. $a_n = \dfrac{n}{\sqrt{n^2 + 1}}$

37. $a_n = \dfrac{3^n}{4^n}$

38. $a_n = \dfrac{(n-2)!}{n!}$

39. $a_n = f^{(n-1)}(2), f(x) = \ln x$

40. $a_n = \dfrac{n^2}{2n+1} - \dfrac{n^2}{2n-1}$

41. $a_n = 3 - \dfrac{1}{2^n}$

42. $a_n = \dfrac{n!}{n^n}$

43. $a_n = \left(1 + \dfrac{k}{n}\right)^n$

44. $a_n = 2^{1/n}$

45. $a_n = \dfrac{n^p}{e^n}(p > 0)$

46. $a_n = n\sin\dfrac{1}{n}$

In Exercises 47–56, determine if the sequence with the given nth term is monotonic.

47. $a_n = 4 - \dfrac{1}{n}$

48. $a_n = \dfrac{4n}{n+1}$

49. $a_n = \dfrac{\cos n}{n}$

50. $a_n = ne^{-n/2}$

51. $a_n = (-1)^n\left(\dfrac{1}{n}\right)$

52. $a_n = \left(-\dfrac{2}{3}\right)^n$

53. $a_n = \left(\dfrac{2}{3}\right)^n$

54. $a_n = \left(\dfrac{3}{2}\right)^n$

55. $a_n = \sin\dfrac{n\pi}{6}$

56. $a_n = \dfrac{n}{2^{n+2}}$

In Exercises 57–60, use Theorem 10.5 to show that the sequence with the given nth term converges, and find the limit.

57. $a_n = 5 + \dfrac{1}{n}$

58. $a_n = 3 - \dfrac{4}{n}$

59. $a_n = \dfrac{1}{3}\left(1 - \dfrac{1}{3^n}\right)$

60. $a_n = 4 + \dfrac{1}{2^n}$

61. Consider the sequence $\{A_n\}$, whose nth term is given by

$$A_n = P\left(1 + \dfrac{r}{12}\right)^n$$

where P is the principal, A_n is the amount at compound interest after n months, and r is the annual percentage rate.
(a) Is $\{A_n\}$ a convergent sequence?
(b) Find the first ten terms of the sequence if $P = \$9,000$ and $r = 0.115$.

62. A deposit of $100 is made each month in an account that earns 12-percent interest compounded monthly. The balance in the account after n months is given by

$$A_n = 100(101)[(1.01)^n - 1].$$

(a) Compute the first 6 terms of the sequence $\{A_n\}$.

(b) Find the balance after 5 years by computing the 60th term of the sequence.

(c) Find the balance after 20 years by computing the 240th term of the sequence.

63. A government program that currently costs taxpayers $2.5 billion per year is to be cut back by 20 percent per year.

(a) Write an expression for the amount budgeted for this program after n years.

(b) Compute the budgets for the first 4 years.

(c) Determine the convergence or divergence of the sequence of reduced budgets. If the sequence converges, find its limit.

64. If the average price of a new car increases $5\frac{1}{2}$ percent per year and the average price is currently $11,000, then the average price after n years is

$$P_n = \$11,000(1.055)^n.$$

Compute the average price for the first 5 years of increases.

65. Consider an idealized population with the characteristic that each member of the population produces 1 offspring at the end of every time period. If each member has a life span of 3 time periods and the population begins with 10 newborn members, then the following table gives the population during the first 5 time periods.

Age bracket	Time period 1	2	3	4	5
0–1	10	10	20	40	70
1–2		10	10	20	40
2–3			10	10	20
Total	10	20	40	70	130

The sequence for the total population has the property that

$$S_n = S_{n-1} + S_{n-2} + S_{n-3}, \quad n > 3.$$

Find the total population during the next 5 time periods.

66. Consider the sequence $\{x_n\}$ defined by

$$x_n = x_{n-1} - \frac{f(x_{n-1})}{f'(x_{n-1})}.$$

If $f(x) = x^2 + x - 1$ and $x_1 = 0.5$, find the next three terms of the sequence.

67. Compute the first 6 terms of the sequence $\{a_n\} = \{\sqrt[n]{n}\}$. If the sequence converges, find its limit.

68. Show that $\{r^n/(1 - r)\}$ diverges if $|r| \geq 1$, and converges to 0 if $|r| < 1$.

69. Prove that if $\{s_n\}$ converges to L and $L > 0$, then there exists a number N such that $s_n > 0$ for $n > N$.

70. If $\{s_n\}$ is a convergent sequence, show that

$$\lim_{n \to \infty} s_{n-1} = \lim_{n \to \infty} s_n.$$

71. In the study of the progeny of rabbits, **Fibonacci** (ca. 1175–ca. 1250) encountered the now famous sequence bearing his name. It is defined recursively by

$$a_{n+2} = a_n + a_{n+1}, \quad \text{where } a_1 = 1 \text{ and } a_2 = 1.$$

(a) Write out the first 12 terms of the sequence.

(b) Write out the first 10 terms of the sequence defined by

$$b_n = \frac{a_{n+1}}{a_n}, \quad \text{for } n \geq 1.$$

(c) Using the definition of part (b) show that

$$b_n = 1 + \frac{1}{b_{n-1}}.$$

(d) If

$$\lim_{n \to \infty} b_n = \rho$$

use the results of part (b) and Exercise 70 to show that

$$\rho = 1 + \frac{1}{\rho}.$$

Solve this equation for ρ. (ρ is called the *golden ratio*.)

72. Complete the proof of Theorem 10.5.

10.2 Series and Convergence

Infinite series ▪ Convergent and divergent series ▪ Telescoping series ▪ Geometric series ▪ *n*th-Term Test for Divergence

One important application of infinite sequences is in representing infinite summations. Informally, if $\{a_n\}$ is an infinite sequence, then

$$\sum_{n=1}^{\infty} a_n = a_1 + a_2 + a_3 + \cdots + a_n + \cdots$$

is called an **infinite series** (or simply a **series**). The numbers $a_1, a_2, a_3, \ldots$ are called the **terms** of the series. For some series it is convenient to begin the index at $n = 0$. As a typesetting convention, it is common to represent an infinite series as simply $\Sigma \, a_n$. In such cases the starting point for the index ($n = 0$ or $n = 1$) must be taken from the context of the statement.

To find the sum of an infinite series, we consider the following **sequence of partial sums.**

$$S_1 = a_1$$
$$S_2 = a_1 + a_2$$
$$S_3 = a_1 + a_2 + a_3$$
$$\vdots$$
$$S_n = a_1 + a_2 + a_3 + \cdots + a_n$$

If this sequence converges, then we say that the series also converges and has the sum indicated in the following definition.

DEFINITION OF CONVERGENT AND DIVERGENT SERIES

For the infinite series $\Sigma \, a_n$, the **nth partial sum** is given by

$$S_n = a_1 + a_2 + \cdots + a_n.$$

If the sequence of partial sums $\{S_n\}$ converges to S, then we say that the series $\Sigma \, a_n$ **converges.** We call S the **sum of the series** and write

$$S = a_1 + a_2 + \cdots + a_n + \cdots.$$

If $\{S_n\}$ diverges, then we say that the series also **diverges.**

As you study this chapter, you will see that there are two basic questions involving infinite series. Does a series converge or does it diverge? If a series converges, what is its sum? These questions are not always easy to answer, especially the second one. We begin with some simple examples.

EXAMPLE 1 Convergent and divergent series

(a) The series

$$\sum_{n=1}^{\infty} \frac{1}{2^n} = \frac{1}{2} + \frac{1}{4} + \frac{1}{8} + \frac{1}{16} + \cdots$$

has the following partial sums.

$$S_1 = \frac{1}{2}$$

$$S_2 = \frac{1}{2} + \frac{1}{4} = \frac{3}{4}$$

$$S_3 = \frac{1}{2} + \frac{1}{4} + \frac{1}{8} = \frac{7}{8}$$

$$\vdots$$

$$S_n = \frac{1}{2} + \frac{1}{4} + \frac{1}{8} + \cdots + \frac{1}{2^n} = \frac{2^n - 1}{2^n}$$

Since

$$\lim_{n \to \infty} \frac{2^n - 1}{2^n} = 1$$

we conclude that the series converges and its sum is 1.

(b) The nth partial sum of the series

$$\sum_{n=1}^{\infty} \left(\frac{1}{n} - \frac{1}{n+1} \right) = \left(1 - \frac{1}{2} \right) + \left(\frac{1}{2} - \frac{1}{3} \right) + \left(\frac{1}{3} - \frac{1}{4} \right) + \cdots$$

is given by

$$S_n = 1 - \frac{1}{n+1}$$

and since the limit of S_n is 1, the series converges and its sum is 1.

(c) The series

$$\sum_{n=1}^{\infty} 1 = 1 + 1 + 1 + 1 + \cdots$$

diverges because $S_n = n$ and the sequence of partial sums diverges.

The series in part (b) of Example 1 is called a **telescoping series.** That is, it is of the form

$$(b_1 - b_2) + (b_2 - b_3) + (b_3 - b_4) + (b_4 - b_5) + \cdots .$$

Note that b_2 is cancelled by the second term, b_3 is cancelled by the third term, and so on. Since the nth partial sum of this series is $S_n = b_1 - b_n$, it follows that a telescoping series will converge if and only if b_n approaches a finite number as $n \to \infty$. Moreover, if the series converges, then its sum is

$$S = b_1 - \lim_{n \to \infty} b_n.$$

In Example 2, we find the limit of a series by writing the series in *telescoping form*.

EXAMPLE 2   Writing a series in telescoping form

Find the sum of the following series.

$$\sum_{n=1}^{\infty} \frac{2}{4n^2 - 1}$$

SOLUTION

Using partial fractions, we find that

$$a_n = \frac{2}{4n^2 - 1} = \frac{2}{(2n - 1)(2n + 1)} = \frac{1}{2n - 1} - \frac{1}{2n + 1}.$$

From this telescoping form, we see that the nth partial sum is

$$S_n = \left(\frac{1}{1} - \frac{1}{3}\right) + \left(\frac{1}{3} - \frac{1}{5}\right) + \cdots + \left(\frac{1}{2n - 1} - \frac{1}{2n + 1}\right)$$

$$= 1 - \frac{1}{2n + 1}.$$

Thus, the series converges and its sum is 1. That is,

$$\sum_{n=1}^{\infty} \frac{2}{4n^2 - 1} = \lim_{n \to \infty} S_n = \lim_{n \to \infty} \left(1 - \frac{1}{2n + 1}\right) = 1. \qquad \blacksquare$$

Geometric series

The series given in part (a) of Example 1 is called a **geometric series.** This important type of series is defined as follows.

DEFINITION OF GEOMETRIC SERIES	The series given by $$\sum_{n=0}^{\infty} ar^n = a + ar + ar^2 + \cdots + ar^n + \cdots, \quad a \neq 0$$ is called a **geometric series** with ratio r.

The conditions for the convergence or divergence of a geometric series are given in the following theorem.

THEOREM 10.6 CONVERGENCE OF A GEOMETRIC SERIES	A geometric series with ratio r diverges if $|r| \geq 1$. If $0 < |r| < 1$, then the series converges to the sum $$\sum_{n=0}^{\infty} ar^n = \frac{a}{1 - r}, \quad 0 < |r| < 1.$$

PROOF   It is easy to see that the series diverges if $r = \pm 1$. If $r \neq \pm 1$, then

$$S_n = a + ar + ar^2 + \cdots + ar^{n-1}.$$

Multiplication by r yields

$$rS_n = ar + ar^2 + ar^3 + \cdots + ar^n.$$

By subtracting the second equation from the first, we obtain $S_n - rS_n = a - ar^n$. Therefore, $S_n(1 - r) = a(1 - r^n)$, and the nth partial sum is

$$S_n = \frac{a}{1 - r}(1 - r^n).$$

Now, since $0 < |r| < 1$, it follows that $r^n \to 0$ as $n \to \infty$, and we obtain

$$\lim_{n\to\infty} S_n = \lim_{n\to\infty}\left[\frac{a}{1-r}(1-r^n)\right] = \frac{a}{1-r}[\lim_{n\to\infty}(1-r^n)] = \frac{a}{1-r}$$

which means the series *converges* and its sum is $a/(1 - r)$. We leave it to you to show that the series diverges if $|r| > 1$.

EXAMPLE 3   A convergent geometric series

The geometric series

$$\sum_{n=0}^{\infty} \frac{3}{2^n} = \sum_{n=0}^{\infty} 3\left(\frac{1}{2}\right)^n = 3(1) + 3\left(\frac{1}{2}\right) + 3\left(\frac{1}{2}\right)^2 + \cdots$$

has a ratio of $r = \frac{1}{2}$ with $a = 3$. Since $0 < |r| < 1$, the series converges and its sum is

$$S = \frac{a}{1 - r} = \frac{3}{1 - (1/2)} = 6.$$

▫

EXAMPLE 4   A divergent geometric series

The geometric series

$$\sum_{n=0}^{\infty} \left(\frac{3}{2}\right)^n = 1 + \frac{3}{2} + \frac{9}{4} + \frac{27}{8} + \cdots$$

has a ratio of $r = \frac{3}{2}$. Since $|r| \geq 1$, the series diverges.

▫

The formula for the sum of a geometric series can be used to write a repeating decimal as the ratio of two integers, as demonstrated in the next example.

EXAMPLE 5   A geometric series for a repeating decimal

Use a geometric series to express $0.080\overline{08}$ as the ratio of two integers.

SOLUTION

For

$$0.080808 \ldots = \frac{8}{10^2} + \frac{8}{10^4} + \frac{8}{10^6} + \frac{8}{10^8} + \cdots = \sum_{n=0}^{\infty} \left(\frac{8}{10^2}\right)\left(\frac{1}{10^2}\right)^n$$

we have $a = 8/10^2$ and $r = 1/10^2$. Thus,

$$0.080808 \ldots = \frac{a}{1 - r} = \frac{8/10^2}{1 - (1/10^2)} = \frac{8}{99}.$$

Try dividing 8 by 99 on a calculator to see that it produces $0.080808 \ldots$.

The convergence of a series is not affected by removing a finite number of terms from the beginning of the series. For instance, the geometric series

$$\sum_{n=4}^{\infty} \left(\frac{1}{2}\right)^n \quad \text{and} \quad \sum_{n=0}^{\infty} \left(\frac{1}{2}\right)^n$$

both converge. Furthermore, because the sum of the second series is $a/(1 - r) = 2$, we can conclude that the sum of the first series is

$$S = 2 - \left[\left(\frac{1}{2}\right)^0 + \left(\frac{1}{2}\right)^1 + \left(\frac{1}{2}\right)^2 + \left(\frac{1}{2}\right)^3\right]$$

$$= 2 - \frac{15}{8} = \frac{1}{8}.$$

Properties of series

The following properties are direct consequences of the corresponding properties of limits of sequences.

THEOREM 10.7 **PROPERTIES OF INFINITE SERIES**	If $\Sigma\, a_n = A$, $\Sigma\, b_n = B$, and c is a real number, then the following series converge to the indicated sums. 1. $\displaystyle\sum_{n=1}^{\infty} ca_n = cA$ 2. $\displaystyle\sum_{n=1}^{\infty} (a_n + b_n) = A + B$ 3. $\displaystyle\sum_{n=1}^{\infty} (a_n - b_n) = A - B$

The following theorem tells us that if a series converges, then the limit of its nth term must be zero.

THEOREM 10.8 **LIMIT OF nTH TERM OF A** **CONVERGENT SERIES**	If the series $\Sigma\, a_n$ converges, then the sequence $\{a_n\}$ converges to 0.

PROOF   Assume that

$$\sum_{n=1}^{\infty} a_n = \lim_{n \to \infty} S_n = L.$$

Then, because $S_n = S_{n-1} + a_n$ and

$$\lim_{n \to \infty} S_n = \lim_{n \to \infty} S_{n-1} = L$$

it follows that

$$L = \lim_{n \to \infty} S_n = \lim_{n \to \infty} (S_{n-1} + a_n) = \lim_{n \to \infty} S_{n-1} + \lim_{n \to \infty} a_n = L + \lim_{n \to \infty} a_n$$

which implies that $\{a_n\}$ converges to 0.

The contrapositive of Theorem 10.8 provides a useful test for *divergence*. This ***n*th-Term Test for Divergence** tells us that if the limit of the *n*th term of a series does *not* converge to 0, then the series must diverge.

THEOREM 10.9 ***n*TH-TERM TEST FOR DIVERGENCE**	If the sequence $\{a_n\}$ does not converge to 0, then the series $\Sigma\, a_n$ diverges.

REMARK   Be sure you see that this theorem does *not* state that the series $\Sigma\, a_n$ converges if $\{a_n\}$ converges to 0.

EXAMPLE 6   Using the *n*th-Term Test for Divergence

(a) For the series

$$\sum_{n=0}^{\infty} 2^n$$

we have

$$\lim_{n \to \infty} 2^n = \infty.$$

Thus, the limit of the *n*th term is not zero, and we conclude that the series *diverges*.

(b) For the series

$$\sum_{n=1}^{\infty} \frac{n!}{2n! + 1}$$

we have

$$\lim_{n \to \infty} \frac{n!}{2n! + 1} = \frac{1}{2}.$$

Thus, the limit of the *n*th term is not zero, and we conclude that the series *diverges*.

(c) For the series

$$\sum_{n=1}^{\infty} \frac{1}{n}$$

we have

$$\lim_{n \to \infty} \frac{1}{n} = 0.$$

Because the limit of the nth term is zero, the nth-Term Test for Divergence does *not* apply and we draw no conclusions about convergence or divergence. (In the next section, we will see that this particular series diverges.)

Applications

EXAMPLE 7   An application of geometric series

A ball is dropped from a height of 6 feet and begins bouncing as shown in Figure 10.5. The height of each bounce is three-fourths the height of the previous bounce. Find the total vertical distance traveled by the ball.

SOLUTION

When the ball hits the ground for the first time, it has traveled a distance of $D_1 = 6$. For subsequent bounces, we let D_n be the distance traveled up *and* down. For example, D_2 and D_3 are as follows.

$$D_2 = \underbrace{6\left(\frac{3}{4}\right)}_{\text{up}} + \underbrace{6\left(\frac{3}{4}\right)}_{\text{down}} = 12\left(\frac{3}{4}\right)$$

$$D_3 = \underbrace{6\left(\frac{3}{4}\right)\left(\frac{3}{4}\right)}_{\text{up}} + \underbrace{6\left(\frac{3}{4}\right)\left(\frac{3}{4}\right)}_{\text{down}} = 12\left(\frac{3}{4}\right)^2$$

By continuing this process, we have a total vertical distance of

$$D = 6 + 12\left(\frac{3}{4}\right) + 12\left(\frac{3}{4}\right)^2 + 12\left(\frac{3}{4}\right)^3 + \cdots.$$

To find the sum of this series, we rewrite the first term in geometric form as $6 = -6 + 12 = -6 + 12\left(\frac{3}{4}\right)^0$. Then we obtain

$$D = -6 + 12\left(\frac{3}{4}\right)^0 + 12\left(\frac{3}{4}\right)^1 + 12\left(\frac{3}{4}\right)^2 + 12\left(\frac{3}{4}\right)^3 + \cdots$$

$$= -6 + \sum_{n=0}^{\infty} 12\left(\frac{3}{4}\right)^n = -6 + \frac{12}{1 - (3/4)} = -6 + 48 = 42 \text{ ft.}$$

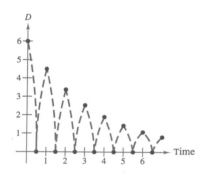

FIGURE 10.5

EXERCISES for Section 10.2

In Exercises 1–6, find the first five terms of the sequence of partial sums.

1. $1 + \dfrac{1}{4} + \dfrac{1}{9} + \dfrac{1}{16} + \dfrac{1}{25} + \cdots$

2. $\dfrac{1}{2 \cdot 3} + \dfrac{2}{3 \cdot 4} + \dfrac{3}{4 \cdot 5} + \dfrac{4}{5 \cdot 6} + \dfrac{5}{6 \cdot 7} + \cdots$

3. $3 - \dfrac{9}{2} + \dfrac{27}{4} - \dfrac{81}{8} + \dfrac{243}{16} - \cdots$

4. $\dfrac{1}{1} + \dfrac{1}{3} + \dfrac{1}{5} + \dfrac{1}{7} + \dfrac{1}{9} + \dfrac{1}{11} + \cdots$

5. $\displaystyle\sum_{n=1}^{\infty} \dfrac{3}{2^{n-1}}$

6. $\displaystyle\sum_{n=1}^{\infty} \dfrac{(-1)^{n+1}}{n!}$

In Exercises 7–18, verify that the infinite series diverges.

7. $\dfrac{1}{2} + \dfrac{2}{3} + \dfrac{3}{4} + \dfrac{4}{5} + \cdots$

8. $3 - \dfrac{9}{2} + \dfrac{27}{4} - \dfrac{81}{8} + \dfrac{243}{16} - \cdots$

9. $\displaystyle\sum_{n=1}^{\infty} \dfrac{3n}{n + 1} = \dfrac{3}{2} + \dfrac{6}{3} + \dfrac{9}{4} + \dfrac{12}{5} + \cdots$

10. $\displaystyle\sum_{n=1}^{\infty} \dfrac{n}{2n + 3} = \dfrac{1}{5} + \dfrac{2}{7} + \dfrac{3}{9} + \dfrac{4}{11} + \cdots$

11. $\displaystyle\sum_{n=1}^{\infty} \dfrac{n^2}{n^2 + 1}$

12. $\displaystyle\sum_{n=1}^{\infty} \dfrac{n}{\sqrt{n^2 + 1}}$

13. $\displaystyle\sum_{n=0}^{\infty} 3\left(\dfrac{3}{2}\right)^n$

14. $\displaystyle\sum_{n=0}^{\infty} \left(\dfrac{4}{3}\right)^n$

15. $\displaystyle\sum_{n=0}^{\infty} 1000(1.055)^n$

16. $\displaystyle\sum_{n=0}^{\infty} 2(-1.03)^n$

17. $\displaystyle\sum_{n=1}^{\infty} \dfrac{2^n + 1}{2^{n+1}}$

18. $\displaystyle\sum_{n=1}^{\infty} \dfrac{n!}{2^n}$

In Exercises 19–24, verify that the infinite series converges.

19. $2 + \dfrac{3}{2} + \dfrac{9}{8} + \dfrac{27}{32} + \dfrac{81}{128} + \cdots$

20. $2 - 1 + \dfrac{1}{2} - \dfrac{1}{4} + \dfrac{1}{8} - \cdots$

21. $\displaystyle\sum_{n=0}^{\infty} (0.9)^n = 1 + 0.9 + 0.81 + 0.729 + \cdots$

22. $\displaystyle\sum_{n=0}^{\infty} (-0.6)^n = 1 - 0.6 + 0.36 - 0.216 + \cdots$

23. $\displaystyle\sum_{n=1}^{\infty} \dfrac{1}{n(n + 1)}$ (Use partial fractions.)

24. $\displaystyle\sum_{n=1}^{\infty} \dfrac{1}{n(n + 2)}$ (Use partial fractions.)

In Exercises 25–40, find the sum of the convergent series.

25. $\displaystyle\sum_{n=0}^{\infty} \left(\dfrac{1}{2}\right)^n$

26. $\displaystyle\sum_{n=0}^{\infty} 2\left(\dfrac{2}{3}\right)^n$

27. $\displaystyle\sum_{n=0}^{\infty} \left(\dfrac{-1}{2}\right)^n$

28. $\displaystyle\sum_{n=0}^{\infty} 2\left(\dfrac{-2}{3}\right)^n$

29. $1 + 0.1 + 0.01 + 0.001 + \cdots$

30. $8 + 6 + \dfrac{9}{2} + \dfrac{27}{8} + \cdots$

31. $3 - 1 + \dfrac{1}{3} - \dfrac{1}{9} + \cdots$

32. $4 - 2 + 1 - \dfrac{1}{2} + \cdots$

33. $\displaystyle\sum_{n=4}^{\infty} 3\left(\dfrac{5}{8}\right)^n$

34. $\displaystyle\sum_{n=5}^{\infty} 2\left(\dfrac{-3}{4}\right)^n$

35. $\displaystyle\sum_{n=2}^{\infty} \dfrac{1}{n^2 - 1}$

36. $\displaystyle\sum_{n=1}^{\infty} \dfrac{1}{n(n + 1)}$

37. $\displaystyle\sum_{n=1}^{\infty} \dfrac{4}{n(n + 2)}$

38. $\displaystyle\sum_{n=1}^{\infty} \dfrac{1}{(2n + 1)(2n + 3)}$

39. $\displaystyle\sum_{n=0}^{\infty} \left(\dfrac{1}{2^n} - \dfrac{1}{3^n}\right)$

40. $\displaystyle\sum_{n=1}^{\infty} [(0.7)^n + (0.9)^n]$

In Exercises 41–44, express the repeated decimal as a geometric series, and write its sum as the ratio of two integers.

41. $0.6\overline{66}$

42. $0.23\overline{23}$

43. $0.075\overline{75}$

44. $0.215\overline{15}$

In Exercises 45–56, determine the convergence or divergence of the series.

45. $\displaystyle\sum_{n=1}^{\infty} \dfrac{n + 10}{10n + 1}$

46. $\displaystyle\sum_{n=0}^{\infty} \dfrac{4}{2^n}$

47. $\displaystyle\sum_{n=1}^{\infty} \left(\dfrac{1}{n} - \dfrac{1}{n + 2}\right)$

48. $\displaystyle\sum_{n=1}^{\infty} \dfrac{n + 1}{2n - 1}$

49. $\displaystyle\sum_{n=1}^{\infty} \dfrac{3n - 1}{2n + 1}$

50. $\displaystyle\sum_{n=0}^{\infty} \dfrac{1}{4^n}$

51. $\displaystyle\sum_{n=0}^{\infty} (1.075)^n$

52. $\displaystyle\sum_{n=1}^{\infty} \dfrac{2^n}{100}$

53. $\displaystyle\sum_{n=2}^{\infty} \dfrac{n}{\ln n}$

54. $\displaystyle\sum_{n=1}^{\infty} \dfrac{2^n}{n^2}$

55. $\displaystyle\sum_{n=1}^{\infty} \left(1 + \dfrac{k}{n}\right)^n$

56. $\displaystyle\sum_{n=1}^{\infty} \dfrac{1}{n(n + 3)}$

57. A company producing a new product estimates the annual sales to be 8000 units. Suppose that in any given year 10 percent of the units (regardless of age) will become inoperative. How many units will be in use after n years?

58. Repeat Exercise 57 with the assumption that 25 percent of the units will become inoperative each year.

59. A ball is dropped from a height of 16 feet. Each time it drops h feet, it rebounds vertically $0.81h$ feet. Find the total distance traveled by the ball.

60. Find the total time it takes for the ball in Exercise 59 to come to rest.

61. Find the fraction of the total area of the square that is eventually shaded if the pattern of shading shown in the figure is continued. (Note that each side of the shaded corner squares is one-fourth that of the square in which it is placed.)

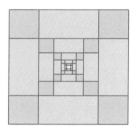

FIGURE FOR 61

In Exercises 62–66, use the formula for the nth partial sum of a geometric series:

$$\sum_{i=0}^{n-1} ar^i = \frac{a(1 - r^n)}{1 - r}.$$

62. A deposit of \$100 is made each month for 5 years in an account that pays 10-percent interest compounded monthly. What is the balance A in the account at the end of the 5 years?

$$A = 100\left(1 + \frac{0.10}{12}\right) + \cdots + 100\left(1 + \frac{0.10}{12}\right)^{60}$$

63. A deposit of \$50 is made each month in an account that pays 12-percent interest compounded monthly. What is the balance in the account at the end of 10 years?

64. A deposit of P dollars is made each month for t years in an account that pays interest at an annual rate of r, compounded monthly. Let $N = 12t$ be the total number of deposits, and show that the balance in the account after t years is

$$A = P\left[\left(1 + \frac{r}{12}\right)^N - 1\right]\left(1 + \frac{12}{r}\right).$$

65. Use the formula in Exercise 64 to find the amount in an account earning 9-percent interest compounded monthly after deposits of \$50 have been made monthly for 40 years.

66. The number of direct ancestors a person has had is given by

$$2 + 2^2 + 2^3 + 2^4 + \cdots + 2^n + \cdots.$$

This formula is valid *provided* the person has had no common ancestors. [A common ancestor is one to whom you are related in more than one way. For example, one of your great-grandmothers on your father's side might also be one of your great-grandmothers on your mother's side (see figure). How many of your direct ancestors have lived since the year 1 A.D.? Assume that the average time between generations was 30 years (resulting in 66 generations) so that the total is given by

$$2 + 2^2 + 2^3 + 2^4 + \cdots + 2^{66}.$$

Considering your total, is it reasonable to assume that you have had no common ancestors in the past 2000 years?

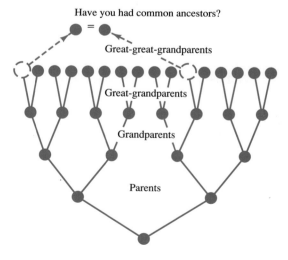

FIGURE FOR 66

67. Prove that $0.75 = 0.749999\ldots$.

68. Prove that every decimal with a repeating pattern of digits is a rational number.

69. Show that the series

$$\sum_{n=1}^{\infty} a_n$$

can be written in the telescoping form

$$\sum_{n=1}^{\infty} [(c - S_{n-1}) - (c - S_n)]$$

where $S_0 = 0$ and S_n is the nth partial sum.

70. Let $\Sigma\, a_n$ be a convergent series, and let

$$R_N = a_{N+1} + a_{N+2} + \cdots$$

be the remainder of the series after the first N terms. Prove that

$$\lim_{N \to \infty} R_N = 0.$$

71. Find two divergent series $\Sigma\, a_n$ and $\Sigma\, b_n$ such that $\Sigma\, (a_n + b_n)$ converges.

72. Given two infinite series $\Sigma\, a_n$ and $\Sigma\, b_n$ such that $\Sigma\, a_n$ converges and $\Sigma\, b_n$ diverges, prove that $\Sigma\, (a_n + b_n)$ diverges.

10.3   The Integral Test and *p*-Series

The Integral Test ▪ *p*-Series ▪ Harmonic series

In this and the following section, we look at several convergence tests that apply to series with *positive* terms.

Since a definite integral is defined as the limit of a sum, it seems logical to expect that we might be able to use integrals to test for convergence or divergence of an infinite series. This is indeed the case, and we show how this is done in the proof of the following theorem.

THEOREM 10.10
INTEGRAL TEST

If f is positive, continuous, and decreasing for $x \geq 1$ and $a_n = f(n)$, then

$$\sum_{n=1}^{\infty} a_n \quad \text{and} \quad \int_1^{\infty} f(x)\, dx$$

either both converge or both diverge.

PROOF   We begin by partitioning the interval $[1, n]$ into $n - 1$ unit intervals, as shown in Figure 10.6. The total area of the inscribed rectangles is given by

$$\sum_{i=2}^{n} f(i) = f(2) + f(3) + \cdots + f(n). \qquad \text{Inscribed area}$$

Similarly, the total area of the circumscribed rectangles is

$$\sum_{i=1}^{n-1} f(i) = f(1) + f(2) + \cdots + f(n-1). \qquad \text{Circumscribed area}$$

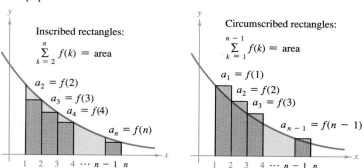

FIGURE 10.6

The exact area under the graph of f from $x = 1$ to $x = n$, $\int_1^n f(x)\, dx$, lies between the inscribed and circumscribed areas, and we have

$$\sum_{i=2}^{n} f(i) \leq \int_1^n f(x)\, dx \leq \sum_{i=1}^{n-1} f(i).$$

Using the nth partial sum, $S_n = f(1) + f(2) + \cdots + f(n)$, we can write this inequality as

$$S_n - f(1) \leq \int_1^n f(x)\, dx \leq S_{n-1}.$$

Now, assuming that $\int_1^\infty f(x)\, dx$ converges to L, it follows that for $n \geq 1$

$$S_n - f(1) \leq L \quad \Longrightarrow \quad S_n \leq L + f(1).$$

Consequently, $\{S_n\}$ is bounded and monotonic, and by Theorem 10.5 it converges. Thus, $\Sigma\, a_n$ converges.

For the other direction of the proof, we assume the improper integral diverges. Then $\int_1^n f(x)\, dx$ approaches infinity as $n \to \infty$, and the inequality

$$S_{n-1} \geq \int_1^n f(x)\, dx$$

implies that $\{S_n\}$ diverges. Thus, $\Sigma\, a_n$ diverges.

We noted earlier that the convergence or divergence of $\Sigma\, a_n$ is not affected by deleting the first N terms. Similarly, if the conditions for the Integral Test are satisfied for all $x \geq N > 1$, we simply use the integral $\int_N^\infty f(x)\, dx$ to test for convergence or divergence. (This is illustrated in Example 4.)

EXAMPLE 1   Using the Integral Test

Apply the Integral Test to the series

$$\sum_{n=1}^{\infty} \frac{n}{n^2 + 1}.$$

SOLUTION

Since $f(x) = x/(x^2 + 1)$ satisfies the conditions for the Integral Test (check this), we integrate to obtain

$$\int_1^\infty \frac{x}{x^2 + 1}\, dx = \frac{1}{2} \int_1^\infty \frac{2x}{x^2 + 1}\, dx$$

$$= \frac{1}{2} \lim_{b \to \infty} \left[\ln(x^2 + 1) \right]_1^b$$

$$= \frac{1}{2} \lim_{b \to \infty} [\ln(b^2 + 1) - \ln 2] = \infty.$$

Thus, the series *diverges*.

EXAMPLE 2   Using the Integral Test

Apply the Integral Test to the series

$$\sum_{n=1}^{\infty} \frac{1}{n^2 + 1}.$$

SOLUTION

Since $f(x) = 1/(x^2 + 1)$ satisfies the conditions for the Integral Test, we integrate to obtain

$$\int_1^{\infty} \frac{1}{x^2 + 1} \, dx = \lim_{b \to \infty} \left[\arctan x \right]_1^b$$

$$= \lim_{b \to \infty} [\arctan b - \arctan 1] = \frac{\pi}{4}.$$

Thus, the series *converges*.

p-Series

In the remainder of this section, we investigate a second type of series that has a simple arithmetic test for convergence or divergence.

DEFINITION OF *p*-SERIES	A series of the form $$\sum_{n=1}^{\infty} \frac{1}{n^p} = \frac{1}{1^p} + \frac{1}{2^p} + \frac{1}{3^p} + \cdots$$ is called a **p-series,** where p is a positive constant. For $p = 1$, the series $$\sum_{n=1}^{\infty} \frac{1}{n} = 1 + \frac{1}{2} + \frac{1}{3} + \cdots$$ is called the **harmonic series.**

REMARK   A **general harmonic series** is of the form $\sum [1/(an + b)]$. In music, strings of the same material, diameter, and tension, whose lengths form a harmonic series, produce harmonic tones.

The Integral Test is convenient for establishing the convergence or divergence of *p*-series. This is shown in the proof of Theorem 10.11.

THEOREM 10.11 **CONVERGENCE OF *p*-SERIES**	The *p*-series $$\sum_{n=1}^{\infty} \frac{1}{n^p} = \frac{1}{1^p} + \frac{1}{2^p} + \frac{1}{3^p} + \frac{1}{4^p} + \cdots$$ 1. converges if $p > 1$, and 2. diverges if $0 < p \leq 1$.

PROOF   The proof follows from the Integral Test and from Theorem 9.3, which states that

$$\int_1^\infty \frac{1}{x^p} \, dx$$

converges if $p > 1$, and diverges if $0 < p \le 1$.

EXAMPLE 3   Convergent and divergent p-series

From Theorem 10.11, it follows that the *harmonic series*

$$\sum_{n=1}^\infty \frac{1}{n} = \frac{1}{1} + \frac{1}{2} + \frac{1}{3} + \cdots \qquad p = 1$$

diverges.
From Theorem 10.11, it follows that the p-series

$$\sum_{n=1}^\infty \frac{1}{n^2} = \frac{1}{1^2} + \frac{1}{2^2} + \frac{1}{3^2} + \cdots \qquad p = 2$$

converges.

The sum of the series in part (b) of Example 3 can be shown to be $\pi^2/6$. (This was shown by Leonhard Euler, but the proof is too difficult to present here.) Be sure you see that the Integral Test does not tell us that the sum of the series is given by the value of the integral. For instance, the sum of the series $\Sigma \, (1/n^2)$ is given by

$$\sum_{n=1}^\infty \frac{1}{n^2} = \frac{\pi^2}{6} \approx 1.645$$

but the value of the corresponding improper integral is

$$\int_1^\infty \frac{1}{x^2} \, dx = 1.$$

EXAMPLE 4   A comparison to the harmonic series

Determine whether the following series converges or diverges.

$$\sum_{n=2}^\infty \frac{1}{n \ln n}$$

SOLUTION

This series is similar to the divergent harmonic series. If its terms were larger than those of the harmonic series, we would expect it to diverge. However, since its terms are smaller, we are not sure what to expect. Using the Integral Test, with $f(x) = 1/(x \ln x)$, we see that the series diverges.

$$\int_2^\infty \frac{1}{x \ln x} \, dx = \int_2^\infty \frac{1/x}{\ln x} \, dx$$

$$= \lim_{b \to \infty} \left[\ln (\ln x) \right]_2^b$$

$$= \lim_{b \to \infty} \left[\ln (\ln b) - \ln (\ln 2) \right] = \infty$$

EXERCISES for Section 10.3

In Exercises 1–10, determine the convergence or divergence of the given series using the Integral Test.

1. $\displaystyle\sum_{n=1}^\infty \frac{1}{n+1}$

2. $\displaystyle\sum_{n=1}^\infty e^{-n}$

3. $\displaystyle\sum_{n=1}^\infty ne^{-n}$

4. $\displaystyle\sum_{n=1}^\infty e^{-n} \cos n$

5. $\dfrac{1}{2} + \dfrac{1}{5} + \dfrac{1}{10} + \dfrac{1}{17} + \dfrac{1}{26} + \cdots$

6. $\dfrac{1}{3} + \dfrac{1}{5} + \dfrac{1}{7} + \dfrac{1}{9} + \dfrac{1}{11} + \cdots$

7. $\dfrac{\ln 2}{2} + \dfrac{\ln 3}{3} + \dfrac{\ln 4}{4} + \dfrac{\ln 5}{5} + \dfrac{\ln 6}{6} + \cdots$

8. $\dfrac{1}{4} + \dfrac{2}{7} + \dfrac{3}{12} + \cdots + \dfrac{n}{n^2+3} + \cdots$

9. $\displaystyle\sum_{n=1}^\infty \frac{n^{k-1}}{n^k + c}$, $\;k$ is a positive integer

10. $\displaystyle\sum_{n=1}^\infty n^k e^{-n}$, $\;k$ is a positive integer

In Exercises 11–20, determine the convergence or divergence of the given p-series.

11. $\displaystyle\sum_{n=1}^\infty \frac{1}{n^3}$

12. $\displaystyle\sum_{n=1}^\infty \frac{1}{n^{1/3}}$

13. $\displaystyle\sum_{n=1}^\infty \frac{1}{\sqrt[5]{n}}$

14. $\displaystyle\sum_{n=1}^\infty \frac{1}{n^{4/3}}$

15. $1 + \dfrac{1}{\sqrt{2}} + \dfrac{1}{\sqrt{3}} + \dfrac{1}{\sqrt{4}} + \cdots$

16. $1 + \dfrac{1}{4} + \dfrac{1}{9} + \dfrac{1}{16} + \dfrac{1}{25} + \cdots$

17. $1 + \dfrac{1}{2\sqrt{2}} + \dfrac{1}{3\sqrt{3}} + \dfrac{1}{4\sqrt{4}} + \dfrac{1}{5\sqrt{5}} + \cdots$

18. $1 + \dfrac{1}{\sqrt[3]{4}} + \dfrac{1}{\sqrt[3]{9}} + \dfrac{1}{\sqrt[3]{16}} + \dfrac{1}{\sqrt[3]{25}} + \cdots$

19. $\displaystyle\sum_{n=1}^\infty \frac{1}{n^{1.04}}$

20. $\displaystyle\sum_{n=1}^\infty \frac{1}{n^\pi}$

In Exercises 21–32, determine the convergence or divergence of the given series.

21. $\displaystyle\sum_{n=1}^\infty \frac{1}{2n-1}$

22. $\displaystyle\sum_{n=2}^\infty \frac{1}{n\sqrt{n^2-1}}$

23. $\displaystyle\sum_{n=1}^\infty \frac{1}{n\sqrt[4]{n}}$

24. $3\displaystyle\sum_{n=1}^\infty \frac{1}{n^{0.95}}$

25. $\displaystyle\sum_{n=0}^\infty \left(\frac{2}{3}\right)^n$

26. $\displaystyle\sum_{n=1}^\infty \frac{n}{\sqrt{n^2+1}}$

27. $\displaystyle\sum_{n=0}^\infty (1.075)^n$

28. $\displaystyle\sum_{n=1}^\infty \left(\frac{1}{n^2} - \frac{1}{n^3}\right)$

29. $\displaystyle\sum_{n=1}^\infty \left(1 + \frac{1}{n}\right)^n$

30. $\displaystyle\sum_{n=2}^\infty \ln n$

31. $\displaystyle\sum_{n=2}^\infty \frac{1}{n(\ln n)^3}$

32. $\displaystyle\sum_{n=2}^\infty \frac{\ln n}{n^3}$

In Exercises 33 and 34, find the positive values of p for which the series converges.

33. $\displaystyle\sum_{n=2}^\infty \frac{1}{n(\ln n)^p}$

34. $\displaystyle\sum_{n=2}^\infty \frac{\ln n}{n^p}$

35. Use a computer to find the smallest value of k such that

$$\sum_{n=1}^k \frac{1}{n} > 4.$$

36. The **Riemann zeta function** for real numbers is

$$\zeta(x) = \sum_{n=1}^\infty n^{-x}.$$

Find the domain of this function.

In the remaining exercises of this set use the following result. Let f be a positive, continuous, and decreasing function for $x \geq 1$, such that $a_n = f(n)$. If the series

$$\sum_{n=1}^\infty a_n$$

converges to S, then the remainder $R_N = S - S_N$ is bounded by

$$0 \leq R_N \leq \int_N^\infty f(x) \, dx.$$

In Exercises 37–42, approximate the sum of the convergent series using the indicated number of terms. Include an estimate of the maximum error for your approximation.

37. $\displaystyle\sum_{n=1}^{\infty} \frac{1}{n^4}$, six terms

38. $\displaystyle\sum_{n=1}^{\infty} \frac{1}{n^5}$, four terms

39. $\displaystyle\sum_{n=1}^{\infty} \frac{1}{n^2 + 1}$, ten terms

40. $\displaystyle\sum_{n=1}^{\infty} \frac{1}{(n + 1)[\ln{(n + 1)}]^3}$, ten terms

41. $\displaystyle\sum_{n=1}^{\infty} ne^{-n^2}$, four terms

42. $\displaystyle\sum_{n=1}^{\infty} e^{-n}$, four terms

In Exercises 43–46, find N so that $R_N \leq 0.001$ for the given convergent series.

43. $\displaystyle\sum_{n=1}^{\infty} \frac{1}{n^4}$

44. $\displaystyle\sum_{n=1}^{\infty} \frac{1}{n^{3/2}}$

45. $\displaystyle\sum_{n=1}^{\infty} e^{-5n}$

46. $\displaystyle\sum_{n=1}^{\infty} \frac{1}{n^2 + 1}$

10.4 Comparisons of Series

Direct Comparison Test ▪ Limit Comparison Test

In the previous two sections we developed tests for convergence or divergence of special types of series. In each case the terms of the series were quite simple, and the series *had to possess* special characteristics in order for the convergence tests to be applied. The slightest deviation from these special characteristics could make a test nonapplicable. For example, notice in the following pairs that the second series cannot be tested by the same convergence test as the first series even though it is similar to the first.

1. $\displaystyle\sum_{n=0}^{\infty} \frac{1}{2^n}$ is geometric, but $\displaystyle\sum_{n=0}^{\infty} \frac{n}{2^n}$ is not.

2. $\displaystyle\sum_{n=1}^{\infty} \frac{1}{n^3}$ is a *p*-series, but $\displaystyle\sum_{n=1}^{\infty} \frac{1}{n^3 + 1}$ is not.

3. $a_n = \dfrac{n}{(n^2 + 3)^2}$ is easily integrated, but $b_n = \dfrac{n^2}{(n^2 + 3)^2}$ is not.

In this section we discuss two additional tests for positive term series. These two tests greatly expand the variety of series we are able to test for convergence or divergence by allowing us to *compare* one series having similar but more complicated terms to a simpler series whose convergence or divergence is known.

THEOREM 10.12
DIRECT COMPARISON TEST

Let $0 \leq a_n \leq b_n$ for all n.

1. If $\displaystyle\sum_{n=1}^{\infty} b_n$ converges, then $\displaystyle\sum_{n=1}^{\infty} a_n$ converges.

2. If $\displaystyle\sum_{n=1}^{\infty} a_n$ diverges, then $\displaystyle\sum_{n=1}^{\infty} b_n$ diverges.

PROOF   To prove the first property, let

$$L = \sum_{n=1}^{\infty} b_n$$

and

$$S_n = a_1 + a_2 + \cdots + a_n.$$

Since $0 \le a_n \le b_n$, we know that the sequence

$$S_1, S_2, S_3, \ldots$$

is nondecreasing and bounded above by L; hence it must converge. Moreover, since

$$\lim_{n \to \infty} S_n = \sum_{n=1}^{\infty} a_n$$

it follows that $\Sigma \, a_n$ converges. The second property is logically equivalent to the first.

REMARK   As stated, the Direct Comparison Test requires that $0 \le a_n \le b_n$ for all n. Since the convergence of a series is not dependent on its first several terms, we could modify the test to require only that $0 \le a_n \le b_n$ for all n greater than some integer N.

EXAMPLE 1   Using the Direct Comparison Test

Determine the convergence or divergence of

$$\sum_{n=1}^{\infty} \frac{1}{2 + 3^n}.$$

SOLUTION

This series resembles

$$\sum_{n=1}^{\infty} \frac{1}{3^n}. \qquad \text{Convergent geometric series}$$

Term-by-term comparison yields

$$a_n = \frac{1}{2 + 3^n} < \frac{1}{3^n} = b_n, \quad n \ge 1.$$

Thus, it follows by the Direct Comparison Test that the series converges.

EXAMPLE 2   Using the Direct Comparison Test

Determine the convergence or divergence of

$$\sum_{n=1}^{\infty} \frac{1}{2 + \sqrt{n}}.$$

SOLUTION

This series resembles

$$\sum_{n=1}^{\infty} \frac{1}{n^{1/2}}. \qquad \text{Divergent } p\text{-series}$$

Term-by-term comparison yields

$$\frac{1}{2 + \sqrt{n}} \le \frac{1}{\sqrt{n}}, \quad n \ge 1$$

which *does not* meet the requirements for divergence. (Remember that if term-by-term comparison reveals a series that is *smaller* than a divergent series, then the Direct Comparison Test tells us nothing.) Still expecting the series to diverge, we compare the given series to

$$\sum_{n=1}^{\infty} \frac{1}{n}. \qquad \text{Divergent harmonic series}$$

In this case, term-by-term comparison yields

$$a_n = \frac{1}{n} \le \frac{1}{2 + \sqrt{n}} = b_n, \quad n \ge 4$$

and, by the Direct Comparison Test, the given series diverges.   ▭

REMARK   To verify the last inequality in Example 2, try showing that $2 + \sqrt{n} \le n$ whenever $n \ge 4$.

When using the Direct Comparison Test, remember that $0 \le a_n \le b_n$ for both parts of the theorem. Informally, the test says the following about two series with nonnegative terms.

1. If the "larger" series $\Sigma\, b_n$ converges, then the "smaller" series $\Sigma\, a_n$ must also converge.
2. If the "smaller" series $\Sigma\, a_n$ diverges, then the "larger" series $\Sigma\, b_n$ must also diverge.

Limit Comparison Test

Often a given series closely resembles a *p*-series or a geometric series, yet we cannot establish the term-by-term comparison necessary to apply the Direct Comparison Test. Under these circumstances we may be able to apply a second comparison test, called the **Limit Comparison Test.**

THEOREM 10.13 **LIMIT COMPARISON TEST**	Suppose $a_n > 0$, $b_n > 0$, and $$\lim_{n \to \infty} \left(\frac{a_n}{b_n}\right) = L$$ where L is *finite and positive*. Then the two series $\Sigma \, a_n$ and $\Sigma \, b_n$ either both converge or both diverge.

PROOF   Since $a_n > 0$, $b_n > 0$, and $(a_n/b_n) \to L$ as $n \to \infty$, there exists $N > 0$ such that for $n \geq N$, $0 < (a_n/b_n) < (L + 1)$. This implies that

$$0 < a_n < (L + 1)b_n.$$

Hence, by the Direct Comparison Test, the convergence of $\Sigma \, b_n$ implies the convergence of $\Sigma \, a_n$. Similarly, the fact that $(b_n/a_n) \to (1/L)$ as $n \to \infty$ can be used to show that the convergence of $\Sigma \, a_n$ implies the convergence of $\Sigma \, b_n$.

REMARK   As with the Direct Comparison Test, the Limit Comparison Test could be modified to require only that a_n and b_n be positive for all n greater than some integer N.

The versatility of the Limit Comparison Test is seen in the next example, where we show that a general harmonic series diverges.

EXAMPLE 3   Using the Limit Comparison Test

Show that the following general harmonic series diverges.

$$\sum_{n=1}^{\infty} \frac{1}{an + b}, \quad a > 0$$

SOLUTION

By comparison to

$$\sum_{n=1}^{\infty} \frac{1}{n} \qquad \text{Divergent harmonic series}$$

we have

$$\lim_{n \to \infty} \frac{1/(an + b)}{1/n} = \lim_{n \to \infty} \frac{n}{an + b} = \frac{1}{a}.$$

Since this limit is greater than 0, we conclude from the Limit Comparison Test that the given series diverges. ▭

The Limit Comparison Test works well for comparing "messy" algebraic series to a p-series. In choosing an appropriate p-series, we must choose one with an nth term of the same magnitude as the nth term of the given series.

Given series	*Comparison series*	*Conclusion*
$\displaystyle\sum_{n=1}^{\infty} \frac{1}{3n^2 - 4n + 5}$	$\displaystyle\sum_{n=1}^{\infty} \frac{1}{n^2}$	Both series converge.
$\displaystyle\sum_{n=1}^{\infty} \frac{1}{\sqrt{3n - 2}}$	$\displaystyle\sum_{n=1}^{\infty} \frac{1}{\sqrt{n}}$	Both series diverge.
$\displaystyle\sum_{n=1}^{\infty} \frac{n^2 - 10}{4n^5 + n^3}$	$\displaystyle\sum_{n=1}^{\infty} \frac{n^2}{n^5} = \sum_{n=1}^{\infty} \frac{1}{n^3}$	Both series converge.
$\displaystyle\sum_{n=1}^{\infty} \frac{\sqrt{n}}{\sqrt{n^3 + 1}}$	$\displaystyle\sum_{n=1}^{\infty} \frac{n^{1/2}}{n^{3/2}} = \sum_{n=1}^{\infty} \frac{1}{n}$	Both series diverge.

In other words, when choosing a series for comparison, we disregard all but the *highest powers of n* in both the numerator and the denominator.

EXAMPLE 4   Using the Limit Comparison Test

Determine the convergence or divergence of

$$\sum_{n=1}^{\infty} \frac{4\sqrt{n} - 1}{n^2 + 2\sqrt{n}}.$$

SOLUTION

Disregarding all but the highest powers of n in the numerator and the denominator, we compare the series to

$$\sum_{n=1}^{\infty} \frac{\sqrt{n}}{n^2} = \sum_{n=1}^{\infty} \frac{1}{n^{3/2}}. \qquad \text{\small Convergent } p\text{-series}$$

Since

$$\lim_{n\to\infty} \frac{a_n}{b_n} = \lim_{n\to\infty} \left(\frac{4\sqrt{n} - 1}{n^2 + 2\sqrt{n}}\right)\left(\frac{n^{3/2}}{1}\right)$$

$$= \lim_{n\to\infty} \frac{4n^2 - n^{3/2}}{n^2 + 2\sqrt{n}} = 4$$

we conclude by the Limit Comparison Test that the given series converges.

EXAMPLE 5   Using the Limit Comparison Test

Determine the convergence or divergence of

$$\sum_{n=1}^{\infty} \frac{n2^n + 5}{4n^3 + 3n}.$$

SOLUTION

A reasonable comparison would be to the series

$$\sum_{n=1}^{\infty} \frac{2^n}{n^2}.$$

Note that this series diverges by the nth-Term Test, since

$$\lim_{n\to\infty} \frac{2^n}{n^2} \neq 0.$$

From the limit

$$\lim_{n\to\infty} \frac{a_n}{b_n} = \lim_{n\to\infty} \left(\frac{n2^n + 5}{4n^3 + 3n}\right)\left(\frac{n^2}{2^n}\right)$$

$$= \lim_{n\to\infty} \frac{1 + [5/(2^n n)]}{4 + (3/n^2)} = \frac{1}{4}$$

we conclude that the given series diverges.

EXERCISES for Section 10.4

In Exercises 1–12, use the Direct Comparison Test to determine the convergence or divergence of the series.

1. $\displaystyle\sum_{n=1}^{\infty} \frac{1}{n^2 + 1}$

2. $\displaystyle\sum_{n=1}^{\infty} \frac{1}{3n^2 + 2}$

3. $\displaystyle\sum_{n=2}^{\infty} \frac{1}{n - 1}$

4. $\displaystyle\sum_{n=2}^{\infty} \frac{1}{\sqrt{n} - 1}$

5. $\displaystyle\sum_{n=0}^{\infty} \frac{1}{3^n + 1}$

6. $\displaystyle\sum_{n=0}^{\infty} \frac{2^n}{3^n + 5}$

7. $\displaystyle\sum_{n=2}^{\infty} \frac{\ln n}{n + 1}$

8. $\displaystyle\sum_{n=1}^{\infty} \frac{1}{\sqrt{n^3 + 1}}$

9. $\displaystyle\sum_{n=0}^{\infty} \frac{1}{n!}$

10. $\displaystyle\sum_{n=1}^{\infty} \frac{1}{3\sqrt[4]{n} - 1}$

11. $\displaystyle\sum_{n=0}^{\infty} e^{-n^2}$

12. $\displaystyle\sum_{n=0}^{\infty} \frac{4^n}{3^n - 1}$

In Exercises 13–26, use the Limit Comparison Test to determine the convergence or divergence of the series.

13. $\displaystyle\sum_{n=1}^{\infty} \frac{n}{n^2 + 1}$

14. $\displaystyle\sum_{n=2}^{\infty} \frac{1}{\sqrt{n^2 - 1}}$

15. $\displaystyle\sum_{n=0}^{\infty} \frac{1}{\sqrt{n^2 + 1}}$

16. $\displaystyle\sum_{n=1}^{\infty} \frac{1}{2^n - 5}$

17. $\displaystyle\sum_{n=1}^{\infty} \frac{2n^2 - 1}{3n^5 + 2n + 1}$

18. $\displaystyle\sum_{n=1}^{\infty} \frac{5n - 3}{n^2 - 2n + 5}$

19. $\displaystyle\sum_{n=1}^{\infty} \frac{n + 3}{n(n + 2)}$

20. $\displaystyle\sum_{n=1}^{\infty} \frac{1}{n(n^2 + 1)}$

21. $\displaystyle\sum_{n=1}^{\infty} \frac{1}{n\sqrt{n^2 + 1}}$

22. $\displaystyle\sum_{n=1}^{\infty} \frac{n}{(n + 1)2^{n-1}}$

23. $\displaystyle\sum_{n=1}^{\infty} \frac{n^{k-1}}{n^k + 1}, \ k > 2$

24. $\displaystyle\sum_{n=1}^{\infty} \frac{1}{n + \sqrt{n^2 + 1}}$

25. $\displaystyle\sum_{n=1}^{\infty} \sin \frac{1}{n}$

26. $\displaystyle\sum_{n=1}^{\infty} \tan \frac{1}{n}$

In Exercises 27–34, test for convergence or divergence, using each of the seven tests at least once.
(a) nth-Term Test
(b) Geometric Series Test
(c) p-Series Test
(d) telescoping series
(e) Integral Test
(f) Direct Comparison Test
(g) Limit Comparison Test

27. $\displaystyle\sum_{n=1}^{\infty} \frac{\sqrt{n}}{n}$

28. $\displaystyle\sum_{n=0}^{\infty} 5\left(\frac{-1}{5}\right)^n$

29. $\displaystyle\sum_{n=1}^{\infty} \frac{1}{3^n + 2}$

30. $\displaystyle\sum_{n=4}^{\infty} \frac{1}{3n^2 - 2n - 15}$

31. $\displaystyle\sum_{n=1}^{\infty} \frac{n}{2n + 3}$

32. $\displaystyle\sum_{n=1}^{\infty} \left(\frac{1}{n + 1} - \frac{1}{n + 2}\right)$

33. $\displaystyle\sum_{n=1}^{\infty} \frac{n}{(n^2 + 1)^2}$

34. $\displaystyle\sum_{n=1}^{\infty} \frac{3}{n(n + 3)}$

35. Use the Limit Comparison Test with the harmonic series to show that the series $\Sigma \, a_n$ (where $0 < a_n < a_{n-1}$) diverges if

$$\lim_{n\to\infty} na_n \neq 0.$$

36. Prove that if $P(n)$ and $Q(n)$ are polynomials of degree j and k, respectively, then the series

$$\sum_{n=1}^{\infty} \frac{P(n)}{Q(n)}$$

converges if $j < k - 1$, and diverges if $j \geq k - 1$.

In Exercises 37–40, use the polynomial test as given in Exercise 36 to determine whether the series converges or diverges.

37. $\dfrac{1}{2} + \dfrac{2}{5} + \dfrac{3}{10} + \dfrac{4}{17} + \dfrac{5}{26} + \cdots$

38. $\dfrac{1}{3} + \dfrac{1}{8} + \dfrac{1}{15} + \dfrac{1}{24} + \dfrac{1}{35} + \cdots$

39. $\displaystyle\sum_{n=1}^{\infty} \frac{1}{n^3 + 1}$ **40.** $\displaystyle\sum_{n=1}^{\infty} \frac{n^2}{n^3 + 1}$

In Exercises 41 and 42, use the divergence test used in Exercise 35 to show that each series diverges.

41. $\displaystyle\sum_{n=1}^{\infty} \frac{n^3}{5n^4 + 3}$ **42.** $\displaystyle\sum_{n=2}^{\infty} \frac{1}{\ln n}$

43. Prove that if the nonnegative series

$$\sum_{n=1}^{\infty} a_n \quad \text{and} \quad \sum_{n=1}^{\infty} b_n$$

converge, then so does the series

$$\sum_{n=1}^{\infty} a_n b_n.$$

44. Use the result of Exercise 43 to prove that if the nonnegative series

$$\sum_{n=1}^{\infty} a_n$$

converges, then so does the series

$$\sum_{n=1}^{\infty} a_n^2.$$

10.5 Alternating Series

Alternating series ▪ Alternating series remainder ▪ Absolute and conditional convergence

Most of the results we have studied so far have been restricted to series with positive terms. In this and the following section we consider series that contain both positive and negative terms. The simplest such series is an **alternating series,** whose terms alternate in sign. For example, the geometric series

$$\sum_{n=0}^{\infty} \left(-\frac{1}{2}\right)^n = \sum_{n=0}^{\infty} (-1)^n \frac{1}{2^n} = 1 - \frac{1}{2} + \frac{1}{4} - \frac{1}{8} + \frac{1}{16} - \cdots$$

is an *alternating geometric series* with $r = -\frac{1}{2}$.

Alternating series occur in two ways:

$$\sum_{n=1}^{\infty} (-1)^n a_n = -a_1 + a_2 - a_3 + a_4 - \cdots, \quad a_n > 0$$

and

$$\sum_{n=1}^{\infty} (-1)^{n-1} a_n = a_1 - a_2 + a_3 - a_4 + \cdots, \quad a_n > 0.$$

In one case the *odd* terms are negative, and in the other case the *even* terms are negative. The conditions for the convergence of an alternating series are given in the following theorem.

THEOREM 10.14 ALTERNATING SERIES TEST	If $a_n > 0$, then the alternating series $$\sum_{n=1}^{\infty} (-1)^n a_n$$ and $$\sum_{n=1}^{\infty} (-1)^{n-1} a_n$$ converge, provided that the following two conditions are met. 1. $a_{n+1} \leq a_n$, for all n 2. $\lim\limits_{n \to \infty} a_n = 0$

PROOF   The proof follows the same pattern for either form of the alternating series. In this proof, we use the form

$$\sum_{n=1}^{\infty} (-1)^{n-1} a_n = a_1 - a_2 + a_3 - a_4 + \cdots.$$

For this series, the partial sum (where $2n$ is even)

$$S_{2n} = (a_1 - a_2) + (a_3 - a_4) + (a_5 - a_6) + \cdots + (a_{2n-1} - a_{2n})$$

has all nonnegative terms, and therefore $\{S_{2n}\}$ is a nondecreasing sequence. But we can also write

$$S_{2n} = a_1 - (a_2 - a_3) - (a_4 - a_5) - \cdots - (a_{2n-2} - a_{2n-1}) - a_{2n}$$

which implies that $S_{2n} \leq a_1$ for every integer n. Thus $\{S_{2n}\}$ is a bounded, nondecreasing sequence that converges to some value L. Now, since $S_{2n-1} = S_{2n} - a_{2n}$ and $a_{2n} \to 0$, we have

$$\lim_{n \to \infty} S_{2n-1} = \lim_{n \to \infty} S_{2n} - \lim_{n \to \infty} a_{2n} = L - \lim_{n \to \infty} a_{2n} = L.$$

Since both S_{2n} and S_{2n-1} converge to the same limit L, it follows that $\{S_n\}$ also converges to L. Consequently, the given alternating series converges.

REMARK   The first condition in the Alternating Series Test can be modified to require only that $0 < a_{n+1} \leq a_n$ for all n greater than some integer N.

EXAMPLE 1   Using the Alternating Series Test

Determine the convergence or divergence of

$$\sum_{n=1}^{\infty} \frac{n}{(-2)^{n-1}} = \frac{1}{1} - \frac{2}{2} + \frac{3}{4} - \frac{4}{8} + \cdots.$$

SOLUTION

To apply the Alternating Series Test, we note that for $n \geq 1$,

$$\frac{1}{2} \leq \frac{n}{n+1}$$

which implies that

$$\frac{2^{n-1}}{2^n} \leq \frac{n}{n+1}$$

$$(n+1) \, 2^{n-1} \leq n2^n$$

$$\frac{n+1}{2^n} \leq \frac{n}{2^{n-1}}.$$

Hence, $a_{n+1} = (n+1)/2^n \leq n/2^{n-1} = a_n$ for all n. Furthermore, by L'Hôpital's Rule,

$$\lim_{x \to \infty} \frac{x}{2^{x-1}} = \lim_{x \to \infty} \frac{1}{2^{x-1}(\ln 2)} = 0 \quad \Longrightarrow \quad \lim_{n \to \infty} \frac{n}{2^{n-1}} = 0.$$

Therefore, by the Alternating Series Test, the given series converges. ▭

EXAMPLE 2   Using the Alternating Series Test

Determine whether the following series converge or diverge.

(a) $\displaystyle\sum_{n=1}^{\infty} \frac{(-1)^n n}{\ln 2n}$ (b) $\displaystyle\sum_{n=1}^{\infty} (-1)^{n+1}\left(\frac{3n+2}{4n^2-3}\right)$

SOLUTION

(a) By L'Hôpital's Rule, we have

$$\lim_{x \to \infty} \frac{x}{\ln 2x} = \lim_{x \to \infty} \frac{1}{1/x} = \lim_{x \to \infty} x = \infty.$$

Thus, $\{a_n\}$ does not converge to zero, and the Alternating Series Test does not apply. However, by the nth-Term Test for Divergence, we can conclude that the series diverges.

(b) Sometimes it is convenient to use differentiation to establish that $a_{n+1} \leq a_n$. In this case, we let

$$f(x) = \frac{3x+2}{4x^2-3}.$$

Then the derivative

$$f'(x) = \frac{-12x^2 - 16x - 9}{(4x^2-3)^2}$$

is always negative. Hence, f is a decreasing function, and it follows that $a_{n+1} \leq a_n$ for $n \geq 1$. Furthermore, since

$$\lim_{n \to \infty} \frac{3n+2}{4n^2-3} = 0$$

the series converges by the Alternating Series Test. ▭

For a convergent alternating series, the partial sum S_N can be a useful approximation for the sum S of the series. Just how close S_N is to S is stated in the following theorem.

THEOREM 10.15
ALTERNATING SERIES REMAINDER

If a convergent alternating series satisfies the condition $a_{n+1} \leq a_n$, then the absolute value of the remainder R_N involved in approximating the sum S by S_N is less than (or equal to) the first neglected term. That is,

$$|S - S_N| = |R_N| \leq a_{N+1}.$$

PROOF

The series obtained by deleting the first N terms of the given series satisfies the conditions of the Alternating Series Test and has a sum of R_N.

$$R_N = S - S_N = \sum_{n=1}^{\infty} (-1)^{n-1} a_n - \sum_{n=1}^{N} (-1)^{n-1} a_n$$
$$= (-1)^N a_{N+1} + (-1)^{N+1} a_{N+2} + (-1)^{N+2} a_{N+3} + \cdots$$
$$= (-1)^N (a_{N+1} - a_{N+2} + a_{N+3} - \cdots)$$
$$|R_N| = a_{N+1} - a_{N+2} + a_{N+3} - a_{N+4} + a_{N+5} - \cdots$$
$$= a_{N+1} - (a_{N+2} - a_{N+3}) - (a_{N+4} - a_{N+5}) - \cdots$$
$$\leq a_{N+1}$$

Consequently, $|S - S_N| = |R_N| \leq a_{N+1}$, which establishes the theorem.

EXAMPLE 3   Approximating the sum of an alternating series

Approximate the sum of the following series by its first six terms.

$$\sum_{n=1}^{\infty} (-1)^{n-1} \left(\frac{1}{n!} \right) = \frac{1}{1!} - \frac{1}{2!} + \frac{1}{3!} - \frac{1}{4!} + \frac{1}{5!} - \frac{1}{6!} + \cdots$$

SOLUTION

The Alternating Series Test establishes that the series converges because

$$\frac{1}{(n + 1)!} \leq \frac{1}{n!} \quad \text{and} \quad \lim_{n \to \infty} \frac{1}{n!} = 0.$$

Now, the sum of the first six terms is

$$S_6 = 1 - \frac{1}{2} + \frac{1}{6} - \frac{1}{24} + \frac{1}{120} - \frac{1}{720} \approx 0.63194$$

and, by the Alternating Series Remainder, we have

$$|S - S_6| = |R_6| \leq a_7 = \frac{1}{5040} \approx 0.0002.$$

Therefore, the sum S lies between $0.63194 - 0.0002$ and $0.63194 + 0.0002$, and we have

$$0.63174 \leq S \leq 0.63214.$$

(Later, we will see that the actual sum is $(e - 1)/e \approx 0.63212$.)

Absolute and conditional convergence

Occasionally, a series may have both positive and negative terms and not be an alternating series. For instance, the series

$$\sum_{n=1}^{\infty} \frac{\sin n}{n^2} = \frac{\sin 1}{1} + \frac{\sin 2}{4} + \frac{\sin 3}{9} + \cdots$$

has both positive and negative terms, yet it is not an alternating series. One way to obtain some information about the convergence of this series is to investigate the convergence of the series

$$\sum_{n=1}^{\infty} \left| \frac{\sin n}{n^2} \right|.$$

By direct comparison, we have $|\sin n| \leq 1$ for all n, so

$$\left| \frac{\sin n}{n^2} \right| \leq \frac{1}{n^2}, \quad n \geq 1.$$

Thus, by the Direct Comparison Test, the series $\Sigma |(\sin n)/n^2|$ converges. But the question still is "Does the original series converge?" The next theorem tells us that the answer is yes.

**THEOREM 10.16
ABSOLUTE CONVERGENCE**

If the series $\Sigma |a_n|$ converges, then the series Σa_n also converges.

PROOF Since $0 \leq a_n + |a_n| \leq 2|a_n|$ for all n, the series

$$\sum_{n=1}^{\infty} (a_n + |a_n|)$$

converges by comparison to the convergent series

$$\sum_{n=1}^{\infty} 2|a_n|.$$

Furthermore, since $a_n = (a_n + |a_n|) - |a_n|$, we can write

$$\sum_{n=1}^{\infty} a_n = \sum_{n=1}^{\infty} (a_n + |a_n|) - \sum_{n=1}^{\infty} |a_n|$$

where both series on the right converge. Hence it follows that Σa_n converges.

The converse of Theorem 10.16 is not true. For instance, the **alternating harmonic series**

$$\sum_{n=1}^{\infty} \frac{(-1)^{n+1}}{n} = \frac{1}{1} - \frac{1}{2} + \frac{1}{3} - \frac{1}{4} + \cdots$$

converges by the Alternating Series Test. Yet the harmonic series diverges. We call this type of convergence **conditional.**

DEFINITION OF ABSOLUTE AND CONDITIONAL CONVERGENCE

1. $\Sigma \, a_n$ is **absolutely convergent** if $\Sigma \, |a_n|$ converges.
2. $\Sigma \, a_n$ is **conditionally convergent** if $\Sigma \, a_n$ converges but $\Sigma \, |a_n|$ diverges.

EXAMPLE 4   Absolute and conditional convergence

Determine whether the following series are convergent or divergent. If convergent, classify the series as absolutely or conditionally convergent.

(a) $\displaystyle \sum_{n=1}^{\infty} \frac{(-1)^{n(n+1)/2}}{3^n} = -\frac{1}{3} - \frac{1}{9} + \frac{1}{27} + \frac{1}{81} - \cdots$

(b) $\displaystyle \sum_{n=1}^{\infty} \frac{(-1)^n}{\ln(n+1)} = -\frac{1}{\ln 2} + \frac{1}{\ln 3} - \frac{1}{\ln 4} + \frac{1}{\ln 5} - \cdots$

(c) $\displaystyle \sum_{n=0}^{\infty} \frac{(-1)^n n!}{2^n} = \frac{0!}{2^0} - \frac{1!}{2^1} + \frac{2!}{2^2} - \frac{3!}{2^3} + \cdots$

(d) $\displaystyle \sum_{n=1}^{\infty} \frac{(-1)^n}{\sqrt{n}} = -\frac{1}{\sqrt{1}} + \frac{1}{\sqrt{2}} - \frac{1}{\sqrt{3}} + \frac{1}{\sqrt{4}} - \cdots$

SOLUTION

(a) This is *not* an alternating series, since the signs change in pairs. However, we note that

$$\sum_{n=1}^{\infty} \left| \frac{(-1)^{n(n+1)/2}}{3^n} \right| = \sum_{n=1}^{\infty} \frac{1}{3^n}$$

is a convergent geometric series with $r = \frac{1}{3} < 1$. Consequently, by Theorem 10.16, we can conclude that the given series is *absolutely* convergent, hence convergent.

(b) In this case, the Alternating Series Test indicates that the given series converges. However, the series

$$\sum_{n=1}^{\infty} \left| \frac{(-1)^n}{\ln(n+1)} \right| = \frac{1}{\ln 2} + \frac{1}{\ln 3} + \frac{1}{\ln 4} + \cdots$$

diverges by direct comparison with the terms of the harmonic series. Therefore, we conclude that the given series is *conditionally* convergent.

(c) This is an alternating series, but the Alternating Series Test does not apply because the limit of the *n*th term is not zero. However, by the *n*th-Term Test for Divergence, the series diverges.

(d) In this case, the Alternating Series Test tells us that the series converges. Moreover, since the series

$$\sum_{n=1}^{\infty} \left| \frac{(-1)^n}{\sqrt{n}} \right| = \frac{1}{\sqrt{1}} + \frac{1}{\sqrt{2}} + \frac{1}{\sqrt{3}} + \frac{1}{\sqrt{4}} + \cdots$$

diverges, we conclude that the given series is *conditionally* convergent.

___

EXERCISES for Section 10.5

In Exercises 1–22, use the Alternating Series Test to determine the convergence or divergence of the series.

1. $\displaystyle\sum_{n=1}^{\infty} \frac{(-1)^{n+1}}{n}$

2. $\displaystyle\sum_{n=1}^{\infty} \frac{(-1)^{n+1} n}{2n - 1}$

3. $\displaystyle\sum_{n=1}^{\infty} \frac{(-1)^{n+1}}{2n - 1}$

4. $\displaystyle\sum_{n=2}^{\infty} \frac{(-1)^n}{\ln n}$

5. $\displaystyle\sum_{n=1}^{\infty} \frac{(-1)^n n^2}{n^2 + 1}$

6. $\displaystyle\sum_{n=1}^{\infty} \frac{(-1)^{n+1} n}{n^2 + 1}$

7. $\displaystyle\sum_{n=1}^{\infty} \frac{(-1)^n}{\sqrt{n}}$

8. $\displaystyle\sum_{n=1}^{\infty} \frac{(-1)^{n+1} n^3}{n^3 + 6}$

9. $\displaystyle\sum_{n=1}^{\infty} \frac{(-1)^{n+1}(n + 1)}{\ln (n + 1)}$

10. $\displaystyle\sum_{n=1}^{\infty} \frac{(-1)^{n+1} \ln (n + 1)}{n + 1}$

11. $\displaystyle\sum_{n=0}^{\infty} (-1)^n e^{-n}$

12. $\displaystyle\sum_{n=1}^{\infty} \frac{(-1)^{n+1} \sqrt{n + 2}}{\sqrt{n(n + 2)}}$

13. $\displaystyle\sum_{n=1}^{\infty} \sin \frac{(2n - 1)\pi}{2}$

14. $\displaystyle\sum_{n=1}^{\infty} \cos n\pi$

15. $\displaystyle\sum_{n=1}^{\infty} \frac{1}{n} \sin \frac{(2n - 1)\pi}{2}$

16. $\displaystyle\sum_{n=1}^{\infty} \frac{1}{n} \cos n\pi$

17. $\displaystyle\sum_{n=0}^{\infty} \frac{(-1)^n}{n!}$

18. $\displaystyle\sum_{n=0}^{\infty} \frac{(-1)^n}{(2n)!}$

19. $\displaystyle\sum_{n=1}^{\infty} \frac{(-1)^{n+1} \sqrt{n}}{n + 2}$

20. $\displaystyle\sum_{n=1}^{\infty} \frac{(-1)^{n+1} \sqrt{n}}{\sqrt[3]{n}}$

21. $\displaystyle\sum_{n=1}^{\infty} \frac{2(-1)^{n+1}}{e^n - e^{-n}} = \sum_{n=1}^{\infty} (-1)^{n+1} \operatorname{csch} n$

22. $\displaystyle\sum_{n=1}^{\infty} \frac{2(-1)^{n+1}}{e^n + e^{-n}} = \sum_{n=1}^{\infty} (-1)^{n+1} \operatorname{sech} n$

In Exercises 23–38, examine the series for conditional convergence or absolute convergence.

23. $\displaystyle\sum_{n=1}^{\infty} \frac{(-1)^{n+1}}{(n + 1)^2}$

24. $\displaystyle\sum_{n=1}^{\infty} \frac{(-1)^{n+1}}{n + 1}$

25. $\displaystyle\sum_{n=1}^{\infty} \frac{(-1)^{n+1}}{\sqrt{n}}$

26. $\displaystyle\sum_{n=1}^{\infty} \frac{(-1)^{n+1}}{n\sqrt{n}}$

27. $\displaystyle\sum_{n=2}^{\infty} \frac{(-1)^n}{\ln n}$

28. $\displaystyle\sum_{n=0}^{\infty} (-1)^n e^{-n^2}$

29. $\displaystyle\sum_{n=2}^{\infty} \frac{(-1)^n n}{n^3 - 1}$

30. $\displaystyle\sum_{n=1}^{\infty} \frac{(-1)^{n+1}}{n^{1.5}}$

31. $\displaystyle\sum_{n=0}^{\infty} \frac{(-1)^n}{(2n + 1)!}$

32. $\displaystyle\sum_{n=0}^{\infty} \frac{(-1)^n}{\sqrt[3]{n + 1}}$

33. $\displaystyle\sum_{n=0}^{\infty} \frac{\cos n\pi}{n + 1}$

34. $\displaystyle\sum_{n=1}^{\infty} (-1)^{n+1} \arctan n$

35. $\displaystyle\sum_{n=1}^{\infty} \frac{\cos n}{n^2}$

36. $\displaystyle\sum_{n=1}^{\infty} \frac{\cos n}{n\sqrt{n}}$

37. $\displaystyle\sum_{n=1}^{\infty} \frac{\sin [(2n - 1)\pi/2]}{\sqrt{n}}$

38. $\displaystyle\sum_{n=1}^{\infty} \frac{\sin [(2n - 1)\pi/2]}{n}$

In Exercises 39–46, approximate the sum of the series with an error of less than 0.001. (Use Theorem 10.15.)

39. $\displaystyle\sum_{n=1}^{\infty} \frac{(-1)^{n+1}}{2n^3 - 1}$

40. $\displaystyle\sum_{n=1}^{\infty} \frac{(-1)^{n+1}}{n^4}$

41. $\displaystyle\sum_{n=0}^{\infty} \frac{(-1)^n}{n!}$ $\left(\text{the sum is } \dfrac{1}{e}\right)$

42. $\displaystyle\sum_{n=0}^{\infty} \frac{(-1)^n}{2^n n!}$ $\left(\text{the sum is } \dfrac{1}{\sqrt{e}}\right)$

43. $\displaystyle\sum_{n=0}^{\infty} \frac{(-1)^n}{(2n + 1)!}$ (the sum is $\sin 1$)

44. $\displaystyle\sum_{n=0}^{\infty} \frac{(-1)^n}{(2n)!}$ (the sum is $\cos 1$)

45. $\displaystyle\sum_{n=1}^{\infty} \frac{(-1)^{n+1}}{n2^n}$ $\left(\text{the sum is } \ln \dfrac{3}{2}\right)$

46. $\displaystyle\sum_{n=1}^{\infty} \frac{(-1)^{n+1}}{n4^n}$ $\left(\text{the sum is } \ln \dfrac{5}{4}\right)$

In Exercises 47 and 48, find the number of terms necessary to approximate the sum of the series with an error less than 0.001. (Use Theorem 10.15.)

47. $\displaystyle\sum_{n=0}^{\infty} \frac{(-1)^n}{2n + 1}$ $\left(\text{the sum is } \dfrac{\pi}{4}\right)$

48. $\displaystyle\sum_{n=1}^{\infty} \frac{(-1)^{n+1}}{n^2}$ $\left(\text{the sum is } \dfrac{\pi^2}{12}\right)$

49. Prove that the alternating *p*-series

$$\sum_{n=1}^{\infty} (-1)^n \left(\frac{1}{n^p}\right)$$

converges if $p > 0$.

50. Prove that if $\Sigma |a_n|$ converges, then Σa_n^2 converges.

51. Find all values of x for which the series

$$\sum_{n=1}^{\infty} \frac{x^n}{n}$$

(a) converges absolutely and
(b) converges conditionally.

52. Determine the error in the following argument that $0 = 1$.

$$
\begin{aligned}
0 &= 0 + 0 + 0 + \cdots \\
&= (1 - 1) + (1 - 1) + (1 - 1) + \cdots \\
&= 1 + (-1 + 1) + (-1 + 1) + \cdots \\
&= 1 + 0 + 0 + \cdots \\
&= 1
\end{aligned}
$$

10.6 The Ratio and Root Tests

Ratio Test ▪ Root Test ▪ Summary of tests for convergence and divergence

We begin this section with a test for absolute convergence—the **Ratio Test.**

THEOREM 10.17
RATIO TEST

Let Σa_n be a series with nonzero terms.

1. Σa_n converges if

$$\lim_{n \to \infty} \left| \frac{a_{n+1}}{a_n} \right| < 1. \qquad \text{Absolute convergence}$$

2. Σa_n diverges if

$$\lim_{n \to \infty} \left| \frac{a_{n+1}}{a_n} \right| > 1.$$

3. The Ratio Test is inconclusive if

$$\lim_{n \to \infty} \left| \frac{a_{n+1}}{a_n} \right| = 1.$$

PROOF   For the proof of the first part, we let

$$\lim_{n \to \infty} \left| \frac{a_{n+1}}{a_n} \right| = r < 1$$

and choose R such that $0 \le r < R < 1$. By the definition of the limit of a sequence, there exists some $N > 0$ such that $|a_{n+1}/a_n| < R$ for all $n > N$. Therefore, we can write the following inequalities.

$$
\begin{aligned}
|a_{N+1}| &< |a_N|R \\
|a_{N+2}| &< |a_{N+1}|R < |a_N|R^2 \\
|a_{N+3}| &< |a_{N+2}|R < |a_{N+1}|R^2 < |a_N|R^3 \\
&\;\;\vdots
\end{aligned}
$$

The geometric series

$$\sum_{n=1}^{\infty} a_N R^n = a_N R + a_N R^2 + \cdots + a_N R^n + \cdots$$

converges, so, by the Direct Comparison Test, the series

$$\sum_{n=1}^{\infty} |a_{N+n}| = |a_{N+1}| + |a_{N+2}| + \cdots + |a_{N+n}| + \cdots$$

also converges. This in turn implies that the series $\Sigma \, |a_n|$ converges, since discarding a finite number of terms ($n = N - 1$) does not affect convergence. Consequently, by Theorem 10.16, the series $\Sigma \, a_n$ converges.

The proof of the second part is similar, except that we choose R such that

$$\lim_{n \to \infty} \left| \frac{a_{n+1}}{a_n} \right| = r > R > 1$$

and show that there exists some $M > 0$ such that $|a_{M+n}| > |a_M| R^n$.

The fact that the Ratio Test fails to give us any useful information when $|a_{n+1}/a_n| \to 1$ can be seen by comparing the two series

$$\sum_{n=1}^{\infty} \frac{1}{n} \quad \text{and} \quad \sum_{n=1}^{\infty} \frac{1}{n^2}.$$

The first series diverges and the second one converges, but in both cases

$$\lim_{n \to \infty} \left| \frac{a_{n+1}}{a_n} \right| = 1.$$

Although the Ratio Test is not a cure for all ills related to tests for convergence, it is particularly useful for series that *converge rapidly*. Series involving factorials or exponentials are frequently of this type.

EXAMPLE 1 Using the Ratio Test

Determine the convergence or divergence of

$$\sum_{n=0}^{\infty} \frac{2^n}{n!}.$$

SOLUTION

Since $a_n = 2^n/n!$, we have

$$\lim_{n \to \infty} \left| \frac{a_{n+1}}{a_n} \right| = \lim_{n \to \infty} \left[\frac{2^{n+1}}{(n+1)!} \div \frac{2^n}{n!} \right] = \lim_{n \to \infty} \left[\frac{2^{n+1}}{(n+1)!} \cdot \frac{n!}{2^n} \right]$$

$$= \lim_{n \to \infty} \frac{2}{n+1} = 0.$$

Therefore, the series converges. ▭

EXAMPLE 2 Using the Ratio Test

Determine whether the following series converge or diverge.

(a) $\displaystyle\sum_{n=0}^{\infty} \frac{n^2 2^{n+1}}{3^n}$ (b) $\displaystyle\sum_{n=0}^{\infty} \frac{n^n}{n!}$

SOLUTION

(a) Since

$$\lim_{n \to \infty} \left| \frac{a_{n+1}}{a_n} \right| = \lim_{n \to \infty} \left[(n+1)^2 \left(\frac{2^{n+2}}{3^{n+1}} \right) \left(\frac{3^n}{n^2 2^{n+1}} \right) \right]$$

$$= \lim_{n \to \infty} \frac{2(n+1)^2}{3n^2} = \frac{2}{3} < 1$$

we conclude by the Ratio Test that the series converges.

(b) Since

$$\lim_{n \to \infty} \left| \frac{a_{n+1}}{a_n} \right| = \lim_{n \to \infty} \left[\frac{(n+1)^{n+1}}{(n+1)!} \left(\frac{n!}{n^n} \right) \right]$$

$$= \lim_{n \to \infty} \left[\frac{(n+1)^{n+1}}{(n+1)} \left(\frac{1}{n^n} \right) \right]$$

$$= \lim_{n \to \infty} \frac{(n+1)^n}{n^n} = \lim_{n \to \infty} \left(1 + \frac{1}{n} \right)^n = e > 1$$

we conclude by the Ratio Test that the series diverges. ▭

EXAMPLE 3 A failure of the Ratio Test

Determine the convergence or divergence of

$$\sum_{n=1}^{\infty} (-1)^n \frac{\sqrt{n}}{n+1}.$$

SOLUTION

Using the Ratio Test, we have

$$\lim_{n \to \infty} \left| \frac{a_{n+1}}{a_n} \right| = \lim_{n \to \infty} \left[\left(\frac{\sqrt{n+1}}{n+2} \right) \left(\frac{n+1}{\sqrt{n}} \right) \right]$$

$$= \lim_{n \to \infty} \left[\sqrt{\frac{n+1}{n}} \left(\frac{n+1}{n+2} \right) \right]$$

$$= \sqrt{1}(1) = 1.$$

Thus, as the Ratio Test gives us no useful information, we use the Alternating Series Test. To show that $a_{n+1} \le a_n$, we let $f(x) = \sqrt{x}/(x+1)$. Then the derivative is

$$f'(x) = \frac{-x+1}{2\sqrt{x}(x+1)^2}$$

and since this is negative for $x > 1$, we know that f is a decreasing function. Also, by L'Hôpital's Rule,

$$\lim_{x \to \infty} \frac{\sqrt{x}}{x+1} = \lim_{x \to \infty} \frac{1/(2\sqrt{x})}{1} = \lim_{x \to \infty} \frac{1}{2\sqrt{x}} = 0.$$

Therefore, the series converges. ▭

REMARK   Note that the series in Example 3 is *conditionally convergent*, since $\Sigma\,|a_n|$ diverges (by the Limit Comparison Test with $\Sigma\,1/\sqrt{n}$), but $\Sigma\,a_n$ converges.

The Root Test

The next test for convergence or divergence of series works especially well for series involving an *n*th power. The proof of this theorem is similar to that given for the Ratio Test, and we leave it as an exercise (see Exercise 51).

THEOREM 10.18 **ROOT TEST**	1. $\Sigma\,a_n$ converges if $$\lim_{n\to\infty} \sqrt[n]{|a_n|} < 1.$$ 2. $\Sigma\,a_n$ diverges if $$\lim_{n\to\infty} \sqrt[n]{|a_n|} > 1.$$ 3. The Root Test is inconclusive if $$\lim_{n\to\infty} \sqrt[n]{|a_n|} = 1.$$

EXAMPLE 4   Using the Root Test

Determine the convergence or divergence of

$$\sum_{n=1}^{\infty} \frac{e^{2n}}{n^n}.$$

SOLUTION

Since

$$\lim_{n\to\infty} \sqrt[n]{|a_n|} = \lim_{n\to\infty} \sqrt[n]{\frac{e^{2n}}{n^n}} = \lim_{n\to\infty} \frac{e^{2n/n}}{n^{n/n}}$$

$$= \lim_{n\to\infty} \frac{e^2}{n} = 0 < 1$$

we conclude by the Root Test that the series converges.

EXAMPLE 5   Using the Root Test

Determine the convergence or divergence of

$$\sum_{n=1}^{\infty} \frac{n^3}{3^n}.$$

SOLUTION

First, we have

$$\lim_{n \to \infty} \sqrt[n]{|a_n|} = \lim_{n \to \infty} \sqrt[n]{\frac{n^3}{3^n}} = \lim_{n \to \infty} \frac{n^{3/n}}{3}.$$

Since the limit in the numerator yields the indeterminate form ∞^0, we apply L'Hôpital's Rule as follows.

$$y = \lim_{x \to \infty} x^{3/x}$$

$$\ln y = \ln \left(\lim_{x \to \infty} x^{3/x} \right)$$

$$= \lim_{x \to \infty} (\ln x^{3/x})$$

$$= \lim_{x \to \infty} \left(\frac{3}{x} \ln x \right)$$

$$= \lim_{x \to \infty} \frac{3 \ln x}{x}$$

$$= \lim_{x \to \infty} \frac{3/x}{1} = 0 \qquad \text{L'Hôpital's Rule}$$

Now, since $\ln y = 0$, we conclude that

$$y = \lim_{x \to \infty} x^{3/x} = 1$$

from which it follows that

$$\lim_{n \to \infty} \frac{n^{3/n}}{3} = \frac{1}{3} < 1.$$

Thus, by the Root Test, the given series converges. ▭

In the last five sections we have discussed ten different tests for determining the convergence or divergence of an infinite series. Skill in choosing and applying the various tests will come only with practice. We summarize in Table 10.2 the various tests we have studied. Below is a useful checklist for choosing an appropriate test.

1. Does the nth term approach zero? If not, the series diverges.
2. Is the series one of the special types—geometric, p-series, telescoping, alternating?
3. Can the Integral, Root, or Ratio Test be applied?
4. Can the series be compared favorably to one of the special types?

In some instances more than one test is applicable. However, your objective should be to learn to choose the most efficient way to test a series.

TABLE 10.2 Summary of Tests for Series

Test	Series	Converges	Diverges	Comment
nth-Term	$\sum\limits_{n=1}^{\infty} a_n$		$\lim\limits_{n\to\infty} a_n \neq 0$	This test cannot be used to show convergence.
Geometric Series	$\sum\limits_{n=0}^{\infty} ar^n$	$|r| < 1$	$|r| \geq 1$	Sum: $S = \dfrac{a}{1-r}$
Telescoping Series	$\sum\limits_{n=1}^{\infty} (b_n - b_{n+1})$	$\lim\limits_{n\to\infty} b_n = L$		Sum: $S = b_1 - L$
p-Series	$\sum\limits_{n=1}^{\infty} \dfrac{1}{n^p}$	$p > 1$	$p \leq 1$	
Alternating Series	$\sum\limits_{n=1}^{\infty} (-1)^{n-1}a_n$	$0 < a_{n+1} \leq a_n$ and $\lim\limits_{n\to\infty} a_n = 0$		Remainder: $|R_N| \leq a_{N+1}$
Integral (f is continuous, positive, and decreasing)	$\sum\limits_{n=1}^{\infty} a_n,\ a_n = f(n) \geq 0$	$\int_1^{\infty} f(x)\,dx$ converges	$\int_1^{\infty} f(x)\,dx$ diverges	Remainder: $0 < R_N < \int_N^{\infty} f(x)\,dx$
Root	$\sum\limits_{n=1}^{\infty} a_n$	$\lim\limits_{n\to\infty} \sqrt[n]{|a_n|} < 1$	$\lim\limits_{n\to\infty} \sqrt[n]{|a_n|} > 1$	Test is inconclusive if $\lim\limits_{n\to\infty} \sqrt[n]{|a_n|} = 1$.
Ratio	$\sum\limits_{n=1}^{\infty} a_n$	$\lim\limits_{n\to\infty} \left|\dfrac{a_{n+1}}{a_n}\right| < 1$	$\lim\limits_{n\to\infty} \left|\dfrac{a_{n+1}}{a_n}\right| > 1$	Test is inconclusive if $\lim\limits_{n\to\infty} \left|\dfrac{a_{n+1}}{a_n}\right| = 1$.
Direct Comparison $(a_n, b_n > 0)$	$\sum\limits_{n=1}^{\infty} a_n$	$0 \leq a_n \leq b_n$ and $\sum\limits_{n=1}^{\infty} b_n$ converges	$0 \leq b_n \leq a_n$ and $\sum\limits_{n=1}^{\infty} b_n$ diverges	
Limit Comparison $(a_n, b_n > 0)$	$\sum\limits_{n=1}^{\infty} a_n$	$\lim\limits_{n\to\infty} \dfrac{a_n}{b_n} = L > 0$ and $\sum\limits_{n=1}^{\infty} b_n$ converges	$\lim\limits_{n\to\infty} \dfrac{a_n}{b_n} = L > 0$ and $\sum\limits_{n=1}^{\infty} b_n$ diverges	

EXERCISES for Section 10.6

In Exercises 1–20, use the Ratio Test to test for convergence or divergence of the series.

1. $\sum\limits_{n=0}^{\infty} \dfrac{n!}{3^n}$

2. $\sum\limits_{n=1}^{\infty} n\left(\dfrac{2}{3}\right)^n$

3. $\sum\limits_{n=0}^{\infty} \dfrac{3^n}{n!}$

4. $\sum\limits_{n=1}^{\infty} n\left(\dfrac{3}{2}\right)^n$

5. $\sum\limits_{n=1}^{\infty} \dfrac{n}{2^n}$

6. $\sum\limits_{n=1}^{\infty} \dfrac{n^2}{2^n}$

7. $\sum\limits_{n=1}^{\infty} \dfrac{2^n}{n^2}$

8. $\sum\limits_{n=1}^{\infty} \dfrac{(-1)^{n+1}(n+2)}{n(n+1)}$

9. $\sum\limits_{n=0}^{\infty} \dfrac{(-1)^n 2^n}{n!}$

10. $\sum\limits_{n=1}^{\infty} \dfrac{(-1)^{n-1}(3/2)^n}{n^2}$

11. $\sum_{n=0}^{\infty} \dfrac{n!}{n3^n}$ **12.** $\sum_{n=1}^{\infty} \dfrac{(2n)!}{n^5}$

13. $\sum_{n=0}^{\infty} \dfrac{4^n}{n!}$ **14.** $\sum_{n=1}^{\infty} \dfrac{n^n}{n!}$

15. $\sum_{n=0}^{\infty} \dfrac{3^n}{(n+1)^n}$ **16.** $\sum_{n=0}^{\infty} \dfrac{(n!)^2}{(3n)!}$

17. $\sum_{n=0}^{\infty} \dfrac{4^n}{3^n+1}$ **18.** $\sum_{n=0}^{\infty} \dfrac{(-1)^n 2^{4n}}{(2n+1)!}$

19. $\sum_{n=0}^{\infty} \dfrac{(-1)^{n+1} n!}{1 \cdot 3 \cdot 5 \cdots (2n+1)}$

20. $\sum_{n=1}^{\infty} \dfrac{(-1)^n 2 \cdot 4 \cdot 6 \cdots (2n)}{2 \cdot 5 \cdot 8 \cdots (3n-1)}$

In Exercises 21–30, use the Root Test to test for convergence or divergence of the series.

21. $\sum_{n=1}^{\infty} \left(\dfrac{n}{2n+1}\right)^n$ **22.** $\sum_{n=1}^{\infty} \left(\dfrac{2n}{n+1}\right)^n$

23. $\sum_{n=2}^{\infty} \dfrac{(-1)^n}{(\ln n)^n}$ **24.** $\sum_{n=1}^{\infty} \left(\dfrac{-2n}{3n+1}\right)^{3n}$

25. $\sum_{n=1}^{\infty} (2\sqrt[n]{n}+1)^n$ **26.** $\sum_{n=1}^{\infty} (\sqrt[n]{n}-1)^n$

27. $\sum_{n=2}^{\infty} \left(\dfrac{\ln n}{n}\right)^n$ **28.** $\sum_{n=0}^{\infty} e^{-n}$

29. $\dfrac{1}{(\ln 3)^3} + \dfrac{1}{(\ln 4)^4} + \dfrac{1}{(\ln 5)^5} + \dfrac{1}{(\ln 6)^6} + \cdots$

30. $1 + \dfrac{2}{3} + \dfrac{3}{3^2} + \dfrac{4}{3^3} + \dfrac{5}{3^4} + \dfrac{6}{3^5} + \cdots$

In Exercises 31–50, test for convergence or divergence using any appropriate test from this chapter. Identify the test used.

31. $\sum_{n=1}^{\infty} \dfrac{(-1)^{n+1} 5}{n}$ **32.** $\sum_{n=1}^{\infty} \dfrac{5}{n}$

33. $\sum_{n=1}^{\infty} \dfrac{3}{n\sqrt{n}}$ **34.** $\sum_{n=1}^{\infty} \left(\dfrac{\pi}{4}\right)^n$

35. $\sum_{n=1}^{\infty} \dfrac{2n}{n+1}$ **36.** $\sum_{n=1}^{\infty} \dfrac{n}{2n^2+1}$

37. $\sum_{n=1}^{\infty} \dfrac{(-1)^n 3^{n-2}}{2^n}$ **38.** $\sum_{n=1}^{\infty} \dfrac{10}{3\sqrt{n^3}}$

39. $\sum_{n=1}^{\infty} \dfrac{10n+3}{n2^n}$ **40.** $\sum_{n=1}^{\infty} \dfrac{2^n}{4n^2-1}$

41. $\sum_{n=1}^{\infty} (-1)^n \ln\left(\dfrac{n+2}{n}\right)$ **42.** $\sum_{n=1}^{\infty} \dfrac{1}{\sqrt{n}+2}$

43. $\sum_{n=1}^{\infty} \dfrac{\cos n}{2^n}$ **44.** $\sum_{n=2}^{\infty} \dfrac{(-1)^n}{n \ln n}$

45. $\sum_{n=1}^{\infty} \dfrac{n7^n}{n!}$ **46.** $\sum_{n=1}^{\infty} \dfrac{\ln n}{n^2}$

47. $\sum_{n=1}^{\infty} \dfrac{(-1)^n 3^{n-1}}{n!}$ **48.** $\sum_{n=1}^{\infty} \dfrac{(-1)^n 3^n}{n2^n}$

49. $\sum_{n=1}^{\infty} \dfrac{(-3)^n}{3 \cdot 5 \cdot 7 \cdots (2n+1)}$

50. $\sum_{n=1}^{\infty} \dfrac{3 \cdot 5 \cdot 7 \cdots (2n+1)}{18^n(2n-1)n!}$

51. Prove Theorem 10.18. [Hint for part 1: If the limit equals $r < 1$, choose a real number R such that $r < R < 1$. By the definition of the limit there exists some $N > 0$ such that $\sqrt[n]{|a_n|} < R$ for $n > N$.]

52. Show that the Ratio Test is inconclusive for a p-series.

10.7 Taylor Polynomials and Approximations

Polynomial approximations of elementary functions ▪ Taylor and Maclaurin polynomials ▪ Remainder of a Taylor polynomial

The remaining four sections of this chapter discuss Taylor polynomials and Taylor series.

Polynomial approximations of elementary functions

The goal of this section is to show how special types of polynomials can be used as approximations for other elementary functions. To find a polynomial function P that approximates another function f, we begin by choosing a number c in the domain of f at which we require that f and P have the same value. That is,

$$P(c) = f(c).\qquad \text{Graphs of } f \text{ and } P \text{ pass through } (c, f(c))$$

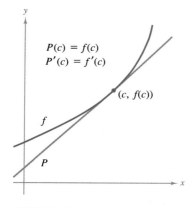

$P(c) = f(c)$
$P'(c) = f'(c)$

$(c, f(c))$

f

P

FIGURE 10.7

We say that the approximating polynomial is **expanded about c** or **centered at c.** Geometrically, the requirement that $P(c) = f(c)$ means that the graph of P passes through the point $(c, f(c))$. Of course, there are many polynomials whose graph passes through the point $(c, f(c))$. Our task is to find a polynomial whose graph resembles the graph of f near this point. One way to do this is to impose the additional requirement that the slope of the polynomial function be the same as the slope of the graph of f at the point $(c, f(c))$. That is, we require that

$$P'(c) = f'(c). \qquad \text{\small Graphs of } f \text{ and } P \text{ have same slope at } (c, f(c))$$

With these two requirements, we can obtain a simple linear approximation of f, as shown in Figure 10.7. This procedure is demonstrated in Example 1.

EXAMPLE 1 First-degree polynomial approximation of $f(x) = e^x$

For the function $f(x) = e^x$, find a first-degree polynomial function

$$P_1(x) = a_1 x + a_0$$

whose value and slope agree with the value and slope of f at $x = 0$.

SOLUTION

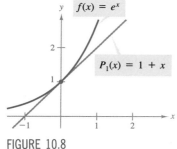

$f(x) = e^x$

$P_1(x) = 1 + x$

FIGURE 10.8

Since $f(x) = e^x$ and $f'(x) = e^x$, the value and slope of f, at $x = 0$, are given by

$$f(0) = e^0 = 1 \quad \text{and} \quad f'(0) = e^0 = 1.$$

Now, since $P_1(x) = a_1 x + a_0$, we can impose the condition that $P_1(0) = f(0)$ to conclude that $a_0 = 1$. Moreover, since $P_1'(x) = a_1$, we can use the condition that $P_1'(0) = f'(0)$ to conclude that $a_1 = 1$. Therefore, we have

$$P_1(x) = x + 1.$$

Figure 10.8 shows the graphs of $P_1(x) = x + 1$ and $f(x) = e^x$. ▭

From Figure 10.8 we can see that at points near $(0, 1)$, the graph of $P_1(x) = x + 1$ is reasonably close to the graph of $f(x) = e^x$. However, as we move away from $(0, 1)$, the graphs move farther from each other and the approximation is not good. To improve the approximation, we can impose yet another requirement—that the values of the second derivatives of P and f agree when $x = 0$. The polynomial of least degree that satisfies all three requirements $P_2(0) = f(0)$, $P_2'(0) = f'(0)$, and $P_2''(0) = f''(0)$ can be shown to be

$$P_2(x) = 1 + x + \frac{1}{2}x^2.$$

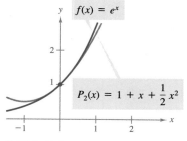

FIGURE 10.9

Moreover, from Figure 10.9, we can see that P_2 is a better approximation to f than P. If we continue this pattern, requiring that the values of $P_n(x)$ and its first n derivatives match those of $f(x) = e^x$ at $x = 0$, we obtain the following.

$$P_n(x) = 1 + x + \frac{1}{2}x^2 + \frac{1}{3!}x^3 + \cdots + \frac{1}{n!}x^n \approx e^x$$

EXAMPLE 2 Third-degree polynomial approximation of $f(x) = e^x$

Construct a table comparing the values of the polynomial

$$P_3(x) = 1 + x + \frac{1}{2}x^2 + \frac{1}{3!}x^3$$

to $f(x) = e^x$ for several values of x near 0.

SOLUTION

Using a calculator, we obtain the results shown in Table 10.3. Note that for $x = 0$, the two functions have the same value, but that as x moves farther away from 0, the accuracy of the approximating polynomial $P_3(x)$ decreases.

TABLE 10.3

x	-1.0	-0.2	-0.1	0.0	0.1	0.2	1.0
e^x	0.367879	0.818731	0.904837	1.000000	1.105171	1.221403	2.718282
$P_3(x)$	0.333333	0.818667	0.904833	1.000000	1.105167	1.221333	2.666667

Taylor and Maclaurin polynomials

The polynomial approximation for $f(x) = e^x$ given in Example 2 is expanded about $c = 0$. For expansions about an arbitrary value of c, it is convenient to write the polynomial in the form

$$P_n(x) = a_0 + a_1(x - c) + a_2(x - c)^2 + a_3(x - c)^3 + \cdots + a_n(x - c)^n.$$

In this form, repeated differentiation produces

$$P_n'(x) = a_1 + 2a_2(x - c) + 3a_3(x - c)^2 + \cdots + na_n(x - c)^{n-1}$$
$$P_n''(x) = 2a_2 + 2 \cdot 3a_3(x - c) + \cdots + n(n - 1)a_n(x - c)^{n-2}$$
$$P_n'''(x) = 2 \cdot 3a_3 + \cdots + n(n - 1)(n - 2)a_n(x - e)^{n-3}$$
$$\vdots$$
$$P_n^{(n)}(x) = n(n - 1)(n - 2) \cdots (2)(1)a_n.$$

Brook Taylor

Letting $x = c$, we then obtain

$$P_n(c) = a_0, \quad P_n'(c) = a_1, \quad P_n''(c) = 2a_2, \quad \ldots, \quad P_n^{(n)}(c) = n!a_n$$

and because the value of f and its first n derivatives must agree with the value of P_n and its first n derivatives at $x = c$, it follows that

$$f(c) = a_0, \qquad f'(c) = a_1, \qquad \frac{f''(c)}{2!} = a_2, \qquad \ldots, \qquad \frac{f^{(n)}(c)}{n!} = a_n.$$

With these coefficients we obtain the following definition of **Taylor polynomials,** named after the English mathematician Brook Taylor (1685–1731). Although Taylor was not the first to seek polynomial approximations of transcendental functions, his published account in 1715 was the first comprehensive work on the subject.

DEFINITION OF nTH TAYLOR POLYNOMIAL AND MACLAURIN POLYNOMIAL

If f has n derivatives at c, then the polynomial

$$P_n(x) = f(c) + f'(c)(x - c) + \frac{f''(c)}{2!}(x - c)^2 + \cdots + \frac{f^{(n)}(c)}{n!}(x - c)^n$$

is called the **nth Taylor polynomial for f at c.** If $c = 0$, then

$$P_n(x) = f(0) + f'(0)x + \frac{f''(0)}{2!}x^2 + \frac{f'''(0)}{3!}x^3 + \cdots + \frac{f^{(n)}(0)}{n!}x^n$$

is called the **nth Maclaurin polynomial for f.**

REMARK   The nth Maclaurin polynomial for f is named after the English mathematician, Colin Maclaurin (1698–1746).

EXAMPLE 3   A Maclaurin polynomial for $f(x) = e^x$

From our discussion earlier in this section, the nth Maclaurin polynomial for $f(x) = e^x$ is given by

$$P_n(x) = 1 + x + \frac{1}{2}x^2 + \frac{1}{3!}x^3 + \cdots + \frac{1}{n!}x^n.$$ ▭

EXAMPLE 4   Finding Taylor polynomials for $\ln x$

Find the Taylor polynomials P_0, P_1, P_2, P_3, and P_4 for $f(x) = \ln x$ centered at $c = 1$.

SOLUTION

Expanding about $c = 1$ yields the following.

$$f(x) = \ln x \qquad f(1) = 0$$

$$f'(x) = \frac{1}{x} \qquad f'(1) = 1$$

$$f''(x) = -\frac{1}{x^2} \qquad f''(1) = -1$$

$$f'''(x) = \frac{2!}{x^3} \qquad f'''(1) = 2$$

$$f^{(4)}(x) = -\frac{3!}{x^4} \qquad f^{(4)}(1) = -6$$

Therefore, the Taylor polynomials are as follows.

$$P_0(x) = f(1) = 0$$

$$P_1(x) = P_0(x) + f'(1)(x - 1) = (x - 1)$$

$$P_2(x) = P_1(x) + \frac{f''(1)}{2!}(x - 1)^2 = (x - 1) - \frac{1}{2}(x - 1)^2$$

$$P_3(x) = P_2(x) + \frac{f'''(1)}{3!}(x - 1)^3 = (x - 1) - \frac{1}{2}(x - 1)^2 + \frac{1}{3}(x - 1)^3$$

$$P_4(x) = P_3(x) + \frac{f^{(4)}(1)}{4!}(x - 1)^4$$

$$= (x - 1) - \frac{1}{2}(x - 1)^2 + \frac{1}{3}(x - 1)^3 - \frac{1}{4}(x - 1)^4$$

Figure 10.10 compares the graphs of P_1, P_2, P_3, and P_4 to the graph of f. Note that near $x = 1$ the graphs are nearly indistinguishable.   ▭ ⋅

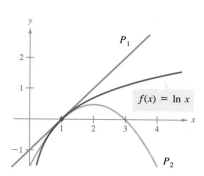

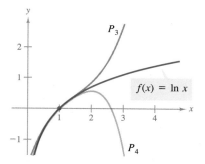

FIGURE 10.10

EXAMPLE 5   Finding Maclaurin polynomials for cos x

Find the Maclaurin polynomials P_0, P_2, P_4, and P_6 for $f(x) = \cos x$. Use $P_6(x)$ to approximate the value of cos (0.1).

SOLUTION

Expanding about $c = 0$ yields the following.

$$f(x) = \cos x \qquad f(0) = 1$$

$$f'(x) = -\sin x \qquad f'(0) = 0$$

$$f''(x) = -\cos x \qquad f''(0) = -1$$

$$f'''(x) = \sin x \qquad f'''(0) = 0$$

Through repeated differentiation, we can see that the pattern 1, 0, −1, 0 continues, and we obtain the following Maclaurin polynomials.

$$P_0(x) = f(0) = 1$$

$$P_2(x) = P_0(x) + f'(0)x + \frac{f''(0)}{2!}x^2 = 1 - \frac{1}{2!}x^2$$

$$P_4(x) = P_2(x) + \frac{f'''(0)}{3!}x^3 + \frac{f^{(4)}(0)}{4!}x^4 = 1 - \frac{1}{2!}x^2 + \frac{1}{4!}x^4$$

$$P_6(x) = P_4(x) + \frac{f^{(5)}(0)}{5!}x^5 + \frac{f^{(6)}(0)}{6!}x^6 = 1 - \frac{1}{2!}x^2 + \frac{1}{4!}x^4 - \frac{1}{6!}x^6$$

Using $P_6(x)$, we obtain the approximation $\cos(0.1) \approx 0.995004165$, which coincides with the calculator value to nine decimal places. ▭

Note in Example 5 that the Maclaurin polynomials for $\cos x$ have only even powers of x. Similarly, the Maclaurin polynomials for $g(x) = \sin x$ have only odd powers of x (see Exercise 5). This is not generally true for the Taylor polynomials for $\sin x$ and $\cos x$ expanded about $c \neq 0$, as we will see in the next example.

EXAMPLE 6 Finding a Taylor polynomial for sin x

Find the third Taylor polynomial for $f(x) = \sin x$, expanded about $c = \pi/6$.

SOLUTION

Expanding about $c = \pi/6$ yields the following.

$$f(x) = \sin x \qquad f\left(\frac{\pi}{6}\right) = \frac{1}{2}$$

$$f'(x) = \cos x \qquad f'\left(\frac{\pi}{6}\right) = \frac{\sqrt{3}}{2}$$

$$f''(x) = -\sin x \qquad f''\left(\frac{\pi}{6}\right) = -\frac{1}{2}$$

$$f'''(x) = -\cos x \qquad f'''\left(\frac{\pi}{6}\right) = -\frac{\sqrt{3}}{2}$$

Thus, the third Taylor polynomial for $f(x) = \sin x$, expanded about $c = \pi/6$, is as follows.

$$P_3(x) = f\left(\frac{\pi}{6}\right) + f'\left(\frac{\pi}{6}\right)\left(x - \frac{\pi}{6}\right) + \frac{f''\left(\frac{\pi}{6}\right)}{2!}\left(x - \frac{\pi}{6}\right)^2 + \frac{f'''\left(\frac{\pi}{6}\right)}{3!}\left(x - \frac{\pi}{6}\right)^3$$

$$= \frac{1}{2} + \frac{\sqrt{3}}{2}\left(x - \frac{\pi}{6}\right) - \frac{1}{2 \cdot 2!}\left(x - \frac{\pi}{6}\right)^2 - \frac{\sqrt{3}}{2 \cdot 3!}\left(x - \frac{\pi}{6}\right)^3 \qquad ▭$$

To approximate the value of a function at a specific point, we can use Taylor or Maclaurin polynomials. For instance, to approximate the value of $\ln(1.1)$, we can use Taylor polynomials for $f(x) = \ln x$ expanded about $c = 1$, as in Example 4. Or we can use Maclaurin polynomials, as shown in the next example.

EXAMPLE 7   Approximation using Maclaurin polynomials

Use a fourth Maclaurin polynomial to approximate the value of ln (1.1).

SOLUTION

Since 1.1 is closer to 1 than to 0, we consider Maclaurin polynomials for the function $g(x) = \ln (1 + x)$.

$$
\begin{aligned}
g(x) &= \ln (1 + x) & g(0) &= 0 \\
g'(x) &= (1 + x)^{-1} & g'(0) &= 1 \\
g''(x) &= -(1 + x)^{-2} & g''(0) &= -1 \\
g'''(x) &= 2(1 + x)^{-3} & g'''(0) &= 2 \\
g^{(4)}(x) &= -6(1 + x)^{-4} & g^{(4)}(0) &= -6
\end{aligned}
$$

Note that we get the same coefficients as we did in Example 4. Therefore, the fourth Maclaurin polynomial for $g(x) = \ln (1 + x)$ is

$$
P_4(x) = g(0) + g'(0)x + \frac{g''(0)}{2!}x^2 + \frac{g'''(0)}{3!}x^3 + \frac{g^{(4)}(0)}{4!}x^4
$$

$$
= x - \frac{1}{2}x^2 + \frac{1}{3}x^3 - \frac{1}{4}x^4.
$$

Consequently,

$$
\ln (1.1) = \ln (1 + 0.1) \approx P_4(0.1)
$$

$$
= (0.1) - \frac{1}{2}(0.1)^2 + \frac{1}{3}(0.1)^3 - \frac{1}{4}(0.1)^4
$$

$$
\approx 0.0953083.
$$

Check to see that the fourth Taylor polynomial (from Example 4), evaluated at $x = 1.1$, yields the same result. ▭

   Table 10.4 illustrates the accuracy of the Taylor polynomial approximation to the calculator value of ln (1.1). We can see that as n becomes larger, $P_n(1.1)$ is closer to the calculator value of 0.0953102.

TABLE 10.4   Approximations of ln (1.1), Using Taylor Polynomials

n	1	2	3	4
$P_n(1.1)$	0.1000000	0.0950000	0.0953333	0.0953083

   On the other hand, Table 10.5 illustrates that as we move away from the expansion point $c = 1$, the accuracy of the approximation decreases.

TABLE 10.5   Fourth Taylor Polynomial Approximations of ln x

x	1.0	1.1	1.5	1.75	2.0
ln x	0.0000000	0.0953102	0.4054651	0.5596158	0.6931472
$P_4(x)$	0.0000000	0.0953083	0.4010417	0.5302734	0.5833333

Tables 10.4 and 10.5 illustrate two very important points about the accuracy of Taylor (or Maclaurin) polynomials for use in approximations.

1. The approximation is usually better at x-values close to c than at x-values far from c.
2. The approximation is usually better for higher degree Taylor (or Maclaurin) polynomials than for those of lower degree.

The remainder of a Taylor polynomial

An approximation technique is of little value without some idea of its accuracy. To measure the accuracy of approximating a functional value $f(x)$ by the Taylor polynomial $P_n(x)$, we use the concept of a **remainder,** $R_n(x)$, defined as follows.

$$f(x) = P_n(x) + R_n(x)$$

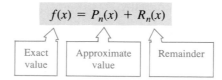

Exact value	Approximate value	Remainder

Thus, $R_n(x) = f(x) - P_n(x)$, and we call the absolute value of $R_n(x)$ the **error** associated with the approximation. That is,

$$\text{error} = |R_n(x)| = |f(x) - P_n(x)|.$$

The next theorem gives a general procedure for estimating the remainder associated with a Taylor polynomial. This important theorem is called **Taylor's Theorem** and the remainder given in the theorem is called the **Lagrange form of the remainder.**

THEOREM 10.19
TAYLOR'S THEOREM

If a function f is differentiable through order $n + 1$ in an interval I containing c, then for each x in I, there exists z between x and c such that

$$f(x) = f(c) + f'(c)(x - c) + \frac{f''(c)}{2!}(x - c)^2 + \cdots + \frac{f^{(n)}(c)}{n!}(x - c)^n + R_n(x)$$

where

$$R_n(x) = \frac{f^{(n+1)}(z)}{(n + 1)!}(x - c)^{n+1}.$$

PROOF   To find $R_n(x)$ we fix x in I $(x \neq c)$ and write

$$R_n(x) = f(x) - P_n(x)$$

where $P_n(x)$ is the nth Taylor polynomial for $f(x)$. Then we let g be a function of t defined by

$$g(t) = f(x) - f(t) - f'(t)(x - t) - \cdots - \frac{f^{(n)}(t)}{n!}(x - t)^n - R_n(x)\frac{(x - t)^{n+1}}{(x - c)^{n+1}}.$$

The reason for defining g this way is that differentiation with respect to t has a telescoping effect. For example, we have

$$\frac{d}{dt}[-f(t) - f'(t)(x - t)] = -f'(t) + f'(t) - f''(t)(x - t)$$

$$= -f''(t)(x - t).$$

The result is that the derivative $g'(t)$ simplifies to

$$g'(t) = -\frac{f^{(n+1)}(t)}{n!}(x - t)^n + (n + 1)R_n(x)\frac{(x - t)^n}{(x - c)^{n+1}}$$

for all t between c and x. Moreover, for a fixed x

$$g(c) = f(x) - [P_n(x) + R_n(x)] = f(x) - f(x) = 0$$

and

$$g(x) = f(x) - f(x) - 0 - \cdots - 0 = f(x) - f(x) = 0.$$

Therefore, g satisfies the conditions of Rolle's Theorem, and it follows that there is a number z between c and x such that $g'(z) = 0$. Substituting z for t in the equation for $g'(t)$ and then solving for $R_n(x)$, we obtain

$$g'(z) = -\frac{f^{(n+1)}(z)}{n!}(x - z)^n + (n + 1)R_n(x)\frac{(x - z)^n}{(x - c)^{n+1}} = 0$$

$$R_n(x) = \frac{f^{(n+1)}(z)}{(n + 1)!}(x - c)^{n+1}.$$

Finally, since $g(c) = 0$, we have

$$0 = f(x) - f(c) - f'(c)(x - c) - \cdots - \frac{f^{(n)}(c)}{n!}(x - c)^n - R_n(x)$$

$$f(x) = f(c) + f'(c)(x - c) + \cdots + \frac{f^{(n)}(c)}{n!}(x - c)^n + R_n(x).$$

When applying Taylor's Theorem, we do not expect to be able to find the exact value of z. (If we could do that, an approximation usually would not have been necessary.) Rather, we try to find bounds for $f^{(n+1)}(z)$ from which we are able to tell how large the remainder $R_n(x)$ is. This is demonstrated in Example 8.

EXAMPLE 8 Determining the accuracy of an approximation

The third Maclaurin polynomial for $\sin x$ is given by

$$P_3(x) = x - \frac{x^3}{3!}.$$

Use Taylor's Theorem to approximate $\sin (0.1)$ by $P_3(0.1)$ and determine the accuracy of the approximation.

SOLUTION

Using Taylor's Theorem, we have

$$\sin x = x - \frac{x^3}{3!} + R_3(x) = x - \frac{x^3}{3!} + \frac{f^{(4)}(z)}{4!}x^4$$

where $0 < z < 0.1$. Therefore,

$$\sin (0.1) \approx 0.1 - \frac{(0.1)^3}{3!} \approx 0.1 - 0.000167 = 0.099833.$$

Since $f^{(4)}(z) = \sin z$ and the sine function is increasing on the interval $[0, 0.1]$, it follows that $0 < \sin z < 1$ and we have

$$0 \le R_3(0.1) = \frac{\sin z}{4!}(0.1)^4 < \frac{0.0001}{4!} \approx 0.000004$$

and we conclude that

$$0.099833 \le \sin (0.1) \le 0.099833 + R_3(x)$$
$$0.099833 \le \sin (0.1) \le 0.099837.$$

▭

EXAMPLE 9 Approximating a functional value to a desired accuracy

Determine the degree of the Taylor polynomial $P_n(x)$ expanded about $c = 1$ that should be used to approximate $\ln (1.2)$ so that the error is less than 0.001.

SOLUTION

Following the pattern of Example 4, we see that the $(n + 1)$st derivative of $f(x) = \ln x$ is given by

$$f^{(n+1)}(x) = (-1)^{n+1}\frac{n!}{x^{n+1}}.$$

Using Taylor's Theorem, we know that the error $|R_n(1.2)|$ is given by

$$|R_n(1.2)| = \left| \frac{f^{(n+1)}(z)}{(n + 1)!}(1.2 - 1)^{n+1} \right| = \frac{n!}{z^{n+1}}\left(\frac{1}{(n + 1)!}\right)(0.2)^{n+1}$$

$$= \frac{(0.2)^{n+1}}{z^{n+1}(n + 1)}$$

where $1 < z < 1.2$. In this interval, $|R_n(1.2)|$ is largest when $z = 1$; thus we are seeking a value of n such that

$$\frac{(0.2)^{n+1}}{(1)^{n+1}(n+1)} < 0.001 \quad \Longrightarrow \quad 1000 < (n+1)5^{n+1}.$$

By trial and error, we can determine that the smallest value of n satisfying this inequality is $n = 3$. Thus, we would need the third Taylor polynomial to achieve the desired accuracy in approximating $\ln(1.2)$. ▭

EXERCISES for Section 10.7

In Exercises 1–14, find the Maclaurin polynomial of degree n for the given function.

1. $f(x) = e^{-x}$, $n = 3$
2. $f(x) = e^{-x}$, $n = 5$
3. $f(x) = e^{2x}$, $n = 4$
4. $f(x) = e^{3x}$, $n = 4$
5. $f(x) = \sin x$, $n = 5$
6. $f(x) = \sin \pi x$, $n = 3$
7. $f(x) = xe^x$, $n = 4$
8. $f(x) = x^2 e^{-x}$, $n = 4$
9. $f(x) = \dfrac{1}{x+1}$, $n = 4$
10. $f(x) = \dfrac{1}{x^2+1}$, $n = 4$
11. $f(x) = \tan x$, $n = 3$
12. $f(x) = \sec x$, $n = 2$
13. $f(x) = 2 - 3x^3 + x^4$, $n = 4$
14. $f(x) = 3 - 2x^2 + x^3$, $n = 3$

In Exercises 15–18, find the Taylor polynomial of degree n centered at c.

15. $f(x) = \dfrac{1}{x}$, $n = 4$, $c = 1$
16. $f(x) = \sqrt{x}$, $n = 4$, $c = 4$
17. $f(x) = \ln x$, $n = 4$, $c = 1$
18. $f(x) = x^2 \cos x$, $n = 2$, $c = \pi$

19. Use the Maclaurin polynomials $P_1(x)$, $P_3(x)$, and $P_5(x)$ for $f(x) = \sin x$ to complete the following table. (See Exercise 5.)

x	0	0.25	0.50	0.75	1.00
$\sin x$	0	0.2474	0.4794	0.6816	0.8415
$P_1(x)$					
$P_3(x)$					
$P_5(x)$					

20. Use the Taylor polynomials $P_1(x)$ and $P_4(x)$ for $f(x) = \ln x$ centered at 1 to complete the following table. (See Exercise 17.)

x	1.00	1.25	1.50	1.75	2.00
$\ln x$	0	0.2231	0.4055	0.5596	0.6931
$P_1(x)$					
$P_4(x)$					

In Exercises 21–24, approximate the function at the given value of x, using the polynomial found in the indicated exercise.

21. $f(x) = e^{-x}$, $f\left(\dfrac{1}{2}\right)$, Exercise 1
22. $f(x) = x^2 e^{-x}$, $f\left(\dfrac{1}{4}\right)$, Exercise 8
23. $f(x) = x^2 \cos x$, $f\left(\dfrac{7\pi}{8}\right)$, Exercise 18
24. $f(x) = \sqrt{x}$, $f(5)$, Exercise 16

25. Compare the Maclaurin polynomials of degree 4 for the functions

$$f(x) = e^x \quad \text{and} \quad g(x) = xe^x$$

What is the relationship between them?

26. Differentiate the Maclaurin polynomial of degree 5 for $f(x) = \sin x$ and compare the result with the Maclaurin polynomial of degree 4 for $g(x) = \cos x$.

In Exercises 27 and 28, determine the values of x for which the given function can be replaced by the Taylor Polynomial if the error cannot exceed 0.001.

27. $f(x) = e^x \approx 1 + x + \dfrac{x^2}{2!} + \dfrac{x^3}{3!}$, $x < 0$

28. $f(x) = \sin x \approx x - \dfrac{x^3}{3!}$

29. Estimate the error in approximating e by the fifth-degree polynomial

$$1 + x + \frac{x^2}{2!} + \frac{x^3}{3!} + \frac{x^4}{4!} + \frac{x^5}{5!}.$$

30. What degree Maclaurin polynomial for $\ln(x + 1)$ should be used to guarantee finding $\ln 1.5$ if the error cannot exceed 0.0001?

In Exercises 31 and 32,
(a) Find the Taylor polynomial $P_3(x)$ of degree three for for $f(x)$.
(b) Complete the accompanying table for $f(x)$ and $P_3(x)$.
(c) Sketch the graphs of $f(x)$ and $P_3(x)$ on the same axes.

31. $f(x) = \arcsin x$ **32.** $f(x) = \arctan x$

x	-1	-0.75	-0.50	-0.25
$f(x)$				
$P_3(x)$				

x	0	0.25	0.50	0.75	1
$f(x)$					
$P_3(x)$					

In Exercises 33 and 34, use a computer or graphics calculator to sketch the graph of the given function over the specified interval and sketch the graph of each of the Maclaurin polynomials. Place all of the graphs on the same set of coordinate axes.

33. $f(x) = \sin x, \quad [-\pi, \pi]$
    (a) $P_1(x)$ (b) $P_3(x)$
    (c) $P_5(x)$ (d) $P_7(x)$

34. $f(x) = e^x, \quad [-1, 1]$
    (a) $P_1(x)$ (b) $P_3(x)$
    (c) $P_5(x)$ (d) $P_7(x)$

35. Prove that if f is an odd function then the nth Maclaurin polynomial for the function will contain only terms with odd powers of x.

36. Prove that if f is an even function then the nth Maclaurin polynomial for the function will contain only terms with even powers of x.

37. Let $P_n(x)$ be the nth Taylor polynomial for f at c. Prove that $P_n(c) = f(c)$ and $P^{(k)}(c) = f^{(k)}(c)$ for $1 \le k \le n$.

10.8   Power Series

Power series ▪ Radius of convergence ▪ Interval of convergence ▪ Endpoint convergence ▪ Differentiation and integration of power series

In Section 10.7, we looked at the problem of approximating functions by Taylor polynomials. For instance, the function $f(x) = e^x$ can be *approximated* by its third-degree Taylor polynomial as follows.

$$e^x \approx 1 + x + \frac{x^2}{2} + \frac{x^3}{3!}$$

In the last three sections of this chapter, we will see that several important types of functions, including $f(x) = e^x$, can be represented *exactly* by a **power series**. For example, the power series representation for e^x is

$$e^x = 1 + x + \frac{x^2}{2} + \frac{x^3}{3!} + \cdots + \frac{x^n}{n!} + \cdots .$$

For each real number x, we will show that the infinite series on the right converges to the number e^x. Before doing that, we look at some preliminary results dealing with power series—beginning with the following definition.

DEFINITION OF POWER SERIES

If x is a variable, then an infinite series of the form

$$\sum_{n=0}^{\infty} a_n x^n = a_0 + a_1 x + a_2 x^2 + a_3 x^3 + \cdots + a_n x^n + \cdots$$

is called a **power series**. More generally, we call a series of the form

$$\sum_{n=0}^{\infty} a_n (x - c)^n = a_0 + a_1 (x - c) + a_2 (x - c)^2 + \cdots + a_n (x - c)^n + \cdots$$

a **power series centered at c,** where c is a constant.

REMARK   To simplify the notation for power series, we agree that $(x - c)^0 = 1$, even if $x = c$.

EXAMPLE 1   Power series

(a) The following power series is centered at 0.

$$\sum_{n=0}^{\infty} \frac{x^n}{n!} = 1 + x + \frac{x^2}{2} + \frac{x^3}{3!} + \cdots$$

(b) The following power series is centered at 1.

$$\sum_{n=1}^{\infty} \frac{1}{n}(x - 1)^n = (x - 1) + \frac{1}{2}(x - 1)^2 + \frac{1}{3}(x - 1)^3 + \cdots \qquad \blacksquare$$

A power series in x can be viewed as a function of x

$$f(x) = \sum_{n=0}^{\infty} a_n (x - c)^n$$

where the *domain of f* is the set of all x for which the power series converges. The determination of the domain of a power series is our primary concern in this section. Of course, every power series converges at its center c, since

$$f(c) = \sum_{n=0}^{\infty} a_n (c - c)^n$$

$$= a_0(1) + 0 + 0 + \cdots + 0 + \cdots$$

$$= a_0.$$

Thus, c always lies in the domain of f. The following important theorem (which we state without proof) tells us that the domain of a power series can take three basic forms: a single point, an interval centered at c, or the entire real line.

THEOREM 10.20 CONVERGENCE OF A POWER SERIES	For a power series centered at c, precisely one of the following is true.

For a power series centered at c, precisely one of the following is true.

1. The series converges only at c.
2. There exists a real number $R > 0$ such that the series converges (absolutely) for $|x - c| < R$, and diverges for $|x - c| > R$.
3. The series converges for all x.

The number R is called the **radius of convergence** of the power series. If the series converges only at c, then we say that the radius of convergence is $R = 0$, and if the series converges for all x, then we say that the radius of convergence is $R = \infty$. The set of all values of x for which the power series converges is called the **interval of convergence** of the power series.

To determine the radius of convergence of a power series, we use the Ratio Test, as shown in Examples 2, 3, and 4.

EXAMPLE 2   Finding the radius of convergence

Find the radius of convergence of the power series

$$\sum_{n=0}^{\infty} n!x^n.$$

SOLUTION

For $x = 0$, we obtain

$$f(0) = \sum_{n=0}^{\infty} n!0^n = 1 + 0 + 0 + \cdots = 1.$$

For any fixed value of x such that $|x| > 0$, let $u_n = n!x^n$. Then

$$\lim_{n \to \infty} \left| \frac{u_{n+1}}{u_n} \right| = \lim_{n \to \infty} \left| \frac{(n + 1)!x^{n+1}}{n!x^n} \right| = |x| \lim_{n \to \infty} (n + 1) = \infty.$$

Therefore, by the Ratio Test, we conclude that the series diverges for $|x| > 0$, and converges only at its center, 0. Hence, the radius of convergence is $R = 0$.

EXAMPLE 3   Finding the radius of convergence

Find the radius of convergence of the power series

$$\sum_{n=0}^{\infty} 3(x - 2)^n.$$

SOLUTION

For $x \neq 2$, we let $u_n = 3(x - 2)^n$. Then,

$$\lim_{n \to \infty} \left| \frac{u_{n+1}}{u_n} \right| = \lim_{n \to \infty} \left| \frac{3(x - 2)^{n+1}}{3(x - 2)^n} \right| = \lim_{n \to \infty} |x - 2| = |x - 2|.$$

By the Ratio Test, it follows that the series converges if $|x - 2| < 1$ and diverges if $|x - 2| > 1$. Therefore, the radius of convergence of the series is $R = 1$. ▭

EXAMPLE 4 Finding the radius of convergence

Find the radius of convergence of the power series

$$\sum_{n=0}^{\infty} \frac{(-1)^n x^{2n+1}}{(2n + 1)!}.$$

SOLUTION

If we let $u_n = (-1)^n x^{2n+1} / (2n + 1)!$, then

$$\lim_{n \to \infty} \left| \frac{u_{n+1}}{u_n} \right| = \lim_{n \to \infty} \left| \frac{[(-1)^{n+1} x^{2n+3}] / (2n + 3)!}{[(-1)^n x^{2n+1}] / (2n + 1)!} \right|$$

$$= \lim_{n \to \infty} \frac{x^2}{(2n + 3)(2n + 2)}.$$

For any *fixed* value for x, this limit is 0, and by the Ratio Test, the series converges for all x. Therefore, the radius of convergence is $R = \infty$. ▭

Endpoint convergence

Note that for a power series whose radius of convergence is a finite number R, Theorem 10.20 says nothing about the convergence of the *endpoints* of the interval of convergence. Each endpoint must be tested separately for convergence or divergence. As a result, the interval of convergence of a power series can take any one of the six forms shown in Figure 10.11.

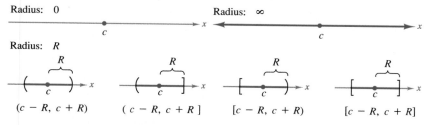

Intervals of Convergence

FIGURE 10.11

EXAMPLE 5   Finding the interval of convergence

Find the interval of convergence of the power series

$$\sum_{n=1}^{\infty} \frac{x^n}{n}.$$

SOLUTION

Letting $u_n = x^n/n$, we have

$$\lim_{n \to \infty} \left| \frac{u_{n+1}}{u_n} \right| = \lim_{n \to \infty} \left| \frac{x^{n+1}/(n+1)}{x^n/n} \right| = \lim_{n \to \infty} \left| \frac{nx}{n+1} \right| = |x|.$$

Therefore, by the Ratio Test, the radius of convergence is $R = 1$. Moreover, because the series is centered at 0, it will converge in the interval $(-1, 1)$, and we proceed to test for convergence at the endpoints. When $x = 1$, we obtain the *divergent* harmonic series

$$\sum_{n=1}^{\infty} \frac{1}{n} = \frac{1}{1} + \frac{1}{2} + \frac{1}{3} + \cdots .$$ Diverges when $x = 1$

When $x = -1$, we obtain the *convergent* alternating harmonic series

$$\sum_{n=1}^{\infty} \frac{(-1)^n}{n} = -1 + \frac{1}{2} - \frac{1}{3} + \frac{1}{4} - \cdots .$$ Converges when $x = -1$

Therefore, the interval of convergence for the given series is $[-1, 1)$, as shown in Figure 10.12.   ▭

Interval: $[-1, 1)$
Radius: $R = 1$

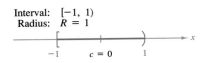

FIGURE 10.12

EXAMPLE 6   Finding the interval of convergence

Find the interval of convergence of the power series

$$\sum_{n=0}^{\infty} \frac{(-1)^n(x+1)^n}{2^n}.$$

SOLUTION

Since

$$\lim_{n \to \infty} \left| \frac{u_{n+1}}{u_n} \right| = \lim_{n \to \infty} \left| \frac{(-1)^{n+1}(x+1)^{n+1}/2^{n+1}}{(-1)^n(x+1)^n/2^n} \right|$$

$$= \lim_{n \to \infty} \left| \frac{2^n(x+1)}{2^{n+1}} \right| = \left| \frac{x+1}{2} \right|$$

the radius of convergence is $R = 2$. Since the series is centered at $x = -1$, it will converge in the interval $(-3, 1)$. Furthermore, at the endpoints we have

$$\sum_{n=0}^{\infty} \frac{(-1)^n(-2)^n}{2^n} = \sum_{n=0}^{\infty} \frac{2^n}{2^n} = \sum_{n=0}^{\infty} 1$$ Diverges when $x = -3$

Interval: $(-3, 1)$
Radius: $R = 2$

FIGURE 10.13

and

$$\sum_{n=0}^{\infty} \frac{(-1)^n (2)^n}{2^n} = \sum_{n=0}^{\infty} (-1)^n \qquad \text{Diverges when } x = 1$$

both of which diverge. Thus, the interval of convergence is $(-3, 1)$, as shown in Figure 10.13.

EXAMPLE 7   Finding the interval of convergence

Find the interval of convergence of the power series

$$\sum_{n=1}^{\infty} \frac{x^n}{n^2}.$$

SOLUTION

Since

$$\lim_{n \to \infty} \left| \frac{u_{n+1}}{u_n} \right| = \lim_{n \to \infty} \left| \frac{x^{n+1}/(n+1)^2}{x^n/n^2} \right| = \lim_{n \to \infty} \left| \frac{n^2 x}{(n+1)^2} \right| = |x|$$

the radius of convergence is $R = 1$. Moreover, since the series is centered at $x = 0$, it will converge in the interval $(-1, 1)$ and we proceed to test for convergence at the endpoints. When $x = 1$, we obtain the *convergent p*-series

$$\sum_{n=1}^{\infty} \frac{1}{n^2} = \frac{1}{1^2} + \frac{1}{2^2} + \frac{1}{3^2} + \frac{1}{4^2} + \cdots. \qquad \text{Converges when } x = 1$$

When $x = -1$, we obtain the *convergent* alternating series

$$\sum_{n=1}^{\infty} \frac{(-1)^n}{n^2} = -\frac{1}{1^2} + \frac{1}{2^2} - \frac{1}{3^2} + \frac{1}{4^2} - \cdots. \qquad \text{Converges when } x = -1$$

Therefore, the interval of convergence for the given series is $[-1, 1]$.

Differentiation and integration of power series

Power series representation of functions has played an important role in the development of calculus. In fact, much of Newton's work with differentiation and integration was done in the context of power series—especially his work with complicated algebraic functions and transcendental functions. Euler, Lagrange, Leibniz, and the Bernoullis all used power series extensively in calculus. One of the earliest mathematicians to work with power series was a Scotsman, James Gregory (1638–1675). He developed a power series method for interpolating table values—a method that was later used by Brook Taylor in the development of Taylor polynomials and Taylor series.

Once we have defined a function with a power series, it is natural to wonder how we can determine the characteristics of the function. Is it continuous? Differentiable? Integrable? Theorem 10.21, which we state without proof, answers these questions.

James Gregory

<table>
<tr><td>

**THEOREM 10.21
PROPERTIES OF A FUNCTION
DEFINED BY A POWER SERIES**

</td><td>

If the function given by

$$f(x) = \sum_{n=0}^{\infty} a_n(x - c)^n = a_0 + a_1(x - c) + a_2(x - c)^2 + a_3(x - c)^3 + \cdots$$

has a radius of convergence of $R > 0$, then on the interval $(c - R, c + R)$ f is continuous, differentiable, and integrable. Moreover, the derivative and antiderivative of f are as follows.

1. $f'(x) = a_1 + 2a_2(x - c) + 3a_3(x - c)^2 + \cdots$

2. $\int f(x)\, dx = C + a_0(x - c) + a_1\dfrac{(x - c)^2}{2} + a_2\dfrac{(x - c)^3}{3} + \cdots$

</td></tr>
</table>

REMARK Theorem 10.21 tells us that, in many ways, a function defined by a power series behaves like a polynomial. It is continuous in its interval of convergence, and both its derivative and antiderivative can be determined by differentiating and integrating each term of the given power series.

The *radius of convergence* for the series obtained by differentiating or integrating a power series is the same as for the original power series. For example, suppose the power series

$$f(x) = a_0 + a_1(x - c) + a_2(x - c)^2 + a_3(x - c)^3 + \cdots$$

has a finite radius of convergence of $R \neq 0$. Then, by the Ratio Test, we know that

$$\lim_{n \to \infty} \left| \frac{a_{n+1}(x - c)^{n+1}}{a_n(x - c)^n} \right| = \left| \frac{x - c}{R} \right|.$$

For the series

$$f'(x) = a_1 + 2a_2(x - c) + 3a_3(x - c)^2 + \cdots$$

it follows that the radius of convergence is also R since

$$\lim_{n \to \infty} \left| \frac{(n + 1)a_{n+1}(x - c)^n}{na_n(x - c)^{n-1}} \right| = \lim_{n \to \infty} \left(\frac{n + 1}{n} \right) \left| \frac{a_{n+1}(x - c)}{a_n} \right|$$

$$= \left| \frac{x - c}{R} \right|.$$

However, the *interval of convergence* may differ due to the behavior at the endpoints. This is demonstrated in Example 8.

EXAMPLE 8 Intervals of convergence for $f(x)$, $f'(x)$, and $\int f(x)\, dx$

Find the intervals of convergence of (a) $\int f(x)\, dx$, (b) $f(x)$, and (c) $f'(x)$, where

$$f(x) = \sum_{n=1}^{\infty} \frac{x^n}{n} = x + \frac{x^2}{2} + \frac{x^3}{3} + \cdots.$$

SOLUTION

By Theorem 10.21 we have

$$f'(x) = \sum_{n=1}^{\infty} x^{n-1} = 1 + x + x^2 + x^3 + \cdots$$

and

$$\int f(x) \, dx = C + \sum_{n=1}^{\infty} \frac{x^{n+1}}{n(n+1)} = C + \frac{x^2}{1 \cdot 2} + \frac{x^3}{2 \cdot 3} + \frac{x^4}{3 \cdot 4} + \cdots .$$

By the Ratio Test we know that each series has a radius of convergence of $R = 1$. Now, considering the interval $(-1, 1)$, we have the following.

(a) For $\int f(x) \, dx$, the series

$$\sum_{n=1}^{\infty} \frac{x^{n+1}}{n(n+1)} \qquad \text{Interval of convergence: } [-1, 1]$$

converges for $x = \pm 1$, and its interval of convergence is $[-1, 1]$.

(b) For $f(x)$, the series

$$\sum_{n=1}^{\infty} \frac{x^n}{n} \qquad \text{Interval of convergence: } [-1, 1)$$

converges for $x = -1$ and diverges for $x = 1$, and its interval of convergence is $[-1, 1)$.

(c) For $f'(x)$, the series

$$\sum_{n=1}^{\infty} x^{n-1} \qquad \text{Interval of convergence: } (-1, 1)$$

diverges for $x = \pm 1$ and its interval of convergence is $(-1, 1)$. ▭

REMARK From Example 8, it appears that of the three series, the one for the derivative, $f'(x)$, is the least likely to converge at the endpoints. In fact, it can be shown that if the series for $f'(x)$ converges at the endpoints $x = c \pm R$, then the series for $f(x)$ will also converge there (see Exercise 38).

EXERCISES for Section 10.8

In Exercises 1–6, find the radius of convergence of the power series.

1. $\displaystyle\sum_{n=0}^{\infty} (-1)^n \frac{x^n}{n+1}$

2. $\displaystyle\sum_{n=0}^{\infty} (4x)^n$

3. $\displaystyle\sum_{n=1}^{\infty} \frac{(2x)^n}{n^2}$

4. $\displaystyle\sum_{n=0}^{\infty} \frac{(-1)^n x^n}{2^n}$

5. $\displaystyle\sum_{n=0}^{\infty} \frac{(2x)^n}{n!}$

6. $\displaystyle\sum_{n=0}^{\infty} \frac{(2n)! x^n}{n!}$

In Exercises 7–30, find the interval of convergence of the power series. (Be sure to include a check for convergence at the endpoints of the interval.)

7. $\displaystyle\sum_{n=0}^{\infty} \left(\frac{x}{2}\right)^n$

8. $\displaystyle\sum_{n=0}^{\infty} \left(\frac{x}{k}\right)^n$

9. $\displaystyle\sum_{n=1}^{\infty} \frac{(-1)^n x^n}{n}$

10. $\displaystyle\sum_{n=0}^{\infty} (-1)^{n+1} n x^n$

11. $\displaystyle\sum_{n=0}^{\infty} \frac{x^n}{n!}$

12. $\displaystyle\sum_{n=0}^{\infty} \frac{(3x)^n}{(2n)!}$

13. $\displaystyle\sum_{n=0}^{\infty} (2n)! \left(\frac{x}{2}\right)^n$

14. $\displaystyle\sum_{n=0}^{\infty} \frac{(-1)^n x^n}{(n+1)(n+2)}$

15. $\displaystyle\sum_{n=1}^{\infty} \frac{(-1)^{n+1} x^n}{4^n}$

16. $\displaystyle\sum_{n=0}^{\infty} \frac{(-1)^n n! (x-4)^n}{3^n}$

17. $\displaystyle\sum_{n=1}^{\infty} \frac{(-1)^{n+1}(x-5)^n}{n5^n}$

18. $\displaystyle\sum_{n=0}^{\infty} \frac{(x-2)^{n+1}}{(n+1)3^{n+1}}$

19. $\displaystyle\sum_{n=0}^{\infty} \frac{(-1)^{n+1}(x-1)^{n+1}}{n+1}$

20. $\displaystyle\sum_{n=1}^{\infty} \frac{(-1)^{n+1}(x-c)^n}{nc^n}$

21. $\displaystyle\sum_{n=1}^{\infty} \frac{(x-c)^{n-1}}{c^{n-1}}, \ 0 < c$

22. $\displaystyle\sum_{n=1}^{\infty} \frac{(-1)^{n+1}x^{2n-1}}{2n-1}$

23. $\displaystyle\sum_{n=1}^{\infty} \frac{n}{n+1}(-2x)^{n-1}$

24. $\displaystyle\sum_{n=0}^{\infty} \frac{(-1)^n x^{2n}}{n!}$

25. $\displaystyle\sum_{n=0}^{\infty} \frac{x^{2n+1}}{(2n+1)!}$

26. $\displaystyle\sum_{n=1}^{\infty} \frac{n! x^n}{(2n)!}$

27. $\displaystyle\sum_{n=1}^{\infty} \frac{k(k+1)(k+2) \cdots (k+n-1)x^n}{n!}, \ k \geq 1$

28. $\displaystyle\sum_{n=1}^{\infty} \left(\frac{2 \cdot 4 \cdot 6 \cdots 2n}{3 \cdot 5 \cdot 7 \cdots (2n+1)} \right) x^{2n+1}$

29. $\displaystyle\sum_{n=1}^{\infty} \frac{(-1)^{n+1} 3 \cdot 7 \cdot 11 \cdots (4n-1)(x-3)^n}{4^n}$

30. $\displaystyle\sum_{n=1}^{\infty} \frac{n!(x-c)^n}{1 \cdot 3 \cdot 5 \cdots (2n-1)}$

In Exercises 31–34, find the interval of convergence of (a) $f(x)$, (b) $f'(x)$, (c) $f''(x)$, and (d) $\int f(x)\, dx$. Include a check for convergence at the endpoints.

31. $f(x) = \displaystyle\sum_{n=0}^{\infty} \left(\frac{x}{2} \right)^n$

32. $f(x) = \displaystyle\sum_{n=1}^{\infty} \frac{(-1)^{n+1}(x-5)^n}{n5^n}$

33. $f(x) = \displaystyle\sum_{n=0}^{\infty} \frac{(-1)^{n+1}(x-1)^{n+1}}{n+1}$

34. $f(x) = \displaystyle\sum_{n=1}^{\infty} \frac{(-1)^{n+1}(x-1)^n}{n}$

35. Let

$$f(x) = \sum_{n=0}^{\infty} \frac{(-1)^n x^{2n+1}}{(2n+1)!}$$

and

$$g(x) = \sum_{n=0}^{\infty} \frac{(-1)^n x^{2n}}{(2n)!}.$$

(a) Find the interval of convergence of f and g.
(b) Show that $f'(x) = g(x)$.
(c) Show that $g'(x) = -f(x)$.
(d) Identify the functions f and g.

36. Let

$$f(x) = \sum_{n=0}^{\infty} \frac{x^{2n+1}}{(2n+1)!}$$

and

$$g(x) = \sum_{n=0}^{\infty} \frac{x^{2n}}{(2n)!}.$$

(a) Find the interval of convergence of f and g.
(b) Show that $f'(x) = g(x)$.
(c) Show that $g'(x) = f(x)$.
(d) Identify the functions f and g. (See Section 8.7.)

37. Let

$$f(x) = \sum_{n=0}^{\infty} \frac{x^n}{n!}.$$

(a) Find the interval of convergence of f.
(b) Show that $f'(x) = f(x)$.
(c) Show that $f(0) = 1$.
(d) Identify the function f.

38. Prove that if the series for f' converges at the endpoints of its interval of convergence, then the series for f does also.

10.9 Representation of Functions by Power Series

Geometric power series ▪ Operations with power series

Joseph Fourier

In this section and the next we look at several techniques for finding a power series to represent a given function. Some of the early work in representing functions by power series was done by the French mathematician Joseph Fourier (1768–1830). Fourier's work is important in the history of calculus, partly because it forced eighteenth century mathematicians to question the then prevailing narrow concept of a function. Both Cauchy and Dirichlet were motivated by Fourier's work with series, and in 1837 Dirichlet published the general definition of function that is used today.

Geometric power series

Consider the function given by $f(x) = 1/(1-x)$. The form of f closely resembles the sum of a geometric series

$$\sum_{n=0}^{\infty} ar^n = \frac{a}{1-r}, \quad |r| < 1.$$

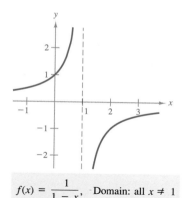

$f(x) = \dfrac{1}{1 - x}$,   Domain: all $x \neq 1$

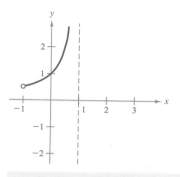

$f(x) = \displaystyle\sum_{n=0}^{\infty} x^n$,   Domain: $-1 < x < 1$

FIGURE 10.14

In other words, if we let $a = 1$ and $r = x$, then a power series representation for $1/(1 - x)$, centered at 0, is

$$\frac{1}{1 - x} = \sum_{n=0}^{\infty} x^n = 1 + x + x^2 + x^3 + \cdots, \quad |x| < 1.$$

$(-1, 1)$, whereas f is defined for all $x \neq 1$, as shown in Figure 10.14. To represent f on another interval, we must develop a different series. For instance, to obtain the power series centered at -1, we write

$$\frac{1}{1 - x} = \frac{1}{2 - (x + 1)}$$

$$= \frac{1/2}{1 - [(x + 1)/2]} = \frac{a}{1 - r}$$

which implies that $a = \frac{1}{2}$ and $r = (x + 1)/2$. Thus, for $|x + 1| < 2$, we have

$$\frac{1}{1 - x} = \sum_{n=0}^{\infty} \left(\frac{1}{2}\right)\left(\frac{x + 1}{2}\right)^n$$

$$= \frac{1}{2}\left[1 + \frac{(x + 1)}{2} + \frac{(x + 1)^2}{4} + \frac{(x + 1)^3}{8} + \cdots\right], \quad |x + 1| < 2$$

which converges on the interval $(-3, 1)$.

EXAMPLE 1 Finding a geometric power series centered at 0

Find a power series for $f(x) = 4/(x + 2)$, centered at 0.

SOLUTION

Writing $f(x)$ in the form $a/(1 - r)$, we have

$$\frac{4}{2 + x} = \frac{2}{1 - (-x/2)} = \frac{a}{1 - r}$$

which implies that $a = 2$ and $r = -x/2$. Therefore, the power series for $f(x)$ is given by

$$\frac{4}{x + 2} = \sum_{n=0}^{\infty} ar^n$$

$$= \sum_{n=0}^{\infty} 2\left(-\frac{x}{2}\right)^n$$

$$= 2\left(1 - \frac{x}{2} + \frac{x^2}{4} - \frac{x^3}{8} + \cdots\right).$$

This power series converges when $|-x/2| < 1$, which implies that the interval of convergence is $(-2, 2)$.

Another way to determine a power series for a rational function such as the one in Example 1 is to use long division. For instance, by dividing $2 + x$ into 4, we obtain the following.

$$
\begin{array}{r}
2 - \ x + \frac{1}{2}x^2 - \frac{1}{4}x^3 + \cdots \\
2 + x \overline{) 4 } \\
\underline{4 + 2x} \\
- \ 2x \\
\underline{- \ 2x - \ x^2} \\
x^2 \\
\underline{x^2 + \frac{1}{2}x^3} \\
- \frac{1}{2}x^3 \\
\underline{- \frac{1}{2}x^3 - \frac{1}{4}x^4}
\end{array}
$$

EXAMPLE 2 Finding a geometric power series centered at 1

Find a power series for $f(x) = 1/x$, centered at 1.

SOLUTION

Writing $f(x)$ in the form $a/(1 - r)$, we have

$$
\frac{1}{x} = \frac{1}{1 - (-x + 1)} = \frac{a}{1 - r}
$$

which implies that $a = 1$ and $r = 1 - x = -(x - 1)$. Therefore, the power series for $f(x)$ is given by

$$
\frac{1}{x} = \sum_{n=0}^{\infty} ar^n
$$

$$
= \sum_{n=0}^{\infty} [-(x - 1)]^n
$$

$$
= 1 - (x - 1) + (x - 1)^2 - (x - 1)^3 + \cdots .
$$

This power series converges when $|x - 1| < 1$, which implies that the interval of convergence is $(0, 2)$. ▭

Operations with power series

The versatility of geometric power series will be shown later in this section, following a discussion of power series operations. These operations, used with differentiation and integration, give us a means for developing power series for a variety of elementary functions. For instance, we will find power series for functions as diverse as

$$
f(x) = \frac{3x - 1}{x^2 - 1}, \qquad f(x) = \ln x, \qquad \text{and} \qquad f(x) = \arctan x.
$$

We summarize the basic operations with power series in the following theorem. For simplicity, we state the results for a series centered at 0.

**THEOREM 10.22
OPERATIONS WITH
POWER SERIES**

Given $f(x) = \Sigma a_n x^n$ and $g(x) = \Sigma b_n x^n$, the following properties are true.

1. $f(kx) = \displaystyle\sum_{n=0}^{\infty} a_n k^n x^n$

2. $f(x^N) = \displaystyle\sum_{n=0}^{\infty} a_n x^{nN}$

3. $f(x) \pm g(x) = \displaystyle\sum_{n=0}^{\infty} (a_n \pm b_n)x^n$

4. $f(x)g(x) = \left(\displaystyle\sum_{n=0}^{\infty} a_n x^n\right)\left(\displaystyle\sum_{n=0}^{\infty} b_n x^n\right)$

The operations described in Theorem 10.22 can change the interval of convergence for the resulting series. For example, in the following addition, the interval of convergence for the sum is the *intersection* of the intervals of convergence of the two original series.

$$\underbrace{\sum_{n=0}^{\infty} x^n}_{(-1, 1)} + \underbrace{\sum_{n=0}^{\infty} \left(\frac{x}{2}\right)^n}_{(-2, 2)} = \underbrace{\sum_{n=0}^{\infty} \left(1 + \frac{1}{2^n}\right)x^n}_{(-1, 1)}$$

$$(-1, 1) \quad \cap \quad (-2, 2) \quad = \quad (-1, 1)$$

EXAMPLE 3 Adding two power series

Find a power series, centered at 0, for

$$f(x) = \frac{3x - 1}{x^2 - 1}.$$

SOLUTION

Using partial fractions, we can write $f(x)$ as

$$\frac{3x - 1}{x^2 - 1} = \frac{2}{x + 1} + \frac{1}{x - 1}.$$

Now, using two geometric power series, we write

$$\frac{2}{x + 1} = \frac{2}{1 - (-x)} = \sum_{n=0}^{\infty} 2(-1)^n x^n, \quad |x| < 1$$

and

$$\frac{1}{x - 1} = \frac{-1}{1 - x} = -\sum_{n=0}^{\infty} x^n, \quad |x| < 1.$$

Finally, adding these two power series produces

$$\frac{3x - 1}{x^2 - 1} = \sum_{n=0}^{\infty} [2(-1)^n - 1]x^n = 1 - 3x + x^2 - 3x^3 + x^4 - \cdots.$$

The interval of convergence for this power series is $(-1, 1)$. ▭

EXAMPLE 4 Finding a power series by integration

Find a power series for $f(x) = \ln x$, centered at 1.

SOLUTION

From Example 2 we know that

$$\frac{1}{x} = \sum_{n=0}^{\infty} (-1)^n (x - 1)^n. \qquad (0, 2)$$

By integration, we have

$$\ln x = \int \frac{1}{x}\, dx + C = C + \sum_{n=0}^{\infty} (-1)^n \frac{(x-1)^{n+1}}{n+1}.$$

To determine C, we let $x = 1$ and conclude that $C = 0$. Therefore,

$$\ln x = \sum_{n=0}^{\infty} (-1)^n \frac{(x-1)^{n+1}}{n+1}$$

$$= \frac{(x-1)}{1} - \frac{(x-1)^2}{2} + \frac{(x-1)^3}{3} - \frac{(x-1)^4}{4} + \cdots. \qquad (0, 2]$$

Note that the series converges at $x = 2$. This is consistent with our observation in the previous section that integration of a power series may alter the convergence at the endpoints of the interval of convergence. ◻

In Section 10.7, we approximated $\ln (1.1)$ using the fourth Taylor polynomial for the natural logarithmic function.

$$\ln x \approx (x - 1) - \frac{(x-1)^2}{2} + \frac{(x-1)^3}{3} - \frac{(x-1)^4}{4}$$

$$\ln (1.1) \approx (0.1) - \frac{1}{2}(0.1)^2 + \frac{1}{3}(0.1)^3 - \frac{1}{4}(0.1)^4 \approx 0.0953083$$

We now know from Example 4 that this polynomial represents the first four terms of the power series for $\ln x$. Moreover, using the Alternating Series Remainder, we can determine that the error in this approximation is less than

$$|R_3| \leq |a_4| = \frac{1}{5}(0.1)^5 \approx 0.000002.$$

During the seventeenth and eighteenth centuries, mathematical tables for logarithms and values of other transcendental functions were computed in this manner. The use for such numerical techniques is far from outdated, since it is precisely by such means that modern calculating devices are programmed to evaluate transcendental functions.

EXAMPLE 5 Finding a power series by integration

Find a power series for $f(x) = \arctan x$, centered at 0.

SOLUTION

Since $D_x[\arctan x] = 1/(1 + x^2)$, we use the series

$$f(x) = \frac{1}{1 + x} = \sum_{n=0}^{\infty} (-1)^n x^n. \qquad (-1, 1)$$

Now, substituting x^2 for x, we obtain

$$f(x^2) = \frac{1}{1 + x^2} = \sum_{n=0}^{\infty} (-1)^n x^{2n}.$$

Finally, by integrating, we obtain

$$\arctan x = \int \frac{1}{1 + x^2} \, dx + C$$

$$= C + \sum_{n=0}^{\infty} (-1)^n \frac{x^{2n+1}}{2n + 1}$$

$$= \sum_{n=0}^{\infty} (-1)^n \frac{x^{2n+1}}{2n + 1} \qquad \text{Let } x = 0, \text{ then } C = 0.$$

$$= x - \frac{x^3}{3} + \frac{x^5}{5} - \frac{x^7}{7} + \cdots. \qquad (-1, 1)$$

It can be shown that the power series developed for arctan x in Example 5 also converges (to arctan x) for $x = \pm 1$. For instance, when $x = 1$, we can write

$$\arctan 1 = 1 - \frac{1}{3} + \frac{1}{5} - \frac{1}{7} + \cdots = \frac{\pi}{4}.$$

However, this series (developed by James Gregory in 1671) does not give us a practical way of approximating π because it converges so slowly that hundreds of terms would have to be used to obtain reasonable accuracy. Example 6 shows how to use *two* different arctangent series to obtain a very good approximation of π using only a few terms. This approximation was developed by John Machin in 1706.

Series that can be used to approximate π have interested mathematicians for the past 300 years. An amazing series for approximating $1/\pi$ was developed by the Indian mathematician Srinivasa Ramanujan (1887–1920) in 1914. Each successive term of Ramanujan's series adds roughly eight more correct digits to the value of $1/\pi$.

Srinivasa Ramanujan

EXAMPLE 6 Approximating π with a series

Use the trigonometric identity

$$4 \arctan \frac{1}{5} - \arctan \frac{1}{239} = \frac{\pi}{4}$$

to approximate the number π.

SOLUTION

By using only five terms from each of the series for arctan $(1/5)$ and arctan $(1/239)$, we obtain

$$4\left(4 \arctan \frac{1}{5} - \arctan \frac{1}{239}\right) \approx 3.141593$$

which agrees with the decimal representation of π with an error less than 0.000001.

EXERCISES for Section 10.9

In Exercises 1–14, find a power series for the given function, centered at c, and determine the interval of convergence.

1. $f(x) = \dfrac{1}{2 - x}$, $c = 0$

2. $f(x) = \dfrac{3}{4 - x}$, $c = 0$

3. $f(x) = \dfrac{1}{2 - x}$, $c = 5$

4. $f(x) = \dfrac{3}{4 - x}$, $c = -2$

5. $f(x) = \dfrac{3}{2x - 1}$, $c = 0$

6. $f(x) = \dfrac{3}{2x - 1}$, $c = 2$

7. $f(x) = \dfrac{1}{2x - 5}$, $c = -3$

8. $f(x) = \dfrac{1}{2x - 5}$, $c = 0$

9. $f(x) = \dfrac{3}{x + 2}$, $c = 0$

10. $f(x) = \dfrac{4}{3x + 2}$, $c = 2$

11. $f(x) = \dfrac{3x}{x^2 + x - 2}$, $c = 0$

12. $f(x) = \dfrac{4x - 7}{2x^2 + 3x - 2}$, $c = 0$

13. $f(x) = \dfrac{2}{1 - x^2}$, $c = 0$

14. $f(x) = \dfrac{4}{4 + x^2}$, $c = 0$

In Exercises 15–22, use the power series

$$\frac{1}{1 + x} = \sum_{n=0}^{\infty} (-1)^n x^n$$

to determine a power series representation, centered at 0, for the given function. List the interval of convergence with your solution.

15. $f(x) = -\dfrac{1}{(x + 1)^2}$

16. $f(x) = \dfrac{2}{(x + 1)^3}$

17. $f(x) = \ln (x + 1)$

18. $f(x) = \ln (x^2 + 1)$

19. $f(x) = \dfrac{1}{4x^2 + 1}$

20. $f(x) = \arctan 2x$

21. $f(x) = \ln \sqrt{\dfrac{1 + x}{1 - x}}$

22. $f(x) = \dfrac{1}{x^2 - x + 1} = \dfrac{x + 1}{x^3 + 1}$

23. Complete the following table to demonstrate the inequalities

$$x - \frac{x^2}{2} \leq \ln (x + 1) \leq x - \frac{x^2}{2} + \frac{x^3}{3}.$$

x	0.0	0.2	0.4	0.6	0.8	1.0
$x - \dfrac{x^2}{2}$						
$\ln (x + 1)$						
$x - \dfrac{x^2}{2} + \dfrac{x^3}{3}$						

24. Complete a table for the same values of x as in Exercise 23 to demonstrate the inequalities

$$x - \frac{x^2}{2} + \frac{x^3}{3} - \frac{x^4}{4} \leq \ln (x + 1)$$

$$\leq x - \frac{x^2}{2} + \frac{x^3}{3} - \frac{x^4}{4} + \frac{x^5}{5}.$$

In Exercises 25–28, use the series for $f(x) = \arctan x$ to approximate the given value, using $R_N \leq 0.001$.

25. $\arctan \dfrac{1}{4}$

26. $\displaystyle\int_0^{3/4} \arctan x^2 \, dx$

27. $\displaystyle\int_0^{1/2} \dfrac{\arctan x^2}{x} \, dx$

28. $\displaystyle\int_0^{1/2} x^2 \arctan x \, dx$

In Exercises 29 and 30, (a) verify the given equation. [Hint: Use trigonometric identities for the tangent of the sum of two angles and the tangent of a double angle.] (b) Use the equation and the series for the arctangent to approximate π to two-decimal-place accuracy.

29. $2 \arctan \dfrac{1}{2} - \arctan \dfrac{1}{7} = \dfrac{\pi}{4}$

30. $2 \arctan \dfrac{2}{3} - \arctan \dfrac{7}{17} = \dfrac{\pi}{4}$

In Exercises 31 and 32, use a computer or graphics calculator to verify the inequality graphically on the interval $0 \le x \le 1$.

31. $x - \dfrac{x^2}{2} \le \ln(x + 1) \le x - \dfrac{x^2}{2} + \dfrac{x^3}{3}$

32. $x - \dfrac{x^2}{2} + \dfrac{x^3}{3} - \dfrac{x^4}{4} \le \ln(x + 1)$

$$\le x - \dfrac{x^2}{2} + \dfrac{x^3}{3} - \dfrac{x^4}{4} + \dfrac{x^5}{5}$$

33. Use long division to develop the power series for $1/(1 + x)$.

10.10 Taylor and Maclaurin Series

Taylor series and Maclaurin series ▪ Binomial series

Colin Maclaurin

In Section 10.9 we derived power series for several functions using geometric series with term-by-term differentiation or integration. In this section we develop a *general* procedure for deriving the power series for a function that has derivatives of all orders.

The development of power series to represent functions is due to the combined work of many seventeenth and eighteenth century mathematicians. Gregory, Newton, John and James Bernoulli, Leibniz, Euler, Lagrange, Wallis, and Fourier all contributed to this work. However, the two names that are most commonly associated with power series are Brook Taylor (1685–1731) and Colin Maclaurin (1698–1746).

We begin with a theorem that gives us the form that *every* (convergent) power series must take.

THEOREM 10.23
THE FORM OF A CONVERGENT POWER SERIES

If f is represented by a power series $f(x) = \Sigma \, a_n(x - c)^n$ for all x in an open interval I containing c, then $a_n = f^{(n)}(c)/n!$ and

$$f(x) = f(c) + f'(c)(x - c) + \frac{f''(c)}{2!}(x - c)^2 + \cdots + \frac{f^{(n)}(c)}{n!}(x - c)^n + \cdots.$$

PROOF

Suppose the power series $\Sigma a_n(x - c)^n$ has a radius of convergence R. Then, by Theorem 10.21, we know that the nth derivative of f exists for $|x - c| < R$, and by successive differentiation we obtain the following.

$$f^{(0)}(x) = a_0 + a_1(x - c) + a_2(x - c)^2 + a_3(x - c)^3 + a_4(x - c)^4 + \cdots$$
$$f^{(1)}(x) = a_1 + 2a_2(x - c) + 3a_3(x - c)^2 + 4a_4(x - c)^3 + \cdots$$
$$f^{(2)}(x) = 2a_2 + 3!a_3(x - c) + 4 \cdot 3a_4(x - c)^2 + \cdots$$
$$f^{(3)}(x) = 3!a_3 + 4!a_4(x - c) + \cdots$$
$$\vdots$$
$$\vdots$$
$$f^{(n)}(x) = n!a_n + (n + 1)!a_{n+1}(x - c) + \cdots$$

Now, evaluating each of these derivatives at $x = c$ yields

$$f^{(0)}(c) = 0!a_0, \qquad f^{(1)}(c) = 1!a_1, \qquad f^{(2)}(c) = 2!a_2, \qquad f^{(3)}(c) = 3!a_3$$

and, in general, $f^{(n)}(c) = n!a_n$. By solving for a_n we find that the coefficients of the power series representation of $f(x)$ are

$$a_n = \frac{f^{(n)}(c)}{n!}.$$

Notice that the coefficients of the power series in Theorem 10.23 are precisely the coefficients of the Taylor polynomials for $f(x)$ at c as defined in Section 10.7. For this reason, the series is called the **Taylor series** for $f(x)$ at c.

DEFINITION OF TAYLOR AND MACLAURIN SERIES

If a function f has derivatives of all orders at $x = c$, then the series

$$\sum_{n=0}^{\infty} \frac{f^{(n)}(c)}{n!}(x - c)^n = f(c) + f'(c)(x - c) + \cdots + \frac{f^{(n)}(c)}{n!}(x - c)^n + \cdots$$

is called the **Taylor series for $f(x)$ at c.** Moreover, if $c = 0$, then this series is also called the **Maclaurin series for f.**

If we know the pattern for the coefficients of the Taylor polynomials for a function, then we can extend the pattern easily to form the corresponding Taylor series. For instance, in Example 4 of Section 10.7, we found the fourth Taylor polynomial for $\ln x$, centered at 1, to be

$$P_4(x) = (x - 1) - \frac{1}{2}(x - 1)^2 + \frac{1}{3}(x - 1)^3 - \frac{1}{4}(x - 1)^4.$$

Following this pattern, we obtain the Taylor series for $\ln x$ centered at 1

$$(x - 1) - \frac{1}{2}(x - 1)^2 + \cdots + \frac{(-1)^{n+1}}{n}(x - 1)^n + \cdots.$$

EXAMPLE 1   Forming a power series

Use the function $f(x) = \sin x$ to form the Maclaurin series

$$\sum_{n=0}^{\infty} \frac{f^{(n)}(0)}{n!}x^n = f(0) + f'(0)x + \frac{f''(0)}{2!}x^2 + \frac{f^{(3)}(0)}{3!}x^3 + \frac{f^{(4)}(0)}{4!}x^4 + \cdots$$

and determine the interval of convergence.

SOLUTION

Successive differentiation of $f(x)$ yields

$$
\begin{aligned}
f(x) &= \sin x & f(0) &= 0 \\
f'(x) &= \cos x & f'(0) &= 1 \\
f''(x) &= -\sin x & f''(0) &= 0 \\
f^{(3)}(x) &= -\cos x & f^{(3)}(0) &= -1 \\
f^{(4)}(x) &= \sin x & f^{(4)}(0) &= 0 \\
f^{(5)}(x) &= \cos x & f^{(5)}(0) &= 1
\end{aligned}
$$

and so on. The pattern repeats after the third derivative. Hence, the power series is as follows.

$$
\sum_{n=0}^{\infty} \frac{f^{(n)}(0)}{n!} x^n = f(0) + f'(0)x + \frac{f''(0)}{2!}x^2 + \frac{f^{(3)}(0)}{3!}x^3 + \frac{f^{(4)}(0)}{4!}x^4 + \cdots
$$

$$
\sum_{n=0}^{\infty} \frac{(-1)^n x^{2n+1}}{(2n+1)!} = x - \frac{x^3}{3!} + \frac{x^5}{5!} - \frac{x^7}{7!} + \cdots
$$

By the Ratio Test, we can conclude that this series converges for all x. ▭

Notice that we did not conclude that the power series in Example 1 converges to $\sin x$ for all x. We simply concluded that the power series converges to some function, but we are not sure what function it is. This is a subtle, but important, point in dealing with Taylor or Maclaurin series. To persuade yourself that the series

$$
f(c) + f'(c)(x - c) + \frac{f''(c)}{2!}(x - c)^2 + \cdots + \frac{f^{(n)}(c)}{n!}(x - c)^n + \cdots
$$

might converge to a function other than f, remember that the derivatives are being evaluated at a single point. It can easily happen that another function will agree with the values of $f^{(n)}(x)$ when $x = c$ and disagree at other x-values. For instance, if we formed the power series (centered at 0) for the function

$$
f(x) = \begin{cases} -1, & x < -\dfrac{\pi}{2} \\[2mm] \sin x, & |x| \le \dfrac{\pi}{2} \\[2mm] 1, & x > \dfrac{\pi}{2} \end{cases} \qquad \text{(See Figure 10.15.)}
$$

we would obtain the same series we obtained in Example 1. We know that the series converges for all x, and yet it obviously cannot converge to both $f(x)$ and $\sin x$ for all x.

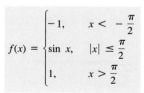

$$
f(x) = \begin{cases} -1, & x < -\dfrac{\pi}{2} \\[2mm] \sin x, & |x| \le \dfrac{\pi}{2} \\[2mm] 1, & x > \dfrac{\pi}{2} \end{cases}
$$

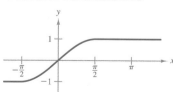

FIGURE 10.15

REMARK Don't confuse this observation with the result given in Theorem 10.23. That theorem says that *if a power series converges to* $f(x)$, then the series must be a Taylor series. The theorem does not say that every series formed with the Taylor coefficients $a_n = f^{(n)}(c)/n!$ will converge to $f(x)$.

To determine whether a power series formed with Taylor coefficients actually converges to $f(x)$, it is helpful to consider the series as the sum of the nth Taylor polynomial for $f(x)$ and a remainder term.

$$\underbrace{f(c) + \cdots + \frac{f^{(n)}(c)}{n!}(x - c)^n}_{P_n(x)} + \underbrace{\frac{f^{(n+1)}(c)}{(n + 1)!}(x - c)^{n+1} + \cdots}_{R_n(x)}$$

Recall that Taylor's Theorem, given in Section 10.7, tells us that the remainder for a Taylor polynomial is given by

$$R_n(x) = \frac{f^{(n+1)}(z)}{(n + 1)!}(x - c)^{n+1}$$

where z lies between x and c. The problem we now want to tackle is that of finding those values of x for which the Taylor polynomials $P_n(x)$ actually converge to $f(x)$ as $n \to \infty$. In other words, we want to find the values of x for which

$$
\begin{aligned}
f(x) &= \lim_{n \to \infty} P_n(x) \\
&= \lim_{n \to \infty} \sum_{k=0}^{n} \frac{f^{(k)}(c)}{k!}(x - c)^k \\
&= \sum_{k=0}^{\infty} \frac{f^{(k)}(c)}{k!}(x - c)^k.
\end{aligned}
$$

The conditions that guarantee this convergence are given in Theorem 10.24.

**THEOREM 10.24
CONVERGENCE OF
TAYLOR SERIES**

If a function f has derivatives of all orders in an interval I centered at c, then the equality

$$f(x) = \sum_{n=0}^{\infty} \frac{f^{(n)}(c)}{n!}(x - c)^n$$

holds if and only if

$$\lim_{n \to \infty} R_n(x) = 0$$

for every x in I.

PROOF   For a Taylor series, the nth partial sum coincides with the nth Taylor polynomial. That is, $S_n(x) = P_n(x)$. Moreover, since

$$P_n(x) = f(x) - R_n(x)$$

it follows that

$$\lim_{n \to \infty} S_n(x) = \lim_{n \to \infty} P_n(x) = \lim_{n \to \infty} [f(x) - R_n(x)] = f(x) - \lim_{n \to \infty} R_n(x).$$

Hence, for a given x, the Taylor series (the sequence of partial sums) converges to $f(x)$ if and only if $R_n(x) \to 0$ as $n \to \infty$.

REMARK   Stated another way, this theorem says that a power series formed with the Taylor coefficients $a_n = f^{(n)}(c)/n!$ converges to the function from which it was derived at precisely those values for which the remainder approaches zero as $n \to \infty$.

In Example 2, we take another look at the series formed in Example 1. We derived that series from the sine function and we also concluded that the series converges to some function on the entire real line. We now show that the series actually converges to sin x.

EXAMPLE 2   A convergent Maclaurin series

Show that the Maclaurin series for $f(x) = \sin x$ converges to sin x for all x.

SOLUTION

Using the result in Example 1, we need to show that

$$\sin x = x - \frac{x^3}{3!} + \frac{x^5}{5!} - \frac{x^7}{7!} + \cdots + \frac{(-1)^n x^{2n+1}}{(2n + 1)!} + \cdots$$

is true for all x. Since

$$f^{(n+1)}(x) = \pm\sin x \qquad \text{or} \qquad f^{(n+1)}(x) = \pm\cos x$$

we know that $\left| f^{(n+1)}(z) \right| \leq 1$ for every real number z. Therefore, for any fixed x, we can apply Taylor's Theorem (Theorem 10.19) to conclude that

$$0 \leq |R_n(x)| \leq \left| \frac{f^{(n+1)}(z)}{(n + 1)!} x^{n+1} \right| \leq \frac{|x|^{n+1}}{(n + 1)!}.$$

Now, from our discussion in Section 10.1 regarding the relative rate of convergence of exponential and factorial sequences, it follows that for a fixed x

$$\lim_{n \to \infty} \frac{|x|^{n+1}}{(n + 1)!} = 0.$$

Finally, by the Squeeze Theorem, it follows that for all x, $R_n(x) \to 0$ as $n \to \infty$. Hence, by Theorem 10.24, the Maclaurin series for sin x converges to sin x for all x.   ▭

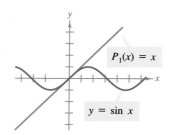

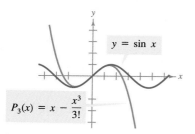

Figure 10.16 visually illustrates the convergence of the Maclaurin series for sin x by comparing the graphs of the Maclaurin polynomials $P_1(x)$, $P_3(x)$, $P_5(x)$, and $P_7(x)$ with the graph of the sine function. Notice that as the degree of the polynomial increases, its graph more closely resembles that of the sine function.

We summarize the steps for finding a Taylor series for $f(x)$ at c as follows.

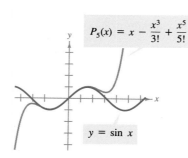

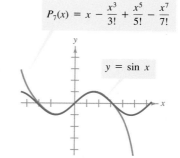

FIGURE 10.16

GUIDELINES FOR FINDING A TAYLOR SERIES

1. Differentiate $f(x)$ several times and evaluate each derivative at c.

$$f(c), f'(c), f''(c), f'''(c), \ldots, f^{(n)}(c), \ldots$$

Try to recognize a pattern for these numbers.

2. Use the sequence developed in the first step to form the Taylor coefficients $a_n = f^{(n)}(c)/n!$, and determine the interval of convergence for the resulting power series

$$f(c) + f'(c)(x - c) + \frac{f''(c)}{2!}(x - c)^2 + \cdots + \frac{f^{(n)}(c)}{n!}(x - c)^n + \cdots.$$

3. Within this interval of convergence, determine whether the series converges to $f(x)$.

The direct determination of Taylor or Maclaurin coefficients using successive differentiation can be difficult, and the next example illustrates a shortcut for finding the coefficients indirectly—using the coefficients of a known Taylor or Maclaurin series.

EXAMPLE 3 Maclaurin series for a composite function

Find the Maclaurin series for $f(x) = \sin x^2$.

SOLUTION

To find the coefficients for this Maclaurin series *directly* we must calculate successive derivatives of $f(x) = \sin x^2$. By calculating just the first two,

$$f'(x) = 2x \cos x^2 \qquad \text{and} \qquad f''(x) = -4x^2 \sin x^2 + 2 \cos x^2$$

we recognize this to be cumbersome. Fortunately there is an alternative. Suppose we first consider the Maclaurin series for $\sin x$ found in Example 1.

$$g(x) = \sin x$$

$$= x - \frac{x^3}{3!} + \frac{x^5}{5!} - \frac{x^7}{7!} + \cdots$$

Now, since $\sin x^2 = g(x^2)$, we can apply Theorem 10.22 and substitute x^2 for x in the series for $\sin x$ to obtain

$$\sin x^2 = g(x^2)$$

$$= x^2 - \frac{x^6}{3!} + \frac{x^{10}}{5!} - \frac{x^{14}}{7!} + \cdots .$$

Be sure to understand the point illustrated in Example 3. Since direct computation of Taylor or Maclaurin coefficients can be tedious, the most practical way to find a Taylor or Maclaurin series is to develop power series for a *basic list* of elementary functions. From this list, we can determine power series for other functions by the operations of addition, subtraction, multiplication, division, differentiation, integration, or composition with known power series. Before presenting the basic list for elementary functions, we develop one additional formula—for a function of the form $f(x) = (1 + x)^k$. This produces the **binomial series.**

EXAMPLE 4 Binomial series

Find the Maclaurin series for $f(x) = (1 + x)^k$ and determine its radius of convergence.

SOLUTION

By successive differentiation, we have

$$f(x) = (1 + x)^k \qquad\qquad f(0) = 1$$
$$f'(x) = k(1 + x)^{k-1} \qquad\qquad f'(0) = k$$
$$f''(x) = k(k - 1)(1 + x)^{k-2} \qquad\qquad f''(0) = k(k - 1)$$
$$f'''(x) = k(k - 1)(k - 2)(1 + x)^{k-3} \qquad f'''(0) = k(k - 1)(k - 2)$$
$$\vdots \qquad\qquad\qquad\qquad \vdots$$
$$f^{(n)}(x) = k \cdots (k - n + 1)(1 + x)^{k-n} \quad f^{(n)}(0) = k(k - 1) \cdots (k - n + 1)$$

which produces the series

$$1 + kx + \frac{k(k - 1)x^2}{2} + \cdots + \frac{k(k - 1) \cdots (k - n + 1)x^n}{n!} + \cdots .$$

By the Ratio Test, we have $a_{n+1}/a_n \to 1$ as $n \to \infty$. This implies that the radius of convergence is $R = 1$, and thus the series converges to *some* function in the interval $(-1, 1)$.

Note that in Example 4 we showed that the Taylor series for $(1 + x)^k$ converges to *some* function in the interval $(-1, 1)$. However, we did not show that the series actually converges to $(1 + x)^k$. To do this, we could show that the remainder $R_n(x)$ converges to 0, as we did in Example 2. Another way to show that the Taylor series for f converges to $f(x)$ is to use the series to form a differential equation. Then, the verification of convergence to $f(x)$ can be obtained by solving the differential equation. For example, if we let y be the binomial series obtained in Example 4,

$$y = 1 + kx + \frac{k(k - 1)x^2}{2} + \cdots + \frac{k(k - 1) \cdots (k - n + 1)x^n}{n!} + \cdots$$

then we can derive the differential equation

$$y' + xy' = ky. \qquad \text{Initial condition } y = 1 \text{ if } x = 0$$

In the following list, we provide the power series for several elementary functions with the corresponding intervals of convergence.

POWER SERIES FOR ELEMENTARY FUNCTIONS

Interval of Convergence

$$\frac{1}{x} = 1 - (x - 1) + (x - 1)^2 - (x - 1)^3 + (x - 1)^4 - \cdots + (-1)^n(x - 1)^n + \cdots, \qquad 0 < x < 2$$

$$\frac{1}{1 + x} = 1 - x + x^2 - x^3 + x^4 - x^5 + \cdots + (-1)^n x^n + \cdots, \qquad -1 < x < 1$$

$$\ln x = (x - 1) - \frac{(x - 1)^2}{2} + \frac{(x - 1)^3}{3} - \frac{(x - 1)^4}{4} + \cdots + \frac{(-1)^{n-1}(x - 1)^n}{n} + \cdots, \qquad 0 < x \leq 2$$

$$e^x = 1 + x + \frac{x^2}{2!} + \frac{x^3}{3!} + \frac{x^4}{4!} + \frac{x^5}{5!} + \cdots + \frac{x^n}{n!} + \cdots, \qquad -\infty < x < \infty$$

$$\sin x = x - \frac{x^3}{3!} + \frac{x^5}{5!} - \frac{x^7}{7!} + \frac{x^9}{9!} - \cdots + \frac{(-1)^n x^{2n+1}}{(2n + 1)!} + \cdots, \qquad -\infty < x < \infty$$

$$\cos x = 1 - \frac{x^2}{2!} + \frac{x^4}{4!} - \frac{x^6}{6!} + \frac{x^8}{8!} - \cdots + \frac{(-1)^n x^{2n}}{(2n)!} + \cdots, \qquad -\infty < x < \infty$$

$$\arctan x = x - \frac{x^3}{3} + \frac{x^5}{5} - \frac{x^7}{7} + \frac{x^9}{9} - \cdots + \frac{(-1)^n x^{2n+1}}{2n + 1} + \cdots, \qquad -1 \leq x \leq 1$$

$$\arcsin x = x + \frac{x^3}{2 \cdot 3} + \frac{1 \cdot 3 x^5}{2 \cdot 4 \cdot 5} + \frac{1 \cdot 3 \cdot 5 x^7}{2 \cdot 4 \cdot 6 \cdot 7} + \cdots + \frac{(2n)! x^{2n+1}}{(2^n n!)^2(2n + 1)} + \cdots, \qquad -1 \leq x \leq 1$$

$$(1 + x)^k = 1 + kx + \frac{k(k - 1)x^2}{2!} + \frac{k(k - 1)(k - 2)x^3}{3!} + \frac{k(k - 1)(k - 2)(k - 3)x^4}{4!} + \cdots, \qquad -1 < x < 1*$$

*The convergence at $x = \pm 1$ depends on the value k.

REMARK The binomial series is valid for noninteger values of k. Moreover, if k happens to be a positive integer, then the binomial series reduces to a simple binomial expansion.

In the next three examples we show how to use the basic list of power series.

EXAMPLE 5 Deriving a new power series from a given series

Find the power series for $f(x) = \cos \sqrt{x}$.

SOLUTION

Using the power series

$$\cos x = 1 - \frac{x^2}{2!} + \frac{x^4}{4!} - \frac{x^6}{6!} + \frac{x^8}{8!} - \cdots$$

we replace x by $\sqrt{x}$ to obtain the series

$$\cos \sqrt{x} = 1 - \frac{x}{2!} + \frac{x^2}{4!} - \frac{x^3}{6!} + \frac{x^4}{8!} - \cdots$$

which converges for all x in the domain of $\cos \sqrt{x}$, that is, for $x \geq 0$. ▭

EXAMPLE 6 Using the binomial series

Find the power series for $g(x) = \sqrt[3]{1 + x}$.

SOLUTION

Using the binomial series

$$(1 + x)^k = 1 + kx + \frac{k(k - 1)x^2}{2!} + \frac{k(k - 1)(k - 2)x^3}{3!} + \cdots$$

we let $k = \frac{1}{3}$ and write

$$(1 + x)^{1/3} = 1 + \frac{x}{3} - \frac{2x^2}{3^2 2!} + \frac{2 \cdot 5x^3}{3^3 3!} - \frac{2 \cdot 5 \cdot 8x^4}{3^4 4!} + \cdots$$

which converges for $-1 \leq x \leq 1$. ▭

EXAMPLE 7 A power series for $\sin^2 x$

Find the power series for $f(x) = \sin^2 x$.

SOLUTION

Instead of squaring the power series for sin x, we write

$$\sin^2 x = \frac{1 - \cos 2x}{2} = \frac{1}{2} - \frac{\cos 2x}{2}.$$

Now, using the series for cos x, we proceed as follows.

$$\cos x = 1 - \frac{x^2}{2!} + \frac{x^4}{4!} - \frac{x^6}{6!} + \frac{x^8}{8!} - \cdots$$

$$\cos 2x = 1 - \frac{2^2}{2!}x^2 + \frac{2^4}{4!}x^4 - \frac{2^6}{6!}x^6 + \frac{2^8}{8!}x^8 - \cdots$$

$$-\frac{1}{2}\cos 2x = -\frac{1}{2} + \frac{2}{2!}x^2 - \frac{2^3}{4!}x^4 + \frac{2^5}{6!}x^6 - \frac{2^7}{8!}x^8 + \cdots$$

$$\sin^2 x = \frac{1}{2} - \frac{1}{2}\cos 2x = \frac{1}{2} - \frac{1}{2} + \frac{2}{2!}x^2 - \frac{2^3}{4!}x^4 + \frac{2^5}{6!}x^6 - \frac{2^7}{8!}x^8 + \cdots$$

$$= \frac{2}{2!}x^2 - \frac{2^3}{4!}x^4 + \frac{2^5}{6!}x^6 - \frac{2^7}{8!}x^8 + \cdots$$

This series converges for $-\infty < x < \infty$. ▭

As mentioned in the previous section, power series can be used to obtain tables of values of transcendental functions. They are also useful for estimating the value of definite integrals for which antiderivatives cannot be found. The next example demonstrates this use.

EXAMPLE 8 Power series approximation of a definite integral

Use a power series to approximate

$$\int_0^1 e^{-x^2}\, dx$$

with an error of less than 0.01.

SOLUTION

Replacing x with $-x^2$ in the series for e^x, we have

$$e^{-x^2} = 1 - x^2 + \frac{x^4}{2!} - \frac{x^6}{3!} + \frac{x^8}{4!} - \cdots$$

$$\int_0^1 e^{-x^2}\, dx = \left[x - \frac{x^3}{3} + \frac{x^5}{5 \cdot 2!} - \frac{x^7}{7 \cdot 3!} + \frac{x^9}{9 \cdot 4!} - \cdots \right]_0^1$$

$$= 1 - \frac{1}{3} + \frac{1}{10} - \frac{1}{42} + \frac{1}{216} - \cdots.$$

Summing the first *four* terms, we have

$$\int_0^1 e^{-x^2} \, dx \approx 0.74$$

which, by the Alternating Series Test, has an error of less than $\frac{1}{216} \approx 0.005$.

EXERCISES for Section 10.10

In Exercises 1–10, find the Taylor series (centered at c) for the given function.

1. $f(x) = e^{2x}$, $c = 0$ **2.** $f(x) = e^{-2x}$, $c = 0$

3. $f(x) = \cos x$, $c = \dfrac{\pi}{4}$ **4.** $f(x) = \sin x$, $c = \dfrac{\pi}{4}$

5. $f(x) = \ln x$, $c = 1$ **6.** $f(x) = e^x$, $c = 1$

7. $f(x) = \sin 2x$, $c = 0$

8. $f(x) = \tan x$, $c = 0$ (first three nonzero terms)

9. $f(x) = \sec x$, $c = 0$ (first three nonzero terms)

10. $f(x) = \ln (x^2 + 1)$, $c = 0$

In Exercises 11–16, use the binomial series to find the Maclaurin series for the given function.

11. $f(x) = \dfrac{1}{(1 + x)^2}$ **12.** $f(x) = \dfrac{1}{\sqrt{1 - x}}$

13. $f(x) = \dfrac{1}{\sqrt{4 + x^2}}$ **14.** $f(x) = \sqrt{1 + x}$

15. $f(x) = \sqrt{1 + x^2}$ **16.** $f(x) = \sqrt{1 + x^3}$

In Exercises 17–30, find the Maclaurin series for the given function from a power series in the table of power series for elementary functions.

17. $f(x) = e^{x^2/2}$ **18.** $g(x) = e^{-3x}$

19. $g(x) = \sin 2x$ **20.** $h(x) = x \cos x$

21. $f(x) = \cos \sqrt{x}$ **22.** $f(x) = \sin x^2$

23. $g(x) = \dfrac{\sin x}{x}$ **24.** $f(x) = \dfrac{\arcsin x}{x}$

25. $f(x) = \dfrac{1}{2}(e^x - e^{-x}) = \sinh x$

26. $f(x) = \dfrac{1}{2}(e^x + e^{-x}) = \cosh x$

27. $g(x) = \dfrac{1}{2i}(e^{ix} - e^{-ix}) = \sin x$

     [Hint: $i^2 = -1$]

28. $g(x) = \dfrac{1}{2}(e^{ix} + e^{-ix}) = \cos x$

     [Hint: $i^2 = -1$]

29. $f(x) = \cos^2 x$

     $\left[\text{Hint: } \cos^2 x = \dfrac{1}{2}(1 + \cos 2x)\right]$

30. $f(x) = \sinh^{-1} x = \ln (x + \sqrt{x^2 + 1})$

$\left[\text{Hint: Integrate the series for } \dfrac{1}{\sqrt{x^2 + 1}}.\right]$

In Exercises 31–38, use power series to approximate the given integral with an error of less than 0.0001.

31. $\displaystyle\int_0^{\pi/2} \dfrac{\sin x}{x} \, dx$ **32.** $\displaystyle\int_0^1 \cos x^2 \, dx$

33. $\displaystyle\int_0^{\pi/2} \sqrt{x} \cos x \, dx$ **34.** $\displaystyle\int_0^{1/2} \dfrac{\ln (x + 1)}{x} \, dx$

35. $\displaystyle\int_0^{1/2} \dfrac{\arctan x}{x} \, dx$ **36.** $\displaystyle\int_1^2 e^{-x^2} \, dx$

37. $\displaystyle\int_{0.1}^{0.3} \sqrt{1 + x^3} \, dx$ **38.** $\displaystyle\int_{0.5}^1 \cos \sqrt{x} \, dx$

In Exercises 39 and 40, find a Maclaurin series for $f(x)$ defined by the given integral.

39. $\displaystyle\int_0^x (e^{-t^2} - 1) \, dt$ **40.** $\displaystyle\int_0^x \sqrt{1 + t^3} \, dt$

In Exercises 41 and 42, use a computer to approximate the given probability with an error of less than 0.0001 for the standard normal probability density function

$$P(a < x < b) = \dfrac{1}{\sqrt{2\pi}} \int_a^b e^{-x^2/2} \, dx.$$

41. $P(0 < x < 1)$ **42.** $P(1 < x < 2)$

43. Show that

$$\sum_{n=0}^{\infty} (-1)^n \left[\dfrac{1}{(2n + 1)!}\right] = \sin 1.$$

44. Show that

$$\sum_{n=1}^{\infty} (-1)^{n-1} \left(\dfrac{1}{n!}\right) = \dfrac{e - 1}{e}.$$

45. As discussed in the Chapter 10 Application, a projectile fired from the ground follows the trajectory given by

$$y = \left(\tan\theta - \frac{g}{kv_0\cos\theta}\right)x - \frac{g}{k^2}\ln\left(1 - \frac{kx}{v_0\cos\theta}\right)$$

where v_0 is the initial speed, θ is the angle of projection, g is the acceleration due to gravity, and k is the drag factor caused by the air resistance. Using the power series representation

$$\ln(1 + x) = x - \frac{x^2}{2} + \frac{x^3}{3} - \frac{x^4}{4} + \cdots,$$

$$-1 < x < 1$$

verify that the trajectory can be rewritten as

$$y = (\tan\theta)x + \frac{gx^2}{2v_0{}^2\cos^2\theta} + \frac{kgx^3}{3v_0{}^3\cos^3\theta}$$

$$+ \frac{k^2gx^4}{4v_0{}^4\cos^4\theta} + \cdots.$$

46. Use the result of Exercise 45 to determine the series for the path of a projectile projected from ground level at an angle of $\theta = 60°$ with an initial speed of $v_0 = 64$ feet per second and a drag factor of $k = \frac{1}{16}$.

47. Consider the function f defined by

$$f(x) = \begin{cases} e^{-1/x^2}, & x \neq 0 \\ 0, & x = 0 \end{cases}.$$

(a) Sketch a graph of the function.

(b) Use the alternate form of the definition of the derivative (Section 3.1) and L'Hôpital's Rule to show that $f'(0) = 0$. (By continuing this process, it can be shown that $f^{(n)}(0) = 0$ for $n > 1$.)

(c) Using the result of part (b), find the Maclaurin series for f. Does the series converge to f?

REVIEW EXERCISES for Chapter 10

In Exercises 1 and 2, find the general term of the sequence.

1. $1, \dfrac{1}{2}, \dfrac{1}{6}, \dfrac{1}{24}, \dfrac{1}{120}, \ldots$

2. $\dfrac{1}{2}, \dfrac{2}{5}, \dfrac{3}{10}, \dfrac{4}{17}, \ldots$

In Exercises 3–10, determine the convergence or divergence of the sequence with the given general term.

3. $a_n = \dfrac{n+1}{n^2}$

4. $a_n = \dfrac{1}{\sqrt{n}}$

5. $a_n = \dfrac{n^3}{n^2+1}$

6. $a_n = \dfrac{n}{\ln n}$

7. $a_n = \sqrt{n+1} - \sqrt{n}$

8. $a_n = \left(1 + \dfrac{1}{2n}\right)^n$

9. $a_n = \dfrac{\sin\sqrt{n}}{\sqrt{n}}$

10. $a_n = (b^n + c^n)^{1/n}$
(b and c are positive real numbers)

In Exercises 11–14, find the first five terms of the sequence of partial sums for the given series.

11. $\displaystyle\sum_{n=0}^{\infty} \left(\frac{3}{2}\right)^n$

12. $\displaystyle\sum_{n=1}^{\infty} \frac{(-1)^{n+1}}{2n}$

13. $\displaystyle\sum_{n=1}^{\infty} \frac{(-1)^{n+1}}{(2n)!}$

14. $\displaystyle\sum_{n=1}^{\infty} \frac{1}{n(n+1)}$

In Exercises 15–18, find the sum of the given series.

15. $\displaystyle\sum_{n=0}^{\infty} \left(\frac{2}{3}\right)^n$

16. $\displaystyle\sum_{n=0}^{\infty} \frac{2^{n+2}}{3^n}$

17. $\displaystyle\sum_{n=0}^{\infty} \left(\frac{1}{2^n} - \frac{1}{3^n}\right)$

18. $\displaystyle\sum_{n=0}^{\infty} \left[\left(\frac{2}{3}\right)^n - \frac{1}{(n+1)(n+2)}\right]$

In Exercises 19 and 20, express the repeating decimal as the ratio of two integers.

19. $0.090909\ldots$

20. $0.923076923076\ldots$

In Exercises 21–32, determine the convergence or divergence of the given series.

21. $\displaystyle\sum_{n=1}^{\infty} \frac{2^n}{n^3}$

22. $\displaystyle\sum_{n=1}^{\infty} \frac{1}{\sqrt{n^2+2n}}$

23. $\displaystyle\sum_{n=1}^{\infty} \frac{1}{\sqrt{n^3+2n}}$

24. $\displaystyle\sum_{n=1}^{\infty} \frac{n+1}{n(n+2)}$

25. $\displaystyle\sum_{n=1}^{\infty} \frac{1}{(n^3+2n)^{1/3}}$

26. $\displaystyle\sum_{n=1}^{\infty} \frac{n!}{e^n}$

27. $\displaystyle\sum_{n=1}^{\infty} \frac{(-1)^n n}{\ln n}$

28. $\displaystyle\sum_{n=1}^{\infty} \frac{(-1)^n\sqrt{n}}{n+1}$

29. $\displaystyle\sum_{n=1}^{\infty} \frac{(-1)^n 1 \cdot 3 \cdot 5 \cdots (2n-1)}{2 \cdot 4 \cdot 6 \cdots (2n)(2n+1)}$

30. $\displaystyle\sum_{n=1}^{\infty} \frac{1 \cdot 3 \cdot 5 \cdots (2n-1)}{2 \cdot 5 \cdot 8 \cdots (3n-1)}$

31. $\displaystyle\sum_{n=1}^{\infty} \left(\frac{1}{n^2} - \frac{1}{n}\right)$ **32.** $\displaystyle\sum_{n=1}^{\infty} \left(\frac{1}{n^2} - \frac{1}{2^n}\right)$

In Exercises 33–36, find the interval of convergence of the power series.

33. $\displaystyle\sum_{n=0}^{\infty} \frac{(-1)^n(x-2)^n}{(n+1)^2}$ **34.** $\displaystyle\sum_{n=0}^{\infty} (2x)^n$

35. $\displaystyle\sum_{n=0}^{\infty} n!(x-2)^n$ **36.** $\displaystyle\sum_{n=0}^{\infty} \frac{(x-2)^n}{2^n}$

In Exercises 37–42, find the power series for $f(x)$ centered at c.

37. $f(x) = \sin x$, $c = \dfrac{3\pi}{4}$ **38.** $f(x) = \sqrt{x}$, $c = 4$

39. $f(x) = 3^x$, $c = 0$

40. $f(x) = \csc x$, $c = \dfrac{\pi}{2}$ (first three terms)

41. $f(x) = \dfrac{1}{x}$, $c = -1$ **42.** $f(x) = \cos x$, $c = -\dfrac{\pi}{4}$

In Exercises 43–46, find the series representation of the function defined by the given integral.

43. $\displaystyle\int_0^x \frac{\sin t}{t}\, dt$ **44.** $\displaystyle\int_0^x \cos \frac{\sqrt{t}}{2}\, dt$

45. $\displaystyle\int_0^x \frac{\ln(t+1)}{t}\, dt$ **46.** $\displaystyle\int_0^x \frac{e^t - 1}{t}\, dt$

In Exercises 47–50, use a Taylor polynomial to approximate the given value with an error of less than 0.001.

47. $\sin 95°$ **48.** $\cos(0.75)$
49. $\ln(1.75)$ **50.** $e^{-0.25}$

In Exercises 51 and 52, show that the function defined by the series is a solution to the specified equation.

51. $y = \displaystyle\sum_{n=0}^{\infty} (-1)^n \frac{x^{2n}}{4^n(n!)^2}$ Bessel function
$x^2 y'' + xy' + x^2 y = 0$

52. $y = \displaystyle\sum_{n=0}^{\infty} \frac{(-3)^n x^{2n}}{2^n n!}$
$y'' + 3xy' + 3y = 0$

53. Sketch the graphs of $f(x) = e^x$ and its fifth Taylor polynomial on the same axes.

54. A ball is dropped from a height of 8 feet. Each time it drops h feet, it rebounds $0.7h$ feet. Find the total distance traveled by the ball.

Chapter 11 Application

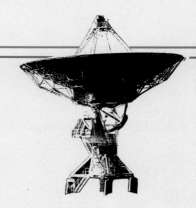

A communications satellite receiving dish, because of the distance involved, may receive signal strength from a satellite on the order of only a few billionths of a watt. To increase the strength of the received signal, receiving terminal designs use the reflective property of a parabola.

Communications Receiving Station

To form a model for a satellite-signal receiving dish, imagine that the parabola given by

$$x^2 = 4py$$

is revolved about the y-axis. Each satellite signal that enters the receiving dish, parallel to the y-axis, is reflected to the *focus* of the parabola, p units up from its vertex. For example, the parabola given by

$$x^2 = 20y$$

has its vertex at the origin, and since $4p = 20$, the focus is at $(0, 5)$.

Larger dishes receive stronger signals, but they are more expensive than smaller dishes. Part of the cost of constructing a receiving dish is related to its surface area. To find the surface area, we use the formula for the area of a surface of revolution. For example, if a segment of the parabola $x^2 = 20y$ is revolved about the y-axis to form a dish of radius r, the surface area is

$$2\pi \int_0^r x\sqrt{1 + \left(\frac{x}{10}\right)^2}\, dx = \frac{\pi}{15}[(100 + r^2)^{3/2} - 1000].$$

This means that the surface area is roughly proportional to the cube of the radius of the dish.

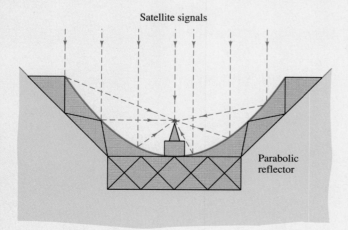

Satellite signals

Parabolic reflector

Chapter Overview

This chapter covers **conic sections**—classic curves that were discovered by the ancient Greeks. We will study conic sections from two points of view: as the intersection of a plane and a double-napped cone, and as the graph of a second-degree equation of the form

$$Ax^2 + Bxy + Cy^2 + Dx + Ey + F = 0.$$

There are four basic types of conics. The first type, circles, was discussed in Chapter 1. The other three basic conics are discussed in Sections 11.1, 11.2, and 11.3.

Conic	Standard Form
Circle:	$(x - h)^2 + (y - k)^2 = r^2$
Parabola:	$(x - h)^2 = 4p(y - k)$
Ellipse:	$\dfrac{(x - h)^2}{a^2} + \dfrac{(y - k)^2}{b^2} = 1$
Hyperbola:	$\dfrac{(x - h)^2}{a^2} - \dfrac{(y - k)^2}{b^2} = 1$

Conics whose axes are horizontal or vertical have simpler equations than do conics whose axes are not horizontal or vertical, and in the first three sections, we restrict our discussion to the simpler type. Then, in Section 11.4, we look at a technique for rotating the coordinate axes so that they are parallel to the axes of a given conic.

Conic sections occur in many applications, from planetary orbits to bridge construction. Throughout the chapter we will describe several of these.

See Exercise 56, Section 11.1

Conic Sections

11.1 Parabolas

Conic sections ▪ Parabolas ▪ Standard equation of a parabola ▪ Applications

Hypatia

The Greeks discovered conic sections sometime between 600 and 300 B.C. By the beginning of the Alexandrian period, enough was known about conics for Apollonius (262–190 B.C.) to produce an eight-volume work on the subject. Later, toward the end of the Alexandrian period, Hypatia (370–415 A.D.) wrote a textbook entitled *On the Conics of Apollonius*. Her death marked the end of major mathematical discoveries in Europe for several hundred years.

The early Greeks were largely concerned with the geometrical properties of conics. It was not until 1900 years later, in the early seventeenth century, that the broader applicability of conics became apparent, and conics then played a prominent role in the development of calculus.

Each **conic section** (or simply **conic**) can be described as the intersection of a plane and a double-napped cone. Notice from Figure 11.1 that in the formation of the four basic conics, the intersecting plane does not pass through the vertex of the cone. When the plane does pass through the vertex, we call the resulting figure a **degenerate conic,** as shown in Figure 11.2.

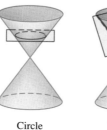

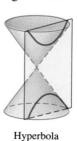

| Circle | Parabola | Ellipse | Hyperbola |

FIGURE 11.1   Conic Sections

| Point | Line | Two intersecting lines |

FIGURE 11.2   Degenerate Conics

645

There are several ways to study conics. We could begin as the Greeks did by defining the conics in terms of the intersections of planes and cones, or we could define them algebraically in terms of the general second-degree equation

$$Ax^2 + Bxy + Cy^2 + Dx + Ey + F = 0$$ General second-degree equation

a procedure to be discussed in Section 11.4. However, a third approach, in which each of the conics is defined as a **locus** (collection) of points satisfying a certain geometric property, suits our needs best. For example, in Section 1.2, we defined a circle as the collection of all points (x, y) that are equidistant from a fixed point (h, k). This locus definition easily produced the standard equation of a circle,

$$(x - h)^2 + (y - k)^2 = r^2.$$ Standard equation of a circle

Parabolas

In this and the following two sections, we give similar definitions to the other three types of conics. We will also identify practical geometric properties used in the construction of objects such as bridges, searchlights, telescopes, and radar detectors. Later, in multivariable calculus, we will often use conics in examples.

DEFINITION OF A PARABOLA	A **parabola** is the set of all points (x, y) that are equidistant from a fixed line (**directrix**) and a fixed point (**focus**) not on the line.

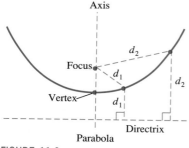

FIGURE 11.3

The midpoint between the focus and the directrix is called the **vertex,** and the line passing through the focus and the vertex is called the **axis** of the parabola. Note in Figure 11.3 that a parabola is symmetric with respect to its axis.

Using the definition of a parabola, we derive the following theorem, which gives the **standard form** of the equation of a parabola whose directrix is parallel to the x-axis or to the y-axis.

THEOREM 11.1 STANDARD EQUATION OF A PARABOLA	The **standard form** of the equation of a parabola with vertex (h, k) and directrix $y = k - p$ is $$(x - h)^2 = 4p(y - k).$$ Vertical axis For directrix $x = h - p$, the equation is $$(y - k)^2 = 4p(x - h).$$ Horizontal axis The focus lies on the axis p units (*directed distance*) from the vertex.

PROOF We prove only the case for which the directrix is parallel to the x-axis and the focus lies above the vertex, as shown in Figure 11.4(a). If (x, y) is any point on the parabola, then by definition (x, y) is equidistant from the focus $(h, k + p)$ and the directrix $y = k - p$, and we have the following.

$$\sqrt{(x - h)^2 + [y - (k + p)]^2} = |y - (k - p)|$$
$$(x - h)^2 + [y - (k + p)]^2 = [y - (k - p)]^2$$
$$(x - h)^2 + y^2 - 2y(k + p) + (k + p)^2 = y^2 - 2y(k - p) + (k - p)^2$$
$$(x - h)^2 - 2py + 2pk = 2py - 2pk$$
$$(x - h)^2 = 4p(y - k)$$

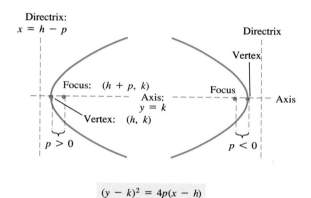

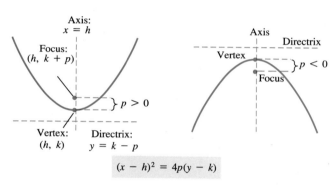

(a) Vertical axis: $p > 0$ (b) Vertical axis: $p < 0$ (c) Horizontal axis: $p > 0$ (d) Horizontal axis: $p < 0$

Parabolic Orientations

FIGURE 11.4

EXAMPLE 1 Finding the standard equation for a parabola

Find the standard form of the equation of the parabola with vertex $(2, 1)$ and focus $(2, 4)$.

SOLUTION

Since the axis of the parabola is vertical, we consider the equation

$$(x - h)^2 = 4p(y - k)$$

where $h = 2$, $k = 1$, and $p = 3$. Thus the standard form is

$$(x - 2)^2 = 12(y - 1).$$

The graph of this parabola is shown in Figure 11.5. ▬

REMARK By expanding the standard equation in Example 1, we obtain the more common quadratic form $y = \frac{1}{12}(x^2 - 4x + 16)$.

FIGURE 11.5

$(x - 2)^2 = 12(y - 1)$

EXAMPLE 2 Finding the vertex and focus

Find the focus of the parabola given by $y = -\frac{1}{2}x^2 - x + \frac{1}{2}$.

SOLUTION

To find the focus, we convert to standard form by completing the square.

$$y = \frac{1}{2}(1 - 2x - x^2)$$

$$2y = 1 - (x^2 + 2x + \quad)$$

$$2y = 2 - (x^2 + 2x + 1)$$

$$(x + 1)^2 = -2(y - 1) \qquad \text{Standard form}$$

Comparing this equation to $(x - h)^2 = 4p(y - k)$, we conclude that

$$h = -1, \quad k = 1, \quad \text{and} \quad p = -\frac{1}{2}.$$

Since p is negative, the parabola opens downward, as shown in Figure 11.6. Therefore, the focus of the parabola is

$$(h, k + p) = \left(-1, \frac{1}{2}\right). \qquad \text{Focus}$$

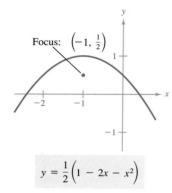

Focus: $\left(-1, \frac{1}{2}\right)$

$$y = \frac{1}{2}\left(1 - 2x - x^2\right)$$

FIGURE 11.6

If the vertex of a parabola is at the origin, then the standard form

$$(x - h)^2 = 4p(y - k) \qquad \text{Vertex at } (h, k)$$

simplifies to

$$x^2 = 4py \qquad \text{Vertex at origin}$$

as demonstrated in the next example.

EXAMPLE 3 Vertex at the origin

Find the standard equation of the parabola with vertex at the origin and focus at (2, 0).

SOLUTION

The axis of the parabola is horizontal, passing through (0, 0) and (2, 0), as shown in Figure 11.7. Thus, we consider the standard form

$$y^2 = 4px$$

where $h = k = 0$ and $p = 2$. Therefore, the equation is

$$y^2 = 8x.$$

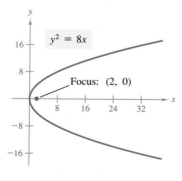

$y^2 = 8x$

Focus: (2, 0)

FIGURE 11.7

A line segment that passes through the focus of a parabola and has endpoints on the parabola is called a **focal chord.** The specific focal chord perpendicular to the axis of the parabola is called the **latus rectum.** In the next example we determine the length of the latus rectum and the length of the corresponding intercepted arc.

EXAMPLE 4 Focal chord length and arc length

Find the length of the latus rectum of the parabola given by $x^2 = 4py$. Then find the length of the parabolic arc intercepted by the latus rectum.

SOLUTION

Since the latus rectum passes through the focus $(0, p)$ and is perpendicular to the y-axis, the coordinates of its endpoints are $(-x, p)$ and (x, p). Substituting p for y in the equation of the parabola, we have

$$x^2 = 4p(p) = 4p^2 \implies x = \pm 2p.$$

Thus, the endpoints of the latus rectum are $(-2p, p)$ and $(2p, p)$, and we conclude that its length is $4p$, as shown in Figure 11.8. In contrast, the length of the intercepted arc is given by the following.

$$
\begin{aligned}
s &= \int_{-2p}^{2p} \sqrt{1 + (y')^2} \, dx = 2\int_0^{2p} \sqrt{1 + \left(\frac{x}{2p}\right)^2} \, dx \\
&= \frac{1}{p} \int_0^{2p} \sqrt{4p^2 + x^2} \, dx \\
&= \frac{1}{2p} \left[x\sqrt{4p^2 + x^2} + 4p^2 \ln \left| x + \sqrt{4p^2 + x^2} \right| \right]_0^{2p} \\
&= \frac{1}{2p} [2p\sqrt{8p^2} + 4p^2 \ln (2p + \sqrt{8p^2}) - 4p^2 \ln (2p)] \\
&= 2p[\sqrt{2} + \ln (1 + \sqrt{2})] \approx 4.59p
\end{aligned}
$$

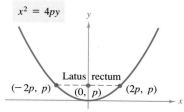

$x^2 = 4py$

$(-2p, p)$ Latus rectum $(2p, p)$
$(0, p)$

Length of latus rectum: $4p$
Arc length: $4.59p$

Latus Rectum and Intercepted Arc

FIGURE 11.8

Applications

One widely used property of a parabola is its reflective property. In physics, a surface is called **reflective** if the tangent line at any point on the surface makes equal angles with an incoming ray and the resulting outgoing ray. The angle corresponding to the incoming ray is called the **angle of incidence,** and the angle corresponding to the outgoing ray is called the **angle of reflection.** One example of a reflective surface is a flat mirror. Another type of reflective surface is that formed by revolving a parabola about its axis. A special property of parabolic reflectors is that they allow us to direct all incoming rays parallel to the axis through the focus of the parabola—this is the principle behind the design of the parabolic mirrors used in reflecting telescopes. Conversely, all light rays emanating from the focus of a parabolic reflector used in a flashlight are parallel, as shown in Figure 11.9.

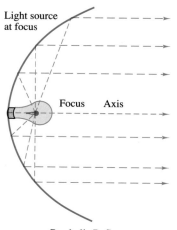

Light source at focus

Focus   Axis

Parabolic Reflector:
Light is reflected
in parallel rays.

FIGURE 11.9

THEOREM 11.2
REFLECTIVE PROPERTY
OF A PARABOLA

The tangent line to a parabola at the point P makes equal angles with the following two lines:

1. the line passing through P and the focus and
2. the line passing through P parallel to the axis of the parabola.

PROOF

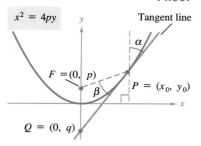

FIGURE 11.10

We consider the point $P = (x_0, y_0)$ on the parabola given by $x^2 = 4py$, as shown in Figure 11.10. Using differentiation, we can determine the tangent line at P to have a slope of $x_0/2p$, which implies that the equation of the tangent line is

$$y - y_0 = \frac{x_0}{2p}(x - x_0). \qquad \text{Tangent line at } P$$

If P lies at the vertex of the parabola, then the other two lines referred to in the theorem coincide and the theorem is trivially valid. Thus, we assume P is a point other than the vertex. This implies that the tangent line intersects the y-axis at a point $Q = (0, q)$, different from P. Using Figure 11.10, we can show that the angles α and β are equal by showing that the triangle with vertices at F, P, and Q is isosceles. Using the equation of the tangent line at $x = 0$, we see that

$$q = y_0 - \frac{x_0^2}{2p} = \frac{x_0^2}{4p} - \frac{x_0^2}{2p} = -\frac{x_0^2}{4p}$$

which implies that the length of $\overline{FQ}$ is

$$p - q = p + \frac{x_0^2}{4p}. \qquad \text{Length of } \overline{FQ}$$

Moreover, using the distance formula, we find the length of $\overline{FP}$ to be

$$\sqrt{x_0^2 + (y_0 - p)^2} = \sqrt{x_0^2 + \left(\frac{x_0^2}{4p} - p\right)^2} = \sqrt{\left(\frac{x_0^2}{4p} + p\right)^2}$$

$$= \frac{x_0^2}{4p} + p. \qquad \text{Length of } \overline{FP}$$

Therefore, $\triangle FQP$ is isosceles, and we conclude that $\alpha = \beta$.

EXERCISES for Section 11.1

In Exercises 1–6, match the equation with the correct graph. [The graphs are labeled (a)–(f).]

1. $y^2 = 4x$
2. $x^2 = -2y$
3. $x^2 = 8y$
4. $y^2 = -12x$
5. $(y - 1)^2 = 4(x - 2)$
6. $(x + 3)^2 = -2(y - 2)$

(a)

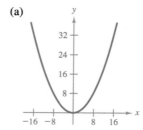

(b)

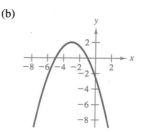

(c)

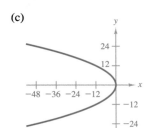

(d)

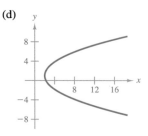

(e)

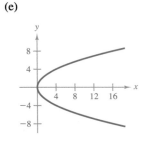

(f)

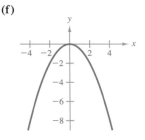

In Exercises 7–26, find the vertex, focus, and directrix of the parabola, and sketch its graph.

7. $y = 4x^2$

8. $y = 2x^2$

9. $y^2 = -6x$

10. $y^2 = 3x$

11. $x^2 + 8y = 0$

12. $x + y^2 = 0$

13. $(x - 1)^2 + 8(y + 2) = 0$

14. $(x + 3) + (y - 2)^2 = 0$

15. $\left(y + \dfrac{1}{2}\right)^2 = 2(x - 5)$

16. $\left(x + \dfrac{1}{2}\right)^2 - 4(y - 3) = 0$

17. $y = \dfrac{1}{4}(x^2 - 2x + 5)$

18. $y = -\dfrac{1}{6}(x^2 + 4x - 2)$

19. $4x - y^2 - 2y - 33 = 0$

20. $y^2 + x + y = 0$

21. $y^2 + 6y + 8x + 25 = 0$

22. $x^2 - 2x + 8y + 9 = 0$

23. $y^2 - 4y - 4x = 0$

24. $y^2 - 4x - 4 = 0$

25. $x^2 + 4x + 4y - 4 = 0$

26. $y^2 + 4y + 8x - 12 = 0$

In Exercises 27–40, find an equation of the specified parabola.

27. Vertex: $(0, 0)$
Focus: $\left(0, -\dfrac{3}{2}\right)$

28. Vertex: $(0, 0)$
Focus: $(2, 0)$

29. Vertex: $(3, 2)$
Focus: $(1, 2)$

30. Vertex: $(-1, 2)$
Focus: $(-1, 0)$

31. Vertex: $(0, -4)$
Directrix: $y = 2$

32. Vertex: $(-2, 1)$
Directrix: $x = 1$

33. Focus: $(0, 0)$
Directrix: $y = 4$

34. Focus: $(2, 2)$
Directrix: $x = -2$

35. Axis parallel to y-axis, graph passes through $(0, 3)$, $(3, 4)$, and $(4, 11)$

36. Axis parallel to x-axis, graph passes through $(4, -2)$, $(0, 0)$, and $(3, -3)$

37.

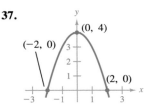

38.

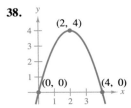

39. Directrix $y = 1$, length of latus rectum is 8, opens downward

40. Directrix $y = -2$, endpoints of latus rectum $(0, 2)$ and $(8, 2)$

41. The filament of a flashlight bulb is $\frac{3}{8}$ inch from the vertex of the parabolic reflector and is located at the focus. Find an equation of a cross section of the reflector. (Assume that it is directed toward the right and the vertex is at the origin.)

42. The receiver in a parabolic television dish antenna is 3 feet from the vertex and is located at the focus. Find an equation of a cross section of the reflector. (Assume that the dish is directed upward and the vertex is at the origin.)

In Exercises 43 and 44, find an equation of the tangent line to the parabola at the specified point.

Parabola	*Point*
43. $x^2 = 2y$	$(4, 8)$
44. $y^2 = -8x$	$(-2, 4)$

In Exercises 45 and 46, find the coordinates of the centroid of the region bounded by the graphs of the given equations.

45. $y^2 = 4x$, $x = 1$

46. $x^2 = \dfrac{1}{2}y$, $y = 8$

47. Find the point on the graph of $x^2 = 8y$ that is closest to the focus of the parabola.

48. Find equations of the tangent lines to $y^2 = 2x$ that pass through the point $(-4, 1)$.

In Exercises 49–52, find the arc length of the parabola over the given interval.

Function	*Interval*
49. $x^2 + 8y = 0$	$0 \le x \le 4$
50. $x + y^2 = 0$	$0 \le y \le 2$
51. $4x - y^2 = 0$	$0 \le y \le 4$
52. $x^2 - 2y = 0$	$0 \le x \le 1$

53. A cable of a parabolic suspension bridge is suspended between two towers that are 400 feet apart and 50 feet above the roadway (see figure). The cable touches the roadway midway between the towers. Find the length of the cable.

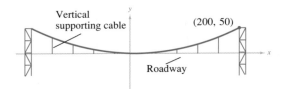

FIGURE FOR 53

54. An earth satellite in a circular orbit 100 miles high has a speed of approximately 17,470 miles per hour. If this speed is multiplied by $\sqrt{2}$, then the satellite will have the minimum velocity necessary to escape the earth's gravitational force and will follow a parabolic path with the center of the earth as the focus (see figure). Find an equation for the resulting parabolic path. (Assume that the radius of the earth is 4000 miles.)

55. Water is flowing from a horizontal pipe 48 feet above the ground with a horizontal velocity of 10 feet per second. The falling stream of water forms a parabola whose vertex, $(0, 48)$, is at the end of the pipe (see figure). Where does the water hit the ground?

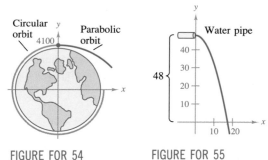

FIGURE FOR 54   FIGURE FOR 55

56. As discussed in the Chapter 11 Application, a satellite-signal receiving dish is formed by revolving the parabola given by the graph of $x^2 = 20y$ about the y-axis. If the radius of the dish is r feet, verify that the surface area of the dish is given by

$$2\pi \int_0^r x\sqrt{1 + \left(\frac{x}{10}\right)^2}\,dx = \frac{\pi}{15}[(100 + r^2)^{3/2} - 1000].$$

57. Find the surface area of the satellite dish in Exercise 56 if its radius is (a) $r = 6$ feet and (b) $r = 10$ feet.

In Exercises 58 and 59, use a computer or graphics calculator to sketch the graph of the parabola and find any x-intercepts.

58. $x^2 - 6.2x - 2.5y + 7.61 = 0$

59. $y^2 - 4y + 4.1x - 4.2 = 0$

60. Prove that the area enclosed by a parabola and a line parallel to its directrix is two-thirds the area of the circumscribed rectangle.

61. Find an equation of the tangent line to the parabola $y = ax^2$ at $x = x_0$. Prove that the x-intercept of this tangent line is $(x_0/2, 0)$.

62. Prove that any two distinct tangent lines to a parabola intersect.

63. Demonstrate the result of Exercise 62 by finding the point of intersection of the tangent lines to the parabola $x^2 - 4x - 4y = 0$ at the points $(0, 0)$ and $(6, 3)$.

64. Prove that if any two tangent lines to a parabola intersect at right angles, then their point of intersection must lie on the directrix.

65. Demonstrate the result of Exercise 64 by showing that the tangent lines to the parabola $x^2 - 4x - 4y + 8 = 0$ at the points $(-2, 5)$ and $\left(3, \frac{5}{4}\right)$ intersect at right angles and that the point of intersection lies on the directrix.

66. Prove that two tangent lines to a parabola intersect at right angles if and only if the focus of the parabola lies on the line segment connecting the two points of tangency.

11.2 Ellipses

Ellipses ▪ Standard equation of an ellipse ▪ Eccentricity

Nicholas Copernicus

More than a thousand years after the close of the Alexandrian period of Greek mathematics, Western civilization finally began a Renaissance of mathematical and scientific discovery. One of the principal figures in this rebirth was the Polish astronomer Nicholas Copernicus (1473–1543). In his work *On the Revolutions of the Heavenly Spheres*, Copernicus claimed that all of the planets, including the earth, revolved about the sun in circular orbits. Although some of Copernicus's claims were invalid, the controversy set off by his heliocentric theory motivated astronomers to search for a mathematical model to explain the observed movements of the sun and planets. The first to find the correct model was the German astronomer Johannes Kepler (1571–1630). Kepler discovered that the planets move about the sun in elliptical orbits, with the sun not as the center but as a focal point of the orbit.

The use of ellipses to explain the movement of the planets is only one of many practical and aesthetic uses. As in the previous section, we begin our study of this second type of conic by defining it as a locus of points. Now, however, we use *two* focal points rather than one.

| DEFINITION OF AN ELLIPSE | An **ellipse** is the set of all points (x, y) the sum of whose distances from two distinct fixed points (**foci**) is constant. (See Figure 11.11.) |

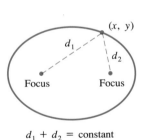

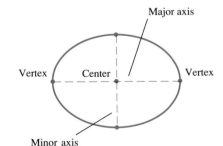

FIGURE 11.11 $d_1 + d_2 = $ constant

The line through the foci intersects the ellipse at two points called the **vertices.** The chord joining the vertices is called the **major axis,** and its midpoint is called the **center** of the ellipse. The chord perpendicular to the major axis at the center is called the **minor axis** of the ellipse.

REMARK   You can visualize the definition of an ellipse by imagining two thumbtacks placed at the foci, as shown in Figure 11.12. If the ends of a fixed length of string are fastened to the thumbtacks and the string is drawn taut with a pencil, the path traced by the pencil will be an ellipse.

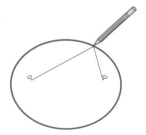

FIGURE 11.12

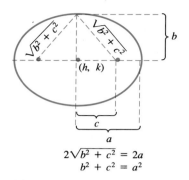

$$2\sqrt{b^2 + c^2} = 2a$$
$$b^2 + c^2 = a^2$$

FIGURE 11.13

To derive the standard form of the equation of an ellipse, consider the ellipse in Figure 11.13 with the following points.

> center: (h, k) vertices: $(h \pm a, k)$ foci: $(h \pm c, k)$

The sum of the distances from any point on the ellipse to the two foci is constant. At a vertex, this constant sum is

$$(a + c) + (a - c) = 2a \qquad \text{Length of major axis}$$

or the length of the major axis. Now, for *any* point (x, y) on the ellipse, the sum of the distances between (x, y) and the two foci must also be $2a$:

$$\sqrt{[x - (h - c)]^2 + (y - k)^2} + \sqrt{[x - (h + c)]^2 + (y - k)^2} = 2a$$

which, after expanding and regrouping, reduces to

$$(a^2 - c^2)(x - h)^2 + a^2(y - k)^2 = a^2(a^2 - c^2).$$

Finally, in Figure 11.13 we can see that $b^2 = a^2 - c^2$, which implies that the equation of the ellipse is

$$b^2(x - h)^2 + a^2(y - k)^2 = a^2 b^2$$
$$\frac{(x - h)^2}{a^2} + \frac{(y - k)^2}{b^2} = 1.$$

Choosing a vertical major axis would have produced a similar equation. Both results are summarized in the following theorem.

THEOREM 11.3
STANDARD EQUATION
OF AN ELLIPSE

The standard form of the equation of an ellipse, with center (h, k) and major and minor axes of lengths $2a$ and $2b$, where $a > b$, is

$$\frac{(x - h)^2}{a^2} + \frac{(y - k)^2}{b^2} = 1 \qquad \text{Major axis is horizontal}$$

$$\frac{(x - h)^2}{b^2} + \frac{(y - k)^2}{a^2} = 1. \qquad \text{Major axis is vertical}$$

The foci lie on the major axis, c units from the center, with $c^2 = a^2 - b^2$.

Figure 11.14 shows an ellipse's vertical and horizontal orientations.

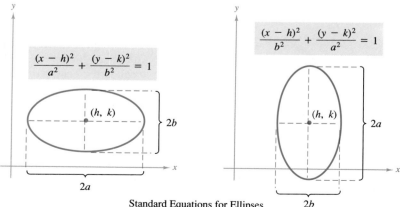

FIGURE 11.14

Standard Equations for Ellipses

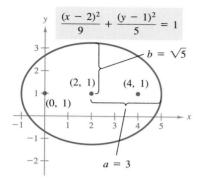

$$\frac{(x - 2)^2}{9} + \frac{(y - 1)^2}{5} = 1$$

FIGURE 11.15

EXAMPLE 1 Finding the standard equation for an ellipse

Find the standard form of the equation of the ellipse having foci at (0, 1) and (4, 1), with a major axis of length 6, as shown in Figure 11.15.

SOLUTION

Since the foci occur at (0, 1) and (4, 1), the center of the ellipse is (2, 1). This implies that the distance from the center to one of the foci is $c = 2$, and since $2a = 6$ we know that $a = 3$. Now, using $c^2 = a^2 - b^2$, we have

$$b = \sqrt{a^2 - c^2} = \sqrt{9 - 4} = \sqrt{5}.$$

Since the major axis is parallel to the x-axis, the standard equation is

$$\frac{(x - 2)^2}{9} + \frac{(y - 1)^2}{5} = 1.$$

EXAMPLE 2 Completing the square to find the standard equation

Find the center, vertices, and foci of the ellipse given by

$$4x^2 + y^2 - 8x + 4y - 8 = 0.$$

SOLUTION

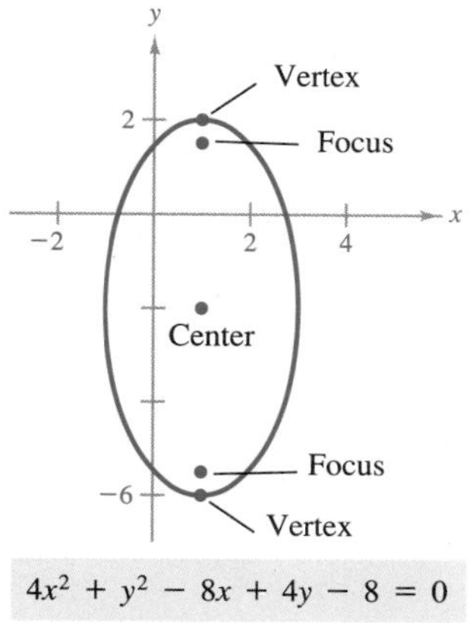

By completing the square, we can write the given equation in standard form.

$$4x^2 + y^2 - 8x + 4y - 8 = 0$$
$$4(x^2 - 2x + 1) + (y^2 + 4y + 4) = 8 + 4 + 4$$
$$4(x - 1)^2 + (y + 2)^2 = 16$$
$$\frac{(x - 1)^2}{4} + \frac{(y + 2)^2}{16} = 1$$

Thus, the major axis is parallel to the y-axis, where $h = 1$, $k = -2$, $a = 4$, $b = 2$, and $c = \sqrt{16 - 4} = 2\sqrt{3}$. Therefore, we obtain the following.

center: (1, −2)   vertices: (1, −6)   foci: $(1, -2 - 2\sqrt{3})$
                              (1, 2) $(1, -2 + 2\sqrt{3})$

The graph of the ellipse is shown in Figure 11.16.

$4x^2 + y^2 - 8x + 4y - 8 = 0$

FIGURE 11.16

REMARK If the constant term $F = -8$ in the equation in Example 2 had been greater than or equal to 8, we would have obtained one of the following degenerate cases.

1. Single point, (1, −2): $\dfrac{(x - 1)^2}{4} + \dfrac{(y + 2)^2}{16} = 0$ $F = 8$

2. No solution points: $\dfrac{(x - 1)^2}{4} + \dfrac{(y + 2)^2}{16} < 0$ $F > 8$

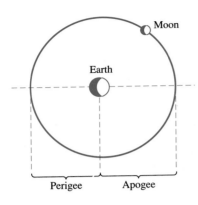

FIGURE 11.17

EXAMPLE 3 An application involving an elliptical orbit

The moon orbits the earth in an elliptical path with the center of the earth at one focus, as shown in Figure 11.17. The major and minor axes of the orbit have lengths of 768,806 kilometers and 767,746 kilometers, respectively. Find the greatest and least distances (the apogee and perigee) from the earth's center to the moon's center.

SOLUTION

Since $2a = 768,806$ and $2b = 767,746$, we have $a = 384,403$, $b = 383,873$, and

$$c = \sqrt{a^2 - b^2} \approx 20,179.$$

Therefore, the greatest distance between the center of the earth and the center of the moon is

$$a + c \approx 404,582 \text{ km}$$

and the least distance is

$$a - c \approx 364,224 \text{ km.}$$

EXAMPLE 4 Finding the area of an ellipse

Find the area of an ellipse whose major and minor axes have lengths of $2a$ and $2b$, respectively.

SOLUTION

For simplicity, we choose an ellipse centered at the origin

$$\frac{x^2}{a^2} + \frac{y^2}{b^2} = 1.$$

Then, using symmetry, we can find the area of the entire region lying within the ellipse by finding the area of the region in the first quadrant and multiplying by 4, as indicated in Figure 11.18. In the first quadrant, we have

$$y = \frac{b}{a} \sqrt{a^2 - x^2}$$

which implies that the entire area is

$$A = 4 \int_0^a \frac{b}{a} \sqrt{a^2 - x^2} \, dx.$$

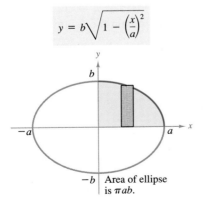

FIGURE 11.18

Using the trigonometric substitution $x = a \sin \theta$, we have

$$
\begin{aligned}
A &= \frac{4b}{a} \int_0^{\pi/2} a^2 \cos^2 \theta \, d\theta \\
&= 4ab \int_0^{\pi/2} \frac{1 + \cos 2\theta}{2} \, d\theta \\
&= 2ab \left[\theta + \frac{\sin 2\theta}{2} \right]_0^{\pi/2} \\
&= 2ab \left(\frac{\pi}{2} \right) = \pi ab.
\end{aligned}
$$

REMARK Note that if $a = b$, then the formula for the area of an ellipse reduces to the area of a circle.

In the previous section we looked at a reflective property of parabolas. Ellipses have a similar reflective property. You are asked to prove the following theorem in Exercise 60.

THEOREM 11.4
REFLECTIVE PROPERTY
OF AN ELLIPSE

The tangent line to an ellipse at the point P makes equal angles with the lines through P and the foci.

Eccentricity

One of the reasons that astronomers had difficulty in detecting that the orbits of the planets are ellipses is that the foci of the planetary orbits are relatively close to the center of the sun, making the orbits nearly circular. To measure the ovalness of an ellipse, we use the concept of **eccentricity.**

DEFINITION OF ECCENTRICITY
OF AN ELLIPSE

The **eccentricity** e of an ellipse is given by the ratio

$$
e = \frac{c}{a}.
$$

To see how this ratio is used to describe the shape of an ellipse, note that since the foci of an ellipse are located along the major axis between the vertices and the center, it follows that

$$
0 < c < a.
$$

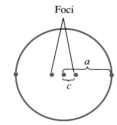

(a) $\dfrac{c}{a}$ is small.

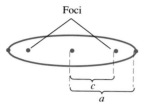

(b) $\dfrac{c}{a}$ is close to 1.

FIGURE 11.19

For an ellipse that is nearly circular, the foci are close to the center and the ratio c/a is small, as shown in Figure 11.19. For an elongated ellipse, the foci are close to the vertices and the ratio is close to 1. Note that $0 < e < 1$ for every ellipse.

The orbit of the moon has an eccentricity of $e = 0.0549$, and the eccentricities of the nine planetary orbits are as follows.

Mercury:	$e = 0.2056$	Saturn:	$e = 0.0543$
Venus:	$e = 0.0068$	Uranus:	$e = 0.0460$
Earth:	$e = 0.0167$	Neptune:	$e = 0.0082$
Mars:	$e = 0.0934$	Pluto:	$e = 0.2481$
Jupiter:	$e = 0.0484$		

In Example 4, we found that the formula for the *area* of an ellipse is $A = \pi ab$. However, it is not so simple to find the *circumference* of an ellipse. The next example shows how to use eccentricity to set up an "elliptic integral" for the circumference of an ellipse.

EXAMPLE 5   Finding the circumference of an ellipse

Show that the circumference of the ellipse given by $(x^2/a^2) + (y^2/b^2) = 1$ is

$$4a \int_0^{\pi/2} \sqrt{1 - e^2 \sin^2 \theta}\, d\theta. \qquad e = c/a$$

SOLUTION

Since the given ellipse is symmetric with respect to both the x-axis and the y-axis, we know that its circumference C is four times the arc length of $y = (b/a)\sqrt{a^2 - x^2}$ in the first quadrant. Thus

$$C = 4 \int_0^a \sqrt{1 + (y')^2}\, dx = 4 \int_0^a \sqrt{1 + \frac{b^2 x^2}{a^2(a^2 - x^2)}}\, dx.$$

Using the trigonometric substitution $x = a \sin \theta$, we obtain

$$C = 4 \int_0^{\pi/2} \sqrt{1 + \frac{b^2 \sin^2 \theta}{a^2 \cos^2 \theta}}\,(a \cos \theta)\, d\theta$$

$$= 4 \int_0^{\pi/2} \sqrt{a^2 \cos^2 \theta + b^2 \sin^2 \theta}\, d\theta$$

$$= 4 \int_0^{\pi/2} \sqrt{a^2(1 - \sin^2 \theta) + b^2 \sin^2 \theta}\, d\theta$$

$$= 4 \int_0^{\pi/2} \sqrt{a^2 - (a^2 - b^2) \sin^2 \theta}\, d\theta.$$

Since $e^2 = c^2/a^2 = (a^2 - b^2)/a^2$, we can rewrite this integral as

$$C = 4a \int_0^{\pi/2} \sqrt{1 - e^2 \sin^2 \theta}\, d\theta. \qquad \blacksquare$$

A great deal of time has been devoted to the study of elliptic integrals. Such integrals generally do not have elementary antiderivatives. To find the circumference of an ellipse, we usually resort to an approximation technique, as illustrated in the following example.

EXAMPLE 6 Approximating the value of an elliptic integral

SOLUTION

Use the elliptic integral in Example 5 to approximate the circumference of the ellipse

$$\frac{x^2}{25} + \frac{y^2}{16} = 1.$$

SOLUTION

Since $e^2 = c^2/a^2 = (a^2 - b^2)/a^2 = 9/25$, we have

$$C = (4)(5) \int_0^{\pi/2} \sqrt{1 - \frac{9 \sin^2 \theta}{25}} \, d\theta.$$

Applying Simpson's Rule with $n = 4$ produces

$$C \approx 20\left(\frac{\pi}{6}\right)\left(\frac{1}{4}\right)[1 + 4(0.9733) + 2(0.9055) + 4(0.8323) + 0.8] \approx 28.36.$$

EXERCISES for Section 11.2

In Exercises 1–6, match the equation with the correct graph. [The graphs are labeled (a)–(f).]

1. $\dfrac{x^2}{1} + \dfrac{y^2}{9} = 1$ **2.** $\dfrac{x^2}{9} + \dfrac{y^2}{1} = 1$

3. $\dfrac{x^2}{9} + \dfrac{y^2}{4} = 1$ **4.** $\dfrac{x^2}{9} + \dfrac{y^2}{9} = 1$

5. $\dfrac{(x - 2)^2}{16} + \dfrac{(y + 1)^2}{4} = 1$

6. $\dfrac{(x + 2)^2}{4} + \dfrac{(y + 2)^2}{25} = 1$

(c)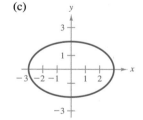

(d)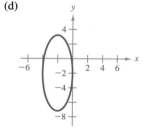

(e)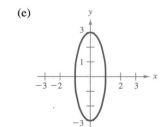

(f)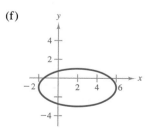

(a)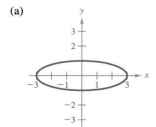

(b)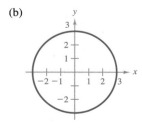

In Exercises 7–26, find the center, foci, vertices, and eccentricity of the ellipse, and sketch its graph.

7. $\dfrac{x^2}{25} + \dfrac{y^2}{16} = 1$

8. $\dfrac{x^2}{144} + \dfrac{y^2}{169} = 1$

9. $\dfrac{x^2}{16} + \dfrac{y^2}{25} = 1$

10. $\dfrac{x^2}{169} + \dfrac{y^2}{144} = 1$

11. $\dfrac{x^2}{9} + \dfrac{y^2}{5} = 1$

12. $\dfrac{x^2}{28} + \dfrac{y^2}{64} = 1$

13. $x^2 + 4y^2 = 4$

14. $5x^2 + 3y^2 = 15$

15. $3x^2 + 2y^2 = 6$

16. $5x^2 + 7y^2 = 70$

17. $4x^2 + y^2 = 1$

18. $16x^2 + 25y^2 = 1$

19. $\dfrac{(x-1)^2}{9} + \dfrac{(y-5)^2}{25} = 1$

20. $(x+2)^2 + 4(y+4)^2 = 1$

21. $9x^2 + 4y^2 + 36x - 24y + 36 = 0$

22. $9x^2 + 4y^2 - 36x + 8y + 31 = 0$

23. $16x^2 + 25y^2 - 32x + 50y + 31 = 0$

24. $9x^2 + 25y^2 - 36x - 50y + 61 = 0$

25. $12x^2 + 20y^2 - 12x + 40y - 37 = 0$

26. $36x^2 + 9y^2 + 48x - 36y + 43 = 0$

In Exercises 27–36, find an equation for the specified ellipse.

27. Center: $(0, 0)$
Focus: $(2, 0)$
Vertex: $(3, 0)$

28. Center: $(0, 0)$
Vertex: $(2, 0)$
Minor axis length: 3

29. Vertices: $(\pm 5, 0)$
Eccentricity: $\frac{3}{5}$

30. Vertices: $(0, \pm 8)$
Eccentricity: $\frac{1}{2}$

31. Vertices: $(3, 1)$, $(3, 9)$
Minor axis length: 6

32. Vertices: $(0, 2)$, $(4, 2)$
Minor axis length: 2

33. Foci: $(0, \pm 5)$
Major axis length: 14

34. Foci: $(\pm 2, 0)$
Major axis length: 8

35. Center: $(0, 0)$
Major axis: horizontal
Points on ellipse:
$(3, 1)$, $(4, 0)$

36. Center: $(1, 2)$
Major axis: vertical
Points on ellipse:
$(1, 6)$, $(3, 2)$

In Exercises 37 and 38, use the following definition. The **latera recta** (plural of **latus rectum**) of an ellipse are the chords passing through the foci and perpendicular to the major axis.

37. Show that the length of a latus rectum of an ellipse is $2b^2/a$.

38. With the result of Exercise 37, sketch the graph of each of the following, making use of the endpoints of the latera recta.

(a) $\dfrac{x^2}{4} + \dfrac{y^2}{1} = 1$

(b) $5x^2 + 3y^2 = 15$

In Exercises 39 and 40, determine the points at which dy/dx is zero or undefined to locate the endpoints of the major and minor axes of the ellipse.

39. $16x^2 + 9y^2 + 96x + 36y + 36 = 0$

40. $9x^2 + 4y^2 + 36x - 24y + 36 = 0$

In Exercises 41 and 42, consider a particle traveling clockwise on the given elliptical path. The particle leaves the orbit at the indicated point and travels in a straight line tangent to the ellipse. At what point will the particle cross the y-axis?

41. $\dfrac{x^2}{100} + \dfrac{y^2}{25} = 1$, $(-8, 3)$

42. $\dfrac{x^2}{16} + \dfrac{y^2}{25} = 1$, $\left(3, \dfrac{5\sqrt{7}}{4}\right)$

In Exercises 43–46, use Simpson's Rule with $n = 8$ to approximate the elliptic integral representing the circumference of the given ellipse.

43. $\dfrac{x^2}{9} + \dfrac{y^2}{16} = 1$

44. $\dfrac{x^2}{9} + \dfrac{y^2}{1} = 1$

45. $\dfrac{x^2}{3} + \dfrac{y^2}{2} = 1$

46. $\dfrac{(x-3)^2}{4} + \dfrac{(y+1)^2}{3} = 1$

In Exercises 47 and 48, find (a) the area of the region bounded by the given ellipse, (b) the volume and surface area of the solid generated by revolving the region about its major axis (prolate spheroid), and (c) the volume and surface area of the solid generated by revolving the region about its minor axis (oblate spheroid).

47. $\dfrac{x^2}{4} + \dfrac{y^2}{1} = 1$

48. $\dfrac{x^2}{16} + \dfrac{y^2}{9} = 1$

49. Find the dimensions of the rectangle of maximum area that can be inscribed in the ellipse

$$\dfrac{x^2}{a^2} + \dfrac{y^2}{b^2} = 1.$$

(Assume that the sides of the rectangle are parallel to the coordinate axes.)

50. A solid has an elliptical base given by

$$\dfrac{x^2}{25} + \dfrac{y^2}{16} = 1.$$

Cross sections perpendicular to the major axis are isosceles triangles of height 6. Find the volume of the solid.

51. Show that the tangent line to the ellipse

$$\frac{x^2}{a^2} + \frac{y^2}{b^2} = 1$$

at the point (x_0, y_0) is given by

$$\frac{x_0}{a^2}x + \frac{y_0}{b^2}y = 1.$$

52. The equation of an ellipse with its center at the origin can be written as

$$\frac{x^2}{a^2} + \frac{y^2}{a^2(1 - e^2)} = 1.$$

Show that as $e \to 0$, with a remaining fixed, the ellipse approaches a circle.

53. Show that the eccentricity of the ellipse

$$\frac{x^2}{a^2} + \frac{y^2}{b^2} = 1$$

is identical to the eccentricity of

$$\frac{(tx)^2}{a^2} + \frac{(ty)^2}{b^2} = 1$$

for any real t. Give a geometrical explanation of this result.

54. A line segment 9 inches long moves so that one endpoint is always on the y-axis and the other always on the x-axis. Find the equation of the curve traced by a point on the line segment 6 inches from the endpoint that is on the y-axis.

55. A fireplace arch is to be constructed in the shape of a semiellipse (see figure). The opening is to be 2 feet high at the center and 5 feet wide along the base. To sketch the outline of the fireplace, the contractor uses a 5-foot string tied to two thumbtacks. Where should the thumbtacks be placed?

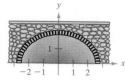

FIGURE FOR 55

56. The length of half of the major axis of the earth's orbit is 14,957,000 kilometers, and the eccentricity is 0.0167. Find the least and greatest distances of the earth from the sun.

57. If the distances to the apogee and the perigee of an elliptical orbit of an earth satellite are measured from the center of the earth, show that the eccentricity of the orbit is given by

$$e = \frac{A - P}{A + P}$$

where A and P are the apogee and perigee distances, respectively.

58. The first artificial satellite to orbit the earth was Sputnik I (launched by Russia in 1957). Its highest point above the earth's surface was 583 miles, and its lowest point was 132 miles. Find the eccentricity of its orbit.

59. Probably the most famous of all comets, **Halley's Comet,** has an elliptical orbit with the sun at the focus. Its maximum distance from the sun is approximately 35.34 au (astronomical unit $\approx 92.956 \times 10^6$ miles), and its minimum distance (perihelion) is approximately 0.59 au. Find the eccentricity of the orbit.

60. Prove that the tangent line to an ellipse at a point P makes equal angles with the lines through P and the foci (see figure). [Hint: (a) Find the slope of the tangent line at P, (b) find the slopes of the lines through P and each focus, and (c) use the formula for the tangent of the angle between two lines.]

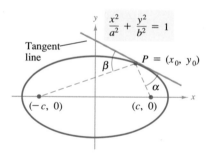

Reflective property of an ellipse: $\alpha = \beta$

FIGURE FOR 60

In Exercises 61 and 62, use a computer or graphics calculator to sketch the graph of the ellipse.

61. $x^2 + 2y^2 - 3x + 4y + 0.25 = 0$

62. $2x^2 + y^2 + 4.8x - 6.4y + 3.12 = 0$

11.3 Hyperbolas

Hyperbola ▪ Standard equation of a hyperbola ▪ Asymptotes ▪ Eccentricity ▪ Applications

The definition of a hyperbola is similar to that of an ellipse. For an ellipse the *sum* of the distances between the foci and a point on the ellipse is fixed, whereas for a hyperbola the *difference* of these distances is fixed.

DEFINITION OF A HYPERBOLA	A **hyperbola** is the set of all points (x, y) the difference of whose distances from two distinct fixed points (foci) is constant. (See Figure 11.20.)

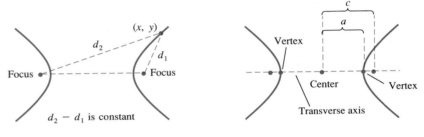

FIGURE 11.20

The line through the two foci intersects a hyperbola at two points called the **vertices.** The line segment connecting the vertices is called the **transverse axis,** and the midpoint of the transverse axis is called the **center** of the hyperbola. One distinguishing feature of hyperbolas is that their graphs have two separate *branches*.

The development of the standard form of the equation of a hyperbola is similar to that for an ellipse, and we list the following theorem without proof.

THEOREM 11.5 **STANDARD EQUATION** **OF A HYPERBOLA**	The standard form of the equation of a hyperbola with center at (h, k) is $\dfrac{(x - h)^2}{a^2} - \dfrac{(y - k)^2}{b^2} = 1$ Transverse axis is *horizontal* $\dfrac{(y - k)^2}{a^2} - \dfrac{(x - h)^2}{b^2} = 1.$ Transverse axis is *vertical* The vertices are a units from the center, and the foci are c units from the center. Moreover, $b^2 = c^2 - a^2$.

Figure 11.21 shows both the horizontal and vertical orientations for a hyperbola.

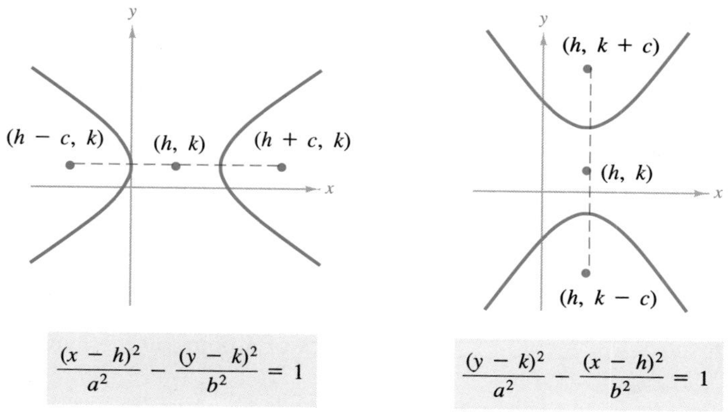

$$\frac{(x - h)^2}{a^2} - \frac{(y - k)^2}{b^2} = 1 \qquad\qquad \frac{(y - k)^2}{a^2} - \frac{(x - h)^2}{b^2} = 1$$

FIGURE 11.21

Standard Equations for Hyperbolas

EXAMPLE 1 Finding the standard equation for a hyperbola

Find the standard form of the equation of the hyperbola with foci at $(-1, 2)$ and $(5, 2)$ and vertices at $(0, 2)$ and $(4, 2)$.

SOLUTION

By the Midpoint Formula, the center of the hyperbola occurs at the point $(2, 2)$. Furthermore, $c = 3$ and $a = 2$, and it follows that

$$b^2 = 3^2 - 2^2 = 9 - 4 = 5.$$

Thus, the equation of the hyperbola is

$$\frac{(x - 2)^2}{4} - \frac{(y - 2)^2}{5} = 1.$$

Figure 11.22 shows the graph of the hyperbola. ▭

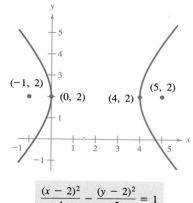

$$\frac{(x - 2)^2}{4} - \frac{(y - 2)^2}{5} = 1$$

FIGURE 11.22

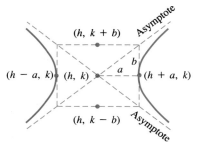

FIGURE 11.23

An important aid in sketching the graph of a hyperbola is the determination of its **asymptotes,** as shown in Figure 11.23. Each hyperbola has two asymptotes that intersect at the center of the hyperbola. The asymptotes pass through the vertices of a rectangle of dimension $2a$ by $2b$, with its center at (h, k). The line segment of length $2b$ joining $(h, k + b)$ and $(h, k - b)$ is referred to as the **conjugate axis** of the hyperbola. The following theorem identifies the equation for the asymptotes.

THEOREM 11.6
ASYMPTOTES OF A HYPERBOLA

For a *horizontal* transverse axis, the equations of the asymptotes are

$$y = k + \frac{b}{a}(x - h) \qquad \text{and} \qquad y = k - \frac{b}{a}(x - h).$$

For a *vertical* transverse axis, the equations of the asymptotes are

$$y = k + \frac{a}{b}(x - h) \qquad \text{and} \qquad y = k - \frac{a}{b}(x - h).$$

PROOF   To simplify the algebra, we assume that the center of the hyperbola is $(h, k) = (0, 0)$. Solving for y in the equation

$$\frac{x^2}{a^2} - \frac{y^2}{b^2} = 1$$

produces

$$y = \pm \frac{b}{a} \sqrt{x^2 - a^2}.$$

If $y = (b/a)x$ is to be an asymptote of the hyperbola, then the difference of the y-values of a point on the hyperbola and a point on the asymptote

$$\frac{b}{a} \sqrt{x^2 - a^2} - \frac{b}{a} x$$

must approach zero as $x \to \infty$. To prove this, observe that

$$\lim_{x \to \infty} \left[\frac{b}{a} \sqrt{x^2 - a^2} - \frac{b}{a} x \right] = \lim_{x \to \infty} \left[\frac{b}{a} \left(\sqrt{x^2 - a^2} - x \right) \right]$$

$$= \lim_{x \to \infty} \left[\frac{b}{a} \left(\frac{x^2 - a^2 - x^2}{\sqrt{x^2 - a^2} + x} \right) \right]$$

$$= \lim_{x \to \infty} \left[\frac{-ab}{\sqrt{x^2 - a^2} + x} \right]$$

$$= 0.$$

The asymptotic behavior of the other three portions of the hyperbola can be established similarly.

REMARK   In Figure 11.23 we can see that the asymptotes coincide with the diagonals of the rectangle with dimensions $2a$ and $2b$, centered at (h, k). This provides us with a quick means of sketching the asymptotes, which in turn aids in sketching the hyperbola.

EXAMPLE 2   Using asymptotes to sketch a hyperbola

Sketch the graph of the hyperbola whose equation is

$$4x^2 - y^2 = 16.$$

SOLUTION

Rewriting this equation in standard form produces

$$\frac{4x^2}{16} - \frac{y^2}{16} = \frac{16}{16}$$

$$\frac{x^2}{2^2} - \frac{y^2}{4^2} = 1.$$

From this, we conclude that the transverse axis is horizontal and the vertices occur at $(-2, 0)$ and $(2, 0)$. Moreover, the ends of the conjugate axis occur at $(0, -4)$ and $(0, 4)$, and we are able to sketch the rectangle shown in Figure 11.24. Finally, by drawing the asymptotes through the corners of this rectangle, we complete the sketch shown in Figure 11.25.

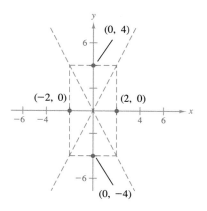

FIGURE 11.24

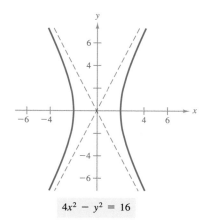

$4x^2 - y^2 = 16$

FIGURE 11.25

EXAMPLE 3 Finding the asymptotes of a hyperbola

Sketch the hyperbola given by

$$4x^2 - 3y^2 + 8x + 16 = 0$$

and find the equations of its asymptotes.

SOLUTION

Rewriting this equation in standard form produces

$$4x^2 - 3y^2 + 8x + 16 = 0$$
$$4(x^2 + 2x) - 3y^2 = -16$$
$$-4(x^2 + 2x + 1) + 3y^2 = 16 - 4$$
$$-4(x + 1)^2 + 3y^2 = 12$$
$$\frac{y^2}{4} - \frac{(x + 1)^2}{3} = 1.$$

From this equation we conclude that the hyperbola is centered at $(-1, 0)$ and has vertices at $(-1, 2)$ and $(-1, -2)$, and the ends of the conjugate axis occur at $(-1 - \sqrt{3}, 0)$ and $(-1 + \sqrt{3}, 0)$. To sketch the graph of the hyperbola, we draw a rectangle through these four points. The asymptotes are the lines passing through the corners of the rectangle, as shown in Figure 11.26. Finally, using $a = 2$ and $b = \sqrt{3}$, we apply Theorem 11.6 to conclude that the equations of the asymptotes are

$$y = \frac{2}{\sqrt{3}}(x + 1) \quad \text{and} \quad y = -\frac{2}{\sqrt{3}}(x + 1).$$

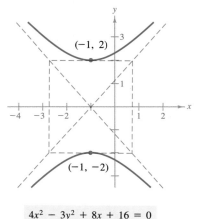

$4x^2 - 3y^2 + 8x + 16 = 0$

FIGURE 11.26

REMARK If the constant term in the equation in Example 3 had been $F = -4$ instead of 16, then we would have obtained the following degenerate case.

Two intersecting lines: $\dfrac{y^2}{4} - \dfrac{(x + 1)^2}{3} = 0$

EXAMPLE 4 Using asymptotes to find the standard equation

Find the standard form of the equation of the hyperbola having vertices at $(3, -5)$ and $(3, 1)$ and asymptotes $y = 2x - 8$ and $y = -2x + 4$, as shown in Figure 11.27.

SOLUTION

By the Midpoint Formula the center of the hyperbola is at $(3, -2)$. Furthermore, the hyperbola has a vertical transverse axis with $a = 3$. By Theorem 11.6 the asymptotes have equations whose slopes are

$$m_1 = \frac{a}{b} \quad \text{and} \quad m_2 = -\frac{a}{b}.$$

From the given equations of the asymptotes, we know that

$$\frac{a}{b} = 2 \quad \text{and} \quad -\frac{a}{b} = -2$$

and since $a = 3$, we conclude that $b = \frac{3}{2}$. Therefore, the standard equation is

$$\frac{(y + 2)^2}{9} - \frac{(x - 3)^2}{9/4} = 1.$$

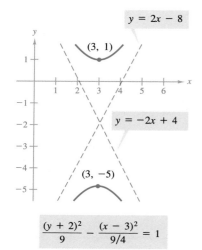

$\dfrac{(y + 2)^2}{9} - \dfrac{(x - 3)^2}{9/4} = 1$

FIGURE 11.27

As with ellipses, the **eccentricity** of a hyperbola is $e = c/a$. Because $c > a$, it follows that $e > 1$. If the eccentricity is large, then the branches of the hyperbola are nearly flat. If the eccentricity is close to 1, then the branches of the hyperbola are more pointed, as shown in Figure 11.28.

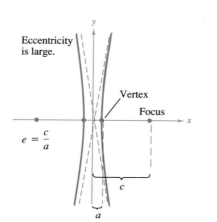
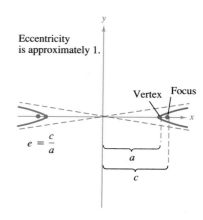

FIGURE 11.28

DEFINITION OF ECCENTRICITY OF A HYPERBOLA	The eccentricity e of a hyperbola is given by the ratio $$e = \frac{c}{a}.$$

Applications

The following application was developed during World War II. It shows how the properties of hyperbolas can be used in radar or other detection systems.

EXAMPLE 5   An application involving hyperbolas

Two microphones, 1 mile apart, record an explosion. Microphone A received the sound 2 seconds before microphone B. Where did the explosion come from?

SOLUTION

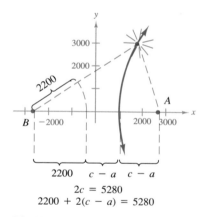

$$2c = 5280$$
$$2200 + 2(c - a) = 5280$$

FIGURE 11.29

Assuming that sound travels at 1100 feet per second, we know that the explosion took place 2200 feet farther from B than from A, as shown in Figure 11.29. The locus of all points that are 2200 feet closer to A than to B is one branch of the hyperbola $(x^2/a^2) - (y^2/b^2) = 1$, where

$$c = \frac{5280}{2} = 2640 \qquad \text{and} \qquad a = \frac{2200}{2} = 1100.$$

Thus, $b^2 = c^2 - a^2 = 5{,}759{,}600$, and we conclude that the explosion occurred somewhere on the right branch of the hyperbola given by

$$\frac{x^2}{1{,}210{,}000} - \frac{y^2}{5{,}759{,}600} = 1.$$

In Example 5 we were able to determine only the hyperbola on which the explosion occurred, but not the exact location of the explosion. If, however, we had received the sound at a third position, C, then two other hyperbolas would be determined. The exact location of the explosion would be the point at which these three hyperbolas intersect.

Another interesting application of conics involves the orbits of comets in our solar system. Of the 610 comets identified prior to 1970, 245 have elliptical orbits, 295 have parabolic orbits, and 70 have hyperbolic orbits. The first woman to be credited with detecting a new comet was the English astronomer Caroline Herschel (1750–1848). During her long life, Caroline Herschel discovered a total of eight new comets.

Caroline Herschel

The center of the sun is a focus of each orbit, and each orbit has a vertex at the point at which the comet is closest to the sun, as shown in Figure 11.30. Undoubtedly, many comets with parabolic or hyperbolic orbits have not been identified. We only get to see such comets *once*. Only comets with elliptical orbits such as Halley's Comet remain in our solar system.

The type of orbit for a comet can be determined as follows.

1. Ellipse: $v < \sqrt{2GM/p}$

2. Parabola: $v = \sqrt{2GM/p}$

3. Hyperbola: $v > \sqrt{2GM/p}$

In these three formulas, p is the distance between the vertex and the focus of the comet's orbit (in meters), v is the velocity of the comet at the vertex (in meters per second), $M \approx 1.991 \times 10^{30}$ kilograms is the mass of the sun, and $G \approx 6.67 \times 10^{-11}$ cubic meters per kilogram-second squared is the gravitational constant.

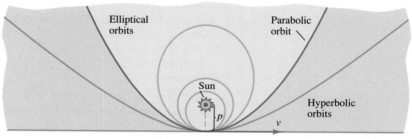

FIGURE 11.30

EXERCISES for Section 11.3

In Exercises 1–6, match the equation with the correct graph. [The graphs are labeled (a)–(f).]

1. $\dfrac{x^2}{9} - \dfrac{y^2}{4} = 1$

2. $\dfrac{y^2}{9} - \dfrac{x^2}{4} = 1$

3. $\dfrac{y^2}{1} - \dfrac{x^2}{16} = 1$

4. $\dfrac{y^2}{16} - \dfrac{x^2}{1} = 1$

5. $\dfrac{(x - 2)^2}{9} - \dfrac{y^2}{4} = 1$

6. $\dfrac{(x + 1)^2}{16} - \dfrac{(y - 3)^2}{9} = 1$

(c)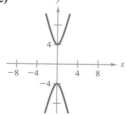

(d)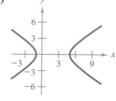

(a)

(b)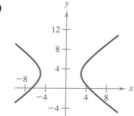

(e)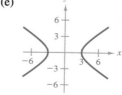

(f)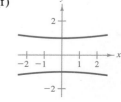

In Exercises 7–26, find the center, vertices, and foci of the hyperbola and sketch its graph, using asymptotes as an aid.

7. $x^2 - y^2 = 1$

8. $\dfrac{x^2}{9} - \dfrac{y^2}{16} = 1$

9. $y^2 - \dfrac{x^2}{4} = 1$

10. $\dfrac{y^2}{9} - x^2 = 1$

11. $\dfrac{y^2}{25} - \dfrac{x^2}{144} = 1$

12. $\dfrac{x^2}{36} - \dfrac{y^2}{4} = 1$

13. $2x^2 - 3y^2 = 6$

14. $3y^2 = 5x^2 + 15$

15. $5y^2 = 4x^2 + 20$

16. $7x^2 - 3y^2 = 21$

17. $\dfrac{(x - 1)^2}{4} - (y + 2)^2 = 1$

18. $\dfrac{(x + 1)^2}{144} - \dfrac{(y - 4)^2}{25} = 1$

19. $(y + 6)^2 - (x - 2)^2 = 1$

20. $4(y - 1)^2 - 9(x + 3)^2 = 1$

21. $9x^2 - y^2 - 36x - 6y + 18 = 0$

22. $x^2 - 9y^2 + 36y - 72 = 0$

23. $9y^2 - x^2 + 2x + 54y + 62 = 0$

24. $16y^2 - x^2 + 2x + 64y + 63 = 0$

25. $x^2 - 9y^2 + 2x - 54y - 80 = 0$

26. $9x^2 - y^2 + 54x + 10y + 55 = 0$

In Exercises 27–36, find an equation for the hyperbola.

27. Center: (0, 0)
Vertex: (0, 2)
Focus: (0, 4)

28. Center: (0, 0)
Vertex: (3, 0)
Focus: (5, 0)

29. Vertices: (±1, 0)
Asymptotes: $y = \pm 3x$

30. Vertices: (0, ±3)
Asymptotes: $y = \pm 3x$

31. Vertices: (0, 2), (6, 2)
Asymptotes: $y = \frac{2}{3}x$
$y = 4 - \frac{2}{3}x$

32. Vertices: (2, ±3)
Foci: (2, ±5)

33. Vertices: (2, ±3)
Point on graph: (0, 5)

34. Focus: (10, 0)
Asymptotes: $y = \pm \frac{3}{4}x$

35. For any point on the hyperbola, the difference of its distances from the points (2, 2) and (10, 2) is 6.

36. For any point on the hyperbola, the difference of its distances from the points (−3, 0) and (−3, 3) is 2.

In Exercises 37 and 38, find equations for (a) the tangent lines and (b) the normal lines to the hyperbola for the given values of x.

37. $\dfrac{x^2}{9} - y^2 = 1$, $x = 6$

38. $\dfrac{y^2}{4} - \dfrac{x^2}{2} = 1$, $x = 4$

In Exercises 39 and 40, the region bounded by the graphs of the given equations revolves about the x-axis. Find the volume and surface area of the resulting solid.

39. $x^2 - y^2 = 1$, $y = 0$, $x = 2$

40. $\dfrac{x^2}{16} - \dfrac{y^2}{9} = 1$, $y = 0$, $x = 5$

41. LORAN (long distance radio navigation) for aircraft and ships uses synchronized pulses transmitted by widely separated transmitting stations. These pulses travel at the speed of light (186,000 miles per second). The difference in the times of arrival of these pulses at an aircraft or ship is constant on a hyperbola having the two transmitting stations as foci. Assume that two stations, 300 miles apart, are positioned on the rectangular coordinate system at points with coordinates (−150, 0) and (150, 0) and that a ship is traveling on a path with coordinates $(x, 75)$ (see figure). Find the x-coordinate of the position of the ship if the time difference between the pulses from the transmitting stations is 1000 microseconds (0.001 second).

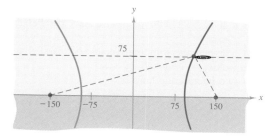

FIGURE FOR 41

42. A rifle, positioned at point $(-c, 0)$, is fired at a target positioned at point $(c, 0)$. A person hears the sound of the rifle and the sound of the bullet hitting the target at the same time. Show that the person is positioned on one branch of the hyperbola given by

$$\dfrac{x^2}{c^2 v_s^2 / v_m^2} - \dfrac{y^2}{c^2(v_m^2 - v_s^2)/v_m^2} = 1$$

where v_m is the muzzle velocity of the rifle and the speed of sound is $v_s = 1100$ feet per second.

43. Show that the equation of the tangent line to

$$\dfrac{x^2}{a^2} - \dfrac{y^2}{b^2} = 1$$

at the point (x_0, y_0) is

$$\dfrac{x_0}{a^2}x - \dfrac{y_0}{b^2}y = 1.$$

44. Show that the ellipse

$$\dfrac{x^2}{a^2} + \dfrac{2y^2}{b^2} = 1$$

and the hyperbola

$$\dfrac{x^2}{a^2 - b^2} - \dfrac{2y^2}{b^2} = 1$$

intersect at right angles.

In Exercises 45 and 46, use the following general form of the equation of the conics.

$$Ax^2 + Cy^2 + Dx + Ey + F = 0$$

45. Prove that the graph of the equation is one of the following (except in degenerate cases).

Conic	Condition
(a) Circle	$A = C$
(b) Parabola	$A = 0$ or $C = 0$
	(but not both)
(c) Ellipse	$AC > 0$
(d) Hyperbola	$AC < 0$

46. If $AC < 0$, determine the relationships among A, C, D, E, and F so that the graph of the equation is
(a) a hyperbola with transverse axis parallel to the x-axis,
(b) a hyperbola with transverse axis parallel to the y-axis,
(c) two intersecting lines.

In Exercises 47–56, use Exercise 45 to classify the graph of each equation as a circle, a parabola, an ellipse, or a hyperbola.

47. $x^2 + 4y^2 - 6x + 16y + 21 = 0$
48. $4x^2 - y^2 - 4x - 3 = 0$
49. $y^2 - 4y - 4x = 0$
50. $25x^2 - 10x - 200y - 119 = 0$
51. $4x^2 + 4y^2 - 16y + 15 = 0$
52. $y^2 - 4y = x + 5$
53. $9x^2 + 9y^2 - 36x + 6y + 34 = 0$
54. $2x(x - y) = y(3 - y - 2x)$
55. $3(x - 1)^2 = 6 + 2(y + 1)^2$
56. $9(x + 3)^2 = 36 - 4(y - 2)^2$

In Exercises 57 and 58, use a computer or graphics calculator to sketch the graph of the hyperbola.

57. $3x^2 - 2y^2 - 6x - 12y - 27 = 0$
58. $3y^2 - x^2 + 6x - 12y = 0$

11.4   Rotation and the General Second-Degree Equation

General second-degree equations ▪ Rotation of axes ▪ Invariants under rotation ▪ Discriminant

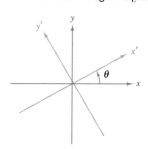

Rotated: x'-axis
            y'-axis

FIGURE 11.31

In previous sections we have shown that the equation of a conic with axes parallel to one of the coordinate axes has a standard form that can be written in the general form

$$Ax^2 + Cy^2 + Dx + Ey + F = 0. \qquad \text{Horizontal or vertical axes}$$

In this section we study the equations of conics whose axes are rotated so that they are not parallel to either the x-axis or the y-axis. The general equation for such conics contains an *xy term*.

$$Ax^2 + Bxy + Cy^2 + Dx + Ey + F = 0 \qquad \text{Equation in } xy\text{-plane}$$

To eliminate this xy term, we use a procedure called **rotation of axes.** We want to rotate the x- and y-axes until they are parallel to the axes of the conic. (We denote the rotated axes as the x'-axis and the y'-axis, as shown in Figure 11.31.) After the rotation has been accomplished, the equation of the conic in the new $x'y'$-plane will have the form

$$A'(x')^2 + C'(y')^2 + D'x' + E'y' + F' = 0. \qquad \text{Equation in } x'y'\text{-plane}$$

Since this equation has no $x'y'$ term, we can obtain a standard form by completing the square.

The following theorem identifies how much to rotate the axes to eliminate an xy term, and also the equations for determining the new coefficients A', C', D', E', and F'.

**THEOREM 11.7
ROTATION OF AXES TO
ELIMINATE AN *XY* TERM**

$Ax^2 + Bxy + Cy^2 + Dx + Ey + F = 0$ can be rewritten as

$$A'(x')^2 + C'(y')^2 + D'x' + E'y' + F' = 0$$

by rotating the coordinate axes through an angle θ, where

$$\cot 2\theta = \frac{A - C}{B}.$$

The coefficients of the new equation are obtained by making the substitutions

$$x = x' \cos \theta - y' \sin \theta$$
$$y = x' \sin \theta + y' \cos \theta.$$

PROOF

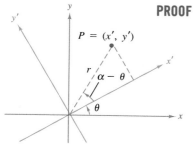

Rotated: $x' = r \cos (\alpha - \theta)$
          $y' = r \sin (\alpha - \theta)$

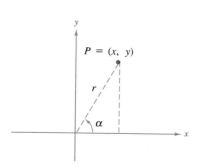

Original: $x = r \cos \alpha$
           $y = r \sin \alpha$

FIGURE 11.32

To discover how the coordinates in the xy-system are related to the coordinates in the $x'y'$-system, we choose a point $P = (x, y)$ in the original system and attempt to find its coordinates (x', y') in the rotated system. In either system the distance r between the point P and the origin is the same, thus the equations for x, y, x', and y' are those given in Figure 11.32. Using the formulas for the sine and cosine of the difference of two angles, we obtain

$$x' = r \cos (\alpha - \theta) = r(\cos \alpha \cos \theta + \sin \alpha \sin \theta)$$
$$= r \cos \alpha \cos \theta + r \sin \alpha \sin \theta = x \cos \theta + y \sin \theta$$
$$y' = r \sin (\alpha - \theta) = r(\sin \alpha \cos \theta - \cos \alpha \sin \theta)$$
$$= r \sin \alpha \cos \theta - r \cos \alpha \sin \theta = y \cos \theta - x \sin \theta.$$

Solving this system for x and y yields

$$x = x' \cos \theta - y' \sin \theta \quad \text{and} \quad y = x' \sin \theta + y' \cos \theta.$$

Finally, by substituting these values for x and y into the original equation and collecting terms, we obtain the following.

$$A' = A \cos^2 \theta + B \cos \theta \sin \theta + C \sin^2 \theta$$
$$C' = A \sin^2 \theta - B \cos \theta \sin \theta + C \cos^2 \theta$$
$$D' = D \cos \theta + E \sin \theta$$
$$E' = -D \sin \theta + E \cos \theta$$
$$F' = F$$

Now, in order to eliminate the $x'y'$ term, we must select θ so that $B' = 0$ as follows.

$$B' = 2(C - A) \sin \theta \cos \theta + B(\cos^2 \theta - \sin^2 \theta)$$
$$= (C - A) \sin 2\theta + B \cos 2\theta$$
$$= B(\sin 2\theta)\left(\frac{C - A}{B} + \cot 2\theta\right) = 0, \quad \sin 2\theta \neq 0$$

If $B = 0$, no rotation is necessary since the xy term is not present in the original equation. If $B \neq 0$, then the only way to make $B' = 0$ is to let

$$\cot 2\theta = \frac{A - C}{B}, \quad B \neq 0.$$

Thus, we have established the desired results.

EXAMPLE 1   Rotation of a hyperbola

Write the equation $xy - 1 = 0$ in standard form.

SOLUTION

Since $A = 0$, $B = 1$, and $C = 0$, we have (for $0 < \theta < \pi/2$)

$$\cot 2\theta = \frac{A - C}{B} = 0 \implies 2\theta = \frac{\pi}{2} \implies \theta = \frac{\pi}{4}.$$

Therefore, the equation in the $x'y'$-system is given by making the following substitutions.

$$x = x' \cos \frac{\pi}{4} - y' \sin \frac{\pi}{4} = x'\left(\frac{\sqrt{2}}{2}\right) - y'\left(\frac{\sqrt{2}}{2}\right) = \frac{x' - y'}{\sqrt{2}}$$

$$y = x' \sin \frac{\pi}{4} + y' \cos \frac{\pi}{4} = x'\left(\frac{\sqrt{2}}{2}\right) + y'\left(\frac{\sqrt{2}}{2}\right) = \frac{x' + y'}{\sqrt{2}}$$

Substituting these expressions into the equation $xy - 1 = 0$ produces

$$\left(\frac{x' - y'}{\sqrt{2}}\right)\left(\frac{x' + y'}{\sqrt{2}}\right) - 1 = 0$$

$$\frac{(x')^2 - (y')^2}{2} - 1 = 0$$

$$\frac{(x')^2}{(\sqrt{2})^2} - \frac{(y')^2}{(\sqrt{2})^2} = 1. \qquad \text{Standard form}$$

This is the equation of a hyperbola centered at the origin with vertices at $(\pm\sqrt{2}, 0)$ in the $x'y'$-system as shown in Figure 11.33. To find the coordinates of the vertices in the xy-system, we substituted the $x'y'$-coordinates into the equations

$$x = \frac{x' - y'}{\sqrt{2}} \qquad \text{and} \qquad y = \frac{x' + y'}{\sqrt{2}}.$$

From these substitutions, we find the vertices to be $(1, 1)$ and $(-1, -1)$ in the xy-system. Note also that the asymptotes of the hyperbola have equations $y' = \pm x'$, which correspond to the original x- and y-axes.

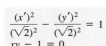

$$\frac{(x')^2}{(\sqrt{2})^2} - \frac{(y')^2}{(\sqrt{2})^2} = 1$$
$$xy - 1 = 0$$

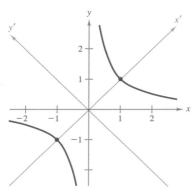

Vertices:
$(\sqrt{2}, 0)$, $(-\sqrt{2}, 0)$ in $x'y'$-system
$(1, 1)$, $(-1, -1)$ in xy-system

FIGURE 11.33

Remember that the substitutions

$$x = x' \cos \theta - y' \sin \theta \qquad \text{and} \qquad y = x' \sin \theta + y' \cos \theta$$

were developed to eliminate the $x'y'$ term in the rotated system. You can use this as a check on your work. In other words, if your final equation contains an $x'y'$ term, you know that you made a mistake.

EXAMPLE 2   Rotation of an ellipse

Sketch the graph of $7x^2 - 6\sqrt{3}xy + 13y^2 - 16 = 0$.

SOLUTION

Since $A = 7$, $B = -6\sqrt{3}$, and $C = 13$, we have (for $0 < \theta < \pi/2$)

$$\cot 2\theta = \frac{A - C}{B} = \frac{7 - 13}{-6\sqrt{3}} = \frac{1}{\sqrt{3}} \implies \theta = \frac{\pi}{6}.$$

Therefore, the equation in the $x'y'$-system is given by making the following substitutions.

$$x = x' \cos \frac{\pi}{6} - y' \sin \frac{\pi}{6} = x'\left(\frac{\sqrt{3}}{2}\right) - y'\left(\frac{1}{2}\right) = \frac{\sqrt{3}x' - y'}{2}$$

$$y = x' \sin \frac{\pi}{6} + y' \cos \frac{\pi}{6} = x'\left(\frac{1}{2}\right) + y'\left(\frac{\sqrt{3}}{2}\right) = \frac{x' + \sqrt{3}y'}{2}$$

Substituting these expressions into the original equation produces

$$7\left(\frac{\sqrt{3}x' - y'}{2}\right)^2 - 6\sqrt{3}\left(\frac{\sqrt{3}x' - y'}{2}\right)\left(\frac{x' + \sqrt{3}y'}{2}\right)$$

$$+ 13\left(\frac{x' + \sqrt{3}y'}{2}\right)^2 = 16$$

which simplifies to

$$4(x')^2 + 16(y')^2 = 16$$

$$\frac{(x')^2}{4} + \frac{(y')^2}{1} = 1. \qquad \text{Standard form}$$

This is the equation of an ellipse centered at the origin with vertices at $(\pm 2, 0)$ in the $x'y'$-system, as shown in Figure 11.34.   ▭

In constructing Examples 1 and 2 we carefully chose the equations so that θ would be one of the common angles 30°, 45°, and so forth. Of course, many second-degree equations do not yield such common solutions to the equation $\cot 2\theta = (A - C)/B$. Example 3 illustrates such a case.

EXAMPLE 3   Rotation of a parabola

Sketch the graph of $x^2 - 4xy + 4y^2 + 5\sqrt{5}y + 1 = 0$.

SOLUTION

Since $A = 1$, $B = -4$, and $C = 4$, we have

$$\cot 2\theta = \frac{A - C}{B} = \frac{1 - 4}{-4} = \frac{3}{4}.$$

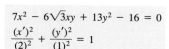

$7x^2 - 6\sqrt{3}xy + 13y^2 - 16 = 0$

$\dfrac{(x')^2}{(2)^2} + \dfrac{(y')^2}{(1)^2} = 1$

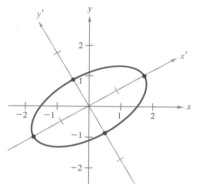

Vertices:

$(\pm 2, 0)$, $(0, \pm 1)$ in $x'y'$-system

$(\pm\sqrt{3}, \pm 1)$, $\left(\pm\frac{1}{2}, \mp\frac{\sqrt{3}}{2}\right)$ in xy-system

FIGURE 11.34

Using the trigonometric identity $\cot 2\theta = (\cot^2 \theta - 1)/(2 \cot \theta)$ produces

$$\cot 2\theta = \frac{3}{4} = \frac{\cot^2 \theta - 1}{2 \cot \theta}$$

from which we obtain the equation

$$6 \cot \theta = 4 \cot^2 \theta - 4 \quad \Longrightarrow \quad 4 \cot^2 \theta - 6 \cot \theta - 4 = 0$$
$$(2 \cot \theta - 4)(2 \cot \theta + 1) = 0.$$

Considering $0 < \theta < \pi/2$, we have $2 \cot \theta = 4$. Thus,

$$\cot \theta = 2 \quad \Longrightarrow \quad \theta \approx 26.6°.$$

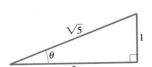

FIGURE 11.35

From the triangle in Figure 11.35 we obtain $\sin \theta = 1/\sqrt{5}$ and $\cos \theta = 2/\sqrt{5}$. Consequently, we obtain the following.

$$x = x' \cos \theta - y' \sin \theta = x'\left(\frac{2}{\sqrt{5}}\right) - y'\left(\frac{1}{\sqrt{5}}\right) = \frac{2x' - y'}{\sqrt{5}}$$

$$y = x' \sin \theta + y' \cos \theta = x'\left(\frac{1}{\sqrt{5}}\right) + y'\left(\frac{2}{\sqrt{5}}\right) = \frac{x' + 2y'}{\sqrt{5}}$$

Substituting these expressions into the original equation, we have

$$\left(\frac{2x' - y'}{\sqrt{5}}\right)^2 - 4\left(\frac{2x' - y'}{\sqrt{5}}\right)\left(\frac{x' + 2y'}{\sqrt{5}}\right) + 4\left(\frac{x' + 2y'}{\sqrt{5}}\right)^2$$
$$+ 5\sqrt{5}\left(\frac{x' + 2y'}{\sqrt{5}}\right) + 1 = 0$$

which simplifies to

$$5(y')^2 + 5x' + 10y' + 1 = 0.$$

By completing the square, we can obtain the standard form

$$5(y' + 1)^2 = -5x' + 4$$
$$(y' + 1)^2 = (-1)\left(x' - \frac{4}{5}\right). \qquad \text{Standard form}$$

The graph of the equation is a parabola with its vertex at $\left(\frac{4}{5}, -1\right)$ and its axis parallel to the x'-axis in the $x'y'$-system, as shown in Figure 11.36. ▭

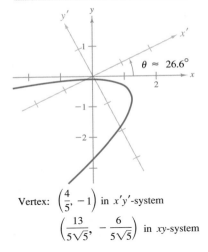

$x^2 - 4xy + 4y^2 + 5\sqrt{5}y + 1 = 0$
$(y' + 1)^2 = 4\left(-\frac{1}{4}\right)\left(x' - \frac{4}{5}\right)$

$\theta \approx 26.6°$

Vertex: $\left(\frac{4}{5}, -1\right)$ in $x'y'$-system

$\left(\frac{13}{5\sqrt{5}}, -\frac{6}{5\sqrt{5}}\right)$ in xy-system

FIGURE 11.36

Invariants under rotation

In Theorem 11.7 note that the constant term $F' = F$ is the same in both equations. We say that it is **invariant under rotation**. Theorem 11.8 lists some other rotation invariants. We leave the proof of this theorem as an exercise (see Exercise 32).

THEOREM 11.8 **ROTATION INVARIANTS**	The rotation of coordinate axes through an angle θ that transforms the equation $Ax^2 + Bxy + Cy^2 + Dx + Ey + F = 0$ into the form $A'(x')^2 + C'(y')^2 + D'x' + E'y' + F' = 0$ has the following rotation invariants.
	1. $F = F'$ 2. $A + C = A' + C'$ 3. $B^2 - 4AC = (B')^2 - 4A'C'$

We can use the results of this theorem to classify the graph of a second-degree equation *with* an *xy* term in much the same way we did for second-degree equations *without* an *xy* term (see Exercise 45 in Section 11.3). Note that since $B' = 0$, the invariant $B^2 - 4AC$ reduces to

$$B^2 - 4AC = -4A'C'. \quad \text{Discriminant}$$

We call this quantity the **discriminant** of the equation

$$Ax^2 + Bxy + Cy^2 + Dx + Ey + F = 0.$$

Now, because the sign of $A'C'$ determines the type of graph for the equation

$$A'(x')^2 + C'(y')^2 + D'x' + E'y' + F' = 0$$

we can conclude that the sign of $B^2 - 4AC$ must determine the type of graph for the original equation. This result is stated in Theorem 11.9.

THEOREM 11.9
CLASSIFICATION OF CONICS
BY THE DISCRIMINANT

The graph of the equation $Ax^2 + Bxy + Cy^2 + Dx + Ey + F = 0$ is, except in degenerate cases, determined by its discriminant as follows.

1. Ellipse or circle: $B^2 - 4AC < 0$
2. Parabola: $B^2 - 4AC = 0$
3. Hyperbola: $B^2 - 4AC > 0$

EXAMPLE 4 Using the discriminant

Classify the graph of each of the following equations.
(a) $4xy - 9 = 0$
(b) $2x^2 - 3xy + 2y^2 - 2x = 0$
(c) $x^2 - 6xy + 9y^2 - 2y + 1 = 0$
(d) $3x^2 + 8xy + 4y^2 - 7 = 0$

SOLUTION

(a) The graph is a hyperbola because

$$B^2 - 4AC = 16 - 0 > 0.$$

(b) The graph is a circle or an ellipse because

$$B^2 - 4AC = 9 - 16 < 0.$$

(c) The graph is a parabola because

$$B^2 - 4AC = 36 - 36 = 0.$$

(d) The graph is a hyperbola because

$$B^2 - 4AC = 64 - 48 > 0.$$

EXERCISES for Section 11.4

In Exercises 1–16, rotate the axes to eliminate the xy term. Sketch the graph of the resulting equation, showing both sets of axes.

1. $xy + 1 = 0$
2. $xy - 4 = 0$
3. $9x^2 + 24xy + 16y^2 + 90x - 130y = 0$
4. $9x^2 + 24xy + 16y^2 + 80x - 60y = 0$
5. $x^2 - 10xy + y^2 + 1 = 0$
6. $xy + x - 2y + 3 = 0$
7. $xy - 2y - 4x = 0$
8. $2x^2 - 3xy - 2y^2 + 10 = 0$
9. $5x^2 - 2xy + 5y^2 - 12 = 0$
10. $13x^2 + 6\sqrt{3}xy + 7y^2 - 16 = 0$
11. $3x^2 - 2\sqrt{3}xy + y^2 + 2x + 2\sqrt{3}y = 0$
12. $16x^2 - 24xy + 9y^2 - 60x - 80y + 100 = 0$
13. $17x^2 + 32xy - 7y^2 = 75$
14. $40x^2 + 36xy + 25y^2 = 52$
15. $32x^2 + 50xy + 7y^2 = 52$
16. $4x^2 - 12xy + 9y^2 + (4\sqrt{13} - 12)x - (6\sqrt{13} + 8)y = 91$

In Exercises 17–24, use the discriminant to determine if the graph of the equation is a parabola, ellipse, or hyperbola.

17. $16x^2 - 24xy + 9y^2 - 30x - 40y = 0$
18. $x^2 - 4xy - 2y^2 - 6 = 0$
19. $13x^2 - 8xy + 7y^2 - 45 = 0$
20. $2x^2 + 4xy + 5y^2 + 3x - 4y - 20 = 0$
21. $x^2 - 6xy - 5y^2 + 4x - 22 = 0$
22. $36x^2 - 60xy + 25y^2 + 9y = 0$
23. $x^2 + 4xy + 4y^2 - 5x - y - 3 = 0$
24. $x^2 + xy + 4y^2 + x + y - 4 = 0$

In Exercises 25–28, sketch the graph (if possible) of the degenerate conic.

25. $y^2 - 4x^2 = 0$
26. $x^2 + y^2 - 2x + 6y + 10 = 0$
27. $5x^2 - 2xy + 5y^2 = 0$
28. $x^2 - 10xy + y^2 = 0$

In Exercises 29 and 30, use a computer or graphics calculator to sketch the graph of the conic.

29. $x^2 + xy + y^2 = 10$
30. $x^2 - 4xy + 2y^2 = 6$

31. Show that the equation $x^2 + y^2 = r^2$ is invariant under rotation of axes.
32. Prove Theorem 11.8.
33. Consider $Ax^2 + Bxy + Cy^2 = 1$, where $B^2 - 4AC < 0$. Show that the area of this ellipse is

$$\pi ab = \frac{2\pi}{\sqrt{4AC - B^2}}$$

where $2a$ and $2b$ are the lengths of the major and minor axes, respectively.

REVIEW EXERCISES for Chapter 11

In Exercises 1–10, match the equation with the correct graph. [The graphs are labeled (a)–(j).]

1. $4x^2 + y^2 = 4$
2. $x^2 = 4y$
3. $4x^2 - y^2 = 4$
4. $y^2 = -4x$
5. $x^2 + 4y^2 = 4$
6. $y^2 - 4x^2 = 4$
7. $x^2 = -6y$
8. $x^2 + 5y^2 = 10$
9. $x^2 - 5y^2 = -5$
10. $y^2 - 8x = 0$

(a)

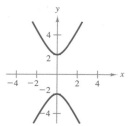

(b)

(c)

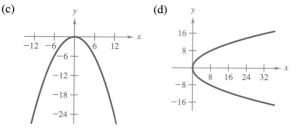

(d)

(e)

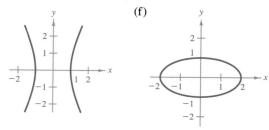

(f)

(g)

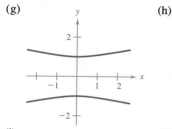

(h)

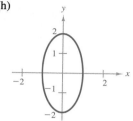

(i)

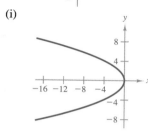

(j)

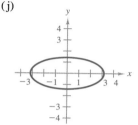

In Exercises 11–22, analyze each equation and sketch its graph.

11. $16x^2 + 16y^2 - 16x + 24y - 3 = 0$
12. $y^2 - 12y - 8x + 20 = 0$
13. $3x^2 - 2y^2 + 24x + 12y + 24 = 0$
14. $4x^2 + y^2 - 16x + 15 = 0$
15. $3x^2 + 2y^2 - 12x + 12y + 29 = 0$
16. $4x^2 - 4y^2 - 4x + 8y - 11 = 0$
17. $x^2 - 6x + 2y + 9 = 0$
18. $x^2 + y^2 - 2x - 4y + 5 = 0$
19. $x^2 + y^2 + 2xy + 2\sqrt{2}x - 2\sqrt{2}y + 2 = 0$
20. $9x^2 + 4xy + 6y^2 - 20 = 0$
21. $4x^2 + 9y^2 - 8x + 9y + 4 = 0$
22. $9x^2 + 4y^2 - 36x + 8y + 31 = 0$

In Exercises 23–32, find an equation of the specified conic.

23. Hyperbola; vertices $(0, \pm1)$, foci $(0, \pm3)$
24. Hyperbola; vertices $(\pm2, 2)$, foci $(\pm4, 2)$
25. Ellipse; foci $(0, 0)$ and $(4, 0)$, sum of the distances from a point on the ellipse to the foci is 10
26. Parabola; vertex $(4, 2)$, focus $(4, 0)$
27. Parabola; vertex $(0, 0)$, focus $(1, 1)$
28. Ellipse; vertices $(2, 0)$ and $(2, 4)$, foci $(2, 1)$ and $(2, 3)$
29. Ellipse; points on graph $(1, 2)$ and $(2, 0)$, center at $(0, 0)$
30. Hyperbola; foci $(\pm4, 0)$, asymptotes $y = \pm2x$
31. Hyperbola; foci $(\pm4, 0)$, absolute value of the difference of the distances from a point on the hyperbola to the foci is 4
32. Parabola; vertex $(0, 2)$, point on graph $(-1, 0)$

33. Find the equation of the line tangent to the parabola $y = x^2 - 2x + 2$ and perpendicular to the line $y = x - 2$.

34. Find the equations of the lines tangent to the ellipse $x^2 + 4y^2 - 4x - 8y - 24 = 0$ and parallel to the line $x - 2y - 12 = 0$.

35. Find a so that the hyperbola given by
$$\frac{x^2}{a^2} - \frac{y^2}{4} = 1$$
is tangent to the line $2x - y - 4 = 0$.

36. A large parabolic antenna is described as the surface formed by revolving the parabola $y = x^2/200$ on the interval $[0, 100]$ about the y-axis. The receiving and transmitting equipment is positioned at the focus.
(a) Find the focus.
(b) Find the antenna's surface area.

In Exercises 37–39, consider the region bounded by the ellipse
$$\frac{x^2}{a^2} + \frac{y^2}{b^2} = 1, \text{ eccentricity } e = c/a.$$

37. Show that the area of the region is πab.
38. Show that the solid (oblate spheroid) generated by revolving the region about the minor axis of the ellipse has a volume of $V = 4\pi a^2 b/3$ and a surface area of
$$S = 2\pi a^2 + \pi\left(\frac{b^2}{e}\right) \ln\left(\frac{1 + e}{1 - e}\right).$$

39. Show that the solid (prolate spheroid) generated by revolving the region about the major axis of the ellipse has a volume of $V = 4\pi ab^2/3$ and a surface area of
$$S = 2\pi b^2 + 2\pi\left(\frac{ab}{e}\right) \arcsin e.$$

40. Use the results of Exercises 38 and 39 to find the volume and surface areas of the prolate and oblate spheroids
$$\frac{x^2}{9} + \frac{y^2}{4} = 1.$$

41. Approximate, to two decimal places, the perimeter of the ellipse
$$\frac{x^2}{9} + \frac{y^2}{4} = 1.$$

In Exercises 42–45, consider a firetruck with a water tank 16 feet long whose vertical cross sections are ellipses described by the equation
$$\frac{x^2}{16} + \frac{y^2}{9} = 1.$$

42. Find the volume of the tank.
43. Find the depth of the water in the tank if it is $\frac{3}{4}$ full (by volume) and the truck is on level ground.
44. Find the force on the end of the tank, full of water.
45. Approximate, to two decimal places, the surface area of the tank.

Chapter 12 Application

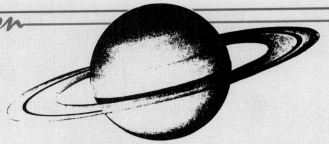

Saturn is the sixth planet from the sun. It is the second largest planet in our solar system, with a volume 815 times that of the earth—only Jupiter is larger. The Voyager I and II flights to Saturn determined that a Saturnian day is short—only 10 hours, 39 minutes, and 24 seconds.

The Orbit of Saturn

The planets in our solar system travel in elliptical orbits with the sun as a focus. In polar coordinates, the equation for the orbit of Saturn is

$$r = \frac{1.4228}{1 - 0.0543 \cos \theta}$$

where r is measured in billions of kilometers. The eccentricity of Saturn's orbit is $e = 0.0543$.

The point on the orbit that is farthest from the sun is called the *aphelion* and occurs when $\theta = 0$. This point is

$$r = \frac{1.4228}{1 - 0.0543 \cos 0} \approx 1.5045 \times 10^9 \text{ kilometers}$$

from the sun.

The point on the orbit that is closest to the sun is called the *perihelion* and occurs when $\theta = \pi$. This point is

$$r = \frac{1.4228}{1 - 0.0543 \cos \pi} \approx 1.3495 \times 10^9 \text{ kilometers}$$

from the sun.

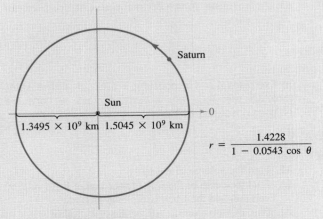

Elliptical Orbit of Saturn

See Exercise 53, Section 12.6.

Chapter Overview

The chapter is divided into two parts—**parametric equations** (Sections 12.1 and 12.2) and the **polar coordinate system** (Sections 12.3–12.6).

In Section 12.1, we show how to use a parameter to represent a curve in the plane. For instance, the set of equations

$$x = t \quad \text{and} \quad y = t^2$$

uses the parameter t to represent the curve whose rectangular equation is $y = x^2$. Parametric representation of a curve is especially useful for studying motion in the plane. In Section 12.2, we discuss the calculus of parametric equations—the parametric form of the derivative, arc length in parametric form, and the area of a surface of revolution.

In Sections 12.3 and 12.4, we introduce the polar coordinate system and discuss techniques for sketching the graph of a polar equation. Then, in Section 12.5, we look at the problem of finding the area of a region bounded by a polar graph, and finding the arc length of a polar graph. Finally, in Section 12.6, we look at polar equations for conics and Kepler's Laws, which can be used to describe the orbits of the planets about the sun.

Plane Curves, Parametric Equations, and Polar Coordinates

12.1 Plane Curves and Parametric Equations

Plane curves ▪ Parametric equations ▪ Curve sketching ▪ Eliminating the parameter ▪ Finding parametric equations ▪ The brachistochrone problem

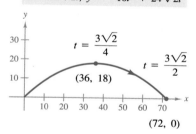

Rectangular equation:
$$y = -\frac{x^2}{72} + x$$
Parametric equations:
$$x = 24\sqrt{2}t, \ y = -16t^2 + 24\sqrt{2}t$$

Two variables for position
One variable for time

FIGURE 12.1

Until now, we have been representing a graph by a single equation involving the *two* variables x and y. In this section we look at situations in which it is useful to introduce a *third* variable to represent a curve in the plane.

To see the usefulness of this procedure, consider the path followed by an object that is propelled into the air at an angle of 45°. If the initial velocity of the object is 48 feet per second, it follows the parabolic path given by

$$y = -\frac{x^2}{72} + x \qquad \text{Rectangular equation}$$

as shown in Figure 12.1. However, this equation does not tell the whole story. Although it does tell us *where* the object has been, it doesn't tell us *when* the object was at a given point (x, y). To determine this time, we introduce a third variable t, called a **parameter.** By writing both x and y as functions of t, we obtain the **parametric equations**

$$x = 24\sqrt{2}t \qquad \text{and} \qquad y = -16t^2 + 24\sqrt{2}t. \qquad \text{Parametric equations}$$

Now, from this set of equations, we can determine that at time $t = 0$, the object is at the point $(0, 0)$. Similarly, at time $t = 1$, the object is at the point $(24\sqrt{2}, 24\sqrt{2} - 16)$, and so on.*

For this particular motion problem, x and y are continuous functions of t, and we call the resulting path a **plane curve.**

DEFINITION OF A PLANE CURVE

If f and g are continuous functions of t on an interval I, then the set of ordered pairs $(f(t), g(t))$ is called a **plane curve** C. The equations $x = f(t)$ and $y = g(t)$ are called **parametric equations** for C, and t is called the **parameter.**

---

*We will discuss a method for determining this particular set of parametric equations—the equations of motion—later in the text.

The set of points $(x, y) = (f(t), g(t))$ in the plane is called the **graph of the curve** C. For simplicity, we will not distinguish between a curve and its graph. When sketching a curve represented by a pair of parametric equations, we still plot points in the xy-plane. Each set of coordinates (x, y) is determined from a value chosen for the parameter t. By plotting the resulting points in the order of *increasing* values of t, we trace the curve out in a specific direction. This is called the **orientation** of the curve.

EXAMPLE 1   Sketching a curve

Sketch the curve described by the parametric equations

$$x = t^2 - 4 \quad \text{and} \quad y = \frac{t}{2}, \quad -2 \le t \le 3.$$

SOLUTION

For values of t on the given interval, the parametric equations yield the points (x, y) shown in Table 12.1.

TABLE 12.1

t	-2	-1	0	1	2	3
x	0	-3	-4	-3	0	5
y	-1	$-\frac{1}{2}$	0	$\frac{1}{2}$	1	$\frac{3}{2}$

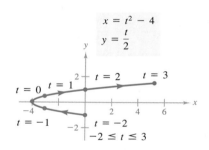

FIGURE 12.2

By plotting these points in the order of increasing t and using the continuity of f and g, we obtain the curve C shown in Figure 12.2. Note that the arrows on the curve indicate its orientation as t increases from -2 to 3.

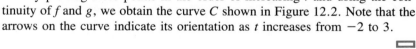

REMARK   By the vertical line test, we can see that the graph shown in Figure 12.2 does not define y as a function of x. This points out one benefit of parametric equations—they can be used to represent graphs that are more general than graphs of functions.

It often happens that two different sets of parametric equations have the same graph. For example, the set of parametric equations

$$x = 4t^2 - 4 \quad \text{and} \quad y = t, \quad -1 \le t \le \frac{3}{2}$$

has the same graph as the set given in Example 1. However, comparing the values of t in Figures 12.2 and 12.3, we see that this second graph is traced out more *rapidly* (considering t as time) than the first graph. Thus, in applications, different parametric representations can be used to represent various *speeds* at which objects travel along a given path.

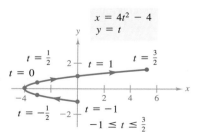

FIGURE 12.3

Eliminating the parameter

In Example 1 we used a simple point-plotting method to sketch the given curve. This tedious process can sometimes be simplified by finding a rectangular equation (in x and y) that has the same graph. We call this process **eliminating the parameter.**

Parametric equations	→	Solve for t in one equation	→	Substitute into second equation	→	Rectangular equation

$x = t^2 - 4$ $\qquad\qquad$ $t = 2y$ $\qquad\qquad$ $x = (2y)^2 - 4$ $\qquad\qquad$ $x = 4y^2 - 4$
$y = t/2$

Now, we recognize that the equation $x = 4y^2 - 4$ represents a parabola with a horizontal axis and vertex at $(-4, 0)$.

The range of x and y implied by the parametric equations may be altered by the change to rectangular form. In such instances the domain of the rectangular equation must be adjusted so that its graph matches the graph of the parametric equations. Such a situation is demonstrated in the next example.

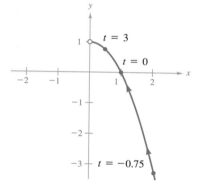

Parametric equations:
$$x = \frac{1}{\sqrt{t + 1}}, \quad y = \frac{t}{t + 1}$$

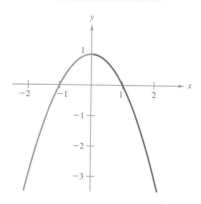

Rectangular equation:
$$y = 1 - x^2$$

FIGURE 12.4

EXAMPLE 2 Adjusting the domain after eliminating the parameter

Sketch the curve represented by the equations

$$x = \frac{1}{\sqrt{t + 1}} \qquad \text{and} \qquad y = \frac{t}{t + 1}$$

by eliminating the parameter and adjusting the domain of the resulting rectangular equation.

SOLUTION

Solving for t in the equation for x, we have

$$x = \frac{1}{\sqrt{t + 1}} \quad \Longrightarrow \quad x^2 = \frac{1}{t + 1}$$

which implies that

$$t = \frac{1 - x^2}{x^2}.$$

Now, substituting into the equation for y, we obtain

$$y = \frac{t}{t + 1} = \frac{(1 - x^2)/x^2}{[(1 - x^2)/x^2] + 1} = 1 - x^2.$$

The rectangular equation, $y = 1 - x^2$, is defined for all values of x, but from the parametric equation for x we see that the curve is defined only when $-1 < t$. This implies that we should restrict the domain of x to positive values, as shown in Figure 12.4.

It is not necessary for the parameter in a set of parametric equations to represent time. Our next example uses an *angle* as the parameter. In this example we use *trigonometric identities* to eliminate the parameter.

EXAMPLE 3   Using a trigonometric identity to eliminate a parameter

Sketch the curve represented by

$$x = 3 \cos \theta \quad \text{and} \quad y = 4 \sin \theta, \quad 0 \leq \theta \leq 2\pi$$

by eliminating the parameter and finding the corresponding rectangular equation.

SOLUTION

We begin by solving for $\cos \theta$ and $\sin \theta$ in the given equations.

$$\cos \theta = \frac{x}{3} \quad \text{and} \quad \sin \theta = \frac{y}{4} \qquad \text{Solve for } \sin \theta \text{ and } \cos \theta$$

Now we make use of the identity $\sin^2 \theta + \cos^2 \theta = 1$ to form an equation involving only x and y.

$$\cos^2 \theta + \sin^2 \theta = 1 \qquad \text{Trigonometric identity}$$

$$\cos^2 \theta + \sin^2 \theta = \left(\frac{x}{3}\right)^2 + \left(\frac{y}{4}\right)^2 = 1 \qquad \text{Substitute}$$

$$\frac{x^2}{9} + \frac{y^2}{16} = 1 \qquad \text{Rectangular equation}$$

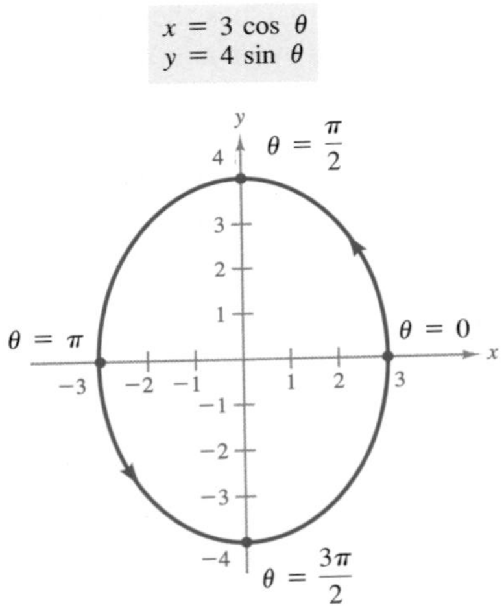

$x = 3 \cos \theta$
$y = 4 \sin \theta$

FIGURE 12.5

From this rectangular equation we see that the graph is an ellipse centered at $(0, 0)$, with vertices at $(0, 4)$ and $(0, -4)$ and minor axis of length $2b = 6$, as shown in Figure 12.5. Note that the elliptic curve is traced out *counterclockwise* as θ varies from 0 to 2π. ∎

In Examples 2 and 3 it is important to realize that eliminating the parameter is primarily an *aid to curve sketching*. If the parametric equations represent the path of a moving object, the graph alone is not sufficient to describe the object's motion. We still need the parametric equations to tell us the *position*, *direction*, and *speed* at a given time.

Finding parametric equations for a graph

We have been looking at techniques for sketching the graph represented by a set of parametric equations. We now look at the reverse problem. How can we determine a set of parametric equations for a given graph or a given physical description? From the discussion following Example 1, we know that such a representation is not unique. This is demonstrated further in the following example, in which we find two different parametric representations for a given graph.

EXAMPLE 4   Finding parametric equations for a given graph

Find a set of parametric equations to represent the graph of $y = 1 - x^2$, using the following parameters.

(a) $t = x$ (b) the slope $m = dy/dx$ at the point (x, y)

SOLUTION

(a) Letting $x = t$, we obtain the parametric equations

$$x = t \quad \text{and} \quad y = 1 - x^2 = 1 - t^2.$$

(b) To express x and y in terms of the parameter m, we proceed as follows.

$$m = \frac{dy}{dx} = -2x \qquad \text{Differentiate } y = 1 - x^2$$

$$x = -\frac{m}{2} \qquad \text{Solve for } x$$

$$y = 1 - x^2 = 1 - \left(-\frac{m}{2}\right)^2 = 1 - \frac{m^2}{4} \qquad \text{Substitute}$$

Thus, the parametric equations are

$$x = -\frac{m}{2} \quad \text{and} \quad y = 1 - \frac{m^2}{4}.$$

In Figure 12.6, note that the resulting curve has a right-to-left orientation as determined by the direction of increasing values of slope m. For part (a) the curve would have the opposite orientation.   ▭

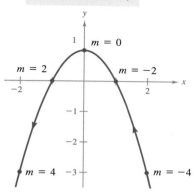

$$x = -\frac{m}{2}, \; y = 1 - \frac{m^2}{4}$$

$m = 0$

$m = 2$ $m = -2$

$m = 4$ $m = -4$

Graph of $y = 1 - x^2$

FIGURE 12.6

EXAMPLE 5   Parametric equations for a cycloid

Determine the curve traced by a point P on the circumference of a circle of radius a rolling along a straight line in a plane. Such a curve is called a **cycloid.**

SOLUTION

We let the parameter θ be the measure of the circle's rotation, and we let the point $P = (x, y)$ begin at the origin. When $\theta = 0$, P is at the origin; when $\theta = \pi$, P is at a maximum point $(\pi a, 2a)$; and when $\theta = 2\pi$, P is back on the x-axis at $(2\pi a, 0)$. From Figure 12.7 we see that $\angle APC = 180° - \theta$. Hence,

$$\sin \theta = \sin (180° - \theta) = \sin (\angle APC) = \frac{AC}{a} = \frac{BD}{a}$$

$$\cos \theta = -\cos (180° - \theta) = -\cos (\angle APC) = \frac{AP}{-a}$$

which implies that

$$AP = -a \cos \theta \quad \text{and} \quad BD = a \sin \theta.$$

Now, since the circle rolls along the *x*-axis, we know that $OD = \overset{\frown}{PD} = a\theta$. Furthermore, since $BA = DC = a$, we have

$$x = OD - BD = a\theta - a \sin \theta$$
$$y = BA + AP = a - a \cos \theta.$$

Therefore, the parametric equations are

$$x = a(\theta - \sin \theta) \qquad \text{and} \qquad y = a(1 - \cos \theta).$$

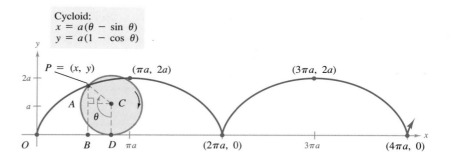

FIGURE 12.7

The cycloid in Figure 12.7 has sharp corners at the values $x = 2n\pi a$. Between these points, the cycloid is called **smooth.**

DEFINITION OF A SMOOTH CURVE	A curve C represented by $x = f(t)$ and $y = g(t)$ on an interval I is called **smooth** if f' and g' are continuous on I and not simultaneously zero, except possibly at the endpoints of I. The curve C is called **piecewise smooth** if it is smooth on each subinterval of some partition of I.

The brachistochrone problem

The type of curve described in Example 5 is related to one of the most famous pairs of problems in the history of calculus. The first problem began with Galileo's discovery that the time required to complete a full swing of a given pendulum is *approximately* the same, whether it makes a large movement at high speeds or a small movement at lower speeds. Late in his life, Galileo (1564–1642) realized that he could use this principle to construct a clock. However, he was not able to conquer the mechanics of actual construction. Christian Huygens (1629–1695) was the first to design and construct a working model. In his work with pendulums, Huygens realized that a pendulum does not take exactly the same time to complete swings of varying lengths. (This doesn't affect a pendulum clock, since the length of the circular arc is kept constant by giving the pendulum a slight boost each time it passes its lowest point.) But, in studying the problem, Huygens discovered that a ball rolling back and forth on an inverted cycloid does complete each cycle in exactly the same time.

The second problem, posed by John Bernoulli in 1696, is called the **brachistochrone problem**—in Greek *brachys* means short and *chronos* means time. The problem was to determine the path down which a particle will slide from point *A* to point *B* in the *shortest time*. Several mathematicians

James Bernoulli

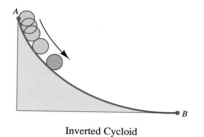

Inverted Cycloid

FIGURE 12.8

took up the challenge, and the following year the problem was solved by Newton, Leibniz, L'Hôpital, John Bernoulli, and James Bernoulli. As it turns out, the solution is not a straight line from A to B, but an inverted cycloid passing through the points A and B, as shown in Figure 12.8. The amazing part of the solution is that a particle starting at rest at *any* other point C of the cycloid between A and B will take exactly the same time to reach B.

James Bernoulli (1654–1705), also called Jacques, was the older brother of John. He was one of several accomplished mathematicians of the Swiss Bernoulli family. James's mathematical accomplishments have given him a prominent place in the early development of calculus.

EXERCISES for Section 12.1

In Exercises 1–30, sketch the curve represented by the parametric equations (indicate the orientation of the curve), and find the corresponding rectangular equation by eliminating the parameter.

1. $x = 3t - 1$
$y = 2t + 1$

2. $x = 3 - 2t$
$y = 2 + 3t$

3. $x = \sqrt{t}$
$y = 1 - t$

4. $x = \sqrt[3]{t}$
$y = 1 - t$

5. $x = t + 1$
$y = t^2$

6. $x = t + 1$
$y = t^3$

7. $x = t^3$
$y = \dfrac{t^2}{2}$

8. $x = 1 + \dfrac{1}{t}$
$y = t - 1$

9. $x = t - 1$
$y = \dfrac{t}{t - 1}$

10. $x = t^2 + t$
$y = t^2 - t$

11. $x = 2t$
$y = |t - 2|$

12. $x = |t - 1|$
$y = t + 2$

13. $x = \sec \theta$
$y = \cos \theta$

14. $x = \tan^2 \theta$
$y = \sec^2 \theta$

15. $x = 3 \cos \theta$
$y = 3 \sin \theta$

16. $x = \cos \theta$
$y = 3 \sin \theta$

17. $x = 4 \sin 2\theta$
$y = 2 \cos 2\theta$

18. $x = \cos \theta$
$y = 2 \sin 2\theta$

19. $x = \cos \theta$
$y = 2 \sin^2 \theta$

20. $x = 4 \cos^2 \theta$
$y = 2 \sin \theta$

21. $x = 4 + 2 \cos \theta$
$y = -1 + \sin \theta$

22. $x = 4 + 2 \cos \theta$
$y = -1 + 2 \sin \theta$

23. $x = 4 + 2 \cos \theta$
$y = -1 + 4 \sin \theta$

24. $x = \sec \theta$
$y = \tan \theta$

25. $x = 4 \sec \theta$
$y = 3 \tan \theta$

26. $x = \cos^3 \theta$
$y = \sin^3 \theta$

27. $x = t^3$
$y = 3 \ln t$

28. $x = e^{2t}$
$y = e^t$

29. $x = e^{-t}$
$y = e^{3t}$

30. $x = \ln 2t$
$y = t^2$

In Exercises 31 and 32, determine how the plane curves differ from each other.

31. (a) $x = t$
$y = 2t + 1$

(b) $x = \cos \theta$
$y = 2 \cos \theta + 1$

(c) $x = e^{-t}$
$y = 2e^{-t} + 1$

(d) $x = e^t$
$y = 2e^t + 1$

32. (a) $x = 2 \cos \theta$
$y = 2 \sin \theta$

(b) $x = \dfrac{\sqrt{4t^2 - 1}}{|t|}$
$y = \dfrac{1}{t}$

(c) $x = \sqrt{t}$
$y = \sqrt{4 - t}$

(d) $x = -\sqrt{4 - e^{2t}}$
$y = e^t$

In Exercises 33–38, sketch the curve represented by the parametric equations.

33. Cycloid: $x = 2(\theta - \sin \theta)$
$y = 2(1 - \cos \theta)$

34. Cycloid: $x = \theta + \sin \theta$
$y = 1 - \cos \theta$

35. Witch of Agnesi: $x = 2 \cot \theta$
$y = 2 \sin^2 \theta$

36. Curtate cycloid: $x = 2\theta - \sin \theta$
$y = 2 - \cos \theta$

37. Prolate cycloid: $x = \theta - \dfrac{3}{2} \sin \theta$
$y = 1 - \dfrac{3}{2} \cos \theta$

38. Folium of Descartes: $x = \dfrac{3t}{1 + t^3}$
$y = \dfrac{3t^2}{1 + t^3}$

In Exercises 39–44, eliminate the parameter and obtain the standard form of the rectangular equation of the curve.

39. Line through (x_1, y_1) and (x_2, y_2):
$$x = x_1 + t(x_2 - x_1)$$
$$y = y_1 + t(y_2 - y_1)$$
40. Circle: $x = r \cos \theta$
$$y = r \sin \theta$$
41. Circle: $x = h + r \cos \theta$
$$y = k + r \sin \theta$$
42. Ellipse: $x = h + a \cos \theta$
$$y = k + b \sin \theta$$
43. Hyperbola: $x = h + a \sec \theta$
$$y = k + b \tan \theta$$
44. Hyperbola: $x = h + a\sqrt{t + 1}$
$$y = k + b\sqrt{t}$$

In Exercises 45–52, find a set of parametric equations for the given graph.

45. Line: passes through $(0, 0)$ and $(5, -2)$
46. Line: passes through $(1, 4)$ and $(5, -2)$
47. Circle: center at $(2, 1)$ and radius of 4
48. Circle: center at $(-3, 1)$ and radius of 3
49. Ellipse: vertices at $(\pm 5, 0)$ and foci at $(\pm 4, 0)$
50. Ellipse: vertices at $(4, 7)$ and $(4, -3)$ and foci at $(4, 5)$ and $(4, -1)$
51. Hyperbola: vertices at $(\pm 4, 0)$ and foci at $(\pm 5, 0)$
52. Hyperbola: vertices at $(0, \pm 1)$ and foci at $(0, \pm 2)$

In Exercises 53 and 54, find two different sets of parametric equations for the given rectangular equation.

53. $y = x^3$ **54.** $y = x^2$

55. A wheel of radius a rolls along a straight line without slipping. The curve traced by a point P that is b units from the center ($b < a$) is called a **curtate cycloid** (see figure). Use the angle θ shown in the figure to find a set of parametric equations for the curve.

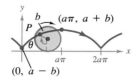

FIGURE FOR 55

56. A circle of radius 1 rolls around the outside of a circle of radius 2 without slipping. The curve traced by a point on the circumference of the smaller circle is called an **epicycloid** (see figure). Use the angle θ shown in the figure to find a set of parametric equations for the curve.

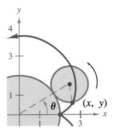

FIGURE FOR 56

In Exercises 57 and 58, sketch the **Lissajous curve**. A Lissajous curve can be represented by

$$x = a \cos nt \quad \text{and} \quad y = b \sin mt$$

where n/m is a rational number.

57. $x = 2 \cos \dfrac{t}{2}$ **58.** $x = \cos 2t$
$\quad\ \ y = \sin t$ $\quad\ \ y = \sin t$

In Exercises 59–62, use a computer or graphics calculator to sketch the curve described by the parametric equations over the specified interval.

Equations	*Interval*
59. $x = 4\sqrt{t} - 1$	$0 \le t \le 4$
$y = 4t^2 - t^3$	
60. $x = 3 \cos^3 \theta$	$0 \le \theta \le 2\pi$
$y = 3 \sin^3 \theta$	
61. $x = 2\theta - 4 \sin \theta$	$-\pi \le \theta \le 3\pi$
$y = 2 - 4 \cos \theta$	
62. $x = 3(\theta + \sin \theta)$	$-\pi \le \theta \le 3\pi$
$y = 3(1 - \cos \theta)$	

12.2 Parametric Equations and Calculus

Slope and tangent lines ▪ Arc length ▪ Area of a surface of revolution

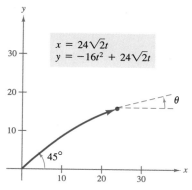

$$x = 24\sqrt{2}\,t$$
$$y = -16t^2 + 24\sqrt{2}\,t$$

FIGURE 12.9

Now that we can represent a curve in the plane by a set of parametric equations, it is natural to ask how to use calculus to study curves. To begin, let's take another look at the projectile represented by the parametric equations

$$x = 24\sqrt{2}\,t \quad \text{and} \quad y = -16t^2 + 24\sqrt{2}\,t$$

as shown in Figure 12.9. From the previous section we know that these parametric equations enable us to locate the position of the projectile at a given time. We also know that the object is initially projected at an angle of 45°. But, how can we find the angle θ representing the object's direction at some other time t? The following theorem answers this question by giving us a formula for the slope of the tangent line as a function of t.

**THEOREM 12.1
PARAMETRIC FORM
OF THE DERIVATIVE**

If a smooth curve C is given by the equations $x = f(t)$ and $y = g(t)$, then the slope of C at (x, y) is

$$\frac{dy}{dx} = \frac{dy/dt}{dx/dt}, \quad \frac{dx}{dt} \neq 0.$$

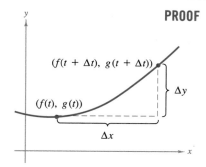

FIGURE 12.10

PROOF   In Figure 12.10 consider $\Delta t > 0$ and let

$$\Delta y = g(t + \Delta t) - g(t)$$

and

$$\Delta x = f(t + \Delta t) - f(t).$$

Since

$$\frac{dy}{dx} = \lim_{\Delta x \to 0} \frac{\Delta y}{\Delta x}$$

and $\Delta x \to 0$ as $\Delta t \to 0$, we can write

$$\frac{dy}{dx} = \lim_{\Delta t \to 0} \frac{g(t + \Delta t) - g(t)}{f(t + \Delta t) - f(t)}.$$

Dividing both the numerator and denominator by Δt, we can use the differentiability of f and g to conclude that

$$\frac{dy}{dx} = \lim_{\Delta t \to 0} \frac{[g(t + \Delta t) - g(t)]/\Delta t}{[f(t + \Delta t) - f(t)]/\Delta t}$$

$$= \frac{\displaystyle\lim_{\Delta t \to 0} \frac{g(t + \Delta t) - g(t)}{\Delta t}}{\displaystyle\lim_{\Delta t \to 0} \frac{f(t + \Delta t) - f(t)}{\Delta t}} = \frac{g'(t)}{f'(t)} = \frac{dy/dt}{dx/dt}.$$

Since dy/dx is a function of t, we can use Theorem 12.1 repeatedly to find *higher-order* derivatives. For instance,

$$\frac{d^2y}{dx^2} = \frac{d}{dx}\left[\frac{dy}{dx}\right] = \frac{\frac{d}{dt}\left[\frac{dy}{dx}\right]}{dx/dt} \qquad \text{Second derivative}$$

$$\frac{d^3y}{dx^3} = \frac{d}{dx}\left[\frac{d^2y}{dx^2}\right] = \frac{\frac{d}{dt}\left[\frac{d^2y}{dx^2}\right]}{dx/dt}. \qquad \text{Third derivative}$$

EXAMPLE 1   Finding slope and concavity

For the curve given by

$$x = \sqrt{t} \qquad \text{and} \qquad y = \frac{1}{4}(t^2 - 4)$$

find the slope and concavity at the point (2, 3).

SOLUTION

Since

$$\frac{dy}{dx} = \frac{dy/dt}{dx/dt} = \frac{(1/2)t}{(1/2)t^{-1/2}} = t^{3/2}$$

we have

$$\frac{d^2y}{dx^2} = \frac{\frac{d}{dt}\left[\frac{dy}{dx}\right]}{dx/dt} = \frac{(3/2)t^{1/2}}{(1/2)t^{-1/2}} = 3t.$$

At $(x, y) = (2, 3)$, it follows that $t = 4$, and the slope is

$$\frac{dy}{dx} = (4)^{3/2} = 8.$$

Moreover, when $t = 4$, the second derivative is

$$\frac{d^2y}{dx^2} = 3(4) = 12 > 0$$

and we conclude that the graph is concave upward at (2, 3), as shown in Figure 12.11.

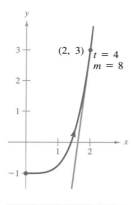

$x = \sqrt{t},\ y = \dfrac{1}{4}(t^2 - 4)$

FIGURE 12.11

Since the parametric equations $x = f(t)$ and $y = g(t)$ need not define y as a function of x, it follows that a curve defined parametrically can loop and cross itself in the plane. At such points the curve may have more than one tangent line, as shown in the next example.

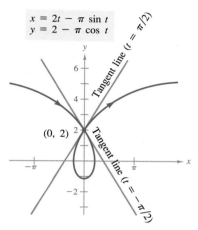

$$x = 2t - \pi \sin t$$
$$y = 2 - \pi \cos t$$

FIGURE 12.12

EXAMPLE 2   A curve with two tangent lines at a point

The **prolate cycloid** given by

$$x = 2t - \pi \sin t \qquad \text{and} \qquad y = 2 - \pi \cos t$$

crosses itself at the point $(0, 2)$, as shown in Figure 12.12. Find the equations of both tangent lines at this point.

SOLUTION

Since $x = 0$ and $y = 2$ when $t = \pm \pi/2$, and since

$$\frac{dy}{dx} = \frac{dy/dt}{dx/dt} = \frac{\pi \sin t}{2 - \pi \cos t}$$

we have $dy/dx = -\pi/2$ when $t = -\pi/2$ and $dy/dx = \pi/2$ when $t = \pi/2$. Therefore, the two tangent lines at $(0, 2)$ are

$$y - 2 = -\left(\frac{\pi}{2}\right)x \qquad \text{Tangent line when } t = -\frac{\pi}{2}$$

$$y - 2 = \left(\frac{\pi}{2}\right)x. \qquad \text{Tangent line when } t = \frac{\pi}{2}$$

If $dy/dt = 0$ and $dx/dt \neq 0$ when $t = t_0$, then the curve represented by $x = f(t)$ and $y = g(t)$ has a *horizontal* tangent at $(f(t_0), g(t_0))$. For instance, in Example 2, the given curve has a horizontal tangent at the point $(0, 2 - \pi)$ (when $t = 0$). Similarly, if $dx/dt = 0$ and $dy/dt \neq 0$ when $t = t_0$, then the curve represented by $x = f(t)$ and $y = g(t)$ has a *vertical* tangent at $(f(t_0), g(t_0))$.

Arc length

We have some idea how parametric equations can be used to describe the path of a particle moving in the plane. We now develop a formula for determining the *distance* traveled by the particle along its path.

Recall from Section 6.4 that the formula for the arc length of a curve C given by $y = h(x)$ over the interval $[x_0, x_1]$ is

$$s = \int_{x_0}^{x_1} \sqrt{1 + [h'(x)]^2} \, dx = \int_{x_0}^{x_1} \sqrt{1 + \left(\frac{dy}{dx}\right)^2} \, dx.$$

If C is represented by the parametric equations $x = f(t)$ and $y = g(t)$, $a \leq t \leq b$, and if $dx/dt = f'(t) > 0$, then we can write

$$s = \int_{x_0}^{x_1} \sqrt{1 + \left(\frac{dy}{dx}\right)^2} \, dx = \int_{x_0}^{x_1} \sqrt{1 + \left(\frac{dy/dt}{dx/dt}\right)^2} \, dx$$

$$= \int_a^b \sqrt{\frac{(dx/dt)^2 + (dy/dt)^2}{(dx/dt)^2}} \frac{dx}{dt} \, dt$$

$$= \int_a^b \sqrt{\left(\frac{dx}{dt}\right)^2 + \left(\frac{dy}{dt}\right)^2} \, dt = \int_a^b \sqrt{[f'(t)]^2 + [g'(t)]^2} \, dt.$$

This formula for arc length can be shown to be valid for *any* smooth curve *C* that does not intersect itself.

THEOREM 12.2 **ARC LENGTH IN** **PARAMETRIC FORM**	If a smooth curve *C* is given by $x = f(t)$ and $y = g(t)$ such that *C* does not intersect itself on the interval $a \le t \le b$ (except possibly at the endpoints), then the arc length of *C* over the interval is given by $$s = \int_a^b \sqrt{\left(\frac{dx}{dt}\right)^2 + \left(\frac{dy}{dt}\right)^2}\, dt = \int_a^b \sqrt{[f'(t)]^2 + [g'(t)]^2}\, dt.$$

In the previous section we saw that if a circle rolls along a line, then a point on its circumference will trace a path called a cycloid. If the circle rolls around the circumference of another circle, then the path of the point is called an **epicycloid.** In the next example, we find the arc length of an epicycloid.

EXAMPLE 3 Finding arc length

A circle of radius 1 rolls around the circumference of a larger circle of radius 4, as shown in Figure 12.13. The epicycloid traced by a point on the circumference of the smaller circle is given by

$$x = 5 \cos t - \cos 5t \qquad \text{and} \qquad y = 5 \sin t - \sin 5t.$$

Find the distance traveled by the point in one complete trip about the larger circle.

SOLUTION

Before applying Theorem 12.2, we note from Figure 12.13 that the curve has sharp points when $t = 0$ and $t = \pi/2$. Between these two points, dx/dt and dy/dt are not simultaneously zero. Thus, the portion of the curve generated from $t = 0$ to $t = \pi/2$ is smooth. To find the total distance traveled by the point, we find the arc length of that portion lying in the first quadrant and multiply by 4.

$$s = 4 \int_0^{\pi/2} \sqrt{\left(\frac{dx}{dt}\right)^2 + \left(\frac{dy}{dt}\right)^2}\, dt$$

$$= 4 \int_0^{\pi/2} \sqrt{(-5 \sin t + 5 \sin 5t)^2 + (5 \cos t - 5 \cos 5t)^2}\, dt$$

$$= 20 \int_0^{\pi/2} \sqrt{2 - 2 \sin t \sin 5t - 2 \cos t \cos 5t}\, dt$$

$$= 20 \int_0^{\pi/2} \sqrt{2 - 2 \cos 4t}\, dt = 20 \int_0^{\pi/2} \sqrt{4 \sin^2 2t}\, dt$$

$$= 40 \int_0^{\pi/2} \sin 2t\, dt = -20 \left[\cos 2t \right]_0^{\pi/2} = 40$$

For the epicycloid shown in Figure 12.13, an arc length of 40 seems about right because the circumference of a circle of radius 6 is $2\pi r = 12\pi \approx 37.7$.

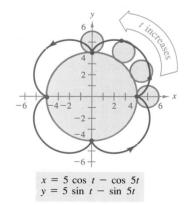

$x = 5 \cos t - \cos 5t$
$y = 5 \sin t - \sin 5t$

FIGURE 12.13

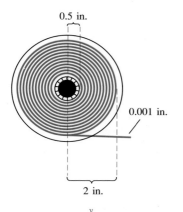

0.5 in.

0.001 in.

2 in.

FIGURE 12.14

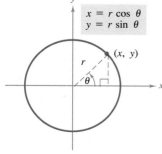

$x = r \cos \theta$
$y = r \sin \theta$

(x, y)

EXAMPLE 4   An application involving arc length

A recording tape 0.001 inch thick is wound around a reel whose inner radius is 0.5 inch and outer radius is 2 inches, as shown in Figure 12.14. How much tape is required to fill the reel?

SOLUTION

To create a model for this problem, we assume that as the tape is wound around the reel its distance r from the center increases linearly at the rate of 0.001 inch per revolution, or

$$r = (0.001)\frac{\theta}{2\pi} = \frac{\theta}{2000\pi}, \quad 1000\pi \le \theta \le 4000\pi$$

where θ is measured in radians. Now, we can determine the coordinates of the point (x, y) corresponding to a given radius to be

$$x = r \cos \theta$$

and

$$y = r \sin \theta.$$

Substituting for r, we obtain the parametric equations

$$x = \left(\frac{\theta}{2000\pi}\right) \cos \theta$$

and

$$y = \left(\frac{\theta}{2000\pi}\right) \sin \theta.$$

Now, we use the arc length formula to determine the total length of the tape to be

$$s = \int_{1000\pi}^{4000\pi} \sqrt{\left(\frac{dx}{d\theta}\right)^2 + \left(\frac{dy}{d\theta}\right)^2} \, d\theta$$

$$= \frac{1}{2000\pi} \int_{1000\pi}^{4000\pi} \sqrt{(-\theta \sin \theta + \cos \theta)^2 + (\theta \cos \theta + \sin \theta)^2} \, d\theta$$

$$= \frac{1}{2000\pi} \int_{1000\pi}^{4000\pi} \sqrt{\theta^2 + 1} \, d\theta \qquad \text{Integration tables, Formula 26}$$

$$= \frac{1}{2000\pi}\left(\frac{1}{2}\right)\left[\theta\sqrt{\theta^2 + 1} + \ln \left| \theta + \sqrt{\theta^2 + 1} \right| \right]_{1000\pi}^{4000\pi}$$

$$\approx 11{,}781 \text{ in.} \approx 982 \text{ ft.}$$

REMARK   The graph of $r = a\theta$ is called the **Spiral of Archimedes.**

Area of a surface of revolution

As we did for arc length, we can use the formula for the area of a surface of revolution in rectangular form to develop a formula for surface area in parametric form.

THEOREM 12.3 **AREA OF A SURFACE** **OF REVOLUTION**	If a smooth curve C given by $x = f(t)$ and $y = g(t)$ does not cross itself on an interval $a \leq t \leq b$, then the area S of the surface of revolution formed by revolving C about the coordinate axes is given by the following. 1. $S = 2\pi \int_a^b g(t) \sqrt{\left(\dfrac{dx}{dt}\right)^2 + \left(\dfrac{dy}{dt}\right)^2}\, dt$ Revolution about x-axis: $g(t) \geq 0$ 2. $S = 2\pi \int_a^b f(t) \sqrt{\left(\dfrac{dx}{dt}\right)^2 + \left(\dfrac{dy}{dt}\right)^2}\, dt$ Revolution about y-axis: $f(t) \geq 0$

EXAMPLE 5   Finding the area of a surface of revolution

Let C be the arc of the circle $x^2 + y^2 = 9$ from $(3, 0)$ to $(3/2, 3\sqrt{3}/2)$, as shown in Figure 12.15. Find the area of the surface formed by revolving C about the x-axis.

SOLUTION

We can represent C parametrically by the equations

$$x = 3 \cos t \quad \text{and} \quad y = 3 \sin t, \quad 0 \leq t \leq \pi/3.$$

(Note that we determine the interval for t by observing that $t = 0$ when $x = 3$ and $t = \pi/3$ when $x = 3/2$.) On this interval, C is smooth and y is nonnegative, and we can apply Theorem 12.3 to obtain a surface area of

$$S = 2\pi \int_0^{\pi/3} (3 \sin t)\sqrt{(-3 \sin t)^2 + (3 \cos t)^2}\, dt$$

$$= 6\pi \int_0^{\pi/3} \sin t \sqrt{9(\sin^2 t + \cos^2 t)}\, dt$$

$$= 6\pi \int_0^{\pi/3} 3 \sin t\, dt$$

$$= -18\pi \left[\cos t\right]_0^{\pi/3}$$

$$= -18\pi \left[\frac{1}{2} - 1\right]$$

$$= 9\pi.$$

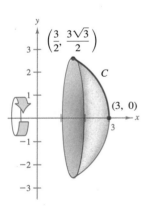

Surface of Revolution

FIGURE 12.15

EXERCISES for Section 12.2

In Exercises 1–10, find dy/dx and d^2y/dx^2, and evaluate each at the specified value of the parameter.

Parametric equations	*Point*
1. $x = 2t$, $y = 3t - 1$	$t = 3$
2. $x = \sqrt{t}$, $y = 3t - 1$	$t = 1$
3. $x = t + 1$, $y = t^2 + 3t$	$t = -1$
4. $x = t^2 + 3t$, $y = t + 1$	$t = 0$
5. $x = 2 \cos \theta$, $y = 2 \sin \theta$	$\theta = \dfrac{\pi}{4}$
6. $x = \cos \theta$, $y = 3 \sin \theta$	$\theta = 0$
7. $x = 2 + \sec \theta$, $y = 1 + 2 \tan \theta$	$\theta = \dfrac{\pi}{6}$
8. $x = \sqrt{t}$, $y = \sqrt{t - 1}$	$t = 2$
9. $x = \cos^3 \theta$, $y = \sin^3 \theta$	$\theta = \dfrac{\pi}{4}$
10. $x = \theta - \sin \theta$, $y = 1 - \cos \theta$	$\theta = \pi$

In Exercises 11–16, find an equation of the tangent line to the curve at the specified value of the parameter.

Parametric equations	*Point*
11. $x = 2t$, $y = t^2 - 1$	$t = 2$
12. $x = t - 1$, $y = \dfrac{1}{t} + 1$	$t = 1$
13. $x = t^2 - t + 2$, $y = t^3 - 3t$	$t = -1$
14. $x = 4 \cos \theta$, $y = 3 \sin \theta$	$\theta = \dfrac{3\pi}{4}$
15. $x = 2 \cot \theta$, $y = 2 \sin^2 \theta$	$\theta = \dfrac{\pi}{4}$
16. $x = 2 - 3 \cos \theta$, $y = 3 + 2 \sin \theta$	$\theta = \dfrac{5\pi}{3}$

In Exercises 17–28, find all points (if any) of horizontal and vertical tangency.

17. $x = 1 - t$
    $y = t^2$

18. $x = t + 1$
    $y = t^2 + 3t$

19. $x = 1 - t$
    $y = t^3 - 3t$

20. $x = t^2 - t + 2$
    $y = t^3 - 3t$

21. $x = 3 \cos \theta$
    $y = 3 \sin \theta$

22. $x = \cos \theta$
    $y = 2 \sin 2\theta$

23. $x = 4 + 2 \cos \theta$
    $y = -1 + \sin \theta$

24. $x = 4 \cos^2 \theta$
    $y = 2 \sin \theta$

25. $x = \sec \theta$
    $y = \tan \theta$

26. $x = \cos^2 \theta$
    $y = \cos \theta$

27. $x = 2\theta$
    $y = 2(1 - \cos \theta)$

28. Involute of a circle:
    $x = \cos \theta + \theta \sin \theta$
    $y = \sin \theta - \theta \cos \theta$

In Exercises 29–34, find the arc length of the given curve.

Parametric equations	*Interval*
29. $x = e^{-t} \cos t$, $y = e^{-t} \sin t$	$0 \le t \le \dfrac{\pi}{2}$
30. $x = t^2$, $y = 4t^3 - 1$	$-1 \le t \le 1$
31. $x = t^2$, $y = 2t$	$0 \le t \le 2$
32. $x = \arcsin t$, $y = \ln \sqrt{1 - t^2}$	$0 \le t \le \dfrac{1}{2}$
33. $x = \sqrt{t}$, $y = 3t - 1$	$0 \le t \le 1$
34. $x = t$, $y = \dfrac{t^5}{10} + \dfrac{1}{6t^3}$	$1 \le t \le 2$

In Exercises 35–38, find the arc length of the given curve.

35. Perimeter of a hypocycloid:
    $x = a \cos^3 \theta$, $y = a \sin^3 \theta$, $0 \le \theta \le 2\pi$

36. Circumference of a circle:
    $x = a \cos \theta$, $y = a \sin \theta$, $0 \le \theta \le 2\pi$

37. Length of one arch of a cycloid:
    $x = a(\theta - \sin \theta)$, $y = a(1 - \cos \theta)$, $0 \le \theta \le 2\pi$

38. Circumference of an ellipse:
    $x = 3 \cos \theta$, $y = 4 \sin \theta$, $0 \le \theta < 2\pi$
    (Use Simpson's Rule with $n = 6$.)

In Exercises 39–44, find the area of the surface generated by revolving the curve about the given axis.

39. $x = t$, $y = 2t$, $0 \le t \le 4$
    (a) x-axis   (b) y-axis

40. $x = t$, $y = 4 - 2t$, $0 \le t \le 2$
    (a) x-axis   (b) y-axis

41. $x = 4 \cos \theta$, $y = 4 \sin \theta$, $0 \le \theta \le \dfrac{\pi}{2}$
    y-axis

42. $x = t^3$, $y = t + 2$, $1 \le t \le 2$
    y-axis

43. $x = a \cos^3 \theta$, $y = a \sin^3 \theta$, $0 \le \theta \le \pi$
    x-axis

44. $x = a \cos \theta$, $y = b \sin \theta$, $0 \le \theta \le 2\pi$
    (a) x-axis   (b) y-axis

45. A portion of a sphere is removed by a circular cone with its vertex at the center of the sphere. Find the surface area removed from the sphere if the vertex of the cone forms an angle 2θ.

46. Use integration by substitution to show that if y is a continuous function of x on the interval $a \le x \le b$, where $x = f(t)$ and $y = g(t)$, then

$$\int_a^b y \, dx = \int_{t_1}^{t_2} g(t) f'(t) \, dt,$$

where $f(t_1) = a$, $f(t_2) = b$, and both g and f' are continuous on $[t_1, t_2]$.

In Exercises 47–50, find the area of the indicated region. (Use the result of Exercise 46.)

47. $x = \cos^3 \theta$
$y = \sin^3 \theta$

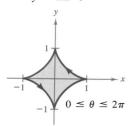

$0 \le \theta \le 2\pi$

48. $x = a \cos \theta$
$y = b \sin \theta$

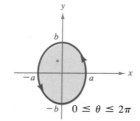

$0 \le \theta \le 2\pi$

49. $x = 2 \sin^2 \theta$
$y = 2 \sin^2 \theta \tan \theta$

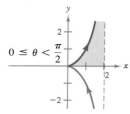

$0 \le \theta < \dfrac{\pi}{2}$

50. $x = 2 \cot \theta$
$y = 2 \sin^2 \theta$

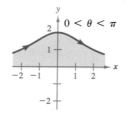

$0 < \theta < \pi$

In Exercises 51 and 52, find the centroid of the region bounded by the graph of the parametric equations and the coordinate axes. (Use the result of Exercise 46.)

51. $x = \sqrt{t}$
$y = 4 - t$

52. $x = \sqrt{4 - t}$
$y = \sqrt{t}$

In Exercises 53 and 54, find the volume of the solid formed by revolving the region bounded by the graphs of the given equations about the x-axis. (Use the result of Exercise 46.)

53. $x = 3 \cos \theta$
$y = 3 \sin \theta$

54. $x = \cos \theta$
$y = 3 \sin \theta$

55. Given the parametric equations

$$x = \frac{4t}{1 + t^3} \quad \text{and} \quad y = \frac{4t^2}{1 + t^3}$$

use a computer or graphics calculator to do the following.
(a) Sketch the curve described by the parametric equations.
(b) Find the points of horizontal tangency to the curve.
(c) Use Simpson's Rule with $n = 10$ to approximate the arc length of the closed loop. [Hint: Because of symmetry we need only integrate for $0 \le t \le 1$.]

12.3 Polar Coordinates and Polar Graphs

Polar coordinates ▪ Coordinate conversion ▪ Graphs of polar equations ▪ Symmetry

So far, we have been representing plane curves as collections of points (x, y) in the rectangular coordinate system, where x and y represent the directed distances from the coordinate axes to the point (x, y). The corresponding equations for these curves have been in either rectangular or parametric form. In this section we introduce a third system called the **polar coordinate system.**

To form the polar coordinate system in the plane, we fix a point O, called the **pole** (or **origin**), and construct from O an initial ray called the **polar axis,** as shown in Figure 12.16. Then each point P in the plane can be assigned **polar coordinates** (r, θ), as follows.

Polar Coordinates

FIGURE 12.16

$r = $ *directed distance* from O to P

$\theta = $ *directed angle*, counterclockwise from the polar axis to segment $\overline{OP}$

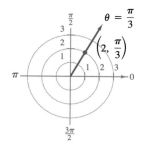

(a)

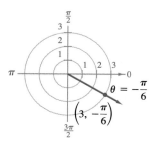

(b)

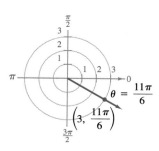

(c)

FIGURE 12.17

In the polar coordinate system, it is convenient to locate points with respect to a grid of concentric circles intersected by **radial lines** through the pole. This procedure is shown in the following example.

EXAMPLE 1 Plotting points in the polar coordinate system

(a) The point $(r, \theta) = (2, \pi/3)$ lies at the intersection of a circle of radius $r = 2$ and a ray that is the terminal side of the angle $\theta = \pi/3$, as shown in Figure 12.17(a).
(b) The point $(r, \theta) = (3, -\pi/6)$ lies in the fourth quadrant, 3 units from the pole. Note that we measure negative angles *clockwise*, as shown in Figure 12.17(b).
(c) The point $(r, \theta) = (3, 11\pi/6)$ coincides with the point $(3, -\pi/6)$, as shown in Figure 12.17(c). ▭

With rectangular coordinates, each point (x, y) has a unique representation. This is not true with polar coordinates. For instance, the coordinates (r, θ) and $(r, 2\pi + \theta)$ represent the same point, as illustrated in Example 1. Another way to obtain multiple representations of a point is to use negative values for r. Since r is a *directed distance*, the coordinates (r, θ) and $(-r, \theta + \pi)$ represent the same point. In general, the point (r, θ) can be written as

$$(r, \theta) = (r, \theta + 2n\pi) \qquad \text{or} \qquad (r, \theta) = (-r, \theta + (2n + 1)\pi)$$

where n is any integer. Moreover, the pole is represented by $(0, \theta)$, where θ is any angle.

EXAMPLE 2 Multiple representations of points

Plot the point $(3, -3\pi/4)$ and find three additional polar coordinate representations of this point, using $-2\pi < \theta < 2\pi$.

SOLUTION

The point is shown in Figure 12.18. Three other representations are as follows.

$$\left(3, -\frac{3\pi}{4} + 2\pi\right) = \left(3, \frac{5\pi}{4}\right) \qquad \text{Add } 2\pi \text{ to } \theta$$

$$\left(-3, -\frac{3\pi}{4} - \pi\right) = \left(-3, -\frac{7\pi}{4}\right) \qquad \text{Replace } r \text{ by } -r \text{ and subtract } \pi \text{ from } \theta$$

$$\left(-3, -\frac{3\pi}{4} + \pi\right) = \left(-3, \frac{\pi}{4}\right) \qquad \text{Replace } r \text{ by } -r \text{ and add } \pi \text{ to } \theta \quad ▭$$

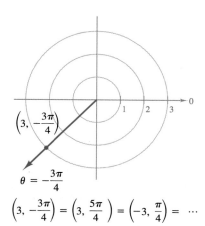

$$\left(3, -\frac{3\pi}{4}\right) = \left(3, \frac{5\pi}{4}\right) = \left(-3, \frac{\pi}{4}\right) = \cdots$$

FIGURE 12.18

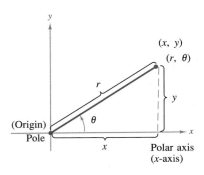

FIGURE 12.19

Relating Polar
and Rectangular Coordinates

To establish the relationship between polar and rectangular coordinates, we let the polar axis coincide with the positive x-axis and the pole with the origin, as shown in Figure 12.19. Since (x, y) lies on a circle of radius r, it follows that $r^2 = x^2 + y^2$. Moreover, for $r > 0$, the definition of the trigonometric functions implies that

$$\tan \theta = \frac{y}{x}, \qquad \cos \theta = \frac{x}{r}, \qquad \text{and} \qquad \sin \theta = \frac{y}{r}.$$

If $r < 0$, we can show that the same relationships hold. For example, consider the point (r, θ), where $r < 0$. Then, since $(-r, \theta + \pi)$ represents the same point and $-r > 0$, we have $-\sin \theta = \sin (\theta + \pi) = -y/r$, which implies that $\sin \theta = y/r$.

These relationships allow us to convert *coordinates* or *equations* from one system to the other, as indicated in the following theorem.

THEOREM 12.4
COORDINATE CONVERSION

The polar coordinates (r, θ) are related to the rectangular coordinates (x, y) as follows.

$$x = r \cos \theta \qquad \text{and} \qquad y = r \sin \theta$$

$$\tan \theta = \frac{y}{x} \qquad \text{and} \qquad r^2 = x^2 + y^2$$

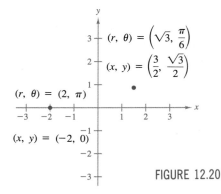

FIGURE 12.20

EXAMPLE 3 Polar-to-rectangular conversion

(a) For the point $(r, \theta) = (2, \pi)$, we have

$$x = r \cos \theta = 2 \cos \pi = -2 \qquad \text{and} \qquad y = r \sin \theta = 2 \sin \pi = 0.$$

Thus, the rectangular coordinates are $(x, y) = (-2, 0)$, as shown in Figure 12.20.

(b) For the point $(r, \theta) = (\sqrt{3}, \pi/6)$, we have

$$x = \sqrt{3} \cos \frac{\pi}{6} = \sqrt{3}\left(\frac{\sqrt{3}}{2}\right) = \frac{3}{2}$$

and

$$y = \sqrt{3} \sin \frac{\pi}{6} = \sqrt{3}\left(\frac{1}{2}\right) = \frac{\sqrt{3}}{2}.$$

Thus, the rectangular coordinates are $(x, y) = (3/2, \sqrt{3}/2)$, as shown in Figure 12.20.

EXAMPLE 4   Rectangular-to-polar conversion

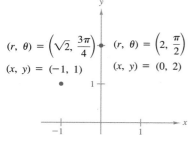

$(r, \theta) = \left(\sqrt{2}, \frac{3\pi}{4}\right)$ $(r, \theta) = \left(2, \frac{\pi}{2}\right)$

$(x, y) = (-1, 1)$ $(x, y) = (0, 2)$

FIGURE 12.21

(a) For the second quadrant point $(x, y) = (-1, 1)$, we have

$$\tan \theta = \frac{y}{x} = -1 \quad \Longrightarrow \quad \theta = \frac{3\pi}{4}.$$

Since our choice of θ lies in the same quadrant as (x, y) we use positive r.

$$r = \sqrt{x^2 + y^2} = \sqrt{(-1)^2 + (1)^2} = \sqrt{2}$$

Thus, *one* set of polar coordinates is $(r, \theta) = (\sqrt{2}, 3\pi/4)$, as shown in Figure 12.21.

(b) Since the point $(x, y) = (0, 2)$ lies on the positive y-axis, we choose $\theta = \pi/2$ and $r = 2$, and one set of polar coordinates is $(r, \theta) = (2, \pi/2)$, as shown in Figure 12.21.

By comparing Examples 3 and 4, we see that point conversion from the polar to the rectangular system is straightforward, whereas point conversion from the rectangular to the polar system is more involved. For equations, the opposite is true. To convert a rectangular equation to polar form, we simply replace x by $r \cos \theta$ and y by $r \sin \theta$. For instance, the rectangular equation $y = x^2$ has the polar form

$$r \sin \theta = (r \cos \theta)^2.$$

On the other hand, to convert a polar equation to rectangular form can require considerable ingenuity.

Graphs of polar equations

Curve sketching in polar coordinates is similar to that for rectangular coordinates. We rely heavily on point-by-point plotting, aided by intercepts and symmetry. Converting to rectangular coordinates and then making the sketch is also an option, as demonstrated in the next example.

EXAMPLE 5 Graphing polar equations

Describe the graphs of the following polar equations and find the corresponding rectangular equations.

(a) $r = 2$ (b) $\theta = \dfrac{\pi}{3}$ (c) $r = \sec \theta$

SOLUTION

(a) The graph of the polar equation $r = 2$ consists of all points that are 2 units from the pole. In other words, this graph is a circle centered at the origin and having a radius of 2, as shown in Figure 12.22(a). We can confirm this by converting to rectangular coordinates, using the relationship $r^2 = x^2 + y^2$.

$$r = 2 \quad \Longrightarrow \quad r^2 = 2^2 \quad \Longrightarrow \quad x^2 + y^2 = 2^2$$

Polar equation   Rectangular equation

(b) The graph of the polar equation $\theta = \pi/3$ consists of all points on the line that makes an angle of $\pi/3$ with the positive x-axis, as shown in Figure 12.22(b). To convert to rectangular form, we make use of the relationship $\tan \theta = y/x$.

$$\theta = \frac{\pi}{3} \quad \Longrightarrow \quad \tan \theta = \sqrt{3} \quad \Longrightarrow \quad y = \sqrt{3}x$$

Polar equation   Rectangular equation

(c) The graph of the polar equation $r = \sec \theta$ is not evident by simple inspection, so we convert to rectangular form by using the relationship $r \cos \theta = x$.

$$r = \sec \theta \quad \Longrightarrow \quad r \cos \theta = 1 \quad \Longrightarrow \quad x = 1$$

Polar equation   Rectangular equation

Now we see that the graph is a vertical line, as shown in Figure 12.22(c). ▢

Curve sketching by converting to rectangular form is not always convenient, and in the next example we demonstrate a straightforward point-plotting technique.

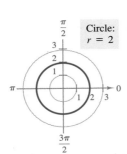

(a)

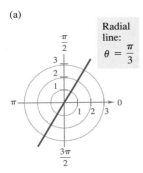

(b)

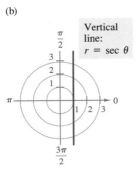

(c)

FIGURE 12.22

EXAMPLE 6 Graphing a polar equation by point-plotting

Sketch the graph of the polar equation

$$r = 4 \sin \theta.$$

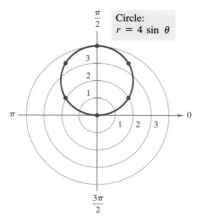

FIGURE 12.23

SOLUTION

Since the sine function is periodic, we can get a full range of r values by considering values of θ in the interval $0 \le \theta \le 2\pi$, as shown in Table 12.2.

TABLE 12.2

θ	0	$\dfrac{\pi}{6}$	$\dfrac{\pi}{3}$	$\dfrac{\pi}{2}$	$\dfrac{2\pi}{3}$	$\dfrac{5\pi}{6}$	π	$\dfrac{7\pi}{6}$	$\dfrac{3\pi}{2}$	$\dfrac{11\pi}{6}$	2π
r	0	2	$2\sqrt{3}$	4	$2\sqrt{3}$	2	0	-2	-4	-2	0

By plotting these points as shown in Figure 12.23, it appears that the graph is a circle of radius 2 whose center is at the point $(x, y) = (0, 2)$. (In Exercise 33 you are asked to confirm this by converting to rectangular form.)

Symmetry

When making a table of points, it is often helpful to use at least one θ-value from each of the four quadrants. However, in Figure 12.23 we see that no points are plotted in Quadrants III and IV because r is negative for these values of θ. Thus, on the interval from $\theta = 0$ to $\theta = 2\pi$ the curve is traced twice. Moreover, we also see that this curve is *symmetric with respect to the line $\theta = \pi/2$*. Had we known about this symmetry and retracing ahead of time we could have plotted fewer points.

Symmetry with respect to the line $\theta = \pi/2$ is one of three important types of symmetry to consider in polar curve sketching. (See Figure 12.24.)

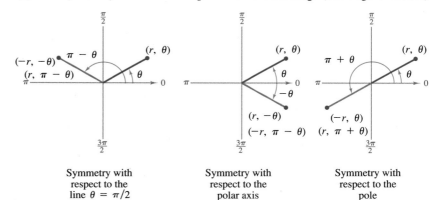

FIGURE 12.24

The following theorem lists a test for each of the three types.

THEOREM 12.5 **TESTS FOR SYMMETRY IN** **POLAR COORDINATES**	The graph of a polar equation is symmetric with respect to the following if the indicated substitution produces an equivalent equation. 1. *The line $\theta = \pi/2$:* Replace (r, θ) by $(r, \pi - \theta)$ or $(-r, -\theta)$. 2. *The polar axis:* Replace (r, θ) by $(r, -\theta)$ or $(-r, \pi - \theta)$. 3. *The pole:* Replace (r, θ) by $(r, \pi + \theta)$ or $(-r, \theta)$.

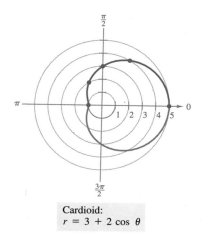

Cardioid:
$r = 3 + 2 \cos \theta$

FIGURE 12.25

EXAMPLE 7   Using symmetry to sketch the graph of a polar equation

Use symmetry to sketch the graph of $r = 3 + 2 \cos \theta$.

SOLUTION

Since the cosine is an even function, we replace (r, θ) by $(r, -\theta)$.

$$r = 3 + 2 \cos (-\theta) \quad \Longrightarrow \quad r = 3 + 2 \cos \theta$$

Since the substitution produces an equivalent equation, the curve is symmetric with respect to the polar axis. This means that we need only use θ-values from the first two quadrants, as shown in Table 12.3.

TABLE 12.3

θ	0	$\dfrac{\pi}{3}$	$\dfrac{\pi}{2}$	$\dfrac{2\pi}{3}$	π
r	5	4	3	2	1

Plotting these points and using polar axis symmetry produces the graph shown in Figure 12.25. ▭

The equations discussed in Examples 6 and 7 are of the following form.

$$r = 4 \sin \theta = f(\sin \theta) \qquad \text{\small r is a function of $\sin \theta$}$$
$$r = 3 + 2 \cos \theta = g(\cos \theta) \qquad \text{\small r is a function of $\cos \theta$}$$

From Figures 12.23 and 12.25 we see that the graphs of these equations are symmetric with respect to the line $\theta = \pi/2$ and the polar axis, respectively. This observation can be generalized to form the following *quick test for symmetry.*

1. The graph of $r = f(\sin \theta)$ is symmetric with respect to the line $\theta = \pi/2$.
2. The graph of $r = g(\cos \theta)$ is symmetric with respect to the polar axis.

When applying Theorem 12.5 note that we are using only two of the many possible polar representations of the point (r, θ). Consequently, the conditions of Theorem 12.5 are *sufficient* to guarantee symmetry, but they are not necessary (that is, a graph can have symmetry without satisfying the conditions of the theorem). For instance, Figure 12.26 shows the graph of $r = \theta + 2\pi$ to be symmetric with respect to the line $\theta = \pi/2$. Yet the test fails to indicate symmetry because neither of the following replacements yields an equivalent equation.

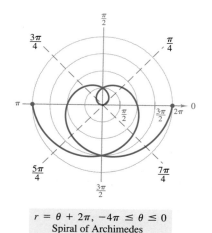

$r = \theta + 2\pi, \ -4\pi \le \theta \le 0$
Spiral of Archimedes

FIGURE 12.26

Original equation	*Replacement*	*New equation*
$r = \theta + 2\pi$	(r, θ) by $(-r, -\theta)$	$-r = -\theta + 2\pi$
$r = \theta + 2\pi$	(r, θ) by $(r, \pi - \theta)$	$r = -\theta + 3\pi$

EXERCISES for Section 12.3

In Exercises 1–8, the polar coordinates of a point are given. Plot the point and find the corresponding rectangular coordinates.

1. $\left(4, \dfrac{3\pi}{6}\right)$

2. $\left(4, \dfrac{3\pi}{2}\right)$

3. $\left(-1, \dfrac{5\pi}{4}\right)$

4. $(0, -\pi)$

5. $\left(4, -\dfrac{\pi}{3}\right)$

6. $\left(-1, -\dfrac{3\pi}{4}\right)$

7. $(\sqrt{2}, 2.36)$

8. $(-3, -1.57)$

In Exercises 9–16, the rectangular coordinates of a point are given. Find two sets of polar coordinates for the point, using $0 \le \theta < 2\pi$.

9. $(1, 1)$

10. $(0, -5)$

11. $(-3, 4)$

12. $(3, -1)$

13. $(-\sqrt{3}, -\sqrt{3})$

14. $(-2, 0)$

15. $(4, 6)$

16. $(5, 12)$

In Exercises 17–32, convert the given rectangular equation to polar form.

17. $x^2 + y^2 = 9$

18. $x^2 + y^2 = a^2$

19. $x^2 + y^2 - 2ax = 0$

20. $x^2 + y^2 - 2ay = 0$

21. $y = 4$

22. $y = b$

23. $x = 10$

24. $x = a$

25. $3x - y + 2 = 0$

26. $4x + 7y - 2 = 0$

27. $xy = 4$

28. $y = x$

29. $y^2 = 9x$

30. $y^2 - 8x - 16 = 0$

31. $x^2 - 4ay - 4a^2 = 0$

32. $(x^2 + y^2)^2 - 9(x^2 - y^2) = 0$

In Exercises 33–44, convert the given polar equation to rectangular form.

33. $r = 4 \sin \theta$

34. $r = 4 \cos \theta$

35. $\theta = \dfrac{\pi}{6}$

36. $r = 4$

37. $r = 2 \csc \theta$

38. $r^2 = \sin 2\theta$

39. $r = 1 - 2 \sin \theta$

40. $r = \dfrac{1}{1 - \cos \theta}$

41. $r = \dfrac{6}{2 - 3 \sin \theta}$

42. $r = 2 \sin 3\theta$

43. $r^2 = 4 \sin \theta$

44. $r = \dfrac{6}{2 \cos \theta - 3 \sin \theta}$

In Exercises 45–54, sketch the graph of the polar equation. Indicate any symmetry possessed by the graph.

45. $r = 5$

46. $r = -2$

47. $r = \theta$

48. $r = \dfrac{\theta}{\pi}$

49. $r = \sin \theta$

50. $r = 3 \cos \theta$

51. $r = 2 \sec \theta$

52. $r = 3 \csc \theta$

53. $r = 4(2 + \sin \theta)$

54. $r = 2 + \cos \theta$

55. Convert the equation

$$r = 2(h \cos \theta + k \sin \theta)$$

to rectangular form and verify that it is the equation of a circle. Find the radius and the rectangular coordinates of the center of the circle.

56. Verify that the distance between the two points (r_1, θ_1) and (r_2, θ_2) in polar coordinates is given by

$$d = \sqrt{r_1^2 + r_2^2 - 2r_1 r_2 \cos(\theta_1 - \theta_2)}.$$

12.4 Tangent Lines and Curve Sketching in Polar Coordinates

Relative extrema of r ▪ Tangent lines to polar graphs ▪ Special polar graphs

We have seen that polar curve sketching can be simplified by considering the periodic nature and symmetry of the graph. We begin this section by using calculus to develop two additional curve sketching aids.

If r is a differentiable function of θ, then we can determine the **relative extrema of r** using the procedures discussed in Chapter 4. However, since the distance between a point (r, θ) and the pole is $|r|$, it is possible that both the relative maxima and relative minima of r yield points whose distance from the pole is a relative maximum. We illustrate this with an example.

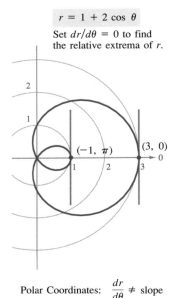

$r = 1 + 2 \cos \theta$

Set $dr/d\theta = 0$ to find
the relative extrema of r.

$(-1, \pi)$ $(3, 0)$

Polar Coordinates: $\dfrac{dr}{d\theta} \neq$ slope

FIGURE 12.27

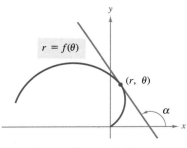

$r = f(\theta)$

(r, θ)

α

Tangent Line to Polar Curve

FIGURE 12.28

EXAMPLE 1   Relative extrema of r

Find the relative extrema of the function given by $r = 1 + 2 \cos \theta$ and locate the corresponding points on the graph of the function.

SOLUTION

By setting the derivative equal to zero, we have

$$\frac{dr}{d\theta} = -2 \sin \theta = 0$$

and we conclude that on the interval $0 \leq \theta < 2\pi$ there are two critical numbers

$$\theta = 0 \qquad \text{and} \qquad \theta = \pi.$$

The corresponding points $(3, 0)$ and $(-1, \pi)$ are shown in Figure 12.27. The graph shown in this figure is called a **limaçon**.

Slope and tangent lines

Be sure you see that $dr/d\theta$ does not represent the slope of a polar graph. To find the slope of a tangent line to a polar graph, consider a differentiable function given by $r = f(\theta)$. If α is the angle from the polar axis to the tangent line at the point (r, θ), then $dy/dx = \tan \alpha$, as shown in Figure 12.28. To convert to polar form, we use the parametric equations

$$x = r \cos \theta = f(\theta) \cos \theta \qquad \text{and} \qquad y = r \sin \theta = f(\theta) \sin \theta.$$

Using the parametric form of dy/dx given in Theorem 12.1, we have

$$\frac{dy}{dx} = \frac{dy/d\theta}{dx/d\theta} = \frac{f(\theta) \cos \theta + f'(\theta) \sin \theta}{-f(\theta) \sin \theta + f'(\theta) \cos \theta}$$

which establishes the following theorem.

THEOREM 12.6 **SLOPE IN POLAR FORM**	If f is a differentiable function of θ, then the *slope* of the tangent line to the graph of $r = f(\theta)$ at the point (r, θ) is $$\frac{dy}{dx} = \frac{dy/d\theta}{dx/d\theta} = \frac{f(\theta) \cos \theta + f'(\theta) \sin \theta}{-f(\theta) \sin \theta + f'(\theta) \cos \theta}$$ provided $dx/d\theta \neq 0$ at (r, θ).

From Theorem 12.6, we can make the following observations.

1. Solutions to $dy/d\theta = 0$ yield horizontal tangents, provided $dx/d\theta \neq 0$.
2. Solutions to $dx/d\theta = 0$ yield vertical tangents, provided $dy/d\theta \neq 0$.

If $dy/d\theta$ and $dx/d\theta$ are *simultaneously* zero, then no conclusions can be drawn about tangent lines.

EXAMPLE 2   Finding horizontal and vertical tangents to polar curves

Find the horizontal and vertical tangents to the graph of $r = 2(1 - \cos\theta)$.

SOLUTION

Using $y = r\sin\theta$, we differentiate and set $dy/d\theta$ equal to zero.

$$y = r\sin\theta = 2(1 - \cos\theta)\sin\theta$$

$$\frac{dy}{d\theta} = 2[(1 - \cos\theta)(\cos\theta) + \sin\theta(\sin\theta)]$$

$$= 2(1 + \cos\theta - 2\cos^2\theta) = -2(2\cos\theta + 1)(\cos\theta - 1) = 0$$

Thus, $\cos\theta = -\frac{1}{2}$ and $\cos\theta = 1$, and we conclude that $dy/d\theta = 0$ when $\theta = 2\pi/3$, $4\pi/3$, and 0. Similarly, using $x = r\cos\theta$, we have

$$x = r\cos\theta = 2\cos\theta - 2\cos^2\theta$$

$$\frac{dx}{d\theta} = -2\sin\theta + 4\cos\theta\sin\theta$$

$$= 2\sin\theta(2\cos\theta - 1) = 0.$$

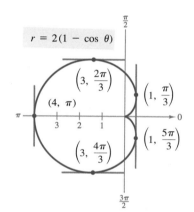

$r = 2(1 - \cos\theta)$

$\left(3, \frac{2\pi}{3}\right)$

$(4, \pi)$

$\left(1, \frac{\pi}{3}\right)$

$\left(3, \frac{4\pi}{3}\right)$

$\left(1, \frac{5\pi}{3}\right)$

FIGURE 12.29

Thus, $\sin\theta = 0$ or $\cos\theta = \frac{1}{2}$, and we conclude that $dx/d\theta = 0$ when $\theta = 0$, π, $\pi/3$, or $5\pi/3$. Finally, we eliminate $\theta = 0$ because both derivatives are zero there, and conclude that the graph, shown in Figure 12.29, has horizontal and vertical tangents at the following points.

$$\left(3, \frac{2\pi}{3}\right), \left(3, \frac{4\pi}{3}\right) \qquad \text{Horizontal tangents}$$

$$\left(1, \frac{\pi}{3}\right), \left(1, \frac{5\pi}{3}\right), (4, \pi) \qquad \text{Vertical tangents}$$

The graph is called a **cardioid.** ▭

Theorem 12.6 has an important consequence. Suppose that the graph of $r = f(\theta)$ passes through the pole when $\theta = \alpha$ and $f'(\alpha) \neq 0$. Then the formula for dy/dx simplifies as follows.

$$\frac{dy}{dx} = \frac{f'(\alpha)\sin\alpha + f(\alpha)\cos\alpha}{f'(\alpha)\cos\alpha - f(\alpha)\sin\alpha} = \frac{f'(\alpha)\sin\alpha + 0}{f'(\alpha)\cos\alpha - 0} = \frac{\sin\alpha}{\cos\alpha} = \tan\alpha$$

This means that the line $\theta = \alpha$ is tangent to the graph at the pole, $(0, \alpha)$. We summarize this result in the following theorem.

THEOREM 12.7
TANGENT LINES AT THE POLE

If $f(\alpha) = 0$ and $f'(\alpha) \neq 0$, then the line $\theta = \alpha$ is tangent at the pole to the graph of $r = f(\theta)$.

Theorem 12.7 is useful because it tells us that the zeros of $r = f(\theta)$ can be used to find the tangent lines at the pole. Note that since a polar curve can cross the pole more than once, it can have more than one tangent line at the pole. Study the next example carefully to see how the various aids to polar curve sketching are combined.

EXAMPLE 3   Sketching a polar graph

Sketch the graph of $r = 2 \cos 3\theta$.

SOLUTION

The graph is symmetric with respect to the polar axis because

$$\cos 3\theta = \cos(-3\theta).$$

Setting $dr/d\theta$ equal to zero, we have

$$\frac{dr}{d\theta} = -6 \sin 3\theta = 0.$$

Thus, in the interval $0 \le \theta \le \pi$, the extrema of r occur when $\theta = 0$, $\pi/3$, and $2\pi/3$. Moreover, by setting r equal to zero, we have

$$2 \cos 3\theta = 0$$

and in the interval $0 \le \theta \le \pi$ the tangent lines to the pole occur when $\theta = \pi/6$, $\pi/2$, and $5\pi/6$. Finally, by plotting a few points, we obtain the graph shown in Figure 12.30. It is interesting to note in this figure that both the upper and lower halves of the graph are traced as θ increases from 0 to π. Were we to continue plotting points as θ increases from π to 2π, the graph would be traced a second time. The graph shown in this figure is called a **rose curve.**

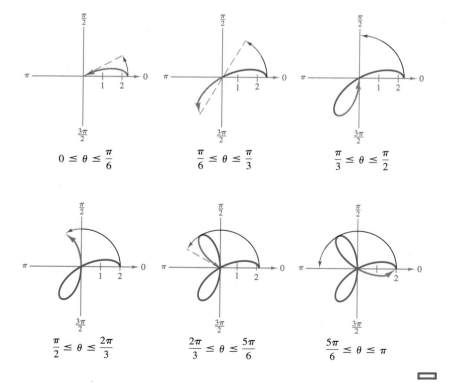

FIGURE 12.30

Special polar graphs

Several important types of graphs have equations that are simpler in polar form than in rectangular form. For example, the polar equation of a circle having a radius of a and centered at the origin is simply $r = a$. Later in the text you will come to appreciate this benefit. For now, we summarize some other types of graphs that have simpler equations in polar form. (Conics are considered in Section 12.6.)

Limaçons

$r = a \pm b \cos \theta$
$r = a \pm b \sin \theta$
$(0 < a,\ 0 < b)$

Rose Curves

n petals if n is odd
$2n$ petals if n is even
$(n \geq 2)$

Circles and Lemniscates

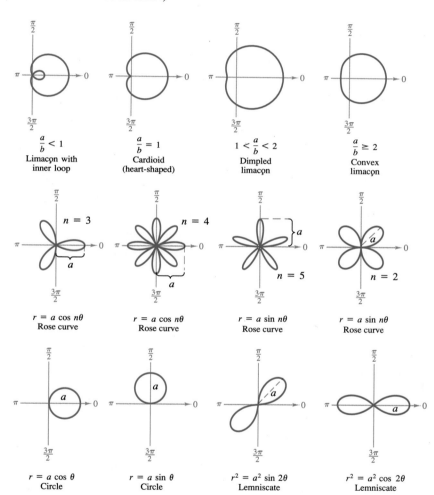

| $\dfrac{a}{b} < 1$ | $\dfrac{a}{b} = 1$ | $1 < \dfrac{a}{b} < 2$ | $\dfrac{a}{b} \geq 2$ |
| Limaçon with inner loop | Cardioid (heart-shaped) | Dimpled limaçon | Convex limaçon |

| $n = 3$ | $n = 4$ | $n = 5$ | $n = 2$ |
| $r = a \cos n\theta$ Rose curve | $r = a \cos n\theta$ Rose curve | $r = a \sin n\theta$ Rose curve | $r = a \sin n\theta$ Rose curve |

| $r = a \cos \theta$ Circle | $r = a \sin \theta$ Circle | $r^2 = a^2 \sin 2\theta$ Lemniscate | $r^2 = a^2 \cos 2\theta$ Lemniscate |

EXAMPLE 4 Sketching a polar graph

Sketch the graph of $r = 3 \cos 2\theta$.

SOLUTION

Type of curve: Rose curve with $2n = 4$ petals
Symmetry: With respect to the polar axis and the line $\theta = \pi/2$
Extrema of r: $(3, 0)$, $(-3,\ \pi/2)$, $(3,\ \pi)$, and $(-3,\ 3\pi/2)$
Tangents at pole: $r = 0$ when $\theta = \pi/4,\ 3\pi/4$

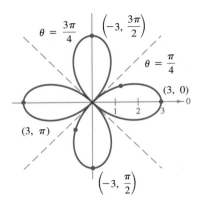

$\theta = \dfrac{3\pi}{4}$ $\left(-3, \dfrac{3\pi}{2}\right)$

$\theta = \dfrac{\pi}{4}$

$(3, 0)$

$(3, \pi)$

$\left(-3, \dfrac{\pi}{2}\right)$

$r = 3 \cos 2\theta$

FIGURE 12.31

By using this information and the additional points shown in Table 12.4, we obtain the graph shown in Figure 12.31.

TABLE 12.4

θ	0	$\dfrac{\pi}{6}$	$\dfrac{\pi}{4}$	$\dfrac{\pi}{3}$
r	3	$\dfrac{3}{2}$	0	$-\dfrac{3}{2}$

EXAMPLE 5   Sketching a polar graph

Sketch the graph of $r^2 = 9 \sin 2\theta$.

SOLUTION

Type of curve: Lemniscate
Symmetry: With respect to the pole
Extrema of r: $r = \pm 3$ when $\theta = \pi/4$
Tangents at pole: $r = 0$ when $\theta = 0,\ \pi/2$

If $\sin 2\theta < 0$, then this equation has no solution points. Thus, we restrict the values of θ to those for which $\sin 2\theta \geq 0$. That is,

$$0 \leq \theta \leq \frac{\pi}{2}$$

or

$$\pi \leq \theta \leq \frac{3\pi}{2}.$$

Moreover, using symmetry, we need only consider the first of these two intervals, In Table 12.5 we find a few additional points, and the graph is shown in Figure 12.32.

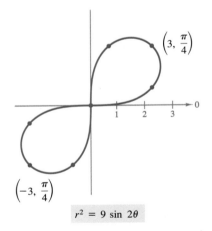

$\left(3, \dfrac{\pi}{4}\right)$

$\left(-3, \dfrac{\pi}{4}\right)$

$r^2 = 9 \sin 2\theta$

FIGURE 12.32

TABLE 12.5

θ	0	$\dfrac{\pi}{12}$	$\dfrac{\pi}{4}$	$\dfrac{5\pi}{12}$	$\dfrac{\pi}{2}$
$r = \pm 3\sqrt{\sin 2\theta}$	0	$\pm\dfrac{3}{\sqrt{2}}$	± 3	$\pm\dfrac{3}{\sqrt{2}}$	0

EXERCISES for Section 12.4

In Exercises 1–6, find the relative extrema of r.

1. $r = 5 \cos 3\theta$

2. $r = \cos \theta + \sqrt{3} \sin \theta$

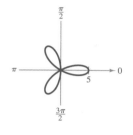

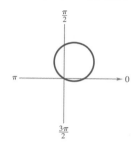

3. $r = 5 + 3 \sin \theta$

4. $r^2 = 9 \sin 2\theta$

5. $r = \dfrac{6}{1 - \cos \theta}$

6. $r = 2 \sin \theta \cos^2 \theta$

In Exercises 7–16, find dy/dx and the slope of the graph of the polar curve at the given value of θ.

Polar equation	Point
7. $r = 3(1 - \cos \theta)$	$\theta = \dfrac{\pi}{2}$
8. $r = 2(1 - \sin \theta)$	$\theta = \dfrac{\pi}{6}$
9. $r = 2 + 3 \sin \theta$	$\theta = \dfrac{\pi}{2}$
10. $r = 3 - 2 \cos \theta$	$\theta = 0$
11. $r = \theta$	$\theta = \pi$
12. $r = 3 \cos \theta$	$\theta = \dfrac{\pi}{4}$
13. $r = 3 \sin \theta$	$\theta = \dfrac{\pi}{3}$
14. $r = 4$	$\theta = \dfrac{\pi}{4}$
15. $r = 2 \sec \theta$	$\theta = \dfrac{\pi}{4}$
16. $r = \dfrac{6}{2 \sin \theta - 3 \cos \theta}$	$\theta = \pi$

In Exercises 17 and 18, find the points of horizontal and vertical tangency (if any) to the polar curve.

17. $r = 1 + \sin \theta$

18. $r = a \sin \theta$

In Exercises 19 and 20, find the points of horizontal tangency (if any) to the polar curve.

19. $r = 2 \csc \theta + 3$

20. $r = a \sin \theta \cos^2 \theta$

In Exercises 21–54, identify and sketch the graph of the given equation. In each case, find the tangents at the pole.

21. $r = 4$

22. $r = -2$

23. $r = 3 \sin \theta$

24. $r = 3 \cos \theta$

25. $r = 3(1 - \cos \theta)$

26. $r = 2(1 - \sin \theta)$

27. $r = 4(1 + \sin \theta)$

28. $r = 1 + \cos \theta$

29. $r = 2 + 3 \sin \theta$

30. $r = 4 + 5 \cos \theta$

31. $r = 3 - 4 \cos \theta$

32. $r = 2(1 - 2 \sin \theta)$

33. $r = 3 - 2 \cos \theta$

34. $r = 5 - 4 \sin \theta$

35. $r = 2 + \sin \theta$

36. $r = 4 + 3 \cos \theta$

37. $r = 2 \cos 3\theta$

38. $r = -\sin 5\theta$

39. $r = 3 \sin 2\theta$

40. $r = 3 \cos 2\theta$

41. $r = 2 \sec \theta$

42. $r = 3 \csc \theta$

43. $r = \dfrac{3}{\sin \theta - 2 \cos \theta}$

44. $r = \dfrac{6}{2 \sin \theta - 3 \cos \theta}$

45. $r^2 = 4 \cos 2\theta$

46. $r^2 = 4 \sin \theta$

47. $r^2 = 4 \sin 2\theta$

48. $r^2 = \cos 3\theta$

49. $r = \sin \theta \cos^2 \theta$

50. $r^2 = \dfrac{1}{\theta}$

51. $r = 2\theta$

52. $r = \dfrac{1}{\theta}$

53. $r = 2 \cos \left(\dfrac{3\theta}{2} \right)$

54. $r = 3 \sin \left(\dfrac{5\theta}{2} \right)$

In Exercises 55–58, sketch the graph of the given equation and show that the indicated line is an asymptote to the graph.

Polar equation	Asymptote
55. $r = 2 - \sec \theta$	$x = -1$
56. $r = 2 + \csc \theta$	$y = 1$
57. $r = \dfrac{2}{\theta}$	$y = 2$
58. $r = 2 \cos 2\theta \sec \theta$	$x = -2$

59. Verify that if the curve whose polar equation is $r = f(\theta)$ is rotated about the pole through an angle ϕ, then an equation for the rotated curve is $r = f(\theta - \phi)$.

60. If the polar form of an equation for a curve is $r = f(\sin \theta)$, show that the form becomes
(a) $r = f(-\cos \theta)$ if the curve is rotated counterclockwise $\pi/2$ radians about the pole.
(b) $r = f(-\sin \theta)$ if the curve is rotated counterclockwise π radians about the pole.
(c) $r = f(\cos \theta)$ if the curve is rotated counterclockwise $3\pi/2$ radians about the pole.

In Exercises 61–64, use the results of Exercises 59 and 60.

61. Write an equation for the limaçon

$$r = 2 - \sin \theta$$

after it has been rotated by the given amount.

(a) $\dfrac{\pi}{4}$ (b) $\dfrac{\pi}{2}$ (c) π (d) $\dfrac{3\pi}{2}$

62. Write an equation for the rose curve

$$r = 2 \sin 2\theta$$

after it has been rotated by the given amount.

(a) $\dfrac{\pi}{6}$ (b) $\dfrac{\pi}{2}$ (c) $\dfrac{2\pi}{3}$ (d) π

63. Sketch the graphs of the equations.

(a) $r = 1 - \sin \theta$ (b) $r = 1 - \sin\left(\theta - \dfrac{\pi}{4}\right)$

64. Sketch the graphs of the equations.

(a) $r = 3 \sec \theta$

(b) $r = 3 \sec\left(\theta - \dfrac{\pi}{4}\right)$

(c) $r = 3 \sec\left(\theta + \dfrac{\pi}{3}\right)$

In Exercises 65–68, use a computer or graphics calculator to sketch the graph of the polar equation and find all points of horizontal tangency.

65. $r = 4 \sin \theta \cos^2 \theta$

66. $r = 3 \cos 2\theta \sec \theta$

67. $r = 2 \csc \theta + 5$

68. $r = 2 \cos(3\theta - 2)$

12.5 Area and Arc Length in Polar Coordinates

Area of a polar region ▪ Points of intersection of polar graphs ▪ Arc length in polar form ▪ Area of a surface of revolution

Area of Sector: $\dfrac{1}{2}\theta r^2$

FIGURE 12.33

The development of a formula for the area of a polar region parallels that for the area of regions in the rectangular coordinate system, but uses *sectors* of a circle instead of rectangles as the basic element of area. In Figure 12.33 note that the area of a circular sector of radius r is given by

$$\text{area of circular sector} = \frac{1}{2}\theta r^2$$

provided θ is measured in radians.

Consider the function given by $r = f(\theta)$, where f is continuous and nonnegative in the interval $[\alpha, \beta]$. The region bounded by the graph of f and the radial lines $\theta = \alpha$ and $\theta = \beta$ is shown in Figure 12.34. To find the area

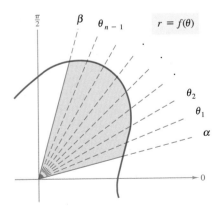

Partition $[\alpha, \beta]$ into n equal subintervals.

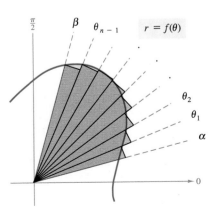

Approximate area by n sectors.

FIGURE 12.34

of this region, we partition the interval $[\alpha, \beta]$ into n equal subintervals,

$$\alpha = \theta_0 < \theta_1 < \theta_2 < \cdots < \theta_{n-1} < \theta_n = \beta.$$

Then we approximate the area of the region by the sum of the areas of the n sectors.

$$\text{radius of } i\text{th sector} = f(\theta_i)$$

$$\text{central angle of } i\text{th sector} = \frac{\beta - \alpha}{n} = \Delta\theta$$

$$A \approx \sum_{i=1}^{n} \left(\frac{1}{2}\right) \Delta\theta f(\theta_i)^2$$

Taking the limit as $n \to \infty$, we have

$$A = \lim_{n \to \infty} \frac{1}{2} \sum_{i=1}^{n} f(\theta_i)^2 \, \Delta\theta = \frac{1}{2} \int_{\alpha}^{\beta} [f(\theta)]^2 \, d\theta$$

which leads to the following theorem.

THEOREM 12.8 **AREA IN POLAR COORDINATES**	If f is continuous and nonnegative on the interval $[\alpha, \beta]$, then the area of the region bounded by the graph of $r = f(\theta)$ between the radial lines $\theta = \alpha$ and $\theta = \beta$ is given by $$A = \frac{1}{2} \int_{\alpha}^{\beta} [f(\theta)]^2 \, d\theta = \frac{1}{2} \int_{\alpha}^{\beta} r^2 \, d\theta.$$

REMARK   We can use the same formula to find the area of a region bounded by the graph of a continuous *nonpositive* function. However, the formula is not necessarily valid if f takes on both positive *and* negative values in the interval $[\alpha, \beta]$.

Sometimes the most difficult part of finding the area of a polar region is determining the limits of integration. A good sketch of the region helps.

EXAMPLE 1   Finding the area of a polar region

Find the area of *one petal* of the rose curve given by

$$r = 3 \cos 3\theta.$$

SOLUTION

In Figure 12.35 we see that the right petal is traced as θ increases from $-\pi/6$ to $\pi/6$. Thus, the area is

$$A = \frac{1}{2} \int_{\alpha}^{\beta} r^2 \, d\theta = \frac{1}{2} \int_{-\pi/6}^{\pi/6} (3 \cos 3\theta)^2 \, d\theta$$

$$= \frac{9}{2} \int_{-\pi/6}^{\pi/6} \frac{1 + \cos 6\theta}{2} \, d\theta$$

$$= \frac{9}{4} \left[\theta + \frac{\sin 6\theta}{6} \right]_{-\pi/6}^{\pi/6} = \frac{9}{4} \left[\frac{\pi}{6} + \frac{\pi}{6} \right] = \frac{3\pi}{4}.$$

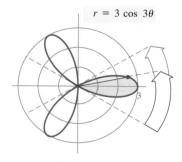

$r = 3 \cos 3\theta$

FIGURE 12.35

Suppose you were asked to find the area of the region lying inside all three petals of the rose curve in Example 1. You could not simply integrate between 0 and 2π. In doing this you would obtain $9\pi/2$, which is twice the area of the three petals—the duplication occurs because the rose curve is traced *twice* as θ increases from 0 to 2π. To avoid this type of error, be sure that the graph of $r = f(\theta)$ does not retrace itself between your chosen limits of integration.

EXAMPLE 2 Finding the area bounded by a single curve

Find the area of the region lying between the inner and outer loops of the limaçon $r = 1 - 2 \sin \theta$.

SOLUTION

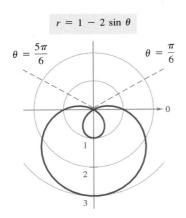

$r = 1 - 2 \sin \theta$

$\theta = \dfrac{5\pi}{6}$ $\theta = \dfrac{\pi}{6}$

FIGURE 12.36

In Figure 12.36 we see that the inner loop is traced as θ increases from $\pi/6$ to $5\pi/6$. Thus, the area of the region lying inside the *inner loop* is given by

$$
\begin{aligned}
A_1 &= \frac{1}{2} \int_{\pi/6}^{5\pi/6} (1 - 2 \sin \theta)^2 \, d\theta \\
&= \frac{1}{2} \int_{\pi/6}^{5\pi/6} (1 - 4 \sin \theta + 4 \sin^2 \theta) \, d\theta \\
&= \frac{1}{2} \int_{\pi/6}^{5\pi/6} \left[1 - 4 \sin \theta + 4 \left(\frac{1 - \cos 2\theta}{2} \right) \right] d\theta \\
&= \frac{1}{2} \int_{\pi/6}^{5\pi/6} (3 - 4 \sin \theta - 2 \cos 2\theta) \, d\theta \\
&= \frac{1}{2} \left[3\theta + 4 \cos \theta - \sin 2\theta \right]_{\pi/6}^{5\pi/6} \\
&= \frac{1}{2} [2\pi - 3\sqrt{3}] = \pi - \frac{3\sqrt{3}}{2}.
\end{aligned}
$$

In a similar way, we can integrate from $5\pi/6$ to $13\pi/6$ to find that the area of the region lying inside the *outer loop* is $A_2 = 2\pi + (3\sqrt{3}/2)$. Finally, to find the area of the region lying between the two loops, we subtract the smaller area from the larger to obtain

$$
\begin{aligned}
A &= A_2 - A_1 \\
&= \left(2\pi + \frac{3\sqrt{3}}{2} \right) - \left(\pi - \frac{3\sqrt{3}}{2} \right) \\
&= \pi + 3\sqrt{3} \approx 8.34.
\end{aligned}
$$

Points of intersection

Because a point may be represented in different ways in polar coordinates, care must be taken in determining the points of intersection of two polar graphs. For example, consider the points of intersection of the graphs of $r = 1 - 2 \cos \theta$ and $r = 1$, as shown in Figure 12.37. If, as with rectangular equations, we attempted to find the points of intersection by solving the two equations simultaneously, we would obtain the following.

$$1 = 1 - 2 \cos \theta \quad \Longrightarrow \quad \cos \theta = 0 \quad \Longrightarrow \quad \theta = \frac{\pi}{2}, \frac{3\pi}{2}$$

The corresponding points of intersection are $(1, \pi/2)$ and $(1, 3\pi/2)$. However, from Figure 12.37 we see that there is a *third* point of intersection that did not show up when we solved the two polar equations simultaneously. (This is one reason we stress sketching a graph when finding the area of a polar region.) The reason we did not find the third point is that it does not occur with the same coordinates in the two graphs. On the graph of $r = 1$, the point occurs with the coordinates $(1, \pi)$, but on the graph of $r = 1 - 2 \cos \theta$, the point occurs with coordinates $(-1, 0)$.

We can compare the problem of finding points of intersection of two polar graphs with that of finding collision points of two satellites in intersecting orbits about the earth, as shown in Figure 12.38. The satellites will not collide as long as they reach the points of intersection at different times (θ-values). A collision will occur only at the points of intersection that are "simultaneous points"—those reached at the same time (θ-value).

REMARK   Because the pole can be represented by $(0, \theta)$, where θ is *any* angle, you should check separately for the pole when hunting for points of intersection.

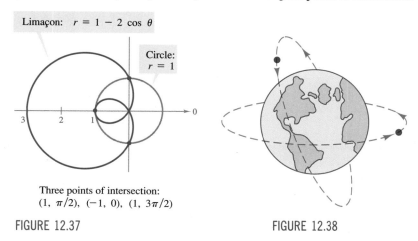

Limaçon: $r = 1 - 2 \cos \theta$

Circle: $r = 1$

Three points of intersection:
$(1, \pi/2), \ (-1, 0), \ (1, 3\pi/2)$

FIGURE 12.37   FIGURE 12.38

EXAMPLE 3   Finding the area of a region between two curves

Find the area of the region common to the two regions bounded by the circle $r = -6 \cos \theta$ and the cardioid $r = 2 - 2 \cos \theta$.

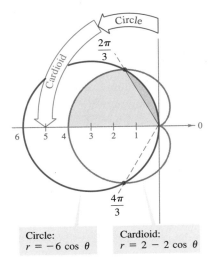

Circle:
$r = -6 \cos \theta$

Cardioid:
$r = 2 - 2 \cos \theta$

FIGURE 12.39

SOLUTION

By symmetry we can work with the upper half-plane, so we begin by dividing the region into two parts, as shown in Figure 12.39. The gray shaded region lies between the circle and the radial line $\theta = 2\pi/3$. Since the circle has coordinates $(0, \pi/2)$ at the pole, we integrate between $\pi/2$ and $2\pi/3$ to obtain the area of this region. The colored region lies between the radial lines $\theta = 2\pi/3$ and $\theta = \pi$ and the cardioid. Thus, we can find the area of this second region by integrating between $2\pi/3$ and π. The sum of these two integrals gives the area of the common region lying *above* the polar axis.

$$
\underbrace{\phantom{\frac{1}{2} \int_{\pi/2}^{2\pi/3} (-6 \cos \theta)^2 \, d\theta}}_{\substack{\text{Region between circle} \\ \text{and radial line } \theta = 2\pi/3}} \qquad \underbrace{\phantom{\frac{1}{2} \int_{2\pi/3}^{\pi} (2 - 2 \cos \theta)^2 \, d\theta}}_{\substack{\text{Region between cardioid and} \\ \text{radial lines } \theta = 2\pi/3 \text{ and } \theta = \pi}}
$$

$$
\frac{A}{2} = \frac{1}{2} \int_{\pi/2}^{2\pi/3} (-6 \cos \theta)^2 \, d\theta + \frac{1}{2} \int_{2\pi/3}^{\pi} (2 - 2 \cos \theta)^2 \, d\theta
$$

$$
= 18 \int_{\pi/2}^{2\pi/3} \cos^2 \theta \, d\theta + \frac{1}{2} \int_{2\pi/3}^{\pi} (4 - 8 \cos \theta + 4 \cos^2 \theta) \, d\theta
$$

$$
= 9 \int_{\pi/2}^{2\pi/3} (1 + \cos 2\theta) \, d\theta + \int_{2\pi/3}^{\pi} (3 - 4 \cos \theta + \cos 2\theta) \, d\theta
$$

$$
= 9 \left[\theta + \frac{\sin 2\theta}{2} \right]_{\pi/2}^{2\pi/3} + \left[3\theta - 4 \sin \theta + \frac{\sin 2\theta}{2} \right]_{2\pi/3}^{\pi}
$$

$$
= 9 \left(\frac{2\pi}{3} - \frac{\sqrt{3}}{4} - \frac{\pi}{2} \right) + \left(3\pi - 2\pi + 2\sqrt{3} + \frac{\sqrt{3}}{4} \right)
$$

$$
= \frac{5\pi}{2}
$$

Finally, multiplying by 2, we conclude that the total area is 5π. ▭

Arc length

The formula for the length of a polar arc can be obtained from the arc length formula for a curve described by parametric equations.

THEOREM 12.9 **ARC LENGTH OF A POLAR CURVE**	Let f be a function whose derivative is continuous in an interval $\alpha \le \theta \le \beta$. The length of the graph of $r = f(\theta)$ from $\theta = \alpha$ to $\theta = \beta$ is $$s = \int_{\alpha}^{\beta} \sqrt{[f(\theta)]^2 + [f'(\theta)]^2} \, d\theta.$$

PROOF Considering the parametric form

$$
x = r \cos \theta = f(\theta) \cos \theta \qquad \text{and} \qquad y = r \sin \theta = f(\theta) \sin \theta
$$

we have

$$
\frac{dx}{d\theta} = f'(\theta) \cos \theta - f(\theta) \sin \theta \qquad \text{and} \qquad \frac{dy}{d\theta} = f'(\theta) \sin \theta + f(\theta) \cos \theta
$$

and it follows that

$$\left(\frac{dx}{d\theta}\right)^2 + \left(\frac{dy}{d\theta}\right)^2 = [f(\theta)]^2 + [f'(\theta)]^2.$$

Consequently, we can apply Theorem 12.2, using θ in place of t, to obtain the desired result.

EXAMPLE 4   Finding the length of a polar curve

Find the length of the arc from $\theta = 0$ to $\theta = 2\pi$ for the cardioid

$$r = f(\theta) = 2 - 2 \cos \theta$$

as shown in Figure 12.40.

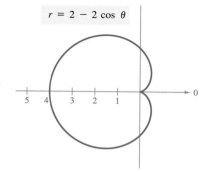

$r = 2 - 2 \cos \theta$

FIGURE 12.40

SOLUTION

Since $f'(\theta) = 2 \sin \theta$, we have

$$\begin{aligned}
s &= \int_\alpha^\beta \sqrt{[f(\theta)]^2 + [f'(\theta)]^2} \; d\theta \\
&= \int_0^{2\pi} \sqrt{(2 - 2 \cos \theta)^2 + (2 \sin \theta)^2} \; d\theta \\
&= 2\sqrt{2} \int_0^{2\pi} \sqrt{1 - \cos \theta} \; d\theta \\
&= 2\sqrt{2} \int_0^{2\pi} \sqrt{2 \sin^2 \frac{\theta}{2}} \; d\theta \\
&= 4 \int_0^{2\pi} \sin \frac{\theta}{2} \; d\theta \\
&= -8 \cos \frac{\theta}{2} \Big]_0^{2\pi} \\
&= 8 + 8 \\
&= 16.
\end{aligned}$$

REMARK   Using Figure 12.40 we can determine the reasonableness of this answer by comparing it to the circumference of a circle. For example, a circle of radius $\frac{5}{2}$ has a circumference of $5\pi \approx 15.7$.

Area of a surface of revolution

The polar coordinate version of the formulas for the area of a surface of revolution can be obtained from the parametric versions given in Theorem 12.3, using the equations $x = r \cos \theta$ and $y = r \sin \theta$.

THEOREM 12.10
AREA OF A SURFACE
OF REVOLUTION

Let f be a function whose derivative is continuous in an interval $\alpha \le \theta \le \beta$. The area of the surface formed by revolving the graph of $r = f(\theta)$ from $\theta = \alpha$ to $\theta = \beta$ about the indicated line is as follows.

1. $S = 2\pi \displaystyle\int_{\alpha}^{\beta} f(\theta) \sin \theta \sqrt{[f(\theta)]^2 + [f'(\theta)]^2} \, d\theta$ About polar axis

2. $S = 2\pi \displaystyle\int_{\alpha}^{\beta} f(\theta) \cos \theta \sqrt{[f(\theta)]^2 + [f'(\theta)]^2} \, d\theta$ About line $\theta = \dfrac{\pi}{2}$

REMARK When using Theorem 12.9 or 12.10, check to see that the graph of $r = f(\theta)$ is traced only once on the interval $\alpha \le \theta \le \beta$. For example, the circle given by $r = \sin \theta$ is traced once on the interval $0 \le \theta \le \pi$.

EXAMPLE 5 Finding the area of a surface of revolution

Find the area of the surface formed by revolving the circle given by

$$r = f(\theta) = \cos \theta$$

about the line $\theta = \pi/2$, as shown in Figure 12.41.

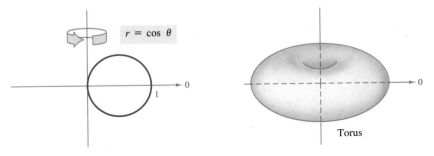

FIGURE 12.41

SOLUTION

We use the second formula given in Theorem 12.10 with $f'(\theta) = -\sin \theta$. Moreover, since the circle is traced once as θ increases from 0 to π, we have the following integral.

$$S = 2\pi \int_{\alpha}^{\beta} f(\theta) \cos \theta \sqrt{[f(\theta)^2 + [f'(\theta)]^2} \, d\theta$$

$$= 2\pi \int_{0}^{\pi} \cos \theta \, (\cos \theta) \sqrt{\cos^2 \theta + \sin^2 \theta} \, d\theta$$

$$= 2\pi \int_{0}^{\pi} \cos^2 \theta \, d\theta = \pi \int_{0}^{\pi} (1 + \cos 2\theta) \, d\theta$$

$$= \pi \left[\theta + \frac{\sin 2\theta}{2} \right]_{0}^{\pi} = \pi^2$$

EXERCISES for Section 12.5

In Exercises 1–12, find the points of intersection of the graphs of the given equations.

1. $r = 1 + \cos \theta$
$r = 1 - \cos \theta$

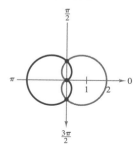

3. $r = 1 + \cos \theta$
$r = 1 - \sin \theta$

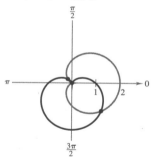

5. $r = 4 - 5 \sin \theta$
$r = 3 \sin \theta$

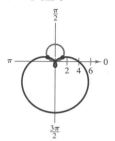

7. $r = \dfrac{\theta}{2}, \quad r = 2$

9. $r = 4 \sin 2\theta, \quad r = 2$

10. $r = 3 + \sin \theta, \quad r = 2 \csc \theta$

11. $r = 2 + 3 \cos \theta, \quad r = \dfrac{\sec \theta}{2}$

12. $r = 3(1 - \cos \theta), \quad r = \dfrac{6}{1 - \cos \theta}$

2. $r = 3(1 + \sin \theta)$
$r = 3(1 - \sin \theta)$

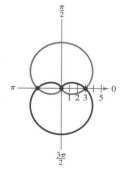

4. $r = 2 - 3 \cos \theta$
$r = \cos \theta$

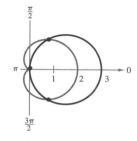

6. $r = 1 + \cos \theta$
$r = 3 \cos \theta$

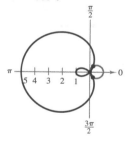

8. $\theta = \dfrac{\pi}{4}, \quad r = 2$

In Exercises 13–32, find the area of the given region.

13. One petal of $r = 2 \cos 3\theta$
14. One petal of $r = 4 \sin 2\theta$
15. One petal of $r = \cos 2\theta$
16. One petal of $r = \cos 5\theta$
17. Interior of $r = 1 - \sin \theta$
18. Interior of $r = 1 - \sin \theta$ (above the polar axis)
19. Inner loop of $r = 1 + 2 \cos \theta$
20. Inner loop of $r = 3 + 4 \sin \theta$
21. Between the loops of $r = 1 + 2 \cos \theta$
22. Between the loops of $r = 2(1 + 2 \sin \theta)$
23. Common interior of $r = 4 \sin 2\theta$ and $r = 2$
24. Common interior of $r = 3(1 + \sin \theta)$ and $r = 3(1 - \sin \theta)$
25. Common interior of $r = 3 - 2 \sin \theta$ and $r = -3 + 2 \sin \theta$
26. Common interior of $r = 3 - 2 \sin \theta$ and $r = 3 - 2 \cos \theta$
27. Common interior of $r = 4 \sin \theta$ and $r = 2$
28. Inside $r = 3 \sin \theta$ and outside $r = 2 - \sin \theta$
29. Inside $r = a(1 + \cos \theta)$ and outside $r = a \cos \theta$
30. Inside $r = 2a \cos \theta$ and outside $r = a$
31. Common interior of $r = a(1 + \cos \theta)$ and $r = a \sin \theta$
32. Region bounded by the graphs of

$$r = \frac{ab}{a \sin \theta + b \cos \theta}, \quad \theta = 0, \quad \text{and} \quad \theta = \frac{\pi}{2}$$

In Exercises 33–40, find the length of the given graph over the indicated interval.

33. $r = a, \quad 0 \le \theta \le 2\pi$
34. $r = a \cos \theta, \quad 0 \le \theta \le 2\pi$
35. $r = 1 + \sin \theta, \quad 0 \le \theta \le 2\pi$
36. $r = 5(1 + \cos \theta), \quad 0 \le \theta \le 2\pi$

37. $r = 2\theta, \quad 0 \le \theta \le \dfrac{\pi}{2}$

38. $r = \sec \theta, \quad 0 \le \theta \le \dfrac{\pi}{3}$

39. $r = \dfrac{1}{\theta}, \quad \pi \le \theta \le 2\pi$

40. $r = e^{\theta}, \quad 0 \le \theta \le \pi$

In Exercises 41–46, find the area of the surface formed by revolving the indicated curve about the given line.

41. $r = 2 \cos \theta, \quad 0 \le \theta \le \dfrac{\pi}{2}$

revolved about the polar axis

42. $r = a \cos \theta, \quad 0 \le \theta \le \dfrac{\pi}{2}$

revolved about the line $\theta = \dfrac{\pi}{2}$

43. $r = e^{a\theta}, \quad 0 \le \theta \le \dfrac{\pi}{2}$

revolved about the line $\theta = \dfrac{\pi}{2}$

44. $r = a(1 + \cos \theta), \quad 0 \le \theta \le \pi$
revolved about the polar axis

45. $r = 4 \cos 2\theta, \quad 0 \le \theta \le \dfrac{\pi}{4}$

revolved about the polar axis
[Hint: Use Simpson's Rule with $n = 4$.]

46. $r = \theta, \quad 0 \le \theta \le \pi$
revolved about the polar axis
[Hint: Use Simpson's Rule with $n = 4$.]

47. Find the surface area of the torus generated by revolving the circle given by $r = a$ about the line $r = b \sec \theta$ where $b > a > 0$. [Hint: Derive the integral that yields the surface area for revolving the graph of $r = f(\theta)$ about the line $r = b \sec \theta$.]

48. The graph of the polar equation $r = a\theta$ is called **Archimedes' spiral**. Let A_n be the area between turns $n - 1$ and n of the spiral for $n \ge 2$ (see figure).

(a) Complete the following table.

n	2	3	4	5	6	7
A_n						

[Hint: Note that

$$A_n = \frac{1}{2} \int_{(n-1)(2\pi)}^{n(2\pi)} (a\theta)^2 \, d\theta - \frac{1}{2} \int_{(n-2)(2\pi)}^{(n-1)(2\pi)} (a\theta)^2 \, d\theta.]$$

(b) Verify that $A_n = (n - 1)A_2$ for $n \ge 3$.

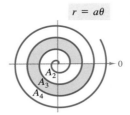

FIGURE FOR 48

In Exercises 49 and 50, use a computer or graphics calculator to sketch the graph of the equation. Use Simpson's Rule with $n = 12$ to find the arc length of the curve and the area of the region enclosed.

49. $r = \sin (3 \cos \theta)$ **50.** $r = 2 \sin (2 \cos \theta)$

12.6 Polar Equations for Conics and Kepler's Laws

Polar equations for conics ▪ Kepler's Laws

In Chapter 11 we saw that the rectangular equations of ellipses and hyperbolas take simple forms when the origin lies at their *center*. As it happens, there are many important applications of conics in which it is more convenient to use one of the *foci* as the reference point (the origin) for the coordinate system. For example, the sun lies at a focus of the earth's orbit. Similarly, the light source of a parabolic reflector lies at its focus. In this section we will see that polar equations of conics take simple forms if one of the foci lies at the pole.

The following theorem uses the concept of *eccentricity*, as defined in Chapter 11, to classify the three basic types of conics. A proof of this theorem is given in Appendix A.

THEOREM 12.11
CLASSIFICATION OF CONICS
BY ECCENTRICITY

The locus of a point in the plane whose distance from a fixed point (*focus*) has a constant ratio to its distance from a fixed line (*directrix*) is a conic. The constant ratio e is the *eccentricity* of the conic.

1. The conic is an ellipse if $0 < e < 1$.
2. The conic is a parabola if $e = 1$.
3. The conic is a hyperbola if $e > 1$.

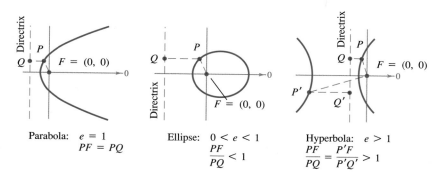

FIGURE 12.42

In Figure 12.42 note that for each type of conic, the pole corresponds to the fixed point (focus) given in the definition. The benefit of this location is seen in the proof of the following theorem.

THEOREM 12.12 **POLAR EQUATIONS FOR CONICS**	The graph of a polar equation of the form $$r = \frac{ed}{1 \pm e \cos \theta} \quad \text{or} \quad r = \frac{ed}{1 \pm e \sin \theta}$$ is a conic, where $e > 0$ is the eccentricity and $|d|$ is the distance between the focus at the pole and its corresponding directrix.

PROOF

We give a proof for $r = ed/(1 + e \cos \theta)$ with $d > 0$. The proofs of the other cases are similar. In Figure 12.43 consider a vertical directrix, d units to the right of the focus $F = (0, 0)$. If $P = (r, \theta)$ is a point on the graph of $r = ed/(1 + e \cos \theta)$, then the distance between P and the directrix is

$$PQ = |d - x| = |d - r \cos \theta|$$

$$= \left| d - \left(\frac{ed}{1 + e \cos \theta} \right) \cos \theta \right|$$

$$= \left| d \left(1 - \frac{e \cos \theta}{1 + e \cos \theta} \right) \right|$$

$$= \left| \frac{d}{1 + e \cos \theta} \right| = \left| \frac{r}{e} \right|.$$

Moreover, since the distance between P and the pole is simply $PF = |r|$, the ratio of PF to PQ is

$$\frac{PF}{PQ} = \frac{|r|}{|r/e|} = |e| = e$$

and, by Theorem 12.11, the graph of the equation must be a conic.

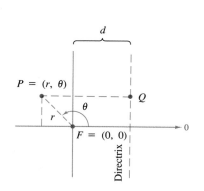

FIGURE 12.43

The four types of equations indicated in Theorem 12.12 can be classified as follows, where $d > 0$.

POLAR EQUATIONS FOR CONICS

1. Horizontal directrix above the pole:
$$r = \frac{ed}{1 + e \sin \theta}$$

2. Horizontal directrix below the pole:
$$r = \frac{ed}{1 - e \sin \theta}$$

3. Vertical directrix to the right of the pole:
$$r = \frac{ed}{1 + e \cos \theta}$$

4. Vertical directrix to the left of the pole:
$$r = \frac{ed}{1 - e \cos \theta}$$

Figure 12.44 illustrates these four possibilities for a parabola.

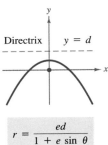

$$r = \frac{ed}{1 + e \sin \theta}$$

(1)

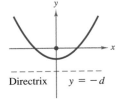

$$r = \frac{ed}{1 - e \sin \theta}$$

(2)

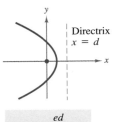

$$r = \frac{ed}{1 + e \cos \theta}$$

(3)

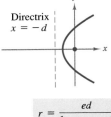

$$r = \frac{ed}{1 - e \cos \theta}$$

(4)

FIGURE 12.44

EXAMPLE 1   Determining a conic from its equation

Sketch the graph of the conic given by

$$r = \frac{15}{3 - 2 \cos \theta}.$$

SOLUTION

To determine the type of conic, we rewrite the equation as

$$r = \frac{15}{3 - 2 \cos \theta} = \frac{5}{1 - (2/3) \cos \theta}.$$

From this form we conclude that the graph is an ellipse with $e = \frac{2}{3}$. We sketch the upper half of the ellipse by plotting points from $\theta = 0$ to $\theta = \pi$, as shown in Figure 12.45. Then, using symmetry with respect to the polar axis, we sketch the lower half.

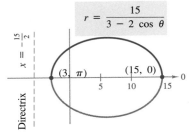

FIGURE 12.45

For the ellipse in Figure 12.45, the major axis is horizontal and the vertices lie at $(15, 0)$ and $(3, \pi)$. Thus, the length of the *major* axis is $2a = 18$. To find the length of the *minor* axis, we use the equations $e = c/a$ and $b^2 = a^2 - c^2$ and conclude that

$$b^2 = a^2 - c^2 = a^2 - (ea)^2 = a^2(1 - e^2). \qquad \text{Ellipse}$$

Since $e = \frac{2}{3}$, we have $b^2 = 9^2[1 - (\frac{2}{3})^2] = 45$, which implies that $b = \sqrt{45} = 3\sqrt{5}$. Thus, the length of the minor axis is $2b = 6\sqrt{5}$. A similar analysis for hyperbolas yields

$$b^2 = c^2 - a^2 = (ea)^2 - a^2 = a^2(e^2 - 1). \qquad \text{Hyperbola}$$

EXAMPLE 2   Sketching a conic from its polar equation

Sketch the graph of the polar equation

$$r = \frac{32}{3 + 5 \sin \theta}.$$

SOLUTION

Dividing the numerator and denominator by 3 produces

$$r = \frac{32/3}{1 + (5/3) \sin \theta}.$$

Since $e = \frac{5}{3} > 1$, the graph is a hyperbola. The transverse axis of the hyperbola lies on the line $\theta = \pi/2$, and the vertices occur at

$$\left(4, \frac{\pi}{2}\right) \qquad \text{and} \qquad \left(-16, \frac{3\pi}{2}\right).$$

Since the length of the transverse axis is 12, we see that $a = 6$. To find b, we write

$$b^2 = a^2(e^2 - 1) = 6^2\left[\left(\frac{5}{3}\right)^2 - 1\right] = 64.$$

Therefore, $b = 8$. Finally, we use a and b to determine the asymptotes of the hyperbola and obtain the sketch shown in Figure 12.46.   ▭

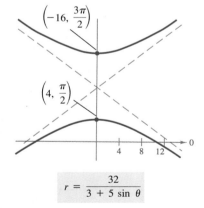

$$r = \frac{32}{3 + 5 \sin \theta}$$

FIGURE 12.46

EXAMPLE 3   Finding the polar equation for a conic

Find a polar equation for the parabola whose focus is the pole and directrix is the line $y = 3$.

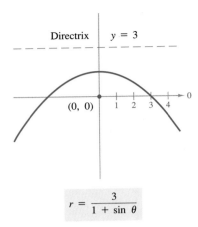

$$r = \frac{3}{1 + \sin \theta}$$

FIGURE 12.47

Johannes Kepler

SOLUTION

From Figure 12.47 we see that the directrix is horizontal, and we choose an equation of the form

$$r = \frac{ed}{1 + e \sin \theta}.$$

Furthermore, since the eccentricity of a parabola is $e = 1$ and the distance between the pole and the directrix is $d = 3$, we have the equation

$$r = \frac{3}{1 + \sin \theta}.$$

Kepler's Laws

Kepler's Laws, named after the German astronomer Johannes Kepler (1571–1630), can be used to describe the orbits of the planets about the sun.

1. Each planet moves in an elliptical orbit with the sun as a focus.
2. The ray from the sun to the planet sweeps out equal areas of the ellipse in equal times.
3. The square of the period is proportional to the cube of the mean distance between the planet and the sun.*

Although Kepler derived these laws empirically, they were later validated by Newton. In fact, Newton was able to show that each law can be deduced from a set of universal laws of motion and gravitation that govern the movement of all heavenly bodies, including comets and satellites. This is illustrated in the next example, involving the comet named after the English mathematician and physicist Edmund Halley (1656–1742).

EXAMPLE 4   An application

Halley's comet has an elliptical orbit with an eccentricity of $e \approx 0.97$. The length of the major axis of the orbit is approximately 36.18 astronomical units. (An astronomical unit is defined to be the mean distance between the earth and the sun, 93 million miles.) Find a polar equation for the orbit. How close does Halley's comet come to the sun?

*If the earth is used as a reference with a period of 1 year and a distance of 1 astronomical unit, the proportionality constant is 1. For example, since Mars has a mean distance to the sun of $D = 1.523$ AU, its period P is given by $D^3 = P^2$. Thus, the period for Mars is $P = 1.88$ years.

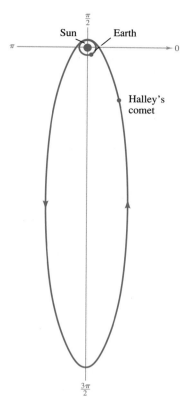

FIGURE 12.48

SOLUTION

Using a vertical axis, we choose an equation of the form $r = ed/(1 + e \sin \theta)$. Since the vertices of the ellipse occur when $\theta = \pi/2$ and $\theta = 3\pi/2$, we can determine the length of the major axis to be the sum of the r-values of the vertices, as shown in Figure 12.48. That is,

$$2a = \frac{0.97d}{1 + 0.97} + \frac{0.97d}{1 - 0.97} \approx 32.83d \approx 36.18.$$

Thus, $d \approx 1.102$ and $ed \approx (0.97)(1.102) \approx 1.069$. Using this value in the equation produces

$$r = \frac{1.069}{1 + 0.97 \sin \theta}$$

where r is measured in astronomical units. To find the closest point to the sun (the focus), we write $c = ea \approx (0.97)(18.09) \approx 17.55$. Since c is the distance between the focus and the center, the closest point is

$$a - c \approx 18.09 - 17.55$$
$$\approx 0.54 \text{ AU}$$
$$\approx 50,000,000 \text{ mi.}$$

Kepler's Second Law states that as a planet moves about the sun, a ray from the sun to the planet sweeps out equal areas in equal times. This law can also be applied to comets or asteroids with elliptical orbits. For example, Figure 12.49 shows the orbit of the asteroid Apollo about the sun. Applying Kepler's Second Law to this asteroid, we know that the closer it is to the sun, the greater its velocity, since a short ray must be moving quickly to sweep out as much area as a long ray.

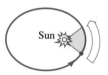

FIGURE 12.49

EXAMPLE 5 An application to elliptical orbits

The asteroid Apollo has a period of 478 earth days, and its orbit is approximated by the ellipse

$$r = \frac{1}{1 + (5/9) \cos \theta} = \frac{9}{9 + 5 \cos \theta}$$

where r is measured in astronomical units. How long does it take Apollo to move from the position given by $\theta = -\pi/2$ to $\theta = \pi/2$, as shown in Figure 12.50?

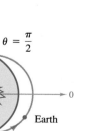

FIGURE 12.50

SOLUTION

We begin by finding the area swept out as θ increases from $-\pi/2$ to $\pi/2$.

$$A = \frac{1}{2}\int_{\alpha}^{\beta} r^2\, d\theta = \frac{1}{2}\int_{-\pi/2}^{\pi/2}\left(\frac{9}{9 + 5\cos\theta}\right)^2 d\theta$$

Using the substitution $u = \tan(\theta/2)$ as discussed in Section 9.6, we can obtain

$$A = \frac{81}{112}\left[\frac{-5\sin\theta}{9 + 5\cos\theta} + \frac{18}{\sqrt{56}}\arctan\frac{\sqrt{56}\tan(\theta/2)}{14}\right]_{-\pi/2}^{\pi/2} \approx 0.90429.$$

Now, since the major axis of the ellipse has length $2a = 81/28$ and the eccentricity is $e = 5/9$, we can determine that $b = a\sqrt{1 - e^2} = 9/\sqrt{56}$. Thus, the area of the ellipse is

$$\text{area of ellipse} = \pi ab = \pi\left(\frac{81}{56}\right)\left(\frac{9}{\sqrt{56}}\right) \approx 5.46507.$$

Since the time required to complete the orbit is 478 days, we can apply Kepler's Second Law to conclude that the time t required to move from the position $\theta = -\pi/2$ to $\theta = \pi/2$ is given by

$$\frac{t}{478} = \frac{\text{area of elliptical segment}}{\text{area of ellipse}} \approx \frac{0.90429}{5.46507}$$

which implies that

$$t = 478\left(\frac{0.90429}{5.46507}\right) \approx 79 \text{ days.}$$

EXERCISES for Section 12.6

In Exercises 1–16, sketch the graph of the equation and identify the curve.

1. $r = \dfrac{2}{1 - \cos\theta}$

2. $r = \dfrac{4}{1 + \sin\theta}$

3. $r = \dfrac{5}{1 + \sin\theta}$

4. $r = \dfrac{6}{1 + \cos\theta}$

5. $r = \dfrac{2}{2 - \cos\theta}$

6. $r = \dfrac{3}{3 + 2\sin\theta}$

7. $r(2 + \sin\theta) = 4$

8. $r(3 - 2\cos\theta) = 6$

9. $r = \dfrac{-1}{1 - \sin\theta}$

10. $r = \dfrac{-3}{2 + 4\sin\theta}$

11. $r = \dfrac{5}{-1 + 2\cos\theta}$

12. $r = \dfrac{3}{-4 + 2\cos\theta}$

13. $r = \dfrac{3}{2 - 6\cos\theta}$

14. $r = \dfrac{4}{1 - 2\cos\theta}$

15. $r = \dfrac{3}{2 + 6\sin\theta}$

16. $r = \dfrac{4}{1 + 2\cos\theta}$

In Exercises 17–28, find a polar equation for the specified conic. In each case consider the focus to be at the pole. (For convenience, the equation for the directrix is given in rectangular form.)

Conic	Eccentricity	Directrix
17. Ellipse	$e = \dfrac{1}{2}$	$y = 1$
18. Ellipse	$e = \dfrac{3}{4}$	$y = -2$
19. Parabola	$e = 1$	$x = -1$
20. Parabola	$e = 1$	$y = 1$
21. Hyperbola	$e = 2$	$x = 1$
22. Hyperbola	$e = \dfrac{3}{2}$	$x = -1$

23. Parabola: vertex at $\left(1, -\dfrac{\pi}{2}\right)$

24. Parabola: directrix is $x = 8$

25. Ellipse: vertices at $\left(6, \dfrac{\pi}{2}\right)$ and $\left(2, \dfrac{3\pi}{2}\right)$

26. Ellipse: vertex at $(2, 0)$, directrix is $x = 6$

27. Hyperbola: vertex at $\left(2, \dfrac{3\pi}{2}\right)$, directrix is $y = -3$

28. Hyperbola: vertices at $(2, 0)$ and $(6, 0)$

29. Show that the polar equation for the ellipse

$$\frac{x^2}{a^2} + \frac{y^2}{b^2} = 1$$

is

$$r^2 = \frac{b^2}{1 - e^2 \cos^2 \theta}.$$

30. Show that the polar equation for the hyperbola

$$\frac{x^2}{a^2} - \frac{y^2}{b^2} = 1$$

is

$$r^2 = \frac{-b^2}{1 - e^2 \cos^2 \theta}.$$

In Exercises 31–34, use the results of Exercises 29 and 30 to find the polar equation of the given conic.

31. Ellipse: focus at $(4, 0)$, vertices at $(5, 0)$, $(5, \pi)$
32. Hyperbola: focus at $(5, 0)$, vertices at $(4, 0)$, $(4, \pi)$
33. $\dfrac{x^2}{9} - \dfrac{y^2}{16} = 1$ **34.** $\dfrac{x^2}{4} + y^2 = 1$

In Exercises 35–38, sketch the graph of the rotated conic.

35. $r = \dfrac{2}{1 - \cos(\theta - \pi/4)}$ (See Exercise 1)

36. $r = \dfrac{4}{1 + \sin(\theta - \pi/3)}$ (See Exercise 2)

37. $r = \dfrac{4}{2 + \sin(\theta + \pi/6)}$ (See Exercise 7)

38. $r = \dfrac{4}{1 + 2\cos(\theta + 2\pi/3)}$ (See Exercise 16)

In Exercises 39 and 40, use Simpson's Rule (with $n = 6$) to find the area of the region bounded by the graph of the given polar equation.

39. $r = \dfrac{3}{2 - \cos \theta}$ **40.** $r = \dfrac{2}{3 - 2\sin \theta}$

41. Show that the graphs of the equations

$$r = \frac{ed}{1 + \sin \theta} \quad \text{and} \quad r = \frac{ed}{1 - \sin \theta}$$

intersect at right angles.

42. Prove that the tangent of the angle ψ $(0 \le \psi \le \pi/2)$ between the radial line and the tangent line at the point (r, θ) on the graph of $r = f(\theta)$ (see figure) is given by

$$\tan \psi = \frac{r}{dr/d\theta}.$$

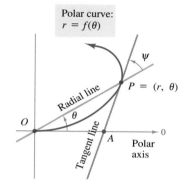

Angle Between Radial
and Tangent Line   FIGURE FOR 42

In Exercises 43–48, use the result of Exercise 42 to find the angle ψ between the radial and tangent lines to the graph at the given point.

Polar equation	Point
43. $r = 2(1 - \cos \theta)$	$\theta = \pi$
44. $r = 3(1 - \cos \theta)$	$\theta = \dfrac{3\pi}{4}$
45. $r = 2 \cos 3\theta$	$\theta = \dfrac{\pi}{6}$
46. $r = 4 \sin 2\theta$	$\theta = \dfrac{\pi}{6}$
47. $r = \dfrac{6}{1 - \cos \theta}$	$\theta = \dfrac{2\pi}{3}$
48. $r = 5$	$\theta = \dfrac{\pi}{6}$

49. The planets travel in elliptical orbits with the sun as a focus. If this focus is at the pole, then the major axis lies on the polar axis, and the length of the major axis is $2a$ (see figure). Show that the polar equation of the orbit is given by

$$r = \frac{(1 - e^2)a}{1 - e \cos \theta}$$

where e is the eccentricity.

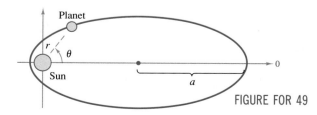

FIGURE FOR 49

50. Use the result of Exercise 49 to show that the minimum distance (*perihelion distance*) from the sun to the planet is $r = a(1 - e)$ and the maximum distance (*aphelion distance*) is $r = a(1 + e)$.

In Exercises 51–53, use the results of Exercises 49 and 50 to find the polar equation of the orbit of the planet and the perihelion and aphelion distances.

51. Earth: $a = 92.957 \times 10^6$ miles
$\qquad e = 0.0167$

52. Pluto: $a = 3.666 \times 10^9$ miles
$\qquad e = 0.2481$

53. Saturn: $a = 1.427 \times 10^9$ kilometers
$\qquad e = 0.0543$

54. Use a computer or graphics calculator to sketch the graph of

$$r = \frac{2e}{1 + e \cos \theta}$$

on the same set of coordinate axes for each of the following eccentricities.

(a) $e = 1$ (b) $e = 0.5$

(c) $e = 0.7$ (d) $e = 1.5$

REVIEW EXERCISES for Chapter 12

In Exercises 1–10, (a) find dy/dx and all points of horizontal tangency, (b) eliminate the parameter where possible, and (c) sketch the curve represented by the parametric equations.

1. $x = 1 + 4t$, $y = 2 - 3t$
2. $x = e^t$, $y = e^{-t}$
3. $x = 3 + 2 \cos \theta$, $y = 2 + 5 \sin \theta$
4. $x = t^2 - 3t + 2$, $y = t^3 - 3t^2 + 2$
5. $x = \dfrac{1}{t}$, $y = 2t + 3$
6. $x = 2t - 1$, $y = \dfrac{1}{t^2 - 2t}$
7. $x = \dfrac{1}{2t + 1}$, $y = \dfrac{2t(t + 1)}{2t + 1}$
8. $x = \cot \theta$, $y = \sin 2\theta$
9. $x = \cos^3 \theta$, $y = 4 \sin^3 \theta$
10. $x = 2\theta - \sin \theta$, $y = 2 - \cos \theta$

In Exercises 11 and 12, find a parametric representation of the conic.

11. Ellipse: center at $(-3, 4)$, horizontal major axis of length 8, and minor axis of length 6
12. Hyperbola: vertices at $(0, \pm 4)$ and foci at $(0, \pm 5)$

13. Eliminate the parameter from

$$x = a(\theta - \sin \theta)$$

and

$$y = a(1 - \cos \theta)$$

to show that the rectangular equation of a cycloid is

$$x = a \arccos \left(\frac{a - y}{a} \right) \pm \sqrt{2ay - y^2}.$$

14. The **involute of a circle** is described by the endpoint P of a string that is held taut as it is unwound from a spool that does not turn (see figure). Show that a parametric representation of the involute is given by

$$x = r(\cos \theta + \theta \sin \theta)$$

and

$$y = r(\sin \theta - \theta \cos \theta).$$

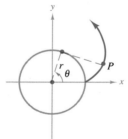

FIGURE FOR 14

In Exercises 15 and 16, find the length of the curve represented by the parametric equations over the given interval.

15. $x = r(\cos \theta + \theta \sin \theta)$, $y = r(\sin \theta - \theta \cos \theta)$,
$\qquad 0 \le \theta \le \pi$
16. $x = 6 \cos \theta$, $y = 6 \sin \theta$, $0 \le \theta \le \pi$

In Exercises 17–32, sketch the graph of the equation.

17. $r = -2(1 + \cos \theta)$ **18.** $r = 3 - 4 \cos \theta$
19. $r = 4 - 3 \cos \theta$ **20.** $r = \cos 5\theta$
21. $r = -3 \cos 2\theta$ **22.** $r^2 = \cos 2\theta$
23. $r = 4$ **24.** $r = 2\theta$

25. $r = -\sec \theta$

26. $r = 3 \csc \theta$

27. $r^2 = 4 \sin^2 2\theta$

28. $r = 2 \sin \theta \cos^2 \theta$

29. $r = \dfrac{2}{1 - \sin \theta}$

30. $r = \dfrac{4}{5 - 3 \cos \theta}$

31. $r = 4 \cos 2\theta \sec \theta$

32. $r = 4(\sec \theta - \cos \theta)$

In Exercises 33–38, convert the given polar equation to rectangular form.

33. $r = 3 \cos \theta$

34. $r = 4 \sec \left(\theta - \dfrac{\pi}{3} \right)$

35. $r = -2(1 + \cos \theta)$

36. $r = 1 + \tan \theta$

37. $r = 4 \cos 2\theta \sec \theta$

38. $\theta = \dfrac{3\pi}{4}$

In Exercises 39–42, convert the given rectangular equation to polar form.

39. $(x^2 + y^2)^2 = ax^2 y$

40. $x^2 + y^2 - 4x = 0$

41. $x^2 + y^2 = a^2 \left(\arctan \dfrac{y}{x} \right)^2$

42. $(x^2 + y^2) \left(\arctan \dfrac{y}{x} \right)^2 = a^2$

In Exercises 43–48, find a polar equation for the given graph.

43. Parabola: focus at the pole, vertex at $(2, \pi)$

44. Ellipse: a focus at the pole, vertices at $(5, 0)$ and $(1, \pi)$

45. Hyperbola: a focus at the pole, vertices at $(1, 0)$ and $(7, 0)$

46. Circle: center $(0, 5)$, passing through the origin

47. Line: intercepts at $(3, 0)$ and $(0, 4)$

48. Line: through the origin, slope $\sqrt{3}$

In Exercises 49 and 50, (a) find the tangents at the pole, (b) find all points of horizontal and vertical tangency, and (c) sketch the graph of the equation.

49. $r = 1 - 2 \cos \theta$

50. $r^2 = 4 \sin 2\theta$

In Exercises 51 and 52, show that the graphs of the given polar equations are orthogonal at the points of intersection.

51. $r = 1 + \cos \theta, r = 1 - \cos \theta$

52. $r = a \sin \theta, r = a \cos \theta$

In Exercises 53–60, find the area of the given region.

53. Interior of $r = 2 + \cos \theta$

54. Interior of $r = 5(1 - \sin \theta)$

55. Interior of $r = \sin \theta \cos^2 \theta$

56. Interior of $r = 4 \sin 3\theta$

57. Interior of $r^2 = a^2 \sin 2\theta$

58. Common interior of $r = a$ and $r^2 = 2a^2 \sin 2\theta$

59. Common interior of $r = 4 \cos \theta$ and $r = 2$

60. Region bounded by the polar axis and $r = e^\theta$ for $0 \le \theta \le \pi$

In Exercises 61 and 62, find the perimeter of the curve represented by the polar equation.

61. $r = a(1 - \cos \theta)$

62. $r = a \cos 2\theta$

63. Find the angle between the circle $r = 3 \sin \theta$ and the limaçon $r = 4 - 5 \sin \theta$ at the point of intersection $(3/2, \pi/6)$.

Chapter 13 Application

Jackie Joyner-Kersee, winner of the heptathlon in the 1988 summer Olympics, is shown throwing a shot-put. The heptathlon is a series of seven events for women: 100-meter hurdles, high jump, shot-put, 200-meter dash, long jump, javelin throw, and 800-meter run. The comparable series for men is called the decathlon and consists of ten events.

Throwing a Shot-Put a Maximum Distance

The path of a shot-put thrown at an angle θ is

$$\mathbf{r}(t) = (v_0 \cos \theta)t\,\mathbf{i} + \left[h + (v_0 \sin \theta)t - \frac{1}{2}gt^2\right]\mathbf{j}$$

where v_0 is the initial speed, h is the initial height, t is the time in seconds, and $g \approx 32$ ft/sec^2 is the acceleration due to gravity. (This formula neglects air resistance.) The shot-put will remain in the air for a total of

$$t = \frac{v_0 \sin \theta + \sqrt{v_0{}^2 \sin^2 \theta + 2gh}}{g} \text{ seconds}$$

and will travel a horizontal distance of

$$\frac{v_0{}^2 \cos \theta}{g}\left(\sin \theta + \sqrt{\sin^2 \theta + \frac{2gh}{v_0{}^2}}\right) \text{ feet.}$$

For fixed values of h and v_0, this distance is a maximum when the angle θ satisfies the equation

$$\cos (2\theta) = \frac{gh}{v_0{}^2 + gh}.$$

For instance, if $h = 7$ feet and $v_0 = 40$ ft/sec, then $\theta \approx 41.5°$.

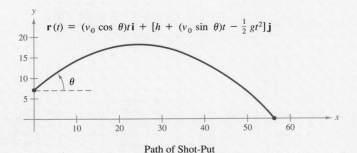

Path of Shot-Put

See Exercise 35, Section 13.4.

Chapter Overview

This chapter begins with a description of vectors in the plane. For instance, a typical vector in the plane can be represented as

$$\mathbf{v} = \langle v_1, v_2 \rangle = v_1\mathbf{i} + v_2\mathbf{j}.$$

Vectors have many applications in geometry, physics, engineering, and other areas such as economics. In this chapter, we concentrate primarily on the geometrical and physical applications of vectors. For instance, vectors can be used to describe forces or the velocity of moving objects.

Section 13.2 continues the study of vectors in the plane by describing an important type of product—the *dot product* of two vectors. The dot product can be used to find the angle between two vectors and to find the projection of one vector onto another vector.

In Section 13.3 we introduce the concept of a **vector-valued function** denoted by

$$\mathbf{r}(t) = f(t)\mathbf{i} + g(t)\mathbf{j}.$$

Vector-valued functions can be used to study curves in the plane. In Section 13.4, we show that these functions can also be used to study the motion of an object along a curve. Specifically, the first derivative of $\mathbf{r}(t)$, $\mathbf{r}'(t) = f'(t)\mathbf{i} + g'(t)\mathbf{j}$, gives the velocity vector for the object at time t, and the second derivative of $\mathbf{r}(t)$, $\mathbf{r}''(t) = f''(t)\mathbf{i} + g''(t)\mathbf{j}$, gives the acceleration vector of the object at time t.

In Section 13.6, we show how to find the arc length of a curve in the plane, and compare the use of an arbitrary parameter with the use of the arc length parameter. Finally, we introduce the concept of **curvature** to measure how much a curve "curves."

Vectors and Curves in the Plane

13

13.1 Vectors in the Plane

Component form of a vector ▪ Length of a vector ▪ Vector operations ▪ Standard unit vectors ▪ Applications of vectors

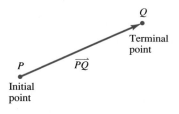

FIGURE 13.1

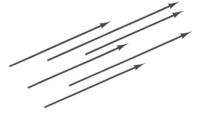

Equivalent Directed
Line Segments

FIGURE 13.2

Many quantities in geometry and physics, such as area, volume, temperature, mass, and time, can be characterized by a single real number scaled to an appropriate unit of measure. We call these **scalar quantities,** and the real number associated with each is called a **scalar.**

Other quantities, such as force and velocity, involve both magnitude and direction and cannot be characterized completely by a single real number. To represent such a quantity, we use a **directed line segment,** as shown in Figure 13.1. The directed line segment $\overrightarrow{PQ}$ has **initial point** P and **terminal point** Q, and we denote its **length** by $\|\overrightarrow{PQ}\|$. Two directed line segments that have the same length and direction are called **equivalent.** For example, the directed line segments in Figure 13.2 are all equivalent. We call the set of all directed line segments that are equivalent to a given directed line segment $\overrightarrow{PQ}$ a **vector in the plane** and write $\mathbf{v} = \overrightarrow{PQ}$.

REMARK We denote vectors by lowercase, boldface letters such as $\mathbf{u}$, $\mathbf{v}$, and $\mathbf{w}$.

Be sure you see that a vector in the plane can be represented by many different directed line segments. This is illustrated in the following example.

EXAMPLE 1 Vector representation by directed line segments

Let $\mathbf{v}$ be represented by the directed line segment from $(0, 0)$ to $(3, 2)$, and let $\mathbf{u}$ be represented by the directed line segment from $(1, 2)$ to $(4, 4)$. Show that $\mathbf{v} = \mathbf{u}$.

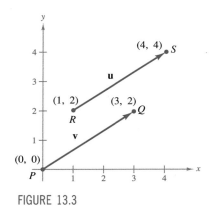

FIGURE 13.3

SOLUTION

Let $P = (0, 0)$, $Q = (3, 2)$, $R = (1, 2)$, and $S = (4, 4)$, as shown in Figure 13.3. We can show that $\overrightarrow{PQ}$ and $\overrightarrow{RS}$ have the *same length* by using the Distance Formula in the plane.

$$\|\overrightarrow{PQ}\| = \sqrt{(3 - 0)^2 + (2 - 0)^2} = \sqrt{13}$$

$$\|\overrightarrow{RS}\| = \sqrt{(4 - 1)^2 + (4 - 2)^2} = \sqrt{13}$$

Moreover, both line segments have the *same direction*, since they both are directed toward the upper right on lines having the same slope.

$$\text{slope of } \overrightarrow{PQ} = \frac{2 - 0}{3 - 0} = \frac{2}{3} \quad \text{and} \quad \text{slope of } \overrightarrow{RS} = \frac{4 - 2}{4 - 1} = \frac{2}{3}$$

Thus, $\overrightarrow{PQ}$ and $\overrightarrow{RS}$ have the same length and direction, and we conclude that $\mathbf{v} = \mathbf{u}$. ▭

The directed line segment whose initial point is the origin is often the most convenient representative of a set of equivalent directed line segments such as those shown in Figure 13.3. We say that this representation of $\mathbf{v}$ is in **standard position.** A directed line segment whose initial point is at the origin can be uniquely represented by the coordinates of its terminal point $Q = (v_1, v_2)$. We call this the **component form of a vector** and write $\mathbf{v} = \langle v_1, v_2 \rangle$.

DEFINITION OF COMPONENT FORM OF A VECTOR IN THE PLANE	If $\mathbf{v}$ is a vector in the plane whose initial point is the origin and whose terminal point is (v_1, v_2), then the **component form of v** is given by $$\mathbf{v} = \langle v_1, v_2 \rangle.$$ The coordinates v_1 and v_2 are called the **components of v.** If both the initial point and the terminal point lie at the origin, then $\mathbf{v}$ is called the **zero vector** and is denoted by $\mathbf{0} = \langle 0, 0 \rangle$.

This definition implies that two vectors $\mathbf{u} = \langle u_1, u_2 \rangle$ and $\mathbf{v} = \langle v_1, v_2 \rangle$ are **equal** if and only if $u_1 = v_1$ and $u_2 = v_2$.

To convert directed line segments to component form or vice versa, we use the following procedures.

1. If $P = (p_1, p_2)$ and $Q = (q_1, q_2)$, then the component form of the vector $\mathbf{v}$ represented by $\overrightarrow{PQ}$ is $\langle v_1, v_2 \rangle = \langle q_1 - p_1, q_2 - p_2 \rangle$. Moreover, the **length** of $\mathbf{v}$ is given by

$$\|\mathbf{v}\| = \sqrt{(q_1 - p_1)^2 + (q_2 - p_2)^2} = \sqrt{v_1^2 + v_2^2}. \qquad \text{Length of a vector}$$

2. If $\mathbf{v} = \langle v_1, v_2 \rangle$, then $\mathbf{v}$ can be represented by the directed line segment, in standard position, from $P = (0, 0)$ to $Q = (v_1, v_2)$.

The length of $\mathbf{v}$ is also called the **norm of v.** If $\|\mathbf{v}\| = 1$, then $\mathbf{v}$ is called a **unit vector.** Moreover, $\|\mathbf{v}\| = 0$ if and only if $\mathbf{v}$ is the zero vector $\mathbf{0}$.

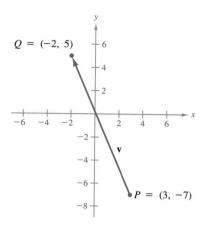

Component form of **v**:
v = ⟨−5, 12⟩

FIGURE 13.4

William Hamilton

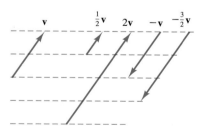

Scalar Multiplication of **v**

FIGURE 13.5

EXAMPLE 2 Finding the component form and length of a vector

Find the component form and length of the vector **v** that has initial point $(3, -7)$ and terminal point $(-2, 5)$.

SOLUTION

We let $P = (3, -7) = (p_1, p_2)$ and $Q = (-2, 5) = (q_1, q_2)$. Then the components of $\mathbf{v} = \langle v_1, v_2 \rangle$ are given by

$$v_1 = q_1 - p_1 = -2 - 3 = -5$$
$$v_2 = q_2 - p_2 = 5 - (-7) = 12.$$

Thus, as shown in Figure 13.4, $\mathbf{v} = \langle -5, 12 \rangle$, and the length of **v** is

$$\|\mathbf{v}\| = \sqrt{(-5)^2 + 12^2} = \sqrt{169} = 13$$

Vector operations

Vectors are relative newcomers to mathematics. Some of the earliest work with vectors was done by the Irish mathematician William Rowan Hamilton (1805–1865). Hamilton spent many years developing a system of vector-like quantities that he called quaternions. This work paved the way for the development of the modern notion of a vector. Although Hamilton was convinced of the benefits of quaternions, the operations he defined did not produce good models for physical phenomena. It wasn't until the latter half of the nineteenth century that the Scottish physicist James Maxwell (1831–1879) restructured Hamilton's quaternions in a form useful for representing physical quantities such as force, velocity, and acceleration.

The two basic vector operations are **vector addition** and **scalar multiplication.** Geometrically, the product of a vector **v** and a scalar k is the vector that is k times as long as **v**. If k is positive, then $k\mathbf{v}$ has the same direction as **v**, and if k is negative, then $k\mathbf{v}$ has the direction opposite that of **v**, as shown in Figure 13.5.

To add two vectors geometrically, we position them (without changing their magnitude or direction) so that the initial point of one coincides with the terminal point of the other. The sum $\mathbf{u} + \mathbf{v}$ (or $\mathbf{v} + \mathbf{u}$), called the **resultant vector,** is formed by joining the initial point of the first vector with the terminal point of the second, as shown in Figure 13.6. The vector $\mathbf{u} + \mathbf{v}$ is the diagonal of a parallelogram having **u** and **v** as its adjacent sides.

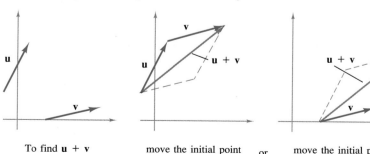

To find **u** + **v**   move the initial point   or   move the initial point
                                of **v** to the terminal   of **u** to the terminal
                                point of **u**   point of **v**.

FIGURE 13.6

Vector addition and scalar multiplication can also be defined using components of vectors.

DEFINITION OF VECTOR ADDITION AND SCALAR MULTIPLICATION	For vectors $\mathbf{u} = \langle u_1, u_2 \rangle$ and $\mathbf{v} = \langle v_1, v_2 \rangle$ and scalar k, we define the following operations. 1. The **vector sum** of $\mathbf{u}$ and $\mathbf{v}$ is the vector $\mathbf{u} + \mathbf{v} = \langle u_1 + v_1, u_2 + v_2 \rangle$. 2. The **scalar multiple** of k and $\mathbf{u}$ is the vector $k\mathbf{u} = \langle ku_1, ku_2 \rangle$. 3. The **negative** of $\mathbf{v}$ is the vector $-\mathbf{v} = (-1)\mathbf{v} = \langle -v_1, -v_2 \rangle$. 4. The **difference** of $\mathbf{u}$ and $\mathbf{v}$ is $\mathbf{u} - \mathbf{v} = \mathbf{u} + (-\mathbf{v}) = \langle u_1 - v_1, u_2 - v_2 \rangle$.

In Figure 13.7, we show that the geometric and algebraic definitions of vector addition and scalar multiplication are the same. Note that to represent $\mathbf{u} - \mathbf{v}$ graphically, we use directed line segments with the *same* initial points. The difference $\mathbf{u} - \mathbf{v}$ is the vector from the terminal point of $\mathbf{v}$ to the terminal point of $\mathbf{u}$.

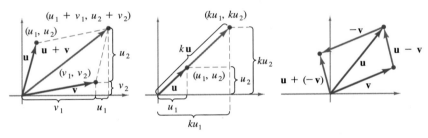

FIGURE 13.7   Vector Addition   Scalar Multiplication   Vector Subtraction

EXAMPLE 3   Vector operations

Given the vectors $\mathbf{v} = \langle -2, 5 \rangle$ and $\mathbf{w} = \langle 3, 4 \rangle$, find the following vectors.

(a) $\dfrac{1}{2}\mathbf{v}$ (b) $\mathbf{w} - \mathbf{v}$ (c) $\mathbf{v} + 2\mathbf{w}$

SOLUTION

(a) Since $\mathbf{v} = \langle -2, 5 \rangle$, we have

$$\frac{1}{2}\mathbf{v} = \left\langle \frac{1}{2}(-2), \frac{1}{2}(5) \right\rangle = \left\langle -1, \frac{5}{2} \right\rangle.$$

(b) $\mathbf{w} - \mathbf{v} = \langle w_1 - v_1, w_2 - v_2 \rangle = \langle 3 - (-2), 4 - 5 \rangle = \langle 5, -1 \rangle$

(c) Since $2\mathbf{w} = \langle 6, 8 \rangle$, it follows that

$$\mathbf{v} + 2\mathbf{w} = \langle -2, 5 \rangle + \langle 6, 8 \rangle = \langle -2 + 6, 5 + 8 \rangle = \langle 4, 13 \rangle. \quad \blacksquare$$

Vector addition and scalar multiplication share many properties of ordinary arithmetic, as shown in the following theorem.

THEOREM 13.1 **PROPERTIES OF VECTOR** **ADDITION AND SCALAR** **MULTIPLICATION**	Let **u**, **v**, and **w** be vectors in the plane, and let c and d be scalars.

1. $\mathbf{u} + \mathbf{v} = \mathbf{v} + \mathbf{u}$ Commutative property
2. $(\mathbf{u} + \mathbf{v}) + \mathbf{w} = \mathbf{u} + (\mathbf{v} + \mathbf{w})$ Associative property
3. $\mathbf{u} + \mathbf{0} = \mathbf{u}$
4. $\mathbf{u} + (-\mathbf{u}) = \mathbf{0}$
5. $c(d\mathbf{u}) = (cd)\mathbf{u}$
6. $(c + d)\mathbf{u} = c\mathbf{u} + d\mathbf{u}$ Distributive property
7. $c(\mathbf{u} + \mathbf{v}) = c\mathbf{u} + c\mathbf{v}$ Distributive property
8. $1(\mathbf{u}) = \mathbf{u},\ 0(\mathbf{u}) = \mathbf{0}$

PROOF We prove only the second and sixth properties and leave the others to you. Note that the proofs rely on the corresponding properties of real numbers. For example, the proof of the *associative property* of vector addition uses the associative property of addition of real numbers, as follows.

$$
\begin{aligned}
(\mathbf{u} + \mathbf{v}) + \mathbf{w} &= [\langle u_1, u_2\rangle + \langle v_1, v_2\rangle] + \langle w_1, w_2\rangle \\
&= \langle u_1 + v_1, u_2 + v_2\rangle + \langle w_1, w_2\rangle \\
&= \langle (u_1 + v_1) + w_1, (u_2 + v_2) + w_2\rangle \\
&= \langle u_1 + (v_1 + w_1), u_2 + (v_2 + w_2)\rangle \\
&= \langle u_1, u_2\rangle + \langle v_1 + w_1, v_2 + w_2\rangle = \mathbf{u} + (\mathbf{v} + \mathbf{w})
\end{aligned}
$$

Similarly, the proof of the following distributive property depends on the distributive property of real numbers.

$$
\begin{aligned}
(c + d)\mathbf{u} &= (c + d)\langle u_1, u_2\rangle \\
&= \langle (c + d)u_1, (c + d)u_2\rangle \\
&= \langle cu_1 + du_1, cu_2 + du_2\rangle \\
&= \langle cu_1, cu_2\rangle + \langle du_1, du_2\rangle = c\mathbf{u} + d\mathbf{u}
\end{aligned}
$$

Emmy Noether

Any set of vectors (with an accompanying set of scalars) that satisfies the eight properties given in Theorem 13.1 is called a **vector space**. The eight properties are called the *vector space axioms*. Thus, this theorem tells us that the set of vectors in the plane (with the set of real numbers) forms a vector space. You will study vector spaces in detail if you take a course in *linear algebra*.

One person who contributed to our knowledge of axiomatic systems was the German mathematician Emmy Noether (1882–1935). Noether is generally recognized as the leading woman mathematician in recent history.

THEOREM 13.2 **LENGTH OF A SCALAR MULTIPLE**	Let **v** be a vector and c be a scalar. Then

$$\|c\mathbf{v}\| = |c|\,\|\mathbf{v}\|$$

where $|c|$ is the absolute value of c.

PROOF Since $c\mathbf{v} = \langle cv_1, cv_2 \rangle$, it follows that

$$\|c\mathbf{v}\| = \|\langle cv_1, cv_2 \rangle\| = \sqrt{(cv_1)^2 + (cv_2)^2} = \sqrt{c^2 v_1^2 + c^2 v_2^2}$$
$$= \sqrt{c^2 (v_1^2 + v_2^2)} = |c|\sqrt{v_1^2 + v_2^2} = |c|\,\|\mathbf{v}\|.$$

In many applications of vectors it is useful to find a unit vector that has the same direction as a given vector. The following theorem gives us a procedure for doing this.

THEOREM 13.3 **UNIT VECTOR IN THE** **DIRECTION OF v**	If $\mathbf{v}$ is a nonzero vector in the plane, then the vector $$\mathbf{u} = \frac{\mathbf{v}}{\|\mathbf{v}\|} = \frac{1}{\|\mathbf{v}\|}\mathbf{v}$$ has length 1 and the same direction as $\mathbf{v}$.

PROOF Since $1/\|\mathbf{v}\|$ is positive and

$$\mathbf{u} = \left(\frac{1}{\|\mathbf{v}\|}\right)\mathbf{v}$$

we conclude that $\mathbf{u}$ has the same direction as $\mathbf{v}$. To see that $\|\mathbf{u}\| = 1$, note that

$$\|\mathbf{u}\| = \left\|\left(\frac{1}{\|\mathbf{v}\|}\right)\mathbf{v}\right\| = \left|\frac{1}{\|\mathbf{v}\|}\right|\|\mathbf{v}\| = \frac{1}{\|\mathbf{v}\|}\|\mathbf{v}\| = 1.$$

In Theorem 13.3, we call $\mathbf{u}$ a **unit vector in the direction of v.** The process of multiplying $\mathbf{v}$ by $1/\|\mathbf{v}\|$ to get a unit vector is called **normalization of v.**

EXAMPLE 4 Finding a unit vector

Find a unit vector in the direction of $\mathbf{v} = \langle -2, 5 \rangle$ and verify that the result has length 1.

SOLUTION

From Theorem 13.3, the unit vector in the direction of $\mathbf{v}$ is

$$\frac{\mathbf{v}}{\|\mathbf{v}\|} = \frac{\langle -2, 5 \rangle}{\sqrt{(-2)^2 + (5)^2}} = \frac{1}{\sqrt{29}}\langle -2, 5 \rangle = \left\langle \frac{-2}{\sqrt{29}}, \frac{5}{\sqrt{29}} \right\rangle.$$

This vector has length 1, since

$$\sqrt{\left(\frac{-2}{\sqrt{29}}\right)^2 + \left(\frac{5}{\sqrt{29}}\right)^2} = \sqrt{\frac{4}{29} + \frac{25}{29}} = \sqrt{\frac{29}{29}} = 1.$$

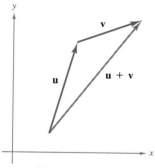

Triangle Inequality   FIGURE 13.8

Generally, the length of the sum of two vectors is not equal to the sum of their lengths. To see this, consider the vectors **u** and **v** as shown in Figure 13.8. By considering **u** and **v** as two sides of a triangle, we see that the length of the third side is $\|\mathbf{u} + \mathbf{v}\|$, and we have

$$\|\mathbf{u} + \mathbf{v}\| \le \|\mathbf{u}\| + \|\mathbf{v}\|.$$

Equality occurs only if the vectors **u** and **v** have the *same* direction. We call this result the **Triangle Inequality** for vectors. (You are asked to prove this in Exercise 40, Section 13.2.)

THEOREM 13.4
TRIANGLE INEQUALITY

If **u** and **v** are vectors in the plane, then

$$\|\mathbf{u} + \mathbf{v}\| \le \|\mathbf{u}\| + \|\mathbf{v}\|.$$

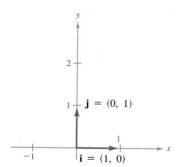

Standard Unit Vectors **i** and **j**

FIGURE 13.9

Standard unit vectors

The unit vectors $\langle 1, 0 \rangle$ and $\langle 0, 1 \rangle$ are called the **standard unit vectors** in the plane and are denoted by

$$\mathbf{i} = \langle 1, 0 \rangle \qquad \text{and} \qquad \mathbf{j} = \langle 0, 1 \rangle$$

as shown in Figure 13.9. These vectors can be used to represent any vector

$$\mathbf{v} = \langle v_1, v_2 \rangle = \langle v_1, 0 \rangle + \langle 0, v_2 \rangle = v_1\langle 1, 0 \rangle + v_2\langle 0, 1 \rangle = v_1\mathbf{i} + v_2\mathbf{j}.$$

We call $\mathbf{v} = v_1\mathbf{i} + v_2\mathbf{j}$ a **linear combination** of **i** and **j**. The scalars v_1 and v_2 are called the **horizontal** and **vertical components of v,** respectively.

EXAMPLE 5 Representing a vector as a linear combination of unit vectors

---

Let **u** be the vector with initial point $(2, -5)$ and terminal point $(-1, 3)$, and let $\mathbf{v} = 2\mathbf{i} - \mathbf{j}$. Write each of the following vectors as a linear combination of the standard unit vectors **i** and **j**.

(a) **u**   (b) $\mathbf{w} = 2\mathbf{u} - 3\mathbf{v}$

SOLUTION

(a) For the vector **u**, we have

$$\mathbf{u} = \langle q_1 - p_1, q_2 - p_2 \rangle = \langle -1 - 2, 3 - (-5) \rangle$$
$$= \langle -3, 8 \rangle = -3\mathbf{i} + 8\mathbf{j}.$$

(b) For the vector **w**, we have

$$\mathbf{w} = 2\mathbf{u} - 3\mathbf{v} = 2(-3\mathbf{i} + 8\mathbf{j}) - 3(2\mathbf{i} - \mathbf{j})$$
$$= -6\mathbf{i} + 16\mathbf{j} - 6\mathbf{i} + 3\mathbf{j} = -12\mathbf{i} + 19\mathbf{j}.$$ ▭

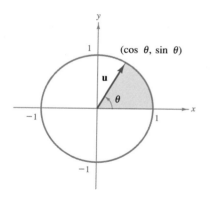

FIGURE 13.10

If **u** is a unit vector such that θ is the angle (measured counterclockwise) from the positive x-axis to **u**, then the terminal point of **u** lies on the unit circle, and we have

$$\mathbf{u} = \langle \cos \theta, \sin \theta \rangle = \cos \theta \, \mathbf{i} + \sin \theta \, \mathbf{j}$$

as shown in Figure 13.10. Moreover, it follows that any other nonzero vector **v** making an angle θ with the positive x-axis has the same direction as **u**, and we can write

$$\mathbf{v} = \|\mathbf{v}\| \langle \cos \theta, \sin \theta \rangle = \|\mathbf{v}\| \cos \theta \, \mathbf{i} + \|\mathbf{v}\| \sin \theta \, \mathbf{j}.$$

For instance, the vector **v** of length 3 making an angle of 30° with the positive x-axis is given by

$$\mathbf{v} = 3 \cos \frac{\pi}{6} \mathbf{i} + 3 \sin \frac{\pi}{6} \mathbf{j}$$

$$= \frac{3\sqrt{3}}{2}\mathbf{i} + \frac{3}{2}\mathbf{j}.$$

Applications of vectors

A host of applications dealing with vectors are found in physics and engineering. We conclude this section by considering two examples.

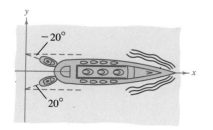

FIGURE 13.11

EXAMPLE 6   An application involving force

Two tugboats are pushing an ocean liner, as shown in Figure 13.11. Each boat is exerting a force of 400 pounds. What is the resultant force on the ocean liner?

SOLUTION

From Figure 13.11, we represent the forces exerted by the first and second tugboats as

$$\mathbf{F}_1 = 400 \langle \cos 20°, \sin 20° \rangle = 400 \cos (20°) \, \mathbf{i} + 400 \sin (20°) \, \mathbf{j}$$
$$\mathbf{F}_2 = 400 \langle \cos (-20°), \sin (-20°) \rangle = 400 \cos (20°) \, \mathbf{i} - 400 \sin (20°) \, \mathbf{j}.$$

To obtain the resultant force on the ocean liner, we add these two forces.

$\mathbf{F} = \mathbf{F}_1 + \mathbf{F}_2$
$\quad = 400 \cos (20°) \, \mathbf{i} + 400 \sin (20°) \, \mathbf{j} + 400 \cos (20°) \, \mathbf{i} - 400 \sin (20°) \, \mathbf{j}$
$\quad = 800 \cos (20°) \, \mathbf{i} \approx 752\mathbf{i}$

Thus, the resultant force on the ocean liner is approximately 752 pounds in the direction of the positive x-axis.

EXAMPLE 7   An application involving velocity

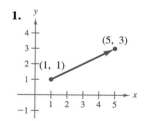
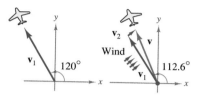

FIGURE 13.12

An airplane is traveling at a fixed altitude with a negligible wind factor. The plane is headed N 30° W at a speed of 500 miles per hour, as shown in Figure 13.12. As the plane reaches a certain point, it encounters wind with a velocity of 70 miles per hour in the direction E 45° N. What is the resultant speed and direction of the plane?

SOLUTION

Using Figure 13.12, we can represent the velocity of the plane by the vector

$$\mathbf{v}_1 = 500\langle\cos(120°), \sin(120°)\rangle = 500\cos(120°)\,\mathbf{i} + 500\sin(120°)\,\mathbf{j}.$$

The velocity of the wind is represented by the vector

$$\mathbf{v}_2 = 70\langle\cos(45°), \sin(45°)\rangle = 70\cos(45°)\,\mathbf{i} + 70\sin(45°)\,\mathbf{j}.$$

The resultant velocity of the plane is

$$\begin{aligned}
\mathbf{v} &= \mathbf{v}_1 + \mathbf{v}_2 \\
&= 500\cos(120°)\,\mathbf{i} + 500\sin(120°)\,\mathbf{j} + 70\cos(45°)\,\mathbf{i} + 70\sin(45°)\,\mathbf{j} \\
&= [500\cos(120°) + 70\cos(45°)]\,\mathbf{i} + [500\sin(120°) + 70\sin(45°)]\,\mathbf{j} \\
&\approx -200.5\mathbf{i} + 482.5\mathbf{j}.
\end{aligned}$$

Now, to find the speed and direction, we write $\mathbf{v} = \|\mathbf{v}\|(\cos\theta\,\mathbf{i} + \sin\theta\,\mathbf{j})$. Since

$$\|\mathbf{v}\| \approx \sqrt{(-200.5)^2 + (482.5)^2} \approx 522.5$$

we have

$$\mathbf{v} \approx 522.5\left[\frac{-200.5}{522.5}\mathbf{i} + \frac{482.5}{522.5}\mathbf{j}\right] \approx 522.5[-0.3837\mathbf{i} + 0.9234\mathbf{j}]$$
$$\approx 522.5[\cos(112.6°)\,\mathbf{i} + \sin(112.6°)\,\mathbf{j}].$$

Therefore, the speed of the plane, as altered by the wind, is approximately 522.5 miles per hour in a flight path that makes an angle of 112.6° with the positive *x*-axis. ▭

EXERCISES for Section 13.1

In Exercises 1–4, (a) find the component form of the vector **v** and (b) sketch the vector with its initial point at the origin.

1.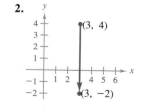

2.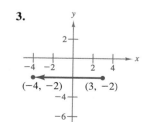

3.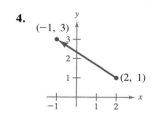

4.

In Exercises 5–12, the initial and terminal points of a vector **v** are given. (a) Sketch the given directed line segment, (b) write the vector in component form, and (c) sketch the vector with its initial point at the origin.

Initial Point	Terminal Point
5. $(1, 2)$	$(5, 5)$
6. $(3, -5)$	$(4, 7)$
7. $(10, 2)$	$(6, -1)$
8. $(0, -4)$	$(-5, -1)$
9. $(6, 2)$	$(6, 6)$
10. $(7, -1)$	$(-3, -1)$
11. $\left(\frac{3}{2}, \frac{4}{3}\right)$	$\left(\frac{1}{2}, 3\right)$
12. $(0.12, 0.60)$	$(0.84, 1.25)$

In Exercises 13 and 14, sketch the scalar multiple of **v**.

13. $\mathbf{v} = \langle 2, 3 \rangle$

    (a) $2\mathbf{v}$ (b) $-3\mathbf{v}$

    (c) $\frac{7}{2}\mathbf{v}$ (d) $\frac{2}{3}\mathbf{v}$

14. $\mathbf{v} = \langle -1, 5 \rangle$

    (a) $4\mathbf{v}$ (b) $-\frac{1}{2}\mathbf{v}$

    (c) $0\mathbf{v}$ (d) $-6\mathbf{v}$

In Exercises 15–20, find the vector **v** and illustrate the indicated vector operations geometrically, where $\mathbf{u} = \langle 2, -1 \rangle$ and $\mathbf{w} = \langle 1, 2 \rangle$.

15. $\mathbf{v} = \frac{3}{2}\mathbf{u}$ **16.** $\mathbf{v} = \mathbf{u} + \mathbf{w}$

17. $\mathbf{v} = \mathbf{u} + 2\mathbf{w}$ **18.** $\mathbf{v} = -\mathbf{u} + \mathbf{w}$

19. $\mathbf{v} = \frac{1}{2}(3\mathbf{u} + \mathbf{w})$ **20.** $\mathbf{v} = \mathbf{u} - 2\mathbf{w}$

In Exercises 21–26, find a and b such that $\mathbf{v} = a\mathbf{u} + b\mathbf{w}$, where $\mathbf{u} = \langle 1, 2 \rangle$ and $\mathbf{w} = \langle 1, -1 \rangle$.

21. $\mathbf{v} = \langle 2, 1 \rangle$ **22.** $\mathbf{v} = \langle 0, 3 \rangle$
23. $\mathbf{v} = \langle 3, 0 \rangle$ **24.** $\mathbf{v} = \langle 3, 3 \rangle$
25. $\mathbf{v} = \langle 1, 1 \rangle$ **26.** $\mathbf{v} = \langle -1, 7 \rangle$

In Exercises 27 and 28, the vector **v** and its initial point are given. Find the terminal point.

27. $\mathbf{v} = \langle -1, 3 \rangle$, initial point $(4, 2)$
28. $\mathbf{v} = \langle 4, -9 \rangle$, initial point $(3, 2)$

In Exercises 29–34, find the magnitude of **v**.

29. $\mathbf{v} = \langle 4, 3 \rangle$ **30.** $\mathbf{v} = \langle 12, -5 \rangle$
31. $\mathbf{v} = 6\mathbf{i} - 5\mathbf{j}$ **32.** $\mathbf{v} = -10\mathbf{i} + 3\mathbf{j}$
33. $\mathbf{v} = 4\mathbf{j}$ **34.** $\mathbf{v} = \mathbf{i} - \mathbf{j}$

In Exercises 35 and 36, find the following.

  (a) $\|\mathbf{u}\|$ (b) $\|\mathbf{v}\|$ (c) $\|\mathbf{u} + \mathbf{v}\|$

  (d) $\left\|\dfrac{\mathbf{u}}{\|\mathbf{u}\|}\right\|$ (e) $\left\|\dfrac{\mathbf{v}}{\|\mathbf{v}\|}\right\|$ (f) $\left\|\dfrac{\mathbf{u} + \mathbf{v}}{\|\mathbf{u} + \mathbf{v}\|}\right\|$

35. $\mathbf{u} = \left\langle 1, \frac{1}{2} \right\rangle$, $\mathbf{v} = \langle 2, 3 \rangle$
36. $\mathbf{u} = \langle 2, -4 \rangle$, $\mathbf{v} = \langle 5, 5 \rangle$

In Exercises 37 and 38, demonstrate the triangle inequality using the vectors **u** and **v**.

37. $\mathbf{u} = \langle 2, 1 \rangle$, $\mathbf{v} = \langle 5, 4 \rangle$
38. $\mathbf{u} = \langle -3, 2 \rangle$, $\mathbf{v} = \langle 1, -2 \rangle$

In Exercises 39–42, find the vector **v** with the given magnitude and the same direction as **u**.

	Magnitude	Direction
39.	$\|\mathbf{v}\| = 4$	$\mathbf{u} = \langle 1, 1 \rangle$
40.	$\|\mathbf{v}\| = 4$	$\mathbf{u} = \langle -1, 1 \rangle$
41.	$\|\mathbf{v}\| = 2$	$\mathbf{u} = \langle \sqrt{3}, 3 \rangle$
42.	$\|\mathbf{v}\| = 3$	$\mathbf{u} = \langle 0, 3 \rangle$

In Exercises 43–46, find a unit vector (a) parallel to and (b) normal to the graph of $f(x)$ at the indicated point.

	Graph	Point
43.	$f(x) = x^3$	$(1, 1)$
44.	$f(x) = x^3$	$(-2, -8)$
45.	$f(x) = \sqrt{25 - x^2}$	$(3, 4)$
46.	$f(x) = \tan x$	$\left(\frac{\pi}{4}, 1\right)$

In Exercises 47–50, find the component form of **v** given its magnitude and the angle it makes with the positive x-axis.

	Magnitude	Angle
47.	$\|\mathbf{v}\| = 3$	$\theta = 0°$
48.	$\|\mathbf{v}\| = 1$	$\theta = 45°$
49.	$\|\mathbf{v}\| = 2$	$\theta = 150°$
50.	$\|\mathbf{v}\| = 1$	$\theta = 3.5°$

In Exercises 51–54, find the component form of $\mathbf{u} + \mathbf{v}$ given the magnitudes of **u** and **v** and the angles **u** and **v** make with the positive x-axis.

51. $\|\mathbf{u}\| = 1$, $\theta = 0°$; $\|\mathbf{v}\| = 3$, $\theta = 45°$
52. $\|\mathbf{u}\| = 4$, $\theta = 0°$; $\|\mathbf{v}\| = 2$, $\theta = 60°$
53. $\|\mathbf{u}\| = 2$, $\theta = 4$; $\|\mathbf{v}\| = 1$, $\theta = 2$
54. $\|\mathbf{u}\| = 5$, $\theta = -0.5$; $\|\mathbf{v}\| = 5$, $\theta = 0.5$

In Exercises 55 and 56, find the component form of **v** given the magnitudes of **u** and **u** + **v** and the angles **u** and **u** + **v** make with the positive *x*-axis.

55. $\|\mathbf{u}\| = 1$, $\theta = 45°$; $\|\mathbf{u} + \mathbf{v}\| = \sqrt{2}$, $\theta = 90°$
56. $\|\mathbf{u}\| = 4$, $\theta = 30°$; $\|\mathbf{u} + \mathbf{v}\| = 6$, $\theta = 120°$

57. A force of 150 pounds in a direction 30° above the horizontal is applied to a bolt. Find the horizontal and vertical components of the force (see figure).

58. A 25-pound weight is suspended from the ceiling by a rope 5 feet long. Determine the magnitude of the horizontal force required to hold the weight 1 foot from the vertical (see figure).

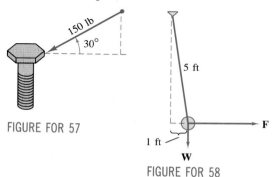

FIGURE FOR 57

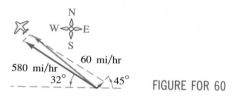

FIGURE FOR 60

61. A plane flies at a constant ground speed of 450 miles per hour due east and encounters a 50 mile per hour wind from the northwest. Find the air speed and compass direction that will allow the plane to maintain its ground speed and eastward direction.

62. A ball is thrown into the air with an initial velocity of 80 feet per second and at an angle of 50° with the horizontal. Find the vertical and horizontal components of the initial velocity.

63. Three vertices of a parallelogram are (1, 2), (3, 1), and (8, 4). Find the three possible fourth vertices (see figure).

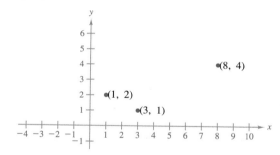

FIGURE FOR 63

64. Use vectors to find the points of trisection of the line segment with endpoints (1, 2) and (7, 5).

65. Prove that the vectors

$$\mathbf{u} = (\cos \theta)\mathbf{i} - (\sin \theta)\mathbf{j}$$

and

$$\mathbf{v} = (\sin \theta)\mathbf{i} + (\cos \theta)\mathbf{j}$$

are unit vectors for any angle θ.

66. Using vectors, prove that the line segment joining the midpoints of two sides of a triangle is parallel to and one-half the length of the third side.

67. Using vectors, prove that the diagonals of a parallelogram bisect each other.

FIGURE FOR 58

59. To carry a 100-pound cylindrical weight, two men lift on the ends of short ropes tied to an eyelet on the top center of the cylinder. If one rope makes a 20° angle away from the vertical and the other a 30° angle, find the following.
(a) the tension in each rope if the resultant force is vertical
(b) the vertical component of each man's force (see figure)

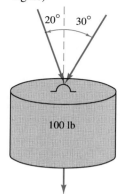

FIGURE FOR 59

60. An airplane is headed 32° north of west. Its speed with respect to the air is 580 miles per hour. The wind at the plane's altitude is from the southwest at 60 miles per hour. What is the true direction of the plane, and what is its speed with respect to the ground? (See figure.)

13.2 The Dot Product of Two Vectors

The dot product ▪ Angle between two vectors ▪ Projections and vector components ▪ Applications

So far we have studied two operations with vectors—vector addition and multiplication by a scalar—each of which yields another vector. In this section we introduce a third vector operation, called the **dot product.** This product yields a scalar, rather than a vector.

DEFINITION OF DOT PRODUCT

The **dot product** of $\mathbf{u} = \langle u_1, u_2 \rangle$ and $\mathbf{v} = \langle v_1, v_2 \rangle$ is

$$\mathbf{u} \cdot \mathbf{v} = u_1 v_1 + u_2 v_2.$$

REMARK The dot product of two vectors is also called the **inner product** (or scalar product) of two vectors.

The properties listed in the next theorem follow readily from the definition of dot product.

**THEOREM 13.5
PROPERTIES OF THE
DOT PRODUCT**

If $\mathbf{u}$, $\mathbf{v}$, and $\mathbf{w}$ are vectors in the plane and c is a scalar, then the following properties are true.

1. $\mathbf{u} \cdot \mathbf{v} = \mathbf{v} \cdot \mathbf{u}$ Commutative property
2. $\mathbf{u} \cdot (\mathbf{v} + \mathbf{w}) = \mathbf{u} \cdot \mathbf{v} + \mathbf{u} \cdot \mathbf{w}$ Distributive property
3. $c(\mathbf{u} \cdot \mathbf{v}) = (c\mathbf{u}) \cdot \mathbf{v} = \mathbf{u} \cdot (c\mathbf{v})$
4. $\mathbf{0} \cdot \mathbf{v} = 0$
5. $\mathbf{v} \cdot \mathbf{v} = \|\mathbf{v}\|^2$

PROOF We prove the first and fifth properties and leave the remaining proofs as an exercise (see Exercise 41). To prove the first property, let $\mathbf{u} = \langle u_1, u_2 \rangle$ and $\mathbf{v} = \langle v_1, v_2 \rangle$. Then

$$\mathbf{u} \cdot \mathbf{v} = u_1 v_1 + u_2 v_2$$
$$= v_1 u_1 + v_2 u_2$$
$$= \mathbf{v} \cdot \mathbf{u}.$$

For the fifth property, we let $\mathbf{v} = \langle v_1, v_2 \rangle$. Then

$$\mathbf{v} \cdot \mathbf{v} = v_1^2 + v_2^2$$
$$= (\sqrt{v_1^2 + v_2^2})^2$$
$$= \|\mathbf{v}\|^2.$$

EXAMPLE 1   Finding dot products

Given $\mathbf{u} = \langle 2, -2 \rangle$, $\mathbf{v} = \langle 5, 8 \rangle$, and $\mathbf{w} = \langle -4, 3 \rangle$, find each of the following.

(a) $\mathbf{u} \cdot \mathbf{v}$ (b) $(\mathbf{u} \cdot \mathbf{v})\mathbf{w}$ (c) $\mathbf{u} \cdot (2\mathbf{v})$ (d) $\|\mathbf{w}\|^2$

SOLUTION

(a) $\mathbf{u} \cdot \mathbf{v} = \langle 2, -2 \rangle \cdot \langle 5, 8 \rangle = 2(5) + (-2)(8) = -6$
(b) $(\mathbf{u} \cdot \mathbf{v})\mathbf{w} = -6\langle -4, 3 \rangle = \langle 24, -18 \rangle$
(c) $\mathbf{u} \cdot (2\mathbf{v}) = 2(\mathbf{u} \cdot \mathbf{v}) = 2(-6) = -12$
(d) $\|\mathbf{w}\|^2 = \mathbf{w} \cdot \mathbf{w} = \langle -4, 3 \rangle \cdot \langle -4, 3 \rangle = (-4)(-4) + (3)(3) = 25$ ▭

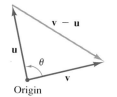

Angle Between Two Vectors

FIGURE 13.13

The **angle between two nonzero vectors** is the angle θ, $0 \le \theta \le \pi$, between their respective standard position vectors, as shown in Figure 13.13. The next theorem shows how to find this angle using the dot product. (Note that we do not define the angle between the zero vector and another vector.)

THEOREM 13.6
ANGLE BETWEEN TWO VECTORS

If θ is the angle between two nonzero vectors $\mathbf{u}$ and $\mathbf{v}$, then

$$\cos \theta = \frac{\mathbf{u} \cdot \mathbf{v}}{\|\mathbf{u}\| \, \|\mathbf{v}\|}.$$

PROOF   Consider the triangle determined by vectors $\mathbf{u}$, $\mathbf{v}$, and $\mathbf{v} - \mathbf{u}$, as shown in Figure 13.13. By the Law of Cosines, we have

$$\|\mathbf{v} - \mathbf{u}\|^2 = \|\mathbf{u}\|^2 + \|\mathbf{v}\|^2 - 2\|\mathbf{u}\| \, \|\mathbf{v}\| \cos \theta.$$

Using the properties of the dot product, we can rewrite the left side as

$$\|\mathbf{v} - \mathbf{u}\|^2 = (\mathbf{v} - \mathbf{u}) \cdot (\mathbf{v} - \mathbf{u}) = (\mathbf{v} - \mathbf{u}) \cdot \mathbf{v} - (\mathbf{v} - \mathbf{u}) \cdot \mathbf{u}$$
$$= \mathbf{v} \cdot \mathbf{v} - \mathbf{u} \cdot \mathbf{v} - \mathbf{v} \cdot \mathbf{u} + \mathbf{u} \cdot \mathbf{u} = \|\mathbf{v}\|^2 - 2\mathbf{u} \cdot \mathbf{v} + \|\mathbf{u}\|^2$$

and substitution back into the Law of Cosines yields

$$\|\mathbf{v}\|^2 - 2\mathbf{u} \cdot \mathbf{v} + \|\mathbf{u}\|^2 = \|\mathbf{u}\|^2 + \|\mathbf{v}\|^2 - 2\|\mathbf{u}\| \, \|\mathbf{v}\| \cos \theta$$
$$-2\mathbf{u} \cdot \mathbf{v} = -2\|\mathbf{u}\| \, \|\mathbf{v}\| \cos \theta$$
$$\cos \theta = \frac{\mathbf{u} \cdot \mathbf{v}}{\|\mathbf{u}\| \, \|\mathbf{v}\|}.$$

If the angle between two vectors is known, then rewriting Theorem 13.6 in the form

$$\mathbf{u} \cdot \mathbf{v} = \|\mathbf{u}\| \, \|\mathbf{v}\| \cos \theta$$

produces an alternate way to calculate the dot product. This form also shows us that because $\|\mathbf{u}\|$ and $\|\mathbf{v}\|$ are always positive, $\mathbf{u} \cdot \mathbf{v}$ and $\cos \theta$ will always have the same sign.

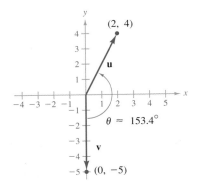

FIGURE 13.14

EXAMPLE 2 Finding the angle between two vectors

Find the angle θ between the vectors $\mathbf{u} = \langle 2, 4 \rangle$ and $\mathbf{v} = \langle 0, -5 \rangle$.

SOLUTION

Using the formula for the angle between two vectors, we have

$$\cos \theta = \frac{\mathbf{u} \cdot \mathbf{v}}{\|\mathbf{u}\| \, \|\mathbf{v}\|} = \frac{0 - 20}{\sqrt{20}\sqrt{25}} = \frac{-20}{10\sqrt{5}} = -\frac{2}{\sqrt{5}}.$$

Therefore, the angle between $\mathbf{u}$ and $\mathbf{v}$ is

$$\theta = \arccos\left(-\frac{2}{\sqrt{5}}\right) \approx 2.68 \text{ radians} \approx 153.4°.$$

(See Figure 13.14.)

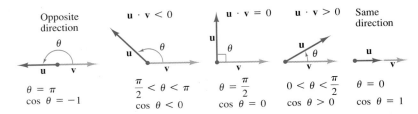

FIGURE 13.15

Figure 13.15 shows the five possible orientations of two vectors.

If the angle between two vectors is $\theta = \pi/2$, then we call the two vectors **orthogonal** (perpendicular). Since $\cos(\pi/2) = 0$, we have the following definition.

DEFINITION OF ORTHOGONAL VECTORS	The vectors $\mathbf{u}$ and $\mathbf{v}$ are **orthogonal** if $$\mathbf{u} \cdot \mathbf{v} = 0.$$ We consider the zero vector to be orthogonal to every vector $\mathbf{u}$ since $\mathbf{0} \cdot \mathbf{u} = 0$.

We say that two nonzero vectors are **parallel** if they have the same or opposite directions. From their coordinate representations, it is easy to determine whether two vectors are parallel. Specifically, two nonzero vectors $\mathbf{u}$ and $\mathbf{v}$ have the *same direction* if and only if $\mathbf{u} = k\mathbf{v}$, where $k > 0$. Similarly, $\mathbf{u}$ and $\mathbf{v}$ have *opposite directions* if and only if $\mathbf{u} = k\mathbf{v}$, where $k < 0$.

EXAMPLE 3 Orthogonal and parallel vectors

(a) Determine whether $\mathbf{u} = \langle 3, -1 \rangle$ and $\mathbf{v} = \langle 2, 6 \rangle$ are orthogonal, parallel, or neither.
(b) Determine whether $\mathbf{u} = \langle -1, 2 \rangle$ and $\mathbf{v} = \langle 2, -4 \rangle$ are orthogonal, parallel, or neither.

SOLUTION

(a) Since

$$\cos \theta = \frac{\mathbf{u} \cdot \mathbf{v}}{\|\mathbf{u}\| \|\mathbf{v}\|} = \frac{6 - 6}{\sqrt{10}\sqrt{40}} = \frac{0}{20} = 0$$

it follows that $\mathbf{u} \cdot \mathbf{v} = 0$, which means that $\mathbf{u}$ and $\mathbf{v}$ are *orthogonal*. (See Figure 13.16(a).)

(b) The cosine of the angle between $\mathbf{u}$ and $\mathbf{v}$ is

$$\cos \theta = \frac{\mathbf{u} \cdot \mathbf{v}}{\|\mathbf{u}\| \|\mathbf{v}\|} = \frac{-2 - 8}{\sqrt{5}\sqrt{20}} = \frac{-10}{\sqrt{100}} = -1.$$

Therefore, the angle between $\mathbf{u}$ and $\mathbf{v}$ is $\theta = \pi$, and $\mathbf{u}$ and $\mathbf{v}$ are *parallel*. (See Figure 13.16(b).)

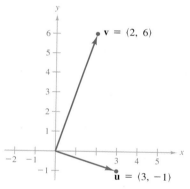

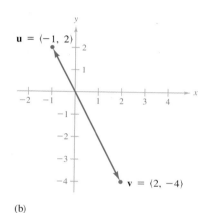

FIGURE 13.16   (a)   (b)

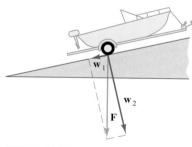

FIGURE 13.17

Projections and vector components

We already have seen applications in which two vectors are added to produce a resultant vector. Many applications in physics and engineering require the reverse problem—decomposing a given vector into the sum of two **vector components.** To see the usefulness of this procedure, we look at a physical example.

Consider a boat on an inclined ramp, as shown in Figure 13.17. The force $\mathbf{F}$ due to gravity pulls the boat *down* the ramp and *against* the ramp. These two forces, $\mathbf{w}_1$ and $\mathbf{w}_2$, are orthogonal, and we call them the vector components of $\mathbf{F}$ and write

$$\mathbf{F} = \mathbf{w}_1 + \mathbf{w}_2.$$

The forces $\mathbf{w}_1$ and $\mathbf{w}_2$ help us analyze the effect of gravity on the boat. For example, $\mathbf{w}_1$ indicates the force necessary to keep the boat from rolling down the ramp. On the other hand, $\mathbf{w}_2$ indicates the force that the tires must withstand.

DEFINITION OF PROJECTION AND VECTOR COMPONENTS

Let $\mathbf{u}$ and $\mathbf{v}$ be nonzero vectors. Moreover, let $\mathbf{u} = \mathbf{w}_1 + \mathbf{w}_2$, where $\mathbf{w}_1$ is parallel to $\mathbf{v}$ and $\mathbf{w}_2$ is orthogonal to $\mathbf{v}$, as shown in Figure 13.18.

1. $\mathbf{w}_1$ is called the **projection of u onto v** or the **vector component of u along v**, and is denoted by $\mathbf{w}_1 = \text{proj}_{\mathbf{v}}\mathbf{u}$.
2. $\mathbf{w}_2 = \mathbf{u} - \mathbf{w}_1$ is called the **vector component of u orthogonal to v.**

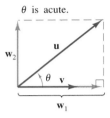
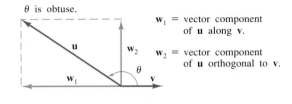

θ is acute.　　　　θ is obtuse.

$\mathbf{w}_1$ = vector component of $\mathbf{u}$ along $\mathbf{v}$.

$\mathbf{w}_2$ = vector component of $\mathbf{u}$ orthogonal to $\mathbf{v}$.

FIGURE 13.18　　　　$\mathbf{w}_1 = \text{proj}_{\mathbf{v}}\,\mathbf{u} = \text{projection of } \mathbf{u} \text{ onto } \mathbf{v}.$

From this definition we see that it is easy to find the vector component $\mathbf{w}_2$ once we have found the projection of $\mathbf{u}$ onto $\mathbf{v}$. To find the projection, we use the dot product, as indicated in the next theorem.

THEOREM 13.7 PROJECTION USING THE DOT PRODUCT

If $\mathbf{u}$ and $\mathbf{v}$ are nonzero vectors, then the projection of $\mathbf{u}$ onto $\mathbf{v}$ is given by

$$\text{proj}_{\mathbf{v}}\mathbf{u} = \left(\frac{\mathbf{u} \cdot \mathbf{v}}{\|\mathbf{v}\|^2}\right)\mathbf{v}.$$

PROOF　From Figure 13.18, we let $\mathbf{w}_1 = \text{proj}_{\mathbf{v}}\mathbf{u}$. Since $\mathbf{w}_1$ is a scalar multiple of $\mathbf{v}$, we can write

$$\mathbf{u} = \mathbf{w}_1 + \mathbf{w}_2 = c\mathbf{v} + \mathbf{w}_2.$$

Now, taking the dot product of both sides with $\mathbf{v}$, we have

$$\mathbf{u} \cdot \mathbf{v} = (c\mathbf{v} + \mathbf{w}_2) \cdot \mathbf{v} = c\mathbf{v} \cdot \mathbf{v} + \mathbf{w}_2 \cdot \mathbf{v} = c\|\mathbf{v}\|^2 + \mathbf{w}_2 \cdot \mathbf{v}.$$

Since $\mathbf{w}_2$ and $\mathbf{v}$ are orthogonal, $\mathbf{w}_2 \cdot \mathbf{v} = 0$ and we have

$$\mathbf{u} \cdot \mathbf{v} = c\|\mathbf{v}\|^2 \quad \Longrightarrow \quad c = \frac{\mathbf{u} \cdot \mathbf{v}}{\|\mathbf{v}\|^2}.$$

Therefore, we can write $\mathbf{w}_1$ as

$$\mathbf{w}_1 = \left(\frac{\mathbf{u} \cdot \mathbf{v}}{\|\mathbf{v}\|^2}\right)\mathbf{v}.$$

The projection of $\mathbf{u}$ onto $\mathbf{v}$ can be written as a scalar multiple of a unit vector in the direction of $\mathbf{v}$. That is,

$$\left(\frac{\mathbf{u} \cdot \mathbf{v}}{\|\mathbf{v}\|^2}\right)\mathbf{v} = \left(\frac{\mathbf{u} \cdot \mathbf{v}}{\|\mathbf{v}\|}\right)\frac{\mathbf{v}}{\|\mathbf{v}\|} = (k)\frac{\mathbf{v}}{\|\mathbf{v}\|}.$$

We call k the **component of u in the direction of v.** Thus,

$$k = \frac{\mathbf{u} \cdot \mathbf{v}}{\|\mathbf{v}\|} = \|\mathbf{u}\| \cos \theta. \qquad \text{Component of } \mathbf{u} \text{ in the direction of } \mathbf{v}$$

REMARK   Be sure you see the distinction between the terms component and vector component. For example, using the standard unit vectors with $\mathbf{u} = u_1 \mathbf{i} + u_2 \mathbf{j}$, we have

u_1 is the *component* of $\mathbf{u}$ in the direction of $\mathbf{i}$

and

$u_1 \mathbf{i}$ is the *vector component* in the direction of $\mathbf{i}$.

EXAMPLE 4   Decomposing a vector into vector components

Find the projection of $\mathbf{u}$ onto $\mathbf{v}$ and the vector component of $\mathbf{u}$ orthogonal to $\mathbf{v}$ for the vectors $\mathbf{u} = 3\mathbf{i} - 5\mathbf{j}$ and $\mathbf{v} = 6\mathbf{i} + 2\mathbf{j}$.

SOLUTION

The projection of $\mathbf{u}$ onto $\mathbf{v}$ is

$$\mathbf{w}_1 = \left(\frac{\mathbf{u} \cdot \mathbf{v}}{\|\mathbf{v}\|^2}\right)\mathbf{v} = \left(\frac{8}{40}\right)(6\mathbf{i} + 2\mathbf{j}) = \frac{6}{5}\mathbf{i} + \frac{2}{5}\mathbf{j}.$$

(See Figure 13.19.) The vector component of $\mathbf{u}$ orthogonal to $\mathbf{v}$ is the vector

$$\mathbf{w}_2 = \mathbf{u} - \mathbf{w}_1 = (3\mathbf{i} - 5\mathbf{j}) - \left(\frac{6}{5}\mathbf{i} + \frac{2}{5}\mathbf{j}\right) = \frac{9}{5}\mathbf{i} - \frac{27}{5}\mathbf{j}. \qquad \blacksquare$$

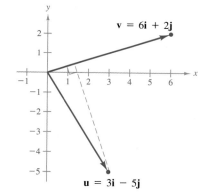

FIGURE 13.19

Now let's return to the problem involving the boat and the ramp to see how we can use vector components in a physical problem.

EXAMPLE 5   An application using a vector component

A 600-pound boat sits on a ramp inclined at 30°, as shown in Figure 13.20. What force is required to keep the boat from rolling down the ramp?

SOLUTION

Since the force due to gravity is vertical and downward, we represent the gravitational force by the vector

$$\mathbf{F} = -600\mathbf{j}. \qquad \text{Force due to gravity}$$

To find the force required to keep the boat from rolling down the ramp, we project $\mathbf{F}$ onto a unit vector $\mathbf{v}$ in the direction of the ramp, as follows.

$$\mathbf{v} = \cos 30° \, \mathbf{i} + \sin 30° \, \mathbf{j} = \frac{\sqrt{3}}{2}\mathbf{i} + \frac{1}{2}\mathbf{j} \qquad \text{Unit vector along ramp}$$

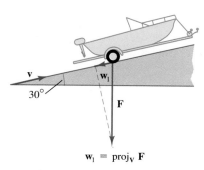

FIGURE 13.20

Therefore, the projection of **F** onto **v** is given by

$$\mathbf{w}_1 = \text{proj}_\mathbf{v} \mathbf{F} = \left(\frac{\mathbf{F} \cdot \mathbf{v}}{\|\mathbf{v}\|^2} \right) \mathbf{v} = (\mathbf{F} \cdot \mathbf{v}) \mathbf{v} = (-600) \left(\frac{1}{2} \right) \mathbf{v}$$

$$= -300 \left(\frac{\sqrt{3}}{2} \mathbf{i} + \frac{1}{2} \mathbf{j} \right).$$

The magnitude of this force is 300, and therefore a force of 300 pounds is required to keep the boat from rolling down the ramp. ▬

Work = $\|\mathbf{F}\| \, \|\overrightarrow{PQ}\|$

(a) Force acts along the line of motion.

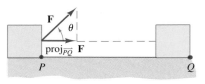

Work = $\|\text{proj}_{\overrightarrow{PQ}} \mathbf{F}\| \, \|\overrightarrow{PQ}\|$

(b) Force acts at angle θ with the line of motion.

FIGURE 13.21

Work

The work W done by a constant force **F** acting along the lines of motion of an object is given by

$$W = (\text{magnitude of force})(\text{distance}) = \|\mathbf{F}\| \, \|\overrightarrow{PQ}\|$$

as shown in Figure 13.21(a). If the constant force **F** is not directed along the line of motion, then we can see from Figure 13.21(b) that the work W done by the force is

$$W = \|\text{proj}_{\overrightarrow{PQ}} \mathbf{F}\| \, \|\overrightarrow{PQ}\| = (\cos \theta) \|\mathbf{F}\| \, \|\overrightarrow{PQ}\| = \mathbf{F} \cdot \overrightarrow{PQ}.$$

We summarize this notion of work in the following definition.

DEFINITION OF WORK

The work W done by a constant force **F** as its point of application moves along the vector $\overrightarrow{PQ}$ is given by either of the following.

1. $W = \|\text{proj}_{\overrightarrow{PQ}} \mathbf{F}\| \, \|\overrightarrow{PQ}\|$ Projection form

2. $W = \mathbf{F} \cdot \overrightarrow{PQ}$ Dot product form

---

EXAMPLE 6   An application involving work

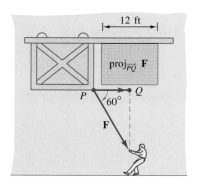

FIGURE 13.22

To close a sliding door, a person pulls on a rope with a constant force of 50 pounds at a constant angle of 60°, as shown in Figure 13.22. Find the work done in moving the door 12 feet to its closed position.

SOLUTION

Using a projection, we calculate the work as follows.

$$W = \|\text{proj}_{\overrightarrow{PQ}} \mathbf{F}\| \, \|\overrightarrow{PQ}\|$$

$$= \cos (60°) \|\mathbf{F}\| \, \|\overrightarrow{PQ}\|$$

$$= \frac{1}{2}(50)(12) = 300 \text{ ft-lb}$$ ▬

EXERCISES for Section 13.2

In Exercises 1–4, find (a) $\mathbf{u} \cdot \mathbf{v}$,　(b) $\mathbf{u} \cdot \mathbf{u}$,　(c) $\|\mathbf{u}\|^2$, (d) $(\mathbf{u} \cdot \mathbf{v})\mathbf{v}$, and　(e) $\mathbf{u} \cdot 2\mathbf{v}$.

1. $\mathbf{u} = \langle 3, 4 \rangle$, $\mathbf{v} = \langle 2, -3 \rangle$
2. $\mathbf{u} = \langle 5, 12 \rangle$, $\mathbf{v} = \langle -3, 2 \rangle$
3. $\mathbf{u} = \mathbf{i} - 4\mathbf{j}$, $\mathbf{v} = \frac{1}{2}\mathbf{i} + 3\mathbf{j}$
4. $\mathbf{u} = \mathbf{i}$, $\mathbf{v} = \mathbf{i}$

In Exercises 5–12, find the angle θ between the given vectors.

5. $\mathbf{u} = \langle 1, 1 \rangle$, $\mathbf{v} = \langle 2, -2 \rangle$
6. $\mathbf{u} = \langle 3, 1 \rangle$, $\mathbf{v} = \langle 2, -1 \rangle$
7. $\mathbf{u} = \langle 1, 1 \rangle$, $\mathbf{v} = \langle 3, -1 \rangle$
8. $\mathbf{u} = \langle 1, 2 \rangle$, $\mathbf{v} = \langle 2, -1 \rangle$
9. $\mathbf{u} = \langle 2, -3 \rangle$, $\mathbf{v} = \langle -9, -6 \rangle$
10. $\mathbf{u} = \langle -1, 2 \rangle$, $\mathbf{v} = \langle 4, 6 \rangle$
11. $\mathbf{u} = \cos \frac{\pi}{6}\mathbf{i} + \sin \frac{\pi}{6}\mathbf{j}$

$\mathbf{v} = \cos \frac{3\pi}{4}\mathbf{i} + \sin \frac{3\pi}{4}\mathbf{j}$

12. $\mathbf{u} = \cos \frac{2\pi}{3}\mathbf{i} + \sin \frac{2\pi}{3}\mathbf{j}$

$\mathbf{v} = \cos \frac{\pi}{12}\mathbf{i} + \sin \frac{\pi}{12}\mathbf{j}$

In Exercises 13–16, find the positive angle θ between the given vector and the positive x-axis.

13. $\mathbf{u} = \mathbf{i} - \mathbf{j}$ 　　　**14.** $\mathbf{u} = \sqrt{3}\mathbf{i} + \mathbf{j}$
15. $\mathbf{u} = \langle 3, 2 \rangle$ 　　　**16.** $\mathbf{u} = \langle 12, 5 \rangle$

In Exercises 17–24, determine whether $\mathbf{u}$ and $\mathbf{v}$ are orthogonal, parallel, or neither.

17. $\mathbf{u} = \langle 4, 0 \rangle$, $\mathbf{v} = \langle 1, 1 \rangle$
18. $\mathbf{u} = \langle 2, -4 \rangle$, $\mathbf{v} = \langle 2, 1 \rangle$
19. $\mathbf{u} = \langle 2, 18 \rangle$, $\mathbf{v} = \left\langle \frac{3}{2}, -\frac{1}{6} \right\rangle$
20. $\mathbf{u} = \langle 0, 4 \rangle$, $\mathbf{v} = \langle \sqrt{3}, 1 \rangle$
21. $\mathbf{u} = \langle 6, -4 \rangle$, $\mathbf{v} = \langle -3, 2 \rangle$
22. $\mathbf{u} = \left\langle -\frac{1}{3}, \frac{2}{3} \right\rangle$, $\mathbf{v} = \langle 2, -4 \rangle$
23. $\mathbf{u} = \mathbf{i} - 3\mathbf{j}$, $\mathbf{v} = 3\mathbf{i} - \mathbf{j}$
24. $\mathbf{u} = 4\mathbf{i} + 3\mathbf{j}$, $\mathbf{v} = \frac{1}{3}\mathbf{i} - \frac{2}{3}\mathbf{j}$

In Exercises 25–30, (a) find the projection of $\mathbf{u}$ onto $\mathbf{v}$, and (b) find the vector component of $\mathbf{u}$ orthogonal to $\mathbf{v}$.

25. $\mathbf{u} = \langle 2, 3 \rangle$, $\mathbf{v} = \langle 5, 1 \rangle$
26. $\mathbf{u} = \langle 1, -2 \rangle$, $\mathbf{v} = \langle 1, 3 \rangle$
27. $\mathbf{u} = \langle 1, 1 \rangle$, $\mathbf{v} = \langle 5, 0 \rangle$
28. $\mathbf{u} = \langle 2, -3 \rangle$, $\mathbf{v} = \langle 5, -1 \rangle$
29. $\mathbf{u} = \langle 2, -3 \rangle$, $\mathbf{v} = \langle 3, 2 \rangle$
30. $\mathbf{u} = \langle \sqrt{3}, 1 \rangle$, $\mathbf{v} = \langle -\sqrt{3}, -1 \rangle$

31. A truck with a gross weight of 32,000 pounds is parked on a 15° slope (see figure). Assuming the only force to overcome is that due to gravity, find the following.
(a) the force required to keep the truck from rolling down the hill
(b) the force perpendicular to the hill

32. Rework Exercise 31 for a truck that is parked on a 16° slope.

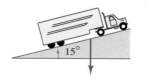

Weight = 32,000 lb

FIGURE FOR 31

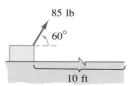

85 lb
60°
10 ft

FIGURE FOR 33

33. An object is dragged 10 feet across a floor, using a force of 85 pounds. Find the work done if the direction of the force is 60° above the horizontal (see figure).

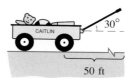

30°
50 ft

FIGURE FOR 34

34. A toy wagon is pulled by exerting a force of 15 pounds on a handle that makes a 30° angle with the horizontal. Find the work done in pulling the wagon 50 feet (see figure).

35. What is known about θ, the angle between two vectors $\mathbf{u}$ and $\mathbf{v}$, if
(a) $\mathbf{u} \cdot \mathbf{v} = 0$ 　　(b) $\mathbf{u} \cdot \mathbf{v} > 0$ 　　(c) $\mathbf{u} \cdot \mathbf{v} < 0$?

36. What can be said about the vectors $\mathbf{u}$ and $\mathbf{v}$ if
(a) the projection of $\mathbf{u}$ onto $\mathbf{v}$ equals $\mathbf{u}$?
(b) the projection of $\mathbf{u}$ onto $\mathbf{v}$ equals $\mathbf{0}$?

37. Use vectors to prove that the diagonals of a rhombus are perpendicular.

38. Prove that

$$\|\mathbf{u} - \mathbf{v}\|^2 = \|\mathbf{u}\|^2 + \|\mathbf{v}\|^2 - 2\mathbf{u} \cdot \mathbf{v}.$$

39. Prove the **Cauchy-Schwarz Inequality**

$$|\mathbf{u} \cdot \mathbf{v}| \le \|\mathbf{u}\| \, \|\mathbf{v}\|.$$

40. Prove the triangle inequality

$$\|\mathbf{u} + \mathbf{v}\| \le \|\mathbf{u}\| + \|\mathbf{v}\|.$$

[Hint: Use the result of Exercise 39 and the fact that $\|\mathbf{u} + \mathbf{v}\|^2 = \|\mathbf{u} + \mathbf{v}\| \cdot \|\mathbf{u} + \mathbf{v}\|.$]

41. Prove properties 2, 3, and 4 of Theorem 13.5.

42. Prove that if $\mathbf{u}$ is orthogonal to $\mathbf{v}$ and $\mathbf{w}$, then $\mathbf{u}$ is orthogonal to $c\mathbf{v} + d\mathbf{w}$ for any scalars c and d.

13.3 Vector-Valued Functions

Vector-valued functions ▪ Limits and continuity ▪ Differentiation of vector-valued functions ▪ Integration of vector-valued functions

We begin this section by looking at a new type of function, called a **vector-valued function,** that maps real numbers onto vectors.

DEFINITION OF A VECTOR-VALUED FUNCTION	A function of the form $$\mathbf{r}(t) = f(t)\mathbf{i} + g(t)\mathbf{j}$$ is called a **vector-valued function,** where the **component functions,** f and g, are real-valued functions of the parameter t.

REMARK Note the distinction between the vector-valued function $\mathbf{r}$ and the real-valued functions f and g. All are functions of the real variable t, but $\mathbf{r}(t)$ is a vector, whereas $f(t)$ and $g(t)$ are real numbers.

Unless stated otherwise, we consider the **domain** of the vector-valued function $\mathbf{r}$ to be the intersection of the domains of the component functions f and g. For instance, the domain of the vector-valued function

$$\mathbf{r}(t) = (\ln t)\,\mathbf{i} + \sqrt{1 - t}\,\mathbf{j}$$

is the interval $(0, 1]$.

Recall from Section 12.1, that a *plane curve* is defined to be the set of ordered pairs $(f(t), g(t))$ satisfying the parametric equations

$$x = f(t) \quad \text{and} \quad y = g(t)$$

where f and g are continuous functions of t on an interval I. Vector-valued functions can be used to represent plane curves. For instance, by letting the parameter t represent time, we can use a vector-valued function to represent *motion* along a curve. Or, in the more general case, we can use a vector-valued function to *trace the graph* of a plane curve. In either case, the terminal point of the position vector $\mathbf{r}(t)$ coincides with the point (x, y) on the curve given by the parametric equations, as shown in Figure 13.23. The arrowhead on the curve indicates the curve's *orientation* by pointing in the direction of increasing values of t.

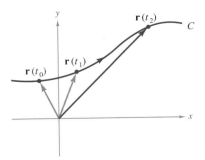

Curve C is traced out by the terminal point of position vector $\mathbf{r}(t)$.

FIGURE 13.23

EXAMPLE 1 A plane curve represented by a vector-valued function

Sketch the plane curve represented by the vector-valued function

$$\mathbf{r}(t) = 2 \cos t \, \mathbf{i} - 3 \sin t \, \mathbf{j}, \qquad 0 \le t \le 2\pi.$$

SOLUTION

From the position vector $\mathbf{r}(t)$, we obtain the parametric equations

$$x = 2 \cos t \quad \text{and} \quad y = -3 \sin t.$$

Solving for $\cos t$ and $\sin t$ and using the identity $\cos^2 t + \sin^2 t = 1$ produces the rectangular equation

$$\frac{x^2}{2^2} + \frac{y^2}{3^2} = 1.$$

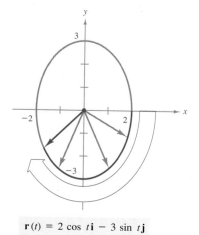

$\mathbf{r}(t) = 2 \cos t \mathbf{i} - 3 \sin t \mathbf{j}$

FIGURE 13.24

The graph of this rectangular equation is the ellipse shown in Figure 13.24. Note that the curve has a *clockwise* orientation. That is, as t increases from 0 to 2π, the position vector $\mathbf{r}(t)$ moves clockwise, and its terminal point traces the ellipse. ▭

In Example 1 we were given a vector-valued function and asked to sketch the corresponding curve. In the next example, we look at the reverse problem—finding a vector-valued function to represent a given curve. Of course, if the curve is described parametrically, then representation by a vector-valued function is straightforward. For instance, to represent the line given by

$$x = 2 + t \quad \text{and} \quad y = 3t$$

we simply use the vector-valued function given by

$$\mathbf{r}(t) = (2 + t)\mathbf{i} + 3t\mathbf{j}.$$

If a set of parametric equations for the curve C is not given, then the problem of representing C by a vector-valued function boils down to finding a set of parametric equations.

EXAMPLE 2 Representing a plane curve by a vector-valued function

Represent the parabola given by $y = x^2 + 1$ by a vector-valued function.

SOLUTION

Although there are many ways to choose the parameter t, a natural choice is to let $x = t$. Then $y = t^2 + 1$ and we have

$$\mathbf{r}(t) = t\mathbf{i} + (t^2 + 1)\mathbf{j}.$$

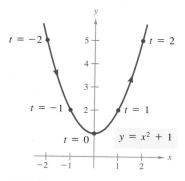

FIGURE 13.25

Note in Figure 13.25 the orientation produced by this particular choice of parameter. Had we chosen $x = -t$ as the parameter, we would have oriented the curve in the opposite direction. ▭

Limits and continuity

Before we can see the real power of representing curves by vector-valued functions, we must do some more preliminary work. It turns out that many techniques and definitions used in the calculus of real-valued functions can be applied to vector-valued functions. For instance, we can add and subtract vector-valued functions, multiply a vector-valued function by a scalar, take the limit of a vector-valued function, differentiate a vector-valued function, and so on. The basic approach is to capitalize on the linearity of vector operations by extending the definitions on a component-by-component basis. For example, to add two vector-valued functions, we write

$$\mathbf{r}_1(t) + \mathbf{r}_2(t) = [f_1(t)\mathbf{i} + g_1(t)\mathbf{j}] + [f_2(t)\mathbf{i} + g_2(t)\mathbf{j}]$$
$$= [f_1(t) + f_2(t)]\mathbf{i} + [g_1(t) + g_2(t)]\mathbf{j}.$$

Similarly, to multiply a vector-valued function by a scalar, we write

$$c\mathbf{r}(t) = c[f_1(t)\mathbf{i} + g_1(t)\mathbf{j}]$$
$$= cf_1(t)\mathbf{i} + cg_1(t)\mathbf{j}.$$

This component-by-component extension of operations with real-valued functions to vector-valued functions is further illustrated in the following definition of the limit of a vector-valued function.

DEFINITION OF THE LIMIT OF A VECTOR-VALUED FUNCTION	If $\mathbf{r}$ is a vector-valued function such that $$\mathbf{r}(t) = f(t)\mathbf{i} + g(t)\mathbf{j}$$ then $$\lim_{t \to a} \mathbf{r}(t) = \left[\lim_{t \to a} f(t)\right]\mathbf{i} + \left[\lim_{t \to a} g(t)\right]\mathbf{j}$$ provided f and g have limits as $t \to a$.

If $\mathbf{r}(t)$ approaches the vector $\mathbf{L}$, as $t \to a$, then the length of the vector $\mathbf{r}(t) - \mathbf{L}$ approaches zero. That is,

$$\|\mathbf{r}(t) - \mathbf{L}\| \to 0 \qquad \text{as} \qquad t \to a.$$

This is illustrated graphically in Figure 13.26. With this definition of the limit of a vector-valued function, we can develop vector versions of most of the limit theorems given in Chapter 2. For example, the limit of the sum of two vector-valued functions is the sum of their individual limits. Also, we can use the orientation of the curve given by $\mathbf{r}(t)$ to define one-sided limits of vector-valued functions. In the following definition, we extend the notion of continuity to vector-valued functions.

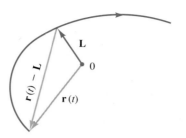
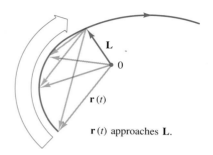

r(t) approaches **L**.

FIGURE 13.26

DEFINITION OF CONTINUITY OF A VECTOR-VALUED FUNCTION

A vector-valued function **r** is **continuous at the point** given by $t = a$ if the limit of **r**(t) exists as $t \to a$ and

$$\lim_{t \to a} \mathbf{r}(t) = \mathbf{r}(a).$$

A vector-valued function **r** is **continuous on an interval** I if it is continuous at every point in the interval.

From this definition, it follows that a vector-valued function is continuous at $t = a$ if and only if each of its component functions is continuous at $t = a$.

EXAMPLE 3   Continuity of vector-valued functions

Discuss the continuity of the vector-valued function given by

$$\mathbf{r}(t) = \tan t \, \mathbf{i} + \sqrt{t + 1} \, \mathbf{j}$$

at $t = 0$ and $t = -2$.

SOLUTION

As t approaches 0, the limit is

$$\lim_{t \to 0} \mathbf{r}(t) = \left[\lim_{t \to 0} \tan t \right] \mathbf{i} + \left[\lim_{t \to 0} \sqrt{t + 1} \right] \mathbf{j}$$

$$= 0\mathbf{i} + \mathbf{j} = \mathbf{j}.$$

Since $\mathbf{r}(0) = (\tan 0)\,\mathbf{i} + \sqrt{1}\,\mathbf{j} = 0\mathbf{i} + \mathbf{j}$, we conclude that **r** is continuous at $t = 0$. At $t = -2$, the component $g(t) = \sqrt{t + 1}$ is not defined; hence $\mathbf{r}(-2)$ is not defined and **r** is not continuous at $t = -2$.   ▭

Differentiation of vector-valued functions

In Sections 13.4 and 13.5, we will look at several important applications involving the calculus of vector-valued functions. In preparation for that, we devote the remainder of this section to the mechanics of differentiation and integration of vector-valued functions.

The definition of the derivative of a vector-valued function parallels that given for real-valued functions.

DEFINITION OF THE DERIVATIVE OF A VECTOR-VALUED FUNCTION

The **derivative of a vector-valued function r** is defined by

$$\mathbf{r}'(t) = \lim_{\Delta t \to 0} \frac{\mathbf{r}(t + \Delta t) - \mathbf{r}(t)}{\Delta t}$$

for all t for which the limit exists.

Differentiation of vector-valued functions can be done on a *component-by-component basis*. To see why this is true, consider the function given by $\mathbf{r}(t) = f(t)\mathbf{i} + g(t)\mathbf{j}$. Applying the definition of the derivative produces

$$\mathbf{r}'(t) = \lim_{\Delta t \to 0} \frac{\mathbf{r}(t + \Delta t) - \mathbf{r}(t)}{\Delta t}$$

$$= \lim_{\Delta t \to 0} \frac{f(t + \Delta t)\mathbf{i} + g(t + \Delta t)\mathbf{j} - f(t)\mathbf{i} - g(t)\mathbf{j}}{\Delta t}$$

$$= \lim_{\Delta t \to 0} \left[\left(\frac{f(t + \Delta t) - f(t)}{\Delta t} \right)\mathbf{i} + \left(\frac{g(t + \Delta t) - g(t)}{\Delta t} \right)\mathbf{j} \right]$$

$$= f'(t)\mathbf{i} + g'(t)\mathbf{j}.$$

We list this important result in the following theorem. Note that the derivative of the vector-valued function **r** is itself a vector-valued function. (A geometric interpretation of the derivative of a vector-valued function will be given in the next section.)

THEOREM 13.8 DIFFERENTIATION OF VECTOR-VALUED FUNCTIONS

If $\mathbf{r}(t) = f(t)\mathbf{i} + g(t)\mathbf{j}$ where f and g are differentiable functions of t, then

$$\mathbf{r}'(t) = f'(t)\mathbf{i} + g'(t)\mathbf{j}.$$

REMARK It is occasionally convenient to use Leibniz or operator notation for the derivative of a vector-valued function. Thus, for the derivative of $\mathbf{r}(t) = f(t)\mathbf{i} + g(t)\mathbf{j}$, we can write

$$\mathbf{r}'(t), \qquad \frac{d\mathbf{r}}{dt}, \qquad \frac{d}{dt}[\mathbf{r}(t)], \qquad \text{or} \qquad D_t[\mathbf{r}(t)].$$

EXAMPLE 4   Differentiation of vector-valued functions

Find the derivative of each of the following vector-valued functions.

(a) $\mathbf{r}(t) = t^2\mathbf{i} - 4\mathbf{j}$ (b) $\mathbf{r}(t) = \dfrac{1}{t}\mathbf{i} + \ln t\,\mathbf{j}$

SOLUTION

Differentiating on a component-by-component basis produces the following.

(a) $\mathbf{r}'(t) = 2t\mathbf{i} - 0\mathbf{j} = 2t\mathbf{i}$

(b) $\mathbf{r}'(t) = -\dfrac{1}{t^2}\mathbf{i} + \dfrac{1}{t}\mathbf{j}$

Higher-order derivatives of vector-valued functions are obtained by successive differentiation of each component function, as demonstrated in Example 5.

EXAMPLE 5   Differentiation of vector-valued functions

For the vector-valued function given by $\mathbf{r}(t) = e^{2t}\mathbf{i} - \sin t\,\mathbf{j}$, find the following.

(a) $\mathbf{r}'(t)$,   (b) $\mathbf{r}''(t)$,   (c) $\mathbf{r}'(t) \cdot \mathbf{r}''(t)$

SOLUTION

(a) $\mathbf{r}'(t) = 2e^{2t}\mathbf{i} - \cos t\,\mathbf{j}$
(b) $\mathbf{r}''(t) = 4e^{2t}\mathbf{i} + \sin t\,\mathbf{j}$
(c) $\mathbf{r}'(t) \cdot \mathbf{r}''(t) = 8e^{4t} - \cos t \sin t$

We call the curve represented by $\mathbf{r}(t) = f(t)\mathbf{i} + g(t)\mathbf{j}$ **smooth** on an open interval I if f' and g' are continuous on I and $\mathbf{r}'(t) \neq \mathbf{0}$ for any value of t in the interval I.

EXAMPLE 6   Finding intervals on which a curve is smooth

Find the intervals on which the epicycloid C given by

$$\mathbf{r}(t) = (5 \cos t - \cos 5t)\mathbf{i} + (5 \sin t - \sin 5t)\mathbf{j}, \qquad 0 \leq t \leq 2\pi$$

is smooth.

SOLUTION

The derivative of **r** is

$$\mathbf{r}'(t) = (-5 \sin t + 5 \sin 5t)\mathbf{i} + (5 \cos t - 5 \cos 5t)\mathbf{j}.$$

In the interval $[0, 2\pi]$, the only values of t for which $\mathbf{r}'(t) = 0\mathbf{i} + 0\mathbf{j}$ are $t = 0,\ \pi/2,\ \pi,\ 3\pi/2,$ and 2π. Therefore, we conclude that C is smooth in the intervals

$$\left(0, \frac{\pi}{2}\right), \quad \left(\frac{\pi}{2}, \pi\right), \quad \left(\pi, \frac{3\pi}{2}\right), \quad \text{and} \quad \left(\frac{3\pi}{2}, 2\pi\right)$$

as shown in Figure 13.27.

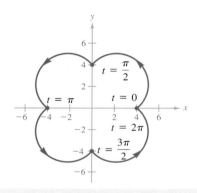

$$\mathbf{r}(t) = (5 \cos t - \cos 5t)\mathbf{i} + (5 \sin t - \sin 5t)\mathbf{j}$$

FIGURE 13.27

Most of the differentiation rules in Chapter 3 have counterparts for vector-valued functions. We list several in the following theorem. Note that the theorem contains two versions of "product rules." Property 3 gives the derivative of the product of a real-valued function f and vector-valued function **r**, and Property 4 gives the derivative of the dot product of two vector-valued functions.

THEOREM 13.9
PROPERTIES OF THE DERIVATIVE OF VECTOR-VALUED FUNCTIONS

1. $D_t[c\,\mathbf{r}(t)] = c\,\mathbf{r}'(t)$
2. $D_t[\mathbf{r}(t) \pm \mathbf{u}(t)] = \mathbf{r}'(t) \pm \mathbf{u}'(t)$
3. $D_t[f(t)\mathbf{r}(t)] = f(t)\mathbf{r}'(t) + f'(t)\mathbf{r}(t)$
4. $D_t[\mathbf{r}(t) \cdot \mathbf{u}(t)] = \mathbf{r}(t) \cdot \mathbf{u}'(t) + \mathbf{r}'(t) \cdot \mathbf{u}(t)$
5. $D_t[\mathbf{r}(f(t))] = \mathbf{r}'(f(t))f'(t)$
6. If $\mathbf{r}(t) \cdot \mathbf{r}(t) = c$, then $\mathbf{r}(t) \cdot \mathbf{r}'(t) = 0$.

PROOF We prove the fourth property and leave the proofs of the other properties as exercises (see Exercises 51–55). Let

$$\mathbf{r}(t) = f_1(t)\mathbf{i} + g_1(t)\mathbf{j} \quad \text{and} \quad \mathbf{u}(t) = f_2(t)\mathbf{i} + g_2(t)\mathbf{j}$$

where f_i and g_i are differentiable functions of t. Then,

$$\mathbf{r}(t) \cdot \mathbf{u}(t) = f_1(t)f_2(t) + g_1(t)g_2(t)$$

and it follows that

$$\begin{aligned} D_t[\mathbf{r}(t) \cdot \mathbf{u}(t)] &= f_1(t)f_2{}'(t) + f_1{}'(t)f_2(t) + g_1(t)g_2{}'(t) + g_1{}'(t)g_2(t) \\ &= [f_1(t)f_2{}'(t) + g_1(t)g_2{}'(t)] + [f_1{}'(t)f_2(t) + g_1{}'(t)g_2(t)] \\ &= \mathbf{r}(t) \cdot \mathbf{u}'(t) + \mathbf{r}'(t) \cdot \mathbf{u}(t). \end{aligned}$$

EXAMPLE 7   Using properties of the derivative

For the vector-valued functions given by

$$\mathbf{r}(t) = \frac{1}{t}\mathbf{i} - \mathbf{j} \qquad \text{and} \qquad \mathbf{u}(t) = t^2\mathbf{i} - 2t\,\mathbf{j}$$

find $D_t[\mathbf{r}(t) \cdot \mathbf{u}(t)]$.

SOLUTION

Since

$$\mathbf{r}'(t) = -\frac{1}{t^2}\mathbf{i} \qquad \text{and} \qquad \mathbf{u}'(t) = 2t\mathbf{i} - 2\mathbf{j}$$

we have

$$\begin{aligned} D_t[\mathbf{r}(t) \cdot \mathbf{u}(t)] &= \mathbf{r}(t) \cdot \mathbf{u}'(t) + \mathbf{r}'(t) \cdot \mathbf{u}(t) \\ &= \left[\frac{1}{t}\mathbf{i} - \mathbf{j}\right] \cdot [2t\mathbf{i} - 2\mathbf{j}] + \left[-\frac{1}{t^2}\mathbf{i}\right] \cdot [t^2\mathbf{i} - 2t\,\mathbf{j}] \\ &= 2 + 2 + (-1) = 3. \end{aligned}$$

REMARK   Try reworking Example 7 by first forming the dot product and then differentiating to see that you obtain the same answer.

Integration of vector-valued functions

The following definition is a natural consequence of the definition of the derivative of a vector-valued function.

DEFINITION OF INTEGRATION OF A VECTOR-VALUED FUNCTION

If $\mathbf{r}(t) = f(t)\mathbf{i} + g(t)\mathbf{j}$ where f and g are continuous on $[a, b]$, then the **indefinite integral (antiderivative)** of $\mathbf{r}$ is

$$\int \mathbf{r}(t)\,dt = \left[\int f(t)\,dt\right]\mathbf{i} + \left[\int g(t)\,dt\right]\mathbf{j}$$

and its **definite integral** over the interval $a \le t \le b$ is

$$\int_a^b \mathbf{r}(t)\,dt = \left[\int_a^b f(t)\,dt\right]\mathbf{i} + \left[\int_a^b g(t)\,dt\right]\mathbf{j}.$$

The antiderivative of a vector-valued function is a family of vector-valued functions all differing by a constant vector $\mathbf{C}$. For instance, if $\mathbf{r}(t) = f(t)\mathbf{i} + g(t)\mathbf{j}$, then for the indefinite integral $\int \mathbf{r}(t)\, dt$, we obtain two constants of integration

$$\int f(t)\, dt = F(t) + C_1 \qquad \text{and} \qquad \int g(t)\, dt = G(t) + C_2$$

where $F'(t) = f(t)$ and $G'(t) = g(t)$. These two *scalar* constants produce one *vector* constant of integration,

$$\begin{aligned} \int \mathbf{r}(t)\, dt &= [F(t) + C_1]\mathbf{i} + [G(t) + C_2]\mathbf{j} \\ &= [F(t)\mathbf{i} + G(t)\mathbf{j}] + [C_1\mathbf{i} + C_2\mathbf{j}] \\ &= \mathbf{R}(t) + \mathbf{C} \end{aligned}$$

where $\mathbf{R}'(t) = \mathbf{r}(t)$.

EXAMPLE 8   Integration of a vector-valued function

Evaluate the indefinite integral

$$\int (t\mathbf{i} + 3\mathbf{j})\, dt.$$

SOLUTION

Integrating on a component-by-component basis produces

$$\int (t\mathbf{i} + 3\mathbf{j})\, dt = \frac{t^2}{2}\mathbf{i} + 3t\mathbf{j} + \mathbf{C}.$$

EXAMPLE 9   Definite integral of a vector-valued function

Evaluate the definite integral

$$\int_0^1 \mathbf{r}(t)\, dt = \int_0^1 \left(\sqrt[3]{t}\,\mathbf{i} + \frac{1}{t+1}\mathbf{j} \right) dt.$$

SOLUTION

$$\begin{aligned} \int_0^1 \mathbf{r}(t)\, dt &= \int_0^1 \left(\sqrt[3]{t}\,\mathbf{i} + \frac{1}{t+1}\,\mathbf{j} \right) dt \\ &= \left[\int_0^1 \sqrt[3]{t}\, dt \right]\mathbf{i} + \left[\int_0^1 \frac{1}{t+1}\, dt \right]\mathbf{j} \\ &= \left[\left(\frac{3}{4}\right)t^{4/3} \right]_0^1 \mathbf{i} + \left[\ln |t+1| \right]_0^1 \mathbf{j} \\ &= \frac{3}{4}\mathbf{i} + (\ln 2)\,\mathbf{j} \end{aligned}$$

As with real-valued functions, we can narrow the family of antiderivatives of a vector-valued function $\mathbf{r}'$ down to a single antiderivative by imposing an initial condition on the vector-valued function $\mathbf{r}$. This is demonstrated in the next example.

EXAMPLE 10 Finding the antiderivative of a vector-valued function

Find the antiderivative of

$$\mathbf{r}'(t) = \cos 2t \, \mathbf{i} - 2 \sin t \, \mathbf{j}$$

that satisfies the initial condition $\mathbf{r}(0) = 3\mathbf{i} - 2\mathbf{j}$.

SOLUTION

$$\mathbf{r}(t) = \int \mathbf{r}'(t) \, dt$$

$$= \left(\int \cos 2t \, dt \right) \mathbf{i} - \left(2 \int \sin t \, dt \right) \mathbf{j}$$

$$= \left(\frac{1}{2} \sin 2t + C_1 \right) \mathbf{i} + (2 \cos t + C_2) \mathbf{j}$$

Letting $t = 0$ and using the fact that $\mathbf{r}(0) = 3\mathbf{i} - 2\mathbf{j}$, we have

$$\mathbf{r}(0) = (0 + C_1)\mathbf{i} + (2 + C_2)\mathbf{j} = 3\mathbf{i} - 2\mathbf{j}.$$

Equating corresponding components produces $C_1 = 3$ and $C_2 = -4$. Thus, the antiderivative that satisfies the given initial condition is

$$\mathbf{r}(t) = \left(\frac{1}{2} \sin 2t + 3 \right) \mathbf{i} + (2 \cos t - 4)\mathbf{j}.$$

EXERCISES for Section 13.3

In Exercises 1–4, find the domain of the given vector-valued function.

1. $\mathbf{r}(t) = 5t\,\mathbf{i} - \dfrac{1}{t}\mathbf{j}$

2. $\mathbf{r}(t) = \sqrt{4 - t^2}\,\mathbf{i} + t^2\,\mathbf{j}$

3. $\mathbf{r}(t) = e^t\mathbf{i} + \ln t\,\mathbf{j}$

4. $\mathbf{r}(t) = \dfrac{1}{t - 3}\mathbf{i} + \dfrac{1}{t - 5}\mathbf{j}$

In Exercises 5 and 6, find $\|\mathbf{r}(t)\|$.

5. $\mathbf{r}(t) = \sin \pi t\,\mathbf{i} + \cos \pi t\,\mathbf{j}$

6. $\mathbf{r}(t) = \sqrt{t}\,\mathbf{i} + 3t\,\mathbf{j}$

In Exercises 7–10, sketch the curve represented by the vector-valued function and give the orientation of the curve.

7. $\mathbf{r}(t) = 3t\,\mathbf{i} + (t - 1)\,\mathbf{j}$

8. $\mathbf{r}(t) = 2 \cos t\,\mathbf{i} + 2 \sin t\,\mathbf{j}$

9. $\mathbf{r}(t) = t\mathbf{i} + t^2\,\mathbf{j}$

10. $\mathbf{r}(t) = t\mathbf{i} + \dfrac{1}{t}\mathbf{j}$

In Exercises 11–16, evaluate the limit.

11. $\displaystyle\lim_{t \to 3} \left(t\mathbf{i} + \frac{t^2 - 9}{t^2 - 3t}\mathbf{j} \right)$

12. $\displaystyle\lim_{t \to 0} \left(e^t\mathbf{i} + \frac{\sin t}{t}\mathbf{j} \right)$

13. $\lim\limits_{t \to 0} \left(\dfrac{1 - \cos t}{t} \mathbf{i} + t^2 \mathbf{j} \right)$

14. $\lim\limits_{t \to 1} \left(\sqrt{t}\,\mathbf{i} + \dfrac{\ln t}{t^2 - 1} \mathbf{j} \right)$

15. $\lim\limits_{t \to \infty} \left(\dfrac{2}{t^2} \mathbf{i} + e^{-2t} \mathbf{j} \right)$

16. $\lim\limits_{t \to \infty} \left(\dfrac{3t^2}{t^2 + 1} \mathbf{i} + \dfrac{5}{t} \mathbf{j} \right)$

In Exercises 17–20, determine the interval(s) on which the vector-valued function is continuous.

17. $\mathbf{r}(t) = t\mathbf{i} + \dfrac{1}{t} \mathbf{j}$

18. $\mathbf{r}(t) = \sqrt{t}\,\mathbf{i} + \sqrt{t - 1}\,\mathbf{j}$

19. $\mathbf{r}(t) = \ln t\,\mathbf{i} + e^t \mathbf{j}$

20. $\mathbf{r}(t) = \arccos t\,\mathbf{i} + t^2 \mathbf{j}$

In Exercises 21–24, (a) sketch the plane curve represented by the vector-valued function, and (b) sketch the vectors $\mathbf{r}(t_0)$ and $\mathbf{r}'(t_0)$ for the specified value of t_0. Position the vectors so that the initial point of $\mathbf{r}(t_0)$ is at the origin and the initial point of $\mathbf{r}'(t_0)$ is at the terminal point of $\mathbf{r}(t_0)$.

21. $\mathbf{r}(t) = t^2 \mathbf{i} + t\mathbf{j}$ $t_0 = 2$

22. $\mathbf{r}(t) = t\mathbf{i} + t^3 \mathbf{j}$ $t_0 = 1$

23. $\mathbf{r}(t) = \cos t\,\mathbf{i} + \sin t\,\mathbf{j}$ $t_0 = \dfrac{\pi}{2}$

24. $\mathbf{r}(t) = t^2 \mathbf{i} + \dfrac{1}{t} \mathbf{j}$ $t_0 = 2$

In Exercises 25–30, find $\mathbf{r}'(t)$ and $\mathbf{r}''(t)$.

25. $\mathbf{r}(t) = 3t\mathbf{i} + (t - 1)\mathbf{j}$

26. $\mathbf{r}(t) = \dfrac{1}{t} \mathbf{i} + \dfrac{t + 1}{t - 1} \mathbf{j}$

27. $\mathbf{r}(t) = \langle a \cos t, a \sin t \rangle$

28. $\mathbf{r}(t) = \langle \sec t, \tan t \rangle$

29. $\mathbf{r}(t) = \langle t - \sin t, 1 - \cos t \rangle$

30. $\mathbf{r}(t) = \langle \cot t, 2 \sin t \cos t \rangle$

In Exercises 31 and 32, find the following.
(a) $\mathbf{r}'(t)$ (b) $D_t[\mathbf{r}(t) \cdot \mathbf{u}(t)]$
(c) $D_t[3\mathbf{r}(t) - \mathbf{u}(t)]$ (d) $D_t[\|\mathbf{r}(t)\|]$

31. $\mathbf{r}(t) = 3t\mathbf{i} + 4t\mathbf{j}, \ \mathbf{u}(t) = 4t\mathbf{i} + t^2 \mathbf{j}$

32. $\mathbf{r}(t) = \sin t\,\mathbf{i} + \cos t\,\mathbf{j}, \ \mathbf{u}(t) = \cos t\,\mathbf{i} - \sin t\,\mathbf{j}$

In Exercises 33–38, find the open interval(s) on which the curve given by the vector-valued function is smooth.

33. $\mathbf{r}(t) = t^2 \mathbf{i} + t^3 \mathbf{j}$

34. $\mathbf{r}(t) = \dfrac{1}{t - 1} \mathbf{i} + 3t\mathbf{j}$

35. $\mathbf{r}(\theta) = \langle 2 \cos^3 \theta, 3 \sin^3 \theta \rangle$

36. $\mathbf{r}(\theta) = \langle \theta + \sin \theta, 1 - \cos \theta \rangle$

37. $\mathbf{r}(\theta) = \langle \theta - 2 \sin \theta, 1 - 2 \cos \theta \rangle$

38. $\mathbf{r}(t) = \left\langle \dfrac{3t}{1 + t^3}, \dfrac{3t^2}{1 + t^3} \right\rangle$

In Exercises 39–42, evaluate the indefinite integral.

39. $\displaystyle\int (6t^2 \mathbf{i} + 3\mathbf{j})\, dt$

40. $\displaystyle\int \left(\dfrac{1}{t} \mathbf{i} + e^t \mathbf{j} \right) dt$

41. $\displaystyle\int (4 \sin t\,\mathbf{i} + 3 \cos t\,\mathbf{j})\, dt$

42. $\displaystyle\int (te^{-t^2} \mathbf{i} + t\mathbf{j})\, dt$

In Exercises 43–46, evaluate the definite integral.

43. $\displaystyle\int_0^1 (6t\mathbf{i} - 3t\mathbf{j})\, dt$

44. $\displaystyle\int_0^1 (\sqrt{t}\,\mathbf{i} + \sqrt{t + 1}\,\mathbf{j})\, dt$

45. $\displaystyle\int_0^{\pi/2} (3 \cos t\,\mathbf{i} + 3 \sin t\,\mathbf{j})\, dt$

46. $\displaystyle\int_0^3 (e^t \mathbf{i} + te^t \mathbf{j})\, dt$

In Exercises 47–50, find $\mathbf{r}(t)$ for the given conditions.

47. $\mathbf{r}'(t) = 4e^{2t} \mathbf{i} + 3e^t \mathbf{j}$
$\mathbf{r}(0) = 2\mathbf{i}$

48. $\mathbf{r}'(t) = 2t\mathbf{i} + \sqrt{t}\,\mathbf{j}$
$\mathbf{r}(0) = \mathbf{i} + \mathbf{j}$

49. $\mathbf{r}''(t) = -32\mathbf{j}$
$\mathbf{r}'(0) = 600\sqrt{3}\,\mathbf{i} + 600\mathbf{j}, \quad \mathbf{r}(0) = \mathbf{0}$

50. $\mathbf{r}''(t) = -4 \cos t\,\mathbf{i} - 3 \sin t\,\mathbf{j}$
$\mathbf{r}'(0) = 3\mathbf{j}, \quad \mathbf{r}(0) = 4\mathbf{i}$

In Exercises 51–55, prove the given property. In each case assume that $\mathbf{r}$ and $\mathbf{u}$ are differentiable vector-valued functions of t, f is a differentiable real-valued function of t, and c is a scalar.

51. $D_t[c\mathbf{r}(t)] = c\mathbf{r}'(t)$

52. $D_t[\mathbf{r}(t) \pm \mathbf{u}(t)] = \mathbf{r}'(t) \pm \mathbf{u}'(t)$

53. $D_t[f(t)\mathbf{r}(t)] = f(t)\mathbf{r}'(t) + f'(t)\mathbf{r}(t)$

54. $D_t[\mathbf{r}(f(t))] = \mathbf{r}'(f(t))f'(t)$

55. If $\mathbf{r}(t) \cdot \mathbf{r}(t) = c$, then $\mathbf{r}(t) \cdot \mathbf{r}'(t) = 0$.

13.4 Velocity and Acceleration

Velocity ▪ Acceleration ▪ Projectile motion

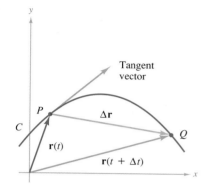

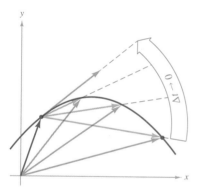

As $\Delta t \to 0$, $\dfrac{\Delta \mathbf{r}}{\Delta t}$ approaches tangent vector.

FIGURE 13.28

We are now ready to combine our work on parametric equations, curves, vectors, and vector-valued functions to form a model for motion along a curve. We begin by looking at the motion of an object in the plane.

As an object moves along a curve in the plane, the coordinates x and y of its center of mass are each functions of time t. Rather than using f and g to represent these two functions, it is convenient to write $x = x(t)$ and $y = y(t)$. Thus, the position vector $\mathbf{r}(t)$ takes the form

$$\mathbf{r}(t) = x(t)\mathbf{i} + y(t)\mathbf{j}. \qquad \text{Position vector}$$

The beauty of this vector model for representing motion is that we can use the first and second derivatives of the vector-valued function $\mathbf{r}$ to find the object's velocity and acceleration.

To find the velocity and acceleration vectors at a given time t, we consider a point $Q = (x(t + \Delta t), y(t + \Delta t))$ that is approaching the point $P = (x(t), y(t))$ along the curve C given by

$$x = x(t) \qquad \text{and} \qquad y = y(t)$$

as shown in Figure 13.28. As $\Delta t \to 0$, the direction of the vector $\overrightarrow{PQ}$ (denoted by $\Delta \mathbf{r}$) approaches the *direction of motion* at the time t, and we write

$$\Delta \mathbf{r} = \mathbf{r}(t + \Delta t) - \mathbf{r}(t)$$

$$\frac{\Delta \mathbf{r}}{\Delta t} = \frac{\mathbf{r}(t + \Delta t) - \mathbf{r}(t)}{\Delta t}$$

$$\lim_{\Delta t \to 0} \frac{\Delta \mathbf{r}}{\Delta t} = \lim_{\Delta t \to 0} \frac{\mathbf{r}(t + \Delta t) - \mathbf{r}(t)}{\Delta t}.$$

We define this limit, if it exists, as the **tangent vector** to the curve at the point P. Note that this is the same limit used to define $\mathbf{r}'(t)$. Thus, the direction of $\mathbf{r}'(t)$ gives us the direction of motion at the time t. Moreover, the magnitude of the vector $\mathbf{r}'(t)$,

$$\|\mathbf{r}'(t)\| = \|x'(t)\mathbf{i} + y'(t)\mathbf{j}\| = \sqrt{[x'(t)]^2 + [y'(t)]^2},$$

gives the **speed** of the object at time t. Similarly, we can use $\mathbf{r}''(t)$ to represent acceleration, as indicated in the following definition.

DEFINITION OF VELOCITY AND ACCELERATION	If x and y are twice differentiable functions of t, and $\mathbf{r}$ is a vector-valued function given by $\mathbf{r}(t) = x(t)\mathbf{i} + y(t)\mathbf{j}$, then the velocity vector, acceleration vector, and speed at time t are as follows.

$$\text{velocity} = \mathbf{v}(t) = \mathbf{r}'(t) = x'(t)\mathbf{i} + y'(t)\mathbf{j}$$

$$\text{acceleration} = \mathbf{a}(t) = \mathbf{r}''(t) = x''(t)\mathbf{i} + y''(t)\mathbf{j}$$

$$\text{speed} = \|\mathbf{v}(t)\| = \|\mathbf{r}'(t)\| = \sqrt{[x'(t)]^2 + [y'(t)]^2}$$

EXAMPLE 1 Finding velocity and acceleration along a plane curve

Find the velocity vector, speed, and acceleration vector of a particle that moves along the plane curve C described by

$$\mathbf{r}(t) = 2 \sin \frac{t}{2} \mathbf{i} + 2 \cos \frac{t}{2} \mathbf{j}.$$

SOLUTION

The velocity vector is given by

$$\mathbf{v}(t) = \mathbf{r}'(t)$$

$$= \cos \frac{t}{2} \mathbf{i} - \sin \frac{t}{2} \mathbf{j}$$

and the speed (at any time) is

$$\|\mathbf{r}'(t)\| = \sqrt{\cos^2 \frac{t}{2} + \sin^2 \frac{t}{2}} = 1.$$

Finally, the acceleration vector is given by

$$\mathbf{a}(t) = \mathbf{r}''(t)$$

$$= -\frac{1}{2} \sin \frac{t}{2} \mathbf{i} - \frac{1}{2} \cos \frac{t}{2} \mathbf{j}.$$

The parametric equations for the curve in Example 1 are $x = 2 \sin (t/2)$ and $y = 2 \cos (t/2)$. By eliminating the parameter t, we can obtain the rectangular equation

$$x^2 + y^2 = 4. \qquad \text{Rectangular equation}$$

Thus, the curve is a circle of radius 2 centered at the origin, as shown in Figure 13.29. Since the velocity vector $\mathbf{v}(t) = \cos (t/2) \mathbf{i} - \sin (t/2) \mathbf{j}$ has a constant magnitude but a changing direction as t increases, the particle moves around the circle at a constant speed.

REMARK It is interesting to note that the velocity and acceleration vectors in Example 1 are orthogonal at any point in time. This is characteristic of motion at a constant speed. (See Exercise 37.)

Circle: $x^2 + y^2 = 4$

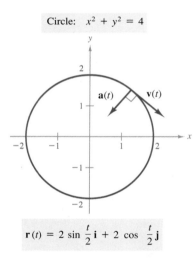

$$\mathbf{r}(t) = 2 \sin \frac{t}{2} \mathbf{i} + 2 \cos \frac{t}{2} \mathbf{j}$$

FIGURE 13.29

EXAMPLE 2 Finding velocity and acceleration along a plane curve

Sketch the path of an object moving along the plane curve given by

$$\mathbf{r}(t) = (t^2 - 4)\mathbf{i} + t\mathbf{j}$$

and find the velocity and acceleration vectors when $t = 0$ and $t = 2$.

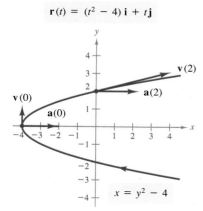

$$\mathbf{r}(t) = (t^2 - 4)\,\mathbf{i} + t\mathbf{j}$$

$\mathbf{v}(2)$

$\mathbf{a}(2)$

$\mathbf{v}(0)$

$\mathbf{a}(0)$

$x = y^2 - 4$

FIGURE 13.30

SOLUTION

Using the parametric equations $x = t^2 - 4$ and $y = t$, we see that the curve is a parabola given by $x = y^2 - 4$, as shown in Figure 13.30. The velocity vector (at any time) is given by

$$\mathbf{v}(t) = \mathbf{r}'(t) = 2t\mathbf{i} + \mathbf{j}$$

and the acceleration vector (at any time) is given by

$$\mathbf{a}(t) = \mathbf{r}''(t) = 2\mathbf{i}.$$

Therefore, when $t = 0$, the velocity and acceleration vectors are given by

$$\mathbf{v}(0) = 2(0)\mathbf{i} + \mathbf{j} = \mathbf{j} \qquad \text{and} \qquad \mathbf{a}(0) = 2\mathbf{i}$$

and when $t = 2$, the velocity and acceleration vectors are given by

$$\mathbf{v}(2) = 2(2)\mathbf{i} + \mathbf{j} = 4\mathbf{i} + \mathbf{j} \qquad \text{and} \qquad \mathbf{a}(2) = 2\mathbf{i}. \qquad \blacksquare$$

So far in this section we have concentrated on finding the velocity and acceleration by differentiating the position function. Many practical applications involve the reverse problem—finding the position function for a given velocity or acceleration. This is demonstrated in the next example.

EXAMPLE 3 Finding the position function by integration

An object starts from rest at the point $P = (2, 0)$ and moves with an acceleration of

$$\mathbf{a}(t) = \mathbf{i} + 2\mathbf{j}$$

where $\|\mathbf{a}(t)\|$ is measured in feet per second per second. Find the location of the object after $t = 2$ seconds.

SOLUTION

From the description of the object's motion, we deduce the following *initial conditions*. Since the object starts from rest, we have

$$\mathbf{v}(0) = \mathbf{0}.$$

Moreover, since the object starts at the point $(x, y) = (2, 0)$, we have

$$\mathbf{r}(0) = x(0)\mathbf{i} + y(0)\mathbf{j} = 2\mathbf{i} + 0\mathbf{j} = 2\mathbf{i}.$$

Now, to find the position function, we integrate twice, each time using one of the initial conditions to solve for the constant of integration. The velocity vector is

$$\mathbf{v}(t) = \int \mathbf{a}(t)\,dt = \int (\mathbf{i} + 2\mathbf{j})\,dt = t\mathbf{i} + 2t\mathbf{j} + \mathbf{C}$$

where $\mathbf{C} = C_1\mathbf{i} + C_2\mathbf{j}$. Letting $t = 0$ and applying the initial condition $\mathbf{v}(0) = \mathbf{0}$, we have

$$\mathbf{v}(0) = C_1\mathbf{i} + C_2\mathbf{j} = \mathbf{0} \quad \Longrightarrow \quad C_1 = C_2 = 0.$$

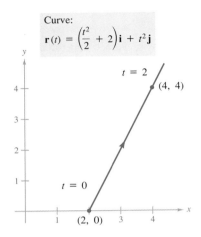

Curve:
$$\mathbf{r}(t) = \left(\frac{t^2}{2} + 2\right)\mathbf{i} + t^2\mathbf{j}$$

FIGURE 13.31

Thus, the *velocity* at any time is

$$\mathbf{v}(t) = t\mathbf{i} + 2t\mathbf{j}. \qquad \text{Velocity vector}$$

Now, integrating once more produces

$$\mathbf{r}(t) = \int \mathbf{v}(t)\,dt = \int (t\mathbf{i} + 2t\mathbf{j})\,dt = \frac{t^2}{2}\mathbf{i} + t^2\mathbf{j} + \mathbf{C}$$

where $\mathbf{C} = C_3\mathbf{i} + C_4\mathbf{j}$. Letting $t = 0$ and applying the initial condition $\mathbf{r}(0) = 2\mathbf{i}$, we have

$$\mathbf{r}(0) = C_3\mathbf{i} + C_4\mathbf{j} = 2\mathbf{i} \quad \Longrightarrow \quad C_3 = 2, \quad C_4 = 0.$$

Thus, the *position* vector is

$$\mathbf{r}(t) = \left(\frac{t^2}{2} + 2\right)\mathbf{i} + t^2\mathbf{j}. \qquad \text{Position vector}$$

The location of the object after 2 seconds is given by $\mathbf{r}(2) = 4\mathbf{i} + 4\mathbf{j}$, as shown in Figure 13.31. ▢

Motion of a projectile

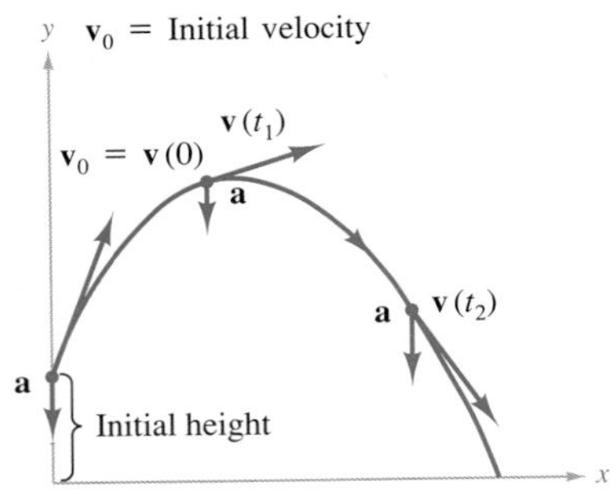

When we introduced parametric equations in Chapter 12, we began with a description of a projectile moving on a parabolic path. We now return to this problem to see how the path can be derived using a vector model.

We assume that gravity is the only force acting on the projectile after it is launched. Hence, the motion occurs in a vertical plane, which we represent by the xy-coordinate system with the origin as a point on the earth's surface, as shown in Figure 13.32. For a projectile of mass m, the force due to gravity is

$$\mathbf{F} = -mg\,\mathbf{j} \qquad \text{Force due to gravity}$$

where the gravitational constant is $g = 32$ feet per second per second, or 9.81 meters per second per second. By **Newton's Second Law of Motion,** this same force produces an acceleration $\mathbf{a} = \mathbf{a}(t)$, and satisfies the equation $\mathbf{F} = m\mathbf{a}$. Consequently, the acceleration of the projectile is given by

$$m\mathbf{a} = -mg\,\mathbf{j}$$
$$\mathbf{a} = -g\,\mathbf{j}. \qquad \text{Acceleration of projectile}$$

With this acceleration, we can derive the position function for the path of a projectile, as shown in the next example.

Parabolic Path of a Projectile

FIGURE 13.32

EXAMPLE 4 Derivation of the position function for a projectile

A projectile of mass m is launched from an initial position $\mathbf{r}_0$ with an initial velocity $\mathbf{v}_0$. Find its position vector as a function of time.

SOLUTION

We begin with the acceleration $\mathbf{a}(t) = -g\,\mathbf{j}$ and integrate twice to obtain

$$\mathbf{v}(t) = \int \mathbf{a}(t)\,dt = \int -g\,\mathbf{j}\,dt = -gt\,\mathbf{j} + \mathbf{C}_1$$

$$\mathbf{r}(t) = \int \mathbf{v}(t)\,dt = \int (-gt\,\mathbf{j} + \mathbf{C}_1)\,dt = -\frac{1}{2}gt^2\,\mathbf{j} + \mathbf{C}_1 t + \mathbf{C}_2$$

where $\mathbf{C}_1$ and $\mathbf{C}_2$ are constant vectors. Since $\mathbf{v}(0) = \mathbf{v}_0$ and $\mathbf{r}(0) = \mathbf{r}_0$, we can solve for C_1 and C_2 as follows.

$$\mathbf{v}(0) = 0\mathbf{j} + \mathbf{C}_1 = \mathbf{v}_0 \qquad \Longrightarrow \qquad \mathbf{C}_1 = \mathbf{v}_0$$
$$\mathbf{r}(0) = 0\mathbf{j} + 0\mathbf{C}_1 + \mathbf{C}_2 = \mathbf{r}_0 \qquad \Longrightarrow \qquad \mathbf{C}_2 = \mathbf{r}_0$$

Therefore, the position vector is given by

$$\mathbf{r}(t) = -\frac{1}{2}gt^2\,\mathbf{j} + t\mathbf{v}_0 + \mathbf{r}_0.$$

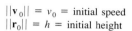

$\|\mathbf{v}_0\| = v_0 =$ initial speed
$\|\mathbf{r}_0\| = h =$ initial height

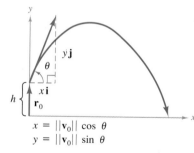

$x = \|\mathbf{v}_0\| \cos\theta$
$y = \|\mathbf{v}_0\| \sin\theta$

Initial Conditions for Projectile

FIGURE 13.33

In many projectile problems the constant vectors $\mathbf{r}_0$ and $\mathbf{v}_0$ are not given explicitly. Often we are given the initial height h, the initial speed v_0, and the angle θ at which the projectile is launched, as shown in Figure 13.33. From the given height, we deduce that

$$h = \|\mathbf{r}_0\| \qquad \Longrightarrow \qquad \mathbf{r}_0 = h\,\mathbf{j}.$$

Similarly, since the speed gives the magnitude of the initial velocity, we have $v_0 = \|\mathbf{v}_0\|$ and we can write

$$\mathbf{v}_0 = x\mathbf{i} + y\mathbf{j} = (\|\mathbf{v}_0\| \cos\theta)\,\mathbf{i} + (\|\mathbf{v}_0\| \sin\theta)\,\mathbf{j}$$
$$= v_0 \cos\theta\,\mathbf{i} + v_0 \sin\theta\,\mathbf{j}.$$

Thus, it follows that the position vector can be written in the form

$$\mathbf{r}(t) = -\frac{1}{2}gt^2\,\mathbf{j} + t\mathbf{v}_0 + \mathbf{r}_0 \qquad \text{Position vector}$$

$$= -\frac{1}{2}gt^2\,\mathbf{j} + tv_0 \cos\theta\,\mathbf{i} + tv_0 \sin\theta\,\mathbf{j} + h\mathbf{j}$$

$$= (v_0 \cos\theta)t\,\mathbf{i} + \left[h + (v_0 \sin\theta)t - \frac{1}{2}gt^2\right]\mathbf{j}$$

which is the result listed in the following theorem.

THEOREM 13.10
POSITION FUNCTION
FOR A PROJECTILE

The path of a projectile launched from an initial height h with initial speed v_0 and angle of elevation θ is described by the vector function

$$\mathbf{r}(t) = (v_0 \cos\theta)t\,\mathbf{i} + \left[h + (v_0 \sin\theta)t - \frac{1}{2}gt^2\right]\mathbf{j}$$

where g is the gravitational constant.

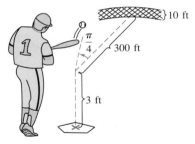

FIGURE 13.34

EXAMPLE 5 Describing the path of a projectile

A baseball is hit 3 feet above ground at 100 feet per second and at an angle of $\pi/4$ with respect to the ground, as shown in Figure 13.34. Find the maximum height reached by the baseball. Will it clear a 10-foot-high fence located 300 feet from home plate?

SOLUTION

We are given $h = 3$, $v_0 = 100$, and $\theta = \pi/4$. Thus, using $g = 32$ feet per second per second produces

$$\mathbf{r}(t) = \left(100 \cos \frac{\pi}{4}\right) t\mathbf{i} + \left[3 + \left(100 \sin \frac{\pi}{4}\right) t - 16t^2\right]\mathbf{j}$$
$$= (50\sqrt{2}t)\mathbf{i} + [3 + 50\sqrt{2}t - 16t^2]\mathbf{j}$$
$$\mathbf{v}(t) = \mathbf{r}'(t) = 50\sqrt{2}\mathbf{i} + (50\sqrt{2} - 32t)\mathbf{j}.$$

The maximum height occurs when $y'(t) = 50\sqrt{2} - 32t = 0$, which implies that

$$t = \frac{25\sqrt{2}}{16} \approx 2.21 \text{ sec.}$$

Hence, the maximum height reached by the ball is

$$y = 3 + 50\sqrt{2}\left(\frac{25\sqrt{2}}{16}\right) - 16\left(\frac{25\sqrt{2}}{16}\right)^2 = \frac{649}{8} \approx 81 \text{ ft.}$$

The ball is 300 feet from where it was hit when

$$300 = x(t) = 50\sqrt{2}t \quad \Longrightarrow \quad t = 3\sqrt{2}.$$

At this time the height of the ball is

$$y = 3 + 50\sqrt{2}(3\sqrt{2}) - 16(3\sqrt{2})^2 = 303 - 288 = 15 \text{ ft.}$$

Therefore, the ball clears the 10-foot fence for a home run. ▭

EXERCISES for Section 13.4

In Exercises 1–8, the position function $\mathbf{r}$ describes the path of an object moving in the xy-plane. Sketch a graph of the path and sketch the velocity and acceleration vectors at the given point.

Position function	*Point*
1. $\mathbf{r}(t) = 3t\mathbf{i} + (t - 1)\mathbf{j}$	$(3, 0)$
2. $\mathbf{r}(t) = (6 - t)\mathbf{i} + t\mathbf{j}$	$(3, 3)$
3. $\mathbf{r}(t) = t^2\mathbf{i} + t\mathbf{j}$	$(4, 2)$
4. $\mathbf{r}(t) = t^3\mathbf{i} + t^2\mathbf{j}$	$(1, 1)$
5. $\mathbf{r}(t) = 2 \cos t\,\mathbf{i} + 2 \sin t\,\mathbf{j}$	$(\sqrt{2}, \sqrt{2})$
6. $\mathbf{r}(t) = 2 \cos t\,\mathbf{i} + 3 \sin t\,\mathbf{j}$	$(2, 0)$
7. $\mathbf{r}(t) = \langle t - \sin t, 1 - \cos t \rangle$	$(\pi, 2)$
8. $\mathbf{r}(t) = \langle e^{-t}, e^t \rangle$	$(1, 1)$

In Exercises 9–12, use the given acceleration function to find the velocity and position functions. Then find the position function at time $t = 2$.

9. $\mathbf{a}(t) = \mathbf{i} + \mathbf{j}$, $\mathbf{v}(0) = \mathbf{0}$, $\mathbf{r}(0) = \mathbf{0}$
10. $\mathbf{a}(t) = t\mathbf{i}$, $\mathbf{v}(0) = 5\mathbf{j}$, $\mathbf{r}(0) = \mathbf{0}$
11. $\mathbf{a}(t) = t\mathbf{i} + t\mathbf{j}$, $\mathbf{v}(1) = 5\mathbf{j}$, $\mathbf{r}(1) = \mathbf{0}$
12. $\mathbf{a}(t) = \mathbf{0}$, $\mathbf{v}(3) = 4\mathbf{i} + 3\mathbf{j}$, $\mathbf{r}(3) = 10\mathbf{i} + 4\mathbf{j}$

13. A baseball player at second base throws the ball 90 feet to the player at first base. The ball is thrown at 50 miles per hour at an angle of 15° with the horizontal. How far will the ball drop by the time it reaches first base? (Neglect air resistance.)

14. The quarterback of a football team releases a pass at a height of 7 feet above the playing field and the football is caught by a receiver 30 yards downfield at a height of 4 feet. The pass is released at an angle of 35° with the horizontal.
  (a) Find the speed of the football when it is released. (Neglect air resistance.)
  (b) Find the maximum height of the ball.
  (c) Find the time the receiver has to position himself after the quarterback releases the ball.

15. A baseball, hit 3 feet above the ground, leaves the bat at an angle of 45° and is caught by an outfielder 300 feet from home plate. What was the initial speed of the ball, and how high did it rise if it was caught 3 feet above the ground?

16. The nozzle of a hose discharges water with a speed of 40 feet per second. Determine how high the water rises if the hose makes an angle of 60° with the ground.

17. A child standing 20 feet from the base of a silo attempts to throw a ball into an opening 40 feet from the point of release (see figure). Find the minimum initial speed and the corresponding angle at which the ball must be thrown to go into the opening.

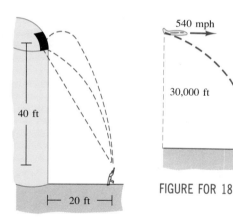

FIGURE FOR 18

FIGURE FOR 17

18. A bomber is flying at an altitude of 30,000 feet with a speed of 540 miles per hour (792 feet per second) (see figure). When should the bomb be released to hit the target? (Neglect air resistance, and give your answer in terms of the angle of depression from the plane to the target.) What is the speed of the bomb at the time of impact?

19. Find the angle at which an object must be thrown to obtain (a) the maximum range and (b) the maximum height.

20. A shot fired from a gun with a muzzle velocity of 1200 feet per second is to hit a target 3000 feet away. Neglecting air resistance, determine the minimum angle of elevation of the gun.

21. A projectile is fired from ground level at an angle of 10° with the horizontal. Find the minimum initial velocity necessary if the projectile is to have a range of 100 feet.

22. Eliminate the parameter t from the position function for the motion of a projectile to show that the rectangular equation is

$$y = -\frac{16 \sec^2 \theta}{v_0{}^2}x^2 + (\tan \theta)x + h.$$

23. The path of a ball is given by the rectangular equation

$$y = x - 0.005x^2.$$

Use the result of Exercise 22 to find the position function. Then find the speed and direction of the ball at the point when it has traveled 60 feet horizontally.

In Exercises 24–26, consider the motion of a particle on the circumference of a rolling circle. As the circle rolls, it generates the cycloid

$$\mathbf{r}(t) = b(\omega t - \sin \omega t)\mathbf{i} + b(1 - \cos \omega t)\mathbf{j}$$

where ω is the constant angular velocity of the circle.

24. Find the velocity and acceleration vectors of the particle.

25. Use the results of Exercise 24 to determine the times that the speed of the particle will be (a) zero and (b) maximum.

26. Find the maximum speed of a point on the circumference of an automobile wheel of radius 1 foot when the automobile is traveling 55 miles per hour. Compare this speed with the speed of the automobile.

In Exercises 27–30, consider a particle moving on a circular path of radius b described by

$$\mathbf{r}(t) = b \cos \omega t \, \mathbf{i} + b \sin \omega t \, \mathbf{j}$$

where $\omega = d\theta/dt$ is the constant angular velocity.

27. Find the velocity vector and show that it is orthogonal to $\mathbf{r}(t)$.

28. Show that the speed of the particle is $b\omega$.

29. Find the acceleration vector and show that its direction is always toward the center of the circle.

30. Show that the magnitude of the acceleration vector is $\omega^2 b$.

In Exercises 31 and 32, use the results of Exercises 27–30.

31. A stone weighing 1 pound is attached to a 2-foot string and is whirled horizontally (see figure). The string will break under a force of 10 pounds. Find the maximum velocity the stone can attain without breaking the string. (Use $\mathbf{F} = m\mathbf{a}$, where $m = 1/32$.)

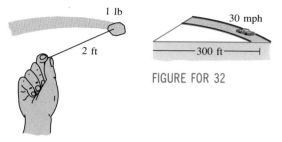

1 lb

2 ft

30 mph

300 ft

FIGURE FOR 32

FIGURE FOR 31

32. A 3000-pound automobile is negotiating a circular interchange of radius 300 feet at 30 miles per hour (see figure). Assuming the roadway to be level, find the force between the tires and the road so that the car stays on the circular path and does not skid. (Use $\mathbf{F} = m\mathbf{a}$, where $m = 3000/32$.)

33. In Exercise 32 find the angle at which the roadway should be banked so that no lateral friction force is exerted on the tires of the automobile.

34. Use a computer or graphics calculator to sketch a graph of the path of a projectile launched from an initial height of 10 feet with an initial speed of 120 feet per second at an angle of elevation of 42°. Find the maximum height and the range of the projectile.

35. In the Chapter 13 Application, the path of a shot-put thrown at an angle θ is given by

$$\mathbf{r}(t) = (v_0 \cos \theta)t\mathbf{i} + \left[h + (v_0 \sin \theta)t - \frac{1}{2}gt^2 \right]\mathbf{j}$$

where v_0 is the initial speed, h is the initial height, t is the time in seconds, and g is the acceleration due to gravity. Verify that the shot-put will remain in the air for a total of

$$t = \frac{v_0 \sin \theta + \sqrt{v_0^2 \sin^2 \theta + 2gh}}{g} \text{ seconds}$$

and will travel a horizontal distance of

$$\frac{v_0^2 \cos \theta}{g}\left(\sin \theta + \sqrt{\sin^2 \theta + \frac{2gh}{v_0^2}} \right) \text{ feet.}$$

36. A shot-put is thrown from a height of $h = 6$ feet with an initial speed of $v_0 = 45$ feet per second. Find the total time of travel and the total horizontal distance traveled if the shot-put is thrown at an angle of $\theta = 42.5°$ with the horizontal.

37. Prove that if an object is traveling at a constant speed, then its velocity and acceleration vectors are orthogonal.

38. Prove that an object moving in a straight line at a constant speed has an acceleration of zero.

13.5 Tangent Vectors and Normal Vectors

Unit tangent vector ▪ Principal unit normal vector ▪ Components of acceleration

In the previous section we saw that the velocity vector points in the direction of motion. This observation leads to the following definition, which applies to any smooth curve—not just those for which the parameter represents time.

DEFINITION OF UNIT TANGENT VECTOR

Let C be a smooth curve represented by $\mathbf{r}$ on an open interval I. If $\mathbf{r}'(t) \neq \mathbf{0}$, then the **unit tangent vector** $\mathbf{T}(t)$ at t is defined to be

$$\mathbf{T}(t) = \frac{\mathbf{r}'(t)}{\|\mathbf{r}'(t)\|}.$$

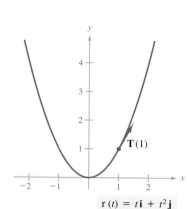

FIGURE 13.35

$r(t) = t\mathbf{i} + t^2\mathbf{j}$

EXAMPLE 1 Finding the unit tangent vector

Find the unit tangent vector to the curve given by $\mathbf{r}(t) = t\mathbf{i} + t^2\mathbf{j}$ when $t = 1$.

SOLUTION

The derivative of $\mathbf{r}(t)$ is $\mathbf{r}'(t) = \mathbf{i} + 2t\mathbf{j}$. Thus, the unit tangent vector is

$$\mathbf{T}(t) = \frac{\mathbf{r}'(t)}{\|\mathbf{r}'(t)\|} = \frac{1}{\sqrt{1 + 4t^2}}(\mathbf{i} + 2t\mathbf{j}).$$

When $t = 1$, the unit tangent vector is

$$\mathbf{T}(1) = \frac{1}{\sqrt{5}}(\mathbf{i} + 2\mathbf{j})$$

as shown in Figure 13.35.

The **tangent line to a curve** at a point is defined to be the line passing through the point and parallel to the unit tangent vector.

In Figure 13.35, there are two vectors that are orthogonal to the tangent vector $\mathbf{T}(t)$. One of these is the vector $\mathbf{T}'(t)$. This follows from Property 6 of Theorem 13.9. That is,

$$\mathbf{T}(t) \cdot \mathbf{T}(t) = \|\mathbf{T}(t)\|^2 = 1 \quad \Longrightarrow \quad \mathbf{T}(t) \cdot \mathbf{T}'(t) = 0.$$

By normalizing the vector $\mathbf{T}'(t)$, we obtain a special vector called the **principal unit normal vector,** as indicated in the following definition.

DEFINITION OF PRINCIPAL UNIT NORMAL VECTOR

Let C be a smooth curve represented by $\mathbf{r}$ on an open interval I. If $\mathbf{T}'(t) \neq \mathbf{0}$, then the **principal unit normal vector** at t is defined to be

$$\mathbf{N}(t) = \frac{\mathbf{T}'(t)}{\|\mathbf{T}'(t)\|}.$$

EXAMPLE 2 Finding the principal unit normal vector

Find $\mathbf{N}(t)$ and $\mathbf{N}(1)$ for the curve represented by

$$\mathbf{r}(t) = 3t\mathbf{i} + 2t^2\mathbf{j}.$$

SOLUTION

By differentiating, we obtain

$$\mathbf{r}'(t) = 3\mathbf{i} + 4t\mathbf{j} \quad \text{and} \quad \|\mathbf{r}'(t)\| = \sqrt{9 + 16t^2}$$

which implies that the unit tangent vector is

$$\mathbf{T}(t) = \frac{\mathbf{r}'(t)}{\|\mathbf{r}'(t)\|} = \frac{1}{\sqrt{9 + 16t^2}}(3\mathbf{i} + 4t\mathbf{j}).$$

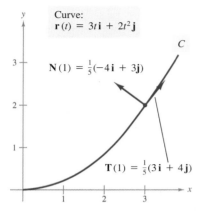

Principal unit normal vector points toward the concave side of the curve.

FIGURE 13.36

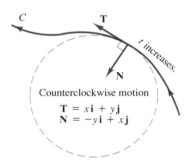

Counterclockwise motion
$T = x\mathbf{i} + y\mathbf{j}$
$N = -y\mathbf{i} + x\mathbf{j}$

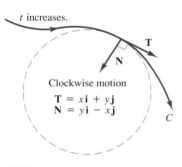

t increases.

Clockwise motion
$T = x\mathbf{i} + y\mathbf{j}$
$N = y\mathbf{i} - x\mathbf{j}$

FIGURE 13.37

Using Theorem 13.9, we differentiate $\mathbf{T}(t)$ with respect to t to obtain

$$\mathbf{T}'(t) = \frac{1}{\sqrt{9 + 16t^2}}(4\mathbf{j}) - \frac{16t}{(9 + 16t^2)^{3/2}}(3\mathbf{i} + 4t\mathbf{j})$$

$$= \frac{12}{(9 + 16t^2)^{3/2}}(-4t\mathbf{i} + 3\mathbf{j})$$

$$\|\mathbf{T}'(t)\| = 12\sqrt{\frac{9 + 16t^2}{(9 + 16t^2)^3}} = \frac{12}{9 + 16t^2}.$$

Therefore, the principal unit normal vector is

$$\mathbf{N}(t) = \frac{\mathbf{T}'(t)}{\|\mathbf{T}'(t)\|} = \frac{1}{\sqrt{9 + 16t^2}}(-4t\mathbf{i} + 3\mathbf{j}).$$

When $t = 1$, the principal unit normal vector is

$$\mathbf{N}(1) = \frac{1}{5}(-4\mathbf{i} + 3\mathbf{j})$$

as shown in Figure 13.36.

The principal unit normal vector can be difficult to evaluate algebraically. For curves in the plane, we can simplify the algebra by finding

$$\mathbf{T}(t) = x(t)\mathbf{i} + y(t)\mathbf{j}$$

and observing that $\mathbf{N}(t)$ must be either

$$\mathbf{N}_1(t) = y(t)\mathbf{i} - x(t)\mathbf{j} \quad \text{or} \quad \mathbf{N}_2(t) = -y(t)\mathbf{i} + x(t)\mathbf{j}.$$

Since $\sqrt{[x(t)]^2 + [y(t)]^2} = 1$, it follows that both $\mathbf{N}_1(t)$ and $\mathbf{N}_2(t)$ are unit normal vectors. The *principal* unit normal vector $\mathbf{N}$ is the one that points toward the concave side of the curve (see Exercise 25). With respect to the closest circular approximation at a point on the curve, if the motion of the object is *counterclockwise*, then we choose $\mathbf{N}(t) = -y(t)\mathbf{i} + x(t)\mathbf{j}$, as indicated in Figure 13.37. If the motion is *clockwise*, we choose $\mathbf{N}(t) = y(t)\mathbf{i} - x(t)\mathbf{j}$.

For instance, in Example 2, the motion is counterclockwise at the point $(3, 2)$. Hence, the unit normal vector at that point is

$$\mathbf{N} = \frac{1}{5}(-4\mathbf{i} + 3\mathbf{j}).$$

This "short-cut" for finding the unit normal vector is further demonstrated in Example 3.

EXAMPLE 3 Finding the principal unit normal vector

Find the principal unit normal vector for the circle given by

$$\mathbf{r}(t) = 2 \cos t\, \mathbf{i} + 2 \sin t\, \mathbf{j}.$$

Curve:
$\mathbf{r}(t) = 2 \cos t \, \mathbf{i} + 2 \sin t \, \mathbf{j}$

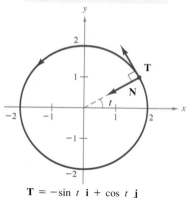

$\mathbf{T} = -\sin t \, \mathbf{i} + \cos t \, \mathbf{j}$
$\mathbf{N} = -\cos t \, \mathbf{i} - \sin t \, \mathbf{j}$

FIGURE 13.38

SOLUTION

The principal unit tangent vector is

$$\mathbf{T}(t) = \frac{1}{2}(-2 \sin t \, \mathbf{i} + 2 \cos t \, \mathbf{j})$$

$$= -\sin t \, \mathbf{i} + \cos t \, \mathbf{j}.$$

Thus, $\mathbf{N}'(t)$ is either

$$\mathbf{N}_1(t) = \cos t \, \mathbf{i} + \sin t \, \mathbf{j} \quad \text{or} \quad \mathbf{N}_2(t) = -\cos t \, \mathbf{i} - \sin t \, \mathbf{j}.$$

Of these two vectors, $\mathbf{N}_1(t)$ points away from the origin and $\mathbf{N}_2(t)$ points toward the origin—toward the direction the curve is turning. Hence, the principal unit normal vector must be $\mathbf{N}_2(t)$, as shown in Figure 13.38. ▬

EXAMPLE 4 Path of a projectile

Use the position vector

$$\mathbf{r}(t) = (v_0 \cos \theta)t \, \mathbf{i} + [(v_0 \sin \theta)t - 16t^2] \, \mathbf{j}$$

where $v_0 = 20$ feet per second and $\theta = \arcsin \frac{4}{5}$ to find $\mathbf{T}(t)$ and $\mathbf{N}(t)$ when $t = \frac{1}{2}$ and $t = \frac{3}{4}$.

SOLUTION

Since $\theta = \arcsin \frac{4}{5}$, $\sin \theta = \frac{4}{5}$ and $\cos \theta = \frac{3}{5}$. Hence, we have

$$\mathbf{r}(t) = 12t \, \mathbf{i} + (16t - 16t^2) \, \mathbf{j}$$
$$\mathbf{r}'(t) = 12 \, \mathbf{i} + (16 - 32t) \, \mathbf{j} = 4[3 \, \mathbf{i} + 4(1 - 2t) \, \mathbf{j}]$$
$$\|\mathbf{r}'(t)\| = 4\sqrt{9 + 16(1 - 2t)^2}.$$

Therefore, the principal unit tangent vector is

$$\mathbf{T}(t) = \frac{\mathbf{r}'(t)}{\|\mathbf{r}'(t)\|} = \frac{3 \, \mathbf{i} + 4(1 - 2t) \, \mathbf{j}}{\sqrt{9 + 16(1 - 2t)^2}}.$$

From Figure 13.39, we can see that the principal unit normal vector should point downward. Hence, $\mathbf{N}(t)$ is given by

$$\mathbf{N}(t) = \frac{4(1 - 2t) \, \mathbf{i} - 3 \, \mathbf{j}}{\sqrt{9 + 16(1 - 2t)^2}}.$$

When $t = \frac{1}{2}$, we have

$$\mathbf{T} = \frac{3 \, \mathbf{i}}{\sqrt{9}} = \mathbf{i} \quad \text{and} \quad \mathbf{N} = -\frac{3 \, \mathbf{j}}{\sqrt{9}} = -\mathbf{j}$$

and when $t = \frac{3}{4}$, we have

$$\mathbf{T} = \frac{3 \, \mathbf{i} - 2 \, \mathbf{j}}{\sqrt{13}} \quad \text{and} \quad \mathbf{N} = \frac{-2 \, \mathbf{i} - 3 \, \mathbf{j}}{\sqrt{13}}. \quad ▬$$

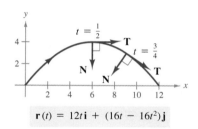

$\mathbf{r}(t) = 12t \, \mathbf{i} + (16t - 16t^2) \, \mathbf{j}$

FIGURE 13.39

Tangential and normal components of acceleration

We now return to the problem of describing the motion of an object along a curve. In the previous section we observed that for an object traveling at a *constant speed*, the velocity and acceleration vectors are perpendicular. This seems reasonable, since the speed would not be constant if any acceleration were acting in the direction of motion. (You can verify this observation by noting that $\mathbf{r}''(t) \cdot \mathbf{r}'(t) = 0$ if $\|\mathbf{r}'(t)\|$ is a constant. See Property 6 of Theorem 13.9.)

However, for an object traveling at a *variable speed*, the velocity and acceleration vectors are not necessarily perpendicular. For instance, we saw that the acceleration vector for a projectile always points down, regardless of the direction of motion.

In general, part of the acceleration (the tangential component) acts in the direction of motion, and part (the normal component) acts perpendicular to the direction of motion. In order to determine these two components, we use the unit vectors $\mathbf{T}(t)$ and $\mathbf{N}(t)$, which serve in much the same way as do $\mathbf{i}$ and $\mathbf{j}$ in representing vectors in the plane.

Recall that the projection of a vector $\mathbf{u}$ onto a *unit* vector $\mathbf{v}$ is given by $(\mathbf{u} \cdot \mathbf{v})\mathbf{v}$. Thus, at any time t, the projections of the acceleration vectors onto the unit vectors $\mathbf{T}(t)$ and $\mathbf{N}(t)$ are given by

$$(\mathbf{a} \cdot \mathbf{T})\mathbf{T} \quad \text{and} \quad (\mathbf{a} \cdot \mathbf{N})\mathbf{N}$$

respectively, as shown in Figure 13.40. We call the two dot products $\mathbf{a} \cdot \mathbf{T}$ and $\mathbf{a} \cdot \mathbf{N}$ the **tangential** and **normal components of acceleration.** (The normal component of acceleration is also referred to as the **centripetal** component of acceleration.)

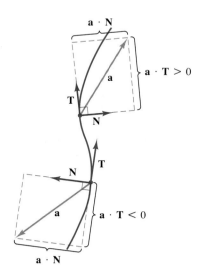

The tangential and normal components of acceleration are obtained by projecting $\mathbf{a}$ onto $\mathbf{T}$ and $\mathbf{N}$.

FIGURE 13.40

DEFINITION OF TANGENTIAL AND NORMAL COMPONENTS OF ACCELERATION	Let $\mathbf{r}(t)$ be the position vector for a smooth curve C, and let the acceleration vector be represented by $$\mathbf{a}(t) = a_{\mathbf{T}}\mathbf{T} + a_{\mathbf{N}}\mathbf{N} = (\mathbf{a} \cdot \mathbf{T})\mathbf{T} + (\mathbf{a} \cdot \mathbf{N})\mathbf{N}.$$ We call $a_{\mathbf{T}} = \mathbf{a} \cdot \mathbf{T}$ the **tangential component of acceleration,** and we call $a_{\mathbf{N}} = \mathbf{a} \cdot \mathbf{N}$ the **normal component of acceleration.**

EXAMPLE 5 The tangential and normal components of acceleration

Find the tangential and normal components of acceleration when $t = 2$ for the following position vector, and sketch the result.

$$\mathbf{r}(t) = 4t\,\mathbf{i} + (2 \ln t - t^2)\,\mathbf{j}, \qquad t > 0$$

SOLUTION

We begin by computing $\mathbf{T}(t)$ and $\mathbf{N}(t)$ as follows.

$$\mathbf{r}'(t) = 4\mathbf{i} + \left(\frac{2 - 2t^2}{t}\right)\mathbf{j}$$

$$\|\mathbf{r}'(t)\| = \sqrt{16 + \frac{4 - 8t^2 + 4t^4}{t^2}} = \frac{2(1 + t^2)}{t}$$

$$\mathbf{T}(t) = \frac{\mathbf{r}'(t)}{\|\mathbf{r}'(t)\|} = \frac{2t}{1 + t^2}\mathbf{i} + \frac{1 - t^2}{1 + t^2}\mathbf{j}$$

$$\mathbf{N}(t) = \frac{1 - t^2}{1 + t^2}\mathbf{i} - \frac{2t}{1 + t^2}\mathbf{j}$$

Furthermore, since the acceleration vector is given by

$$\mathbf{a}(t) = \mathbf{r}''(t)$$
$$= -2\left(\frac{1 + t^2}{t^2}\right)\mathbf{j}$$

it follows that when $t = 2$, we have

$$\mathbf{a} = -\frac{5}{2}\mathbf{j}, \quad \mathbf{T} = \frac{4}{5}\mathbf{i} - \frac{3}{5}\mathbf{j}, \quad \text{and} \quad \mathbf{N} = -\frac{3}{5}\mathbf{i} - \frac{4}{5}\mathbf{j}.$$

Therefore, when $t = 2$, the tangential and normal components of acceleration are given by

$$\mathbf{a} \cdot \mathbf{T} = \frac{3}{2}$$

and

$$\mathbf{a} \cdot \mathbf{N} = 2$$

as shown in Figure 13.41.

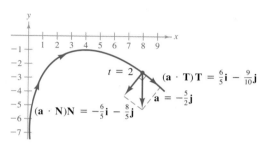

$$\mathbf{r}(t) = 4t\mathbf{i} + (2 \ln t - t^2)\mathbf{j}$$

FIGURE 13.41

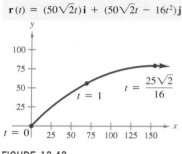

$$\mathbf{r}(t) = (50\sqrt{2}t)\mathbf{i} + (50\sqrt{2}t - 16t^2)\mathbf{j}$$

FIGURE 13.42

EXAMPLE 6 An application

The position vector for the projectile shown in Figure 13.42 is given by

$$\mathbf{r}(t) = (50\sqrt{2}t)\mathbf{i} + (50\sqrt{2}t - 16t^2)\mathbf{j}.$$

Find the tangential component of acceleration when $t = 0$, 1, and $25\sqrt{2}/16$.

SOLUTION

$$\mathbf{r}'(t) = 50\sqrt{2}\mathbf{i} + (50\sqrt{2} - 32t)\mathbf{j}$$
$$\|\mathbf{r}'(t)\| = 2\sqrt{50^2 - 16(50)\sqrt{2}t + 16^2t^2}$$
$$\mathbf{T}(t) = \frac{\mathbf{r}'(t)}{\|\mathbf{r}'(t)\|} = \frac{50\sqrt{2}\mathbf{i} + (50\sqrt{2} - 32t)\mathbf{j}}{2\sqrt{50^2 - 16(50)\sqrt{2}t + 16^2t^2}}$$
$$\mathbf{a}(t) = \mathbf{r}''(t) = -32\mathbf{j}$$

The tangential component of acceleration is

$$a_{\mathbf{T}} = \mathbf{a} \cdot \mathbf{T} = \frac{-32(50\sqrt{2} - 32t)}{2\sqrt{50^2 - 16(50)\sqrt{2}t + 16^2t^2}}.$$

At the specified times, we have

$$t = 0: \quad a_{\mathbf{T}} = \frac{-32(50\sqrt{2})}{100} = -16\sqrt{2} \approx -22.6$$

$$t = 1: \quad a_{\mathbf{T}} = \frac{-32(50\sqrt{2} - 32)}{2\sqrt{50^2 - 16(50)\sqrt{2} + 16^2}} \approx -15.4$$

$$t = \frac{25\sqrt{2}}{16}: \quad a_{\mathbf{T}} = \frac{-32(50\sqrt{2} - 50\sqrt{2})}{50\sqrt{2}} = 0.$$

We can see from Figure 13.42 that, at the maximum height, the tangential component is zero. This is reasonable because the direction of motion is horizontal at that point and the tangential component of acceleration is equal to the horizontal component of acceleration. ▭

EXERCISES for Section 13.5

In Exercises 1–14, find $\mathbf{T}(t)$, $\mathbf{N}(t)$, $a_{\mathbf{T}}$, and $a_{\mathbf{N}}$ at the given time t.

Function	Time
1. $\mathbf{r}(t) = 4t\mathbf{i}$	$t = 2$
2. $\mathbf{r}(t) = 4t\mathbf{i} - 2t\mathbf{j}$	$t = 1$
3. $\mathbf{r}(t) = 4t^2\mathbf{i}$	$t = 4$
4. $\mathbf{r}(t) = t^2\mathbf{i} + \mathbf{j}$	$t = 0$
5. $\mathbf{r}(t) = t\mathbf{i} + \dfrac{1}{t}\mathbf{j}$	$t = 1$
6. $\mathbf{r}(t) = t\mathbf{i} + t^2\mathbf{j}$	$t = 1$
7. $\mathbf{r}(t) = 4\cos 2\pi t\mathbf{i} + 4\sin 2\pi t\mathbf{j}$	$t = \dfrac{1}{8}$

Function	Time
8. $\mathbf{r}(t) = a\cos \omega t\mathbf{i} + a\sin \omega t\mathbf{j}$	$t = t_0$
9. $\mathbf{r}(t) = 2\cos \pi t\mathbf{i} + 2\sin \pi t\mathbf{j}$	$t = \dfrac{1}{3}$
10. $\mathbf{r}(t) = a\cos \omega t\mathbf{i} + b\sin \omega t\mathbf{j}$	$t = 0$
11. $\mathbf{r}(t) = e^t\cos t\mathbf{i} + e^t\sin t\mathbf{j}$	$t = \dfrac{\pi}{2}$
12. $\mathbf{r}(t) = \langle \omega t - \sin \omega t, 1 - \cos \omega t\rangle$	$t = t_0$
13. $\mathbf{r}(t) = \langle\cos \omega t + \omega t \sin \omega t,$ $\sin \omega t - \omega t \cos \omega t\rangle$	$t = t_0$
14. $\mathbf{r}(t) = \langle t \ln t, t - 1\rangle$	$t = t_0$

In Exercises 15 and 16, sketch the graph of the curve represented by the vector-valued function, and at the point on the curve determined by $\mathbf{r}(t_0)$ sketch the vectors $\mathbf{T}$ and $\mathbf{N}$. Note that $\mathbf{N}$ points toward the concave side of the curve.

15. $\mathbf{r}(t) = t\mathbf{i} + \dfrac{1}{t}\mathbf{j}$ $\qquad\qquad t_0 = 2$

16. $\mathbf{r}(t) = 2\cos t\,\mathbf{i} + 2\sin t\,\mathbf{j}$ $\qquad t_0 = \dfrac{\pi}{4}$

17. Find the tangential and normal components of acceleration for a projectile fired at an angle θ with the horizontal at an initial speed of v_0. What are the components when the projectile is at its maximum height?

18. A plane flying at an altitude of 30,000 feet and with a speed of 540 miles per hour (792 feet per second) releases a bomb. Find the tangential and normal components of acceleration acting on the bomb.

19. An object is spinning at a constant speed on the end of a string, according to the position function given in Exercise 10.
   (a) If the angular velocity ω is doubled, how is the centripetal component of acceleration changed?
   (b) If the angular velocity is unchanged but the length of the string is halved, how is the centripetal component of acceleration changed?

20. An object of mass m moves at a constant speed v in a circular path of radius r. The force required to produce the centripetal component of acceleration is called the centripetal force and is given by $F = mv^2/r$. Newton's Law of Universal Gravitation is given by $F = GMm/d^2$, where d is the distance between the centers of the two bodies of mass M and m. Use this to show that the speed required for circular motion is $v = \sqrt{GM/r}$.

In Exercises 21–24, use the result of Exercise 20 to find the speed necessary for *the given circular orbit* around the earth. Let $GM = 9.56 \times 10^4$ cubic miles per second per second, and assume the radius of the earth is 4000 miles.

21. A space shuttle 100 miles above the surface of the earth

22. A space shuttle 200 miles above the surface of the earth

23. A heat capacity mapping satellite 385 miles above the surface of the earth (see figure).

FIGURE FOR 23

24. A SYNCOM satellite r miles above the surface of the earth in a geosynchronous orbit. [The satellite completes one orbit per sidereal day (23 hours, 56 minutes), and thus appears to remain stationary above a point on the earth.]

25. Prove that for a plane curve, the principal unit normal vector $\mathbf{N}$ points toward the concave side of the curve.

13.6 Arc Length and Curvature

Arc length ▪ Arc length parameter ▪ Curvature

In Section 12.2 we saw that the arc length of a smooth curve C given by the parametric equations $x = x(t)$ and $y = y(t)$, $a \le t \le b$, is

$$s = \int_a^b \sqrt{[x'(t)]^2 + [y'(t)]^2}\, dt.$$

In vector form, where C is given by $\mathbf{r}(t) = x(t)\mathbf{i} + y(t)\mathbf{j}$, we can rewrite this equation for arc length as

$$s = \int_a^b \|\mathbf{r}'(t)\|\, dt$$

as summarized in the following theorem.

THEOREM 13.11
ARC LENGTH OF A PLANE CURVE

If C is a smooth curve given by $\mathbf{r}(t) = x(t)\mathbf{i} + y(t)\mathbf{j}$, on an interval $[a, b]$, then the arc length of C on the interval is

$$s = \int_a^b \|\mathbf{r}'(t)\| \, dt.$$

$\mathbf{r}(t) = 2 \cos t \, \mathbf{i} + 3 \sin t \, \mathbf{j}$

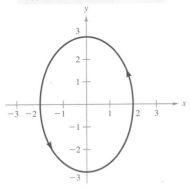

FIGURE 13.43

EXAMPLE 1   Finding the arc length of a plane curve

Find the arc length of the ellipse given by

$$\mathbf{r}(t) = 2 \cos t \, \mathbf{i} + 3 \sin t \, \mathbf{j} \qquad 0 \le t \le 2\pi$$

as shown in Figure 13.43.

SOLUTION

Using the formula for arc length, we obtain

$$\begin{aligned}
s &= \int_0^{2\pi} \|\mathbf{r}'(t)\| \, dt \\
&= \int_0^{2\pi} \sqrt{[x'(t)]^2 + [y'(t)]^2} \, dt \\
&= \int_0^{2\pi} \sqrt{[-2 \sin t]^2 + [3 \cos t]^2} \, dt \\
&= \int_0^{2\pi} \sqrt{4 + 5 \cos^2 t} \, dt \qquad \text{Simpson's Rule} \\
&\approx 15.87.
\end{aligned}$$

REMARK   A rough approximation formula for the circumference of the ellipse given in Example 1 is $2\pi\sqrt{2^2 + 3^2}/\sqrt{2} \approx 16.01$. The approximation derived in Example 1 is more accurate.

Arc length parameter

We have seen that curves can be represented by vector-valued functions in different ways, depending on the choice of parameter. For *motion* along a curve, the convenient parameter is time, t. However, for studying the *geometric properties* of a curve, the convenient parameter is often arc length, s. We define the **arc length parameter** in terms of the following arc length function.

DEFINITION OF ARC
LENGTH FUNCTION

Let C be a smooth curve given by $\mathbf{r}(t)$ defined on the closed interval $[a, b]$. For $a \le t \le b$, the **arc length function** is given by

$$s(t) = \int_a^t \|\mathbf{r}'(\tau)\| \, d\tau.$$

(τ is the lowercase Greek letter tau.)

REMARK   Note that the arc length function is *nonnegative*.

By using the definition of the arc length function and the Second Fundamental Theorem of Calculus, we can conclude that $ds/dt = \|\mathbf{r}'(t)\|$, as stated in the following theorem.

THEOREM 13.12 **THE DERIVATIVE OF THE** **ARC LENGTH FUNCTION**	Let C be a smooth curve given by $\mathbf{r}(t)$ on the closed interval $[a, b]$. The derivative of the arc length function s for this curve is given by $$\frac{ds}{dt} = \|\mathbf{r}'(t)\|.$$ In differential form, we can write $ds = \|\mathbf{r}'(t)\|\, dt$.

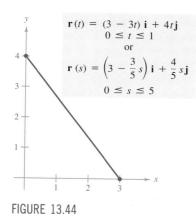

$\mathbf{r}(t) = (3 - 3t)\,\mathbf{i} + 4t\,\mathbf{j}$
$0 \le t \le 1$
or
$\mathbf{r}(s) = \left(3 - \dfrac{3}{5}s\right)\mathbf{i} + \dfrac{4}{5}s\,\mathbf{j}$
$0 \le s \le 5$

FIGURE 13.44

EXAMPLE 2   Finding the arc length function for a line

Find the arc length function $s(t)$ for the line segment given by

$$\mathbf{r}(t) = (3 - 3t)\mathbf{i} + 4t\,\mathbf{j}, \quad 0 \le t \le 1$$

and express $\mathbf{r}$ as a function of the parameter s. (See Figure 13.44.)

SOLUTION

Since $\mathbf{r}'(t) = -3\mathbf{i} + 4\mathbf{j}$, and $\|\mathbf{r}'(t)\| = \sqrt{3^2 + 4^2} = 5$, we have

$$s(t) = \int_0^t \|\mathbf{r}'(\tau)\|\, d\tau = \int_0^t 5\, d\tau = 5t.$$

Now, using $s = 5t$ (or $t = s/5$), we can rewrite $\mathbf{r}$ using the arc length parameter as follows.

$$\mathbf{r}(s) = \left(3 - \frac{3}{5}s\right)\mathbf{i} + \frac{4}{5}s\,\mathbf{j}, \quad 0 \le s \le 5.$$

One of the advantages of writing a vector-valued function in terms of the arc length parameter is that $\|\mathbf{r}'(s)\| = 1$. For instance, in Example 2, we have

$$\|\mathbf{r}'(s)\| = \sqrt{\left(-\frac{3}{5}\right)^2 + \left(\frac{4}{5}\right)^2} = 1.$$

Thus, for a smooth curve C, represented by $\mathbf{r}(s)$, where s is the arc length parameter, the arc length between a and b is

$$\text{length of arc} = \int_a^b \|\mathbf{r}'(s)\|\, ds = \int_a^b ds = b - a$$

$$= \text{length of interval}.$$

Furthermore, if t is *any* parameter such that $\|\mathbf{r}'(t)\| = 1$, then t must be the arc length parameter. We summarize these results in the following theorem, which we state without proof.

THEOREM 13.13 **ARC LENGTH PARAMETER**	If C is a smooth curve given by $\mathbf{r}(s) = x(s)\mathbf{i} + y(s)\mathbf{j}$, where s is the arc length parameter, then $$\|\mathbf{r}'(s)\| = 1.$$ Moreover, if t is *any* parameter for the vector-valued function $\mathbf{r}$ such that $\|\mathbf{r}'(t)\| = 1$, then t must be the arc length parameter.

Curvature

An important use of the arc length parameter is to find **curvature**—the measure of how sharply a curve bends. For instance, in Figure 13.45, the curve bends more sharply at P than at Q, and we say that the curvature is greater at P than at Q. To measure the bending at a point on a curve, we calculate the norm of the rate of change of the unit tangent vector $\mathbf{T}$ with respect to the arc length s, as indicated in Figure 13.46.

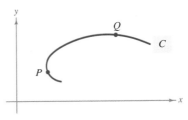

Curvature at P is greater than at Q.

FIGURE 13.45

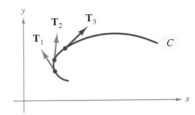

$\mathbf{T}$ changes as s changes.

FIGURE 13.46

DEFINITION OF CURVATURE	Let C be a smooth curve given by $\mathbf{r}(s)$, where s is the arc length parameter. The **curvature** at s is given by $$K = \left\|\frac{d\mathbf{T}}{ds}\right\| = \|\mathbf{T}'(s)\|.$$

Since a straight line doesn't "curve," its curvature should be zero. We demonstrate this fact in Example 3.

EXAMPLE 3 Curvature of a straight line

Find the curvature of the line given by

$$\mathbf{r}(s) = \left(3 - \frac{3}{5}s\right)\mathbf{i} + \frac{4}{5}s\,\mathbf{j}.$$

SOLUTION

To begin, we note that $\mathbf{r}$ is written in terms of the arc length parameter (see Example 2). Since $\mathbf{r}'(s) = -\frac{3}{5}\mathbf{i} + \frac{4}{5}\mathbf{j}$, and $\|\mathbf{r}'(s)\| = 1$, we have

$$\mathbf{T}(s) = \frac{\mathbf{r}'(s)}{\|\mathbf{r}'(s)\|} = -\frac{3}{5}\mathbf{i} + \frac{4}{5}\mathbf{j}.$$

Therefore, the curvature is

$$K = \|\mathbf{T}'(s)\| = \|0\mathbf{i} + 0\mathbf{j}\| = 0$$

at every point on the line. ▭

A circle has the same curvature at any point. Moreover, the curvature and the radius of the circle are inversely related. That is, a circle with a large radius has a small curvature, and a circle with a small radius has a large curvature. This inverse relationship is made explicit in the following example.

EXAMPLE 4   Finding the curvature of a circle

Show that the curvature of a circle of radius r is $K = 1/r$.

SOLUTION

Without loss of generality we consider the circle to be centered at the origin. Let (x, y) be any point on the circle and let s be the length of the arc from $(r, 0)$ to (x, y), as shown in Figure 13.47. By letting θ be the central angle of the circle, we can represent the circle by

$$\mathbf{r}(\theta) = r \cos \theta \, \mathbf{i} + r \sin \theta \, \mathbf{j}.$$

Using the formula for the length of a circular arc $s = r\theta$, we can rewrite $\mathbf{r}(\theta)$ in terms of the arc length parameter as follows.

$$\mathbf{r}(s) = r \cos \frac{s}{r} \, \mathbf{i} + r \sin \frac{s}{r} \, \mathbf{j}$$

Thus, the unit tangent vector is

$$\mathbf{T}(s) = \frac{\mathbf{r}'(s)}{\|\mathbf{r}'(s)\|} = -\sin \frac{s}{r} \, \mathbf{i} + \cos \frac{s}{r} \, \mathbf{j} \qquad \|\mathbf{r}'(s)\| = 1$$

and the curvature is given by

$$K = \|\mathbf{T}'(s)\| = \left\| -\frac{1}{r} \cos \frac{s}{r} \, \mathbf{i} - \frac{1}{r} \sin \frac{s}{r} \, \mathbf{j} \right\| = \frac{1}{r}$$

at every point on the circle. ▭

In Examples 3 and 4, we found the curvature by applying the definition directly. This requires that the curve be written in terms of the arc length parameter s. The following useful theorem gives another formula for finding the curvature of a curve written in terms of an arbitrary parameter t. We leave the proof of this theorem as an exercise. (See Exercise 53.)

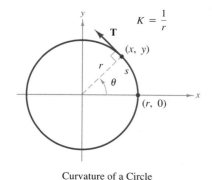

Curvature of a Circle

FIGURE 13.47

THEOREM 13.14
FORMULA FOR CURVATURE

If C is a smooth curve given by $\mathbf{r}(t)$, then the curvature of C at t is given by

$$K = \frac{\|\mathbf{T}'(t)\|}{\|\mathbf{r}'(t)\|}.$$

Since $\|\mathbf{r}'(t)\| = ds/dt$, this theorem implies that curvature is the ratio of the rate of change in the tangent vector $\mathbf{T}$ to the rate of change in arc length. This seems reasonable if we consider that as Δt approaches zero, we can write

$$\frac{\mathbf{T}'(t)}{ds/dt} \approx \frac{[\mathbf{T}(t + \Delta t) - \mathbf{T}(t)]/\Delta t}{[s(t + \Delta t) - s(t)]/\Delta t} = \frac{\mathbf{T}(t + \Delta t) - \mathbf{T}(t)}{s(t + \Delta t) - s(t)} = \frac{\Delta \mathbf{T}}{\Delta s}.$$

In other words, for a given Δs, the greater the length of $\Delta \mathbf{T}$, the more the curve bends at t, as shown in Figure 13.48.

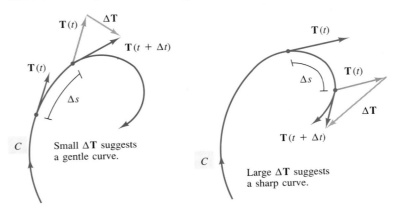

FIGURE 13.48

Let C be a curve with curvature K at the point P. A circle passing through the point P with radius $r = 1/K$ is called the **circle of curvature** if the circle lies on the concave side of the curve and shares a common tangent line with the curve at the point P. We call r the **radius of curvature** at P, and the center of the circle is called the **center of curvature.**

The circle of curvature gives us a nice way to graphically estimate the curvature K at a point P on a curve. Using a compass, we sketch a circle that snuggles up against the concave side of the curve at the point P, as shown in Figure 13.49. If the circle has a radius of r, then we estimate the curvature to be $K = 1/r$.

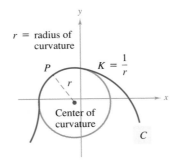

FIGURE 13.49   Circle of Curvature

To calculate curvature for a plane curve given by $y = f(x)$, we use the following theorem, which we state without proof.

THEOREM 13.15 **CURVATURE IN RECTANGULAR** **COORDINATES**	If C is the graph of a twice differentiable function given by $y = f(x)$, then the curvature at the point (x, y) is given by $$K = \frac{|y''|}{[1 + (y')^2]^{3/2}}.$$

EXAMPLE 5   Finding curvature in rectangular coordinates

Find the curvature of the parabola given by

$$y = x - \frac{1}{4}x^2$$

at $x = 2$ and $x = 4$. Sketch the circle of curvature at $(2, 1)$.

SOLUTION

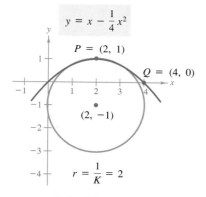

$y = x - \frac{1}{4}x^2$

$P = (2, 1)$

$Q = (4, 0)$

$(2, -1)$

$r = \frac{1}{K} = 2$

Circle of Curvature

FIGURE 13.50

	At x = 2	At x = 4
$y' = 1 - \dfrac{x}{2}$	$y' = 0$	$y' = -1$
$y'' = -\dfrac{1}{2}$	$y'' = -\dfrac{1}{2}$	$y'' = -\dfrac{1}{2}$
$K = \dfrac{|y''|}{[1 + (y')^2]^{3/2}}$	$K = \dfrac{1}{2}$	$K = \dfrac{1}{2^{5/2}} \approx 0.177$

Note in Figure 13.50 that at the point $(2, 1)$ the circle of curvature has a radius of $r = 1/K = 2$. ▭

Arc length and curvature are closely related to the tangential and normal components of acceleration. The tangential component of acceleration is the rate of change of the speed, which in turn is the rate of change of the arc length. This component is negative as a moving object slows down and positive as it speeds up—regardless of whether the object is turning or traveling on a straight line. Thus, the tangential component is solely a function of the arc length and is independent of the curvature. On the other hand, the normal component of acceleration is a function of *both* speed and curvature. This component measures the acceleration acting perpendicular to the direction of motion. To see why the normal component is affected by both speed and curvature, imagine that you are driving a car around a turn, as shown in Figure 13.51. If your speed is high and the turn is sharp, you feel yourself thrown against the car door. By lowering your speed *or* taking a more gentle turn, you are able to lessen this sideways thrust.

The next theorem explicitly states the relationships among speed, curvature, and the components of acceleration.

FIGURE 13.51

**THEOREM 13.16
ACCELERATION, SPEED,
AND CURVATURE**

If $\mathbf{r}(t)$ is the position vector for a smooth curve C, then the acceleration vector is given by

$$\mathbf{a}(t) = \frac{d^2s}{dt^2}\mathbf{T} + K\left(\frac{ds}{dt}\right)^2\mathbf{N}$$

where K is the curvature of C and ds/dt is the speed.

PROOF Since $\mathbf{T} = \mathbf{r}'/\|\mathbf{r}'\|$, it follows that

$$\mathbf{r}' = \|\mathbf{r}'\|\mathbf{T}.$$

By differentiating, we obtain

$$
\begin{aligned}
\mathbf{a} = \mathbf{r}'' &= \frac{d}{dt}\left[\|\mathbf{r}'\|\right]\mathbf{T} + \|\mathbf{r}'\|\mathbf{T}' \\
&= \frac{d}{dt}\left[\frac{ds}{dt}\right]\mathbf{T} + \frac{ds}{dt}\mathbf{T}'\left(\frac{\|\mathbf{T}'\|}{\|\mathbf{T}'\|}\right) \\
&= \frac{d^2s}{dt^2}\mathbf{T} + \frac{ds}{dt}\|\mathbf{T}'\|\mathbf{N} \\
&= \frac{d^2s}{dt^2}\mathbf{T} + \frac{ds}{dt}(\|\mathbf{r}'\|K)\mathbf{N} \\
&= \frac{d^2s}{dt^2}\mathbf{T} + K\left(\frac{ds}{dt}\right)^2\mathbf{N}.
\end{aligned}
$$

There are many applications in physics and engineering dynamics that involve the relationships among speed, arc length, curvature, and acceleration. One such application concerns frictional force.

Suppose a moving object with mass m is in contact with a stationary object. The total force required to produce an acceleration $\mathbf{a}$ along a given path is

$$\mathbf{F} = m\mathbf{a} = m\left(\frac{d^2s}{dt^2}\right)\mathbf{T} + mK\left(\frac{ds}{dt}\right)^2\mathbf{N}.$$

The portion of this total force that is supplied by the stationary object is called the **force of friction.** For example, if a car is rounding a turn, the roadway exerts a frictional force that keeps the car from sliding off the road. If the car is not sliding, then the frictional force is perpendicular to the direction of motion and has magnitude equal to the normal component of acceleration, as shown in Figure 13.52. The potential frictional force of a road around a turn can be increased by banking the roadway at an angle.

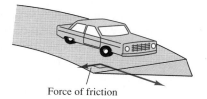

Force of friction

FIGURE 13.52 Force of friction is perpendicular to the direction of motion.

EXAMPLE 6 An application of the normal component of acceleration

A 360-kilogram Go-cart is driven at a speed of 60 kilometers per hour around a circular race track of radius 12 meters. To keep the car from skidding off course, what frictional force must the track surface exert on the tires?

SOLUTION

The frictional force must equal the normal component of acceleration. For this circular path we know that the curvature is $K = \frac{1}{12}$. Therefore, the normal component is

$$a_\mathbf{N} = mK\left(\frac{ds}{dt}\right)^2 = (360 \text{ kg})\left(\frac{1}{12 \text{ m}}\right)\left(\frac{60{,}000 \text{ m}}{3600 \text{ sec}}\right)^2 \approx 8333 \text{ (kg)(m)/sec}^2.$$

We summarize several of the formulas dealing with vector-valued functions as follows.

SUMMARY OF VELOCITY, ACCELERATION, AND CURVATURE

Let C be a curve given by the position function $\mathbf{r}(t) = x(t)\mathbf{i} + y(t)\mathbf{j}$, and let s be the arc length parameter.

Velocity vector, speed, and acceleration vector:

$$\mathbf{v}(t) = \mathbf{r}'(t)$$

$$\|\mathbf{v}(t)\| = \frac{ds}{dt}$$

$$\mathbf{a}(t) = \mathbf{r}''(t) = a_\mathbf{T}\mathbf{T}(t) + a_\mathbf{N}\mathbf{N}(t)$$

Unit tangent vector and principal unit normal vector:

$$\mathbf{T}(t) = \frac{\mathbf{r}'(t)}{\|\mathbf{r}'(t)\|}$$

$$\mathbf{N}(t) = \frac{\mathbf{T}'(t)}{\|\mathbf{T}'(t)\|}$$

Components of acceleration:

$$a_\mathbf{T} = \mathbf{a} \cdot \mathbf{T} = \frac{d^2s}{dt^2}$$

$$a_\mathbf{N} = \mathbf{a} \cdot \mathbf{N} = K\left(\frac{ds}{dt}\right)^2$$

Formulas for curvature:

$$K = \|\mathbf{T}'(s)\| = \|\mathbf{r}''(s)\| \qquad s \text{ is arc length parameter}$$

$$K = \frac{\|\mathbf{T}'(t)\|}{\|\mathbf{r}'(t)\|} \qquad t \text{ is general parameter}$$

$$K = \frac{|y''|}{[1 + (y')^2]^{3/2}} \qquad C \text{ given by } y = f(x)$$

$$K = \frac{|x'y'' - y'x''|}{[(x')^2 + (y')^2]^{3/2}} \qquad C \text{ given by } x = x(t), y = y(t)$$

EXERCISES for Section 13.6

In Exercises 1–6, sketch the given curve and find its length over the indicated interval.

Function	Interval
1. $\mathbf{r}(t) = t\mathbf{i} + 3t\mathbf{j}$	$[0, 4]$
2. $\mathbf{r}(t) = t\mathbf{i} + t^2\mathbf{j}$	$[0, 4]$
3. $\mathbf{r}(t) = a\cos^3 t\,\mathbf{i} + a\sin^3 t\,\mathbf{j}$	$[0, 2\pi]$
4. $\mathbf{r}(t) = a\cos t\,\mathbf{i} + a\sin t\,\mathbf{j}$	$[0, 2\pi]$
5. $\mathbf{r}(t) = \langle \sin t - t\cos t,\ \cos t + t\sin t \rangle$	$\left[0, \dfrac{\pi}{2}\right]$
6. $\mathbf{r}(t) = \langle t^{3/2},\ t \rangle$	$[0, 4]$

In Exercises 7 and 8, approximate the length of the plane curve over the indicated interval by using Simpson's Rule with $n = 4$.

Function	Interval
7. $\mathbf{r}(t) = t^2\mathbf{i} + \ln t\,\mathbf{j}$	$[1, 3]$
8. $\mathbf{r}(t) = t^2\mathbf{i} + \dfrac{1}{t}\mathbf{j}$	$[1, 3]$

In Exercises 9–14, find the curvature and radius of curvature of the plane curve at the indicated point.

Function	Point
9. $y = 3x - 2$	$x = a$
10. $y = mx + b$	$x = a$
11. $y = 2x^2 + 3$	$x = -1$
12. $y = x + \dfrac{1}{x}$	$x = 1$
13. $y = \sqrt{a^2 - x^2}$	$x = 0$
14. $y = \dfrac{3}{4}\sqrt{16 - x^2}$	$x = 0$

In Exercises 15–18, sketch the graph of the function and sketch the circle of curvature to the graph at the specified value of x.

Function	Point
15. $y = \sin x$	$x = \dfrac{\pi}{2}$
16. $y = \ln x$	$x = 1$
17. $y = e^x$	$x = 0$
18. $y = \dfrac{1}{3}x^3$	$x = 1$

In Exercises 19–24, find the curvature K of the given curve at the indicated point.

Function	Point
19. $\mathbf{r}(t) = 4t\mathbf{i}$	$t = 2$
20. $\mathbf{r}(t) = 4t\mathbf{i} - 2t\mathbf{j}$	$t = 1$
21. $\mathbf{r}(t) = 4t^2\mathbf{i}$	$t = 4$
22. $\mathbf{r}(t) = t^2\mathbf{i} + \mathbf{j}$	$t = 0$
23. $\mathbf{r}(t) = t\mathbf{i} + \dfrac{1}{t}\mathbf{j}$	$t = 1$
24. $\mathbf{r}(t) = t\mathbf{i} + t^2\mathbf{j}$	$t = 1$

In Exercises 25–30, find the curvature K of the given curve.

25. $\mathbf{r}(t) = 2\cos \pi t\,\mathbf{i} + 2\sin \pi t\,\mathbf{j}$
26. $\mathbf{r}(t) = a\cos \omega t\,\mathbf{i} + a\sin \omega t\,\mathbf{j}$
27. $\mathbf{r}(t) = e^t\cos t\,\mathbf{i} + e^t\sin t\,\mathbf{j}$
28. $\mathbf{r}(t) = a\cos \omega t\,\mathbf{i} + b\sin \omega t\,\mathbf{j}$
29. $\mathbf{r}(t) = \langle \cos \omega t + \omega t\sin \omega t,\ \sin \omega t - \omega t\cos \omega t \rangle$
30. $\mathbf{r}(t) = \langle a(\omega t - \sin \omega t),\ a(1 - \cos \omega t) \rangle$

In Exercises 31–34, (a) find the point on the curve at which the curvature K is a maximum and (b) find the limit of K as $x \to \infty$.

31. $y = (x - 1)^2 + 3$ **32.** $y = x^3$
33. $y = x^{2/3}$ **34.** $y = \ln x$

In Exercises 35 and 36, use Theorem 13.16 to find a_T and a_N for the curve given by the vector-valued function.

35. $\mathbf{r}(t) = 3t^2\mathbf{i} + (3t - t^3)\mathbf{j}$
36. $\mathbf{r}(t) = t^2\mathbf{i} + 2t\mathbf{j}$

37. Find the circle of curvature of the graph of $y = (x + 1)/x$ at the point $(1, 2)$.

38. Find all the points on the graph of $y = (x - 1)^3 + 3$ at which the curvature is zero.

39. Show that the curvature is greatest at the endpoints of the major axis and least at the endpoints of the minor axis for the ellipse given by

$$x^2 + 4y^2 = 4.$$

40. Find all a and b such that the two curves given by

$$y_1 = ax(b - x)$$

and

$$y_2 = \frac{x}{x + 2}$$

intersect at only one point and have a common tangent line and equal curvature at the point. Sketch a graph for each set of values for a and b.

41. The smaller the curvature in a bend of a road, the faster a car can travel. Assume that the maximum speed around a turn is inversely proportional to the square root of the curvature. A car moving on the path $y = \frac{1}{3}x^3$ (x and y measured in miles) can safely go 30 miles per hour at $\left(1, \frac{1}{3}\right)$. How fast can it go at $\left(\frac{3}{2}, \frac{9}{8}\right)$?

42. The curve C is given by the polar equation $r = f(\theta)$. Show that the curvature K at the point (r, θ) is

$$K = \frac{|2(r')^2 - rr'' + r^2|}{[(r')^2 + r^2]^{3/2}}.$$

[Hint: Represent the curve by $\mathbf{r}(\theta) = r \cos \theta \, \mathbf{i} + r \sin \theta \, \mathbf{j}$.]

In Exercises 43–46, use the result of Exercise 42 to find the curvature of the polar curve.

43. $r = 1 + \sin \theta$ **44.** $r = a \sin \theta$

45. $r = \theta$ **46.** $r = e^\theta$

47. Given the polar curve $r = e^{a\theta}$, find the curvature K and determine the limit of K as (a) $\theta \to \infty$, and (b) $a \to \infty$.

48. Show that the formula for the curvature of a polar curve $r = f(\theta)$ given in Exercise 42 reduces to

$$K = \frac{2}{|r'|}$$

for the curvature *at the pole*.

In Exercises 49 and 50, use the result of Exercise 48 to find the curvature of the rose curve at the pole.

49. $r = 4 \sin 2\theta$ **50.** $r = 6 \cos 3\theta$

51. For a smooth curve given by the parametric equations $x = f(t)$ and $y = g(t)$, prove that the curvature is

$$K = \frac{|f'(t)g''(t) - g'(t)f''(t)|}{([f'(t)]^2 + [g'(t)]^2)^{3/2}}.$$

52. Use a computer or calculator and Simpson's Rule with $n = 12$ to approximate the length of the curve

$$\mathbf{r}(t) = t\mathbf{i} + \sin t \, \mathbf{j}$$

over the interval $0 \le t \le 2$.

53. Prove Theorem 13.14.

REVIEW EXERCISES for Chapter 13

In Exercises 1 and 2, let $\mathbf{u} = \overrightarrow{PQ}$ and $\mathbf{v} = \overrightarrow{PR}$, and find (a) the component forms of $\mathbf{u}$ and $\mathbf{v}$, (b) the magnitude of $\mathbf{v}$, (c) $\mathbf{u} \cdot \mathbf{v}$, (d) $2\mathbf{u} + \mathbf{v}$, (e) the vector component of $\mathbf{u}$ in the direction of $\mathbf{v}$, and (f) the vector component of $\mathbf{u}$ orthogonal to $\mathbf{v}$.

1. $P = (1, 2)$, $Q = (4, 1)$, $R = (5, 4)$
2. $P = (-2, -1)$, $Q = (5, -1)$, $R = (2, 4)$

In Exercises 3 and 4, find the angle θ between the vectors $\mathbf{u}$ and $\mathbf{v}$.

3. $\mathbf{u} = \left\langle 5 \cos \dfrac{3\pi}{4}, 5 \sin \dfrac{3\pi}{4} \right\rangle$,

    $\mathbf{v} = \left\langle 2 \cos \dfrac{2\pi}{3}, 2 \sin \dfrac{2\pi}{3} \right\rangle$

4. $\mathbf{u} = \langle -12, 5 \rangle$, $\mathbf{v} = \langle 2, 8 \rangle$

In Exercises 5 and 6, find $\mathbf{u}$.

5. The angle, measured counterclockwise, from $\mathbf{u}$ to the positive x-axis is 135°, and $\|\mathbf{u}\| = 4$.

6. The angle between $\mathbf{u}$ and the positive x-axis is 180°, and $\|\mathbf{u}\| = 8$.

In Exercises 7–10, determine whether $\mathbf{u}$ and $\mathbf{v}$ are orthogonal, parallel, or neither.

7. $\mathbf{u} = \langle 8, -12 \rangle$, $\mathbf{v} = \langle -2, 3 \rangle$
8. $\mathbf{u} = \langle -22, 11 \rangle$, $\mathbf{v} = \langle 2, 4 \rangle$
9. $\mathbf{u} = \langle 6, 27 \rangle$, $\mathbf{v} = \langle 9, -2 \rangle$
10. $\mathbf{u} = \langle -17, 20 \rangle$, $\mathbf{v} = \langle 3, 2 \rangle$

In Exercises 11 and 12, sketch the plane curve represented by the vector-valued function.

11. $\mathbf{r}(t) = 4\mathbf{i} + t\mathbf{j}$
12. $\mathbf{r}(t) = t\mathbf{i} + 3 \cos t\,\mathbf{j}$

In Exercises 13 and 14, (a) find the domain of $\mathbf{r}$ and (b) determine the values of t for which the function is discontinuous.

13. $\mathbf{r}(t) = \sqrt{t}\,\mathbf{i} + \dfrac{1}{t-4}\mathbf{j}$
14. $\mathbf{r}(t) = \langle t^2, \sqrt{25-t^2}\rangle$

In Exercises 15 and 16, find the indicated limit.

15. $\displaystyle\lim_{t\to 3^-} (t^2\mathbf{i} + \sqrt{9-t^2}\,\mathbf{j})$
16. $\displaystyle\lim_{t\to\infty} \left(\dfrac{t^2+3}{2t^2}\mathbf{i} + \dfrac{1}{t}\mathbf{j}\right)$

In Exercises 17 and 18, find the following.
(a) $\mathbf{r}'(t)$
(b) $D_t[\mathbf{u}(t) - 2\mathbf{r}(t)]$
(c) $D_t[\mathbf{r}(t) \cdot \mathbf{u}(t)]$
(d) $D_t[\|\mathbf{r}(t)\|]$

17. $\mathbf{r}(t) = 3t\mathbf{i} + (t-1)\mathbf{j}$
$\mathbf{u}(t) = \sqrt{t}\,\mathbf{i} + 4t\mathbf{j}$
18. $\mathbf{r}(t) = \cos t\,\mathbf{i} + \sin t\,\mathbf{j}$
$\mathbf{u}(t) = -\sin t\,\mathbf{i} + \cos t\,\mathbf{j}$

In Exercises 19 and 20, find the indefinite integral.

19. $\displaystyle\int (\cos t\,\mathbf{i} + t\cos t\,\mathbf{j})\,dt$
20. $\displaystyle\int (3t^2\mathbf{i} + t\ln t\,\mathbf{j})\,dt$

In Exercises 21–26, find the velocity, speed, and acceleration at any time t. Then, find $\mathbf{a}\cdot\mathbf{T}$, $\mathbf{a}\cdot\mathbf{N}$, and the curvature at time t.

21. $\mathbf{r}(t) = (1+4t)\mathbf{i} + (2-3t)\mathbf{j}$
22. $\mathbf{r}(t) = 5t\mathbf{i}$
23. $\mathbf{r}(t) = 2(t+1)\mathbf{i} + \dfrac{2}{t+1}\mathbf{j}$
24. $\mathbf{r}(t) = t\mathbf{i} + \sqrt{t}\,\mathbf{j}$
25. $\mathbf{r}(t) = e^t\mathbf{i} + e^{-t}\mathbf{j}$
26. $\mathbf{r}(t) = t\cos t\,\mathbf{i} + t\sin t\,\mathbf{j}$

In Exercises 27 and 28, find the length of the curve over the indicated interval.

27. $\mathbf{r}(t) = 12t\mathbf{i} + 2t^2\mathbf{j}$, $0 \le t \le 4$
28. $\mathbf{r}(t) = e^t \sin t\,\mathbf{i} + e^t \cos t\,\mathbf{j}$, $0 \le t \le \pi$

29. Find the largest weight W that can be supported by the structure in the accompanying figure if the strut can withstand a maximum compression of 500 pounds.

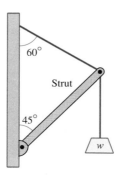

FIGURE FOR 29

30. Use vectors to prove that the midpoints of the sides of *any* quadrilateral are the vertices of a parallelogram.

31. A projectile is fired from ground level at an angle of elevation of 30°. Find the range of the projectile if the initial velocity is 75 feet per second.

32. The center of a truckbed is 6 feet below and 4 feet horizontally from the end of a horizontal conveyor that is discharging gravel (see figure). Determine the speed ds/dt at which the conveyor belt should be moving so that the gravel falls onto the center of the truck bed.

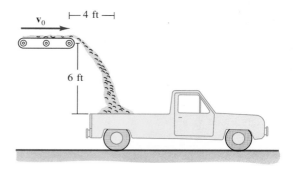

FIGURE FOR 32

33. Find the speed necessary for a satellite to maintain a circular orbit 600 miles above the surface of the earth.

34. An automobile is in a circular traffic exchange and its speed is twice that posted. By what factor is the centripetal force increased over that which would occur at the posted speed?

35. A civil engineer designs a highway as indicated in the accompanying figure. *BC* is an arc of a circle. *AB* and *CD* are straight lines tangent to the circular arc. Criticize the design.

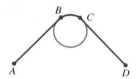

FIGURE FOR 35

36. A highway has an exit ramp that begins at the origin of a coordinate system and follows the curve given by

$$y = \frac{1}{32}x^{5/2}$$

to the point (4, 1) (see figure). Then it follows a circular path whose curvature is that given by the curve at (4, 1). What is the radius of the circular arc?

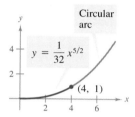

FIGURE FOR 36

Derricks and cranes are used on construction sites to raise and lower objects. The tension on the *load line* is primarily due to the weight of the object being supported. Sometimes ropes are attached to the load line to guide the object's rise or descent. Pulling on a "guide rope" from ground level increases the tension on the load line.

Vector Analysis of a System

Vectors are commonly used in engineering to analyze the way in which several forces interact. The following example describes the interaction of three forces—the weight of an automobile, the force exerted by a person pulling on a rope, and the resultant force on a cable.

At a loading dock, a 3600-pound automobile is supported by a cable. A rope, attached to the cable, is pulled to position the automobile on the dock. The angle between the rope and the horizontal is 35°, and the angle between the cable and the vertical is 2°. (The side of the ship makes an angle of 45° with the plane determined by the cable and the rope.) To determine the tension on the cable and the tension on the rope, we set up the following system of vectors.

$$\mathbf{F}_1 = -3600\mathbf{k} \qquad \text{Automobile}$$
$$\mathbf{F}_2 = T_2(0.5792\mathbf{i} + 0.5792\mathbf{j} - 0.5736\mathbf{k}) \qquad \text{Rope}$$
$$\mathbf{F}_3 = T_3(-0.0247\mathbf{i} - 0.0247\mathbf{j} + 0.9994\mathbf{k}) \qquad \text{Cable}$$

Because these three vectors represent a state of equilibrium, we can write $\mathbf{F}_1 + \mathbf{F}_2 + \mathbf{F}_3 = \mathbf{0}$. Solving this system, we can determine that the tension on the cable is $T_3 \approx 3692.6$ pounds, and the tension on the rope is $T_2 \approx 157.3$ pounds.

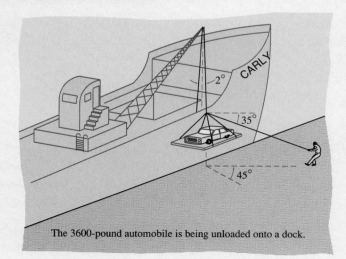

The 3600-pound automobile is being unloaded onto a dock.

Chapter Overview

The chapter begins by introducing the three-dimensional coordinate system that we use as a model for space. Most of the concepts and formulas used in the plane have straightforward generalizations in space. For instance, the distance between the two points (x_1, y_1, z_1) and (x_2, y_2, z_2) in space is given by

$$d = \sqrt{(x_2 - x_1)^2 + (y_2 - y_1)^2 + (z_2 - z_1)^2}.$$

(Note the similarity between this formula and the formula for the distance between two points in the plane.)

A typical vector in space can be represented as

$$\mathbf{v} = \langle v_1, v_2, v_3 \rangle = v_1\mathbf{i} + v_2\mathbf{j} + v_3\mathbf{k}.$$

Section 14.1 gives a quick summary of the vector properties that have straightforward generalizations from two to three dimensions. For instance, the dot product of two vectors in space, $\mathbf{v} = \langle v_1, v_2, v_3 \rangle$ and $\mathbf{u} = \langle u_1, u_2, u_3 \rangle$, is given by

$$\mathbf{v} \cdot \mathbf{u} = v_1u_1 + v_2u_2 + v_3u_3.$$

In Chapter 14 you will encounter some concepts involving three dimensions that either have no two-dimensional counterparts, or are not straightforward generalizations of concepts in the plane. For instance, in Section 14.2 we describe a type of product, called the *cross product* of two vectors, that is defined only for vectors in space.

The next two sections in the chapter discuss the geometry of space. Equations of lines and planes are discussed in Section 14.3, and equations of surfaces are discussed in Section 14.4.

Sections 14.5 and 14.6 describe vector-valued functions and curves in three-dimensional space.

See Exercise 91, Section 14.1.

Solid Analytic Geometry and Vectors in Space

14.1 Space Coordinates and Vectors in Space

Coordinates in space ▪ Solid analytic geometry ▪ Vectors in space ▪ Applications ▪ Direction cosines

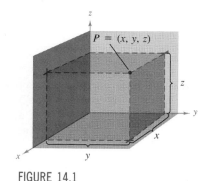

FIGURE 14.1

In Chapter 1 we described the Cartesian plane as the plane determined by two real number lines (the x- and y-axes) intersecting at a right angle. These axes, together with their point of intersection (the origin), allowed us to develop a two-dimensional coordinate system for identifying points in the plane and for discussing topics in plane analytic geometry. To identify a point in space, we need to introduce a third dimension to our model. The geometry of this three-dimensional model is called **solid analytic geometry.**

We construct a **three-dimensional coordinate system** by passing a z-axis perpendicular to both the x- and y-axes at the origin. Figure 14.1 shows the positive portion of each coordinate axis. Taken as pairs, the axes determine three **coordinate planes** called the **xy-plane,** the **xz-plane,** and the **yz-plane.** These three coordinate planes separate three-space into eight **octants.** The first octant is the one for which all three coordinates are positive. In this three-dimensional system, a point P in space is determined by an ordered triple (x, y, z), where

x = directed distance from yz-plane to P

y = directed distance from xz-plane to P

z = directed distance from xy-plane to P.

Several points are shown in Figure 14.2.

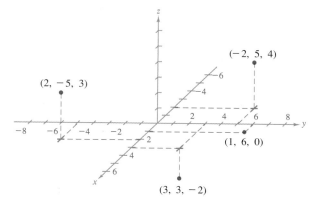

FIGURE 14.2

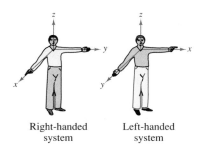

Right-handed Left-handed
system system

FIGURE 14.3

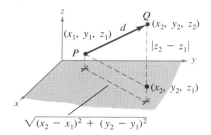

FIGURE 14.4

A three-dimensional coordinate system can have either a **left-handed** or a **right-handed** orientation. To determine the orientation of a system, imagine that you are standing at the origin, with your arms pointing in the direction of the positive *x*- and *y*-axes, and the *z*-axis pointing up, as shown in Figure 14.3. The system is right-handed or left-handed, depending on which hand points along the *x*-axis. In this text we work exclusively with right-handed systems.

Many of the formulas established for the two-dimensional coordinate system can be extended to three dimensions. For example, to find the distance between two points in space, we use the Pythagorean Theorem twice, as shown in Figure 14.4, to obtain the formula for the distance between the points (x_1, y_1, z_1) and (x_2, y_2, z_2).

$$d = \sqrt{(x_2 - x_1)^2 + (y_2 - y_1)^2 + (z_2 - z_1)^2}$$ Distance Formula

EXAMPLE 1 Finding the distance between two points in space

Find the distance between $(2, -1, 3)$ and $(1, 0, -2)$.

SOLUTION

By the Distance Formula, we have
$$d = \sqrt{(1 - 2)^2 + (0 + 1)^2 + (-2 - 3)^2}$$
$$= \sqrt{1 + 1 + 25} = \sqrt{27} = 3\sqrt{3}.$$

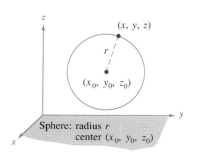

Sphere: radius *r*
center (x_0, y_0, z_0)

FIGURE 14.5

A **sphere** with center at (x_0, y_0, z_0) and radius *r* is defined to be the set of all points (x, y, z) such that the distance between (x, y, z) and (x_0, y_0, z_0) is *r*. We can use the Distance Formula to find the **standard equation of a sphere** of radius *r*, centered at (x_0, y_0, z_0). If (x, y, z) is an arbitrary point on the sphere, then the equation of the sphere is

$$(x - x_0)^2 + (y - y_0)^2 + (z - z_0)^2 = r^2$$ Equation of sphere

as shown in Figure 14.5. Moreover, the midpoint of the line segment joining the points (x_1, y_1, z_1) and (x_2, y_2, z_2) has coordinates

$$\left(\frac{x_1 + x_2}{2}, \frac{y_1 + y_2}{2}, \frac{z_1 + z_2}{2} \right).$$ Midpoint Rule

The following example makes use of both of these formulas.

EXAMPLE 2 Finding the equation of a sphere

Find the standard equation for the sphere that has the points $(5, -2, 3)$ and $(0, 4, -3)$ as endpoints of a diameter.

SOLUTION

By the Midpoint Rule, the center of the sphere is

$$\left(\frac{5+0}{2}, \frac{-2+4}{2}, \frac{3-3}{2}\right) = \left(\frac{5}{2}, 1, 0\right).$$

By the Distance Formula, the radius is

$$r = \sqrt{\left(0 - \frac{5}{2}\right)^2 + (4-1)^2 + (-3-0)^2} = \sqrt{\frac{97}{4}} = \frac{\sqrt{97}}{2}.$$

Therefore, the standard equation of the sphere is

$$\left(x - \frac{5}{2}\right)^2 + (y-1)^2 + (z-0)^2 = \frac{97}{4}.$$

Vectors in space

Physical forces and velocities are not confined to the plane, and hence it is natural to extend our discussion of vectors in the plane to vectors in space. In space, we denote vectors by ordered triples

$$\mathbf{v} = \langle v_1, v_2, v_3 \rangle$$

with the **zero vector** denoted by $\mathbf{0} = \langle 0, 0, 0 \rangle$. Using the unit vectors $\mathbf{i} = \langle 1, 0, 0 \rangle$, $\mathbf{j} = \langle 0, 1, 0 \rangle$, and $\mathbf{k} = \langle 0, 0, 1 \rangle$ in the directions of the positive x-, y-, and z-axes, respectively, the **standard unit vector notation** for $\mathbf{v}$ is

$$\mathbf{v} = v_1 \mathbf{i} + v_2 \mathbf{j} + v_3 \mathbf{k}$$

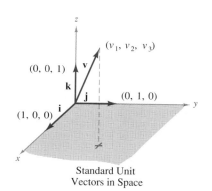

Standard Unit
Vectors in Space

FIGURE 14.6

as shown in Figure 14.6. If $\mathbf{v}$ is represented by the directed line segment from $P = (p_1, p_2, p_3)$ to $Q = (q_1, q_2, q_3)$, as shown in Figure 14.7, then the component form of $\mathbf{v}$ is given by subtracting the coordinates of the initial point from the coordinates of the terminal point as follows.

$$\mathbf{v} = \langle v_1, v_2, v_3 \rangle = \langle q_1 - p_1, q_2 - p_2, q_3 - p_3 \rangle$$

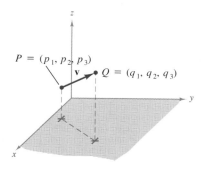

$$\mathbf{v} = \langle q_1 - p_1, q_2 - p_2, q_3 - p_3 \rangle$$

FIGURE 14.7

EXAMPLE 3   Standard unit vector notation

(a) Write the vector $\mathbf{v} = 4\mathbf{i} - 5\mathbf{k}$ in component form.
(b) Find the terminal point of the vector $\mathbf{v} = 7\mathbf{i} - \mathbf{j} + 3\mathbf{k}$, given that the initial point is $P = (-2, 3, 5)$.

SOLUTION

(a) Since $\mathbf{j}$ is missing, its component is zero and $\mathbf{v} = 4\mathbf{i} - 5\mathbf{k} = \langle 4, 0, -5 \rangle$.
(b) We seek $Q = (q_1, q_2, q_3)$ such that $\mathbf{v} = \overrightarrow{PQ} = 7\mathbf{i} - \mathbf{j} + 3\mathbf{k}$. This implies that

$$q_1 - (-2) = 7, \qquad q_2 - 3 = -1, \qquad q_3 - 5 = 3$$

with solutions $q_1 = 5$, $q_2 = 2$, and $q_3 = 8$. Therefore, $Q = (5, 2, 8)$.

In the following list, we summarize the basic definitions and operations of vectors in space.

VECTORS IN SPACE

Let $\mathbf{u} = \langle u_1, u_2, u_3 \rangle$ and $\mathbf{v} = \langle v_1, v_2, v_3 \rangle$ be vectors in space, and let c be a scalar.

1. *Equality of Vectors:* $\mathbf{u} = \mathbf{v}$ if and only if $u_1 = v_1$, $u_2 = v_2$, and $u_3 = v_3$.
2. *Length:* $\|\mathbf{v}\| = \sqrt{v_1^2 + v_2^2 + v_3^2}$
3. *Unit Vector in the Direction of* $\mathbf{v}$: $\dfrac{\mathbf{v}}{\|\mathbf{v}\|} = \left(\dfrac{1}{\|\mathbf{v}\|}\right)\langle v_1, v_2, v_3 \rangle$
4. *Vector Addition:* $\mathbf{v} + \mathbf{u} = \langle v_1 + u_1, v_2 + u_2, v_3 + u_3 \rangle$
5. *Scalar Multiplication:* $c\mathbf{v} = \langle cv_1, cv_2, cv_3 \rangle$
6. *Parallel Vectors:* Two nonzero vectors $\mathbf{u}$ and $\mathbf{v}$ are parallel if there is some scalar c such that $\mathbf{u} = c\mathbf{v}$.
7. *Dot Product:* $\mathbf{u} \cdot \mathbf{v} = u_1v_1 + u_2v_2 + u_3v_3$
8. *Orthogonal Vectors:* $\mathbf{u}$ and $\mathbf{v}$ are orthogonal if $\mathbf{u} \cdot \mathbf{v} = 0$.
9. *Angle Between Two Vectors:* The angle θ between two nonzero vectors is given by

$$\cos \theta = \frac{\mathbf{u} \cdot \mathbf{v}}{\|\mathbf{u}\| \, \|\mathbf{v}\|}.$$

10. *Triangle Inequality:* $\|\mathbf{u} + \mathbf{v}\| \le \|\mathbf{u}\| + \|\mathbf{v}\|$
11. *Projection:* The projection of $\mathbf{u}$ onto the nonzero vector $\mathbf{v}$ is given by

$$\text{proj}_\mathbf{v} \, \mathbf{u} = \frac{\mathbf{u} \cdot \mathbf{v}}{\|\mathbf{v}\|^2}\mathbf{v}.$$

REMARK   The properties of vector addition, scalar multiplication, vector length, and dot products given in Theorems 13.1, 13.2, and 13.5 are valid also for vectors in space. We leave the proof to you.

EXAMPLE 4   Finding the component form of a vector in space

Find the component form and length of the vector $\mathbf{v}$ having initial point $(-2, 3, 1)$ and terminal point $(0, -4, 4)$. Then find a unit vector in the direction of $\mathbf{v}$.

SOLUTION

The component form of $\mathbf{v}$ is

$$\begin{aligned}
\mathbf{v} &= \langle q_1 - p_1, q_2 - p_2, q_3 - p_3 \rangle \\
&= \langle 0 - (-2), -4 - 3, 4 - 1 \rangle \\
&= \langle 2, -7, 3 \rangle
\end{aligned}$$

which implies that its length is

$$\|\mathbf{v}\| = \sqrt{(2)^2 + (-7)^2 + (3)^2} = \sqrt{62}.$$

Now, using this length, we find that the unit vector in the direction of $\mathbf{v}$ is

$$\mathbf{u} = \frac{\mathbf{v}}{\|\mathbf{v}\|} = \frac{1}{\sqrt{62}}\langle 2, -7, 3 \rangle.$$

EXAMPLE 5 Parallel vectors

Vector **w** has initial point $(2, -1, 3)$ and terminal point $(-4, 7, 5)$. Which of the following vectors is parallel to **w**?

(a) $\mathbf{u} = \langle 3, -4, -1 \rangle$ (b) $\mathbf{v} = \langle 12, -16, 4 \rangle$

SOLUTION

First, we determine **w** to have the component form

$$\mathbf{w} = \langle -4 - 2, 7 - (-1), 5 - 3 \rangle = \langle -6, 8, 2 \rangle.$$

(a) Since $\mathbf{u} = \langle 3, -4, -1 \rangle = -\frac{1}{2}\langle -6, 8, 2 \rangle = -\frac{1}{2}\mathbf{w}$, we conclude that **u** *is* parallel to **w**.

(b) In this case, we want to find a scalar c such that

$$\langle 12, -16, 4 \rangle = c\langle -6, 8, 2 \rangle.$$

However, when we equate corresponding components, we get $c = -2$ for the first two components and $c = 2$ for the third. Hence, the equation has no solution, and we conclude that the vectors are *not* parallel. ▭

EXAMPLE 6 Using vectors to determine collinear points

Determine whether the points $P = (1, -2, 3)$, $Q = (2, 1, 0)$, and $R = (4, 7, -6)$ lie on the same line.

SOLUTION

We can solve this problem by finding the component forms of the vectors $\overrightarrow{PQ}$ and $\overrightarrow{PR}$. Since these two vectors have a common initial point, it follows that P, Q, and R lie on the same line if and only if $\overrightarrow{PQ}$ and $\overrightarrow{PR}$ are parallel.

$$\overrightarrow{PQ} = \langle 2 - 1, 1 - (-2), 0 - 3 \rangle = \langle 1, 3, -3 \rangle$$
$$\overrightarrow{PR} = \langle 4 - 1, 7 - (-2), -6 - 3 \rangle = \langle 3, 9, -9 \rangle$$

Since $\overrightarrow{PR} = 3\overrightarrow{PQ}$, we conclude that these two vectors are parallel, and therefore the three given points are collinear, as shown in Figure 14.8. ▭

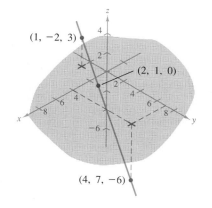

FIGURE 14.8

EXAMPLE 7 Finding the angle between two vectors

For $\mathbf{u} = \langle 3, -1, 2 \rangle$, $\mathbf{v} = \langle -4, 0, 2 \rangle$, $\mathbf{w} = \langle 1, -1, -2 \rangle$, and $\mathbf{z} = \langle 2, 0, -1 \rangle$, find the angle between the following pairs of vectors.

(a) **u** and **v**   (b) **u** and **w**   (c) **v** and **z**

SOLUTION

(a) $\cos \theta = \dfrac{\mathbf{u} \cdot \mathbf{v}}{\|\mathbf{u}\| \, \|\mathbf{v}\|} = \dfrac{-12 + 4}{\sqrt{14}\sqrt{20}} = \dfrac{-8}{2\sqrt{14}\sqrt{5}} = \dfrac{-4}{\sqrt{70}}$

Since $\mathbf{u} \cdot \mathbf{v} < 0$, $\theta = \arccos \dfrac{-4}{\sqrt{70}} \approx 2.069$ radians.

(b) $\cos \theta = \dfrac{\mathbf{u} \cdot \mathbf{w}}{\|\mathbf{u}\| \, \|\mathbf{w}\|} = \dfrac{3 + 1 - 4}{\sqrt{14}\sqrt{6}} = \dfrac{0}{\sqrt{84}} = 0$

Since $\mathbf{u} \cdot \mathbf{w} = 0$, $\mathbf{u}$ and $\mathbf{w}$ are *orthogonal* vectors, and furthermore, $\theta = \pi/2$.

(c) $\cos \theta = \dfrac{\mathbf{v} \cdot \mathbf{z}}{\|\mathbf{v}\| \, \|\mathbf{z}\|} = \dfrac{-8 + 0 - 2}{\sqrt{20}\sqrt{5}} = \dfrac{-10}{\sqrt{100}} = -1$

Consequently, $\theta = \pi$.

EXAMPLE 8   Decomposing a vector into vector components

Find the projection of **u** onto **v** and the vector component of **u** orthogonal to **v** for the vectors

$$\mathbf{u} = 3\mathbf{i} - 5\mathbf{j} + 2\mathbf{k} \qquad \text{and} \qquad \mathbf{v} = 7\mathbf{i} + \mathbf{j} - 2\mathbf{k}.$$

SOLUTION

The projection of **u** onto **v** is

$$\text{proj}_{\mathbf{v}}\mathbf{u} = \left(\dfrac{\mathbf{u} \cdot \mathbf{v}}{\|\mathbf{v}\|^2}\right)\mathbf{v} = \left(\dfrac{12}{54}\right)(7\mathbf{i} + \mathbf{j} - 2\mathbf{k}) = \dfrac{14}{9}\mathbf{i} + \dfrac{2}{9}\mathbf{j} - \dfrac{4}{9}\mathbf{k}.$$

The vector component of **u** orthogonal to **v** is the vector

$$\mathbf{w} = \mathbf{u} - \text{proj}_{\mathbf{v}}\mathbf{u} = (3\mathbf{i} - 5\mathbf{j} + 2\mathbf{k}) - \left(\dfrac{14}{9}\mathbf{i} + \dfrac{2}{9}\mathbf{j} - \dfrac{4}{9}\mathbf{k}\right)$$

$$= \dfrac{13}{9}\mathbf{i} - \dfrac{47}{9}\mathbf{j} + \dfrac{22}{9}\mathbf{k}.$$

(See Figure 14.9.)

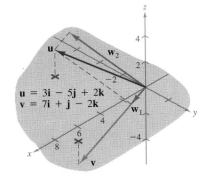

FIGURE 14.9

EXAMPLE 9   An application involving force

An object weighing 120 pounds is supported by a tripod, as shown in Figure 14.10. Represent the force exerted on each leg of the tripod as a vector. (Assume that the weight is distributed equally on each of the three legs.)

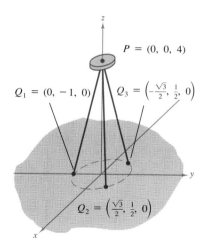

FIGURE 14.10

SOLUTION

We let the vectors $\mathbf{F}_1$, $\mathbf{F}_2$, and $\mathbf{F}_3$ represent the force exerted on the three legs. From Figure 14.10, we can determine the direction of $\mathbf{F}_1$, $\mathbf{F}_2$, and $\mathbf{F}_3$ to be given by

$$\overrightarrow{PQ_1} = \langle 0 - 0, -1 - 0, 0 - 4 \rangle = \langle 0, -1, -4 \rangle$$

$$\overrightarrow{PQ_2} = \left\langle \frac{\sqrt{3}}{2} - 0, \frac{1}{2} - 0, 0 - 4 \right\rangle = \left\langle \frac{\sqrt{3}}{2}, \frac{1}{2}, -4 \right\rangle$$

$$\overrightarrow{PQ_3} = \left\langle -\frac{\sqrt{3}}{2} - 0, \frac{1}{2} - 0, 0 - 4 \right\rangle = \left\langle -\frac{\sqrt{3}}{2}, \frac{1}{2}, -4 \right\rangle.$$

Because each leg has the same length, and the total force is distributed equally on each leg, we know that $\|\mathbf{F}_1\| = \|\mathbf{F}_2\| = \|\mathbf{F}_3\|$ and hence there exists a constant c such that

$$\mathbf{F}_1 = c\langle 0, -1, -4 \rangle, \quad \mathbf{F}_2 = c\left\langle \frac{\sqrt{3}}{2}, \frac{1}{2}, -4 \right\rangle, \quad \mathbf{F}_3 = c\left\langle -\frac{\sqrt{3}}{2}, \frac{1}{2}, -4 \right\rangle.$$

Now, we let the total force exerted by the object be given by $\mathbf{F} = -120\mathbf{k}$, and using the fact that

$$\mathbf{F} = \mathbf{F}_1 + \mathbf{F}_2 + \mathbf{F}_3$$

we can conclude that $\mathbf{F}_1$, $\mathbf{F}_2$, and $\mathbf{F}_3$ each has a vertical component of -40. This implies that $c(-4) = -40$ and $c = 10$. Therefore, we have

$$\mathbf{F}_1 = \langle 0, -10, -40 \rangle, \quad \mathbf{F}_2 = \langle 5\sqrt{3}, 5, -40 \rangle, \quad \mathbf{F}_3 = \langle -5\sqrt{3}, 5, -40 \rangle.$$

Direction cosines

For a vector in the plane, we found it convenient to measure direction in terms of the angle, measured counterclockwise, *from* the positive x-axis to the vector. In space it is more convenient to measure direction in terms of the angles *between* the vector $\mathbf{v}$ and the three unit vectors $\mathbf{i}$, $\mathbf{j}$, and $\mathbf{k}$, as shown in Figure 14.11. We call the angles α, β, and γ the **direction angles of v**, and we call $\cos \alpha$, $\cos \beta$, and $\cos \gamma$ the **direction cosines of v**. Since

$$\mathbf{v} \cdot \mathbf{i} = \langle v_1, v_2, v_3 \rangle \cdot \langle 1, 0, 0 \rangle = v_1$$

it follows that $\cos \alpha = v_1/\|\mathbf{v}\|$. By similar reasoning with the unit vectors $\mathbf{j}$ and $\mathbf{k}$, we have

$$\cos \alpha = \frac{v_1}{\|\mathbf{v}\|}, \quad \cos \beta = \frac{v_2}{\|\mathbf{v}\|}, \quad \text{and} \quad \cos \gamma = \frac{v_3}{\|\mathbf{v}\|}.$$

Consequently, any nonzero vector $\mathbf{v}$ in space has the normalized form

$$\frac{\mathbf{v}}{\|\mathbf{v}\|} = \frac{v_1}{\|\mathbf{v}\|}\mathbf{i} + \frac{v_2}{\|\mathbf{v}\|}\mathbf{j} + \frac{v_3}{\|\mathbf{v}\|}\mathbf{k} = \cos \alpha\, \mathbf{i} + \cos \beta\, \mathbf{j} + \cos \gamma\, \mathbf{k}$$

and since $\mathbf{v}/\|\mathbf{v}\|$ is a unit vector, it follows that

$$\cos^2 \alpha + \cos^2 \beta + \cos^2 \gamma = 1.$$

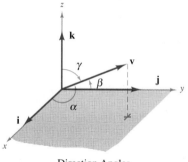

Direction Angles

FIGURE 14.11

EXAMPLE 10 Finding direction angles

Find the direction cosines and angles for the vector $\mathbf{v} = 2\mathbf{i} + 3\mathbf{j} + 4\mathbf{k}$, and show that $\cos^2 \alpha + \cos^2 \beta + \cos^2 \gamma = 1$.

SOLUTION

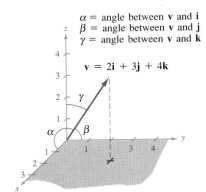

α = angle between $\mathbf{v}$ and $\mathbf{i}$
β = angle between $\mathbf{v}$ and $\mathbf{j}$
γ = angle between $\mathbf{v}$ and $\mathbf{k}$

Since $\|\mathbf{v}\| = \sqrt{2^2 + 3^2 + 4^2} = \sqrt{29}$, we have

$$\cos \alpha = \frac{v_1}{\|\mathbf{v}\|} = \frac{2}{\sqrt{29}} \quad \Longrightarrow \quad \alpha \approx 68.2°$$

$$\cos \beta = \frac{v_2}{\|\mathbf{v}\|} = \frac{3}{\sqrt{29}} \quad \Longrightarrow \quad \beta \approx 56.1°$$

$$\cos \gamma = \frac{v_3}{\|\mathbf{v}\|} = \frac{4}{\sqrt{29}} \quad \Longrightarrow \quad \gamma \approx 42.0°.$$

Furthermore, the sum of the squares of the direction cosines is

$$\cos^2 \alpha + \cos^2 \beta + \cos^2 \gamma = \frac{4}{29} + \frac{9}{29} + \frac{16}{29} = \frac{29}{29} = 1.$$

Direction Angles of $\mathbf{v}$

FIGURE 14.12

(See Figure 14.12.)

EXERCISES for Section 14.1

In Exercises 1–4, plot the points on the same three-dimensional coordinate system.

1. (a) $(2, 1, 3)$ (b) $(-1, 2, 1)$

2. (a) $(3, -2, 5)$ (b) $\left(\frac{3}{2}, 4, -2\right)$

3. (a) $(5, -2, 2)$ (b) $(5, -2, -2)$
4. (a) $(0, 4, -5)$ (b) $(4, 0, 5)$

In Exercises 5–8, find the lengths of the sides of the triangle with the indicated vertices, and determine whether the triangle is a right triangle, an isosceles triangle, or neither.

5. $(0, 0, 0)$, $(2, 2, 1)$, $(2, -4, 4)$
6. $(5, 3, 4)$, $(7, 1, 3)$, $(3, 5, 3)$
7. $(1, -3, -2)$, $(5, -1, 2)$, $(-1, 1, 2)$
8. $(5, 0, 0)$, $(0, 2, 0)$, $(0, 0, -3)$

In Exercises 9 and 10, find the coordinates of the midpoint of the line segment joining the given points.

9. $(5, -9, 7)$ and $(-2, 3, 3)$
10. $(4, 0, -6)$ and $(8, 8, 20)$

In Exercises 11–14, find the general form of the equation of the sphere.

11. Center $(0, 2, 5)$, radius 2
12. Center $(4, -1, 1)$, radius 5
13. Endpoints of a diameter are $(2, 0, 0)$ and $(0, 6, 0)$
14. Center $(-2, 1, 1)$, tangent to the xy-coordinate plane

In Exercises 15–18, find the center and radius of the sphere.

15. $x^2 + y^2 + z^2 - 2x + 6y + 8z + 1 = 0$
16. $x^2 + y^2 + z^2 + 9x - 2y + 10z + 19 = 0$
17. $9x^2 + 9y^2 + 9z^2 - 6x + 18y + 1 = 0$
18. $4x^2 + 4y^2 + 4z^2 - 4x - 32y + 8z + 33 = 0$

In Exercises 19–22, (a) find the component form of the vector **v** and (b) sketch the vector with its initial point at the origin.

19. **20.**

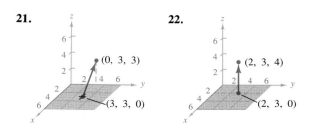

21. **22.**

In Exercises 23 and 24, the initial and terminal points of a vector **v** are given. (a) Sketch the directed line segment, (b) find the component form of the vector, and (c) sketch the vector with its initial point at the origin.

23. Initial point $(-1, 2, 3)$, terminal point $(3, 3, 4)$
24. Initial point $(2, -1, -2)$, terminal point $(-4, 3, 7)$

In Exercises 25 and 26, sketch each scalar multiple of **v**.

25. $\mathbf{v} = \langle 1, 2, 2 \rangle$
   (a) $2\mathbf{v}$ (b) $-\mathbf{v}$
   (c) $\frac{3}{2}\mathbf{v}$ (d) $0\mathbf{v}$

26. $\mathbf{v} = \langle 2, -2, 1 \rangle$
   (a) $-\mathbf{v}$ (b) $2\mathbf{v}$
   (c) $\frac{1}{2}\mathbf{v}$ (d) $\frac{5}{2}\mathbf{v}$

In Exercises 27–32, find the indicated vector, given $\mathbf{u} = \langle 1, 2, 3 \rangle$, $\mathbf{v} = \langle 2, 2, -1 \rangle$, and $\mathbf{w} = \langle 4, 0, -4 \rangle$.

27. $\mathbf{u} - \mathbf{v}$ **28.** $\mathbf{u} - \mathbf{v} + 2\mathbf{w}$
29. $2\mathbf{u} + 4\mathbf{v} - \mathbf{w}$ **30.** $5\mathbf{u} - 3\mathbf{v} - \frac{1}{2}\mathbf{w}$
31. **z**, where $2\mathbf{z} - 3\mathbf{u} = \mathbf{w}$
32. **z**, where $2\mathbf{u} + \mathbf{v} - \mathbf{w} + 3\mathbf{z} = \mathbf{0}$

In Exercises 33–36, determine which of the vectors are parallel to **z**.

33. $\mathbf{z} = \langle 3, 2, -5 \rangle$
   (a) $\langle -6, -4, 10 \rangle$ (b) $\left\langle 2, \frac{4}{3}, -\frac{10}{3} \right\rangle$
   (c) $\langle 6, 4, 10 \rangle$ (d) $\langle 1, -4, 2 \rangle$

34. $\mathbf{z} = \frac{1}{2}\mathbf{i} - \frac{2}{3}\mathbf{j} + \frac{3}{4}\mathbf{k}$
   (a) $6\mathbf{i} - 4\mathbf{j} + 9\mathbf{k}$ (b) $-\mathbf{i} + \frac{4}{3}\mathbf{j} - \frac{3}{2}\mathbf{k}$
   (c) $12\mathbf{i} + 9\mathbf{k}$ (d) $\frac{3}{4}\mathbf{i} - \mathbf{j} + \frac{9}{8}\mathbf{k}$

35. **z** has initial point $(1, -1, 3)$ and terminal point $(-2, 3, 5)$
   (a) $-6\mathbf{i} + 8\mathbf{j} + 4\mathbf{k}$ (b) $4\mathbf{j} + 2\mathbf{k}$

36. **z** has initial point $(3, 2, -1)$ and terminal point $(-1, -3, 5)$
   (a) $\langle 0, 5, -6 \rangle$ (b) $\langle 8, 10, -12 \rangle$

In Exercises 37–40, use vectors to determine whether the given points lie on a straight line.

37. $(0, -2, -5)$, $(3, 4, 4)$, $(2, 2, 1)$
38. $(1, -1, 5)$, $(0, -1, 6)$, $(3, -1, 3)$
39. $(1, 2, 4)$, $(2, 5, 0)$, $(0, 1, 5)$
40. $(0, 0, 0)$, $(1, 3, -2)$, $(2, -6, 4)$

In Exercises 41 and 42, use vectors to show that the given points form the vertices of a parallelogram.

41. $(2, 9, 1)$, $(3, 11, 4)$, $(0, 10, 2)$, $(1, 12, 5)$
42. $(1, 1, -3)$, $(9, -1, 2)$, $(11, 2, 1)$, $(3, 4, -4)$

In Exercises 43 and 44, the vector **v** and its initial point are given. Find the terminal point.

43. $\mathbf{v} = \langle 3, -5, 6 \rangle$, initial point $(0, 6, 2)$
44. $\mathbf{v} = \left\langle 0, \frac{1}{2}, -\frac{1}{3} \right\rangle$, initial point $\left(3, 0, -\frac{2}{3} \right)$

In Exercises 45–50, find the magnitude of **v**.

45. $\mathbf{v} = \langle 0, 0, 0 \rangle$ **46.** $\mathbf{v} = \langle 1, 0, 3 \rangle$
47. **v** has $(1, -3, 4)$ and $(1, 0, -1)$ as its initial and terminal points, respectively.
48. **v** has $(0, -1, 0)$ and $(1, 2, -2)$ as its initial and terminal points, respectively.
49. $\mathbf{v} = \mathbf{i} - 2\mathbf{j} - 3\mathbf{k}$ **50.** $\mathbf{v} = -4\mathbf{i} + 3\mathbf{j} + 7\mathbf{k}$

In Exercises 51–54, find a unit vector (a) in the direction of **u** and (b) in the direction opposite that of **u**.

51. $\mathbf{u} = \langle 2, -1, 2 \rangle$ **52.** $\mathbf{u} = \langle 6, 0, 8 \rangle$
53. $\mathbf{u} = \langle 3, 2, -5 \rangle$ **54.** $\mathbf{u} = \langle 8, 0, 0 \rangle$

In Exercises 55–58, determine the values of c that satisfy the given equation. Let $\mathbf{u} = \mathbf{i} + 2\mathbf{j} + 3\mathbf{k}$ and $\mathbf{v} = 2\mathbf{i} + 2\mathbf{j} - \mathbf{k}$.

55. $\|c\mathbf{u}\| = 1$ **56.** $\|c\mathbf{v}\| = 1$

57. $\|c\mathbf{v}\| = 5$ **58.** $\|c\mathbf{u}\| = 3$

In Exercises 59–62, use vectors to find the point that lies two-thirds of the way from P to Q.

59. $P = (4, 3, 0)$, $Q = (1, -3, 3)$
60. $P = (-2, 1, 6)$, $Q = (6, 1, 4)$
61. $P = (1, 2, 5)$, $Q = (6, 8, 2)$
62. $P = (-9, -8, 5)$, $Q = (12, 3, -1)$

In Exercises 63 and 64, write the component form of $\mathbf{v}$ and sketch it.

63. $\mathbf{v}$ lies in the yz-plane, has magnitude 2, and makes an angle of 30° with the positive y-axis.
64. $\mathbf{v}$ lies in the xz-plane, has magnitude 5, and makes an angle of 45° with the positive z-axis.

In Exercises 65 and 66, find (a) $\mathbf{u} \cdot \mathbf{v}$, (b) $\mathbf{u} \cdot \mathbf{u}$, (c) $\|\mathbf{u}\|^2$, (d) $(\mathbf{u} \cdot \mathbf{v})\mathbf{v}$, and (e) $\mathbf{u} \cdot (2\mathbf{v})$.

65. $\mathbf{u} = 2\mathbf{i} - \mathbf{j} + \mathbf{k}$, $\mathbf{v} = \mathbf{i} - \mathbf{k}$
66. $\mathbf{u} = 2\mathbf{i} + \mathbf{j} - 2\mathbf{k}$, $\mathbf{v} = \mathbf{i} - 3\mathbf{j} + 2\mathbf{k}$

In Exercises 67–70, find the angle θ between the given vectors.

67. $\mathbf{u} = \langle 1, 1, 1 \rangle$, $\mathbf{v} = \langle 2, 1, -1 \rangle$
68. $\mathbf{u} = 2\mathbf{i} + 3\mathbf{j} + \mathbf{k}$, $\mathbf{v} = -3\mathbf{i} + 2\mathbf{j}$
69. $\mathbf{u} = 3\mathbf{i} + 4\mathbf{j}$, $\mathbf{v} = -2\mathbf{j} + 3\mathbf{k}$
70. $\mathbf{u} = 2\mathbf{i} - 3\mathbf{j} + \mathbf{k}$, $\mathbf{v} = \mathbf{i} - 2\mathbf{j} + \mathbf{k}$

In Exercises 71–74, determine whether $\mathbf{u}$ and $\mathbf{v}$ are orthogonal, parallel, or neither.

71. $\mathbf{u} = \mathbf{j} + 6\mathbf{k}$, $\mathbf{v} = \mathbf{i} - 2\mathbf{j} - \mathbf{k}$
72. $\mathbf{u} = -2\mathbf{i} + 3\mathbf{j} - \mathbf{k}$, $\mathbf{v} = 2\mathbf{i} + \mathbf{j} - \mathbf{k}$
73. $\mathbf{u} = \langle 2, -3, 1 \rangle$, $\mathbf{v} = \langle -1, -1, -1 \rangle$
74. $\mathbf{u} = \langle \cos\theta, \sin\theta, -1 \rangle$
      $\mathbf{v} = \langle \sin\theta, -\cos\theta, 0 \rangle$

In Exercises 75–80, (a) find the projection of $\mathbf{u}$ onto $\mathbf{v}$, and (b) find the vector component of $\mathbf{u}$ orthogonal to $\mathbf{v}$.

75. $\mathbf{u} = \langle 2, 1, 2 \rangle$, $\mathbf{v} = \langle 0, 3, 4 \rangle$
76. $\mathbf{u} = \langle 0, 4, 1 \rangle$, $\mathbf{v} = \langle 0, 2, 3 \rangle$
77. $\mathbf{u} = \langle 1, 1, 1 \rangle$, $\mathbf{v} = \langle -2, -1, 1 \rangle$
78. $\mathbf{u} = \langle -2, -1, 1 \rangle$, $\mathbf{v} = \langle 1, 1, 1 \rangle$
79. $\mathbf{u} = \langle 5, -4, 3 \rangle$, $\mathbf{v} = \langle 1, 0, 0 \rangle$
80. $\mathbf{u} = \langle 5, -4, 3 \rangle$, $\mathbf{v} = \langle 0, 1, 0 \rangle$

In Exercises 81–84, find the direction cosines of $\mathbf{u}$ and demonstrate that the sum of the squares of the direction cosines is 1.

81. $\mathbf{u} = \mathbf{i} + 2\mathbf{j} + 2\mathbf{k}$ **82.** $\mathbf{u} = 3\mathbf{i} - \mathbf{j} + 5\mathbf{k}$
83. $\mathbf{u} = \langle 0, 6, -4 \rangle$ **84.** $\mathbf{u} = \langle a, b, c \rangle$

85. Let $\mathbf{u} = \mathbf{i} + \mathbf{j}$, $\mathbf{v} = \mathbf{j} + \mathbf{k}$, and $\mathbf{w} = a\mathbf{u} + b\mathbf{v}$.
(a) Sketch $\mathbf{u}$ and $\mathbf{v}$.
(b) If $\mathbf{w} = \mathbf{0}$, show that a and b must both be zero.
(c) Find a and b such that $\mathbf{w} = \mathbf{i} + 2\mathbf{j} + \mathbf{k}$.
(d) Show that no choice of a and b yields $\mathbf{w} = \mathbf{i} + 2\mathbf{j} + 3\mathbf{k}$.

86. The initial and terminal points of the vector $\mathbf{v}$ are (x_1, y_1, z_1) and (x, y, z), respectively. Describe the set of all points (x, y, z) such that $\|\mathbf{v}\| = 4$.

87. Find the component form of the unit vector $\mathbf{v}$ representing the diagonal of the cube (see figure).

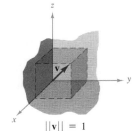

$\|\mathbf{v}\| = 1$ FIGURE FOR 87

88. The guy wire to a 100-foot tower has a tension of 550 pounds. Using the distances shown in the accompanying figure, write the component form of the vector $\mathbf{F}$ representing the tension in the wire.

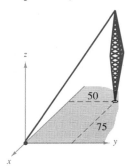

FIGURE FOR 88

89. The lights in an auditorium are 25-pound disks of radius 18 inches. Each disk is supported by three equally spaced 48-inch wires to the ceiling (see figure). Find the tension in each wire.

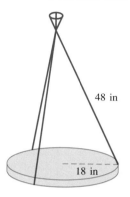

48 in

18 in

FIGURE FOR 89

90. Repeat Exercise 89 if the radius of each light fixture is r_0 inches. Determine the limit of the tension of each wire as $r_0 \to 48^-$.

91. Refer to the Chapter 14 Application. Use the fact that $\mathbf{F}_1 + \mathbf{F}_2 + \mathbf{F}_3 = \mathbf{0}$ to determine T_2 and T_3 for the following vectors.

$$\mathbf{F}_1 = -3600\mathbf{k}$$
$$\mathbf{F}_2 = T_2(0.5792\mathbf{i} + 0.5792\mathbf{j} - 0.5736\mathbf{k})$$
$$\mathbf{F}_3 = T_3(-0.0247\mathbf{i} - 0.0247\mathbf{j} + 0.9994\mathbf{k})$$

14.2   The Cross Product of Two Vectors in Space

The cross product ▪ Triple scalar product ▪ Applications

Many applications in physics, engineering, and geometry involve finding a vector in space that is orthogonal to two given vectors. In this section we look at a product that will yield such a vector. It is called the **cross product,** and it is most conveniently defined and calculated using the standard unit vector form.

DEFINITION OF CROSS PRODUCT OF TWO VECTORS IN SPACE	Let $\mathbf{u} = u_1\mathbf{i} + u_2\mathbf{j} + u_3\mathbf{k}$ and $\mathbf{v} = v_1\mathbf{i} + v_2\mathbf{j} + v_3\mathbf{k}$ be vectors in space. The **cross product** of $\mathbf{u}$ and $\mathbf{v}$ is the vector $$\mathbf{u} \times \mathbf{v} = (u_2v_3 - u_3v_2)\mathbf{i} - (u_1v_3 - u_3v_1)\mathbf{j} + (u_1v_2 - u_2v_1)\mathbf{k}.$$

REMARK   Be sure you see that this definition applies only to three-dimensional vectors. We do not define the cross product of two-dimensional vectors.

A convenient way to calculate $\mathbf{u} \times \mathbf{v}$ is to use the following *determinant form* with cofactor expansion. (This 3×3 determinant form is used simply to help remember the formula for the cross product—it is technically not a determinant since the entries of the corresponding matrix are not all real numbers.)

$$\mathbf{u} \times \mathbf{v} = \begin{vmatrix} \mathbf{i} & \mathbf{j} & \mathbf{k} \\ u_1 & u_2 & u_3 \\ v_1 & v_2 & v_3 \end{vmatrix} \xleftarrow{\hspace{1cm}} \text{Put "}\mathbf{u}\text{" in Row 2.}$$
$$\xleftarrow{\hspace{1cm}} \text{Put "}\mathbf{v}\text{" in Row 3.}$$

$$= \begin{vmatrix} \mathbf{i} & \mathbf{j} & \mathbf{k} \\ u_1 & u_2 & u_3 \\ v_1 & v_2 & v_3 \end{vmatrix} \mathbf{i} - \begin{vmatrix} \mathbf{i} & \mathbf{j} & \mathbf{k} \\ u_1 & u_2 & u_3 \\ v_1 & v_2 & v_3 \end{vmatrix} \mathbf{j} + \begin{vmatrix} \mathbf{i} & \mathbf{j} & \mathbf{k} \\ u_1 & u_2 & u_3 \\ v_1 & v_2 & v_3 \end{vmatrix} \mathbf{k}$$

$$= \begin{vmatrix} u_2 & u_3 \\ v_2 & v_3 \end{vmatrix} \mathbf{i} - \begin{vmatrix} u_1 & u_3 \\ v_1 & v_3 \end{vmatrix} \mathbf{j} + \begin{vmatrix} u_1 & u_2 \\ v_1 & v_2 \end{vmatrix} \mathbf{k}$$

Note the minus sign in front of the **j**-component. Each of the three 2×2 determinants can be evaluated by using the following diagonal pattern.

$$\begin{vmatrix} a & b \\ c & d \end{vmatrix} = ad - bc.$$

EXAMPLE 1   Finding the cross product

Given $\mathbf{u} = \mathbf{i} - 2\mathbf{j} + \mathbf{k}$ and $\mathbf{v} = 3\mathbf{i} + \mathbf{j} - 2\mathbf{k}$, find the following.

(a) $\mathbf{u} \times \mathbf{v}$ (b) $\mathbf{v} \times \mathbf{u}$ (c) $\mathbf{v} \times \mathbf{v}$

SOLUTION

(a) $\mathbf{u} \times \mathbf{v} = \begin{vmatrix} \mathbf{i} & \mathbf{j} & \mathbf{k} \\ 1 & -2 & 1 \\ 3 & 1 & -2 \end{vmatrix} = \begin{vmatrix} -2 & 1 \\ 1 & -2 \end{vmatrix} \mathbf{i} - \begin{vmatrix} 1 & 1 \\ 3 & -2 \end{vmatrix} \mathbf{j} + \begin{vmatrix} 1 & -2 \\ 3 & 1 \end{vmatrix} \mathbf{k}$

$$= (4 - 1)\mathbf{i} - (-2 - 3)\mathbf{j} + (1 + 6)\mathbf{k}$$
$$= 3\mathbf{i} + 5\mathbf{j} + 7\mathbf{k}$$

(b) $\mathbf{v} \times \mathbf{u} = \begin{vmatrix} \mathbf{i} & \mathbf{j} & \mathbf{k} \\ 3 & 1 & -2 \\ 1 & -2 & 1 \end{vmatrix} = \begin{vmatrix} 1 & -2 \\ -2 & 1 \end{vmatrix} \mathbf{i} - \begin{vmatrix} 3 & -2 \\ 1 & 1 \end{vmatrix} \mathbf{j} + \begin{vmatrix} 3 & 1 \\ 1 & -2 \end{vmatrix} \mathbf{k}$

$$= (1 - 4)\mathbf{i} - (3 + 2)\mathbf{j} + (-6 - 1)\mathbf{k}$$
$$= -3\mathbf{i} - 5\mathbf{j} - 7\mathbf{k}$$

Note that this result is the negative of that in part (a).

(c) $\mathbf{v} \times \mathbf{v} = \begin{vmatrix} \mathbf{i} & \mathbf{j} & \mathbf{k} \\ 3 & 1 & -2 \\ 3 & 1 & -2 \end{vmatrix} = 0$

The results obtained in Example 1 suggest some interesting *algebraic* properties of the cross product. For instance,

$$\mathbf{u} \times \mathbf{v} = -(\mathbf{v} \times \mathbf{u}) \quad \text{and} \quad \mathbf{v} \times \mathbf{v} = \mathbf{0}.$$

These properties, and several others, are given in the following theorem.

**THEOREM 14.1
ALGEBRAIC PROPERTIES
OF THE CROSS PRODUCT**

If **u**, **v**, and **w** are vectors in space and c is any scalar, then the following properties are true.

1. $\mathbf{u} \times \mathbf{v} = -(\mathbf{v} \times \mathbf{u})$
2. $\mathbf{u} \times (\mathbf{v} + \mathbf{w}) = (\mathbf{u} \times \mathbf{v}) + (\mathbf{u} \times \mathbf{w})$
3. $c(\mathbf{u} \times \mathbf{v}) = (c\mathbf{u}) \times \mathbf{v} = \mathbf{u} \times (c\mathbf{v})$
4. $\mathbf{u} \times \mathbf{0} = \mathbf{0} \times \mathbf{u} = \mathbf{0}$
5. $\mathbf{u} \times \mathbf{u} = \mathbf{0}$
6. $\mathbf{u} \cdot (\mathbf{v} \times \mathbf{w}) = (\mathbf{u} \times \mathbf{v}) \cdot \mathbf{w}$

PROOF

Each of the six properties can be proved by writing the vectors in component form and then applying the definition of the cross product. For instance, to prove the first property, we let $\mathbf{u} = u_1\mathbf{i} + u_2\mathbf{j} + u_3\mathbf{k}$ and $\mathbf{v} = v_1\mathbf{i} + v_2\mathbf{j} + v_3\mathbf{k}$. Then,

$$\mathbf{u} \times \mathbf{v} = (u_2v_3 - u_3v_2)\mathbf{i} - (u_1v_3 - u_3v_1)\mathbf{j} + (u_1v_2 - u_2v_1)\mathbf{k}$$

and

$$\mathbf{v} \times \mathbf{u} = (v_2u_3 - v_3u_2)\mathbf{i} - (v_1u_3 - v_3u_1)\mathbf{j} + (v_1u_2 - v_2u_1)\mathbf{k}$$

which implies that $\mathbf{u} \times \mathbf{v} = -(\mathbf{v} \times \mathbf{u})$. Proofs of the remaining properties are left as exercises (see Exercises 35–39).

Note that the first property listed in Theorem 14.1 indicates that the cross product is *not commutative*. In particular, this property indicates that the vectors $\mathbf{u} \times \mathbf{v}$ and $\mathbf{v} \times \mathbf{u}$ have equal lengths but opposite directions. The following theorem lists some other *geometric* properties of the cross product of two vectors.

**THEOREM 14.2
GEOMETRIC PROPERTIES
OF THE CROSS PRODUCT**

If **u** and **v** are nonzero vectors in space, and θ is the angle between **u** and **v**, then the following properties are true.

1. $\mathbf{u} \times \mathbf{v}$ is orthogonal to both **u** and **v**.
2. $\|\mathbf{u} \times \mathbf{v}\| = \|\mathbf{u}\| \|\mathbf{v}\| \sin \theta$.
3. $\mathbf{u} \times \mathbf{v} = \mathbf{0}$ if and only if **u** and **v** are scalar multiples of each other.
4. $\|\mathbf{u} \times \mathbf{v}\| = $ area of parallelogram having **u** and **v** as adjacent sides.

PROOF

We prove the second and fourth properties and leave the proofs of the first and third properties as exercises (see Exercises 40 and 41). For the second property, we note that since $\cos \theta = (\mathbf{u} \cdot \mathbf{v})/(\|\mathbf{u}\| \|\mathbf{v}\|)$, it follows that

$$\|\mathbf{u}\| \|\mathbf{v}\| \sin \theta = \|\mathbf{u}\| \|\mathbf{v}\|\sqrt{1 - \cos^2 \theta}$$

$$= \|\mathbf{u}\| \|\mathbf{v}\| \sqrt{1 - \frac{(\mathbf{u} \cdot \mathbf{v})^2}{\|\mathbf{u}\|^2\|\mathbf{v}\|^2}} = \sqrt{\|\mathbf{u}\|^2\|\mathbf{v}\|^2 - (\mathbf{u} \cdot \mathbf{v})^2}$$

$$= \sqrt{(u_1{}^2 + u_2{}^2 + u_3{}^2)(v_1{}^2 + v_2{}^2 + v_3{}^2) - (u_1v_1 + u_2v_2 + u_3v_3)^2}$$

$$= \sqrt{(u_2v_3 - u_3v_2)^2 + (u_1v_3 - u_3v_1)^2 + (u_1v_2 - u_2v_1)^2}$$

$$= \|\mathbf{u} \times \mathbf{v}\|.$$

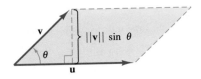

FIGURE 14.13

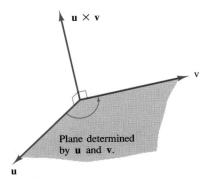

FIGURE 14.14

Right-handed Systems

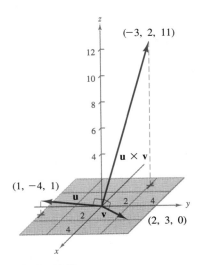

FIGURE 14.15

To prove the fourth property, we use Figure 14.13, which is a parallelogram having **v** and **u** as adjacent sides. Since the height of the parallelogram is $\|\mathbf{v}\| \sin \theta$, the area is

$$\text{area} = (\text{base})(\text{height}) = \|\mathbf{u}\| \|\mathbf{v}\| \sin \theta = \|\mathbf{u} \times \mathbf{v}\|.$$

REMARK It follows from the first two properties listed in Theorem 14.2 that if **n** is a unit vector orthogonal to both **u** and **v**, then

$$\mathbf{u} \times \mathbf{v} = \pm(\|\mathbf{u}\| \|\mathbf{v}\| \sin \theta)\mathbf{n}.$$

Both $\mathbf{u} \times \mathbf{v}$ and $\mathbf{v} \times \mathbf{u}$ are perpendicular to the plane determined by **u** and **v**. One way to remember the orientation of the vectors **u**, **v**, and $\mathbf{u} \times \mathbf{v}$ is to compare them with the unit vectors **i**, **j**, and $\mathbf{k} = \mathbf{i} \times \mathbf{j}$, as shown in Figure 14.14. The three vectors **u**, **v**, and $\mathbf{u} \times \mathbf{v}$ form a *right-handed system*. whereas the three vectors **u**, **v**, and $\mathbf{v} \times \mathbf{u}$ form a *left-handed system*. Using a right-handed system (or the determinant form of the cross product), we can verify each of the following.

$\mathbf{i} \times \mathbf{j} = \mathbf{k}$	$\mathbf{j} \times \mathbf{k} = \mathbf{i}$	$\mathbf{k} \times \mathbf{i} = \mathbf{j}$
$\mathbf{j} \times \mathbf{i} = -\mathbf{k}$	$\mathbf{k} \times \mathbf{j} = -\mathbf{i}$	$\mathbf{i} \times \mathbf{k} = -\mathbf{j}$
$\mathbf{i} \times \mathbf{i} = 0$	$\mathbf{j} \times \mathbf{j} = 0$	$\mathbf{k} \times \mathbf{k} = 0$

EXAMPLE 2 Applications of the cross product

Find a unit vector that is orthogonal to both

$$\mathbf{u} = \mathbf{i} - 4\mathbf{j} + \mathbf{k} \qquad \text{and} \qquad \mathbf{v} = 2\mathbf{i} + 3\mathbf{j}.$$

SOLUTION

The cross product $\mathbf{u} \times \mathbf{v}$, as shown in Figure 14.15, is orthogonal to both **u** and **v**.

$$\mathbf{u} \times \mathbf{v} = \begin{vmatrix} \mathbf{i} & \mathbf{j} & \mathbf{k} \\ 1 & -4 & 1 \\ 2 & 3 & 0 \end{vmatrix} = -3\mathbf{i} + 2\mathbf{j} + 11\mathbf{k}$$

Since $\|\mathbf{u} \times \mathbf{v}\| = \sqrt{(-3)^2 + 2^2 + 11^2} = \sqrt{134}$, a unit vector orthogonal to both **u** and **v** is

$$\frac{\mathbf{u} \times \mathbf{v}}{\|\mathbf{u} \times \mathbf{v}\|} = -\frac{3}{\sqrt{134}}\mathbf{i} + \frac{2}{\sqrt{134}}\mathbf{j} + \frac{11}{\sqrt{134}}\mathbf{k}. \qquad \blacksquare$$

REMARK In Example 2, note that we could have used the cross product $\mathbf{v} \times \mathbf{u}$ to form a unit vector that is orthogonal to both **u** and **v**. With that choice, we would have obtained the negative of the unit vector found in the example.

EXAMPLE 3   An application of the cross product

Show that the quadrilateral with vertices at the following points is a parallelogram, and find its area.

$$A = (5, 2, 0) \qquad B = (2, 6, 1)$$
$$C = (2, 4, 7) \qquad D = (5, 0, 6)$$

SOLUTION

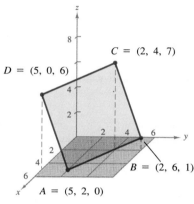

FIGURE 14.16

From Figure 14.16 we see that the sides of the quadrilateral correspond to the following four vectors.

$$\overrightarrow{AB} = -3\mathbf{i} + 4\mathbf{j} + \mathbf{k} \qquad \overrightarrow{CD} = 3\mathbf{i} - 4\mathbf{j} - \mathbf{k} = -\overrightarrow{AB}$$
$$\overrightarrow{AD} = 0\mathbf{i} - 2\mathbf{j} + 6\mathbf{k} \qquad \overrightarrow{CB} = 0\mathbf{i} + 2\mathbf{j} - 6\mathbf{k} = -\overrightarrow{AD}$$

Thus, $\overrightarrow{AB}$ is parallel to $\overrightarrow{CD}$ and $\overrightarrow{AD}$ is parallel to $\overrightarrow{CB}$, and we conclude that the quadrilateral is a parallelogram with $\overrightarrow{AB}$ and $\overrightarrow{AD}$ as adjacent sides. Moreover, since

$$\overrightarrow{AB} \times \overrightarrow{AD} = \begin{vmatrix} \mathbf{i} & \mathbf{j} & \mathbf{k} \\ -3 & 4 & 1 \\ 0 & -2 & 6 \end{vmatrix} = 26\mathbf{i} + 18\mathbf{j} + 6\mathbf{k}$$

the area of the parallelogram is

$$\|\overrightarrow{AB} \times \overrightarrow{AD}\| = \sqrt{1036} \approx 32.19.$$

In physics, the cross product can be used to measure **torque**—the **moment M of a force F about a point P,** as shown in Figure 14.17. If the point of application of the force is Q, then the moment of $\mathbf{F}$ about P is given by

$$\mathbf{M} = \overrightarrow{PQ} \times \mathbf{F}. \qquad \text{Moment of F about } P$$

The magnitude of the moment $\mathbf{M}$ measures the tendency of the vector $\overrightarrow{PQ}$ to rotate counterclockwise (using the right-hand rule) about an axis directed along the vector $\mathbf{M}$.

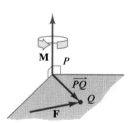

FIGURE 14.17

EXAMPLE 4   An application of the cross product

A vertical force of 50 pounds is applied to the end of a 1-foot lever that is attached to an axle at point P, as shown in Figure 14.18. Find the moment of this force about the point P when $\theta = 60°$.

SOLUTION

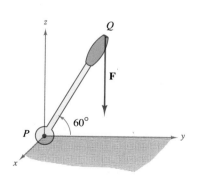

FIGURE 14.18

If we represent the 50-pound force as $\mathbf{F} = -50\mathbf{k}$ and the lever as

$$\overrightarrow{PQ} = \cos(60°)\,\mathbf{j} + \sin(60°)\,\mathbf{k} = \frac{1}{2}\mathbf{j} + \frac{\sqrt{3}}{2}\mathbf{k}$$

the moment of **F** about *P* is given by

$$\mathbf{M} = \overrightarrow{PQ} \times \mathbf{F} = \begin{vmatrix} \mathbf{i} & \mathbf{j} & \mathbf{k} \\ 0 & \dfrac{1}{2} & \dfrac{\sqrt{3}}{2} \\ 0 & 0 & -50 \end{vmatrix} = -25\mathbf{i}$$

and the magnitude of this moment is 25 foot-pounds.

REMARK In Example 4, note that the moment (the tendency of the lever to rotate about its axle) is dependent upon the angle θ. When $\theta = \pi/2$, the moment is 0, and the moment is greatest when $\theta = 0$.

The triple scalar product

For vectors **u**, **v**, and **w** in space, the dot product of **u** and $\mathbf{v} \times \mathbf{w}$

$$\mathbf{u} \cdot (\mathbf{v} \times \mathbf{w})$$

is called the **triple scalar product.** This product has a practical geometric interpretation and it can be evaluated by the determinant given in the following theorem. We leave the proof of this theorem as an exercise (see Exercise 44).

THEOREM 14.3 **THE TRIPLE SCALAR PRODUCT**	For $\mathbf{u} = u_1\mathbf{i} + u_2\mathbf{j} + u_3\mathbf{k}$, $\mathbf{v} = v_1\mathbf{i} + v_2\mathbf{j} + v_3\mathbf{k}$, and $\mathbf{w} = w_1\mathbf{i} + w_2\mathbf{j} + w_3\mathbf{k}$, the triple scalar product is given by $$\mathbf{u} \cdot (\mathbf{v} \times \mathbf{w}) = \begin{vmatrix} u_1 & u_2 & u_3 \\ v_1 & v_2 & v_3 \\ w_1 & w_2 & w_3 \end{vmatrix}.$$

REMARK From the properties of determinants we know that the value of a determinant is multiplied by -1 if two rows are interchanged. After two such interchanges, the value of the determinant will be unchanged. Thus, the following triple scalar products are equivalent.

$$\mathbf{u} \cdot (\mathbf{v} \times \mathbf{w}) = \mathbf{v} \cdot (\mathbf{w} \times \mathbf{u}) = \mathbf{w} \cdot (\mathbf{u} \times \mathbf{v})$$

If the vectors **u**, **v**, and **w** do not lie in the same plane, then the triple scalar product $\mathbf{u} \cdot (\mathbf{v} \times \mathbf{w})$ can be used to determine the volume of the parallelepiped with **u**, **v**, and **w** as adjacent sides, as shown in Figure 14.19. This is established in the following theorem.

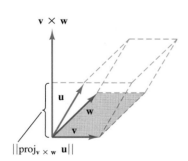

$\mathbf{v} \times \mathbf{w}$

$\|\text{proj}_{\mathbf{v} \times \mathbf{w}}\, \mathbf{u}\|$

Area of base $= \|\mathbf{v} \times \mathbf{w}\|$

Volume of Parallelepiped $= |\mathbf{u} \cdot (\mathbf{v} \times \mathbf{w})|$

FIGURE 14.19

THEOREM 14.4 **GEOMETRIC PROPERTY OF** **TRIPLE SCALAR PRODUCT**	The volume V of a parallelepiped with vectors **u**, **v**, and **w** as adjacent sides is given by $$V = |\mathbf{u} \cdot (\mathbf{v} \times \mathbf{w})|.$$

PROOF   In Figure 14.19, we note that

$$\|\mathbf{v} \times \mathbf{w}\| = \text{area of base}$$

$$\|\text{proj}_{\mathbf{v} \times \mathbf{w}}\mathbf{u}\| = \text{height of parallelepiped.}$$

Therefore, the volume is

$$V = (\text{height})(\text{area of base}) = \|\text{proj}_{\mathbf{v} \times \mathbf{w}}\mathbf{u}\| \, \|\mathbf{v} \times \mathbf{w}\|$$

$$= \left| \frac{\mathbf{u} \cdot (\mathbf{v} \times \mathbf{w})}{\|\mathbf{v} \times \mathbf{w}\|} \right| \, \|\mathbf{v} \times \mathbf{w}\|$$

$$= |\mathbf{u} \cdot (\mathbf{v} \times \mathbf{w})|.$$

EXAMPLE 5   Volume by the triple scalar product

Find the volume of the parallelepiped having $\mathbf{u} = 3\mathbf{i} - 5\mathbf{j} + \mathbf{k}$, $\mathbf{v} = 2\mathbf{j} - 2\mathbf{k}$, and $\mathbf{w} = 3\mathbf{i} + \mathbf{j} + \mathbf{k}$ as adjacent edges.

SOLUTION

By Theorem 14.4, we have

$$V = |\mathbf{u} \cdot (\mathbf{v} \times \mathbf{w})| = \begin{vmatrix} 3 & -5 & 1 \\ 0 & 2 & -2 \\ 3 & 1 & 1 \end{vmatrix} = 36.$$

A natural consequence of Theorem 14.4 is that the volume of the parallelepiped is zero if and only if the three vectors are coplanar. This gives us the following test.

Test for Coplanar Vectors: If the vectors $\mathbf{u} = \langle u_1, u_2, u_3 \rangle$, $\mathbf{v} = \langle v_1, v_2, v_3 \rangle$, and $\mathbf{w} = \langle w_1, w_2, w_3 \rangle$ have the same initial point, then they lie in the same plane if and only if

$$\mathbf{u} \cdot (\mathbf{v} \times \mathbf{w}) = \begin{vmatrix} u_1 & u_2 & u_3 \\ v_1 & v_2 & v_3 \\ w_1 & w_2 & w_3 \end{vmatrix} = 0.$$

EXERCISES for Section 14.2

In Exercises 1–6, find the cross product of the given unit vectors and sketch your result.

1. $\mathbf{j} \times \mathbf{i}$ **2.** $\mathbf{i} \times \mathbf{j}$
3. $\mathbf{j} \times \mathbf{k}$ **4.** $\mathbf{k} \times \mathbf{j}$
5. $\mathbf{i} \times \mathbf{k}$ **6.** $\mathbf{k} \times \mathbf{i}$

In Exercises 7–14, find $\mathbf{u} \times \mathbf{v}$ and show that it is orthogonal to both $\mathbf{u}$ and $\mathbf{v}$.

7. $\mathbf{u} = \langle 2, -3, 1 \rangle$, $\mathbf{v} = \langle 1, -2, 1 \rangle$
8. $\mathbf{u} = \langle -1, 1, 2 \rangle$, $\mathbf{v} = \langle 0, 1, 0 \rangle$
9. $\mathbf{u} = \langle 12, -3, 0 \rangle$, $\mathbf{v} = \langle -2, 5, 0 \rangle$

10. $\mathbf{u} = \langle -10, 0, 6 \rangle$, $\mathbf{v} = \langle 7, 0, 0 \rangle$

11. $\mathbf{u} = \mathbf{i} + \mathbf{j} + \mathbf{k}$, $\mathbf{v} = 2\mathbf{i} + \mathbf{j} - \mathbf{k}$

12. $\mathbf{u} = \mathbf{j} + 6\mathbf{k}$, $\mathbf{v} = \mathbf{i} - 2\mathbf{j} + \mathbf{k}$

13. $\mathbf{u} = -3\mathbf{i} + 2\mathbf{j} - 5\mathbf{k}$, $\mathbf{v} = \frac{1}{2}\mathbf{i} - \frac{3}{4}\mathbf{j} + \frac{1}{10}\mathbf{k}$

14. $\mathbf{u} = \frac{2}{3}\mathbf{k}$, $\mathbf{v} = \frac{1}{2}\mathbf{i} + 6\mathbf{k}$

In Exercises 15–18, find the area of the parallelogram that has the given vectors as adjacent sides.

15. $\mathbf{u} = \mathbf{j}$, $\mathbf{v} = \mathbf{j} + \mathbf{k}$

16. $\mathbf{u} = \mathbf{i} + \mathbf{j} + \mathbf{k}$, $\mathbf{v} = \mathbf{j} + \mathbf{k}$

17. $\mathbf{u} = \langle 3, 2, -1 \rangle$, $\mathbf{v} = \langle 1, 2, 3 \rangle$

18. $\mathbf{u} = \langle 2, -1, 0 \rangle$, $\mathbf{v} = \langle -1, 2, 0 \rangle$

In Exercises 19 and 20, find the area of the parallelogram with the given vertices.

19. $(1, 1, 1)$, $(2, 3, 4)$, $(6, 5, 2)$, $(7, 7, 5)$

20. $(2, -1, 1)$, $(5, 1, 4)$, $(0, 1, 1)$, $(3, 3, 4)$

In Exercises 21–24, find the area of the triangle with the given vertices. ($\frac{1}{2}\|\mathbf{u} \times \mathbf{v}\|$ is the area of the triangle having $\mathbf{u}$ and $\mathbf{v}$ as adjacent sides.)

21. $(0, 0, 0)$, $(1, 2, 3)$, $(-3, 0, 0)$

22. $(2, -3, 4)$, $(0, 1, 2)$, $(-1, 2, 0)$

23. $(1, 3, 5)$, $(3, 3, 0)$, $(-2, 0, 5)$

24. $(1, 2, 0)$, $(-2, 1, 0)$, $(0, 0, 0)$

In Exercises 25–28, find $\mathbf{u} \cdot (\mathbf{v} \times \mathbf{w})$.

25. $\mathbf{u} = \mathbf{i}$, $\mathbf{v} = \mathbf{j}$, $\mathbf{w} = \mathbf{k}$

26. $\mathbf{u} = \langle 1, 1, 1 \rangle$, $\mathbf{v} = \langle 2, 1, 0 \rangle$, $\mathbf{w} = \langle 0, 0, 1 \rangle$

27. $\mathbf{u} = \langle 2, 0, 1 \rangle$, $\mathbf{v} = \langle 0, 3, 0 \rangle$, $\mathbf{w} = \langle 0, 0, 1 \rangle$

28. $\mathbf{u} = \langle 2, 0, 0 \rangle$, $\mathbf{v} = \langle 1, 1, 1 \rangle$, $\mathbf{w} = \langle 0, 2, 2 \rangle$

In Exercises 29 and 30, use the triple scalar product to find the volume of the parallelepiped having adjacent edges $\mathbf{u}$, $\mathbf{v}$, and $\mathbf{w}$.

29. $\mathbf{u} = \mathbf{i} + \mathbf{j}$, $\mathbf{v} = \mathbf{j} + \mathbf{k}$, $\mathbf{w} = \mathbf{i} + \mathbf{k}$
(See figure.)

30. $\mathbf{u} = \langle 1, 3, 1 \rangle$, $\mathbf{v} = \langle 0, 5, 5 \rangle$, $\mathbf{w} = \langle 4, 0, 4 \rangle$
(See figure.)

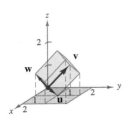

FIGURE FOR 29

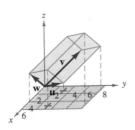

FIGURE FOR 30

In Exercises 31 and 32, find the volume of the parallelepiped with the given vertices.

31. $(0, 0, 0)$, $(3, 0, 0)$, $(0, 5, 1)$, $(3, 5, 1)$, $(2, 0, 5)$, $(5, 0, 5)$, $(2, 5, 6)$, $(5, 5, 6)$
(See figure.)

32. $(0, 0, 0)$, $(1, 1, 0)$, $(1, 0, 2)$, $(0, 1, 1)$, $(2, 1, 2)$, $(1, 1, 3)$, $(1, 2, 1)$, $(2, 2, 3)$
(See figure.)

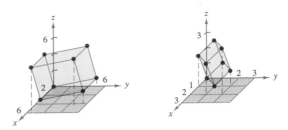

FIGURE FOR 31   **FIGURE FOR 32**

33. A child applies the brakes on a bicycle by applying a downward force of 20 pounds on the pedal when the crank makes a 40° angle with the horizontal (see figure). Find the torque at P if the crank is 6 inches in length.

34. Both the magnitude and direction of the force on a crankshaft change as the crankshaft rotates. Find the torque on the crankshaft using the position and data shown in the accompanying figure.

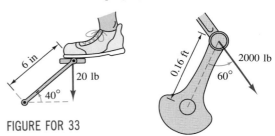

FIGURE FOR 33

FIGURE FOR 34

In Exercises 35–43, prove the property of the cross product.

35. $\mathbf{u} \times (\mathbf{v} + \mathbf{w}) = (\mathbf{u} \times \mathbf{v}) + (\mathbf{u} \times \mathbf{w})$

36. $(c\mathbf{u}) \times \mathbf{v} = c(\mathbf{u} \times \mathbf{v})$

37. $\mathbf{u} \times \mathbf{0} = \mathbf{0} \times \mathbf{u} = \mathbf{0}$

38. $\mathbf{u} \times \mathbf{u} = \mathbf{0}$

39. $\mathbf{u} \cdot (\mathbf{v} \times \mathbf{w}) = (\mathbf{u} \times \mathbf{v}) \cdot \mathbf{w}$

40. $\mathbf{u} \times \mathbf{v}$ is orthogonal to both $\mathbf{u}$ and $\mathbf{v}$

41. $\mathbf{u} \times \mathbf{v} = \mathbf{0}$ if and only if $\mathbf{u}$ and $\mathbf{v}$ are scalar multiples of each other.

42. $\|\mathbf{u} \times \mathbf{v}\| = \|\mathbf{u}\| \|\mathbf{v}\|$ if $\mathbf{u}$ and $\mathbf{v}$ are orthogonal.

43. $\mathbf{u} \times (\mathbf{v} \times \mathbf{w}) = (\mathbf{u} \cdot \mathbf{w})\mathbf{v} - (\mathbf{u} \cdot \mathbf{v})\mathbf{w}$

44. Prove Theorem 14.3.

14.3 Lines and Planes in Space

Lines in space ▪ Planes in space ▪ Distance between points, lines, and planes

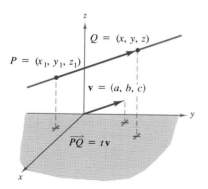

Line has direction vector **v**.

FIGURE 14.20

In the plane we use *slope* to determine an equation of a line. In space it is more convenient to use *vectors* to determine the equation of a line.

In Figure 14.20, consider the line L through the point $P = (x_1, y_1, z_1)$ and parallel to the vector $\mathbf{v} = \langle a, b, c \rangle$. We call $\mathbf{v}$ the **direction vector** for the line L, and a, b, and c the **direction numbers.** One way of describing the line L is to say that it consists of all points $Q = (x, y, z)$ for which the vector $\overrightarrow{PQ}$ is parallel to $\mathbf{v}$. This means that $\overrightarrow{PQ}$ is a scalar multiple of $\mathbf{v}$, and we write $\overrightarrow{PQ} = t\mathbf{v}$, where t is a scalar. In component form, we have

$$\overrightarrow{PQ} = \langle x - x_1, y - y_1, z - z_1 \rangle$$
$$= \langle at, bt, ct \rangle = t\mathbf{v}.$$

By equating corresponding components, we obtain the equations

$$x = x_1 + at, \qquad y = y_1 + bt, \qquad \text{and} \qquad z = z_1 + ct.$$

These are called **parametric equations** of a line in space.

THEOREM 14.5
PARAMETRIC EQUATIONS
OF A LINE IN SPACE

A line L parallel to the vector $\mathbf{v} = \langle a, b, c \rangle$ and passing through the point $P = (x_1, y_1, z_1)$ is represented by the **parametric equations**

$$x = x_1 + at, \qquad y = y_1 + bt, \qquad \text{and} \qquad z = z_1 + ct.$$

If the direction numbers a, b, and c are all nonzero, then we can eliminate the parameter t and obtain the following **symmetric equations** for a line.

$$\frac{x - x_1}{a} = \frac{y - y_1}{b} = \frac{z - z_1}{c} \qquad \text{Symmetric equations}$$

EXAMPLE 1 Finding parametric and symmetric equations for a line

---

Find a set of parametric equations and a set of symmetric equations for the line L that passes through the point $(1, -2, 4)$ and is parallel to $\mathbf{v} = \langle 2, 4, -4 \rangle$. (See Figure 14.21.)

SOLUTION

For the coordinates $x_1 = 1$, $y_1 = -2$, and $z_1 = 4$ and direction numbers $a = 2$, $b = 4$, and $c = -4$, a set of parametric equations for the line L is

$$x = 1 + 2t, \qquad y = -2 + 4t, \qquad \text{and} \qquad z = 4 - 4t.$$

Since a, b, and c are all nonzero, a set of symmetric equations is

$$\frac{x - 1}{2} = \frac{y + 2}{4} = \frac{z - 4}{-4}.$$

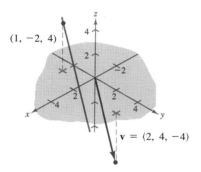

FIGURE 14.21

Neither the parametric equations nor the symmetric equations for a given line are unique. For instance, in Example 1, by letting $t = 1$ in the parametric equations we obtain the point $(3, 2, 0)$. Using this point with the direction numbers $a = 2$, $b = 4$, and $c = -4$, we have the parametric equations

$$x = 3 + 2t, \qquad y = 2 + 4t, \qquad \text{and} \qquad z = -4t.$$

EXAMPLE 2 Finding parametric equations for a line through two points

Find a set of parametric equations for the line that passes through the points $(-2, 1, 0)$ and $(1, 3, 5)$.

SOLUTION

We begin by letting $P = (-2, 1, 0)$ and $Q = (1, 3, 5)$. Then a direction vector for the line passing through P and Q is given by

$$\mathbf{v} = \overrightarrow{PQ} = \langle 1 - (-2), 3 - 1, 5 - 0 \rangle = \langle 3, 2, 5 \rangle = \langle a, b, c \rangle.$$

Now, using the direction numbers $a = 3$, $b = 2$, and $c = 5$, with the point $P = (-2, 1, 0)$, we obtain the parametric equations

$$x = -2 + 3t, \qquad y = 1 + 2t, \qquad \text{and} \qquad z = 5t. \qquad \blacksquare$$

Planes in space

We have seen how an equation for a line in space can be obtained from a point on the line and a vector *parallel* to it. We now look at how to obtain an equation for a plane in space from a point in the plane and a vector *normal* (perpendicular) to it.

Consider the plane containing the point $P = (x_1, y_1, z_1)$ having a nonzero normal vector $\mathbf{n} = \langle a, b, c \rangle$, as shown in Figure 14.22. This plane consists of all points $Q = (x, y, z)$ for which vector $\overrightarrow{PQ}$ is orthogonal to $\mathbf{n}$. Using the dot product, we have

$$\mathbf{n} \cdot \overrightarrow{PQ} = 0$$
$$\langle a, b, c \rangle \cdot \langle x - x_1, y - y_1, z - z_1 \rangle = 0$$
$$a(x - x_1) + b(y - y_1) + c(z - z_1) = 0.$$

The third equation of the plane is said to be in **standard form.**

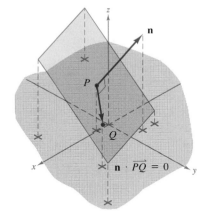

FIGURE 14.22

THEOREM 14.6 **STANDARD EQUATION OF** **A PLANE IN SPACE**	The plane containing the point (x_1, y_1, z_1) and having a normal vector $\mathbf{n} = \langle a, b, c \rangle$ can be represented, in **standard form,** by the equation $a(x - x_1) + b(y - y_1) + c(z - z_1) = 0.$

By regrouping terms, we obtain the **general form** of the equation of a plane in space,

$$ax + by + cz + d = 0.$$

Given this general form, it is easy to find a normal vector to the plane. Simply use the coefficients of x, y, and z and write $\mathbf{n} = \langle a, b, c \rangle$.

EXAMPLE 3   Finding an equation of a plane in three-space

Find the general equation of the plane containing the points $(2, 1, 1)$, $(0, 4, 1)$, and $(-2, 1, 4)$.

SOLUTION

To apply Theorem 14.6 we need a point in the plane and a vector that is normal to the plane. There are three choices for the point, but no normal vector is given. To obtain a normal vector, we use the cross product of vectors $\mathbf{u}$ and $\mathbf{v}$ extending from the point $(2, 1, 1)$ to the points $(0, 4, 1)$ and $(-2, 1, 4)$, as shown in Figure 14.23. We have

$$\mathbf{u} = \langle 0 - 2, 4 - 1, 1 - 1 \rangle = \langle -2, 3, 0 \rangle$$
$$\mathbf{v} = \langle -2 - 2, 1 - 1, 4 - 1 \rangle = \langle -4, 0, 3 \rangle$$

and it follows that

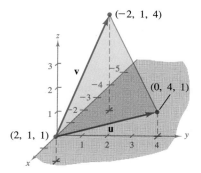

Plane determined by $\mathbf{u}$ and $\mathbf{v}$.

FIGURE 14.23

$$\mathbf{n} = \mathbf{u} \times \mathbf{v} = \begin{vmatrix} \mathbf{i} & \mathbf{j} & \mathbf{k} \\ -2 & 3 & 0 \\ -4 & 0 & 3 \end{vmatrix} = 9\mathbf{i} + 6\mathbf{j} + 12\mathbf{k} = \langle a, b, c \rangle$$

is normal to the given plane. Using the direction numbers for $\mathbf{n}$ and the point $(x_1, y_1, z_1) = (2, 1, 1)$, we determine an equation of the plane to be

$$a(x - x_1) + b(y - y_1) + c(z - z_1) = 0$$
$$9(x - 2) + 6(y - 1) + 12(z - 1) = 0$$
$$9x + 6y + 12z - 36 = 0$$
$$3x + 2y + 4z - 12 = 0.$$

REMARK   In Example 3, check that each of the three points satisfies the equation $3x + 2y + 4z - 12 = 0$.

Two distinct planes in three-space either are parallel or intersect in a line. If they intersect we can determine the angle between them from the angle between their normal vectors, as shown in Figure 14.24. Specifically, if vectors $\mathbf{n}_1$ and $\mathbf{n}_2$ are normal to two intersecting planes, then the angle θ between the normal vectors is equal to the angle between the two planes and is given by

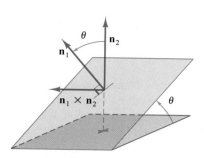

Angle Between Two Planes

FIGURE 14.24

$$\cos \theta = \frac{|\mathbf{n}_1 \cdot \mathbf{n}_2|}{\|\mathbf{n}_1\| \, \|\mathbf{n}_2\|}.$$

Consequently, two planes with normal vectors $\mathbf{n}_1$ and $\mathbf{n}_2$ are

1. *perpendicular* if $\mathbf{n}_1 \cdot \mathbf{n}_2 = 0$
2. *parallel* if $\mathbf{n}_1$ is a scalar multiple of $\mathbf{n}_2$.

EXAMPLE 4 Finding the line of intersection of two planes

Find the angle between the two planes given by

$$x - 2y + z = 0 \qquad \text{and} \qquad 2x + 3y - 2z = 0$$

and find parametric equations for their line of intersection.

SOLUTION

The normal vectors for the planes are $\mathbf{n}_1 = \langle 1, -2, 1 \rangle$ and $\mathbf{n}_2 = \langle 2, 3, -2 \rangle$. Consequently, the angle between the two planes is determined as follows.

$$\cos \theta = \frac{|\mathbf{n}_1 \cdot \mathbf{n}_2|}{\|\mathbf{n}_1\| \, \|\mathbf{n}_2\|} = \frac{|-6|}{\sqrt{6} \, \sqrt{17}} = \frac{6}{\sqrt{102}} \approx 0.59409$$

$$\theta \approx 53.55°$$

We can find the line of intersection of the two planes by simultaneously solving the two linear equations representing the planes.

$$\begin{array}{ll} x - 2y + z = 0 \\ 2x + 3y - 2z = 0 \end{array} \implies \begin{array}{ll} -2x + 4y - 2z = 0 \\ \underline{2x + 3y - 2z = 0} \\ 7y - 4z = 0 \end{array} \implies y = \frac{4z}{7}$$

Now, substituting $y = 4z/7$ back into one of the original equations, we find that $x = z/7$. Finally, by letting $t = z/7$, we have the parametric equations

$$x = t, \qquad y = 4t, \qquad \text{and} \qquad z = 7t$$

which indicate that 1, 4, and 7 are direction numbers for the line of intersection. ▬

Note that the direction numbers in Example 4 can be obtained from the cross product of the two normal vectors as follows.

$$\mathbf{n}_1 \times \mathbf{n}_2 = \begin{vmatrix} \mathbf{i} & \mathbf{j} & \mathbf{k} \\ 1 & -2 & 1 \\ 2 & 3 & -2 \end{vmatrix} = \mathbf{i} + 4\mathbf{j} + 7\mathbf{k}$$

This means that the line of intersection of the two planes is parallel to the cross product of their normal vectors.

Sketching planes in space

If a plane in space intersects one of the coordinate planes, we call the line of intersection the **trace** of the given plane in the coordinate plane. To sketch a plane in space, it is helpful to find its points of intersection with the coordinate axes and its traces in the coordinate planes. For example, consider the plane given by

$$3x + 2y + 4z = 12. \qquad \text{Equation of plane}$$

We find the xy-trace by letting $z = 0$ and sketching the line

$$3x + 2y = 12 \qquad \text{\small \textit{xy}-trace}$$

in the xy-plane. This line intersects the x-axis at $(4, 0, 0)$ and the y-axis at $(0, 6, 0)$. In Figure 14.25, we continue this process by finding the yz-trace and the xz-trace, and then shading in the triangular region lying in the first octant.

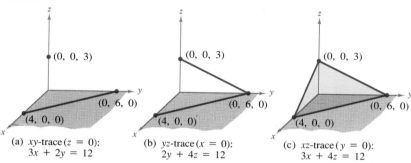

FIGURE 14.25

(a) xy-trace $(z = 0)$:
$3x + 2y = 12$

(b) yz-trace $(x = 0)$:
$2y + 4z = 12$

(c) xz-trace $(y = 0)$:
$3x + 4z = 12$

Traces of the Plane: $3x + 2y + 4z = 12$

If the equation of a plane has a missing variable such as $2x + z = 1$, then the plane must be *parallel to the axis* represented by the missing variable, as shown in Figure 14.26. If two variables are missing from the equation of a plane, then it is *parallel to the coordinate plane* represented by the missing variables, as shown in Figure 14.27.

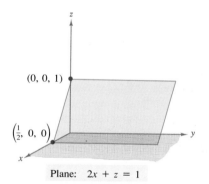

Plane: $2x + z = 1$

FIGURE 14.26

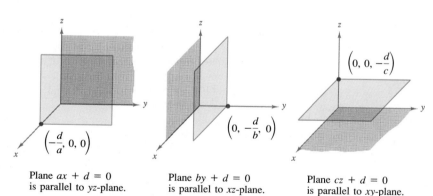

FIGURE 14.27

Plane $ax + d = 0$
is parallel to yz-plane.

Plane $by + d = 0$
is parallel to xz-plane.

Plane $cz + d = 0$
is parallel to xy-plane.

Distances between points, planes, and lines

We conclude this section with two basic distance problems in space.

1. Finding the distance between a point and a plane
2. Finding the distance between a point and a line

The solutions to these problems illustrate the versatility and usefulness of vectors in coordinate geometry. In the first problem we use the *dot product* of two vectors, and in the second problem we use the *cross product*.

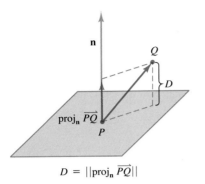

$$D = \|\text{proj}_{\mathbf{n}} \, \overrightarrow{PQ}\|$$

By the distance between a point Q and a plane, we mean the length of the shortest line segment connecting Q to the plane. If we knew which point in the plane was closest to Q, we could simply apply the formula for the distance between two points in space. However, this closest point is not easy to find, and so we use a different approach. In Figure 14.28, let D be the distance between Q and the given plane. Then if P is *any* point on the plane, we can find the distance by projecting the vector $\overrightarrow{PQ}$ onto the normal vector $\mathbf{n}$. The length of this projection gives us the desired distance. This result is stated in the following theorem.

FIGURE 14.28

THEOREM 14.7
DISTANCE BETWEEN A
POINT AND A PLANE

The distance between a plane and a point Q (not on the plane) is

$$D = \|\text{proj}_{\mathbf{n}} \, \overrightarrow{PQ}\| = \frac{|\overrightarrow{PQ} \cdot \mathbf{n}|}{\|\mathbf{n}\|}$$

where P is a point on the plane and $\mathbf{n}$ is normal to the plane.

To find a point in the plane given by $ax + by + cz + d = 0$, where $a \neq 0$, let $y = 0$ and $z = 0$. Then from the equation $ax + d = 0$, you can conclude that the point $(-d/a, 0, 0)$ lies in the plane.

EXAMPLE 5   Finding the distance between a point and a plane

Find the distance between the point $Q = (1, 5, -4)$ and the plane given by $3x - y + 2z = 6$.

SOLUTION

We know that $\mathbf{n} = \langle 3, -1, 2 \rangle$ is normal to the given plane. To find a point in the plane, we let $y = 0$ and $z = 0$ and obtain the point $P = (2, 0, 0)$. Now, the vector from P to Q is given by

$$\overrightarrow{PQ} = \langle 1 - 2, 5 - 0, -4 - 0 \rangle = \langle -1, 5, -4 \rangle.$$

Finally, using the distance formula given in Theorem 14.7, we have

$$D = \frac{|\overrightarrow{PQ} \cdot \mathbf{n}|}{\|\mathbf{n}\|} = \frac{|\langle -1, 5, -4 \rangle \cdot \langle 3, -1, 2 \rangle|}{\sqrt{9 + 1 + 4}} = \frac{|-3 - 5 - 8|}{\sqrt{14}}$$

$$= \frac{16}{\sqrt{14}}.$$

REMARK   The choice of the point P in Example 5 is arbitrary. Try choosing a different point to verify that you obtain the same distance.

From Theorem 14.7, we can determine that the distance between the point $Q = (x_0, y_0, z_0)$ and the plane given by $ax + by + cz + d = 0$ is

$$D = \frac{|a(x_0 - x_1) + b(y_0 - y_1) + c(z_0 - z_1)|}{\sqrt{a^2 + b^2 + c^2}}$$

$$D = \frac{|ax_0 + by_0 + cz_0 + d|}{\sqrt{a^2 + b^2 + c^2}} \qquad \text{Distance between point and plane}$$

where $P = (x_1, y_1, z_1)$ is a point on the plane and $d = -(ax_1 + by_1 + cz_1)$. (Note the similarity between this formula and the formula for the distance between a point and a line in the plane, as given in Exercises 69–74 in Section 1.4.) We demonstrate the use of this alternate formula for the distance between a point and a plane in the following example.

EXAMPLE 6 Finding the distance between two parallel planes

Find the distance between the two parallel planes given by

$$3x - y + 2z - 6 = 0 \qquad \text{and} \qquad 6x - 2y + 4z + 4 = 0$$

as shown in Figure 14.29.

SOLUTION

We begin by choosing a point in the first plane, say $(x_0, y_0, z_0) = (2, 0, 0)$. Then from the second plane, we determine that $a = 6$, $b = -2$, $c = 4$, and $d = 4$, and conclude that the distance is

$$D = \frac{|ax_0 + by_0 + cz_0 + d|}{\sqrt{a^2 + b^2 + c^2}}$$

$$= \frac{|6(2) + (-2)(0) + (4)(0) + 4|}{\sqrt{6^2 + (-2)^2 + 4^2}} = \frac{16}{\sqrt{56}} = \frac{8}{\sqrt{14}}.$$

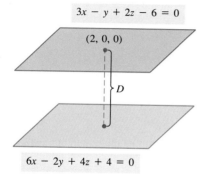

$3x - y + 2z - 6 = 0$

$(2, 0, 0)$

D

$6x - 2y + 4z + 4 = 0$

FIGURE 14.29

The formula for the distance between a point and a line in space resembles that for the distance between a point and a plane. For this distance, we replace the dot product by the cross product and replace the normal vector **n** by a direction vector for the given line. The validity of this procedure is shown in the proof of the following theorem.

THEOREM 14.8
DISTANCE BETWEEN A POINT
AND A LINE IN SPACE

The distance between a point Q and a line in space is given by

$$D = \frac{\|\overrightarrow{PQ} \times \mathbf{u}\|}{\|\mathbf{u}\|}$$

where **u** is the direction vector for the line and P is any point on the line.

PROOF In Figure 14.30, we let D be the distance between the point Q and the given line. Then

$$D = \|\overrightarrow{PQ}\| \sin \theta$$

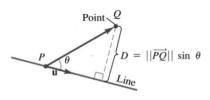

$D = \|\overrightarrow{PQ}\| \sin \theta$

Distance Between a Point and a Line

FIGURE 14.30

where θ is the angle between $\mathbf{u}$ and $\overrightarrow{PQ}$. By Theorem 14.2, we have

$$\|\mathbf{u}\| \|\overrightarrow{PQ}\| \sin \theta = \|\mathbf{u} \times \overrightarrow{PQ}\| = \|\overrightarrow{PQ} \times \mathbf{u}\|.$$

Consequently,

$$D = \|\overrightarrow{PQ}\| \sin \theta = \frac{\|\overrightarrow{PQ} \times \mathbf{u}\|}{\|\mathbf{u}\|}.$$

EXAMPLE 7 Finding the distance between a point and a line

Find the distance between the point $Q = (3, -1, 4)$ and the line given by

$$x = -2 + 3t, \qquad y = -2t, \qquad \text{and} \qquad z = 1 + 4t.$$

SOLUTION

Using the direction numbers 3, -2, and 4, we find the direction vector for the line to be

$$\mathbf{u} = \langle 3, -2, 4 \rangle. \qquad \text{Direction vector for line}$$

To find a point on the line, we let $t = 0$ and obtain

$$P = (-2, 0, 1). \qquad \text{Point on the line}$$

Thus,

$$\overrightarrow{PQ} = \langle 3 - (-2), -1 - 0, 4 - 1 \rangle = \langle 5, -1, 3 \rangle$$

and we form the cross product

$$\overrightarrow{PQ} \times \mathbf{u} = \begin{vmatrix} \mathbf{i} & \mathbf{j} & \mathbf{k} \\ 5 & -1 & 3 \\ 3 & -2 & 4 \end{vmatrix} = 2\mathbf{i} - 11\mathbf{j} - 7\mathbf{k} = \langle 2, -11, -7 \rangle.$$

Finally, using Theorem 14.8, we find the distance to be

$$D = \frac{\|\overrightarrow{PQ} \times \mathbf{u}\|}{\|\mathbf{u}\|} = \frac{\sqrt{174}}{\sqrt{29}} = \sqrt{6}.$$

EXERCISES for Section 14.3

In Exercises 1–10, find a set of (a) parametric equations and (b) symmetric equations of the specified line. (For each line, express the direction numbers as integers.)

1. The line passes through the origin and is parallel to $\mathbf{v} = \langle 1, 2, 3 \rangle$.
2. The line passes through the origin and is parallel to $\mathbf{v} = \langle -2, \frac{5}{2}, 1 \rangle$.

3. The line passes through the point $(-2, 0, 3)$ and is parallel to $\mathbf{v} = 2\mathbf{i} + 4\mathbf{j} - 2\mathbf{k}$.
4. The line passes through the point $(-2, 0, 3)$ and is parallel to $\mathbf{v} = 6\mathbf{i} + 3\mathbf{j}$.
5. The line passes through the points $(5, -3, -2)$ and $\left(-\frac{2}{3}, \frac{2}{3}, 1\right)$.
6. The line passes through the points $(1, 0, 1)$ and $(1, 3, -2)$.

7. The line passes through the point $(1, 0, 1)$ and is parallel to the line given by

$$x = 3 + 3t$$
$$y = 5 - 2t$$
$$z = -7 + t.$$

8. The line passes through the point $(-3, 5, 4)$ and is parallel to the line given by

$$\frac{x - 1}{3} = \frac{y + 1}{-2} = z - 3.$$

9. The line passes through the point $(2, 3, 4)$ and is parallel to the xz-plane and the yz-plane.

10. The line passes through the point $(2, 3, 4)$ and is perpendicular to the plane given by $3x + 2y - z = 6$.

In Exercises 11 and 12, determine which of the points lie on the line L.

11. The line L passes through the point $(-2, 3, 1)$ and is parallel to the vector $\mathbf{v} = 4\mathbf{i} - \mathbf{k}$.
  (a) $(2, 3, 0)$ (b) $(-6, 3, 2)$
  (c) $(2, 1, 0)$ (d) $(10, 3, -2)$
  (e) $(6, 3, -2)$

12. The line L passes through the points $(2, 0, -3)$ and $(4, 2, -2)$.

  (a) $(4, 1, -2)$ (b) $\left(3, 1, -\frac{5}{2}\right)$

  (c) $\left(\frac{5}{2}, \frac{1}{2}, -\frac{11}{4}\right)$ (d) $(-1, -3, -4)$

  (e) $(0, -2, -4)$

In Exercises 13–16, determine whether the lines intersect, and if so, find the point of intersection and the cosine of the angle of intersection.

13. $\begin{aligned}x &= 4t + 2 \\ y &= 3 \\ z &= -t + 1\end{aligned}$ $\begin{aligned}x &= 2s + 2 \\ y &= 2s + 3 \\ z &= s + 1\end{aligned}$

14. $\begin{aligned}x &= -3t + 1 \\ y &= 4t + 1 \\ z &= 2t + 4\end{aligned}$ $\begin{aligned}x &= 3s + 1 \\ y &= 2s + 4 \\ z &= -s + 1\end{aligned}$

15. $\dfrac{x}{3} = \dfrac{y - 2}{-1} = z + 1$

    $\dfrac{x - 1}{4} = y + 2 = \dfrac{z + 3}{-3}$

16. $\dfrac{x - 2}{-3} = \dfrac{y - 2}{6} = z - 3$

    $\dfrac{x - 3}{2} = y + 5 = \dfrac{z + 2}{4}$

In Exercises 17–32, find the equation of the specified plane.

17. The plane passes through the point $(2, 1, 2)$ and has normal vector $\mathbf{n} = \mathbf{i}$.

18. The plane passes through the point $(1, 0, -3)$ and has normal vector $\mathbf{n} = \mathbf{k}$.

19. The plane passes through the point $(3, 2, 2)$ and has normal vector $\mathbf{n} = 2\mathbf{i} + 3\mathbf{j} - \mathbf{k}$.

20. The plane passes through the point $(3, 2, 2)$ and is perpendicular to the line given by

$$\frac{x - 1}{4} = y + 2 = \frac{z + 3}{-3}.$$

21. The plane passes through the points $(0, 0, 0)$, $(1, 2, 3)$, and $(-2, 3, 3)$.

22. The plane passes through the points $(1, 2, -3)$, $(2, 3, 1)$, and $(0, -2, -1)$.

23. The plane passes through the points $(1, 2, 3)$, $(3, 2, 1)$, and $(-1, -2, 2)$.

24. The plane passes through the point $(1, 2, 3)$ and is parallel to the yz-plane.

25. The plane passes through the point $(1, 2, 3)$ and is parallel to the xy-plane.

26. The plane contains the y-axis and makes an angle of $\pi/6$ with the positive x-axis.

27. The plane contains lines given by

$$\frac{x - 1}{-2} = y - 4 = z$$

$$\frac{x - 2}{-3} = \frac{y - 1}{4} = \frac{z - 2}{-1}.$$

28. The plane passes through the point $(2, 2, 1)$ and contains the line given by

$$\frac{x}{2} = \frac{y - 4}{-1} = z.$$

29. The plane passes through the points $(2, 2, 1)$ and $(-1, 1, -1)$ and is perpendicular to the plane $2x - 3y + z = 3$.

30. The plane passes through the points $(3, 2, 1)$ and $(3, 1, -5)$ and is perpendicular to the plane $6x + 7y + 2z = 10$.

31. The plane passes through the points $(1, -2, -1)$ and $(2, 5, 6)$ and is parallel to the x-axis.

32. The plane passes through the points $(4, 2, 1)$ and $(-3, 5, 7)$ and is parallel to the z-axis.

In Exercises 33–40, determine whether the planes are parallel, orthogonal, or neither. If they are neither parallel nor orthogonal, find the angle of intersection.

33. $5x - 3y + z = 4$, $x + 4y + 7z = 1$
34. $3x + y - 4z = 3$, $-9x - 3y + 12z = 4$

35. $x - 3y + 6z = 4$, $5x + y - z = 4$
36. $3x + 2y - z = 7$, $x - 4y + 2z = 0$
37. $x - 5y - z = 1$, $5x - 25y - 5z = -3$
38. $2x - z = 1$, $4x + y + 8z = 10$
39. $x + 3y + z = 7$, $x - 5z = 0$
40. $2x + y = 3$, $x - 5z = 0$

In Exercises 41–48, mark the intercepts and sketch the graph of the plane.

41. $4x + 2y + 6z = 12$ **42.** $3x + 6y + 2z = 6$
43. $2x - y + 3z = 4$ **44.** $2x - y + z = 4$
45. $y + z = 5$ **46.** $x + 2y = 4$
47. $2x + y - z = 6$ **48.** $x - 3z = 3$

In Exercises 49 and 50, find a set of parametric equations for the line of intersection of the planes.

49. $3x + 2y - z = 7$, $x - 4y + 2z = 0$
50. $x - 3y + 6z = 4$, $5x + y - z = 4$

In Exercises 51–54, find the point of intersection (if any) of the plane and the line. Also determine whether the line lies in the plane.

51. $2x - 2y + z = 12$
$$x - \frac{1}{2} = \frac{y + (3/2)}{-1} = \frac{z + 1}{2}$$

52. $2x + 3y = -5$
$$\frac{x - 1}{4} = \frac{y}{2} = \frac{z - 3}{6}$$

53. $2x + 3y = 10$
$$\frac{x - 1}{3} = \frac{y + 1}{-2} = z - 3$$

54. $5x + 3y = 17$
$$\frac{x - 4}{2} = \frac{y + 1}{-3} = \frac{z + 2}{5}$$

In Exercises 55 and 56, find the distance between the point and the line.

55. $(10, 3, -2)$; $x = 4t - 2$, $y = 3$, $z = -t + 1$
56. $(4, 1, -2)$; $x = 2t + 2$, $y = 2t$, $z = t - 3$

In Exercises 57 and 58, find the distance between the point and the plane.

57. $(0, 0, 0)$, $2x + 3y + z = 12$
58. $(1, 2, 3)$, $2x - y + z = 4$

In Exercises 59 and 60, find the distance between the planes.

59. $x - 3y + 4z = 10$, $x - 3y + 4z = 6$
60. $2x - 4z = 4$, $2x - 4z = 10$

In Exercises 61 and 62, find the distance between the two skew lines (lines that are neither parallel nor intersecting).

61. $x = \dfrac{y}{2} = \dfrac{z}{3}$, $\dfrac{x - 1}{-1} = y - 4 = z + 1$

62. $x = 3t$ $x = 4s + 1$
     $y = -t + 2$ $y = s - 2$
     $z = t - 1$ $z = -3s - 3$

63. Use a computer to sketch the graphs of the two intersecting lines

     $x = 2t + 3$ $x = -2s + 7$
     $y = 5t - 2$ $y = s + 8$
     $z = -t + 1$ $z = 2s - 1$

and find the point of intersection.

64. Use a computer to sketch the graph of the plane $2.1x - 4.7y - z + 3 = 0$.

65. If a_1, b_1, c_1 and a_2, b_2, c_2 are two sets of direction numbers for the same line, show that there exists a scalar d such that $a_1 = a_2d$, $b_1 = b_2d$, and $c_1 = c_2d$.

14.4 Surfaces in Space

Cylindrical surfaces ▪ Quadric surfaces ▪ Surfaces of revolution

In the first five sections of this chapter, we introduced the vector portion of the preliminary work necessary to study vector calculus and the calculus of space. In this and the next section we complete this preliminary development.

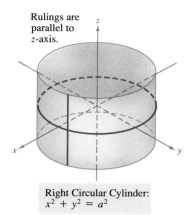

Right Circular Cylinder:
$x^2 + y^2 = a^2$

FIGURE 14.31

We begin with the classification of several important surfaces in space. So far, we have studied two special types of surfaces.

1. Spheres: $(x - x_0)^2 + (y - y_0)^2 + (z - z_0)^2 = r^2$ Section 14.1
2. Planes: $ax + by + cz + d = 0$ Section 14.3

A third type of surface in space is called a **cylindrical surface,** or, more simply, a **cylinder.** To see how we define a cylinder, consider the familiar right circular cylinder shown in Figure 14.31. We can imagine that this cylinder is generated by a vertical line moving around the circle $x^2 + y^2 = a^2$ in the xy-plane. We call this circle a **generating curve** for the cylinder, as indicated in the following definition.

DEFINITION OF A CYLINDER

Let C be a curve in a plane and L be a line not in a parallel plane. The set of all lines parallel to L and intersecting C is called a **cylinder.** C is called the **generating curve** (or **directrix**) of the cylinder, and the parallel lines are called **rulings.**

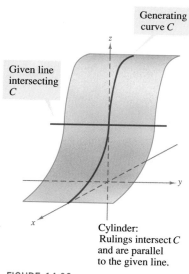

Cylinder:
Rulings intersect C
and are parallel
to the given line.

FIGURE 14.32

REMARK   Without loss of generality, we can assume that C lies in one of the three coordinate planes. Moreover, in this text we restrict our discussion to *right* cylinders— cylinders whose rulings are perpendicular to the coordinate plane containing C, as shown in Figure 14.32.

For the right circular cylinder shown in Figure 14.31, the equation of the generating curve is

$$x^2 + y^2 = a^2.$$ Equation of generating curve in xy-plane

To find an equation for the cylinder, note that we can generate any one of the rulings by fixing the values of x and y and then allowing z to take on all real values. In this sense the value of z is arbitrary and is therefore not included in the equation. In other words, the equation of this cylinder is simply the equation of its generating curve:

$$x^2 + y^2 = a^2.$$ Equation of cylinder in space

This result is generalized in the following theorem.

THEOREM 14.9
EQUATIONS OF CYLINDERS

In space, the graph of an equation in two of the three variables x, y, and z is a cylinder whose rulings are parallel to the axis of the missing variable.

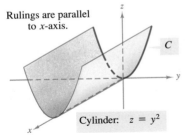

Generating curve C lies in yz-plane.

Rulings are parallel to x-axis.

C

Cylinder: $z = y^2$

FIGURE 14.33

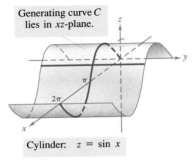

Generating curve C lies in xz-plane.

Cylinder: $z = \sin x$

Rulings are parallel to y-axis.

FIGURE 14.34

EXAMPLE 1 Sketching a cylinder

Sketch the surfaces represented by the following equations.

(a) $z = y^2$
(b) $z = \sin x, \quad 0 \le x \le 2\pi$

SOLUTION

(a) The graph is a cylinder whose generating curve, $z = y^2$, is a parabola in the yz-plane. The rulings of the cylinder are parallel to the x-axis, as shown in Figure 14.33.
(b) The graph is a cylinder generated by the sine curve in the xz-plane. The rulings are parallel to the y-axis, as shown in Figure 14.34. ▭

The intersection of a surface with a plane is called the **trace of the surface** in the plane. To visualize a surface in space, it is helpful to determine its traces in some well-chosen planes. This will become clear as we study the fourth common type of surface—a **quadric surface.** Quadric surfaces are three-dimensional analogues of conic sections, as represented by the general second-degree equation

$$Ax^2 + Bxy + Cy^2 + Dx + Ey + F = 0. \qquad \text{Conic section in } xy\text{-plane}$$

The general second-degree equation in three variables follows a similar pattern.

DEFINITION OF A QUADRIC SURFACE	In space, the graph of a second-degree equation of the form $$Ax^2 + By^2 + Cz^2 + Dxy + Exz + Fyz + Gx + Hy + Iz + J = 0$$ is called a **quadric surface.**

There are six basic types of quadric surfaces: **ellipsoid, hyperboloid of one sheet, hyperboloid of two sheets, elliptic cone, elliptic paraboloid,** and **hyperbolic paraboloid.** From the definition of a quadric surface, we can see that the traces taken in each of the coordinate planes are conic sections. These traces, together with the **standard form** of the equation of each quadric surface, are shown in Table 14.1.

TABLE 14.1   Quadric Surfaces

	Ellipsoid $$\frac{x^2}{a^2} + \frac{y^2}{b^2} + \frac{z^2}{c^2} = 1$$ *Trace* *Plane* Ellipse Parallel to *xy*-plane Ellipse Parallel to *xz*-plane Ellipse Parallel to *yz*-plane The surface is a sphere if $a = b = c \neq 0$.	
	Hyperboloid of One Sheet $$\frac{x^2}{a^2} + \frac{y^2}{b^2} - \frac{z^2}{c^2} = 1$$ *Trace* *Plane* Ellipse Parallel to *xy*-plane Hyperbola Parallel to *xz*-plane Hyperbola Parallel to *yz*-plane The axis of the hyperboloid corresponds to the variable whose coefficient is negative.	
	Hyperboloid of Two Sheets $$\frac{z^2}{c^2} - \frac{x^2}{a^2} - \frac{y^2}{b^2} = 1$$ *Trace* *Plane* Ellipse Parallel to *xy*-plane Hyperbola Parallel to *xz*-plane Hyperbola Parallel to *yz*-plane The axis of the hyperboloid corresponds to the variable whose coefficient is positive. There is no trace in the coordinate plane perpendicular to this axis.	

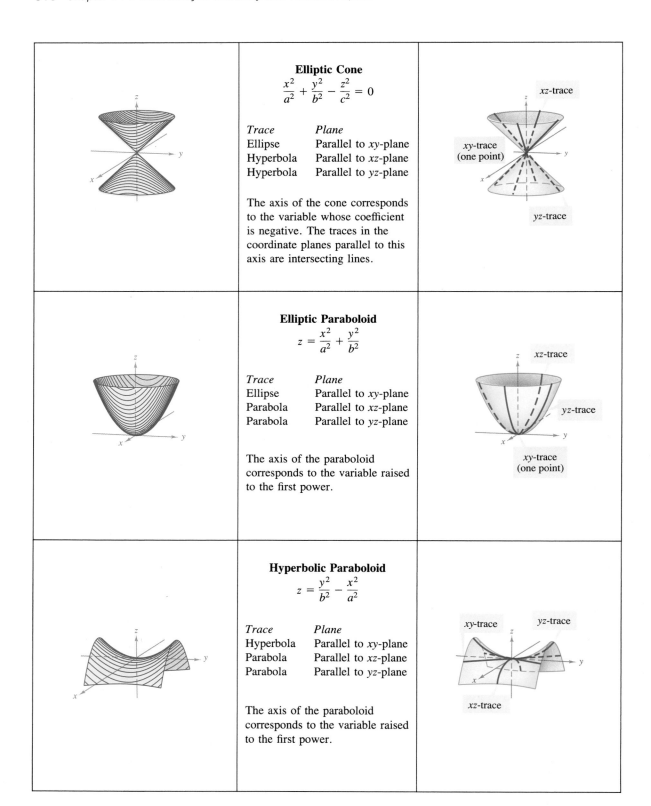

Elliptic Cone

$$\frac{x^2}{a^2} + \frac{y^2}{b^2} - \frac{z^2}{c^2} = 0$$

Trace	Plane
Ellipse	Parallel to xy-plane
Hyperbola	Parallel to xz-plane
Hyperbola	Parallel to yz-plane

The axis of the cone corresponds to the variable whose coefficient is negative. The traces in the coordinate planes parallel to this axis are intersecting lines.

xz-trace

xy-trace (one point)

yz-trace

Elliptic Paraboloid

$$z = \frac{x^2}{a^2} + \frac{y^2}{b^2}$$

Trace	Plane
Ellipse	Parallel to xy-plane
Parabola	Parallel to xz-plane
Parabola	Parallel to yz-plane

The axis of the paraboloid corresponds to the variable raised to the first power.

xz-trace

yz-trace

xy-trace (one point)

Hyperbolic Paraboloid

$$z = \frac{y^2}{b^2} - \frac{x^2}{a^2}$$

Trace	Plane
Hyperbola	Parallel to xy-plane
Parabola	Parallel to xz-plane
Parabola	Parallel to yz-plane

The axis of the paraboloid corresponds to the variable raised to the first power.

xy-trace

yz-trace

xz-trace

In Table 14.1 only one of several orientations of each quadric surface is shown. If the surface is oriented along a different axis, then its standard equation will change accordingly, as illustrated in the next two examples. The fact that the two types of paraboloids have one variable raised to the first power can be helpful in classifying quadric surfaces. The other four types of basic quadric surfaces have equations that are of *second degree* in all three variables.

EXAMPLE 2 Sketching a quadric surface

Describe and sketch the surface given by $4x^2 - 3y^2 + 12z^2 + 12 = 0$.

SOLUTION

We express the given equation in standard form as follows.

$$\frac{x^2}{-3} + \frac{y^2}{4} - z^2 - 1 = 0 \qquad \text{Divide by } -12$$

$$\frac{y^2}{4} - \frac{x^2}{3} - \frac{z^2}{1} = 1 \qquad \text{Standard form}$$

From Table 14.1 we conclude that the surface is a hyperboloid of two sheets with the y-axis as its axis. To help sketch the graph of this surface, we find the following traces.

xy-trace: $\quad \dfrac{y^2}{4} - \dfrac{x^2}{3} = 1 \qquad$ Hyperbola
$(z = 0)$

xz-trace: $\quad \dfrac{x^2}{3} + \dfrac{z^2}{1} = -1 \qquad$ No trace
$(y = 0)$

yz-trace: $\quad \dfrac{y^2}{4} - \dfrac{z^2}{1} = 1 \qquad$ Hyperbola
$(x = 0)$

The graph is shown in Figure 14.35.   ▭

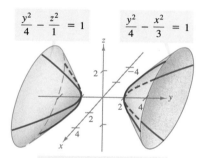

$$\frac{y^2}{4} - \frac{z^2}{1} = 1 \qquad \frac{y^2}{4} - \frac{x^2}{3} = 1$$

Hyperboloid of Two Sheets:
$$-\frac{x^2}{3} + \frac{y^2}{4} - z^2 = 1$$

FIGURE 14.35

EXAMPLE 3 Sketching a quadric surface

Classify and sketch the surface given by $x - y^2 - 4z^2 = 0$.

SOLUTION

Since x is raised only to the first power, the surface will be a paraboloid, and in this case its axis is the x-axis. In the standard form, we have

$$x = y^2 + 4z^2.$$

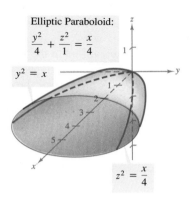

Elliptic Paraboloid:
$$\frac{y^2}{4} + \frac{z^2}{1} = \frac{x}{4}$$

$y^2 = x$

$z^2 = \frac{x}{4}$

FIGURE 14.36

Some convenient traces are as follows.

xy-trace: $(z = 0)$	$x = y^2$	Parabola
xz-trace: $(y = 0)$	$x = 4z^2$	Parabola
parallel to yz-plane: $(x = 4)$	$\dfrac{y^2}{4} + \dfrac{z^2}{1} = 1$	Ellipse

Thus, the surface is an *elliptic* paraboloid, as shown in Figure 14.36. ▭

For a quadric surface not centered at the origin, we can form the standard equation by completing the square, as we did for conics. A translation of axes then makes sketching easier. This is demonstrated in the next example.

EXAMPLE 4 A quadric surface not centered at the origin

Classify and sketch the surface given by

$$x^2 + 2y^2 + z^2 - 4x + 4y - 2z + 3 = 0.$$

SOLUTION

Completing the square for each variable produces

$$(x^2 - 4x +) + 2(y^2 + 2y +) + (z^2 - 2z +) = -3$$
$$(x^2 - 4x + 4) + 2(y^2 + 2y + 1) + (z^2 - 2z + 1) = -3 + 4 + 2 + 1$$
$$(x - 2)^2 + 2(y + 1)^2 + (z - 1)^2 = 4$$
$$\frac{(x - 2)^2}{4} + \frac{(y + 1)^2}{2} + \frac{(z - 1)^2}{4} = 1.$$

For this equation, we see that the quadric surface is centered at $(2, -1, 1)$, and if we let

$$x' = x - 2, \qquad y' = y + 1, \qquad \text{and} \qquad z' = z - 1$$

we obtain the standard form

$$\frac{(x')^2}{4} + \frac{(y')^2}{2} + \frac{(z')^2}{4} = 1$$

of an ellipsoid. Its graph is shown in Figure 14.37. ▭

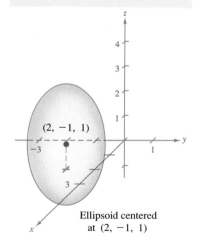

$$\frac{(x - 2)^2}{4} + \frac{(y + 1)^2}{2} + \frac{(z - 1)^2}{4} = 1$$

$(2, -1, 1)$

Ellipsoid centered
at $(2, -1, 1)$

FIGURE 14.37

Surfaces of revolution

The fifth special type of surface we will study is called a **surface of revolution.** In Section 6.4, we looked at a method for finding the *area* of such a surface. We now look at a procedure for finding its *equation.*

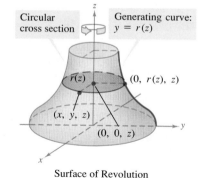

Surface of Revolution

FIGURE 14.38

Consider the graph of the **radius function**

$$y = r(z) \qquad \text{Generating curve}$$

in the yz-plane. If this graph is revolved about the z-axis, it forms a surface of revolution, as shown in Figure 14.38. The trace of the surface in the plane $z = z_0$ is a circle whose radius is $r(z_0)$ and whose equation is

$$x^2 + y^2 = [r(z_0)]^2. \qquad \text{Circular trace in plane: } z = z_0$$

Replacing z_0 by z produces an equation that is valid for all values of z. In a similar manner we can obtain equations for surfaces of revolution for the other two axes, and we summarize the results in the following theorem.

THEOREM 14.10
SURFACE OF REVOLUTION

If the graph of a radius function r is revolved about one of the coordinate axes, then the equation of the resulting surface of revolution has one of the following forms.

1. Revolved about the x-axis: $\quad y^2 + z^2 = [r(x)]^2$
2. Revolved about the y-axis: $\quad x^2 + z^2 = [r(y)]^2$
3. Revolved about the z-axis: $\quad x^2 + y^2 = [r(z)]^2$

EXAMPLE 5 Finding an equation for a surface of revolution

(a) An equation for the surface of revolution formed by revolving the graph of $y = 1/z$ about the z-axis is

$$x^2 + y^2 = [r(z)]^2 = \left(\frac{1}{z}\right)^2.$$

(b) To find an equation for the surface formed by revolving the graph of $x^2 = y^3$ about the y-axis, we solve for x in terms of y to obtain

$$x = y^{3/2} = r(y). \qquad \text{Radius function}$$

Thus, the equation for this surface is

$$x^2 + z^2 = [r(y)]^2 = [y^{3/2}]^2 = y^3.$$

The graph is shown in Figure 14.39.

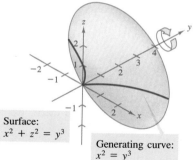

Surface:
$x^2 + z^2 = y^3$

Generating curve:
$x^2 = y^3$

FIGURE 14.39

The generating curve for a surface of revolution is not unique. For instance, the surface $x^2 + z^2 = e^{-2y}$ can be formed by revolving either the graph of $x = e^{-y}$ about the y-axis or the graph of $z = e^{-y}$ about the y-axis, as shown in Figure 14.40.

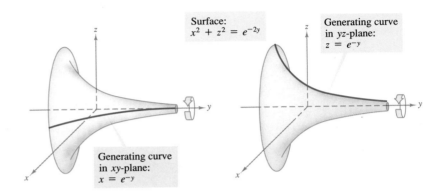

Surface:
$x^2 + z^2 = e^{-2y}$

Generating curve in yz-plane:
$z = e^{-y}$

Generating curve in xy-plane:
$x = e^{-y}$

FIGURE 14.40

EXAMPLE 6   Finding a generating curve for a surface of revolution

Find a generating curve and the axis of revolution for the surface given by $x^2 + 3y^2 + z^2 = 9$.

SOLUTION

From Theorem 14.10 we know that the equation has one of the following forms.

$$x^2 + y^2 = [r(z)]^2$$
$$y^2 + z^2 = [r(x)]^2$$
$$x^2 + z^2 = [r(y)]^2$$

Since the coefficients of x^2 and z^2 are equal, we choose the third form and write

$$x^2 + z^2 = 9 - 3y^2.$$

We conclude that the y-axis is the axis of revolution. We can choose a generating curve from either of the following traces.

$$x^2 = 9 - 3y^2 \qquad \text{Trace in } xy\text{-plane}$$
$$z^2 = 9 - 3y^2 \qquad \text{Trace in } yz\text{-plane}$$

For example, using the first trace, we choose the semiellipse given by

$$x = \sqrt{9 - 3y^2}. \qquad \text{Generating curve}$$

The graph of this surface is shown in Figure 14.41.

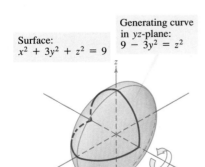

Surface:
$x^2 + 3y^2 + z^2 = 9$

Generating curve in yz-plane:
$9 - 3y^2 = z^2$

Generating curve in xy-plane:
$9 - 3y^2 = x^2$

FIGURE 14.41

EXERCISES for Section 14.4

In Exercises 1–8, match the equation with its graph.
[The graphs are labeled (a)–(h).]

1. $\dfrac{x^2}{9} + \dfrac{y^2}{16} + \dfrac{z^2}{9} = 1$

2. $15x^2 - 4y^2 + 15z^2 = -4$

3. $4x^2 - y^2 + 4z^2 = 4$ **4.** $y^2 = 4x^2 + 9z^2$

5. $4x^2 - 4y + z^2 = 0$ **6.** $12z = -3y^2 + 4x^2$

7. $4x^2 - y^2 + 4z = 0$ **8.** $x^2 + y^2 + z^2 = 9$

(a)
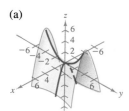

(b)
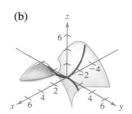

(c)
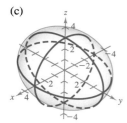

(d)
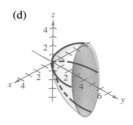

(e)
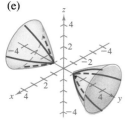

(f)
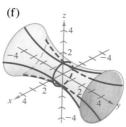

(g)
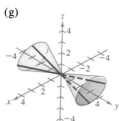

(h)
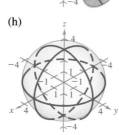

In Exercises 9–18, describe and sketch each surface.

9. $z = 3$ **10.** $x = 4$

11. $y^2 + z^2 = 9$ **12.** $x^2 + z^2 = 16$

13. $x^2 - y = 0$ **14.** $y^2 + z = 4$

15. $4x^2 + y^2 = 4$ **16.** $z - \sin y = 0$

17. $y^2 - z^2 = 4$ **18.** $z - e^y = 0$

In Exercises 19–34, identify and sketch the given quadric surface.

19. $x^2 + \dfrac{y^2}{4} + z^2 = 1$ **20.** $\dfrac{x^2}{9} + \dfrac{y^2}{16} + \dfrac{z^2}{16} = 1$

21. $16x^2 - y^2 + 16z^2 = 4$ **22.** $9x^2 + 4y^2 - 8z^2 = 72$

23. $x^2 - y + z^2 = 0$ **24.** $z = 4x^2 + y^2$

25. $x^2 - y^2 + z = 0$ **26.** $z^2 - x^2 - \dfrac{y^2}{4} = 1$

27. $4x^2 - y^2 + 4z^2 = -16$

28. $z^2 = x^2 + \dfrac{y^2}{4}$

29. $z^2 = x^2 + 4y^2$ **30.** $4y = x^2 + z^2$

31. $3z = -y^2 + x^2$ **32.** $z^2 = 2x^2 + 2y^2$

33. $16x^2 + 9y^2 + 16z^2 - 32x - 36y + 36 = 0$

34. $4x^2 + y^2 - 4z^2 - 16x - 6y - 16z + 9 = 0$

In Exercises 35–40, sketch the region bounded by the graphs of the equations.

35. $z = 2\sqrt{x^2 + y^2}$, $z = 2$

36. $z = \sqrt{4 - x^2}$, $y = \sqrt{4 - x^2}$, $x = 0$, $y = 0$, $z = 0$

37. $x^2 + y^2 = 1$, $x + z = 2$, $z = 0$

38. $x^2 + y^2 + z^2 = 4$, $z = \sqrt{x^2 + y^2}$, $z = 0$

39. $z = \sqrt{4 - x^2 - y^2}$, $y = 2z$, $z = 0$

40. $z = \sqrt{x^2 + y^2}$, $z = 4 - x^2 - y^2$

In Exercises 41–46, find an equation for the surface of revolution generated by revolving the given curve about the specified axis.

41. $z^2 = 4y$ in the yz-plane about the y-axis

42. $z = 2y$ in the yz-plane about the y-axis

43. $z = 2y$ in the yz-plane about the z-axis

44. $2z = \sqrt{4 - x^2}$ in the xz-plane about the x-axis

45. $xy = 2$ in the xy-plane about the x-axis

46. $z = \ln y$ in the yz-plane about the z-axis

In Exercises 47 and 48, find an equation of a generating curve given the equation of its surface of revolution.

47. $x^2 + y^2 - 2z = 0$ **48.** $x^2 + z^2 = \sin^2 y$

In Exercises 49 and 50, analyze the trace when the surface given by

$$z = \dfrac{x^2}{2} + \dfrac{y^2}{4}$$

is intersected by the given planes.

49. Find the length of the major and minor axes and the coordinates of the foci of the ellipse generated when the surface is intersected by the planes given by (a) $z = 2$ and (b) $z = 8$.

50. Find the coordinates of the focus of the parabola formed when the surface is intersected by the planes given by (a) $y = 4$ and (b) $x = 2$.

55. Because of its rotation, the earth is an oblate ellipsoid rather than a sphere. The equatorial radius is 3963 miles and the polar radius is 3942 miles. Find an equation of the ellipsoid. (Assume the center of the earth is at the origin and the trace formed by the plane $z = 0$ corresponds to the equator.)

In Exercises 51–54, use a computer to sketch the graph of the surface.

51. $z = 2 \sin x$

52. $z = 5e^{-x/3} \sin y$

53. $z = x^2 + 0.5y^2$

54. $z = \dfrac{y^2}{7} - \dfrac{x^2}{2}$

14.5 Curves and Vector-Valued Functions in Space

Space curves ▪ Vector-valued functions in space ▪ Arc length of a space curve

In Section 13.3, we showed how to represent a *plane curve* by a vector-valued function of the form

$$\mathbf{r}(t) = f(t)\mathbf{i} + g(t)\mathbf{j}$$

where f and g are continuous functions of t on an interval I. This definition can be extended to three-dimensional space in a natural way. That is, we can represent a curve in space by a vector-valued function of the form

$$\mathbf{r}(t) = f(t)\mathbf{i} + g(t)\mathbf{j} + h(t)\mathbf{k}$$

where f, g, and h are continuous functions of t on an interval I.

EXAMPLE 1 A space curve represented by a vector-valued function

Sketch the space curve represented by the vector-valued function

$$\mathbf{r}(t) = 4 \cos t\, \mathbf{i} + 4 \sin t\, \mathbf{j} + t\mathbf{k}, \quad 0 \le t \le 4\pi.$$

SOLUTION

From the first two parametric equations, $x = 4 \cos t$ and $y = 4 \sin t$, we obtain

$$x^2 + y^2 = 16.$$

This means that the curve lies on a right circular cylinder of radius 4, centered about the z-axis. To locate the curve on this cylinder, we use the third parametric equation $z = t$. Then, as t increases from 0 to 4π, the point (x, y, z) spirals up the cylinder to produce the **helix** shown in Figure 14.42.

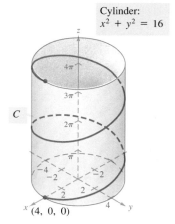

Cylinder:
$x^2 + y^2 = 16$

Curve C is a helix.

FIGURE 14.42

In the next example, we look at a space curve that is defined as the intersection of two surfaces in space.

EXAMPLE 2   Representing a space curve by a vector-valued function

Sketch the curve C represented by the intersection of the semiellipsoid

$$\frac{x^2}{12} + \frac{y^2}{24} + \frac{z^2}{4} = 1, \quad z \geq 0$$

and the parabolic cylinder $y = x^2$. Then, find a vector-valued function to represent C.

SOLUTION

The intersection of the two surfaces is shown in Figure 14.43. A natural choice of parameter is $x = t$. For this choice, we use the given equation $y = x^2$ to obtain $y = t^2$. Then, it follows that

$$\frac{z^2}{4} = 1 - \frac{x^2}{12} - \frac{y^2}{24} = 1 - \frac{t^2}{12} - \frac{t^4}{24} = \frac{24 - 2t^2 - t^4}{24}.$$

Since the curve lies above the xy-plane, we choose the positive square root for z and obtain the following parametric equations

$$x = t, \quad y = t^2, \quad \text{and} \quad z = \sqrt{\frac{24 - 2t^2 - t^4}{6}}.$$

The resulting vector-valued function is

$$\mathbf{r}(t) = t\mathbf{i} + t^2\mathbf{j} + \sqrt{\frac{24 - 2t^2 - t^4}{6}}\,\mathbf{k}.$$

For the points $(-2, 4, 0)$ and $(2, 4, 0)$ shown in Figure 14.43, we see that the curve is traced as t increases from -2 to 2.

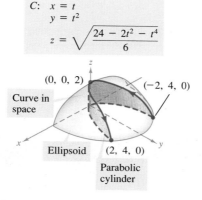

FIGURE 14.43

Vector-valued functions in space

The properties of vector-valued functions in space are similar to those for vector-valued functions in the plane. We summarize several of these in the following list.

PROPERTIES OF VECTOR-VALUED FUNCTIONS IN SPACE

Let $\mathbf{r}(t) = f(t)\mathbf{i} + g(t)\mathbf{j} + h(t)\mathbf{k}$ be a vector-valued function in space.

1. *Limit:* The **limit** of $\mathbf{r}(t)$ as $t \to a$ is given by

$$\lim_{t \to a} \mathbf{r}(t) = \left[\lim_{t \to a} f(t)\right]\mathbf{i} + \left[\lim_{t \to a} g(t)\right]\mathbf{j} + \left[\lim_{t \to a} h(t)\right]\mathbf{k}.$$

2. *Continuity:* A vector-valued function $\mathbf{r}$ is **continuous at the point** given by $t = a$ if the limit of $\mathbf{r}(t)$ exists as $t \to a$ and

$$\lim_{t \to a} \mathbf{r}(t) = \mathbf{r}(a).$$

A vector-valued function $\mathbf{r}$ is **continuous on an interval** I if it is continuous at every point in the interval. (Note that, as is true for vector-valued functions in the plane, a vector-valued function in space is continuous at $t = a$ if and only if each of its component functions is continuous at $t = a$.)

3. *Smooth Curve in Space:* The curve C represented by $\mathbf{r}(t)$ is called **smooth** on the interval I if f', g', and h' are continuous on I and $\mathbf{r}'(t) \neq \mathbf{0}$ for any value of t in the interval I.

4. *Derivative:* If f, g, and h are differentiable functions of t, then the derivative of $\mathbf{r}$ is given by

$$\mathbf{r}'(t) = f'(t)\mathbf{i} + g'(t)\mathbf{j} + h'(t)\mathbf{k}.$$

5. *Integration:* If f, g, and h are continuous functions of t on the interval $[a, b]$, then the indefinite integral (or antiderivative) of $\mathbf{r}$ is

$$\int \mathbf{r}(t)\, dt = \left[\int f(t)\, dt\right]\mathbf{i} + \left[\int g(t)\, dt\right]\mathbf{j} + \left[\int h(t)\, dt\right]\mathbf{k}$$

and its definite integral over the interval $a \leq t \leq b$ is

$$\int_a^b \mathbf{r}(t)\, dt = \left[\int_a^b f(t)\, dt\right]\mathbf{i} + \left[\int_a^b g(t)\, dt\right]\mathbf{j} + \left[\int_a^b h(t)\, dt\right]\mathbf{k}.$$

EXAMPLE 3   Finding the derivative of a vector-valued function

Find the derivative of

$$\mathbf{r}(t) = t^2\,\mathbf{i} + \sin t\,\mathbf{j} + e^{2t}\,\mathbf{k}.$$

SOLUTION

The derivative of $\mathbf{r}$ is

$$\mathbf{r}'(t) = 2t\,\mathbf{i} + \cos t\,\mathbf{j} + 2e^{2t}\,\mathbf{k}.$$

Each of the properties in Theorem 13.9 dealing with properties of derivatives of vector-valued functions in the plane is also true for vector-valued functions in space. For instance, the derivative of the dot product of two vector-valued functions is given by

$$D_t[\mathbf{r}(t) \cdot \mathbf{u}(t)] = \mathbf{r}(t) \cdot \mathbf{u}'(t) + \mathbf{r}'(t) \cdot \mathbf{u}(t).$$

Also, if the dot product of a vector-valued function with itself is constant, then the dot product of the function with its derivative must be zero. That is, if $\mathbf{r}(t) \cdot \mathbf{r}(t) = c$, then

$$\mathbf{r}(t) \cdot \mathbf{r}'(t) = 0.$$

An additional property of derivatives of vector-valued functions that applies only in space is as follows.

$$D_t[\mathbf{r}(t) \times \mathbf{u}(t)] = \mathbf{r}(t) \times \mathbf{u}'(t) + \mathbf{r}'(t) \times \mathbf{u}(t)$$

(See Exercise 76.)

EXAMPLE 4   Using properties of the derivative

For the vector-valued function given by

$$\mathbf{r}(t) = t^2\mathbf{i} - 2t\mathbf{j} + \mathbf{k}$$

find $D_t[\mathbf{r}(t) \times \mathbf{r}'(t)]$.

SOLUTION

Since $\mathbf{r}'(t) = 2t\mathbf{i} - 2\mathbf{j}$ and $\mathbf{r}''(t) = 2\mathbf{i}$, we have

$$D_t[\mathbf{r}(t) \times \mathbf{r}'(t)] = \mathbf{r}(t) \times \mathbf{r}''(t) + \mathbf{r}'(t) \times \mathbf{r}'(t)$$

$$= \begin{vmatrix} \mathbf{i} & \mathbf{j} & \mathbf{k} \\ t^2 & -2t & 1 \\ 2 & 0 & 0 \end{vmatrix} + 0$$

$$= \begin{vmatrix} -2t & 1 \\ 0 & 0 \end{vmatrix}\mathbf{i} - \begin{vmatrix} t^2 & 1 \\ 2 & 0 \end{vmatrix}\mathbf{j} + \begin{vmatrix} t^2 & -2t \\ 2 & 0 \end{vmatrix}\mathbf{k}$$

$$= 0\mathbf{i} - (-2)\mathbf{j} + 4t\mathbf{k}$$

$$= 2\mathbf{j} + 4t\mathbf{k}.$$

REMARK   Try reworking Example 4 by first forming the cross product and then differentiating to see that you obtain the same answer.

Arc length of a space curve

In Section 13.6 we saw that the arc length of a smooth *plane* curve C given by $\mathbf{r}(t) = x(t)\mathbf{i} + y(t)\mathbf{j}$ is

$$s = \int_a^b \|\mathbf{r}'(t)\| \, dt.$$

This formula has a natural extension to a smooth curve in *space*, as stated in the following theorem.

THEOREM 14.11
ARC LENGTH OF A SPACE CURVE

If C is a smooth curve given by $\mathbf{r}(t) = x(t)\mathbf{i} + y(t)\mathbf{j} + z(t)\mathbf{k}$, on an interval $[a, b]$, then the arc length of C on the interval is

$$s = \int_a^b \sqrt{[x'(t)]^2 + [y'(t)]^2 + [z'(t)]^2}\, dt = \int_a^b \|\mathbf{r}'(t)\|\, dt.$$

EXAMPLE 5 Finding the arc length of a curve in space

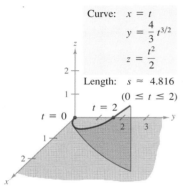

Curve: $x = t$
$y = \dfrac{4}{3} t^{3/2}$
$z = \dfrac{t^2}{2}$

Length: $s \approx 4.816$

$(0 \le t \le 2)$

$t = 2$

$t = 0$

FIGURE 14.44

Find the arc length of the curve given by

$$\mathbf{r}(t) = t\mathbf{i} + \frac{4}{3}t^{3/2}\mathbf{j} + \frac{1}{2}t^2\mathbf{k}$$

from $t = 0$ to $t = 2$, as shown in Figure 14.44.

SOLUTION

Using $x(t) = t$, $y(t) = \frac{4}{3}t^{3/2}$, and $z(t) = \frac{1}{2}t^2$, we obtain $x'(t) = 1$, $y'(t) = 2t^{1/2}$, and $z'(t) = t$. Thus, the arc length from $t = 0$ to $t = 2$ is given by

$$s = \int_a^b \|\mathbf{r}'(t)\|\, dt$$

$$= \int_0^2 \sqrt{[x'(t)]^2 + [y'(t)]^2 + [z'(t)]^2}\, dt$$

$$= \int_0^2 \sqrt{1 + 4t + t^2}\, dt$$

$$= \int_0^2 \sqrt{(t + 2)^2 - 3}\, dt \qquad \text{Integration Formula 26}$$

$$= \left[\frac{t + 2}{2}\sqrt{(t + 2)^2 - 3} - \frac{3}{2}\ln\left|(t + 2) + \sqrt{(t + 2)^2 - 3}\right| \right]_0^2$$

$$= 2\sqrt{13} - \frac{3}{2}\ln(4 + \sqrt{13}) - 1 + \frac{3}{2}\ln 3$$

$$\approx 4.816. \qquad \blacksquare$$

REMARK The formula for arc length given in Theorem 14.11 is independent of the parameter used to represent C. To illustrate this independence, try using the vector-valued function

$$\mathbf{r}(t) = t^2\mathbf{i} + \frac{4}{3}t^3\mathbf{j} + \frac{1}{2}t^4\mathbf{k}$$

to represent the curve given in Example 5. Then find the arc length from $t = 0$ to $t = \sqrt{2}$ and compare the result to that found in Example 5.

EXAMPLE 6 Finding the arc length of a helix

Find the length of one turn of the helix given by

$$\mathbf{r}(t) = b \cos t \, \mathbf{i} + b \sin t \, \mathbf{j} + \sqrt{1 - b^2} \, t \mathbf{k}$$

as shown in Figure 14.45.

SOLUTION

Since

$$\mathbf{r}'(t) = -b \sin t \, \mathbf{i} + b \cos t \, \mathbf{j} + \sqrt{1 - b^2} \, t \, \mathbf{k}$$

the arc length of one turn is

$$s = \int_0^{2\pi} \|\mathbf{r}'(t)\| \, dt = \int_0^{2\pi} \sqrt{b^2(\sin^2 t + \cos^2 t) + (1 - b^2)} \, dt$$

$$= \int_0^{2\pi} dt = 2\pi.$$

Curve:
$\mathbf{r}(t) = b \cos t \mathbf{i} + b \sin t \mathbf{j} + \sqrt{1 - b^2} t \mathbf{k}$

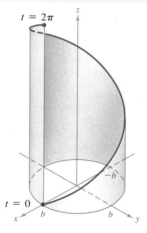

FIGURE 14.45   One Turn of a Helix

EXERCISES for Section 14.5

In Exercises 1–8, find the domain of the given vector-valued function.

1. $\mathbf{r}(t) = 5t\mathbf{i} - 4t\mathbf{j} - \dfrac{1}{t}\mathbf{k}$

2. $\mathbf{r}(t) = \sqrt{4 - t^2}\mathbf{i} + t^2\mathbf{j} - 6t\mathbf{k}$

3. $\mathbf{r}(t) = \ln t \, \mathbf{i} - e^t\mathbf{j} - t\mathbf{k}$

4. $\mathbf{r}(t) = \sin t \, \mathbf{i} + 4 \cos t \, \mathbf{j} + t\mathbf{k}$

5. $\mathbf{r}(t) = \mathbf{F}(t) + \mathbf{G}(t)$ where

$$\mathbf{F}(t) = \cos t \, \mathbf{i} - \sin t \, \mathbf{j} + \sqrt{t}\mathbf{k}$$

and

$$\mathbf{G}(t) = \cos t \, \mathbf{i} + \sin t \, \mathbf{j}$$

6. $\mathbf{r}(t) = \mathbf{F}(t) - \mathbf{G}(t)$ where

$\mathbf{F}(t) = \ln t\,\mathbf{i} + 5t\,\mathbf{j} - 3t^2\,\mathbf{k}$

and

$\mathbf{G}(t) = \mathbf{i} + 4t\,\mathbf{j} - 3t^2\,\mathbf{k}$

7. $\mathbf{r}(t) = \mathbf{F}(t) \times \mathbf{G}(t)$ where

$\mathbf{F}(t) = \sin t\,\mathbf{i} + \cos t\,\mathbf{j}$

and

$\mathbf{G}(t) = \sin t\,\mathbf{j} + \cos t\,\mathbf{k}$

8. $\mathbf{r}(t) = \mathbf{F}(t) \times \mathbf{G}(t)$ where

$\mathbf{F}(t) = t^3\,\mathbf{i} - t\,\mathbf{j} + t\,\mathbf{k}$

and

$\mathbf{G}(t) = \sqrt[3]{t}\,\mathbf{i} + \dfrac{1}{t+1}\,\mathbf{j} + (t+2)\,\mathbf{k}$

In Exercises 9 and 10, find $\|\mathbf{r}(t)\|$.

9. $\mathbf{r}(t) = \sin 3t\,\mathbf{i} + \cos 3t\,\mathbf{j} + t\,\mathbf{k}$
10. $\mathbf{r}(t) = \sqrt{t}\,\mathbf{i} + 3t\,\mathbf{j} - 4t\,\mathbf{k}$

In Exercises 11–16, sketch the curve represented by the vector-valued function and give the orientation of the curve.

11. $\mathbf{r}(t) = 2 \cos t\,\mathbf{i} + 2 \sin t\,\mathbf{j} + t\,\mathbf{k}$
12. $\mathbf{r}(t) = 3 \cos t\,\mathbf{i} + 4 \sin t\,\mathbf{j} + \dfrac{t}{2}\,\mathbf{k}$
13. $\mathbf{r}(t) = 2 \sin t\,\mathbf{i} + 2 \cos t\,\mathbf{j} + e^{-t}\,\mathbf{k}$
14. $\mathbf{r}(t) = t\,\mathbf{i} + t^2\,\mathbf{j} + \dfrac{3t}{2}\,\mathbf{k}$
15. $\mathbf{r}(t) = \left\langle t, t^2, \dfrac{2}{3}t^3 \right\rangle$
16. $\mathbf{r}(t) = \langle \cos t + t \sin t, \sin t - t \cos t, t \rangle$

In Exercises 17–24, sketch the space curve represented by the intersection of the given surfaces, and represent the curve by a vector-valued function using the given parameter.

Surfaces	*Parameter*
17. $z = x^2 + y^2$, $x + y = 0$	$x = t$
18. $z = x^2 + y^2$, $z = 4$	$x = 2 \cos t$
19. $x^2 + y^2 = 4$, $z = x^2$	$x = 2 \sin t$
20. $4x^2 + y^2 + 4z^2 = 16$, $x = y^2$	$y = t$
21. $x^2 + y^2 + z^2 = 4$, $x + z = 2$	$x = 1 + \sin t$
22. $x^2 + y^2 + z^2 = 10$, $x + y = 4$	$x = 2 + \sin t$
23. $x^2 + z^2 = 4$, $y^2 + z^2 = 4$ (first octant)	$x = t$
24. $x^2 + y^2 + z^2 = 16$, $xy = 4$ (first octant)	$x = t$

In Exercises 25–30, evaluate the limit.

25. $\displaystyle\lim_{t \to 2} \left(t\,\mathbf{i} + \dfrac{t^2 - 4}{t^2 - 2t}\,\mathbf{j} + \dfrac{1}{t}\,\mathbf{k} \right)$
26. $\displaystyle\lim_{t \to 0} \left(e^t\,\mathbf{i} + \dfrac{\sin t}{t}\,\mathbf{j} + e^{-t}\,\mathbf{k} \right)$
27. $\displaystyle\lim_{t \to 0} \left(t^2\,\mathbf{i} + 3t\,\mathbf{j} + \dfrac{1 - \cos t}{t}\,\mathbf{k} \right)$
28. $\displaystyle\lim_{t \to 1} \left(\sqrt{t}\,\mathbf{i} + \dfrac{\ln t}{t^2 - 1}\,\mathbf{j} + 2t^2\,\mathbf{k} \right)$
29. $\displaystyle\lim_{t \to 0} \left(\dfrac{1}{t}\,\mathbf{i} + \cos t\,\mathbf{j} + \sin t\,\mathbf{k} \right)$
30. $\displaystyle\lim_{t \to \infty} \left(e^{-t}\,\mathbf{i} + \dfrac{1}{t}\,\mathbf{j} + \dfrac{t}{t^2 + 1}\,\mathbf{k} \right)$

In Exercises 31–34, determine the intervals on which the vector-valued function is continuous.

31. $\mathbf{r}(t) = t\,\mathbf{i} + \arcsin t\,\mathbf{j} + (t - 1)\,\mathbf{k}$
32. $\mathbf{r}(t) = \sin t\,\mathbf{i} + \cos t\,\mathbf{j} + \ln t\,\mathbf{k}$
33. $\mathbf{r}(t) = \langle e^{-t}, t^2, \tan t \rangle$
34. $\mathbf{r}(t) = \langle 8, \sqrt{t}, \sqrt[3]{t} \rangle$

In Exercises 35 and 36, (a) sketch the space curve represented by the vector-valued function, and (b) sketch the vectors $\mathbf{r}(t_0)$ and $\mathbf{r}'(t_0)$ for the specified value of t_0. Position the vectors so that the initial point of $\mathbf{r}(t_0)$ is at the origin and the initial point of $\mathbf{r}'(t_0)$ is at the terminal point of $\mathbf{r}(t_0)$.

35. $\mathbf{r}(t) = 2 \cos t\,\mathbf{i} + 2 \sin t\,\mathbf{j} + t\,\mathbf{k}$, $\quad t_0 = \dfrac{3\pi}{2}$

36. $\mathbf{r}(t) = t\,\mathbf{i} + t^2\,\mathbf{j} + \dfrac{3}{2}t\,\mathbf{k}$, $\quad t_0 = 2$

In Exercises 37–44, find $\mathbf{r}'(t)$.

37. $\mathbf{r}(t) = 6t\,\mathbf{i} - 7t^2\,\mathbf{j} + t^3\,\mathbf{k}$
38. $\mathbf{r}(t) = \dfrac{1}{t}\,\mathbf{i} + 16t\,\mathbf{j} + \dfrac{t^2}{2}\,\mathbf{k}$
39. $\mathbf{r}(t) = a \cos^3 t\,\mathbf{i} + a \sin^3 t\,\mathbf{j} + \mathbf{k}$
40. $\mathbf{r}(t) = \sqrt{t}\,\mathbf{i} + t\sqrt{t}\,\mathbf{j} + \ln t\,\mathbf{k}$
41. $\mathbf{r}(t) = e^{-t}\,\mathbf{i} + 4\,\mathbf{j}$
42. $\mathbf{r}(t) = \langle \sin t - t \cos t, \cos t + t \sin t, t^2 \rangle$
43. $\mathbf{r}(t) = \langle t \sin t, t \cos t, t \rangle$
44. $\mathbf{r}(t) = \langle \arcsin t, \arccos t, 0 \rangle$

In Exercises 45 and 46, find the following.
(a) $\mathbf{r}'(t)$ (b) $\mathbf{r}''(t)$
(c) $D_t[\mathbf{r}(t) \cdot \mathbf{u}(t)]$ (d) $D_t[3\mathbf{r}(t) - \mathbf{u}(t)]$
(e) $D_t[\mathbf{r}(t) \times \mathbf{u}(t)]$ (f) $D_t[\|\mathbf{r}(t)\|]$

45. $\mathbf{r}(t) = t\,\mathbf{i} + 3t\,\mathbf{j} + t^2\,\mathbf{k}$
$\mathbf{u}(t) = 4t\,\mathbf{i} + t^2\,\mathbf{j} + t^3\,\mathbf{k}$

46. $\mathbf{r}(t) = t^2\mathbf{i} + \sin t\,\mathbf{j} + \cos t\,\mathbf{k}$

$\mathbf{u}(t) = \dfrac{1}{t^2}\mathbf{i} + \sin t\,\mathbf{j} + \cos t\,\mathbf{k}$

In Exercises 47–54, evaluate the indefinite integral.

47. $\displaystyle\int (2t\mathbf{i} + \mathbf{j} + \mathbf{k})\,dt$

48. $\displaystyle\int (3t^2\mathbf{i} + 4t\mathbf{j} - 8t^3\mathbf{k})\,dt$

49. $\displaystyle\int \left(\dfrac{1}{t}\mathbf{i} + \mathbf{j} - t^{3/2}\mathbf{k}\right)dt$

50. $\displaystyle\int [(2t - 1)\mathbf{i} + 4t^3\mathbf{j} + 3\sqrt{t}\mathbf{k}]\,dt$

51. $\displaystyle\int [e^t\mathbf{i} + \sin t\,\mathbf{j} + \cos t\,\mathbf{k}]\,dt$

52. $\displaystyle\int \left[\ln t\,\mathbf{i} + \dfrac{1}{t}\mathbf{j} + \mathbf{k}\right]dt$

53. $\displaystyle\int \left[\sec^2 t\,\mathbf{i} + \dfrac{1}{1 + t^2}\mathbf{j}\right]dt$

54. $\displaystyle\int [e^{-t}\sin t\,\mathbf{i} + e^{-t}\cos t\,\mathbf{j}]\,dt$

In Exercises 55–58, find $\mathbf{r}(t)$ for the given conditions.

55. $\mathbf{r}'(t) = 2t\mathbf{j} + \sqrt{t}\mathbf{k}$
$\mathbf{r}(0) = \mathbf{i} + \mathbf{j}$

56. $\mathbf{r}''(t) = -4\cos t\,\mathbf{j} - 3\sin 5\,\mathbf{k}$
$\mathbf{r}'(0) = 3\mathbf{k}$
$\mathbf{r}(0) = 4\mathbf{j}$

57. $\mathbf{r}'(t) = te^{-t^2}\mathbf{i} - e^{-t}\mathbf{j} + \mathbf{k}$
$\mathbf{r}(0) = \dfrac{1}{2}\mathbf{i} - \mathbf{j} + \mathbf{k}$

58. $\mathbf{r}'(t) = \dfrac{1}{1 + t^2}\mathbf{i} + \dfrac{1}{t^2}\mathbf{j} + \dfrac{1}{t}\mathbf{k}$
$\mathbf{r}(1) = 2\mathbf{i}$

In Exercises 59–62, evaluate the definite integral.

59. $\displaystyle\int_0^1 (8t\mathbf{i} + t\mathbf{j} - \mathbf{k})\,dt$

60. $\displaystyle\int_{-1}^1 (t\mathbf{i} + t^3\mathbf{j} + \sqrt[3]{t}\mathbf{k})\,dt$

61. $\displaystyle\int_0^{\pi/2} [(a\cos t)\mathbf{i} + (a\sin t)\mathbf{j} + \mathbf{k}]\,dt$

62. $\displaystyle\int_0^3 (e^t\mathbf{i} + te^t\mathbf{k})\,dt$

In Exercises 63–68, sketch the given curve and find its length over the indicated interval.

Function	Interval
63. $\mathbf{r}(t) = t\mathbf{i} + 3t\mathbf{j}$	$[0, 4]$
64. $\mathbf{r}(t) = t\mathbf{i} + t^2\mathbf{k}$	$[0, 4]$

Function	Interval
65. $\mathbf{r}(t) = a\cos t\,\mathbf{i} + a\sin t\,\mathbf{j} + bt\mathbf{k}$	$[0, 2\pi]$
66. $\mathbf{r}(t) = a\cos^3 t\,\mathbf{i} + a\sin^3 t\,\mathbf{j}$	$[0, 2\pi]$
67. $\mathbf{r}(t) = \langle \sin t - t\cos t, \cos t + t\sin t, t^2\rangle$	$\left[0, \dfrac{\pi}{2}\right]$
68. $\mathbf{r}(t) = \langle 4t, 3\cos t, 3\sin t\rangle$	$\left[0, \dfrac{\pi}{2}\right]$

In Exercises 69 and 70, approximate the length of the space curve over the indicated interval by using Simpson's Rule with $n = 4$.

69. $\mathbf{r}(t) = t^2\mathbf{i} + t\mathbf{j} + \ln t\,\mathbf{k}, \quad 1 \le t \le 3$

70. $\mathbf{r}(t) = \sin \pi t\,\mathbf{i} + \cos \pi t\,\mathbf{j} + t^3\mathbf{k}, \quad 0 \le t \le 2$

71. Consider the helix represented by the parametric equations $x(t) = 2\cos t$, $y(t) = 2\sin t$, and $z(t) = t$.
(a) Express the length of the arc s on the helix as a function of t by evaluating the integral

$$s = \int_0^t \sqrt{[x'(\tau)]^2 + [y'(\tau)]^2 + [z'(\tau)]^2}\,d\tau.$$

(b) Solve for t in the relationship derived in part (a), and substitute the result into the original set of parametric equations. This yields a parametrization of the curve in terms of the arc length parameter s.

72. Repeat Exercise 71 for the curve represented by the parametric equations

$x(t) = \sin t - t\cos t$
$y(t) = \cos t + t\sin t$
$z(t) = t^2.$

In Exercises 73 and 74, use a computer to sketch the curve represented by the vector-valued function.

73. $\mathbf{r}(t) = t\mathbf{i} + 0.6t^2\mathbf{j} + 2t\mathbf{k}$

74. $\mathbf{r}(t) = 2\sin t\,\mathbf{i} + 2\cos t\,\mathbf{j} + 2\sin^2 t\,\mathbf{k}$

75. Use a computer or calculator with $n = 12$ and Simpson's Rule to approximate the length of the curve given by

$$\mathbf{r}(t) = t\mathbf{i} + \dfrac{t^2}{5}\mathbf{j} + \dfrac{t^3}{10}\mathbf{k}$$

over the interval $0 \le t \le 2$.

76. Assuming that $\mathbf{r}(t)$ and $\mathbf{u}(t)$ are differentiable vector-valued functions of t, prove each of the following.
(a) If $\mathbf{r}(t) \cdot \mathbf{r}(t)$ is a constant, then $\mathbf{r}(t) \cdot \mathbf{r}'(t) = 0$.
(b) $D_t[\mathbf{r}(t) \times \mathbf{u}(t)] = \mathbf{r}(t) \times \mathbf{u}'(t) + \mathbf{r}'(t) \times \mathbf{u}(t)$

77. Prove that if $\mathbf{r}$ is a vector-valued function that is continuous at c, then $\|\mathbf{r}\|$ is continuous at c.

78. Verify that the converse of Exercise 77 is not true by finding a vector-valued function $\mathbf{r}$ such that $\|\mathbf{r}\|$ is continuous at c but $\mathbf{r}$ is not continuous at c.

14.6 Tangent Vectors, Normal Vectors, and Curvature in Space

Velocity and acceleration ▪ Tangent vectors and normal vectors ▪ Tangential and normal components of acceleration ▪ Curvature

In the last section of this chapter, we extend our work with plane curves and motion in the plane to space curves and motion in space. We begin by defining the velocity and acceleration vectors for an object moving in space.

DEFINITION OF VELOCITY AND ACCELERATION	If x, y, and z are twice differentiable functions of t and $\mathbf{r}$ is a vector-valued function given by $\mathbf{r}(t) = x(t)\mathbf{i} + y(t)\mathbf{j} + z(t)\mathbf{k}$, then the velocity vector, acceleration vector, and speed at time t are as follows.

$$\text{velocity} = \mathbf{v}(t) \quad = \mathbf{r}'(t) \quad = x'(t)\mathbf{i} + y'(t)\mathbf{j} + z'(t)\mathbf{k}$$
$$\text{acceleration} = \mathbf{a}(t) \quad = \mathbf{r}''(t) \quad = x''(t)\mathbf{i} + y''(t)\mathbf{j} + z''(t)\mathbf{k}$$
$$\text{speed} = \|\mathbf{v}(t)\| = \|\mathbf{r}'(t)\| = \sqrt{[x'(t)]^2 + [y'(t)]^2 + [z'(t)]^2}$$

EXAMPLE 1   Sketching velocity and acceleration vectors in space

Sketch the path of an object moving along the space curve C given by

$$\mathbf{r}(t) = t\mathbf{i} + t^3\mathbf{j} + 3t\mathbf{k}, \qquad t \geq 0$$

and find the velocity and acceleration vectors when $t = 1$.

SOLUTION

Using the parametric equations $x = t$ and $y = t^3$, we see that the path of the object lies on the cubic cylinder given by $y = x^3$. Moreover, since $z = 3t$, we see that the object starts at $(0, 0, 0)$ and moves upward as t increases, as shown in Figure 14.46. Since $\mathbf{r}(t) = t\mathbf{i} + t^3\mathbf{j} + 3t\mathbf{k}$, we have

$$\mathbf{v}(t) = \mathbf{r}'(t) = \mathbf{i} + 3t^2\mathbf{j} + 3\mathbf{k}$$

and

$$\mathbf{a}(t) = \mathbf{r}''(t) = 6t\mathbf{j}.$$

Therefore, the velocity and acceleration vectors when $t = 1$ are

$$\mathbf{v}(1) = \mathbf{r}'(1) = \mathbf{i} + 3\mathbf{j} + 3\mathbf{k}$$

and

$$\mathbf{a}(1) = \mathbf{r}''(1) = 6\mathbf{j}.$$

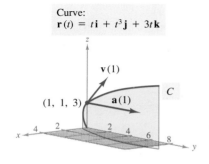

Curve:
$\mathbf{r}(t) = t\mathbf{i} + t^3\mathbf{j} + 3t\mathbf{k}$

FIGURE 14.46

Tangent vectors and normal vectors

In space, the definitions of the unit tangent vector and the principal unit normal vector are similar to those given for curves in the plane.

**UNIT TANGENT VECTOR AND
PRINCIPAL UNIT NORMAL VECTOR**

Let C be a smooth curve represented by $\mathbf{r}$ on an open interval I. If $\mathbf{r}'(t) \neq \mathbf{0}$, then the **unit tangent vector** $\mathbf{T}(t)$ at t is defined to be

$$\mathbf{T}(t) = \frac{\mathbf{r}'(t)}{\|\mathbf{r}'(t)\|}.$$

The **principal unit normal vector** $\mathbf{N}(t)$ at t is defined to be

$$\mathbf{N}(t) = \frac{\mathbf{T}'(t)}{\|\mathbf{T}'(t)\|}.$$

As in the plane, the **tangent line to a curve** at a point is defined to be the line passing through the point and parallel to the unit tangent vector. In Example 2, we use the unit tangent vector to find the tangent line at a point on a helix.

EXAMPLE 2   Finding the tangent line at a point on a curve

Find $\mathbf{T}(t)$ and then find a set of parametric equations for the tangent line to the helix given by $\mathbf{r}(t) = 2 \cos t\,\mathbf{i} + 2 \sin t\,\mathbf{j} + t\mathbf{k}$ at the point corresponding to $t = \pi/4$.

SOLUTION

The derivative of $\mathbf{r}(t)$ is $\mathbf{r}'(t) = -2 \sin t\,\mathbf{i} + 2 \cos t\,\mathbf{j} + \mathbf{k}$, which implies that $\|\mathbf{r}'(t)\| = \sqrt{4 \sin^2 t + 4 \cos^2 t + 1} = \sqrt{5}$. Therefore, the unit tangent vector is

$$\mathbf{T}(t) = \frac{\mathbf{r}'(t)}{\|\mathbf{r}'(t)\|} = \frac{1}{\sqrt{5}}(-2 \sin t\,\mathbf{i} + 2 \cos t\,\mathbf{j} + \mathbf{k}).$$

When $t = \pi/4$, the unit tangent vector is

$$\mathbf{T}\left(\frac{\pi}{4}\right) = \frac{1}{\sqrt{5}}\left(-2\frac{\sqrt{2}}{2}\mathbf{i} + 2\frac{\sqrt{2}}{2}\mathbf{j} + \mathbf{k}\right) = \frac{1}{\sqrt{5}}(-\sqrt{2}\mathbf{i} + \sqrt{2}\mathbf{j} + \mathbf{k}).$$

Using the direction numbers $a = -\sqrt{2}$, $b = \sqrt{2}$, and $c = 1$, and the point $(x_1, y_1, z_1) = (\sqrt{2}, \sqrt{2}, \pi/4)$, we obtain the following parametric equations (given with parameter s).

$$x = x_1 + as = \sqrt{2} - \sqrt{2}s$$
$$y = y_1 + bs = \sqrt{2} + \sqrt{2}s$$
$$z = z_1 + cs = \frac{\pi}{4} + s$$

This tangent line is shown in Figure 14.47.

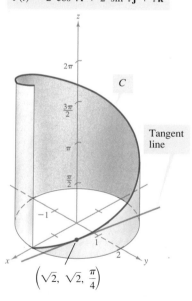

Curve:
$r(t) = 2 \cos t\mathbf{i} + 2 \sin t\mathbf{j} + t\mathbf{k}$

C

Tangent line

$\left(\sqrt{2}, \sqrt{2}, \frac{\pi}{4}\right)$

FIGURE 14.47

EXAMPLE 3   Finding the principal unit normal vector

Find the principal unit normal vector for the helix given by

$$\mathbf{r}(t) = 2 \cos t\,\mathbf{i} + 2 \sin t\,\mathbf{j} + t\mathbf{k}.$$

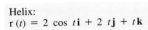

Helix:
$r(t) = 2 \cos t\mathbf{i} + 2 t\mathbf{j} + t\mathbf{k}$

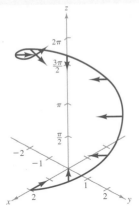

$\mathbf{N}(t)$ points toward the z-axis.

FIGURE 14.48

SOLUTION

From Example 2, we know that the unit tangent vector is

$$\mathbf{T}(t) = \frac{1}{\sqrt{5}}(-2 \sin t\,\mathbf{i} + 2 \cos t\,\mathbf{j} + \mathbf{k}).$$

Thus, $\mathbf{T}'(t)$ is given by

$$\mathbf{T}'(t) = \frac{1}{\sqrt{5}}(-2 \cos t\,\mathbf{i} - 2 \sin t\,\mathbf{j}).$$

Since $\|\mathbf{T}'(t)\| = 2/\sqrt{5}$, it follows that the principal unit normal vector is

$$\mathbf{N}(t) = \frac{\mathbf{T}'(t)}{\|\mathbf{T}'(t)\|} = \frac{1}{2}(-2 \cos t\,\mathbf{i} - 2 \sin t\,\mathbf{j}) = -\cos t\,\mathbf{i} - \sin t\,\mathbf{j}.$$

Note that this vector points toward the z-axis, as shown in Figure 14.48.

Tangential and normal components of acceleration

The following theorem is important because it tells us that the acceleration vector must lie in the plane determined by the unit tangent vector and principal unit normal vector.

THEOREM 14.12 **ACCELERATION VECTOR**	If $\mathbf{r}(t)$ is the position vector for a smooth curve C, then the acceleration vector $\mathbf{a}(t)$ lies in the plane determined by $\mathbf{T}(t)$ and $\mathbf{N}(t)$.

PROOF To simplify the notation, we write $\mathbf{T}$ for $\mathbf{T}(t)$, $\mathbf{T}'$ for $\mathbf{T}'(t)$, and so on. Since $\mathbf{T} = \mathbf{r}'/\|\mathbf{r}'\| = \mathbf{v}/\|\mathbf{v}\|$, it follows that

$$\mathbf{v} = \|\mathbf{v}\|\mathbf{T}.$$

By differentiating, we obtain

$$\mathbf{a} = \mathbf{v}' = D_t[\|\mathbf{v}\|]\mathbf{T} + \|\mathbf{v}\|\mathbf{T}'$$

$$= D_t[\|\mathbf{v}\|]\mathbf{T} + \|\mathbf{v}\|\mathbf{T}'\left(\frac{\|\mathbf{T}'\|}{\|\mathbf{T}'\|}\right)$$

$$= D_t[\|\mathbf{v}\|]\mathbf{T} + \|\mathbf{v}\|\,\|\mathbf{T}'\|\mathbf{N}.$$

Since $\mathbf{a}$ is written as a linear combination of $\mathbf{T}$ and $\mathbf{N}$, it must lie in the plane determined by $\mathbf{T}$ and $\mathbf{N}$.

With this theorem, we can now extend the results described in Section 13.5 to curves in space. For instance, we call the coefficients of $\mathbf{T}$ and $\mathbf{N}$ given in Theorem 14.12, the **tangential** and **normal components of acceleration**, and they are denoted by $a_{\mathbf{T}} = D_t[\|\mathbf{v}\|]$ and $a_{\mathbf{N}} = \|\mathbf{v}\|\,\|\mathbf{T}'\|$. Thus, we can write

$$\mathbf{a}(t) = a_{\mathbf{T}}\mathbf{T}(t) + a_{\mathbf{N}}\mathbf{N}(t).$$

The following theorem gives some convenient formulas for $a_{\mathbf{N}}$ and $a_{\mathbf{T}}$.

**THEOREM 14.13
TANGENTIAL AND NORMAL
COMPONENTS OF ACCELERATION**

If $\mathbf{r}(t)$ is the position vector for a smooth curve C in space, then the tangential and normal components of acceleration are as follows.

$$a_{\mathbf{T}} = \mathbf{a} \cdot \mathbf{T} = \frac{\mathbf{v} \cdot \mathbf{a}}{\|\mathbf{v}\|}$$

$$a_{\mathbf{N}} = \mathbf{a} \cdot \mathbf{N} = \frac{\|\mathbf{v} \times \mathbf{a}\|}{\|\mathbf{v}\|} = \sqrt{\|\mathbf{a}\|^2 - a_{\mathbf{T}}^2}$$

PROOF

Since $\mathbf{a}$ lies in the plane of $\mathbf{T}$ and $\mathbf{N}$, it follows that for any time t, the component of the projection of the acceleration vector onto $\mathbf{T}$ is given by $a_{\mathbf{T}} = \mathbf{a} \cdot \mathbf{T}$, and onto $\mathbf{N}$ is given by $a_{\mathbf{N}} = \mathbf{a} \cdot \mathbf{N}$. Moreover, since $\mathbf{a} = \mathbf{v}'$ and $\mathbf{T} = \mathbf{v}/\|\mathbf{v}\|$, we have

$$a_{\mathbf{T}} = \mathbf{a} \cdot \mathbf{T} = \mathbf{T} \cdot \mathbf{a} = \frac{\mathbf{v}}{\|\mathbf{v}\|} \cdot \mathbf{a} = \frac{\mathbf{v} \cdot \mathbf{a}}{\|\mathbf{v}\|}.$$

Furthermore, using $\mathbf{a} = a_{\mathbf{T}}\mathbf{T} + a_{\mathbf{N}}\mathbf{N}$, $\mathbf{T} \times \mathbf{T} = \mathbf{0}$, and $\|\mathbf{T} \times \mathbf{N}\| = 1$, we obtain the following.

$$\mathbf{v} \times \mathbf{a} = \|\mathbf{v}\|\mathbf{T} \times (a_{\mathbf{T}}\mathbf{T} + a_{\mathbf{N}}\mathbf{N}) = \|\mathbf{v}\|a_{\mathbf{T}}(\mathbf{T} \times \mathbf{T}) + \|\mathbf{v}\|a_{\mathbf{N}}(\mathbf{T} \times \mathbf{N})$$

$$= \|\mathbf{v}\|a_{\mathbf{N}}(\mathbf{T} \times \mathbf{N})$$

$$\|\mathbf{v} \times \mathbf{a}\| = \|\mathbf{v}\|a_{\mathbf{N}}\|\mathbf{T} \times \mathbf{N}\| = \|\mathbf{v}\|a_{\mathbf{N}}$$

$$a_{\mathbf{N}} = \mathbf{a} \cdot \mathbf{N} = \frac{\|\mathbf{v} \times \mathbf{a}\|}{\|\mathbf{v}\|}$$

Moreover, since

$$\|\mathbf{a}\|^2 = \mathbf{a} \cdot \mathbf{a} = (a_{\mathbf{T}}\mathbf{T} + a_{\mathbf{N}}\mathbf{N}) \cdot (a_{\mathbf{T}}\mathbf{T} + a_{\mathbf{N}}\mathbf{N})$$

$$= a_{\mathbf{T}}^2\|\mathbf{T}\|^2 + 2a_{\mathbf{T}}a_{\mathbf{N}}\mathbf{T} \cdot \mathbf{N} + a_{\mathbf{N}}^2\|\mathbf{N}\|^2$$

$$= a_{\mathbf{T}}^2\|\mathbf{T}\|^2 + a_{\mathbf{N}}^2\|\mathbf{N}\|^2 = a_{\mathbf{T}}^2 + a_{\mathbf{N}}^2$$

it follows that $\|\mathbf{a}\| = \sqrt{a_{\mathbf{T}}^2 + a_{\mathbf{N}}^2}$, and because $a_{\mathbf{N}} \geq 0$, we have

$$a_{\mathbf{N}} = \sqrt{a_{\mathbf{N}}^2} = \sqrt{a_{\mathbf{N}}^2 + a_{\mathbf{T}}^2 - a_{\mathbf{T}}^2} = \sqrt{\|\mathbf{a}\|^2 - a_{\mathbf{T}}^2}.$$

EXAMPLE 4   Finding the tangential and normal components of acceleration

Find the tangential and normal components of acceleration for the position function given by $\mathbf{r}(t) = 3t\mathbf{i} - t\mathbf{j} + t^2\mathbf{k}$.

SOLUTION

We begin by finding the velocity, speed, and acceleration.

$$\mathbf{v}(t) = \mathbf{r}'(t) = 3\mathbf{i} - \mathbf{j} + 2t\mathbf{k}$$

$$\|\mathbf{v}(t)\| = \sqrt{9 + 1 + 4t^2} = \sqrt{10 + 4t^2}$$

$$\mathbf{a}(t) = \mathbf{r}''(t) = 2\mathbf{k}$$

By Theorem 14.13, the tangential component of acceleration is

$$a_T = \frac{\mathbf{v} \cdot \mathbf{a}}{\|\mathbf{v}\|} = \frac{4t}{\sqrt{10 + 4t^2}}$$

and since

$$\mathbf{v} \times \mathbf{a} = \begin{vmatrix} \mathbf{i} & \mathbf{j} & \mathbf{k} \\ 3 & -1 & 2t \\ 0 & 0 & 2 \end{vmatrix} = -2\mathbf{i} - 6\mathbf{j}$$

the normal component of acceleration is

$$a_N = \frac{\|\mathbf{v} \times \mathbf{a}\|}{\|\mathbf{v}\|} = \frac{\sqrt{4 + 36}}{\sqrt{10 + 4t^2}} = \frac{2\sqrt{10}}{\sqrt{10 + 4t^2}}.$$

REMARK   In Example 4, we could have used the alternate formula for a_N as follows.

$$a_N = \sqrt{\|\mathbf{a}\|^2 - a_T^2} = \sqrt{(2)^2 - \frac{16t^2}{10 + 4t^2}} = \frac{2\sqrt{10}}{\sqrt{10 + 4t^2}}$$

EXAMPLE 5   Finding a_T and a_N for a circular helix

Find the tangential and normal components of acceleration for the helix given by

$$\mathbf{r}(t) = b \cos t\, \mathbf{i} + b \sin t\, \mathbf{j} + ct\mathbf{k}, \quad b > 0.$$

SOLUTION

$$\mathbf{v}(t) = \mathbf{r}'(t) = -b \sin t\, \mathbf{i} + b \cos t\, \mathbf{j} + c\, \mathbf{k}$$
$$\|\mathbf{v}(t)\| = \sqrt{b^2 \sin^2 t + b^2 \cos^2 t + c^2} = \sqrt{b^2 + c^2}$$
$$\mathbf{a}(t) = \mathbf{r}''(t) = -b \cos t\, \mathbf{i} - b \sin t\, \mathbf{j}$$

By Theorem 14.13, the tangential component of acceleration is

$$a_T = \frac{\mathbf{v} \cdot \mathbf{a}}{\|\mathbf{v}\|} = b^2 \sin t \cos t - b^2 \sin t \cos t + 0 = 0.$$

Moreover, since

$$\|\mathbf{a}(t)\| = \sqrt{b^2 \cos^2 t + b^2 \sin^2 t} = b$$

we can use the alternate formula for the normal component of acceleration to obtain

$$a_N = \sqrt{\|\mathbf{a}(t)\|^2 - a_T^2} = \sqrt{b^2 - 0^2} = b.$$

Note that the normal component of acceleration is equal to the magnitude of the acceleration. In other words, since the speed is constant, the acceleration is perpendicular to the velocity.

Curvature

In space, we define the arc length function the same way as in the plane. That is, if C is a smooth curve given by $\mathbf{r}(t)$ defined on the closed interval $a \leq t \leq b$, the **arc length function** is given by

$$s(t) = \int_a^t \|\mathbf{r}'(\tau)\| \, d\tau.$$

Moreover, by using the definition of the arc length function and the Second Fundamental Theorem of Calculus, we can conclude that

$$\frac{ds}{dt} = \|\mathbf{r}'(t)\|$$

or in differential form, $ds = \|\mathbf{r}'(t)\| \, dt$.

As in the plane, the **curvature of a smooth curve** in space can be written in terms of the arc length parameter s

$$K = \left\|\frac{d\mathbf{T}}{ds}\right\| = \|\mathbf{T}'(s)\| \qquad \text{Curvature}$$

or in terms of an arbitrary parameter t

$$K = \frac{\|\mathbf{T}'(t)\|}{\|\mathbf{r}'(t)\|} = \frac{\|\mathbf{r}'(t) \times \mathbf{r}''(t)\|}{\|\mathbf{r}'(t)\|^3}.$$

You are asked to verify the cross product formula for curvature in Exercise 35.

EXAMPLE 6 Finding the curvature of a space curve

Find the curvature of the curve given by $\mathbf{r}(t) = 2t\mathbf{i} + t^2\mathbf{j} - \frac{1}{3}t^3\mathbf{k}$.

SOLUTION

It is not apparent whether this parameter is arc length, so we use the formula $K = \|\mathbf{T}'(t)\|/\|\mathbf{r}'(t)\|$. Since $\mathbf{r}'(t) = 2\mathbf{i} + 2t\mathbf{j} - t^2\mathbf{k}$, it follows that $\|\mathbf{r}'(t)\| = \sqrt{4 + 4t^2 + t^4} = t^2 + 2$, and we have

$$\mathbf{T}(t) = \frac{\mathbf{r}'(t)}{\|\mathbf{r}'(t)\|} = \frac{2\mathbf{i} + 2t\mathbf{j} - t^2\mathbf{k}}{t^2 + 2}$$

$$\mathbf{T}'(t) = \frac{(t^2 + 2)(2\mathbf{j} - 2t\mathbf{k}) - (2t)(2\mathbf{i} + 2t\mathbf{j} - t^2\mathbf{k})}{(t^2 + 2)^2}$$

$$= \frac{-4t\mathbf{i} + (4 - 2t^2)\mathbf{j} - 4t\mathbf{k}}{(t^2 + 2)^2}$$

$$\|\mathbf{T}'(t)\| = \frac{\sqrt{16t^2 + 16 - 16t^2 + 4t^4 + 16t^2}}{(t^2 + 2)^2} = \frac{2(t^2 + 2)}{(t^2 + 2)^2} = \frac{2}{t^2 + 2}.$$

Therefore,

$$K = \frac{\|\mathbf{T}'(t)\|}{\|\mathbf{r}'(t)\|} = \frac{2}{(t^2 + 2)^2}.$$

Theorem 13.16, given in Section 13.6, is also valid in space. That is, if $\mathbf{r}(t)$ is the position vector for a smooth curve C in space, then the acceleration vector is given by

$$\mathbf{a}(t) = \frac{d^2s}{dt^2}\,\mathbf{T} + K\left(\frac{ds}{dt}\right)^2 \mathbf{N}$$

where K is the curvature of C and ds/dt is the speed. We use this result in the next example.

EXAMPLE 7 Finding the tangential and normal components of acceleration

Find $a_\mathbf{T}$ and $a_\mathbf{N}$ for the curve given by $\mathbf{r}(t) = 2t\mathbf{i} + t^2\mathbf{j} - \frac{1}{3}t^3\mathbf{k}$.

SOLUTION

From Example 6 we know that

$$\frac{ds}{dt} = \|\mathbf{r}'(t)\| = t^2 + 2 \qquad \text{and} \qquad K = \frac{2}{(t^2 + 2)^2}.$$

Therefore,

$$a_\mathbf{T} = \frac{d^2s}{dt^2} = 2t$$

and

$$a_\mathbf{N} = K\left(\frac{ds}{dt}\right)^2 = \frac{2}{(t^2 + 2)^2}(t^2 + 2)^2 = 2.$$

EXERCISES for Section 14.6

In Exercises 1–8, the position function $\mathbf{r}$ describes the path of an object moving in space. Find the velocity, speed, and acceleration of the object.

1. $\mathbf{r}(t) = t\mathbf{i} + (2t - 5)\mathbf{j} + 3t\mathbf{k}$
2. $\mathbf{r}(t) = 4t\mathbf{i} + 4t\mathbf{j} + 2t\mathbf{k}$
3. $\mathbf{r}(t) = t\mathbf{i} + t^2\mathbf{j} + \dfrac{t^2}{2}\mathbf{k}$
4. $\mathbf{r}(t) = t\mathbf{i} + 3t\mathbf{j} + \dfrac{t^2}{2}\mathbf{k}$
5. $\mathbf{r}(t) = t\mathbf{i} + t\mathbf{j} + \sqrt{9 - t^2}\,\mathbf{k}$
6. $\mathbf{r}(t) = t^2\mathbf{i} + t\mathbf{j} + 2t^{3/2}\mathbf{k}$
7. $\mathbf{r}(t) = \langle 4t, 3\cos t, 3\sin t\rangle$
8. $\mathbf{r}(t) = \langle e^t \cos t, e^t \sin t, e^t\rangle$

In Exercises 9–12, use the given acceleration function to find the velocity and position functions. Then find the position at time $t = 2$.

9. $\mathbf{a}(t) = \mathbf{i} + \mathbf{j} + \mathbf{k}$, $\mathbf{v}(0) = \mathbf{0}$, $\mathbf{r}(0) = \mathbf{0}$
10. $\mathbf{a}(t) = \mathbf{i} + \mathbf{k}$, $\mathbf{v}(0) = 5\mathbf{j}$, $\mathbf{r}(0) = \mathbf{0}$

11. $\mathbf{a}(t) = t\mathbf{j} + t\mathbf{k}$, $\mathbf{v}(1) = 5\mathbf{j}$, $\mathbf{r}(1) = \mathbf{0}$
12. $\mathbf{a}(t) = -\cos t\,\mathbf{i} - \sin t\,\mathbf{j}$, $\mathbf{v}(0) = \mathbf{j} + \mathbf{k}$, $\mathbf{r}(0) = \mathbf{i}$

In Exercises 13–20, find a set of parametric equations for the line tangent to the space curve at the given point.

Function	*Point*
13. $\mathbf{r}(t) = t\mathbf{i} + t^2\mathbf{j} + t\mathbf{k}$	$(0, 0, 0)$
14. $\mathbf{r}(t) = t\mathbf{i} + t^2\mathbf{j} + \dfrac{2}{3}\mathbf{k}$	$\left(1, 1, \dfrac{2}{3}\right)$
15. $\mathbf{r}(t) = 2\cos t\,\mathbf{i} + 2\sin t\,\mathbf{j} + t\mathbf{k}$	$(2, 0, 0)$
16. $\mathbf{r}(t) = 3\cos t\,\mathbf{i} + 4\sin t\,\mathbf{j} + \dfrac{t}{2}\mathbf{k}$	$\left(0, 4, \dfrac{\pi}{4}\right)$
17. $\mathbf{r}(t) = \left\langle t, t^2, \dfrac{2}{3}t^3\right\rangle$	$(3, 9, 18)$
18. $\mathbf{r}(t) = \langle t, t, \sqrt{4 - t^2}\rangle$	$(1, 1, \sqrt{3})$
19. $\mathbf{r}(t) = \langle 2\cos t, 2\sin t, 4\rangle$	$(\sqrt{2}, \sqrt{2}, 4)$
20. $\mathbf{r}(t) = \langle 2\sin t, 2\cos t, 4\sin^2 t\rangle$	$(1, \sqrt{3}, 1)$

In Exercises 21–26, find $\mathbf{T}(t)$, $\mathbf{N}(t)$, $a_{\mathbf{T}}$, and $a_{\mathbf{N}}$ at the given time t.

Function	Time
21. $\mathbf{r}(t) = 4\mathbf{i} + t\mathbf{j} + t^2\mathbf{k}$	$t = 1$
22. $\mathbf{r}(t) = 4t\mathbf{i} - 4t\mathbf{j} + 2t\mathbf{k}$	$t = 2$
23. $\mathbf{r}(t) = t\mathbf{i} + t^2\mathbf{j} + \dfrac{t^2}{2}\mathbf{k}$	$t = 1$
24. $\mathbf{r}(t) = t\mathbf{i} + 3t^2\mathbf{j} + \dfrac{t^2}{2}\mathbf{k}$	$t = 2$
25. $\mathbf{r}(t) = 4t\mathbf{i} + 3\cos t\,\mathbf{j} + 3\sin t\,\mathbf{k}$	$t = \dfrac{\pi}{2}$
26. $\mathbf{r}(t) = e^t\cos t\,\mathbf{i} + e^t\sin t\,\mathbf{j} + e^t\mathbf{k}$	$t = 0$

In Exercises 27 and 28, sketch the space curve represented by the vector-valued function, and at the point on the curve determined by $\mathbf{r}(t_0)$ sketch the vectors $\mathbf{T}$, $\mathbf{N}$, and $\mathbf{B} = \mathbf{T} \times \mathbf{N}$. The vector $\mathbf{B}$ is called the **unit binormal vector** to the curve (see figure).

27. $\mathbf{r}(t) = 2\cos t\,\mathbf{i} + 2\sin t\,\mathbf{j} + \dfrac{t}{2}\mathbf{k}, \quad t_0 = \dfrac{\pi}{2}$

28. $\mathbf{r}(t) = t\mathbf{i} + t\mathbf{j} + \dfrac{t^2}{2}\mathbf{k}, \quad t_0 = 2$

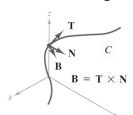

Vectors $\mathbf{T}$, $\mathbf{N}$, and $\mathbf{B}$
are mutually orthogonal.

FIGURE FOR 27 and 28

In Exercises 29–34, find the curvature K of the given curve.

29. $\mathbf{r}(t) = 4\mathbf{i} + t\mathbf{j} + \dfrac{t^2}{2}\mathbf{k}$

30. $\mathbf{r}(t) = 4t\mathbf{i} - 4t\mathbf{j} + 2t\mathbf{k}$

31. $\mathbf{r}(t) = t\mathbf{i} - t^2\mathbf{j} + \dfrac{t^2}{2}\mathbf{k}$

32. $\mathbf{r}(t) = t\mathbf{i} + 3t^2\mathbf{j} + \dfrac{t^2}{2}\mathbf{k}$

33. $\mathbf{r}(t) = 4t\mathbf{i} + 3\cos t\,\mathbf{j} + 3\sin t\,\mathbf{k}$

34. $\mathbf{r}(t) = e^t\cos t\,\mathbf{i} + e^t\sin t\,\mathbf{j} + e^t\mathbf{k}$

35. Use the definition of curvature in space, $K = \|\mathbf{T}'(s)\| = \|\mathbf{r}''(s)\|$, to verify the following alternate formula for curvature.

$$K = \frac{\|\mathbf{r}'(t) \times \mathbf{r}''(t)\|}{\|\mathbf{r}'(t)\|^3}$$

In Exercises 36–38, use the result of Exercise 35 to find the curvature K of the given curve.

36. $\mathbf{r}(t) = t^2\mathbf{i} + 2t\mathbf{j} + t^2\mathbf{k}$

37. $\mathbf{r}(t) = 3t^2\mathbf{i} + (3t - t^3)\mathbf{j} - t\mathbf{k}$

38. $\mathbf{r}(t) = \sin t\,\mathbf{i} + \cos t\,\mathbf{j} + 4\mathbf{k}$

In Exercises 39 and 40, use the formula

$$\mathbf{a}(t) = \frac{d^2 s}{dt^2}\mathbf{T} + K\left(\frac{ds}{dt}\right)^2\mathbf{N}$$

to find $a_{\mathbf{T}}$ and $a_{\mathbf{N}}$ for the curve given by the vector-valued function.

39. $\mathbf{r}(t) = 3t^2\mathbf{i} + (3t - t^3)\mathbf{j} - t\mathbf{k}$

40. $\mathbf{r}(t) = t^2\mathbf{i} + 2t\mathbf{j} + t^2\mathbf{k}$

REVIEW EXERCISES for Chapter 14

In Exercises 1 and 2, let $\mathbf{u} = \overrightarrow{PQ}$ and $\mathbf{v} = \overrightarrow{PR}$, and find (a) the component forms of $\mathbf{u}$ and $\mathbf{v}$, (b) $\mathbf{u} \cdot \mathbf{v}$, (c) $\mathbf{u} \times \mathbf{v}$, (d) an equation of the plane containing P, Q, and R, and (e) a set of parametric equations of the line through P and Q.

1. $P = (5, 0, 0)$, $Q = (4, 4, 0)$, $R = (2, 0, 6)$
2. $P = (2, -1, 3)$, $Q = (0, 5, 1)$, $R = (5, 5, 0)$

In Exercises 3 and 4, find the angle θ between the vectors $\mathbf{u}$ and $\mathbf{v}$.

3. $\mathbf{u} = \langle 10, -5, 15 \rangle$, $\mathbf{v} = \langle -2, 1, -3 \rangle$
4. $\mathbf{u} = \langle 1, 0, -3 \rangle$, $\mathbf{v} = \langle 2, -2, 1 \rangle$

In Exercises 5–8, find $\mathbf{u}$.

5. The angle, measured counterclockwise, from $\mathbf{u}$ to the positive x-axis is $135°$, and $\|\mathbf{u}\| = 4$.

6. The angle between $\mathbf{u}$ and the positive x-axis is $180°$, and $\|\mathbf{u}\| = 8$.

7. $\mathbf{u}$ is perpendicular to the plane $x - 3y + 4z = 0$, and $\|\mathbf{u}\| = 3$.

8. $\mathbf{u}$ is a unit vector perpendicular to the lines

$$\begin{array}{ll} x = 4 - t & x = -3 + 7t \\ y = 3 + 2t & y = -2 + t \\ z = 1 + 5t & z = 1 + 2t \end{array}$$

In Exercises 9–14, let $\mathbf{u} = \langle 3, -2, 1 \rangle$, $\mathbf{v} = \langle 2, -4, -3 \rangle$, and $\mathbf{w} = \langle -1, 2, 2 \rangle$.

9. Find $\|\mathbf{u}\|$.

10. Find the angle between $\mathbf{u}$ and $\mathbf{v}$.

11. Show that $\mathbf{u} \cdot \mathbf{u} = \|\mathbf{u}\|^2$.

12. Determine a unit vector perpendicular to the plane containing $\mathbf{v}$ and $\mathbf{w}$.

13. Show that $\mathbf{u} \cdot (\mathbf{v} + \mathbf{w}) = \mathbf{u} \cdot \mathbf{v} + \mathbf{u} \cdot \mathbf{w}$.

14. Show that

$$\mathbf{u} \times (\mathbf{v} + \mathbf{w}) = (\mathbf{u} \times \mathbf{v}) + (\mathbf{u} \times \mathbf{w}).$$

In Exercises 15–18, find (a) a set of parametric equations, and (b) a set of symmetric equations for the given line.

15. The line passes through $(1, 2, 3)$ and is perpendicular to the xz-coordinate plane.

16. The line passes through $(1, 2, 3)$ and is parallel to the line given by $x = y = z$.

17. The line is the intersection of the planes given by

$$3x - 3y - 7z = -4$$
$$x - y + 2z = 3.$$

18. The line passes through the point $(0, 1, 4)$ and is perpendicular to $\mathbf{u} = \langle 2, -5, 1 \rangle$ and $\mathbf{v} = \langle -3, 1, 4 \rangle$.

In Exercises 19–22, find an equation of the plane.

19. The plane passes through the point $(1, 2, 3)$ and is orthogonal to the line given by $x = y = z$.

20. The plane passes through the point $(4, 2, 1)$ and is parallel to the yz-coordinate plane.

21. The plane contains the lines

$$\frac{x - 1}{-2} = y = z + 1$$

$$\frac{x + 1}{-2} = y - 1 = z - 2.$$

22. The plane passes through the points $(-3, -4, 2)$, $(-3, 4, 1)$, and $(1, 1, -2)$.

In Exercises 23 and 24, find the distance from the point to the plane.

23. $(1, 0, 2)$, $2x - 3y + 6z = 6$

24. $(0, 0, 0)$, $4x - 7y + z = 2$

In Exercises 25 and 26, find the distance between the parallel planes.

25. $5x - 3y + z = 2$, $5x - 3y + z = -3$

26. $3x + 2y - z = -1$, $6x + 4y - 2z = 1$

In Exercises 27–30, find the distance between the lines.

27. $\begin{aligned} x &= 4 \\ y &= -2t \\ z &= 1 + t \end{aligned} \qquad \begin{aligned} x &= -1 - s \\ y &= 2 + s \\ z &= 3 \end{aligned}$

28. $\begin{aligned} x &= 4 + t \\ y &= 3 - t \\ z &= 7 + 3t \end{aligned} \qquad \begin{aligned} x &= -3 - 5s \\ y &= 7 + 2s \\ z &= -5 - 6s \end{aligned}$

29. $\dfrac{x}{1} = \dfrac{y}{2} = \dfrac{z}{3}$

$$\frac{x + 1}{-1} = \frac{y}{3} = \frac{z + 2}{2}$$

30. $\dfrac{x - 2}{1} = \dfrac{y + 3}{-2} = \dfrac{z - 1}{4}$

$$\frac{x}{3} = \frac{y - 2}{1} = \frac{z - 3}{1}$$

In Exercises 31–42, sketch the graph of the specified surface.

31. $x + 2y + 3z = 6$ **32.** $y = z^2$

33. $x^2 + z^2 = 4$

34. $x^2 + y^2 + z^2 - 2x + 4y - 6z + 5 = 0$

35. $16x^2 + 16y^2 - 9z^2 = 0$

36. $\dfrac{x^2}{16} + \dfrac{y^2}{9} + z^2 = 1$

37. $y = \dfrac{1}{2}z$ **38.** $\dfrac{x^2}{16} + \dfrac{y^2}{9} - z^2 = 1$

39. $\dfrac{x^2}{16} - \dfrac{y^2}{9} + z^2 = -1$ **40.** $\dfrac{x^2}{25} + \dfrac{y^2}{4} - \dfrac{z^2}{100} = 1$

41. $y^2 - 4x^2 = z$ **42.** $y = \cos z$

In Exercises 43–46, sketch the space curve represented by the vector function.

43. $\mathbf{r}(t) = \mathbf{i} + t\mathbf{j} + t^2\mathbf{k}$

44. $\mathbf{r}(t) = 2t\mathbf{i} + t\mathbf{j} + t^2\mathbf{k}$

45. $\mathbf{r}(t) = 2 \cos t \, \mathbf{i} + t\mathbf{j} + 2 \sin t \, \mathbf{k}$

46. $\mathbf{r}(t) = \mathbf{i} + \sin t \, \mathbf{j} + \mathbf{k}$

In Exercises 47 and 48, sketch the space curve represented by the intersection of the given surfaces. Find a vector-valued function for the space curve using the indicated parameter.

47. $z = x^2 + y^2$, $x + y = 0$, $t = x$

48. $x^2 + z^2 = 4$, $x - y = 0$, $t = x$

In Exercises 49 and 50, find the indicated limit.

49. $\displaystyle\lim_{t \to 2} (t^2\mathbf{i} + \sqrt{4 - t^2}\,\mathbf{j} + \mathbf{k})$

50. $\displaystyle\lim_{t \to 0} \left(\frac{\sin 2t}{t}\mathbf{i} + e^{-t}\mathbf{j} + e^t\mathbf{k} \right)$

In Exercises 51–54, (a) find the domain of **r** and (b) determine the values of t for which the function is discontinuous.

51. $\mathbf{r}(t) = t\mathbf{i} + \csc t\,\mathbf{k}$

52. $\mathbf{r}(t) = \sqrt{t}\,\mathbf{i} + \dfrac{1}{t-4}\mathbf{j} + \mathbf{k}$

53. $\mathbf{r}(t) = \ln t\,\mathbf{i} + t\mathbf{j} + t\mathbf{k}$

54. $\mathbf{r}(t) = (2t+1)\mathbf{i} + t^2\mathbf{j} + t\mathbf{k}$

In Exercises 55 and 56, find the following.
(a) $\mathbf{r}'(t)$ (b) $\mathbf{r}''(t)$
(c) $D_t[\mathbf{r}(t) \cdot \mathbf{u}(t)]$ (d) $D_t[\mathbf{u}(t) - 2\mathbf{r}(t)]$
(e) $D_t[\|\mathbf{r}(t)\|]$, $t > 0$ (f) $D_t[\mathbf{r}(t) \times \mathbf{u}(t)]$

55. $\mathbf{r}(t) = 3t\mathbf{i} + (t-1)\mathbf{j}$

$\mathbf{u}(t) = t\mathbf{i} + t^2\mathbf{j} + \dfrac{2}{3}t^3\mathbf{k}$

56. $\mathbf{r}(t) = \sin t\,\mathbf{i} + \cos t\,\mathbf{j} + t\mathbf{k}$

$\mathbf{u}(t) = \sin t\,\mathbf{i} + \cos t\,\mathbf{j} + \dfrac{1}{t}\mathbf{k}$

In Exercises 57–60, evaluate the indefinite integral.

57. $\displaystyle\int (\cos t\,\mathbf{i} + t\cos t\,\mathbf{j})\,dt$

58. $\displaystyle\int (\ln t\,\mathbf{i} + t\ln t\,\mathbf{j} + \mathbf{k})\,dt$

59. $\displaystyle\int \|\cos t\,\mathbf{i} + \sin t\,\mathbf{j} + t\mathbf{k}\|\,dt$

60. $\displaystyle\int (t\mathbf{j} + t^2\mathbf{k}) \times (\mathbf{i} + t\mathbf{j} + t\mathbf{k})\,dt$

In Exercises 61 and 62, find a set of parametric equations for the line tangent to the space curve at the indicated point.

61. $\mathbf{r}(t) = 2\cos t\,\mathbf{i} + 2\sin t\,\mathbf{j} + t\mathbf{k}$, $t = \dfrac{3\pi}{4}$

62. $\mathbf{r}(t) = t\mathbf{i} + t^2\mathbf{j} + \dfrac{2}{3}t^3\mathbf{k}$, $t = 3$

In Exercises 63–70, find the velocity, speed, and acceleration at time t. Then, find $\mathbf{a} \cdot \mathbf{T}$, $\mathbf{a} \cdot \mathbf{N}$, and the curvature at time t.

63. $\mathbf{r}(t) = (1 + 4t)\mathbf{i} + (2 - 3t)\mathbf{j} + \mathbf{k}$

64. $\mathbf{r}(t) = 5t\mathbf{k}$

65. $\mathbf{r}(t) = 2(t+1)\mathbf{j} + \dfrac{2}{t+1}\mathbf{k}$

66. $\mathbf{r}(t) = t\mathbf{i} + \sqrt{t}\,\mathbf{k}$

67. $\mathbf{r}(t) = e^t\mathbf{i} + e^{-t}\mathbf{j} + t\mathbf{k}$

68. $\mathbf{r}(t) = t\cos t\,\mathbf{i} + t\sin t\,\mathbf{j} - \mathbf{k}$

69. $\mathbf{r}(t) = t\mathbf{i} + t^2\mathbf{j} + \dfrac{1}{2}t^2\mathbf{k}$

70. $\mathbf{r}(t) = (t-1)\mathbf{i} + t\mathbf{j} + \dfrac{1}{t}\mathbf{k}$

In Exercises 71 and 72, find the length of the space curve over the indicated interval.

71. $\mathbf{r}(t) = \dfrac{1}{2}t\mathbf{i} + \sin t\,\mathbf{j} + \cos t\,\mathbf{k}$, $0 \le t \le \pi$

72. $\mathbf{r}(t) = e^t\sin t\,\mathbf{i} + e^t\cos t\,\mathbf{k}$, $0 \le t \le \pi$

Chapter 15 Application

Computer-aided-design (CAD) systems are used by engineers to generate wire-frame layouts of overall body designs. This wire-frame layout shows the initial design for an engineering project.

Computer Graphics

Early applications of computer graphics in engineering and science had to rely on main-frame computers and expensive software. Today, however, advances in computer technology make the use of computer graphics practical for even small businesses.

Computer graphics takes many forms. In engineering, a common form involves *computer aided design* (CAD). In a typical CAD system, an object's dimensions are specified to the computer, after which the computer creates various three-dimensional renderings of the object.

In scientific applications, computer graphics often involves a computer generated graph in two or three dimensions. For example, the computer generated surface shown below represents the relative brightness observed for the *Whirlpool Nebula*. For this particular graph, the observed brightness z is a function of two variables,

$$z = f(x, y)$$

where the point (x, y) is in the domain of the function f.

Computer-generated surface showing the observed brightness for the *Whirlpool Nebula*. Note that the surface reveals two distinct galaxies in the nebula. [Photograph: Los Alamos National Laboratory.]

Chapter Overview

This chapter begins our study of "multivariable calculus." In Section 15.1, we introduce the concept of a function of several variables. For instance, the function given by

$$f(x, y) = 3xy - y^2 + 4$$

has two (independent) variables, and the function given by

$$f(x, y, z) = \sin xy + \cos xz$$

has three (independent) variables.

In this chapter, we extend many of the ideas of calculus of a single variable to calculus of several variables. For instance, in Section 15.1, we discuss the graph of a function of two variables. Then, in Section 15.2, we discuss limits and continuity in the context of functions of two or three variables.

In Section 15.3–15.6, we discuss several ideas related to derivatives of functions of several variables. These include partial derivatives, higher-order partial derivatives, differentials, Chain Rules, and gradients.

In Sections 15.7–15.9, we look at several applications involving partial derivatives and gradients. For instance, in Section 15.7, we show how to find the tangent plane and normal line at a point on a surface. In Section 15.8, we discuss extrema of functions of two variables, and in the next section we look at several applications involving extrema of functions of two variables.

The last section in the chapter shows how to use Lagrange Multipliers to solve optimization problems involving constraints.

Functions of Several Variables

15.1 Introduction to Functions of Several Variables

Functions of several variables ▪ Surfaces ▪ Level curves ▪ Level surfaces ▪ Computer graphics

Although we have dealt so far only with functions of a single (independent) variable, many familiar quantities are functions of two or more variables. For instance, the work done by a force ($W = FD$) and the volume of a right circular cylinder ($V = \pi r^2 h$) are both functions of two variables. The volume of a rectangular solid ($V = lwh$) is a function of three variables. The notation for a function of two or more variables is similar to that for a function of a single variable. Here are two examples:

$$z = f(x, y) = x^2 + xy \qquad \text{and} \qquad w = f(x, y, z) = x + 2y - 3z.$$

$$\underbrace{}_{\text{2 variables}} \qquad\qquad \underbrace{}_{\text{3 variables}}$$

DEFINITION OF A FUNCTION OF TWO VARIABLES	Let D be a set of ordered pairs of real numbers. If to each ordered pair (x, y) in D there corresponds a real number $f(x, y)$, then f is called a **function of x and y.** The set D is the **domain** of f, and the corresponding set of values for $f(x, y)$ is the **range** of f.

For the function given by $z = f(x, y)$, we call x and y the **independent variables** and z the **dependent variable.**

Similar definitions can be given for functions of three, four, or n variables, where the domains consist of ordered triples (x_1, x_2, x_3), quadruples (x_1, x_2, x_3, x_4), and n-tuples $(x_1, x_2, \ldots, x_n)$, respectively. In all cases, the range is a set of real numbers. We will limit our discussions in this chapter to functions of two and three variables.

As with functions of one variable, we usually use *equations* to describe functions of several variables, and unless otherwise restricted, we assume the domain to be the set of all points for which the equation is defined. For instance, the domain of the function given by

$$f(x, y) = x^2 + y^2$$

is assumed to be the entire *xy*-plane. In Example 1, we look at two functions whose domains are restricted to regions in the *xy*-plane or in space.

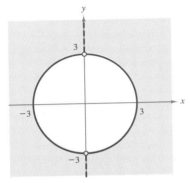

Domain of
$$f(x, y) = \frac{\sqrt{x^2 + y^2 - 9}}{x}$$

FIGURE 15.1

EXAMPLE 1   Finding the domain of functions of several variables

Find the domains for the following functions.

(a) $f(x, y) = \dfrac{\sqrt{x^2 + y^2 - 9}}{x}$ (b) $g(x, y, z) = \dfrac{x}{\sqrt{9 - x^2 - y^2 - z^2}}$

SOLUTION

(a) The function f is defined for all points (x, y) such that $x \neq 0$ and $x^2 + y^2 \geq 9$. Thus, the domain is the set of all points lying on or outside the circle $x^2 + y^2 = 9$, *except* those on the y-axis, as shown in Figure 15.1.

(b) The function g is defined for all points (x, y, z) such that $x^2 + y^2 + z^2 < 9$. Consequently, the domain is the set of all points (x, y, z) lying inside a sphere of radius 3 that is centered at the origin.   ▭

Functions of several variables can be combined in the same ways as functions of a single variable. For two variables, we have the following.

$$(f \pm g)(x, y) = f(x, y) \pm g(x, y) \qquad \text{Sum or difference}$$
$$(fg)(x, y) = f(x, y)g(x, y) \qquad \text{Product}$$
$$\frac{f}{g}(x, y) = \frac{f(x, y)}{g(x, y)}, \quad g(x, y) \neq 0 \qquad \text{Quotient}$$

The **composite** function given by $(g \circ h)(x, y)$ is defined only if h is a function of x and y and g is a function of a single variable. Then

$$(g \circ h)(x, y) = g(h(x, y)) \qquad \text{Composition}$$

for all (x, y) in the domain of h such that $h(x, y)$ is in the domain of g. For example, the function given by

$$f(x, y) = \sqrt{16 - 4x^2 - y^2}$$

can be viewed as the composite of the function of two variables given by $h(x, y) = 16 - 4x^2 - y^2$ and the function of a single variable given by $g(u) = \sqrt{u}$.

A function that can be expressed as a sum of functions of the form $cx^m y^n$ (where c is a real number and m and n are nonnegative integers) is called a **polynomial function** of two variables. For instance, the functions given by

$$f(x, y) = x^2 + y^2 - 2xy + x + 2 \qquad \text{and} \qquad g(x, y) = 3xy^2 + x - 2$$

are polynomial functions of two variables. A **rational function** is the quotient of two polynomial functions. Similar terminology is used for functions of more than two variables.

The graph of a function of two variables

As with functions of a single variable, we can learn a lot about the behavior of a function of two variables by sketching its graph. The **graph** of a function f of two variables is the set of all points (x, y, z) for which $z = f(x, y)$ and

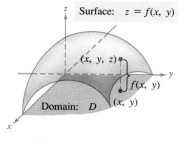

FIGURE 15.2

(x, y) is in the domain of f. This graph can be interpreted geometrically as a *surface in space*, and we already have done much of the preliminary work necessary for sketching such surfaces in Sections 14.3 and 14.4. In Figure 15.2, note that the graph of $z = f(x, y)$ is a surface whose projection onto the xy-plane is D, the domain of f. Consequently, to each point (x, y) in D there corresponds a point (x, y, z) on the surface, and, conversely, to each point (x, y, z) on the surface there corresponds a point (x, y) in D.

EXAMPLE 2   Sketching the graph of a function of two variables

Sketch the graph of $f(x, y) = \sqrt{16 - 4x^2 - y^2}$. What is the range of f?

SOLUTION

The domain D implied by the equation for f is the set of all points (x, y) such that $16 - 4x^2 - y^2 \geq 0$. Thus, D is the set of all points lying on or inside the ellipse given by

$$\frac{x^2}{4} + \frac{y^2}{16} = 1.$$

The range of f is all values $z = f(x, y)$ such that $0 \leq z \leq \sqrt{16} = 4$. A point (x, y, z) is on the graph of f if and only if

$$z = \sqrt{16 - 4x^2 - y^2}$$
$$z^2 = 16 - 4x^2 - y^2$$
$$4x^2 + y^2 + z^2 = 16$$
$$\frac{x^2}{4} + \frac{y^2}{16} + \frac{z^2}{16} = 1, \quad 0 \leq z \leq 4.$$

From our work in Section 14.4, we know that the graph of f is the upper half of an ellipsoid, as shown in Figure 15.3. ▭

Surface: $z = \sqrt{16 - 4x^2 - y^2}$

Trace in plane $z = 2$

Range

Domain

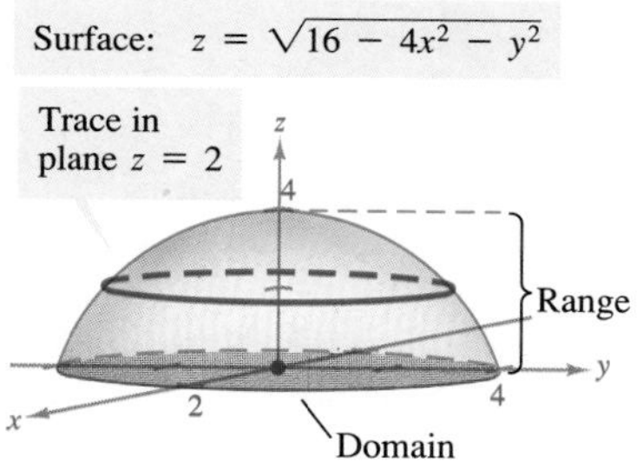

Graph of $f(x, y) = \sqrt{16 - 4x^2 - y^2}$

FIGURE 15.3

REMARK   To sketch a surface in space, it is helpful to use traces in planes parallel to the coordinate planes, as shown in Figure 15.3. For example, to find the trace of the surface in the plane $z = 2$, we substitute $z = 2$ in the equation $z = \sqrt{16 - 4x^2 - y^2}$ and obtain

$$2 = \sqrt{16 - 4x^2 - y^2} \quad \Longrightarrow \quad \frac{x^2}{3} + \frac{y^2}{12} = 1.$$

Thus, the trace is an ellipse centered at the point $(0, 0, 2)$ with major and minor axes of lengths $4\sqrt{3}$ and $2\sqrt{3}$, respectively.

Level curves

A second way to visualize a function of two variables is as a **scalar field** in which the scalar $z = f(x, y)$ is assigned to the point (x, y). A scalar field can be characterized by **level curves** (or **contour lines**) along which the value of $f(x, y)$ is constant. For instance, the weather map in Figure 15.4 shows level curves of equal pressure called **isobars.** In weather maps for which the level curves represent points of equal temperature, the level curves are called

isotherms, as shown in Figure 15.5. Another common use of level curves is in representing electric potential fields. In this type of map the level curves are called **equipotential lines.**

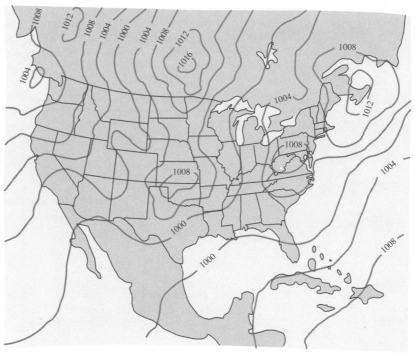

FIGURE 15.4   Level curves show lines of equal pressure (isobars) measured in millibars.

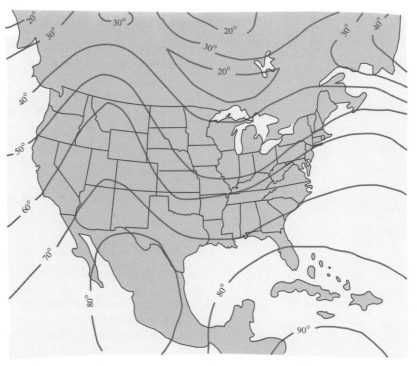

FIGURE 15.5   Level curves show lines of equal temperature (isotherms) measured in degrees Fahrenheit.

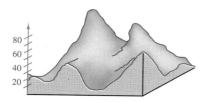

FIGURE 15.6

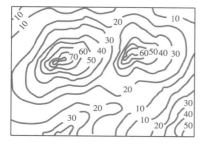

Topographic Map of Two Peaks. One is over 70 units high and the other is over 50 units high.

FIGURE 15.7

Contour maps are commonly used to show regions on the earth's surface, with the level curves representing the height above sea level. This type of map is called a **topographic map.** For example, the two mountain peaks in Figure 15.6 are shown on the topographic map in Figure 15.7.

A contour map depicts the variation of z with respect to x and y by the spacing between level curves. Much space between level curves indicates that z is changing slowly, whereas little space indicates a rapid change in z. Furthermore, to give a good three-dimensional illusion in a contour map it is important to choose c-values that are *evenly spaced.*

EXAMPLE 3 Sketching a contour map

The hemisphere given by $f(x, y) = \sqrt{64 - x^2 - y^2}$ is shown in Figure 15.8. Sketch a contour map for this surface using level curves corresponding to $c = 0, 1, 2, \ldots, 8$.

SOLUTION

For each value of c the equation given by $f(x, y) = c$ is a circle (or point) in the xy-plane. For example, when $c_1 = 0$ the level curve is

$$x^2 + y^2 = 64 \qquad \text{Circle of radius 8}$$

which is a circle of radius 8. Figure 15.9 shows the nine level curves for the hemisphere.

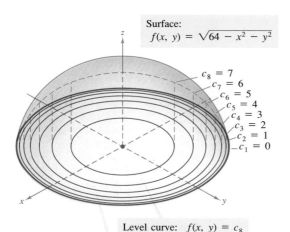

FIGURE 15.8

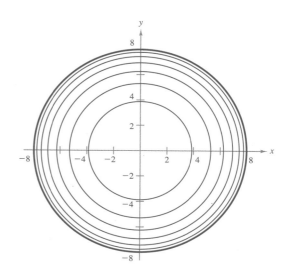

FIGURE 15.9

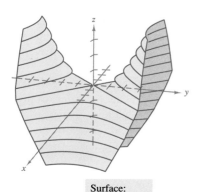

Surface:
$z = y^2 - x^2$

FIGURE 15.10

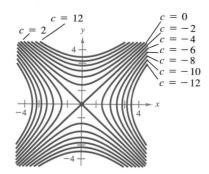

Hyperbolic Level Curves
(at increments of 2)

FIGURE 15.11

EXAMPLE 4   Sketching a contour map

The hyperbolic paraboloid given by

$$z = y^2 - x^2$$

is shown in Figure 15.10. Sketch a contour map for this surface.

SOLUTION

For each value of c, we let $f(x, y) = c$ and sketch the resulting level curve in the xy-plane. For this function, each of the level curves ($c \neq 0$) is a hyperbola whose asymptotes are the lines $y = \pm x$. If $c < 0$, the transverse axis is horizontal. For instance, the level curve for $c = -4$ is given by

$$\frac{x^2}{2^2} - \frac{y^2}{2^2} = 1.$$ Hyperbola with horizontal transverse axis

If $c > 0$, the transverse axis is vertical. For instance, the level curve for $c = 4$ is given by

$$\frac{y^2}{2^2} - \frac{x^2}{2^2} = 1.$$ Hyperbola with vertical transverse axis

If $c = 0$, the level curve is the degenerate conic representing the intersecting asymptotes, as shown in Figure 15.11.   ▭

One example of a function of two variables used in economics is the **Cobb-Douglas production function.** This function is used as a model to represent the number of units produced by varying amounts of labor and capital. If x measures the units of labor and y measures the units of capital, then the number of units produced is given by

$$f(x, y) = Cx^a y^{1-a}$$

where C is constant and $0 < a < 1$. We illustrate a property of this function in the next example.

EXAMPLE 5   An application from economics

Suppose that a manufacturer estimates a particular production function to be $f(x, y) = 100x^{0.6}y^{0.4}$, where x is the number of units of labor and y is the number of units of capital. Compare the production level when $x = 1000$ and $y = 500$ to the production level when $x = 2000$ and $y = 1000$.

SOLUTION

When $x = 1000$ and $y = 500$, the production level is

$$f(1000, 500) = 100(1000^{0.6})(500^{0.4}) \approx 100(63.10)(12.01) \approx 75{,}786.$$

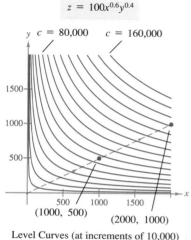

$$z = 100x^{0.6}y^{0.4}$$

Level Curves (at increments of 10,000)

FIGURE 15.12

When $x = 2000$ and $y = 1000$, the production level is

$$f(2000, 1000) = 100(2000^{0.6})(1000^{0.4}) \approx 100(95.64)(15.85) \approx 151,572.$$

The level curves of the surface $z = f(x, y)$ are shown in Figure 15.12. Note that by doubling *both* x and y we doubled the production level. In Exercise 58 you are asked to show that this is characteristic of the Cobb-Douglas production function. ▭

Level surfaces for a function of three variables

The concept of a level curve can be extended by one dimension to define a **level surface.** If f is a function of three variables and c is a constant, then the graph of the equation $f(x, y, z) = c$ is called a **level surface** of the function f, as shown in Figure 15.13. At every point on a given level surface, the function f has a constant value, $f(x, y, z) = c$. For example, if the function $f(x, y, z)$ represents the temperature at (x, y, z) in a region of space, then the level surfaces of equal temperature are called **isothermal surfaces,** as shown in Figure 15.14.

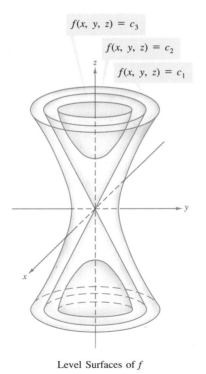

$f(x, y, z) = c_3$

$f(x, y, z) = c_2$

$f(x, y, z) = c_1$

Level Surfaces of f

FIGURE 15.13

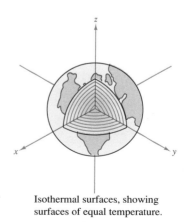

Isothermal surfaces, showing surfaces of equal temperature.

FIGURE 15.14

Mary Fairfax Somerville

One person to write about the problem of creating geometrical models for functions of several variables was the English mathematician Mary Fairfax Somerville (1780–1872). Somerville's most well-known book, *The Mechanics of the Heavens*, was published in 1831.

EXAMPLE 6 Level surfaces

Describe the level surfaces of the function

$$f(x, y, z) = 4x^2 + y^2 + z^2.$$

SOLUTION

Each level surface has an equation of the form $4x^2 + y^2 + z^2 = c$. Therefore, the level surfaces are ellipsoids (whose cross sections parallel to the yz-plane are circles). As c increases, the radii of the circular cross sections increase according to the square root of c. For example, the level surfaces corresponding to the values $c = 4$ and $c = 16$ are as follows.

$$\frac{x^2}{1} + \frac{y^2}{4} + \frac{z^2}{4} = 1 \qquad \text{Level surface for } c = 4$$

$$\frac{x^2}{4} + \frac{y^2}{16} + \frac{z^2}{16} = 1 \qquad \text{Level surface for } c = 16$$

Two of these level surfaces are shown in Figure 15.15.

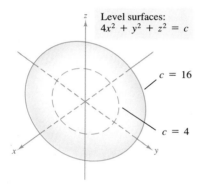

FIGURE 15.15

Computer graphics

The problem of sketching the graph of a surface in space can be simplified by using a computer. Although there are several types of computer graphics plotting routines, most use some form of trace analysis to give the illusion of three dimensions. To use such a plotting routine, you need to enter the equation for the surface, the region in the xy-plane over which the surface is to be plotted, the scale for the z-axis, and the number of traces to be taken. For instance, to sketch the surface given by

$$f(x, y) = (x^2 + y^2)e^{1-x^2-y^2}$$

we might choose the following bounds for x, y, and z.

$$-3 \le x \le 3 \qquad \text{Bounds for } x$$
$$-3 \le y \le 3 \qquad \text{Bounds for } y$$
$$0 \le z \le 1 \qquad \text{Bounds for } z$$

Figure 15.16 shows a computer-generated sketch of this surface using 26 traces taken parallel to the yz-plane. To heighten the three-dimensional effect, the program uses a "hidden line" routine. That is, it begins by plotting the traces in the foreground (those corresponding to the largest x-values), and then, as each new trace is plotted, the program determines whether all or only part of the next trace should be shown.

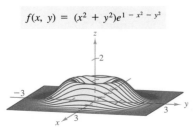

FIGURE 15.16

In Table 15.1 we show a variety of surfaces that were plotted by a computer.

TABLE 15.1

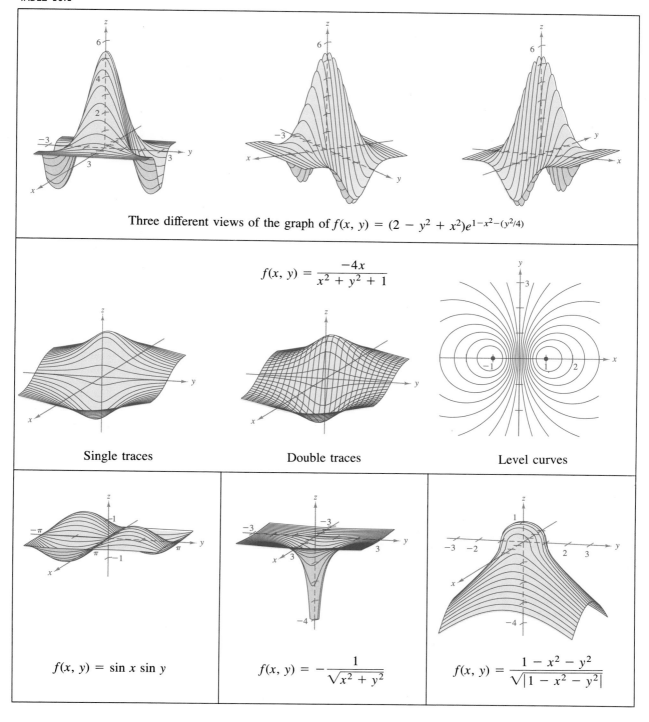

Three different views of the graph of $f(x, y) = (2 - y^2 + x^2)e^{1-x^2-(y^2/4)}$

$$f(x, y) = \frac{-4x}{x^2 + y^2 + 1}$$

Single traces   Double traces   Level curves

$f(x, y) = \sin x \sin y$ $f(x, y) = -\dfrac{1}{\sqrt{x^2 + y^2}}$ $f(x, y) = \dfrac{1 - x^2 - y^2}{\sqrt{|1 - x^2 - y^2|}}$

EXERCISES for Section 15.1

In Exercises 1–10, evaluate the given function at the indicated point.

1. $f(x, y) = \dfrac{x}{y}$

   (a) $(3, 2)$ (b) $(-1, 4)$
   (c) $(30, 5)$ (d) $(5, y)$
   (e) $(x, 2)$ (f) $(5, t)$

2. $f(x, y) = 4 - x^2 - 4y^2$

   (a) $(0, 0)$ (b) $(0, 1)$
   (c) $(2, 3)$ (d) $(1, y)$
   (e) $(x, 0)$ (f) $(t, 1)$

3. $f(x, y) = xe^y$

   (a) $(5, 0)$ (b) $(3, 2)$
   (c) $(2, -1)$ (d) $(5, y)$
   (e) $(x, 2)$ (f) (t, t)

4. $g(x, y) = \ln |x + y|$

   (a) $(2, 3)$ (b) $(5, 6)$
   (c) $(e, 0)$ (d) $(0, 1)$
   (e) $(2, -3)$ (f) (e, e)

5. $h(x, y, z) = \dfrac{xy}{z}$

   (a) $(2, 3, 9)$ (b) $(1, 0, 1)$

6. $f(x, y, z) = \sqrt{x + y + z}$

   (a) $(0, 5, 4)$ (b) $(6, 8, -3)$

7. $f(x, y) = x \sin y$

   (a) $\left(2, \dfrac{\pi}{4}\right)$ (b) $(3, 1)$

8. $V(r, h) = \pi r^2 h$

   (a) $(3, 10)$ (b) $(5, 2)$

9. $f(x, y) = \displaystyle\int_x^y (2t - 3)\, dt$

   (a) $(0, 4)$ (b) $(1, 4)$

10. $g(x, y) = \displaystyle\int_x^y \dfrac{1}{t}\, dt$

   (a) $(4, 1)$ (b) $(6, 3)$

In Exercises 11–22, describe the region R in the xy-coordinate plane that corresponds to the domain of the given function, and find the range of the function.

11. $f(x, y) = \sqrt{4 - x^2 - y^2}$
12. $f(x, y) = \sqrt{4 - x^2 - 4y^2}$
13. $f(x, y) = \arcsin (x + y)$
14. $f(x, y) = \arccos (y/x)$
15. $z = \dfrac{x + y}{xy}$ **16.** $z = \dfrac{xy}{x - y}$
17. $f(x, y) = \ln (4 - x - y)$
18. $f(x, y) = \ln (4 - xy)$
19. $f(x, y) = e^{x/y}$ **20.** $f(x, y) = x^2 + y^2$
21. $g(x, y) = \dfrac{1}{xy}$ **22.** $g(x, y) = x\sqrt{y}$

In Exercises 23–28, match the graph of the given surface with one of the contour maps. [The contour maps are labeled (a)–(f).]

23. $f(x, y) = e^{1 - x^2 - y^2}$

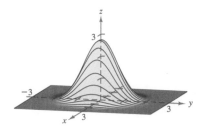

24. $f(x, y) = e^{1 - x^2 + y^2}$

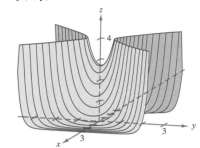

25. $f(x, y) = |y|^{1 + |x|}$

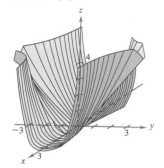

26. $f(x, y) = -\dfrac{x}{y^2 - x}$

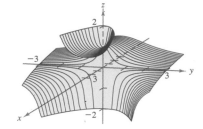

27. $f(x, y) = \ln |y - x^2|$

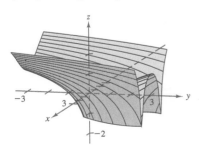

28. $f(x, y) = \cos\left(\dfrac{x^2 + 2y^2}{4}\right)$

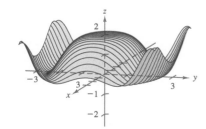

(a)

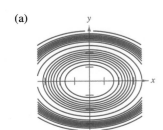

(b)

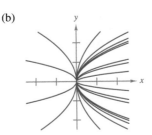

(c)

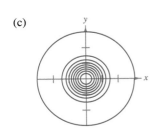

(d)

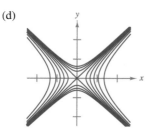

(e)

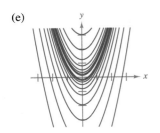

(f)

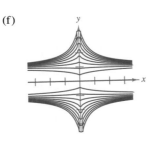

In Exercises 29–36, sketch the graph of the surface specified by the function.

29. $z = 4 - x^2 - y^2$

30. $z = \sqrt{x^2 + y^2}$

31. $f(x, y) = y^2$

32. $z = y^2 - x^2 + 1$

33. $f(x, y) = 6 - 2x - 3y$

34. $f(x, y) = \dfrac{1}{12}\sqrt{144 - 16x^2 - 9y^2}$

35. $f(x, y) = e^{-x}$

36. $f(x, y) = \begin{cases} xy, & x \geq 0, \ y \geq 0 \\ 0, & x < 0 \text{ or } y < 0 \end{cases}$

37. (a) Sketch the graph of the surface given by $f(x, y) = x^2 + y^2$.
   (b) On the surface of part (a), sketch the graphs of $z = f(1, y)$ and $z = f(x, 1)$.

38. (a) Sketch the graph of the surface given by $f(x, y) = xy$ in the first octant.
   (b) On the surface of part (a), sketch the graph of $z = f(x, x)$.

In Exercises 39–48, describe the level curves for each function. Sketch the level curves for the given c-values.

39. $f(x, y) = \sqrt{25 - x^2 - y^2}$ $c = 0, 1, 2, 3, 4, 5$

40. $f(x, y) = x^2 + y^2$ $c = 0, 2, 4, 6, 8$

41. $f(x, y) = xy$ $c = \pm 1, \pm 2, \ldots, \pm 6$

42. $f(x, y) = 6 - 2x - 3y$ $c = 0, 2, 4, 6, 8, 10$

43. $f(x, y) = \dfrac{x}{x^2 + y^2}$ $c = \pm\dfrac{1}{2}, \pm 1, \pm\dfrac{3}{2}, \pm 2$

44. $f(x, y) = \arctan\dfrac{y}{x}$ $c = 0, \pm\dfrac{\pi}{6}, \pm\dfrac{\pi}{3}, \pm\dfrac{5\pi}{6}$

45. $f(x, y) = \ln(x - y)$ $c = 0, \pm\dfrac{1}{2}, \pm 1, \pm\dfrac{3}{2}, \pm 2$

46. $f(x, y) = \dfrac{x + y}{x - y}$ $c = 0, \pm 1, \pm 2, \pm 3$

47. $f(x, y) = e^{xy}$ $c = 1, 2, 3, 4, \dfrac{1}{2}, \dfrac{1}{3}, \dfrac{1}{4}$

48. $f(x, y) = \cos(x + y)$ $c = 0, \pm\dfrac{1}{2}, \pm 1$

In Exercises 49–54, sketch the level surface $f(x, y, z) = c$ for the given function at the specified value of c.

49. $f(x, y, z) = x - 2y + 3z$ $c = 6$

50. $f(x, y, z) = 4x + y + 2z$ $c = 4$

51. $f(x, y, z) = x^2 + y^2 + z^2$ $c = 9$

52. $f(x, y, z) = x^2 + y^2 - z$ $c = 1$

53. $f(x, y, z) = 4x^2 + 4y^2 - z^2$ $c = 0$

54. $f(x, y, z) = \sin x - z$ $c = 0$

55. The **Doyle Log Rule** is one of several methods used to determine the lumber yield of a log (in board-feet) in terms of its diameter d (in inches) and its length L (in feet). The number of board-feet is given by

$$N(d, L) = \left(\dfrac{d - 4}{4}\right)^2 L.$$

(a) Find the number of board-feet of lumber in a log 22 inches in diameter and 12 feet in length.

(b) Find $N(30, 12)$.

56. A principal of $1000 is deposited in a savings account that earns an interest rate of r (expressed as a decimal), compounded continuously. The amount $A(r, t)$ after t years is given by

$$A(r, t) = 1000e^{rt}.$$

Use this function of two variables to complete the following table.

Rate	Number of years			
	5	10	15	20
0.08				
0.10				
0.12				
0.14				

57. *Queuing model* The average length of time that a customer waits in line for service is given by

$$W(x, y) = \frac{1}{x - y}, \quad y < x$$

where y is the average arrival rate, expressed as the number of customers per unit of time, and x is the average service rate, expressed in the same units. Evaluate W at the following points.

(a) $(15, 10)$ (b) $(12, 9)$

(c) $(12, 6)$ (d) $(4, 2)$

58. Use the Cobb-Douglas production function (see Example 5) to show that if the number of units of labor and the number of units of capital are both doubled, then the production level is also doubled.

59. The temperature T (in degrees Celsius) at any point (x, y) in a circular steel plate of radius 10 feet is

$$T = 600 - 0.75x^2 - 0.75y^2$$

where x and y are measured in feet. Sketch some of the isothermal curves.

60. According to the Ideal Gas Law, $PV = kT$, where P is pressure, V is volume, T is temperature (in degrees Kelvin), and k is a constant of proportionality. A tank contains 2600 cubic inches of nitrogen at a pressure of 20 pounds per square inch and a temperature of $300° K$.

(a) Determine k.

(b) Express P as a function of V and T and describe the level curves.

61. A rectangular box with an open top has a length of x feet, width of y feet, and height of z feet. Express the cost C of constructing the box as a function of x, y, and z if it costs $0.75 per square foot to build the base and $0.40 per square foot to build the sides.

62. A propane tank is constructed by welding hemispheres to the ends of a right circular cylinder. Write the volume V of the tank as a function of r and l, where r is the radius of the cylinder and hemispheres and l is the length of the cylinder.

63. Meteorologists measure the atmospheric pressure in units called millibars. From these observations they create weather maps on which the curves of equal atmospheric pressure are called isobars (see figure). State which of A, B, and C is the location of (a) highest pressure, (b) lowest pressure, and (c) highest wind velocity if the closer the isobars the higher the wind speed.

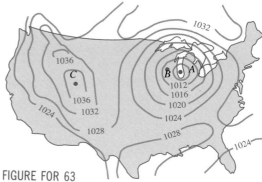

FIGURE FOR 63

64. The acidity of rainwater is measured in units called pH. A pH of 7 is neutral, smaller values are increasingly acidic, and larger values are increasingly alkaline. The accompanying map shows the curves of equal pH and gives evidence that downwind of heavily industrialized areas the acidity has been increasing. Using the level curves on the map, determine the direction of the prevailing winds in the northeastern United States.

FIGURE FOR 64

In Exercises 65 and 66, use a computer or graphics calculator to sketch several level curves of the function.

65. $f(x, y) = 4x^2 + y^2$ **66.** $f(x, y) = \dfrac{y^2}{7} - \dfrac{x^2}{5}$

15.2 Limits and Continuity

Neighborhoods in the plane ▪ Limit of a function of two variables ▪ Continuity of a function of two variables ▪ Continuity of a function of three variables

Sonya Kovalevsky

In this section we discuss limits and continuity as applied to functions of two or three variables. We begin with functions of two variables. Then, at the end of the section, we indicate how the concepts can be extended to cover functions of three variables.

Neighborhoods in the plane

Before discussing limits and continuity for functions of two variables we need to define some preliminary terms. Much of this terminology was introduced by the German mathematician Karl Weierstrass (1815–1897), whose rigorous approach to limits and other topics in calculus gained him the reputation as the "father of modern analysis." Weierstrass was a gifted teacher. One of his best known students was the Russian mathematician Sonya Kovalevsky (1850–1891), who applied many of Weierstrass's techniques to problems in mathematical physics and became one of the first women to gain acceptance as a research mathematician.

We begin our discussion of the limit of a function of two variables by defining a two-dimensional analog to an interval on the real line. Using the formula for the distance $\delta > 0$ between two points (x, y) and (x_0, y_0) in the plane, we define the **δ-neighborhood** about (x_0, y_0) to be the **disc** centered at (x_0, y_0) with radius δ:

$$\{(x, y): \sqrt{(x - x_0)^2 + (y - y_0)^2} < \delta\}$$

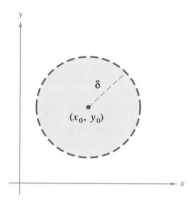

Open Disc

FIGURE 15.17

as shown in Figure 15.17. When this formula contains the *less than* inequality, $<$, the disc is called **open,** and when it contains the *less than or equal to* inequality, $\leq$, the disc is called **closed.** This corresponds to the use of $<$ and $\leq$ to define open and closed intervals.

A point (x_0, y_0) in a plane region R is an **interior point** of R if there exists a δ-neighborhood about (x_0, y_0) that lies entirely in R, as shown in Figure 15.18. If every point in R is an interior point, then we call R an **open region.** A point (x_0, y_0) is a **boundary point** of R if every open disc centered at (x_0, y_0) contains points inside R *and* points outside R. By definition, a region must contain its interior points, but it need not contain its boundary points. If a region contains all its boundary points, then we say that the region is **closed.** A region that contains some but not all its boundary points is neither open nor closed.

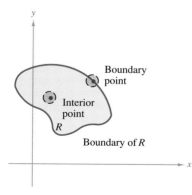

Boundary and Interior
Points of a Region R

Limit of a function of two variables

We are now ready to define the limit of a function of two variables.

FIGURE 15.18

DEFINITION OF THE LIMIT OF A FUNCTION OF TWO VARIABLES	Let f be a function of two variables defined, except possibly at (x_0, y_0), on an open disc centered at (x_0, y_0), and let L be a real number. Then $$\lim_{(x,y) \to (x_0,y_0)} f(x, y) = L$$ if for each $\varepsilon > 0$, there corresponds a $\delta > 0$ such that $$|f(x, y) - L| < \varepsilon \quad \text{whenever} \quad 0 < \sqrt{(x - x_0)^2 + (y - y_0)^2} < \delta.$$

"For any (x, y) in the circle of radius δ, the value $f(x, y)$ lies between $L + \varepsilon$ and $L - \varepsilon$."

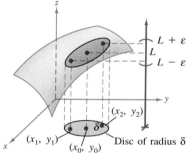

FIGURE 15.19

REMARK Graphically, this definition of a limit implies that for any point (x, y) in the disc of radius δ, the value $f(x, y)$ lies between $L + \varepsilon$ and $L - \varepsilon$, as shown in Figure 15.19.

The definition of the limit of a function of two variables is similar to the definition of the limit of a function of a single variable, yet there is a critical difference. To determine whether a function of a single variable possesses a limit, we need only test the approach from two directions—from the left and from the right. If the function approaches the same limit from the right and from the left, we can conclude that the limit exists. However, for a function of two variables, when we write

$$(x, y) \to (x_0, y_0)$$

we mean that the point (x, y) is allowed to approach (x_0, y_0) from any "direction." If the value of

$$\lim_{(x,y) \to (x_0,y_0)} f(x, y)$$

is not the same for all possible approaches, or **paths,** to (x_0, y_0), then the limit does not exist.

EXAMPLE 1 Verifying a limit by the definition

Show that $\displaystyle\lim_{(x,y) \to (a,b)} x = a.$

SOLUTION

Let $f(x, y) = x$ and $L = a$. We need to show that for each $\varepsilon > 0$, there exists a δ-neighborhood about (a, b) such that

$$|f(x, y) - L| = |x - a| < \varepsilon$$

whenever $(x, y) \neq (a, b)$ lies in the neighborhood. We first observe that from

$$0 < \sqrt{(x - a)^2 + (y - b)^2} < \delta$$

it follows

$$|f(x, y) - a| = |x - a| = \sqrt{(x - a)^2} \leq \sqrt{(x - a)^2 + (y - b)^2} < \delta.$$

Thus, we can choose $\delta = \varepsilon$, and the limit is verified. ▭

Limits of functions of several variables have the same properties regarding sums, differences, products, and quotients as do limits of functions of a single variable. (See Theorem 2.3 in Section 2.1.) We make use of these properties in our next example.

EXAMPLE 2   Verifying a limit

Evaluate the following limits.

(a) $\displaystyle\lim_{(x,y)\to(1,2)} \frac{5x^2y}{x^2 + y^2}$ (b) $\displaystyle\lim_{(x,y)\to(0,0)} \frac{5x^2y}{x^2 + y^2}$

SOLUTION

(a) By using the properties of limits of products and sums, we obtain

$$\lim_{(x,y)\to(1,2)} 5x^2y = 5(1^2)(2) = 10$$

and

$$\lim_{(x,y)\to(1,2)} (x^2 + y^2) = (1^2 + 2^2) = 5.$$

Since the limit of a quotient is equal to the quotient of the limits, we have

$$\lim_{(x,y)\to(1,2)} \frac{5x^2y}{x^2 + y^2} = \frac{10}{5} = 2.$$

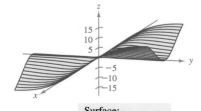

Surface:

$$f(x, y) = \frac{5x^2y}{x^2 + y^2}$$

FIGURE 15.20

(b) In this case, the limits of the numerator and of the denominator are both 0, and so we cannot determine the existence (or nonexistence) of a limit by taking the limits of the numerator and denominator separately and then dividing. However, from the graph of f in Figure 15.20, it seems reasonable that the limit might be zero. Thus, we try applying the definition to $L = 0$. First, we note that

$$|y| \le \sqrt{x^2 + y^2} \quad \text{and} \quad \frac{x^2}{x^2 + y^2} \le 1.$$

Then, in a δ-neighborhood about $(0, 0)$, we have $0 < \sqrt{x^2 + y^2} < \delta$, and it follows that, for $(x, y) \neq (0, 0)$,

$$|f(x, y) - 0| = \left|\frac{5x^2y}{x^2 + y^2}\right| = 5|y| \left(\frac{x^2}{x^2 + y^2}\right)$$
$$\le 5|y| \le 5\sqrt{x^2 + y^2} < 5\delta.$$

Thus, we choose $\delta = \varepsilon/5$ and conclude that

$$\lim_{(x,y)\to(0,0)} \frac{5x^2y}{x^2 + y^2} = 0.$$

The next example describes a limit that does not exist because the function approaches different values along different paths.

EXAMPLE 3   A limit that does not exist

Show that the following limit does not exist.

$$\lim_{(x,y) \to (0,0)} \left(\frac{x^2 - y^2}{x^2 + y^2} \right)^2$$

SOLUTION

The domain of the function given by

$$f(x, y) = \left(\frac{x^2 - y^2}{x^2 + y^2} \right)^2$$

consists of all points in the xy-plane except for the point $(0, 0)$. To show that the limit as (x, y) approaches $(0, 0)$ does not exist, we consider approaching along two different paths, as shown in Figure 15.21. Along the x-axis, every point is of the form $(x, 0)$, and the limit along this approach is

$$\lim_{(x,0) \to (0,0)} \left(\frac{x^2 - 0^2}{x^2 + 0^2} \right)^2 = \lim_{(x,0) \to (0,0)} (1)^2 = 1.$$

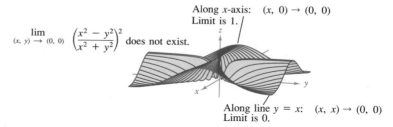

$$\lim_{(x, y) \to (0, 0)} \left(\frac{x^2 - y^2}{x^2 + y^2} \right)^2 \text{ does not exist.}$$

Along x-axis: $(x, 0) \to (0, 0)$
Limit is 1.

Along line $y = x$: $(x, x) \to (0, 0)$
Limit is 0.

FIGURE 15.21

However, if (x, y) approaches $(0, 0)$ along the line $y = x$, we obtain

$$\lim_{(x,x) \to (0,0)} \left(\frac{x^2 - x^2}{x^2 + x^2} \right)^2 = \lim_{(x,x) \to (0,0)} \left(\frac{0}{2x^2} \right)^2 = 0.$$

This means that in any open disc centered at $(0, 0)$ there are points (x, y) at which f takes on the value 1, and other points at which f takes on the value 0. For instance, $f(x, y) = 1$ at the points

$$(1, 0), \quad (0.1, 0), \quad (0.01, 0), \quad \text{and} \quad (0.001, 0)$$

and $f(x, y) = 0$ at the points

$$(1, 1), \quad (0.1, 0.1), \quad (0.01, 0.01), \quad \text{and} \quad (0.001, 0.001).$$

Hence, f does not have a limit as $(x, y) \to (0, 0)$.

In Example 3 we were able to conclude that the limit did not exist because we found two approaches that yielded different limits. Be sure you see that if the two approaches had yielded the same limit, we still could not conclude that the limit exists. To form such a conclusion, we must show that the limit is the same along *all* possible approaches.

Continuity of a function of two variables

Notice in Example 2 part (a) that the limit of $f(x, y) = 5x^2y/(x^2 + y^2)$ as $(x, y) \rightarrow (1, 2)$ can be evaluated by direct substitution. That is, the limit is $f(1, 2) = 2$. In such cases, we say that f is **continuous** at the point $(1, 2)$. We define continuity of a function of two variables as follows.

DEFINITION OF CONTINUITY OF A FUNCTION OF TWO VARIABLES

A function f of two variables is **continuous at a point** (x_0, y_0) in an open region R if $f(x_0, y_0)$ is defined and is equal to the limit of $f(x, y)$ as (x, y) approaches (x_0, y_0). That is,

$$\lim_{(x,y) \rightarrow (x_0,y_0)} f(x, y) = f(x_0, y_0).$$

The function f is **continuous in the open region** R if it is continuous at every point in R.

REMARK   This definition of continuity can be extended to *boundary points* of the open region R by considering a special type of limit in which (x, y) is allowed to approach (x_0, y_0) along paths lying in the region R. This notion is similar to that of one-sided limits, as discussed in Chapter 2.

In Example 2(b), the function $f(x, y) = 5x^2y/(x^2 + y^2)$ is not continuous at $(0, 0)$. However, since the limit at this point exists, we can remove the discontinuity of defining f at $(0, 0)$ to be equal to its limit there. Such a discontinuity is called **removable.** In Example 3, the function $f(x, y) = [(x^2 - y^2)/(x^2 + y^2)]^2$ is also discontinuous at $(0, 0)$, but this discontinuity is **nonremovable.**

THEOREM 15.1 PROPERTIES OF CONTINUOUS FUNCTIONS OF TWO VARIABLES

If k is a real number and f and g are continuous at (x_0, y_0), then the following functions are continuous at (x_0, y_0).

1. Scalar multiple: kf 2. Sum and difference: $f \pm g$
3. Product: fg 4. Quotient: f/g, if $g(x_0, y_0) \neq 0$

Theorem 15.1 establishes the continuity of *polynomial* and *rational* functions at every point in their domains. Furthermore, the continuity of other types of functions can be extended naturally from one to two variables, as indicated in the following example.

EXAMPLE 4 Functions that are continuous at every point in the plane

The functions given by

$$f(x, y) = \sin xy \qquad \text{and} \qquad g(x, y) = (\cos y^2)e^{-\sqrt{x^2+y^2}}$$

are both continuous at every point in the plane, as indicated in Figures 15.22 and 15.23.

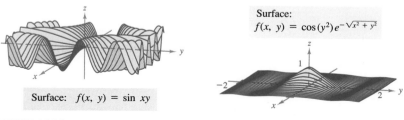

Surface: $f(x, y) = \cos(y^2)e^{-\sqrt{x^2 + y^2}}$

Surface: $f(x, y) = \sin xy$

FIGURE 15.22

FIGURE 15.23

THEOREM 15.2
CONTINUITY OF A COMPOSITE FUNCTION

If h is continuous at (x_0, y_0) and g is continuous at $h(x_0, y_0)$, then the composite function given by $(g \circ h)(x, y) = g(h(x, y))$ is continuous at (x_0, y_0). That is,

$$\lim_{(x,y)\to(x_0,y_0)} g(h(x, y)) = g(h(x_0, y_0)).$$

REMARK   Note in Theorem 15.2 that h is a function of two variables and g is a function of one variable.

EXAMPLE 5 Testing for continuity

Discuss the continuity of the following functions.

(a) $f(x, y) = \dfrac{x - 2y}{x^2 + y^2}$ (b) $g(x, y) = \dfrac{2}{y - x^2}$

SOLUTION

(a) Since a rational function is continuous at every point in its domain, we conclude that f is continuous at each point in the xy-plane except at $(0, 0)$, as shown in Figure 15.24.

(b) The function given by $g(x, y) = 2/(y - x^2)$ is continuous except at the points at which the denominator is zero, $y - x^2 = 0$. Thus, we conclude that the function is continuous at all points except those lying on the parabola $y = x^2$. Inside this parabola, we have $y > x^2$, and the surface represented by the function lies above the xy-plane, as shown in Figure 15.25. Outside the parabola, $y < x^2$, and the surface lies below the xy-plane.

$$g(x, y) = \frac{2}{y - x^2}$$

Discontinuous on
parabola $y = x^2$

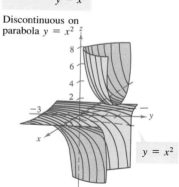

$y = x^2$

FIGURE 15.25

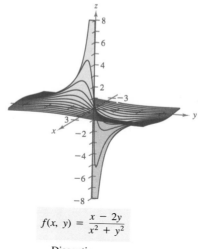

$$f(x, y) = \frac{x - 2y}{x^2 + y^2}$$

Discontinuous
at $(0, 0)$

FIGURE 15.24

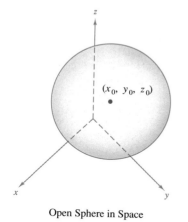

(x_0, y_0, z_0)

Open Sphere in Space

FIGURE 15.26

Continuity of a function of three variables

The preceding definitions of limits and continuity can be extended to functions of three variables by considering points (x, y, z) within the *open sphere*

$$(x - x_0)^2 + (y - y_0)^2 + (z - z_0)^2 = \delta^2.$$

The radius of this sphere is δ and the sphere is centered at (x_0, y_0, z_0), as shown in Figure 15.26. At point (x_0, y_0, z_0) in a region R in space is called an **interior point** of R if there exists a δ-sphere about (x_0, y_0, z_0) that lies entirely in R. If every point in R is an interior point, then we say that R is **open**.

DEFINITION OF CONTINUITY OF A FUNCTION OF THREE VARIABLES	A function f of three variables is **continuous at a point** (x_0, y_0, z_0) in an open region R if $f(x_0, y_0, z_0)$ is defined and equal to the limit of $f(x, y, z)$ as (x, y, z) approaches (x_0, y_0, z_0). That is, $$\lim_{(x,y,z) \to (x_0,y_0,z_0)} f(x, y, z) = f(x_0, y_0, z_0).$$ The function f is **continuous in the open region** R if it is continuous at every point in R.

EXAMPLE 6   Testing for continuity of a function of three variables

The function

$$f(x, y, z) = \frac{1}{x^2 + y^2 - z}$$

is continuous at each point in space except at the points on the paraboloid given by $z = x^2 + y^2$. ▭

EXERCISES for Section 15.2

In Exercises 1–4, find the indicated limit by using the limits

$$\lim_{(x,y)\to(a,b)} f(x, y) = 5 \quad \text{and} \quad \lim_{(x,y)\to(a,b)} g(x, y) = 3.$$

1. $\displaystyle\lim_{(x,y)\to(a,b)} [f(x, y) - g(x, y)]$

2. $\displaystyle\lim_{(x,y)\to(a,b)} \left[\frac{4f(x, y)}{g(x, y)}\right]$

3. $\displaystyle\lim_{(x,y)\to(a,b)} [f(x, y)g(x, y)]$

4. $\displaystyle\lim_{(x,y)\to(a,b)} \left[\frac{f(x, y) - g(x, y)}{f(x, y)}\right]$

In Exercises 5–14, find the indicated limit and discuss the continuity of the function.

5. $\displaystyle\lim_{(x,y)\to(2,1)} (x + 3y^2)$

6. $\displaystyle\lim_{(x,y)\to(0,0)} (5x + 3xy + y + 1)$

7. $\displaystyle\lim_{(x,y)\to(2,4)} \frac{x + y}{x - y}$

8. $\displaystyle\lim_{(x,y)\to(1,1)} \frac{x}{\sqrt{x + y}}$

9. $\displaystyle\lim_{(x,y)\to(0,1)} \frac{\arcsin (x/y)}{1 + xy}$

10. $\displaystyle\lim_{(x,y)\to(\pi/4,2)} y \sin xy$

11. $\displaystyle\lim_{(x,y)\to(0,0)} e^{xy}$

12. $\displaystyle\lim_{(x,y)\to(1,1)} \frac{xy}{x^2 + y^2}$

13. $\displaystyle\lim_{(x,y,z)\to(1,2,5)} \sqrt{x + y + z}$

14. $\displaystyle\lim_{(x,y,z)\to(2,0,1)} xe^{yz}$

In Exercises 15 and 16, find the limit of $f(x, y)$ (if it exists) as $(x, y) \to (0, 0)$.

15. $f(x, y) = -\dfrac{xy^2}{x^2 + y^4}$

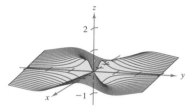

16. $f(x, y) = \dfrac{x^2}{(x^2 + 1)(y^2 + 1)}$

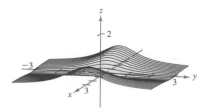

In Exercises 17–24, discuss the continuity of the function and evaluate the indicated limit, if it exists.

17. $\displaystyle\lim_{(x,y)\to(0,0)} \frac{\sin (x^2 + y^2)}{x^2 + y^2}$

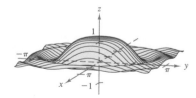

18. $\displaystyle\lim_{(x,y)\to(0,0)} e^{xy}$

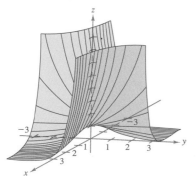

19. $\displaystyle\lim_{(x,y)\to(0,0)} \frac{xy}{x^2 + y^2}$

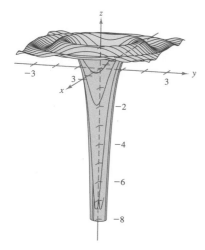

20. $\displaystyle\lim_{(x,y)\to(0,0)} \left[1 - \frac{\cos(x^2 + y^2)}{x^2 + y^2}\right]$

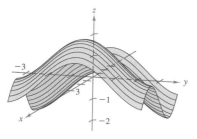

21. $\displaystyle\lim_{(x,y)\to(0,0)} (\sin x + \cos y)$

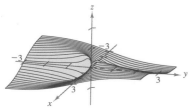

22. $\displaystyle\lim_{(x,y)\to(0,0)} \frac{y}{x^2 + y^2}$

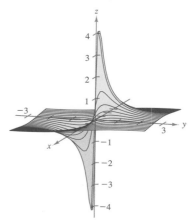

23. $\displaystyle\lim_{(x,y)\to(0,0)} \ln(x^2 + y^2)$

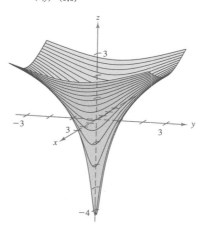

24. $\displaystyle\lim_{(x,y)\to(0,0)} \frac{2x - y^2}{2x^2 + y}$

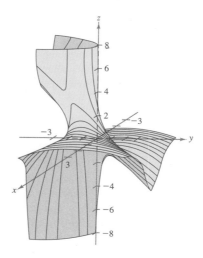

In Exercises 25–28, find the limit by using polar coordinates. [Hint: Let $x = r \cos \theta$ and $y = r \sin \theta$, and note that $(x, y) \to (0, 0)$ is equivalent to $r \to 0$.]

25. $\displaystyle \lim_{(x,y)\to(0,0)} \frac{\sin (x^2 + y^2)}{x^2 + y^2}$ **26.** $\displaystyle \lim_{(x,y)\to(0,0)} \frac{xy^2}{x^2 + y^2}$

27. $\displaystyle \lim_{(x,y)\to(0,0)} \frac{x^3 + y^3}{x^2 + y^2}$ **28.** $\displaystyle \lim_{(x,y)\to(0,0)} \frac{x^2 y^2}{x^2 + y^2}$

In Exercises 29–32, discuss the continuity of the function.

29. $f(x, y, z) = \dfrac{1}{\sqrt{x^2 + y^2 + z^2}}$

30. $f(x, y, z) = \dfrac{z}{x^2 + y^2 - 4}$

31. $f(x, y, z) = \dfrac{\sin z}{e^x + e^y}$

32. $f(x, y, z) = xy \sin z$

In Exercises 33–36, discuss the continuity of the composite function $f \circ g$.

33. $f(t) = t^2$, $g(x, y) = 3x - 2y$

34. $f(t) = \dfrac{1}{t}$, $g(x, y) = x^2 + y^2$

35. $f(t) = \dfrac{1}{t}$, $g(x, y) = 3x - 2y$

36. $f(t) = \dfrac{1}{4 - t}$, $g(x, y) = x^2 + y^2$

In Exercises 37–40, find the following limits.

(a) $\displaystyle \lim_{\Delta x \to 0} \frac{f(x + \Delta x, y) - f(x, y)}{\Delta x}$

(b) $\displaystyle \lim_{\Delta y \to 0} \frac{f(x, y + \Delta y) - f(x, y)}{\Delta y}$

37. $f(x, y) = x^2 - 4y$
38. $f(x, y) = x^2 + y^2$
39. $f(x, y) = 2x + xy - 3y$
40. $f(x, y) = \sqrt{y}(y + 1)$

In Exercises 41 and 42, use a computer to find the limit

$$\lim_{(x,y)\to(0,0)} f(x, y) \quad \text{(if it exists)}$$

along the specified path.

41. $f(x, y) = \dfrac{0.2x^2 y}{x^4 + 4y^2}$

(a) $y = x$ (b) $y = x^2$

42. $f(x, y) = 2x + 3y \sin \dfrac{1}{x}$

(a) $y = 0$ (b) $y = x$

43. Prove that

$$\lim_{(x,y)\to(a,b)} [f(x, y) + g(x, y)] = L_1 + L_2$$

where $f(x, y)$ approaches L_1 and $g(x, y)$ approaches L_2 as $(x, y) \to (a, b)$.

44. Prove that if f is continuous and $f(a, b) < 0$, then there exists a δ-neighborhood about (a, b) such that $f(x, y) < 0$ for every point (x, y) in the neighborhood.

15.3 Partial Derivatives

Partial derivatives ▪ Higher-order partial derivatives

Jean Le Rond d'Alembert

In applications of functions of several variables, the question often arises, "How will the function be affected by a change in one of its independent variables?" We can answer this by considering the independent variables one at a time. For example, to determine the effect of a catalyst in an experiment, a chemist could conduct the experiment several times using varying amounts of the catalyst, while keeping constant other variables such as temperature and pressure. We follow a similar procedure to determine the rate of change of a function f with respect to one of its several independent variables. That is, we take the derivative of f with respect to one independent variable at a time, while holding the others constant. This process is called **partial differentiation,** and the result is referred to as the **partial derivative** of f with respect to the chosen independent variable.

The introduction of partial derivatives followed Newton's and Leibniz's work in calculus by several years. Between 1730 and 1760, Leonhard Euler and Jean Le Rond d'Alembert (1717–1783) separately published several papers on dynamics, in which they established much of the theory of partial derivatives. These papers used functions of two or more variables to study problems involving equilibrium, fluid motion, and vibrating strings.

DEFINITION OF PARTIAL DERIVATIVES OF A FUNCTION OF TWO VARIABLES

If $z = f(x, y)$, then the **first partial derivatives** of f with respect to x and to y are the functions f_x and f_y defined by

$$f_x(x, y) = \lim_{\Delta x \to 0} \frac{f(x + \Delta x, y) - f(x, y)}{\Delta x}$$

$$f_y(x, y) = \lim_{\Delta y \to 0} \frac{f(x, y + \Delta y) - f(x, y)}{\Delta y}$$

provided the limits exist.

This definition indicates that if $z = f(x, y)$, then to find f_x we *consider y constant* and differentiate with respect to x. Similarly, to find f_y, we *consider x constant* and differentiate with respect to y.

EXAMPLE 1   Finding partial derivatives

Find f_x and f_y for $f(x, y) = 3x - x^2y^2 + 2x^3y$.

SOLUTION

Considering y to be constant and differentiating with respect to x, we have

$$f_x(x, y) = 3 - 2xy^2 + 6x^2y.$$

Considering x to be constant and differentiating with respect to y, we have

$$f_y(x, y) = -2x^2y + 2x^3.$$ ▬

There are several notations for first partial derivatives. We list the common ones here, along with the notation for a partial derivative evaluated at the point (a, b).

NOTATION FOR FIRST PARTIAL DERIVATIVES

For $z = f(x, y)$, the partial derivatives f_x and f_y are denoted by

$$\frac{\partial}{\partial x} f(x, y) = f_x(x, y) = z_x = \frac{\partial z}{\partial x}$$

and $\dfrac{\partial}{\partial y} f(x, y) = f_y(x, y) = z_y = \dfrac{\partial z}{\partial y}.$

The first partials evaluated at the point (a, b) are denoted by

$$\frac{\partial z}{\partial x}\bigg|_{(a,b)} = f_x(a, b) \qquad \text{and} \qquad \frac{\partial z}{\partial y}\bigg|_{(a,b)} = f_y(a, b).$$

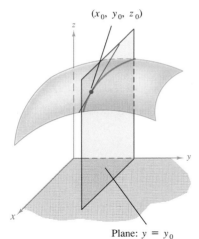

(x_0, y_0, z_0)

Plane: $y = y_0$

$\dfrac{\partial f}{\partial x} = $ (slope in x-direction)

FIGURE 15.27

EXAMPLE 2   Finding and evaluating partial derivatives

For $f(x, y) = xe^{x^2y}$, find f_x and f_y, and evaluate each at the point $(1, \ln 2)$.

SOLUTION

Since

$$f_x(x, y) = xe^{x^2y}(2xy) + e^{x^2y}$$

the partial derivative of f with respect to x at $(1, \ln 2)$ is

$$f_x(1, \ln 2) = e^{\ln 2}(2 \ln 2) + e^{\ln 2} = 4 \ln 2 + 2.$$

Since

$$f_y(x, y) = xe^{x^2y}(x^2) = x^3e^{x^2y}$$

the partial derivative of f with respect to y at $(1, \ln 2)$ is

$$f_y(1, \ln 2) = e^{\ln 2} = 2.$$

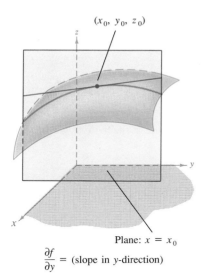

(x_0, y_0, z_0)

Plane: $x = x_0$

$\dfrac{\partial f}{\partial y} = $ (slope in y-direction)

FIGURE 15.28

The partial derivatives of a function of two variables, $z = f(x, y)$, have a useful geometric interpretation. If $y = y_0$, then $z = f(x, y_0)$ represents the curve formed by intersecting the surface $z = f(x, y)$ with the plane $y = y_0$, as shown in Figure 15.27. Therefore,

$$f_x(x_0, y_0) = \lim_{\Delta x \to 0} \frac{f(x_0 + \Delta x, y_0) - f(x_0, y_0)}{\Delta x}$$

represents the slope of this curve at the point $(x_0, y_0, f(x_0, y_0))$. (Note that both the curve and the tangent line lie in the plane $y = y_0$.) Similarly,

$$f_y(x_0, y_0) = \lim_{\Delta y \to 0} \frac{f(x_0, y_0 + \Delta y) - f(x_0, y_0)}{\Delta y}$$

represents the slope of the curve given by the intersection of $z = f(x, y)$ and the plane $x = x_0$ at $(x_0, y_0, f(x_0, y_0))$, as shown in Figure 15.28.

Informally, we can say that the values of $\partial f/\partial x$ and $\partial f/\partial y$ at the point (x_0, y_0, z_0) denote the **slope of the surface in the x and y directions,** respectively.

EXAMPLE 3   Finding the slope of a surface in the x and y directions

Find the slope of the surface given by

$$f(x, y) = -\frac{x^2}{2} - y^2 + \frac{25}{8}$$

at the point $\left(\frac{1}{2}, 1, 2\right)$ in the x direction and in the y direction.

SOLUTION

In the x direction, the slope is given by

$$f_x(x, y) = -x \quad \Longrightarrow \quad f_x\left(\frac{1}{2}, 1\right) = -\frac{1}{2}. \qquad \text{Figure 15.29(a)}$$

In the y direction, the slope is given by

$$f_y(x, y) = -2y \quad \Longrightarrow \quad f_y\left(\frac{1}{2}, 1\right) = -2. \qquad \text{Figure 15.29(b)}$$

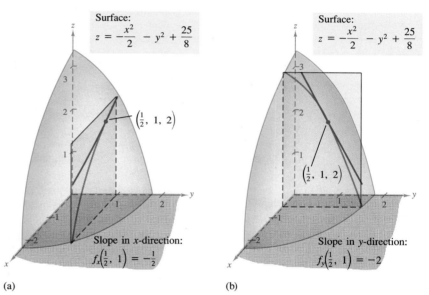

Surface:
$$z = -\frac{x^2}{2} - y^2 + \frac{25}{8}$$

$\left(\frac{1}{2}, 1, 2\right)$

Slope in x-direction:
$$f_x\left(\frac{1}{2}, 1\right) = -\frac{1}{2}$$

Surface:
$$z = -\frac{x^2}{2} - y^2 + \frac{25}{8}$$

$\left(\frac{1}{2}, 1, 2\right)$

Slope in y-direction:
$$f_y\left(\frac{1}{2}, 1\right) = -2$$

FIGURE 15.29   (a)   (b)

No matter how many variables are involved, partial derivatives of several variables can be interpreted as *rates of change*. This is illustrated in Example 4.

EXAMPLE 4   Using partial derivatives to find rates of change

The area of a parallelogram with adjacent sides a and b and included angle θ is given by $A = ab \sin \theta$, as shown in Figure 15.30.

(a) Find the rate of change of A with respect to a when $a = 10$, $b = 20$, and $\theta = \pi/6$.

(b) Find the rate of change of A with respect to θ when $a = 10$, $b = 20$, and $\theta = \pi/6$.

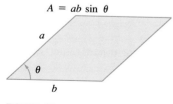

$A = ab \sin \theta$

a

θ

b

FIGURE 15.30

SOLUTION

(a) To find the rate of change of the area with respect to a, we hold b and θ constant and differentiate with respect to a to obtain

$$\frac{\partial A}{\partial a} = b \sin \theta.$$

When $a = 10$, $b = 20$, and $\theta = \pi/6$, we have

$$\partial A/\partial a = 20 \sin (\pi/6) = 10.$$

(b) To find the rate of change of the area with respect to θ, we hold a and b constant and differentiate with respect to θ to obtain

$$\frac{\partial A}{\partial \theta} = ab \cos \theta.$$

When $a = 10$, $b = 20$, and $\theta = \pi/6$, we have

$$\partial A/\partial \theta = 200 \cos (\pi/6) = 100\sqrt{3}.$$

Partial derivatives of a function of three or more variables

The concept of a partial derivative can be extended naturally to functions of three or more variables. For instance, if $w = f(x, y, z)$, then there are three partial derivatives, each of which is formed by considering two of the variables to be constant. That is, to define the partial derivative of w with respect to x, we consider y and z to be constant and write

$$\frac{\partial w}{\partial x} = f_x(x, y, z) = \lim_{\Delta x \to 0} \frac{f(x + \Delta x, y, z) - f(x, y, z)}{\Delta x}.$$

To define the partial derivative of w with respect to y, we consider x and z to be constant and write

$$\frac{\partial w}{\partial y} = f_y(x, y, z) = \lim_{\Delta y \to 0} \frac{f(x, y + \Delta y, z) - f(x, y, z)}{\Delta y}.$$

To define the partial derivative of w with respect to z, we consider x and y to be constant and write

$$\frac{\partial w}{\partial z} = f_z(x, y, z) = \lim_{\Delta z \to 0} \frac{f(x, y, z + \Delta z) - f(x, y, z)}{\Delta z}.$$

In general, if $w = f(x_1, x_2, \ldots, x_n)$, then there are n partial derivatives denoted by

$$\frac{\partial w}{\partial x_k} = f_{x_k}(x_1, x_2, \ldots, x_n), \quad k = 1, 2, \ldots, n.$$

To find the partial derivative with respect to one of the variables, we consider the others to be constant and differentiate with respect to the given variable.

EXAMPLE 5   Finding partial derivatives

Find the indicated partial derivatives.
(a) For $f(x, y, z) = xy + yz^2 + xz$, find $f_z(x, y, z)$.
(b) For $f(x, y, z) = z \sin (xy^2 + 2z)$, find $f_z(x, y, z)$.
(c) For $f(x, y, z, w) = (x + y + z)/w$, find $f_w(x, y, z, w)$.

SOLUTION

(a) To find the partial derivative of f with respect to z, we consider x and y to be constant and obtain

$$\frac{\partial}{\partial z}[xy + yz^2 + xz] = 2yz + x.$$

(b) To find the partial derivative of f with respect to z, we consider x and y to be constant and, using the Product Rule, we obtain

$$\frac{\partial}{\partial z}[z \sin (xy^2 + 2z)] = (z)\frac{\partial}{\partial z}[\sin (xy^2 + 2z)] + \sin (xy^2 + 2z)\frac{\partial}{\partial z}[z]$$

$$= (z)[\cos (xy^2 + 2z)](2) + \sin (xy^2 + 2z)$$

$$= 2z \cos (xy^2 + 2z) + \sin (xy^2 + 2z).$$

(c) To find the partial derivative of f with respect to w, we consider x, y, and z to be constant and obtain

$$\frac{\partial}{\partial w}\left[\frac{x + y + z}{w}\right] = -\frac{x + y + z}{w^2}.$$ ▭

Higher-order partial derivatives

As is true for ordinary derivatives, it is possible to take second, third, and higher partial derivatives of a function of several variables, provided such derivatives exist. We denote higher-order partial derivatives by the order in which the differentiation occurs. For instance, the function $z = f(x, y)$ has the following second partial derivatives.

1. Differentiate twice with respect to x:

$$\frac{\partial}{\partial x}\left(\frac{\partial f}{\partial x}\right) = \frac{\partial^2 f}{\partial x^2} = f_{xx}.$$

2. Differentiate twice with respect to y:

$$\frac{\partial}{\partial y}\left(\frac{\partial f}{\partial y}\right) = \frac{\partial^2 f}{\partial y^2} = f_{yy}.$$

3. Differentiate first with respect to x and then with respect to y:

$$\frac{\partial}{\partial y}\left(\frac{\partial f}{\partial x}\right) = \frac{\partial^2 f}{\partial y \partial x} = f_{xy}.$$

4. Differentiate first with respect to y and then with respect to x:

$$\frac{\partial}{\partial x}\left(\frac{\partial f}{\partial y}\right) = \frac{\partial^2 f}{\partial x \partial y} = f_{yx}.$$

The third and fourth cases are called **mixed partial derivatives.** Note that the two types of notation for mixed partials have different conventions for indicating the order of differentiation. For instance, the partial

$$\frac{\partial}{\partial y}\left(\frac{\partial f}{\partial x}\right) = \frac{\partial^2 f}{\partial y \partial x} \qquad \text{Right-to-left order}$$

indicates differentiation with respect to x first, but the partial

$$(f_y)_x = f_{yx} \qquad \text{Left-to-right order}$$

indicates differentiation with respect to y first. You can remember this by observing that in both notations, you differentiate first with respect to the variable "nearest" f.

EXAMPLE 6 Finding second partial derivatives

Find the second partial derivatives of $f(x, y) = 3xy^2 - 2y + 5x^2y^2$, and determine the value of $f_{xy}(-1, 2)$.

SOLUTION

We begin by finding the first partial derivatives with respect to x and y.

$$f_x(x, y) = 3y^2 + 10xy^2 \qquad \text{and} \qquad f_y(x, y) = 6xy - 2 + 10x^2y$$

Then, by differentiating each of these with respect to x and y, we have

$$f_{xx}(x, y) = 10y^2 \qquad \text{and} \qquad f_{yy}(x, y) = 6x + 10x^2$$
$$f_{xy}(x, y) = 6y + 20xy \qquad \text{and} \qquad f_{yx}(x, y) = 6y + 20xy.$$

Finally, $f_{xy}(-1, 2) = 12 - 40 = -28$. ▭

Notice in Example 6 that the two mixed partials are equal. This is often the case, as indicated in the next theorem, which we list without proof.

THEOREM 15.3 **EQUALITY OF MIXED PARTIAL** **DERIVATIVES**	If f is a function of x and y such that f, f_x, f_y, f_{xy}, and f_{yx} are continuous on an open region R, then for every (x, y) in R, $f_{xy}(x, y) = f_{yx}(x, y)$.

Theorem 15.3 also applies to a function f of *three or more variables* so long as f and all of its first and second partial derivatives are continuous. For example, if $w = f(x, y, z)$ and f and all of its first and second partial derivatives

are continuous in an open region R, then at each point in R the order of differentiation in the mixed second partial derivatives is irrelevant. That is,

$$f_{xy}(x, y, z) = f_{yx}(x, y, z)$$
$$f_{xz}(x, y, z) = f_{zx}(x, y, z)$$
$$f_{yz}(x, y, z) = f_{zy}(x, y, z).$$

Moreover, if the third partial derivatives of f are also continuous, then the order of differentiation of the mixed third partial derivatives is irrelevant. For example, $f_{xzz} = f_{zxz} = f_{zzx}$, as demonstrated in the next example.

EXAMPLE 7 Finding higher-order partial derivatives

Show that $f_{xz} = f_{zx}$ and $f_{xzz} = f_{zxz} = f_{zzx}$ for the function given by

$$f(x, y, z) = ye^x + x \ln z.$$

SOLUTION

First partials:

$$f_x(x, y, z) = ye^x + \ln z, \qquad f_z(x, y, z) = \frac{x}{z}$$

Second partials:

$$f_{xz}(x, y, z) = \frac{1}{z}, \qquad f_{zx}(x, y, z) = \frac{1}{z}, \qquad f_{zz}(x, y, z) = -\frac{x}{z^2}$$

(Note that the first two are equal.)

Third partials:

$$f_{xzz}(x, y, z) = -\frac{1}{z^2}, \qquad f_{zxz}(x, y, z) = -\frac{1}{z^2}, \qquad f_{zzx}(x, y, z) = -\frac{1}{z^2}$$

(Note that all three are equal.)

EXERCISES for Section 15.3

In Exercises 1–22, find the first partial derivatives with respect to x and with respect to y.

1. $f(x, y) = 2x - 3y + 5$ **2.** $f(x, y) = x^2 - 3y^2 + 7$

3. $f(x, y) = xy$ **4.** $f(x, y) = \dfrac{x}{y}$

5. $z = x\sqrt{y}$ **6.** $z = x^2 - 3xy + y^2$

7. $z = x^2 e^{2y}$ **8.** $z = xe^{x/y}$

9. $z = \ln(x^2 + y^2)$ **10.** $z = \ln\sqrt{xy}$

11. $z = \ln\dfrac{x + y}{x - y}$ **12.** $z = \dfrac{x^2}{2y} + \dfrac{4y^2}{x}$

13. $h(x, y) = e^{-(x^2 + y^2)}$

14. $g(x, y) = \ln\sqrt{x^2 + y^2}$

15. $f(x, y) = \sqrt{x^2 + y^2}$ **16.** $f(x, y) = \dfrac{xy}{x^2 + y^2}$

17. $z = \sin(2x - y)$ **18.** $z = \sin 3x \cos 3y$

19. $z = e^y \sin xy$ **20.** $z = \cos(x^2 + y^2)$

21. $f(x, y) = \displaystyle\int_x^y (t^2 - 1)\, dt$

22. $f(x, y) = \displaystyle\int_x^y (2t + 1)\, dt + \int_y^x (2t - 1)\, dt$

In Exercises 23–26, evaluate f_x and f_y at the indicated point.

23. $f(x, y) = \arctan \dfrac{y}{x}$, $(2, -2)$

24. $f(x, y) = \arcsin xy$, $(1, 0)$

25. $f(x, y) = \dfrac{xy}{x - y}$, $(2, -2)$

26. $f(x, y) = \dfrac{4xy}{\sqrt{x^2 + y^2}}$, $(1, 0)$

In Exercises 27–32, find the first partial derivatives with respect to x, y, and z.

27. $w = \sqrt{x^2 + y^2 + z^2}$ **28.** $w = \dfrac{xy}{x + y + z}$

29. $F(x, y, z) = \ln \sqrt{x^2 + y^2 + z^2}$

30. $G(x, y, z) = \dfrac{1}{\sqrt{1 - x^2 - y^2 - z^2}}$

31. $H(x, y, z) = \sin (x + 2y + 3z)$
32. $f(x, y, z) = 3x^2y - 5xyz + 10yz^2$

In Exercises 33–40, find the second partial derivatives

$$\dfrac{\partial^2 z}{\partial x^2}, \dfrac{\partial^2 z}{\partial y^2}, \dfrac{\partial^2 z}{\partial y \partial x}, \text{ and } \dfrac{\partial^2 z}{\partial x \partial y}.$$

33. $z = x^2 - 2xy + 3y^2$ **34.** $z = x^4 - 3x^2y^2 + y^4$
35. $z = e^x \tan y$ **36.** $z = 2e^{xy^2}$

37. $z = \arctan \dfrac{y}{x}$ **38.** $z = \sin (x - 2y)$

39. $z = \sqrt{x^2 + y^2}$ **40.** $z = \dfrac{xy}{x - y}$

In Exercises 41–46, show that

$$\dfrac{\partial^2 z}{\partial y \partial x} = \dfrac{\partial^2 z}{\partial x \partial y}.$$

41. $z = x^3 + 3x^2y$ **42.** $z = \ln (x - y)$
43. $z = x \sec y$ **44.** $z = \sqrt{9 - x^2 - y^2}$
45. $z = xe^{-y^2}$ **46.** $z = xe^y + ye^x$

In Exercises 47–50, show that the mixed partials f_{xyy}, f_{yxy}, and f_{yyx} are equal.

47. $f(x, y, z) = xyz$
48. $f(x, y, z) = x^2 - 3xy + 4yz + z^3$
49. $f(x, y, z) = e^{-x} \sin yz$ **50.** $f(x, y, z) = \dfrac{x}{y + z}$

In Exercises 51–54, show that each function satisfies **Laplace's equation**

$$\dfrac{\partial^2 z}{\partial x^2} + \dfrac{\partial^2 z}{\partial y^2} = 0.$$

51. $z = 5xy$ **52.** $z = \dfrac{1}{2}(e^y - e^{-y}) \sin x$

53. $z = e^x \sin y$ **54.** $z = \arctan \dfrac{y}{x}$

In Exercises 55 and 56, show that the function satisfies the **wave equation**

$$\dfrac{\partial^2 z}{\partial t^2} = c^2 \dfrac{\partial^2 z}{\partial x^2}.$$

55. $z = \sin (x - ct)$ **56.** $z = \sin \omega ct \sin \omega x$

In Exercises 57 and 58, show that the function satisfies the **heat equation**

$$\dfrac{\partial z}{\partial t} = c^2 \dfrac{\partial^2 z}{\partial x^2}.$$

57. $z = e^{-t} \cos \dfrac{x}{c}$ **58.** $z = e^{-t} \sin \dfrac{x}{c}$

In Exercises 59–62, use the limit definition of partial derivatives to find $f_x(x, y)$ and $f_y(x, y)$.

59. $f(x, y) = 2x + 3y$ **60.** $f(x, y) = \dfrac{1}{x + y}$

61. $f(x, y) = \sqrt{x + y}$
62. $f(x, y) = x^2 - 2xy + y^2$

In Exercises 63–66, sketch the curve formed by the intersection of the given surface and plane. Find the slope of the curve at the given point.

Surface	Plane	Point
63. $z = \sqrt{49 - x^2 - y^2}$	$x = 2$	$(2, 3, 6)$
64. $z = x^2 + 4y^2$	$y = 1$	$(2, 1, 8)$
65. $z = 9x^2 - y^2$	$y = 3$	$(1, 3, 0)$
66. $z = 9x^2 - y^2$	$x = 1$	$(1, 3, 0)$

67. A company manufactures two types of wood-burning stoves: a freestanding model and a fireplace-insert model. The cost function for producing x freestanding and y fireplace-insert stoves is

$$C = 32\sqrt{xy} + 175x + 205y + 1050.$$

Find the marginal costs ($\partial C/\partial x$ and $\partial C/\partial y$) when $x = 80$ and $y = 20$.

68. Let $x = 1000$ and $y = 500$ in the Cobb-Douglas production function

$$f(x, y) = 100x^{0.6}y^{0.4}.$$

(a) Find the marginal productivity of labor, $\partial f/\partial x$.
(b) Find the marginal productivity of capital, $\partial f/\partial y$.

69. Let N be the number of applicants to a university, p the charge for food and housing at the university, and t the tuition. Suppose that N is a function of p and t such that $\partial N/\partial p < 0$ and $\partial N/\partial t < 0$. How would you interpret the fact that both partials are negative?

70. The range of a projectile fired at an angle θ above the horizontal with velocity v_0 is

$$R = \frac{v_0^2 \sin 2\theta}{32}.$$

Evaluate $\partial R/\partial v_0$ and $\partial R/\partial \theta$ when $v_0 = 2000$ feet per second and $\theta = 5°$.

71. The temperature at any point (x, y) in a steel plate is given by

$$T = 500 - 0.6x^2 - 1.5y^2$$

where x and y are measured in feet. At the point $(2, 3)$, find the rate of change of the temperature with respect to the distance moved along the plate in the directions of the x- and y-axes, respectively.

72. According to the Ideal Gas Law, $PV = kT$, where P is pressure, V is volume, T is temperature, and k is a constant of proportionality. Find (a) $\partial P/\partial T$ and (b) $\partial V/\partial P$.

73. Consider the function defined by

$$f(x, y) = \begin{cases} \dfrac{xy(x^2 - y^2)}{x^2 + y^2}, & (x, y) \neq (0, 0) \\ 0, & (x, y) = (0, 0). \end{cases}$$

(a) Find $f_x(x, y)$ and $f_y(x, y)$ for $(x, y) \neq (0, 0)$.

(b) Use the definition of partial derivatives to find $f_x(0, 0)$ and $f_y(0, 0)$.

$$\left[\text{Hint:} \quad f_x(0, 0) = \lim_{\Delta x \to 0} \frac{f(\Delta x, 0) - f(0, 0)}{\Delta x}. \right]$$

(c) Use the definition of partial derivatives to find $f_{xy}(0, 0)$ and $f_{yx}(0, 0)$.

(d) Using Theorem 15.3 and the result of part (c), what can be said about at least one of the functions f, f_x, f_y, f_{xy}, and f_{yx}?

15.4 Differentials

Increments ▪ Differentials ▪ Differentiability ▪ Approximation by differentials

In this section we generalize the concepts of increments and differentials to functions of two or more variables. Recall from Section 4.9 that for $y = f(x)$ we defined the differential of y to be $dy = f'(x)\,dx$. For a function of two variables, given by $z = f(x, y)$, we use similar terminology. We call Δx and Δy the **increments of x and y,** and the **increment of z** is given by

$$\Delta z = f(x + \Delta x, y + \Delta y) - f(x, y). \qquad \text{Increment of } z$$

The differentials dx, dy, and dz are defined as follows.

DEFINITION OF TOTAL DIFFERENTIAL

If $z = f(x, y)$ and Δx and Δy are increments of x and y, then the **differentials** of the independent variables x and y are

$$dx = \Delta x \quad \text{and} \quad dy = \Delta y$$

and the **total differential** of the dependent variable z is

$$dz = \frac{\partial z}{\partial x}\,dx + \frac{\partial z}{\partial y}\,dy = f_x(x, y)\,dx + f_y(x, y)\,dy.$$

This definition can be extended to a function of three or more variables. For instance, if $w = f(x, y, z, u)$, then $dx = \Delta x$, $dy = \Delta y$, $dz = \Delta z$, $du = \Delta u$, and the total differential of w is

$$dw = \frac{\partial w}{\partial x}\,dx + \frac{\partial w}{\partial y}\,dy + \frac{\partial w}{\partial z}\,dz + \frac{\partial w}{\partial u}\,du.$$

EXAMPLE 1 Finding the total differential

(a) The total differential dz for $z = 2x \sin y - 3x^2y^2$ is

$$dz = \frac{\partial z}{\partial x}\, dx + \frac{\partial z}{\partial y}\, dy$$
$$= (2 \sin y - 6xy^2)\, dx + (2x \cos y - 6x^2y)\, dy.$$

(b) The total differential dw for $w = x^2 + y^2 + z^2$ is

$$dz = \frac{\partial w}{\partial x}\, dx + \frac{\partial w}{\partial y}\, dy + \frac{\partial w}{\partial z}\, dz$$
$$= 2x\, dx + 2y\, dy + 2z\, dz.$$

In Section 4.9 we saw that for a *differentiable* function given by $y = f(x)$, we can use the differential $dy = f'(x)\, dx$ as an approximation (for small Δx) to the value $\Delta y = f(x + \Delta x) - f(x)$. When a similar approximation is possible for a function of two variables, we say that it is **differentiable.** This is stated explicitly in the following definition.

DEFINITION OF DIFFERENTIABILITY	A function f given by $z = f(x, y)$ is **differentiable** at (x_0, y_0) if Δz can be expressed in the form $$\Delta z = f_x(x_0, y_0)\Delta x + f_y(x_0, y_0)\Delta y + \varepsilon_1 \Delta x + \varepsilon_2 \Delta y$$ where both ε_1 and $\varepsilon_2 \to 0$ as $(\Delta x, \Delta y) \to (0, 0)$. The function f is said to be **differentiable in a region R** if it is differentiable at each point of R.

EXAMPLE 2 Showing that a function is differentiable

Show that the function given by $f(x, y) = x^2 + 3y$ is differentiable at every point in the plane.

SOLUTION

Letting $z = f(x, y)$, the increment of z at an arbitrary point (x, y) in the plane is

$$\Delta z = f(x + \Delta x, y + \Delta y) - f(x, y)$$
$$= (x^2 + 2x\Delta x + \Delta x^2) + 3(y + \Delta y) - (x^2 + 3y)$$
$$= 2x\Delta x + \Delta x^2 + 3\Delta y$$
$$= 2x(\Delta x) + 3\Delta y + \Delta x(\Delta x) + 0(\Delta y)$$
$$= f_x(x, y)\Delta x + f_y(x, y)\Delta y + \varepsilon_1 \Delta x + \varepsilon_2 \Delta y$$

where $\varepsilon_1 = \Delta x$ and $\varepsilon_2 = 0$. Since $\varepsilon_1 \to 0$ and $\varepsilon_2 \to 0$ as $(\Delta x, \Delta y) \to (0, 0)$, it follows that f is differentiable at every point in the plane.

Be sure you see that the term "differentiable" is used differently for functions of two variables than for one variable. A function of one variable is differentiable at a point if its derivative exists at the point. However, for a function of two variables, the existence of the partial derivatives f_x and f_y does not guarantee that the function is differentiable (see Example 5). In the following theorem, we present a *sufficient* condition for differentiability of a function of two variables.

THEOREM 15.4
SUFFICIENT CONDITION FOR DIFFERENTIABILITY

If f is a function of x and y, where f, f_x, and f_y are continuous in an open region R, then f is differentiable on R.

PROOF

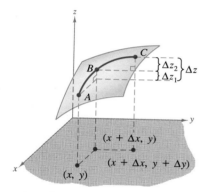

$\Delta z = f(x + \Delta x, y + \Delta y) - f(x, y)$

FIGURE 15.31

Let S be the surface defined by $z = f(x, y)$, where f, f_x, and f_y are continuous at (x, y). Let A, B, and C be points on surface S, as shown in Figure 15.31. From this figure we can see that the change in f from point A to point C is given by

$$\Delta z = f(x + \Delta x, y + \Delta y) - f(x, y)$$
$$= [f(x + \Delta x, y) - f(x, y)] + [f(x + \Delta x, y + \Delta y) - f(x + \Delta x, y)]$$
$$= \Delta z_1 + \Delta z_2.$$

Between A and B, y is fixed and x changes. Hence, by the Mean Value Theorem, there is a value x_1 between x and $x + \Delta x$ such that

$$\Delta z_1 = f(x + \Delta x, y) - f(x, y) = f_x(x_1, y)\Delta x.$$

Similarly, between B and C, x is fixed and y changes, and there is a value y_1 between y and $y + \Delta y$ such that

$$\Delta z_2 = f(x + \Delta x, y + \Delta y) - f(x + \Delta x, y) = f_y(x + \Delta x, y_1)\Delta y.$$

By combining these two results, we can write

$$\Delta z = \Delta z_1 + \Delta z_2 = f_x(x_1, y)\Delta x + f_y(x + \Delta x, y_1)\Delta y.$$

Now, if we define ε_1 and ε_2 by

$$\varepsilon_1 = f_x(x_1, y) - f_x(x, y) \qquad \text{and} \qquad \varepsilon_2 = f_y(x + \Delta x, y_1) - f_y(x, y)$$

it follows that

$$\Delta z = \Delta z_1 + \Delta z_2 = [\varepsilon_1 + f_x(x, y)]\Delta x + [\varepsilon_2 + f_y(x, y)]\Delta y$$
$$= [f_x(x, y)\Delta x + f_y(x, y)\Delta y] + \varepsilon_1\Delta x + \varepsilon_2\Delta y.$$

Then, by the continuity of f_x and f_y and the fact that $x \leq x_1 \leq x + \Delta x$ and $y \leq y_1 \leq y + \Delta y$, it follows that $\varepsilon_1 \to 0$ and $\varepsilon_2 \to 0$ as $\Delta x \to 0$ and $\Delta y \to 0$. Therefore, by definition, f is differentiable.

Theorem 15.4 tells us that we can choose $(x + \Delta x, y + \Delta y)$ close enough to (x, y) to make $\varepsilon_1\Delta x$ and $\varepsilon_2\Delta y$ insignificant. In other words, for small Δx and Δy, we can use the approximation

$$\Delta z \approx dz.$$

This approximation is illustrated graphically in Figure 15.32. Recall that the partial derivatives $\partial z/\partial x$ and $\partial z/\partial y$ can be interpreted as the slope of the surface in the x and y directions, respectively. This means that

$$dz = \frac{\partial z}{\partial x}\Delta x + \frac{\partial z}{\partial y}\Delta y$$

represents the change in height of a plane that is tangent to the surface at the point $(x, y, f(x, y))$. Since a plane in space is represented by a linear equation in the variables x, y, and z, we call the approximation of Δz by dz a **linear approximation.** We will say more about this geometrical interpretation in Section 15.7.

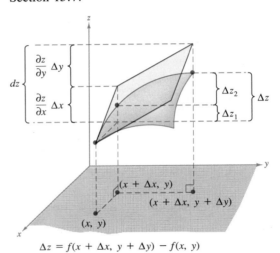

FIGURE 15.32

$$\Delta z = f(x + \Delta x, y + \Delta y) - f(x, y)$$

EXAMPLE 3   Using a differential as an approximation

Use the differential dz to approximate the change in $z = \sqrt{4 - x^2 - y^2}$ as (x, y) moves from the point $(1, 1)$ to $(1.01, 0.97)$. Compare this approximation to the exact change in z.

SOLUTION

Letting $(x, y) = (1, 1)$ and $(x + \Delta x, y + \Delta y) = (1.01, 0.97)$ produces

$$dx = \Delta x = 0.01 \qquad \text{and} \qquad dy = \Delta y = -0.03.$$

Thus, the change in z can be approximated by

$$\Delta z \approx dz = \frac{\partial z}{\partial x}\, dx + \frac{\partial z}{\partial y}\, dy = \frac{-x}{\sqrt{4 - x^2 - y^2}}\,\Delta x + \frac{-y}{\sqrt{4 - x^2 - y^2}}\,\Delta y.$$

When $x = 1$ and $y = 1$, we have

$$\Delta z \approx -\frac{1}{\sqrt{2}}(0.01) - \frac{1}{\sqrt{2}}(-0.03) = \frac{0.02}{\sqrt{2}} = \sqrt{2}(0.01) \approx 0.0141.$$

In Figure 15.33 we see that the exact change corresponds to the difference

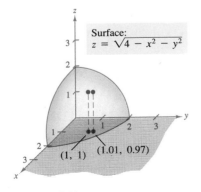

Surface:
$z = \sqrt{4 - x^2 - y^2}$

$(1, 1)$ $(1.01, 0.97)$

FIGURE 15.33

in the heights of two points on the surface of a hemisphere. This difference is given by

$$\Delta z = f(1.01, 0.97) - f(1, 1)$$
$$= \sqrt{4 - (1.01)^2 - (0.97)^2} - \sqrt{4 - 1^2 - 1^2} \approx 0.0137.$$ ▭

A function of three variables, $w = f(x, y, z)$, is called **differentiable** at (x, y, z) provided that $\Delta w = f(x + \Delta x, y + \Delta y, z + \Delta z) - f(x, y, z)$ can be expressed in the form

$$\Delta w = f_x \Delta x + f_y \Delta y + f_z \Delta z + \varepsilon_1 \Delta x + \varepsilon_2 \Delta y + \varepsilon_3 \Delta z$$

where ε_1, ε_2, and $\varepsilon_3 \to 0$ as $(\Delta x, \Delta y, \Delta z) \to (0, 0, 0)$. With this definition of differentiability, Theorem 15.4 has the following extension for functions of three variables. If f is a function of x, y, and z, where $f, f_x, f_y,$ and f_z are continuous in an open region R, then f is differentiable on R.

In Section 4.9, we used differentials to approximate the propagated error introduced by an error in measurement. We demonstrate this application in the following example involving a differential approximation with a function of three variables.

EXAMPLE 4   An application of differentials

The possible error involved in measuring each dimension of a rectangular box is ± 0.1 millimeter. The dimensions of the box are $x = 50$ centimeters, $y = 20$ centimeters, and $z = 15$ centimeters, as shown in Figure 15.34. Use dV to estimate the propagated error and the relative error in the calculated volume of the box.

SOLUTION

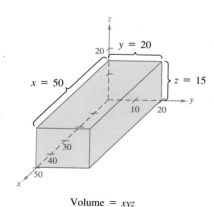

Volume = xyz

FIGURE 15.34

The volume of the box is given by

$$V = xyz \qquad \text{Volume of box}$$

and thus

$$dV = \frac{\partial V}{\partial x}\, dx + \frac{\partial V}{\partial y}\, dy + \frac{\partial V}{\partial z}\, dz = yz\, dx + xz\, dy + xy\, dz.$$

Now, since 0.1 millimeter = 0.01 centimeter, we have $dx = dy = dz = \pm 0.01$, and the propagated error is approximately

$$dV = (20)(15)(\pm 0.01) + (50)(15)(\pm 0.01) + (50)(20)(\pm 0.01)$$
$$= 300(\pm 0.01) + 750(\pm 0.01) + 1000(\pm 0.01)$$
$$= 2050(\pm 0.01) = \pm 20.5 \text{ cm}^3.$$

Since the measured volume is

$$V = (50)(20)(15) = 15{,}000 \text{ cm}^3$$

the relative error, $\Delta V/V$, is approximately

$$\frac{\Delta V}{V} \approx \frac{dV}{V} = \frac{20.5}{15{,}000} \approx 0.14\%.$$ ▭

As is true for a function of a single variable, if a function in two or more variables is differentiable at some point in its domain, then it is also continuous there. This is established in the proof of Theorem 15.5.

THEOREM 15.5
DIFFERENTIABILITY IMPLIES
CONTINUITY

If a function of x and y is differentiable at (x_0, y_0), then it is continuous at (x_0, y_0).

PROOF   Let f be differentiable at (x_0, y_0), where $z = f(x, y)$. Then

$$\Delta z = [f_x(x_0, y_0) + \varepsilon_1]\Delta x + [f_y(x_0, y_0) + \varepsilon_2]\Delta y$$

where both ε_1 and $\varepsilon_2 \to 0$ as $(\Delta x, \Delta y) \to (0, 0)$. However, by definition, we know that Δz is given by

$$\Delta z = f(x_0 + \Delta x, y_0 + \Delta y) - f(x_0, y_0).$$

Thus, letting $x = x_0 + \Delta x$ and $y = y_0 + \Delta y$, we obtain

$$f(x, y) - f(x_0, y_0) = [f_x(x_0, y_0) + \varepsilon_1]\Delta x + [f_y(x_0, y_0) + \varepsilon_2]\Delta y$$
$$= [f_x(x_0, y_0) + \varepsilon_1](x - x_0) + [f_y(x_0, y_0) + \varepsilon_2](y - y_0).$$

Now, taking the limit as $(x, y) \to (x_0, y_0)$, we have

$$\lim_{(x,y)\to(x_0,y_0)} f(x, y) = f(x_0, y_0)$$

which means that f is continuous at (x_0, y_0).

We already pointed out that the existence of f_x and f_y is not sufficient to guarantee differentiability. The next example uses Theorem 15.5 to validate this claim.

EXAMPLE 5   A function that is not differentiable

Show that $f_x(0, 0)$ and $f_y(0, 0)$ both exist, but that f is not differentiable at $(0, 0)$ where f is defined as

$$f(x, y) = \begin{cases} \dfrac{-3xy}{x^2 + y^2}, & \text{if } (x, y) \neq (0, 0) \\ 0, & \text{if } (x, y) = (0, 0). \end{cases}$$

SOLUTION

Using Theorem 15.5 we can show that f is not differentiable at $(0, 0)$ by showing that it is not continuous at this point. To see that f is not continuous at $(0, 0)$, we look at the values of $f(x, y)$ along two different approaches to $(0, 0)$, as shown in Figure 15.35. Along the line $y = x$, the limit is

$$\lim_{(x,x)\to(0,0)} f(x, y) = \lim_{(x,x)\to(0,0)} \frac{-3x^2}{2x^2} = -\frac{3}{2}$$

whereas along $y = -x$ we have

$$\lim_{(x,-x)\to(0,0)} f(x, y) = \lim_{(x,-x)\to(0,0)} \frac{3x^2}{2x^2} = \frac{3}{2}.$$

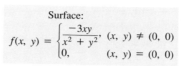

Surface:
$$f(x, y) = \begin{cases} \dfrac{-3xy}{x^2 + y^2}, & (x, y) \neq (0, 0) \\ 0, & (x, y) = (0, 0) \end{cases}$$

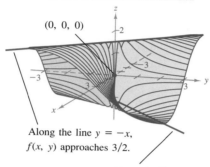

$(0, 0, 0)$

Along the line $y = -x$,
$f(x, y)$ approaches $3/2$.

Along the line $y = x$,
$f(x, y)$ approaches $-3/2$.

FIGURE 15.35

Thus, the limit of $f(x, y)$ as $(x, y) \to (0, 0)$ does not exist, and we conclude that f is not continuous at $(0, 0)$. Therefore, by Theorem 15.5, we know that f is not differentiable at $(0, 0)$. On the other hand, by the definition of the partial derivatives f_x and f_y, we have

$$f_x(0, 0) = \lim_{\Delta x \to 0} \frac{f(\Delta x, 0) - f(0, 0)}{\Delta x} = \lim_{\Delta x \to 0} \frac{0 - 0}{\Delta x} = 0$$

and

$$f_y(0, 0) = \lim_{\Delta y \to 0} \frac{f(0, \Delta y) - f(0, 0)}{\Delta y} = \lim_{\Delta y \to 0} \frac{0 - 0}{\Delta y} = 0.$$

Thus, the partial derivatives at $(0, 0)$ exist.   ▭

EXERCISES for Section 15.4

In Exercises 1–10, find the total differential.

1. $z = 3x^2y^3$

2. $z = \dfrac{x^2}{y}$

3. $z = \dfrac{-1}{x^2 + y^2}$

4. $z = e^x \sin y$

5. $z = x \cos y - y \cos x$

6. $z = \dfrac{1}{2}(e^{x^2+y^2} - e^{-x^2-y^2})$

7. $w = 2z^3y \sin x$

8. $w = e^x \cos y + z$

9. $w = \dfrac{x + y}{z - 2y}$

10. $w = x^2yz^2 + \sin yz$

In Exercises 11–16, (a) evaluate $f(1, 2)$ and $f(1.05, 2.1)$ and calculate Δz and (b) use the total differential dz to approximate Δz.

11. $f(x, y) = 9 - x^2 - y^2$

12. $f(x, y) = \sqrt{x^2 + y^2}$

13. $f(x, y) = x \sin y$

14. $f(x, y) = xy$

15. $f(x, y) = 3x - 4y$

16. $f(x, y) = \dfrac{x}{y}$

In Exercises 17–20, show that the given function is differentiable by finding values for ε_1 and ε_2 (see the definition of differentiability), and verify that both ε_1 and $\varepsilon_2 \to 0$ as $(\Delta x, \Delta y) \to (0, 0)$.

17. $f(x, y) = x^2 - 2x + y$ **18.** $f(x, y) = x^2 + y^2$

19. $f(x, y) = x^2 y$ **20.** $f(x, y) = 5x - 10y + y^3$

21. The radius r and height h of a right-circular cylinder are measured with a possible error of 4 percent and 2 percent, respectively. Approximate the maximum possible percentage error in measuring the volume.

22. The centripetal acceleration of a particle moving in a circle is $a = v^2/r$, where v is the velocity and r is the radius of the circle. Approximate the maximum percentage error in measuring the acceleration due to errors of 2 percent in v and 1 percent in r.

23. Electrical power P is given by $P = E^2/R$, where E is voltage and R is resistance. Approximate the maximum percentage error in calculating power if 200 volts is applied to a 4000-ohm resistor and the possible percentage errors in measuring E and R are 2 percent and 3 percent, respectively.

24. The total resistance R of two resistors connected in parallel is given by

$$\frac{1}{R} = \frac{1}{R_1} + \frac{1}{R_2}.$$

Approximate the change in R if R_1 is increased from 10 ohms to 10.5 ohms and R_2 is decreased from 15 ohms to 13 ohms.

25. The period T of a pendulum of length L is given by $T = 2\pi\sqrt{L/g}$, where g is the acceleration due to gravity. A pendulum is moved from the Canal Zone, where $g = 32.09$ feet per second per second, to Greenland, where $g = 32.24$ feet per second per second. Because of the change in temperature, the length of the pendulum changes from 2.5 feet to 2.48 feet. Approximate the change in the period of the pendulum.

26. A triangle is measured and two adjacent sides are found to be 3 and 4 inches long, with an included angle of $\pi/4$. The possible errors in measurement are $\frac{1}{16}$ inch in the sides and 0.02 radian in the angle. Approximate the maximum possible error in the computation of the area.

27. The inductance L (in micro-henrys) of a straight non-magnetic wire in free space is given by

$$L = 0.00021\left(\ln \frac{2h}{r} - 0.75\right)$$

where h is the length of the wire in millimeters and r is the radius of a circular cross section. Approximate L when $r = 2 \pm \frac{1}{16}$ millimeters and $h = 100 \pm \frac{1}{100}$ millimeters.

28. A right-circular cone of height $h = 6$ and radius $r = 3$ is constructed, and in the process errors Δr and Δh are made in the radius and height, respectively. Complete the accompanying table to show the relationship between ΔV and dV for the given errors.

Δr	Δh	dV	ΔV	$\Delta V - dV$
0.1	0.1			
0.1	−0.1			
0.001	0.002			
−0.0001	0.0002			

In Exercises 29 and 30, use the given function to prove that (a) $f_x(0, 0)$ and $f_y(0, 0)$ exist, and (b) f is not differentiable at $(0, 0)$.

29. $f(x, y) = \begin{cases} \dfrac{3x^2 y}{x^4 + y^2}, & (x, y) \neq (0, 0) \\ 0, & (x, y) = (0, 0) \end{cases}$

30. $f(x, y) = \begin{cases} \dfrac{2x^2 y^2}{x^4 + y^4}, & (x, y) \neq (0, 0) \\ 0, & (x, y) = (0, 0) \end{cases}$

15.5 Chain Rules for Functions of Several Variables

Chain Rules for functions of several variables ▪ Implicit partial differentiation

The work with differentials in the previous section provides the basis for the extension of the Chain Rule to functions of two variables. In this extension, we consider two cases. The first case involves w as a function of x and y, where x and y are functions of a single independent variable t.

THEOREM 15.6 CHAIN RULE: ONE INDEPENDENT VARIABLE	Let $w = f(x, y)$, where f is a differentiable function of x and y. If $x = g(t)$ and $y = h(t)$, where g and h are differentiable functions of t, then w is a differentiable function of t, and $$\frac{dw}{dt} = \frac{\partial w}{\partial x}\frac{dx}{dt} + \frac{\partial w}{\partial y}\frac{dy}{dt}.$$

PROOF Since g and h are differentiable functions of t, we know that both Δx and Δy approach zero as Δt approaches zero. Moreover, since f is a differentiable function of x and y, we know that

$$\Delta w = \frac{\partial w}{\partial x}\Delta x + \frac{\partial w}{\partial y}\Delta y + \varepsilon_1\Delta x + \varepsilon_2\Delta y$$

where both ε_1 and $\varepsilon_2 \to 0$ as $(\Delta x, \Delta y) \to (0, 0)$. Thus, for $\Delta t \neq 0$, we have

$$\frac{\Delta w}{\Delta t} = \frac{\partial w}{\partial x}\frac{\Delta x}{\Delta t} + \frac{\partial w}{\partial y}\frac{\Delta y}{\Delta t} + \varepsilon_1\frac{\Delta x}{\Delta t} + \varepsilon_2\frac{\Delta y}{\Delta t}$$

from which it follows that

$$\frac{dw}{dt} = \lim_{\Delta t \to 0}\frac{\Delta w}{\Delta t} = \frac{\partial w}{\partial x}\frac{dx}{dt} + \frac{\partial w}{\partial y}\frac{dy}{dt} + 0\left(\frac{dx}{dt}\right) + 0\left(\frac{dy}{dt}\right)$$

$$= \frac{\partial w}{\partial x}\frac{dx}{dt} + \frac{\partial w}{\partial y}\frac{dy}{dt}.$$

EXAMPLE 1   Using the Chain Rule with one independent variable

Let $w = x^2y - y^2$, where $x = \sin t$ and $y = e^t$. Find dw/dt when $t = 0$.

SOLUTION

By the Chain Rule for one independent variable, we have

$$\frac{dw}{dt} = \frac{\partial w}{\partial x}\frac{dx}{dt} + \frac{\partial w}{\partial y}\frac{dy}{dt} = 2xy(\cos t) + (x^2 - 2y)e^t.$$

When $t = 0$, $x = 0$, and $y = 1$, it follows that

$$\frac{dw}{dt} = 0 - 2 = -2.$$
▬

The Chain Rules presented in this section provide alternate techniques for solving many problems in single-variable calculus. For instance, in Example 1, we could have used single-variable techniques to find dw/dt by first writing w as a function of t,

$$w = x^2y - y^2 = (\sin t)^2(e^t) - (e^t)^2 = e^t \sin^2 t - e^{2t}$$

and then differentiating as usual.

The Chain Rule in Theorem 15.6 can be extended to any number of variables. For example, if each x_i is a differentiable function of a single variable t, then for $w = f(x_1, x_2, \ldots, x_n)$ we have

$$\frac{dw}{dt} = \frac{\partial w}{\partial x_1}\frac{dx_1}{dt} + \frac{\partial w}{\partial x_2}\frac{dx_2}{dt} + \cdots + \frac{\partial w}{\partial x_n}\frac{dx_n}{dt}.$$

In the next example, we apply this Chain Rule to a function of four variables.

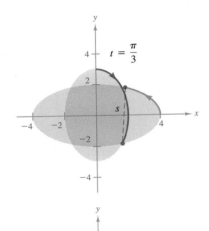

EXAMPLE 2   An application of a Chain Rule to related rates

Two objects are traveling in elliptical paths given by the following parametric equations.

$$x_1 = 4 \cos t \qquad \text{and} \qquad y_1 = 2 \sin t \qquad \text{First object}$$
$$x_2 = 2 \sin 2t \qquad \text{and} \qquad y_2 = 3 \cos 2t \qquad \text{Second object}$$

At what rate is the distance between the two objects changing when $t = \pi$?

SOLUTION

From Figure 15.36 we see that the distance s between the two objects is given by

$$s = \sqrt{(x_2 - x_1)^2 + (y_2 - y_1)^2}$$

and when $t = \pi$, we have $x_1 = -4$, $y_1 = 0$, $x_2 = 0$, $y_2 = 3$, and

$$s = \sqrt{(0 + 4)^2 + (3 - 0)^2} = 5.$$

When $t = \pi$, the partial derivatives of s are as follows.

$$\frac{\partial s}{\partial x_1} = \frac{-(x_2 - x_1)}{\sqrt{(x_2 - x_1)^2 + (y_2 - y_1)^2}} = -\frac{1}{5}(0 + 4) = -\frac{4}{5}$$

$$\frac{\partial s}{\partial y_1} = \frac{-(y_2 - y_1)}{\sqrt{(x_2 - x_1)^2 + (y_2 - y_1)^2}} = -\frac{1}{5}(3 - 0) = -\frac{3}{5}$$

$$\frac{\partial s}{\partial x_2} = \frac{(x_2 - x_1)}{\sqrt{(x_2 - x_1)^2 + (y_2 - y_1)^2}} = \frac{1}{5}(0 + 4) = \frac{4}{5}$$

$$\frac{\partial s}{\partial y_2} = \frac{(y_2 - y_1)}{\sqrt{(x_2 - x_1)^2 + (y_2 - y_1)^2}} = \frac{1}{5}(3 - 0) = \frac{3}{5}$$

Moreover, when $t = \pi$, the derivatives of x_1, y_1, x_2, and y_2 are

$$\frac{dx_1}{dt} = -4 \sin t = 0 \qquad \frac{dy_1}{dt} = 2 \cos t = -2$$

$$\frac{dx_2}{dt} = 4 \cos 2t = 4 \qquad \frac{dy_2}{dt} = -6 \sin 2t = 0.$$

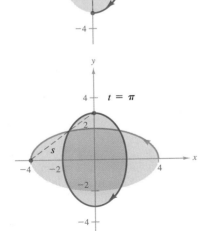

FIGURE 15.36

Therefore, using the appropriate Chain Rule, we know that the distance is changing at the rate of

$$\frac{ds}{dt} = \frac{\partial s}{\partial x_1}\frac{dx_1}{dt} + \frac{\partial s}{\partial y_1}\frac{dy_1}{dt} + \frac{\partial s}{\partial x_2}\frac{dx_2}{dt} + \frac{\partial s}{\partial y_2}\frac{dy_2}{dt}$$

$$= \left(-\frac{4}{5}\right)(0) + \left(-\frac{3}{5}\right)(-2) + \left(\frac{4}{5}\right)(4) + \left(\frac{3}{5}\right)(0)$$

$$= \frac{22}{5}.$$

In Example 2, note that s is the function of four *intermediate* variables, x_1, y_1, x_2, and y_2, which in turn are each functions of a single variable t. Another type of composite function is one in which the intermediate variables are themselves functions of more than one variable. For instance, if

$$w = f(x, y)$$

where

$$x = g(s, t) \qquad \text{and} \qquad y = h(s, t)$$

then it follows that w is a function of s and t, and we can consider the partial derivatives of w with respect to s and t. One way to find these partial derivatives is to write w as a function of s and t explicitly by substituting the equations $x = g(s, t)$ and $y = h(s, t)$ into the equation $w = f(x, y)$. Then, we can find the partial derivatives in the usual way, as demonstrated in the next example.

EXAMPLE 3   Finding partial derivatives by substitution

Find $\partial w/\partial s$ and $\partial w/\partial t$ for $w = 2xy$, where $x = s^2 + t^2$ and $y = s/t$.

SOLUTION

We begin by substituting $x = s^2 + t^2$ and $y = s/t$ into the equation $w = 2xy$ to obtain

$$w = 2xy = 2(s^2 + t^2)\left(\frac{s}{t}\right) = 2\left(\frac{s^3}{t} + st\right).$$

Then, to find $\partial w/\partial s$, we hold t constant and differentiate with respect to s.

$$\frac{\partial w}{\partial s} = 2\left(\frac{3s^2}{t} + t\right) = \frac{6s^2 + 2t^2}{t}$$

Similarly, to find $\partial w/\partial t$, we hold s constant and differentiate with respect to t to obtain

$$\frac{\partial w}{\partial t} = 2\left(-\frac{s^3}{t^2} + s\right) = 2\left(\frac{-s^3 + st^2}{t^2}\right) = \frac{2st^2 - 2s^3}{t^2}.$$

Theorem 15.7 given an alternate method for finding the partial derivatives in Example 3—without explicitly writing w as a function of s and t.

**THEOREM 15.7
CHAIN RULE: TWO INDEPENDENT
VARIABLES**

Let $w = f(x, y)$, where f is a differentiable function of x and y. If $x = g(s, t)$ and $y = h(s, t)$ such that the first partials $\partial x/\partial s$, $\partial x/\partial t$, $\partial y/\partial s$, and $\partial y/\partial t$ all exist, then $\partial w/\partial s$ and $\partial w/\partial t$ exist and are given by

$$\frac{\partial w}{\partial s} = \frac{\partial w}{\partial x}\frac{\partial x}{\partial s} + \frac{\partial w}{\partial y}\frac{\partial y}{\partial s}$$

and

$$\frac{\partial w}{\partial t} = \frac{\partial w}{\partial x}\frac{\partial x}{\partial t} + \frac{\partial w}{\partial y}\frac{\partial y}{\partial t}.$$

PROOF

For $\partial w/\partial s$ we hold t constant, which means that g and h are differentiable functions of s. Thus, we can apply Theorem 15.6 to obtain the desired result. Similarly, for $\partial w/\partial t$ we hold s constant, which means that g and h are differentiable functions of t alone, and apply Theorem 15.6.

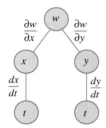

Chain Rule
One Independent Variable

FIGURE 15.37

The "chains" mentioned in Theorems 15.6 and 15.7 can be represented schematically, as shown in Figures 15.37 and 15.38.

Now, we can use the Chain Rule given in Theorem 15.7 to find the same partial derivatives we found in Example 3.

EXAMPLE 4 Using the Chain Rule with two independent variables

Use the Chain Rule to find $\partial w/\partial s$ and $\partial w/\partial t$ for $w = 2xy$, where $x = s^2 + t^2$ and $y = s/t$.

SOLUTION

If we hold t fixed, then, by Theorem 15.7, we have

$$\frac{\partial w}{\partial s} = \frac{\partial w}{\partial x}\frac{\partial x}{\partial s} + \frac{\partial w}{\partial y}\frac{\partial y}{\partial s} = 2y(2s) + 2x\left(\frac{1}{t}\right)$$

$$= 4\left(\frac{s^2}{t}\right) + \frac{2s^2 + 2t^2}{t} = \frac{6s^2 + 2t^2}{t}.$$

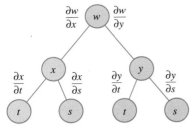

Chain Rule
Two Independent Variables

FIGURE 15.38

Similarly, holding s fixed gives us

$$\frac{\partial w}{\partial t} = \frac{\partial w}{\partial x}\frac{\partial x}{\partial t} + \frac{\partial w}{\partial y}\frac{\partial y}{\partial t} = 2y(2t) + 2x\left(\frac{-s}{t^2}\right)$$

$$= 4s - \frac{2s^3 + 2st^2}{t^2} = \frac{4st^2 - 2s^3 - 2st^2}{t^2}$$

$$= \frac{2st^2 - 2s^3}{t^2}.$$

The Chain Rule in Theorem 15.7 can also be extended to any number of variables. For example, if w is a differentiable function of the n variables

$x_1, x_2, \ldots, x_n$ where each x_i is a differentiable function of the m variables $t_1, t_2, \ldots, t_m$, then for $w = f(x_1, x_2, \ldots, x_n)$ we have the following.

$$\frac{\partial w}{\partial t_1} = \frac{\partial w}{\partial x_1}\frac{\partial x_1}{\partial t_1} + \frac{\partial w}{\partial x_2}\frac{\partial x_2}{\partial t_1} + \cdots + \frac{\partial w}{\partial x_n}\frac{\partial x_n}{\partial t_1}$$

$$\frac{\partial w}{\partial t_2} = \frac{\partial w}{\partial x_1}\frac{\partial x_1}{\partial t_2} + \frac{\partial w}{\partial x_2}\frac{\partial x_2}{\partial t_2} + \cdots + \frac{\partial w}{\partial x_n}\frac{\partial x_n}{\partial t_2}$$

$$\vdots$$

$$\frac{\partial w}{\partial t_m} = \frac{\partial w}{\partial x_1}\frac{\partial x_1}{\partial t_m} + \frac{\partial w}{\partial x_2}\frac{\partial x_2}{\partial t_m} + \cdots + \frac{\partial w}{\partial x_n}\frac{\partial x_n}{\partial t_m}$$

EXAMPLE 5 Using the Chain Rule for a function of three variables

Find $\partial w/\partial s$ and $\partial w/\partial t$ when $s = 1$ and $t = 2\pi$ for the function given by

$$w = xy + yz + xz$$

where $x = s \cos t$, $y = s \sin t$, and $z = t$.

SOLUTION

By extending the result of Theorem 15.7, we have

$$\frac{\partial w}{\partial s} = \frac{\partial w}{\partial x}\frac{\partial x}{\partial s} + \frac{\partial w}{\partial y}\frac{\partial y}{\partial s} + \frac{\partial w}{\partial z}\frac{\partial z}{\partial s}$$

$$= (y + z)(\cos t) + (x + z)(\sin t) + (y + x)(0).$$

When $s = 1$ and $t = 2\pi$, we have $x = 1$, $y = 0$, and $z = 2\pi$. Therefore,

$$\frac{\partial w}{\partial s} = 2\pi(1) + (1 + 2\pi)(0) + 0 = 2\pi.$$

Furthermore,

$$\frac{\partial w}{\partial t} = \frac{\partial w}{\partial x}\frac{\partial x}{\partial t} + \frac{\partial w}{\partial y}\frac{\partial y}{\partial t} + \frac{\partial w}{\partial z}\frac{\partial z}{\partial t}$$

$$= (y + z)(-s \sin t) + (x + z)(s \cos t) + (y + x)(1)$$

and for $s = 1$ and $t = 2\pi$ it follows that

$$\frac{\partial w}{\partial t} = (0 + 2\pi)(0) + (1 + 2\pi)(1) + (0 + 1)(1) = 2 + 2\pi. \qquad \blacksquare$$

Implicit partial differentiation

We conclude this section with an application of the Chain Rule to determine the derivative of a function defined *implicitly*. Suppose that x and y are related by the equation $F(x, y) = 0$, where it is assumed that $y = f(x)$ is a differentiable function of x. To find dy/dx, we could use the techniques discussed in Section 3.6. However, we will see that the Chain Rule provides a convenient alternative. If we consider the function given by

$$w = F(x, y) = F(x, f(x))$$

then we can apply Theorem 15.6 to obtain

$$\frac{dw}{dx} = F_x(x, y) \frac{dx}{dx} + F_y(x, y) \frac{dy}{dx}.$$

Since $w = F(x, y) = 0$ for all x in the domain of f, we know that $dw/dx = 0$ and we have

$$F_x(x, y) \frac{dx}{dx} + F_y(x, y) \frac{dy}{dx} = 0.$$

Now, if $F_y(x, y) \neq 0$, we can conclude that

$$\frac{dy}{dx} = -\frac{F_x(x, y)}{F_y(x, y)}.$$

A similar procedure can be used to find the partial derivatives of functions of several variables that are defined implicitly.

**THEOREM 15.8
CHAIN RULE: IMPLICIT
DIFFERENTIATION**

If the equation $F(x, y) = 0$ defines y implicitly as a differentiable function of x, then

$$\frac{dy}{dx} = -\frac{F_x(x, y)}{F_y(x, y)}, \quad F_y(x, y) \neq 0.$$

If the equation $F(x, y, z) = 0$ defines z implicitly as a differentiable function of x and y, then

$$\frac{\partial z}{\partial x} = -\frac{F_x(x, y, z)}{F_z(x, y, z)} \quad \text{and} \quad \frac{\partial z}{\partial y} = -\frac{F_y(x, y, z)}{F_z(x, y, z)}, \quad F_z(x, y, z) \neq 0.$$

This theorem can be extended to differentiable functions defined implicitly with any number of variables.

EXAMPLE 6   Finding a derivative implicitly

Find dy/dx, given $y^3 + y^2 - 5y - x^2 + 4 = 0$.

SOLUTION

We define a function F by

$$F(x, y) = y^3 + y^2 - 5y - x^2 + 4.$$

Then, using Theorem 15.8, we have

$$F_x(x, y) = -2x$$
$$F_y(x, y) = 3y^2 + 2y - 5$$

and it follows that

$$\frac{dy}{dx} = -\frac{F_x(x, y)}{F_y(x, y)} = \frac{-(-2x)}{3y^2 + 2y - 5} = \frac{2x}{3y^2 + 2y - 5}.$$

REMARK   Compare the solution in Example 6 to the solution in Example 2 of Section 3.6.

EXAMPLE 7   Finding partial derivatives implicitly

Find $\partial z/\partial x$ and $\partial z/\partial y$, given $3x^2z - x^2y^2 + 2z^3 + 3yz - 5 = 0$.

SOLUTION

To apply Theorem 15.8 we let $F(x, y, z) = 3x^2z - x^2y^2 + 2z^3 + 3yz - 5$. Then

$$F_x(x, y, z) = 6xz - 2xy^2$$
$$F_y(x, y, z) = -2x^2y + 3z$$
$$F_z(x, y, z) = 3x^2 + 6z^2 + 3y$$

and we obtain

$$\frac{\partial z}{\partial x} = -\frac{F_x}{F_z} = \frac{2xy^2 - 6xz}{3x^2 + 6z^2 + 3y}$$

$$\frac{\partial z}{\partial y} = -\frac{F_y}{F_z} = \frac{2x^2y - 3z}{3x^2 + 6z^2 + 3y}.$$

EXERCISES for Section 15.5

In Exercises 1–6, find dw/dt using the appropriate Chain Rule.

1. $w = x^2 + y^2$
   $x = e^t, y = e^{-t}$

2. $w = \sqrt{x^2 + y^2}$
   $x = \sin t, y = e^t$

3. $w = x \sec y$
   $x = e^t, y = \pi - t$

4. $w = \ln \dfrac{y}{x}$
   $x = \cos t, y = \sin t$

5. $w = x^2 + y^2 + z^2$
   $x = e^t \cos t, y = e^t \sin t, z = e^t$

6. $w = xy \cos z$
   $x = t, y = t^2, z = \arccos t$

In Exercises 7–10, find $\partial w/\partial s$ and $\partial w/\partial t$ using the appropriate Chain Rule, and evaluate each partial derivative at the indicated values of s and t.

Function	Point
7. $w = x^2 + y^2$ $x = s + t, y = s - t$	$s = 2, t = -1$
8. $w = y^3 - 3x^2y$ $x = e^s, y = e^t$	$s = 0, t = 1$
9. $w = x^2 - y^2$ $x = s \cos t, y = s \sin t$	$s = 3, t = \dfrac{\pi}{4}$
10. $w = \sin(2x + 3y)$ $x = s + t, y = s - t$	$s = 0, t = \dfrac{\pi}{2}$

In Exercises 11–14, find dw/dt (a) by the appropriate Chain Rule and (b) by converting w to a function of t before differentiating.

11. $w = xy$
$x = 2 \sin t, y = \cos t$

12. $w = \cos (x - y)$
$x = t^2, y = 1$

13. $w = xy + xz + yz$
$x = t - 1, y = t^2 - 1, z = t$

14. $w = xyz$
$x = t^2, y = 2t, z = e^{-t}$

In Exercises 15–18, find $\partial w/\partial r$ and $\partial w/\partial \theta$ (a) by the appropriate Chain Rule and (b) by converting w to a function of r and θ before differentiating.

15. $w = x^2 - 2xy + y^2$
$x = r + \theta, y = r - \theta$

16. $w = \sqrt{4 - 2x^2 - 2y^2}$
$x = r \cos \theta, y = r \sin \theta$

17. $w = \arctan \dfrac{y}{x}$
$x = r \cos \theta, y = r \sin \theta$

18. $w = \dfrac{xy}{z}$
$x = r + \theta, y = r - \theta, z = \theta^2$

In Exercises 19–22, differentiate implicitly to find the first partial derivatives of z.

19. $x^2 + y^2 + z^2 = 25$

20. $xz + yz + xy = 0$

21. $\tan (x + y) + \tan (y + z) = 1$

22. $z = e^x \sin (y + z)$

In Exercises 23 and 24, differentiate implicitly to find all first and second partial derivatives of z.

23. $x^2 + 2yz + z^2 = 1$

24. $x + \sin (y + z) = 0$

In Exercises 25 and 26, differentiate implicitly to find the first partial derivatives of w.

25. $xyz + xzw - yzw + w^2 = 5$

26. $x^2 + y^2 + z^2 + 6xw - 8w^2 = 5$

27. The dimensions of a rectangular chamber are increasing at the following rates: length 3 feet per minute, width 2 feet per minute, and depth $\frac{1}{2}$ foot per minute. What are the rates of change of the volume and the surface area when the length, width, and depth are 10 feet, 6 feet, and 4 feet, respectively?

28. The radius of a right-circular cylinder is increasing at the rate of 6 inches per minute, and the height is decreasing at the rate of 4 inches per minute. What is the rate of change of the volume and surface area when the radius is 12 inches and the height is 36 inches?

29. Repeat Exercise 28 for a right-circular cone.

30. The two radii of the frustrum of a right-circular cone are increasing at the rate of 4 centimeters per minute, and the height is increasing at the rate of 12 centimeters per minute (see figure). Find the rate at which the volume and surface area are changing when the two radii are 15 centimeters and 25 centimeters and the height is 10 centimeters.

31. An annular cylinder has an inside radius of r_1 and an outside radius of r_2 (see figure). The moment of inertia is given by

$$I = \frac{1}{2}m(r_1^2 + r_2^2)$$

where m is the mass. Find the rate at which I is changing at the instant the radii are 6 centimeters and 8 centimeters, respectively, if the two radii are increasing at the rate of 2 centimeters per second.

FIGURE FOR 30   FIGURE FOR 31

32. The Ideal Gas Law is $pV = RT$, where R is a constant. If p and V are functions of time, find dT/dt, the rate at which the temperature changes with respect to time.

In Exercises 33–37, the given function is **homogeneous of degree *n***, which means that

$$f(tx, ty) = t^n f(x, y).$$

Find the degree of the given function, and show that

$$xf_x(x, y) + yf_y(x, y) = nf(x, y).$$

33. $f(x, y) = x^3 - 3xy^2 + y^3$

34. $f(x, y) = \dfrac{xy}{\sqrt{x^2 + y^2}}$

35. $f(x, y) = e^{x/y}$

36. $f(x, y) = 2x^3 - 3xy^2$

37. $f(x, y) = \dfrac{x^2}{\sqrt{x^2 + y^2}}$

38. Show that if $f(x, y)$ is homogeneous of degree n, then

$$xf_x(x, y) + yf_y(x, y) = nf(x, y).$$

[Hint: Let $g(t) = f(tx, ty) = t^n f(x, y)$. Find $g'(t)$ and then let $t = 1$.]

39. Show that

$$\frac{\partial w}{\partial u} + \frac{\partial w}{\partial v} = 0$$

for $w = f(x, y)$, $x = u - v$, and $y = v - u$.

40. Demonstrate the result of Exercise 39 for the function given by

$$w = (x - y) \sin (y - x).$$

41. Consider the function $w = f(x, y)$ where $x = r \cos \theta$ and $y = r \sin \theta$. Prove the following.

(a) $\dfrac{\partial w}{\partial x} = \dfrac{\partial w}{\partial r} \cos \theta - \dfrac{\partial w}{\partial \theta} \dfrac{\sin \theta}{r}$

$\dfrac{\partial w}{\partial y} = \dfrac{\partial w}{\partial r} \sin \theta + \dfrac{\partial w}{\partial \theta} \dfrac{\cos \theta}{r}$

(b) $\left(\dfrac{\partial w}{\partial x}\right)^2 + \left(\dfrac{\partial w}{\partial y}\right)^2 = \left(\dfrac{\partial w}{\partial r}\right)^2 + \left(\dfrac{1}{r^2}\right)\left(\dfrac{\partial w}{\partial \theta}\right)^2$

42. Demonstrate the result of Exercise 41(b) for the function given by

$$w = \arctan \frac{y}{x}.$$

43. Given the functions $u(x, y)$ and $v(x, y)$, show that the **Cauchy-Riemann differential equations**

$$\frac{\partial u}{\partial x} = \frac{\partial v}{\partial y} \quad \text{and} \quad \frac{\partial u}{\partial y} = -\frac{\partial v}{\partial x}$$

can be written in polar coordinate form as

$$\frac{\partial u}{\partial r} = \frac{1}{r} \frac{\partial v}{\partial \theta} \quad \text{and} \quad \frac{\partial v}{\partial r} = -\frac{1}{r} \frac{\partial u}{\partial \theta}.$$

44. Demonstrate the result of Exercise 43 for the functions

$$u = \ln \sqrt{x^2 + y^2} \quad \text{and} \quad v = \arctan \frac{y}{x}.$$

45. Show that

$$u(x, t) = \frac{1}{2}[f(x - ct) + f(x + ct)]$$

is a solution to the one-dimensional wave equation

$$\frac{\partial^2 u}{\partial t^2} = c^2 \frac{\partial^2 u}{\partial x^2}.$$

[This equation describes the small transverse vibration of an elastic string like those on certain musical instruments.]

15.6 Directional Derivatives and Gradients

Directional derivative ▪ Gradient ▪ Applications of the gradient ▪ Functions of three variables

Suppose you were standing on the hillside pictured in Figure 15.39 and wanted to determine the hill's incline toward the z-axis. If the hill were represented by $z = f(x, y)$, then you would already know how to determine the slope in two different directions—the slope in the y direction would be given by the partial derivative $f_y(x, y)$, and the slope in the x direction would be given by the partial derivative $f_x(x, y)$. In this section, you will see that these two partial derivatives can be used to find the slope in *any* direction.

To determine the slope at a point on a surface, we define a new type of derivative called a **directional derivative.** We begin by letting $z = f(x, y)$ be a *surface* and $P = (x_0, y_0)$ a *point* in the domain of f, as shown in Figure 15.40. We specify *direction* by a unit vector $\mathbf{u} = \cos \theta \, \mathbf{i} + \sin \theta \, \mathbf{j}$, where θ

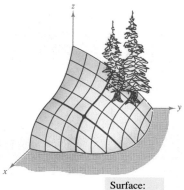

FIGURE 15.39

Surface:
$z = f(x, y)$

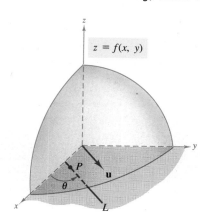

FIGURE 15.40

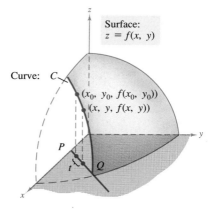

Surface:
$z = f(x, y)$

Curve: C

$(x_0, y_0, f(x_0, y_0))$
$(x, y, f(x, y))$

P

Q

FIGURE 15.41

is the angle the vector makes with the positive x-axis. Now, to find the desired slope, we reduce the problem to two dimensions by intersecting the surface with a vertical plane passing through the point P and parallel to **u**, as shown in Figure 15.41. This vertical plane intersects the surface to form a curve C, and we define the slope of the surface at $(x_0, y_0, f(x_0, y_0))$ in the direction of **u** to be the slope of the curve C at that point.

Informally, we can write the slope of the curve C as a limit that looks much like those used in single-variable calculus. The vertical plane used to form C intersects the xy-plane in a line L, represented by the parametric equations

$$x = x_0 + t \cos \theta \quad \text{and} \quad y = y_0 + t \sin \theta$$

so that for any value of t, the point $Q = (x, y)$ lies on the line L. For each of the points P and Q, there is a corresponding point on the surface,

$$(x_0, y_0, f(x_0, y_0)) \qquad \text{Point above } P$$

and

$$(x, y, f(x, y)). \qquad \text{Point above } Q$$

Moreover, since the distance between P and Q is

$$\sqrt{(x - x_0)^2 + (y - y_0)^2} = \sqrt{(t \cos \theta)^2 + (t \sin \theta)^2} = |t|$$

we can write the slope of the secant line through $(x_0, y_0, f(x_0, y_0))$ and $(x, y, f(x, y))$ as

$$\frac{f(x, y) - f(x_0, y_0)}{t} = \frac{f(x_0 + t \cos \theta, y_0 + t \sin \theta) - f(x_0, y_0)}{t}.$$

Finally, by letting t approach zero, we arrive at the following definition.

DEFINITION OF DIRECTIONAL DERIVATIVE

Let f be a function of two variables x and y and let $\mathbf{u} = \cos \theta \, \mathbf{i} + \sin \theta \, \mathbf{j}$ be a unit vector. Then the **directional derivative of f in the direction of u,** denoted by $D_\mathbf{u} f$, is

$$D_\mathbf{u} f(x, y) = \lim_{t \to 0} \frac{f(x + t \cos \theta, y + t \sin \theta) - f(x, y)}{t}.$$

Calculating directional derivatives by this definition is similar to finding the derivative of a function of one variable by the limit process (given in Section 3.1). A simpler "working" formula for finding directional derivatives involves the partial derivatives f_x and f_y.

THEOREM 15.9 DIRECTIONAL DERIVATIVE

If f is a differentiable function of x and y, then the directional derivative of f in the direction of the unit vector $\mathbf{u} = \cos \theta \, \mathbf{i} + \sin \theta \, \mathbf{j}$ is

$$D_\mathbf{u} f(x, y) = f_x(x, y) \cos \theta + f_y(x, y) \sin \theta.$$

PROOF   For a fixed point (x_0, y_0), we let $x = x_0 + t \cos \theta$ and $y = y_0 + t \sin \theta$. Then we let $g(t) = f(x, y)$. Since f is differentiable, we can apply the Chain Rule given in Theorem 15.7 to obtain

$$g'(t) = f_x(x, y) \frac{\partial x}{\partial t} + f_y(x, y) \frac{\partial y}{\partial t}$$

$$= f_x(x, y) \cos \theta + f_y(x, y) \sin \theta.$$

If $t = 0$, then $x = x_0$ and $y = y_0$, so

$$g'(0) = f_x(x_0, y_0) \cos \theta + f_y(x_0, y_0) \sin \theta.$$

By definition, it is also true that

$$g'(0) = \lim_{t \to 0} \frac{g(t) - g(0)}{t}$$

$$= \lim_{t \to 0} \frac{f(x_0 + t \cos \theta, y_0 + t \sin \theta) - f(x_0, y_0)}{t}.$$

Consequently, $D_{\mathbf{u}} f(x_0, y_0) = f_x(x_0, y_0) \cos \theta + f_y(x_0, y_0) \sin \theta.$

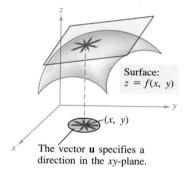

The vector **u** specifies a direction in the *xy*-plane.

FIGURE 15.42

Be sure to see that there are infinitely many directional derivatives to a surface at a given point—one for each direction specified by **u**, as indicated in Figure 15.42. Two of these turn out to be the partial derivatives f_x and f_y.

1. Direction of positive *x*-axis $(\theta = 0)$: $\mathbf{u} = \cos 0 \, \mathbf{i} + \sin 0 \, \mathbf{j} = \mathbf{i}$

$$D_{\mathbf{i}} f(x, y) = f_x(x, y) \cos 0 + f_y(x, y) \sin 0 = f_x(x, y)$$

2. Direction of positive *y*-axis $\left(\theta = \dfrac{\pi}{2}\right)$: $\mathbf{u} = \cos \dfrac{\pi}{2} \mathbf{i} + \sin \dfrac{\pi}{2} \mathbf{j} = \mathbf{j}$

$$D_{\mathbf{j}} f(x, y) = f_x(x, y) \cos \frac{\pi}{2} + f_y(x, y) \sin \frac{\pi}{2} = f_y(x, y)$$

EXAMPLE 1   Finding a directional derivative

Find the directional derivative of $f(x, y) = 4 - x^2 - \frac{1}{4}y^2$ at $(1, 2)$ in the direction of $\mathbf{u} = \cos (\pi/3) \, \mathbf{i} + \sin (\pi/3) \, \mathbf{j}$.

SOLUTION

$$D_{\mathbf{u}} f(x, y) = f_x(x, y) \cos \theta + f_y(x, y) \sin \theta$$

$$= (-2x) \cos \theta + \left(-\frac{y}{2}\right) \sin \theta$$

Evaluating at $\theta = \pi/3$, $x = 1$, and $y = 2$ produces

$$D_{\mathbf{u}} f(1, 2) = (-2)\left(\frac{1}{2}\right) + (-1)\left(\frac{\sqrt{3}}{2}\right) = -1 - \frac{\sqrt{3}}{2} \approx -1.866. \quad \blacksquare$$

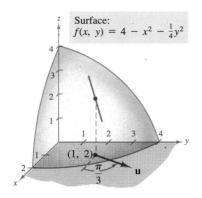

Surface:
$f(x, y) = 4 - x^2 - \frac{1}{4}y^2$

FIGURE 15.43

REMARK   Note in Figure 15.43 that we can interpret the directional derivative as giving the slope of the surface at the point (1, 2, 2) in the direction of the unit vector **u**.

We have been specifying direction by a unit vector **u**. If the direction is given by a vector whose length is not 1, then we must normalize the vector before applying the formula in Theorem 15.9.

EXAMPLE 2   Finding a directional derivative

Find the directional derivative of $f(x, y) = x^2 \sin 2y$ at $(1, \pi/4)$ in the direction of $\mathbf{v} = 3\mathbf{i} - 4\mathbf{j}$.

SOLUTION

We begin by finding a unit vector in the direction of **v**.

$$\mathbf{u} = \frac{\mathbf{v}}{\|\mathbf{v}\|} = \frac{3}{5}\mathbf{i} - \frac{4}{5}\mathbf{j} = \cos\theta\,\mathbf{i} + \sin\theta\,\mathbf{j}$$

Now, using this unit vector, we have

$$D_{\mathbf{u}} f(x, y) = (2x \sin 2y)(\cos\theta) + (2x^2 \cos 2y)(\sin\theta)$$

$$D_{\mathbf{u}} f\left(1, \frac{\pi}{4}\right) = \left(2 \sin\frac{\pi}{2}\right)\left(\frac{3}{5}\right) + \left(2 \cos\frac{\pi}{2}\right)\left(-\frac{4}{5}\right)$$

$$= (2)\left(\frac{3}{5}\right) + (0)\left(-\frac{4}{5}\right) = \frac{6}{5}.$$

The gradient of a function of two variables

The directional derivative $D_{\mathbf{u}} f(x, y)$ can be expressed as the dot product of the unit vector

$$\mathbf{u} = \cos\theta\,\mathbf{i} + \sin\theta\,\mathbf{j}$$

and the vector

$$f_x(x, y)\mathbf{i} + f_y(x, y)\mathbf{j}.$$

This latter vector is important and has many uses. We call it the **gradient of f.**

DEFINITION OF GRADIENT OF A FUNCTION OF TWO VARIABLES	If $z = f(x, y)$, then the **gradient of f**, denoted by $\nabla f(x, y)$, is the vector $$\nabla f(x, y) = f_x(x, y)\mathbf{i} + f_y(x, y)\mathbf{j}.$$ We read ∇f as "del f." Another notation for the gradient is **grad** $f(x, y)$.

REMARK   We do not assign a value to the symbol ∇ by itself. It is an operator in the same sense that d/dx is an operator. When ∇ operates on $f(x, y)$, it produces the vector $\nabla f(x, y)$.

Since the gradient of f is a vector, we can write the directional derivative of f in the direction of **u** as

$$D_{\mathbf{u}}f(x, y) = [f_x(x, y)\mathbf{i} + f_y(x, y)\mathbf{j}] \cdot [\cos\theta\,\mathbf{i} + \sin\theta\,\mathbf{j}].$$

In other words, the directional derivative is the dot product of the gradient and the direction vector. We summarize this useful result in the following theorem.

THEOREM 15.10 **ALTERNATE FORM OF THE** **DIRECTIONAL DERIVATIVE**	If f is a differentiable function of x and y, then the directional derivative of f in the direction of the unit vector **u** is $$D_{\mathbf{u}}f(x, y) = \nabla f(x, y) \cdot \mathbf{u}.$$

EXAMPLE 3 Using $\nabla f(x, y)$ to find a directional derivative

Find the directional derivative of $f(x, y) = 3x^2 - 2y^2$ at $(-1, 3)$ in the direction from $P = (-1, 3)$ to $Q = (1, -2)$.

SOLUTION

A vector in the specified direction is

$$\overrightarrow{PQ} = \mathbf{v} = (1 + 1)\mathbf{i} + (-2 - 3)\mathbf{j} = 2\mathbf{i} - 5\mathbf{j}$$

and a unit vector in this direction is

$$\mathbf{u} = \frac{\mathbf{v}}{\|\mathbf{v}\|} = \frac{2}{\sqrt{29}}\mathbf{i} - \frac{5}{\sqrt{29}}\mathbf{j}.$$

Since $\nabla f(x, y) = f_x(x, y)\mathbf{i} + f_y(x, y)\mathbf{j} = 6x\mathbf{i} - 4y\mathbf{j}$, the gradient at $(-1, 3)$ is

$$\nabla f(-1, 3) = -6\mathbf{i} - 12\mathbf{j}.$$

Consequently, at $(-1, 3)$ the directional derivative is

$$D_{\mathbf{u}}f(-1, 3) = \nabla f(-1, 3) \cdot \mathbf{u} = (-6\mathbf{i} - 12\mathbf{j}) \cdot \left(\frac{2}{\sqrt{29}}\mathbf{i} - \frac{5}{\sqrt{29}}\mathbf{j}\right)$$

$$= \frac{-12}{\sqrt{29}} + \frac{60}{\sqrt{29}} = \frac{48}{\sqrt{29}}.$$ ▬

We have already seen that there are many directional derivatives at the point (x, y) on a surface. In many applications we would like to know in which direction to move so that $f(x, y)$ increases most rapidly. We call this the direction of steepest ascent, and it is given by the gradient, as stated in the following theorem.

THEOREM 15.11
PROPERTIES OF THE GRADIENT

Let f be differentiable at the point (x, y).

1. If $\nabla f(x, y) = \mathbf{0}$, then $D_{\mathbf{u}}f(x, y) = 0$ for all $\mathbf{u}$.
2. The direction of *maximum* increase of f is given by $\nabla f(x, y)$. The maximum value of $D_{\mathbf{u}}f(x, y)$ is $\|\nabla f(x, y)\|$.
3. The direction of *minimum* increase of f is given by $-\nabla f(x, y)$. The minimum value of $D_{\mathbf{u}}f(x, y)$ is $-\|\nabla f(x, y)\|$.

PROOF

If $\nabla f(x, y) = \mathbf{0}$, then for any direction (any $\mathbf{u}$), we have

$$D_{\mathbf{u}}f(x, y) = \nabla f(x, y) \cdot \mathbf{u} = (0\mathbf{i} + 0\mathbf{j}) \cdot (\cos \theta\, \mathbf{i} + \sin \theta\, \mathbf{j}) = 0.$$

If $\nabla f(x, y) \neq \mathbf{0}$, then let ϕ be the angle between $\nabla f(x, y)$ and a unit vector $\mathbf{u}$. Using the dot product, we can apply Theorem 13.6 to conclude that

$$D_{\mathbf{u}}f(x, y) = \nabla f(x, y) \cdot \mathbf{u} = \|\nabla f(x, y)\|\,\|\mathbf{u}\| \cos \phi = \|\nabla f(x, y)\| \cos \phi$$

and it follows that the maximum value $D_{\mathbf{u}}f(x, y)$ will occur when $\cos \phi = 1$. Thus, $\phi = 0$, and the maximum value for the directional derivative occurs when $\mathbf{u}$ has the same direction as $\nabla f(x, y)$. Moreover, this largest value for $D_{\mathbf{u}}f(x, y)$ is precisely

$$\|\nabla f(x, y)\| \cos \phi = \|\nabla f(x, y)\|.$$

Similarly, the minimum value of $D_{\mathbf{u}}f(x, y)$ can be obtained by letting $\phi = \pi$ so that $\mathbf{u}$ points in the direction opposite that of $\nabla f(x, y)$, as indicated in Figure 15.44.

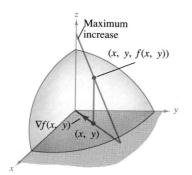

FIGURE 15.44

To visualize one of the properties of the gradient, imagine a skier coming down a mountainside. If $f(x, y)$ denotes the altitude of the skier, then $-\nabla f(x, y)$ indicates the *compass direction* the skier should take to ski the path of steepest descent. (Remember that the gradient indicates direction in the xy-plane and does not itself point up or down the mountainside.)

As another illustration of the gradient, consider the temperature $T(x, y)$ at any point (x, y) on a flat metal plate. In this case, $\nabla T(x, y)$ gives the direction of greatest temperature increase at point (x, y), as illustrated in the next example.

EXAMPLE 4   Finding the direction of maximum increase

The temperature in degrees Celsius on the surface of a metal plate is

$$T(x, y) = 20 - 4x^2 - y^2$$

where x and y are measured in inches. In what direction from $(2, -3)$ does the temperature increase most rapidly? What is this rate of increase?

SOLUTION

The gradient is

$$\nabla T(x, y) = T_x(x, y)\mathbf{i} + T_y(x, y)\mathbf{j} = -8x\mathbf{i} - 2y\mathbf{j}.$$

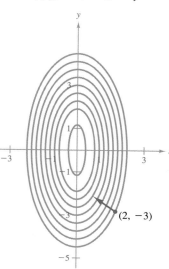

Level curves:
$T(x, y) = 20 - 4x^2 - y^2$

$(2, -3)$

Direction of Most Rapid Increase
in Temperature at $(2, -3)$

FIGURE 15.45

It follows that the direction of maximum increase is given by

$$\nabla T(2, -3) = -16\mathbf{i} + 6\mathbf{j}$$

as shown in Figure 15.45, and the rate of increase is

$$\|\nabla T(2, -3)\| = \sqrt{256 + 36} = \sqrt{292} \approx 17.09° \text{ per inch.}$$

The solution presented in Example 4 can be misleading. Although the gradient points in the direction of maximum temperature increase, it does not necessarily point toward the hottest spot on the plate. In other words, the gradient provides a local solution to finding an increase relative to the temperature at the point $(2, -3)$. *Once we leave that position, the direction of maximum increase may change.*

EXAMPLE 5   Finding the path of a heat-seeking particle

A heat-seeking particle is located at the point $(2, -3)$ on a metal plate whose temperature at (x, y) is $T(x, y) = 20 - 4x^2 - y^2$. Find the path of the particle as it continuously moves in the direction of maximum temperature increase.

SOLUTION

We let the path be represented by the position function

$$\mathbf{r}(t) = x(t)\mathbf{i} + y(t)\mathbf{j}.$$

A tangent vector at each point $(x(t), y(t))$ is given by

$$\mathbf{r}'(t) = \frac{dx}{dt}\mathbf{i} + \frac{dy}{dt}\mathbf{j}.$$

Since the particle seeks maximum temperature increase, the directions of $\mathbf{r}'(t)$ and $\nabla T(x, y) = -8x\mathbf{i} - 2y\mathbf{j}$ are the same at each point of the path. Thus,

$$-8x = \frac{dx}{dt} \quad \text{and} \quad -2y = \frac{dy}{dt}.$$

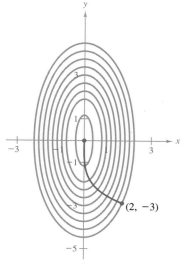

Level curves:
$T(x, y) = 20 - 4x^2 - y^2$

$(2, -3)$

Path Followed by
Heat-Seeking Particle

FIGURE 15.46

These differential equations represent exponential growth (see Theorem 7.14), and the solutions are

$$x(t) = C_1 e^{-8t} \quad \text{and} \quad y(t) = C_2 e^{-2t}.$$

Now, since the particle starts at $(2, -3)$, it follows that $2 = x(0) = C_1$ and $-3 = y(0) = C_2$. Thus, the path is represented by

$$\mathbf{r}(t) = x(t)\mathbf{i} + y(t)\mathbf{j} = 2e^{-8t}\mathbf{i} - 3e^{-2t}\mathbf{j}.$$

Eliminating the parameter t produces

$$x = 2e^{-8t} = \frac{2}{81}(-3e^{-2t})^4 = \frac{2}{81}y^4 \quad \Longrightarrow \quad x = \frac{2}{81}y^4.$$

The path is shown in Figure 15.46.

In Figure 15.46, the path of the particle (determined by the gradient at each point) appears to be orthogonal to each of the level curves. This becomes clear when we consider that the temperature $T(x, y)$ is constant along a given level curve. Hence, at any point (x, y) on the curve, the rate of change of T in the direction of a unit tangent vector $\mathbf{u}$ is 0, and we can write

$$\nabla f(x, y) \cdot \mathbf{u} = D_{\mathbf{u}} T(x, y) = 0. \qquad \text{\footnotesize \textbf{u} is a unit tangent vector}$$

Since the dot product of $\nabla f(x, y)$ and $\mathbf{u}$ is zero, we conclude that they must be orthogonal. This result is stated in the following theorem.

THEOREM 15.12 **GRADIENT IS NORMAL** **TO LEVEL CURVES**	If f is differentiable at (x_0, y_0), and $\nabla f(x_0, y_0) \neq \mathbf{0}$, then $\nabla f(x_0, y_0)$ is normal to the level curve through (x_0, y_0).

EXAMPLE 6   Finding a normal vector to a level curve

Sketch the level curve corresponding to $c = 0$ for the function given by

$$f(x, y) = y - \sin x$$

and find a normal vector at several points on the curve.

SOLUTION

The level curve for $c = 0$ is given by

$$0 = y - \sin x \quad \Longrightarrow \quad y = \sin x$$

as shown in Figure 15.47. Since the gradient vector of f at (x, y) is

$$\nabla f(x, y) = f_x(x, y)\mathbf{i} + f_y(x, y)\mathbf{j} = -\cos x\,\mathbf{i} + \mathbf{j}$$

we can use Theorem 15.12 to conclude that $\nabla f(x, y)$ is normal to the level curve at the point (x, y). Some gradient vectors are

$$\nabla f(-\pi, 0) = \mathbf{i} + \mathbf{j},$$
$$\nabla f\left(-\frac{2\pi}{3}, -\frac{\sqrt{3}}{2}\right) = \frac{1}{2}\mathbf{i} + \mathbf{j},$$
$$\nabla f\left(-\frac{\pi}{2}, -1\right) = \mathbf{j}.$$

Several others are shown in Figure 15.47.

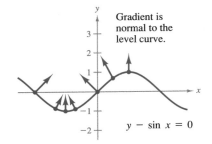

Gradient is normal to the level curve.

$y - \sin x = 0$

FIGURE 15.47

Functions of three variables

The definitions of the directional derivative and the gradient can be extended naturally to functions of three or more variables. As often happens, some of the geometrical interpretation is lost in the generalization from functions of two variables to three variables. For example, we do not interpret the directional derivative of a function of three variables to represent slope.

We list the definitions and properties of the directional derivative and the gradient of a function of three variables in the following summary.

DIRECTIONAL DERIVATIVE AND GRADIENT FOR A FUNCTION OF THREE VARIABLES

Let f be a function of x, y, and z, with continuous first partial derivatives. The **directional derivative of f** in the direction of a unit vector $\mathbf{u} = a\mathbf{i} + b\mathbf{j} + c\mathbf{k}$ is given by

$$D_{\mathbf{u}} f(x, y, z) = af_x(x, y, z) + bf_y(x, y, z) + cf_z(x, y, z).$$

The **gradient of f** is defined to be

$$\nabla f(x, y, z) = f_x(x, y, z)\mathbf{i} + f_y(x, y, z)\mathbf{j} + f_z(x, y, z)\mathbf{k}.$$

Properties of the gradient are as follows.

1. $D_{\mathbf{u}} f(x, y, z) = \nabla f(x, y, z) \cdot \mathbf{u}$
2. If $\nabla f(x, y, z) = \mathbf{0}$, then $D_{\mathbf{u}} f(x, y, z) = 0$ for all $\mathbf{u}$.
3. The direction of *maximum* increase of f is given by $\nabla f(x, y, z)$. The maximum value of $D_{\mathbf{u}} f(x, y, z)$ is $\|\nabla f(x, y, z)\|$.
4. The direction of *minimum* increase of f is given by $-\nabla f(x, y, z)$. The minimum value of $D_{\mathbf{u}} f(x, y, z)$ is $-\|\nabla f(x, y, z)\|$.

EXAMPLE 7   Finding the gradient for a function of three variables

Find $\nabla f(x, y, z)$ for the function given by

$$f(x, y, z) = x^2 + y^2 - 4z$$

and find the direction of maximum increase of f at the point $(2, -1, 1)$.

SOLUTION

The gradient vector is given by

$$\nabla f(x, y, z) = f_x(x, y, z)\mathbf{i} + f_y(x, y, z)\mathbf{j} + f_z(x, y, z)\mathbf{k}$$
$$= 2x\mathbf{i} + 2y\mathbf{j} - 4\mathbf{k}.$$

Hence, it follows that the direction of maximum increase at $(2, -1, 1)$ is

$$\nabla f(2, -1, 1) = 4\mathbf{i} - 2\mathbf{j} - 4\mathbf{k}.$$

EXERCISES for Section 15.6

In Exercises 1–12, find the directional derivative of the function at P in the direction of $\mathbf{v}$.

1. $f(x, y) = 3x - 4xy + 5y$,

$P = (1, 2)$, $\mathbf{v} = \dfrac{1}{2}(\mathbf{i} + \sqrt{3}\mathbf{j})$

2. $f(x, y) = x^2 - y^2$,

$P = (4, 3)$, $\mathbf{v} = \dfrac{\sqrt{2}}{2}(\mathbf{i} + \mathbf{j})$

3. $f(x, y) = xy$,

$P = (2, 3)$, $\mathbf{v} = \mathbf{i} + \mathbf{j}$

4. $f(x, y) = \dfrac{x}{y}$,

$P = (1, 1)$, $\mathbf{v} = -\mathbf{j}$

5. $g(x, y) = \sqrt{x^2 + y^2}$,

$P = (3, 4)$, $\mathbf{v} = 3\mathbf{i} - 4\mathbf{j}$

6. $g(x, y) = \arcsin xy$,

$P = (1, 0)$, $\mathbf{v} = \mathbf{i} + 5\mathbf{j}$

7. $h(x, y) = e^x \sin y,$
$$P = \left(1, \frac{\pi}{2}\right), \mathbf{v} = -\mathbf{i}$$

8. $h(x, y) = e^{-(x^2+y^2)},$
$P = (0, 0), \mathbf{v} = \mathbf{i} + \mathbf{j}$

9. $f(x, y, z) = xy + yz + xz,$
$P = (1, 1, 1), \mathbf{v} = 2\mathbf{i} + \mathbf{j} - \mathbf{k}$

10. $f(x, y, z) = x^2 + y^2 + z^2,$
$P = (1, 2, -1), \mathbf{v} = \mathbf{i} - 2\mathbf{j} + 3\mathbf{k}$

11. $h(x, y, z) = x \arctan yz,$
$P = (4, 1, 1), \mathbf{v} = \langle 1, 2, -1 \rangle$

12. $h(x, y, z) = xyz,$
$P = (2, 1, 1), \mathbf{v} = \langle 2, 1, 2 \rangle$

In Exercises 13–16, find the directional derivative of the function in the direction $\mathbf{u} = \cos \theta \, \mathbf{i} + \sin \theta \, \mathbf{j}$.

13. $f(x, y) = x^2 + y^2, \theta = \dfrac{\pi}{4}$

14. $f(x, y) = \dfrac{y}{x + y}, \theta = -\dfrac{\pi}{6}$

15. $f(x, y) = \sin (2x - y), \theta = -\dfrac{\pi}{3}$

16. $g(x, y) = xe^y, \theta = \dfrac{2\pi}{3}$

In Exercises 17–20, find the directional derivative of the given function at the point P in the direction of Q.

17. $f(x, y) = x^2 + 4y^2,$
$P = (3, 1), Q = (1, -1)$

18. $f(x, y) = \cos (x + y),$
$$P = (0, \pi), Q = \left(\frac{\pi}{2}, 0\right)$$

19. $h(x, y, z) = \ln (x + y + z),$
$P = (1, 0, 0), Q = (4, 3, 1)$

20. $g(x, y, z) = xye^z,$
$P = (2, 4, 0), Q = (0, 0, 0)$

In Exercises 21–30, find the gradient of the function and the maximum value of the directional derivative at the indicated point.

Function	Point
21. $f(x, y) = x^2 - 3xy + y^2$	$(4, 2)$
22. $f(x, y) = y\sqrt{x}$	$(4, 2)$
23. $h(x, y) = x \tan y$	$\left(2, \dfrac{\pi}{4}\right)$
24. $h(x, y) = y \cos (x - y)$	$\left(0, \dfrac{\pi}{3}\right)$
25. $g(x, y) = \ln \sqrt[3]{x^2 + y^2}$	$(1, 2)$
26. $g(x, y) = ye^{-x^2}$	$(0, 5)$
27. $f(x, y, z) = \sqrt{x^2 + y^2 + z^2}$	$(1, 4, 2)$

Function	Point
28. $f(x, y, z) = xe^{yz}$	$(2, 0, -4)$
29. $w = \dfrac{1}{\sqrt{1 - x^2 - y^2 - z^2}}$	$(0, 0, 0)$
30. $w = xy^2z^2$	$(2, 1, 1)$

In Exercises 31–38, use the function given by

$$f(x, y) = 3 - \frac{x}{3} - \frac{y}{2}.$$

31. Sketch the graph of f in the first octant and plot the point $(3, 2, 1)$.

32. Find $D_{\mathbf{u}} f(3, 2)$ where $\mathbf{u} = \cos \theta \, \mathbf{i} + \sin \theta \, \mathbf{j}$.
    (a) $\theta = \dfrac{\pi}{4}$ (b) $\theta = \dfrac{2\pi}{3}$

33. Find $D_{\mathbf{u}} f(3, 2)$ where $\mathbf{u} = \cos \theta \, \mathbf{i} + \sin \theta \, \mathbf{j}$.
    (a) $\theta = \dfrac{4\pi}{3}$ (b) $\theta = -\dfrac{\pi}{6}$

34. Find $D_{\mathbf{u}} f(3, 2)$ where $\mathbf{u} = \mathbf{v}/\|\mathbf{v}\|$.
    (a) $\mathbf{v} = \mathbf{i} + \mathbf{j}$ (b) $\mathbf{v} = -3\mathbf{i} - 4\mathbf{j}$

35. Find $D_{\mathbf{u}} f(3, 2)$ where $\mathbf{u} = \mathbf{v}/\|\mathbf{v}\|$.
    (a) $\mathbf{v}$ is the vector from $(1, 2)$ to $(-2, 6)$
    (b) $\mathbf{v}$ is the vector from $(3, 2)$ to $(4, 5)$

36. Find $\nabla f(x, y)$.

37. Find the maximum value of the directional derivative at $(3, 2)$.

38. Find a unit vector $\mathbf{u}$ orthogonal to $\nabla f(3, 2)$ and calculate $D_{\mathbf{u}} f(3, 2)$. Discuss the geometric meaning of the result.

In Exercises 39–42, use the function given by

$$f(x, y) = 9 - x^2 - y^2.$$

39. Sketch the graph of f in the first octant and plot the point $(1, 2, 4)$ on the surface.

40. Find $D_{\mathbf{u}} f(1, 2)$ where $\mathbf{u} = \cos \theta \, \mathbf{i} + \sin \theta \, \mathbf{j}$.
    (a) $\theta = -\dfrac{\pi}{4}$ (b) $\theta = \dfrac{\pi}{3}$

41. Find $\nabla f(1, 2)$ and $\|\nabla f(1, 2)\|$.

42. Find a unit vector $\mathbf{u}$ orthogonal to $\nabla f(1, 2)$, and calculate $D_{\mathbf{u}} f(1, 2)$. Discuss the geometric meaning of the result.

In Exercises 43–46, find a normal vector to the level curve $f(x, y) = c$ at P.

43. $f(x, y) = x^2 + y^2$
    $c = 25, P = (3, 4)$

44. $f(x, y) = 6 - 2x - 3y$
    $c = 6, P = (0, 0)$

45. $f(x, y) = \dfrac{x}{x^2 + y^2}$
    $c = \dfrac{1}{2}, P = (1, 1)$

46. $f(x, y) = xy$
    $c = -3, P = (-1, 3)$

In Exercises 47–50, use the gradient to find a unit normal vector to the graph of the equation at the indicated point. Sketch your results.

Equation	Point
47. $4x^2 - y = 6$	$(2, 10)$
48. $3x^2 - 2y^2 = 1$	$(1, 1)$
49. $9x^2 + 4y^2 = 40$	$(2, -1)$
50. $xe^y - y = 5$	$(5, 0)$

51. The temperature at the point (x, y) on a metal plate is given by

$$T = \frac{x}{x^2 + y^2}.$$

Find the direction of greatest increase in heat from the point $(3, 4)$.

52. The surface of a mountain is described by the equation

$$h(x, y) = 4000 - 0.001x^2 - 0.004y^2.$$

Suppose that a mountain climber is at the point $(500, 300, 3390)$. In what direction should the climber move in order to ascend at the greatest rate?

In Exercises 53 and 54, find the path followed by a heat-seeking particle placed at point P on a metal plate with a temperature field given by $T(x, y)$.

53. $T(x, y) = 400 - 2x^2 - y^2$, $P = (10, 10)$
54. $T(x, y) = 50 - x^2 - 2y^2$, $P = (4, 3)$

55. Let P be a point on the ellipse given by

$$\frac{x^2}{a^2} + \frac{y^2}{b^2} = 1$$

where $\mathbf{T}$ is a tangent vector to the ellipse at P and $f(x, y) = d_1 + d_2$ is the sum of the distances from the foci to P (see figure). Show that $T \cdot \nabla f(x, y)$ is zero, and give a geometric interpretation of this result.

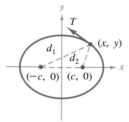

FIGURE FOR 55

15.7 Tangent Planes and Normal Lines

Equations of surfaces ▪ Tangent plane to a surface ▪ Normal line to a surface ▪ Angle of inclination of a plane

So far we have represented surfaces in space primarily by equations of the form

$$z = f(x, y). \qquad \text{Equation of a surface } S$$

In the development to follow, however, it is convenient to use the more general representation $F(x, y, z) = 0$. For a surface S given by $z = f(x, y)$ we can easily convert to the general form by defining F as

$$F(x, y, z) = f(x, y) - z.$$

Then, since $z = f(x, y)$, we have $f(x, y) - z = 0$, which means that we can consider S to be the level surface of F given by

$$F(x, y, z) = 0. \qquad \text{Alternative equation for surface } S$$

For instance, for the surface given by $z = x^2 + 2y^2 + 4$ we let

$$F(x, y, z) = x^2 + 2y^2 + 4 - z.$$

Then the surface can be written as $F(x, y, z) = 0$.

Tangent plane and normal line to a surface

We have seen many examples of the usefulness of normal lines in applications involving curves. Normal lines are equally important in analyzing surfaces and solids. For example, consider the collision of two billiard balls. When a

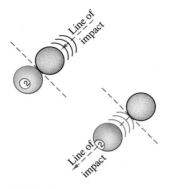

FIGURE 15.48

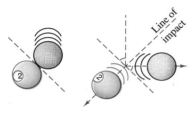

FIGURE 15.49

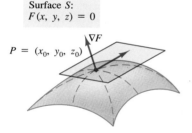

Tangent Plane to Surface *S* at *P*

FIGURE 15.50

stationary ball is struck at a point *P* on its surface, it moves along the **line of impact** determined by *P* and the center of the ball. The impact can occur in *two* ways. If the cue ball is moving along the line of impact, it stops dead and imparts all of its momentum to the stationary ball, as shown in Figure 15.48. This kind of shot requires precision, since the line of impact must coincide exactly with the direction of the cue ball. More often, the cue ball is deflected to one side or the other and retains part of its momentum. That part of the momentum that is transferred to the stationary ball occurs along the line of impact, *regardless* of the direction of the cue ball, as shown in Figure 15.49. We call this line of impact the **normal line** to the surface of the ball at the point *P*.

In the process of finding a normal line to a surface, we are also able to solve the problem of finding a **tangent plane** to the surface. Let *S* be a surface given by $F(x, y, z) = 0$, and let $P = (x_0, y_0, z_0)$ be a point on *S*. Let *C* be a curve on *S* through *P* that is defined by the vector-valued function $\mathbf{r}(t) = x(t)\mathbf{i} + y(t)\mathbf{j} + z(t)\mathbf{k}$. Then, for all *t*,

$$F(x(t), y(t), z(t)) = 0.$$

If *F* is differentiable and $x'(t)$, $y'(t)$, and $z'(t)$ all exist, it follows from the Chain Rule that

$$0 = F'(t) = F_x(x, y, z)x'(t) + F_y(x, y, z)y'(t) + F_z(x, y, z)z'(t).$$

At (x_0, y_0, z_0), the equivalent vector form is

$$0 = \nabla F(x_0, y_0, z_0) \cdot \mathbf{r}'(t_0) = \text{(gradient)} \cdot \text{(tangent vector)}.$$

This result means that the gradient at *P* is orthogonal to the tangent vector of every curve on *S* through *P*. Thus, all tangent lines at *P* lie in a plane that is normal to $\nabla F(x_0, y_0, z_0)$ and contains *P*, as shown in Figure 15.50. We call this plane the **tangent plane to *S* at *P*** and we call the line passing through *P* in the direction of $\nabla F(x_0, y_0, z_0)$ the **normal line to *S* at *P*.**

DEFINITION OF TANGENT PLANE AND NORMAL LINE	Let *F* be differentiable at the point $P = (x_0, y_0, z_0)$ on the surface *S* given by $F(x, y, z) = 0$ such that $\nabla F(x_0, y_0, z_0) \neq \mathbf{0}$. 1. The plane through *P* that is normal to $\nabla F(x_0, y_0, z_0)$ is called the **tangent plane to *S* at *P*.** 2. The line through *P* having the direction of $\nabla F(x_0, y_0, z_0)$ is called the **normal line to *S* at *P*.**

REMARK   In the remainder of this section we will assume $\nabla F(x_0, y_0, z_0)$ to be nonzero unless we state otherwise.

To find an equation for the tangent plane to *S* at (x_0, y_0, z_0), we let (x, y, z) be an arbitrary point in the tangent plane. Then the vector

$$\mathbf{v} = (x - x_0)\mathbf{i} + (y - y_0)\mathbf{j} + (z - z_0)\mathbf{k}$$

lies in the tangent plane. Since $\nabla F(x_0, y_0, z_0)$ is normal to the tangent plane at (x_0, y_0, z_0), it must be orthogonal to every vector in the tangent plane and we have

$$\nabla F(x_0, y_0, z_0) \cdot \mathbf{v} = 0$$

which leads to the result in the following theorem.

THEOREM 15.13
EQUATION OF TANGENT PLANE

If F is differentiable at (x_0, y_0, z_0), then an equation of the tangent plane to the surface given by $F(x, y, z) = 0$ at (x_0, y_0, z_0) is

$$F_x(x_0, y_0, z_0)(x - x_0) + F_y(x_0, y_0, z_0)(y - y_0) + F_z(x_0, y_0, z_0)(z - z_0) = 0.$$

EXAMPLE 1   Finding an equation of a tangent plane

Find an equation of the tangent plane to the hyperboloid given by

$$z^2 - 2x^2 - 2y^2 - 12 = 0$$

at the point $(1, -1, 4)$.

SOLUTION

Considering

$$F(x, y, z) = z^2 - 2x^2 - 2y^2 - 12 = 0$$

we have

$$F_x(x, y, z) = -4x \qquad F_y(x, y, z) = -4y \qquad F_z(x, y, z) = 2z$$

and at the point $(1, -1, 4)$ the partial derivatives are

$$F_x(1, -1, 4) = -4 \qquad F_y(1, -1, 4) = 4 \qquad F_z(1, -1, 4) = 8.$$

Therefore, an equation of the tangent plane at $(1, -1, 4)$ is

$$-4(x - 1) + 4(y + 1) + 8(z - 4) = 0$$
$$-4x + 4y + 8z - 24 = 0$$
$$x - y - 2z + 6 = 0.$$

Figure 15.51 shows a portion of the hyperboloid and tangent plane.   ▭

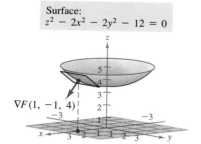

Surface:
$z^2 - 2x^2 - 2y^2 - 12 = 0$

$\nabla F(1, -1, 4)$

Tangent Plane to Surface

FIGURE 15.51

To find the equation of the tangent plane at a point on a surface given by $z = f(x, y)$, we define the function F by

$$F(x, y, z) = f(x, y) - z.$$

Then S is given by the level surface $F(x, y, z) = 0$, and by Theorem 15.13 an equation of the tangent plane to S at the point (x_0, y_0, z_0) is

$$f_x(x_0, y_0)(x - x_0) + f_y(x_0, y_0)(y - y_0) - (z - z_0) = 0. \qquad \text{Equation of tangent plane}$$

This form of the tangent plane equation is demonstrated in the next example.

EXAMPLE 2   Finding an equation for the tangent plane to $z = f(x, y)$

Find the equation of the tangent plane to the paraboloid

$$z = \frac{x^2 + 4y^2}{10}$$

at the point $(2, -2, 2)$.

SOLUTION

From $z = f(x, y) = (x^2 + 4y^2)/10$, we obtain

$$f_x(x, y) = \frac{x}{5} \qquad f_y(x, y) = \frac{4y}{5}$$

$$f_x(2, -2) = \frac{2}{5} \qquad f_y(2, -2) = -\frac{8}{5}.$$

Therefore, an equation of the tangent plane at $(2, -2, 2)$ is

$$f_x(2, -2)(x - 2) + f_y(2, -2)(y + 2) - (z - 2) = 0$$

$$\frac{2}{5}(x - 2) - \frac{8}{5}(y + 2) - (z - 2) = 0$$

$$2x - 8y - 5z - 10 = 0.$$

This tangent plane is shown in Figure 15.52.

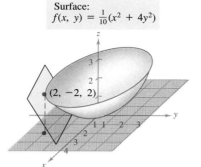

Surface:
$f(x, y) = \frac{1}{10}(x^2 + 4y^2)$

$(2, -2, 2)$

Tangent Plane to Surface

FIGURE 15.52

REMARK   You don't need to memorize the alternate formula used for the tangent plane equation in Example 2, since it can be derived easily from the level surface equation $F(x, y, z) = f(x, y) - z = 0$.

In Section 15.4, we saw that the total differential

$$dz = f_x(x, y)\, dx + f_y(x, y)\, dy$$

can be used to approximate the increment Δz. (see Figure 15.32 in Section 15.4.) We now give a geometric interpretation of this approximation in terms of the tangent plane to a surface. The tangent plane to the surface $z = f(x, y)$ at (x_0, y_0, z_0) is given by

$$f_x(x_0, y_0)(x - x_0) + f_y(x_0, y_0)(y - y_0) - (z - z_0) = 0$$

or equivalently,

$$z = z_0 + f_x(x_0, y_0)(x - x_0) + f_y(x_0, y_0)(y - y_0).$$

Thus, at (x_0, y_0) the *tangent plane* has height z_0 and at $(x_0 + dx, y_0 + dy)$ it has height

$$z = z_0 + f_x(x_0, y_0)(x_0 + dx - x_0) + f_y(x_0, y_0)(y_0 + dy - y_0)$$

$$= z_0 + f_x(x_0, y_0)\, dx + f_y(x_0, y_0)\, dy.$$

That is, $z = z_0 + dz$. We can see from Figure 15.53 that for points near (x_0, y_0, z_0) the increments $dx = \Delta x$ and $dy = \Delta y$ are small and $dz \approx \Delta z$.

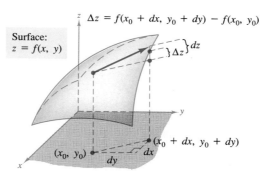

FIGURE 15.53

When dx and dy are small, $dz \approx \Delta z$.

The gradient $\nabla F(x, y, z)$ gives a convenient way to find equations of normal lines, as shown in Example 3.

EXAMPLE 3 Finding an equation for a normal line to a surface

Find a set of symmetric equations for the normal line to the surface given by $xyz = 12$ at the point $(2, -2, -3)$.

SOLUTION

We let

$$F(x, y, z) = xyz - 12.$$

Then, the gradient is given by

$$\nabla F(x, y, z) = F_x(x, y, z)\mathbf{i} + F_y(x, y, z)\mathbf{j} + F_z(x, y, z)\mathbf{k}$$
$$= yz\mathbf{i} + xz\mathbf{j} + xy\mathbf{k}$$

and at the point $(2, -2, -3)$, we have

$$\nabla F(2, -2, -3) = (-2)(-3)\mathbf{i} + (2)(-3)\mathbf{j} + (2)(-2)\mathbf{k}$$
$$= 6\mathbf{i} - 6\mathbf{j} - 4\mathbf{k}.$$

Therefore, the normal line at $(2, -2, -3)$ has direction numbers 6, −6, and −4. The corresponding set of symmetric equations is

$$\frac{x - 2}{6} = \frac{y + 2}{-6} = \frac{z + 3}{-4}.$$

Knowing that the gradient $\nabla F(x, y, z)$ is normal to the surface given by $F(x, y, z) = 0$ allows us to solve a variety of problems dealing with surfaces and curves in space. A typical problem is shown in the next example.

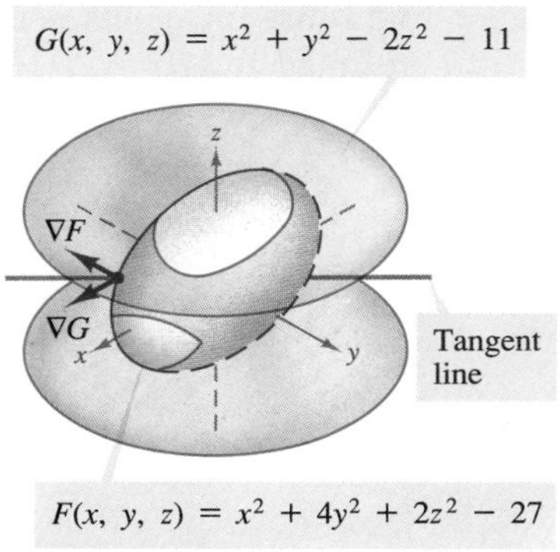

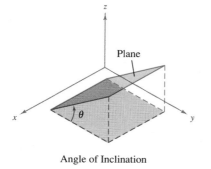

$G(x, y, z) = x^2 + y^2 - 2z^2 - 11$

$F(x, y, z) = x^2 + 4y^2 + 2z^2 - 27$

FIGURE 15.54

EXAMPLE 4 Finding the equation of a tangent line to a curve

Find symmetric equations for the tangent line to the curve of intersection of the ellipsoid given by

$$x^2 + 4y^2 + 2z^2 = 27 \qquad \text{Ellipsoid}$$

and the hyperboloid given by

$$x^2 + y^2 - 2z^2 = 11 \qquad \text{Hyperboloid}$$

at the point $(3, -2, 1)$, as shown in Figure 15.54.

SOLUTION

To find the equation of the tangent line, we first find the gradients to both surfaces at the point $(3, -2, 1)$. The cross product of these two gradients is a vector that is tangent to both surfaces at the point $(3, -2, 1)$. For the ellipsoid, we let $F(x, y, z) = x^2 + 4y^2 + 2z^2 - 27$ and obtain

$$\nabla F(x, y, z) = 2x\mathbf{i} + 8y\mathbf{j} + 4z\mathbf{k}$$
$$\nabla F(3, -2, 1) = 6\mathbf{i} - 16\mathbf{j} + 4\mathbf{k}.$$

For the hyperboloid, we let $G(x, y, z) = x^2 + y^2 - 2z^2 - 11$ and obtain

$$\nabla G(x, y, z) = 2x\mathbf{i} + 2y\mathbf{j} - 4z\mathbf{k}$$
$$\nabla G(3, -2, 1) = 6\mathbf{i} - 4\mathbf{j} - 4\mathbf{k}.$$

The cross product of these two gradient vectors is

$$\nabla F(3, -2, 1) \times \nabla G(3, -2, 1) = \begin{vmatrix} \mathbf{i} & \mathbf{j} & \mathbf{k} \\ 6 & -16 & 4 \\ 6 & -4 & -4 \end{vmatrix} = 8(10\mathbf{i} + 6\mathbf{j} + 9\mathbf{k}).$$

Hence, the direction numbers 10, 6, and 9 give the direction of the required tangent line. Symmetric equations for this tangent line at $(3, -2, 1)$ are

$$\frac{x - 3}{10} = \frac{y + 2}{6} = \frac{z - 1}{9}.$$

The angle of inclination of a plane

Another use of the gradient $\nabla F(x, y, z)$ is to determine the angle of inclination of the tangent plane to a surface. The **angle of inclination** of a plane is defined to be the angle θ, $0 \le \theta \le \pi/2$, between the given plane and the xy-plane, as shown in Figure 15.55. (The angle of inclination of a horizontal plane is defined to be zero.) Since the vector $\mathbf{k}$ is normal to the xy-plane, we can use the formula for the cosine of the angle between two planes (given in Section 14.3 to conclude that the angle of inclination of a plane with normal vector $\mathbf{n}$ is given by

$$\cos \theta = \frac{|\mathbf{n} \cdot \mathbf{k}|}{\|\mathbf{n}\| \|\mathbf{k}\|} = \frac{|\mathbf{n} \cdot \mathbf{k}|}{\|\mathbf{n}\|}. \qquad \text{Angle of inclination of a plane}$$

Angle of Inclination

FIGURE 15.55

EXAMPLE 5   Finding the angle of inclination of a tangent plane

Find the angle of inclination of the tangent plane to the ellipsoid given by

$$\frac{x^2}{12} + \frac{y^2}{12} + \frac{z^2}{3} = 1$$

at the point $(2, 2, 1)$.

SOLUTION

If we let

$$F(x, y, z) = \frac{x^2}{12} + \frac{y^2}{12} + \frac{z^2}{3} - 1$$

then the gradient of F at the point $(2, 2, 1)$ is given by

$$\nabla F(x, y, z) = \frac{x}{6}\mathbf{i} + \frac{y}{6}\mathbf{j} + \frac{2z}{3}\mathbf{k}$$

$$\nabla F(2, 2, 1) = \frac{1}{3}\mathbf{i} + \frac{1}{3}\mathbf{j} + \frac{2}{3}\mathbf{k}.$$

Now, since $\nabla F(2, 2, 1)$ is normal to the tangent plane and $\mathbf{k}$ is normal to the xy-plane, it follows that the angle of inclination of the tangent plane is given by

$$\cos\theta = \frac{|\nabla F(2, 2, 1) \cdot \mathbf{k}|}{\|\nabla F(2, 2, 1)\|} = \frac{2/3}{\sqrt{(1/3)^2 + (1/3)^2 + (2/3)^2}} = \sqrt{2/3}$$

which implies that $\theta = \arccos\sqrt{2/3} \approx 35.3°$, as shown in Figure 15.56.

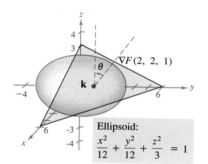

FIGURE 15.56

Ellipsoid:
$$\frac{x^2}{12} + \frac{y^2}{12} + \frac{z^2}{3} = 1$$

A special case of the procedure shown in Example 5 is worth noting. The angle of inclination θ of the tangent plane to the surface $z = f(x, y)$ at (x_0, y_0, z_0) is given by

$$\cos\theta = \frac{1}{\sqrt{[f_x(x_0, y_0)]^2 + [f_y(x_0, y_0)]^2 + 1}}.$$

Alternate formula for angle of inclination

(See Exercise 50.)

A comparison of the gradients $\nabla f(x, y)$ and $\nabla F(x, y, z)$

We conclude this section with a comparison of the gradients $\nabla f(x, y)$ and $\nabla F(x, y, z)$. In the previous section we discovered that the gradient of a function f of two variables is normal to the *level curves* of f. Specifically, in Theorem 15.12, we stated that if f is differentiable at (x_0, y_0) and $\nabla f(x_0, y_0) \neq \mathbf{0}$, then $\nabla f(x_0, y_0)$ is normal to the level curve through (x_0, y_0). Having developed normal lines to surfaces, we can now extend this result to a function of three variables.

THEOREM 15.14 GRADIENT IS NORMAL TO LEVEL SURFACES	If F is differentiable at (x_0, y_0, z_0) and $\nabla F(x_0, y_0, z_0) \neq \mathbf{0}$, then $\nabla F(x_0, y_0, z_0)$ is normal to the level surface through (x_0, y_0, z_0).

When working with the gradients $\nabla f(x, y)$ and $\nabla F(x, y, z)$, be sure that you remember that $\nabla f(x, y)$ is a vector in the xy-plane and $\nabla F(x, y, z)$ is a vector in space.

EXERCISES for Section 15.7

In Exercises 1–10, find a unit normal vector to the given surface at the indicated point. [Hint: Normalize the gradient vector $\nabla F(x, y, z)$.]

Surface	Point
1. $x + y + z = 4$	$(2, 0, 2)$
2. $x^2 + y^2 + z^2 = 11$	$(3, 1, 1)$
3. $z = \sqrt{x^2 + y^2}$	$(3, 4, 5)$
4. $z = x^3$	$(2, 1, 8)$
5. $x^2 y^4 - z = 0$	$(1, 2, 16)$
6. $x^2 + 3y + z^3 = 9$	$(2, -1, 2)$
7. $z - x \sin y = 4$	$\left(6, \dfrac{\pi}{6}, 7\right)$
8. $z e^{x^2 - y^2} - 3 = 0$	$(2, 2, 3)$
9. $\ln\left(\dfrac{x}{y - z}\right) = 0$	$(1, 4, 3)$
10. $\sin(x - y) - z = 2$	$\left(\dfrac{\pi}{3}, \dfrac{\pi}{6}, -\dfrac{3}{2}\right)$

In Exercises 11–24, find an equation of the tangent plane to the given surface at the indicated point.

Surface	Point
11. $f(x, y) = 25 - x^2 - y^2$	$(3, 1, 15)$
12. $f(x, y) = \sqrt{x^2 + y^2}$	$(3, 4, 5)$
13. $f(x, y) = \dfrac{y}{x}$	$(1, 2, 2)$
14. $f(x, y) = 2 - \dfrac{2}{3}x - y$	$(3, -1, 1)$
15. $g(x, y) = x^2 - y^2$	$(5, 4, 9)$
16. $g(x, y) = \arctan \dfrac{y}{x}$	$(1, 0, 0)$
17. $z = e^x (\sin y + 1)$	$\left(0, \dfrac{\pi}{2}, 2\right)$
18. $z = x^3 - 3xy + y^3$	$(1, 2, 3)$
19. $h(x, y) = \ln \sqrt{x^2 + y^2}$	$(3, 4, \ln 5)$
20. $h(x, y) = \cos y$	$\left(5, \dfrac{\pi}{4}, \dfrac{\sqrt{2}}{2}\right)$
21. $x^2 + 4y^2 + z^2 = 36$	$(2, -2, 4)$

Surface	Point
22. $x^2 + 2z^2 = y^2$	$(1, 3, -2)$
23. $xy^2 + 3x - z^2 = 4$	$(2, 1, -2)$
24. $y = x(2z - 1)$	$(4, 4, 1)$

In Exercises 25–30, find an equation for the tangent plane and find symmetric equations for the normal line to the given surface at the indicated point.

Surface	Point
25. $x^2 + y^2 + z = 9$	$(1, 2, 4)$
26. $x^2 + y^2 + z^2 = 9$	$(1, 2, 2)$
27. $xy - z = 0$	$(-2, -3, 6)$
28. $x^2 + y^2 - z^2 = 0$	$(5, 12, 13)$
29. $z = \arctan \dfrac{y}{x}$	$\left(1, 1, \dfrac{\pi}{4}\right)$
30. $xyz = 10$	$(1, 2, 5)$

In Exercises 31–34, find the angle of inclination θ of the tangent plane to the given surface at the indicated point.

Surface	Point
31. $3x^2 + 2y^2 - z = 15$	$(2, 2, 5)$
32. $xy - z^2 = 0$	$(2, 2, 2)$
33. $x^2 - y^2 + z = 0$	$(1, 2, 3)$
34. $x^2 + y^2 = 5$	$(2, 1, 3)$

In Exercises 35–40, (a) find symmetric equations of the tangent line to the curve of intersection of the given surfaces at the indicated point and (b) find the cosine of the angle between the gradient vectors at this point. State whether the surfaces are orthogonal at the point of intersection.

Surfaces	Point
35. $x^2 + y^2 = 5$ $z = x$	$(2, 1, 2)$

Surfaces	Point
36. $z = x^2 + y^2$	$(2, -1, 5)$
$z = 4 - y$	
37. $x^2 + z^2 = 25$	$(3, 3, 4)$
$y^2 + z^2 = 25$	
38. $z = \sqrt{x^2 + y^2}$	$(3, 4, 5)$
$2x + y + 2z = 20$	
39. $x^2 + y^2 + z^2 = 6$	$(2, 1, 1)$
$x - y - z = 0$	
40. $z = x^2 + y^2$	$(1, 2, 5)$
$x + y + 6z = 33$	

In Exercises 41 and 42, find the point on the surface where the tangent plane is horizontal.

41. $z = 3 - x^2 - y^2 + 6y$
42. $z = 3x^2 + 2y^2 - 3x + 4y - 5$

In Exercises 43 and 44, find the path of a heat-seeking particle in the temperature field T, starting at the specified point.

43. $T(x, y, z) = 400 - 2x^2 - y^2 - 4z^2$, $(4, 3, 10)$
44. $T(x, y, z) = 100 - 3x - y - z^2$, $(2, 2, 5)$

In Exercises 45 and 46, show that the tangent plane to the quadric surface at the point (x_0, y_0, z_0) can be written in the given form.

45. Ellipsoid: $\dfrac{x^2}{a^2} + \dfrac{y^2}{b^2} + \dfrac{z^2}{c^2} = 1$

Plane: $\dfrac{x_0 x}{a^2} + \dfrac{y_0 y}{b^2} + \dfrac{z_0 z}{c^2} = 1$

46. Hyperboloid: $\dfrac{x^2}{a^2} + \dfrac{y^2}{b^2} - \dfrac{z^2}{c^2} = 1$

Plane: $\dfrac{x_0 x}{a^2} + \dfrac{y_0 y}{b^2} - \dfrac{z_0 z}{c^2} = 1$

47. Show that any tangent plane to the cone $z^2 = a^2 x^2 + b^2 y^2$ passes through the origin.
48. Show that any line normal to a sphere passes through the center of the sphere.
49. Prove Theorem 15.14.
50. Prove that the angle of inclination θ of the tangent plane to the surface $z = f(x, y)$ at the point (x_0, y_0, z_0) is given by

$$\cos \theta = \frac{1}{\sqrt{[f_x(x_0, y_0)]^2 + [f_y(x_0, y_0)]^2 + 1}}.$$

15.8 Extrema of Functions of Two Variables

Absolute extrema ▪ Relative extrema ▪ Second-Partials Test

Karl Weierstrass

In Chapter 4 we studied techniques for finding the extreme values of a function of a single variable. In this section we extend these techniques to functions of two variables. For example, in Theorem 15.15 we extend the Extreme Value Theorem for a function of a single variable to a function of two variables. This theorem is difficult to prove in both the single- and two-variable cases. Although the theorem had been used by earlier mathematicians, the first person to provide a rigorous proof was the German mathematician Karl Weierstrass (1815–1897). Weierstrass also provided rigorous justifications for many other mathematical results already in common use, and we are indebted to him for much of the solid logical foundation upon which modern calculus is built.

The values $f(a, b)$ and $f(c, d)$ such that $f(a, b) \le f(x, y) \le f(c, d)$ for all (x, y) in R are called the **absolute minimum** and **absolute maximum** of f in the region R, as shown in Figure 15.57. As in single-variable calculus, we distinguish between absolute extrema and **relative extrema.** Several relative extrema are shown in Figure 15.58.

Recall from Section 15.2 that a region in the plane is *closed* if it contains all of its boundary points. The Extreme Value Theorem deals with a region in the plane that is both closed and *bounded*. A region in the plane is called **bounded** if it is a subregion of a closed disc in the plane.

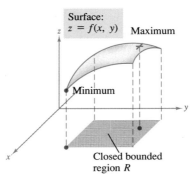

R contains point(s) at which $f(x, y)$ is minimum and point(s) at which $f(x, y)$ is maximum.

Absolute Extrema

FIGURE 15.57

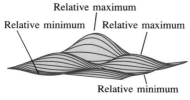

Relative Extrema

FIGURE 15.58

**THEOREM 15.15
EXTREME VALUE THEOREM**

Let f be a continuous function of two variables x and y defined on a closed bounded region R in the xy-plane.

1. There is at least one point in R where f takes on a minimum value.
2. There is at least one point in R where f takes on a maximum value.

DEFINITION OF RELATIVE EXTREMA

Let f be a function defined on a region R containing (x_0, y_0).

1. $f(x_0, y_0)$ is a **relative minimum** of f if $f(x, y) \geq f(x_0, y_0)$ for all (x, y) in an *open* disc containing (x_0, y_0).
2. $f(x_0, y_0)$ is a **relative maximum** of f if $f(x, y) \leq f(x_0, y_0)$ for all (x, y) in an *open* disc containing (x_0, y_0).

To say that $z_0 = f(x_0, y_0)$ is a relative maximum of f means that the point (x_0, y_0, z_0) is at least as high as all nearby points on the graph of $z = f(x, y)$. Similarly, $z_0 = f(x_0, y_0)$ is a relative minimum if (x_0, y_0, z_0) is at least as low as all nearby points on the graph.

To locate relative extrema of f, we investigate the points at which the gradient of f is zero or undefined. We call such points **critical points** of f.

DEFINITION OF CRITICAL POINT

Let f be defined on an open region R containing (x_0, y_0). We call (x_0, y_0) a **critical point** of f if one of the following is true.

1. $f_x(x_0, y_0) = 0$ and $f_y(x_0, y_0) = 0$.
2. $f_x(x_0, y_0)$ or $f_y(x_0, y_0)$ does not exist.

Recall from Theorem 15.11 that if f is differentiable and

$$\nabla f(x_0, y_0) = f_x(x_0, y_0)\mathbf{i} + f_y(x_0, y_0)\mathbf{j} = 0\mathbf{i} + 0\mathbf{j}$$

then every directional derivative at (x_0, y_0) must be zero. In Exercise 35 you are asked to show that this implies that the function has a horizontal tangent plane at the point (x_0, y_0), as shown in Figure 15.59. It appears that such a point is a likely location of a relative extremum.

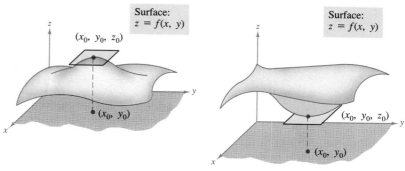

FIGURE 15.59   Relative Maximum   Relative Minimum

**THEOREM 15.16
RELATIVE EXTREMA OCCUR
ONLY AT CRITICAL POINTS**

If $f(x_0, y_0)$ is a relative extremum of f on an open region R, then (x_0, y_0) is a critical point of f.

PROOF   If either f_x or f_y does not exist at (x_0, y_0), then (x_0, y_0) is a critical point. Thus, we assume both first partial derivatives exist at (x_0, y_0), and we consider the function of one variable, $g(x) = f(x, y_0)$. By hypothesis, it has a relative extremum at x_0 and is differentiable there. Hence,

$$g'(x_0) = \lim_{\Delta x \to 0} \frac{f(x_0 + \Delta x, y_0) - f(x_0, y_0)}{\Delta x} = f_x(x_0, y_0) = 0.$$

Similarly, $h(y) = f(x_0, y)$ has a local extremum at y_0, and being differentiable there, it satisfies the equation

$$h'(y_0) = \lim_{\Delta y \to 0} \frac{f(x_0, y_0 + \Delta y) - f(x_0, y_0)}{\Delta y} = f_y(x_0, y_0) = 0.$$

Thus, $f_x(x_0, y_0) = f_y(x_0, y_0) = 0$, and (x_0, y_0) is a critical point of f.

EXAMPLE 1   Finding a relative extremum

Determine the relative extrema of

$$f(x, y) = 2x^2 + y^2 + 8x - 6y + 20.$$

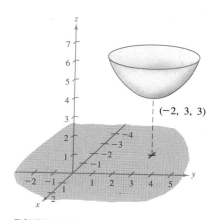

FIGURE 15.60

SOLUTION

We begin by finding the critical points of f. Since

$$f_x(x, y) = 4x + 8 \quad \text{and} \quad f_y(x, y) = 2y - 6$$

are defined for all x and y, the only critical points are those for which both first partial derivatives are zero. To locate these points, let $f_x(x, y)$ and $f_y(x, y)$ be zero, and solve the system of equations

$$4x + 8 = 0 \quad \text{and} \quad 2y - 6 = 0$$

to obtain the critical point $(-2, 3)$. By completing the square, we can conclude that for all $(x, y) \neq (-2, 3)$,

$$f(x, y) = 2(x + 2)^2 + (y - 3)^2 + 3 > 3.$$

Therefore, a relative *minimum* of f occurs at $(-2, 3)$. The value of the relative minimum is $f(-2, 3) = 3$, as shown in Figure 15.60. ▭

Example 1 shows a relative minimum occurring at one type of critical point—the type for which both $f_x(x, y)$ and $f_y(x, y)$ are zero. In the next example we look at a relative maximum that occurs at the other type of critical point—the type for which either $f_x(x, y)$ or $f_y(x, y)$ is undefined.

EXAMPLE 2   Finding a relative extremum

Determine the relative extrema of

$$f(x, y) = 1 - (x^2 + y^2)^{1/3}.$$

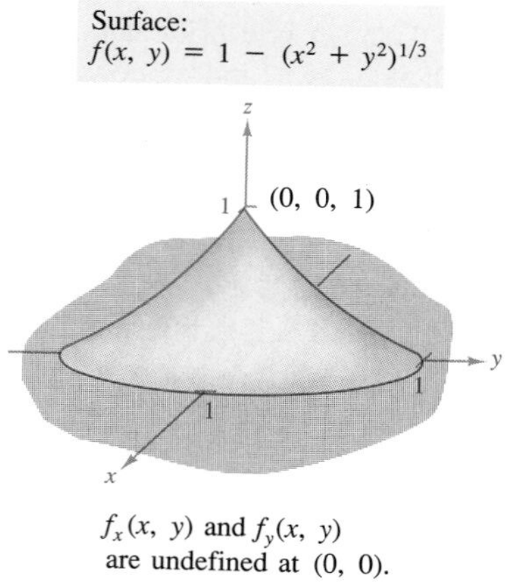

SOLUTION

Since

$$f_x(x, y) = -\frac{2x}{3(x^2 + y^2)^{2/3}} \quad \text{and} \quad f_y(x, y) = -\frac{2y}{3(x^2 + y^2)^{2/3}}$$

we see that both partial derivatives are defined for all points in the xy-plane except for $(0, 0)$. Moreover, this is the only critical point, since the partial derivatives cannot both be zero unless both x and y are zero. In Figure 15.61 we see that $f(0, 0)$ is 1. For all other (x, y) it is clear that

$$f(x, y) = 1 - (x^2 + y^2)^{1/3} < 1.$$

Therefore, $f(0, 0)$ is a relative *maximum* of f. ▭

Surface:
$f(x, y) = 1 - (x^2 + y^2)^{1/3}$

$f_x(x, y)$ and $f_y(x, y)$
are undefined at $(0, 0)$.

FIGURE 15.61

REMARK   In Example 2, $f_x(x, y) = 0$ for every point on the y-axis other than $(0, 0)$. However, since $f_y(x, y)$ is nonzero, these are not critical points. Remember that *one* of the partials must be undefined or *both* must be zero in order to yield a critical point.

Theorem 15.16 tells us that to find relative extrema we need only examine values of $f(x, y)$ at critical points. However, as is true for a function of one variable, the critical points of a function of two variables do not always yield

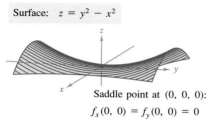

Surface: $z = y^2 - x^2$

Saddle point at $(0, 0, 0)$:
$f_x(0, 0) = f_y(0, 0) = 0$

FIGURE 15.62

relative maxima or minima. Some critical points yield **saddle points,** which are neither relative maxima nor relative minima. For example, the saddle point shown in Figure 15.62 is not a relative extremum, since in any open disc centered at $(0, 0)$ the function takes on both negative values (along the *x*-axis) *and* positive values (along the *y*-axis).

For the functions in Examples 1 and 2, it is relatively easy to determine the relative extrema, because each function was either given or able to be written in completed square form. For more complicated functions, algebraic arguments are not so fruitful, and we rely on the more analytic means presented in the following Second-Partials Test. This is the two-variable counterpart of the Second-Derivative Test for functions of one variable. The proof of this theorem is best left to a course in advanced calculus.

THEOREM 15.17
SECOND-PARTIALS TEST

Let f have continuous first and second partial derivatives on an open region containing a point (a, b) for which $f_x(a, b) = 0$ and $f_y(a, b) = 0$. To test for relative extrema of f, we define the quantity

$$d = f_{xx}(a, b)f_{yy}(a, b) - [f_{xy}(a, b)]^2.$$

1. If $d > 0$ and $f_{xx}(a, b) > 0$, then $f(a, b)$ is a **relative minimum.**
2. If $d > 0$ and $f_{xx}(a, b) < 0$, then $f(a, b)$ is a **relative maximum.**
3. If $d < 0$, then $(a, b, f(a, b))$ is a **saddle point.**
4. The test gives no information if $d = 0$.

REMARK   If $d > 0$, then $f_{xx}(a, b)$ and $f_{yy}(a, b)$ must have the same signs. This means that $f_{xx}(a, b)$ can be replaced by $f_{yy}(a, b)$ in the first two parts of the test.

A convenient device for remembering the formula for d in the Second-Partials Test is given by the 2×2 determinant

$$d = \begin{vmatrix} f_{xx}(a, b) & f_{xy}(a, b) \\ f_{yx}(a, b) & f_{yy}(a, b) \end{vmatrix}$$

where $f_{xy}(a, b) = f_{yx}(a, b)$ by Theorem 15.3.

EXAMPLE 3   Using the Second-Partials Test

Find the relative extrema of $f(x, y) = -x^3 + 4xy - 2y^2 + 1$.

SOLUTION

We begin by finding the critical points of f. Since

$$f_x(x, y) = -3x^2 + 4y \quad \text{and} \quad f_y(x, y) = 4x - 4y$$

are defined for all x and y, the only critical points are those for which both first partial derivatives are zero. To locate these points, we let $f_x(x, y)$ and $f_y(x, y)$ be zero and obtain the following system of equations.

$$-3x^2 + 4y = 0$$
$$4x - 4y = 0$$

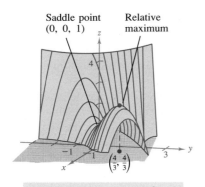

$f(x, y) = -x^3 + 4xy - 2y^2 + 1$

FIGURE 15.63

From the second equation we see that $x = y$, and, by substitution into the first equation, we obtain two solutions: $y = x = 0$ and $y = x = \frac{4}{3}$. Since

$$f_{xx}(x, y) = -6x, \qquad f_{yy}(x, y) = -4, \qquad \text{and} \qquad f_{xy}(x, y) = 4$$

it follows that for the critical point $(0, 0)$,

$$d = f_{xx}(0, 0)f_{yy}(0, 0) - [f_{xy}(0, 0)]^2 = 0 - 16 < 0$$

and, by the Second-Partials Test, we conclude that $(0, 0, 1)$ is a saddle point of f. Furthermore, for the critical point $\left(\frac{4}{3}, \frac{4}{3}\right)$,

$$d = f_{xx}\left(\frac{4}{3}, \frac{4}{3}\right)f_{yy}\left(\frac{4}{3}, \frac{4}{3}\right) - \left[f_{xy}\left(\frac{4}{3}, \frac{4}{3}\right)\right]^2 = -8(-4) - 16 = 16 > 0$$

and since

$$f_{xx}\left(\frac{4}{3}, \frac{4}{3}\right) = -8 < 0$$

we conclude that $f\left(\frac{4}{3}, \frac{4}{3}\right)$ is a relative maximum, as shown in Figure 15.63.

The Second-Partials Test can fail to find relative extrema in two ways. If either of the first partial derivatives is undefined, then we cannot use the test. Also, if

$$d = f_{xx}(a, b)f_{yy}(a, b) - [f_{xy}(a, b)]^2 = 0$$

the test fails. In such cases, we must rely on a sketch or some other approach, as demonstrated in the next example.

EXAMPLE 4   Failure of the Second-Partials Test

Find the relative extrema of $f(x, y) = x^2y^2$.

SOLUTION

Since

$$f_x(x, y) = 2xy^2 \qquad \text{and} \qquad f_y(x, y) = 2x^2y$$

we see that both partial derivatives are zero if $x = 0$ *or* $y = 0$. That is, every point along the x- or y-axis is a critical point. Now, since

$$f_{xx}(x, y) = 2y^2, \qquad f_{yy}(x, y) = 2x^2, \qquad \text{and} \qquad f_{xy}(x, y) = 4xy$$

we see that if either $x = 0$ or $y = 0$, then

$$d = f_{xx}(x, y)f_{yy}(x, y) - [f_{xy}(x, y)]^2 = 4x^2y^2 - 16x^2y^2 = -12x^2y^2 = 0.$$

Thus, the Second-Partials Test fails. However, since $f(x, y) = 0$ for every point along the x- or y-axis, and since $f(x, y) = x^2y^2 > 0$ for all other points, we can conclude that each of these critical points yields an absolute minimum, as shown in Figure 15.64.

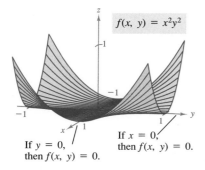

$f(x, y) = x^2y^2$

If $y = 0$, then $f(x, y) = 0$.

If $x = 0$, then $f(x, y) = 0$.

FIGURE 15.64

Absolute extrema of a function can occur in two ways. First, some relative extrema also happen to be absolute extrema. For instance, in Example 1, $f(-2, 3)$ is an absolute minimum of the function. (On the other hand, the relative maximum found in Example 3 is not an absolute maximum of the function.) Second, absolute extrema can occur at a boundary point of the domain. This is illustrated in the next example.

EXAMPLE 5 Finding absolute extrema

Find the absolute extrema of the function $f(x, y) = \sin xy$ on the closed region given by $0 \leq x \leq \pi$ and $0 \leq y \leq 1$.

SOLUTION

From the partial derivatives

$$f_x(x, y) = y \cos xy \qquad \text{and} \qquad f_y(x, y) = x \cos xy$$

we see that each point lying on the hyperbola given by $xy = \pi/2$ is a critical point. Moreover, these points each yield the value $f(x, y) = \sin (\pi/2) = 1$, which we know is the absolute maximum, as shown in Figure 15.65. The only other critical point of f *lying in the given region* is $(0, 0)$. It yields an absolute minimum of 0, since

$$0 \leq xy \leq \pi \quad \Longrightarrow \quad 0 \leq \sin xy \leq 1.$$

To hunt for other absolute extrema, we consider the four boundaries of the region formed by taking traces with the vertical planes $x = 0$, $x = \pi$, $y = 0$, and $y = 1$. Having done this, we find that $\sin xy = 0$ at all points on the x-axis, the y-axis, and at the point $(\pi, 1)$. Each of these points yields an absolute minimum for the surface, as shown in Figure 15.65. ▭

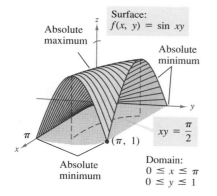

FIGURE 15.65

The concepts of relative extrema and critical points can be extended to functions of three or more variables. If all first partial derivatives of $w = f(x_1, x_2, x_3, \ldots, x_n)$ exist, then it can be shown that a relative maximum or minimum can occur at $(x_1, x_2, x_3, \ldots, x_n)$ only if every first partial derivative is zero at that point. This means the critical points are obtained by solving the following system of equations.

$$f_{x_1}(x_1, x_2, x_3, \ldots, x_n) = 0$$
$$f_{x_2}(x_1, x_2, x_3, \ldots, x_n) = 0$$
$$\vdots$$
$$f_{x_n}(x_1, x_2, x_3, \ldots, x_n) = 0$$

The extension of Theorem 15.17 to three or more variables is also possible, although we will not consider such an extension in this text.

EXERCISES for Section 15.8

In Exercises 1–18, examine each function for relative extrema and saddle points.

1. $f(x, y) = 2x^2 + 2xy + y^2 + 2x - 3$
2. $f(x, y) = -x^2 - 5y^2 + 8x - 10y - 13$
3. $f(x, y) = -5x^2 + 4xy - y^2 + 16x + 10$
4. $f(x, y) = x^2 + 6xy + 10y^2 - 4y + 4$
5. $z = 2x^2 + 3y^2 - 4x - 12y + 13$
6. $z = -3x^2 - 2y^2 + 3x - 4y + 5$
7. $h(x, y) = x^2 - y^2 - 2x - 4y - 4$
8. $h(x, y) = x^2 - 3xy - y^2$
9. $g(x, y) = xy$
10. $g(x, y) = 120x + 120y - xy - x^2 - y^2$
11. $f(x, y) = x^3 - 3xy + y^3$
12. $f(x, y) = 4xy - x^4 - y^4$

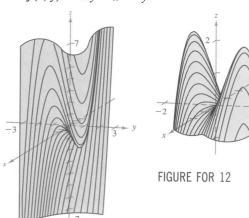

FIGURE FOR 12

FIGURE FOR 11

13. $z = \dfrac{-4x}{x^2 + y^2 + 1}$
14. $f(x, y) = y^3 - 3yx^2 - 3y^2 - 3x^2 + 1$

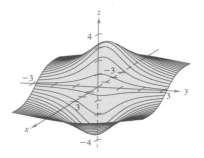

FIGURE FOR 13

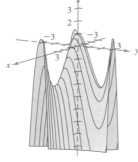

FIGURE FOR 14

15. $z = (x^2 + 4y^2)e^{1-x^2-y^2}$

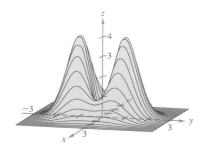

16. $z = e^{-x} \sin y$

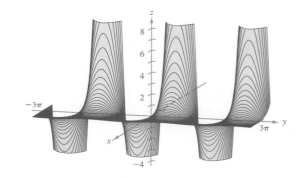

17. $z = \left(\dfrac{1}{2} - x^2 + y^2\right)e^{1-x^2-y^2}$

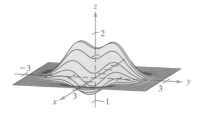

18. $z = e^{-(x^2+y^2)}$

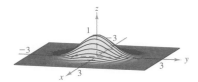

In Exercises 19–22, find the absolute extrema of the function over the region R.

19. $f(x, y) = x^2 + xy$,
   $R = \{(x, y): |x| \leq 2, |y| \leq 1\}$
20. $f(x, y) = x^2 + 2xy + y^2$,
   $R = \{(x, y): |x| \leq 2, |y| \leq 1\}$
21. $f(x, y) = x^2 + 2xy + y^2$,
   $R = \{(x, y): x^2 + y^2 \leq 8\}$
22. $f(x, y) = x^2 - 4xy$,
   $R = \{(x, y): 0 \leq x \leq 4, 0 \leq y \leq \sqrt{x}\}$

In Exercises 23–28, find the critical points and test for relative extrema. List the critical points for which the Second-Partials Test fails.

23. $f(x, y) = x^3 + y^3$
24. $f(x, y) = x^3 + y^3 - 3x^2 + 6y^2 + 3x + 12y + 7$

25. $f(x, y) = (x - 1)^2(y + 4)^2$
26. $f(x, y) = \sqrt{(x - 1)^2 + (y + 2)^2}$
27. $f(x, y) = x^{2/3} + y^{2/3}$
28. $f(x, y) = (x^2 + y^2)^{2/3}$

In Exercises 29–34, find the critical points of the given function.

29. $f(x, y, z) = x^2 + (y - 3)^2 + (z + 1)^2$
30. $f(x, y, z) = 4 - [x(y - 1)(z + 2)]^2$
31. $f(x, y, z) = x^2 + y^2 + 2xz - 4yz + 10z$
32. $f(x, y, z) = x^2 - y^2 + yz - x^2z$
33. $f(x, y, z) = (x - 1)^2(1 - z) + y(z - y)$
34. $f(x, y, z) = (x + z - 3)^2 + y^2 - z^2 + 2z(5 - 2y)$

35. Prove that if f is a differentiable function such that $\nabla f(x_0, y_0) = \mathbf{0}$, then the tangent plane at (x_0, y_0) is horizontal.

15.9   Applications of Extrema of Functions of Two Variables

Applied optimization problems ▪ The method of least squares

In this section we survey a few of the many applications of extrema of functions of two (or more) variables.

EXAMPLE 1   Finding maximum volume

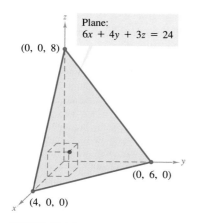

Plane:
$6x + 4y + 3z = 24$

$(0, 0, 8)$

$(0, 6, 0)$

$(4, 0, 0)$

FIGURE 15.66

A rectangular box is resting on the xy-plane with one vertex at the origin. Find the maximum volume of the box if its vertex opposite the origin lies in the plane $6x + 4y + 3z = 24$, as shown in Figure 15.66.

SOLUTION

Since one vertex of the box lies in the plane $6x + 4y + 3z = 24$, we have

$$z = \frac{1}{3}(24 - 6x - 4y)$$

and we can write the volume, xyz, of the box as a function of two variables.

$$V(x, y) = (x)(y)\left[\frac{1}{3}(24 - 6x - 4y)\right] = \frac{1}{3}(24xy - 6x^2y - 4xy^2)$$

By setting the first partial derivatives equal to zero,

$$V_x(x, y) = \frac{1}{3}(24y - 12xy - 4y^2) = \frac{y}{3}(24 - 12x - 4y) = 0$$

$$V_y(x, y) = \frac{1}{3}(24x - 6x^2 - 8xy) = \frac{x}{3}(24 - 6x - 8y) = 0$$

we obtain the critical points $(0, 0)$ and $\left(\frac{4}{3}, 2\right)$. At $(0, 0)$ the volume is zero, so we apply the Second-Partials Test to the point $\left(\frac{4}{3}, 2\right)$.

$$V_{xx}(x, y) = -4y, \qquad V_{yy}(x, y) = \frac{-8x}{3}, \qquad V_{xy}(x, y) = \frac{1}{3}(24 - 12x - 8y)$$

Since

$$V_{xx}\left(\frac{4}{3}, 2\right)V_{yy}\left(\frac{4}{3}, 2\right) - \left[V_{xy}\left(\frac{4}{3}, 2\right)\right]^2 = (-8)\left(-\frac{32}{9}\right) - \left(-\frac{8}{3}\right)^2$$

$$= \frac{64}{3} > 0$$

and

$$V_{xx}\left(\frac{4}{3}, 2\right) = -8 < 0$$

we conclude from the Second-Partials Test that the maximum volume is

$$V\left(\frac{4}{3}, 2\right) = \frac{1}{3}\left[24\left(\frac{4}{3}\right)(2) - 6\left(\frac{4}{3}\right)^2(2) - 4\left(\frac{4}{3}\right)(2^2)\right] = \frac{64}{9} \text{ cubic units.}$$

(Note that the volume is zero at the boundary points of the triangular domain of *V*.) ▭

REMARK   In many applied problems the domain of the function to be optimized is a closed bounded region. To find minimum or maximum points, you must not only test critical points, but also consider the value of the function at points on the boundary.

In Section 4.10 we looked at several applications of extrema in economics and business. In practice, such applications often involve more than one independent variable. For instance, a company may produce several models of one type of product. The price per unit and profit per unit are usually different for each model. Moreover, the demand for each model is often a function of the prices of the other models (as well as its own price). The next example illustrates an application involving two products.

EXAMPLE 2   Finding the maximum profit

The profit obtained by producing *x* units of product A and *y* units of product B is approximated by the model

$$P(x, y) = 8x + 10y - (0.001)(x^2 + xy + y^2) - 10,000.$$

Find the production level that produces a maximum profit.

SOLUTION

We have

$$P_x(x, y) = 8 - (0.001)(2x + y)$$

and

$$P_y(x, y) = 10 - (0.001)(x + 2y).$$

By setting these partial derivatives equal to zero, we obtain the following system of equations.

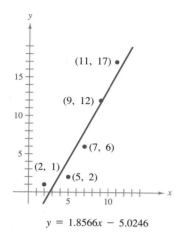

$$y = 1.8566x - 5.0246$$

FIGURE 15.67

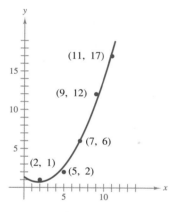

$$y = 0.1996x^2 - 0.7281x + 1.3749$$

FIGURE 15.68

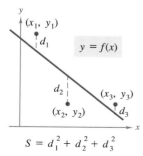

$$S = d_1^2 + d_2^2 + d_3^2$$

FIGURE 15.69

$$8 - (0.001)(2x + y) = 0 \quad \Longrightarrow \quad 2x + y = 8000$$
$$10 - (0.001)(x + 2y) = 0 \quad \Longrightarrow \quad x + 2y = 10{,}000$$

Solving this system, we have $x = 2000$ and $y = 4000$. The second partial derivatives of P are

$$P_{xx}(2000, 4000) = -0.002$$
$$P_{yy}(2000, 4000) = -0.002$$
$$P_{xy}(2000, 4000) = -0.001.$$

Moreover, since $P_{xx} < 0$ and

$$P_{xx}(2000, 4000)P_{yy}(2000, 4000) - [P_{xy}(2000, 4000)]^2$$
$$= (-0.002)^2 - (-0.001)^2 > 0$$

we conclude that the production level of $x = 2000$ units and $y = 4000$ units yields a *maximum* profit.   ▭

REMARK   In Example 2 we assumed that the plant is able to produce the required number of units to yield a maximum profit. In actual practice, the production would be bounded by physical constraints. We will consider such constrained optimization problems in the next section.

The method of least squares

We have spent some time discussing **mathematical models.** For instance, in Example 2 we gave a quadratic model for profit. We now look at one method for obtaining such a model. In constructing a model to represent a particular phenomenon, the goals are simplicity and accuracy. Of course, these goals often conflict. For instance, a simple linear model for the points in Figure 15.67 is

$$y = 1.8566x - 5.0246.$$

However, Figure 15.68 shows that by choosing the slightly more complicated quadratic model

$$y = 0.1996x^2 - 0.7281x + 1.3749$$

we can achieve greater accuracy.

As a measure of how well the model $y = f(x)$ fits the collection of points

$$\{(x_1, y_1), (x_2, y_2), (x_3, y_3), \ldots, (x_n, y_n)\}$$

we add the squares of the differences between the actual y-values and the values given by the model to obtain the **sum of the squared errors**

$$S = \sum_{i=1}^{n} [f(x_i) - y_i]^2.$$

Graphically, S can be interpreted as the sum of the squares of the vertical distances between the graph of f and the given points in the plane, as shown in Figure 15.69. If the model is perfect, then $S = 0$. However, when perfection

Adrien-Marie Legendre

is not feasible, we settle for a model that minimizes S. Statisticians call the *linear model* that minimizes S the **least squares regression line.** The proof that this line actually minimizes S involves the minimum of a function of two variables.

The method of least squares was introduced by the French mathematician Adrien-Marie Legendre (1752–1833). Legendre is best known for his work in geometry. In fact, his text *Elements of Geometry* was so popular in the United States that it continued to be used for 33 editions, spanning a period of more than 100 years.

**THEOREM 15.18
LEAST SQUARES
REGRESSION LINE**

The **least squares regression line** for $\{(x_1, y_1), (x_2, y_2), \ldots, (x_n, y_n)\}$ is given by $f(x) = ax + b$, where

$$a = \frac{n \sum\limits_{i=1}^{n} x_i y_i - \sum\limits_{i=1}^{n} x_i \sum\limits_{i=1}^{n} y_i}{n \sum\limits_{i=1}^{n} x_i^2 - \left(\sum\limits_{i=1}^{n} x_i\right)^2} \quad \text{and} \quad b = \frac{1}{n}\left(\sum\limits_{i=1}^{n} y_i - a \sum\limits_{i=1}^{n} x_i\right).$$

PROOF   We begin by letting $S(a, b)$ represent the sum of the squared errors for the model $f(x) = ax + b$ and the given set of points. That is,

$$S(a, b) = \sum_{i=1}^{n} [f(x_i) - y_i]^2 = \sum_{i=1}^{n} [ax_i + b - y_i]^2$$

where the points (x_i, y_i) represent constants. Because S is a function of a and b, we can use the methods discussed in the previous section to find the minimum value of S. Specifically, the first partial derivatives of S are

$$S_a(a, b) = \sum_{i=1}^{n} 2x_i[ax_i + b - y_i]$$

$$= 2a \sum_{i=1}^{n} x_i^2 + 2b \sum_{i=1}^{n} x_i - 2 \sum_{i=1}^{n} x_i y_i$$

$$S_b(a, b) = \sum_{i=1}^{n} 2[ax_i + b - y_i]$$

$$= 2a \sum_{i=1}^{n} x_i + 2nb - 2 \sum_{i=1}^{n} y_i.$$

By setting these two partial derivatives equal to zero, we obtain the values for a and b that are listed in the theorem. We leave it to you to apply the Second-Partials Test (see Exercise 22) to verify that these values of a and b yield a minimum.

If the x-values are symmetrically spaced about the y-axis, then $\Sigma\, x_i = 0$ and the formulas for a and b simplify to

$$a = \frac{\displaystyle\sum_{i=1}^{n} x_i y_i}{\displaystyle\sum_{i=1}^{n} x_i^2} \quad\text{and}\quad b = \frac{1}{n}\sum_{i=1}^{n} y_i.$$

This simplification is often possible with a translation of the x-values. For instance, if the x-values in a data collection consisted of the years 1980, 1981, 1982, 1983, and 1984, we could let 1982 be represented by 0.

EXAMPLE 3   Finding the least squares regression line

Find the least squares regression line for the following points.

$$(-3, 0), \quad (-1, 1) \quad (0, 2), \quad (2, 3)$$

SOLUTION

Table 15.2 shows the calculations involved in finding the least squares regression line using $n = 4$.

TABLE 15.2

x	y	xy	x^2
-3	0	0	9
-1	1	-1	1
0	2	0	0
2	3	6	4
$\displaystyle\sum_{i=1}^{n} x_i = -2$	$\displaystyle\sum_{i=1}^{n} y_i = 6$	$\displaystyle\sum_{i=1}^{n} x_i y_i = 5$	$\displaystyle\sum_{i=1}^{n} x_i^2 = 14$

Now, applying Theorem 15.18, we have

$$a = \frac{n\displaystyle\sum_{i=1}^{n} x_i y_i - \sum_{i=1}^{n} x_i \sum_{i=1}^{n} y_i}{n\displaystyle\sum_{i=1}^{n} x_i^2 - \left(\sum_{i=1}^{n} x_i\right)^2} = \frac{4(5) - (-2)(6)}{4(14) - (-2)^2} = \frac{8}{13}$$

and

$$b = \frac{1}{n}\left(\sum_{i=1}^{n} y_i - a\sum_{i=1}^{n} x_i\right) = \frac{1}{4}\left(6 - \frac{8}{13}(-2)\right) = \frac{47}{26}.$$

Thus, the least squares regression line is

$$y = \frac{8}{13}x + \frac{47}{26}$$

as shown in Figure 15.70.

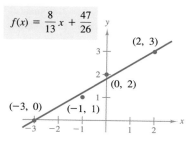

$f(x) = \dfrac{8}{13}x + \dfrac{47}{26}$

FIGURE 15.70

EXERCISES for Section 15.9

In Exercises 1 and 2, find the minimum distance from the point to the plane given by $2x + 3y + z = 12$.

1. $(0, 0, 0)$

2. $(1, 2, 3)$

In Exercises 3 and 4, find the minimum distance from the given point to the paraboloid given by $z = x^2 + y^2$.

3. $(5, 5, 0)$

4. $(5, 0, 0)$

In Exercises 5–8, find three positive numbers x, y, and z satisfying the given conditions.

5. The sum is 30 and the product is maximum.

6. The sum is 32 and $P = xy^2z$ is maximum.

7. The sum is 30 and the sum of the squares is minimum.

8. The sum is 1 and the sum of the squares is minimum.

9. The sum of the length and the girth (perimeter of a cross section) of packages carried by parcel post cannot exceed 108 inches. Find the dimensions of the rectangular package of largest volume that may be sent by parcel post.

10. Repeat Exercise 9 if the sum of the two perimeters of the two cross sections shown in the accompanying figure cannot exceed 108 inches.

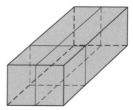

11. A water line is to be built from point P to point S and must pass through regions where construction costs differ (see figure). Find x and y so that the total cost C will be minimum if the cost per mile in dollars is $3k$ from P to Q, $2k$ from Q to R, and k from R to S.

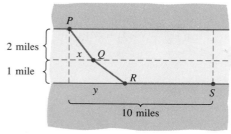

12. The material for constructing the base of an open box costs 1.5 times as much as the material for constructing the sides. For a fixed amount of money C, find the dimensions of the box of largest volume that can be made.

13. The volume of an ellipsoid

$$\frac{x^2}{a^2} + \frac{y^2}{b^2} + \frac{z^2}{c^2} = 1$$

is $4\pi abc/3$. For fixed $a + b + c$, show that the ellipsoid of maximum volume is a sphere.

14. Show that the rectangular box of maximum volume inscribed in a sphere of radius r is a cube.

15. Show that the rectangular box of given volume and minimum surface area is a cube.

16. A trough with trapezoidal cross sections is formed by turning up the edges of a 10-inch-wide sheet of aluminum. Find the cross section of maximum area.

17. Repeat Exercise 16 for a sheet of aluminum that is w inches wide.

18. A company manufactures two products. The total revenue from x_1 units of product 1 and x_2 units of product 2 is

$$R = -5x_1{}^2 - 8x_2{}^2 - 2x_1x_2 + 42x_1 + 102x_2.$$

Find x_1 and x_2 so as to maximize the revenue.

19. A retail outlet sells two competitive products, the prices of which are p_1 and p_2. Find p_1 and p_2 so as to maximize total revenue, where

$$R = 500p_1 + 800p_2 + 1.5p_1p_2 - 1.5p_1{}^2 - p_2{}^2.$$

20. A corporation manufactures a product at two locations. The cost of producing x_1 units at location 1 is

$$C_1 = 0.02x_1{}^2 + 4x_1 + 500$$

and the cost of producing x_2 units at location 2 is

$$C_2 = 0.05x_2 + 4x_2 + 275.$$

If the product sells for $15 per unit, find the quantity that should be produced at each location to maximize the profit, $P = 15(x_1 + x_2) - C_1 - C_2$.

21. Find a system of equations whose solution yields the coefficients a, b, and c for the least squares regression quadratic $y = ax^2 + bx + c$ for the points

$$(x_1, y_1), (x_2, y_2), \ldots, (x_n, y_n)$$

by minimizing the sum

$$S(a, b, c) = \sum_{i=1}^{n} (y_i - ax_i{}^2 - bx_i - c)^2.$$

22. Use the Second-Partials Test to verify that the formulas for a and b given in Theorem 15.18 yield a minimum.

[Hint: Use the fact that $n\sum_{i=1}^{n} x_i{}^2 \geq \left[\sum_{i=1}^{n} x_i\right]^2$.]

In Exercises 23–26, (a) find the least squares regression line and (b) calculate S, the sum of the squared errors.

23.

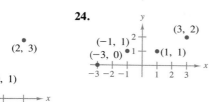

24.

25.

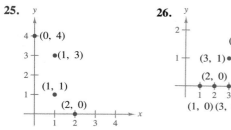

26.

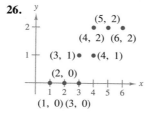

In Exercises 27–30, find the least squares regression line for the given points.

27. (0, 0), (1, 1), (3, 4), (4, 2), (5, 5)
28. (1, 0), (3, 3), (5, 6)
29. (0, 6), (4, 3), (5, 0), (8, −4), (10, −5)
30. (5, 2), (0, 0), (2, 1), (7, 4), (10, 6), (12, 6)

In Exercises 31–34, use the result of Exercise 21 to find the least squares regression quadratic for the given points. Then plot the points and sketch the graph of the least squares quadratic.

31. (−2, 0), (−1, 0), (0, 1), (1, 2), (2, 5)
32. (−4, 5), (−2, 6), (2, 6), (4, 2)
33. (0, 0), (2, 2), (3, 6), (4, 12)
34. (0, 10), (1, 9), (2, 6), (3, 0)

35. A store manager wants to know the demand for a certain product as a function of price. The daily sales for three different prices of the product are given in the following table.

Price (x)	$1.00	$1.25	$1.50
Demand (y)	450	375	330

(a) Find the least squares regression line for these data.
(b) Estimate the demand when the price is $1.40.

36. After contamination with a carcinogen, people in different geographic regions were assigned an exposure index, which represented the degree of contamination. Using the following data, find a least squares regression line to estimate the mortality per 100,000 people for a given exposure.

Exposure (x)	2.67	1.35	3.93	5.14	7.43
Mortality (y)	135.2	118.5	167.3	197.6	204.7

37. An agronomist used four test plots to determine the relationship between the wheat yield (in bushels per acre) and the amount of fertilizer (in hundreds of pounds per acre). The results are in the following table.

Fertilizer (x)	1.0	1.5	2.0	2.5
Yield (y)	32	41	48	53

Find the least squares regression line for this data, and estimate the yield for a fertilizer application of 160 pounds per acre.

38. The following table gives the world population in billions for five different years.

Year (x)	1960	1970	1975	1980	1985
Population (y)	3.0	3.7	4.1	4.5	4.8

Let $x = 0$ represent the year 1975.
(a) Find the least squares regression quadratic for these data.
(b) Use this quadratic to estimate the world population for the year 1990.

39. After developing a new turbocharger for an automobile engine, the following experimental data were obtained for speed in miles per hour at 2-second intervals. Fit a least squares quadratic to the data.

Time (x)	0	2	4	6	8	10
Speed (y)	0	15	30	50	65	70

15.10   Lagrange Multipliers

Lagrange multipliers ▪ Constrained optimization problems

Many optimization problems have restrictions or **constraints** on the values that can be used to produce the optimal solution. Such constraints tend to complicate optimization problems because the optimal solution can occur easily at a boundary point of the domain. In this section we look at an ingenious technique for solving such problems. It is called the **Method of Lagrange Multipliers,** after the French mathematician Joseph Louis Lagrange (1736–1813). Lagrange first introduced the method in his famous paper on mechanics, written when he was nineteen years old.

We begin with a straightforward constrained optimization problem. Suppose we want to find the rectangle of maximum area that can be inscribed in the ellipse given by

$$\frac{x^2}{3^2} + \frac{y^2}{4^2} = 1.$$

Let (x, y) be the vertex of the rectangle in the first quadrant, as shown in Figure 15.71. Then, since the rectangle has sides of length $2x$ and $2y$, its area is given by

$$f(x, y) = 4xy. \qquad \text{Objective function}$$

We want to find x and y such that $f(x, y)$ is a maximum. Our choice of (x, y) is restricted to points that lie on the ellipse in the first quadrant.

$$\frac{x^2}{3^2} + \frac{y^2}{4^2} = 1 \qquad \text{Constraint}$$

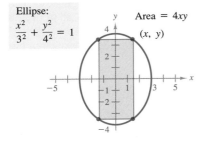

Ellipse:
$\dfrac{x^2}{3^2} + \dfrac{y^2}{4^2} = 1$

Area = $4xy$

FIGURE 15.71

To see how this problem can be solved by Lagrange multipliers, we consider the constraint equation to be a fixed level curve of

$$g(x, y) = \frac{x^2}{3^2} + \frac{y^2}{4^2}.$$

Now, the level curves of f represent a family of hyperbolas

$$f(x, y) = k$$

and in this family the level curves that meet the given constraint correspond to the hyperbolas that intersect the ellipse. Moreover, to maximize $f(x, y)$, we want to find the hyperbola that just barely satisfies the constraint. The level curve that does this is the one that is *tangent* to the ellipse, as shown in Figure 15.72.

To find the appropriate hyperbola, we use the fact that two curves are tangent at a point if and only if their gradient vectors are parallel. This means that $\nabla f(x, y)$ must be a scalar multiple of $\nabla g(x, y)$ at the point of tangency. In the context of constrained optimization problems, we denote this scalar by λ and write

$$\nabla f(x, y) = \lambda \nabla g(x, y).$$

The scalar λ is called a **Lagrange multiplier.** Theorem 15.19 gives the necessary conditions for the existence of such multipliers.

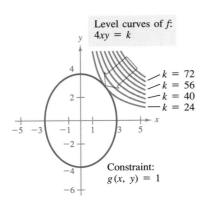

Level curves of f:
$4xy = k$

$k = 72$
$k = 56$
$k = 40$
$k = 24$

Constraint:
$g(x, y) = 1$

FIGURE 15.72

THEOREM 15.19 LAGRANGE'S THEOREM	Let f and g have continuous first partial derivatives such that f has an extremum at a point (x_0, y_0) on the smooth constraint curve $g(x, y) = c$. If $\nabla g(x_0, y_0) \neq 0$, then there is a real number λ such that $$\nabla f(x_0, y_0) = \lambda \nabla g(x_0, y_0).$$

PROOF   To begin, we represent the smooth curve given by $g(x, y) = c$ by the vector-valued function

$$\mathbf{r}(t) = x(t)\mathbf{i} + y(t)\mathbf{j}, \quad \mathbf{r}'(t) \neq 0$$

where x' and y' are continuous on an open interval I. If we define the function h by $h(t) = f(x(t), y(t))$, then since $f(x_0, y_0)$ is an extreme value of f, we know that

$$h(t_0) = f(x(t_0), y(t_0)) = f(x_0, y_0)$$

is an extreme value of h. This implies that $h'(t_0) = 0$, and, by the Chain Rule,

$$h'(t_0) = f_x(x_0, y_0)x'(t_0) + f_y(x_0, y_0)y'(t_0) = \nabla f(x_0, y_0) \cdot \mathbf{r}'(t_0) = 0.$$

Thereofre $\nabla f(x_0, y_0)$ is orthogonal to $\mathbf{r}'(t_0)$. Moreover, by Theorem 15.12, $\nabla g(x_0, y_0)$ is also orthogonal to $\mathbf{r}'(t_0)$. Consequently, the gradients $\nabla f(x_0, y_0)$ and $\nabla g(x_0, y_0)$ are parallel, and there must exist a scalar λ such that

$$\nabla f(x_0, y_0) = \lambda \nabla g(x_0, y_0).$$

Lagrange's Theorem can be shown to be true for functions of three variables, using a singular argument with level surfaces and Theorem 15.14.

The method of Lagrange Multipliers uses Theorem 15.19 to find the extreme values of a function f subject to a constraint.

METHOD OF LAGRANGE MULTIPLIERS	Let f and g satisfy the hypothesis of Lagrange's Theorem, and let f have a minimum or maximum subject to the constraint $g(x, y) = c$. To find the minimum or maximum of f, use the following steps.

1. Simultaneously solve the equations $\nabla f(x, y) = \lambda \nabla g(x, y)$ and $g(x, y) = c$ by solving the following system of equations.

$$f_x(x, y) = \lambda g_x(x, y)$$
$$f_y(x, y) = \lambda g_y(x, y)$$
$$g(x, y) = c$$

2. Evaluate f at each solution point obtained in the first step and at each endpoint (if any) of the constraint curve. The largest value yields the maximum of f subject to the constraint $g(x, y) = c$, and the smallest value yields the minimum of f subject to the constraint $g(x, y) = c$.

EXAMPLE 1 Using a Lagrange multiplier with one constraint

Find the maximum value of

$$f(x, y) = 4xy, \quad x > 0, y > 0$$

subject to the constraint $(x^2/3^2) + (y^2/4^2) = 1$.

SOLUTION

First, we let

$$g(x, y) = \frac{x^2}{3^2} + \frac{y^2}{4^2} - 1.$$

Then, since $\nabla f(x, y) = 4y\mathbf{i} + 4x\mathbf{j}$ and $\lambda\nabla g(x, y) = (2\lambda x/9)\mathbf{i} + (\lambda y/8)\mathbf{j}$, we obtain the following system of equations.

$$4y = \frac{2}{9}\lambda x \qquad f_x(x, y) = \lambda g_x(x, y)$$

$$4x = \frac{1}{8}\lambda y \qquad f_y(x, y) = \lambda g_y(x, y)$$

$$\frac{x^2}{3^2} + \frac{y^2}{4^2} = 1 \qquad \text{Constraint}$$

From the first equation we obtain $\lambda = 18y/x$, and substitution into the second equation produces

$$4x = \frac{1}{8}\left(\frac{18y}{x}\right)y \quad \Longrightarrow \quad x^2 = \frac{9}{16}y^2.$$

Now, substituting this value for x^2 into the third equation produces

$$\frac{1}{9}\left(\frac{9}{16}y^2\right) + \frac{1}{16}y^2 = 1 \quad \Longrightarrow \quad y^2 = 8.$$

Thus, $y = \pm 2\sqrt{2}$. Since it is required that $y > 0$, we choose the positive value and find that

$$16x^2 - 9(8) = 0 \quad \Longrightarrow \quad x^2 = \frac{9}{2} \quad \Longrightarrow \quad x = \frac{3}{\sqrt{2}}.$$

Thus, the maximum of f is

$$f\left(\frac{3}{\sqrt{2}}, 2\sqrt{2}\right) = 4\left(\frac{3}{\sqrt{2}}\right)(2\sqrt{2}) = 24. \qquad \blacksquare$$

EXAMPLE 2 A business application

The Cobb-Douglas production function (see Example 5, Section 15.1) for a particular manufacturer is given by

$$f(x, y) = 100x^{3/4}y^{1/4} \qquad \text{Objective function}$$

where x represents the units of labor (at \$150 per unit) and y represents the units of capital (at \$250 per unit). The total cost of labor and capital is limited to \$50,000. Find the maximum production level for this manufacturer.

SOLUTION

From the limit on the cost of labor and capital we have the constraint

$$150x + 250y = 50,000. \qquad \text{Constraint}$$

Using Lagrange's method, we let $g(x, y) = 150x + 250y - 50,000$ and obtain $\nabla f(x, y) = 75x^{-1/4}y^{1/4}\mathbf{i} + 25x^{3/4}y^{-3/4}\mathbf{j}$ and $\lambda\nabla g(x, y) = 150\lambda\mathbf{i} + 250\lambda\mathbf{j}$. This gives rise to the following system of equations.

$$75x^{-1/4}y^{1/4} = 150\lambda \qquad f_x(x, y) = \lambda g_x(x, y)$$

$$25x^{3/4}y^{-3/4} = 250\lambda \qquad f_y(x, y) = \lambda g_y(x, y)$$

$$150x + 250y = 50,000 \qquad \text{Constraint}$$

By solving for λ in the first equation

$$\lambda = \frac{75x^{-1/4}y^{1/4}}{150} = \frac{x^{-1/4}y^{1/4}}{2}$$

and substituting into the second equation, we obtain

$$25x^{3/4}y^{-3/4} = 250\left(\frac{x^{-1/4}y^{1/4}}{2}\right) \qquad \text{Multiply by } x^{1/4}y^{3/4}$$

$$25x = 125y.$$

Therefore, $x = 5y$, and, by substituting into the third equation, we have

$$150(5y) + 250y = 50,000$$

$$1000y = 50,000$$

$$y = 50 \text{ units of capital}$$

$$x = 250 \text{ units of labor.}$$

Finally, the maximum production is

$$f(250, 50) = 100(250)^{3/4}(50)^{1/4} \approx 16,719 \text{ product units.} \qquad \blacksquare$$

Economists call the Lagrange multiplier obtained in a production function the **marginal productivity of money.** For instance, in Example 2 the marginal productivity of money at $x = 250$ and $y = 50$ is

$$\lambda = \frac{x^{-1/4}y^{1/4}}{2} = \frac{(250)^{-1/4}(50)^{1/4}}{2} \approx 0.334$$

which means that for each additional dollar spent on production, 0.334 additional units of the product can be produced.

EXAMPLE 3 Using Lagrange multipliers with functions of three variables

Find the minimum value of

$$f(x, y, z) = 2x^2 + y^2 + 3z^2 \qquad \text{Objective function}$$

subject to the constraint $2x - 3y - 4z = 49$.

SOLUTION

Let $g(x, y, z) = 2x - 3y - 4z - 49$. Then, since $\nabla f(x, y, z) = 4x\mathbf{i} + 2y\mathbf{j} + 6z\mathbf{k}$ and $\lambda\nabla g(x, y, z) = 2\lambda\mathbf{i} - 3\lambda\mathbf{j} - 4\lambda\mathbf{k}$, we obtain the following system of equations.

$$
\begin{aligned}
4x &= 2\lambda & & f_x(x, y, z) = \lambda g_x(x, y, z) \\
2y &= -3\lambda & & f_y(x, y, z) = \lambda g_y(x, y, z) \\
6z &= -4\lambda & & f_z(x, y, z) = \lambda g_z(x, y, z) \\
2x - 3y - 4z &= 49 & & \text{Constraint}
\end{aligned}
$$

The solution of this system is $x = 3$, $y = -9$, and $z = -4$. Therefore, the minimum value of f is

$$f(3, -9, -4) = 2(3)^2 + (-9)^2 + 3(-4)^2 = 147. \qquad \blacksquare$$

At the beginning of this section we gave a graphical interpretation of constrained optimization problems in two variables. In three variables, the interpretation is similar, except that we use level surfaces instead of level curves. For instance, in Example 3, the level surfaces of f are ellipsoids centered at the origin, and the constraint $2x - 3y - 4z = 49$ is a plane. The minimum value of f is represented by the ellipsoid that is tangent to the constraint plane, as shown in Figure 15.73.

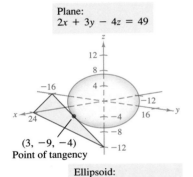

Plane:
$2x + 3y - 4z = 49$

$(3, -9, -4)$
Point of tangency

Ellipsoid:
$2x^2 + y^2 + 3z^2 = 147$

FIGURE 15.73

The method of Lagrange multipliers with two constraints

For optimization problems involving *two* constraint functions g and h, we introduce a second Lagrange multiplier, μ (the lowercase Greek letter *mu*), and solve the equation

$$\nabla f = \lambda\nabla g + \mu\nabla h$$

as illustrated in the next example.

EXAMPLE 4 Optimization with two constraints

Let

$$T(x, y, z) = 20 + 2x + 2y + z^2 \qquad \text{Objective function}$$

represent the temperature at each point on the sphere $x^2 + y^2 + z^2 = 11$. Find the extreme temperatures on the curve formed by the intersection of the plane $x + y + z = 3$ and the sphere.

SOLUTION

In this case we have two constraints.

$$g(x, y, z) = x^2 + y^2 + z^2 - 11 \qquad \text{Constraint 1}$$
$$h(x, y, z) = x + y + z - 3 \qquad \text{Constraint 2}$$

Since $\nabla T(x, y, z) = 2\mathbf{i} + 2\mathbf{j} + 2z\mathbf{k}$, $\lambda\nabla g(x, y, z) = 2\lambda x\mathbf{i} + 2\lambda y\mathbf{j} + 2\lambda z\mathbf{k}$, and $\mu\nabla h(x, y, z) = \mu\mathbf{i} + \mu\mathbf{j} + \mu\mathbf{k}$, we have the following system of equations.

$$2 = 2\lambda x + \mu \qquad T_x(x, y, z) = \lambda g_x(x, y, z) + \mu h_x(x, y, z)$$
$$2 = 2\lambda y + \mu \qquad T_y(x, y, z) = \lambda g_y(x, y, z) + \mu h_y(x, y, z)$$
$$2z = 2\lambda z + \mu \qquad T_z(x, y, z) = \lambda g_z(x, y, z) + \mu h_z(x, y, z)$$
$$x^2 + y^2 + z^2 = 11 \qquad \text{Constraint 1}$$
$$x + y + z = 3 \qquad \text{Constraint 2}$$

By subtracting the second equation from the first, we obtain the following system.

$$\lambda(x - y) = 0$$
$$2z(1 - \lambda) - \mu = 0$$
$$x^2 + y^2 + z^2 = 11$$
$$x + y + z = 3$$

From the first equation we conclude that $\lambda = 0$ or $x = y$. We consider these two cases separately. If $\lambda = 0$, then since $2 = 2\lambda x + \mu$, we have $\mu = 2$, and, by substituting into the second equation we have $z = 1$. The last two equations now become

$$x^2 + y^2 = 10 \qquad \text{and} \qquad x + y = 2$$

and, by substitution, we obtain

$$x^2 + (2 - x)^2 - 10 = 0$$
$$2x^2 - 4x - 6 = 0$$
$$x^2 - 2x - 3 = 0$$

with solutions $x = 3$ and $x = -1$. The corresponding y-values are $y = -1$ and $y = 3$, respectively. Thus, if $\lambda = 0$, the critical points are $(3, -1, 1)$ and $(-1, 3, 1)$. If $\lambda \neq 0$, then $x = y$, and the last two equations become

$$2x^2 + z^2 = 11 \qquad \text{and} \qquad 2x + z = 3$$

and, by substitution, we obtain

$$2x^2 + (3 - 2x)^2 = 11 \implies 3x^2 - 6x - 1 = 0.$$

The solutions are $x = (3 \pm 2\sqrt{3})/3$, and the corresponding values of z are $z = (3 \mp 4\sqrt{3})/3$, which yield two additional critical points. Finally, to find the optimal solutions, we compare the temperatures at the following four critical points.

$$T(3, -1, 1) = T(-1, 3, 1) = 25$$

$$T\left(\frac{3 - 2\sqrt{3}}{3}, \frac{3 - 2\sqrt{3}}{3}, \frac{3 + 4\sqrt{3}}{3}\right) = \frac{91}{3} \approx 30.33$$

$$T\left(\frac{3 + 2\sqrt{3}}{3}, \frac{3 + 2\sqrt{3}}{3}, \frac{3 - 4\sqrt{3}}{3}\right) = \frac{91}{3} \approx 30.33$$

Thus, $T = 25$ is the minimum temperature and $T = \frac{91}{3}$ is the maximum temperature on the curve. ▭

The system of equations that arises in the method of Lagrange multipliers is not, in general, a linear system, and the solution often requires ingenuity.

EXERCISES for Section 15.10

In Exercises 1–16, use Lagrange multipliers to find the indicated extrema. In each case, assume that x, y, and z are positive.

1. Maximize $f(x, y) = xy$
   Constraint: $x + y = 10$
2. Maximize $f(x, y) = xy$
   Constraint: $2x + y = 4$
3. Minimize $f(x, y) = x^2 + y^2$
   Constraint: $x + y - 4 = 0$
4. Minimize $f(x, y) = x^2 + y^2$
   Constraint: $2x - 4y + 5 = 0$
5. Maximize $f(x, y) = x^2 - y^2$
   Constraint: $y - x^2 = 0$
6. Maximize $f(x, y) = x^2 - y^2$
   Constraint: $x - 2y + 6 = 0$
7. Maximize $f(x, y) = 2x + 2xy + y$
   Constraint: $2x + y = 100$
8. Maximize $f(x, y) = 3x + y + 10$
   Constraint: $x^2y = 6$
9. Maximize $f(x, y) = \sqrt{6 - x^2 - y^2}$
   Constraint: $x + y - 2 = 0$
10. Minimize $f(x, y) = \sqrt{x^2 + y^2}$
    Constraint: $2x + 4y - 15 = 0$
11. Maximize $f(x, y) = e^{xy}$
    Constraint: $x^2 + y^2 - 8 = 0$
12. Minimize $f(x, y) = 2x + y$
    Constraint: $xy = 32$
13. Minimize $f(x, y, z) = x^2 + y^2 + z^2$
    Constraint: $x + y + z - 6 = 0$
14. Maximize $f(x, y, z) = xyz$
    Constraint: $x + y + z - 6 = 0$
15. Minimize $f(x, y, z) = x^2 + y^2 + z^2$
    Constraint: $x + y + z = 1$
16. Minimize $f(x, y) = x^2 - 8x + y^2 - 12y + 48$
    Constraint: $x + y = 8$

In Exercises 17–20, use Lagrange multipliers to find the indicated extrema of f subject to two constraints. In each case, assume x, y, and z are positive.

17. Maximize $f(x, y, z) = xyz$
    Constraints: $x + y + z = 32$
    $\qquad\qquad x - y + z = 0$
18. Minimize $f(x, y, z) = x^2 + y^2 + z^2$
    Constraints: $x + 2z = 4$
    $\qquad\qquad x + y = 8$
19. Maximize $f(x, y, z) = xy + yz$
    Constraints: $x + 2y = 6$
    $\qquad\qquad x - 3z = 0$
20. Maximize $f(x, y, z) = xyz$
    Constraints: $x^2 + z^2 = 5$
    $\qquad\qquad x - 2y = 0$

21. Find the dimensions of the rectangular package of largest volume subject to the constraint that the sum of the length and the girth cannot exceed 108 inches. (Maximize $V = xyz$ subject to the constraint $x + 2y + 2z = 108$.)
22. The material for the base of an open box costs 1.5 times as much as the material for the sides. Find the dimensions of the box of largest volume that can be made for a fixed cost C. (Maximize $V = xyz$ subject to $1.5xy + 2xz + 2yz = C$.)
23. A cargo container (in the shape of a rectangular solid) must have a volume of 480 cubic feet. Use Lagrange multipliers to find the dimensions of the container of this size that has minimum cost if the bottom will cost $5 per square foot to construct and the sides and top will cost $3 per square foot to construct.
24. Use Lagrange multipliers to find the dimensions of the right circular cylinder with volume V_0 cubic units and minimum surface area.

25. Use Lagrange multipliers to find the dimensions of the rectangular box of maximum volume that can be inscribed (with edges parallel to the coordinate axes) in the ellipsoid

$$\frac{x^2}{a^2} + \frac{y^2}{b^2} + \frac{z^2}{c^2} = 1.$$

26. Use Lagrange multipliers to prove that the product of three positive numbers, x, y, and z, whose sum has the constant value S, is maximum when the three numbers are equal. Use this result to prove that

$$\sqrt[3]{xyz} \le \frac{x + y + z}{3}, \quad x, y, z > 0.$$

27. When light waves traveling in a transparent medium strike the surface of a second transparent medium, they tend to "bend" in order to follow the path of minimum time. This tendency is called refraction and is described by **Snell's Law of Refraction,**

$$\frac{\sin \theta_1}{v_1} = \frac{\sin \theta_2}{v_2}$$

where θ_1 and θ_2 are the magnitudes of the angles shown in the figure, and v_1 and v_2 are the velocities of light in the two mediums. Use Lagrange multipliers to derive this law using the constraint $x + y = a$.

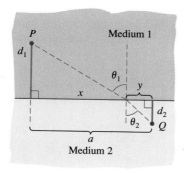

FIGURE FOR 27

28. Let

$$T(x, y, z) = 100 + x^2 + y^2$$

represent the temperature at each point on the sphere $x^2 + y^2 + z^2 = 50$. Find the maximum temperature on the curve formed by the intersection of the sphere and the plane $x - z = 0$.

In Exercises 29–32, find the minimum distance from the curve or surface to the specified point. [Hint: In Exercise 29, minimize $f(x, y) = x^2 + y^2$ subject to the constraint $2x + 3y = -1$.]

Curve	Point
29. Line: $2x + 3y = -1$	$(0, 0)$
30. Circle: $(x - 4)^2 + y^2 = 4$	$(0, 10)$

Surface	Point
31. Plane: $x + y + z = 1$	$(2, 1, 1)$
32. Cone: $z = \sqrt{x^2 + y^2}$	$(4, 0, 0)$

In Exercises 33 and 34, find the maximum production level if the total cost of labor (at $48 per unit) and capital (at $36 per unit) is limited to $100,000.

33. $P(x, y) = 100x^{0.25}y^{0.75}$
34. $P(x, y) = 100x^{0.6}y^{0.4}$

In Exercises 35 and 36, find the minimum cost of producing 20,000 product units, where x is the number of units of labor (at $48 per unit) and y is the number of units of capital (at $36 per unit).

35. $P(x, y) = 100x^{0.25}y^{0.75}$
36. $P(x, y) = 100x^{0.6}y^{0.4}$

In Exercises 37 and 38, find the highest point on the curve of intersection of the given surfaces.

37. Sphere: $x^2 + y^2 + z^2 = 36$
      Plane: $2x + y - z = 2$
38. Cone: $x^2 + y^2 - z^2 = 0$
      Plane: $x + 2z = 4$

REVIEW EXERCISES for Chapter 15

In Exercises 1–4, discuss the continuity of the function and evaluate the limit, if it exists.

1. $\displaystyle\lim_{(x,y)\to(1,1)} \frac{xy}{x^2 + y^2}$

2. $\displaystyle\lim_{(x,y)\to(1,1)} \frac{xy}{x^2 - y^2}$

3. $\displaystyle\lim_{(x,y)\to(0,0)} \frac{-4x^2y}{x^4 + y^2}$

4. $\displaystyle\lim_{(x,y)\to(0,0)} \frac{y + xe^{-y^2}}{1 + x^2}$

In Exercises 5–8, sketch several level curves for the given function.

5. $f(x, y) = e^{x^2+y^2}$

6. $f(x, y) = \ln xy$

7. $f(x, y) = x^2 - y^2$

8. $f(x, y) = \dfrac{x}{x + y}$

In Exercises 9–18, find all first partial derivatives.

9. $f(x, y) = e^x \cos y$

10. $f(x, y) = \dfrac{xy}{x + y}$

11. $z = xe^y + ye^x$

12. $z = \ln(x^2 + y^2 + 1)$

13. $g(x, y) = \dfrac{xy}{x^2 + y^2}$

14. $w = \sqrt{x^2 + y^2 + z^2}$

15. $f(x, y, z) = z \arctan \dfrac{y}{x}$

16. $f(x, y, z) = \dfrac{1}{\sqrt{1 - x^2 - y^2 - z^2}}$

17. $u(x, t) = ce^{-n^2 t} \sin nx$

18. $u(x, t) = c \sin(akx) \cos kt$

In Exercises 19 and 20, find $\partial z/\partial x$ and $\partial z/\partial y$.

19. $x^2y - 2xyz - xz - z^2 = 0$

20. $xz^2 - y \sin z = 0$

In Exercises 21–24, find all second partial derivatives and verify that the second mixed partials are equal.

21. $f(x, y) = 3x^2 - xy + 2y^3$

22. $h(x, y) = \dfrac{x}{x + y}$

23. $h(x, y) = x \sin y + y \cos x$

24. $g(x, y) = \cos(x - 2y)$

In Exercises 25–28, show that the function satisfies the **Laplace equation**

$$\frac{\partial^2 z}{\partial x^2} + \frac{\partial^2 z}{\partial y^2} = 0.$$

25. $z = x^2 - y^2$

26. $z = x^3 - 3xy^2$

27. $z = \dfrac{y}{x^2 + y^2}$

28. $z = e^x \sin y$

In Exercises 29 and 30, find dz.

29. $z = x \sin \dfrac{y}{x}$

30. $z = \dfrac{xy}{\sqrt{x^2 + y^2}}$

In Exercises 31 and 32, find the indicated derivatives (a) by the Chain Rule and (b) by substitution before differentiating.

31. $u = x^2 + y^2 + z^2,\ \dfrac{\partial u}{\partial r},\ \dfrac{\partial u}{\partial t}$

$\quad x = r \cos t,\ y = r \sin t,\ z = t$

32. $u = y^2 - x,\ \dfrac{du}{dt}$

$\quad x = \cos t,\ y = \sin t$

In Exercises 33–36, find the directional derivative at the given point in the direction of **v**.

33. $f(x, y) = x^2y$
$\quad \mathbf{v} = \mathbf{i} - \mathbf{j},\ (2, 1)$

34. $f(x, y) = \dfrac{1}{4}y^2 - x^2$
$\quad \mathbf{v} = 2\mathbf{i} + \mathbf{j},\ (1, 4)$

35. $f(x, y, z) = y^2 + xz$
$\quad \mathbf{v} = 2\mathbf{i} - \mathbf{j} + 2\mathbf{k},\ (1, 2, 2)$

36. $f(x, y, z) = 6x^2 + 3xy - 4y^2z$
$\quad \mathbf{v} = \mathbf{i} + \mathbf{j} - \mathbf{k},\ (1, 0, 1)$

In Exercises 37–40, find the gradient and the maximum value of the directional derivative of the function at the specified point.

	Function	Point
37.	$z = \dfrac{y}{x^2 + y^2}$	$(1, 1)$
38.	$z = \dfrac{x^2}{x - y}$	$(2, 1)$
39.	$z = e^{-x} \cos y$	$\left(0, \dfrac{\pi}{4}\right)$
40.	$z = x^2y$	$(2, 1)$

In Exercises 41–44, find an equation of the tangent plane and equations for the normal line to the given surface at the specified point.

	Surface	Point
41.	$f(x, y) = x^2y$	$(2, 1, 4)$
42.	$f(x, y) = \sqrt{25 - y^2}$	$(2, 3, 4)$
43.	$z = -9 + 4x - 6y - x^2 - y^2$	$(2, -3, 4)$
44.	$z = \sqrt{9 - x^2 - y^2}$	$(1, 2, 2)$

In Exercises 45 and 46, find symmetric equations of the tangent line to the curve of intersection of the given surfaces at the indicated point.

Surfaces	Point
45. $z = x^2 - y^2$, $z = 3$	$(2, 1, 3)$
46. $z = 25 - y^2$, $y = x$	$(4, 4, 9)$

In Exercises 47–50, locate and classify any extrema of the function.

47. $f(x, y) = x^3 - 3xy + y^2$

48. $f(x, y) = 2x^2 + 6xy + 9y^2 + 8x + 14$

49. $f(x, y) = xy + \dfrac{1}{x} + \dfrac{1}{y}$

50. $z = 50(x + y) - (0.1x^3 + 20x + 150)$
$$- (0.05y^3 + 20.6y + 125)$$

In Exercises 51 and 52, locate and classify any extrema of the function by using Lagrange multipliers.

51. $z = x^2y$

Constraint: $x + 2y = 2$

52. $w = xy + yz + xz$

Constraint: $x + y + z = 1$

53. The legs of a right triangle are measured to be 5 inches and 12 inches, with a possible error of $\frac{1}{16}$ inch. Approximate the maximum possible error in computing the length of the hypotenuse. Approximate the maximum percentage error.

54. To determine the height of a tower, the angle of elevation to the top of the tower was measured from a point $100 \pm \frac{1}{2}$ foot from the base. The angle is measured at $33°$, with a possible error of $1°$. Assuming the ground to be horizontal, approximate the maximum error in determining the height of the tower.

55. The volume of a right circular cone is $V = \frac{1}{3}\pi r^2 h$. Find the approximate error in the volume due to possible errors of $\frac{1}{8}$ inch in the measured values of r and h, if these are found to be 2 and 5 inches, respectively.

56. Approximate the error in the lateral surface area of the cone of Exercise 55. (The lateral surface area is given by $A = \pi r \sqrt{r^2 + h^2}$.)

57. A corporation manufactures a product at two locations. The cost functions for producing x_1 units at location 1 and x_2 units at location 2 are given by

$$C_1 = 0.05x_1^2 + 15x_1 + 5400$$
$$C_2 = 0.03x_2^2 + 15x_2 + 6100$$

and the total revenue function is

$$R = [225 - 0.4(x_1 + x_2)](x_1 + x_2).$$

Find the production levels at the two locations that will maximize the profit $P(x_1, x_2) = R - C_1 - C_2$.

58. A manufacturer has an order for 1000 units that can be produced at two locations. Let x_1 and x_2 be the number of units produced at the two locations. Find the number that should be produced at each to meet the order and minimize cost, if the cost function is

$$C = 0.25x_1^2 + 10x_1 + 0.15x_2^2 + 12x_2.$$

59. The production function for a manufacturer is

$$f(x, y) = 4x + xy + 2y.$$

Assume that the total amount available for labor and capital is $2000, and that units of labor and capital cost $20 and $4, respectively. Find the maximum production level for this manufacturer.

60. Show that a triangle is equilateral if the product of the sines of its angles is maximum.

Chapter 16 Application

Stars & Stripes, the winner of the 1987 America's Cup competition, employed a concentrated technological effort in sailboat design. Computers played a significant role in the victory. [*Scientific American*, August, 1987.]

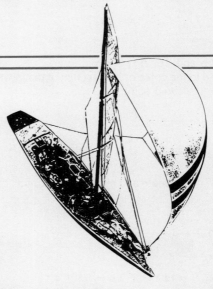

Center of Pressure on a Sail

The center of pressure on a sail is that point (x_p, y_p) at which the total aerodynamic force may be assumed to act. If the sail is represented by a plane region R, then the center of pressure is given by

$$x_p = \frac{\iint\limits_R xy \, dA}{\iint\limits_R y \, dA} \quad \text{and} \quad y_p = \frac{\iint\limits_R y^2 \, dA}{\iint\limits_R y \, dA}.$$

For example, to compute the center of pressure on a triangular sail with vertices at $(0, 0)$, $(2, 1)$, and $(0, 5)$, we evaluate the following double integrals.

$$\iint\limits_R y \, dA = \int_0^2 \int_{x/2}^{-2x+5} y \, dy \, dx = 10$$

$$\iint\limits_R xy \, dA = \int_0^2 \int_{x/2}^{-2x+5} xy \, dy \, dx = \frac{35}{6}$$

$$\iint\limits_R y^2 \, dA = \int_0^2 \int_{x/2}^{-2x+5} y^2 \, dy \, dx = \frac{155}{6}$$

Thus, the center of pressure occurs when

$$x_p = \frac{35/6}{10} = \frac{7}{12} \quad \text{and} \quad y_p = \frac{155/6}{10} = \frac{31}{12}.$$

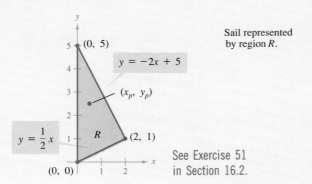

Sail represented by region R.

$y = -2x + 5$

(x_p, y_p)

$y = \frac{1}{2}x$

R

$(2, 1)$

$(0, 0)$

See Exercise 51 in Section 16.2.

Chapter Overview

In this chapter we introduce the concepts of **double integrals** over regions in the plane and **triple integrals** over regions in space.

Double integrals are used to evaluate the area of a region in the plane and the volume of a solid region under a surface. For example, the volume of the solid region lying between the paraboloid $z = 1 - x^2 - y^2$ and the xy-plane is given by the double integral

$$\text{volume} = \iint\limits_R (1 - x^2 - y^2) \, dA$$

where R represents the region in the xy-plane bounded by the unit circle $x^2 + y^2 = 1$. Double integrals are also used to find the moments, the mass, the center of mass, and the radii of gyration of a planar lamina. Triple integrals have analogous uses—they are used to evaluate the volume of a region in space, as well as the moments, the mass, the center of mass, and the radii of gyration of a solid region in space.

The techniques used to evaluate double or triple integrals involve conversions to **iterated integrals.** For example, the double integral listed above can be written as the following iterated integral.

$$\text{volume} = \int_{-1}^{1} \int_{-\sqrt{1-x^2}}^{\sqrt{1-x^2}} (1 - x^2 - y^2) \, dy \, dx$$

Occasionally, double or triple integrals are more easily evaluated by a **change of variables** to polar coordinates, spherical coordinates, or cylindrical coordinates. The chapter closes by considering other ways of changing variables using Jacobians.

Multiple Integration

16.1 Iterated Integrals and Area in the Plane

Iterated integrals ▪ Area of a plane region

In Chapters 16 and 17 we survey several applications of integration involving functions of several variables. Chapter 16 is much like Chapter 6 in that we will use integration to find plane areas, volumes, surface areas, moments, and centers of mass.

In Chapter 15 we saw that it is meaningful to differentiate functions of several variables with respect to one variable while holding the other variables constant. We can *integrate* functions of several variables by a similar procedure. For example, if we are given the partial derivative $f_x(x, y) = 2xy$, then, by considering y constant, we can integrate with respect to x to obtain

$$f(x, y) = \int 2xy \, dx = y \int 2x \, dx = y(x^2) + C(y) = x^2 y + C(y).$$

Note that the "constant" of integration, $C(y)$, is actually a function of y. In other words, by integrating with respect to x, we are able to recover $f(x, y)$ only partially. The total recovery of a function of x and y from its partial derivatives is a topic we will study in Chapter 17. For now, we are more concerned with extending definite integrals to functions of several variables. For instance, by considering y constant, we can apply the Fundamental Theorem of Calculus to evaluate

$$\int_1^{2y} 2xy \, dx = x^2 y \Big]_1^{2y} = (2y)^2 y - (1)^2 y = 4y^3 - y.$$

| x is the variable of integration and y is fixed. | Replace x by the limits of integration. | The result is a function of y. |

Similarly, we can integrate with respect to y by holding x fixed. We summarize both procedures as follows.

$$\int_{h_1(y)}^{h_2(y)} f_x(x, y) \, dx = f(x, y) \Big]_{h_1(y)}^{h_2(y)} = f(h_2(y), y) - f(h_1(y), y)$$

$$\int_{g_1(x)}^{g_2(x)} f_y(x, y) \, dy = f(x, y) \Big]_{g_1(x)}^{g_2(x)} = f(x, g_2(x)) - f(x, g_1(x))$$

Note that the variable of integration cannot appear in either limit of integration. For instance, it makes no sense to write $\int_0^x y \, dx$.

EXAMPLE 1 Integrating with respect to y

Evaluate the integral

$$\int_1^x (2x^2y^{-2} + 2y) \, dy.$$

SOLUTION

Considering x to be constant and integrating with respect to y produces

$$\int_1^x (2x^2y^{-2} + 2y) \, dy = \left[\frac{-2x^2}{y} + y^2 \right]_1^x$$

$$= \left(\frac{-2x^2}{x} + x^2 \right) - \left(\frac{-2x^2}{1} + 1 \right) = 3x^2 - 2x - 1.$$

Notice in Example 1 that the integral defines a function of x and can *itself* be integrated, as shown in the next example.

EXAMPLE 2 The integral of an integral

Evaluate the integral

$$\int_1^2 \left[\int_1^x (2x^2y^{-2} + 2y) \, dy \right] dx.$$

SOLUTION

Using the results from Example 1, we have

$$\int_1^2 \left[\int_1^x (2x^2y^{-2} + 2y) \, dy \right] dx = \int_1^2 (3x^2 - 2x - 1) \, dx$$

$$= \left[x^3 - x^2 - x \right]_1^2 = 2 - (-1) = 3.$$

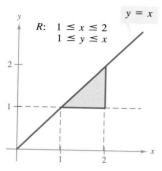

Region of Integration for

$$\int_1^2 \int_1^x f(x,\ y)\ dy\ dx$$

FIGURE 16.1

Iterated integrals

Integrals of the form

$$\int_a^b \left[\int_{g_1(x)}^{g_2(x)} f(x,\ y)\ dy \right] dx \qquad \text{and} \qquad \int_c^d \left[\int_{h_1(y)}^{h_2(y)} f(x,\ y)\ dx \right] dy$$

are called **iterated integrals.** We normally omit the brackets and write simply

$$\int_a^b \int_{g_1(x)}^{g_2(x)} f(x,\ y)\ dy\ dx \qquad \text{and} \qquad \int_c^d \int_{h_1(y)}^{h_2(y)} f(x,\ y)\ dx\ dy.$$

The **inside limits of integration** can be variable with respect to the outer variable of integration. However, the **outside limits of integration** must be constant with respect to both variables of integration. After performing the inside integration, we are left with a standard definite integral, and the second integration produces a real number. The limits of integration for an iterated integral identify two sets of boundary intervals for the variables. For instance, in Example 2, the outside limits indicate that x lies in the interval $1 \leq x \leq 2$ and the inside limits indicate that y lies in the interval $1 \leq y \leq x$. Together, these two intervals determine the **region of integration R** of the iterated integral, as shown in Figure 16.1.

Because an iterated integral is just a special type of definite integral—one in which the integrand is also an integral—we can use the properties of definite integrals to evaluate iterated integrals.

The area of a plane region

In the remainder of this section we take a new look at an old problem—that of finding the area of a plane region. Consider the plane region R bounded by $a \leq x \leq b$ and $g_1(x) \leq y \leq g_2(x)$, as shown in Figure 16.2. The area of R is given by the definite integral

$$\int_a^b [g_2(x) - g_1(x)]\ dx. \qquad \text{Area of } R$$

Using the Fundamental Theorem of Calculus, we can rewrite the integrand $g_2(x) - g_1(x)$ as a definite integral. Specifically, if we consider x to be fixed and let y vary from $g_1(x)$ to $g_2(x)$, we can write

$$\int_{g_1(x)}^{g_2(x)} dy = y \Big]_{g_1(x)}^{g_2(x)} = g_2(x) - g_1(x).$$

Combining these two integrals, we can write the area of the region R as an iterated integral

$$\int_a^b \int_{g_1(x)}^{g_2(x)} dy\ dx = \int_a^b y \Big]_{g_1(x)}^{g_2(x)} dx = \int_a^b [g_2(x) - g_1(x)]\ dx. \qquad \text{Area of } R$$

Region is bounded by
$$a \leq x \leq b$$
$$g_1(x) \leq y \leq g_2(x)$$

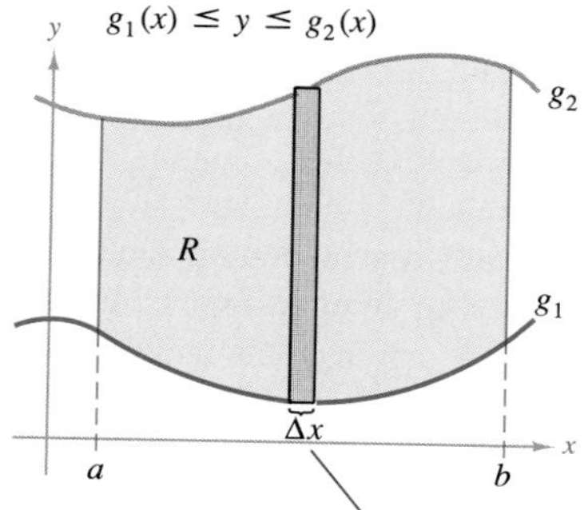

$$\text{Area} = \int_a^b \int_{g_1(x)}^{g_2(x)} dy\ dx$$

Vertically Simple Region

FIGURE 16.2

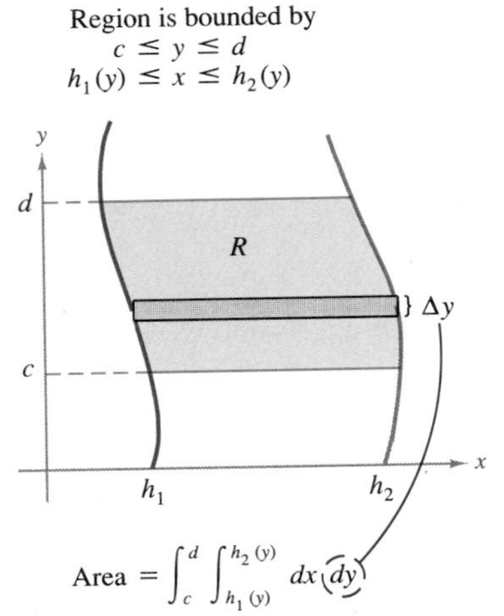

Region is bounded by
$$c \le y \le d$$
$$h_1(y) \le x \le h_2(y)$$

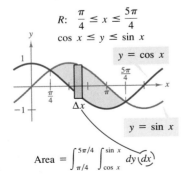

$$\text{Area} = \int_c^d \int_{h_1(y)}^{h_2(y)} dx\, dy$$

Horizontally Simple Region

Placing a representative rectangle in the region R helps determine both the order and the limits of integration. A vertical rectangle implies the order $dy\,dx$, with the inside limits corresponding to the upper and lower bounds of the rectangle, as shown in Figure 16.2. We call this type of region **vertically simple,** because the outside limits of integration represent the vertical lines $x = a$ and $x = b$. Similarly, a horizontal rectangle implies the order $dx\,dy$, with the inside limits determined by the left and right bounds of the rectangle, as shown in Figure 16.3. This type of region is called **horizontally simple,** because the outside limits represent the horizontal lines $y = c$ and $y = d$. We summarize the iterated integrals for these two types of simple regions as follows.

FIGURE 16.3

AREA OF A REGION IN THE PLANE

1. If R is defined by $a \le x \le b$ and $g_1(x) \le y \le g_2(x)$, where g_1 and g_2 are continuous on $[a, b]$, then the area of R is given by

$$A = \int_a^b \int_{g_1(x)}^{g_2(x)} dy\, dx. \qquad \text{Figure 16.2}$$

2. If R is defined by $c \le y \le d$ and $h_1(y) \le x \le h_2(y)$, where h_1 and h_2 are continuous on $[c, d]$, then the area of R is given by

$$A = \int_c^d \int_{h_1(y)}^{h_2(y)} dx\, dy. \qquad \text{Figure 16.3}$$

EXAMPLE 3 Finding area by an iterated integral

Use an iterated integral to find the area of the region bounded by the graphs of $f(x) = \sin x$ and $g(x) = \cos x$ between $x = \pi/4$ and $x = 5\pi/4$.

SOLUTION

Since f and g are given as functions of x, a vertical representative rectangle is convenient, and we choose $dy\,dx$ as the order of integration, as shown in Figure 16.4. The outside limits of integration are $\pi/4 \le x \le 5\pi/4$. Moreover, since the rectangle is bounded above by $f(x) = \sin x$ and below by $g(x) = \cos x$, we have

$$\text{area of } R = \int_{\pi/4}^{5\pi/4} \int_{\cos x}^{\sin x} dy\, dx = \int_{\pi/4}^{5\pi/4} \Big] y \Big]_{\cos x}^{\sin x} dx$$

$$= \int_{\pi/4}^{5\pi/4} (\sin x - \cos x)\, dx$$

$$= \Big[-\cos x - \sin x \Big]_{\pi/4}^{5\pi/4} = 2\sqrt{2}.$$

$$R: \frac{\pi}{4} \le x \le \frac{5\pi}{4}$$
$$\cos x \le y \le \sin x$$

$$y = \cos x$$

$$y = \sin x$$

$$\text{Area} = \int_{\pi/4}^{5\pi/4} \int_{\cos x}^{\sin x} dy\, dx$$

FIGURE 16.4

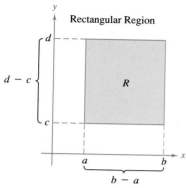

Rectangular Region

$$\text{Area} = \int_a^b \int_c^d dy\ dx = (b - a)(d - c)$$

FIGURE 16.5

In Figure 16.4, note that the region of integration of an iterated integral need not have any straight lines as boundaries. If all four limits of integration happen to be constants, then the region of integration is rectangular. For instance, the region of integration corresponding to the iterated integral

$$\int_a^b \int_c^d dy\ dx$$

is the rectangle shown in Figure 16.5. Try evaluating this integral to find that the area of the rectangle is $(b - a)(d - c)$.

One order of integration will often produce a simpler integration problem than the other order. For instance, try reworking Example 3 with the order $dx\ dy$—you may be surprised to see that the task is formidable! However, if you succeed, you will see that the answer is the same. In other words, the order of integration affects the ease of integration, but not the value of the integral. This is demonstrated further in the next example.

EXAMPLE 4   Comparing different orders of integration

Sketch the region whose area is represented by the integral

$$\int_0^2 \int_{y^2}^4 dx\ dy.$$

Then find another iterated integral using the order $dy\ dx$ to represent the same area and show that both integrals yield the same value.

SOLUTION

From the given limits of integration, we know that

$$y^2 \le x \le 4 \qquad \text{Inner limits of integration}$$

which means that the region R is bounded on the left by the parabola $x = y^2$ and on the right by the line $x = 4$. Furthermore, since

$$0 \le y \le 2 \qquad \text{Outer limits of integration}$$

we know that R is bounded below by the y-axis, as shown in Figure 16.6(a). Now, to change the order of integration to $dy\ dx$, we place a vertical rectangle in the region, as shown in Figure 16.6(b). From this we see that the constant bounds $0 \le x \le 4$ serve as the outer limits of integration. Solving for y in the equation $x = y^2$, we conclude that the inner bounds are $0 \le y \le \sqrt{x}$. Therefore, the area of the region can also be represented by

$$\int_0^4 \int_0^{\sqrt{x}} dy\ dx.$$

Evaluating these two integrals, we see that each yields the same value.

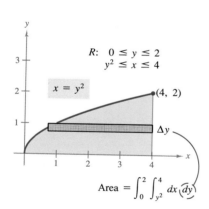

$R:$ $0 \le y \le 2$
$y^2 \le x \le 4$

$x = y^2$

(4, 2)

Δy

$$\text{Area} = \int_0^2 \int_{y^2}^4 dx\ dy$$

(a)

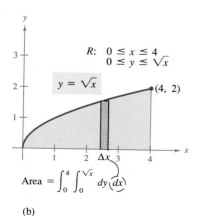

$R:$ $0 \le x \le 4$
$0 \le y \le \sqrt{x}$

$y = \sqrt{x}$

(4, 2)

Δx

$$\text{Area} = \int_0^4 \int_0^{\sqrt{x}} dy\ dx$$

(b)

FIGURE 16.6

$$\int_0^2 \int_{y^2}^4 dx\ dy = \int_0^2 x \Big]_{y^2}^4 dy = \int_0^2 (4 - y^2)\ dy = \left[4y - \frac{y^3}{3} \right]_0^2 = \frac{16}{3}$$

$$\int_0^4 \int_0^{\sqrt{x}} dy\ dx = \int_0^4 y \Big]_0^{\sqrt{x}} dx = \int_0^4 \sqrt{x}\ dx = \frac{2}{3} x^{3/2} \Big]_0^4 = \frac{16}{3}$$

It is not always possible to calculate the area of a region with a single iterated integral. In such cases we divide the region into subregions such that the area of each subregion can be calculated by an iterated integral. The total area is then the sum of the iterated integrals.

EXAMPLE 5   An area represented by two iterated integrals

Find the area of the region R that lies below the parabola $y = 4x - x^2$, above the x-axis, and above the line $y = -3x + 6$, as shown in Figure 16.7.

SOLUTION

We begin by dividing R into the two subregions R_1 and R_2 shown in Figure 16.7. In both regions it is convenient to use vertical rectangles, and we have

$$\text{area} = \int_1^2 \int_{-3x+6}^{4x-x^2} dy\, dx + \int_2^4 \int_0^{4x-x^2} dy\, dx$$

$$= \int_1^2 (4x - x^2 + 3x - 6)\, dx + \int_2^4 (4x - x^2)\, dx$$

$$= \left[\frac{7x^2}{2} - \frac{x^3}{3} - 6x \right]_1^2 + \left[2x^2 - \frac{x^3}{3} \right]_2^4$$

$$= \left(14 - \frac{8}{3} - 12 - \frac{7}{2} + \frac{1}{3} + 6 \right) + \left(32 - \frac{64}{3} - 8 + \frac{8}{3} \right)$$

$$= \frac{15}{2}.$$

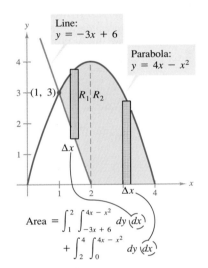

FIGURE 16.7

REMARK   In Examples 3, 4, and 5, be sure you see the benefit of sketching the region of integration. We strongly recommend that you develop the habit of making sketches to help determine the limits of integration for all iterated integrals in this chapter.

At this point you may be wondering why we need iterated integrals. After all, we already know how to use conventional integration to find the area of a region in the plane. (For instance, compare the solution of Example 3 in this section with that given in Example 3 in Section 6.1.) The need for iterated integrals will become clear in the next section. In this section we have chosen to give primary atention to procedures for finding the limits of integration of the region of an iterated integral, and the following exercise set is designed to develop skill in this important procedure.

EXERCISES for Section 16.1

In Exercises 1–10, evaluate the given integral.

1. $\int_0^x (2x - y)\,dy$

2. $\int_x^{x^2} \frac{y}{x}\,dy$

3. $\int_1^{2y} \frac{y}{x}\,dx$

4. $\int_0^{\cos y} y\,dx$

5. $\int_0^{\sqrt{4-x^2}} x^2 y\,dy$

6. $\int_{x^2}^{\sqrt{x}} (x^2 + y^2)\,dy$

7. $\int_{e^y}^{y} \frac{y \ln x}{x}\,dx$

8. $\int_{-\sqrt{1-y^2}}^{\sqrt{1-y^2}} (x^2 + y^2)\,dx$

9. $\int_0^{x^3} ye^{-y/x}\,dy$

10. $\int_y^{\pi/2} \sin^3 x \cos y\,dx$

In Exercises 11–20, evaluate the given iterated integral.

11. $\int_0^1 \int_0^2 (x + y)\,dy\,dx$

12. $\int_0^1 \int_0^x \sqrt{1 - x^2}\,dy\,dx$

13. $\int_1^2 \int_0^4 (x^2 - 2y^2 + 1)\,dx\,dy$

14. $\int_0^1 \int_y^{2y} (1 + 2x^2 + 2y^2)\,dx\,dy$

15. $\int_0^1 \int_0^{\sqrt{1-y^2}} (x + y)\,dx\,dy$

16. $\int_0^2 \int_{3y^2-6y}^{2y-y^2} 3y\,dx\,dy$

17. $\int_0^2 \int_0^{\sqrt{4-y^2}} \frac{2}{\sqrt{4 - y^2}}\,dx\,dy$

18. $\int_0^{\pi/2} \int_0^{2\cos\theta} r\,dr\,d\theta$

19. $\int_0^{\pi/2} \int_0^{\sin\theta} \theta r\,dr\,d\theta$

20. $\int_0^{\pi/4} \int_0^{\cos\theta} 3r^2 \sin\theta\,dr\,d\theta$

In Exercises 21–24, evaluate the improper iterated integral.

21. $\int_1^\infty \int_0^{1/x} y\,dy\,dx$

22. $\int_0^3 \int_0^\infty \frac{x^2}{1 + y^2}\,dy\,dx$

23. $\int_0^\infty \int_0^\infty xy\,e^{-(x^2+y^2)}\,dx\,dy$

24. $\int_1^\infty \int_1^\infty \frac{1}{xy}\,dx\,dy$

In Exercises 25–28, sketch the region R of integration and switch the order of integration.

25. $\int_0^4 \int_0^y f(x, y)\,dx\,dy$

26. $\int_0^4 \int_{\sqrt{y}}^2 f(x, y)\,dx\,dy$

27. $\int_{-\pi/2}^{\pi/2} \int_0^{\cos x} f(x, y)\,dy\,dx$

28. $\int_{-1}^1 \int_{x^2}^1 f(x, y)\,dy\,dx$

In Exercises 29–36, sketch the region R whose area is given by the iterated integral. Then switch the order of integration and show that both orders yield the same area.

29. $\int_0^1 \int_0^2 dy\,dx$

30. $\int_1^2 \int_2^4 dx\,dy$

31. $\int_0^1 \int_{-\sqrt{1-y^2}}^{\sqrt{1-y^2}} dx\,dy$

32. $\int_0^2 \int_0^x dy\,dx + \int_2^4 \int_0^{4-x} dy\,dx$

33. $\int_0^2 \int_{x/2}^1 dy\,dx$

34. $\int_0^4 \int_{\sqrt{x}}^2 dy\,dx$

35. $\int_0^1 \int_{y^2}^{\sqrt[3]{y}} dx\,dy$

36. $\int_{-2}^2 \int_0^{4-y^2} dx\,dy$

In Exercises 37–42, use an iterated integral to find the area of the specified region.

37.

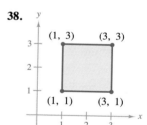

38.

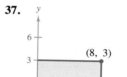

39.

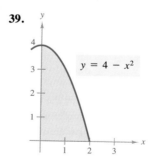

40.

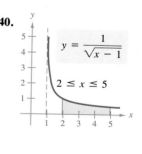

41.

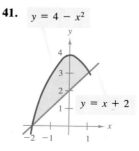

42.

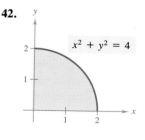

In Exercises 43–48, use an iterated integral to find the area of the region bounded by the graphs of the given equations.

43. $\sqrt{x} + \sqrt{y} = 2$, $x = 0$, $y = 0$
44. $y = x^{3/2}$, $y = x$
45. $2x - 3y = 0$, $x + y = 5$, $y = 0$
46. $xy = 9$, $y = x$, $y = 0$, $x = 9$
47. $\dfrac{x^2}{a^2} + \dfrac{y^2}{b^2} = 1$
48. $y = x$, $y = 2x$, $x = 2$

In Exercises 49–52, evaluate the iterated integral. Note that it is necessary to switch the order of integration.

49. $\displaystyle\int_0^2 \int_x^2 x\sqrt{1 + y^3}\, dy\, dx$ **50.** $\displaystyle\int_0^2 \int_x^2 e^{-y^2}\, dy\, dx$

51. $\displaystyle\int_0^1 \int_y^1 \sin x^2\, dx\, dy$ **52.** $\displaystyle\int_0^2 \int_{y^2}^4 \sqrt{x}\, \sin x\, dx\, dy$

In Exercises 53 and 54, use a computer to approximate the double integral.

53. $\displaystyle\int_0^2 \int_0^{4-x^2} e^{xy}\, dy\, dx$ **54.** $\displaystyle\int_0^1 \int_y^{2y} \sin(x + y)\, dx\, dy$

16.2 Double Integrals and Volume

Double integral ■ Volume of a solid region ■ Properties of double integrals ■ Evaluation of double integrals

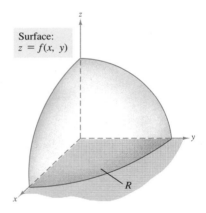

Surface:
$z = f(x, y)$

R

FIGURE 16.8

The essence of the definite integral over an *interval* is that it uses a limit process to assign measure to quantities such as area, volume, arc length, and mass. In this section we will see that a similar process can be used to define the **double integral** of a function of two variables over a *region in the plane*. Later, we will define other types of integrals—triple integrals over solid regions, line integrals over curves, and surface integrals over surfaces. In each case, we follow the pattern established in Section 5.2. That is, we partition the region into subregions, evaluate the function at a point in each subregion, and take the limit of the resulting Riemann sum. To help visualize the process for double integrals, we use a geometric model—the volume of the solid region lying under a surface.

Consider a continuous function f such that $f(x, y) \geq 0$ for all (x, y) in a region R in the xy-plane. Our goal is to find the volume of the solid region lying between the surface given by $z = f(x, y)$ and the xy-plane, as shown in Figure 16.8. To begin, we superimpose a rectangular grid over the region, as shown in Figure 16.9. The rectangles lying entirely within R form an **inner partition**, whose **norm** $\|\Delta\|$ is defined to be the length of the longest diagonal of the n rectangles. Next, we choose a point (x_i, y_i) in each rectangle and form the rectangular prism whose height is $f(x_i, y_i)$, as shown in Figure 16.10. Since the area of the ith rectangle is $\Delta A_i = \Delta x_i \Delta y_i$, it follows that the volume of the ith prism is

$$f(x_i, y_i)\Delta A_i = f(x_i, y_i)\Delta x_i \Delta y_i \qquad \text{Volume of } i\text{th prism}$$

and we can approximate the volume of the solid region by the Riemann sum of the volumes of all n prisms,

$$\sum_{i=1}^{n} f(x_i, y_i)\Delta x_i \Delta y_i \qquad \text{Riemann sum}$$

as shown in Figure 16.11. This approximation can be improved by tightening the mesh of the grid to form smaller and smaller rectangles. That is, we take the limit as $\|\Delta\| \to 0$. If the limit of this sum exists, then we have

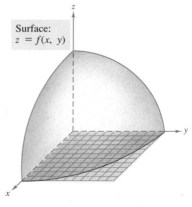

Surface:
$z = f(x, y)$

Inner Partition of Region R

FIGURE 16.9

$$\text{volume} = \lim_{\|\Delta\| \to 0} \sum_{i=1}^{n} f(x_i, y_i)\Delta x_i \Delta y_i.$$

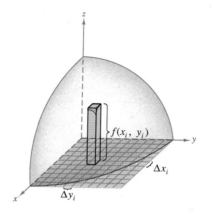

$f(x_i, y_i)$

Δx_i

Δy_i

FIGURE 16.10

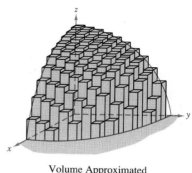

Volume Approximated
by Rectangular Prisms

FIGURE 16.11

Example 1 shows how to use a sum of the form

$$\sum_{i=1}^{n} f(x_i, y_i)\Delta x_i \Delta y_i$$

to approximate the volume of a solid.

EXAMPLE 1 Approximating the volume of a solid

Approximate the volume of the solid lying between the paraboloid

$$f(x, y) = 4 - x^2 - 2y^2$$

and the square region R given by $0 \le x \le 1$, $0 \le y \le 1$. Use a partition made up of squares whose edges have a length of $\frac{1}{4}$.

SOLUTION

We form the specified partition of R and choose the centers of the subregions as the points at which to evaluate $f(x, y)$.

$\left(\frac{1}{8}, \frac{1}{8}\right)$ $\left(\frac{1}{8}, \frac{3}{8}\right)$ $\left(\frac{1}{8}, \frac{5}{8}\right)$ $\left(\frac{1}{8}, \frac{7}{8}\right)$

$\left(\frac{3}{8}, \frac{1}{8}\right)$ $\left(\frac{3}{8}, \frac{3}{8}\right)$ $\left(\frac{3}{8}, \frac{5}{8}\right)$ $\left(\frac{3}{8}, \frac{7}{8}\right)$

$\left(\frac{5}{8}, \frac{1}{8}\right)$ $\left(\frac{5}{8}, \frac{3}{8}\right)$ $\left(\frac{5}{8}, \frac{5}{8}\right)$ $\left(\frac{5}{8}, \frac{7}{8}\right)$

$\left(\frac{7}{8}, \frac{1}{8}\right)$ $\left(\frac{7}{8}, \frac{3}{8}\right)$ $\left(\frac{7}{8}, \frac{5}{8}\right)$ $\left(\frac{7}{8}, \frac{7}{8}\right)$

Since the area of each square is $\Delta x_i \Delta y_i = \frac{1}{16}$, we approximate the volume by the sum

$$\sum_{i=1}^{16} f(x_i, y_i)\Delta x_i \Delta y_i = \sum_{i=1}^{16} (4 - x_i^2 - 2y_i^2)\left(\frac{1}{16}\right) \approx 3.016.$$

This approximation is shown graphically in Figure 16.12. (With a grid of squares with sides of length $\frac{1}{10}$, the approximation would be 3.003.) ▭

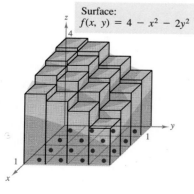

Surface:
$f(x, y) = 4 - x^2 - 2y^2$

FIGURE 16.12

Using the limit of a Riemann sum to define volume is a special case of using the limit to define a **double integral.** In the general case we do not require that the function be positive or continuous.

DEFINITION OF DOUBLE INTEGRAL

If f is defined on a closed, bounded region R in the xy-plane, then the **double integral of f over R** is given by

$$\iint\limits_{R} f(x, y)\, dA = \lim_{\|\Delta\| \to 0} \sum_{i=1}^{n} f(x_i, y_i)\Delta x_i \Delta y_i$$

provided the limit exists. If this limit exists, then we say that f is **integrable** over R.

REMARK Having defined a double integral, we will occasionally refer to a definite integral as a **single integral.**

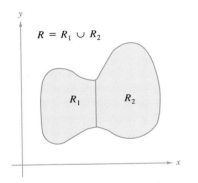

R_1 and R_2 are nonoverlapping.

FIGURE 16.13

The definition of the double integral of f on the region R is qualified by the phrase "provided the limit exists." A precise statement of the conditions necessary for this limit to exist is not appropriate in an introductory calculus course. However, we can give some sufficient conditions that cover almost all of the uses of double integrals in engineering and physics. One condition deals with the plane region R. For the remainder of this chapter, we will assume that R can be written as the union of a finite number of nonoverlapping subregions that are vertically or horizontally simple. (Two regions are nonoverlapping if their intersection is a set to which we assign an area of 0, as shown in Figure 16.13. For example, the area of a line segment is 0.) If f is continuous on such a region R, then the double integral of f over R exists.

Double integrals share many properties of single integrals, and we summarize some of these properties in the following theorem.

THEOREM 16.1 PROPERTIES OF DOUBLE INTEGRALS

Let f and g be continuous over a closed, bounded plane region R, and let c be a constant.

1. $\displaystyle\iint\limits_{R} cf(x, y)\, dA = c \iint\limits_{R} f(x, y)\, dA$

2. $\displaystyle\iint\limits_{R} [f(x, y) \pm g(x, y)]\, dA = \iint\limits_{R} f(x, y)\, dA \pm \iint\limits_{R} g(x, y)\, dA$

3. $\displaystyle\iint\limits_{R} f(x, y)\, dA \geq 0 \text{ if } f(x, y) \geq 0$

4. $\displaystyle\iint\limits_{R} f(x, y)\, dA \geq \iint\limits_{R} g(x, y)\, dA \text{ if } f(x, y) \geq g(x, y)$

5. $\displaystyle\iint\limits_{R} f(x, y)\, dA = \iint\limits_{R_1} f(x, y)\, dA + \iint\limits_{R_2} f(x, y)\, dA$

where R is the union of two nonoverlapping subregions R_1 and R_2.

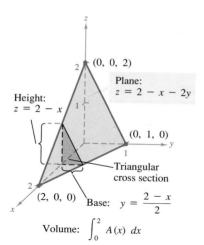

Volume: $\int_0^2 A(x)\,dx$

FIGURE 16.14

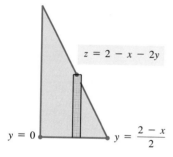

Triangular Cross Section

FIGURE 16.15

Evaluation of double integrals

To evaluate a double integral, we use an iterated integral. To see how this is done, we return to our geometric model of a double integral as the volume of a solid.

Consider the solid region bounded by the plane $z = f(x, y) = 2 - x - 2y$ and the three coordinate planes, as shown in Figure 16.14. Each vertical cross section taken parallel to the yz-plane is a triangular region whose base has a length of $y = (2 - x)/2$ and whose height is $z = 2 - x$. This implies that for a fixed value of x, the area of the triangular cross section is

$$A(x) = \frac{1}{2}(\text{base})(\text{height}) = \frac{1}{2}\left(\frac{2-x}{2}\right)(2-x) = \frac{(2-x)^2}{4}.$$

By the formula for the volume of a solid with known cross sections (Section 6.2), the volume of the solid is

$$\text{volume} = \int_a^b A(x)\,dx = \int_0^2 \frac{(2-x)^2}{4}\,dx = -\frac{(2-x)^3}{12}\bigg]_0^2 = \frac{8}{12} = \frac{2}{3}.$$

This procedure works no matter how $A(x)$ is obtained. In particular, we can find $A(x)$ by integration, as indicated in Figure 16.15. That is, we consider x to be constant, and integrate $z = 2 - x - 2y$ from 0 to $(2 - x)/2$ to obtain

$$A(x) = \int_0^{(2-x)/2} (2 - x - 2y)\,dy$$

$$= (2-x)y - y^2\bigg]_0^{(2-x)/2} = \frac{(2-x)^2}{4}.$$

Finally, combining these results, we have the *iterated integral*

$$\text{volume} = \iint_R f(x, y)\,dA = \int_0^2 \int_0^{(2-x)/2} (2 - x - 2y)\,dy\,dx.$$

To better understand this procedure, it helps to imagine the integration as two sweeping motions. For the inner integration, a vertical line sweeps out the area of a cross section. For the outer integration, the triangular cross section sweeps out the volume, as shown in Figure 16.16.

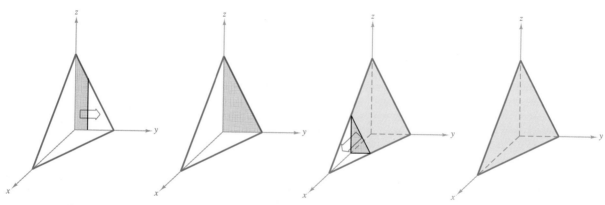

FIGURE 16.16

Integrate with respect to y to obtain the area of the cross section.

Integrate with respect to x to obtain the volume of the solid.

The following theorem was proved by the Italian mathematician Guido Fubini (1879–1943). The theorem states that if R is a vertically or horizontally simple region and f is continuous on R, then the double integral of f on R is equal to an iterated integral.

THEOREM 16.2
FUBINI'S THEOREM

Let f be continuous on a plane region R.

1. If R is defined by $a \le x \le b$ and $g_1(x) \le y \le g_2(x)$, where g_1 and g_2 are continuous on $[a, b]$, then

$$\iint_R f(x, y) \, dA = \int_a^b \int_{g_1(x)}^{g_2(x)} f(x, y) \, dy \, dx.$$

2. If R is defined by $c \le y \le d$ and $h_1(y) \le x \le h_2(y)$, where h_1 and h_2 are continuous on $[c, d]$, then

$$\iint_R f(x, y) \, dA = \int_c^d \int_{h_1(y)}^{h_2(y)} f(x, y) \, dx \, dy.$$

EXAMPLE 2   Evaluating a double integral as an iterated integral

Evaluate

$$\iint_R (4 - x^2 - 2y^2) \, dA$$

where R is the region given by $0 \le x \le 1$, $0 \le y \le 1$.

SOLUTION

Since the region R is a simple square, it is both vertically and horizontally simple, and we can use either order of integration. Suppose we choose $dy \, dx$ by placing a vertical representative rectangle in the region, as shown in Figure 16.17. This produces the following.

$$\iint_R (4 - x^2 - 2y^2) \, dA = \int_0^1 \int_0^1 (4 - x^2 - 2y^2) \, dy \, dx$$

$$= \int_0^1 \left[(4 - x^2)y - \frac{2y^3}{3} \right]_0^1 dx$$

$$= \int_0^1 \left(\frac{10}{3} - x^2 \right) dx = \left[\frac{10}{3}x - \frac{x^3}{3} \right]_0^1 = 3 \ \blacksquare$$

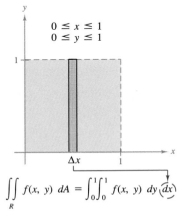

$$0 \le x \le 1$$
$$0 \le y \le 1$$

$$\iint_R f(x, y) \, dA = \int_0^1 \int_0^1 f(x, y) \, dy \, dx$$

FIGURE 16.17

REMARK   Note that the double integral evaluated in Example 2 represents the volume of the solid region approximated in Example 1. Note also that the approximation obtained in Example 1 is quite good (3.016 vs. 3.000), even though we used a partition consisting of only sixteen squares. We obtained so little error because we used the centers of the square subregions as the points in the approximation. This is comparable to the Midpoint Rule approximation of a single integral.

The difficulty of evaluating a single integral $\int_a^b f(x)\,dx$ depends largely on the function f, and not on the interval $[a, b]$. This is a major difference between single and double integrals. In the next example we integrate the same function we did in Examples 1 and 2. Notice that a change in the region R produces a much more difficult integration problem.

EXAMPLE 3   Finding volume by a double integral

Find the volume of the solid region bounded by the paraboloid $z = 4 - x^2 - 2y^2$ and the xy-plane.

SOLUTION

By letting $z = 0$, we see that the base of the region in the xy-plane is the ellipse $x^2 + 2y^2 = 4$, as shown in Figure 16.18. This plane region is both vertically and horizontally simple, and we choose the order $dy\,dx$.

Variable bounds for y: $-\sqrt{\dfrac{(4 - x^2)}{2}} \le y \le \sqrt{\dfrac{(4 - x^2)}{2}}$

Constant bounds for x: $-2 \le x \le 2$

Therefore, the volume is given by

$$V = \int_{-2}^{2} \int_{-\sqrt{(4-x^2)/2}}^{\sqrt{(4-x^2)/2}} (4 - x^2 - 2y^2)\,dy\,dx$$

$$= \int_{-2}^{2} \left[(4 - x^2)y - \frac{2y^3}{3} \right]_{-\sqrt{(4-x^2)/2}}^{\sqrt{(4-x^2)/2}} dx$$

$$= \frac{4}{3\sqrt{2}} \int_{-2}^{2} (4 - x^2)^{3/2}\,dx$$

$$= \frac{4}{3\sqrt{2}} \int_{-\pi/2}^{\pi/2} 16 \cos^4 \theta\,d\theta \qquad\qquad x = 2 \sin \theta$$

$$= \frac{64}{3\sqrt{2}}(2) \int_{0}^{\pi/2} \cos^4 \theta\,d\theta = \frac{128}{3\sqrt{2}}\left(\frac{3\pi}{16}\right) = 4\sqrt{2}\,\pi. \qquad \text{Wallis's Formula}$$

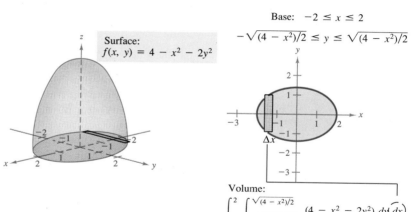

Surface:
$f(x, y) = 4 - x^2 - 2y^2$

Base: $-2 \le x \le 2$

$-\sqrt{(4 - x^2)/2} \le y \le \sqrt{(4 - x^2)/2}$

Volume:

$$\int_{-2}^{2} \int_{-\sqrt{(4 - x^2)/2}}^{\sqrt{(4 - x^2)/2}} (4 - x^2 - 2y^2)\,dy\,dx$$

FIGURE 16.18

REMARK In Example 3, note the usefulness of Wallis's Formula. You may want to review this formula in Section 9.3.

In Examples 2 and 3, the order of integration was optional, because the regions were both vertically and horizontally simple. Moreover, had we used the order $dx\,dy$, we would have obtained integrals of comparable difficulty. There are, however, some occasions in which one order of integration is much more convenient than the other. Example 4 shows such a case.

EXAMPLE 4 Comparing different orders of integration

Find the volume of the solid region R bounded by the surface $f(x,\,y) = e^{-x^2}$ and the planes $y = 0$, $y = x$, and $x = 1$, as shown in Figure 16.19.

SOLUTION

The base of R in the xy-plane is bounded by the lines $y = 0$, $x = 1$, and $y = x$. The two possible orders of integration are given in Figure 16.20.

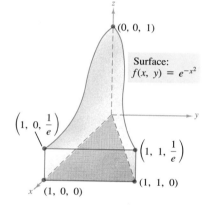

Surface: $f(x,\,y) = e^{-x^2}$

$(0,\,0,\,1)$

$\left(1,\,0,\,\dfrac{1}{e}\right)$

$\left(1,\,1,\,\dfrac{1}{e}\right)$

$(1,\,1,\,0)$

$(1,\,0,\,0)$

Base is bounded by
$y = 0$, $y = x$, $x = 1$.

FIGURE 16.19

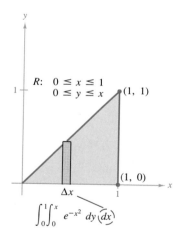

$R:\ \ 0 \le x \le 1$
$\ \ \ \ \ \ 0 \le y \le x$ $\ (1,\,1)$

$(1,\,0)$

Δx

$\displaystyle\int_0^1\int_0^x e^{-x^2}\,dy\,(dx)$

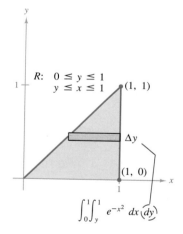

$R:\ \ 0 \le y \le 1$
$\ \ \ \ \ \ y \le x \le 1$ $\ (1,\,1)$

Δy

$(1,\,0)$

$\displaystyle\int_0^1\int_y^1 e^{-x^2}\,dx\,(dy)$

FIGURE 16.20

By setting up the corresponding iterated integrals, we discover that the order $dx\,dy$ requires the antiderivative $\int e^{-x^2}\,dx$, which we know is not an elementary function. On the other hand, the order $dy\,dx$ produces the integral

$$\int_0^1\int_0^x e^{-x^2}\,dy\,dx = \int_0^1 e^{-x^2}y\,\Big]_0^x\,dx = \int_0^1 xe^{-x^2}\,dx$$

$$= -\frac{1}{2}e^{-x^2}\Big]_0^1 = -\frac{1}{2}\left(\frac{1}{e} - 1\right)$$

$$= \frac{e - 1}{2e} \approx 0.316.$$

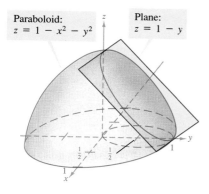

Paraboloid:
$z = 1 - x^2 - y^2$

Plane:
$z = 1 - y$

R:
$0 \le y \le 1$
$-\sqrt{y - y^2} \le x \le \sqrt{y - y^2}$

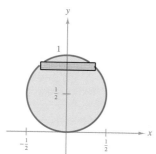

FIGURE 16.21

EXAMPLE 5   Finding volume of a region bounded by two surfaces

Find the volume of the solid region R bounded above by the paraboloid $z = 1 - x^2 - y^2$ and below by the plane $z = 1 - y$, as shown in Figure 16.21.

SOLUTION

Equating z-values, we find that the intersection of the two surfaces occurs on the right-circular cylinder given by

$$1 - y = 1 - x^2 - y^2 \quad \Longrightarrow \quad x^2 = y - y^2.$$

Since the volume of R is the difference between the volume under the paraboloid and the volume under the plane, we have

$$\text{volume} = \int_0^1 \int_{-\sqrt{y-y^2}}^{\sqrt{y-y^2}} (1 - x^2 - y^2) \, dx \, dy - \int_0^1 \int_{-\sqrt{y-y^2}}^{\sqrt{y-y^2}} (1 - y) \, dx \, dy$$

$$= \int_0^1 \int_{-\sqrt{y-y^2}}^{\sqrt{y-y^2}} (y - y^2 - x^2) \, dx \, dy$$

$$= \int_0^1 \left[(y - y^2)x - \frac{x^3}{3} \right]_{-\sqrt{y-y^2}}^{\sqrt{y-y^2}} dy = \frac{4}{3} \int_0^1 (y - y^2)^{3/2} \, dy$$

$$= \left(\frac{4}{3} \right)\left(\frac{1}{8} \right) \int_0^1 [1 - (2y - 1)^2]^{3/2} \, dy$$

$$= \frac{1}{6} \int_{-\pi/2}^{\pi/2} \frac{\cos^4 \theta}{2} \, d\theta \qquad\qquad 2y - 1 = \sin \theta$$

$$= \frac{1}{6} \int_0^{\pi/2} \cos^4 \theta \, d\theta = \left(\frac{1}{6} \right)\left(\frac{3\pi}{16} \right) = \frac{\pi}{32}. \qquad \text{Wallis's Formula} \quad \square$$

EXERCISES for Section 16.2

In Exercises 1–6, sketch the region R and evaluate the double integral

$$\iint_R f(x, y) \, dA.$$

1. $\displaystyle\int_0^2 \int_0^1 (1 + 2x + 2y) \, dy \, dx$

2. $\displaystyle\int_0^\pi \int_0^{\pi/2} \sin^2 x \cos^2 y \, dy \, dx$

3. $\displaystyle\int_0^6 \int_{y/2}^3 (x + y) \, dx \, dy$

4. $\displaystyle\int_0^1 \int_y^{\sqrt{y}} x^2 y^2 \, dx \, dy$

5. $\displaystyle\int_{-a}^a \int_{-\sqrt{a^2-x^2}}^{\sqrt{a^2-x^2}} (x + y) \, dy \, dx$

6. $\displaystyle\int_0^1 \int_{y-1}^0 e^{x+y} \, dx \, dy + \int_0^1 \int_0^{1-y} e^{x+y} \, dx \, dy$

In Exercises 7–12, set up the integral for both orders of integration, and use the more convenient order to evaluate the integral over the region R.

7. $\displaystyle\iint_R xy \, dA$

R: rectangle with vertices $(0, 0)$, $(0, 5)$, $(3, 5)$, $(3, 0)$

8. $\displaystyle\iint_R \sin x \sin y \, dA$

R: rectangle with vertices $(-\pi, 0)$, $(\pi, 0)$, $(\pi, \pi/2)$, $(-\pi, \pi/2)$

9. $\displaystyle\iint_R \frac{y}{x^2 + y^2} \, dA$

R: triangle bounded by $y = x$, $y = 2x$, $x = 2$

10. $\displaystyle\iint_R \frac{y}{1 + x^2}\, dA$

R: region bounded by $y = 0$, $y = \sqrt{x}$, $x = 4$

11. $\displaystyle\iint_R x\, dA$

R: sector of a circle in the first quadrant bounded by $y = \sqrt{25 - x^2}$, $3x - 4y = 0$, $y = 0$

12. $\displaystyle\iint_R (x^2 + y^2)\, dA$

R: semicircle bounded by $y = \sqrt{4 - x^2}$, $y = 0$

In Exercises 13–24, use a double integral to find the volume of the specified solid.

13.

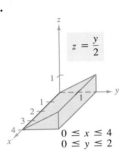

14.

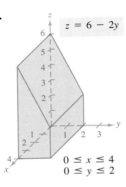

15.

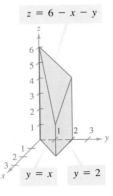

16.

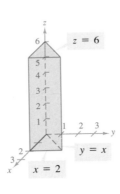

17.

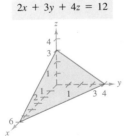

18.

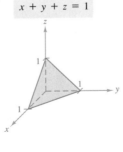

19.

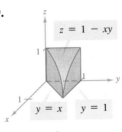

20.

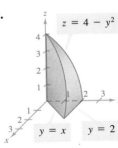

21.

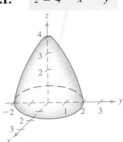

22.

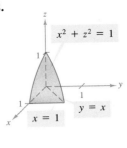

23. Improper integral

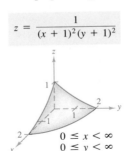

24. Improper integral

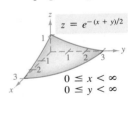

In Exercises 25–32, use a double integral to find the volume of the solid bounded by the graphs of the given equations.

25. $z = xy$, $z = 0$, $y = x$, $x = 1$ (first octant)
26. $y = 0$, $z = 0$, $y = x$, $z = x$, $x = 0$, $x = 5$
27. $z = 0$, $z = x^2$, $x = 0$, $x = 2$, $y = 0$, $y = 4$
28. $x^2 + y^2 + z^2 = r^2$
29. $x^2 + z^2 = 1$, $y^2 + z^2 = 1$ (first octant)
30. $y = 1 - x^2$, $z = 1 - x^2$ (first octant)
31. $z = x + y$, $x^2 + y^2 = 4$ (first octant)
32. $z = \dfrac{1}{1 + y^2}$, $x = 0$, $x = 2$, $y \geq 0$

In Exercises 33–36, use Wallis's Formula as an aid to finding the volume of the solid bounded by the graph of the equations.

33. $z = 4 - x^2 - y^2$, $z = 0$
34. $x^2 = 9 - y$, $z^2 = 9 - y$ (first octant)

35. $z = x^2 + y^2$, $x^2 + y^2 = 4$, $z = 0$

36. $z = \sin^2 x$, $z = 0$, $0 \le x \le \pi$, $0 \le y \le 5$

In Exercises 37–40, find the average value of $f(x, y)$ over the region R where

$$\text{average} = \frac{1}{A} \iint\limits_{R} f(x, y) \, dA$$

and where A is the area of R.

37. $f(x, y) = x$

R: rectangle with vertices $(0, 0)$, $(4, 0)$, $(4, 2)$, $(0, 2)$

38. $f(x, y) = xy$

R: rectangle with vertices $(0, 0)$, $(4, 0)$, $(4, 2)$, $(0, 2)$

39. $f(x, y) = x^2 + y^2$

R: square with vertices $(0, 0)$, $(2, 0)$, $(2, 2)$, $(0, 2)$

40. $f(x, y) = e^{x+y}$

R: triangle with vertices $(0, 0)$, $(0, 1)$, $(1, 1)$

41. For a particular company, the Cobb-Douglas production function is

$$f(x, y) = 100x^{0.6}y^{0.4}.$$

Estimate the average production level if the number of units of labor varies between 200 and 250 and the number of units of capital varies between 300 and 325.

42. A firm's profit in marketing two products is given by

$$P = 192x + 576y - x^2 - 5y^2 - 2xy - 5000$$

where x and y represent the number of units of each product. Estimate the average weekly profit if x varies between 40 and 50 units and y varies between 45 and 60 units.

In Exercises 43–46, evaluate the iterated integral. Note that it is necessary to switch the order of integration.

43. $\displaystyle\int_0^1 \int_{y/2}^{1/2} e^{-x^2} \, dx \, dy$

44. $\displaystyle\int_0^1 \int_0^{\arccos y} \sin x \sqrt{1 + \sin^2 x} \, dx \, dy$

45. $\displaystyle\int_0^{\ln 10} \int_{e^x}^{10} \frac{1}{\ln y} \, dy \, dx$

46. $\displaystyle\int_0^2 \int_{x^2}^4 \sqrt{y} \cos y \, dy \, dx$

In Exercises 47–50, approximate the integral

$$\iint\limits_{R} f(x, y) \, dA$$

by dividing the rectangle R with vertices $(0, 0)$, $(4, 0)$, $(4, 2)$, and $(0, 2)$ into 8 equal squares and finding the sum

$$\sum_{i=1}^{8} f(x_i, y_i) \Delta x_i \Delta y_i$$

where (x_i, y_i) is the center of the ith square. Evaluate the double integral and compare it to the approximation.

47. $\displaystyle\int_0^4 \int_0^2 (x + y) \, dy \, dx$

48. $\displaystyle\int_0^4 \int_0^2 xy \, dy \, dx$

49. $\displaystyle\int_0^4 \int_0^2 (x^2 + y^2) \, dy \, dx$

50. $\displaystyle\int_0^4 \int_0^2 \frac{1}{(x + 1)(y + 1)} \, dy \, dx$

51. Refer to the Chapter 16 Application. Perform the required integrations to verify that the center of pressure on a triangular sail with vertices at $(0, 0)$, $(2, 1)$, and $(0, 5)$ is given by $(x_p, y_p) = \left(\frac{7}{12}, \frac{31}{12}\right)$ where

$$x_p = \frac{\displaystyle\iint\limits_{R} xy \, dA}{\displaystyle\iint\limits_{R} y \, dA} \quad \text{and} \quad y_p = \frac{\displaystyle\iint\limits_{R} y^2 \, dA}{\displaystyle\iint\limits_{R} y \, dA}.$$

52. Repeat Exercise 51 if the sail has vertices at $(0, 0)$, $(3, 1)$, and $(0, 7)$.

53. If f is a continuous function such that $0 \le f(x, y) \le 1$ over a region R of area 1, prove that

$$0 \le \iint\limits_{R} f(x, y) \, dA \le 1.$$

54. Find the volume of the solid in the first octant bounded by the coordinate planes and the plane

$$\frac{x}{a} + \frac{y}{b} + \frac{z}{c} = 1$$

where $a > 0$, $b > 0$, and $c > 0$.

In Exercises 55 and 56, use a computer to approximate the volume of the solid bounded by the graphs of the given equations.

55. $z = \dfrac{2}{1 + x^2 + y^2}$, $z = 0$, $y = 0$, $x = 0$, $y = -0.5x + 1$

56. $z = \ln(1 + x + y)$, $z = 0$, $y = 0$, $x = 0$, $x = 4 - \sqrt{y}$

16.3 Change of Variables: Polar Coordinates

Double integrals in polar coordinates ▪ Change of variables to polar form

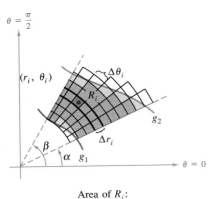

Area of R_i:
$\Delta A_i = r_i \Delta r_i \Delta \theta_i$

FIGURE 16.22

Some double integrals are *much* easier to evaluate in polar form than in rectangular form. This is especially true for regions such as circles, cardioids, and petal curves, and for integrands that involve the sum $x^2 + y^2$. We begin with the definition of a double integral in the polar coordinate system.

Consider a polar region R bounded by the graphs of $r = g_1(\theta)$ and $r = g_2(\theta)$ and by the lines $\theta = \alpha$ and $\theta = \beta$. We begin by superimposing on the region a polar grid made of rays and circular arcs, as shown in Figure 16.22. The polar sectors R_i lying entirely within R form an **inner polar partition,** whose **norm** $\|\Delta\|$ is defined to be the length of the longest diagonal of the n polar sectors. Recall from the development of area in polar coordinates (Section 12.5) that if we choose a point (r_i, θ_i) in R_i such that r_i is the average radius of R_i, then the area of R_i is

$$\Delta A_i = r_i \Delta r_i \Delta \theta_i \qquad \text{Area of } R_i$$

where θ is measured in radians. (See Exercise 35.) Now, if f is a continuous function of r and θ on the region R, then by taking the limit, as $\|\Delta\| \to 0$, of the sum

$$\sum_{i=1}^{n} f(r_i, \theta_i) r_i \Delta r_i \Delta \theta_i$$

it can be shown that we obtain the following polar form of a double integral.

DOUBLE INTEGRAL IN POLAR COORDINATES

If f is a continuous function of r and θ on a closed bounded plane region R, then the **double integral of f over R** in polar coordinates is given by

$$\iint_R f(r, \theta) \, dA = \lim_{\|\Delta\| \to 0} \sum_{i=1}^{n} f(r_i, \theta_i) r_i \Delta r_i \Delta \theta_i = \iint_R f(r, \theta) r \, dr \, d\theta.$$

As with rectangular coordinates, we restrict the region R to two basic types, which we call ***r*-simple** and ***θ*-simple** regions, as in Figure 16.23.

Fixed bounds for θ:
$\theta_1 \leq \theta \leq \theta_2$
Variable bounds for r:
$g_1(\theta) \leq r \leq g_2(\theta)$

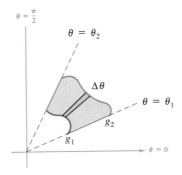

Variable bounds for θ:
$h_1(r) \leq \theta \leq h_2(r)$
Fixed bounds for r:
$r_1 \leq r \leq r_2$

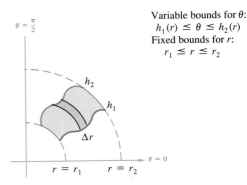

FIGURE 16.23

r-Simple region

θ-Simple region

We evaluate a double integral in polar form by an iterated integral, as stated in the following polar form of Fubini's Theorem.

THEOREM 16.3 **POLAR FORM OF** **FUBINI'S THEOREM**	Let f be continuous on a plane region R. 1. If R is defined by $\theta_1 \leq \theta \leq \theta_2$ and $g_1(\theta) \leq r \leq g_2(\theta)$, where g_1 and g_2 are continuous on $[\theta_1, \theta_2]$, then $$\iint\limits_R f(r, \theta)\, dA = \int_{\theta_1}^{\theta_2} \int_{g_1(\theta)}^{g_2(\theta)} f(r, \theta) r\, dr\, d\theta.$$ 2. If R is defined by $r_1 \leq r \leq r_2$ and $h_1(r) \leq \theta \leq h_2(r)$, where h_1 and h_2 are continuous on $[r_1, r_2]$, then $$\iint\limits_R f(r, \theta)\, dA = \int_{r_1}^{r_2} \int_{h_1(r)}^{h_2(r)} f(r, \theta) r\, d\theta\, dr.$$

REMARK   Note that the integrand contains a factor r, which arose from the area of the ith polar sector $\Delta A_i = r_i \Delta r_i \Delta \theta_i$.

EXAMPLE 1   Evaluating a double polar integral

Evaluate

$$\iint\limits_R \sin \theta\, dA$$

where R is the first-quadrant region lying inside the circle given by $r = 4 \cos \theta$ and outside the circle given by $r = 2$.

SOLUTION

From Figure 16.24 we see that R is an r-simple region, and we sketch a representative sector whose bounds yield the following limits of integration.

$$0 \leq \theta \leq \frac{\pi}{3} \qquad \text{Fixed bounds on } \theta$$

$$2 \leq r \leq 4 \cos \theta \qquad \text{Variable bounds on } r$$

Thus, we obtain

$$\iint\limits_R \sin \theta\, dA = \int_0^{\pi/3} \int_2^{4\cos\theta} (\sin \theta) r\, dr\, d\theta = \int_0^{\pi/3} (\sin \theta) \frac{r^2}{2} \bigg]_2^{4\cos\theta} d\theta$$

$$= \frac{1}{2} \int_0^{\pi/3} (\sin \theta)(16 \cos^2 \theta - 4)\, d\theta$$

$$= 2 \int_0^{\pi/3} [4 \cos^2 \theta\, (\sin \theta) - \sin \theta]\, d\theta$$

$$= 2 \left[-\frac{4 \cos^3 \theta}{3} + \cos \theta \right]_0^{\pi/3} = \frac{4}{3}. \qquad \blacksquare$$

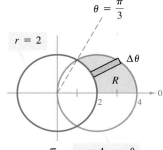

$\theta = \dfrac{\pi}{3}$

$r = 2$

$\Delta\theta$

R

$r = 4 \cos \theta$

$R: \ 0 \leq \theta \leq \dfrac{\pi}{3}$

$2 \leq r \leq 4 \cos \theta$

FIGURE 16.24

EXAMPLE 2 Finding area of polar regions

Use a double integral to find the area enclosed by the graph of $r = 3 \cos 3\theta$.

SOLUTION

We consider R to be one petal of the curve shown in Figure 16.25. This region is r-simple, and the boundaries are as follows.

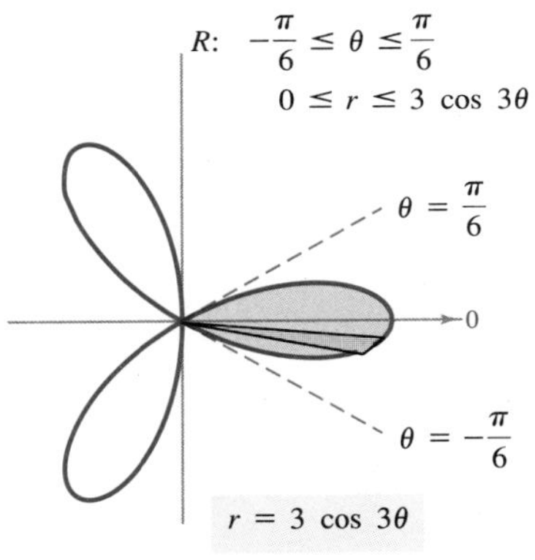

$$-\frac{\pi}{6} \le \theta \le \frac{\pi}{6} \qquad \text{Fixed bounds on } \theta$$

$$0 \le r \le 3 \cos 3\theta \qquad \text{Variable bounds on } r$$

Thus, the area of one petal is

$$\frac{1}{3}A = \iint_R dA = \int_{-\pi/6}^{\pi/6} \int_0^{3\cos 3\theta} r \, dr \, d\theta = \int_{-\pi/6}^{\pi/6} \frac{r^2}{2} \Big]_0^{3\cos 3\theta} d\theta$$

$$= \frac{9}{2} \int_{-\pi/6}^{\pi/6} \cos^2 3\theta \, d\theta$$

$$= \frac{9}{4} \int_{-\pi/6}^{\pi/6} (1 + \cos 6\theta) \, d\theta$$

$$= \frac{9}{4} \left[\theta + \frac{1}{6} \sin 6\theta \right]_{-\pi/6}^{\pi/6} = \frac{3\pi}{4}.$$

Therefore, the total area is $A = 9\pi/4$.

R: $-\dfrac{\pi}{6} \le \theta \le \dfrac{\pi}{6}$
$0 \le r \le 3 \cos 3\theta$

$\theta = \dfrac{\pi}{6}$

$\theta = -\dfrac{\pi}{6}$

$r = 3 \cos 3\theta$

FIGURE 16.25

Change of variables to polar form

We now come to the reason for introducing double integrals in polar form. With sufficient restrictions on a function f and a region R, we can make the following change of variables.

$$x = r \cos \theta, \qquad y = r \sin \theta$$
$$r^2 = x^2 + y^2, \qquad dA = r \, dr \, d\theta$$

This change of variables converts a double integral in rectangular coordinates to a double integral in polar coordinates, as stated in the following theorem. We postpone a discussion of the proof of this theorem until Section 16.8.

THEOREM 16.4 **CHANGE OF VARIABLES** **TO POLAR FORM**	Let R be a plane region consisting of all points $(x, y) = (r \cos \theta, r \sin \theta)$ satisfying the condition $$0 \le g_1(\theta) \le r \le g_2(\theta), \quad \theta_1 \le \theta \le \theta_2$$ where $0 < (\theta_2 - \theta_1) \le 2\pi$. If g_1 and g_2 are continuous on $[\theta_1, \theta_2]$ and f is continuous on R, then $$\iint_R f(x, y) \, dA = \int_{\theta_1}^{\theta_2} \int_{g_1(\theta)}^{g_2(\theta)} f(r \cos \theta, r \sin \theta) r \, dr \, d\theta.$$

If $z = f(x, y)$ is nonnegative on R, then the integral in Theorem 16.4 can be interpreted as the *volume* of the solid region between the graph of f and the region R. We use this interpretation in our next example.

EXAMPLE 3   Change of variables to polar coordinates

Use polar coordinates to find the volume of the solid region bounded above by the hemisphere $z = \sqrt{16 - x^2 - y^2}$ and below by the circular region R given by $x^2 + y^2 = 4$, as shown in Figure 16.26.

SOLUTION

From Figure 16.26 we can see that R has the bounds

$$-\sqrt{4 - y^2} \leq x \leq \sqrt{4 - y^2}, \quad -2 \leq y \leq 2$$

and that $0 \leq z \leq \sqrt{16 - x^2 - y^2}$. In polar coordinates the bounds are

$$0 \leq r \leq 2 \quad \text{and} \quad 0 \leq \theta \leq 2\pi$$

with height $z = \sqrt{16 - x^2 - y^2} = \sqrt{16 - r^2}$. Consequently, the volume V is given by

$$V = \iint_R f(x, y)\, dA = \int_0^{2\pi} \int_0^2 \sqrt{16 - r^2}\, r\, dr\, d\theta$$

$$= -\frac{1}{3} \int_0^{2\pi} (16 - r^2)^{3/2} \Big]_0^2 d\theta$$

$$= -\frac{1}{3} \int_0^{2\pi} (24\sqrt{3} - 64)\, d\theta$$

$$= -\frac{8}{3}(3\sqrt{3} - 8)\theta \Big]_0^{2\pi}$$

$$= \frac{16\pi}{3}(8 - 3\sqrt{3}) \approx 46.98.$$

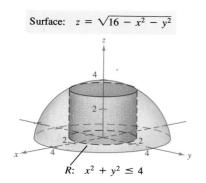

Surface: $z = \sqrt{16 - x^2 - y^2}$

R: $x^2 + y^2 \leq 4$

FIGURE 16.26

REMARK   To see the remarkable benefit of polar coordinates in Example 3, you should try to evaluate the corresponding rectangular double integral

$$\int_{-2}^{2} \int_{-\sqrt{4-y^2}}^{\sqrt{4-y^2}} \sqrt{16 - x^2 - y^2}\, dx\, dy.$$

EXAMPLE 4   Change of variables to polar coordinates

Let R be the annular region lying between the two circles $x^2 + y^2 = 1$ and $x^2 + y^2 = 5$, as shown in Figure 16.27. Evaluate the integral

$$\iint_R (x^2 + y)\, dA.$$

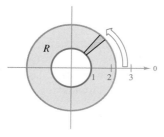

R: $1 \leq r \leq \sqrt{5}$
    $0 \leq \theta \leq 2\pi$

r-Simple region:.
$dA = r \, dr \, d\theta$

FIGURE 16.27

SOLUTION

The polar boundaries are $1 \leq r \leq \sqrt{5}$ and $0 \leq \theta \leq 2\pi$. Thus, we have

$$\iint_R (x^2 + y) \, dA = \int_0^{2\pi} \int_1^{\sqrt{5}} (r^2 \cos^2 \theta + r \sin \theta) \, r \, dr \, d\theta.$$

For the sake of illustration, we change the order of integration, as indicated in Figure 16.28, and obtain

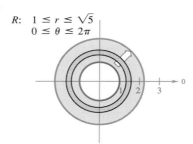

R: $1 \leq r \leq \sqrt{5}$
    $0 \leq \theta \leq 2\pi$

FIGURE 16.28

θ-Simple region:
$dA = r \, d\theta \, dr$

$$\iint_R (x^2 + y) \, dA = \int_1^{\sqrt{5}} \int_0^{2\pi} r^2(r \cos^2 \theta + \sin \theta) \, d\theta \, dr$$

$$= \int_1^{\sqrt{5}} r^2 \left[\int_0^{2\pi} \left(\frac{r}{2} + \frac{r \cos 2\theta}{2} + \sin \theta \right) d\theta \right] dr$$

$$= \int_1^{\sqrt{5}} r^2 \left[\frac{r\theta}{2} + \frac{r \sin 2\theta}{4} - \cos \theta \right]_0^{2\pi} dr$$

$$= \int_1^{\sqrt{5}} r^2(\pi r) \, dr = \pi \frac{r^4}{4} \Big]_1^{\sqrt{5}} = 6\pi.$$

Try integrating with $dA = r \, dr \, d\theta$ to see which order is more convenient.

___

EXERCISES for Section 16.3

In Exercises 1–6, evaluate the double integral $\iint_R f(r, \theta) \, dA$, and sketch the region R.

1. $\int_0^{2\pi} \int_0^6 3r^2 \sin \theta \, dr \, d\theta$

2. $\int_0^{\pi/4} \int_0^4 r^2 \sin \theta \cos \theta \, dr \, d\theta$

3. $\int_0^{\pi/2} \int_2^3 \sqrt{9 - r^2} \, r \, dr \, d\theta$

4. $\int_0^{\pi/2} \int_0^3 re^{-r^2} \, dr \, d\theta$

5. $\int_0^{\pi/2} \int_0^{1+\sin \theta} \theta \, dr \, d\theta$

6. $\int_0^{\pi/2} \int_0^{1-\cos \theta} \sin \theta \, dr \, d\theta$

In Exercises 7–12, use a double integral to find the area of the indicated region.

7.

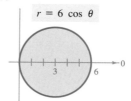

$r = 6 \cos \theta$

8.

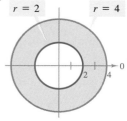

$r = 2$ $\qquad$ $r = 4$

9.

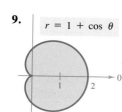

$r = 1 + \cos \theta$

10.

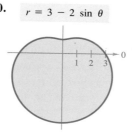

$r = 3 - 2 \sin \theta$

11.

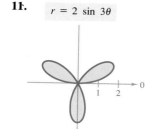

$r = 2 \sin 3\theta$

12.

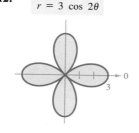

$r = 3 \cos 2\theta$

In Exercises 13–18, evaluate the double integral by changing to polar coordinates.

13. $\int_0^a \int_0^{\sqrt{a^2-y^2}} y \, dx \, dy$

14. $\int_0^a \int_0^{\sqrt{a^2-x^2}} x \, dy \, dx$

15. $\int_0^3 \int_0^{\sqrt{9-x^2}} \arctan \frac{y}{x} \, dy \, dx$

16. $\int_0^2 \int_y^{\sqrt{8-y^2}} \sqrt{x^2 + y^2} \, dx \, dy$

17. $\int_0^2 \int_0^{\sqrt{2x-x^2}} xy \, dy \, dx$

18. $\int_0^4 \int_0^{\sqrt{4y-y^2}} x^2 \, dx \, dy$

In Exercises 19 and 20, combine the sum of the two double integrals into a single double integral by using polar coordinates. Evaluate the resulting double integral.

19. $\int_0^2 \int_0^x \sqrt{x^2 + y^2} \, dy \, dx$
$+ \int_2^{2\sqrt{2}} \int_0^{\sqrt{8-x^2}} \sqrt{x^2 + y^2} \, dy \, dx$

20. $\int_0^{5\sqrt{2}/2} \int_0^x xy \, dy \, dx + \int_{5\sqrt{2}/2}^5 \int_0^{\sqrt{25-x^2}} xy \, dy \, dx$

In Exercises 21–24, use polar coordinates to evaluate the double integral $\iint\limits_R f(x, y) \, dA$.

21. $f(x, y) = x + y$
$R: x^2 + y^2 \leq 4, \, 0 \leq x, \, 0 \leq y$

22. $f(x, y) = e^{-(x^2+y^2)}$
$R: x^2 + y^2 \leq 4, \, 0 \leq x, \, 0 \leq y$

23. $f(x, y) = \arctan \frac{y}{x}$
$R: x^2 + y^2 \leq 1, \, 0 \leq x, \, 0 \leq y$

24. $f(x, y) = 9 - x^2 - y^2$
$R: x^2 + y^2 \leq 9, \, 0 \leq x, \, 0 \leq y$

In Exercises 25–30, use a double integral in polar coordinates to find the volume of the solid bounded by the graphs of the equations.

25. $z = xy, \, x^2 + y^2 = 1$ (first octant)

26. $z = x^2 + y^2 + 1, \, z = 0, \, x^2 + y^2 = 4$

27. $z = \sqrt{x^2 + y^2}, \, z = 0, \, x^2 + y^2 = 25$

28. $z = \sqrt{x^2 + y^2}, \, z = 0,$
$x^2 + y^2 \geq 4, \, x^2 + y^2 \leq 16$

29. Inside the hemisphere $z = \sqrt{16 - x^2 - y^2}$ and inside the cylinder $x^2 + y^2 - 4x = 0$

30. Inside the hemisphere $z = \sqrt{16 - x^2 - y^2}$ and outside the cylinder $x^2 + y^2 = 1$

31. Find a so that the volume inside the hemisphere $z = \sqrt{16 - x^2 - y^2}$ and outside the cylinder $x^2 + y^2 = a^2$ is one-half the volume of the hemisphere.

32. Use a double integral in polar coordinates to find the volume of a sphere of radius a.

33. The integral

$$I = \int_{-\infty}^{\infty} e^{-x^2/2} \, dx$$

is important in the study of normal distributions. Use polar coordinates to evaluate I by finding the double integral

$$I^2 = \left(\int_{-\infty}^{\infty} e^{-x^2/2} \, dx \right) \left(\int_{-\infty}^{\infty} e^{-y^2/2} \, dy \right)$$
$$= \int_{-\infty}^{\infty} \int_{-\infty}^{\infty} e^{-(x^2+y^2)/2} \, dA.$$

34. Using Exercise 33 as a model, evaluate the integral

$$\int_{-\infty}^{\infty} e^{-x^2} \, dx.$$

35. Show that the area of the polar sector R (see figure) is

area $= r \, \Delta r \, \Delta \theta$

where $r = (r_1 + r_2)/2$ is the average radius of R.

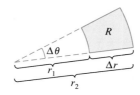

16.4 Center of Mass and Moments of Inertia

Mass ▪ Moments ▪ Center of mass ▪ Moments of inertia

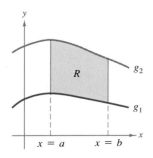

Lamina of Constant Density

FIGURE 16.29

In Section 6.7 we discussed several applications of integration involving a lamina of *constant* density ρ. For example, if the lamina corresponding to the region R, as shown in Figure 16.29, has a constant density ρ, then the mass of the lamina is given by

$$
\begin{aligned}
\text{mass} &= \rho A \\
&= \rho \int_a^b [g_2(x) - g_1(x)] \, dx \\
&= \rho \iint_R dA \\
&= \iint_R \rho \, dA. \qquad \text{Constant density}
\end{aligned}
$$

The use of a double integral suggests a natural extension of the formula for finding the mass of a lamina of *variable* density, where the density at (x, y) is given by the **density function** ρ.

DEFINITION OF MASS OF A PLANAR LAMINA OF VARIABLE DENSITY	If ρ is a continuous density function on the lamina corresponding to a plane region R, then the mass m of the lamina is given by $$m = \iint_R \rho(x, y) \, dA. \qquad \text{Variable density}$$

REMARK Density is normally expressed as mass per unit volume. However, for a planar lamina we consider density as mass per unit surface area.

EXAMPLE 1 Finding the mass of a planar lamina

Find the mass of the triangular lamina with vertices $(0, 0)$, $(0, 3)$, and $(2, 3)$, given that the density at (x, y) is

$$\rho(x, y) = 2x + y. \qquad \text{Density function}$$

SOLUTION

As shown in Figure 16.30, region R has the boundaries $x = 0$, $y = 3$, and $y = 3x/2$ (or $x = 2y/3$). Therefore, the mass of the lamina is

$$
\begin{aligned}
m &= \iint_R (2x + y) \, dA = \int_0^3 \int_0^{2y/3} (2x + y) \, dx \, dy = \int_0^3 \left[x^2 + xy \right]_0^{2y/3} dy \\
&= \frac{10}{9} \int_0^3 y^2 \, dy = \frac{10}{9} \frac{y^3}{3} \Big]_0^3 = 10.
\end{aligned}
$$

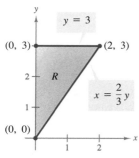

Lamina of Variable Density

FIGURE 16.30

REMARK   Note in Figure 16.30 that we shade a plane lamina of variable density so that the darkest shading corresponds to the densest part.

EXAMPLE 2   Finding mass by polar coordinates

Find the mass of the lamina corresponding to the first-quadrant portion of the circle $x^2 + y^2 = 4$, where the density at the point (x, y) is proportional to the distance between the point and the origin, as shown in Figure 16.31.

SOLUTION

At any point (x, y) the density of the lamina is

$$\rho(x, y) = k\sqrt{(x - 0)^2 + (y - 0)^2} = k\sqrt{x^2 + y^2}.$$

Since $0 \le x \le 2$ and $0 \le y \le \sqrt{4 - x^2}$, the mass is given by

$$m = \iint_R k\sqrt{x^2 + y^2}\, dA = \int_0^2 \int_0^{\sqrt{4-x^2}} k\sqrt{x^2 + y^2}\, dy\, dx.$$

To simplify the integration, we change to polar coordinates, using the bounds $0 \le \theta \le \pi/2$ and $0 \le r \le 2$. Thus, we have

$$m = \iint_R k\sqrt{x^2 + y^2}\, dA = \int_0^{\pi/2} \int_0^2 k\sqrt{r^2}\, r\, dr\, d\theta$$

$$= \int_0^{\pi/2} \int_0^2 kr^2\, dr\, d\theta$$

$$= \int_0^{\pi/2} \frac{kr^3}{3} \Big]_0^2 \, d\theta$$

$$= \frac{8k}{3} \int_0^{\pi/2} d\theta = \frac{8k}{3} \theta \Big]_0^{\pi/2} = \frac{4\pi k}{3}.$$

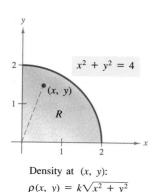

Density at (x, y):
$\rho(x, y) = k\sqrt{x^2 + y^2}$

FIGURE 16.31

Moments

For a lamina of variable density, moments of mass are defined in a manner similar to that used for the uniform density case. For a partition Δ of a lamina corresponding to a plane region R, we consider the ith rectangle R_i of area ΔA_i, as shown in Figure 16.32. Assuming that the mass of R_i is concentrated at one of its interior points (x_i, y_i), it follows that the moment of mass of R_i with respect to the x-axis is approximated by

$$(\text{mass})(y_i) \approx [\rho(x_i, y_i)\Delta A_i](y_i).$$

Similarly, the moment of mass with respect to the y-axis is approximated by

$$(\text{mass})(x_i) \approx [\rho(x_i, y_i)\Delta A_i](x_i).$$

By forming the Riemann sum of all such products and taking the limit as the norm of Δ approaches zero, we obtain the following formulas for moments of mass with respect to the x- and y-axes.

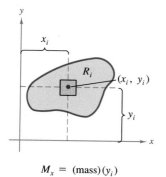

$M_x = (\text{mass})(y_i)$
$M_y = (\text{mass})(x_i)$

FIGURE 16.32

DEFINITION OF MOMENTS AND CENTER OF MASS OF A VARIABLE DENSITY PLANAR LAMINA

If ρ is a continuous density function on the lamina corresponding to a plane region R, then the **moments of mass** with respect to the x- and y-axes are

$$M_x = \iint_R y\rho(x, y)\, dA \quad \text{and} \quad M_y = \iint_R x\rho(x, y)\, dA.$$

Furthermore, if m is the mass of the lamina, then the **center of mass** is

$$(\bar{x}, \bar{y}) = \left(\frac{M_y}{m}, \frac{M_x}{m}\right).$$

REMARK If R represents a simple plane region rather than a lamina, the point $(\bar{x}, \bar{y})$ is called the **centroid** of the region.

Sometimes we can tell that either $\bar{x} = 0$ or $\bar{y} = 0$ by the form of ρ and the symmetry of the region R, as illustrated in the next example.

EXAMPLE 3 Finding the center of mass

Find the center of mass of the lamina corresponding to the parabolic region $0 \leq y \leq 4 - x^2$, where the density at the point (x, y) is proportional to the distance between (x, y) and the x-axis, as shown in Figure 16.33.

SOLUTION

Since the lamina is symmetric with respect to the y-axis and $\rho(x, y) = ky$, the center of mass will lie on the y-axis. Thus, $\bar{x} = 0$. To find $\bar{y}$, we first find the mass of the lamina.

$$\text{mass} = \int_{-2}^{2}\int_{0}^{4-x^2} ky\, dy\, dx = \frac{k}{2}\int_{-2}^{2} y^2\Big]_{0}^{4-x^2} dx$$

$$= \frac{k}{2}\int_{-2}^{2} (16 - 8x^2 + x^4)\, dx$$

$$= \frac{k}{2}\left[16x - \frac{8x^3}{3} + \frac{x^5}{5}\right]_{-2}^{2}$$

$$= k\left(32 - \frac{64}{3} + \frac{32}{5}\right) = \frac{256k}{15}$$

Next, we find the moment about the x-axis.

$$M_x = \int_{-2}^{2}\int_{0}^{4-x^2} (y)(ky)\, dy\, dx = \frac{k}{3}\int_{-2}^{2} y^3\Big]_{0}^{4-x^2} dx$$

$$= \frac{k}{3}\int_{-2}^{2} (64 - 48x^2 + 12x^4 - x^6)\, dx$$

$$= \frac{k}{3}\left[64x - 16x^3 + \frac{12x^5}{5} - \frac{x^7}{7}\right]_{-2}^{2}$$

$$= \frac{4096k}{105}$$

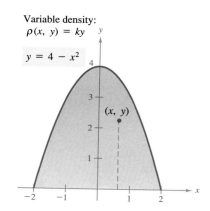

Variable density:
$\rho(x, y) = ky$

$y = 4 - x^2$

(x, y)

FIGURE 16.33

Thus,

$$\bar{y} = \frac{M_x}{m} = \frac{4096k/105}{256k/15}$$

$$= \frac{16}{7}$$

and the center of mass is $\left(0, \frac{16}{7}\right)$.

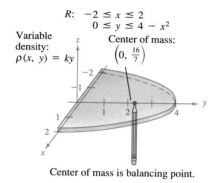

$R: \quad -2 \leq x \leq 2$
$\quad\quad 0 \leq y \leq 4 - x^2$

Variable density:
$\rho(x, y) = ky$

Center of mass:
$\left(0, \frac{16}{7}\right)$

Center of mass is balancing point.

FIGURE 16.34

Although we think of the moments M_x and M_y as measuring the tendency to rotate about the x- or y-axis, the calculation of moments is usually an intermediate step toward a more tangible goal. The use of the moments M_x and M_y in Example 3 is typical—to find the center of mass. Determination of the center of mass is useful in a variety of applications that allow us to treat a lamina as if its mass were concentrated at just one point. Intuitively, we can think of the center of mass as the balancing point of the lamina. For instance, the lamina in Example 3 should balance on the point of a pencil placed at $\left(0, \frac{16}{7}\right)$, as shown in Figure 16.34.

Moments of inertia

The moments M_x and M_y used in determining the center of mass of a lamina are sometimes referred to as the **first moments** about the x- and y-axes. In each case the moment is the product of a mass times a distance.

$$M_x = \iint\limits_{R} \underbrace{(y)}_{\substack{\text{distance to} \\ x\text{-axis}}}\underbrace{\rho(x, y)}_{\text{mass}}\, dA, \qquad M_y = \iint\limits_{R} \underbrace{(x)}_{\substack{\text{distance to} \\ y\text{-axis}}}\underbrace{\rho(x, y)}_{\text{mass}}\, dA$$

We now look at another type of moment—the **second moment,** or the **moment of inertia** of a lamina about a line. In the same way that mass is a measure of the tendency of matter to resist a change in straight-line motion, the moment of inertia about a line is a *measure of the tendency of matter to resist a change in rotational motion.* For example, if a particle of mass m has a distance of d from a fixed line, then its moment of inertia about the line is defined to be

$$I = md^2 = (\text{mass})(\text{distance})^2.$$

As with moments of mass, we can generalize this concept to obtain the moments of inertia about the x- and y-axes of a lamina of variable density. We denote these second moments by I_x and I_y, and in each case the moment is the product of a mass times the square of a distance.

$$I_x = \iint\limits_{R} \underbrace{(y^2)}_{\substack{\text{square of distance} \\ \text{to } x\text{-axis}}}\underbrace{\rho(x, y)}_{\text{mass}}\, dA, \qquad I_y = \iint\limits_{R} \underbrace{(x^2)}_{\substack{\text{square of distance} \\ \text{to } y\text{-axis}}}\underbrace{\rho(x, y)}_{\text{mass}}\, dA$$

The sum of the moments I_x and I_y is called the **polar moment of inertia** and is denoted by I_0. For a lamina in the xy-plane, I_0 represents the moment of inertia of the lamina about the z-axis. The term polar moment of inertia stems from the fact that the square of the polar distance r is used in the calculation.

$$I_0 = \iint\limits_{R} (x^2 + y^2)\rho(x, y)\, dA = \iint\limits_{R} r^2\rho(x, y)\, dA$$

EXAMPLE 4   Finding the moment of inertia

Find the moment of inertia about the x-axis of the lamina in Example 3.

SOLUTION

From the definition of moment of inertia we have

$$I_x = \int_{-2}^{2}\int_{0}^{4-x^2} y^2(ky)\, dy\, dx = \frac{k}{4}\int_{-2}^{2} y^4 \Big]_{0}^{4-x^2} dx$$

$$= \frac{k}{4}\int_{-2}^{2} (256 - 256x^2 + 96x^4 - 16x^6 + x^8)\, dx$$

$$= \frac{k}{4}\left[256x - \frac{256x^3}{3} + \frac{96x^5}{5} - \frac{16x^7}{7} + \frac{x^9}{9}\right]_{-2}^{2} = \frac{32{,}768k}{315}.$$ ▭

The moment of inertia I of a revolving lamina can be used to measure its kinetic energy. For example, suppose a planar lamina is revolving about a line with an **angular speed** of ω radians per second, as shown in Figure 16.35. The kinetic energy of the revolving lamina is given by

$$E = \frac{1}{2}I\omega^2. \qquad \text{\small Kinetic energy for rotational motion}$$

On the other hand, the kinetic energy of a mass m moving in a straight line at a velocity v is given by

$$E = \frac{1}{2}mv^2. \qquad \text{\small Kinetic energy for linear motion}$$

Thus, the kinetic energy of a mass moving in a straight line is proportional to its mass, but the kinetic energy of a mass revolving about an axis is proportional to its moment of inertia.

The **radius of gyration** $\bar{\bar{r}}$ of a revolving mass m with moment of inertia I is defined to be

$$\bar{\bar{r}} = \sqrt{\frac{I}{m}}. \qquad \text{\small Radius of gyration}$$

If the entire mass were located at a distance $\bar{\bar{r}}$ from its axis of revolution, it would have the same moment of inertia and, consequently, the same kinetic energy. For instance, the radius of gyration of the lamina in Example 4 about the x-axis is given by

$$\bar{\bar{y}} = \sqrt{\frac{I_x}{m}} = \sqrt{\frac{32{,}768k/315}{256k/15}} = \sqrt{\frac{128}{21}} \approx 2.47.$$

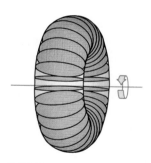

FIGURE 16.35

EXAMPLE 5 Finding the radius of gyration

Find the radius of gyration about the y-axis for the lamina corresponding to the region

$$R: 0 \leq y \leq \sin x, \quad 0 \leq x \leq \pi$$

where the density at (x, y) is given by $\rho(x, y) = x$.

SOLUTION

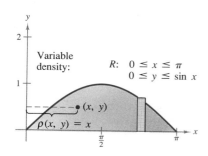

FIGURE 16.36

The region R is shown in Figure 16.36. The mass is given by

$$m = \int_0^\pi \int_0^{\sin x} x \, dy \, dx = \int_0^\pi xy \Big]_0^{\sin x} dx$$

$$= \int_0^\pi x \sin x \, dx = \Big[-x \cos x + \sin x \Big]_0^\pi = \pi.$$

The moment of inertia about the y-axis is

$$I_y = \int_0^\pi \int_0^{\sin x} x^3 \, dy \, dx = \int_0^\pi x^3 y \Big]_0^{\sin x} dx = \int_0^\pi x^3 \sin x \, dx$$

$$= \Big[(3x^2 - 6)(\sin x) - (x^3 - 6x)(\cos x) \Big]_0^\pi = \pi^3 - 6\pi.$$

Thus, the radius of gyration about the y-axis is

$$\bar{\bar{x}} = \sqrt{\frac{I_y}{m}} = \sqrt{\frac{\pi^3 - 6\pi}{\pi}} = \sqrt{\pi^2 - 6} \approx 1.97.$$

EXERCISES for Section 16.4

In Exercises 1–4, find the mass and center of mass for the lamina of specified density.

1. R: rectangle with vertices $(0, 0)$, $(a, 0)$, $(0, b)$, (a, b)
   (a) $\rho = k$ (b) $\rho = ky$
2. R: rectangle with vertices $(0, 0)$, $(a, 0)$, $(0, b)$, (a, b)
   (a) $\rho = kxy$ (b) $\rho = k(x^2 + y^2)$
3. R: triangle with vertices $(0, 0)$, $(b/2, h)$, $(b, 0)$
   (a) $\rho = k$ (b) $\rho = ky$
4. R: triangle with vertices $(0, 0)$, $(0, a)$, $(a, 0)$
   (a) $\rho = k$ (b) $\rho = x^2 + y^2$

In Exercises 5–20, find the mass and center of mass of the lamina bounded by the graphs of the given equations and of the specified density. [Hint: Some of the integrals are simpler in polar coordinates.]

5. $y = \sqrt{a^2 - x^2}$, $y = 0$
   (a) $\rho = k$ (b) $\rho = k(a - y)y$

6. $x^2 + y^2 = a^2$, $0 \leq x, 0 \leq y$
   (a) $\rho = k$ (b) $\rho = k(x^2 + y^2)$
7. $y = \sqrt{x}$, $y = 0$, $x = 4$; $\rho = kxy$
8. $y = x^2$, $y = 0$, $x = 4$; $\rho = kx$
9. $y = e^{-x}$, $y = 0$, $x = 0$, $x = 2$; $\rho = ky$
10. $y = \ln x$, $y = 0$, $x = 1$, $x = e$; $\rho = k/x$
11. $y = \dfrac{1}{1 + x^2}$, $y = 0$, $x = -1$, $x = 1$; $\rho = k$
12. $xy = 4$, $x = 1$, $x = 4$; $\rho = kx^2$
13. $x = 16 - y^2$, $x = 0$; $\rho = kx$
14. $y = 9 - x^2$, $y = 0$; $\rho = ky^2$
15. $y = \sin \dfrac{\pi x}{L}$, $y = 0$, $x = 0$, $x = L$; $\rho = ky$
16. $y = \cos \dfrac{\pi x}{L}$, $y = 0$, $x = 0$, $x = \dfrac{L}{2}$; $\rho = k$
17. $y = \sqrt{a^2 - x^2}$, $0 \leq y \leq x$; $\rho = k$
18. $y = \sqrt{a^2 - x^2}$, $y = 0$, $y = x$; $\rho = k\sqrt{x^2 + y^2}$
19. $r = 2 \cos 3\theta$, $-\dfrac{\pi}{6} \leq \theta \leq \dfrac{\pi}{6}$; $\rho = k$ [Hint: Use Simpson's Rule with $n = 6$.]
20. $r = 1 + \cos \theta$; $\rho = k$

In Exercises 21–26, verify the given moment(s) of inertia and find $\bar{\bar{x}}$ and $\bar{\bar{y}}$. Assume each lamina has a density of $\rho = 1$. (These regions are common shapes used in engineering.)

21. Rectangle

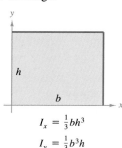

$$I_x = \tfrac{1}{3}bh^3$$
$$I_y = \tfrac{1}{3}b^3h$$

22. Right Triangle

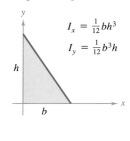

$$I_x = \tfrac{1}{12}bh^3$$
$$I_y = \tfrac{1}{12}b^3h$$

23. Circle

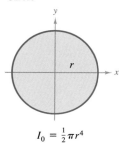

$$I_0 = \tfrac{1}{2}\pi r^4$$

24. Semicircle

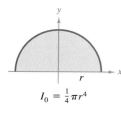

$$I_0 = \tfrac{1}{4}\pi r^4$$

25. Quarter Circle

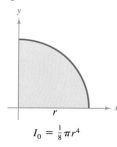

$$I_0 = \tfrac{1}{8}\pi r^4$$

26. Ellipse

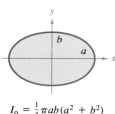

$$I_0 = \tfrac{1}{4}\pi ab(a^2 + b^2)$$

In Exercises 27–34, find I_x, I_y, I_0, $\bar{\bar{x}}$, and $\bar{\bar{y}}$ for the lamina bounded by the graphs of the given equations.

27. $y = 0$, $y = b$, $x = 0$, $x = a$; $\rho = ky$
28. $y = \sqrt{a^2 - x^2}$, $y = 0$; $\rho = ky$
29. $y = 4 - x^2$, $y = 0$, $x > 0$; $\rho = kx$
30. $y = x$, $y = x^2$; $\rho = kxy$
31. $y = \sqrt{x}$, $y = 0$, $x = 4$; $\rho = kxy$
32. $y = x^2$, $y^2 = x$; $\rho = x^2 + y^2$
33. $y = x^2$, $y^2 = x$; $\rho = kx$
34. $y = x^3$, $y = 4x$; $\rho = ky$

In Exercises 35–40, find the moment of inertia I of the lamina bounded by the graphs of the given equations about the line.

35. $x^2 + y^2 = b^2$; $\rho = k$, line: $x = a$ $(a > b)$
36. $y = 0$, $y = 2$, $x = 0$, $x = 4$; $\rho = k$, line: $x = 6$
37. $y = \sqrt{x}$, $y = 0$, $x = 4$; $\rho = kx$, line: $x = 6$
38. $y = \sqrt{a^2 - x^2}$, $y = 0$; $\rho = ky$, line: $y = a$
39. $y = \sqrt{a^2 - x^2}$, $y = 0$, $0 \le x$; $\rho = k(a - y)$, line: $y = a$
40. $y = 4 - x^2$, $y = 0$; $\rho = k$, line: $y = 2$

41. Use a computer to approximate the center of mass of the lamina with density $\rho = \sqrt{x}$ and bounded by the graphs of $y = \ln x$, $y = 0$, $x = 1$, and $x = 10$.

42. Use a computer to approximate the moment of inertia of the lamina of density $\rho = 1.6$, bounded by the graph of $x^2 + y^2 = 2$ about the line $x = 3$.

43. Prove the following Theorem of Pappus.

Let R be a region in a plane and let L be a line in the same plane such that L does not intersect the interior of R. If r is the distance between the centroid of R and the line, then the volume V of the solid of revolution formed by revolving R about the line is given by

$$V = 2\pi rA$$

where A is the area of R.

16.5 Surface Area

Surface area

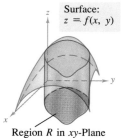

Region R in xy-Plane

FIGURE 16.37

At this point we know a great deal about the solid region lying between a surface and a closed and bounded region R in the xy-plane, as shown in Figure 16.37. For example, we know how to find the following.

1. The extrema of f on R (Section 15.8)
2. The area of the base R of the solid (Section 16.1)
3. The volume of the solid (Section 16.2)
4. The centroid of the base R (Section 16.4)

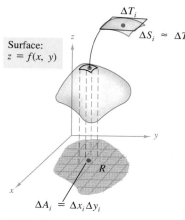

Surface:
$z = f(x, y)$

$\Delta S_i \approx \Delta T_i$

R

$\Delta A_i = \Delta x_i \Delta y_i$

FIGURE 16.38

In this section we show how to find the upper **surface area** of the solid. In later sections we will find the centroid (center of mass) of the solid (Section 16.6) and the lateral surface area (Section 17.2).

We begin with a surface given by $z = f(x, y)$ defined over a region R. To find the surface area, we construct an inner partition of R consisting of n rectangles, where the area of the ith rectangle R_i is $\Delta A_i = \Delta x_i \Delta y_i$, as shown in Figure 16.38. For a point (x_i, y_i) in R_i, there corresponds a point (x_i, y_i, z_i) on the surface S at which we construct a tangent plane T_i. From Section 15.7 we know that the angle of inclination of this tangent plane is given by

$$\cos \theta_i = \frac{1}{\sqrt{1 + [f_x(x_i, y_i)]^2 + [f_y(x_i, y_i)]^2}}$$

or

$$\sec \theta_i = \sqrt{1 + [f_x(x_i, y_i)]^2 + [f_y(x_i, y_i)]^2}.$$

From trigonometry we know that the portion of the tangent plane T_i that lies directly above R_i has an area of $\Delta T_i = \sec \theta_i \, \Delta A_i$ (see Exercise 32). Moreover, we can use the area of this small section of the tangent plane to approximate the area of the portion of the surface that lies directly above R_i, as follows.

$$\Delta S_i \approx \Delta T_i = \sec \theta_i \, \Delta A_i = \sqrt{1 + [f_x(x_i, y_i)]^2 + [f_y(x_i, y_i)]^2} \, \Delta A_i$$

The total area of S can be approximated by the Riemann sum

$$\text{surface area of } S \approx \sum_{i=1}^{n} \Delta S_i \approx \sum_{i=1}^{n} \sqrt{1 + [f_x(x_i, y_i)]^2 + [f_y(x_i, y_i)]^2} \, \Delta A_i.$$

Finally, by taking the limit as $\|\Delta\|$ approaches zero, we have the following double integral formula for surface area.

DEFINITION OF SURFACE AREA

If f and its first partial derivatives are continuous on the closed region R in the xy-plane, then the **area of the surface** $z = f(x, y)$ over R is given by

$$\text{surface area} = \iint\limits_{R} dS = \iint\limits_{R} \sqrt{1 + [f_x(x, y)]^2 + [f_y(x, y)]^2} \, dA.$$

As an aid to remembering the double integral for surface area, it is helpful to note its similarity to the integral for arc length.

Length on x-axis: $\displaystyle\int_a^b dx$

Arc length in xy-plane: $\displaystyle\int_a^b ds = \int_a^b \sqrt{1 + [f'(x)]^2} \, dx$

Area in xy-plane: $\displaystyle\iint\limits_{R} dA$

Surface area in space: $\displaystyle\iint\limits_{R} dS = \iint\limits_{R} \sqrt{1 + [f_x(x, y)]^2 + [f_y(x, y)]^2} \, dA$

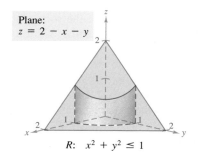

Plane:
$z = 2 - x - y$

R: $x^2 + y^2 \leq 1$

FIGURE 16.39

Like integrals for arc length, integrals for surface area are often very difficult to evaluate. However, one type that is easily evaluated is demonstrated in the next example.

EXAMPLE 1   The surface area of a plane region

Find the surface area of that portion of the plane $z = 2 - x - y$ that lies above the circle $x^2 + y^2 = 1$ in the first quadrant, as shown in Figure 16.39.

SOLUTION

Since $f_x(x, y) = -1$ and $f_y(x, y) = -1$, the surface area is given by

$$S = \iint_R \sqrt{1 + [f_x(x, y)]^2 + [f_y(x, y)]^2} \, dA$$

$$= \iint_R \sqrt{3} \, dA = \sqrt{3} \iint_R dA.$$

Now, since the integral on the right is simply $\sqrt{3}$ times the area of the region R, we have

$$S = \sqrt{3}(\text{area of } R)$$

$$= \sqrt{3}\left(\frac{\pi}{4}\right) = \frac{\sqrt{3}\pi}{4}.$$

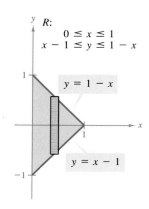

Surface:
$f(x, y) = 1 - x^2 + y$

(0, 1, 2)

R:
$0 \leq x \leq 1$
$x - 1 \leq y \leq 1 - x$

$y = 1 - x$

$y = x - 1$

FIGURE 16.40

EXAMPLE 2   Finding surface area

Find the area of that portion of the surface $f(x, y) = 1 - x^2 + y$ that lies above the triangular region with vertices $(1, 0, 0)$, $(0, -1, 0)$, and $(0, 1, 0)$, as shown in Figure 16.40.

SOLUTION

Since $f_x(x, y) = -2x$ and $f_y(x, y) = 1$, the surface area is given by

$$S = \iint_R \sqrt{1 + [f_x(x, y)]^2 + [f_y(x, y)]^2} \, dA$$

$$= \iint_R \sqrt{1 + 4x^2 + 1} \, dA.$$

From Figure 16.40 we see that the bounds for R are

$$0 \leq x \leq 1 \qquad \text{and} \qquad x - 1 \leq y \leq 1 - x.$$

Thus, the integral becomes

$$S = \int_0^1 \int_{x-1}^{1-x} \sqrt{2 + 4x^2} \, dy \, dx$$

$$= \int_0^1 y\sqrt{2 + 4x^2} \bigg]_{x-1}^{1-x} dx$$

$$= \int_0^1 (2\sqrt{2 + 4x^2} - 2x\sqrt{2 + 4x^2}) \, dx$$

$$= \left[x\sqrt{2 + 4x^2} + \ln (2x + \sqrt{2 + 4x^2}) - \frac{(2 + 4x^2)^{3/2}}{6} \right]_0^1$$

$$= \sqrt{6} + \ln (2 + \sqrt{6}) - \sqrt{6} - \ln \sqrt{2} + \frac{1}{3}\sqrt{2}$$

$$\approx 1.618.$$

Many integrals for surface area can be simplified by making a change of variables to polar coordinates, as demonstrated in the next three examples.

EXAMPLE 3   Change of variables to polar coordinates

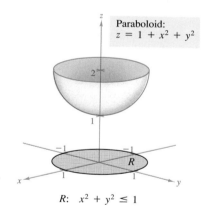

Paraboloid:
$z = 1 + x^2 + y^2$

$R:$ $x^2 + y^2 \le 1$

FIGURE 16.41

Find the surface area of the paraboloid $z = 1 + x^2 + y^2$ that lies above the unit circle, as shown in Figure 16.41.

SOLUTION

Since $f_x(x, y) = 2x$ and $f_y(x, y) = 2y$, we have

$$S = \iint_R \sqrt{1 + [f_x(x, y)]^2 + [f_y(x, y)]^2} \, dA$$

$$= \iint_R \sqrt{1 + 4x^2 + 4y^2} \, dA.$$

We convert to polar coordinates by letting $x = r \cos \theta$ and $y = r \sin \theta$. Then, since the region R is bounded by

$$0 \le r \le 1 \quad \text{and} \quad 0 \le \theta \le 2\pi$$

we have

$$S = \int_0^{2\pi} \int_0^1 \sqrt{1 + 4r^2} \, r \, dr \, d\theta = \int_0^{2\pi} \frac{1}{12}(1 + 4r^2)^{3/2} \bigg]_0^1 d\theta$$

$$= \int_0^{2\pi} \frac{5\sqrt{5} - 1}{12} \, d\theta$$

$$= \frac{\pi(5\sqrt{5} - 1)}{6} \approx 5.33.$$

Hemisphere:
$f(x, y) = \sqrt{25 - x^2 - y^2}$

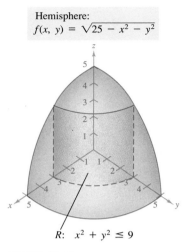

R: $x^2 + y^2 \leq 9$

FIGURE 16.42

EXAMPLE 4 Finding surface area

Find the surface area S of that portion of the hemisphere $f(x, y) = \sqrt{25 - x^2 - y^2}$ that lies above the region R bounded by the circle $x^2 + y^2 = 9$, as shown in Figure 16.42.

SOLUTION

The first partial derivatives of f are

$$f_x(x, y) = \frac{-x}{\sqrt{25 - x^2 - y^2}} \quad \text{and} \quad f_y(x, y) = \frac{-y}{\sqrt{25 - x^2 - y^2}}$$

and from the formula for surface area, we have

$$dS = \sqrt{1 + [f_x(x, y)]^2 + [f_y(x, y)]^2} \, dA = \frac{5}{\sqrt{25 - x^2 - y^2}} \, dA.$$

Therefore, the required surface area is

$$S = \iint_R \frac{5}{\sqrt{25 - x^2 - y^2}} \, dA.$$

Now, we convert to polar coordinates by letting $x = r \cos \theta$ and $y = r \sin \theta$. Then, since the region R is bounded by

$$0 \leq r \leq 3 \quad \text{and} \quad 0 \leq \theta \leq 2\pi$$

we obtain

$$S = \int_0^{2\pi} \int_0^3 \frac{5}{\sqrt{25 - r^2}} r \, dr \, d\theta = 5 \int_0^{2\pi} \left[-\sqrt{25 - r^2} \right]_0^3 d\theta$$

$$= 5 \int_0^{2\pi} d\theta$$

$$= 10\pi. \qquad \blacksquare\!\square$$

REMARK The procedure used in Example 4 can be extended to find the surface area of a sphere by using the region R bounded by the circle $x^2 + y^2 = a^2$, where $0 < a < 5$. The surface area of that portion of the hemisphere lying above the circular region can be shown to be $10\pi(5 - \sqrt{25 - a^2})$. By taking the limit as a approaches 5 and doubling the result, we obtain a total area of 100π. (The surface area of a sphere of radius r is $S = 4\pi r^2$.)

We can use Simpson's Rule or the Trapezoidal Rule to approximate the value of a double integral, *provided* we can get through the first integration. This is demonstrated in the next example.

Paraboloid:
$f(x, y) = 2 - x^2 - y^2$

R: $-1 \leq x \leq 1$
$-1 \leq y \leq 1$

FIGURE 16.43

EXAMPLE 5 Approximating surface area by Simpson's Rule

Find the area of the surface of the paraboloid $f(x, y) = 2 - x^2 - y^2$ that lies above the square region bounded by $-1 \leq x \leq 1$ and $-1 \leq y \leq 1$, as shown in Figure 16.43.

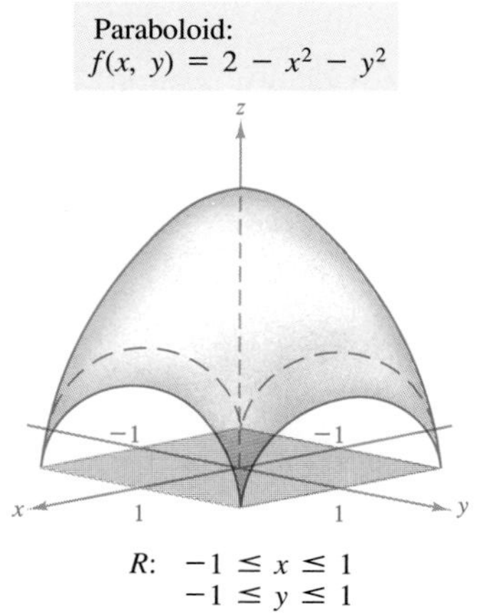

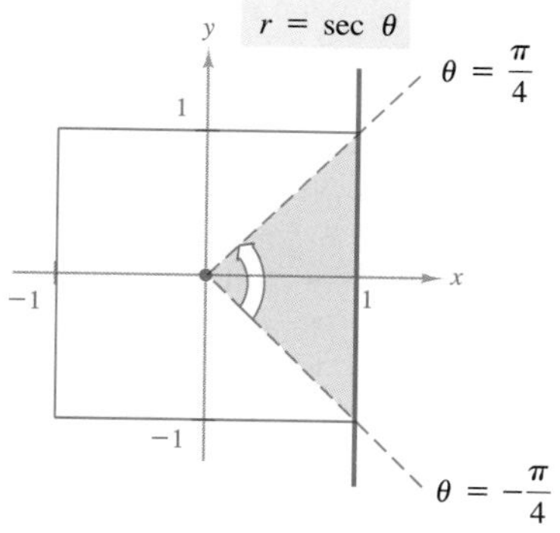

FIGURE 16.44

SOLUTION

Considering the partial derivatives $f_x(x, y) = -2x$ and $f_y(x, y) = -2y$, we have a surface area of

$$S = \iint_R \sqrt{1 + [f_x(x, y)]^2 + [f_y(x, y)]^2}\, dA = \iint_R \sqrt{1 + 4x^2 + 4y^2}\, dA.$$

In polar coordinates, the line $x = 1$ is given by $r \cos \theta = 1$ or $r = \sec \theta$, and we determine from Figure 16.44 that one-fourth of the region R is bounded by

$$0 \le r \le \sec \theta \qquad \text{and} \qquad -\frac{\pi}{4} \le \theta \le \frac{\pi}{4}.$$

Letting $x = r \cos \theta$ and $y = r \sin \theta$ produces

$$\frac{1}{4}S = \int_{-\pi/4}^{\pi/4} \int_0^{\sec \theta} \sqrt{1 + 4r^2}\, r\, dr\, d\theta$$

$$= \int_{-\pi/4}^{\pi/4} \frac{1}{12}(1 + 4r^2)^{3/2} \Big]_0^{\sec \theta} d\theta$$

$$= \frac{1}{12} \int_{-\pi/4}^{\pi/4} [(1 + 4 \sec^2 \theta)^{3/2} - 1]\, d\theta.$$

Now, using Simpson's Rule with $n = 10$, we approximate this *single* integral to be

$$S = \frac{1}{3} \int_{-\pi/4}^{\pi/4} [(1 + 4 \sec^2 \theta)^{3/2} - 1]\, d\theta \approx 7.45.$$

EXERCISES for Section 16.5

In Exercises 1–18, find the area of the surface given by $z = f(x, y)$ over the region R. [Hint: Some of the integrals are simpler in polar coordinates.]

1. $f(x, y) = 2x + 2y$
  R: triangle with vertices $(0, 0, 0)$, $(2, 0, 0)$, $(0, 2, 0)$
2. $f(x, y) = 10 + 2x - 3y$
  R: square with vertices $(0, 0, 0)$, $(2, 0, 0)$, $(0, 2, 0)$, $(2, 2, 0)$
3. $f(x, y) = 8 + 2x + 2y$
  $R = \{(x, y): x^2 + y^2 \le 4\}$
4. $f(x, y) = 10 + 2x - 3y$
  $R = \{(x, y): x^2 + y^2 \le 9\}$
5. $f(x, y) = 9 - x^2$
  R: square with vertices $(0, 0, 0)$, $(3, 0, 0)$, $(0, 3, 0)$, $(3, 3, 0)$
6. $f(x, y) = y^2$
  R: square with vertices $(0, 0, 0)$, $(3, 0, 0)$, $(0, 3, 0)$, $(3, 3, 0)$
7. $f(x, y) = 2y + x^2$
  R: triangle with vertices $(0, 0, 0)$, $(1, 0, 0)$, $(1, 1, 0)$

8. $f(x, y) = 2x + y^2$
  R: triangle with vertices $(0, 0, 0)$, $(2, 0, 0)$, $(0, 2, 0)$
9. $f(x, y) = 2 + x^{3/2}$
  R: quadrangle with vertices $(0, 0, 0)$, $(0, 4, 0)$, $(3, 4, 0)$, $(3, 0, 0)$
10. $f(x, y) = 2 + \frac{2}{3}x^{3/2}$
  $R = \{(x, y): 0 \le x \le 1, 0 \le y \le 1 - x\}$
11. $f(x, y) = \ln |\sec x|$
  $R = \{(x, y): 0 \le x \le \frac{\pi}{4}, 0 \le y \le \tan x\}$
12. $f(x, y) = 4 + x^2 - y^2$
  $R = \{(x, y): x^2 + y^2 \le 1\}$
13. $f(x, y) = 4 - x^2 - y^2$
  $R = \{(x, y): 0 \le f(x, y)\}$
14. $f(x, y) = x^2 + y^2$
  $R = \{(x, y): 0 \le f(x, y) \le 16\}$
15. $f(x, y) = \sqrt{x^2 + y^2}$
  $R = \{(x, y): 0 \le f(x, y) \le 1\}$
16. $f(x, y) = xy$
  $R = \{(x, y): x^2 + y^2 \le 16\}$

17. $f(x, y) = \sqrt{a^2 - x^2 - y^2}$
$R = \{(x, y): x^2 + y^2 \leq b^2, b < a\}$
18. $f(x, y) = \sqrt{a^2 - x^2 - y^2}$
$R = \{(x, y): x^2 + y^2 \leq a^2\}$

19. Find the surface area of the solid of intersection of the cylinders $x^2 + z^2 = 1$ and $y^2 + z^2 = 1$ (see figure).

20. Show that the surface area of the cone

$$z = k\sqrt{x^2 + y^2}, \quad k > 0$$

over the circular region $x^2 + y^2 \leq r^2$ in the xy-plane is $\pi r^2 \sqrt{k^2 + 1}$, as shown in the accompanying figure.

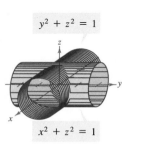

$y^2 + z^2 = 1$

$x^2 + z^2 = 1$

FIGURE FOR 19

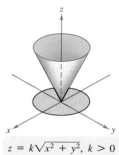

$z = k\sqrt{x^2 + y^2}, k > 0$

FIGURE FOR 20

In Exercises 21–26, set up the double integral that gives the area of the surface on the graph of f over the region R.

21. $f(x, y) = x^3 - 3xy + y^3$
R: square with vertices $(1, 1, 0)$, $(-1, 1, 0)$,
$(-1, -1, 0)$, $(1, -1, 0)$
22. $f(x, y) = e^{-x} \sin y$
$R = \{(x, y): 0 \leq x \leq 4, 0 \leq y \leq x\}$
23. $f(x, y) = e^{-x} \sin y$
$R = \{(x, y): x^2 + y^2 \leq 4\}$
24. $f(x, y) = x^2 - 3xy - y^2$
$R = \{(x, y): 0 \leq x \leq 4, 0 \leq y \leq x\}$
25. $f(x, y) = e^{xy}$
$R = \{(x, y): 0 \leq x \leq 4, 0 \leq y \leq 10\}$

26. $f(x, y) = \cos (x^2 + y^2)$
$$R = \left\{(x, y): x^2 + y^2 \leq \frac{\pi}{2}\right\}$$

In Exercises 27–30, approximate the double integral that gives the surface area on the graph of f over the region $R = \{(x, y): 0 \leq x \leq 1, 0 \leq y \leq 1\}$. Use Simpson's Rule with $n = 4$ on the second integral.

27. $f(x, y) = e^x$ **28.** $f(x, y) = \dfrac{2}{5}y^{5/2}$

29. $f(x, y) = 4 - x^2 - y^2$

30. $f(x, y) = \dfrac{2}{3}x^{3/2} + \cos x$

31. A company produces a spherical object whose radius is 25 cm. A hole, whose radius is 4 cm, is drilled through the center of the object. Find (a) the volume of the object, and (b) the outer surface area of the object.

32. The angle between a plane P and the xy-plane is θ, where $0 \leq \theta < \pi/2$. The projection of a rectangular region in P onto the xy-plane is a rectangle whose sides have lengths of Δx and Δy, as shown in the accompanying figure. Prove that the area of the rectangular region in P is $\sec \theta \, \Delta x \, \Delta y$.

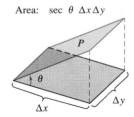

Area: $\sec \theta \, \Delta x \Delta y$

P

θ

Δx Δy

Area in xy-plane: $\Delta x \Delta y$ **FIGURE FOR 32**

33. Use a computer to approximate the area of the surface of the paraboloid $f(x, y) = 4 - x^2 - y^2$ that lies above the square region bounded by $-2 \leq x \leq 2$ and $-2 \leq y \leq 2$.

16.6 Triple Integrals and Applications

Triple integral ▪ Evaluation by iterated integrals ▪ Volume of a solid region ▪ Center of mass ▪ Moments of inertia

The procedure used to define a **triple integral** follows that used for double integrals. First, we assume that f is a continuous function of three variables, defined over a bounded solid region Q. Next we encompass Q with a network of boxes and form the **inner partition** consisting of all boxes lying entirely within Q, as shown in Figure 16.45. The volume of the ith box is

$$\Delta V_i = \Delta x_i \Delta y_i \Delta z_i \qquad \text{Volume of } i\text{th box}$$

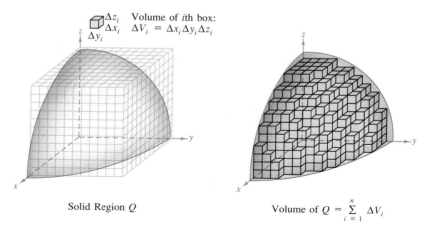

Δz_i Volume of *i*th box:
Δx_i $\Delta V_i = \Delta x_i \Delta y_i \Delta z_i$
Δy_i

FIGURE 16.45   Solid Region Q Volume of $Q \approx \sum\limits_{i=1}^{n} \Delta V_i$

and we define the **norm** $\|\Delta\|$ of the partition to be the length of the longest diagonal of the n boxes in the partition. Then we choose a point (x_i, y_i, z_i) in each box and form the Riemann sum

$$\sum_{i=1}^{n} f(x_i, y_i, z_i) \Delta V_i.$$

Finally, by taking the limit as $\|\Delta\|$ approaches zero, we obtain the following definition.

DEFINITION OF TRIPLE INTEGRAL

If f is continuous over a bounded solid region Q, then the **triple integral of f over Q** is defined to be

$$\iiint\limits_{Q} f(x, y, z) \, dV = \lim_{\|\Delta\| \to 0} \sum_{i=1}^{n} f(x_i, y_i, z_i) \Delta V_i$$

provided the limit exists. The **volume** of the solid region Q is given by

$$\text{volume of } Q = \iiint\limits_{Q} dV.$$

Each of the properties of double integrals in Theorem 16.1 can be restated in terms of triple integrals.

1. $\displaystyle\iiint\limits_{Q} cf(x, y, z) \, dV = c \iiint\limits_{Q} f(x, y, z) \, dV$

2. $\displaystyle\iiint\limits_{Q} [f(x, y, z) \pm g(x, y, z)] \, dV$

$$= \iiint\limits_{Q} f(x, y, z) \, dV \pm \iiint\limits_{Q} g(x, y, z) \, dV$$

3. $\displaystyle\iiint\limits_{Q} f(x, y, z) \, dV = \iiint\limits_{Q_1} f(x, y, z) \, dV + \iiint\limits_{Q_2} f(x, y, z) \, dV$

where Q is the union of two nonoverlapping solid subregions Q_1 and Q_2.

If the solid region Q is simple, the triple integral $\iiint f(x, y, z) \, dV$ can be evaluated with an iterated integral using one of the six orders of integration:

$$dx \, dy \, dz \qquad dy \, dx \, dz \qquad dz \, dx \, dy$$
$$dx \, dz \, dy \qquad dy \, dz \, dx \qquad dz \, dy \, dx$$

In the following version of Fubini's Theorem, we describe a region that is considered simple with respect to the order $dz \, dy \, dx$. Similar descriptions can be given for the other five orders.

THEOREM 16.5
EVALUATION BY ITERATED
INTEGRALS

Let f be continuous on a solid region Q defined by

$$a \le x \le b, \qquad h_1(x) \le y \le h_2(x), \qquad g_1(x, y) \le z \le g_2(x, y)$$

where h_1, h_2, g_1, and g_2 are continuous functions. Then,

$$\iiint_Q f(x, y, z) \, dV = \int_a^b \int_{h_1(x)}^{h_2(x)} \int_{g_1(x,y)}^{g_2(x,y)} f(x, y, z) \, dz \, dy \, dx.$$

To evaluate a triple iterated integral in the order $dz \, dy \, dx$, we hold *both* x and y constant for the innermost integration, and then hold x constant for the second integration. This is demonstrated in the first example.

EXAMPLE 1   Evaluating a triple iterated integral

Evaluate the iterated integral

$$\int_0^2 \int_0^x \int_0^{x+y} e^x(y + 2z) \, dz \, dy \, dx.$$

SOLUTION

Holding x and y constant, we have

$$\int_0^2 \int_0^x \int_0^{x+y} e^x(y + 2z) \, dz \, dy \, dx = \int_0^2 \int_0^x e^x(yz + z^2) \Big]_0^{x+y} dy \, dx$$

$$= \int_0^2 \int_0^x e^x(x^2 + 3xy + 2y^2) \, dy \, dx.$$

Now, holding x constant, we have

$$\int_0^2 \int_0^x e^x(x^2 + 3xy + 2y^2) \, dy \, dx = \int_0^2 \left[e^x \left(x^2 y + \frac{3xy^2}{2} + \frac{2y^3}{3} \right) \right]_0^x dx$$

$$= \frac{19}{6} \int_0^2 x^3 e^x \, dx$$

$$= \frac{19}{6} \left[e^x(x^3 - 3x^2 + 6x - 6) \right]_0^2$$

$$= 19 \left(\frac{e^2}{3} + 1 \right).$$

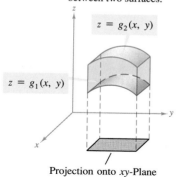

Solid region Q lies between two surfaces.

$z = g_2(x, y)$

$z = g_1(x, y)$

Projection onto xy-Plane

FIGURE 16.46

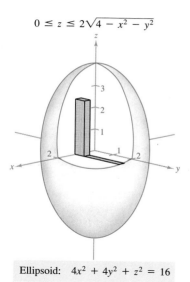

$0 \leq z \leq 2\sqrt{4 - x^2 - y^2}$

Ellipsoid: $4x^2 + 4y^2 + z^2 = 16$

FIGURE 16.47

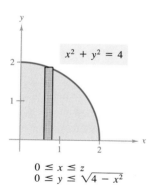

$x^2 + y^2 = 4$

$0 \leq x \leq z$
$0 \leq y \leq \sqrt{4 - x^2}$

FIGURE 16.48

In Example 1, we demonstrated the integration order $dz\, dy\, dx$. For other orders, we follow a similar procedure. For instance, to evaluate a triple iterated integral in the order $dx\, dy\, dz$ we hold both y and z constant for the innermost integration and integrate with respect to x. Then, for the second integration we hold z constant and integrate with respect to y. Finally, for the third integration, we integrate with respect to z.

To find the limits for a particular order of integration, it is generally advisable to first determine the innermost limits, which may be functions of the outer two variables. Then, by projecting the solid Q onto the coordinate plane of the outer two variables, we can determine their limits of integration by the methods used for double integrals. For instance, to evaluate

$$\iiint\limits_{Q} f(x, y, z) \, dz \, dy \, dx$$

first determine the limits for z, and then the integral has the form

$$\iint \left[\int_{g_1(x,y)}^{g_2(x,y)} f(x, y, z) \, dz \right] dy \, dx.$$

Now, by projecting the solid Q onto the xy-plane, we can determine the limits for x and y as we did for double integrals, as in Figure 16.46.

EXAMPLE 2   Finding volume by triple integrals

Find the volume of the ellipsoidal solid given by $4x^2 + 4y^2 + z^2 = 16$.

SOLUTION

Since x, y, and z play similar roles in the equation, the order of integration is probably immaterial, and we arbitrarily choose $dz \, dy \, dx$. Moreover, we can simplify the calculation by considering only that portion of the ellipsoid lying in the first octant, as shown in Figure 16.47. From the order $dz \, dy \, dx$, we first determine the bounds for z:

$$0 \leq z \leq 2\sqrt{4 - x^2 - y^2}$$

From Figure 16.48 we see that the boundaries for y and x are $0 \leq x \leq 2$ and $0 \leq y \leq \sqrt{4 - x^2}$, so the volume of the ellipsoid is

$$V = \iiint\limits_{Q} dV = 8 \int_0^2 \int_0^{\sqrt{4-x^2}} \int_0^{2\sqrt{4-x^2-y^2}} dz \, dy \, dx$$

$$= 8 \int_0^2 \int_0^{\sqrt{4-x^2}} z \Big]_0^{2\sqrt{4-x^2-y^2}} dy \, dx$$

$$= 16 \int_0^2 \int_0^{\sqrt{4-x^2}} \sqrt{(4 - x^2) - y^2} \, dy \, dx$$

$$= 8 \int_0^2 \left[y\sqrt{4 - x^2 - y^2} + (4 - x^2) \arcsin\left(\frac{y}{\sqrt{4 - x^2}} \right) \right]_0^{\sqrt{4-x^2}} dx$$

$$= 8 \int_0^2 (4 - x^2)\left(\frac{\pi}{2}\right) dx = 4\pi \left[4x - \frac{x^3}{3} \right]_0^2 = \frac{64\pi}{3}.$$

Example 2 is unusual in that each of the six possible orders of integration produces integrals of comparable difficulty. Try setting up some other possible orders of integration to find the volume of the ellipsoid. For instance, the order *dx dy dz* yields the integral

$$V = 8 \int_0^4 \int_0^{\sqrt{16-z^2}/2} \int_0^{\sqrt{16-4y^2-z^2}/2} dx \, dy \, dz.$$

If you solve this integral you will obtain the same volume obtained in Example 2. This is always the case—the order of integration does not affect the value of an integral. However, the order of integration often does affect the complexity of the integral. In Example 3 the given order of integration is not convenient and we change the order to simplify the problem.

EXAMPLE 3 Changing the order of integration

Evaluate

$$\int_0^{\sqrt{\pi/2}} \int_x^{\sqrt{\pi/2}} \int_1^3 \sin y^2 \, dz \, dy \, dx.$$

SOLUTION

Note that after one integration in the given order, we would encounter the integral $2 \int \sin (y^2) \, dy$, which is not an elementary function. To avoid this problem, we change the order of integration to *dz dx dy*, so that *y* is the outer variable.

The solid region *Q* is given by

$$0 \le x \le \sqrt{\frac{\pi}{2}}, \qquad x \le y \le \sqrt{\frac{\pi}{2}}, \qquad 1 \le z \le 3$$

as shown in Figure 16.49, and the projection of *Q* in the *xy*-plane yields the bounds

$$0 \le y \le \sqrt{\frac{\pi}{2}} \qquad \text{and} \qquad 0 \le x \le y.$$

Therefore, we have

$$\begin{aligned} V = \iiint_Q dV &= \int_0^{\sqrt{\pi/2}} \int_0^y \int_1^3 \sin (y^2) \, dz \, dx \, dy \\ &= \int_0^{\sqrt{\pi/2}} \int_0^y z \sin (y^2) \Big]_1^3 \, dx \, dy \\ &= 2 \int_0^{\sqrt{\pi/2}} \int_0^y \sin (y^2) \, dx \, dy \\ &= 2 \int_0^{\sqrt{\pi/2}} x \sin (y^2) \Big]_0^y \, dy \\ &= 2 \int_0^{\sqrt{\pi/2}} y \sin (y^2) \, dy = -\cos (y^2) \Big]_0^{\sqrt{\pi/2}} = 1. \quad \blacksquare \end{aligned}$$

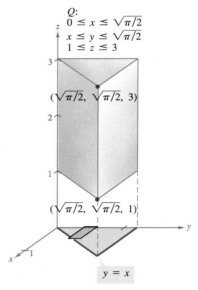

Q:
$0 \le x \le \sqrt{\pi/2}$
$x \le y \le \sqrt{\pi/2}$
$1 \le z \le 3$

$(\sqrt{\pi/2}, \sqrt{\pi/2}, 3)$

$(\sqrt{\pi/2}, \sqrt{\pi/2}, 1)$

$y = x$

FIGURE 16.49

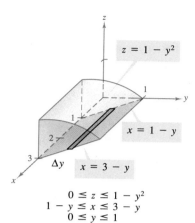

$$z = 1 - y^2$$
$$x = 1 - y$$
$$x = 3 - y$$
$$\Delta y$$

$$0 \le z \le 1 - y^2$$
$$1 - y \le x \le 3 - y$$
$$0 \le y \le 1$$

FIGURE 16.50

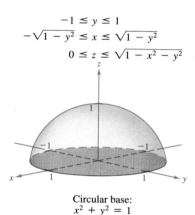

$$-1 \le y \le 1$$
$$-\sqrt{1 - y^2} \le x \le \sqrt{1 - y^2}$$
$$0 \le z \le \sqrt{1 - x^2 - y^2}$$

Circular base:
$$x^2 + y^2 = 1$$

FIGURE 16.51

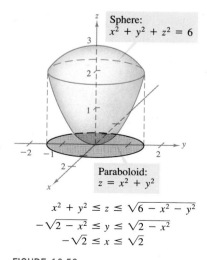

Sphere:
$$x^2 + y^2 + z^2 = 6$$

Paraboloid:
$$z = x^2 + y^2$$

$$x^2 + y^2 \le z \le \sqrt{6 - x^2 - y^2}$$
$$-\sqrt{2 - x^2} \le y \le \sqrt{2 - x^2}$$
$$-\sqrt{2} \le x \le \sqrt{2}$$

FIGURE 16.52

EXAMPLE 4   Determining the limits of integration

Set up a triple integral for the volume of each of the following solid regions.
(a) The region in the first octant bounded above by the cylinder $z = 1 - y^2$ and lying between the vertical planes $x + y = 1$ and $x + y = 3$.
(b) The upper hemisphere given by $z = \sqrt{1 - x^2 - y^2}$.
(c) The region bounded below by the paraboloid $z = x^2 + y^2$ and above by the sphere $x^2 + y^2 + z^2 = 6$.

SOLUTION

(a) From Figure 16.50, we see that the solid is bounded below by the xy-plane ($z = 0$) and above by the cylinder $z = 1 - y^2$. Therefore, we have
$$0 \le z \le 1 - y^2.$$

Now, by projecting the region onto the xy-plane, we obtain a parallelogram. Since two sides of the parallelogram are parallel to the x-axis, we set up the bounds
$$1 - y \le x \le 3 - y \qquad \text{and} \qquad 0 \le y \le 1.$$

Therefore, the volume of the region is given by
$$V = \iiint_Q dV = \int_0^1 \int_{1-y}^{3-y} \int_0^{1-y^2} dz \, dx \, dy.$$

(b) For the upper hemisphere given by $z = \sqrt{1 - x^2 - y^2}$, we have
$$0 \le z \le \sqrt{1 - x^2 - y^2}.$$

In Figure 16.51, we see that the projection of the hemisphere onto the xy-plane is the circle given by $x^2 + y^2 = 1$, and we can use either order $dx\,dy$ or $dy\,dx$. Choosing the first, we have
$$-\sqrt{1 - y^2} \le x \le \sqrt{1 - y^2} \qquad \text{and} \qquad -1 \le y \le 1$$

which implies that the volume of the region is given by
$$V = \iiint_Q dV = \int_{-1}^1 \int_{-\sqrt{1-y^2}}^{\sqrt{1-y^2}} \int_0^{\sqrt{1-x^2-y^2}} dz \, dx \, dy.$$

(c) For the region bounded below by the paraboloid $z = x^2 + y^2$ and above by the sphere $x^2 + y^2 + z^2 = 6$ we have
$$x^2 + y^2 \le z \le \sqrt{6 - x^2 - y^2}.$$

The sphere and the paraboloid intersect when $z = 2$. Moreover, we see from Figure 16.52 that the projection of the solid region onto the xy-plane is the circle given by $x^2 + y^2 = 2$. Using the order $dy\,dx$, we have
$$-\sqrt{2 - x^2} \le y \le \sqrt{2 - x^2} \qquad \text{and} \qquad -\sqrt{2} \le x \le \sqrt{2}$$

which implies that the volume of the region is given by
$$V = \iiint_Q dV = \int_{-\sqrt{2}}^{\sqrt{2}} \int_{-\sqrt{2-x^2}}^{\sqrt{2-x^2}} \int_{x^2+y^2}^{\sqrt{6-x^2-y^2}} dz \, dy \, dx.$$

Center of mass and moments of inertia

In the remainder of this section we look at two important applications of triple integrals. We consider a solid region Q whose density at (x, y, z) is given by the **density function** ρ. The **center of mass** of a solid region Q of mass m is given by $(\bar{x}, \bar{y}, \bar{z})$, where

$$m = \iiint\limits_{Q} \rho(x, y, z)\, dV \qquad M_{xz} = \iiint\limits_{Q} y\rho(x, y, z)\, dV$$

$$M_{yz} = \iiint\limits_{Q} x\rho(x, y, z)\, dV \qquad M_{xy} = \iiint\limits_{Q} z\rho(x, y, z)\, dV$$

and

$$\bar{x} = \frac{M_{yz}}{m}, \qquad \bar{y} = \frac{M_{xz}}{m}, \qquad \bar{z} = \frac{M_{xy}}{m}.$$

The quantities M_{yz}, M_{xz}, and M_{xy} are called the **first moments** of the region Q about the yz-, xz-, and xy-planes, respectively.

The first moments for solid regions are taken about a plane, whereas the second moments for solids are taken about a line. The **second moments** (or **moments of inertia**) about the x-, y-, and z-axes are as follows.

$$I_x = \iiint\limits_{Q} (y^2 + z^2)\rho(x, y, z)\, dV \qquad \text{Moment of inertia about } x\text{-axis}$$

$$I_y = \iiint\limits_{Q} (x^2 + z^2)\rho(x, y, z)\, dV \qquad \text{Moment of inertia about } y\text{-axis}$$

$$I_z = \iiint\limits_{Q} (x^2 + y^2)\rho(x, y, z)\, dV \qquad \text{Moment of inertia about } z\text{-axis}$$

For problems requiring the calculation of all three moments, considerable effort can be saved by applying the additive property of triple integrals and writing

$$I_x = I_{xz} + I_{xy}, \qquad I_y = I_{yz} + I_{xy}, \qquad \text{and} \qquad I_z = I_{yz} + I_{xz}$$

where I_{xy}, I_{xz}, and I_{yz} are as follows.

$$I_{xy} = \iiint\limits_{Q} z^2\rho(x, y, z)\, dV$$

$$I_{xz} = \iiint\limits_{Q} y^2\rho(x, y, z)\, dV$$

$$I_{yz} = \iiint\limits_{Q} x^2\rho(x, y, z)\, dV$$

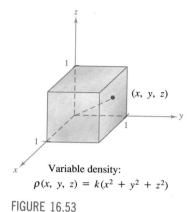

Variable density:
$$\rho(x, y, z) = k(x^2 + y^2 + z^2)$$

FIGURE 16.53

EXAMPLE 5 Finding the center of mass of a solid region

Find the center of mass of the unit cube shown in Figure 16.53, given that the density at the point (x, y, z) is proportional to the square of its distance from the origin.

SOLUTION

Since the density at (x, y, z) is proportional to the square of the distance between $(0, 0, 0)$ and (x, y, z), we have

$$\rho(x, y, z) = k(x^2 + y^2 + z^2).$$

Now, using this density function, we begin by finding the mass of the cube. Because of the symmetry of the region, the order of integration is irrelevant and we have

$$
\begin{aligned}
m &= \int_0^1 \int_0^1 \int_0^1 k(x^2 + y^2 + z^2)\, dz\, dy\, dx \\
&= k \int_0^1 \int_0^1 \left[(x^2 + y^2)z + \frac{z^3}{3} \right]_0^1 dy\, dx \\
&= k \int_0^1 \int_0^1 \left(x^2 + y^2 + \frac{1}{3} \right) dy\, dx \\
&= k \int_0^1 \left[\left(x^2 + \frac{1}{3} \right)y + \frac{y^3}{3} \right]_0^1 dx \\
&= k \int_0^1 \left(x^2 + \frac{2}{3} \right) dx = k \left[\frac{x^3}{3} + \frac{2x}{3} \right]_0^1 = k.
\end{aligned}
$$

The first moment about the yz-plane is

$$
\begin{aligned}
M_{yz} &= k \int_0^1 \int_0^1 \int_0^1 x(x^2 + y^2 + z^2)\, dz\, dy\, dx \\
&= k \int_0^1 x \left[\int_0^1 \int_0^1 (x^2 + y^2 + z^2)\, dz\, dy \right] dx.
\end{aligned}
$$

Note that x can be factored out of the two inner integrals, since it is constant with respect to y and z. After factoring, the two inner integrals are the same as for the mass m; hence we have

$$
M_{yz} = k \int_0^1 x \left(x^2 + \frac{2}{3} \right) dx = k \left[\frac{x^4}{4} + \frac{x^2}{3} \right]_0^1 = \frac{7k}{12}.
$$

Therefore,

$$
\bar{x} = \frac{M_{yz}}{m} = \frac{7k/12}{k} = \frac{7}{12}.
$$

Finally, from the nature of ρ and the symmetry of x, y, and z in this solid region, we have $\bar{x} = \bar{y} = \bar{z}$, and the center of mass is $\left(\frac{7}{12}, \frac{7}{12}, \frac{7}{12} \right)$.

$-2 \le x \le 2$

$-\sqrt{4 - x^2} \le y \le \sqrt{4 - x^2}$

$0 \le z \le \sqrt{4 - x^2 - y^2}$

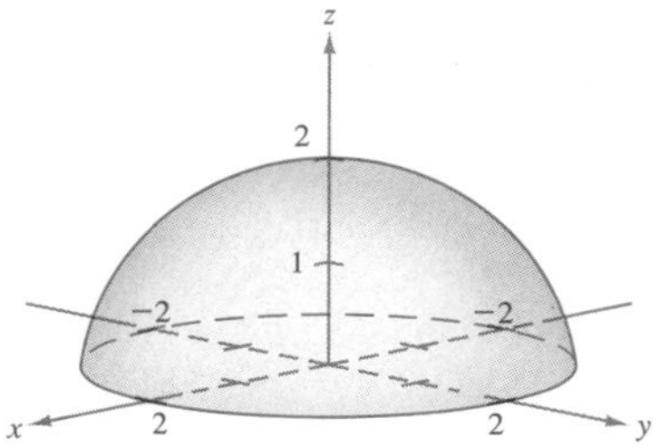

Circular base:
$x^2 + y^2 = 4$

Variable density:
$\rho(x, y, z) = kz$

FIGURE 16.54

EXAMPLE 6   Moments of inertia for a solid region

Find the moments of inertia about the x- and y-axes for the solid region lying between the hemisphere $z = \sqrt{4 - x^2 - y^2}$ and the xy-plane, given that the density at (x, y, z) is proportional to the distance between (x, y, z) and the xy-plane.

SOLUTION

The density of the region is given by $\rho(x, y, z) = kz$. Considering the symmetry of this problem, we know that $I_x = I_y$, and we need to compute only one moment, say I_x. From Figure 16.54 we choose the order $dz\,dy\,dx$ and write

$$I_x = \iiint_Q (y^2 + z^2)\rho(x, y, z)\,dV$$

$$= \int_{-2}^{2} \int_{-\sqrt{4-x^2}}^{\sqrt{4-x^2}} \int_{0}^{\sqrt{4-x^2-y^2}} (y^2 + z^2)(kz)\,dz\,dy\,dx$$

$$= k \int_{-2}^{2} \int_{-\sqrt{4-x^2}}^{\sqrt{4-x^2}} \left[\frac{y^2 z^2}{2} + \frac{z^4}{4}\right]_0^{\sqrt{4-x^2-y^2}}\,dy\,dx$$

$$= k \int_{-2}^{2} \int_{-\sqrt{4-x^2}}^{\sqrt{4-x^2}} \left[\frac{y^2(4 - x^2 - y^2)}{2} + \frac{(4 - x^2 - y^2)^2}{4}\right]\,dy\,dx$$

$$= \frac{k}{4} \int_{-2}^{2} \int_{-\sqrt{4-x^2}}^{\sqrt{4-x^2}} [(4 - x^2)^2 - y^4]\,dy\,dx$$

$$= \frac{k}{4} \int_{-2}^{2} \left[(4 - x^2)^2 y - \frac{y^5}{5}\right]_{-\sqrt{4-x^2}}^{\sqrt{4-x^2}}\,dx$$

$$= \frac{k}{4} \int_{-2}^{2} \frac{8}{5}(4 - x^2)^{5/2}\,dx = \frac{4k}{5} \int_{0}^{2} (4 - x^2)^{5/2}\,dx \qquad x = 2\sin\theta$$

$$= \frac{4k}{5} \int_{0}^{\pi/2} 64\cos^6\theta\,d\theta = \left(\frac{256k}{5}\right)\left(\frac{5\pi}{32}\right) = 8k\pi. \qquad \text{Wallis's Formula}$$

Thus, $I_x = 8k\pi = I_y$.

EXERCISES for Section 16.6

In Exercises 1–10, evaluate the triple integral.

1. $\displaystyle\int_0^3 \int_0^2 \int_0^1 (x + y + z)\,dx\,dy\,dz$

2. $\displaystyle\int_{-1}^1 \int_{-1}^1 \int_{-1}^1 x^2 y^2 z^2\,dx\,dy\,dz$

3. $\displaystyle\int_0^1 \int_0^x \int_0^{xy} x\,dz\,dy\,dx$

4. $\displaystyle\int_0^4 \int_0^\pi \int_0^{1-x} x\sin y\,dz\,dy\,dx$

5. $\displaystyle\int_1^4 \int_0^1 \int_0^x 2ze^{-x^2}\,dy\,dx\,dz$

6. $\displaystyle\int_1^4 \int_1^{e^2} \int_0^{1/xz} \ln z\,dy\,dz\,dx$

7. $\displaystyle\int_0^9 \int_0^{y/3} \int_0^{\sqrt{y^2-9x^2}} z\,dz\,dx\,dy$

8. $\displaystyle\int_0^{\sqrt{2}} \int_0^{\sqrt{2-x^2}} \int_{2x^2+y^2}^{4-y^2} y\,dz\,dy\,dx$

9. $\displaystyle\int_0^2 \int_{-\sqrt{4-x^2}}^{\sqrt{4-x^2}} \int_0^{x^2} x\,dz\,dy\,dx$

10. $\displaystyle\int_0^{\pi/2} \int_0^{y/2} \int_0^{1/y} \sin y\,dz\,dx\,dy$

In Exercises 11–14, sketch the solid whose volume is given by the triple integral and rewrite the integral with the specified order of integration.

11. $\int_0^4 \int_0^{(4-x)/2} \int_0^{(12-3x-6y)/4} dz\, dy\, dx$

Rewrite using the order $dy\, dx\, dz$.

12. $\int_0^4 \int_0^{\sqrt{16-x^2}} \int_0^{10-x-y} dz\, dy\, dx$

Rewrite using the order $dz\, dx\, dy$.

13. $\int_0^1 \int_y^1 \int_0^{\sqrt{1-y^2}} dz\, dx\, dy$

Rewrite using the order $dz\, dy\, dx$.

14. $\int_0^2 \int_{2x}^4 \int_0^{\sqrt{y^2-4x^2}} dz\, dy\, dx$

Rewrite using the order $dx\, dy\, dz$.

In Exercises 15 and 16, list the six possible orders of integration for the triple integral

$$\iiint_Q xyz\, dV$$

over the solid Q.

15. $Q = \{(x, y, z): 0 \le x \le 1, 0 \le y \le x, 0 \le z \le 3\}$

16. $Q =$
$\{(x, y, z): 0 \le x \le 2, x^2 \le y \le 4, 0 \le z \le 2 - x\}$

In Exercises 17–22, use a triple integral to find the volume of the solid bounded by the graphs of the given equations.

17. $x = 4 - y^2$, $z = 0$, $z = x$
18. $z = xy$, $z = 0$, $x = 0$, $x = 1$, $y = 0$, $y = 1$
19. $x^2 + y^2 + z^2 = r^2$
20. $z = 9 - x^2 - y^2$, $z = 0$
21. $z = 4 - x^2$, $y = 4 - x^2$ (first octant)
22. $z = 9 - x^2$, $y = -x + 2$, $y = 0$, $z = 0$, $x \ge 0$

In Exercises 23–26, find the mass and the indicated coordinates of the center of mass of the solid of specified density bounded by the graphs of the equations.

23. Find $\bar{x}$ using $\rho(x, y, z) = k$.
Q: $2x + 3y + 6z = 12$, $x = 0$, $y = 0$, $z = 0$

24. Find $\bar{y}$ using $\rho(x, y, z) = ky$.
Q: $2x + 3y + 6z = 12$, $x = 0$, $y = 0$, $z = 0$

25. Find $\bar{z}$ using $\rho(x, y, z) = kx$.
Q: $z = 4 - x$, $z = 0$, $y = 0$, $y = 4$

26. Find $\bar{y}$ using $\rho(x, y, z) = k$.
Q: $\dfrac{x}{a} + \dfrac{y}{b} + \dfrac{z}{c} = 1$ $(a, b, c > 0)$, $x = 0$, $y = 0$, $z = 0$

In Exercises 27 and 28, find the mass and the center of mass of the solid bounded by the graphs of the given equations.

27. $x = 0$, $x = b$, $y = 0$, $y = b$, $z = 0$, $z = b$,
$\rho(x, y, z) = kxy$

28. $x = 0$, $x = a$, $y = 0$, $y = b$, $z = 0$, $z = c$,
$\rho(x, y, z) = kz$

In Exercises 29–32, find the centroid of the solid region bounded by the graphs of the given equations. (Assume uniform density and find the center of mass.)

29. $z = \dfrac{h}{r}\sqrt{x^2 + y^2}$, $z = h$

30. $y = \sqrt{4 - x^2}$, $y = 0$, $z = y$, $z = 0$

31. $z = \sqrt{4^2 - x^2 - y^2}$, $z = 0$

32. $z = \dfrac{1}{y^2 + 1}$, $z = 0$, $x = -2$, $x = 2$, $y = 0$, $y = 1$

In Exercises 33–36, find I_x, I_y, and I_z for the indicated solid of specified density.

33. (a) $\rho = k$
(b) $\rho = kxyz$

34. (a) $\rho(x, y, z) = k$
(b) $\rho(x, y, z) = k(x^2 + y^2)$

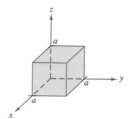

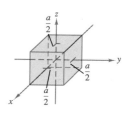

Cube Centered at Origin

35. (a) $\rho(x, y, z) = k$
(b) $\rho = ky$

36. (a) $\rho = kz$
(b) $\rho = k(4 - z)$

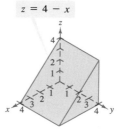

$z = 4 - x$

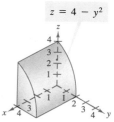

$z = 4 - y^2$

In Exercises 37 and 38, verify the moments of inertia for the solids of uniform density.

37. $I_x = I_z = \dfrac{1}{12}m(3a^2 + L^2)$

$I_y = \dfrac{1}{2}ma^2$

[Hint: Use Wallis's Formula.]

38. $I_x = \dfrac{1}{12}m(a^2 + b^2)$

$I_y = \dfrac{1}{12}m(b^2 + c^2)$

$I_z = \dfrac{1}{12}m(a^2 + c^2)$

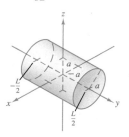

Right Circular Cylinder

FIGURE FOR 37

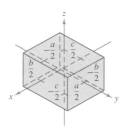

FIGURE FOR 38

In Exercises 39 and 40, set up a triple integral that gives the moment of inertia about the z-axis of the solid Q of density ρ.

39. $Q =$
$\{(x, y, z): -1 \le x \le 1, -1 \le y \le 1, 0 \le z \le 1 - x\}$
$\rho = \sqrt{x^2 + y^2 + z^2}$

40. $Q = \{(x, y, z): x^2 + y^2 \le 1, 0 \le z \le 4 - x^2 - y^2\}$
$\rho = kx^2$

In Exercises 41 and 42, use a computer to approximate the triple integral.

41. $\displaystyle\int_0^2 \int_0^{\sqrt{4-x^2}} \int_1^4 \dfrac{x^2 \sin y}{z}\, dz\, dy\, dx$

42. $\displaystyle\int_0^3 \int_0^{2-(2y/3)} \int_0^{6-2y-3z} ze^{-x^2y^2}\, dx\, dz\, dy$

16.7 Cylindrical and Spherical Coordinates

Cylindrical coordinates ▪ Spherical coordinates

We have seen that in the plane some graphs are easier to represent in polar coordinates than in rectangular coordinates. In this section we will see that a similar situation exists for surfaces in space. We will look at two new space coordinate systems. The first is an extension of polar coordinates to space and is called the **cylindrical coordinate system.**

THE CYLINDRICAL COORDINATE SYSTEM

In a **cylindrical coordinate system,** a point P in space is represented by an ordered triple (r, θ, z).

1. (r, θ) is a polar representation of the projection of P in the xy-plane.
2. z is the directed distance from (r, θ) to P.

Cylindrical coordinates:
$r^2 = x^2 + y^2$
$\tan \theta = \dfrac{y}{x}$
$z = z$

Rectangular coordinates:
$x = r \cos \theta$
$y = r \sin \theta$
$z = z$

$P \bullet (x, y, z)$
(r, θ, z)

FIGURE 16.55

To convert from rectangular to cylindrical coordinates (or vice versa), we follow the conversion guidelines for polar coordinates, as illustrated in Figure 16.55.

Cylindrical to rectangular:

$x = r \cos \theta, y = r \sin \theta, z = z$

Rectangular to cylindrical:

$r^2 = x^2 + y^2, \tan \theta = \dfrac{y}{x}, z = z$

We call the point $(0, 0, 0)$ the **pole.** Moreover, since the representation of a point in the polar coordinate system is not unique, it follows that the representation in the cylindrical coordinate system is also not unique.

EXAMPLE 1   Changing from cylindrical to rectangular coordinates

Express the point $(r, \theta, z) = (4, 5\pi/6, 3)$ in rectangular coordinates.

SOLUTION

Using the *cylindrical-to-rectangular* conversion equations produces

$$x = 4 \cos \frac{5\pi}{6} = 4\left(-\frac{\sqrt{3}}{2}\right) = -2\sqrt{3}$$

$$y = 4 \sin \frac{5\pi}{6} = 4\left(\frac{1}{2}\right) = 2$$

$$z = 3.$$

Thus, the point is $(-2\sqrt{3}, 2, 3)$ in rectangular coordinates, as shown in Figure 16.56.

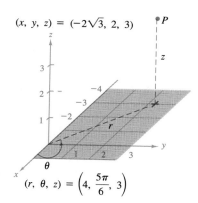

$(x, y, z) = (-2\sqrt{3}, 2, 3)$

$(r, \theta, z) = \left(4, \dfrac{5\pi}{6}, 3\right)$

FIGURE 16.56

EXAMPLE 2   Changing from rectangular to cylindrical coordinates

Express the point $(x, y, z) = (1, \sqrt{3}, 2)$ in cylindrical coordinates.

SOLUTION

Using the rectangular-to-cylindrical conversion equations produces

$$r = \pm\sqrt{1 + 3} = \pm 2$$

$$\tan \theta = \sqrt{3} \quad\Longrightarrow\quad \theta = \arctan (\sqrt{3}) + n\pi = \frac{\pi}{3} + n\pi$$

$$z = 2.$$

We have two choices for r and infinitely many choices for θ. As shown in Figure 16.57, two convenient representations of the point are

$$\left(2, \frac{\pi}{3}, 2\right) \qquad r > 0 \text{ and } \theta \text{ in Quadrant I}$$

$$\left(-2, \frac{4\pi}{3}, 2\right). \qquad r < 0 \text{ and } \theta \text{ in Quadrant III}$$

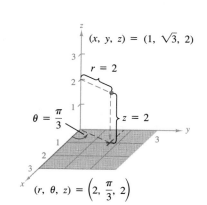

$(x, y, z) = (1, \sqrt{3}, 2)$

$r = 2$

$\theta = \dfrac{\pi}{3}$

$z = 2$

$(r, \theta, z) = \left(2, \dfrac{\pi}{3}, 2\right)$

FIGURE 16.57

Cylindrical coordinates are especially convenient for representing cylindrical surfaces and surfaces of revolution with the z-axis as the axis of symmetry, as shown in Figure 16.58.

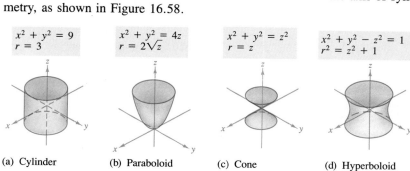

$x^2 + y^2 = 9$
$r = 3$

$x^2 + y^2 = 4z$
$r = 2\sqrt{z}$

$x^2 + y^2 = z^2$
$r = z$

$x^2 + y^2 - z^2 = 1$
$r^2 = z^2 + 1$

FIGURE 16.58   (a) Cylinder   (b) Paraboloid   (c) Cone   (d) Hyperboloid

Vertical planes containing the z-axis and horizontal planes also have simple cylindrical coordinate equations, as shown in Figure 16.59.

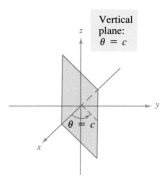
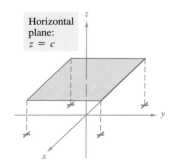

FIGURE 16.59

EXAMPLE 3   Rectangular-to-cylindrical conversion

Find equations in cylindrical coordinates for the surfaces whose rectangular equations are as follows.

(a) $x^2 + y^2 = 4z^2$ (b) $y^2 = x$

SOLUTION

(a) From the previous section we know that the graph of $x^2 + y^2 = 4z^2$ is a cone with its axis along the z-axis, as shown in Figure 16.60. If we replace $x^2 + y^2$ by r^2, the equation in cylindrical coordinates is

$$r^2 = 4z^2 \quad \Longrightarrow \quad r = \pm 2z.$$

(b) The graph of the surface $y^2 = x$ is a parabolic cylinder with rulings parallel to the z-axis, as shown in Figure 16.61. By replacing y^2 by $r^2 \sin^2 \theta$ and x by $r \cos \theta$, we obtain the following equation in cylindrical coordinates.

$$r^2 \sin^2 \theta = r \cos \theta$$
$$r(r \sin^2 \theta - \cos \theta) = 0$$

By temporarily discarding the possibility that $r = 0$, we obtain

$$r \sin^2 \theta - \cos \theta = 0$$
$$r = \frac{\cos \theta}{\sin^2 \theta} = \csc \theta \cot \theta.$$

Note that this equation includes a point for which $r = 0$, so nothing was lost by discarding the factor r.   ▭

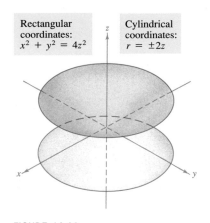

Rectangular coordinates:
$x^2 + y^2 = 4z^2$

Cylindrical coordinates:
$r = \pm 2z$

FIGURE 16.60

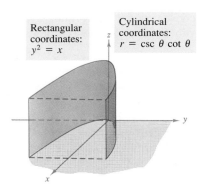

Rectangular coordinates:
$y^2 = x$

Cylindrical coordinates:
$r = \csc \theta \cot \theta$

FIGURE 16.61

EXAMPLE 4 Cylindrical-to-rectangular conversion

Find a rectangular equation for the graph represented by the cylindrical equation $r^2 \cos 2\theta + z^2 + 1 = 0$.

SOLUTION

If we replace $\cos 2\theta$ by $\cos^2 \theta - \sin^2 \theta$, we obtain

$$r^2(\cos^2 \theta - \sin^2 \theta) + z^2 + 1 = 0$$
$$r^2 \cos^2 \theta - r^2 \sin^2 \theta + z^2 = -1.$$

Now, by replacing $r \cos \theta$ by x and $r \sin \theta$ by y, we have $x^2 - y^2 + z^2 = -1$, or

$$y^2 - x^2 - z^2 = 1$$

which is a hyperboloid of two sheets whose axis lies along the y-axis, as shown in Figure 16.62.

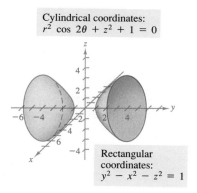

Cylindrical coordinates:
$r^2 \cos 2\theta + z^2 + 1 = 0$

Rectangular coordinates:
$y^2 - x^2 - z^2 = 1$

FIGURE 16.62

Spherical coordinates

In the **spherical coordinate system,** each point is represented by an ordered triple—the first coordinate is a distance, and the second and third coordinates are angles. This system is similar to the latitude-longitude system used to identify points on the surface of the earth. For example, the point on the surface of the earth whose latitude is 40° North (of the equator) and whose longitude is 80° West (of the prime meridian) is shown in Figure 16.63. Assuming that the earth is spherical with a radius of 4000 miles, we would label this point as

$$(4000, -80°, 50°).$$

radius   80° clockwise from prime meridian   50° down from North Pole

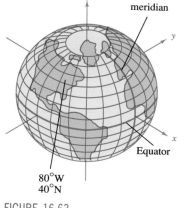

Prime meridian

Equator

80°W
40°N

FIGURE 16.63

THE SPHERICAL COORDINATE SYSTEM

In a **spherical coordinate system,** a point P in space is represented by an ordered triple (ρ, θ, ϕ).

1. ρ is the distance between P and the origin, $\rho \geq 0$.
2. θ is the same angle used in cylindrical coordinates, $0 \leq \theta < 2\pi$.
3. ϕ is the angle *between* the positive z-axis and the line segment $\overrightarrow{OP}$, $0 \leq \phi \leq \pi$.

REMARK ρ is the lowercase Greek letter rho, and ϕ is the lowercase Greek letter phi.

The relationship between the rectangular and the spherical coordinates is illustrated in Figure 16.64. To convert from one system to the other, we use the following conversion equations.

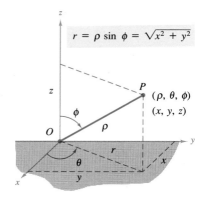

$r = \rho \sin \phi = \sqrt{x^2 + y^2}$

Spherical Coordinates

FIGURE 16.64

Spherical to rectangular:

$$x = \rho \sin \phi \cos \theta, \qquad y = \rho \sin \phi \sin \theta, \qquad z = \rho \cos \phi$$

Rectangular to spherical:

$$\rho^2 = x^2 + y^2 + z^2, \qquad \tan \theta = \frac{y}{x}, \qquad \phi = \arccos \left(\frac{z}{\sqrt{x^2 + y^2 + z^2}} \right)$$

The spherical coordinate system is useful primarily for surfaces in space that have a *point* or *center* of symmetry. For example, Figure 16.65 shows three surfaces with simple spherical equations.

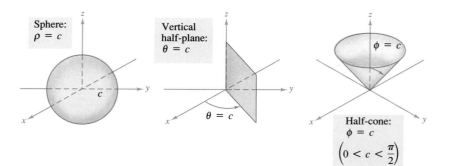

FIGURE 16.65

EXAMPLE 5 Rectangular-to-spherical conversion

Find an equation in spherical coordinates for the surfaces represented by the following rectangular coordinate equations.

(a) Cone: $x^2 + y^2 = z^2$ (b) Sphere: $x^2 + y^2 + z^2 - 4z = 0$

SOLUTION

(a) Making the appropriate replacements for x, y, and z in the given equation yields the following.

$$x^2 + y^2 = z^2$$
$$\rho^2 \sin^2 \phi \cos^2 \theta + \rho^2 \sin^2 \phi \sin^2 \theta = \rho^2 \cos^2 \phi$$
$$\rho^2 \sin^2 \phi(\cos^2 \theta + \sin^2 \theta) = \rho^2 \cos^2 \phi$$
$$\rho^2 \sin^2 \phi = \rho^2 \cos^2 \phi$$
$$\frac{\sin^2 \phi}{\cos^2 \phi} = 1 \qquad \rho > 0$$
$$\tan^2 \phi = 1$$

Thus, $\phi = \pi/4$ or $\phi = 3\pi/4$. The equation $\phi = \pi/4$ represents the *upper* half-cone, and the equation $\phi = 3\pi/4$ represents the *lower* half-cone.

(b) Since $\rho^2 = x^2 + y^2 + z^2$ and $z = \rho \cos \phi$, the given equation has the following spherical form.

$$\rho^2 - 4\rho \cos \phi = 0$$
$$\rho(\rho - 4 \cos \phi) = 0$$

Temporarily discarding the possibility that $\rho = 0$, we have the spherical equation

$$\rho - 4 \cos \phi = 0 \qquad \text{or} \qquad \rho = 4 \cos \phi.$$

Note that the solution set for this equation includes a point for which $\rho = 0$, so nothing is lost by discarding the factor ρ. The sphere represented by the equation $\rho = 4 \cos \phi$ is shown in Figure 16.66. ▭

Rectangular:
$x^2 + y^2 + z^2 - 4z = 0$

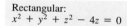

Spherical:
$\rho = 4 \cos \phi$

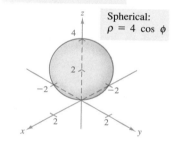

FIGURE 16.66

EXERCISES for Section 16.7

In Exercises 1–6, convert the given point from rectangular to cylindrical coordinates.

1. $(0, 5, 1)$ **2.** $(2\sqrt{2}, -2\sqrt{2}, 4)$
3. $(1, \sqrt{3}, 4)$ **4.** $(\sqrt{3}, -1, 2)$
5. $(2, -2, -4)$ **6.** $(-3, 2, -1)$

In Exercises 7–12, convert the given point from cylindrical to rectangular coordinates.

7. $(5, 0, 2)$ **8.** $\left(4, \frac{\pi}{2}, -2\right)$

9. $\left(2, \frac{\pi}{3}, 2\right)$ **10.** $\left(3, -\frac{\pi}{4}, 1\right)$

11. $\left(4, \frac{7\pi}{6}, 3\right)$ **12.** $\left(1, \frac{3\pi}{2}, 1\right)$

In Exercises 13–18, convert the given point from rectangular to spherical coordinates.

13. $(4, 0, 0)$ **14.** $(1, 1, 1)$
15. $(-2, 2\sqrt{3}, 4)$ **16.** $(2, 2, 4\sqrt{2})$
17. $(\sqrt{3}, 1, 2\sqrt{3})$ **18.** $(-4, 0, 0)$

In Exercises 19–24, convert the given point from spherical to rectangular coordinates.

19. $\left(4, \frac{\pi}{6}, \frac{\pi}{4}\right)$ **20.** $\left(12, \frac{3\pi}{4}, \frac{\pi}{9}\right)$

21. $\left(12, -\frac{\pi}{4}, 0\right)$ **22.** $\left(9, \frac{\pi}{4}, \pi\right)$

23. $\left(5, \frac{\pi}{4}, \frac{3\pi}{4}\right)$ **24.** $\left(6, \pi, \frac{\pi}{2}\right)$

In Exercises 25–30, convert the given point from cylindrical to spherical coordinates.

25. $\left(4, \dfrac{\pi}{4}, 0\right)$ **26.** $\left(2, \dfrac{2\pi}{3}, -2\right)$

27. $\left(4, -\dfrac{\pi}{6}, 6\right)$ **28.** $\left(-4, \dfrac{\pi}{3}, 4\right)$

29. $(12, \pi, 5)$ **30.** $\left(4, \dfrac{\pi}{2}, 3\right)$

In Exercises 31–36, convert the given point from spherical to cylindrical coordinates.

31. $\left(10, \dfrac{\pi}{6}, \dfrac{\pi}{2}\right)$ **32.** $\left(4, \dfrac{\pi}{18}, \dfrac{\pi}{2}\right)$

33. $\left(6, -\dfrac{\pi}{6}, \dfrac{\pi}{3}\right)$ **34.** $\left(5, -\dfrac{5\pi}{6}, \pi\right)$

35. $\left(8, \dfrac{7\pi}{6}, \dfrac{\pi}{6}\right)$ **36.** $\left(7, \dfrac{\pi}{4}, \dfrac{3\pi}{4}\right)$

In Exercises 37–42, match the equation (expressed in terms of cylindrical or spherical coordinates) to the correct graph. [The graphs are labeled (a)–(f).]

37. $r = 5$ **38.** $\theta = \dfrac{\pi}{4}$

39. $\rho = 5$ **40.** $\phi = \dfrac{\pi}{4}$

41. $r^2 = z$ **42.** $\rho = 4 \sec \phi$

(a)
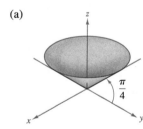

(b)
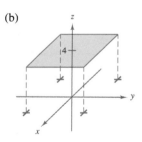

(c)
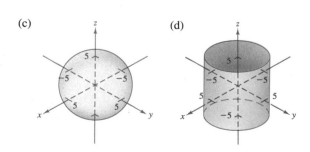

(d)

(e)
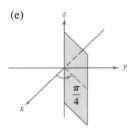

(f)
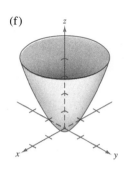

In Exercises 43–50, find an equation in rectangular coordinates for the equation in cylindrical coordinates, and sketch its graph.

43. $r = 2$ **44.** $z = 2$

45. $\theta = \dfrac{\pi}{6}$ **46.** $r = \dfrac{z}{2}$

47. $r = 2 \sin \theta$ **48.** $r = 2 \cos \theta$
49. $r^2 + z^2 = 4$ **50.** $z = r^2 \sin^2 \theta$

In Exercises 51–58, find an equation in rectangular coordinates for the equation in spherical coordinates and sketch its graph.

51. $\rho = 2$ **52.** $\theta = \dfrac{3\pi}{4}$

53. $\phi = \dfrac{\pi}{6}$ **54.** $\phi = \dfrac{\pi}{2}$

55. $\rho = 4 \cos \phi$ **56.** $\rho = 2 \sec \phi$
57. $\rho = \csc \phi$ **58.** $\rho = 4 \csc \phi \sec \theta$

In Exercises 59–66, find an equation of the given surface in (a) cylindrical coordinates and (b) spherical coordinates.

59. $x^2 + y^2 + z^2 = 16$ **60.** $4(x^2 + y^2) = z^2$
61. $x^2 + y^2 + z^2 - 2z = 0$
62. $x^2 + y^2 = z$
63. $x^2 + y^2 = 4y$ **64.** $x^2 + y^2 = 16$
65. $x^2 - y^2 = 9$ **66.** $y = 4$

In Exercises 67 and 68, sketch the solid that has the given description in cylindrical coordinates.

67. $0 \leq \theta \leq 2\pi,\ 0 \leq r \leq a,\ r \leq z \leq a$
68. $0 \leq \theta \leq 2\pi,\ 2 \leq r \leq 4,\ z^2 \leq -r^2 + 6r - 8$

In Exercises 69 and 70, sketch the solid that has the given description in spherical coordinates.

69. $0 \leq \theta \leq 2\pi,\ 0 \leq \phi \leq \pi/6,\ 0 \leq \rho \leq a \sec \phi$
70. $0 \leq \theta \leq 2\pi,\ \pi/4 \leq \phi \leq \pi/2,\ 0 \leq \rho \leq 1$

In Exercises 71–74, find inequalities that describe the solid, and state the coordinate system used. Position the solid on the coordinate system you choose so that the inequalities are as simple as possible.

71. cube with each edge 10 centimeters long

72. cylindrical shell 8 feet long with an inside diameter of 0.75 inches and an outside diameter of 1.25 inches

73. spherical shell with inside and outside radii of 4 inches and 6 inches, respectively

74. the solid that remains after a hole 1 inch in diameter is drilled through the center of a sphere 6 inches in diameter

75. Los Angeles is located at 34.05° North latitude and 118.24° West longitude, and Rio de Janeiro, Brazil is located at 22.90° South latitude and 43.22° West longitude (see figure). Assume that the earth is spherical with a radius of 4000 miles.

(a) Find the spherical coordinates for the location of each city.

(b) Find the rectangular coordinates for the location of each city.

(c) Find the angle (in radians) between the vectors from the center of the earth to each city.

(d) Find the great-circle distance s between the cities. [Hint: $s = r\theta$]

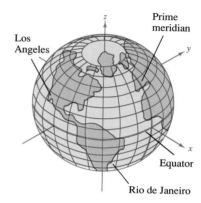

FIGURE FOR 75

76. Repeat Exercise 75 for the cities of Boston located at 42.36° North latitude and 71.06° West longitude, and Honolulu located at 21.31° North latitude and 157.86° West longitude.

16.8 Triple Integrals in Cylindrical and Spherical Coordinates

Cylindrical coordinates ▪ Spherical coordinates

Pierre Simon de Laplace

Many common solid regions such as spheres, ellipsoids, cones, and paraboloids can yield difficult triple integrals in rectangular coordinates. In fact, it is precisely this difficulty that led to the introduction of nonrectangular coordinate systems. In this section, we show how *cylindrical* and *spherical* coordinates can be used to evaluate triple integrals. One of the first to use such a system was the French mathematician Pierre Simon de Laplace (1749–1827). Laplace has been called the "Newton of France," and he published many important works in mechanics, differential equations, and probability.

Triple integrals in cylindrical coordinates

We begin by looking at triple integrals in cylindrical coordinates. You will see that this development is similar to that given for double integrals in polar coordinates earlier in this chapter. Recall from Section 16.7 that the rectangular conversion equations for cylindrical coordinates are

$$x = r\cos\theta, \qquad y = r\sin\theta, \qquad z = z.$$

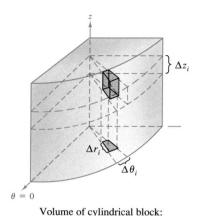

Volume of cylindrical block:
$$\Delta V_i = r_i \Delta r_i \Delta \theta_i \Delta z_i$$

FIGURE 16.67

In this coordinate system, the simplest solid region is a cylindrical block bounded by

$$r_1 \leq r \leq r_2$$
$$\theta_1 \leq \theta \leq \theta_2$$

and

$$z_1 \leq z \leq z_2$$

as shown in Figure 16.67.

To obtain the cylindrical coordinate form of a triple integral, we let f be a continuous function of r, θ, and z defined over a bounded solid region Q. To begin, we encompass the solid by a network of cylindrical blocks and form the **inner partition** Δ, consisting of all blocks lying entirely within Q. The **norm** $\|\Delta\|$ of the partition is the length of the longest diagonal of the n blocks in Δ. If we choose a point (r_i, θ_i, z_i) in the ith block such that r_i is the average radius of the block, then the block's volume is

$$\Delta V_i = (\text{area of base})(\text{height}) = (r_i \Delta \theta_i \Delta r_i)\Delta z_i.$$

By forming the sum

$$\sum_{i=1}^{n} f(r_i, \theta_i, z_i) r_i \Delta r_i \Delta \theta_i \Delta z_i$$

and taking the limit as $\|\Delta\| \to 0$, it can be shown that we obtain the following cylindrical coordinate form of a triple integral.

TRIPLE INTEGRAL IN CYLINDRICAL COORDINATES

If f is a continuous function of r, θ, and z on a bounded solid region Q, then in cylindrical coordinates the **triple integral of f over Q** is

$$\iiint_Q f(r, \theta, z)\, dV = \lim_{\|\Delta\| \to 0} \sum_{i=1}^{n} f(r_i, \theta_i, z_i) r_i \Delta r_i \Delta \theta_i \Delta z_i.$$

We can evaluate a triple integral in cylindrical coordinates, provided that the solid region Q is simple with respect to one of the six possible orders of integration. For example, if the region is bounded above and below by $h_1(r, \theta) \leq z \leq h_2(r, \theta)$, then we can write

$$\iiint_Q f(r, \theta, z)\, dV = \iint_R \left[\int_{h_1(r,\theta)}^{h_2(r,\theta)} f(r, \theta, z)\, dz\right] r\, dr\, d\theta$$

where the double integral over R is evaluated in polar coordinates. (That is, R is a plane region that is either r-simple or θ-simple, as discussed in Section 16.3.) Now, if R is r-simple, then the iterated form is

$$\int_{\theta_1}^{\theta_2} \int_{g_1(\theta)}^{g_2(\theta)} \int_{h_1(r,\theta)}^{h_2(r,\theta)} f(r, \theta, z) r\, dz\, dr\, d\theta.$$

To visualize a particular order of integration, it helps to view the iterated integral in terms of three sweeping motions—each adding another dimension to the solid. For instance, in the order $dr\, d\theta\, dz$, the first integration occurs

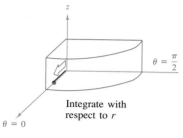

Integrate with
respect to r

$\theta = 0$

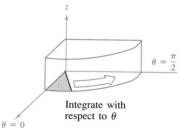

Integrate with
respect to θ

$\theta = 0$

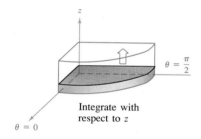

Integrate with
respect to z

$\theta = 0$

FIGURE 16.68

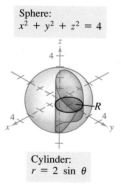

Sphere:
$x^2 + y^2 + z^2 = 4$

Cylinder:
$r = 2 \sin \theta$

FIGURE 16.69

in the r direction as a point sweeps out a ray. Then, as θ increases, the line sweeps out a sector. Finally, as z increases, the sector sweeps out a solid wedge, as shown in Figure 16.68.

EXAMPLE 1   Finding volume by cylindrical coordinates

Find the volume of the solid region Q cut from the sphere $x^2 + y^2 + z^2 = 4$ by the cylinder $r = 2 \sin \theta$, as shown in Figure 16.69.

SOLUTION

Since $x^2 + y^2 + z^2 = r^2 + z^2 = 4$, the bounds on z are

$$-\sqrt{4 - r^2} \le z \le \sqrt{4 - r^2}.$$

We let R be the circular projection of the solid onto the $r\theta$-plane. Then the bounds on R are given by

$$0 \le r \le 2 \sin \theta \qquad \text{and} \qquad 0 \le \theta \le \pi.$$

Thus, the volume of Q is

$$
\begin{aligned}
V &= \int_0^{\pi} \int_0^{2 \sin \theta} \int_{-\sqrt{4-r^2}}^{\sqrt{4-r^2}} r \, dz \, dr \, d\theta \\
&= 2 \int_0^{\pi/2} \int_0^{2 \sin \theta} 2r\sqrt{4 - r^2} \, dr \, d\theta \\
&= 2 \int_0^{\pi/2} -\frac{2}{3}(4 - r^2)^{3/2} \Big]_0^{2 \sin \theta} d\theta \\
&= \frac{4}{3} \int_0^{\pi/2} [8 - 8 \cos^3 \theta] \, d\theta \\
&= \frac{32}{3} \int_0^{\pi/2} [1 - (\cos \theta)(1 - \sin^2 \theta)] \, d\theta \\
&= \frac{32}{3} \Big[\theta - \sin \theta + \frac{\sin^3 \theta}{3} \Big]_0^{\pi/2} = \frac{16}{9}(3\pi - 4).
\end{aligned}
$$

EXAMPLE 2   Finding mass by cylindrical coordinates

Find the mass of the ellipsoidal solid Q given by $4x^2 + 4y^2 + z^2 = 16$, lying above the xy-plane. The density at a point in the solid is proportional to the distance between the point and the xy-plane.

SOLUTION

The density function $\rho(r, \theta, z) = kz$. The bounds on z are

$$0 \le z \le \sqrt{16 - 4x^2 - 4y^2} = 2\sqrt{4 - r^2}$$

$0 \leq z \leq 2\sqrt{4 - r^2}$

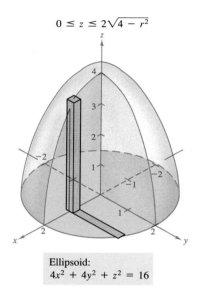

Ellipsoid:
$4x^2 + 4y^2 + z^2 = 16$

FIGURE 16.70

where $0 \leq r \leq 2$ and $0 \leq \theta \leq 2\pi$, as shown in Figure 16.70. Thus, the mass of the solid is

$$
\begin{aligned}
m &= \int_0^{2\pi} \int_0^2 \int_0^{\sqrt{16-4r^2}} kzr \, dz \, dr \, d\theta \\
&= \frac{k}{2} \int_0^{2\pi} \int_0^2 z^2 r \Big]_0^{\sqrt{16-4r^2}} dr \, d\theta \\
&= \frac{k}{2} \int_0^{2\pi} \int_0^2 (16r - 4r^3) \, dr \, d\theta \\
&= \frac{k}{2} \int_0^{2\pi} \left[8r^2 - r^4 \right]_0^2 d\theta \\
&= 8k \int_0^{2\pi} d\theta = 16\pi k.
\end{aligned}
$$

In Examples 1 and 2 we used the integration order $dz \, dr \, d\theta$. In the next example a different order of integration is more convenient.

EXAMPLE 3   Finding a moment of inertia

Find the moment of inertia about the axis of symmetry of the solid bounded by the paraboloid $z = x^2 + y^2$ and the plane $z = 4$, as shown in Figure 16.71. The density at each point is proportional to the distance between the point and the z-axis.

SOLUTION

Since the z-axis is the axis of symmetry, and $\rho(x, y, z) = k\sqrt{x^2 + y^2}$, it follows that

$$
I_z = \iiint_Q k(x^2 + y^2)\sqrt{x^2 + y^2} \, dV.
$$

In cylindrical coordinates, $0 \leq r \leq \sqrt{x^2 + y^2} = \sqrt{z}$. Therefore, we have

$$
\begin{aligned}
I_z &= k \int_0^4 \int_0^{2\pi} \int_0^{\sqrt{z}} r^2(r)r \, dr \, d\theta \, dz \\
&= k \int_0^4 \int_0^{2\pi} \frac{r^5}{5} \Big]_0^{\sqrt{z}} d\theta \, dz \\
&= k \int_0^4 \int_0^{2\pi} \frac{z^{5/2}}{5} \, d\theta \, dz \\
&= k \int_0^4 \frac{z^{5/2}}{5}(2\pi) \, dz \\
&= k \left[\left(\frac{2\pi}{5}\right)\left(\frac{2}{7}\right) z^{7/2} \right]_0^4 = \frac{512k\pi}{35}.
\end{aligned}
$$

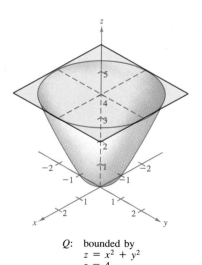

Q: bounded by
$z = x^2 + y^2$
$z = 4$

FIGURE 16.71

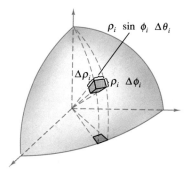

$\rho_i \sin \phi_i \, \Delta\theta_i$

$\Delta\rho_i$ $\rho_i \, \Delta\phi_i$

Spherical block:
$\Delta V_i \approx \rho_i^2 \sin \phi_i \, \Delta\rho_i \, \Delta\theta_i \, \Delta\phi_i$

FIGURE 16.72

Triple integrals in spherical coordinates

Recall from Section 16.7 that the rectangular conversion equations for spherical coordinates are

$$x = \rho \sin \phi \cos \theta$$
$$y = \rho \sin \phi \sin \theta$$

and

$$z = \rho \cos \phi.$$

For solids in spherical coordinates, the fundamental element of volume is a spherical block bounded by $\rho_1 \leq \rho \leq \rho_2$, $\theta_1 \leq \theta \leq \theta_2$, and $\phi_1 \leq \phi \leq \phi_2$, as shown in Figure 16.72. If (ρ, θ, ϕ) is a point in the interior of such a block, then the volume of the block can be approximated as follows.

$$\Delta V \approx \rho^2 \sin \phi \, \Delta\rho \, \Delta\theta \, \Delta\phi$$

Using the usual inner partition-summation-limit process, we can develop the following version of a triple integral in spherical coordinates.

TRIPLE INTEGRAL IN SPHERICAL COORDINATES

If f is a continuous function of ρ, θ, and ϕ on a bounded solid region Q, then in spherical coordinates the **triple integral of f over Q** is

$$\iiint_Q f(\rho, \, \theta, \, \phi) \, dV = \lim_{\|\Delta\| \to 0} \sum_{i=1}^{n} f(\rho_i, \, \theta_i, \, \phi_i)\rho_i^2 \sin \phi_i \, \Delta\rho_i \Delta\theta_i \Delta\phi_i.$$

REMARK   The Greek letter ρ used in spherical coordinates is not related to density. Rather, it is the three-dimensional analogue of the r used in polar coordinates. For problems involving spherical coordinates and a density function, we will use a different symbol to denote density.

Like triple integrals in cylindrical coordinates, triple integrals in spherical coordinates are evaluated with iterated integrals, as demonstrated in the next two examples.

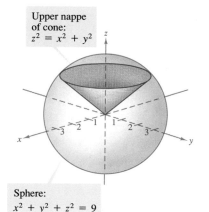

Upper nappe
of cone:
$z^2 = x^2 + y^2$

Sphere:
$x^2 + y^2 + z^2 = 9$

FIGURE 16.73

EXAMPLE 4   Finding volume in spherical coordinates

Find the volume of the solid region Q bounded below by the upper nappe of the cone $z^2 = x^2 + y^2$ and above by the sphere $x^2 + y^2 + z^2 = 9$, as shown in Figure 16.73.

SOLUTION

In spherical coordinates, the equation of the sphere is

$$\rho^2 = x^2 + y^2 + z^2 = 9 \quad \Longrightarrow \quad \rho = 3.$$

Furthermore, the sphere and cone intersect when

$$(x^2 + y^2) + z^2 = (z^2) + z^2 = 9 \quad \Longrightarrow \quad z = \frac{3}{\sqrt{2}}$$

and since $z = \rho \cos \phi$, it follows that

$$\left(\frac{3}{\sqrt{2}}\right)\left(\frac{1}{3}\right) = \cos \phi \quad \Longrightarrow \quad \phi = \frac{\pi}{4}.$$

Consequently, we can use the integration order $d\rho\, d\phi\, d\theta$, where

$$0 \le \rho \le 3, \qquad 0 \le \phi \le \pi/4, \qquad \text{and} \qquad 0 \le \theta \le 2\pi$$

as shown in Figure 16.74.

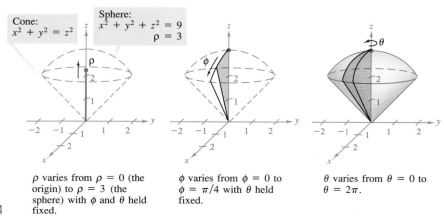

FIGURE 16.74

ρ varies from $\rho = 0$ (the origin) to $\rho = 3$ (the sphere) with ϕ and θ held fixed.

ϕ varies from $\phi = 0$ to $\phi = \pi/4$ with θ held fixed.

θ varies from $\theta = 0$ to $\theta = 2\pi$.

Finally, the volume is

$$V = \iiint_Q dV = \int_0^{2\pi} \int_0^{\pi/4} \int_0^3 \rho^2 \sin \phi \, d\rho \, d\phi \, d\theta$$

$$= \int_0^{2\pi} \int_0^{\pi/4} 9 \sin \phi \, d\phi \, d\theta = 9 \int_0^{2\pi} \left. -\cos \phi \right]_0^{\pi/4} d\theta$$

$$= 9 \int_0^{2\pi} \left(1 - \frac{\sqrt{2}}{2}\right) d\theta = 9\pi(2 - \sqrt{2}) \approx 16.56. \quad \blacksquare$$

The next example demonstrates the use of a different order of integration to find the center of mass of the spherical sector described in Example 4.

EXAMPLE 5 Finding the center of mass of a solid region

Find the center of mass of the solid region Q of uniform density, bounded below by the upper nappe of the cone $z^2 = x^2 + y^2$ and above by the sphere $x^2 + y^2 + z^2 = 9$.

SOLUTION

Since the density is uniform, we consider the density at the point (x, y, z) to be k. By symmetry, the center of mass lies on the z-axis, and we need only calculate $\bar{z} = M_{xy}/m$, where $m = kV = 9k\pi(2 - \sqrt{2})$ from Example 4. Now, since $z = \rho \cos \phi$, it follows that

$$M_{xy} = \iiint_Q kz \, dV = k \int_0^3 \int_0^{2\pi} \int_0^{\pi/4} (\rho \cos \phi)\rho^2 \sin \phi \, d\phi \, d\theta \, d\rho$$

$$= k \int_0^3 \int_0^{2\pi} \rho^3 \frac{\sin^2 \phi}{2}\Big]_0^{\pi/4} d\theta \, d\rho$$

$$= \frac{k}{4} \int_0^3 \int_0^{2\pi} \rho^3 \, d\theta \, d\rho = \frac{k\pi}{2} \int_0^3 \rho^3 \, d\rho = \frac{81k\pi}{8}.$$

Therefore, we have

$$\bar{z} = \frac{M_{xy}}{m} = \frac{81k\pi/8}{9k\pi(2 - \sqrt{2})} = \frac{9(2 + \sqrt{2})}{16} \approx 1.92$$

and the center of mass is approximately $(0, 0, 1.92)$.

EXERCISES for Section 16.8

In Exercises 1–8, evaluate the triple integral.

1. $\int_0^4 \int_0^{\pi/2} \int_0^2 r \cos \theta \, dr \, d\theta \, dz$

2. $\int_0^{\pi/4} \int_0^2 \int_0^{2-r} rz \, dz \, dr \, d\theta$

3. $\int_0^{\pi/2} \int_0^{2\cos^2 \theta} \int_0^{4-r^2} r \sin \theta \, dz \, dr \, d\theta$

4. $\int_0^4 \int_0^z \int_0^{\pi/2} re^r \, d\theta \, dr \, dz$

5. $\int_0^\pi \int_0^{\pi/2} \int_0^2 e^{-\rho^3}\rho^2 \, d\rho \, d\theta \, d\phi$

6. $\int_0^{\pi/2} \int_0^\pi \int_0^{\sin \theta} (2 \cos \phi)\rho^2 \, d\rho \, d\theta \, d\phi$

7. $\int_0^{2\pi} \int_0^{\pi/4} \int_0^{\cos \phi} \rho^2 \sin \phi \, d\rho \, d\theta \, d\phi$

8. $\int_0^{\pi/4} \int_0^{\pi/4} \int_0^{\cos \theta} \rho^2 \sin \phi \cos \phi \, d\rho \, d\theta \, d\phi$

In Exercises 9–12, sketch the solid region whose volume is given by the integral and evaluate the integral.

9. $\int_0^{\pi/2} \int_0^3 \int_0^{e^{-r^2}} r \, dz \, dr \, d\theta$

10. $\int_0^{2\pi} \int_0^{\sqrt{3}} \int_0^{3-r^2} r \, dz \, dr \, d\theta$

11. $4 \int_0^{\pi/2} \int_{\pi/6}^{\pi/2} \int_0^4 \rho^2 \sin \phi \, d\rho \, d\phi \, d\theta$

12. $\int_0^{2\pi} \int_0^\pi \int_2^5 \rho^2 \sin \phi \, d\rho \, d\phi \, d\theta$

In Exercises 13–16, convert the integral from rectangular coordinates to both cylindrical and spherical coordinates and evaluate the simplest integral.

13. $\int_{-2}^2 \int_{-\sqrt{4-x^2}}^{\sqrt{4-x^2}} \int_{x^2+y^2}^4 x \, dz \, dy \, dx$

14. $\int_0^2 \int_0^{\sqrt{4-x^2}} \int_0^{\sqrt{16-x^2-y^2}} \sqrt{x^2 + y^2} \, dz \, dy \, dx$

15. $\int_{-a}^a \int_{-\sqrt{a^2-x^2}}^{\sqrt{a^2-x^2}} \int_a^{a+\sqrt{a^2-x^2-y^2}} x \, dz \, dy \, dx$

16. $\int_0^1 \int_0^{\sqrt{1-x^2}} \int_0^{\sqrt{1-x^2-y^2}} \sqrt{x^2 + y^2 + z^2} \, dz \, dy \, dx$

In Exercises 17–22, use cylindrical coordinates to find the indicated characteristic of the cone (see figure).

17. Find the volume of the cone.

18. Find the centroid of the cone.

19. Find the center of mass of the cone, assuming that its density at any point is proportional to the distance between the point and the axis of the cone.

20. Find the center of mass of the cone, assuming that its density at any point is proportional to the distance between the point and the base.

21. Assume that the cone has uniform density, and show that the moment of inertia about the z-axis is

$$I_z = \frac{3}{10}mr_0^2.$$

22. Assume that the density is given by

$$\rho(x, y, z) = k(x^2 + y^2)$$

and find the moment of inertia about the z-axis.

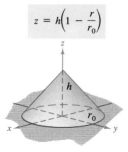

$$z = h\left(1 - \frac{r}{r_0}\right)$$

FIGURE FOR 17–22

In Exercises 23 and 24, use cylindrical coordinates to verify the given formula for the moment of inertia of the solid of uniform density.

23. Cylindrical shell: $I_z = \dfrac{1}{2}m(a^2 + b^2)$

$$0 < a \le r \le b, \, 0 \le z \le h$$

24. Right circular cylinder: $I_z = \dfrac{3}{2}ma^2$

$$r = 2a \sin \theta, \, 0 \le z \le h$$

In Exercises 25 and 26, use cylindrical coordinates to find the volume of the solid.

25. Solid inside both $x^2 + y^2 + z^2 = a^2$

$$\text{and } \left(x - \frac{a}{2}\right)^2 + y^2 = \left(\frac{a}{2}\right)^2$$

26. Solid inside $x^2 + y^2 + z^2 = 16$

and outside $z = \sqrt{x^2 + y^2}$

In Exercises 27 and 28, use spherical coordinates to find the volume of the solid.

27. The torus given by $\rho = 4 \sin \phi$

28. The solid between the spheres $x^2 + y^2 + z^2 = a^2$ and $x^2 + y^2 + z^2 = b^2$, $b > a$, and inside the cone $z^2 = x^2 + y^2$

In Exercises 29 and 30, use spherical coordinates to find the mass of the solid.

29. Sphere: $x^2 + y^2 + z^2 = a^2$
The density at any point is proportional to the distance between the point and the origin.

30. Sphere: $x^2 + y^2 + z^2 = a^2$
The density at any point is proportional to the distance of the point from the z-axis.

In Exercises 31 and 32, use spherical coordinates to find the center of mass of the solid of uniform density.

31. Hemispherical solid of radius r
32. Solid lying between two concentric hemispheres of radii r and R, where $r < R$

In Exercises 33 and 34, use spherical coordinates to find the moment of inertia about the z-axis of the solid of uniform density.

33. Solid bounded by the hemisphere $\rho = \cos \phi$, $\pi/4 \le \phi \le \pi/2$ and the cone $0 \le \phi \le \pi/4$
34. Solid lying between two concentric hemispheres of radii r and R, where $r < R$

16.9 Change of Variables: Jacobians

Jacobians ▪ Change of variables for double integrals

For the single integral $\int_a^b f(x) \, dx$, we can change variables by letting $x = g(u)$, so that $dx = g'(u) \, du$, and obtain

$$\int_a^b f(x) \, dx = \int_c^d f(g(u))g'(u) \, du$$

where $a = g(c)$ and $b = g(d)$. Note that the change of variables process

Carl Gustav Jacobi

introduces an additional factor $g'(u)$ into the integrand. This also occurs in the case of double integrals

$$\iint\limits_{R} f(x, y)\, dA = \iint\limits_{S} f(g(u, v), h(u, v)) \underbrace{\left| \frac{\partial x}{\partial u} \frac{\partial y}{\partial v} - \frac{\partial y}{\partial u} \frac{\partial x}{\partial v} \right|}_{\text{Jacobian}} du\, dv$$

where the change of variables $x = g(u, v)$ and $y = h(u, v)$ introduces a factor called the **Jacobian** of x and y with respect to u and v, named after the German mathematician Carl Gustav Jacobi (1804–1851). Jacobi is known for his work in many areas of mathematics, but his interest in integration stemmed from the problem of finding the circumference of an ellipse. In defining the Jacobian, it is convenient to use the following determinant notation.

DEFINITION OF THE JACOBIAN

If $x = g(u, v)$ and $y = h(u, v)$, then the **Jacobian** of x and y with respect to u and v, denoted by $\partial(x, y)/\partial(u, v)$, is

$$\frac{\partial(x, y)}{\partial(u, v)} = \begin{vmatrix} \dfrac{\partial x}{\partial u} & \dfrac{\partial x}{\partial v} \\ \dfrac{\partial y}{\partial u} & \dfrac{\partial y}{\partial v} \end{vmatrix} = \frac{\partial x}{\partial u} \frac{\partial y}{\partial v} - \frac{\partial y}{\partial u} \frac{\partial x}{\partial v}.$$

EXAMPLE 1 Finding the Jacobian for rectangular-to-polar conversion

Find the Jacobian for the change of variables defined by

$$x = r \cos \theta \quad \text{and} \quad y = r \sin \theta.$$

SOLUTION

From the definition of a Jacobian, we obtain

$$\frac{\partial(x, y)}{\partial(r, \theta)} = \begin{vmatrix} \dfrac{\partial x}{\partial r} & \dfrac{\partial x}{\partial \theta} \\ \dfrac{\partial y}{\partial r} & \dfrac{\partial y}{\partial \theta} \end{vmatrix} = \begin{vmatrix} \cos \theta & -r \sin \theta \\ \sin \theta & r \cos \theta \end{vmatrix} = r \cos^2 \theta + r \sin^2 \theta = r.$$

Example 1 points out that the change of variables from rectangular to polar coordinates for a double integral can be written as

$$\iint\limits_{R} f(x, y)\, dA = \iint\limits_{S} f(r \cos \theta, r \sin \theta) r\, dr\, d\theta, \qquad r > 0$$

$$= \iint\limits_{S} f(r \cos \theta, r \sin \theta) \left| \frac{\partial(x, y)}{\partial(r, \theta)} \right| dr\, d\theta$$

where S is the region in the $r\theta$-plane that corresponds to the region R in the xy-plane, as shown in Figure 16.75.

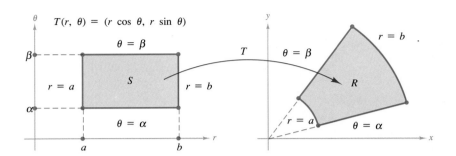

FIGURE 16.75

In general, we consider a change of variables that is given by a one-to-one **transformation** T from a region S in the uv-plane to a region R in the xy-plane given by

$$T(u, v) = (x, y) = (g(u, v), h(u, v))$$

where g and h have continuous first partial derivatives in the region S. In Figure 16.76, note that the point (u, v) lies in S and the point (x, y) lies in R.

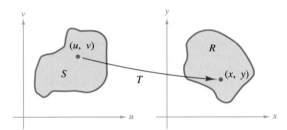

FIGURE 16.76

The type of transformation we usually seek is one for which the region S is simpler than the region R, as demonstrated in Example 2.

EXAMPLE 2   Finding a change of variables to simplify a region

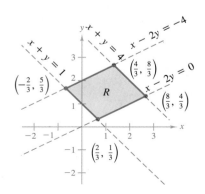

Region R in xy-Plane

FIGURE 16.77

Let R be the region bounded by the lines $x - 2y = 0$, $x - 2y = -4$, $x + y = 4$, and $x + y = 1$, as shown in Figure 16.77. Find a transformation T from a region S to R such that S is a rectangular region (with sides parallel to the u- or v-axis).

SOLUTION

To begin, we let $u = x + y$ and $v = x - 2y$. Solving this system of equations for x and y produces $T(u, v) = (x, y)$, where

$$x = \frac{1}{3}(2u + v) \quad \text{and} \quad y = \frac{1}{3}(u - v).$$

Moreover, the four boundaries for R in the xy-plane give rise to the following boundaries for S in the uv-plane as follows.

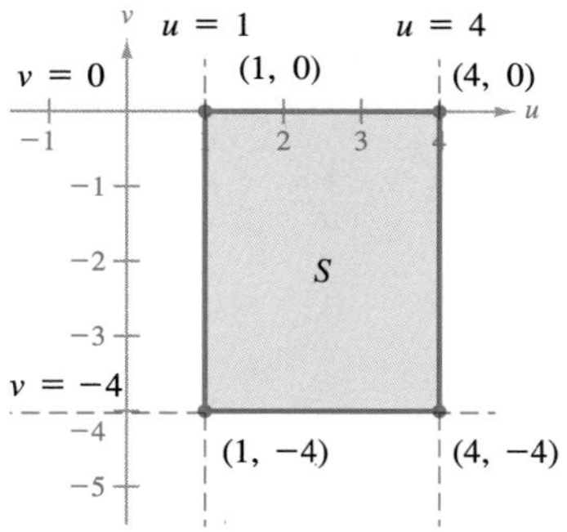

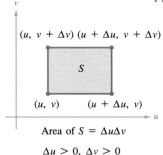

Region S in uv-Plane

FIGURE 16.78

Bounds in xy-plane	Bounds in uv-plane
$x + y = 1$ ⟹	$u = 1$
$x + y = 4$ ⟹	$u = 4$
$x - 2y = 0$ ⟹	$v = 0$
$x - 2y = -4$ ⟹	$v = -4$

The region S is shown in Figure 16.78. Note that the transformation T maps the vertices of the region S onto the vertices of the region R. For instance,

$$T(1, 0) = \left(\frac{1}{3}[2(1) + 0], \frac{1}{3}[1 - 0]\right) = \left(\frac{2}{3}, \frac{1}{3}\right).$$

Change of variables for double integrals

We now look at how a change of variables can be used for a double integral.

THEOREM 16.6
CHANGE OF VARIABLES
FOR DOUBLE INTEGRALS

Let R and S be regions in the xy- and uv-planes that are related by the equations $x = g(u, v)$ and $y = h(u, v)$ such that each point in R is the image of a unique point in S. If f is continuous on R and g and h have continuous partial derivatives on S and $\partial(x, y)/\partial(u, v)$ is nonzero on S, then

$$\iint_R f(x, y)\, dA = \iint_S f(g(u, v), h(u, v))\left|\frac{\partial(x, y)}{\partial(u, v)}\right| du\, dv.$$

PROOF

We will discuss a proof for the case in which S is a rectangular region in the uv-plane with vertices (u, v), $(u + \Delta u, v)$, $(u + \Delta u, v + \Delta v)$, and $(u, v + \Delta v)$, as shown in Figure 16.79. The images of these vertices in the xy-plane are as follows.

$$M = (g(u, v), h(u, v))$$
$$N = (g(u + \Delta u, v), h(u + \Delta u, v))$$
$$P = (g(u + \Delta u, v + \Delta v), h(u + \Delta u, v + \Delta v))$$
$$Q = (g(u, v + \Delta v), h(u, v + \Delta v))$$

If Δu and Δv are small, then the continuity of g and h implies that R is approximately a parallelogram determined by the vectors $\overrightarrow{MN}$ and $\overrightarrow{MQ}$, as shown in Figure 16.80. Thus, the area ΔA of R is given by

$$\Delta A \approx \|\overrightarrow{MN} \times \overrightarrow{MQ}\|.$$

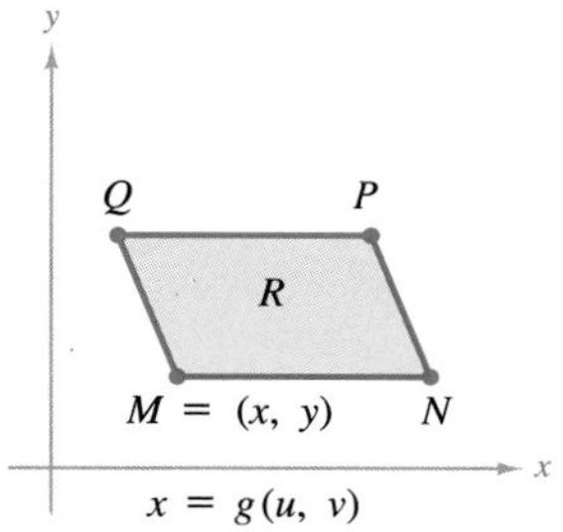

Moreover, for small Δu and Δv, the partial derivatives of g and h with respect to u can be approximated by

$$g_u(u, v) \approx \frac{g(u + \Delta u, v) - g(u, v)}{\Delta u}$$

and $h_u(u, v) \approx \dfrac{h(u + \Delta u, v) - h(u, v)}{\Delta u}.$

Area of $S = \Delta u \Delta v$

$\Delta u > 0$, $\Delta v > 0$

FIGURE 16.79

$x = g(u, v)$
$y = h(u, v)$

FIGURE 16.80

Consequently,

$$\overrightarrow{MN} = [g(u + \Delta u, v) - g(u, v)]\mathbf{i} + [h(u + \Delta u, v) - h(u, v)]\mathbf{j}$$

$$\approx [g_u(u, v)\Delta u]\mathbf{i} + [h_u(u, v)\Delta u]\mathbf{j} = \frac{\partial x}{\partial u}\Delta u\,\mathbf{i} + \frac{\partial y}{\partial u}\Delta u\,\mathbf{j}.$$

Similarly, we can approximate $\overrightarrow{MQ}$ as

$$\overrightarrow{MQ} \approx \frac{\partial x}{\partial v}\Delta v\,\mathbf{i} + \frac{\partial y}{\partial v}\Delta v\,\mathbf{j}$$

which implies that

$$\overrightarrow{MN} \times \overrightarrow{MQ} \approx \begin{vmatrix} \mathbf{i} & \mathbf{j} & \mathbf{k} \\ \dfrac{\partial x}{\partial u}\Delta u & \dfrac{\partial y}{\partial u}\Delta u & 0 \\ \dfrac{\partial x}{\partial v}\Delta v & \dfrac{\partial y}{\partial v}\Delta v & 0 \end{vmatrix} = \begin{vmatrix} \dfrac{\partial x}{\partial u} & \dfrac{\partial y}{\partial u} \\ \dfrac{\partial x}{\partial v} & \dfrac{\partial y}{\partial v} \end{vmatrix} \Delta u \Delta v\,\mathbf{k}.$$

It follows that, in Jacobian notation,

$$\Delta A \approx \|\overrightarrow{MN} \times \overrightarrow{MQ}\| \approx \left|\frac{\partial(x, y)}{\partial(u, v)}\right| \Delta u \Delta v.$$

Finally, since this approximation improves as Δu and Δv approach 0, the limiting case can be written as

$$dA \approx \|\overrightarrow{MN} \times \overrightarrow{MQ}\| \approx \left|\frac{\partial(x, y)}{\partial(u, v)}\right| du\,dv.$$

The next two examples in this section show how a change of variables can simplify the integration process. The simplification occurs in *two* ways. We can make a change of variables to simplify either the *region R*, the *integrand f(x, y)*, or both.

EXAMPLE 3   Using a change of variables to simplify a region

Let R be the region bounded by the lines $x - 2y = 0$, $x - 2y = -4$, $x + y = 4$, and $x + y = 1$, as shown in Figure 16.77. Evaluate the double integral

$$\iint\limits_R 3xy\,dA.$$

SOLUTION

From Example 2, we use the following change of variables.

$$y = \frac{1}{3}(u - v) \quad \text{and} \quad x = \frac{1}{3}(2u + v)$$

Thus, the Jacobian is

$$\frac{\partial(x, y)}{\partial(u, v)} = \begin{vmatrix} \dfrac{\partial x}{\partial u} & \dfrac{\partial x}{\partial v} \\ \dfrac{\partial y}{\partial u} & \dfrac{\partial y}{\partial v} \end{vmatrix} = \begin{vmatrix} \dfrac{2}{3} & \dfrac{1}{3} \\ \dfrac{1}{3} & -\dfrac{1}{3} \end{vmatrix} = -\frac{2}{9} - \frac{1}{9} = -\frac{1}{3}.$$

Therefore, by Theorem 16.6, we obtain

$$\iint_R 3xy \, dA = \iint_S 3\left[\frac{1}{3}(2u + v)\frac{1}{3}(u - v)\right]\left|\frac{\partial(x, y)}{\partial(u, v)}\right| dv \, du$$

$$= \int_1^4 \int_{-4}^0 \frac{1}{9}(2u^2 - uv - v^2) \, dv \, du$$

$$= \frac{1}{9} \int_1^4 \left[2u^2v - \frac{uv^2}{2} - \frac{v^3}{3}\right]_{-4}^0 du$$

$$= \frac{1}{9} \int_1^4 \left(8u^2 + 8u - \frac{64}{3}\right) du = \frac{1}{9}\left[\frac{8u^3}{3} + 4u^2 - \frac{64}{3}u\right]_1^4$$

$$= \frac{164}{9}.$$

EXAMPLE 4 Using a change of variables to simplify an integrand

Let R be the region bounded by the square with vertices $(0, 1)$, $(1, 2)$, $(2, 1)$, and $(1, 0)$. Evaluate the integral

$$\iint_R (x + y)^2 \sin^2 (x - y) \, dA.$$

SOLUTION

First, we note that the sides of R lie on the lines $x + y = 1$, $x - y = 1$, $x + y = 3$, and $x - y = 1$, as shown in Figure 16.81. Letting $u = x + y$ and $v = x - y$, we find the bounds for region S in the uv-plane to be

$$1 \le u \le 3$$
$$-1 \le v \le 1$$

as shown in Figure 16.82. Solving for x and y in terms of u and v, we have

$$x = \frac{1}{2}(u + v) \qquad y = \frac{1}{2}(u - v).$$

Thus, the Jacobian is

$$\frac{\partial(x, y)}{\partial(u, y)} = \begin{vmatrix} \dfrac{1}{2} & \dfrac{1}{2} \\ \dfrac{1}{2} & -\dfrac{1}{2} \end{vmatrix} = -\frac{1}{4} - \frac{1}{4} = -\frac{1}{2}.$$

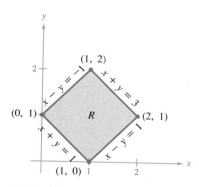

FIGURE 16.81

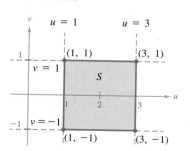

FIGURE 16.82

By Theorem 16.6, it follows that

$$\iint\limits_{R} (x + y)^2 \sin^2 (x - y) \, dA = \int_{-1}^{1} \int_{1}^{3} u^2 \sin^2 v \left(\frac{1}{2}\right) du \, dv$$

$$= \frac{1}{2} \int_{-1}^{1} (\sin^2 v) \frac{u^3}{3}\Big]_{1}^{3} dv$$

$$= \frac{13}{3} \int_{-1}^{1} \sin^2 v \, dv$$

$$= \frac{13}{6} \int_{-1}^{1} (1 - \cos 2v) \, dv$$

$$= \frac{13}{6}\left[v - \frac{1}{2} \sin 2v\right]_{-1}^{1}$$

$$= \frac{13}{6}\left[2 - \frac{1}{2} \sin 2 + \frac{1}{2} \sin (-2)\right]$$

$$= \frac{13}{6}[2 - \sin 2] \approx 2.363. \qquad \blacksquare$$

In each of the change-of-variable examples we have looked at so far, the region S has been a rectangle with sides parallel to the u- or v-axis. Occasionally, a change of variables can be used for other types of regions. For instance, Figure 16.83 shows a change of variables from a circular region S to an elliptical region R.

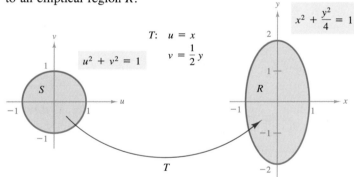

FIGURE 16.83

EXERCISES for Section 16.9

In Exercises 1–8, find the Jacobian $\partial(x, y)/\partial(u, v)$ for the indicated change of variables.

1. $x = -\dfrac{1}{2}(u - v), \ y = \dfrac{1}{2}(u + v)$

2. $x = au + bv, \ y = cu + dv$

3. $x = u - v^2, \ y = u + v$

4. $x = u - uv, \ y = uv$

5. $x = u \cos \theta - v \sin \theta, \ y = u \sin \theta + v \cos \theta$

6. $x = u + a, \ y = v + a$

7. $x = e^u \sin v, \ y = e^u \cos v$

8. $x = \dfrac{u}{v}, \ y = u + v$

In Exercises 9 and 10, sketch the image S in the uv-plane of the region R in the xy-plane using the given transformations.

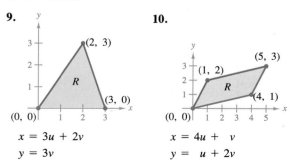

9.

$x = 3u + 2v$

$y = 3v$

10.

$x = 4u + v$

$y = u + 2v$

In Exercises 11–16, use the indicated change of variables to evaluate the double integral.

11. $\displaystyle\iint\limits_{R} 48xy \, dx \, dy$

$x = \dfrac{1}{2}(u + v)$

$y = \dfrac{1}{2}(u - v)$

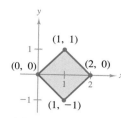

12. $\displaystyle\iint\limits_{R} 4(x^2 + y^2) \, dx \, dy$

$x = \dfrac{1}{2}(u + v)$

$y = \dfrac{1}{2}(u - v)$

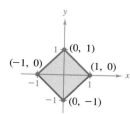

13. $\displaystyle\iint\limits_{R} 4(x + y)e^{x-y} \, dy \, dx$

$x = \dfrac{1}{2}(u + v)$

$y = \dfrac{1}{2}(u - v)$

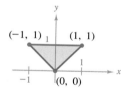

14. $\displaystyle\iint\limits_{R} y(x - y) \, dx \, dy$

$x = u + v$

$y = u$

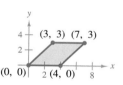

15. $\displaystyle\iint\limits_{R} \dfrac{\sqrt{x + y}}{x} \, dy \, dx$

$x = u, \ y = uv$

R: triangle with vertices $(0, 0)$, $(4, 0)$, $(4, 4)$

16. $\displaystyle\iint\limits_{R} y \sin xy \, dy \, dx$

$x = \dfrac{u}{v}, \ y = v$

R: region lying between the graphs of $xy = 1$, $xy = 4$, $y = 1$, $y = 4$

In Exercises 17–22, use a change of variables to find the volume of the solid region lying below the surface $z = f(x, y)$ and above the plane region R.

17. $f(x, y) = (x + y)e^{x-y}$

R: region bounded by the square with vertices $(4, 0)$, $(6, 2)$, $(4, 4)$, $(2, 2)$

18. $f(x, y) = (x + y)^2 \sin^2 (x - y)$

R: region bounded by the square with vertices $(\pi, 0)$, $(3\pi/2, \pi/2)$, (π, π), $(\pi/2, \pi/2)$

19. $f(x, y) = \sqrt{(x - y)(x + 4y)}$

R: region bounded by the parallelogram with vertices $(0, 0)$, $(1, 1)$, $(5, 0)$, $(4, -1)$

20. $f(x, y) = (3x + 2y)^2\sqrt{2y - x}$

R: region bounded by the parallelogram with vertices $(0, 0)$, $(-2, 3)$, $(2, 5)$, $(4, 2)$

21. $f(x, y) = \sqrt{x + y}$

R: region bounded by the triangle with vertices $(0, 0)$, $(a, 0)$, $(0, a)$ where $0 \le a$

22. $f(x, y) = \dfrac{xy}{1 + x^2y^2}$

R: region bounded by the graphs of $xy = 1$, $xy = 4$, $x = 1$, $x = 4$ [Hint: Let $x = u$, $y = v/u$.]

23. Consider the region R in the xy-plane described by

$$\frac{x^2}{a^2} + \frac{y^2}{b^2} = 1$$

and the transformation $x = au$ and $y = bv$.

(a) Sketch the graph of the region R and its image S under the specified transformation.

(b) Find $\partial(x, y)/\partial(u, v)$.

24. Use the result of Exercise 23 to find the volume of each dome-shaped solid lying below the surface $z = f(x, y)$ and above the elliptical region R. [Hint: After making the change of variables specified by the results of Exercise 23, make a second change of variables to polar coordinates.]

(a) $f(x, y) = 16 - x^2 - y^2$

R: $\dfrac{x^2}{16} + \dfrac{y^2}{9} \le 1$

(b) $f(x, y) = A \cos\left(\dfrac{\pi}{2}\sqrt{\dfrac{x^2}{a^2} + \dfrac{y^2}{b^2}}\right)$

R: $\dfrac{x^2}{a^2} + \dfrac{y^2}{b^2} \le 1$

In Exercises 25–28, find the Jacobian $\partial(x, y, z)/\partial(u, v, w)$ for the indicated change of variables. If $x = f(u, v, w)$, $y = g(u, v, w)$, and $z = h(u, v, w)$, then the Jacobian of x, y, and z with respect to u, v, and w, is

$$\frac{\partial(x, y, z)}{\partial(u, v, w)} = \begin{vmatrix} \dfrac{\partial x}{\partial u} & \dfrac{\partial x}{\partial v} & \dfrac{\partial x}{\partial w} \\[2mm] \dfrac{\partial y}{\partial u} & \dfrac{\partial y}{\partial v} & \dfrac{\partial y}{\partial w} \\[2mm] \dfrac{\partial z}{\partial u} & \dfrac{\partial z}{\partial v} & \dfrac{\partial z}{\partial w} \end{vmatrix}.$$

25. $x = u(1 - v)$, $y = uv(1 - w)$, $z = uvw$

26. $x = 4u - v$, $y = 4v - w$, $z = u + w$

27. *Spherical coordinates* $\ x = \rho \sin \phi \cos \theta$, $y = \rho \sin \phi \sin \theta$, $z = \rho \cos \phi$

28. *Cylindrical coordinates* $\ x = r \cos \theta$, $y = r \sin \theta$, $z = z$

REVIEW EXERCISES for Chapter 16

In Exercises 1–14, evaluate the multiple integral. Change coordinate systems when that makes the integration easier.

1. $\int_0^1 \int_0^{1+x} (3x + 2y) \, dy \, dx$ **2.** $\int_0^2 \int_{x^2}^{2x} (x^2 + 2y) \, dy \, dx$

3. $\int_0^3 \int_0^{\sqrt{9-x^2}} 4x \, dy \, dx$ **4.** $\int_0^{\sqrt{3}} \int_{2-\sqrt{4-y^2}}^{2+\sqrt{4-y^2}} dx \, dy$

5. $\int_{-2}^4 \int_{y^2/4}^{(4+y)/2} (x - y) \, dx \, dy$

6. $\int_{-2}^2 \int_0^{4-y^2} (8x - 2y^2) \, dx \, dy$

7. $\int_0^h \int_0^x \sqrt{x^2 + y^2} \, dy \, dx$ **8.** $\int_0^4 \int_0^{\sqrt{16-y^2}} (x^2 + y^2) \, dx \, dy$

9. $\int_{-3}^3 \int_{-\sqrt{9-x^2}}^{\sqrt{9-x^2}} \int_{x^2+y^2}^9 \sqrt{x^2 + y^2} \, dz \, dy \, dx$

10. $\int_{-2}^2 \int_{-\sqrt{4-x^2}}^{\sqrt{4-x^2}} \int_0^{(x^2+y^2)/2} (x^2 + y^2) \, dz \, dy \, dx$

11. $\int_{-1}^1 \int_{-\sqrt{1-x^2}}^{\sqrt{1-x^2}} \int_{-\sqrt{1-x^2-y^2}}^{\sqrt{1-x^2-y^2}} (x^2 + y^2) \, dz \, dy \, dx$

12. $\int_0^5 \int_0^{\sqrt{25-x^2}} \int_0^{\sqrt{25-x^2-y^2}} \frac{1}{\sqrt{x^2 + y^2 + z^2}} \, dz \, dy \, dx$

13. $\int_0^a \int_0^b \int_0^c (x^2 + y^2 + z^2) \, dx \, dy \, dz$

14. $\int_0^2 \int_0^{\sqrt{4-x^2}} \int_0^{\sqrt{4-x^2-y^2}} xyz \, dz \, dy \, dx$

In Exercises 15–22, write the limits to the double integral

$$\iint_R f(x, y) \, dA$$

for both orders of integration. Compute the area of R by letting $f(x, y) = 1$ and integrating.

15. Triangle: vertices $(0, 0)$, $(3, 0)$, $(0, 1)$
16. Triangle: vertices $(0, 0)$, $(3, 0)$, $(2, 2)$
17. The larger area between the graphs of $x^2 + y^2 = 25$, $x = 3$
18. Region bounded by the graphs of $y = 6x - x^2$, $y = x^2 - 2x$
19. Region enclosed by the graph of $y^2 = x^2 - x^4$
20. Region bounded by the graphs of $x = y^2 + 1$, $x = 0$, $y = 0$, $y = 2$
21. Region bounded by the graphs of $x = y + 3$, $x = y^2 + 1$
22. Region bounded by the graphs of $x = -y$, $x = 2y - y^2$

In Exercises 23 and 24, convert the given point from rectangular to (a) cylindrical coordinates and (b) spherical coordinates.

23. $(-2\sqrt{2}, 2\sqrt{2}, 2)$ **24.** $\left(\frac{\sqrt{3}}{4}, \frac{3}{4}, \frac{3\sqrt{3}}{2}\right)$

In Exercises 25 and 26, find an equation of the given surface in (a) cylindrical coordinates and (b) spherical coordinates.

25. $x^2 - y^2 = 2z$ **26.** $x^2 + y^2 + z^2 = 16$

In Exercises 27 and 28, find an equation in rectangular coordinates for the equation in cylindrical coordinates.

27. $r^2(\cos^2 \theta - \sin^2 \theta) + z^2 + 1$
28. $r^2 = 16z$

In Exercises 29 and 30, find an equation in rectangular coordinates for the equation in spherical coordinates.

29. $\rho = \csc \phi$ **30.** $\rho = 5$

In Exercises 31–36, use an appropriate multiple integral and coordinate system to find the volume of the solid.

31. Solid bounded by the graphs of $z = x^2 - y + 4$, $z = 0$, $x = 0$, $x = 4$
32. Solid bounded by the graphs of $z = x + y$, $z = 0$, $x = 0$, $x = 3$, $y = x$
33. Solid bounded by the graphs of $z = 0$, $z = h$, outside the cylinder $x^2 + y^2 = 1$ and inside the hyperboloid $x^2 + y^2 - z^2 = 1$
34. Solid that remains after drilling a hole of radius b through the center of a sphere of radius R ($b < R$)
35. Solid inside the graphs of $r = 2 \cos \theta$ and $r^2 + z^2 = 4$
36. Solid inside the graphs of $r^2 + z = 16$ and $r = 2 \sin \theta$

In Exercises 37 and 38, find the mass and center of mass of the lamina bounded by the graphs of the equations of specified density.

37. $y = 2x$, $y = 2x^3$, (first quadrant)
    (a) $\rho = kxy$ (b) $\rho = k(x^2 + y^2)$

38. $y = \dfrac{h}{2}\left(2 - \dfrac{x}{L} - \dfrac{x^2}{L^2}\right)$, (first quadrant)
    $\rho = k$

In Exercises 39–42, find the center of mass of the solid of uniform density bounded by the graphs of the given equations.

39. Solid inside the hemisphere $\rho = \cos \phi$, $\pi/4 \le \phi \le \pi/2$ and outside the cone $\phi = \pi/4$

40. Wedge: $x^2 + y^2 = a^2$, $z = cy$ $(0 < c)$, $0 \le y$, $0 \le z$

41. $x^2 + y^2 + z^2 = a^2$ (first octant)

42. $x^2 + y^2 + z^2 = 25$, $z = 4$ (the larger solid)

In Exercises 43 and 44, find the area of the surface on the function $f(x, y)$ over the region R.

43. $f(x, y) = 16 - x^2 - y^2$
    $R = \{(x, y): x^2 + y^2 \le 16\}$

44. $f(x, y) = 16 - x - y^2$
    $R = \{(x, y): 0 \le x \le 2, 0 \le y \le x\}$

In Exercises 45 and 46, find the moment of inertia I_z of the solid of specified density.

45. The solid of uniform density inside the paraboloid $z = 16 - x^2 - y^2$ and outside the cylinder $x^2 + y^2 = 9$, $z \ge 0$.

46. $x^2 + y^2 + z^2 = a^2$, density is proportional to the distance from the center

47. Give a geometrical interpretation of
$$\int_0^{2\pi} \int_{-\pi/2}^{\pi/2} \int_0^{6 \sin \phi} \rho^2 \sin \phi \, d\rho \, d\phi \, d\theta.$$

Chapter 17 Application

The Earth's magnetic field is very weak compared to that of even a toy magnet. Thus, it was difficult for early investigators to study the earth's magnetism. The first important work on the subject was performed by William Gilbert (1540–1603).

The Earth's Magnetic Field

A cross section of the earth's magnetic field can be represented as a vector field in which the center of the earth is located at the origin and the positive *y*-axis points in the direction of the magnetic north pole. The equation for this field is

$$\mathbf{F}(x, y) = M(x, y)\,\mathbf{i} + N(x, y)\,\mathbf{j}$$

$$= \frac{m}{(x^2 + y^2)^{5/2}}[3xy\,\mathbf{i} + (2y^2 - x^2)\,\mathbf{j}]$$

where *m* is the magnetic moment of the earth.

This vector field is conservative because

$$\frac{\partial M}{\partial y} = \frac{(3x^3 - 12xy^2)m}{(x^2 + y^2)^{7/2}} = \frac{\partial N}{\partial x}.$$

Many other important types of vector fields are also conservative (gravitational fields, electric force fields), which means that the work done in moving a particle from one position to another is independent of the path along which the particle is moved.

This computer generated graph indicates the flow of the earth's magnetic field.

See Exercise 71, Section 17.1.

Chapter Overview

This chapter combines our work with vectors and our work with integral calculus. The chapter begins by introducing **vector fields** in the plane and in space. For instance, a vector field in the plane takes the form

$$\mathbf{F}(x, y) = M(x, y)\,\mathbf{i} + N(x, y)\,\mathbf{j}.$$

Section 17.2 introduces three forms of **line integrals.**

$$\int_C f(x, y)\, ds$$

$$\int_C \mathbf{F}(x, y) \cdot d\mathbf{r}$$

$$\int_C M\, dx + N\, dy$$

Line integrals can be used to calculate the mass of a spring, as well as the work done by a force that moves an object in a force field.

Line integrals can be difficult to evaluate directly, and Sections 17.3 and 17.4 present ways of simplifying the process. In Section 17.3, we present the Fundamental Theorem of Line Integrals, which is roughly the line integral equivalent of the Fundamental Theorem of Calculus. Green's Theorem, which is discussed in Section 17.4, gives conditions under which a line integral can be evaluated as a double integral.

The remainder of the chapter introduces the concept of a **surface integral** and presents two important theorems for working with surface integrals—the Divergence Theorem and Stokes's Theorem.

Vector Analysis

17.1 Vector Fields

Vector fields ▪ Conservative vector fields ▪ Curl of a vector field ▪ Divergence of a vector field

In Chapter 13 we studied vector-valued functions—functions that assign a vector to a *real number*. There we saw that vector-valued functions of a real number are useful in representing curves and motion along a curve. In this chapter we will study two other types of vector-valued functions—functions that assign a vector to a *point in the plane* or a *point in space*. Such functions are called **vector fields,** and they are useful in representing various types of **force fields** and **velocity fields.**

DEFINITION OF A VECTOR FIELD

Let M and N be functions of two variables x and y, defined on a plane region R. The function $\mathbf{F}$ defined by

$$\mathbf{F}(x, y) = M\mathbf{i} + N\mathbf{j} \qquad \text{Plane}$$

is called a **vector field over R.**

Let M, N, and P be functions of three variables x, y, and z, defined on a solid region Q in space. The function $\mathbf{F}$ defined by

$$\mathbf{F}(x, y, z) = M\mathbf{i} + N\mathbf{j} + P\mathbf{k} \qquad \text{Space}$$

is called a **vector field over Q.**

From this definition we see that the *gradient* is one example of a vector field, since it can be written in the form

$$\nabla f(x, y) = f_x(x, y)\mathbf{i} + f_y(x, y)\mathbf{j}$$

or

$$\nabla f(x, y, z) = f_x(x, y, z)\mathbf{i} + f_y(x, y, z)\mathbf{j} + f_z(x, y, z)\mathbf{k}.$$

Some common *physical* examples of vector fields are **velocity fields, gravitational fields,** and **electric force fields.**

Velocity field

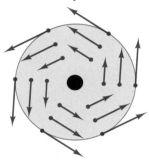

Rotating Wheel

FIGURE 17.1

Air-flow Vector Field

FIGURE 17.2

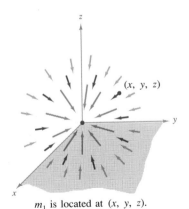

m_1 is located at (x, y, z).
m_2 is located at $(0, 0, 0)$.

Gravitational Force Field

FIGURE 17.3

1. *Velocity fields* describe the motion of a system of particles in the plane or in space. For instance, Figure 17.1 shows the vector field determined by a wheel rotating on an axle. Notice that the velocity vectors are determined by the location of their initial points—the farther a point is from the axle, the greater its velocity. Velocity fields are also determined by the flow of liquids through a container or by the flow of air currents around a moving object, as shown in Figure 17.2.

2. *Gravitational fields* are defined by **Newton's Law of Gravitation,** which states that the force of attraction exerted on a particle of mass m_1 located at (x, y, z) by a particle of mass m_2 located at $(0, 0, 0)$ is given by

$$\mathbf{F}(x, y, z) = \frac{-Gm_1m_2}{x^2 + y^2 + z^2}\mathbf{u}$$

where G is the gravitational constant and $\mathbf{u}$ is the unit vector in the direction from the origin to (x, y, z). In Figure 17.3 we can see that the gravitational field $\mathbf{F}$ has the properties that $\mathbf{F}(x, y, z)$ always points toward the origin, and that the magnitude of $\mathbf{F}(x, y, z)$ is the same at all points equidistant from the origin. A vector field with these two properties is called a **central force field.** Using the position vector $\mathbf{r} = x\mathbf{i} + y\mathbf{j} + z\mathbf{k}$ for the point (x, y, z), we can express the gravitational field $\mathbf{F}$ as

$$\mathbf{F}(x, y, z) = \frac{-Gm_1m_2}{\|\mathbf{r}\|^2} \cdot \frac{\mathbf{r}}{\|\mathbf{r}\|} = \frac{-Gm_1m_2}{\|\mathbf{r}\|^2}\mathbf{u}.$$

3. *Electric force fields* are defined by **Coulomb's Law,** which states that the force exerted on a particle with electric charge q_1 located at (x, y, z) by a particle with electric charge q_2 located at $(0, 0, 0)$ is given by

$$\mathbf{F}(x, y, z) = \frac{cq_1q_2}{\|\mathbf{r}\|^2}\mathbf{u}$$

where $\mathbf{r} = x\mathbf{i} + y\mathbf{j} + z\mathbf{k}$, $\mathbf{u} = \mathbf{r}/\|\mathbf{r}\|$, and c is a constant that depends on the choice of units for $\|\mathbf{r}\|$, q_1, and q_2.

Note that an electric force field has the same form as a gravitational field. That is, $\mathbf{F}(x, y, z) = (k/\|\mathbf{r}\|^2)\mathbf{u}$. We call such a force field an **inverse square field.**

| **DEFINITION OF INVERSE SQUARE FIELD** | Let $\mathbf{r}(t) = x(t)\mathbf{i} + y(t)\mathbf{j} + z(t)\mathbf{k}$ be the position vector. The vector field $\mathbf{F}$ is an **inverse square field** if $$\mathbf{F}(x, y, z) = \frac{k}{\|\mathbf{r}\|^2}\mathbf{u}$$ where k is a real number and $\mathbf{u} = \mathbf{r}/\|\mathbf{r}\|$ is a unit vector in the direction of $\mathbf{r}$. |

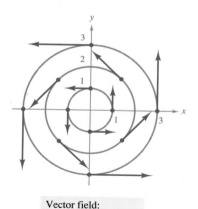

Vector field:
$\mathbf{F}(x, y) = -y\mathbf{i} + x\mathbf{j}$

FIGURE 17.4

EXAMPLE 1 Sketching a vector field

Sketch some vectors in the vector field given by $\mathbf{F}(x, y) = -y\mathbf{i} + x\mathbf{j}$.

SOLUTION

We could plot vectors at several random points in the plane. However, it is more enlightening to plot vectors of equal magnitude. This corresponds to finding level curves in scalar fields. In this case, vectors of equal magnitude lie on circles given by

$$\|\mathbf{F}\| = \sqrt{x^2 + y^2} = c \quad \Longrightarrow \quad x^2 + y^2 = c^2.$$

Now, we choose a value for c^2 and plot several vectors on the resulting circle. For instance, the following vectors occur on the unit circle.

Point: (1, 0)   (0, 1)   (−1, 0)   (0, −1)
Vector: $\mathbf{F}(1, 0) = \mathbf{j}$ $\mathbf{F}(0, 1) = -\mathbf{i}$ $\mathbf{F}(-1, 0) = -\mathbf{j}$ $\mathbf{F}(0, -1) = \mathbf{i}$

These and several other vectors in the vector field are shown in Figure 17.4. Note in the figure that this vector field is similar to that given by the rotating wheel shown in Figure 17.1. ▭

EXAMPLE 2 Sketching a vector field

Sketch some vectors in the vector field given by $\mathbf{F}(x, y) = 2x\mathbf{i} + y\mathbf{j}$.

SOLUTION

For this vector field, vectors of equal length lie on ellipses given by

$$\|\mathbf{F}\| = \sqrt{(2x)^2 + (y)^2} = c \quad \Longrightarrow \quad 4x^2 + y^2 = c^2.$$

For $c = 1$, we sketch several vectors $2x\mathbf{i} + y\mathbf{j}$ of magnitude 1 at points on the ellipse given by

$$4x^2 + y^2 = 1.$$

For $c = 2$, we sketch several vectors $2x\mathbf{i} + y\mathbf{j}$ of magnitude 2 at points on the ellipse given by

$$4x^2 + y^2 = 4.$$

These vectors are shown in Figure 17.5. ▭

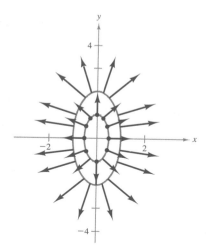

Vector field:
$\mathbf{F}(x, y) = 2x\mathbf{i} + y\mathbf{j}$

FIGURE 17.5

EXAMPLE 3 Sketching a velocity field

Sketch some vectors in the velocity field given by

$$\mathbf{v}(x, y, z) = (25 - x^2 - y^2)\mathbf{k}$$

where $x^2 + y^2 \leq 25$.

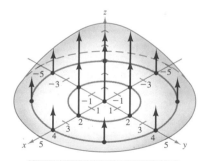

Velocity field:
$\mathbf{v}(x, y, z) = (25 - x^2 - y^2)\mathbf{k}$

FIGURE 17.6

SOLUTION

We can imagine that $\mathbf{v}$ describes the velocity of a liquid flowing through a tube of radius 5. Vectors near the z-axis are longer than those near the edge of the tube. For instance, at the point $(0, 0, 0)$, the velocity vector is

$$\mathbf{v}(0, 0, 0) = 25\mathbf{k}$$

while at the point $(0, 4, 0)$, the velocity vector is

$$\mathbf{v}(0, 4, 0) = (25 - 0^2 - 4^2)\mathbf{k} = 9\mathbf{k}.$$

Figure 17.6 shows these and several other vectors for the velocity field.

Conservative vector fields

Notice in Figure 17.5 that all the vectors appear to be normal to the level curve from which they emanate. Since this is a property of gradients, it is natural to ask whether the vector field given by $\mathbf{F}(x, y) = 2x\mathbf{i} + y\mathbf{j}$ is the *gradient* for some differentiable function f. Not every vector field is the gradient of a differentiable function—we call those that are, **conservative** vector fields.

DEFINITION OF CONSERVATIVE VECTOR FIELD

A vector field $\mathbf{F}$ is called **conservative** if there exists a differentiable function f such that

$$\mathbf{F} = \nabla f.$$

The function f is called the **potential function** for $\mathbf{F}$.

EXAMPLE 4   Conservative vector fields

(a) The vector field given by $\mathbf{F}(x, y) = 2x\mathbf{i} + y\mathbf{j}$ is conservative. To see this, consider the potential function $f(x, y) = x^2 + \frac{1}{2}y^2$. Since

$$\nabla f = 2x\mathbf{i} + y\mathbf{j} = \mathbf{F}$$

it follows that $\mathbf{F}$ is conservative.

(b) Every inverse square field is conservative. To see this, let

$$\mathbf{F}(x, y, z) = \frac{k}{\|\mathbf{r}\|^2}\mathbf{u} \quad \text{and} \quad f(x, y, z) = \frac{-k}{\sqrt{x^2 + y^2 + z^2}}$$

where $\mathbf{u} = \mathbf{r}/\|\mathbf{r}\|$. Since

$$\nabla f = \frac{kx}{(x^2 + y^2 + z^2)^{3/2}}\mathbf{i} + \frac{ky}{(x^2 + y^2 + z^2)^{3/2}}\mathbf{j} + \frac{kz}{(x^2 + y^2 + z^2)^{3/2}}\mathbf{k}$$

$$= \frac{k}{x^2 + y^2 + z^2}\left(\frac{x\mathbf{i} + y\mathbf{j} + z\mathbf{k}}{\sqrt{x^2 + y^2 + z^2}}\right) = \frac{k}{\|\mathbf{r}\|^2}\frac{\mathbf{r}}{\|\mathbf{r}\|} = \frac{k}{\|\mathbf{r}\|^2}\mathbf{u}$$

it follows that $\mathbf{F}$ is conservative.

As seen in Example 4, many important vector fields, including gravitational fields, magnetic fields, and electric force fields, are conservative. Most of the terminology in this chapter comes from physics. For example, the term "conservative" is derived from the classic physical law regarding the conservation of energy. This law states that the sum of the kinetic energy and the potential energy of a particle moving in a conservative force field is constant. (The kinetic energy of a particle is the energy due to its motion, and the potential energy is the energy due to its position in the force field.)

The following important theorem gives a necessary and sufficient condition for a vector field *in the plane* to be conservative.

THEOREM 17.1 **TEST FOR CONSERVATIVE VECTOR FIELD IN THE PLANE**	Let M and N have continuous first partial derivatives on an open disc R. The vector field given by $\mathbf{F}(x, y) = M\mathbf{i} + N\mathbf{j}$ is conservative if and only if $$\frac{\partial N}{\partial x} = \frac{\partial M}{\partial y}.$$

PROOF   To prove that the given condition is necessary for $\mathbf{F}$ to be conservative, we suppose there exists a potential function f such that

$$\mathbf{F}(x, y) = \nabla f(x, y) = M\mathbf{i} + N\mathbf{j}.$$

Then, we have

$$f_x(x, y) = M \quad \Longrightarrow \quad f_{xy}(x, y) = \frac{\partial M}{\partial y}$$

$$f_y(x, y) = N \quad \Longrightarrow \quad f_{yx}(x, y) = \frac{\partial N}{\partial x}$$

and, by the equivalence of the mixed partials f_{xy} and f_{yx}, we conclude that $\partial N/\partial x = \partial M/\partial y$ for all (x, y) in R. We prove the sufficiency of the condition in Section 17.4.

REMARK   Theorem 17.1 requires that the domain of $\mathbf{F}$ be an open disc. If R is simply an open region, then the given condition is necessary but not sufficient to produce a conservative vector field.

EXAMPLE 5   Testing for conservative vector fields in the plane

(a) The vector field given by $\mathbf{F}(x, y) = x^2 y\mathbf{i} + xy\mathbf{j}$ *is not* conservative, since

$$\frac{\partial M}{\partial y} = \frac{\partial}{\partial y}[x^2 y] = x^2 \quad \text{and} \quad \frac{\partial N}{\partial x} = \frac{\partial}{\partial x}[xy] = y.$$

(b) The vector field given by $\mathbf{F}(x, y) = 2x\mathbf{i} + y\mathbf{j}$ *is* conservative, since

$$\frac{\partial M}{\partial y} = \frac{\partial}{\partial y}[2x] = 0 \quad \text{and} \quad \frac{\partial N}{\partial x} = \frac{\partial}{\partial x}[y] = 0. \qquad \blacksquare$$

Theorem 17.1 tells us whether a vector field is conservative. It does not tell us how to find a potential function for **F**. The problem is comparable to antidifferentiation, and sometimes you will be able to find a potential function by simple inspection. For instance, as we saw in Example 4,

$$f(x, y) = x^2 + \frac{1}{2}y^2$$

has the property that $\nabla f(x, y) = 2x\mathbf{i} + y\mathbf{j}$. We will discuss techniques for finding potential functions in detail in Chapter 18 (in the section on exact differential equations). For now, we outline the procedure by way of an example.

EXAMPLE 6   Finding a potential function for $\mathbf{F}(x, y)$

Find a potential function for

$$\mathbf{F}(x, y) = 2xy\mathbf{i} + (x^2 - y)\mathbf{j}.$$

SOLUTION

From Theorem 17.1 it follows that **F** is conservative, since

$$\frac{\partial}{\partial y}[2xy] = 2x \qquad \text{and} \qquad \frac{\partial}{\partial x}[x^2 - y] = 2x.$$

Now, if f is a function such that $\nabla f(x, y) = f_x(x, y)\mathbf{i} + f_y(x, y)\mathbf{j}$, then we have

$$f_x(x, y) = 2xy \qquad \text{and} \qquad f_y(x, y) = x^2 - y.$$

To reconstruct the function f from these two partial derivatives, we integrate $f_x(x, y)$ with respect to x and $f_y(x, y)$ with respect to y, as follows.

$$f(x, y) = \int f_x(x, y)\, dx = \int 2xy\, dx = x^2y + g(y) + K$$

$$f(x, y) = \int f_y(x, y)\, dy = \int (x^2 - y)\, dy = x^2y - \frac{y^2}{2} + h(x) + K$$

Now, we reconcile the differences between these two expressions for $f(x, y)$. That is, since $g(y) = -y^2/2$, it follows that $h(x) = 0$, and we have

$$f(x, y) = x^2y + g(y) + K = x^2y - \frac{y^2}{2} + K.$$

REMARK   Note that the solution in Example 6 is comparable to that given by an indefinite integral. That is, the solution represents a family of potential functions, any two of which differ by a constant. To find a unique solution we would have to be given an initial condition satisfied by the vector field.

Curl of a vector field

Theorem 17.1 has a counterpart for vector fields in space. Before stating that result, we define the **curl of a vector field** in space.

DEFINITION OF CURL OF A VECTOR FIELD	The **curl** of $\mathbf{F}(x, y, z) = M\mathbf{i} + N\mathbf{j} + P\mathbf{k}$ is

$$\text{curl } \mathbf{F}(x, y, z) = \nabla \times \mathbf{F}(x, y, z)$$

$$= \left(\frac{\partial P}{\partial y} - \frac{\partial N}{\partial z}\right)\mathbf{i} - \left(\frac{\partial P}{\partial x} - \frac{\partial M}{\partial z}\right)\mathbf{j} + \left(\frac{\partial N}{\partial x} - \frac{\partial M}{\partial y}\right)\mathbf{k}.$$

REMARK   If **curl F** = **0**, then we say that **F** is **irrotational.**

The cross-product notation used for curl comes from viewing the gradient ∇f as the result of the **differential operator** ∇ acting on the function f. In this context, we can use the following determinant form as an aid in remembering the formula for curl.

$$\text{curl } \mathbf{F}(x, y, z) = \nabla \times \mathbf{F}(x, y, z) = \begin{vmatrix} \mathbf{i} & \mathbf{j} & \mathbf{k} \\ \dfrac{\partial}{\partial x} & \dfrac{\partial}{\partial y} & \dfrac{\partial}{\partial z} \\ M & N & P \end{vmatrix}$$

$$= \left(\frac{\partial P}{\partial y} - \frac{\partial N}{\partial z}\right)\mathbf{i} - \left(\frac{\partial P}{\partial x} - \frac{\partial M}{\partial z}\right)\mathbf{j} + \left(\frac{\partial N}{\partial x} - \frac{\partial M}{\partial y}\right)\mathbf{k}$$

EXAMPLE 7   Finding the curl of a vector field

Find **curl F** for the vector field given by

$$\mathbf{F}(x, y, z) = 2xy\mathbf{i} + (x^2 + z^2)\mathbf{j} + 2zy\mathbf{k}.$$

SOLUTION

The curl of **F** is given by

$$\nabla \times \mathbf{F}(x, y, z) = \begin{vmatrix} \mathbf{i} & \mathbf{j} & \mathbf{k} \\ \dfrac{\partial}{\partial x} & \dfrac{\partial}{\partial y} & \dfrac{\partial}{\partial z} \\ 2xy & x^2 + z^2 & 2zy \end{vmatrix}$$

$$= \begin{vmatrix} \dfrac{\partial}{\partial y} & \dfrac{\partial}{\partial z} \\ x^2 + z^2 & 2zy \end{vmatrix}\mathbf{i} - \begin{vmatrix} \dfrac{\partial}{\partial x} & \dfrac{\partial}{\partial z} \\ 2xy & 2zy \end{vmatrix}\mathbf{j} + \begin{vmatrix} \dfrac{\partial}{\partial x} & \dfrac{\partial}{\partial y} \\ 2xy & x^2 + z^2 \end{vmatrix}\mathbf{k}$$

$$= (2z - 2z)\mathbf{i} - (0 - 0)\mathbf{j} + (2x - 2x)\mathbf{k}$$

$$= \mathbf{0}.$$

Later in this chapter we will assign a physical interpretation to the curl of a vector field. But for now, the primary use we make of curl is in the following test for conservative vector fields in space. The test states that for a vector field whose domain is all of three-dimensional space (or an open sphere), the curl is zero at every point in the domain if and only if **F** is conservative. The proof is similar to that given in Theorem 17.1.

THEOREM 17.2
TEST FOR CONSERVATIVE
VECTOR FIELD IN SPACE

Suppose M, N, and P have continuous first partial derivatives in an open sphere Q in space. The vector field given by $\mathbf{F}(x, y, z) = M\mathbf{i} + N\mathbf{j} + P\mathbf{k}$ is conservative if and only if

$$\text{curl } \mathbf{F}(x, y, z) = \mathbf{0}.$$

That is, $\mathbf{F}$ is conservative if and only if

$$\frac{\partial P}{\partial y} = \frac{\partial N}{\partial z}, \qquad \frac{\partial P}{\partial x} = \frac{\partial M}{\partial z}, \qquad \text{and} \qquad \frac{\partial N}{\partial x} = \frac{\partial M}{\partial y}.$$

From Theorem 17.2, we can see that the vector field given in Example 7 is conservative.

EXAMPLE 8   Testing for conservative vector fields in space

Find the curl of the vector field $\mathbf{F}(x, y, z) = x^3y^2z\mathbf{i} + x^2z\mathbf{j} + x^2y\mathbf{k}$ and show that it is not conservative.

SOLUTION

By the definition of curl, we have

$$\text{curl } \mathbf{F}(x, y, z) = \begin{vmatrix} \mathbf{i} & \mathbf{j} & \mathbf{k} \\ \dfrac{\partial}{\partial x} & \dfrac{\partial}{\partial y} & \dfrac{\partial}{\partial z} \\ x^3y^2z & x^2z & x^2y \end{vmatrix}$$

$$= \begin{vmatrix} \dfrac{\partial}{\partial y} & \dfrac{\partial}{\partial z} \\ x^2z & x^2y \end{vmatrix}\mathbf{i} - \begin{vmatrix} \dfrac{\partial}{\partial x} & \dfrac{\partial}{\partial z} \\ x^3y^2z & x^2y \end{vmatrix}\mathbf{j} + \begin{vmatrix} \dfrac{\partial}{\partial x} & \dfrac{\partial}{\partial y} \\ x^3y^2z & x^2z \end{vmatrix}\mathbf{k}$$

$$= (x^2 - x^2)\mathbf{i} - (2xy - x^3y^2)\mathbf{j} + (2xz - 2x^3yz)\mathbf{k}$$

$$= (x^3y^2 - 2xy)\mathbf{j} + (2xz - 2x^3yz)\mathbf{k}.$$

Since $\text{curl } \mathbf{F} \neq \mathbf{0}$, we conclude by Theorem 17.2 that $\mathbf{F}$ is not conservative. ▭

For vector fields in space that pass the test for being conservative, we can find a potential function by following the same pattern used in the plane (as demonstrated in Example 6).

EXAMPLE 9   Finding a potential function for $\mathbf{F}(x, y, z)$

Find a potential function for $\mathbf{F}(x, y, z) = 2xy\mathbf{i} + (x^2 + z^2)\mathbf{j} + 2zy\mathbf{k}$.

SOLUTION

From Example 7, we know that the vector field given by $\mathbf{F}$ is conservative. Now, if f is a function such that $\mathbf{F}(x, y, z) = \nabla f(x, y, z)$, then

$$f_x(x, y, z) = 2xy, \qquad f_y(x, y, z) = x^2 + z^2, \qquad \text{and} \qquad f_z(x, y, z) = 2zy$$

and by integrating with respect to x, y, and z separately, we obtain

$$f(x, y, z) = \int M\, dx = \int 2xy\, dx = x^2 y + g(y, z) + K$$

$$f(x, y, z) = \int N\, dy = \int (x^2 + z^2)\, dy = x^2 y + z^2 y + h(x, z) + K$$

$$f(x, y, z) = \int P\, dz = \int 2zy\, dz = z^2 y + k(x, y) + K.$$

Comparing these three versions of $f(x, y, z)$, we conclude that

$$g(y, z) = z^2 y, \qquad h(x, z) = 0, \qquad \text{and} \qquad k(x, y) = x^2 y.$$

Therefore, $f(x, y, z)$ is given by

$$f(x, y, z) = x^2 y + z^2 y + K.$$

REMARK Examples 6 and 9 are illustrations of a type of problem called *recovering a function from its gradient*, and we will discuss other methods for solving this problem in Chapter 18. One popular method gives an interplay between successive "partial integrations" and partial differentiations.

Divergence of a vector field

We have seen that the curl of a vector field $\mathbf{F}$ is itself a vector field. Another important function defined on a vector field is **divergence,** which is a scalar function. This function can be viewed as a type of derivative of $\mathbf{F}$ in that, for vector fields representing velocities of moving particles, the divergence measures the rate of particle flow per unit volume at a point. (More will be said about this in Section 17.6.)

DEFINITION OF DIVERGENCE OF A VECTOR FIELD	The **divergence** of $\mathbf{F}(x, y) = M\mathbf{i} + N\mathbf{j}$ is $$\operatorname{div} \mathbf{F}(x, y) = \nabla \cdot \mathbf{F}(x, y) = \frac{\partial M}{\partial x} + \frac{\partial N}{\partial y}. \qquad \text{Plane}$$ The **divergence** of $\mathbf{F}(x, y, z) = M\mathbf{i} + N\mathbf{j} + P\mathbf{k}$ is $$\operatorname{div} \mathbf{F}(x, y, z) = \nabla \cdot \mathbf{F}(x, y, z) = \frac{\partial M}{\partial x} + \frac{\partial N}{\partial y} + \frac{\partial P}{\partial z}. \qquad \text{Space}$$

REMARK If $\operatorname{div} \mathbf{F} = 0$, then $\mathbf{F}$ is said to be **divergence free** or **solenoidal.**

The dot-product notation used for divergence comes from considering ∇ as a **differential operator,** as follows.

$$\nabla \cdot \mathbf{F}(x, y, z) = \left[\left(\frac{\partial}{\partial x}\right)\mathbf{i} + \left(\frac{\partial}{\partial y}\right)\mathbf{j} + \left(\frac{\partial}{\partial z}\right)\mathbf{k}\right] \cdot [M\mathbf{i} + N\mathbf{j} + P\mathbf{k}]$$

$$= \frac{\partial M}{\partial x} + \frac{\partial N}{\partial y} + \frac{\partial P}{\partial z}$$

EXAMPLE 10   Finding the divergence of a vector field

Find the divergence at $(2, 1, -1)$ for the vector field

$$\mathbf{F}(x, y, z) = x^3y^2z\,\mathbf{i} + x^2z\,\mathbf{j} + x^2y\,\mathbf{k}.$$

SOLUTION

By definition, we have

$$\text{div } \mathbf{F}(x, y, z) = \frac{\partial}{\partial x}[x^3y^2z] + \frac{\partial}{\partial y}[x^2z] + \frac{\partial}{\partial z}[x^2y] = 3x^2y^2z.$$

At the point $(2, 1, -1)$, we have

$$\text{div } \mathbf{F}(2, 1, -1) = 3(2^2)(1^2)(-1) = -12. \qquad \blacksquare$$

There are many important properties of the divergence and curl of a vector field $\mathbf{F}$ (see Exercises 59–65). One that we will use often is described in the following theorem. You are asked to supply a proof for this theorem in Exercise 66.

THEOREM 17.3
RELATIONSHIP BETWEEN
DIVERGENCE AND CURL

If the second partial derivatives of a function f are continuous, then

$$\text{div } (\mathbf{curl\ F}) = 0.$$

EXERCISES for Section 17.1

In Exercises 1–14, sketch several representative vectors in the given vector field.

1. $\mathbf{F}(x, y) = \mathbf{i} + \mathbf{j}$
2. $\mathbf{F}(x, y) = 2\mathbf{i}$
3. $\mathbf{F}(x, y) = x\mathbf{j}$
4. $\mathbf{F}(x, y) = y\mathbf{i}$
5. $\mathbf{F}(x, y) = x\mathbf{i} + y\mathbf{j}$
6. $\mathbf{F}(x, y) = -x\mathbf{i} + y\mathbf{j}$
7. $\mathbf{F}(x, y) = x\mathbf{i} + 3y\mathbf{j}$
8. $\mathbf{F}(x, y) = y\mathbf{i} - x\mathbf{j}$
9. $\mathbf{F}(x, y) = \dfrac{x}{\sqrt{x^2 + y^2}}\mathbf{i} + \dfrac{y}{\sqrt{x^2 + y^2}}\mathbf{j}$
10. $\mathbf{F}(x, y) = 4x\mathbf{i} + y\mathbf{j}$
11. $\mathbf{F}(x, y, z) = 3y\mathbf{j}$
12. $\mathbf{F}(x, y) = \mathbf{i} + (x^2 + y^2)\mathbf{j}$

13. $\mathbf{F}(x, y, z) = \mathbf{i} + \mathbf{j} + \mathbf{k}$
14. $\mathbf{F}(x, y, z) = x\mathbf{i} + y\mathbf{j} + z\mathbf{k}$

In Exercises 15–20, find the gradient vector field for the given scalar function. (That is, find the conservative vector field for the given potential function.)

15. $f(x, y) = 5x^2 + 3xy + 10y^2$
16. $f(x, y) = \sin 3x \cos 4y$
17. $f(x, y, z) = z - ye^{x^2}$
18. $f(x, y, z) = \dfrac{y}{z} + \dfrac{z}{x} - \dfrac{xz}{y}$

19. $g(x, y, z) = xy \ln (x + y)$
20. $g(x, y, z) = x \arcsin yz$

In Exercises 21–28, determine whether the vector field is conservative. If it is, find a potential function for the vector field.

21. $\mathbf{F}(x, y) = 2xy\mathbf{i} + x^2\mathbf{j}$

22. $\mathbf{F}(x, y) = \dfrac{1}{y^2}(y\mathbf{i} - 2x\mathbf{j})$

23. $\mathbf{F}(x, y) = xe^{x^2y}(2y\mathbf{i} + x\mathbf{j})$
24. $\mathbf{F}(x, y) = 2xy^3\mathbf{i} + 3y^2x^2\mathbf{j}$

25. $\mathbf{F}(x, y) = \dfrac{x\mathbf{i} + y\mathbf{j}}{x^2 + y^2}$

26. $\mathbf{F}(x, y) = \dfrac{2y}{x}\mathbf{i} - \dfrac{x^2}{y^2}\mathbf{j}$

27. $\mathbf{F}(x, y) = e^x(\cos y\, \mathbf{i} + \sin y\, \mathbf{j})$

28. $\mathbf{F}(x, y) = \dfrac{2x\mathbf{i} + 2y\mathbf{j}}{(x^2 + y^2)^2}$

In Exercises 29–32, find the curl of the vector field $\mathbf{F}$ at the indicated point.

Vector field	Point
29. $\mathbf{F}(x, y, z) = xyz\mathbf{i} + y\mathbf{j} + z\mathbf{k}$	$(1, 2, 1)$
30. $\mathbf{F}(x, y, z) = x^2z\mathbf{i} - 2xz\mathbf{j} + yz\mathbf{k}$	$(2, -1, 3)$
31. $\mathbf{F}(x, y, z) = e^x \sin y\, \mathbf{i} - e^x \cos y\, \mathbf{j}$	$(0, 0, 3)$
32. $\mathbf{F}(x, y, z) = e^{-xyz}(\mathbf{i} + \mathbf{j} + \mathbf{k})$	$(3, 2, 0)$

In Exercises 33–36, find the curl of the vector field $\mathbf{F}$.

33. $\mathbf{F}(x, y, z) = \arctan \dfrac{x}{y}\mathbf{i} + \ln \sqrt{x^2 + y^2}\, \mathbf{j} + \mathbf{k}$

34. $\mathbf{F}(x, y, z) = \dfrac{yz}{y - z}\mathbf{i} + \dfrac{xz}{x - z}\mathbf{j} + \dfrac{xy}{x - y}\mathbf{k}$

35. $\mathbf{F}(x, y, z) = \sin (x - y)\, \mathbf{i} + \sin (y - z)\, \mathbf{j}$
$$+ \sin (z - x)\mathbf{k}$$

36. $\mathbf{F}(x, y, z) = \sqrt{x^2 + y^2 + z^2}(\mathbf{i} + \mathbf{j} + \mathbf{k})$

In Exercises 37–42, determine whether the vector field $\mathbf{F}$ is conservative. If it is, find a potential function for the vector field.

37. $\mathbf{F}(x, y, z) = \sin y\, \mathbf{i} - x \cos y\, \mathbf{j} + \mathbf{k}$
38. $\mathbf{F}(x, y, z) = e^z(y\mathbf{i} + x\mathbf{j} + \mathbf{k})$
39. $\mathbf{F}(x, y, z) = e^z(y\mathbf{i} + x\mathbf{j} + xy\mathbf{k})$
40. $\mathbf{F}(x, y, z) = 3x^2y^2z\mathbf{i} + 2x^3yz\mathbf{j} + x^3y^2\mathbf{k}$

41. $\mathbf{F}(x, y, z) = \dfrac{1}{y}\mathbf{i} - \dfrac{x}{y^2}\mathbf{j} + (2z - 1)\mathbf{k}$

42. $\mathbf{F}(x, y, z) = \dfrac{x}{x^2 + y^2}\mathbf{i} + \dfrac{y}{x^2 + y^2}\mathbf{j} + \mathbf{k}$

In Exercises 43 and 44, find **curl** $(\mathbf{F} \times \mathbf{G})$.

43. $\mathbf{F}(x, y, z) = \mathbf{i} + 2x\mathbf{j} + 3y\mathbf{k}$
$\quad \mathbf{G}(x, y, z) = x\mathbf{i} - y\mathbf{j} + z\mathbf{k}$
44. $\mathbf{F}(x, y, z) = x\mathbf{i} - z\mathbf{k}$
$\quad \mathbf{G}(x, y, z) = x^2\mathbf{i} + y\mathbf{j} + z^2\mathbf{k}$

In Exercises 45 and 46, find

$$\mathbf{curl}\ (\mathbf{curl}\ \mathbf{F}) = \nabla \times (\nabla \times \mathbf{F}).$$

45. $\mathbf{F}(x, y, z) = xyz\mathbf{i} + y\mathbf{j} + z\mathbf{k}$
46. $\mathbf{F}(x, y, z) = x^2z\mathbf{i} - 2xz\mathbf{j} + yz\mathbf{k}$

In Exercises 47–50, find the divergence of the vector field $\mathbf{F}$.

47. $\mathbf{F}(x, y, z) = 6x^2\mathbf{i} - xy^2\mathbf{j}$
48. $\mathbf{F}(x, y, z) = xe^x\mathbf{i} + ye^y\mathbf{j}$
49. $\mathbf{F}(x, y, z) = \sin x\, \mathbf{i} + \cos y\, \mathbf{j} + z^2\mathbf{k}$
50. $\mathbf{F}(x, y, z) = \ln (x^2 + y^2)\, \mathbf{i} + xy\mathbf{j} + \ln (y^2 + z^2)\, \mathbf{k}$

In Exercises 51–54, find the divergence of the vector field $\mathbf{F}$ at the indicated point.

51. $\mathbf{F}(x, y, z) = xyz\mathbf{i} + y\mathbf{j} + z\mathbf{k}, (1, 2, 1)$
52. $\mathbf{F}(x, y, z) = x^2z\mathbf{i} - 2xz\mathbf{j} + yz\mathbf{k}, (2, -1, 3)$
53. $\mathbf{F}(x, y, z) = e^x \sin y\, \mathbf{i} - e^x \cos y\, \mathbf{j}, (0, 0, 3)$
54. $\mathbf{F}(x, y, z) = e^{-xyz}(\mathbf{i} + \mathbf{j} + \mathbf{k}), (3, 2, 0)$

In Exercises 55 and 56, find div $(\mathbf{F} \times \mathbf{G})$.

55. $\mathbf{F}(x, y, z) = \mathbf{i} + 2x\mathbf{j} + 3y\mathbf{k}$
$\quad \mathbf{G}(x, y, z) = x\mathbf{i} - y\mathbf{j} + z\mathbf{k}$
56. $\mathbf{F}(x, y, z) = x\mathbf{i} - z\mathbf{k}$
$\quad \mathbf{G}(x, y, z) = x^2\mathbf{i} + y\mathbf{j} + z^2\mathbf{k}$

In Exercises 57 and 58, find

$$\text{div}\ (\mathbf{curl}\ \mathbf{F}) = \nabla \cdot (\nabla \times \mathbf{F}).$$

57. $\mathbf{F}(x, y, z) = xyz\mathbf{i} + y\mathbf{j} + z\mathbf{k}$
58. $\mathbf{F}(x, y, z) = x^2z\mathbf{i} - 2xz\mathbf{j} + yz\mathbf{k}$

In Exercises 59–65, prove the given property for vector fields $\mathbf{F}$ and $\mathbf{G}$ and scalar function f. (Assume the required partial derivatives are continuous.)

59. $\mathbf{curl}(\mathbf{F} + \mathbf{G}) = \mathbf{curl}\ \mathbf{F} + \mathbf{curl}\ \mathbf{G}$
60. $\mathbf{curl}(\nabla f) = \nabla \times (\nabla f) = \mathbf{0}$
61. $\text{div}(\mathbf{F} + \mathbf{G}) = \text{div}\ \mathbf{F} + \text{div}\ \mathbf{G}$
62. $\text{div}(\mathbf{F} \times \mathbf{G}) = (\mathbf{curl}\ \mathbf{F}) \cdot \mathbf{G} - \mathbf{F} \cdot (\mathbf{curl}\ \mathbf{G})$
63. $\nabla \times [\nabla f + (\nabla \times \mathbf{F})] = \nabla \times (\nabla \times \mathbf{F})$
64. $\nabla \times (f\mathbf{F}) = f(\nabla \times \mathbf{F}) + (\nabla f) \times \mathbf{F}$
65. $\text{div}(f\mathbf{F}) = f \text{ div } \mathbf{F} + \nabla f \cdot \mathbf{F}$

66. Prove Theorem 17.3.

In Exercises 67–70, let $\mathbf{F}(x, y, z) = x\mathbf{i} + y\mathbf{j} + z\mathbf{k}$, and $f(x, y, z) = \|\mathbf{F}(x, y, z)\|$.

67. Show that $\nabla(\ln f) = \dfrac{\mathbf{F}}{f^2}$.

68. Show that $\nabla\left(\dfrac{1}{f}\right) = -\dfrac{\mathbf{F}}{f^3}$.

69. Show that $\nabla f^n = nf^{n-2}\mathbf{F}$.

70. The **Laplacian** is the differential operator

$$\nabla^2 = \nabla \cdot \nabla = \frac{\partial^2}{\partial x^2} + \frac{\partial^2}{\partial y^2} + \frac{\partial^2}{\partial z^2}$$

and Laplace's equation is

$$\nabla^2 w = \frac{\partial^2 w}{\partial x^2} + \frac{\partial^2 w}{\partial y^2} + \frac{\partial^2 w}{\partial z^2} = 0.$$

Any function that satisfies this equation is called **harmonic.** Show that the function $1/f$ is harmonic.

71. Refer to the Chapter 17 Application. A cross section of the earth's magnetic field can be represented as a vector field in which the center of the earth is located at the origin and the positive y-axis points in the direction of the magnetic north pole. The equation for this field is

$$\mathbf{F}(x, y) = M(x, y)\mathbf{i} + N(x, y)\mathbf{j}$$
$$= \frac{m}{(x^2 + y^2)^{5/2}}[3xy\,\mathbf{i} + (2y^2 - x^2)\mathbf{j}]$$

where m is the magnetic moment of the earth. Show that this vector field is conservative.

17.2 Line Integrals

Piecewise smooth curves ▪ Line integrals ▪ Line integrals of vector fields ▪ Line integrals in differential form

Josiah Willard Gibbs

A classical property of gravitational fields is that, subject to certain physical constraints, the work done by gravity on an object moving between two points in the field is independent of the path taken by the object. One of the constraints is that the **path** must be a piecewise smooth curve. Recall that a curve C given by

$$\mathbf{r}(t) = x(t)\mathbf{i} + y(t)\mathbf{j} + z(t)\mathbf{k}, \quad a \le t \le b$$

is **smooth** if dx/dt, dy/dt, and dz/dt are continuous on $[a, b]$ and not simultaneously zero on (a, b). Moreover, a curve C is **piecewise smooth** if the interval $[a, b]$ can be partitioned into a finite number of subintervals, on each of which C is smooth. The parametric representation of a smooth curve is called a **smooth parameterization.**

Many physicists and mathematicians have contributed to the theory and applications described in this chapter—Newton, Gauss, Laplace, Hamilton, Maxwell, and many others. However, the use of vector analysis to describe the results is primarily due to the American mathematical physicist Josiah Willard Gibbs (1839–1903).

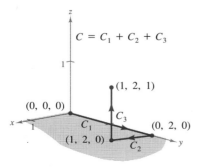

FIGURE 17.7

EXAMPLE 1 Finding a piecewise smooth parameterization

Find a piecewise smooth parameterization of the curve C shown in Figure 17.7.

SOLUTION

Since C consists of three smooth line segments C_1, C_2, and C_3, we construct a parameterization for each and piece them together by making the last t-value in C_i correspond to the first t-value in C_{i+1}, as follows.

$$C_1: x(t) = 0, \qquad y(t) = 2t, \qquad z(t) = 0, \qquad 0 \le t \le 1$$
$$C_2: x(t) = t - 1, \qquad y(t) = 2, \qquad z(t) = 0, \qquad 1 \le t \le 2$$
$$C_3: x(t) = 1, \qquad y(t) = 2, \qquad z(t) = t - 2, \quad 2 \le t \le 3$$

Therefore, C is given by

$$\mathbf{r}(t) = \begin{cases} 2t\,\mathbf{j}, & 0 \le t \le 1 \\ (t - 1)\mathbf{i} + 2\mathbf{j}, & 1 \le t \le 2 \\ \mathbf{i} + 2\mathbf{j} + (t - 2)\mathbf{k}, & 2 \le t \le 3. \end{cases}$$

Recall that the parameterization of a curve induces an **orientation** to the curve. For instance, in Example 1 the curve is oriented so that the positive direction is from $(0, 0, 0)$, following the curve to $(1, 2, 1)$. Try finding a parameterization that induces the opposite orientation.

Line integrals

At this point in the text, we have looked at various types of integrals. For a single integral $\int_a^b f(x)\,dx$, we integrate over the interval $[a, b]$. Similarly, for a double integral $\iint_R f(x, y)\,dA$, we integrate over the region R in the plane. We now introduce a new type of integral called a **line integral** for which we integrate over a piecewise smooth curve C. (The terminology is somewhat unfortunate—this type of integral might be better described as a "curve integral.")

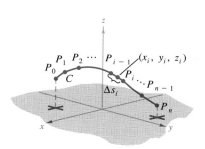

Partition of Curve C

FIGURE 17.8

To introduce the concept of a line integral, we consider the mass of a wire of finite length, given by a curve C in space. The density (mass per unit length) of the wire at the point (x, y, z) is given by $f(x, y, z)$. Suppose we partition the curve C by the points $P_0, P_1, \ldots, P_n$, producing n subarcs, as shown in Figure 17.8. The length of the ith subarc is given by Δs_i. Next, suppose that we choose a point (x_i, y_i, z_i) in each subarc. If the length of each subarc is small, then the total mass of the wire can be approximated by the sum

$$\text{Mass of wire} \approx \sum_{i=1}^{n} f(x_i, y_i, z_i)\Delta s_i.$$

Now, if we let $\|\Delta\|$ denote the length of the longest subarc and let $\|\Delta\|$ approach zero, it seems reasonable that the limit of this sum approaches the mass of the wire. This leads to the following definition.

DEFINITION OF LINE INTEGRAL

If f is defined in a region containing a smooth curve C of finite length, then the **line integral of f along C** is given by

$$\int_C f(x, y)\,ds = \lim_{\|\Delta\| \to 0} \sum_{i=1}^{n} f(x_i, y_i)\Delta s_i \qquad \text{Plane}$$

or

$$\int_C f(x, y, z)\,ds = \lim_{\|\Delta\| \to 0} \sum_{i=1}^{n} f(x_i, y_i, z_i)\Delta s_i \qquad \text{Space}$$

provided this limit exists.

As with the integrals discussed in Chapter 16, evaluation of a line integral is best accomplished by converting to a definite integral. It can be shown that if f is *continuous*, then the above limit exists and is the same for all parameterizations of C that are oriented in the same direction.

To evaluate a line integral over a plane curve C given by $\mathbf{r}(t) = x(t)\mathbf{i} + y(t)\mathbf{j}$, we use the fact that

$$ds = \|\mathbf{r}'(t)\|\, dt = \sqrt{[x'(t)]^2 + [y'(t)]^2}\, dt.$$

A similar formula holds for a space curve, as indicated in the following theorem.

THEOREM 17.4
EVALUATION OF A LINE INTEGRAL AS A DEFINITE INTEGRAL

Let f be continuous in a region containing a smooth curve C. If C is given by $\mathbf{r}(t) = x(t)\mathbf{i} + y(t)\mathbf{j}$, where $a \le t \le b$, then

$$\int_C f(x, y)\, ds = \int_a^b f(x(t), y(t)) \sqrt{[x'(t)]^2 + [y'(t)]^2}\, dt.$$

If C is given by $\mathbf{r}(t) = x(t)\mathbf{i} + y(t)\mathbf{j} + z(t)\mathbf{k}$, where $a \le t \le b$, then

$$\int_C f(x, y, z)\, ds = \int_a^b f(x(t), y(t), z(t)) \sqrt{[x'(t)]^2 + [y'(t)]^2 + [z'(t)]^2}\, dt.$$

REMARK   Note that if $f(x, y, z) = 1$, then the line integral gives the arc length of the curve C, as defined in Section 13.6. That is,

$$\int_C 1\, ds = \int_a^b \|\mathbf{r}'(t)\|\, dt = \text{length of curve } C.$$

EXAMPLE 2   Evaluating a line integral

Evaluate

$$\int_C (x^2 - y + 3z)\, ds$$

where C is the line segment shown in Figure 17.9.

SOLUTION

Using a parametric form of the equation of a line, we have

$$x = (1 - 0)t = t, \qquad y = (2 - 0)t = 2t, \qquad \text{and} \qquad z = (1 - 0)t = t$$

where $0 \le t \le 1$. Hence, $x'(t) = 1$, $y'(t) = 2$, and $z'(t) = 1$, which implies that

$$\sqrt{[x'(t)]^2 + [y'(t)]^2 + [z'(t)]^2} = \sqrt{1^2 + 2^2 + 1^2} = \sqrt{6}.$$

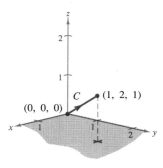

FIGURE 17.9

Thus, the line integral takes the following form.

$$\int_C (x^2 - y + 3z)\, ds = \int_0^1 (t^2 - 2t + 3t)\sqrt{6}\, dt$$

$$= \sqrt{6} \int_0^1 (t^2 + t)\, dt$$

$$= \sqrt{6} \left[\frac{t^3}{3} + \frac{t^2}{2} \right]_0^1 = \frac{5\sqrt{6}}{6} \qquad \blacksquare$$

Suppose C is a path composed of smooth curves $C_1, C_2, \ldots, C_n$. If f is continuous on C, then it can be shown that

$$\int_C f(x, y)\, ds = \int_{C_1} f(x, y)\, ds + \int_{C_2} f(x, y)\, ds + \cdots + \int_{C_n} f(x, y)\, ds.$$

We use this property in our next example.

EXAMPLE 3   Evaluating a line integral over a path

Evaluate

$$\int_C x\, ds$$

where C is the piecewise smooth curve shown in Figure 17.10.

SOLUTION

We first integrate up the line $y = x$, using the following parameterization.

$$C_1: x = t, \qquad y = t, \quad 0 \le t \le 1.$$

For this curve, we have $\mathbf{r}(t) = t\mathbf{i} + t\mathbf{j}$, which implies that $x'(t) = 1$ and $y'(t) = 1$. Thus, $\sqrt{[x'(t)]^2 + [y'(t)]^2} = \sqrt{2}$, and we have

$$\int_{C_1} x\, ds = \int_0^1 t\sqrt{2}\, dt = \frac{\sqrt{2}}{2} t^2 \Big]_0^1 = \frac{\sqrt{2}}{2}.$$

Next, we integrate down the parabola $y = x^2$, using the parameterization

$$C_2: x = 1 - t, \qquad y = (1 - t)^2, \quad 0 \le t \le 1.$$

For this curve, we have $\mathbf{r}(t) = (1 - t)\mathbf{i} + (1 - t)^2\mathbf{j}$, which implies that $x'(t) = -1$ and $y'(t) = -2(1 - t)$. Thus,

$$\sqrt{[x'(t)]^2 + [y'(t)]^2} = \sqrt{1 + 4(1 - t)^2},$$

and we have

$$\int_{C_2} x\, ds = \int_0^1 (1 - t)\sqrt{1 + 4(1 - t)^2}\, dt$$

$$= -\frac{1}{8} \left[\frac{2}{3}[1 + 4(1 - t)^2]^{3/2} \right]_0^1 = \frac{1}{12}(5^{3/2} - 1).$$

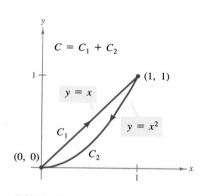

$C = C_1 + C_2$

$y = x$

$(1, 1)$

C_1

$y = x^2$

$(0, 0)$ C_2

FIGURE 17.10

Consequently,

$$\int_C x \, ds = \int_{C_1} x \, ds + \int_{C_2} x \, ds = \frac{\sqrt{2}}{2} + \frac{1}{12}(5^{3/2} - 1) \approx 1.56.$$ ▭

For parameterizations given by $\mathbf{r}(t) = x(t)\mathbf{i} + y(t)\mathbf{j} + z(t)\mathbf{k}$, it is helpful to remember the form of ds as

$$ds = \|\mathbf{r}'(t)\| \, dt = \sqrt{[x'(t)]^2 + [y'(t)]^2 + [z'(t)]^2} \, dt.$$

This is demonstrated in the next example.

EXAMPLE 4 Evaluating a line integral

Evaluate

$$\int_C (x + 2) \, ds$$

where C is the curve represented by

$$\mathbf{r}(t) = t\mathbf{i} + \frac{4}{3}t^{3/2}\mathbf{j} + \frac{1}{2}t^2\mathbf{k}, \quad 0 \le t \le 2.$$

SOLUTION

Since $\mathbf{r}'(t) = \mathbf{i} + 2t^{1/2}\mathbf{j} + t\mathbf{k}$, and

$$\|\mathbf{r}'(t)\| = \sqrt{[x'(t)]^2 + [y'(t)]^2 + [z'(t)]^2} = \sqrt{1 + 4t + t^2}$$

it follows that

$$\int_C (x + 2) \, ds = \int_0^2 (t + 2)\sqrt{1 + 4t + t^2} \, dt$$

$$= \frac{1}{2}\int_0^2 2(t + 2)(1 + 4t + t^2)^{1/2} \, dt$$

$$= \frac{1}{3}\left[(1 + 4t + t^2)^{3/2}\right]_0^2 = \frac{1}{3}(13\sqrt{13} - 1) \approx 15.29.$$ ▭

In the next example we compute the mass of a wire in the shape of a helix.

EXAMPLE 5 Finding the mass of a wire

Find the mass of a spring in the shape of the circular helix

$$\mathbf{r}(t) = \frac{1}{\sqrt{2}}(\cos t \, \mathbf{i} + \sin t \, \mathbf{j} + t\mathbf{k}), \quad 0 \le t \le 6\pi$$

where the density of the wire is $\rho(x, y, z) = 1 + z$, as shown in Figure 17.11.

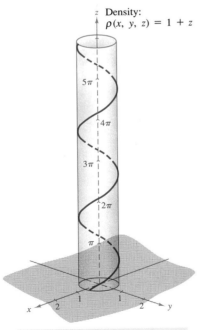

Density:
$\rho(x, y, z) = 1 + z$

$\mathbf{r}(t) = \dfrac{1}{\sqrt{2}}(\cos t \, \mathbf{i} + \sin t \, \mathbf{j} + t\mathbf{k})$

FIGURE 17.11

SOLUTION

Since

$$\|\mathbf{r}'(t)\| = \frac{1}{\sqrt{2}}\sqrt{(-\sin t)^2 + (\cos t)^2 + (1)^2} = 1$$

it follows that the mass of the spring is

$$
\begin{aligned}
\text{mass} = \int_C (1 + z)\, ds &= \int_0^{6\pi} \left(1 + \frac{t}{\sqrt{2}}\right) dt \\
&= \left[t + \frac{t^2}{2\sqrt{2}} \right]_0^{6\pi} \\
&= 6\pi \left(1 + \frac{3\pi}{\sqrt{2}}\right) \approx 144.47.
\end{aligned}
$$

Line integrals of vector fields

One of the most important physical applications of line integrals is that of finding the **work** done on an object moving in a force field. For example, Figure 17.12 shows an inverse square field similar to the gravitational field of the sun. Note that the magnitude of the force along a circular path about the center is constant, whereas the magnitude of the force along a parabolic path varies from point to point.

To see how a line integral can be used to find work done in a force field **F**, consider an object moving along a path C in the field, as shown in Figure 17.13. To determine the work done by the force, we need consider only that part of the force that is acting in the same direction as that in which the object is moving (or the opposite direction). This means that at each point on C, we consider the projection $\mathbf{F} \cdot \mathbf{T}$ of the force vector $\mathbf{F}$ onto the unit tangent vector $\mathbf{T}$. On a small subarc of length Δs_i, the increment of work is

$$\Delta W_i = (\text{force})(\text{distance}) \approx [\mathbf{F}(x_i, y_i, z_i) \cdot \mathbf{T}(x_i, y_i, z_i)]\Delta s_i$$

where (x_i, y_i, z_i) is a point in the ith subarc. Consequently, the total work done is given by the following integral.

$$W = \int_C \mathbf{F}(x, y, z) \cdot \mathbf{T}(x, y, z)\, ds$$

Inverse square force field **F**

Vectors along a parabolic
path in the force field **F**

FIGURE 17.12

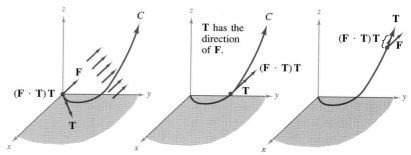

FIGURE 17.13

At each point of C, the force in the direction of motion is $(\mathbf{F} \cdot \mathbf{T})\mathbf{T}$.

This line integral appears in other contexts and is the basis of the following definition of the **line integral of a vector field.** Note in the definition that

$$\mathbf{F} \cdot \mathbf{T} \, ds = \mathbf{F} \cdot \frac{\mathbf{r}'(t)}{\|\mathbf{r}'(t)\|} \|\mathbf{r}'(t)\| \, dt = \mathbf{F} \cdot \mathbf{r}'(t) \, dt = \mathbf{F} \cdot d\mathbf{r}.$$

DEFINITION OF LINE INTEGRAL OF A VECTOR FIELD

Let **F** be a continuous vector field defined on a smooth curve C given by $\mathbf{r}(t)$. The **line integral** of **F** on C is given by

$$\int_C \mathbf{F} \cdot d\mathbf{r} = \int_C \mathbf{F} \cdot \mathbf{T} \, ds = \int_a^b \mathbf{F}(x(t), y(t), z(t)) \cdot \mathbf{r}'(t) \, dt.$$

EXAMPLE 6   Work done by a force

Find the work done by the force field

$$\mathbf{F}(x, y, z) = x\mathbf{i} - xy\mathbf{j} + z^2\mathbf{k}$$

in moving a particle along the helix given by $\mathbf{r}(t) = \cos t \, \mathbf{i} + \sin t \, \mathbf{j} + t\mathbf{k}$ from the point $(1, 0, 0)$ to $(-1, 0, 3\pi)$.

SOLUTION

Since $\mathbf{r}(t) = x(t)\mathbf{i} + y(t)\mathbf{j} + z(t)\mathbf{k} = \cos t \, \mathbf{i} + \sin t \, \mathbf{j} + t\mathbf{k}$, we have

$$\mathbf{r}'(t) = -\sin t \, \mathbf{i} + \cos t \, \mathbf{j} + \mathbf{k}$$
$$\mathbf{F}(x(t), y(t), z(t)) = \cos t \, \mathbf{i} - \cos t \sin t \, \mathbf{j} + t^2\mathbf{k}.$$

Therefore, the work is

$$W = \int_C \mathbf{F} \cdot d\mathbf{r} = \int_a^b \mathbf{F}(x(t), y(t), z(t)) \cdot \mathbf{r}'(t) \, dt$$

$$= \int_0^{3\pi} (\cos t \, \mathbf{i} - \cos t \sin t \, \mathbf{j} + t^2\mathbf{k}) \cdot (-\sin t \, \mathbf{i} + \cos t \, \mathbf{j} + \mathbf{k}) \, dt$$

$$= \int_0^{3\pi} (-\sin t \cos t - \sin t \cos^2 t + t^2) \, dt$$

$$= \left[\frac{\cos^2 t}{2} + \frac{\cos^3 t}{3} + \frac{t^3}{3} \right]_0^{3\pi} = 9\pi^3 - \frac{2}{3}.$$

For line integrals of vector functions, the orientation of the curve C is important. If the orientation of the curve is reversed, then the unit tangent vector $\mathbf{T}(t)$ is changed to $-\mathbf{T}(t)$, and we have

$$\int_{-C} \mathbf{F} \cdot d\mathbf{r} = -\int_C \mathbf{F} \cdot d\mathbf{r}.$$

Moreover, it can be shown that line integrals do not depend on the specific parameterization of the curve C. These ideas are illustrated in the next example.

EXAMPLE 7   Orientation and parameterization of a curve

C_1: $\mathbf{r}_1(t) = (4 - t)\mathbf{i} + (4t - t^2)\mathbf{j}$
C_2: $\mathbf{r}_2(t) = t\mathbf{i} + (4t - t^2)\mathbf{j}$

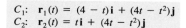

Let $\mathbf{F}(x, y) = y\mathbf{i} + x^2\mathbf{j}$ and evaluate the line integral $\int_C \mathbf{F} \cdot d\mathbf{r}$ for each of the following parabolic curves. (See Figure 17.14.)

(a) C_1: $\mathbf{r}_1(t) = (4 - t)\mathbf{i} + (4t - t^2)\mathbf{j}$, $0 \le t \le 3$
(b) C_2: $\mathbf{r}_2(t) = t\mathbf{i} + (4t - t^2)\mathbf{j}$, $1 \le t \le 4$

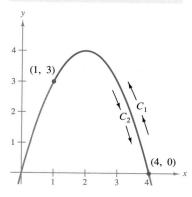

FIGURE 17.14

SOLUTION

(a) Since $\mathbf{r}_1'(t) = -\mathbf{i} + (4 - 2t)\mathbf{j}$ and $\mathbf{F}(x(t), y(t)) = (4t - t^2)\mathbf{i} + (4 - t)^2\mathbf{j}$, we have

$$\int_{C_1} \mathbf{F} \cdot d\mathbf{r} = \int_0^3 [(4t - t^2)\mathbf{i} + (4 - t)^2\mathbf{j}] \cdot [-\mathbf{i} + (4 - 2t)\mathbf{j}]\, dt$$

$$= \int_0^3 (-4t + t^2 + 64 - 64t + 20t^2 - 2t^3)\, dt$$

$$= \int_0^3 (-2t^3 + 21t^2 - 68t + 64)\, dt$$

$$= \left[-\frac{t^4}{2} + 7t^3 - 34t^2 + 64t\right]_0^3$$

$$= -\frac{81}{2} + 189 - 306 + 192 = \frac{69}{2}.$$

(b) Since $\mathbf{r}_2'(t) = \mathbf{i} + (4 - 2t)\mathbf{j}$ and $\mathbf{F}(x(t), y(t)) = (4t - t^2)\mathbf{i} + t^2\mathbf{j}$, we have

$$\int_{C_2} \mathbf{F} \cdot d\mathbf{r} = \int_1^4 [(4t - t^2)\mathbf{i} + t^2\mathbf{j}] \cdot [\mathbf{i} + (4 - 2t)\mathbf{j}]\, dt$$

$$= \int_1^4 (4t - t^2 + 4t^2 - 2t^3)\, dt$$

$$= \int_1^4 (-2t^3 + 3t^2 + 4t)\, dt$$

$$= \left[-\frac{t^4}{2} + t^3 + 2t^2\right]_1^4$$

$$= -\frac{69}{2}.$$

Note that the answer in part (b) is the negative of that found in part (a).

A second commonly used form of line integral is derived from the vector field notation used in the previous section. If **F** is a vector field of the form $\mathbf{F}(x, y) = M\mathbf{i} + N\mathbf{j}$, and C is given by $\mathbf{r}(t) = x(t)\mathbf{i} + y(t)\mathbf{j}$, then $\mathbf{F} \cdot d\mathbf{r}$ is often written as $M\, dx + N\, dy$, as follows.

$$\int_C \mathbf{F} \cdot d\mathbf{r} = \int_C \mathbf{F} \cdot \frac{d\mathbf{r}}{dt}\, dt = \int_a^b (M\mathbf{i} + N\mathbf{j}) \cdot (x'(t)\mathbf{i} + y'(t)\mathbf{j})\, dt$$

$$= \int_a^b \left(M\frac{dx}{dt} + N\frac{dy}{dt} \right) dt = \int_C (M\, dx + N\, dy)$$

This **differential form** can be extended to three variables. We often omit the parentheses and write

$$\int_C M\, dx + N\, dy \qquad \text{and} \qquad \int_C M\, dx + N\, dy + P\, dz.$$

Notice how this differential notation is used in the next example.

EXAMPLE 8 Evaluating a line integral in differential form

Let C be the circle of radius 3 given by

$$\mathbf{r}(t) = 3 \cos t\, \mathbf{i} + 3 \sin t\, \mathbf{j}, \quad 0 \leq t \leq 2\pi$$

and evaluate the line integral

$$\int_C y^3\, dx + (x^3 + 3xy^2)\, dy.$$

SOLUTION

Since $x = 3 \cos t$ and $y = 3 \sin t$, we have $dx = -3 \sin t\, dt$ and $dy = 3 \cos t\, dt$. Thus, the line integral is

$$\int_C M\, dx + N\, dy$$

$$= \int_C y^3\, dx + (x^3 + 3xy^2)\, dy$$

$$= \int_0^{2\pi} [(27 \sin^3 t)(-3 \sin t) + (27 \cos^3 t + 81 \cos t \sin^2 t)(3 \cos t)]\, dt$$

$$= 81 \int_0^{2\pi} (\cos^4 t - \sin^4 t + 3 \cos^2 t \sin^2 t)\, dt$$

$$= 81 \int_0^{2\pi} \left(\cos^2 t - \sin^2 t + \frac{3}{4} \sin^2 2t \right) dt$$

$$= 81 \int_0^{2\pi} \left[\cos 2t + \frac{3}{4}\left(\frac{1 - \cos 4t}{2} \right) \right] dt$$

$$= 81 \left[\frac{\sin 2t}{2} + \frac{3}{8}t - \frac{3 \sin 4t}{32} \right]_0^{2\pi} = \frac{243\pi}{4}.$$

For curves represented by $y = g(x)$, $a \leq x \leq b$, we can let $x = t$ and obtain the parametric form

$$x = t \quad \text{and} \quad y = g(t), \quad a \leq t \leq b.$$

Since $dx = dt$ for this form, we have the option of evaluating the line integral in the variable x or t. This is demonstrated in the next example.

EXAMPLE 9 Evaluating a line integral in differential form

Evaluate

$$\int_C y \, dx + x^2 \, dy$$

where C is the parabolic arc given by $y = 4x - x^2$ from $(4, 0)$ to $(1, 3)$, as shown in Figure 17.15.

SOLUTION

Rather than converting to the parameter t, we simply retain the variable x and write

$$y = 4x - x^2 \quad \Longrightarrow \quad dy = (4 - 2x) \, dx.$$

Then, in the direction from $(4, 0)$ to $(1, 3)$, the line integral is

$$\int_C y \, dx + x^2 \, dy = \int_4^1 [(4x - x^2) \, dx + x^2(4 - 2x) \, dx]$$

$$= \int_4^1 [4x + 3x^2 - 2x^3] \, dx$$

$$= \left[2x^2 + x^3 - \frac{x^4}{2} \right]_4^1 = \frac{69}{2}.$$

(See Example 7.)

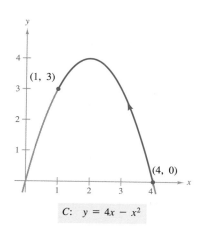

C: $y = 4x - x^2$

FIGURE 17.15

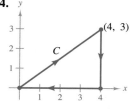

EXERCISES for Section 17.2

In Exercises 1–6, find a piecewise smooth parameter-ization of the path C.

1. $x^2 + y^2 = 9$

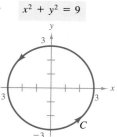

2. $\dfrac{x^2}{16} + \dfrac{y^2}{9} = 1$

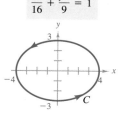

3.

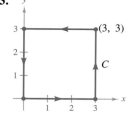

4.

5.

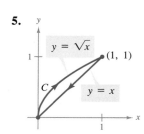

6.

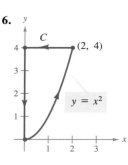

In Exercises 7–10, evaluate the line integral over the specified path.

7. $\displaystyle\int_C (x - y)\, ds$

C: $\mathbf{r}(t) = 4t\mathbf{i} + 3t\mathbf{j}$
$0 \le t \le 2$

8. $\displaystyle\int_C 4xy\, ds$

C: $\mathbf{r}(t) = t\mathbf{i} + (1 - t)\mathbf{j}$
$0 \le t \le 1$

9. $\displaystyle\int_C (x^2 + y^2 + z^2)\, ds$

C: $\mathbf{r}(t) = \sin t\,\mathbf{i} + \cos t\,\mathbf{j} + 8t\mathbf{k}$
$0 \le t \le \pi/2$

10. $\displaystyle\int_C 8xyz\, ds$

C: $\mathbf{r}(t) = 3\mathbf{i} + 12t\mathbf{j} + 5t\mathbf{k}$
$0 \le t \le 2$

In Exercises 11–14, evaluate

$$\int_C (x^2 + y^2)\, ds$$

along the given path.

11. C: x-axis from $x = 0$ to $x = 3$
12. C: y-axis from $y = 1$ to $y = 10$
13. C: counterclockwise around the circle $x^2 + y^2 = 1$ from $(1, 0)$ to $(0, 1)$
14. C: counterclockwise around the circle $x^2 + y^2 = 4$ from $(2, 0)$ to $(0, 2)$

In Exercises 15–18, evaluate

$$\int_C (x + 4\sqrt{y})\, ds$$

along the given path.

15. C: line from $(0, 0)$ to $(1, 1)$
16. C: line from $(0, 0)$ to $(3, 9)$
17. C: counterclockwise around the triangle with vertices $(0, 0)$, $(1, 0)$ and $(0, 1)$
18. C: counterclockwise around the square with vertices $(0, 0)$, $(1, 0)$ $(1, 1)$ and $(0, 1)$

In Exercises 19–24, evaluate

$$\int_C \mathbf{F} \cdot d\mathbf{r}$$

where C is represented by $\mathbf{r}(t)$.

19. $\mathbf{F}(x, y) = xy\mathbf{i} + y\mathbf{j}$
C: $\mathbf{r}(t) = 4t\mathbf{i} + t\mathbf{j}$,
$0 \le t \le 1$

20. $\mathbf{F}(x, y) = xy\mathbf{i} + y\mathbf{j}$
C: $\mathbf{r}(t) = 4 \cos t\,\mathbf{i} + 4 \sin t\,\mathbf{j}$,
$0 \le t \le \dfrac{\pi}{2}$

21. $\mathbf{F}(x, y) = 3x\mathbf{i} + 4y\mathbf{j}$
C: $\mathbf{r}(t) = 2 \cos t\,\mathbf{i} + 2 \sin t\,\mathbf{j}$,
$0 \le t \le \dfrac{\pi}{2}$

22. $\mathbf{F}(x, y) = 3x\mathbf{i} + 4y\mathbf{j}$
C: $\mathbf{r}(t) = t\mathbf{i} + \sqrt{4 - t^2}\,\mathbf{j}$,
$-2 \le t \le 2$

23. $\mathbf{F}(x, y, z) = x^2y\mathbf{i} + (x - z)\mathbf{j} + xyz\mathbf{k}$
C: $\mathbf{r}(t) = t\mathbf{i} + t^2\mathbf{j} + 2\mathbf{k}$,
$0 \le t \le 1$

24. $\mathbf{F}(x, y, z) = x^2\mathbf{i} + y^2\mathbf{j} + z^2\mathbf{k}$
C: $\mathbf{r}(t) = \sin t\,\mathbf{i} + \cos t\,\mathbf{j} + t^2\mathbf{k}$,
$0 \le t \le \dfrac{\pi}{2}$

In Exercises 25–28, evaluate the line integral over the path C given by $x = 2t$, $y = 10t$, where $0 \le t \le 1$.

25. $\displaystyle\int_C (x + 3y^2)\, dy$

26. $\displaystyle\int_C (x + 3y^2)\, dx$

27. $\displaystyle\int_C xy\, dx + y\, dy$

28. $\displaystyle\int_C (y - 3x)\, dx + x^2\, dy$

In Exercises 29–34, evaluate the integral

$$\int_C (2x - y)\, dx + (x + 3y)\, dy$$

along the given path.

29. C: x-axis from $x = 0$ to $x = 5$

30. C: y-axis from $y = 0$ to $y = 2$

31. C: line segments from $(0, 0)$ to $(3, 0)$ and from $(3, 0)$ to $(3, 3)$

32. C: line segments from $(0, 0)$ to $(0, -3)$ and from $(0, -3)$ to $(2, -3)$

33. C: parabolic path $x = t$, $y = 2t^2$, from $(0, 0)$ to $(2, 8)$

34. C: elliptic path $x = 4 \sin t$, $y = 3 \cos t$, from $(0, 3)$ to $(4, 0)$

In Exercises 35 and 36, find the total mass of two coils of a spring with density ρ in the shape of the circular helix

$$\mathbf{r}(t) = 3 \cos t\, \mathbf{i} + 3 \sin t\, \mathbf{j} + 2t \mathbf{k}.$$

35. $\rho(x, y, z) = \dfrac{1}{2}(x^2 + y^2 + z^2)$

36. $\rho(x, y, z) = 2$

In Exercises 37–42, find the work done by the force field $\mathbf{F}$ on an object moving along the specified path.

37. $\mathbf{F}(x, y) = -x\mathbf{i} - 2y\mathbf{j}$
C: $y = x^3$ from $(0, 0)$ to $(2, 8)$

38. $\mathbf{F}(x, y) = x^2\mathbf{i} - xy\mathbf{j}$
C: $x = \cos^3 t$, $y = \sin^3 t$ from $(1, 0)$ to $(0, 1)$

39. $\mathbf{F}(x, y) = 2x\mathbf{i} + y\mathbf{j}$
C: counterclockwise around the triangle whose vertices are $(0, 0)$, $(1, 0)$, and $(1, 1)$

40. $\mathbf{F}(x, y) = -y\mathbf{i} - x\mathbf{j}$
C: counterclockwise along the semicircle $y = \sqrt{4 - x^2}$ from $(2, 0)$ to $(-2, 0)$

41. $\mathbf{F}(x, y, z) = x\mathbf{i} + y\mathbf{j} - 5z\mathbf{k}$
C: $\mathbf{r}(t) = 2 \cos t\, \mathbf{i} + 2 \sin t\, \mathbf{j} + t\mathbf{k}$,
$0 \le t \le 2\pi$

42. $\mathbf{F}(x, y, z) = yz\mathbf{i} + xz\mathbf{j} + xy\mathbf{k}$
C: line from $(0, 0, 0)$ to $(5, 3, 2)$

In Exercises 43–46, demonstrate the property that

$$\int_C \mathbf{F} \cdot d\mathbf{r} = 0$$

regardless of the initial and terminal points of C, if the tangent vector $\mathbf{r}'(t)$ is orthogonal to the force vector $\mathbf{F}$.

43. $\mathbf{F}(x, y) = y\mathbf{i} - x\mathbf{j}$
C: $\mathbf{r}(t) = t\mathbf{i} - 2t\mathbf{j}$

44. $\mathbf{F}(x, y) = -3y\mathbf{i} + x\mathbf{j}$
C: $\mathbf{r}(t) = t\mathbf{i} + t^3\mathbf{j}$

45. $\mathbf{F}(x, y) = (x^3 - 2x^2)\mathbf{i} + \left(x - \dfrac{y}{2}\right)\mathbf{j}$
C: $\mathbf{r}(t) = t\mathbf{i} + t^2\mathbf{j}$

46. $\mathbf{F}(x, y) = x\mathbf{i} + y\mathbf{j}$
C: $\mathbf{r}(t) = 3 \sin t\, \mathbf{i} + 3 \cos t\, \mathbf{j}$

In Exercises 47–54, find the area of the lateral surface (see figure) over the curve C in the xy-plane and under the surface $z = f(x, y)$, where

$$\text{lateral surface area} = \int_C f(x, y)\, ds.$$

47. $f(x, y) = h$
C: line from $(0, 0)$ to $(3, 4)$

48. $f(x, y) = y$
C: line from $(0, 0)$ to $(4, 4)$

49. $f(x, y) = xy$
C: $x^2 + y^2 = 1$ from $(1, 0)$ to $(0, 1)$

50. $f(x, y) = x + y$
C: $x^2 + y^2 = 1$ from $(1, 0)$ to $(0, 1)$

51. $f(x, y) = h$
C: $y = 1 - x^2$ from $(1, 0)$ to $(0, 1)$

52. $f(x, y) = y + 1$
C: $y = 1 - x^2$ from $(1, 0)$ to $(0, 1)$

53. $f(x, y) = xy$
C: $y = 1 - x^2$ from $(1, 0)$ to $(0, 1)$

54. $f(x, y) = x^2 - y^2 + 4$
C: $x^2 + y^2 = 4$

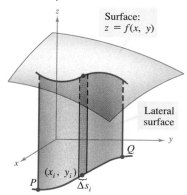

C: Curve in xy-plane

FIGURE FOR 47–54

55. Find the work done by a person weighing 150 pounds walking exactly one revolution up a circular helical staircase of radius 3 feet if the person rises 10 feet.

56. The roof on a building has a height above the floor given by $z = 20 + \frac{1}{4}x$, and one of the walls follows a path given by $y = x^{3/2}$. Find the surface area of the wall if $0 \le x \le 40$.

In Exercises 57 and 58, use a computer to approximate the integral

$$\int_C \mathbf{F} \cdot d\mathbf{r}$$

where C is represented by $\mathbf{r}(t)$.

57. $\mathbf{F}(x, y, z) = x^2z\mathbf{i} + 6y\mathbf{j} + yz^2\mathbf{k}$
$C: \mathbf{r}(t) = t\mathbf{i} + t^2\mathbf{j} + \ln t\,\mathbf{k}$
$\quad 1 \le t \le 3$

58. $\mathbf{F}(x, y, z) = \dfrac{x\mathbf{i} + y\mathbf{j} + z\mathbf{k}}{\sqrt{x^2 + y^2 + z^2}}$
$C: \mathbf{r}(t) = t\mathbf{i} + t\mathbf{j} + e^t\mathbf{k}$
$\quad 0 \le t \le 2$

17.3 Conservative Vector Fields and Independence of Path

Fundamental Theorem of Line Integrals ▪ Independence of path ▪ Conservation of energy

We began the previous section by stating that in a gravitational field the work done by gravity on an object moving between two points in the field is independent of the path taken by the object. In this section we will prove an important generalization of this result. The generalization is called the **Fundamental Theorem of Line Integrals.**

We begin with an example in which the line integral of a *conservative vector field* is evaluated over three different paths.

EXAMPLE 1 Line integral of a conservative vector field over different paths

Find the work done by the force field $\mathbf{F}(x, y) = 4xy\mathbf{i} + 2x^2\mathbf{j}$ in moving a particle from $(0, 0)$ to $(1, 1)$ along the following paths. (See Figure 17.16.)

(a) $C_1: y = x$ (b) $C_2: x = y^2$ (c) $C_3: y = x^3$

SOLUTION

(a) Let $\mathbf{r}(t) = t\mathbf{i} + t\mathbf{j}$ for $0 \le t \le 1$, so that
$$d\mathbf{r} = (\mathbf{i} + \mathbf{j})\,dt \quad \text{and} \quad \mathbf{F}(x, y) = 4t^2\mathbf{i} + 2t^2\mathbf{j}.$$

Then, the work done is
$$W = \int_{C_1} \mathbf{F} \cdot d\mathbf{r} = \int_0^1 (4t^2\mathbf{i} + 2t^2\mathbf{j}) \cdot (\mathbf{i} + \mathbf{j})\,dt$$
$$= \int_0^1 6t^2\,dt = 2t^3 \Big]_0^1 = 2.$$

(b) Let $\mathbf{r}(t) = t\mathbf{i} + \sqrt{t}\mathbf{j}$ for $0 \le t \le 1$, so that
$$d\mathbf{r} = \left(\mathbf{i} + \frac{1}{2\sqrt{t}}\mathbf{j}\right)dt \quad \text{and} \quad \mathbf{F}(x, y) = 4t^{3/2}\mathbf{i} + 2t^2\mathbf{j}.$$

Then, the work done is
$$W = \int_{C_2} \mathbf{F} \cdot d\mathbf{r} = \int_0^1 5t^{3/2}\,dt = 2t^{5/2} \Big]_0^1 = 2.$$

(c) Let $\mathbf{r}(t) = \frac{1}{2}t\mathbf{i} + \frac{1}{8}t^3\mathbf{j}$ for $0 \le t \le 2$, so that
$$d\mathbf{r} = \left(\frac{1}{2}\mathbf{i} + \frac{3}{8}t^2\mathbf{j}\right)dt \quad \text{and} \quad \mathbf{F}(x, y) = \frac{1}{4}t^4\mathbf{i} + \frac{1}{2}t^2\mathbf{j}.$$

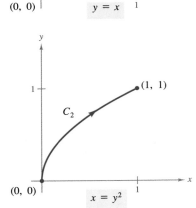

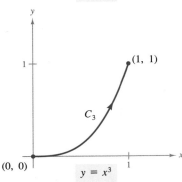

FIGURE 17.16

Then, the work done is

$$W = \int_{C_3} \mathbf{F} \cdot d\mathbf{r} = \int_0^2 \frac{5}{16}t^4 \, dt = \frac{1}{16}t^5 \Big]_0^2 = 2.$$

In Example 1(a), note that the vector field $\mathbf{F}(x, y) = 4xy\mathbf{i} + 2x^2\mathbf{j}$ is conservative because $\mathbf{F}(x, y) = \nabla f(x, y)$, where $f(x, y) = 2x^2y$. In such cases, the following theorem states that the value of $\int_C \mathbf{F} \cdot d\mathbf{r}$ is given by

$$\int_C \mathbf{F} \cdot d\mathbf{r} = f(x(1), y(1)) - f(x(0), y(0)) = 2 - 0 = 2.$$

THEOREM 17.5
FUNDAMENTAL THEOREM
OF LINE INTEGRALS

Let C be a piecewise smooth curve lying in an open region R and given by

$$\mathbf{r}(t) = x(t)\mathbf{i} + y(t)\mathbf{j}, \quad a \le t \le b.$$

If $\mathbf{F}(x, y) = M\mathbf{i} + N\mathbf{j}$ is conservative in R, and M and N are continuous in R, then

$$\int_C \mathbf{F} \cdot d\mathbf{r} = \int_C \nabla f \cdot d\mathbf{r} = f(x(b), y(b)) - f(x(a), y(a))$$

where f is a potential function of $\mathbf{F}$. That is, $\mathbf{F}(x, y) = \nabla f(x, y)$.

PROOF   We provide a proof only for a smooth curve. For piecewise smooth curves, the procedure is carried out separately on each smooth portion. Since $\mathbf{F}(x, y) = \nabla f(x, y) = f_x(x, y)\mathbf{i} + f_y(x, y)\mathbf{j}$, it follows that

$$\int_C \mathbf{F} \cdot d\mathbf{r} = \int_a^b \mathbf{F} \cdot \frac{d\mathbf{r}}{dt} \, dt = \int_a^b \left[f_x(x, y) \frac{dx}{dt} + f_y(x, y) \frac{dy}{dt} \right] dt$$

and, by the Chain Rule (Theorem 15.6), we have

$$\int_C \mathbf{F} \cdot d\mathbf{r} = \int_a^b \frac{d}{dt}[f(x(t), y(t))] \, dt = f(x(b), y(b)) - f(x(a), y(a)).$$

The last step is an application of the Fundamental Theorem of Calculus.

In space, the Fundamental Theorem of Line Integrals takes the following form. Let C be a piecewise smooth curve lying in an open region Q and given by $\mathbf{r}(t) = x(t)\mathbf{i} + y(t)\mathbf{j} + z(t)\mathbf{k}, \; a \le t \le b$. If $\mathbf{F}(x, y, z) = M\mathbf{i} + N\mathbf{j} + P\mathbf{k}$ is conservative and M, N, and P are continuous, then

$$\int_C \mathbf{F} \cdot d\mathbf{r} = \int_C \nabla f \cdot d\mathbf{r} = f(x(b), y(b), z(b)) - f(x(a), y(a), z(a))$$

where $\mathbf{F}(x, y, z) = \nabla f(x, y, z)$.

The Fundamental Theorem of Line Integrals states that if the vector field $\mathbf{F}$ is conservative, then the line integral between any two points is simply the difference in the values of the *potential* function f at these points.

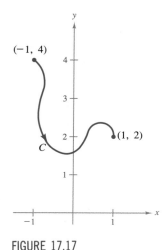

FIGURE 17.17

EXAMPLE 2   Using the Fundamental Theorem of Line Integrals

Evaluate

$$\int_C \mathbf{F} \cdot d\mathbf{r}$$

where C is a piecewise smooth curve from $(-1, 4)$ to $(1, 2)$ and $\mathbf{F}(x, y) = 2xy\mathbf{i} + (x^2 - y)\mathbf{j}$. (See Figure 17.17.)

SOLUTION

From Example 6 of Section 17.1 we know that $\mathbf{F}$ is the gradient of f where

$$f(x, y) = x^2 y - \frac{1}{2}y^2 + K.$$

Consequently, $\mathbf{F}$ is conservative, and by the Fundamental Theorem ·of Line Integrals, it follows that

$$\int_C \mathbf{F} \cdot d\mathbf{r} = f(1, 2) - f(-1, 4)$$

$$= \left[1^2(2) - \frac{1}{2}(2^2) \right] - \left[(-1)^2(4) - \frac{1}{2}(4^2) \right] = 4.$$

Note that it is unnecessary to include K as part of f, since it is cancelled by subtraction.   ▭

Independence of path

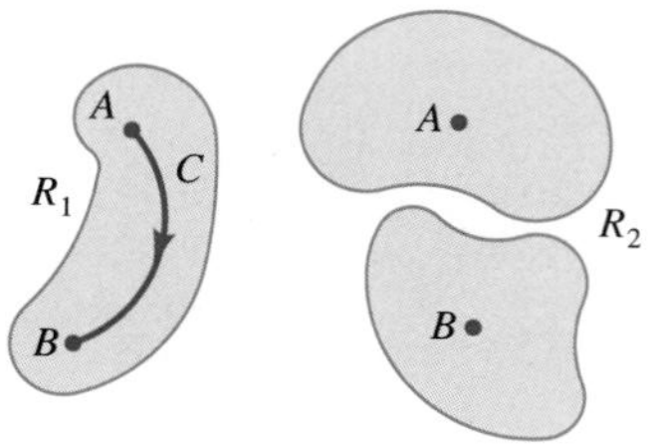

From the Fundamental Theorem of Line Integrals it is clear that if $\mathbf{F}$ is continuous and conservative in an open region R, then the value of $\int_C \mathbf{F} \cdot d\mathbf{r}$ is the same for every piecewise smooth curve C from one fixed point in R to another fixed point in R. We describe this result by saying that the line integral $\int_C \mathbf{F} \cdot d\mathbf{r}$ is **independent of path** in the region R. In open regions that are *connected*, we will be able to show that the path independence of $\int_C \mathbf{F} \cdot d\mathbf{r}$ is equivalent to the condition that $\mathbf{F}$ is conservative. We say a region in the plane (or in space) is **connected** if any two points in the region can be joined by a piecewise smooth curve lying entirely within the region, as shown in Figure 17.18.

R_1 is connected. R_2 is not connected.

FIGURE 17.18

THEOREM 17.6 **INDEPENDENCE OF PATH AND CONSERVATIVE VECTOR FIELDS**	If $\mathbf{F}$ is continuous on an open connected region, then the line integral $$\int_C \mathbf{F} \cdot d\mathbf{r}$$ is independent of path if and only if $\mathbf{F}$ is conservative.

PROOF

If **F** is conservative, then by the Fundamental Theorem of Line Integrals the line integral is independent of path. We establish the converse for a plane region R. Let $\mathbf{F}(x, y) = M\mathbf{i} + N\mathbf{j}$, and let (x_0, y_0) be a fixed point in R. If (x, y) is any point in R, then we choose a piecewise smooth curve C running from (x_0, y_0) to (x, y), and define f by

$$f(x, y) = \int_C \mathbf{F} \cdot d\mathbf{r} = \int_C M\ dx + N\ dy.$$

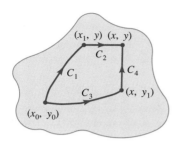

FIGURE 17.19

The existence of C in R is guaranteed by the fact that R is connected. We will show that f is a potential function of **F** by considering two different paths between (x_0, y_0) and (x, y). For the *first* path, we choose (x_1, y) in R such that $x \neq x_1$. This is possible since R is open. Then we choose C_1 and C_2, as shown in Figure 17.19, and using the independence of path, we write

$$f(x, y) = \int_C M\ dx + N\ dy = \int_{C_1} M\ dx + N\ dy + \int_{C_2} M\ dx + N\ dy.$$

Since the first integral does not depend on x, and $dy = 0$ in the second integral, we have

$$f(x, y) = g(y) + \int_{C_2} M\ dx$$

and it follows that the partial derivative of f with respect to x is $f_x(x, y) = M$. For the *second* path, we choose a point (x, y_1), and following a similar reasoning we conclude that $f_y(x, y) = N$. Therefore,

$$\begin{aligned} \nabla f(x, y) &= f_x(x, y)\mathbf{i} + f_y(x, y)\mathbf{j} \\ &= M\mathbf{i} + N\mathbf{j} = \mathbf{F}(x, y) \end{aligned}$$

and it follows that **F** is conservative.

EXAMPLE 3   Finding work in a conservative force field

For the force field given by $\mathbf{F}(x, y, z) = e^x \cos y\ \mathbf{i} - e^x \sin y\ \mathbf{j} + 2\mathbf{k}$, show that $\int_C \mathbf{F} \cdot d\mathbf{r}$ is independent of path, and calculate the work done by **F** on an object moving along a curve C from $(0, \pi/2, 1)$ to $(1, \pi, 3)$.

SOLUTION

Writing the force field in the form $\mathbf{F}(x, y, z) = M\mathbf{i} + N\mathbf{j} + P\mathbf{k}$, we have $M = e^x \cos y$, $N = -e^x \sin y$, and $P = 2$, and it follows that

$$\frac{\partial P}{\partial y} = 0 = \frac{\partial N}{\partial z}$$

$$\frac{\partial P}{\partial x} = 0 = \frac{\partial M}{\partial z}$$

$$\frac{\partial N}{\partial x} = -e^x \sin y = \frac{\partial M}{\partial y}.$$

Hence, **F** is conservative. If f is a potential function of **F**, then

$$f_x(x, y, z) = e^x \cos y$$
$$f_y(x, y, z) = -e^x \sin y$$
$$f_z(x, y, z) = 2.$$

By integrating with respect to x, y, and z separately, we obtain

$$f(x, y, z) = \int f_x(x, y, z)\, dx = \int e^x \cos y\, dx = e^x \cos y + g(y, z) + K$$

$$f(x, y, z) = \int f_y(x, y, z)\, dy = \int -e^x \sin y\, dy = e^x \cos y + h(x, z) + K$$

$$f(x, y, z) = \int f_z(x, y, z)\, dz = \int 2\, dz = 2z + k(x, y) + K.$$

By comparing these three versions of $f(x, y, z)$, we conclude that

$$f(x, y, z) = e^x \cos y + 2z + K.$$

Therefore, the work done by **F** along *any* curve C from $(0,\ \pi/2,\ 1)$ to $(1,\ \pi,\ 3)$ is

$$W = \int_C \mathbf{F} \cdot d\mathbf{r} = \left[e^x \cos y + 2z \right]_{(0,\pi/2,1)}^{(1,\pi,3)}$$

$$= (-e + 6) - (0 + 2) = 4 - e.$$

How much work would be done if the object in Example 3 moved from the point $(0,\ \pi/2,\ 1)$ to $(1,\ \pi,\ 3)$ and then back to the starting point $(0,\ \pi/2,\ 1)$? The Fundamental Theorem tells us that there is zero work done. Remember that by definition work can be negative. Hence, by the time the object gets back to its starting point, the amount of work that registers positively is canceled out by the amount of work that registers negatively. We say that a curve that has the same initial and terminal point is **closed,** as shown in Figure 17.20. By the Fundamental Theorem, we conclude that if **F** is continuous and conservative on an open region R, then the line integral over every closed curve R is zero. We summarize this result in the following theorem.

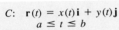

C: $\mathbf{r}(t) = x(t)\mathbf{i} + y(t)\mathbf{j}$
  $a \le t \le b$

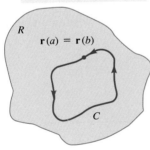

Closed Curve

FIGURE 17.20

THEOREM 17.7
EQUIVALENT CONDITIONS

Let $\mathbf{F}(x, y, z) = M\mathbf{i} + N\mathbf{j} + P\mathbf{k}$ have continuous first partial derivatives in an open connected region R, and let C be a piecewise smooth curve in R. The following conditions are equivalent.

1. **F** is conservative. That is, $\mathbf{F} = \nabla f$ for some function f.

2. $\displaystyle\int_C \mathbf{F} \cdot d\mathbf{r}$ is independent of path.

3. $\displaystyle\int_C \mathbf{F} \cdot d\mathbf{r} = 0$ for every *closed* curve C in R.

Theorem 17.7 tells us that we have some options when evaluating a line integral involving a conservative vector field. We can use a potential function, or it might be more convenient to choose a particularly simple path, such as a straight line.

EXAMPLE 4   Evaluating a line integral

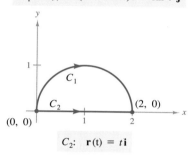

C_1: $\mathbf{r}(t) = (1 - \cos t)\mathbf{i} + \sin t\,\mathbf{j}$

C_2: $\mathbf{r}(t) = t\mathbf{i}$

FIGURE 17.21

Evaluate

$$\int_{C_1} \mathbf{F} \cdot d\mathbf{r}$$

where $\mathbf{F}(x, y) = (y^3 + 1)\mathbf{i} + (3xy^2 + 1)\mathbf{j}$ and C_1 is the semicircular path from (0, 0) to (2, 0), as shown in Figure 17.21.

SOLUTION

We have the following three options.
(a) We can use the method of the previous section to evaluate the line integral along the *given curve*. To do this, we use the parameterization

$$\mathbf{r}(t) = (1 - \cos t)\mathbf{i} + \sin t\,\mathbf{j}, \quad 0 \leq t \leq \pi$$

so that

$$d\mathbf{r} = \mathbf{r}'(t)\,dt = (\sin t\,\mathbf{i} + \cos t\,\mathbf{j})\,dt$$

and

$$\mathbf{F}(x, y) = (\sin^3 t + 1)\mathbf{i} + [1 + 3\sin^2 t - 3\cos t\,\sin^2 t]\,\mathbf{j}.$$

Thus,

$$\int_{C_1} \mathbf{F} \cdot d\mathbf{r} = \int_0^\pi [\sin t + \sin^4 t + \cos t + 3\sin^2 t\,\cos t - 3\cos^2 t\,\sin^2 t]\,dt.$$

The sight of this integral dampens our enthusiasm for this option.
(b) We can try to find a *potential function* and evaluate the line integral by the Fundamental Theorem. For example, if f is a potential function of $\mathbf{F}$, then $f_x(x, y) = y^3 + 1$ and $f_y(x, y) = 3xy^2 + 1$, and we have

$$f(x, y) = \int f_x(x, y)\,dx = \int (y^3 + 1)\,dx = xy^3 + x + g(y) + K$$

$$f(x, y) = \int f_y(x, y)\,dy = \int (3xy^2 + 1)\,dy = xy^3 + y + h(x) + K.$$

Thus, $f(x, y) = xy^3 + x + y + K$, and, by the Fundamental Theorem, we have

$$W = \int_{C_1} \mathbf{F} \cdot d\mathbf{r} = f(2, 0) - f(0, 0) = 2.$$

(c) Knowing that **F** is conservative, we have a third option. Since the value of the line integral is independent of path, we can replace the semicircular path with a *simpler path*. Suppose we choose the straight line path C_2 from (0, 0) to (2, 0). Then, we have

$$\mathbf{r}(t) = t\mathbf{i}, \quad 0 \le t \le 2.$$

Thus, $d\mathbf{r} = \mathbf{i}\, dt$ and $\mathbf{F}(x, y) = (y^3 + 1)\mathbf{i} + (3xy^2 + 1)\mathbf{j} = \mathbf{i} + \mathbf{j}$, so that

$$\int_{C_1} \mathbf{F} \cdot d\mathbf{r} = \int_{C_2} \mathbf{F} \cdot d\mathbf{r} = \int_0^2 1 \, dt = t \Big]_0^2 = 2.$$

Of the three options, obviously the third one is the easiest. ▭

Conservation of energy

Michael Faraday

In 1840, the English physicist Michael Faraday (1791–1867) wrote, "Nowhere is there a pure creation or production of power without a corresponding exhaustion of something to supply it." This statement represents the first formulation of one of the most important laws of physics—the **Law of Conservation of Energy.** Several philosophers of science have considered this law as the greatest generalization ever conceived by humankind. Many physicists have contributed to our knowledge of this law. Two early and influential ones were James Prescott Joule (1818–1889) and Hermann Ludwig Helmholtz (1821–1894). In modern terminology, we state the law as follows: *In a conservative force field, the sum of the potential and kinetic energies of an object remains constant from point to point.*

We can use the Fundamental Theorem of Line Integrals to derive this law. From physics, the **kinetic energy** of a particle of mass m and speed v is $k = \frac{1}{2}mv^2$. The **potential energy** p of a particle at point (x, y, z) in a conservative vector field **F** is defined as $p(x, y, z) = -f(x, y, z)$, where f is the potential function for **F**. Consequently, the work done by **F** along a smooth curve C from A to B is

$$W = \int_C \mathbf{F} \cdot d\mathbf{r} = f(x, y, z) \Big]_A^B = -p(x, y, z) \Big]_A^B = p(A) - p(B)$$

as indicated in Figure 17.22. In other words, work W is equal to the difference in the potential energies of A and B. Now, suppose $\mathbf{r}(t)$ is the position vector

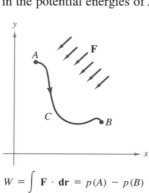

$$W = \int_C \mathbf{F} \cdot d\mathbf{r} = p(A) - p(B)$$

FIGURE 17.22

for a particle moving along C from $A = \mathbf{r}(a)$ to $B = \mathbf{r}(b)$. At any time t, the particle's velocity, acceleration, and speed are $\mathbf{v}(t) = \mathbf{r}'(t)$, $\mathbf{a}(t) = \mathbf{r}''(t)$, and $v(t) = \|\mathbf{v}(t)\|$, respectively. Thus, by Newton's Second Law of Motion, $\mathbf{F} = m\,\mathbf{a}(t) = m(\mathbf{v}'(t))$, and the work done by $\mathbf{F}$ is

$$
\begin{aligned}
W = \int_C \mathbf{F} \cdot d\mathbf{r} &= \int_a^b \mathbf{F} \cdot \mathbf{r}'(t)\, dt \\
&= \int_a^b \mathbf{F} \cdot \mathbf{v}(t)\, dt = \int_a^b [m\,\mathbf{v}'(t)] \cdot \mathbf{v}(t)\, dt \\
&= \int_a^b m[\mathbf{v}'(t) \cdot \mathbf{v}(t)]\, dt \\
&= \frac{m}{2} \int_a^b \frac{d}{dt}[\mathbf{v}(t) \cdot \mathbf{v}(t)]\, dt \\
&= \frac{m}{2} \int_a^b \frac{d}{dt}[\|\mathbf{v}(t)\|^2]\, dt = \frac{m}{2}\|\mathbf{v}(t)\|^2 \Big]_a^b \\
&= \frac{m}{2}[v(t)]^2 \Big]_a^b = \frac{1}{2}m[v(b)]^2 - \frac{1}{2}m[v(a)]^2 \\
&= k(B) - k(A).
\end{aligned}
$$

Equating these two results for W, we obtain

$$
p(A) - p(B) = k(B) - k(A) \quad\Longrightarrow\quad p(A) + k(A) = p(B) + k(B)
$$

which implies that the sum of the potential and kinetic energies remains constant from point to point.

EXERCISES for Section 17.3

In Exercises 1–4, show that the value of $\int_C \mathbf{F} \cdot d\mathbf{r}$ is the same for the given parametric representations of C.

1. $\mathbf{F}(x, y) = x^2\mathbf{i} + xy\mathbf{j}$
   (a) $\mathbf{r}_1(t) = t\mathbf{i} + t^2\mathbf{j}$, $0 \le t \le 1$

   (b) $r_2(\theta) = \sin\theta\,\mathbf{i} + \sin^2\theta\,\mathbf{j}$, $0 \le \theta \le \dfrac{\pi}{2}$

2. $\mathbf{F}(x, y) = (x^2 + y^2)\mathbf{i} - x\mathbf{j}$
   (a) $\mathbf{r}_1(t) = t\mathbf{i} + \sqrt{t}\,\mathbf{j}$, $0 \le t \le 4$
   (b) $\mathbf{r}_2(w) = w^2\mathbf{i} + w\mathbf{j}$, $0 \le w \le 2$

3. $\mathbf{F}(x, y) = y\mathbf{i} - x\mathbf{j}$

   (a) $\mathbf{r}_1(\theta) = \sec\theta\,\mathbf{i} + \tan\theta\,\mathbf{j}$, $0 \le \theta \le \dfrac{\pi}{3}$

   (b) $\mathbf{r}_2(t) = \sqrt{t+1}\,\mathbf{i} + \sqrt{t}\,\mathbf{j}$, $0 \le t \le 3$

4. $\mathbf{F}(x, y) = y\mathbf{i} - x^2\mathbf{j}$
   (a) $\mathbf{r}_1(t) = (2 + t)\mathbf{i} + (3 - t)\mathbf{j}$, $0 \le t \le 3$
   (b) $\mathbf{r}_2(w) = (2 + \ln w)\,\mathbf{i} + (3 - \ln w)\,\mathbf{j}$, $1 \le w \le e^3$

In Exercises 5–18, find the value of the line integral $\int_C \mathbf{F} \cdot d\mathbf{r}$. [Hint: If the integral is independent of path, the integration may be easier on an alternate path.]

5. $\mathbf{F}(x, y) = 2xy\mathbf{i} + x^2\mathbf{j}$
   (a) $\mathbf{r}_1(t) = t\mathbf{i} + t^2\mathbf{j}$, $0 \le t \le 1$
   (b) $\mathbf{r}_2(t) = t\mathbf{i} + t^3\mathbf{j}$, $0 \le t \le 1$
6. $\mathbf{F}(x, y) = ye^{xy}\mathbf{i} + xe^{xy}\mathbf{j}$

   (a) $\mathbf{r}_1(t) = t\mathbf{i} - \dfrac{3}{2}(t - 2)\mathbf{j}$, $0 \le t \le 2$

   (b) line segments from $(0, 3)$ to $(0, 0)$, and then from $(0, 0)$ to $(2, 0)$
7. $\mathbf{F}(x, y) = y\mathbf{i} - x\mathbf{j}$
   (a) $\mathbf{r}_1(t) = t\mathbf{i} + t\mathbf{j}$, $0 \le t \le 1$
   (b) $\mathbf{r}_2(t) = t\mathbf{i} + t^2\mathbf{j}$, $0 \le t \le 1$
   (c) $\mathbf{r}_3(t) = t\mathbf{i} + t^3\mathbf{j}$, $0 \le t \le 1$
8. $\mathbf{F}(x, y) = xy^2\mathbf{i} + 2x^2y\mathbf{j}$

   (a) $\mathbf{r}_1(t) = t\mathbf{i} + \dfrac{1}{t}\mathbf{j}$, $1 \le t \le 3$

   (b) $\mathbf{r}_2(t) = (t + 1)\mathbf{i} - \dfrac{1}{3}(t - 3)\mathbf{j}$, $0 \le t \le 2$

9. $\displaystyle\int_C y^2\,dx + 2xy\,dy$

(a)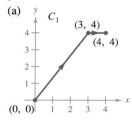

(b) $y = \sqrt{1 - x^2}$

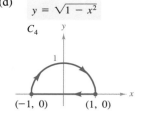

(c)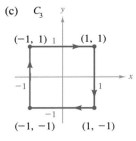

(d) $y = \sqrt{1 - x^2}$

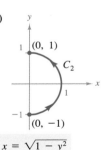

10. $\displaystyle\int_C (2x - 3y + 1)\,dx - (3x + y - 5)\,dy$

(a)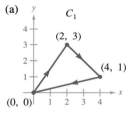

(b)

(c)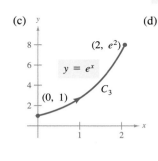

(d) $x = \sqrt{1 - y^2}$

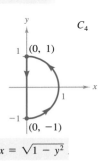

$x = \sqrt{1 - y^2}$

11. $\displaystyle\int_C 2xy\,dx + (x^2 + y^2)\,dy$

(a) C: ellipse $\dfrac{x^2}{25} + \dfrac{y^2}{16} = 1$ from $(5, 0)$ to $(0, 4)$

(b) C: parabola $y = 4 - x^2$ from $(2, 0)$ to $(0, 4)$

12. $\displaystyle\int_C (x^2 + y^2)\,dx + 2xy\,dy$

(a) $\mathbf{r}_1(t) = t^3\mathbf{i} + t^2\mathbf{j},\ 0 \le t \le 2$

(b) $\mathbf{r}_2(t) = 2\cos t\,\mathbf{i} + 2\sin t\,\mathbf{j},\ 0 \le t \le \dfrac{\pi}{2}$

13. $\mathbf{F}(x, y, z) = yz\,\mathbf{i} + xz\,\mathbf{j} + xy\,\mathbf{k}$
    (a) $\mathbf{r}_1(t) = t\mathbf{i} + 2\mathbf{j} + t\mathbf{k},\ 0 \le t \le 4$
    (b) $\mathbf{r}_2(t) = t^2\mathbf{i} + t\mathbf{j} + t^2\mathbf{k},\ 0 \le t \le 2$

14. $\mathbf{F}(x, y, z) = \mathbf{i} + z\mathbf{j} + y\mathbf{k}$
    (a) $\mathbf{r}_1(t) = \cos t\,\mathbf{i} + \sin t\,\mathbf{j} + t^2\mathbf{k},\ 0 \le t \le \pi$
    (b) $\mathbf{r}_2(t) = (1 - 2t)\mathbf{i} + \pi^2 t\mathbf{k},\ 0 \le t \le 1$

15. $\mathbf{F}(x, y, z) = (2y + x)\mathbf{i} + (x^2 - z)\mathbf{j} + (2y - 4z)\mathbf{k}$
    (a) $\mathbf{r}_1(t) = t\mathbf{i} + t^2\mathbf{j} + \mathbf{k},\ 0 \le t \le 1$
    (b) $\mathbf{r}_2(t) = t\mathbf{i} + t\mathbf{j} + (2t - 1)^2\mathbf{k},\ 0 \le t \le 1$

16. $\mathbf{F}(x, y, z) = -y\mathbf{i} + x\mathbf{j} + 3xz^2\mathbf{k}$
    (a) $\mathbf{r}_1(t) = \cos t\,\mathbf{i} + \sin t\,\mathbf{j} + t\mathbf{k},\ 0 \le t \le \pi$
    (b) $\mathbf{r}_2(t) = (1 - 2t)\mathbf{i} + \pi t\mathbf{k},\ 0 \le t \le 1$

17. $\mathbf{F}(x, y, z) = e^z(y\mathbf{i} + x\mathbf{j} + xy\mathbf{k})$
    (a) $\mathbf{r}_1(t) = 4\cos t\,\mathbf{i} + 4\sin t\,\mathbf{j} + 3\mathbf{k},\ 0 \le t \le \pi$
    (b) $\mathbf{r}_2(t) = (4 - 8t)\mathbf{i} + 3\mathbf{k},\ 0 \le t \le 1$

18. $\mathbf{F}(x, y, z) = y\sin z\,\mathbf{i} + x\sin z\,\mathbf{j} + xy\cos z\,\mathbf{k}$
    (a) $\mathbf{r}_1(t) = t^2\mathbf{i} + t^2\mathbf{j},\ 0 \le t \le 2$
    (b) $\mathbf{r}_2(t) = 4t\mathbf{i} + 4t\mathbf{j},\ 0 \le t \le 1$

In Exercises 19–28, evaluate the given line integral using the Fundamental Theorem of Line Integrals.

19. $\displaystyle\int_C (y\mathbf{i} + x\mathbf{j}) \cdot d\mathbf{r}$

    C: smooth curve from $(0, 0)$ to $(3, 8)$

20. $\displaystyle\int_C [2(x + y)\mathbf{i} + 2(x + y)\mathbf{j}] \cdot d\mathbf{r}$

    C: smooth curve from $(-1, 1)$ to $(3, 2)$

21. $\displaystyle\int_{(0,-\pi)}^{(3\pi/2,\,\pi/2)} \cos x \sin y\,dx + \sin x \cos y\,dy$

22. $\displaystyle\int_{(1,1)}^{(2\sqrt{3},2)} \frac{y\,dx - x\,dy}{x^2 + y^2}$

23. $\displaystyle\int_C e^x \sin y\,dx + e^x \cos y\,dy$

    C: cycloid $x = \theta - \sin\theta,\ y = 1 - \cos\theta$ from $(0, 0)$ to $(2\pi, 0)$

24. $\displaystyle\int_C \frac{2x}{(x^2 + y^2)^2}\,dx + \frac{2y}{(x^2 + y^2)^2}\,dy$

    C: circle $(x - 4)^2 + (y - 5)^2 = 9$
    clockwise from $(7, 5)$ to $(1, 5)$

25. $\displaystyle\int_C (z + 2y)\,dx + (2x - z)\,dy + (x - y)\,dz$

    (a) C: line segment from $(0, 0, 0)$ to $(1, 1, 1)$
    (b) C: line segments from $(0, 0, 0)$ to $(0, 0, 1)$ to $(1, 1, 1)$
    (c) C: line segments from $(0, 0, 0)$ to $(1, 0, 0)$ to $(1, 1, 0)$ to $(1, 1, 1)$

26. Repeat Exercise 25 using the integral

$$\int_C zy\,dx + xz\,dy + xy\,dz.$$

27. $\displaystyle\int_{(0,0,0)}^{(\pi/2,3,4)} -\sin x \, dx + z \, dy + y \, dz$

28. $\displaystyle\int_{(0,0,0)}^{(3,4,0)} 6x \, dx - 4z \, dy - (4y - 20z) \, dz$

In Exercises 29 and 30, find the work done by the force field **F** in moving an object from P to Q.

29. $\mathbf{F}(x, y) = 9x^2y^2\mathbf{i} + (6x^3y - 1)\mathbf{j}$
$P = (0, 0), \, Q = (5, 9)$

30. $\mathbf{F}(x, y) = \dfrac{2x}{y}\mathbf{i} - \dfrac{x^2}{y^2}\mathbf{j}$
$P = (-1, 1), \, Q = (3, 2)$

31. A stone weighing 1 pound is attached to the end of a 2-foot string and is whirled horizontally with one end held fixed. It makes one revolution per second. Find the work done by the force **F** that keeps the stone moving in a circular path. [Hint: Use force = (mass)(centripetal acceleration).]

32. If $\mathbf{F}(x, y, z) = a_1\mathbf{i} + a_2\mathbf{j} + a_3\mathbf{k}$ is a constant force vector field, show that the work done in moving a particle along any path from P to Q is
$$W = \mathbf{F} \cdot \overrightarrow{PQ}.$$

33. The kinetic energy of an object moving through a conservative force field is decreasing at a rate of 10 units per minute. At what rate is the potential energy changing?

34. Let $\mathbf{F}(x, y) = \dfrac{y}{x^2 + y^2}\mathbf{i} - \dfrac{x}{x^2 + y^2}\mathbf{j}$.

(a) Show that
$$\frac{\partial N}{\partial x} = \frac{\partial M}{\partial y}$$
where
$$M = \frac{y}{x^2 + y^2} \text{ and }$$
$$N = \frac{-x}{x^2 + y^2}.$$

(b) If $\mathbf{r}(t) = \cos t\,\mathbf{i} + \sin t\,\mathbf{j}$, for $0 \le t \le \pi$, find $\int_C \mathbf{F} \cdot d\mathbf{r}$.

(c) If $\mathbf{r}(t) = \cos t\,\mathbf{i} - \sin t\,\mathbf{j}$, for $0 \le t \le \pi$, find $\int_C \mathbf{F} \cdot d\mathbf{r}$.

(d) If $\mathbf{r}(t) = \cos t\,\mathbf{i} + \sin t\,\mathbf{j}$, for $0 \le t \le 2\pi$, find $\int_C \mathbf{F} \cdot d\mathbf{r}$. Why doesn't this contradict Theorem 17.7?

17.4 Green's Theorem

Green's Theorem ▪ Applications of Green's Theorem ▪ Alternate forms of Green's Theorem

In this section we discuss a theorem named after the English mathematician George Green (1793–1841). Taken alone, the theorem is fascinating enough—it states that the value of a double integral over a *simply connected* plane region R is determined by the value of a line integral around the boundary of R. But the result becomes even more impressive when we see some of the loose ends this theorem allows us to collect.

A plane curve C given by
$$\mathbf{r}(t) = x(t)\mathbf{i} + y(t)\mathbf{j}, \quad a \le t \le b$$
is **simple** if it does not cross itself. A plane region R is **simply connected** if its boundary consists of *one* simple closed curve, as shown in Figure 17.23.

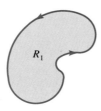

Simply Connected

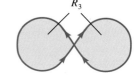

Not Simply Connected

FIGURE 17.23

THEOREM 17.8 GREEN'S THEOREM	Let R be a simply connected region with a piecewise smooth boundary C, oriented counterclockwise (that is, C is traversed *once* so that the region R always lies to the *left*). If $M, N, \partial M/\partial y$ and $\partial N/\partial x$ are all continuous in an open region containing R, then $$\int_C M\,dx + N\,dy = \iint_R \left(\frac{\partial N}{\partial x} - \frac{\partial M}{\partial y}\right)\,dA.$$

PROOF

We give a proof only for a region that is both vertically simple and horizontally simple. Thus, R is described by the two forms

$$R: a \le x \le b, f_1(x) \le y \le f_2(x) \qquad \text{Figure 17.24}$$

and

$$R: c \le y \le d, g_1(y) \le x \le g_2(y). \qquad \text{Figure 17.25}$$

The line integral $\int_C M\,dx$ can be written as

$$\int_C M\,dx = \int_{C_1} M\,dx + \int_{C_2} M\,dx = \int_a^b M(x, f_1(x))\,dx + \int_b^a M(x, f_2(x))\,dx$$

$$= \int_a^b [M(x, f_1(x)) - M(x, f_2(x))]\,dx.$$

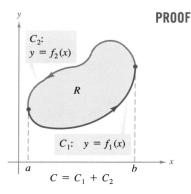

$C = C_1 + C_2$

R is vertically simple.

FIGURE 17.24

On the other hand, we have

$$\iint_R \frac{\partial M}{\partial y}\,dA = \int_a^b \int_{f_1(x)}^{f_2(x)} \frac{\partial M}{\partial y}\,dy\,dx = \int_a^b M(x, y)\Big]_{f_1(x)}^{f_2(x)}\,dx$$

$$= \int_a^b [M(x, f_2(x)) - M(x, f_1(x))]\,dx.$$

Consequently,

$$\int_C M\,dx = -\iint_R \frac{\partial M}{\partial y}\,dA.$$

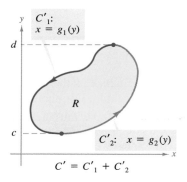

$C' = C'_1 + C'_2$

R is horizontally simple.

FIGURE 17.25

Similarly, we can use $g_1(y)$ and $g_2(y)$ to show that

$$\int_C N\,dy = \iint_R \frac{\partial N}{\partial x}\,dA.$$

Thus, we obtain

$$\int_C M\,dx + N\,dy = -\iint_R \frac{\partial M}{\partial y}\,dA + \iint_R \frac{\partial N}{\partial x}\,dA = \iint_R \left(\frac{\partial N}{\partial x} - \frac{\partial M}{\partial y}\right)\,dA.$$

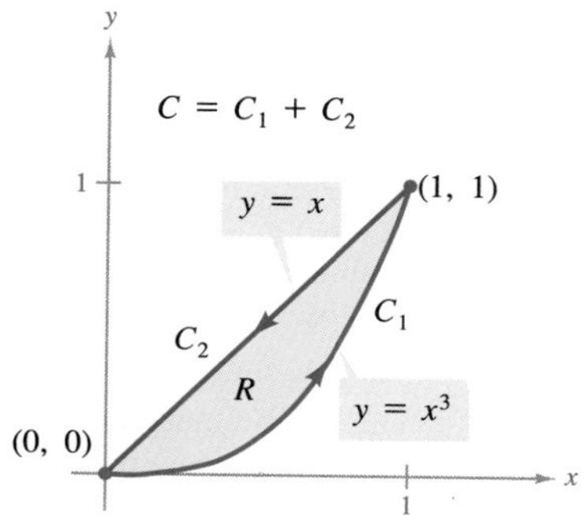

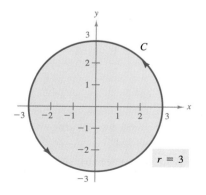

FIGURE 17.26

EXAMPLE 1 Using Green's Theorem

Use Green's Theorem to evaluate the line integral

$$\int_C y^3 \, dx + (x^3 + 3xy^2) \, dy$$

where C is the path from $(0, 0)$ to $(1, 1)$ along the graph of $y = x^3$ and from $(1, 1)$ to $(0, 0)$ along the graph of $y = x$, as shown in Figure 17.26.

SOLUTION

Since $M = y^3$ and $N = x^3 + 3xy^2$, it follows that

$$\frac{\partial N}{\partial x} = 3x^2 + 3y^2 \qquad \text{and} \qquad \frac{\partial M}{\partial y} = 3y^2.$$

Applying Green's Theorem, we then have

$$\int_C y^3 \, dx + (x^3 + 3xy^2) \, dy = \iint_R \left(\frac{\partial N}{\partial x} - \frac{\partial M}{\partial y} \right) dA$$

$$= \int_0^1 \int_{x^3}^x [(3x^2 + 3y^2) - 3y^2] \, dy \, dx$$

$$= \int_0^1 \int_{x^3}^x 3x^2 \, dy \, dx$$

$$= \int_0^1 3x^2 y \Big]_{x^3}^x \, dx$$

$$= \int_0^1 (3x^3 - 3x^5) \, dx$$

$$= \left[\frac{3x^4}{4} - \frac{x^6}{2} \right]_0^1 = \frac{1}{4}.$$

EXAMPLE 2 Using Green's Theorem to calculate work

While subject to the force $\mathbf{F}(x, y) = y^3 \mathbf{i} + (x^3 + 3xy^2) \mathbf{j}$, a particle travels once around the circle of radius 3 shown in Figure 17.27. Use Green's Theorem to find the work done by $\mathbf{F}$.

SOLUTION

From Example 1 we know by Green's Theorem that

$$\int_C y^3 \, dx + (x^3 + 3xy^2) \, dy = \iint_R 3x^2 \, dA.$$

FIGURE 17.27

In polar coordinates, using $x = r \cos \theta$ and $dA = r \, dr \, d\theta$, the work done is

$$W = \iint_R 3x^2 \, dA = \int_0^{2\pi} \int_0^3 3(r \cos \theta)^2 r \, dr \, d\theta$$

$$= 3 \int_0^{2\pi} \int_0^3 r^3 \cos^2 \theta \, dr \, d\theta$$

$$= 3 \int_0^{2\pi} \frac{r^4}{4} \cos^2 \theta \Big]_0^3 \, d\theta$$

$$= 3 \int_0^{2\pi} \frac{81}{4} \cos^2 \theta \, d\theta = \frac{243}{8} \int_0^{2\pi} (1 + \cos 2\theta) \, d\theta$$

$$= \frac{243}{8} \left[\theta + \frac{\sin 2\theta}{2} \right]_0^{2\pi} = \frac{243\pi}{4}.$$

Although a few more steps would be necessary, the line integrals in Examples 1 and 2 could have been evaluated directly, without the aid of Green's Theorem. (See Example 8 in Section 17.2.) In the next example, the advantage of Green's Theorem is more obvious, since direct evaluation would require four separate line integrals.

EXAMPLE 3   Using Green's Theorem for a piecewise smooth curve

Evaluate

$$\int_C (\arctan x + y^2) \, dx + (\ln y - x^2) \, dy$$

where C is the path enclosing the annular region shown in Figure 17.28.

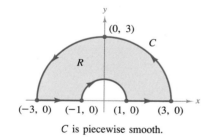

FIGURE 17.28

SOLUTION

In polar coordinates, R is given by $1 \le r \le 3$ for $0 \le \theta \le \pi$. Moreover,

$$\frac{\partial N}{\partial x} - \frac{\partial M}{\partial y} = -2x - 2y = -2(r \cos \theta + r \sin \theta).$$

Thus, by Green's Theorem,

$$\int_C (\arctan x + y^2) \, dx + (\ln y - x^2) \, dy = \iint_R -2(x + y) \, dA$$

$$= \int_0^\pi \int_1^3 -2r(\cos \theta + \sin \theta)r \, dr \, d\theta$$

$$= \int_0^\pi -2(\cos \theta + \sin \theta) \frac{r^3}{3} \Big]_1^3 \, d\theta$$

$$= \int_0^\pi \left(-\frac{52}{3} \right) (\cos \theta + \sin \theta) \, d\theta$$

$$= -\frac{52}{3} \Big[\sin \theta - \cos \theta \Big]_0^\pi = -\frac{104}{3}.$$

In Examples 1–3 we used Green's Theorem to evaluate line integrals as double integrals. We can also use the theorem to evaluate double integrals as line integrals. One useful application occurs when $\partial N/\partial x - \partial M/\partial y = 1$.

$$\int_C M\,dx + N\,dy = \iint_R 1\,dA = \text{area of region } R$$

Among the many choices for M and N satisfying the stated condition, we choose $M = -y/2$ and $N = x/2$ to obtain the following line integral for the area of region R.

THEOREM 17.9
LINE INTEGRAL FOR AREA

If R is a plane region bounded by a piecewise smooth simple closed curve C, then the area of R is given by

$$A = \frac{1}{2}\int_C x\,dy - y\,dx.$$

EXAMPLE 4   Finding area by a line integral

Use a line integral to find the area of the ellipse

$$\frac{x^2}{a^2} + \frac{y^2}{b^2} = 1.$$

SOLUTION

Using Figure 17.29, we can induce a counterclockwise orientation to this elliptical path by letting

$$x = a\cos t \quad\text{and}\quad y = b\sin t, \quad 0 \le t \le 2\pi.$$

Therefore, we have

$$A = \frac{1}{2}\int_C x\,dy - y\,dx = \frac{1}{2}\int_0^{2\pi} [(a\cos t)(b\cos t)\,dt - (b\sin t)(-a\sin t)\,dt]$$

$$= \frac{ab}{2}\int_0^{2\pi} (\cos^2 t + \sin^2 t)\,dt = \frac{ab}{2}t\Big]_0^{2\pi}$$

$$= \pi ab.$$

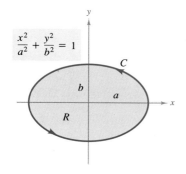

FIGURE 17.29

Green's Theorem can be extended to cover some regions that are not simply connected. This is demonstrated in the next example.

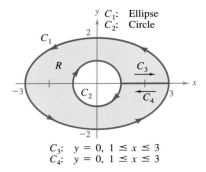

C_1: Ellipse
C_2: Circle

C_3: $y = 0$, $1 \leq x \leq 3$
C_4: $y = 0$, $1 \leq x \leq 3$

FIGURE 17.30

EXAMPLE 5 Green's Theorem extended to a region with a hole

Let R be the region inside the ellipse $(x^2/9) + (y^2/4) = 1$ and outside the circle $x^2 + y^2 = 1$. Evaluate the line integral

$$\int_C 2xy \, dx + (x^2 + 2x) \, dy$$

where $C = C_1 + C_2$ is the boundary of R, as shown in Figure 17.30.

SOLUTION

To begin, we introduce the line segments C_3 and C_4, as shown in Figure 17.30. Note that because the curves C_3 and C_4 have opposite orientations, the line integrals over them cancel. Furthermore, we can apply Green's Theorem to the region R using the boundary $C_1 + C_4 + C_2 + C_3$ to obtain

$$\int_C 2xy \, dx + (x^2 + 2x) \, dy = \iint_R \left(\frac{\partial N}{\partial x} - \frac{\partial M}{\partial y} \right) dA = \iint_R (2x + 2 - 2x) \, dA$$

$$= 2 \iint_R dA = 2(\text{area of } R)$$

$$= 2[\pi(3)(2) - \pi(1^2)] = 10\pi.$$

In Section 17.1 we listed a necessary and sufficient condition for conservative vector fields. There, we proved only one direction of the proof. We now outline the other direction, using Green's Theorem. Let $\mathbf{F}(x, y) = M\mathbf{i} + N\mathbf{j}$ be defined on an open disc R. We want to show that if M and N have continuous first partial derivatives and

$$\frac{\partial M}{\partial y} = \frac{\partial N}{\partial x}$$

then $\mathbf{F}$ is conservative. Suppose that C is a closed path forming the boundary of a connected region lying in R. Then, using the fact that $\partial M/\partial y = \partial N/\partial x$, we can apply Green's Theorem to conclude that

$$\int_C \mathbf{F} \cdot d\mathbf{r} = \int_C M \, dx + N \, dy = \iint_R \left(\frac{\partial N}{\partial x} - \frac{\partial M}{dy} \right) dA = 0.$$

This, in turn, is equivalent to showing that $\mathbf{F}$ is conservative (see Theorem 17.7).

Alternate forms of Green's Theorem

We conclude this section with the derivation of two vector forms of Green's Theorem for regions in the plane. The extension of these vector forms to three dimensions is the basis for the discussion in the remaining sections of this chapter. If $\mathbf{F}$ is a vector field in the plane, we can write

$$\mathbf{F}(x, y) = M\mathbf{i} + N\mathbf{j} + 0\mathbf{k}$$

so that the curl of **F**, as described in Section 17.1, is given by

$$\text{curl } \mathbf{F} = \nabla \times \mathbf{F} = \begin{vmatrix} \mathbf{i} & \mathbf{j} & \mathbf{k} \\ \dfrac{\partial}{\partial x} & \dfrac{\partial}{\partial y} & \dfrac{\partial}{\partial z} \\ M & N & 0 \end{vmatrix} = -\frac{\partial N}{\partial z}\mathbf{i} + \frac{\partial M}{\partial z}\mathbf{j} + \left(\frac{\partial N}{\partial x} - \frac{\partial M}{\partial y}\right)\mathbf{k}.$$

Consequently,

$$(\text{curl } \mathbf{F}) \cdot \mathbf{k} = \left[-\frac{\partial N}{\partial z}\mathbf{i} + \frac{\partial M}{\partial z}\mathbf{j} + \left(\frac{\partial N}{\partial x} - \frac{\partial M}{\partial y}\right)\mathbf{k}\right] \cdot \mathbf{k} = \frac{\partial N}{\partial x} - \frac{\partial M}{\partial y}.$$

With appropriate conditions on **F**, C, and R, we can write Green's Theorem in the vector form

$$\int_C \mathbf{F} \cdot d\mathbf{r} = \iint_R \left(\frac{\partial N}{\partial x} - \frac{\partial M}{\partial y}\right) dA = \iint_R (\text{curl } \mathbf{F}) \cdot \mathbf{k} \, dA.$$

The extension of this vector form of Green's Theorem to surfaces in space produces **Stokes's Theorem,** discussed in Section 17.7.

For the second vector form of Green's Theorem, we assume the same conditions for **F**, C, and R. Using the arc length parameter s for C, we have $\mathbf{r}(s) = x(s)\mathbf{i} + y(s)\mathbf{j}$. Thus, a unit tangent vector **T** to curve C is given by

$$\mathbf{r}'(s) = \mathbf{T} = x'(s)\mathbf{i} + y'(s)\mathbf{j}.$$

From Figure 17.31 we can see that the *outward* unit normal vector **N** can then be written as

$$\mathbf{N} = y'(s)\mathbf{i} - x'(s)\mathbf{j}.$$

Consequently, for $\mathbf{F}(x, y) = M\mathbf{i} + N\mathbf{j}$, we can apply Green's Theorem to obtain

$$\begin{aligned} \int_C \mathbf{F} \cdot \mathbf{N} \, ds &= \int_a^b (M\mathbf{i} + N\mathbf{j}) \cdot (y'(s)\mathbf{i} - x'(s)\mathbf{j}) \, ds \\ &= \int_a^b \left(M\frac{dy}{ds} - N\frac{dx}{ds}\right) ds \\ &= \int_C M \, dy - N \, dx \\ &= \int_C -N \, dx + M \, dy \\ &= \iint_R \left(\frac{\partial M}{\partial x} + \frac{\partial N}{\partial y}\right) dA \\ &= \iint_R \text{div } \mathbf{F} \, dA. \end{aligned}$$

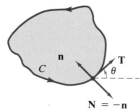

$$\mathbf{T} = \cos\theta\,\mathbf{i} + \sin\theta\,\mathbf{j}$$
$$\mathbf{n} = \cos\left(\theta + \frac{\pi}{2}\right)\mathbf{i} + \sin\left(\theta + \frac{\pi}{2}\right)\mathbf{j}$$
$$= -\sin\theta\,\mathbf{i} + \cos\theta\,\mathbf{j}$$
$$\mathbf{N} = \sin\theta\,\mathbf{i} - \cos\theta\,\mathbf{j}$$

FIGURE 17.31

The extension of this form to three dimensions is called the **Divergence Theorem,** discussed in Section 17.6. The physical interpretation of divergence and curl will be discussed in Sections 17.6 and 17.7.

EXERCISES for Section 17.4

In Exercises 1–4, verify Green's Theorem by evaluating both integrals

$$\int_C y^2 \, dx + x^2 \, dy = \int\int_R \left(\frac{\partial N}{\partial x} - \frac{\partial M}{\partial y} \right) dA$$

for the given path.

1. C: boundary of the square with vertices $(0, 0)$, $(4, 0)$, $(4, 4)$, $(0, 4)$
2. C: boundary of the triangle with vertices $(0, 0)$, $(4, 0)$, $(4, 4)$
3. C: boundary of the region lying between the graphs of $y = x$ and $y = x^2/4$
4. C: $x^2 + y^2 = 1$

In Exercises 5–8, use Green's Theorem to evaluate the integral

$$\int_C (y - x) \, dx + (2x - y) \, dy$$

for the given path.

5. C: boundary of region lying between the graphs of $y = x$ and $y = x^2 - x$
6. C: $x = 2 \cos \theta$, $y = \sin \theta$
7. C: boundary of the region lying inside the rectangle bounded by $x = -5$, $x = 5$, $y = -3$, $y = 3$ and outside the square bounded by $x = -1$, $x = 1$, $y = -1$, $y = 1$
8. C: boundary of the region lying inside the circle $x^2 + y^2 = 16$ and outside the circle $x^2 + y^2 = 1$

In Exercises 9–18, use Green's Theorem to evaluate the line integral.

9. $\int_C 2xy \, dx + (x + y) \, dy$

C: boundary of the region lying between the graphs of $y = 0$ and $y = 4 - x^2$

10. $\int_C y^2 \, dx + xy \, dy$

C: boundary of the region lying between the graphs of $y = 0$, $y = \sqrt{x}$, and $x = 4$

11. $\int_C (x^2 - y^2) \, dx + 2xy \, dy$

C: $x^2 + y^2 = a^2$

12. $\int_C (x^2 - y^2) \, dx + 2xy \, dy$

C: $r = 1 + \cos \theta$

13. $\int_C 2 \arctan \frac{y}{x} \, dx + \ln (x^2 + y^2) \, dy$

C: $x = 4 + 2 \cos \theta$, $y = 4 + \sin \theta$

14. $\int_C e^x \sin 2y \, dx + 2e^x \cos 2y \, dy$

C: $x^2 + y^2 = a^2$

15. $\int_C \sin x \cos y \, dx + (xy + \cos x \sin y) \, dy$

C: boundary of the region lying between the graphs of $y = x$ and $y = \sqrt{x}$

16. $\int_C (e^{-x^2/2} - y) \, dx + (e^{-y^2/2} + x) \, dy$

C: boundary of the region lying between the graphs of the circle $x = 5 \cos \theta$, $y = 5 \sin \theta$ and the ellipse $x = 2 \cos \theta$, $y = \sin \theta$

17. $\int_C xy \, dx + (x + y) \, dy$

C: boundary of the region lying between the graphs of $x^2 + y^2 = 1$ and $x^2 + y^2 = 9$

18. $\int_C 3x^2 e^y \, dx + e^y \, dy$

C: boundary of the region lying between the squares with vertices $(1, 1)$, $(-1, 1)$, $(-1, -1)$, $(1, -1)$, and $(2, 2)$, $(-2, 2)$, $(-2, -2)$, $(2, -2)$

In Exercises 19–22, use Green's Theorem to calculate the work done by the force $\mathbf{F}$ in moving a particle around the closed path C.

19. $\mathbf{F}(x, y) = xy\mathbf{i} + (x + y)\mathbf{j}$
C: $x^2 + y^2 = 4$
20. $\mathbf{F}(x, y) = (e^x - 3y)\mathbf{i} + (e^y + 6x)\mathbf{j}$
C: $r = 2 \cos \theta$
21. $\mathbf{F}(x, y) = (x^{3/2} - 3y)\mathbf{i} + (6x + 5\sqrt{y})\mathbf{j}$
C: boundary of the triangle with vertices $(0, 0)$, $(5, 0)$, $(0, 5)$
22. $\mathbf{F}(x, y) = (3x^2 + y)\mathbf{i} + 4xy^2\mathbf{j}$
C: boundary of the region bounded by the graphs of $y = \sqrt{x}$, $y = 0$, $x = 4$

In Exercises 23–26, use a line integral to find the area of the region R.

23. R: region bounded by the graph of $x^2 + y^2 = a^2$
24. R: triangle bounded by the graphs of $x = 0$, $2x - 3y = 0$, $x + 3y = 9$
25. R: region bounded by the graphs of $y = 2x + 1$, $y = 4 - x^2$

26. *R*: region inside the loop of the folium of Descartes bounded by the graph of

$$x = \frac{3t}{t^3 + 1}, \quad y = \frac{3t^2}{t^3 + 1}$$

In Exercises 27 and 28, use Green's Theorem to verify the line integral formulas.

27. The centroid of the region having area *A* bounded by the simple closed path *C* is

$$\bar{x} = \frac{1}{2A} \int_C x^2 \, dy, \quad \bar{y} = -\frac{1}{2A} \int_C y^2 \, dx.$$

28. The area of a plane region bounded by the simple closed path *C* given in polar coordinates is

$$A = \frac{1}{2} \int_C r^2 \, d\theta.$$

In Exercises 29–32, use the result of Exercise 27 to find the centroid of the region.

29. *R*: region bounded by the graphs of $y = 0$ and $y = 4 - x^2$

30. *R*: region bounded by the graphs of $y = \sqrt{a^2 - x^2}$ and $y = 0$

31. *R*: region bounded by the graphs of $y = x^3$, $y = x$, $0 \le x \le 1$

32. *R*: triangle with vertices $(-a, 0)$, $(a, 0)$, (b, c), where $-a \le b \le a$

In Exercises 33–36, use the result of Exercise 28 to find the area of the region bounded by the graphs of the polar equations.

33. $r = a(1 - \cos \theta)$ **34.** $r = a \cos 3\theta$

35. $r = 1 + 2 \cos \theta$ (inner loop)

36. $r = \dfrac{3}{2 - \cos \theta}$

37. Let

$$I = \int_C \frac{y \, dx - x \, dy}{x^2 + y^2}$$

where *C* is a circle. Show that $I = 0$ if *C* does not contain the origin. What is *I* if *C* does contain the origin?

38. (a) Let *C* be the line segment joining (x_1, y_1) and (x_2, y_2). Show that

$$\int_C -y \, dx + x \, dy = x_1 y_2 - x_2 y_1.$$

(b) Let (x_1, y_1), (x_2, y_2), . . . , (x_n, y_n) be the vertices of a polygon. Prove that the area enclosed is

$$\frac{1}{2}[(x_1 y_2 - x_2 y_1) + (x_2 y_3 - x_3 y_2) + \cdots$$
$$+ (x_{n-1} y_n - x_n y_{n-1}) + (x_n y_1 - x_1 y_n)].$$

In Exercises 39 and 40, find the area enclosed by the polygon with the given vertices.

39. Pentagon: $(0, 0)$, $(2, 0)$, $(3, 2)$, $(1, 4)$, $(-1, 1)$
40. Hexagon: $(0, 0)$, $(2, 0)$, $(3, 2)$, $(2, 4)$, $(0, 3)$, $(-1, 1)$

41. Use Green's Theorem to prove that

$$\int_C f(x) \, dx + g(y) \, dy = 0$$

if *f* and *g* are differentiable functions and *C* is a piecewise smooth, simple closed path.

42. Let $\mathbf{F} = M\mathbf{i} + N\mathbf{j}$, where *M* and *N* have continuous first partial derivatives in a simply connected region *R*. Prove that if *C* is simple, smooth, and closed, and $N_x = M_y$ then

$$\int_C \mathbf{F} \cdot d\mathbf{r} = 0.$$

In Exercises 43 and 44, prove the identity where *R* is a simply connected region with boundary *C*. Also assume that the required partial derivatives of the scalar functions *f* and *g* are continuous. The expressions $D_{\mathbf{N}}f$ and $D_{\mathbf{N}}g$ are the derivatives in the direction of the outward normal vector **N** of *C* and are defined by

$$D_{\mathbf{N}}f = \nabla f \cdot \mathbf{N}, \quad D_{\mathbf{N}}g = \nabla g \cdot \mathbf{N}.$$

43. Green's first identity:

$$\iint_R (f \nabla^2 g + \nabla f \cdot \nabla g) \, dA = \int_C f D_{\mathbf{N}}g \, ds$$

[Hint: Use the alternate form of Green's Theorem and the property

$$\text{div} \, (f\mathbf{G}) = f \, \text{div} \, \mathbf{G} + \nabla f \cdot \mathbf{G}.]$$

44. Green's second identity:

$$\iint_R (f \nabla^2 g - g \nabla^2 f) \, dA = \int_C (f D_{\mathbf{N}}g - g D_{\mathbf{N}}f) \, ds$$

[Hint: Use Exercise 43 twice.]

In Exercises 45 and 46, verify Green's Theorem by using a computer to approximate both integrals

$$\int_C xe^y \, dx + e^x \, dy = \iint_R \left(\frac{\partial N}{\partial x} - \frac{\partial M}{\partial y} \right) dA$$

for the given path.

45. C: $x^2 + y^2 = 4$

46. C: boundary of the region lying between the graphs of $y = x$ and $y = x^3$.

17.5 Surface Integrals

Surface integrals ▪ Orientation of a surface ▪ Flux integrals

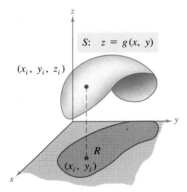

S: $z = g(x, y)$

(x_i, y_i, z_i)

R

(x_i, y_i)

Scalar function f
assigns a number
to each point of S.

FIGURE 17.32

The remainder of this chapter deals primarily with **surface integrals.** The name comes from the fact that instead of integrating over plane regions bounded by curves, we will integrate over regions in space bounded by surfaces.

Let S be a surface given by $z = g(x, y)$ and R its projection on the xy-plane, as shown in Figure 17.32. Suppose g, g_x, and g_y are continuous at all points in R, and f is defined on S. Employing the procedure used to find surface area in Section 16.5, we evaluate f at (x_i, y_i, z_i) and form the sum

$$\sum_{i=1}^{n} f(x_i, y_i, z_i)\Delta S_i$$

where $\Delta S_i \approx \sqrt{1 + [g_x(x_i, y_i)]^2 + [g_y(x_i, y_i)]^2}\Delta A_i$. Providing the limit as $\|\Delta\|$ approaches zero exists, we define the **surface integral of f over S** to be

$$\iint_S f(x, y, z) \, dS = \lim_{\|\Delta\| \to 0} \sum_{i=1}^{n} f(x_i, y_i, z_i)\Delta S_i.$$

This integral can be evaluated by the double integral given in Theorem 17.10.

THEOREM 17.10
EVALUATING A
SURFACE INTEGRAL

Let S be a surface with equation $z = g(x, y)$ and R its projection on the xy-plane. If g, g_x, and g_y are continuous in R and f is continuous on S, then the surface integral of f over S is

$$\iint_S f(x, y, z) \, dS = \iint_R f(x, y, g(x, y)) \sqrt{1 + [g_x(x, y)]^2 + [g_y(x, y)]^2} \, dA.$$

For surfaces described by functions of x and z (or y and z), we make the following adjustments to Theorem 17.10. If S is the graph of $y = g(x, z)$ and R is its projection onto the xz-plane, then

$$\iint_S f(x, y, z) \, dS = \iint_R f(x, g(x, z), z)\sqrt{1 + [g_x(x, z)]^2 + [g_z(x, z)]^2} \, dA.$$

If S is the graph of $x = g(y, z)$ and R is its projection onto the yz-plane, then

$$\iint_S f(x, y, z) \, dS = \iint_R f(g(y, z), y, z)\sqrt{1 + [g_y(y, z)]^2 + [g_z(y, z)]^2} \, dA.$$

EXAMPLE 1 Evaluating a surface integral

Evaluate the surface integral

$$\iint\limits_S (y^2 + 2yz)\, dS$$

where S is the first-octant portion of the plane $2x + y + 2z = 6$.

SOLUTION

We can write S as $z = \frac{1}{2}(6 - 2x - y) = g(x, y)$, so that $g_x(x, y) = -1$ and $g_y(x, y) = -\frac{1}{2}$ and obtain

$$\sqrt{1 + [g_x(x, y)]^2 + [g_y(x, y)]^2} = \sqrt{1 + 1 + \frac{1}{4}} = \frac{3}{2}.$$

Using Figure 17.33 and Theorem 17.10, we obtain

$$\iint\limits_S (y^2 + 2yz)\, dS = \iint\limits_R f(x, y, g(x, y))\sqrt{1 + [g_x(x, y)]^2 + [g_y(x, y)]^2}\, dA$$

$$= \iint\limits_R \left[y^2 + 2y\left(\frac{1}{2}\right)(6 - 2x - y)\right]\left(\frac{3}{2}\right) dA$$

$$= 3\int_0^3 \int_0^{2(3-x)} y(3 - x)\, dy\, dx = 6\int_0^3 (3 - x)^3\, dx$$

$$= -\frac{3}{2}(3 - x)^4\bigg]_0^3 = \frac{243}{2}.$$

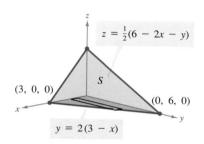

$z = \frac{1}{2}(6 - 2x - y)$

$(3, 0, 0)$

S

$(0, 6, 0)$

$y = 2(3 - x)$

FIGURE 17.33

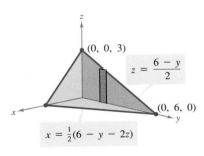

$(0, 0, 3)$

$z = \dfrac{6 - y}{2}$

$(0, 6, 0)$

$x = \frac{1}{2}(6 - y - 2z)$

FIGURE 17.34

One alternate solution to Example 1 would be to project S onto the yz-plane, as shown in Figure 17.34. Then, $x = \frac{1}{2}(6 - y - 2z)$, and

$$\sqrt{1 + [g_y(y, z)]^2 + [g_z(y, z)]^2} = \sqrt{1 + \frac{1}{4} + 1} = \frac{3}{2}.$$

Thus, the surface integral is

$$\iint\limits_S (y^2 + 2yz)\, dS = \iint\limits_R f(g(y, z), y, z)\sqrt{1 + [g_y(y, z)]^2 + [g_x(y, z)]^2}\, dA$$

$$= \int_0^6 \int_0^{(6-y)/2} [y^2 + 2yz]\left(\frac{3}{2}\right) dz\, dy$$

$$= \frac{3}{8}\int_0^6 (36y - y^3)\, dy = \frac{243}{2}.$$

In Example 1 we could have projected the surface S onto any one of the three coordinate planes. In the next example the surface is a cylinder centered about the z-axis, and we have the option of projecting S onto either the xz-plane or the xy-plane.

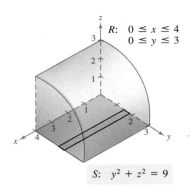

$R:\ 0 \le x \le 4$
$\quad 0 \le y \le 3$

$S:\ y^2 + z^2 = 9$

FIGURE 17.35

EXAMPLE 2 Evaluating a surface integral

Evaluate the surface integral

$$\iint_S (x + z)\, dS$$

where S is the first-octant portion of the cylinder $z^2 + y^2 = 9$ between $x = 0$ and $x = 4$, as shown in Figure 17.35.

SOLUTION

We project onto the xy-plane, so that $z = g(x, y) = \sqrt{9 - y^2}$, and obtain

$$\sqrt{1 + [g_x(x, y)]^2 + [g_y(x, y)]^2} = \sqrt{1 + \left(\frac{-y}{\sqrt{9 - y^2}}\right)^2} = \frac{3}{\sqrt{9 - y^2}}.$$

Theorem 17.10 does not apply directly because g_y is not continuous when $y = 3$. However, we can apply the theorem for $0 \le b < 3$ and then take the limit as b approaches 3, as follows.

$$\iint_S (x + z)\, dS = \lim_{b \to 3^-} \int_0^b \int_0^4 (x + \sqrt{9 - y^2}) \frac{3}{\sqrt{9 - y^2}}\, dx\, dy$$

$$= \lim_{b \to 3^-} 3 \int_0^b \int_0^4 \left(\frac{x}{\sqrt{9 - y^2}} + 1\right) dx\, dy$$

$$= \lim_{b \to 3^-} 3 \int_0^b \left[\frac{x^2}{2\sqrt{9 - y^2}} + x\right]_0^4 dy$$

$$= \lim_{b \to 3^-} 3 \int_0^b \left(\frac{8}{\sqrt{9 - y^2}} + 4\right) dy$$

$$= \lim_{b \to 3^-} 3 \left[4y + 8 \arcsin \frac{y}{3}\right]_0^b$$

$$= \lim_{b \to 3^-} 3 \left(4b + 8 \arcsin \frac{b}{3}\right)$$

$$= 36 + 24\left(\frac{\pi}{2}\right) = 36 + 12\pi \qquad \square$$

If the function f defined on the surface S is simply $f(x, y, z) = 1$, then the surface integral yields the *surface area* of S.

$$\text{area of surface} = \iint_S 1\, dS$$

On the other hand, if S is a lamina of variable density and $\rho(x, y, z)$ is the density at the point (x, y, z), then the *mass* of the lamina is given by

$$\text{mass of lamina} = \iint_S \rho(x, y, z)\, dS.$$

We illustrate this application in our next example.

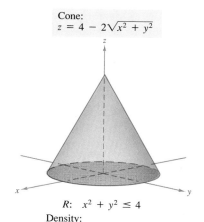

Cone:
$z = 4 - 2\sqrt{x^2 + y^2}$

$R: \; x^2 + y^2 \le 4$

Density:
$\rho(x, y, z) = k\sqrt{x^2 + y^2}$

FIGURE 17.36

EXAMPLE 3 Finding the mass of a surface lamina

A cone-shaped surface lamina S is given by $z = 4 - 2\sqrt{x^2 + y^2}, 0 \le z \le 4$, as shown in Figure 17.36. At each point on S the density is proportional to the distance between the point and the z-axis. Find the mass m of the lamina.

SOLUTION

Projecting S onto the xy-plane produces the following.

$$S: z = 4 - 2\sqrt{x^2 + y^2} = g(x, y), \quad 0 \le z \le 4$$
$$R: x^2 + y^2 \le 4$$
$$\text{Density: } \rho(x, y, z) = k\sqrt{x^2 + y^2}$$

Using a surface integral, we find the mass to be

$$m = \iint_S \rho(x, y, z) \, dS = \iint_R k\sqrt{x^2 + y^2}\sqrt{1 + [g_x(x, y)]^2 + [g_y(x, y)]^2} \, dA$$

$$= k \iint_R \sqrt{x^2 + y^2}\sqrt{1 + \frac{4x^2}{x^2 + y^2} + \frac{4y^2}{x^2 + y^2}} \, dA$$

$$= k \iint_R \sqrt{5}\sqrt{x^2 + y^2} \, dA$$

$$= k \int_0^{2\pi} \int_0^2 (\sqrt{5}r)r \, dr \, d\theta \qquad \text{Polar coordinates}$$

$$= \frac{\sqrt{5}k}{3} \int_0^{2\pi} r^3 \Big]_0^2 \, d\theta$$

$$= \frac{8\sqrt{5}k}{3} \Big[\theta\Big]_0^{2\pi}$$

$$= \frac{16\sqrt{5}k\pi}{3}.$$

∎

Orientation of a surface

In using a line integral to find work, we developed the vector form of a line integral

$$\int_C \mathbf{F} \cdot \mathbf{T} \, ds$$

where the unit tangent vector $\mathbf{T}$ points in the direction of positive orientation along C. For every smooth curve, the unit tangent vector is continuous along C. In a similar way, we use unit normal vectors to induce an orientation to a surface S in space. We say that a surface is **orientable** if a unit normal vector $\mathbf{N}$ can be defined at every nonboundary point of S in such a way that the normal vectors vary continuously over the surface S. If this is possible, we call S an **oriented surface.**

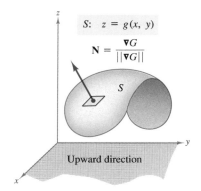

$$S: \ z = g(x, y)$$

$$\mathbf{N} = \frac{\nabla G}{\|\nabla G\|}$$

S

Upward direction

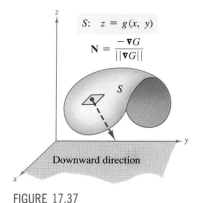

$$S: \ z = g(x, y)$$

$$\mathbf{N} = \frac{-\nabla G}{\|\nabla G\|}$$

S

Downward direction

FIGURE 17.37

An orientable surface S has two distinct sides. Orienting the surface consists of selecting one of the two possible unit normal vectors. If S is a closed surface such as a sphere, it is customary to choose the unit normal vector $\mathbf{N}$ to be the one that points outward from the sphere.

Most common surfaces, such as spheres, paraboloids, ellipses, and planes, are orientable. (See Exercise 31 for an example of a surface that is *not* orientable.) Moreover, for an orientable surface, the gradient vector provides a convenient way to find a unit normal vector.

For an orientable surface S given by $z = g(x, y)$, we let $G(x, y, z) = z - g(x, y)$. Then, S can be oriented by either of the following unit normal vectors

$$\mathbf{N} = \frac{\nabla G(x, y, z)}{\|\nabla G(x, y, z)\|} = \frac{-g_x(x, y)\,\mathbf{i} - g_y(x, y)\,\mathbf{j} + \mathbf{k}}{\sqrt{1 + [g_x(x, y)]^2 + [g_y(x, y)]^2}} \qquad \begin{array}{l}\text{Upward unit} \\ \text{normal}\end{array}$$

$$\mathbf{N} = \frac{-\nabla G(x, y, z)}{\|\nabla G(x, y, z)\|} = \frac{g_x(x, y)\,\mathbf{i} + g_y(x, y)\,\mathbf{j} - \mathbf{k}}{\sqrt{1 + [g_x(x, y)]^2 + [g_y(x, y)]^2}} \qquad \begin{array}{l}\text{Downward unit} \\ \text{normal}\end{array}$$

as shown in Figure 17.37. Similarly, we can use the gradient vectors

$$\nabla G(x, y, z), \quad G(x, y, z) = y - g(x, z)$$

and

$$\nabla G(x, y, z), \quad G(x, y, z) = x - g(y, z)$$

to orient surfaces given by $y = g(x, z)$ or $x = g(y, z)$.

Flux integrals

One of the principal applications involving the vector form of a surface integral relates to the flow of a fluid through a surface S. Suppose an oriented surface S is submerged in a fluid having a continuous velocity field $\mathbf{F}$. Let ΔS be the area of a small patch of the surface S over which $\mathbf{F}$ is nearly constant. Then the amount of fluid crossing this region per unit of time is approximated by the volume of the column of height $\mathbf{F} \cdot \mathbf{N}$, as shown in Figure 17.38. That is,

$$\Delta V = (\text{height})(\text{area of base})$$
$$= (\mathbf{F} \cdot \mathbf{N})\Delta S.$$

Velocity field $\mathbf{F}$ has direction of fluid flow.

FIGURE 17.38

Consequently, the volume of fluid crossing surface S per unit of time (called the **flux of F across** S) is given by the surface integral in the following definition.

DEFINITION OF FLUX INTEGRAL	Let $\mathbf{F}(x, y, z) = M\mathbf{i} + N\mathbf{j} + P\mathbf{k}$, where $M, N,$ and P have continuous first partial derivatives on the surface S oriented by a unit normal vector $\mathbf{N}$. The **flux integral of F across** S is given by $$\iint_S \mathbf{F} \cdot \mathbf{N} \, dS.$$

Geometrically, a flux integral is the surface integral over S of the *normal component* of $\mathbf{F}$. If $\rho(x, y, z)$ is the density of the fluid at (x, y, z), then the flux integral

$$\iint_S \rho\mathbf{F} \cdot \mathbf{N} \, dS$$

represents the *mass* of the fluid flowing across S per unit of time.

EXAMPLE 4   Using a flux integral to find the rate of mass flow

Let S be that portion of the paraboloid $z = 4 - x^2 - y^2$ lying above the xy-plane, oriented by an upward unit normal vector, as shown in Figure 17.39. A fluid of constant density ρ is flowing through the surface S according to the velocity field $\mathbf{F}(x, y, z) = x\mathbf{i} + y\mathbf{j} + z\mathbf{k}$. Find the rate of mass flow through S.

SOLUTION

Projecting S onto the xy-plane, we have the following.

$$S: z = 4 - x^2 - y^2 \quad \Longrightarrow \quad G(x, y, z) = z - 4 + x^2 + y^2$$
$$R: x^2 + y^2 \leq 4$$

Unit normal: $\mathbf{N} = \dfrac{\nabla G(x, y, z)}{\|\nabla G(x, y, z)\|} = \dfrac{2x\mathbf{i} + 2y\mathbf{j} + \mathbf{k}}{\sqrt{1 + 4x^2 + 4y^2}}$

$dS: \sqrt{1 + [g_x(x, y)]^2 + [g_y(x, y)]^2} \, dA = \sqrt{1 + 4x^2 + 4y^2} \, dA$

Therefore, the rate of mass flow through S is

$$\iint_S \rho\mathbf{F} \cdot \mathbf{N} \, dS = \rho \iint_R \left(\frac{2x^2 + 2y^2 + z}{\sqrt{1 + 4x^2 + 4y^2}} \right) \sqrt{1 + 4x^2 + 4y^2} \, dA$$

$$= \rho \iint_R [2x^2 + 2y^2 + (4 - x^2 - y^2)] \, dA$$

$$= \rho \iint_R [4 + x^2 + y^2] \, dA$$

$$= \rho \int_0^{2\pi} \int_0^2 [4 + r^2]r \, dr \, d\theta \qquad \text{Polar coordinates}$$

$$= \rho \int_0^{2\pi} 12 \, d\theta = 24\pi\rho.$$

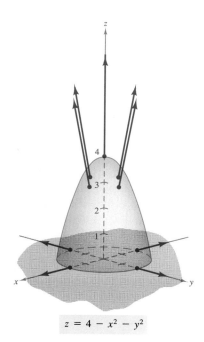

$z = 4 - x^2 - y^2$

FIGURE 17.39

Notice the cancellation that occurs in the product

$$\mathbf{N}\,dS = \frac{\nabla G(x,\, y,\, z)}{\|\nabla G(x,\, y,\, z)\|}\,dS$$

$$= \frac{\nabla G(x,\, y,\, z)}{\sqrt{[g_x(x,\, y,\, z)]^2 + [g_y(x,\, y,\, z)]^2 + 1}}\,\sqrt{[g_x(x,\, y,\, z)]^2 + [g_y(x,\, y,\, z)]^2 + 1}\,dA$$

$$= \nabla G(x,\, y,\, z)\,dA.$$

By generalizing this result, we can write flux integrals in the following simplified form.

THEOREM 17.11
EVALUATING A FLUX INTEGRAL

Let S be an oriented surface given by $z = g(x,\, y)$ and let R be its projection on the xy-plane.

$$\iint_S \mathbf{F} \cdot \mathbf{N}\,dS = \iint_R \mathbf{F} \cdot (-g_x(x,\, y)\,\mathbf{i} - g_y(x,\, y)\,\mathbf{j} + \mathbf{k})\,dA \qquad \text{Oriented upward}$$

$$\iint_S \mathbf{F} \cdot \mathbf{N}\,dS = \iint_R \mathbf{F} \cdot (g_x(x,\, y)\,\mathbf{i} + g_y(x,\, y)\,\mathbf{j} - \mathbf{k})\,dA \qquad \text{Oriented downward}$$

EXAMPLE 5 Finding the flux of an inverse square field

Find the flux over the sphere S given by $x^2 + y^2 + z^2 = a^2$, where $\mathbf{F}$ is an inverse square field given by

$$\mathbf{F}(\mathbf{r}) = \frac{q}{\|\mathbf{r}\|^2}\,\frac{\mathbf{r}}{\|\mathbf{r}\|} = \frac{q\,\mathbf{r}}{\|\mathbf{r}\|^3}.$$

Assume S is oriented outward, as shown in Figure 17.40.

SOLUTION

We need only compute the flux on the upper hemisphere

$$z = g(x,\, y) = \sqrt{a^2 - x^2 - y^2}$$

because the vector field $\mathbf{F}$ and the surface S are both symmetric with respect to the origin. The projection of this hemisphere onto the xy-plane is the circular region R given by $x^2 + y^2 \le a^2$. Moreover, since the partial derivatives of g

$$g_x(x,\, y) = \frac{-x}{\sqrt{a^2 - x^2 - y^2}} \qquad \text{and} \qquad g_y(x,\, y) = \frac{-y}{\sqrt{a^2 - x^2 - y^2}}$$

are not continuous on the boundary of R, we consider a smaller circular region R_b given by $x^2 + y^2 \le b^2$, where $0 < b < a$. Finally, using

$$\mathbf{F}(x,\, y,\, z) = \frac{q(x\,\mathbf{i} + y\,\mathbf{j} + z\,\mathbf{k})}{(x^2 + y^2 + z^2)^{3/2}} = \frac{q(x\,\mathbf{i} + y\,\mathbf{j} + z\,\mathbf{k})}{a^3}$$

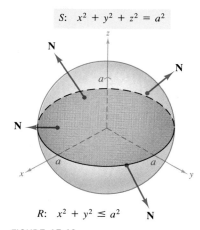

S: $x^2 + y^2 + z^2 = a^2$

R: $x^2 + y^2 \le a^2$

FIGURE 17.40

and

$$-g_x(x, y)\mathbf{i} - g_y(x, y)\mathbf{j} + \mathbf{k} = \frac{x\mathbf{i} + y\mathbf{j}}{\sqrt{a^2 - x^2 - y^2}} + \mathbf{k}$$

we obtain the following flux over the upper hemisphere.

$$\begin{aligned}
\text{flux} &= \lim_{b \to a^-} \iint_{R_b} \frac{q(x\mathbf{i} + y\mathbf{j} + z\mathbf{k})}{a^3} \cdot \left(\frac{x\mathbf{i} + y\mathbf{j}}{\sqrt{a^2 - x^2 - y^2}} + \mathbf{k} \right) dA \\
&= \lim_{b \to a^-} q \iint_{R_b} \left(\frac{x^2 + y^2}{a^3\sqrt{a^2 - x^2 - y^2}} + \frac{z}{a^3} \right) dA \\
&= \lim_{b \to a^-} q \iint_{R_b} \left(\frac{x^2 + y^2}{a^3\sqrt{a^2 - x^2 - y^2}} + \frac{\sqrt{a^2 - x^2 - y^2}}{a^3} \right) dA \\
&= \lim_{b \to a^-} q \iint_{R_b} \frac{a^2}{a^3\sqrt{a^2 - x^2 - y^2}} \, dA \\
&= \lim_{b \to a^-} q \int_0^{2\pi} \int_0^b \frac{1}{a\sqrt{a^2 - r^2}} \, r \, dr \, d\theta \qquad \text{Polar coordinates} \\
&= \lim_{b \to a^-} \frac{q}{a} \int_0^{2\pi} \left. -\sqrt{a^2 - r^2} \right]_0^b \, d\theta \\
&= \lim_{b \to a^-} \left[\frac{q}{a} (a - \sqrt{a^2 - b^2})2\pi \right] = 2\pi q
\end{aligned}$$

Therefore, by Theorem 17.11, the flux over the entire sphere is given by

$$\iint_S \mathbf{F} \cdot \mathbf{N} \, dS = 4\pi q.$$ ▭

The result in Example 5 shows that the flux across a sphere S in an inverse square field is independent of the radius of S. In particular, if $\mathbf{E}$ is an electric field, then the result in Example 5, along with Coulomb's Law, yields one of the basic laws of electrostatics, known as **Gauss's Law:**

$$\iint_S \mathbf{E} \cdot \mathbf{N} \, dS = 4\pi q \qquad \text{Gauss's Law}$$

where q is a point charge located at the center of the sphere. Gauss's Law is valid for more general closed surfaces, and relates the flux out of the surface to the total charge q inside the surface.

EXERCISES for Section 17.5

In Exercises 1–4, evaluate

$$\iint_S (x - 2y + z) \, dS.$$

1. $S: z = 4 - x, 0 \le x \le 4, 0 \le y \le 4$

2. $S: z = 10 - 2x + 2y, 0 \le x \le 2, 0 \le y \le 4$

3. $S: z = 10, x^2 + y^2 \le 1$

4. $S: z = \frac{2}{3}x^{3/2}, 0 \le x \le 1, 0 \le y \le x$

In Exercises 5–8, evaluate

$$\iint_S xy \, dS.$$

5. $S: z = 6 - x - 2y$ (first octant)
6. $S: z = xy, \, 0 \le x \le 2, \, 0 \le y \le 2$
7. $S: z = 9 - x^2, \, 0 \le x \le 2, \, 0 \le y \le x$
8. $S: z = h, \, 0 \le x \le 2, \, 0 \le y \le \sqrt{4 - x^2}$

In Exercises 9–14, evaluate

$$\iint_S f(x, y, z) \, dS.$$

9. $f(x, y, z) = x^2 + y^2 + z^2$
$S: z = x + 2, \, x^2 + y^2 \le 1$
10. $f(x, y, z) = \dfrac{xy}{z}$
$S: z = x^2 + y^2, \, 4 \le x^2 + y^2 \le 16$
11. $f(x, y, z) = \sqrt{x^2 + y^2 + z^2}$
$S: z = \sqrt{x^2 + y^2}, \, x^2 + y^2 \le 4$
12. $f(x, y, z) = \sqrt{x^2 + y^2 + z^2}$
$S: z = \sqrt{x^2 + y^2}, \, (x - 1)^2 + y^2 \le 1$
13. $f(x, y, z) = x^2 + y^2 + z^2$
$S: x^2 + y^2 = 9, \, 0 \le x \le 3, \, 0 \le y \le 3, \, 0 \le z \le 9$
14. $f(x, y, z) = x^2 + y^2 + z^2$
$S: x^2 + y^2 = 9, \, 0 \le x \le 3, \, 0 \le z \le x$

In Exercises 15–20, find the flux of **F** through S,

$$\iint_S \mathbf{F} \cdot \mathbf{N} \, dS$$

where **N** is the upper unit normal vector to S.

15. $\mathbf{F}(x, y, z) = 3z\mathbf{i} - 4\mathbf{j} + y\mathbf{k}$
$S: x + y + z = 1$ (first octant)
16. $\mathbf{F}(x, y, z) = x\mathbf{i} + y\mathbf{j}$
$S: 2x + 3y + z = 6$ (first octant)
17. $\mathbf{F}(x, y, z) = x\mathbf{i} + y\mathbf{j} + z\mathbf{k}$
$S: z = 9 - x^2 - y^2, \, 0 \le z$
18. $\mathbf{F}(x, y, z) = x\mathbf{i} + y\mathbf{j} + z\mathbf{k}$
$S: x^2 + y^2 + z^2 = 16$ (first octant)
19. $\mathbf{F}(x, y, z) = 4\mathbf{i} - 3\mathbf{j} + 5\mathbf{k}$
$S: z = x^2 + y^2, \, x^2 + y^2 \le 4$
20. $\mathbf{F}(x, y, z) = x\mathbf{i} + y\mathbf{j} - 2z\mathbf{k}$
$S: z = \sqrt{a^2 - x^2 - y^2}$

In Exercises 21 and 22, find the flux of **F** over the closed surface. (Let **N** be the outward unit normal vector of each surface.)

21. $\mathbf{F}(x, y, z) = 4xy\mathbf{i} + z^2\mathbf{j} + yz\mathbf{k}$
S: unit cube bounded by $x = 0, \, x = 1, \, y = 0, \, y = 1,$
$z = 0, \, z = 1$

22. $\mathbf{F}(x, y, z) = (x + y)\mathbf{i} + y\mathbf{j} + z\mathbf{k}$
$S: z = 1 - x^2 - y^2, \, z = 0$

In Exercises 23 and 24, find the mass of the surface lamina S of density ρ.

23. $S: 2x + 3y + 6z = 12$ (first octant), $\rho(x, y, z) = x^2 + y^2$
24. $S: z = \sqrt{a^2 - x^2 - y^2}, \, \rho(x, y, z) = kz$

In Exercises 25 and 26, use the following formulas for the moments of inertia about the coordinate axes of a surface lamina of density ρ.

$$I_x = \iint_S (y^2 + z^2)\rho(x, y, z) \, dS$$

$$I_y = \iint_S (x^2 + z^2)\rho(x, y, z) \, dS$$

$$I_z = \iint_S (x^2 + y^2)\rho(x, y, z) \, dS$$

25. Show that the moment of inertia of a conical shell about its axis is $\frac{1}{2}ma^2$, where m is the mass and a is the radius.
26. Show that the moment of inertia of a spherical shell of uniform density about its diameter is $\frac{2}{3}ma^2$, where m is the mass and a is the radius.

In Exercises 27 and 28, find I_z for the given lamina with uniform density of 1.

27. $x^2 + y^2 = a^2, \, 0 \le z \le h$
28. $z = x^2 + y^2, \, 0 \le z \le h$

In Exercises 29 and 30, find the rate of mass flow of a fluid of density ρ through the surface S oriented upwards if the velocity field is given by $\mathbf{F}(x, y, z) = 0.5z\mathbf{k}$.

29. $S: z = 16 - x^2 - y^2, \, z \ge 0$
30. $S: z = \sqrt{16 - x^2 - y^2}$

31. The surface shown in the accompanying figure is called a Möbius strip. Explain why this surface is not orientable.
32. Is the surface shown in the accompanying figure orientable?

FIGURE FOR 31   FIGURE FOR 32

In Exercises 33 and 34, use a computer to approximate

$$\iint_S (x^2 - 2xy)\, dS$$

over the indicated surface S.

33. S: $z = 10 - x^2 - y^2$, $0 \le x \le 2$, $0 \le y \le 2$

34. S: $z = \cos x$, $0 \le x \le \dfrac{\pi}{2}$, $0 \le y \le x$

35. Let $\mathbf{E} = yz\mathbf{i} + xz\mathbf{j} + xy\mathbf{k}$ be an electrostatic field. Use Gauss's Law to find the total charge enclosed by the closed surface consisting of the hemisphere $z = \sqrt{1 - x^2 - y^2}$ and its circular base in the xy-plane.

17.6 Divergence Theorem

Divergence Theorem ▪ Flux and the Divergence Theorem

Carl Friedrich Gauss

At the end of Section 17.4, we pointed out that Green's Theorem can be extended to three dimensions. This important extension is called the **Divergence Theorem** or **Gauss's Theorem,** after the famous German mathematician Carl Friedrich Gauss (1777–1855). Gauss is recognized, with Newton and Archimedes, as one of the three greatest mathematicians in history. One of his many contributions to mathematics was made at the age of twenty-two, when, as part of his doctoral dissertation, he proved the *Fundamental Theorem of Algebra*.

Recall from Section 17.4 that an alternate form of Green's Theorem is

$$\int_C \mathbf{F} \cdot \mathbf{N}\, ds = \iint_R \left(\frac{\partial M}{\partial x} + \frac{\partial N}{\partial y} \right) dA = \iint_R \operatorname{div} \mathbf{F}\, dA.$$

In an analogous way, the Divergence Theorem gives the relationship between a triple integral over a solid region Q and a surface integral over the surface of Q. In the statement of the theorem, the surface S is **closed** in the sense that it forms the complete boundary of the solid Q. Regions bounded by spheres, ellipsoids, cubes, tetrahedrons, or some combination of these surfaces are typical examples of closed surfaces. We assume that Q is a solid region on which a triple integral can be evaluated, and that the closed surface S is oriented by *outward* unit normal vectors, as shown in Figure 17.41. With these restrictions on S and Q, we state the following theorem.

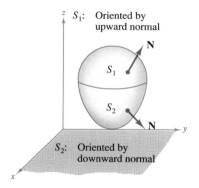

S_1: Oriented by upward normal

S_2: Oriented by downward normal

FIGURE 17.41

Semiellipsoid Topped by a Hemisphere

THEOREM 17.12 THE DIVERGENCE THEOREM	Let Q be a solid region bounded by a closed surface S oriented by a unit normal vector directed outward from Q. If $\mathbf{F}$ is a vector field whose component functions have continuous partial derivatives in Q, then

$$\iint_S \mathbf{F} \cdot \mathbf{N} \, dS = \iiint_Q \operatorname{div} \mathbf{F} \, dV.$$

PROOF   If we let $\mathbf{F}(x, y, z) = M\mathbf{i} + N\mathbf{j} + P\mathbf{k}$, the theorem takes the form

$$\iint_S (M\mathbf{i} \cdot \mathbf{N} + N\mathbf{j} \cdot \mathbf{N} + P\mathbf{k} \cdot \mathbf{N}) \, dS = \iiint_Q \left(\frac{\partial M}{\partial x} + \frac{\partial N}{\partial y} + \frac{\partial P}{\partial z} \right) dV.$$

We can prove this by showing that the following three equations are valid.

$$\iint_S M\mathbf{i} \cdot \mathbf{N} \, dS = \iiint_Q \frac{\partial M}{\partial x} \, dV$$

$$\iint_S N\mathbf{j} \cdot \mathbf{N} \, dS = \iiint_Q \frac{\partial N}{\partial y} \, dV$$

$$\iint_S P\mathbf{k} \cdot \mathbf{N} \, dS = \iiint_Q \frac{\partial P}{\partial z} \, dV$$

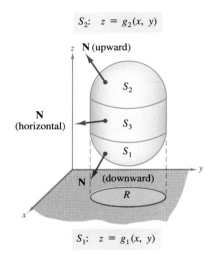

S_2: $z = g_2(x, y)$

z $\mathbf{N}$ (upward)

S_2

$\mathbf{N}$ (horizontal) S_3

S_1

$\mathbf{N}$ (downward)

R

x y

S_1: $z = g_1(x, y)$

FIGURE 17.42

Since the verifications of the three equations are similar, we will only discuss the third. We restrict the proof to a **simple solid** region with upper surface $z = g_2(x, y)$ and lower surface $z = g_1(x, y)$, whose projections on the xy-plane coincide and form region R. If Q has a lateral surface like S_3 in Figure 17.42, then a normal vector is horizontal, which implies that $P\mathbf{k} \cdot \mathbf{N} = 0$. Consequently, we have

$$\iint_S P\mathbf{k} \cdot \mathbf{N} \, dS = \iint_{S_1} P\mathbf{k} \cdot \mathbf{N} \, dS + \iint_{S_2} P\mathbf{k} \cdot \mathbf{N} \, dS + 0.$$

On the upper surface S_2, the outward normal vector is upward, whereas on the lower surface S_1, the outward normal vector is downward. Therefore, by Theorem 17.11, we have the following.

$$\iint_{S_1} P\mathbf{k} \cdot \mathbf{N} \, dS = \iint_R P(x, y, g_1(x, y))\mathbf{k} \cdot \left(\frac{\partial g_1}{\partial x}\mathbf{i} + \frac{\partial g_1}{\partial y}\mathbf{j} - \mathbf{k} \right) dA$$

$$= -\iint_R P(x, y, g_1(x, y)) \, dA$$

$$\iint_{S_2} P\mathbf{k} \cdot \mathbf{N} \, dS = \iint_R P(x, y, g_2(x, y))\mathbf{k} \cdot \left(-\frac{\partial g_2}{\partial x}\mathbf{i} - \frac{\partial g_2}{\partial y}\mathbf{j} + \mathbf{k} \right) dA$$

$$= \iint_R P(x, y, g_2(x, y)) \, dA$$

Adding these results, we obtain

$$\iint_S P\mathbf{k} \cdot \mathbf{N} \, dS = \iint_R [P(x, y, g_2(x, y)) - P(x, y, g_1(x, y))] \, dA$$

$$= \iint_R \left[\int_{g_1(x,y)}^{g_2(x,y)} \frac{\partial P}{\partial z} \, dz \right] dA = \iiint_Q \frac{\partial P}{\partial z} \, dV.$$

EXAMPLE 1   Using the Divergence Theorem

Let Q be the solid region bounded by the coordinate planes and the plane $2x + 2y + z = 6$, and let $\mathbf{F} = x\mathbf{i} + y^2\mathbf{j} + z\mathbf{k}$. Find

$$\iint_S \mathbf{F} \cdot \mathbf{N} \, dS$$

where S is the surface of Q.

SOLUTION

From Figure 17.43 we see that Q is bounded by four subsurfaces, and we would need four surface integrals to evaluate

$$\iint_S \mathbf{F} \cdot \mathbf{N} \, dS.$$

However, by the Divergence Theorem, we need only one triple integral. Since

$$\operatorname{div} \mathbf{F} = \frac{\partial M}{\partial x} + \frac{\partial N}{\partial y} + \frac{\partial P}{\partial z} = 1 + 2y + 1$$

we have

$$\iint_S \mathbf{F} \cdot \mathbf{N} \, dS = \iiint_Q \operatorname{div} \mathbf{F} \, dV$$

$$= \int_0^3 \int_0^{3-y} \int_0^{6-2x-2y} (1 + 2y + 1) \, dz \, dx \, dy$$

$$= \int_0^3 \int_0^{3-y} (2z + 2yz) \Big]_0^{6-2x-2y} \, dx \, dy$$

$$= \int_0^3 \int_0^{3-y} (12 - 4x + 8y - 4xy - 4y^2) \, dx \, dy$$

$$= \int_0^3 \left[12x - 2x^2 + 8xy - 2x^2y - 4xy^2 \right]_0^{3-y} \, dy$$

$$= \int_0^3 (18 + 6y - 10y^2 + 2y^3) \, dy$$

$$= \left[18y + 3y^2 - \frac{10y^3}{3} + \frac{y^4}{2} \right]_0^3 = \frac{63}{2}.$$

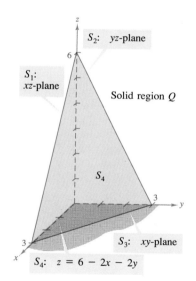

S_2: yz-plane

S_1: xz-plane

Solid region Q

S_4

S_3: xy-plane

S_4: $z = 6 - 2x - 2y$

FIGURE 17.43

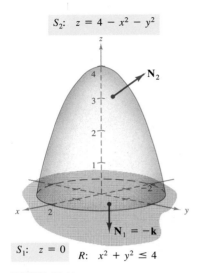

S_2: $z = 4 - x^2 - y^2$

N_2

$N_1 = -k$

S_1: $z = 0$ R: $x^2 + y^2 \le 4$

FIGURE 17.44

EXAMPLE 2   Verifying the Divergence Theorem

Let Q be the solid region between the paraboloid $z = 4 - x^2 - y^2$ and the xy-plane. Verify the Divergence Theorem for $\mathbf{F}(x, y, z) = 2z\mathbf{i} + x\mathbf{j} + y^2\mathbf{k}$.

SOLUTION

From Figure 17.44 we see that the outward normal vector for the surface S_1 is $\mathbf{N}_1 = -\mathbf{k}$, whereas the outward normal vector for the surface S_2 is

$$\mathbf{N}_2 = \frac{2x\mathbf{i} + 2y\mathbf{j} + \mathbf{k}}{\sqrt{4x^2 + 4y^2 + 1}}.$$

Thus, by Theorem 17.11, we have

$$\iint_S \mathbf{F} \cdot \mathbf{N} \, dS$$

$$= \iint_{S_1} \mathbf{F} \cdot (-\mathbf{k}) \, dS + \iint_{S_2} \mathbf{F} \cdot \mathbf{N}_2 \, dS$$

$$= \iint_R -y^2 \, dA + \iint_R \mathbf{F} \cdot (2x\mathbf{i} + 2y\mathbf{j} + \mathbf{k}) \, dA$$

$$= -\int_{-2}^{2} \int_{-\sqrt{4-y^2}}^{\sqrt{4-y^2}} y^2 \, dx \, dy + \int_{-2}^{2} \int_{-\sqrt{4-y^2}}^{\sqrt{4-y^2}} (4xz + 2xy + y^2) \, dx \, dy$$

$$= \int_{-2}^{2} \int_{-\sqrt{4-y^2}}^{\sqrt{4-y^2}} [4x(4 - x^2 - y^2) + 2xy] \, dx \, dy$$

$$= \int_{-2}^{2} \left[8x^2 - x^4 - 2x^2y^2 + x^2y \right]_{-\sqrt{4-y^2}}^{\sqrt{4-y^2}} dy = \int_{-2}^{2} 0 \, dy = 0.$$

On the other hand, because

$$\operatorname{div} \mathbf{F} = \frac{\partial}{\partial x}[2z] + \frac{\partial}{\partial y}[x] + \frac{\partial}{\partial z}[y^2] = 0$$

we obtain the equivalent result

$$\iiint_Q 0 \, dV = 0. \qquad \blacksquare$$

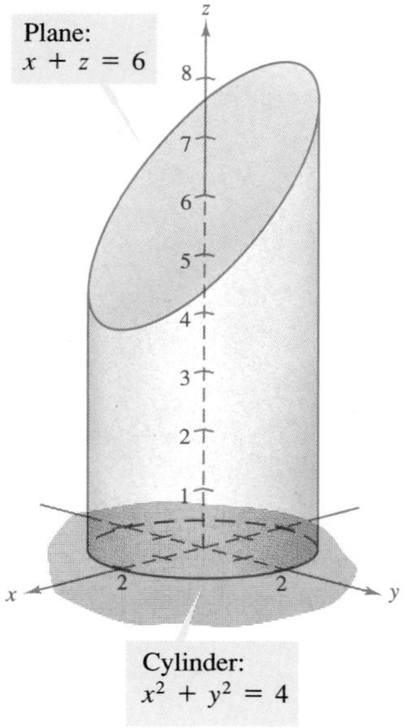

EXAMPLE 3   Using the Divergence Theorem

Plane: $x + z = 6$

Cylinder: $x^2 + y^2 = 4$

FIGURE 17.45

Let Q be the solid bounded by the cylinder $x^2 + y^2 = 4$, the plane $x + z = 6$, and the xy-plane, as shown in Figure 17.45. Find

$$\iint_S \mathbf{F} \cdot \mathbf{N} \, dS$$

where S is the surface of Q and $\mathbf{F}(x, y, z) = (x^2 + \sin z)\mathbf{i} + (xy + \cos z)\mathbf{j} + e^y\mathbf{k}$.

SOLUTION

Direct evaluation of this surface integral would be difficult. However, by the Divergence Theorem, we have

$$\iint_S \mathbf{F} \cdot \mathbf{N} \, dS = \iiint_Q \text{div } \mathbf{F} \, dV = \iiint_Q (2x + x + 0) \, dV$$

$$= \int_0^{2\pi} \int_0^2 \int_0^{6 - r\cos\theta} (3r\cos\theta)r \, dz \, dr \, d\theta \qquad \text{Cylindrical coordinates}$$

$$= \int_0^{2\pi} \int_0^2 (18r^2\cos\theta - 3r^3\cos^2\theta) \, dr \, d\theta$$

$$= \int_0^{2\pi} (48\cos\theta - 12\cos^2\theta) \, d\theta$$

$$= \left[48\sin\theta - 6\left(\theta + \frac{1}{2}\sin 2\theta\right) \right]_0^{2\pi} = -12\pi. \qquad \blacksquare$$

Flux and the Divergence Theorem

To help understand the Divergence Theorem, we consider the two sides of the equation

$$\iint_S \mathbf{F} \cdot \mathbf{N} \, dS = \iiint_Q \text{div } \mathbf{F} \, dV.$$

We know from Section 17.5 that the flux integral on the left determines the total fluid flow across the surface S per unit of time. This can be approximated by summing the fluid flow across small patches of the surface. The triple integral on the right measures this same fluid flow across S, but from a very different perspective—namely, by calculating the flow of fluid into (or out of) small *cubes* of volume ΔV_i. The flux of the ith cube is approximately

$$\text{flux of } i\text{th cube} \approx \text{div } \mathbf{F}(x_i, y_i, z_i)\Delta V_i$$

for some point (x_i, y_i, z_i) in the ith cube. Note that for a cube in the interior of Q, the gain (or loss) of fluid through any one of its six sides is offset by a corresponding loss (or gain) through one of the sides of an adjacent cube. After summing over all the cubes in Q, the only fluid flow that is not cancelled by adjoining cubes is that on the outside edges of the cubes on the boundary. Thus, the sum

$$\sum_{i=1}^{n} \text{div } \mathbf{F}(x_i, y_i, z_i)\Delta V_i$$

approximates the total flux into (or out of) Q, and therefore through the surface S.

To see what is meant by the divergence of $\mathbf{F}$ at a point, we consider ΔV_α to be the volume of a small sphere S_α of radius α and center (x_0, y_0, z_0), contained in region Q, as shown in Figure 17.46. Applying the Divergence Theorem to S_α produces

$$\text{flux of } \mathbf{F} \text{ across } S_\alpha = \iiint_{Q_\alpha} \text{div } \mathbf{F} \, dV \approx \text{div } \mathbf{F}(x_0, y_0, z_0)\Delta V_\alpha$$

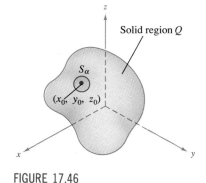

FIGURE 17.46

where Q_α is the interior of S_α. Consequently, we have

$$\text{div } \mathbf{F}(x_0, y_0, z_0) \approx \frac{\text{flux of } \mathbf{F} \text{ across } S_\alpha}{\Delta V_\alpha}$$

and, by taking the limit as $\alpha \to 0$, we obtain the divergence of $\mathbf{F}$ at the point (x_0, y_0, z_0).

$$\text{div } \mathbf{F}(x_0, y_0, z_0) = \lim_{\alpha \to 0} \frac{\text{flux of } \mathbf{F} \text{ across } S_\alpha}{\Delta V_\alpha} = \frac{\text{flux per unit volume}}{\text{at } (x_0, y_0, z_0)}$$

The point (x_0, y_0, z_0) in a vector field is classified as a source, a sink, or incompressible, as follows.

1. **Source,** if div $\mathbf{F} > 0$. [See Figure 17.47(a).]
2. **Sink,** if div $\mathbf{F} < 0$. [See Figure 17.47(b).]
3. **Incompressible,** if div $\mathbf{F} = 0$. [See Figure 17.47(c).]

FIGURE 17.47   (a) Source: div $\mathbf{F} > 0$ (b) Sink: div $\mathbf{F} < 0$ (c) Incompressible: div $\mathbf{F} = 0$

EXAMPLE 4   Calculating flux by the Divergence Theorem

Let Q be the region bounded by the sphere $x^2 + y^2 + z^2 = 4$. Find the outward flux of the vector field $\mathbf{F}(x, y, z) = 2x^3\mathbf{i} + 2y^3\mathbf{j} + 2z^3\mathbf{k}$ through the sphere.

SOLUTION

By the Divergence Theorem, we have

$$\begin{aligned}
\text{flux across } S &= \iint_S \mathbf{F} \cdot \mathbf{N} \, dS = \iiint_Q \text{div } \mathbf{F} \, dV \\
&= \iiint_Q 6(x^2 + y^2 + z^2) \, dV \\
&= 6 \int_0^2 \int_0^\pi \int_0^{2\pi} \rho^4 \sin\phi \, d\theta \, d\phi \, d\rho \qquad \text{\small Spherical coordinates} \\
&= 6 \int_0^2 \int_0^\pi 2\pi\rho^4 \sin\phi \, d\phi \, d\rho \\
&= 12\pi \int_0^2 2\rho^4 \, d\rho = 24\pi\left(\frac{32}{5}\right) = \frac{768\pi}{5}.
\end{aligned}$$

EXERCISES for Section 17.6

In Exercises 1–4, verify the Divergence Theorem by evaluating

$$\iint\limits_{S} \mathbf{F} \cdot \mathbf{N}\, dS$$

as a surface integral and as a triple integral.

1. $\mathbf{F}(x, y, z) = 2x\mathbf{i} - 2y\mathbf{j} + z^2\mathbf{k}$
   S: cube bounded by the planes $x = 0$, $x = a$, $y = 0$, $y = a$, $z = 0$, $z = a$
2. $\mathbf{F}(x, y, z) = 2x\mathbf{i} - 2y\mathbf{j} + z^2\mathbf{k}$
   S: cylinder $x^2 + y^2 = 1$, $0 \le z \le h$
3. $\mathbf{F}(x, y, z) = (2x - y)\mathbf{i} - (2y - z)\mathbf{j} + z\mathbf{k}$
   S: surface bounded by the plane $2x + 4y + 2z = 12$ and the coordinate planes
4. $\mathbf{F}(x, y, z) = xy\mathbf{i} + z\mathbf{j} + (x + y)\mathbf{k}$
   S: surface bounded by the planes $y = 4$, $z = 4 - x$ and the coordinate planes

In Exercises 5–14, use the Divergence Theorem to evaluate

$$\iint\limits_{S} \mathbf{F} \cdot \mathbf{N}\, dS$$

and find the outward flux of $\mathbf{F}$ through the surface of the solid bounded by the graphs of the equations.

5. $\mathbf{F}(x, y, z) = x^2\mathbf{i} + y^2\mathbf{j} + z^2\mathbf{k}$
   S: $x = 0$, $x = a$, $y = 0$, $y = a$, $z = 0$, $z = a$
6. $\mathbf{F}(x, y, z) = x^2z\mathbf{i} - y\mathbf{j} + xyz\mathbf{k}$
   S: $x = 0$, $x = a$, $y = 0$, $y = a$, $z = 0$, $z = a$
7. $\mathbf{F}(x, y, z) = x^2\mathbf{i} - 2xy\mathbf{j} + xyz^2\mathbf{k}$
   S: $z = \sqrt{a^2 - x^2 - y^2}$, $z = 0$
8. $\mathbf{F}(x, y, z) = xy\mathbf{i} + yz\mathbf{j} - yz\mathbf{k}$
   S: $z = \sqrt{a^2 - x^2 - y^2}$, $z = 0$
9. $\mathbf{F}(x, y, z) = x\mathbf{i} + y\mathbf{j} + z\mathbf{k}$
   S: $x^2 + y^2 + z^2 = 4$
10. $\mathbf{F}(x, y, z) = xyz\mathbf{j}$
    S: $x^2 + y^2 = 9$, $z = 0$, $z = 4$
11. $\mathbf{F}(x, y, z) = x\mathbf{i} + y^2\mathbf{j} - z\mathbf{k}$
    S: $x^2 + y^2 = 9$, $z = 0$, $z = 4$
12. $\mathbf{F}(x, y, z) = (xy^2 + \cos z)\mathbf{i} + (x^2y + \sin z)\mathbf{j} + e^z\mathbf{k}$
    S: $z = \sqrt{x^2 + y^2}$, $z = 4$
13. $\mathbf{F}(x, y, z) = x^3\mathbf{i} + x^2y\mathbf{j} + x^2e^y\mathbf{k}$
    S: $z = 4 - y$, $z = 0$, $x = 0$, $x = 6$, $y = 0$
14. $\mathbf{F}(x, y, z) = xe^z\mathbf{i} + ye^z\mathbf{j} + e^z\mathbf{k}$
    S: $z = 4 - y$, $z = 0$, $x = 0$, $x = 6$, $y = 0$

15. Use the Divergence Theorem to show that the volume of the solid bounded by a surface S is

$$\iint\limits_{S} x\, dy\, dz = \iint\limits_{S} y\, dz\, dx = \iint\limits_{S} z\, dx\, dy.$$

16. Verify the result of Exercise 15 for the cube bounded by $x = 0$, $x = a$, $y = 0$, $y = a$, $z = 0$, and $z = a$.

In Exercises 17 and 18, evaluate

$$\iint\limits_{S} \operatorname{curl} \mathbf{F} \cdot \mathbf{N}\, dS$$

where S is the closed surface of the solid bounded by the graphs of $x = 4$, $z = 9 - y^2$, and the coordinate planes.

17. $\mathbf{F}(x, y, z) = (4xy + z^2)\mathbf{i} + (2x^2 + 6yz)\mathbf{j} + 2xz\mathbf{k}$
18. $\mathbf{F}(x, y, z) = xy\cos z\,\mathbf{i} + yz\sin x\,\mathbf{j} + xyz\mathbf{k}$

19. Show that

$$\iint\limits_{S} \operatorname{curl} \mathbf{F} \cdot \mathbf{N}\, dS = 0$$

for any closed surface S.

20. For the constant vector field given by

$$\mathbf{F}(x, y, z) = a_1\mathbf{i} + a_2\mathbf{j} + a_3\mathbf{k}$$

show that

$$\iint\limits_{S} \mathbf{F} \cdot \mathbf{N}\, dS = 0$$

where S is any closed surface.

21. Given the vector field

$$\mathbf{F}(x, y, z) = x\mathbf{i} + y\mathbf{j} + z\mathbf{k}$$

show that

$$\iint\limits_{S} \mathbf{F} \cdot \mathbf{N}\, dS = 3V$$

where V is the volume of the solid bounded by the closed surface S.

22. Given the vector field

$$\mathbf{F}(x, y, z) = x\mathbf{i} + y\mathbf{j} + z\mathbf{k}$$

show that

$$\frac{1}{\|\mathbf{F}\|}\iint\limits_{S} \mathbf{F} \cdot \mathbf{N}\, dS = \frac{3}{\|\mathbf{F}\|}\iiint\limits_{Q} dV.$$

In Exercises 23 and 24, prove the identity, assuming that Q, S, and $\mathbf{N}$ meet the conditions of the Divergence Theorem and that the required partial derivatives of the scalar functions f and g are continuous. The expressions $D_{\mathbf{N}}f$ and $D_{\mathbf{N}}g$ are the derivatives in the direction of the vector $\mathbf{N}$ and are defined by

$$D_{\mathbf{N}}f = \nabla f \cdot \mathbf{N}, \quad D_{\mathbf{N}}g = \nabla g \cdot \mathbf{N}.$$

23. $\displaystyle\iiint_Q (f\nabla^2 g + \nabla f \cdot \nabla g)\, dV = \iint_S f\, D_{\mathbf{N}}g \, dS$

[Hint: Use div $(f\mathbf{G}) = f$ div $\mathbf{G} + \nabla f \cdot \mathbf{G}$.]

24. $\displaystyle\iiint_Q (f\nabla^2 g - g\nabla^2 f)\, dV = \iint_S (f\, D_{\mathbf{N}}g - g\, D_{\mathbf{N}}f)\, dS$

[Hint: Use Exercise 23 twice.]

17.7 Stokes's Theorem

Stokes's Theorem ▪ Physical interpretation of curl

George Gabriel Stokes

A second higher-dimension analogue of Green's Theorem is called **Stokes's Theorem,** after the English mathematical physicist George Gabriel Stokes (1819–1903). Stokes was part of a group of English mathematical physicists referred to as the Cambridge School, which included William Thompson (Lord Kelvin) and James Clerk Maxwell. In addition to making contributions to physics, Stokes worked with infinite series and differential equations, as well as with the integration results presented in this section.

Stokes's Theorem gives the relationship between a surface integral over an oriented surface S and a line integral along a closed space curve C forming the boundary of S, as shown in Figure 17.48. The positive direction along C is counterclockwise relative to the normal vector $\mathbf{N}$. That is, if you imagine grasping the normal vector $\mathbf{N}$ with your right hand, with your thumb pointing in the direction of $\mathbf{N}$, then your fingers will point toward the positive direction of C, as shown in Figure 17.49.

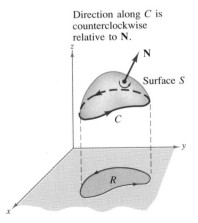

FIGURE 17.48

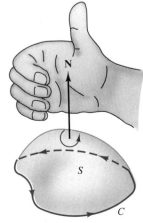

FIGURE 17.49

THEOREM 17.13 STOKES'S THEOREM	Let S be an oriented surface with unit normal vector $\mathbf{N}$, bounded by a piecewise smooth, simple closed curve C. If $\mathbf{F}$ is a vector field whose component functions have continuous partial derivatives on an open region containing S and C, then $$\int_C \mathbf{F} \cdot d\mathbf{r} = \iint_S (\text{curl } \mathbf{F}) \cdot \mathbf{N} \, dS.$$

REMARK The line integral may be expressed in the differential form $\int_C M \, dx + N \, dy + P \, dz$ or in the vector form $\int_C \mathbf{F} \cdot \mathbf{T} \, ds$.

EXAMPLE 1 Using Stokes's Theorem

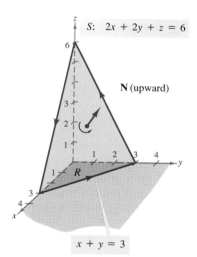

S: $2x + 2y + z = 6$

$\mathbf{N}$ (upward)

R

$x + y = 3$

FIGURE 17.50

Let C be the oriented triangle lying in the plane $2x + 2y + z = 6$, as shown in Figure 17.50. Evaluate

$$\int_C \mathbf{F} \cdot d\mathbf{r}$$

where $\mathbf{F}(x, y, z) = -y^2 \mathbf{i} + z \mathbf{j} + x \mathbf{k}$.

SOLUTION

Using Stokes's Theorem, we begin by finding the curl of $\mathbf{F}$.

$$\text{curl } \mathbf{F} = \begin{vmatrix} \mathbf{i} & \mathbf{j} & \mathbf{k} \\ \dfrac{\partial}{\partial x} & \dfrac{\partial}{\partial y} & \dfrac{\partial}{\partial z} \\ -y^2 & z & x \end{vmatrix} = -\mathbf{i} - \mathbf{j} + 2y\mathbf{k}$$

Considering $z = 6 - 2x - 2y = g(x, y)$ and replacing $\mathbf{F}$ by **curl F**, we can use Theorem 17.11 for an upward normal vector to obtain

$$\int_C \mathbf{F} \cdot d\mathbf{r} = \iint_S (\text{curl } \mathbf{F}) \cdot \mathbf{N} \, dS$$

$$= \iint_R (-\mathbf{i} - \mathbf{j} + 2y\mathbf{k}) \cdot [-g_x(x, y)\mathbf{i} - g_y(x, y)\mathbf{j} + \mathbf{k}] \, dA$$

$$= \iint_R (-\mathbf{i} - \mathbf{j} + 2y\mathbf{k}) \cdot (2\mathbf{i} + 2\mathbf{j} + \mathbf{k}) \, dA$$

$$= \int_0^3 \int_0^{3-y} (2y - 4) \, dx \, dy = \int_0^3 (-2y^2 + 10y - 12) \, dy$$

$$= \left[-\frac{2y^3}{3} + 5y^2 - 12y \right]_0^3 = -9.$$

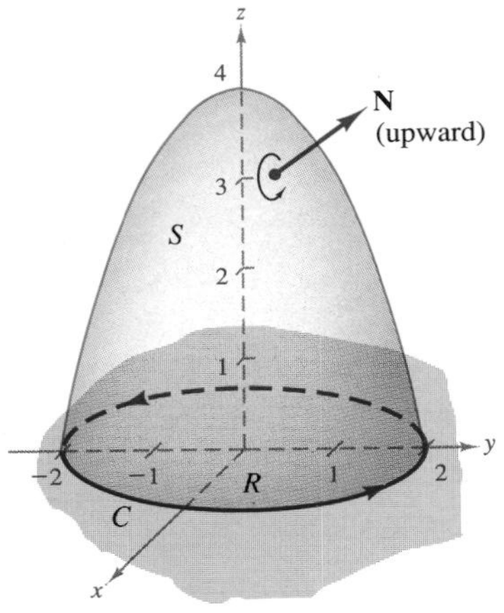

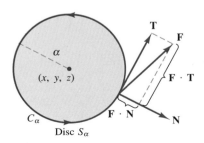

S: $z = 4 - x^2 - y^2$

R: $x^2 + y^2 \le 4$

FIGURE 17.51

EXAMPLE 2 Verifying Stokes's Theorem

Verify Stokes's Theorem for $\mathbf{F}(x, y, z) = 2z\mathbf{i} + x\mathbf{j} + y^2\mathbf{k}$, where S is the surface of the paraboloid $z = 4 - x^2 - y^2$ and C is the trace of S in the xy-plane, as shown in Figure 17.51.

SOLUTION

As a *surface integral*, we have $z = g(x, y) = 4 - x^2 - y^2$ and

$$\mathbf{curl\ F} = \begin{vmatrix} \mathbf{i} & \mathbf{j} & \mathbf{k} \\ \dfrac{\partial}{\partial x} & \dfrac{\partial}{\partial y} & \dfrac{\partial}{\partial z} \\ 2z & x & y^2 \end{vmatrix} = 2y\mathbf{i} + 2\mathbf{j} + \mathbf{k}.$$

By Theorem 17.11 for an upward normal vector $\mathbf{N}$, we obtain

$$\iint_S (\mathbf{curl\ F}) \cdot \mathbf{N}\, dS = \iint_R (2y\mathbf{i} + 2\mathbf{j} + \mathbf{k}) \cdot (2x\mathbf{i} + 2y\mathbf{j} + \mathbf{k})\, dA$$

$$= \int_{-2}^{2} \int_{-\sqrt{4-y^2}}^{\sqrt{4-y^2}} (4xy + 4y + 1)\, dx\, dy$$

$$= \int_{-2}^{2} (8y\sqrt{4 - y^2} + 2\sqrt{4 - y^2})\, dy$$

$$= \left[-\frac{8}{3}(4 - y^2)^{3/2} + y\sqrt{4 - y^2} + 4 \arcsin \frac{y}{2} \right]_{-2}^{2}$$

$$= 4\pi.$$

As a *line integral*, we parameterize C by

$$\mathbf{r}(t) = 2 \cos t\, \mathbf{i} + 2 \sin t\, \mathbf{j} + 0\mathbf{k}, \quad 0 \le t \le 2\pi.$$

For $\mathbf{F}(x, y, z) = 2z\mathbf{i} + x\mathbf{j} + y^2\mathbf{k}$, we obtain

$$\int_C \mathbf{F} \cdot d\mathbf{r} = \int_C M\, dx + N\, dy + P\, dz = \int_C 2z\, dx + x\, dy + y^2\, dz$$

$$= \int_0^{2\pi} [0 + 2 \cos t(2 \cos t) + 0]\, dt$$

$$= 2 \int_0^{2\pi} (1 + \cos 2t)\, dt = 2 \left[t + \frac{1}{2} \sin 2t \right]_0^{2\pi} = 4\pi. \quad \blacksquare$$

Physical interpretation of curl

Stokes's Theorem provides insight into a physical interpretation of curl. In a vector field $\mathbf{F}$, let S_α be a *small* circular disc of radius α, centered at (x, y, z) and with boundary C_α, as shown in Figure 17.52. At each point on the circle C_α, $\mathbf{F}$ has a normal component $\mathbf{F} \cdot \mathbf{N}$ and a tangential component $\mathbf{F} \cdot \mathbf{T}$. The more closely $\mathbf{F}$ and $\mathbf{T}$ are aligned, the greater the value of $\mathbf{F} \cdot \mathbf{T}$.

FIGURE 17.52

Thus, the fluid tends to move along the circle rather than across it. Consequently, we say that the line integral around C_α measures the **circulation of F around C_α**. That is,

$$\int_{C_\alpha} \mathbf{F} \cdot \mathbf{T} \, ds = \text{circulation of } \mathbf{F} \text{ around } C_\alpha.$$

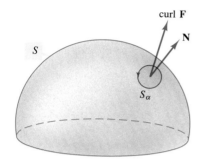

curl F

N

S

S_α

FIGURE 17.53

Now consider a small disc S_α to be centered at some point (x, y, z) on the surface S, as shown in Figure 17.53. On such a small disc, **curl F** is nearly constant, since it varies little from its value at (x, y, z). Moreover, **curl F** $\cdot$ **N** is also nearly constant on S_α, since all unit normals to S_α are about the same. Consequently, Stokes's Theorem yields

$$\int_{C_\alpha} \mathbf{F} \cdot \mathbf{T} \, ds = \iint_{S_\alpha} (\textbf{curl F}) \cdot \mathbf{N} \, dS$$

$$\approx (\textbf{curl F}) \cdot \mathbf{N} \iint_{S_\alpha} dS$$

$$\approx (\textbf{curl F}) \cdot \mathbf{N}(\pi\alpha^2).$$

Therefore,

$$(\textbf{curl F}) \cdot \mathbf{N} \approx \frac{\displaystyle\int_{C_\alpha} \mathbf{F} \cdot \mathbf{T} \, ds}{\pi\alpha^2}$$

$$= \frac{\text{circulation of } \mathbf{F} \text{ around } C_\alpha}{\text{area of disc } S_\alpha}$$

$$= \text{rate of circulation.}$$

Assuming conditions are such that the approximation improves over smaller and smaller discs ($\alpha \to 0$), it follows that

$$(\textbf{curl F}) \cdot \mathbf{N} = \lim_{\alpha \to 0} \frac{1}{\pi\alpha^2} \int_{C_\alpha} \mathbf{F} \cdot \mathbf{T} \, ds$$

which is referred to as the **rotation of F about N**. That is,

curl $\mathbf{F}(x, y, z) \cdot \mathbf{N} = $ rotation of **F** about **N** at (x, y, z).

In this case, the rotation of **F** is maximum when **curl F** and **N** have the same direction. Normally, this tendency to rotate will vary from point to point on the surface S, and Stokes's Theorem says that the collective measure of this *rotational* tendency taken over the entire surface S (surface integral) is equal to the tendency of the fluid to *circulate* around the boundary C (line integral).

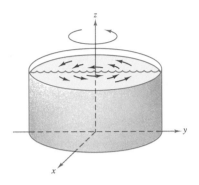

FIGURE 17.54

EXAMPLE 3 An application of curl

A liquid is swirling around in a circular container of radius 2, so that its motion is described by the velocity field

$$\mathbf{F}(x, y, z) = -y\sqrt{x^2 + y^2}\,\mathbf{i} + x\sqrt{x^2 + y^2}\,\mathbf{j}$$

as shown in Figure 17.54. Find

$$\iint_S (\mathbf{curl\ F}) \cdot \mathbf{N}\ dS$$

where S is the upper surface of the cylindrical tank.

SOLUTION

The curl of $\mathbf{F}$ is given by

$$\mathbf{curl\ F} = \begin{vmatrix} \mathbf{i} & \mathbf{j} & \mathbf{k} \\ \dfrac{\partial}{\partial x} & \dfrac{\partial}{\partial y} & \dfrac{\partial}{\partial z} \\ -y\sqrt{x^2 + y^2} & x\sqrt{x^2 + y^2} & 0 \end{vmatrix} = 3\sqrt{x^2 + y^2}\,\mathbf{k}.$$

Letting $\mathbf{N} = \mathbf{k}$, we have

$$\iint_S (\mathbf{curl\ F}) \cdot \mathbf{N}\ dS = \iint_R 3\sqrt{x^2 + y^2}\ dA$$

$$= \int_0^{2\pi}\int_0^2 (3r)r\ dr\ d\theta$$

$$= \int_0^{2\pi} 8\ d\theta = 16\pi. \qquad \blacksquare$$

If $\mathbf{curl\ F} = \mathbf{0}$ throughout region Q, then the rotation of $\mathbf{F}$ about each unit normal $\mathbf{N}$ is zero. That is, $\mathbf{F}$ is irrotational. From earlier work, we know that this is a characteristic of conservative vector fields.

EXERCISES for Section 17.7

In Exercises 1–6, find the curl of the vector field $\mathbf{F}$.

1. $\mathbf{F}(x, y, z) = (2y - z)\mathbf{i} + xyz\,\mathbf{j} + e^z\mathbf{k}$
2. $\mathbf{F}(x, y, z) = z^2\mathbf{i} + y^2\mathbf{j} + x^2\mathbf{k}$
3. $\mathbf{F}(x, y, z) = 2z\mathbf{i} - 4x^2\mathbf{j} + \arctan x\ \mathbf{k}$
4. $\mathbf{F}(x, y, z) = x\sin y\ \mathbf{i} - y\cos x\,\mathbf{j} + yz^2\mathbf{k}$
5. $\mathbf{F}(x, y, z) = e^{x^2+y^2}\mathbf{i} + e^{y^2+z^2}\mathbf{j} + xyz\,\mathbf{k}$
6. $\mathbf{F}(x, y, z) = \arcsin y\ \mathbf{i} + \sqrt{1 - x^2}\,\mathbf{j} + y^2\mathbf{k}$

In Exercises 7–10, verify Stokes's Theorem by evaluating

$$\int_C \mathbf{F} \cdot \mathbf{T}\ dS$$

as a line integral and as a double integral.

7. $\mathbf{F}(x, y, z) = (-y + z)\mathbf{i} + (x - z)\mathbf{j} + (x - y)\mathbf{k}$
   $S:\ z = \sqrt{1 - x^2 - y^2}$

8. $\mathbf{F}(x, y, z) = (-y + z)\mathbf{i} + (x - z)\mathbf{j} + (x - y)\mathbf{k}$
  $S: z = 4 - x^2 - y^2, 0 \le z$

9. $\mathbf{F}(x, y, z) = xyz\mathbf{i} + y\mathbf{j} + z\mathbf{k}$
  $S: 3x + 4y + 2z = 12$ (first octant)

10. $\mathbf{F}(x, y, z) = z^2\mathbf{i} + x^2\mathbf{j} + y^2\mathbf{k}$
  $S: z = x^2, 0 \le x \le a, 0 \le y \le a$

In Exercises 11–20, use Stokes's Theorem to evaluate

$$\int_C \mathbf{F} \cdot d\mathbf{r}.$$

11. $\mathbf{F}(x, y, z) = 2y\mathbf{i} + 3z\mathbf{j} + x\mathbf{k}$
  C: triangle with vertices $(0, 0, 0)$, $(0, 2, 0)$, $(1, 1, 1)$

12. $\mathbf{F}(x, y, z) = \arctan \dfrac{x}{y}\mathbf{i} + \ln \sqrt{x^2 + y^2}\,\mathbf{j} + \mathbf{k}$
  C: triangle with vertices $(0, 0, 0)$, $(1, 1, 1)$, $(0, 0, 2)$

13. $\mathbf{F}(x, y, z) = z^2\mathbf{i} + x^2\mathbf{j} + y^2\mathbf{k}$
  $S: z = 4 - x^2 - y^2, 0 \le z$

14. $\mathbf{F}(x, y, z) = 4xz\mathbf{i} + y\mathbf{j} + 4xy\mathbf{k}$
  $S: z = 4 - x^2 - y^2, 0 \le z$

15. $\mathbf{F}(x, y, z) = z^2\mathbf{i} + y\mathbf{j} + xz\mathbf{k}$
  $S: z = \sqrt{4 - x^2 - y^2}$

16. $\mathbf{F}(x, y, z) = x^2\mathbf{i} + z^2\mathbf{j} - xyz\mathbf{k}$
  $S: z = \sqrt{4 - x^2 - y^2}$

17. $\mathbf{F}(x, y, z) = -\ln \sqrt{x^2 + y^2}\,\mathbf{i} + \arctan \dfrac{x}{y}\mathbf{j} + \mathbf{k}$
  $S: z = 9 - 2x - 3y$ over one petal of $r = 2 \sin 2\theta$ in the first octant

18. $\mathbf{F}(x, y, z) = yz\mathbf{i} + (2 - 3y)\mathbf{j} + (x^2 + y^2)\mathbf{k}$
  S: the first octant portion of $x^2 + z^2 = 16$ over $x^2 + y^2 = 16$

19. $\mathbf{F}(x, y, z) = xyz\mathbf{i} + y\mathbf{j} + z\mathbf{k}$
  $S: z = x^2, 0 \le x \le a, 0 \le y \le a$

20. $\mathbf{F}(x, y, z) = xyz\mathbf{i} + y\mathbf{j} + z\mathbf{k}$
  S: the first octant portion of $z = x^2$ over $x^2 + y^2 = a^2$

In Exercises 21 and 22, the motion of a liquid in a cylindrical container of radius 1 is described by the velocity field $\mathbf{F}(x, y, z)$. Find

$$\iint_S (\text{curl }\mathbf{F}) \cdot \mathbf{N}\, dS$$

where S is the upper surface of the cylindrical tank.

21. $\mathbf{F}(x, y, z) = \mathbf{i} + \mathbf{j} - 2\mathbf{k}$
22. $\mathbf{F}(x, y, z) = -y\mathbf{i} + x\mathbf{j}$

23. Let f and g be scalar functions with continuous partial derivatives, and let C and S satisfy the conditions of Stokes's Theorem. Verify each of the following identities.

(a) $\displaystyle\int_C (f\nabla g) \cdot d\mathbf{r} = \iint_S (\nabla f \times \nabla g) \cdot \mathbf{N}\, dS$

(b) $\displaystyle\int_C (f\nabla f) \cdot d\mathbf{r} = 0$

(c) $\displaystyle\int_C (f\nabla g + g\nabla f) \cdot d\mathbf{r} = 0$

24. Demonstrate the result of Exercise 23 for the functions $f(x, y, z) = xyz$ and $g(x, y, z) = z$. Let S be the hemisphere $z = \sqrt{4 - x^2 - y^2}$.

REVIEW EXERCISES for Chapter 17

In Exercises 1 and 2, find a three-dimensional vector field that has the potential function f.

1. $f(x, y, z) = 8x^2 + xy + z^2$
2. $f(x, y, z) = x^2 e^{yz}$

In Exercises 3–10, determine if $\mathbf{F}$ is conservative. If it is, find the potential function f.

3. $\mathbf{F}(x, y) = \dfrac{1}{y}\mathbf{i} - \dfrac{y}{x^2}\mathbf{j}$

4. $\mathbf{F}(x, y) = -\dfrac{y}{x^2}\mathbf{i} + \dfrac{1}{x}\mathbf{j}$

5. $\mathbf{F}(x, y) = (6xy^2 - 3x^2)\mathbf{i} + (6x^2y + 3y^2 - 7)\mathbf{j}$
6. $\mathbf{F}(x, y) = (-2y^3 \sin 2x)\mathbf{i} + 3y^2(1 + \cos 2x)\mathbf{j}$
7. $\mathbf{F}(x, y, z) = (4xy + z)\mathbf{i} + (2x^2 + 6y)\mathbf{j} + 2z\mathbf{k}$
8. $\mathbf{F}(x, y, z) = (4xy + z^2)\mathbf{i} + (2x^2 + 6yz)\mathbf{j} + 2xz\mathbf{k}$

9. $\mathbf{F}(x, y, z) = \dfrac{yz\mathbf{i} - xz\mathbf{j} - xy\mathbf{k}}{y^2z^2}$

10. $\mathbf{F}(x, y, z) = \sin z\,(y\mathbf{i} + x\mathbf{j} + \mathbf{k})$

In Exercises 11–18, find (a) the divergence and (b) the curl of the vector field $\mathbf{F}$.

11. $\mathbf{F}(x, y, z) = x^2\mathbf{i} + y^2\mathbf{j} + z^2\mathbf{k}$
12. $\mathbf{F}(x, y, z) = xy^2\mathbf{j} - zx^2\mathbf{k}$
13. $\mathbf{F}(x, y, z) = (\cos y + y \cos x)\mathbf{i}$
  $+ (\sin x - x \sin y)\mathbf{j} + xyz\mathbf{k}$
14. $\mathbf{F}(x, y, z) = (3x - y)\mathbf{i} + (y - 2z)\mathbf{j} + (z - 3x)\mathbf{k}$
15. $\mathbf{F}(x, y, z) = \arcsin x\,\mathbf{i} + xy^2\mathbf{j} + yz^2\mathbf{k}$
16. $\mathbf{F}(x, y, z) = (x^2 - y)\mathbf{i} - (x + \sin^2 y)\mathbf{j}$
17. $\mathbf{F}(x, y, z) = \ln (x^2 + y^2)\mathbf{i} + \ln (x^2 + y^2)\mathbf{j} + z\mathbf{k}$
18. $\mathbf{F}(x, y, z) = \dfrac{z}{x}\mathbf{i} + \dfrac{z}{y}\mathbf{j} + z^2\mathbf{k}$

19. Evaluate

$$\int_C (x^2 + y^2)\, ds$$

for the following curves.
(a) C: line segment from $(-1, -1)$ to $(2, 2)$
(b) C: $x^2 + y^2 = 16$, one revolution counterclockwise, starting at $(4, 0)$

20. Evaluate

$$\int_C xy\, ds$$

for the following curves.
(a) C: line segment from $(0, 0)$ to $(5, 4)$
(b) C: counterclockwise around the triangle with vertices $(0, 0)$, $(4, 0)$, $(0, 2)$

21. Evaluate

$$\int_C (x^2 + y^2)\, ds$$

for the following curve.
C: $\mathbf{r}(t) = (\cos t + t \sin t)\,\mathbf{i} + (\sin t - t \cos t)\,\mathbf{j}$,
    $0 \le t \le 2\pi$

22. Evaluate

$$\int_C x\, ds$$

for the following curve.
C: $\mathbf{r}(t) = (t - \sin t)\,\mathbf{i} + (1 - \cos t)\,\mathbf{j}$, $0 \le t \le 2\pi$

23. Evaluate

$$\int_C (2x - y)\, dx + (x + 3y)\, dy$$

for the following curves.
(a) C: line segment from $(0, 0)$ to $(2, -3)$
(b) C: counterclockwise around the circle $x = 3 \cos t$, $y = 3 \sin t$

24. Evaluate

$$\int_C (2x - y)\, dx + (x + 3y)\, dy$$

for the following curve.
C: $\mathbf{r}(t) = (\cos t + t \sin t)\,\mathbf{i} + (\sin t - t \sin t)\,\mathbf{j}$,
    $0 \le t \le \pi/2$

In Exercises 25–30, evaluate

$$\int_C \mathbf{F} \cdot d\mathbf{r}.$$

25. $\mathbf{F}(x, y) = xy\,\mathbf{i} + x^2\,\mathbf{j}$
    C: $\mathbf{r}(t) = t^2\,\mathbf{i} + t^3\,\mathbf{j}$, $0 \le t \le 1$
26. $\mathbf{F}(x, y) = (x - y)\,\mathbf{i} + (x + y)\,\mathbf{j}$
    C: $\mathbf{r}(t) = 4 \cos t\,\mathbf{i} + 3 \sin t\,\mathbf{j}$, $0 \le t \le 2\pi$
27. $\mathbf{F}(x, y, z) = x\,\mathbf{i} + y\,\mathbf{j} + z\,\mathbf{k}$
    C: $\mathbf{r}(t) = 2 \cos t\,\mathbf{i} + 2 \sin t\,\mathbf{j} + t\,\mathbf{k}$, $0 \le t \le 2\pi$
28. $\mathbf{F}(x, y, z) = (2y - z)\,\mathbf{i} + (z - x)\,\mathbf{j} + (x - y)\,\mathbf{k}$
    C: curve of intersection of $x^2 + z^2 = 4$ and $y^2 + z^2 = 4$ from $(2, 2, 0)$ to $(0, 0, 2)$
29. $\mathbf{F}(x, y, z) = (y - z)\,\mathbf{i} + (z - x)\,\mathbf{j} + (x - y)\,\mathbf{k}$
    C: curve of intersection of $z = x^2 + y^2$ and $x + y = 0$ from $(-2, 2, 8)$ to $(2, -2, 8)$
30. $\mathbf{F}(x, y, z) = (x^2 - z)\,\mathbf{i} + (y^2 + z)\,\mathbf{j} + x\,\mathbf{k}$
    C: curve of intersection of $z = x^2$ and $x^2 + y^2 = 4$ from $(0, -2, 0)$ to $(0, 2, 0)$

In Exercises 31 and 32, use the Fundamental Theorem of Line Integrals to evaluate the given integral.

31. $\displaystyle \int_{(0,0,0)}^{(1,4,3)} 2xyz\, dx + x^2z\, dy + x^2y\, dz$

32. $\displaystyle \int_{(0,0,1)}^{(4,4,4)} y\, dx + x\, dy + \frac{1}{z}\, dz$

In Exercises 33–38, use Green's Theorem to evaluate each line integral.

33. $\displaystyle \int_C y\, dx + 2x\, dy$

   C: boundary of the square with vertices $(0, 0)$, $(0, 2)$, $(2, 0)$, $(2, 2)$

34. $\displaystyle \int_C xy\, dx + (x^2 + y^2)\, dy$

   C: boundary of the square with vertices $(0, 0)$, $(0, 2)$, $(2, 0)$, $(2, 2)$

35. $\displaystyle \int_C xy^2\, dx + x^2y\, dy$

   C: $x = 4 \cos t$, $y = 2 \sin t$

36. $\displaystyle \int_C (x^2 - y^2)\, dx + 2xy\, dy$

   C: $x^2 + y^2 = a^2$

37. $\int_C xy \, dx + x^2 \, dy$

    C: boundary of the region between the graphs of $y = x^2$ and $y = x$

38. $\int_C y^2 \, dx + x^{2/3} \, dy$

    C: $x^{2/3} + y^{2/3} = 1$

In Exercises 39 and 40, verify the Divergence Theorem by evaluating

$$\int_S \mathbf{F} \cdot \mathbf{N} \, dS$$

as a surface integral and as a triple integral.

39. $\mathbf{F}(x, y, z) = x^2 \mathbf{i} + xy \mathbf{j} + z \mathbf{k}$

    Q: solid region bounded by the coordinate planes and the plane $2x + 3y + 4z = 12$

40. $\mathbf{F}(x, y, z) = x \mathbf{i} + y \mathbf{j} + z \mathbf{k}$

    Q: solid region bounded by the coordinate planes and the plane $2x + 3y + 4z = 12$

In Exercises 41 and 42, verify Stokes's Theorem by evaluating

$$\int_C \mathbf{F} \cdot d\mathbf{r}$$

as a line integral and as a double integral.

41. $\mathbf{F}(x, y, z) = (\cos y + y \cos x) \mathbf{i}$
                     $+ (\sin x - x \sin y) \mathbf{j} + xyz \mathbf{k}$

    S: portion of $z = y^2$ over the square in the xy-plane with vertices $(0, 0)$, $(a, 0)$, (a, a), $(0, a)$

42. $\mathbf{F}(x, y, z) = (x - z) \mathbf{i} + (y - z) \mathbf{j} + x^2 \mathbf{k}$

    S: first octant portion of the plane $3x + y + 2z = 12$

43. Consider the integral

$$\int_C \mathbf{F} \cdot d\mathbf{r}$$

where

$$\mathbf{F}(x, y) = \frac{-y}{x^2 + y^2} \mathbf{i} + \frac{x}{x^2 + y^2} \mathbf{j}$$

and

$$\mathbf{r}(t) = \cos t \, \mathbf{i} + \sin t \, \mathbf{j}.$$

Since $N_x = M_y$ and C is a circle, it is expected that the line integral will have a value of zero. However, upon direct integration we obtain

$$\int_C \mathbf{F} \cdot d\mathbf{r} = 2\pi.$$

Which is correct, and why?

Chapter 18 Application

Descent speeds vary among parachutists. A speed of approximately 16 miles per hour is common. This is the speed that would be attained by a person jumping (without a parachute) from a height of 8.6 feet. The parachutist must know proper landing techniques to avoid injury at such a landing speed.

Velocity of a Parachutist

The fall of a parachutist is given by the second-order linear differential equation

$$\frac{w}{g}\frac{d^2y}{dt^2} - k\frac{dy}{dt} = w$$

where w is the weight of the parachutist, y is the height at time t, g is the acceleration due to gravity, and k measures the drag-factor of the parachute.

For instance, if the parachute is opened at 2000 feet, $y(0) = 2000$, and at that time the velocity is -100 ft/sec, $y'(0) = -100$, then for a 160-pound parachutist, using $k = 8$, the differential equation is

$$-5y'' - 8y' = 160.$$

Using the given initial conditions, the solution of this equation is

$$y = 1950 + 50e^{-1.6t} - 20t.$$

From this solution we can determine that the parachutist will reach the ground about 97.5 seconds after opening the parachute, and will be traveling at a speed of about 20 ft/sec (≈ 13.6 mi/hr) at impact.

Chapter Overview

In this chapter we look at techniques for solving differential equations. A **differential equation** is an equation involving a function and one or more of its derivatives. For example,

$$\frac{dy}{dx} - xy = 0$$

is a first-order differential equation, and

$$y'' + 4y' - 8y = 0$$

is a second-order differential equation.

The chapter begins by looking at the *form* of the general solution of a differential equation and methods for using **initial conditions** to find a particular solution of a differential equation.

Then, in the next three sections, we look at various methods for *solving* first-order differential equations. In Sections 18.5 and 18.6, we study methods for solving second-order linear differential equations and show how these methods can be extended to higher-order linear differential equations.

Finally, in the last section, we discuss methods of approximating the solution of a differential equation by means of a series.

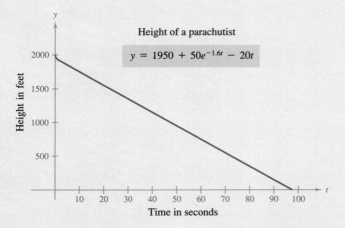

Height of a parachutist

$$y = 1950 + 50e^{-1.6t} - 20t$$

See Exercise 31, Section 18.6.

Differential Equations

18

18.1 Definitions and Basic Concepts

Type and order ▪ General and particular solutions

Several times in the text we have identified physical phenomena that can be described by differential equations. For example, we saw that problems involving radioactive decay, population growth, chemical reactions, Newton's Law of Cooling, and gravitational force can be formulated in terms of differential equations.

A **differential equation** is an equation involving a function and one or more of its derivatives. If the function has only one independent variable, the equation is called an **ordinary differential equation.** For instance,

$$\frac{d^2y}{dx^2} + 3\frac{dy}{dx} - 2y = 0$$

is an ordinary differential equation in which the dependent variable $y = f(x)$ is a twice differentiable function of x. A differential equation involving a function of several variables and its partial derivatives is called a **partial differential equation.** In this chapter we restrict our discussion to ordinary differential equations.

In addition to **type** (ordinary or partial), differential equations are classified by order. The **order** of a differential equation is determined by the highest-order derivative in the equation.

EXAMPLE 1 Classifying differential equations

Equation	Type	Order
(a) $y''' + 4y = 2$	Ordinary	3
(b) $\dfrac{d^2s}{dt^2} = -32$	Ordinary	2
(c) $(y')^2 - 3y = e^x$	Ordinary	1
(d) $\dfrac{\partial^2 u}{\partial x^2} + \dfrac{\partial^2 u}{\partial y^2} = 0$	Partial	2
(e) $y - \sin y' = 0$	Ordinary	1

A function $y = f(x)$ is called a **solution** of a differential equation if the equation is satisfied when y and its derivatives are replaced by $f(x)$ and its derivatives, respectively. For example, differentiation and substitution would show that $y = e^{-2x}$ is a solution of the differential equation

$$y' + 2y = 0.$$

It can be shown that every solution of this differential equation is of the form $y = Ce^{-2x}$, where C is any real number, and we call $y = Ce^{-2x}$ the **general solution.** (Some differential equations have **singular solutions** that cannot be written as special cases of the general solution. However, we will not consider such solutions in this text.)

In Section 5.1, Example 8, we saw that the second-order differential equation $s''(t) = -32$ has the general solution

$$s(t) = -16t^2 + C_1 t + C_2$$

which contains two arbitrary constants. It can be shown that a differential equation of order n has a general solution with n arbitrary constants.

EXAMPLE 2   Verifying solutions

Determine whether or not the given functions are solutions to the differential equation $y'' - y = 0$.

(a) $y = \sin x$ (b) $y = e^{2x}$ (c) $y = 4e^{-x}$ (d) $y = Ce^x$

SOLUTION

(a) Since $y = \sin x$, $y' = \cos x$, and $y'' = -\sin x$, it follows that

$$y'' - y = -\sin x - \sin x = -2 \sin x \neq 0.$$

Hence, $y = \sin x$ is *not* a solution.

(b) Since $y = e^{2x}$, $y' = 2e^{2x}$, and $y'' = 4e^{2x}$, it follows that

$$y'' - y = 4e^{2x} - e^{2x} = 3e^{2x} \neq 0.$$

Hence, $y = e^{2x}$ is *not* a solution.

(c) Since $y = 4e^{-x}$, $y' = -4e^{-x}$, and $y'' = 4e^{-x}$, it follows that

$$y'' - y = 4e^{-x} - 4e^{-x} = 0.$$

Hence, $y = 4e^{-x}$ *is* a solution.

(d) Since $y = Ce^x$, $y' = Ce^x$, and $y'' = Ce^x$, it follows that

$$y'' - y = Ce^x - Ce^x = 0.$$

Hence, $y = Ce^x$ *is* a solution for any value of C.   ⊏⊐

Later in this chapter we will see that the general solution of the differential equation in Example 2 is $y = C_1 e^x + C_2 e^{-x}$. A **particular solution** of a differential equation is any solution that is obtained by assigning specific values to the constants in the general solution.

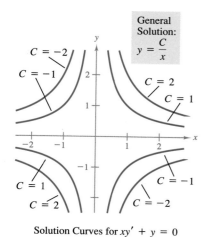

General Solution:

$$y = \frac{C}{x}$$

Solution Curves for $xy' + y = 0$

FIGURE 18.1

Geometrically, the general solution of a first-order differential equation represents a family of curves known as **solution curves,** one for each value assigned to the arbitrary constant. For instance, we can easily verify that every function of the form $y = C/x$ is a solution of the differential equation $xy' + y = 0$. Figure 18.1 shows some of the solution curves corresponding to different values of C.

Particular solutions of a differential equation are obtained from **initial conditions** that give the value of the dependent variable or one of its derivatives for a particular value of the independent variable. The term "initial condition" stems from the fact that, often in problems involving time, the value of the dependent variable or one of its derivatives is known at the *initial* time $t = 0$. For instance, the second-order differential equation $s''(t) = -32$ having the general solution

$$s(t) = -16t^2 + C_1 t + C_2$$

might have the following initial conditions.

$$s(0) = 80, \qquad s'(0) = 64 \qquad \text{Initial conditions}$$

In this case, the initial conditions yield the particular solution

$$s(t) = -16t^2 + 64t + 80.$$

EXAMPLE 3 Finding a particular solution

For the differential equation $xy' - 3y = 0$, verify that $y = Cx^3$ is a solution and find the particular solution determined by the initial condition $y = 2$ when $x = -3$.

SOLUTION

We know that $y = Cx^3$ is a solution because $y' = 3Cx^2$ and

$$xy' - 3y = x(3Cx^2) - 3(Cx^3) = 0.$$

Furthermore, the initial condition $y = 2$ when $x = -3$ yields

$$y = Cx^3 \quad \Longrightarrow \quad 2 = C(-3)^3 \quad \Longrightarrow \quad C = -\frac{2}{27}$$

and we conclude that the particular solution is $y = -2x^3/27$.

REMARK   To determine a particular solution, the number of initial conditions must match the number of constants in the general solution.

In the remainder of this chapter we will discuss procedures for finding the general solution to several classes of differential equations. Each section will include practical applications of that particular type of differential equation. We will focus mainly on first- and second-order equations and linear equations with constant coefficients. By a **linear differential equation with constant coefficients** $a_0, a_1, a_2, \ldots, a_n$, we mean an equation of the form

$$a_n y^{(n)} + a_{n-1} y^{(n-1)} + \cdots + a_2 y'' + a_1 y' + a_0 y = F(x)$$

which is linear in y and its derivatives.

EXERCISES for Section 18.1

In Exercises 1–12, classify the differential equation according to type and order.

1. $\dfrac{dy}{dx} + 3xy = x^2$

2. $y'' + 2y' + y = 1$

3. $\dfrac{d^2x}{dt^2} + 2\dfrac{dx}{dt} - 4x = e^t$

4. $\dfrac{d^2u}{dt^2} + \dfrac{du}{dt} = \sec t$

5. $y^{(4)} + 3(y')^2 - 4y = 0$

6. $x^2y'' + 3xy' = 0$

7. $(y'')^2 + 3y' - 4y = 0$

8. $\dfrac{\partial u}{\partial t} = C^2 \dfrac{\partial^2 u}{\partial x^2}$

9. $\dfrac{\partial u}{\partial t} + \dfrac{\partial u}{\partial y} = 2u$

10. $\dfrac{\partial^2 u}{\partial x \partial y} = \dfrac{\partial u}{\partial y}$

11. $\dfrac{d^2y}{dx^2} = \sqrt{1 + \left(\dfrac{dy}{dx}\right)^2}$

12. $\sqrt{\dfrac{d^2y}{dx^2}} = \dfrac{dy}{dx}$

In Exercises 13–18, verify that the given equation represents a solution of the differential equation.

13. $y = Ce^{4x}$, $\dfrac{dy}{dx} = 4y$

14. $x^2 + y^2 = Cy$, $y' = \dfrac{2xy}{x^2 - y^2}$

15. $y = C_1 \cos x + C_2 \sin x$, $y'' + y = 0$

16. $y = C_1 e^{-x} \cos x + C_2 e^{-x} \sin x$, $y'' + 2y' + 2y = 0$

17. $u = e^{-t} \sin bx$, $b^2 \dfrac{\partial u}{\partial t} = \dfrac{\partial^2 u}{\partial x^2}$

18. $u = \dfrac{y}{x^2 + y^2}$, $\dfrac{\partial^2 u}{\partial x^2} + \dfrac{\partial^2 u}{\partial y^2} = 0$

In Exercises 19–24, determine whether the given function is a solution to the differential equation

$$y^{(4)} - 16y = 0.$$

19. $y = 3 \cos x$

20. $y = 3 \cos 2x$

21. $y = e^{-2x}$

22. $y = 5 \ln x$

23. $y = C_1 e^{2x} + C_2 e^{-2x} + C_3 \sin 2x + C_4 \cos 2x$

24. $y = 5e^{-2x} + 3 \cos 2x$

In Exercises 25–30, determine if the function is a solution to the differential equation

$$x \dfrac{\partial u}{\partial x} - y \dfrac{\partial u}{\partial y} = 0.$$

25. $u = e^{x+y}$

26. $u = 5$

27. $u = x^2 y^2$

28. $u = \sin xy$

29. $u = (xy)^n$

30. $u = x^2 + y^2$

In Exercises 31 and 32, some of the curves corresponding to different values of C in the general solution to the

differential equation are given. Find the particular solution that passes through the point indicated on each graph.

31. $y^2 = Cx^3$, $2xy' - 3y = 0$

32. $2x^2 - y^2 = C$, $yy' - 2x = 0$

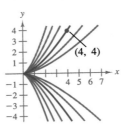

FIGURE FOR 31

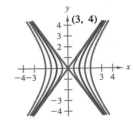

FIGURE FOR 32

In Exercises 33–38, verify that the general solutions satisfy the differential equation. Then find the particular solution satisfying the given initial condition.

33. $y = Ce^{-2x}$
$y' + 2y = 0$
$y = 3$ when $x = 0$

34. $2x^2 + 3y^2 = C$
$2x + 3yy' = 0$
$y = 2$ when $x = 1$

35. $y = C_1 \sin 3x + C_2 \cos 3x$
$y'' + 9y = 0$
$y = 2$ and $y' = 1$ when $x = \dfrac{\pi}{6}$

36. $y = C_1 + C_2 \ln x$
$xy'' + y' = 0$
$y = 0$ and $y' = \dfrac{1}{2}$ when $x = 2$

37. $y = C_1 x + C_2 x^3$
$x^2 y'' - 3xy' + 3y = 0$
$y = 0$ and $y' = 4$ when $x = 2$

38. $y = e^{2x/3}(C_1 + C_2 x)$
$9y'' - 12y' + 4y = 0$
$y = 4$ when $x = 0$ and $y = 0$ when $x = 3$

In Exercises 39 and 40, the general solution to the differential equation is given. Sketch the graph of the particular solutions for the given values of C.

39. $4yy' - x = 0$
$2y^2 - x^2 = C$
$C = 0$, $C = \pm 1$, $C = \pm 4$

40. $yy' + x = 0$
$x^2 + y^2 = C$
$C = 0$, $C = 1$, $C = 4$

In Exercises 41–48, use integration to find a general solution to the differential equation.

41. $\dfrac{dy}{dx} = 3x^2$

42. $\dfrac{dy}{dx} = \dfrac{1}{1 + x^2}$

43. $\dfrac{dy}{dx} = \dfrac{x - 2}{x}$

44. $\dfrac{dy}{dx} = x \cos x$

45. $\dfrac{dy}{dx} = e^x \sin 2x$

46. $\dfrac{dy}{dx} = \tan^2 x$

47. $\dfrac{dy}{dx} = x\sqrt{x - 3}$

48. $\dfrac{dy}{dx} = xe^x$

49. The limiting capacity of the habitat for a particular wildlife herd is L. The growth rate dN/dt of the herd is proportional to the unused opportunities for growth, as described by the differential equation

$$\frac{dN}{dt} = k(L - N).$$

The general solution to this differential equation is

$$N = L - Ce^{-kt}.$$

Suppose 100 animals are released into a tract of land that can support 750 of these animals. After 2 years the herd has grown to 160 animals.
  (a) Find the population function in terms of the time t in years.
  (b) Verify that the function of part (a) is a solution to the given differential equation.
  (c) Sketch the graph of this population function.

50. The rate of growth of an investment is proportional to the amount of the investment at any time t. That is,

$$\frac{dA}{dt} = kA.$$

  (a) Show that $A = Ce^{kt}$ is a solution to this differential equation.
  (b) Find the particular solution to this differential equation if the initial investment is \$1000, and 10 years later the amount is \$3320.12.

18.2 Separation of Variables in First-Order Equations

Separation of variables ▪ Homogeneous differential equations ▪ Applications

In this section we begin studying techniques for solving specific kinds of ordinary differential equations. We start with a procedure for solving a first-order differential equation that can be written in the form

$$M(x) + N(y)\frac{dy}{dx} = 0$$

where M is a continuous function of x alone and N is a continuous function of y alone. For this type of equation, all x terms can be collected with dx and all y terms with dy, and a solution can be obtained by integration. Such equations are said to be **separable,** and the solution procedure is called **separation of variables.** The steps are as follows.

SEPARATION OF VARIABLES

1. Express the equation in the differential form

$$M(x)\, dx + N(y)\, dy = 0 \qquad \text{or} \qquad M(x)\, dx = -N(y)\, dy.$$

2. Integrate to obtain the general solution

$$\int M(x)\, dx + \int N(y)\, dy = C$$

or

$$\int M(x)\, dx = -\int N(y)\, dy + C.$$

EXAMPLE 1 Separation of variables

Find the general solution of

$$(x^2 + 4)\frac{dy}{dx} = xy.$$

SOLUTION

To begin, we note that $y = 0$ is a solution. To find other solutions, we assume $y \neq 0$ and we separate variables as follows.

$$(x^2 + 4)\, dy = xy\, dx \qquad\qquad \text{Differential form}$$

$$\frac{dy}{y} = \frac{x}{x^2 + 4}\, dx \qquad\qquad \text{Separate variables}$$

Now we integrate to obtain

$$\int \frac{dy}{y} = \int \frac{x}{x^2 + 4}\, dx \qquad\qquad \text{Integrate}$$

$$\ln |y| = \frac{1}{2} \ln (x^2 + 4) + C_1$$

$$= \ln \sqrt{x^2 + 4} + C_1$$

$$|y| = e^{C_1}\sqrt{x^2 + 4}$$

$$y = \pm e^{C_1}\sqrt{x^2 + 4}.$$

Because $y = 0$ is also a solution, we can write the general solution as

$$y = C\sqrt{x^2 + 4}. \qquad\qquad \text{General solution} \qquad \rule[0.3ex]{1.2em}{0.4ex}$$

REMARK   We encourage you to *check your solutions* throughout this chapter. For instance, in Example 1 you can check the solution $y = C\sqrt{x^2 + 4}$ by differentiating and substituting into the original equation.

In some cases it is not feasible to write the general solution in the explicit form $y = f(x)$. The next example is a case in point; implicit differentiation can be used to verify this solution.

EXAMPLE 2 Finding a particular solution by separation of variables

Given the initial condition $y(0) = 1$, find the particular solution to the equation

$$xy\, dx + e^{-x^2}(y^2 - 1)\, dy = 0.$$

SOLUTION

Note that $y = 0$ is a solution of the differential equation. But this solution does not satisfy the initial condition; hence, we assume $y \neq 0$. To separate variables we must rid the first term of y and the second of e^{-x^2}. Thus, we multiply by e^{x^2}/y and obtain the following.

$$\left(\frac{e^{x^2}}{y}\right)xy\ dx + \left(\frac{e^{x^2}}{y}\right)e^{-x^2}(y^2 - 1)\ dy = 0, \quad y \neq 0$$

$$xe^{x^2}\ dx + \left(y - \frac{1}{y}\right)dy = 0$$

$$\int xe^{x^2}\ dx + \int \left(y - \frac{1}{y}\right)dy = 0$$

$$\frac{1}{2}e^{x^2} + \frac{y^2}{2} - \ln|y| = C_1 \quad \cdot$$

$$e^{x^2} + y^2 - \ln y^2 = 2C_1 = C$$

From the given initial condition, we use the fact that $y = 1$ when $x = 0$ to obtain $1 + 1 + 0 = 2 = C$. Thus, the particular solution has the implicit form

$$e^{x^2} + y^2 - \ln y^2 = 2.$$

▭

EXAMPLE 3 Finding a particular solution curve

Find the equation of the curve that passes through the point $(1, 3)$ and has a slope of y/x^2 at the point (x, y), as shown in Figure 18.2.

SOLUTION

Since the slope of the curve is given by y/x^2, we have

$$\frac{dy}{dx} = \frac{y}{x^2}$$

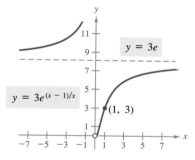

$y = 3e$

$y = 3e^{(x-1)/x}$

(1, 3)

FIGURE 18.2

with the initial condition $y(1) = 3$. Separating variables and integrating produces

$$\int \frac{dy}{y} = \int \frac{dx}{x^2}, \quad y \neq 0$$

$$\ln|y| = -\frac{1}{x} + C_1$$

$$y = e^{-(1/x)+C_1} = Ce^{-1/x}.$$

Since $y = 3$ when $x = 1$, it follows that $3 = Ce^{-1}$ and $C = 3e$. Therefore, the equation of the specified curve is

$$y = (3e)e^{-1/x} = 3e^{(x-1)/x}.$$

▭

Homogeneous differential equations

Some differential equations that are not separable in x and y can be made separable by a change of variables. This is true for differential equations of the form $y' = f(x, y)$, where f is a **homogeneous function.**

DEFINITION OF HOMOGENEOUS FUNCTION	The function given by $z = f(x, y)$ is said to be **homogeneous of degree n** if $$f(tx, ty) = t^n f(x, y)$$ where n is a real number.

EXAMPLE 4   Verifying homogeneous functions

(a) $f(x, y) = x^2 y - 4x^3 + 3xy^2$ is a homogeneous function of degree 3 because

$$
\begin{aligned}
f(tx, ty) &= (tx)^2(ty) - 4(tx)^3 + 3(tx)(ty)^2 \\
&= t^3(x^2 y) - t^3(4x^3) + t^3(3xy^2) \\
&= t^3(x^2 y - 4x^3 + 3xy^2) \\
&= t^3 f(x, y).
\end{aligned}
$$

(b) $f(x, y) = xe^{x/y} + y \sin(y/x)$ is a homogeneous function of degree 1 because

$$
f(tx, ty) = txe^{tx/ty} + ty \sin \frac{ty}{tx} = t\left(xe^{x/y} + y \sin \frac{y}{x} \right) = tf(x, y).
$$

(c) $f(x, y) = x + y^2$ is not a homogeneous function because

$$
f(tx, ty) = tx + t^2 y^2 = t(x + ty^2) \neq t^n(x + y^2).
$$

(d) $f(x, y) = x/y$ is homogeneous of degree zero because

$$
f(tx, ty) = \frac{tx}{ty} = t^0 \frac{x}{y}.
$$

DEFINITION OF HOMOGENEOUS DIFFERENTIAL EQUATION	A **homogeneous differential equation** is an equation of the form $$M(x, y)\, dx + N(x, y)\, dy = 0$$ where M and N are homogeneous functions of the same degree.

To solve a homogeneous differential equation by the method of separation of variables, we use the following change of variables theorem.

THEOREM 18.1 CHANGE OF VARIABLES FOR HOMOGENEOUS EQUATIONS	If $M(x, y)\, dx + N(x, y)\, dy = 0$ is homogeneous, then it can be transformed into a differential equation whose variables are separable by the substitution $$y = vx$$ where v is a differentiable function of x.

PROOF   Let $y = vx$. Then $dy = v\, dx + x\, dv$, and, by substitution,

$$M(x, y)\, dx + N(x, y)\, dy = M(x, \cdot vx)\, dx + N(x, vx)(v\, dx + x\, dv) = 0.$$

Since M and N are homogeneous of degree n, it follows that

$$x^n M(1, v)\, dx + x^n N(1, v)(v\, dx + x\, dv) = 0$$
$$M(1, v)\, dx + N(1, v)v\, dx = -N(1, v)x\, dv$$
$$[M(1, v) + vN(1, v)]\, dx = -N(1, v)x\, dv$$
$$\frac{dx}{x} + \frac{N(1, v)}{M(1, v) + vN(1, v)}\, dv = 0$$

provided no denominators are zero. Moreover, if v is a solution of this differential equation, we can reverse the steps to show that $y = vx$ is a solution of the given homogeneous differential equation.

EXAMPLE 5   Solving a homogeneous differential equation

Find the general solution to the differential equation

$$(x^2 - y^2)\, dx + 3xy\, dy = 0.$$

SOLUTION

Since $(x^2 - y^2)$ and $3xy$ are both homogeneous of degree 2, we let $y = vx$ to obtain $dy = x\, dv + v\, dx$. Then, by substitution, we have

$$(x^2 - v^2x^2)\, dx + 3x(vx)\overbrace{(x\, dv + v\, dx)}^{dy} = 0$$
$$(x^2 + 2v^2x^2)\, dx + 3x^3v\, dv = 0$$
$$x^2(1 + 2v^2)\, dx + x^2(3vx)\, dv = 0.$$

Dividing by x^2 and separating variables produces

$$(1 + 2v^2)\, dx = -3vx\, dv$$
$$\int \frac{dx}{x} = \int \frac{-3v}{1 + 2v^2}\, dv$$
$$\ln|x| = -\frac{3}{4} \ln(1 + 2v^2) + C_1$$
$$4 \ln|x| = -3 \ln(1 + 2v^2) + \ln|C|$$
$$\ln x^4 = \ln|C(1 + 2v^2)^{-3}|$$
$$x^4 = C(1 + 2v^2)^{-3}.$$

Substituting for v, we obtain the following general solution.

$$\left(1 + 2\frac{y^2}{x^2}\right)^3 x^4 = C$$
$$(x^2 + 2y^2)^3 = Cx^2$$

Applications

In the remainder of this section we look at two applications involving differential equations. We begin with one involving exponential growth.

EXAMPLE 6 Application to food preservation

In the preservation of food, cane sugar is broken down (inverted) into two simpler sugars: glucose and fructose. In dilute solutions, the inversion rate is proportional to the concentration $y(t)$ of unaltered sugar. If the concentration is $\frac{1}{50}$ when $t = 0$ and $\frac{1}{200}$ after 3 hours, find the concentration of unaltered sugar after 6 hours and after 12 hours.

SOLUTION

Since the rate of inversion is proportional to $y(t)$, we have the differential equation

$$\frac{dy}{dt} = ky.$$

Separating the variables and integrating produces

$$\int \frac{1}{y} \, dy = \int k \, dt$$

$$\ln |y| = kt + C_1$$

$$y = Ce^{kt}.$$

From the given conditions, we obtain the following.

$$y(0) = \frac{1}{50} \quad \Longrightarrow \quad C = \frac{1}{50}$$

$$y(3) = \frac{1}{200} \quad \Longrightarrow \quad \frac{1}{200} = \left(\frac{1}{50}\right)e^{3k} \quad \Longrightarrow \quad k = -\frac{\ln 4}{3}$$

Therefore, the concentration of unaltered sugar is given by

$$y(t) = \frac{1}{50}e^{-(\ln 4)t/3}$$

$$= \frac{1}{50}(4^{-t/3}).$$

When $t = 6$ and $t = 12$, we have the following concentrations of unaltered sugar.

$$y(6) = \frac{1}{50}(4^{-2}) = \frac{1}{800} \qquad \text{After 6 hours}$$

$$y(12) = \frac{1}{50}(4^{-4}) = \frac{1}{12,800} \qquad \text{After 12 hours}$$

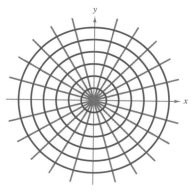

Each line $y = Kx$ is an orthogonal trajectory to the family of circles.

FIGURE 18.3

A common problem in electrostatics, thermodynamics, and hydrodynamics involves finding a family of curves, each of which is orthogonal to all members of a given family of curves. For example, Figure 18.3 shows a family of circles

$$x^2 + y^2 = C \qquad \text{Family of circles}$$

each of which intersects the lines in the family

$$y = Kx \qquad \text{Family of lines}$$

at right angles. Two such families of curves are said to be **mutually orthogonal,** and each curve in one of the families is called an **orthogonal trajectory** of the other family. In electrostatics, lines of force are orthogonal to the *equipotential curves.* In thermodynamics, the flow of heat across a plane surface is orthogonal to the *isothermal curves.* In hydrodynamics, the flow (stream) lines are orthogonal trajectories of the *velocity potential curves.*

EXAMPLE 7 Finding orthogonal trajectories

Describe the orthogonal trajectories for the family of curves given by $y = C/x$ for $C \neq 0$. Sketch several members of each family.

SOLUTION

First, we solve the given equation for C and write $xy = C$. Then, differentiating implicitly with respect to x, we obtain the differential equation

$$xy' + y = 0 \quad \Longrightarrow \quad \frac{dy}{dx} = -\frac{y}{x}. \qquad \text{Slope of given family}$$

Since y' represents the slope of the given family of curves at (x, y), it follows that the orthogonal family has the negative reciprocal slope, x/y, and we write

$$\frac{dy}{dx} = \frac{x}{y}. \qquad \text{Slope of orthogonal family}$$

Now we can find the orthogonal family by separating variables and integrating.

$$\int y \, dy = \int x \, dx$$

$$\frac{y^2}{2} = \frac{x^2}{2} + C_1$$

Therefore, each orthogonal trajectory is a hyperbola given by

$$\frac{y^2}{K} - \frac{x^2}{K} = 1$$

$$2C_1 = K \neq 0.$$

The centers are at the origin, and the transverse axes are vertical for $K > 0$ and horizontal for $K < 0$. Several trajectories are shown in Figure 18.4.

Given family:
$xy = C$

Orthogonal family:
$\dfrac{y^2}{K} - \dfrac{x^2}{K} = 1$

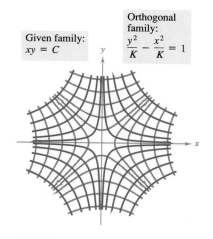

FIGURE 18.4

EXERCISES for Section 18.2

In Exercises 1–6, find the general solution of the given differential equation.

1. $\dfrac{dy}{dx} = \dfrac{x}{y}$

2. $\dfrac{dy}{dx} = \dfrac{x^2 + 2}{3y^2}$

3. $(2 + x)y' = 3y$

4. $xy' = y$

5. $yy' = \sin x$

6. $\sqrt{1 - 4x^2}\,y' = 1$

In Exercises 7–14, find the particular solution of the differential equation that satisfies the given initial condition.

Differential equation	Initial condition
7. $yy' - e^x = 0$	$y(0) = 4$
8. $\sqrt{x} + \sqrt{y}\,y' = 0$	$y(1) = 4$
9. $y(x + 1) + y' = 0$	$y(-2) = 1$
10. $xyy' - \ln x = 0$	$y(1) = 0$
11. $(1 + x^2)y' - (1 + y^2) = 0$	$y(0) = \sqrt{3}$
12. $\sqrt{1 - x^2}\,y' - \sqrt{1 - y^2} = 0$	$y(0) = 1$
13. $dP - kP\,dt = 0$	$P(0) = P_0$
14. $dT + k(T - 70)\,dt = 0$	$T(0) = 140$

In Exercises 15–20, determine whether the function is homogeneous, and if so, determine the degree.

15. $f(x, y) = x^3 - 4xy^2 + y^3$

16. $f(x, y) = \dfrac{xy}{\sqrt{x^2 + y^2}}$

17. $f(x, y) = 2 \ln xy$

18. $f(x, y) = \tan(x + y)$

19. $f(x, y) = 2 \ln \dfrac{x}{y}$

20. $f(x, y) = \tan \dfrac{y}{x}$

In Exercises 21–26, solve the homogeneous differential equation.

21. $y' = \dfrac{x + y}{2x}$

22. $y' = \dfrac{2x + y}{y}$

23. $y' = \dfrac{x - y}{x + y}$

24. $y' = \dfrac{x^2 + y^2}{2xy}$

25. $y' = \dfrac{xy}{x^2 - y^2}$

26. $y' = \dfrac{3x + 2y}{x}$

In Exercises 27–30, find the particular solution that satisfies the given initial condition.

Differential equation	Initial condition
27. $x\,dy - (2xe^{-y/x} + y)\,dx = 0$	$y(1) = 0$
28. $-y^2\,dx + x(x + y)\,dy = 0$	$y(1) = 1$
29. $\left(x \sec \dfrac{y}{x} + y\right) dx - x\,dy = 0$	$y(1) = 0$
30. $(y - \sqrt{x^2 - y^2})\,dx - x\,dy = 0$	$y(1) = 0$

In Exercises 31–36, find the orthogonal trajectories of the given family and sketch several members of each family.

31. $x^2 + y^2 = C$

32. $2x^2 - y^2 = C$

33. $x^2 = Cy$

34. $y^2 = 2Cx$

35. $y^2 = Cx^3$

36. $y = Ce^x$

In Exercises 37 and 38, find an equation for the curve that passes through the given point and has the specified slope.

Point	Slope
37. $(1, 1)$	$y' = -\dfrac{9x}{16y}$
38. $(8, 2)$	$y' = \dfrac{2y}{3x}$

39. The amount A of an investment P increases at a rate proportional to A at any instant of time t.
 (a) Find an equation for A as a function of t.
 (b) If the initial investment is \$1000.00 and the rate is 11 percent, find the amount after 10 years.
 (c) Find the time necessary to double the investment if the rate is 11 percent.

40. The rate of growth of a population of fruit flies is proportional to the size of the population at any instant. If there were 180 flies after the second day of the experiment and 300 after the fourth day, how many flies were there in the original population?

41. The rate of decomposition of radioactive radium is proportional to the amount present at a given instant. Find the percentage that remains of a present amount after 25 years, if the half-life of radioactive radium is 1600 years.

42. In a chemical reaction, a certain compound changes into another compound at a rate proportional to the unchanged amount. If initially there was 20 grams of the original compound, and there is 16 grams after 1 hour, when will 75 percent of the compound be changed?

43. Newton's Law of Cooling states that the rate of change in the temperature of an object is proportional to the difference between its temperature and the temperature of the surrounding air. Suppose a room is kept at a constant temperature of 70°, and an object cools from 350° to 150° in 45 minutes. How long will it take for the object to cool to a temperature of 80°?

44. If s gives the position of an object moving in a straight line, the velocity and acceleration are given by

$$v = \frac{ds}{dt}, \qquad a = \frac{dv}{dt} = \frac{ds}{dt}\frac{dv}{ds} = v\frac{dv}{ds}.$$

Suppose that a boat weighing 640 pounds has a motor that exerts a force of 60 pounds. If the resistance (in pounds) to the motion of the boat is 3 times the velocity (in feet per second) and if the boat starts from rest, find its maximum speed. Use the equation

$$\text{force} = F = ma = mv\frac{dv}{ds} = \frac{w}{32}v\frac{dv}{ds}$$

where w is the weight of the boat.

18.3 Exact First-Order Equations

Exact differential equations ▪ Test for exactness ▪ Integrating factors

In this section we introduce a method for solving the first-order differential equation

$$M(x, y)\, dx + N(x, y)\, dy = 0$$

for the special case in which this equation represents the exact differential of a function $z = f(x, y)$.

DEFINITION OF AN EXACT DIFFERENTIAL EQUATION

The equation

$$M(x, y)\, dx + N(x, y)\, dy = 0$$

is an **exact differential equation** if there exists a function f of two variables x and y having continuous partial derivatives such that

$$f_x(x, y) = M(x, y) \qquad \text{and} \qquad f_y(x, y) = N(x, y).$$

The general solution to the equation is $f(x, y) = C$.

From Section 15.3, we know that if f has continuous second partials, then

$$\frac{\partial M}{\partial y} = \frac{\partial^2 f}{\partial y \partial x} = \frac{\partial^2 f}{\partial x \partial y} = \frac{\partial N}{\partial x}.$$

This suggests the following test for exactness.

THEOREM 18.2 TEST FOR EXACTNESS

Let M and N have continuous partial derivatives on an open disc R. The differential equation $M(x, y)\, dx + N(x, y)\, dy = 0$ is exact if and only if

$$\frac{\partial M}{\partial y} = \frac{\partial N}{\partial x}.$$

Exactness is a fragile condition in the sense that seemingly minor alterations in an exact equation can destroy its exactness. This is demonstrated in the following example.

EXAMPLE 1   Testing for exactness

(a) The differential equation

$$(xy^2 + x)\, dx + yx^2\, dy = 0$$

is exact because

$$\overbrace{\frac{\partial}{\partial y}[xy^2 + x]}^{\frac{\partial M}{\partial y}} = 2xy = \overbrace{\frac{\partial}{\partial x}[yx^2]}^{\frac{\partial N}{\partial x}}.$$

But the equation $(y^2 + 1)\, dx + xy\, dy = 0$ is not exact, even though it is obtained by dividing both sides of the first equation by x.

(b) The differential equation

$$\cos y\, dx + (y^2 - x \sin y)\, dy = 0$$

is exact because

$$\overbrace{\frac{\partial}{\partial y}[\cos y]}^{\frac{\partial M}{\partial y}} = -\sin y = \overbrace{\frac{\partial}{\partial x}[y^2 - x \sin y]}^{\frac{\partial N}{\partial x}}.$$

But the equation $\cos y\, dx + (y^2 + x \sin y)\, dy = 0$ is not exact, even though it differs from the first equation only by a single sign. ▭

REMARK   Every differential equation of the form $M(x)\, dx + N(y)\, dy = 0$ is exact. In other words, a separable variables equation is actually a special type of exact equation.

Note that the test for exactness of $M(x, y)\, dx + N(x, y)\, dy = 0$ is the same as the test for determining whether $\mathbf{F}(x, y) = M(x, y)\mathbf{i} + N(x, y)\mathbf{j}$ is the gradient of a potential function (Theorem 17.1). This means that a general solution $f(x, y) = C$ to an exact differential equation can be found by the method used to find a potential function for a conservative vector field. We demonstrate this procedure further in the next two examples.

EXAMPLE 2   Solving an exact differential equation

Show that the differential equation

$$(2xy - 3x^2)\, dx + (x^2 - 2y)\, dy = 0$$

is exact, and find its general solution.

SOLUTION

The given differential equation is exact because

$$\underbrace{\frac{\partial}{\partial y}[2xy - 3x^2]}_{\frac{\partial M}{\partial y}} = 2x = \underbrace{\frac{\partial}{\partial x}[x^2 - 2y]}_{\frac{\partial N}{\partial x}}.$$

Thus, the general solution, $f(x, y) = C$, is given by

$$f(x, y) = \int M(x, y) \, dx$$
$$= \int (2xy - 3x^2) \, dx = x^2y - x^3 + g(y).$$

In Section 17.1 we determined $g(y)$ by integrating $N(x, y)$ with respect to y and reconciling the two expressions for $f(x, y)$. An alternate method is to partially differentiate this version of $f(x, y)$ with respect to y and compare the result to $N(x, y)$. In other words,

$$f_y(x, y) = \frac{\partial}{\partial y}[x^2y - x^3 + g(y)] = x^2 + g'(y) = \overbrace{x^2 - 2y}^{N(x, y)}.$$

$$g'(y) = -2y$$

Thus, $g'(y) = -2y$, and it follows that $g(y) = -y^2 + C_1$. Therefore,

$$f(x, y) = x^2y - x^3 - y^2 + C_1$$

and the general solution is

$$x^2y - x^3 - y^2 = C.$$

■

EXAMPLE 3   Solving an exact differential equation

Find the particular solution of

$$(\cos x - x \sin x + y^2) \, dx + 2xy \, dy = 0$$

that satisfies the initial condition $y = 1$ when $x = \pi$.

SOLUTION

The given equation is exact because

$$\underbrace{\frac{\partial}{\partial y}[\cos x - x \sin x + y^2]}_{\frac{\partial M}{\partial y}} = 2y = \underbrace{\frac{\partial}{\partial x}[2xy]}_{\frac{\partial N}{\partial x}}.$$

In this case, $N(x, y)$ is simpler than $M(x, y)$, and we proceed as follows.

$$f(x, y) = \int N(x, y) \, dy = \int 2xy \, dy = xy^2 + g(x)$$

$$f_x(x, y) = \frac{\partial}{\partial x}[xy^2 + g(x)] = y^2 + g'(x) = \overbrace{\cos x - x \sin x + y^2}^{M(x,\, y)}$$

$$\boxed{g'(x) = \cos x - x \sin x}$$

Thus, $g'(x) = \cos x - x \sin x$ and

$$g(x) = \int (\cos x - x \sin x) \, dx = x \cos x + C_1$$

which implies $f(x, y) = xy^2 + x \cos x + C_1$, and the general solution is

$$xy^2 + x \cos x = C.$$

Applying the given initial condition produces $\pi(1)^2 + \pi \cos \pi = C$, which implies that $C = 0$. Hence, the particular solution is

$$xy^2 + x \cos x = 0.$$

In Example 3, note that if $z = f(x, y) = xy^2 + x \cos x$, then the total differential of z is given by

$$dz = f_x(x, y) \, dx + f_y(x, y) \, dy = (\cos x - x \sin x + y^2) \, dx + 2xy \, dy$$
$$= M(x, y) \, dx + N(x, y) \, dy.$$

In other words, we call $M \, dx + N \, dy = 0$ an *exact* differential equation because $M \, dx + N \, dy$ is exactly the differential of $f(x, y)$.

Integrating factors

If the differential equation $M(x, y) \, dx + N(x, y) \, dy = 0$ is not exact, it may be possible to make it exact by multiplying by an appropriate factor $u(x, y)$, called an **integrating factor** for the differential equation. For instance, if the differential equation

$$2y \, dx + x \, dy = 0 \qquad \text{Not an exact equation}$$

is multiplied by the integrating factor $u(x, y) = x$, the resulting equation

$$2xy \, dx + x^2 \, dy = 0 \qquad \text{Exact equation}$$

is exact, and the left side is the total differential of $x^2 y$. Similarly, if the equation

$$y \, dx - x \, dy = 0 \qquad \text{Not an exact equation}$$

is multiplied by the integrating factor $u(x, y) = 1/y^2$, the resulting equation

$$\frac{1}{y} \, dx - \frac{x}{y^2} \, dy = 0 \qquad \text{Exact equation}$$

is exact.

Finding an integrating factor can be difficult. However, there are two classes of differential equations whose integrating factors can be found routinely—namely, those that possess integrating factors that are functions of either x alone or y alone. The following theorem, which we list without proof, outlines a procedure for finding these two special categories of integrating factors.

**THEOREM 18.3
INTEGRATING FACTORS**

Consider the differential equation $M(x, y) \, dx + N(x, y) \, dy = 0$.

1. If

$$\frac{1}{N(x, y)}[M_y(x, y) - N_x(x, y)] = h(x)$$

is a function of x alone, then $e^{\int h(x) \, dx}$ is an integrating factor.

2. If

$$\frac{1}{M(x, y)}[N_x(x, y) - M_y(x, y)] = k(y)$$

is a function of y alone, then $e^{\int k(y) \, dy}$ is an integrating factor.

REMARK If either $h(x)$ or $k(y)$ is constant, Theorem 18.3 still applies. As an aid to remembering the formulas, note that the subtracted partial derivative identifies both the denominator and the variable for the integrating factor.

EXAMPLE 4 Finding an integrating factor

Find the general solution of the differential equation

$$(y^2 - x) \, dx + 2y \, dy = 0.$$

SOLUTION

The given equation is not exact because $M_y(x, y) = 2y$ and $N_x(x, y) = 0$. However, since

$$\frac{M_y(x, y) - N_x(x, y)}{N(x, y)} = \frac{2y - 0}{2y} = 1 = h(x)$$

it follows that $e^{\int h(x) \, dx} = e^{\int dx} = e^x$ is an integrating factor. Multiplying the given differential equation by e^x, we obtain the exact differential equation

$$(y^2 e^x - x e^x) \, dx + 2y e^x \, dy = 0$$

whose solution is obtained as follows.

$$f(x, y) = \int N(x, y) \, dy = \int 2y e^x \, dy = y^2 e^x + g(x)$$

$$f_x(x, y) = y^2 e^x + g'(x) = \overbrace{y^2 e^x - x e^x}^{M(x, y)}$$

$$g'(x) = -x e^x$$

Therefore, $g'(x) = -xe^x$ and $g(x) = -xe^x + e^x + C_1$, which implies that

$$f(x, y) = y^2 e^x - xe^x + e^x + C_1.$$

The general solution is

$$y^2 e^x - xe^x + e^x = C \implies y^2 - x + 1 = Ce^{-x}.$$

In the next example we show how a differential equation can aid in sketching a force field given by $\mathbf{F}(x, y) = M(x, y)\mathbf{i} + N(x, y)\mathbf{j}$.

EXAMPLE 5 An application to force fields

Sketch the force field given by

$$\mathbf{F}(x, y) = \frac{2y}{\sqrt{x^2 + y^2}}\mathbf{i} - \frac{y^2 - x}{\sqrt{x^2 + y^2}}\mathbf{j}$$

by finding and sketching the family of curves tangent to $\mathbf{F}$.

SOLUTION

At the point (x, y) in the plane, the vector $\mathbf{F}(x, y)$ has a slope of

$$\frac{dy}{dx} = \frac{-(y^2 - x)/\sqrt{x^2 + y^2}}{2y/\sqrt{x^2 + y^2}} = \frac{-(y^2 - x)}{2y}$$

which, in differential form, is

$$2y\, dy = -(y^2 - x)\, dx$$
$$(y^2 - x)\, dx + 2y\, dy = 0.$$

From Example 4 we know that the general solution of this differential equation is $y^2 - x + 1 = Ce^{-x}$, or

$$y^2 = x - 1 + Ce^{-x}.$$

Figure 18.5 shows several representative curves from this family. Note that the force vector at (x, y) is tangent to the curve passing through (x, y). ▭

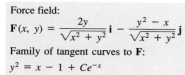

Force field:
$$\mathbf{F}(x, y) = \frac{2y}{\sqrt{x^2 + y^2}}\mathbf{i} - \frac{y^2 - x}{\sqrt{x^2 + y^2}}\mathbf{j}$$
Family of tangent curves to $\mathbf{F}$:
$$y^2 = x - 1 + Ce^{-x}$$

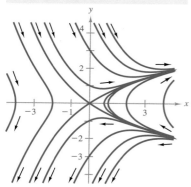

FIGURE 18.5

EXERCISES for Section 18.3

In Exercises 1–10, determine whether the differential equation is exact, and if so, find the general solution.

1. $(2x - 3y)\, dx + (2y - 3x)\, dy = 0$

2. $ye^x\, dx + e^x\, dy = 0$

3. $(3y^2 + 10xy^2)\, dx + (6xy - 2 + 10x^2y)\, dy = 0$

4. $2\cos(2x - y)\, dx - \cos(2x - y)\, dy = 0$

5. $(4x^3 - 6xy^2)\, dx + (4y^3 - 6xy)\, dy = 0$

6. $2y^2 e^{xy^2}\, dx + 2xye^{xy^2}\, dy = 0$

7. $\dfrac{1}{x^2 + y^2}(x\, dy - y\, dx) = 0$

8. $e^{-(x^2+y^2)}(x\, dx + y\, dy) = 0$

9. $\dfrac{1}{(x - y)^2}(y^2\, dx + x^2\, dy) = 0$

10. $e^y \cos xy\, [y\, dx + (x + \tan xy)\, dy] = 0$

In Exercises 11–16, find the particular solution that satisfies the given initial condition.

11. $\dfrac{y}{x-1} dx + [\ln (x-1) + 2y] dy = 0$

$y(2) = 4$

12. $\dfrac{1}{\sqrt{x^2 + y^2}}(x\, dx + y\, dy) = 0$

$y(4) = 3$

13. $\dfrac{1}{x^2 + y^2}(x\, dx + y\, dy) = 0$

$y(0) = 4$

14. $e^{3x}(\sin 3y\, dx + \cos 3y\, dy) = 0$

$y(0) = \pi$

15. $(2x \tan y + 5)\, dx + (x^2 \sec^2 y)\, dy = 0$

$y(0) = 0$

16. $(x^2 + y^2)\, dx + 2xy\, dy = 0$

$y(3) = 1$

In Exercises 17–26, find the integrating factor that is a function of x or y alone and use it to find the general solution of the differential equation.

17. $y\, dx - (x + 6y^2)\, dy = 0$
18. $(2x^3 + y)\, dx - x\, dy = 0$
19. $(5x^2 - y)\, dx + x\, dy = 0$
20. $(5x^2 - y^2)\, dx + 2y\, dy = 0$
21. $(x + y)\, dx + \tan x\, dy = 0$
22. $(2x^2y - 1)\, dx + x^3\, dy = 0$
23. $y^2\, dx + (xy - 1)\, dy = 0$
24. $(x^2 + 2x + y)\, dx + 2\, dy = 0$
25. $2y\, dx + (x - \sin \sqrt{y})\, dy = 0$
26. $(-2y^3 + 1)\, dx + (3xy^2 + x^3)\, dy = 0$

In Exercises 27–30, use the given integrating factor to find the general solution of the differential equation.

27. $(4x^2y + 2y^2)\, dx + (3x^3 + 4xy)\, dy = 0$,

$u(x, y) = xy^2$

28. $(3y^2 + 5x^2y)\, dx + (3xy + 2x^3)\, dy = 0$,

$u(x, y) = x^2y$

29. $(-y^5 + x^2y)\, dx + (2xy^4 - 2x^3)\, dy = 0$,

$u(x, y) = x^{-2}y^{-3}$

30. $-y^3\, dx + (xy^2 - x^2)\, dy = 0$,

$u(x, y) = x^{-2}y^{-2}$

31. Show that each of the following are integrating factors for the differential equation

$y\, dx - x\, dy = 0.$

(a) $\dfrac{1}{x^2}$

(b) $\dfrac{1}{y^2}$

(c) $\dfrac{1}{xy}$

(d) $\dfrac{1}{x^2 + y^2}$

32. Show that the differential equation

$(axy^2 + by)\, dx + (bx^2y + ax)\, dy = 0$

is exact only if $a = b$. If $a \neq b$, show that $x^m y^n$ is an integrating factor, where

$$m = -\frac{2b + a}{a + b}, \qquad n = -\frac{2a + b}{a + b}.$$

In Exercises 33–36, sketch the family of tangent curves to the given force field.

33. $\mathbf{F}(x, y) = \dfrac{y}{\sqrt{x^2 + y^2}}\mathbf{i} - \dfrac{x}{\sqrt{x^2 + y^2}}\mathbf{j}$

34. $\mathbf{F}(x, y) = \dfrac{x}{\sqrt{x^2 + y^2}}\mathbf{i} - \dfrac{y}{\sqrt{x^2 + y^2}}\mathbf{j}$

35. $\mathbf{F}(x, y) = 4x^2y\,\mathbf{i} - \left(2xy^2 + \dfrac{x}{y^2}\right)\mathbf{j}$

36. $\mathbf{F}(x, y) = (1 + x^2)\mathbf{i} - 2xy\,\mathbf{j}$

In Exercises 37 and 38, find an equation for the curve passing through the given point with the specified slope.

Point	Slope
37. $(2, 1)$	$\dfrac{dy}{dx} = \dfrac{y - x}{3y - x}$
38. $(0, 2)$	$\dfrac{dy}{dx} = \dfrac{-2xy}{x^2 + y^2}$

39. If $y = C(x)$ represents the cost of producing x units in a manufacturing process, then the **elasticity of cost** is defined to be

$$E(x) = \frac{\text{marginal cost}}{\text{average cost}} = \frac{C'(x)}{C(x)/x} = \frac{x\, dy}{y\, dx}.$$

Find the cost function if the elasticity function is

$$E(x) = \frac{20x - y}{2y - 10x}$$

where $C(100) = 500$ and $100 \leq x$.

18.4 First-Order Linear Differential Equations

First-order linear differential equations ▪ Bernoulli equations ▪ Applications

In this section we will see how integrating factors help to solve a very important class of first-order differential equations—first-order *linear* differential equations.

DEFINITION OF FIRST-ORDER LINEAR DIFFERENTIAL EQUATION	A first-order linear differential equation is an equation of the form $$\frac{dy}{dx} + P(x)y = Q(x)$$ where P and Q are continuous functions of x.

REMARK   The first-order linear differential equation given in this definition is said to be in **standard form**.

To solve a first-order linear differential equation, we use an integrating factor $u(x)$, which converts the left side into the derivative of the product $u(x)y$. That is, we need a factor $u(x)$ such that

$$u(x)\frac{dy}{dx} + u(x)P(x)y = \frac{d[u(x)y]}{dx}.$$

By expanding the right side and using prime notation, we find that

$$u(x)y' + u(x)P(x)y = u(x)y' + yu'(x)$$
$$u(x)P(x)y = yu'(x)$$
$$P(x) = \frac{u'(x)}{u(x)}.$$

Integration with respect to x yields

$$\ln|u(x)| = \int P(x)\,dx + C_1 \quad \Longrightarrow \quad u(x) = Ce^{\int P(x)\,dx}.$$

Since we don't need the most general integrating factor, we let $C = 1$. Multiplying the original equation $y' + P(x)y = Q(x)$ by $u(x) = e^{\int P(x)\,dx}$ produces

$$y'e^{\int P(x)\,dx} + yP(x)e^{\int P(x)\,dx} = Q(x)e^{\int P(x)\,dx}$$
$$\frac{d}{dx}\Big[ye^{\int P(x)\,dx}\Big] = Q(x)e^{\int P(x)\,dx}.$$

The general solution is given by

$$ye^{\int P(x)\,dx} = \int Q(x)e^{\int P(x)\,dx}\,dx + C.$$

THEOREM 18.4
SOLUTION OF A FIRST-ORDER
LINEAR DIFFERENTIAL EQUATION

An integrating factor for the first-order differential equation $y' + P(x)y = Q(x)$ is $u(x) = e^{\int P(x)\,dx}$. The solution of the differential equation is

$$ye^{\int P(x)\,dx} = \int Q(x)e^{\int P(x)\,dx}\,dx + C.$$

REMARK Rather than memorizing this formula, just remember that multiplication by the integrating factor $e^{\int P(x)\,dx}$ converts the left side of the differential equation into the derivative of the product $ye^{\int P(x)\,dx}$.

EXAMPLE 1 Solving a first-order linear differential equation

Find the general solution to $xy' - 2y = x^2$.

SOLUTION

The *standard form* of the given equation is

$$y' - \left(\frac{2}{x}\right)y = x.$$

Thus, $P(x) = -2/x$, and we have

$$\int P(x)\,dx = -\int \frac{2}{x}\,dx = -\ln x^2$$

$$e^{\int P(x)\,dx} = e^{-\ln x^2} = \frac{1}{x^2}.$$

Therefore, multiplying both sides of the standard form by $1/x^2$ yields

$$\frac{y'}{x^2} - \frac{2y}{x^3} = \frac{1}{x}$$

$$\frac{d}{dx}\left[\frac{y}{x^2}\right] = \frac{1}{x}$$

$$\frac{y}{x^2} = \int \frac{1}{x}\,dx = \ln|x| + C$$

$$y = x^2 \ln|x| + Cx^2. \qquad \text{General solution}$$

EXAMPLE 2 Solving a first-order linear differential equation

Find the general solution

$$y' - y \tan t = 1.$$

SOLUTION

Since $P(t) = -\tan t$, we have

$$\int P(t)\, dt = -\int \tan t\, dt = \ln|\cos t|$$
$$e^{\int P(t)\, dt} = e^{\ln|\cos t|} = |\cos t|.$$

A quick check would show that $\cos t$ is also an integrating factor. Thus, multiplying $y' - y \tan t = 1$ by $\cos t$ produces

$$\frac{d}{dt}[y \cos t] = \cos t$$

$$y \cos t = \int \cos t\, dt = \sin t + C$$

$$y = \tan t + C \sec t. \qquad \text{General solution}$$

Bernoulli equations

A well-known *nonlinear* equation that reduces to a linear one with an appropriate substitution is the **Bernoulli equation,** named after James Bernoulli (1654–1705).

$$y' + P(x)y = Q(x)y^n \qquad \text{Bernoulli equation}$$

This equation is linear if $n = 0$, and has separable variables if $n = 1$. Thus, in the following development, we assume that $n \neq 0$ and $n \neq 1$. We begin by multiplying by y^{-n} and $(1 - n)$ to obtain

$$y^{-n}y' + P(x)y^{1-n} = Q(x)$$
$$(1 - n)y^{-n}y' + (1 - n)P(x)y^{1-n} = (1 - n)Q(x)$$
$$\frac{d}{dx}[y^{1-n}] + (1 - n)P(x)y^{1-n} = (1 - n)Q(x)$$

which is a linear equation in the variable y^{1-n}. Letting $z = y^{1-n}$ produces the linear equation

$$\frac{dz}{dx} + (1 - n)P(x)z = (1 - n)Q(x).$$

Finally, by Theorem 18.4, the *general solution to the Bernoulli equation* is

$$y^{1-n}e^{\int(1-n)P(x)\,dx} = \int (1 - n)Q(x)e^{\int(1-n)P(x)\,dx}\, dx + C.$$

EXAMPLE 3   Solving a Bernoulli equation

Find the general solution of $y' + xy = xe^{-x^2}y^{-3}$.

SOLUTION

For this Bernoulli equation, $n = -3$, and we use the substitution

$$z = y^{1-n} = y^4 \quad \Longrightarrow \quad z' = 4y^3 y'.$$

Multiplying the original equation by $4y^3$ produces

$$4y^3 y' + 4xy^4 = 4xe^{-x^2}$$

$$z' + 4xz = 4xe^{-x^2}. \qquad \text{Linear equation: } z' + P(x)z = Q(x)$$

Now, since this equation is linear in z, we have $P(x) = 4x$ and

$$\int P(x)\, dx = \int 4x\, dx = 2x^2$$

which implies that e^{2x^2} is an integrating factor. Multiplying by this factor produces

$$\frac{d}{dx}\left[ze^{2x^2}\right] = 4xe^{x^2}$$

$$ze^{2x^2} = \int 4xe^{x^2}\, dx = 2e^{x^2} + C$$

$$z = 2e^{-x^2} + Ce^{-2x^2}.$$

Finally, by substituting $z = y^4$, the general solution is

$$y^4 = 2e^{-x^2} + Ce^{-2x^2}. \qquad \text{General solution}$$

So far we have studied several types of first-order differential equations. Of these, the separable variables case is usually the simplest, and solution by an integrating factor is usually a last resort. The following summary lists the various types we have studied.

SUMMARY OF FIRST-ORDER DIFFERENTIAL EQUATIONS

Method	Form of equation
1. Separable variables:	$M(x)\, dx + N(y)\, dy = 0$
2. Homogeneous:	$M(x, y)\, dx + N(x, y)\, dy = 0$, where M and N are nth-degree homogeneous.
3. Exact:	$M(x, y)\, dx + N(x, y)\, dy = 0$, where $\partial M/\partial y = \partial N/\partial x$.
4. Integrating factor:	$u(x, y)M(x, y)\, dx + u(x, y)N(x, y)\, dy = 0$ is exact.
5. Linear:	$y' + P(x)y = Q(x)$
6. Bernoulli equation:	$y' + P(x)y = Q(x)y^n$

Applications

One type of problem that can be described in terms of a differential equation involves chemical mixtures, as illustrated in the next example.

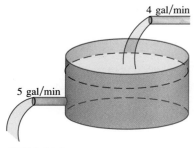

4 gal/min

5 gal/min

FIGURE 18.6

EXAMPLE 4 A mixture application

A tank contains 50 gallons of a solution composed of 90 percent water and 10 percent alcohol. A second solution containing 50 percent water and 50 percent alcohol is added to the tank at the rate of 4 gallons per minute. As the second solution is being added, the tank is being drained at the rate of 5 gallons per minute, as shown in Figure 18.6. Assuming the solution in the tank is stirred constantly, how much alcohol is in the tank after 10 minutes?

SOLUTION

Let y be the number of gallons of alcohol in the tank at any time t. We know that $y = 5$ when $t = 0$. Since the number of gallons of solution in the tank at any time is $50 - t$ and since the tank loses 5 gallons of solution per minute, it must lose

$$\left(\frac{5}{50 - t}\right) y$$

gallons of alcohol per minute. Furthermore, since the tank is gaining 2 gallons of alcohol per minute, the rate of change of alcohol in the tank is given by

$$\frac{dy}{dt} = 2 - \left(\frac{5}{50 - t}\right) y \quad \Longrightarrow \quad \frac{dy}{dt} + \left(\frac{5}{50 - t}\right) y = 2.$$

To solve this linear equation, we let $P(t) = 5/(50 - t)$ and obtain

$$\int P(t)\, dt = \int \frac{5}{50 - t}\, dt = -5 \ln |50 - t|.$$

Since $t < 50$, we can drop the absolute value signs and conclude that

$$e^{\int P(t)\, dt} = e^{-5 \ln (50-t)} = \frac{1}{(50 - t)^5}.$$

Thus, the general solution is

$$\frac{y}{(50 - t)^5} = \int \frac{2}{(50 - t)^5}\, dt = \frac{1}{2(50 - t)^4} + C$$

$$y = \frac{50 - t}{2} + C(50 - t)^5.$$

Since $y = 5$ when $t = 0$, we have

$$5 = \frac{50}{2} + C(50)^5 \quad \Longrightarrow \quad -\frac{20}{50^5} = C$$

which means that the particular solution is

$$y = \frac{50 - t}{2} - 20\left(\frac{50 - t}{50}\right)^5.$$

Finally, when $t = 10$, the amount of alcohol in the tank is

$$y = \frac{50 - 10}{2} - 20\left(\frac{50 - 10}{50}\right)^5 = 13.45 \text{ gal}$$

which represents a solution containing 33.6 percent alcohol.

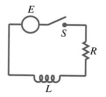

E

S

R

L

FIGURE 18.7

A simple electric circuit consists of electric current I (in amperes), a resistance R (in ohms), an inductance L (in henrys), and a constant electromotive force E (in volts), as shown in Figure 18.7. According to Kirchhoff's Second Law, if the switch S is closed when $t = 0$, then the applied electromotive force (voltage) is equal to the sum of the voltage drops in the rest of the circuit. This in turn means that the current I satisfies the differential equation

$$L\frac{dI}{dt} + RI = E.$$

EXAMPLE 5　An electric circuit application

Find the current I as a function of time t (in seconds), given that I satisfies the differential equation $L(dI/dt) + RI = \sin 2t$, where R and L are nonzero constants.

SOLUTION

In standard form the given linear equation is

$$\frac{dI}{dt} + \frac{R}{L}I = \frac{1}{L}\sin 2t.$$

Thus, we let $P(t) = R/L$, so that $e^{\int P(t)dt} = e^{(R/L)t}$, and, by Theorem 18.4,

$$Ie^{(R/L)t} = \frac{1}{L}\int e^{(R/L)t}\sin 2t\, dt$$

$$= \frac{1}{4L^2 + R^2}e^{(R/L)t}(R\sin 2t - 2L\cos 2t) + C.$$

Thus, the general solution is

$$I = e^{-(R/L)t}\left[\frac{1}{4L^2 + R^2}e^{(R/L)t}(R\sin 2t - 2L\cos 2t) + C\right]$$

$$I = \frac{1}{4L^2 + R^2}(R\sin 2t - 2L\cos 2t) + Ce^{-(R/L)t}.　\blacksquare$$

In most falling-body problems discussed so far in the text, we have neglected air resistance. In the next example we include this factor, and we assume that the air resistance on the object falling in the earth's atmosphere is proportional to its velocity v. If g is the gravitational constant, then the downward force F on a falling object of mass m is given by the difference $mg - kv$. But by Newton's Second Law of Motion, we know that $F = ma = m(dv/dt)$, which yields the following differential equation.

$$m\frac{dv}{dt} = mg - kv \quad\Longrightarrow\quad \frac{dv}{dt} + \frac{k}{m}v = g$$

EXAMPLE 6 A falling object with air resistance

An object of mass m is dropped from a hovering helicopter. Find its velocity as a function of time t, assuming the air resistance is proportional to the velocity of the object.

SOLUTION

The velocity v satisfies the equation

$$\frac{dv}{dt} + \frac{kv}{m} = g$$

where g is the gravitational constant and k is the constant of proportionality. Letting $b = k/m$, we can *separate variables* to obtain

$$dv = (g - bv) \, dt$$

$$\int \frac{dv}{g - bv} = \int dt$$

$$-\frac{1}{b} \ln |g - bv| = t + C_1$$

$$\ln |g - bv| = -bt - bC_1$$

$$g - bv = Ce^{-bt}.$$

Since the object was dropped, $v = 0$ when $t = 0$; thus $g = C$, and it follows that

$$-bv = -g + ge^{-bt} \quad \Longrightarrow \quad v = \frac{g - ge^{-bt}}{b} = \frac{mg}{k}(1 - e^{-kt/m}). \quad \blacksquare$$

REMARK Note in Example 6 that the velocity approaches a limit of mg/k. This limit is due to the air resistance. For falling-body problems in which the air resistance is neglected, the velocity increases without bound.

EXERCISES for Section 18.4

In Exercises 1–10, solve the first-order linear differential equation.

1. $\dfrac{dy}{dx} + \left(\dfrac{1}{x}\right)y = 3x + 4$

2. $\dfrac{dy}{dx} + \left(\dfrac{2}{x}\right)y = 3x + 1$

3. $\dfrac{dy}{dx} = e^x - y$

4. $y' + 2y = \sin x$

5. $y' - y = \cos x$

6. $y' + 2xy = 2x$

7. $(3y + \sin 2x) \, dx - dy = 0$

8. $(y - 1) \sin x \, dx - dy = 0$

9. $(x - 1)y' + y = x^2 - 1$

10. $y' + 5y = e^{5x}$

In Exercises 11–16, find the particular solution that satisfies the given boundary condition.

11. $y' \cos^2 x + y - 1 = 0$, $y(0) = 5$

12. $x^3y' + 2y = e^{1/x^2}$, $y(1) = e$

13. $y' + y \tan x = \sec x + \cos x$, $y(0) = 1$

14. $y' + y \sec x = \sec x$, $y(0) = 4$

15. $y' + \left(\dfrac{1}{x}\right)y = 0$, $y(2) = 2$

16. $y' + (2x - 1)y = 0$, $y(1) = 2$

In Exercises 17–22, solve the Bernoulli differential equation.

17. $y' + 3x^2 y = x^2 y^3$

18. $y' + 2xy = xy^2$

19. $y' + \left(\dfrac{1}{x}\right)y = xy^2$

20. $y' + \left(\dfrac{1}{x}\right)y = x\sqrt{y}$

21. $y' - y = x^3 \sqrt[3]{y}$

22. $yy' - 2y^2 = e^x$

In Exercises 23 and 24, use the differential equation for electrical circuits given by

$$L\frac{dI}{dt} + RI = E.$$

23. Solve the differential equation given a constant voltage E_0.

24. Use the result of Exercise 23 to find the equation for the current if $I(0) = 0$, $E_0 = 110$ volts, $R = 550$ ohms, and $L = 4$ henrys. When does the current reach 90 percent of its limiting value?

25. When predicting population growth, demographers must consider birth and death rates as well as the net change caused by the difference between the rates of immigration and emigration. Let P be the population at time t and N be the net increase per unit time due to the difference between immigration and emigration. Thus, the rate of growth of the population is given by

$$\frac{dP}{dt} = kP + N, \quad N \text{ is constant.}$$

Solve this differential equation to find P as a function of time if at time $t = 0$ the size of the population is P_0.

26. A large corporation starts at time $t = 0$ to invest part of its receipts at a rate of P dollars per year in a fund for future corporate expansion. Assume that the fund earns r percent interest per year compounded continuously. Thus, the rate of growth of the amount A in the fund is given by

$$\frac{dA}{dt} = rA + P$$

where $A = 0$ when $t = 0$. Solve this differential equation for A as a function of t.

In Exercises 27 and 28, use the result of Exercise 26.

27. Find A if
(a) $P = \$100{,}000$, $r = 12\%$, and $t = 5$ years
(b) $P = \$250{,}000$, $r = 15\%$, and $t = 10$ years

28. Find t if the corporation needs $\$800{,}000$ and it can invest $\$75{,}000$ per year in a fund earning 13% interest compounded continuously.

29. Glucose is added intravenously to the bloodstream at the rate of q units per minute, and the body removes glucose from the bloodstream at a rate proportional to the amount present. Assume $Q(t)$ is the amount of glucose in the bloodstream at time t.
(a) Determine the differential equation describing the rate of change with respect to time of glucose in the bloodstream.
(b) Solve the differential equation, letting $Q = Q_0$ when $t = 0$.
(c) Find the limit of $Q(t)$ as $t \to \infty$.

30. The management at a certain factory has found that the maximum number of units a worker can produce in a day is 30. The rate of increase in the number of units N produced with respect to time t in days by a new employee is proportional to $30 - N$.
(a) Determine the differential equation describing the rate of change of performance with respect to time.
(b) Solve the differential equation of part (a).
(c) Find the particular solution for a new employee who produced 10 units on the first day at the factory and 19 units on the twentieth day. This is called the *learning curve* of the employee.

In Exercises 31–34, consider a tank that at time $t = 0$ contains v_0 gallons of a solution, of which, by weight, q_0 pounds is a soluble concentrate. Another solution containing q_1 pounds of the concentrate per gallon is running into the tank at the rate of r_1 gallons per minute. The solution in the tank is kept well stirred and is withdrawn at the rate of r_2 gallons per minute.

31. If Q is the amount of concentrate in the solution at any time t and $Q = q_0$, show that

$$\frac{dQ}{dt} + \frac{r_2 Q}{v_0 + (r_1 - r_2)t} = q_1 r_1.$$

32. A 200-gallon tank is full of a solution containing 25 pounds of concentrate. Starting at time $t = 0$, distilled water is admitted to the tank at the rate of 10 gallons per minute, and the well-stirred solution is withdrawn at the same rate.
(a) Find the amount of the concentrate in the solution as a function of t.
(b) Find the time when the amount of concentrate in the tank reaches 15 pounds.

33. Repeat Exercise 32, assuming that the solution entering the tank contains 0.05 pound of concentrate per gallon.

34. A 200-gallon tank is half full of distilled water. At time $t = 0$, a solution containing 0.5 pound of concentrate per gallon enters the tank at the rate of 5 gallons per minute, and the well-stirred mixture is withdrawn at the rate of 3 gallons per minute.
(a) At what time will the tank be full?
(b) At the time the tank is full, how many pounds of concentrate will it contain?

In Exercises 35–50, solve the first-order differential equation by any appropriate method.

35. $\dfrac{dy}{dx} = \dfrac{e^{2x+y}}{e^{x-y}}$

36. $\dfrac{dy}{dx} = \dfrac{x+1}{y(y+2)}$

37. $(1 + y^2)\, dx + (2xy + y + 2)\, dy = 0$

38. $(1 + 2e^{2x+y})\, dx + e^{2x+y}\, dy = 0$

39. $y \cos x - \cos x + \dfrac{dy}{dx} = 0$

40. $(x + 1)\dfrac{dy}{dx} = e^x - y$

41. $(x^2 + \cos y)\dfrac{dy}{dx} = -2xy$

42. $y' = 2x\sqrt{1 - y^2}$

43. $(3y^2 + 4xy)\, dx + (2xy + x^2)\, dy = 0$

44. $(x + y)\, dx - x\, dy = 0$

45. $(2y - e^x)\, dx + x\, dy = 0$

46. $(y^2 + xy)\, dx - x^2\, dy = 0$

47. $(x^2y^4 - 1)\, dx + x^3y^3\, dy = 0$

48. $y\, dx + (3x + 4y)\, dy = 0$

49. $3y\, dx - (x^2 + 3x + y^2)\, dy = 0$

50. $x\, dx + (y + e^y)(x^2 + 1)\, dy = 0$

18.5 Second-Order Homogeneous Linear Equations

Second-order linear differential equations ▪ Higher-order linear differential equations ▪ Applications

In this and the following section we discuss methods for solving higher-order linear differential equations.

DEFINITION OF A LINEAR DIFFERENTIAL EQUATION OF ORDER *n*	Let $g_1, g_2, \ldots, g_n$ and f be functions of x with a common domain. An equation of the form $$y^{(n)} + g_1(x)y^{(n-1)} + g_2(x)y^{(n-2)} + \cdots + g_{n-1}(x)y' + g_n(x)y = f(x)$$ is called a **linear differential equation of order *n*.** If $f(x) = 0$, the equation is **homogeneous;** otherwise, it is **nonhomogeneous.**

REMARK Note that this use of the term *homogeneous* differs from that in Section 18.2.

We will study homogeneous equations in this section, and leave the nonhomogeneous case for the next section.

We begin with a definition of linear independence. We say that the functions $y_1, y_2, \ldots, y_n$ are **linearly independent** if the *only* solution of the equation

$$C_1y_1 + C_2y_2 + \cdots + C_ny_n = 0$$

is the trivial one, $C_1 = C_2 = \cdots = C_n = 0$. Otherwise, this set of functions is called **linearly dependent.** For example, the functions $y_1(x) = \sin x$ and $y_2 = x$ are linearly independent because the only values of C_1 and C_2 for which

$$C_1 \sin x + C_2x = 0 \qquad \textit{for all } x$$

are $C_1 = 0$ and $C_2 = 0$. It can be shown that two functions form a linearly dependent set if and only if one is a constant multiple of the other. For example, $y_1(x) = x$ and $y_2(x) = 3x$ are linearly dependent because $C_1x + C_2(3x) = 0$ has the nonzero solutions $C_1 = -3$ and $C_2 = 1$.

The following theorem points out the importance of linear independence in constructing the general solution of a second-order linear homogeneous differential equation with constant coefficients.

THEOREM 18.5 **GENERAL SOLUTION AS A LINEAR** **COMBINATION OF LINEARLY** **INDEPENDENT SOLUTIONS**	If y_1 and y_2 are linearly independent solutions of the differential equation $y'' + ay' + by = 0$, then the general solution is $$y = C_1y_1 + C_2y_2$$ where C_1 and C_2 are constants.

PROOF   We prove only one direction of the theorem. If y_1 and y_2 are solutions, then we obtain the following system of equations.

$$y_1''(x) + ay_1'(x) + by_1(x) = 0$$
$$y_2''(x) + ay_2'(x) + by_2(x) = 0$$

Multiplying the first equation by C_1, the second by C_2, and adding, we obtain

$$[C_1y_1''(x) + C_2y_2''(x)] + a[C_1y_1'(x) + C_2y_2'(x)] + b[C_1y_1(x) + C_2y_2(x)] = 0$$

which means that $y = C_1y_1 + C_2y_2$ is a solution, as desired. The proof that all solutions are of this form is best left to a full course on differential equations.

This theorem tells us that if we can find two linearly independent solutions, then we can obtain the general solution by forming a **linear combination** of the two solutions.

To find two linearly independent solutions, we note that the nature of the equation $y'' + ay' + by = 0$ suggests that it may have solutions of the form $y = e^{mx}$. If so, then $y' = me^{mx}$ and $y'' = m^2e^{mx}$. Thus, by substitution, $y = e^{mx}$ is a solution if and only if

$$y'' + ay' + by = 0$$
$$m^2e^{mx} + ame^{mx} + be^{mx} = 0$$
$$e^{mx}(m^2 + am + b) = 0.$$

Since e^{mx} is never zero, $y = e^{mx}$ is a solution if and only if

$$m^2 + am + b = 0. \qquad \text{Characteristic equation}$$

This equation is called the **characteristic equation** of the differential equation $y'' + ay' + by = 0$. Note that the characteristic equation can be determined from its differential equation simply by replacing y'' by m^2, y' by m, and y by 1.

EXAMPLE 1   Characteristric equation with distinct real roots

Solve the differential equation

$$y'' - 4y = 0.$$

SOLUTION

In this case the characteristic equation is

$$m^2 - 4 = 0 \qquad \text{Characteristic equation}$$

so $m = \pm 2$. Thus, $y_1 = e^{m_1 x} = e^{2x}$ and $y_2 = e^{m_2 x} = e^{-2x}$ are particular solutions of the given differential equation. Furthermore, since these two solutions are linearly independent, we can apply Theorem 18.5 to conclude that the general solution is

$$y = C_1 e^{2x} + C_2 e^{-2x}. \qquad \text{General solution}$$ ▭

The characteristic equation in Example 1 has two distinct real roots. From algebra we know that this is only one of *three* possibilities for quadratic equations. In general, the quadratic equation $m^2 + am + b = 0$ has roots

$$m_1 = \frac{-a + \sqrt{a^2 - 4b}}{2} \qquad \text{and} \qquad m_2 = \frac{-a - \sqrt{a^2 - 4b}}{2}$$

which fall into one of three cases.

1. Two distinct real roots, $m_1 \neq m_2$
2. Two equal real roots, $m_1 = m_2$
3. Two complex conjugate roots, $m_1 = \alpha + \beta i$ and $m_2 = \alpha - \beta i$

In terms of the differential equation $y'' + ay' + by = 0$, these three cases correspond to three different types of general solutions.

THEOREM 18.6
SOLUTIONS OF
$y'' + ay' + by = 0$

Distinct Real Roots: If $m_1 \neq m_2$ are distinct real roots of the characteristic equation, then the general solution is

$$y = C_1 e^{m_1 x} + C_2 e^{m_2 x}.$$

Equal Real Roots: If $m_1 = m_2$ are equal real roots of the characteristic equation, then the general solution is

$$y = C_1 e^{m_1 x} + C_2 x e^{m_1 x} = (C_1 + C_2 x) e^{m_1 x}.$$

Complex Roots: If $m_1 = \alpha + \beta i$ and $m_2 = \alpha - \beta i$ are complex roots of the characteristic equation, then the general solution is

$$y = C_1 e^{\alpha x} \cos \beta x + C_2 e^{\alpha x} \sin \beta x.$$

EXAMPLE 2   Characteristic equation with complex roots

Find the general solution of the differential equation $y'' + 6y' + 12y = 0$.

SOLUTION

The characteristic equation

$$m^2 + 6m + 12 = 0$$

has two complex roots, as follows.

$$m = \frac{-6 \pm \sqrt{36 - 48}}{2}$$

$$= \frac{-6 \pm \sqrt{-12}}{2}$$

$$= -3 \pm \sqrt{-3} = -3 \pm \sqrt{3}\,i$$

Thus, $\alpha = -3$ and $\beta = \sqrt{3}$, and the general solution is

$$y = C_1 e^{-3x} \cos \sqrt{3}x + C_2 e^{-3x} \sin \sqrt{3}x.$$ ▭

REMARK   In Example 2, note that although the characteristic equation has two *complex* roots, the solution of the differential equation is *real*.

EXAMPLE 3   Characteristic equation with repeated roots

Solve the differential equation $y'' + 4y' + 4y = 0$, subject to the initial conditions $y(0) = 2$ and $y'(0) = 1$.

SOLUTION

The characteristic equation

$$m^2 + 4m + 4 = (m + 2)^2 = 0$$

has two equal real roots given by $m = -2$. Thus, the general solution is

$$y = C_1 e^{-2x} + C_2 x e^{-2x}. \qquad \text{General solution}$$

Now, since $y = 2$ when $x = 0$, we have

$$2 = C_1(1) + C_2(0)(1) = C_1.$$

Furthermore, since $y' = 1$ when $x = 0$, we have

$$y' = -2C_1 e^{-2x} + C_2(-2xe^{-2x} + e^{-2x})$$

$$1 = -2(2)(1) + C_2[-2(0)(1) + 1]$$

$$5 = C_2.$$

Therefore, the solution is

$$y = 2e^{-2x} + 5xe^{-2x}. \qquad \text{Particular solution}$$ ▭

Higher-order linear differential equations

For higher-order homogeneous linear differential equations, we find the general solution in much the same way as we do for second-order equations. That is, we begin by determining the *n* roots of the characteristic equation, and then, based on these *n* roots, we form a linearly independent collection of *n* solutions. The major difference is that with equations of third or higher order, roots of the characteristic equation may occur more than twice. When this happens, the linearly independent solutions are formed by multiplying by increasing powers of *x*. We demonstrate this in the next two examples.

EXAMPLE 4   Solving a third-order equation

Find the general solution of $y''' + 3y'' + 3y' + y = 0$.

SOLUTION

The characteristic equation is

$$m^3 + 3m^2 + 3m + 1 = (m + 1)^3 = 0.$$

Since the root $m = -1$ occurs three times, the general solution is

$$y = C_1 e^{-x} + C_2 x e^{-x} + C_3 x^2 e^{-x}. \qquad \text{General solution}$$

EXAMPLE 5   Solving a fourth-order equation

Find the general solution of $y^{(4)} + 2y'' + y = 0$.

SOLUTION

The characteristic equation is as follows.

$$m^4 + 2m^2 + 1 = (m^2 + 1)^2 = 0$$
$$m = \pm i$$

Since the roots $m_1 = \alpha + \beta i = 0 + i$ and $m_2 = \alpha - \beta i = 0 - i$ each occur twice, the general solution is

$$y = C_1 \cos x + C_2 \sin x + C_3 x \cos x + C_4 x \sin x. \qquad \text{General solution}$$

Applications

One of many applications of linear differential equations is that of describing the motion of an oscillating spring. According to Hooke's Law (Section 6.5), a spring that is stretched (or compressed) *y* units from its natural length 1 tends to *restore* itself to its natural length by a force *F* that is proportional to *y*. That is, $F(y) = -ky$, where *k* is called the **spring constant** and indicates the stiffness of a given spring.

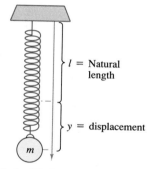

l = Natural length

y = displacement

m

Spring Displacement

FIGURE 18.8

Suppose a rigid object of mass m is attached to the end of the spring and causes a displacement, as shown in Figure 18.8. Assume that the mass of the spring is negligible compared to m. Now suppose the object is pulled down and released. The resulting oscillations are a product of two opposing forces—the spring force $F(y) = -ky$ and the weight mg of the object. Under such conditions, we can use a differential equation to find the position y of the object as a function of time t. According to Newton's Second Law of Motion, the force acting on the weight is $F = ma$, where $a = d^2y/dt^2$ is the acceleration. Assuming the motion is **undamped**—that is, there are no other external forces acting on the object—it follows that $m(d^2y/dt^2) = -ky$, and we have

$$\frac{d^2y}{dt^2} + \left(\frac{k}{m}\right)y = 0. \qquad \text{Undamped motion of a spring}$$

We use this equation in our next example.

EXAMPLE 6   Undamped motion of a spring

Suppose a 4-pound weight stretches a spring 8 inches from its natural length. If the weight is pulled down an additional 6 inches and released with an initial upward velocity of 8 feet per second, find a formula for the position of the weight as a function of time t.

SOLUTION

By Hooke's Law, $4 = k\left(\frac{2}{3}\right)$, so $k = 6$. Moreover, since the weight w is given by mg, it follows that $m = w/g = \frac{4}{32} = \frac{1}{8}$. Hence, the resulting differential equation for this undamped motion is

$$\frac{d^2y}{dt^2} + 48y = 0.$$

Since the characteristic equation $m^2 + 48 = 0$ has complex roots $m = 0 \pm 4\sqrt{3}\,i$, the general solution is

$$y = C_1 e^0 \cos 4\sqrt{3}\,t + C_2 e^0 \sin 4\sqrt{3}\,t = C_1 \cos 4\sqrt{3}\,t + C_2 \sin 4\sqrt{3}\,t.$$

Using the initial conditions, we have

$$\frac{1}{2} = C_1(1) + C_2(0) \implies C_1 = \frac{1}{2} \qquad y(0) = \frac{1}{2}$$

$$y'(t) = -4\sqrt{3}\,C_1 \sin 4\sqrt{3}\,t + 4\sqrt{3}\,C_2 \cos 4\sqrt{3}\,t$$

$$8 = -4\sqrt{3}\left(\frac{1}{2}\right)(0) + 4\sqrt{3}\,C_2(1) \implies C_2 = \frac{2\sqrt{3}}{3}. \qquad y'(0) = 8$$

Consequently, the position at time t is given by

$$y = \frac{1}{2}\cos 4\sqrt{3}\,t + \frac{2\sqrt{3}}{3}\sin 4\sqrt{3}\,t.$$

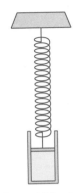

Damped Vibration

FIGURE 18.9

Suppose the object in Figure 18.9 experiences an additional damping or frictional force that is proportional to its velocity. A case in point would be the damping force due to friction and movement through a fluid. With this damping force, $-p(dy/dt)$, considered, the differential equation for the oscillation is

$$m\frac{d^2y}{dt^2} = -ky - p\frac{dy}{dt}$$

or, in standard linear form,

$$\frac{d^2y}{dt^2} + \frac{p}{m}\left(\frac{dy}{dt}\right) + \frac{k}{m}y = 0. \quad \text{Damped motion of a spring}$$

EXERCISES for Section 18.5

In Exercises 1–26, find the general solution of the linear differential equation.

1. $y'' - y' = 0$
2. $y'' + 2y' = 0$
3. $y'' - y' - 6y = 0$
4. $y'' + 6y' + 5y = 0$
5. $2y'' + 3y' - 2y = 0$
6. $16y'' - 16y' + 3y = 0$
7. $y'' + 6y' + 9y = 0$
8. $y'' - 10y' + 25y = 0$
9. $16y'' - 8y' + y = 0$
10. $9y'' - 12y' + 4y = 0$
11. $y'' + y = 0$
12. $y'' + 4y = 0$
13. $y'' - 9y = 0$
14. $y'' - 2y = 0$
15. $y'' - 2y' + 4y = 0$
16. $y'' - 4y' + 21y = 0$
17. $y'' - 3y' + y = 0$
18. $3y'' + 4y' - y = 0$
19. $9y'' - 12y' + 11y = 0$
20. $2y'' - 6y' + 7y = 0$
21. $y^{(4)} - y = 0$
22. $y^{(4)} - y'' = 0$
23. $y''' - 6y'' + 11y' - 6y = 0$
24. $y''' - y'' - y' + y = 0$
25. $y''' - 3y'' + 7y' - 5y = 0$
26. $y''' - 3y'' + 3y' - y = 0$

In Exercises 27 and 28, find the particular solution to the linear differential equation.

27. $y'' - y' - 30y = 0$
    $y(0) = 1,\ y'(0) = -4$
28. $y'' + 2y' + 3y = 0$
    $y(0) = 2,\ y'(0) = 1$

In Exercises 29–34, describe the motion of a 32-pound weight suspended on a spring. Assume that the weight stretches the spring $\frac{2}{3}$ foot from its natural position.

29. The weight is pulled $\frac{1}{2}$ foot below the equilibrium position and released.
30. The weight is raised $\frac{2}{3}$ foot above the equilibrium position and released.
31. The weight is raised $\frac{2}{3}$ foot above the equilibrium position and started off with a downward velocity of $\frac{1}{2}$ foot per second.

32. The weight is pulled $\frac{1}{2}$ foot below the equilibrium position and started off with an upward velocity of $\frac{1}{2}$ foot per second.
33. The weight is pulled $\frac{1}{2}$ foot below the equilibrium position and released. The motion takes place in a medium that furnishes a damping force of magnitude $\frac{1}{8}$ speed at all times.
34. The weight is pulled $\frac{1}{2}$ foot below the equilibrium position and released. The motion takes place in a medium that furnishes a damping force of magnitude $\frac{1}{4}|v|$ at all times.

35. If the characteristic equation of the differential equation

    $$y'' + ay' + by = 0$$

    has two equal real roots given by $m = r$, show that

    $$y = C_1 e^{rx} + C_2 x e^{rx}$$

    is a solution.

36. If the characteristic equation of the differential equation

    $$y'' + ay' + by = 0$$

    has complex roots given by $m_1 = \alpha + \beta i$ and $m_2 = \alpha - \beta i$, show that

    $$y = C_1 e^{\alpha x} \cos \beta x + C_2 e^{\alpha x} \sin \beta x$$

    is a solution.

The *Wronskian* of two differentiable functions f and g, denoted by $W(f, g)$, is defined to be the function given by the determinant

$$W(f, g) = \begin{vmatrix} f & g \\ f' & g' \end{vmatrix}.$$

The functions f and g are linearly independent if there exists at least one value of x for which $W(f, g) \neq 0$. In Exercises 37–40, use the Wronskian to verify the linear independence of the two functions.

37. $y_1 = e^{ax}, y_2 = e^{bx} \quad (a \neq b)$

38. $y_1 = e^{ax}, y_2 = xe^{ax}$

39. $y_1 = e^{ax} \sin bx, y_2 = e^{ax} \cos bx, b \neq 0$

40. $y_1 = x, y_2 = x^2$

18.6 Second-Order Nonhomogeneous Linear Equations

Nonhomogeneous equations ■ Method of undetermined coefficients ■ Variation of parameters

Sophie Germain

Many of the early contributors to calculus were interested in forming mathematical models for vibrating strings and membranes, oscillating springs, and elasticity. One of these was the French mathematician Sophie Germain (1776–1831), who in 1816 was awarded a prize by the French Academy for a paper entitled *Memoir on the Vibrations of Elastic Plates*.

In the previous section we represented damped oscillations of a spring by the *homogeneous* second-order linear equation

$$\frac{d^2y}{dt^2} + \frac{p}{m}\left(\frac{dy}{dt}\right) + \frac{k}{m}y = 0. \qquad \text{Free motion}$$

We call this type of oscillation **free** because it is determined solely by the spring and gravity and is free of the action of other external forces. If such a system is also subject to an external periodic force such as $a \sin bt$, caused by vibrations at the opposite end of the spring, then we call the motion **forced,** and it is characterized by the *nonhomogeneous* equation

$$\frac{d^2y}{dt^2} + \frac{p}{m}\left(\frac{dy}{dt}\right) + \frac{k}{m}y = a \sin bt. \qquad \text{Forced motion}$$

In this section we describe two methods for finding the general solution of a nonhomogeneous linear differential equation. In both methods, the first step is to find the general solution, called y_h, of the corresponding homogeneous equation. Having done that, we try to find a particular solution, y_p, of the nonhomogeneous equation. By combining these two results, we can conclude that the general solution of the nonhomogeneous equation is $y = y_h + y_p$, as stated in the following theorem.

**THEOREM 18.7
GENERAL SOLUTION OF
NONHOMOGENEOUS
LINEAR EQUATION**

Let $y'' + ay' + by = F(x)$ be a second-order nonhomogeneous linear differential equation. If y_p is a particular solution to this equation and y_h is the general solution of the corresponding homogeneous equation, then

$$y = y_h + y_p$$

is the general solution of the nonhomogeneous equation.

PROOF The sum $y = y_h + y_p$ is a solution to the nonhomogeneous equation because

$$y'' + ay' + by = (y_h'' + y_p'') + a(y_h' + y_p') + b(y_h + y_p)$$
$$= (y_h'' + ay_h' + by_h) + (y_p'' + ay_p' + by_p)$$
$$= 0 + F(x) = F(x).$$

Conversely, if y is any solution to the nonhomogeneous equation, then $y - y_p$ is a solution to the homogeneous equation because

$$(y - y_p)'' + a(y - y_p)' + b(y - y_p)$$

$$= (y'' - y_p'') + a(y' - y_p') + b(y - y_p)$$
$$= (y'' + ay' + by) - (y_p'' + ay_p' + by_p)$$
$$= F(x) - F(x) = 0.$$

Consequently, $y - y_p$ is of the form $y - y_p = y_h$, which in turn means that $y = y_p + y_h$ is the general solution of the nonhomogeneous equation.

Method of undetermined coefficients

Since we already have the tools for finding y_h, we focus on ways to find the particular solution y_p. If $F(x)$ in $y'' + ay' + by = F(x)$ consists of sums or products of

$$x^n, \quad e^{mx}, \quad \cos \beta x, \quad \sin \beta x$$

then we can find a particular solution y_p by the method of **undetermined coefficients.** The gist of the method is to guess that the solution y_p is a generalized form of $F(x)$. Here are some examples.

1. If $F(x) = 3x^2$, choose $y_p = Ax^2 + Bx + C$.
2. If $F(x) = 4xe^x$, choose $y_p = Axe^x + Be^x$.
3. If $F(x) = x + \sin 2x$, choose $y_p = (Ax + B) + C \sin 2x + D \cos 2x$.

Then, by substitution, we determine the coefficients for this generalized solution. We illustrate this method in the next four examples.

EXAMPLE 1   Method of undetermined coefficients

Find the general solution of the equation $y'' - 2y' - 3y = 2 \sin x$.

SOLUTION

To find y_h, we solve the characteristic equation

$$m^2 - 2m - 3 = (m + 1)(m - 3) = 0 \quad \Longrightarrow \quad m = -1 \text{ and } m = 3.$$

Thus, $y_h = C_1 e^{-x} + C_2 e^{3x}$. Next, we let y_p be a generalized form of $2 \sin x$.

$$y_p = A \cos x + B \sin x$$
$$y_p' = -A \sin x + B \cos x$$
$$y_p'' = -A \cos x - B \sin x.$$

Substitution into the given equation yields

$$-A \cos x - B \sin x + 2A \sin x - 2B \cos x - 3A \cos x - 3B \sin x = 2 \sin x$$
$$(-4A - 2B) \cos x + (2A - 4B) \sin x = 2 \sin x.$$

Consequently, y_p is a solution, provided the coefficients of like terms are equal. Thus, we obtain the system

$$-4A - 2B = 0 \quad \text{and} \quad 2A - 4B = 2$$

with solutions $A = \frac{1}{5}$ and $B = -\frac{2}{5}$. Therefore, the general solution is

$$y = y_h + y_p = C_1 e^{-x} + C_2 e^{3x} + \frac{1}{5} \cos x - \frac{2}{5} \sin x. \qquad \blacksquare$$

In Example 1, the form of the homogeneous solution $y_h = C_1 e^{-x} + C_2 e^{3x}$ has no overlap with the function $F(x)$ in the equation $y'' + ay' + by = F(x)$. However, suppose the given differential equation in Example 1 were of the form

$$y'' - 2y' - 3y = e^{-x}.$$

Now, it would make no sense to guess the particular solution to be $y = Ae^{-x}$, since we know that this solution yields zero. In such cases, we alter the guess by multiplying by the lowest power of x that removes the duplication. For this particular problem, we would guess $y_p = Axe^{-x}$. This is demonstrated further in the next example.

EXAMPLE 2   Method of undetermined coefficients

Find the general solution of $y'' - 2y' = x + 2e^x$.

SOLUTION

The characteristic equation $m^2 - 2m = 0$ has solutions $m = 0$ and $m = 2$. Thus,

$$y_h = C_1 + C_2 e^{2x}.$$

Since $F(x) = x + 2e^x$, our first choice for y_p would be $(A + Bx) + Ce^x$. However, since y_h *already* contains a constant term C_1, we multiply the *polynomial part* by x and use

$$y_p = Ax + Bx^2 + Ce^x$$
$$y_p' = A + 2Bx + Ce^x$$
$$y_p'' = 2B + Ce^x.$$

Substituting into the differential equation, we have

$$(2B + Ce^x) - 2(A + 2Bx + Ce^x) = x + 2e^x$$
$$(2B - 2A) - 4Bx - Ce^x = x + 2e^x.$$

Equating coefficients of like terms yields the system

$$2B - 2A = 0, \quad -4B = 1, \quad -C = 2$$

with solutions $A = B = -\frac{1}{4}$ and $C = -2$. Therefore,

$$y_p = -\frac{1}{4}x - \frac{1}{4}x^2 - 2e^x$$

and the general solution is

$$y = C_1 + C_2e^{2x} - \frac{1}{4}x - \frac{1}{4}x^2 - 2e^x.$$

◻

In Example 2 the polynomial part of the initial guess $(A + Bx) + Ce^x$ for y_p overlapped by a constant term with $y_h = C_1 + C_2e^{2x}$, and it was necessary to multiply the polynomial part by a power of x that removed the overlap. In the next example we further illustrate some choices for y_p that eliminate overlap with y_h. Remember that in all cases the first guess for y_p should match the types of functions occurring in $F(x)$.

EXAMPLE 3   Choosing the form of the particular solution

Determine a suitable choice for y_p for the following.

$y'' + ay' + by = F(x)$	y_h
(a) $y'' = x^2$	$C_1 + C_2x$
(b) $y'' + 2y' + 10y = 4 \sin 3x$	$C_1e^{-x} \cos 3x + C_2e^{-x} \sin 3x$
(c) $y'' - 4y' + 4 = e^{2x}$	$C_1e^{2x} + C_2xe^{2x}$

SOLUTION

(a) Since $F(x) = x^2$, the normal choice for y_p would be

$$A + Bx + Cx^2.$$

However, since $y_h = C_1 + C_2x$ already contains a linear term, we multiply by x^2 to obtain

$$y_p = Ax^2 + Bx^3 + Cx^4.$$

(b) Since $F(x) = 4 \sin 3x$ and since each term in y_h contains a factor of e^{-x}, we simply let

$$y_p = A \cos 3x + B \sin 3x.$$

(c) Since $F(x) = e^{2x}$, the normal choice for y_p would be Ae^{2x}. However, since

$$y_h = C_1e^{2x} + C_2xe^{2x}$$

already contains an xe^{2x} term, we multiply by x^2 to get

$$y_p = Ax^2e^{2x}.$$

◻

We can also use the method of undetermined coefficients for nonhomogeneous equations of higher order. The procedure is demonstrated in the next example.

EXAMPLE 4 Undetermined coefficients with a third-order equation

Find the general solution of $y''' + 3y'' + 3y' + y = x$.

SOLUTION

From Example 4 in the previous section we know that the homogeneous solution is

$$y_h = C_1 e^{-x} + C_2 x e^{-x} + C_3 x^2 e^{-x}.$$

Since $F(x) = x$, we let $y_p = A + Bx$ and obtain $y_p' = B$, and $y_p'' = y_p''' = 0$. Thus, by substitution, we have

$$(0) + 3(0) + 3(B) + (A + Bx) = (3B + A) + Bx = x.$$

Thus, $B = 1$ and $A = -3$ and the general solution is

$$y = C_1 e^{-x} + C_2 x e^{-x} + C_3 x^2 e^{-x} - 3 + x.$$

Variation of parameters

The method of undetermined coefficients works well if $F(x)$ is made up of polynomials or functions whose successive derivatives have a cyclic pattern. For functions like $1/x$ and $\tan x$, which do not possess such characteristics, we use a more general method called **variation of parameters.** In this method we assume that y_p has the same *form* as y_h, except that the constants in y_h are replaced by variables.

VARIATION OF PARAMETERS

To find the general solution to the equation $y'' + ay' + by = F(x)$:

1. Find $y_h = C_1 y_1 + C_2 y_2$.
2. Replace the constants by variables to form $y_p = u_1 y_1 + u_2 y_2$.
3. Solve the following system for u_1' and u_2'.

$$u_1' y_1 + u_2' y_2 = 0$$
$$u_1' y_1' + u_2' y_2' = F(x)$$

4. Integrate to find u_1 and u_2. The general solution is $y = y_h + y_p$.

EXAMPLE 5 Variation of parameters

Solve the differential equation

$$y'' - 2y' + y = \frac{e^x}{2x}, \quad x > 0.$$

SOLUTION

The characteristic equation $m^2 - 2m + 1 = (m - 1)^2 = 0$ has one solution, $m = 1$. Thus, the homogeneous solution is

$$y_h = C_1 y_1 + C_2 y_2 = C_1 e^x + C_2 x e^x.$$

Replacing C_1 and C_2 by u_1 and u_2 produces

$$y_p = u_1 y_1 + u_2 y_2 = u_1 e^x + u_2 x e^x.$$

The resulting system of equations is

$$u_1' e^x + u_2' x e^x = 0$$

$$u_1' e^x + u_2' (x e^x + e^x) = \frac{e^x}{2x}.$$

Subtracting the second equation from the first produces $u_2' = \frac{1}{2}x$. Then, by substitution in the first equation, we have $u_1' = -\frac{1}{2}$. Finally, integration yields

$$u_1 = -\int \frac{1}{2} dx = -\frac{x}{2} \quad \text{and} \quad u_2 = \frac{1}{2} \int \frac{1}{x} dx = \frac{1}{2} \ln x = \ln \sqrt{x}.$$

From this result it follows that a particular solution is

$$y_p = -\frac{1}{2} x e^x + (\ln \sqrt{x}) x e^x$$

and the general solution is

$$y = C_1 e^x + C_2 x e^x - \frac{1}{2} x e^x + x e^x \ln \sqrt{x}.$$

EXAMPLE 6   Variation of parameters

Solve the differential equation $y'' + y = \tan x$.

SOLUTION

Since the characteristic equation $m^2 + 1 = 0$ has solutions $m = \pm i$, the homogeneous solution is

$$y_h = C_1 \cos x + C_2 \sin x.$$

Replacing C_1 and C_2 by u_1 and u_2, we have

$$y_p = u_1 \cos x + u_2 \sin x.$$

The resulting system of equations is

$$u_1' \cos x + u_2' \sin x = 0$$
$$-u_1' \sin x + u_2' \cos x = \tan x.$$

Multiplying the first equation by $\sin x$ and the second by $\cos x$, we have

$$u_1' \sin x \cos x + u_2' \sin^2 x = 0$$
$$-u_1' \sin x \cos x + u_2' \cos^2 x = \sin x.$$

Adding these two equations, we have $u_2' = \sin x$, which implies that

$$u_1' = -\frac{\sin^2 x}{\cos x} = \frac{\cos^2 x - 1}{\cos x} = \cos x - \sec x.$$

Integration yields

$$u_1 = \int (\cos x - \sec x) \, dx = \sin x - \ln|\sec x + \tan x|$$

$$u_2 = \int \sin x \, dx = -\cos x$$

so that

$$y_p = \sin x \cos x - \cos x \ln|\sec x + \tan x| - \sin x \cos x$$
$$= -\cos x \ln|\sec x + \tan x|$$

and the general solution is

$$y = y_h + y_p = C_1 \cos x + C_2 \sin x - \cos x \ln|\sec x + \tan x|. \quad \blacksquare$$

EXERCISES for Section 18.6

In Exercises 1–16, solve the differential equation by the method of undetermined coefficients.

1. $y'' - 3y' + 2y = 2x$

2. $y'' - 2y' - 3y = x^2 - 1$

3. $y'' + y = x^3$
  $y(0) = 1, \ y'(0) = 0$

4. $y'' + 4y = 4$
  $y(0) = 1, \ y'(0) = 6$

5. $y'' + 2y' = 2e^x$

6. $y'' - 9y = 5e^{3x}$

7. $y'' - 10y' + 25y = 5 + 6e^x$

8. $16y'' - 8y' + y = 4(x + e^x)$

9. $y'' + y' = 2 \sin x$
  $y(0) = 0, \ y'(0) = -3$

10. $y'' + y' - 2y = 3 \cos 2x$
  $y(0) = -1, \ y'(0) = 2$

11. $y'' + 9y = \sin 3x$

12. $y'' + 4y' + 5y = \sin x + \cos x$

13. $y''' - 3y' + 2y = 2e^{-2x}$

14. $y''' - y'' = 4x^2$
  $y(0) = 1, \ y'(0) = 1, \ y''(0) = 1$

15. $y' - 4y = xe^x - xe^{4x}$
  $y(0) = \dfrac{1}{3}$

16. $y' + 2y = \sin x$
  $y\left(\dfrac{\pi}{2}\right) = \dfrac{2}{5}$

In Exercises 17–22, solve the differential equation by the method of variation of parameters.

17. $y'' + y = \sec x$

18. $y'' + y = \sec x \tan x$

19. $y'' + 4y = \csc 2x$

20. $y'' - 4y' + 4y = x^2 e^{2x}$

21. $y'' - 2y' + y = e^x \ln x$

22. $y'' - 4y' + 4y = \dfrac{e^{2x}}{x}$

In Exercises 23 and 24, use the electrical circuit differential equation

$$\frac{d^2q}{dt^2} + \left(\frac{R}{L}\right)\frac{dq}{dt} + \left(\frac{1}{LC}\right)q = \left(\frac{1}{L}\right)E(t)$$

where R is the resistance (in ohms), C is the capacitance (in farads), L is the inductance (in henrys), $E(t)$ is the electromotive force (in volts), and q is the charge on the capacitor C (in coulombs). Find the charge q as a function of time for the electrical circuit described. Assume that $q(0) = 0$ and $q'(0) = 0$.

23. $R = 20, \ C = 0.02, \ L = 2$
  $E(t) = 12 \sin 5t$

24. $R = 20, \ C = 0.02, \ L = 1$
  $E(t) = 10 \sin 5t$

In Exercises 25–28, use the given information to find the particular solution to the differential equation

$$\frac{w}{g}y''(t) + by'(t) + ky(t) = \frac{w}{g}F(t)$$

for the oscillating motion of an object on the end of a spring. In the equation, y is the displacement from the equilibrium (positive direction is downward) measured in feet, and t is time in seconds (see figure). The constant w is the weight of the object, g is the acceleration due to gravity, b measures the magnitude of the resistance to the motion, k is the spring constant from Hooke's Law, and $F(t)$ is the acceleration imposed on the system.

25. $\dfrac{24}{32}y'' + 48y = \dfrac{24}{32}(48 \sin 4t)$

$y(0) = \dfrac{1}{4}, \ y'(0) = 0$

26. $\dfrac{2}{32}y'' + 4y = \dfrac{2}{32}(4 \sin 8t)$

$y(0) = \dfrac{1}{4}, \ y'(0) = 0$

27. $\dfrac{2}{32}y'' + y' + 4y = \dfrac{2}{32}(4 \sin 8t)$

$y(0) = \dfrac{1}{4}, \ y'(0) = -3$

28. $\dfrac{4}{32}y'' + \dfrac{1}{2}y' + \dfrac{25}{2}y = 0$

$y(0) = \dfrac{1}{2}, \ y'(0) = -4$

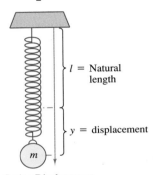

l = Natural length

y = displacement

m

Spring Displacement

FIGURE FOR 25–28

29. Rewrite y_h in the solution for Exercise 25 by using the identity

$$a \cos \omega t + b \sin \omega t = \sqrt{a^2 + b^2} \sin(\omega t + \phi)$$

where $\phi = \arctan a/b$.

30. Find $y(t)$ as $t \to \infty$ in Exercise 28. When there is no external acceleration imposed on the system, what is the ultimate effect of the retarding force?

31. As discussed in the Chapter 18 Application, the fall of a parachutist is described by the second-order linear differential equation

$$\frac{w}{g}\frac{d^2y}{dt^2} - k\frac{dy}{dt} = w$$

where w is the weight of the parachutist, y is the height at time t, g is the acceleration due to gravity, and k measures the drag factor of the parachute. If the parachute is opened at 2000 feet, $y(0) = 2000$, and at that time the velocity is $y'(0) = -100$ feet per second, then for a 160-pound parachutist, using $k = 8$, the differential equation is

$$-5y'' - 8y' = 160.$$

Using the given initial conditions, verify that the solution of the differential equation is

$$y = 1950 + 50e^{-1.6t} - 20t.$$

32. Repeat Exercise 31 for a parachutist who weighs 192 pounds and has a parachute with a drag factor of $k = 9$.

18.7 Series Solutions of Differential Equations

Power series solution to a differential equation ▪ Approximation by Taylor series

We conclude this chapter by showing how power series can be used to solve certain types of differential equations. We will limit our discussion to the statement and demonstration of the method and omit the theoretical development.

We begin with the general **power series solution** method. Recall from Chapter 10 that a power series represents a function f on an interval of convergence, and that we can successively differentiate the power series to obtain a series for f', f'', and so on.

$$f(x) = a_0 + a_1x + a_2x^2 + a_3x^3 + \cdots = \sum_{n=0}^{\infty} a_nx^n$$

$$f'(x) = a_1 + 2a_2x + 3a_3x^2 + 4a_4x^3 + \cdots = \sum_{n=0}^{\infty} na_nx^{n-1}$$

$$f''(x) = 2a_2 + 6a_3x + 12a_4x^2 + 20a_5x^3 + \cdots = \sum_{n=0}^{\infty} n(n-1)a_nx^{n-2}$$

These properties are used in the *power series* solution method demonstrated in the first two examples. We begin with a simple differential equation that can be solved by other methods.

EXAMPLE 1 Power series solution

Use a power series to find the general solution to the differential equation $y' - 2y = 0$.

SOLUTION

Assume

$$y = \sum_{n=0}^{\infty} a_n x^n$$

is a solution. Then

$$y' = \sum_{n=0}^{\infty} n a_n x^{n-1}.$$

Substituting for y' and $-2y$, we obtain the following series form for the given differential equation.

$$y' - 2y = \sum_{n=0}^{\infty} n a_n x^{n-1} - 2 \sum_{n=0}^{\infty} a_n x^n = 0$$

$$\sum_{n=0}^{\infty} n a_n x^{n-1} = 2 \sum_{n=0}^{\infty} a_n x^n$$

Next we adjust the summation indices so that x^n appears in each series. In this case we need only to replace n by $n + 1$ in the left-hand series to obtain

$$\sum_{n=-1}^{\infty} (n + 1) a_{n+1} x^n = 2 \sum_{n=0}^{\infty} a_n x^n.$$

Now, by equating coefficients of like terms, we obtain the **recursion formula** $(n + 1) a_{n+1} = 2a_n$, which implies that

$$a_{n+1} = \frac{2a_n}{n + 1}, \quad n \geq 0.$$

This formula generates the following results in terms of a_0.

$$a_1 = 2a_0$$

$$a_2 = \frac{2a_1}{2} = \frac{2^2 a_0}{2}$$

$$a_3 = \frac{2a_2}{3} = \frac{2^3 a_0}{2 \cdot 3} = \frac{2^3 a_0}{3!}$$

$$a_4 = \frac{2a_3}{4} = \frac{2^4 a_0}{2 \cdot 3 \cdot 4} = \frac{2^4 a_0}{4!}$$

$$\vdots$$

$$a_n = \frac{2^n a_0}{n!}$$

Using these values as the coefficients for the *solution* series, we have

$$y = \sum_{n=0}^{\infty} \frac{2^n a_0}{n!} x^n = a_0 \sum_{n=0}^{\infty} \frac{(2x)^n}{n!} = a_0 e^{2x}.$$ ▬

The next example *cannot* be solved by any of the methods discussed in previous sections of this chapter.

EXAMPLE 2   Power series solution

Use a power series to solve the differential equation $y'' + xy' + y = 0$.

SOLUTION

Assume $\sum_{n=0}^{\infty} a_n x^n$ is a solution. Then we have

$$y' = \sum_{n=0}^{\infty} n a_n x^{n-1}, \qquad xy' = \sum_{n=0}^{\infty} n a_n x^n, \qquad y'' = \sum_{n=0}^{\infty} n(n-1) a_n x^{n-2}.$$

Substituting for y'', xy', and y in the given differential equation, we obtain the following series.

$$\sum_{n=0}^{\infty} n(n-1) a_n x^{n-2} + \sum_{n=0}^{\infty} n a_n x^n + \sum_{n=0}^{\infty} a_n x^n = 0$$

$$\sum_{n=0}^{\infty} n(n-1) a_n x^{n-2} = -\sum_{n=0}^{\infty} (n+1) a_n x^n$$

To obtain equal powers of x, we adjust the summation indices by replacing n by $n + 2$ in the left-hand sum, to obtain

$$\sum_{n=-2}^{\infty} (n+2)(n+1) a_{n+2} x^n = -\sum_{n=0}^{\infty} (n+1) a_n x^n.$$

Now, by equating coefficients, we have $(n+2)(n+1) a_{n+2} = -(n+1) a_n$, from which we get the recursion formula

$$a_{n+2} = -\frac{(n+1)}{(n+2)(n+1)} a_n = -\frac{a_n}{n+2}, \quad n \ge 0$$

and the coefficients of the solution series are as follows.

$$a_2 = -\frac{a_0}{2} \qquad\qquad\qquad a_3 = -\frac{a_1}{3}$$

$$a_4 = -\frac{a_2}{4} = \frac{a_0}{2 \cdot 4} \qquad\qquad a_5 = -\frac{a_3}{5} = \frac{a_1}{3 \cdot 5}$$

$$a_6 = -\frac{a_4}{6} = -\frac{a_0}{2 \cdot 4 \cdot 6} \qquad\qquad a_7 = -\frac{a_5}{7} = -\frac{a_1}{3 \cdot 5 \cdot 7}$$

$$\vdots \qquad\qquad\qquad\qquad \vdots$$

$$a_{2k} = \frac{(-1)^k a_0}{2 \cdot 4 \cdot 6 \cdots (2k)} = \frac{(-1)^k a_0}{2^k (k!)} \qquad a_{2k+1} = \frac{(-1)^k a_1}{3 \cdot 5 \cdot 7 \cdots (2k+1)}$$

Thus, we can represent the general solution as the sum of two series—one for the even-powered terms with coefficients in terms of a_0 and one for the odd-powered terms with coefficients in terms of a_1.

$$y = a_0\left(1 - \frac{x^2}{2} + \frac{x^4}{2 \cdot 4} - \cdots\right) + a_1\left(x - \frac{x^3}{3} + \frac{x^5}{3 \cdot 5} - \cdots\right)$$

$$= a_0 \sum_{k=0}^{\infty} \frac{(-1)^k x^{2k}}{2^k (k!)} + a_1 \sum_{k=0}^{\infty} \frac{(-1)^k x^{2k+1}}{3 \cdot 5 \cdot 7 \cdots (2k+1)}$$

Note that the solution has two arbitrary constants, a_0 and a_1, as we would expect in the general solution of a second-order differential equation. ▭

Approximation by Taylor series

A second type of series solution method involves a differential equation *with initial conditions* and makes use of Taylor series, as given in Section 10.10.

EXAMPLE 3 Approximation by Taylor series

Use a Taylor series to find the series solution of

$$y' = y^2 - x$$

given the initial condition $y = 1$ when $x = 0$. Then, use the first six terms of this series solution to approximate values of y for $0 \le x \le 1$.

SOLUTION

Recall from Section 10.10 that, for $c = 0$,

$$y = y(0) + y'(0)x + \frac{y''(0)}{2!}x^2 + \frac{y'''(0)}{3!}x^3 + \cdots.$$

Since $y(0) = 1$ and $y' = y^2 - x$, we obtain the following.

$$y(0) = 1$$

$$y' = y^2 - x \qquad\qquad y'(0) = 1$$
$$y'' = 2yy' - 1 \qquad\qquad y''(0) = 2 - 1 = 1$$
$$y''' = 2yy'' + 2(y')^2 \qquad\qquad y'''(0) = 2 + 2 = 4$$
$$y^{(4)} = 2yy''' + 6y'y'' \qquad\qquad y^{(4)}(0) = 8 + 6 = 14$$
$$y^{(5)} = 2yy^{(4)} + 8y'y''' + 6(y'')^2 \qquad y^{(5)}(0) = 28 + 32 + 6 = 66$$

Therefore, we can approximate the values of the solution from the series

$$y = y(0) + y'(0)x + \frac{y''(0)}{2!}x^2 + \frac{y'''(0)}{3!}x^3 + \frac{y^{(4)}(0)}{4!}x^4 + \frac{y^{(5)}(0)}{5!}x^5 + \cdots$$

$$= 1 + x + \frac{1}{2}x^2 + \frac{4}{3!}x^3 + \frac{14}{4!}x^4 + \frac{66}{5!}x^5 + \cdots.$$

Using the first six terms of this series, we compute several values for y in the interval $0 \leq x \leq 1$, as shown in Table 18.1.

TABLE 18.1

x	0.0	0.1	0.2	0.3	0.4	0.5	0.6	0.7	0.8	0.9	1.0
y	1.00000	1.1057	1.2264	1.3691	1.5432	1.7620	2.0424	2.4062	2.8805	3.4985	4.3000

EXERCISES for Section 18.7

In Exercises 1–12, use power series to solve the differential equation.

1. $y' - y = 0$

2. $y' - ky = 0$

3. $y'' - 9y = 0$

4. $y'' - k^2y = 0$

5. $y'' + 4y = 0$

6. $y'' + k^2y = 0$

7. $y' + 3xy = 0$

8. $y' - 2xy = 0$

9. $y'' - xy' = 0$

10. $y'' - xy' - y = 0$

11. $(x^2 + 4)y'' + y = 0$

12. $y'' + x^2y = 0$

In Exercises 13–16, use Taylor's Theorem to find the series solution to the differential equation with the specified initial conditions. Use n terms of the series to approximate y for the given value of x.

13. $y' + (2x - 1)y = 0$, $y(0) = 2$,

$n = 5$, $x = \dfrac{1}{2}$

14. $y' - 2xy = 0$, $y(0) = 1$,

$n = 4$, $x = 1$

15. $y'' - 2xy = 0$, $y(0) = 1$, $y'(0) = -3$,

$n = 6$, $x = \dfrac{1}{4}$

16. $y'' - 2xy' + y = 0$, $y(0) = 1$, $y'(0) = 2$,

$n = 8$, $x = \dfrac{1}{2}$

In Exercises 17–20, verify that the series converges to the given function on the indicated interval. [Hint: Use the given differential equation.]

17. $\displaystyle\sum_{n=0}^{\infty} \frac{x^n}{n!} = e^x$, $(-\infty, \infty)$

Differential Equation: $y' - y = 0$

18. $\displaystyle\sum_{n=0}^{\infty} \frac{(-1)^n x^{2n}}{(2n)!} = \cos x$, $(-\infty, \infty)$

Differential Equation: $y'' + y = 0$

19. $\displaystyle\sum_{n=0}^{\infty} \frac{(-1)^n x^{2n+1}}{2n + 1} = \arctan x$, $(-1, 1)$

Differential Equation: $(x^2 + 1)y'' + 2xy' = 0$

20. $\displaystyle\sum_{n=0}^{\infty} \frac{(2n)!\,x^{2n+1}}{(2^n n!)^2(2n + 1)} = \arcsin x$, $(-1, 1)$

Differential Equation: $(1 - x^2)y'' - xy' = 0$

REVIEW EXERCISES for Chapter 18

In Exercises 1–30, find the general solution of the first-order differential equation.

1. $\dfrac{dy}{dx} - \dfrac{y}{x} = 2 + \sqrt{x}$

2. $\dfrac{dy}{dx} + xy = 2y$

3. $y' - \dfrac{2y}{x} = \dfrac{y'}{x}$

4. $\dfrac{dy}{dx} - 3x^2y = e^{x^3}$

5. $\dfrac{dy}{dx} - \dfrac{y}{x} = \dfrac{x}{y}$

6. $\dfrac{dy}{dx} - \dfrac{3y}{x^2} = \dfrac{1}{x^2}$

7. $y' - 2y = e^x$

8. $y' + \dfrac{2y}{x} = -x^9y^5$

9. $(10x + 8y + 2)\,dx + (8x + 5y + 2)\,dy = 0$

10. $(y + x^3 + xy^2)\,dx - x\,dy = 0$

11. $(2x - 2y^3 + y)\,dx + (x - 6xy^2)\,dy = 0$

12. $3x^2y^2\,dx + (2x^3y + x^3y^4)\,dy = 0$

13. $x\,dy = (x + y + 2)\,dx$

14. $ye^{xy}\,dx + xe^{xy}\,dy = 0$

15. $(1 + y)\ln(1 + y)\,dx + dy = 0$

16. $(2x + y - 3)\,dx + (x - 3y + 1)\,dy = 0$

17. $dy = (y\tan x + 2e^x)\,dx$

18. $y\,dx - (x + \sqrt{xy})\,dy = 0$

19. $(x - y - 5)\,dx - (x + 3y - 2)\,dy = 0$

20. $y' = x^2y^2 - 9x^2$

21. $2xy' - y = x^3 - x$

22. $y' = 2x\sqrt{1 - y^2}$

23. $x + yy' = \sqrt{x^2 + y^2}$

24. $xy' + y = \sin x$

25. $yy' + y^2 = 1 + x^2$

26. $2x\,dx + 2y\,dy = (x^2 + y^2)\,dx$

27. $(1 + x^2)\,dy = (1 + y^2)\,dx$

28. $x^3yy' = x^4 + 3x^2y^2 + y^4$

29. $xy' - ay = bx^4$

30. $y' = y + 2x(y - e^x)$

In Exercises 31–36, find the general solution of the second-order differential equation.

31. $y'' + y = x^3 + x$

32. $y'' + 2y = e^{2x} + x$

33. $y'' + y = 2\cos x$

34. $y'' + 5y' + 4y = x^2 + \sin 2x$

35. $y'' - 2y' + y = 2xe^x$

36. $y'' + 2y' + y = \dfrac{1}{x^2e^x}$

In Exercises 37 and 38, find the orthogonal trajectories of the given family and sketch several members of each family.

37. $(x - C)^2 + y^2 = C^2$

38. $y - 2x = C$

In Exercises 39 and 40, find the series solution to the differential equation.

39. $(x - 4)y' + y = 0$

40. $y'' + 3xy' - 3y = 0$

APPENDIX A

Proofs of Selected Theorems

THEOREM 2.3
PROPERTIES OF LIMITS
(Properties 3, 4, and 5)
(page 58)

Let b and c be real numbers, n a positive integer, and let f and g be functions whose limits exist as $x \to c$. Then the following properties are true:

3. $\lim\limits_{x \to c} f(x)g(x) = \left[\lim\limits_{x \to c} f(x) \right]\left[\lim\limits_{x \to c} g(x) \right]$

4. $\lim\limits_{x \to c} \dfrac{f(x)}{g(x)} = \dfrac{\lim\limits_{x \to c} f(x)}{\lim\limits_{x \to c} g(x)}$, provided $\lim\limits_{x \to c} g(x) \neq 0$

5. $\lim\limits_{x \to c} [f(x)]^n = \left[\lim\limits_{x \to c} f(x) \right]^n$

PROOF

Let

$$\lim_{x \to c} f(x) = L \qquad \text{and} \qquad \lim_{x \to c} g(x) = K.$$

To prove Property 3, we write

$$f(x)g(x) = [f(x) - L][g(x) - K] + [Lg(x) + Kf(x)] - LK.$$

Since the limit of $f(x)$ is L, and the limit of $g(x)$ is K, we have

$$\lim_{x \to c} [f(x) - L] = 0 \qquad \text{and} \qquad \lim_{x \to c} [g(x) - K] = 0.$$

Let $0 < \varepsilon < 1$. Then there exists $\delta > 0$ such that if $0 < |x - c| < \delta$, then

$$|f(x) - L - 0| < \varepsilon \qquad \text{and} \qquad |g(x) - K - 0| < \varepsilon,$$

which implies that

$$|[f(x) - L][g(x) - K] - 0| = |f(x) - L|\,|g(x) - K| < \varepsilon\varepsilon < \varepsilon.$$

Hence,

$$\lim_{x \to c} [f(x) - L][g(x) - K] = 0.$$

Furthermore, by Property 1 of Theorem 2.3, we have

$$\lim_{x \to c} Lg(x) = LK \quad \text{and} \quad \lim_{x \to c} Kf(x) = KL.$$

Finally, by Property 2 of Theorem 2.3, we obtain

$$\lim_{x \to c} f(x)g(x) = \lim_{x \to c} [f(x) - L][g(x) - K] + \lim_{x \to c} Lg(x) + \lim_{x \to c} Kf(x) - \lim_{x \to c} LK$$

$$= 0 + LK + KL - LK$$

$$= LK.$$

To prove Property 4, we use Property 3 and the limit of a reciprocal function and write

$$\lim_{x \to c} \frac{f(x)}{g(x)} = \lim_{x \to c} \left[f(x) \frac{1}{g(x)} \right] = \left[\lim_{x \to c} f(x) \right] \left[\lim_{x \to c} \frac{1}{g(x)} \right] = \frac{L}{K}.$$

Finally, the proof of Property 5 can be obtained by a straightforward application of mathematical induction coupled with Property 3.

THEOREM 2.6
LIMIT OF A FUNCTION INVOLVING A RADICAL (page 60)

If $c > 0$ and n is *any* positive integer, or if $c \le 0$ and n is an *odd* positive integer, then

$$\lim_{x \to c} \sqrt[n]{x} = \sqrt[n]{c}.$$

PROOF Consider the case for which $c > 0$ and n is any positive integer. For a given $\varepsilon > 0$, we need to find $\delta > 0$ such that

$$|\sqrt[n]{x} - \sqrt[n]{c}| < \varepsilon \quad \text{whenever} \quad 0 < |x - c| < \delta,$$

which is the same as saying

$$-\varepsilon < \sqrt[n]{x} - \sqrt[n]{c} < \varepsilon \quad \text{whenever} \quad -\delta < x - c < \delta.$$

We assume $\varepsilon < \sqrt[n]{c}$, which implies that $0 < \sqrt[n]{c} - \varepsilon < \sqrt[n]{c}$. Now, we let δ be the smaller of the two numbers.

$$c - \left(\sqrt[n]{c} - \varepsilon \right)^n \quad \text{and} \quad \left(\sqrt[n]{c} + \varepsilon \right)^n - c$$

Then we have

$$-\delta < x - c \quad\quad < \delta$$
$$-[c - \left(\sqrt[n]{c} - \varepsilon \right)^n] < x - c \quad\quad < \left(\sqrt[n]{c} + \varepsilon \right)^n - c$$
$$\left(\sqrt[n]{c} - \varepsilon \right)^n - c < x - c \quad\quad < \left(\sqrt[n]{c} + \varepsilon \right)^n - c$$
$$\left(\sqrt[n]{c} - \varepsilon \right)^n < x \quad\quad < \left(\sqrt[n]{c} + \varepsilon \right)^n$$
$$\sqrt[n]{c} - \varepsilon < \sqrt[n]{x} \quad\quad < \sqrt[n]{c} + \varepsilon$$
$$-\varepsilon < \sqrt[n]{x} - \sqrt[n]{c} < \varepsilon$$

THEOREM 2.7
LIMIT OF A COMPOSITE FUNCTION
(page 60)

If f and g are functions such that

$$\lim_{x \to c} g(x) = L \qquad \text{and} \qquad \lim_{x \to L} f(x) = f(L)$$

then

$$\lim_{x \to c} f(g(x)) = f(L).$$

PROOF

For a given $\varepsilon > 0$, we must find $\delta > 0$ such that

$$\left| f(g(x)) - f(L) \right| < \varepsilon \qquad \text{whenever} \qquad 0 < |x - c| < \delta.$$

Since the limit of $f(x)$ as $x \to L$ is $f(L)$, we know there exists $\delta_1 > 0$ such that

$$\left| f(u) - f(L) \right| < \varepsilon \qquad \text{whenever} \qquad 0 < |u - L| < \delta_1.$$

Moreover, since the limit of $g(x)$ as $x \to c$ is L, we know there exists $\delta > 0$ such that

$$\left| g(x) - L \right| < \delta_1 \qquad \text{whenever} \qquad 0 < |x - c| < \delta.$$

Finally, letting $u = g(x)$, we have

$$\left| f(g(x)) - f(L) \right| < \varepsilon \qquad \text{whenever} \qquad 0 < |x - c| < \delta.$$

THEOREM 2.10
CONTINUITY OF A COMPOSITE
FUNCTION   (page 73)

If g is continuous at c and f is continuous at $g(c)$, then the composite function given by $f \circ g(x) = f(g(x))$ is continuous at c.

PROOF

Since g is continuous at c and f is continuous at $g(c)$, we can write

$$\lim_{x \to c} g(x) = g(c) \qquad \text{and} \qquad \lim_{u \to g(c)} f(u) = f(g(c)),$$

and by applying the theorem on the limit of a composite function, we can conclude that

$$\lim_{x \to c} f(g(x)) = f(g(c)).$$

THEOREM 2.12
VERTICAL ASYMPTOTES
(page 79)

Let f and g be continuous on an open interval containing c. If $f(c) \neq 0$, $g(c) = 0$, and there exists an open interval containing c such that $g(x) \neq 0$ for all $x \neq c$ in the interval, then the graph of the function given by

$$F(x) = \frac{f(x)}{g(x)}$$

has a vertical asymptote at $x = c$.

PROOF   We consider the case for which $f(c) > 0$, and there exists $b > c$ such that $c < x < b$ implies $g(x) > 0$. Then for $M > 0$, we choose δ_1 such that

$$0 < x - c < \delta_1 \qquad \text{implies that} \qquad \frac{f(c)}{2} < f(x) < \frac{3f(c)}{2}$$

and δ_2 such that

$$0 < x - c < \delta_2 \qquad \text{implies that} \qquad 0 < g(x) < \frac{f(c)}{2M}.$$

Now we let δ be the smaller of δ_1 and δ_2. Then, it follows that

$$0 < x - c < \delta \qquad \text{implies that} \qquad \frac{f(x)}{g(x)} > \frac{f(c)}{2}\left(\frac{2M}{f(c)}\right) = M.$$

Therefore, it follows that

$$\lim_{x \to c^+} \frac{f(x)}{g(x)} = \infty$$

and the line $x = c$ is a vertical asymptote of the graph of F.

THEOREM 3.9
THE CHAIN RULE   **(page 129)**

If $y = f(u)$ is a differentiable function of u, and $u = g(x)$ is a differentiable function of x, then $y = f(g(x))$ is a differentiable function of x, and

$$\frac{dy}{dx} = \frac{dy}{du} \cdot \frac{du}{dx}.$$

PROOF   In Section 3.5, we let $F(x) = f(g(x))$ and used the alternate form of the derivative to show that $F'(c) = f'(g(c))g'(c)$, provided $g(x) \neq g(c)$ for values of x other than c. We now consider a more general proof. We begin by considering the derivative of f.

$$f'(x) = \lim_{\Delta x \to 0} \frac{f(x + \Delta x) - f(x)}{\Delta x} = \lim_{\Delta x \to 0} \frac{\Delta y}{\Delta x}.$$

For a fixed value of x, we define a function η such that

$$\eta(\Delta x) = \begin{cases} 0, & \Delta x = 0 \\ \dfrac{\Delta y}{\Delta x} - f'(x), & \Delta x \neq 0. \end{cases}$$

Since the limit of $\eta(\Delta x)$ as $\Delta x \to 0$ doesn't depend upon the value of $\eta(0)$, we have

$$\lim_{\Delta x \to 0} \eta(\Delta x) = \lim_{\Delta x \to 0} \left[\frac{\Delta y}{\Delta x} - f'(x)\right] = 0$$

and we can conclude that η is continuous at 0. Moreover, since $\Delta y = 0$ when $\Delta x = 0$, the equation

$$\Delta y = \Delta x \eta(\Delta x) + \Delta x f'(x)$$

is valid whether Δx is a zero or not. Now, by letting $\Delta u = g(x + \Delta x) - g(x)$, we use the continuity of g to conclude that

$$\lim_{\Delta x \to 0} \Delta u = \lim_{\Delta x \to 0} [g(x + \Delta x) - g(x)] = 0,$$

which implies that

$$\lim_{\Delta x \to 0} \eta(\Delta u) = 0.$$

Finally,

$$\Delta y = \Delta u \eta(\Delta u) + \Delta u f'(u) \implies \frac{\Delta y}{\Delta x} = \frac{\Delta u}{\Delta x} \eta(\Delta u) + \frac{\Delta u}{\Delta x} f'(u), \quad \Delta x \neq 0$$

and taking the limit as $\Delta x \to 0$, we have

$$\frac{dy}{dx} = \frac{du}{dx}\left[\lim_{\Delta x \to 0} \eta(\Delta u)\right] + \frac{du}{dx}f'(u) = \frac{dy}{dx}(0) + \frac{du}{dx}f'(u)$$

$$= \frac{du}{dx}f'(u) = \frac{du}{dx} \cdot \frac{dy}{du}.$$

EQUIVALENCE OF CONCAVITY DEFINITIONS   (page 175)

Let f be a function that is differentiable on some open interval containing c.

The graph of f is concave *upward* at $f(c)$ if and only if the graph of f lies *above* the tangent line to the graph of f at $(c, f(c))$ on some open interval containing c.

The graph of f is concave *downward* at $f(c)$ if and only if the graph of f lies *below* the tangent line to the graph of f at $(c, f(c))$ on some open interval containing c.

PROOF

There are many parts to this proof, but they are all similar. We give only one part here, and ask you to supply the remaining parts. Assume f is concave upward at c, then by definition, there exists an interval (a, b) containing c such that f' is increasing on (a, b). The equation of the tangent line to the graph of f at c is given by

$$g(x) = f(c) + f'(c)(x - c).$$

If x is in the open interval (c, b), then the directed distance from the point $(x, f(x))$ (on the graph of f) to the point $(x, g(x))$ (on the tangent line) is given by

$$d = f(x) - [f(c) + f'(c)(x - c)]$$
$$= f(x) - f(c) - f'(c)(x - c).$$

Moreover, by the Mean Value Theorem there exists a number z in (c, x) such that

$$f'(z) = \frac{[f(x) - f(c)]}{(x - c)}.$$

Thus, we have

$$d = f(x) - f(c) - f'(c)(x - c)$$
$$= f'(z)(x - c) - f'(c)(x - c)$$
$$= [f'(z) - f'(c)](x - c).$$

The second factor $(x - c)$ is positive because $c < x$. Moreover, since f' is increasing, it follows that the first factor $[f'(z) - f'(c)]$ is also positive. Therefore, $d > 0$ and we conclude that the graph of f lies above the tangent line. If x is in the open interval (a, c), a similar argument can be given.

THEOREM 4.10
LIMITS AT INFINITY
(page 184)

If r is a positive rational number, and c is any real number, then

$$\lim_{x \to \infty} \left(\frac{c}{x^r}\right) = 0.$$

Furthermore, if x^r is defined when $x < 0$, then

$$\lim_{x \to -\infty} \left(\frac{c}{x^r}\right) = 0.$$

PROOF   We begin by proving that

$$\lim_{x \to \infty} \frac{1}{x} = 0.$$

For $\varepsilon > 0$, let $M = 1/\varepsilon$. Then, for $x > M$, we have

$$x > M = \frac{1}{\varepsilon} \implies \frac{1}{x} < \varepsilon \implies \left|\frac{1}{x} - 0\right| < \varepsilon.$$

Therefore, by the definition of a limit at infinity, we conclude that the limit of $1/x$ as $x \to \infty$ is 0. Now, using this result, and letting $r = m/n$, we can write the following.

$$\lim_{x \to \infty} \left(\frac{c}{x^r}\right) = \lim_{x \to \infty} \left(\frac{c}{x^{m/n}}\right)$$
$$= c \left[\lim_{x \to \infty} \left(\frac{1}{\sqrt[n]{x}}\right)^m\right]$$
$$= c \left[\lim_{x \to \infty} \sqrt[n]{\frac{1}{x}}\right]^m$$
$$= c \left[\sqrt[n]{\lim_{x \to \infty} \frac{1}{x}}\right]^m$$
$$= c \left[\sqrt[n]{0}\right]^m$$
$$= 0$$

The proof of the second part of the theorem is similar.

THEOREM 5.4
SUMMATION FORMULAS
(page 240)

1. $\displaystyle\sum_{i=1}^{n} c = cn$

2. $\displaystyle\sum_{i=1}^{n} i = \frac{n(n+1)}{2}$

3. $\displaystyle\sum_{i=1}^{n} i^2 = \frac{n(n+1)(2n+1)}{6}$

4. $\displaystyle\sum_{i=1}^{n} i^3 = \frac{n^2(n+1)^2}{4}$

PROOF

The proof of part 1 is straightforward. By adding c to itself n times, we obtain a sum of nc.

To prove part 2, we write the sum in increasing and decreasing order and add corresponding terms as follows.

$$\sum_{i=1}^{n} i = \quad 1 \quad + \quad 2 \quad + \quad 3 \quad + \cdots + \quad (n-1) \quad + \quad n$$

$$\downarrow \qquad \downarrow \qquad \downarrow \qquad\qquad \downarrow \qquad \downarrow$$

$$\sum_{i=1}^{n} i = \quad n \quad + (n-1) + (n-2) + \cdots + \quad 2 \quad + \quad 1$$

$$\downarrow \qquad \downarrow \qquad \downarrow \qquad\qquad \downarrow \qquad \downarrow$$

$$2\sum_{i=1}^{n} i = (n+1) + (n+1) + (n+1) + \cdots + (n+1) + (n+1)$$

$$\underbrace{\hspace{9cm}}_{n \text{ terms}}$$

Therefore,

$$\sum_{i=1}^{n} i = \frac{n(n+1)}{2}.$$

To prove part 3, we use mathematical induction. First, if $n = 1$, the result is true because

$$\sum_{i=1}^{1} i^2 = 1^2 = 1 = \frac{1(1+1)(2+1)}{6}.$$

Now, assuming the result is true for $n = k$, we show that it is true for $n = k + 1$, as follows.

$$\sum_{i=1}^{k+1} i^2 = \sum_{i=1}^{k} i^2 + (k+1)^2$$

$$= \frac{k(k+1)(2k+1)}{6} + (k+1)^2$$

$$= \frac{k+1}{6}(2k^2 + k + 6k + 6)$$

$$= \frac{k+1}{6}[(2k+3)(k+2)]$$

$$= \frac{(k+1)(k+2)[2(k+1)+1]}{6}$$

Part 4 can be proved using a similar argument with mathematical induction.

THEOREM 5.10
PRESERVATION OF INEQUALITY
(page 258)

1. If f is integrable and nonnegative on the closed interval $[a, b]$, then

$$0 \leq \int_a^b f(x)dx.$$

2. If f and g are integrable on the closed interval $[a, b]$, and $f(x) \leq g(x)$ for every x in $[a, b]$, then

$$\int_a^b f(x)dx \leq \int_a^b g(x)dx.$$

PROOF

To prove part 1, we suppose, on the contrary, that

$$\int_a^b f(x)dx = I < 0.$$

Then, let $a = x_0 < x_1 < x_2 < \cdots < x_n = b$ be a partition of $[a, b]$, and let

$$R = \sum_{i=1}^{n} f(c_i)\Delta x_i$$

be a Riemann sum. Since $f(x) \geq 0$, it follows that $R \geq 0$. Now, for $\|\Delta\|$ sufficiently small, we have $|R - I| < -I/2$, which implies that

$$\sum_{i=1}^{n} f(c_i)\Delta x_i = R < I - \frac{I}{2} < 0,$$

which is not possible. From this contradiction, we can conclude that

$$0 \leq \int_a^b f(x)dx.$$

To prove the second part of the theorem, we note that $f(x) \leq g(x)$ implies that $g(x) - f(x) \geq 0$. Hence, we can apply the result of part 1 to conclude that

$$0 \leq \int_a^b (g(x) - f(x))dx$$

$$0 \leq \int_a^b g(x)dx - \int_a^b f(x)dx$$

$$\int_a^b f(x)dx \leq \int_a^b g(x)dx.$$

THEOREM 7.6 CONTINUITY AND DIFFERENTIABILITY OF INVERSE FUNCTIONS (page 375)	1. If f is continuous on its domain, then f^{-1} is continuous on its domain. 2. If f is increasing on its domain, then f^{-1} is increasing on its domain. 3. If f is decreasing on its domain, then f^{-1} is decreasing on its domain. 4. If f is differentiable at c and $f'(c) \neq 0$, then f^{-1} is differentiable at $f(c)$.

PROOF

To prove part 1, assume that a is in an open interval in the domain of f^{-1}, and let $f^{-1}(a) = c$. (A similar proof using one-sided limits can be given for the endpoints.) For $\varepsilon > 0$, let $\delta_1 = a - f(c - \varepsilon)$ and $\delta_2 = f(c + \varepsilon) - a$. If y is in the interval

$$(a - \delta_1, a + \delta_2) = (f(c - \varepsilon), \quad f(c + \varepsilon))$$

then by the one-to-one nature of f and f^{-1}, it follows that $f^{-1}(y)$ is in the interval $(c - \varepsilon, c + \varepsilon)$. By taking δ to be the minimum of δ_1 and δ_2, it follows that y is in $(a - \delta, a + \delta)$, and since $f^{-1}(y)$ is in $(c - \varepsilon, c + \varepsilon)$, we can conclude that

$$\lim_{y \to a} f^{-1}(y) = f^{-1}(a),$$

which implies that f^{-1} is continuous at a.

To prove part 2, let y_1 and y_2 be in the domain of f^{-1}, with $y_1 < y_2$. Then, there exist x_1 and x_2 in the domain of f such that

$$f(x_1) = y_1 < y_2 = f(x_2).$$

Since f is increasing, $f(x_1) < f(x_2)$ holds precisely when $x_1 < x_2$. Therefore,

$$f^{-1}(y_1) = x_1 < x_2 = f^{-1}(y_2),$$

which implies that f^{-1} is increasing. (Part 3 can be proved in a similar way.) Finally, to prove part 4, we consider the limit

$$(f^{-1})'(a) = \lim_{y \to a} \frac{f^{-1}(y) - f^{-1}(a)}{y - a}$$

where a is in the domain of f^{-1} and $f^{-1}(a) = c$. Since f is differentiable, f is continuous, and so is f^{-1}. Thus, $y \to a$ implies that $x \to c$, and we have

$$(f^{-1})'(a) = \lim_{x \to c} \frac{x - c}{f(x) - f(c)}$$

$$= \lim_{x \to c} \frac{1}{\dfrac{f(x) - f(c)}{x - c}}$$

$$= \frac{1}{\lim_{x \to c} \dfrac{f(x) - f(c)}{x - c}}$$

$$= \frac{1}{f'(c)}.$$

Hence, $(f^{-1})'(a)$ exists, and f^{-1} is differentiable at $f(c)$.

THE EXTENDED MEAN VALUE THEOREM (page 407)	If f and g are differentiable on an open interval (a, b) and continuous on $[a, b]$ such that $g(x) \neq 0$ for any x in (a, b), then there exists a point c in (a, b) such that $$\frac{f'(c)}{g'(c)} = \frac{f(b) - f(a)}{g(b) - g(a)}.$$

PROOF We can assume that $g(a) \neq g(b)$, since otherwise by Rolle's Theorem it would follow that $g'(x) = 0$ for some x in (a, b). Now, we define $h(x)$ to be

$$h(x) = f(x) - \left(\frac{f(b) - f(a)}{g(b) - g(a)} \right) g(x).$$

Then

$$h(a) = f(a) - \left(\frac{f(b) - f(a)}{g(b) - g(a)} \right) g(a) = \frac{f(a)g(b) - f(b)g(a)}{g(b) - g(a)}$$

and

$$h(b) = f(b) - \left(\frac{f(b) - f(a)}{g(b) - g(a)} \right) g(b) = \frac{f(a)g(b) - f(b)g(a)}{g(b) - g(a)}$$

and by Rolle's Theorem there exists a point c in (a, b) such that

$$h'(c) = f'(c) - \frac{f(b) - f(a)}{g(b) - g(a)} g'(c) = 0,$$

which implies that

$$\frac{f'(c)}{g'(c)} = \frac{f(b) - f(a)}{g(b) - g(a)}.$$

We can use the Extended Mean Value Theorem to prove L'Hôpital's Rule. Of the several different cases of this rule, we illustrate the proof of only one case and leave the remaining cases for you to prove.

THEOREM 7.15 L'HÔPITAL'S RULE (page 407)	Let f and g be functions that are differentiable on an open interval (a, b) containing c, except possibly at c itself. If the limit of $f(x)/g(x)$ as x approaches c produces the indeterminate form $0/0$ or ∞/∞, then $$\lim_{x \to c} \frac{f(x)}{g(x)} = \lim_{x \to c} \frac{f'(x)}{g'(x)},$$ provided the limit on the right exists (or is infinite).

PROOF　We consider the case for which

$$\lim_{x \to a^+} f'(x) = 0 \qquad \text{and} \qquad \lim_{x \to a^+} g'(x) = 0.$$

Moreover, we assume that

$$\lim_{x \to a^+} \frac{f'(x)}{g'(x)} = L.$$

Now, we choose b such that both f and g are differentiable in the interval (a, b) and define the following two new functions:

$$F(x) = \begin{cases} f(x), & x \neq a \\ 0, & x = a \end{cases} \qquad \text{and} \qquad G(x) = \begin{cases} g(x), & x \neq a \\ 0, & x = a. \end{cases}$$

It follows that F and G are continuous on $[a, b]$, and we can apply the Extended Mean Value Theorem to conclude that there exists a number c in (a, b) such that

$$\frac{F'(c)}{G'(c)} = \frac{F(b) - F(a)}{G(b) - G(a)} = \frac{F(b) - 0}{G(b) - 0} = \frac{F(b)}{G(b)}.$$

Finally, by letting b approach a, we also force c to approach a, and it follows that

$$\lim_{b \to a^+} \frac{f(b)}{g(b)} = \lim_{b \to a^+} \frac{F(b)}{G(b)} = \lim_{c \to a^+} \frac{F'(c)}{G'(c)} = \lim_{c \to a^+} \frac{f'(c)}{g'(c)}$$

and we conclude that

$$\lim_{x \to a^+} \frac{f(x)}{g(x)} = \lim_{x \to a^+} \frac{f'(x)}{g'(x)}.$$

**THEOREM 12.11
CLASSIFICATION OF CONICS BY
ECCENTRICITY　(page 716)**

The locus of a point in the plane whose distance from a fixed point (*focus*) has a constant ratio to its distance from a fixed line (*directrix*) is a conic. The constant ratio e is the *eccentricity* of the conic.

1. The conic is an ellipse if $0 < e < 1$.
2. The conic is a parabola if $e = 1$.
3. The conic is a hyperbola if $e > 1$.

PROOF　If $e = 1$, then by definition, the conic must be a parabola. If $e \neq 1$, then we consider the focus F to lie at the origin and the directrix $x = d$ to lie to the right of the origin, as shown in Figure A.1. For the point $P = (r, \theta) = (x, y)$, we have $|PF| = r$ and $|PQ| = d - r \cos \theta$. Given that $e = |PF|/|PQ|$, it follows that

$$|PF| = |PQ|e \quad \Longrightarrow \quad r = e(d - r \cos \theta).$$

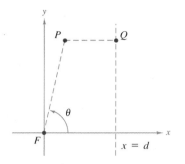

FIGURE A.1

By converting to rectangular coordinates and squaring both sides, we obtain

$$x^2 + y^2 = e^2(d - x)^2 = e^2(d^2 - 2dx + x^2).$$

Completing the square produces

$$\left(x + \frac{e^2d}{1 - e^2}\right)^2 + \frac{y^2}{1 - e^2} = \frac{e^2d^2}{(1 - e^2)^2}.$$

If $e < 1$, then this equation represents an ellipse. If $e > 1$, then $1 - e^2 < 0$, and the equation represents a hyperbola.

APPENDIX B

Basic Differentiation Rules for Elementary Functions

1. $\dfrac{d}{dx}[cu] = cu'$

2. $\dfrac{d}{dx}[u \pm v] = u' \pm v'$

3. $\dfrac{d}{dx}[uv] = uv' + vu'$

4. $\dfrac{d}{dx}\left[\dfrac{u}{v}\right] = \dfrac{vu' - uv'}{v^2}$

5. $\dfrac{d}{dx}[c] = 0$

6. $\dfrac{d}{dx}[u^n] = nu^{n-1}u'$

7. $\dfrac{d}{dx}[x] = 1$

8. $\dfrac{d}{dx}[|u|] = \dfrac{u}{|u|}(u'),\ u \neq 0$

9. $\dfrac{d}{dx}[\ln u] = \dfrac{u'}{u}$

10. $\dfrac{d}{dx}[e^u] = e^u u'$

11. $\dfrac{d}{dx}[\sin u] = (\cos u)u'$

12. $\dfrac{d}{dx}[\cos u] = -(\sin u)u'$

13. $\dfrac{d}{dx}[\tan u] = (\sec^2 u)u'$

14. $\dfrac{d}{dx}[\cot u] = -(\csc^2 u)u'$

15. $\dfrac{d}{dx}[\sec u] = (\sec u \tan u)u'$

16. $\dfrac{d}{dx}[\csc u] = -(\csc u \cot u)u'$

17. $\dfrac{d}{dx}[\arcsin u] = \dfrac{u'}{\sqrt{1 - u^2}}$

18. $\dfrac{d}{dx}[\arccos u] = \dfrac{-u'}{\sqrt{1 - u^2}}$

19. $\dfrac{d}{dx}[\arctan u] = \dfrac{u'}{1 + u^2}$

20. $\dfrac{d}{dx}[\operatorname{arccot} u] = \dfrac{-u'}{1 + u^2}$

21. $\dfrac{d}{dx}[\operatorname{arcsec} u] = \dfrac{u'}{|u|\sqrt{u^2 - 1}}$

22. $\dfrac{d}{dx}[\operatorname{arccsc} u] = \dfrac{-u'}{|u|\sqrt{u^2 - 1}}$

APPENDIX C

Integration Tables

Forms Involving u^n

1. $\displaystyle\int u^n\, du = \frac{u^{n+1}}{n+1} + C,\ n \neq -1$

2. $\displaystyle\int \frac{1}{u}\, du = \ln |u| + C$

Forms Involving $a + bu$

3. $\displaystyle\int \frac{u}{a+bu}\, du = \frac{1}{b^2}(bu - a\ln|a+bu|) + C$

4. $\displaystyle\int \frac{u}{(a+bu)^2}\, du = \frac{1}{b^2}\left(\frac{a}{a+bu} + \ln|a+bu|\right) + C$

5. $\displaystyle\int \frac{u}{(a+bu)^n}\, du = \frac{1}{b^2}\left[\frac{-1}{(n-2)(a+bu)^{n-2}} + \frac{a}{(n-1)(a+bu)^{n-1}}\right] + C,\ n \neq 1,2$

6. $\displaystyle\int \frac{u^2}{a+bu}\, du = \frac{1}{b^3}\left(-\frac{bu}{2}(2a - bu) + a^2\ln|a+bu|\right) + C$

7. $\displaystyle\int \frac{u^2}{(a+bu)^2}\, du = \frac{1}{b^3}\left(bu - \frac{a^2}{a+bu} - 2a\ln|a+bu|\right) + C$

8. $\displaystyle\int \frac{u^2}{(a+bu)^3}\, du = \frac{1}{b^3}\left[\frac{2a}{a+bu} - \frac{a^2}{2(a+bu)^2} + \ln|a+bu|\right] + C$

9. $\displaystyle\int \frac{u^2}{(a+bu)^n}\, du = \frac{1}{b^3}\left[\frac{-1}{(n-3)(a+bu)^{n-3}} + \frac{2a}{(n-2)(a+bu)^{n-2}} - \frac{a^2}{(n-1)(a+bu)^{n-1}}\right] + C,$
$$n \neq 1, 2, 3$$

10. $\displaystyle\int \frac{1}{u(a+bu)}\, du = \frac{1}{a}\ln\left|\frac{u}{a+bu}\right| + C$

11. $\displaystyle\int \frac{1}{u(a+bu)^2}\, du = \frac{1}{a}\left(\frac{1}{a+bu} + \frac{1}{a}\ln\left|\frac{u}{a+bu}\right|\right) + C$

12. $\displaystyle\int \frac{1}{u^2(a+bu)}\, du = -\frac{1}{a}\left(\frac{1}{u} + \frac{b}{a}\ln\left|\frac{u}{a+bu}\right|\right) + C$

13. $\displaystyle\int \frac{1}{u^2(a + bu)^2}\, du = -\frac{1}{a^2}\left[\frac{a + 2bu}{u(a + bu)} + \frac{2b}{a}\ln\left|\frac{u}{a + bu}\right|\right] + C$

Forms Involving $a + bu + cu^2$, $b^2 \neq 4ac$

14. $\displaystyle\int \frac{1}{a + bu + cu^2}\, du = \begin{cases} \dfrac{2}{\sqrt{4ac - b^2}}\arctan \dfrac{2cu + b}{\sqrt{4ac - b^2}} + C, & b^2 < 4ac \\[3mm] \dfrac{1}{\sqrt{b^2 - 4ac}}\ln\left|\dfrac{2cu + b - \sqrt{b^2 - 4ac}}{2cu + b + \sqrt{b^2 - 4ac}}\right| + C, & b^2 > 4ac \end{cases}$

15. $\displaystyle\int \frac{u}{a + bu + cu^2}\, du = \frac{1}{2c}\left(\ln|a + bu + cu^2| - b\int \frac{1}{a + bu + cu^2}\, du\right)$

Forms Involving $\sqrt{a + bu}$

16. $\displaystyle\int u^n\sqrt{a + bu}\, du = \frac{2}{b(2n + 3)}\left[u^n(a + bu)^{3/2} - na\int u^{n-1}\sqrt{a + bu}\, du\right]$

17. $\displaystyle\int \frac{1}{u\sqrt{a + bu}}\, du = \begin{cases} \dfrac{1}{\sqrt{a}}\ln\left|\dfrac{\sqrt{a + bu} - \sqrt{a}}{\sqrt{a + bu} + \sqrt{a}}\right| + C, & 0 < a \\[3mm] \dfrac{2}{\sqrt{-a}}\arctan \sqrt{\dfrac{a + bu}{-a}} + C, & a < 0 \end{cases}$

18. $\displaystyle\int \frac{1}{u^n\sqrt{a + bu}}\, du = \frac{-1}{a(n - 1)}\left[\frac{\sqrt{a + bu}}{u^{n-1}} + \frac{(2n - 3)b}{2}\int \frac{1}{u^{n-1}\sqrt{a + bu}}\, du\right], \quad n \neq 1$

19. $\displaystyle\int \frac{\sqrt{a + bu}}{u}\, du = 2\sqrt{a + bu} + a\int \frac{1}{u\sqrt{a + bu}}\, du$

20. $\displaystyle\int \frac{\sqrt{a + bu}}{u^n}\, du = \frac{-1}{a(n - 1)}\left[\frac{(a + bu)^{3/2}}{u^{n-1}} + \frac{(2n - 5)b}{2}\int \frac{\sqrt{a + bu}}{u^{n-1}}\, du\right], \quad n \neq 1$

21. $\displaystyle\int \frac{u}{\sqrt{a + bu}}\, du = \frac{-2(2a - bu)}{3b^2}\sqrt{a + bu} + C$

22. $\displaystyle\int \frac{u^n}{\sqrt{a + bu}}\, du = \frac{2}{(2n + 1)b}\left(u^n\sqrt{a + bu} - na\int \frac{u^{n-1}}{\sqrt{a + bu}}\, du\right)$

Forms Involving $a^2 \pm u^2$, $0 < a$

23. $\displaystyle\int \frac{1}{a^2 + u^2}\, du = \frac{1}{a}\arctan \frac{u}{a} + C$

24. $\displaystyle\int \frac{1}{u^2 - a^2}\, du = -\int \frac{1}{a^2 - u^2}\, du = \frac{1}{2a}\ln\left|\frac{u - a}{u + a}\right| + C$

25. $\displaystyle\int \frac{1}{(a^2 \pm u^2)^n}\, du = \frac{1}{2a^2(n - 1)}\left[\frac{u}{(a^2 \pm u^2)^{n-1}} + (2n - 3)\int \frac{1}{(a^2 \pm u^2)^{n-1}}\, du\right], \quad n \neq 1$

Forms Involving $\sqrt{u^2 \pm a^2}$, $0 < a$

26. $\displaystyle\int \sqrt{u^2 \pm a^2}\, du = \frac{1}{2}\left(u\sqrt{u^2 \pm a^2} \pm a^2\ln|u + \sqrt{u^2 \pm a^2}|\right) + C$

27. $\displaystyle\int u^2\sqrt{u^2 \pm a^2}\, du = \frac{1}{8}\left[u(2u^2 \pm a^2)\sqrt{u^2 \pm a^2} - a^4 \ln \left|u + \sqrt{u^2 \pm a^2}\right|\right] + C$

28. $\displaystyle\int \frac{\sqrt{u^2 + a^2}}{u}\, du = \sqrt{u^2 + a^2} - a \ln \left|\frac{a + \sqrt{u^2 + a^2}}{u}\right| + C$

29. $\displaystyle\int \frac{\sqrt{u^2 - a^2}}{u}\, du = \sqrt{u^2 - a^2} - a \operatorname{arcsec} \frac{|u|}{a} + C$

30. $\displaystyle\int \frac{\sqrt{u^2 \pm a^2}}{u^2}\, du = \frac{-\sqrt{u^2 \pm a^2}}{u} + \ln \left|u + \sqrt{u^2 \pm a^2}\right| + C$

31. $\displaystyle\int \frac{1}{\sqrt{u^2 \pm a^2}}\, du = \ln \left|u + \sqrt{u^2 \pm a^2}\right| + C$

32. $\displaystyle\int \frac{1}{u\sqrt{u^2 + a^2}}\, du = \frac{-1}{a} \ln \left|\frac{a + \sqrt{u^2 + a^2}}{u}\right| + C$

33. $\displaystyle\int \frac{1}{u\sqrt{u^2 - a^2}}\, du = \frac{1}{a} \operatorname{arcsec} \frac{|u|}{a} + C$

34. $\displaystyle\int \frac{u^2}{\sqrt{u^2 \pm a^2}}\, du = \frac{1}{2}\left(u\sqrt{u^2 \pm a^2} \mp a^2 \ln \left|u + \sqrt{u^2 \pm a^2}\right|\right) + C$

35. $\displaystyle\int \frac{1}{u^2\sqrt{u^2 \pm a^2}}\, du = \mp \frac{\sqrt{u^2 \pm a^2}}{a^2 u} + C$

36. $\displaystyle\int \frac{1}{(u^2 \pm a^2)^{3/2}}\, du = \frac{\pm u}{a^2\sqrt{u^2 \pm a^2}} + C$

Forms Involving $\sqrt{a^2 - u^2},\ 0 < a$

37. $\displaystyle\int \sqrt{a^2 - u^2}\, du = \frac{1}{2}\left(u\sqrt{a^2 - u^2} + a^2 \arcsin \frac{u}{a}\right) + C$

38. $\displaystyle\int u^2\sqrt{a^2 - u^2}\, du = \frac{1}{8}\left[u(2u^2 - a^2)\sqrt{a^2 - u^2} + a^4 \arcsin \frac{u}{a}\right] + C$

39. $\displaystyle\int \frac{\sqrt{a^2 - u^2}}{u}\, du = \sqrt{a^2 - u^2} - a \ln \left|\frac{a + \sqrt{a^2 - u^2}}{u}\right| + C$

40. $\displaystyle\int \frac{\sqrt{a^2 - u^2}}{u^2}\, du = \frac{-\sqrt{a^2 - u^2}}{u} - \arcsin \frac{u}{a} + C$

41. $\displaystyle\int \frac{1}{\sqrt{a^2 - u^2}}\, du = \arcsin \frac{u}{a} + C$

42. $\displaystyle\int \frac{1}{u\sqrt{a^2 - u^2}}\, du = \frac{-1}{a} \ln \left|\frac{a + \sqrt{a^2 - u^2}}{u}\right| + C$

43. $\displaystyle\int \frac{u^2}{\sqrt{a^2 - u^2}}\, du = \frac{1}{2}\left(-u\sqrt{a^2 - u^2} + a^2 \arcsin \frac{u}{a}\right) + C$

44. $\displaystyle\int \frac{1}{u^2\sqrt{a^2 - u^2}}\, du = \frac{-\sqrt{a^2 - u^2}}{a^2 u} + C$

45. $\displaystyle\int \frac{1}{(a^2 - u^2)^{3/2}}\, du = \frac{u}{a^2\sqrt{a^2 - u^2}} + C$

Forms Involving sin u or cos u

46. $\displaystyle\int \sin u\ du = -\cos u + C$

47. $\displaystyle\int \cos u\ du = \sin u + C$

48. $\displaystyle\int \sin^2 u\ du = \frac{1}{2}(u - \sin u \cos u) + C$

49. $\displaystyle\int \cos^2 u\ du = \frac{1}{2}(u + \sin u \cos u) + C$

50. $\displaystyle\int \sin^n u\ du = -\frac{\sin^{n-1} u \cos u}{n} + \frac{n-1}{n}\int \sin^{n-2} u\ du$

51. $\displaystyle\int \cos^n u\ du = \frac{\cos^{n-1} u \sin u}{n} + \frac{n-1}{n}\int \cos^{n-2} u\ du$

52. $\displaystyle\int u \sin u\ du = \sin u - u \cos u + C$

53. $\displaystyle\int u \cos u\ du = \cos u + u \sin u + C$

54. $\displaystyle\int u^n \sin u\ du = -u^n \cos u + n\int u^{n-1} \cos u\ du$

55. $\displaystyle\int u^n \cos u\ du = u^n \sin u - n\int u^{n-1} \sin u\ du$

56. $\displaystyle\int \frac{1}{1 \pm \sin u}\ du = \tan u \mp \sec u + C$

57. $\displaystyle\int \frac{1}{1 \pm \cos u}\ du = -\cot u \pm \csc u + C$

58. $\displaystyle\int \frac{1}{\sin u \cos u}\ du = \ln |\tan u| + C$

Forms Involving tan u, cot u, sec u, csc u

59. $\displaystyle\int \tan u\ du = -\ln |\cos u| + C$

60. $\displaystyle\int \cot u\ du = \ln |\sin u| + C$

61. $\displaystyle\int \sec u\ du = \ln |\sec u + \tan u| + C$

62. $\displaystyle\int \csc u\ du = \ln |\csc u - \cot u| + C$

63. $\displaystyle\int \tan^2 u\ du = -u + \tan u + C$

64. $\displaystyle\int \cot^2 u\ du = -u - \cot u + C$

65. $\displaystyle\int \sec^2 u\ du = \tan u + C$

66. $\displaystyle\int \csc^2 u \, du = -\cot u + C$

67. $\displaystyle\int \tan^n u \, du = \frac{\tan^{n-1} u}{n-1} - \int \tan^{n-2} u \, du, \, n \neq 1$

68. $\displaystyle\int \cot^n u \, du = -\frac{\cot^{n-1} u}{n-1} - \int (\cot^{n-2} u) \, du, \, n \neq 1$

69. $\displaystyle\int \sec^n u \, du = \frac{\sec^{n-2} u \tan u}{n-1} + \frac{n-2}{n-1} \int \sec^{n-2} u \, du, \, n \neq 1$

70. $\displaystyle\int \csc^n u \, du = -\frac{\csc^{n-2} u \cot u}{n-1} + \frac{n-2}{n-1} \int \csc^{n-2} u \, du, \, n \neq 1$

71. $\displaystyle\int \frac{1}{1 \pm \tan u} \, du = \frac{1}{2}(u \pm \ln |\cos u \pm \sin u|) + C$

72. $\displaystyle\int \frac{1}{1 \pm \cot u} \, du = \frac{1}{2}(u \mp \ln |\sin u \pm \cos u|) + C$

73. $\displaystyle\int \frac{1}{1 \pm \sec u} \, du = u + \cot u \mp \csc u + C$

74. $\displaystyle\int \frac{1}{1 \pm \csc u} \, du = u - \tan u \pm \sec u + C$

Forms Involving Inverse Trigonometric Functions

75. $\displaystyle\int \arcsin u \, du = u \arcsin u + \sqrt{1 - u^2} + C$

76. $\displaystyle\int \arccos u \, du = u \arccos u - \sqrt{1 - u^2} + C$

77. $\displaystyle\int \arctan u \, du = u \arctan u - \ln \sqrt{1 + u^2} + C$

78. $\displaystyle\int \text{arccot } u \, du = u \, \text{arccot } u + \ln \sqrt{1 + u^2} + C$

79. $\displaystyle\int \text{arcsec } u \, du = u \, \text{arcsec } u - \ln |u + \sqrt{u^2 - 1}| + C$

80. $\displaystyle\int \text{arccsc } u \, du = u \, \text{arccsc } u + \ln |u + \sqrt{u^2 - 1}| + C$

Forms Involving e^u

81. $\displaystyle\int e^u \, du = e^u + C$

82. $\displaystyle\int u e^u \, du = (u - 1)e^u + C$

83. $\displaystyle\int u^n e^u \, du = u^n e^u - n \int u^{n-1} e^u \, du$

84. $\displaystyle\int \frac{1}{1 + e^u} \, du = u - \ln (1 + e^u) + C$

85. $\displaystyle\int e^{au} \sin bu \, du = \frac{e^{au}}{a^2 + b^2}(a \sin bu - b \cos bu) + C$

86. $\displaystyle\int e^{au} \cos bu \, du = \frac{e^{au}}{a^2 + b^2}(a \cos bu + b \sin bu) + C$

Forms Involving ln u

87. $\int \ln u \; du = u(-1 + \ln u) + C$

88. $\int u \ln u \; du = \dfrac{u^2}{4}(-1 + 2 \ln u) + C$

89. $\int u^n \ln u \; du = \dfrac{u^{n+1}}{(n+1)^2}[-1 + (n+1) \ln u] + C, \; n \neq -1$

90. $\int (\ln u)^2 \; du = u[2 - 2 \ln u + (\ln u)^2] + C$

91. $\int (\ln u)^n \; du = u(\ln u)^n - n \int (\ln u)^{n-1} \; du$

Answers to Odd-Numbered Exercises

Chapter 1

Section 1.1

1. Rational **3.** Irrational **5.** Rational
7. Rational **9.** Rational **11.** $\frac{4}{11}$ **13.** $\frac{11}{37}$
15. (a) True **(b)** False **(c)** True
  (d) False **(e)** False **(f)** False
17.

Interval Notation	Set Notation	Graph
$[-2, 0)$	$\{x: -2 \le x < 0\}$	
$(-\infty, -4]$	$\{x: x \le -4\}$	
$[3, 11/2]$	$\{x: 3 \le x \le 11/2\}$	
$(-1, 7)$	$\{x: -1 < x < 7\}$	

19. $x \ge 12$

21. $x < -\frac{1}{2}$

23. $x \ge \frac{1}{2}$

25. $-\frac{1}{2} < x < \frac{7}{2}$

27. $x < -4$

29. $x > 6$

31. $-1 < x < 1$

33. $x \ge 13, \quad x \le -7$

35. $a - b < x < a + b$

37. $-3 < x < 2$

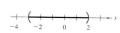

39. $0 < x < 3$ **41.** $-3 \le x \le 1$

43. $-3 \le x \le 2$ **45.** $4, -4, 4$

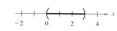

47. (a) $-51, 51, 51$ **(b)** $51, -51, 51$
49. 1 **51. (a)** 14 **(b)** 10
53. $|x| \le 2$ **55.** $|x - 2| > 2$
57. (a) $|x - 12| \le 10$ **(b)** $|x - 12| \ge 10$
59. $r > 12.5\%$ **61.** $m < 24{,}062.5$ miles
63. $x \le 41$ or $x \ge 59$
65. (a) $\frac{355}{113} > \pi$ **(b)** $\frac{22}{7} > \pi$

Section 1.2

1. $d = 2\sqrt{5}$ **3.** $d = 2\sqrt{10}$

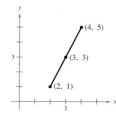

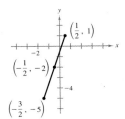

5. $d = \sqrt{8 - 2\sqrt{3}}$

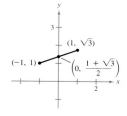

7. Right triangle,
$d_1 = \sqrt{45}, d_2 = \sqrt{5},$
$d_3 = \sqrt{50}$
$(d_1)^2 + (d_2)^2 = (d_3)^2$

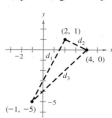

9. Rhombus; length of
each side is $\sqrt{5}$

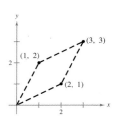

45. $\left(x - \frac{1}{2}\right)^2 + \left(y - \frac{1}{2}\right)^2 = 2$

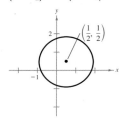

11. Collinear, since
$d_1 + d_2 = d_3$

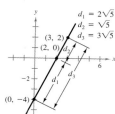

13. Not collinear, since
$d_1 + d_2 > d_3$

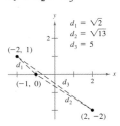

47. $\left(x + \frac{1}{2}\right)^2 + \left(y + \frac{5}{4}\right)^2 = \frac{9}{4}$

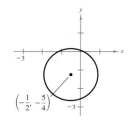

49. $x^2 + y^2 = 26{,}000^2$
51. $(x - 1)^2 + (y + 1)^2 = 25$
53.   **55.**

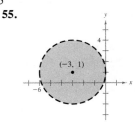

15. $x = \pm 3$ **17.** $y = \pm\sqrt{55}$
19. $3x - 2y - 1 = 0$
21. $\left(\dfrac{3x_1 + x_2}{4}, \dfrac{3y_1 + y_2}{4}\right), \left(\dfrac{x_1 + x_2}{2}, \dfrac{y_1 + y_2}{2}\right),$
$\left(\dfrac{x_1 + 3x_2}{4}, \dfrac{y_1 + 3y_2}{4}\right)$

23. (a) $\left(x + \frac{5}{2}\right)^2 - \frac{25}{4}$ **(b)** $(x + 4)^2 - 9$ **25.** d
26. c   **27.** b   **28.** e   **29.** a   **30.** f
31. $x^2 + y^2 - 9 = 0$
33. $x^2 + y^2 - 4x + 2y - 11 = 0$
35. $x^2 + y^2 + 2x - 4y = 0$
37. $x^2 + y^2 - 6x - 4y + 3 = 0$
39. $x^2 + y^2 - 6x - 8y = 0$
41. $(x - 1)^2 + (y + 3)^2 = 4$

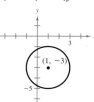

43. $(x - 1)^2 + (y + 3)^2 = 0$

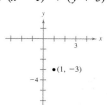

Section 1.3

1. c   **2.** d   **3.** b   **4.** f   **5.** a   **6.** e
7. $(0, -3), \left(\frac{3}{2}, 0\right)$ **9.** $(0, -2), (-2, 0), (1, 0)$
11. $(0, 0), (-3, 0), (3, 0)$ **13.** $\left(0, \frac{1}{2}\right), (1, 0)$
15. $(0, 0)$ **17.** Symmetric with respect to the y-axis
19. Symmetric with respect to the y-axis
21. Symmetric with respect to the x-axis
23. Symmetric with respect to the origin
25. Symmetric with respect to the origin
27. $y = x$ **29.** $y = x + 3$

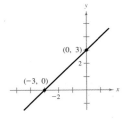

31. $y = -3x + 2$

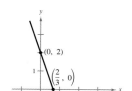

33. $y = \frac{1}{2}x - 4$

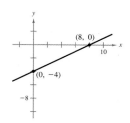

65. (a)

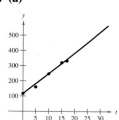

67. (a)

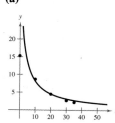

35. $y = 1 - x^2$

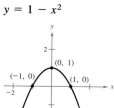

37. $y = -2x^2 + x + 1$

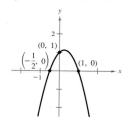

(b) 1995, 471.4

69. $(-1.2164, 0)$, $(0, 2)$

(b) 1995, 1.9%

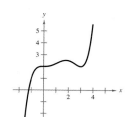

39. $y = x^3 + 2$

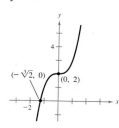

41. $x^2 + 4y^2 = 4$

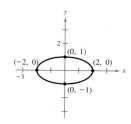

Section 1.4

1. $m = 1$ **3.** $m = 0$ **5.** $m = -3$
7. $m = 3$ **9.** $m = 0$

43. $y = (x + 2)^2$

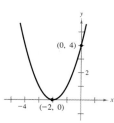

45. $y = \dfrac{1}{x}$

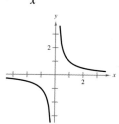

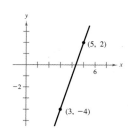

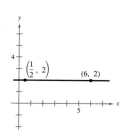

11. $m = -\frac{2}{3}$ **13.** $(0, 1)$, $(1, 1)$, $(3, 1)$

47. $(1, 1)$ **49.** $(5, 2)$ **51.** $(-1, -2)$, $(2, 1)$
53. $(-1, -1)$, $(0, 0)$, $(1, 1)$
55. $(-1, -5)$, $(0, -1)$, $(2, 1)$ **57.** $x \approx 192$ units
59. $(1, 2)$ is not on the graph
    $(1, -1)$ is on the graph
    $(4, 5)$ is on the graph
61. $\left(1, \frac{1}{5}\right)$ is on the graph
    $\left(2, \frac{1}{2}\right)$ is on the graph
    $(-1, -2)$ is not on the graph
63. (a) $k = 4$ **(b)** $k = -\frac{1}{8}$
    (c) $k = $ any real number   **(d)** $k = 1$

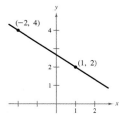

15. $(0, 10)$, $(2, 4)$, $(3, 1)$ **17.** $m = -\frac{1}{5}$, $(0, 4)$
19. m is undefined, no y-intercept

21. $2x - y - 3 = 0$

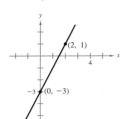

23. $3x + y = 0$

25. $y + 2 = 0$

27. $3x - 4y + 12 = 0$

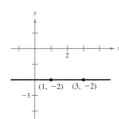

29. $2x - 3y = 0$

31. $4x - y + 2 = 0$

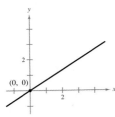

33. $x - 3 = 0$

35. $3x + 2y - 6 = 0$

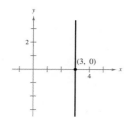

37. $12x + 3y + 2 = 0$ **39.** $x + y - 3 = 0$
41. **(a)** $2x - y - 3 = 0$ **(b)** $x + 2y - 4 = 0$
43. **(a)** $40x + 24y - 53 = 0$ **(b)** $24x - 40y + 9 = 0$
45. **(a)** $x - 2 = 0$ **(b)** $y - 5 = 0$
47.

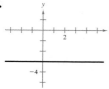

49.

51.

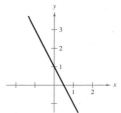

53. $2x - y = 0$ **55.** Not collinear, since $m_1 \neq m_2$
57. $\left(0, \dfrac{-a^2 + b^2 + c^2}{2c}\right)$ **59.** $\left(b, \dfrac{a^2 - b^2}{c}\right)$
61. $5F - 9C - 160 = 0$ **63.** $C = 0.25x + 95$
65. $y = 875 - 175t$ where $0 \leq t \leq 5$
67. **(a)** $x = \frac{1}{15}(1130 - p)$ **(b)** 45 units
    (c) 49 units
69. 2   **71.** $\dfrac{5\sqrt{2}}{2}$ **73.** $2\sqrt{2}$

Section 1.5

1. **(a)** -3 **(b)** -9 **(c)** $2b - 3$ **(d)** $2x - 5$
3. **(a)** 1   **(b)** 3   **(c)** $\sqrt{c + 3}$
    (d) $\sqrt{x + \Delta x + 3}$
5. **(a)** 1   **(b)** -1 **(c)** 1   **(d)** $\dfrac{|x - 1|}{x - 1}$
7. $3 + \Delta x$ **9.** $3x^2 + 3x\Delta x + (\Delta x)^2$
11. $\dfrac{-1}{\sqrt{x - 1}(1 + \sqrt{x - 1})}$
13. $f(x) = 4 - x$
    Domain: $(-\infty, \infty)$
    Range: $(-\infty, \infty)$

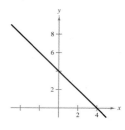

15. $f(x) = 4 - x^2$
    Domain: $(-\infty, \infty)$
    Range: $(-\infty, 4]$

17. $h(x) = \sqrt{x - 1}$
    Domain: $[1, \infty)$
    Range: $[0, \infty)$

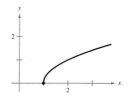

19. $f(x) = \sqrt{9 - x^2}$
Domain: $[-3, 3]$
Range: $[0, 3]$

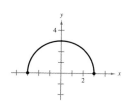

21. $f(x) = |x - 2|$
Domain: $(-\infty, \infty)$
Range: $[0, \infty)$

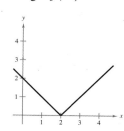

45. $(f \circ g)(x) = \dfrac{x + 1}{x}$
Domain: $(-\infty, 0), (0, \infty)$
$(g \circ f)(x) = \dfrac{1}{x + 1}$
Domain: $(-\infty, -1), (-1, \infty)$

47. $x = \pm 3$ **49.** $\frac{10}{7}$ **51.** Even **53.** Odd

59. $R(x) = 4 - \dfrac{x^2}{2}, \quad r(x) = 2$

61. $h(x) = x^2, \quad p(x) = x$

63. $A = xy = x\left(\dfrac{100 - 2x}{2}\right) = x(50 - x)$

65. $V = x(12 - 2x)^2$ **67.** $V = 108x^2 - 4x^3$

69. $T = \dfrac{\sqrt{x^2 + 4}}{2} + \dfrac{\sqrt{x^2 - 6x + 10}}{4}$

71. (a)

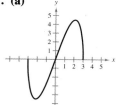

(b) $x = 0, \pm 3$
(c) $[-3, 3]$

23. y is a function of x **25.** y is not a function of x
27. y is a function of x **29.** y is not a function of x
31. y is a function of x **33.** y is a function of x
35. y is not a function of x
37. (a) $y = \sqrt{x} + 2$ **(b)** $y = -\sqrt{x}$

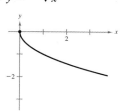

(c) $y = \sqrt{x - 2}$ **(d)** $y = \sqrt{x + 3}$

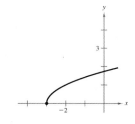

(e) $y = \sqrt{x - 4}$ **(f)** $y = 2\sqrt{x}$

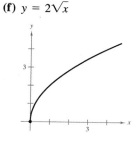

39. (a) $y = (x + 3)^2$ **(b)** $y = x^2 + 3$
41. (a) 0   **(b)** 0   **(c)** -1 **(d)** $\sqrt{15}$
   (e) $\sqrt{x^2 - 1}$ **(f)** $x - 1$
43. $(f \circ g)(x) = x$
Domain: $[0, \infty)$
$(g \circ f)(x) = |x|$
Domain: $(-\infty, \infty)$

Review Exercises for Chapter 1

1. $-1 \le x \le 5$

3. $x < -5, \quad x > -1$

5. $\frac{27}{16}$ **7.** $(-1, 3), (3, 2), (1, 1)$
9. Center: $(-3, 1)$ **11.** Point: $(-3, 1)$
   Radius: 3

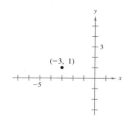

13. $c = -21$ **15.** $x^2 + y^2 - 2x - 4y - 4 = 0$
   (a) on the circle
   (b) inside the circle
   (c) outside the circle
   (d) inside the circle

17.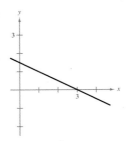

19.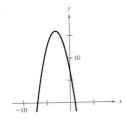

21. The points are not collinear.

23.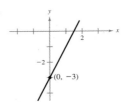

25.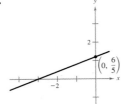

27. (a) $7x - 16y + 78 = 0$ **29.** $(-4, 5)$ **31.** $(4, 1)$
(b) $5x - 3y + 22 = 0$
(c) $y + 2x = 0$
(d) $x + 2 = 0$

33. $v = 850a + 300,000$ **35.** $s = 6x^2$
Domain: $\{a: a \geq 0\}$ Domain: $\{x: x \geq 0\}$

37. $d = 45t$ **39.** $P(x) = 500x - x^2$
Domain: $\{t: t \geq 0\}$

41. Function   **43.** Not a function

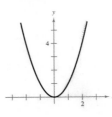

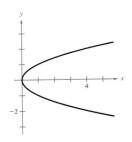

45. Function

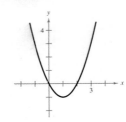

47. (a) $-x^2 + 2x + 2$
(b) $-x^2 - 2x$
(c) $-2x^3 - x^2 + 2x + 1$
(d) $\dfrac{1 - x^2}{2x + 1}$
(e) $-4x^2 - 4x$
(f) $3 - 2x^2$

49. (a)

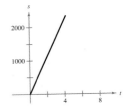

(b) Speed of the plane is 560 mi/hr

51. $C = 0.30x + 150$

Chapter 2

Section 2.1

1.

x	1.9	1.99	1.999
$f(x)$	0.3448	0.3344	0.3334

x	2.001	2.01	2.1
$f(x)$	0.3332	0.3322	0.3226

$$\lim_{x \to 2} \frac{x - 2}{x^2 - x - 2} \approx 0.3333 \ \left(\text{Actual limit is } \tfrac{1}{3}.\right)$$

3.

x	-0.1	-0.01	-0.001
$f(x)$	0.2911	0.2889	0.2887

x	0.001	0.01	0.1
$f(x)$	0.2887	0.2884	0.2683

$$\lim_{x \to 0} \frac{\sqrt{x + 3} - \sqrt{3}}{x} \approx 0.2887 \ \left(\text{Actual limit is } \frac{1}{2\sqrt{3}}.\right)$$

5.

x	2.9	2.99	2.999
$f(x)$	-0.0641	-0.0627	-0.0625

x	3.001	3.01	3.1
$f(x)$	-0.0625	-0.0623	-0.0610

$$\lim_{x \to 3} \frac{[1/(x + 1)] - (1/4)}{x - 3} \approx -0.0625$$
$$\left(\text{Actual limit is } -\tfrac{1}{16}.\right)$$

7. 1   **9.** 2   **11.** Limit does not exist.
13. 4   **15.** −1   **17.** −4   **19.** 2   **21.** 1
23. $\frac{1}{2}$ **25.** −2
27. (a) 15   **(b)** 5   **(c)** 6   **(d)** $\frac{2}{3}$
29. (a) 64   **(b)** 2   **(c)** 12   **(d)** 8
31. $\lim\limits_{x \to 4} f(x) = \frac{1}{6}$

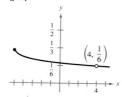

33. No. The value of $f(x)$ at $x = 2$ has no bearing on the behavior of f as x approaches 2.

35. $f(x) = -\dfrac{1}{x}, \quad g(x) = \dfrac{1}{x}$

Section 2.2

1. (a) 1   **(b)** 3   **3. (a)** 2   **(b)** 0   **5.** −2
7. $\frac{1}{6}$ **9.** 12   **11.** 2   **13.** $2x - 2$ **15.** $\frac{1}{10}$
17. $\dfrac{3}{2}$ **19.** $\dfrac{\sqrt{3}}{6}$ **21.** $-\dfrac{1}{4}$

23.

x	−0.1	−0.01	−0.001
$f(x)$	0.358	0.354	0.354

x	0.001	0.01	0.1
$f(x)$	0.354	0.353	0.349

$\lim\limits_{x \to 0} \dfrac{\sqrt{x + 2} - \sqrt{2}}{x} \approx 0.354 \left(\text{Actual limit is } \dfrac{1}{2\sqrt{2}}.\right)$

25.

x	−0.1	−0.01	−0.001
$f(x)$	−0.263	−0.251	−0.250

x	0.001	0.01	0.1
$f(x)$	−0.250	−0.249	−0.238

$\lim\limits_{x \to 0} \dfrac{[1/(2 + x)] - (1/2)}{x} \approx -0.250$

$\left(\text{Actual limit is } -\frac{1}{4}.\right)$

27. (a) 1   **(b)** 1   **(c)** 1
29. (a) 0   **(b)** 0   **(c)** 0
31. (a) 3   **(b)** −3   **(c)** Limit does not exist.

33. $\frac{1}{10}$ **35.** Limit does not exist.   **37.** 2
39. $-\dfrac{1}{x^2}$ **41.** Limit does not exist.
43. Limit does not exist.   **45.** 2   **47.** −2
49. 1   **51.** −16 ft/sec   **53.** $\lim\limits_{x \to -4} f(x) = 3$

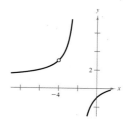

Section 2.3

1. Continuous for all real x
3. Discontinuous at $x = -1$
5. Discontinuous at $x = 1$
7. Continuous for all real x
9. Nonremovable discontinuity at $x = 1$
11. Continuous for all real x
13. Removable discontinuity at $x = -2$; Nonremovable discontinuity at $x = 5$
15. Continuous for all real x
17. Nonremovable discontinuity at $x = 2$
19. Nonremovable discontinuity at $x = -2$
21. Continuous for all real x
23. Nonremovable discontinuities at each integer
25. Continuous for all real x
27. Continuous for all real x
29. Continuous for all real x
31. Removable discontinuity at $x = 4$

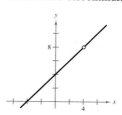

33. Nonremovable discontinuity at each integer

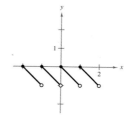

35. $(-\infty, -6), (-6, 6), (6, \infty)$ **37.** $(-\infty, \infty)$

41. **(a)** $(0.6, 0.7)$ **(b)** $(0.68, 0.69)$

43. $f(3) = 11$ **45.** $f(2) = 4$ **47.** $a = 2$

49. Yes, f is continuous on $[-1, 1]$.

51. $C = \begin{cases} 1.04, & 0 < t \leq 2 \\ 1.04 + 0.36[[t - 1]], & t > 2, \ t \text{ is not an integer} \\ 1.04 + 0.36(t - 1), & t > 2, \ t \text{ is an integer} \end{cases}$

Nonremovable discontinuity at each integer greater than 2

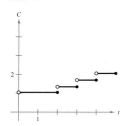

53. Discontinuous at $x = 3$

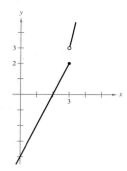

Section 2.4

1. $\lim\limits_{x \to -2^+} \dfrac{1}{(x + 2)^2} = \infty$ $\lim\limits_{x \to -2^-} \dfrac{1}{(x + 2)^2} = \infty$

3. $\lim\limits_{x \to -3^+} \dfrac{1}{x^2 - 9} = -\infty$ $\lim\limits_{x \to -3^-} \dfrac{1}{x^2 - 9} = \infty$

5. $\lim\limits_{x \to -3^+} \dfrac{x^3}{x^2 - 9} = \infty$ $\lim\limits_{x \to -3^-} \dfrac{x^3}{x^2 - 9} = -\infty$

7. $x = -1, \ x = 2$ **9.** $x = 0$

11. $x = -2, \ x = 1$ **13.** $x = \pm 2$ **15.** $x = 0$

17. $x = -2, \ x = 1$

19. Removable discontinuity at $x = -1$

21. Vertical asymptote at $x = -1$ **23.** $-\infty$

25. ∞ **27.** $-\infty$ **29.** $\frac{1}{2}$ **31.** ∞ **33.** ∞

35. ∞ **37.** $-\infty$

39. **(a)** \$14,117.65   **(b)** \$80,000

(c) \$720,000   **(d)** ∞

41. **(a)** $\frac{7}{12}$ ft/sec   **(b)** $\frac{3}{2}$ ft/sec   **(c)** ∞

43. $-\infty$

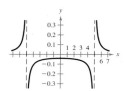

Section 2.5

1. $\lim\limits_{x \to 2} (3x + 2) = 8$ Let $\delta = \dfrac{0.01}{3} \approx 0.0033$.

3. $\lim\limits_{x \to 2} (x^2 - 3) = 1$

Assume $1 < x < 3$ and let $\delta = \dfrac{0.01}{5} = 0.002$.

5. $\lim\limits_{x \to 2} \dfrac{x^2 - 3x + 2}{x - 2} = 1$ Let $\delta = 0.01$.

7. $\lim\limits_{x \to 2} (x + 3) = 5$ Let $\delta = \varepsilon$.

9. $\lim\limits_{x \to 6} 3 = 3$ Any δ will work.

11. $\lim\limits_{x \to 0} \sqrt[3]{x} = 0$ Let $\delta = \varepsilon^3$.

13. $\lim\limits_{x \to 0} x^2 = 0$ Let $\delta = \sqrt{\varepsilon}$.

15. $\lim\limits_{x \to 2} \dfrac{x^2 + x - 6}{x - 2} = 5$ Let $\delta = \varepsilon$.

17. $\lim\limits_{x \to 2} \dfrac{1}{x} = \dfrac{1}{2}$ Let $\delta = 2\varepsilon$.

19. $\lim\limits_{x \to 2} (x^2 - 2) = 2$ Let $\delta = \dfrac{\varepsilon}{5}$.

21. $\lim\limits_{x \to 0^+} \sqrt{x} = 0$ Let $\delta = \varepsilon^2$.

23. $\lim\limits_{x \to -1^+} \dfrac{1}{x + 1} = \infty$ Let $\delta = \dfrac{1}{M}$.

25. $\lim\limits_{x \to 2} \dfrac{1}{(x - 2)^2} = \infty$ Let $\delta = \dfrac{1}{\sqrt{M}}$.

27. $\lim\limits_{x \to 3} x^2 = f(3) = 9$ **29.** 4

Review Exercises for Chapter 2

1. 7 **3.** 77 **5.** $\frac{10}{3}$ **7.** $-\frac{1}{4}$ **9.** -1

11. 3 **13.** $-\infty$ **15.** $-\infty$ **17.** $\frac{1}{3}$ **19.** 0

21.

x	1.1	1.01	1.001	1.0001
$f(x)$	0.5680	0.5764	0.5773	0.5773

$\lim\limits_{x \to 1^+} \dfrac{\sqrt{2x + 1} - \sqrt{3}}{x - 1} \approx 0.577$

23. $\dfrac{1}{\sqrt{3}}$ **25.** False   **27.** False   **29.** True

31. Nonremovable discontinuity at each integer
33. Removable discontinuity at $x = 1$
35. Nonremovable discontinuity at $x = 2$
37. Nonremovable discontinuity at $x = -1$
39. $c = -\frac{1}{2}$
41. Nonremovable discontinuity every 6 months

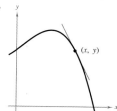

Chapter 3

Section 3.1

1.

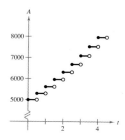

3. (a) $m = 0$
     (b) $m = -3$

5. $f'(x) = 0$ **7.** $f'(x) = -5$

9. $f'(x) = 4x + 1$ **11.** $f'(x) = -\dfrac{1}{(x-1)^2}$

13. $f'(t) = 3t^2 - 12$
15. $f'(x) = 2x$
Tangent line:
$y = 4x - 3$

17. $f'(x) = 3x^2$
Tangent line:
$y = 12x - 16$

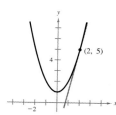

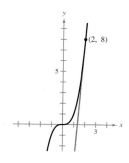

19. $f'(x) = \dfrac{1}{2\sqrt{x+1}}$
Tangent line: $4y = x + 5$

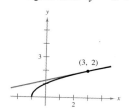

21. $f(x) = x^2 - 1$
$$f'(2) = \lim_{x \to 2} \frac{f(x) - f(2)}{x - 2}$$
$$= \lim_{x \to 2} \frac{(x^2 - 1) - 3}{x - 2}$$
$$= \lim_{x \to 2} (x + 2) = 4$$

23. $f(x) = x^3 + 2x^2 + 1$
$$f'(-2) = \lim_{x \to -2} \frac{f(x) - f(-2)}{x + 2}$$
$$= \lim_{x \to -2} \frac{(x^3 + 2x^2 + 1) - 1}{x + 2}$$
$$= \lim_{x \to -2} x^2 = 4$$

25. $f(x) = (x - 1)^{2/3}$
$$f'(1) = \lim_{x \to 1} \frac{f(x) - f(1)}{x - 1}$$
$$= \lim_{x \to 1} \frac{(x - 1)^{2/3} - 0}{x - 1}$$
$$= \lim_{x \to 1} \frac{1}{(x - 1)^{1/3}}$$
Limit does not exist.
f is not differentiable at $x = 1$.

27. $(-\infty, -3), (-3, \infty)$
29. $(-\infty, -1), (-1, \infty)$
31. $(-\infty, 3), (3, \infty)$ **33.** $(1, \infty)$
35. $(-\infty, 0), (0, \infty)$
37. f is not differentiable at $x = 1$.
39. $f'(1) = 0$
41. $y = 3x - 2$ **43.** $y = 2x + 1$
     $y = 3x + 2$ $y = -2x + 9$

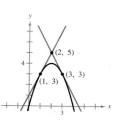

45. (a) 3   (b) -3 **47.** True   **49.** True

51.

x	−2	−1.5	−1	−0.5
f(x)	−2	−0.84375	−0.25	−0.03125
f'(x)	3	1.6875	0.75	0.1875

x	0	0.5	1	1.5	2
f(x)	0	0.03125	0.25	0.84375	2
f'(x)	0	0.1875	0.75	1.6875	3

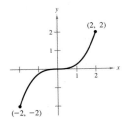

Section 3.2

1. Average rate: 2
Instantaneous rates:
$f'(1) = 2$
$f'(2) = 2$

3. Average rate: $-\frac{1}{4}$
Instantaneous rates:
$f'(0) = -1$
$f'(3) = -\frac{1}{16}$

5. Average rate: 4.1
Instantaneous rates:
$f'(2) = 4$
$f'(2.1) = 4.2$

7. (a) −48 ft/sec
(b) $s'(1) = -32$ ft/sec
$s'(2) = -64$ ft/sec
(c) $t = \dfrac{15\sqrt{6}}{4} \approx 9.2$ sec
(d) −293.9 ft/sec

9. $v(5) = 224$ ft/sec, $v(10) = 64$ ft/sec

11. $v\left(\dfrac{5\sqrt{6}}{2}\right) = -80\sqrt{6}$ ft/sec ≈ -195.96 ft/sec

13. 740 ft

15.

t	0	1	2	3	4
s(t)	0	57.75	99	123.75	132
v(t)	66	49.5	33	16.5	0
a(t)	−16.5	−16.5	−16.5	−16.5	−16.5

17. [0, 1], 57.75 ft/sec
[1, 2], 41.25 ft/sec
[2, 3], 24.75 ft/sec
[3, 4], 8.25 ft/sec

19. 2x

21. $\dfrac{2}{x^2}$

23. 0   **25.** −$1.91, −$1.93
27. (a) 83.67   **(b)** 40.79

29. $\dfrac{dF}{dr} = -2K\dfrac{m_1 m_2}{r^3}$
Force decreases as
the distance between
the particles increases.

31. $t \approx 3.48$ sec
$v(3.48) = -84.4$ ft/sec

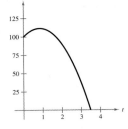

Section 3.3

1. (a) $\frac{1}{2}$ **(b)** $\frac{3}{2}$ **(c)** 2 **(d)** 3 **3.** 0
5. 1 **7.** 2x **9.** −4t + 3 **11.** $3t^2 - 2$
13. −1 **15.** 0 **17.** 4 **19.** $2x + \dfrac{4}{x^2}$
21. $3x^2 - 3 + \dfrac{8}{x^5}$ **23.** $\dfrac{x^3 - 8}{x^3}$ **25.** $3x^2 + 1$
27. $\dfrac{4}{5x^{1/5}}$ **29.** $\dfrac{1}{3x^{2/3}} + \dfrac{1}{5x^{4/5}}$

	Function	Rewrite	Derivative	Simplify
31.	$y = \dfrac{1}{3x^3}$	$y = \dfrac{1}{3}x^{-3}$	$y' = -x^{-4}$	$y' = -\dfrac{1}{x^4}$
33.	$y = \dfrac{1}{(3x)^3}$	$y = \dfrac{1}{27}x^{-3}$	$y' = -\dfrac{1}{9}x^{-4}$	$y' = -\dfrac{1}{9x^4}$
35.	$y = \dfrac{\sqrt{x}}{x}$	$y = x^{-1/2}$	$y' = -\dfrac{1}{2}x^{-3/2}$	$y' = -\dfrac{1}{2x^{3/2}}$

37. $2x + y - 2 = 0$
39. $(0, 2)$, $\left(\sqrt{\frac{3}{2}}, -\frac{1}{4}\right)$, $\left(-\sqrt{\frac{3}{2}}, -\frac{1}{4}\right)$
41. No horizontal tangents
43. $y = 4x - 4$ $y = 2x - 1$

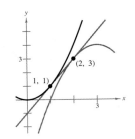
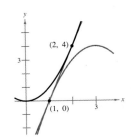

45. 8
47. (a) 10,000 **(b)** 4,000 **(c)** 0 **(d)** $-8,000$
49. (a) $\frac{4}{9}$ **(b)** $\frac{1}{3}$ **(c)** 0 **(d)** $-\frac{5}{9}$
51. -44.1 ft/sec **53.** -5.4 ft/sec^2
Approximately $\frac{1}{6}$ the acceleration due to gravity on earth.
55. True **57.** (0.1104, 0.1353)
(1.8408, -10.4862)

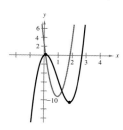

Section 3.4

1. $f'(x) = 2x^2$, $f'(0) = 0$
3. $f'(x) = 5[x^{-2}(1) + (x + 3)(-2x^{-3})]$
$$= -\frac{5(x + 6)}{x^3}$$
$f'(1) = -35$
5. $f'(x) = (x^3 - 3x)(4x + 3) + (2x^2 + 3x + 5)(3x^2 - 3)$
$= 10x^4 + 12x^3 - 3x^2 - 18x - 15$
$f'(0) = -15$
7. $f'(x) = (x^5 - 3x)\left(-\frac{2}{x^3}\right) + \left(\frac{1}{x^2}\right)(5x^4 - 3)$
$$= 3x^2 + \frac{3}{x^2}$$
$f'(-1) = 6$
9. $\dfrac{(2x - 3)(3) - (3x - 2)(2)}{(2x - 3)^2} = -\dfrac{5}{(2x - 3)^2}$
11. $\dfrac{(x^2 - 1)(-2 - 2x) - (3 - 2x - x^2)(2x)}{(x^2 - 1)^2} = \dfrac{2}{(x + 1)^2}$
13. $\dfrac{\sqrt{x}(1) - (x + 1)[1/(2\sqrt{x})]}{x} = \dfrac{x - 1}{2x^{3/2}}$
15. $\dfrac{(t^2 + 2t + 2)(1) - (t + 1)(2t + 2)}{(t^2 + 2t + 2)^2} = \dfrac{-t^2 - 2t}{(t^2 + 2t + 2)^2}$
17. $6s^2(s^3 - 2)$
19. $\left[\dfrac{x + 1}{x + 2}\right](2) + (2x - 5)\left[\dfrac{(x + 2)(1) - (x + 1)(1)}{(x + 2)^2}\right]$
$$= \dfrac{2x^2 + 8x - 1}{(x + 2)^2}$$
21. $15x^4 - 48x^3 - 33x^2 - 32x - 20$
23. $\dfrac{(c^2 + x^2)(-2x) - (c^2 - x^2)(2x)}{(c^2 + x^2)^2} = -\dfrac{4xc^2}{(c^2 + x^2)^2}$

Function	Rewrite	Derivative	Simplify
25. $y = \dfrac{x^2 + 2x}{x}$	$y = x + 2$	$y' = 1$	$y' = 1$
27. $y = \dfrac{7}{3x^3}$	$y = \dfrac{7}{3}x^{-3}$	$y' = -7x^{-4}$	$y' = -\dfrac{7}{x^4}$
29. $y = \dfrac{3x^2 - 5}{7}$	$y = \dfrac{1}{7}(3x^2 - 5)$	$y' = \dfrac{1}{7}(6x)$	$y' = \dfrac{6x}{7}$

31. $\dfrac{3}{\sqrt{x}}$ **33.** $\dfrac{2}{(x - 1)^3}$ **35.** $y = -x + 4$
37. $y = -x - 2$ **39.** (0, 0), (2, 4)
41. (a) -0.48 **(b)** 0.12 **(c)** 0.0149 **43.** 31.55
45. $a(2) = 4.069$

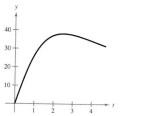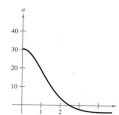

Section 3.5

$y = f(g(x))$	$u = g(x)$	$y = f(u)$
1. $y = (6x - 5)^4$	$u = 6x - 5$	$y = u^4$
3. $y = \sqrt{x^2 - 1}$	$u = x^2 - 1$	$y = \sqrt{u}$
5. $y = (x^2 - 3x + 4)^6$	$u = x^2 - 3x + 4$	$y = u^6$

7. $6(2x - 7)^2$
9. $12(9x - 4)^3(9) = 108(9x - 4)^3$
11. $-\dfrac{1}{(x - 2)^2}$ **13.** $-2(t - 3)^{-3}(1) = -\dfrac{2}{(t - 3)^3}$
15. $-3(x^3 - 4)^{-2}(3x^2) = -\dfrac{9x^2}{(x^3 - 4)^2}$
17. $x^2[4(x - 2)^3(1)] + (x - 2)^4(2x) = 2x(x - 2)^3(3x - 2)$
19. $\dfrac{1}{2}(1 - t)^{-1/2}(-1) = -\dfrac{1}{2\sqrt{1 - t}}$
21. $\dfrac{1}{2}(t^2 + 2t - 1)^{-1/2}(2t + 2) = \dfrac{t + 1}{\sqrt{t^2 + 2t - 1}}$
23. $\dfrac{1}{3}(9x^2 + 4)^{-2/3}(18x) = \dfrac{6x}{(9x^2 + 4)^{2/3}}$
25. $(4 - x^2)^{-1/2}(-2x) = -\dfrac{2x}{\sqrt{4 - x^2}}$
27. $\dfrac{2}{3}(9 - x^2)^{-1/3}(-2x) = -\dfrac{4x}{3(9 - x^2)^{1/3}}$
29. $-\dfrac{1}{2(x + 2)^{3/2}}$
31. $x\left(\dfrac{1}{2}\right)(1 - x^2)^{-1/2}(-2x) + (1 - x^2)^{1/2}(1) = \dfrac{1 - 2x^2}{\sqrt{1 - x^2}}$
33. $\dfrac{(x^2 + 1)^{1/2}(1) - x(1/2)(x^2 + 1)^{-1/2}(2x)}{x^2 + 1} = \dfrac{1}{(x^2 + 1)^{3/2}}$

35. $\dfrac{(x^2 + 1)(1/2)(x^{-1/2}) - (\sqrt{x} + 1)(2x)}{(x^2 + 1)^2} = \dfrac{1 - 3x^2 - 4x^{3/2}}{2\sqrt{x}(x^2 + 1)^2}$

37. $\dfrac{(t - 1)(3) - (3t + 2)}{(t - 1)^2} = -\dfrac{5}{(t - 1)^2}$

39. $3t^2\left(-\dfrac{1}{2}\right)(t^2 + 2t - 1)^{-3/2}(2t + 2) + (t^2 + 2t - 1)^{-1/2}(6t)$

$\qquad = \dfrac{3t(t^2 + 3t - 2)}{(t^2 + 2t - 1)^{3/2}}$

41. $\dfrac{1}{2}\left(\dfrac{x + 1}{x}\right)^{-1/2}\left(\dfrac{x - (x + 1)}{x^2}\right) = -\dfrac{1}{2x^{3/2}\sqrt{x + 1}}$

43. $-\dfrac{2}{3}\left[(2 - t)\dfrac{1}{2\sqrt{1 + t}} + (-1)\sqrt{1 + t}\right] = \dfrac{t}{\sqrt{1 + t}}$

45. $9x - 5y - 2 = 0$ **47.** $12(5x^2 - 1)(x^2 - 1)$

49. $\dfrac{3}{4(x^2 + x + 1)^{3/2}}$ **53.** $(3x^2 + 1)\left(\dfrac{x^3 + x}{|x^3 + x|}\right)$

55. (a) 1.461 **(b)** -1.016

Section 3.6

1. $-\dfrac{x}{y}, -\dfrac{3\sqrt{7}}{7}$ **3.** $-\dfrac{y}{x}, -\dfrac{1}{4}$ **5.** $-\sqrt{\dfrac{y}{x}}, -\dfrac{5}{4}$

7. $\dfrac{y - 3x^2}{2y - x}, \dfrac{1}{2}$ **9.** $\dfrac{18x}{(x^2 + 9)^2 y}$, undefined

11. $\dfrac{1 - 3x^2 y^3}{3x^3 y^2 - 1}, -1$ **13.** $-\sqrt[3]{\dfrac{y}{x}}, -\dfrac{1}{2}$

15. $\dfrac{4xy - 3x^2 - 3y^2}{2x(3y - x)}, -\dfrac{15}{28}$ **17.** $-\dfrac{1}{2}$ **19.** 0

21. $y' = -\dfrac{x}{y}$ **23.** $y' = -\dfrac{9x}{16y}$

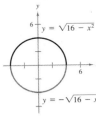

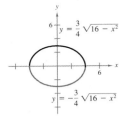

25. $\dfrac{10}{x^3}$ **27.** $-\dfrac{16}{y^3}$ **29.** $\dfrac{3x}{4y}$

31. (a) At $(4, 3)$:
  Tangent line: $4x + 3y - 25 = 0$
  Normal line: $3x - 4y = 0$
(b) At $(-3, 4)$:
  Tangent line: $3x - 4y + 25 = 0$
  Normal line: $4x + 3y = 0$
33. Horizontal tangents: $(-4, 0), (-4, 10)$
  Vertical tangents: $(0, 5), (-8, 5)$

35. At $(1, 2)$:
  Slope of ellipse: -1
  Slope of parabola: 1

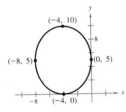

At $(1, -2)$:
  Slope of ellipse: 1
  Slope of parabola: -1

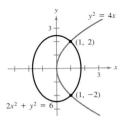

37. At $(\sqrt{2}, -\sqrt{2})$:
  Slope of line: -1
  Slope of circle: 1
At $(-\sqrt{2}, \sqrt{2})$:
  Slope of line: -1
  Slope of circle: 1

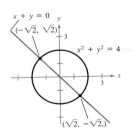

39. $(x - 1 - 2\sqrt{2})^2 + (y - 2 + 2\sqrt{2})^2 = 16$
  $(x - 1 + 2\sqrt{2})^2 + (y - 2 - 2\sqrt{2})^2 = 16$
41. $x + 2y - 6 = 0$

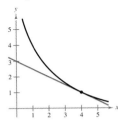

Section 3.7

1. (a) $\frac{3}{4}$ **(b)** 20 **3. (a)** $-\frac{5}{8}$ **(b)** $\frac{3}{2}$
5. (a) 24π in.^{2}/min   **(b)** 96π in.^{2}/min
7. If dr/dt is constant, dA/dt is proportional to r.

9. (a) $\dfrac{5}{\pi}$ ft/min   **(b)** $\dfrac{5}{4\pi}$ ft/min

11. $\dfrac{8}{405\pi}$ ft/min

13. (a) 9 cm³/sec   (b) 900 cm³/sec

15. (a) 0 cm/min   (b) 12 cm/min

17. (a) $\frac{8}{25}$ cm/min   (b) 0 cm/min

     (c) $-\frac{8}{25}$ cm/min   (d) -0.0039 cm/min

19. (a) 24.6%   (b) $\frac{1}{64}$ ft/min

21. (a) $-\frac{7}{12}$ ft/sec   (b) $-\frac{3}{2}$ ft/sec   (c) $-\frac{48}{7}$ ft/sec

23. 21.96 ft²/sec

25. (a) -750 mi/hr   (b) 20 minutes

27. $-\dfrac{28}{\sqrt{10}} \approx -8.85$ ft/sec

29. (a) $\frac{25}{3}$ ft/sec   (b) $\frac{10}{3}$ ft/sec

33. $v^{0.3}\left(1.3p\dfrac{dv}{dt} + v\dfrac{dp}{dt}\right) = 0$

Review Exercises for Chapter 3

1. $3x(x - 2)$ **3.** $\dfrac{x + 1}{2x^{3/2}}$ **5.** $-\dfrac{4}{3t^3}$

7. $\dfrac{3x^2}{2\sqrt{x^3 + 1}}$ **9.** $2(6x^3 - 9x^2 + 16x - 7)$

11. $s(s^2 - 1)^{3/2}(8s^3 - 3s + 25)$ **13.** $\dfrac{2x(2 - x)}{(x - 1)^2}$

15. $-\dfrac{x^2 + 1}{(x^2 - 1)^2}$ **17.** $32x - 128x^3$

19. $\dfrac{6x}{(4 - 3x^2)^2}$ **21.** $\dfrac{x + 2}{(x + 1)^{3/2}}$ **23.** $\dfrac{5}{6(t + 1)^{1/6}}$

25. $\dfrac{9}{(x^2 + 9)^{3/2}}$ **27.** $\dfrac{2(t + 2)}{(1 - t)^4}$

29. $\dfrac{2(6x^3 - 15x^2 - 18x + 5)}{(x^2 + 1)^3}$ **31.** $-\dfrac{2x + 3y}{3(x + y^2)}$

33. $\dfrac{2y\sqrt{x} - y\sqrt{y}}{2x\sqrt{y} - x\sqrt{x}}$ **35.** $\dfrac{x}{y}$

37. Tangent line: $3x - y + 7 = 0$
Normal line: $x + 3y - 1 = 0$

39. Tangent line: $x + 2y - 10 = 0$
Normal line: $2x - y = 0$

41. Tangent line: $2x - 3y - 3 = 0$
Normal line: $3x + 2y - 11 = 0$

43. (a) $(0, -1)$, $\left(-2, \dfrac{7}{3}\right)$ (b) $(-3, 2)$, $\left(1, -\dfrac{2}{3}\right)$

     (c) $\left(-1 + \sqrt{2}, \dfrac{2[1 - 2\sqrt{2}]}{3}\right)$, $\left(-1 - \sqrt{2}, \dfrac{2[1 + 2\sqrt{2}]}{3}\right)$

45. $f'(x) = -\dfrac{2}{x^3}$ **47.** $f'(x) = \dfrac{1}{2\sqrt{x + 2}}$

49. $v(t) = 1 - \dfrac{1}{(t + 1)^2}$, $a(t) = \dfrac{2}{(t + 1)^3}$

51. (a) -18.667 (b) -7.284 (c) -3.240

     (d) -0.747 **53.** 56 ft/sec

55. (a)

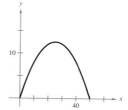

   (b) 50

   (c) $x = 25$

   (d) $y' = 1 - 0.04x$

x	0	10	25	30	50
y'	1	0.6	0	-0.2	-1

   (e) $y'(25) = 0$

59. (a) $2\sqrt{2}$ units/sec   (b) 4 units/sec

   (c) 8 units/sec   **61.** $\frac{2}{25}$ ft/min

65. (a)

   (b) No

   (c) No

Chapter 4

Section 4.1

1. $f'(0) = 0$ **3.** $f'(4) = 0$

5. $f'(-2)$ is undefined

7. Minimum: $(2, 2)$
Maximum: $(-1, 8)$

9. Minimum: $(0, 0)$ and $(3, 0)$
Maximum: $\left(\frac{3}{2}, \frac{9}{4}\right)$

11. Minimum: $(-1, -4)$ and $(2, -4)$
Maximum: $(0, 0)$ and $(3, 0)$

13. Minimum: $(0, 0)$ **15.** Minimum: $(1, 1)$
Maximum: $(-1, 5)$ Maximum: $(4, 4)$

17. Minimum: $(1, -1)$
Maximum: $\left(0, -\frac{1}{2}\right)$

19. f is bounded on $[1, 2]$ **21.** (a) Yes
but not bounded on $(0, 2]$.   (b) No

23. (a) No
(b) Yes

25. (a) Minimum: $(0, -3)$
Maximum: $(2, 1)$
(b) Minimum: $(0, -3)$
(c) Maximum: $(2, 1)$
(d) No extrema

27. Maximum: $|f''(0)| = 2$
29. Maximum: $\left|f''\left(\sqrt[3]{-10 + \sqrt{108}}\right)\right| \approx 1.47$
31. Maximum: $\left|f^{(4)}\left(\frac{1}{2}\right)\right| = 360$
33. Maximum: $|f^{(4)}(0)| = \frac{56}{81}$
35. Maximum: $P(12) = 72$
37. 0.4398

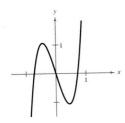

Section 4.2

1. $f(0) = f(2) = 0$
f is not differentiable on $(0, 2)$

3. $f'(1) = 0$
5. $[1, 2]$, $f'\left(\dfrac{6 - \sqrt{3}}{3}\right) = 0$
$[2, 3]$, $f'\left(\dfrac{6 + \sqrt{3}}{3}\right) = 0$

7. Not differentiable at $x = 0$
9. Not differentiable at $x = 0$
11. $f'(-2 + \sqrt{5}) = 0$
13. $f'\left(-\frac{1}{2}\right) = -1$
15. $f'\left(\dfrac{8}{27}\right) = 1$
17. $f'\left(\dfrac{-2 + \sqrt{6}}{2}\right) = \dfrac{2}{3}$
19. $f'\left(\dfrac{\sqrt{3}}{3}\right) = 1$
21. (a) $f(1) = f(2) = 64$
(b) Velocity $= 0$ for some t in $[1, 2]$
23. (a) -48 ft/sec
(b) $t = \frac{3}{2}$ sec
25. $f(x)$ is not continuous on $[2, 6]$
33. (a) $x - 4y + 3 = 0$
(b) $c = 4$, $x - 4y + 4 = 0$

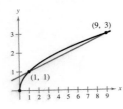

Section 4.3

1. Increasing on $(3, \infty)$
Decreasing on $(-\infty, 3)$

3. Increasing on $(-\infty, -2)$ and $(2, \infty)$
Decreasing on $(-2, 2)$
5. Increasing on $(-\infty, 0)$
Decreasing on $(0, \infty)$
7. Critical number: $x = 1$
Increasing on $(-\infty, 1)$
Decreasing on $(1, \infty)$
Relative maximum: $(1, 5)$
9. Critical number: $x = 3$
Increasing on $(3, \infty)$
Decreasing on $(-\infty, 3)$
Relative minimum: $(3, -9)$
11. Critical numbers: $x = -2, 1$
Increasing on $(-\infty, -2)$ and $(1, \infty)$
Decreasing on $(-2, 1)$
Relative maximum: $(-2, 20)$
Relative minimum: $(1, -7)$
13. Critical numbers: $x = 0, 4$
Increasing on $(-\infty, 0)$ and $(4, \infty)$
Decreasing on $(0, 4)$
Relative maximum: $(0, 15)$
Relative minimum: $(4, -17)$
15. Critical number: $x = 1$
Increasing on $(1, \infty)$
Decreasing on $(-\infty, 1)$
Relative minimum: $(1, 0)$
17. Critical number: $x = 0$
Increasing on $(-\infty, \infty)$
No relative extrema
19. Critical numbers: $x = -1, 1$
Discontinuity: $x = 0$
Increasing on $(-\infty, -1)$ and $(1, \infty)$
Decreasing on $(-1, 0)$ and $(0, 1)$
Relative maximum: $(-1, -2)$
Relative minimum: $(1, 2)$
21. Critical number: $x = 0$
Discontinuities: $x = -3, 3$
Increasing on $(-\infty, -3)$ and $(-3, 0)$
Decreasing on $(0, 3)$ and $(3, \infty)$
Relative maximum: $(0, 0)$
23. Critical numbers: $x = -1, 1$
Increasing on $(-\infty, -1)$ and $(1, \infty)$
Decreasing on $(-1, 1)$
Relative maximum: $\left(-1, \frac{4}{5}\right)$
Relative minimum: $\left(1, -\frac{4}{5}\right)$
25. Critical numbers: $x = -3, 1$
Discontinuity: $x = -1$
Increasing on $(-\infty, -3)$ and $(1, \infty)$
Decreasing on $(-3, -1)$ and $(-1, 1)$
Relative maximum: $(-3, -8)$
Relative minimum: $(1, 0)$
27. (a) Not monotonic
(b) Strictly monotonic
(c) Strictly monotonic
29. Moving upward when $0 < t < 3$

Moving downward when $3 < t < 6$
Maximum height: $s(3) = 144$ ft

31. $r = \dfrac{2R}{3}$

33. Increasing when $0 < t < 84.3388$ minutes
Decreasing when $84.3388 < t < 120$ minutes

35. Increasing when $6.02 < t < 14$ days
Decreasing when $0 < t < 6.02$ days

37. $T = 10°$ **39.** $f(x) = -\frac{1}{2}x^3 + \frac{3}{2}x^2$

41. $g'(0) < 0$ **43.** $g'(-6) < 0$ **45.** $g'(0) > 0$

47. (a)

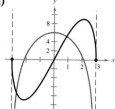

(b) Critical numbers: $x = \pm\dfrac{3\sqrt{2}}{2}$

(c) $f' > (0)$ on $\left(-\dfrac{3\sqrt{2}}{2}, \dfrac{3\sqrt{2}}{2}\right)$

$f' < (0)$ on $\left(-3, -\dfrac{3\sqrt{2}}{2}\right), \left(\dfrac{3\sqrt{2}}{2}, 3\right)$

Section 4.4

1. Concave upward: $(-\infty, \infty)$

3. Concave upward: $(-\infty, 1)$
Concave downward: $(1, \infty)$

5. Concave upward: $(-\infty, -1), (1, \infty)$
Concave downward: $(-1, 1)$

7. Relative maximum: $(3, 9)$

9. Relative minimum: $(5, 0)$

11. Relative maximum: $(0, 3)$
Relative minimum: $(2, -1)$

13. Relative minimum: $(3, -25)$

15. Relative minimum: $(0, -3)$

17. Relative maximum: $(-2, -4)$
Relative minimum: $(2, 4)$

19. Relative maximum: $(-2, 16)$
Relative minimum: $(2, -16)$
Point of inflection: $(0, 0)$

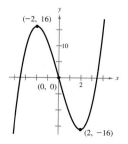

21. Point of inflection: $(2, 0)$

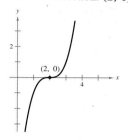

23. Relative minima: $(\pm2, -4)$
Relative maximum: $(0, 0)$
Points of inflection: $\left(\pm\dfrac{2}{\sqrt{3}}, -\dfrac{20}{9}\right)$

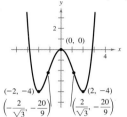

25. Relative minimum: $(1, -27)$
Points of inflection: $(2, -16), (4, 0)$

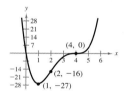

27. Relative minima: $(-1, 2), (1, 2)$

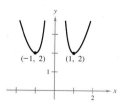

29. Relative minimum: $(-2, -2)$

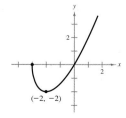

31. Point of inflection: (0, 0)

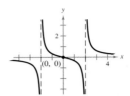

33. Point of inflection: (2, 0)

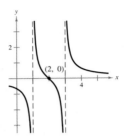

35. (a) **(b)**

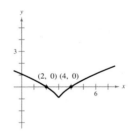 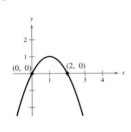

37. **39.**

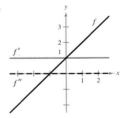 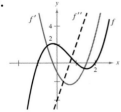

41. $f(x) = (x - c)^n$ has a point of inflection at $(c, 0)$ if n is odd and $n \geq 3$.

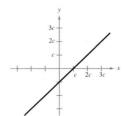 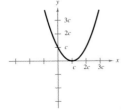

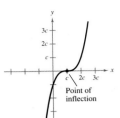 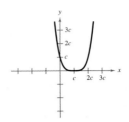

Point of inflection

43. $f(x) = \frac{1}{2}x^3 - 6x^2 + \frac{45}{2}x - 24$
45. Relative extrema: (2, 32), (6, 0)
Point of inflection: (4, 16)
47. $x = \left(\dfrac{15 - \sqrt{33}}{16}\right) L \approx 0.578 L$
49. $x = 100$ units
51. (a) 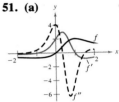 **(b)** Critical numbers: $x = -1, 1$
Relative minimum: $\left(-1, -\frac{2}{3}\right)$
Relative maximum: (1, 2)

(c) Points of inflection:
$(-1.8794, -0.5863)$, $(0.3473, 0.8982)$,
$(1.5321, 1.6881)$

Section 4.5

1. h **2.** c **3.** e **4.** a **5.** d **6.** g
7. b **8.** f **9.** $\frac{2}{3}$ **11.** 0
13. Limit does not exist.
15. Limit does not exist. **17.** 5 **19.** -1
21. 2 **23.** 1 **25.** 0 **27.** $-\frac{1}{2}$
29. **31.**

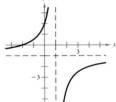 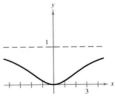

33. **35.**

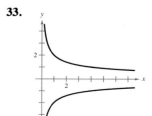 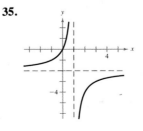

37.

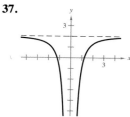

39.

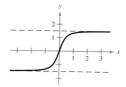

41.

x	1	10	10^2	10^3
$f(x)$	2.000	0.348	0.101	0.032

x	10^4	10^5	10^6
$f(x)$	0.010	0.0032	0.001

$$\lim_{x \to \infty} \frac{x+1}{x\sqrt{x}} = 0$$

43.

x	1	10	10^2	10^3
$f(x)$	-0.236	-0.025	-0.003	-0.000

x	10^4	10^5	10^6
$f(x)$	-0.000	-0.000	-0.000

$$\lim_{x \to \infty} (2x - \sqrt{4x^2 + 1}) = 0$$

45. 0.5   **47.** 100%

49. Horizontal asymptotes: $y = \pm\frac{3}{2}$

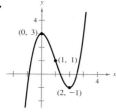

Section 4.6

1.

3.

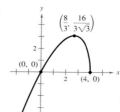

5.

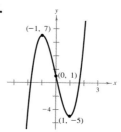

7.

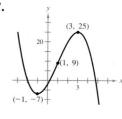

9.

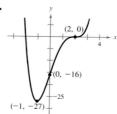

11.

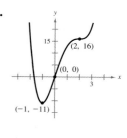

13.

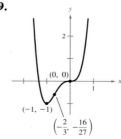

15.

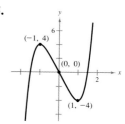

17.

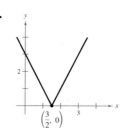

19.

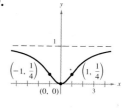

21.

23.

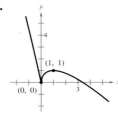

25.

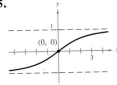

27.

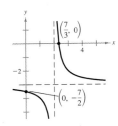

29.

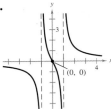

31.

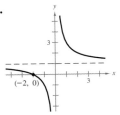

33.

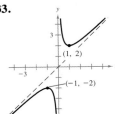

35.

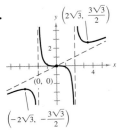

37. $a < 0$ and $b^2 < 3ac$ **39.** $a < 0$ and $b^2 = 3ac$
41. $a < 0$ and $b^2 > 3ac$ **43.**

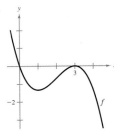

45.

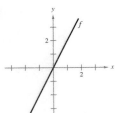

47.

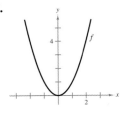

Section 4.7

1. 55 and 55   **3.** $\sqrt{192}$ and $\sqrt{192}$
5. 1 and 1   **7.** $l = w = 25$ ft

9. $l = w = 8$ ft   **11.** $\left(\dfrac{7}{2}, \sqrt{\dfrac{7}{2}} \right)$

13. $x = \dfrac{Q}{2}$ **15.** $600m \times 300m$

17. $V = 128$ when $x = 2$

19. $\dfrac{5 - \sqrt{7}}{6}$ ft $\times \dfrac{1 + \sqrt{7}}{3}$ ft $\times \dfrac{4 + \sqrt{7}}{3}$ ft

21. Rectangular portion: $\dfrac{16}{\pi + 4}$ ft $\times \dfrac{32}{\pi + 4}$ ft

23. $x = 3$, $y = \dfrac{3}{2}$ **25.** $(0, 0), (4, 0), (0, 6)$

27. Height: $\dfrac{5\sqrt{2}}{2}$; width: $5\sqrt{2}$

29. Bases: r and $2r$; altitude: $\dfrac{\sqrt{3}\,r}{2}$

31. $r \approx 1.51$, $h = 2r \approx 3.02$

33. 18 in. $\times$ 18 in. $\times$ 36 in.   **35.** $\dfrac{32\pi r^3}{81}$

37. $r = \sqrt[3]{\dfrac{9}{\pi}} \approx 1.42$ in.

39. Side of square: $\dfrac{10\sqrt{3}}{9 + 4\sqrt{3}}$

Side of triangle: $\dfrac{30}{9 + 4\sqrt{3}}$

41. (a) $\dfrac{10\pi}{\pi + 3 + 2\sqrt{2}} \approx 3.5$ ft   (b) 10 ft

43. $w = 8\sqrt{3}$, $h = 8\sqrt{6}$
45. 1 mile from the nearest point on the coast
47. 14.1421×7.071

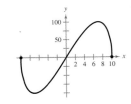

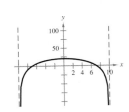

Section 4.8

1.

n	x_n	$f(x_n)$	$f'(x_n)$	$\dfrac{f(x_n)}{f'(x_n)}$	$x_n - \dfrac{f(x_n)}{f'(x_n)}$
1	1.700	−0.110	3.400	−0.032	1.732

3.

n	x_n	$f(x_n)$	$f'(x_n)$	$\dfrac{f(x_n)}{f'(x_n)}$	$x_n - \dfrac{f(x_n)}{f'(x_n)}$
1	1	1.000	9.000	0.111	0.889

5. 0.682   **7.** 1.146   **9.** 3.317   **11.** 0.569
13. 1.379   **15.** $f'(x_1) = 0$
17. $1 = x_1 = x_3 = \ldots$ **19.** $x_{i+1} = \dfrac{x_i^2 + a}{2x_i}$
$\quad\quad 0 = x_2 = x_4 = \ldots$
21. 2.646   **23.** 1.565   **27.** (1.939, 0.240)
29. $x \approx 1.563$ mi
31. $x \approx 11.803$

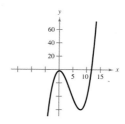

Section 4.9

1. $6x\,dx$ **3.** $12x^2\,dx$ **5.** $-\dfrac{3}{(2x-1)^2}\,dx$

7. $\dfrac{1}{2\sqrt{x}}\,dx$ **9.** $\dfrac{1-2x^2}{\sqrt{1-x^2}}\,dx$

11.

$dx = \Delta x$	dy	Δy	$\Delta y - dy$	$\dfrac{dy}{\Delta y}$
1.000	1.000	1.000	0.000	1.000
0.500	0.500	0.500	0.000	1.000
0.100	0.100	0.100	0.000	1.000
0.010	0.010	0.010	0.000	1.000
0.001	0.001	0.001	0.000	1.000

13.

$dx = \Delta x$	dy	Δy	$\Delta y - dy$	$\dfrac{dy}{\Delta y}$
1.000	4.000	5.000	1.000	0.800
0.500	2.000	2.250	0.250	0.889
0.100	0.400	0.410	0.010	0.976
0.010	0.040	0.040	0.000	1.000
0.001	0.004	0.004	0.000	1.000

15.

$dx = \Delta x$	dy	Δy	$\Delta y - dy$	$\dfrac{dy}{\Delta y}$
1.000	80.000	211.000	131.000	0.379
0.500	40.000	65.656	25.656	0.609
0.100	8.000	8.841	0.841	0.905
0.010	0.800	0.808	0.008	0.990
0.001	0.080	0.080	0.000	1.000

17. (a) $dA = 2x\Delta x,\ \Delta A = 2x\Delta x + (\Delta x)^2$
(b)

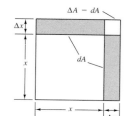

(c) $\Delta A - dA = (\Delta x)^2$
19. $\pm 7\pi$ in.^{2} **21. (a)** $\frac{2}{3}\%$ **(b)** 1.25%
23. (a) $\pm 2.88\pi$ in.^{3} **(b)** $\pm 0.96\pi$ in.^{2}
(c) 1%, $\frac{2}{3}\%$
25. (a) $\frac{1}{4}\%$ **(b)** 216 sec = 3.6 min

Section 4.10

1. 4500   **3.** 300   **5.** 200   **7.** 200
9. $x = 30$ **11.** $x = 1500$ **13.** 20   **15.** 200
$\quad p = 60$ $\quad\quad p = 35$
17. Line should run from power station to point across the river $3/(2\sqrt{7})$ mile downstream.
19. 8%   **21.** $x = 3$
23. (a) $p = \$68$ **(b)** $\overline{C} \approx \$50.88$
25. $x \approx 40$ units   **27.** $\$30,000$
29. $\eta = -\frac{17}{3}$, elastic   **31.** $\eta = -\frac{1}{2}$, inelastic

Review Exercises for Chapter 4

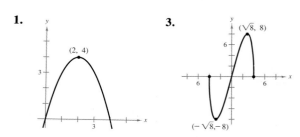

5.

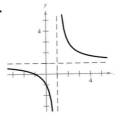

7.

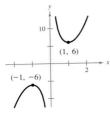

9.

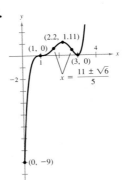

11.

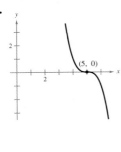

13.

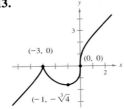

15.

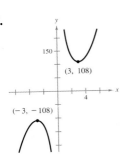

17.

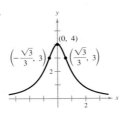

19.

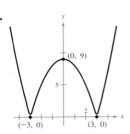

21.

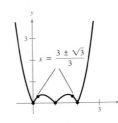

23.

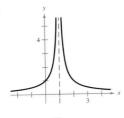

25. Maximum: $(1, 3)$
Minimum: $(1, 1)$
27. $f'\left(\dfrac{-2 + \sqrt{85}}{3}\right) = -\dfrac{1}{17}$

29. $f'\left(\dfrac{2744}{729}\right) = \dfrac{3}{7}$ **31.** $f'(2) = \dfrac{5}{4}$

33. No, f is discontinuous at $x = 0$

35. $c = \dfrac{x_1 + x_2}{2}$ **37.** $t \approx 4.92 \approx 4{:}55$ P.M.
$d \approx 64$ mi

39. \$48 **41.** $(0, 0), (5, 0), (0, 10)$ **43.** 120

47. 14.05 ft **49.** $x = \sqrt{\dfrac{2Qs}{r}}$ **51.** -0.347

53. -0.453 **55.** **(a)** 2% **(b)** 3%

Chapter 5

Section 5.1

Given	Rewrite	Integrate	Simplify
1. $\displaystyle\int \sqrt[3]{x}\,dx$	$\displaystyle\int x^{1/3}\,dx$	$\dfrac{x^{4/3}}{4/3} + C$	$\dfrac{3}{4}x^{4/3} + C$
3. $\displaystyle\int \dfrac{1}{x\sqrt{x}}\,dx$	$\displaystyle\int x^{-3/2}\,dx$	$\dfrac{x^{-1/2}}{-1/2} + C$	$-\dfrac{2}{\sqrt{x}} + C$
5. $\displaystyle\int \dfrac{1}{2x^3}\,dx$	$\dfrac{1}{2}\displaystyle\int x^{-3}\,dx$	$\dfrac{1}{2}\left(\dfrac{x^{-2}}{-2}\right) + C$	$-\dfrac{1}{4x^2} + C$

7. $\frac{1}{4}x^4 + 2x + C$ **9.** $\frac{2}{5}x^{5/2} + x^2 + x + C$

11. $\dfrac{3}{5}x^{5/3} + C$ **13.** $-\dfrac{1}{2x^2} + C$ **15.** $-\dfrac{1}{4x} + C$

17. $\frac{2}{15}x^{1/2}(3x^2 + 5x + 15) + C$

19. $x^3 + \dfrac{1}{2}x^2 - 2x + C$ **21.** $t - \dfrac{2}{t} + C$

23. $\frac{2}{7}y^{7/2} + C$ **25.** $x + C$
27. $y = x^2 - x + 1$ **29.** $y = x^3 - x + 2$
31. $f(x) = x^2 + x + 4$
33. $f(x) = -4x^{1/2} + 3x = -4\sqrt{x} + 3x$
35. $s(t) = -16t^2 + 1600, t = 10$ sec
37. $v_0 \approx 187.617$ ft/sec

39. **(a)** $\dfrac{1 + \sqrt{17}}{2} \approx 2.562$ sec
(b) $-16\sqrt{17} \approx -65.970$ ft/sec
41. **(a)** $\frac{154}{39} \approx 3.95$ ft/sec² **(b)** $\frac{1859}{3} \approx 619.67$ ft
43. **(a)** 300 ft **(b)** 60 ft/sec ≈ 41 mi/hr
49. $R = x\left(100 - \frac{5}{2}x\right), p = 100 - \frac{5}{2}x$

Section 5.2

1. 35 **3.** $\dfrac{158}{85}$ **5.** $4c$ **7.** 238 **9.** $\displaystyle\sum_{i=1}^{9} \dfrac{1}{3i}$

11. $\displaystyle\sum_{j=1}^{8} \left[2\left(\dfrac{j}{8}\right) + 3\right]$ **13.** $\dfrac{1}{6}\displaystyle\sum_{k=1}^{6}\left[\left(\dfrac{k}{6}\right)^2 + 2\right]$

15. $\dfrac{2}{n}\displaystyle\sum_{i=1}^{n}\left[\left(\dfrac{2i}{n}\right)^3 - \left(\dfrac{2i}{n}\right)\right]$ **17.** $\dfrac{3}{n}\displaystyle\sum_{i=1}^{n}\left[2\left(1 + \dfrac{3i}{n}\right)^2\right]$

19. 420 **21.** 2470 **23.** $\dfrac{1015}{n^3}$

25. $\frac{8}{3}$ **27.** $\frac{81}{4}$ **29.** 9

31. $\lim\limits_{n\to\infty} \dfrac{1}{6}\left[\dfrac{2n^3 - 3n^2 + n}{n^3}\right] = \dfrac{1}{3}$

33. $\lim\limits_{n\to\infty} \left[8\left(\dfrac{n^2 + n}{n^2}\right)\right] = 8$

35. $\lim\limits_{n\to\infty} 2\left[\dfrac{10n^4 + 13n^3 + 4n^2}{n^4}\right] = 20$

37. $S \approx 0.768$ **39.** $S \approx 0.746$ **41.** $S \approx 0.859$
$s \approx 0.518$ $s \approx 0.646$ $s \approx 0.659$

43.

n	5	10	50	100
$s(n)$	1.6	1.8	1.96	1.98
$S(n)$	2.4	2.2	2.04	2.02

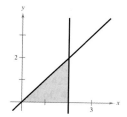

45. $A = 2$

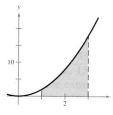

47. $A = \frac{7}{3}$

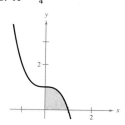

49. $A = \frac{52}{3}$

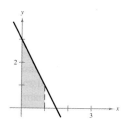

51. $A = \frac{3}{4}$

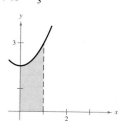

53. $A = \frac{2}{3}$

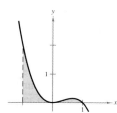

55. $A = 6$

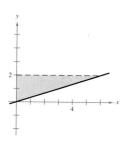

57. $\frac{69}{8}$ **59.** 0.673

61. $N(5) \approx 167$, $N(10) \approx 250$, $N(25) \approx 400$

$\lim\limits_{t\to\infty} \dfrac{10(5 + 3t)}{1 + 0.04t} = 750$

63.

n	4	8	12	16	20
$s(n)$	4.1463	4.7650	4.9623	5.0587	5.1156
$S(n)$	6.1463	5.7650	5.6290	5.5587	5.5156

Section 5.3

1. $\displaystyle\int_0^5 3\,dx$ **3.** $\displaystyle\int_{-4}^4 (4 - |x|)\,dx$

5. $\displaystyle\int_{-2}^2 (4 - x^2)\,dx$ **7.** $\displaystyle\int_0^2 y^3\,dy$

9. $\displaystyle\int_0^2 \sqrt{x + 1}\,dx$

11. $A = 12$

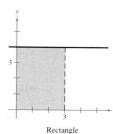

Rectangle

13. $A = 8$

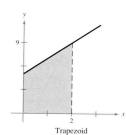

Triangle

15. $A = 14$

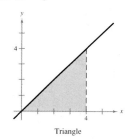

Trapezoid

17. $A = 1$

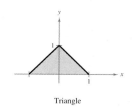

Triangle

19. $A = \dfrac{9\pi}{2}$

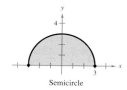

Semicircle

21. (a) 13   **(b)** -10 **(c)** 0   **(d)** 30
23. (a) 8   **(b)** -12 **(c)** -4 **(d)** 30
25. 36   **27.** 0   **29.** $\dfrac{10}{3}$ **31.** $\dfrac{4\sqrt{2}}{3}$
33. $\displaystyle\int_{-1}^{5} (3x + 10)\, dx$ **35.** $\displaystyle\int_{0}^{3} \sqrt{x^2 + 4}\, dx$
39. Not integrable because there are an infinite number of discontinuities.

Section 5.4

1. 1   **3.** $-\dfrac{5}{2}$ **5.** $-\dfrac{10}{3}$ **7.** $\dfrac{1}{3}$ **9.** $\dfrac{1}{2}$
11. 36   **13.** -4 **15.** $\dfrac{2}{3}$ **17.** $-\dfrac{1}{18}$
19. $-\dfrac{27}{20}$ **21.** 1   **23.** 4   **25.** $\dfrac{1}{6}$ **27.** $\dfrac{8}{5}$
29. 6   **31.** 10   **33.** 6   **35.** $\sqrt[3]{2}$
37. 1, 3

39. Average $= \dfrac{8}{3}$

$x = \pm\dfrac{2\sqrt{3}}{3} \approx \pm 1.155$

41. Average $= -\dfrac{2}{3}$

$x = \left(1 + \dfrac{1}{\sqrt{3}}\right)^2$

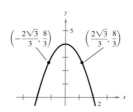

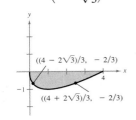

43. $\dfrac{1}{2}x^2 + 2x$ **45.** $\dfrac{3}{4}x^{4/3} - 12$ **47.** $1 - \dfrac{1}{x}$
49. $x^2 - 2x + 5$ **51.** $\sqrt{x^4 + 1}$
53. 0.5318 liter   **55. (a)** $64.4°$ **(b)** $68.6°$

Section 5.5

$\displaystyle\int f(g(x))g'(x)\, dx$	$u = g(x)$	$du = g'(x)\, dx$
1. $\displaystyle\int (5x^2 + 1)^2(10x)\, dx$	$5x^2 + 1$	$10x\, dx$
3. $\displaystyle\int \dfrac{x}{\sqrt{x^2 + 1}}\, dx$	$x^2 + 1$	$2x\, dx$

5. $\dfrac{(1 + 2x)^5}{5} + C$ **7.** $\dfrac{2}{3}(9 - x^2)^{3/2} + C$
9. $\dfrac{(x^3 - 1)^5}{15} + C$ **11.** $-\dfrac{15}{8}(1 - x^2)^{4/3} + C$
13. $-\dfrac{1}{3(1 + x^3)} + C$ **15.** $-4\sqrt{16 - x^2} + C$
17. $-\dfrac{1}{2(x^2 + 2x - 3)} + C$ **19.** $-\dfrac{1}{4}\left(1 + \dfrac{1}{t}\right)^4 + C$
21. $\sqrt{2x} + C$ **23.** $\dfrac{2}{5}\sqrt{x}(x^2 + 5x + 35) + C$

25. $\dfrac{1}{4}t^4 - t^2 + C$ **27.** $\dfrac{2}{5}y^{3/2}(15 - y) + C$
29. $2\left[\dfrac{1}{5}(x + 2)^{5/2} - \dfrac{2}{3}(x + 2)^{3/2}\right] + C$
$\qquad = \dfrac{2}{15}(x + 2)^{3/2}(3x - 4) + C$
31. $-2\left[\dfrac{1}{3}(1 - x)^{3/2} - \dfrac{2}{5}(1 - x)^{5/2} + \dfrac{1}{7}(1 - x)^{7/2}\right] + C$
$\qquad = -\dfrac{2}{105}(1 - x)^{3/2}(15x^2 + 12x + 8) + C$
33. $\dfrac{1}{4}\left[\dfrac{1}{5}(2x - 1)^{5/2} + \dfrac{2}{3}(2x - 1)^{3/2} - 3(2x - 1)^{1/2}\right] + C$
$\qquad = \dfrac{\sqrt{2x - 1}}{15}(3x^2 + 2x - 13) + C$
35. $-x - 1 - 2\sqrt{x + 1} + C$
$\qquad$ or $-(x + 2\sqrt{x + 1}) + C_1$
37. $\dfrac{1}{2}\left[\dfrac{1}{3}(2x + 1)^{3/2} - (2x - 1)\right] + C$
$\qquad = \dfrac{1}{3}\sqrt{2x + 1}(x - 1) + C$
39. 0   **41.** 2   **43.** $\dfrac{1}{2}$ **45.** $\dfrac{4}{15}$ **47.** $\dfrac{144}{5}$
49. $\dfrac{1209}{28}$ **51. (a)** $\dfrac{8}{3}$ **(b)** $\dfrac{16}{3}$ **(c)** $-\dfrac{8}{3}$ **(d)** 8
53. (a) $\dfrac{3}{2}[\sqrt{16t + 9} - 3]$
   (b) $\dfrac{3}{2}[\sqrt{1609} - 3] \approx 55.67$ lb

Section 5.6

	Exact	Trapezoidal	Simpson's
1.	2.6667	2.7500	2.6667
3.	4.0000	4.2500	4.0000
5.	4.0000	4.0625	4.0000
7.	12.6667	12.6640	12.6667
9.	0.1667	0.1676	0.1667

	Trapezoidal	Simpson's
11.	1.683	1.622
13.	3.41	3.22
15.	0.342	0.372
17.	2.200	2.210
19.	2.352	2.438
21.	0.500	0.000
23.	0.010	0.001
25.	$n = 366$	$n = 26$
27.	$n = 130$	$n = 12$

29. 3.14159   **31.** 89,250 ft^2
33. (a) 9,920 ft^2 **(b)** $10,413\frac{1}{3}$ ft^2

35.

n	Trapezoidal Rule	Simpson's Rule
4	15.6055	15.4845
8	15.5010	15.4662
12	15.4814	15.4657
16	15.4745	15.4657
20	15.4713	15.4657

Review Exercises for Chapter 5

1. $x^{2/3} + C$ **3.** $\frac{2}{3}x^3 + \frac{1}{2}x^2 - x + C$

5. $\frac{2\sqrt{x}}{15}(15 + 10x + 3x^2) + C$

7. $\frac{2}{3}\sqrt{x^3 + 3} + C$ **9.** $\frac{1}{7}x^7 + \frac{3}{5}x^5 + x^3 + x + C$

11. $\frac{1}{8}(x^2 + 1)^4 + C$ **13.** $-\frac{1}{4(x^2 + 1)^2} + C$

15. $2\left[\frac{1}{7}(x + 5)^{7/2} - 2(x + 5)^{5/2} + \frac{25}{3}(x + 5)^{3/2}\right] + C$

$$= \frac{2(x + 5)^{3/2}}{21}(3x^2 - 12x + 40) + C$$

17. $\frac{1}{2}x^2 - \frac{1}{x} + C$

19. (a) $\sum_{i=1}^{10} (2i - 1)$ (b) $\sum_{i=1}^{n} i^3$ (c) $\sum_{i=1}^{10} (4i + 2)$

21. 16   **23.** 0   **25.** 2   **27.** $\frac{422}{5}$

29. $\frac{28\pi}{15}$ **31.** $y = 2 - x^2$ **33.** 240 ft/sec

35. (a) 3 sec   (b) 144 ft   (c) $\frac{3}{2}$ sec   (d) 108 ft

37. (a) $S = \frac{5mb^2}{8}$, $s = \frac{3mb^2}{8}$

   (b) $S(n) = \frac{mb^2(n + 1)}{2n}$

   $s(n) = \frac{mb^2(n - 1)}{2n}$

   (c) $\frac{1}{2}mb^2$ (d) $\frac{1}{2}mb^2$

39. 6

41. $\frac{10}{3}$

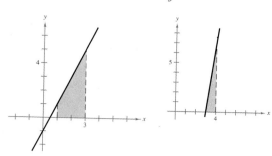

43. $\frac{1}{4}$

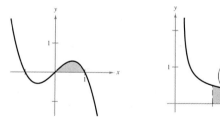

45. Average $= \frac{2}{5}$, $x = \frac{29}{4}$

47. Average $= 2$, $x = 2$ **49.** 0.254

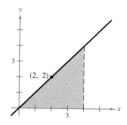

51. (a) $\frac{20,650}{M}$ (b) $\frac{43,150}{M}$

53. (a) $0.025 = 2.5\%$ (b) $0.736 = 73.6\%$

Chapter 6

Section 6.1

1. $A = 36$ **3.** $A = 9$ **5.** $A = \frac{3}{2}$

7. $A = \frac{32}{3}$ **9.** $A = \frac{9}{2}$

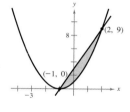

11. $A = 1$ **13.** $A = \frac{500}{27}$

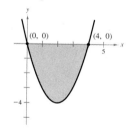

15. $A = 2$ **17.** $A = \frac{3}{2}$

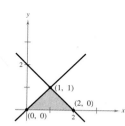

19. $A = \frac{64}{3}$

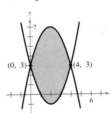

21. $A = \frac{9}{2}$

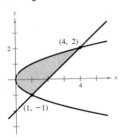

23. $A = 6$

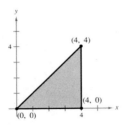

25. $A = 8$

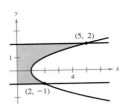

27. $A = \frac{1}{2}ac$

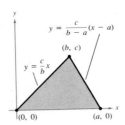

29. $b = 9\left(1 - \dfrac{1}{\sqrt[3]{4}}\right) \approx 3.330$

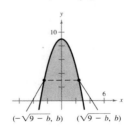

31. $x^4 - 2x^2 + 1 \le 1 - x^2$ on $[-1, 1]$

$$A = \int_{-1}^{1} [(1 - x^2) - (x^4 - 2x^2 + 1)]\, dx = \frac{4}{15}$$

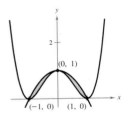

33. $A = \displaystyle\int_{-2}^{1} [x^3 - (3x - 2)]\, dx = \frac{27}{4}$

35. 1.760   **37.** $\frac{1}{6}$

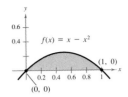

39. \$1.625 billion   **41.** Consumer surplus $= 1600$
Producer surplus $= 400$

43. Consumer surplus $= 50,000$ **45.** 4.773
Producer surplus $= 25,497$

Section 6.2

1. $\dfrac{\pi}{3}$ **3.** $\dfrac{16\pi}{3}$ **5.** $\dfrac{15\pi}{2}$ **7.** $\dfrac{2\pi}{35}$ **9.** 8π

11. $\dfrac{\pi}{4}$

13. **(a)** 8π **(b)** $\dfrac{128\pi}{5}$ **(c)** $\dfrac{256\pi}{15}$ **(d)** $\dfrac{192\pi}{5}$

15. **(a)** $\dfrac{32\pi}{3}$ **(b)** $\dfrac{64\pi}{3}$ **17.** $\dfrac{128\pi}{15}$ **19.** $\dfrac{3\pi}{4}$

21. 18π **23.** $\dfrac{363\pi}{64}$ **25.** $\dfrac{208\pi}{3}$

27. $\dfrac{384\pi}{5}$ **29.** $\dfrac{512\pi}{15}$ **31.** 60π **33.** 18π

37. $\pi r^2 h\left(1 - \dfrac{h}{H} + \dfrac{h^2}{3H^2}\right)$

41. $\dfrac{\pi}{30}$ **43.** 9.293

45. **(a)** $\dfrac{128}{3}$ **(b)** $\dfrac{32\sqrt{3}}{3}$ **(c)** $\dfrac{16\pi}{3}$ **(d)** $\dfrac{32}{3}$

47. **(a)** $\dfrac{1}{10}$ **(b)** $\dfrac{\pi}{80}$ **(c)** $\dfrac{\sqrt{3}}{40}$ **(d)** $\dfrac{3}{80}$ **(e)** $\dfrac{\pi}{20}$

49. $\dfrac{2r^3}{3}$ **51.** 122.92

Section 6.3

1. $\dfrac{16\pi}{3}$ **3.** $\dfrac{8\pi}{3}$ **5.** $\dfrac{128\pi}{5}$ **7.** 8π **9.** $\dfrac{16\pi}{3}$

11. 16π **13.** 64π **15.** $\dfrac{8\pi}{3}$ **17.** $\dfrac{192\pi}{5}$ **19.** $\dfrac{\pi}{2}$

21. (a) $\dfrac{128\pi}{7}$ (b) $\dfrac{64\pi}{5}$ (c) $\dfrac{96\pi}{5}$ (d) $\dfrac{320\pi}{7}$

23. (a) $\dfrac{\pi a^3}{15}$ (b) $\dfrac{\pi a^3}{15}$ (c) $\dfrac{4\pi a^3}{15}$

25. Diameter $= 2\sqrt{4 - 2\sqrt{3}} \approx 1.464$

29. 2.97 (Shell), 3.01 (Disc)

31. 122,313 ft^3 **33.** $4\pi^2$ **35.** 186.055

Section 6.4

1. 13 **3.** $\frac{2}{3}(\sqrt{8} - 1) \approx 1.219$ **5.** $\frac{33}{16}$

7. $\frac{779}{240}$ **9.** $s = \displaystyle\int_{-2}^{1} \sqrt{2 + 4x + 4x^2}\, dx$

11. $s = \displaystyle\int_{0}^{2} \sqrt{1 + 4x^2}\, dx$ **13.** $\displaystyle\int_{1}^{2} \dfrac{\sqrt{y^6 + 4}}{y^3}\, dy$

15. 2.1517 **17.** 8.6101 **19.** $\frac{2}{3}$

21. 3 arcsin $\frac{2}{3} \approx 2.1892$

23. $\dfrac{\pi}{9}(82\sqrt{82} - 1) \approx 258.85$ **25.** $\dfrac{47\pi}{16}$

27. $\dfrac{\pi}{27}(145\sqrt{145} - 10\sqrt{10}) \approx 199.48$

29. $\pi r \sqrt{r^2 + h^2}$

31. $6\pi(3 - \sqrt{5}) \approx 14.40$

33. Surface area $= \dfrac{\pi}{27}$ ft$^2 \approx 16.8$ in.^{2} **37.** 50.86

Amount of glass $= \dfrac{\pi}{27}\left(\dfrac{0.015}{12}\right)$

≈ 0.00015 ft^3

≈ 0.25 in.^{3}

Section 6.5

1. 1000 ft · lb **3.** 300 ft · lb

5. 30.625 in · lb $\approx$ 2.55 ft · lb

7. 360 in · lb = 30 ft · lb

9. 180 in · lb = 15 ft · lb

11. (a) 2496 ft · lb (b) 9984 ft · lb

13. 431,308.8π ft · lb **15.** 20,217.6π ft · lb

17. 2995.2π ft · lb **19.** 2457π ft · lb

21. (a) 761.905 mi · ton $\approx$ 8,046,000,000 ft · lb

(b) 1,454.545 mi · ton $\approx$ 15,360,000,000 ft · lb

23. (a) 2.93×10^4 mi · tons $\approx 3.10 \times 10^{11}$ ft · lb

(b) 3.38×10^4 mi · tons $\approx 3.57 \times 10^{11}$ ft · lb

25. $\dfrac{3k}{4}$ **27.** 337.5 ft · lb **29.** 300 ft · lb

31. 168.75 ft · lb **33.** 7987.5 ft · lb

35. 2000 ln $\frac{3}{2} \approx 810.93$ ft · lb **37.** 10,202.6 ft · lb

Section 6.6

1. 936 lb **3.** 748.8 lb **5.** 1123.2 lb

7. 748.8 lb **9.** 1064.96 lb **11.** 748.8 lb

13. 15,163.2 lb **15.** 2814 lb **17.** 3376.8 lb

19. 94.5 lb **23.** 960 lb

25. 9,984 lb (on the wall at the shallow end)

39,936 lb (on the wall at the deep end)

46,592 lb (on the side walls)

27. 6483 lb

Section 6.7

1. $\bar{x} = -\frac{6}{7}$ **3.** $\bar{x} = 12$ **5.** $\bar{x} = 17$

7. $(\bar{x}, \bar{y}) = \left(\dfrac{10}{9}, -\dfrac{1}{9}\right)$ **9.** $(\bar{x}, \bar{y}) = \left(-\dfrac{7}{8}, -\dfrac{7}{16}\right)$

11. $(\bar{x}, \bar{y}) = \left(\dfrac{4 + 3\pi}{4 + \pi}, 0\right)$ **13.** $(\bar{x}, \bar{y}) = \left(0, \dfrac{135}{34}\right)$

15. $(\bar{x}, \bar{y}) = \left(0, \dfrac{2 + 3\pi}{2 + \pi}\right)$

17. $M_x = 4\rho$, $M_y = \dfrac{64\rho}{5}$, $(\bar{x}, \bar{y}) = \left(\dfrac{12}{5}, \dfrac{3}{4}\right)$

19. $M_x = \dfrac{\rho}{12}$, $M_y = \dfrac{\rho}{15}$, $(\bar{x}, \bar{y}) = \left(\dfrac{2}{5}, \dfrac{1}{2}\right)$

21. $M_x = \dfrac{99\rho}{5}$, $M_y = \dfrac{27\rho}{4}$, $(\bar{x}, \bar{y}) = \left(\dfrac{3}{2}, \dfrac{22}{5}\right)$

23. $M_x = 0$, $M_y = \dfrac{256\rho}{15}$, $(\bar{x}\ \bar{y}) = \left(\dfrac{8}{5}, 0\right)$

25. $M_x = \dfrac{27\rho}{4}$, $M_y = -\dfrac{27\rho}{10}$, $(\bar{x}, \bar{y}) = \left(-\dfrac{3}{5}, \dfrac{3}{2}\right)$

27. $M_x = \dfrac{192\rho}{7}$, $M_y = 96\rho$, $(\bar{x}, \bar{y}) = \left(5, \dfrac{10}{7}\right)$

29. $(\bar{x}, \bar{y}) = \left(0, \dfrac{4}{3\pi}\right)$ **31.** $(\bar{x}, \bar{y}) = \left(\dfrac{1}{2}, \dfrac{2}{5}\right)$

33. $(\bar{x}, \bar{y}) = \left(\dfrac{12}{7}, \dfrac{26}{7}\right)$ **35.** $(\bar{x}, \bar{y}) = \left(\dfrac{b}{3}, \dfrac{c}{3}\right)$

37. $\bar{y} = 16.5$ ft

39. $V = 160\pi^2 \approx 1579.14$

41. $V = \dfrac{128\pi}{3}$ **43.** $(\bar{x}, \bar{y}) = \left(0, \dfrac{2r}{\pi}\right)$

Review Exercises for Chapter 6

1. $A = \dfrac{4}{5}$

3. $A = \dfrac{1}{4}$

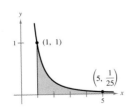

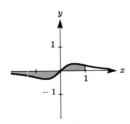

5. $A = \frac{4}{3}$

7. $A = \frac{1}{2}$

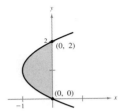

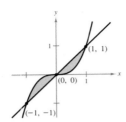

9. $A = \frac{512}{3}$

11. $A = \frac{14}{3}$

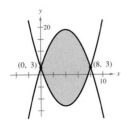

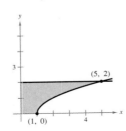

13. $A = \frac{1}{6}$

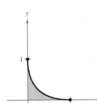

15. (a) $\dfrac{64\pi}{3}$

(b) $\dfrac{128\pi}{3}$

(c) $\dfrac{64\pi}{3}$

(d) $\dfrac{160\pi}{3}$

17. (a) 64π (b) 48π

19. $\dfrac{\pi}{2}$

21. (a) $\dfrac{512\pi}{15}$

(b) 64π

23. 50 in · lb ≈ 4.167 ft · lb

25. 104,000π ft · lb ≈ 163.4 ft · ton

27. 250 ft · lb

29. 72,800 lb (on side walls)

62,400 lb (on wall at deep end)

15,600 lb (on wall at shallow end)

31. 4992π lb   **33.** $(\bar{x}, \bar{y}) = \left(\dfrac{a}{5}, \dfrac{a}{5}\right)$

35. $(\bar{x}, \bar{y}) = \left(0, \dfrac{2a^2}{5}\right)$

37. $s = \displaystyle\int_0^{\sqrt{3}} \dfrac{4}{\sqrt{4 - x^2}}\, dx$ **39.** 15π

41. $\dfrac{4}{15}$ **43.** $\dfrac{32\pi}{105}$ **45.** $\dfrac{8}{15}(1 + 6\sqrt{3}) \approx 6.076$

Chapter 7

Section 7.1

1. (a) 125   (b) 9   (c) $\frac{1}{9}$ (d) $\frac{1}{3}$

3. (a) 5^5 (b) $\frac{1}{5}$ (c) $\frac{1}{5}$ (d) 2^2

5. (a) e^6 (b) e^{12} (c) $\dfrac{1}{e^6}$ (d) e^2

7. $x = 4$ **9.** $x = -2$ **11.** $x = 2$

13. $x = 16$ **15.** $x = -\frac{5}{2}$

17. (a) $\dfrac{271{,}801}{99{,}990} = 2.7182818281 \ldots ,$

$e = 2.7182818284 \ldots$

(b) $\dfrac{299}{110} = 2.718181818 \ldots ,$

$e = 2.7182818284 \ldots$

19.

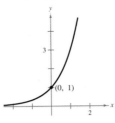

21.

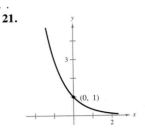

23.

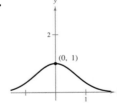

25.

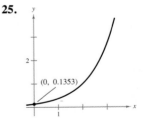

27.

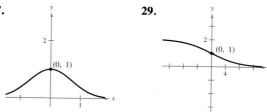

29.

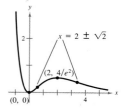

31. (a) $2593.74   **(b)** $2653.30   **(c)** $2707.04
(d) $2717.91   **(e)** $2718.28
33. (a) $88,692.04   **(b)** $30,119.42
(c) $9,071.80   **(d)** $247.88
35. (a) $849.53   **(b)** $421.12
37. (a) 0.154   **(b)** 0.487   **(c)** 0.811
39. (a) 850   **(b)**

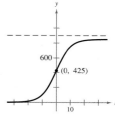

41. (a) 0.731   **(b)** 0.83

Section 7.2

1. 3   **3.** 1   **5.** -2

7. $2e^{2x}$ **9.** $2(x - 1)e^{-2x+x^2}$ **11.** $-\dfrac{e^{1/x}}{x^2}$

13. $\dfrac{e^{\sqrt{x}}}{2\sqrt{x}}$ **15.** $e^{3x}(3x + 4)$ **17.** $\dfrac{e^{x^2}(2x^2 - 1)}{x^2}$

19. $3(e^{-x} + e^x)^2(e^x - e^{-x})$ **21.** $\dfrac{-2(e^x - e^{-x})}{(e^x + e^{-x})^2}$

23. xe^x **25.** $\dfrac{10 - e^y}{xe^y + 3}$ **27.** $6(3e^{3x} + 2e^{-2x})$

29. $32(x + 1)e^{4x}$
31. Point of inflection: $(0, 1)$

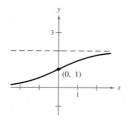

33. Relative minimum: $(0, 0)$
Relative maximum: $(2, 4e^{-2})$
Points of inflection: $(2 \pm \sqrt{2}, (6 \pm 4\sqrt{2})e^{-(2\pm\sqrt{2})})$

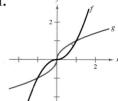

35. $y = x + 1$ **37.** $A = \sqrt{2}e^{-1/2}$
39. $V'(20) = 0.073$ million ft^3/yr
$V'(60) = 0.040$ million ft^3/yr
41. $P'(3) = 0.038$ $P'(10) = 0.017$
43. $\dfrac{e^2 - 1}{2e^2}$ **45.** $\dfrac{e}{3}(e^2 - 1)$ **47.** $\dfrac{1}{1 + e^{-x}} + C$

49. $\dfrac{1}{2a}e^{ax^2} + C$ **51.** $\dfrac{e}{3}(e^2 - 1)$

53. $-\dfrac{1}{3}(1 + e^{-x})^3 + C$ **55.** $-\dfrac{2}{3}(1 - e^x)^{3/2} + C$

57. $2\sqrt{e^x - e^{-x}} + C$ **59.** $-\dfrac{5}{2}e^{-2x} + e^{-x} + C$

61. 4   **63.** $f(x) = \dfrac{1}{2}(e^x + e^{-x})$

65. $e^5 - 1 \approx 147.41$ **67.** $1 - e^{-1} \approx 0.632$

69. $\dfrac{\pi}{2}(e^2 - 1)$ **73. (a)** 0.212   **(b)** 0.035

75. 0.3413   **77.** 19.0162

Section 7.3

1.

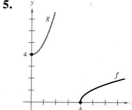

3.

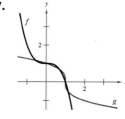

5.

7.

9. $f^{-1}(x) = \dfrac{x + 3}{2}$

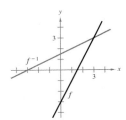

11. $f^{-1}(x) = x^{1/5}$

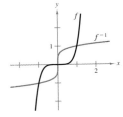

13. $f^{-1}(x) = x^2,\ x \geq 0$

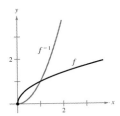

15. $f^{-1}(x) = \sqrt{4 - x^2},\ 0 \leq x \leq 2$

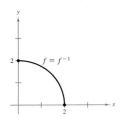

17. $f^{-1}(x) = x^3 + 1$

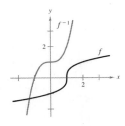

19. $f^{-1}(x) = x^{3/2},\ x \geq 0$

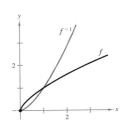

21. $f^{-1}(x) = \dfrac{\sqrt{7x}}{\sqrt{1 - x^2}},\ -1 < x < 1$

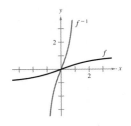

23.

x	1	2	3	4
$f^{-1}(x)$	0	1	2	4

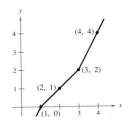

25. 32   **27.** 600   **29.** Inverse exists
31. Inverse does not exist   **33.** Inverse exists
35. Inverse exists   **37.** Inverse exists
39. $f'(x) = 2(x - 4) > 0$ on $(4, \infty)$

41. $f'(x) = -\dfrac{8}{x^3} < 0$ on $(0, \infty)$

43. $f'\left(\tfrac{1}{2}\right) = \tfrac{3}{4},\ (f^{-1})'\left(\tfrac{1}{8}\right) = \tfrac{4}{3}$
45. $f'(5) = \tfrac{1}{2},\ (f^{-1})'(1) = 2$
47. Not continuous at $x = 0$

49. $f^{-1}(x) = \dfrac{-8x}{1 + \sqrt{1 + 16x^2}}$

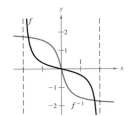

Section 7.4

1. (a) $\log_2 8 = 3$ **(b)** $\log_3 \tfrac{1}{3} = -1$
3. (a) $10^{-2} = 0.01$ **(b)** $\left(\tfrac{1}{2}\right)^{-3} = 8$
5. (a) $e^{0.6931...} = 2$ **(b)** $e^{2.128...} = 8.4$
7. (a) $x = 3$ **(b)** $x = -1$
9. (a) $x = \tfrac{1}{3}$ **(b)** $x = \tfrac{1}{16}$
11. (a) $x = \tfrac{1}{9}$ **(b)** $x = 3$
13. (a) $x = -1, 2$ **(b)** $x = \tfrac{1}{3}$
15.

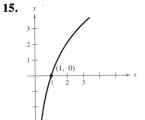

17.

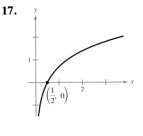

19.
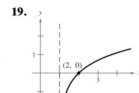

21.
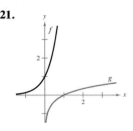

23.
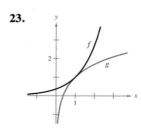

25. x^2 **27.** $5x + 2$ **29.** $\sqrt{x}$
31. (a) 1.7917 **(b)** -0.4055
(c) 4.3944 **(d)** 0.5493
33. $\ln 2 - \ln 3$ **35.** $\ln x + \ln y - \ln z$
37. $\frac{3}{2} \ln 2$ **39.** $3[\ln(x + 1) + \ln(x - 1) - 3 \ln x]$

41. $\ln z + 2 \ln (z - 1)$ **43.** $\ln \dfrac{x - 2}{x + 2}$

45. $\ln \sqrt[3]{\dfrac{x(x + 3)^2}{x^2 - 1}}$ **47.** $\ln \dfrac{9}{\sqrt{x^2 + 1}}$

49. $x = 4$ **51.** $x = 1$
53. $x = \ln 4 - 1 \approx 0.386$
55. $x = \dfrac{\ln 5 - \ln 6}{0.11} \approx -1.657$

57. $x = \dfrac{\ln 15}{\ln 25} \approx 0.841$ **59.** $t = \dfrac{\ln 2}{\ln 1.07} \approx 10.245$

61. (a) $t \approx 6.642$ yrs. **(b)** $t \approx 6.330$ yrs.
(c) $t \approx 6.302$ yrs. **(d)** $t \approx 6.301$ yrs.

63.

r	2%	4%	6%	8%	10%	12%
t (years)	54.93	27.47	18.31	13.73	10.99	9.16

65.

x	y	$\dfrac{\ln x}{\ln y}$	$\ln \dfrac{x}{y}$	$\ln x - \ln y$
1	2	0	-0.6931	-0.6931
3	4	0.7925	-0.2877	-0.2877
10	5	1.4307	0.6931	0.6931
4	0.5	-2.000	2.0794	2.0794

67.
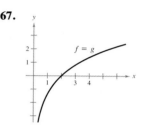

69. (a) 1.771 **(b)** 0.712
(c) -3.322 **(d)** -0.431

Section 7.5

1. 3 **3.** 2 **5.** $\dfrac{2}{x}$ **7.** $\dfrac{2(x^3 - 1)}{x(x^3 - 4)}$

9. $\dfrac{4(\ln x)^3}{x}$ **11.** $\dfrac{2x^2 - 1}{x(x^2 - 1)}$ **13.** $\dfrac{1 - x^2}{x(x^2 + 1)}$

15. $\dfrac{1 - 2 \ln x}{x^3}$ **17.** $\dfrac{2}{x \ln x^2} = \dfrac{1}{x \ln x}$

19. $\dfrac{1}{1 - x^2}$ **21.** $\dfrac{-4}{x(x^2 + 4)}$ **23.** $\dfrac{\sqrt{x^2 + 1}}{x^2}$

25. $(\ln 4)4^x$ **27.** $(\ln 5)5^{x-2}$

29. $x2^x(x \ln 2 + 2)$ **31.** $\dfrac{1}{x(\ln 3)}$

33. $\dfrac{x - 2}{(\ln 2)x(x - 1)}$ **35.** $\dfrac{x}{(\ln 5)(x^2 - 1)}$

37. $\dfrac{2x}{x^2 - 1}$ **39.** $\dfrac{2x^2 - 1}{\sqrt{x^2 - 1}}$

41. $\dfrac{3x^3 - 15x^2 + 8x}{2(x - 1)^3 \sqrt{3x - 2}}$

43. $\dfrac{(2x^2 + 2x - 1)\sqrt{x - 1}}{(x + 1)^{3/2}}$

45. $2(1 - \ln x)x^{(2/x)-2}$

47. $(x - 2)^{x+1}\left[\dfrac{x + 1}{x - 2} + \ln(x - 2)\right]$

49. $xy'' + y' = x\left(\dfrac{-2}{x^2}\right) + \dfrac{2}{x} = 0$ **51.** $\dfrac{2xy}{3 - 2y^2}$

53. $5x - y - 2 = 0$
55. Relative minimum: $\left(1, \frac{1}{2}\right)$

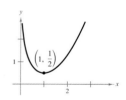

57. Relative minimum: $(e^{-1}, -e^{-1})$

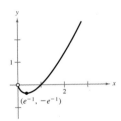

59. Relative minimum: (e, e)

Point of inflection: $\left(e^2, \dfrac{e^2}{2}\right)$

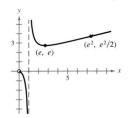

61. 0.567 **63.** $c = e - 1$
65. (a) 0 **(b)** $-\sqrt{3}$

Section 7.6

1. $\ln |x + 1| + C$ **3.** $-\frac{1}{2} \ln |3 - 2x| + C$

5. $\ln \sqrt{x^2 + 1} + C$ **7.** $\dfrac{x^2}{2} - 4 \ln |x| + C$

9. $\frac{1}{4}$ **11.** $\frac{7}{3}$ **13.** $-\ln 3$
15. $2\sqrt{x + 1} + C$
17. $\frac{1}{3} \ln |x^3 + 3x^2 + 9x| + C$
19. $3 \ln |1 + x^{1/3}| + C$
21. $2[\sqrt{x} - \ln(1 + \sqrt{x})] + C$
23. $x + 6\sqrt{x} + 18 \ln |\sqrt{x} - 3| + C$
25. $-\frac{2}{3} \ln |1 - x\sqrt{x}| + C$

27. $\ln |x - 1| + \dfrac{1}{2(x - 1)^2} + C$ **29.** $\dfrac{3^x}{\ln 3} + C$

31. $\dfrac{7}{\ln 4}$ **33.** $\dfrac{1}{2 \ln 5} 5^{x^2} + C$

35. $\frac{15}{2} + 8 \ln 2 \approx 13.045$ square units

37. $\pi \ln 4$ **39.** $\dfrac{\pi}{4}(32 \ln 4 - 3)$ **41.** $\dfrac{26}{\ln 3}$

43. $2000 \ln \frac{3}{2} \approx 810.93$ ft · lb
45. $P(t) = 1000[12 \ln |1 + 0.25t| + 1]$
$P(3) \approx 7715$

47. $\dfrac{10}{\ln 2} \ln \dfrac{4}{3} \approx 4.15$ min

Section 7.7

1. $y = \frac{1}{2}e^{0.4605t}$ **3.** $y = 0.6687e^{0.4024t}$

5. Time to double: 5.78 years
Amount after 10 years: \$3,320.12
Amount after 25 years: \$20,085.54
7. Annual rate: 8.94%
Amount after 10 years: \$1,833.67
Amount after 25 years: \$7,009.86
9. Annual rate: 9.50%
Time to double: 7.30 years
Amount after 25 years: \$5,375.51
11. (a) $N = 30(1 - e^{-0.0502t})$ **(b)** 36 days
13. (a) $S = 30e^{-1.7918/t}$ **(b)** 20,965 units
(c)

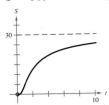

15. 900 **17.** 6015
19. Amount after 1,000 years: 6.52 grams
Amount after 10,000 years: 0.14 gram
21. Initial quantity: 6.70 grams
Amount after 1,000 years: 5.94 grams
23. Initial quantity: 2.16 grams
Amount after 10,000 years: 1.63 grams
25. 95.81% **27.** 15,683 years **29.** 22.35°
31. 11.75° **33.** 527.06 mm Hg

Section 7.8

1. 3 **3.** 0 **5.** 2
7. $n = 1:0$, $n = 2:\frac{1}{2}$, $n \geq 3:\infty$ **9.** 0
11. $\frac{3}{2}$ **13.** ∞ **15.** 0 **17.** $-\frac{3}{2}$ **19.** 1
21. 0 **23.** 1 **25.** 1 **27.** 0 **29.** 0 **31.** 0
33. Limit is not of the form $0/0$ or ∞/∞.

35.

x	10	10^2	10^4	10^6	10^8	10^{10}
$\dfrac{(\ln x)^4}{x}$	2.811	4.498	0.720	0.036	0.001	0.000

39. (a) 0 **(b)** ∞ **41.** $v = 32t + v_0$
43. Horizontal asymptote: $y = 1$
Relative maximum: $(e, e^{1/e})$

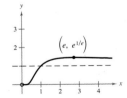

47. Horizontal asymptote: $y = 0$
Relative maximum: $(1, 2/e)$

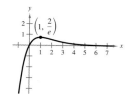

49. $\lim\limits_{k \to 0^+} \dfrac{x^k - 1}{k} = \ln x$

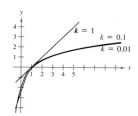

Review Exercises for Chapter 7

1. $f^{-1}(x) = 2x + 6$ **3.** $f^{-1}(x) = x^2 - 1,\ x \geq 0$

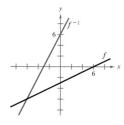

5. $f^{-1}(x) = \sqrt{x + 5}$

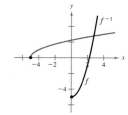

7. $f^{-1}(x) = \sqrt{\dfrac{x}{2}} + 4,\ x \geq 4$ **9.** $x = 3$

11. $x = 5 \pm \sqrt{7}$ **13.** $\dfrac{1}{2x}$

15. $\dfrac{1 + 2\ln x}{2\sqrt{\ln x}}$ **17.** $\dfrac{-y}{x(2y + \ln x)}$

19. $y(1 + \ln x)$ **21.** $\dfrac{x}{(a + bx)^2}$ **23.** $\dfrac{1}{x(a + bx)}$

25. $-2x$ **27.** $xe^x(x + 2)$ **29.** $\dfrac{e^{2x} - e^{-2x}}{\sqrt{e^{2x} + e^{-2x}}}$

31. $3^{x-1}\ln 3$ **33.** $\dfrac{y - ye^x - e^y}{e^x + xe^y - x}$ **35.** ax^{a-1}

37. $x^x(1 + \ln x)$ **39.** $\frac{1}{7}\ln |7x - 2| + C$

41. $\ln \big|\ln |3x|\big| + C$ **43.** $\dfrac{x^2}{2} + 3\ln |x| + C$

45. $\frac{1}{3}\ln |x^3 - 1| + C$ **47.** $2\sqrt{\ln x} + C$

49. $-\frac{1}{6}e^{-3x^2} + C$ **51.** $\dfrac{e^{4x} - 3e^{2x} - 3}{3e^x} + C$

53. $\ln |e^x - 1| + C$ **55.** $-e^{-x^2/2} + C$

57. $3 + \ln 4$ **59.** $\ln \frac{5}{3}$

61. $2[1 - e^{-4}] \approx 1.963$

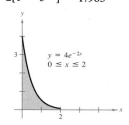

63. average $= \dfrac{2}{5}\ln \dfrac{3}{2} \approx 0.162$,

$x = 1 + \dfrac{5}{2\ln (3/2)} \approx 7.166$

65. $\frac{1}{2}(1 - e^{-16}) \approx 0.500$

67. (a) \$525.64   **(b)** \$824.36   **(c)** \$74,206.58

69. \$3,499.38

71. (a) 27.73 yrs   **(b)** 43.94 yrs

73. (a) -24.26% **(b)** -14.72%

75. (a) 0.3935   **(b)** 0.7769   **(c)** 0.2492
     (d) 0.9502

77. (a) 0.60   **(b)** 0.85

81. $v(t) = \dfrac{1}{k}[32 - (20k + 32)e^{kt}]$

83. $s(t) = \dfrac{32}{k}t - \dfrac{32}{k^2}(1 - e^{kt}) + s_0$ **85.** 0

87. ∞ **89.** 1   **91.** $1000e^{0.09} \approx 1094.17$

Chapter 8

Section 8.1

1. (a) $396°, -324°$ **(b)** $240°, -480°$

3. (a) $\dfrac{19\pi}{9}, -\dfrac{17\pi}{9}$ **(b)** $\dfrac{10\pi}{3}, -\dfrac{2\pi}{3}$

5. (a) $\dfrac{\pi}{6}$ **(b)** $\dfrac{5\pi}{6}$ **(c)** $\dfrac{7\pi}{4}$ **(d)** $\dfrac{2\pi}{3}$

7. (a) $270°$ **(b)** $210°$ **(c)** $-105°$ **(d)** $20°$

9.

r	8 ft	15 in.	85 cm	24 in.	$\dfrac{12{,}963}{\pi}$ mi
s	12 ft	24 in.	63.75π cm	96 in.	8642 mi
θ	1.5	1.6	$\dfrac{3\pi}{4}$	4	$\dfrac{2\pi}{3}$

11. (a) $\dfrac{5\pi}{12}$ (b) 7.8125π in.

13. (a) $\sin\theta = \frac{4}{5}$, $\csc\theta = \frac{5}{4}$
$\cos\theta = \frac{3}{5}$, $\sec\theta = \frac{5}{3}$
$\tan\theta = \frac{4}{3}$, $\cot\theta = \frac{3}{4}$
(b) $\sin\theta = -\frac{15}{17}$, $\csc\theta = -\frac{17}{15}$
$\cos\theta = \frac{8}{17}$, $\sec\theta = \frac{17}{8}$
$\tan\theta = -\frac{15}{8}$, $\cot\theta = -\frac{8}{15}$

15. Quadrant III **17.** 2 **19.** $\frac{4}{3}$ **21.** $\frac{17}{15}$

23. (a) $\sin 60° = \dfrac{\sqrt{3}}{2}$ (b) $\sin\dfrac{2\pi}{3} = \dfrac{\sqrt{3}}{2}$

$\cos 60° = \dfrac{1}{2}$ $\cos\dfrac{2\pi}{3} = -\dfrac{1}{2}$

$\tan 60° = \sqrt{3}$ $\tan\dfrac{2\pi}{3} = -\sqrt{3}$

(c) $\sin\dfrac{\pi}{4} = \dfrac{\sqrt{2}}{2}$ (d) $\sin\dfrac{5\pi}{4} = -\dfrac{\sqrt{2}}{2}$

$\cos\dfrac{\pi}{4} = \dfrac{\sqrt{2}}{2}$ $\cos\dfrac{5\pi}{4} = -\dfrac{\sqrt{2}}{2}$

$\tan\dfrac{\pi}{4} = 1$ $\tan\dfrac{5\pi}{4} = 1$

25. (a) $\sin 225° = -\dfrac{\sqrt{2}}{2}$ (b) $\sin(-225°) = \dfrac{\sqrt{2}}{2}$

$\cos 225° = -\dfrac{\sqrt{2}}{2}$ $\cos(-225°) = -\dfrac{\sqrt{2}}{2}$

$\tan 225° = 1$ $\tan(-225°) = -1$

(c) $\sin 300° = -\dfrac{\sqrt{3}}{2}$ (d) $\sin 330° = -\dfrac{1}{2}$

$\cos 300° = \dfrac{1}{2}$ $\cos 330° = \dfrac{\sqrt{3}}{2}$

$\tan 300° = -\sqrt{3}$ $\tan 330° = -\dfrac{\sqrt{3}}{3}$

27. (a) 0.1736 **29.** (a) 0.3640
(b) 5.759 (b) 0.3640

31. (a) $\theta = \dfrac{\pi}{4}, \dfrac{7\pi}{4}$ **33.** (a) $\theta = \dfrac{\pi}{4}, \dfrac{5\pi}{4}$

(b) $\theta = \dfrac{3\pi}{4}, \dfrac{5\pi}{4}$ (b) $\theta = \dfrac{5\pi}{6}, \dfrac{11\pi}{6}$

35. $\theta = \dfrac{\pi}{4}, \dfrac{3\pi}{4}, \dfrac{5\pi}{4}, \dfrac{7\pi}{4}$ **37.** $\theta = 0, \dfrac{\pi}{4}, \pi, \dfrac{5\pi}{4}$

39. $\theta = \dfrac{\pi}{3}, \dfrac{5\pi}{3}$ **41.** $\theta = 0, \dfrac{\pi}{2}, \pi$

43. $y = \dfrac{100\sqrt{3}}{3}$

45. $x = \dfrac{25\sqrt{3}}{3}$

47. $20\sin 75° \approx 19.32$ ft
49. $150\cot 4° \approx 2145.1$ ft

Section 8.2

1. Period: π Amplitude: 2
3. Period: 4 Amplitude: $\frac{5}{2}$
5. Period: 6π Amplitude: 2
7. Period: $\dfrac{\pi}{5}$ Amplitude: 2
9. Period $\frac{1}{2}$ Amplitude: 3
11. Period: $\dfrac{\pi}{2}$ **13.** Period: $\dfrac{2\pi}{5}$

15. **17.**

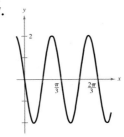

19. **21.**

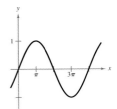

23. **25.**

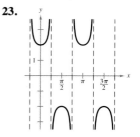
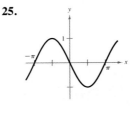

27.

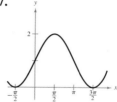

29.

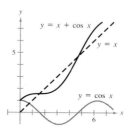

31.

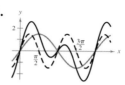

33. $\frac{1}{5}$ **35.** 0   **37.** 0   **39.** 1   **41.** 1
43. $\frac{2}{3}$ **45.** 0   **47.** Limit does not exist.
49. ∞ **51.** Continuous for all real x
53. Nonremovable discontinuities at integer multiples
of $\pi/2$
55. Continuous for all real x
57. (a)   **(b)**

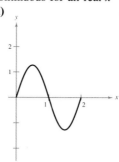

59. $\lim_{x \to 0} f(x) = \frac{5}{2}$

61.

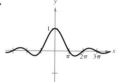

x	± 0.1	± 0.01	± 0.001
$\dfrac{\sin x}{x}$	0.9983	1.0000	1.0000

63.

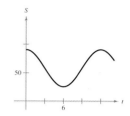

January, November, December

Section 8.3

1. $2x + \frac{1}{2} \sin x$
3. $-\dfrac{1}{x^2} - 3 \cos x$ **5.** $\dfrac{2}{\sqrt{x}} - 3 \sin x$
7. $t(t \cos t + 2 \sin t)$ **9.** $-\dfrac{t \sin t + \cos t}{t^2}$
11. $-1 + \sec^2 x = \tan^2 x$
13. $-5x \csc x \cot x + 5 \csc x$
$= 5 \csc x(1 - x \cot x)$
15. $\csc \theta \cot \theta - \cos \theta = \cos \theta \cot^2 \theta$
17. $t^2 \cos t + 2t \sin t - 2t \sin t + 2 \cos t - 2 \cos t$
$= t^2 \cos t$
19. $\pi \cos 2\pi x$ **21.** $\dfrac{-2 \csc x \cot x}{(1 - \csc x)^2}$
23. $-3 \sin 3x$ **25.** $12 \sec^2 4x$ **27.** $\pi \cos \pi x$
29. $\frac{1}{4} \sin 2x$ **31.** $\frac{1}{2} \sin 4x$ **33.** $\frac{1}{2} \cot x \sqrt{\sin x}$
35. $6 \sec^3 2x \tan 2x$ **37.** $\csc x$ **39.** $2e^x \cos x$
41. $\sec^2 x e^{\tan x}$ **43.** $\sec x \csc x$
45. $\dfrac{\cos x}{2 \sin 2y}$, undefined   **47.** $-\dfrac{x^2}{x^2 + 1}$, 0
49. $\dfrac{\cot y}{x}$, $\dfrac{1}{2\sqrt{3}}$ **53. (a)** 1   **(b)** 2
55. $4x - 2y = 2 - \pi$ **57.** $\frac{2}{3}$ **59.** -2
61. $\frac{1}{2}$ **63.** 1

65.

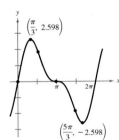

$\left(\frac{\pi}{3}, 2.598\right)$

$\left(\frac{5\pi}{3}, -2.598\right)$

67.

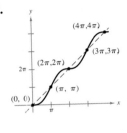

$(4\pi, 4\pi)$

$(3\pi, 3\pi)$

$(2\pi, 2\pi)$

(π, π)

$(0, 0)$

69. $-2.03, 0, 2.03$

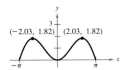

$(-2.03, 1.82)$ $(2.03, 1.82)$

71. **(a)** $y = 1/4$ in. $v = 4$ in./sec
(c) Period: $\pi/6$ Frequency: $6/\pi$

73. $\theta = \arctan k$, $F = \dfrac{kW}{\sqrt{k^2 + 1}}$

75. $\theta = \dfrac{2\pi}{3}(3 - \sqrt{6}) \approx 66$

77. **(a)** $\frac{1}{2}$ rad/min **(b)** $\frac{3}{2}$ rad/min
(c) 1.87 rad/min

79. **(a)** 0 ft/sec **(b)** 10π ft/sec **(c)** $10\sqrt{3}\pi$ ft/sec

81. $rg \sec^2 \theta \dfrac{d\theta}{dt} = 2v\dfrac{dv}{dt}$

83. **(a)**

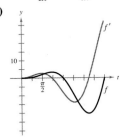

f'

f

(b) Critical numbers: $x = 2.2889, 5.0870$
(c) $f' > 0$ on $(0, 2.2889)$, $(5.0870, 2\pi)$
$f' < 0$ on $(2.2889, 5.0870)$

85. **(a)**

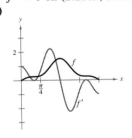

f

f'

(b) Critical numbers: $x = 0.5236, 1.5708, 2.6180$
(c) $f' > 0$ on $(0, 0.5236)$, $(0.5236, 1.5708)$
$f' < 0$ on $(1.5708, 2.6180)$, $(2.6180, \pi)$

Section 8.4

1. $-2 \cos x + 3 \sin x + C$ **3.** $t + \csc t + C$
5. $\tan \theta + \cos \theta + C$ **7.** $-\frac{1}{2} \cos 2x + C$
9. $\dfrac{1}{2} \sin x^2 + C$ **11.** $2 \tan\left(\dfrac{x}{2}\right) + C$
13. $\frac{1}{2} \tan^2 x + C$ or $\frac{1}{2} \sec^2 x + C_1$
15. $-\cot x - x + C$ **17.** $\frac{1}{5} \tan^5 x + C$
19. $\dfrac{1}{\pi} \ln |\sin \pi x| + C$
21. $-\frac{1}{2} \ln |\csc 2x + \cot 2x| + C$
23. $\ln |\tan x| + C$ **25.** $\ln |\sec x - 1| + C$
27. $\ln |1 + \sin t| + C$ **29.** $\ln |\theta - \sin \theta| + C$
31. $\sin e^x + C$ **33.** $\ln |\cos e^{-x}| + C$
35. $x - \frac{1}{4} \cos 4x + C$
37. $\dfrac{3\sqrt{3}}{4}$ **39.** $2(\sqrt{3} - 1)$ **41.** $\dfrac{1}{2}$
43. $-1 + \sec 1$ **45.** 2 **47.** 4
49. $2\left(\dfrac{2\pi}{3} - \ln (2 + \sqrt{3})\right) \approx 1.5549$
51. 2π **53.** π **55.** 3.829
57. **(a)** 102.352 thousand units
(b) 102.352 thousand units
(c) 74.5 thousand units
59. **(a)** 1.273 amps **(b)** 1.382 amps **(c)** 0 amps
61. $\frac{1}{2} \sin^2 x + C_1 = -\frac{1}{2} \cos^2 x + C_2$
67. **(a)** 0.957 **(b)** 0.978
69. **(a)** 0.334 **(b)** 0.305
71. **(a)** 0.194 **(b)** 0.186
73. 3.4624

Section 8.5

1. $\dfrac{\pi}{6}$ **3.** $\dfrac{\pi}{3}$ **5.** $\dfrac{\pi}{6}$ **7.** $\dfrac{\pi}{4}$
9. $\arccos\left(\frac{1}{1.269}\right) \approx 0.663$
11. **(a)** $\frac{1}{2}$ **(b)** $\cos 2\theta = 1 - 2 \sin^2 \theta = \frac{1}{2}$

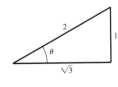

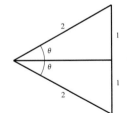

13. (a) $\frac{3}{5}$ **(b)** $\frac{5}{3}$

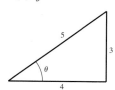

15. (a) $\cot\left[\arcsin\left(-\frac{1}{2}\right)\right] = -\cot\left[\arcsin\left(\frac{1}{2}\right)\right] = -\sqrt{3}$

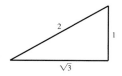

(b) $-\frac{13}{5}$

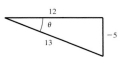

17. x **19.** $\sqrt{1-4x^2}$ **21.** $\dfrac{\sqrt{x^2-1}}{x}$

23. $\dfrac{\sqrt{x^2-9}}{3}$ **25.** $\dfrac{\sqrt{x^2+2}}{x}$

27. $\arcsin\left(\dfrac{9}{\sqrt{x^2+81}}\right)$

31.

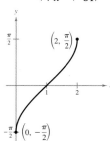

33.

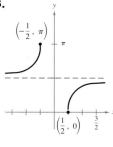

35. $x = \frac{1}{3}\left[\sin\left(\frac{1}{2}\right) + \pi\right] \approx 1.207$ **37.** $x = \frac{1}{3}$

39. $\dfrac{2}{\sqrt{1-4x^2}}$ **41.** $\dfrac{2}{\sqrt{2x-x^2}}$ **43.** $-\dfrac{3}{\sqrt{4-x^2}}$

45. $\dfrac{5}{1+25x^2}$ **47.** $\dfrac{1}{|x|\sqrt{x^2-1}}$ **49.** 0

51. $-\dfrac{t}{\sqrt{1-t^2}}$ **53.** $\dfrac{1}{2+3t^2}$ **55.** $\dfrac{1}{1-x^4}$

57. $\arcsin x$ **59.** $(0, 0)$

61. Relative maximum: $(1.272, -0.606)$
Relative minimum: $(-1.272, 3.747)$

63. $(0.7862, 0.6662)$ **65.** $\dfrac{\sqrt{3}}{40}$ rad/sec

Section 8.6

1. $\dfrac{\pi}{18}$ **3.** $\dfrac{\pi}{6}$ **5.** $\text{arcsec } 2x + C$

7. $\frac{1}{2}x^2 - \frac{1}{2}\ln(x^2+1) + C$ **9.** $\arcsin(x+1) + C$

11. $\frac{1}{2}\arctan t^2 + C$ **13.** $\frac{1}{2}(\arctan x)^2 + C$

15. $\dfrac{\pi^2}{32} \approx 0.308$ **17.** $\dfrac{\sqrt{3}-2}{2} \approx -0.134$

19. $\arcsin e^x + C$ **21.** $\dfrac{1}{3}\arctan\left(\dfrac{x-3}{3}\right) + C$

23. $2\arctan\sqrt{x} + C$ **25.** $\dfrac{\pi}{4}$ **27.** $\dfrac{\pi}{2}$

29. $\ln|x^2+6x+13| - 3\arctan\left(\dfrac{x+3}{2}\right) + C$

31. $\arcsin\left(\dfrac{x+2}{2}\right) + C$

33. $-\sqrt{-x^2-4x} + C$ **35.** $4 - 2\sqrt{3} + \dfrac{\pi}{6} \approx 1.059$

37. $\frac{1}{2}\arctan(x^2+1) + C$
39. $\frac{1}{4}\arcsin(4x-2) + C$
41. $2(\sqrt{x-1} - \arctan\sqrt{x-1}) + C$

43. $2\sqrt{e^t-3} - 2\sqrt{3}\arctan\left(\dfrac{\sqrt{e^t-3}}{\sqrt{3}}\right) + C$

45. $\dfrac{\pi}{4}$ **47.** $\dfrac{\pi}{8}$

49. $\sqrt{\dfrac{32}{k}}\tan\left[\arctan\left(500\sqrt{\dfrac{k}{32}}\right) - \sqrt{32k}\,t\right]$

51. (b) 3.14159 **53. (a)** $\arcsin x + C$
(b) $-\sqrt{1-x^2} + C$
(c) Does not fit Basic Formulas

55. (a) $\frac{2}{3}(x-1)^{3/2} + C$
(b) $\frac{2}{15}(x-1)^{3/2}(3x+2) + C$
(c) $\frac{2}{3}\sqrt{x-1}(x+2) + C$

Section 8.7

1. (a) 10.018 **3. (a)** $\frac{4}{3}$ **5. (a)** 1.317
(b) -0.964 **(b)** $\frac{13}{12}$ **(b)** 0.962

7. $\left(\dfrac{e^x-e^{-x}}{e^x+e^{-x}}\right)^2 + \left(\dfrac{2}{e^x+e^{-x}}\right)^2 = 1$

9. $\left(\dfrac{e^x-e^{-x}}{2}\right)\left(\dfrac{e^y+e^{-y}}{2}\right) + \left(\dfrac{e^x+e^{-x}}{2}\right)\left(\dfrac{e^y-e^{-y}}{2}\right)$

$= \dfrac{e^{(x+y)} - e^{-(x+y)}}{2} = \sinh(x+y)$

11. $\left(\dfrac{e^x-e^{-x}}{2}\right)\left[3 + 4\left(\dfrac{e^x-e^{-x}}{2}\right)^2\right]$

$= \dfrac{e^{3x} - e^{-3x}}{2} = \sinh(3x)$

13. $-2x\cosh(1-x^2)$ **15.** $\coth x$
17. $\text{csch } x$ **19.** $\sinh^2 x$ **21.** $\text{sech } x$

23. $\frac{y}{x}[\cosh x + x(\sinh x) \ln x]$ **25.** $-2e^{-2x}$

27. $\frac{3}{\sqrt{9x^2 - 1}}$ **29.** $|\sec x|$ **31.** $2 \sec 2x$

33. $2 \sinh^{-1} (2x)$ **35.** $-\frac{1}{2} \cosh (1 - 2x) + C$

37. $\frac{1}{3} \cosh^3 (x - 1) + C$ **39.** $\ln |\sinh x| + C$

41. $-\coth \frac{x^2}{2} + C$ **43.** $\operatorname{csch} \frac{1}{x} + C$

45. $\frac{1}{5} \ln 3$ **47.** $\frac{\pi}{4}$ **49.** $\frac{1}{2} \arctan x^2 + C$

51. $-\operatorname{csch}^{-1}(e^x) + C = -\ln \left(\frac{1 + \sqrt{1 + e^{2x}}}{e^x} \right) + C$

53. $2 \sinh^{-1} \sqrt{x} + C = 2 \ln (\sqrt{x} + \sqrt{1 + x}) + C$

55. $-\ln \left[\frac{1 + \sqrt{(x - 1)^2 + 1}}{|x - 1|} \right] + C$

57. $\frac{1}{2\sqrt{6}} \ln \left| \frac{\sqrt{2}(x + 1) + \sqrt{3}}{\sqrt{2}(x + 1) - \sqrt{3}} \right| + C$

59. $\frac{1}{4} \arcsin \left(\frac{4x - 1}{9} \right) + C$

61. $-\frac{x^2}{2} - 4x - \frac{10}{3} \ln \left| \frac{x - 5}{x + 1} \right| + C$

63. Point of inflection: $(0, 0)$

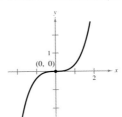

(0, 0)

65. $y''' - y' = a \cosh x - a \cosh x = 0$

67. $-\frac{\sqrt{a^2 - x^2}}{x}$ **69.** $\frac{52}{31}$ kg   **71.** $2a \sinh \frac{b}{a}$

23. $-\frac{e^y + 2x \sin x^2}{xe^y}$

25. $2 \csc^2 x \cot x$

27. $\frac{-x^2 \cos x + 2x \sin x + 2 \cos x}{x^3}$

29. $\frac{8x}{(1 - 4x^2)^{3/2}}$ **31.** $\arctan (\sin x) + C$

33. $\frac{1}{4}(\arctan 2x)^2 + C$ **35.** $\frac{\tan^{n+1} x}{n + 1} + C$

37. $-\frac{1}{2} \ln (1 + e^{-2x}) + C$

39. $\frac{1}{2} \arctan (e^{2x}) + C$ **41.** $-\frac{1}{2} \ln |\cos x^2| + C$

43. $\frac{1}{2} \arcsin x^2 + C$ **45.** $\frac{1}{2} \ln (16 + x^2) + C$

47. $\frac{1}{4} \left[\arctan \frac{x}{2} \right]^2 + C$

49. $4 \arcsin \frac{x}{2} + \sqrt{4 - x^2} + C$

53. $2\pi x + 4y = \pi^2$ **55.** 1.122

57. $3(3^{2/3} + 2^{2/3})^{3/2}$ ft ≈ 21.07 ft

59. $\frac{32}{L} \sqrt{A^2 + B^2}$ **61.** $x = \frac{3\pi}{4}, \frac{7\pi}{4}$

63. (a) $C \approx \$9.17$
    (b) $C \approx \$3.14$ Savings $\approx \$6.03$

65. $\frac{\pi^2}{4}$ **67.** $2x \sinh (x^2 + 1)$ **69.** $\frac{2}{4 - x^2}$

71. $-\cosh \frac{1}{x} + C$ **73.** $\frac{1}{2} \sinh^{-1} (2x) + C$

77. $\lim_{x \to 3^+} f(x) = -\infty$

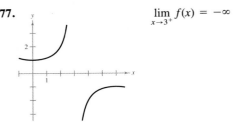

Review Exercises for Chapter 8

1. $\frac{x \cos x - 2 \sin x}{x^3}$ **3.** $-x \sec^2 x - \tan x$

5. $\cos 4x + 1$ **7.** $\frac{1}{2}(1 - \cos 2x) = \sin^2 x$

9. $-\frac{3}{\sqrt{x}} \csc^3 \sqrt{x} \cot \sqrt{x}$ **11.** $-\frac{\sec^2 \sqrt{1 - x}}{2\sqrt{1 - x}}$

13. $(1 - x^2)^{-3/2}$ **15.** $\frac{x}{|x|\sqrt{x^2 - 1}} + \operatorname{arcsec} x$

17. $(\arcsin x)^2$ **19.** $\frac{1}{2} + x \arctan x$

21. $\frac{1}{\cos y} = \frac{1}{\sqrt{1 - (x - 2)^2}}$

Chapter 9

Section 9.1

1. $\frac{1}{15}(3x - 2)^5 + C$ **3.** $-\frac{1}{5}(-2x + 5)^{5/2} + C$

5. $\frac{1}{2}v^2 - \frac{1}{6(3v - 1)^2} + C$

7. $-\frac{1}{3} \ln |-t^3 + 9t + 1| + C$

9. $\frac{1}{2}x^2 + x + \ln |x - 1| + C$

11. $\frac{1}{3} \ln \left| \frac{3x - 1}{3x + 1} \right| + C$ **13.** $-\frac{1}{2} \cos t^2 + C$

15. $e^{\sin x} + C$ **17.** $-e^{-t} + 2t + e^t + C$

19. $\frac{1}{3}\sec 3x + C$ **21.** $2\ln(1 + e^x) + C$

23. $-\cot x - \csc x + C$

25. $\ln(t^2 + 4) - \frac{1}{2}\arctan\frac{t}{2} + C$

27. $3\arctan t + C$ **29.** $\frac{1}{2}\operatorname{arcsec}\frac{|x|}{2} + C$

31. $-\frac{1}{2}\arcsin(2t - 1) + C$ **33.** $-\frac{1}{\pi}\csc(\pi x) + C$

35. $\frac{1}{2}\arcsin t^2 + C$ **37.** $\frac{1}{2}\arctan\frac{\tan x}{2} + C$

39. $\frac{1}{2}\ln\left|\cos\frac{2}{t}\right| + C$

41. $\frac{x}{15}(12x^4 + 20x^2 + 15) + C$

43. $\frac{1}{2}x^2 + 3x + 3\ln|x| - \frac{1}{x} + C$

45. $\frac{1}{2}e^{2x} + 2e^x + x + C$ **47.** $3\arcsin\frac{x - 3}{3} + C$

49. $\frac{1}{4}\arctan\frac{2x + 1}{8} + C$ **51.** $\frac{1}{2}(1 - e^{-1}) \approx 0.316$

53. $\frac{1}{2}$ **55.** 4   **57.** $\frac{1}{2}\ln 2$

59. $\frac{1}{2}\left(\operatorname{arcsec} 4 - \frac{\pi}{3}\right) \approx 0.135$ **61.** $\frac{4}{3}$

63. $a = \frac{1}{2}$ **65.** $b = \sqrt{\ln\left(\frac{3\pi}{3\pi - 4}\right)} \approx 0.743$

67. $\frac{2}{\arcsin(4/5)} \approx 2.157$ **69.** 1.0320

29. $x\arcsin 2x + \frac{1}{2}\sqrt{1 - 4x^2} + C$

31. $x\arctan x - \frac{1}{2}\ln(1 + x^2) + C$

33. $\frac{1}{5}e^{2x}(2\sin x - \cos x) + C$

35. $-\frac{\pi}{2}$ **37.** $\frac{e[\sin(1) - \cos(1)] + 1}{2} \approx 0.909$

39. $\frac{\pi}{2} - 1$ **41.** $\frac{e^{2x}}{4}(2x^2 - 2x + 1) + C$

43. $(3x^2 - 6)\sin x - (x^3 - 6x)\cos x + C$

45. $x\tan x + \ln|\cos x| + C$

47. $\frac{2}{5}(2x - 3)^{3/2}(x + 1) + C$

49. $\frac{1}{3}\sqrt{4 + x^2}(x^2 - 8) + C$

57. $\frac{x^4}{16}(4\ln x - 1) + C$

59. $\frac{e^{2x}}{13}(2\cos 3x + 3\sin 3x) + C$

61. $1 - \frac{5}{e^4} \approx 0.908$ **63.** $\frac{\pi}{1 + \pi^2}\left(\frac{1}{e} + 1\right) \approx 0.395$

65. (a) 1   (b) $\pi(e - 2) \approx 2.257$

     (c) $\frac{(e^2 + 1)\pi}{2} \approx 13.177$

     (d) $\left(\frac{e^2 + 1}{4}, \frac{e - 2}{2}\right) \approx (2.097, 0.359)$

67. (a) $3.2(\ln 2) - 0.2 \approx 2.018$

     (b) $12.8(\ln 4) - 7.2(\ln 3) - 1.8 \approx 8.035$

69. \$771,721.44   **71.** $\frac{8h}{(n\pi)^2}\sin\left(\frac{n\pi}{2}\right)$

73. 125.5 mi

75. For any integrable function,   **77.** 5.739
$\int f(x)\,dx = C + \int f(x)\,dx$,
but this does not imply that $C = 0$.

Section 9.2

1. $\frac{e^{2x}}{4}(2x - 1) + C$ **3.** $\frac{1}{2}e^{x^2} + C$

5. $-\frac{1}{4e^{2x}}(2x + 1) + C$

7. $e^x(x^3 - 3x^2 + 6x - 6) + C$

9. $\frac{x^4}{16}(4\ln x - 1) + C$

11. $\frac{1}{4}[2(t^2 - 1)\ln|t + 1| - t^2 + 2t] + C$

13. $x(\ln x)^2 - 2x\ln x + 2x + C$

15. $\frac{(\ln x)^3}{3} + C$ **17.** $\frac{e^{2x}}{4(2x + 1)} + C$

19. $(x - 1)^2 e^x + C$ **21.** $\frac{2(x - 1)^{3/2}}{15}(3x + 2) + C$

23. $\frac{2}{3}x^2(2 + 3x)^{1/2} - \frac{8}{27}x(2 + 3x)^{3/2} + \frac{16}{405}(2 + 3x)^{5/2}$
     $= \frac{2}{405}(27x^2 - 24x + 32)\sqrt{2 + 3x} + C$

25. $x\sin x + \cos x + C$

27. $x\tan x + \ln|\cos x| + C$

Section 9.3

1. $-\frac{1}{4}\cos^4 x + C$ **3.** $\frac{1}{12}\sin^6 2x + C$

5. $-\frac{1}{3}\cos^3 x + \frac{2}{5}\cos^5 x - \frac{1}{7}\cos^7 x + C$

7. $\frac{1}{12}(6x + \sin 6x) + C$

9. $\frac{1}{32\pi}(12\pi x - 8\sin 2\pi x + \sin 4\pi x) + C$

11. $\frac{1}{32}(4x - \sin 4x) + C$

13. $\frac{1}{8}(2x^2 - 2x\sin 2x - \cos 2x) + C$

15. $\frac{1}{3}\ln|\sec 3x + \tan 3x| + C$

17. $\frac{1}{15}\tan 5x(3 + \tan^2 5x) + C$

19. $\frac{1}{2\pi}(\sec\pi x\tan\pi x + \ln|\sec\pi x + \tan\pi x|) + C$

21. $-\frac{\tan^2(1 - x)}{2} - \ln|\cos(1 - x)| + C$

23. $\tan^4\left(\frac{x}{4}\right) - 2\tan^2\left(\frac{x}{4}\right) - 4\ln\left|\cos\frac{x}{4}\right| + C$

25. $\frac{1}{2}\tan^2 x + C$ **27.** $\frac{\tan^3 x}{3} + C$

29. $\frac{1}{5\pi}\sec^5 \pi x + C$ **31.** $\frac{\sec^6 4x}{24} + C$

33. $\frac{1}{3}\sec^3 x + C$ **35.** $\frac{1}{9}\sec^3 3x - \frac{1}{3}\sec 3x + C$

37. $\frac{1}{4}[\ln|\csc^2(2x)| - \cot^2(2x)] + C$

39. $-\cot\theta - \frac{1}{3}\cot^3\theta + C$

41. $\ln|\csc t - \cot t| + \cos t + C$

43. $-\frac{1}{10}(\cos 5x + 5\cos x) + C$

45. $\frac{1}{8}(2\sin 2\theta - \sin 4\theta) + C$

47. $\ln|\csc x - \cot x| + \cos x + C$

49. $t - 2\tan t + C$ **51.** π **53.** $\frac{1}{2}(1 - \ln 2)$

55. $\ln 2$ **57.** $\frac{3\sqrt{2}}{10}$ **59.** $\frac{4}{3}$

61. $\frac{\tan^6 3x}{18} + \frac{\tan^4 3x}{12} + C_1 = \frac{\sec^6 3x}{18} - \frac{\sec^4 3x}{12} + C_2$

63. $\frac{1}{2}$ **65.** $2\pi\left(1 - \frac{\pi}{4}\right) \approx 1.348$

67. (a) $\frac{\pi^2}{2}$ **(b)** $(\bar{x}, \bar{y}) = \left(\frac{\pi}{2}, \frac{\pi}{8}\right)$

79. $-\frac{1}{15}\cos x(3\sin^4 x + 4\sin^2 x + 8) + C$

81. $\frac{1}{6}\cos^5 x\sin x + \frac{5}{24}\cos^3 x\sin x$
$\quad\quad\quad\quad + \frac{5}{16}\cos x\sin x + \frac{5}{16}x + C$

83. $\frac{1}{8\pi}(-2\cos^3 \pi x\sin \pi x + \cos \pi x\sin \pi x + \pi x) + C$

87. 8.586

Section 9.4

1. $\frac{x}{25\sqrt{25 - x^2}} + C$

3. $5\ln\left|\frac{5 - \sqrt{25 - x^2}}{x}\right| + \sqrt{25 - x^2} + C$

5. $\ln|x + \sqrt{x^2 - 4}| + C$

7. $\frac{1}{15}(x^2 - 4)^{3/2}(3x^2 + 8) + C$ **9.** $\frac{1}{3}(1 + x^2)^{3/2} + C$

11. $\frac{1}{2}\left(\arctan x + \frac{x}{1 + x^2}\right) + C$ **13.** $\sqrt{x^2 + 9} + C$

15. 2π **17.** $\ln|x + \sqrt{x^2 - 9}| + C$

19. $\sqrt{3} - \frac{\pi}{3} \approx 0.685$ **21.** $-\frac{(1 - x^2)^{3/2}}{3x^3} + C$

23. $-\frac{1}{3}\ln\left|\frac{\sqrt{4x^2 + 9} + 3}{2x}\right| + C$

25. $-\frac{1}{\sqrt{x^2 + 3}} + C$ **27.** $\frac{1}{3}(1 + e^{2x})^{3/2} + C$

29. $\frac{1}{3}(x^2 + 2x + 2)^{3/2} + C$

31. $\sqrt{x^2 + 4x + 8}$
$\quad\quad - 2\ln|\sqrt{x^2 + 4x + 8} + (x + 2)| + C$

33. $\frac{1}{2}(x - 15)\sqrt{x^2 + 10x + 9}$
$\quad\quad + 33\ln|\sqrt{x^2 + 10x + 9} + (x + 5)| + C$

35. $\frac{1}{2}(\arcsin e^x + e^x\sqrt{1 - e^{2x}}) + C$

37. $\frac{1}{4}\left(\frac{x}{x^2 + 2} + \frac{1}{\sqrt{2}}\arctan\frac{x}{\sqrt{2}}\right) + C$

39. $\frac{1}{6}(3x\sqrt{9x^2 + 4} + 4\ln|3x + \sqrt{9x^2 + 4}|) + C$

41. $\frac{1}{2}(x\sqrt{x^2 - 1} + \ln|x + \sqrt{x^2 - 1}|) + C$

43. $x\,\text{arcsec}\,2x - \frac{1}{2}\ln|2x + \sqrt{4x^2 - 1}| + C$

45. $9(2 - \sqrt{2})$ **47.** $1 - \frac{\sqrt{3}\pi}{6} \approx 0.093$

49. 187.2π lb   **51.** πr^2 **53.** $6\pi^2$

55. $\ln\left[\frac{5(\sqrt{2} + 1)}{\sqrt{26} + 1}\right] + \sqrt{26} - \sqrt{2} \approx 4.367$

57. $100\sqrt{2} + 50\ln\left(\frac{\sqrt{2} + 1}{\sqrt{2} - 1}\right) \approx 229.559$

59. $\frac{\pi}{32}[102\sqrt{2} - \ln(3 + 2\sqrt{2})] \approx 13.989$

61. 121.3 lb   **63.** $(\bar{x}, \bar{y}) \approx (5.30, 1.43)$

Section 9.5

1. $\frac{1}{2}\ln\left|\frac{x - 1}{x + 1}\right| + C$ **3.** $\ln\left|\frac{x - 1}{x + 2}\right| + C$

5. $\frac{3}{2}\ln|2x - 1| - 2\ln|x + 1| + C$

7. $5\ln|x - 2| - \ln|x + 2| - 3\ln|x| + C$

9. $x^2 + \frac{3}{2}\ln|x - 4| - \frac{1}{2}\ln|x + 2| + C$

11. $\frac{1}{x} + \ln|x^4 + x^3| + C$

13. $\frac{1}{2}x^2 + 3x + 6\ln|x - 1| - \frac{4}{x - 1} - \frac{1}{2(x - 1)^2} + C$

15. $3\ln|x - 3| - \frac{9}{x - 3} + C$

17. $\ln\left|\frac{x^2 + 1}{x}\right| + C$

19. $\frac{1}{6}\left[\ln\left|\frac{x - 2}{x + 2}\right| + \sqrt{2}\arctan\left(\frac{x}{\sqrt{2}}\right)\right] + C$

21. $\frac{1}{16}\ln\left|\frac{4x^2 - 1}{4x^2 + 1}\right| + C$

23. $\frac{\sqrt{2}}{2}\arctan\frac{x}{\sqrt{2}} - \frac{1}{2(x^2 + 2)} + C$

25. $\ln|x + 1| + \sqrt{2}\arctan\left(\frac{x - 1}{\sqrt{2}}\right) + C$

27. $4\ln|x - 2| + \ln(x^2 + 4) + \frac{1}{2}\arctan\frac{x}{2} + C$

29. $\ln|x - 2| + \frac{1}{2}\ln|x^2 + x + 1|$
$\quad\quad\quad - \sqrt{3}\arctan\left(\frac{2x + 1}{\sqrt{3}}\right) + C$

31. $\ln 2$ **33.** $\frac{1}{2}\ln\left(\frac{8}{5}\right) - \frac{\pi}{4} + \arctan(2) \approx 0.557$

35. $\ln 2 + \arctan 2 - \arctan 3$

37. $\ln\left|\frac{\cos x}{\cos x - 1}\right| + C$ **39.** $\ln\left|\frac{-1 + \sin x}{2 + \sin x}\right| + C$

41. $\dfrac{1}{5}\ln\left|\dfrac{e^x-1}{e^x+4}\right|+C$ **47.** $6-\dfrac{7}{4}\ln 7\approx 2.595$

49. $2\pi\left[\arctan 3-\dfrac{3}{10}\right]\approx 5.963$

51. $x=\dfrac{n[e^{(n+1)kt}-1]}{n+e^{(n+1)kt}}$

53. (a) As $t\to\infty$, $x(t)\to y_0$ **55.** 14.666

  (b) $y_0>z_0$: As $t\to\infty$, $x(t)\to z_0$

    $y_0=z_0$: As $t\to\infty$, $x(t)\to y_0=z_0$

Section 9.6

1. $-\frac{1}{2}x(2-x)+\ln|1+x|+C$

3. $-\dfrac{\sqrt{1-x^2}}{x}+C$ **5.** $\frac{1}{9}x^3(-1+3\ln x)+C$

7. $\dfrac{\sqrt{x^2-4}}{4x}+C$ **9.** $\frac{1}{2}e^{x^2}+C$

11. $\dfrac{2}{9}\left(\ln|1-3x|+\dfrac{1}{1-3x}\right)+C$

13. $e^x\arccos(e^x)-\sqrt{1-e^{2x}}+C$

15. $\frac{1}{16}x^4(4\ln|x|-1)+C$

17. $\dfrac{1}{27}\left(3x-\dfrac{25}{3x-5}+10\ln|3x-5|\right)+C$

19. $\frac{1}{2}(x^2+\cot x^2+\csc x^2)+C$

21. $\arctan(\sin x)+C$ **23.** $x-\frac{1}{2}\ln(1+e^{2x})+C$

25. $\dfrac{\sqrt{2}}{2}\arctan\left(\dfrac{1+\sin\theta}{\sqrt{2}}\right)+C$

27. $-\dfrac{\sqrt{2+9x^2}}{2x}+C$

29. $\frac{1}{2}(e^x\sqrt{e^{2x}+1}+\ln|e^x+\sqrt{e^{2x}+1}|)+C$

31. $\frac{1}{16}(6x-3\sin 2x\cos 2x-2\sin^3 2x\cos 2x)+C$

33. $-2(\cot\sqrt{x}+\csc\sqrt{x})+C$

35. $(t^4-12t^2+24)\sin t+(4t^3-24t)\cos t+C$

37. $\frac{1}{2}[(x^2+1)\text{ arcsec}(x^2+1)$

$\qquad -\ln|(x^2+1)+\sqrt{x^4+2x^2}|]+C$

39. $\frac{1}{4}(2\ln|x|-3\ln|3+2\ln|x||)+C$

41. $\sqrt{2-2x-x^2}$

$\qquad -\sqrt{3}\ln\left|\dfrac{\sqrt{3}+\sqrt{2-2x-x^2}}{x+1}\right|+C$

43. $\frac{1}{2}\arctan(x^2-3)+C$

45. $\frac{1}{2}\ln|(x^2-3)+\sqrt{x^4-6x^2+5}|+C$

47. $-\frac{1}{3}\sqrt{4-x^2}(x^2+8)+C$ **49.** $-\dfrac{2\sqrt{1-x}}{\sqrt{x}}+C$

51. $\dfrac{2}{1+e^x}-\dfrac{1}{2(1+e^x)^2}+\ln|1+e^x|+C$

59. $\dfrac{1}{\sqrt{5}}\ln\left|\dfrac{2\tan(\theta/2)-3-\sqrt{5}}{2\tan(\theta/2)-3+\sqrt{5}}\right|+C$

61. $\ln 2$ **63.** $\frac{1}{2}\ln(3-2\cos\theta)+C$

65. $2\sin\sqrt{\theta}+C$ **67.** $-\csc\theta+C$

69. 1919.145 ft · lb

Section 9.7

1. 4   **3.** 6   **5.** 1   **7.** Diverges   **9.** 2

11. 1   **13.** $\dfrac{1}{2}$ **15.** π **17.** $\dfrac{\pi}{4}$

19. Diverges   **21.** Diverges   **23.** 6

25. $-\frac{1}{4}$ **27.** Diverges   **29.** $\ln(2+\sqrt{3})$

31. 0   **33.** $p>1$ **37.** Diverges

39. Converges   **41.** Converges   **43.** Diverges

45. Converges   **47.** (a) 1   (b) $\dfrac{\pi}{3}$ (c) Diverges

49. 6   **51.** (a) $\Gamma(1)=1$

    (b) $\Gamma(2)=1$

    (c) $\Gamma(3)=2$

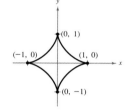

53. (b) 51.05%   **55.** (a) \$743,997.58

  (c) 7   (b) \$795,584.54

    (c) \$858,333.33

57. $k\dfrac{\sqrt{a^2+1}-1}{a^2\sqrt{a^2+1}}$

Review Exercises for Chapter 9

1. $x+\frac{9}{8}\ln|x-3|-\frac{25}{8}\ln|x+5|+C$

3. $\tan\theta+\sec\theta+C$

5. $\dfrac{e^{2x}}{13}(2\sin 3x-3\cos 3x)+C$

7. $-\dfrac{1}{x}(1+\ln 2x)+C$

9. $\dfrac{1}{2}\left(4\arcsin\dfrac{x}{2}+x\sqrt{4-x^2}\right)+C$

11. $\dfrac{3\sqrt{4-x^2}}{x}+C$

13. $\dfrac{2}{3}\left[\tan^3\left(\dfrac{x}{2}\right)+3\tan\left(\dfrac{x}{2}\right)\right]+C$

15. $\dfrac{3}{2}\ln\left|\dfrac{x-3}{x+3}\right|+C$

17. $\frac{1}{4}[6\ln|x-1|-\ln(x^2+1)+6\arctan x]+C$

19. $\dfrac{3}{2}\ln(x^2+1)-\dfrac{1}{2(x^2+1)}+C$

21. $16\arcsin\dfrac{x}{4}+C$ **23.** $\dfrac{1}{2}\arctan\dfrac{e^x}{2}+C$

25. $\frac{1}{2} \ln |x^2 + 4x + 8| - \arctan \left(\dfrac{x+2}{2}\right) + C$

27. $\frac{1}{8}(\sin 2\theta - 2\theta \cos 2\theta) + C$

29. $\frac{1}{2}(2\theta - \cos 2\theta) + C$

31. $\frac{4}{3}[x^{3/4} - 3x^{1/4} + 3 \arctan (x^{1/4})] + C$

33. $2\sqrt{1 - \cos x} + C$

35. $x \ln |x^2 + x| - 2x + \ln |x + 1| + C$

37. $\sin x \ln (\sin x) - \sin x + C$

39. $x - \dfrac{1}{2(x^2 + 1)} + C$　**41.** $-\dfrac{\sqrt{4 + x^2}}{4x} + C$

43. $\frac{1}{3}\sqrt{4 + x^2}(x^2 - 8) + C$

45. (a) $e - 1 \approx 1.72$　**(b)** 1
(c) $\frac{1}{2}(e - 1) \approx 0.86$　**(d)** 1.46 (Simpson's Rule)

47. 3.82　**49.** $(\bar{x}, \bar{y}) = \left(0, \dfrac{4}{3\pi}\right)$

51. $0.015846 < \displaystyle\int_2^\infty \dfrac{1}{x^5 - 1}\, dx < 0.015851$

Month	6	7	8
Amount	$9,530.06	$9,621.39	$9,713.59

Month	9	10
Amount	$9,806.68	$9,900.66

63. (a) $\$2,500,000,000(0.8)^n$
(b)

Year	1	2
Budget	$2,000,000,000	$1,600,000,000

Year	3	4
Budget	$1,280,000,000	$1,024,000,000

(c) Converges to 0

65. $S_6 = 240$, $S_7 = 440$, $S_8 = 810$,
$S_9 = 1490$, $S_{10} = 2740$

67. $1, 1.4142, 1.4422, 1.4142, 1.3797, 1.3480$
Converges to 1

71. (a) $1, 1, 2, 3, 5, 8, 13, 21, 34, 55, 89, 144$
(b) $1, 2, 1.5, 1.6667, 1.6, 1.6250, 1.6154, 1.6190,$
$1.6176, 1.6182$
(d) $\rho = \dfrac{1 + \sqrt{5}}{2} \approx 1.6180$

Chapter 10

Section 10.1

1. $2, 4, 8, 16, 32$　**3.** $-\frac{1}{2}, \frac{1}{4}, -\frac{1}{8}, \frac{1}{16}, -\frac{1}{32}$

5. $3, \frac{9}{2}, \frac{27}{6}, \frac{81}{24}, \frac{243}{120}$　**7.** $-1, -\frac{1}{4}, \frac{1}{9}, \frac{1}{16}, -\frac{1}{25}$

9. $3n - 2$　**11.** $n^2 - 2$　**13.** $\dfrac{n+1}{n+2}$

15. $\dfrac{(-1)^{n-1}}{2^{n-2}}$　**17.** $\dfrac{n+1}{n}$　**19.** $\dfrac{n}{(n+1)(n+2)}$

21. $\dfrac{2^n n!}{(2n)!}$　**23.** $(-1)^{n(n-1)/2}$　**25.** Converges to 1

27. Diverges　**29.** Converges to $\frac{3}{2}$　**31.** Diverges

33. Converges to 0　**35.** Diverges

37. Converges to 0　**39.** Diverges

41. Converges to 3　**43.** Converges to e^k

45. Converges to 0　**47.** Monotonic

49. Not monotonic　**51.** Not monotonic

53. Monotonic　**55.** Not monotonic　**57.** 5

59. $\frac{1}{3}$

61. (a) No
(b)

Month	1	2	3
Amount	$9,086.25	$9,173.33	$9,261.24

Month	4	5
Amount	$9,349.99	$9,439.60

Section 10.2

1. $1, 1.25, 1.361, 1.424, 1.464$

3. $3, -1.5, 5.25, -4.875, 10.3125$

5. $3, 4.5, 5.25, 5.625, 5.8125$

7. $\lim\limits_{n\to\infty} a_n = 1 \neq 0$　**9.** $\lim\limits_{n\to\infty} a_n = 3 \neq 0$

11. $\lim\limits_{n\to\infty} a_n = 1 \neq 0$　**13.** Geometric series: $r = \frac{3}{2} > 1$

15. Geometric series: $r = 1.055 > 1$

17. $\lim\limits_{n\to\infty} a_n = \frac{1}{2} \neq 0$　**19.** Geometric series: $r = \frac{3}{4} < 1$

21. Geometric series: $r = 0.9 < 1$

23. Telescoping series: $a_n = \dfrac{1}{n} - \dfrac{1}{n+1}$, converges to 1

25. 2　**27.** $\frac{2}{3}$　**29.** $\frac{10}{9}$　**31.** $\frac{9}{4}$　**33.** $5\left(\frac{5}{8}\right)^3 = \frac{625}{512}$

35. $\frac{3}{4}$　**37.** 3　**39.** $\frac{1}{2}$　**41.** $\displaystyle\sum_{n=0}^{\infty} \frac{3}{5}(0.1)^n = \frac{2}{3}$

43. $\displaystyle\sum_{n=0}^{\infty} \frac{3}{40}(0.01)^n = \frac{5}{66}$　**45.** Diverges

47. Converges　**49.** Diverges　**51.** Diverges

53. Diverges   **55.** Diverges   **57.** $80,000(1 - 0.9^n)$
59. 152.42 ft   **61.** $\frac{1}{3}$ **63.** $11,616.95
65. $235,821.51

71. $\sum\limits_{n=0}^{\infty} 1,\ \sum\limits_{n=0}^{\infty} (-1)$ (Answer is not unique.)

Section 10.3

1. Diverges   **3.** Converges   **5.** Converges
7. Diverges   **9.** Diverges   **11.** Converges
13. Diverges   **15.** Diverges   **17.** Converges
19. Converges   **21.** Diverges   **23.** Converges
25. Converges   **27.** Diverges   **29.** Diverges
31. Converges   **33.** $p > 1$ **35.** 31
37. $R_6 \approx 0.0015,\ S_6 \approx 1.0811$
39. $R_{10} \approx 0.0997,\ S_{10} \approx 0.9818$
41. $R_4 \approx 5.6 \times 10^{-8},\ S_4 \approx 0.4049$
43. $N \geq 7$ **45.** $N \geq 2$

Section 10.4

1. Converges   **3.** Diverges   **5.** Converges
7. Diverges   **9.** Converges   **11.** Converges
13. Diverges   **15.** Diverges   **17.** Converges
19. Diverges   **21.** Converges   **23.** Diverges
25. Diverges   **27.** Diverges; p-Series Test
29. Converges; Comparison Test with $\sum\limits_{n=1}^{\infty} \left(\frac{1}{3}\right)^n$
31. Diverges; nth-Term Test
33. Converges; Integral Test   **37.** Diverges
39. Converges

Section 10.5

1. Converges   **3.** Converges   **5.** Diverges
7. Converges   **9.** Diverges   **11.** Converges
13. Diverges   **15.** Converges   **17.** Converges
19. Converges   **21.** Converges
23. Converges absolutely
25. Converges conditionally
27. Converges conditionally
29. Converges absolutely   **31.** Converges absolutely
33. Converges conditionally
35. Converges absolutely
37. Converges conditionally   **39.** 0.947
41. 0.368   **43.** 0.842   **45.** 0.406
47. At least 499   **51.** **(a)** $-1 < x < 1$
                                 (b) $-1 \leq x < 1$

Section 10.6

1. Diverges   **3.** Converges   **5.** Converges
7. Diverges   **9.** Converges   **11.** Diverges
13. Converges   **15.** Converges   **17.** Diverges
19. Converges   **21.** Converges   **23.** Converges
25. Diverges   **27.** Converges   **29.** Converges
31. Converges; Alternating Series Test
33. Converges; p-Series Test
35. Diverges; nth-Term Test
37. Diverges; Ratio Test
39. Converges; Limit Comparison Test with $b_n = \dfrac{1}{2^n}$
41. Converges; Alternating Series Test
43. Converges; Comparison Test with $b_n = \dfrac{1}{2^n}$
45. Converges; Ratio Test
47. Converges; Ratio Test   **49.** Converges; Ratio Test

Section 10.7

1. $1 - x + \frac{1}{2}x^2 - \frac{1}{6}x^3$
3. $1 + 2x + 2x^2 + \frac{4}{3}x^3 + \frac{2}{3}x^4$
5. $x - \frac{1}{6}x^3 + \frac{1}{120}x^5$.   **7.** $x + x^2 + \frac{1}{2}x^3 + \frac{1}{6}x^4$
9. $1 - x + x^2 - x^3 + x^4$ **11.** $x + \frac{1}{3}x^3$
13. $2 - 3x^3 + x^4$
15. $1 - (x - 1) + (x - 1)^2 - (x - 1)^3 + (x - 1)^4$
17. $(x - 1) - \frac{1}{2}(x - 1)^2 + \frac{1}{3}(x - 1)^3 - \frac{1}{4}(x - 1)^4$

19.

x	$\sin x$	$P_1(x)$	$P_3(x)$	$P_5(x)$
0	0	0	0	0
0.25	0.2474	0.25	0.2474	0.2474
0.50	0.4794	0.50	0.4792	0.4794
0.75	0.6816	0.75	0.6797	0.6817
1.00	0.8415	1.00	0.8333	0.8417

21. 0.6042   **23.** -6.7954
25. For e^x: $P_4(x) = 1 + x + \frac{1}{2}x^2 + \frac{1}{6}x^3 + \frac{1}{24}x^4$
      For xe^x: $Q_4(x) = x + x^2 + \frac{1}{2}x^3 + \frac{1}{6}x^4$
                $Q_4(x) = xP_4(x) - \frac{1}{24}x^5$
27. $-0.3936 < x < 0$ **29.** 0.0038
31. **(a)** $P_3(x) = x + \frac{1}{6}x^3$

(b)

x	-1	-0.75	-0.50	-0.25
$f(x)$	-1.571	-0.848	-0.524	-0.253
$P_3(x)$	-1.167	-0.820	-0.521	-0.253

x	0	0.25	0.50	0.75	1
$f(x)$	0	0.253	0.524	0.848	1.571
$P_3(x)$	0	0.253	0.521	0.820	1.167

(c)

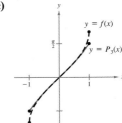

33.

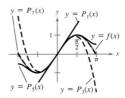

Section 10.8

1. $R = 1$ **3.** $R = \frac{1}{2}$ **5.** $R = \infty$
7. $(-2, 2)$ **9.** $(-1, 1]$ **11.** $(-\infty, \infty)$
13. $x = 0$ **15.** $(-4, 4)$ **17.** $(0, 10]$
19. $(0, 2]$ **21.** $(0, 2c)$ **23.** $\left(-\frac{1}{2}, \frac{1}{2}\right)$
25. $(-\infty, \infty)$ **27.** $(-1, 1)$ **29.** $x = 3$
31. (a) $(-2, 2)$ **33. (a)** $(0, 2]$
    (b) $(-2, 2)$ **(b)** $(0, 2)$
    (c) $(-2, 2)$ **(c)** $(0, 2)$
    (d) $[-2, 2)$ **(d)** $[0, 2]$
35. (a) For $f(x)$: $(-\infty, \infty)$ **37. (a)** $(-\infty, \infty)$
    For $g(x)$: $(-\infty, \infty)$ **(d)** $f(x) = e^x$
    (d) $f(x) = \sin x$
        $g(x) = \cos x$

Section 10.9

1. $\sum_{n=0}^{\infty} \frac{x^n}{2^{n+1}}$, $(-2, 2)$ **3.** $\sum_{n=0}^{\infty} \frac{(x-5)^n}{(-3)^{n+1}}$, $(2, 8)$

5. $-3 \sum_{n=0}^{\infty} (2x)^n$, $\left(-\frac{1}{2}, \frac{1}{2}\right)$

7. $-\frac{1}{11} \sum_{n=0}^{\infty} \left[\frac{2}{11}(x + 3)\right]^n$, $\left(-\frac{17}{2}, \frac{5}{2}\right)$

9. $\frac{3}{2} \sum_{n=0}^{\infty} \left(\frac{x}{-2}\right)^n$, $(-2, 2)$

11. $\sum_{n=0}^{\infty} \left[\frac{1}{(-2)^n} - 1\right] x^n$, $(-1, 1)$

13. $2 \sum_{n=0}^{\infty} x^{2n}$, $(-1, 1)$

15. $\sum_{n=1}^{\infty} n(-1)^n x^{n-1}$, $(-1, 1)$

17. $\sum_{n=0}^{\infty} \frac{(-1)^n x^{n+1}}{n + 1}$, $(-1, 1]$

19. $\sum_{n=0}^{\infty} (-1)^n (2x)^{2n}$, $\left(-\frac{1}{2}, \frac{1}{2}\right)$

21. $\sum_{n=0}^{\infty} \frac{x^{2n+1}}{2n + 1}$, $(-1, 1)$

23.

x	$x - \dfrac{x^2}{2}$	$\ln(x + 1)$	$x - \dfrac{x^2}{2} + \dfrac{x^3}{3}$
0.0	0.000	0.000	0.000
0.2	0.180	0.182	0.183
0.4	0.320	0.336	0.341
0.6	0.420	0.470	0.492
0.8	0.480	0.588	0.651
1.0	0.500	0.693	0.833

25. 0.245 **27.** 0.125 **29.** 3.14

31.

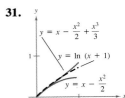

33. $\sum_{n=0}^{\infty} (-1)^n x^n$

Section 10.10

1. $\sum_{n=0}^{\infty} \frac{(2x)^n}{n!}$ **3.** $\frac{\sqrt{2}}{2} \sum_{n=0}^{\infty} \frac{(-1)^{n(n+1)/2}}{n!} \left(x - \frac{\pi}{4}\right)^n$

5. $\sum_{n=0}^{\infty} \frac{(-1)^n (x - 1)^{n+1}}{n + 1}$ **7.** $\sum_{n=0}^{\infty} \frac{(-1)^n (2x)^{2n+1}}{(2n + 1)!}$

9. $1 + \frac{x^2}{2!} + \frac{5x^4}{4!} + \cdots$

11. $\displaystyle\sum_{n=0}^{\infty} (-1)^n(n+1)x^n$

13. $\displaystyle\frac{1}{2}\left[1 + \sum_{n=1}^{\infty} \frac{(-1)^n 1 \cdot 3 \cdot 5 \cdots (2n-1)x^{2n}}{2^{3n}n!}\right]$

15. $\displaystyle 1 + \frac{x^2}{2} + \sum_{n=2}^{\infty} \frac{(-1)^{n+1} 1 \cdot 3 \cdot 5 \cdots (2n-3)x^{2n}}{2^n n!}$

17. $\displaystyle 1 + \frac{x^2}{2} + \frac{x^4}{2^2 2!} + \frac{x^6}{2^3 3!} + \cdots$

19. $\displaystyle\sum_{n=0}^{\infty} \frac{(-1)^n(2x)^{2n+1}}{(2n+1)!}$ **21.** $\displaystyle\sum_{n=0}^{\infty} \frac{(-1)^n x^n}{(2n)!}$

23. $\displaystyle\sum_{n=0}^{\infty} \frac{(-1)^n x^{2n}}{(2n+1)!}$ **25.** $\displaystyle\sum_{n=0}^{\infty} \frac{x^{2n+1}}{(2n+1)!}$

27. $\displaystyle\sum_{n=0}^{\infty} \frac{(-1)^n x^{2n+1}}{(2n+1)!}$

29. $\displaystyle\frac{1}{2}\left[1 + \sum_{n=0}^{\infty} \frac{(-1)^n(2x)^{2n}}{(2n)!}\right]$ **31.** 1.3708

33. 0.7040   **35.** 0.4872   **37.** 0.2010

39. $\displaystyle\sum_{n=0}^{\infty} \frac{(-1)^{n+1}x^{2n+3}}{(2n+3)(n+1)!}$ **41.** 0.3413

47. (a)

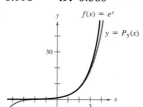

(c) $\displaystyle\sum_{n=0}^{\infty} 0\, x^n = 0 \neq f(x)$

Review Exercises for Chapter 10

1. $a_n = \dfrac{1}{n!}$ **3.** Converges to 0   **5.** Diverges

7. Converges to 0   **9.** Converges to 0

11. 1, 2.5, 4.75, 8.125, 13.1875

13. 0.5, 0.45833, 0.45972, 0.45970, 0.45970

15. 3   **17.** $\frac{1}{2}$ **19.** $\frac{1}{11}$ **21.** Diverges

23. Converges   **25.** Diverges   **27.** Diverges

29. Converges   **31.** Diverges   **33.** [1, 3]

35. Converges only at $x = 2$

37. $\displaystyle\frac{\sqrt{2}}{2}\sum_{n=0}^{\infty} \frac{(-1)^{n(n+1)/2}}{n!}\left(x - \frac{3\pi}{4}\right)^n$

39. $\displaystyle\sum_{n=0}^{\infty} \frac{(x \ln 3)^n}{n!}$ **41.** $\displaystyle -\sum_{n=0}^{\infty} (x+1)^n$

43. $\displaystyle\sum_{n=0}^{\infty} \frac{(-1)^n x^{2n+1}}{(2n+1)(2n+1)!}$ **45.** $\displaystyle\sum_{n=0}^{\infty} \frac{(-1)^n x^{n+1}}{(n+1)^2}$

47. 0.996   **49.** 0.560

53.

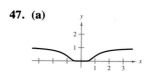

Chapter 11

Section 11.1

1. e   **2.** f   **3.** a   **4.** c   **5.** d   **6.** b

7. Vertex: (0, 0)
Focus: $\left(0, \frac{1}{16}\right)$
Directrix: $y = -\frac{1}{16}$

9. Vertex: (0, 0)
Focus: $\left(-\frac{3}{2}, 0\right)$
Directrix: $x = \frac{3}{2}$

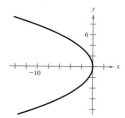

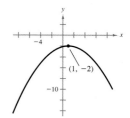

11. Vertex: (0, 0)
Focus: (0, −2)
Directrix: $y = 2$

13. Vertex: (1, −2)
Focus: (1, −4)
Directrix: $y = 0$

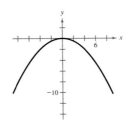

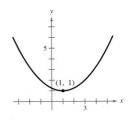

15. Vertex: $\left(5, -\frac{1}{2}\right)$
Focus: $\left(\frac{11}{2}, -\frac{1}{2}\right)$
Directrix: $x = \frac{9}{2}$

17. Vertex: (1, 1)
Focus: (1, 2)
Directrix: $y = 0$

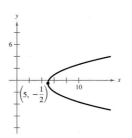

19. Vertex: $(8, -1)$
Focus: $(9, -1)$
Directrix: $x = 7$

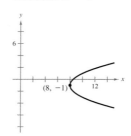

21. Vertex: $(-2, -3)$
Focus: $(-4, -3)$
Directrix: $x = 0$

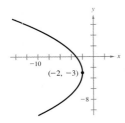

23. Vertex: $(-1, 2)$
Focus: $(0, 2)$
Directrix: $x = -2$

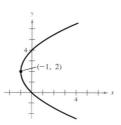

25. Vertex: $(-2, 2)$
Focus: $(-2, 1)$
Directrix: $y = 3$

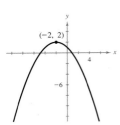

27. $x^2 + 6y = 0$ **29.** $y^2 - 4y + 8x - 20 = 0$
31. $x^2 + 24y + 96 = 0$ **33.** $x^2 = -16y$
35. $5x^2 - 14x - 3y + 9 = 0$
37. $x^2 + y - 4 = 0$ **39.** $(x - h)^2 = -8(y + 1)$
41. $3x - 2y^2 = 0$ **43.** $4x - y - 8 = 0$
45. $\left(\frac{3}{5}, 0\right)$ **47.** $(0, 0)$
49. $2[\sqrt{2} + \ln(\sqrt{2} + 1)]$
51. $2\sqrt{5} + \ln(\sqrt{5} + 2)$
53. $100\left[\sqrt{5} + 4\ln\left(\frac{1 + \sqrt{5}}{2}\right)\right] \approx 416.1$ ft
55. $10\sqrt{3}$ ft from end of pipe
57. **(a)** 122.7 ft^2
(b) 382.9 ft^2
59. $(1.02439, 0)$ **61.** $y = 2ax_0x - ax_0^2$

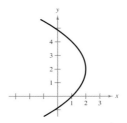

63. Tangent lines: $x + y = 0$, $2x - y - 9 = 0$
Point of intersection: $(3, -3)$
65. Tangent lines: $2x + y - 1 = 0$, $2x - 4y - 1 = 0$
Point of intersection (on the directrix): $\left(\frac{1}{2}, 0\right)$

Section 11.2

1. e **2.** a **3.** c **4.** b **5.** f **6.** d
7. Center: $(0, 0)$
Foci: $(\pm 3, 0)$
Vertices: $(\pm 5, 0)$
$e = \frac{3}{5}$

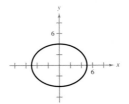

9. Center: $(0, 0)$
Foci: $(0, \pm 3)$
Vertices: $(0, \pm 5)$
$e = \frac{3}{5}$

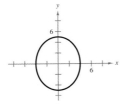

11. Center: $(0, 0)$
Foci: $(\pm 2, 0)$
Vertices: $(\pm 3, 0)$
$e = \frac{2}{3}$

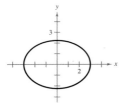

13. Center: $(0, 0)$
Foci: $(\pm\sqrt{3}, 0)$
Vertices: $(\pm 2, 0)$
$e = \dfrac{\sqrt{3}}{2}$

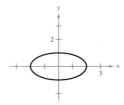

15. Center: $(0, 0)$
Foci: $(0, \pm 1)$
Vertices: $(0, \pm\sqrt{3})$
$e = \dfrac{\sqrt{3}}{3}$

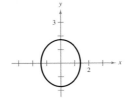

17. Center: $(0, 0)$

Foci: $\left(0, \pm\dfrac{\sqrt{3}}{2}\right)$

Vertices: $(0, \pm1)$

$e = \dfrac{\sqrt{3}}{2}$

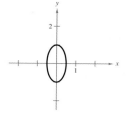

19. Center: $(1, 5)$

Foci: $(1, 9), (1, 1)$

Vertices: $(1, 10), (1, 0)$

$e = \frac{4}{5}$

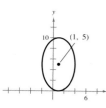

21. Center: $(-2, 3)$

Foci: $(-2, 3 \pm \sqrt{5})$

Vertices: $(-2, 6), (-2, 0)$

$e = \dfrac{\sqrt{5}}{3}$

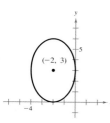

23. Center: $(1, -1)$

Foci: $\left(1 \pm \dfrac{3\sqrt{10}}{20}, -1\right)$

Vertices: $\left(1 \pm \dfrac{\sqrt{10}}{4}, -1\right)$

$e = \frac{3}{5}$

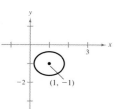

25. Center: $\left(\dfrac{1}{2}, -1\right)$

Foci: $\left(\dfrac{1}{2} \pm \sqrt{2}, -1\right)$

Vertices: $\left(\dfrac{1}{2} \pm \sqrt{5}, -1\right)$

$e = \dfrac{\sqrt{10}}{5}$

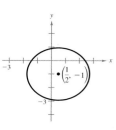

27. $\dfrac{x^2}{9} + \dfrac{y^2}{5} = 1$ **29.** $\dfrac{x^2}{25} + \dfrac{y^2}{16} = 1$

31. $\dfrac{(x - 3)^2}{9} + \dfrac{(y - 5)^2}{16} = 1$ **33.** $\dfrac{x^2}{24} + \dfrac{y^2}{49} = 1$

35. $\dfrac{x^2}{16} + \dfrac{7y^2}{16} = 1$

39. Minor axis: $(-6, -2), (0, -2)$

Major axis: $(-3, -6), (-3, 2)$

41. $\left(0, \frac{25}{3}\right)$ **43.** 22.10 **45.** 9.91

47. **(a)** 2π

(b) Volume $= \dfrac{8\pi}{3}$

Surface area $= \dfrac{2\pi(9 + 4\sqrt{3}\pi)}{9} \approx 21.48$

(c) Volume $= \dfrac{16\pi}{3}$

Surface area $= \dfrac{4\pi[6 + \sqrt{3}\ln(2 + \sqrt{3})]}{3} \approx 34.69$

49. $\sqrt{2}a \times \sqrt{2}b$ **55.** 1.5 ft from center

59. $e \approx 0.96716$

61.

Section 11.3

1. e **2.** a **3.** f **4.** c **5.** d **6.** b

7. Center: $(0, 0)$

Vertices: $(\pm1, 0)$

Foci: $(\pm\sqrt{2}, 0)$

9. Center: $(0, 0)$

Vertices: $(0, \pm1)$

Foci: $(0, \pm\sqrt{5})$

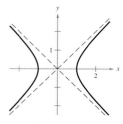
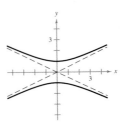

11. Center: $(0, 0)$

Vertices: $(0, \pm5)$

Foci: $(0, \pm13)$

13. Center: $(0, 0)$

Vertices: $(\pm\sqrt{3}, 0)$

Foci: $(\pm\sqrt{5}, 0)$

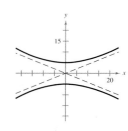
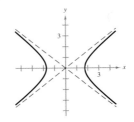

15. Center: (0, 0)
Vertices: (0, ±2)
Foci: (0, ±3)

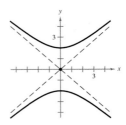

17. Center: (1, −2)
Vertices: (−1, −2), (3, −2)
Foci: (1 ± $\sqrt{5}$, −2)

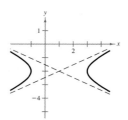

19. Center: (2, −6)
Vertices: (2, −5), (2, −7)
Foci: (2, −6 ± $\sqrt{2}$)

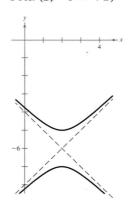

21. Center: (2, −3)
Vertices: (1, −3), (3, −3)
Foci: (2 ± $\sqrt{10}$, −3)

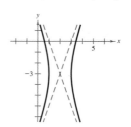

23. Center: (1, −3)
Vertices: (1, −3 ± $\sqrt{2}$)
Foci: (1, −3 ± 2$\sqrt{5}$)

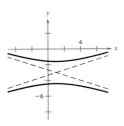

25. Degenerate hyperbola
Graph is two lines intersecting
at (−1, −3).

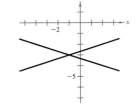

27. $\dfrac{y^2}{4} - \dfrac{x^2}{12} = 1$

29. $\dfrac{x^2}{1} - \dfrac{y^2}{9} = 1$ **31.** $\dfrac{(x-3)^2}{9} - \dfrac{(y-2)^2}{4} = 1$

33. $\dfrac{y^2}{9} - \dfrac{(x-2)^2}{9/4} = 1$

35. $\dfrac{(x-6)^2}{9} - \dfrac{(y-2)^2}{7} = 1$

37. **(a)** At (6, $\sqrt{3}$): $2x - 3\sqrt{3}y - 3 = 0$
At (6, −$\sqrt{3}$): $2x + 3\sqrt{3}y - 3 = 0$
(b) At (6, $\sqrt{3}$): $9x + 2\sqrt{3}y - 60 = 0$
At (6, −$\sqrt{3}$): $9x - 2\sqrt{3}y - 60 = 0$

39. Volume = $\dfrac{4\pi}{3}$

Surface area =
$\pi\left[2\sqrt{7} - 1 + \dfrac{\sqrt{2}}{2}\ln\left(2\dfrac{\sqrt{2}+1}{\sqrt{2}+\sqrt{7}}\right)\right] \approx 11.66$

41. $x \approx 110.3$ mi   **47.** Ellipse   **49.** Parabola
51. Circle   **53.** Circle   **55.** Hyperbola
57.

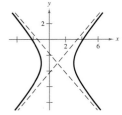

Section 11.4

1. $\dfrac{(y')^2}{2} - \dfrac{(x')^2}{2} = 1$

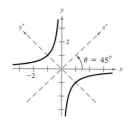

3. $y' = \dfrac{(x')^2}{6} - \dfrac{x'}{3}$

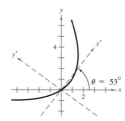

13. $\dfrac{(x')^2}{3} - \dfrac{(y')^2}{5} = 1$

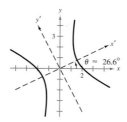

15. $\dfrac{(x')^2}{1.096} - \dfrac{(y')^2}{6.153} = 1$

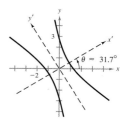

5. $\dfrac{(x')^2}{1/4} - \dfrac{(y')^2}{1/6} = 1$

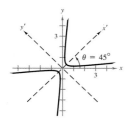

17. Parabola   **19.** Ellipse   **21.** Hyperbola

23. Parabola   **25.** Two lines

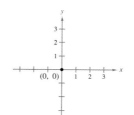

7. $\dfrac{(x' - 3\sqrt{2})^2}{16} - \dfrac{(y' - \sqrt{2})^2}{16} = 1$

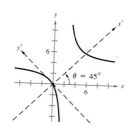

27. Point

29.

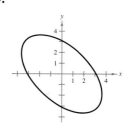

Review Exercises for Chapter 11

1. h   **2.** a   **3.** e   **4.** i   **5.** f   **6.** b

7. c   **8.** j   **9.** g   **10.** d

11. Circle
Center: $\left(\frac{1}{2}, -\frac{3}{4}\right)$
Radius: 1

13. Hyperbola
Center: $(-4, 3)$
Vertices: $(-4 \pm \sqrt{2}, 3)$

9. $\dfrac{(x')^2}{3} + \dfrac{(y')^2}{2} = 1$

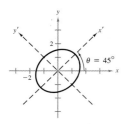

11. $x' = -(y')^2$

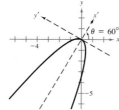

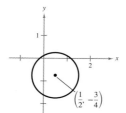

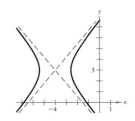

15. Ellipse
Center: $(2, -3)$
Vertices: $\left(2, -3 \pm \dfrac{\sqrt{2}}{2}\right)$

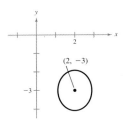

17. Parabola
Vertex: $(3, 0)$

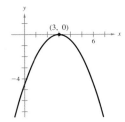

19. Parabola
Vertex: $\left(-\dfrac{\sqrt{2}}{4}, \dfrac{\sqrt{2}}{4}\right)$

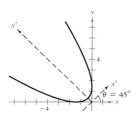

21. Ellipse
Center: $\left(1, -\dfrac{1}{2}\right)$
Vertices: $\left(1 \pm \dfrac{3}{4}, -\dfrac{1}{2}\right)$

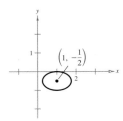

23. $\dfrac{y^2}{1} - \dfrac{x^2}{8} = 1$ **25.** $\dfrac{(x-2)^2}{25} + \dfrac{y^2}{21} = 1$

27. $x^2 - 2xy + y^2 - 8x - 8y = 0$

29. $\dfrac{x^2}{4} + \dfrac{y^2}{16/3} = 1$ **31.** $\dfrac{x^2}{4} - \dfrac{y^2}{12} = 1$

33. $4x + 4y - 7 = 0$ **35.** $a = \sqrt{5}$

41. 15.87   **43.** 4.212 ft   **45.** 428.68 ft²

Chapter 12

Section 12.1

1. $2x - 3y + 5 = 0$ **3.** $y = 1 - x^2,\ x \geq 0$

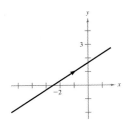
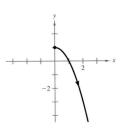

5. $y = (x - 1)^2$ **7.** $y = \dfrac{1}{2}x^{2/3}$

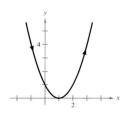
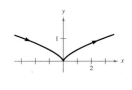

9. $y = \dfrac{x + 1}{x}$ **11.** $y = \dfrac{|x - 4|}{2}$

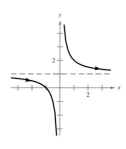
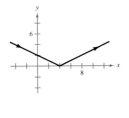

13. $y = \dfrac{1}{x},\ |x| \geq 1$ **15.** $x^2 + y^2 = 9$

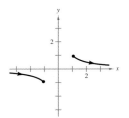
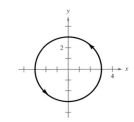

17. $\dfrac{x^2}{16} + \dfrac{y^2}{4} = 1$

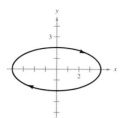

19. $y = 2 - 2x^2,\ -1 \le x \le 1$

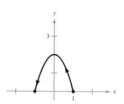

21. $\dfrac{(x-4)^2}{4} + \dfrac{(y+1)^2}{1} = 1$

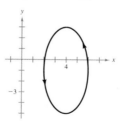

23. $\dfrac{(x-4)^2}{4} + \dfrac{(y+1)^2}{16} = 1$

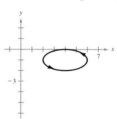

25. $\dfrac{x^2}{16} - \dfrac{y^2}{9} = 1$

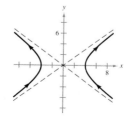

27. $y = \ln x$

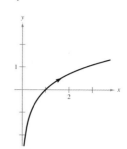

29. $y = \dfrac{1}{x^3},\ x > 0$

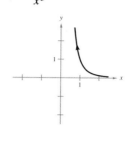

31. Each curve represents a portion of the line $y = 2x + 1$.

	Domain	**Orientation**
(a)	$-\infty < x < \infty$	Up
(b)	$-1 \le x \le 1$	Oscillates
(c)	$0 < x < \infty$	Down
(d)	$0 < x < \infty$	Up

33.   **35.**

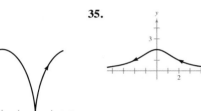

37.

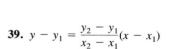

39. $y - y_1 = \dfrac{y_2 - y_1}{x_2 - x_1}(x - x_1)$

41. $(x - h)^2 + (y - k)^2 = r^2$

43. $\dfrac{(x - h)^2}{a^2} - \dfrac{(y - k)^2}{b^2} = 1$

45. $x = 5t$
$y = -2t$
(Solution is not unique.)

47. $x = 2 + 4 \cos \theta$
$y = 1 + 4 \sin \theta$
(Solution is not unique.)

49. $x = 5 \cos \theta$
$y = 3 \sin \theta$
(Solution is not unique.)

51. $x = 4 \sec \theta$
$y = 3 \tan \theta$
(Solution is not unique.)

53. $x = t$
$y = t^3$
$x = \tan t$
$y = \tan^3 t$
(Solution is not unique.)

55. $x = a\theta - b \sin \theta$
$y = a - b \cos \theta$

57.

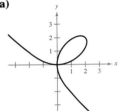

59.

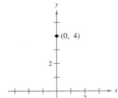

61.

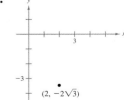

19. Horizontal: $(0, -2)$, $(2, 2)$
Vertical: None
21. Horizontal: $(0, 3)$, $(0, -3)$
Vertical: $(3, 0)$, $(-3, 0)$
23. Horizontal: $(4, 0)$, $(4, -2)$
Vertical: $(2, -1)$, $(6, -1)$
25. Horizontal: None
Vertical: $(1, 0)$, $(-1, 0)$
27. Horizontal: $(4n\pi, 0)$, $(2[2n - 1]\pi, 4)$
Vertical: None
29. $\sqrt{2}(1 - e^{-\pi/2}) \approx 1.12$
31. $2\sqrt{5} + \ln (2 + \sqrt{5}) \approx 5.916$
33. $\frac{1}{12}[\ln (\sqrt{37} + 6) + 6\sqrt{37}] \approx 3.249$ **35.** $6a$
37. $8a$ **39.** (a) $32\pi\sqrt{5}$ (b) $16\pi\sqrt{5}$
41. 32π **43.** $\dfrac{12\pi a^2}{5}$ **45.** $2\pi r^2(1 - \cos \theta)$
47. $\dfrac{3\pi}{8}$ **49.** $\dfrac{3\pi}{2}$ **51.** $\left(\dfrac{3}{4}, \dfrac{8}{5}\right)$ **53.** 36π
55. (a)

(b) $(0, 0)$, $(1.6799, 2.1165)$
(c) 6.557

Section 12.2

1. $\dfrac{dy}{dx} = \dfrac{3}{2}$, $\dfrac{d^2y}{dx^2} = 0$

3. At $t = -1$, $\dfrac{dy}{dx} = 2t + 3$, $\dfrac{d^2y}{dx^2} = 2$
$\dfrac{dy}{dx} = 1$, $\dfrac{d^2y}{dx^2} = 2$

5. At $\theta = \dfrac{\pi}{4}$, $\dfrac{dy}{dx} = -\cot \theta$, $\dfrac{d^2y}{dx^2} = -\dfrac{\csc^3 \theta}{2}$
$\dfrac{dy}{dx} = -1$, $\dfrac{d^2y}{dx^2} = -\sqrt{2}$

7. At $\theta = \dfrac{\pi}{6}$, $\dfrac{dy}{dx} = 2 \csc \theta$, $\dfrac{d^2y}{dx^2} = -2 \cot^3 \theta$
$\dfrac{dy}{dx} = 4$, $\dfrac{d^2y}{dx^2} = -6\sqrt{3}$

9. At $\theta = \dfrac{\pi}{4}$, $\dfrac{dy}{dx} = -\tan \theta$, $\dfrac{d^2y}{dx^2} = \dfrac{\sec^4 \theta \csc \theta}{3}$
$\dfrac{dy}{dx} = -1$, $\dfrac{d^2y}{dx^2} = \dfrac{4\sqrt{2}}{3}$

11. $2x - y - 5 = 0$ **13.** $y = 2$
15. $x + 2y - 4 = 0$ **17.** Horizontal: $(1, 0)$
Vertical: None

Section 12.3

1.

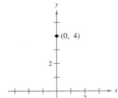

3.

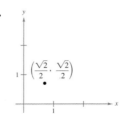

5.

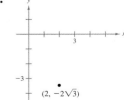

7.

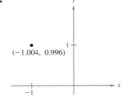

9. $\left(\sqrt{2}, \dfrac{\pi}{4}\right), \left(-\sqrt{2}, \dfrac{5\pi}{4}\right)$

11. $(5, 2.214), (-5, 5.356)$

13. $\left(\sqrt{6}, \dfrac{5\pi}{4}\right), \left(-\sqrt{6}, \dfrac{\pi}{4}\right)$

15. $(2\sqrt{13}, 0.983), (-2\sqrt{13}, 4.124)$ **17.** $r = 3$

19. $r = 2a \cos \theta$ **21.** $r = 4 \csc \theta$

23. $r = 10 \sec \theta$ **25.** $r = -\dfrac{2}{3 \cos \theta - \sin \theta}$

27. $r^2 = 8 \csc 2\theta$ **29.** $r = 9 \csc^2 \theta \cos \theta$

31. $r = \dfrac{2a}{1 - \sin \theta}, \quad r = \dfrac{-2a}{1 + \sin \theta}$

33. $x^2 + y^2 - 4y = 0$ **35.** $\sqrt{3}x - 3y = 0$

37. $y = 2$ **39.** $(x^2 + y^2 + 2y)^2 = x^2 + y^2$

41. $4x^2 - 5y^2 - 36y - 36 = 0$

43. $(x^2 + y^2)^3 = 16y^2$

45. Symmetry: Pole
Polar axis
$\theta = \dfrac{\pi}{2}$

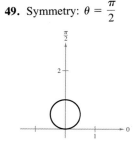

47. Symmetry: $\theta = \dfrac{\pi}{2}$

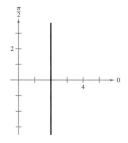

49. Symmetry: $\theta = \dfrac{\pi}{2}$

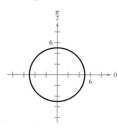

51. Symmetry: Polar axis   **53.** Symmetry: $\theta = \dfrac{\pi}{2}$

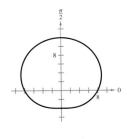

55. $(x - h)^2 + (y - k)^2 = h^2 + k^2$
Center: (h, k)
Radius: $\sqrt{h^2 + k^2}$

Section 12.4

1. Relative maxima:
$(5, 0), \left(5, \dfrac{2\pi}{3}\right), \left(5, \dfrac{4\pi}{3}\right)$
Relative minima:
$\left(0, \dfrac{\pi}{6}\right), \left(0, \dfrac{\pi}{2}\right), \left(0, \dfrac{5\pi}{6}\right), \left(0, \dfrac{7\pi}{6}\right), \left(0, \dfrac{3\pi}{2}\right),$
$\left(0, \dfrac{11\pi}{6}\right)$

3. Relative maximum: $\left(8, \dfrac{\pi}{2}\right)$
Relative minimum: $\left(2, \dfrac{3\pi}{2}\right)$

5. Relative minimum: $(3, \pi)$

7. $\dfrac{(1 + 2 \cos \theta)(1 - \cos \theta)}{\sin \theta (2 \cos \theta - 1)}, -1$

9. $\dfrac{2 \cos \theta (3 \sin \theta + 1)}{3 \cos 2\theta - 2 \sin \theta}, 0$

11. $\dfrac{\sin \theta + \theta \cos \theta}{\cos \theta - \theta \sin \theta}, \pi$ **13.** $\dfrac{2 \sin \theta \cos \theta}{1 - 2 \sin^2 \theta}, -\sqrt{3}$

15. Graph of $r = 2 \sec \theta$ is a vertical line.
Therefore, $\dfrac{dy}{dx}$ is undefined.

17. Horizontal: $\left(2, \dfrac{\pi}{2}\right), \left(\dfrac{1}{2}, \dfrac{7\pi}{6}\right), \left(\dfrac{1}{2}, \dfrac{11\pi}{6}\right)$
Vertical: $\left(\dfrac{3}{2}, \dfrac{\pi}{6}\right), \left(\dfrac{3}{2}, \dfrac{5\pi}{6}\right)$

19. $\left(5, \dfrac{\pi}{2}\right), \left(1, \dfrac{3\pi}{2}\right)$

21.

23. $\theta = 0$

25.

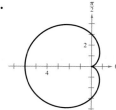

27.

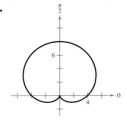

41.

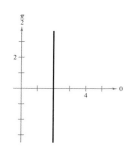

43.

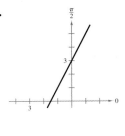

29. $\theta = \arcsin\left(-\frac{2}{3}\right)$
$\theta = \pi + \arcsin\left(\frac{2}{3}\right)$

31. $\theta = \arccos\left(\frac{3}{4}\right)$
$\theta = 2\pi - \arccos\left(\frac{3}{4}\right)$

45. $\theta = \dfrac{\pi}{4}, \dfrac{3\pi}{4}$

47. $\theta = 0, \dfrac{\pi}{2}$

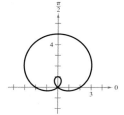

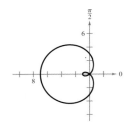

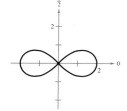

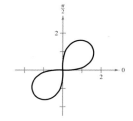

49. $\theta = 0, \dfrac{\pi}{2}$

51. $\theta = 0$

33.

35.

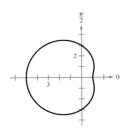

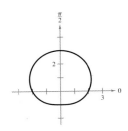

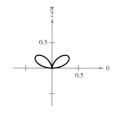

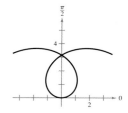

37. $\theta = \dfrac{\pi}{6}, \dfrac{\pi}{2}, \dfrac{5\pi}{6}$

39. $\theta = 0, \dfrac{\pi}{2}$

53. $\theta = \dfrac{\pi}{3}, \pi, \dfrac{5\pi}{3}$

55.

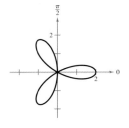

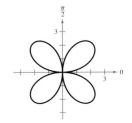

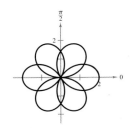

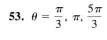

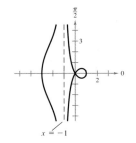

$x = -1$

57.

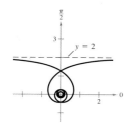

61. (a) $r = 2 - \sin\left[\theta - \left(\dfrac{\pi}{4}\right)\right]$

$= 2 - \dfrac{\sqrt{2}(\sin\theta - \cos\theta)}{2}$

(b) $r = 2 + \cos\theta$
(c) $r = 2 + \sin\theta$ **(d)** $r = 2 - \cos\theta$

63. (a)   **(b)**

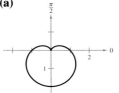   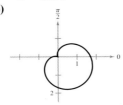

65. $(0, 0)$, $(1.4142, 0.7854)$, **67.** $(7, 1.5708)$, $(3, 4.7124)$
$(1.4142, 2.3562)$

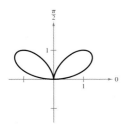   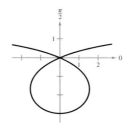

Section 12.5

1. $\left(1, \dfrac{\pi}{2}\right)$, $\left(1, \dfrac{3\pi}{2}\right)$, $(0, 0)$

3. $\left(\dfrac{2 - \sqrt{2}}{2}, \dfrac{3\pi}{4}\right)$, $\left(\dfrac{2 + \sqrt{2}}{2}, \dfrac{7\pi}{4}\right)$, $(0, 0)$

5. $\left(\dfrac{3}{2}, \dfrac{\pi}{6}\right)$, $\left(\dfrac{3}{2}, \dfrac{5\pi}{6}\right)$, $(0, 0)$ **7.** $(2, 4)$, $(-2, -4)$

9. $\left(2, \dfrac{\pi}{12}\right)$, $\left(2, \dfrac{5\pi}{12}\right)$, $\left(2, \dfrac{7\pi}{12}\right)$, $\left(2, \dfrac{11\pi}{12}\right)$,
$\left(2, \dfrac{13\pi}{12}\right)$, $\left(2, \dfrac{17\pi}{12}\right)$, $\left(2, \dfrac{19\pi}{12}\right)$, $\left(2, \dfrac{23\pi}{12}\right)$

11. $(0.581, \pm0.535)$, $(2.581, \pm1.376)$ **13.** $\dfrac{\pi}{3}$

15. $\dfrac{\pi}{8}$ **17.** $\dfrac{3\pi}{2}$ **19.** $\dfrac{2\pi - 3\sqrt{3}}{2}$

21. $\pi + 3\sqrt{3}$ **23.** $\dfrac{4}{3}(4\pi - 3\sqrt{3})$

25. $11\pi - 24$ **27.** $\dfrac{2}{3}(4\pi - 3\sqrt{3})$ **29.** $\dfrac{5\pi a^2}{4}$

31. $\dfrac{a^2}{2}(\pi - 2)$ **33.** $2\pi a$ **35.** 8

37. $\dfrac{\pi}{4}\sqrt{\pi^2 + 4} + \ln\left(\dfrac{\pi + \sqrt{\pi^2 + 4}}{2}\right)$

39. $\ln\left(\dfrac{\sqrt{4\pi^2 + 1} + 2\pi}{\sqrt{\pi^2 + 1} + \pi}\right) + \dfrac{2\sqrt{\pi^2 + 1} - \sqrt{4\pi^2 + 1}}{2\pi}$

41. 4π **43.** $\dfrac{2\pi\sqrt{1 + a^2}}{1 + 4a^2}(e^{\pi a} - 2a)$ **45.** 21.88

47. $4\pi^2 ab$ **49.** $A \approx 0.667$, $s \approx 4.439$

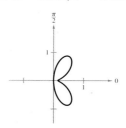

Section 12.6

1. Parabola   **3.** Parabola

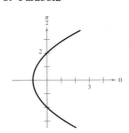   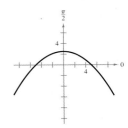

5. Ellipse   **7.** Ellipse

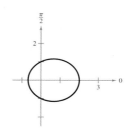   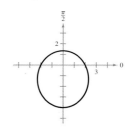

9. Parabola

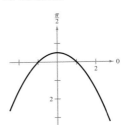

11. Hyperbola

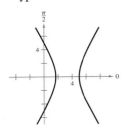

53. $r = \dfrac{1,422,792,504.77}{1 - 0.0543 \cos \theta}$

Perihelion: 1,349,513,900 mi

Aphelion: 1,504,486,100 mi

Review Exercises for Chapter 12

1. (a) $\dfrac{dy}{dx} = -\dfrac{3}{4}$

Horizontal tangents: None

(b) $y = \dfrac{-3x + 11}{4}$

(c)

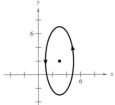

13. Hyperbola

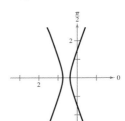

15. Hyperbola

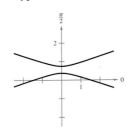

17. $r = \dfrac{1}{2 + \sin \theta}$

19. $r = \dfrac{1}{1 - \cos \theta}$

21. $r = \dfrac{2}{1 + 2 \cos \theta}$

23. $r = \dfrac{2}{1 - \sin \theta}$

25. $r = \dfrac{6}{2 - \sin \theta}$

27. $r = \dfrac{6}{1 - 2 \sin \theta}$

31. $r^2 = \dfrac{9}{1 - \left(\frac{16}{25}\right) \cos^2 \theta}$

33. $r^2 = \dfrac{-16}{1 - \left(\frac{25}{9}\right) \cos^2 \theta}$

3. (a) $\dfrac{dy}{dx} = -\dfrac{5}{2} \cot \theta$

Horizontal tangents: $(3, 7)$, $(3, -3)$

(b) $\dfrac{(x - 3)^2}{4} + \dfrac{(y - 2)^2}{25} = 1$

(c)

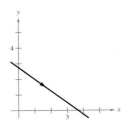

35.

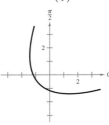

37.

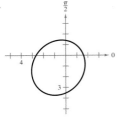

5. (a) $\dfrac{dy}{dx} = -2t^2$

Horizontal tangents: None

(b) $y = 3 + \dfrac{2}{x}$

(c)

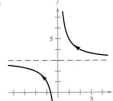

39. 10.87   **43.** $\dfrac{\pi}{2}$ **45.** 0   **47.** $\dfrac{\pi}{3}$

51. $r = \dfrac{92,931,075.2223}{1 - 0.0167 \cos \theta}$

Perihelion: 91,404,618 mi

Aphelion: 94,509,382 mi

7. (a) $\dfrac{dy}{dx} = -(2t^2 + 2t + 1)$

  Horizontal tangents: None

(b) $y = \dfrac{1 - x^2}{2x}$

(c)

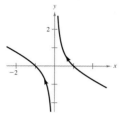

9. (a) $\dfrac{dy}{dx} = -4 \tan \theta$

  Horizontal tangents: $(\pm 1, 0)$

(b) $x^{2/3} + (y/4)^{2/3} = 1$

(c)

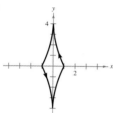

11. $x = -3 + 4 \cos \theta$
  $y = 4 + 3 \sin \theta$

15. $\dfrac{\pi^2 r}{2}$

17. Cardioid

19. Limaçon

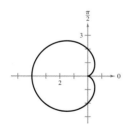

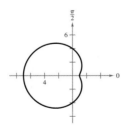

21. Rose curve

23. Circle

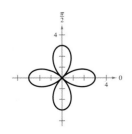

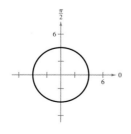

25. Line

27. Rose curve

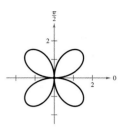

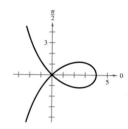

29. Parabola

31. Strophoid

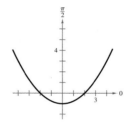

33. $x^2 + y^2 - 3x = 0$

35. $(x^2 + y^2 + 2x)^2 = 4(x^2 + y^2)$

37. $y^2 = x^2\left(\dfrac{4 - x}{4 + x}\right)$ **39.** $r = a \cos^2 \theta \sin \theta$

41. $r^2 = a^2\theta^2$ **43.** $r = \dfrac{4}{1 - \cos \theta}$

45. $r = \dfrac{7}{3 + 4 \cos \theta}$

47. $r = \dfrac{12}{4 \cos \theta + 3 \sin \theta}$

49. (a) $\pm \dfrac{\pi}{3}$

(b) Vertical: $(-1, 0)$, $(3, \pi)$, $\left(\frac{1}{2}, \pm 1.318\right)$
  Horizontal: $(-0.686, \pm 0.568)$, $(2.186, \pm 2.206)$

(c)

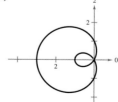

53. $\dfrac{9\pi}{2}$ **55.** $\dfrac{\pi}{32}$ **57.** a^2 **59.** $\dfrac{8\pi - 6\sqrt{3}}{3}$

61. $8a$ **63.** $\arctan\left(\dfrac{2\sqrt{3}}{3}\right)$

Chapter 13

Section 13.1

1. $\langle 4, 2 \rangle$

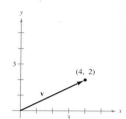

3. $\langle -7, 0 \rangle$

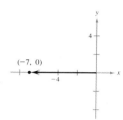

5. $\langle 4, 3 \rangle$

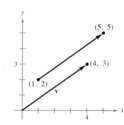

7. $\langle -4, -3 \rangle$

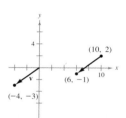

9. $\langle 0, 4 \rangle$

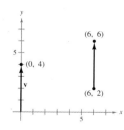

11. $\langle -1, \frac{5}{3} \rangle$

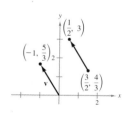

13. (a) $\langle 4, 6 \rangle$

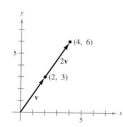

(b) $\langle -6, -9 \rangle$

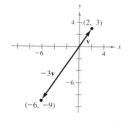

(c) $\langle 7, \frac{21}{2} \rangle$

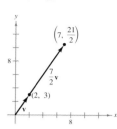

(d) $\langle \frac{4}{3}, 2 \rangle$

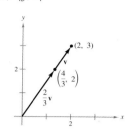

15. $\langle 3, -\frac{3}{2} \rangle$

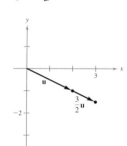

17. $\langle 4, 3 \rangle$

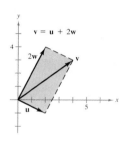

19. $\langle \frac{7}{2}, -\frac{1}{2} \rangle$

21. $a = 1, b = 1$

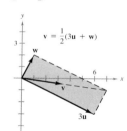

23. $a = 1, b = 2$ **25.** $a = \frac{2}{3}, b = \frac{1}{3}$

27. $(3, 5)$ **29.** 5 **31.** $\sqrt{61}$ **33.** 4

35. (a) $\dfrac{\sqrt{5}}{2}$ **(b)** $\sqrt{13}$ **(c)** $\dfrac{\sqrt{85}}{2}$

(d) 1 **(e)** 1 **(f)** 1

39. $\langle 2\sqrt{2}, 2\sqrt{2} \rangle$ **41.** $\langle 1, \sqrt{3} \rangle$

43. (a) $\pm \dfrac{1}{\sqrt{10}} \langle 1, 3 \rangle$ **45. (a)** $\pm \dfrac{1}{5} \langle -4, 3 \rangle$

(b) $\pm \dfrac{1}{\sqrt{10}} \langle 3, -1 \rangle$ **(b)** $\pm \dfrac{1}{5} \langle 3, 4 \rangle$

47. $\langle 3, 0 \rangle$ **49.** $\langle -\sqrt{3}, 1 \rangle$

51. $\left(\dfrac{3 + \sqrt{2}}{\sqrt{2}} \right) \mathbf{i} + \left(\dfrac{3}{\sqrt{2}} \right) \mathbf{j}$

53. $(2 \cos 4 + \cos 2) \mathbf{i} + (2 \sin 4 + \sin 2) \mathbf{j}$

55. $-\dfrac{\sqrt{2}}{2} \mathbf{i} + \dfrac{\sqrt{2}}{2} \mathbf{j}$

57. $F_H = -75\sqrt{3}$ lb, $F_V = -75$ lb

59. Tensions: 44.65 lb and 65.27 lb
Vertical components: 38.67 lb and 61.33 lb
61. N 85.13° E   **63.** $(-4, -1)$, $(6, 5)$, $(10, 3)$
416.15 mi/hr

Section 13.2

1. (a) -6 **(b)** 25   **(c)** 25
(d) $\langle -12, 18 \rangle$ **(e)** -12
3. (a) $-\frac{23}{2}$ **(b)** 17   **(c)** 17
(d) $\langle -\frac{23}{4}, -\frac{69}{2} \rangle$ **(e)** -23
5. $\frac{\pi}{2}$ **7.** $\arccos \dfrac{1}{\sqrt{5}} \approx 63.4°$ **9.** $\dfrac{\pi}{2}$
11. $\dfrac{7\pi}{12}$ **13.** $\dfrac{\pi}{4}$ **15.** $\arccos \dfrac{3}{\sqrt{13}} \approx 33.7°$
17. Neither   **19.** Orthogonal   **21.** Parallel
23. Neither   **25. (a)** $\langle \frac{5}{2}, \frac{1}{2} \rangle$ **(b)** $\langle -\frac{1}{2}, \frac{5}{2} \rangle$
27. (a) $\langle 1, 0 \rangle$ **(b)** $\langle 0, 1 \rangle$
29. (a) $\langle 0, 0 \rangle$ **(b)** $\langle 2, -3 \rangle$
31. (a) 8282.2 lb   **(b)** 30,909.6 lb   **33.** 425 ft · lb
35. (a) $\theta = \dfrac{\pi}{2}$ **(b)** $0 < \theta < \dfrac{\pi}{2}$ **(c)** $\dfrac{\pi}{2} < \theta < \pi$

Section 13.3

1. $(-\infty, 0)$, $(0, \infty)$ **3.** $(0, \infty)$ **5.** 1
7.   **9.**

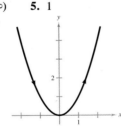

11. $3\mathbf{i} + 2\mathbf{j}$ **13.** 0   **15.** 0
17. $(-\infty, 0)$, $(0, \infty)$ **19.** $(0, \infty)$
21. $\mathbf{r}(2) = 4\mathbf{i} + 2\mathbf{j}$ **23.** $\mathbf{r}\left(\dfrac{\pi}{2}\right) = \mathbf{j}$
$\mathbf{r}'(2) = 4\mathbf{i} + \mathbf{j}$
$\mathbf{r}'\left(\dfrac{\pi}{2}\right) = -\mathbf{i}$

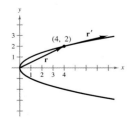

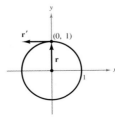

25. $\mathbf{r}'(t) = 3\mathbf{i} + \mathbf{j}$, $\mathbf{r}''(t) = \mathbf{0}$
27. $\mathbf{r}'(t) = \langle -a \sin t, a \cos t \rangle$,
$\mathbf{r}''(t) = \langle -a \cos t, -a \sin t \rangle$
29. $\mathbf{r}'(t) = \langle 1 - \cos t, \sin t \rangle$, $\mathbf{r}''(t) = \langle \sin t, \cos t \rangle$
31. (a) $3\mathbf{i} + 4\mathbf{j}$ **(b)** $12t(2 + t)$
(c) $5\mathbf{i} + (12 - 2t)\mathbf{j}$ **(d)** 5
33. $(-\infty, 0)$, $(0, \infty)$ **35.** $\left(\dfrac{n\pi}{2}, \dfrac{(n + 1)\pi}{2}\right)$
37. $(-\infty, \infty)$ **39.** $2t^3\mathbf{i} + 3t\mathbf{j} + \mathbf{C}$
41. $-4 \cos t\,\mathbf{i} + 3 \sin t\,\mathbf{j} + \mathbf{C}$ **43.** $3\mathbf{i} - \frac{3}{2}\mathbf{j}$
45. $-3\mathbf{i} - 3\mathbf{j}$ **47.** $2e^{2t}\mathbf{i} + 3(e^t - 1)\mathbf{j}$
49. $600\sqrt{3}\,\mathbf{i} + (-16t^2 + 600t)\mathbf{j}$

Section 13.4

1. $\mathbf{v}(1) = 3\mathbf{i} + \mathbf{j}$, $\mathbf{a}(1) = \mathbf{0}$

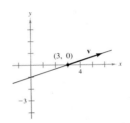

3. $\mathbf{v}(2) = 4\mathbf{i} + \mathbf{j}$, $\mathbf{a}(2) = 2\mathbf{i}$

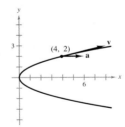

5. $\mathbf{v}\left(\dfrac{\pi}{4}\right) = -\sqrt{2}\,\mathbf{i} + \sqrt{2}\,\mathbf{j}$,
$\mathbf{a}\left(\dfrac{\pi}{4}\right) = -\sqrt{2}\,\mathbf{i} - \sqrt{2}\,\mathbf{j}$

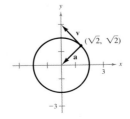

7. $\mathbf{v}(\pi) = 2\mathbf{i}, \quad \mathbf{a}(\pi) = -\mathbf{j}$

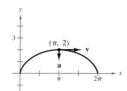

9. $\mathbf{v}(t) = t\mathbf{i} + t\mathbf{j}$
$\mathbf{r}(t) = \frac{1}{2}t^2\mathbf{i} + \frac{1}{2}t^2\mathbf{j}$
$\mathbf{r}(2) = 2\mathbf{i} + 2\mathbf{j}$

11. $\mathbf{v}(t) = \left(\frac{t^2}{2} - \frac{1}{2}\right)\mathbf{i} + \left(\frac{t^2}{2} + \frac{9}{2}\right)\mathbf{j}$
$\mathbf{r}(t) = \left(\frac{t^3}{6} - \frac{t}{2} + \frac{1}{3}\right)\mathbf{i} + \left(\frac{t^3}{6} + \frac{9}{2}t - \frac{14}{3}\right)\mathbf{j}$
$\mathbf{r}(2) = \frac{2}{3}\mathbf{i} + \frac{17}{3}\mathbf{j}$

13. 1.7 ft   **15.** $v_0 = 40\sqrt{6}$ ft/sec, 78 ft
17. $v_0 = 4\sqrt{170}$ ft/sec, $\theta \approx 76°$

19. (a) $\theta = \dfrac{\pi}{4}$ (b) $\theta = \dfrac{\pi}{2}$

21. 96.7 ft/sec
23. $\mathbf{r}(t) = 40\sqrt{2}t\mathbf{i} + (40\sqrt{2}t - 16t^2)\mathbf{j}$
Speed: $8\sqrt{58}$ ft/sec
Direction: $8\sqrt{2}(5\mathbf{i} + 2\mathbf{j})$

25. (a) $\omega t = 0, 2\pi, 4\pi, \ldots$ (b) $\omega t = \pi, 3\pi, \ldots$
27. $\mathbf{v}(t) = -b\omega \sin \omega t\, \mathbf{i} + b\omega \cos \omega t\, \mathbf{j}, \quad \mathbf{v}(t) \cdot \mathbf{r}(t) = 0$
29. $\mathbf{a}(t) = -b\omega^2[\cos \omega t\, \mathbf{i} + \sin \omega t\, \mathbf{j}] = -\omega^2 \mathbf{r}(t)$
31. $8\sqrt{10}$ ft/sec   **33.** 11.4°

Section 13.5

1. $\mathbf{T} = \mathbf{i}$
$\mathbf{N} = \mathbf{0}$
$\mathbf{a} \cdot \mathbf{T} = 0$
$\mathbf{a} \cdot \mathbf{N} = 0$

3. $\mathbf{T} = \mathbf{i}$
$\mathbf{N} = \mathbf{0}$
$\mathbf{a} \cdot \mathbf{T} = 8$
$\mathbf{a} \cdot \mathbf{N} = 0$

5. $\mathbf{T} = \dfrac{\sqrt{2}}{2}(\mathbf{i} - \mathbf{j})$
$\mathbf{N} = \dfrac{\sqrt{2}}{2}(\mathbf{i} + \mathbf{j})$
$\mathbf{a} \cdot \mathbf{T} = -\sqrt{2}$
$\mathbf{a} \cdot \mathbf{N} = \sqrt{2}$

7. $\mathbf{T} = \dfrac{\sqrt{2}}{2}(-\mathbf{i} + \mathbf{j})$
$\mathbf{N} = -\dfrac{\sqrt{2}}{2}(\mathbf{i} + \mathbf{j})$
$\mathbf{a} \cdot \mathbf{T} = 0$
$\mathbf{a} \cdot \mathbf{N} = 16\pi^2$

9. $\mathbf{T} = \frac{1}{2}(-\sqrt{3}\mathbf{i} + \mathbf{j})$
$\mathbf{N} = -\frac{1}{2}(\mathbf{i} + \sqrt{3}\mathbf{j})$
$\mathbf{a} \cdot \mathbf{T} = 0$
$\mathbf{a} \cdot \mathbf{N} = 2\pi^2$

11. $\mathbf{T} = \dfrac{\sqrt{2}}{2}(-\mathbf{i} + \mathbf{j})$
$\mathbf{N} = -\dfrac{\sqrt{2}}{2}(\mathbf{i} + \mathbf{j})$
$\mathbf{a} \cdot \mathbf{T} = \sqrt{2}e^{\pi/2}$
$\mathbf{a} \cdot \mathbf{N} = \sqrt{2}e^{\pi/2}$

13. $\mathbf{T} = (\cos \omega t_0)\mathbf{i} + (\sin \omega t_0)\mathbf{j}$
$\mathbf{N} = (-\sin \omega t_0)\mathbf{i} + (\cos \omega t_0)\mathbf{j}$
$\mathbf{a} \cdot \mathbf{T} = \omega^2$
$\mathbf{a} \cdot \mathbf{N} = \omega^3 t_0$

15. $\mathbf{r}(2) = 2\mathbf{i} + \frac{1}{2}\mathbf{j}$
$\mathbf{T}(2) = \dfrac{\sqrt{17}}{17}(4\mathbf{i} - \mathbf{j})$
$\mathbf{N}(2) = \dfrac{\sqrt{17}}{17}(\mathbf{i} + 4\mathbf{j})$

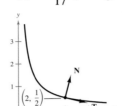

17. $\mathbf{a} \cdot \mathbf{T} = \dfrac{-32(v_0 \sin \theta - 32t)}{\sqrt{v_0^2 \cos^2 \theta + (v_0 \sin \theta - 32t)^2}}$
$\mathbf{a} \cdot \mathbf{N} = \dfrac{32v_0 \cos \theta}{\sqrt{v_0^2 \cos^2 \theta + (v_0 \sin \theta - 32t)^2}}$
At maximum height $\mathbf{a} \cdot \mathbf{T} = 0$ and $\mathbf{a} \cdot \mathbf{N} = 32$.

19. (a) Centripetal component is quadrupled.
(b) Centripetal component is halved.
21. 4.83 mi/sec   **23.** 4.67 mi/sec

Section 13.6

1. $4\sqrt{10}$

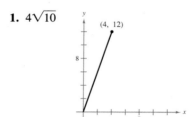

3. $6a$

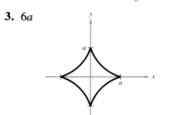

5. $\dfrac{\pi^2}{8}$

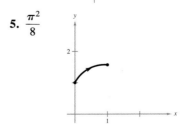

7. 8.111   **9.** $K = 0$, undefined
11. $K = \dfrac{4}{(17)^{3/2}}, \dfrac{1}{K} = \dfrac{(17)^{3/2}}{4}$ **13.** $K = \dfrac{1}{a}, \dfrac{1}{K} = a$

15. $\left(x - \dfrac{\pi}{2}\right)^2 + y^2 = 1$

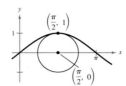

17. $(x + 2)^2 + (y - 3)^2 = 8$

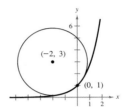

19. 0   **21.** 0   **23.** $\dfrac{\sqrt{2}}{2}$ **25.** $\dfrac{1}{2}$

27. $\dfrac{\sqrt{2}}{2}e^{-t}$ **29.** $\dfrac{1}{\omega t}$ **31.** (a) $(1, 3)$ (b) 0

33. (a) $K \to \infty$ as $x \to 0$ (b) 0

35. $a_{\mathbf{T}} = 6t$, $a_{\mathbf{N}} = 6$

37. $(x - 1)^2 + \left(y - \dfrac{5}{2}\right)^2 = \dfrac{1}{4}$ **41.** 56.27 mi/hr

43. $\dfrac{3}{2\sqrt{2}(1 + \sin \theta)}$ **45.** $\dfrac{2 + \theta^2}{(1 + \theta^2)^{3/2}}$

47. (a) 0   (b) 0   **49.** $\dfrac{1}{4}$

Review Exercises for Chapter 13

1. (a) $\mathbf{u} = \langle 3, -1 \rangle$, $\mathbf{v} = \langle 4, 2 \rangle$ (b) $2\sqrt{5}$
   (c) 10   (d) $\langle 10, 0 \rangle$ (e) $\langle 2, 1 \rangle$ (f) $\langle 1, -2 \rangle$

3. $\theta = \arccos\left(\dfrac{\sqrt{2} + \sqrt{6}}{4}\right) = \dfrac{\pi}{12}$

5. $4\cos 135° \, \mathbf{i} + 4\sin 135° \, \mathbf{j} = -2\sqrt{2}\mathbf{i} + 2\sqrt{2}\mathbf{j}$

7. Parallel   **9.** Orthogonal

11.

13. (a) $[0, 4)$, $(4, \infty)$
   (b) Discontinuous when $t = 4$ **15.** $9\mathbf{i}$

17. (a) $3\mathbf{i} + \mathbf{j}$ (b) $\left(\dfrac{1}{2\sqrt{t}} - 6\right)\mathbf{i} + 2\mathbf{j}$

   (c) $\dfrac{9}{2}\sqrt{t} + 8t - 4$ (d) $\dfrac{10t - 1}{\sqrt{10t^2 - 2t + 1}}$

19. $\sin t \, \mathbf{i} + (t \sin t + \cos t)\mathbf{j} + \mathbf{C}$

21. $\mathbf{v} = 4\mathbf{i} - 3\mathbf{j}$
   $\|\mathbf{v}\| = 5$
   $\mathbf{a} = \mathbf{0}$
   $\mathbf{a} \cdot \mathbf{T} = 0$
   $\mathbf{a} \cdot \mathbf{N} = 0$
   $K = 0$

23. $\mathbf{v} = 2\mathbf{i} - \dfrac{2}{(t + 1)^2}\mathbf{j}$
   $\|\mathbf{v}\| = \dfrac{2\sqrt{(t + 1)^4 + 1}}{(t + 1)^2}$
   $\mathbf{a} = \dfrac{4}{(t + 1)^3}\mathbf{j}$
   $\mathbf{a} \cdot \mathbf{T} = \dfrac{-4}{(t + 1)^3\sqrt{(t + 1)^4 + 1}}$
   $\mathbf{a} \cdot \mathbf{N} = \dfrac{4}{(t + 1)\sqrt{(t + 1)^4 + 1}}$
   $K = \dfrac{(t + 1)^3}{[(t + 1)^4 + 1]^{3/2}}$

25. $\mathbf{v} = e^t\mathbf{i} - e^{-t}\mathbf{j}$
   $\|\mathbf{v}\| = \sqrt{e^{2t} + e^{-2t}}$
   $\mathbf{a} = e^t\mathbf{i} + e^{-t}\mathbf{j}$
   $\mathbf{a} \cdot \mathbf{T} = \dfrac{e^{2t} - e^{-2t}}{\sqrt{e^{2t} + e^{-2t}}}$
   $\mathbf{a} \cdot \mathbf{N} = \dfrac{2}{\sqrt{e^{2t} + e^{-2t}}}$
   $K = \dfrac{2}{(e^{2t} + e^{-2t})^{3/2}}$

27. $2(20 + 9 \ln 3) \approx 59.775$

29. $\dfrac{250\sqrt{2}}{3}(3 + \sqrt{3}) \approx 557.68$ lb

31. 152 ft   **33.** 4.56 mi/sec

Chapter 14

Section 14.1

1.

3.

5. 3, $3\sqrt{5}$, 6, Right triangle
7. 6, 6, $2\sqrt{10}$, Isosceles triangle   **9.** $\left(\dfrac{3}{2}, -3, 5\right)$
11. $x^2 + y^2 + z^2 - 4y - 10z + 25 = 0$
13. $x^2 + y^2 + z^2 - 2x - 6y = 0$
15. Center: $(1, -3, -4)$ Radius: 5
17. Center: $\left(\dfrac{1}{3}, -1, 0\right)$ Radius: 1

19. (a) $\langle -2, 2, 2 \rangle$

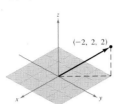

21. (a) $\langle -3, 0, 3 \rangle$

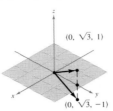

23. $\langle 4, 1, 1 \rangle$

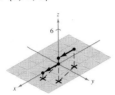

25. (a)

(b)

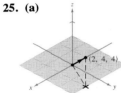

(c)

(d)

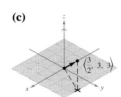

27. $\langle -1, 0, 4 \rangle$ **29.** $\langle 6, 12, 6 \rangle$
31. $\langle \frac{7}{2}, 3, \frac{5}{2} \rangle$ **33.** a and b   **35.** a
37. Collinear   **39.** Not collinear   **43.** $(3, 1, 8)$
45. 0   **47.** $\sqrt{34}$ **49.** $\sqrt{14}$
51. (a) $\frac{1}{3}\langle 2, -1, 2 \rangle$ **(b)** $-\frac{1}{3}\langle 2, -1, 2 \rangle$
53. (a) $\dfrac{1}{\sqrt{38}}\langle 3, 2, -5 \rangle$ **(b)** $-\dfrac{1}{\sqrt{38}}\langle 3, 2, -5 \rangle$
55. $\pm\dfrac{\sqrt{14}}{14}$ **57.** $\pm\dfrac{5}{3}$ **59.** $(2, -1, 2)$
61. $\left(\frac{13}{3}, 6, 3 \right)$

63. $\langle 0, \sqrt{3}, \pm 1 \rangle$

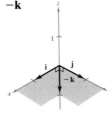

65. (a) 1   **(b)** 6   **(c)** 6   **(d)** $\mathbf{i} - \mathbf{k}$ **(e)** 2
67. $\arccos \dfrac{\sqrt{2}}{3} \approx 61.9°$
69. $\arccos \left(-\dfrac{8\sqrt{13}}{65} \right) \approx 116.3°$
71. Neither   **73.** Orthogonal
75. (a) $\langle 0, \frac{33}{25}, \frac{44}{25} \rangle$ **(b)** $\langle 2, -\frac{8}{25}, \frac{6}{25} \rangle$
77. (a) $\langle \frac{2}{3}, \frac{1}{3}, -\frac{1}{3} \rangle$ **(b)** $\langle \frac{1}{3}, \frac{2}{3}, \frac{4}{3} \rangle$
79. (a) $\langle 5, 0, 0 \rangle$ **(b)** $\langle 0, -4, 3 \rangle$
81. $\cos \alpha = \frac{1}{3}, \cos \beta = \frac{2}{3}, \cos \gamma = \frac{2}{3}$
83. $\cos \alpha = 0, \cos \beta = \dfrac{3}{\sqrt{13}}, \cos \gamma = -\dfrac{2}{\sqrt{13}}$
85. (c) $a = b = 1$

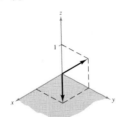

87. $\dfrac{\sqrt{3}}{3}\langle 1, 1, 1 \rangle$ **89.** 8.99 lb
91. $T_2 = 157.5$
$T_3 = 3692.5$

Section 14.2

1. $-\mathbf{k}$

3. $\mathbf{i}$

5. $-\mathbf{j}$

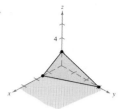

7. $\langle -1, -1, -1 \rangle$
9. $\langle 0, 0, 54 \rangle$
11. $\langle -2, 3, -1 \rangle$

41.

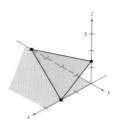

43.

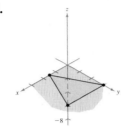

13. $\langle -\frac{71}{20}, -\frac{11}{5}, \frac{5}{4} \rangle$ **15.** 1   **17.** $6\sqrt{5}$
19. $2\sqrt{83}$ **21.** $\dfrac{3\sqrt{13}}{2}$ **23.** $\dfrac{9\sqrt{6}}{2}$ **25.** 1
27. 6   **29.** 2   **31.** 75
33. $10 \cos 40° \approx 7.66 \text{ ft} \cdot \text{lb}$

45.

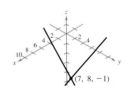

47.

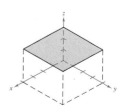

Section 14.3

Parametric Equations	Symmetric Equations	Direction Numbers
1. $x = t$ $y = 2t$ $z = 3t$	$x = \dfrac{y}{2} = \dfrac{z}{3}$	1, 2, 3
3. $x = -2 + 2t$ $y = 4t$ $z = 3 - 2t$	$\dfrac{x+2}{2} = \dfrac{y}{4}$ $= \dfrac{z-3}{-2}$	2, 4, −2
5. $x = 5 + 17t$ $y = -3 - 11t$ $z = -2 - 9t$	$\dfrac{x-5}{17} = \dfrac{y+3}{-11}$ $= \dfrac{z+2}{-9}$	17, −11, −9
7. $x = 1 + 3t$ $y = -2t$ $z = 1 + t$	$\dfrac{x-1}{3} = \dfrac{y}{-2}$ $= \dfrac{z-1}{1}$	3, −2, 1
9. $x = 2$ $y = 3$ $z = 4 + t$	None	0, 0, 1

11. a, b, d   **13.** $(2, 3, 1)$, $\cos \theta = \dfrac{7\sqrt{17}}{51}$
15. Nonintersecting   **17.** $x - 2 = 0$
19. $2x + 3y - z = 10$ **21.** $3x + 9y - 7z = 0$
23. $4x - 3y + 4z = 10$ **25.** $z = 3$
27. $x + y + z = 5$ **29.** $7x + y - 11z = 5$
31. $y - z = -1$ **33.** Orthogonal   **35.** 83.5°
37. Parallel   **39.** 76.3°

49. $x = 2$ **51.** $(2, -3, 2)$
   $y = 1 + t$ **53.** Nonintersecting
   $z = 1 + 2t$
55. 0   **57.** $\dfrac{6\sqrt{14}}{7}$ **59.** $\dfrac{2\sqrt{26}}{13}$ **61.** $\dfrac{10\sqrt{26}}{13}$
63. $(7, 8, -1)$

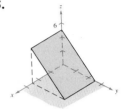

Section 14.4

1. c   **2.** e   **3.** f   **4.** g   **5.** d   **6.** b
7. a   **8.** h
9. Plane   **11.** Right circular cylinder

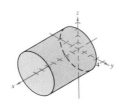

13. Parabolic cylinder

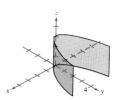

15. Elliptic cylinder

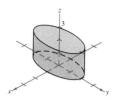

29. Elliptic cone

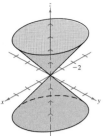

31. Hyperbolic paraboloid

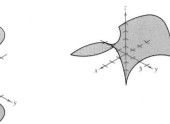

17. Hyperbolic cylinder

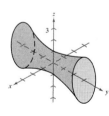

19. Ellipsoid

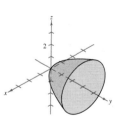

33. Ellipsoid

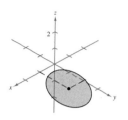

35.

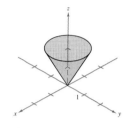

37.

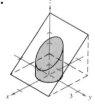

39.

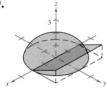

21. Hyperboloid of one sheet

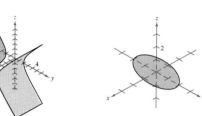

23. Elliptic paraboloid

41. $x^2 + z^2 = 4y$ **43.** $4x^2 + 4y^2 = z^2$

45. $y^2 + z^2 = \dfrac{4}{x^2}$ **47.** $y = \sqrt{2z}$

49. (a) Major axis: $4\sqrt{2}$ **(b)** Major axis: $8\sqrt{2}$
     Minor axis: 4   Minor axis: 8
     Foci: $(0, \pm 2, 2)$ Foci: $(0, \pm 4, 8)$

51.

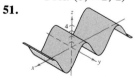

53.

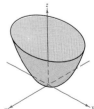

25. Hyperbolic paraboloid

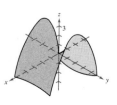

27. Hyperboloid of two sheets

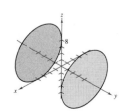

55. $\dfrac{x^2}{3963^2} + \dfrac{y^2}{3963^2} + \dfrac{z^2}{3942^2} = 1$

Section 14.5

1. $(-\infty, 0), (0, \infty)$ **3.** $(0, \infty)$ **5.** $[0, \infty)$
7. $(-\infty, \infty)$ **9.** $\sqrt{1 + t^2}$
11.   **13.**

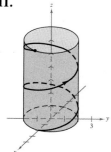   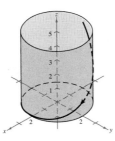

15.

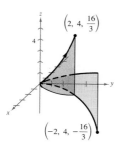

17. $x = t, y = -t, z = 2t^2$

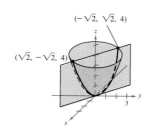

19. $x = 2 \sin t, y = 2 \cos t, z = 4 \sin^2 t$

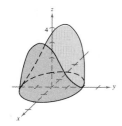

21. $x = 1 + \sin t, y = \pm\sqrt{2} \cos t, z = 1 - \sin t$

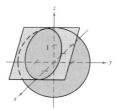

23. $x = t, y = t, z = \sqrt{4 - t^2}$

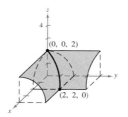

25. $2\mathbf{i} + 2\mathbf{j} + \frac{1}{2}\mathbf{k}$ **27. 0**
29. Limit does not exist.   **31.** $[-1, 1]$
33. $\left(-\dfrac{\pi}{2} + n\pi, \dfrac{\pi}{2} + n\pi\right)$
35. $\mathbf{r}\left(\dfrac{3\pi}{2}\right) = -2\mathbf{j} + \left(\dfrac{3\pi}{2}\right)\mathbf{k}, \mathbf{r}'\left(\dfrac{3\pi}{2}\right) = 2\mathbf{i} + \mathbf{k}$

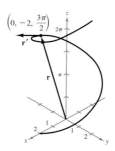

37. $6\mathbf{i} - 14t\mathbf{j} + 3t^2\mathbf{k}$
39. $-3a \cos^2 t \sin t\,\mathbf{i} + 3a \sin^2 t \cos t\,\mathbf{j}$
41. $-e^{-t}\mathbf{i}$ **43.** $\langle \sin t + t \cos t, \cos t - t \sin t, 1 \rangle$
45. (a) $\mathbf{i} + 3\mathbf{j} + 2t\mathbf{k}$ **(b)** $2\mathbf{k}$
    (c) $8t + 9t^2 + 5t^4$
    (d) $-\mathbf{i} + (9 - 2t)\mathbf{j} + (6t - 3t^2)\mathbf{k}$
    (e) $8t^3\mathbf{i} + (12t^2 - 4t^3)\mathbf{j} + (3t^2 - 24t)\mathbf{k}$
    (f) $\dfrac{10 + 2t^2}{\sqrt{10 + t^2}}$
47. $t^2\mathbf{i} + t\mathbf{j} + t\mathbf{k} + \mathbf{C}$
49. $\ln t\,\mathbf{i} + t\mathbf{j} - \frac{2}{5}t^{5/2}\mathbf{k} + \mathbf{C}$
51. $e^t\mathbf{i} - \cos t\,\mathbf{j} + \sin t\,\mathbf{k} + \mathbf{C}$

53. $\tan t\,\mathbf{i} + \arctan t\,\mathbf{j} + \mathbf{C}$

55. $\mathbf{r}(t) = \mathbf{i} + (t^2 + 1)\mathbf{j} + \frac{2}{3}t^{3/2}\mathbf{k}$

57. $\left(\dfrac{2 - e^{-t^2}}{2}\right)\mathbf{i} + (e^{-t} - 2)\mathbf{j} + (t + 1)\mathbf{k}$

59. $4\mathbf{i} + \dfrac{1}{2}\mathbf{j} - \mathbf{k}$ **61.** $a\mathbf{i} + a\mathbf{j} + \dfrac{\pi}{2}\mathbf{k}$

63. $4\sqrt{10}$

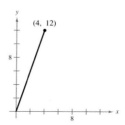

65. $2\pi\sqrt{a^2 + b^2}$

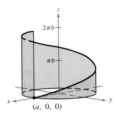

67. $\dfrac{\sqrt{5}\pi^2}{8}$

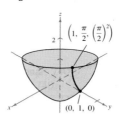

69. 8.37

71. (a) $s = \sqrt{5}t$

 (b) $\mathbf{r}(s) = 2\cos\dfrac{s}{\sqrt{5}}\mathbf{i} + 2\sin\dfrac{s}{\sqrt{5}}\mathbf{j} + \dfrac{s}{\sqrt{5}}\mathbf{k}$

73.   **75.** 2.412

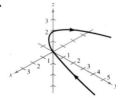

Section 14.6

1. $\mathbf{v}(t) = \mathbf{i} + 2\mathbf{j} + 3\mathbf{k}$ **3.** $\mathbf{v}(t) = \mathbf{i} + 2t\mathbf{j} + t\mathbf{k}$
$\ \ \ s(t) = \sqrt{14}$ $\ \ \ s(t) = \sqrt{1 + 5t^2}$
$\ \ \ \mathbf{a}(t) = \mathbf{0}$ $\ \ \ \mathbf{a}(t) = 2\mathbf{j} + \mathbf{k}$

5. $\mathbf{v}(t) = \mathbf{i} + \mathbf{j} - \dfrac{t}{\sqrt{9 - t^2}}\mathbf{k}$

$\ \ \ s(t) = \sqrt{\dfrac{18 - t^2}{9 - t^2}}$

$\ \ \ \mathbf{a}(t) = \dfrac{-9}{(9 - t^2)^{3/2}}\mathbf{k}$

7. $\mathbf{v}(t) = 4\mathbf{i} - 3\sin t\,\mathbf{j} + 3\cos t\,\mathbf{k}$
$\ \ \ s(t) = 5$
$\ \ \ \mathbf{a}(t) = -3\cos t\,\mathbf{j} - 3\sin t\,\mathbf{k}$

9. $\mathbf{v}(t) = t(\mathbf{i} + \mathbf{j} + \mathbf{k})$
$\ \ \ \mathbf{r}(t) = \dfrac{t^2}{2}(\mathbf{i} + \mathbf{j} + \mathbf{k})$
$\ \ \ \mathbf{r}(2) = 2(\mathbf{i} + \mathbf{j} + \mathbf{k})$

11. $\mathbf{v}(t) = \left(\dfrac{t^2}{2} + \dfrac{9}{2}\right)\mathbf{j} + \left(\dfrac{t^2}{2} - \dfrac{1}{2}\right)\mathbf{k}$

$\ \ \ \mathbf{r}(t) = \left(\dfrac{t^3}{6} + \dfrac{9}{2}t - \dfrac{14}{3}\right)\mathbf{j} + \left(\dfrac{t^3}{6} - \dfrac{1}{2}t + \dfrac{1}{3}\right)\mathbf{k}$

$\ \ \ \mathbf{r}(2) = \dfrac{17}{3}\mathbf{j} + \dfrac{2}{3}\mathbf{k}$

13. $\mathbf{T}(0) = \dfrac{\sqrt{2}}{2}(\mathbf{i} + \mathbf{k})$, $x = t$, $y = 0$, $z = t$

15. $\mathbf{T}(0) = \dfrac{\sqrt{5}}{5}(2\mathbf{j} + \mathbf{k})$, $x = 2$, $y = 2t$, $z = t$

17. $\mathbf{T}(3) = \frac{1}{19}\langle 1, 6, 18\rangle$
$\ \ \ x = 3 + t$, $y = 9 + 6t$, $z = 18 + 18t$

19. $\mathbf{T}\left(\dfrac{\pi}{4}\right) = \dfrac{1}{2}\langle -\sqrt{2}, \sqrt{2}, 0\rangle$

21. $\mathbf{T} = \dfrac{\sqrt{5}}{5}(\mathbf{j} + 2\mathbf{k})$

$\ \ \ \mathbf{N} = \dfrac{\sqrt{5}}{5}(-2\mathbf{j} + \mathbf{k})$

$\ \ \ \mathbf{a}\cdot\mathbf{T} = \dfrac{4\sqrt{5}}{5}$

$\ \ \ \mathbf{a}\cdot\mathbf{N} = \dfrac{2\sqrt{5}}{5}$

23. $\mathbf{T} = \dfrac{\sqrt{6}}{6}(\mathbf{i} + 2\mathbf{j} + \mathbf{k})$ **25.** $\mathbf{T} = \frac{1}{5}(4\mathbf{i} - 3\mathbf{j})$
$\ \ \ \mathbf{N} = \dfrac{\sqrt{30}}{30}(-5\mathbf{i} + 2\mathbf{j} + \mathbf{k})$ $\ \ \ \mathbf{N} = -\mathbf{k}$
             $\ \ \ \mathbf{a}\cdot\mathbf{T} = 0$
$\ \ \ \mathbf{a}\cdot\mathbf{T} = \dfrac{5\sqrt{6}}{6}$ $\ \ \ \mathbf{a}\cdot\mathbf{N} = 3$

$\ \ \ \mathbf{a}\cdot\mathbf{N} = \dfrac{\sqrt{30}}{3}$

27. $\mathbf{r}\left(\dfrac{\pi}{2}\right) = 2\mathbf{j} + \dfrac{\pi}{4}\mathbf{k}$

$\mathbf{T}\left(\dfrac{\pi}{2}\right) = \dfrac{\sqrt{17}}{17}(-4\mathbf{i} + \mathbf{k})$

$\mathbf{N}\left(\dfrac{\pi}{2}\right) = -\mathbf{j}$

$\mathbf{B}\left(\dfrac{\pi}{2}\right) = \dfrac{\sqrt{17}}{17}(\mathbf{i} + 4\mathbf{k})$

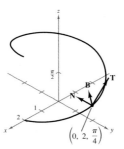

29. $\dfrac{1}{(1 + t^2)^{3/2}}$ **31.** $\dfrac{\sqrt{5}}{(1 + 5t^2)^{3/2}}$ **33.** $\dfrac{3}{25}$

37. $\dfrac{6\sqrt{9t^4 + 19t^2 + 10}}{(9t^4 + 18t^2 + 10)^{3/2}}$ **39.** $a_\mathbf{T} = \dfrac{18t(t^2 + 1)}{\sqrt{9t^4 + 18t^2 + 10}}$

$a_\mathbf{N} = \dfrac{6\sqrt{9t^4 + 19t^2 + 10}}{\sqrt{9t^4 + 18t^2 + 10}}$

Review Exercises for Chapter 14

1. (a) $\mathbf{u} = -\mathbf{i} + 4\mathbf{j}$, $\mathbf{v} = -3\mathbf{i} + 6\mathbf{j}$ (b) 3
   (c) $24\mathbf{i} + 6\mathbf{j} + 12\mathbf{k}$ (d) $4x + y + 2z = 20$
   (e) $x = 4 - t,\ y = 4 + 4t,\ z = 0$
3. π **5.** $-2\sqrt{2}\mathbf{i} + 2\sqrt{2}\mathbf{j}$
7. $\dfrac{3}{\sqrt{26}}\mathbf{i} - \dfrac{9}{\sqrt{26}}\mathbf{j} + \dfrac{12}{\sqrt{26}}\mathbf{k}$ **9.** $\sqrt{14}$
15. (a) $x = 1,\ y = 2 + t,\ z = 3$ (b) None
17. (a) $x = t,\ y = -1 + t,\ z = 1$
   (b) $x = y + 1,\ z = 1$
19. $x + y + z = 6$ **21.** $x + 2y = 1$ **23.** $\frac{8}{7}$
25. $\dfrac{\sqrt{35}}{7}$ **27.** $\dfrac{\sqrt{6}}{6}$ **29.** $\dfrac{\sqrt{3}}{3}$
31.

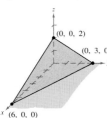

33.

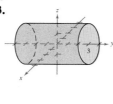

35.

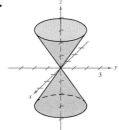

37.

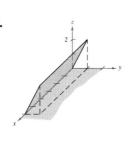

39.

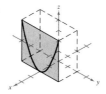

41.

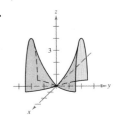

43.

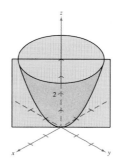

45.

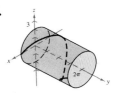

47. $x = t,\ y = -t,\ z = 2t^2$

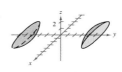

49. $4\mathbf{i} + \mathbf{k}$
51. (a) All reals except $n\pi$, n is an integer
   (b) $t = n\pi$, n is an integer
53. (a) $(0, \infty)$ (b) Continuous
55. (a) $3\mathbf{i} + \mathbf{j}$ (b) $\mathbf{0}$ (c) $4t + 3t^2$
   (d) $-5\mathbf{i} + (2t - 2)\mathbf{j} + 2t^2\mathbf{k}$
   (e) $\dfrac{10t - 1}{\sqrt{10t^2 - 2t + 1}}$
   (f) $\left(\frac{8}{3}t^3 - 2t^2\right)\mathbf{i} - 8t^3\mathbf{j} + (9t^2 - 2t + 1)\mathbf{k}$
57. $\sin t\,\mathbf{i} + (t \sin t + \cos t)\mathbf{j} + \mathbf{C}$
59. $\frac{1}{2}(t\sqrt{1 + t^2} + \ln|t + \sqrt{1 + t^2}|) + C$
61. $x = -\sqrt{2} - \sqrt{2}t,\quad y = \sqrt{2} - \sqrt{2}t,\quad z = \dfrac{3\pi}{4} + t$
63. $\mathbf{v} = 4\mathbf{i} - 3\mathbf{j}$
   $\|\mathbf{v}\| = 5$
   $\mathbf{a} = \mathbf{0}$
   $\mathbf{a} \cdot \mathbf{T} = 0$
   $\mathbf{a} \cdot \mathbf{N} = 0$
   $K = 0$

65. $\mathbf{v} = 2\mathbf{j} - \dfrac{2}{(t + 1)^2}\mathbf{k}$

$\|\mathbf{v}\| = \dfrac{2\sqrt{(t + 1)^4 + 1}}{(t + 1)^2}$

$\mathbf{a} = \dfrac{4}{(t + 1)^3}\mathbf{k}$

$\mathbf{a} \cdot \mathbf{T} = \dfrac{-4}{(t + 1)^3\sqrt{(t + 1)^4 + 1}}$

$\mathbf{a} \cdot \mathbf{N} = \dfrac{4}{(t + 1)\sqrt{(t + 1)^4 + 1}}$

$K = \dfrac{(t + 1)^3}{[(t + 1)^4 + 1]^{3/2}}$

67. $\mathbf{v} = e^t\mathbf{i} - e^{-t}\mathbf{j} + \mathbf{k}$

$\|\mathbf{v}\| = \sqrt{e^{2t} + e^{-2t} + 1}$

$\mathbf{a} = e^t\mathbf{i} + e^{-t}\mathbf{j}$

$\mathbf{a} \cdot \mathbf{T} = \dfrac{e^{2t} - e^{-2t}}{\sqrt{e^{2t} + e^{-2t} + 1}}$

$\mathbf{a} \cdot \mathbf{N} = \dfrac{\sqrt{e^{2t} + e^{-2t} + 4}}{\sqrt{e^{2t} + e^{-2t} + 1}}$

$K = \dfrac{\sqrt{e^{2t} + e^{-2t} + 4}}{(e^{2t} + e^{-2t} + 1)^{3/2}}$

69. $\mathbf{v} = \mathbf{i} + 2t\mathbf{j} + t\mathbf{k}$

$\|\mathbf{v}\| = \sqrt{1 + 5t^2}$

$\mathbf{a} = 2\mathbf{j} + \mathbf{k}$

$\mathbf{a} \cdot \mathbf{T} = \dfrac{5t}{\sqrt{1 + 5t^2}}$

$\mathbf{a} \cdot \mathbf{N} = \dfrac{\sqrt{5}}{\sqrt{1 + 5t^2}}$

$K = \dfrac{\sqrt{5}}{(1 + 5t^2)^{3/2}}$

71. $\dfrac{\sqrt{5}\pi}{2}$

Chapter 15

Section 15.1

1. (a) $3/2$ **(b)** $-1/4$ **(c)** 6 **(d)** $5/y$
(e) $x/2$ **(f)** $5/t$
3. (a) 5 **(b)** $3e^2$ **(c)** $2/e$ **(d)** $5e^y$
(e) xe^2 **(f)** te^t
5. (a) $2/3$ **(b)** 0 **7. (a)** $\sqrt{2}$ **(b)** $3\sin 1$
9. (a) 4 **(b)** 6
11. Domain: $\{(x, y): x^2 + y^2 \le 4\}$
Range: $0 \le z \le 2$
13. Domain: $\{(x, y): -1 \le x + y \le 1\}$
Range: $-\dfrac{\pi}{2} \le z \le \dfrac{\pi}{2}$

15. Domain: $\{(x, y): x \ne 0, y \ne 0\}$
Range: All real numbers
17. Domain: $\{(x, y): y < -x + 4\}$
Range: All real numbers
19. Domain: $\{(x, y): y \ne 0\}$
Range: $0 < z$
21. Domain: $\{(x, y): x \ne 0, y \ne 0\}$
Range: $0 < |z|$
23. c **24.** d **25.** f **26.** b **27.** e **28.** a
29.

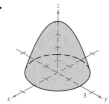

31.

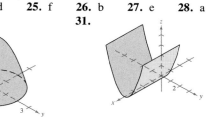

33.

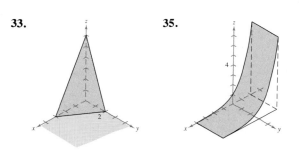

35.

37.

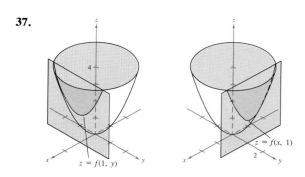

$z = f(1, y)$ $z = f(x, 1)$

39. Circles centered at $(0, 0)$
Radius ≤ 5

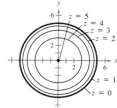

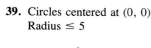

41. Hyperbolas: $xy = c$

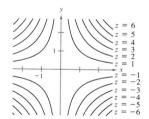

43. Circles passing through $(0, 0)$

Centered at $\left(\dfrac{1}{2c}, 0\right)$

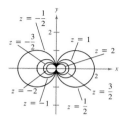

45. Lines with a slope of 1 passing through Quadrant IV

$y = x - e^c$

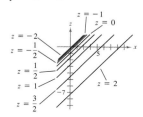

47. Hyperbolas centered at $(0, 0)$

Asymptotes: x- and y-axes

$xy = \ln c$

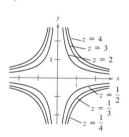

49.

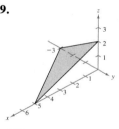

51.

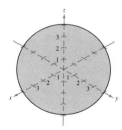

53.

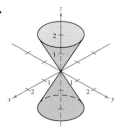

55. (a) 243 board-feet

(b) 507 board-feet

57. (a) 12 minutes

(b) 20 minutes

(c) 10 minutes

(d) 30 minutes

59.

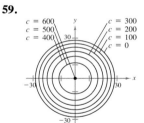

61. $C = 0.75xy + 0.80(xz + yz)$

63. (a) C

(b) A

(c) B

65.

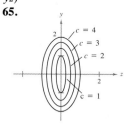

Section 15.2

1. 2   **3.** 15   **5.** 5, Continuous

7. -3, Continuous for $x \ne y$

9. 0, Continuous for $xy \ne -1$, $y \ne 0$, $\left|\dfrac{x}{y}\right| \le 1$

11. 1, Continuous

13. $2\sqrt{2}$, Continuous for $x + y + z \ge 0$

15. Limit does not exist

17. Continuous except at (0, 0); 1
19. Continuous except at (0, 0); limit does not exist.
21. Continuous, 1
23. Continuous except at (0, 0); $-\infty$ **25.** 1   **27.** 0
29. Continuous except at (0, 0, 0)   **31.** Continuous
33. Continuous   **35.** Continuous for $y \neq \dfrac{3x}{2}$
37. (a) $2x$ **(b)** -4 **39. (a)** $2 + y$ **(b)** $x - 3$
41. (a) 0   **(b)** 0.04

Section 15.3

1. $f_x(x, y) = 2$
$f_y(x, y) = -3$

3. $f_x(x, y) = y$
$f_y(x, y) = x$

5. $\dfrac{\partial z}{\partial x} = \sqrt{y}$
$\dfrac{\partial z}{\partial y} = \dfrac{x}{2\sqrt{y}}$

7. $\dfrac{\partial z}{\partial x} = 2xe^{2y}$
$\dfrac{\partial z}{\partial y} = 2x^2 e^{2y}$

9. $\dfrac{\partial z}{\partial x} = \dfrac{2x}{x^2 + y^2}$
$\dfrac{\partial z}{\partial y} = \dfrac{2y}{x^2 + y^2}$

11. $\dfrac{\partial z}{\partial x} = \dfrac{-2y}{x^2 - y^2}$
$\dfrac{\partial z}{\partial y} = \dfrac{2x}{x^2 - y^2}$

13. $h_x(x, y) = -2xe^{-(x^2+y^2)}$
$h_y(x, y) = -2ye^{-(x^2+y^2)}$

15. $f_x(x, y) = \dfrac{x}{\sqrt{x^2 + y^2}}$
$f_y(x, y) = \dfrac{y}{\sqrt{x^2 + y^2}}$

17. $\dfrac{\partial z}{\partial x} = 2 \cos (2x - y)$
$\dfrac{\partial z}{\partial y} = -\cos (2x - y)$

19. $\dfrac{\partial z}{\partial x} = ye^y \cos xy$
$\dfrac{\partial z}{\partial y} = e^y(x \cos xy + \sin xy)$

21. $f_x(x, y) = 1 - x^2$
$f_y(x, y) = y^2 - 1$

23. $\dfrac{\partial z}{\partial x} = \dfrac{1}{4}$
$\dfrac{\partial z}{\partial y} = \dfrac{1}{4}$

25. $\dfrac{\partial z}{\partial x} = -\dfrac{1}{4}$
$\dfrac{\partial z}{\partial y} = \dfrac{1}{4}$

27. $\dfrac{\partial w}{\partial x} = \dfrac{x}{\sqrt{x^2 + y^2 + z^2}}$
$\dfrac{\partial w}{\partial y} = \dfrac{y}{\sqrt{x^2 + y^2 + z^2}}$
$\dfrac{\partial w}{\partial z} = \dfrac{z}{\sqrt{x^2 + y^2 + z^2}}$

29. $F_x(x, y, z) = \dfrac{x}{x^2 + y^2 + z^2}$
$F_y(x, y, z) = \dfrac{y}{x^2 + y^2 + z^2}$
$F_z(x, y, z) = \dfrac{z}{x^2 + y^2 + z^2}$

31. $H_x(x, y, z) = \cos (x + 2y + 3z)$
$H_y(x, y, z) = 2 \cos (x + 2y + 3z)$
$H_z(x, y, z) = 3 \cos (x + 2y + 3z)$

33. $\dfrac{\partial^2 z}{\partial x^2} = 2$
$\dfrac{\partial^2 z}{\partial y^2} = 6$
$\dfrac{\partial^2 z}{\partial y \partial x} = \dfrac{\partial^2 z}{\partial x \partial y} = -2$

35. $\dfrac{\partial^2 z}{\partial x^2} = e^x \tan y$
$\dfrac{\partial^2 z}{\partial y^2} = 2e^x \sec^2 y \tan y$
$\dfrac{\partial^2 z}{\partial y \partial x} = \dfrac{\partial^2 z}{\partial x \partial y} = e^x \sec^2 y$

37. $\dfrac{\partial^2 z}{\partial x^2} = \dfrac{2xy}{(x^2 + y^2)^2}$
$\dfrac{\partial^2 z}{\partial y^2} = \dfrac{-2xy}{(x^2 + y^2)^2}$
$\dfrac{\partial^2 z}{\partial y \partial x} = \dfrac{\partial^2 z}{\partial x \partial y} = \dfrac{y^2 - x^2}{(x^2 + y^2)^2}$

39. $\dfrac{\partial^2 z}{\partial x^2} = \dfrac{y^2}{(x^2 + y^2)^{3/2}}$
$\dfrac{\partial^2 z}{\partial y^2} = \dfrac{x^2}{(x^2 + y^2)^{3/2}}$
$\dfrac{\partial^2 z}{\partial y \partial x} = \dfrac{\partial^2 z}{\partial x \partial y} = \dfrac{-xy}{(x^2 + y^2)^{3/2}}$

41. $\dfrac{\partial^2 z}{\partial y \partial x} = \dfrac{\partial^2 z}{\partial x \partial y} = 6x$

43. $\dfrac{\partial^2 z}{\partial y \partial x} = \dfrac{\partial^2 z}{\partial x \partial y} = \sec y \tan y$

45. $\dfrac{\partial^2 z}{\partial y \partial x} = \dfrac{\partial^2 z}{\partial x \partial y} = -2ye^{-y^2}$

47. $f_{xyy}(x, y, z) = f_{yxy}(x, y, z)$
$\qquad = f_{yyx}(x, y, z) = 0$

49. $f_{xyy}(x, y, z) = f_{yxy}(x, y, z)$
$\qquad = f_{yyx}(x, y, z) = z^2 e^{-x} \sin yz$

51. $\dfrac{\partial^2 z}{\partial x^2} + \dfrac{\partial^2 z}{\partial y^2} = 0 + 0 = 0$

53. $\dfrac{\partial^2 z}{\partial x^2} + \dfrac{\partial^2 z}{\partial y^2} = e^x \sin y - e^x \sin y = 0$

55. $\dfrac{\partial^2 z}{\partial t^2} = -c^2 \sin (x - ct) = c^2 \dfrac{\partial^2 z}{\partial x^2}$

57. $\dfrac{\partial z}{\partial t} = -e^{-t} \cos \dfrac{x}{c} = c^2 \dfrac{\partial^2 z}{\partial x^2}$

59. $f_x(x, y) = 2$
$f_y(x, y) = 3$

61. $f_x(x, y) = \dfrac{1}{2\sqrt{x + y}}$
$f_y(x, y) = \dfrac{1}{2\sqrt{x + y}}$

63. $-\frac{1}{2}$ **65.** 18

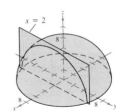

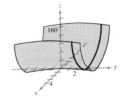

67. $\dfrac{\partial C}{\partial x} = 183, \quad \dfrac{\partial C}{\partial y} = 237$

69. An increase in either price will cause a decrease in demand.

71. $\dfrac{\partial T}{\partial x} = -2.4°/\text{ft}$

$\dfrac{\partial T}{\partial y} = -9°/\text{ft}$

73. **(a)** $f_x(x, y) = \dfrac{y(x^4 + 4x^2y^2 - y^4)}{(x^2 + y^2)^2}$

$f_y(x, y) = \dfrac{x(x^4 - 4x^2y^2 - y^4)}{(x^2 + y^2)^2}$

(b) $f_x(0, 0) = f_y(0, 0) = 0$

(c) $f_{xy}(0, 0) = -1, f_{yx}(0, 0) = 1$

(d) Discontinuous

Section 15.4

1. $dz = 6xy^3 \, dx + 9x^2y^2 \, dy$

3. $dz = \dfrac{2}{(x^2 + y^2)^2}(x \, dx + y \, dy)$

5. $dz = (\cos y + y \sin x) \, dx - (x \sin y + \cos x) \, dy$

7. $dw = 2z^3 \, y \cos x \, dx + 2z^3 \sin x \, dy + 6z^2y \sin x \, dz$

9. $dw = \dfrac{1}{z - 2y} \, dx + \dfrac{z + 2x}{(z - 2y)^2} \, dy - \dfrac{x + y}{(z - 2y)^2} \, dz$

11. **(a)** -0.5125 **(b)** -0.5

13. **(a)** -0.00293 **(b)** 0.00385

15. **(a)** -0.25 **(b)** -0.25

17. $\varepsilon_1 = \Delta x$

$\varepsilon_2 = 0$

19. $\varepsilon_1 = y\Delta x$

$\varepsilon_2 = 2x\Delta x + (\Delta x)^2$

(Answer is not unique.)

21. 10% **23.** 7% **25.** -0.0111 sec

27. $L \approx 8.096 \times 10^{-4} \pm 6.6 \times 10^{-6}$ micro-henrys

Section 15.5

1. $2(e^{2t} - e^{-2t})$

3. $e^t \sec(\pi - t)[1 - \tan(\pi - t)]$ **5.** $4e^{2t}$

7. $\dfrac{\partial w}{\partial s} = 4s, 8, \quad \dfrac{\partial w}{\partial t} = 4t, -4$

9. $\dfrac{\partial w}{\partial s} = 2s \cos 2t, 0, \quad \dfrac{\partial w}{\partial t} = -2s^2 \sin 2t, -18$

11. $2 \cos 2t$ **13.** $3(2t^2 - 1)$

15. $\dfrac{\partial w}{\partial r} = 0$ **17.** $\dfrac{\partial w}{\partial r} = 0$

$\dfrac{\partial w}{\partial \theta} = 4(x - y)$ $\dfrac{\partial w}{\partial \theta} = 1$

19. $\dfrac{\partial z}{\partial x} = \dfrac{-x}{z}$ **21.** $\dfrac{\partial z}{\partial x} = \dfrac{-\sec^2(x + y)}{\sec^2(y + z)}$

$\dfrac{\partial z}{\partial y} = \dfrac{-y}{z}$ $\dfrac{\partial z}{\partial y} = -1 - \dfrac{\sec^2(x + y)}{\sec^2(y + z)}$

23. $\dfrac{\partial z}{\partial x} = \dfrac{-x}{y + z}$

$\dfrac{\partial z}{\partial y} = \dfrac{-z}{y + z}$

$\dfrac{\partial^2 z}{\partial x^2} = -\dfrac{(y + z)^2 + x^2}{(y + z)^3}$

$\dfrac{\partial^2 z}{\partial y^2} = \dfrac{z(2y + z)}{(y + z)^3}$

$\dfrac{\partial^2 z}{\partial x \partial y} = \dfrac{\partial^2 z}{\partial y \partial x} = \dfrac{xy}{(y + x)^3}$

25. $\dfrac{\partial w}{\partial z} = \dfrac{yw - xy - xw}{xz - yz + 2w}$

$\dfrac{\partial w}{\partial y} = \dfrac{-xz + zw}{xz - yz + 2w}$

$\dfrac{\partial w}{\partial x} = \dfrac{-yz - zw}{xz - yz + 2w}$

27. $\dfrac{dV}{dt} = 182 \text{ ft}^3/\text{min}$

$\dfrac{dS}{dt} = 132 \text{ ft}^2/\text{min}$

29. $\dfrac{dV}{dt} = 1536\pi \text{ in.}^3/\text{min}$

$\dfrac{dS}{dt} = \dfrac{36\pi}{5}(20 + 9\sqrt{10}) \text{ in.}^2/\text{min}$

(Surface area includes base.)

31. $28m \text{ cm}^2/\text{sec}$ **33.** 3 **35.** 0 **37.** 1

Section 15.6

1. $\dfrac{\sqrt{3} - 5}{2}$ **3.** $\dfrac{5\sqrt{2}}{2}$ **5.** $-\dfrac{7}{25}$ **7.** $-e$

9. $\dfrac{2\sqrt{6}}{3}$ **11.** $\dfrac{(8 + \pi)\sqrt{6}}{24}$ **13.** $\sqrt{2}(x + y)$

15. $\left(\dfrac{2 + \sqrt{3}}{2}\right) \cos(2x - y)$ **17.** $-7\sqrt{2}$

19. $\dfrac{7\sqrt{19}}{19}$ **21.** $(2x - 3y)\mathbf{i} + (2y - 3x)\mathbf{j}, 2\sqrt{17}$

23. $\tan y \, \mathbf{i} + x \sec^2 y \, \mathbf{j}, \sqrt{17}$

25. $\dfrac{2}{3(x^2 + y^2)}(x\mathbf{i} + y\mathbf{j}), \dfrac{2\sqrt{5}}{15}$

27. $\dfrac{x\mathbf{i} + y\mathbf{j} + z\mathbf{k}}{\sqrt{x^2 + y^2 + z^2}}, 1$

29. $\dfrac{x\mathbf{i} + y\mathbf{j} + z\mathbf{k}}{(1 - x^2 - y^2 - z^2)^{3/2}}, 0$

31.

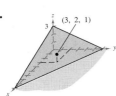

33. (a) $\dfrac{2 + 3\sqrt{3}}{12}$

(b) $\dfrac{3 - 2\sqrt{3}}{12}$

7. $\dfrac{\sqrt{113}}{113}(-\mathbf{i} - 6\sqrt{3}\mathbf{j} + 2\mathbf{k})$ **9.** $\dfrac{\sqrt{3}}{3}(\mathbf{i} - \mathbf{j} + \mathbf{k})$

11. $6x + 2y + z = 35$ **13.** $2x - y + z = 2$
15. $10x - 8y - z = 9$ **17.** $2x - z = -2$
19. $3x + 4y - 25z = 25(1 - \ln 5)$
21. $x - 4y + 2z = 18$ **23.** $x + y + z = 1$
25. $2x + 4y + z = 14$

$$\dfrac{x - 1}{2} = \dfrac{y - 2}{4} = \dfrac{z - 4}{1}$$

27. $3x + 2y + z = -6$

$$\dfrac{x + 2}{3} = \dfrac{y + 3}{2} = \dfrac{z - 6}{1}$$

29. $x - y + 2z = \dfrac{\pi}{2}$

$$\dfrac{x - 1}{1} = \dfrac{y - 1}{-1} = \dfrac{z - (\pi/4)}{2}$$

31. $86.0°$ **33.** $77.4°$

35. (a) $\dfrac{x - 2}{1} = \dfrac{y - 1}{-2} = \dfrac{z - 2}{1}$

(b) $\dfrac{\sqrt{10}}{5}$, Not orthogonal

37. (a) $\dfrac{x - 3}{4} = \dfrac{y - 3}{4} = \dfrac{z - 4}{-3}$

(b) $\dfrac{16}{25}$, Not orthogonal

35. (a) $-\dfrac{1}{5}$ **(b)** $-\dfrac{11\sqrt{10}}{60}$ **37.** $\dfrac{\sqrt{13}}{6}$

39.

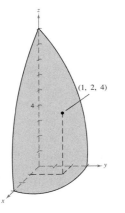

41. $-2\mathbf{i} - 4\mathbf{j}, 2\sqrt{5}$

39. (a) $\dfrac{y - 1}{1} = \dfrac{z - 1}{-1}, x = 2$

(b) 0, Orthogonal

41. $(0, 3, 12)$ **43.** $x = 4e^{-4t}, y = 3e^{-2t}, z = 10e^{-8t}$

43. $6\mathbf{i} + 8\mathbf{j}$ **45.** $-\dfrac{1}{2}\mathbf{j}$

47. $\dfrac{\sqrt{257}}{257}(16\mathbf{i} - \mathbf{j})$ **49.** $\dfrac{\sqrt{85}}{85}(9\mathbf{i} - 2\mathbf{j})$

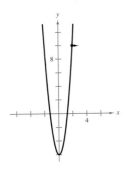

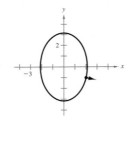

Section 15.8

1. Relative minimum: $(-1, 1, -4)$
3. Relative maximum: $(8, 16, 74)$
5. Relative minimum: $(1, 2, -1)$
7. Saddle point: $(1, -2, -1)$
9. Saddle point: $(0, 0, 0)$
11. Saddle point: $(0, 0, 0)$
 Relative minimum: $(1, 1, -1)$
13. Relative maximum: $(-1, 0, 2)$
 Relative minimum: $(1, 0, -2)$
15. Relative maxima: $(0, \pm 1, 4)$
 Relative minimum: $(0, 0, 0)$
 Saddle points: $(\pm 1, 0, 1)$

51. $\dfrac{1}{625}(7\mathbf{i} - 24\mathbf{j})$ **53.** $y^2 = 10x$

Section 15.7

1. $\dfrac{\sqrt{3}}{3}(\mathbf{i} + \mathbf{j} + \mathbf{k})$ **3.** $\dfrac{\sqrt{2}}{10}(3\mathbf{i} + 4\mathbf{j} - 5\mathbf{k})$

5. $\dfrac{\sqrt{2,049}}{2,049}(32\mathbf{i} + 32\mathbf{j} - \mathbf{k})$

17. Relative maxima: $\left(0, \pm\dfrac{\sqrt{2}}{2}, \sqrt{e}\right)$

 Relative minima: $\left(\pm\dfrac{\sqrt{6}}{2}, 0, -\dfrac{\sqrt{e}}{e}\right)$

 Saddle point: $\left(0, 0, \dfrac{e}{2}\right)$

19. Absolute maxima: $(2, 1, 6)$, $(-2, -1, 6)$
Absolute minima: $\left(-\frac{1}{2}, 1, -\frac{1}{4}\right)$, $\left(\frac{1}{2}, -1, -\frac{1}{4}\right)$
21. Absolute maxima: $(2, 2, 16)$, $(-2, -2, 16)$
Absolute minima: $(x, -x, 0)$, $|x| \le 2$
23. Saddle point: $(0, 0, 0)$
Test fails
25. Absolute minima: $(1, a, 0)$, $(b, -4, 0)$
Test fails
27. Absolute minimum: $(0, 0, 0)$
Test fails
29. Relative minimum: $(0, 3, -1)$
31. Relative minimum: $(-1, 2, 1)$
33. Relative maxima: $\left(1 \pm \dfrac{\sqrt{2}}{2}, \dfrac{1}{2}, 1\right)$
Relative minimum: $(1, 0, 0)$

Section 15.9

1. $\dfrac{6\sqrt{14}}{7}$ **3.** 6 **5.** 10, 10, 10
7. 10, 10, 10 **9.** $36 \times 18 \times 18$ in.
11. $x = \dfrac{\sqrt{2}}{2} \approx 0.707$ mi
$y = \dfrac{3\sqrt{2} + 2\sqrt{3}}{6} \approx 1.284$ mi
17. Each edge of $w/3$ inches is turned up $60°$ from the horizontal.
19. $p_1 = \$586.67$, $p_2 = \$840.00$
21. $a \displaystyle\sum_{i=1}^{n} x_i^4 + b \sum_{i=1}^{n} x_i^3 + c \sum_{i=1}^{n} x_i^2 = \sum_{i=1}^{n} x_i^2 y_i$
$a \displaystyle\sum_{i=1}^{n} x_i^3 + b \sum_{i=1}^{n} x_i^2 + c \sum_{i=1}^{n} x_i = \sum_{i=1}^{n} x_i y_i$
$a \displaystyle\sum_{i=1}^{n} x_i^2 + b \sum_{i=1}^{n} x_i + cn = \sum_{i=1}^{n} y_i$
23. $y = \frac{3}{4}x + \frac{4}{3}$, $\frac{1}{6}$ **25.** $y = -2x + 4$, 2
27. $y = \frac{37}{43}x + \frac{7}{43}$ **29.** $y = -\frac{175}{148}x + \frac{945}{148}$
31. $y = \frac{3}{7}x^2 + \frac{6}{5}x + \frac{26}{35}$ **33.** $y = x^2 - x$

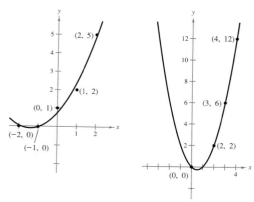

35. (a) $y = -240x + 685$ **37.** $y = 14x + 19$
(b) $y = 349$ 41.4 bushels/acre
39. $y = -\frac{25}{112}x^2 + \frac{541}{56}x - \frac{25}{14}$

Section 15.10

1. $f(5, 5) = 25$ **3.** $f(2, 2) = 8$
5. $f\left(\dfrac{\sqrt{2}}{2}, \dfrac{1}{2}\right) = \dfrac{1}{4}$ **7.** $f(25, 50) = 2600$
9. $f(1, 1) = 2$ **11.** $f(2, 2) = e^4$
13. $f(2, 2, 2) = 12$ **15.** $f\left(\frac{1}{3}, \frac{1}{3}, \frac{1}{3}\right) = \frac{1}{3}$
17. $f(8, 16, 8) = 1024$ **19.** $f\left(3, \frac{3}{2}, 1\right) = 6$
21. $36 \times 18 \times 18$ in.
23. $\sqrt[3]{360} \times \sqrt[3]{360} \times \frac{4}{3}\sqrt[3]{360}$ ft
25. $\dfrac{2\sqrt{3}a}{3} \times \dfrac{2\sqrt{3}b}{3} \times \dfrac{2\sqrt{3}c}{3}$ **29.** $\dfrac{\sqrt{13}}{13}$
31. $\sqrt{3}$ **33.** $P\left(\frac{3125}{6}, \frac{6250}{3}\right) = 147{,}314$
35. $x = 50\sqrt{2}$
$y = 200\sqrt{2}$
cost $= \$13{,}576.45$
37. $x = \dfrac{10 + 2\sqrt{265}}{15}$
$y = \dfrac{5 + \sqrt{265}}{15}$
$z = \dfrac{-1 + \sqrt{265}}{3}$

Review Exercises for Chapter 15

1. Continuous except at $(0, 0)$
Limit: $\frac{1}{2}$
3. Continuous except at $(0, 0)$
Limit does not exist
5.

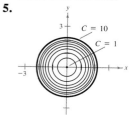

7.

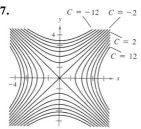

9. $f_x(x, y) = e^x \cos y$
$f_y(x, y) = -e^x \sin y$
11. $f_x(x, y) = e^y + ye^x$
$f_y(x, y) = xe^y + e^x$
13. $g_x(x, y) = \dfrac{y(y^2 - x^2)}{(x^2 + y^2)^2}$
$g_y(x, y) = \dfrac{x(x^2 - y^2)}{(x^2 + y^2)^2}$
15. $f_x(x, y, z) = \dfrac{-yz}{x^2 + y^2}$
$f_y(x, y, z) = \dfrac{xz}{x^2 + y^2}$
$f_z(x, y, z) = \arctan \dfrac{y}{x}$

17. $u_x(x, t) = cne^{-n^2t} \cos nx$
$u_t(x, t) = -cn^2e^{-n^2t} \sin nx$

19. $\dfrac{\partial z}{\partial x} = -\dfrac{2yz + z - 2xy}{2xy + x + 2z}$

$\dfrac{\partial z}{\partial y} = \dfrac{x^2 - 2xz}{2xy + x + 2z}$

21. $f_{xx}(x, y) = 6$
$f_{yy}(x, y) = 12y$
$f_{xy}(x, y) = f_{yx}(x, y) = -1$

23. $h_{xx}(x, y) = -y \cos x$
$h_{yy}(x, y) = -x \sin y$
$h_{xy}(x, y) = h_{yx}(x, y) = \cos y - \sin x$

29. $\left(\sin \dfrac{y}{x} - \dfrac{y}{x} \cos \dfrac{y}{x} \right) dx + \left(\cos \dfrac{y}{x} \right) dy$

31. $\dfrac{\partial u}{\partial r} = 2r$ **33.** 0 **35.** $\frac{2}{3}$

$\dfrac{\partial u}{\partial t} = 2t$

37. $\left\langle -\dfrac{1}{2}, 0 \right\rangle, \dfrac{1}{2}$ **39.** $\left\langle -\dfrac{\sqrt{2}}{2}, -\dfrac{\sqrt{2}}{2} \right\rangle, 1$

41. $4x + 4y - z = 8$
$\dfrac{x - 2}{4} = \dfrac{y - 1}{4} = \dfrac{z - 4}{-1}$

43. Tangent plane: $z = 4$
Normal line: $x = 2, y = -3, z = 4 + t$

45. $\dfrac{x - 2}{1} = \dfrac{y - 1}{2}, z = 3$

47. Relative minimum: $\left(\frac{3}{2}, \frac{9}{4}, -\frac{27}{16} \right)$
Saddle point: $(0, 0, 0)$

49. Relative minimum: $(1, 1, 3)$

51. Relative maximum: $\left(\frac{4}{3}, \frac{1}{3}, \frac{16}{27} \right)$
Relative minimum: $(0, 1, 0)$

53. 0.082 in., 0.63% **55.** π in.^{3}

57. $x_1 = 94$ **59.** $49.4, 253$
$x_2 = 157$

25. $\displaystyle\int_0^4 \int_0^y f(x, y)\, dx\, dy = \int_0^4 \int_x^4 f(x, y)\, dy\, dx$

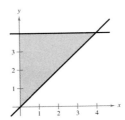

27. $\displaystyle\int_{-\pi/2}^{\pi/2} \int_0^{\cos x} f(x, y)\, dy\, dx = \int_0^1 \int_{-\arccos y}^{\arccos y} f(x, y)\, dx\, dy$

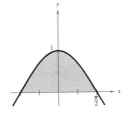

29. $\displaystyle\int_0^1 \int_0^2 dy\, dx = \int_0^2 \int_0^1 dx\, dy = 2$

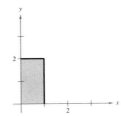

31. $\displaystyle\int_0^1 \int_{-\sqrt{1-y^2}}^{\sqrt{1-y^2}} dx\, dy = \int_{-1}^1 \int_0^{\sqrt{1-x^2}} dy\, dx = \dfrac{\pi}{2}$

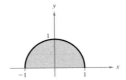

Chapter 16

Section 16.1

1. $\dfrac{3x^2}{2}$ **3.** $y \ln (2y)$ **5.** $\dfrac{4x^2 - x^4}{2}$

7. $\dfrac{y}{2}[(\ln y)^2 - y^2]$ **9.** $x^2(1 - e^{-x^2} - x^2 e^{-x^2})$

11. 3 **13.** $\frac{20}{3}$ **15.** $\frac{2}{3}$ **17.** 4

19. $\dfrac{\pi^2}{32} + \dfrac{1}{8}$ **21.** $\dfrac{1}{2}$ **23.** $\dfrac{1}{4}$

33. $\int_0^2 \int_{x/2}^1 dy\,dx = \int_0^1 \int_0^{2y} dx\,dy = 1$

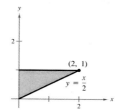

35. $\int_0^1 \int_{y^2}^{\sqrt[3]{y}} dx\,dy = \int_0^1 \int_{x^3}^{\sqrt{x}} dy\,dx = \dfrac{5}{12}$

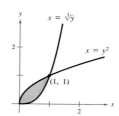

37. 24   **39.** $\frac{16}{3}$ **41.** $\frac{9}{2}$ **43.** $\frac{8}{3}$ **45.** 5
47. πab **49.** $\frac{26}{9}$ **51.** $\frac{1}{2}(1 - \cos 1) \approx 0.230$
53. 20.5648

Section 16.2

1. 8   **3.** 36

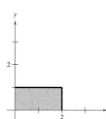   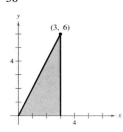

5. 0

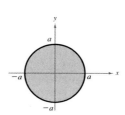

7. $\int_0^3 \int_0^5 xy\,dy\,dx = \dfrac{225}{4}$

$\int_0^5 \int_0^3 xy\,dx\,dy = \dfrac{225}{4}$

9. $\int_0^2 \int_x^{2x} \dfrac{y}{x^2 + y^2}\,dy\,dx = \ln\dfrac{5}{2}$

$\int_0^2 \int_{y/2}^y \dfrac{y}{x^2 + y^2}\,dx\,dy + \int_2^4 \int_{y/2}^2 \dfrac{y}{x^2 + y^2}\,dx\,dy = \ln\dfrac{5}{2}$

11. $\int_0^3 \int_{4y/3}^{\sqrt{25-y^2}} x\,dx\,dy = 25$

$\int_0^4 \int_0^{3x/4} x\,dy\,dx + \int_4^5 \int_0^{\sqrt{25-x^2}} x\,dy\,dx = 25$

13. 4   **15.** 8   **17.** 12   **19.** $\frac{3}{8}$ **21.** 8π
23. 1   **25.** $\frac{1}{8}$ **27.** $\frac{32}{3}$ **29.** $\frac{2}{3}$ **31.** $\frac{16}{3}$
33. 8π **35.** 8π **37.** 2   **39.** $\frac{8}{3}$
41. 25,645.24   **43.** $1 - e^{-1/4} \approx 0.221$ **45.** 9
47. 24 (Approximation is exact.)
49. Approximation: 52   **55.** 1.2315
Exact: $\frac{160}{3}$

Section 16.3

1. 0   **3.** $\dfrac{5\sqrt{5}\,\pi}{6}$

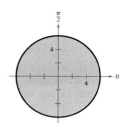   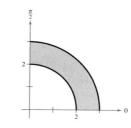

5. $\dfrac{\pi^2}{8} + 1$ **7.** 9π

9. $\dfrac{3\pi}{2}$ **11.** π **13.** $\dfrac{a^3}{3}$ **15.** $\dfrac{9\pi^2}{16}$ **17.** $\dfrac{2}{3}$
19. $\dfrac{4\sqrt{2}\,\pi}{3}$ **21.** $\dfrac{16}{3}$ **23.** $\dfrac{\pi^2}{16}$ **25.** $\dfrac{1}{8}$
27. $\dfrac{250\pi}{3}$ **29.** $\dfrac{64}{9}(3\pi - 4)$
31. $2\sqrt{4 - 2\sqrt[3]{2}}$ **33.** $\sqrt{2}\pi$

Section 16.4

1. (a) $m = kab$, $\left(\dfrac{a}{2}, \dfrac{b}{2}\right)$ **(b)** $m = \dfrac{kab^2}{2}$, $\left(\dfrac{a}{2}, \dfrac{2b}{3}\right)$

3. (a) $m = \dfrac{kbh}{2}$, $\left(\dfrac{b}{2}, \dfrac{h}{3}\right)$ **(b)** $m = \dfrac{kh^2b}{6}$, $\left(\dfrac{b}{2}, \dfrac{h}{2}\right)$

5. (a) $m = \dfrac{k\pi a^2}{2}$, $\left(0, \dfrac{4a}{3\pi}\right)$ **(b)** $m = \dfrac{ka^4}{24}(16 - 3\pi)$

$\left(0, \dfrac{a}{5}\left[\dfrac{15\pi - 32}{16 - 3\pi}\right]\right)$

7. $m = \dfrac{32k}{3}$, $\left(3, \dfrac{8}{7}\right)$ **9.** $m = \dfrac{k}{4}(1 - e^{-4})$

$\left(\dfrac{e^4 - 5}{2(e^4 - 1)}, \dfrac{4}{9}\left[\dfrac{e^6 - 1}{e^6 - e^2}\right]\right)$

11. $m = \dfrac{k\pi}{2}$, $\left(0, \dfrac{\pi + 2}{4\pi}\right)$ **13.** $m = \dfrac{8129k}{15}$, $\left(\dfrac{64}{7}, 0\right)$

15. $m = \dfrac{kL}{4}$, $\left(\dfrac{L}{2}, \dfrac{16}{9\pi}\right)$

17. $m = \dfrac{k\pi a^2}{8}$, $\left(\dfrac{4\sqrt{2}a}{3\pi}, \dfrac{4a(2 - \sqrt{2})}{3\pi}\right)$

19. $m = \dfrac{k\pi}{3}$, $(1.12, 0)$

21. $\bar{\bar{x}} = \dfrac{\sqrt{3}b}{3}$, $\bar{\bar{y}} = \dfrac{\sqrt{3}h}{3}$

23. $\bar{\bar{x}} = \dfrac{r}{2}$, $\bar{\bar{y}} = \dfrac{r}{2}$ **25.** $\bar{\bar{x}} = \dfrac{r}{2}$, $\bar{\bar{y}} = \dfrac{r}{2}$

27. $I_x = \dfrac{kab^4}{4}$ **29.** $I_x = \dfrac{32k}{3}$

$I_y = \dfrac{kb^2a^3}{6}$ $I_y = \dfrac{16k}{3}$

$I_o = \dfrac{3kab^4 + 2ka^3b^2}{12}$ $I_o = 16k$

$\bar{\bar{x}} = \dfrac{\sqrt{3}a}{3}$ $\bar{\bar{x}} = \dfrac{2\sqrt{3}}{3}$

$\bar{\bar{y}} = \dfrac{\sqrt{2}b}{2}$ $\bar{\bar{y}} = \dfrac{2\sqrt{6}}{3}$

31. $I_x = 16k$ **33.** $I_x = \dfrac{3k}{56}$

$I_y = \dfrac{512k}{5}$ $I_y = \dfrac{k}{18}$

$I_o = \dfrac{592k}{5}$ $I_o = \dfrac{55k}{504}$

$\bar{\bar{x}} = \dfrac{4\sqrt{15}}{5}$ $\bar{\bar{x}} = \dfrac{\sqrt{30}}{9}$

$\bar{\bar{y}} = \dfrac{\sqrt{6}}{2}$ $\bar{\bar{y}} = \dfrac{\sqrt{70}}{14}$

35. $\dfrac{k\pi b^2}{4}(b^2 + 4a^2)$ **37.** $\dfrac{42{,}752k}{315}$

39. $a^5\left(\dfrac{7\pi}{16} - \dfrac{17}{15}\right)k$

41. $m = 34.9326$, $(6.8938, 0.9332)$

Section 16.5

1. 6   **3.** 12π **5.** $\dfrac{3}{4}[6\sqrt{37} + \ln(\sqrt{37} + 6)]$

7. $\dfrac{27 - 5\sqrt{5}}{12}$ **9.** $\dfrac{4}{27}(31\sqrt{31} - 8)$

11. $\sqrt{2} - 1$ **13.** $\dfrac{\pi}{6}(17\sqrt{17} - 1)$ **15.** $\sqrt{2}\pi$

17. $2\pi a(a - \sqrt{a^2 - b^2})$ **19.** 16

21. $\displaystyle\int_{-1}^{1}\int_{-1}^{1} \sqrt{1 + 9(x^2 - y)^2 + 9(y^2 - x)^2}\, dy\, dx$

23. $\displaystyle\int_{-2}^{2}\int_{-\sqrt{4-x^2}}^{\sqrt{4-x^2}} \sqrt{1 + e^{-2x}}\, dy\, dx$

25. $\displaystyle\int_{0}^{4}\int_{0}^{10} \sqrt{1 + e^{2xy}(x^2 + y^2)}\, dy\, dx$

27. 2.00349   **29.** 1.86173

31. (a) $406\pi\sqrt{609}$ cm³   **(b)** $50\pi\sqrt{609}$ cm²

33. 52.0182

Section 16.6

1. 18   **3.** $\dfrac{1}{10}$ **5.** $\dfrac{15}{2}\left(1 - \dfrac{1}{e}\right)$

7. $\dfrac{729}{4}$ **9.** $\dfrac{128}{15}$

11. $\displaystyle\int_{0}^{3}\int_{0}^{(12-4z)/3}\int_{0}^{(12-4z-3x)/6} dy\, dx\, dz$

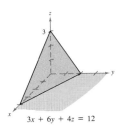

$3x + 6y + 4z = 12$

13. $\displaystyle\int_{0}^{1}\int_{0}^{x}\int_{0}^{\sqrt{1-y^2}} dz\, dy\, dx$

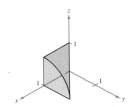

15. $\int_0^1 \int_0^x \int_0^3 xyz \; dz \; dy \; dx \quad \int_0^1 \int_y^1 \int_0^3 xyz \; dz \; dx \; dy$

$\int_0^1 \int_0^3 \int_0^x xyz \; dy \; dz \; dx \quad \int_0^3 \int_0^1 \int_0^x xyz \; dy \; dx \; dz$

$\int_0^3 \int_0^1 \int_y^1 xyz \; dx \; dy \; dz \quad \int_0^1 \int_0^3 \int_y^1 xyz \; dx \; dz \; dy$

17. $\dfrac{256}{15}$ **19.** $\dfrac{4\pi r^3}{3}$ **21.** $\dfrac{256}{15}$ **23.** $m = 8k$

$\bar{x} = \dfrac{3}{2}$

25. $m = \dfrac{128k}{3}$ **27.** $m = \dfrac{kb^5}{4}$ **29.** $\left(0, 0, \dfrac{3h}{4}\right)$

$\bar{z} = 1$ $\left(\dfrac{2b}{3}, \dfrac{2b}{3}, \dfrac{b}{2}\right)$

31. $\left(0, 0, \dfrac{3}{2}\right)$

33. (a) $I_x = \dfrac{2ka^5}{3}$ (b) $I_x = \dfrac{ka^8}{8}$

$I_y = \dfrac{2ka^5}{3}$ $I_y = \dfrac{ka^8}{8}$

$I_z = \dfrac{2ka^5}{3}$ $I_z = \dfrac{ka^8}{8}$

35. (a) $I_x = 256k$ (b) $I_x = \dfrac{2048k}{3}$

$I_y = \dfrac{512k}{3}$ $I_y = \dfrac{1024k}{3}$

$I_z = 256k$ $I_z = \dfrac{2048k}{3}$

39. $\int_{-1}^1 \int_{-1}^1 \int_0^{1-x} (x^2 + y^2)\sqrt{x^2 + y^2 + z^2} \; dz \; dy \; dx$

41. 2.44166

Section 16.7

1. $\left(5, \dfrac{\pi}{2}, 1\right)$ **3.** $\left(2, \dfrac{\pi}{3}, 4\right)$

5. $\left(2\sqrt{2}, -\dfrac{\pi}{4}, -4\right)$ **7.** $(5, 0, 2)$

9. $(1, \sqrt{3}, 2)$ **11.** $(-2\sqrt{3}, -2, 3)$

13. $\left(4, 0, \dfrac{\pi}{2}\right)$ **15.** $\left(4\sqrt{2}, \dfrac{2\pi}{3}, \dfrac{\pi}{4}\right)$

17. $\left(4, \dfrac{\pi}{6}, \dfrac{\pi}{6}\right)$ **19.** $(\sqrt{6}, \sqrt{2}, 2\sqrt{2})$

21. $(0, 0, 12)$ **23.** $\left(\dfrac{5}{2}, \dfrac{5}{2}, -\dfrac{5\sqrt{2}}{2}\right)$

25. $\left(4, \dfrac{\pi}{4}, \dfrac{\pi}{2}\right)$ **27.** $\left(2\sqrt{13}, -\dfrac{\pi}{6}, \arccos\left[\dfrac{3}{\sqrt{13}}\right]\right)$

29. $\left(13, \pi, \arccos\left[\dfrac{5}{13}\right]\right)$ **31.** $\left(10, \dfrac{\pi}{6}, 0\right)$

33. $\left(3\sqrt{3}, -\dfrac{\pi}{6}, 3\right)$ **35.** $\left(4, \dfrac{7\pi}{6}, 4\sqrt{3}\right)$ **37.** d

38. e   **39.** c   **40.** a   **41.** f   **42.** b

43. $x^2 + y^2 = 4$ **45.** $x - \sqrt{3}y = 0$

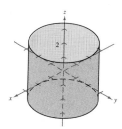

47. $x^2 + y^2 - 2y = 0$ **49.** $x^2 + y^2 + z^2 = 4$

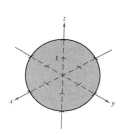

51. $x^2 + y^2 + z^2 = 4$ **53.** $3x^2 + 3y^2 - z^2 = 0$

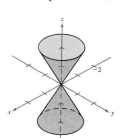

55. $x^2 + y^2 + (z - 2)^2 = 4$ **57.** $x^2 + y^2 = 1$

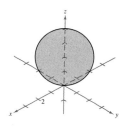
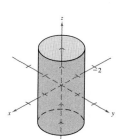

59. (a) $r^2 + z^2 = 16$ (b) $\rho = 4$

61. (a) $r^2 + (z - 1)^2 = 1$ (b) $\rho = 2\cos\phi$

63. (a) $r = 4\sin\theta$ (b) $\rho = \dfrac{4\sin\theta}{\sin\phi} = 4\sin\theta\csc\phi$

65. (a) $r^2 = \dfrac{9}{\cos^2 \theta - \sin^2 \theta}$

(b) $\rho^2 = \dfrac{9 \csc^2 \phi}{\cos^2 \theta - \sin^2 \theta}$

67.

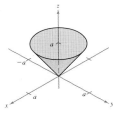

69.

71. Rectangular: $0 \le x \le 10$
$0 \le y \le 10$
$0 \le z \le 10$

73. Spherical: $4 \le \rho \le 6$

75. (a) Los Angeles: $(4000, -118.24°, 55.95°)$
Rio de Janeiro: $(4000, -43.22°, 112.90°)$
(b) Los Angeles: $(-1568.2, -2919.7, 2239.7)$
Rio de Janeiro: $(2685.2, -2523.3, -1556.5)$
(c) $91.18° \approx 1.59$ radians **(d)** 6366 miles

Section 16.8

1. 8 **3.** $\dfrac{52}{45}$ **5.** $\dfrac{\pi^2}{6}(1 - e^{-8})$ **7.** 0

9. $\dfrac{\pi}{4}(1 - e^{-9})$ **11.** $\dfrac{64\sqrt{3}\,\pi}{3}$

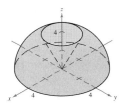

13. Cylindrical: $\displaystyle\int_0^{2\pi}\int_0^2\int_{r^2}^4 r^2 \cos \theta \, dz \, dr \, d\theta$

Spherical: $\displaystyle\int_0^{2\pi}\int_0^{\pi/2}\int_0^{4 \sec \phi} \rho^3 \sin^2 \phi \cos \theta \, d\rho \, d\phi \, d\theta, \ 0$

15. Cylindrical: $\displaystyle\int_0^{2\pi}\int_0^a\int_a^{a+\sqrt{a^2-r^2}} r^2 \cos \theta \, dz \, dr \, d\theta$

Spherical: $\displaystyle\int_0^{\pi/4}\int_0^{2\pi}\int_{a \sec \phi}^{2a \cos \phi} \rho^3 \sin^2 \phi \cos \theta \, d\rho \, d\theta \, d\phi, \ 0$

17. $\dfrac{\pi r_0^2 h}{3}$ **19.** $\left(0, 0, \dfrac{h}{5}\right)$ **25.** $\dfrac{4}{3}a^3\left(\dfrac{\pi}{2} - \dfrac{2}{3}\right)$

27. $16\pi^2$ **29.** $k\pi a^4$ **31.** $\left(0, 0, \dfrac{3r}{8}\right)$ **33.** $\dfrac{k\pi}{96}$

Section 16.9

1. $-\dfrac{1}{2}$ **3.** $1 + 2v$ **5.** 1 **7.** $-e^{2u}$

9.

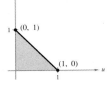

11. 0
13. $2(1 - e^{-2})$
15. $\dfrac{32}{9}(2\sqrt{2} - 1)$

17. $12(e^4 - 1)$ **19.** $\dfrac{100}{9}$ **21.** $\dfrac{2}{5}a^{5/2}$

23. (a)

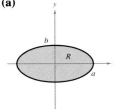
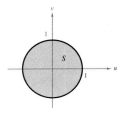

(b) ab

25. u^2v **27.** $\rho^2 \sin \phi$

Review Exercises for Chapter 16

1. $\dfrac{29}{6}$ **3.** 36 **5.** $\dfrac{27}{5}$

7. $\dfrac{h^3}{6}[\ln (\sqrt{2} + 1) + \sqrt{2}]$ **9.** $\dfrac{324\pi}{5}$

11. $\dfrac{8\pi}{15}$ **13.** $\dfrac{abc}{3}(a^2 + b^2 + c^2)$

15. $\displaystyle\int_0^3\int_0^{(3-x)/3} dy \, dx = \int_0^1\int_0^{3-3y} dx \, dy = \dfrac{3}{2}$

17. $\displaystyle\int_{-5}^3\int_{-\sqrt{25-x^2}}^{\sqrt{25-x^2}} dy \, dx$

$= \displaystyle\int_{-5}^{-4}\int_{-\sqrt{25-y^2}}^{\sqrt{25-y^2}} dx \, dy + \int_{-4}^4\int_{-\sqrt{25-y^2}}^3 dx \, dy + \int_4^5\int_{-\sqrt{25-y^2}}^{\sqrt{25-y^2}} dx \, dy$

$= \dfrac{25\pi}{2} + 12 + 25 \arcsin \dfrac{3}{5} \approx 67.36$

19. $4\displaystyle\int_0^1\int_0^{x\sqrt{1-x^2}} dy \, dx = 4\int_0^{1/2}\int_{\sqrt{(1-\sqrt{1-4y^2})/2}}^{\sqrt{(1+\sqrt{1-4y^2})/2}} dx \, dy = \dfrac{4}{3}$

21. $\displaystyle\int_2^5\int_{x-3}^{\sqrt{x-1}} dy \, dx + 2\int_1^2\int_0^{\sqrt{x-1}} dy \, dx = \int_{-1}^2\int_{y^2+1}^{y+3} dx \, dy = \dfrac{9}{2}$

23. (a) $\left(4, \dfrac{3\pi}{4}, 2\right)$ **(b)** $\left(2\sqrt{5}, \dfrac{3\pi}{4}, \arccos\left[\dfrac{\sqrt{5}}{5}\right]\right)$

25. (a) $r^2 \cos 2\theta = 2z$
(b) $\rho = 2 \sec 2\theta \cos \phi \csc^2 \phi$

27. $x^2 - y^2 + z^2 = 1$ **29.** $x^2 + y^2 = 1$

31. $\dfrac{3296}{15}$ **33.** $\dfrac{\pi h^3}{3}$ **35.** $\dfrac{16}{3}\left(\dfrac{\pi}{2} - \dfrac{2}{3}\right)$

37. (a) $m = \dfrac{k}{4}, \left(\dfrac{32}{45}, \dfrac{64}{55}\right)$ **(b)** $m = \dfrac{17k}{30}, \left(\dfrac{936}{1309}, \dfrac{784}{663}\right)$

39. $\left(0, 0, \dfrac{1}{4}\right)$ **41.** $\left(\dfrac{3a}{8}, \dfrac{3a}{8}, \dfrac{3a}{8}\right)$

43. $\dfrac{\pi}{6}(65\sqrt{65} - 1)$ **45.** $\dfrac{833k\pi}{3}$

47. Volume of a torus:
Formed by a circle of radius 3
Centered at (0, 3, 0)
Revolved about the z-axis

Chapter 17

Section 17.1

1.

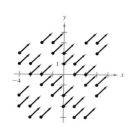

3.

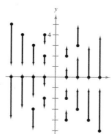

5.

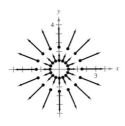

7.

9.

11.

13.

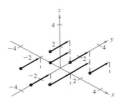

15. $(10x + 3y)\mathbf{i} + (3x + 20y)\mathbf{j}$

17. $-2xye^{x^2}\mathbf{i} - e^{x^2}\mathbf{j} + \mathbf{k}$

19. $\left[\dfrac{xy}{x + y} + y\ln(x + y)\right]\mathbf{i} + \left[\dfrac{xy}{x + y} + x\ln(x + y)\right]\mathbf{j}$

21. Conservative: $f(x, y) = x^2y + K$

23. Conservative: $f(x, y) = e^{x^2y} + K$

25. Conservative: $f(x, y) = \frac{1}{2}\ln(x^2 + y^2) + K$

27. Not conservative   **29.** $2\mathbf{j} - \mathbf{k}$ **31.** $-2\mathbf{k}$

33. $\dfrac{2x\,\mathbf{k}}{x^2 + y^2}$

35. $\cos(y - z)\mathbf{i} + \cos(z - x)\mathbf{j} + \cos(x - y)\mathbf{k}$

37. Not conservative

39. Conservative: $f(x, y, z) = xye^z + K$

41. Conservative: $f(x, y, z) = \dfrac{x}{y} + z^2 - z + K$

43. $6x\mathbf{j} - 3y\mathbf{k}$ **45.** $z\mathbf{j} + y\mathbf{k}$

47. $12x - 2xy$ **49.** $\cos x - \sin y + 2z$

51. 4   **53.** 0   **55.** $2z + 3x$ **57.** 0

Section 17.2

1. $3\cos t\,\mathbf{i} + 3\sin t\,\mathbf{j},\ 0 \le t \le 2\pi$

3. $\mathbf{r}(t) = \begin{cases} t\mathbf{i}, & 0 \le t \le 3 \\ 3\mathbf{i} + (t - 3)\mathbf{j}, & 3 \le t \le 6 \\ (9 - t)\mathbf{i} + 3\mathbf{j}, & 6 \le t \le 9 \\ (12 - t)\mathbf{j}, & 9 \le t \le 12 \end{cases}$

5. $\mathbf{r}(t) = \begin{cases} t\mathbf{i} + \sqrt{t}\,\mathbf{j}, & 0 \le t \le 1 \\ (2 - t)\mathbf{i} + (2 - t)\mathbf{j}, & 1 \le t \le 2 \end{cases}$

7. 10   **9.** $\dfrac{\sqrt{65}\,\pi}{6}(3 + 16\pi^2)$ **11.** 9

13. $\dfrac{\pi}{2}$ **15.** $\dfrac{19\sqrt{2}}{6}$ **17.** $\dfrac{19}{6}(1 + \sqrt{2})$

19. $\frac{35}{6}$ **21.** 2   **23.** $-\frac{17}{15}$ **25.** 1010

27. $\frac{190}{3}$ **29.** 25   **31.** $\frac{63}{2}$ **33.** $\frac{316}{3}$

35. $\dfrac{2\pi\sqrt{13}}{3}(27 + 64\pi^2) \approx 4{,}973.8$ **37.** -66

39. 0   **41.** $-10\pi^2$ **47.** $5h$ **49.** $\frac{1}{2}$

51. $\dfrac{h}{4}[2\sqrt{5} + \ln(2 + \sqrt{5})]$ **53.** $\dfrac{1}{120}(25\sqrt{5} - 11)$

55. 1500 ft · lb   **57.** 249.49

Section 17.3

1. $\frac{11}{15}$ **3.** $-\ln(2 + \sqrt{3}) = -\frac{1}{2}\ln(7 + 4\sqrt{3})$
5. 1 **7. (a)** 0 **(b)** $-\frac{1}{3}$ **(c)** $-\frac{1}{2}$
9. (a) 64 **(b)** 0 **(c)** 0 **(d)** 0
11. (a) $\frac{64}{3}$ **(b)** $\frac{64}{3}$ **13. (a)** 32 **(b)** 32
15. (a) $\frac{2}{3}$ **(b)** $\frac{17}{6}$ **17. (a)** 0 **(b)** 0
19. 24 **21.** -1 **23.** 0
25. (a) 2 **(b)** 2 **(c)** 2 **27.** 11
29. 30,366 **31.** 0 **33.** Increasing: 10 units/min

Section 17.4

1. 0 **3.** $\frac{32}{15}$ **5.** $\frac{4}{3}$ **7.** 56 **9.** $\frac{32}{3}$
11. $\dfrac{16a^3}{3}$ **13.** 0 **15.** $\dfrac{1}{12}$ **17.** 8π **19.** 4π
21. $\frac{225}{2}$ **23.** πa^2 **25.** $\frac{32}{3}$ **29.** $\left(0, \frac{8}{5}\right)$
31. $\left(\dfrac{8}{15}, \dfrac{8}{21}\right)$ **33.** $\dfrac{3\pi a^2}{2}$ **35.** $\pi - \dfrac{3\sqrt{3}}{2}$
37. $I = -2\pi$ when C is a circle that contains the origin.
39. $\frac{19}{2}$ **45.** 19.99

Section 17.5

1. 0 **3.** 10π **5.** $\dfrac{27\sqrt{6}}{2}$ **7.** $\dfrac{391\sqrt{17} + 1}{240}$
9. $\dfrac{19\sqrt{2}\pi}{4}$ **11.** $\dfrac{32\pi}{3}$ **13.** 486π **15.** $-\dfrac{4}{3}$
17. $\dfrac{243\pi}{2}$ **19.** 20π **21.** $\dfrac{5}{2}$ **23.** $\dfrac{364}{3}$
27. $2\pi a^3 h$ **29.** $64\pi\rho$
31. If a normal vector at a point P on the surface is moved around the Mobius strip once, it will point in the opposite direction.
33. -11.47 **35.** 0

Section 17.6

1. a^4 **3.** 18 **5.** $3a^4$ **7.** 0 **9.** 32π
11. 0 **13.** 2304 **17.** 0

Section 17.7

1. $-xy\mathbf{i} - \mathbf{j} + (yz - 2)\mathbf{k}$
3. $\left(2 - \dfrac{1}{1 + x^2}\right)\mathbf{j} - 8x\mathbf{k}$
5. $z(x - 2e^{y^2+z^2})\mathbf{i} - yz\mathbf{j} - 2ye^{x^2+y^2}\mathbf{k}$ **7.** 2π

9. 0 **11.** 1 **13.** 0 **15.** 0 **17.** $\frac{8}{3}$
19. $\dfrac{a^5}{4}$ **21.** 0

Review Exercises for Chapter 17

1. $(16x + y)\mathbf{i} + x\mathbf{j} + 2z\mathbf{k}$ **3.** Not conservative
5. Conservative: $f(x, y) = 3x^2y^2 - x^3 + y^3 - 7y + K$
7. Not conservative
9. Conservative: $f(x, y, z) = \dfrac{x}{yz} + K$
11. (a) div $\mathbf{F} = 2x + 2y + 2z$ **(b)** curl $\mathbf{F} = \mathbf{0}$
13. (a) div $\mathbf{F} = -y \sin x - x \cos y + xy$
(b) curl $\mathbf{F} = xz\mathbf{i} - yz\mathbf{j}$
15. (a) div $\mathbf{F} = \dfrac{1}{\sqrt{1 - x^2}} + 2xy + 2yz$
(b) curl $\mathbf{F} = z^2\mathbf{i} + y^2\mathbf{k}$
17. (a) div $\mathbf{F} = \dfrac{2x + 2y}{x^2 + y^2} + 1$
(b) curl $\mathbf{F} = \dfrac{2x - 2y}{x^2 + y^2}\mathbf{k}$
19. (a) $6\sqrt{2}$ **(b)** 128π **21.** $2\pi^2(1 + 2\pi^2)$
23. (a) $\frac{35}{2}$ **(b)** 18π **25.** $\frac{5}{7}$ **27.** $2\pi^2$
29. $\frac{64}{3}$ **31.** 12 **33.** 4 **35.** 0 **37.** $\frac{1}{12}$
39. 66 **41.** $-\dfrac{2a^6}{5}$
43. $\displaystyle\int_C \mathbf{F} \cdot d\mathbf{r} = 2\pi$
$\mathbf{F}$ is undefined at the point $(0, 0)$.

Chapter 18

Section 18.1

Type	Order
1. Ordinary	1
3. Ordinary	2
5. Ordinary	4
7. Ordinary	2
9. Partial	1
11. Ordinary	2

19. Not a solution **21.** Solution **23.** Solution
25. Not a solution **27.** Solution **29.** Solution
31. $4y^2 = x^3$ **33.** $y = 3e^{-2x}$
35. $y = 2 \sin 3x - \frac{1}{3}\cos 3x$ **37.** $y = -2x + \frac{1}{2}x^3$

39.

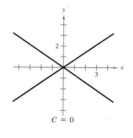

$C = 0$

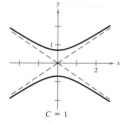

$C = 1$

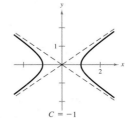

$C = -1$

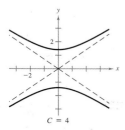

$C = 4$

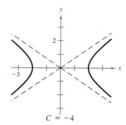

$C = -4$

41. $y = x^3 + C$ **43.** $y = x - \ln x^2 + C$

45. $y = \dfrac{e^x}{5}(\sin 2x - 2 \cos 2x) + C$

47. $y = \frac{2}{5}(x - 3)^{5/2} + 2(x - 3)^{3/2} + C$

49. (a) $N = 750 - 650e^{-(t/2)\ln(65/59)}$

$= 750 - 650e^{-0.0484t}$

(c)

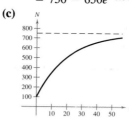

21. $x = C(x - y)^2$ **23.** $y^2 + 2xy - x^2 = C$

25. $y = Ce^{-x^2/2y^2}$ **27.** $e^{y/x} = 1 + \ln x^2$

29. $x = e^{\sin(y/x)}$

31. Circles: $x^2 + y^2 = C$

Lines: $y = Kx$

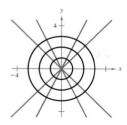

33. Parabolas: $x^2 = Cy$

Ellipses: $x^2 + 2y^2 = K$

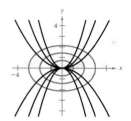

35. Curves: $y^2 = Cx^3$

Ellipses: $2x^2 + 3y^2 = K$

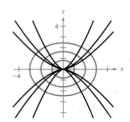

37. $9x^2 + 16y^2 = 25$

39. (a) $A = Pe^{kt}$ **(b)** $A = \$3,004.17$

(c) $t = 6.3$ years

41. 98.9% of original amount   **43.** $t = 119.7$ min

Section 18.2

1. $y^2 - x^2 = C$ **3.** $y = C(x + 2)^3$

5. $y^2 = C - 2 \cos x$ **7.** $y^2 = 2e^x + 14$

9. $y = e^{-(x^2+2x)/2}$ **11.** $\dfrac{y - x}{1 + xy} = \sqrt{3}$

13. $P = P_0 e^{kt}$ **15.** Homogeneous of degree 3

17. Not homogeneous   **19.** Homogeneous of degree 0

Section 18.3

1. $x^2 - 3xy + y^2 = C$

3. $3xy^2 + 5x^2y^2 - 2y = C$ **5.** Not exact

7. $\arctan \dfrac{x}{y} = C$ **9.** Not exact

11. $y \ln(x - 1) + y^2 = 16$ **13.** $x^2 + y^2 = 16$

15. $x^2 \tan y + 5x = 0$

17. Integrating factor: $\dfrac{1}{y^2}$ **19.** Integrating factor: $\dfrac{1}{x^2}$

$\dfrac{x}{y} - 6y = C$ $\dfrac{y}{x} + 5x = C$

21. Integrating factor: $\cos x$
$y \sin x + x \sin x + \cos x = C$

23. Integrating factor: $\dfrac{1}{y}$ **25.** Integrating factor: $\dfrac{1}{\sqrt{y}}$

$xy - \ln y = C$ $x\sqrt{y} + \cos \sqrt{y} = C$

27. Integrating factor: xy^2
$x^4y^3 + x^2y^4 = C$

29. Integrating factor: $\dfrac{1}{x^2y^3}$ **33.** $x^2 + y^2 = C$

$\dfrac{y^2}{x} + \dfrac{x}{y^2} = C$

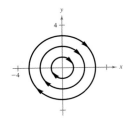

35. $2x^2y^4 + x^2 = C$ **37.** $x^2 - 2xy + 3y^2 = 3$

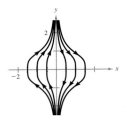

39. $C = \dfrac{5(x^2 + \sqrt{x^4 - 1,000,000x})}{x}$

Section 18.4

1. $y = x^2 + 2x + \dfrac{C}{x}$ **3.** $y = \dfrac{1}{2}e^x + Ce^{-x}$

5. $y = \dfrac{1}{2}(\sin x - \cos x) + Ce^x$

7. $y = -\dfrac{1}{13}(3 \sin 2x + 2 \cos 2x) + Ce^{3x}$

9. $y = \dfrac{x^3 - 3x + C}{3(x - 1)}$ **11.** $y = 1 + 4e^{-\tan x}$

13. $y = \sin x + (x + 1) \cos x$ **15.** $xy = 4$

17. $\dfrac{1}{y^2} = Ce^{2x^3} + \dfrac{1}{3}$ **19.** $y = \dfrac{1}{Cx - x^2}$

21. $y^{2/3} = Ce^{2x/3} - \dfrac{1}{4}(4x^3 + 18x^2 + 54x + 81)$

23. $y = \dfrac{E_0}{R} + Ce^{-Rt/L}$ **25.** $P = -\dfrac{N}{k} + \left(\dfrac{N}{k} + P_0\right)e^{kt}$

27. (a) $\$685,099.00$ (b) $\$5,802,815.12$

29. (a) $\dfrac{dQ}{dt} = q - kQ$ (b) $Q = \dfrac{q}{k} + \left(Q_0 - \dfrac{q}{k}\right)e^{-kt}$

   (c) $\dfrac{q}{k}$

33. (a) $Q = 10 + 15e^{-0.05t}$
   (b) $t = -20 \ln \dfrac{1}{3} \approx 21.97$ min

35. $2e^x + e^{-2y} = C$ **37.** $x + xy^2 + \dfrac{1}{2}y^2 + 2y = C$

39. $y = Ce^{-\sin x} + 1$ **41.** $x^2y + \sin y = C$

43. $x^3y^2 + x^4y = C$ **45.** $y = \dfrac{e^x(x - 1) + C}{x^2}$

47. $x^4y^4 - 2x^2 = C$ **49.** $3 \arctan \dfrac{x}{y} - y = C$

Section 18.5

1. $y = C_1 + C_2e^x$ **3.** $y = C_1e^{3x} + C_2e^{-2x}$

5. $y = C_1e^{x/2} + C_2e^{-2x}$ **7.** $y = C_1e^{-3x} + C_2xe^{-3x}$

9. $y = C_1e^{x/4} + C_2xe^{x/4}$

11. $y = C_1 \sin x + C_2 \cos x$

13. $y = C_1e^{3x} + C_2e^{-3x}$

15. $y = e^x(C_1 \sin \sqrt{3}x + C_2 \cos \sqrt{3}x)$

17. $y = C_1e^{(3+\sqrt{5})x/2} + C_2e^{(3-\sqrt{5})x/2}$

19. $y = e^{2x/3}\left(C_1 \sin \dfrac{\sqrt{7}x}{3} + C_2 \cos \dfrac{\sqrt{7}x}{3}\right)$

21. $y = C_1e^x + C_2e^{-x} + C_3 \sin x + C_4 \cos x$

23. $y = C_1e^x + C_2e^{2x} + C_3e^{3x}$

25. $y = C_1e^x + e^x(C_2 \sin 2x + C_3 \cos 2x)$

27. $y = \dfrac{1}{11}(e^{6x} + 10e^{-5x})$ **29.** $y = \dfrac{1}{2} \cos 4\sqrt{3}t$

31. $y = \dfrac{2}{3} \cos 4\sqrt{3}t - \dfrac{\sqrt{3}}{24} \sin 4\sqrt{3}t$

33. $y = \dfrac{e^{-t/16}}{2}\left(\cos \dfrac{\sqrt{12,287}t}{16} + \dfrac{\sqrt{12,287}}{12,287} \sin \dfrac{\sqrt{12,287}t}{16}\right)$

Section 18.6

1. $y = C_1e^x + C_2e^{2x} + x + \dfrac{3}{2}$

3. $y = \cos x + 6 \sin x + x^3 - 6x$

5. $y = C_1 + C_2e^{-2x} + \dfrac{2}{3}e^x$

7. $y = (C_1 + C_2x)e^{5x} + \dfrac{3}{8}e^x + \dfrac{1}{5}$

9. $y = -1 + 2e^{-x} - \cos x - \sin x$

11. $y = \left(C_1 - \dfrac{x}{6}\right) \cos 3x + C_2 \sin 3x$

13. $y = C_1e^x + C_2xe^x + \left(C_3 + \dfrac{2x}{9}\right)e^{-2x}$

15. $y = \left(\dfrac{4}{9} - \dfrac{1}{2}x^2\right)e^{4x} - \dfrac{1}{9}(1 + 3x)e^x$

17. $y = (C_1 + \ln |\cos x|) \cos x + (C_2 + x) \sin x$

19. $y = \left(C_1 - \dfrac{x}{2}\right) \cos 2x$

   $+ \left(C_2 + \dfrac{1}{4} \ln |\sin 2x|\right) \sin 2x$

21. $y = (C_1 + C_2 x)e^x + \dfrac{x^2 e^x}{4}(\ln x^2 - 3)$

23. $q = \frac{3}{25}(e^{-5t} + 5te^{-5t} - \cos 5t)$

25. $y = \frac{1}{4}\cos 8t - \frac{1}{2}\sin 8t + \sin 4t$

27. $y = \left(\frac{9}{32} - \frac{3}{4}t\right)e^{-8t} - \frac{1}{32}\cos 8t$

29. $y = \dfrac{\sqrt{5}}{4}\sin\left(8t - \arctan\dfrac{1}{2}\right)$

$\quad = \dfrac{\sqrt{5}}{4}\sin(8t - 0.4636)$

Section 18.7

1. $y = a_0 \displaystyle\sum_{n=0}^{\infty} \dfrac{x^n}{n!} = a_0 e^x$

3. $y = a_0 \displaystyle\sum_{k=0}^{\infty} \dfrac{(3x)^{2k}}{(2k)!} + \dfrac{a_1}{3}\sum_{k=0}^{\infty}\dfrac{(3x)^{2k+1}}{(2k+1)!}$

$\quad = C_0 \displaystyle\sum_{k=0}^{\infty}\dfrac{(3x)^k}{k!} + C_1 \sum_{k=0}^{\infty}\dfrac{(-3x)^k}{k!}$

$\quad = C_0 e^{3x} + C_1 e^{-3x}$

where $C_0 + C_1 = a_0$ and $C_0 - C_1 = \frac{1}{3}a_1$.

5. $y = a_0 \displaystyle\sum_{k=0}^{\infty}\dfrac{(-1)^k(2x)^{2k}}{(2k)!} + \dfrac{a_1}{4}\sum_{k=0}^{\infty}\dfrac{(-1)^k(2x)^{2k+1}}{(2k+1)!}$

$\quad = C_0 \cos 2x + C_1 \sin 2x$

7. $y = a_0 \displaystyle\sum_{k=0}^{\infty}\dfrac{(-3)^k}{2^k k!}x^{2k}$

$\quad + a_1 \displaystyle\sum_{k=0}^{\infty}\dfrac{(-3)^k}{1 \cdot 3 \cdot 5 \cdots (2k+1)}x^{2k+1}$

9. $y = a_1 \displaystyle\sum_{k=0}^{\infty}\dfrac{x^{2k+1}}{2^k(k!)(2k+1)}$

11. $y = a_0\left(1 - \dfrac{x^2}{8} + \dfrac{x^4}{128} - \cdots\right)$

$\quad + a_1\left(x - \dfrac{x^3}{24} + \dfrac{7x^5}{1920} - \cdots\right)$

13. $y = 2 + \dfrac{2x}{1!} - \dfrac{2x^2}{2!} - \dfrac{10x^3}{3!} + \dfrac{2x^4}{4!} + \cdots,$

when $x = \dfrac{1}{2}$, $y \approx 2.547$

15. $y = 1 - \dfrac{3x}{1!} + \dfrac{2x^3}{3!} - \dfrac{12x^4}{4!} + \dfrac{16x^6}{6!} - \dfrac{120x^7}{7!} + \cdots,$

when $x = \dfrac{1}{4}$, $y \approx 0.253$

Review Exercises for Chapter 18

1. $y = x \ln x^2 + 2x^{3/2} + Cx$ **3.** $y = C(1 - x)^2$

5. $y^2 = x^2 \ln x^2 + Cx^2$ **7.** $y = Ce^{2x} - e^x$

9. $5x^2 + 8xy + 2x + \frac{5}{2}y^2 + 2y = C$

11. $xy - 2xy^3 + x^2 = C$

13. $y = x \ln|x| - 2 + Cx$

15. $\ln|1 + y| = Ce^{-x}$

17. $y = e^x(1 + \tan x) + C \sec x$

19. $x^2 - 2xy - 10x - 3y^2 + 4y = C$

21. $y = \frac{1}{5}x^3 - x + C\sqrt{x}$ **23.** $y^2 = 2Cx + C^2$

25. $y^2 = x^2 - x + \frac{3}{2} + Ce^{-2x}$

27. $\dfrac{y - x}{1 + xy} = C$ **29.** $y = \dfrac{bx^4}{4 - a} + Cx^a$

31. $y = C_1 \sin x + C_2 \cos x - 5x + x^3$

33. $y = (C_1 + x)\sin x + C_2 \cos x$

35. $y = \left(C_1 + C_2 x + \dfrac{x^3}{3}\right)e^x$

37. Family of circles: $x^2 + (y - K)^2 = K^2$

39. $y = a_0 \displaystyle\sum_{n=0}^{\infty}\dfrac{x^n}{4^n}$

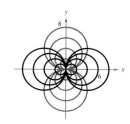

Index

NOBODY SAID CALCULUS WAS EASY. . .
BUT TURNING ON YOUR VCR IS!

Increasing your success in mastering calculus is now as easy as turning on your VCR, thanks to **The Calculus Series**, a comprehensive and compelling new video program created by Video Tutorial Service (VTS). These twelve videotapes (six per semester for the first two semesters) will enable you to:

- Review calculus thoroughly at your own pace and convenience.
- Build your understanding of challenging textbook material using problems and step-by-step solutions.
- Master information covered in complicated or missed lectures.

A SPECIAL OFFER TO STUDENTS:

Any indiviudal student using **Calculus with Analytic Geometry**, Alternate Fourth Edition, by Larson and Hostetler is entitled to substantial discounts off **The Calculus Series'** retail price of $49.95 per videotape:

- Purchase Tape I of either or both semesters' six-part package for only $14.95 each—**save 70%**!
- Purchase Tapes II-VI of either or both packages for $24.95 each—**save 50%**!
- Purchase each six-part package for only $124.70—*a $25 savings over these already discounted prices*—and receive a comprehensive workbook and answer key at no cost.
- Receive a $5.00 rebate for each tape returned in usable condition within a year of its purchase.

Prices subject to change. For more information or rush orders call 800 USAMATH.

Clip and mail to: **Video Tutorial Service, Inc., 1840 52nd Street, Brooklyn, NY 11204**

_____ YES! Send me the complete **Calculus Series** video program for:

_____ Semester I ($124.70) _____ Semester II ($124.70) _____ Both Semesters

_____ Send me Tape I for:

_____ Semester I ($14.95) _____ Semester II ($14.95)

_____ Send me information about purchasing individual tapes.

Name _____

Address _____

City _____ State _____ Zip _____

Phone # _____ _____

Payment option: _____ Check or M.O. _____ Visa _____ MC _____ AMEX

Credit Card #: _____ Exp. Date: _____

Text Purchased: **Calculus with Analytic Geometry**, Alternate Fourth Edition

ALGEBRA

Factors and Zeros of Polynomials

Let $p(x) = a_n x^n + a_{n-1} x^{n-1} + \cdots + a_1 x + a_0$ be a polynomial. If $p(a) = 0$, then a is a *zero* of the polynomial and a solution of the equation $p(x) = 0$. Furthermore, $(x - a)$ is a *factor* of the polynomial.

Fundamental Theorem of Algebra

An nth degree polynomial has n (not necessarily distinct) zeros. Although all of these zeros may be imaginary, a real polynomial of odd degree must have at least one real zero.

Quadratic Formula

If $p(x) = ax^2 + bx + c$, and $0 \le b^2 - 4ac$, then the real zeros of p are $x = (-b \pm \sqrt{b^2 - 4ac})/2a$.

Special Factors

$$x^2 - a^2 = (x - a)(x + a)$$
$$x^3 + a^3 = (x + a)(x^2 - ax + a^2)$$

$$x^3 - a^3 = (x - a)(x^2 + ax + a^2)$$
$$x^4 - a^4 = (x^2 - a^2)(x^2 + a^2)$$

Binomial Theorem

$$(x + y)^2 = x^2 + 2xy + y^2$$
$$(x + y)^3 = x^3 + 3x^2 y + 3xy^2 + y^3$$
$$(x + y)^4 = x^4 + 4x^3 y + 6x^2 y^2 + 4xy^3 + y^4$$
$$(x + y)^n = x^n + nx^{n-1} y + \frac{n(n - 1)}{2!} x^{n-2} y^2 + \cdots + nxy^{n-1} + y^n$$
$$(x - y)^n = x^n - nx^{n-1} y + \frac{n(n - 1)}{2!} x^{n-2} y^2 - \cdots \pm nxy^{n-1} \mp y^n$$

$$(x - y)^2 = x^2 - 2xy + y^2$$
$$(x - y)^3 = x^3 - 3x^2 y + 3xy^2 - y^3$$
$$(x - y)^4 = x^4 - 4x^3 y + 6x^2 y^2 - 4xy^3 + y^4$$

Rational Zero Theorem

If $p(x) = a_n x^n + a_{n-1} x^{n-1} + \cdots + a_1 x + a_0$ has integer coefficients, then every *rational zero* of p is of the form $x = r/s$, where r is a factor of a_0 and s is a factor of a_n.

Factoring by Grouping

$$acx^3 + adx^2 + bcx + bd = ax^2(cx + d) + b(cs + d) = (ax^2 + b)(cx + d)$$

Arithmetic Operations

$$ab + ac = a(b + c)$$

$$\frac{a}{b} + \frac{c}{d} = \frac{ad + bc}{bd}$$

$$\frac{a + b}{c} = \frac{a}{c} + \frac{b}{c}$$

$$\frac{\left(\dfrac{a}{b}\right)}{\left(\dfrac{c}{d}\right)} = \left(\frac{a}{b}\right)\left(\frac{d}{c}\right) = \frac{ad}{bc}$$

$$\frac{\left(\dfrac{a}{b}\right)}{c} = \frac{a}{bc}$$

$$\frac{a}{\left(\dfrac{b}{c}\right)} = \frac{ac}{b}$$

$$a\left(\frac{b}{c}\right) = \frac{ab}{c}$$

$$\frac{a - b}{c - d} = \frac{b - a}{d - c}$$

$$\frac{ab + ac}{a} = b + c$$

Exponents and Radicals

$$a^0 = 1, \quad a \ne 0 \qquad (ab)^x = a^x b^x \qquad a^x a^y = a^{x+y} \qquad \sqrt{a} = a^{1/2} \qquad \frac{a^x}{a^y} = a^{x-y} \qquad \sqrt[n]{a} = a^{1/n}$$

$$\left(\frac{a}{b}\right)^x = \frac{a^x}{b^x} \qquad \sqrt[n]{a^m} = a^{m/n} \qquad a^{-x} = \frac{1}{a^x} \qquad \sqrt[n]{ab} = \sqrt[n]{a}\sqrt[n]{b} \qquad (a^x)^y = a^{xy} \qquad \sqrt[n]{\frac{a}{b}} = \frac{\sqrt[n]{a}}{\sqrt[n]{b}}$$